DeGarmo's MATERIALS AND PROCESSES IN MANUFACTURING

INTERNATIONAL STUDENT VERSION

ELEVENTH EDITION

J T. Black
Auburn University-Emeritus

Ronald A. Kohser
Missouri University of Science & Technology

WILEY

John Wiley & Sons, Inc.

Founded in 1807, John Wiley & Sons, Inc. has been a valued source of knowledge and understanding for more than 200 years, helping people around the world meet their needs and fulfill their aspirations. Our company is built on a foundation of principles that include responsibility to the communities we serve and where we live and work. In 2008, we launched a Corporate Citizenship Initiative, a global effort to address the environmental, social, economic, and ethical challenges we face in our business. Among the issues we are addressing are carbon impact, paper specifications and procurement, ethical conduct within our business and among our vendors, and community and charitable support. For more information, please visit our website: www.wiley.com/go/citizenship.

ISBN: 978-0-470-87375-5

Printed in Asia

10 9 8 7 6 5 4 3 2 1

PREFACE

It's a world of manufactured goods. Whether we like it or not, we all live in a technological society. Every day we come in contact with hundreds of manufactured items, made from every possible material. From the bedroom to the kitchen, to the workplace, we use appliances, phones, cars, trains, and planes, TVs, cell phones, VCRs, DVD's, furniture, clothing, sports equipment, books and more! These goods are manufactured in factories all over the world using manufacturing processes.

Basically, manufacturing is a *value-adding* activity, where the conversion of materials into products adds value to the original material. Thus, the objective of a company engaged in manufacturing is to add value and to do so in the most efficient manner, with the least amount of waste in terms of time, material, money, space, and labor. To minimize waste and increase productivity, the processes and operations need to be properly selected and arranged to permit smooth and controlled flow of material through the factory and provide for product variety. Meeting these goals requires an engineer who can design and operate an efficient manufacturing system. Here are the trends that are impacting the manufacturing world.

- **Manufacturing is a global activity**

  Manufacturing is a global activity with companies sending work to other countries (China, Taiwan, Mexico) to take advantage of low-cost labor. Many US companies have plants in other countries and foreign companies have built plants in the United States, to be nearer their marketplace. Automobile manufacturers from all around the globe and their suppliers use just about every process described in this book and some that we do not describe, often because they are closely held secrets.

- **It's a digital world**

  Information technology and computers are growing exponentially, doubling in power every year. Every manufacturing company has ready access to world-wide digital technology. Products can be built by suppliers anywhere in the world working using a common set of digital information. Designs can be emailed to manufacturers who can rapidly produce a prototype in metal or plastic in a day.

- **Lean manufacturing is widely practiced**

  Most (over 60%) manufacturing companies have restructured their factories (their manufacturing systems) to become lean producers, making goods of superior quality, cheaper, faster in a flexible way (i.e., they are more responsive to the customers). Almost every plant is doing something to make itself leaner. Many of them have adopted some version of the Toyota Production System. More importantly, these manufacturing factories are designed with the internal customer (the workforce) in mind, so things like ergonomics and safety are key design requirements. So while this book is all about materials and processes for making the products, the design of the factory cannot be ignored when it comes to making the external customer happy with the product and the internal customer satisfied with the employer.

- **New products and materials need new processes**

  The number and variety of products and the materials from which they are made continues to proliferate, while production quantities (lot sizes) have become smaller. Existing processes must be modified to be more flexible, and new processes must be developed.

- **Customers expect great quality**

  Consumers want better quality and reliability, so the methods, processes, and people responsible for the quality must be continually improved. The trend toward (improving) zero defects and continuous improvement requires continual changes to the manufacturing system.

- **Rapid product development is required**

Finally, the effort to reduce the *time-to-market* for new products is continuing. Many companies are taking wholistic or system wide perspectives, including concurrent engineering efforts to bring product design and manufacturing closer to the customer. There are two key aspects here. First, products are designed to be easier to manufacture and assemble (called *design for manufacture/assembly*). Second, the manufacturing system design is flexible (able to rapidly assimilate new products), so the company can be competitive in the global marketplace.

History of the Text

E. Paul DeGarmo was a mechanical engineering professor at the University of California, Berkley when he wrote the first edition of Materials and Processes in Manufacturing, published by Macmillan in 1957. The book quickly became the emulated standard for introductory texts in manufacturing. Second, third, and fourth editions followed in 1962, 1969, and 1974. DeGarmo had begun teaching at Berkeley in 1937, after earning his M.S. in mechanical engineering from California Institute of Technology. DeGarmo was a founder of the Department of Industrial Engineering (now Industrial Engineering and Operations Research) and served as its chair from 1956–1960. He was also assistant dean of the College of Engineering for three years while continuing his teaching responsibilities.

Dr. DeGarmo observed that engineering education had begun to place more emphasis on the underlying sciences at the expense of hands on experience. Most of his students were coming to college with little familiarity with materials, machine tools, and manufacturing methods that their predecessors had acquired through the old "shop" classes. If these engineers and technicians were to successfully convert their ideas into reality, they needed a foundation in materials and processes, with emphasis on their opportunities and their limitations. He sought to provide a text that could be used in either a one-or two-semester course designed to meet these objectives. The materials sections were written with an emphasis on use and application. Processes and machine tools were described in terms of what they could do, how they do it, and their relative advantages and limitations, including economic considerations. Recognizing that many students would be encountering the material for the first time, clear description was accompanied by numerous visual illustrations.

Paul's efforts were well received, and the book quickly became the standard text in many schools and curricula. As materials and processes evolved, advances were incorporated into subsequent editions. Computer usage, quality control, and automation were added to the text, along with other topics, so that it continued to provide state-of-the-art instruction in both materials and processes. As competing books entered the market, their subject material and organization tended to mimic the DeGarmo text.

Paul DeGarmo retired from active teaching in 1971, but he continued his research, writing, and consulting for many years. In 1977, after the publication of the fourth edition of *Materials and Processes in Manufacturing*, he received a letter from Ron Kohser, then an assistant professor at the University of Missouri-Rolla who had many suggestions regarding the materials chapters. DeGarmo asked Ron to rewrite those chapters for the upcoming fifth edition. After the 5th edition DeGarmo decided he was really going to retire and after a national search, recruited J T. Black, then a Professor at Ohio State, to co-author the book with Dr. Kohser.

For the sixth through tenth editions (published in 1984 and 1988 by Macmillan, 1997 by Prentice Hall and 2003 and 2008 by John Wiley & Sons), Ron Kohser and J T. Black have shared the responsibility for the text. The chapters on engineering materials, casting, forming, powder metallurgy, additive manufacturing, joining and non-destructive testing have been written or revised by Ron Kohser. J T. Black has responsibility for the introduction and chapters on material removal, metrology, surface finishing, quality control, manufacturing systems design, and lean engineering.

DeGarmo died in 2000, three weeks short of his 93rd birthday. His wife Mary died in 1995; he is survived by his sons, David and Richard, and many grandchildren. For the

mass production system changed to the *"Economy of Scope"*, featuring flexibility, small lots, superior quality, minimum-inventory and short throughput times.

Later chapters provide an introduction to surface engineering, measurements and quality control. Engineers need to know how to determine process capability and if they get involved in six sigma projects, to know what sigma really measures. There is also introductory material on surface integrity, since so many processes produce the finished surface and residual stresses in the components.

With each new edition, new and emerging technology is incorporated, and existing technologies are updated to accurately reflect current capabilities. Through its 50-plus year history and 10 previous editions, the DeGarmo text was often the first introductory book to incorporate processes such as friction-stir welding, microwave heating and sintering, and machining dynamics.

Somewhat open-ended case studies have been incorporated throughout the text. These have been designed to make students aware of the great importance of properly coordinating design, material selection, and manufacturing to produce cost competitive, reliable products.

The text is intended for use by engineering (mechanical, lean, manufacturing, and industrial) and engineering technology students, in both two-and four-year undergraduate degree programs. In addition, the book is also used by engineers and technologists in other disciplines concerned with design and manufacturing (such as aerospace and electronics). Factory personnel will find this book to be a valuable reference that concisely presents the various production alternatives and the advantages and limitations of each. Additional or more in-depth information on specific materials or processes can be found in the expanded list of references that accompanies the text.

Supplements

For instructors adopting the text for use in their course, an *instructor solutions manual* is available through the book website: www.wiley.com/go/global/degarmo. Also available on the website is a set of *PowerPoint lecture slides* created by Philip Appel.

Two additional chapters, as well as three Advanced Topic sections, are available on the book website. These chapters cover: measurement and inspection, nondestructive inspection and testing, lean engineering, quality engineering, and the enterprise (production system). The registration card attached on the inside front cover provides information on how to access and download this material. If the registration card is missing, access can be purchased directly on the website www.wiley.com/go/global/degarmo, by clicking on "student companion site" and then on the links to the chapter titles.

Acknowledgments

The authors wish to acknowledge the multitude of assistance, information, and illustrations that have been provided by a variety of industries, professional organizations, and trade associations. The text has become known for the large number of clear and helpful photos and illustrations that have been graciously provided by a variety of sources. In some cases, equipment is photographed or depicted without safety guards, so as to show important details, and personnel are not wearing certain items of safety apparel that would be worn during normal operation.

Over the many editions, there have been hundreds of reviewers, faculty, and students who have made suggestions and corrections to the text. We continue to be grateful for the time and interest that they have put into this book. For this edition we benefited from the comments of the following reviewers:

Jerald Brevick, The Ohio State University; Zezhong Chen, Concordia University; Emmanuel Enemuoh, University of Minnesota; Ronald Huston, University of Cincinnati; Thenkurussi Kesavadas, University at Buffalo, *The State University of New York*; Shuting Lei, Kansas State University; Lee Gearhart, University at Buffalo, *The State University of New York*; ZJ Pei, Kansas State University; Christine Corum,

10th edition, which coincided with the 50th anniversary of the text, we honored our mentor with a change in the title to include his name—*DeGarmo's Materials and Processes in Manufacturing*. We recognize Paul for his insight and leadership and are forever indebted to him for selecting us to carry on the tradition of his book for this, the 11th edition!

Purpose of the Book

The purpose of this book is to provide basic information on materials, manufacturing processes and systems to engineers and technicians. The materials section focuses on properties and behavior. Thus, aspects of smelting and refining (or other material production processes) are presented only as they affect manufacturing and manufactured products. In terms of the processes used to manufacture items (converting materials into products), this text seeks to provide a descriptive introduction to a wide variety of options, emphasizing how each process works and its relative advantages and limitations. Our goal is to present this material in a way that can be understood by individuals seeing it for the very first time. This is not a graduate text where the objective is to thoroughly understand and optimize manufacturing processes. Mathematical models and analytical equations are used only when they enhance the basic understanding of the material. So, while the text is an introductory text, we do attempt to incorporate new and emerging technologies like direct-digital-and micro-manufacturing processes as they are introduced into usage.

Organization of the Book

E. Paul DeGarmo wanted a book that explained to engineers how the things they designed are made. *DeGarmo's Materials and Processes in Manufacturing* is still being written to provide a broad, basic introduction to the fundamentals of manufacturing. The book begins with a survey of engineering materials, the "stuff" that manufacturing begins with, and seeks to provide the basic information that can be used to match the properties of a material to the service requirements of a component. A variety of engineering materials are presented, along with their properties and means of modifying them. The materials section can be used in curricula that lack preparatory courses in metallurgy, materials science, or strength of materials, or where the student has not yet been exposed to those topics. In addition, various chapters in this section can be used as supplements to a basic materials course, providing additional information on topics such as heat treatment, plastics, composites, and material selection.

Following the materials chapters are sections on casting, forming, powder metallurgy, material removal, and joining. Each section begins with a presentation of the fundamentals on which those processes are based. The introductions are followed by a discussion of the various process alternatives, which can be selected to operate individually or be combined into an integrated system.

The chapter on rapid prototyping, which had been moved to a web-based supplement in the 10th edition, has been restored to the print text, significantly expanded, and renamed Additive Processes: Rapid Prototyping and Direct-Digital Manufacturing, to incorporate the aspects of rapid prototyping, rapid tooling, and direct-digital manufacturing, and provide updated information on many recent advances in this area.

Reflecting the growing role of plastics, ceramics and composites, the chapter on the processes used with these materials has also been expanded.

New to this edition is an Advanced Topic section on lean engineering. The lean engineer works to transform the mass production system into a lean production system. To achieve lean production, the final assembly line is converted to a mixed model delivery system so that the demand for subassemblies and components is made constant. The conveyor type flow lines are dismantled and converted into U-shaped manufacturing cells also capable of one-piece flow. The subassembly and manufacturing cells are linked to the final assembly by a pull system called Kanban (visible record) to form an integrated production and inventory control system. Hence, economy of scale of the

Purdue University; Allen Yi, The Ohio State University; Stephen Oneyear, North Carolina State; Roger Wright, Rennselaer Polytechnic Institute.

The authors would also like to acknowledge the contributions of Dr. Elliot Stern for the dynamics of machining section in Chapter 20, Dr. Memberu Lulu for inputs to the quality chapter, Dr. Lewis Payton for writing the micro manufacturing chapter, Dr. Subbu Subramanium for inputs to the abrasive chapter, Dr. David Cochran for his contributions in lean engineering and system design, and Mr. Chris Huskamp of the Boeing Company for valuable assistance with the chapter on additive manufacturing.

As always, our wives have played a major role in preparing the manuscript. Carol Black and Barb Kohser have endured being "textbook widows" during the time when the book was being were written. Not only did they provide loving support, but Carol also provided hours of expert proofreading, typing, and editing as the manuscript was prepared.

About the Authors

J T. Black received his Ph.D. from Mechanical and Industrial Engineering, University of Illinois, Urbana in 1969, an M.S. in Industrial Engineering from West Virginia University in 1963 and his B.S. in Industrial Engineering, Lehigh University in 1960. J T. is Professor Emeritus from Industrial and Systems Engineering at Auburn University. He was the Chairman and a Professor of Industrial and Systems Engineering at The University of Alabama-Huntsville. He also taught at The Ohio State University, the University of Rhode Island, the University of Vermont, and the University of Illinois. He taught his first processes class in 1960 at West Virginia University. J T. is a Fellow in the American Society of Mechanical Engineers, the Institute of Industrial Engineering and the Society of Manufacturing Engineers. J loves to write music (mostly down home country) and poetry, play tennis in the backyard and show his champion pug dog VBo.

Ron Kohser received his Ph.D. from Lehigh University Institute for Metal Forming in 1975. Ron is currently in his 37th year on the faculty of Missouri University of Science & Technology (formerly the University of Missouri-Rolla), where he is a Professor of Metallurgical Engineering and Dean's Teaching Scholar. While maintaining a full commitment to classroom instruction, he has served as department chair and Associate Dean for Undergraduate Instruction. He currently teaches courses in Metallurgy for Engineers, Introduction to Manufacturing Processes, and Material Selection, Fabrication and Failure Analysis. In addition to his academic responsibilities, Ron and his wife Barb operate *A Miner Indulgence,* a bed-and-breakfast in Rolla, Missouri, and they enjoy showing their three collector cars.

CONTENTS

CHAPTER 1

INTRODUCTION TO DEGARMO'S MATERIALS AND PROCESSES IN MANUFACTURING

1.1 MATERIALS, MANUFACTURING, AND THE STANDARD OF LIVING

Manufacturing is critical to a country's economic welfare and standard of living because the standard of living in any society is determined, primarily, by the *goods* and *services* that are available to its people. Manufacturing companies contribute about 20% of the GNP, employ about 18% of the workforce, and account for 40% of the exports of the United States. In most cases, materials are utilized in the form of manufactured goods. **Manufacturing** and **assembly** represent the organized activities that convert raw materials into salable goods. The manufactured goods are typically divided into two classes: producer goods and consumer goods. **Producer goods** are those goods manufactured for other companies to use to manufacture either producer or consumer goods. **Consumer goods** are those purchased directly by the consumer or the general public. For example, someone has to build the machine tool (a lathe) that produces (using machining processes) the large rolls that are sold to the rolling mill factory to be used to roll the sheets of steel that are then formed (using dies) into body panels of your car. Similarly, many service industries depend heavily on the use of manufactured products, just as the agricultural industry is heavily dependent on the use of large farming machines for efficient production.

Processes convert materials from one form to another adding value to them. The more efficiently materials can be produced and converted into the desired products that function with the prescribed quality, the greater will be the companies' productivity and the better will be the standard of living of the employees.

The history of man has been linked to his ability to work with tools and materials, beginning with the Stone Age and ranging through the eras of copper and bronze, the Iron Age, and recently the age of steel. While ferrous materials still dominate the manufacturing world, we are entering the age of tailor-made plastics, composite materials, and exotic alloys.

A good example of this progression is shown in Figure 1-1. The goal of the manufacturer of any product or service is to continually improve. For a given product or service, this improvement process usually follows an S-shaped curve, as shown in Figure 1-1a, often called a product life-cycle curve. After the initial invention/creation and development, a period of rapid growth in performance occurs, with relatively few resources required. However, each improvement becomes progressively more difficult. For a delta

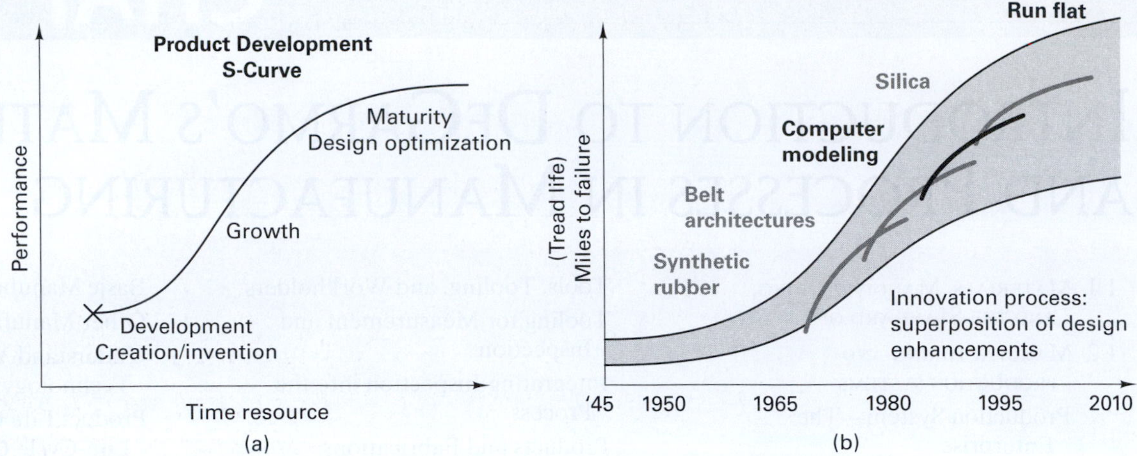

FIGURE 1-1 (a) A product development curve usually has an "S"-shape. (b) Example of the S-curve for the radial tire. *(Courtesy of Bart Thomas, Michelin)*

gain, more money and time and ingenuity are required. Finally, the product or service enters the maturity phase, during which additional performance gains become very costly.

For example, in the automobile tire industry, Figure 1-1b shows the evolution of radial tire performance from its birth in 1946 to the present. Growth in performance is actually the superposition of many different improvements in material, processes, and design.

These innovations, known as **sustaining technology,** serve to continually bring more value to the consumer of existing products and services. In general, sustaining manufacturing technology is the backbone of American industry and the ever-increasing productivity metric.

Although materials are no longer used only in their natural state, there is obviously an absolute limit to the amounts of many materials available here on earth. Therefore, as the variety of man-made materials continues to increase, resources must be used efficiently and recycled whenever possible. Of course, recycling only postpones the exhaustion date.

Like materials, processes have also proliferated greatly in the past 50 years, with new processes being developed to handle the new materials more efficiently and with less waste. A good example is the laser, invented around 1960, which now finds many uses in machining, measurement, inspection, heat treating, welding, and more. New developments in manufacturing technology often account for improvements in productivity. Even when the technology is proprietary, the competition often gains access to it, usually quite quickly.

Starting with the product design, materials, labor, and equipment are interactive factors in manufacturing that must be combined properly (integrated) to achieve low cost, superior quality, and on-time delivery. Figure 1-2 shows a breakdown of costs for a

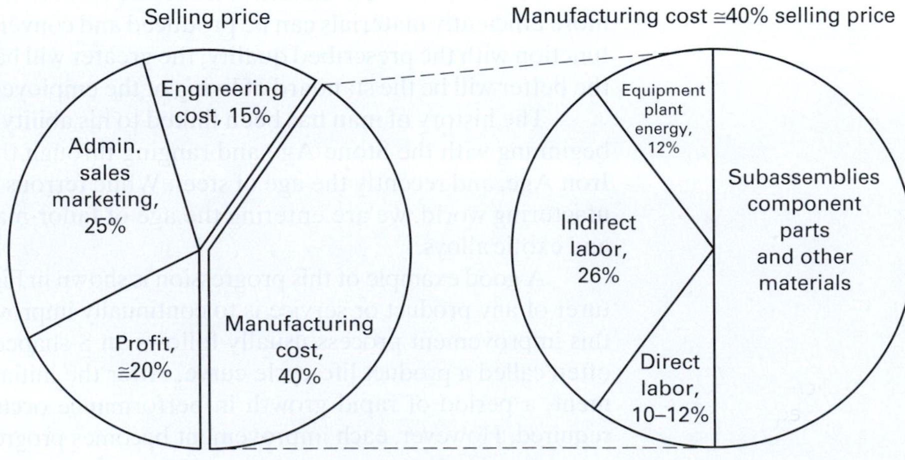

FIGURE 1-2 Manufacturing cost is the largest part of the selling price, usually around 40%. The largest part of the manufacturing cost is materials, usually 50%.

product (like a car). Typically about 40% of the selling price of a product is the **manufacturing cost.** Because the selling price determines how much the customer is willing to pay, maintaining the profit often depends on reducing manufacturing cost. The internal customers who really make the product are called direct labor. They are usually the targets of automation, but typically they account for only about 10% of the manufacturing cost, even though they are the main element in increasing productivity. In Chapter 2, a manufacturing strategy is presented that attacks the materials cost, indirect costs, and general administration costs, in addition to labor costs. The materials costs include the cost of storing and handling the materials within the plant. The strategy depends on a new factory design and is called **lean production.**

Referring again to the total expenses shown in Figure 1-2 (selling price less profit), about 68% of dollars are spent on people, but only 5 to 10% on director labor, the breakdown for the rest being about 15% for engineers and 25% for marketing, sales, and general management people. The average labor cost in manufacturing in the United States is around $15 per hour for hourly workers (2010). Reductions in direct labor will have only marginal effects on the total people costs. The optimal combination of factors for producing a small quantity of a given product may be very inefficient for a larger quantity of the same product. Consequently, a systems approach, taking all the factors into account, must be used. This requires a sound and broad understanding on the part of the decision makers on the value of materials, processes, and equipment to the company, and their customers, accompanied by an understanding of the manufacturing systems. Materials, processes, and manufacturing systems are what this book is all about.

1.2 MANUFACTURING AND PRODUCTION SYSTEMS

Manufacturing is the economic term for making goods and services available to satisfy human wants. Manufacturing implies creating value by applying useful mental or physical labor. The *manufacturing processes* are collected together to form a *manufacturing system* (MS). The manufacturing system is a complex arrangement of physical elements characterized by measurable parameters (Figure 1-3). The manufacturing system takes inputs and produces products for the external customer.

The entire company is often referred to as the enterprise or the production system. The production services the manufacturing system, as shown in Figure 1-4. In this book, a production system will refer to the total company and will include within it the manufacturing system. The production system includes the manufacturing system plus all the other functional areas of the plant for information, design, analysis, and control. These subsystems are connected by various means to each other to produce either goods or services or both.

Goods refers to material things. **Services** are nonmaterial things that we buy to satisfy our wants, needs, or desires. Service production systems include transportation, banking, finance, savings and loan, insurance, utilities, health care, education, communication, entertainment, sporting events, and so forth. They are useful labors that do not directly produce a product. Manufacturing has the responsibility for designing

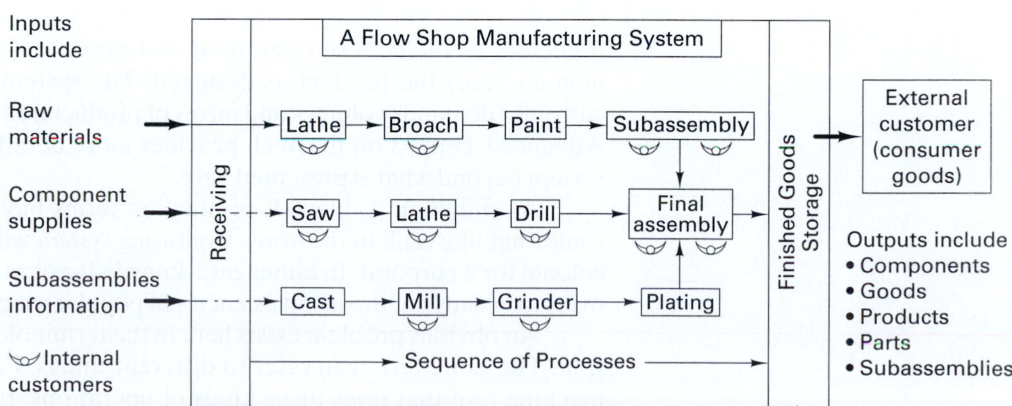

FIGURE 1-3 The manufacturing system design (aka the factory design) is composed of machines, tooling, material handling equipment, and people.

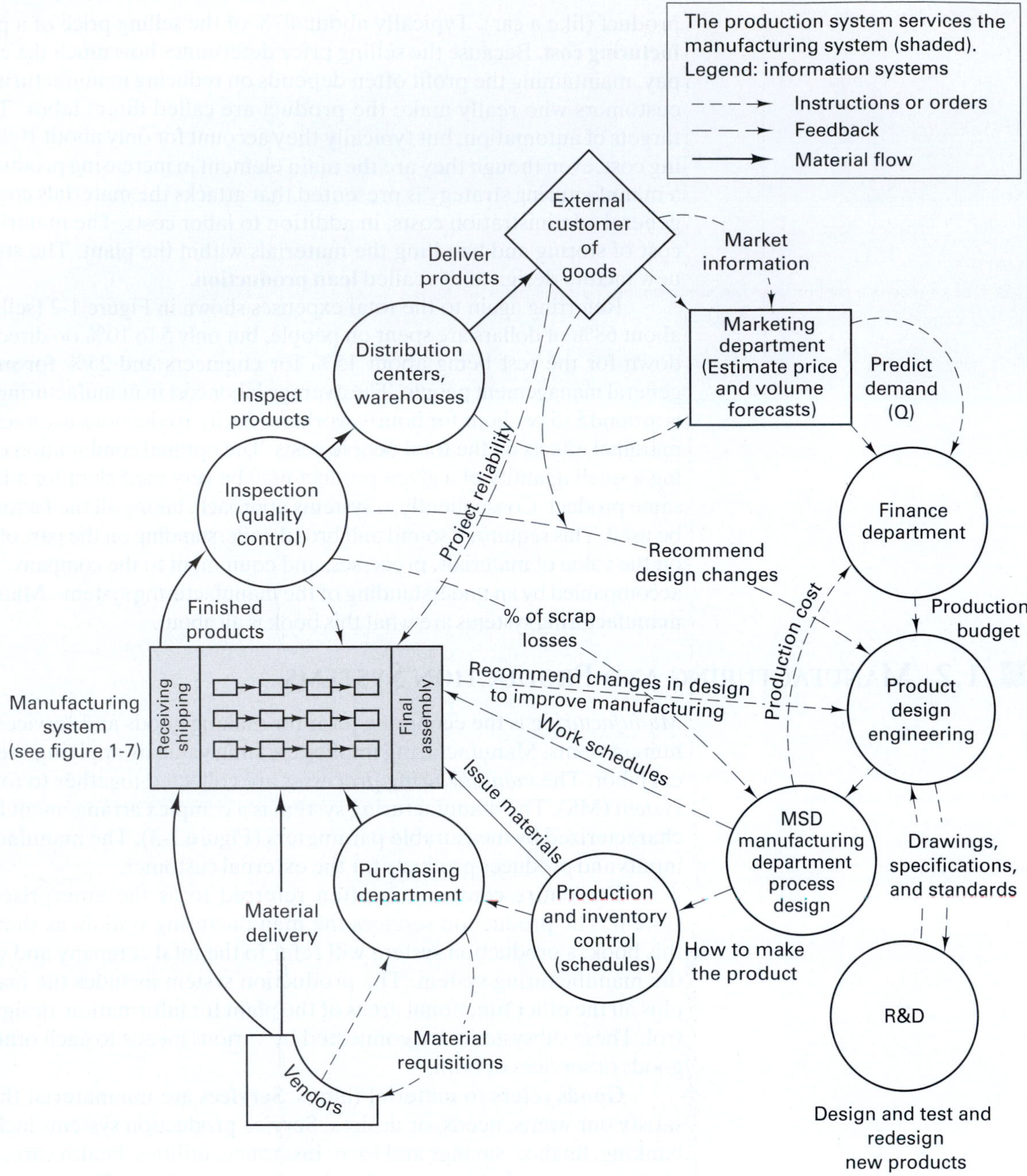

FIGURE 1-4 The production system includes and services the manufacturing system. The functional departments are connected by formal and informal information systems, designed to service the manufacturing that produces the goods.

processes (sequences of operations and processes) and systems to create (make or manufacture) the product as designed. The system must exhibit flexibility to meet customer demand (volumes and mixes of products) as well as changes in product design. Advanced Topic 3 on the Web provides more detailed discussions of the production system beyond what is presented here.

As shown in Table 1-1, production terms have a definite rank of importance, somewhat like rank in the army. Confusing *system* with *section* is similar to mistaking a colonel for a corporal. In either case, knowledge of rank is necessary. The terms tend to overlap because of the inconsistencies of popular usage.

An obvious problem exists here in the terminology of manufacturing and production. The same term can refer to different things. For example, *drill* can refer to the machine tool that does these kinds of operations; the operation itself, which can be

TABLE 1-1 Production Terms for Manufacturing Production Systems

Term	Meaning	Examples
Production system; the enterprise	All aspects of workers, machines, and information, considered collectively, needed to manufacture parts or products; integration of all units of the system is critical.	Company that makes engines, assembly plant, glassmaking factory, foundry; sometimes called the enterprise or the business.
Manufacturing system (sequence of operations, collection of processes) or factory	The collection of manufacturing processes and operations resulting in specific end products; an arrangement or layout of many processes, materials-handling equipment, and operators.	Rolling steel plates, manufacturing of automobiles, series of connected operations or processes, a job shop, a flow shop, a continuous process.
Machine or machine tool or manufacturing process	A specific piece of equipment designed to accomplish specific processes, often called a *machine tool;* machine tools linked together to make a manufacturing system.	Spot welding, milling machine, lathe, drill press, forge, drop hammer, die caster, punch press, grinder, etc.
Job (sometimes called a *station*; a collection of tasks)	A collection of operations done on machines or a collection of tasks performed by one worker at one location on the assembly line.	Operation of machines, inspection, final assembly; e.g., forklift driver has the job of moving materials.
Operation (sometimes called a *process*)	A specific action or treatment, often done on a machine, the collection of which makes up the job of a worker.	Drill, ream, bend, solder, turn, face, mill extrude, inspect, load.
Tools or tooling	Refers to the implements used to hold, cut, shape, or deform the work materials; called *cutting tools* if referring to machining; can refer to *jigs* and *fixtures* in workholding and *punches* and *dies* in metal forming.	Grinding wheel, drill bit, end milling cutter, die, mold, clamp, three-jaw chuck, fixture.

done on many different kinds of machines; or the cutting tool, which exists in many different forms. It is therefore important to use modifiers whenever possible: "Use the *radial* drill *press* to drill a hole with a 1-in.-diameter spade drill." The emphasis of this book will be directed toward the understanding of the processes, machines, and tools required for manufacturing and how they interact with the materials being processed. In the last section of the book, an introduction to systems aspects is presented.

PRODUCTION SYSTEM—THE ENTERPRISE

The highest-ranking term in the hierarchy is **production system.** A production system includes people, money, equipment, materials and supplies, markets, management, and the manufacturing system. In fact, all aspects of commerce (manufacturing, sales, advertising, profit, and distribution) are involved. Table 1-2 provides a partial list of

TABLE 1-2 Partial List of Production Systems for Producer and Consumer Goods

Aerospace and airplanes	Foods (canned, dairy, meats, etc.)
Appliances	Footwear
Automotive (cars, trucks, vans, wagons, etc.)	Furniture
Beverages	Glass
Building supplies (hardware)	Hospital suppliers
Cement and asphalt	Leather and fur goods
Ceramics	Machines
Chemicals and allied industries	Marine engineering
Clothing (garments)	Metals (steel, aluminum, etc.)
Construction	Natural resources (oil, coal, forest, pulp and paper)
Construction materials (brick, block, panels)	Publishing and printing (books, CDs, newspapers)
Drugs, soaps, cosmetics	Restaurants
Electrical and microelectronics	Retail (food, department stores, etc.)
Energy (power, gas, electric)	Ship building
Engineering	Textiles
Equipment and machinery (agricultural, construction and electrical products, electronics, household products, industrial machine tools, office equipment, computers, power generators)	Tire and rubber
	Tobacco
	Transportation vehicles (railroad, airline, truck, bus)
	Vehicles (bikes, cycles, ATVs, snowmobiles)

TABLE 1-3 Types of Service Industries

Advertising and marketing
Communication (telephone, computer networks)
Education
Entertainment (radio, TV, movies, plays)
Equipment and furniture rental
Financial (banks, investment companies, loan companies)
Health care
Insurance
Transportation and car rental
Travel (hotel, motel, cruise lines)

production systems. Another term for them is "industries" as in the "aerospace industry." Further discussion on the enterprise is found in Advanced Topic 3, on the Web.

Much of the information given for manufacturing systems is relevant to the service system. Most require a service production system [SPS] for proper product sales. This is particularly true in industries, such as the food (restaurant) industry, in which customer service is as important as quality and on-time delivery. Table 1-3 provides a short list of service industries.

MANUFACTURING SYSTEMS

A collection of operations and processes used to obtain a desired product(s) or component(s) is called a **manufacturing system.** The manufacturing system is therefore the design or arrangement of the manufacturing processes in the factory. Control of a system applies to overall control of the whole, not merely of the individual processes or equipment. The entire manufacturing system must be controlled in order to schedule and control the factory—all its inputs, inventory levels, product quality, output rates, and so forth. Designs or layouts of factories are discussed in Chapter 2.

MANUFACTURING PROCESSES

A **manufacturing process** converts unfinished materials to finished products, often using machines or machine tools. For example, injection molding, die casting, progressive stamping, milling, arc welding, painting, assembling, testing, pasteurizing, homogenizing, and annealing are commonly called processes or manufacturing processes. The term *process* can also refer to a sequence of steps, processes, or operations for production of goods and services, as shown in Figure 1-5, which shows the processes to manufacture an Olympic-type medal.

A **machine tool** is an assembly of related mechanisms on a frame or bed that together produce a desired result. Generally, motors, controls, and auxiliary devices are included. Cutting tools and workholding devices are considered separately.

A machine tool may do a single process (e.g., cutoff saw) or multiple processes, or it may manufacture an entire component. Machine sizes vary from a tabletop drill press to a 1000-ton forging press.

JOB AND STATION

In the classical manufacturing system, a **job** is the total of the work or duties a worker performs. A **station** is a location or area where a production worker performs tasks or his job.

A job is a group of related operations and tasks performed at one station or series of stations in cells. For example, the job at a final assembly station may consist of four tasks:

1. Attach carburetor.
2. Connect gas line.
3. Connect vacuum line.
4. Connect accelerator rod.

The job of a turret lathe (a semiautomatic machine) operator may include the following operations and tasks: load, start, index and stop, unload, inspect. The operator's job may also include setting up the machine (i.e., getting ready for manufacturing). Other machine operations include drilling, reaming, facing, turning, chamfering, and knurling. The operator can run more than one machine or service at more than one station.

The terms *job* and *station* have been carried over to unmanned machines. A job is a group of related operations generally performed at one station, and a station is a position or location in a machine (or process) where specific operations are performed. A simple machine may have only one station. Complex machines can be composed of

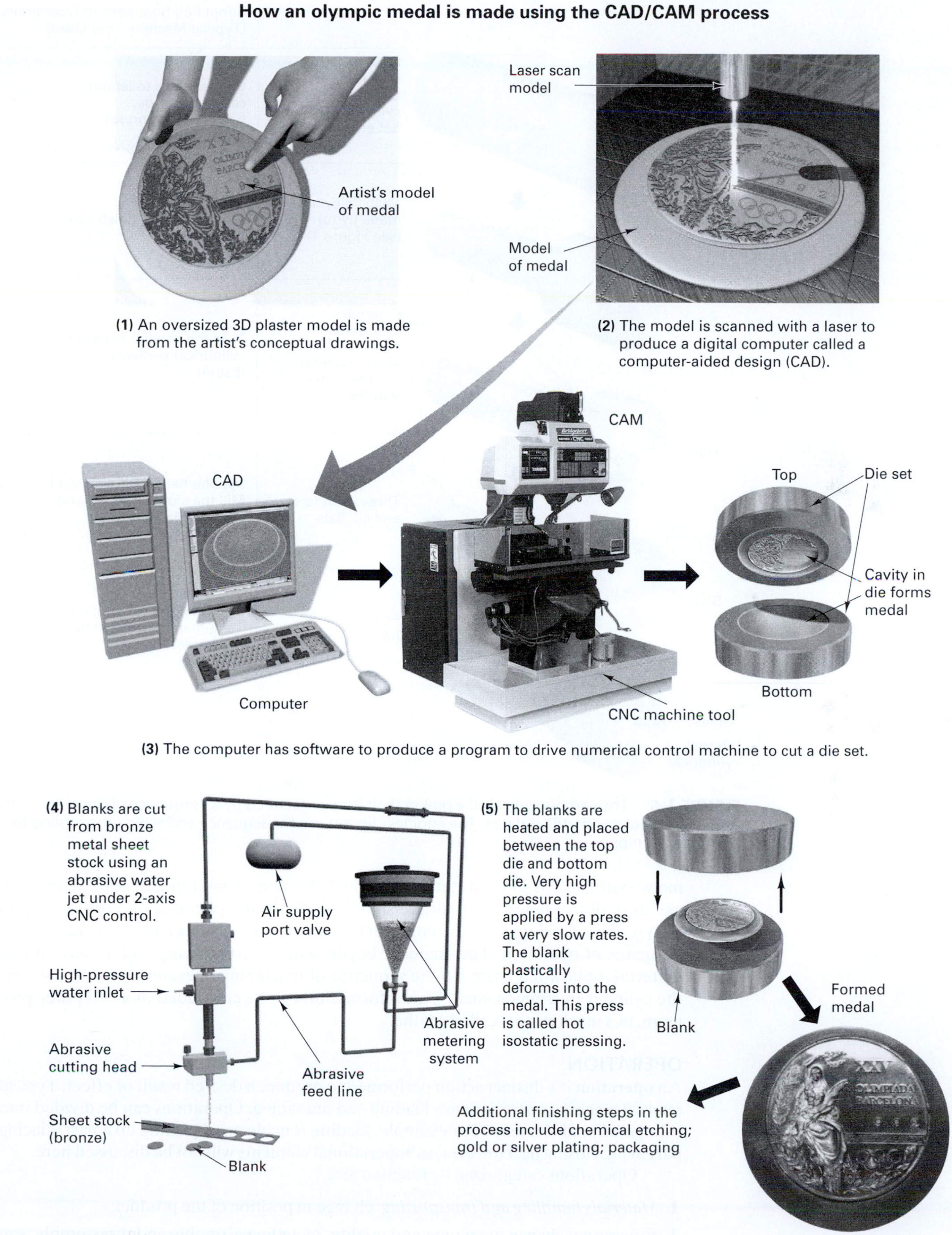

How an olympic medal is made using the CAD/CAM process
Artist's model of medal
Laser scan model
Model of medal
(1) An oversized 3D plaster model is made from the artist's conceptual drawings.
(2) The model is scanned with a laser to produce a digital computer called a computer-aided design (CAD).
CAM
CAD
Top
Die set
Cavity in die forms medal
Bottom
Computer
CNC machine tool
(3) The computer has software to produce a program to drive numerical control machine to cut a die set.
(4) Blanks are cut from bronze metal sheet stock using an abrasive water jet under 2-axis CNC control.
Air supply port valve
High-pressure water inlet
Abrasive metering system
Abrasive cutting head
Abrasive feed line
Sheet stock (bronze)
Blank
(5) The blanks are heated and placed between the top die and bottom die. Very high pressure is applied by a press at very slow rates. The blank plastically deforms into the medal. This press is called hot isostatic pressing.
Blank
Formed medal
Additional finishing steps in the process include chemical etching; gold or silver plating; packaging

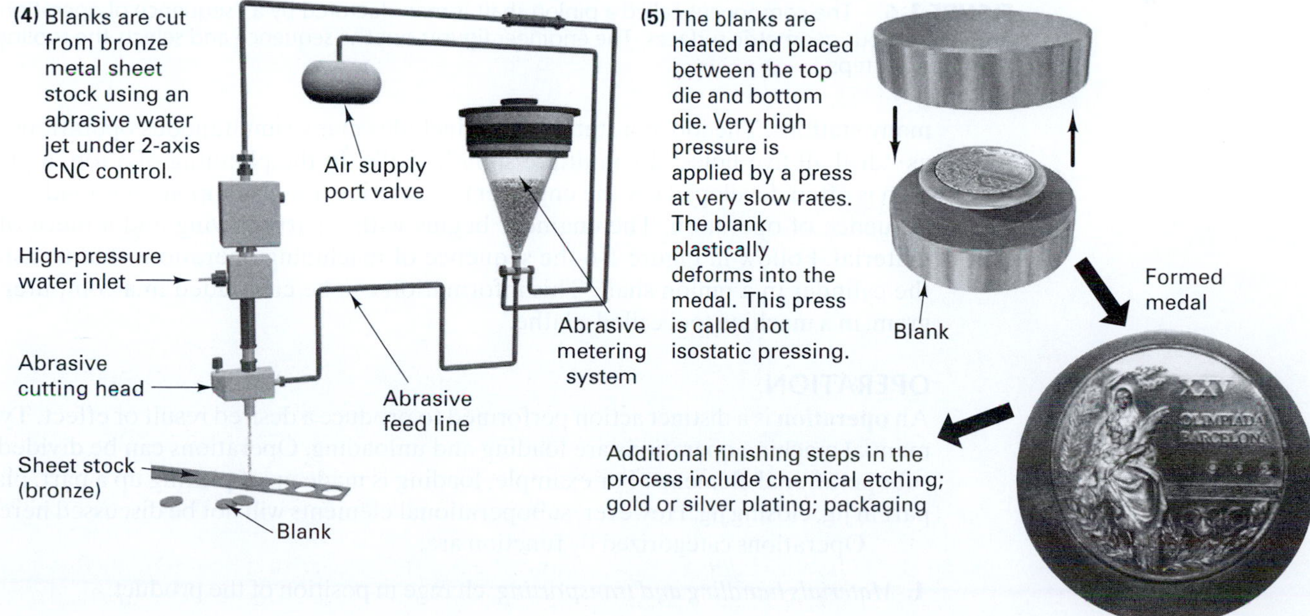

FIGURE 1-5 The manufacturing process for making Olympic medals has many steps or operations, beginning with design and including die making. *(Courtesy J T. Black)*

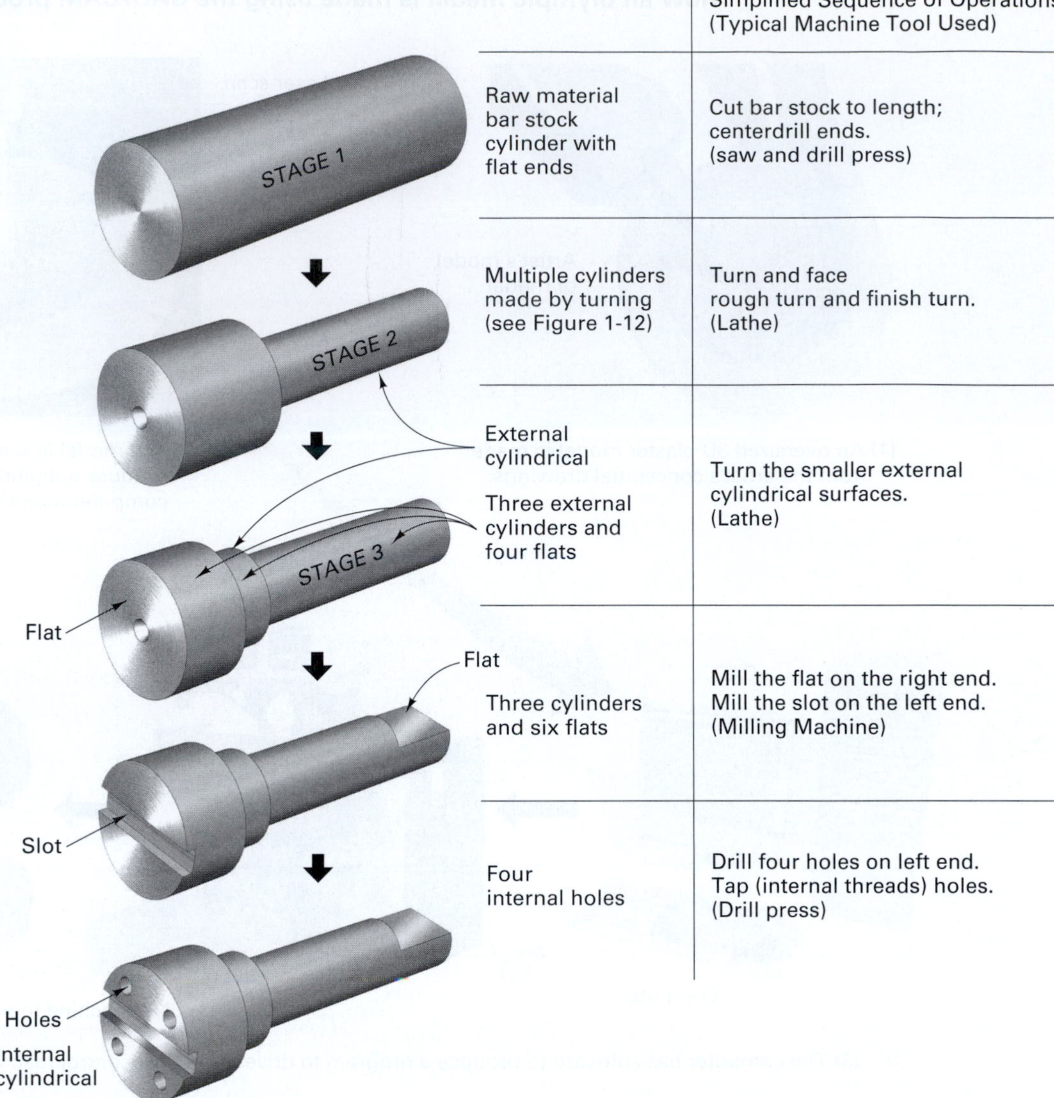

Simplified Sequence of Operations
(Typical Machine Tool Used)
Raw material
bar stock
cylinder with
flat ends
STAGE 1
Cut bar stock to length;
centerdrill ends.
(saw and drill press)
Multiple cylinders
made by turning
(see Figure 1-12)
STAGE 2
Turn and face
rough turn and finish turn.
(Lathe)
External
cylindrical
Three external
cylinders and
four flats
STAGE 3
Turn the smaller external
cylindrical surfaces.
(Lathe)
Flat
Flat
Three cylinders
and six flats
Mill the flat on the right end.
Mill the slot on the left end.
(Milling Machine)
Slot
Four
internal holes
Drill four holes on left end.
Tap (internal threads) holes.
(Drill press)
Holes
Internal
cylindrical

FIGURE 1-6 The component called a pinion shaft is manufactured by a "sequence of operations" to produce various geometric surfaces. The engineer figures out the sequence and selects the tooling to perform the steps.

many stations. The job at a station often includes many simultaneous operations, such as "drill all five holes" by multiple spindle drills. In the planning of a job, a process plan is often developed (by the engineer) to describe how a component is made using a sequence of operation. The engineer begins with a part drawing and a piece of raw material. Follow in Figure 1-6 the sequence of machining operations that transforms the cylinder in a pinion shaft. This information can be embedded in a computer program, in a machine tool called a lathe.

OPERATION

An **operation** is a distinct action performed to produce a desired result or effect. Typical manual machine operations are loading and unloading. Operations can be divided into suboperational elements. For example, loading is made up of picking up a part, placing part in jig, closing jig. However, suboperational elements will not be discussed here.

Operations categorized by function are:

1. *Materials handling and transporting:* change in position of the product.
2. *Processing:* change in volume and quality, including assembly and disassembly; can include packaging.

3. *Packaging:* special processing; may be temporary or permanent for shipping.
4. *Inspecting and testing:* comparison to .the standard or check of process behavior
5. *Storing:* time lapses without further operations.

These basic operations may occur more than once in some processes, or they may sometimes be omitted. *Remember, it is the manufacturing processes that change the value and quality of the materials.* Defective processes produce poor quality or scrap. Other operations may be necessary but do not, in general, add value, whereas operations performed by machines that do material processing usually do add value.

TREATMENTS

Treatments operate continuously on the workpiece. They usually alter or modify the product-in-process without tool contact. Heat treating, curing, galvanizing, plating, finishing, (chemical) cleaning, and painting are examples of treatments. Treatments usually add value to the part.

These processes are difficult to include in manufacturing cells because they often have long cycle times, are hazardous to the workers' health, or are unpleasant to be around because of high heat or chemicals. They are often done in large tanks or furnaces or rooms. The cycle time for these processes may dictate the cycle times for the entire system. These operations also tend to be material specific. Many manufactured products are given decorative and protective surface treatments that control the finished appearance. A customer may not buy a new vehicle because it has a visible defect in the chrome bumper, although this defect will not alter the operation of the car.

TOOLS, TOOLING, AND WORKHOLDERS

The lowest mechanism in the production term rank is the **tool.** Tools are used to hold, cut, shape, or form the unfinished product. Common hand tools include the saw, hammer, screwdriver, chisel, punch, sandpaper, drill, clamp, file, torch, and grindstone.

Basically, machines are mechanized versions of such hand tools and are called cutting tools. Some examples of tools for cutting are drill bits, reamers, single-point turning tools, milling cutters, saw blades, broaches, and grinding wheels. Noncutting tools for forming include extrusion dies, punches, and molds.

Tools also include workholders, jigs, and fixtures. These tools and cutting tools are generally referred to as the **tooling,** which usually must be considered (purchased) separate from machine tools. Cutting tools wear and fail and must be periodically replaced before parts are ruined. The workholding devices must be able to locate and secure the workpieces during processing in a repeatable, mistake-proof way.

TOOLING FOR MEASUREMENT AND INSPECTION

Measuring tools and instruments are also important for manufacturing. Common examples of measuring tools are rulers, calipers, micrometers, and gages. Precision devices that use laser optics or vision systems coupled with sophisticated electronics are becoming commonplace. Vision systems and coordinate measuring machines are becoming critical elements for achieving superior quality.

INTEGRATING INSPECTION INTO THE PROCESS

The integration of the **inspection** process into the manufacturing process or the manufacturing system is a critical step toward building products of superior quality. An example will help. Compare an electric typewriter with a computer that does word processing. The electric typewriter is flexible. It types whatever words are wanted in whatever order. It can type in Pica, Elite, or Orator, but the font (disk or ball that has the appropriate type size on it) has to be changed according to the size and face of type wanted. The computer can do all of this but can also, through its software, set italics; set bold, dark type; vary the spacing to justify the right margin; plus many other functions. It checks immediately for incorrect spelling and other defects like repeated words. The software system provides a signal to the hardware to flash the word so that the operator will know something is wrong and can make an immediate correction. If the system were designed to prevent the typist from typing repeated words, then this would be a *poka-yoke,* a term meaning defect prevention. Defect prevention is better than immediate defect detection and correction. Ultimately, the system should be able to forecast the probability of a defect, correcting the problem at the source. This means that the typist would have to be removed from the process loop, perhaps by having the system type out what it is told (convert oral to written directly). Poka-yoke devices and source inspection techniques are keys to designing manufacturing systems that produce superior-quality products at low cost.

PRODUCTS AND FABRICATIONS

In manufacturing, material things (goods) are made to satisfy human wants. **Products** result from manufacture. Manufacture also includes conversion processes such as refining, smelting, and mining.

Products can be manufactured by fabricating or by processing. **Fabricating** is the manufacture of a product from pieces such as parts, components, or assemblies. Individual products or parts can also be fabricated. Separable discrete items such as tires, nails, spoons, screws, refrigerators, or hinges are fabricated.

Processing is also used to refer to the manufacture of a product by continuous means, or by a continuous series of operations, for a specific purpose. Continuous items such as steel strip, beverages, breakfast foods, tubing, chemicals, and petroleum are "processed." Many processed products are marketed as discrete items, such as bottles of beer, bolts of cloth, spools of wire, and sacks of flour.

Separable discrete products, both piece parts and assemblies, are fabricated in a plant, factory, or mill, for instance, a textile or rolling mill. Products that flow (liquids, gases, grains, or powders) are processed in a *plant* or *refinery*. The *continuous-process industries* such as petroleum and chemical plants are sometimes called processing industries or flow industries.

To a lesser extent, the terms *fabricating industries* and *manufacturing industries* are used when referring to fabricators or manufacturers of large products composed of many parts, such as a car, a plane, or a tractor. Manufacturing often includes continuous-process treatments such as electroplating, heating, demagnetizing, and extrusion forming.

Construction or building is making goods by means other than manufacturing or processing in factories. Construction is a form of project manufacturing of useful goods like houses, highways, and buildings. The public may not consider construction as manufacturing because the work is not usually done in a plant or factory, but it can be. There is a company in Delaware that can build a custom house of any design in its factory, truck it to the building site, and assemble it on a foundation in two or three weeks.

Agriculture, fisheries, and commercial fishing produce real goods from useful labor. Lumbering is similar to both agriculture and mining in some respects, and mining should be considered processing. Processes that convert the raw materials from agriculture, fishing, lumbering, and mining into other usable and consumable products are also forms of manufacturing.

WORKPIECE AND ITS CONFIGURATION

In the manufacturing of goods, the primary objective is to produce a component having a desired geometry, size, and finish. Every component has a shape that is bounded by various types of surfaces of certain sizes that are spaced and arranged relative to each other. Consequently, a component is manufactured by producing the surfaces that bound the shape. Surfaces may be:

1. Plane or flat.
2. Cylindrical (external or internal).
3. Conical (external or internal).
4. Irregular (curved or warped).

Figure 1-6 illustrates how a shape can be analyzed and broken up into these basic bounding surfaces. Parts are manufactured by using a set or sequence of processes that will either (1) remove portions of a rough block of material (bar stock, casting, forging) so as to produce and leave the desired bounding surface or (2) cause material to form into a stable configuration that has the required bounding surfaces (casting, forging). Consequently, in designing an object, the designer specifies the shape, size, and arrangement of the bounding surface. The part design must be analyzed to determine what materials will provide the desired properties, including mating to other components, and what processes can best be employed to obtain the end product at the most reasonable cost. This is often the job of the manufacturing engineer.

ROLES OF ENGINEERS IN MANUFACTURING

Many engineers have as their function the designing of products. The products are brought into reality through the processing or fabrication of materials. In this capacity designers are a key factor in the material selection and manufacturing procedure. A **design engineer,** better than any other person, should know what the design is to accomplish, what assumptions can be made about service loads and requirements, what service environment the product must withstand, and what appearance the final product is to have. To meet these requirements, the material(s) to be used must be selected and specified. In most cases, to utilize the material and to enable the product to have the desired form, the designer knows that certain manufacturing processes will have to be employed. In many instances, the selection of a specific material may dictate what processing must be used. On the other hand, when certain processes must be used, the design may have to be modified in order for the process to be utilized effectively and economically. Certain dimensional sizes can dictate the processing, and some processes require certain sizes for the parts going into them. In converting the design into reality, many decisions must be made. In most instances, they can be made most effectively at the design stage. It is thus apparent that design engineers are a vital factor in the manufacturing process, and it is indeed a blessing to the company if they can *design for manufacturing,* that is, design the product so that it can be manufactured and/or assembled economically (i.e., at low unit cost). Design for manufacturing uses the knowledge of manufacturing processes, and so the design and manufacturing engineers should work together to integrate design and manufacturing activities.

Manufacturing engineers select and coordinate specific processes and equipment to be used or supervise and manage their use. Some design special tooling is used so that standard machines can be utilized in producing specific products. These engineers must have a broad knowledge of manufacturing processes and material behavior so that desired operations can be done effectively and efficiently without overloading or damaging machines and without adversely affecting the materials being processed. Although it is not obvious, the most hostile environment the material may ever encounter in its lifetime is the processing environment.

Industrial and lean engineers are responsible for manufacturing systems design (or layout) of factories. They must take into account the interrelationships of the factory design and the properties of the materials that the machines are going to process as well as the interreaction of the materials and processes. The choice of machines and equipment used in manufacturing and their arrangement in the factory are key design tasks.

The **lean engineer** has expertise in cell design, setup reduction (tool design), integrated quality control devices (poka-yokes and decouplers) and reliability (maintenance of machines and people) for the lean production system. See Advanced Topic 1 online for discussion of cell design and lean engineering.

Materials engineers devote their major efforts to developing new and better materials. They, too, must be concerned with how these materials can be processed and with the effects that the processing will have on the properties of the materials. Although their roles may be quite different, it is apparent that a large proportion of engineers must concern themselves with the interrelationships of materials and manufacturing processes.

As an example of the close interrelationship of design, materials selection, and the selection and use of manufacturing processes, consider the common desk stapler. Suppose that this item is sold at the retail store for $20. The wholesale outlet sold the stapler for $16 and the manufacturer probably received about $10 for it. Staplers typically consist of 10 to 12 parts and some rivets and pins. Thus, the manufacturer had to produce and assemble the 10 parts for about $1 per part. Only by giving a great deal of attention to design, selection of materials, selection of processes, selection of equipment used for manufacturing (tooling), and utilization of personnel could such a result be achieved.

The stapler is a relatively simple product, yet the problems involved in its manufacture are typical of those that manufacturing industries must deal with. The elements of design, materials, and processes are all closely related, each having its effect on the performance of the device and the other elements. For example, suppose the designer calls for the component that holds the staples to be a metal part. Will it be a machined part rather than a formed part? Entirely different processes and materials need to be specified depending on the choice. Or, if a part is to be changed from metal to plastic, then a whole new set of fundamentally different materials and processes would need to come into play. Such changes would also have a significant impact on cost as well as the service (useful life) of the product.

CHANGING WORLD COMPETITION

In recent years, major changes in the world of goods manufacturing have taken place. Three of these are:

1. Worldwide competition for global products and their manufacture.
2. High-tech manufacturing or advanced technology.
3. New manufacturing systems designs, strategies, and management.

Worldwide (global) competition is a fact of manufacturing life, and it will get stronger in the future. The goods you buy today may have been made anywhere in the world. For many U.S. companies, suppliers in China, India, and Mexico are not uncommon.

The second aspect, advanced manufacturing technology, usually refers to new machine tools or processes controlled by computers. Companies that produce such machine tools, though small, can have an enormous impact on factory productivity. Improved processes lead to better components and more durable goods. However, the new technology is often purchased from companies that have developed the technology, so this approach is important but may not provide a unique competitive advantage if your competitors can also buy the technology, provided that they have the capital. Some companies develop their own unique process technology and try to keep it proprietary as long as they can.

The third change and perhaps the real key to success in manufacturing is to implement lean manufacturing system design that can deliver, on time to the customer, superquality goods at the lowest possible cost in a flexible way. Lean production is an effort to reduce waste and improve markedly the methodology by which goods are produced rather than simply upgrading the manufacturing process technology.

Manufacturing system design is discussed extensively in Chapter 2 and Advanced Topic 1 of the book, and it is strongly recommended that students examine this material closely after they have gained a working knowledge of materials and processes. The next section provides a brief discussion of manufacturing system designs.

MANUFACTURING SYSTEM DESIGNS

Five manufacturing system designs can be identified: the job shop, the flow shop, the linked-cell shop, the project shop, and the continuous process. See Figure 1-7. The continuous process deals with liquids and/or gases (such as an oil refinery) rather than solids or discrete parts and is used mostly by the chemical engineer.

The most common of these layouts is the **job shop,** characterized by large varieties of components, general-purpose machines, and a functional layout (Figure 1-8). This means that machines are collected by function (all lathes together, all broaches together, all grinding machines together) and the parts are routed around the shop in small lots to the various machines. The layout of the factory shows the multiple paths through the shop and a detail on one of the seven broaching machine tools. The material is moved from machine to machine in carts or containers and is called the lot or batch.

(a) **Job shop**

Receiving
Lathes
Grinders
Heat-treating
Saws
Drill process
Presses (sheet metal)
Milling machines
Painting
Assembly
Storage
C
B
A

Job shop makes components for subassembly using a functional layout.

(b) **Project shop**

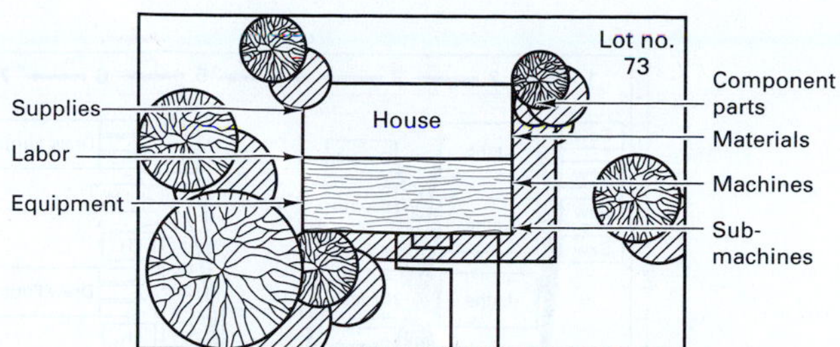

(c) **Flow shop**

STA 4
STA 3
STA 2
STA 1
Moving assembly line
Product
Rework line for detects
Work is devided equally into the stations
STA 10
REPAIR
Subassembly feeder lines
STA 14
STA 13
STA 12
STA 11
Finished Product

Flow shop uses line balancing to achieve one piece flow.

(d) **Continuous process**

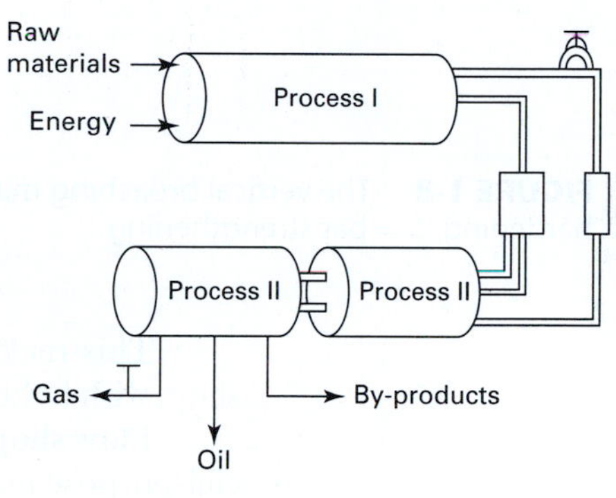

Continuous process systems make products that can flow like gas and oil.

(e) **Linked-cell**

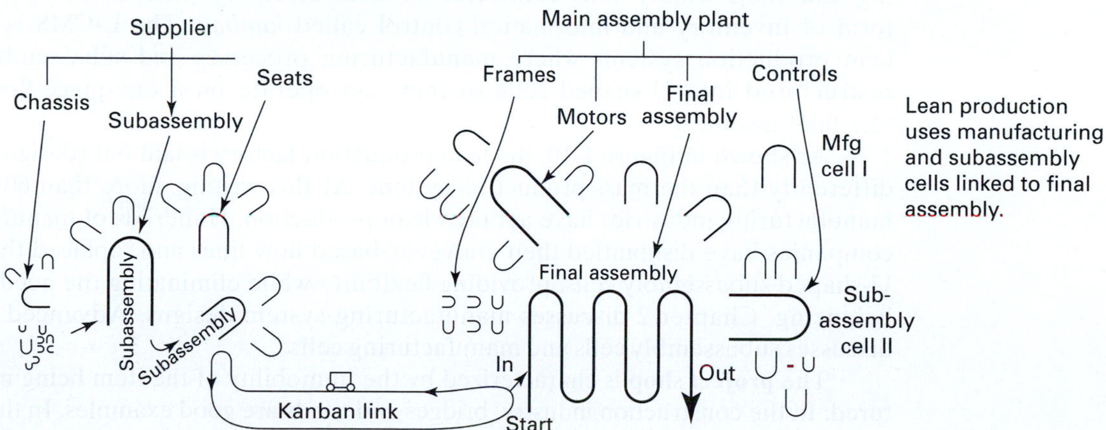

Lean production uses manufacturing and subassembly cells linked to final assembly.

FIGURE 1-7 Schematic layouts of factory designs: (a) functional or job shop, (b) fixed location or project shop, (c) flow shop or assembly line, (d) continuous process or lean shop or linked-cell design.

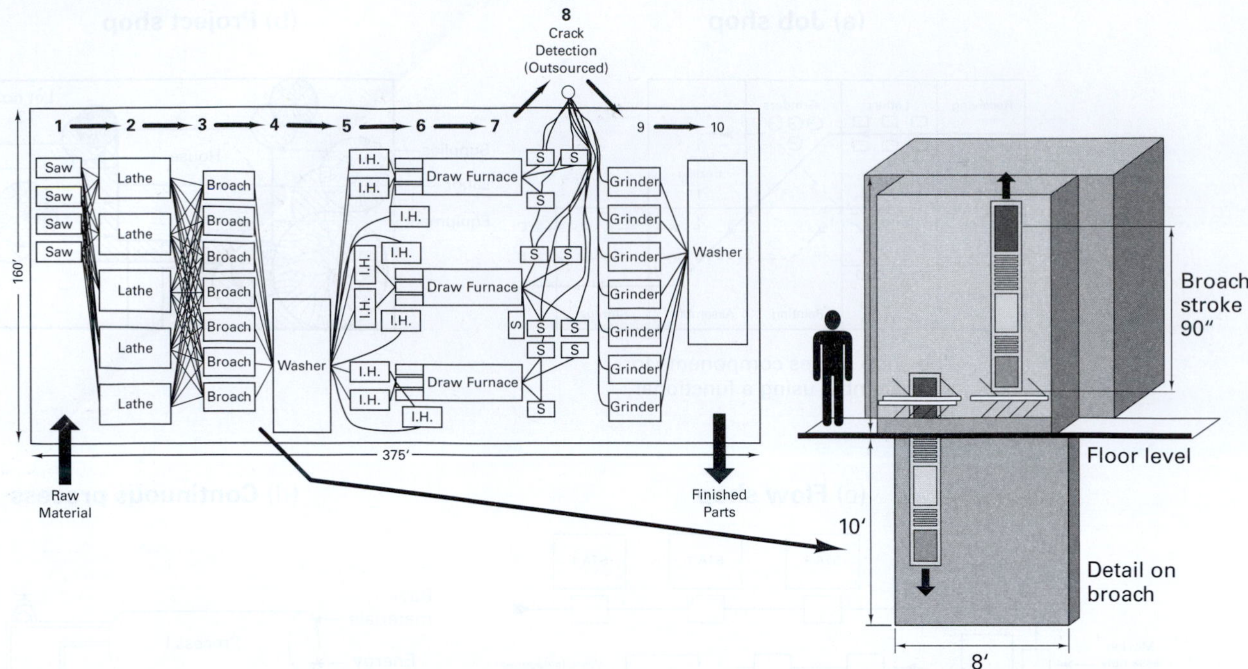

FIGURE 1-8 The vertical broaching machine is one of seven machines in this production job shop. IH = induction hardening, S = bar strengthening.

This rack bar machining area is functionally designed so it operates like a job shop, with lathes, broaches, and grinders lined up.

Flow shops are characterized by larger volumes of the same part or assembly, special-purpose machines and equipment, less variety, less flexibility, and more mechanization. Flow shop layouts are typically either continuous or interrupted and can be for manufacturing or assembly, as shown in Figure 1-9. If continuous, a production line is built that basically runs one large-volume complex item in great quantity and nothing else. The common light bulb is made this way. A transfer line producing an engine block is another typical example. If interrupted, the line manufactures large lots but is periodically "changed over" to run a similar but different component.

The **linked-cell manufacturing system (L-CMS)** is composed of manufacturing and subassembly cells connected to final assembly (linked) using a unique form of inventory and information control called *kanban*. The L-CMS is used in lean production systems where manufacturing processes and subassemblies are restructured into U-shaped cells so they can operate on a one-piece-flow basis, like final assembly.

As shown in Figure 1-10, the lean production factory is laid out (designed) very differently than the mass production system. At this writing, more than 60% of all manufacturing industries have adopted lean production. Hundreds of manufacturing companies have dismantled their conveyor-based flow lines and replaced them with U-shaped subassembly cells, providing flexibility while eliminating the need for line balancing. Chapter 2 discusses manufacturing system designs. Advanced Topic 1 discusses subassembly cells and manufacturing cells.

The **project shop** is characterized by the immobility of the item being manufactured. In the construction industry, bridges and roads are good examples. In the manufacture of goods, large airplanes, ships, large machine tools, and locomotives are manufactured in project shops. It is necessary that the workers, machines, and materials come to the site. The number of end items is not very large, and therefore the lot sizes of the components going into the end item are not large. Thus, the job shop usually supplies parts and subassemblies to the project shop in small lots.

Continuous processes are used to manufacture liquids, oils, gases, and powders. These manufacturing systems are usually large plants producing goods for other

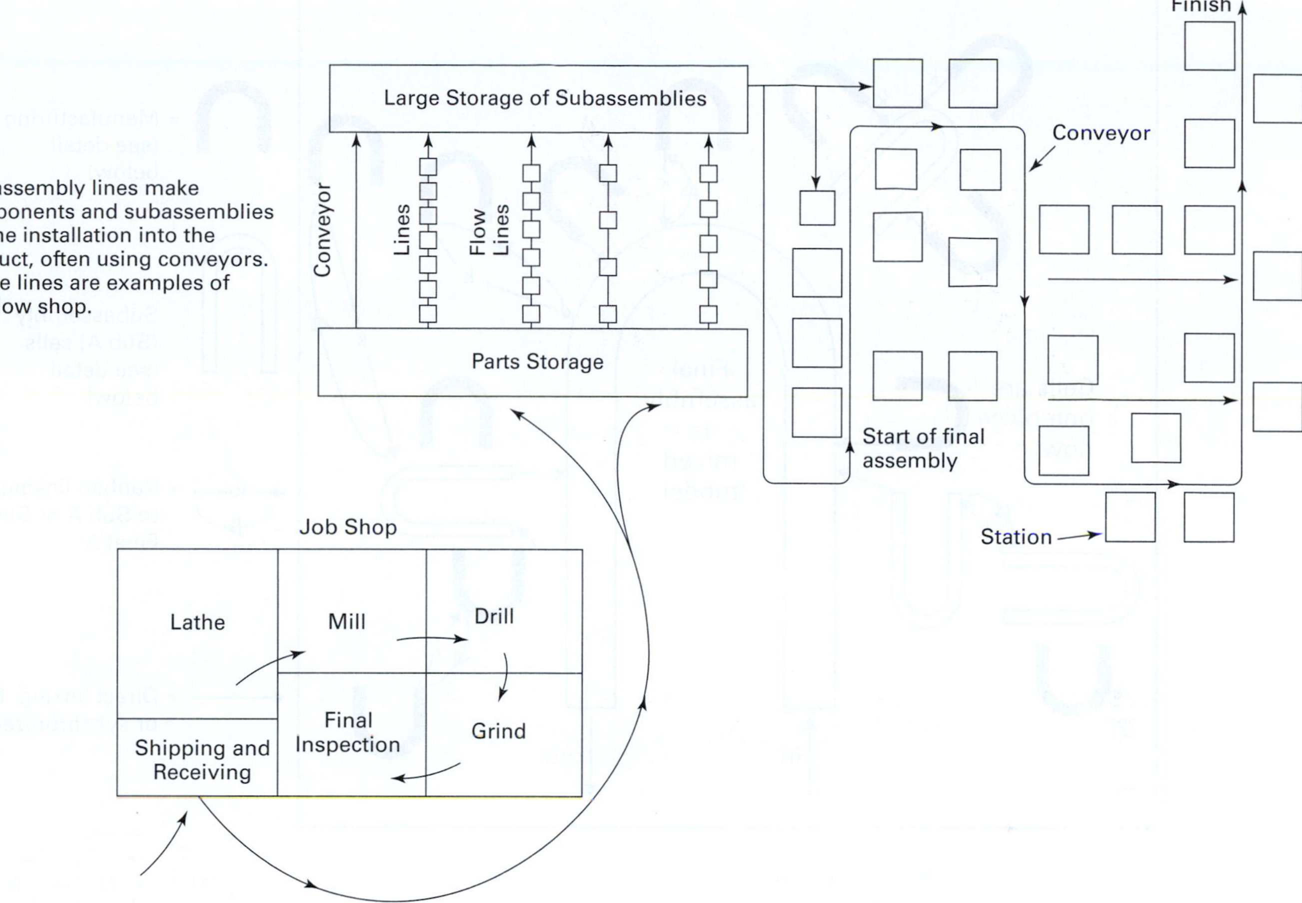

FIGURE 1-9 Flow shops and lines are common in the mass production system. Final assembly is usually a moving assembly line. The product travels through stations in a specific amount of time. The work needed to assemble the product is distributed into the stations, called division of labor. The moving assembly line for cars is an example of the flow shop.

producers or mass-producing canned or bottled goods for consumers. The manufacturing engineer in these factories is often a chemical engineer.

Naturally, there are many hybrid forms of these manufacturing systems, but the job shop is the most common system. Because of its design, the job shop has been shown to be the least cost-efficient of all the systems. Component parts in a typical job shop spend only 5% of their time in machines and the rest of the time waiting or being moved from one functional area to the next. Once the part is on the machine, it is actually being processed (i.e., having value added to it by the changing of its shape) only about 30 to 40% of the time. The rest of the time parts are being loaded, unloaded, inspected, and so on. The advent of **numerical control** machines increased the percentage of time that the machine is making chips because tool movements are programmed and the machines can automatically change tools or load or unload parts.

However, there are a number of trends that are forcing manufacturing management to consider means by which the job shop system itself can be redesigned to improve its overall efficiency. These trends have forced manufacturing companies to convert their batch-oriented job shops into linked-cell manufacturing systems, with the manufacturing and subassembly cells structured around specific products.

Another way to identify families of products with a similar set of manufacturing processes is called group technology. **Group technology (GT)** can be used to restructure the factory floor. GT is a concept whereby similar parts are grouped together into part families. Parts of similar size and shape can often be processed through a similar set of processes. A part family based on manufacturing would have the same set or sequences of manufacturing processes. The machine tools needed to process the part family are

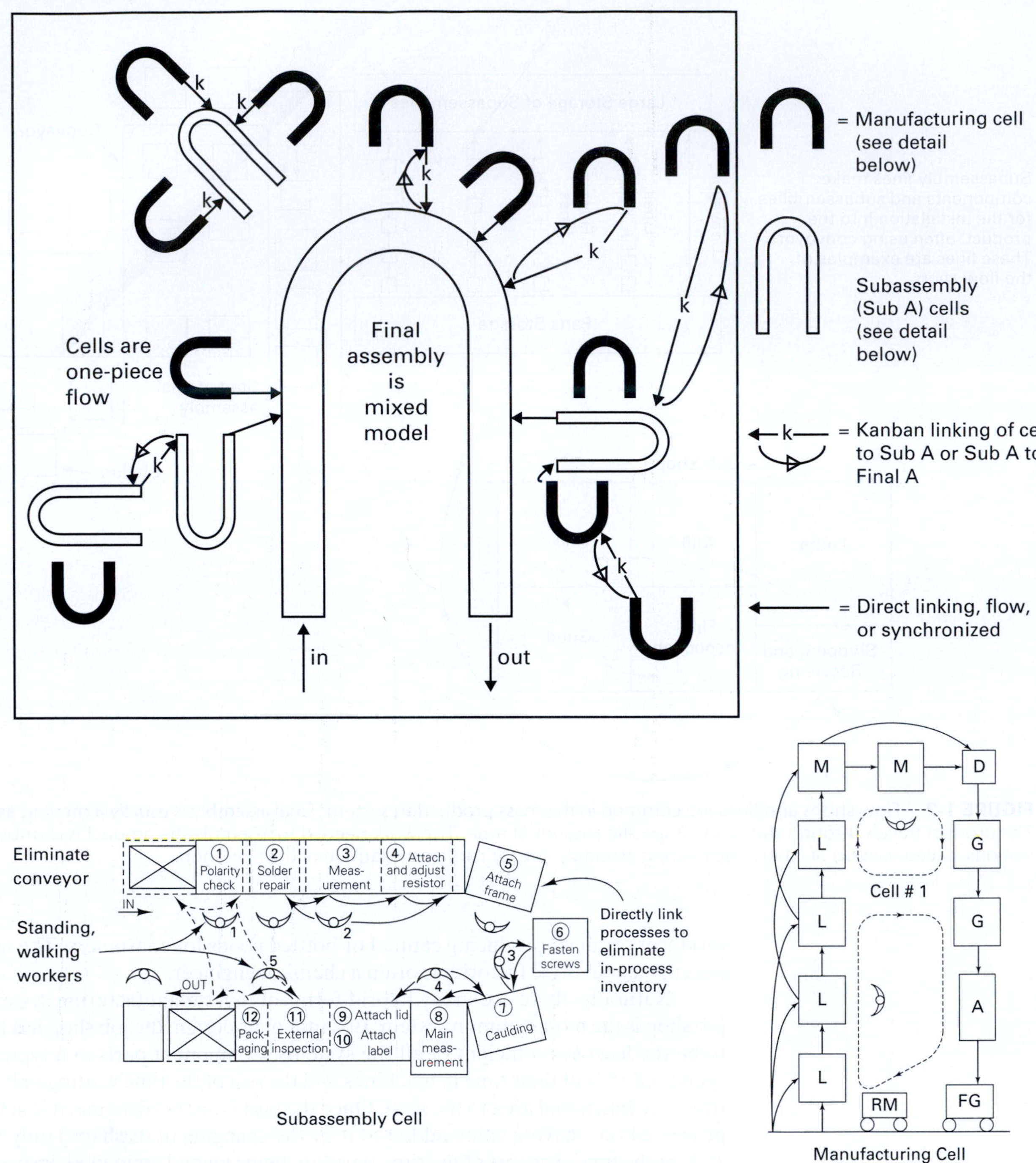

FIGURE 1-10 The linked-cell manufacturing system for lean production has subassembly and manufacturing cells connected to final assembly by kanban links. The traditional subassembly lines can be redesigned into U-shaped cells as part of the conversion of mass production to lean production.

gathered into a cell. Thus, with GT, job shops can be restructured into cells, each cell specializing in a particular family of parts. The parts are handled less, machine setup time is shorter, in-process inventory is lower, and the time needed for parts to get through the manufacturing system (called the throughput time) is greatly reduced.

BASIC MANUFACTURING PROCESSES

It is the manufacturing processes that create or add value to a product. The manufacturing processes can be classified as:

- Casting, foundry, or molding processes.
- Forming or metalworking processes.
- Machining (material removal) processes.
- Joining and assembly.
- Surface treatments (finishing).
- Rapid prototyping.
- Heat treating.
- Other.

These classifications are not mutually exclusive. For example, some finishing processes involve a small amount of metal removal or metal forming. A laser can be used either for joining or for metal removal or heat treating. Occasionally, we have a process such as shearing, which is really metal cutting but is viewed as a (sheet) metal-forming process. Assembly may involve processes other than joining. The categories of process types are far from perfect.

Casting and **molding** processes are widely used to produce parts that often require other follow-on processes, such as **machining.** Casting uses molten metal to fill a cavity. The metal retains the desired shape of the mold cavity after solidification. An important advantage of casting and molding is that, in a single step, materials can be converted from a crude form into a desired shape. In most cases, a secondary advantage is that excess or scrap material can easily be recycled. Figure 1-11 illustrates schematically some of the basic steps in the *lost-wax casting process,* one of many processes used in the foundry industry.

Casting processes are commonly classified into two types: permanent mold (a mold can be used repeatedly) or nonpermanent mold (a new mold must be prepared for each casting made). Molding processes for plastics and composites are included in the chapters on forming processes.

Forming and **shearing** operations typically utilize material (metal or plastics) that has been previously cast or molded. In many cases, the materials pass through a series of forming or shearing operations, so the form of the material for a specific operation may be the result of all the prior operations. The basic purpose of forming and shearing is to modify the shape and size and/or physical properties of the material.

Metal-forming and shearing operations are done both "hot" and "cold," a reference to the temperature of the material at the time it is being processed with respect to the temperature at which this material can recrystallize (i.e., grow new grain structure). Figure 1-12 shows the process by which the fender of a car is made using a series of metalforming processes.

Metal cutting, machining, or metal removal processes refer to the removal of certain selected areas from a part in order to obtain a desired shape or finish. Chips are formed by interaction of a cutting tool with the material being machined. Figure 1-13 shows a chip being formed by a single-point cutting tool in a machine tool called a lathe. The manufacturing engineer may be called upon to specify the cutting parameters such as cutting speed, feed, or depth of cut (DOC). The engineer may also have to select the cutting tools for the job.

Cutting tools used to perform the basic turning on the lathe are shown in Figure 1-14. The cutting tools are mounted in machine tools, which provide the required movements of the tool with respect to the work (or vice versa) to accomplish the process desired. In recent years many new machining processes have been developed.

The seven basic machining processes are shaping, drilling, turning, milling, sawing, broaching, and abrasive machining. Each of these basic processes is extensively discussed. Historically, eight basic types of machine tools have been developed to accomplish the basic processes. These machine tools are called shapers (and planers), drill presses, lathes, boring machines, milling machines, saws, broaches, and grinders. Most of these machine tools are capable of performing more than one of the basic machining processes. Shortly after numerical control was invented, machining centers

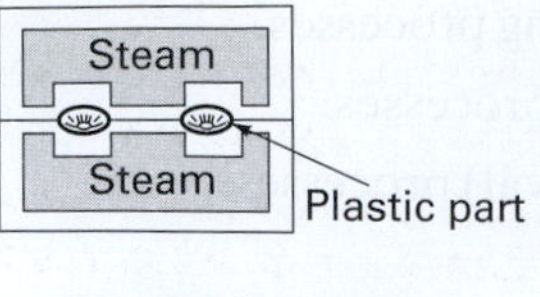

To make the foam parts, metal molds are used. Beads of polystyrene are heated and expanded in the mold to get parts.

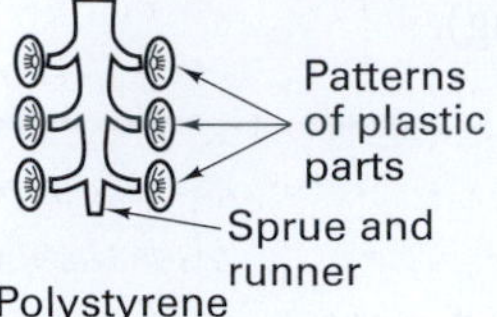

A pattern containing a sprue, runners, risers, and parts is made from single or multiple pieces of foamed polystyrene plastic.

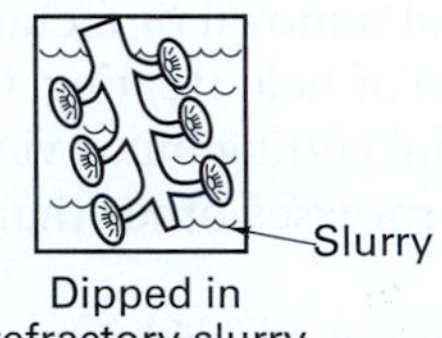

The polystyrene pattern is dipped in a ceramic slurry, which wets the surface and forms a coating about 0.005 inch thick.

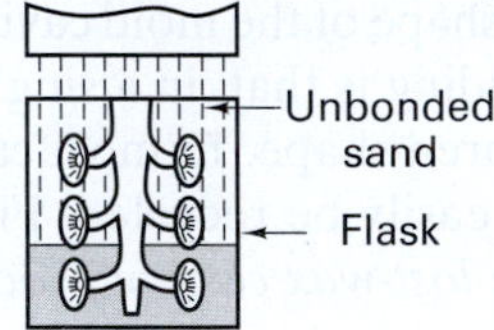

The coated pattern is placed in a flask and surrounded with loose, unbonded sand.

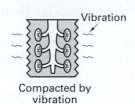

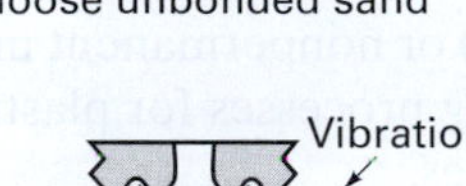

The flask is vibrated so that the loose sand is compacted around the pattern.

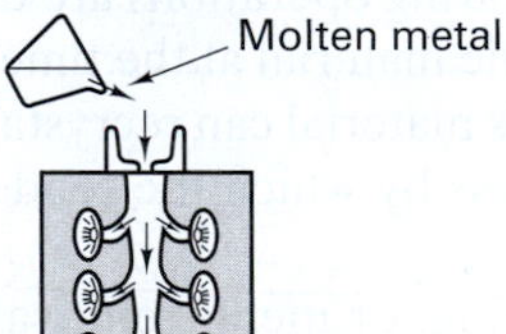

During the pouring of molten metal, the hot metal vaporizes the pattern and fills the resulting cavity.

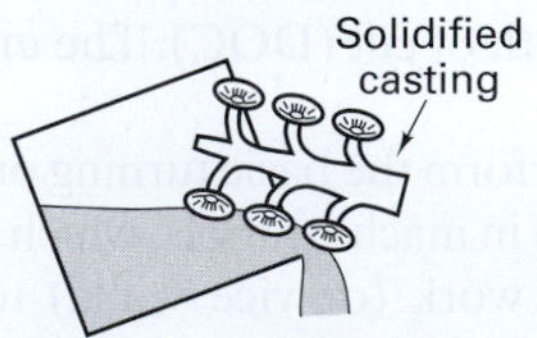

The solidified casting is removed from flask and the loose sand reclaimed.

FIGURE 1-11 Schematic of the lost-foam casting process.

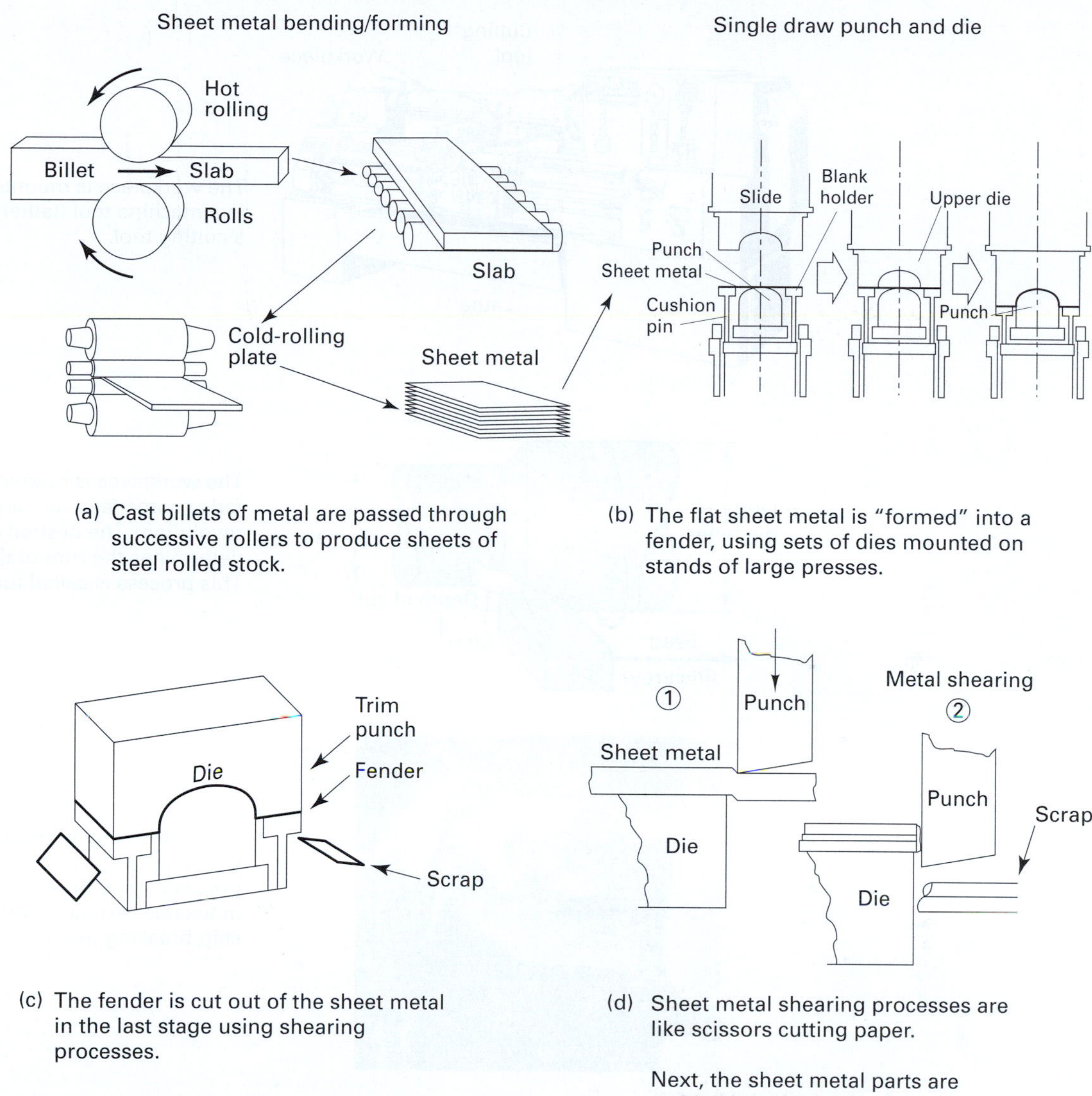

FIGURE 1-12 The forming process used to make a fender for a car.

were developed that could combine many of the basic processes, plus other related processes, into a single machine tool with a single workpiece setup.

Aside from the chip-making processes, there are processes wherein metal is removed by chemical, electrical, electrochemical, or thermal sources. Generally speaking, these nontraditional processes have evolved to fill a specific need when conventional processes were too expensive or too slow when machining very hard materials. One of the first uses of a laser was to machine holes in ultra-high-strength metals. Lasers are being used today to drill tiny holes in turbine blades for jet engines. Because of its ability to produce components with great precision and accuracy, metal cutting, using machine tools, is recognized as having great value-adding capability.

In recent years a new family of processes has emerged called *rapid prototyping* or *rapid manufacturing* or *free-form manufacturing* (see Chapter 19). These additive-type processes produce first, or prototype, components directly from the software using specialized machines driven by computer-aided design packages. The prototypes can be field tested and modifications to the design quickly implemented. Early versions of these machines produced only nonmetallic components, but modern machines can

The Machining Process
(turning on a lathe)

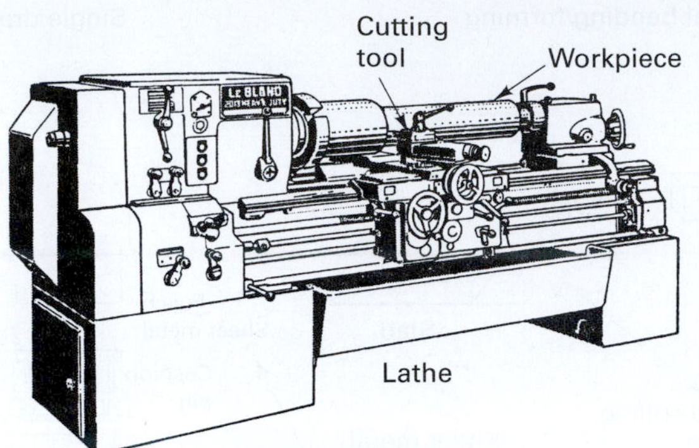

The workpiece is mounted in a workholding device in a machine tool (lathe) and is cut (machined) with a cutting tool.

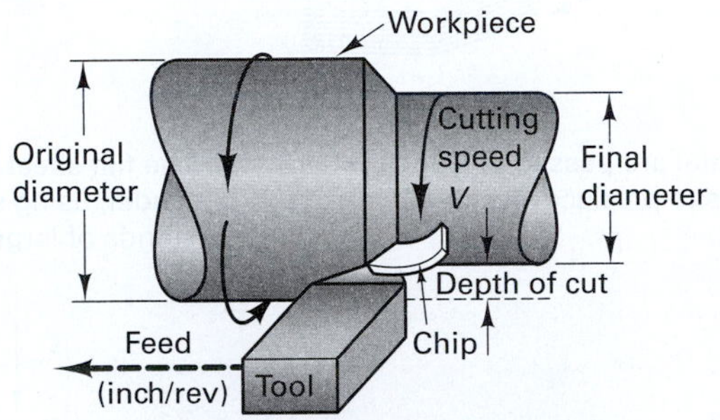

The workpiece is rotated while the tool is fed at some feed rate (inches per revolution). The desired cutting speed *V* determines the rpm of the workpiece. This process is called turning.

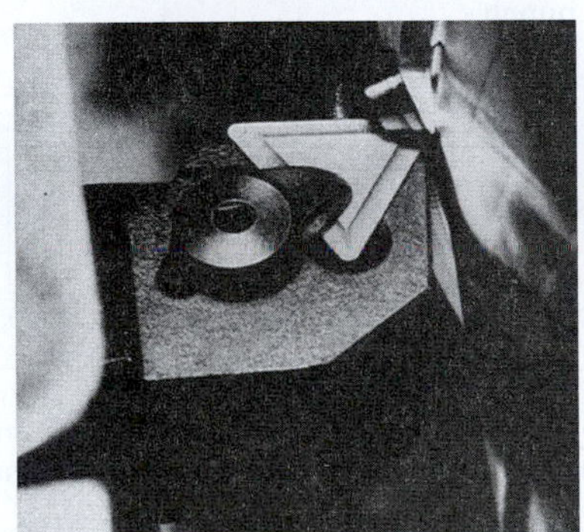

The cutting tool interacts with the workpiece to form a chip by a shearing process. The tool shown here is an indexable carbide insert tool with a chip-breaking groove.

FIGURE 1-13 Single-point metal-cutting process (turning) produces a chip while creating a new surface on the workpiece. *(Courtesy J T. Black)*

make metal parts. In contrast, the machining processes are recognized as having great value-adding capability, that is, the ability to produce components with great precision and accuracy. Companies have sprung up; you can send your CAD drawing over the Internet and a prototype is made in hours.

Perhaps the largest collection of processes, in terms of both diversity and quantity, are the **joining processes,** which include the following:

1. Mechanical fastening.
2. Soldering and brazing.
3. Welding.
4. Press, shrink, or snap fittings.
5. Adhesive bonding.
6. Assembly processes.

Many of these joining processes are often found in the assembly area of the plant. Figure 1-14 provides one example where all but welding are used in the sequence of

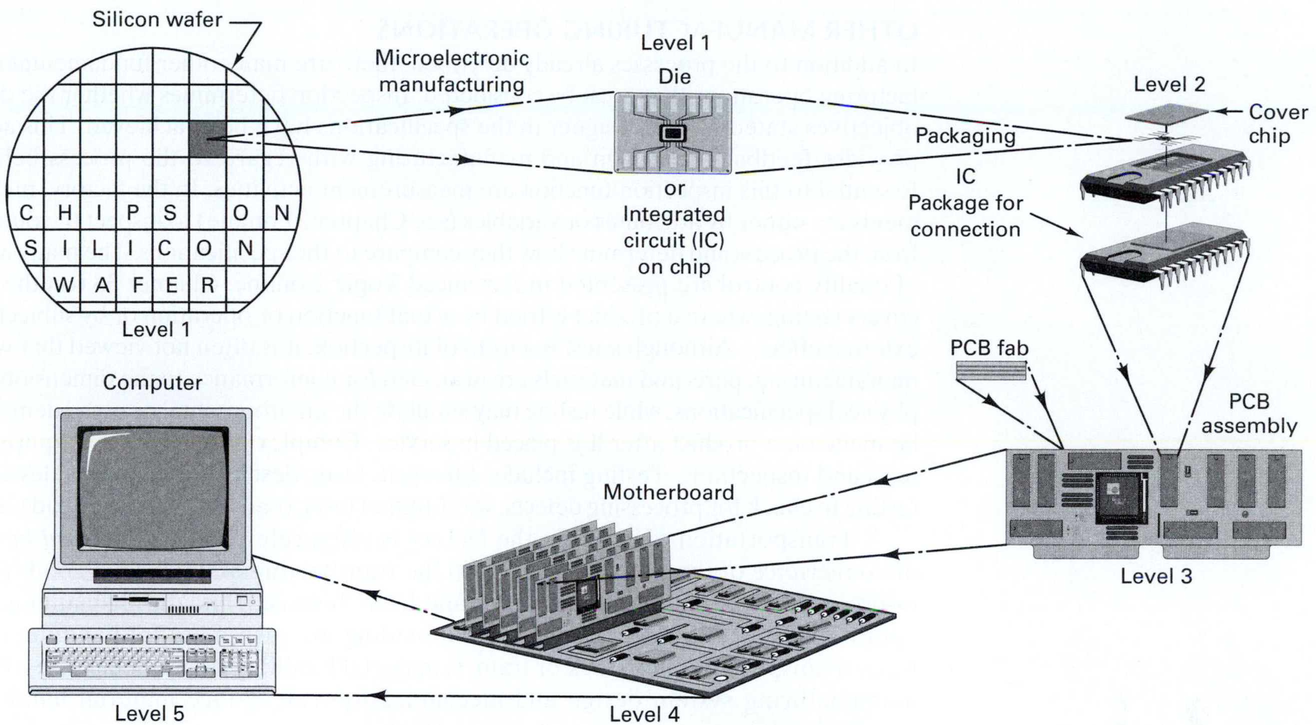

FIGURE 1-14 How an electronic product is made.

operations to produce a computer. Starting in the upper left corner, microelectronic fabrication methods produce entire integrated circuits (ICs) of solid-state (no moving parts) components, with wiring and connections, on a single piece of semiconductor material, usually single-crystalline silicon. Arrays of ICs are produced on thin, round disks of semiconductor material called wafers. Once the semiconductor on the wafer has been fabricated, the finished wafer is cut up into individual ICs, or chips. Next, at level 2, these chips are individually housed with connectors or leads making up "dies" that are placed into "packages" using adhesives. The packages provide protection from the elements and a connection between the die and another subassembly called the printed circuit boards (PCBs). At level 3, IC packages, along with other discrete components (e.g., resistors, capacitors, etc.), are soldered onto PCBs and then assembled with even larger circuits on PCBs. This is sometimes referred to as electronic assembly. Electronic packages at this level are called cards or printed wiring assemblies (PWAs). Next, series of cards are combined on a back-panel PCB, also known as a motherboard or simply a board. This level of packaging is sometimes referred to as card-on-board packaging. Ultimately, card-on-board assemblies are put into housings using mechanical fasteners and snap fitting and finally integrated with power supplies and other electronic peripherals through the use of cables to produce final commercial products. See Chapter 35 for more details on electronic manufacturing.

Finishing processes are yet another class of processes typically employed for cleaning, removing burrs left by machining, or providing protective and/or decorative surfaces on workpieces. Surface treatments include chemical and mechanical cleaning, deburring, painting, plating, buffing, galvanizing, and anodizing.

Heat treatment is the heating and cooling of a metal for the specific purpose of altering its metallurgical and mechanical properties. Because changing and controlling these properties is so important in the processing and performance of metals, heat treatment is a very important manufacturing process. Each type of metal reacts differently to heat treatment. Consequently, a designer should know not only how a selected metal can be altered by heat treatment but, equally important, *how a selected metal will react, favorably or unfavorably, to any heating or cooling that may be incidental to the manufacturing processes.*

OTHER MANUFACTURING OPERATIONS

In addition to the processes already described, there are many other fundamental manufacturing operations that must be considered. Inspection determines whether the desired objectives stated by the designer in the specifications have been achieved. This activity provides feedback to design and manufacturing with regard to the process behavior. Essential to this inspection function are measurement activities. In the factory, measurements are either by attributes or variables (see Chapter 37 online) to inspect the outcomes from the process and determine how they compare to the specifications. The many aspects of quality control are presented in Advanced Topic 2 online. Chapter 38 (on the Web) covers testing, where a product is tried by actual function or operation or by subjection to external effects. Although a test is a form of inspection, it is often not viewed that way. In manufacturing, parts and materials are inspected for conformance to the dimensional and physical specifications, while testing may simulate the environmental or usage demands to be made on a product after it is placed in service. Complex processes may require many tests and inspections. Testing includes life-cycle tests, destructive tests, nondestructive testing to check for processing defects, wind-tunnel tests, road tests, and overload tests.

Transportation of goods in the factory is often referred to as *material handling* or *conveyance* of the goods and refers to the transporting of unfinished goods (work-in-process) in the plant and supplies to and from, between, and during manufacturing operations. Loading, positioning, and unloading are also material-handling operations. Transportation, by truck or train, is material handling between factories. Proper manufacturing system design and mechanization can reduce material handling in countless ways.

Automatic material handling is a critical part of continuous automatic manufacturing. The word *automation* is derived from automatic material handling. Material handling, a fundamental operation done by people and by conveyors and loaders, often includes positioning the workpiece within the machine by indexing, shuttle bars, slides, and clamps. In recent years, wire-guided automated guided vehicles (AGVs) and automatic storage and retrieval systems (AS/RSs) have been developed in an attempt to replace forklift trucks on the factory floor. Another form of material handling, the mechanized removal of waste (chips, trimming, and cutoffs), can be more difficult than handling the product. Chip removal must be done before a tangle of scrap chips damages tooling or creates defective workpieces.

Most texts on manufacturing processes do not mention **packaging,** yet the packaging is often the first thing the customer sees. Also, packaging often maintains the product's quality between completion and use. (The term *packaging* is also used in electronics manufacturing to refer to placing microelectronic chips in containers for mounting on circuit boards.) Packaging can also prepare the product for delivery to the user. It varies from filling ampules with antibiotics to steel-strapping aluminum ingots into palletized loads. A product may require several packaging operations. For example, Hershey Kisses are (1) individually wrapped in foil, (2) placed in bags, (3) put into boxes, and (4) placed in shipping cartons.

Weighing, filling, sealing, and labeling are packaging operations that are highly automated in many industries. When possible, the cartons or wrappings are formed from material on rolls in the packaging machine. Packaging is a specialty combining elements of product design (styling), material handling, and quality control. Some packages cost more than their contents (e.g., cosmetics and razor blades).

During **storage,** nothing happens intentionally to the product or part except the passage of time. Part or product deterioration on the shelf is called **shelf life,** meaning that items can rust, age, rot, spoil, embrittle, corrode, creep, and otherwise change in state or structure, while supposedly nothing is happening to them. Storage is detrimental, wasting the company's time and money. The best strategy is to keep the product moving with as little storage as possible. Storage during processing must be *eliminated*, not automated or computerized. Companies should avoid investing heavily in large automated systems that do not alter the bottom line. Have the outputs improved with respect to the inputs, or has storage simply increased the costs (indirectly) without improving either the quality or the throughput time?

TABLE 1-4 Characterizing a Process Technology

Mechanics (statics and dynamics of the process)
How does the process work?
What are the process mechanics (statics, dynamics, friction)?
What physically happens, and what makes it happen? (Understand the physics.)

Economics or costs
What are the tooling costs, the engineering costs?
Which costs are short term, which long term?
What are the setup costs?

Time spans
How long does it take to set up the process initially?
What is the throughput time?
How can these times be shortened?
How long does it take to run a part once it is set up (cycle time)?
What process parameters affect the cycle time?

Constraints
What are the process limits?
What cannot be done?
What constrains this process (sizes, speeds, forces, volumes, power, cost)?
What is very hard to do within an acceptable time/cost frame?

Uncertainties and process reliability
What can go wrong?
How can this machine fail?
What do people worry about with this process?
Is this a reliable, stable process?

Skills
What operator skills are critical?
What is not done automatically?
How long does it take to learn to do this process?

Flexibility
Can this process be adapted easily for new parts of a new design or material?
How does the process react to changes in part design and demand?
What changes are easy to do?

Process capability
What are the accuracy and precision of the process?
What tolerances does the process meet? (What is the process capability?)
How repeatable are those tolerances?

By not storing a product, the company avoids having to (1) remember where the product is stored, (2) retrieve it, (3) worry about its deteriorating, or (4) pay storage (including labor) costs. Storage is the biggest waste of all and should be eliminated at every opportunity.

UNDERSTAND YOUR PROCESS TECHNOLOGY

Understanding the process technology of the company is very important for everyone in the company. Manufacturing technology affects the design of the product and the manufacturing system, the way in which the manufacturing system can be controlled, the types of people employed, and the materials that can be processed. Table 1-4 outlines the factors that characterize a process technology. Take a process you are familiar with and think about these factors. One valid criticism of American companies is that their managers seem to have an aversion to understanding their companies' manufacturing technologies. Failure to understand the company business (i.e., its fundamental process technology) can lead to the failure of the company.

The way to overcome technological aversion is to run the process and study the technology. Only someone who has run a drill press can understand the sensitive relationship between feed rate and drill torque and thrust. All processes have these "know-how" features. Those who run the processes must be part of the decision making for the factory. The CEO who takes a vacation working on the plant floor and learning the processes will be well on the way to being the head of a successful company.

PRODUCT LIFE CYCLE AND LIFE-CYCLE COST

Manufacturing systems are dynamic and change with time. There is a general, traditional relationship between a product's life cycle and the kind of manufacturing system used to make the product. Figure 1-15 simplifies the **product life cycle** into these steps, again using an S-shaped curve.

1. *Startup*. New product or new company, low volume, small company.
2. *Rapid growth*. Products become standardized and volume increases rapidly. Company's ability to meet demand stresses its capacity.
3. *Maturation*. Standard designs emerge. Process development is very important.
4. *Commodity*. Long-life, standard-of-the-industry type of product or
5. *Decline*. Product is slowly replaced by improved products.

The maturation of a product in the marketplace generally leads to fewer competitors, with competition based more on price and on-time delivery than on unique product features. As the competitive focus shifts during the different stages of the product life cycle, the requirements placed on manufacturing—cost, quality, flexibility, and delivery dependability—also change. The stage of the product life cycle affects the product design stability, the length of the product development cycle, the frequency of

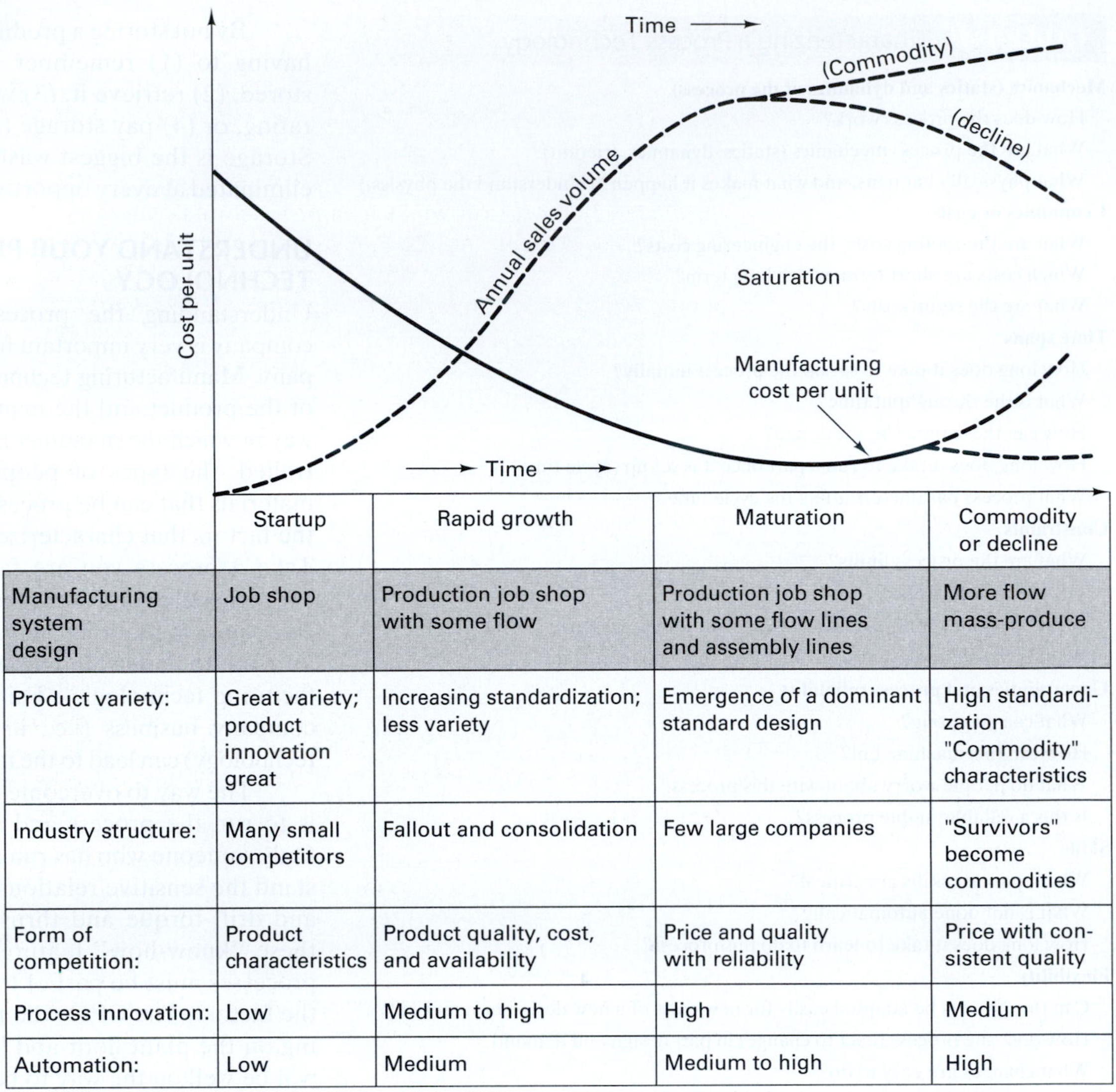

	Startup	Rapid growth	Maturation	Commodity or decline
Manufacturing system design	Job shop	Production job shop with some flow	Production job shop with some flow lines and assembly lines	More flow mass-produce
Product variety:	Great variety; product innovation great	Increasing standardization; less variety	Emergence of a dominant standard design	High standardization "Commodity" characteristics
Industry structure:	Many small competitors	Fallout and consolidation	Few large companies	"Survivors" become commodities
Form of competition:	Product characteristics	Product quality, cost, and availability	Price and quality with reliability	Price with consistent quality
Process innovation:	Low	Medium to high	High	Medium
Automation:	Low	Medium	Medium to high	High

FIGURE 1-15 Product life-cycle costs change with the classic manufacturing system designs.

engineering change orders, and the commonality of components—all of which have implications for manufacturing process technology.

During the design phase of the product, much of the cost of manufacturing and assembly is determined. Assembly of the product is inherently integrative as it focuses on pairs and groups of parts.

It is crucial to achieve this integration during the design phase because about 70% of the life-cycle cost of a product is determined when it is designed. Design choices determine materials; fabrication methods; assembly methods; and, to a lesser degree, material-handling options, inspection techniques, and other aspects of the production system. Manufacturing engineers and internal customers can influence only a small part of the overall cost if they are presented with a finished design that does not reflect their concerns. Therefore, all aspects of production should be included if product designs are to result in real functional integration.

Life-cycle costs include the costs of all the materials, manufacture, use, repair, and disposal of a product. Early design decisions determine about 60% of the cost, and all activities up to the start of full-scale development determine about 75%. Later decisions can make only minor changes to the ultimate total unless the design of the manufacturing system is changed.

In short, the concept of product life-cycle provides a framework for thinking about the product's evolution through time and the kind of market segments that are likely to develop at various times. Analysis of life-cycle costs shows that the design of the manufacturing system determines the cost per unit, which generally decreases over time with process improvements and increased volumes. For additional discussion on reliability and maintainability of manufacturing equipment, see the Society of Automotive Engineers SAE publication M-110.2.

The linked-cell manufacturing system design discussed in Chapter 2 and Advanced Topic 1 (known as lean production or the **Toyota Production System**) has transformed the automobile industry and many other industries to be able to make a large variety of products in small volumes with very short throughput times. Thousands of companies have implemented lean to reduce waste, decrease cost per unit significantly while maintaining flexibility and making smooth transitions. This is a new business model affecting product development, design, purchasing, marketing, customer service and all other aspects of the company.

Low-cost manufacturing does not just happen. There is a close, interdependent relationship among the design of a product, the selection of materials, the selection of processes and equipment, the design of the processes, and tooling selection and design. Each of these steps must be carefully considered, planned, and coordinated before manufacturing starts.

Some of the steps involved in getting the product from the original idea stage to daily manufacturing are discussed in more detail in Chapter 10. The steps are closely related to each other. For example, the design of the tooling is dependent on the design of the parts to be produced. It is often possible to simplify the tooling if certain changes are made in the design of the parts or the design of the manufacturing systems. Similarly, the material selection will affect the design of the tooling or the processes selected. Can the design be altered so that it can be produced with tooling already on hand and thus avoid the purchase of new equipment? Close coordination of all the various phases of design and manufacture is essential if economy is to result.

With the advent of computers and computer-controlled machines, the integration of the design function and the manufacturing function through the computer is a reality. This is usually called CAD/CAM (computed-aided design/computer-aided manufacturing). The key is a common database from which detailed drawings can be made for the designer and the manufacturer and from which programs can be generated to make all the tooling. In addition, extensive computer-aided testing and inspection (CATI) of the manufactured parts is taking place. There is no doubt that this trend will continue at ever-accelerating rates as computers become cheaper and smarter, but at this time, the computers necessary to accomplish complete computer-integrated manufacturing (CIM) are expensive and the software very complex. Implementing CIM requires a lot of manpower as well.

COMPARISONS OF MANUFACTURING SYSTEM DESIGN

When designing a manufacturing system, two customers must be taken into consideration: the external customer who buys the product and the internal customer who makes the product. The external customer is likely to be global and demand greater variety with superior quality and reliability. The internal customer is often empowered to make critical decisions about how to make the products. The Toyota Motor Company is making vehicles in 25 countries. Their truck plant in Indiana has the capacity to make 150,000 vehicles per year (creating 2300 new jobs), using the Toyota Production System (TPS). An appreciation of the complexity of the manufacturing system design problem is shown in Figure 1-16, where the choices between the system designs are reflected against the number of different products, or parts being made in the system, often called *variety.* Clearly, there are many choices regarding which method (or system) to use to make the goods. A manufacturer never really knows how large or diverse a market will be. If a diverse and specialized market emerges, a company with a focused flow-line system may be too inflexible to meet the varying demand. If a large but homogeneous market develops, a manufacturer with a flexible system may find production costs too high and the flexibility unexploitable. Another general relationship between manufacturing system designs and production volumes is shown in Figure 1-17.

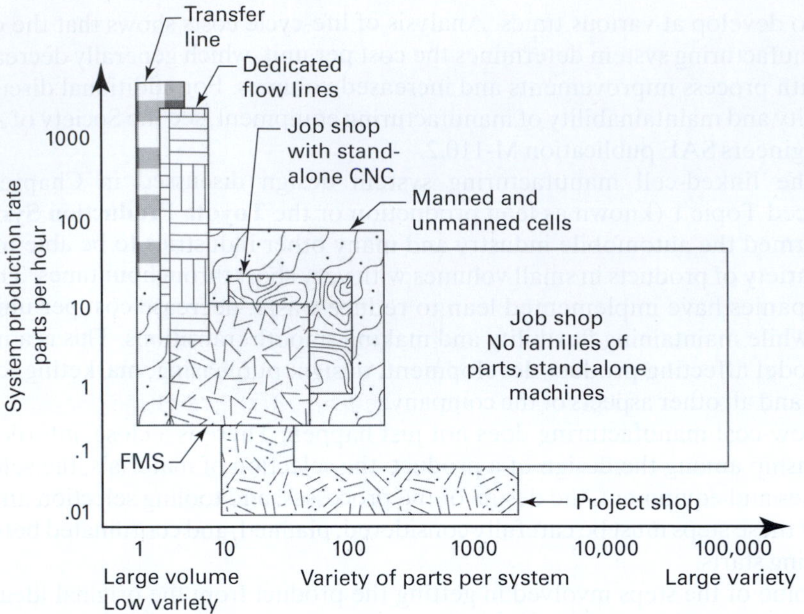

This part variety-production rate matrix shows examples of particular manufacturing system designs. This matrix was developed by Black based on real factory data. Notice there is a large amount of overlap in the middle of the matrix, so the manufacturing engineer has many choices regarding which method or system to use to make the goods. This book will show the connection between the process and the manufacturing system used to produce the products, turning raw materials into finished goods.

FIGURE 1-16 Different manufacturing system designs produce goods at different production rates.

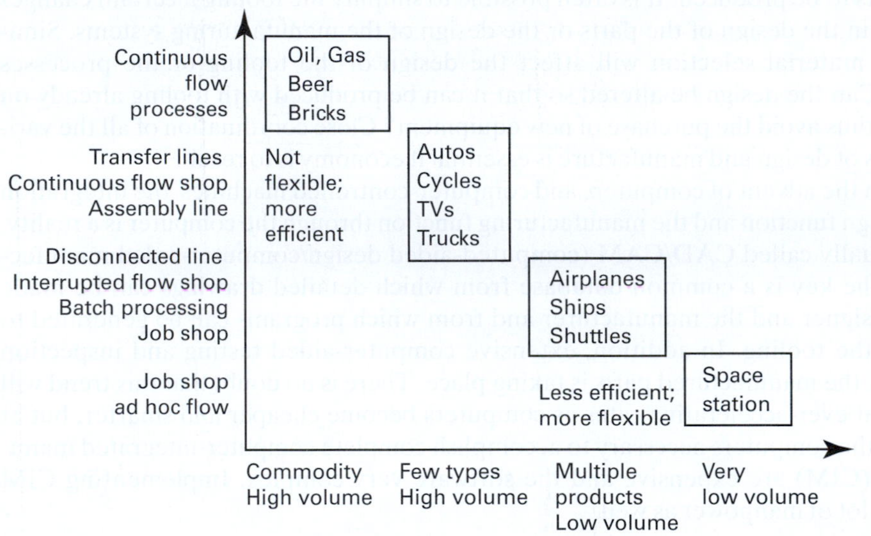

The figure shows in a general way the relationship between manufacturing systems and production volumes. The upper left represents systems with low flexibility but high efficiency compared to the lower right, where volumes are low and so is efficiency. Where a particular company lies in this matrix is determined by many forces, not all of which are controllable. The job of manufacturing and industrial engineers is to design and implement a system which can achieve low unit cost, superior quality, with on-time delivery in a flexible way.

FIGURE 1-17 This figure shows in a general way the relationship between manufacturing systems and production volumes.

NEW MANUFACTURING SYSTEMS

The manufacturing process technology described in this text is available worldwide. Many countries have about the same level of process development when it comes to manufacturing technology. Much of the technology existing in the world today was developed in the United States, Germany, France, and Japan. More recently Taiwan, Korea, and China have been making great inroads into American markets, particularly in the automotive and electronics industries. Many companies have developed and promoted a different kind of manufacturing system design. This new manufacturing system, called lean manufacturing, will take its place with the American Armory System and the Ford System for mass production. This new manufacturing system, developed by the Toyota Motor Company, has been successfully adopted by many American companies.

For lean production to work, units with no defects (100% good) must flow rhythmically to subsequent processes without interruption. In order to accomplish this, an integrated quality control (IQC) program has to be developed. The responsibility for quality has been given to manufacturing. All the employees are inspectors and are empowered to make it right the first time. There is a companywide attitude toward constant quality improvement. Make quality easy to see, stop the line when something goes wrong, and inspect things 100% if necessary to prevent defects from occurring. The results of this system are astonishing in terms of quality, low cost, and on-time delivery of goods to the customer.

The most important factor in economical and successful manufacturing is the manner in which the resources—labor, materials, and capital—are organized and managed so as to provide effective coordination, responsibility, and control. Part of the success of lean production can be attributed to a different management approach. This approach is characterized by a holistic attitude toward people.

The real secret of successful manufacturing lies in designing a manufacturing system in which everyone who works in the system understands how the system works and how goods are controlled, with the decision making placed at the correct level. The engineers also must possess a broad fundamental knowledge of design, metallurgy, processing, economics, accounting, and human relations. In the manufacturing game, low-cost mass production is the result of teamwork within an integrated manufacturing/production system. This is the key to producing superior quality at less cost with on-time delivery.

Key Words

assembly
casting
construction
consumer goods
continuous process
design engineer
fabricating
finishing process
flow shop
forming
goods
group technology (GT)
heat treatment
inspection
job
job shop
joining process
lean engineer
lean production
linked-cell manufacturing system (L-CMS)
machine tool
machining
manufacturing
manufacturing cost
manufacturing engineer
manufacturing process
manufacturing system
materials engineer
molding
numerical control
operation
packaging
processing
producer goods
product
product life cycle
production system
project shop
shearing
shelf life
station
storage
sustaining technology
tooling
tools
Toyota Production System
treatments

Review Questions

1. What role does manufacturing play relative to the standard of living of a country?
2. Aren't all goods really consumer goods, depending on how you define the customer? Discuss.
3. Is the Subway sandwich shop an example of a job shop, flow shop, or project shop?
4. How does a system differ from a process? From a machine tool? From a job? From an operation?
5. Is a cutting tool the same thing as a machine tool?
6. What are the major classifications of basic manufacturing processes?
7. Casting is often used to produce a complex-shaped part to be made from a hard-to-machine metal. How else could the part be made?
8. In the lost-wax casting process, what happens to the foam?
9. In making a gold medal, what do we mean by a "relief image" cut into the die?
10. How is a railroad station like a station on an assembly line?
11. Because no work is being done on a part when it is in storage, it does not cost you anything. True or false? Explain.
12. What forming processes are used to make a paper clip?
13. What is tooling in a manufacturing system?
14. It is acknowledged that chip-type machining is basically an inefficient process. Yet it is probably used more than any other to produce desired shapes. Why?
15. Compare Figure 1-1 and Figure 1-15. What are the stages of the product life cycle for a computer?
16. In a modern safety razor with three or four blades that sells for $1, what do you think the cost of the blades might be?
17. List three purposes of packaging operations.
18. *Assembly* is defined as "the putting together of all the different parts to make a complete machine." Think of (and describe) an assembly process. Is making a club sandwich an assembly process? What about carving a turkey? Is this an assembly process?
19. What are the physical elements in a manufacturing system?
20. In the production system, who usually figures out how to make the product?
21. In Figure 1-8, what do the lines connecting the processes represent?
22. Characterize the process of squeezing toothpaste from a tube (extrusion of toothpaste) using Table 1-4 as a guideline. See the index for help on extrusion.
23. What difficulties would result if production planning and scheduling were omitted from the procedure outlined in Chapter 9 for making a product in a job shop?

24. It has been said that low-cost products are more likely to be more carefully designed than high-priced items. Do you think this is true? Why or why not?
25. Proprietary processes are closely held or guarded company secrets. The chemical makeup of a lubricant for an extrusion process is a good example. Give another example of a proprietary process.
26. If the rolls for the cold-rolling mill that produces the sheet metal used in your car cost $300,000 to $400,000, how is it that your car can still cost less than $20,000?
27. Make a list of service systems, giving an example of each.
28. What is the fundamental difference between a service systems and a manufacturing system?
29. In the process of buying a calf, raising it to a cow, and disassembling it into "cuts" of meat for sale, where is the "value added"?
30. What kind of process is powder metallurgy: casting or forming?
31. In view of Figure 1-2, who really determines the selling price per unit?
32. What costs make up manufacturing cost (sometimes called factory cost)?
33. What are major phases of a product life cycle?
34. How many different manufacturing systems might be used to make a component with annual projected sales of 16,000 parts per year with 10 to 12 different models (varieties)?
35. In general, as the annual volume for a product increases, the unit cost decreases. Explain.

PROBLEMS

1. The Toyota truck plant in Indiana produces 150,000 trucks per year. The plant runs one eight-hour shift, 300 days per year, and makes 500 trucks per day. About 1300 people work on the final assembly line. Each car has about 20 labor hours per car in it.
   a. Assuming the truck sells for $16,000 and workers earn $30 per hour in wages and benefits, what percentage of the cost of the truck is in direct labor?
   b. What is the production rate of the final assembly line?
2. Suppose you wanted to redesign a stapler to have fewer components. (You should be able to find a stapler at a local discount store.) How much did it cost? How many parts does it have? Make up a "new parts" list and indicate which parts would have to be redesigned and which parts would be eliminated. Estimate the manufacturing cost of the stapler assuming that manufacturing costs are 40% of the selling price. What are the disadvantages of your new stapler design versus the old stapler?
3. A company is considering making automobile bumpers from aluminum instead of from steel. List some of the factors it would have to consider in arriving at its decision.
4. Many companies are critically examining the relationship of product design to manufacturing and assembly. Why do they call this concurrent engineering?
5. We can analogize your university to a manufacturing system that produces graduates. Assuming that it takes four years to get a college degree and that each course really adds value to the student's knowledge base, what percentage of the four years is "value adding" (percentage of time in class plus two hours of preparation for each hour in class)?
6. What are the major process steps in the assembly of an automobile?

www.wiley.com/go/global/degarmo

Chapter 1 CASE STUDY

Famous Manufacturing Engineers

Manufacturing engineering is that engineering function charged with the responsibility of interpreting product design in terms of manufacturing requirements and process capability. Specifically, the manufacturing engineer may:

- Determine how the product is to be made in terms of specific manufacturing processes.
- Design workholding and work transporting tooling or containers.
- Select the tools (including the tool materials) that will machine or form the work materials.
- Select, design, and specify devices and instruments that inspect products that have been manufactured to determine their quality.
- Design and evaluate the performance of the manufacturing system.
- Perform all these functions (and many more) related to the actual making of the product at the most reasonable cost per unit without sacrifice of the functional requirements or the users' service life.

There's no great glory in being a great manufacturing engineer (MfE). If you want to be a manufacturing engineer, you had better be ready to get your hands dirty. Of course, there are exceptions. There have been some very famous manufacturing engineers. For example:

- John Wilkinson of Bersham, England built a boring mill in 1775 to bore the cast iron cylinders for James Watt's steam engine. How good was this machine?
- Eli Whitney was said to have invented the cotton gin, a machine to separate seeds from cotton. His machine was patented but was so simple, anyone could make one. He was credited with "interchangeability"—but we know Thomas Jefferson observed interchangeability in France in 1785 and probably the French gunsmith LeBlanc is the real inventor here. Jefferson tried to bring the idea to America and Whitney certainly did. He took 10 muskets to Congress, disassembled them, and scattered the pieces. Interchangeable parts permitted them to be reassembled. He was given a contract for 2000 guns to be made in two years. But what is the rest of his story?
- Joe Brown started a business in Rhode Island in 1833 making lathes and small tools as well as timepieces (watchmaker). Lucian Sharp joined the company in 1848 and developed a pocket sheet metal gage in 1877 and a 1-inch micrometer, and in 1862 developed the universal milling machine.
- At age 16, Sam Colt sailed to Calcutta on the Brig "Curve." He whittled a wood model of a revolver on this voyage. He saved his money and had models of a gun built in Hartford by Anson Chase, for which he got a patent. He set up a factory in New Jersey—but he could not sell his guns to the Army because they were too complicated. He sold to the Texas Rangers and the Florida Frontiersmen, but he had to close the plant. In 1846, the Mexican war broke out. General Zachary Taylor and Captain Sam Walters wanted to buy guns. Colt had none but accepted orders for 1000 guns and constructed a model (Walker Colt); he arranged to have them made at Whitney's (now 40-year-old) plant in Whitneyville. Here he learned about mass production methods. In 1848, he rented a plant in Hartford, Connecticut, and the Colt legend spread. In 1853 he had built one of the world's largest arms plant in Connecticut, which had 1400 machine tools. Colt helped start the careers of
    - E. K. Root, mechanic and superintendent, paying him a salary of $25,000 in the 1800s. Abolished hand work—jigs and fixtures.
    - Francis Pratt and Amos Whitney—famous machine tool builders.
    - William Gleason—gear manufacturer
    - E. P. Bullard—invented the Mult–An–Matic Multiple spindle machine, which cut the time to make a flywheel from 18 minutes to slightly over 1 minute. Sold this to Ford.
    - Christopher Sponer.
    - E. J. Kingsbury—invented a drilling machine to drill holes through toy wheel hubs that had a spring-loaded cam that enabled the head to sense the condition of the casting and modify feed rate automatically.

Now here are some more names from the past of famous and not-so-famous manufacturing, mechanical, and industrial engineers. Relate them to the development of manufacturing processes or manufacturing system designs.

- Eli Whitney
- Henry Ford
- Charles Sorenson
- Sam Colt
- John Parsons
- Eiji Toyoda
- Elisha Root
- John Hall
- Thomas Blanchard
- Fred Taylor
- Taiichi Ohno
- Ambrose Swasey

CHAPTER 2

MANUFACTURING SYSTEMS DESIGN

2.1 INTRODUCTION

In a factory, *manufacturing processes* are assembled together to form a *manufacturing system* to produce a desired set of goods. The manufacturing system takes specific inputs and materials, adds value through processes, and transforms the inputs into products for the customer. It is important to distinguish between the manufacturing system and the production system, which is also known as the enterprise system, or the whole company. The production system includes the manufacturing system.

As shown in Figure 2-1, the **production system** services the manufacturing system, using all the other functional areas of the plant for information, design, analysis, and control. These subsystems are connected to each other to produce goods or services, or both.

A production system includes all aspects of the business, including design engineering, manufacturing engineering, sales, advertising, production and inventory control (scheduling and distribution), and, most important, the manufacturing system. The enterprise and all its functional areas are discussed in Advanced Topic 3 on the Web.

2.2 MANUFACTURING SYSTEMS

A sequence of processes and people that actually produce the desired product(s) is called the **manufacturing system.** In Figure 2-2, the manufacturing system is defined as the *complex arrangement of physical elements characterized (and controlled) by measurable parameters* (Black, 1991). The relationship among the elements determines how well the system can run or be controlled. The control of a system refers to the entire manufacturing system (which means the control of the operators in a harmonious way relative to the system's objectives), not merely the individual processes or equipment. The entire manufacturing system must be under daily control to enable the management of material movement, people and processes (scheduling), inventory levels, product quality, production rates, throughput, and, of course, cost.

As shown in Figure 2-2, inputs to the manufacturing system include materials, information, and energy. The system is a complex set of elements that includes machines (or machine tools), people, materials-handling equipment, and tooling. Workers are the internal customers. They process materials within the system, which gain value as the material progresses from process to machine. Manufacturing system outputs may be finished or semifinished goods. Semifinished goods serve as inputs to some other process at other locations. Manufacturing systems are dynamic, meaning that they must be designed to adapt constantly to change. Many of the inputs cannot be fully controlled by management, and the effect of disturbances must be counteracted by manipulating the controllable inputs or the system itself. Controlling the input

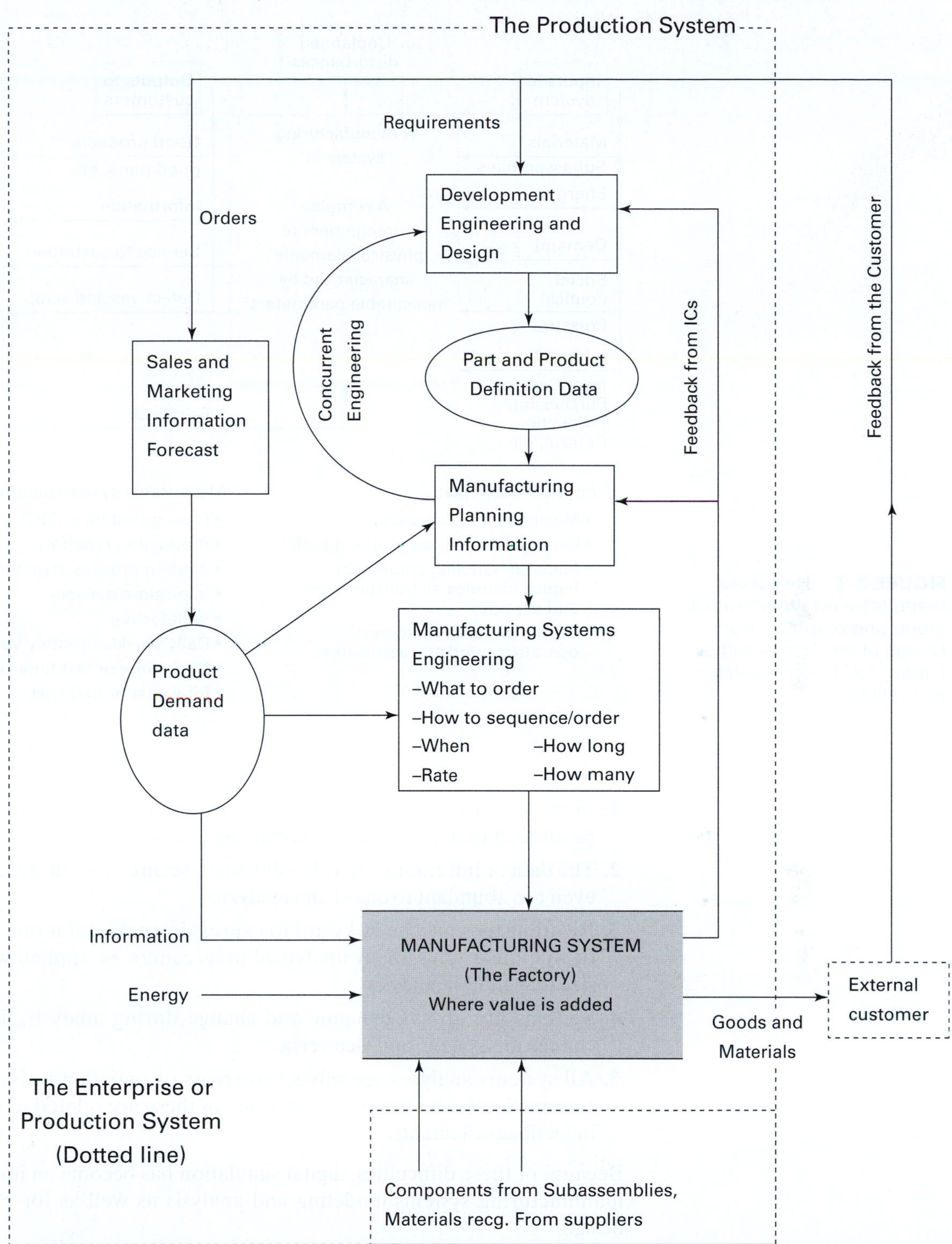

FIGURE 2-1 The manufacturing system (shaded) is the heart of the company. It lies within and is served by the production system or the enterprise.

material availability and/or predicting demand fluctuations may be difficult. A national economic decline or recession can cause shifts in the business environment that can seriously change any of these inputs. In manufacturing systems, not all inputs are fully controllable. To understand how manufacturing systems work and be able to design manufacturing systems, computer modeling (simulation) and analysis are used. However, modeling and analysis are difficult because

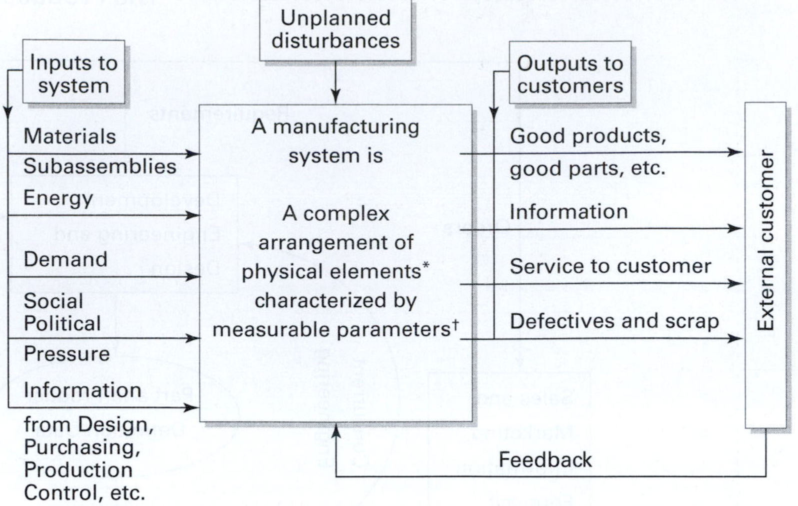

* Physical elements:

- Machines for processing
- Tooling (fixtures, dies, cutting tools)
- Material handling equipment (which includes all transportation and storage)
- People (internal customers) operators, workers, associates

† Measurable system parameters:

- Throughput time (TPT)
- Production rate (PR)
- Work-in-process inventory
- % on-time delivery
- % defective
- Daily/weekly/monthly volume
- Cycle time or takt time (TT)
- Total cost or unit cost

FIGURE 2-2 Here is our manufacturing system with its inputs and outputs. *(From* Design of the Factory with a Future, *1991, McGraw-Hill, by J T. Black)*

1. In the absence of a system design, the manufacturing systems can be very complex, be difficult to define, and have conflicting goals.
2. The data or information may be difficult to secure, inaccurate, conflicting, missing, or even too abundant to digest and analyze.
3. Relationships may be awkward to express in analytical terms, and interactions may be nonlinear; thus, many analytical tools cannot be applied with accuracy. System size may inhibit analysis.
4. Systems are always dynamic and change during analysis. The environment can change the system, and vice versa.
5. All systems analyses are subject to errors of omission (missing information) and commission (extra information). Some of these are related to breakdowns or delays in feedback elements.

Because of these difficulties, digital simulation has become an important technique for manufacturing systems modeling and analysis as well as for manufacturing system design.

2.3 CONTROL OF THE MANUFACTURING SYSTEM

In general, a manufacturing system should be an integrated whole, composed of integrated subsystems, each of which interacts with the entire system. The critical control functions are production rate and mix control, inventory control, quality control, and machine tool control (reliability).While the system may have a number of objectives or goals, the users of the system may seek to optimize the whole. Optimizing bits and pieces does not optimize the entire system. System control functions require information gathering, communication capabilities, and decision-making processes that are integral parts of the manufacturing system.

2.4 CLASSIFICATION OF MANUFACTURING SYSTEMS

Manufacturing industries vary by the products that they make or assemble. While almost all factories are different, there are five basic **manufacturing system designs (MSDs):** four classic (or hybrid combinations thereof) and one new manufacturing system design that is rapidly gaining acceptance in almost all of these industries. The classical systems here are the *job shop,* the *flow shop,* the *project shop,* and the *continuous process.* The *lean shop* is a new kind of system. The *assembly line* is a form of the flow shop, and a system that has *batch flow* might also be added as another manufacturing system. Figure 2-3 shows schematics of the four classical systems along with the linked-cell manufacturing system, or the lean shop.

Table 2-1 lists examples of five types of manufacturing systems. These lists are not meant to be complete, just informative as there are hundreds of examples of each of the basic system designs.

TABLE 2-1 Types and Examples of Manufacturing Systems

Type of Manufacturing System	Examples	
	Service[a]	Product[b]
Job shop	Auto repair	Machine shop
	Hospital	Metal fabrication
	Restaurant	Custom jewelry
	University	FMS
Flow shop or flow line	X-ray	TV factory
	Cafeteria	Auto assembly line
	College registration	
	Car wash	
Lean shop or linked-cell	Fast-food restaurant (KFC, Wendy's)	Product families
	Food court at mall	Family of turned parts
	10-minute oil change	Composite part families
		Design families
Project shop	Producing a movie	Locomotive assembly
	Broadway play	Bridge construction
	TV show	House construction
Continuous process	Telephone company	Oil refinery
	Phone company	Chemical plant

[a] Customer receives a service or perishable product.
[b] Products can be for customers or for other companies.

JOB SHOP

The job shop's distinguishing feature is its functional design. In the **job shop,** a variety of products are manufactured, which results in small manufacturing lot sizes, often one of a kind. Job shop manufacturing is commonly done to specific customer order, but, in truth, many job shops produce to fill finished-goods inventories. Because the plant must perform a wide variety of manufacturing processes, general-purpose production equipment is required. Workers must have relatively high skill levels to perform a range of different work assignments, often due to a lack of work standardization, defects, and variety in goods produced. Job shop products include space vehicles, aircraft, machine tools, special tools, and equipment. Figure 2-4 depicts the functionally arranged job shop. Production machines are grouped according to the general type of manufacturing process. The lathes are in one department, drill presses in another, plastic molding in still another, and so on. The advantage of this layout is its ability to make a wide variety of products. Each different part requiring its own unique sequence of operations can be routed through the respective departments in the proper order. In the job shop, **process planning** consists of determining the *sequence* of individual manufacturing processes and operations needed to produce the parts. An overview of the product is developed with a process flow chart, which shows the various levels of the product, subassemblies, and components. Figure 2-5 shows a process flow chart and a **bill of materials (BOM),** which lists all the parts and components in a product. The **route sheet** and the **operations sheet** are the documents that specify the process sequence through the job shop and the sequence of operations to be performed at specific machines. Route sheets are used as the production control device to define the path of the material through the manufacturing system; see Figure 2-6, for example. Forklifts and handcarts are used to move materials from one machine to the next. As the company grows, the job shop evolves into a production job shop making products in large lots or batches.

The route sheet lists manufacturing operations and associated machine tools for each workpiece. The route sheet travels with the parts, which move in batches (or lots) between the processes. When a cart of parts has to be moved from one point in the job shop to another, the route sheet provides routing (travel) information, telling the material handler which machine in which department the parts must go to next. Now look at Figure 2-4, which shows the path through the job shop that the punch would take from start to finish as described in the route sheet.

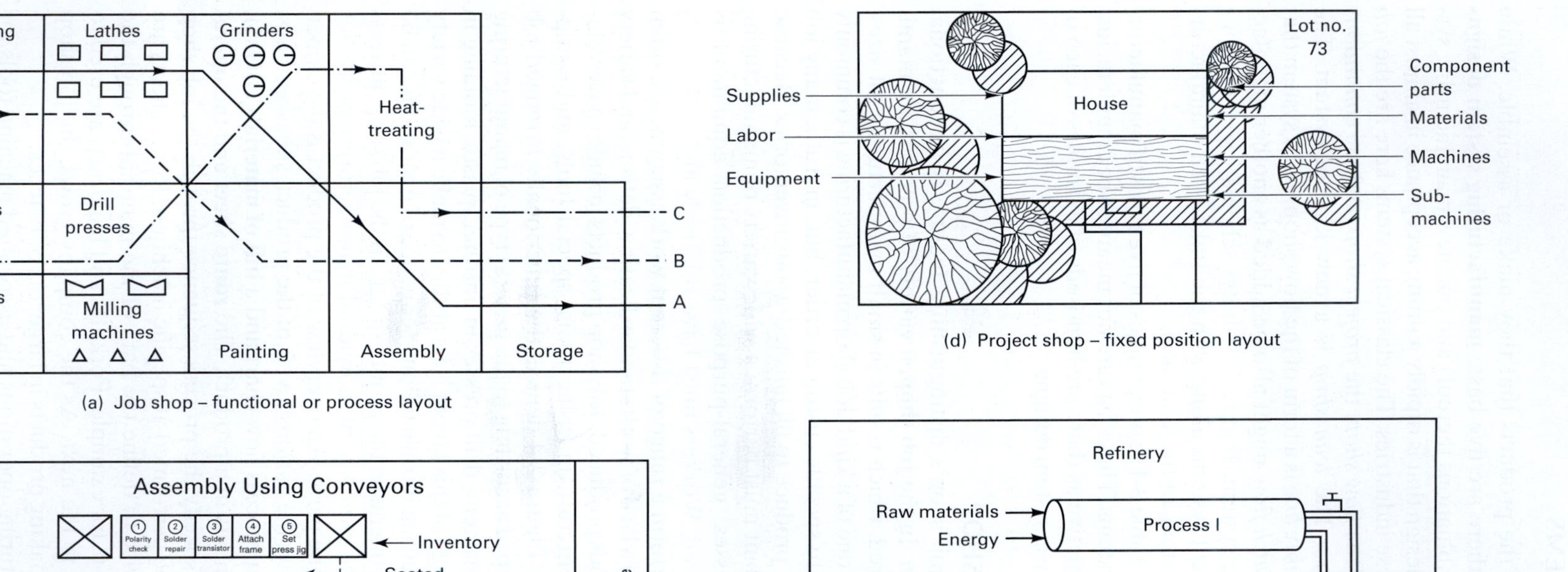

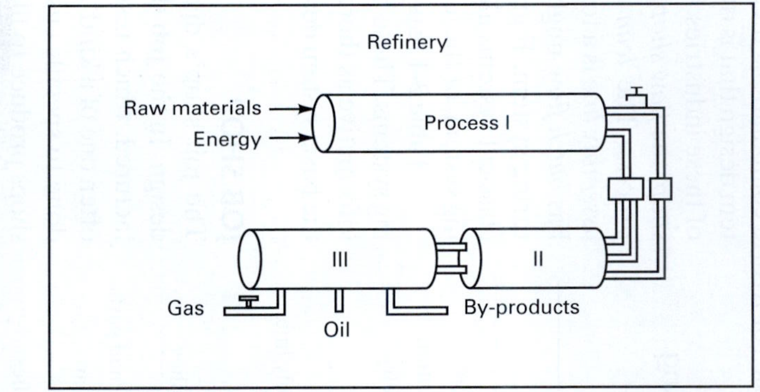

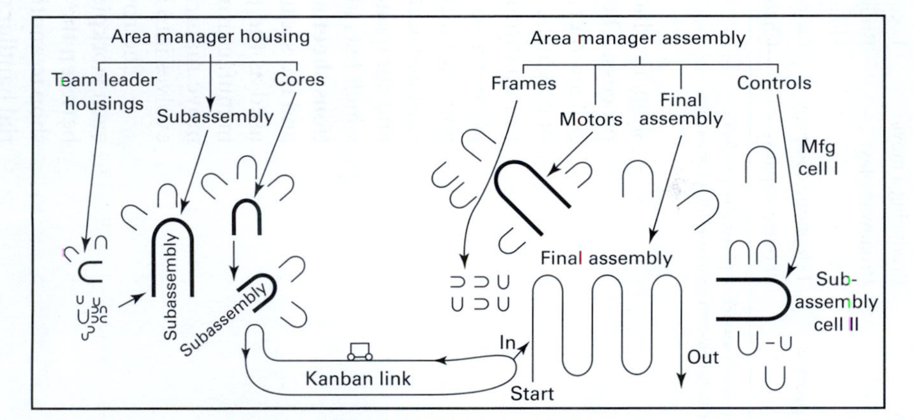

FIGURE 2-3 Schematic layouts of five manufacturing systems: (a) job shop (functional or process layout); (b) flow shop (line or product layout); (c) linked-cell layout, or the lean production system; (d) project shop (fixed-position layout); and (e) continuous process.

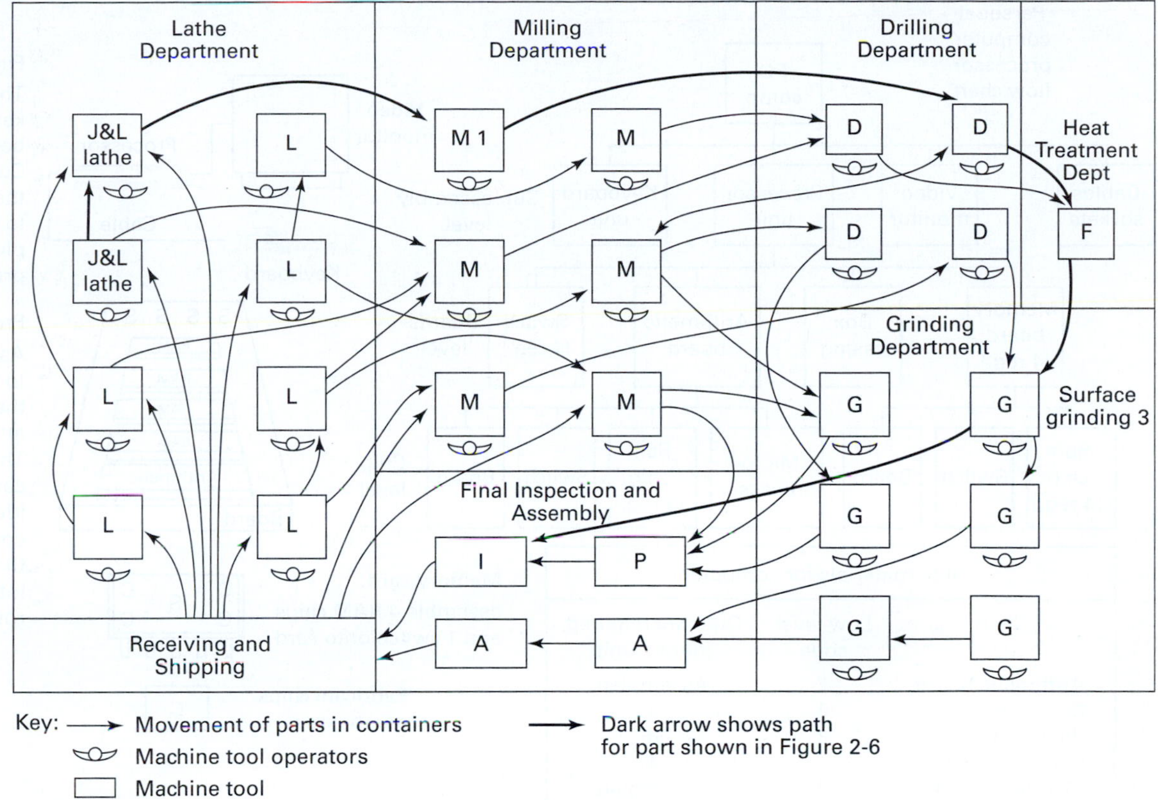

FIGURE 2-4 Schematic layout of a job shop where processes are gathered functionally into areas or departments. Each square block represents a manufacturing process. Sometimes called the "spaghetti design."

The operations sheet describes what machining or assembly operations are done to the parts at particular machines. The operations performed on an engine lathe to make the part shown in Figure 2-7 are shown. Note that the details of speed, feed, and depth of cut are specified (actually, recommended, because the machinist may change them). Over time, many different people may plan the same part; therefore, there can be many different process sequences and many different routes through the factory for the same or similar parts.

Process Planning for the Job Shop. The first step in planning is to determine the basic job requirements that must be satisfied. These are usually determined by analysis of the drawings and the job orders. They involve consideration and determination of the following:

1. Size and shape of the geometric components of the workpiece.
2. Tolerances, as applied by the designer.
3. Material from which the part is to be made.
4. Properties of material being machined (hardness).
5. Number of pieces to be produced (see the section in this chapter on quantity versus process and case studies on economic analysis).
6. Machine tools available for this workpiece.

Such an analysis for the threaded shaft shown in Figure 2-7 would be as follows:

1. a. Two concentric and adjacent cylinders having diameters of 0.877/0.873 and 0.501/0.499, respectively, and lengths of 2 in. and $1\frac{1}{2}$ in.
   b. Three parallel plain surfaces forming the ends of the cylinders.
   c. A 45° × $^1/_8$-in. bevel on the outer end of the $^7/_8$-in. cylinder.
   d. A $^7/_8$-in. NF-2 thread cut the entire length of the $^7/_8$-in. cylinder.

Personal computer processor flow chart

PC comp

Cables sockets | Video monitor | Processor unit | Keyboard unit | Sub-assembly level

Memory board (4 req.) | Box casing | Arithmetic board | Switch (4 req.) | Boards level

Ram chip (4 req.) | Switch | Board | Micro-processor | Ram chip (2 req.) | Switch | Board | Chip level

Video monitor

Processor

Cable

Keyboard

Final assembly

The video unit and the keyboard unit have been preassembled. Connect the sockets on the end of their cables to the corresponding plugs at the rear of the processor unit.

S S S S

RAM

RAM

RAM

RAM

Arithmetic

Board

Processor subassembly

Assemble one switch(s) to the inside of each of the 4 plug connections at the back of the box. Then fit 4 memory cards boards into the 4 identical rows of connectors. Finally, fit the arithmetic cards into the front connector row.

Memory card; assemble 4 RAM chips and 1 switch onto card

C S C
C C

Fabricate chips and boards

C

Bill of materials for computer

Item	Low-level code	Quantity required for one unit
Arithmetic board	2	Assembled
Box casing	2	1
Ram chip	3	16
Keyboard unit	1	1
Memory board	2	Assembled
Processor unit	1	Assembled
Ram chip	3	2
Switch	3	9
Video unit	1	1
Board type X	3	4
Board type Y	3	1
Microprocessor	3	1

FIGURE 2-5 The process flow chart and the bill of materials (BOM) for a personal computer. See Chapter 34.

2. The tightest tolerance is 0.002 in., and the angular tolerance on the bevel is $\pm 1°$.
3. The material is AISI1340 cold-rolled steel, BHN 200.
4. The job order calls for 25 parts.

Now the engineer can draw a number of conclusions regarding the processing of the part. First, because concentric, external, cylindrical surfaces are involved, turning operations are required and the piece should be made on some type of lathe. Second, because 25 pieces are to be made, the use of an engine lathe or a computer numerical control (CNC) lathe would be preferred over an automatic screw machine where the setup time will likely be too long to justify this small a lot size. As the company's numerical control (NC) lathe is in use, an engine lathe will be used. Third, because the maximum required diameter is approximately $^7/_8$ in., 1-in.-diameter cold-rolled stock will be satisfactory; it will provide about $^1/_{16}$ in. of material for rough and finish turning of the large diameter. From this information, the operations sheet(s) for a particular engine lathe is (are) prepared.

The engineer prepares the operations sheet shown in Figure 2-7 listing, in sequence, the operations required for machining the threaded shaft shown in the part drawing. A single operation sheet lists the operations that are done in sequence on a single machine. This sheet is for an engine lathe (see Chapter 22).

Operations sheets vary greatly as to details. The simpler types often list only the required operations and the machines to be used. Speeds and feeds may be left to the discretion of the operator, particularly when skilled workers and small quantities are involved. However, it is common practice for complete details to be given regarding

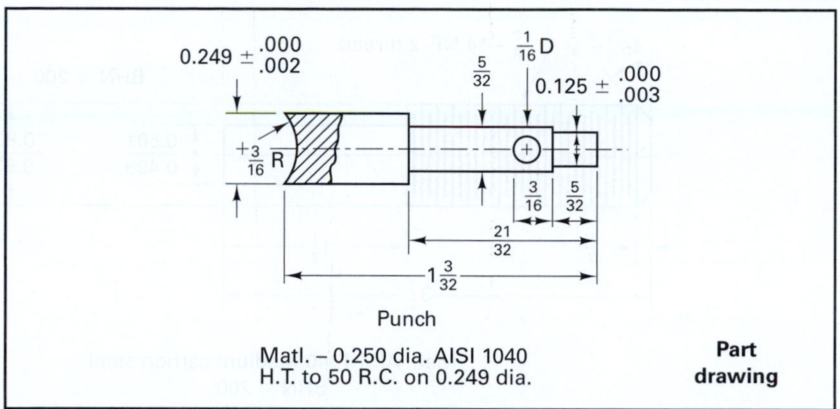

DARIC INDUSTRIES

ROUTING SHEET

NAME OF PART: Punch		PART NO.: 2	
QUANTITY: 1,000		MATERIAL: SAE 1040	
OPERATION NUMBER	DESCRIPTION OF OPERATION	EQUIPMENT OR MACHINES See Fig 41-4	TOOLING
1	Turn, $\frac{5}{32}$, 0.125, and 0.249 diameters	J & L turret lathe 12	#642 box tools
2	Cut off to $1\frac{3}{32}$ length	"	#6 cutoff in cross turret
3	Mill $\frac{3}{16}$ radius	#1 Milwaukee	Special jaws in vise $\frac{3}{16}$ form cutter × 4″ D
4	Drill $\frac{1}{16}$ hole	Turret Drill 4	
5	Heat treat. 1,700° F for 30 minutes, oil quench	Atmosphere furnace	
6	Grind (cutting edges)	Surface grinder 3	$\frac{3}{16}$ end radius on wheel
7	Check hardness	Rockwell tester	

FIGURE 2-6 Part drawing for a punch used in a progressive die set (above) with the route sheet (traveler) for making the punch. The route sheet is used in the job shop (layout in Figure 2-4) to tell the material handler (forklift truck driver) where to take the containers of parts after each operation is completed. The manufacturing engineer designs the process plan or sequence of operations to make the product.

tools, speeds, and often the time allowed for completing each operation. Such data are necessary if the work is to be done on NC machines, and experience has shown that these preplanning steps are advantageous when ordinary machine tools are used.

The selection of speeds and feeds required to manufacture the part is discussed in Chapters 20–24 and will depend on many factors such as tool material, workpiece

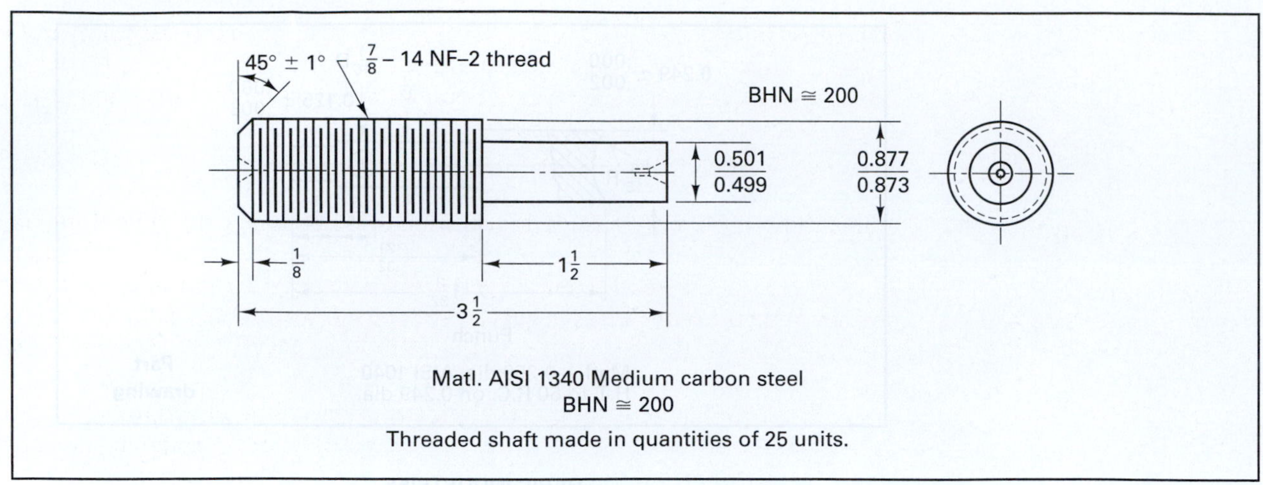

DEBLAKOS INDUSTRIES

OPERATION SHEET

PART NAME: Threaded Shaft 1340 Cold Rolled Steel

Part No. 7358-267-10

OPER. NO.	NAME OF OPERATION	MACH. TOOL	CUTTING TOOL	CUTTING SPEED ft/min	CUTTING SPEED rpm	FEED ipr	DEPTH OF CUT Inches	REMARKS
10	Face end of bar	Engine Lathe		120	458	Hand		Use 3-jaw Universal chuck
20	Center drill end	·	Combination center drill		750	Hand		
30	Cut off to $3\frac{9}{16}$ length	·	Parting tool	120	458	Hand		To prevent chattering, keep overhang of work and tool at a minimum and feed steadily. Use lubricant.
40	Face to length	·	RH facing tool (small radius point)	120	458	Hand	(R) $\frac{1}{8}$ max. (F) .005	Before replacing part in 3-jaw chuck, scribe a line marking the $3\frac{1}{2}$ inch length.
50	Center drill end	·	Combination center drill		750	Hand		
60	Place between centers, turn $\frac{.501}{.499}$ diameter, and face shoulder	·	RH turning tool (small radius point)	120 160	(R) 458 (F) 611	(R) .0089 (F) .0029	(R) .081 (3) (F) .007	
70	Remove and replace end for end and turn $\frac{.877}{.873}$ diameter	·	RH tools (R) (small radius point) (F) Round nose tool	120 160	(R) 458 (F) 611	(R) .0089 (F) .0029	(R) .057 (F) .005	
80	Produce 45°-chamfer	·	RH round nose tool	120	458	Hand	(R) $\frac{1}{8}$ max. (F) .005	
90	Cut $\frac{7}{8}$ - 14 NF-2 thread	·	Threading tool	60	208		(R) .004 (F) .001	(1) Swivel compound rest to 30 degrees. (2) Set tool with thread gage. (3) When tool touches outside diamter of work set cross slide to zero. (4) Depth of cut for roughing = .004. (5) Engage thread dial indicator on any line. (6) Depth of cut for finishing = .001 Use compound rest.
	Remove burrs and sharp edges	·	Hand file					

Date ____________ ____________

FIGURE 2-7 Part drawing (above) and operations sheet, which provides details of the recommended manufacturing process steps.

material, and depth of cut. Tables of machining data will give suggested values for the turning and facing operations for either high-speed-steel or carbide tools. The tables are segregated by workpiece material (medium-carbon-alloy steels, wrought or cold worked) and then by process—in this case, turning. For these materials, additional tables for drilling and threading would have to be referenced. The depth of cut dictates the speed and feed selection. The depth of cut for a roughing pass for operation 60 was 0.081 in., and the finish pass was 0.007 in. Therefore, for operation 60, three roughing cuts and one finishing cut are used. Looking ahead to Chapter 22, could you estimate or determine the cutting time for one of these roughing cuts?

While the operation sheet does not specifically say so, high-speed-steel tools are being used throughout. If the job were being done on an NC lathe, it is likely that all the cutting tools would be carbides, which would change all the cutting parameters. Notice also that the job as described for the engine lathe required two setups, a three-jaw chuck to get the part to length and produce the centers and a between-centers setup to complete the part. It was necessary to stop the lathe to invert the part. Do you think this part could be manufactured efficiently in an NC lathe? How would you orient the part in the NC lathe?

Speed and feed recommendations are often computerized to be compatible with computer-aided manufacturing and computer-aided process planning systems, but the manufacturing engineer is still responsible for the final process plan. As noted earlier, the cutting time, as computed from the machining parameters, represents only 30 to 40% of the total time needed to complete the part. Referring again to Figure 2-6, the operator will need to pick up the part after it is cut off at operation 30, stop the lathe, open up the chuck, take out the piece of metal in the chuck, scribe a line on the part marking the desired $3\frac{1}{2}$-in. length, place the part back in the chuck, change the cutoff tool to a facing tool, adjust the facing tool to the right height, and move the tool to the proper position for the desired facing cut in operation 40. All these operations take time, and someone must estimate how much time is required for such noncutting operations if an accurate estimate of total time to make the part is to be obtained.

When an operation is machine controlled, as in making a lathe cut of a certain length with power feed, the required time can be determined by simple mathematics. For example, if a cut is 10 in. in length and the turning speed is 200 rpm, with a feed of 0.005 in. per revolution, the cutting time required will be 10 minutes. See Chapter 22 for details of this calculation. Procedures are available for determining the time required for people-controlled machining elements, such as moving the carriage of a lathe by hand, back from the end of one cut to the starting point of a following cut. Such determinations of time estimates are generally considered the job of the industrial engineer, and space does not permit us to cover this material, but the techniques are well established. Actual time studies, accumulated data from past operations, or some type of motion-time data, such as MTM (methods-time-measurement), can be employed for estimating such times. Each can provide accurate results that can be used for establishing standard times for use in planning. Various handbooks and books on machine shop estimating contain tables of average times for a wide variety of elemental operations for use in estimating and setting standards. However, such data should be used with great caution, even for planning purposes, and they should never be used as a basis of wage payment. The conditions under which they were obtained may have been very different from those for which a standard is being set.

Quantity versus Process and Material Alternatives. Most processes are not equally suitable and economical for producing a wide range of quantities for a given product. Consequently, the quantity to be produced should be considered, and the product design should be adjusted to the process that actually is to be used before the design is finalized. As an example, consider the part shown in Figure 2-8. Assume that, functionally, a brass alloy, a heat-treated aluminum alloy, or stainless steel would be suitable materials. What material and process would be most economical if 10, 100, or 1000 parts were to be made?

If only 10 parts were to be made, lathe turning, milling the flat, milling the slot, and drilling and tapping the holes would be very economical. The part could be machined out of bar stock. Casting would require the making of a pattern, which would be about as costly to produce as the part itself. It is likely that a suitable piece of stainless steel, brass, or heat-treated aluminum alloy would be available in the correct diameter and finish so that the largest diameter would not need any additional machining. Brass may

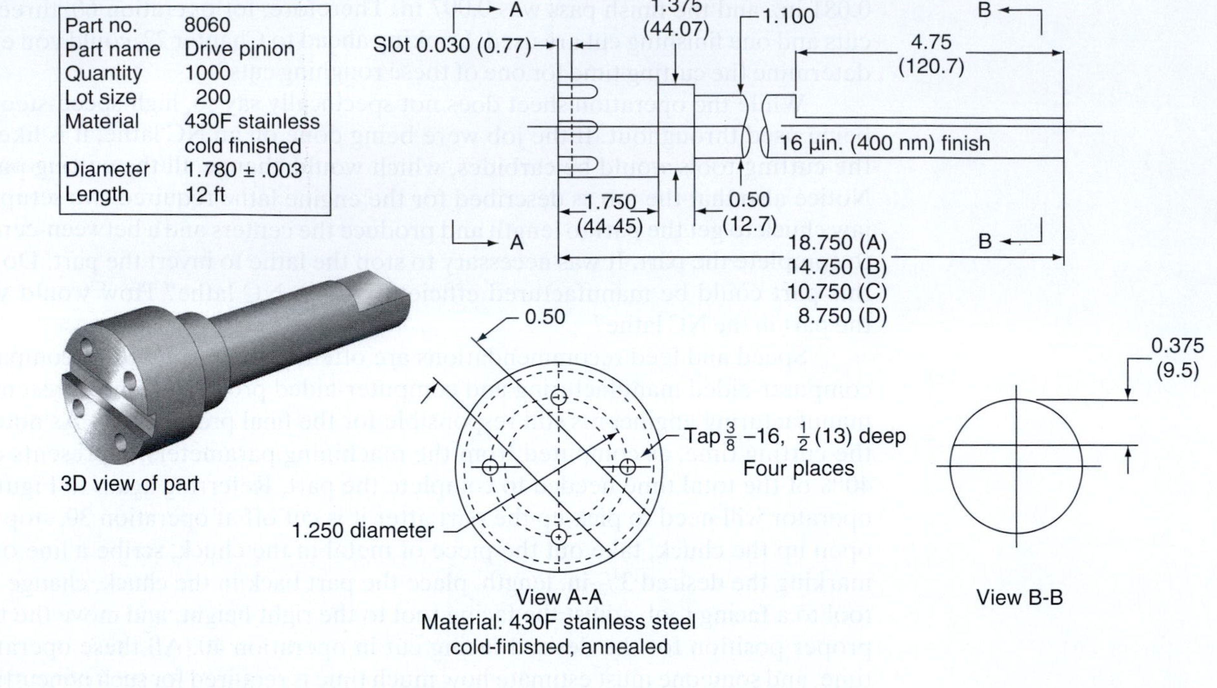

Part no. 8060 Ordering quantity 1000

Part name Drive Pinion Lot requirement 200

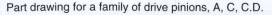

Material 430F Stainless steel, 1.780 ± 0.003 in. cold - finished 12-ft bars = 1000 pieces

Unit material cost $ 22.47

Workstation	Operation no.	Description of operations (list tools and gages)	Setup hour	Cycle hour/ 100 units	Unit estimate	Labor rate	Labor + overhead rate	Cost for labor + overhead rate
Engine lathe # 137	10	Face A-A end 0.05 center drill. A-A and rough turn has cast off to length 18.750	3.2	10.067	0.117	18.35	1.70	3.65
Engine lathe # 227	11	Center drill B-B end finish turn 1.100 turn 1.735 dirn	3.2	8.067	0.095	18.35	1.70	2.96
Vertical drill # 357	20	End mill 0.50 stat with 1/2 HSS and mill (collet future)	1.8	7.850	0.088	19.65	1.85	3.20
Horizontal drill # 469	30	Slab mill 4.75 x 3/8 (resting vise HSS tool)	1.3	1.500	0.022	19.65	1.80	0.78
NC rarret drill press # 474	40	Drill 3/8 hcles-4x tap 3/8-16 (collet fixture)	0.66	5.245	0.056	17.40	2.15	2.10
Cylindrical Grinder # 67	50	Grind shaft to 16gm - 1.10	1.0	10.067	0.110	19.65	1.80	3.89

$ 16.58 + 22.47 = 39.051

Estimated mig cost per unit

FIGURE 2-8 (Top) Part drawing with 3D view of part. (Bottom) Process planning sheet based on manufacturing the pinion in a job shop.

be slightly more expensive than the other materials in low volumes. However, brass material with sawing, turning, milling, and drilling most likely would be the best combination. For 10 parts, the excess cost of brass over stainless steel would not be great, and this combination would require no special consideration on the part of the designer.

For a quantity of 100 parts, an effort to minimize machining costs may be worthwhile. Forging discussed in Chapter 16 might be the most economical process, followed by machining. Stainless steel would be cheaper than any of the other permissible materials. Although the design requirements for forging this simple shape would be minimal, the designer would want to consider them, particularly as to whether the part should be forged, as this will reduce machining time. (Can stainless steel be forged?)

For 1000 parts, entirely different solutions become feasible. The use of an aluminum extrusion, with the individual units being sawed off, might be the most economical solution, assuming that the lead time required obtaining a special extrusion die was not a detriment. Aluminum can be cut at much higher speeds than stainless steel. What material would you recommend for the part now? How many feet of extrusion would be required, including sawing allowance? How much should be spent on the die so that the per-piece cost would not be great and probably would be more than offset by the savings in machining costs? If extrusion is used, the designer should make sure that any tolerances specified are well within commercial extrusion tolerances.

So what did the manufacturing engineer (MfE) decide to do for this part? Figure 2-8 shows the process plan developed for this part using stainless steel bars that were purchased in 12-ft lengths with diameters of 1.70 ± 0.003 in. Follow the processing steps (operations 10–50) to understand how the pinion was made. Notice that the process plan is used to obtain a cost estimate for this part. The total cost estimate takes into account all of the one-time or fixed costs associated with the component plus the labor and material costs (variable costs) required to make the component. Estimating the cost of making a product prior to ever making the first one is usually the job of the industrial engineer (IE) or MfE, and there are many good reference texts available for this work.

This simple example clearly illustrates how quantity can interact with material and process selection and how the selection of the process may require special considerations and design revisions on the part of the designer. Obviously, if the dimensional tolerances were changed, entirely different solutions might result. When more complex products are involved, these relations become more complicated, but they also are usually more important and require detailed consideration by the designer.

The production job shop becomes extremely difficult to manage as it grows, resulting in long product throughput times and very large in-process inventory levels. In the job shop, parts typically spend 95% of the time waiting (delay) or being transported and only 5% of the time on the machine. Thus, the time spent actually adding value may only be 2 or 3% of the total time available. The job shop typically builds large volumes of products but still builds lots or batches, usually medium-sized lots of 50 to 200 units. The lots may be produced only once, or they may be produced at regular intervals. The purpose of batch production is often to satisfy continuous customer demand for an item. This system usually operates in the following manner: Because the production rate can exceed the customer demand rate, the shop builds an inventory of the item, then changes over the machines to produce other products to fill other orders. This involves tearing down the setups on many machines and resetting them for new products. When the stock of the first item becomes depleted, production is repeated to build the inventory again. See Advanced Topic 3 on the Web for a discussion on inventory control.

Some machine tools are designed for higher production rates. For example, automatic lathes capable of holding many cutting tools can have shorter processing times than engine lathes. The machine tools are often equipped with specially designed workholding devices, jigs, and fixtures, which increase process output rate, precision, accuracy, and repeatability.

Industrial equipment, furniture, textbooks, and components for many assembled consumer products (household appliances, lawn mowers, and so on) are made in production job shops. Such systems are called machine shops, foundries, plastic-molding factories, and pressworking shops. Because the job shop has for years been the dominant factory design, most service companies are also job shops, organized functionally. Does your school have separate engineering or technology departments? Are you routed to different departments for processing? This is how the job shop works.

FLOW SHOP

The **flow shop** has a product-oriented layout composed mainly of flow lines. When the volume gets very large, especially in an assembly line, this is called **mass production,** shown schematically in Figure 2-9. This kind of system can have (very) high production rates. Specialized equipment, dedicated to the manufacture of a particular product, is used. The entire plant may be designed exclusively to produce the particular product or family of products, using special-purpose rather than general-purpose equipment. The investment in specialized machines and specialized tooling is high. Many production skills are transferred from the operator to the machines so that the manual labor skill level in a flow shop tends to be lower than in a production job shop. Items are made to "flow" through a sequence of operations by material-handling devices (conveyors, moving belts, and transfer devices).The items move through the operations one at a time. The time the item spends at each station or location is fixed or equal (balanced). Figure 2-10 shows a layout of an assembly line or flow line requiring line balancing.

Figure 2-11 shows an example of an automated transfer machine for the assembly of engine blocks at the rate of 100 per hour. All the machines are specially designed and built to perform specific tasks and are not capable of making any other products. Consequently, to be economical, such machines must be operated for considerable periods of time to spread the cost of the initial investment over many units. These machines and systems, although highly efficient, can be utilized only to make products in very large volume, hence the term *mass production*. Changes in product design are often avoided or delayed because it would be too costly to change the process or scrap the machines.

However, as we have already noted, products manufactured to meet the demands of free-economy, mass-consumption markets need to incorporate changes in design for improved product performance as well as style changes. Therefore, hard automation systems need to be as flexible as possible while retaining the ability to mass-produce.

The incorporation of **programmable logic controllers (PLCs)** and feedback control devices have made these machines more flexible. Modern PLCs have the functional sophistication to perform virtually any control task. These devices are rugged, reliable, easy to program, and economically competitive with alternative control devices. PLCs have replaced conventional hard-wired relay panels in many applications because they are easy to reprogram. Relay panels have the advantage of being well understood by maintenance people and are invulnerable to electronic noise, but construction time is long and tedious. PLCs allow for mathematical algorithms to be included in the closed-loop control system and are being widely used for single-axis, point-to-point control as typically required in straight-line machining, robot handling, and robot-assembly applications. They do not at this time challenge the computer numerical controls used on multi-axis contouring machines. However, PLCs are used for monitoring temperature, pressure, and voltage on such machines. PLCs are used on transfer lines to handle complex material movement problems, gaging, automatic tool setting, online tool wear compensation, and automatic inspection, giving these systems flexibility that they never had before.

The transfer line has been combined with CNC machines to form **flexible manufacturing systems (FMSs).**These systems are discussed in Chapter 25.

In the flow-line manufacturing system, the processing and assembly facilities are arranged in accordance with the product's sequence of operations; see Figure 2-11. Workstations or machines are arranged in line with only one workstation of a type, except where duplicates are needed for balancing the time products take at each station. The line is organized by the processing sequence needed to make a single product or a regular mix of products. A hybrid form of the flow line produces batches of products moving through clusters of workstations or processes organized by product flow. In most cases, the setup times to change from one product to another are long and often complicated.

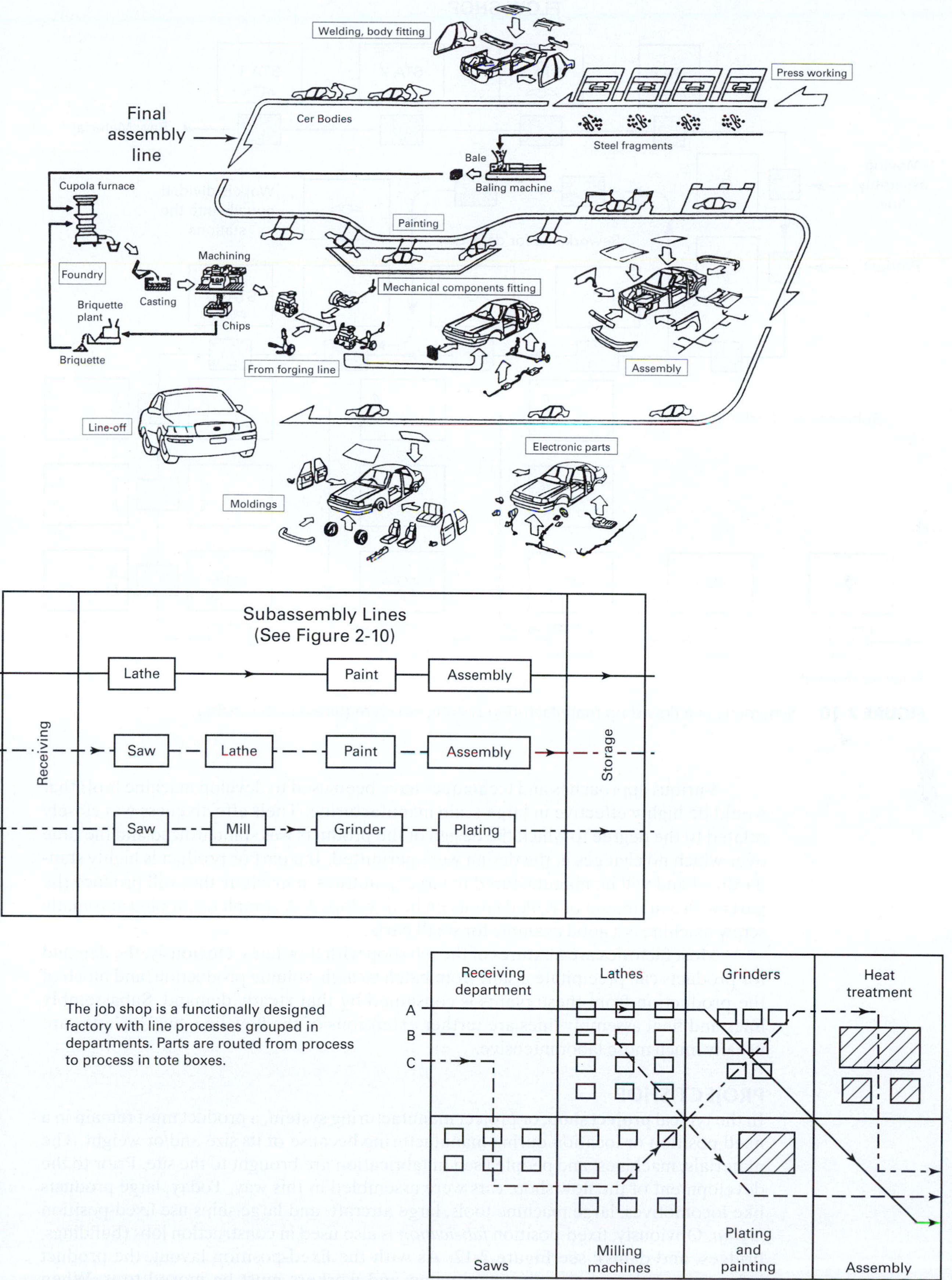

FIGURE 2-9 The mass production system produces large volumes at low unit cost, and personifies the economy of scale where large fixed costs are spread out over many units.

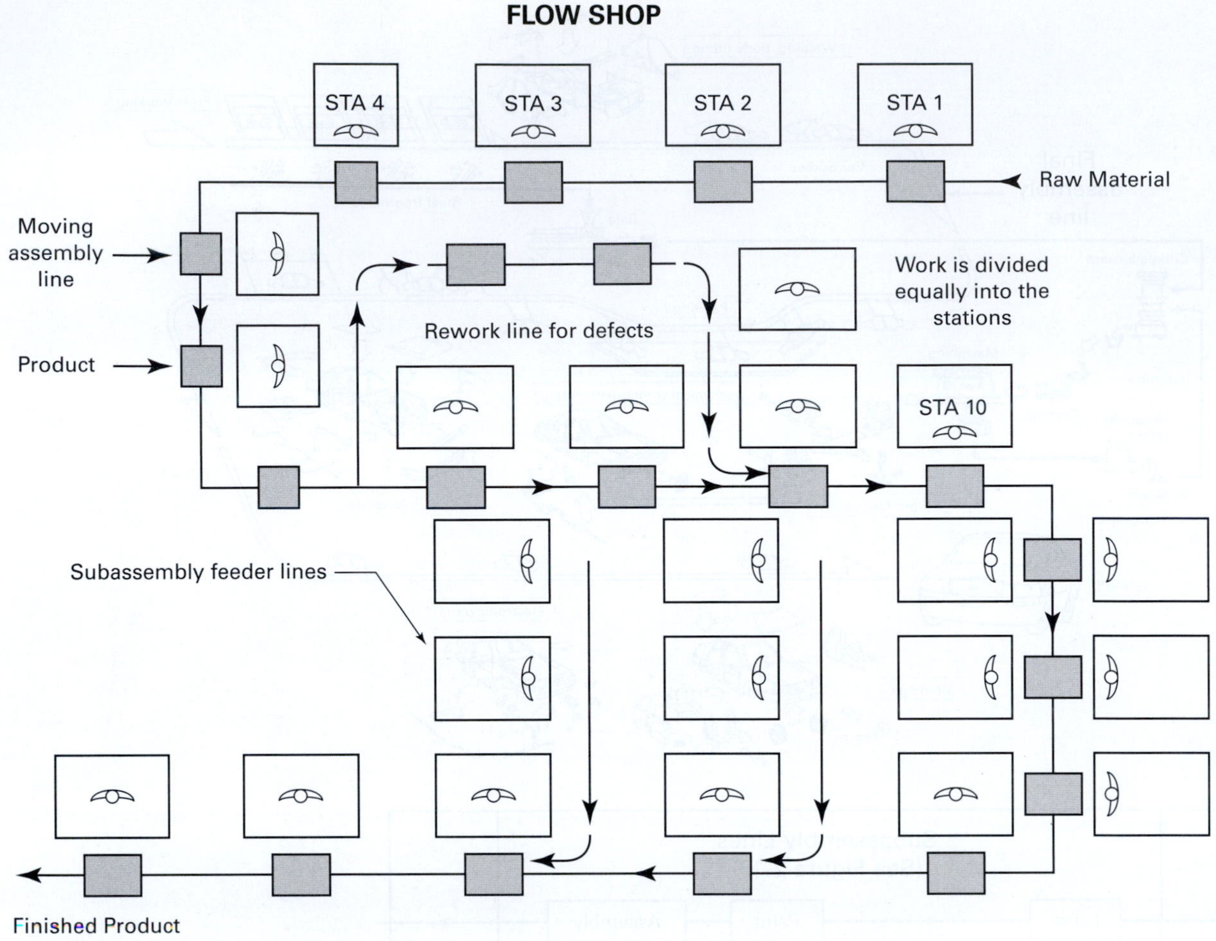

FIGURE 2-10 Schematic of a flow shop manufacturing system, which requires line balancing.

Various approaches and techniques have been used to develop machine tools that would be highly effective in large-scale manufacturing. Their effectiveness was closely related to the degree to which the design of the products was standardized and the time over which no changes in the design were permitted. If a part or product is highly standardized and will be manufactured in large quantities, a machine that will produce the parts with a minimum of skilled labor can be developed. A completely tooled automatic screw machine is a good example for small parts.

Most factories are mixtures of the job shop with flow lines. Obviously, the demand for products can precipitate a shift from batch to high-volume production, and much of the production from these plants is consumed by that steady demand. Subassembly lines and final assembly lines are further extensions of the flow line, but the latter are usually much more labor intensive.

PROJECT SHOP

In the typical **project shop,** or project manufacturing system, a product must remain in a fixed position or location during manufacturing because of its size and/or weight. The materials, machines, and people used in fabrication are brought to the site. Prior to the development of the flow shop, cars were assembled in this way. Today, large products like locomotives, large machine tools, large aircraft, and large ships use fixed-position *layout.* Obviously, fixed-position *fabrication* is also used in construction jobs (buildings, bridges, and dams); see Figure 2-12. As with the fixed-position layout, the product is large, and the construction equipment and workers must be moved to it. When the job is completed, the equipment is removed from the construction site. The project shop invariably has job shop/flow shop elements manufacturing all the

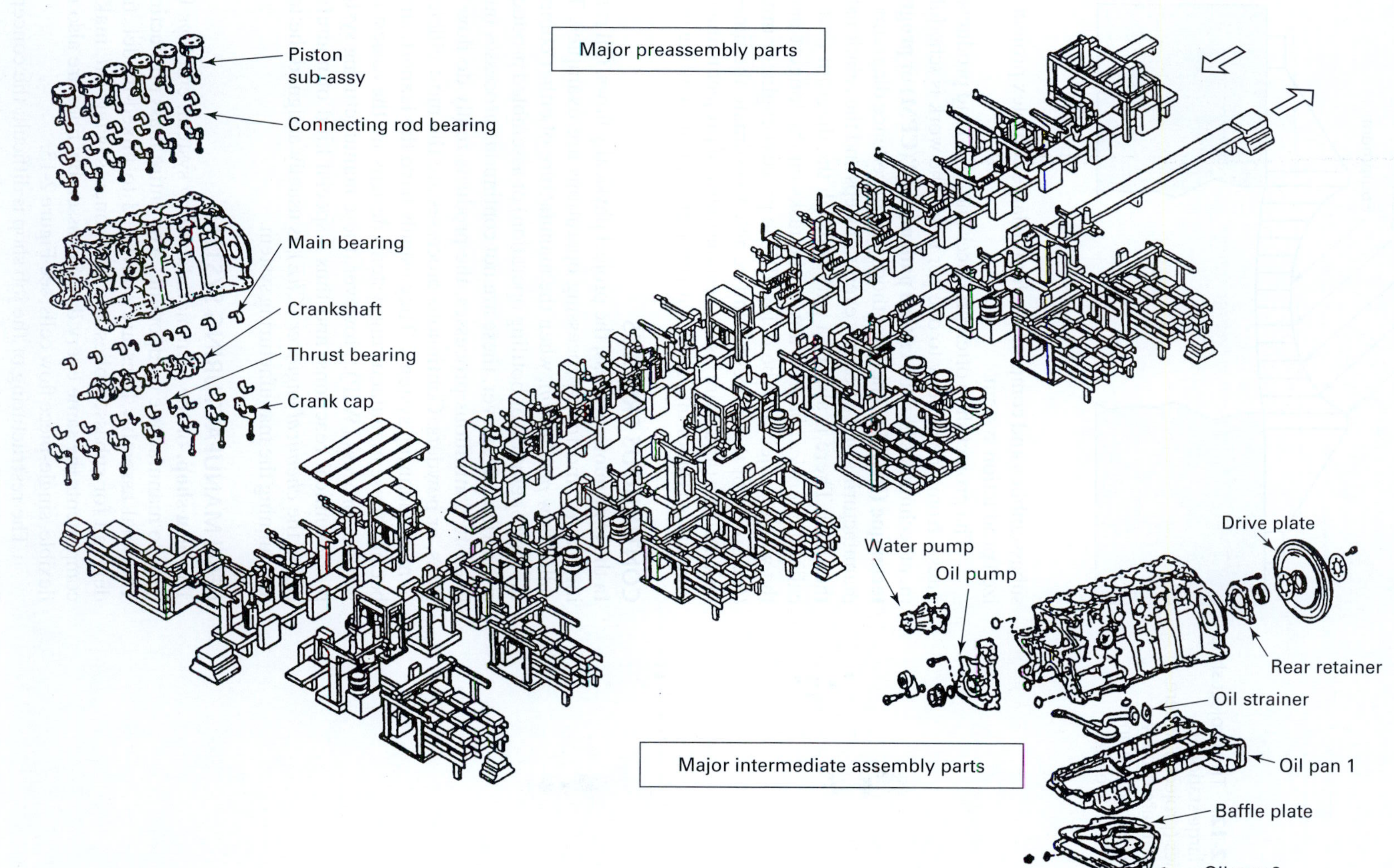

FIGURE 2-11 Transfer line for assembly of an engine is an example of a very automated flow line. *(From* Toyota Technical Review, *Vol. 44, no. 1, 1994)*

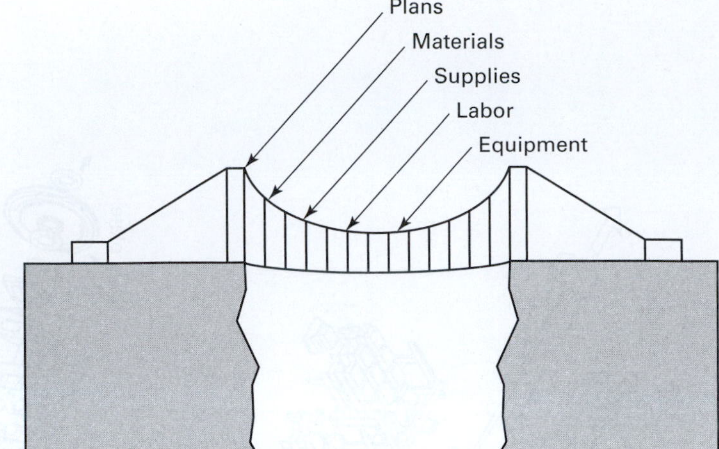

FIGURE 2-12 The project shop involves large stationary assemblies (projects), where components are usually fabricated elsewhere and transported to the site.

subassemblies and components for the large complex project and thus has a functionalized production system.

The project shop often produces one-of-a-kind products with very low production rates—from one per day to one per year. The work is scheduled using project management techniques like the **critical path method (CPM)** or **program evaluation and review technique (PERT).** These methods use precedence diagrams that show the sequence of manufacturing and assembly events or steps and the relationship (precedence) between the steps. There is always a path through the diagram that consumes the most time, called the *critical path.* If any of the tasks on the longest path are delayed, it is likely the whole project will be delayed. The process engineer must know the relationships (the order) of the processes and be able to estimate the times needed to perform each task. All this (and more) is the responsibility of project management. Project shop manufacturing is labor intensive, with projects typically costing in the millions of dollars.

CONTINUOUS PROCESS

In the **continuous process,** the project physically flows. Oil refineries, chemical processing plants, and food-processing operations are examples. This system is sometimes called flow *production*, when the manufacture of either complex single parts (such as a canning operation or bottling operation) or assembled products (such as television sets) is described. However, these are not continuous processes but rather high-volume flow lines. In continuous processes, the products really do flow because they are liquids, gases, or powders. Continuous processes are the most efficient but least flexible kinds of manufacturing systems. They usually have the leanest, simplest production systems because these manufacturing systems designs are the easiest to control, having the least work-in-process (WIP). However, these manufacturing systems usually involve complex chemical reactions, and thus a special kind of manufacturing engineer (MfE), called the *chemical engineer (ChE),* is usually assigned the task of designing, building, and running the manufacturing system.

LEAN MANUFACTURING SYSTEM

The **lean shop,** or the lean manufacturing system, employs U-shaped cells or parallel rows to manufacture components. The entire mass production factory is reconfigured. The final assembly lines are converted to mixed-model, final assembly so that the demand for subassemblies and components is *leveled*, making the daily demand for components the same every day. Subassembly lines are also reconfigured into volume-flexible, single-piece flow cells; see Figure 2-13.

The restructuring of the job shop is difficult; the concept is shown in Figure 2-14. In the cells, the machine tools are upgraded to be single-cycle automatics so parts can be loaded into the machine and the machining cycle started. The operator moves off to the next machine in the sequence, carrying the part from the previous process with him, so

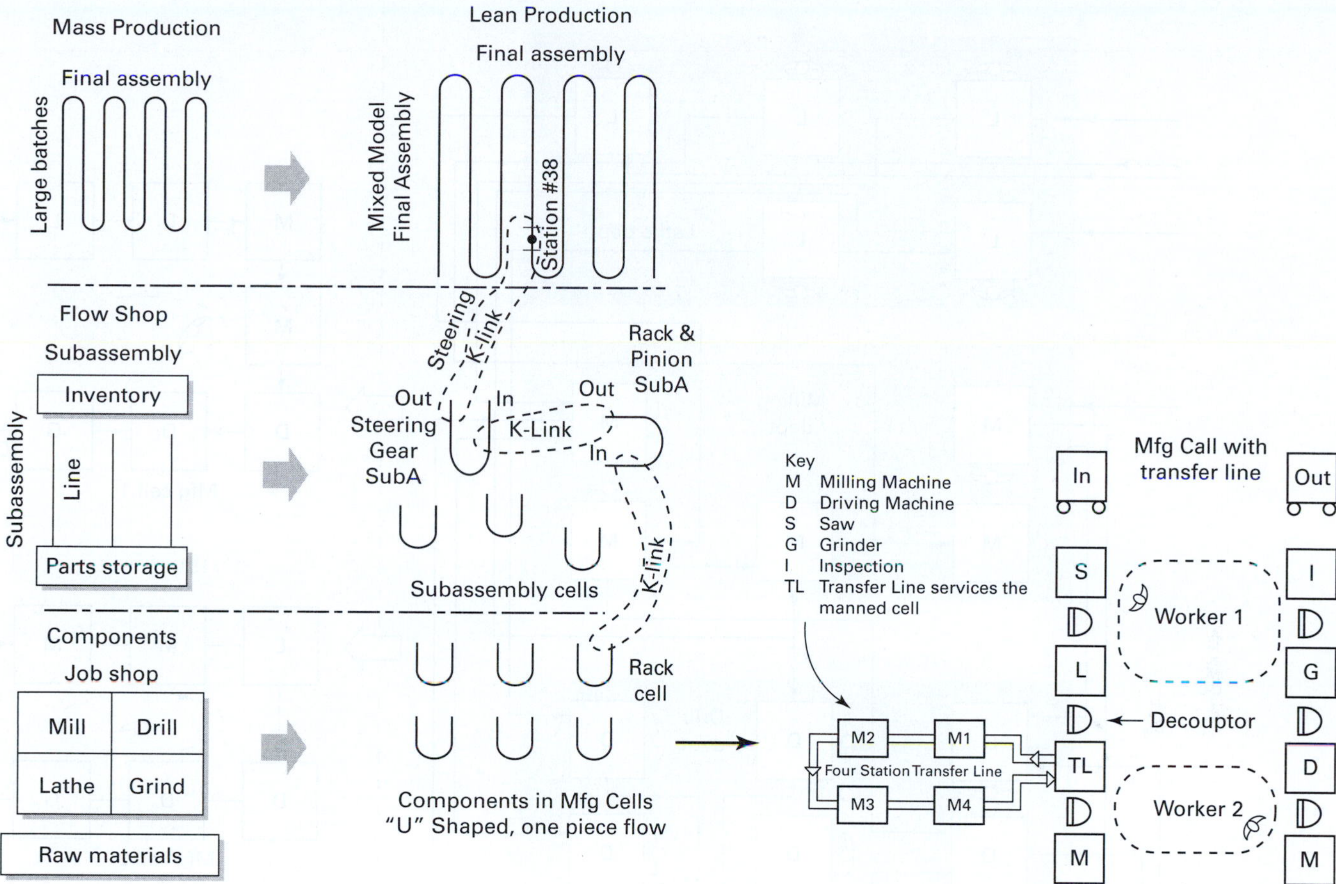

FIGURE 2-13 The manufacturing system design on the left, called "mass production," produces large volumes at low unit cost. It can be restructured into a lean manufacturing system design to achieve single-piece flow, on the requirements of a L-CMS design. On the right is a manufacturing cell with a small transfer serving one station in the cell.

the lean shop employs standing, walking multiprocess workers. System design requirements are used to design the cells. In a **linked-cell system,** the key proprietary aspects are the U-shaped manufacturing and assembly cells. In the cells, the axiomatic design concept of decoupling is employed to separate (decouple) the processing times for individual machines from the cycle time for the cell, enabling the lead time for a batch of parts to be independent of the processing times for individual machines. This effect of this is to take all the variation out of the supply chain lead times, so scheduling of the supply system becomes so simple that the supply chain can be operated by a pull system of production control *(kanban).* Using *kanban,* the inventory levels can be dropped, which decreases in turn, the throughput time for the manufacturing system.

Workers in the cells are also multifunctional; each worker can operate more than one kind of process and also perform inspection and machine maintenance duties according to a standard work pattern. Cells eliminate the job shop concept of one person–one machine and thereby greatly increase worker productivity and utilization. The restriction of the cell to a family of parts makes reduction of setup in the cell possible. The general approach to setup reduction is discussed in Chapter 27 as part of the cell design strategy.

In some cells **decouplers** are placed between the processes, operations, or machines to connect the movement of parts between operators. A decoupler physically holds one unit, decouples the variability in processing time between the machines, and enables the separation of the worker from the machine. Decouplers can provide flexibility, part transportation, inspection for defect prevention (poka-yoke) and quality control, and process delay for the manufacturing cell.

Here is how an inspection decoupler might work. The part is removed from a process and placed in a decoupler. The decoupler inspects the part for a critical dimension

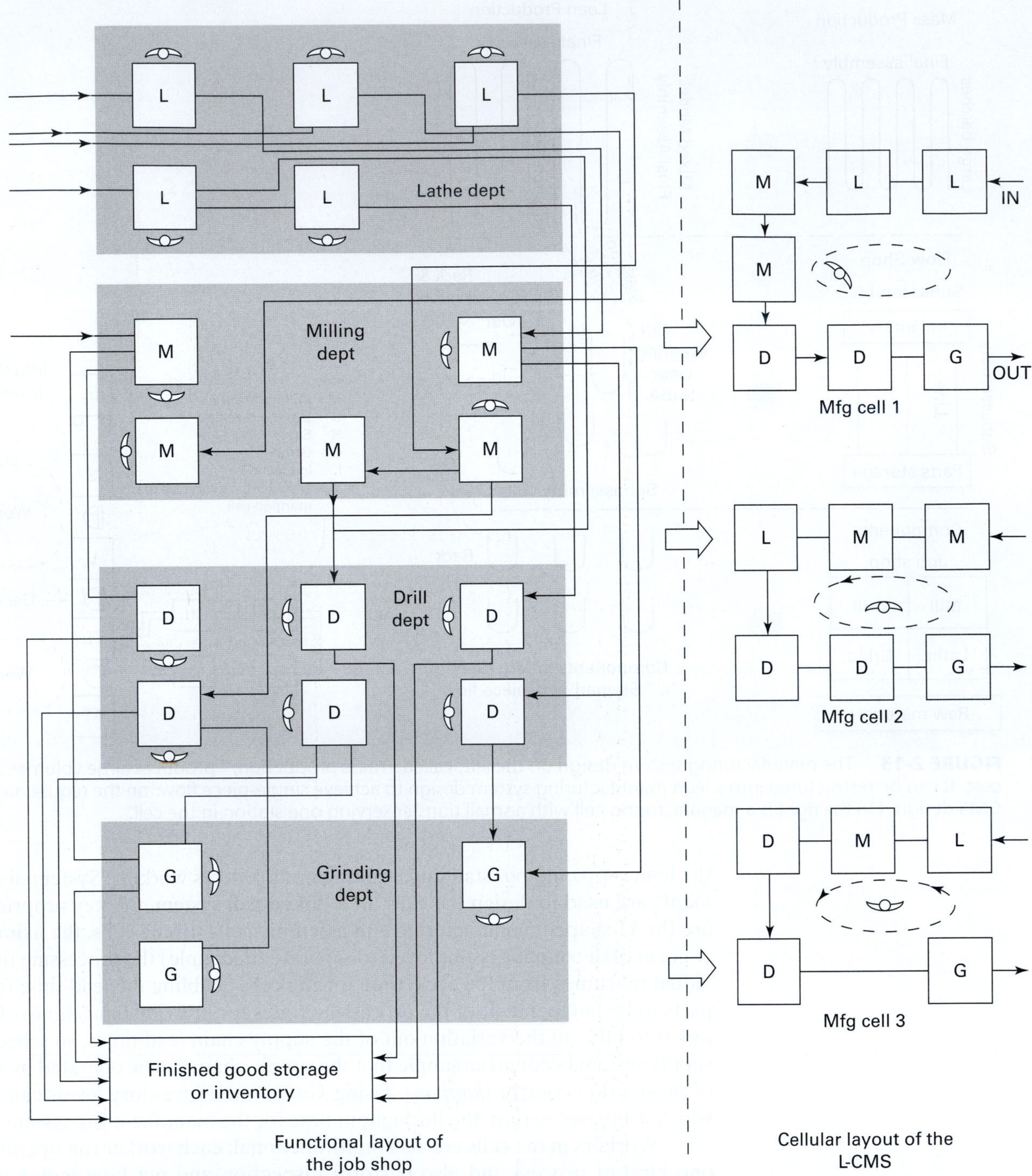

FIGURE 2-14 The job shop portion of the plant requires a systems level conversion to reconfigure it into manufacturing cells, operated by standing and walking workers.

and turns on a light if a bad (oversized) part is detected. A process delay decoupler delays the part movement to allow the part to cool down, heat up, cure, or whatever is necessary, for a period of time greater than the cycle time for the cell. Decouplers are vital parts of manned and unmanned cells and will be discussed further in Advanced Topic 1.

The system design ensures that the right mix and quantity of parts are made according to an averaged customer demand. This system is robust in that it ensures that the right quantity and mix are made, even though there is variation or disturbance in the manufacturing process or other operations in the manufacturing system. If process variation or disturbances to the system occur from outside the manufacturing system's

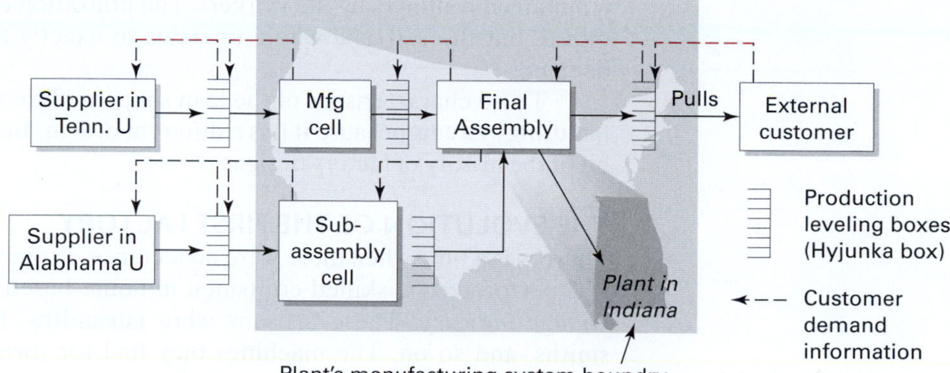

FIGURE 2-15 The linked-cell manufacturing pulls the goods from the suppliers through the assembly to the external customer.

boundary (i.e., incoming parts are defective), the system design is robust enough to handle these disturbances due to its feedback control design.

A typical linked-cell system is shown in Figure 2-15 with the customer demand information illustrated by dashed lines.

A specific level of inventory is established after each producing cell and is held in the *kanban* links. This set-point level is sometimes called the **standard work-in-process (SWIP)** inventory. The SWIP defines the minimum inventory necessary for the system to produce the right quantity and mix to the external customer and to be able to compensate for disturbances and variation to the system.

2.5 SUMMARY OF FACTORY DESIGNS

Let us summarize the characteristics of the lean shop versus the other basic factory designs; see Table 2-2. The cell makes parts one at a time in a flexible design. Cell capacity (the cycle time) can be altered quickly to respond to changes in customer demand by increasing or decreasing operators. The cycle time does not depend on the machining times.

Families of parts with similar designs, flexible workholding devices, and tool changers in programmable machines allow rapid changeover from one component to another. Rapid change are over means that quick or one-touch setup is employed, often like flipping a light switch. This leads to small lot production. Significant inventory reductions between the cells are possible and the inventory level can be directly controlled by the user. Quality is controlled within the cell, and the equipment within the cell is

TABLE 2-2 Characteristics of Basic Manufacturing Systems

Characteristics	Job Shop	Flow Shop	Project Shop	Continuous Process	Lean Shop
Types of machines	Flexible General purpose	Single purpose Single function	General purpose Mobile, manual Automation	Specialized, high technology	Simple, customized Single-cycle automatic
Design of processes	Functional or process Process	Product flow layout	Project or fixed-position layout	Product	Linked U-shaped cells
Setup time	Long, variable frequent	Long and complex	Variable, every job different	Skill level varies	Multifunctional, multipurpose
Workers	Single function, highly skilled: one worker–one machine	One function, lower-skilled: one worker–one machine	Specialized, highly skilled	Skill level varies	Multifunctional, multiprocess
Inventories (WIP)	Large inventory to provide for large variety	Large to provide buffer storage	Variable, usually large	Very small	Small
Lot sizes	Small to medium	Large lot	Small lot	Very large	Small
Manufacturing lead time	Long, variable	Short, constant time	Long, variable lead	Very fast, constant	Short, constant

maintained routinely by the workers. The utilization of individual machines is not optimized, but the cell as a whole operates to exactly meet the pace of the customer's demand.

These characteristics of the lean shop are different from that of the other manufacturing systems because it has a different design. In the next section, we will review a bit of the history of factory designs.

THE EVOLUTION OF THE FIRST FACTORY

There was a time when there were no factories. From 1700 to 1850, most manufacturing was performed by skilled craftsmen in home-based workshops, what historians call *cottage industry*. These artisans were gunsmiths, blacksmiths, toolmakers, silversmiths, and so on. The machines they had for forming and cutting materials were manually powered.

Amber & Amber (1962) in their yardstick for automation pointed out that the first step in mechanizing and automating the factory was to *power the machines*. The need to power the machine tools efficiently created the need to gather the machine tools into one location where the power source was concentrated. By this time, the steel and cast iron to build machine tools were becoming economically available. Railroads were built to transport goods. These were the technologies that influenced the factory design and are now called *enabling technologies;* see Table 2-3. A functional design evolved because of the *method* needed to drive or power the machines.

The first factories were built next to rivers, and the machines were powered by flowing water driving waterwheels, which drove overhead shafts that ran into the factory. See Figure 2-16. Machines of like type were set, all in a line, underneath the appropriate power shaft (i.e., the shaft that turned at the speed needed to drive this type of machine). Huge leather belts were used to take power off the shaft to the machines. Thus, a factory design of collected like processes evolved, and it came to be known as the job shop. So, all the lathes were collected under their own power shaft, likewise all the milling machines and all the drill presses. Later, the waterwheel was replaced by a steam engine, which allowed the factory to be built somewhere other than next to a river. Eventually, large electric motors were used, and later individual electric motors for each machine replaced the large steam and electric engines. Nevertheless, the job shop design was replicated, and the functional design held. Many early factories were in the northeastern part of the United States and were manufacturing guns. Filing jigs were used by laborers to create interchangeable parts. In time, this design became

TABLE 2-3 The Evolution of Manufacturing System Designs[a]

	First Factory Revolution	Second Factory Revolution	Third Factory Revolution	Fourth Factory Revolution
Time period	1840–1910	1910–1970	1960–2010	2000–???
Manufacturing system design	Job shop	Flow shop	Lean shop	Global Information Technology
Layout	Functional layout	Product layout	Linked-cells to Mixed Model Final Assembly line	Integrated supply chain
Enabling technologies	Power for machines, steel production, and railroad for transportation	Moving final assembly line, standardization leading to true interchangeability, and automatic material handling	U-shaped cells, *kanban,* and rapid die exchange	Virtual reality/simulation, 3D design using low-cost, and very high performance computers, digital technology[c]
Historical company name	Whitney, Colt, and Remington	Singer, Ford	Toyota Motor Company[b], Hewlett-Packard, Omark, Harley Davidson, Honda	Boeing, Lockheed, Electric Boat, and Mercedes
Economics	Economy of collected technology	Economy of scale, high volume > low unit cost	Economy of scope, wide variety at low unit cost	Economy of global manufacturing

[a] MSD (manufacturing system design), including machine tools, material handling equipment, tooling and people
[b] System developed by Taiichi Ohno, who called it the Toyota Production System (TPS)
[c] Single-source digital product definition and 3D design, a manufacturing system simulated using virtual reality

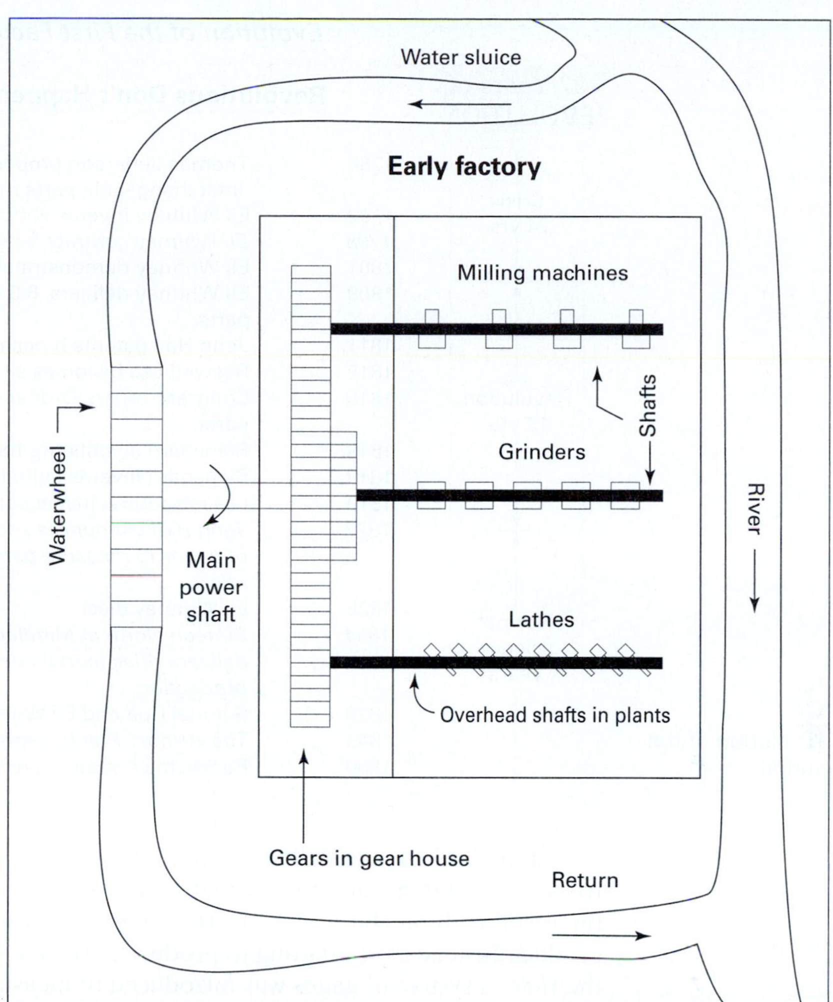

FIGURE 2-16 Early factories used water power to drive the machines.

known as the **American Armory System.** People from all over the world came to America to observe the American Armory System, and this functionally designed system was duplicated around the factory world as a key driver in the first factory revolution.

FACTORY REVOLUTIONS EVOLVE

Revolutions do not happen overnight but require many simultaneous events. An evolution in thought, philosophy, and the mindset of people in the factory, as well as the redesign of the factory, are needed to revolutionize the factory setup. Even though manufacturing is the leading wealth producer in the world, historians have often neglected the technical aspects of history.

The model used here is based on Thomas Kuhn's book, *The Structure of Scientific Revolutions*. He said there are three phases in any revolution: the crisis stage, the revolution stage, and the normal science stage, as shown in Figure 2-17. Some people have portrayed this last stage as the diffusion of science. The first factory revolution can be considered a crisis due to the problems associated with craft production such as producing products in which filing was used to fit products together. This crisis was highlighted in the United States by a congressional mandate to have interchangeable parts for muskets in the field in the late 18th century.

Eli Whitney received a contract from Congress to produce 4000 muskets in $1\frac{1}{2}$ years with interchangeable parts. In wartime there was a need for part interchangeability so that particular musket parts could be replaced in the field, instead of replacing the entire musket. The goal was to reduce replacement time and cost. At that time, muskets consisted of parts that were all craft-made, filed-to-fit products.

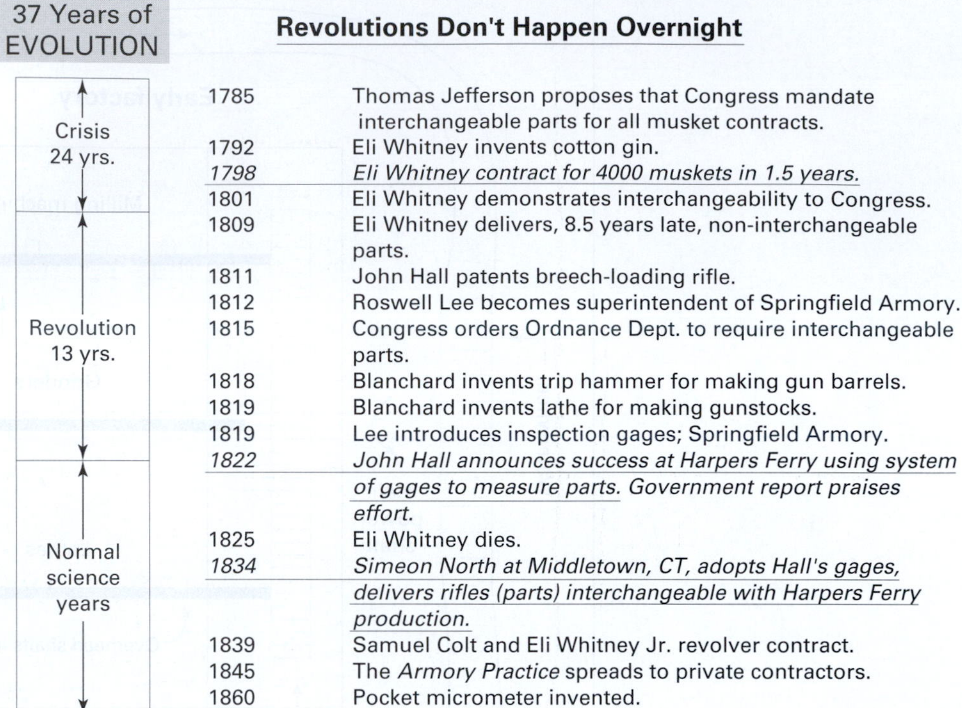

Evolution of the First Factory Revolution
37 Years of EVOLUTION
Revolutions Don't Happen Overnight
Crisis 24 yrs.
Revolution 13 yrs.
Normal science years
1785 Thomas Jefferson proposes that Congress mandate interchangeable parts for all musket contracts.
1792 Eli Whitney invents cotton gin.
1798 Eli Whitney contract for 4000 muskets in 1.5 years.
1801 Eli Whitney demonstrates interchangeability to Congress.
1809 Eli Whitney delivers, 8.5 years late, non-interchangeable parts.
1811 John Hall patents breech-loading rifle.
1812 Roswell Lee becomes superintendent of Springfield Armory.
1815 Congress orders Ordnance Dept. to require interchangeable parts.
1818 Blanchard invents trip hammer for making gun barrels.
1819 Blanchard invents lathe for making gunstocks.
1819 Lee introduces inspection gages; Springfield Armory.
1822 John Hall announces success at Harpers Ferry using system of gages to measure parts. Government report praises effort.
1825 Eli Whitney dies.
1834 Simeon North at Middletown, CT, adopts Hall's gages, delivers rifles (parts) interchangeable with Harpers Ferry production.
1839 Samuel Colt and Eli Whitney Jr. revolver contract.
1845 The Armory Practice spreads to private contractors.
1860 Pocket micrometer invented.

FIGURE 2-17 Evolution of the first factory revolution. *(David Cochran)*

Part interchangeability was a manufacturing challenge for the armory system from the beginning, and the ultimate success in achieving part interchangeability was the result of John Hall's work in 1822. His innovation was the ability to measure the product. Instead of attempting to produce parts to a master part, as was the practice at the time, a system of gages was introduced to measure the part. As a result, the engineers and designers had to be able to draw the parts with specifications in terms of their geometric dimensions and tolerances. The entire need for drawings, tolerances, and specifications was the result of part measurement to achieve part interchangeability.

Many people viewed part interchangeability as adding additional manufacturing cost and concluded that it was unnecessary. However, standard measurements and part interchangeability were key enabling technologies for the development of the moving assembly line during the second factory revolution.

After this evolution occurred, the practice of using gages to achieve part interchangeability was then moved to Middletown, Connecticut, from Harper's Ferry, which had adopted Hall's system of gages. A gun could then be assembled from parts made from the Harper's Ferry and Middletown locations. By 1845, this armory practice spread to private contractors. Soon, the ability to do gaging became ubiquitous throughout the country. In 1860, the pocket micrometer was invented, and the science of measurement spread throughout the world.

As a system design, the job shop still exists. Modern-day versions of the job shop produce large volumes of goods in batches or lots of 50 to 200 pieces. Processes are still grouped functionally but the walls are removed. Between the machines, filling the aisles, are tote boxes filled with components in various stages of completion. While it may be difficult to actually count the inventory, these designs have thousands of parts on the floor at any one time. However, the production system for the early job shop was minimal and most decisions about how to make the products were made by the operators on the shop floor.

EVOLUTION OF THE SECOND MSD—THE FLOW SHOP

The second factory revolution led to the development of mass production, shown in Figure 2-18. Henry Ford defined the concept of economies of scale with mass production.

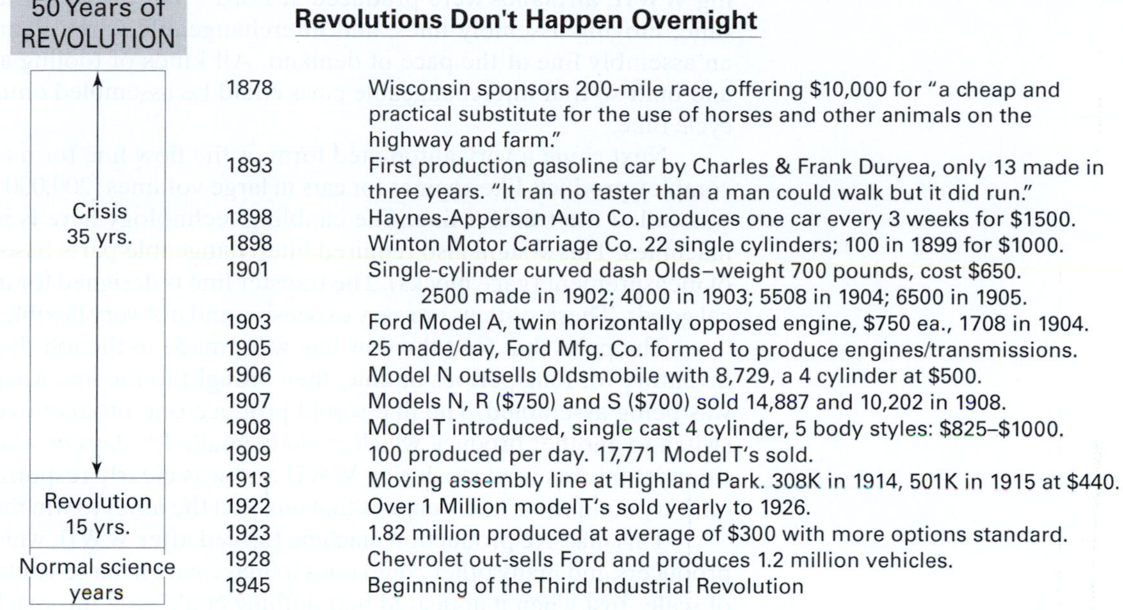

FIGURE 2-18 Evolution of the second factory revolution. *(David Cochran)*

His primary "mass production" tools were the extreme division of labor, the moving assembly line, the shortage chaser to control minimum and maximum stock levels, and the overarching reliance of assembly cycle time predictability with interchangeable parts.

Ford is credited with many product design innovations, as well as manufacturing innovations. He developed the process for engine blocks. One reason for the Model T's success was a single cast engine block instead of four cylinders bolted together. The use of this manufacturing process innovation decreased the car's weight and increased power.

However, the enabling technology for mass production was interchangeable parts based on exact standards of measurement. Ford insisted that every product meet specifications. He was a stickler about keeping all gages calibrated in the factory. In other words, the second factory revolution was founded on the first factory revolution concept of part interchangeability. Flow-line manufacturing began in the 1900s for small items and evolved to the moving **assembly line** at the Ford Motor Company around 1913. This methodology was developed by Ford production engineers led by Charles Sorenson. Today's moving assembly line for automobile production has hundreds of stations where the car is assembled. This requires the work at each station to be balanced where tasks at each station take about the same amount of time. This is called **line balancing.** The moving assembly line makes cars one at a time, in what is now called single piece flow.

Just as in the 1800s, people throughout the world came to observe how this system worked, and the new design methodology was again spread around the world. For many companies, a hybrid system evolved, which included a mixture of job shop and flow shop, with the components made in the job shop feeding the assembly line. This design permitted companies to manufacture large volumes of identical products at low unit cost.

Mass production relied on the first factory revolution for part interchangeability, while producing products in a fixed cycle time with moving assembly lines. To produce at a fixed cycle time, division of labor was used, and unskilled workers replaced the craftsmen in the factory (see Adam Smith's *Wealth of Nations*). With the division of labor, instead of assembling an entire transmission, the workers performed the same small set of tasks on each transmission. As a result, labor turnover in the factory increased dramatically, so Ford introduced the "five-dollar day," salary for all his workers, an exorbitant amount of money in those days. In fact, the five-dollar day created the economic system that enabled the emergence of the middle class. The workers were able to buy the products they produced in high-volume, mass-production factories.

Between 1928 and 1945, the Ford factory system diffused to other products. During WWII, airplanes were produced at Ford's Willow Run Factory using fixed cycle time, moving assembly lines, and interchangeable parts. Planes were moved down an assembly line at the pace of demand. All kinds of tooling and jigs were designed and built so that interchangeable parts could be assembled onto the aircraft in a fixed cycle time.

Next came a very automated form of the flow line for machining or assembling complex products like engines for cars in large volumes (200,000 to 400,000 per year). It was called the **transfer line.** The enabling technology here is repeat cycle automatic machines. This system also required interchangeable parts based on precise standards of measurement (gage blocks). The transfer line is designed for large volumes of identical goods. These systems are very expensive and not very flexible.

The parts that feed the flow line were made in the job shop in large lots, held in inventory for long periods of time, then brought to the line where a particular product was being assembled. The line would produce one product over a long run and then switch to another product, which could be made for days or weeks. This mass-production system was in place during WWII and was clearly responsible for producing the military equipment and weapons that enabled the allies to win the war.

This massive production machine thrived after WWII, which enabled automobile producers and many other companies to make cars in large volumes using the economy of scale. Just when it appeared that nothing could stop this machine, a new player, the Toyota Motor Company, evolved a new manufacturing system that truly changed the manufacturing world. This system is based on a different system design, which ushered in the third factory revolution, characterized by global companies producing for worldwide markets.

THE THIRD MSD—LEAN PRODUCTION

The latest MSD is sometimes called **lean production,** the **Toyota Production System (TPS), just-in-time (JIT) manufacturing,** or many other names. A new manufacturing system design brings one or more new companies to the forefront of the industrial world (this time, Toyota). See Figure 2-19 for details of this revolution. Each new design employs some unique enabling technologies, in this case, manufacturing and assembly

Third Factory Revolution at Toyota in Japan (Ohno, 1988)

Revolutions Don't Happen Overnight

21 Years of REVOLUTION	Year	Event
Crisis 3 yrs.	1945	Need to rebuild wide variety of products in low volume after World War II. Only had six presses, requiring frequent and fast changeover.
Revolution 18 yrs.	1948	Withdrawal by subsequent processes.
	1949	Intermediate warehouses abolished.
	1950	In-line cells. Horseshoe or U-shaped machine layout replaces job shop.
	1950	Machining and assembly lines balanced.
	1953	Supermarket system in machine shop.
	1955	Assembly and body plants linked.
	1955	Main plant assembly line production system adopts visual control, line stop, and mixed load. Automation to *autonomation.*
	1958	Warehouse withdrawal slips abolished.
	1961	*Andon* installed, Motomachi assembly plant.
	1962	15-minute main plant setups.
	1962	Kanban adopted company-wide.
	1962	Full work control of machines *pokayoke.*
	1965	Kanban adopted for ordering outside parts for 100% of supply system; began teaching system to affiliates.
	1966	First autonomated line Kamigo plant.
Diffusion & normal science	1971	Main office and Motomachi setups reach 3 minutes.
	1971	Body indication system at Motomachi Crown line.
	1981	Publication of *Toyota Production System* in English and infusion in United States.
	1990	Publication of the *Machine That Changed the World.*

FIGURE 2-19 Third factory revolution at Toyota in Japan. *(David Cochran)*

TABLE 2-4 Factory Revolutions Are Driven by New Manufacturing System Designs (MSDs)

Zero	1700–1850 Craft/cottage production–hand powered tools	No MSD
First Factory Revolution	1840–1910 First factory revolution (American Armory System) * Creation of factories with powered machines * Mechanization/Interchangeable parts	Job shop (functional layout)
Second Factory Revolution	1910-1970 * Second factory revolution (the Ford system) * Assembly line–flow shop product layout * Economy of scale yields mass production era * Automation (automatic material handling)	Flow shop (product layout)
Third Factory Revolution	1960–2010 (estimated) * Third factory revolution (lean production) * U-shaped manufacturing and assembly cells * One-piece flow with mixed-model final assembly * Flexibility in customer demand and product design * Integrated control functions (*kanban*)	Lean shop (linked-cell layout)

cells that produce defect-free goods. The new system has now been adopted by more than 60% of American manufacturing companies and has been disseminated around the industrial world.

Black's Theory of Factory Revolutions as outlined in Table 2-4, proposes that we are now 40-plus years into the factory revolution. This factory revolution is not based on computers, hardware, or a particular process, but once again on the design of the manufacturing system—the complex arrangement of physical elements characterized by measurable parameters. Again, people throughout the world went to observe the new design, but this time they went to Japan to try to understand how this tiny nation became a giant in the global manufacturing arena.

After World War II, the Japanese were confronted with different requirements of manufacturing than that of the Americans. The United States had almost an infinite capacity to produce. There were rows of stamping presses in the factories, a surplus of resources, pent-up demand, and many people with money, so all the United States had to do was to produce. In Japan, there were very few presses and very little money. One of the first concepts of the Toyota Production System was that all parts had to be good, because there was no excess capacity and no dealers in the United States to fix the defects. The right mix of cars of perfect quality had to be made with limited resources, and they had to be exactly right the first time in spite of any variations or disturbances to the system.

The result was a new manufacturing system design, and just as a new MSD pushed Colt and Remington to the forefront in the first factory revolution and Ford and Singer in the second factory revolution, the development of this new manufacturing system design vaulted Toyota into world leadership. Black defines the new physical system design as linked-cell (Black, 1991) or L-CMS for linked-cell manufacturing system. Toyota called it the Toyota Production System (TPS). Schonberger called it the JIT/TQC system or World Class Manufacturing (WCM) system. In 1990 it was finally given a name that would become universal: lean production. This term was coined by John Krafcik, an engineer in the International Motor Vehicle program at the Massachusetts Institute of Technology (Womack et al., 1991).

What was different about this system design was the development of manufacturing and assembly cells linked to final assembly by a unique material control system, producing a functionally integrated system for inventory and production control. In cells, processes are grouped according to the sequence of operations needed to make a product. This design uses one-piece flow like the flow shop, but is designed for flexibility.

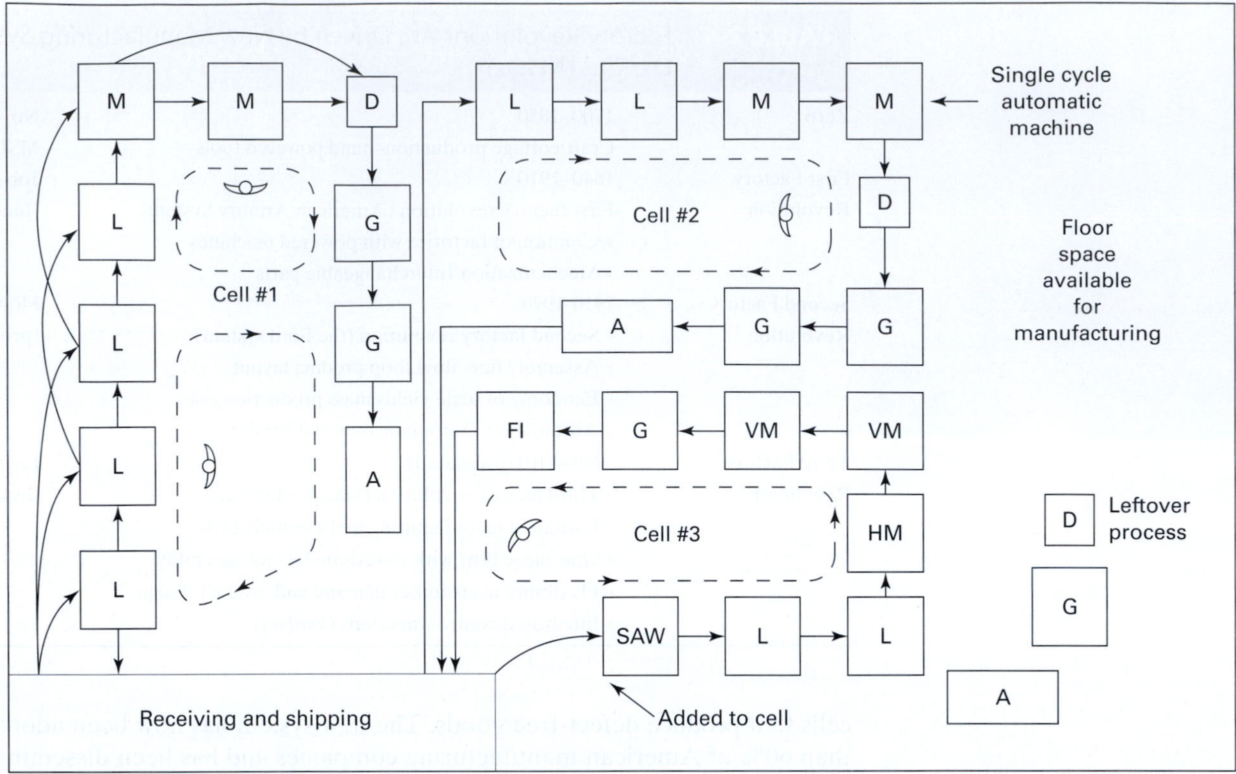

FIGURE 2-20 The functionally designed job shop can be restructured into manufacturing cells to process families of components at production rates that match part consumption.

The cell is designed in a U-shape or in parallel rows so that the workers can readily rebalance the line and change the output rate while moving from machine to machine loading and unloading parts. Figure 2-20 shows how the job shop in Figure 2-4 can be rearranged into manned cells. Cell 3 has one worker who can make a walking loop around the cell in 60 seconds. The machines in the cell have been upgraded to single-cycle automatic capability so they can complete the desired processing untended, turning themselves off when done with a machining cycle. The operator comes to a machine, unloads a part, checks the part, loads a new part into the machine, and starts the machining cycle again. The cell usually includes all the processing needed for a complete part or subassembly and may even include assembly steps.

To form a linked-cell manufacturing system, the first step is to restructure portions of the job shop, converting it in stages into manned cells. At the same time, the linear flow lines in subassembly are also reconfigured into U-shaped cells, which operate much like the manufacturing cells. The long setup times typical in flow lines must be vigorously attacked and reduced so that the flow lines can be changed quickly from making one product to another. The need to perform line-balancing tasks is eliminated through design. The standing, walking workers are capable of performing multiples of operations. More details are given in Advanced Topic 1.

SUMMARY ON MANUFACTURING CELLS

Product designers can easily see how parts are made in the manufacturing cells because all the operations and processes are together. Because quality-control techniques are also integrated into the cells, the designer knows exactly the cell's process capability. The designer can easily configure the future designs to be made in the cell. This is truly designing for manufacturing. CNC machining centers can do the same sequence of steps but are not as flexible as a cell composed of multiple, simple machines. Cellular layouts facilitate the integration of critical production functions while maintaining flexibility in producing superior-quality families of components. The cells facilitate

single-piece-flow (SPF) and **volume flexibility.** SPF is the movement of one part at a time between machines by the multiprocess operators. In the main, each machine executes a step in the sequence of processes or operations. The outcome of that step is checked before the part is advanced to the next step. Volume flexibility is achieved by the separation of man's activities from the operations that machines do better. Output per hour can be changed by the rapid reallocation of operations to workers.

Key Words

American Armory System
assembly line
bill of materials (BOM)
continuous process
critical path method (CPM)
decouplers
flexible manufacturing system (FMS)
flow shop
job shop
Ford Production System
just-in-time (JIT) manufacturing
lean production
lean shop
line balancing
linked-cell system
manufacturing system
manufacturing system design (MSD)
mass production
operations sheet
part interchangeability
process planning
production system
program evaluation and review technique (PERT)
programmable logic controller (PLC)
project shop
route sheet
single-piece-flow (SPF)
standard work-in-process (SWIP)
Toyota Production System (TPS)
transfer line
volume flexibility

Review Questions

1. What are the major functional elements or departmental areas of the production system? (See Chapter 44 on the Web for help.)
2. What is a route sheet? Who uses it?
3. What is the function of a route sheet?
4. Find an example of a route sheet other than the one in the book.
5. What are other names for a route sheet?
6. What is a process flow chart? How is it related to the bill of materials?
7. What is an operations sheet? How is it related to the route sheet?
8. How does the design of the product influence the design of the manufacturing system, including assembly and the production system?
9. Explain the difference between a route sheet and an operations sheet (or a process planning sheet).
10. What does the study of ergonomics entail? (This was not discussed in Chapter 2.)
11. In project shop manufacturing, what is the critical path? (Not discussed in Chapter 2.)
12. How do the job shop, flow shop, and lean shop manufacturing systems perform in terms of quality, cost, delivery, and flexibility?
13. What are the possibilities of incorporating lean manufacturing concepts into a high-volume transfer line for machining?
14. How did the Ford thinking build on the first factory revolution?

Problems

1. Discuss this statement: "Software can be as costly to design and develop as hardware and will require long production runs to recover, even though these costs may be hidden in the overhead costs."
2. Table 2-1 lists some examples of service job shops. Compare your college to a manufacturing/production system, using the definition of a manufacturing system given in the chapter. Who is the internal customer in the academic job shop? What or who are the products in the academic job shop?
3. Outline your critical path through the academic job shop.
4. Explain how function dictates design with respect to the design of footwear. Use examples of different kinds of footwear (shoes, sandals, high heels, boots, etc.) to emphasize your points. For example, cowboy boots have pointed toes so that they slip into the stirrups easily and high heels to keep the foot in the stirrup.
5. Most companies, when computing or estimating costs for a job, will add in an overhead cost, often tying that cost to some direct cost, such as direct labor, through the academic job shop. How would you calculate the cost per unit of a product to include overhead?
6. What is the impact of minimizing the unit cost of each operation:

   a. On machine design?
   b. On the workers?
   c. On the factory as a system?

www.wiley.com/go/global/degarmo

Chapter 2 CASE STUDY

Jury Duty for an Engineer

Katrin S. is suing the PogiBear Snowmobile Company and an engineer for PippenCat Components for $750,000 over her friend's death. He was killed while racing his snowmobile through the woods in the upper peninsula of Michigan. Her lawyer, Ken, claims that her friend was killed because a tie-rod broke, causing him to lose control and crash into a tree, breaking his neck. While it was impossible to determine whether the tie-rod broke before the crash or as a result of the crash, the following evidence has been put forth.

The tie-rod was originally designed and made entirely out of low-carbon steel (heat treated by case hardening) in three pieces, as shown in Figure CS-2 These tie-rods were subcontracted by PogiBear to PippenCat Components. PippenCat Components changed the material of the sleeves from steel to a heat-treated aluminum having the same *ultimate tensile strength (UTS)*, value as the steel. They did this because aluminum sleeves were easier to thread than steel sleeves. It was further found that threads on one of the tie-rod bolts were not as completely formed as they should have been. The sleeve of the tie-rod in question was split open (fractured) and one of the tie-rod bolts was bent. Katrin's lawyer further claimed that the tie-rod was not assembled properly. He claimed that one rod was screwed into the sleeve too far and the other not far enough, thereby giving it insufficient thread engagement. The engineer for PippenCat testified that these tie-rods are hand assembled and checked only for overall length and that such a misassembly was possible. In his summary, Ken, Katrin's lawyer, stated that the failure was due to a combination of material change, manufacturing error, and bad assembly—all combining to result in a failure of the tie-rod.

A design engineer for PogiBear testified that the tie-rods were "way overdesigned" and would not fail even with slightly small threads or misassembly. PogiBear's lawyer then claimed that the accident was caused by driver failure and that the tie-rod broke upon impact of the snowmobile with the tree. One of the men racing with Katrin's friend claimed that her friend's snowmobile had veered sharply just before he crashed, but under cross examination he admitted that they had all been drinking that night because it was so cold (he guessed –20° to –30°F). Because this accident had taken place more than 5 years ago, he could not remember how much they had had to drink.

You are a member of the jury and have now been sequestered to decide if PogiBear and PippenCat are guilty of negligence resulting in death. The rest of the jury, knowing you are an engineer, has asked for your opinion. What do you think? Who is really to blame for this accident? What actually caused the accident?

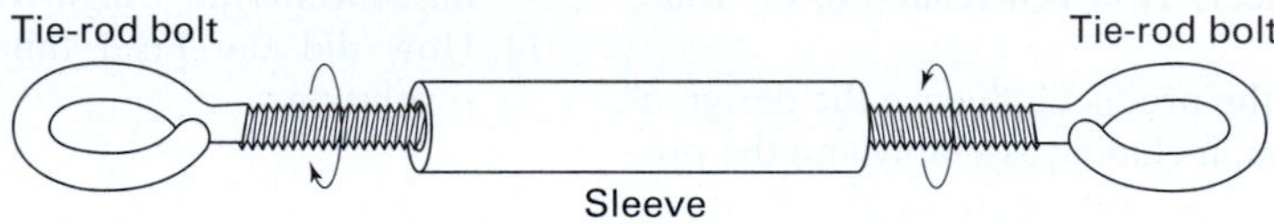

CHAPTER 3

PROPERTIES OF MATERIALS

3.1 INTRODUCTION

The history of man has been intimately linked to the materials that have shaped his world, so much so that we have associated periods of time with the dominant material, such as the Stone Age, Bronze Age, and Iron Age. Stone was used in its natural state, but bronze and iron were made possible by advances in processing. Each contributed to the comfort, productivity, safety, and security of everyday living. One replaced the other when new advantages and capabilities were realized, iron being lighter and stronger than bronze for example. Many refer to the close of the 20th century as the Silicon Age. The multitude of devices that have been made possible by the transistor and computer chip [ultra-fast computers, cell phones, global positioning system (GPS) units, etc.] have revolutionized virtually every aspect of our lives. From the manufacturing perspective, however, the current era lacks a materials designation. We now use an extremely wide array of materials, falling into the categories of metals, ceramics, polymers, and a myriad of combinations known as composites.

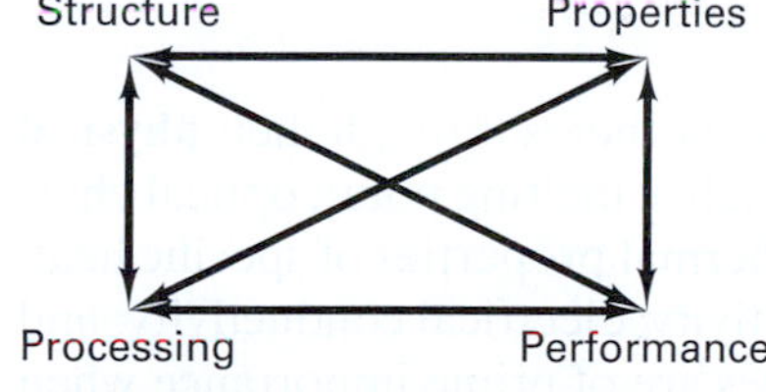

FIGURE 3-1 The interdependent relationships between structure, properties, processing, and performance.

With each of these materials, the ultimate desire is to convert it into some form of useful product. **Manufacturing** has been described as the various activities that are performed to convert "stuff" into "things." Successful products begin with appropriate materials. You wouldn't build an airplane out of lead, or an automobile out of concrete—you need to start with the right stuff. But "stuff" rarely comes in the right shape, size, and quantity for the desired use. Parts and components must be produced by subjecting materials to one or more processes (often a series of operations) that alter their shape, their properties, or both. Much of a manufacturing education relates to understanding: (1) the **structure** of materials, (2) the **properties** of materials, (3) the **processing** of materials, and (4) the **performance** of materials, as well as the interrelations between these four factors, as illustrated in Figure 3-1.

This chapter will begin to address the properties of engineering materials. Chapters 4 and 5 will discuss the subject of structure and begin to provide the whys behind various properties. Chapter 6 introduces the possibility of controlling and modifying structure to produce desired properties. Many engineering materials do not have a single set of properties, offering instead a range or spectrum of possibilities. Taking advantage of this range, we might want to intentionally make a material weak and ductile for easy shaping (making forming loads low, extending tool life, and preventing cracking or fracture), and then, once the shape has been produced, make the material strong for enhanced performance during use.

When selecting a material for a product or application, it is important to ensure that its properties will be adequate for the anticipated operating conditions. The various requirements of each part or component must first be estimated or determined. These requirements typically include mechanical characteristics (strength, rigidity, resistance to fracture, the ability to withstand vibrations or impacts) and physical characteristics (weight, electrical properties, appearance), as well as features relating to the service environment (ability to operate under extremes of temperature or resist corrosion). Candidate materials must possess the desired properties within their range of possibilities.

To help evaluate the properties of engineering materials, a variety of standard tests have been developed, and data from these tests have been tabulated and made readily available. Proper use of these data, however, requires sound engineering judgment. It is important to consider which of the evaluated properties are significant, under what conditions the test values were determined, and what cautions or restrictions should be placed on their use. Only by being familiar with the various test procedures, their capabilities, and their limitations can one determine if the resulting data are applicable to a particular problem.

METALLIC AND NONMETALLIC MATERIALS

While engineering materials are often grouped as metals, ceramics, polymers and composites, a more simplistic distinction might be to separate into metallic and nonmetallic. The common **metallic** materials include iron, copper, aluminum, magnesium, nickel, titanium, lead, tin, and zinc, as well as the many alloys of these metals, including steel, brass, and bronze. They possess the metallic properties of luster, high thermal conductivity, and high electrical conductivity; they are relatively ductile; and some have good magnetic properties. Some common **nonmetals** are wood, brick, concrete, glass, rubber, and plastics. Their properties vary widely, but they generally tend to be weaker, less ductile, and less dense than the metals, with poor electrical and thermal conductivities.

Although metals have traditionally been the more important of the two groups, the nonmetallic materials have become increasingly important in modern manufacturing. Advanced ceramics, composite materials, and engineered plastics have emerged in a number of applications. In many cases, metals and nonmetals are viewed as competing materials, with selection being based on how well each is capable of providing the required properties. Where both perform adequately, total cost often becomes the deciding factor, where total cost includes both the cost of the material and the cost of fabricating the desired component. Factors such as product lifetime, environmental impact, energy requirements, and recyclability are also considered.

PHYSICAL AND MECHANICAL PROPERTIES

A common means of distinguishing one material from another is through their **physical properties.** These include such features as density (weight); melting point; optical characteristics (transparency, opaqueness, or color); the thermal properties of specific heat, coefficient of thermal expansion, and thermal conductivity; electrical conductivity; and magnetic properties. In some cases, physical properties are of prime importance when selecting a material, and several will be discussed in more detail near the end of this chapter.

More often, however, material selection is dominated by the properties that describe how a material responds to applied loads or forces. These **mechanical properties** are usually determined by subjecting prepared specimens to standard test conditions. When using the obtained results, however, it is important to remember that they apply only to the specific conditions that were employed in the test. The actual service conditions of engineered products rarely duplicate the conditions of laboratory testing, so considerable caution should be exercised.

STRESS AND STRAIN

When a force or load is applied to a material, it deforms or distorts (becomes *strained*), and internal reactive forces *(stresses)* are transmitted through the solid. For example, if a weight, *W*, is suspended from a bar of uniform cross section and

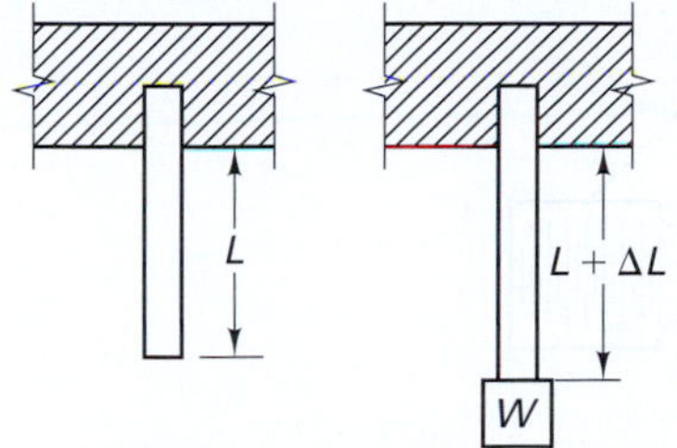

FIGURE 3-2 Tension loading and the resultant elongation.

length L, as in Figure 3-2, the bar will elongate by an amount ΔL. For a given weight, the magnitude of the **elongation,** ΔL, depends on the original length of the bar. The amount of elongation per unit length, expressed as $e = \Delta L/L$, is called the **unit strain.** Although the ratio is that of a length to another length and is therefore dimensionless, **strain** is usually expressed in terms of millimeters per meter, inches per inch, or simply as a percentage.

Application of the force also produces reactive stresses, which serve to transmit the load through the bar and on to its supports. **Stress** is defined as the force or load being transmitted divided by the cross-sectional area transmitting the load. Thus, in Figure 3-2, the stress is $S = W/A$, where A is the cross-sectional area of the supporting bar. Stress is normally expressed in megapascals in SI units (where a Pascal is one Newton per square meter) or pounds per square inch in the English system.

In Figure 3-2, the weight tends to stretch or lengthen the bar, so the strain is known as a **tensile strain** and the stress as a **tensile stress.** Other types of loadings produce other types of stresses and strains (Figure 3-3). Compressive forces tend to shorten the material and produce *compressive stresses and strains*. *Shear stresses and strains* result when two opposing forces acting on a body are offset with respect to one another.

3.2 STATIC PROPERTIES

When the forces that are applied to a material are constant, or nearly so, they are said to be *static*. Because static or steady loadings are observed in many applications, it is important to characterize the behavior of materials under these conditions. For design engineers, the strength of a material may be of primary concern, along with the amount of elastic stretching or deflection that may be experienced when the product is under load. Manufacturing engineers, wanting to shape products with mechanical forces, need to know the stresses necessary to effect permanent deformation. At the same time, they want to perform this deformation without inducing cracking or fracture.

As a result, a number of standardized tests have been developed to evaluate the **static properties** of engineering materials. Individual test results can be used to determine if a given material or batch of material has the necessary properties to meet specified requirements. The results of multiple tests can provide the materials characterization information that is used when selecting materials for various applications. In all cases, it is important to determine that the conditions for the product being considered are indeed similar to those of the standard testing. Even when the service conditions differ, however, the results of standard tests may still be helpful in qualitatively rating and comparing various materials.

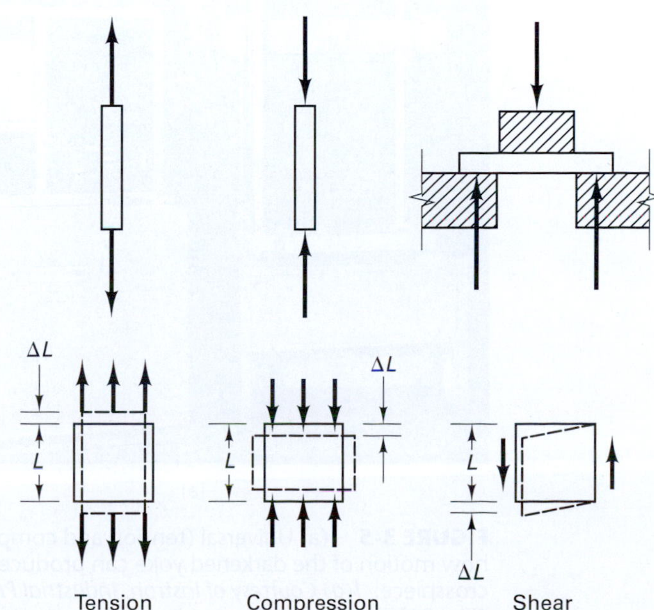

FIGURE 3-3 Examples of tension, compression, and shear loading—and their response.

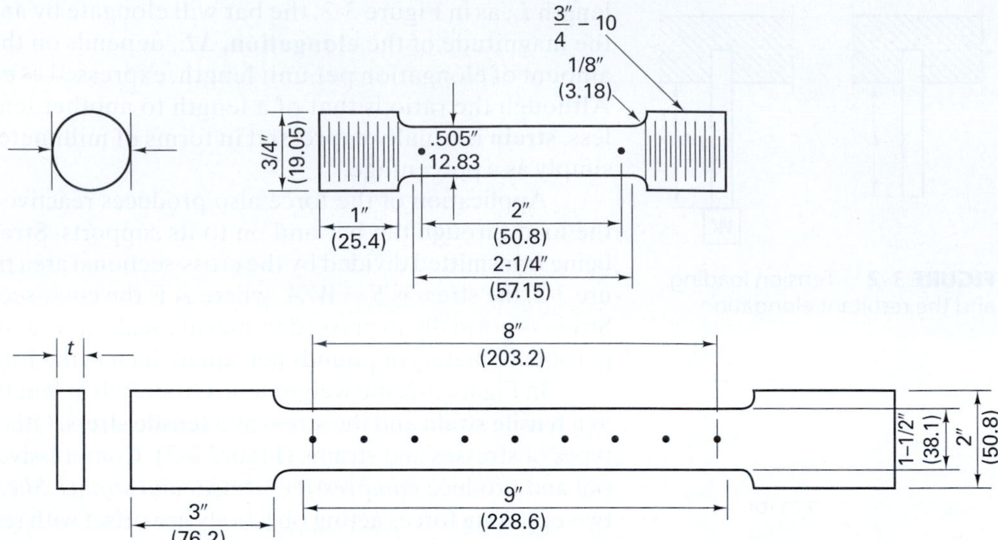

FIGURE 3-4 Two common types of standard tensile test specimens: (a) round; (b) flat. Dimensions are in inches, with millimeters in parentheses.

TENSILE TEST

The most common of the static **tests** is the **uniaxial tensile test.** The test begins with the preparation of a standard specimen with prescribed geometry, like the round and flat specimens described in Figure 3-4. The standard specimens ensure meaningful and reproducible results and have been designed to produce uniform uniaxial tension in the central portion while ensuring reduced stresses in the enlarged ends or shoulders that are placed in moving grips.

Strength Properties. The standard specimen is then inserted into a testing machine like the one shown in Figure 3-5. A tensile force or load, W, is applied and measured by the testing machine, while the elongation or stretch (ΔL) of a specified length **(gage length)** is simultaneously monitored. A plot of the coordinated load–elongation data produces a curve similar to that of Figure 3-6. Because the loads will differ for different-size

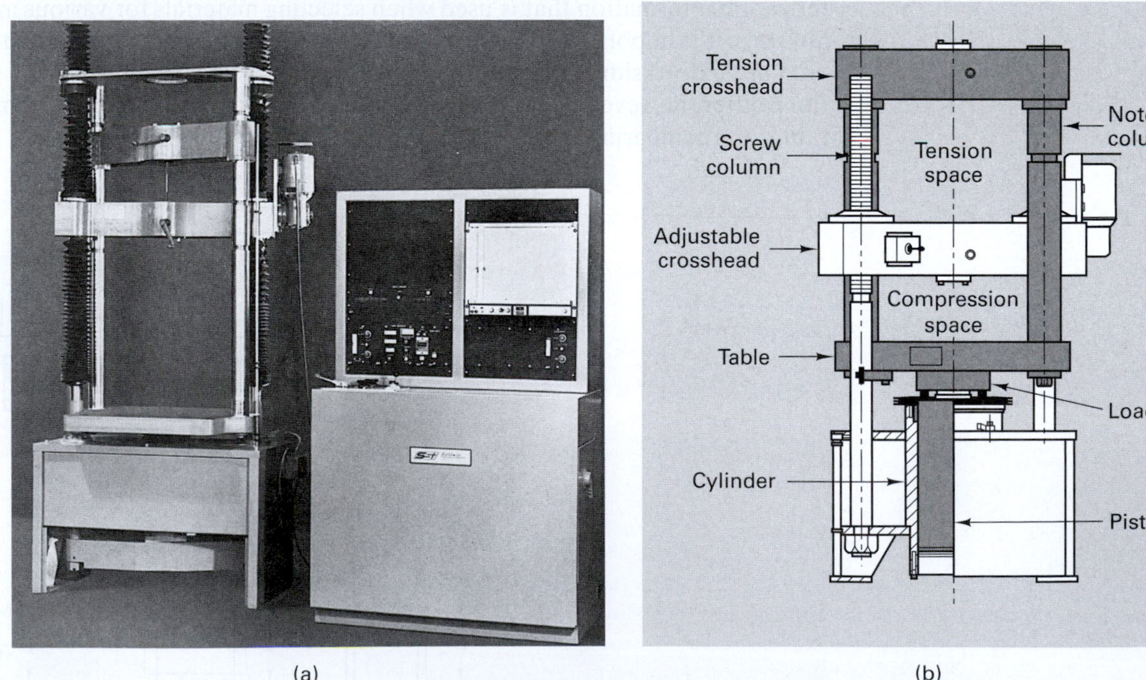

FIGURE 3-5 (a) Universal (tension and compression) testing machine; (b) schematic of the load frame showing how motion of the darkened yoke can produce tension or compression with respect to the stationary (white) crosspiece. *[(a) Courtesy of Instron, Industrial Products Group, Grove City, PA; (b) Courtesy of Satec Systems Inc., Grove City, PA]*

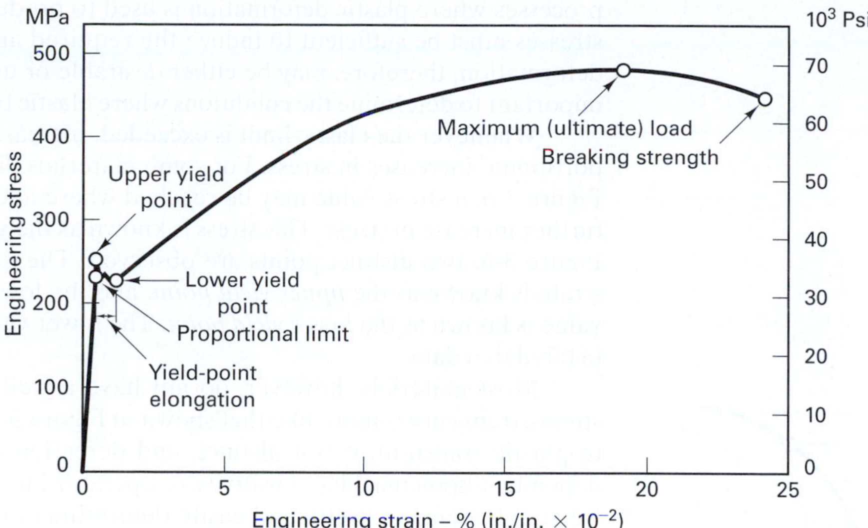

FIGURE 3-6 Engineering stress-strain diagram for a low-carbon steel.

specimens and the amount of elongation will vary with different gage lengths, it is important to remove these geometric or size effects if we are to produce data that are characteristic of a given material and not a particular specimen. If the load is divided by the *original* cross-sectional area, A_o, and the elongation is divided by the *original* gage length, L_o, the size effects are eliminated and the resulting plot becomes known as an **engineering stress-engineering strain curve** (see Figure 3-6). This is simply a load–elongation plot with the scales of both axes modified to remove the effects of specimen size.

In Figure 3-6 it can be noted that the initial response is linear. Up to a certain point, the stress and strain are directly proportional to one another. The stress at which this proportionality ceases is known as the **proportional limit.** Below this value, the material obeys **Hooke's law,** which states that the strain is directly proportional to the stress. The proportionality constant, or ratio of stress to strain, is known as **Young's modulus** or the **modulus of elasticity.** This is an inherent property of a given material[1] and is of considerable engineering importance. As a measure of **stiffness,** it indicates the ability of a material to resist deflection or stretching when loaded and is commonly designated by the symbol E.

Up to a certain stress, if the load is removed, the specimen will return to its original length. The response is elastic or recoverable, like the stretching and relaxation of a rubber band. The uppermost stress for which this behavior is observed is known as the **elastic limit.** For most materials the elastic limit and proportional limit are almost identical, with the elastic limit being slightly higher. Neither quantity should be assigned great engineering significance, however, because the determined values are often dependent on the sensitivity and precision of the test equipment.

The amount of energy that a material can absorb while in the elastic range is called the **resilience.** The area under a load–elongation curve is the product of a force and a distance, and is therefore a measure of the energy absorbed by the specimen. If the area is determined up to the elastic limit, the absorbed energy will be elastic (or potential) energy and is regained when the specimen is unloaded. If the same determination is performed on an engineering stress-engineering strain diagram, the area beneath the elastic region corresponds to an energy per unit volume, and is known as the **modulus of resilience.**

Elongation beyond the elastic limit becomes unrecoverable and is known as **plastic deformation.** When the load is removed, only the elastic stretching will be recovered, and the specimen will retain a permanent change in shape (in this case, an increase in length). For most components, the onset of plastic flow represents failure, because the part dimensions will now be outside of allowable tolerances. In manufacturing

[1] The modulus of elasticity is determined by the binding forces between the atoms. Since these forces cannot be changed, the elastic modulus is characteristic of a specific material and is not alterable by the structure modifications that can be induced by processing.

processes where plastic deformation is used to produce the desired shape, the applied stresses must be sufficient to induce the required amount of plastic flow. Permanent deformation, therefore, may be either desirable or undesirable, but in either case, it is important to determine the conditions where elastic behavior transitions to plastic flow.

Whenever the elastic limit is exceeded, increases in strain no longer require proportionate increases in stress. For some materials, like the low-carbon steel tested in Figure 3-6, a stress value may be reached where additional strain occurs without any further increase in stress. This stress is known as the **yield point,** or *yield-point stress*. In Figure 3-6, two distinct points are observed. The highest stress preceding extensive strain is known as the *upper yield point,* and the lower, relatively constant, "run-out" value is known as the *lower yield point*. The lower value is the one that usually appears in tabulated data.

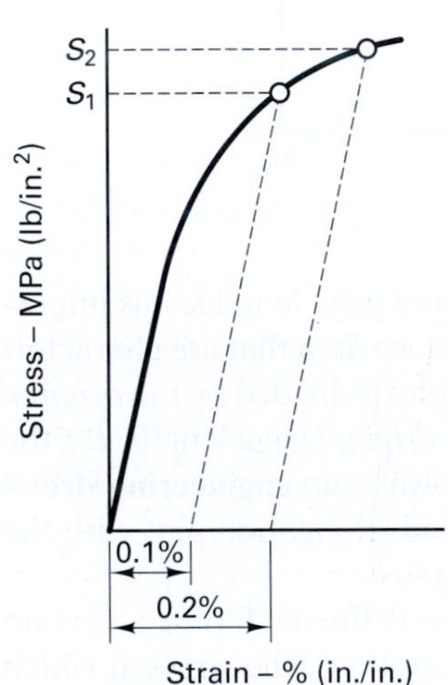

FIGURE 3-7 Stress-strain diagram for a material not having a well-defined yield point, showing the offset method for determining yield strength. S_1 is the 0.1% offset yield strength; S_2 is the 0.2% offset yield strength.

Most materials, however, do not have a well-defined yield point, and exhibit stress-strain curves more like that shown in Figure 3-7. For these materials, the elastic-to-plastic transition is not distinct, and detection of plastic deformation would be dependent upon machine sensitivity or operator interpretation. To solve this dilemma, we simply define a useful and easily determined property known as the **offset yield strength.** *Offset yield strength does not describe the onset of plastic deformation,* but instead defines the stress required to produce a specified, acceptable, amount of permanent strain. If this strain, or "offset," is specified to be 0.2% (a common value), we simply determine the stress required to plastically deform a 1-in. length to a final length of 1.002 in. (a 0.2% strain). If the applied stresses are then kept below this 0.2% offset yield strength value, the user can be guaranteed that any resulting plastic deformation will be less than 0.2% of the original dimension.

Offset yield strength is determined by drawing a line parallel to the elastic line, but displaced by the offset strain, and reporting the stress where the constructed line intersects the actual stress-strain curve. Figure 3-7 shows the determination of both 0.1% offset and 0.2% offset yield strength values, S_1 and S_2, respectively. The intersection values are reproducible and are independent of equipment sensitivity. It should be noted that the offset yield strength values are meaningless unless they are reported in conjunction with the amount of offset strain used in their determination. While 0.2% is a common offset for many mechanical products (and is generally assumed unless another number is specified), applications that cannot tolerate that amount of deformation may specify offset values of 0.1% or even 0.02%. It is important, therefore, to verify that any tabulated data being used was determined under the desired conditions.

As shown in Figure 3-6, the load (or engineering stress) required to produce additional plastic deformation continues to increase. Because the material is deforming, this load is the product of the material strength times the cross-sectional area. During tensile deformation, the specimen is continually increasing in length. The cross-sectional area, therefore, must be decreasing, but the overall load-bearing ability of the specimen continues to increase! For this to occur, the material must be getting stronger. The mechanism for this phenomenon will be discussed in Chapter 4, where we will learn that the strength of a metal continues to increase with increased deformation.

During the plastic deformation portion of a tensile test, the weakest location of the specimen is continually undergoing deformation and becoming stronger. As each weakest location strengthens, another location assumes that status and deforms. As a consequence, the specimen deforms and strengthens uniformly, maintaining its original cylindrical or rectangular geometry. As plastic deformation progresses, however, the additional increments of strength decrease in magnitude, and a point is reached where the decrease in area cancels the increase in strength. When this occurs, the load-bearing ability peaks, and the force required to continue straining the specimen begins to decrease, as seen in the Figure 3-6. The stress at which the load-bearing ability peaks is known as the ultimate strength, tensile strength, or **ultimate tensile strength** of the material. The weakest location in the test specimen at that time continues to be the weakest location by virtue of the decrease in area, and further deformation becomes localized. This localized reduction in cross-sectional area is known as **necking,** and is shown in Figure 3-8.

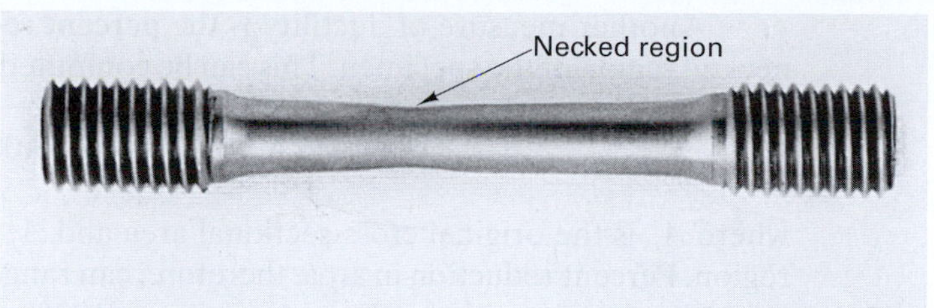

FIGURE 3-8 A standard 0.505-in.-diameter tensile specimen showing a necked region that has developed prior to failure. *(E. Paul DeGarmo)*

If the straining is continued, necking becomes intensified and the tensile specimen will ultimately fracture. The stress at which fracture occurs is known as the **breaking strength** or **fracture strength.** For ductile materials, necking precedes fracture, and the breaking strength is less than the ultimate tensile strength. For a brittle material, fracture usually terminates the stress-strain curve before necking, and often before the onset of plastic flow.

Ductility and Brittleness. When evaluating the suitability of a material for certain manufacturing processes or its appropriateness for a given application, the amount of plasticity that precedes fracture, or the **ductility,** can often be a significant property. For metal deformation processes, the greater the ductility, the more a material can be deformed without fracture. Ductility also plays a key role in toughness, a property that will be described shortly.

One of the simplest ways to evaluate ductility is to determine the **percent elongation** of a tensile test specimen at the time of fracture. As shown in Figure 3-9, ductile materials do not elongate uniformly when loaded beyond necking. If the percent change of the entire 8-in. gage length were computed, the elongation would be 31%. However, if only the center 2-in. segment is considered, the elongation of that portion is 60%. A valid comparison of material behavior, therefore, requires similar specimens with the same standard gage length.

In many cases, material "failure" is defined as the onset of localized deformation or necking. Consider a sheet of metal being formed into an automobile body panel. If we are to assure uniform strength and corrosion resistance in the final panel, the operation must be performed in such a way as to maintain uniform sheet thickness. For this application, a more meaningful measure of material ductility would be the **uniform elongation** or the *percent elongation prior to the onset of necking*. This value can be determined by constructing a line parallel to the elastic portion of the diagram, passing through the point of highest force or stress. The intercept where the line crosses the strain axis denotes the available uniform elongation. Because the additional deformation that occurs after necking is not considered, uniform elongation is always less than the total elongation at fracture (the generally reported elongation value).

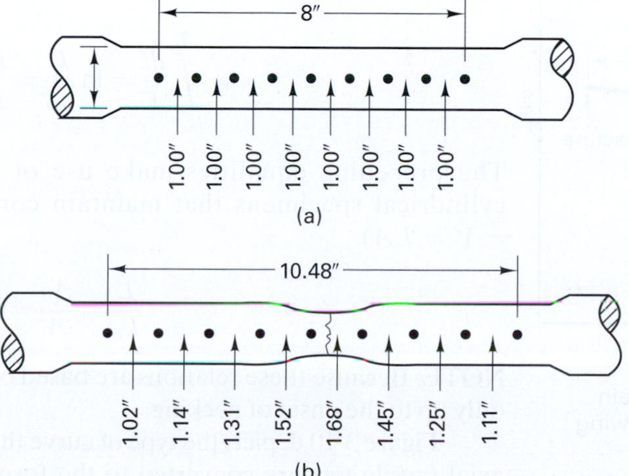

FIGURE 3-9 Final elongation in various segments of a tensile test specimen: (a) original geometry; (b) shape after fracture.

Another measure of ductility is the **percent reduction in area** that occurs in the necked region of the specimen. This can be computed as

$$\text{R.A.} = \frac{A_o - A_f}{A_o} \times 100\%$$

where A_o is the original cross-sectional area and A_f is the smallest area in the necked region. Percent reduction in area, therefore, can range from 0% (for a brittle glass specimen that breaks with no change in area) to 100% (for extremely plastic soft bubble gum that pinches down to a point before fracture).

When materials fail with little or no ductility, they are said to be **brittle.** Brittleness, however, is simply the lack of ductility, and should not be confused with a lack of strength. Strong materials can be brittle, and brittle materials can be strong.

Toughness. **Toughness,** or *modulus of toughness*, is the work per unit volume required to fracture a material. The tensile test can provide one measure of this property, because toughness corresponds to the total area under the stress-strain curve from test initiation to fracture, and thereby encompasses both strength and ductility. Caution should be exercised when using toughness data, however, because the work or energy to fracture can vary markedly with different conditions of testing. Variations in the temperature or the speed of loading can significantly alter both the stress-strain curve and the toughness.

In most cases, toughness is associated with impact or shock loadings, and the values obtained from high-speed (dynamic) impact tests often fail to correlate with those obtained from the relatively slow-speed (static) tensile test.

True Stress–True Strain Curves. The stress-strain curve in Figure 3-6 is a plot of **engineering stress,** S, versus **engineering strain,** e, where S is computed as the applied load divided by the original cross-sectional area, A_o, and e is the elongation, ΔL, divided by the original gage length, L_o. As the test progresses, the cross section of the test specimen changes continually, first in a uniform manner and then nonuniformly after necking begins. The actual stress should be computed based on the instantaneous cross-sectional area, A, not the original, A_o. Because the area is decreasing, the actual or true stress will be greater than the engineering stress plotted in Figure 3-6. **True stress,** σ, can be computed by taking simultaneous readings of the load, W, and the minimum specimen diameter. The actual area can then be computed, and true stress can be determined as

$$\sigma = \frac{W}{A}$$

The determination of **true strain** is a bit more complex. In place of the change in length divided by the original length that was used to compute engineering strain, true strain is defined as the summation of the incremental strains that occur throughout the test. For a specimen that has been stretched from length L_o to length L, the *true*, *natural*, or *logarithmic strain*, would be:

$$\varepsilon = \int_{L_o}^{L} \frac{d\ell}{\ell} = \ln\frac{L}{L_o} = \frac{D_o^2}{D^2} = 2\ln\frac{D_o}{D}$$

The preceding equalities make use of the following relationships for cylindrical specimens that maintain constant volume (i.e., $V_o = L_oA_o = V = LA$)

$$\frac{L}{L_o} = \frac{A_o}{A} = \frac{D_o^2}{D^2}$$

NOTE: Because these relations are based on cylindrical geometry, they apply only up to the onset of necking.

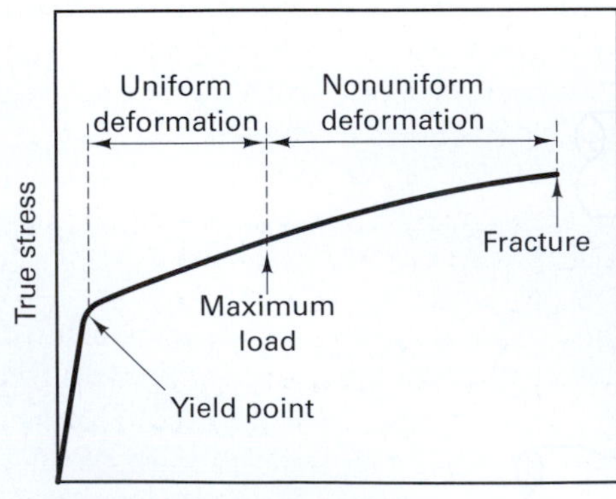

FIGURE 3-10 True stress–true strain curve for an engineering metal, showing true stress continually increasing throughout the test.

Figure 3-10 depicts the type of curve that results when the data from a uniaxial tensile test are converted to the form of true stress versus true strain.

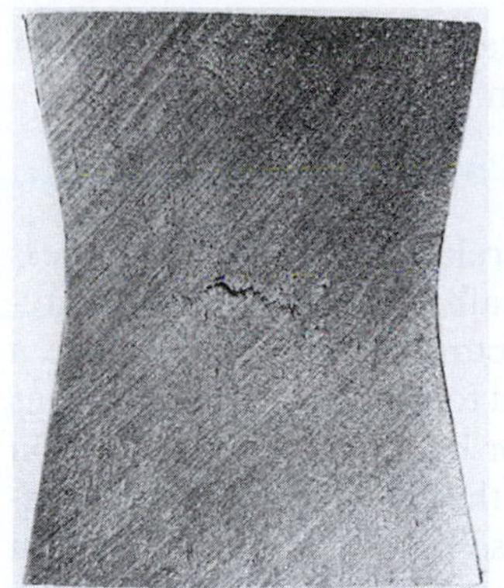

FIGURE 3-11 Section of a tensile test specimen stopped just prior to failure, showing a crack already started in the necked region, which is experiencing triaxial tension. *(Photo by E. R. Parker, courtesy E. Paul DeGarmo)*

Because the true stress is a measure of the material strength at any point during the test, it will continue to rise even after necking. Data beyond the onset of necking should be used with extreme caution, however, because the geometry of the neck transforms the stress state from uniaxial tension (stretching in one direction with compensating contractions in the other two) to triaxial tension, in which the material is stretched or restrained in all three directions. Because of the triaxial tension, voids or cracks (Figure 3-11) tend to form in the necked region and serve as a precursor to final fracture. Measurements of the external diameter no longer reflect the true load-bearing area, and the data are further distorted.

Strain Hardening and the Strain-Hardening Exponent. Figure 3-12 is a true stress–true strain diagram, which has been modified to show how a ductile metal (such as steel) will behave when subjected to slow loading and unloading. Loading and unloading within the elastic region will result in simply cycling up and down the linear portion of the curve between points O and A. However, if the initial loading is carried through point B (in the plastic region), unloading will follow the path BeC, which is approximately parallel to the line OA, and the specimen will exhibit a permanent elongation of the amount OC. Upon reloading from point C, elastic behavior is again observed as the stress follows the line CfD, a slightly different path from that of unloading. Point D is now the yield point or yield stress for the material in its partially deformed state. A comparison of points A and D reveals that plastic deformation has made the material stronger. If the test were again interrupted at point E, we would find a new, even higher-yield stress. Thus, within the region of plastic deformation, each of the points along the true stress–true strain curve represents the yield stress for the material at the corresponding value of strain.

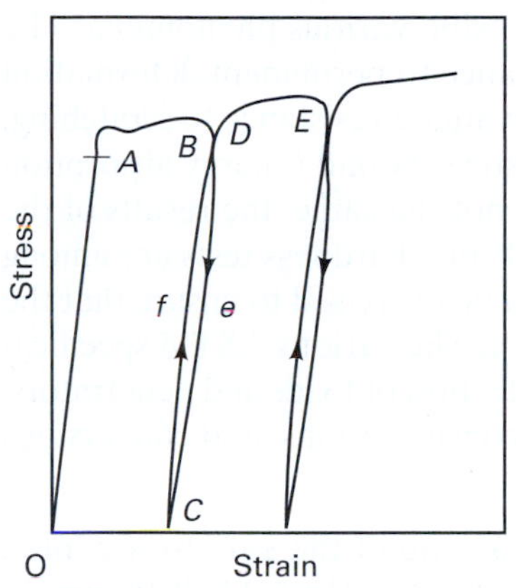

FIGURE 3-12 Stress-strain diagram obtained by unloading and reloading a specimen.

When metals are plastically deformed, they become harder and stronger, a phenomenon known as **strain hardening.** Therefore, if a stress induces plastic flow, an even greater stress will be required to continue the deformation. In Chapter 4 we will discuss the atomic-scale features that are responsible for this phenomenon.

Various materials strain harden at different rates; that is, for a given amount of deformation different materials will exhibit different increases in strength. One method of describing this behavior is to mathematically fit the plastic region of the true stress–true strain curve to the equation

$$\sigma = K\varepsilon^n$$

and determine the best-fit value of n, the **strain-hardening exponent.**[2] As shown in Figure 3-13, a material with a high value of n will have a significant increase in strength with a small amount of deformation. A material with a small n value will show little change in strength with plastic deformation.

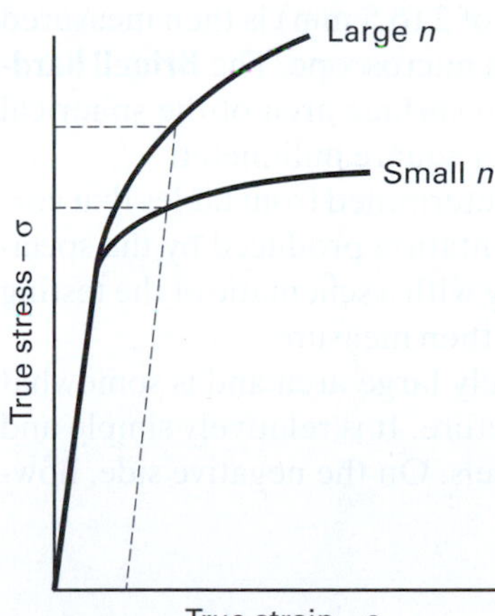

FIGURE 3-13 True stress–true strain curves for metals with large and small strain hardening. Metals with larger n values experience larger amounts of strengthening for a given strain.

Damping Capacity. In Figure 3-12 the unloading and reloading of the specimen follow slightly different paths. The area between the two curves is proportional to the amount of energy that is converted from mechanical form to heat and is therefore absorbed by the material. When this area is large, the material is said to exhibit good **damping capacity** and is able to absorb mechanical vibrations or damp them out quickly. This is an important property in applications such as crankshafts and machinery bases. Gray cast iron is used in many applications because of its high damping capacity. Materials with low damping capacity, such as brass and steel, readily transmit both sound and vibrations.

Rate Considerations. The rate or speed at which a tensile test is conducted can have a significant effect on the various properties. **Strain rate** sensitivity varies widely for the engineering materials. Plastics and polymers are very sensitive to testing speed, as are metals with low melting points, such as lead and zinc. Those materials that are sensitive to speed variations exhibit higher strengths and lower ductility when speed is increased.

[2] Taking the logarithm of both sides of the equation yields $\log \sigma = \log K + n \log \varepsilon$, which has the same form as the equation $y = mx + b$ if y is $\log \sigma$ and x is $\log \varepsilon$. This is the equation of a straight line with slope m and intercept b. Therefore, if the true stress–true strain data were plotted on a log-log scale with stress (σ) on the y-axis and strain (ε) on the x, the slope of the data in the plastic region would be n.

It is important to recognize that standard testing selects a standard speed, which may or may not correlate with the conditions of product application.

COMPRESSION TESTS

When a material is subjected to compressive loadings, the relationships between stress and strain are similar to those for a tension test. Up to a certain value of stress, the material behaves elastically. Beyond this value, plastic flow occurs. In general, however, a compression test is more difficult to conduct than a standard tensile test. Test specimens must have larger cross-sectional areas to resist bending or buckling. As deformation proceeds, the material strengthens by strain hardening and the cross section of the specimen increases, combining to produce a substantial increase in required load. Friction between the testing machine surfaces and the ends of the test specimen will alter the results if not properly considered. The type of service for which the material is intended, however, should be the primary factor in determining whether the testing should be performed in tension or **compression.**

HARDNESS TESTING

The wear resistance and strength of a material can also be evaluated by assessing its "hardness." **Hardness** is actually a hard-to-define property of engineering materials, and a number of different tests have been developed using various phenomena. The most common of the hardness tests are based on resistance to permanent deformation in the form of penetration or indentation. Other tests evaluate resistance to scratching, wear resistance, resistance to cutting or drilling, or elastic rebound (energy absorption under impact loading). Because these phenomena are not the same, the results of the various tests often do not correlate with one another. While hardness tests are among the easiest to perform on the shop floor, caution should be exercised to ensure that the selected test clearly evaluates the phenomena of interest. The various ASTM specifications[3] provide details regarding sample preparation, selection of loads and penetrators, minimum sample thicknesses, spacing and near-edge considerations, and conversions between scales.

Brinell Hardness Test. The **Brinell hardness test** was one of the earliest accepted methods of measuring hardness. A tungsten carbide or hardened steel ball 10 mm. in diameter is pressed into the flat surface of a material by a standard load of 500 or 3000 kg. The load is maintained for a period of time to permit sufficient plastic deformation to occur to support the applied load (10 to 15 seconds for iron or steel and up to 30 seconds for softer metals), and the load and ball are then removed. The diameter of the resulting spherical indentation (usually in the range of 2 to 5 mm) is then measured to an accuracy of 0.05 mm using a special grid or traveling microscope. The **Brinell hardness number (BHN)** is equal to the load divided by the surface area of the spherical indentation when the units are expressed as kilograms per square millimeter.

In actual practice, the Brinell hardness number is determined from tables that correlate the Brinell number with the diameter of the indentation produced by the specified load. Figure 3-14 shows a typical Brinell tester, along with a schematic of the testing procedure, which is actually a two-step operation—load then measure.

The Brinell test measures hardness over a relatively large area and is somewhat indifferent to small-scale variations in the material structure. It is relatively simple and easy to conduct and is used extensively on irons and steels. On the negative side, however, the Brinell test has the following limitations:

1. It cannot be used on very hard or very soft materials.
2. The results may not be valid for thin specimens. It is best if the thickness of material is at least 10 times the depth of the indentation. Some standards specify the minimum hardnesses for which the tests on thin specimens will be considered valid.
3. The test is not valid for case-hardened surfaces.

[3] ASTM hardness testing specifications include E3, E10, E18, E103, E140, and E384.

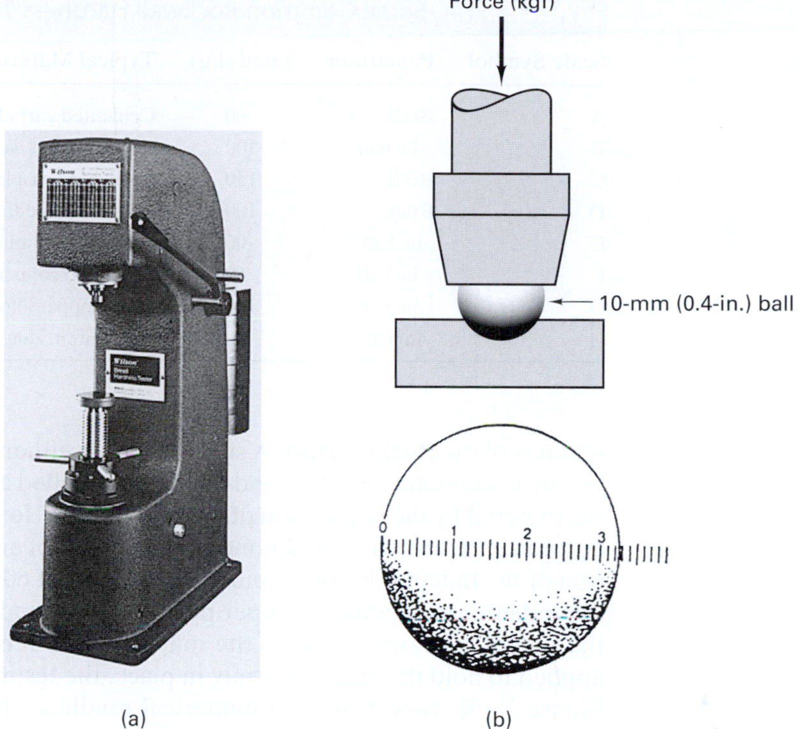

FIGURE 3-14 (a) Brinell hardness tester; (b) Brinell test sequence showing loading and measurement of the indentation under magnification with a scale calibrated in millimeters. *(Courtesy of Wilson Hardness, an Instron Company, Norwood, MA)*

4. The test must be conducted far enough from the edge of the material so that no edge bulging occurs.
5. The substantial indentation may be objectionable on finished parts.
6. The edge or rim of the indentation may not be clearly defined or may be difficult to see.

Portable testers are available for use on pieces that are too large to be brought to a benchtop machine.

The Rockwell Test. The **Rockwell hardness test** is the most widely used hardness test, and is similar to the Brinell test, with the hardness value again being determined by an indentation or penetration produced by a static load. Figure 3-15a shows the key

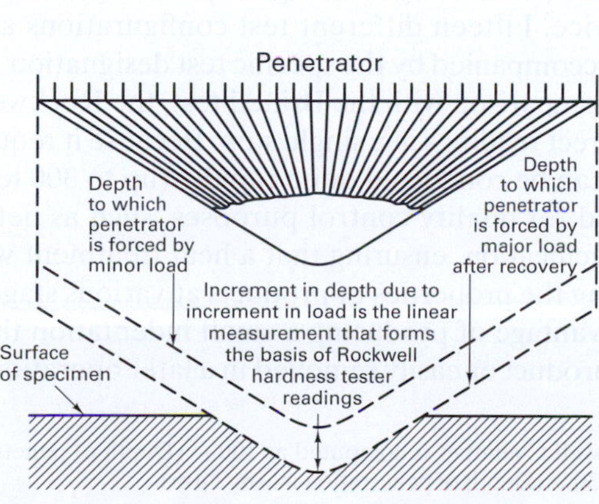

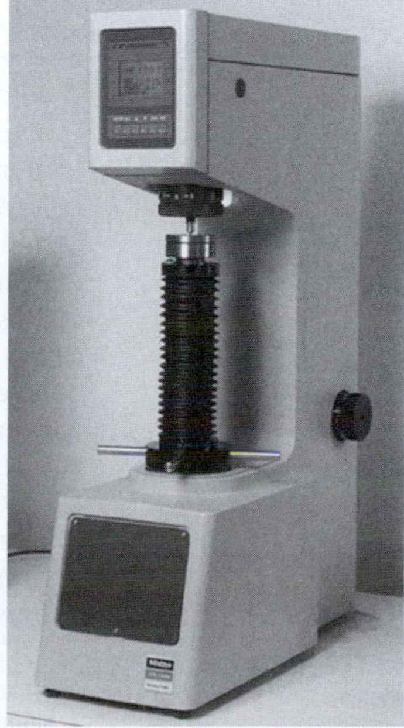

FIGURE 3-15 (a) Operating principle of the Rockwell hardness tester; (b) typical Rockwell hardness tester with digital readout. *(Courtesy of Mitutoyo America Corporation, Aurora, IL)*

TABLE 3-1 Some Common Rockwell Hardness Tests

Scale Symbol	Penetrator	Load (kg)	Typical Materials
A	Brale	60	Cemented carbides, thin steel, shallow case-hardened steel
B	$\frac{1}{16}$-in. ball	100	Copper alloys, soft steels, aluminum alloys, malleable iron
C	Brale	150	Steel, hard cast irons, titanium, deep case-hardened steel
D	Brale	100	Thin steel, medium case-hardened steel
E	$\frac{1}{8}$-in. ball	100	Cast iron, aluminum, magnesium
F	$\frac{1}{16}$-in. ball	60	Annealed coppers, thin soft sheet metals
G	$\frac{1}{16}$-in. ball	150	Hard copper alloys, malleable irons
H	$\frac{1}{8}$-in ball	60	Aluminum, zinc, lead

features of the Rockwell test. A small indenter, either a hardened steel ball of 1/16, 1/8, $\frac{1}{4}$ or $\frac{1}{2}$-in. in diameter, or a diamond-tipped cone called a **brale,** is first seated firmly against the material by the application of a 10-kg "minor" load. This causes a slight elastic penetration into the surface and removes the effects of any surface irregularities. The location of the indenter is noted, and "major" load of 60, 100, or 150 kg. is then applied to the indenter to produce a deeper penetration by inducing plastic deformation. When the indenter ceases to move, the major load is removed. With the minor load still applied to hold the indenter firmly in place, the testing machine, like the one shown in Figure 3-15b, now displays a numerical reading. This Rockwell hardness number is really an indication of the distance of indenter travel or the *depth* of the plastic or permanent penetration that was produced by the major load, with each unit representing a penetration depth of 2 μm.

To accommodate a wide range of materials with a wide range of strength, there are 15 different Rockwell test scales, each having a specified major load and indenter geometry. Table 3-1 provides a partial listing of Rockwell scales, which are designated by letters, and some typical materials for which they are used. Because of the different scales, a Rockwell hardness number must be accompanied by the letter corresponding to the particular combination of load and indenter used in its determination. The notation R_C60 (or Rockwell C 60), for example, indicates that a 120-degree diamond-tipped brale indenter was used in combination with a major load of 150 kg, and a reading of 60 was obtained. The B and C scales are the most common, with B being used for copper and aluminum and C for steels.[4]

Rockwell tests should not be conducted on thin materials (typically less than 1.5 mm or 1/16 in.), on rough surfaces, or on materials that are not homogeneous, such as gray cast iron. Because of the small size of the indentation, variations in roughness, composition, or structure can greatly influence test results. For thin materials, or where a very shallow indentation is desired (as in the evaluation of surface-hardening treatments such as nitriding or carburizing), the *Rockwell superficial hardness test* is preferred. Operating on the same Rockwell principle, this test employs smaller major and minor loads (15 or 45 kg and 3 kg, respectively), and uses a more sensitive depth-measuring device. Fifteen different test configurations are again available, so test results must be accompanied by the specific test designation.

In comparison with the Brinell test, the Rockwell test offers the attractive advantage of direct readings in a single step. Because it requires little (if any) surface preparation and can be conducted quite rapidly (up to 300 tests per hour or 5 per minute), it is often used for quality control purposes, such as determining if an incoming product meets specification, ensuring that a heat treatment was performed properly, or simply monitoring the properties of products at various stages of manufacture. It has the additional advantage of producing a small indentation that can be easily concealed on the finished product or easily removed in a later operation.

[4] The Rockwell C number is computed as 100 – (Depth of penetration in μm/2 μm), while the Rockwell B number is 130 – (Depth of penetration in μm/2 μm).

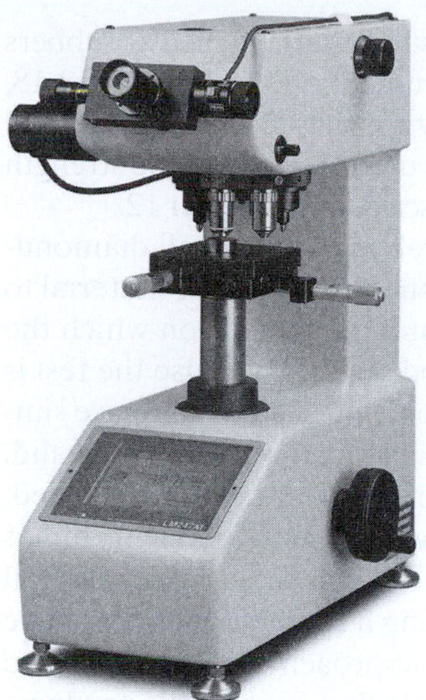

FIGURE 3-16 Microindentation hardness tester. *(Image used with permission from LECO Corporation)*

Vickers Hardness Test (also called Diamond Pyramid Hardness). The **Vickers hardness test** is similar to the Brinell test, but uses a 136-degree square-based diamond-tipped pyramid as the indenter and loads between 1 and 120 kg. Like the Brinell value, the Vickers hardness number is also defined as load divided by the surface area of the indentation expressed in units of kilograms per square millimeter. The advantages of the Vickers approach include the increased accuracy in determining the diagonal of a square impression as opposed to the diameter of a circle, and the assurance that even light loads will produce some plastic deformation. The use of diamond as the indenter material enables the test to evaluate any material and effectively places the hardness of all materials on a single scale.

Like the other indentation or penetration methods, the Vickers test has a number of attractive features: (1) it is simple to conduct, (2) little time is involved, (3) little surface preparation is required, (4) the marks are quite small and are easily hidden or removed, (5) the test can be done on location, (6) it is relatively inexpensive, and (7) it provides results that can be used to evaluate material strength or assess product quality.

Microindentation Hardness. Hardness tests have also been developed for applications where the testing involves a very precise area of material, or where the material or modified surface layer is exceptionally thin. These tests were previously called **microhardness tests,** but the newer **microindentation** term is more appropriate since it is the size of the indentation that is extremely small, not the measured value of hardness.[5] Special machines, such as the one shown in Figure 3-16, have been constructed for this type of testing, which must be performed on specimens with a polished metallographic surface. The location for the test is selected under high magnification. A predetermined load ranging from 25 to 3600 g is then applied through a small diamond-tipped penetrator. In the **Knoop test,** a diamond-shaped indenter with the long diagonal seven times the short diagonal is used, and the length of the indentation is measured under a magnification of about 200 to 400X. Figure 3-17 compares the indenters for the Vickers and Knoop tests, and shows a series of Knoop indentations progressing left-to-right across a surface-hardened steel specimen, from the hardened surface to the unhardened core. The hardness value, known as the **Knoop hardness number,** is again obtained by dividing the load in kilograms by the projected area of the indentation, expressed in square millimeters. A light-load Vickers test can also be used to determine microindentation hardness.

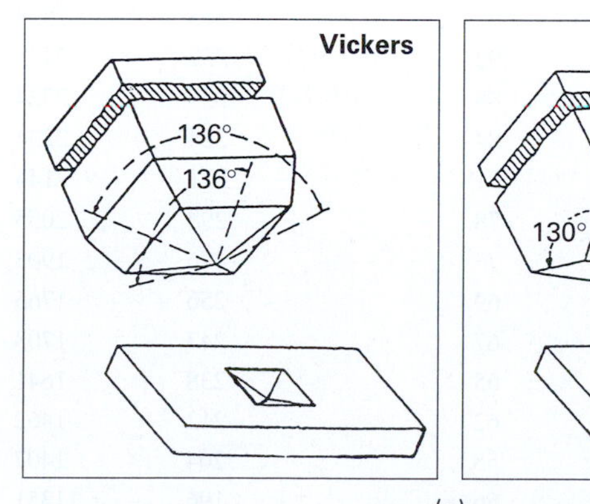

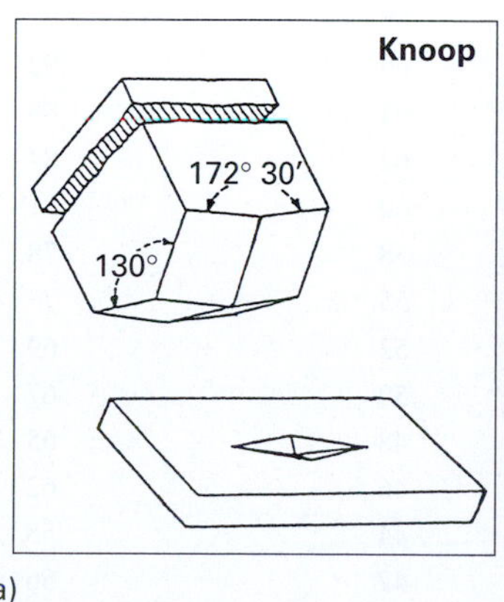

(a)

(b)

FIGURE 3-17 (a) Comparison of the diamond-tipped indenters used in the Vickers and Knoop hardness tests; (b) series of Knoop hardness indentations progressing left-to-right across a surface-hardened steel specimen (hardened surface to unhardened core). *(Courtesy of Buehler)*

[5] The ASTM Standard E384 has been renamed "Standard Test Method for Microindentation Hardness of Materials" to reflect this change in nomenclature.

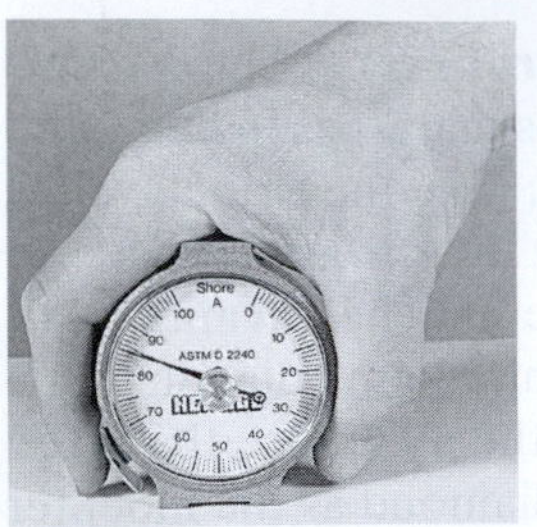

FIGURE 3-18 Durometer hardness tester. *(Courtesy of Newage Testing Instruments, An AMETEK Co.; www.hardnesstesters.com)*

Other Hardness Determinations. When testing soft, elastic materials, such as rubbers and nonrigid plastics, a **durometer** is often be used. This instrument, shown in Figure 3-18, measures the resistance of a material to elastic penetration by a spring-loaded conical steel indenter. No permanent deformation occurs. A similar test is used to evaluate the strength of molding sands used in the foundry industry, and will be described in Chapter 12.

In the **scleroscope test,** hardness is measured by the rebound of a small diamond-tipped "hammer" that is dropped from a fixed height onto the surface of the material to be tested. This test evaluates the resilience of a material, and the surface on which the test is conducted must have a fairly high polish to yield good results. Because the test is based on resilience, scleroscope hardness numbers should only be used to compare similar materials. A comparison between steel and rubber, for example, would not be valid.

Another definition of hardness is the ability of a material to resist being scratched. A crude but useful test that employs this principle is the **file test,** where one determines if a material can be cut by a simple metalworking file. The test can be either a pass–fail test using a single file, or a semiquantitative evaluation using a series of files that have been pretreated to various levels of known hardness. This approach is commonly used by geologists. Ten selected materials are used to create a scale that enables the hardness of rocks and minerals to be classified from 0 to 10.

Relationships among the Various Hardness Tests. Because the various hardness tests often evaluate different phenomena, there are no simple relationships between the different types of hardness numbers. Approximate relationships have been developed, however, by testing the same material on a variety of devices. Table 3-2 presents a correlation of hardness values for plain carbon and low-alloy steels. It may be noted that for Rockwell C numbers above 20, the Brinell values are approximately 10 times the Rockwell number. Also, for Brinell values below 320, the Vickers and Brinell values agree quite closely. Because the relationships among the various tests will differ with material, mechanical processing, and heat treatment, correlations such as Table 3-2 should be used with caution.

TABLE 3-2 Hardness Conversion Table for Steels

Brinell Number	Vickers Number	Rockwell Number C	Rockwell Number B	Scleroscope Number	Tensile Strength ksi	Tensile Strength MPa
	940	68		97	368	2537
757[a]	860	66		92	352	2427
722[a]	800	64		88	337	2324
686[a]	745	62		84	324	2234
660[a]	700	60		81	311	2144
615[a]	655	58		78	298	2055
559[a]	595	55		73	276	1903
500	545	52		69	256	1765
475	510	50		67	247	1703
452	485	48		65	238	1641
431	459	46		62	212	1462
410	435	44		58	204	1407
390	412	42		56	196	1351
370	392	40		53	189	1303
350	370	38	110	51	176	1213
341	350	36	109	48	165	1138
321	327	34	108	45	155	1069
302	305	32	107	43	146	1007

(continued)

Brinell Number	Vickers Number	Rockwell Number C	B	Scleroscope Number	Tensile Strength ksi	MPa
285	287	30	105	40	138	951
277	279	28	104	39	34	924
262	263	26	103	37	128	883
248	248	24	102	36	122	841
228	240	20	98	34	116	800
210	222	17	96	32	107	738
202	213	14	94	30	99	683
192	202	12	92	29	95	655
183	192	9	90	28	91	627
174	182	7	88	26	87	600
166	175	4	86	25	83	572
159	167	2	84	24	80	552
153	162		82	23	76	524
148	156		80	22	74	510
140	148		78	22	71	490
135	142		76	21	68	469
131	137		74	20	66	455
126	132		72	20	64	441
121	121		70		62	427
112	114		66		58	

[a] Tungsten, carbide ball; others, standard ball.

Relationship of Hardness to Tensile Strength. Table 3-2 and Figure 3-19 show a definite relationship between tensile strength and hardness. For plain carbon and low-alloy steels, the tensile strength (in pounds per square inch) can be estimated by multiplying the Brinell hardness number by 500. In this way, an inexpensive and quick hardness test can be used to provide a close approximation of the tensile strength of the steel. For other materials, however, the relationship is different and may even exhibit too much variation to be dependable. The multiplying factor for age-hardened aluminum is about 600, while for soft brass it is around 800.

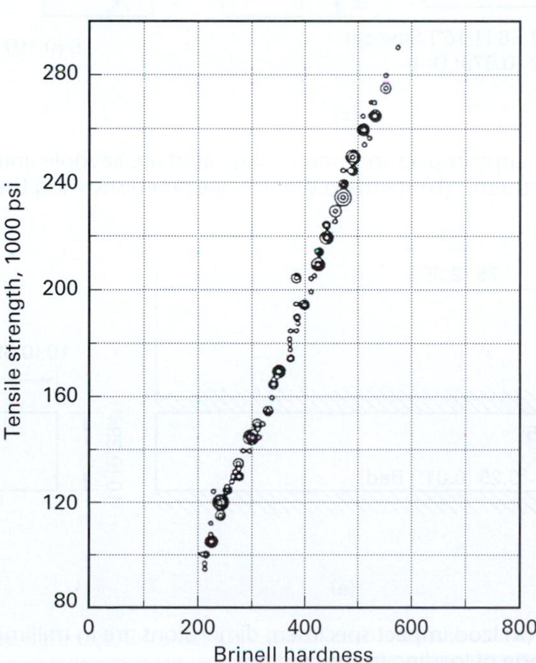

FIGURE 3-19 Relationship of hardness and tensile strength for a group of standard alloy steels. *(Courtesy of ASM International, Materials Park, OH)*

3.3 DYNAMIC PROPERTIES

Products or components can also be subjected to a wide variety of **dynamic loadings.** These may include (1) sudden impacts or loads that change rapidly in magnitude, (2) repeated cycles of loading and unloading, or (3) frequent changes in the mode of loading, such as from tension to compression. To select materials and design for these conditions, we must be able to characterize the mechanical properties of engineering materials under dynamic loadings.

Most dynamic tests subject standard specimens to a well-controlled set of test conditions. The conditions experienced by actual parts, however, rarely duplicate the controlled conditions of the standardized tests. While identical tests on different materials can indeed provide a comparison of material behavior, the assumption that similar results can be expected for similar conditions may not always be true. Because dynamic conditions can vary greatly, the quantitative results of standardized tests should be used with extreme caution, and one should always be aware of test limitations.

IMPACT TEST

Several tests have been developed to evaluate the *toughness* or fracture resistance of a material when it is subjected to a rapidly applied load, or impact. Of the tests that have become common, two basic types have emerged: (1) bending impacts, which include the standard Charpy and Izod tests, and (2) tension impacts.

The bending impact tests utilize specimens that are supported as beams. In the **Charpy test,** shown schematically in Figure 3-20, the standard specimen is a square bar containing a V-, keyhole-, or U-shaped notch. The test specimen is positioned horizontally, supported on the ends, and an impact is applied to the center, behind the notch, to complete a three-point bending. The **Izod test** specimen, while somewhat similar in size and appearance, is supported vertically as a cantilever beam and is impacted on the unsupported end, striking from the side of the notch (Figure 3-21). Impact testers, like

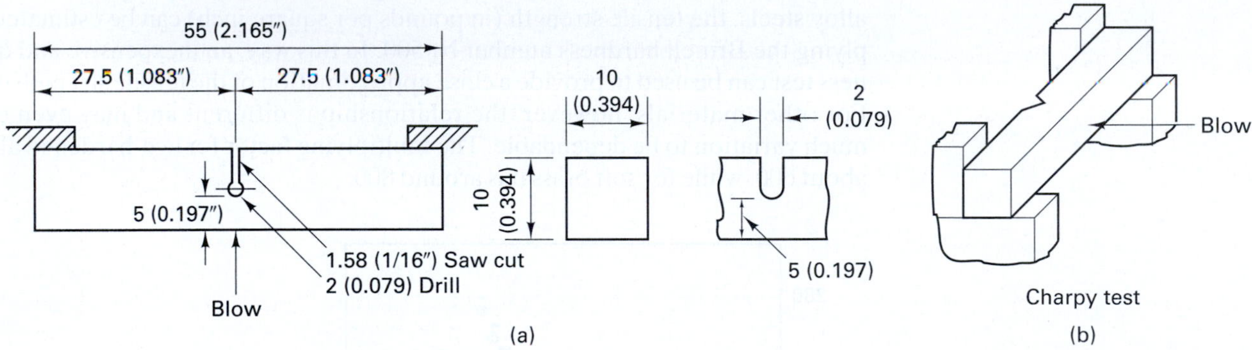

FIGURE 3-20 (a) Standard Charpy impact specimens; illustrated are keyhole and U-notches; dimensions are in millimeters with inches in parentheses; (b) standard V-notch specimen showing the three-point bending type of impact loading.

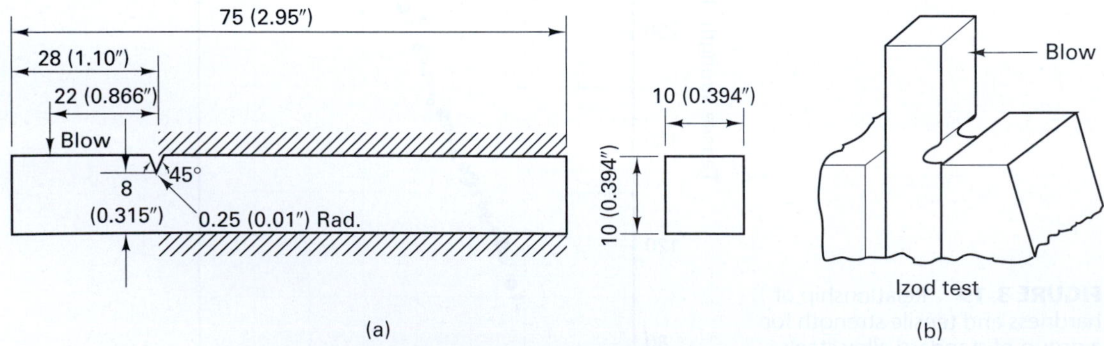

FIGURE 3-21 (a) Izod impact specimen; dimensions are in millimeters with inches in parentheses; (b) cantilever mode of loading in the Izod test.

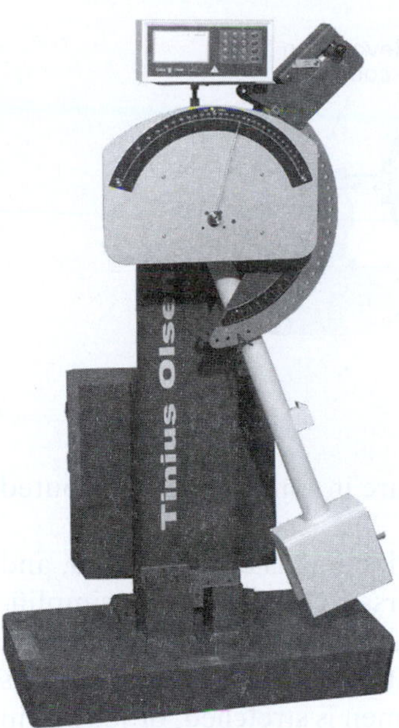

FIGURE 3-22 Impact testing machine. *(Courtesy of Tinius Olsen, Inc., Horsham, PA)*

the one shown in Figure 3-22, supply a predetermined impact energy in the form of a pendulum swinging from a starting height. After breaking or deforming the specimen, the pendulum continues its upward swing with an energy equal to its original minus that absorbed by the impacted specimen. The loss of energy is measured by the angle that the pendulum attains during its upward swing.

The test specimens for bending impacts must be prepared with geometric precision to ensure consistent and reproducible results. Notch profile is extremely critical, because the test measures the energy required to both initiate and propagate a fracture. The effect of notch profile is shown dramatically in Figure 3-23. Here, two specimens have been made from the same piece of steel with the same reduced cross-sectional area. The one with the keyhole notch fractures and absorbs only 43 ft-lb of energy, while the unnotched specimen resists fracture and absorbs 65 ft-lb during the impact.

Caution should also be placed on the use of impact data for design purposes. The test results apply only to standard specimens containing a standard notch that have been subjected to very specific test conditions. Changes in the form of the notch, minor variations in the overall specimen geometry, or faster or slower rates of loading (speed of the pendulum) can all produce significant changes in the results. Under conditions of sharp notches, wide specimens, and rapid loading, many ductile materials lose their energy-absorbing capability and fail in a brittle manner. (For example, the standard impact tests should not be used to evaluate materials for bullet-proof armor, because the velocities of loading are extremely different.)

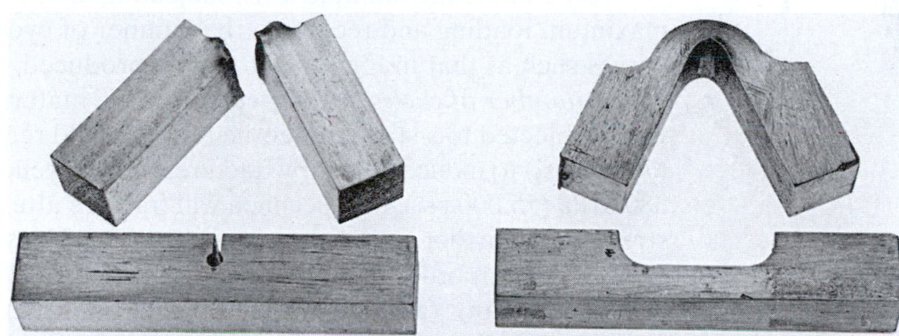

FIGURE 3-23 Notched and unnotched impact specimens before and after testing. Both specimens had the same cross-sectional area, but the notched specimen fractures while the other doesn't. *(E. Paul DeGarmo)*

The results of standard tests, however, can be quite valuable in assessing a material's sensitivity to notches and the multi-axial stresses that exist around a notch. Materials whose properties vary with notch geometry are termed **notch-sensitive.** Good surface finish and the absence of scratches, gouges and defects in workmanship will be key to satisfactory performance. Materials that are **notch-insensitive** can often be used with as-cast or rough-machined surfaces with no risk of premature failure.

Impact testing can also be performed at a variety of temperatures. As will be seen in a later section of this chapter, the evaluation of how fracture resistance changes with temperature, such as a ductile-to-brittle transition, can be crucial to success when selecting engineering materials for low-temperature service.

The **tensile impact test,** illustrated schematically in Figure 3-24, eliminates the use of a notched specimen, thereby avoiding many of the objections inherent in the Charpy and Izod tests. Turned specimens are subjected to uniaxial impact loadings applied through drop weights, modified pendulums, or variable-speed flywheels.

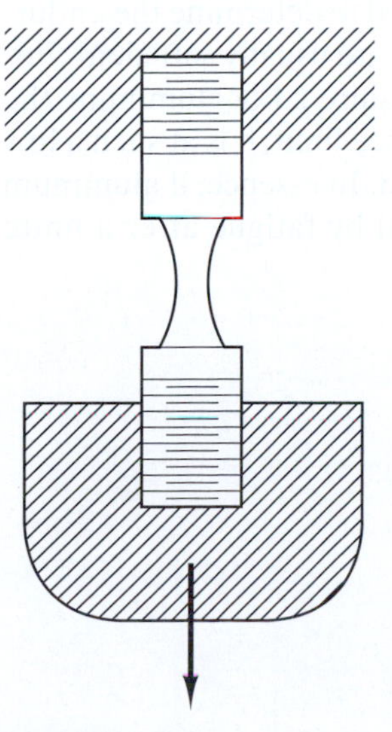

FIGURE 3-24 Tensile impact test.

FATIGUE AND THE ENDURANCE LIMIT

Materials can also fail by fracture if they are subjected to repeated applications of stress, even though the peak stresses have magnitudes less than the ultimate tensile strength and usually less than the yield strength. This phenomenon, known as **fatigue,** can result from either the cyclic repetition of a particular loading cycle or entirely random

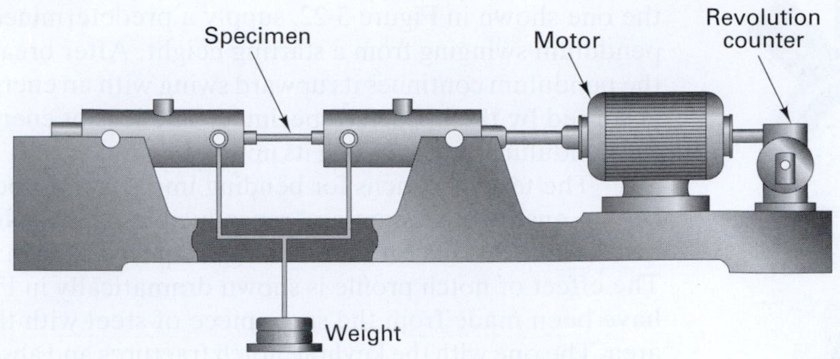

FIGURE 3-25 Schematic diagram of a Moore rotating-beam fatigue machine. *(Adapted from Hayden et al.,* The Structure and Properties of Materials, *Vol. 3, p. 15, Wiley, 1965)*

variations in stress. Almost 90% of all metallic fractures are in some degree attributed to fatigue.

For experimental simplicity, a periodic, sinusoidal loading is often utilized, and conditions of equal-magnitude tension–compression reversals provide further simplification. These conditions can be achieved by placing a placing a cylindrical specimen in a rotating drive and hanging a weight so as to produce elastic bending along the axis, as shown in Figure 3-25. Material at the bottom of the specimen is stretched, or loaded in tension, while material on the top surface is compressed. As the specimen turns, the surface of the specimen experiences a sinusoidal application of tension and compression with each rotation.

By conducting multiple tests, subjecting identical specimens to different levels of maximum loading and recording the number of cycles necessary to achieve fracture, curves such as that in Figure 3-26 can be produced. These curves are known as *stress versus number of cycles*, or ***S–N* curves.** If the material being evaluated in Figure 3-26 were subjected to a standard tensile test, it would require a stress in excess of 480 MPa (70,000 psi) to induce failure by fracture. Under cyclic loading with a peak stress of only 380 MPa (55,000 psi), the specimen will fracture after about 100,000 cycles. If the peak stress were further reduced to 350 MPa (51,000 psi), the fatigue lifetime would be extended by an order of magnitude to approximately 1,000,000 cycles. With a further reduction to any value below 340 MPa (49,000 psi), the specimen would not fail by fatigue, regardless of the number of stress application cycles.

The stress below which the material will not fail regardless of the number of load cycles is known as the **endurance limit** or *endurance strength*, and may be an important criterion in many designs. Above this value, any point on the curve is the **fatigue strength,** the maximum stress that can be sustained for a specified number of loading cycles.

A different number of loading cycles is generally required to determine the endurance limit for different materials. For steels, 10 million cycles are usually sufficient. For several of the nonferrous metals, 500 million cycles may be required. For aluminum, the curve continues to drop such that if aluminum has an endurance limit, it is at such a low value that a cheaper and much weaker material could be used. In essence, if aluminum is used under realistic stresses and cyclic loading, it will fail by fatigue after a finite lifetime.

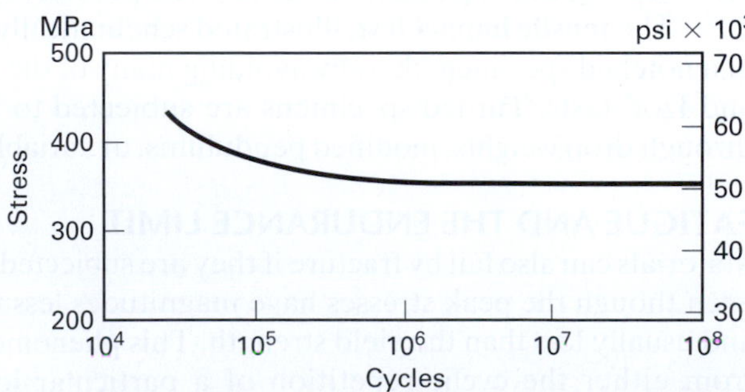

FIGURE 3-26 Typical *S–N* curve for steel showing an endurance limit. Specific numbers will vary with the type of steel and treatment.

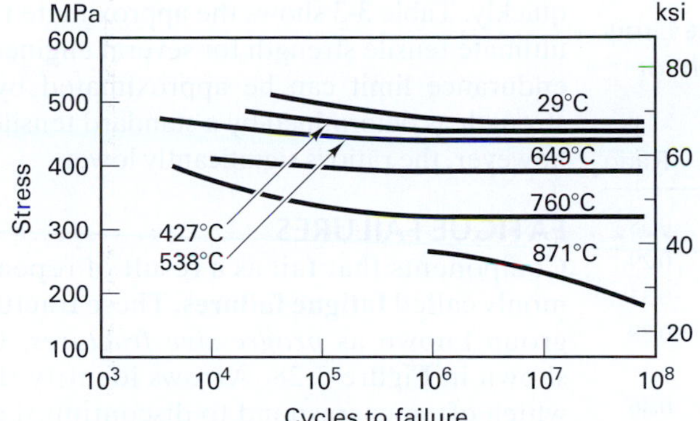

FIGURE 3-27 Fatigue strength of Inconel alloy 625 at various temperatures. *(Courtesy of Huntington Alloy Products Division, The International Nickel Company, Inc., Toronto, Canada)*

The fatigue resistance of an actual product is sensitive to a number of additional factors. One of the most important of these is the presence of stress raisers (or stress concentrators), such as sharp corners, small surface cracks, machining marks, or surface gouges. Data for the *S–N* curves are obtained from polished-surface, "flaw-free" specimens, and the reported lifetime is the cumulative number of cycles required to initiate a fatigue crack and then grow or propagate it to failure. If a part already contains a surface crack or flaw, the number of cycles required for crack initiation can be reduced significantly. In addition, the stress concentrator magnifies the stress experienced at the tip of the crack, accelerating the rate of subsequent crack growth. Great care should be taken to eliminate stress raisers and surface flaws on parts that will be subjected to cyclic loadings. Proper design and good manufacturing practices are often more important than material selection and heat treatment.

Operating temperature can also affect the fatigue performance of a material. Figure 3-27 shows *S–N* curves for Inconel 625 (a high-temperature Ni–Cr–Fe alloy) determined over a range of temperatures. As temperature is increased, the fatigue strength drops significantly. Because most test data are generated at room temperature, caution should be exercised when the product application involves elevated service temperatures.

Fatigue lifetime can also be affected by changes in the environment. When metals are subjected to corrosion during the cyclic loadings, the condition is known as *corrosion fatigue,* and both specimen lifetime and the endurance limit can be significantly reduced. Moreover, the nature of the environmental attack need not be severe. For some materials, tests conducted in air have been shown to have shorter lifetimes than those run in a vacuum, and further lifetime reductions have been observed with increasing levels of humidity. The test results can also be dependent on the frequency of the loading cycles. For slower frequencies, the environment has a longer time to act between loadings. At high frequencies, the environmental effects may be somewhat masked. The application of test data to actual products, therefore, requires considerable caution.

Residual stresses can also alter fatigue behavior. If the specimen surface is in a state of compression, such as that produced from shot peening, carburizing, or burnishing, it is more difficult to initiate a fatigue crack, and lifetime is extended. Conversely, processes that produce residual tension on the surface, such as welding or machining, can significantly reduce the fatigue lifetime of a product.

If the magnitude of the load varies during service, the fatigue response can be extremely complex. For example, consider the wing of a commercial airplane. As the wing vibrates during flight, the wing-fuselage joint is subjected to a large number of low-stress loadings. While large in number, these in-flight loadings may be far less damaging than a few high-stress loadings, like those that occur when the plane impacts the runway during landing. From a different perspective, however, the heavy loads may be sufficient to stretch and blunt a sharp fatigue crack, requiring many additional small-load cycles to "reinitiate" it. Evaluating how materials respond to complex patterns of loading is an area of great importance to design engineers.

Because reliable fatigue data may take a considerable time to generate, we may prefer to estimate fatigue behavior from properties that can be determined more

TABLE 3-3 Ratio of Endurance Limit to Tensile Strength for Various Materials

Material	Ratio
Aluminum	0.38
Beryllium copper (heat-treated)	0.29
Copper, hard	0.33
Magnesium	0.38
Steel	
AISI 1035	0.46
Screw stock	0.44
AISI 4140 normalized	0.54
Wrought iron	0.63

quickly. Table 3-3 shows the approximate ratio of the endurance limit to the ultimate tensile strength for several engineering metals. For many steels the endurance limit can be approximated by 0.5 times the ultimate tensile strength as determined by a standard tensile test. For the nonferrous metals, however, the ratio is significantly lower.

FATIGUE FAILURES

Components that fail as a result of repeated or cyclic loadings are commonly called **fatigue failures.** These fractures form a major part of a larger group known as *progressive fractures.* Consider the fracture surfaces shown in Figure 3-28. Arrows identify the points of fracture initiation, which often correspond to discontinuities in the form of surface cracks, sharp corners, machining marks, or even "metallurgical notches," such as an abrupt change in metal structure. With each repeated application of load, the stress at the tip of the crack exceeds the strength of the material, and the crack grows a very small amount. Crack growth continues with each successive application of load until the remaining cross section is no longer sufficient to withstand the peak stresses. Sudden overload fracture then occurs through the remainder of the material.

The overall fracture surface tends to exhibit two distinct regions: a smooth, relatively flat region where the crack was propagating by cyclic fatigue and a fibrous, irregular, or ragged region that corresponds to the sudden overload tearing. The size of the overload region reflects the area that must be intact in order to support the highest applied load. In Figure 3-28a, the applied load is high and only a small fatigue area is necessary to reduce the specimen to the point of overload fracture. In the b section of the figure, the fatigue fracture propagates about halfway through before sudden overload occurs.

The smooth areas of the fracture often contain a series of parallel ridges radiating outward from the origin of the crack. These ridges may not be visible under normal examination, however. They may be extremely fine, they may have been obliterated by a rubbing action during the compressive stage of repeated loading, or they may be very few in number if the failure occurred after only a few cycles of loading (low-cycle fatigue). Electron microscopy may be required to reveal the ridges, or **fatigue striations,**

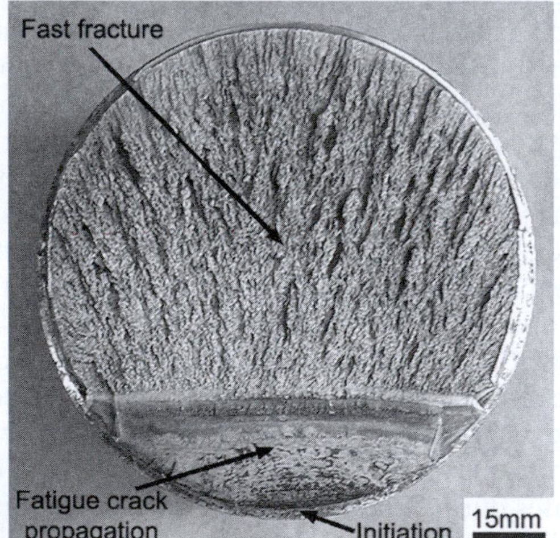

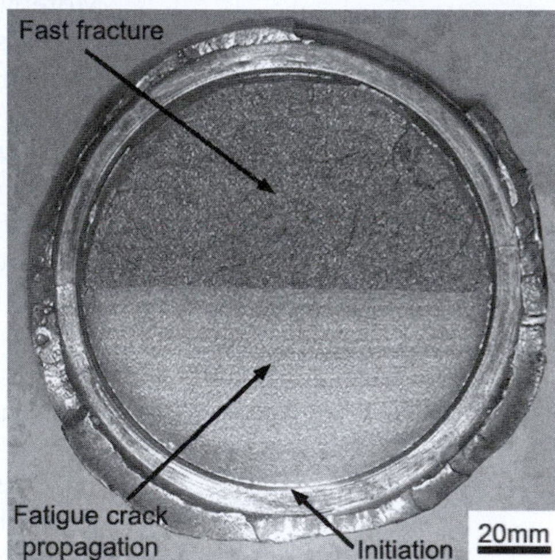

FIGURE 3-28 Fatigue fractures with arrows indicating the points of fracture initiation, the regions of fatigue crack propagation, and the regions of sudden overload or fast fracture. (a) High applied load results in a small fatigue region compared to the area of overload fracture; (b) low applied load results in a large area of fatigue fracture compared to the area of overload fracture. NOTE: The overload area is the minimum area required to carry the applied loads. *(From "Fatigue Crack Propagation," an article publshed in the May 2008 issue of* Advanced Materials & Processes *magazine, Reprinted with permission of ASM International®. All rights reserved. www.asminternational.org)*

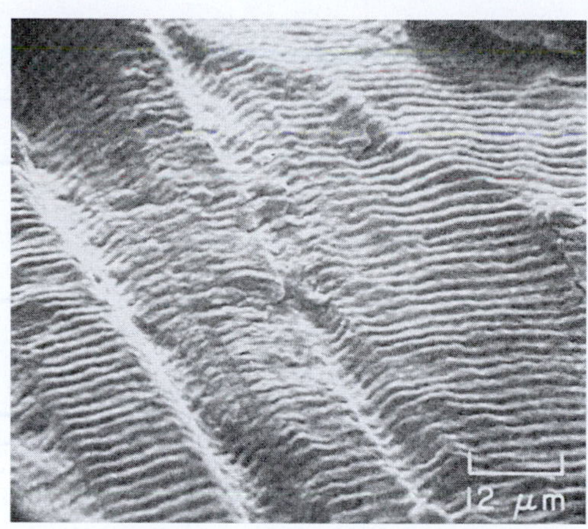

FIGURE 3-29 Fatigue fracture of AISI type 304 stainless steel viewed in a scanning electron microscope at 810×. Well-defined striations are visible. *(From "Interpretation of SEM Fractographs,"* Metals Handbook *Vol. 9, 8th ed. p.70. Reprinted with permission of ASM International®. All rights reserved. www.asminternational.org)*

that are characteristic of fatigue failure. Figure 3-29 shows an example of these markings at high magnification. Larger marks, known as *beach marks* may appear on the fatigued surface, lying parallel to the striations. These can be caused by interruptions to the cyclic loadings, changes in the magnitude of the applied load, and isolated overloads (not sufficient to cause ultimate fracture). *Ratchet marks,* or offset steps, can appear on the fracture surface if multiple fatigue cracks nucleate at different points and grow together.

For some fatigue failures, the overload area may exhibit a crystalline appearance, and the failure is sometimes attributed to the metal having "crystallized." As will be noted in Chapter 4, engineering metals are almost always crystalline materials. The final overload fracture simply propagated along the intercrystalline surfaces (grain boundaries), revealing the already-existing crystalline nature of the material. The conclusion that the material failed because it crystallized is totally erroneous, and the term is a definite misnomer.

Another common error is to classify all progressive-type failures as fatigue failures. Other progressive failure mechanisms, such as creep failure and stress–corrosion cracking, will also produce the characteristic two-region fracture. In addition, the same mechanism can produce fractures with different appearances depending on the magnitude of the load, type of loading (torsion, bending, or tension), temperature, and operating environment. Correct interpretation of a metal failure generally requires far more information than that acquired by a visual examination of the fracture surface.

A final misconception regarding fatigue failures is to assume that the failure is time dependent. The failure of materials under repeated loads below their static strength is primarily a function of the magnitude and number of loading cycles. If the frequency of loading is increased, the time to failure should decrease proportionately. If the time does not change, the failure is dominated by one or more environmental factors, and fatigue is a secondary component.

3.4 TEMPERATURE EFFECTS (BOTH HIGH AND LOW)

The test data used in design and engineering decisions should always be obtained under conditions that simulate those of actual service. A number of engineered structures, such as aircraft, space vehicles, gas turbines, and nuclear power plants, are required to operate under temperatures as low as −130°C (−200°F) or as high as 1250°C (2300°F). To cover these extremes, the designer must consider both the short- and long-range effects of temperature on the mechanical and physical properties of the material being considered. From a manufacturing viewpoint, the effects of temperature are equally important. Numerous manufacturing processes involve heat, and the elevated temperature and processing may alter the material properties in both favorable and unfavorable ways. A material can often be processed successfully, or economically, only because heating or cooling can be used to change its properties.

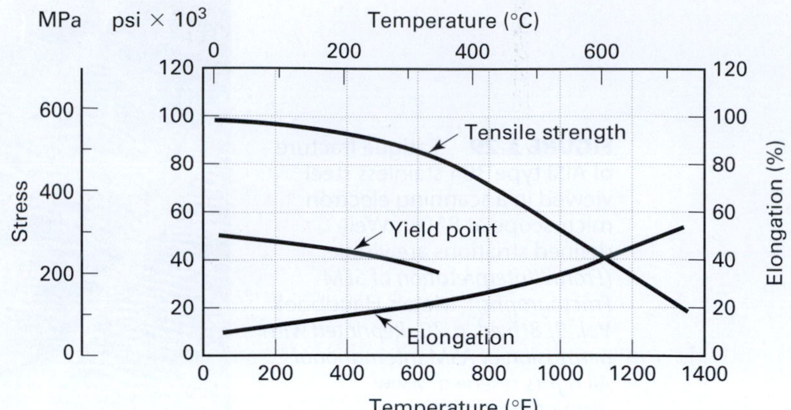

FIGURE 3-30 The effects of temperature on the tensile properties of a medium-carbon steel.

Elevated temperatures can be quite useful in modifying the strength and ductility of a material. Figure 3-30 summarizes the results of tensile tests conducted over a wide range of temperatures using a medium-carbon steel. Similar effects are presented for magnesium in Figure 3-31. As expected, an increase in temperature will typically induce a decrease in strength and hardness and an increase in elongation. For manufacturing operations such as metal forming, heating to elevated temperature may be extremely attractive because the material is now both weaker and more ductile.

Figure 3-32 shows the combined effects of temperature and strain rate (speed of testing) on the ultimate tensile strength of copper. For a given temperature, the **rate of deformation** can also have a strong influence on mechanical properties. Room-temperature

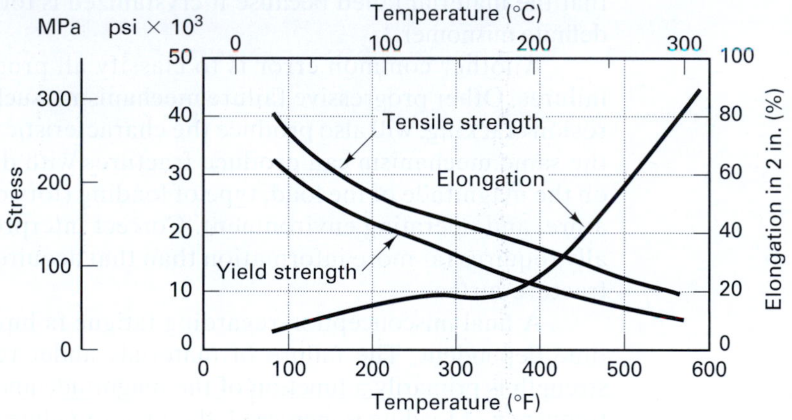

FIGURE 3-31 The effects of temperature on the tensile properties of magnesium.

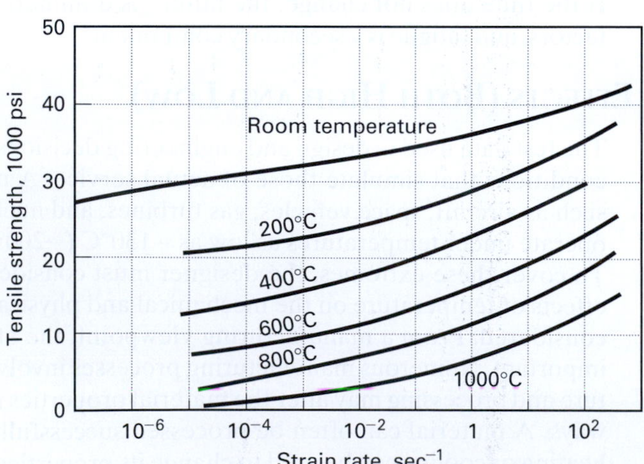

FIGURE 3-32 The effects of temperature and strain rate on the tensile strength of copper. *(From A. Nadai and M. J. Manjoine,* Journal of Applied Mechanics, *Vol. 8, 1941, p. A82, courtesy of ASME)*

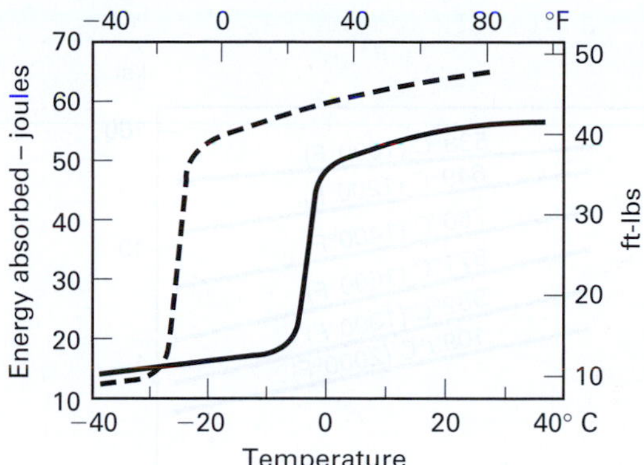

FIGURE 3-33 The effect of temperature on the impact properties of two low-carbon steels.

standard-rate tensile test data will be of little value if the application involves a material being hot-rolled at speeds of 1300 m/min (5000 ft/min).

The effect of temperature on impact properties became the subject of intense study in the 1940s when the increased use of welded-steel construction led to catastrophic failures of ships and other structures while operating in cold environments. Welding produces a monolithic (single-piece) product where cracks can propagate through a joint and continue on to other sections of the structure! Figure 3-33 shows the effect of decreasing temperature on the impact properties of two low-carbon steels. Although similar in form, the two curves are significantly different. The steel indicated by the solid line becomes brittle (requires very little energy to fracture) at temperatures below −4°C (25°F) while the other steel retains good fracture resistance down to −26°C (−15°F). The temperature at which the toughness goes from high-energy absorption to low energy absorption is known as the **ductile-to-brittle transition temperature.** As the temperature changes across this transition, the fracture appearance also changes. At high temperatures, ductile fracture occurs, and the specimen deforms in the region of ultimate fracture. At low temperatures, the fracture is brittle in nature, and the fracture area retains the original square shape. A plot of percent shear in the fracture surface (ductile fracture) versus temperature results in a curve similar to that for toughness, and the 50% point is reported as the **fracture appearance transition temperature (FATT).**

Figure 3-34 shows the ductile-to-brittle transition temperature for steel salvaged from the *R.M.S. Titanic* along with data from currently used ship plate material. While both are quality materials for their era, the *Titanic* steel has a much higher transition temperature and is generally more brittle. The *Titanic* struck an iceberg in salt water. The water temperature at the time of the accident was −2°C, and the results show that the steel would have been quite brittle. All steels tend to exhibit a ductile-to-brittle transition, but the temperature at which it occurs varies with carbon content and alloy. Metals such as aluminum, copper, and some types of stainless steel do not have a ductile-to-brittle transition and can be used at low temperatures with no significant loss of toughness.

Two separate curves are provided for each of the steels in Figure 3-34, reflecting test specimens cut in different orientation with respect to the direction of product rolling. Here, we see that processing features can further affect the properties and performance of a material. Because the performance properties can vary widely with the type of material, chemistry variations within the class of material, and prior processing, special cautions should be taken when selecting materials for low-temperature applications.

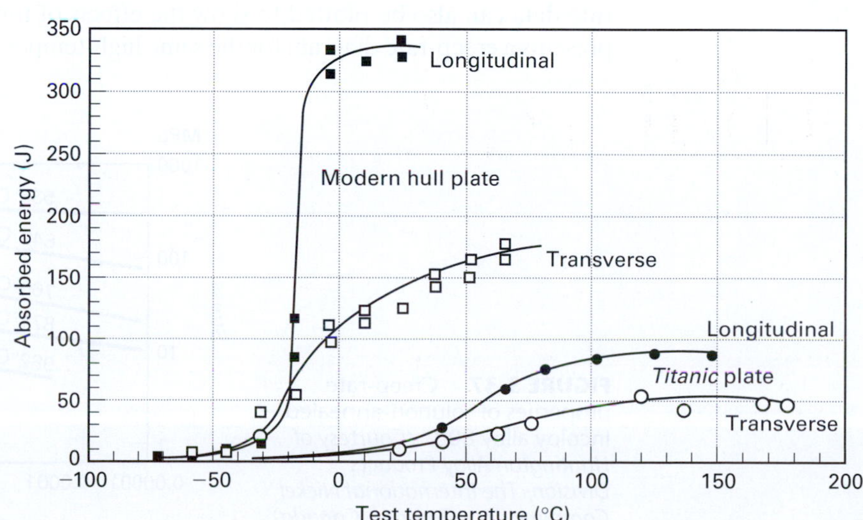

FIGURE 3-34 Notch toughness impact data: steel from the *Titanic* versus modern steel plate for both longitudinal and transverse specimens. *(Courtesy I&SM, September 1999, p. 33, Iron and Steel Society, Warrendale, PA)*

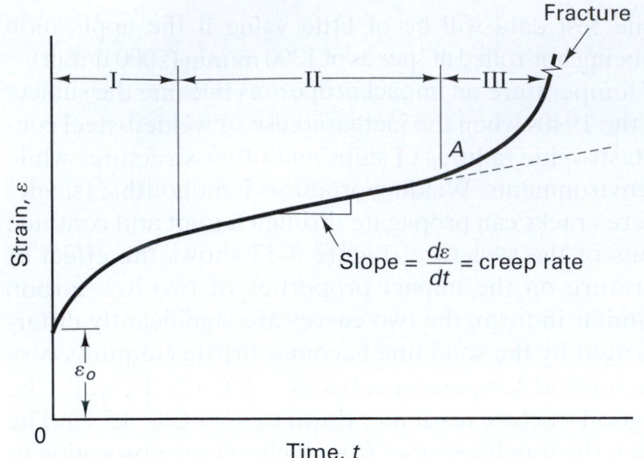

FIGURE 3-35 Creep curve for a single specimen at a fixed elevated temperature, showing the three stages of creep and reported creep rate. Note the nonzero strain at time zero due to the initial application of the load.

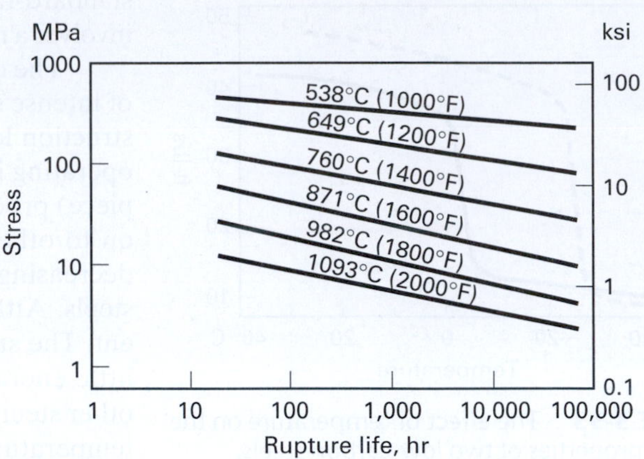

FIGURE 3-36 Stress–rupture diagram of solution-annealed Incoloy alloy 800 (Fe–Ni–Cr alloy). *(Courtesy of Huntington Alloy Products Division, The International Nickel Company, Inc., Toronto, Canada)*

CREEP

Long-term exposure to elevated temperatures can also lead to failure by a phenomenon known as **creep.** If a tensile-type specimen is subjected to a constant load at elevated temperature, it will elongate continuously until rupture occurs, even though the applied stress is below the yield strength of the material at the temperature of testing. While the rate of elongation is often quite small, creep can be an important consideration when designing equipment such as steam or gas turbines, power plant boilers, and other devices that operate under loads or pressures for long periods of time at high temperature.

If a test specimen is subjected to conditions of fixed load and fixed elevated temperature, an elongation-versus-time plot can be generated, similar to the one shown in Figure 3-35. The curve contains three distinct stages: a short-lived initial stage, a rather long second stage where the elongation rate is somewhat linear, and a short-lived third stage leading to fracture. Two significant pieces of engineering data are obtained from this curve: the rate of elongation in the second stage, or **creep rate,** and the total elapsed **time to rupture.** These results are unique to the material being tested and the specific conditions of the test. Tests conducted at higher temperatures or with higher applied loads would exhibit higher creep rates and shorter rupture times.

When creep behavior is a concern, multiple tests are conducted over a range of temperatures and stresses, and the rupture time data are collected into a single **stress–rupture diagram,** like the one shown in Figure 3-36. This simple engineering tool provides an overall picture of material performance at elevated temperature. In a similar manner, creep rate data can also be plotted to show the effects of temperature and stress. Figure 3-37 presents a creep-rate diagram for the same high-temperature nickel-base alloy.

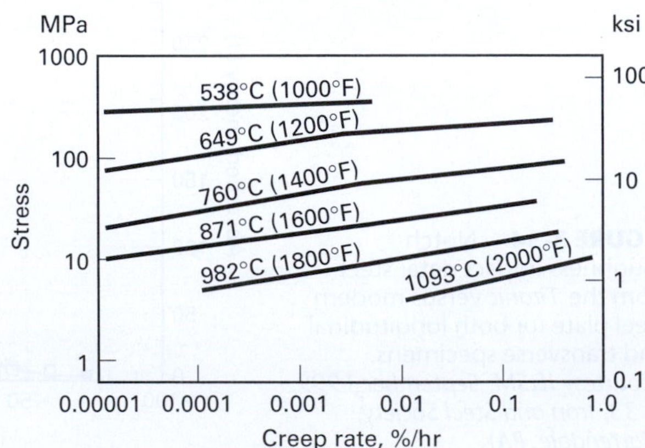

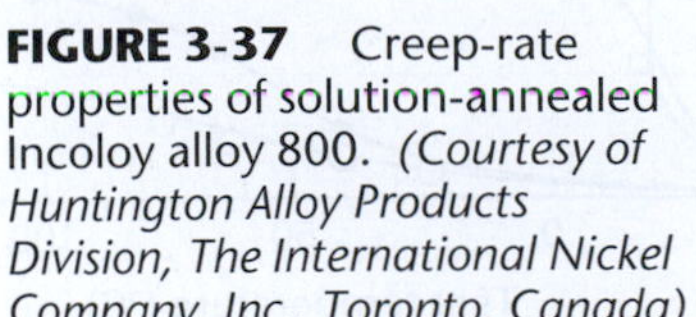

FIGURE 3-37 Creep-rate properties of solution-annealed Incoloy alloy 800. *(Courtesy of Huntington Alloy Products Division, The International Nickel Company, Inc., Toronto, Canada)*

3.5 MACHINABILITY, FORMABILITY, AND WELDABILITY

While it is common to assume that the various "-ability" terms also refer to specific material properties, they actually refer to the way a material responds to specific processing techniques. As a result, they can be quite nebulous. **Machinability,** for example, depends not only on the material being machined but also on the specific machining process, the conditions of that process (such as cutting speed), and the aspects of that process that are of greatest interest. Machinability ratings are frequently based on relative tool life. In certain applications, however, we may be more interested in how easy a metal is to cut, or how it performs under high machining speeds, and less interested in the tool life or the resulting surface finish. For other applications, surface finish or the formation of fine chips may be the most desirable feature. A material with high machinability to one individual may be considered to have poor machinability by a person using a different process or different process conditions.

In a similar manner, **malleability, workability,** and **formability** all refer to a material's suitability for plastic deformation processing. Because a material often behaves differently at different temperatures, a material with good "hot formability" may have poor deformation characteristics at room temperature. Furthermore, materials that flow nicely at low deformation speeds may behave in a brittle manner when loaded at rapid rates. Formability, therefore, needs to be evaluated for a specific combination of material, process, and process conditions, and the results should not be extrapolated or transferred to other processes or process conditions. Likewise, the **weldability** of a material will also depend on the specific welding or joining process and the specific process parameters.

3.6 FRACTURE TOUGHNESS AND THE FRACTURE MECHANICS APPROACH

A discussion of the mechanical properties of materials would not be complete without mention of the many tests and design concepts based on the fracture mechanics approach. Instead of treating test specimens as flaw-free materials, fracture mechanics begins with the premise that *all materials contain flaws or defects of some given size*. These may be **material defects,** such as pores, cracks, or inclusions; **manufacturing defects,** in the form of machining marks, arc strikes, or contact damage to external surfaces; or **design defects,** such as abrupt section changes, excessively small fillet radii, and holes. When the specimen is subjected to loads, the applied stresses are amplified or intensified in the vicinity of these defects, potentially causing accelerated failure or failure under unexpected conditions.

Fracture mechanics seeks to identify the conditions under which a defect will grow or propagate to failure and, if possible, the rate of crack or defect growth. The methods concentrate on three principal quantities: (1) the size of the largest or most critical flaw, usually denoted as a; (2) the applied stress, denoted by σ; and (3) the fracture toughness, a quantity that describes the resistance of a material to fracture or crack growth, which is usually denoted by K with subscripts to signify the conditions of testing. Equations have been developed that relate these three quantities at the onset of crack growth or propagation for various specimen geometries, flaw locations, and flaw orientations. If nondestructive testing or quality control methods have been applied, the size of the largest flaw that could go undetected is often known. By mathematically placing this worst possible flaw in the worst possible location and orientation, and coupling this with the largest applied stress for that location, a designer can determine the value of fracture toughness necessary to prevent that flaw from propagating during service. Specifying any two of the three parameters allows the computation of the third. If the material and stress conditions were defined, the size of the maximum permissible flaw could be computed. Inspection conditions could then be selected to ensure that flaws greater than this magnitude are cause for product rejection. Finally, if a component is found to have a significant flaw and the material is known, the maximum operating stress can be determined that will ensure no further growth of that flaw.

In the past, detection of a flaw or defect was usually cause for rejection of the part (Detection = Rejection). With enhanced methods and sensitivities of inspection, almost every product can now be shown to contain flaws. Fracture mechanics comes to the rescue. According to the philosophy of fracture mechanics, each of the flaws or defects in a material can be either **dormant** or **dynamic.** Dormant defects are those whose size remains unchanged through the lifetime of the part, and are indeed permissible. A major goal of fracture mechanics, therefore, is to define the distinction between dormant and dynamic for the specific conditions of material, part geometry, and applied loading. The basic equation of fracture mechanics assumes the form of $K \geq \alpha\sigma\sqrt{\pi a}$, where K is the **fracture toughness** of the material (a material property); σ is the maximum applied tensile stress; a is the size of the largest or most critical flaw; and α is a dimensionless factor that considers the flaw location, orientation, and shape. The left side of the equation considers the material and the right side describes the usage condition (a combination of flaw and loading). The relationship is usually described as a greater than or equal. When the material number, K, is greater than the usage condition, the flaw is dormant. When equality is reached, the flaw becomes dynamic, and crack growth or fracture occurs. Alternative efforts to prevent material fracture generally involve overdesign, excessive inspection, or the use of premium-quality materials—all of which increase cost and possibly compromise performance.

Fracture mechanics can also be applied to fatigue, which has already been cited as causing as much as 90% of all dynamic failures. The standard method of fatigue testing applies cyclic loads to polished, "flaw-free" specimens, and the reported lifetime includes both crack initiation and crack propagation. In contrast, fracture mechanics focuses on the growth of an already existing flaw. Figure 3-38 shows the **crack growth**

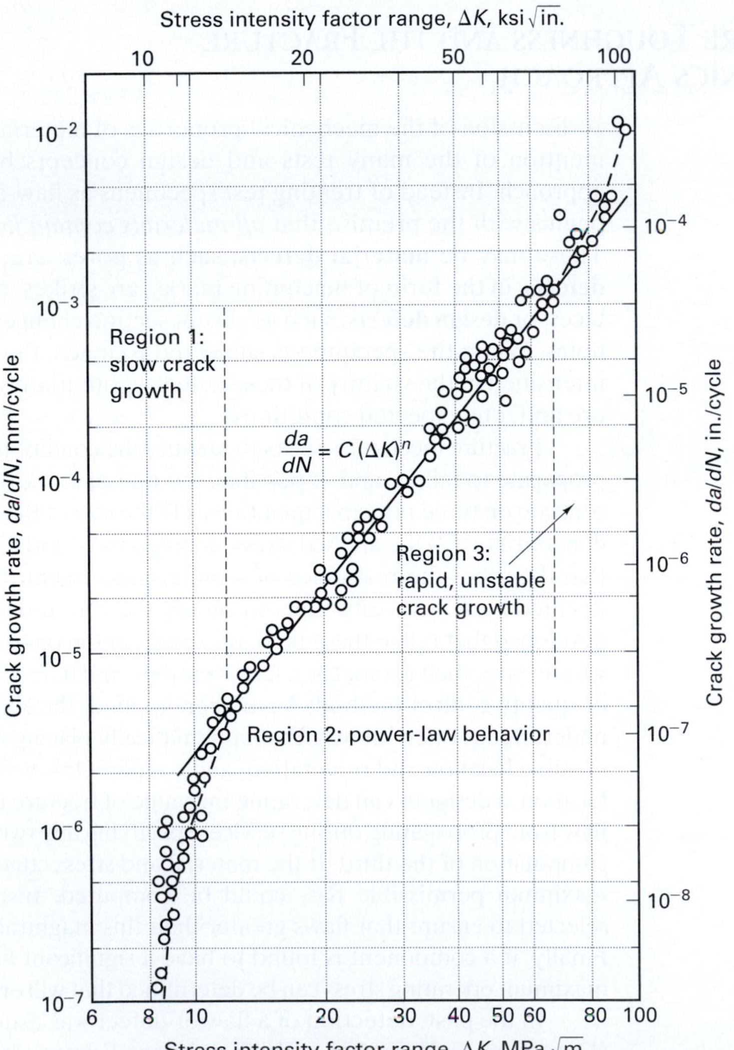

FIGURE 3-38 Plot of the fatigue crack growth rate versus ΔK for a typical steel—the fracture mechanics approach. Similar shape curves are obtained for most engineering metals. *(Courtesy of ASM International, Materials Park, OH)*

rate (change in size per loading cycle denoted as ***da/dN***) plotted as a function of the fracture mechanics parameter, ΔK (where ΔK increases with an increase in either the flaw size and/or the magnitude of applied stress). Because the fracture mechanics approach begins with an existing flaw, it provides a far more realistic guarantee of minimum service life.

Fracture mechanics is a truly integrated blend of design (applied stresses), inspection (flaw-size determination), and materials (fracture toughness). The approach has proven valuable in many areas where fractures could be catastrophic.

3.7 PHYSICAL PROPERTIES

For certain applications, the *physical properties* of a material may be even more important than the mechanical. These include the thermal, electrical, magnetic, and optical characteristics.

We have already seen several ways in which the mechanical properties of materials change with variations in temperature. In addition to these effects, there are some truly *thermal properties* that should be considered. The **heat capacity** or **specific heat** of a material is the amount of energy that must be added to or removed from a given mass of material to produce a 1-degree change in temperature. This property is extremely important in processes such as casting, where heat must be extracted rapidly to promote solidification, or heat treatment, where large quantities of material are heated and cooled. **Thermal conductivity** measures the rate at which heat can be transported through a material. While this may be tabulated separately in reference texts, it is helpful to remember that for metals, thermal conductivity is directly proportional to electrical conductivity. Metals such as copper, gold, and aluminum that possess good electrical conductivity are also good transporters of thermal energy. **Thermal expansion** is another important thermal property. Most materials expand upon heating and contract upon cooling, but the amount of expansion or contraction will vary with the material. For components that are machined at room temperature but put in service at elevated temperatures, or castings that solidify at elevated temperatures and then cool to room temperature, the as-manufactured dimensions must be adjusted to compensate for the subsequent changes.

Electrical conductivity or **electrical resistivity** may also be an important design consideration. These properties will vary not only with the material, but also with the temperature and the way the material has been processed.

The **magnetic response** of materials can be classified as diamagnetic, paramagnetic, ferromagnetic, antiferromagnetic, or ferrimagnetic. These terms refer to the way in which the material responds to an applied magnetic field. Material properties, such as saturation strength, remanence, and magnetic hardness or softness, describe the strength, duration, and nature of this response.

Still other physical properties that may assume importance include *weight* or *density*, *melting* and *boiling points*, and the various *optical properties*, such as the ability to transmit, absorb, or reflect light or other electromagnetic radiation.

3.8 TESTING STANDARDS AND TESTING CONCERNS

When evaluating the mechanical and physical properties of materials, it is important that testing be conducted in a consistent and reproducible manner. **ASTM International,** formerly the American Society of Testing and Materials, maintains and updates many testing standards, and it is important to become familiar with their contents. For example, ASTM specification E370 describes the "Standard Test Methods and Definitions for Mechanical Testing of Steel Products." Tensile testing is described in specifications E8 and E83, impact testing in E23, creep in E139, and penetration hardness in E10. Other specifications describe fracture mechanics testing, as well as procedures to evaluate corrosion resistance, compressive strength, shear strength, torsional properties, and corrosion-fatigue.

In addition, it is important to note not only the material being tested, but also the location from which the specimen was taken and its orientation. Rolled sheet, rolled plate and rolled bars, for example, will have different properties when tested parallel to the direction of rolling (longitudinal) and perpendicular to the rolling direction (transverse). See Figure 3-34 for an example. This variation of properties with direction is known as **anisotropy** and may be crucial to the success or failure of a product.

KEY WORDS

ASTM International
anisotropy
brale
breaking strength
Brinell hardness number (BHN)
Brinell hardness test
brittle
Charpy test
compression
crack growth rate (*da/dN*)
creep
creep rate
damping
design defects
dormant flaw
ductile-to-brittle transition temperature (DBTT)
ductility
durometer
dynamic flaw
dynamic properties
elastic limit
electrical conductivity
electrical resistivity
elongation
endurance limit
engineering strain
engineering stress–engineering strain curve
engineering stress
fatigue
fatigue failure
fatigue strength
fatigue striations
file test
formability
fracture appearance transition temperature (FATT)
fracture strength
fracture toughness
gage length
hardness
heat capacity
Hooke's Law
Izod test
impact test
Knoop hardness
Knoop test
machinability
magnetic response
malleability
manufacturing
manufacturing defects
material defects
mechanical properties
metal
microhardness tests
microindentation hardness testing
modulus of elasticity
modulus of resilience
necking
nonmetal
notch-sensitive
notch-insensitive
offset yield strength
percent elongation
percent reduction in area
performance
physical properties
plastic deformation
processing
properties
proportional limit
rate of deformation
resilience
Rockwell hardness test
S–N curve
scleroscope test
specific heat
static properties
stiffness
strain
strain hardening
strain-hardening exponent
strain rate
stress
stress–rupture diagram
structure
tensile strain
tensile strength
tensile test
tensile impact test
thermal conductivity
thermal expansion
time to rupture (rupture time)
toughness
transition temperature
true strain
true stress
ultimate tensile strength
uniaxial tensile test
uniform elongation
unit strain
Vickers hardness test
weldability
workability
yield point
Young's modulus

REVIEW QUESTIONS

1. What eras in the history of man have been linked to materials?
2. Knowledge of what four aspects is critical to the successful application of a material in an engineering design?
3. Give an example of how we might take advantage of a material that has a range of properties.
4. What are some properties commonly associated with metallic materials?
5. What are some of the more common nonmetallic engineering materials?
6. What are some of the important physical properties of materials?
7. Why should caution be exercised when applying the results from any of the standard mechanical property tests?
8. What are the standard units used to report stress and strain in the English system? In the metric or SI system?
9. What are static properties?
10. What is the most common static test to determine mechanical properties?
11. Why might Young's modulus or stiffness be an important material property?
12. What are some of the tensile test properties that are used to describe or define the elastic-to-plastic transition in a material?
13. Why is it important to specify the "offset" when providing yield strength data?
14. During the plastic deformation portion of a tensile test, a cylindrical specimen first maintains its cylindrical shape (increasing in length and decreasing in diameter) then transitions into a state called "necking." What is the explanation for this behavior?
15. What are two tensile test properties that can be used to describe the ductility of a material?
16. Is a brittle material a weak material? What does *brittleness* mean?
17. What is the toughness of a material?
18. What is the difference between true stress and engineering stress? True strain and engineering strain?
19. Explain how the plastic portion of a true stress–true strain curve can be viewed as a continuous series of yield strength values.
20. What is strain hardening or work hardening? How might this phenomenon be measured or reported? How might it be used in manufacturing?
21. How might tensile test data be misleading for a "strain rate sensitive" material?

22. What are some of the different material characteristics or responses that have been associated with the term *hardness?*
23. What are the similarities and differences between the Brinell and Rockwell hardness tests?
24. Why are there different Rockwell hardness scales?
25. When might a microhardness test be preferred over the more-standard Brinell or Rockwell tests?
26. Why might the various types of hardness tests fail to agree with one another?
27. What is the relationship between penetration hardness and the ultimate tensile strength for steel?
28. Describe several types of dynamic loading.
29. Why should the results of standardized dynamic tests be applied with considerable caution?
30. What are the two most common types of bending impact tests? How are the specimens supported and loaded in each?
31. What aspects or features can significantly alter impact data?
32. What is "notch sensitivity," and how might it be important in the manufacture and performance of a product?
33. Which type of failure accounts for almost 90% of metal failures?
34. What is the endurance limit? What occurs when stresses are above it? Below it?
35. Are the stresses applied during a fatigue test above or below the yield strength (as determined in a tensile test)?
36. What features may significantly alter the fatigue lifetime or fatigue behavior of a material?
37. What relationship can be used to estimate the endurance limit of a steel?
38. What material, design, or manufacturing features can contribute to the initiation of a fatigue crack?
39. How might the relative sizes of the fatigue region and the overload region provide useful information about the design of the product?
40. What are fatigue striations, and why do they form?
41. Why is it important for a designer or engineer to know a material's properties at all possible temperatures of operation?
42. Why should one use caution when using steel at low (below zero Fahrenheit) temperature?
43. How might the orientation of a piece of metal (with respect to its rolling direction) affect properties such as fracture resistance?
44. How might we evaluate the long-term effect of elevated temperature on an engineering material?
45. What is a stress–rupture diagram, and how is one developed?
46. Why are terms such as *machinability, formability,* and *weldability* considered to be poorly defined and therefore quite nebulous?
47. What is the basic premise of the fracture mechanics approach to testing and design?
48. What are some of the types of flaws or defects that might be present in a material?
49. What three principal quantities does fracture mechanics attempt to relate?
50. What is a dormant flaw? A dynamic flaw? How do these features relate to the former "Detection = Rejection" criteria for product inspection?
51. What are the three most common thermal properties of a material, and what do they measure?
52. Describe an engineering application where the density of the selected material would be an important material consideration.
53. Why is it important that property testing be performed in a standardized and reproducible manner?
54. Why is it important to consider the orientation of a test specimen with respect to the overall piece of material?

Problems

1. Select a product or component for which physical properties are more important than mechanical properties.
   a. Describe the product or component and its function.
   b. What are the most important properties or characteristics?
   c. What are the secondary properties or characteristics that would also be desirable?
2. Repeat Problem 1 for a product or component whose dominant required properties are of a static mechanical nature.
3. Repeat Problem 1 for a product or component whose dominant requirements are dynamic mechanical properties.
4. One of the important considerations when selecting a material for an application is to determine the highest and lowest operating temperature along with the companion properties that must be present at each extreme. The ductile-to-brittle transition temperature, discussed in Section 3.4, has been an important factor in a number of failures. An article that summarized the features of 56 catastrophic brittle fractures that made headline news between 1888 and 1956 noted that low temperatures were present in nearly every case. The water temperature at the time of the sinking of the *Titanic* was above the freezing point for salt water but below the transition point for the steel used in construction of the hull of the ship.
   a. Which of the common engineering materials exhibits a ductile-to-brittle transition?
   b. For plain carbon and low-alloy steels, what is a typical value (or range of values) for the transition temperature?
   c. What type of material would you recommend for construction of a small vessel to transport liquid nitrogen within a building or laboratory?
   d. Figure 3-34 summarizes the results of impact testing performed on hull plate from the *R.M.S. Titanic* and similar material produced for modern steel-hulled ships. Why should there be a difference between specimens cut longitudinally (along the rolling direction) and transversely (across the rolling direction)? What advances in steel making have led to the significant improvement in low-temperature impact properties?
5. Several of the property tests described in this chapter produce results that are quite sensitive to the presence or absence of notches or other flaws. The fracture mechanics approach to materials testing incorporates flaws into the tests

and evaluates their performance. The review article mentioned in Problem 4 cites the key role of a flaw or defect in nearly all of the headline-news fractures.

a. What are some of the various "flaws or defects" that might be present in a product? Consider flaws that might be present in the starting material; flaws that might be introduced during manufacture; and flaws that might occur due to shipping, handling, use, maintenance, or repair.
b. What particular properties might be most sensitive to flaws or defects?
c. Discuss the relationship of flaws to the various types of loading (tension versus compression, torsion, shear).
d. Fracture mechanics considers both surface and interior flaws and assigns terms such as "crack initiator," "crack propagator," and "crack arrestor." Briefly discuss why location and orientation may be as important as the physical size of a flaw.

www.wiley.com/go/global/degarmo

Chapter 3 CASE STUDY

Separation of Mixed Materials

Because of the amount of handling that occurs during material production, within warehouses, and during manufacturing operations, along with loading, shipping, and unloading, material mix-ups and mixed materials are not an uncommon occurrence. Mixed materials also occur when industrial scrap is collected, or when discarded products are used as new raw materials through recycling. Assume that you have equipment to perform each of the tests described in this chapter (as well as access to the full spectrum of household and department store items and even a small machine shop). For each of the following material combinations, determine one or more procedures that would permit separation of the mixed materials. Use standard data-source references to help identify distinguishable properties.

1. Steel and aluminum cans that have been submitted for recycling
2. Stainless steel sheets of Type 430 ferritic stainless and Type 316 austenitic stainless.
3. 6061-T6 aluminum and AZ91 magnesium that have become mixed in a batch of machine shop scrap.
4. Transparent bottles of polyethylene and polypropylene (both thermoplastic polymers) that have been collected for recycling.
5. Hot-rolled bars of AISI 1008 and 1040 steel.
6. Hot-rolled bars of AISI 1040 (plain-carbon) steel and 4140 steel (a molybdenum-containing alloy)
7. Mixed plastic consisting of recyclable thermoplastic polyvinylchloride (PVC) and nonrecyclable polyester—as might occur from automotive dashboards, consoles, and other interior components.

CHAPTER 4

NATURE OF METALS AND ALLOYS

4.1 STRUCTURE–PROPERTY–PROCESSING–PERFORMANCE RELATIONSHIPS

The success of many manufactured products depends on the selection of materials whose properties meet the requirements of the application. Primitive cultures were often limited to the naturally occurring materials in their environment. As civilizations developed, the spectrum of construction materials expanded. Materials could now be processed and their properties altered and improved. The alloying or heat treatment of metals, and the firing of ceramics are examples of techniques that can substantially alter the properties of a material. Fewer compromises were required and enhanced design possibilities emerged. Products, in turn, became more sophisticated. While the early successes in altering materials were largely the result of trial and error, we now recognize that the **properties** and **performance** of a material are a direct result of its **structure** and **processing.** If we want to change the properties, we will most likely have to induce changes in the material structure.

Because all materials are composed of the same basic components—particles that include *protons, neutrons,* and *electrons*—it is amazing that so many different materials exist with such widely varying properties. This variation can be explained, however, by the many possible combinations these units can assume in a macroscopic assembly. The subatomic particles combine in different arrangements to form the various elemental *atoms,* each having a nucleus of protons and neutrons surrounded by the proper number of electrons to maintain charge neutrality. The specific arrangement of the electrons surrounding the nucleus affects the electrical, magnetic, thermal, and optical properties as well as the way the atoms bond to one another. Atomic bonding then produces a higher level of structure, which may be in the form of a *molecule*, *crystal*, or *amorphous aggregate*. This structure, along with the imperfections that may be present, has a profound effect on the mechanical properties. The size, shape, and arrangement of multiple crystals, or the mixture of two or more different structures within a material, produce a higher level of structure, known as **microstructure.** Variations in microstructure further affect the material properties.

Because of the ability to control structures through processing, and the ability to develop new structures through techniques such as composite materials, engineers now have at their disposal a wide variety of materials with an almost unlimited range of properties. The specific properties of these materials depend on all levels of structure, from subatomic to macroscopic (Figure 4-1). This chapter will attempt to develop an understanding of the basic structure of engineering materials and how changes in that structure affect their properties and performance.

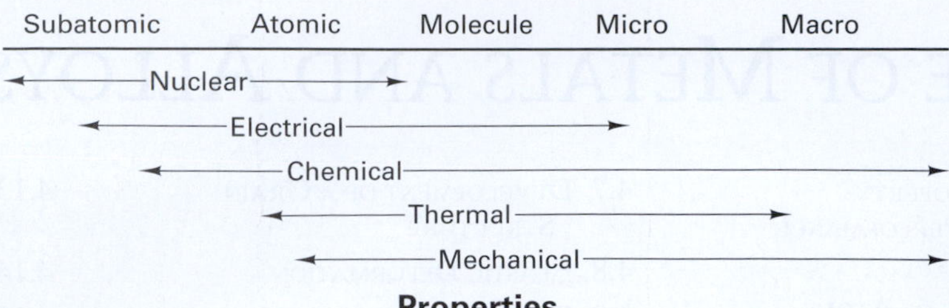

FIGURE 4-1 General relationships between structural level and the various types of engineering properties.

4.2 THE STRUCTURE OF ATOMS

Experiments have revealed that atoms consist of a relatively dense nucleus composed of positively charged protons and neutral particles of nearly identical mass, known as neutrons. Surrounding the nucleus are the negatively charged electrons, which appear in numbers equal to the protons so as to maintain a neutral charge balance. Distinct groupings of these basic particles produce the known elements, ranging from the relatively simple hydrogen atom to the unstable transuranium atoms more than 250 times as heavy. Except for density and specific heat, however, the weight of atoms has very little influence on their engineering properties.

The light electrons that surround the nucleus play an extremely significant role in determining material properties. These electrons are arranged in a characteristic structure consisting of shells and subshells, each of which can contain only a limited number of electrons. The first shell, nearest the nucleus, can contain only 2. The second shell can contain 8, the third, 18, and the fourth, 32. Each shell and subshell is most stable when it is completely filled. For atoms containing electrons in the third shell and beyond, however, relative stability is achieved with eight electrons in the outermost layer or subshell.

If an atom has slightly less than the number of outer-layer electrons required for stability, it will readily accept electrons from another source. It will then have more electrons than protons and becomes a negatively charged atom, or **negative ion.** Depending on the number of additional electrons, ions can have negative charges of 1, 2, 3, or more. Conversely, if an atom has a slight excess of electrons beyond the number required for stability (such as sodium, with one electron in the third shell), it will readily give up the excess electron and become a **positive ion.** The remaining electrons become more strongly attached, so further removal of electrons becomes progressively more difficult.

The number of electrons surrounding the nucleus of a neutral atom is called the **atomic number.** More important, however, are those electrons in the outermost shell or subshell, which are known as **valence electrons.** These are influential in determining chemical properties, electrical conductivity, some mechanical properties, the nature of interatomic bonding, the atom size, and optical characteristics. Elements with similar electron configurations in their outer shells tend to have similar properties.

4.3 ATOMIC BONDING

Atoms are rarely found as free and independent units, but are usually linked or bonded to other atoms in some manner as a result of interatomic attraction. The electron structure of the atoms plays the dominant role in determining the nature of the bond.

Three types of **primary bonds** are generally recognized, the simplest of which is the **ionic bond.** If more than one type of atom is present, the outermost electrons can break free from atoms with excesses in their valence shell, transforming them into positive ions. These electrons then transfer to atoms with deficiencies in their outer shell, converting them into negative ions. The positive and negative ions have an electrostatic attraction for each other, resulting in a strong bonding force. Figure 4-2 presents a crude schematic of the ionic bonding process for sodium and chlorine. Ionized atoms do not

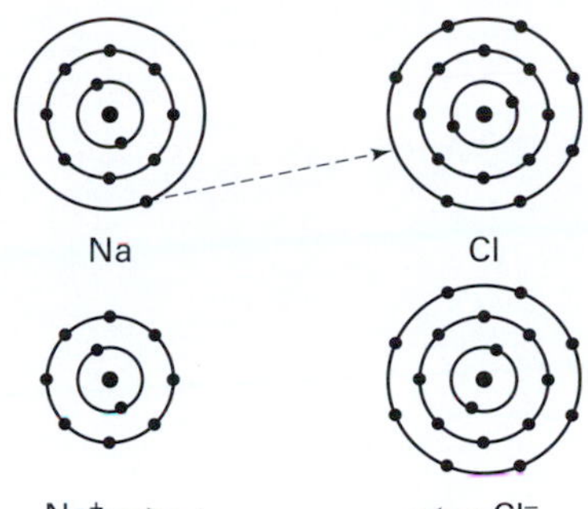

FIGURE 4-2 Ionization of sodium and chlorine, producing stable outer shells by electron transfer.

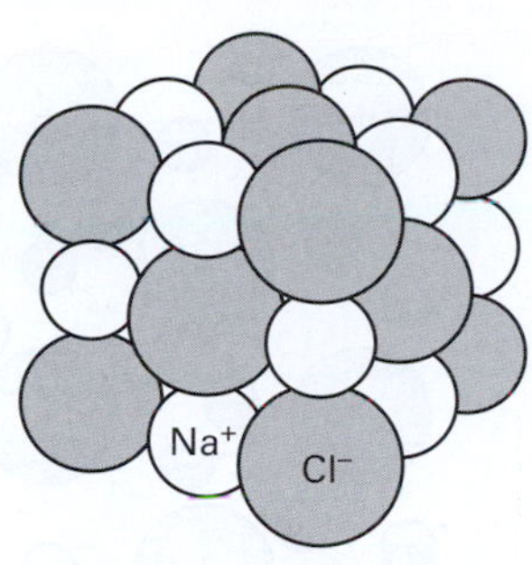

FIGURE 4-3 Three-dimensional structure of the sodium chloride crystal. Note how the various ions are surrounded by ions of the opposite charge.

usually unite in simple pairs, however. All positively charged atoms attract all negatively charged atoms. Therefore, each sodium ion will attempt to surround itself with negative chlorine ions, and each chlorine ion will attempt to surround itself with positive sodium ions. Because the attraction is equal in all directions, the result will be a three-dimensional structure, like the one shown in Figure 4-3. Because charge neutrality must be maintained within the structure, equal numbers of positive and negative charges must be present in each neighborhood. General characteristics of materials joined by ionic bonds include high density, moderate to high strength, high hardness, brittleness, high melting point, and low electrical and thermal conductivities (because all electrons are captive to specific atoms, movement of electrical charge would require movement of entire atoms or ions).

A second type of primary bond is the **covalent bond.** Here, the atoms in the assembly find it impossible to produce completed shells by electron transfer but achieve the same goal through electron sharing. Adjacent atoms share outer-shell electrons so that each achieves a stable electron configuration. The shared (negatively charged) electrons locate between the positive nuclei, forming a positive–negative–positive bonding link. Figure 4-4 illustrates this type of bond for a pair of chlorine atoms, each of which contains seven electrons in the valence shell. The result is a stable two-atom molecule, Cl_2. Stable molecules can also form from the sharing of more than one electron from each atom, as in the case of nitrogen (Figure 4-5a). The atoms in the assembly need not be identical (as in HF, Figure 4-5b), the sharing does not have to be equal, and a single atom can share electrons with more than one other atom. For atoms such as carbon and silicon, with four electrons in the valence shell, one atom may share its valence electrons with each of four neighboring atoms. The resulting structure is a three-dimensional network of bonded atoms, like the one shown in Figure 4-5c, where each atom is the center of a four-atom tetrahedron formed by its four neighbors as shown in Figure 4-5d. Because each atom only wants four neighbors, carbon and silicon materials tend to be light in weight. The covalent bond tends to produce materials with high strength and high melting point. Because atom movement within the three-dimensional structure (plastic deformation) requires the breaking of discrete bonds, covalent materials are characteristically brittle. Electrical conductivity depends on bond strength, ranging from conductive tin (weak covalent bonding), through semiconductive silicon and germanium, to insulating diamond (carbon). Ionic or covalent bonds are commonly found in ceramic and polymeric materials.

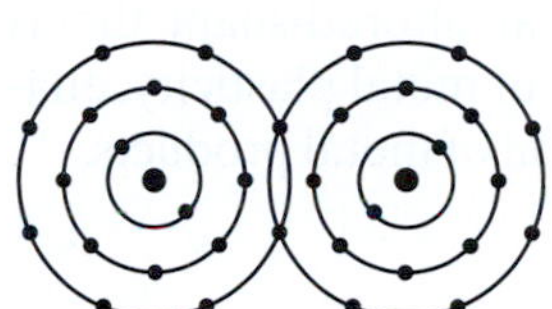

FIGURE 4-4 Formation of a chlorine molecule by the electron sharing of a covalent bond.

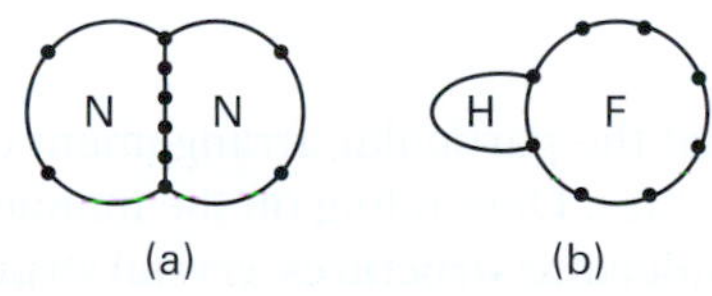

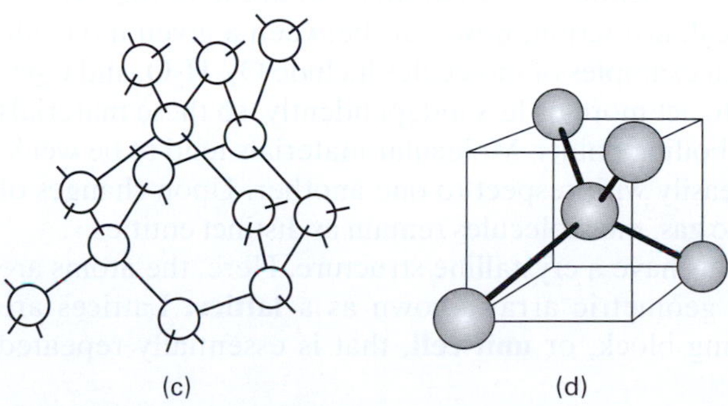

FIGURE 4-5 Examples of covalent bonding in (a) nitrogen molecule, (b) HF, and (c) silicon. Part (d) shows the tetrahedron formed by a silicon atom and its four neighbors.

A third type of primary bond is possible when a complete outer shell cannot be formed by either electron transfer or electron sharing. This bond is known as the **metallic bond** (Figure 4-6). If each of the atoms in an aggregate contains only a few valence electrons (one, two, or three), these electrons can be easily removed to produce "stable" ions. The positive ions (nucleus and inner, nonvalence electrons) then arrange in a three-dimensional periodic array, and are surrounded by wandering, universally shared, valence electrons, sometimes referred to as an electron cloud or electron gas. These highly mobile, free electrons account

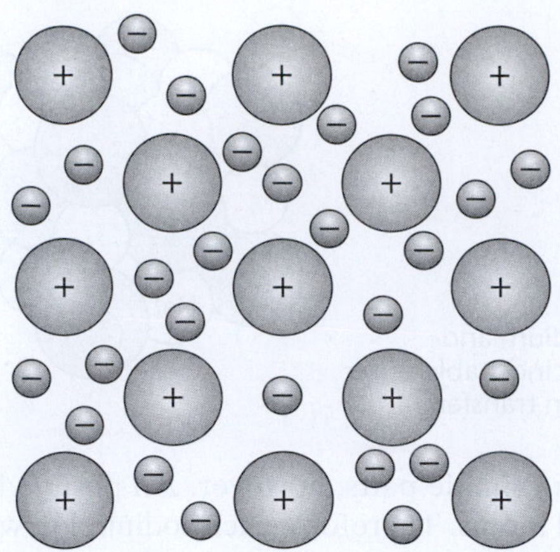

FIGURE 4-6 Schematic of the metallic bond showing the positive ions and free electrons.

for the high electrical and thermal conductivity values as well as the opaque (nontransparent) characteristic observed in metals (the free electrons are able to absorb the various discrete energies of light radiation). They also provide the "cement" required for the positive–negative–positive attractions that result in bonding. Bond strength, and therefore material strength and melting temperature, varies over a wide range. More significant, however, is the observation that the positive ions can now move within the structure without the breaking of discrete bonds. Materials bonded by metallic bonds can be deformed by atom-movement mechanisms and produce an altered-shape that is every bit as strong as the original. This phenomenon is the basis of metal plasticity, enabling the wide variety of forming processes used in the fabrication of metal products.

4.4 SECONDARY BONDS

Weak or **secondary bonds,** known as **van der Waals forces,** can form between molecules that possess a nonsymmetrical distribution of electrical charge. Some molecules, such as hydrogen fluoride and water,[1] can be viewed as electric dipoles. Certain portions of the molecule tend to be more positive or more negative than others (an effect referred to as **polarization**). The negative part of one molecule tends to attract the positive region of another, forming a weak bond. Van der Waals forces contribute to the mechanical properties of a number of molecular polymers, such as polyethylene and polyvinyl chloride (PVC).

4.5 ATOM ARRANGEMENTS IN MATERIALS

As atoms bond together to form aggregates, we find that the particular arrangement of the atoms has a significant effect on the material properties. Depending on the manner of atomic grouping, materials are classified as having **molecular structures, crystal structures,** or **amorphous structures.**

Molecular structures have a distinct number of atoms that are held together by primary bonds. There is only a weak attraction, however, between a given molecule and other similar groupings. Typical examples of molecules include O_2, H_2O, and C_2H_4 (ethylene). Each molecule is free to act more or less independently, so these materials exhibit relatively low melting and boiling points. Molecular materials tend to be weak, because the molecules can move easily with respect to one another. Upon changes of state from solid to liquid or liquid to gas, the molecules remain as distinct entities.

Solid metals and most minerals have a crystalline structure. Here, the atoms are arranged in a three-dimensional geometric array known as a **lattice.** Lattices are describable through a unit building block, or **unit cell,** that is essentially repeated

[1] The H_2O molecule can be viewed as a 109-degree boomerang or elbow with oxygen in the middle and the two hydrogens on the extending arms. The eight valence electrons (six from oxygen and two from hydrogen) associate with oxygen, giving it a negative charge. The hydrogen arms are positive. Therefore, when two or more water molecules are present, the positive hydrogen locations of one molecule are attracted to the negative oxygen location of an adjacent molecule.

throughout space. Crystalline structures will be discussed more fully in the following section.

In an amorphous structure, such as glass, the atoms have a certain degree of local order (arrangement with respect to neighboring atoms), but when viewed as an aggregate, they lack the periodically ordered arrangement that is characteristic of a crystalline solid.

4.6 CRYSTAL STRUCTURES OF METALS

From a manufacturing viewpoint, metals are an extremely important class of materials. They are frequently the materials being processed and often form both the tool and the machinery performing the processing. They are characterized by the metallic bond and possess the distinguishing characteristics of strength, good electrical and thermal conductivity, luster, the ability to be plastically deformed to a fair degree without fracturing, and a relatively high specific gravity or density compared to nonmetals. The fact that some metals possess properties different from the general pattern simply expands their engineering utility.

When metals solidify, the atoms assume a crystalline structure; that is, they arrange themselves in a geometric lattice. Many metals exist in only one lattice form. Some, however, can exist in the solid state in two or more lattice forms, with the particular form depending on the conditions of temperature and pressure. These metals are said to be **allotropic** or **polymorphic** (poly means "more than one"; morph means "structure"), and the change from one lattice form to another is called an *allotropic transformation*. The most notable example of such a metal is iron, where the allotropic change makes it possible for heat-treating procedures that yield a wide range of final properties. It is largely because of its allotropy that iron has become the basis of our most important alloys.

There are 14 basic types of crystal structures or lattices. Fortunately, however, nearly all of the commercially important metals solidify into one of three lattice types: body-centered cubic, face-centered cubic, or hexagonal close-packed. Table 4-1 lists the room temperature structure for a number of common metals. Figure 4-7 compares these three structures to one another, along with the easily visualized, but rarely observed, simple cubic structure.

TABLE 4-1 The Type of Crystal Lattice for Common Metals at Room Temperature

Metal	Lattice Type
Aluminum	Face-centered cubic
Copper	Face-centered cubic
Gold	Face-centered cubic
Iron	Body-centered cubic
Lead	Face-centered cubic
Magnesium	Hexagonal
Silver	Face-centered cubic
Tin	Body-centered tetragonal
Titanium	Hexagonal

To begin our study of crystals, consider the **simple cubic structure** illustrated in Figure 4-7a. This crystal can be constructed by placing single atoms on all corners of a cube, and then linking identical cube units together. If we assume that the atoms are rigid spheres with atomic radii touching one another, computation reveals that only 52% of available space is occupied. Each atom is in direct contact with only six neighbors (plus and minus in each of the x, y, and z cube edge directions). Both of these observations are unfavorable to the metallic bond, where atoms desire both a high number of nearest neighbors and high-efficiency packing.

The largest region of unoccupied space is in the geometric center of the cube, where a sphere of 0.732 times the atom diameter could be inserted.[2] If the cube is expanded to permit the insertion of an entire atom, the **body-centered cubic (BCC)** structure results (Figure 4-7b). Each atom now has eight nearest neighbors, and 68% of the space is occupied. This structure is more favorable to metals and is observed in room temperature iron, chromium, manganese, and the other metals listed in Figure 4-7b.

[2] The diagonal of a cube is equal to the square root of three times the length of the cube edge, and the cube edge is here equal to two atomic radii or one atomic diameter. Thus, the diagonal is equal to 1.732 times the atom diameter and is made up of an atomic radius, open space, and another atomic radius. Because two radii equals one diameter, the open space must be equal in size to 0.732 times the atomic diameter.

	Lattice structure	Unit cell schematic	Unit cells	Number of nearest neighbors	Packing efficiency	Typical metals
a	Simple cubic	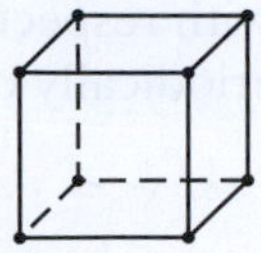	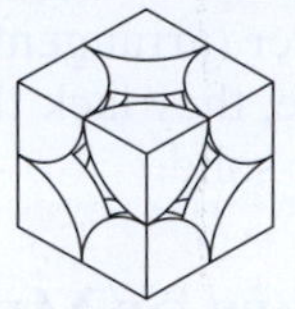	6	52%	None
b	Body-centered cubic	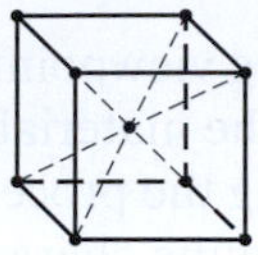		8	68%	Fe, Cr, Mn, Cb, W, Ta, Ti, V, Na, K
c	Face-centered cubic	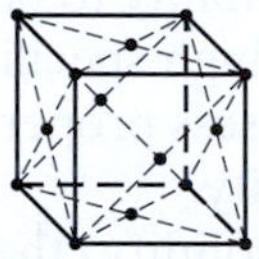		12	74%	Fe, Al, Cu, Ni, Ca, Au, Ag, Pb, Pt
d	Hexagonal close-packed	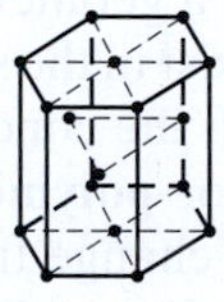	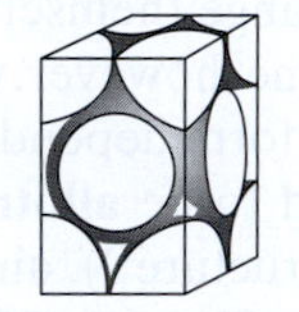	12	74%	Be, Cd, Mg, Zn, Zr

FIGURE 4-7 Comparison of crystal structures: simple cubic, body-centered cubic, face-centered cubic, and hexagonal close-packed.

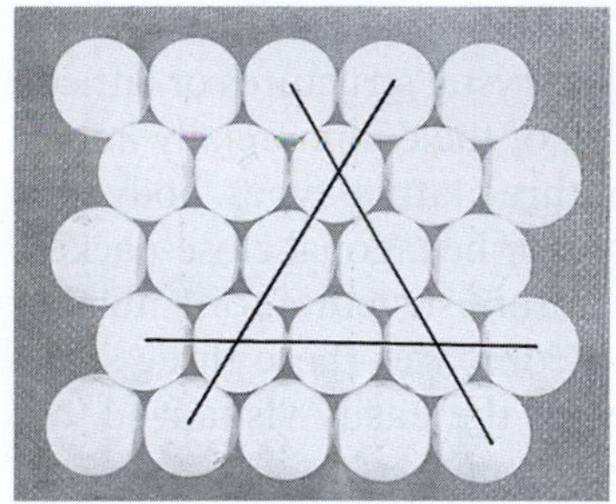

FIGURE 4-8 Close-packed atomic plane showing three directions of atom touching or close-packing. *(Courtesy Ronald Kohser)*

Compared to materials with other structures, body-centered-cubic metals tend to be high strength.

In seeking efficient packing and a large number of adjacent neighbors, consider maximizing the number of spheres in a single layer and then stacking those layers. The layer of maximized packing is known as a **close-packed plane** and exhibits the hexagonal symmetry shown in Figure 4-8. The next layer is positioned with its spheres occupying either the "point-up" or "point-down" triangular recesses in the original layer. Depending on the sequence in which the various layers are stacked, two distinctly different structures can be produced. Both have 12 nearest neighbors (six within the original plane and three from each of the layers above and below) and a 74% efficiency of occupying space.

If the layers are stacked in sets of three (original location, point-up recess of the original layer, and point-down recess of the original layer), rotation of the resulting structure reveals cubic symmetry where an atom has been inserted into the center of each of the six cube faces, like a dice with the number five on each of its sides. This is the **face-centered cubic (FCC)** structure shown in Figure 4-7c. It is the preferred structure for many engineering metals and tends to provide the exceptionally high ductility (ability to be plastically deformed without fracture) that is characteristic of aluminum, copper, silver, gold, and elevated temperature iron.

A stacking sequence of any two alternating layers results in a structure known as **hexagonal close-packed (HCP),** where the individual close-packed planes can be clearly identified (Figure 4-7d). Metals having this structure, such as magnesium and zinc, tend to have poor ductility, fail in a brittle manner, and often require special processing procedures.

■ 4.7 DEVELOPMENT OF A GRAIN STRUCTURE

When a metal solidifies, a small particle of solid forms from the liquid with a lattice structure characteristic of the given material. This particle then acts like a seed or nucleus and grows as other atoms attach themselves. The basic crystalline unit, or unit cell, is repeated, as illustrated in Figure 4-9.

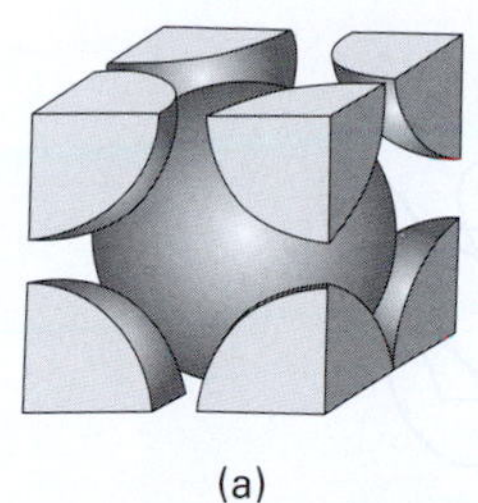

(a)

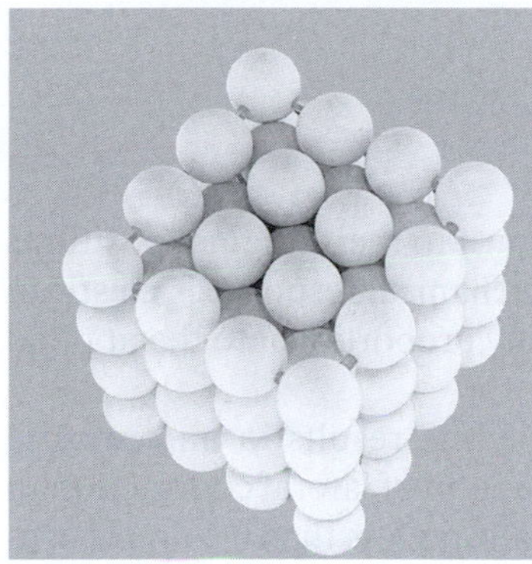

(b)

FIGURE 4-9 Growth of crystals to produce an extended lattice: (a) unit cell; (b) multi-cell aggregate. *(Courtesy Ronald Kohser)*

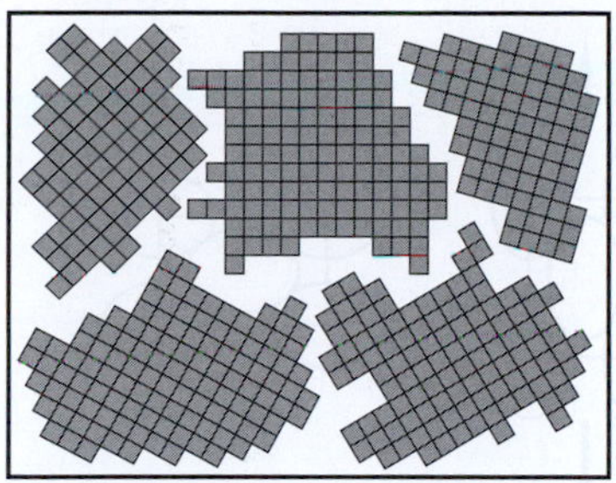

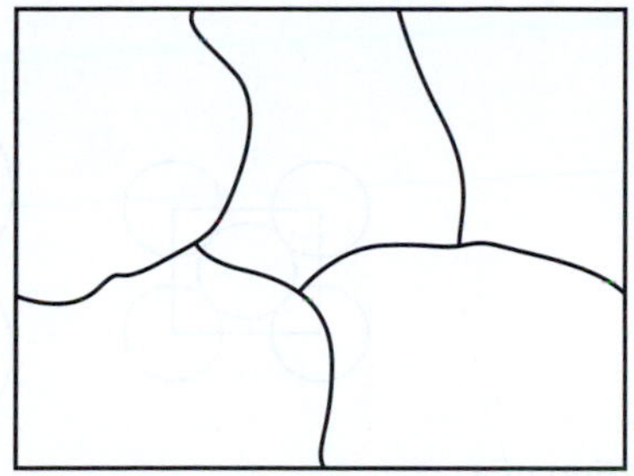

FIGURE 4-10 Schematic representation of the growth of crystals to produce a polycrystalline material.

In actual solidification, many nuclei form independently throughout the liquid and have random orientations with respect to one another. Each then grows until it encounters its neighbors. Because the adjacent lattice structures have different alignments or orientations, growth cannot produce a single continuous structure, and a polycrystalline solid is produced. Figure 4-10 provides a two-dimensional illustration of this phenomenon. The small, continuous regions of solid are known as crystals or **grains,** and the surfaces that divide them (i.e., the surfaces of crystalline discontinuity) are known as **grain boundaries.** The process of solidification is one of crystal **nucleation and growth.**

Grains are the smallest unit of structure in a metal that can be observed with an ordinary light microscope. If a piece of metal is polished to mirror finish with a series of abrasives and then exposed to an attacking chemical for a short time (etched), the grain structure can be revealed. The atoms along the grain boundaries are more loosely bonded and tend to react with the chemical more readily than those that are part of the grain interior. When viewed under reflected light, the attacked boundaries scatter light and appear dark compared to the relatively unaffected (still flat) grains (Figure 4-11). In some cases, the individual grains may be large enough to be seen by the unaided eye, as with some galvanized steels, but usually magnification is required.

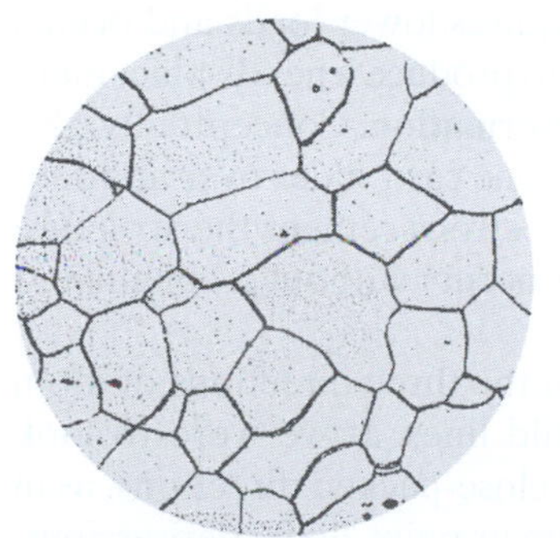

FIGURE 4-11 Photomicrograph of alpha ferrite. *(Courtesy Ronald Kohser)*

The number and size of the grains in a metal vary with the rate of nucleation and the rate of growth. The greater the nucleation rate, the smaller the resulting grains. Conversely, the greater the rate of growth, the larger the grains. Because the resulting **grain structure** will influence certain mechanical and physical properties, it is an important property to control and specify. One means of specification is through the **ASTM grain size number,** defined in ASTM specification E112 as

$$N = 2^{n-1}$$

where N is the number of grains per square inch visible in a prepared specimen at 100× magnification, and n is the ASTM grain size number. Low ASTM numbers mean a few massive grains, while high numbers refer to materials with many small grains.

4.8 ELASTIC DEFORMATION

The mechanical properties of a material are strongly dependent on its crystal structure. An understanding of mechanical behavior, therefore, begins with an understanding of the way crystals react to mechanical loads. Most studies begin with carefully prepared single crystals. Through them, we learn that the mechanical behavior depends on (1) the type of lattice, (2) the interatomic forces (i.e., bond strength), (3) the spacing between adjacent planes of atoms, and (4) the density of the atoms on the various planes.

If the applied loads are relatively low, the crystals respond by simply stretching or compressing the distance between adjacent atoms (Figure 4-12). The basic lattice does not change, and all of the atoms remain in their original positions relative to one another. The applied load serves only to alter the force balance of the atomic bonds, and the atoms assume new equilibrium positions with the applied load as an additional

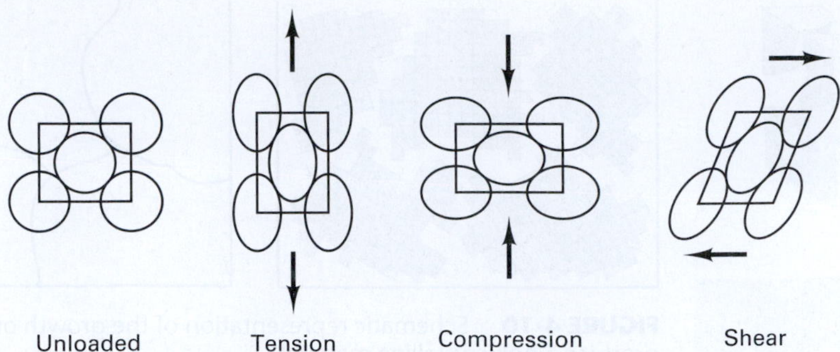

FIGURE 4-12 Distortion of a crystal lattice in response to various elastic loadings.

component of force. If the load is removed, the atoms return to their original positions and the crystal resumes its original size and shape. The mechanical response is **elastic** in nature, and the amount of stretch or compression is directly proportional to the applied load or stress.

Elongation or compression in the direction of loading results in an opposite change of dimensions at right angles to that direction. The ratio of lateral contraction to axial stretching is known as **Poisson's ratio.** This value is always less than 0.5 and is usually about 0.3.

4.9 PLASTIC DEFORMATION

As the magnitude of applied load becomes greater, distortion (or elastic strain) continues to increase, and a point is reached where the atoms either (1) break bonds to produce a fracture or (2) slide over one another in a way that would reduce the load. For metallic materials, the second phenomenon generally requires lower loads and occurs preferentially. The atomic planes shear over one another to produce a net displacement or permanent shift of atom positions, known as **plastic deformation.** Conceptually, this is similar to the distortion of a deck of playing cards when one card slides over another. Because of the metallic bond, where the freely moving electrons cement the structure together, the result is a permanent change in shape that occurs without a concurrent deterioration in properties.

Consider the close-packed plane of Figure 4-8, with the three directions of atom touching (the close-packed directions) identified by bold lines. If we were to look across the top surface of this plane along one of the three close-packed directions, as in Figure 4-13, we see a series of parallel ridges. The point-up or point-down depressions, which become the sites for the next layer of atoms, lie along valleys that parallel the ridges. If the upper layer were to slide in one of the ridge directions, its atoms would simply traverse the valleys and would encounter little resistance. Movement in any other direction would require atoms to climb over the ridges, requiring a greater applied force. Hence, the preference for deformation to occur by movement along close-packed planes in directions of atom touching. If close-packed planes are not available within the crystal structure, plastic deformation tends to occur along planes having the highest atomic density and greatest separation. The rationale for this can be seen in the simplified two-dimensional array of Figure 4-14. Planes A and A' have higher density and greater separation than planes B and B'. In visualizing relative motion, the atoms of B and B' would interfere significantly with one another, whereas planes A and A' do not experience this difficulty.

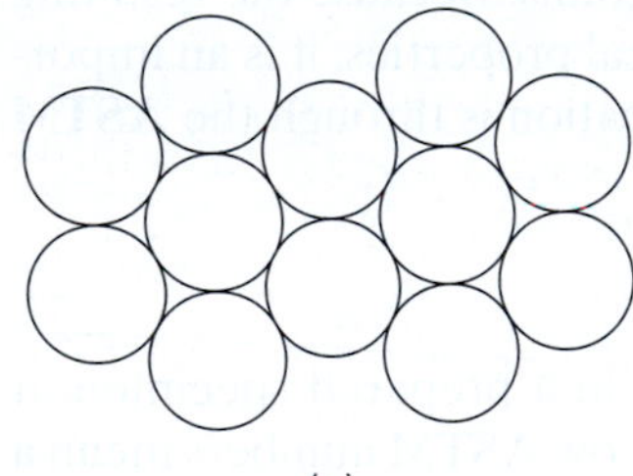

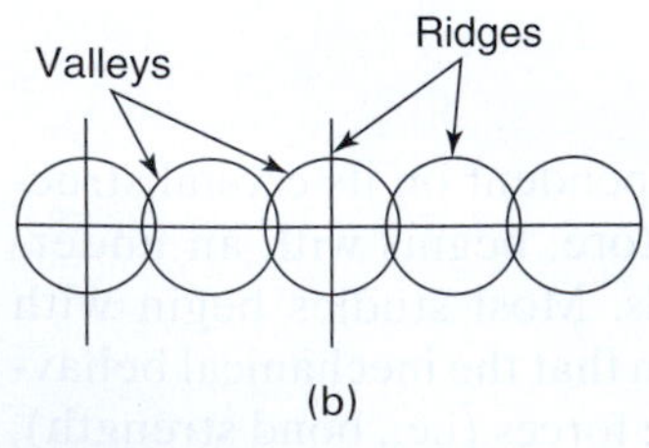

FIGURE 4-13 (a) Close-packed atomic plane viewed from above; (b) view from the side (across the surface) showing the ridges and valleys that lie in directions of close-packing.

Plastic deformation, therefore, tends to occur by the preferential sliding of maximum-density planes (close-packed planes if present) in directions of closest packing. A specific combination of plane and direction is called a **slip system,** and the resulting shear deformation or sliding is known as **slip.** The ability of a metal to deform along a given slip system depends on the ease of shearing along that system and the orientation of the plane with respect to the applied load. Consider a deck of playing cards. The deck will not "deform" when laid flat on the table and pressed from the top, or when stacked on edge and pressed uniformly. The cards will slide over one another, however,

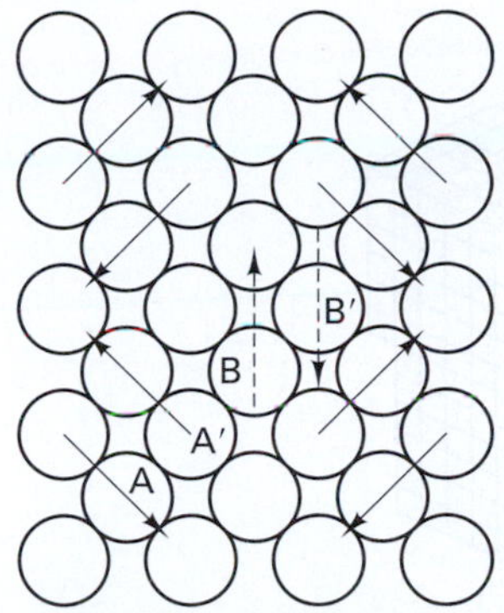

FIGURE 4-14 Simple schematic illustrating the lower deformation resistance of planes with higher atomic density and larger interplanar spacing.

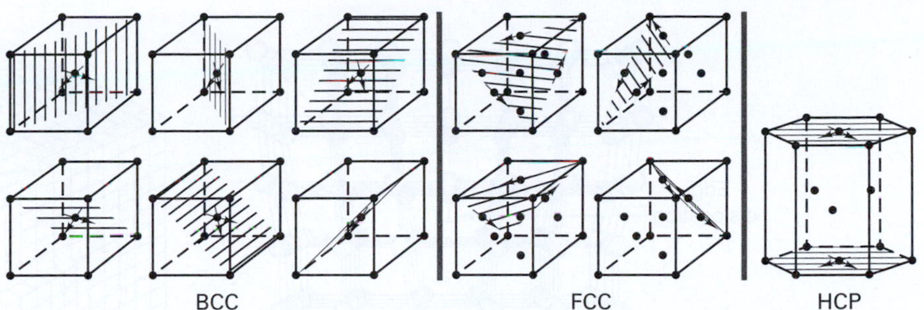

FIGURE 4-15 Slip planes within the BCC, FCC, and HCP crystal structures.

if the deck is skewed with respect to the applied load so as to induce a shear stress along the plane of sliding.

With this understanding, consider the deformation properties of the three most common crystal structures:

1. *Body-centered cubic.* In the BCC structure, there are no close-packed planes. Slip occurs on the most favorable alternatives, which are those planes with the greatest interplanar spacing (six of which are illustrated in Figure 4-15). Within these planes, slip occurs along the directions of closest packing, which are the diagonals through the body of the cube. If each specific combination of plane and direction is considered as a separate slip system, we find that the BCC materials contain 48 attractive ways to slip (plastically deform). The probability that one or more of these systems will be oriented in a favorable manner is great, but the force required to produce deformation is extremely large since there are no close-packed planes. As a result, materials with this structure generally possess high strength with moderate ductility. (Refer to the typical BCC metals in Figure 4-7.)
2. *Face-centered cubic.* In the FCC structure, each unit cell contains four close-packed planes, as illustrated in Figure 4-15. Each of those planes contains three close-packed directions, the diagonals along the cube faces, giving 12 possible means of slip. The probability that one or more of these will be favorably oriented is great, and for this structure, the force required to induce slip is quite low. Metals with the FCC structure are relatively weak and possess excellent ductility, as can be confirmed by a check of the metals listed in Figure 4-7.
3. *Hexagonal close-packed.* The hexagonal lattice also contains close-packed planes, but only one such plane exists within the lattice. Although this plane contains three close-packed directions and the force required to produce slip is again rather low, the probability of favorable orientation to the applied load is small (especially if one considers a polycrystalline aggregate). As a result, metals with the HCP structure tend to have low ductility and are often classified as brittle.

4.10 Dislocation Theory of Slippage

A theoretical calculation of the strength of metals based on the sliding of entire atomic planes over one another predicts yield strengths on the order of 3 million pounds per square inch or 20,000 MPa. The observed strengths in actual testing are typically 100 to 150 times lower than this value. Extremely small laboratory-grown crystals, however, have been shown to exhibit the full theoretical strength.

An explanation can be provided by the fact that plastic deformation does not occur by all of the atoms in one plane slipping simultaneously over all the atoms of an adjacent plane. Instead, deformation is the result of the progressive slippage of a localized disruption known as a **dislocation.** Consider a simple analogy. A carpet has been rolled onto a floor, and we now want to move it a short distance in a given direction. One approach would be to pull on one end and try to "shear the carpet across the floor," simultaneously overcoming the frictional resistance of the entire area of contact. This would require a large force acting over a small distance. An alternative approach might

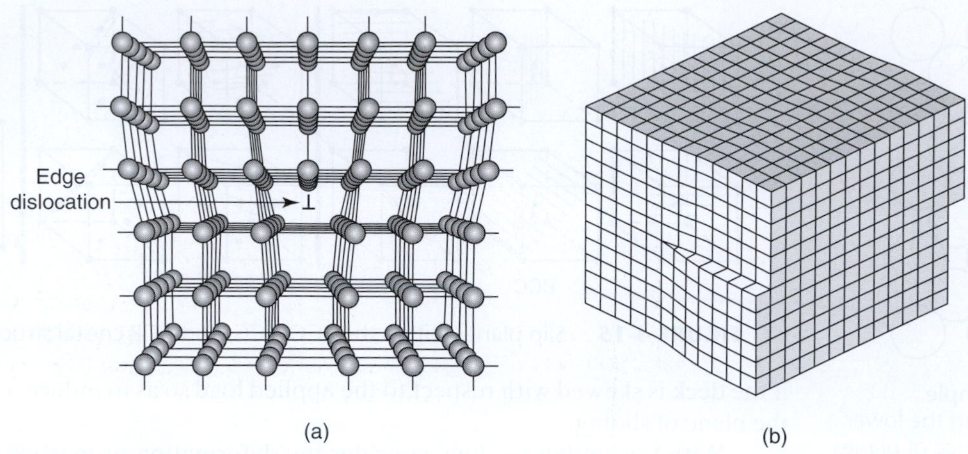

FIGURE 4-16 Schematic representation of (a) edge and (b) screw dislocations. *[(a) From* Elements of Physical Metallurgy, *by A. G. Guy, Addison-Wesley Publishing Co., Inc., Reading, MA, 1959; (b) Adapted from* Materials Science and Engineering, *7th ed., by William D. Callister Jr., John Wiley & Sons, Inc., 2007]*

be to form a wrinkle at one end of the carpet and walk the wrinkle across the floor to produce a net shift in the carpet as a whole—a low-force-over-large distance approach to the same task. In the region of the wrinkle, there is an excess of carpet with respect to the floor beneath it, and the movement of this excess is relatively easy.

Electron microscopes have revealed that metal crystals do not have all of their atoms in perfect arrangement, but rather contain a variety of localized imperfections. Two such imperfections are the **edge dislocation** and **screw dislocation** (Figure 4-16). Edge dislocations are the edges of extra half-planes of atoms. Screw dislocations correspond to partial tearing of the crystal plane. In each case, the dislocation is a disruption to the regular, periodic arrangement of atoms and can be moved about with a rather low applied force. It is the motion of these atomic-scale dislocations under applied load that is responsible for the observed macroscopic plastic deformation.

All metals contain dislocations, usually in abundant quantities. The ease of deformation depends on the ease of making them move. Barriers to dislocation motion, therefore, would tend to increase the overall strength of a metal. These barriers take the form of other crystal imperfections and may be of the point type (missing atoms or **vacancies,** extra atoms or **interstitials,** or **substitution atoms** of a different variety, as shown in Figure 4-17), line type (another *dislocation*), or surface type (*crystal grain boundary* or *free surface*). To increase the strength of a material, we can either remove all defects to create a perfect crystal (nearly impossible) or work to impede the movement of existing dislocations by adding other crystalline defects (the basis of a variety of strengthening mechanisms).

4.11 STRAIN HARDENING OR WORK HARDENING

As noted in our discussion of the tensile test in Chapter 3, most metals become stronger when they are plastically deformed, a phenomenon known as **strain hardening** or **work hardening.** Understanding of this phenomenon can now come from our knowledge of dislocations and a further extension of the carpet analogy. Suppose that this time our goal is to move the carpet diagonally. The best way would be to move a wrinkle in one direction, and then move a second one perpendicular to the first. But suppose that both wrinkles were started simultaneously. We would find that wrinkle 1 would impede the motion of wrinkle 2, and vice versa. In essence, the feature that makes deformation easy can also serve to impede the motion of other, similar dislocations.

In metals, plastic deformation occurs through dislocation movement. As dislocations move, they are more likely to encounter and interact with other dislocations or other crystalline defects, thereby producing resistance to further motion. In addition, mechanisms exist that markedly increase the number of dislocations in a metal during

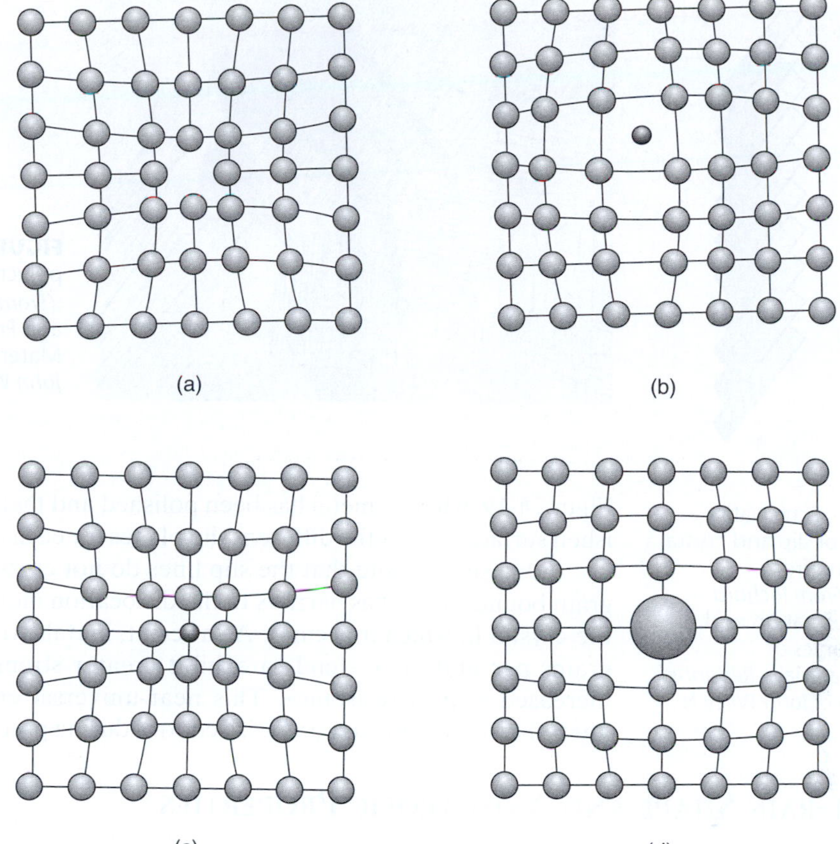

FIGURE 4-17 Two-dimensional schematic showing the various point defects: (a) vacancy; (b) interstitial; (c) smaller-than-host substitutional; (d) larger-than-host substitutional. *(Adapted from* Essentials of Materials Science and Engineering, *2nd ed., by Donald R. Askeland and Pradeep P. Fulay, Cengage Learning, 2009)*

deformation (usually by several orders of magnitude), thereby enhancing the probability of interaction.

The effects of strain hardening become attractive when one considers that mechanical deformation (metal-forming) is frequently used in the shaping of metal products. As the product shape is being formed, the material is simultaneously becoming stronger. Because strength can be increased substantially during deformation, a strain-hardened (deformed), inexpensive metal can often be substituted for a more costly, stronger one that is machined or cast to shape.

Transmission electron microscope studies have confirmed the existence of dislocations and the slippage theory of deformation. By observing the images of individual dislocations in a thin metal section, we can see the increase in the number of dislocations and their interactions during deformation. Macroscopic observations also lend support. When a load is applied to a single metal crystal, deformation begins on the slip system that is most favorably oriented. The net result is often an observable slip and rotation, like that of a skewed deck of cards (Figure 4-18). Dislocation motion becomes more difficult as strain hardening produces increased resistance, and rotation makes the slip system orientation less favorable. Further deformation may then occur on alternative systems that now offer less resistance, a phenomenon known as **cross slip.**

4.12 PLASTIC DEFORMATION IN POLYCRYSTALLINE METALS

Commercial metals are not single crystals, but usually take the form of polycrystalline aggregates. Within each crystal, deformation proceeds in the manner previously described. Because the various grains have different orientations, an applied load will produce different deformations within each of the crystals. This can be seen in

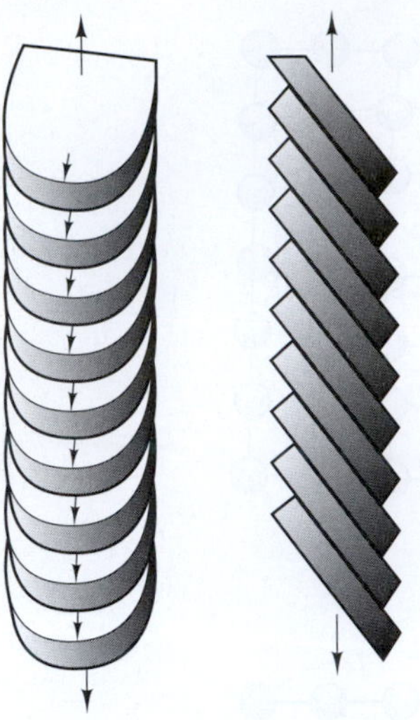

FIGURE 4-18 Schematic representation of slip and crystal rotation resulting from deformation. (*From Richard Hertzberg,* Deformation and Fracture Mechanics of Engineering Materials. *Reprinted with permission of John Wiley & Sons, Inc.*)

FIGURE 4-19 Slip lines in a polycrystalline material. (*From Richard Hertzberg,* Deformation and Fracture Mechanics of Engineering Materials. *Reprinted with permission of John Wiley & Sons, Inc.*)

Figure 4-19, where a metal has been polished and then deformed. The relief of the polished surface reveals the different slip planes for each of the grains.

One should note that the slip lines do not cross from one grain to another. The grain boundaries act as barriers to the dislocation motion (i.e., the defect is confined to the crystal in which it occurs). As a result, metals with a finer grain structure—more grains per unit area—tend to exhibit greater strength and hardness, coupled with increased impact resistance. This near-universal enhancement of properties is an attractive motivation for grain size control during processing.

4.13 GRAIN SHAPE AND ANISOTROPIC PROPERTIES

When a metal is deformed, the grains tend to elongate in the direction of metal flow (Figure 4-20). Accompanying the nonsymmetric structure are directionally varying properties. Mechanical properties (such as strength and ductility), as well as physical properties (such as electrical and magnetic characteristics), may all exhibit directional differences. Properties that vary with direction are said to be **anisotropic.** Properties that are uniform in all directions are **isotropic.**

The directional variation of properties can be harmful or beneficial. By controlling the metal flow in processes such as forging, enhanced strength or fracture resistance can be imparted to certain locations or directions. Caution should be exercised, however, because an improvement in one direction is generally accompanied by a decline in another. Moreover, directional variation in properties may create problems during subsequent processing operations, such as the further forming of rolled metal sheets. For

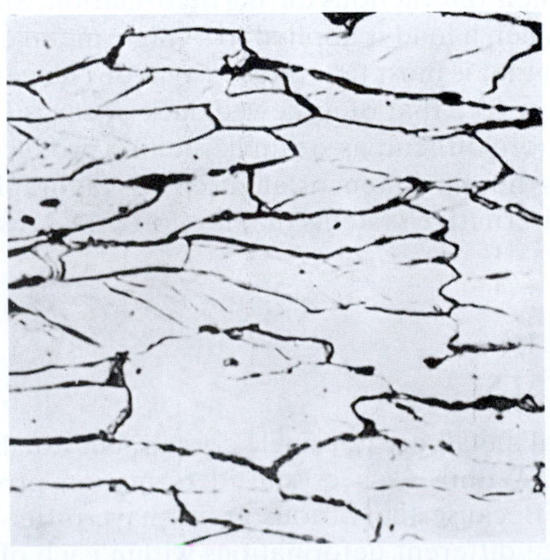

FIGURE 4-20 Deformed grains in a cold-worked 1008 steel after 50% reduction by rolling; (*From* Metals Handbook, *8th ed., 1972. Reprinted with permission of ASM International®. All rights reserved. www.asminternational.org)*

these and other reasons, both the part designer and the part manufacturer should consider the effects of directional property variations.

4.14 FRACTURE OF METALS

When metals are deformed, strength and hardness increase, while ductility decreases. If too much plastic deformation is attempted, the metal may respond by fracture. If plastic deformation precedes the break, the fracture is known as a **ductile fracture.** Fractures can also occur before the onset of plastic deformation. These sudden, catastrophic failures, known as **brittle fractures,** are more common in metals having the BCC or HCP crystal structures. Whether the fracture is ductile or brittle, however, often depends on the specific conditions of material, temperature, state of stress, and rate of loading.

4.15 COLD WORKING, RECRYSTALLIZATION, AND HOT WORKING

During plastic deformation, a portion of the deformation energy is stored within the material in the form of additional dislocations and increased grain boundary surface area.[3] If a deformed polycrystalline metal is subsequently heated to a high enough temperature, the material will seek to lower its energy. New crystals nucleate and grow to consume and replace the original structure (Figure 4-21). This process of reducing the internal energy through the formation of new crystals is known as **recrystallization.**

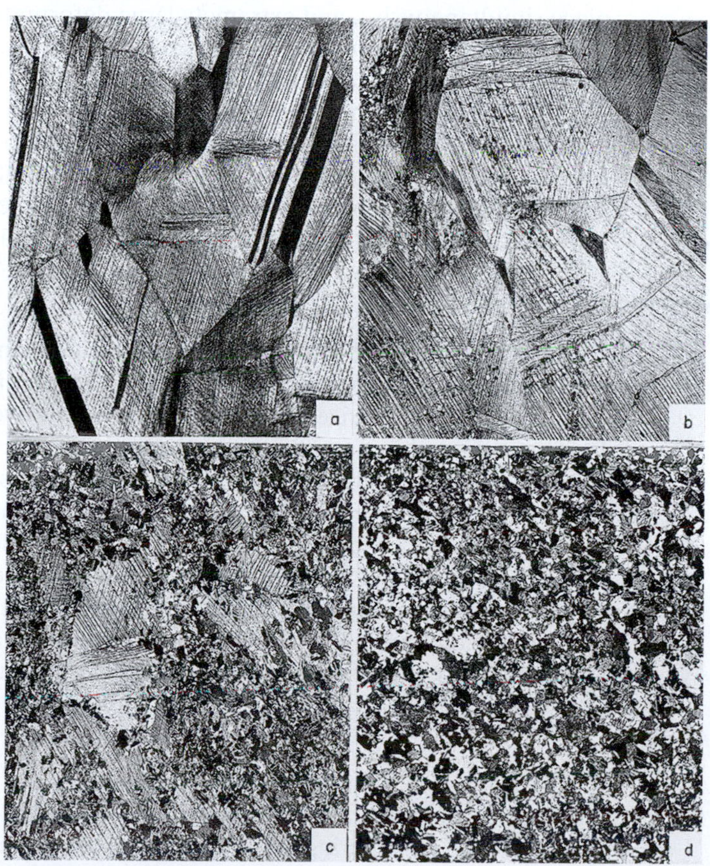

FIGURE 4-21 Recrystallization of 70–30 cartridge brass: (a) cold-worked 33%; (b) heated at 580°C (1075°F) for 3 seconds; (c) 4 seconds; (d) 8 seconds; 45×. *(Courtesy J. E. Burke, General Electric Company, Fairfield, CT)*

[3] A sphere has the least amount of surface area of any shape to contain a given volume of material. When the shape becomes altered from that of a sphere, the surface area must increase. Consider a round balloon filled with air. If the balloon is stretched or flattened into another shape, the rubber balloon is stretched further. When the applied load is removed, the balloon snaps back to its original shape, the one involving the least surface energy. Metals behave in an analogous manner. During deformation, the distortion of the crystals increases the energy of the material. Given the opportunity, the material will try to lower its energy by returning to spherical grains.

TABLE 4-2 The Lowest Recrystallization Temperature of Common Metals

Metal	Temperature [°F(°C)]
Aluminum	300 (150)
Copper	390 (200)
Gold	390 (200)
Iron	840 (450)
Lead	Below room temperature
Magnesium	300 (150)
Nickel	1100 (590)
Silver	390 (200)
Tin	Below room temperature
Zinc	Room temperature

The temperature at which recrystallization occurs is different for each metal and also varies with the amount of prior deformation. The greater the amount of prior deformation, the more stored energy, and the lower the recrystallization temperature. There is a lower limit, however, below which recrystallization will not take place in a reasonable amount of time. Table 4-2 gives the lowest practical recrystallization temperatures for several materials. This is the temperature at which atomic diffusion (atom movement within the solid) becomes significant, and can often be estimated by taking 0.4 times the melting point of the metal when the melting point is expressed as an absolute temperature (Kelvin or Rankine).

When metals are plastically deformed at temperatures below their recrystallization temperature, the process is called **cold working.** The metal strengthens by strain hardening, and the resultant structure consists of distorted grains. As deformation continues, the metal decreases in ductility and may ultimately fracture. It is a common practice, therefore, to recrystallize the material after a certain amount of cold work. Through this **recrystallization anneal,** the structure is replaced by one of new crystals that have never experienced deformation. All strain hardening is lost, but ductility is restored, and the material is now capable of further deformation without the danger of fracture.

If the temperature of deformation is sufficiently above the recrystallization temperature, the deformation process becomes **hot working.** Recrystallization begins as soon as sufficient driving energy is created, (i.e. deformation and recrystallization take place simultaneously), and extremely large deformations are now possible. Because a recrystallized grain structure is constantly forming, the final product will not exhibit the increased strength of strain hardening.

Recrystallization can also be used to control or improve the grain structure of a material. A coarse grain structure can be converted to a more attractive fine grain structure through recrystallization. The material must first be plastically deformed to store sufficient energy to provide the driving force. Subsequent control of the recrystallization process then establishes the more desirable final grain size.

4.16 GRAIN GROWTH

Recrystallization is a continuous process in which a material seeks to lower its overall energy. Ideally, recrystallization will result in a structure of uniform crystals with a comparatively small grain size. If a metal is held at or above its recrystallization temperature for any appreciable time, however, the grains in the recrystallized structure can continue to increase in size. In effect, some of the grains become larger at the expense of their smaller neighbors as the material seeks to further lower its energy by decreasing the amount of grain boundary surface area. Because engineering properties tend to diminish as the size of the grains increase, control of recrystallization is of prime importance. A deformed material should be held at elevated temperature just long enough to complete the recrystallization process. The temperature should then be decreased to stop the process and avoid the property changes that accompany **grain growth.**

4.17 ALLOYS AND ALLOY TYPES

Our discussion thus far has been directed toward the nature and behavior of pure metals. For most manufacturing applications, however, metals are not used in their pure form. Instead, engineering metals tend to be **alloys,** materials composed of two or more different elements, and they tend to exhibit their own characteristic properties.

There are three ways in which a metal might respond to the addition of another element. The first, and probably the simplest, response occurs when the *two materials are insoluble in one another in the solid state*. In this case the base metal and the alloying addition each maintain their individual identities, structures, and properties. The alloy in effect becomes a composite structure, consisting of two types of building blocks in an intimate mechanical mixture.

The second possibility occurs when the *two elements exhibit some degree of solubility in the solid state*. The two materials can form a **solid solution,** where the alloy element dissolves in the base metal. The solutions can be: *substitutional* or *interstitial*. In the substitutional solution, atoms of the alloy element occupy lattice sites normally filled by atoms of the base metal. In an interstitial solution, the alloy element atoms squeeze into the open spaces between the atoms of the base metal lattice.

A third possibility exists where the *elements combine to form intermetallic compounds*. In this case, the atoms of the alloying element interact with the atoms of the base metal in definite proportions and in definite geometric relationships. The bonding is primarily of the nonmetallic variety (i.e., ionic or covalent), and the lattice structures are often quite complex. Because of the type of bonding, **intermetallic compounds** tend to be hard, but brittle, high-strength materials.

Even though alloys are composed of more than one type of atom, their structure is still one of crystalline lattices and grains. Their behavior in response to applied loadings is similar to that of pure metals, with some features reflecting the increased level of structural complexity. Dislocation movement can be further impeded by the presence of unlike atoms. If neighboring grains have different chemistries and/or structures, they may respond differently to the same type and magnitude of load.

4.18 ATOMIC STRUCTURE AND ELECTRICAL PROPERTIES

In addition to mechanical properties, the structure of a material also influences its physical properties, such as its electrical behavior. **Electrical conductivity** refers to the net movement of charge through a material. In metals, the charge carriers are the valence electrons. The more perfect the atomic arrangement, the greater the freedom of electron movement, and the higher the electrical conductivity. Lattice imperfections or irregularities provide impediments to electron transport, and lower conductivity.

The electrical resistance of a metal, therefore, depends largely on two factors: (1) lattice imperfections and (2) temperature. Vacant atomic sites, interstitial atoms, substitutional atoms, dislocations, and grain boundaries all act as disruptions to the regularity of a crystalline lattice. Thermal energy causes the atoms to vibrate about their equilibrium position. These vibrations cause the atoms to be out of position, which further interferes with electron travel. For a metal, electrical conductivity will decrease with an increase in temperature. As the temperature drops, the number and type of crystalline imperfections becomes more of a factor. The best metallic conductors, therefore, are those with fewer defects (such as pure metals with large grain size) at low temperature.

The electrical conductivity of a metal is due to the movement of the free electrons in the metallic bond. For covalently bonded materials, however, bonds must be broken to provide the electrons required for charge transport. Therefore, the electrical properties of these materials are a function of bond strength. Diamond, for instance, has strong bonds and is a strong insulator. Silicon and germanium have weaker bonds that are more easily broken by thermal energy. These materials are known as **intrinsic semiconductors,** because moderate amounts of thermal energy enable them to conduct small amounts of electricity. Continuing down Group IV of the periodic table of elements, we find that tin has such weak bonding that a high number of bonds are broken at room temperature, and the electrical behavior resembles that of a metal.

The electrical conductivity of intrinsic semiconductors can be substantially improved by a process known as **doping.** Silicon and germanium each have four valence electrons and form four covalent bonds. If one of the bonding atoms is replaced with an atom containing five valence electrons, such as phosphorus or arsenic, the four covalent bonds would form, leaving an additional valence electron that is not involved in the bonding process. This extra electron would be free to move about and provide additional conductivity. Materials doped in this manner are known as ***n*-type semiconductors.**

A similar effect can be created by inserting an atom with only three valence electrons, such as aluminum. An electron will be missing from one of the bonds, creating an **electron hole.** When a voltage is applied, a nearby electron can jump into this hole, creating a hole in the location that it vacated. Movement of electron holes is equivalent to a countermovement of electrons and thus provides additional conductivity. Materials containing dopants with three valence electrons are known as ***p*-type semiconductors.** The ability to control the electrical conductivity of semiconductor material is the functional basis of solid-state electronics and circuitry.

In ionically bonded materials, all electrons are captive to atoms (ions). Charge transport, therefore, requires the movement of entire atoms, not electrons. Consider a large block of salt (sodium chloride). It is a good electrical insulator, until it becomes wet, whereupon the ions are free to move in the liquid solution and conductivity is observed.

■ KEY WORDS

allotropic
alloy
amorphous structure
anisotropic
ASTM grain size number
atomic number
body-centered cubic (BCC)
brittle fracture
close-packed planes
cold work
covalent bond
cross slip
crystal structure
dislocation
doping
ductile fracture
edge dislocation
elastic deformation
electrical conductivity
electron hole
extrinsic semiconductor
face-centered cubic (FCC)
grain
grain boundary
grain growth
grain structure
hexagonal close-packed (HCP)
hot work
intermetallic compound
interstitial
intrinsic semiconductor
ionic bond
isotropic
lattice
metallic bond
microstructure
molecular structure
n-type semiconductor
negative ion
nucleation and growth
p-type semiconductor
performance
plastic deformation
Poisson's ratio
polarization
polymorphic
positive ion
primary bond
processing
properties
recrystallization
recrystallization anneal
screw dislocation
secondary bonds
simple cubic structure
slip
slip system
solid solution
strain hardening
structure
substitutional atom
unit cell
vacancy
valence electrons
van der Waals forces
work hardening

■ REVIEW QUESTIONS

1. What enables us to control the properties and performance of engineering materials?
2. What are the next levels of structure that are greater than the atom?
3. What is meant by the term *microstructure?*
4. What is the most stable configuration for an electron shell or subshell?
5. What is an ion and what are the two varieties?
6. What properties or characteristics of a material are influenced by the valence electrons?
7. What are the three types of primary bonds, and what types of atoms do they unite?
8. What are some general characteristics of ionically bonded materials?
9. Where are the bonding electrons located in a covalent bond?
10. What are some general properties and characteristics of covalently bonded materials?
11. Why are the covalently bonded hydrocarbon polymers light in weight?
12. What are some unique property features of materials bonded by metallic bonds?
13. For what common engineering materials are van der Waals forces important?
14. What is the difference between a crystalline material and one with an amorphous structure?
15. What is a lattice? A unit cell?
16. What are some of the general characteristics of metallic materials?
17. What is an allotropic or polymorphic material?
18. Why did we elect to focus on only three of the fourteen basic crystal structures or lattices? What are those three structures?
19. Why is the simple cubic crystal structure not observed in the engineering metals?
20. What is the efficiency of filling space with spheres in the simple cubic structure? Body-centered-cubic structure? Face-centered cubic structure? Hexagonal close-packed structure?
21. What is the dominant characteristic of body-centered-cubic metals? Face-centered-cubic metals? Hexagonal-close-packed metals?
22. What is a grain? A grain boundary?
23. What is the most common means of describing or quantifying the grain size of a solid metal?
24. What is implied by a low ASTM grain size number? A large ASTM grain size number?
25. How does a metallic crystal respond to low applied loads?
26. What is plastic deformation?
27. What is a slip system in a material? What types of planes and directions tend to be preferred?
28. What structural features account for each of the dominant properties cited in Question 21?

29. What is a dislocation? What is the difference between an edge dislocation and a screw dislocation?
30. What role do dislocations play in determining the mechanical properties of a metal?
31. What are some of the common barriers to dislocation movement that can be used to strengthen metals?
32. What are the three major types of point defects in crystalline materials?
33. What is the mechanism (or mechanisms) responsible for the observed deformation strengthening or strain hardening of a metal?
34. Why is a fine grain size often desired in an engineering metal?
35. What is an anisotropic property? Why might anisotropy be a concern?
36. What is the difference between brittle fracture and ductile fracture?
37. How does a metal increase its internal energy during plastic deformation?
38. What is required in order to drive the recrystallization of a cold-worked or deformed material?
39. In what ways can recrystallization be used to enable large amounts of deformation without fear of fracture?
40. How might the lowest recrystallization temperature of a metal be estimated?
41. What is the major distinguishing feature between hot and cold working?
42. How can deformation and recrystallization improve the grain structure of a metal?
43. Why is grain growth usually undesirable?
44. What types of structures can be produced when an alloy element is added to a base metal?
45. As a result of their ionic or covalent bonding, what types of mechanical properties are characteristic of intermetallic compounds?
46. How is electrical charge transported in a metal (electrical conductivity)?
47. What features in a metal structure tend to impede or reduce electrical conductivity?
48. What is the difference between an intrinsic semiconductor and an extrinsic semiconductor?

■ Problems

1. A prepared sample of metal reveals a structure with 64 grains per square inch at 100× magnification.
   a. What is its ASTM grain size number?
   b. Would this material be weaker or stronger than the same metal with an ASTM grain size number of 4? Why?
2. Brass is an alloy of copper with a certain amount of zinc dissolved and dispersed throughout the structure. Based on the material presented in this chapter:
   a. Would you expect brass to be stronger or weaker than pure copper? Why?
   b. Low brass (copper alloy 240) contains 20% zinc. Cartridge brass (copper alloy 260) contains 30% dissolved zinc. Which would you expect to be stronger? Why?
3. It is not uncommon for subsequent processing to expose manufactured products to extreme elevated temperature. Zinc coatings can be applied by immersion into a bath of molten zinc (hot-dip galvanizing). Welding actually melts and resolidifies the crystalline metals. Brazing deposits molten filler metal. How might each of the following structural features, and their associated properties, be altered by an exposure to elevated temperature?
   a. A recrystallized polycrystalline metal.
   b. A cold-worked metal.
   c. A solid-solution alloy, such as brass where zinc atoms dissolve and disperse throughout copper.
4. Polyethylene consists of fibrous molecules of covalently bonded atoms tangled and interacting like the fibers of a cotton ball. Weaker van der Waals forces act between the molecules with a strength that is inversely related to separation distance.
   a. What properties of polyethylene can be attributed to the covalent bonding?
   b. What properties are most likely the result of the weaker van der Waals forces?
   c. If we pull on the ends of a cotton ball, the cotton fibers go from a random arrangement to an array of somewhat aligned fibers. Assuming we get a similar response from deformed polyethylene, how might properties change? Why?

CHAPTER 5

EQUILIBRIUM PHASE DIAGRAMS AND THE IRON–CARBON SYSTEM

5.1 INTRODUCTION

As our study of engineering materials becomes more focused on specific metals and alloys, it is increasingly important that we acquire an understanding of their natural characteristics and properties. What is the basic structure of the material? Is the material uniform throughout, or is it a mixture of two or more distinct components? If there are multiple components, how much of each is present, and what are the different chemistries? Is there a component that may impart undesired properties or characteristics? What will happen if temperature is increased or decreased, pressure is changed, or chemistry is varied? The answers to these and other important questions can be obtained through the use of **equilibrium phase diagrams.**

5.2 PHASES

Before we move to a discussion of equilibrium phase diagrams, it is important that we first develop a working definition of the term *phase*. As a starting definition, a **phase** is simply a form of material possessing a characteristic structure and characteristic properties. Uniformity of chemistry, structure, and properties is assumed throughout a phase. More rigorously, a phase has *a definable structure, a uniform and identifiable chemistry* (also known as **composition**), and distinct *boundaries* or **interfaces** that separate it from other different phases.

A phase can be continuous (like the air in a room) or discontinuous (like grains of salt in a shaker). A phase can be solid, liquid, or gas. In addition, a phase can be a pure substance or a solution, provided that the structure and composition are uniform throughout. Alcohol and water mix in all proportions and will therefore form a single phase when combined. There are no boundaries across which structure and/or chemistry changes. Oil and water, on the other hand, tend to separate into regions with distinct boundaries and must be regarded as two distinct phases. Ice cubes in water are another two-phase system, since there are two distinct structures with interfaces between them.

5.3 EQUILIBRIUM PHASE DIAGRAMS

An *equilibrium phase diagram* is a graphic mapping of the natural tendencies of a material or a material system, assuming that equilibrium has been attained for all possible conditions. There are three primary variables to be considered: *temperature, pressure,* and *composition*. The simplest phase diagram is a **pressure–temperature (*P–T*) diagram**

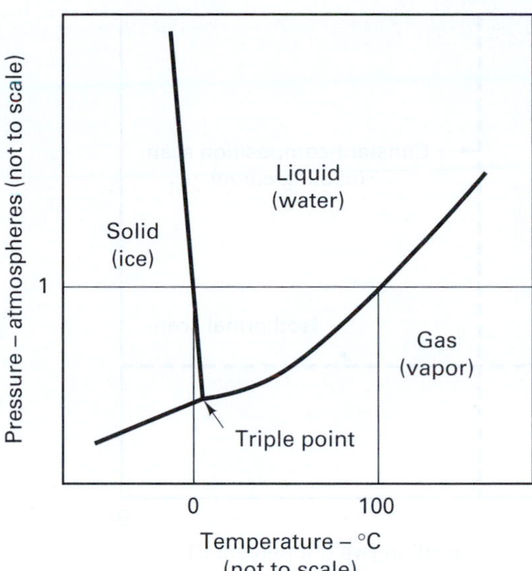

FIGURE 5-1 Pressure–temperature equilibrium phase diagram for water.

for a fixed-composition material. Areas of the diagram are assigned to the various phases, with the boundaries indicating the equilibrium conditions of transition.

As an introduction, consider the pressure–temperature diagram for water, presented as Figure 5-1. With the composition fixed as H_2O, the diagram maps the stable form of water for various conditions of temperature and pressure. If the pressure is held constant and temperature is varied, the region boundaries denote the melting and boiling points. For example, at 1 atmosphere pressure, the diagram shows that water melts at 0°C and boils at 100°C. Still other uses are possible. Locate a temperature where the stable phase is liquid at atmospheric pressure. Maintaining the pressure at one atmosphere, drop the temperature until the material goes from liquid to solid (i.e., ice). Now, maintain that new temperature and begin to decrease the pressure. A transition will be encountered where solid goes directly to gas without melting (sublimation). The combined process just described, known as **freeze drying,** is employed in the manufacture of numerous dehydrated products. With an appropriate phase diagram, process conditions can be determined that might reduce the amount of required cooling and the magnitude of pressure drop required for sublimation. A process operating about the triple point would be most efficient.

TEMPERATURE–COMPOSITION DIAGRAMS

While the *P–T* diagram for water is an excellent introduction to phase diagrams, *P–T* phase diagrams are rarely used for engineering applications. Most engineering processes are conducted at atmospheric pressure, and variations are more likely to occur in temperature and composition as we consider both alloys and impurities. The most useful mapping, therefore, is usually a **temperature–composition phase diagram** at 1 atmosphere pressure. For the remainder of the chapter, this will be the form of phase diagram that will be considered.

For mapping purposes, temperature is placed on the vertical axis and composition on the horizontal. Figure 5-2 shows the axes for mapping the *A–B* system, where the left-hand vertical corresponds to pure material *A*, and the percentage of *B* (usually expressed in weight percent) increases as we move toward pure material *B* at the right side of the diagram. The temperature range often includes only solids and liquids, since few processes involve engineering materials in the gaseous state. Experimental investigations that provide the details of the diagram take the form of either vertical or horizontal scans that seek to locate the various phase transitions.

COOLING CURVES

Considerable information can be obtained from vertical scans through the diagram where a fixed composition material is heated and slowly cooled. By plotting the cooling

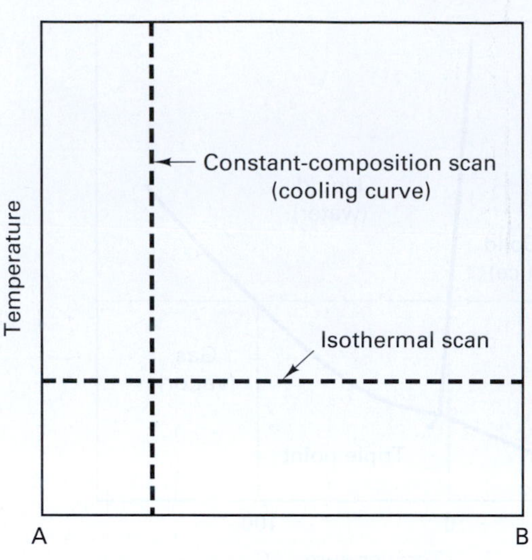

FIGURE 5-2 Mapping axes for a temperature–composition equilibrium phase diagram.

history in the form of a temperature-versus-time plot, known as a **cooling curve,** the transitions in structure will appear as characteristic points, such as slope changes or isothermal (constant-temperature) holds.

Consider the system composed of sodium chloride (common table salt) and water. Five different cooling curves are presented in Figure 5-3. Curve a is for pure water being cooled from the liquid state. A decreasing-temperature line is observed for the liquid where the removal of heat produces a concurrent drop in temperature. When the freezing point of 0°C is reached (point *a*), the material begins to change state and releases heat energy as part of the liquid-to-solid transition. Heat is being continuously extracted from the system, but because its source is now the change in state, there is no companion decrease in temperature. An isothermal or constant-temperature hold (*a–b*) is observed until the solidification is complete. From this point, as heat extraction continues, the newly formed solid experiences a steady drop in temperature. This type of curve is characteristic of pure metals and other substances with a distinct melting point.

Curve b in Figure 5-3 presents the cooling curve for a solution of 10% salt in water. The liquid region undergoes continuous cooling down to point *c*, where the slope abruptly decreases. At this temperature, small particles of ice (i.e., solid) begin to form

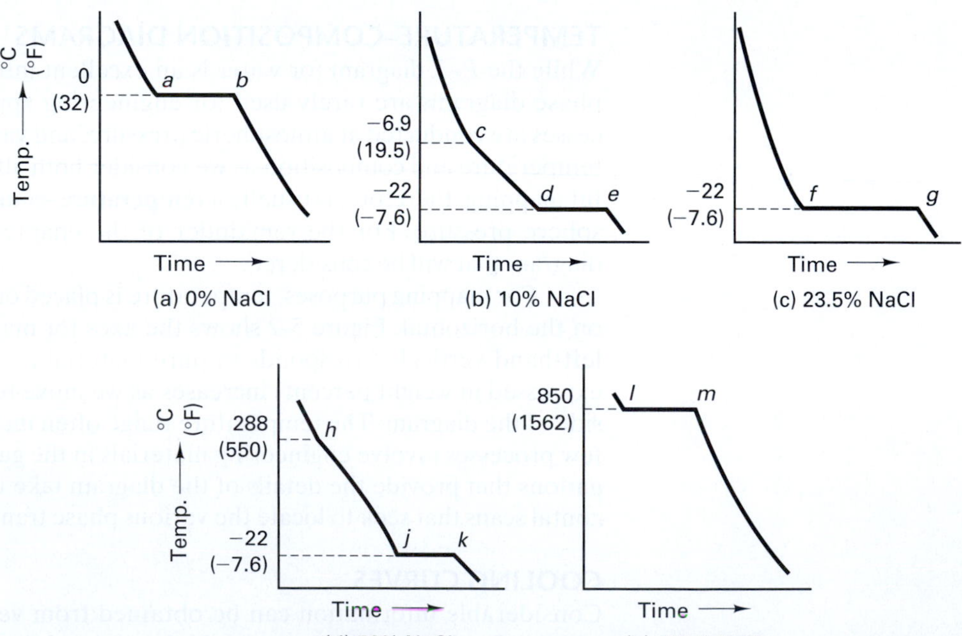

FIGURE 5-3 Cooling curves for five different solutions of salt and water: (a) 0% NaCl; (b) 10% NaCl; (c) 23.5% NaCl; (d) 50% NaCl; (e) 100% NaCl.

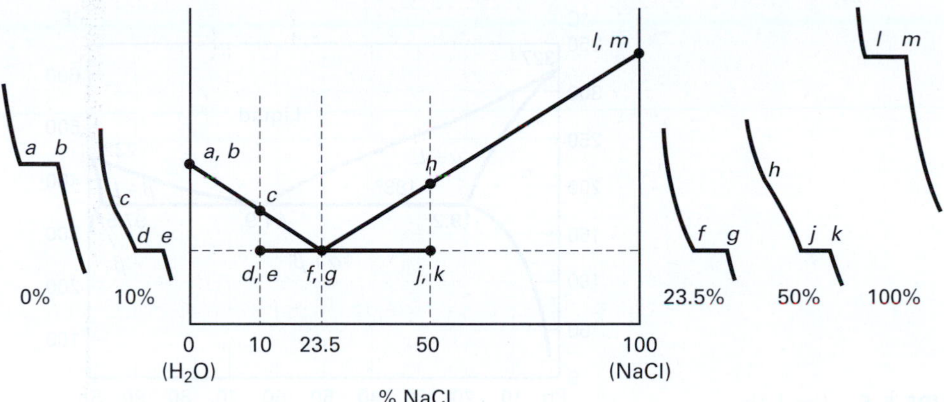

FIGURE 5-4 Partial equilibrium diagram for NaCl and H_2O derived from cooling-curve information.

and the reduced slope is attributed to the energy released in this transition. The formation of these ice particles leaves the remaining solution richer in salt and imparts a lower freezing temperature. Further cooling results in the formation of additional ice, which continues to enrich the remaining liquid and further lowers its freezing point. Instead of possessing a distinct melting point or freezing point, this material is said to have a **freezing range.** When the temperature of point *d* is reached, the remaining liquid undergoes an abrupt reaction and solidifies into an intimate mixture of solid salt and solid water (discussed later), and an isothermal hold is observed. Further extraction of heat produces a drop in the temperature of the fully solidified material.

For a solution of 23.5% salt in water, a distinct freezing point is again observed, as shown in curve c. Compositions with richer salt concentration, such as curve d, show phenomena similar to those in curve b, but with salt being the first solid to form from the liquid. Finally, the curve for pure salt, curve e, exhibits behavior similar to that of pure water.

If the observed transition points are now transferred to a temperature–composition diagram, such as Figure 5-4, we have the beginnings of a map that summarizes the behavior of the system. Line *a–c–f–h–l* denotes the lowest temperature at which the material is totally liquid, and is known as the **liquidus** line. Line *d–f–j* denotes a particular three-phase reaction and will be discussed later. Between the lines, two phases coexist, one being a liquid and the other a solid. The equilibrium phase diagram, therefore, can be viewed as a collective presentation of cooling curve data for an entire range of alloy compositions.

The cooling curve studies have provided some key information regarding the salt-water system, including some insight into the use of salt on highways in the winter. With the addition of salt, the freezing point of water can be lowered from 0°C (32°F) to as low as −22°C (−7.6°F).

SOLUBILITY STUDIES

The observant reader will note that the ends of the diagram still remain undetermined. Both pure materials have a distinct melting point, below which they appear as a pure solid. Can ice retain some salt as a single-phase solid? Can solid salt hold some water and remain a single phase? If so, how much, and does the amount vary with temperature? Completion of the diagram, therefore, requires several horizontal scans to determine any **solubility limits** and their possible variation with temperature.

These isothermal (constant temperature) scans usually require the preparation of specimens over a range of composition and their subsequent examination by X-ray techniques, microscopy, or other methods to determine whether the structure and chemistry are uniform or if the material is a two-phase mixture. As we move away from a pure material, we often encounter a single-phase solid solution, in which one component is dissolved and dispersed throughout the other. If there is a limit to this solubility, there will be line in the phase diagram, known as a **solvus** line, denoting the conditions of saturation where the single-phase solid solution becomes a two-phase mixture.

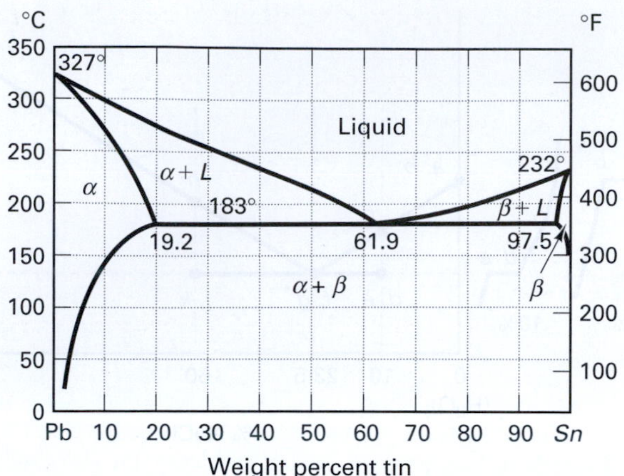

FIGURE 5-5 Lead–tin equilibrium phase diagram.

Figure 5-5 presents the equilibrium phase diagram for the lead–tin system, using the conventional notation in which Greek letters are used to denote the various single-phase solids. The upper portion of the diagram closely resembles the salt-water diagram, but the partial solubility of one material in the other can be observed on both ends of the diagram.[1]

COMPLETE SOLUBILITY IN BOTH LIQUID AND SOLID STATES

Having developed the basic concepts of equilibrium phase diagrams, we now consider a series of examples in which solubility changes. If two materials are completely soluble in each other in both the liquid and solid states, a rather simple diagram results, like the copper–nickel diagram of Figure 5-6. The upper line is the liquidus line, the lowest temperature for which the material is 100% liquid. Above the liquidus, the two materials form a uniform-chemistry liquid solution. The lower line, denoting the highest temperature at which the material is completely solid, is known as a **solidus** line. Below the solidus, the materials form a solid-state solution in which the two types of atoms are uniformly distributed throughout a single crystalline lattice. Between the liquidus and solidus is a freezing range, a two-phase region where liquid and solid solutions coexist.

PARTIAL SOLID SOLUBILITY

Many materials do not exhibit complete solubility in the solid state. Each is often soluble in the other up to a certain limit or saturation point, which varies with temperature. Such a diagram has already been observed for the lead–tin system in Figure 5-5.

At the point of maximum solubility, 183°C, lead can hold up to 19.2 wt% tin in a single-phase solution and tin can hold up to 2.5% lead within its structure and still be

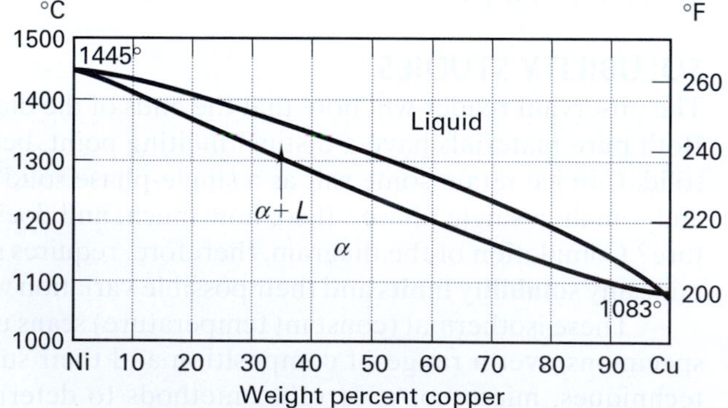

FIGURE 5-6 Copper–nickel equilibrium phase diagram, showing complete solubility in both liquid and solid states.

[1] Lead–tin solders have had a long history in joining electronic components. With the miniaturization of components, and the evolution of the circuit board or chip to ever-smaller features, exposure to the potentially damaging temperatures of the soldering operation became an increasing concern. Figure 5-5 reveals why 60–40 solder (60 wt% tin) became the primary joining material in the lead–tin system. Of all possible alloys, it has the lowest (all liquid) melting temperature.

a single phase. If the temperature is decreased, however, the amount of **solute** that can be held in solution decreases in a continuous manner. If a saturated solution of tin in lead (19.2 wt% tin) is cooled from 183°C, the material will go from a single-phase solution to a two-phase mixture as a tin-rich second phase precipitates from solution. This change in structure can be used to alter and control the properties in a number of engineering alloys.

INSOLUBILITY

If one or both of the components is totally insoluble in the other, the diagrams will also reflect this phenomenon. Figure 5-7 illustrates the case where component *A* is completely insoluble in component *B* in both the liquid and solid states.

UTILIZATION OF DIAGRAMS

Before moving to more complex diagrams, let us first return to a simple phase diagram, such as the one in Figure 5-8, and develop several useful tools. For each condition of temperature and composition (i.e., for each point in the diagram), we would like to obtain three pieces of information:

1. *The phases present.* The stable phases can be determined by simply locating the point of consideration on the temperature–composition mapping and identifying the region of the diagram in which the point appears.
2. *The composition of each phase.* If the point lies in a single-phase region, there is only one component present, and the composition (or chemistry) of the phase is simply the composition of the alloy being considered. If the point lies in a two-phase region, a **tie-line** must be constructed. A tie-line is simply an isothermal (constant-temperature) line drawn through the point of consideration, terminating at the boundaries of the single-phase regions on either side. The compositions where the tie-line intersects the neighboring single-phase regions will be the compositions of those respective phases in the two-phase mixture. For example, consider point *a* in Figure 5-8. The tie-line for this temperature runs from S_2 to L_2. The tie-line intersects the solid phase region at point S_2. Therefore, the solid in the two-phase mixture at point *a* has the composition of point S_2. The other end of the tie-line intersects the liquid region at L_2, so the liquid phase that is present at point *a* will have the composition of point L_2.
3. *The amount of each phase present.* If the point lies in a single-phase region, all of the material, or 100%, must be of that phase. If the point lies in a two-phase region, the relative amounts of the two components can be determined by a **lever-law** calculation using the previously drawn tie-line. Consider the cooling of alloy X in Figure 5-8 in a manner sufficiently slow so as to preserve equilibrium at all temperatures. For

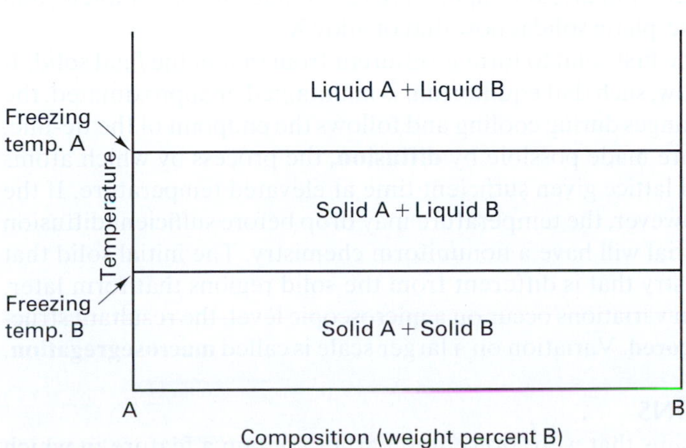

FIGURE 5-7 Equilibrium diagram of two materials that are completely insoluble in each other in both the liquid and solid states.

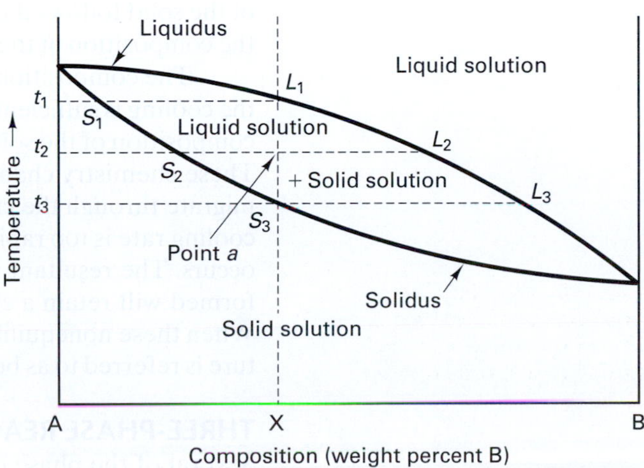

FIGURE 5-8 Equilibrium diagram showing the changes that occur during the cooling of alloy X.

temperatures above t_1, the material is a single-phase liquid. Temperature t_1 is the lowest temperature for which the alloy is 100% liquid. If we draw a tie-line at this temperature, it runs from S_1 to L_1 and lies entirely to the left of composition X. At temperature t_3, the alloy is completely solid, and the tie-line lies completely to the right of composition X. As the alloy cools from temperature t_1 to temperature t_3, the amount of solid goes from 0 to 100% while the segment of the tie-line that lies to the right of composition X also goes from 0 to 100%. Similarly, the amount of liquid goes from 100% to 0 as the segment of the tie-line lying to the left of composition X also goes from 100% to 0. Extrapolating these observations to intermediate temperatures, such as temperature t_2, we predict that the fraction of the tie-line that lies to the left of point a corresponds to the fraction of the material that is liquid. This fraction can be computed as:

$$\%\text{Liquid} = \frac{a - S_2}{L_2 - S_2} \times 100\%$$

where the values of a, S_2, and L_2 are their composition values in weight percent B read from the bottom scale of the diagram. In a similar manner, the fraction of solid corresponds to the fraction of the tie-line that lies to the right of point a.

$$\%\text{Solid} = \frac{L_2 - a}{L_2 - S_2} \times 100\%$$

Each of these mathematical relations could be rigorously derived from the conservation of either A or B atoms, as the material divides into the two different compositions of S_2 and L_2. Because the calculations consider the tie-line as a lever with the fulcrum at the composition line and the component phases at either end, they are called lever-law calculations.

Equilibrium phase diagrams can also be used to provide an overall picture of an alloy system, or to identify the transition points for phase changes in a given alloy. For example, the temperature required to redissolve a second phase or melt an alloy can be easily determined. The various changes that will occur during the slow heating or slow cooling of a material can now be predicted. In fact, most of the questions posed at the beginning of this chapter can now be answered.

SOLIDIFICATION OF ALLOY X

Let us now apply the tools that we have just developed—tie-lines and lever-laws—to follow the solidification of alloy X in Figure 5-8. At temperature t_1, the first minute amount of solid forms with the chemistry of point S_1. As the temperature drops, more solid forms, but the chemistries of both the solid and liquid phases shift to follow the tie-line endpoints. The chemistry of the liquid follows the liquidus line, and the chemistry of the solid follows the solidus. Finally, at temperature t_3, solidification is complete, and the composition of the single-phase solid is now that of alloy X.

The composition of the first solid to form is different from that of the final solid. If the cooling is sufficiently slow, such that equilibrium is maintained or approximated, the composition of the solid changes during cooling and follows the endpoint of the tie-line. These chemistry changes are made possible by **diffusion,** the process by which atoms migrate through the crystal lattice given sufficient time at elevated temperature. If the cooling rate is too rapid, however, the temperature may drop before sufficient diffusion occurs. The resultant material will have a nonuniform chemistry. The initial solid that formed will retain a chemistry that is different from the solid regions that form later. When these nonequilibrium variations occur on a microscopic level, the resultant structure is referred to as being **cored**. Variation on a larger scale is called **macrosegregation**.

THREE-PHASE REACTIONS

Several of the phase diagrams that were presented earlier contain a feature in which phase regions are separated by a horizontal (or constant temperature) line. These lines are further characterized by either a V intersecting from above or an inverted-V intersecting from below. The intersection of the V and the line denotes the location of a **three-phase reaction**.

One common type of three-phase reaction, known as a **eutectic,** has already been observed in Figures 5-4 and 5-5. It is possible to understand these reactions through use of the tie-line and lever-law concepts that have been developed. Refer to the lead–tin diagram of Figure 5-5 and consider any alloy containing between 19.2 and 97.5 wt% tin at a temperature just above the 183°C horizontal line. Tie-line and lever-law computations reveal that the material contains either a lead-rich or tin-rich solid and remaining liquid. At this temperature, any liquid that is present will have a composition of 61.9 wt% tin, regardless of the overall composition of the alloy. If we now focus on this liquid and allow it to cool to just below 183°C, a transition occurs in which the liquid of composition 61.9% tin transforms to a mixture of lead-rich solid with 19.2% tin and tin-rich solid containing 97.5% tin. The three-phase reaction that occurs upon cooling through 183°C can be written as:

$$\text{Liquid}_{61.9\%\ \text{Sn}} \xrightarrow{183^\circ\text{C}} \alpha_{19.2\%\ \text{Sn}} + \beta_{97.5\%\ \text{Sn}}$$

Note the similarity to the very simple chemical reaction in which water dissociates or separates into hydrogen and oxygen: $H_2O \rightarrow H_2 + \frac{1}{2}O_2$. Because the two solids in the lead–tin eutectic reaction have chemistries on either side of the original liquid, a similar separation must have occurred. Any chemical separation requires atom movement, but the distances involved in a eutectic reaction cannot be great. The resulting structure, known as **eutectic structure,** will be an intimate mixture of the two single-phase solids, with a multitude of interphase boundaries.

In the preceding example, the eutectic structure always forms from the same chemistry at the same temperature and has its own characteristic set of physical and mechanical properties. Alloys with the eutectic composition have the lowest melting point of all neighboring alloys, and generally possess relatively high strength. For these reasons, they are often used as casting alloys or as filler material in soldering or brazing operations.

The eutectic reaction can be written in the general form of:

$$\text{liquid} \rightarrow \text{solid}_1 + \text{solid}_2$$

Figure 5-9 summarizes the various types of three-phase reactions that may occur in equilibrium phase diagrams, along with the generic form of the reaction shown below the figures.[2] These include the *eutectic*, **peritectic, monotectic,** and **syntectic** reactions, where the suffix *-ic* denotes that at least one of the three phases in the reaction is a liquid. If the same prefix appears with an *-oid* suffix, the reaction is of a similar form, but all phases involved are solids. Two all-solid reactions are the **eutectoid** and the **peritectoid**. The separation that occurs in the eutectoid reaction produces an extremely fine two-phase mixture. The combination reactions of the peritectic and peritectoid tend to be very sluggish. During actual material processing, these reactions may not reach the completion states predicted by the equilibrium diagram.

INTERMETALLIC COMPOUNDS

A final phase diagram feature occurs in alloy systems where the bonding attraction between the component materials is strong enough to form stable compounds. These compounds appear as single-phase solids in the middle of a diagram. If components *A* and *B* form such a compound, and the compound cannot tolerate any deviation from its fixed atomic ratio, the product is known as a **stoichiometric intermetallic compound** and it appears as a single vertical line in the A-B phase diagram. (Note: An example of this is

[2] To determine the specific form of a three-phase reaction, locate its horizontal line and the V intersecting from either above or below the line. Go above the point of the V and write the phases that are present. Then go below and identify the equilibrium phase or phases. Write the reaction as the phases above the line transform to those below. Apply this method to the diagrams in Figure 5-9 to identify the specific reactions, and compare them to their generic forms presented below the figures, remembering that the Greek letters denote single-phase solids.

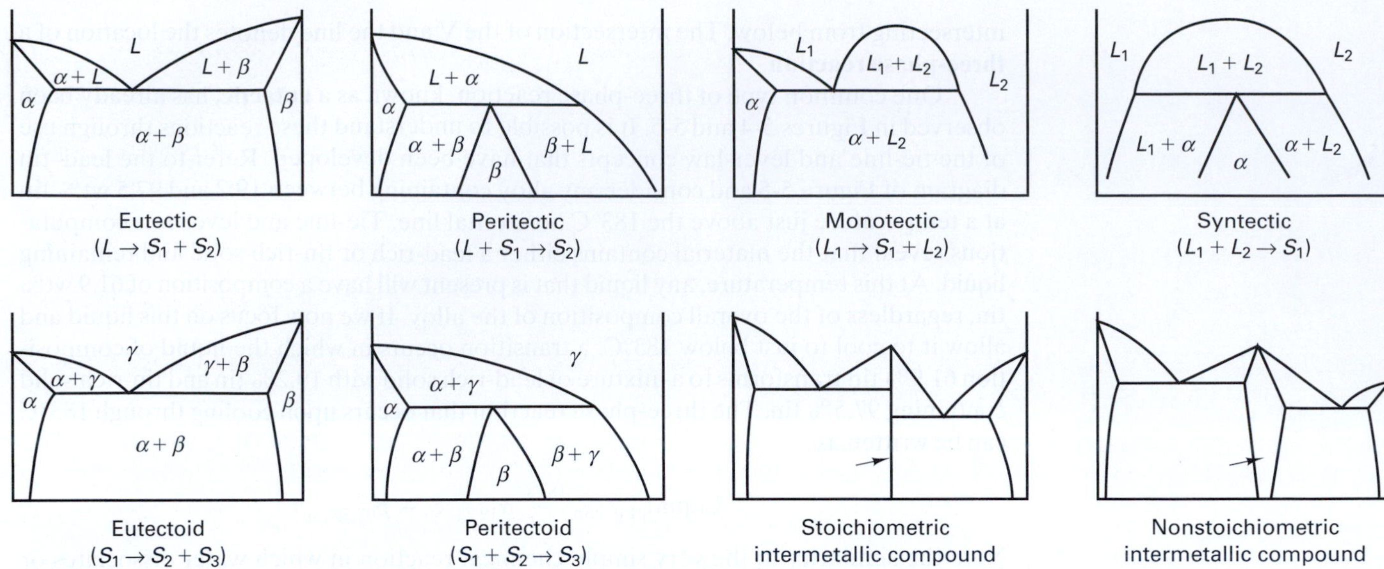

FIGURE 5-9 Schematic summary of three-phase reactions and intermetallic compounds.

the Fe_3C or iron carbide that appears at 6.67 wt.% carbon in the upcoming iron-carbon equilibrium diagram.) If some degree of chemical deviation is tolerable, the vertical line expands into a single-phase region, and the compound is known as a **nonstoichiometric intermetallic compound**. Figure 5-9 shows schematic representations of both stoichiometric and nonstoichiometric compounds.

The single-phase intermetallic compounds appear in the middle of equilibrium diagrams, with locations consistent with whole number atomic ratios, such as AB, A_2B, AB_2, A_3B, AB_3, etc.[3] In general, they tend to be hard, brittle materials, because these properties are a consequence of their ionic or covalent bonding. If they are present in large quantities or lie along grain boundaries in the form of a continuous film, the overall alloy can be extremely brittle. If the same compound is dispersed throughout the alloy in the form of small discrete particles, the result can be a considerable strengthening of the base metal.

COMPLEX DIAGRAMS

The equilibrium diagrams for actual alloy systems may be one of the basic types just discussed or some combination of them. In some cases the diagrams appear to be quite complex and formidable. However, by focusing on a particular composition and analyzing specific points using the tie-line and lever-law concepts, even the most complex diagram can be interpreted and understood. If the properties of the various components are known, phase diagrams can then be used to predict the behavior of the resultant structures.

5.4 IRON–CARBON EQUILIBRIUM DIAGRAM

Steel, composed primarily of iron and carbon, is clearly the most important of the engineering metals. For this reason, the **iron–carbon equilibrium diagram** assumes special importance. The diagram most frequently encountered, however, is not the full iron–carbon diagram but the iron–iron carbide diagram shown in Figure 5-10. Here, a stoichiometric intermetallic compound, Fe_3C, is used to terminate the carbon range at 6.67 wt% carbon. The names of key phases and structures, and the specific notations

[3] The use of "weight percent" along the horizontal axis tends to mask the whole number atomic ratio of intermetallic compounds. Many equilibrium phase diagrams now include a second horizontal scale to reflect "atomic percent." Intermetallic compounds then appear at atomic percents of 25, 33, 50, 67, 75, and similar values that reflect whole number atomic ratios.

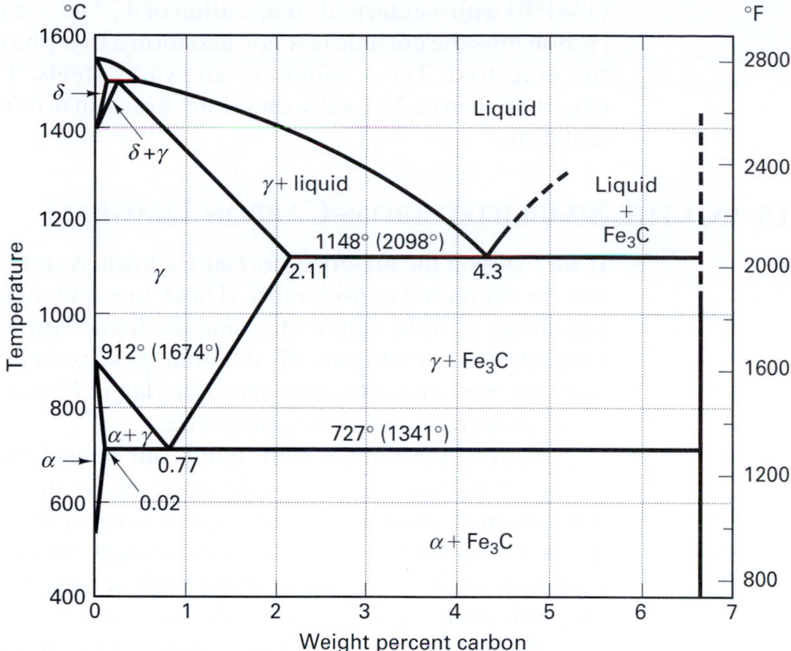

FIGURE 5-10 The iron–carbon equilibrium phase diagram. Single phases are: α, ferrite; γ, austenite; δ, δ–ferrite; Fe_3C, cementite.

used on the diagram, have evolved historically, and will be used in their generally accepted form.

There are four single-phase solids within the diagram. Three of these occur in pure iron, and the fourth is the iron carbide intermetallic that forms at 6.67% carbon. Upon cooling, pure iron solidifies into a body-centered-cubic solid that is stable down to 1394°C (2541°F). Known as **delta-ferrite,** this phase is present only at extremely elevated temperatures and has little engineering importance. From 1394 to 912°C (2541 to 1674°F) pure iron assumes a face-centered-cubic structure known as **austenite** in honor of the famed metallurgist Roberts-Austen of England. Designated by the Greek letter γ, austenite exhibits the high formability that is characteristic of the face-centered-cubic structure and is capable of dissolving more than 2% carbon in single-phase solid solution. Hot forming of steel takes advantage of the low strength, high ductility, and chemical uniformity of austenite. Most of the heat treatments of steel begin by forming the high-temperature austenite structure. *Alpha ferrite,* or more commonly just **ferrite,** is the stable form of iron at temperatures below 912°C (1674°C). This body-centered-cubic structure can hold only 0.02 wt% carbon in solid solution and forces the creation of a two-phase mixture in most steels. Upon further cooling to 770°C (1418°F), iron undergoes a transition from nonmagnetic-to-magnetic. The temperature of this transition is known as the **Curie point,** but because it is not associated with any change in phase (but is an atomic-level transition), it does not appear on the equilibrium phase diagram.

The fourth single phase is the stoichiometric intermetallic compound, Fe_3C, which goes by the name **cementite,** or iron–carbide. Like most intermetallics, it is quite hard and brittle, and care should be exercised in controlling the structures in which it occurs. Alloys with excessive amounts of cementite, or cementite in undesirable form, tend to have brittle characteristics. Because cementite dissociates prior to melting, its exact melting point is unknown, and the liquidus line remains undetermined in the high carbon region of the diagram.

Three distinct three-phase reactions can also be identified. At 1495°C (2723°F), a *peritectic* reaction occurs for alloys with a low weight percentage of carbon. Because of its high temperature and the extensive single-phase austenite region immediately below it, the peritectic reaction rarely assumes any engineering significance. A *eutectic* is observed at 1148°C (2098°F), with the eutectic composition of 4.3% carbon. All alloys containing more than 2.11% carbon will experience the eutectic reaction and are classified by the general term **cast irons.** The final three-phase reaction is a *eutectoid* at 727°C

(1341°F) with a eutectoid composition of 0.77 wt% carbon. Alloys with less than 2.11% carbon miss the eutectic reaction and form a two-phase mixture when they cool through the eutectoid. These alloys are known as **steels.** The point of maximum solubility of carbon in iron, 2.11 wt%, therefore, forms an arbitrary separation between steels and cast irons.

5.5 STEELS AND THE SIMPLIFIED IRON–CARBON DIAGRAM

If we focus on the materials normally known as steel, the phase diagram of Figure 5-10 can be simplified considerably. Those portions near the delta phase (or peritectic) region are of little significance, and the higher carbon region of the eutectic reaction only applies to cast irons. By deleting these segments and focusing on the eutectoid reaction, we can use the simplified diagram of Figure 5-11 to provide an understanding of the properties and processing of steel.

Rather than begin with liquid, our considerations generally begin with high-temperature, face-centered-cubic, single-phase austenite. The key transition will be the conversion of austenite to the two-phase ferrite plus carbide mixture as the temperature drops. Control of this reaction, which arises as a result of the drastically different carbon solubilities of the face-centered and body-centered structures, enables a wide range of properties to be achieved through heat treatment.

To begin to understand these processes, consider a steel of the eutectoid composition, 0.77% carbon, being slow cooled along line x–x' in Figure 5-11. At the upper temperatures, only austenite is present, with the 0.77% carbon being dissolved in solid solution within the face-centered structure. When the steel cools through 727°C (1341°F), several changes occur simultaneously. The iron wants to change crystal structure from the face-centered-cubic austenite to the body-centered-cubic ferrite, but the ferrite can only contain 0.02% carbon in solid solution. The excess carbon is rejected, and forms the carbon-rich intermetallic (Fe_3C) known as cementite. The net reaction at the eutectoid composition and temperature is:

$$\text{Austenite}_{0.77\%\ \text{C;FCC}} \rightarrow \text{Ferrite}_{0.02\%\ \text{C;BCC}} + \text{Cementite}_{6.67\%\ \text{C}}$$

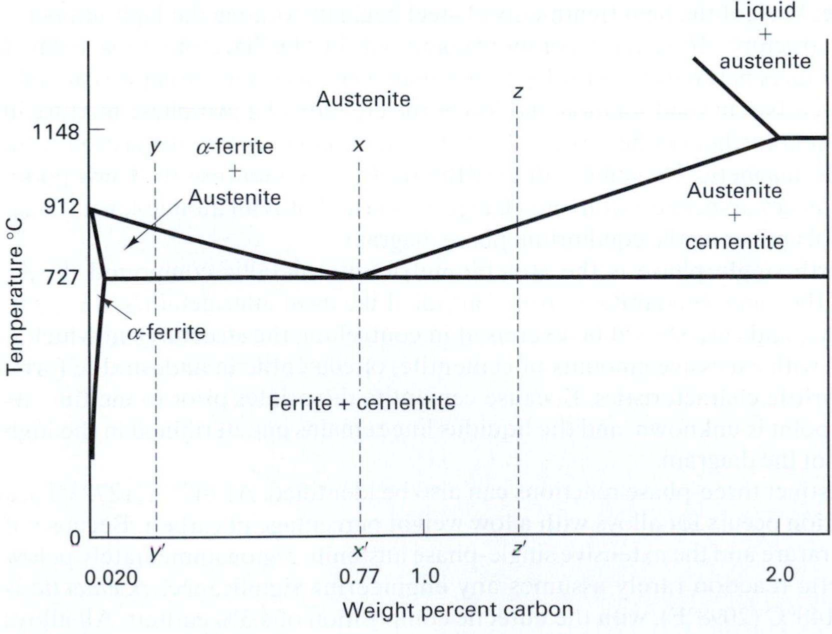

FIGURE 5-11 Simplified iron–carbon phase diagram with labeled regions. Figure 5-10 shows the more-standard Greek letter notation.

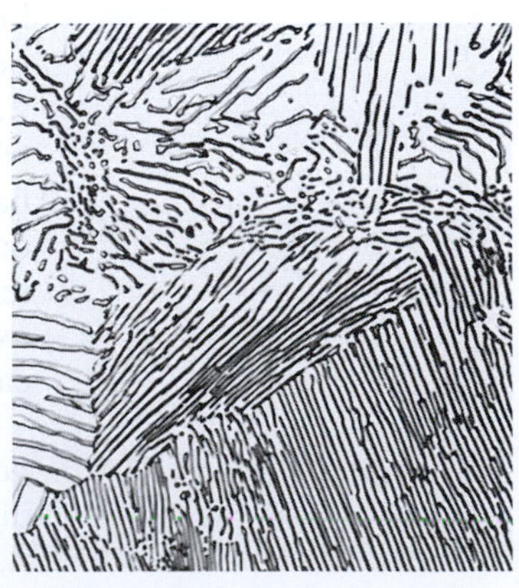

FIGURE 5-12 Pearlite; 1000×. *(Courtesy United States Steel Corporation)*

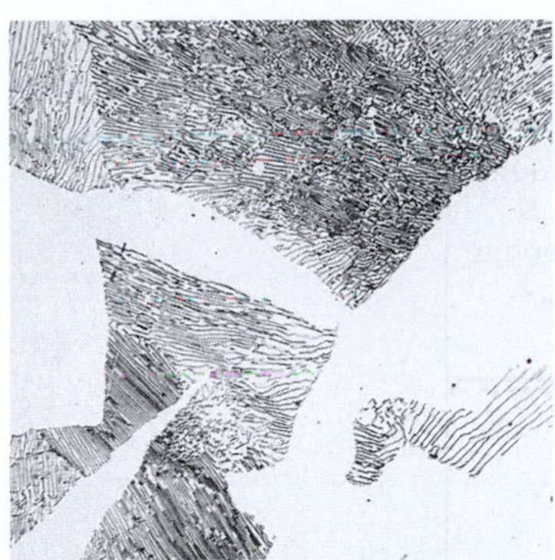

FIGURE 5-13 Photomicrograph of a hypoeutectoid steel showing regions of primary ferrite (white) and pearlite; 500×. *(Courtesy United States Steel Corporation)*

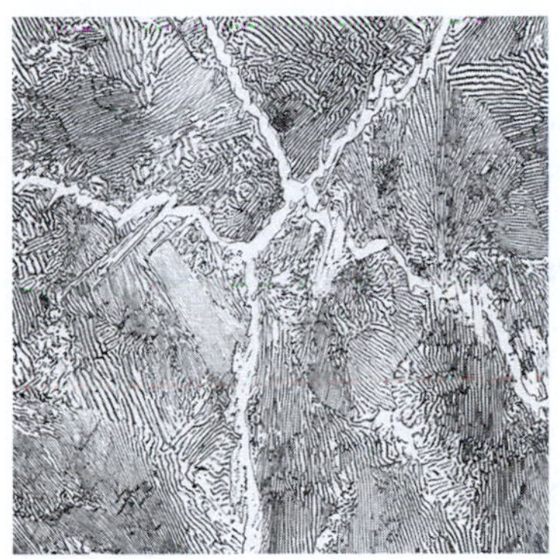

FIGURE 5-14 Photomicrograph of a hypereutectoid steel showing primary cementite along grain boundaries; 500×. *(Courtesy United States Steel Corporation)*

Because the chemical separation occurs entirely within crystalline solids, the resultant structure is a fine mixture of ferrite and cementite. Specimens prepared by polishing and etching in a weak solution of nitric acid and alcohol reveal a lamellar structure composed of alternating layers or plates, as shown in Figure 5-12. Because it always forms from a fixed composition at a fixed temperature, this structure has its own set of characteristic properties (even though it is composed of two distinct phases) and goes by the name **pearlite** because of its metallic luster and resemblance to mother-of-pearl when viewed at low magnification.

Steels having less than the eutectoid amount of carbon (less than 0.77%) are called **hypoeutectoid steels** (*hypo* means "less than"). Consider the cooling of a typical hypoeutectoid alloy along line y–y' in Figure 5-11. At high temperatures the material is entirely austenite. Upon cooling, however, it enters a region where the stable phases are ferrite and austenite. Tie-line and lever-law calculations show that the low-carbon ferrite nucleates and grows, leaving the remaining austenite richer in carbon. At 727°C (1341°F), the remaining austenite will have assumed the eutectoid composition (0.77% carbon), and further cooling transforms it to pearlite. The resulting structure, therefore, is a mixture of *primary* or *proeutectoid ferrite* (ferrite that forms before the eutectoid reaction) and regions of pearlite as shown in Figure 5-13.

Hypereutectoid steels (*hyper* means "greater than") are those that contain more than the eutectoid amount of carbon. When such a steel cools, as along line z–z' in Figure 5-11, the process is similar to the hypoeutectoid case, except that the **primary phase** or proeutectoid phase is now cementite instead of ferrite. As the carbon-rich phase nucleates and grows, the remaining austenite decreases in carbon content, again reaching the eutectoid composition at 727°C (1341°F). This austenite then transforms to pearlite upon slow cooling through the eutectoid temperature. Figure 5-14 is a photomicrograph of the resulting structure, which consists of primary cementite and pearlite. In this case the continuous network of primary cementite (a brittle intermetallic compound) will cause the overall material to be extremely brittle.

It should be noted that the transitions just described are for equilibrium conditions, which can be approximated by slow cooling. Upon slow heating, the transitions will occur in the reverse manner. When the alloys are cooled rapidly, however, entirely different results may be obtained, since sufficient time may not be provided for the normal phase reactions to occur. In these cases, the equilibrium phase diagram is no longer a valid tool for engineering analysis. Because the rapid-cool processes are important in the heat treatment of steels and other metals, their characteristics will be discussed in Chapter 6, and new tools will be introduced to aid our understanding.

5.6 CAST IRONS

As shown in Figure 5-10, iron–carbon alloys with more than 2.11% carbon experience the eutectic reaction during cooling. These alloys are known as cast irons. Being relatively inexpensive, with good fluidity and rather low liquidus (full-melting) temperatures, they are readily cast and occupy an important place in engineering applications.

While we are exploring the iron–carbon equilibrium phase diagram, we should note that most commercial cast irons also contain a significant amount of silicon. Cast irons typically contain 2.0 to 4.0% carbon, 0.5 to 3.0% silicon, less than 1.0% manganese, and less than 0.2% sulfur. The silicon produces several metallurgical effects. By promoting the formation of a tightly adhering surface oxide, the high silicon enhances the oxidation and corrosion resistance of cast irons. Therefore, cast irons generally exhibit a level of corrosion resistance that is superior to most steels.

Because silicon partially substitutes for carbon (both have four valence electrons in their outermost shell), use of the iron-carbon equilibrium phase diagram requires replacing the weight percent carbon scale with a **carbon equivalent.** Several formulations exist to compute this number, with the simplest being the weight percent of carbon plus one-third of the weight percent of silicon:

$$\text{Carbon Equivalent(CE)} = (\text{wt\% Carbon}) + 1/3(\text{wt\% Silicon})$$

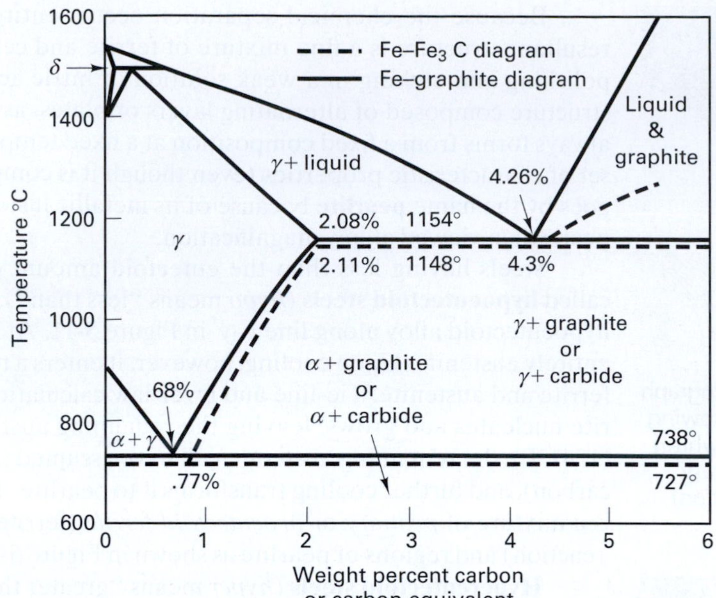

FIGURE 5-15 An iron–carbon diagram showing two possible high-carbon phases. Solid lines denote the iron–graphite system; dashed lines denote iron–cementite (or iron–carbide).

By using carbon equivalent, the two-component iron-carbon diagram can be used to determine melting points and compute microstructures for the three-component iron-carbon-silicon alloys.

Silicon also tends to promote the formation of **graphite** as the carbon-rich phase instead of the Fe_3C intermetallic. The eutectic reaction, therefore, now has two distinct possibilities, as shown in the modified phase diagram of Figure 5-15:

$$\text{Liquid} \rightarrow \text{Austenite} + Fe_3C$$
$$\text{Liquid} \rightarrow \text{Austenite} + \text{Graphite}$$

The final microstructure of cast iron, therefore, will contain either the carbon-rich intermetallic compound, Fe_3C, or pure carbon in the form of graphite. Which occurs depends on the metal chemistry and various other process variables. Graphite is the more stable of the two, and is the true equilibrium structure. Its formation is promoted by slow cooling, high carbon and silicon contents, heavy or thick section sizes, **inoculation** practices, and the presence of sulfur, phosphorus, aluminum, magnesium, antimony, tin, copper, nickel, and cobalt. Cementite (Fe_3C) formation is favored by fast cooling, low carbon and silicon levels, thin sections, and alloy additions of titanium, vanadium, zirconium, chromium, manganese, and molybdenum.

Cast iron is really a generic term that is applied to a variety of metal alloys. Depending on which type of high-carbon phase is present and the form or nature of that phase, the cast iron can be classified as gray, white, malleable, ductile or nodular, or compacted graphite. The ferrous metals, including steels, stainless steels, tool steels, and cast irons, will be presented in more detail in Chapter 6.

■ KEY WORDS

austenite
carbon equivalent
cast iron
cementite
composition
cooling curve
cored structure
Curie point
delta-ferrite
diffusion
equilibrium phase diagram
eutectic
eutectic structure
eutectoid
ferrite
freeze drying
freezing range
graphite
hypereutectoid steel
hypoeutectoid steel
inoculation
interfaces
lever law
liquidus
macrosegregation
monotectic
nonstoichiometric intermetallic compound
pearlite
peritectic
peritectoid
phase
pressure–temperature (P–T) diagram
primary phase
primary phasesolidus
solubility limit
solute
solvus
steel
stoichiometric intermetallic compound
syntectic
temperature–composition phase diagram
three-phase reaction
tie-line

REVIEW QUESTIONS

1. What kind of questions can be answered by equilibrium phase diagrams?
2. What are some features that are useful in defining a phase?
3. Supplement the examples provided in the text with another example of a single phase that is each of the following: continuous, discontinuous, gaseous, and a liquid solution.
4. What is an equilibrium phase diagram?
5. What three primary variables are generally considered in equilibrium phase diagrams?
6. Why is a pressure–temperature phase diagram not that useful for most engineering applications?
7. What is a cooling curve?
8. What features in a cooling curve indicate some form of change in a material's structure? What causes a constant-temperature hold? A slope change?
9. What is a solubility limit, and how might it be determined?
10. In general, how does the solubility of one material in another change as temperature is increased?
11. Describe the conditions of complete solubility, partial solubility, and insolubility.
12. What types of changes occur upon cooling through a liquidus line? A solidus line? A solvus line?
13. What three pieces of information can be obtained for each point in an equilibrium phase diagram?
14. What is a tie-line? For what types of phase diagram regions would it be useful?
15. What points on a tie-line are used to determine the chemistry (or composition) of the component phases?
16. What tool can be used to compute the relative amounts of the component phases in a two-phase mixture? How does this tool work?
17. What is a cored structure? Under what conditions is it produced?
18. What is the difference between a cored structure and macrosegregation?
19. What features in a phase diagram can be used to identify three-phase reactions?
20. What is the general form of a eutectic reaction?
21. What is the general form of the eutectic structure?
22. Why are alloys of eutectic composition attractive for casting and as filler metals in soldering and brazing?
23. What is a stoichiometric intermetallic compound, and how would it appear in a temperature–composition phase diagram? How would a nonstoichiometric intermetallic compound appear?
24. What type of mechanical properties would be expected for intermetallic compounds?
25. In what form(s) might intermetallic compounds be undesirable in an engineering material? In what form(s) might they be attractive?
26. What are the four single phases in the iron–iron carbide diagram? Provide both the phase diagram notation and the assigned name.
27. What features of austenite make it attractive for forming operations? What features make it attractive as a starting structure for many heat treatments?
28. What feature in the iron–carbon diagram is used to distinguish between cast irons and steels?
29. Which of the three-phase reactions in the iron–carbon diagram is most important in understanding the behavior of steels? Write this reaction in terms of the interacting phases and their composition.
30. Describe the relative ability of iron to dissolve carbon in solution when in the form of austenite (the elevated temperature phase) and when in the form of ferrite at room temperature.
31. What is pearlite? Describe its structure.
32. What is a hypoeutectoid steel, and what structure will it assume upon slow cooling? What is a hypereutectoid steel, and how will its structure differ from that of a hypoeutectoid?
33. In addition to iron and carbon, what other element is present in rather large amounts in cast iron?
34. What is carbon equivalent and how is it computed?
35. What are the two possible high-carbon phases in cast irons? What features tend to favor the formation of each?

PROBLEMS

1. Obtain a binary (two-component) phase diagram for a system not discussed in this chapter. Identify the following:
   a. Single phase.
   b. Three phase reaction.
   c. Intermetallic compound.
2. Copper and aluminum are both extremely ductile materials, as evidenced by the manufacture of fine copper wire and aluminum foil. Equal weights of copper and aluminum are melted together to produce an alloy and solidified in a pencil-shaped mold to produce short-length rods approximately $\frac{1}{4}$ in. or 6.5 mm in diameter. These rods appear extremely bright and shiny, almost as if they had been chrome-plated. When dropped on a concrete floor from about waist height, however, the rods shatter into a multitude of pieces, a behavior similar to that observed with glass.
   a. How might you explain this result? [Hint: Use the aluminum-copper phase diagram provided in Figure 6-3 of this text to determine the structure of 50–50 wt% alloy.]
   b. Several of the high-strength aerospace aluminum alloys are aluminum-copper alloys. Explain how the observations above might be useful in providing the desired properties in these alloys.

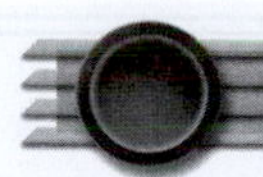

Chapter 5 CASE STUDY

Fish Hooks

The fish hook has been identified as one of the most important tools in the history of man and dates back a number of centuries. Early examples were made from wood, bones, animal horn, seashells, and stone. As technology developed, so did the fish hook. Copper hooks were made in the Far East more than 7000 years ago. Bronze and iron soon followed. Steel has now become the material of preference, but hooks are available in an enormous variety—sizes, shapes and materials—with selection depending on the purpose and preference of the fisherman.

To be successful, fish hooks must be super strong, but not brittle—balancing both strength and flexibility. Because they may find use in both fresh and salt water, and are often placed in storage while wet, they must possess a certain degree of corrosion resistance. The points and barbs must be sharp and retain their sharpness through use and handling.

The simple single hook contains a point (the sharp end that penetrates the fish's mouth), the barbs (backward projections that secure the bait on the hook and subsequently keeps the fish from unhooking), the bend and shank (wire lengths that form the J-portion of the hook), the gap (the specified distance between the legs of the J portion), and the eye (the rounded closure at the end of the hook where the fishing line is tied). Most hooks are currently made from high-carbon steel, alloy steel, or stainless steel. A wide variety of coatings are used to provide corrosion resistance and a full spectrum of appearances and colors.

1. Consider the manufacture of a fish hook beginning with 1080 carbon steel wire. What forming operations would you want to perform before subjecting the material to a quench-and-temper hardening operation? Would it be best to sharpen the point before or after hardening? Would your answer be different if you were to sharpen by chemical removal instead of mechanical grinding?
2. If a stainless steel were to be used, what type of stainless would be best? Consider ferritic, austenitic, and martensitic (described in Chapter 7). What are the advantages and disadvantages of the stainless steel option?
3. A wide spectrum of coatings and surface treatments have been applied to fish hooks. These have included clear lacquer; platings of gold, tin, nickel, and other metals; Teflon; and a variety of coloration treatments involving chemical treatments, thermal oxides, and others. Select several possible surface treatments, and discuss how they might be applied while (a) maintaining the sharpness of the point and barbs and (b) preserving the mechanical balance of strength and flexibility. At what stage of manufacture should each of the proposed surface treatments be performed?
4. The geometry of the part may present some significant problems to heat treatment and surface treatment. For example, many small parts are plated by tumbling in plating baths. Masses of fish hooks would emerge as a tangled mass. Quantities of fish hooks going into a quench tank might experience similar problems. Consider the manufacturing sequences proposed earlier, and discuss how they could be specifically adapted or performed on fish hooks.

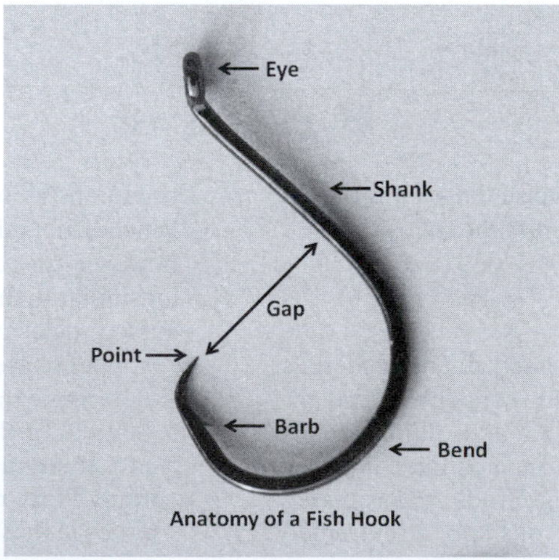

(Wikimedia Commons and Mike Cline)

CHAPTER 6

HEAT TREATMENT

6.1 INTRODUCTION

In the previous chapters, you were introduced to the interrelationships among the structure, properties, processing, and performance of engineering materials. Chapters 4 and 5 considered aspects of structure, while Chapter 3 focused on properties. In this chapter, we begin to incorporate processing as a means of manipulating and controlling the structure and the companion properties of materials.

Many engineering materials can be characterized not by a single set of properties but by an entire spectrum of possibilities that can be selected and varied at will. **Heat treatment** is the term used to describe *the controlled heating and cooling of materials for the purpose of altering their structures and properties.* The same material can be made weak and ductile for ease in manufactur, and then retreated to provide high strength and good fracture resistance for use and application. Because both physical and mechanical properties (such as strength, toughness, machinability, wear resistance, and corrosion resistance) can be altered by heat treatment, and these changes can be induced with no concurrent change in product shape, heat treatment is one of the most important and widely used manufacturing processes.

Technically, the term *heat treatment* applies only to processes where the heating and cooling are performed for the specific purpose of altering properties, but heating and cooling often occur as incidental phases of other manufacturing processes, such as hot forming or welding. The structure and properties of the material will be altered, however, just as though an intentional heat treatment had been performed, and the results can be either beneficial or harmful. For this reason, both the individual who selects material and the person who specifies its processing must be fully aware of the possible changes that can occur during heating or cooling activities. Heat treatment should be fully integrated with other manufacturing processes if effective results are to be obtained. To provide a basic understanding, this chapter will present both the theory of heat treatment and a survey of the more common heat-treatment processes. Because more than 90% of all heat treatment is performed on steel and other ferrous metals, these materials will receive the bulk of our attention.

6.2 PROCESSING HEAT TREATMENTS

The term *heat treatment* is often associated with those thermal processes that increase the strength of a material, but the broader definition permits inclusion of another set of processes that we will call **processing heat treatments.** These are often performed as a means of preparing the material for fabrication. Specific objectives may be the improvement of machining characteristics, the reduction of forming forces, or the restoration of ductility to enable further processing.

EQUILIBRIUM DIAGRAMS AS AIDS

Most of the processing heat treatments involve rather slow cooling or extended times at elevated temperatures. These conditions tend to approximate equilibrium, and the resulting structures, therefore, can be reasonably predicted through the use of an **equilibrium phase diagram** (presented in Chapter 4). These diagrams can be used to determine the temperatures that must be attained to produce a desired starting structure and to describe the changes that will then occur upon subsequent cooling. It should be noted, however, that these diagrams are for true equilibrium conditions, and any departure from equilibrium may lead to substantially different results.

PROCESSING HEAT TREATMENTS FOR STEEL

Because many of the processing heat treatments are applied to plain-carbon and low-alloy steels, they will be presented here with the simplified iron–carbon equilibrium diagram of Figure 5-11 serving as a reference guide. Figure 6-1 shows this diagram with the key transition lines labeled in standard notation. The eutectoid line is designated by the symbol $\mathbf{A_1}$, and $\mathbf{A_3}$ designates the boundary between austenite and ferrite + austenite.[1] The transition from austenite to austenite + cementite is designated as the $\mathbf{A_{cm}}$ line.

A number of process heat-treating operations have been classified under the general term of **annealing.** These may be employed to reduce strength or hardness, remove residual stresses, improve machinability, improve toughness, restore ductility, refine grain size, reduce segregation, stabilize dimensions, or alter the electrical or magnetic properties of the material. By producing a certain desired structure, characteristics can be imparted that will be favorable to subsequent operations (such as machining or forming) or applications. Because of the variety of anneals, it is important to designate the specific treatment, which is usually indicated by a preceding adjective. The specific temperatures, cooling rate, and details of the process will depend on the material being treated and the objectives of the treatment.

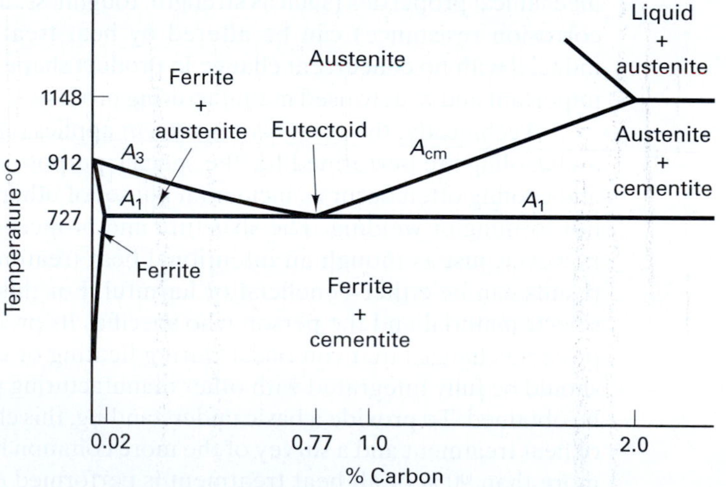

FIGURE 6-1 Simplified iron–carbon phase diagram for steels with transition lines labeled in standard notation as A_1, A_3, and A_{cm}.

[1] Historically, an A_2 line once appeared between the A_1 and A_3. This line designated the magnetic property change known as the Curie point. Because this transition was later shown to be an atomic change, and not a change in phase, the line was deleted from the equilibrium phase diagram without a companion relabeling.

In the **full annealing** process, hypoeutectoid steels (less than 0.77% carbon) are heated to 30 to 60°C (50 to 100°F) above the A_3 temperature, held for sufficient time to convert the structure to homogeneous single-phase austenite of uniform composition and temperature, and then slowly cooled at a controlled rate through the A_1 temperature. Cooling is usually done in the furnace by decreasing the temperature by 10 to 30°C (20 to 50°F) per hour to at least 30°C (50°F) below the A_1 temperature. At this point, all structural changes are complete, and the metal can be removed from the furnace and air cooled to room temperature. The resulting structure is one of coarse pearlite (widely spaced layers or lamellae) with excess ferrite in amounts predicted by the equilibrium phase diagram. In this condition, the steel is quite soft and ductile.

The procedure to full-anneal a hypereutectoid alloy (greater than 0.77% carbon) is basically the same, except that the original heating is only into the austenite plus cementite region (30 to 60°C above the A_1). If the material were to be slow cooled from the all-austenite region, a continuous network of cementite may form on the grain boundaries and make the entire material brittle. When properly annealed, a hypereutectoid steel will have a structure of coarse pearlite with excess cementite in dispersed spheroidal form.

While full anneals produce the softest and weakest properties, they are quite time consuming, and considerable amounts of energy must be spent to maintain the elevated temperatures required during soaking and furnace cooling. When maximum softness and ductility are not required and cost savings are desirable, **normalizing** may be specified. In this process, the steel is heated to 60°C (100°F) above the A_3 (hypoeutectoid) or A_{cm} (hypereutectoid) temperature, held at this temperature to produce uniform austenite, and then removed from the furnace and allowed to cool in still air. The resultant structures and properties will depend on the subsequent cooling rate. Wide variations are possible, depending on the size and geometry of the product, but fine pearlite with excess ferrite or cementite is generally produced.

One should note a key difference between full annealing and normalizing. In the full anneal, the furnace imposes identical cooling conditions at all locations within the metal, which results in identical structures and properties. With normalizing, the cooling will be different at different locations. Properties will vary between surface and interior, and different thickness regions will also have different properties. When subsequent processing involves a substantial amount of machining that may be automated, the added cost of a full anneal may be justified, because it produces a product with uniform machining characteristics at all locations.

If cold working has severely strain hardened a metal, it is often desirable to restore the ductility, either for service or to permit further processing without danger of fracture. This is often achieved through the **recrystallization** process described in Chapter 4. When the material is a low-carbon steel (<0.25% carbon), the specific procedure is known as a **process anneal.** The steel is heated to a temperature slightly below the A_1, held long enough to induce recrystallization of the dominant ferrite phase, and then cooled at a desired rate (usually in still air). Because the entire process is performed at temperatures within the same phase region, the process simply induces a change in phase morphology (size, shape, and distribution). The material is not heated to as high a temperature as in the full anneal or normalizing process, so a process anneal is somewhat cheaper and tends to produce less scaling.

A **stress-relief anneal** may be employed to reduce the **residual stresses** in large steel castings, welded assemblies, and cold-formed products. Parts are heated to temperatures below the A_1 (between 550 and 650°C or 1000 and 1200°F), held for a period of time, and then slow cooled to prevent the creation of additional stresses. Times and temperatures vary with the condition of the component, but the basic microstructure and associated mechanical properties generally remain unchanged.

When high-carbon steels (>0.60% carbon) are to undergo extensive machining or cold-forming, a heat treatment known as **spheroidization** is often employed. Here, the objective is to produce a structure in which all of the cementite is in the form of small spheroids or globules dispersed throughout a ferrite matrix. This can be accomplished by a variety of techniques, including (1) prolonged heating at a temperature just below

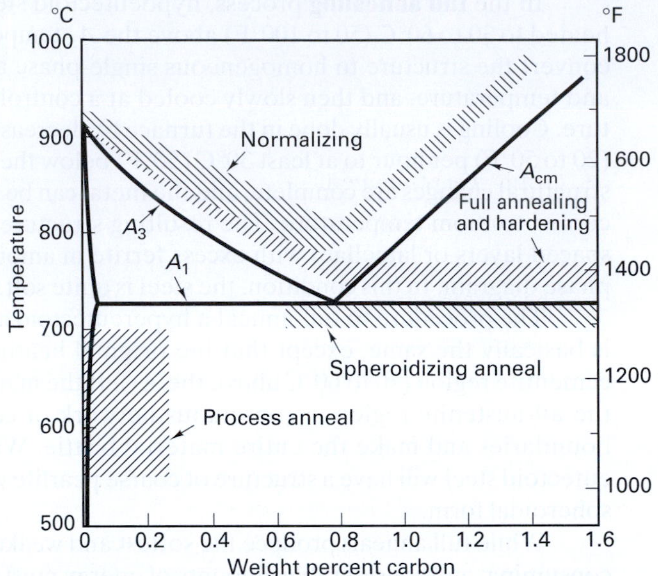

FIGURE 6-2 Graphical summary of the process heat treatments for steels on an equilibrium diagram.

the A_1 followed by relatively slow cooling; (2) prolonged cycling between temperatures slightly above and slightly below the A_1; or (3) in the case of tool or high-alloy steels, heating to 750 to 800°C (1400 to 1500°F) or higher and holding at this temperature for several hours, followed by slow cooling.

Although the selection of a processing heat treatment often depends on the desired objectives, steel composition strongly influences the choice. Process anneals are restricted to low-carbon steels, and spheroidization is a treatment for high-carbon material. Normalizing and full annealing can be applied to all carbon contents, but even here, preferences are noted. Because different cooling rates do not produce a wide variation of properties in low-carbon steels, the air cool of a normalizing treatment often produces acceptable uniformity. For higher carbon contents, different cooling rates can produce wider property variations, and the uniform furnace cooling of a full anneal is often preferred. For plain-carbon steels, the recommended treatments are: 0 to 0.4% carbon—normalize; 0.4 to 0.6% carbon—full anneal; and above 0.6% carbon—spheroidize. Due to the effects of hardenability (to be discussed later in this chapter), the transition between normalizing and full annealing for alloy steels is generally lowered to 0.2% carbon.

Figure 6-2 provides a graphical summary of the process heat treatments.

HEAT TREATMENTS FOR NONFERROUS METALS

Most of the nonferrous metals do not have the significant phase transitions observed in the iron–carbon system, and for them, the process heat treatments do not play such a significant role. Aside from the strengthening treatment of precipitation hardening, which is discussed later, the nonferrous metals are usually heat-treated for three purposes: (1) to produce a uniform, homogeneous structure; (2) to provide stress relief; or (3) to bring about recrystallization. Castings that have been cooled too rapidly can possess a segregated solidification structure known as coring (discussed more fully in Chapter 5). **Homogenization** can be achieved by heating to moderate temperatures and then holding for a sufficient time to allow thorough diffusion to take place. Similarly, heating for several hours at relatively low temperatures can reduce the internal stresses that are often produced by forming, welding, or brazing. Recrystallization (discussed in Chapter 4) is a function of the particular metal, the amount of prior deformation, and the desired recrystallization time. In general, the more a metal has been strained, the lower the recrystallization temperature or the shorter the time. Without prior straining, however, recrystallization will not occur and heating will only produce undesirable grain growth.

6.3 HEAT TREATMENTS USED TO INCREASE STRENGTH

Six major mechanisms are available to increase the strength of metals:

1. Solid-solution strengthening
2. Strain hardening
3. Grain-size refinement
4. Precipitation hardening
5. Dispersion hardening
6. Phase transformations

While all of these may not be applicable to a given metal or alloy, these heat treatments can often play a significant role in inducing or altering the final properties of a product.

In **solid-solution strengthening,** a base metal dissolves other atoms, either as **substitutional solutions,** where the new atoms occupy sites in the host crystal lattice, or as **interstitial solutions,** where the new atoms squeeze into "holes" between the atoms of the base lattice. The amount of strengthening depends on the amount of dissolved solute and the size difference of the atoms involved. Because distortion of the host structure makes dislocation movement more difficult, the greater the size difference, the more effective the addition.

Strain hardening (discussed in Chapter 4) produces an increase in strength by means of plastic deformation under cold-working conditions.

Because grain boundaries act as barriers to dislocation motion, a metal with small grains tends to be stronger than the same metal with larger grains. Thus, **grain-size refinement** can be used to increase strength, except at elevated temperatures, where grain growth can occur and grain boundary diffusion contributes to creep and failure. It is important to note that grain-size refinement is one of the few processes that can improve strength without a companion loss of ductility and toughness.

In **precipitation hardening,** or **age hardening,** strength is obtained from a nonequilibrium structure that is produced by a three-step heat treatment. Details of this method will be provided in Section 6.4.

Strength obtained by dispersing second-phase particles throughout a base material is known as **dispersion hardening.** To be effective, the dispersed particles should be stronger than the matrix, adding strength through both their reinforcing action and the additional interfacial surfaces that present barriers to dislocation movement.

Phase transformation strengthening involves those alloys that can be heated to form a single phase at elevated temperature and subsequently transform to one or more low temperature phases upon cooling. When this feature is used to increase strength, the cooling is usually rapid and the phases that are produced are usually of a nonequilibrium nature.

6.4 STRENGTHENING HEAT TREATMENTS FOR NONFERROUS METALS

All six of the mechanisms just described can be used to increase the strength of nonferrous metals. Solid-solution strengthening can impart strength to single-phase materials. Strain hardening can be quite useful if sufficient ductility is present. Alloys containing eutectic structure exhibit considerable dispersion hardening. Among all of the possibilities, however, the most effective strengthening mechanism for the nonferrous metals tends to be precipitation hardening.

PRECIPITATION OR AGE HARDENING

To be a candidate for precipitation hardening, an alloy system must exhibit solubility that decreases with decreasing temperature, such as the aluminum-rich portion of the aluminum–copper system shown in Figure 6-3 and enlarged in Figure 6-4. Consider the alloy with 4% copper, and use the phase diagram to determine its equilibrium structure. Liquid metal cools through the alpha plus liquid region and solidifies into a single-phase solid (a phase). At 1000°F, the full 4% of copper would be dissolved and distributed

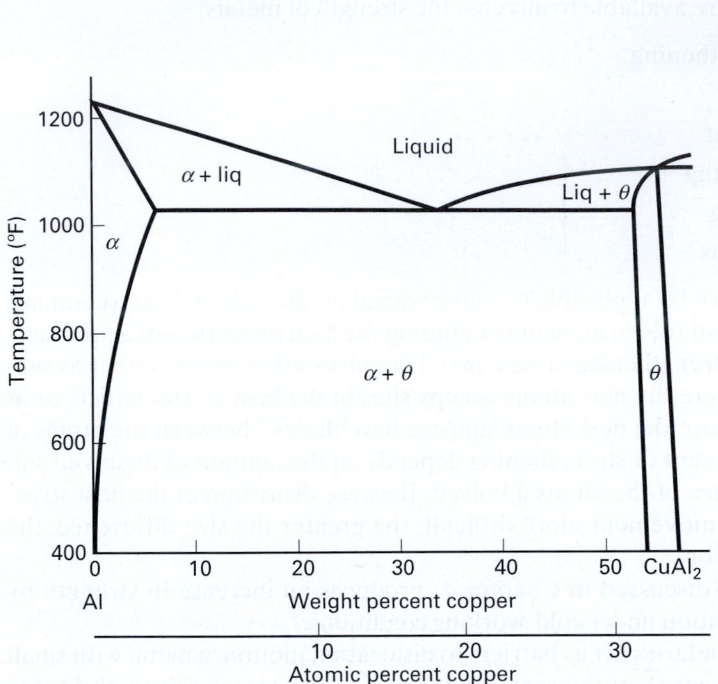

FIGURE 6-3 High-aluminum section of the aluminum–copper equilibrium phase diagram.

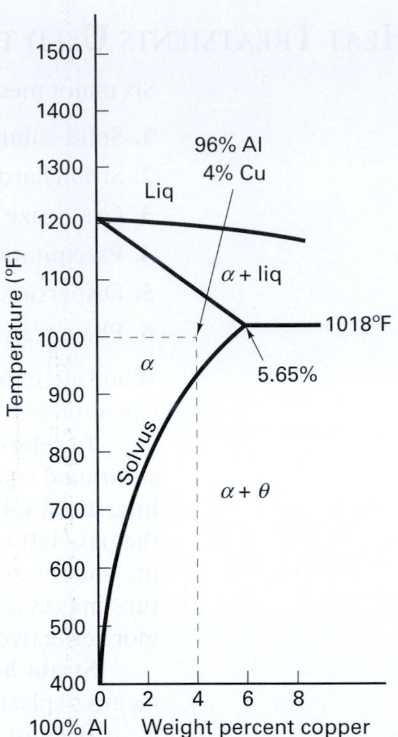

FIGURE 6-4 Enlargement of the solvus-line region of the aluminum–copper equilibrium diagram of Figure 6-3.

throughout the alpha-phase crystals. As the temperature drops, however, the maximum solubility of copper in aluminum decreases from 5.65% at 1018°F to less than 0.2% at room temperature. Upon cooling through the solvus (or solubility limit) line at 930°F, the 4% copper alloy enters a two-phase region, and copper-rich theta-phase precipitates begin to form and grow. (*Note:* Theta phase is actually a hard, brittle intermetallic compound with the chemical formula of $CuAl_2$). The equilibrium structure at room temperature, therefore, would be an aluminum-rich alpha-phase structure with coarse theta-phase precipitates, generally lying along alpha-phase grain boundaries where the nucleation of second-phase particles can benefit from the existing interfacial surface.

Whenever two or more phases are present, the material exhibits dispersion strengthening. Dislocations are confined to their own crystal and cannot cross interfacial boundaries. Therefore, each interface between alpha-phase and the theta-phase precipitate is a strengthening boundary. Take a particle of theta precipitate, cut it into two halves, and separate the segments. Forming the two half-size precipitates has just added two additional interfaces, corresponding to both sides of the cut. If the particle were to be further cut into quarters, eighths, and sixteenths, we would expect strength to increase as we continually add interfacial surface. Ideally, we would like to have millions of ultra-small particles dispersed throughout the alpha-phase structure. When we try to form this more desirable nonequilibrium configuration (nonequilibrium because energy is added each time new interfacial surface is created), we gain an unexpected benefit that adds significant strength. This new nonequilibrium treatment is known as *age hardening* or *precipitation hardening*.

The process of precipitation hardening is actually a three-step sequence. The first step, known as **solution treatment,** erases the room-temperature structure and redissolves any existing precipitate. The metal is heated to a temperature above the solvus and held in the single-phase region for sufficient time to redissolve the second phase and uniformly distribute the solute atoms (in this case, copper).

If the alloy were slow cooled, the second-phase precipitate would nucleate and the material would revert back to a structure similar to equilibrium. To prevent this from

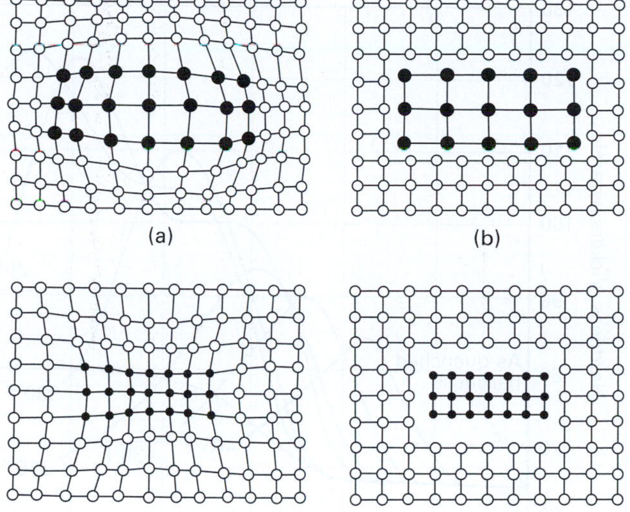

FIGURE 6-5 Two-dimensional illustrations depicting (a) a coherent precipitate cluster where the precipitate atoms are larger than those in the host structure and (b) its companion overaged or discrete second-phase precipitate particle. Parts (c) and (d) show equivalent sketches where the precipitate atoms are smaller than the host.

happening, age hardening alloys are **quenched** from their solution treatment temperature. The rapid-cool quenching, usually in water, suppresses diffusion, trapping the dissolved atoms in place. The result is a room-temperature *supersaturated* solid solution. For the alloy discussed earlier, the alpha phase would now be holding 4% copper in solution at room temperature—far in excess of its equilibrium maximum of <0.2%. In this nonequilibrium quenched condition, the material is often soft and can be easily straightened, formed, or machined.

If the supersaturated material is then reheated to a temperature where atom movement (diffusion) can occur but still within the two-phase region, the alloy will attempt to form its equilibrium two-phase structure as the excess solute atoms precipitate out of the supersaturated matrix. This stage of the process, known as **aging,** is actually a continuous transition. Solute atoms begin to cluster at locations within the parent crystal, still occupying atom sites within the original lattice. Various transitions may then occur, leading ultimately to the formation of distinct second phase particles with their own characteristic chemistry and crystal structure.

A key concept in the aging sequence is that of **coherency** or crystalline continuity. If the clustered solute atoms continue to occupy lattice sites within the parent structure, the crystal planes remain continuous in all directions, and the clusters of solute atoms (which are of different size and possibly different valence from the host material) tend to distort or strain the adjacent lattice for a sizable distance in all directions, as illustrated two-dimensionally in Figures 6-5a and 6-5c. For this reason, each small cluster appears to be much larger with respect to its ability to interfere with dislocation motion (i.e., impart strength). When the clusters reach a certain size, however, the associated strain becomes so great that the clusters can lower their energy by breaking free from the parent structure to form distinct second-phase particles with their own crystal structure and well-defined interphase boundaries, as shown in Figure 6-5b and 6-5d. Coherency is lost and the strengthening reverts to dispersion hardening, where dislocation interference is limited to the actual size of the particle. Strength and hardness decrease, and the material is said to be **overaged.**

Aging can be performed at any temperature within the two-phase region where diffusion is sufficiently rapid to form the desired precipitate in a reasonable time. Figure 6-6 presents a family of aging curves for the 4% copper–96% aluminum alloy. For higher aging temperatures, the peak properties are achieved in a shorter time, but the peak hardness (or strength) is not as great as can be achieved at lower aging temperatures. The higher peak strength at lower aging temperature can be attributed to the combined effect of the increased amount of precipitate that forms at lower temperature (use a lever-law calculation as described in Chapter 5) and the fineness of the precipitate that forms when diffusion is more limited. Actual selection of the aging conditions

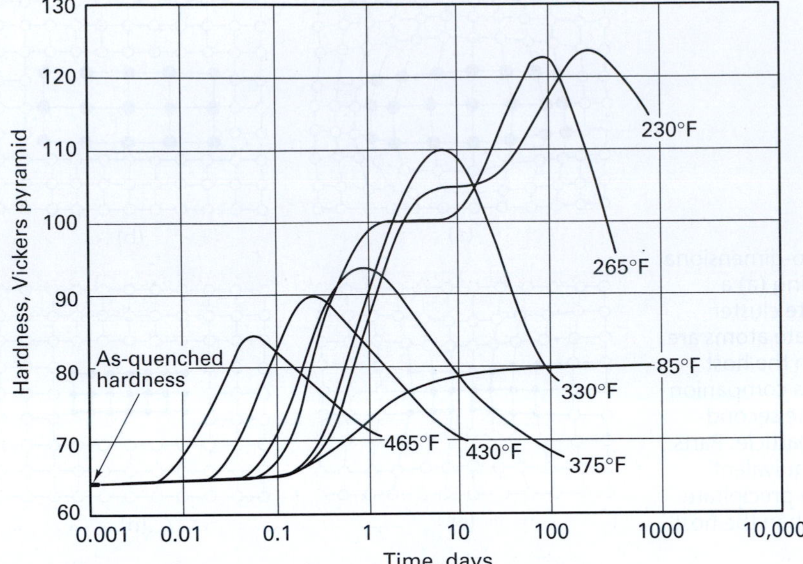

FIGURE 6-6 Aging curves for the Al–4%Cu alloy at various temperatures showing peak strengths and times of attainment. *(Adapted from Journal of the Institute for Metals, Vol. 79, p. 321, 1951)*

(temperature and time) is a decision that is made on the basis of desired strength, available equipment, and production constraints.

The aging step can be used to divide precipitation-hardening materials into two types: (1) **natural aging** materials, where room temperature is sufficient to move the unstable supersaturated solution toward the stable two-phase structure, and (2) **artificial aging** materials, where elevated temperatures are required to provide the necessary diffusion. With natural aging materials, such as aluminum alloy rivets, some form of refrigeration may be required to retain the after-quench condition of softness. Upon removal from the refrigeration, the rivets are easily headed, but progress to full strength after several days at room temperature.

Because artificial aging requires elevated temperature to provide diffusion, the aging process can be stopped at any time by simply dropping the temperature (quenching). Diffusion is halted, and the current structure and properties are "locked-in," provided that the material is not subsequently exposed to elevated temperatures that would reactivate diffusion. When diffusion is possible, the material will always attempt to revert to its equilibrium structure! According to Figure 6-6, if the 4% copper alloy were aged for one day at 375°F and then quenched to prevent overaging, the metal would attain a hardness of 94 Vickers (and the associated strength), and retain these properties throughout its useful lifetime provided subsequent diffusion did not occur. If a higher strength is required, a lower temperature and longer time could be selected.

Precipitation hardening is an extremely effective strengthening mechanism and is responsible for the attractive engineering properties of many aluminum, copper, magnesium, and titanium alloys. In many cases, the strength can more than double that observed upon conventional cooling. While other strengthening methods are traditionally used with steels and cast irons, those methods have been combined with age hardening to produce some of the highest strength ferrous alloys, such as the maraging steels and precipitation hardenable stainless.

6.5 Strengthening Heat Treatments for Steel

Iron-based metals have been heat treated for centuries, and today more than 90% of all heat-treatment operations are performed on steel. The striking changes that resulted from plunging red-hot metal into cold water or some other quenching medium were awe-inspiring to the ancients. Those who performed these acts in the making of swords and armor were looked upon as possessing unusual powers, and much superstition arose regarding the process. Because quality was directly related to the act of quenching, great importance was placed on the quenching medium that was used. Urine, for example, was found to be a superior quenching medium, and that from a red-haired boy was deemed particularly effective, as was that from a 3-year-old goat fed only ferns.

It has only been within the last century that the art of heat treating has begun to turn into a science. One of the major barriers to understanding was the fact that the strengthening treatments were nonequilibrium in nature. Minor variations in cooling often produced major variations in structure and properties.

ISOTHERMAL-TRANSFORMATION (I-T) OR TIME-TEMPERATURE-TRANSFORMATION (T-T-T) DIAGRAMS

A useful aid to understanding nonequilibrium heat-treatment processes is the **isothermal-transformation (I-T)** or **time-temperature-transformation (T-T-T) diagram.** The information in this diagram is obtained by heating thin specimens of a particular steel to produce elevated-temperature uniform-chemistry **austenite,** "instantaneously" quenching to a temperature where austenite is no longer the stable phase, holding for variable periods of time at this new temperature, and observing the resultant structures via metallographic photomicrographs (i.e., optical microscope examination).

For simplicity, consider a carbon steel of eutectoid composition (0.77% carbon) and its T-T-T diagram shown as Figure 6-7. Above the A_1 temperature of 1341°F (727°C), austenite is the stable phase and will persist regardless of the time. Below this temperature, the face-centered austenite would like to transform to body-centered ferrite and carbon-rich cementite. Two factors control the rate of transition: (1) the motivation or driving force for the change and (2) the ability to form the desired products (i.e., the ability to redistribute the atoms through diffusion). The region below 1341°F in Figure 6-7 can be interpreted as follows. Zero time corresponds to a sample "instantaneously" quenched to its new lower temperature. The structure is usually unstable austenite. As time passes (moving horizontally across the diagram), a line is encountered representing the start of transformation and a second line indicating completion of the phase change. At elevated temperatures (just below 1341°F), atom movement within the solid (diffusion) is rapid, but the rather sluggish driving force dominates the kinetics. At a low temperature, the driving force is high but diffusion is quite limited. The kinetics of phase transformation are most rapid at a compromise intermediate temperature, resulting in the characteristic C-curve shape. The portion of the C that extends farthest to the left is known as the *nose* of the T-T-T or I-T diagram.

If the transformation occurs between the A_1 temperature and the nose of the curve, the departure from equilibrium is not very great. The austenite transforms into alternating layers of ferrite and cementite, producing the **pearlite** structure that was introduced with the equilibrium phase diagram description in Chapter 5. Because the diffusion rate is greater at higher temperatures, pearlite produced under those conditions has a larger lamellar spacing (separation distance between similar layers). The pearlite formed near the A_1 temperature is known as *coarse* pearlite, while the closer-spaced structures formed near the nose are called *fine* pearlite. Because the resulting structures and properties are similar to those of the near-equilibrium process heat treatments, the constant-temperature transformation procedure just described is called an **isothermal anneal.**

If the austenite is quenched to a temperature between the nose and the temperature designated as M_s, a different structure is produced. These transformation conditions are a significant departure from equilibrium, and the amount of diffusion required to form the continuous layers within pearlite is no longer available. The metal still has the goal of changing crystal structure from face-centered austenite to body-centered ferrite, with the excess carbon being accommodated in the form of cementite. The resulting structure, however, does not contain cementite layers but rather a dispersion of discrete cementite particles dispersed throughout a matrix of ferrite. Electron microscopy may be required to resolve the carbides in this structure, which is known as **bainite.** Because of the fine dispersion of carbide, it is stronger than fine pearlite, and ductility is retained because the soft ferrite is the continuous matrix.

If austenite is quenched to a temperature below the M_s line, a different type of transformation occurs. The steel still wants to change its crystal structure from face-centered cubic to body-centered cubic, but it can no longer expel the amount of carbon

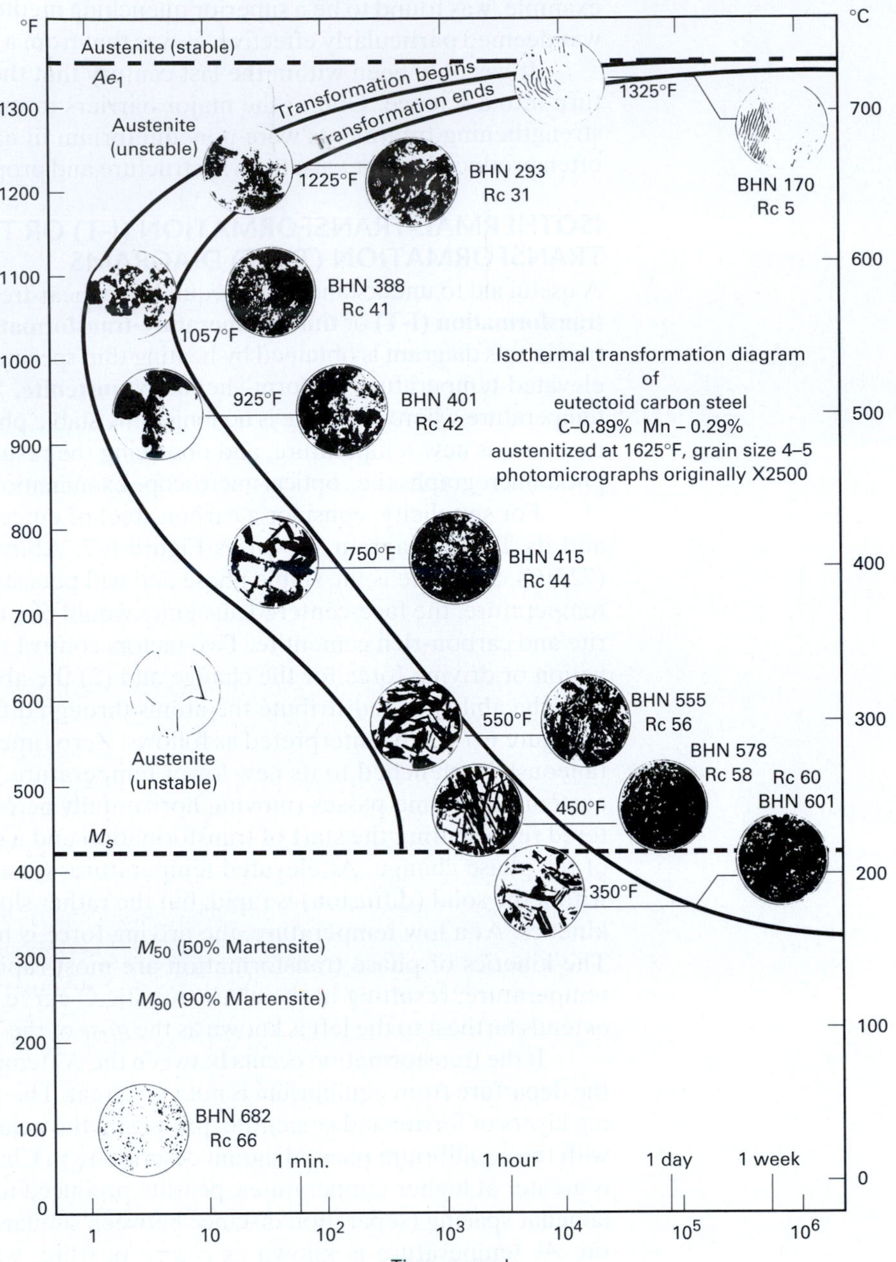

FIGURE 6-7 Isothermal transformation diagram (or T-T-T diagram) for eutectoid composition steel. Structures resulting from transformation at various temperatures are shown as insets. *(Courtesy United States Steel Corporation)*

necessary to form ferrite. Responding to the severe nonequilibrium conditions, it simply undergoes an abrupt change in crystal structure with no significant movement of carbon. The excess carbon becomes trapped, distorting the structure into a body-centered tetragonal crystal lattice (distorted body-centered cubic), with the amount of distortion being proportional to the amount of excess carbon. The new structure, shown in Figure 6-8, is known as **martensite,** and, with sufficient carbon, it is exceptionally strong, hard, and brittle. The highly distorted lattice effectively blocks the dislocation motion that is necessary for metal deformation.

As shown in Figure 6-9, the hardness and strength of steel with the martensitic structure are strong functions of the carbon content. Below 0.10% carbon, martensite is not very strong. Because no diffusion occurs during the transformation, higher-carbon steels form higher-carbon martensite, with an increase in strength and hardness and a concurrent decrease in toughness and ductility. From 0.3 to 0.7% carbon, strength and hardness increase rapidly. Above 0.7% carbon, however, the rise is far less dramatic and may actually be a decline, a feature related to the presence of retained austenite (to be described below).

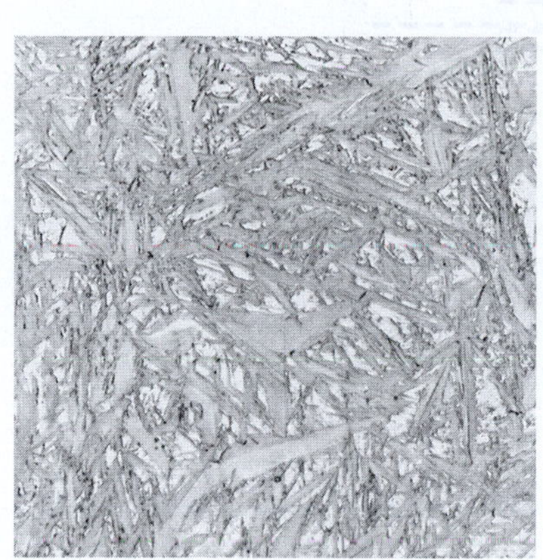

FIGURE 6-8 Photomicrograph of martensite; 1000×. *(Courtesy United States Steel Corporation)*

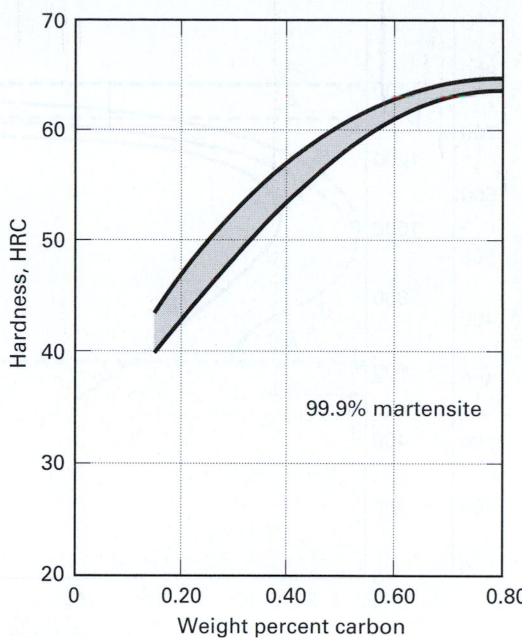

FIGURE 6-9 Effect of carbon on the hardness of martensite.

Unlike the other structure transformations, the *amount* of martensite that forms is not a function of time, but depends only on the lowest temperature that is encountered during the quench. This feature is shown in Figure 6-10 where the amount of martensite is recorded as a function of temperature. Returning to the C-curve of Figure 6-7, there is a temperature designated as M_{50}, where the structure is 50% martensite and 50% untransformed austenite. At the lower M_{90} temperature, the structure has become 90% martensite. If no further cooling were to occur, the untransformed austenite could remain within the structure. This **retained austenite** can cause loss of strength or hardness, dimensional instability, and cracking or brittleness. Because many quenches are to room temperature, retained austenite becomes a significant problem when the martensite finish, or 100% martensite, temperature lies below room temperature. Figure 6-11 presents the

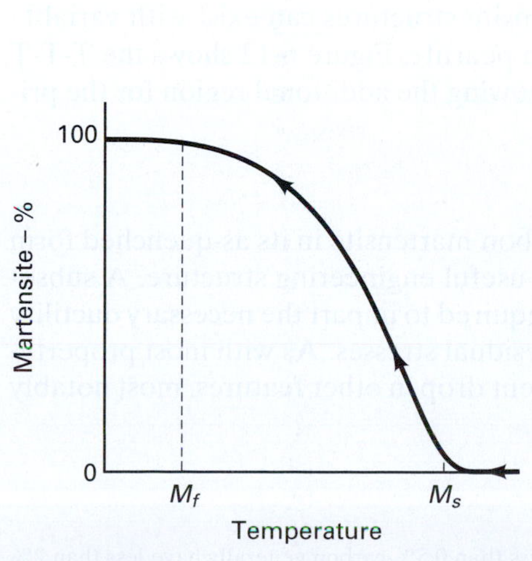

FIGURE 6-10 Schematic representation depicting the amount of martensite formed upon quenching to various temperatures from M_s through M_f.

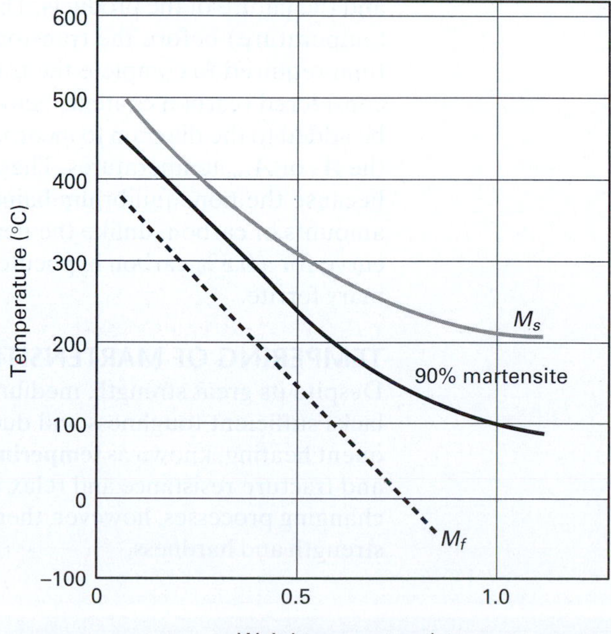

FIGURE 6-11 Variation of M_s and M_f temperatures with carbon content. Note that for high carbon steels, completion of the martensite transformation requires cooling to below room temperature.

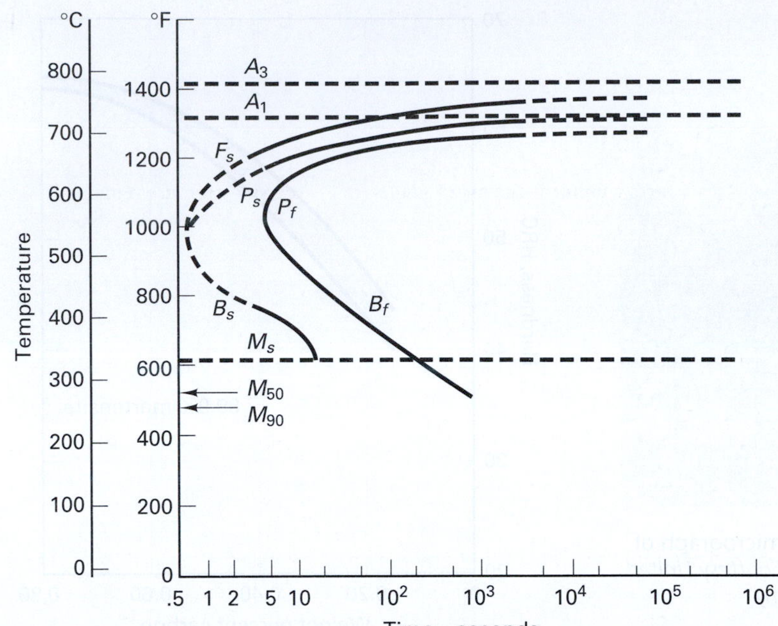

FIGURE 6-12 Isothermal transformation diagram for a hypoeutectoid steel (1050) showing the additional region for primary ferrite.

martensite start and martensite finish temperatures for a range of carbon contents. Higher carbon contents, as well as most alloy additions, decrease all martensite-related temperatures, and materials with these chemistries may require refrigeration or a quench in dry ice or liquid nitrogen to produce full hardness.[2]

It is important to note that all of the transformations that occur below the A_1 temperature are one-way transitions (austenite to something). The steel is simply seeking to change its crystal structure, and the various products are the result of this change. It is impossible, therefore, to convert one transformation product to another without first reheating to above the A_1 temperature to again form the face-centered-cubic austenite.

T-T-T diagrams can be quite useful in determining the kinetics of transformation and the nature of the products. The left-hand curve shows the elapsed time (at constant temperature) before the transformation begins, and the right-hand curve shows the time required to complete the transformation. If hypo- or hypereutectoid steels were considered (carbon contents below or above 0.77%), additional regions would have to be added to the diagram to incorporate the primary equilibrium phases that form below the A_3 or A_{cm} temperatures. These regions would not extend below the nose, however, because the nonequilibrium bainite and martensite structures can exist with variable amounts of carbon, unlike the near-equilibrium pearlite. Figure 6-12 shows the T-T-T curve for a 0.5% carbon hypoeutectoid steel, showing the additional region for the primary ferrite.

TEMPERING OF MARTENSITE

Despite its great strength, medium- or high-carbon martensite in its as-quenched form lacks sufficient toughness and ductility to be a useful engineering structure. A subsequent heating, known as **tempering,** is usually required to impart the necessary ductility and fracture resistance and relax undesirable residual stresses. As with most property-changing processes, however, there is a concurrent drop in other features, most notably strength and hardness.

[2] With a quench to room temperature, carbon steels with less than 0.5% carbon generally have less than 2% retained austenite. As the amount of carbon increases, so does the amount of retained austenite, rising to about 6% at 0.8% carbon and to more than 30% when the carbon content is 1.25%.

TABLE 6-1 Comparison of Age Hardening with the Quench-and-Temper Process

Heat Treatment	Step 1	Step 2	Step 3
Age hardening	*Solution treatment.* Heat into the stable single-phase region (above the solvus) and hold to form a uniform-chemistry single-phase solid solution.	*Quench.* Rapid cool to form a nonequilibrium supersaturated single-phase solid solution (crystal structure remains unchanged, material is soft and ductile).	*Age.* A controlled reheat in the stable two-phase region (below the solvus). The material moves toward the formation of the stable two-phase structure, becoming stronger and harder. The properties can be "frozen in" by dropping the temperature to stop further diffusion.
Quench and temper for steel	*Austenitize.* Heat into the stable single-phase region (above the A_3 or A_{cm}) and hold to form a uniform-chemistry single-phase solid solution (austenite).	*Quench.* Rapid cool to form a nonequilibrium supersaturated single-phase solid solution (crystal structure changes to body-centered martensite, which is hard but brittle).	*Temper.* A controlled reheat in the stable two-phase region (below the A_1). The material moves toward the formation of the stable two-phase structure, becoming weaker but tougher. The properties can be "frozen in" by dropping the temperature to stop further diffusion.

Martensite is a supersaturated solid solution of carbon in alpha ferrite and, therefore, is a metastable structure. When heated into the range of 100 to 700°C (200 to 1300°F), the excess carbon atoms are rejected from solution, and the structure moves toward a mixture of the stable phases of ferrite and cementite. This decomposition of martensite into ferrite and cementite is a time- and temperature-dependent, diffusion-controlled phenomenon with a continuous spectrum of intermediate and transitory structures.

Table 6-1 presents a chart-type comparison of the previously discussed precipitation hardening process and the austenitize-quench-and-temper sequence. Both are nonequilibrium heat treatments that involve three distinct stages. In both, the first step is an elevated temperature soaking designed to erase the prior structure, redissolving material to produce a uniform-chemistry, single-phase starting condition. Both treatments follow this soak with a rapid-cool quench. In precipitation hardening, the purpose of the quench is to prevent nucleation of the second phase, thereby producing a supersaturated solid solution. This material is usually soft, weak, and ductile, with good toughness. Subsequent aging (reheating within the temperatures of the stable two-phase region) allows the material to move toward the formation of the stable two-phase structure and sacrifices toughness and ductility for an increase in strength. When the proper balance is achieved, the temperature is dropped, diffusion ceases, and the current structure and properties are preserved—provided that the material is never subsequently exposed to any elevated temperature that would reactivate diffusion and permit the structure to move further toward equilibrium.

For steels, the quench induces a phase transformation as the material changes from the face-centered-cubic austenite to the distorted body-centered structure known as martensite. The quench product is again a supersaturated, single-phase solid solution, this time of carbon in iron, but the associated properties are the reverse of precipitation hardening. Martensite is strong and hard, but relatively brittle. When the material is tempered (reheated to a temperature within the stable two-phase region, i.e., below the A_1 temperature), strength and hardness are sacrificed for an increase in ductility and toughness. Figure 6-13 shows the final properties of a steel that has been tempered at a variety of temperatures. During tempering, diffusion enables movement *toward* the stable two-phase structure, and a drop in temperature can again halt diffusion and lock in properties. By quenching steel to form martensite and then tempering it at various temperatures, an infinite range of structures and corresponding properties can be produced. This procedure is known as the **quench-and-temper process** and the product, which offers an outstanding combination of strength and toughness, is called **tempered martensite.**

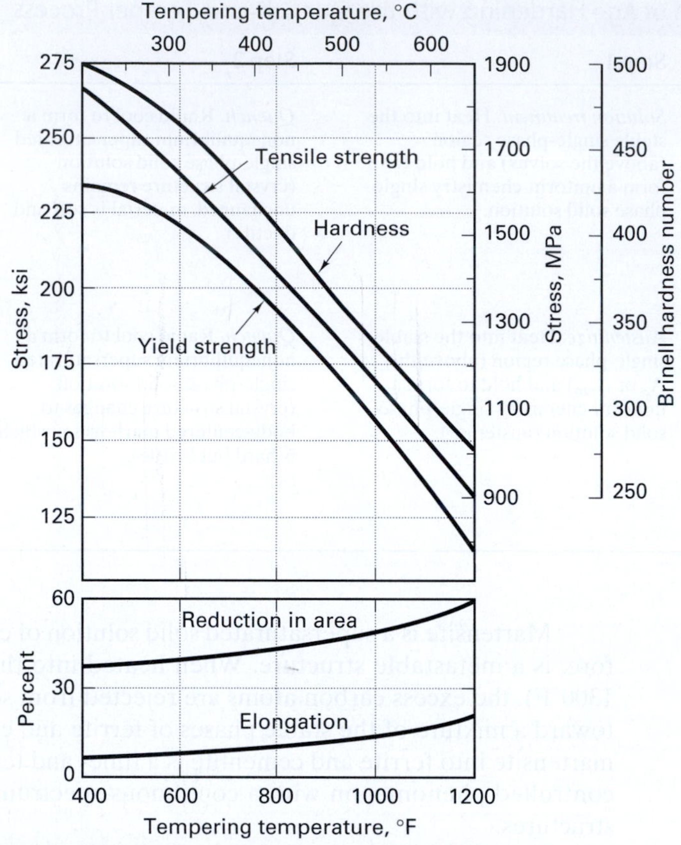

FIGURE 6-13 Properties of an AISI 4140 steel that has been austenitized, oil-quenched, and tempered at various temperatures. *(Adapted from Engineering Properties of Steel, ASM International, Materials Park, OH, 1982)*

CONTINUOUS COOLING TRANSFORMATIONS

While the T-T-T diagrams provide considerable information about the structures obtained through nonequilibrium thermal processing, the assumptions of instantaneous cooling followed by constant temperature transformation rarely match reality. Actual parts generally experience continuous cooling from elevated temperature, and a diagram showing the results of this type of cooling at various rates would be far more useful. What would be the result if the temperature were to be decreased at a rate of 500°F per second, 50°F per second, or 5°F per second?

A **continuous-cooling-transformation (C-C-T) diagram,** like the one shown in Figure 6-14, can provide answers to these questions and numerous others. If the cooling is sufficiently fast (dashed curve A), the structure will be martensite. The slowest cooling rate that will produce a fully martensitic structure is referred to as the "critical cooling rate." Slow cooling (curve D) generally produces coarse pearlite along with a possible primary phase. Intermediate rates usually result in mixed structures, because the time at any one temperature is usually insufficient to complete the transformation. If each structure is regarded as providing a companion set of properties, the wide range of possibilities obtainable through the controlled heating and cooling of steel becomes even more evident.

JOMINY TEST FOR HARDENABILITY

The C-C-T diagram shows that different cooling rates will produce different structures each with their own associated properties. While the details of the C-C-T diagram will change with material chemistry, in all cases there exists a general relation of the following form:

$$\text{Material} + \text{Cooling rate} \rightarrow \text{Structure} \rightarrow \text{Properties}$$

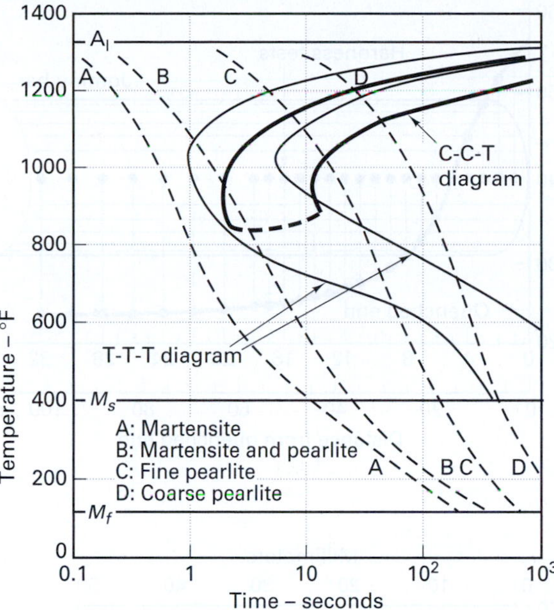

FIGURE 6-14 C-C-T diagram for a eutectoid composition steel (bold), with several superimposed cooling curves and the resultant structures. The lighter curves are the T-T-T transitions for the same steel. *(Courtesy of United States Steel Corporation, Pittsburgh, PA)*

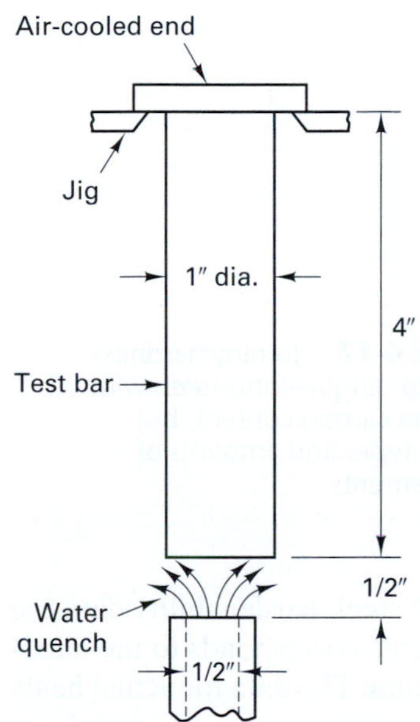

FIGURE 6-15 Schematic diagram of the Jominy hardenability test.

For a given material, the C-C-T diagram (if available) can provide the desired link between cooling rate and structure. Individuals whose focus is on use and application, however, are often more interested in the resulting properties and are less concerned with the structures that produce them. For these individuals, the **Jominy end-quench hardenability test**[3] and associated diagrams provide another useful tool, which further expands our understanding of nonequilibrium heat treatment. In this test, depicted schematically in Figure 6-15, a standard specimen (1 in. diameter and 4 in. long) is heated to austenitizing temperature and then quenched on one end. The quench is standardized by specifying the fluid medium (water at 75°F), the internal nozzle diameter ($\frac{1}{2}$ in.), the rate of water flow (that producing a 2-$\frac{1}{2}$-in. vertical fountain), and the gap between the nozzle and the specimen ($\frac{1}{2}$ in.). Because none of the water contacts the side of the specimen, all cooling is directional, and an entire spectrum of cooling rates is produced along the length of the bar. One end sees a rapid-cool quench, while the other is essentially air-cooled. Because the thermal conductivity of steel does not change over the normal ranges of carbon and alloy additions, a characteristic cooling rate can be assigned to each location along the length of the standard-geometry specimen.

After the test bar has cooled to room temperature, a flat region is ground along opposite sides and Rockwell C hardness readings are taken along the length (every $\frac{1}{16}$ in. for the first inch, every $\frac{1}{8}$ in. for the next inch, and every $\frac{1}{4}$ in. for the remaining length). The resulting data are then plotted as shown in Figure 6-16. The hardness values are taken at specific positions, and because the cooling rate is known for each location within the bar, the hardnesses, therefore, are indirectly correlated with the cooling rate that produced them. Because hardness is also an indicator of strength, we have experimentally linked the entire spectrum of cooling rates to their resultant strengths.

Application of the test then assumes that equivalent cooling conditions will produce equivalent results. If the cooling rate is known for a specific location within a part (from experimentation or theory), the properties at that location can be predicted to be the same as those at the Jominy test bar location with the same cooling rate. Conversely, if specific properties are required, the necessary cooling rate can be determined. Should the cooling rates be restricted by either geometry or processing limitations, various materials can be compared and a satisfactory alloy selected. Figure 6-17 shows the Jominy curves for several engineering steels. Because differences in chemical

[3] This test is described in detail in the following standards: ASTM A255, SAE J406, DIN 50191, and ISO 642.

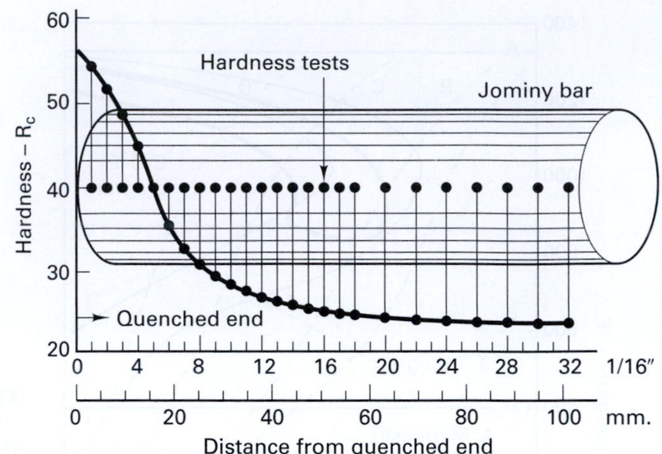

FIGURE 6-16 Typical hardness distribution along a Jominy test specimen.

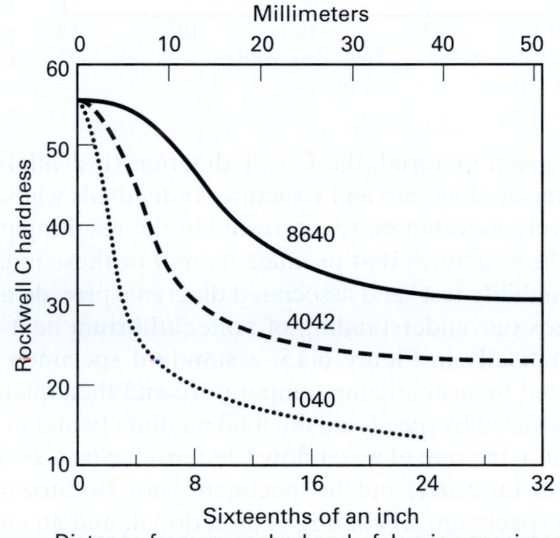

FIGURE 6-17 Jominy hardness curves for engineering steels with the same carbon content, but varying types and amounts of alloy elements.

composition can exist between heats of the same grade of steel, hardenability data are often presented in the form of bands, where the upper curve corresponds to the maximum expected hardness and the lower curve to the minimum. The data for actual heats should then fall between these two extremes.

HARDENABILITY CONSIDERATIONS

Several key effects must be considered as we seek to understand the heat treatment of steel: (1) the effect of carbon content, (2) the effect of alloy additions, and (3) the effect of various quenching conditions. The first two relate to the material being treated and the third to the heat-treatment process.

Hardness is a mechanical property related to strength and is a strong function of the carbon content of a steel and the particular microstructure. With different heat treatments, the same steel can have different hardness values. **Hardenability** is a measure of the depth to which full hardness can be obtained under a normal hardening cycle and is related primarily to the amounts and types of alloying elements. Hardenability, therefore, is a material property dependent upon chemical composition. In Figure 6-17, all of the steels have the same carbon content, but they differ in the type and amounts of alloy elements. The maximum hardness, which occurs on the quenched end, is the same in all cases, but the depth of hardening (or the way hardness varies with distance from the quenched end) varies considerably. Figure 6-18 shows Jominy test results for steels

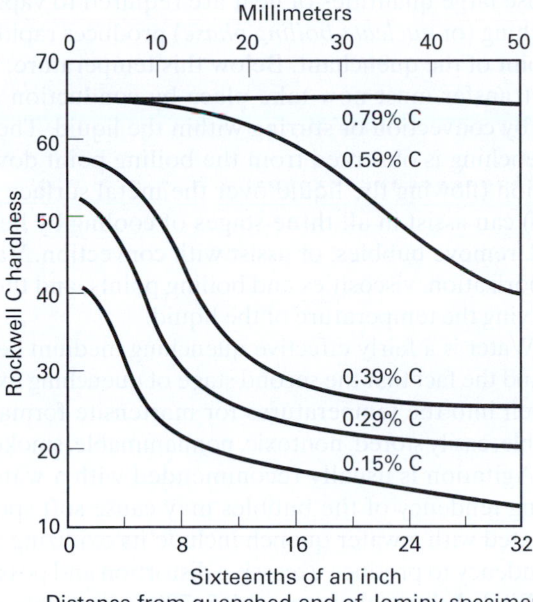

FIGURE 6-18 Jominy hardness curves for engineering steels with identical alloy conditions but variable carbon content.

containing the same alloying elements but variable amounts of carbon. Note the change in peak hardness as carbon content is increased.

The results of a heat-treat operation depend on both the hardenability of the metal and the rate of heat extraction. The primary reason for adding alloy elements to commercial steels is to increase their hardenability, not to improve their strength. Steels with greater hardenability can achieve a desired level of strength or hardness with slower rates of cooling, and, for this reason, can be completely hardened in thicker sections. Slower cooling also serves to reduce the amount of quench-induced distortion and the likelihood of **quench cracking.**

An accurate determination of need is required if steels are to be selected for specific applications. Strength tends to be associated with carbon content, and a general rule is to select the lowest possible level that will meet the specifications. Because heat can be extracted only from the surface of a metal, the size of the piece and the depth of required hardening set the conditions for hardenability and quench. For a given quench condition, different alloys will produce different results. Because alloy additions increase the cost of a material, it is best to select only what is required to ensure compliance with specifications. Money is often wasted by specifying an alloy steel for an application where a plain carbon steel, or a steel with lower alloy content (less costly), would be satisfactory. When greater depth of hardness is required, another alternative is to modify the quench conditions so that a faster cooling rate is achieved. Quench changes may be limited, however, by cracking or warping problems, and other considerations relating to the size, shape, complexity, and desired precision of the part being treated.

QUENCHING AND QUENCH MEDIA

Quenchants are selected to provide the cooling rates required to produce the desired structure and properties (such as hardness, strength and toughness) in the size and shape part being treated. Quench media vary in their effectiveness, and one can best understand the variation by considering the three stages of quenching. Let us begin with a piece of hot metal being inserted into a tank of liquid quenchant. If the temperature of the metal is above the boiling point of the quenchant, the liquid adjacent to the metal will vaporize and form a thin gaseous layer between the metal and the liquid. Cooling is slow through this **vapor jacket** (first stage) because the gas has an insulating effect and heat transfer is largely through radiation. Bubbles soon nucleate, however, and break the jacket. New liquid contacts the hot metal, vaporizes (removing its heat of vaporization from the metal), forms another bubble, and the process continues.

Because large quantities of heat are required to vaporize a liquid, this second stage of quenching (or *nucleate-boiling phase*) produces rapid rates of cooling down to the boiling point of the quenchant. Below this temperature, vaporization can no longer occur. Heat transfer must now take place by conduction across the solid–liquid interface, aided by convection or stirring within the liquid. The slower cooling of this third stage of quenching is observed from the boiling point down to room temperature. Flow or agitation (flowing the liquid over the metal surface or moving the metal through the liquid) can assist in all three stages of cooling by helping to break through the vapor jacket, remove bubbles, or assist with convection. Various liquids offer different heats of vaporization, viscosities and boiling points, and the quenches can be further tailored by varying the temperature of the liquid.

Water is a fairly effective quenching medium because of its high heat of vaporization, and the fact that the second stage of quenching extends down to 100°C (212°F), usually well into the temperatures for martensite formation. Water is also cheap, readily available, easily stored, nontoxic, nonflammable, smokeless, and easy to filter and pump.

Agitation is usually recommended with a water quench, however, because the clinging tendency of the bubbles may cause soft spots on the metal. Other problems associated with a water quench include its oxidizing nature (i.e., its corrosiveness) and the tendency to produce excessive distortion and possible cracking.

While **brine** (salt water with 5–7% sodium chloride or calcium chloride) has a similar heat of vaporization and boiling point to water, it produces more rapid cooling because the salt nucleates bubbles, forcing a quick transition through the vapor jacket stage. Unfortunately, the salt in a brine quench also tends to accelerate corrosion problems for both the pieces being treated and the quenching equipment. Different types of salts can be used, and various degrees of agitation or spraying can be used to adjust the effectiveness of the quench. Alkaline or caustic solutions, involving 5 to 10% sodium or potassium hydroxide, are similar to the brines. (*Note:* Because of all of the dissolved salts, the urines cited in quenching folklore are actually quite similar to the brines of today.)

If a slower cooling rate is desired, *oil quenches* are often utilized. Various oils are available that have high flash points and different degrees of quenching effectiveness. Because the boiling points can be quite high, the transition to third-stage cooling usually precedes the martensite start temperature. The slower cooling through the M_s-to-M_f martensite transformation leads to a milder temperature gradient within the piece, reduced distortion, and reduced likelihood of cracking. Heating the oil actually increases its cooling ability, because the reduced viscosity assists bubble formation and removal. Problems associated with oil quenchants include water contamination, smoke, fumes, spill and disposal problems, and fire hazard. In addition, quench oils tend to be somewhat expensive.

Quite often, there is a need for a quenchant that will cool more rapidly than the oils, but slower than water or brine. To fill this gap, a number of water-based **polymer quench** solutions (also called **synthetic quenchants**) have been developed. Tailored quenchants can be produced by varying the concentrations of the components (such as liquid organic polymers, corrosion inhibitors, and water) and adjusting the operating temperature and amount of agitation. The polymer quenchants provide extremely uniform and reproducible results, are less corrosive than water and brine, are nontoxic, and are less of a fire hazard than oils (no fires, fumes, smoke, or need for air pollution control apparatus). Distortion and cracking are less of a problem because the boiling point can be adjusted to be above the martensite start temperature. In addition, the polymer-rich film that forms initially on the hot metal part serves to modify the cooling rate.

If slow cooling is required, molten salt baths can be employed to provide a medium where the quench goes directly to the third stage of cooling. Still slower cooling can be obtained by cooling in still air, burying the hot material in sand, or a variety of other methods.

High-pressure gas quenching uses a stream of flowing gas to extract heat, and the cooling rates can be adjusted by controlling the gas velocity and pressure. Results are comparable to oil quenching, with far fewer environmental and safety concerns. From an environmental perspective, vegetable oils may also be attractive quenchants. They are biodegradable, offer low toxicity, and are a renewable resource.

DESIGN CONCERNS, RESIDUAL STRESSES, DISTORTION, AND CRACKING

Product design and material selection play important roles in the satisfactory and economical heat treatment of parts. Proper consideration of these factors usually leads to simpler, more economical, and more reliable products. Failure to relate design and materials to heat-treatment procedures usually produces disappointing or variable results and may lead to a variety of service failures.

From the viewpoint of heat treatment, undesirable design features include (1) nonuniform sections or thicknesses, (2) sharp interior corners, and (3) sharp exterior corners. Because these features often find their way into the design of parts, the designer should be aware of their effect on heat treatment. Undesirable results may include nonuniform structure and properties, undesirable residual stresses, cracking, warping, and dimensional changes.

Heat can only be extracted from a piece through its exposed surfaces. Therefore, if the piece to be hardened has a nonuniform cross section, any thin region will cool rapidly and may fully harden, while thick regions may harden only on the surface, if at all. The shape that might be closest to ideal from the viewpoint of quenching would be a doughnut. The uniform cross section with high exposed surface area and absence of sharp corners would be quite attractive. Because most shapes are designed to perform a function, however, compromises are usually necessary.

Residual stresses are the often-complex stresses that are present within a body, independent of any applied load. They can be induced in a number of ways, but the complex dimensional changes that can occur during heat treatment are a primary cause. Thermal expansion during heating and thermal contraction during cooling are well-understood phenomena, but when these occur in a nonuniform manner, the results can be extremely complex. In addition, the various phases and structures that can exist within a material usually possess different densities. Volume expansions or contractions accompany any phase transformation. For example, when austenite transforms to martensite, there is a volume expansion of up to 4%. Transformations to ferrite, pearlite, or other room-temperature structure also involve volume expansions but of a smaller magnitude.

If all of the temperature changes occurred uniformly throughout a part, all of the associated dimensional changes would occur simultaneously and the resultant product would be free of residual stresses. Most of the parts being heat-treated, however, experience nonuniform temperatures during the cooling or quenching operation. Consider a block of hot aluminum being cooled by water sprays from top and bottom. For simplicity, let us model the block as a three-layer sandwich, as in the top sequence of Figure 6-19. At the start of the quench, all layers are uniformly hot. As the water spray begins, the surface layers cool and contract, but the center layer does not experience the quench and remains hot. Because the part is actually one piece, however, the various layers must accommodate each other. The contracting surface layers exert compressive forces on the hot, weak interior, causing it to also contract, but by plastic deformation not thermal contraction. As time passes, the interior now cools and wants to thermally contract but finds itself sandwiched between the cold, strong surface layers. It pulls on the surface layers, placing them in compression, while the surface layers restrict its movement, creating tension in the interior. While the net force is zero (because there is no applied load), counterbalancing tension and compression stresses exist within the product.

Let us now change the material to steel and repeat the sequence. When heated, all three layers become hot, face-centered-cubic austenite, as shown in the bottom sequence of Figure 6-19. Upon quenching, the surface layers transform to martensite (the structure changes to body-centered tetragonal) and *expand!* The expanding surfaces deform the soft, weak (and still hot), austenite center, which then cools and wants to expand as it undergoes the crystal structure change to martensite or another

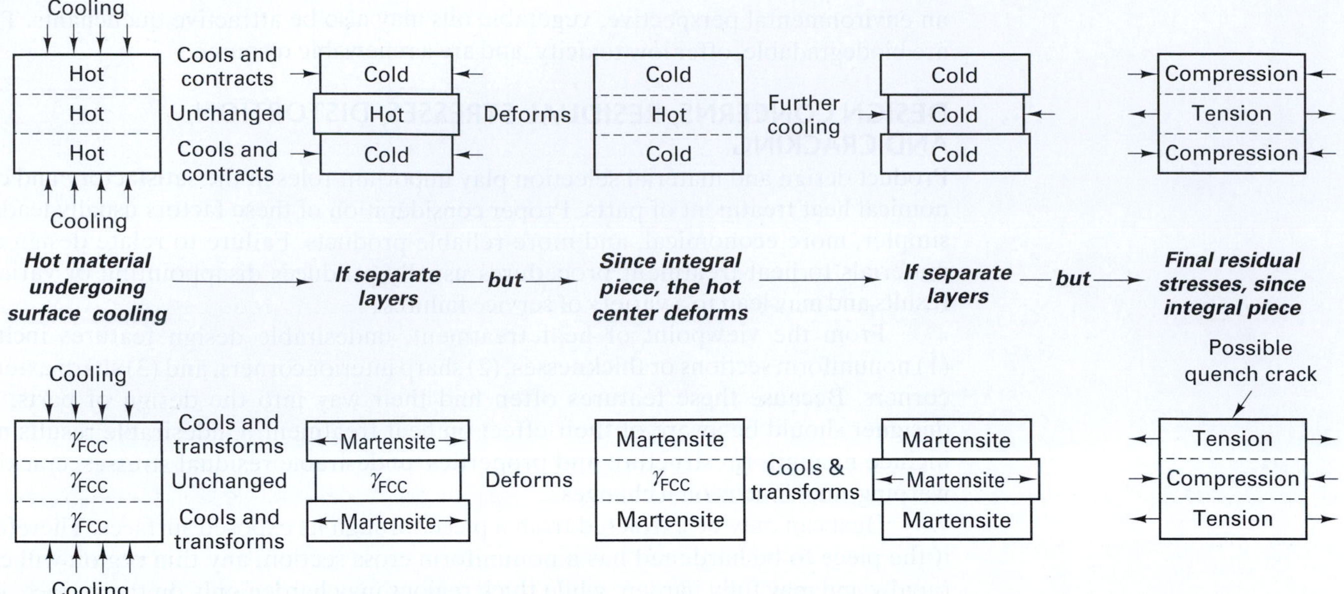

FIGURE 6-19 Three-layer model of a plate undergoing cooling: upper sequence depicts a material such as aluminum that contracts upon cooling while the bottom sequence depicts steel, which expands during the cooling-induced phase transformation.

body-centered, room-temperature product. The hard, strong surface layers hold it back, placing the center in compression, while the expanding center tries to stretch the surface layers, producing surface tension. If the tension at the surface becomes great enough, cracking can result, a phenomenon known as quench cracking. One should note that for the rapid quench conditions just described, aluminum will never quench crack because the surface is in compression, but the steel might, because the residual stresses are reversed. If the cooling conditions were not symmetrical, there might be more contraction or expansion on one side, and the block might warp. With more complex shapes and the resultant nonuniform cooling, the residual stresses induced by heat treatment can be extremely complex.

Various techniques can be employed to minimize or reduce the problems associated with residual stresses. If product cross sections can be made more uniform, temperature differences can be minimized and will not be concentrated at any specific location. If this is not possible, slower cooling may be recommended, coupled with a material that will provide the desired properties after a slower oil or air quench. Because materials with greater hardenability are more expensive, design alternatives should generally precede quench modifications.

When temperature differences and the resultant residual stresses become severe or localized, cracking or distortion problems can be expected. Figure 6-20a shows an example where a sharp interior corner has been placed at a change in cross section. Upon quenching, stresses will concentrate along line A–B, and a crack is almost certain to result. If changes in cross section are required, they should be gradual, as in the redesigned version of Figure 6-20a. Generous fillets at interior corners, radiused exterior corners, and smooth transitions all reduce problems. A material with greater hardenability and a less severe quench would also help. Figure 6-20b shows the cross section of a blanking die that consistently cracked during hardening. Rounding the sharp corners and adding additional holes to provide a more uniform cross section during quenching eliminated the problem.

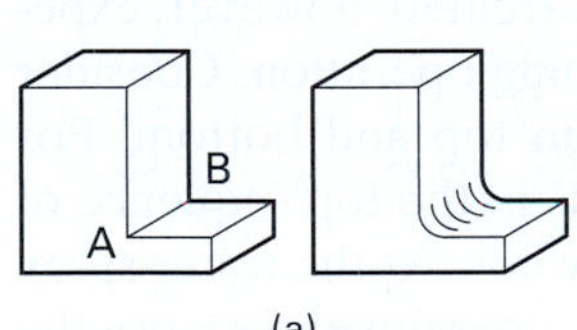

(a)

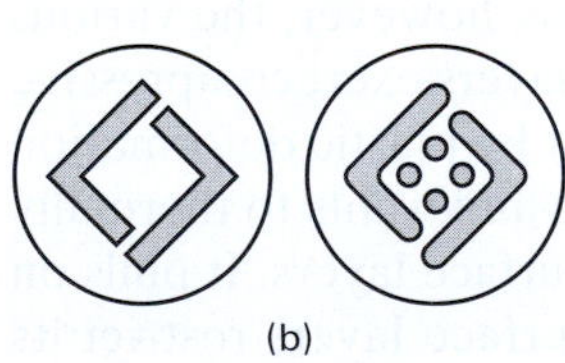

(b)

FIGURE 6-20 (a) Shape containing nonuniform sections and a sharp interior corner that may crack during quenching. This is improved by using a large radius to join the sections. (b) Original design containing sharp corner holes, which can be further modified to produce more uniform sections.

One of the ominous features of poor product and process design is the fact that the residual stresses may not produce immediate failure, but may contribute to failure at a later time. Applied stresses add to the residual stresses already present within the part. Therefore, it is possible for applied stresses that are well within the "safe" designed limit to couple with residual stresses and produce a value sufficient to induce failure.

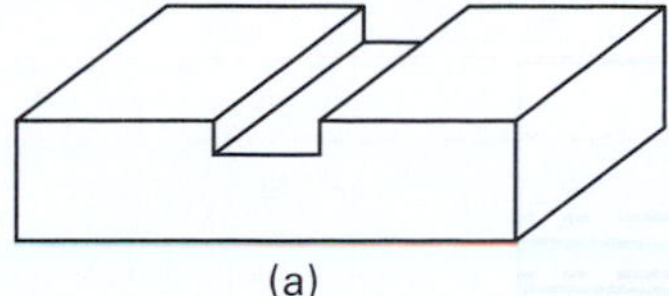

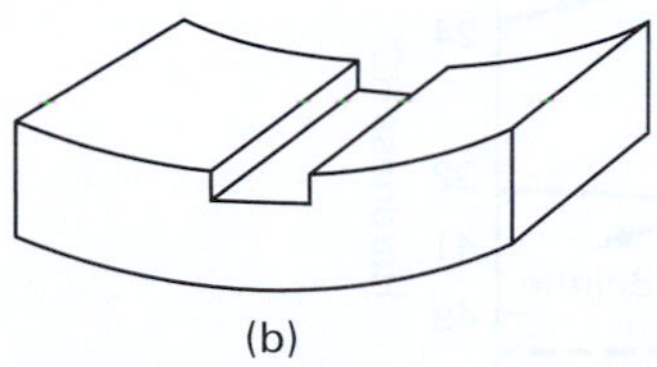

FIGURE 6-21 Warping can occur when subsequent machining upsets the equilibrium balance of residual stresses. (a) Sketch of the desired product, a flat plate with a milled slot. (b) Resulting geometry if the unmachined plate contained the residual stress pattern described in the upper portion of Figure 6-19.

Residual stresses can also accelerate corrosion reactions. As illustrated in Figure 6-21, dimensional changes or warping can occur when subsequent machining or grinding operations upset the equilibrium balance of the residual stresses. After the removal of some material, the remaining piece adjusts its shape to produce a new (sum of forces equal zero) equilibrium balance. When we consider all of the possible difficulties, it is apparent that considerable time and money can be saved if good design, material selection, and heat-treatment practices are employed.

TECHNIQUES TO REDUCE CRACKING AND DISTORTION

Press hardening, or **press quenching**, is one approach to reducing distortion during quenching. Steel parts are austenitized in a furnace and are then robotically transferred to the bottom die of a press. The top die descends and closes, applying a predetermined amount of pressure. A high volume of oil or polymer quenchant is then forced through apertures, flowing over all surfaces of the part. Many quench-hardening presses apply the pressure in pulses, allowing the part to expand or contract during the off-pressure periods. Still others use internal water cooling of the dies, rather than flushing the part with the direct flow of a quenchant.

The steel segment of Figure 6-19 has already introduced the phenomena of residual surface tension and the possibility of quench cracking. Figure 6-22a further illustrates its cause using a T-T-T diagram (a misuse of the diagram, because continuous cooling is employed, but helpful for visualization). The surface and center of a quenched product generally have different cooling rates. As a result, when the surface has cooled and is transforming to martensite with its companion expansion, the center is still hot and remains as soft, untransformed austenite. At a later time, the center cools to the martensite transformation, and it expands. The result is significant tension in the cold, hard surface and possible cracking.

Figure 6-22b depicts two variations of rapid quenching that have been developed to produce strong structures while reducing the likelihood of cracking. A rapid cool must still be employed to prevent transformation to the softer, weaker pearlitic structure, but instead of quenching through the martensite transformation, the component is rapidly quenched into a liquid medium that is now about 15°C (25°F) above the martensite start (M_s) temperature, such as hot oil or molten salt. Holding for a period of time in this bath allows the entire piece to return to a nearly uniform temperature. If the material is held at this temperature for sufficient time, the austenite will transform

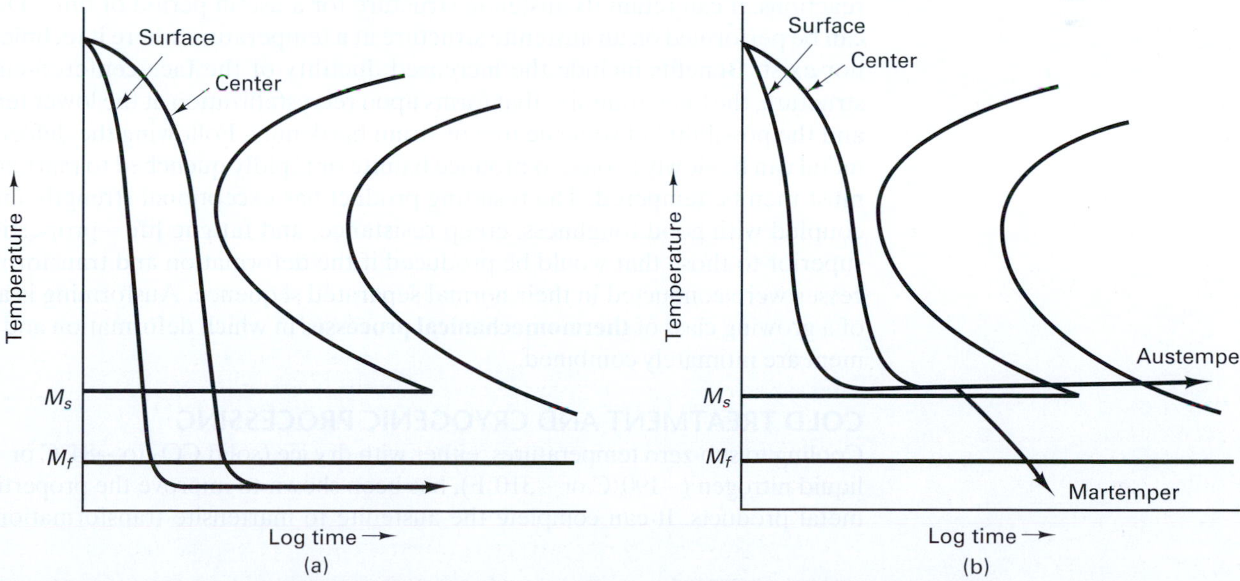

FIGURE 6-22 (a) Schematic representation of the cooling paths of surface and center during a direct quench. (b) The modified cooling paths experienced during the austempering and martempering processes.

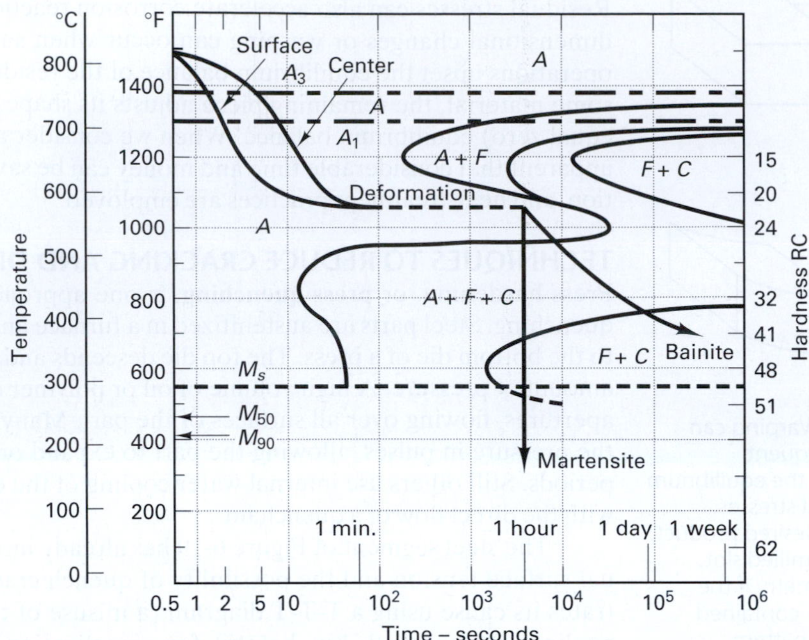

FIGURE 6-23 T-T-T diagram for 4340 steel showing the "bay," along with a schematic of the ausforming process. *(Courtesy of United States Steel Corporation, Pittsburgh, PA)*

to bainite, which usually has sufficient toughness that a subsequent temper is not required. This process is known as **austempering.**[4] If the material is brought to a uniform temperature and then slow cooled, usually in air, through the martensite transformation, the process is known as **martempering,** or *marquenching*. The resulting structure is martensite, which must be tempered the same as the martensite that forms directly upon quenching. In both of these processes, all transformations (and related volume expansions) occur at the same time, thereby eliminating the residual stresses and tendency to crack.

AUSFORMING

A process that is often confused with austempering is **ausforming.** Certain alloys tend to retard the pearlite transformation more than the bainite reaction and produce a T-T-T curve of the shape shown in Figure 6-23. If this material is heated to form austenite and then quenched to the temperature of the "bay" between the pearlite and bainite reactions, it can retain its austenite structure for a useful period of time. Deformation can be performed on an austenite structure at a temperature where it technically should not exist. Benefits include the increased ductility of the face-centered-cubic crystal structure, the finer grain size that forms upon recrystallization at the lower temperature, and the possibility of some degree of strain hardening. Following the deformation, the metal can be slowly cooled to produce bainite or rapidly quenched to martensite, which must then be tempered. The resulting product has exceptional strength and ductility, coupled with good toughness, creep resistance, and fatigue life—properties that are superior to those that would be produced if the deformation and transformation processes were conducted in their normal separated sequence. Ausforming is an example of a growing class of **thermomechanical processes** in which deformation and heat treatment are intimately combined.

COLD TREATMENT AND CRYOGENIC PROCESSING

Cooling to sub-zero temperatures, either with dry ice (solid CO_2 to −84°C or −120°F) or liquid nitrogen (−190°C or −310°F), has been shown to improve the properties of some metal products. It can complete the austenite to martensite transformation when the

[4] Attractive features of bainite include high hardness with good strength, ductility, toughness, and fatigue life. In addition to the reduced distortion and cracking, if the hardness is greater than R_C40, austempering will produce a better combination of properties than conventional quench and temper.

martensite finish temperature is low, as with high-carbon and highly alloyed steels, thereby increasing strength and hardness. Performance improvements have also been noted for a number of nonferrous metals, and the underlying reason appears to be the relaxation or removal of the unfavorable residual stresses. By slow cooling to such extreme temperatures, and then slowly heating back to room temperature, the parts undergo a significant amount of uniform thermal contraction, and this is often sufficient to induce the small amounts of deformation necessary to relax the residual stresses. Numerous parts, such as race-car engine blocks and brake rotors, exhibit reduced wear, longer lifetime, and enhanced performance after **cryogenic processing.** Thermal shock is avoided by slowly feeding the liquid nitrogen and cooling the part in the evaporated gas.

6.6 SURFACE HARDENING OF STEEL

Many products require different properties at different locations. Quite frequently, this variation takes the form of a hard, wear-resistant surface coupled with a tough, fracture-resistant core. The methods developed to produce the varied properties can be classified into three basic groups: (1) selective heating of the surface, (2) altered surface chemistry, and (3) deposition of an additional surface layer. The first two approaches will be discussed in the next sections, while platings and coatings will be described in Chapter 37.

SELECTIVE HEATING TECHNIQUES

If a steel has sufficient carbon to achieve the desired surface hardness, generally greater than 0.3%, the different properties can often be obtained by varying the thermal histories of the various regions. Core properties are set by a bulk treatment, with the surface properties being established by a subsequent surface treatment. Maximum hardness depends on the carbon content of the material, while the depth of that hardness depends on both the depth of heating and the material's hardenability. The various methods generally differ in the way the surface is brought to elevated temperature.

Flame hardening uses an oxy-acetylene flame to raise the surface temperature high enough to reform austenite. The surface is then water quenched[5] to produce martensite and tempered to the desired level of toughness. Heat input is quite rapid, leaving the interior at low temperature and free from any significant change. Considerable flexibility is provided because the rate and depth of heating can be easily varied. Depth of hardening can range from thin skins to more than 6 mm. ($\frac{1}{4}$ in.). Flame hardening is often used on large objects, because alternative methods tend to be limited by both size and shape. Equipment varies from crude handheld torches to fully automated and computerized units.

When using **induction hardening,** the steel part is placed inside a conductor coil, which is then energized with alternating current. The changing magnetic field induces surface currents in the steel, which then heat by electrical resistance. The heating rates can be extremely rapid, and energy efficiency is high. Rapid cooling is then provided by either an immersion quench or water spray using a quench ring that follows the induction coil.

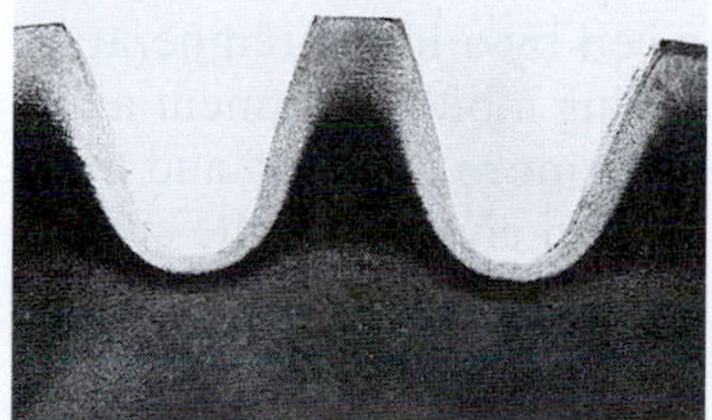

FIGURE 6-24 Section of gear teeth showing induction-hardened surfaces. *(Courtesy of Ajax Tocco Magnethermic Corp, Warren, OH)*

Induction heating is particularly well suited to **surface hardening** because the rate and depth of heating can be controlled directly through the amperage and frequency of the generator (higher frequencies produce shallower heating). Induction hardening is ideal for round bars and cylindrical parts but can also be adapted to more complex geometries. The process offers high quality, good reproducibility, and the possibility of automation. Figure 6-24 shows a partial cross section of an induction-hardened gear, where hardening has been applied to those areas expected to see high wear. Distortion

[5] There is no real danger of surface cracking during the water quench. When the surface is reaustenitized, the soft austenite adjusts to the colder, stronger, underlying material. Upon quenching, the surface austenite expands during transformation. The interior is still cold and restrains the expansion, producing a surface in compression with no tendency toward cracking. This is just the opposite of the conditions that occur during the through-hardening of a furnace-soaked workpiece! The compressive surface stresses delay crack initiation and extend the fatigue life of components.

during hardening is negligible since the dark areas remain cool and rigid throughout the entire process.

Laser-beam hardening has been used to produce hardened surfaces on a wide variety of geometries. An absorptive coating such as zinc or manganese phosphate is often applied to the steel to improve the efficiency of converting light energy into heat. The surface is then scanned with the laser, where beam size, beam intensity, and scanning speed have been selected to obtain the desired amount of heat input and depth of heating. Because of the localized heating of the beam, it is possible for the process to be **autoquenching** (cooled simply by conductive transfer into the underlying metal), but a water or oil quench can also be used. Through laser-beam hardening, a 0.4% carbon steel can attain a surface hardness as high as Rockwell C 65. Typical depth of hardening is between 0.5 and 1 mm (0.02 to 0.04 in.). The process operates at high speeds, produces little distortion, induces compressive residual stresses on the surface, and can be used to harden selected surface areas while leaving the remaining surfaces unaffected. Computer software and automation can be used to control the process parameters, and conventional mirrors and optics can be used to shape and manipulate the beam.

Electron-beam hardening is similar to laser-beam hardening. Here, the heat source is a beam of high-energy electrons rather than a beam of light, with the charged particles being focused and directed by electromagnetic controls. Like laser-beam treating, the process can be readily automated, and production equipment can perform a variety of operations with efficiencies often greater than 90%. Electrons cannot travel in air, however, so the entire operation must be performed in a hard vacuum, which is the major limitation of this process. More information on laser- and electron-beam techniques, as well as other means of heating material, is provided in the welding and joining chapters (Chapters 29 through 33).

Still other surface-heating techniques employ immersion in a pool of molten lead or molten salt (*lead pot* or *salt bath* heating). These processes are attractive for treating complex-shaped products and hardening relatively inaccessible surfaces.

TECHNIQUES INVOLVING ALTERED SURFACE CHEMISTRY

If the steels contain insufficient carbon to achieve the desired surface properties, or the difference in surface and interior properties is too great for a single chemistry material, an alternative approach is to alter the surface chemistry. The most common technique within this category, **carburizing,** involves the diffusion of carbon into the elevated temperature, face-centered-cubic, austenite structure, at temperatures between 800° and 1050°C (1450° and 1950°F). When sufficient carbon has diffused to the desired depth, the parts are then thermally processed. Direct quenching from the carburization treatment is the simplest alternative, and the different carbon contents and cooling rates can often produce the desired variation in properties. Other options include a slow cool from the high temperature carburizing treatment, followed by a lower-temperature reaustenitizing and quenching, or a duplex process involving a bulk treatment and a separate surface heat treatment. These latter processes are more involved and more costl, but produce improved product properties. The carbon content of the surface usually varies from 0.7 to 1.2%, depending on the material, the process, and the desired results. Case depth may range from a few hundredths of a millimeter to 9.5 mm (a few thousandths of an inch to more than $\frac{3}{8}$ in.).

The various carburizing processes differ in the source of the carbon. The most common is *gas carburizing*, where a hot, carbon-containing gas surrounds the parts. The process is fast, is easily controlled, and produces accurate and uniformly modified surfaces. In *pack carburizing*, the steel components are surrounded by a high-carbon solid (such as carbon powder, charcoal, or cast iron turnings) and heated in a furnace. The hot carburizing compound produces CO gas that reacts with the metal, releasing carbon, which is readily absorbed by the hot austenite. A molten bath supplies the carbon in *liquid carburizing*. At one time, liquid-carburizing baths contained cyanide, which supplied both carbon and nitrogen to the surface. Safety and environmental concerns now dictate the use of noncyanide liquid compounds that are generally used to produce thin cases on small parts.

Nitriding provides surface hardness by producing alloy nitrides in special steels that contain nitride-forming elements like aluminum, chromium, molybdenum, or vanadium. The parts are first heat-treated and tempered at 525 to 675°C (1000 to 1250°F). After cleaning and removal of any decarburized surface material, they are heated in an atmosphere containing dissociated ammonia (nitrogen and hydrogen) for 10 to 40 hours at 500 to 625°C (950 to 1150°F). Because the temperatures are below the A_1 temperature, the nitrogen is diffusing into ferrite, not austenite, and subsequent cooling will not induce a phase transformation. Extremely hard cases are formed to a depth of about 0.65 mm (0.025 in.), and distortion is low. No subsequent thermal processing is required. In fact, subsequent heating should be avoided because the thermal expansions and contractions will crack the hard nitrided case. Finish grinding should also be avoided because the nitrided layer is exceptionally thin.

Ionitriding (or **plasma nitriding**) is a plasma process that has emerged as an attractive alternative to the conventional method of nitriding. Parts to be treated are placed in an evacuated "furnace" and a direct current (DC) potential of 500 to 1000 volts is applied between the parts and the furnace walls. Low-pressure nitrogen gas is introduced into the chamber and becomes ionized. The ions are accelerated toward the negatively charged product surface, where they impact and generate sufficient heat to promote inward diffusion. This is the only heat associated with the process; the "furnace" acts only as a vacuum container and electrode. Advantages of the process include shorter cycle times, reduced consumption of gases, significantly reduced energy costs, and reduced space requirements. Product quality is improved over that of conventional nitriding, and the process is applicable to a wider range of materials. **Ion carburizing** is a parallel process in which low-pressure methane is substituted for the low-pressure nitrogen, and carbon diffuses into the surface.

Carbon and nitrogen can be added simultaneously. If ammonia is added to a standard carburizing atmosphere, and the steel is heated to a temperature where the structure is austenite, the process becomes one of **carbonitriding.** The temperature is usually lower than standard carburizing, and the treatment time is somewhat shorter. If CO_2 is added to the ammonia of nitriding, and the process is carried out at a temperature below the A_1, the process is known as **nitrocarburizing.** The resulting surface resists scuffing, and fatigue resistance is improved.

Ion plating and **ion implantation** are other processes that can modify the surface chemistry. In Chapter 34, we will expand on the surface treatment of materials and compare the processes discussed in the preceding sections to various platings, coatings, and other techniques.

6.7 FURNACES

FURNACE TYPES AND FURNACE ATMOSPHERES

To facilitate production heat treatment, many types of furnaces have been developed in a wide range of sizes, each having its characteristic advantages and disadvantages. These furnaces are generally classified by heat source (electricity or combustible fuel) and style (batch or continuous). **Batch furnaces,** where the workpiece remains stationary throughout its treatment, are preferred for large parts or small lots of a particular part or grade of steel. **Continuous furnaces** move the components through the heat-treatment operation at rates selected to be compatible with the other manufacturing operations. Continuous furnaces are used for large production runs where the same or similar parts undergo the same thermal processing. The workpieces are moved through the furnace by some type of transfer mechanism (conveyor belt, walking beam, pusher, roller, or rotary hearth) and often fall into a quench tank to complete the treatment. By incorporating various zones, complex cycles of heating, holding, and quenching/cooling can be conducted in an exact and repeatable manner with low labor cost.

Horizontal batch furnaces, often called *box furnaces* because of their overall shape, are the simplest, most basic furnace. As shown in Figure 6-25, a door is provided on one end to allow the work to be inserted and removed. When large or very long workpieces are to be heated, a *car-bottom box furnace* may be employed, like the one

FIGURE 6-25 Box-type electric heat-treating furnace. *(Courtesy of Lindberg/MPH, a Division of SPX, Riverside, MI)*

FIGURE 6-26 Car-bottom box furnace. *(Commercial Metals Company, Ohio heat treat facility)*

shown in Figure 6-26. Here, the work is loaded onto a refractory-topped flatcar, which can be rolled into and out of the furnace on railway rails.

In a *bell furnace,* the heating elements are contained within a bottomless "bell" that is lowered over the work. An airtight inner shell is often placed over the workpieces to contain a protective atmosphere during the heating and cooling operations. After the work is heated, the furnace unit can be lifted off and transferred to another batch, while the inner shell maintains the protective atmosphere during cooling. If extremely slow cooling is desired, an insulated cover can be placed over the heated shell. An interesting modification of the bell design is the *elevator furnace,* where the bell remains stationary and the workpieces are raised into it on a movable platform that then forms the bottom of the furnace. By placing a quench tank below the furnace, this design enables the workpieces to first be raised into the furnace and then lowered into the quench tank. It is extremely attractive for applications where the work must be quenched as soon as possible after being removed from the heat.

When long, slender parts are positioned horizontally and heated, there is little resistance to sagging or warping. For these types of workpieces, a *vertical pit furnace* may be preferred. These furnaces are usually cylindrical chambers sunk into the floor with a door on top that can be swung aside to allow suspended workpieces to be lowered into the furnace. They can also be used to heat large quantities of small parts by loading them into wire-mesh baskets that are then stacked within the column.

While most continuous furnaces use the straight-chamber design where pieces enter one end and exit the other, the circular, *rotary-hearth* design offers some unique features. Workpieces are placed on the rotating hearth and are heated and soaked as they rotate through the furnace. These furnaces can be placed adjacent to production equipment, such as forge presses or hammers. They occupy minimum floor space, and because loading and unloading can be conducted through the same opening, a single operator can perform both functions and receive a continuous supply of heated materials.

All of the furnaces that have been described can heat in air, but most commercial furnaces can also employ **artificial gas atmospheres.** These are often selected to prevent scaling or tarnishing, to prevent decarburization, or even to provide carbon or nitrogen for surface modification. Many of the artificial atmospheres are generated from either the combustion or decomposition of natural gas, but nitrogen-based atmospheres frequently offer reduced cost, energy savings, increased safety, and environmental attractiveness. Other common atmospheres include argon, dissociated ammonia, dry hydrogen, helium, steam and vacuum.

The heating rates of gas atmosphere furnaces can be significantly increased by incorporating the **fluidized-bed** concept. These furnaces consist of a bed of dry, inert particles, such as aluminum oxide (a ceramic), which are heated and fluidized (suspended) in a stream of upward-flowing gas. Products introduced into the bed become engulfed in the particles, which then radiate uniform heat. Temperature and atmosphere can be altered quickly, and high heat-transfer rates, high thermal efficiency, good temperature uniformity, and low fuel consumption have been observed. Because atmosphere changes can be performed in minutes, a single furnace can be used for nitriding, stress relieving, carburizing, carbonitriding, annealing, and hardening.

When a liquid heating medium is preferred, **salt bath furnaces** or **pot furnaces** are a popular choice. Electrically conductive salt can be heated by passing a current between two electrodes suspended in the bath, which also causes the bath to circulate and maintain uniform temperature. Nonconductive salts can be heated by some form of immersion heater, or the containment vessels can be externally fired. In these furnaces, the molten salt not only serves as a uniform source of heat but can also be selected to prevent decarburization. The immersion isolates the hot metal from air and prevents oxidation and scaling. Parts can be partially immersed for localized or selective heating. A *lead pot* is a similar device, where molten lead replaces salt as the heat transfer medium.

Electrical induction is another popular means of heating electrically conductive materials, such as metal. Small parts can be through-heated and hardened. Long products can be heated and quenched in a continuous manner by passing them through a stationary heating coil or by having a moving coil traverse a stationary part. Localized or selective heating can also be performed at rapid production rates. Flexibility is another attractive feature, because a standard induction unit can be adapted to a wide variety of products simply by changing the induction coil and adjusting the equipment settings. If the power settings are adjusted, the same unit that austenitizes can also be used to temper.

High-temperature **microwave processing** of materials is quickly becoming an industrial reality. Because microwave furnaces generally heat only the objects to be processed, and not the furnace walls and atmospheres, they are extremely energy efficient. Moreover, conventional heating transfers heat through the outer surface of a material to the interior, while microwave heating, like induction, places the energy directly into the volume of the material. Processing times can be reduced up to 90% with a corresponding decrease of up to 80% in energy consumption and reduced environmental impact. Originally applied to the sintering of ceramics, microwave processing has recently expanded to include powder metal binder removal and sintering, melting of metal, brazing, and surface treating.

FURNACE CONTROL

All heat treatment operations should be conducted with rigid control if the desired results are to be obtained in a consistent fashion. Most furnaces are equipped with one or more temperature sensors, which can be coupled to a controller or computer to regulate the temperature and the rate of heating or cooling. It should be remembered, however, that it is the temperature of the workpiece, and not the temperature of the furnace, that controls the result, and it is this temperature that should be monitored.

6.8 HEAT TREATMENT AND ENERGY

Because of the elevated temperatures and the time required at those temperatures, heat treatments can consume considerable amounts of energy. However, if one considers the broader picture, heat treatment may actually prove to be an energy conservation measure. The manufacture of higher-quality, more durable products can often eliminate the need for frequent replacements. Higher strengths may also permit the use of less material in the manufacture of a product, thereby saving additional energy.

Further savings can often be obtained by integrating the manufacturing operations. For example, a direct quench and temper from hot forging may be used to replace the conventional sequence of forge, air cool, reheat, quench, and temper. One should note, however, that the integrated procedure quenches from the conditions of forging, which generally have greater variability in temperature, temperature uniformity, and austenite grain size. If these variations are too great, the additional energy for the reheat and soak may be well justified.

Heat treatment, a $15 to $20 billion a year business in the United States, impacts nearly every industrial market sector. It is both capital intensive (specialized and dedicated equipment) and energy intensive. Industry goals currently include reducing energy consumption, reducing processing times, reducing emissions, increasing furnace life, improving heat transfer during heating and cooling, reducing distortion, and improving uniformity of structure and properties—both within a given part and throughout an entire production quantity.

KEY WORDS

A_1	dispersion hardening	laser-beam hardening	quenching
A_3	equilibrium phase diagram	martempering	recrystallization
A_{cm}	electrical induction	martensite	residual stresses
age hardening	electron-beam hardening	microwave processing	retained austenite
aging	flame hardening	natural aging	salt bath furnace
annealing	fluidized-bed furnaces	nitriding	solid-solution strengthening
artificial aging	full anneal	nitrocarburizing	solution treatment
artificial gas atmospheres	grain-size refinement	normalize	spheroidization
ausforming	hardenability	overaged	strain hardening
austempering	hardness	pearlite	stress-relief anneal
austenite	heat treatment	phase transformation strengthening	substitutional solution
autoquenching	homogenization	plasma nitriding	surface hardening
bainite	induction hardening	polymer quench	synthetic quenchant
batch furnaces	interstitial solution	pot furnace	time-temperature-transformation (T-T-T) diagram
brine	ion carburizing	precipitation hardening	tempered martensite
continuous-cooling-transformation (C-C-T) diagram	ion implantation	press quenching	tempering
carbonitriding	ion plating	process anneal	thermomechanical processing
carburizing	ionitriding	processing heat treatments	vapor jacket
coherency	isothermal anneal	quench-and-temper process	
continuous furnaces	isothermal-transformation (I-T) diagram	quenchant	
cryogenic processing	Jominy end-quench hardenability test	quench cracking	

REVIEW QUESTIONS

1. What is heat treatment?
2. What types of properties can be altered through heat treatment?
3. Why should people performing hot forming or welding be aware of the effects of heat treatment?
4. What is the broad goal of the processing heat treatments? Cite some of the specific objectives that may be sought.
5. Why might equilibrium phase diagrams be useful aids in designing and understanding the processing heat treatments?
6. What are the A_1, A_3, and A_{cm} lines?
7. What are some possible objectives of annealing operations?
8. Why are the hypereutectoid steels not furnace-cooled from the all-austenite region?
9. While full anneals often produce the softest and most ductile structures, what may be some of the objections or undesirable features of these treatments?
10. What is the major process difference between full annealing and normalizing?
11. While normalizing is less expensive than a full anneal, some manufacturers cite cost saving through the use of a full anneal. How is this achieved?
12. What are some of the process heat treatments that can be performed without reaustenitizing the material (heating above the A_1 temperature)?
13. What types of steel would be candidates for a process anneal? Spheroidization?
14. How might steel composition influence the selection of a processing heat treatment?
15. Other than increasing strength, for what three purposes are nonferrous metals often heat-treated?
16. What are the six major mechanisms that can be used to increase the strength of a metal?

17. What is the most effective strengthening mechanism for the nonferrous metals?
18. What is the room-temperature equilibrium or slow-cool structure of an aluminum–4% copper alloy?
19. What are the three steps in an age-hardening treatment? Describe the material structure at the end of each of the stages.
20. What is the difference between a coherent precipitate and a distinct second-phase particle? Why does coherency offer significant strengthening?
21. What is overaging?
22. Describe the various aging responses (maximum attainable strength and time for attaining that strength) that can occur over the range of possible aging temperatures.
23. What is the difference between natural and artificial aging? Which offers more flexibility? Over which does the engineer have more control?
24. Why is it more difficult to understand the nonequilibrium strengthening treatments?
25. What types of heating and cooling conditions are imposed in an I-T or T-T-T diagram? Are they realistic for the processing of commercial items?
26. What are the stable equilibrium phases for steels at temperatures below the A_1 temperature?
27. What are some nonequilibrium structures that appear in the T-T-T diagram for a eutectoid composition steel?
28. Which steel structure is produced by a diffusionless phase change?
29. What is the major factor that influences the strength and hardness of martensite?
30. Most structure changes proceed to completion over time. The martensite transformation is different. What must be done to produce more martensite in a partially transformed structure?
31. Why is retained austenite an undesirable structure in heat-treated steels?
32. What types of steels are more prone to retained austenite?
33. Why are martensitic structures usually tempered before being put into use? What properties increase during tempering? Which ones decrease?
34. In what ways is the quench-and-temper heat treatment similar to age hardening? How are the property changes different in the two processes?
35. What is a C-C-T diagram? Why is it more useful than a T-T-T diagram?
36. What is the "critical cooling rate," and how is it determined?
37. What two features combine to determine the structure and properties of a heat-treated steel?
38. What conditions are used to standardize the quench in the Jominy test?
39. How do the various locations of a Jominy test specimen correlate with cooling rate?
40. What is the assumption that allows the data from a Jominy test to be used to predict the properties of various locations on a manufactured product?
41. What is hardenability? What capabilities are provided by high-hardenability materials?
42. What are the three stages of liquid quenching?
43. What are some of the major advantages and disadvantages of a water quench?
44. Why is an oil quench less likely to produce quench cracks than water or brine?
45. What are some of the attractive qualities of a polymer or synthetic quench?
46. What are some undesirable design features that may be present in parts that are to be heat-treated?
47. What is the cause of thermally induced residual stresses?
48. Why would the residual stresses in steel be different from the residual stresses in an identically processed aluminum part?
49. What are some of the potentially undesirable effects of residual stresses?
50. What causes quench cracking to occur when steel is rapidly cooled?
51. Describe several techniques that utilize simultaneous transformation to reduce residual stresses in steel products.
52. What is thermomechanical processing?
53. What are some of the benefits of ausforming?
54. What are some possible mechanisms for the improvements that have been noted on materials that have been cryogenically processed?
55. What are some of the methods that can be used to selectively alter the surface properties of metal parts?
56. How can the depth of heating be controlled during induction hardening of surfaces?
57. What are some of the attractive features of surface hardening with a laser beam?
58. How can a laser beam be manipulated and focused? How are these operations performed with an electron beam?
59. What is carburizing?
60. Why does a carburized part have to be further heat-treated after the carbon is diffused into the surface? What are the various options?
61. In what ways might ionitriding be more attractive than conventional nitriding or carburizing?
62. For what type of products or product mixes might a batch furnace be preferred to a continuous furnace?
63. What are some possible functions of artificial atmospheres in a heat-treating furnace?
64. How are parts heated in a fluidized-bed furnace? What are some of the attractive features?
65. What are some of the potential benefits of microwave processing or heating?
66. Heat treatments consume energy. In what ways might the heat treatment of metals actually be an energy conservation measure?

■ PROBLEMS

1. This chapter presented four processing-type heat treatments whose primary objective is to soften, weaken, enhance ductility, or promote machinability. Consider each of the following processes as they are applied to steels:
   a. Full annealing
   b. Normalizing
   c. Process annealing
   d. Spheriodizing

Provide information relating to
(1) A basic description of how the process works and what its primary objectives are
(2) Typical materials on which the process is performed
(3) Type of equipment used
(4) Typical times, temperatures, and atmospheres required
(5) Recommended rates of heating and cooling
(6) Typical properties achieved

2. A number of different quenchants were discussed in the chapter, including brine, water, oil, synthetic polymer mixes, and even high-pressure gas flow. Select two of these and investigate the environmental concerns that may accompany their use.
3. It has been noted that *hot oil* is often a more effective quench than *cold oil*. Can you explain this apparent contradiction?
4. Traditional manufacturing generally separates mechanical processing (such as forging, extrusion, presswork, or machining) and thermal processing (heat treatment), and applies them as sequential operations. Ausforming was presented as an example of a thermomechanical process where the mechanical and thermal processes are performed concurrently. When this is done, the resulting structures and properties are often quite different from the traditional. Identify another thermomechanical process and discuss its use and attributes.
5. A number of heat treatments have been devised to harden the surfaces of steel and other engineering metals. Consider the following processes:
   a. Flame hardening
   b. Induction hardening
   c. Laser-beam hardening
   d. Carburizing
   e. Nitriding
   f. Ionitriding

   For each of these processes, provide information relating to:
   (1) A basic description of how the process works
   (2) Typical materials on which the process is performed
   (3) Type of equipment required
   (4) Typical times, temperatures, and atmospheres required
   (5) Typical depth of hardening and reasonable limits
   (6) Hardness achievable
   (7) Subsequent treatments or processes that might be required
   (8) Information relating to distortion and/or stresses
   (9) Ability to use the process to harden selective areas
6. Select one of the lesser-known surface modification techniques, such as ion implantation, boriding, chromizing, or other similar technique, and investigate the nine numbered areas of Problem 5.

www.wiley.com/go/global/degarmo

Chapter 6 CASE STUDY

A Carpenter's Claw Hammer

Carpenter claw hammers are actually a rather sophisticated metallurgical product, because the loadings differ for the various locations. The claw sees static bending, while the eye sustains impacts, and the striking face sees impact contact with potentially hard surfaces. As one might expect, the optimum properties

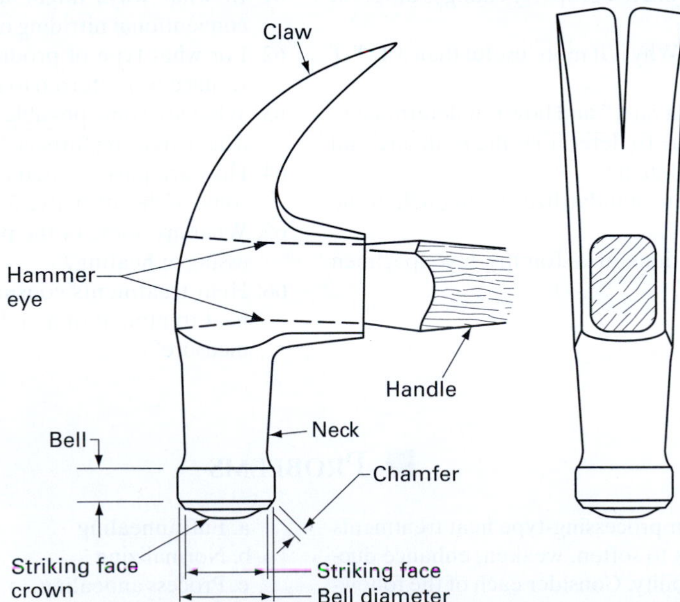

and microstructures vary with location. While hammer handles have been made from a variety of materials, including heat-treated tubular 4140 steel, our problem will focus on the head.

The following information was obtained from the American National Standards Institute (ANSI) Standard B173.1, "American National Standard Safety Requirements for Nail Hammers." This is a voluntary specification (recommendation only) developed as a "guide to aid the manufacturer, the consumer, and the general public."

According to the ANSI specification:

> "Hammerheads shall be forged in one piece from special quality hot rolled carbon steel bars."

While the specification allows for steels ranging from 1045 to 1088, two major manufacturers of high-quality tools have used 1078 steel as their material of choice, so we will go with their selection.

> "The hammer striking face shall be hardened and tempered to a Rockwell hardness of not less than C 40 or more than C 60, and the steel directly behind the striking face shall be a toughened supporting core gradually decreasing in hardness.
> Hammer claws shall be hardened to a Rockwell hardness of not less than C 40 or more than C 55 for a minimum length of $\frac{3}{4}$-inch from the tip end; the remaining length to the base of the V-slot shall be of the same through hardness, or shall contain a toughened core gradually decreasing in hardness to the core center."

While there is no specification for the eye region, many manufacturers prefer for this area to have the greatest toughness (i.e., even softer still—as low as R_C25!).

In summary, we are looking at a single piece of heat-treated steel that preferably exhibits different properties at different locations. For example, one top-quality hammer has a striking face of R_C 55 to 58, coupled with a claw of R_C 46 to 48. Another top-quality hammer has a striking face hardness of R_C 50 to 58, claw tip hardness of R_C 47 to 55, and a hardness in the crotch of the V of R_C 44 to 52. The rim of the striking face is softened to a lower hardness (R_C 41 to 48) to prevent chipping—a characteristic feature of this particular manufacturer.

Fixing our material as the previously used 1078 hot-rolled steel bar, and using forging as our shaping process:

1. What problems might be expected if the material on the striking face were too hard? Too soft? Consider each with respect to possible liability.
2. Describe some heat treatment processes or sequences that could be used to produce a quality product with the property variations described earlier.
3. Discuss the methods of heating, cooling or quenching, target temperatures, etc., that you are proposing to accomplish this task.
4. Finally, how might you duplicate the rim softening being achieved by the cited manufacturer?
5. Inexpensive hammers frequently use a single material and single heat treatment, rendering the properties similar for all locations. What are the major compromises? If these hammers were to be used by a professional carpenter, how might they be deficient?
6. While the ANSI specification calls for a forged product, some hammer heads are shaped by casting. What might be some of the pros and cons of forged versus cast?
7. The ANSI specification calls for the material to be a carbon steel.
   a. Hammer heads have also been made of alloy steel. Discuss the pros and cons of this substitution.
   b. A small, inexpensive hammer was shown to have a head of gray cast iron. What problems might accompany this choice of material? Would one of the other types of cast iron be an acceptable material for a cast hammer head? Discuss.

CHAPTER 7

FERROUS METALS AND ALLOYS

7.1 INTRODUCTION TO HISTORY-DEPENDENT MATERIALS

Engineering materials are available with a wide range of useful properties and characteristics. Some of these are inherent to the particular material, but many others can be varied by controlling the manner of production and the details of processing. Metals are classic examples of such "history-dependent" materials. The final properties are clearly affected by their past processing history. The particular details of the smelting and refining process control the resulting purity and the type and nature of any influential contaminants. The solidification process imparts structural features that may be transmitted to the final product. Preliminary operations such as the rolling of sheet or plate often impart directional variations to properties, and their impact should be considered during subsequent processing and use. Thus, while it is easy to take the attitude that "metals come from warehouses," it is important to recognize that aspects of prior processing can significantly influence further operations as well as the final properties of a product. The breadth of this book does not permit full coverage of the processes and methods involved in the production of engineering metals, but certain aspects will be presented because of their role in affecting subsequent performance.

7.2 FERROUS METALS

In this chapter we will introduce the major **ferrous** (iron-based) **metals** and **alloys,** summarized in Figure 7-1. These materials made possible the Industrial Revolution nearly 150 years ago, and they continue to be the backbone of modern civilization. We see them everywhere in our lives—in the buildings where we work, the cars we drive, the homes in which we live, the cans we open, and the appliances that enhance our standard of living. Numerous varieties have been developed over the years to meet the specific needs of various industries. These developments and improvements have continued, with recent decades seeing the introduction of a number of new varieties and even classes of ferrous metals. According to the American Iron and Steel Institute, more than 50% of the steels made today did not exist as little as 10 years ago. The newer steels are stronger than ever, rolled thinner, are easier to shape, and are more corrosion resistant. As a result, steel still accounts for roughly 60% of the metal used in an average vehicle in North America.

In addition, all steel is recyclable, and this recycling does not involve any loss in material quality. In fact, more steel is recycled each year than all other materials

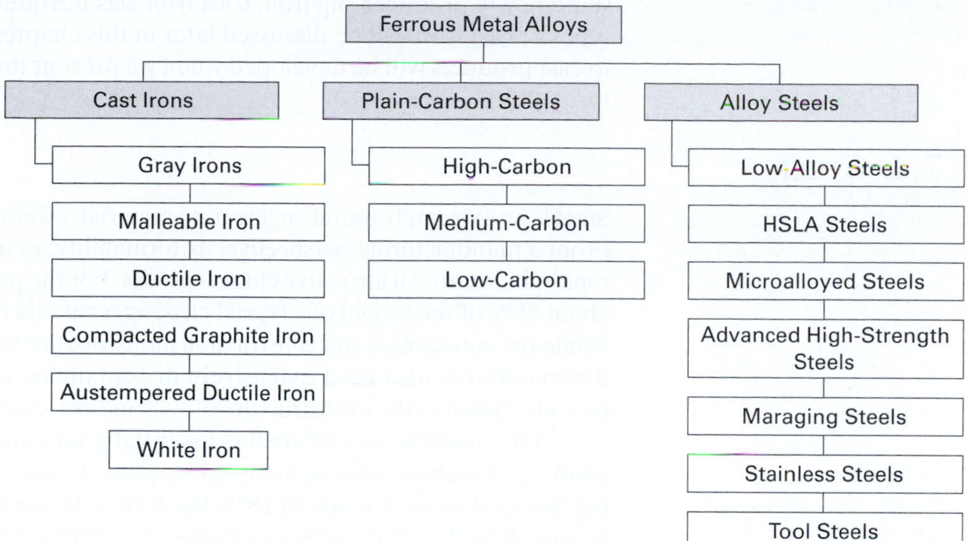

FIGURE 7-1 Classification of common ferrous metals and alloys.

combined, including aluminum, glass, and paper. Because steel is magnetic, it is easily separated and recovered from demolished buildings, junked automobiles, and discarded appliances. In 2006, nearly 70 million tons of steel were recycled in the United States, for an overall recycling rate of nearly 69%. Structural beams and plates are the most recycled products at 97.5%. The recycling rate for steel cans was 60.2%, 90.0% for large appliances, and 103.8% for automobiles. That's correct—more steel was recovered from scrap cars than was used in the production of new vehicles! Moreover, each ton of recycled steel saves more than 4000 pounds of raw materials (including 1400 pounds of coal) and 74% of the energy required to make a ton of new steel.

7.3 IRON

For centuries, **iron** has been the most important of the engineering metals. While iron is the fourth most plentiful element in the earth's crust, it is rarely found in the metallic state. Instead, it occurs in a variety of mineral compounds, known as *ores*, the most attractive of which are iron oxides coupled with companion impurities. To produce metallic iron, the ores are processed in a manner that breaks the iron–oxygen bonds (chemical reducing reactions). Ore, limestone, coke (carbon), and air are continuously introduced into specifically designed furnaces and molten metal is periodically withdrawn.

Within the furnace, other oxides (that were impurities in the original ore) will also be reduced. All of the phosphorus and most of the manganese will enter the molten iron. Oxides of silicon and sulfur compounds are partially reduced, and these elements also become part of the resulting metal. Other contaminant elements, such as calcium, magnesium, and aluminum, are collected in the limestone-based slag and are largely removed from the system. The resulting **pig iron** tends to have roughly the following composition:

Carbon	3.0–4.5%
Manganese	0.15–2.5%
Phosphorus	0.1–2.0%
Silicon	1.0–3.0%
Sulfur	0.05–0.1%

While the bulk of molten pig iron is further processed into steel, a small portion is cast directly into final shape and is classified as cast iron. Most commercial cast irons, however, are produced by recycling scrap iron and steel, with the possible addition of some newly produced pig iron. Cast iron was introduced in Chapter 5, and the various types of cast iron will be discussed later in this chapter. The conversion of this material to cast products will be developed when we present the casting processes in Chapters 11 through 13.

7.4 STEEL

Steel is an extremely useful engineering material, offering strength, rigidity, and durability. From a manufacturing perspective, its formability, joinability, and paintability, as well as repairability, are all attractive characteristics. For the past 20 years, steel has accounted for about 55% of the weight of a typical passenger car and is expected to continue at this level. While the automotive and construction industries are indeed the major consumers of steel, the material is also used extensively in containers, appliances, and machinery, and it provides much of the infrastructure of such industries as oil and gas.

The manufacture of steel is essentially an oxidation process that decreases the amount of carbon, silicon, manganese, phosphorus, and sulfur in a molten mixture of pig iron and/or steel scrap. In 1856, the Kelly–Bessemer process opened up the industry by enabling the manufacture of commercial quantities of steel. The open-hearth process surpassed the Bessemer process in tonnage produced in 1908, and was producing more than 90% of all steel in 1960. Most of our commercial steels are currently produced by a variety of oxygen and electric arc furnaces.

In many of our steelmaking processes, air or oxygen passes over or through the molten metal to drive a variety of exothermic refining reactions. Carbon oxidizes to form gaseous CO or CO_2, which then exits the melt. Other elements, such as silicon and phosphorus, are similarly oxidized and, being lighter than the metal, rise to be collected in a removable slag. At the same time, however, oxygen and other elements from the reaction gases dissolve in the molten metal and may later become a cause for concern.

SOLIDIFICATION CONCERNS

Regardless of the method by which the steel is made, it must undergo a change from liquid to solid before it can become a usable product. The liquid can be converted directly into finish-shape steel castings or solidified into a form suitable for further processing. In most cases, some form of continuous casting produces the feedstock material for subsequent forging or rolling operations.

Prior to solidification, we want to remove as much contamination as possible. The molten metal is first poured from the steelmaking furnaces into containment vessels, known as **ladles.** Historically, the ladles simply served as transfer and pouring containers, but they have recently emerged as the site for additional processing. **Ladle metallurgy** refers to a variety of processes designed to provide final purification and to fine-tune both the chemistry and temperature of the melt. Alloy additions can be made, carbon can be further reduced, dissolved gases can be reduced or removed, and steps can be taken to control subsequent grain size, limit inclusion content, reduce sulfur, and control the shape of any included sulfides. Stirring, degassing, reheating, and the injection of powdered alloys or cored wire can all be performed to increase the cleanliness of the steel and provide for tighter control of the chemistry and properties.

The processed liquid is then poured from these ladles into molds or some form of **continuous caster,** usually through a bottom-pouring process such as the one shown schematically in Figure 7-2. By extracting the metal from the bottom of the ladle, slag and floating matter are not transferred, and a cleaner product results. Figure 7-3a illustrates a typical continuous caster, in which molten metal flows from a ladle, through a tundish, into a bottomless, water-cooled mold, usually made of copper. Cooling is controlled to ensure that the outside has solidified before the metal exits the mold. Direct water sprays further cool the emerging metal to complete the solidification. The newly solidified metal can then be cut to desired length, or, because the cast solid is still hot, it can be bent and fed horizontally through a short reheat furnace or directly to a rolling operation. By varying the size and shape of the mold, products can be cast with a variety of cross sections with names such as slab, bloom, billet and strand. Figure 7-3b depicts

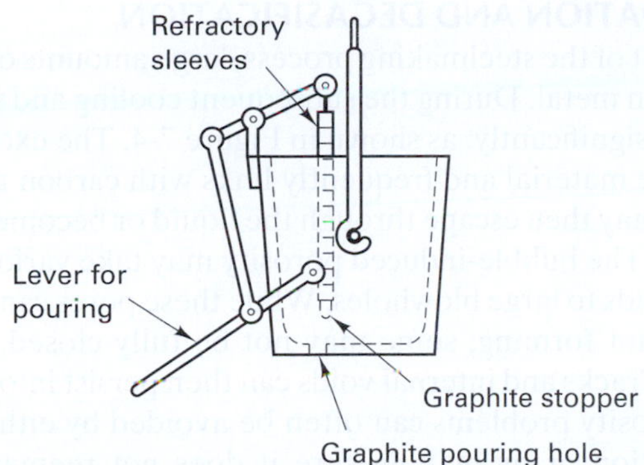

FIGURE 7-2 Diagram of a bottom-pouring ladle.

the simultaneous casting of multiple strands. Compared to the casting of discrete ingots, continuous casting offers significant reduction in cost, energy, and scrap. In addition, the products have improved surfaces, more uniform chemical composition, and fewer oxide inclusions.

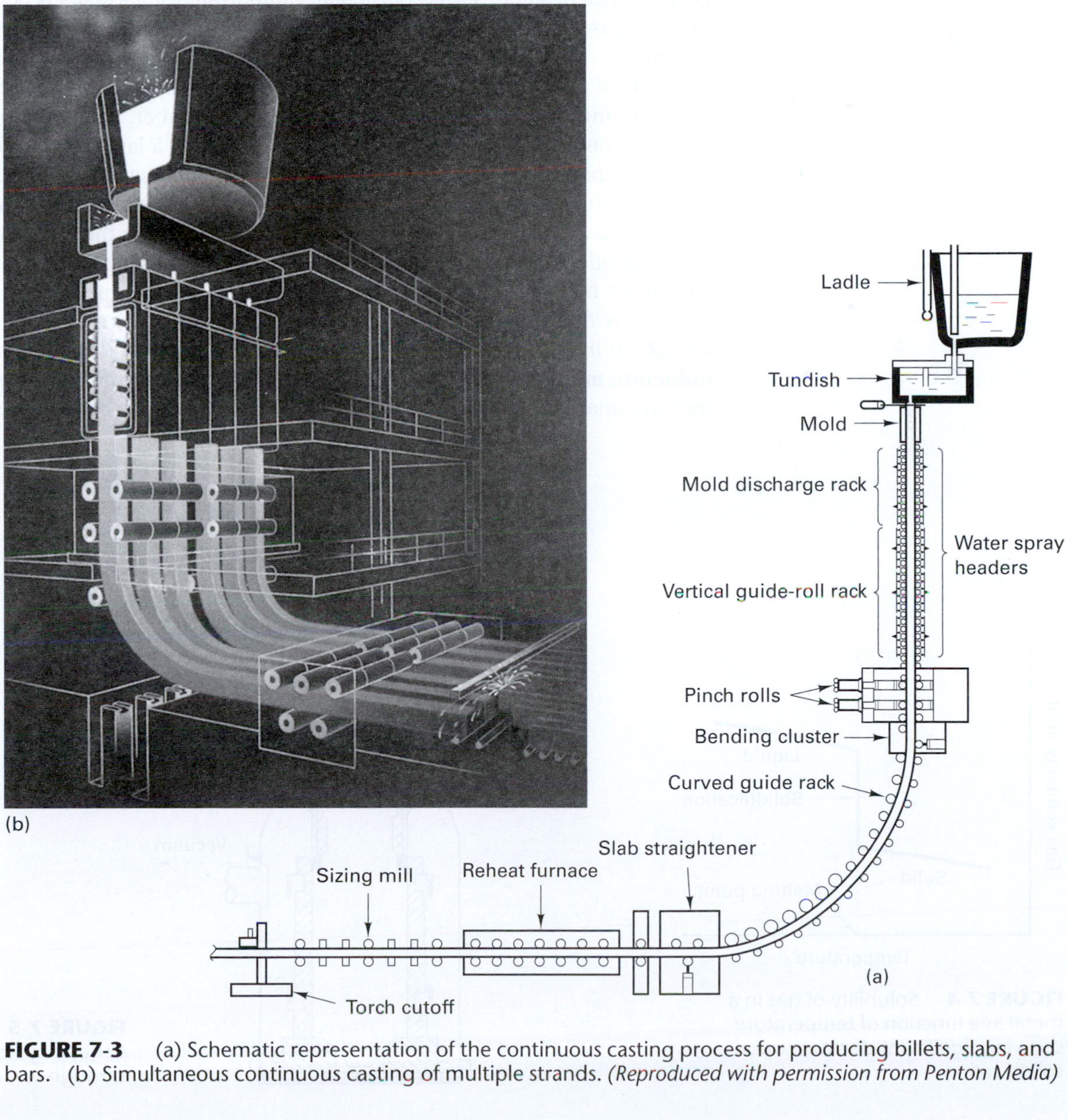

FIGURE 7-3 (a) Schematic representation of the continuous casting process for producing billets, slabs, and bars. (b) Simultaneous continuous casting of multiple strands. *(Reproduced with permission from Penton Media)*

DEOXIDATION AND DEGASIFICATION

As a result of the steelmaking process, large amounts of oxygen can become dissolved in the molten metal. During the subsequent cooling and solidification, the solubility levels decrease significantly, as shown in Figure 7-4. The excess oxygen can no longer be held within the material and frequently links with carbon to produce carbon monoxide gas. This gas may then escape through the liquid or become trapped to produce pores within the solid. The bubble-induced porosity may take various forms ranging from small, dispersed voids to large blowholes. While these pores can often be welded shut during subsequent hot forming, some may not be fully closed, and others may not weld upon closure. Cracks and internal voids can then persist into a finished product.

Porosity problems can often be avoided by either removing the oxygen prior to solidification or by making sure it does not reemerge as a gas. Aluminum, ferromanganese, or ferrosilicon can be added to molten steel to provide a material whose affinity for oxygen is higher than that of carbon. The rejected oxygen then reacts with these **deoxidization** additions to produce solid metal oxides that are either removed from the molten metal when they float to the top or become dispersed throughout the structure.

While deoxidization additions can effectively tie up dissolved oxygen, small amounts of other gases, such as hydrogen and nitrogen, can also have deleterious effects on the performance of steels. This is particularly important for alloy steels because the solubility of these gases tends to be increased by alloy additions, such as vanadium, niobium, and chromium. Alternative **degasification** processes have been devised that reduce the amounts of all dissolved gases. Figure 7-5 illustrates one form of **vacuum degassing,** in which an ingot mold is placed in an evacuated chamber, and a stream of molten metal passes through a vacuum during pouring. By creating a large amount of exposed surface during the pouring operation, the vacuum is able to extract most of the dissolved gas.

An alternative to vacuum degassing is the **consumable-electrode remelting** process, where an already solidified metal electrode replaces the ladle of molten metal. As the electrode is progressively remelted, molten droplets pass through a vacuum, and the extremely high surface area again provides an effective means of gas removal. If the melting is done by an electric arc, the process is known as **vacuum arc remelting (VAR).** If induction heating is used to melt the electrode, the process becomes **vacuum induction melting (VIM).** Both are highly effective in removing dissolved gases, but they are unable to remove any nonmetallic impurities that may be present in the metal.

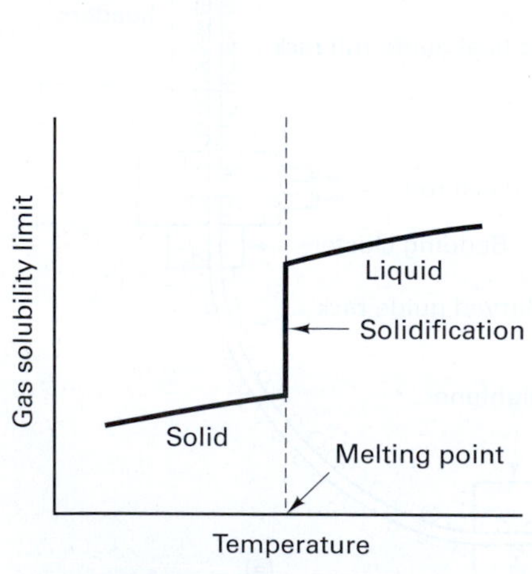

FIGURE 7-4 Solubility of gas in a metal as a function of temperature showing significant decrease upon solidification.

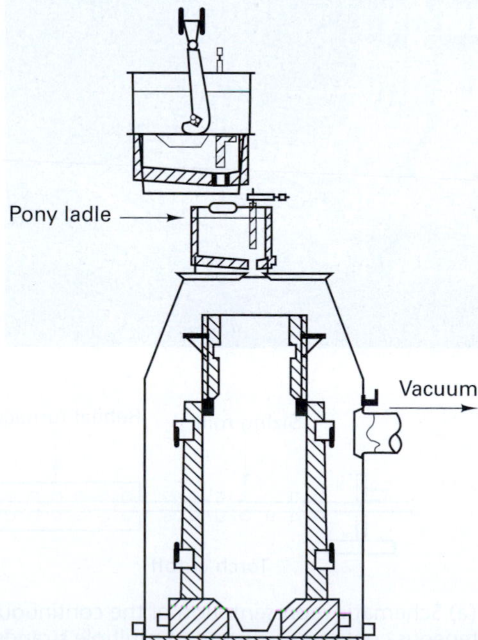

FIGURE 7-5 Method of degassing steel by pouring through a vacuum.

(a)

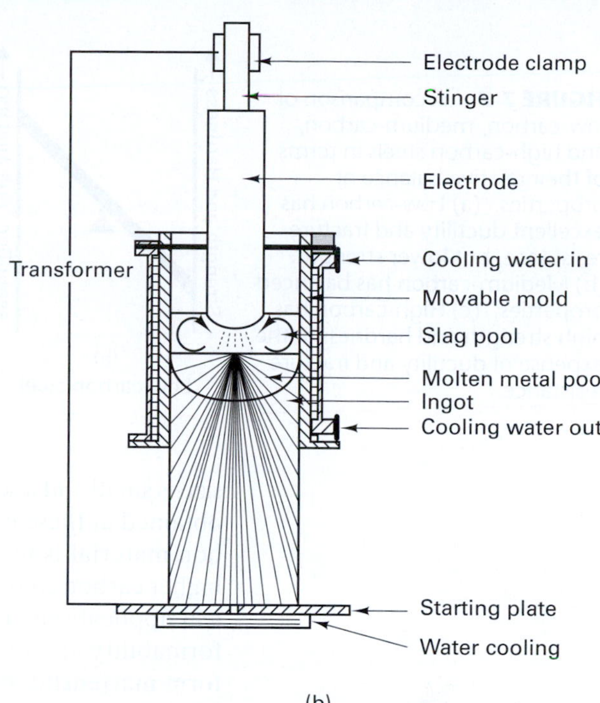

(b)

FIGURE 7-6 (a) Production of an ingot by the electroslag remelting process. (b) Schematic representation of this process showing the starting electrode, melting arc, and resolidified ingot. *(Courtesy Carpenter Technology Corporation, Reading, PA)*

The **electroslag remelting (ESR)** process, shown in Figure 7-6, can be used to produce extremely clean, gas-free metal. A solid electrode is again melted and recast using an electric current, but the entire remelting is conducted under a blanket of molten flux. Nonmetallic impurities float and are collected in the flux, leaving a newly solidified metal structure with much-improved quality. No vacuum is required, because the molten material is confined beneath the flux and the progressive freezing permits easy escape for the rejected gas. This process is simply a large-scale version of the electroslag welding process that will be discussed in Chapter 32.

PLAIN-CARBON STEEL

While theoretically an alloy of only iron and carbon, commercial steel actually contains manganese, phosphorus, sulfur, and silicon in significant and detectable amounts. When these four additional elements are present in their normal percentages and no minimum amount is specified for any other constituent, the product is referred to as **plain-carbon steel.** Strength is primarily a function of carbon content, increasing with increasing carbon, as shown in Table 7-1. Unfortunately, the ductility, toughness, and weldability of plain-carbon steels decrease as the carbon content is increased, and hardenability is quite low. In addition, the properties of ordinary carbon steels are impaired by both high and low temperatures (loss of strength and embrittlement, respectively), and they are subject to corrosion in most environments.

TABLE 7-1 Effect of Carbon on the Strength of Annealed Plain-Carbon Steels[a]

		Minimum Tensile Strength	
Type of Steel	Carbon Content	Mpa	ksi
1020	0.20%	414	60
1030	0.30%	448	65
1040	0.40%	517	75
1050	0.50%	621	90

[a] Data are from ASTM Specification A732.

Plain-carbon steels are generally classed into three subgroups based on their carbon content. **Low-carbon steels** have less than 0.20% carbon and possess good formability (can be strengthened by cold work) and weldability. Their structures are usually ferrite and pearlite, and the material is generally used as it comes from the hot-forming or cold-forming processes, or in the as-welded condition. **Medium-carbon steels** have between 0.20 and 0.50% carbon, and they can be quenched to form martensite or bainite if the section

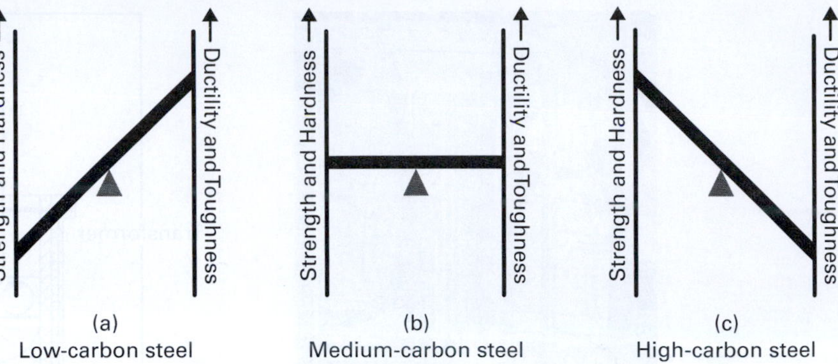

FIGURE 7-7 A comparison of low-carbon, medium-carbon, and high-carbon steels in terms of their relative balance of properties. (a) Low-carbon has excellent ductility and fracture resistance, but lower strength. (b) Medium-carbon has balanced properties. (c) High-carbon has high strength and hardness at the expense of ductility and fracture resistance.

size is small and a severe water or brine quench is used. The best balance of properties is obtained at these carbon levels, where the high toughness and ductility of the low-carbon material is in good compromise with the strength and hardness that come with higher carbon contents. These steels are extremely popular and find numerous mechanical applications. **High-carbon steels** have more than 0.50% carbon. Toughness and formability are low, but hardness and wear resistance are high. Severe quenches can form martensite, but hardenability is still poor. Quench cracking is often a problem when the material is pushed to its limit. Figure 7-7 depicts the characteristic properties of low-, medium- and high-carbon steels using a balance of properties that shows the offsetting characteristics of "strength and hardness" and "ductility and toughness."

Compared to other engineering materials, the plain-carbon steels offer high strength and high stiffness, coupled with reasonable toughness. Unfortunately, they also rust easily and generally require some form of surface protection, such as paint, galvanizing, or other coating. Because the plain-carbon steels are generally the lowest-cost steel material, they are often given first consideration for many applications. Their limitations, however, may become restrictive. When improved performance is required, these steels can often be upgraded by the addition of one or more alloying elements.

ALLOY STEELS

The differentiation between plain-carbon and alloy steel is often somewhat arbitrary. Both contain carbon, manganese, and usually silicon. Copper and boron are possible additions to both classes. Steels containing more than 1.65% manganese, 0.60% silicon, or 0.60% copper are usually designated as **alloy steels.** Also, a steel is considered to be an alloy steel if a definite or minimum amount of other alloying element is specified. The most common alloy elements are chromium, nickel, molybdenum, vanadium, tungsten, cobalt, boron, and copper, as well as manganese, silicon, phosphorus, and sulfur in amounts greater than are normally present. If the steel contains less than 8% of total alloy addition, it is considered to be a **low-alloy steel.** Steels with more than 8% alloying elements are **high-alloy steels.**

In general, alloying elements are added to steels in small percentages (usually less than 5%) to improve strength or hardenability, or in much larger amounts (often up to 20%) to produce special properties such as corrosion resistance or stability at high or low temperatures. Additions of manganese, silicon, or aluminum may be made during the steelmaking process to remove dissolved oxygen from the melt. Manganese, silicon, nickel, and copper add strength by forming solid solutions in ferrite. Chromium, vanadium, molybdenum, tungsten, and other elements increase strength by forming dispersed second-phase carbides. Nickel and copper can be added in small amounts to improve corrosion resistance. Nickel has been shown to impart increased toughness and impact resistance, and molybdenum helps resist embrittlement. Zirconium, cerium, and calcium can also promote increased toughness by controlling the shape of inclusions. Machinability can be enhanced through the formation of manganese sulfides, or by additions of lead, bismuth, selenium, or tellurium. Still other additions can be used to provide ferrite or austenite grain-size control.

TABLE 7-2 Effect of Carbon on the Strength of Quenched-and-Tempered Alloy Steels[a]

		Minimum Tensile Strength	
Type of Steel	Carbon Content	Mpa	ksi
4130	0.30%	1030	150
4330	0.30%	1030	150
8630	0.30%	1030	150
4140	0.40%	1241	180
4340	0.40%	1241	180

[a] Data from ASTM Specification A732.

Compared to carbon steels, the alloy steels can be considerably stronger. For a specified level of strength, the ductility and toughness tend to be higher. For a specified level of ductility or toughness, the strength is higher. In essence, there is an improvement in the overall combination of properties. Mechanical properties at both low and high temperatures can be increased, but all of these improvements come with an increase in cost. Weldability, machinability and formability generally decline.

Selection of an alloy steel still begins with identifying the proper carbon content. Table 7-2 shows the effect of carbon on the strength of quenched-and-tempered alloy steels. The strength values are significantly higher than those of Table 7-1, reflecting the difference between the annealed and quenched-and-tempered microstructures. The 4130 steel has about 1.2% total alloying elements, 4330 has 3.0%, and 8630 has about 1.3%, yet all have the same quenched-and-tempered tensile strength. Strength and hardness depend primarily on carbon content. The primary role of an alloy addition, therefore, is usually to increase **hardenability,** but other effects such as modified toughness or machinability are also possible. The most common hardenability-enhancing elements (in order of decreasing effectiveness) are manganese, molybdenum, chromium, silicon, and nickel. Boron is an extremely powerful hardenability agent. Only a few thousandths of a percent are sufficient to produce a significant effect in low-carbon steels, but the results diminish rapidly with increasing carbon content. Because no carbide formation or ferrite strengthening accompanies the addition, improved machinability and cold-forming characteristics may favor the use of boron in place of other hardenability additions. Small amounts of vanadium can also be quite effective, but the response drops off as the quantity is increased.

Table 7-3 summarizes the primary effects of the common alloying elements in steel. A working knowledge of this information may be useful in selecting an alloy steel to meet a given set of requirements. Alloying elements are often used in combination,

TABLE 7-3 Principal Effects of Alloying Elements in Steel

Element	Percentage	Primary Function
Aluminum	0.95–1.30	Alloying element in nitriding steels
Bismuth	—	Improves machinability
Boron	0.001–0.003	Powerful hardenability agent
Chromium	0.5–2	Increase of hardenability
	4–18	Corrosion resistance
Copper	0.1–0.4	Corrosion resistance
Lead	—	Improved machinability
Manganese	0.25–0.40	Combines with sulfur to prevent brittleness
	>1	Increases hardenability by lowering transformation points and causing transformations to be sluggish
Molybdenum	0.2–5	Stable carbides; inhibits grain growth
Nickel	2–5	Toughener
	12–20	Corrosion resistance
Silicon	0.2–0.7	Increases strength
	2	Spring steels
	Higher percentages	Improves magnetic properties
Sulfur	0.08–0.15	Free-machining properties
Titanium	—	Fixes carbon in inert particles
		Reduces martensitic hardness in chromium steels
Tungsten	—	Hardness at high temperatures
Vanadium	0.15	Stable carbides; increases strength while retaining ductility, Promotes fine grain structure

however, resulting in the immense variety of alloy steels that are commercially available. To provide some degree of simplification, a classification system has been developed and has achieved general acceptance in a variety of industries.

AISI–SAE CLASSIFICATION SYSTEM

The most common classification scheme for alloy steels is the **AISI–SAE identification system.** This system, which classifies alloys by chemistry, was started by the Society of Automotive Engineers (SAE) to provide some standardization for the steels used in the automotive industry. It was later adopted and expanded by the American Iron and Steel Institute (AISI) and has been incorporated into the Universal Numbering System that was developed to include all engineering metals. Both plain-carbon and low-alloy steels are identified by a four-digit number, where the first number indicates the major alloying elements and the second number designates a subgrouping within the major alloy system. These first two digits can be interpreted by looking them up on a list, such as the one presented in Table 7-4. The last two digits of the number indicate the approximate amount of carbon, expressed as "points," where one point is equal to 0.01%. Thus, a 1080 steel would be a plain-carbon steel with 0.80% carbon. Similarly, a 4340 steel would be a Mo–Cr–Ni alloy with 0.40% carbon. Because of the meanings associated with the numbers, the designation is not read as a series of single digits, such as

TABLE 7-4 AISI–SAE Standard Steel Designations and Associated Chemistries

AISI Number	Type	Alloying Elements (%) Mn	Ni	Cr	Mo	V	Other
1xxx	Carbon steels						
10xx	Plain carbon						
11xx	Free cutting (S)						
12xx	Free cutting (S) and (P)						
13xx	High manganese	1.60–1.90					
15xx	High manganese						
2xxx	Nickel steels		3.5–5.0				
3xxx	Nockel–chromium		1.0–3.5	0.5–1.75			
4xxx	Molybdenum						
40xx	Mo				0.15–0.30		
41xx	Mo, Cr			0.40–1.10	0.08–0.35		
43xx	Mo, Cr, Ni		1.65–2.00	0.40–0.90	0.20–0.30		
44xx	Mo				0.35–0.60		
46xx	Mo, Ni (low)		0.70–2.00		0.15–0.30		
47xx	Mo, Cr, Ni		0.90–1.20	0.35–0.55	0.15–0.40		
48xx	Mo, Ni (high)		3.25–3.75		0.20–0.30		
5xxx	Chromium						
50xx				0.20–0.60			
51xx				0.70–1.15			
6xxx	Chromum–vanadium						
61xx				0.50–1.10		0.10–0.15	
8xxx	Ni, Cr,Mo						
81xx			0.20–0.40	0.30–0.55	0.08–0.15		
86xx			0.40–0.70	0.40–0.60	0.15–0.25		
87xx			0.40–0.70	0.40–0.60	0.20–0.30		
88xx			0.40–0.70	0.40–0.60	0.30–0.40		
9xxx	Other						
92xx	High silicon						1.20–2.20Si
93xx	Ni, Cr,Mo		3.00–3.50	1.00–1.40	0.08–0.15		
94xx	Ni, Cr,Mo		0.30–0.60	0.30–0.50	0.08–0.15		

four-three-four-zero, but as a pair of double-digit groupings, such as ten-eighty or forty-three forty.

Letters may also be incorporated into the designation. The letter *B* between the second and third digits indicates that the base metal has been supplemented by the addition of boron. Similarly, an *L* in this position indicates a lead addition for enhanced machinability. A letter prefix may also be employed to designate the process used to produce the steel, such as *E* for electric furnace.

When hardenability is a major requirement, one might consider the H grades of AISI steels, designated by an H suffix attached to the standard designation. The chemistry specifications are somewhat less stringent, but the steel must now meet a hardenability standard. The hardness values obtained for each location on a Jominy test specimen (see Chapter 6) must fall within a predetermined band for that particular type of steel. When the AISI designation is followed by an RH suffix (restricted hardenability), an even narrower range of hardness values is imposed.

Other designation organizations, such as the American Society for Testing and Materials (ASTM) and the U.S. government [military (MIL) and federal], have specification systems based more on specific applications. Acceptance into a given classification is generally determined by physical or mechanical properties rather than the chemistry of the metal. ASTM designations are often used when specifying structural steels.

SELECTING ALLOY STEELS

From the previous discussion it is apparent that two or more alloying elements can often produce similar effects. Thus, when properly heat-treated, steels with substantially different chemical compositions can possess almost identical mechanical properties. Figure 7-8 clearly demonstrates this fact, which becomes particularly important when one realizes that some alloying elements can be very costly and others may be in short supply due to emergencies or political constraints. Overspecification has often been employed to guarantee success despite sloppy manufacturing and heat-treatment

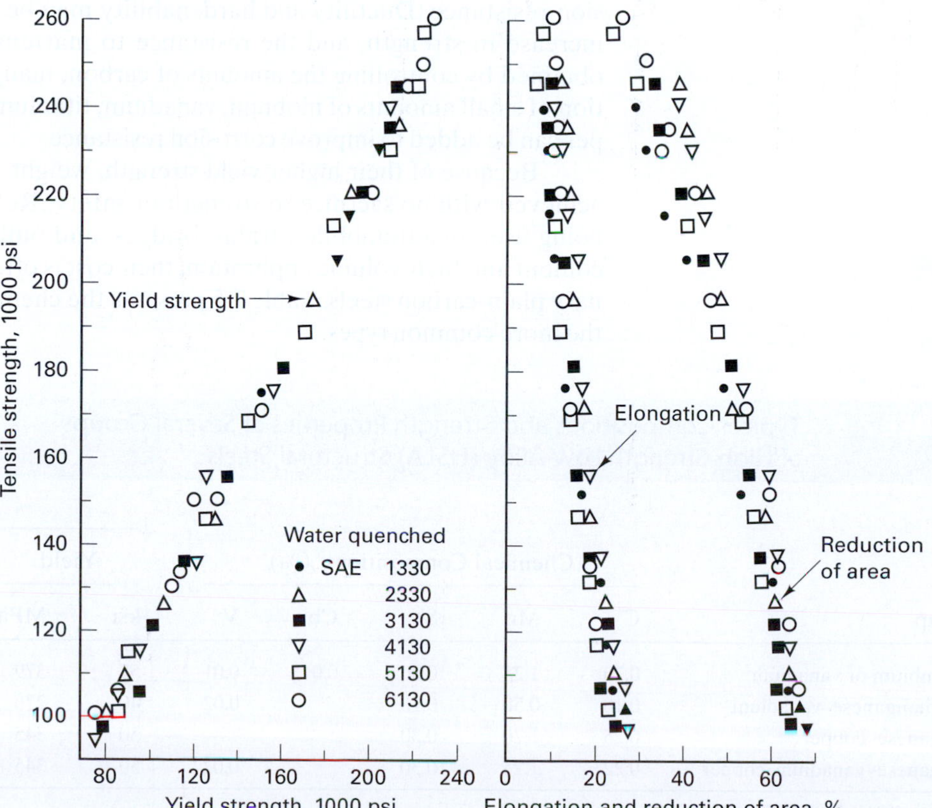

FIGURE 7-8 Relationships between the mechanical properties of a variety of properly heat-treated AISI–SAE alloy steels. *(Courtesy of ASM International, Materials Park, OH)*

practice. The correct steel, however, is usually the least expensive one that can be consistently processed to achieve the desired properties. This usually involves taking advantage of the effects provided by all of the alloy elements.

When selecting alloy steels, it is also important to consider both use and fabrication. For one product, it might be permissible to increase the carbon content to obtain greater strength. For another application, such as one involving assembly by welding, it might be best to keep the carbon content low and use a balanced amount of alloy elements, obtaining the desired strength while minimizing the risk of weld cracking. Steel selection involves defining the required properties, determining the best microstructure to provide those properties (strength can be achieved through alloying, cold work, and heat treatment, as well as combinations thereof), determining the method of part or product manufacture (casting, machining, metal forming, etc.), and then selecting the steel with the best carbon content and hardenability characteristics to facilitate those processes and achieve the desired goals.

HIGH-STRENGTH LOW-ALLOY STRUCTURAL STEELS

Among the general categories of alloy steels are (1) the **constructional alloys,** where the desired properties are typically developed by a separate thermal treatment and the specific alloy elements tend to be selected for their effect on hardenability, and (2) the **high-strength low-alloy (HSLA)** or microalloyed types, which rely largely on chemical composition to develop the desired properties in the as-rolled or normalized condition. The constructional alloys are usually purchased by AISI–SAE identification, which effectively specifies chemistry. The HSLA designations generally focus on product (size and shape) and desired properties. When steels are specified by mechanical properties, the supplier or producer is free to adjust the chemistry (within limits), and substantial cost savings may result. To ensure success, however, it is important that all of the necessary properties be specified.

The HSLA materials provide increased strength-to-weight compared to conventional carbon steels for only a modest increase in cost. They are available in a variety of forms, including sheet, strip, plate, structural shapes, and bars. The dominant property requirements generally are high yield strength, good weldability, and acceptable corrosion resistance. Ductility and hardenability may be somewhat limited, however. The increase in strength, and the resistance to martensite formation in a weld zone, is obtained by controlling the amounts of carbon, manganese, and silicon, with the addition of small amounts of niobium, vanadium, titanium or other alloys. About 0.2% copper can be added to improve corrosion resistance.

Because of their higher yield strength, weight savings of 20 to 30% can often be achieved with no sacrifice to strength or safety. Rolled and welded HSLA steels are being used in automobiles, trains, bridges, and buildings. Because of their low-alloy content and high-volume application, their cost is often little more than that of the ordinary plain-carbon steels. Table 7-5 presents the chemistries and properties of several of the more common types.

TABLE 7-5 Typical Compositions and Strength Properties of Several Groups of High-Strength Low-Alloy (HSLA) Structural Steels

	Chemical Compositions[a] (%)					Strength Properties				
						Yield		Tensile		
Group	C	Mn	Si	Cb	V	ksi	MPa	ksi	MPa	Elongation in 2 in. (%)
Columbium or vanadium	0.20	1.25	0.30	0.01	0.01	55	379	70	483	20
Low manganese–vanadium	0.10	0.50	0.10		0.02	40	276	60	414	35
Manganese–copper	0.25	1.20	0.30			50	345	75	517	20
Manganese–vanadium–copper	0.22	1.25	0.30		0.02	50	345	70	483	22

[a] All have 0.04% P, 0.05% S, and 0.20% Cu.

MICROALLOYED STEELS IN MANUFACTURED PRODUCTS

In terms of both cost and performance, **microalloyed steels** occupy a position between carbon steels and the alloy grades, and are being used increasingly as substitutes for heat-treated steels in the manufacture of small- to medium-sized discrete parts. These low- and medium-carbon steels contain small amounts (0.05 to 0.15%) of alloying elements, such as niobium, vanadium, titanium, molybdenum, zirconium, boron, rare earth elements, or combinations thereof. Many of these additions form alloy carbides—nitrides or carbonitrides—whose primary effect is to provide grain refinement and/or precipitation strengthening. Yield strengths between 500 and 750 MPa (70 and 110 ksi) can be obtained without heat treatment. Weldability can be retained or even improved if the carbon content is simultaneously decreased. In essence, these steels offer maximum strength with minimum carbon, while simultaneously preserving weldability, machinability, and formability. Compared to a quenched-and-tempered alternative, however, ductility and toughness are generally somewhat inferior.

Cold-formed microalloyed steels require less cold work to achieve a desired level of strength, so they tend to have greater residual ductility. Hot-formed products, such as forgings, can often be used in the air-cooled condition. By means of accurate temperature control and controlled-rate cooling directly from the forming operation, mechanical properties can be produced that approximate those of quenched-and-tempered material. Machinability can be enhanced because of the more uniform hardness and the fact that the ferrite–pearlite structure of the microalloyed steel is often more machinable than the ferrite–carbide structure of the quenched-and-tempered variety. Fatigue life and wear resistance can also be superior to those of the heat-treated counterparts.

In applications where the properties are adequate, microalloyed steels can often provide attractive cost savings. Energy savings can be substantial, straightening or stress relieving after heat treatment is no longer necessary, and quench cracking is not a problem. Due to the increase in material strength, the size and weight of finished products can often be reduced. As a result, the cost of a finished forging may be reduced by 5 to 25%.

If these materials are to attain their optimum properties, however, certain precautions must be observed. During the elevated-temperature segments of processing, the material must be heated high enough to place all of the alloys into solution. After forming, the products should be rapidly air cooled to 540 to 600°C (1000 to 1100°F) before dropping into collector boxes. In addition, microalloyed steels tend to through-harden upon air cooling, so products fail to exhibit the lower-strength, higher-toughness interiors that are typical of the quenched-and-tempered materials.

BAKE-HARDENABLE STEEL SHEET

Bake-hardenable steel has assumed a significant role in automotive sheet applications. These low-carbon steels are processed in such a way that they are resistant to aging during normal storage but begin to age during sheet-metal forming. A subsequent exposure to heat during the paint-baking operation completes the aging process and adds an additional 35 to 70 MPa (5 to 10 ksi), raising the final yield strength to approximately 275 MPa (40 ksi). Because the increase in strength occurs after the forming operation, the material offers good formability coupled with improved dent resistance in the final product. In addition, it allows weight savings to be achieved without compromising the attractive features of steel sheet, which include spot weldability, good crash energy absorption, low cost, and full recyclability.

ADVANCED HIGH-STRENGTH STEELS (AHSS)

Traditional methods of producing high-strength steel have included adding carbon and/or alloy elements followed by heat treatment or cold-working to a high level followed by a partial anneal to restore some ductility. As strength increased, however, ductility and toughness decreased and often became a limiting feature. The high-strength low-alloy (HSLA) and microalloyed steels, introduced about 40 years ago, used thermomechanical processing to further increase strength, but were accompanied by an even further decline in ductility. Beginning in the mid-1990s, enhanced thermomechanical processing capabilities and controls have led to the development of a variety of new high-strength steels that go collectively by the name **advanced high-strength steel (AHSS).** Many were developed for weight savings in automotive applications (higher strength enabling reduced size or thickness) while preserving or enhancing energy absorption. As a result, large amounts of low-carbon and HSLA steels are being replaced by the advanced high-strength steels. One group (including the dual-phase and TRIP steels to be discussed later) provided greater formability for already-existing levels of strength. Another group (including complex-phase and martensitic varieties) provided higher levels of strength while retaining current levels of ductility. Because of the improved formability, the AHSS materials can often be stamped or hydroformed into more complex parts. Parts can often be integrated into single pieces, eliminating the cost and time associated with assembly, and the higher strength can provide weight reduction accompanied by improved fatigue and crash performance.

The various types of advanced high-strength steels are primarily ferrite-phase, soft steels with varying amounts of martensite, bainite, or retained austenite that offer high strength with enhanced ductility. Each of the types will be described briefly below:

Dual-phase steels form when material is cooled to a temperature that is above the A_1 but below the A_3 to form a structure that consists of ferrite and high-carbon austenite, and then followed with a rapid-cool quench. During the quench, the ferrite remains unaffected, while the high-carbon austenite transforms to high-carbon martensite. A low- or medium-carbon steel now has a mixed microstructure of a continuous, weak, ductile ferrite matrix combined with islands of high-strength, high-hardness, high-carbon martensite. The dual-phase structure offers strengths that are comparable to HSLA materials, coupled with improved forming characteristics and no loss in weldability. The high work-hardening rates and excellent elongation lead to a high ultimate tensile strength coupled with a low initial yield strength. The high strain-rate sensitivity means that the faster the steel is crushed, the more energy it absorbs—a feature that further enhances the crash resistance of automotive structures. Dual-phase steels also exhibit the bake-hardening effect, a precipitation-induced increase in yield strength when stamping or forming is followed by the elevated temperature of a paint-bake oven.

While the dual-phase steels have structures of ferrite and martensite, **transformation-induced plasticity (TRIP) steels** contain a matrix of ferrite combined with hard martensite or bainite and at least 5 vol% of retained austenite. Because of the hard phase dispersed in the soft ferrite, deformation behavior begins much like the dual-phase steels. At higher strains, however, the retained austenite transforms progressively to martensite, enabling the high work-hardening to persist to greater levels of deformation. At lower levels of carbon, the austenite transformation begins at lower levels of strain, and the extended ductility of the lower-carbon TRIP steels offers significant advantages in operations such as stretch-forming and deep-drawing. At higher carbon levels, the retained austenite is more stable and requires greater strains to induce transformation. If the retained austenite can be carried into a finished part, subsequent deformation (such as a crash) can induce transformation. The conversion of retained austenite to martensite, and the companion high rate of work hardening, can then be used to provide excellent energy absorption.

Complex-phase (CP) steels and **martensitic (Mart) steels** offer even higher strengths with useful capacity for deformation and energy absorption. The CP steels have a microstructure of ferrite and bainite, combined with small amounts of martensite, retained austenite and pearlite, and are strengthened further by grain refinement created by a fine precipitate of niobium, titanium or vanadium carbides or nitrides. The Mart steels are almost entirely martensite, and can have tensile strengths up to 1700 MPa (245 ksi).

Still other types are under development and are making the transition from research into production. These include the *ferritic-bainitic (FB) steels*—also known as stretch-flangeable (SF) and high hole expansion (HHE) because of the improved stretch formability of sheared edges—*twinning-induced plasticity (TWIP) steels, nano steels,* and others. The FB steels have a microstructure of fine ferrite and bainite, coupled with grain refinement. TWIP steels contain between 17 and 24% manganese, making the steel fully austenitic at room temperature. Deformation occurs by twinning inside the grains, with the newly created twin boundaries providing increased strength and a high rate of strain hardening. The result is a high strength (>1000 MPa or 145 ksi)

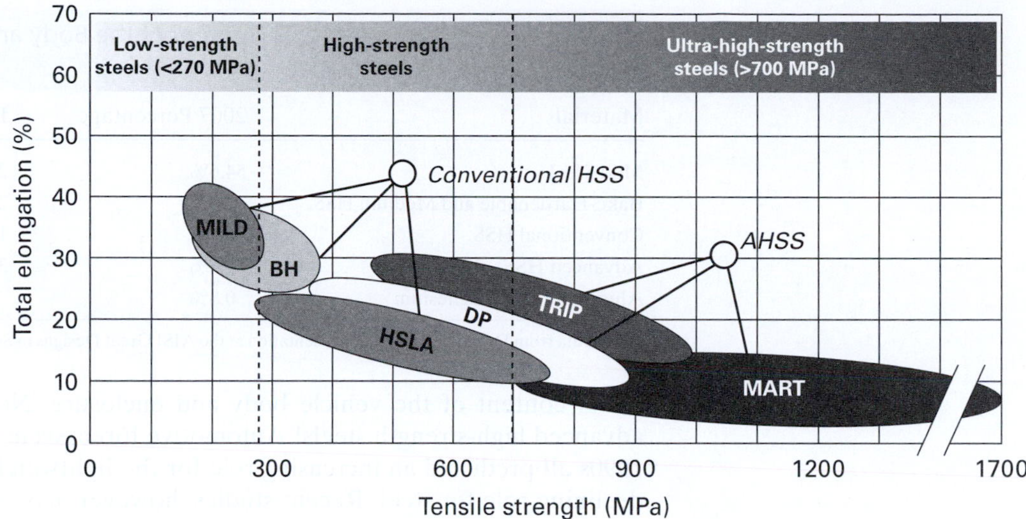

FIGURE 7-9 Relative strength and formability (elongation) of conventional, high-strength low-alloy, and advanced high-strength steels. BH = bake hardenable; DP = dual phase; Mart = martensitic.

combined with extremely high ductility (as high as 70% elongation). Nano steels replace the hard phases that are present in the dual-phase (DP) and TRIP steels with an array of ultra-fine nano-sized precipitates (diameters <10 nm). During sheet forming operations, fractures often initiate at the interface between very soft and very hard phases. Because the interfaces are so small in the nano steels, these fractures are avoided and formability is increased.

Figure 7-9 shows the relative strengths and formability (elongation) of the conventional steels (including mild steels and bake-hardenable steels), carbon-manganese steels, HSLA steels, and the newer AHSS materials. Note how the newer AHSS steels increase strength or formability or both. Also included are the TWIP steels, which occupy a region of very high formability that lies outside of the overlapping band. Some useful distinctions between low-strength steel (ultimate tensile strength, or UTS) below 270 MPa or 40 ksi), high-strength steel, and ultra-high-strength steel (UTS above 700 MPa or 100 ksi) have also been included in this figure.

Table 7-6 shows the changes in the material content of a North American light vehicle from 1975 through 2007 with a projection for 2015. Table 7-7 looks only at the

TABLE 7-6 Material Content of a North American Light Vehicle from 1975 through 2007 with a Projection for 2015

	Material Content in Pounds				
	1975	2005	2007	2015	Change From 1975 to 2015
Mild Steel	2,180	1,751	1,748	1,314	Down 866 lbs.
High Strength Steel	140	324	334	315	Up 175 lbs.
Advanced HSS	—	111	149	403	Up 403 lbs.
Other Steels	65	76	76	77	Up 12 lbs.
Iron	585	290	284	244	Down 341 lbs.
Aluminum	84	307	327	374	Up 290 lbs.
Magnesium	—	9	9	22	Up 22 lbs.
Other Metals	120	150	149	145	Up 25 lbs.
Plastic/Composites	180	335	340	364	Up 184 lbs.
Other Materials	546	629	634	650	Up 104 lbs.
Total Pounds	3,900	3,982	4,050	3,908*	Up 8 lbs.

Note: Data from Drucker Worldwide presentation at the AISI "Great Designs in Steel" Conference, March, 2007.

TABLE 7-7 Metallic Material Content of the Body and Enclosure of a North American Light Vehicle

Material	2007 Percentage	Projected 2015 Percentage
Mild Steel	54.6%	29.0%
Bake-Hardenable and Medium HSS	22.4%	23.5%
Conventional HSS	12.7%	10.2%
Advanced HSS	9.5%	34.8%
Aluminum and Magnesium	0.8%	2.5%

Note: Data from Drucker Worldwide presentation at the AISI Great Designs in Steel Conference, March 2007.

metal content of the vehicle body and enclosure. Note the increased role of the advanced high-strength steels! Automotive forecasts in the 1960s, 1980s, and into the 1990s all predicted an increasing role for the lightweight alternative materials and a declining role for steel. Recent studies, however, have shown that the percentage of steel has actually risen over the years. This has largely been due to the development of new types of steels that are stronger, more energy absorbent, and easy to fabricate into thinner-gage, reduced weight structures. Table 7-8 provides the mechanical properties of some grades of advanced high-strength automotive sheet steels.

TABLE 7-8 Mechanical Properties of Various Grades of Advanced High-Strength Automotive Sheet Steels[a]

Cold-Rolled AHSS Grades

	Min. Yield Str		Min. Tensile Str.	
Type	MPa	ksi	MPa	ksi
Dual-phase				
	250	36	440	64
	290	42	490	71
	340	49	590	85.5
	550	80	690	100
	420	61	780	113
	550	80	980	142
TRIP Steels				
	380	55	590	85.5
	400	58	690	100
	420	61	780	113
Martensitic Steels				
	700	101.5	900	130.5
	860	125	1100	1
	1030	149.5	1300	188.5
	1200	174	1500	217.5

Hot-Rolled AHSS Grades

	Min. Yield Str		Min. Tensile Str.	
Type	MPa	ksi	MPa	ksi
Dual-phase				
	300	43.5	590	85.5
	380	55	780	113
TRIP Steels				
	400	58	590	85.5
	450	65	780	113

[a] Data from society of Automotive Engineers document SAE J2745, July, 2007.

FREE-MACHINING STEELS

The increased use of high-speed, automated machining has spurred the use and development of several varieties of **free-machining steels.** These steels machine readily and form small chips when cut. The smaller chips reduce the length of contact between the chip and cutting tool, thereby reducing the associated friction and heat, as well as required power and wear on the cutting tool. The formation of small chips also reduces the likelihood of chip entanglement in the machine and makes chip removal much easier. On the negative side, free-machining steels often carry a cost premium of 15 to 20% over conventional alloys, but this increase may be easily recovered through higher machining speeds, larger depths of cut, and extended tool life.

Free-machining steels are basically carbon steels that have been modified by an addition of sulfur, lead, bismuth, selenium, tellurium, or phosphorus plus sulfur to enhance machinability. Sulfur combines with manganese to form soft manganese sulfide inclusions. These, in turn, serve as chip-breaking discontinuities within the structure. The inclusions also provide a built-in lubricant that prevents formation of a built-up edge on the cutting tool and imparts an improved cutting geometry (see Chapter 20). In leaded materials, the insoluble lead particles work in much the same way.

The bismuth free-machining steels are an attractive alternative to the previous varieties. Bismuth is more environmentally acceptable (compared to lead), has a reduced tendency to form stringers, and can be more uniformly dispersed because its density is a better match to that of iron. Machinability is improved because the heat generated by cutting is sufficient to form a thin film of liquid bismuth that lasts for only fractions of a microsecond. Tool life is noticeably extended and the machined product is still weldable.

The use of free-machining steels is not without compromise, however. Ductility and impact properties are somewhat reduced compared to the unmodified steels. Copper-based braze joints tend to embrittle when used to join bismuth free-machining steels, and the machining additions reduce the strength of shrink-fit assemblies. If these compromises are objectionable, other methods may be used to enhance machinability. For example, the machinability of steels can be improved by cold working the metal. As the strength and hardness of the metal increase, the metal loses ductility, and subsequent machining produces chips that tear away more readily and fracture into smaller segments.

PRECOATED STEEL SHEET

Traditional sheet metal fabrication involves the shaping of components from bare steel, followed by the finishing (or coating) of these products on a piece-by-piece basis. In this sequence, it is not uncommon for the finishing processes to be the most expensive and time-consuming stages of manufacture because they involve handling, manipulation, and possible curing or drying, as well as adherence to the various Environmental Protection Agency (EPA) and Occupational Safety and Health Administration (OSHA) requirements (environmental and safety and health, respectively).

An alternative to this procedure is to purchase **precoated steel sheet**, where the steel supplier applies the coating when the material is still in the form of a long, continuous strip. Cleaning, pretreatment, coating, and curing can all be performed in a continuous manner, producing a coating that is uniform in thickness and offers improved adhesion. Numerous coatings can be specified, including the entire spectrum of dipped and plated metals (such as aluminum, zinc, and chromium), vinyls, paints, primers, and other polymers or organics. Many of these coatings are specially formulated to endure the rigors of subsequent forming and bending. The continuous sheets can also be printed, striped, or embossed to provide a number of visual effects. Extra caution must be exercised during handling and fabrication to prevent damage to the coating, but the additional effort and expense are often less than the cost of finishing individual pieces.

STEELS FOR ELECTRICAL AND MAGNETIC APPLICATIONS

Soft magnetic materials can be magnetized by relatively low-strength magnetic fields but lose almost all of their magnetism when the applied field is removed. They are widely used in products such as solenoids, transformers, motors, and generators. The most common soft magnetic materials are high-purity iron, low-carbon steels, iron-silicon electrical steels, amorphous ferromagnetic alloys, iron-nickel alloys, and soft ferrites (ceramic material).

In recent years, the **amorphous metals** have shown attractive electrical and magnetic properties. Because the material has no crystal structure, grains, or grain boundaries, (1) the magnetic domains can move freely in response to magnetic fields, (2) the properties are the same in all directions, and (3) corrosion resistance is improved. The high magnetic strength and low hysteresis losses offer the possibility of smaller, lighter-weight magnets. When used to replace silicon steel in power transformer cores, this material has the potential of reducing core losses by as much as 50%.

To exhibit permanent magnetism, materials must remain magnetized when removed from the applied field. While most permanent magnets are ceramic materials or complex metal alloys, cobalt alloy steels (containing up to 36% cobalt) may be specified for electrical equipment where high magnetic densities are required.

MARAGING STEELS

When super-high strength is required from a steel, the **maraging** grades become a very attractive option. These alloys contain between 15 and 25% nickel, plus significant amounts of cobalt, molybdenum, and titanium—all added to a very-low-carbon steel. They can be hot worked at elevated temperatures, machined or cold worked in the air-cooled condition, and then strengthened by a precipitation reaction (aged) to yield strengths in excess of 1725 MPa (250 ksi), with good residual elongation.

Maraging alloys are very useful in applications where ultra-high strength and good toughness are important. (The fracture toughness of maraging steel is considerably higher than that of the conventional high-stength steels.) They can be welded, provided the weldment is followed by the full solution and aging treatment. As might be expected from the large amount of alloy additions (more than 30%) and the multistep thermal processing, maraging steels are quite expensive and should be specified only when their outstanding properties are absolutely required.

STEELS FOR HIGH-TEMPERATURE SERVICE

As a general rule of thumb, plain-carbon steels should not be used at temperatures in excess of about 250°C (500°F). Conventional alloy steels extend this upper limit to around 350°C (650°F). Continued developments in areas such as missiles and jet aircraft, however, have increased the demand for metals that offer good strength characteristics, corrosion resistance, and creep resistance at operating temperatures in excess of 550°C (1000°F).

The high-temperature ferrous alloys tend to be low-carbon materials with less than 0.1% carbon. At their peak operating temperatures, 1000-hour rupture stresses tend to be quite low, often in the neighborhood of 50 MPa (7 ksi). While iron is also a major component of other high-temperature alloys, when the amounts fall below 50% the metal is not generally classified as a ferrous material. High strength at high temperature usually requires the more expensive nonferrous materials that will be discussed in Chapter 8.

7.5 STAINLESS STEELS

Low-carbon steel with the addition of 4 to 6% chromium acquires good resistance to many of the corrosive media encountered in the chemical industry. This behavior is attributed to the formation of a strongly adherent iron chromium oxide on the surface. If more improved corrosion resistance and outstanding appearance are required, materials should be specified that use a superior oxide that forms when the amount of chromium in solution (excluding chromium carbides and other forms where the chromium is no longer available to react with oxygen) exceeds 12%. When damaged, this tough, adherent, transparent, corrosion-resistant oxide (which is only 1 to 2 nm thick) actually heals itself, provided oxygen is present, even in very small amounts. Materials that form this superior protective oxide are known as **stainless steels,** or more specifically, **true stainless steels.**

Several classification schemes have been devised to categorize these alloys. The American Iron and Steel Institute (AISI) groups the metals by chemistry and assigns a three-digit number that identifies the basic family and the particular alloy within that family. In this text, however, we will group these alloys into microstructural families, because it is the basic structure that controls the engineering properties of the metal. Table 7-9 presents the AISI designation scheme for stainless steels and correlates it with the microstructural families.

TABLE 7-9 AISI Designation Scheme for Stainless Steels

Series	Alloys	Structure
200	Chromium, nickel, manganese, or nitrogen	Austenitic
300	Chromium and nickel	Austenitic
400	Chromium and possibly carbon	Ferritic or martensitic
500	Low chromium (<12%) and possibly carbon	Martensitic

Having a body-centered-cubic structure, chromium tends to stabilize the body-centered ferrite structure in a steel. A chromium addition, therefore, increases the temperature range over which ferrite is the stable structure. With sufficient chromium and a low level of carbon, a corrosion-resistant iron alloy can be produced that is ferrite at all temperatures below solidification. These alloys are known as the **ferritic stainless steels.** They possess rather limited ductility and poor toughness, but are readily weldable. No martensite can form in the welds because there is no possibility of forming the face-centered-cubic (FCC) austenite structure that can then transform during cooling. These alloys cannot be heat treated, and poor ductility limits the amount of strengthening by cold work. The primary source of strength is the body-centered-cubic (BCC) crystal structure combined with the effects of solid solution strengthening. Characteristic of the BCC metals, the ferritic stainless steels exhibit a ductile-to-brittle transition as the temperature is reduced. The ferritic alloys are the cheapest type of stainless steel, however, and, as such, they should be given first consideration when a stainless alloy is required.

If increased strength is needed, the **martensitic stainless steels** should be considered. For these alloys, carbon is added and the chromium content is reduced to a level where the material can be austenite (FCC) at high temperature and ferrite (BCC) at low. Upon heating, the carbon will dissolve in the face-centered-cubic austenite, which can then be quenched to trap it in the body-centered martensitic structure. The carbon contents can be varied up to 1.2% to provide a wide range of strengths and hardnesses. Caution should be taken, however, to ensure more than 12% chromium remains in solution. Slow cools may allow the carbon and chromium to react and form chromium carbides. When this occurs, the chromium is no longer available to react with oxygen and form the protective oxide. As a result, the martensitic stainless steels may only exhibit good corrosion resistance when in the martensitic condition (when the chromium is trapped in atomic solution) and may be susceptible to red rust when annealed or normalized for ease of machining or fabrication. The martensitic stainless steels cost about $1\frac{1}{2}$ times as much as the ferritic alloys, with part of the increase being due to the additional heat treatment, which generally consists of an austenitization, quench, stress relief, and temper. They are less corrosion resistant than the other varieties, are the least weldable of the stainless steels, and a ductile-to-brittle transition occurs at low temperatures. The martensitic stainlesses tend to be used in applications such as cutlery, where strength and hardness are the dominant requirements.

Nickel, being face-centered cubic, is an austenite stabilizer, and with sufficient amounts of both chromium and nickel (and low carbon), it is possible to produce a stainless steel in which austenite is the stable structure from elevated to cryogenic temperatures. Known as **austenitic stainless steels,** these alloys may cost two to three times as much as the ferritic variety, but here the added expense is attributed to the cost of the nickel and chromium alloys. (The most widely used austenitic stainless steel, Type 304, is also known as 18-8, because it contains 18% chromium and 8% nickel.) Manganese and nitrogen are also austenite stabilizers and may be substituted for some of the nickel to produce a lower-cost, somewhat lower-quality austenitic stainless steel (the AISI 200-series).

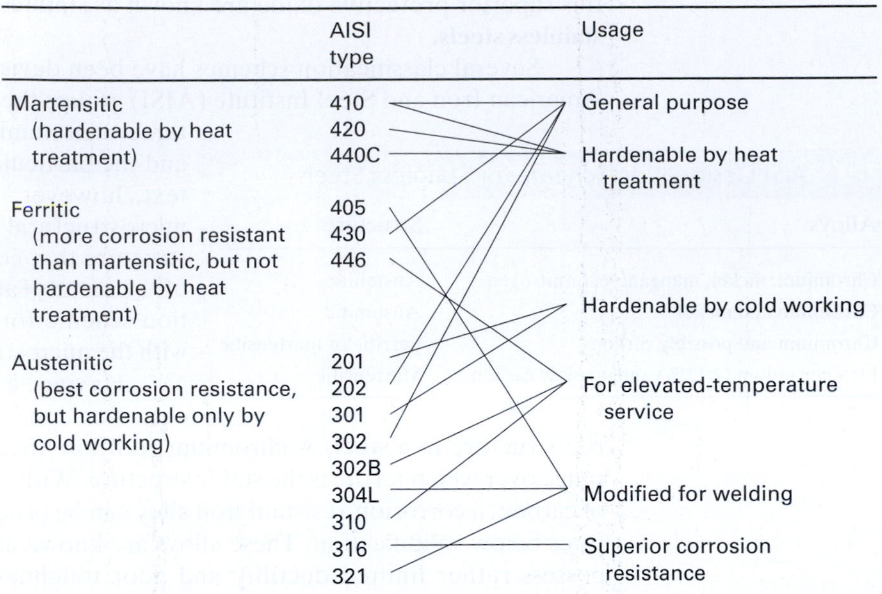

AISI type
Usage
Martensitic (hardenable by heat treatment)
410
420
440C
General purpose
Hardenable by heat treatment
Ferritic (more corrosion resistant than martensitic, but not hardenable by heat treatment)
405
430
446
Hardenable by cold working
Austenitic (best corrosion resistance, but hardenable only by cold working)
201
202
301
302
302B
304L
310
316
321
For elevated-temperature service
Modified for welding
Superior corrosion resistance

FIGURE 7-10 Popular alloys and key properties for different types of stainless steels.

Austenitic stainless steels are easily identified by their nonmagnetic characteristic (the ferritic and martensitic stainlesses are attracted to a magnet). They are highly resistant to corrosion in almost all media (except hydrochloric acid and other halide acids and salts) and may be polished to a mirror finish, thereby combining attractive appearance and corrosion resistance. Formability is outstanding (due to the low yield strength and high elongation that is characteristic of the FCC crystal structure), and these steels strengthen significantly when cold worked. The following table shows the response of the popular 304 alloy to a small amount of cold work:

	Water Quench	Cold Rolled 15%
Yield strength [MPa (ksi)]	260 (38)	805 (117)
Tensile strength [MPA (ksi)]	620 (90)	965 (140)
Elongation in 2 in. (%)	68	11

The austenitic stainless steels offer the best combination of corrosion resistance and toughness of the stainless varieties, are easily welded, and do not embrittle at low temperatures. Because they are also some of the most costly, they should not be specified where the less expensive ferritic or martensitic alloys would be adequate or where a true stainless steel is not required. Figure 7-10 lists some of the popular alloys from each of the three major structural classifications and links them to some associated properties. Table 7-10 shows the basic types and the primary mechanism of strengthening.

TABLE 7-10 Primary Strengthening Mechanism for the Various Types of Stainless Steel

Type of Stainless Steel	Primary Strengthening Mechanism
Ferritic	Solid-solution strengthening
Martensitic	Phase transformation strengthening (martensite)
Austenitic	Cold work (deformation strengthening)

A fourth and special class of stainless steels is the **precipitation-hardening stainless steel.** These alloys are basically martensitic or austenitic types—modified by the addition of alloying elements such as copper, aluminum, and titanium—that permit the precipitation of hard intermetallic compounds at the temperatures used to temper martensite. With the addition of age hardening, these materials are capable of attaining high-strength properties such as a 1790 MPa (260 ksi) yield strength, 1825 MPa (265 ksi) tensile strength, and a 2% elongation. Because the additional alloys and extra processing make the precipitation-hardening alloys some of the most expensive stainless steels, they should be used only when their high-strength feature is absolutely required.

While the four structures just described constitute the bulk of stainless steels, there are also some additional variants. **Duplex stainless steels** contain between 18 and 25% chromium, 4 to 7% nickel, and up to 4% molybdenum, and can be water quenched from a hot-working temperature to produce a microstructure that is approximately half ferrite and half austenite. This mixed structure offers good toughness and a higher yield strength and greater resistance to both stress corrosion cracking and pitting corrosion than either the full-austenitic or full-ferritic grades.

Because stainless steels are difficult to machine because of their work-hardening properties and their tendency to seize during cutting, special free-machining alloys have been produced within each family. Additions of sulfur, phosphorus, or selenium can raise machinability to approximately that of a medium-carbon steel. The free-machining grades are designated by the letters F or Se following the three-digit alloy code.

The preceding discussion has focused on the wrought stainless alloys. Cast stainless steels have structures and properties that are similar to the wrought grades. Most are specified by properties using standards of ASTM or ISO (International Organization for Standardization), while chemistries have been classified by the High Alloy Product Group of the Steel Founders Society of America. The C-series are used primarily to impart corrosion resistance and are used in valves, pumps and fittings. The H-grades (heat-resistant) have been designed to provide useful properties at elevated temperature and are used for furnace parts and turbine components.

Several potential problems are unique to the family of stainless steels. Because the protective oxide provides the excellent corrosion resistance, this feature can be lost whenever the amount of chromium in solution drops below 12%. A localized depletion of chromium can occur when elevated temperatures allow chromium carbides to form along grain boundaries **(sensitization).** To prevent their formation, one can keep the carbon content of stainless steels as low as possible, usually below 0.10%. Another method is to tie up existing carbon with small amounts of stabilizing elements, such as titanium or niobium, that have a stronger affinity for carbon than does chromium. A letter suffix again designates the modification. Rapidly cooling these metals through the carbide-forming range of 480 to 820°C (900 to 1500°F) also works to prevent carbide formation.

Another problem with high-chromium stainless steels is an embrittlement that can occur after long times at elevated temperatures. This is attributed to the formation of a brittle compound that forms at elevated temperature and coats grain boundaries. Known as **sigma phase,** this material then provides a brittle crack path through the metal. Stainless steels used in high-temperature service should be checked periodically to detect and monitor sigma-phase formation.

7.6 TOOL STEELS

Tool steels are high-carbon, high-strength, ferrous alloys that have been modified by alloy additions to provide a desired balance of strength (resistance to deformation), toughness (ability to absorb shock or impact), and wear resistance (ability to resist erosion between the tool steel and contact material) when properly heat-treated. Several classification systems have been developed, some using chemistry as a basis, while others employ hardening method or major mechanical property. The AISI system uses a letter designation to identify basic features such as quenching method, primary application, special alloy, or dominant characteristic. Table 7-11 lists seven basic families of tool steels, the corresponding AISI letter grades, and the associated feature or characteristic. Individual alloys within the letter grades are then listed numerically to produce a letter–number identification system, such as W1, D2, or H13.

Water-hardening tool steels (W-grade) are essentially high-carbon plain-carbon steels. They are the least expensive variety and are used for a wide range of parts that are usually quite small and not subject to severe usage or elevated temperature. Because strength and hardness are functions of the carbon content, a wide range of properties can be achieved through composition variation. Hardenability is low, so these steels must be quenched in water to attain high hardness. They can be used only

TABLE 7-11 Basic Types of Tool Steel and Corresponding AISI Grades

Type	AISI Grade	Significant Characteristic
1.Water-hardening	W	
2. Cold-work	O	Oil-hardening
	A	Air-hardening medium alloy
	D	High-carbon–high-chromium
3. Shock-resisting	S	
4. High-speed	T	Tungsten alloy
	M	Molybdenum alloy
5.Hot-work	H	H1–H19: chromium alloy
		H20–H39: tungsten alloy
		H40–H59: molybdenum alloy
6. Plastic-mold	P	
7. Special-purpose	L	Low alloy
	F	Carbon–tungsten

for relatively thin sections if the full depth of hardness is desired. They are also rather brittle, particularly at higher hardness.

Typical uses of the various plain-carbon steels are as follows:

0.60–0.75% carbon: machine parts, chisels, setscrews, and similar products where medium hardness is required coupled with good toughness and shock resistance

0.75–0.90% carbon: forging dies, hammers, and sledges

0.90–1.10% carbon: general-purpose tooling applications that require good balance of wear resistance and toughness, such as drills, cutters, shear blades, and other heavy-duty cutting edges

1.10–1.30% carbon: small drills, lathe tools, razor blades, and other light-duty applications in which extreme hardness is required without great toughness

In applications where improved toughness is required, small amounts of manganese, silicon, and molybdenum are often added. Vanadium additions of about 0.20% are used to form strong, stable carbides that retain fine grain size during heat treatment. One of the main weaknesses of the plain-carbon tool steels is their loss of hardness at elevated temperature, which can occur with prolonged exposure to temperatures over 150°C (300°F).

When larger parts must be hardened or distortion must be minimized, the **cold-work tool steels** are usually recommended. The alloy additions and higher hardenability of the **oil-** or **air-hardening grades** (O and A designations, respectively) enable hardening by less severe quenches. Tighter dimensional tolerances can be maintained during heat treatment, and the cracking tendency is reduced. The **high-chromium tool steels** (D designation) contain between 10 and 18% chromium and are air-hardening and offer outstanding deep-hardening wear resistance. Blanking, stamping, and cold-forming dies, punches, and other tools for large production runs are all common applications for this class. Because these steels do not have the alloy content necessary to resist softening at elevated temperatures, they should not be used for applications that involve prolonged service at temperatures in excess of 250°C (500°F).

Shock-resisting tool steels (S designation) offer the high toughness needed for impact applications. Low carbon content (approximately 0.5% carbon) is usually specified to ensure the necessary toughness, with carbide-forming alloys providing the necessary abrasion resistance, hardenability, and hot-work characteristics. Applications include parts for pneumatic tooling, chisels, punches, and shear blades.

High-speed tool steels are used for cutting tools and other applications where strength and hardness must be retained at temperatures up to or exceeding red heat (about 760°C or 1400°F). One popular member of the *tungsten high-speed tool steels*

(T designation) is the T1 alloy, which contains 0.7% carbon, 18% tungsten, 4% chromium, and 1% vanadium. It offers a balanced combination of shock resistance and abrasion resistance and is used for a wide variety of cutting applications. The *molybdenum high-speed steels* (M designation) were developed to reduce the amount of tungsten and chromium required to produce the high-speed properties, and M1 has become quite popular for drill bits.

Hot-work tool steels (H designation) were developed to provide strength and hardness during prolonged exposure to elevated temperature. All employ substantial additions of carbide-forming alloys. H1 to H19 are chromium-based alloys with about 5.0% chromium; H20 to H39 are tungsten-based types with 9 to 18% tungsten, coupled with 3 to 4% chromium; and H40 to H59 are molybdenum-based. The chromium types tend to be less expensive than the tungsten or molybdenum alloys.

Other types of tool steels include (1) the *plastic mold steels* (P designation), designed to meet the requirements of zinc die casting and plastic injection molding dies; (2) the *low-alloy special-purpose tool steels* (L designation), such as the L6 extreme toughness variety; and (3) the *carbo–tungsten type* of special-purpose tool steels (F designation), which are water hardening but substantially more wear-resistant than the plain-carbon tool steels.

Most tool steels are wrought materials, but some are designed specifically for fabrication by casting. Powder metallurgy processing has also been used to produce special compositions that are difficult or impossible to produce by wrought or cast methods, or to provide key structural enhancements. By subjecting water-atomized powders to hot-isostatic pressing (HIP), 100% dense billets can be produced with fine grain size and small, uniformly distributed carbide particles. These materials offer superior wear resistance compared to conventional tool steels, combined with useful levels of toughness.

7.7 CAST IRONS

The term **cast iron** applies to an entire family of metals that are alloys of iron, carbon, and silicon and offer a wide variety of properties. Various types of cast iron can be produced, depending on the chemical composition, cooling rate, and the type and amount of inoculants that are used. (Inoculants and inoculation practice will be discussed shortly.) Each of these, however, can be described as having a structure consisting of an iron-based metal matrix (such as ferrite, pearlite, bainite, or martensite) and a high-carbon second phase that is either graphite (pure carbon) or iron carbide (Fe_3C). The basic types of cast iron are gray, white, malleable, ductile or nodular, austempered ductile, compacted graphite, and high-alloy.

TYPES OF CAST IRON

Gray cast iron is the least expensive and most common variety and can be characterized by those features that promote the formation of graphite (discussed at the end of Chapter 5). Typical compositions range from 2.5 to 4.0% carbon, 1.0 to 3.0% silicon, and 0.4 to 1.0% manganese. The microstructure consists of three-dimensional graphite flakes (which form during the eutectic reaction) dispersed in a matrix of ferrite, pearlite, or other iron-based structure that forms from the cooling of austenite. Figure 7-11 presents a typical section through gray cast iron, showing the graphite flakes dispersed throughout the metal matrix. Because the graphite flakes have no appreciable strength, they act essentially as voids in the structure. The pointed edges of the flakes act as pre-existing notches or crack initiation sites, giving the material a characteristic brittle nature. Because a large portion of any fracture follows the graphite flakes, the freshly exposed fracture surfaces have a characteristic gray appearance, and a graphite smudge can usually be obtained if one rubs a finger across the fracture. On a more positive note, the formation of the lower-density graphite reduces the amount of shrinkage that occurs when the liquid goes to solid, making possible the production of more complex iron castings.

The size, shape, and distribution of the graphite flakes have a considerable effect on the overall properties of gray cast iron. When maximum strength is desired,

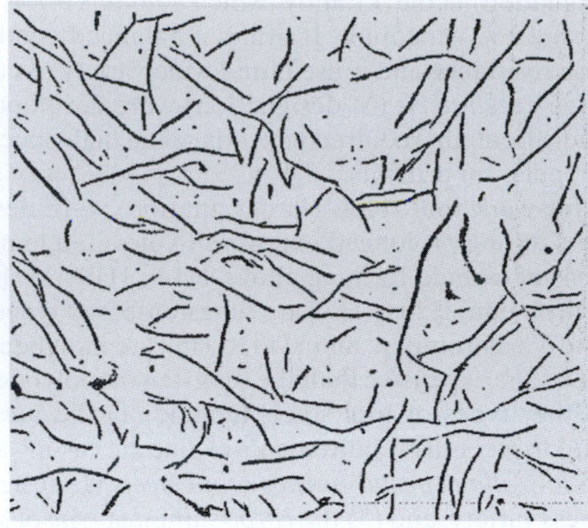

FIGURE 7-11 Photomicrograph showing the distribution of graphite flakes in gray cast iron; unetched, 100×. *(Courtesy Ronald Kohser)*

small, uniformly distributed flakes with a minimum amount of intersection are preferred. A more effective means of controlling strength, however, is through control of the metal matrix structure, which is in turn controlled by the carbon and silicon contents and the cooling rate of the casting. Gray cast iron is normally sold by **class,** with the class number corresponding to the minimum tensile strength in thousands of pounds per square inch.[1] Class 20 iron (minimum tensile strength of 20,000 psi) consists of high-carbon, high-silicon metal with a ferrite matrix. Higher strengths, up to class 40, can be obtained with lower carbon and silicon and a pearlite matrix. To go above class 40, alloying is required to provide solid solution strengthening, and heat-treatment practices must be performed to modify the matrix. Gray cast irons can be obtained up through class 80, but, regardless of strength, the presence of the graphite flakes results in extremely low ductility.

Gray cast irons offer excellent compressive strength (compressive forces do not promote crack propagation, so compressive strength is typically three to four times the tensile strength), excellent machinability (the graphite flakes act to break up the chips and lubricate contact surfaces), good resistance to adhesive wear and galling (graphite flakes self-lubricate), and outstanding sound and vibration damping characteristics (graphite flakes absorb transmitted energy). Table 7-12 compares the relative damping capacities of various engineering metals, and clearly shows the unique characteristic of the high-carbon-equivalent (high-carbon and high-silicon) gray cast irons. This material is 20 to 25 times better than steel and 250 times better than aluminum! High silicon contents promote good corrosion resistance and provide the enhanced fluidity desired for casting operations. For these reasons, coupled with low cost, high thermal conductivity, low rate of thermal expansion, good stiffness, resistance to thermal fatigue, and

TABLE 7-12 Relative Damping Capacity of Various Metals

Material	Damping Capacity[a]
Gray iron (high carbon equivalent)	100–500
Gray iron (low carbon equivalent)	20–100
Ductile iron	5–20
Malleable iron	8–15
White iron	2–4
Steel	4
Aluminum	0.4

[a] Natural log of the ratio of successive amplitudes.

[1] The ASTM class uses numbers reflecting thousands of pounds per square inch, while the International Standards Organization (ISO) assigns class numbers using megapascals.

100% recyclability, gray cast iron is specified for a number of applications, including automotive engine blocks, heads, and cylinder liners; transmission housings; machine tool bases; and large equipment parts that are subjected to compressive loads and vibrations.

White cast iron has all of its excess carbon in the form of iron carbide and receives its name from the white surface that appears when the material is fractured. Features promoting its formation are those that favor cementite over graphite: a low carbon equivalent (1.8 to 3.6% carbon, 0.5 to 1.9% silicon, and 0.25 to 0.8% manganese) and rapid cooling.

Because the large amount of iron carbide dominates the microstructure, white cast iron is very hard and brittle, and finds applications where high abrasion resistance is the overwhelming requirement. For these uses it is also common to pursue the hard, wear-resistant martensite structure (described in Chapter 6) as the metal matrix. In this way, both the metal matrix and the high-carbon second phase contribute to the wear-resistant characteristics of the material.

White cast iron surfaces can also be applied over a base of another material. For example, mill rolls that require extreme wear resistance may have a white cast iron surface on top of a steel interior. By accelerating the cooling rate and controlling chemistry, white iron surfaces or regions can be produced in gray iron castings. Tapered sections or metal chill bars placed in the molding sand provide the accelerated cooling. When regions of white and gray cast iron occur in the same component, there is generally a transition zone containing regions of both white and gray irons, known as the **mottled zone.**

If white cast iron is exposed to an extended heat treatment at temperatures in the range of 900°C (1650°F), the cementite will dissociate into its component elements, and some or all of the carbon will be converted into irregularly shaped clusters of graphite (also referred to as clump or popcorn graphite). The product, known as **malleable cast iron,** has significantly improved ductility compared to gray cast iron because the more favorable graphite shape no longer resembles an internal crack or notch. The rapid cooling required to produce the starting white iron structure, however, restricts the size and thickness of malleable iron products such that most weigh less than 5 kg (10 lb).

Various types of malleable iron can be produced, depending on the type of heat treatment that is employed. If the white iron is heated and held for a prolonged time just below the melting point, the carbon in the cementite converts to graphite (first-stage graphitization). Subsequent slow cooling through the eutectoid reaction causes the carbon-containing austenite to transform to ferrite and more graphite (second-stage graphitization), and the resulting product, known as *ferritic malleable cast iron,* has a structure of irregular particles of graphite dispersed in a ferrite matrix (Figure 7-12). Typical properties of this material are 10% elongation; 35 ksi (240 MPa) yield strength; 50 ksi (345 MPa) tensile strength; and excellent impact strength, corrosion resistance,

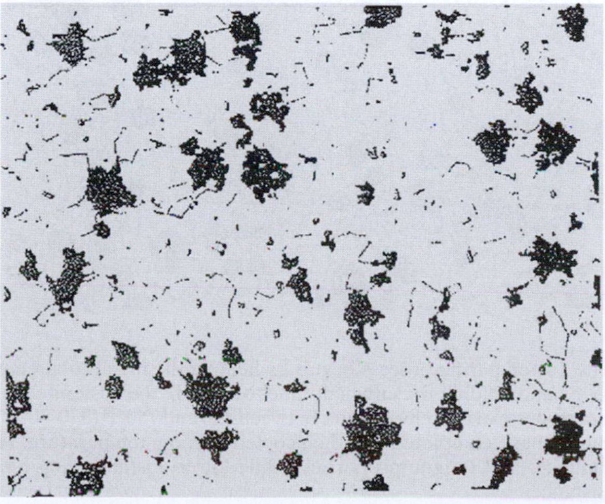

FIGURE 7-12 Photomicrograph of malleable iron showing the irregular graphite clusters, etched to reveal the ferrite matrix, 100×. *(Courtesy Ronald Kohser)*

and machinability. The heat-treatment times, however, are quite lengthy, often involving more than 100 hours at elevated temperature.

If the material is cooled more rapidly through the eutectoid transformation, the carbon in the austenite does not form additional graphite but is retained in a pearlite or martensite matrix. The resulting *pearlitic malleable cast iron* is characterized by higher strength and lower ductility than its ferritic counterpart. Properties range from 1 to 4% elongation, 45 to 85 ksi (310 to 590 MPa) yield strength, and 65 to 105 ksi (450 to 725 MPa) tensile strength, with reduced machinability compared to the ferritic material.

The modified graphite structure of malleable iron provided quite an improvement in properties compared to gray cast iron. However, it would be even more attractive if a similar structure could be obtained directly upon solidification rather than through a prolonged heat treatment at high elevated temperature. If a high-carbon-equivalent cast iron is sufficiently low in sulfur (either by original chemistry or by desulfurization), the addition of certain materials can promote graphite formation and change the morphology (shape) of the graphite product. If ferrosilicon is injected into the melt **(inoculation),** it will promote the formation of graphite. If magnesium (in the form of MgFeSi or MgNi alloy) is also added just prior to solidification, the graphite will form as smooth-surface spheres. The latter addition is known as a **nodulizer,** and the product becomes **ductile** or **nodular cast iron.** One should note that the magnesium nodulizer volatilizes and its effectiveness will diminish or be lost with time, and the graphite structure will transition to the flake form of gray cast iron. This phenomenon is known as **fading.**

Subsequent control of cooling can produce a variety of matrix structures, with ferrite or pearlite being the most common (Figure 7-13). By controlling the matrix structure, properties can be produced that span a wide range from 2 to 18% elongation, 40 to 90 ksi (275 to 620 MPa) yield strength, and 60 to 120 ksi (415 to 825 MPa) tensile strength. The combination of good ductility, high strength, toughness, wear resistance, machinability, low-melting-point castability, and up to a 10% weight reduction compared to steel, makes ductile iron an attractive engineering material. High silicon–molybdenum ductile irons offer excellent high temperature strength and good corrosion resistance. Unfortunately, the costs of a nodulizer, higher-grade melting stock, better furnaces, and the improved process control required for its manufacture combine to place it among the most expensive of the cast irons.

Austempered ductile iron (ADI), or ductile iron that has undergone a special austempering heat treatment to modify and enhance its properties,[2] has emerged as a

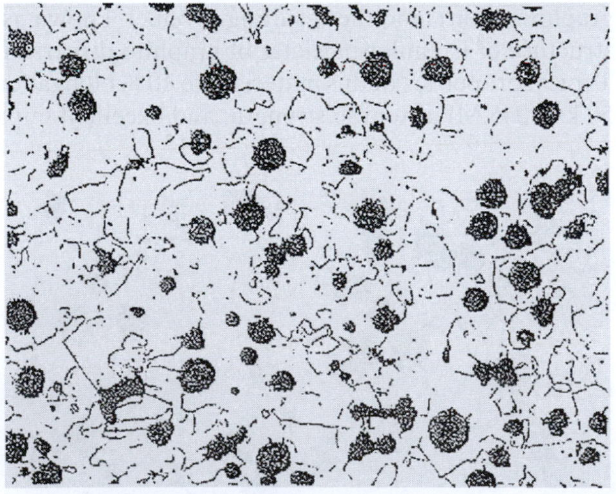

FIGURE 7-13 Ductile cast iron with a ferrite matrix. Note the spheroidal shape of the graphite, 100×. *(Courtesy Ronald Kohser)*

[2] The austempering process begins by heating the metal to a temperature between 1500 and 1750°F (815 to 955°C) and holding for sufficient time to saturate the austenite with carbon. The metal is then rapidly cooled to an austempering temperature between 450 and 750°F (230-400°C), where it is held until all crystal structure changes have completed and then cooled to room temperature. High austempering temperatures give good toughness and fatigue properties, while lower austempering temperatures give better strength and wear resistance.

TABLE 7-13 Typical Mechanical Properties of Malleable, Ductile, and Austempered Ductile Cast Irons

Class or Grade	Minimum Yield Strength ksi	Minimum Yield Strength MPa	Minimum Tensile Strength ksi	Minimum Tensile Strength MPa	Minimum Percentage Elongation	Brinell Hardness Number
Malleable Iron[a]						
M3210	32	224	50	345	10	156 max
M4504	45	310	65	448	4	163–217
M5003	50	345	75	517	3	187–241
M5503	55	379	75	517	3	187–241
M7002	70	483	90	621	2	229–269
M8501	85	586	105	724	1	269–302
Ductile Iron[b]						
60–40–18	40	276	60	414	18	149–187
65–45–12	45	310	65	448	12	170–207
80–55–06	55	379	80	552	6	187–248
100–70–03	70	483	100	689	3	217–269
120–90–02	90	621	120	827	2	240–300
Austempered Ductile Iron[c]						
1	70	500	110	750	11	241–302
2	90	550	130	900	9	269–341
3	110	750	150	1050	7	302–375
4	125	850	175	1200	4	341–444
5	155	1100	200	1400	2	388–477
6	185	1300	230	1600	1	402–512

[a] ASTM Specification A602 (Also SAE J 158).
[b] ASTM Specification A536.
[c] ASTM Specification A897.

significant engineering material over the past 30 years. It combines the ability to cast intricate shapes with strength, fatigue, and wear-resistance properties that are similar to those of heat-treated steel. Compared to conventional as-cast ductile iron, it offers nearly double the strength at the same level of ductility. Compared to steel, it offers reduced cost, an 8 to 10% reduction in density (so strength-to-weight is excellent), and enhanced damping capability, both due to the graphite nodules. Machinability is generally poorer, thermal conductivity is lower, and there is about a 20% drop in elastic modulus. ADI is approximately three times stronger than aluminum, more than two times stiffer, with better fatigue strength and wear resistance. Table 7-13 compares some typical mechanical properties of malleable and ductile irons with the six grades of austempered ductile cast iron that are specified in ASTM Standard A-897.

Compacted graphite cast iron (CGI) is also attracting considerable attention. Produced by a method similar to that used to make ductile iron (a Mg–Ce–Ti addition is made), compacted graphite iron is characterized by a graphite structure that is intermediate to the flake graphite of gray iron and the nodular graphite of ductile iron, and it tends to possess some of the desirable properties and characteristics of each. Table 7-14 shows how the properties of compacted graphite iron bridge the gap between gray and ductile. Strength, stiffness, and ductility are greater than those of gray iron, while castability, machinability, thermal conductivity, and damping capacity all exceed those of ductile. Impact and fatigue properties are good.

ASTM Specification A842 identifies five grades of compacted graphite cast iron—250, 300, 350, 400, and 450—where the numbers correspond to tensile strength in

TABLE 7-14 Typical Properties of Pearlitic Gray, Compacted Graphite, and Ductile Cast Irons

Property	Gray	CGI	Ductile
Tensile strength (MPa)	250	450	750
Elastic modulus (Gpa)	105	145	160
Elongation (%)	0	1.5	5
Thermal conductivity (w/mk)	48	37	28
Relative damping capacity (Gray = 1)	1	0.35	0.22

megapascals. Areas of application tend to be those where the mechanical properties of gray iron are insufficient and those of ductile iron, along with its higher cost, are considered to be overkill. More specific, compacted graphite iron is attractive when the desired properties include high strength, castability, machinability, thermal conductivity, and thermal shock resistance.

THE ROLE OF ALLOYS IN CAST IRONS

Because the effects of alloying elements are often the same regardless of the process used to produce the final shape, much of what was presented earlier for wrought steel also applies to cast irons. When the desired shape is to be made by casting, however, some alloys can be used to enhance process-specific features, such as fluidity and as-solidified properties.

Many cast iron products are used in the as-cast condition, with the only heat treatment being a stress relief or annealing. For these applications, the alloy elements are selected for their ability to alter properties by (1) affecting the formation of graphite or cementite, (2) modifying the morphology of the carbon-rich phase, (3) strengthening the matrix material, or (4) enhancing wear resistance through the formation of alloy carbides. Nickel, for example, promotes graphite formation and tends to promote finer graphite structures. Chromium retards graphite formation and stabilizes cementite. These alloys are frequently used together in a ratio of two or three parts of nickel to one part of chromium. Between 0.5 and 1.0% molybdenum is often added to gray cast iron to impart additional strength, form alloy carbides, and help to control the size of the graphite flakes.

High-alloy cast irons have been designed to provide enhanced corrosion resistance and/or good elevated temperature service. Within this family, the austenitic gray cast irons, which contain about 14% nickel, 5% copper, and 2.5% chromium, offer good corrosion resistance to many acids and alkalis at temperatures up to about 800°C (1500°F). The addition of up to 1% molybdenum and 5% silicon to ductile iron greatly enhances high temperature tensile strength and creep strength. They are attractive for applications involving temperatures between 650 and 875°C (1200 to 1600°F).

7.8 CAST STEELS

If a ferrous casting alloy contains less than about 2.0% carbon, it is considered to be a **cast steel.** (Alloys with more than 2% carbon are *cast irons.*) Cast steels are generally used whenever a cast iron is not adequate for the application. Compared to cast irons, the cast steels offer enhanced stiffness, toughness, and ductility over a wide range of operating temperatures and can be readily welded. They are usually heat-treated to produce a final quenched-and-tempered structure, and the alloy additions are selected to provide the desired hardenability and balance of properties. The enhanced properties come with a price, however, because the cast steels have a higher melting point (more energy to melt and higher cost refractories are necessary), lower fluidities (leading to increased probability of incomplete die or mold filling), and increased shrinkage (because graphite is not formed during solidification). The diverse applications take advantage of the material's structural strength and its ability to contain pressure, resist impacts, withstand elevated temperatures, and resist wear.

7.9 The Role of Processing on Cast Properties

While typical properties have been presented for the various types of cast materials, it should be noted that the properties of all metals are influenced by how they are processed. For cast materials, properties will often vary with the manner of solidification and cooling. Because cast components often have complex geometries, the cooling rate may vary from location to location, with companion variation in properties. To ensure compliance with industry specifications, standard geometry test bars are often cast along with manufactured products so the material can be evaluated and quality can be assured independent of product geometry.

Alloy cast irons and cast steels are usually specified by their ASTM designation numbers, which relate the materials to their mechanical properties and intended service applications. The Society of Automotive Engineers (SAE) also has specifications for cast steels used in the automotive industry.

Key Words

advanced high-strength steel (AHSS)
air-hardenable tool steel
AISI–SAE identification system
alloys
alloy steel
amorphous metals
austempered ductile iron (ADI)
austenitic stainless steel
bake-hardenable steel
cast iron
cast steel
class
cold-work tool steel
compacted graphite cast iron (CGI)
complex-phase (CP) steels
constructional alloys
consumable-electrode remelting
continuous casting
degasification
deoxidation
dual-phase steels
ductile cast iron
duplex stainless steel
electroslag remelting (ESR)
fading
ferritic stainless steel
ferrous metals
free-machining steel
gray cast iron
hardenability
high-alloy cast iron
high-alloy steel
high-carbon steel
high-chromium tool steel
high-speed tool steel
high-strength low-alloy (HSLA) steel
hot-work tool steel
inoculation
iron
ladle metallurgy
ladles
low-alloy steel
low-carbon steel
malleable cast iron
maraging steel
martensitic stainless steel
martensitic (Mart) steels
medium-carbon steel
microalloyed steel
mottled zone
nodular cast iron
nodulizer
oil-hardenable tool steel
pig iron
plain-carbon steel
precipitation-hardenable stainless steel
precoated steel sheet
sensitization
sigma phase
shock-resisting tool steel
soft magnetic materials
stainless steel
steel
tool steel
transformation-induced plasticity (TRIP) steels
true stainless steels
vacuum arc remelting (VAR)
vacuum degassing
vacuum induction melting (VIM)
water-hardenable tool steel
white cast iron

Review Questions

1. Why might it be important to know the prior processing history of an engineering material?
2. What is a ferrous material?
3. How does the recycling of steel compare with the recycling of aluminum?
4. Why is the recycling of steel so attractive compared to the manufacture from virgin ore?
5. When iron ore is reduced to metallic iron, what other elements are generally present in the metal?
6. What properties or characteristics have made steel such an attractive engineering material?
7. How does steel differ from pig iron?
8. What are some of the modification processes that can be performed on a steel during ladle metallurgy operations?
9. What is the advantage of pouring molten metal from the bottom of a ladle?
10. What are some of the attractive economic and processing advantages of continuous casting?
11. What problems are associated with oxygen or other gases that are dissolved in molten steel?
12. What are some of the techniques used to reduce the amount of dissolved oxygen in molten steel?
13. How might other gases, such as nitrogen and hydrogen, be reduced?
14. What are some of the attractive features of electroslag remelting?
15. What is plain-carbon steel?
16. What is considered a low-carbon steel? Medium-carbon? High-carbon?
17. What properties account for the high-volume use of medium-carbon steels?
18. Why should plain-carbon steels be given first consideration for applications requiring steel?
19. What are some of the common alloy elements added to steel?

20. For what different reasons might alloying elements be added to steel?
21. What are some of the alloy elements that tend to form stable carbides within a steel?
22. What alloys are particularly effective in increasing the hardenability of steel?
23. What is the significance of the last two digits in a typical four-digit AISI–SAE steel designation?
24. How are letters incorporated into the AISI–SAE designation system for steel, and what do some of the more common ones mean?
25. What is an H-grade steel, and when should it be considered in a material specification?
26. Why should the proposed fabrication processes enter into the considerations when selecting a steel?
27. How are the final properties usually obtained in the constructional alloy steels? In the HSLA steels?
28. How are HSLA steels specified?
29. What are microalloyed steels?
30. What are some of the potential benefits that may be obtained through the use of microalloyed steels?
31. What is the primary attraction of the bake-hardenable steels?
32. What are advanced high-strength steels (AHSS)?
33. What are the two phases that are present in dual-phase steels?
34. What is the "transformation" that occurs during the deformation of the transformation-induced plasticity (TRIP) steels?
35. What are some other types of advanced high-strength steels (AHHS)?
36. Describe the role of steel in the automotive industry over the past 50 years. What is the projection for the near future?
37. What is the role of chip-breakers in free-machining steels?
38. What are some of the various alloy additions that have been used to improve the machinability of steels?
39. What are some of the compromises associated with the use of free-machining steels?
40. What factors might be used to justify the added expense of precoated steel sheet?
41. Why have the amorphous metals attracted attention as potential materials for magnetic applications?
42. What are maraging steels, and for what conditions might they be required?
43. What are the typical elevated temperature limits of plain-carbon and alloy steels?
44. What is a stainless steel?
45. What feature is responsible for the observed corrosion resistance of stainless steels?
46. Why should ferritic stainless steels be given first consideration when selecting a stainless steel?
47. Which of the major types of stainless steel is likely to contain significant amounts of carbon? Why?
48. Under what conditions might a martensitic stainless steel "rust" when exposed to a hostile environment?
49. What are some of the unique properties of austenitic stainless steels?
50. How can an austenitic stainless steel be easily identified?
51. What two structures are present in a duplex stainless steel?
52. What is sensitization of a stainless steel, and how can it be prevented?
53. What problem is associated with the presence of sigma phase in a stainless steel?
54. What is a tool steel?
55. How does the AISI–SAE designation system for tool steels differ from that for plain-carbon and alloy steels?
56. For what types of applications might an air-hardenable tool steel be attractive?
57. What alloying elements are used to produce the hot-work tool steels?
58. What assets can be provided by the hot-isostatic-presses powder metallurgy tool steels?
59. Describe the microstructure of gray cast iron.
60. Which of the structural units is generally altered to increase the strength of a gray cast iron?
61. What are some of the attractive engineering properties of gray cast iron?
62. What are some of the key limitations to the engineering use of gray cast iron?
63. What is the dominant mechanical property of white cast iron?
64. What structural feature is responsible for the increased ductility and fracture resistance of malleable cast iron?
65. How is malleable cast iron produced?
66. What is unique about shape or form of the graphite in ductile cast iron?
67. What requirements of ductile iron manufacture are responsible for its increased cost over materials such as gray cast iron?
68. What is the purpose of inoculation when making ductile cast iron? Of nodulizing?
69. What is fading? Why should ductile iron be solidified immediately following nodulizing?
70. What are some of the attractive features of austempered ductile cast iron?
71. Compacted graphite iron has a structure and properties intermediate to what two other types of cast irons?
72. What are some of the reasons that alloy additions are made to cast irons that will be used in their as-cast condition?
73. When should a cast steel be used instead of a cast iron?

PROBLEMS

1. Identify at least one easily identified product or component that is currently being produced from each of the following types of cast irons:
   a. Gray cast iron
   b. White cast iron
   c. Malleable cast iron
   d. Ductile cast iron
   e. Compacted graphite cast iron
2. Find an example where one of the types of cast iron has been used in place of a previous material. What feature or features might have prompted the substitution?

Chapter 7 CASE STUDY

The Paper Clip

Sometimes even the simplest of everyday items can have a rather extensive materials and manufacturing history. For example, consider the standard wire paper clip.

By the 13th century, paper had become the standard material for written documents, and there was a need for attaching pieces together when the assembly did not merit permanent binding in the form of a book. The earliest method of fastening paper was to thread a string, strip of cloth, or ribbon through two parallel slits that had been made in the upper left-hand corner of the papers by a sharp knife. The ends could be sealed with wax to ensure that no substitutions had been made. Pins, made of metal or bone were also used as paper fasteners.

Bent-wire paper clips did not become common until the late 1800s. Various designs were introduced, all based on creating two metal surfaces that are pressed together by the elasticity of the metal wire. Various designs were proposed and were marketed with some of the following claims:

Does not catch, mutilate, or tear papers
Does not get tangled with other clips in the box
Holds a thick set of papers
Holds papers securely
Is thinner and takes less space in files
Is easily inserted;
Is light in weight and requires less postage
Is cheap, because it uses less wire

The familiar double-oval "Gem" paper clip was developed around 1900 and has proven to be a satisfactory product for more than a century. For this case study, assume you have been challenged to take the concept and design of the now-standard paper clip and transition it to manufactured reality. The following questions relate to materials selection, process selection, and surface treatments.

1. Material Selection:
   a. What are the necessary mechanical properties?
   b. What are the necessary physical properties?
   c. What would be appropriate families or classes of candidate materials?
   d. What specific metal, alloy, or other material would you recommend?
2. Process Selection:
   a. Because the design is essentially a bent wire product, describe the features you would specify in purchasing your starting material. Would you want to purchase it in the annealed (weakest with greatest ductility) condition? By positioning the last anneal in the wire-making sequence, the final wire can be produced with a specified amount of residual cold work. This wire would be stronger but less ductile than an annealed product. Would some amount of cold work in your starting wire be a desirable asset? Do you have any concerns about surface finish at this stage of manufacture?
   b. In converting the wire to the double-oval paper clip shape, cold work is imparted in the bend regions. The properties will no longer be uniform in the product. Is this acceptable in your final product?
   c. In separating the starting wire or finished clip from the long length, sheared ends will be created. Are there any concerns about these ends?
   d. Is any form of final heat treatment needed to impart the desired final properties?
3. Surface Treatment (Depending on the material selected in Part 1, it may be necessary to apply some form of coating or surface treatment to suppress corrosion and possible rust marks on the paper.)
   a. In addition to corrosion resistance, are there any other surface requirements?
   b. What types of surface treatments would you consider to be possibilities?
   c. Which of the preceding possibilities would you recommend?
   d. At what stage of manufacture do you want to perform the surface treatment and by what method? Give consideration to the following observations:
      - If applied to the starting wire, the coating will have to have sufficient ductility to endure the shaping operation.
      - If applied to the starting wire, the sheared ends will not be coated.
      - If applied to the finished product, the gap between the parallel wires may be bridged or may not be coated.
      - If the coating is applied at elevated temperature, the coating process may anneal the underlying wire.
      - Some coating operations may affect the ductility of the underlying material, such as hydrogen embrittlement.
      - Some processes, such as hot dip immersion in molten metal, apply a thicker-than-desired coating. If this is applied earlier in the wire-making process, will the subsequent wire drawing operations maintain the integrity of the coating, or simply wipe off the deposited surface?
   e. Would stainless steel be a desirable material for paper clip manufacture? Which type or variety would you recommend (ferritic, austenitic, or martensitic)? Would there be any advantage to having a nonmagnetic austenitic paper clip?

CHAPTER 8

NONFERROUS METALS AND ALLOYS

8.1 INTRODUCTION

Nonferrous metals and alloys have assumed increasingly important roles in modern technology. Because of their number and the fact that their properties vary widely, they provide an almost limitless range of properties for the design engineer. While they tend to be more costly than iron or steel, these metals often possess certain properties or combinations of properties that are not available in the ferrous metals, such as:

1. Resistance to corrosion
2. Ease of fabrication
3. High electrical and thermal conductivity
4. Light weight
5. Strength at elevated temperatures
6. Color

Nearly all of the nonferrous alloys possess at least two of the qualities listed, and some possess nearly all. For many applications, specific combinations of these properties are highly desirable. Each year, the average American uses about 65 lb of aluminum, 21 lb of copper, 12 lb of lead, 11 lb of zinc, and 25 lb of various other nonferrous metals. Figure 8-1 groups some of the nonferrous metals by attractive engineering property.

As a whole, the strength of the nonferrous alloys is generally inferior to that of steel, and the modulus of elasticity is usually lower, a fact that places them at a distinct disadvantage when stiffness is required. Ease of fabrication, however, is often attractive. Those alloys with low melting points are easy to cast in sand molds, permanent molds, or dies. Many alloys have high ductility coupled with low yield points, the ideal combination for cold working. Good machinability is also characteristic of many nonferrous alloys. The savings obtained through ease of fabrication can often overcome the higher cost of the nonferrous material and justify its use in place of steel. Weldability is the one fabrication area where the nonferrous alloys tend to be somewhat inferior to

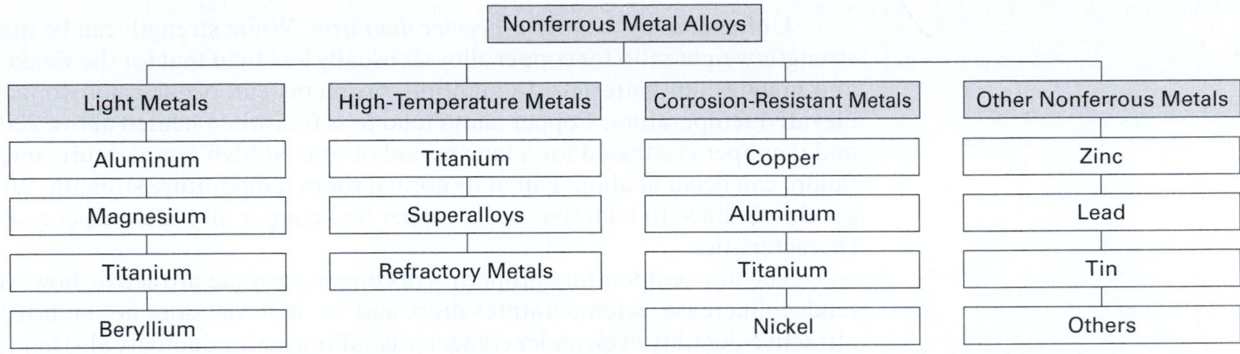

FIGURE 8-1 Some common nonferrous metals and alloys, classified by attractive engineering property.

steel. With modern joining techniques, however, it is generally possible to produce satisfactory weldments in all of the nonferrous metals.

8.2 COPPER AND COPPER ALLOYS

GENERAL PROPERTIES AND CHARACTERISTICS

Copper has been an important engineering metal for more than 6000 years. As a pure metal, it has been the backbone of the electrical industry. It is also the base metal of a number of alloys, generically known as brasses and bronzes. Compared to other engineering materials, copper and copper alloys offer three important properties: (1) *high electrical and thermal conductivity*, (2) *useful strength with high ductility*, and (3) *corrosion resistance* to a wide range of media. Because of its excellent conductivity, about one-third of all copper produced is used in some form of electrical application, such as the items shown in Figure 8-2. Other large areas of use include plumbing, heating, and air conditioning.

Pure copper in its annealed state, has a tensile strength of only about 200 MPa (30 ksi), with an elongation of nearly 60%. Through cold working, the tensile strength can be more than doubled to greater than 450 MPa (65 ksi), with a decrease in elongation to about 5%. Because of its relatively low strength and high ductility, copper is a very desirable metal for applications where extensive forming is required. Because the recrystallization temperature for copper is less than 260°C (500°F), the hardening effects of cold working can be easily removed to establish desired final properties or permit further deformation. Copper and copper alloys also lend themselves nicely to the whole spectrum of fabrication processes, including casting, machining, joining, and surface finishing by either plating or polishing.

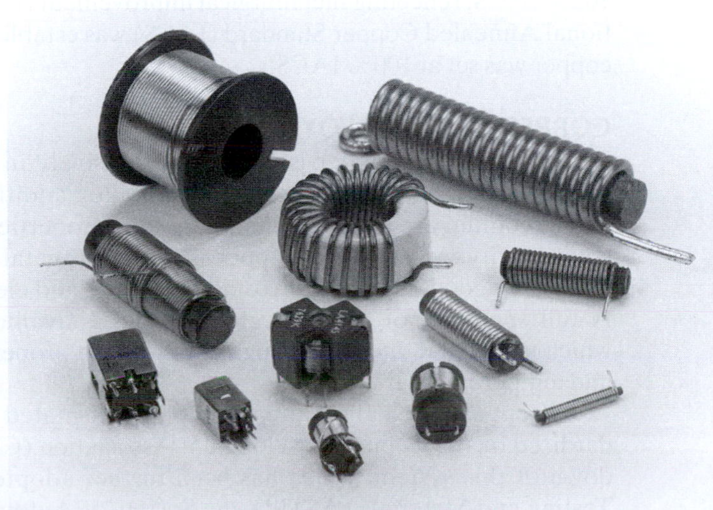

FIGURE 8-2 Copper and copper alloys are used for a variety of electrical applications. *(© David J. Green–electrical/Alamy)*

Unfortunately, copper is *heavier than iron*. While strength can be quite high, the strength/weight ratio for copper alloys is usually less than that for the weaker aluminum and magnesium materials. In addition, problems can occur when copper is used at elevated temperature. Copper alloys tend to soften when heated above 220°C (400°F), and if copper is stressed for a long period of time at high temperature, intercrystalline failure can occur at about half of its normal room temperature strength. While offering good resistance to adhesive wear, copper and copper alloys have poor abrasive wear characteristics.

The low-temperature properties of copper are quite attractive, however. Strength tends to increase as temperatures drop, and the material does not embrittle, retaining attractive ductility even under cryogenic conditions. Conductivity also tends to increase with a drop in temperature.

Copper and copper alloys respond well to strengthening methods, with the strongest alloy being 15 to 20 times stronger than the weakest. Because of the wide range of properties, the material can often be tailored to the specific needs of a design. Elastic stiffness is between 50 and 60% of steel. Additionally, they are nonmagnetic, nonpyrophoric (slivers or particles do not burn in air, i.e., they are nonsparking), nonbiofouling (inhibits marine organism growth), and available in a wide spectrum of colors (including yellow, red, brown, and silver).

The U.S. Environmental Protection Agency (EPA) recently confirmed the antimicrobial properties of more than 300 copper alloys, killing more than 99.9% of bacteria (including staphylococcus and E-coli) within two hours of contamination and delivering continuous antibacterial action even after repeated wear and recontamination. These alloys may well replace stainless steel in applications involving surfaces that are repeatedly touched by multiple individuals, such as door knobs and sink faucet handles. Copper components can also reduce airborne pathogens in heating, ventilation, and air-conditioning systems. Use is expected to expand in hospitals, schools, public buildings, exercise facilities, nursing homes, shopping malls, mass transit facilities, airports, interiors of airplanes and cruise ships, and even private homes.

Another evolving market is the hybrid or electric car, with each using up to 100 lb of copper in the various motors and wiring.

COMMERCIALLY PURE COPPER

Being second only to silver in conductivity, commercially pure copper is used primarily for electrical applications. Refined copper containing between 0.02 and 0.05% oxygen is called **electrolytic tough-pitch (ETP) copper.** It is often used as a base for copper alloys, and may be used for electrical applications such as wire and cable when the highest conductivity is not required. For superior conductivity, additional refining can reduce the oxygen content and produce **oxygen-free high-conductivity (OFHC) copper.** The better grades of conductor copper now have a conductivity rating of about 102% IACS, reflecting metallurgical improvements made since 1913, when the International Annealed Copper Standard (IACS) was established and the conductivity of pure copper was set at 100% IACS.

COPPER-BASED ALLOYS

As a pure metal, copper is not used extensively in manufactured products, except in electrical applications, and even here alloy additions of silver, arsenic, cadmium, and zirconium are used to enhance various properties without significantly impairing conductivity. More often, copper is the base metal for an alloy, where the copper imparts its good ductility, corrosion resistance, and electrical and thermal conductivity. A full spectrum of mechanical properties is available, ranging from pure copper, which is soft and ductile, through alloys whose properties can rival those of quenched and tempered steel.

Copper-based alloys are commonly designated using a system of numbers standardized by the Copper Development Association (CDA). Table 8-1 presents a breakdown of this system, which has been further adopted by the American Society for Testing and Materials (ASTM), the Society of Automotive Engineers (SAE), and the

TABLE 8-1 Designation System for Copper and Copper Alloys (Copper Development Association System)

Wrought Alloys		Cast Alloys	
100–155	Commercial coppers	833–838	Red brasses and leaded red brasses
162–199	High-copper alloys	842–848	Semired brasses and leaded semired brasses
200–299	Copper–zinc alloys (brasses)	852–858	Yellow brasses and leaded yellow brasses
300–399	Copper–zinc–lead alloys (leaded brasses)	861–868	Manganese and leaded manganese bronzes
400–499	Copper–zinc–tin alloys (tin brasses)	872–879	Silicon bronzes and silicon brasses
500–529	Copper–tin alloys (phosphor bronzes)	902–917	Tin bronzes
532–548	Copper–tin–lead alloys (leaded phosphor bronzes)	922–929	Leaded tin bronzes
600–642	Copper–aluminum alloys (aluminum bronzes)	932–945	High-leaded tin bronzes
647–661	Copper–silicon alloys (silicon bronzes)	947–949	Nickel–tin bronzes
667–699	Miscellaneous copper–zinc alloys	952–958	Aluminum bronzes
700–725	Copper–nickel alloys	962–966	Copper nickels
732–799	Copper–nickel–zinc alloys (nickel silvers)	973–978	Leaded nickel bronzes

U.S. government. Alloys numbered from 100 to 199 are mostly copper with less than 2% alloy addition. Numbers 200 to 799 are **wrought alloys,** and the 800 and 900 series are **cast alloys.**[1] When converted to the Unified Numbering System for metals and alloys, the three-digit numbers are converted to five digits by placing two zeros at the end, and the letter *C* as a prefix to denote the copper base.

COPPER–ZINC ALLOYS

Zinc is by far the most popular alloying addition, and the resulting alloys are generally known as some form of **brass.** If the zinc content is less than 36%, the brass is a single-phase solid solution. Because this structure is identified as the alpha phase, these alloys are often called *alpha brasses*. They are quite ductile and formable, with both strength and ductility increasing with the amount of zinc throughout the single-phase region. The alpha brasses can be strengthened significantly by cold working and are commercially available in various degrees of cold-worked strength and hardness. *Cartridge brass,* the 70% copper–30% zinc alloy, offers the best overall combination of strength and ductility. As its name implies, it has become a popular material for sheet-forming operations like deep drawing.

Many applications of these alloys result from the high electrical and thermal conductivity coupled with useful engineering strength. The wide range of colors (red, orange, yellow, silver, and white), enhanced by further variations that can be produced through the addition of a third alloy element, account for a number of decorative uses. Because the plating characteristics are excellent, the material is also a frequently used base for decorative chrome or similar coatings. Another attractive property of alpha brass is its ability to have rubber vulcanized to it without any special treatment except thorough cleaning. As a result, brass is widely used in mechanical rubber goods.

With more than 36% zinc, the copper–zinc alloys enter a two-phase region involving a brittle, zinc-rich phase, and ductility drops markedly. While cold-working properties are rather poor for these high-zinc brasses, deformation can be performed easily at elevated temperature.

Most brasses have good corrosion resistance. In the range of 0 to 40% zinc, the addition of a small amount of tin imparts improved resistance to seawater corrosion. Cartridge brass with tin becomes admiralty brass, and the 40% zinc Muntz metal with a tin addition is called *naval brass.* Brasses with 20 to 36% zinc, however, are subject to a selective corrosion, known as **dezincification,** when exposed to acidic or salt solutions. Brasses with more than 15% zinc often experience **season cracking** or **stress-corrosion cracking.** Both stress and exposure to corrosive media are required for this failure to

[1] The term *wrought* means "shaped or fabricated in the solid state." Key properties for wrought material generally relate to ductility. *Cast* alloys are shaped as a liquid, where the attractive features include low melting point, high fluidity, and good as-solidified strength.

TABLE 8-2 Composition, Properties, and Uses of Some Common Copper–Zinc Alloys

CDA Number	Common Name	Composition (%) Cu	Zn	Sn	Pb	Mn	Condition	Tensile Strength ksi	MPa	Elongation in 2 in. (%)	Typical Uses
220	Commercial bronze	90	10				Soft sheet Hard sheet	38 64	262 441	45 4	Screen wire, hardware, screws. jewelry
240	Low brass	80	20				Spring Annealed sheet Hard	73 47 75	503 324 517	3 47 7	Drawing, architectural work, ornamental
260	Cartridge brass	70	30				Spring Annealed sheet Hard	91 53 76	627 365 524	3 54 7	Munitions, hardware, musical instruments, tubing
270	Yellow brass	65	35				Spring Annealed sheet Hard	92 46 76	634 317 524	3 64 7	Cold forming, radiator cores, springs, screws
280	Muntz metal	60	40				Hot-rolled Cold-rolled	54 80	372 551	45 5	Architectural work; condenser tube
443–445	Admiralty metal	71	28	1			Soft Hard	45 95	310 655	60 5	Condenser tube (salt water), heat exchangers
360	Free-cutting brass	61.5	35.3		3		Soft Hard	47 62	324 427	60 20	Screw-machine parts
675	Manganese bronze	58.5	39	1		0.1	Soft Bars, half hard	65 84	448 579	33 19	Clutch disks, pump rods, valve stems, highstrength propellers

occur (but residual stresses and atmospheric moisture may be sufficient!). As a result, cold-worked brass is usually stress relieved (to remove the residual stresses) before being placed in service.

When high machinability is required, as with automatic screw machine stock, 2 to 3% lead can be added to the brass to ensure the formation of free-breaking chips. Brass casting alloys are quite popular for use in plumbing fixtures and fittings, low-pressure valves, and a variety of decorative hardware. They have good fluidity during pouring and attractive low melting points. An alloy containing between 50 and 55% copper and the remainder zinc is often used as a filler metal in brazing. It is an effective material for joining steel, cast iron, brasses, and copper, producing joints that are nearly as strong as those obtained by welding.

Table 8-2 lists some of the more common copper–zinc alloys and their composition, properties, and typical uses.

COPPER–TIN ALLOYS

Because **tin** is more costly than zinc, alloys of copper and tin, commonly called **tin bronzes,** are usually specified when they offer some form of special property or characteristic. The term **bronze** is often confusing, however, because it can be used to designate any copper alloy where the major alloy addition is not zinc or nickel. To provide clarification, the major alloy addition is usually included in the designation name.

The tin bronzes usually contain less than 12% tin. (Strength continues to increase as tin is added up to about 20%, but the high-tin alloys tend to be brittle.) Tin bronzes offer good strength, toughness, wear resistance, and corrosion resistance. They are often used for bearings, gears, and fittings that are subjected to heavy compressive loads. When the copper–tin alloys are used for bearing applications, up to 10% lead is frequently added.

The most popular wrought alloy is phosphor bronze, which usually contains from 1 to 11% tin along with an addition of phosphorus. Alloy 521 (CDA), with 8% tin, is typical of this class. Hard sheet has a tensile strength of 760 MPa (110 ksi) and an elongation of 3%. Soft sheet has a tensile strength of 380 MPa (55 ksi) and 65% elongation. The material is often specified for pump parts, gears, springs, and bearings.

Alloy 905 is a tin-bronze casting alloy containing 10% tin and 2% zinc. In the as-cast condition, the tensile strength is about 310 MPa (45 ksi), with an elongation of 45%. It has very good resistance to seawater corrosion and is used on ships for pipe fittings, gears, pump parts, bushings, and bearings.

Bronzes can also be made by mixing powders of copper and tin, followed by low-density powder metallurgy processing (described in Chapter 18). The porous product can be used as a filter for high-temperature or corrosive media, or can be infiltrated with oil to produce self-lubricating bearings.

COPPER–NICKEL ALLOYS

Copper and nickel exhibit complete solubility (as shown previously in Figure 5-6), and a wide range of useful alloys has been developed. Key features include high thermal conductivity, high-temperature strength, and corrosion resistance to a range of materials, including seawater. These properties, coupled with a high resistance to stress-corrosion cracking, make the copper–nickel alloys a good choice for heat exchangers, cookware, desalination apparatus, and a wide variety of coinage. **Cupronickels** contain 2 to 30% nickel. **Nickel silvers** contain no silver, but 10 to 30% nickel and at least 5% zinc. The bright silvery luster makes them attractive for ornamental applications, and they are also used for musical instruments. An alloy with 45% nickel is known as **constantan,** and the 67% nickel material is called *Monel*. Monel will be discussed later in the chapter as a nickel alloy.

OTHER COPPER-BASED ALLOYS

The copper alloys discussed previously acquire their strength primarily through solid solution strengthening and cold work. Within the copper-alloy family, alloys containing aluminum, silicon, or beryllium can be strengthened by precipitation hardening.

Aluminum bronze alloys are best known for their combination of high strength and excellent corrosion resistance, and are often considered as cost-effective alternatives to stainless steel and nickel-based alloys. The wrought alloys can be strengthened by solid-solution strengthening, cold work, and the precipitation of iron- or nickel-rich phases.

With less than 8% aluminum, the alloys are very ductile. When aluminum exceeds 9%, however, the ductility drops and the hardness approaches that of steel. Still higher aluminum contents result in brittle, but wear-resistant, materials. By varying the aluminum content and heat treatment, the tensile strength can range from about 415 to 1000 MPa (60 to 145 ksi). Typical applications include marine hardware, power shafts, sleeve bearings, and pump and valve components for handling seawater, sour mine water, and various industrial fluids. Cast alloys are available for applications where casting is the preferred means of manufacture. Because aluminum bronze exhibits large amounts of solidification shrinkage, castings made of this material should be designed with shrinkage in mind.

Silicon bronzes contain up to 4% silicon and 1.5% zinc (higher zinc contents may be used when the material is to be cast). Strength, formability, machinability, and corrosion resistance are all quite good. Tensile strengths range from a soft condition of about 380 MPa (55 ksi) through a maximum that approaches 900 MPa (130 ksi). Uses include boiler, tank, and stove applications, which require a combination of weldability, high strength, and corrosion resistance.

Copper–beryllium alloys, which contain up to $2\frac{1}{2}$% beryllium, can be age hardened to produce the highest strengths of the copper-based metals but are quite expensive to use. When annealed the material has a yield strength of 170 MPa (25 ksi), tensile strength of 480 MPa (70 ksi), and an elongation of 50%. After heat treatment, these properties can rise to 1100 MPa (160 ksi), 1250 MPa (180 ksi), and 5%, respectively. Cold work coupled with age hardening can produce even stronger material. The modulus of elasticity is about 125,000 MPa (8×10^6 psi) and the endurance limit is around 275 MPa (40 ksi). These properties make the material an excellent choice for electrical contact springs, but cost limits application to small components requiring long life and high reliability. Other applications, such as spark-resistant safety tools and spot welding electrodes, utilize a unique combination of properties: (1) the material has the strength of heat-treated steel, but is also (2) nonsparking, nonmagnetic, and electrically and thermally conductive. Concerns over the toxicity of beryllium have created a demand for substitute alloys with similar properties, but no clear alternative has emerged.

LEAD-FREE CASTING ALLOYS

For many years, **lead** has been a common alloy additive to cast copper alloys. It helped to fill and seal the micro-porosity that forms during solidification, thereby providing the pressure tightness required for use with pressurized gases and fluids. The lead also acted as a lubricant and chip-breaker, enhancing the machinability and machined surface finish. Many plumbing components have been made from leaded red and semi-red brass casting alloys.

With increased concern for lead in drinking water, and the introduction of environmental regulations, efforts were made to develop lead-free copper-based casting alloys. Among the most common are the EnviroBrass alloys, which use **bismuth** and **selenium** as substitutes for lead. Bismuth is not known to be toxic for humans and has been used in a popular remedy for upset stomach. Selenium is an essential nutrient for humans. While somewhat lower in ductility, the new alloys have been shown to have mechanical properties, machinability and platability that are quite similar to the traditional leaded materials.

8.3 ALUMINUM AND ALUMINUM ALLOYS

GENERAL PROPERTIES AND CHARACTERISTICS

Although **aluminum** has only been a commercial metal for about 125 years, it now ranks second to steel in both worldwide quantity and expenditure, and is clearly the most important of the nonferrous metals. It has achieved importance in virtually all segments of the economy, with principal uses in transportation, containers and packaging, building construction, electrical applications, consumer durables, and mechanical equipment. We are all familiar with uses such as aluminum cookware, window frames, aluminum siding, and the ever-present aluminum beverage can.

A number of unique and attractive properties account for the engineering significance of aluminum. These include its workability, light weight, corrosion resistance, good electrical and thermal conductivity, optical reflectivity, and a nearly limitless array of available finishes. Aluminum has a specific gravity of 2.7 compared to 7.85 for steel, making aluminum about one-third the weight of steel for an equivalent volume. Cost comparisons are often made on the basis of cost per pound, where aluminum is at a distinct disadvantage, being four to five times more expensive than carbon steel. There are a number of applications, however, where a more appropriate comparison would be based on cost per unit volume. A pound of aluminum produces three times as many same-size parts as a pound of steel, so the cost difference becomes markedly less.

Aluminum can be recycled repeatedly with no loss in quality, and recycling saves 95% of the energy required to produce aluminum from ore. Since the 1980s, the overall reclamation rate for aluminum has been greater than 50%. The aluminum can is the most recycled beverage container in North America, and nearly 90% of all aluminum used in cars is recovered at the end of their useful life.

A serious weakness of aluminum from an engineering viewpoint is its relatively low modulus of elasticity, which is also about one-third that of steel. Under identical loadings, an aluminum component will deflect three times as much as a steel component of the same design. Because the modulus of elasticity cannot be significantly altered by alloying or heat treatment, it is usually necessary to provide stiffness and buckling resistance through design features such as ribs or corrugations. These can be incorporated with relative ease, however, because aluminum adapts easily to the full spectrum of fabrication processes.

COMMERCIALLY PURE ALUMINUM

In its pure state, aluminum is soft, ductile, and not very strong. In the annealed condition, pure aluminum has only about one fifth the strength of hot-rolled structural steel. Commercially pure aluminum, therefore, is used primarily for its physical rather than its mechanical properties.

Electrical-conductor-grade aluminum is used in large quantities and has replaced copper in many applications, such as electrical transmission lines. Commonly designated by the letters EC, this grade contains a minimum of 99.45% aluminum and has an electrical conductivity that is 62% that of copper for the same size wire and 200% that of copper on an equal-weight basis.

ALUMINUMS FOR MECHANICAL APPLICATIONS

For nonelectrical applications, most aluminum is used in the form of alloys. These have much greater strength than pure aluminum, yet retain the advantages of light weight, good conductivity, and corrosion resistance. While usually weaker than steel, some alloys are now available that have tensile properties (except for ductility) that are comparable to those of the HSLA structural grades. Because alloys can be as much as 30 times stronger than pure aluminum, designers can frequently optimize their design and then tailor the material to their specific requirements. Some alloys are specifically designed for casting, while others are intended for the manufacture of wrought products.

On a strength-to-weight basis, most of the aluminum alloys are superior to steel and other structural metals, but wear, creep, and fatigue properties are generally rather poor. Aluminum alloys have a finite fatigue life at all reasonable values of applied stress. In addition, aluminum alloys rapidly lose their strength and dimensions change by creep when temperature is increased. As a result, most aluminum alloys should not be considered for applications involving service temperatures much above 150°C (300°F). At subzero temperatures, however, aluminum is actually stronger than at room temperature with no loss in ductility. Both the adhesive and the abrasive varieties of wear can be extremely damaging to aluminum alloys.

The selection of steel or aluminum for any given component is often a matter of cost, but considerations of light weight, corrosion resistance, low maintenance expense, and high thermal or electrical conductivity may be sufficient to justify the added cost of aluminum. With the drive for lighter, more fuel-efficient vehicles, the fraction of aluminum targeted for transportation applications has risen markedly. In 2001, aluminum passed plastics as a percentage of automotive material content, and in 2006, it passed iron. It is now second only to steel. The average North American automobile now contains nearly 150 kg (325 lb) of aluminum, with a projected growth rate of 3.5 to 4.5 kg (8 to 10 lb) per year. Approximately 80% of automotive aluminum is used in engines, transmissions, wheels, and heat exchangers (radiators). Nearly 70% of all automotive engine blocks are now cast in aluminum. An aluminum space frame, such as the ones shown in Figure 8-3 for the 2005 Ford GT and the 2006 Corvette Z06, can reduce the overall weight of the structure, enhance recyclability, and reduce the number of parts required for the primary body structure. The all-aluminum space frame of the 2006 Z06 Corvette resulted in a 30% reduction in weight from the all-steel design of the previous model.

CORROSION RESISTANCE OF ALUMINUM AND ITS ALLOYS

Pure aluminum is very reactive and forms a tight, adherent oxide coating on the surface as soon as it is exposed to air. This oxide is resistant to many corrosive media and serves as a corrosion-resistant barrier to protect the underlying metal. Like stainless steels, the corrosion resistance of aluminum is actually a property of the oxide, not the metal itself. Because the oxide formation is somewhat retarded when alloys are added, aluminum alloys do not have quite the corrosion resistance of pure aluminum.

(a)

(b)

FIGURE 8-3 (a) The space frame chassis for the 2005 Ford GT is comprised of 35 aluminum extrusions, 7 complex castings, 2 semisolid castings, and various aluminum panels, some superplastically formed. (b) The aluminum frame of the 2006 Corvette Z06 yielded a 30% weight savings compared to the previous steel design. *[(a) Courtesy Ford Motor Company, Dearborn, MI and HydroAluminum North America; (b) Courtesy of General Motors, Detroit, MI]*

The oxide coating also causes difficulty when welding. To produce consistent quality resistance welds, it is usually necessary to remove the tenacious oxide immediately before welding. For fusion welding, special fluxes or protective inert gas atmospheres must be used to prevent material oxidation. While welding aluminum may be more difficult than steel, suitable techniques have been developed to permit the production of high-quality, cost-effective welds with most of the welding processes.

CLASSIFICATION SYSTEM

Aluminum alloys can be divided into two major groups based on the method of fabrication. *Wrought alloys* are those that are shaped as solids and are therefore designed to have attractive forming characteristics, such as low yield strength, high ductility, good fracture resistance, and good strain hardening. *Casting alloys* achieve their shape as they solidify in molds or dies. Attractive features for the casting alloys include low melting point, high fluidity, resistance to hot cracking during and after solidification, and attractive as-solidified structures and properties. Clearly, these properties are distinctly different, and the alloys that have been designed to meet them are also different. As a result, separate classification systems exist for the wrought and cast aluminum alloys.

WROUGHT ALUMINUM ALLOYS

The wrought aluminum alloys are generally identified using the standard four-digit designation system for aluminums. The first digit indicates the major alloy element or elements as described here:

Major Alloying Element	
Aluminum, 99.00% and greater	1xxx
Copper	2xxx
Manganese	3xxx
Silicon	4xxx
Magnesium	5xxx
Magnesium and silicon	6xxx
Zinc	7xxx
Other element	8xxx

For the 1xxx series, the remaining digits indicate the level of purity. For all remaining series, the second digit is usually zero, with nonzero numbers being used to indicate some form of modification or improvement to the original alloy. The last two digits simply indicate the particular alloy within the family. For example, 2024 simply means alloy number 24 within the 2xxx, or aluminum–copper, system.

The four digits of a wrought aluminum designation identify the chemistry of the alloy. Additional information about the processing history of the alloy (i.e., its condition) is then provided through a **temper designation,** in the form of a letter or letter–number suffix using the following system:

-F: as fabricated

-H: strain-hardened

- -H1: strain-hardened by working to desired dimensions; a second digit, 1 through 9, indicates the degree of hardening, 8 being commercially full-hard and 9 extra-hard
- -H2: strain-hardened by cold working, followed by partial annealing
- -H3: strain-hardened and stabilized

-O: annealed

-T: thermally treated (heat treated)

- -T1: cooled from hot working and naturally aged
- -T2: cooled from hot working, cold-worked, and naturally aged
- -T3: solution-heat-treated, cold-worked, and naturally aged
- -T4: solution-heat-treated and naturally aged
- -T5: cooled from hot working and artificially aged
- -T6: solution-heat-treated and artificially aged
- -T7: solution-heat-treated and stabilized
- -T8: solution-heat-treated, cold-worked, and artificially aged
- -T9: solution-heat-treated, artificially aged, and cold-worked
- -T10: cooled from hot working, cold-worked, and artificially aged

-W: solution-heat-treated only

These treatments or tempers can be used to increase strength, hardness, wear resistance, or machinability; stabilize mechanical or physical properties; alter electrical characteristics; reduce residual stresses; or improve corrosion resistance. By identifying both the chemistry and the condition, one can reasonably estimate the mechanical and physical properties.

Wrought alloys can be further divided into two basic types: those that achieve strength by solid-solution strengthening and cold working and those that can be strengthened by an age-hardening heat treatment. Table 8-3 lists some of the common wrought aluminum alloys in each of these families. It can be noted that the work-hardenable alloys (those that cannot be age-hardened) are primarily those in the 1xxx (pure aluminum), 3xxx (aluminum–manganese), and 5xxx (aluminum–magnesium) series. A comparison of the annealed (O-suffix) and cold-worked (H-suffix) conditions reveals the amount of strengthening achievable through strain hardening.

The precipitation-hardenable alloys are found primarily in the 2xxx, 6xxx, and 7xxx series. By comparing the properties in the heat-treated condition to those of the strain-hardened alloys, we see that heat-treatment offers significantly higher strength. Alloy 2017, the original **duralumin,** is probably the oldest age-hardenable aluminum alloy. The 2024 alloy is stronger and has seen considerable use in aircraft applications. An attractive feature of the 2xxx series is the fact that ductility does not significantly decrease during the strengthening heat treatment. Within the 7xxx series are some newer alloys with strengths that approach or exceed those of the high-strength structural steels. Ductility, however, is generally low, and fabrication is more difficult than for the 2xxx-type alloys. Nevertheless, the 7xxx series alloys have also found wide use in aircraft applications. To maintain properties, age-hardened alloys should not be used at temperatures greater than 175°C (350°F). Welding should be performed with considerable caution because the exposure to elevated temperature will significantly diminish the strengthening achieved through either cold working or age hardening.

Because of their two-phase structure, the heat-treatable alloys tend to have poorer corrosion resistance than either pure aluminum or the single-phase work-hardenable alloys. When both high strength and superior corrosion resistance are desired, wrought aluminum is often produced as **Alclad** material. A thin layer of corrosion-resistant aluminum is bonded to one or both surfaces of a high-strength alloy during rolling, and the material is further processed as a composite.

Because only moderate temperatures are required to lower the strength of aluminum alloys, extrusions and forgings are relatively easy to produce and are manufactured in large quantities. Deep drawing and other sheet-metal-forming operations can also be carried out quite easily. In general, the high ductility and low yield strength of the aluminum alloys make them appropriate for almost all forming operations. Good dimensional tolerances and fairly intricate shapes can be produced with relative ease.

The machinability of aluminum-based alloys, however, can vary greatly, and special tools and techniques may be desirable if large amounts of machining are required. Free-machining alloys, such as 2011, have been developed for screw-machine work. These special alloys can be machined at very high speeds and have replaced brass screw machine stock in many applications.

Color anodizing offers an inexpensive and attractive means of surface finishing. A thick aluminum oxide is produced on the surface. Colored dye then penetrates the porous surface and is sealed by immersion into hot water. The result is the colored metallic finish commonly observed on products such as bicycle frames and softball bats.

ALUMINUM CASTING ALLOYS

Although its low melting temperature tends to make it suitable for casting, pure aluminum is seldom cast. Its high shrinkage upon solidification (about 7%) and susceptibility to hot cracking cause considerable difficulty, and scrap is high. By adding small amounts of alloying elements, however, very suitable casting characteristics can be obtained and strength can be increased. Aluminum alloys are cast in considerable quantity by a variety of processes. Many of the most popular alloys contain enough silicon to produce the eutectic reaction, which is characterized by a low melting point and high as-cast

TABLE 8-3 Composition and Properties of Some Wrought Aluminum Alloys in Various Conditions

Designation[a]	Composition (%) Aluminum = Balance					Form Tested	Tensile Strength		Yield Strength[b]		Elongation in 2 in. (%)	Brinell Hardness	Uses and Characteristics
	Cu	Si	Mn	Mg	Others		ksi	MPa	ksi	MPa			
						Work-Hardening Alloys—Not Heat-Treatable							
1100–0	0.12				99 A1	$\frac{1}{16}$-in. sheet	13	90	5	34	35	23	Commercial Al: good forming properties
1100–H14						$\frac{1}{16}$-in. sheet	16	110	14	97	9	32	Good corrosion resistance, low yield strength
110–H18						$\frac{1}{16}$-in. sheet	24	165	21	145	5	44	Cooking utensils; sheet and tubing
3003–0	0.12		1.2			$\frac{1}{16}$-in. sheet	16	110	6	41	30	28	Similar to 1100
3003–H14						$\frac{1}{16}$-in. sheet	22	152	21	145	8	40	Slightly stronger and less ductile
3003–H18						$\frac{1}{16}$-in. sheet	29	200	27	186	4	55	Cooking utensils; sheet-metal work
5052–0				2.5	0.25 Cr	$\frac{1}{16}$-in. sheet	28	193	13	90	25	45	Strongest work-hardening alloy
5052–H32						$\frac{1}{16}$-in. sheet	33	228	28	193	12	60	Highly yield strength and fatigue limit
5052–H36						$\frac{1}{16}$-in. sheet	40	276	35	241	8	73	Highly stressed sheet-metal products
						Precipitation-Hardening Alloys—Heat-Treatable							
2017–0	4.0	0.5	0.7	0.6		$\frac{1}{16}$-in. sheet	26	179	10	69	20	45	Duralumin, original strong alloy
2017–T4						$\frac{1}{16}$-in. sheet	62	428	40	276	20	105	Hardened by quenching and aging
2024–0	4.4		0.6	1.5		$\frac{1}{16}$-in. sheet	27	186	11	76	20	42	Stronger than 2017
2024–T4						$\frac{1}{16}$-in. sheet	64	441	45	290	19	120	Used widely in aircraft construction
2014–0	4.4	0.8	0.8	0.5		$\frac{1}{2}$-in. extruded shapes	27	186	14	97	12	45	Strong alloy for extruded shapes
2014–T6						Forgings	65	448	55	379	10	125	Strong forging alloy
2014–T6						$\frac{1}{16}$-in. sheet	70	483	60	413	8		Higher yield strength than Alclad 2024
Alclad 2014–T6	4.5	1.0	0.8	0.4		$\frac{1}{16}$-in. sheet	63	434	56	386	7		Clad with heat-treatable alloy[c]
7075–0	1.6		0.2	2.5	0.3 Cr, 5.6 Zn	$\frac{1}{16}$-in. sheet	33	228	15	103	17	60	Alloy of highest strength
7075–T6						$\frac{1}{16}$-in. sheet	76	524	67	462	11	150	Lower ductility than 2024
Alclad 7075–T6						$\frac{1}{16}$-in. sheet	76	524	67	462	11		Strongest Alclad product
7075–T6						$\frac{1}{2}$-in. extruded shapes	80	552	70	483	6		Strongest alloy for extrusions
6061–T6	0.28	0.6		1.0	0.20 Cr	$\frac{1}{2}$-in. extruded shapes	42	290	40	276	12	95	Strong, corrosion resistant
6063–T6		0.4		0.7		$\frac{1}{2}$-in. rod extruded	35	241	31	214	12	80	Good forming properties and corrosion resistance
6151–T6		0.9		0.6	0.20 Cr	Forgings	48	331	43	297	17	90	For intricate forgings
2025–T6	4.5	0.8	0.8			Forgings	55	379	30	207	18	100	Good forgeability, lower cost
2018–T6	4			0.7	2 Ni	Forgings	55	379	40	276	10	100	Strong at elevated temperatures; forged pistons
4032–T6	0.9	12.2		1.1	0.9 Ni	Forgings	55	379	46	317	9	115	Forged aircraft pistons
2011–T3	5.5			(0.5 Bi)	0.5 Pb	$\frac{1}{2}$-in. rod	55	379	43	297	15	95	Free cutting, screw-machine products

[a] O, annealed; T, quenched and aged; H, cold-rolled to hard temper.
[b] Yield strength taken at 0.2% permanent set.
[c] Cladding alloy; 1.0 Mg, 0.7 Si, 0.5 Mn.

strength. Silicon also improves the fluidity of the metal, making it easier to produce complex shapes or thin sections, but high silicon also produces an abrasive, difficult-to-cut material. Copper, zinc, and magnesium are other popular alloy additions that permit the formation of age-hardening precipitates.

Table 8-4 lists some of the commercial aluminum casting alloys and uses the three-digit designation system of the Aluminum Association to designate alloy chemistry. The first digit indicates the alloy group as follows:

Major Alloying Element	
Aluminum, 99.00% and greater	1xx.x
Copper	2xx.x
Silicon with copper and/or magnesium	3xx.x
Silicon	4xx.x
Magnesium	5xx.x
Zinc	7xx.x
Tin	8xx.x
Other element	9xx.x

The second and third digits identify the particular alloy or aluminum purity, and the last digit, separated by a decimal point, indicates the product form (e.g., casting or ingot). A letter before the numerical designation indicates a modification of the original alloy, such as a small variation in the amount of an alloying element or impurity.

Aluminum casting alloys have been designed for both properties and process. When the strength requirements are low, as-cast properties are usually adequate. High-strength castings usually require the use of alloys that can subsequently be heat-treated. Sand casting has the fewest process restrictions and the widest range of aluminum casting alloys, with alloy 356 being the most common. The aluminum alloys used for permanent mold casting are designed to have lower coefficients of thermal expansion (or contraction) because the molds offer restraint to the dimensional changes that occur upon cooling. Die-casting alloys require high degrees of fluidity because they are often cast in thin sections. Most of the die-casting alloys are also designed to produce high "as-cast" strength without heat treatment, using the rapid cooling conditions of the die-casting process to promote a fine grain size and fine eutectic structure. Alloy 380 comprises about 85% of aluminum die casting production. Tensile strengths of the aluminum permanent-mold and die-casting alloys can be in excess of 275 MPa (40 ksi).

ALUMINUM–LITHIUM ALLOYS

Lithium is the lightest of all metallic elements, and in the search for aluminum alloys with higher strength, greater stiffness, and lighter weight, aluminum–lithium alloys have emerged. Each percent of lithium reduces the overall weight by 3% and increases stiffness by 6%. The initially developed alloys offered 8 to 10% lower density, 15 to 20% greater stiffness, strengths comparable to those of existing alloys, and good resistance to fatigue crack propagation. Unfortunately, fracture toughness, ductility, and stress-corrosion resistance were poorer than for conventional alloys. The current generation alloys are aluminum–copper–lithium with about 4% copper and no more than 2% lithium. The weight benefits are still sufficient to warrant use in a number of aerospace applications, and the fact that they can be fabricated by conventional processes make them attractive alternatives to the advanced composites.

Because aluminum alloys can comprise as much as 80% of the weight of commercial aircraft, even small percentage reductions can be significant. Improved strength and stiffness can further facilitate weight reduction. Fuel savings over the life of the airplane would more than compensate for any additional manufacturing expense. As an example of potential, the weight of the external liquid-hydrogen tank on the U.S. space shuttle booster rocket was reduced by approximately 3400 kg (7500 lb) by conversion to an aluminum–lithium alloy.

TABLE 8-4 Composition, Properties, and Characteristics of Some Aluminum Casting Alloys

Alloy Designation[a]	Process[b]	Composition (%) (Major Alloys > 1%)						Temper	Tensile Strength		Elongation in 2 in. (%)	Uses and Chracteristics
		Cu	Si	Mg	Zn	Fe	Other		ksi[c]	MPa		
208	S	4.0	3.0		1.0	1.2		F	19	1.31	1.5	General-purposes and castings, can be heat treated
242	S,P	4.0		1.6		1.0	2.0 Ni	T61	40	276	—	Withstands elevated temperatures
295	S	4.5	1.0			1.0		T6	32	221	3.0	Structural castings, heat-treatable
296	P	4.5	2.5			1.2		T6	35	241	2.0	Permanent-mold version of 295
308	P	4.5	5.5		1.0	1.0		F	24	166	—	General-purpose permanent mold
319	S,P	3.5	6.0		1.0	1.0		T6	31	214	1.5	Superior casting characteristics
354	P	1.8	9.0					—	—	—	—	High-strength, aircraft
355	S,P	1.3	5.0					T6	32	221	2.0	High strength and pressure tightness
C355	S,P	1.3	5.0					T61	40	276	3.0	Stronger and more ductile than 355
356	S,P		7.0					T6	30	207	3.0	Excellent castability and impact strength
A356	S,P		7.0					T61	37	255	5.0	Stronger and more ductile than 356
357	S,P		7.0					T6	45	310	3.0	High strength-to-weight castings
359	S,P		9.0					—	—	—	—	High-strength aircraft usage
360	D		9.5			2.0		F	44[d]	303	2.5[d]	Good corrosion resistance and strength
A360	D		9.5			2.0		F	46[d]	317	3.5[d]	Similar to 360
380	D	3.5	8.5		3.0	2.0		F	46[d]	317	2.5[d]	High strength and hardness
A380	D	3.5	8.5		3.0	1.3		F	47[d]	324	3.5[d]	Similar to 380
383	D	1.5	10.5		3.0	1.3		F	45[d]	310	3.5[d]	High strength and hardness
384	D	3.75	11.3		1.0	1.3		F	48[d]	331	2.5	High strength and hardness
413	D	1.0	12.0			2.0		F	43[d]	297	2.5[d]	General-purpose, good castability
A413	D	1.0	12.0			1.3		F	42[d]	290	3.5[d]	Similar to 413
443	D	5.25				2.0		F	33[d]	228	9.0[d]	General-purpose, good castability
B443	S,P	5.25				2.0		F	17	117	3.0	General-purpose casting alloy
514	S			4.0				F	22	152	6.0	High corrosion resistance
518	D			8.0		1.8		F	45[d]	310	5.0[d]	Good corrosion resistance, strength, and toughness
520	S			10.0				T4	42	290	12.0	High strength with good ductility
535	S			6.9				F	35	241	9.0	Good corrosion resistance and machinability
712	S				5.8			F	34	234	4.0	Good properties without heat treatment
713	S,P				7.5	1.1		F	32	221	3.0	Similar to 712
771	S				7.0			T6	42	290	5.0	Aircraft and computer components
850	S,P	1.0					6.3 Sn, 1.0 Ni	T5	16	110	5.0	Bearing alloy

(Note the compatibility with specific casting processes presented in the second column.)

[a] Aluminum Association.

[b] Sand-cast; P, permanent-mold-cast; D, die cast.

[c] Minimum figures unless noted.

[d] Typical values.

ALUMINUM FOAM

A material known as **stabilized aluminum foam** can be made by mixing ceramic particles with molten aluminum and blowing gas into the mixture. The bubbles remain through solidification, yielding a structure that resembles metallic styrofoam. Panels can be cast up to 1.2 m (4 ft) wide, 15.2 m (50 ft) long, and 2.5 to 10.2 cm (1–4 in.) thick, with densities ranging from 2.5 to 20% that of solid aluminum. Originally developed for automotive, aerospace, and military applications, the material has found additional uses in architecture and design. Strength-to-weight is outstanding, and the material offers excellent energy absorption from impacts, crashes, and explosive blasts. The fuel cells of race cars have been shrouded with aluminum foam, and foam fill has been inserted between the front of cars and the driver compartment. Tubular structures can be filled with foam to increase strength, absorb energy, and provide resistance to crushing. Still other applications capitalize on the excellent thermal insulation, vibration damping and sound absorption that result from the numerous trapped air pockets. The metal foams are easily machined and can be joined by adhesive bonding and brazing, as well as laser and gas-tungsten-arc welding.

8.4 MAGNESIUM AND MAGNESIUM ALLOYS

GENERAL PROPERTIES AND CHARACTERISTICS

Magnesium is the lightest of the commercially important metals, having a specific gravity of about 1.74 (two-thirds that of aluminum, one-quarter that of steel, and only slightly higher than fiber-reinforced plastics). Like aluminum, magnesium is relatively weak in the pure state and for engineering purposes is almost always used as an alloy. Even in alloy form, however, the metal is characterized by poor wear, creep, and fatigue properties. It has the highest thermal expansion of all engineering metals. Strength drops rapidly when the temperature exceeds 100°C (200°F), so magnesium should not be considered for elevated-temperature service. Its modulus of elasticity is even less than that of aluminum, being between one-fourth and one-fifth that of steel. Thick sections are required to provide adequate stiffness, but the alloy is so light that it is often possible to use thicker sections for the required rigidity and still have a lighter structure than can be obtained with any other metal. Cost per unit volume is low, so the use of thick sections is generally not prohibitive. Moreover, because a large portion of magnesium components are cast, the thicker sections actually become a desirable feature. Ductility is frequently low, a characteristic of the hexagonal-close-packed (HCP) crystal structure, but some alloys have values exceeding 10%.

On the more positive side, magnesium alloys have a relatively high strength-to-weight ratio, with some commercial alloys attaining strengths as high as 380 MPa (55 ksi). High energy absorption provides good damping of noise and vibration, as well as impact and dent resistance, and electrical and thermal conductivity are relatively good. While many magnesium alloys require enamel or lacquer finishes to impart adequate corrosion resistance, this property has been improved markedly with the development of higher-purity alloys. In the absence of unfavorable galvanic couples, these materials have excellent corrosion resistance. While aluminum alloys are often used for the load-bearing members of mechanical structures, magnesium alloys are best suited for those applications where light weight is the primary consideration and strength is a secondary requirement. Compared to reinforced plastics, the strengths and densities are quite comparable, but magnesium is three to six times more rigid. As a result, magnesium alloys are finding applications in a wide range of markets, including automotive, aerospace, power tools, sporting goods, luggage, and electronic products (where it offers the combination of electromagnetic shielding, light weight, and durability).

MAGNESIUM ALLOYS AND THEIR FABRICATION

Like aluminum, magnesium alloys are classified as either cast or wrought, and some are heat-treatable by age hardening. A designation system for magnesium alloys has been developed by ASTM, identifying both chemical composition and temper, and is presented in specification B93. Two prefix letters designate the two largest alloying metals in order of decreasing amount, using the following format:

A	aluminum	F	iron	M	manganese	R	chromium
B	bismuth	H	thorium	N	nickel	S	silicon
C	copper	K	zirconium	P	lead	T	tin
D	cadmium	L	beryllium	Q	silver	Z	zinc
E	rare earth						

Aluminum is the most common alloying element and, along with zinc, zirconium, and thorium, promotes precipitation hardening. Manganese improves corrosion resistance, and tin improves castability. The two letters are then followed by two or three numbers and a possible suffix letter. The numbers correspond to the rounded-off whole-number percentages of the two main alloy elements and are arranged in the same order as the letters. Thus, the AZ91 alloy would contain approximately 9% aluminum and 1% zinc. A suffix letter is used to denote variations of the same base alloy, such as AZ91A. The temper designation is quite similar to that used with the aluminum alloys. Table 8-5 lists some of the more common magnesium alloys together with their properties and uses.

Sand, permanent mold, die, semisolid, and investment casting are all well developed for magnesium alloys and take advantage of the low melting points and high fluidity. Die casting is clearly the most popular manufacturing process, accounting for 70% of all magnesium castings. Although the magnesium alloys typically cost about twice as much as aluminum, the hot-chamber die-casting process used with magnesium is easier, more economical, and 40 to 50% faster than the cold-chamber process generally required for aluminum. Wall thickness, draft angle, and dimensional tolerances are all

TABLE 8-5 Composition, Properties, and Characteristics of Common Magnesium Alloys

		Composition (%)						Tensile Strength[a]		Yield Strength[a]			
Alloy	Temper	Al	Rare Earths	Mn	Th	Zn	Zr	ksi	MPa	ksi	MPa	Elongation in 2 in. (%)	Uses and Characteristics
AM60A	F	6.0		0.13				30	207	17	117	6	Die castings
AM100A	T4	10.0		0.1				34	234	10	69	6	Sand and permanent-mold castings
AZ31B	F	3.0				1.0		32	221	15	103	6	Sheet, plate, extrusions, forgings
AZ61A	F	6.5				1.0		36	248	16	110	7	Sheet, plate, extrusions forgings
AZ63A	T5	6.0				3.0		34	234	11	76	7	Sand and permanent-mold castings
AZ80A	T5	8.5				0.5		34	234	22	152	2	High-strength forgings, extrusions
AZ81A	T4	7.6				0.7		34	234	11	76	7	Sand and permanent-mold castings
AZ91A	F	9.0				0.7		34	234	23	159	3	Die castings
AZ92A	T4	9.0				2.0		34	234	11	76	6	High-strength sand and permanent-mold castings
EZ33A	T5		3.2			2.6	0.7	20	138	14	97	2	Sand and permanent-mold castings
HK31A	H24				3.2		0.7	33	228	24	166	4	Sheet and plates; castings in T6 temper
HM21A	T5			0.8	2.0			33	228	25	172	3	High-temperature (800°F) sheets, plates, forgings
HZ32A	T5				3.2	2.1		27	186	13	90	4	Sand and permanent-mold castings
ZH62A	T5				1.8	5.7	0.7	35	241	22	152	5	Sand and permanent-mold castings
ZK51A	T5					4.6	0.7	34	234	20	138	5	Sand and permanent-mold castings
ZK60A	T5					5.5	0.45	38	262	20	138	7	Extrusions, forgings

[a] Properties are minimums for the designated temper.

lower than for both aluminum die castings and thermoplastic moldings. Die life is significantly greater than that observed with aluminum. As a result, magnesium die castings compete well with aluminum[2] and often replace plastic injection-molded components when improved stiffness or dimensional stability, or the benefits of electrical or thermal conductivity, are required.

Forming behavior is poor at room temperature, but most conventional processes can be performed when the material is heated to temperatures between 250 and 500°C (480 and 775°F). Because these temperatures are easily attained and generally do not require a protective atmosphere, many formed and drawn magnesium products are manufactured. Magnesium extrusions and sheet metal products have properties similar to the more common wrought aluminum alloys. While slightly heavier than plastics, they offer an order of magnitude or greater improvement in stiffness or rigidity.

The machinability of magnesium alloys is the best of any commercial metal and, in many applications, the savings in machining costs, achieved through deeper cuts, higher cutting speeds and longer tool life, more than compensate for the increased cost of the material. It is necessary, however, to keep the tools sharp and provide adequate cooling for the chips.

Magnesium alloys can be spot welded almost as easily as aluminum, but scratch brushing or chemical cleaning is necessary before forming the weld. Fusion welding is best performed with processes using an inert shielding atmosphere of argon or helium gas.

While heat treatments can be used in increase strength, the added increment achieved by age hardening is far less than observed with aluminum. In fact, the strongest magnesium alloy is only about three times stronger than the weakest. Because of this, designs must be made to accommodate the material, rather than the material being tailored to the design.

Considerable misinformation exists regarding the fire hazards when processing or using magnesium alloys. It is true that magnesium alloys are highly combustible when in a finely divided form, such as powder or fine chips, and this hazard should never be ignored. In the form of sheet, bar, extruded product, or finished castings, however, magnesium alloys rarely present a fire hazard. When the metal is heated above 500°C (950°F), a noncombustible, oxygen-free atmosphere is recommended to suppress burning, which will initiate around 600°C (1100°F). Casting operations often require additional precautions due to the reactivity of magnesium with certain mold materials, such as those containing sand and water.

■ 8.5 ZINC AND ZINC ALLOYS

More than 50% of all metallic *zinc* is used in the **galvanizing** of iron and steel. In this process the iron-based material is coated with a layer of zinc by one of a variety of processes that include direct immersion in a bath of molten metal (hot dipping) and electrolytic plating. The resultant coating provides excellent corrosion resistance, even when the surface is badly scratched or marred, and this corrosion resistance will persist until all of the sacrificial zinc has been depleted.

Zinc is also used as the base metal for a variety of die-casting alloys. For this purpose, zinc offers low cost, a low melting point (only 380°C or 715°F), and the attractive property of not adversely affecting steel dies when in molten metal contact. Unfortunately, pure zinc is almost as heavy as steel and is also rather weak and brittle. Therefore, when alloys are designed for die casting, the alloy elements are usually selected for their ability to increase strength and toughness in the as-cast condition while retaining the low melting point.

The composition and properties of common zinc die-casting alloys are presented in Table 8-6. Alloy AG40A (also known as alloy 903 or Zamak 3) is widely used because of its excellent dimensional stability, and alloy AC41A (also known as alloy 925 or Zamak 5) offers higher strength and better corrosion resistance. As a whole, the zinc

[2] The most common magnesium die casting alloy, AZ91, has the same yield strength and ductility as the most common die cast aluminum, alloy 380.

TABLE 8-6 Composition and Properties of Some Zinc Die-Casting Alloys

Alloy	#3 SAE 903 ASTM AG40A	#5 SAE 925 ASTM AC41A	#7 ASTM AG408	ZA-8			ZA-12			ZA-27		
				S[a]	P	D	S	P	D	S	P	D
Composition[b]												
Aluminum	3.5–4.3	3.5–4.3	3.5–4.3	8.0–8.8			10.5–11.5			25.0–28.0		
Copper	0.25 max	0.75–1.25	0.25 max	0.8–1.3			0.75–1.2			2.0–2.5		
Zinc	balance	balance	balance	balance			balance			balance		
Properties												
Density (g/cc)	6.6	6.6	6.6	6.3			6.0			5.0		
Yield strength (MPa)	221	228	221	200	206	290	214	269	317	372		379
(ksi)	32	33	32	29	30	42	31	39	46	54		55
Tensile strength (MPa)	283	328	283	263	255	374	317	345	400	441		421
(ksi)	41	48	41	38	37	54	46	50	58	64		61
Elongation (% in 2 in.)	10	7	13	2	2	10	3	3	7	6		3
Impact strength (J)	58	65	58	20		42	25		29	47		5
Modulus of elasticity (GPa)	85.5	85.5	85.5	85.5			82.7			77.9		
Machinability[c]	E	E	E	E			VG			G		

[a] S, sand-cast; P, permanent-mold cast; D, die-cast.
[b] Also contains small amounts of Fe, Pb, Cd, Sn, and Ni.
[c] E, excellent; VG, very good; G, good.

die-casting alloys offer a reasonably high strength and impact resistance, along with the ability to be cast to close dimensional limits with extremely thin sections. The dimensions are quite stable, and the products can be finish machined at a minimum of cost. Resistance to surface corrosion is adequate for a number of applications, and the material can be surface finished by a variety of means that include polishing, plating, painting, anodizing, or a chromate conversion coating. Energy costs are low (low melting temperature), tool life is excellent, and the zinc alloys can be efficiently recycled. While the rigidity is low compared to other metals, it is far superior to engineering plastics, and zinc die castings often compete with plastic injection moldings.

The attractiveness of zinc die casting has been further enhanced by the zinc–aluminum casting alloys (ZA-8, ZA-12, and ZA-27,with 8, 12, and 27% aluminum, respectively). Initially developed for sand, permanent mold, and graphite mold casting, these alloys can also be die cast to achieve higher strength (up to 415 MPa or 60 ksi), higher hardness (up to 120 BHN), improved creep resistance and wear resistance, and lighter weight than is possible with any of the conventional alloys. Because of their lower melting and casting costs, these materials are becoming attractive alternatives to the conventional aluminum, brass, and bronze casting alloys, as well as cast iron.

8.6 TITANIUM AND TITANIUM ALLOYS

Titanium is a strong, lightweight, corrosion-resistant metal that has been of commercial importance since about 1950. Because its properties are generally between those of steel and aluminum, its importance has been increasing rapidly. The yield strength of commercially pure titanium is about 210 MPa (30 ksi), but this can be raised to 1300 MPa (190 ksi) or higher through alloying and heat treatment, a strength comparable to that of many heat-treated alloy steels. Density, on the other hand, is approximately 56% that of steel (making strength-to-weight quite attractive), and the modulus of elasticity is a little more than one-half that of steel. Good mechanical properties are retained up to temperatures of 535°C (1000°F), so the metal is often considered to be a high-temperature engineering material. The coefficient of thermal expansion is lower than that of steel and less than half that of aluminum. On the negative side, titanium and its alloys suffer from high cost, fabrication difficulties, a high energy content (they

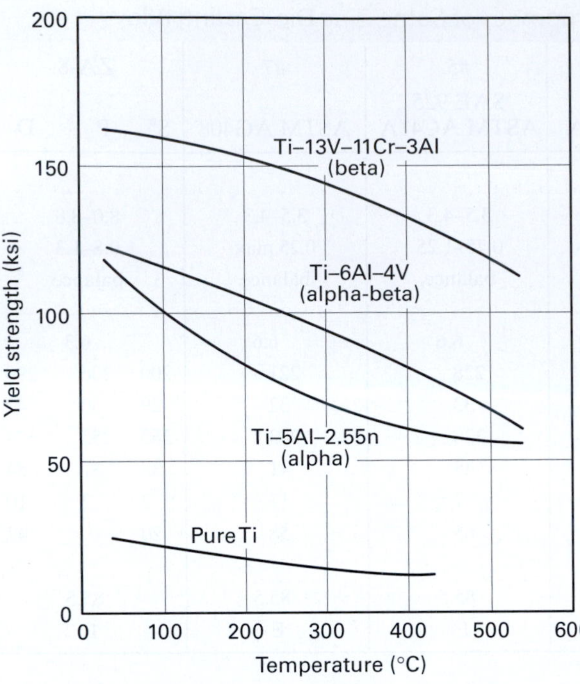

FIGURE 8-4 Strength retention at elevated temperature for various titanium alloys.

require about 10 times as much energy to produce as steel), and a high reactivity at elevated temperatures (above 535°C).

Titanium alloys are designated by major alloy and amount (see ASTM specification B-265) and are generally grouped into three classes based on their microstructural features. These classes are known as *alpha-*, *beta-*, and *alpha-beta-titanium alloys,* where the terms denote the stable phase or phases at room temperature. Alloying elements can be used to stabilize the room-temperature hexagonal-close-packed alpha phase or the elevated-temperature body-centered-cubic beta phase, and heat treatments can be applied to manipulate structure and improve properties. Fabrication can be by casting (generally investment or graphite-mold), forging, rolling, extrusion, or welding, provided that special process modifications and controls are implemented. Advanced processing methods include powder metallurgy, mechanical alloying, rapid-solidification processing (RSP), superplastic forming, diffusion bonding, and hot-isostatic pressing (HIP).

While titanium is an abundant metal, it is difficult to extract from ore, difficult to process, and difficult to fabricate. These difficulties make it significantly more expensive than either steel or aluminum, so its uses relate primarily to its light weight, high strength/weight ratio, good stiffness, good fatigue strength and fracture toughness, good thermal conductivity, excellent corrosion resistance (the result of a thin, tenacious oxide coating), and the retention of mechanical properties at elevated temperatures, as shown for several alloys in Figure 8-4. Titanium is also nontoxic and biocompatible with human tissues and bones. Aluminum, magnesium, and beryllium are the only base metals that are lighter than titanium, and none of these come close in either mechanical performance or elevated temperature properties. Aerospace applications tend to dominate, with titanium comprising up to 40% of the structural weight of high-performance military fighters. Titanium and titanium alloys are also used in such diverse areas as chemical- and electrochemical-processing equipment, food-processing equipment, heat exchangers, marine implements, medical implants, high-performance bicycle and automotive components, and sporting goods. They are often used in place of steel where weight savings are desired and to replace aluminums where high-temperature performance is necessary. Some bonding applications utilize the unique property that titanium wets glass and some ceramics. The titanium–6% aluminum–4% vanadium alloy (Ti-6-4) is the most popular titanium alloy, accounting for nearly 50% of all titanium usage worldwide.

8.7 NICKEL-BASED ALLOYS

Nickel-based alloys are most noted for their outstanding strength and corrosion resistance, particularly at high temperatures, and are available in a wide range of wrought and cast grades. Wrought alloys are generally known by trade names, such as Monel, Hastelloy, Inconel, Incoloy, and others. Cast alloys are generally identified by Alloy Casting Institute or ASTM designations. General characteristics include good formability (FCC crystal structure), good creep resistance, and the retention of strength and ductility at cold or even cryogenic temperatures.

Monel metal, an alloy containing about 67% nickel and 30% copper, has been used for years in the chemical and food-processing industries because of its outstanding corrosion characteristics. In fact, Monel probably has better corrosion resistance to more media than any other commercial alloy. It is particularly resistant to saltwater, sulfuric acid, and even high-velocity, high-temperature steam. For the latter reason, Monel has been used for steam turbine blades. It can be polished to have an excellent appearance, similar to that of stainless steel, and is often used in ornamental trim and household ware. In its most common form, Monel has a tensile strength ranging from 500 to 1200 MPa (70 to 170 ksi), with a companion elongation between 2 and 50%.

Nickel-based alloys have also been used for electrical resistors and heating elements. These materials are primarily nickel–chromium alloys and are known by the trade name **Nichrome.** They have excellent resistance to oxidation while retaining useful strength at red heats. **Invar,** an alloy of nickel and 36% iron, has a near-zero thermal expansion and is used where dimensions cannot change with a change in temperature.

Other nickel-based alloys have been designed to provide good mechanical properties at extremely high temperatures and are generally classified as *superalloys*. These alloys will be discussed along with other, similar materials in the following section.

8.8 SUPERALLOYS, REFRACTORY METALS, AND OTHER MATERIALS DESIGNED FOR HIGH-TEMPERATURE SERVICE

Titanium and titanium alloys have already been cited as being useful in providing strength at elevated temperatures, but the maximum temperature for these materials is approximately 535°C (1000°F). Jet engine, gas turbine, rocket, and nuclear applications often require materials that possess high strength, creep resistance, oxidation and corrosion resistance, and fatigue resistance at temperatures up to and in excess of 1100°C (2000°F). Other application areas include heat exchangers, chemical reaction vessels, and furnace components.

One group of materials offering these properties is the **superalloys,** first developed in the 1940s for use in the elevated-temperature areas of turbojet aircraft. These alloys are based on **nickel, iron and nickel,** or **cobalt,** and have the ability to retain most of their strength even after long exposures to extremely high temperatures. Strength comes from solid-solution strengthening, precipitation hardening, and dispersed alloy carbides or oxides. The nickel-based alloys tend to have higher strengths at room temperature, with yield strengths up to 1200 MPa (175 ksi) and ultimate tensile strengths as high as 1450 MPa (210 ksi). The 1000-hour rupture strengths of the nickel-based alloys at 815°C (1500°F) are also higher than those of the cobalt-based material. Nickel alloys currently comprise approximately 50% of the weight of a jet engine. One of the more common uses of cobalt is as the binder in cemented carbide cutting tools, providing good strength into and above the red-heat range of temperatures. Unfortunately, the density of all of the superalloy metals is significantly greater than iron, so their use is often at the expense of additional weight.

Most of the superalloys are difficult to form or machine, so methods such as electrodischarge, electrochemical, or ultrasonic machining are often used, or the products are made to final shape as investment castings. Powder metallurgy techniques are also used extensively. Because of their ingredients, all of the alloys are quite expensive, and this limits their use to small or critical parts where the cost is not the determining factor.

FIGURE 8-5 Superalloys and refractory metals are needed to withstand the high temperatures of jet engine exhaust. (© *George Impey/Alamy*)

Still other engineering applications require materials whose temperature limits exceed those of the superalloys. Figure 8-5 shows the high temperature exhaust of a jet engine. One reference estimates that the exhaust of future jet engines will reach temperatures in excess of 1425°C (2600°F). Rocket nozzles go well beyond this point. Materials such as TD-nickel (a powder metallurgy nickel alloy containing 2% dispersed thorium oxide) can operate at service temperatures somewhat above 1100°C (2000°F). Going to higher temperatures, we look to the **refractory metals,** which include **niobium, molybdenum, tantalum, rhenium,** and **tungsten**. All have melting points near or in excess of 2500°C (4500°F) and low thermal expansion. They retain a significant fraction of their strength at elevated temperature, resist creep, and can be used at temperatures as high as 1650°C (3000°F) provided that protective ceramic coatings effectively isolate them from gases in their operating environment. Coating technology is quite challenging, however, because the ceramic coatings must (1) have a high melting point, (2) not react with the metal it is protecting, (3) provide a diffusion barrier to oxygen and other gases, and (4) have thermal expansion characteristics that match the underlying metal. While the refractory metals could be used at higher temperatures, the uppermost temperature is currently being set by limitations and restrictions imposed by the coating. In addition to its high-temperature properties, tantalum is resistant to chemical attack from virtually all environments at temperatures below 150°C (300°F) and is frequently used as a corrosion-resistant material.

Table 8-7 presents key properties for several refractory metals. Unfortunately, all are heavier than steel, and several are significantly heavier. In fact, tungsten, with a density about 1.7 times that of lead, is often used in counterbalances and compact flywheels and weights, with other applications as diverse as military projectiles, gyratory compasses, and golf clubs.

TABLE 8-7 Properties of Some Refractory Metals

		Room Temperature				Elevated Temperature [1832°F (1000°C)]	
Metal	Melting Temperature [°F(°C)]	Density (g/cm^3)	Yield Strength (ksi)	Tensil Strength (ksi)	Elongation (%)	Yield Strength (ksi)	Tensile Strength (ksi)
Molybdenum	4730 (2610)	10.22	80	120	10	30	50
Niobium	4480 (2470)	8.57	20	45	25	8	17
Tantalum	5430 (3000)	16.6	35	50	35	24	27
Tungsten	6170 (3410)	19.25	220	300	3	15	66

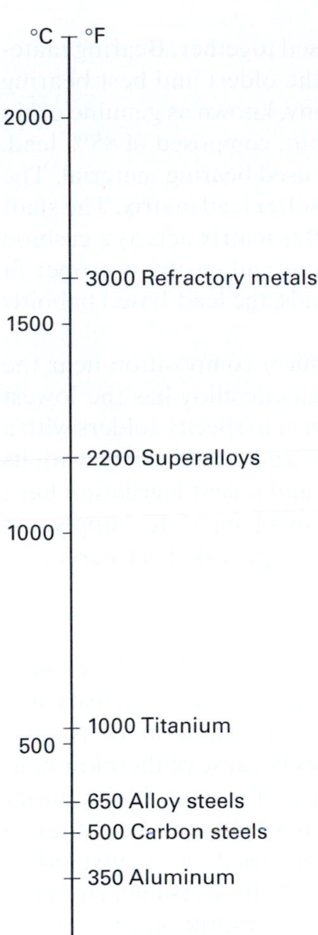

FIGURE 8-6 Temperature scale indicating the upper limit to useful mechanical properties for various engineering metals.

Other materials and technologies that offer promise for high-temperature service include intermetallic compounds, engineered ceramics and ceramic composites, graded materials, and advanced coating systems. The **intermetallic compounds** provide properties that are between metals and ceramics and are excellent candidates for high-temperature applications. They are hard, stiff, creep resistant, and oxidation resistant, with good high-temperature strength that often increases with temperature. The titanium and nickel aluminides offer the additional benefit of being significantly lighter than the superalloys. Unfortunately, the intermetallics are also characterized by poor ductility, poor fracture toughness, and poor fatigue resistance. They are difficult to fabricate using traditional techniques, such as forming and welding. On a positive note, research and development efforts have begun to overcome some of these limitations, and the intermetallics are now appearing in commercial products. The high-temperature ceramics will be discussed in a future chapter.

Figure 8-6 compares the upper limit for useful mechanical properties for a variety of engineering metals, ranging from aluminum through the refractory metals. Figure 8-7 graphically presents the densities of the various metals. Note that the superalloys (cobalt and nickel), and the refractory metals (niobium, molybdenum, tantalum, tungsten, and rhenium) are all heavier than steel. Titanium is the only elevated temperature metal that is also lightweight, and its temperature range is limited.

8.9 LEAD AND TIN, AND THEIR ALLOYS

The dominant properties of *lead* and lead alloys are high density coupled with strength and stiffness values that are among the lowest of the engineering metals. The principal uses of lead as a pure metal include storage batteries, cable cladding, and radiation-absorbing or sound-and vibration-dampening shields. Lead-acid batteries are clearly the dominant product, and more than 60% of U.S. lead consumption is generated from battery recycling. Other applications utilize the properties of good corrosion resistance, low melting point, and the ease of casting or forming. Structural applications are severely limited because lead is susceptible to creep under low loads, even at room temperature. As a pure metal, *tin* is used primarily as a corrosion resistant coating on steel.

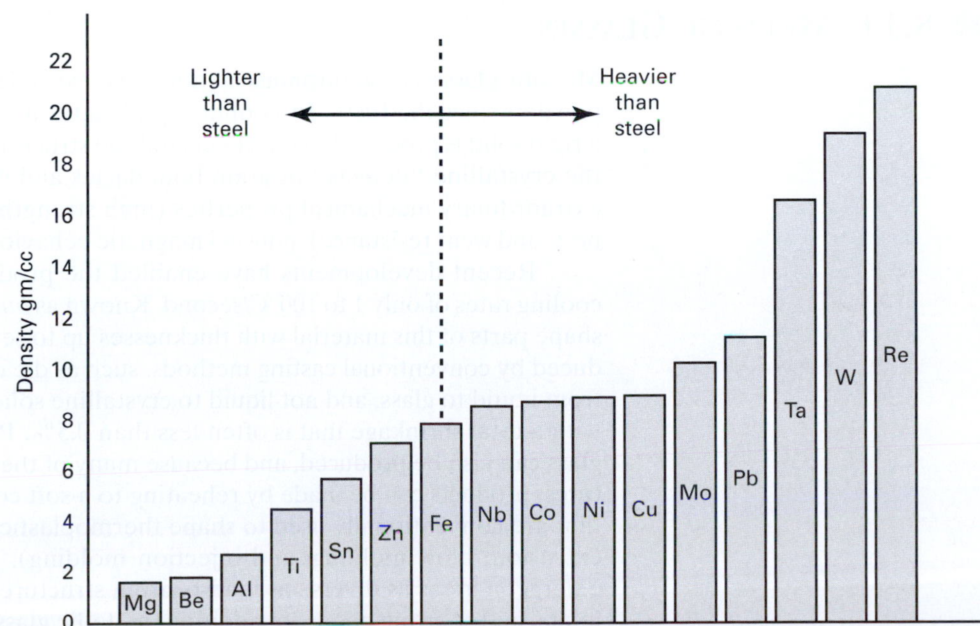

FIGURE 8-7 Densities of the various engineering metals. The elevated-temperature superalloys and refractory metals are all heavier than steel.

In the form of alloys, lead and tin are almost always used together. Bearing material and **solder** are the two most important uses. One of the oldest and best bearing materials is an alloy of 84% tin, 8% copper, and 8% antimony, known as genuine or tin **babbitt.** Because of the high cost of tin, however, lead babbitt, composed of 85% lead, 5% tin, 10% antimony, and 0.5% copper, is a more widely used bearing material. The tin and antimony combine to form hard particles within the softer lead matrix. The shaft rides on the harder particles with low friction, while the softer matrix acts as a cushion that can distort sufficiently to compensate for misalignment and ensure a proper fit between the two surfaces. For slow speeds and moderate loads, the lead-based babbitts have proven to be quite adequate.

Soft solders are basically lead–tin alloys with a chemical composition near the eutectic value of 61.9% tin (see Figure 5-5). While the eutectic alloy has the lowest melting temperature, the high cost of tin has forced many users to specify solders with a lower-than-optimum tin content. A variety of compositions are available, each with its own characteristic melting range. Environmental concerns and recent legislation have prompted a move toward lead-free solders for applications involving water supply and distribution. Additional information on solders and soldering is provided in Chapter 33.

8.10 SOME LESSER-KNOWN METALS AND ALLOYS

Several of the lesser-known metals have achieved importance as a result of their somewhat unique physical and mechanical properties. **Beryllium** combines a density less than aluminum with a stiffness greater than steel, and is transparent to X-rays. **Hafnium, thorium,** and beryllium are used in nuclear reactors because of their low neutron-absorption characteristics. Depleted **uranium,** because of its very high density (19.1 g/cm^3), is useful in special applications where maximum weight must be put into a limited space, such as counterweights or flywheels. **Zirconium** is used for its outstanding corrosion resistance to most acids, chlorides, and organic acids. It offers high strength, good weldability and fatigue resistance, and attractive neutron absorption characteristics. **Rare-earth metals** have been incorporated into magnets that offer increased strength compared to the standard ferrite variety. Neodymium–iron–boron and samarium–cobalt are two common varieties.

While the **precious metals** (*gold, silver*, and the platinum group metals—*platinum, palladium, rhodium, ruthenium, iridium and osmium*) may seem unlikely as engineering materials, they offer outstanding corrosion resistance and electrical conductivity, often under extreme conditions of temperature and environment.

8.11 METALLIC GLASSES

Metallic glasses, or **amorphous metals,** have existed in the form of thin ribbon and fine powders since the 1960s. By cooling liquid metal at a rate that exceeds 10^5 to 10^6°C/sec, a rigid solid is produced that lacks crystalline structure. Because the structure also lacks the crystalline "defects" of grain boundaries and dislocations, the materials exhibit extraordinary mechanical properties (high strength, large elastic strain, good toughness, and wear resistance), unusual magnetic behavior, and high corrosion resistance.

Recent developments have enabled the production of amorphous metal with cooling rates of only 1 to 100°C/second. Known as *bulk metallic glass (BMG),* complex-shape parts of this material with thicknesses up to several centimeters can now be produced by conventional casting methods, such as die casting. Because the material goes from liquid to glass, and not liquid to crystalline solid, precision products can be made with a total shrinkage that is often less than 0.5%. Pellets or powders of bulk metallic glass can also be produced, and because many of the alloys have low melting temperatures, products can be made by reheating to a soft condition and forming by processes that are conventionally used to shape thermoplastic polymers (compression molding, extrusion, blow molding, and injection molding). Applications have just begun to emerge in areas as diverse as load-bearing structures, electronic casings, replacement joints, and sporting goods. In addition, metallic glasses have also been developed that retain their glassy structure at temperatures as high as 870°C (1600°F).

8.12 GRAPHITE

While technically not a metal, **graphite** is an engineering material with considerable potential. It offers properties of both a metal and nonmetal, including good thermal and electrical conductivity, inertness, the ability to withstand high temperature, and lubricity. In addition, it possesses the unique property of increasing in strength as the temperature is elevated. Polycrystralline graphites can have mechanical strengths up to 70 MPa (10 ksi) at room temperature, which double when the temperature reaches 2500°C (4500°F). The material is stable in air at temperatures up to 500°C (930°F) and in vacuum or inert atmospheres up to 3000°C (5430°F).

Large quantities of graphite are used as electrodes in arc furnaces, but other uses are developing rapidly. The addition of small amounts of borides, carbides, nitrides, and silicides greatly lowers the oxidation rate at elevated temperatures and improves the mechanical strength. This makes the material highly suitable for use as rocket-nozzle inserts and as permanent molds for casting various metals, where it costs less than tool steel, requires no heat treating, and has a lower coefficient of thermal expansion. It can be machined quite readily to excellent surface finishes. Graphite fibers have also found extensive use in composite materials. This application will be discussed in Chapter 9.

KEY WORDS

Alclad
aluminum
amorphous metal
babbitt
beryllium
bismuth
brass
bronze
cast alloys
cobalt
color anodizing
constantan
copper
copper–beryllium alloys
cupronickels
dezincification
duralumin
electrolytic tough-pitch (ETP) copper
galvanizing
graphite
hafnium
intermetallic compound
Invar
lead
lithium
magnesium
metallic glass
molybdenum
Monel
Nichrome
nickel
nickel-based alloys
nickel silver
niobium
nonferrous metals and alloys
oxygen-free high-conductivity (OFHC) copper
precious metals
rare-earth metals
refractory metals
rhenium
season cracking
selenium
silicon bronze
solder
stabilized aluminum foam
stress-corrosion cracking
superalloys
tantalum
temper designation
thorium
tin
tin bronzes
titanium
tungsten
uranium
wrought alloys
zinc
zirconium

REVIEW QUESTIONS

1. What types of properties do nonferrous metals possess that may not be available in the ferrous metals?
2. In what respects are the nonferrous metals generally inferior to steel?
3. In what ways might the nonferrous metals offer attractive ease of fabrication?
4. What are the three properties of copper and copper alloys that account for many of their uses and applications?
5. What properties make copper attractive for cold-working processes?
6. What are some of the limiting properties of copper that might restrict its area of application?
7. What properties of copper make it attractive for low-temperature applications?
8. What are some potential applications that would make use of copper's antimicrobial properties?
9. Why does the copper designation system separate wrought and cast alloys? What properties are attractive for each group?
10. What are some of the attractive engineering properties that account for the wide use of the copper–zinc alpha brasses?
11. Why might cold-worked brass require a stress relief prior to being placed in service?
12. Why might the term *bronze* be potentially confusing when used in reference to a copper-based alloy?
13. What are some attractive engineering properties of copper–nickel alloys?
14. Describe the somewhat unique properties available with heat-treated copper–beryllium alloys. What has limited their use in recent years?
15. What alloys have been used to replace lead in copper casting alloys being targeted to drinking water applications?
16. What are some of the attractive engineering properties of aluminum and aluminum alloys?
17. How does aluminum compare to steel in terms of weight? Discuss the merits of comparing cost per unit weight versus cost per unit volume.

18. What is the primary benefit of aluminum recycling compared to making new aluminum from ore?
19. When designing with aluminum, why might there be concern regarding rigidity or stiffness?
20. How does aluminum compare to copper in terms of electrical conductivity?
21. What features might limit the mechanical uses and applications of aluminum and aluminum alloys?
22. What features make aluminum attractive for transportation applications?
23. How is the corrosion-resistance mechanism observed in aluminum and aluminum alloys similar to that observed in stainless steels?
24. How are the wrought alloys distinguished from the cast alloys in the aluminum designation system? Why would these two groups of metals have distinctly different properties?
25. What feature in the wrought aluminum designation scheme is used to denote the condition or structure of a given alloy?
26. What is the primary strengthening mechanism in the high-strength "aircraft-quality" aluminum alloys?
27. What unique combination of properties is offered by the composite Alclad materials?
28. What surface finishing technique is used in the production of numerous metallic-colored aluminum products?
29. Why are aluminum–silicon alloys popular for casting operations?
30. What specific material properties might make an aluminum casting alloy attractive for permanent mold casting? For die casting?
31. What are the attractive features of aluminum–lithium alloys? What has limited their success and expansion?
32. What are some possible applications of aluminum foam?
33. What are some attractive and restrictive properties of magnesium and magnesium alloys?
34. Describe the designation system applied to magnesium alloys.
35. What is the most popular fabrication process applied to magnesium alloys?
36. In what way can ductility be imparted to magnesium alloys so that they can be formed by conventional processes?
37. Under what conditions should magnesium be considered to be a flammable or explosive material?
38. What is the primary application of pure zinc? Of the zinc-based engineering alloys?
39. What are some of the attractive features of the zinc–aluminum casting alloys?
40. What are some of the attractive engineering properties of titanium and titanium alloys?
41. What feature is used to provide the metallurgical classification of titanium alloys?
42. Under what conditions might titanium replace steel? Replace aluminum?
43. What conditions favor the selection and use of nickel-based alloys?
44. What property of Monel alloys dominates most of the applications?
45. What metals or combinations of metals form the bases of the superalloys?
46. What class of metals or alloys must be used when the operating temperatures exceed the limits of the superalloys?
47. Which metals are classified as refractory metals?
48. What are some general characteristics of intermetallic compounds?
49. What temperature is generally considered to be the upper limit for which titanium alloys retain their useful engineering properties? The superalloys? Refractory metals?
50. What is the dominant product for which lead is used?
51. What features makes beryllium a unique, lightweight metal?
52. What are some of the attractive properties of metallic glasses?
53. What unique property of graphite makes it attractive for elevated-temperature applications?

PROBLEMS

1. Your company is considering an expansion of its line of conventional hand tools to include safety tools, capable of being used in areas such as gas leaks where the potential of explosion or fire exists. Conventional irons and steels are pyrophoric (i.e., small slivers or fragments can burn in air, forming sparks if dropped or impacted on a hard surface).

   You are asked to evaluate potential materials and processes that might be used to manufacture a nonsparking pipe wrench. This product is to be produced in the same shape and range of sizes as conventional pipe wrenches and possess all of the same characteristic properties (strength in the handle, hardness in the teeth, fracture resistance, corrosion resistance, etc.). In addition, the new safety wrench must be nonsparking (or nonpyrophoric).

   Your initial review of the nonferrous metals reveals that aluminum is nonpyrophoric, but it lacks the strength and wear resistance needed in the teeth and jaw region of the wrench. Copper is also nonpyrphoric, but it is heavier than steel, which may be unattractive for the larger wrenches. Copper–2% beryllium can be age hardened to provide the strength and hardness properties equivalent to the steel that is currently being used for the jaws of the wrench, but the cost of this material is also quite high. Titanium is difficult to fabricate and may not possess the needed hardness and wear resistance. Mixed materials may create an unattractive galvanic corrosion cell. Both forging and casting appear to be viable means of forming the desired shape. You want to produce a quality product but also wish to make the wrench in the most economical manner possible so that the new line of safety tools is attractive to potential customers.

   Suggest some alternative manufacturing systems (materials coupled with companion methods of fabrication) that could be used to produce the desired wrench. What might be the advantages and disadvantages of each? Which of your alternatives would you recommend to your supervisor?

www.wiley.com/go/global/degarmo

Chapter 8 CASE STUDY

Hip Replacement Prosthetics

Hip replacement surgery is currently among the most common orthopedic procedures, used to relieve arthritis pain or fix severe physical joint damage. In a total hip replacement, both the top of the femur (the "ball") and the acetabulum "socket" in the pelvis are replaced. To be successful, the materials used in the implant components must be (1) biocompatible—accepted by the body with no rejection symptoms or consequences; (2) resistant to corrosion, degradation, and wear—retaining strength and shape, ensuring proper joint function, and not generating any harmful particulate or corrosive debris; and (3) have mechanical properties that duplicate the natural structures that they are intended to replace—strong enough to withstand loads, flexible enough to bear stresses without breaking, and able to move smoothly over one another.

Early attempts at hip replacement used glass and ivory. While glass was biocompatible, it was not sufficiently durable and failed by brittle fracture. Efforts then shifted to various plastics combined with stainless steel. Today's prostheses are available in a variety of materials, often different for each of the various components: (1) the stem, which fits into the thigh bone or femur; (2) the ball, which replaces the spherical head of the femur; and (3) the cup, which replaces the worn-out hip socket. In some designs, the stem and ball are a single piece, while in others, they are separate, allowing for possible variation in materials. Stems are currently made from titanium, cobalt–chromium alloys, and stainless steel. The heads or balls are made from cobalt–chromium metal or ceramic. The cup may be of a singular material or a modular combination. Single-material cups are either polyethylene or metal. Modular cups use a metal shell and a liner (that contacts the ball) of polyethylene, ceramic, or metal. Highly cross-linked, ultra-high-density polyethylene has recently been used in place of standard polyethylene. Various coatings have also been used to impart desired surface characteristics. (The accompanying photo shows a titanium stem, ceramic head and polyethylene acetabular cup.)

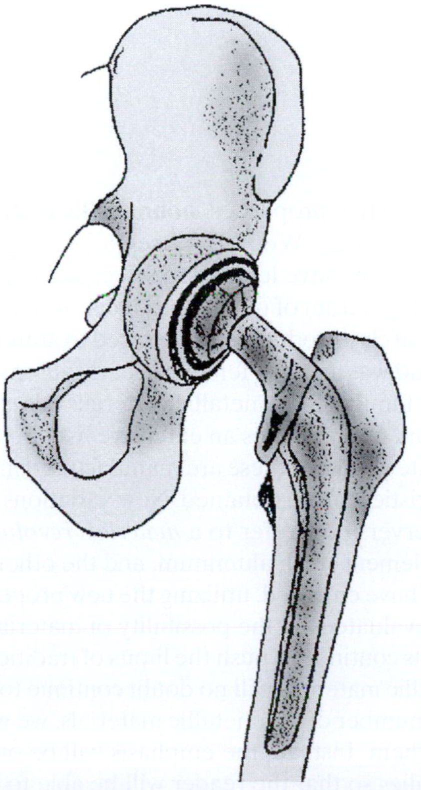

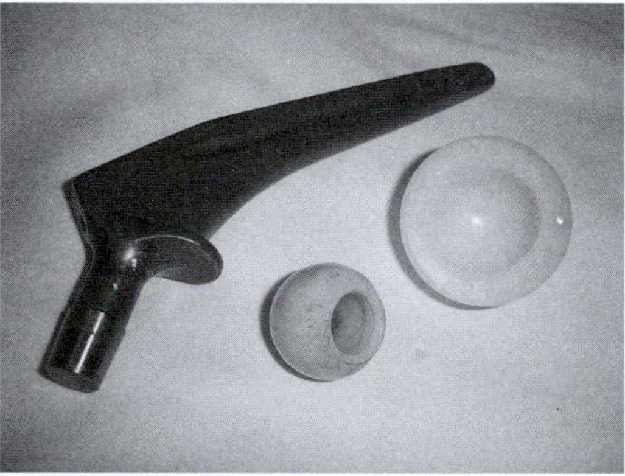

Photo by Nuno Nogueira, from http://en.wikipedia.org/wiki/File:Hip_prosthesis.jpg

1. For each of the materials identified in this case study, determine their properties and discuss their assets and liabilities for use in the various components of a hip prosthesis.
2. There are various pairings between the ball and socket. For each of these, discuss their attractive features and their possible adverse characteristics. Consider interface friction, wear rates, the possibility of wear debris, and any associated consequences.
   a. Metal-on-polyethylene
   b. Meta-on-cross-linked polyethylene
   c. Ceramic-on-ceramic
   d. Metal-on-metal
3. Which of the preceding combinations would you expect to be least expensive? Most expensive?
4. Which of the preceding combinations would you expect to have the shortest lifetime? Greatest lifetime?
5. Is your answer to Question 4 consistent with Question 3?

CHAPTER 9

NONMETALLIC MATERIALS: PLASTICS, ELASTOMERS, CERAMICS, AND COMPOSITES

9.1 INTRODUCTION

Because of their wide range of attractive properties, **nonmetallic materials** have always played a significant role in manufacturing. Wood has been a key engineering material down through the centuries, and artisans have learned to select and use the various types and grades to manufacture a broad spectrum of quality products. Stone and rock continue to be key construction materials, and clay products can be traced to antiquity. Even leather has been a construction material and was used for fenders in early automobiles.

More recently, however, the family of nonmetallic materials has expanded from the natural materials just described, and now includes an extensive list of plastics (polymers), elastomers, ceramics, and composites. Most of these are manufactured materials, so a wide variety of properties and characteristics can be obtained. New variations are being created on a continuous basis. Many observers now refer to a *materials revolution* as these new materials compete with and complement steel, aluminum, and the other more traditional engineering metals. New products have emerged, utilizing the new properties, and existing products are continually being reevaluated for the possibility of material substitution. As the design requirements of products continue to push the limits of traditional materials, the role of the manufactured nonmetallic materials will no doubt continue to expand.

Because of the breadth and number of nonmetallic materials, we will not attempt to provide information about all of them. Instead, the emphasis will be on the basic nature and properties of the various families so that the reader will be able to determine if they may be reasonable candidates for specific products and applications. For detailed information about specific materials within these families, more extensive and dedicated texts, handbooks, and compilations should be consulted.

9.2 PLASTICS

It is difficult to provide a precise definition of the term **plastics.** From a technical viewpoint, the term is applied to engineered materials characterized by large molecules that are built up by the joining of smaller molecules. On a more practical level, these materials are natural or synthetic resins, or their compounds, that can be molded, extruded, cast, or used as thin films or coatings. They offer low density, low tooling costs, good resistance to corrosion and chemicals, cost reduction, and design versatility. From a chemical viewpoint, most are organic substances containing hydrogen, oxygen, carbon, and nitrogen.

In less than a century, we have gone from a world without plastic to a world where its use and applications are limitless. The United States currently produces more plastic than steel, aluminum, and copper combined. Plastics are used to save lives in applications such as artificial organs, shatter-proof glass, and bullet-proof vests. They reduce the weight of cars, provide thermal insulation to our homes, and encapsulate our medicines. They form the base material in products as diverse as shower curtains, contact lenses, and clothing and are used in some of the primary components in televisions, computers, cell phones, and furniture. Even the Statue of Liberty has a plastic coating to protect it from corrosion.

MOLECULAR STRUCTURE OF PLASTICS

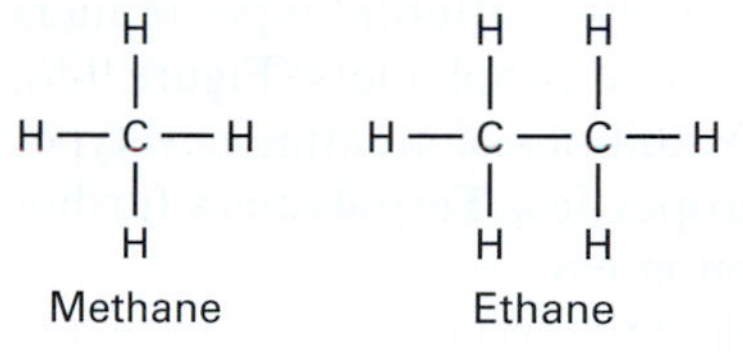

FIGURE 9-1 The linking of carbon and hydrogen to form methane and ethane molecules. Each dash represents a shared electron pair or covalent bond.

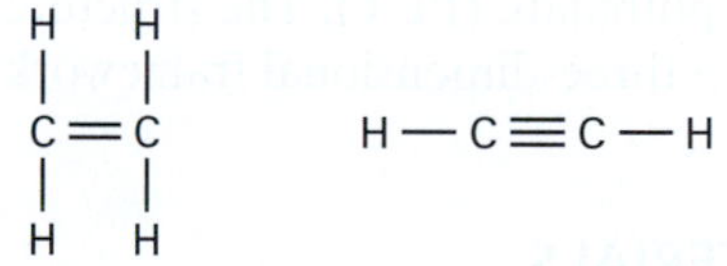

To understand the properties of plastics, it is important to first understand their molecular structure. For simplicity, let us begin with the paraffin-type hydrocarbons, in which carbon and hydrogen combine in the relationship C_nH_{2n+2}. Theoretically, the atoms can link together indefinitely to form very large molecules, extending the series depicted in Figure 9-1. The bonds between the various atoms are all pairs of shared electrons (covalent bonds). Bonding within the molecule, therefore, is quite strong, but the attractive forces between adjacent molecules are much weaker. Because there is no provision for additional atoms to be added to the chain, these molecules are said to be **saturated monomers**.

Carbon and hydrogen can also form molecules where the carbon atoms are held together by double or triple covalent bonds. Ethylene and acetylene are common examples (Figure 9-2). Because these molecules do not have the maximum number of hydrogen atoms, they are said to be **unsaturated monomers** and are important in the polymerization process, where small molecules link to form large ones with the same constituent atoms.

Ethylene Acetylene

FIGURE 9-2 Double and triple covalent bonds exist between the carbon atoms in unsaturated ethylene and acetylene molecules.

In all of the described molecules, four electron pairs surround each carbon atom and one electron pair is shared with each hydrogen atom. Other atoms or structures can be substituted for carbon and hydrogen. Chlorine, fluorine, or even a benzene ring can take the place of hydrogen. Oxygen, silicon, sulfur, or nitrogen can take the place of carbon. Because of these substitutions, a wide range of organic compounds can be created.

ISOMERS

The same kind and number of atoms can also unite in different structural arrangements, known as **isomers,** and these ultimately behave as different compounds with different properties. Figure 9-3 shows an example of this feature, involving propyl and isopropyl alcohol. Isomers can be considered analogous to allotropism or polymorphism in crystalline materials, where the same material possesses different properties because of different crystal structures.

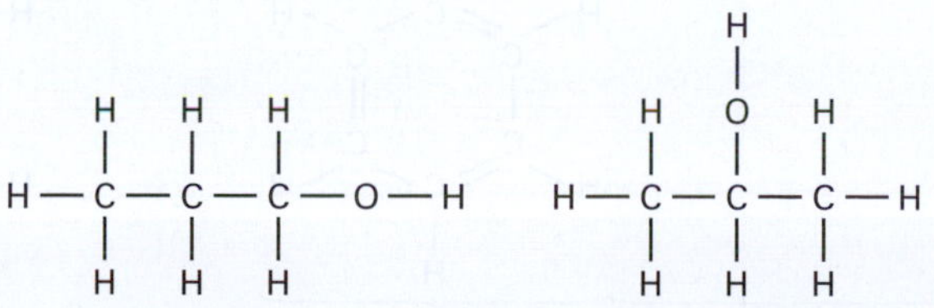

Propyl Alcohol Isopropyl Alcohol

FIGURE 9-3 Linking of eight hydrogen, one oxygen, and three carbon atoms to form two isomers: propyl alcohol and isopropyl alcohol. Note the different locations of the –OH attachment.

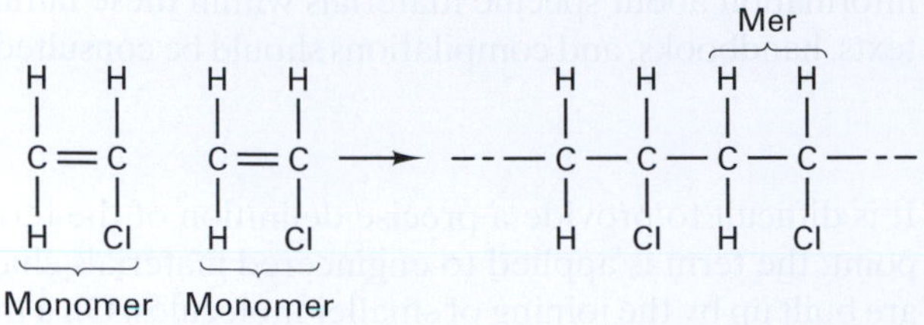

FIGURE 9-4 Addition polymerization—the linking of monomers, in this case, identical ethylene molecules.

FORMING MOLECULES BY POLYMERIZATION

The **polymerization** process, or linking of molecules, occurs by either an **addition** or **condensation** mechanism. Figure 9-4 illustrates polymerization by addition, where a number of basic units **(monomers)** link together to form a large molecule **(polymer)** in which there is a repeated unit **(mer).** Activators or catalysts, such as benzoyl peroxide, initiate and terminate the chain. Thus, the amount of activator relative to the amount of monomer determines the average molecular weight (or average length) of the polymer chain. The average number of mers in the polymer, known as the **degree of polymerization,** ranges from 75 to 750 for most commercial plastics. Chain length controls many of the properties of a plastic. Increasing the chain length tends to increase toughness, creep resistance, melting temperature, melt viscosity, and difficulty in processing.

Copolymers are a special category of polymer where two different types of mers are combined into the same addition chain. The formation of copolymers (Figure 9-5), analogous to alloying in metals, greatly expands the possibilities of creating new types of plastics with improved physical and mechanical properties. **Terpolymers** further extend the possibilities by combining three different monomers.

In contrast to polymerization by addition, where all of the original atoms appear in the product molecule, condensation polymerization occurs when reactive molecules combine with one another to produce polymer plus small, by-product molecules, such as water or alcohol. Heat, pressure, and catalysts are often required to drive the reaction. Figure 9-6a illustrates the reaction between phenol and formaldehyde to form phenol-formaldehyde, otherwise known as Bakelite, first performed in 1910. Figure 9-6b shows the condensation reaction to produce polyethylene terephthalate (PET). The structure of condensation polymers can be either linear chains or a three-dimensional framework in which all atoms are linked by strong, primary bonds.

THERMOSETTING AND THERMOPLASTIC MATERIALS

The terms **thermosetting** and **thermoplastic** refer to the material's response to elevated temperature. Addition polymers (or linear condensation polymers) can be viewed as long chains of bonded carbon atoms with attached pendants of hydrogen, fluorine, chlorine, or benzene rings. All of the bonds within the molecules are strong covalent bonds. The attraction between neighboring molecules is through the much

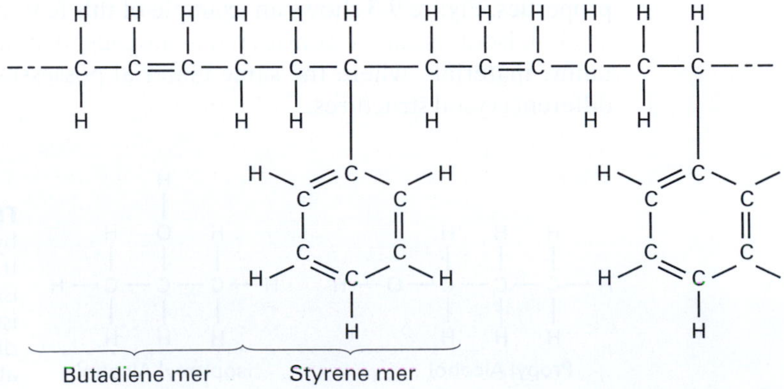

FIGURE 9-5 Addition polymerization with two kinds of mers—here, the copolymerization of butadiene and styrene.

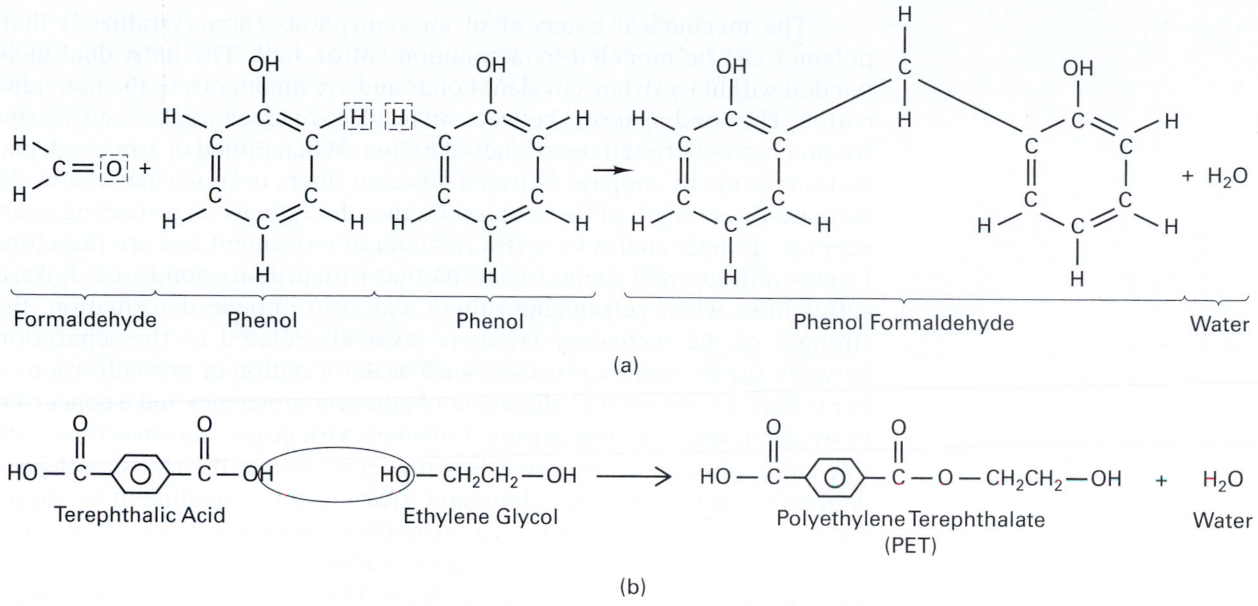

FIGURE 9-6 The formation of (a) phenol–formaldehyde (Bakelite) and (b) polyethylene terephthalate (PET) by condensation polymerization. Note the H_2O or water by-product.

weaker van der Waals forces. For these materials, the intermolecular forces strongly influence the mechanical and physical properties. In general, the linear polymers tend to be flexible and tough. Because the intermolecular bonds are weakened by elevated temperature, plastics of this type soften with increasing temperature, and the individual molecules can slide over each other in a molding process. When the material is cooled, it becomes harder and stronger. The softening and hardening of these thermoplastic or heat-softening materials can be repeated as often as desired, and no chemical change is involved.

Because thermoplastic materials contain molecules of different lengths, they do not have a definite melting temperature, but, instead, soften over a range of temperatures. Above the temperature required for melting, the material can be poured and cast as a liquid or formed by injection molding. When cooled to a temperature where it is fully solid, the material can retain its amorphous structure, but with companion properties that are somewhat rubbery. The application of a force produces both elastic and plastic deformation. Large amounts of permanent deformation are possible and make this range attractive for molding and extrusion. At still lower temperatures, the bonds become stronger and the polymer is stiffer and somewhat leathery. Many commercial polymers, such as polyethylene, have useful strength in this condition. When further cooled below the glass transition temperature, however, the linear polymer retains its amorphous structure but becomes hard, brittle, and glasslike.

Many thermoplastics can partially **crystallize**[1] when cooled below the melting temperature. This should not be confused with the crystal structures discussed previously in this text. When polymers "crystallize," the chains closely align over appreciable distances, with a companion increase in density. In addition, the polymer becomes stiffer, harder, less ductile, and more resistant to solvents and heat. The ability of a polymer to crystallize depends on the complexity of its molecules, the degree of polymerization (length of the chains), the cooling rate, and the amount of deformation during cooling.

[1] It should be noted that the term *crystallize,* when applied to polymers, has a different meaning than when applied to metals and ceramics. Metals and ceramics are crystalline materials, meaning that the atoms occupy sites in a regular, periodic array, known as a lattice. In polymers, it is not the atoms that become aligned, but the molecules. Because van der Waals bonding has a bond strength that is inversely related to the separation distance, the parallel alignment of the crystallized state is a lower energy configuration and is promoted by slow cooling and equilibrium-type processing conditions.

The mechanical behavior of an amorphous (noncrystallized) thermoplastic polymer can be modeled by a common cotton ball. The individual molecules are bonded within by strong covalent bonds and are analogous to the individual fibers of cotton. The bonding forces between molecules are much weaker and are similar to the friction forces between the strands of cotton. When pulled or stretched, plastic deformation occurs by slippage between adjacent fibers or molecular chains. Methods to increase the strength of thermoplastics, therefore, focus on restricting intermolecular slippage. Longer chains have less freedom of movement and are therefore stronger. Connecting adjacent chains to one another with primary bond cross-links, as with the sulfur links when vulcanizing rubber, can also impede deformation. Because the strength of the secondary bonds is inversely related to the separation distance between the molecules, processes such as deformation or crystallization can be used to produce a tight parallel alignment of adjacent molecules and a concurrent increase in strength, stiffness, and density. Polymers with larger side structures, such as chlorine atoms or benzene rings, may be stronger or weaker than those with just hydrogen, depending on whether the dominant effect is the impediment to slippage or the increased separation distance. Branched polymers, where the chains divide in a Y with primary bonds linking all segments, are often weaker because branching reduces the density and close packing of the chains. Physical, mechanical, and electrical properties all vary with these changes in structure.

The four most common thermoplastic polymers are polyethylene (PE), polypropylene (PP), polystyrene (PS), and polyvinylchloride (PVC). Other thermoplastics include polycarbonate (PC), polyethylene terephthalate (PET), polymethylmethacrylate (PMMA), and acrylonitrile-butadiene-styrene (ABS).

In contrast to the thermoplastic polymers, *thermosetting plastics* usually have a highly cross-linked or three-dimensional framework structure in which all atoms are connected by strong, covalent bonds. These materials are generally produced by condensation polymerization where elevated temperature promotes an irreversible reaction, hence the term *thermosetting*. Once set, subsequent heating will not produce the softening observed with the thermoplastics. Instead, thermosetting materials maintain their mechanical properties up to the temperature at which they char or burn. Because deformation requires the breaking of primary bonds, the thermosetting polymers are significantly stronger and more rigid than the thermoplastics. They can resist higher temperatures and have greater dimensional stability, but they also have lower ductility and poorer impact properties. Some common thermosetting polymers include polyurethane (PUR), phenol-formaldehyde or Bakelite (PF), and various epoxies.

As a helpful analogy, thermoplastic polymers are a lot like candle wax. They can be softened or melted by heat and then cooled to assume a solid shape. Thermosets are more like egg whites or bread dough. Heating changes their structure and properties in an irreversible fashion.

Although classification of a polymer as thermosetting or thermoplastic provides insight as to properties and performance, it also has a strong effect on fabrication. For example, thermoplastics can be easily molded. After the hot, soft material has been formed to the desired shape, however, the mold must be cooled so that the plastic will harden and be able to retain its shape upon removal. The repetitive heating and cooling cycles affect mold life, and the time required for the thermal cycles influences productivity. When a part is produced from thermosetting materials, the mold can remain at a constant temperature throughout the entire process, but the setting or curing of the resins now determines the time in the mold. Because the material hardens as a result of the reaction and has strength and rigidity even when hot, product removal can be performed without cooling the mold.

PROPERTIES AND APPLICATIONS

Because there are so many varieties of plastics, and new ones are being developed almost continuously, it is helpful to have knowledge of both the general properties of plastics as well as the unique or specific properties of the various families. General properties of plastics include:

1. *Light weight*. Most plastics have specific gravities between 1.1 and 1.6, compared with about 1.75 for magnesium (the lightest engineering metal).
2. *Corrosion resistance*. Many plastics perform well in hostile, corrosive, or chemical environments. Some are notably resistant to acid corrosion.
3. *Electrical resistance*. Plastics are widely used as insulating materials.
4. *Low thermal conductivity*. Plastics are relatively good thermal insulators.
5. *Variety of optical properties*. Through the incorporation of pigments and dyes, many plastics have an almost unlimited color range, and the color goes throughout, not just on the surface. Both transparent and opaque materials are available.
6. *Formability or ease of fabrication*. Objects can frequently be produced from plastics in a single operation. Raw material can be converted to final shape through such processes as casting, extrusion, and molding. Relatively low temperatures are required for the forming of plastics.
7. *Surface finish*. The same processes that produce the shape also produce excellent surface finish. Additional surface finishing may not be required.
8. *Comparatively low cost*. The low cost of plastics generally applies to both the material itself and the manufacturing process. Plastics frequently offer reduced tool costs and high rates of production.
9. *Low energy content.*

While the attractive features of plastics tend to be in the area of physical properties, the inferior features generally relate to mechanical strength. Plastics can be flexible or rigid, but none of the plastics possess strength properties that approach those of the engineering metals unless they are reinforced in the form of a composite. Their low density allows them to compete effectively on a strength-to-weight (or specific strength) basis, however. Many have low impact strength, although several (such as ABS, high-density polyethylene, and polycarbonate) are exceptions to this rule. Aluminum is nearly 10 times more rigid than a high-rigidity plastic, and steel is 30 times more rigid.

The dimensional stability of plastics tends to be greatly inferior to that of metals, and the coefficient of thermal expansion is rather high. Thermoplastics are quite sensitive to heat, and their strength often drops rapidly as temperatures increase above normal environmental conditions. Thermosetting materials offer good strength retention at elevated temperature but have an upper limit of about 250°C (500°F). Low-temperature properties are generally inferior to those of other materials. While the corrosion resistance of plastics is generally good, they often absorb moisture, and this, in turn, decreases strength. Some thermoplastics can exhibit a 50% drop in tensile strength as the humidity increases from 0 to 100%. Radiation, both ultraviolet and particulate, can markedly alter the properties. Many plastics used in an outdoor environment have ultimately failed due to the cumulative effect of ultraviolet radiation. Plastics are also difficult to repair if broken.

Table 9-1 summarizes the properties of a number of common plastics. By considering the information in this table along with the preceding discussion of general properties, it becomes apparent that plastics are best used in applications that require materials with low to moderate strength, light weight, low electrical and/or thermal conductivity, a wide range of available colors, and ease of fabrication into finished products. No other family of materials can offer this combination of properties. Because of their light weight, attractive appearance, and ease of fabrication, plastics have been selected for many packaging and container applications. This classification includes such items as household appliance housings, clock cases, and exteriors of electronic products, where the primary role is to contain the interior mechanisms. Applications such as insulation on electrical wires and handles for hot articles capitalize on the low electrical and thermal conductivities. Soft, pliable, **foamed plastics** are used extensively as cushioning material. Rigid foams are used inside sheet metal structures to provide compressive strength. Nylon has been used for gears, acrylic for lenses, and polycarbonate for safety helmets and unbreakable windows.

TABLE 9-1 Properties and Major Characteristics of Some Common Types of Plastics

Material	Specific Gravity	Tensile Strength (1000 lb/in.^{2})	Impact Strength Izod (ft-lb/in. of Notch)	Top Working Temperature [°F(°C)]	Dielectric Strength[b] (V/mil)	24-Hour Water Absorption (%)	Special Characteristics[a]: Weatherability	Colorability	Optical Clarity	Chemical Resistance	Common Forms: Injection Molding	Extrusions	Formable Sheet	Film	Fiber	Compression or Transfer Moldings	Castings	Reinforced Plastics	Moldings	Industrial Thermosetting Laminates	Foam
Thermoplastics																					
ABS material	1.02–1.06	4–8	1.3–10.0		300–400	0.2–0.3	0	×		0	•	•	•								
Acetal	1.4	10	1.5	250(121)	1200	0.22		×		0	•	•									
Acrylics	1.12–1.19	5.5–10	0.2–2.3	200(93)	400–530	0.2–0.4	×		×	0	•	•	•		•			•			
Cellulose acetate	1.25–1.50	3–8	0.75–4.0	260(127)	300–600	2.0–6.0		×	×		•	•	•	•	•						
Cellulose acetate				130(54)	250–350	1.8–2.1	×	×	×												
butyrate	1.18–1.24	2–6	0.6–3.2																		
Cellulose propionate	1.19–1.24	1–5	0.8–9	140(60)	300	1.8–2.1		×	×		•	•	•	•							
Chlorinated polyether	1.4	6	3.3	300(149)	400	0.01				×	•	•									
Ethyl cellulose	1.16	3–6	1.8–4.0	150(66)	350	1.6–2.2		×			•	•									
TFE-fluorocarbon	2.1–2.3	1.5–3	2.5–4.0	500(260)	450	0	×			×		•	•	•	•	•					
CFE-fluorocarbon	2.1–2.15	4.5–6	3.5–3.6	390(199)	550	0	×			×	•	•	•	•							
Nylon	1.1–1.2	8–10	2	250(121)	385–470	0.4–5.5		0	0	0	•	•	•	•	•						
Polycarbonate	1.2	9.5	14	250(121)	400	0.15	×	0		×	•	•									
Polyethylene	0.96	4	10	200(93)	440	0.003		0	×	×	•	•	•	•	•						•
Polypropylene	0.9–1.27	3.4–5.3	1.02	230(110)	520–800	0.03		0	×	×	•	•	•	•	•						
Polystyrene	1.05–1.15	5–9	0.3–0.6	190(8)	400–600	<0.2		×	×	0	•	•	•	•							
Modified polystyrene	1.0–1.1	2.5–6	0.25–11.0	212(100)	300–600	0.03–02		×		×	•	•	•	•							
Vinyl	1.16–1.55	1–5.9	0.25–2.0	220(104)	25–500	0.2–1		×	×	×	•	•	•	•	•						•
Thermosetting plastics																					
Epoxy	1.1–1.7	4–13	0.4–1.5	325(163)	500	0.1–0.5	×	×		×							•	•		•	•
Melamine	1.76–1.98	5–8		350(177)	460	0.1		×		0						•		•		•	
Phenolic	1.2–1.45	5–9	0.25–5	300(149)	100–500	0.2–0.6				0						•	•	•		•	•
Polyester (other than molding compounds)	1.06–1.46	4–10	0.18–0.4	300(149)	340–570	0.5		×	×	0				•	•		•	•	•		
Polyester (alkyd, DAP)	1.6–1.75	3.2–8	3.6–8			0.16–0.67										•		•		•	
Silicone	2.0	3–5	0.2–3.0	550(288)	250–350	0.4–0.5										•	•	•		•	•
Urea	1.41–1.80	4–8.5	0.2–0.5	185(85)	300–600	1–3		×	0												

[a] X denotes a principal reason for its use; 0 indicates a secondary reason.
[b] Short-time ASTM test.

There are many applications where only one or two of the properties of plastics are sufficient to justify their use. When special characteristics are desired that are not normally found in the commercial plastics, composite materials can often be designed that use a polymeric matrix. For example, high directional strength may be achieved by incorporating a fabric or fiber reinforcement within a plastic resin. These materials will be discussed in some detail later in this chapter.

COMMON TYPES OR FAMILIES OF PLASTICS

The following is a brief descriptive summary of some of the types of plastics listed in Table 9-1.

THERMOPLASTICS

ABS: contains acrylonitrile, butadiene, and styrene; low weight, good strength, rigid, hard and very tough, even at low temperatures; opaque; resists heat, weather, and chemicals quite well; dimensionally stable but flammable.

Acrylics: hard, brittle (at room temperature); high impact, flexural, tensile, and dielectric strengths; transparent or easily colored; resist weathering and UV light; the most common example is *PMMA (polymethyl methacrylate)*; highest optical clarity, transmitting greater than 90% of light; shatterproof; common trade names include Lucite and Plexiglas; used in automotive tail-light lenses, aircraft windows, and other optical applications; scratches easily, however.

Cellulosics: Commercially important examples include *cellulose acetate*—wide range of colors, good insulating qualities, easily molded, high moisture absorption in most grades; and *cellulose acetate butyrate*—higher impact strength and moisture resistance than cellulose acetate; will withstand rougher usage.

Fluorocarbons: inert to most chemicals (solvents, acids, and bases); high temperature resistance; high strength; low moisture absorption; good weathering; very low coefficients of friction (polyfluoroethylene or Teflon accounts for about 85% of this family); used for nonlubricated bearings and nonstick coatings for cooking utensils and electrical irons.

Polyamides: *Nylons* are the most important member; good strength, toughness, stiffness, abrasion resistance, and chemical resistance; low coefficient of friction; excellent dimensional stability; good heat resistance; used for small gears and bearings, zip fasteners, and as monofilaments for textiles, carpets, fishing line, and ropes. *Aromatic polyamides* or *aramids* form another important subgroup, with the most important member being *Kevlar.* As a reinforcing fiber, it offers the strength of steel at $\frac{1}{5}$ the weight.

Polycarbonates: high strength and outstanding toughness; excellent heat resistance; good dimensional stability; transparent or easily colored; easy to process and readily recyclable; most recognizable as the base material for CDs and DVD discs, but also used for office machine housings, safety helmets, and pump impellers.

Polyesters: can be either thermoplastic or thermosetting depending on the presence or absence of cross-linking,—key example is *polyethylene terephthalate (PET)*; easily processed by injection- or blow-molding or extrusion; commonly used for food packaging, such as soft-drink bottles; polyester fibers are common in wearing apparel.

Polyethylenes: the most common polymer (30% of global plastics consumption); inexpensive, tough, good chemical resistance to acids, bases, and salts; high electrical resistance; low strength; easy to shape and join; easily recycled; reasonably clear in thin-film form; subject to weathering via ultraviolet light; limited resistance to elevated temperatures; flammable; used for grocery bags, milk jugs, and other food containers, tubes, pipes, sheeting, and electrical wire insulation. Variations include low-density polyethylene (LDPE, floats in water), high-density polyethylene (HDPE), and ultra-high molecular weight polyethylene (UHMW).

Polypropylene: lightest of the plastics; inexpensive; stronger, stiffer, and better heat resistance than polyethylene; transparent; reasonable toughness; used for beverage containers, luggage, pipes, and ropes.

Polystyrenes: high dimensional stability and stiffness with low water absorption; best all-around dielectric; excellent thermal insulator; clear, hard, and brittle at room temperature; often used for rigid packaging; can be foamed to produce expanded polystyrene (trade name of Styrofoam); burns readily; softens at about 95°C; *high-impact polystyrenes* contain additions of rubber to improve toughness.

Polyurethane: can be thermoplastic, thermosetting, or elastomeric; the thermoplastic polyurethanes bridge the gap between flexible rubbers and rigid thermoplastics; easy to process; strong, tough, and durable; flexible at low temperatures; cut and tear resistant; resistant to oil, grease, fuels, solvents, and other chemicals; common applications include flexible foam (seat cushions and carpet underlays), coatings, sealants, and adhesives.

Polyvinylchloride (PVC): extensively used general-purpose thermoplastic; strong, light, and durable; good resistance to ultraviolet light (good for outside applications); easily molded or extruded; always used with fillers, plasticizers, and pigments; used largely in building construction (gas and water pipes, window frames).

Vinyls: wide range of types, from thin, rubbery films to rigid forms; tear resistant; good aging properties; good dimensional stability and water resistance in rigid forms; used for floor and wall covering, upholstery fabrics, and lightweight water hose; common trade names include Saran and Tygon.

THERMOSETS

Amino resins: Two primary groups are melamines and urea-formaldehydes. *Melamines*—excellent resistance to heat, water, and many chemicals; full range of translucent and opaque colors; excellent electric-arc resistance; tableware and counter tops (trade name Formica); used in treating paper and cloth to impart water-repellent properties. *Urea–formaldehyde*—properties similar to those of phenolics but available in lighter colors; useful in containers and housings, but not outdoors; used in lighting fixtures because of translucence in thin sections; as a foam, may be used as household insulation.

Epoxies: good strength, toughness, elasticity, chemical resistance, moisture resistance, and dimensional stability; easily compounded to cure at room temperature; used as adhesives, bonding agents, coatings, and in fiber laminates.

Phenolics: oldest of the plastics but still widely used; hard, strong, low cost, and easily molded, but rather brittle; resistant to heat and moisture; dimensionally stable; opaque, but with a range of dark colors; wide variety of forms: sheet, rod, tube, and laminate; uses include molded products, printed circuit boards, counter tops, and as a bonding material in grinding wheels; phenol-formaldehyde goes by the trade name of Bakelite.

Polyesters: can be thermoplastic or thermoset; thermoset polyesters are strong with good resistance to environmental influences; uses include fiber-reinforced polymer-matrix composites (boat and car bodies, pipes, tanks, and construction panels), vents and ducts, textiles, adhesives, coatings, and laminates.

Polyimides: good chemical resistance; high strength and stiffness; good elevated temperature stability (considered to be a high-temperature polymer).

Polyurethanes: the thermosetting polyurethanes are often used as rigid foams where they provide support, rigidity, and thermal insulation.

Silicones: heat and weather resistant; low moisture absorption; chemically inert; high dielectric properties; excellent sealants.

ADDITIVE AGENTS IN PLASTICS

For most uses, additional materials are incorporated into plastics to (1) impart or improve properties, (2) reduce cost, (3) improve moldability, and/or (4) impart color. These **additive agents** are usually classified as *fillers and reinforcements*, *plasticizers*, *lubricants or release agents, coloring agents, stabilizers, antioxidants, flame retardants, conductive compounds,* and *foaming agents.*

Ordinarily, **fillers** comprise a large percentage of the total volume of a molded plastic product. Their primary roles are to improve strength, stiffness, or toughness; reduce shrinkage; reduce weight; or simply serve as an extender, providing cost-saving bulk (often at the expense of reduced moldability). To a large degree, they determine the general properties of a molded plastic. Selection tends to favor materials that are much less expensive than the plastic resin. Some of the most common fillers and their properties are:

1. *Wood flour* (fine sawdust): a general-purpose filler; low cost with fair strength; good moldability
2. *Cloth fibers*: improved impact strength; fair moldability
3. *Macerated cloth*: high impact strength; limited moldability
4. *Glass fibers*: high strength; dimensional stability; translucence
5. *Mica*: excellent electrical properties and low moisture absorption
6. *Calcium carbonate, silica, talc, and clay*: serve primarily as extenders

When fillers are used with a plastic resin, the resin acts as a binder, surrounding the filler material and holding the mass together. The surface of a molded part, therefore, will be almost pure resin with no exposed filler. Cutting or scratching through the shiny surface, however, will expose the less-attractive filler.

Plasticizers can be added in small amounts to reduce viscosity and improve the flow of the plastic during molding or to increase the flexibility of thermoplastic products by reducing the intermolecular contact and strength of the secondary bonds between the polymer chains. When used for molding purposes, the amount of plasticizer is governed by the intricacy of the mold. In general, it should be kept to a minimum because it is likely to affect the stability of the finished product through a gradual aging loss. When used for flexibility, plasticizers should be selected with minimum volatility, so as to impart the desired property for as long as possible. Examples of products where plasticizers improve flexibility include wire and cable coatings, tubes and hoses, a wide variety of sheets and films from construction membranes to food wraps, adhesives, and sealants.

Lubricants and **mold release agents,** such as waxes, silicones, stearates, and soaps, can be used to improve the moldability of plastics and to facilitate removal of parts from the mold. They are also used to keep thin polymer sheets from sticking to each other when stacked or rolled. When applied directly onto mold surfaces between cycles, these materials are called external mold releases. In processes such as injection molding, it may be more economical to incorporate the release as a resin additive than to periodically interrupt the process to spray the mold surfaces. When used in this manner, however, only a minimum amount should be used because the lubricants adversely affect most engineering properties.

Coloring agents can be used to provide almost any color and frequently eliminates the need for secondary coating operations. They may be either **dyes,** which are soluble in the resins, or **pigments,** which are insoluble and impart color simply by their presence as they are dispersed in the polymer matrix. In general, dyes are used for transparent plastics, and pigments for the opaque or translucent ones. Optical brighteners can also be used to enhance appearance. Carbon black can provide both a black color and electrical conductivity.

Heat, light (especially ultraviolet), and oxidation tend to degrade polymers, but **stabilizers** and **antioxidants** can be added to retard these effects. Heat-stabilizing additives seek to prevent changes in the chemical structure caused by elevated temperatures, both during processing and in the final products. UV stabilizers absorb ultraviolet radiation and convert it to heat, preventing photolytic decomposition of the polymer. Polymers can undergo chemical reactions, both during processing and during use, that lead to degradation and loss of favorable properties. The most common of these degradation reactions are based on oxidative free radicals. Antioxidant additives or stabilizers can intercept these radicals and significantly slow or halt the degradation.

TABLE 9-2 Additive Agents in Plastics and Their Purpose

Type	Purpose
Fillers	Enhance mechanical properties, reduce shrinkage, reduce weight, or provide bulk
Plasticizer	Increase flexibility, improve flow during molding, reduce elastic modulus
Lubricant	Improve moldability and extraction from molds
Coloring agents (dyes and pigments)	Impart color
Stabilizers	Retard degradation due to heat or light
Antioxidants	Retard degradation due to oxidation
Flame retardants	Reduce flammability
Conductive/antistatic additives	Impart various degrees of electrical conductivity

Flame retardants can be added when nonflammability is important. Electrically conductive additives allow for the migration of electrical charge, and combat problems associated with the generation and accumulation of static electricity, electrostatic discharges, and electromagnetic or radio-frequency interference. **Conductive polymers** are being used for a number of applications, such as electronics packaging. The antistatic agents can also reduce the attraction of dust and contaminants, making the materials more suitable for use in food or drug applications. Antimicrobial additives can provide long-term protection from both fungus (such as mildew) and bacteria. Fibers can be incorporated to increase strength and stiffness, and metal flakes, fibers, or powders can further modify electrical and magnetic properties. Table 9-2 summarizes the purposes of the various additives.

ORIENTED PLASTICS

Because the intermolecular bond strength increases with reduced separation distance, any processing that aligns the molecules parallel to the applied load can be used to give the long-chain thermoplastics high strength in a given direction.[2] This orientation can be accomplished by various forming processes, such as stretching, rolling, or extrusion. The material is usually heated prior to the orienting process to aid in overcoming the intermolecular forces and is cooled immediately afterward to "freeze" the molecules in the desired orientation.

Orienting may increase the tensile strength by more than 50%, but a 25% increase is more typical. In addition, the elongation may be increased by several hundred percent. If the **oriented plastics** are reheated, however, they tend to deform back toward their original shape, a phenomenon known as **viscoelastic memory.** The various shrink-wrap materials are examples of this effect.

ENGINEERING PLASTICS

The standard polymers tend to be lightweight, corrosion-resistant materials with low strength and low stiffness. They are relatively inexpensive and are readily formed into a wide range of useful shapes, but they are not suitable for use at elevated temperatures.

In contrast, a group of plastics has been developed with improved thermal properties (up to 350°C or 650°F), enhanced impact and stress resistance, high rigidity, superior electrical characteristics, excellent processing properties, and little dimensional change with varying temperature and humidity. These true engineering plastics include the polyamides, polyacetals, polyacrylates, polycarbonates, modified polyphenylene oxides, polybutylene terepthalates, polyketones, polysulfones, polyetherimides, and

[2] The effects of orienting can be observed in the common disposable flexible-walled plastic drinking cup. Start at the top lip. Place a sharp bend in the lip and then tear down the side wall. The material tears easily. Move around the lip about $\frac{1}{2}$ in. and make another side-wall tear—also easy. Now try to tear across the strip that you have created. This tear is much more difficult, because you are tearing across molecules that have been oriented vertically along the cup walls by the cup-forming operation.

liquid crystal polymers. While stabilizers, fibrous reinforcements, and particulate fillers can upgrade the conventional plastics, there is usually an accompanying reduction in other properties. The engineering plastics offer a more balanced set of properties. They are usually produced in small quantities, however, and are often quite expensive.

Materials producers have also developed electroconductive polymers with tailored electrical and electronic properties, and high-crystalline polymers with properties comparable to some metals.

PLASTICS AS ADHESIVES

Polymeric adhesives are used in many industrial applications. They are quite attractive for the bonding of dissimilar materials, such as metals to nonmetals, and have even been used to replace welding or riveting. A wide range of mechanical properties is available through variations in composition and additives, and a variety of curing mechanisms can be used. Examples can be found from the thermoplastics (hot melt glues), thermosets (two-part epoxies), and even elastomers (silicone adhesives). The seven most common structural adhesives are epoxies, urethanes, cyanoacrylates, acrylics, anaerobics, hot melts, and silicones. Selection usually involves consideration of the manufacturing conditions, the substrates to be bonded, the end-use environment, and cost. The various features of adhesive bonding are discussed in greater detail in Chapter 33.

PLASTICS FOR TOOLING

Polymers can also provide inexpensive tooling for applications where pressures, temperatures, and wear requirements are not extreme. Because of their wide range of properties, their ease of conversion into desired shapes, and their excellent properties when loaded in compression, plastics have been widely used in applications such as jigs, fixtures, and a wide variety of forming-die components. Both thermoplastic and thermoset polymers (particularly the cold-setting types) have been used. By using plastics in these applications, costs can be reduced and smaller quantities of products can be economically justified. In addition, the tooling can often be produced in a much shorter time, enabling quicker production.

FOAMED PLASTICS

A number of polymeric materials can be produced in the form of foams that incorporate arrays of voids throughout their structure. Chemical **foaming agents** are added to the resins, often in the form of powders or pellets. When activated by temperature, they generate gas by a decomposition reaction. This gas dissolves in the polymer melt, and when the pressure is dropped, emerges as gas pockets or bubbles.

The resulting foamed materials are extremely versatile, with properties ranging from soft and flexible to hard and rigid. The softer foams are generally used for cushioning in upholstery and automobile seats and in various applications as vibration absorbers. Semirigid foams find use in floatation devices, refrigerator insulation, disposable food trays and containers, building insulation panels, and sound attenuation. Rigid foams have been used as construction materials for boats, airplane components, electronic encapsulation and furniture.

Foamed products can be made by a wide variety of processes and can be made as either discrete products or used as a "foamed-in-place" material. In addition to the sound and vibration attenuation and thermal insulation properties mentioned earlier, foams offer light weight (often as much as 95% air) and the possibility of improved stiffness and reduced cost (less material to make the part).

POLYMER COATINGS

Polymer coatings are used extensively to enhance appearance, but they have also assumed a significant role in providing corrosion protection. The tough, thick coatings must adhere to the substrate; not chip or peel; and resist exposure to heat, moisture, salt and chemicals. Polymer coatings have been replacing chrome and cadmium due to environmental concerns relating to the heavy metals. In addition, polymers provide better resistance to the effects of acid rain.

PLASTICS VERSUS OTHER MATERIALS

Polymeric materials have successfully competed with traditional materials in a number of areas. Plastics have replaced glass in containers and other transparent products. PVC pipe and fittings compete with copper and brass in many plumbing applications. Plastics have even replaced ceramics in areas as diverse as sewer pipe and lavatory facilities.

While plastics and metals are often viewed as competing materials, their engineering properties are really quite different. Many of the attractive features of plastics have already been discussed. In addition to these, we can add (1) the ability to be fabricated with lower tooling costs; (2) the ability to be molded at the same rate as product assembly, thereby reducing inventory; (3) a possible reduction in assembly operations and easier assembly through snap fits, friction welds, or the use of self-tapping fasteners; (4) the ability to reuse manufacturing scrap; and (5) reduced finishing costs.

Metals, on the other hand, are often cheaper and offer faster fabrication speeds and greater impact resistance. They are considerably stronger and more rigid and can withstand traditional paint cure temperatures. In addition, resistance to flames, acids, and various solvents is significantly better. Table 9-3 compares the mechanical properties of selected polymers to annealed commercially pure aluminum and annealed 1040 steel. Note the mechanical superiority of the metals, even though they are being presented in their weakest condition. Table 9-4 compares the cost per pound and elastic modulus of several engineering plastics with values for steel and aluminum. When the size of the part is fixed, cost per cubic inch becomes a more valid comparison, and the figures show plastics to be quite competitive because of their low density.

TABLE 9-3 Property Comparison of Metals and Polymers

Material	Condition	TS (ksi)	E (10^6 psi)	Elongation
Polyethylene	Branched	2	0.025	90–650
Polyethylene	Crystallized	4	0.100	50–800
Polyvinylchloride	Cl-sides	8	0.375	2–40
Polystyrene	Benzene-sides	7	0.500	1–3
Bakelite	Framework	7	1.0	1
Aluminum	Annealed	13	10.0	15–30
1040 steel	Annealed	75	30.0	30

TABLE 9-4 Comparison of Varoius Materials (Modulus and Cost)[a]

Material	Modulus ($\times 10^6$ psi)	$/pound	$/in.^{3}
Aluminum	10.0	1.30	0.132
Steel	30.0	30–50	0.075–0.125
Nylon	0.1	1.80	0.129
ABS	0.3	0.92	0.031
High-density polyethylene	0.1	0.75	0.028
Polycarbonate	0.35	1.45	0.063
Polypropylene	0.2	0.94	0.029
Polystyrene	0.3	0.85	0.026
Epoxy (bisphenol)	0.45	1.06	0.033

[a] Cost figures are 2010 values and are clearly subject to change.

The automotive industry is a good indication of the expanding use of plastics. Polymeric materials now account for more than 250 lb of a typical vehicle, compared to only 25 lb in 1960, 105 in 1970, 195 in 1980, and 229 in 1990. In addition to the traditional application areas of dashboards, interiors, body panels, and trim, plastics are now being used for bumpers, intake manifolds, valve covers, fuel tanks, and fuel lines and fittings. If we include clips and fasteners, there are now more than 1000 plastic parts in a typical automobile.

RECYCLING OF PLASTICS

More than 200 million tons of plastic are manufactured annually around the world—much of it for disposable, low-value items such as food-wrap and product packaging. Unfortunately, plastics are not particularly disposable. The features that make them so durable also make them resistant to environmental and biological processes. Burning them often releases toxic chemicals, while recycling is difficult because there are so many different kinds and each has to be recycled by a different process. Moreover, many conventional plastics are often commingled with organic wastes, such as food scraps, wet paper, and liquids. Expensive cleaning and sanitizing procedures would be required prior to recycling.

Because of the wide variety of types and compositions, all with similar physical properties, the recycling of mixed plastics is far more difficult than the recycling of mixed metals. These materials must be sorted not only on the basis of resin type, but also by type of filler, color, and other additive features. If the various types of resins can be identified and kept separate, many of the thermoplastic materials can be readily recycled into useful products. Packaging is the largest single market for plastics, and there is currently a well-established network to collect and recycle PET (the polyester used in soft-drink

TABLE 9-5 Symbols for Recyclable Plastics and Some Common Uses

Symbol	Abbreviation[a]	Polymer Name	Recycled Uses
1	**PETE** or **PET**	Polyethylene terephthalate	Polyester fibres, thermoformed sheet, strapping, and soft drink bottles
2	**HDPE**	High density polyethylene	Bottles, grocery bags, milk jugs, recycling bins, agricultural pipe, base cups, car stops, playground equipment, and plastic lumber
3	**PVC** or **V**	Polyvinyl chloride	Pipe, fencing, and non-food bottles
4	**LDPE**	Low density polyethylene	Plastic bags, 6 pack rings, various containers, dispensing bottles, wash bottles, tubing, and various molded laboratory equipment
5	**PP**	Polypropylene	Auto parts, industrial fibers, food containers, and dishware
6	**PS**	Polystyrene	Desk accessories, cafeteria trays, plastic utensils, toys, video cassettes and cases, and insulation board and other expanded polystyrene products (e.g., Styrofoam)
7	**OTHER** or **O**	Other plastics, including acrylic, acrylonitrile butadiene styrene, fiberglass, nylon, polycarbonate, and polylactic acid	Bottles, plastic lumber applications

[a] Often found under symbol

bottles) and high-density polyethylene (the plastic used in milk, juice, and water jugs). The properties generally deteriorate with recycling, however, so applications must often be downgraded with reuse. PET is being recycled into new bottles, fiberfill insulation, and carpeting. Recycled polyethylene is used for new containers, plastic bags, and recycling bins. Polystyrene has been recycled into cafeteria trays and the cases for CDs and DVDs. Plastic "lumber," made from low-density polyethylene, offers weather and insect resistance and a reduction in required maintenance (but at higher cost than traditional wood). Polyvinyl chloride (PVC) and polypropylene are also readily recycled. Table 9-5 presents the standard recycling symbols often incorporated onto plastic products to identify the material, along with a list of some typical recycled uses.

Because of the cross-linking or network bonding, thermosets and elastomers cannot be recycled by simple remelting. Thermoset materials can be ground into particulates and used as fillers in other plastic parts. Rubber tires can be ground into chunks or nuggets and used for purposes such as landscape mulch or playground surfaces.

When thermoplastics and thermosets are mixed or separation is difficult, the material is often regarded more as an alternative fuel (competing with coal and oil) than as a resource for recycling into quality products. On an equivalent-weight basis, polystyrene and polyethylene have heat contents greater than fuel oil and far in excess of paper and wood. As a recycling alternative, decomposition processes can be used to break polymers down into useful building blocks. Hydrolysis (exposure to high-pressure steam) and pyrolysis (heating in the absence of oxygen) methods can be used to convert plastics into simple petrochemical materials. Ford Motor Company, for example, is using this approach to convert used nylon carpeting into nylon cylinder-head covers. These processes, however, require good control of the input material. As a result, only about 5% of all plastic (and less than one-third of plastic containers) now finds a second life.

BIODEGRADABLE PLASTICS

While additives like UV and heat stabilizers are often added to plastics to increase durability, biodegradable plastics are designed to break down more quickly in either natural aerobic (composting) or anaerobic (landfill) environments. **Bioplastics** are made from natural materials, such as corn starch. Lactic acid, obtained from corn, can be polymerized to form *polylactide* or *polylactic acid (PLA)*, a thermoplastic that looks and behaves like polyethylene and polypropylene. With a natural filler, such as cellulose, this material has been used in the manufacture of a variety of quickly discarded food containers and packaging items. When discarded, these products absorb water and swell up, breaking into small fragments that are easily processed by microorganisms.

Biodegradable plastics are made using conventional petrochemical polymers and a natural (biodegradable) filler. The degradation of the filler converts the material to a sponge-like consistency that may be more easily broken down over time. These materials may also contain bio-active additives and swelling agents that further accelerate the decay.

There is some debate regarding the benefit of these materials, however. All require a specific environment for biodegradation. In the absence of such an environment (such as natural-material biodegradable plastics placed in an anaerobic landfill), these materials remain extremely durable. When they do degrade, some release potentially harmful greenhouse gases (methane or carbon dioxide) or toxic residues. The bioplastics (all natural materials) and biodegradable plastics (petrochemical based) cannot be easily recycled. When PLA plastics are mixed with PET (polyethylene terephthalate), the entire mix becomes impossible to recycle—thereby impairing existing recycling efforts.

9.3 ELASTOMERS

The term **elastomer,** a contraction of the words *elastic polymer,* refers to a special class of linear polymers that display an exceptionally large amount of elastic deformation when a force is applied. Many can be stretched to several times their original length. Upon release of the stretching force, the deformation is completely recovered as the material quickly returns to its original shape. In addition, the cycle can be repeated numerous times with identical results, as with the stretching of a rubber band.

The elastic properties of most engineering materials are the result of a change in the distance between adjacent atoms (i.e., bond length) when loads are applied. Hooke's Law is commonly obeyed, where twice the force produces twice the stretch. When the applied load is removed, the interatomic forces return all of the atoms to their original position and the elastic deformation is completely recovered.

In the elastomeric polymers, the linear chain-type molecules are twisted or curled, much like a coil spring. When a force is applied, the polymer stretches by uncoiling. When the load is removed, the molecules recoil as the bond angles return to their original, unloaded values, and the material returns to its original size and shape. The relationship between force and stretch, however, does not follow Hooke's Law.

In reality, the behavior of elastomers is a bit more complex. While the chains indeed uncoil when placed under load, they can also slide with respect to one another to produce a small degree of viscous deformation. When the load is removed, the molecules return to their coiled shape, but the viscous deformation is not recovered and there is some permanent change in shape.

By linking the coiled molecules to one another by strong covalent bonds, a process known as **cross-linking,** it is possible to restrict the viscous deformation while retaining the large elastic response. The elasticity or rigidity of the product can be determined by controlling the number of cross-links. Small amounts of cross-linking leave the elastomer soft and flexible, as in a rubber band. Additional cross-linking further restricts the uncoiling, and the material becomes harder, stiffer, and more brittle, like the rubber used in bowling balls. Because the cross-linked bonds can only be destroyed by extremely high temperatures, the engineering elastomers can be tailored to possess a wide range of stable properties and stress-strain characteristics.

If placed under constant strain, however, even highly cross-linked material will exhibit some viscous flow over time. Consider a rubber band stretched between two nails. While the dimensions remain fixed, the force or stress being applied to the nails will continually decrease. This phenomenon is known as **stress relaxation.** The rate of this relaxation depends on the material, the force, and the temperature.

NATURAL RUBBER

Natural **rubber,** the oldest commercial elastomer, is made from latex, a secretion from the inner bark of a tropical tree. In its crude form it is an excellent adhesive, and many cements can be made by dissolving it in suitable solvents. Its use as an engineering material dates from 1839, when Charles Goodyear discovered that it could be vulcanized (cross-linked) by the addition of about 30% sulfur followed by heating to a suitable temperature. The cross-linking restricted the movement of the molecular chains and imparted strength. Subsequent research found that the properties could be further improved by various additives (such as carbon black), which act as stiffeners, tougheners, and antioxidants. Accelerators have been found that speed up the **vulcanization** process. These have enabled a reduction in the amount of sulfur, such that most rubber compounds now contain less than 3% sulfur. Softeners can be added to facilitate processing, and fillers can be used to add bulk.

Rubber (polyisoprene) can now be compounded to provide a wide variety of characteristics, ranging from soft and gummy to extremely hard. When additional strength is required, textile cords or fabrics can be coated with rubber. The fibers carry the load, and the rubber serves as a matrix to join the cords while isolating them from one another to prevent chafing. For severe service, steel wires can be used as the load-bearing medium. Vehicle tires and heavy-duty conveyor belts are examples of this technology.

Natural rubber compounds are outstanding for their high tensile and tear strength; good resilience (ability to recover shape after deformation); good electrical insulation; low internal friction; and resistance to most inorganic acids, salts, and alkalis. However, they have poor resistance to petroleum products, such as oil, gasoline, and naphtha. In addition, they lose their strength at elevated temperatures, so it is advisable that they not be used at temperatures above 80°C (175°F). Unless they are specially compounded, they also deteriorate fairly rapidly in direct sunlight.

ARTIFICIAL ELASTOMERS

To overcome some of these limitations, as well as the uncertainty in the supply and price of natural rubber, a number of synthetic or artificial elastomers have been developed and have come to assume great commercial importance. While some are a bit inferior to natural rubber, others offer distinctly different and, frequently, superior properties. Polyisoprene can also be synthesized chemically and offers properties closest to natural rubber. Styrene-butadiene or polybutadiene is today's largest-tonnage elastomer. It is an oil-derivative, high-volume substitute for natural rubber that has become the standard material for passenger-car tires. For this material, some form of reinforcement is generally required to provide the desired tensile strength, tear resistance and durability. Polychloroprene or Neoprenes (trade name) have properties similar to natural rubber, with improved stiffness and better resistance to oils, ozone, oxidation, and flame. They are used for a wide range of applications, including fuel hoses, automotive hoses and belts, footwear, wet suits, and seals.

Polyurethane can be a thermoplastic, thermoset or elastomer, depending on the degree of cross-linking. With a small amount of cross-linking, the material has elastomeric properties and is frequently used in the form of foams for cushioning in furniture and automobile seats. In its unfoamed condition, it can be molded into products like car bumpers.

Silicone rubbers look and feel like organic rubber but are based on a linear chain of silicon and oxygen atoms (not carbon). Various mixes and blends offer retention of physical properties at elevated temperatures [as hot as 230°C (450°F)]; flexibility at low temperatures [as low as −100°C (−150°F)]; resistance to acids, bases, and other aqueous and organic fluids; resistance to flex fatigue; ability to absorb energy and provide damping; good weatherability; ozone resistance; and availability in a variety of different hardnesses. A number of other artificial elastomers are available and are identified by both chemical and commercial trade names.

Elastomers are often classified as thermosetting elastomers and thermoplastic elastomers. The thermoset materials are formed during the irreversible vulcanization (cross-linking) process, which may be somewhat time consuming. Thermoplastic elastomers do not employ cross-links, but derive their properties from a complex combination of soft and hard component phases, and can be processed into products by all of the conventional thermoplastic polymer processes (injection molding, extrusion, blow molding, thermoforming, and others). They soften at elevated temperatures, which the thermosets easily withstand, but offer good low-temperature flexibility, scrap recyclability, availability in a variety of colors, and high gripping friction. Unfortunately, many are more costly than the conventional rubber materials.

SELECTION OF AN ELASTOMER

Elastomeric materials can be used for a wide range of engineering applications where they impart properties that include shock absorption, noise and vibration control, sealing, corrosion protection, abrasion protection, friction modification, electrical and thermal insulation, waterproofing, and load bearing. Selection of an elastomer for a specific application requires consideration of many factors, including the mechanical and physical service requirements, the operating environment (including temperature), the desired lifetime, the ability to manufacture the product, and cost. There are a number of families, and within each family, there exists a wide range of available properties. Moreover, almost any physical or mechanical property can be altered through additives, which can also be used to enhance processing or reduce cost, and modifications of the processing parameters.

Table 9-6 lists some of the more common artificial elastomers, along with natural rubber for comparison, and gives their properties and some typical uses.

ELASTOMERS FOR TOOLING APPLICATIONS

When an elastomer is confined, it acts like a fluid, transmitting force uniformly in all directions. For this reason, elastomers can be substituted for one-half of a die set in sheet-metal-forming operations. Elastomers are also used to perform bulging and to form reentrant sections that would be impossible to form with rigid dies except through the use of costly multipiece tooling. The engineering elastomers have become increasingly popular as tool materials because they can be compounded to range from very soft to very hard; hold up well under compressive loading; are impervious to oils, solvents, and other similar fluids; and can be made into a desired shape quickly and economically. In addition, the elastomeric tooling will not mark or damage highly polished or prepainted surfaces. The urethanes are currently the most popular elastomer for tooling applications.

9.4 CERAMICS

The first materials used by humans were natural materials such as wood and stone. The discovery that certain clays could be mixed, shaped, and hardened by firing led to what was probably the first man-made material. Many people today, when they hear the word **ceramics,** think of a coffee mug or some other form of pottery, or dinnerware, or even decorative tiles. While traditional ceramic products, such as bricks and pottery, have continued to be key materials throughout history, ceramic materials have also assumed important roles in a number of engineering applications. They are used in aircraft engines and catalytic converters, to make lightweight body armor, as the basis of artificial bones and bio-implants, as the fiber in fiber-optic communications, in a variety of sensor and control devices, and in a multitude of electronic applications that make

TABLE 9-6 Properties and Uses of Some Common Elastomers

Elastomer	Specific Gravity	Durometer Hardness	Tensile Strength (psi) Pure Gum	Black	Elongation (%) Pure Gum	Black	Service Temperature F(C)] Min.	Max.	Resistance to:[a] Oil	Water Swell	Tear	Typical Application
Natural rubber	0.93	20–100	2500	4000	75	650	–65(54)	180(82)	P	G	G	Tires, gaskets, hose
Polyacrylate	1.10	40–100	350	2500	600	400	0(–18)	300(149)	G	P	F	Oil hose, O-rings
EDPM (ethylene propylene)	0.85	30–100	1	3		500	–40(–40)	300(149)	P	G	G	Electric insulation, footwear, hose, belts
Chlorosulfonated polyethylene	1.10	50–90	4	2		400	–65(–54)	250(121)	G	E	G	Tank lining, chemical hose; shoes, soles and heels
Polychloroprene (neoprene)	1.23	20–90	3500	4000	800	550	–50(–46)	225(107)	G	G	G	Wire insulation, belts, hose, gaskets, seals, linings
Polybutadiene	1.93	30–100	1000	3000	800	550	–80(–62)	212(100)	P	P	G	Tires, soles and heels, gaskets, seals
Polyisoprene	0.94	20–100	3000	4000		600	–65(–54)	180(82)	P	G	G	Same as natural rubber
Polysulfide	1.34	20–80	350	00600	600	400	–65(–54)	180(82)	E	G	G	Seals, gaskets, diaphragms, valve disks
SBR (styrene-butadiene)	0.94	40–100	2			1200	–65(–54)	225(107)	P	G	G	Molded mechanical goods, disposable pharmaceutical items
Silicone	1.1	25–90		1200		450	–120(–84)	450(232)	F	E	P	Electric insulation, seals, gaskets, O-rings
Epichlorohydrin	1.27	40–90		2		325	–50(–46)	250(121)	G	G	G	Diaphragms, seals, molded goods, low-temperature parts
Urethane	0.85	62–95	5000		700		–54(–65)	212(100)	E	F	E	Caster wheels, heels, foam padding
Fluoroelastomers	1.65	60–90	1	3		400	–40(–40)	450(232)	E	E	F	O-rings, seals, gaskets, roll coverings

[a] P, poor; F, fair; G, good; E, excellent.

possible many of the devices in our everyday world. Most of these applications utilize their outstanding physical properties, including the ability to withstand high temperatures, resist chemical attack, provide a wide variety of electrical and magnetic properties, and resist wear. In general, ceramics are hard, brittle, high-melting-point materials with low electrical and thermal conductivity, low thermal expansion, good chemical and thermal stability, good creep resistance, high elastic modulus, and high compressive strengths that are retained at elevated temperature. A family of "structural ceramics" has also emerged, and these materials now provide enhanced mechanical properties that make them attractive for many load-bearing applications.

Glass and glass products now account for about half of the ceramic materials market. Advanced ceramic materials (including the structural ceramics, electrical and magnetic ceramics, and fiber-optic material) compose another 20%. Whiteware and porcelain enameled products (such as household appliances) account for about 10% each, while refractories and structural clay products make up most of the difference.

NATURE AND STRUCTURE OF CERAMICS

Ceramic materials are compounds of metallic and nonmetallic elements (often in the form of oxides, carbides, and nitrides) and exist in a wide variety of compositions and forms. Most have crystalline structures, but unlike metals, the bonding electrons are generally captive in strong ionic or covalent bonds. The absence of free electrons makes the ceramic materials poor electrical conductors and results in many being transparent in thin sections. Because of the strength of the primary bonds, most ceramics have high melting temperatures, high rigidity, and high compressive strength.

The crystal structures of ceramic materials can be quite different from those observed in metals. In many ceramics, atoms of significantly different size must be accommodated within the same structure, and the interstitial sites, therefore, become extremely important. Charge neutrality must be maintained throughout ionic structures. Covalent materials must have structures with a limited number of nearest neighbors, set by the number of shared-electron bonds. These features often dictate a less efficient packing, and hence lower densities, than those observed for metallic materials. As with metals, the same chemistry material can often exist in more than one structural arrangement (polymorphism), depending on the conditions of temperature and pressure. Silica (SiO_2) for example, can exist in three forms—quartz, tridymite, and crystobalite.

Ceramic materials can also exist in the form of chains, similar to the linear molecules in plastics. Like the polymeric materials having this structure, the bonds between the chains are not as strong as those within the chains. Consequently, when forces are applied, cleavage or shear can occur between the chains. In other ceramics, the atoms bond in the form of sheets, producing layered structures. Relatively weak bonds exist between the sheets, and these interfacial surfaces become the preferred sites for fracture. Mica is a good example of such a material. A noncrystalline structure is also possible in solid ceramics. This **amorphous** condition is referred to as the *glassy state*, and the materials are known as *glasses*.

Elevated temperatures can be used to decrease the viscosity of glass, allowing the atoms to move as groups and the material to be shaped and formed. When the temperature is dropped, the material again becomes hard and rigid. The crystalline ceramics do not soften, but they can creep at elevated temperature by means of grain boundary sliding. Therefore, when ceramic materials are produced for elevated temperature service, large grain size is generally desired.

CERAMICS ARE BRITTLE, BUT CAN BE TOUGH

Both crystalline and noncrystalline ceramics tend to be brittle. The glass materials have a three-dimensional network of strong primary bonds that impart brittleness. The crystalline materials do contain dislocations, but for ceramic materials, brittle fracture tends to occur at stresses lower than those required to induce plastic deformation.

There is little that can be done to alter the brittle nature of ceramic materials. However, the energy required to induce brittle fracture (the material toughness) can often be increased. **Tempered glass** uses rapid cooling of the surfaces to induce residual surface compression. The surfaces cool, contract, and harden. As the center then cools and tries to contract, it compresses or squeezes the surface. Because fractures initiate on the surface, the applied stresses must first cancel the residual compression before they can become tensile. *Cermet* materials surround particles of brittle ceramic with a continuous matrix of tough, fracture-resistant metal. **Ceramic-ceramic composites** use weak interfaces that separate or delaminate to become crack arrestors or crack diverters, allowing the remaining structure to continue carrying the load.

Stabilization involves compounding or alloying to eliminate crystal structure changes that might occur over a desired temperature range, thereby eliminating the dimensional expansions or contractions that would accompany them. Nonuniform heating or cooling can now occur without the stresses that induce fracture. **Transformation toughening** stops the progress of a crack by a crystal structure change that occurs whenever volume expansion is permitted. Fine grain size, high purity, and high density can be promoted by enhanced processing, and these all act to improve toughness.

CLAY AND WHITEWARE PRODUCTS

Many ceramic products are still based on **clay,** to which various amounts of quartz and feldspar and other materials are added. Selected proportions are mixed with water, shaped, dried, and fired to produce the structural clay products of brick, roof and structural tiles, drainage pipe, and sewer pipe, as well as the **whiteware** products of sanitary ware (toilets, sinks, and bathtubs), dinnerware, decorative floor and wall tile, pottery, and other artware. The terms *earthenware, stoneware, china,* and *porcelain* have been applied to products with increasing firing temperatures and decreasing amounts of residual porosity.

REFRACTORY MATERIALS

Refractory materials are ceramics that have been designed to provide acceptable mechanical or chemical properties at high operating temperatures, generally in excess of 550°C or 1000°F. Most are based on stable oxide compounds, where the coarse oxide particles are bonded by finer refractory material. Various carbides, nitrides, and borides can also be used in refractory applications. Desirable properties include high melting temperature, excellent hot strength, resistance to elevated-temperature chemical attack, low thermal conductivity, low thermal expansion, resistance to creep, and resistance to thermal shock.

Refractory ceramics fall into three distinct chemical classes: *acidic*, *basic,* and *neutral.* Common acidic refractories are based on silica (SiO_2) and alumina (Al_2O_3) and can be compounded to provide high-temperature resistance along with high hardness and good mechanical properties. The insulating tiles on the U.S. space shuttle were made from machinable silica ceramic. Magnesium oxide (MgO) is the core material for most basic refractories. These are generally more expensive than the acidic materials but provide better chemical resistance and are often required in metal-processing applications to provide compatibility with the metal. Neutral refractories, containing chromite (Cr_2O_3), are often used to separate the acidic and basic materials because they tend to attack one another. The combination is often attractive when a basic refractory is necessary on the surface for chemical reasons, and the cheaper, acidic material is used beneath to provide strength and insulation. Refractories that exhibit extraordinary properties for special applications may be based on silicon carbide (abrasion resistance), zircon or zirconia, fused silica (temperatures up to 1450°C or 3000°F), and carbon or graphite (not wetted by most molten metals or slags).

Refractory ceramics are the principal materials in the construction of furnaces, in the lining of ladles and containment vessels, in metal casting molds (like the one shown in Figure 9-7), and in the flues and stacks through which hot gases are conducted. They may take the form of bricks and shaped products, bulk materials (often used as coatings), and insulating ceramic fibers. Insulating bricks, designed with a high degree of porosity, are often used behind surface refractories to provide enhanced thermal insulation. Bulk or monolithic (continuous structure) refractories that do not require firing for their manufacture now account for about 50% of the refractory market. There is a

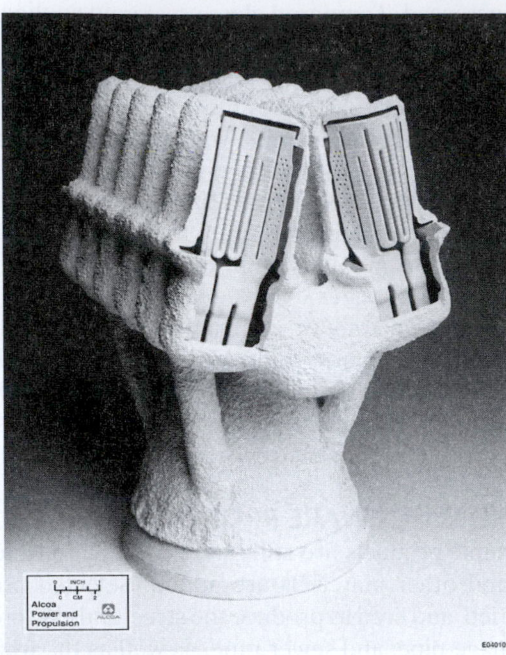

FIGURE 9-7 Ceramic investment-casting mold for casting gas turbine engine rotor blades. The positioned cores create intricate internal cooling passages. *(Photo courtesy of Alcoa Howmet)*

significant energy saving by eliminating the firing operation, the coatings are joint-free, and repairs are easily made.

Figure 9-8 shows a variety of high-strength alumina components.

ABRASIVES

Because of their high hardness, ceramic materials, such as silicon carbide and aluminum oxide (alumina), are often used as the basis for **abrasives** in the following forms: *bonded* (such as grinding wheels), *coated* (like sandpaper), and *loose* (like grit blasting and polishing compounds). Tungsten carbide is often used in wear applications, and manufactured diamond and cubic boron nitride have such phenomenal properties that they are often termed **superabrasives.** Materials used for abrasive applications are discussed in greater detail in Chapter 26.

CERAMICS FOR ELECTRICAL AND MAGNETIC APPLICATIONS

Ceramic materials also offer a variety of useful electrical and magnetic properties. Some ceramics, such as silicon carbide, are used as resistors and high-temperature heating elements for electric furnaces. Others have semiconducting properties and are used for thermistors and rectifiers. Dielectric, piezoelectric, and ferroelectric behavior can

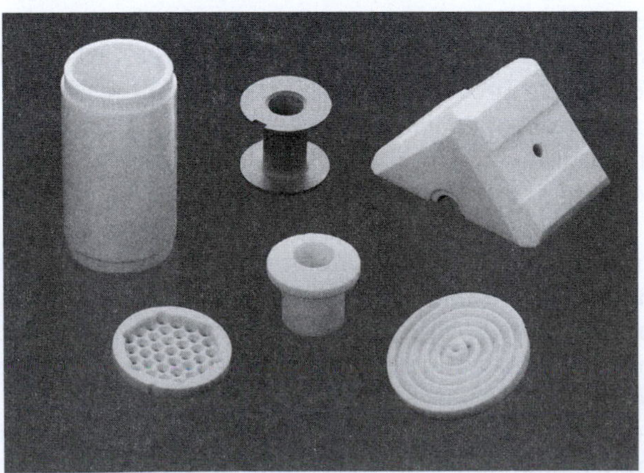

FIGURE 9-8 A variety of high-strength alumina components. *(Supplied by Morgan Technical Ceramics)*

also be utilized in many applications. Barium titanate, for example, is used in capacitors and transducers. High-density clay-based ceramics and aluminum oxide make excellent high-voltage insulators, such as those found on automotive spark plugs. The magnetic ferrites have been used in a number of magnetic applications. Considerable attention has also been directed toward the "high-temperature" ceramic superconductors.

GLASSES

When some molten ceramics are cooled at a rate that exceeds a critical value, the material solidifies into a hard, rigid, noncrystalline (i.e., amorphous) solid, known as a **glass.** Most commercial glasses are based on silica (SiO_2), lime ($CaCO_3$), and sodium carbonate ($NaCO_3$), with additives to alter the structure or reduce the melting point. Various chemistries can be used to optimize or adjust: fluidity of the molten glass, optical properties including color and index of refraction, thermal stability, and resistance to thermal shock.

Glass is soft and moldable when hot, making shaping rather straightforward. When cool and solid, glass is strong in compression, but brittle and weak in tension. In addition, most glasses exhibit excellent resistance to weathering and attack by most chemicals. Traditional applications include automotive and window glass, bottles and other containers, light bulbs, mirrors, lenses, and fiberglass insulation. There are also a wide variety of specialty applications, including glass fiber for fiber-optic communications and glass fiber to reinforce composites, glass cookware, and a variety of glass uses in medical and biological products. Glass and other ceramic fibers have also been used for filtration, where they provide a chemical inertness and the possibility to withstand elevated temperature.

Glass is recyclable, and manufacturers use as much recycled glass or cullet as possible because the material never loses its quality, purity, or clarity. In addition, the use of recycled glass enables lower furnace temperatures to be used, reducing the amount of energy needed to produce a product.

GLASS CERAMICS

These materials are first shaped as a glass and then heat-treated to promote partial devitrification or crystallization of the material, resulting in a structure that contains large amounts of fine-grain-size crystalline material within an amorphous base. Because they were initially formed as a glass, **glass ceramics** do not have the strength-limiting or fracture-inducing porosity that is characteristic of the conventional sintered ceramics. This feature, coupled with the fine grain size, provides strength that is considerably greater than that of the traditional glasses. In addition, the crystalline phase makes the material opaque and helps to retard creep at high temperatures. Because the thermal expansion coefficient is near zero, the material has good resistance to thermal shock. The white Pyroceram (trade name) material commonly found in Corningware is a common example of a glass ceramic.

CERMETS

Cermets are combinations of metals and ceramics (usually oxides, carbides, nitrides, or carbonitrides), united into a single product by the procedures of powder metallurgy. This usually involves pressing mixed powders at pressures ranging from 70 to 280 MPa (10 to 40 ksi) followed by sintering in a controlled-atmosphere furnace at about 1650°C (3000°F). Cermets combine the high hardness and refractory characteristics of ceramics with the toughness and thermal shock resistance of metals. They are used as crucibles, jet engine nozzles, and aircraft brakes and in other applications requiring hardness, strength, and toughness at elevated temperature. Cemented tungsten carbide (tungsten carbide particles cemented in a cobalt binder) has been used in dies and cutting tools for quite some time. The more advanced cermets now enable higher cutting speeds than those achievable with high-speed tool steel, tungsten carbide, or the coated carbides. See Chapter 21.

CEMENTS

Various ceramic materials can harden by chemical reaction, enabling their use as a binder that does not require firing or sintering. Sodium silicate hardens in the presence of carbon dioxide, and is used to produce sand cores in metal casting. Plaster of Paris and Portland cement both harden by hydration reactions.

CERAMIC COATINGS

A wide spectrum of enamels, glazes, and other ceramic coatings has been developed to decorate, seal, and protect substrate materials. Porcelain enamel can be applied to carbon steel in the perforated tubs of washing machines, where the material must withstand the scratching of zippers, buttons and snaps along with the full spectrum of laundry products. Chemical reaction vessels are often glass lined.

CERAMICS FOR MECHANICAL APPLICATIONS: STRUCTURAL AND ADVANCED CERAMICS

Because of the strong ionic or covalent bonding and high shear resistance, ceramic materials tend to have low ductility and high compressive strength. Theoretically, ceramics could also have high tensile strengths. However, because of their high melting points and lack of ductility, most ceramics are processed in the solid state, where products are made from powdered material. After various means of compaction, voids remain between the powder particles, and a portion of these persists through the sintering process (described in Chapter 18). Contamination can also occur on particle surfaces and then become part of the internal structure of the product. As a result, full theoretical density is extremely difficult to achieve, and small cracks, pores, and impurity inclusions tend to be an integral part of most ceramic materials. These act as mechanical stress concentrators. As loads are applied, the effect of these flaws cannot be reduced through plastic flow, and the result is generally a brittle fracture. Applying the principles of fracture mechanics,[3] we find that ceramics are sensitive to very small flaws. Tensile failures typically occur at stress values between 20 and 210 MPa (3 and 30 ksi), more than an order of magnitude less than the corresponding strength in compression.

Because the number, size, shape, and location of the flaws are likely to differ from part to part, ceramic parts produced from identical material by identical methods often fail at very different applied loads. As a result, the mechanical properties of ceramic products tend to follow a statistical spread that is much less predictable than for metals. This feature tends to limit the use of ceramics in critical high-strength applications.

If the various flaws and defects could be eliminated or reduced to very small size, high and consistent tensile strengths could be obtained. Hardness, wear resistance, and strength at elevated temperatures would be attractive properties, along with light weight (specific gravities of 2.3 to 3.85), high stiffness, dimensional stability, low thermal conductivity, corrosion resistance, and chemical inertness. Failure would still occur by brittle fracture, however. Because of the poor thermal conductivity, thermal shock may also be a problem. The cost of these "flaw-free" or "restricted flaw" materials would be rather high. Joining to other engineering materials and machining would be extremely difficult, so products would have to be fabricated to final shape through the use of net-shape processing.

Advanced (also known as *structural* or *engineering*) **ceramics** is an emerging technology with a broad base of current and potential applications. The base materials currently include silicon nitride, silicon carbide, partially stabilized zirconia, transformation toughened zirconia, alumina, sialons, boron carbide, boron nitride, titanium diboride, and ceramic composites (such as ceramic fibers in a glass, glass–ceramic, or ceramic matrix). The materials and products are characterized by high strength and hardness, high fracture

[3] According to the principles of fracture mechanics, fracture will occur in a brittle material when the fracture toughness, K, is equal to a product involving a dimensionless geometric factor α, the applied stress, σ, and the square root of the number π (3.14), times the size of the most critical flaw, a.

$$K = \alpha\sigma(\pi a)^{1/2}$$

When the right-hand side is less than the value of K, the material bears the load without breaking. Fracture occurs when the combination of applied stress and flaw size equals the critical value, K. Because K is a material property, any attempt to increase the load or stress a material can withstand must be achieved by a companion reduction in flaw size.

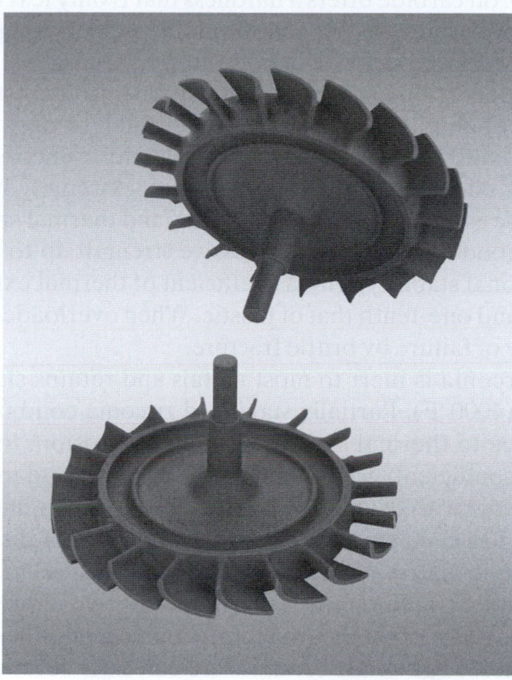

FIGURE 9-9 Gas-turbine rotors made from advanced ceramic silicon nitride. The lightweight material (one-half the weight of stainless steel) offers strength at elevated temperature as well as excellent corrosion resistance and thermal shock. *(Kyocera Industrial Ceramics Corp)*

toughness, resistance to heat and chemicals, fine grain size, and little or no porosity. Applications include a wide variety of wear-resistant parts, including cutting tools, punches, dies, and engine components (such as bearings, seals, and valves), as well as use in heat exchangers, gas turbines, and furnaces. Porous products have been used as substrate material for catalytic converters and as filters for streams of molten metal. Biocompatible ceramics have been used as substitutes for joints and bones, and as dental implants.

Alumina (or aluminum oxide) ceramics are the most common for industrial applications. They are relatively inexpensive and offer high hardness and abrasion resistance, low density, and high electrical resistivity. Alumina is strong in compression and retains useful properties at temperatures as high as 1900°C (3500°F), but it is limited by low toughness, low tensile strength, and susceptibility to thermal shock and attack by highly corrosive media. Due to its high melting point, it is generally processed in a powder form.

Silicon carbide and silicon nitride offer excellent strength, extreme hardness, adequate toughness, high thermal conductivity, low thermal expansion, and good thermal shock resistance, coupled with light weight and corrosion resistance. They work well in high-stress, high-temperature applications, such as turbine blades, and may well replace nickel- or cobalt-based superalloys. Figure 9-9 shows gas-turbine rotors made from injection-molded silicon nitride. They are designed to operate at 1250°C (2300°F), where the material retains more than half of its room-temperature strength and does not require external cooling. Figure 9-10 shows some additional silicon nitride products.

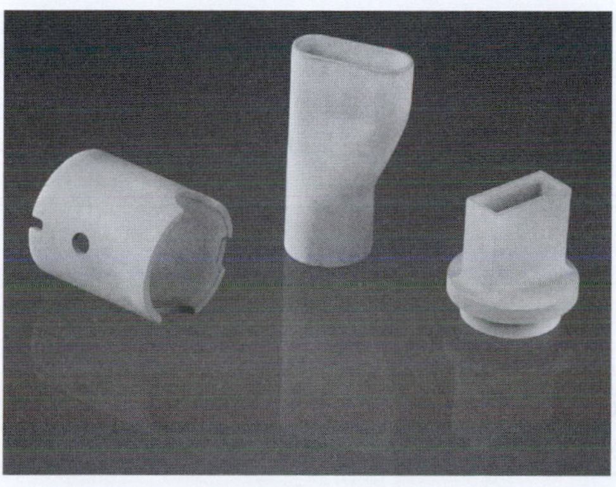

FIGURE 9-10 A variety of components manufactured from silicon nitride. *(Supplied by Morgan Technical Ceramics)*

Boron carbide offers a hardness that is only less than diamond, cubic boron nitride and boron carbide. Its high elastic modulus, high compressive strength, and light weight make it ideal for light weight body armor.

Sialon (a *si*licon–*al*uminum–*o*xygen–*n*itrogen structural ceramic) is really a solid solution of alumina and silicon nitride, and it bridges the gap between them. More aluminum oxide enhances hardness, while more silicon nitride improves toughness. The resulting material is stronger than steel, extremely hard, and as light as aluminum. It has good resistance to corrosion, wear, and thermal shock; is an electrical insulator; and retains good tensile and compressive strength up to 1400°C (2550°F). It has excellent dimensional stability, with a coefficient of thermal expansion that is only one-third that of steel and one-tenth that of plastic. When overloaded, however, it exhibits the ceramic property of failure by brittle fracture.

Zirconia is inert to most metals and retains strength to temperatures well over 2200°C (4000°F). Partially stabilized zirconia combines the zirconia characteristics of resistance to thermal shock, wear, and corrosion; low thermal conductivity; and low friction coefficient, with the enhanced strength and toughness brought about by doping the material with oxides of calcium, yttrium, or magnesium. Transformation-toughened zirconia has even greater toughness as a result of dispersed second phases throughout the ceramic matrix. When a crack approaches the metastable phase, it transforms to a more stable structure, increasing in volume to compress and stop the crack.

The high cost of the structural ceramics continues to be a barrier to their widespread acceptance. High-grade ceramics are currently several times more expensive than their metal counterparts. Even factoring in enhanced lifetime and improved performance, there is still a need to reduce cost. Work continues, however, toward the development of a low-cost, high-strength, high-toughness ceramic with a useful temperature range. Parallel efforts are under way to ensure flaw detection in the range of 10 to 50 mm. If these efforts are successful, ceramics could compete where tool steels, powdered metals, coated materials, and tungsten carbide are now being used. Potential applications include engines, turbochargers, gas turbines, bearings, pump and valve seals, and other products that operate under high-temperature, high-stress environments.

In addition to the attractive properties described earlier, the advanced ceramics are also biocompatible, which make them ideal for medical implant applications. Ceramic-on-ceramic artificial joints now have a predicted life in excess of 20 years. Ceramic materials have also emerged as the material-of-choice in numerous dental procedures.

A ceramic automobile engine has been discussed for a number of years. By allowing higher operating temperatures, engine efficiency could be increased. Sliding friction would be reduced, and there would be no need for cooling. The radiator, water pump, coolant, fan belt and water lines could all be eliminated. The net result would be up to a 30% reduction in fuel consumption. Unfortunately, this is still a dream, because of the inability to produce large, complex-shaped products with few, small-sized flaws.

Table 9-7 provides the mechanical properties of some of today's advanced or structural ceramics.

TABLE 9-7 Properties of Some Advanced or Structural Ceramics

Material	Density (g/cm^3)	Tensile Strength (ksi)	Compressive Strength (ksi)	Modulus of Elasticity (10^6 psi)	Fracture Toughness (ksi$\sqrt{\text{in.}}$)
Al_2O_3	3.98	30	400	56	5
Sialon	3.25	60	500	45	9
SiC	3.1	25	560	60	4
ZrO_2 (partially stabilized)	5.8	65	270	30	10
ZrO_2 (transformation toughened)	5.8	50	250	29	11
Si_3N_4 (hot pressed)	3.2	80	500	45	5

ADVANCED CERAMICS AS CUTTING TOOLS

Because of their high hardness, retention of hardness at elevated temperature, and low reactivity with metals, ceramic materials are attractive for a variety of cutting applications. Silicon carbide is a common abrasive in many grinding wheels. Cobalt-bonded tungsten carbide has been a popular alternative to high-speed tool steels for many tool and die applications. Tool steel and carbide tools are often enhanced by a variety of vapor-deposited ceramic coatings. Thin layers of titanium carbide, titanium nitride, and aluminum oxide can inhibit reactions between the metal being cut and the tool steel or binder phase of the carbide. This results in a significant reduction in friction and wear and enables faster rates of cutting. Silicon nitride, boron carbide, cubic boron nitride, and polycrystalline diamond cutting tools offer even greater tool life, higher cutting speeds, and reduced machine downtime. Cutting tool materials are developed in greater detail in Chapter 21.

With advanced tool materials, cutting speeds can be increased from 60 to 1500 m/min (200 to 5000 ft/min). The use of these ultra-high-speed materials, however, requires companion developments in the machine tools themselves. High-speed spindles must be perfectly balanced, and workholding devices must withstand high centrifugal forces. Chip-removal methods must be able to remove the chips as fast as they are formed.

As environmental regulations become more stringent, dry machining may be pursued as a means of reducing or eliminating coolant- and lubricant-disposal problems. Ceramic materials are currently the best materials for dry operations. Ceramic tools have also been used in the direct machining of materials that once required grinding, a process sometimes called **hard machining.** Figure 9-11 shows the combination of toughness and hardness for a variety of cutting-tool materials.

ULTRA-HIGH-TEMPERATURE CERAMICS

Ultra-high-temperature ceramics can be defined as compounds with melting temperatures in excess of 3000°C (5400°F), most of which are the borides, carbides, or nitrides of the early transition metals (niobium, zirconium, tantalum, and hafnium). Because of their chemical and structural stability, they are materials of interest in applications like the handling of high-temperature molten metals; as electrodes in electric arcs; and in aerospace applications such as hypersonic flight, scramjet propulsion, rocket propulsion, and atmospheric reentry. The boride compounds offer the ceramic properties of high melting point, high elastic modulus, and high hardness, coupled with metal-like values of high electrical and thermal conductivity. The carbides have higher melting temperatures, but lower electrical and thermal conductivities. Both families are significantly lighter than the high-temperature refractory metals.

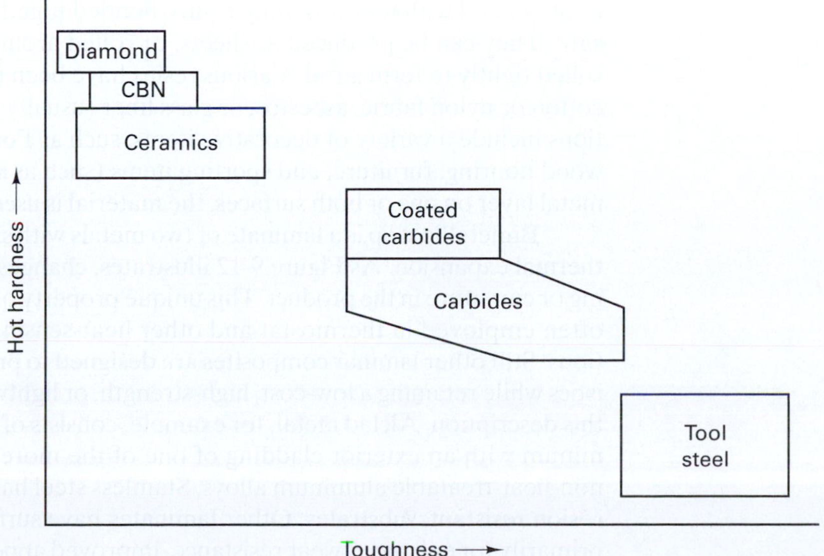

FIGURE 9-11 Graphical mapping of the combined toughness and hardness for a variety of cutting-tool materials. Note the superior hardness of the ceramic materials. (CBN = cubic boron nitride)

9.5 COMPOSITE MATERIALS

A **composite material** is a nonuniform solid consisting of two or more different materials that are mechanically or metallurgically bonded together. Each of the various components retains its identity in the composite and maintains its characteristic structure and properties. There are recognizable interfaces between the component materials. The composite, however, generally possesses characteristic properties (or combinations of properties), such as stiffness, strength, weight, high-temperature performance, corrosion resistance, hardness, and conductivity, which are not possible with the individual components by themselves. Analysis of these properties shows that they depend on (1) the properties of the individual components; (2) the relative amounts of the components; (3) the size, shape, and distribution of the discontinuous components; (4) the orientation of the various components; and (5) the degree of bonding between the components. The materials involved can be natural materials, man-made organics, metals, or ceramics. Hence, a wide range of freedom exists, and composite materials can often be designed to meet a desired set of engineering properties and characteristics.

There are many types of composite materials and several methods of classifying them. One method is based on geometry and consists of three distinct families: laminar or layered composites, particulate composites, and fiber-reinforced composites.

LAMINAR OR LAYERED COMPOSITES

Laminar composites are those having distinct layers of material bonded together in some manner and include thin coatings, thicker protective surfaces, claddings, bimetallics, laminates, sandwiches, and others. They are used to impart properties such as reduced cost, enhanced corrosion resistance or wear resistance, electrical insulation or conductivity, unique expansion characteristics, lighter weight, improved strength, or altered appearance.

Plywood is probably the most common engineering material in this category and is an example of a laminate material. Layers of wood veneer are adhesively bonded with their grain orientations at various angles to one another. Strength and fracture resistance are improved, properties are somewhat uniform within the plane of the sheet, swelling and shrinkage tendencies are minimized, and large pieces are available at reasonable cost. Automotive tires have a series of bonded belts or plies. Safety glass is another laminate in which a layer of polymeric adhesive is placed between two pieces of glass and serves to retain the fragments when the glass is broken. A*r*amid-*al*uminum-*l*aminates (Arall) consist of thin sheets of aluminum bonded with woven adhesive-impregnated aramid fibers. The combination offers light weight coupled with high fracture, impact, and fatigue resistance.

Laminated plastics are made from layers of reinforcing material that have been impregnated with thermosetting resins, bonded together and cured under heat and pressure. They can be produced as sheets, or rolled around a mandrel to produce a tube or rolled tightly to form a rod. Various resins have been used with reinforcements of paper, cotton or nylon fabric, asbestos, or glass fiber (usually in woven form). Common applications include a variety of decorative items, such as Formica countertops, imitation hardwood flooring, furniture, and sporting items (such as snow skis). When combined with a metal layer on one or both surfaces, the material is used for printed circuit boards.

Bimetallic strip is a laminate of two metals with significantly different coefficients of thermal expansion. As Figure 9-12 illustrates, changes in temperature now produce flexing or curvature in the product. This unique property of shape varying with temperature is often employed in thermostat and other heat-sensing or temperature-control applications. Still other laminar composites are designed to provide enhanced surface characteristics while retaining a low-cost, high-strength, or lightweight core. Many clad materials fit this description. Alclad metal, for example, consists of high strength, age-hardenable aluminum with an exterior cladding of one of the more corrosion-resistant, single-phase, non-heat-treatable aluminum alloys. Stainless steel has been applied to cheaper, less corrosion resistant, substrates. Other laminates have surface layers that have been selected primarily for enhanced wear resistance, improved appearance, or electrical conductivity.

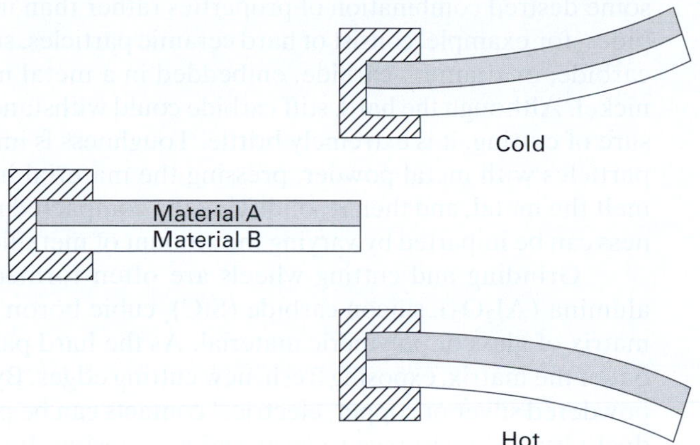

FIGURE 9-12 Schematic of a bimetallic strip where material A has the greater coefficient of thermal expansion. Note the response to cold and hot temperatures.

In 1964, the value of the silver in a U.S. quarter exceeded 25 cents, and a less-expensive replacement was needed. The replacement needed to be accepted in coin-operated vending machines, however, and these machines assessed both size and weight. The replacement quarter, therefore, had to have the same overall density as the original silver coin. The solution involved creating a laminar composite of metal alloys that are heavier than silver (copper) and lighter than silver (nickel). By placing the high-nickel alloy on the surface, the brightness and appearance of the previous coin could be retained.

Sandwich material is a laminar structure composed of a thick, low-density core placed between thin, high-density surfaces. Corrugated cardboard is an example of a sandwich structure. Other engineering sandwiches incorporate cores of a polymer foam or honeycomb structure, to produce a lightweight, high-strength, high-rigidity composite (See Figure 14-20).

It should be noted that the properties of laminar composites are always **anisotropic**—that is, they are not the same in all directions. Because of the variation in structure, properties will always be different in the direction perpendicular to the layers.

PARTICULATE COMPOSITES

Particulate composites consist of discrete particles of one material surrounded by a matrix of another material. Concrete is a classic example, consisting of sand and gravel particles surrounded by hydrated cement. Asphalt consists of mineral aggregate in a matrix of bitumin, a thermoplastic polymer. In both of these examples, the particles are rather coarse. Other particulate composites involve extremely fine particles and include many of the multicomponent powder metallurgy products, specifically those where the dispersed particles do not diffuse into the matrix material.

Dispersion-strengthened materials are particulate composites where a small amount of hard, brittle, small-size particles (typically, oxides or carbides) are dispersed throughout a softer, more ductile metal matrix. Because the dispersed material is not soluble in the matrix, it does not redissolve, overage, or overtemper when the material is heated. Pronounced strengthening can be induced, which decreases only gradually as temperature is increased. Creep resistance, therefore, is improved significantly. Examples of dispersion-strengthened materials include sintered aluminum powder (SAP), which consists of an aluminum matrix strengthened by up to 14% aluminum oxide, and thoria-dispersed (or TD) nickel, a nickel alloy containing 1 to 2 wt% thoria (ThO_2). Silicon carbide dispersed in aluminum has been shown to be a low-cost, easily fabricated material with isotropic properties, and has been used in automotive connecting rods. Because of the metal–ceramic mix and the desire to distribute materials of differing density, the dispersion-strengthened composites are generally produced by powder metallurgy techniques.

Other types of particulate composites are known as **true particulate composites,** and contain large amounts of coarse particles. They are usually designed to produce some desired combination of properties rather than increased strength. Cemented carbides, for example, consist of hard ceramic particles, such as tungsten carbide, tantalum carbide, or titanium carbide, embedded in a metal matrix, which is usually cobalt or nickel. Although the hard, stiff carbide could withstand the high temperatures and pressure of cutting, it is extremely brittle. Toughness is imparted by combining the carbide particles with metal powder, pressing the material into the desired shape, heating to melt the metal, and then resolidifying the compacted material. Varying levels of toughness can be imparted by varying the amount of metallic material in the composite.

Grinding and cutting wheels are often formed by bonding abrasives, such as alumina (Al_2O_3), silicon carbide (SiC), cubic boron nitride (CBN), or diamond, in a matrix of glass or polymeric material. As the hard particles wear, they fracture or pull out of the matrix, exposing fresh, new cutting edges. By combining tungsten powder and powdered silver or copper, electrical contacts can be produced that offer both high conductivity and resistance to wear and arc erosion. Foundry molds and cores are often made from sand (particles) and an organic or inorganic binder (matrix).

Metal-matrix composites of the particulate type have been made by introducing a variety of ceramic or glass particles into aluminum or magnesium matrices. Particulate-toughened ceramics using zirconia and alumina matrices are being used as bearings, bushings, valve seats, die inserts, and cutting tool inserts. Many plastics could be considered to be particulate composites because the additive fillers and extenders are actually dispersed particles. Designation as a particulate composite, however, is usually reserved for polymers where the particles are added for the primary purpose of property modification. One such example is the combination of granite particles in an epoxy matrix that is currently being used in some machine tool bases. This unique material offers high strength and a vibration damping capacity that exceeds that of gray cast iron.

Because of their unique geometry, the properties of particulate composites are usually **isotropic,** that is, uniform in all directions. This may be particularly important in engineering applications.

FIBER-REINFORCED COMPOSITES

The most popular type of composite material is the **fiber-reinforced composite** geometry, where continuous or discontinuous thin fibers of one material are embedded in a matrix of another. The objective is usually to enhance strength, stiffness, fatigue resistance, or strength-to-weight ratio by incorporating strong, stiff, but possibly brittle fibers in a softer, more ductile, matrix. The matrix supports and transmits forces to the fibers, protects them from environments and handling, and provides ductility and toughness as well as the necessary corrosion resistance. The fibers carry most of the load and impart enhanced stiffness. Coupling agents are often incorporated to enhance the bonding between the fibers and the matrix.

Wood and bamboo are two naturally occurring fiber composites, consisting of cellulose fibers in a lignin matrix. Bricks of straw and mud may well have been the first human-made material of this variety, dating back to near 800 B.C. Automobile tires now use fibers of nylon, rayon, aramid (Kevlar), or steel in various numbers and orientations to reinforce the rubber and provide added strength and durability. Steel-reinforced concrete is actually a double composite, consisting of a particulate matrix reinforced with steel fibers.

Glass-fiber-reinforced resins, the first of the modern fibrous composites, were developed shortly after World War II in an attempt to produce lightweight materials with high strength and high stiffness. Glass fibers about 10 mm in diameter are bonded in a variety of polymers, generally epoxy or polyester resins. Between 30 and 60 vol% is made up of fibers of either E-type borosilicate glass (tensile strength of 500 ksi and elastic modulus of 10.5×10^6 psi) or the stronger, stiffer, high-performance S-type magnesia–alumina–silicate glass (with tensile strength of 670 ksi and elastic modulus of 12.4×10^6 psi).[4]

[4] It is important to note that a fiber of material tends to be stronger than the same material in bulk form. The size of any flaw is limited to the diameter of the fiber, and the complete failure of a given fiber does not propagate through the assembly, as would occur in an identical bulk material.

Glass fibers are still the most widely used reinforcement, primarily because of their lower cost and adequate properties for many applications. Current uses of glass-fiber-reinforced plastics include sporting goods, boat hulls, and bathtubs. Limitations of the glass-fiber material are generally related to strength and stiffness. Alternative fibers have been developed for applications requiring enhanced properties. Boron–tungsten fibers (boron deposited on a tungsten core) offer an elastic modulus of 55×10^6 psi with tensile strengths in excess of 400 ksi. Silicon carbide filaments (SiC on tungsten) have an even higher modulus of elasticity.

Graphite (or carbon) and aramid (DuPont tradename of Kevlar) are other popular reinforcing fibers. Graphite fibers can be either the PAN type, produced by the thermal pyrolysis of synthetic organic fibers, primarily polyacrylonitrile, or pitch type, made from petroleum pitch. They have low density and a range of high tensile strengths (600 to 750 ksi) and high elastic moduli (40 to 65×10^6 psi). Graphite's negative thermal-expansion coefficient can also be used to offset the positive values of most matrix materials, leading to composites with low or zero thermal expansion. Kevlar is an organic aramid fiber with a tensile strength up to 650 ksi, elastic modulus of 27×10^6 psi, a density approximately one-half that of aluminum, and good toughness. In addition, it is flame retardant and transparent to radio signals, making it attractive for a number of military and aerospace applications where the service temperature is not excessive.

Ceramic fibers, metal wires, and specially grown whiskers have also been used as reinforcing fibers for high-strength, high-temperature applications. Metal fibers can also be used to provide electrical conductivity or shielding from electromagnetic interference to a lightweight polymeric matrix. With the demand for less expensive, lightweight, environmentally friendly (renewable and compostable) materials, the natural fibers have also assumed an engineering material role. Cotton, hemp, flax, jute, coir (coconut husk), and sisal have found use in various composites.[5] Thermoplastic fibers, such as nylon and polyester, have been used to enhance the toughness and impact strength of the brittle thermoset resins.

Table 9-8 lists some of the key engineering properties for several of the common reinforcing fibers. Because the objectives are often high strength coupled with light weight, or high stiffness coupled with light weight, properties are often reported as **specific strength** and **specific stiffness,** where the strength or stiffness values are divided by density.

The orientation of the fibers within the composite is often key to properties and performance. Sheet-molding compound (SMC), bulk-molding compound (BMC), and fiberglass generally contain short, randomly oriented fibers. Long, unidirectional fibers, either monofilament or multifilament bundles, can be used to produce highly directional properties, with the fiber directions being tailored to the direction of loading. Woven fabrics or tapes can be produced and then layered in various orientations to produce a laminar or plywood-like product. The layered materials can then be stitched together (knitted) to add a third dimension to the weave, and complex three-dimensional shapes can be woven from fibers and later injected with a matrix material.

TABLE 9-8 Properties and Characteristics of Some Common Reinforcing Fibers

Fiber Material	Specific Strength[a] (10^6 in.)	Specific Stiffness[b] (10^6 in.)	Density (lb/in.^{3})	Melting Temperature[c] (°F)
Al_2O_3 whiskers	21.0	434	0.142	3600
Boron	4.7	647	0.085	3690
Ceramic fiber (mullite)	1.1	200	0.110	5430
E-type glass	5.6	114	0.092	<3140
High-strength graphite	7.4	742	0.054	6690
High-modulus graphite	5.0	1430	0.054	6690
Kevlar	10.1	347	0.052	—
SiC whiskers	26.2	608	0.114	4890

[a] Strength divided by density.
[b] Elastic modulus divided by density.
[c] Or maximum temperature of use.

[5] Ford Motor Company is currently using wheat straw to replace 20% of the petroleum-based polypropylene plastic in two internal storage bins of its Flex crossover vehicle and is considering use in center console bins and trays, door trim panels, and other interior components. The wheat straw acts as a weight-reducing filler and simultaneously increases stiffness.

The properties of fiber-reinforced composites depend strongly on several characteristics: (1) the properties of the fiber material; (2) the volume fraction of fibers; (3) the **aspect ratio** of the fibers, that is, the length-to-diameter ratio; (4) the orientation of the fibers; (5) the degree of bonding between the fiber and the matrix; and (6) the properties of the matrix. While more fibers tend to provide greater strength and stiffness, the volume fraction of fibers generally cannot exceed 80% to allow for a continuous matrix. Long, thin fibers (higher aspect ratio) provide greater strength, and a strong bond is usually desired between the fiber and the matrix.

POLYMER-MATRIX FIBER-REINFORCED COMPOSITES

The matrix materials should be strong, tough, and ductile so that they can transmit the loads to the fibers and prevent cracks from propagating through the composite. In addition, the matrix material is often responsible for providing the electrical properties, chemical behavior, and elevated-temperature stability. For polymer-matrix composites, both thermosetting and thermoplastic resins have been used. The thermosets provide high strength and high stiffness, and the low-viscosity, uncured resins readily impregnate the fibers. Popular thermosets include polyesters and vinyls, epoxies, bismaleimides, and polyimides. While epoxies are the most common thermosets for high-strength use, they are often limited to temperatures of 120°C (250°F). Bismaleimides extend this range to 170 to 230°C (350 to 400°F), and the polyimides can be used up to 260 to 320°C (500 to 600°F).

Bulk-molding compound (BMC) is a viscous, putty-like material, consisting of thermosetting resin, 5 to 50% discontinuous fiber reinforcement consisting of chopped strand (usually glass) 0.75 to 12.5mm ($\frac{1}{32}$ to $\frac{1}{2}$ in.) in length, and various additives. It can be shaped by various thermoset molding operations to produce parts with good dimensional precision, good mechanical properties, high strength-to-weight ratio (up to one-third lighter than metal), integral color, and surfaces receptive to powder coating, painting, or other coating techniques.

Sheet-molding compound (SMC) is the sheet equivalent of BMC with 10 to 65% reinforcement fibers that range from 12.5 to 50 mm ($\frac{1}{2}$ to 2 in.) in length. Parts ranging from 30 to 150 cm (12 to 60 in.) in width can be formed from the various thermoset sheet-forming techniques.

From a manufacturing viewpoint, it may be easier and faster to heat and cool a thermoplastic than to cure a thermoset. Moreover, the thermoplastics are tougher, more tolerant to damage, and often recyclable or remoldable. Polyethylene, polystyrene, and nylon are traditional thermoplastic matrix materials. Improved high-temperature and chemical-resistant properties can be achieved with the thermoplastic polyimides, polyphenylene sulfide (PPS), polyether ether ketone (PEEK), and the liquid-crystal polymers. When reinforced with high-strength, high-modulus fibers, these materials can show dramatic improvements in strength, stiffness, toughness, and dimensional stability compared to traditional thermoplastics.

ADVANCED FIBER-REINFORCED COMPOSITES

Advanced composites are materials that have been developed for applications requiring exceptional combinations of strength, stiffness, and light weight. Fiber content often exceeds 50 vol%, and the modulus of elasticity is typically greater than 16×10^6 psi. Superior creep and fatigue resistance, low thermal expansion, low friction and wear, vibration-damping characteristics, and environmental stability are other properties that may also be required in these materials.

There are four basic types of advanced composites where the matrix material is matched to the fiber and the conditions of application:

1. The advanced **polymer-matrix composites** frequently use high-strength, high-modulus fibers of graphite (carbon), aramid (Kevlar), or boron with epoxy, bismaleimide, or polyimide matrix to create a high-strength, lightweight, fatigue-resistant material. Properties can be put in desired locations or orientations at about one-half the weight of aluminum or one-sixth that of steel. Thermal expansion

TABLE 9-9 Properties of Several Polymer-Matrix Fiber-Reinforced Composites (in the Fiber Direction) Compared to Lightweight or Low-Thermal-Expansion Metals

Material	Specific Strength[a] (10^6 in.)	Specific Stiffness[b] (10^6 in.)	Density ($lb/in.^3$)	Thermal Expansion Coefficient [in./(in.-°F)]	Thermal Conductivity [Btu/(hr-ft-°F)]
Boron–epoxy	3.3	457	0.07	2.2	1.1
Glass–epoxy (woven cloth)	0.7	45	0.065	6	0.1
Graphite–epoxy: high modulus (unidirectional)	2.1	700	0.063	−0.5	75
Graphite–epoxy: high strength (unidirectional)	5.4	400	0.056	−0.3	3
Kevlar–epoxy (woven cloth)	1	80	0.5	1	0.5
Aluminum	0.7	100	0.10	13	100
Beryllium	1.1	700	0.07	7.5	120
Invar[c]	0.2	70	0.29	1	6
Titanium	0.8	100	0.16	5	4

[a] Strength divided by density.
[b] Elastic modulus divided by density.
[c] A low-expansion metal containing 36% Ni and 64% Fe.

can be designed to be low or even negative, and the fibers frequently provide or enhance electrical and thermal conductivity. Unfortunately, these materials have a maximum service temperature of about 315°C (600°F) because the polymer matrix loses strength when heated. Table 9-9 compares the properties of some of the common resin-matrix composites with those of several of the lightweight or low-thermal-expansion metals. Typical applications include sporting equipment (tennis rackets, skis, golf clubs, and fishing poles), lightweight armor plate, and a myriad of low-temperature aerospace components.

2. **Metal-matrix composites (MMCs)** can be used for operating temperatures up to 1250°C (2300°F), where the conditions require high strength, high stiffness, good electrical and/or thermal conductivity, exceptional wear resistance, and good ductility and toughness. The ductile matrix material can be aluminum, copper, magnesium, titanium, nickel, superalloy, or even intermetallic compound, and the reinforcing fibers may be graphite, boron carbide, alumina, or silicon carbide. Fine whiskers (tiny needle-like single crystals of 1 to 10 mm in diameter) of sapphire, silicon carbide, and silicon nitride have also been used as the reinforcement, as well as wires of titanium, tungsten, molybdenum, beryllium, and stainless steel. The reinforcing fibers may be either continuous or discontinuous, and typically comprise between 10 and 60 vol% of the composite.

   Compared to the engineering metals, these composites offer higher stiffness and strength (especially at elevated temperatures); a lower coefficient of thermal expansion; better elevated temperature properties; and enhanced resistance to fatigue, abrasion, and wear. Compared to the polymer-matrix composites, they offer higher heat resistance, as well as improved electrical and thermal conductivity. They are nonflammable, do not absorb water or gases, and are corrosion resistant to fuels and solvents. Unfortunately, these materials are quite expensive. The vastly different thermal expansions of the components may lead to debonding, and the assemblies may be prone to degradation through interdiffusion or galvanic corrosion.

   Graphite-reinforced aluminum can be designed to have near-zero thermal expansion in the fiber direction. Aluminum oxide–reinforced aluminum has been used in automotive connecting rods to provide stiffness and fatigue resistance with lighter weight. Aluminum reinforced with silicon carbide has been fabricated into automotive drive shafts, cylinder liners, and brake drums, as well as aircraft wing

TABLE 9-10 Properties of Some Fiber-Reinforced Metal-Matrix Composites.*

Material	Orientation	Specific Str (10^6 in.)	Specific Stiffness (10^6 in.)	Density (lb/in.^{3})
Aluminum with 50 vol % boron fiber	Along fiber	2.26	312	0.096
	Across fiber	0.21	229	
Aluminum with 50 vol % silicon carbide fiber	Along fiber	0.35	437	0.103
	Across fiber	0.15	—	
Titanium-6Al-4V with 35 vol % silicon carbide fiber	Along fiber	1.83	316	0.139
	Across fiber	0.42	—	
Aluminum-Lithium with 60 vol % Al_2O_3 fiber	Along fiber	0.80	304	0.125
	Across fiber	0.22	176	
Magnesium alloy with 38 vol % graphite fiber	Along fiber	1.14	—	0.065
Aluminum with 30 vol % graphite fiber	Along fiber	1.12	258	0.089

*Note the variation of properties with fiber orientation.

panels, all offering significant weight savings. Fiber-reinforced superalloys may well become a preferred material for applications such as turbine blades.

Table 9-10 presents some of the properties of fiber-reinforced metal-matrix composites. Note the significant difference between the properties measured along the fiber direction and those measured across the fibers.

3. **Carbon–carbon composites** (graphite fibers in a graphite or carbon matrix) offer the possibility of a heat-resistant material that could operate at temperatures above 2000°C (3600°F), along with a strength that is 20 times that of conventional graphite, a density that is 30% lighter (1.38 g/cm^3), and a low coefficient of thermal expansion. Not only does this material withstand high temperatures, it actually gets stronger when heated. Companion properties include good toughness, good thermal and electrical conductivity, and resistance to corrosion and abrasion. For temperatures over 540°C (1000°F), however, the composite requires some form of coating to protect it from oxidizing. Various coatings can be used for different temperature ranges. Current applications include the nose cone and leading edges of the space shuttle, aircraft and racing car disc brakes, automotive clutches, aerospace turbines and jet engine components, rocket nozzles, and surgical implants.
4. **Ceramic-matrix composites (CMCs)** offer light weight, high-temperature strength and stiffness, and good dimensional and environmental stability. The matrix provides high-temperature resistance. Glass matrices can operate at temperatures as high as 1500°C (2700°F). The crystalline ceramics—usually based on alumina, silicon carbide, silicon nitride, boron nitride, titanium diboride, or zirconia—can be used at even higher temperatures. The fibers add directional strength; increase fracture toughness; improve thermal shock resistance; and can be incorporated in unwoven, woven, knitted, and braided form. Typical reinforcements include carbon fiber, glass fiber, fibers of the various matrix materials, and ceramic whiskers. Composites with discontinuous fibers tend to be used primarily for wear applications, such as cutting tools, forming dies, and automotive parts such as valve guides. Other applications include lightweight armor plate and radomes. Continuous-fiber ceramic composites are used for applications involving the combination of high temperatures and high stresses and have been shown to fail in a noncatastrophic manner. Application examples include gas-turbine components, high-pressure heat exchangers, and high-temperature filters. Unfortunately, the cost of ceramic–ceramic composites ranges from high to extremely high, so applications are restricted to those where the benefits are quite attractive.

HYBRID COMPOSITES

Hybrid composites involve two or more different types of fibers in a common matrix. The particular combination of fibers is usually selected to balance strength and stiffness, provide dimensional stability, reduce cost, reduce weight, or improve fatigue and fracture resistance. Types of hybrid composites include (1) interply (alternating layers of fibers), (2) intraply (mixed strands in the same layer), (3) interply–intraply, (4) selected placement (where the more costly material is used only where needed), and (5) interply knitting (where plies of one fiber are stitched together with fibers of another type).

DESIGN AND FABRICATION

The design of composite materials involves the selection of the component materials; the determination of the relative amounts of each component; the determination of size shape, distribution, and orientation of the components; and the selection of an appropriate fabrication method. Many of the possible fabrication methods have been specifically developed for use with composite materials. For example, fibrous composites can be manufactured into useful shapes through compression molding, filament winding, pultrusion (where bundles of coated fibers are drawn through a heated die), cloth lamination, and autoclave curing (where pressure and elevated temperature are applied simultaneously). A variety of fiber-containing thermoset resins premixed with fillers and additives (bulk-molding compounds) can be shaped and cured by compression, transfer, and injection molding to produce three-dimensional fiber-reinforced products for numerous applications. Sheets of glass-fiber-reinforced thermoset resin, again with fillers and additives, (sheet-molding compound) can be press formed to provide lightweight, corrosion-resistant products that are similar to those made from sheet metal. The reinforcing fibers can be short and random, directionally oriented, or fully continuous in a specified direction. With a wide spectrum of materials, geometries, and processes, it is now possible to tailor a composite material product for a specific application. As one example, consider the cargo beds for pickup trucks, where composite products offer reduced weight coupled with resistance to dents, scratches, and corrosion.

A significant portion of Chapter 14 is devoted to a more complete description of the fabrication methods that have been developed for composite materials.

ASSETS AND LIMITATIONS

Figure 9-13 graphically presents the strength/weight ratios of various aerospace materials as a function of temperature. The superiority of the various advanced composites over the conventional aerospace metals is clearly evident. The weight of a graphite–epoxy composite I-beam is less than one-fifth that of steel, one-third that of titanium, and one-half that of aluminum. Its ultimate tensile strength equals or exceeds that of the other three materials, and it possesses an almost infinite fatigue life. The greatest limitations of this and other composites are their relative brittleness and the high cost of both materials and fabrication.

While there has been considerable advancement in the field, composite materials remain relatively expensive, and manufacturing with composites can still be quite labor intensive and involve long cycle times. There is a persistent lack of trained designers, established design guidelines and data, information about fabrication costs, and reliable methods of quality control and inspection. It is often difficult to predict the interfacial bond strength, the strength of the composite and its response to impacts, and the probable modes of failure. Defects can involve delaminations, voids, missing layers, contamination, fiber breakage, and (hard-to-detect) improperly cured resin. There is often concern about heat resistance. Many composites with polymeric matrices are sensitive to moisture, acids, chlorides, organic solvents, oils, and ultraviolet radiation, and tend to cure forever, causing continually changing properties. In addition, most composites have limited ability to be repaired if damaged, preventive maintenance procedures are not well established, and recycling is often extremely difficult. Assembly operations with composites generally require the use of industrial adhesives.

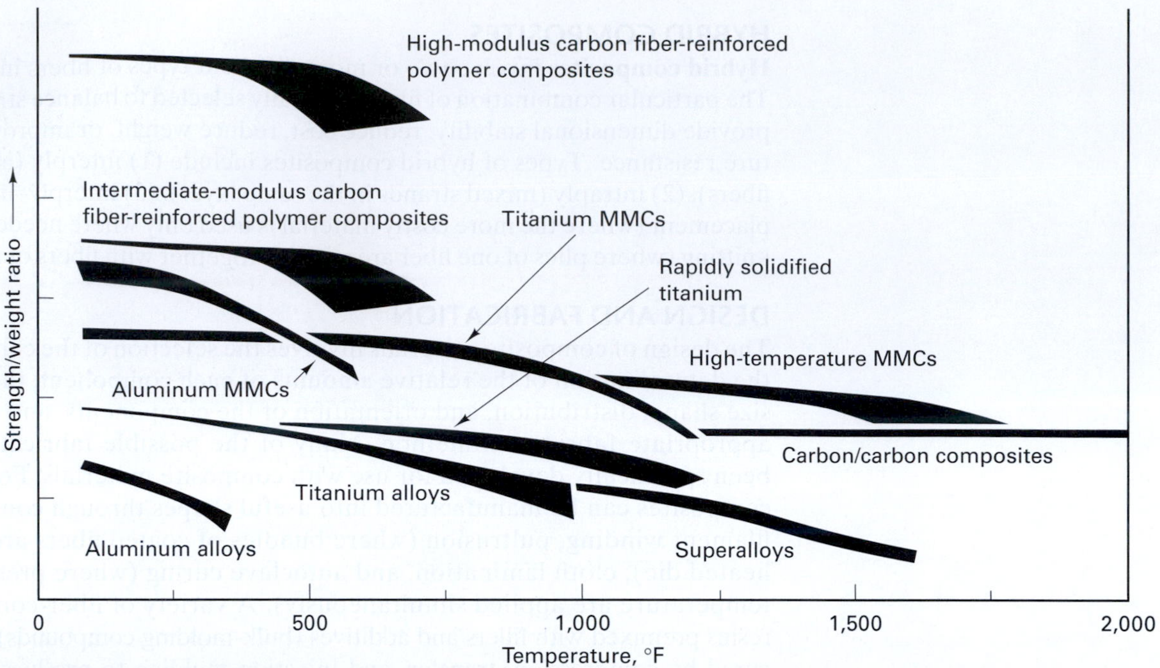

FIGURE 9-13 The strength/weight ratio of various aerospace materials as a function of temperature. Note the superiority of the various fiber-reinforced composites. *(Adapted with permission of DuPont Company, Wilmington, DE)*

On the positive side, the availability of a corrosion-resistant material with strength and stiffness greater than those of steel at only one-fifth the weight may be sufficient to justify some engineering compromises. Reinforcement fibers can be oriented in the direction of maximum stiffness and strength. In addition, products can often be designed to significantly reduce the number of parts (part consolidation), number of fasteners, assembly time, and both tooling cost and overall cost. Adam Aircraft manufactures two styles of high-performance aircraft using carbon-fiber composites. Their planes have 240 structural parts in comparison to more than 5000 for traditional aluminum aircraft.

AREAS OF APPLICATION

Many composite materials are stronger than steel, lighter than aluminum, and stiffer than titanium. They can also possess low thermal conductivity, good heat resistance, good fatigue life, low corrosion rates, and adequate wear resistance. For these reasons, they have become well established in several areas, often ones involving relatively low production rates or small production quantities.

Aerospace applications frequently require light weight, high strength, stiffness, and fatigue resistance. As a result, composites may well account for a considerable fraction of the weight of a current airplane design. Figure 9-14 shows a schematic of the F-22 Raptor fighter airplane. Traditional materials, such as aluminum and steel, make up only about 20 wt% of the F-22 structure. Its higher speed, longer range, greater agility, and reduced detectability are made possible through the use of 42% titanium and 24% composite material. Boeing's new 787 Dreamliner, a 200-seat intercontinental commercial airliner, has a majority of its primary structure, including wings and fuselage, made of polymer-matrix (carbon–epoxy) composites. A titanium–graphite composite will also be used in the wings of this aircraft, which will use 15 to 20% less fuel than current wide-body planes. The Airbus Industries' new A380 wide-body plane also utilizes a high proportion of composite materials (about 16 wt%), and will mark the introduction of a new composite material, known as glass-reinforced aluminum, a laminate composite of alternating layers of aluminum and glass prepreg. The new material, which enables a 25% reduction in the weight of fuselage skin, is more fatigue-resistant than aluminum and less expensive that a full composite.

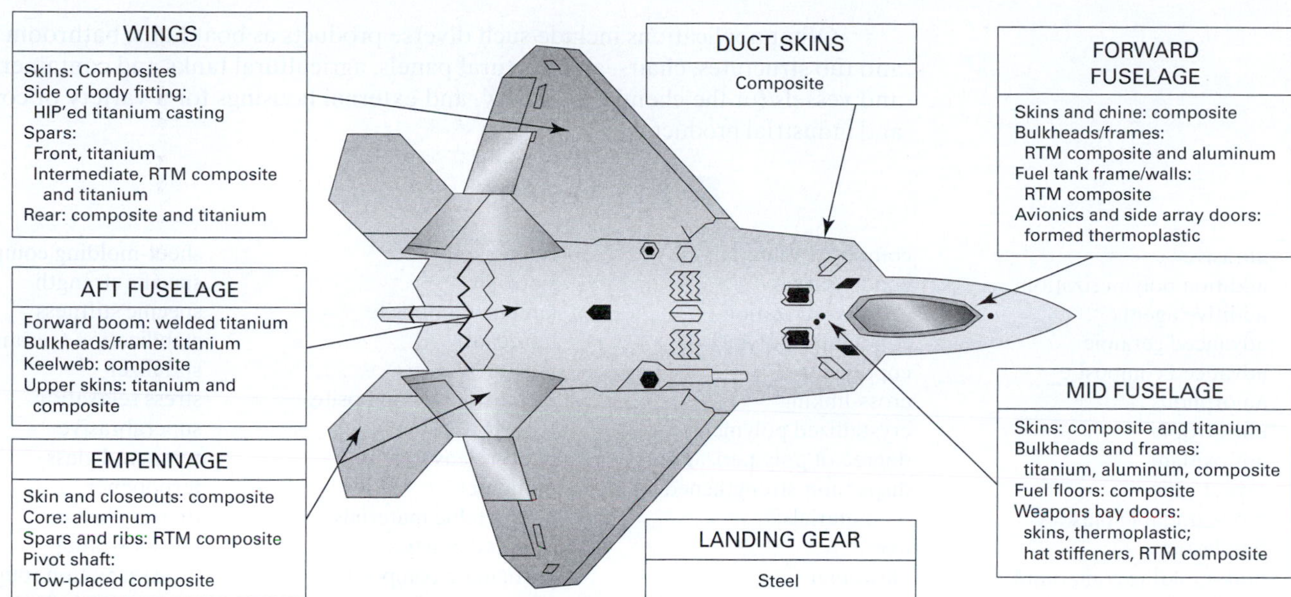

FIGURE 9-14 Schematic diagram showing the materials used in the various sections of the F-22 Raptor fighter airplane. Traditional materials, such as aluminum and steel, comprise only 20 wt%. Titanium accounts for 42%, and 24% is composite material. The plane is capable of flying at Mach 2. (*Note:* RTM is resin-transfer molding.) *(Reprinted with permission of ASM International, Materials Park, OH)*

Sports are highly competitive, and fractions of a second or tenths of a millimeter often decide victories. As a result, both professionals and amateurs are willing to invest in athletic equipment that will improve performance. The materials of choice have evolved from naturally occurring wood, twine, gut, and rubber to a wide variety of high-technology metals, polymers, ceramics, and composites. Golf club shafts, baseball bats, fishing rods, archery bows, tennis rackets, bicycle frames, skis, and snowboards are now available in a wide variety of fibrous composites. Figure 9-15 shows an example of these applications.

In addition to body panels, automotive uses of composite materials include drive shafts, springs, and bumpers. Weight savings compared to existing parts is generally 20 to 25%. Truck manufacturers now use fiber-reinforced composites for cab shells and bodies, oil pans, fan shrouds, instrument panels, and engine covers. Some pickup trucks now have cargo boxes formed from sheet-molding compound.

FIGURE 9-15 Composite materials are often used in sporting goods, like this snowboard, to improve performance through light weight, high stiffness, and high strength, and also to provide attractive styling. *(© Taissiya Shaidarova/iStockphoto)*

Other applications include such diverse products as boat hulls, bathroom shower and tub structures, chairs, architectural panels, agricultural tanks and containers, pipes and vessels for the chemical industry, and external housings for a variety of consumer and industrial products.

KEY WORDS

abrasive
addition polymerization
additive agents
advanced ceramic
advanced composite
amorphous
anisotropic
antioxidant
aspect ratio
biodegradable plastics
bioplastics
bulk-molding compound (BMC)
carbon–carbon composite
ceramic
ceramic–ceramic composite
ceramic-matrix composite
cermet
clay
coloring agent
composite material
condensation polymerization
conductive polymer
copolymer
cross-linking
crystallized polymer
degree of polymerization
dispersion-strengthened material
dye
elastomer
fiber-reinforced composite
filler
flame retardant
foamed plastic
foaming agent
glass
glass ceramic
hard machining
hybrid composite
isomer
isotropic
laminar composite
lubricant
mer
metal-matrix composite (MMC)
mold-release agent
monomer
nonmetallic materials
oriented plastics
particulate composite
pigment
plastic
plasticizer
polymer
polymerization
polymer-matrix composite
refractory material
rubber
saturated monomer
sheet-molding compound
specific strength
specific stiffness
stabilization (ceramic)
stabilizer
stress relaxation
superabrasive
tempered glass
terpolymer
thermoplastic
thermosetting
transformation toughening
true particulate composites
ultra-high-temperature ceramics
unsaturated monomer
viscoelastic memory
vulcanization
whiteware

REVIEW QUESTIONS

1. What are some naturally occurring nonmetallic materials that have been used for engineering applications?
2. What are some material families that would be classified under the general term *nonmetallic engineering materials?*
3. How might plastics be defined from the viewpoints of chemistry, structure, fabrication, and processing?
4. What is the primary type of atomic bonding within polymers?
5. What is the difference between a saturated and an unsaturated molecule?
6. What is an isomer?
7. Describe and differentiate the two means of forming polymers: addition polymerization and condensation polymerization.
8. What is degree of polymerization?
9. Describe and differentiate thermoplastic and thermosetting plastics.
10. Describe the mechanism by which thermoplastic polymers soften under heat and deform under pressure.
11. What does it mean when a polymer "crystallizes"? How is this different from the crystal structures observed in metals and ceramics?
12. What are some of the ways that a thermoplastic polymer can be made stronger?
13. What are the four most common thermoplastic polymers?
14. Why are thermosetting polymers characteristically brittle?
15. How do thermosetting polymers respond to subsequent heating?
16. Describe how thermoplastic or thermosetting characteristics affect productivity during the fabrication of a molded part.
17. What are some attractive engineering properties of polymeric materials?
18. What are some limiting properties of plastics, and in what general area do they fall?
19. What are some environmental conditions that might adversely affect the engineering properties plastics?
20. What are some reasons that additive agents are incorporated into plastics?
21. What are some functions of a filler material in a polymer?
22. What are some of the more common filler materials used in plastics?
23. What is the function of a plasticizer?
24. What is the difference between a dye and a pigment?
25. What is the role of a stabilizer or antioxidant?
26. What is an oriented plastic, and what is the primary engineering benefit?
27. What are some properties and characteristics of the "engineering plastics"?
28. Describe the use of plastic materials as adhesives. In tooling applications.
29. Describe some of the applications for foamed plastics.
30. Provide some examples where plastics have competed with or replaced metals.
31. What are some features of plastics that make them attractive from a manufacturing perspective?
32. In a cost comparison, why might cost per unit volume be a more valid figure than cost per unit weight?
33. How has the use of plastics grown in the automotive industry?

34. Which type of plastic is most easily recycled?
35. Why is the recycling of mixed plastics more difficult than the recycling of mixed metals?
36. What are some recycling alternatives for thermosetting polymers?
37. What are some of the approaches to producing a biodegradable plastic?
38. What is the unique mechanical property of elastomeric materials, and what structural feature is responsible for it?
39. How can cross-linking be used to control the engineering properties of elastomers?
40. What are some of the materials that can be added to natural rubber, and for what purpose?
41. What are some of the limitations of natural rubber?
42. What are some of the attractive features of the silicone rubbers?
43. What factors should be considered when selecting an elastomer for a specific application?
44. What are some outstanding physical properties of ceramic materials?
45. Which class of ceramic material accounts for nearly half of the ceramic materials market?
46. Why are the crystal structures of ceramics frequently more complex than those observed for metals?
47. What is the common name given for ceramic material in the noncrystalline, or amorphous, state?
48. What are some of the ways that toughness can be imparted to ceramic materials?
49. What is the dominant property of refractory ceramics?
50. What are some common applications of refractory ceramics?
51. What is the dominant property of ceramic abrasives?
52. How are glass products formed or shaped?
53. What are some of the specialty applications of glass?
54. How is a glass ceramic different from a glass?
55. What are cermets, and what properties or combination of properties do they offer?
56. Why do most ceramic materials fail to possess their theoretically high tensile strength?
57. Why do the mechanical properties of ceramics generally show a wider statistical spread than the same properties of metals?
58. If all significant flaws or defects could be eliminated from the structural ceramics, what properties might be present and what features might still limit their possible applications?
59. What are some specific materials that are classified as structural ceramics?
60. What are some attractive and limiting properties of sialon (one of the structural ceramics)?
61. What are some potential applications of the advanced ceramic materials? What is limiting their use?
62. What are some ceramic materials that are currently being used for cutting-tool applications, and what features or properties make them attractive?
63. What properties are desired in ultra-high-temperature ceramics, and what are some possible applications?
64. What is a composite material?
65. What are the basic features of a composite material that influence and determine its properties?
66. What are the three primary geometries of composite materials?
67. Give several examples of laminar composites.
68. What feature in a bimetallic strip makes its shape sensitive to temperature?
69. What are some reasons for creating clad composites?
70. What is the attractive aspect of the strength that is induced by the particles in a dispersion-strengthened particulate composite material?
71. Provide some examples of true particulate composites.
72. Which of the three primary composite geometries is most likely to possess isotropic properties?
73. What are the primary roles of the matrix in a fiber-reinforced composite? Of the fibers?
74. What are some of the more popular fiber materials used in fiber-reinforced composite materials?
75. What is specific strength? Specific stiffness?
76. What are some possible fiber orientations or arrangements in a fiber-reinforced composite material?
77. What are some features that influence the properties of fiber-reinforced composites?
78. What is bulk-molding compound? Sheet-molding compound?
79. What are the attractive features of a thermosetting polymer matrix in a polymer-matrix fiber-reinforced composite? A thermoplastic matrix?
80. What are "advanced composites"?
81. In what ways are metal-matrix composites superior to straight engineering metals? To polymer-matrix composites?
82. Consider the data in Table 9-10. How significant are the property variations between directions along the fiber and across the fiber?
83. What are the primary application conditions for the carbon–carbon composites?
84. What features might be imparted by the fibers in a ceramic-matrix composite?
85. What are hybrid composites?
86. What are some of the significant attractive properties of the fiber-reinforced composites?
87. What are some of the limitations that may restrict the use of composite materials in engineering applications?

Problems

1. a. One of Leonardo da Vinci's sketchbooks contains a crude sketch of an underwater boat (or submarine). Da Vinci did not attempt to develop or refine this sketch further, possibly because he recognized that the engineering materials of his day (wood, stone, and leather) were inadequate for the task. What properties would be required for the body of a submersible vehicle? What materials might you consider?
   b. Another of da Vinci's sketches bears a crude resemblance to a helicopter—a flying machine. What properties would be desirable in a material that would be used for this type of application?
   c. Try to identify a possible engineering product that would require a material with properties that do not exist among today's engineering materials. For your application, what

are the demanding features or requirements? If a material were to be developed for this application, from what family or group do you think it would emerge? Why?

2. Select a product (or component of a product) that can reasonably be made from materials from two or more of the basic materials families (metals, polymers, ceramics, and composites).
   a. Briefly describe the function of the product or component.
   b. What properties would be required for this product or component to perform its function?
   c. What two materials groups might provide reasonable candidates for your product?
   d. Select a candidate material from the first of your two families and describe its characteristics. In what ways does it meet your requirements? How might it fall short of the needs?
   e. Repeat Part d for a candidate material from the second material family.
   f. Compare the two materials to one another. Which of the two would you prefer? Why?
3. Coatings have been applied to cutting tool materials since the early 1970s. Desirable properties of these coatings include high-temperature stability, chemical stability, low coefficient of friction, high hardness for edge retention, and good resistance to abrasive wear. Consider the ceramic material coatings of titanium carbide (TiC), titanium nitride (TiN), and aluminum oxide (Al_2O_3), and compare them with respect to the conditions required for deposition and the performance of the resulting coatings.
4. Ceramic engines continue to constitute an area of considerable interest and are frequently discussed in the popular literature. If perfected, they would allow higher operating temperatures with a companion increase in engine efficiency. In addition, they would lower sliding friction and permit the elimination of radiators, fan belts, cooling system pumps, coolant lines, and coolant. The net result would be reduced weight and a more compact design. Estimated fuel savings could amount to 30% or more.
   a. What are the primary limitations to the successful manufacture of such a product?
   b. What types of ceramic materials would you consider to be appropriate?
   c. What methods of fabrication could produce a product of the required size and shape?
   d. What types of special material properties or special processing might be required?
5. Material recyclability has become an important requirement in many manufactured products.
   a. Consider each of the four major materials groups (metals, polymers, ceramics, and composites) and evaluate each for recyclability. What properties or characteristics tend to limit or restrict recyclability?
   b. Which materials within each group are currently being recycled in large or reasonable quantities?
   c. Europe has recently legislated extensive recycling of automobiles and electronic products. How might this legislation change the material make-up of these products?
   d. Consider a typical family automobile and discuss how the factors of (1) recyclability, (2) fuel economy, and (3) energy required to produce materials and convert them to products might favor or oppose various candidate engineering materials.

Chapter 9 CASE STUDY

Lightweight Armor

There are many approaches to producing lightweight armor, and many different materials or combinations of materials have been employed. As an example, consider a ceramic-material-based personnel armor $\frac{1}{3}$ in. thick that can stop an armor-piercing projectile that is capable of penetrating 2 in. of steel. A hard ceramic plate is bonded to a fiberglass plate with a glue similar to rubber cement. When a bullet strikes the ceramic, it is hard enough to shatter the bullet into small pieces, while the ceramic itself also shatters. Both of these actions absorb some of the energy of the projectile. The fiberglass then catches all of the broken pieces (both the bullet and ceramic). While the individual will sustain a significant impact, the result is not likely to be fatal.

One of the ceramic materials developed for armor usage was boron carbide. It is extremely hard and one-third the weight of steel. Other armor materials use woven Kevlar, alumina (aluminum oxide ceramic), and various metals—often combined in some form of composite material.

Personnel armor involves both body armor and helmets. Comfort, freedom of movement, and weight are all important design considerations. Lightweight armor on a military vehicle would enable faster traveling speed, greater range on a tank of fuel, or the ability to carry more munitions, supplies, or personnel while providing crew safety and survivability. Weight is even more important on aircraft armor, where the various applications include pilot and crew seats as well as panels, tiles, and flooring units for both helicopters and fixed-wing aircraft. The armor on the various sizes and types of naval vessels must also endure the corrosive conditions imposed by salt water. The various applications just described will likely see a range of projectile sizes, types and incoming velocities.

1. Use the Internet and other sources to research lightweight armor, identifying several types. For each, note the materials being employed and the design features being employed to produce the desired result.
2. Discuss briefly how the materials and design features might differ for two or more of the applications described above.
3. How have the materials used in military helmets changed over the past 50 or more years?
4. How are the requirements for a football helmet or a motorcycle helmet similar or different from a military helmet? What materials are being used in these applications?

CHAPTER 10

MATERIAL SELECTION

10.1 INTRODUCTION

The objective of manufacturing operations is to make products or components that adequately perform their intended task. Meeting this objective implies the manufacture of components from selected engineering materials, with the required geometrical shape and precision, and with companion material structures and properties that are optimized for the intended service environment. The ideal product is one that will just meet all requirements. Anything better will usually incur added cost through excess or higher-grade materials, enhanced processing, or improved properties that may not be necessary. Anything worse will likely cause product failure, dissatisfied customers, and the possibility of unemployment.

It was not that long ago that each of the materials groups had its own well-defined uses and markets. Metals were specified when strength, toughness, and durability were the primary requirements. Ceramics were generally limited to low-value applications where heat or chemical resistance was required and any loadings were compressive. Glass was used for its optical transparency, and plastics were relegated to low-value applications where low cost and light weight were attractive features and performance properties were secondary.

Such clear delineations no longer exist. Many of the metal alloys in use today did not exist as little as 30 years ago, and the common alloys that have been in use for a century or more have been much improved due to advances in metallurgy and production processes. New on the scene are amorphous metals, dispersion-strengthened alloys produced by powder metallurgy, mechanical alloyed products, and directionally solidified materials. Ceramics, polymers, and composites are now available with specific properties that often transcend the traditional limits and boundaries. Advanced structural materials offer higher strength and stiffness; strength at elevated temperature; light weight; and resistance to corrosion, creep, and fatigue. Still other materials offer enhanced thermal, electrical, optical, magnetic, and chemical properties.

To the inexperienced individual, "wood is wood," but to the carpenter or craftsman, oak is best for one application, while maple excels for another, and yellow pine is preferred for a third. The ninth edition of *Woldman's Engineering Alloys*[1] includes more than 56,000 metal alloys, and that does not consider polymers, ceramics, or composites. Even if we eliminate the obsolete and obscure, we are still left with tens of thousands of options from which to select the "right" or "best" material for the task at hand.

Unfortunately, the availability of so many alternatives has often led to poor materials selection. Money can be wasted in the unnecessary specification of an expensive

[1] J. Frick, Ed., *Woldman's Engineering Alloys*, 9th ed., (Metals Park, OH: ASM International, 2000).

alloy or one that is difficult to fabricate. At other times, these materials may be absolutely necessary, and selection of a cheaper alloy would mean certain failure. It is the responsibility of the design and manufacturing engineer, therefore, to be knowledgeable in the area of engineering materials and to be able to make the best selection among the numerous alternatives.

In addition, it is also important that the **material selection** process be one of constant reevaluation. New materials are continually being developed, others may no longer be available, and prices are always subject to change. Concerns regarding environmental pollution, recycling, and worker health and safety may impose new constraints. Desires for weight reduction, energy savings, or improved corrosion resistance may well motivate a change in engineering material. Pressures from domestic and foreign competition, increased demand for quality and serviceability, or negative customer feedback can all prompt a reevaluation. Finally, the proliferation of manufacturer recalls and product liability actions—many of which are the result of improper material use—has further emphasized the need for constant reevaluation of the engineering materials in a product.

The automotive industry alone consumes approximately 60 million metric tons of engineering materials worldwide every year—including substantial amounts of steel, aluminum, cast iron, copper, glass, lead, polymers, rubber, and zinc. In recent years, the shift toward lighter, more fuel-efficient, or alternative-powered vehicles has led to an increase in the use of the lightweight metals, high-strength steels, plastics, and composites.

A million metric tons of engineering materials go into aerospace applications every year. The principal materials tend to be aluminum, magnesium, titanium, superalloys, polymers, rubber, steel, metal-matrix composites, and polymer-matrix composites. Competition is intense, and materials substitutions are frequent. The use of advanced composite materials in aircraft construction has risen from less than 2% in 1970 to the point where they now account for one-quarter of the weight of the U.S. Air Force's Advanced Tactical Fighter and are already being used in the main fuselage of commercial planes. Titanium is used extensively for applications that include the exterior skins surrounding the engines, as well as the engine frames. The cutaway section of the Rolls Royce Trent 900 jet engine in Figure 10-1a reveals the myriad components—each with its own characteristic shape, precision, stresses, and operating temperatures—that require a variety of engineering materials. Figure 10-1b shows an actual engine in a manner that reveals both its size and complexity. The intake fan diameter is nearly 3 m in diameter (9 ft. 8 in.).

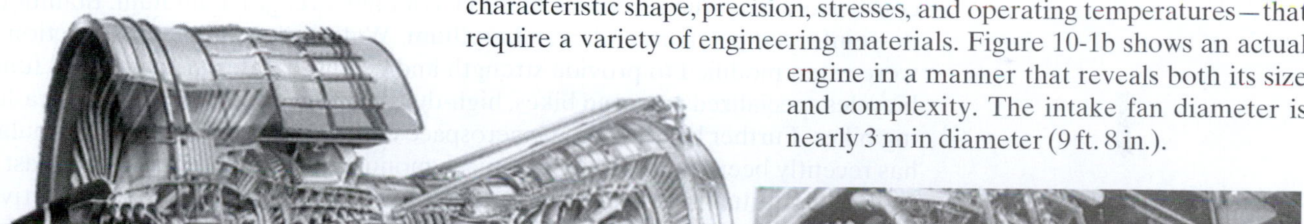

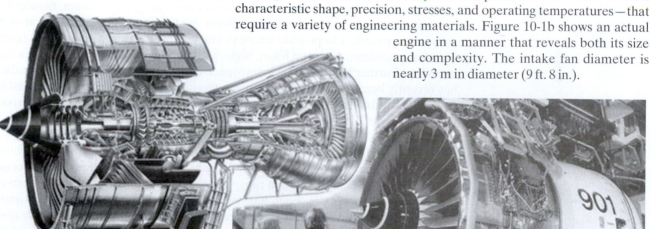

FIGURE 10-1 (a) Cut-away drawing showing the internal design of a Rolls-Royce jet engine. Notice the number and intricacy of the components. Ambient air enters the front and hot exhaust exits the back. A wide range of materials and processes will be used in its construction. (b) A full-size Rolls-Royce engine, showing the size and complexity of the product. *(Courtesy of Rolls-Royce Corporation, London, England)*

(a)

(b)

FIGURE 10-2 (a) A traditional two-wheel bicycle frame (1970s vintage) made from joined segments of metal tubing. (b) A top-of-the-line (Tour de France or triathalon-type) bicycle with one-piece frame, made from fiber-reinforced polymer-matrix composite. *[(a) Courtesy Ronald Kohser; (b) Trek Bicycle]*

The earliest two-wheeled bicycle frames were constructed of wood, with various methods and materials employed at the joints. Then, for nearly a century, the requirements of yield strength, stiffness, and acceptable weight were met by steel tubing, either low-carbon plain-carbon or thinner-walled, higher-strength chrome-moly steel tubing (such as 4130), with either a brazed or welded assembly. In the 1970s a full circle occurred. Where a pair of bicycle builders (the Wright brothers) pioneered aerospace, the aerospace industry returned to revolutionize bicycles. Lightweight frames were constructed from the aerospace materials of high-strength aluminum, titanium, graphite-reinforced polymer, and even beryllium. Wall thickness and cross-section profiles were often modified to provide strength and rigidity. Materials paralleled function as bicycles specialized into road bikes, high-durability mountain bikes, and ultra-light racing bikes. Further building on the aerospace experience, the century-old tubular frame has recently been surpassed by one-piece monocoque frames of either die-cast magnesium or continually wound carbon-fiber epoxy tapes with or without selective metal reinforcements. One top-of-the-line carbon fiber frame now weighs only 2.5 lb! Figure 10-2 compares a traditional tubular frame with one of the newer designs.

Window frames were once made almost exclusively from wood. While wood remains a competitive material, a trip to any building supply will reveal a selection that includes anodized aluminum in a range of colors, as well as frames made from colored vinyl and other polymers. Each has its companion advantages and limitations. Auto bodies have been traditionally fabricated from steel sheet and assembled by resistance spot welding. Designers now select from steel, aluminum, and polymeric sheet molding compounds and may use adhesive bonding to produce the joints.

The vacuum cleaner assembly shown in Figure 10-3, while not a current model, is typical of many engineering products, where a variety of materials are used for the various components. Table 10-1 lists the material changes that were recommended in just one past revision of the appliance. The materials for 12 components were changed completely, and that for a 13th was modified. Eleven different reasons were given for the changes. An increased emphasis on lighter weight has brought about even further changes in both design and materials.

A listing of available engineering materials now includes metals and alloys, ceramics, plastics, elastomers, glasses, concrete, composite materials, and others. It is not surprising, therefore, that a single person might have difficulty making the necessary decisions concerning the materials in even a simple manufactured product.

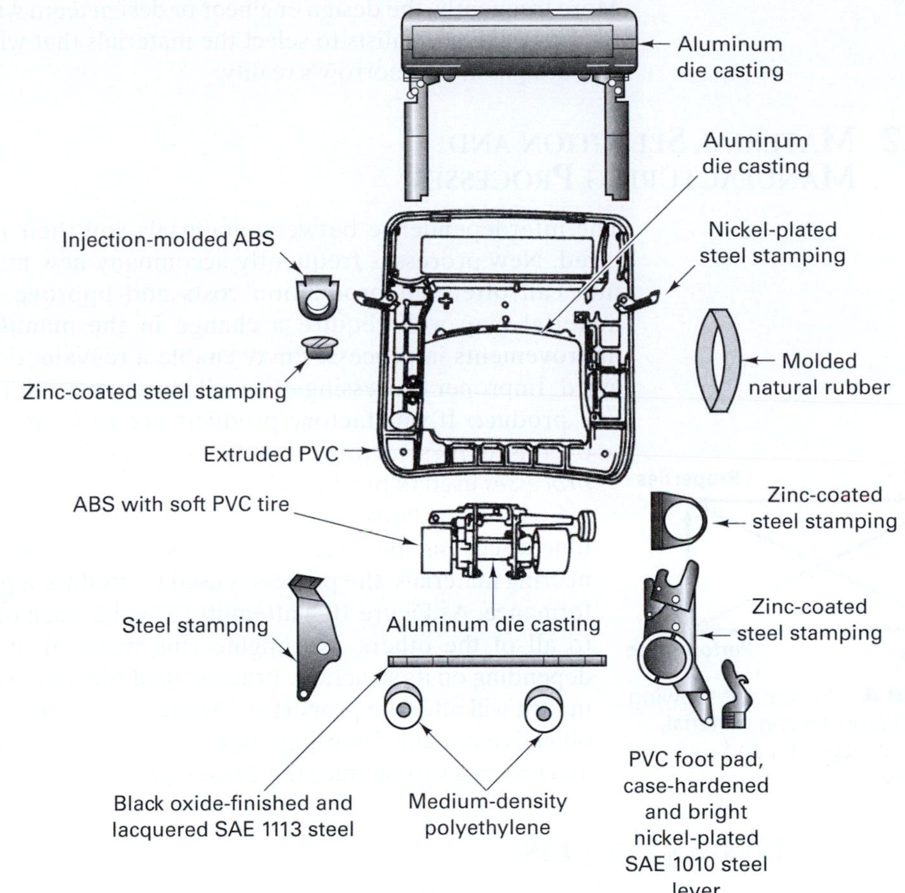

FIGURE 10-3 Materials used in various parts of a vacuum cleaner assembly. *(Courtesy of Advanced Materials and Processes, ASM International)*

TABLE 10-1 Examples of Material Selection and Substitution in the Redesign of a Vacuum Cleaner

Part	Former Material	New Material	Benefits
Bottom plate	Assembly of steel stampings	One-piece aluminum die casting	More convenient servicing
Wheels (carrier and caster)	Molded phenolic	Molded medium-density polyethylene	Reduced noise
Wheel mounting	Screw-machine parts	Preassembled with a cold-headed steel shaft	Simplified replacement, more economical
Agitator brush	Horsehair bristles in a die-cast zinc or aluminum brush back	Nylon bristles stapled to a polyethylene brush back	Nylon bristles last seven times longer and are now cheaper than horsehair
Switch toggle	Bakelite molding	Molded ABS	Breakage eliminated
Handle tube	AISI 1010 lock-seam tubing	Electric seam-welded tubing	Less expensive, better dimensional control
Handle bail	Steel stamping	Die-cast aluminum	Better appearance, allowed lower profile for cleaning under furniture
Motor hood	Molded cellulose acetate (replaced Bakelite)	Molded ABS	Reasonable cost, equal impact strength, much improved heat and moisture resistance: eliminated warpage problems
Extension-tube spring latch	Nickel-plated spring steel, extruded PVC cover	Molded acetal resin	More economical
Crevice tool	Wrapped fiber paper	Molded polyethylene	More flexibility
Rug nozzle	Molded ABS	High-impact styrene	Reduced costs
Hose	PVC-coated wire with a single-ply PVC extruded covering	PVC-coated wire with a two-ply PVC extruded covering separated by a nylon reinforcement	More durability, lower cost
Bellows, cleaning-tool nozzles, cord insulation, bumper strips	Rubber	PVC	More economical, better aging and color, less marking

Source: Metal Progress, by permission.

More frequently, the design engineer or design team will work in conjunction with various materials specialists to select the materials that will be needed to convert conceptual designs into tomorrow's reality.

10.2 MATERIAL SELECTION AND MANUFACTURING PROCESSES

The interdependence between materials and their processing must also be recognized. New processes frequently accompany new materials, and their implementation can often cut production costs and improve product quality. A change in material may well require a change in the manufacturing process. Conversely, improvements in processes may enable a reevaluation of the materials being processed. Improper processing of a well-chosen material can definitely result in a defective product. If satisfactory products are to be made, considerable care must be exercised in selecting *both* the *engineering materials* and the *manufacturing processes* used to produce the product.

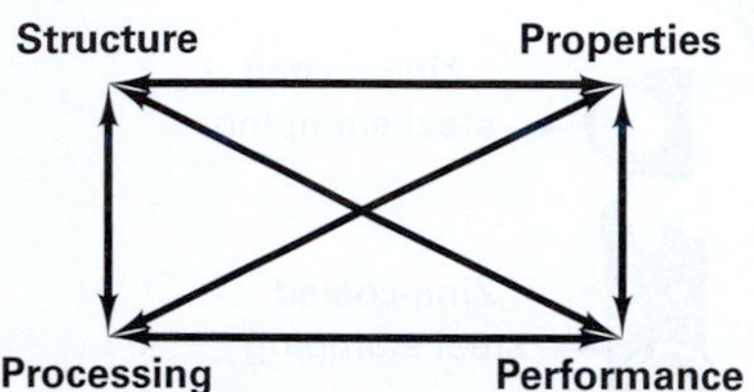

FIGURE 10-4 Schematic showing the interrelation among material, properties, processing, and performance.

Most textbooks on materials and manufacturing processes spend considerable time discussing the interrelationships between the structure and properties of engineering materials, the processes used to produce a product, and the subsequent performance. As Figure 10-4 attempts to depict, each of these aspects is directly related to all of the others. An engineering material may possess different properties depending on its structure. Processing of that material can alter the structure, which in turn will alter the properties. Altered properties certainly alter performance. The objective of manufacturing, therefore, is to devise an optimized system of material and processes to produce the desired product.

10.3 THE DESIGN PROCESS

The first step in the manufacturing process is **design**—starting with a need and determining in rather precise detail what it is that we want to produce, and for each component of the product or assembly, what properties it must possess, what to make it out of, how to make it, how many to make, and what conditions it will see during use.

Design usually takes place in several distinct stages: (1) conceptual, (2) functional, and (3) production. During the **conceptual design stage,** the designer is concerned primarily with the functions that the product is to fulfill. Several concepts are often considered, and a determination is made that the concept is either not practical, or is sound and should be developed further. Here the only concern about materials is that materials exist that could provide the desired properties. If such materials are not available, consideration is given to whether there is a reasonable prospect that new ones could be developed within the limitations of cost and time.

At the **functional,** or *engineering,* **design stage,** a workable design is developed, including a detailed plan for manufacturing. Geometric features are determined and dimensions are specified, along with allowable tolerances. Specific materials are selected for each component. Consideration is given to appearance, cost, reliability, producibility, and serviceability, in addition to the various functional factors. It is important to have a complete understanding of the functions and performance requirements of each component and to perform a thorough materials analysis, selection, and specification. If these decisions are deferred, they may end up being made by individuals who are less knowledgeable about all of the functional aspects of the product.

Often, a **prototype** or working model is constructed to permit a full evaluation of the product. It is possible that the prototype evaluation will show that some changes have to be made in either the design or material before the product can be advanced to production. This should not be taken, however, as an excuse for not doing a thorough job. It is strongly recommended that all prototypes be built with the same materials that will be used in production and, where possible, with the same manufacturing techniques. It is of little value to have a perfectly functioning prototype that cannot be manufactured economically in the desired volume, or one that is substantially different from what the production units will be like.[2]

In the **production design stage,** we look to full production and determine if the proposed solution is compatible with production speeds and quantities. Can the parts be processed economically, and will they be of the desired quality?

As actual manufacturing begins, changes in both the materials and processes may be suggested. In most cases, however, changes made after the tooling and machinery have been placed in production tend to be quite costly. Good up-front material selection and thorough product evaluation can do much to eliminate the need for change.

As production continues, the availability of new materials and new processes may well present possibilities for cost reduction or improved performance. Before adopting new materials, however, the candidates should be evaluated very carefully to ensure that all of their characteristics related to both processing and performance are well established. Remember that it is indeed rare that as much is known about the properties and reliability of a new material as an established one. Numerous product failures and product liability cases have resulted from new materials being substituted before their long-term properties were fully known.

10.4 APPROACHES TO MATERIAL SELECTION

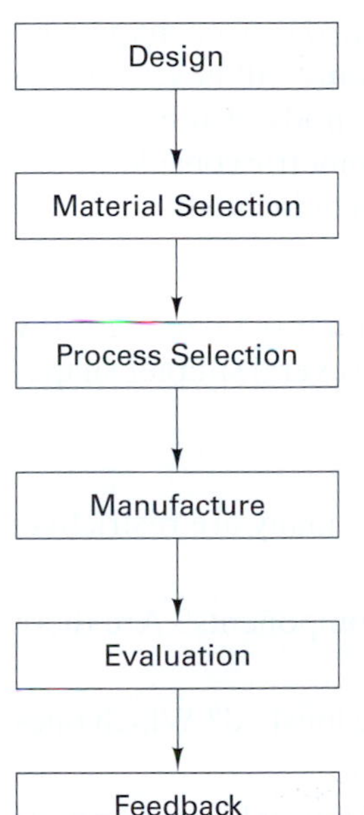

FIGURE 10-5 Sequential flow chart showing activities leading to the production of a part or product.

The selection of an appropriate material and its subsequent conversion into a useful product with desired shape and properties can be a rather complex process. As depicted in Figure 10-5, nearly every engineered item goes through a sequence of activities:

$$\text{Design} \rightarrow \text{Material selection} \rightarrow \text{Process selection} \rightarrow \text{Manufacture or fabrication} \rightarrow \text{Evaluation} \rightarrow \text{Feedback}$$

Feedback can involve possible redesign or modification at one or more of the preceding steps. Numerous engineering decisions must be made along the way, and several iterations may be necessary.

Several methods have been developed for approaching a design and selection problem. The **case-history method** is one of the simplest. Begin by evaluating what has been done in the past in terms of engineering material and method of manufacture, or what a competitor is currently doing. This can yield important information that will serve as a starting base. Then, either duplicate or modify the details of that solution. The basic assumption of this approach is that similar requirements can be met with similar solutions. The information acquired during evaluation may require redesign or modification at one or more of the preceding steps.

The case-history approach is quite useful, and many manufacturers continually examine and evaluate their competitors' products for just this purpose. The real issue here, however, is "how similar is similar." A minor variation in service requirement, such as a different operating temperature or a different corrosive environment, may be sufficient to justify a totally different material. Different production quantities may justify a different manufacturing method. In addition, this approach tends to preclude the use of new materials, new technology, and any manufacturing advances that may have occurred since the formulation of the original solution. It is equally unwise, however, to totally ignore the benefits and insights that can be gained through past experience.

Other design and selection activities may be related to the *modification of an existing product*, generally in an effort to reduce cost, improve quality, or overcome a problem or defect that has been encountered. A customer may have requested a product like

[2] Because of the prohibitive cost of a dedicated die or pattern, as might be required for forging or casting, one-of-a-kind or limited quantity prototype parts are often made by machining or one of the newer rapid-prototype techniques. If the objective is simply to verify dimensional fit and interaction, the prototype material may be selected for compatibility with the prototype process. If performance is to be verified, however, it is best to use the proper material and process. Machining, for example, simply cuts through the material structure imparted in the manufacture of the starting bar or plate. Casting erases all prior structure during melting and establishes a new structure during solidification. Metal forming processes reorient the starting structure by plastic flow. The altered features caused by these processes may lead to altered performance.

the current one but that is capable of operating at higher temperatures, or in an acidic environment, or at higher loads or pressures. Efforts here generally begin with an evaluation of the current product and its present method of manufacture. The most frequent pitfall, however, is to overlook one of the original design requirements and recommend a change that in some way compromises the total performance of the product. Examples of such oversights are provided in Section 10.8, where material substitutions have been recommended to meet a specific objectives.

The safest and most comprehensive approach to part manufacture is to follow the full sequence of design, material selection, and process selection—considering all aspects and all alternatives. This is the approach one would take in the *development of an entirely new product.*

Before any decisions are made, take the time to fully define the needs of the product. What exactly is the "target" that we wish to hit? Develop a clear picture of all of the characteristics necessary for this part to adequately perform its intended function, and do so with no prior biases about material, or method of fabrication. These requirements will fall into three major areas: (1) shape or geometry considerations, (2) property requirements, and (3) manufacturing concerns. By first formulating these requirements, we will be in a better position to evaluate candidate materials and companion methods of fabrication.

GEOMETRIC CONSIDERATIONS

A dimensioned sketch can answer many of the questions about the size, shape, and complexity of a part, and these **geometric,** or *shape*, **considerations** will have a strong influence on decisions relating to the proposed method or methods of fabrication.[3] While many features of part geometry are somewhat obvious, geometric considerations are often more complex than first imagined. Typical questions might include:

1. What is the relative size of the component?
2. How complex is the shape? Are there any axes or planes of symmetry? Are there any uniform cross sections? Could the component be divided into several simpler shapes that might be easier to manufacture?
3. How many dimensions must be specified?
4. How precise must these dimensions be? Are all precise? How many are restrictive, and which ones?
5. How does this component interact geometrically with other components? Are there any restrictions imposed by the interaction?
6. What are the surface finish requirements? Must all surfaces be finished? Which ones do not?
7. How much can each dimension change by wear, corrosion, thermal expansion, or thermal contraction and still have the part function adequately?
8. Could a minor change in part geometry increase the ease of manufacture or improve the performance (fracture resistance, fatigue resistance, etc.) of the part?

Producing the right shape is only part of the desired objective. If the part is to perform adequately, it must also possess the necessary **mechanical** and **physical properties,** as well as the ability to endure anticipated environments for a specified period of time. **Environmental considerations** should include all aspects of shipping, storage, and use. Some key questions include:

[3] Die casting, for example, can be used to produce parts ranging from less than an ounce to more than 100 lb, but the ideal wall thickness should be less than $\frac{5}{16}$ in. Permanent mold casting can produce thicknesses up to 2 in., and there is no limit to the thickness for sand casting. At the same time, dimensional precision and surface finish become progressively worse as we move from die casting, to permanent mold, to sand. Extrusion and rolling can be used to produce long parts with constant cross section. Powder metallurgy parts must be able to be ejected from a compacting die.

MECHANICAL PROPERTIES

1. How will the loads be applied? Tension? Compression? Torsion? Shear?
2. How much static strength is required?
3. If the part is accidentally overloaded, is it permissible to have a sudden brittle fracture, or is plastic deformation and distortion a desirable precursor to failure?
4. How much can the material bend, stretch, twist or compress under load and still function properly?
5. Are any impact loadings anticipated? If so, of what type, magnitude, and velocity?
6. Can you envision vibrations or cyclic loadings? If so, of what type, magnitude, and frequency?
7. Is wear resistance desired? Where? How much? How deep?
8. Will all of the preceding requirements be needed over the entire range of operating temperature? If not, which properties are needed at the lowest extreme? At the highest extreme?

PHYSICAL PROPERTIES (ELECTRICAL, MAGNETIC, THERMAL, AND OPTICAL)

1. Are there any electrical requirements? Conductor? Insulator?
2. Are any magnetic properties desired?
3. Are thermal properties significant? Thermal conductivity? Changes in dimension with change in temperature?
4. Are there any optical requirements?
5. Is weight a significant factor?
6. How important is appearance? Is there a preferred color, texture, or feel?

ENVIRONMENTAL CONSIDERATIONS

1. What are the lowest, highest, and normal temperatures the product will see? Will temperature changes be cyclic? How fast will temperature changes occur?
2. What is the most severe environment that is anticipated as far as corrosion or deterioration of material properties is concerned?
3. What is the *required* service lifetime for the product? The *desired* service life?
4. What is the anticipated level of inspection and maintenance during use?
5. Should the product be manufactured with disassembly, repairability, or **recyclability** in mind?
6. Are there any concerns relating to toxicity, end-of-use recycling or disposal?

MANUFACTURING CONCERNS

A final area of consideration is the variety of factors that will directly influence the method of manufacture. Some of these **manufacturing concerns** are:

1. How many of the components are to be produced? At what rate? (*Note:* One-of-a-kind parts and small quantities are rarely made by processes that require dedicated patterns, molds or dies, because the expense of the tooling is hard to justify. High-volume, high-rate products may require automatable processes.)
2. What is the desired level of quality compared to similar products on the market?
3. What are the quality control and inspection requirements?
4. Are there any assembly (or disassembly) concerns? Any key relationships or restrictions with respect to mating parts?
5. What is the thickest section? The thinnest section? Is the thickness uniform or widely varying?

6. Have standard sizes and shapes been specified wherever possible (both as finished shapes and as starting raw material)? What would be the preferred form of starting material (plate, sheet, foil, bar, rod, wire, powder, ingot)?
7. Has the design addressed the requirements that will facilitate ease of manufacture (machinability, castability, formability, weldability, hardenability)?
8. What is the potential liability if the product should fail?
9. Is there a desire to utilize recycled materials in the manufacture of the product?
10. Is there a desire to minimize the overall energy consumption in the manufacture of the part?

The considerations just mentioned are only a sample of the many questions that must be addressed when precisely defining what it is that we want to produce. While there is a natural tendency to want to jump to an answer—in this case, a material and method of manufacture—time spent determining the various requirements will be well rewarded. Collectively, the requirements direct and restrict material and process selections. It is possible that several families of materials, and numerous members within those families, all appear to be adequate. In this case, selections may become a matter of preference. It is also possible, however, that one or more of the requirements emerges as a dominant restrictor (such as the need for ultra-high strength, superior wear resistance, the ability to function at extreme operating temperatures, or the ability to withstand highly corrosive environments), and selection then becomes focused on those materials offering that specific characteristic.

It is important that *all* factors are listed and *all* service conditions and uses are considered. Many failures and product liability claims have resulted from engineering oversights or failure to consider the entire spectrum of conditions that a product might experience in its lifetime. Consider the failure of several large electric power transformers where fatigue cracks formed at the joint where horizontal cooling fins had been welded to the exterior of the casing. The subsequent loss of cooling oil through the cracks led to overheating and failure of the transformer coils. Because transformers operate under static conditions, fatigue was not considered in the original design and material selection. However, when the horizontal fins were left unsupported during shipping, the resulting vibrations were sufficient to induce the fatal cracks. It is also not uncommon for the most severe corrosion environment to be experienced during shipping or storage as opposed to normal operation. Products can also encounter unusual service conditions. Consider the parts that failed on earthmoving and construction equipment during the construction of the trans-Alaskan oil pipeline. When this equipment was originally designed and the materials were selected, extreme subzero temperatures were not considered as possible operating conditions.

After completing a thorough evaluation of the required properties; it may be helpful to assign a relative importance to the various needs. Some requirements may be **absolutes,** while others may be **relative** or *compromisable.* Absolute requirements are those for which there can be no compromise. The consequence of not meeting them will be certain failure of the product. Materials that fall short of absolute requirements should be automatically eliminated. For example, if a component must possess good electrical conductivity, most plastics and ceramics would not be appropriate. Relative or compromisable properties are those that frequently differentiate "good," "better," and "best," where all would be considered as acceptable.

10.5 ADDITIONAL FACTORS TO CONSIDER

When evaluating candidate materials, an individual is often directed to handbook-type data that have been obtained through standardized materials characterization tests. It is important to note the conditions of these tests in comparison with those of the proposed application. Significant variations in factors such as temperature, rates of loading, or surface finish can lead to major changes in a material's behavior. In addition, one should keep in mind that the handbook values often represent an average or mean and that actual material properties may vary to either side of that value. Where vital information is missing or the data may not be applicable to the proposed use, one is advised to consult with the various materials producers or qualified materials engineers.

At this point, it is probably appropriate to introduce **cost** as an additional factor. Because of competition and marketing pressures, economic considerations are often as important as the technical ones. However, we have chosen to adopt the philosophy that cost should not be considered until a material has been shown to meet the necessary requirements. If acceptable candidates can be identified, cost will certainly become an important part of the selection process, and both material cost and the cost of fabrication should be considered. A more appropriate cost consideration might be **total lifetime cost,** which begins with the starting material, the energy to produce it, and the environmental impact of its production. To this is added the cost of converting it into the desired product, the cost of operating or using the product through its full lifetime, and finally the cost of disposal or recycling.

The final decision often involves some form of compromise between material cost, ease of fabrication, and performance or quality. Numerous questions might be asked, such as:

1. Is the material too expensive to meet the marketing objectives?
2. Is a more expensive material justifiable if it offers improved performance?
3. How much additional expense might be justified to gain ease of fabrication?

In addition, it is important that the appropriate cost figures be considered. Material costs are most often reported in the form of dollars per pound, or some other form of cost per unit weight. If the product has a fixed size, however, material comparisons should probably be based on cost per unit volume. For example, aluminum has a density about one-third that of steel. For products where the size is fixed, one pound of aluminum can be used to produce three times as many parts as one pound of steel. If the per-pound cost of aluminum were less than three times that of steel, aluminum would actually be the cheaper material. Whenever the densities of materials are quite different, as with magnesium and stainless steel, the relative rankings based on cost per pound and cost per cubic inch can be radically different.

Material availability is another important consideration. The material selected may not be available in the size, quantity, or shape desired, or may not be available in any form at all. Diversity and reliability of supply will facilitate competitive pricing and avoid production bottlenecks. If the desired material is not available domestically, import dependence and geopolitics may further influence the ultimate selection. It may be necessary to identify alternative materials that are also feasible candidates for the specific use.

Still other factors to be considered when making material selections include:

1. Are there possible misuses of the product that should be considered? If the product is to be used by the general public, one should definitely anticipate the worst. Screwdrivers are routinely used as chisels and pry bars (different forms of loading from the intended torsional twist). Scissors may be used as wire cutters. Other products are similarly misused.
2. Have there been any failures of this or similar products? If so, what were the identified causes, and have they been addressed in the current product? Failure analysis results should definitely be made available to the designers, who can directly benefit from them.
3. Has the material (or class of materials) being considered established a favorable or unfavorable performance record? Under what conditions was unfavorable performance noted?
4. Has an attempt been made to benefit from material standardization, whereby multiple components are manufactured from the same material or by the same manufacturing process? Although function, reliability, and appearance should not be sacrificed, one should not overlook the potential for savings and simplification that standardization has to offer.

10.6 CONSIDERATION OF THE MANUFACTURING PROCESS

The overall attractiveness of an engineering material depends not only on its physical and mechanical properties but also on one's ability to shape it into useful objects in an economical and timely manner. Without the necessary shape, parts cannot perform, and without economical production, the material will be limited to a few high-value applications. For this reason, material selection should be further refined by considering the possible fabrication processes and the suitability of each "prescreened" material to each of those processes. Familiarity with the various manufacturing alternatives is a necessity, together with a knowledge of the associated limitations, economics, product quality, surface finish, precision, and so on. All processes are not compatible with all materials. Steel, for example, cannot be fabricated by die casting. Titanium can be forged successfully by isothermal techniques but generally not by conventional drop hammers. Wrought alloys cannot be cast, and casting alloys are not attractive for forming. Ceramics cannot be welded or machined, and most cannot be cast.

Certain fabrication processes have distinct ranges of product size, shape, and thickness, and these should be compared with the requirements of the product. Each process has its characteristic precision and surface finish. Because secondary operations, such as machining, grinding, and polishing, all require the handling, positioning, and processing of individual parts, as well as additional tooling, they can add significantly to manufacturing cost. Usually it is best to hit the target with as few operations as possible. Some processes require prior heating or subsequent heat treatment. Still other considerations include production rate, production volume, desired level of automation, and the amount of labor required, especially if it is skilled labor. All of these concerns will be reflected in the cost of fabrication. There may also be additional constraints, such as the need to design a product so that it can be produced with existing equipment or facilities, or with a minimum of lead time, or with a minimal expenditure for dedicated tooling.

It is not uncommon for a certain process to be implied by the geometric details of a component design, such as the presence of cored features in a casting, the magnitude of draft allowances, or the recommended surface finish. The designer often specifies these features prior to consultation with manufacturing experts. It is best, therefore, to consider all possible methods of manufacture and, where appropriate, work with the designer to incorporate changes that would enable a more attractive means of production.

10.7 ULTIMATE OBJECTIVE

The real objective of this activity is to develop a manufacturing system—a combination of material and process (or sequence of processes) that is the best solution for a given product. Figure 10-5 presented a series of activities that move from a well-defined set of needs and objectives through material and process selection to the manufacture and evaluation of a product. Numerous decisions are required and most are judgmental in nature. For example, we may have to select between "good," "better," and "best," where "better" and "best" carry increments of added cost, or make intelligent compromises when all of the requirements cannot be simultaneously met.

While Figure 10-5 depicts the various activities as having a definite, sequential pattern, one should be aware that they are definitely interrelated. Figure 10-6 shows an alternative form, where material selection and process selection have been moved to be parallel instead of sequential. It is not uncommon for one of the two selections to be dominant, and the other to become dependent or secondary. For example, the production of a large quantity of small, intricate parts with thin walls, precise dimensions, and smooth surfaces is an ideal candidate for die casting. Material selection, therefore, may focus on die-castable materials—assuming feasible alternatives are available. In a converse example, highly restrictive material properties, such as the ability to endure extreme elevated temperatures or severe corrosive environments, may significantly limit the possible materials. Fabrication options will tend to be limited to those processes that are compatible with the candidate materials.

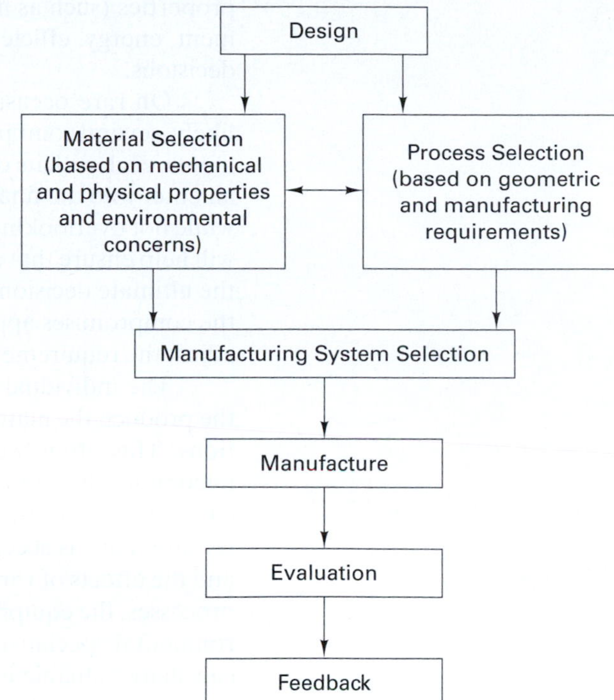

FIGURE 10-6 Alternative flow chart showing parallel selection of material and process.

In both models, decisions in one area generally impose restrictions or limitations in another. As shown in Figure 10-7, selection of a material may limit processes, and selection of a process may limit material. Each material has its own set of performance characteristics, both strengths and limitations. The various fabrication methods impart characteristic properties to the material, and all of these may not be beneficial (consider anisotropy, porosity, or residual stresses). Processes designed to improve certain

Process \ Material	Irons	Steel	Aluminum	Copper	Magnesium	Nickel	Refractory Metals	Titanium	Zinc
Sand Casting	X	X	X	X	X	X			0
Permanent Mold Cast	X	0	X	0	X	0			0
Die Casting			X	0	X				X
Investment Casting		X	X	X	0	0			
Closed-Die Forging		X	0	0	0	0	0	0	
Extrusion		0	X	X	X	0	0	0	
Cold Heading		X	X	X		0			
Stamping, Deep Draw		X	X	X	0	X		0	0
Screw Machine	0	X	X	X	0	X	0	0	0
Powder Metallurgy	X	X	0	X		0	X	0	

Key: X = Routinely performed
0 = Performed with difficulty, caution, or some sacrifice (such as die life)
Blank = Not recommended

FIGURE 10-7 Compatibility chart of materials and processes. Selection of a material may restrict possible processes. Selection of a process may restrict possible materials.

properties (such as heat treatment) may adversely affect others. Economics, environment, energy, efficiency, recycling, inspection, and serviceability all tend to influence decisions.

On rare occasions, a single solution will emerge as the obvious choice. More likely, several combinations of materials and processes will all meet the specific requirements, each with its own strengths and limitations. Compromise, opinion, and judgment all enter into the final decision making, where our desire is to achieve the best solution while not overlooking a major requirement. Listing and ranking the required properties will help ensure that all of the necessary factors were considered and weighed in making the ultimate decision. If no material-process combination meets the requirements, or if the compromises appear to be too severe, it may be necessary to redesign the product, adjust the requirements, or develop new materials or processes.

The individuals making materials and manufacturing decisions must understand the product, the materials, the manufacturing processes, and all of the various interrelations. This often requires multiple perspectives and diverse expertise, and it is not uncommon to find the involvement of an entire team. Design engineers ensure that each of the requirements is met and that any compromise or adjustment in those requirements is acceptable. Materials specialists bring expertise in candidate materials and the effects of various processing. Manufacturing personnel know the capabilities of processes, the equipment available, and the cost of associated tooling. Quality and environmental specialists add their perspective and expertise. Failure analysis personnel can share valuable experience gained from past unsuccessful efforts. Customer representatives or marketing specialists may also be consulted for their opinions. Clear and open communication is vital to the making of sound decisions and compromises.

The design and manufacture of a successful product is an iterative, evolving, and continual process. The failure of a component or product may have revealed deficiencies in design, poor material selection, material defects, manufacturing defects, improper assembly, or improper or unexpected product use. The costs of both material and processing continually change, and these changes may prompt a reevaluation. The availability of new materials, technological advances in processing methods, increased restrictions in environment or energy, or the demand for enhanced performance of an existing product all provide a continuing challenge. Materials availability may also have become an issue. A change in material may well require companion changes in the manufacturing process. Improvements in processing may warrant a reevaluation of the material.

10.8 MATERIALS SUBSTITUTION

As new technology is developed or market pressures arise, it is not uncommon for new materials to be substituted into an existing design or manufacturing system. Quite often, the substitution brings about improved quality, reduced cost, ease of manufacture, simplified assembly, or enhanced performance. When making a **material substitution,** however, it is also possible to overlook certain requirements and cause more harm than good.

Consider the efforts related to the production of lighter-weight, more-fuel-efficient, less emission-producing automobiles. The development of higher-strength steel sheets has provided the opportunity to match or exceed the strength of traditional body panels with thinner-gage material. Forming and fabricating the higher-strength materials will bring new problems, but even if these can be overcome, there are still other issues. Strength was definitely increased, but it is unlikely that corrosion resistance and elastic stiffness (rigidity) were also increased in a proportional manner. Thinner sheets might corrode in a shorter time, and previously unnoticed vibrations could become a significant problem. Measures to retard corrosion and design modifications to reduce vibration may be necessary before the higher-strength material could be effectively substituted.

Aluminum is one-third the weight of steel, but it also has one-third the rigidity. Aluminum sheet has been used to replace steel panels, enabling a 50% reduction in

TABLE 10-2 Material Substitutions to Reduce Weight in an Automobile[a]

New Material	Previous Material	Weight Reduction	New Relative Cost
High-strength steel	Mild steel	10%	100% (no change)
Aluminum	Steel or Cast Iron	40–60%	130–200%
Magnesium	Steel or cast iron	60–75%	150–250%
Magnesium	Aluminum	25–35%	100–150%
Glass fiber Reinforced plastic	Steel	25–35%	100–150%

[a] Data taken from "Automotive Materials in the 21st Century" by William F. Powers, published in *Advanced Materials and Processes,* May 2000.

weight,[4] but the vibration problems associated with the lower elastic modulus has required special consideration.

Aluminum has also been considered as an alternative to cast iron for engine blocks and transmission housings. Corrosion resistance would be enhanced and weight savings would be substantial. If we assume that the mechanical properties are adequate, consideration would also have to be given to the area of noise and vibration. Gray cast iron has excellent damping characteristics and effectively eliminates these undesirable features. Aluminum transmits noise and vibration, and its use in transmission housings would probably require the addition of some form of sound isolation material. When aluminum was first used for engine blocks, the transmitted vibrations required a companion redesign of the engine support system.

Automotive body panels have also been produced from polymeric materials, such as sheet-molding compound. Thermoplastic materials must be cooled, generally in the mold, to impart strength and rigidity. Thermosetting materials must be cured in the desired final shape. As a result, fabrication times tend to be longer than traditional sheet metal forming, and we generally see polymeric panels being used on specialty vehicles with lower production rates.

Polymeric materials have also been used successfully for bumpers, fuel tanks, pumps, and housings. Composite-material drive shafts have been used in place of metal. Cast metal, powder metallurgy products, and composite materials have all been used for connecting rods. Ceramic and reinforced plastic components have been used for engine components. Magnesium is being used for instrument panels and steering wheels. Fiber-reinforced polymer composite has been used to produce the cargo beds for pickup trucks. When making a material substitution in a successful product, however, it is important to first consider all of the design requirements. Approaching a design or material modification as thoroughly as one approaches a new problem may well avoid costly errors.

Table 10-2 summarizes some of the weight-saving material substitutions that have been used on automobiles and calls attention to the fact that many of these substitutions are accompanied by an increase in cost, where total cost incorporates both cost of the material itself and the cost of converting that material into the desired product.

10.9 EFFECT OF PRODUCT LIABILITY ON MATERIALS SELECTION

Product liability actions, court awards, and rising insurance costs have made it imperative that designers and manufacturers employ the very best procedures in selecting and processing materials. Although many individuals feel that the situation has grown to absurd proportions, there have also been many instances where sound procedures were not used in selecting materials and methods of manufacture. In today's business and legal climate, such negligence cannot be tolerated.

[4] While aluminum is one-third the weight of steel, it is also weaker. A thicker gage metal was used in the substitution, resulting in only a 50% savings in weight.

An examination of recent product liability claims has revealed that the five most common causes have been:

1. Failure to know and use the latest and best information about the materials being specified.
2. Failure to foresee, and account for, all reasonable uses of the product.
3. Use of materials for which there were insufficient or uncertain data, particularly with regard to long-term properties.
4. Inadequate and unverified quality control procedures.
5. Material selection made by people who were completely unqualified.

An examination of these faults reveals that there is no good reason for them to exist. Consideration of each, however, is good practice when seeking to ensure the production of a quality product and can greatly reduce the number and magnitude of product liability claims.

10.10 AIDS TO MATERIAL SELECTION

From the discussion in this chapter, it is apparent that those who select materials should have a broad, basic understanding of the nature and properties of materials and their processing characteristics. Providing this background is a primary purpose of this text. The number of engineering materials is so great, however, and the mass of information that is both available and useful is so large, that a single book of this type and size cannot be expected to furnish all that is required. Anyone who does much work in material selection needs to have ready access to many sources of data.

It is almost imperative that one has access to the information contained in the various volumes of *Metals Handbook*, published by ASM International. This multivolume series contains a wealth of information about both engineering metals and associated manufacturing processes. The one-volume *Metals Handbook Desk Edition* provides the highlights of this information in a less voluminous, more concise format. A parallel *ASM Engineered Materials Handbook* series and one-volume *Desk Edition* provide similar information for composites, plastics, adhesives, and ceramics. These resources are also available on computer CD-ROMs and directly via Internet through paid subscription.

ASM also offers a one-volume *ASM Metals Reference Book* that provides extensive data about metals and metalworking in tabular or graphic form. *Smithell's Metal Reference Book* provides nearly 2000 pages of useful information and data. Additional handbooks are available for specific classes of materials, such as titanium alloys, stainless steels, tool steels, plastics, and composites. Various technical magazines often provide annual issues that serve as information databooks. Some of these include *Modern Plastics*, *Industrial Ceramics*, and ASM's *Advanced Materials and Processes*.

Persons selecting materials and processes should also have available several of the handbooks published by various materials organizations, technical societies, and trade associations. These may be material related (such as the Aluminum Association's *Aluminum Standards and Data* and the Copper Development Association's *Standards Handbook: Copper, Brass, and Bronze*), process related (such as the *Steel Castings Handbook* by the Steel Founder's Society of America and the *Heat Treater's Guide* by ASM International), or profession related (such as the *SAE Handbook* by the Society of Automotive Engineers, *ASME Handbook* by the American Society for Mechanical Engineers, and the *Tool and Manufacturing Engineers Handbook* by the Society for Manufacturing Engineers). These may be supplemented further by a variety of supplier-provided information. While the latter is excellent and readily available, the user should recognize that supplier information might not provide a truly objective viewpoint.

It is also important to have accurate information on the cost of various materials. Because these tend to fluctuate, it may be necessary to consult a daily or weekly publication such as the *American Metal Market* newspaper or online service. Costs associated with various processing operations are more difficult to obtain and can vary greatly from one company to another. These costs may be available from within the firm or may have to be estimated from outside sources. A variety of texts and software packages are available.

Each of the preceding references provides focused information about a class of materials or a specific type of process. A number of texts have attempted to achieve integration with a focus on design and material selection. Possibly the most well known are the works of M. F. Ashby of Cambridge University and the associated computer-based software developed and marketed through Granta Design Ltd.(*CES Selector* and *CES EduPack* software). Specific Ashby texts include *Materials Selection in Mechanical Design*, 3rd ed. (2005); *Materials and the Environment—Eco-informed Material Choice* (2009); and *Materials and Design, The Art and Science of Material Selection in Product Design,* 2nd ed. (2009).

With the evolution of high-speed computers with large volumes of searchable memory, materials selection can now be computerized. Most of the textbook and handbook references are now available on CDs, or directly on the Internet, and all of the information in an entire handbook series can be accessed almost instantaneously. Programs have been written to utilize information databases and actually perform materials selection. The various property requirements can be specified, and the entire spectrum of engineering materials can be searched to identify possible candidates. Search parameters can then be tightened or relaxed so as to produce a desired number of candidate materials. In a short period of time, a wide range of materials can be considered, far greater than could be considered in a manual selection. Process simulation packages can then be used to verify the likelihood of producing a successful product.

While the capabilities of computers and computer software are indeed phenomenal, the knowledge and experience of trained individuals should not be overlooked. Experienced personnel should reevaluate the final materials and manufacturing sequence to assure full compliance with the needs of the product.

The appendix titled "Selected References for Additional Study" provides an extensive list of additional resources.

Key Words

absolute requirement
case-history method
conceptual design stage
cost
design
environmental considerations
functional design stage
geometric considerations
manufacturing concerns
material availability
material selection
material substitution
mechanical properties
physical properties
product liability
production design stage
prototype
recyclability
relative requirement
total lifetime cost

Review Questions

1. What is the objective of a manufacturing operation, and what are some of the details in meeting this objective?
2. What are some possible undesirable features of significantly exceeding the requirements of a product?
3. What problems are created by the availability of so many alternative materials from which to choose?
4. In a manufacturing environment, why should the selection and use of engineering materials be a matter of constant reevaluation?
5. How have different materials enabled advances and specializations in bicycle manufacture?
6. Discuss the interrelation between engineering material and the fabrication processes used to produce the desired shape and properties.
7. What is design?
8. What are the three primary stages of product design, and how does the consideration of materials differ in each?
9. What is the benefit of requiring prototype products to be manufactured from the same materials that will be used in production and by the same manufacturing techniques?
10. What sequence of activities is common to nearly every engineered component or product?
11. What are some of the possible pitfalls in the case-history approach to materials selection? Some of the benefits?
12. What is the most frequent pitfall when seeking to improve an existing product?
13. What should be the first step in any materials selection problem?
14. In what ways does the concept of shape or geometry go beyond a dimensioned sketch?

15. How might geometric requirements influence selection of the method of fabrication?
16. How might temperature enter into the specification of mechanical properties?
17. What are "physical properties" of materials?
18. What are some of the important aspects of the service environment to be considered when selecting an engineering material?
19. What are some of the possible manufacturing concerns that should be considered?
20. Why is it important to resist jumping to the answer and first perform a thorough evaluation of product needs and requirements, considering all factors and all service conditions?
21. Give an example where failure to consider a specific requirement has led to product failure.
22. What is the difference between an absolute and relative requirement?
23. What are some possible pitfalls when using handbook data to assist in materials selection?
24. Why might it be appropriate to defer cost considerations until after evaluating the performance capabilities of various engineering materials?
25. Why might it be appropriate to consider "total lifetime cost"?
26. Give an example of a product or component where material cost should be compared on a cost per pound basis. Give a contrasting example where cost per unit volume would be more appropriate.
27. What are some possible considerations relating to materials availability?
28. Provide some examples of reasonable misuse of a product.
29. In what way might failure analysis data be useful in a material selection decision?
30. How might using the same material or the same process for multiple parts result in cost savings?
31. Why should consideration of the various fabrication process possibilities be included in material selection? What aspects of a manufacturing process should be considered?
32. Why might it be better to perform material selection and process selection in a parallel fashion as opposed to sequential?
33. Give an example of where selection of a material may limit processes, and where selection of a process may limit materials.
34. Why is it likely that compromise, opinion, and judgment will all enter into material and process selection?
35. Why is it likely that multiple individuals will be involved in the material and process selection activity?
36. Why should the design and manufacture of a successful product be an iterative, evolving, and continual process?
37. Give an example of an unexpected problem that occurred in a materials substitution.
38. What are some of the most common causes of product liability losses?
39. How have high-speed, high-capacity computers changed materials selection? Have they replaced trained individuals?

PROBLEMS

1. One simple tool that has been developed to assist in materials selection is a rating chart, such as the one shown in Figure 10-A. Absolute properties are identified and must be present for a material to be considered. The various relative properties are weighted as to their significance, and candidate materials are rated on a scale such as 1 to 5 or 1 to 10 with regard to their ability to provide that property. A rating number is then computed by multiplying the property rating by its weighted significance and summing the results. Potential materials can then be compared in a uniform, unbiased manner, and the best candidates can often be identified. In addition, by placing all the requirements on a single sheet of paper, the designer is less likely to overlook a major requirement. Finalist materials should then be reevaluated to assure that no key requirement has been overlooked or excessively compromised.

   Three materials (X, Y, and Z) are available for a certain use. Any material selected must have good weldability. Tensile strength, stiffness, stability, and fatigue strength have also been identified as key requirements. Fatigue strength is considered the most important of these requirements, and stiffness is least important. The three materials can be rated as follows:

	X	Y	Z
Weldability	Excellent	Poor	Good
Tensile strength	Good	Excellent	Fair
Stiffness	Good	Good	Good
Stability	Good	Excellent	Good
Fatigue strength	Fair	Good	Excellent

   Develop a rating chart such as that in Figure 10-A to determine which material you would recommend.
2. The chalk tray on a classroom chalkboard has very few performance requirements. As a result, it can be made from a wide spectrum of materials. Wood, aluminum, and even plastic have been used in this application. Discuss the performance and durability requirements and the pros and cons of the three listed materials. Chalk trays have a continuous cross section, but the processes used to produce such a configuration may vary with material. Discuss how a chalk tray might be mass-produced from each of the three materials classifications. Might this be a candidate for some form of wood by-product similar to particle board? Because the product demands are low, might some form of recycled material be considered?
3. Soda has been sold in a variety of containers, including aluminum cans, polymer (plastic) bottles, and glass bottles. Discuss the pros and cons of each in terms of the following aspects: cost, convenience, durability, weight (both shipping and customer preference), environmental impact and recyclability, affects on flavor and carbonation retention, and others. Is container transparency an asset? Why or why not?
4. Examine the properties of wood, aluminum, and extruded vinyl as they relate to household window frames. Discuss the pros and cons of each, considering cost, ease of manufacture, and aspects of performance—including strength, energy efficiency, thermal expansion and contraction, response to moisture and humidity, durability, rigidity, appearance (the ability to be finished in a variety of colors), ease of

Material	Go-No-Go** screening			Relative rating number (†rating number x * weighting factor)								Material rating number
	Corrosion	Weldability	Brazability	Strength (5)*	Toughness (5)	Stiffness (5)	Stability (5)	Fatigue (4)	As-welded strength (4)	Thermal stresses (3)	Cost (1)	$\frac{\Sigma \text{ rel rating no.}}{\Sigma \text{ rating factors}}$

*Weighting factor = 1 lowest to 5 most important

† Range = 1 poorest to 5 best

**Code = S = satisfactory

U = unsatisfactory

FIGURE 10-A Rating chart for comparing materials for a specific application.

maintenance, property changes with low and high extremes of temperature, and any other factor you feel is important. Which would be your preference for your particular location? Might your preference change if you were located in the dry southwest (Arizona), New England, Alaska, or Hawaii? Can you imagine some means of combining materials to produce windows that might be superior to any single material? Which of the preceding features would apply to residential home siding?

5. Automobile body panels have been made from carbon steel, high-strength steel, aluminum, and various polymer-base molding compounds (both thermoplastic and thermoset). Discuss the key material properties and the relative performance characteristics of each, considering both use and manufacture. For what type of vehicle might you prefer the various materials? Consider low-volume versus high volume production, family versus commercial versus performance, low-cost versus luxury, etc. How might preferences change if recyclability were required?
6. Consider the two-wheel bicycle frame and the variety of materials that have been used in its construction—low-carbon plain-carbon steel, somewhat higher carbon chrome–moly alloy steel (such as 4130), cold-drawn aluminum tubing (strengthened by cold work), age-hardened aluminum tubing (strengthened by the age-hardening heat treatment), titanium alloy, fiber-reinforced composite, and still others. While all bicycle frames experience fatigue loadings, some materials have endurance limits, while others do not. Some can be assembled by conventional welding or brazing. Others require low-temperature joining methods, because exposure to high temperature will compromise material strength. For still others, a one-piece structure (no joints) may be feasible. Select a material, other than steel, and discuss the possible methods of manufacture and concerns you might have. Would your solution be appropriate for high- or low-production bicycles? Would it be good for pleasure bikes? Rugged mountain bikes? Racing bikes? What would be its unique selling features?
7. Go to the local hardware or building supply store and examine a specific class of fastener (nail, screw, bolt, rivet, etc.). Is it available in different grades or classes based on strength or intended use? What are they and how do they differ? How might the materials and methods of manufacture be different for these identified groups? Summarize your findings. What surface treatments, if any, have been applied to the items? How was it applied and for what purpose?
8. The handles of carpenter claw hammers have been made from wood (often hickory), metal, fiberglass, and other composites. Discuss the relative pros and cons of the various candidate materials.
9. Decorative fence posts for a residential home have been made from wood, extruded PVC, recycled polyethylene, decorative concrete, and various metals. Discuss the key material requirements and the pros and cons of the various potential materials.
10. The individual turbine blades used in the exhaust region of jet engines must withstand high temperatures, high stresses, and highly corrosive operating conditions. These demanding conditions severely limit the material possibilities, and most jet engine turbine blades have been manufactured from one of the high-temperature superalloys. The fabrication processes are limited to those that are compatible with both the material and the desired geometry. Through the 1960s and early 1970s, the standard method of production was investment casting, and the resultant product was a polycrystalline solid with thousands of polyhedral crystals. In the mid- to late 1970s, production shifted to unidirectional solidification, where elongated crystals ran the entire length of the blade. More recently, advances have enabled the production of single-crystal turbine blades. Investigate this product to determine how the various material and processing conditions produce products with differing performance characteristics.
11. Identify a product for which you have experienced a shorter-than-expected service life. What do you feel was the cause of the reduced lifetime? Consider the following possibilities: (a) poor design, (b) improper material selection, (c) poor

selection of manufacturing process, (d) insufficient quality control, (e) poor maintenance, and (f) product misuse, as well as others. What changes would you recommend?

12. Hockey sticks are currently available in wood, aluminum, and composites. Wood was the original material, and wooden sticks are still popular today. Aluminum sticks use a tubular (or rectangular tube) aluminum shaft, mated with a traditional wooden blade. Composite sticks have utilized fibers of Kevlar, graphite, or fiberglass in a thermoplastic resin, such as acrylonitrile-butadiene-styrene (ABS).
    a. Compare the three materials in terms of cost, weight, durability, performance, and other important properties.
    b. Describe one or more methods that could be used to manufacture the fiber-reinforced composite sticks. Would you want the reinforcement fibers in the shaft only, blade only, or both shaft and blade? How would you achieve your desired configuration?
    c. Titanium has been proposed as a potentially more attractive alternative. What do you consider to be the pros and cons of titanium? How might it be superior to aluminum?

www.wiley.com/go/global/degarmo

Chapter 10 CASE STUDY

Material Selection

This study is designed to get you to question why parts are made from a particular material and how they could be fabricated to their final shape. For one or more of the products listed, write a brief evaluation that addresses the following questions.

QUESTIONS:

1. What are the normal use or uses of this product or component? What are the normal operating conditions in terms of temperatures, loadings, impacts, corrosive media, etc.? Are there any unusual extremes?
2. What are the major properties or characteristics that the material must possess in order for the product to function?
3. What material (or materials) would you suggest and why?
4. How might you propose to fabricate this product?
5. Would the product require heat treatment? For what purpose? What kind of treatment?
6. Would this product require any surface treatment or coating? For what purpose? What would you recommend?
7. Would there be any concerns relating to environment? Recycling? Product liability?

PRODUCTS:

a. The head of a carpenter's claw hammer
b. Exterior of an office filing cabinet
c. Residential interior doorknob
d. A paper clip
e. Staples for an office stapler
f. A pair of scissors
g. A moderate- to high-quality household cook pot or frying pan
h. Case for a jeweler-quality wristwatch
i. Jet engine turbine blade to operate in the exhaust region of the engine
j. Standard open-end wrench
k. A socket-wrench socket to install and remove spark plugs
l. The frame of a 10-speed bicycle
m. Interior panels of a microwave oven
n. Handle segments of a retractable blade utility knife with internal storage for additional blades
o. The outer skin of an automobile muffler
p. The exterior case for a classroom projector
q. The basket section of a grocery-store shopping cart
r. The body of a child's toy wagon
s. Decorative handle for a kitchen cabinet
t. Automobile radiator
u. The motor housing for a chain saw
v. The blade of a household screwdriver
w. Household dinnerware (knife, fork, and spoon)
x. The blades on a high-quality cutlery set
y. A shut-off valve for a $\frac{1}{2}$ in. household water line
z. The base plate (with heating element) for an electric steam iron
aa. The front sprocket of a 10-speed bicycle
bb. The load-bearing structure of a child's outdoor swing set
cc. The perforated spin tub of a washing machine
dd. A commemorative coin for a corporation's 100th anniversary
ee. The keys for a commercial quality door lock
ff. The exterior canister for an automobile oil filter

CHAPTER 11

FUNDAMENTALS OF CASTING

11.1 INTRODUCTION TO MATERIALS PROCESSING

Almost every manufactured product (or component of a product) goes through a series of activities that include (1) design, or defining what we want to produce; (2) material selection; (3) process selection; (4) manufacture; (5) inspection and evaluation; and (6) feedback. Previous chapters have presented the fundamentals of *materials engineering*, the study of the structure, properties, processing, and performance of engineering materials and the systems interactions between these aspects. Other chapters address the use of heat treatment to achieve desired properties and the use of surface treatments to alter features such as wear or corrosion resistance. In this chapter, we begin a focus on **materials processing,** the science and technology through which a material is converted into a useful shape with structure and properties that are optimized for the proposed service environment. A less technical definition of materials processing might be "whatever must be done to convert stuff into things."

A primary objective of materials processing is the production of a desired shape in the desired quantity. Shape-producing processes can be grouped into five basic "families," as indicated in Figure 11-1. **Casting** processes exploit the properties of a liquid as it flows into and assumes the shape of a prepared container, and then solidifies upon cooling. The **material removal** processes remove selected segments from an initially oversized piece. Traditionally, these processes have often been referred to as **machining,** a term used to describe the mechanical cutting of materials. The more general term, *material removal*, includes a wide variety of techniques, including those based on chemical, thermal, and physical processes. **Deformation** processes exploit the ductility or plasticity of certain materials, mostly metals, and produce the desired shape by mechanically moving or rearranging the solid. **Consolidation** processes build a desired shape by putting smaller pieces together. Included here are welding, brazing, soldering, adhesive bonding, and mechanical fasteners. **Powder metallurgy** is the manufacture of a desired shape from particulate material, a definite form of consolidation, but it can also involve aspects of casting and forming. The newest addition is **direct digital manufacturing** or **additive manufacturing,** which includes a variety of processes developed to directly convert a computer-file "drawing" to a finished product without any intervening patterns, dies, or other tooling.

Each of the five basic families has distinct advantages and limitations, and the various processes within the families have their own unique characteristics. For example, cast products can have extremely complex shapes but also possess structures that are produced by solidification, and, as such, they are subject to such defects as shrinkage and porosity. Material removal processes are capable of outstanding dimensional precision but produce scrap when material is cut away to produce the desired shape.

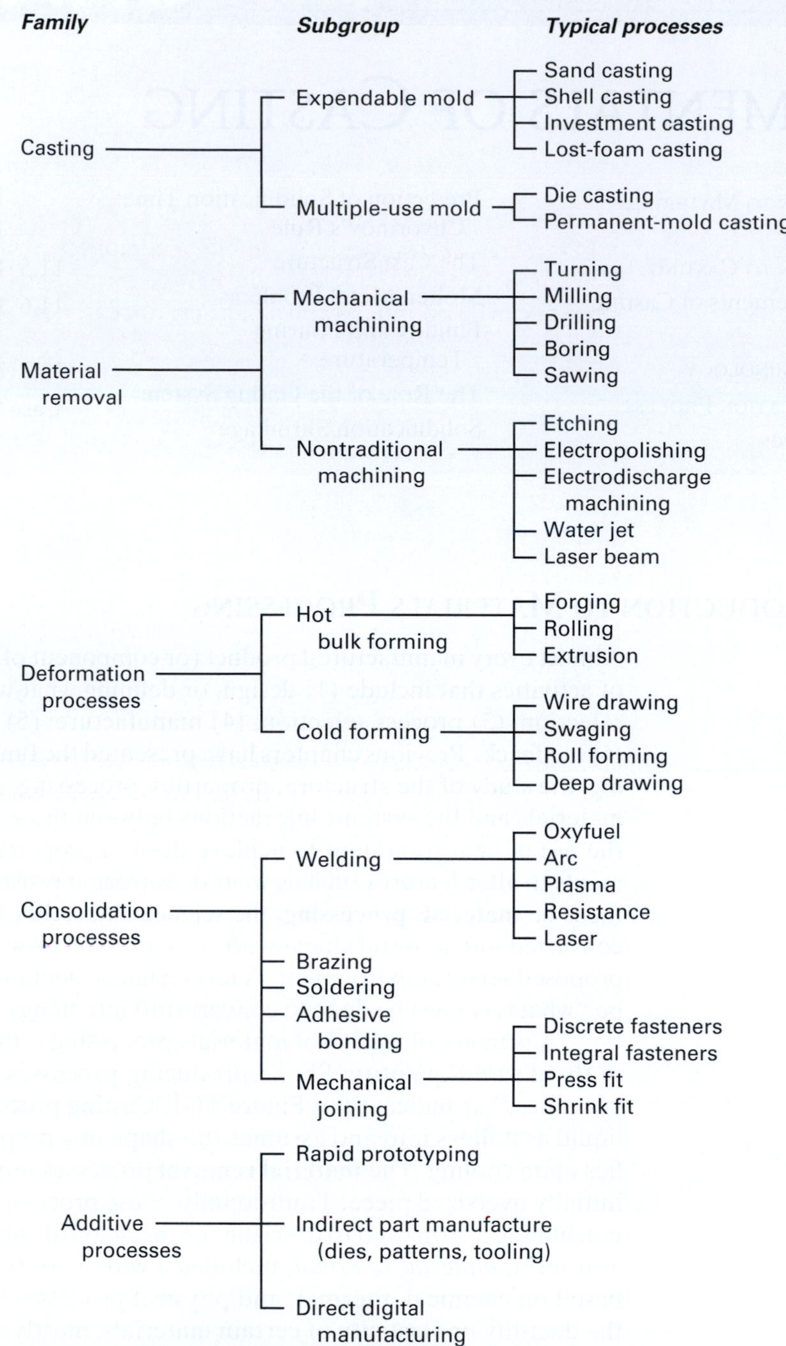

FIGURE 11-1 The five materials processing families, with subgroups and typical processes.

Deformation processes can have high rates of production but generally require powerful equipment and dedicated tools or dies. Complex products can often be assembled from simple shapes, but the joint areas are often affected by the joining process and may possess characteristics different from the original base material. Direct digital manufacturing can produce parts almost on-demand but is limited in the time it takes to produce a part and the range of materials and properties that are available.

When selecting the process or processes to be used in obtaining a desired shape and achieving the desired properties, decisions should be made with the knowledge of all available alternatives and their associated assets and limitations. A large portion of this book is dedicated to presenting the various processes that can be applied to engineering materials. They are grouped according to the basic categories, with powder metallurgy being included at the end of the deformation process section, along with direct digital manufacturing. The emphasis is on process fundamentals, descriptions of the various alternatives, and an assessment of associated assets and limitations. We will begin with a survey of the casting processes.

11.2 INTRODUCTION TO CASTING

In the casting processes, a material is first melted, heated to proper temperature, and sometimes treated to modify its chemical composition. The molten material is then poured into a cavity or mold that holds it in the desired shape during cool-down and solidification. In a single step, simple or complex shapes can be made from any material that can be melted. By proper design and process control, the resistance to working stresses can be optimized and a pleasing appearance can be produced.

Ninety percent of all manufactured goods contain at least one metal casting. Cast parts range in size from a fraction of a centimeter and a fraction of a gram (such as the individual teeth on a zipper), to more than 10 meters and many tons (as in the huge propellers and stern frames of ocean liners). Moreover, the casting processes have distinct advantages when the production involves complex shapes, parts having hollow sections or internal cavities, parts that contain irregular curved surfaces (except those that can be made from thin sheet metal), very large parts, or parts made from metals that are difficult to machine.

It is almost impossible to design a part that cannot be cast by one or more of the commercial casting processes. However, as with all manufacturing techniques, the best results and lowest cost are only achieved if the designer understands the various options and tailors the design to use the most appropriate process in the most efficient manner. The variety of casting processes use different mold materials (sand, metal, or various ceramics) and pouring methods (gravity, vacuum, low pressure, or high pressure). All share the requirement that the material should solidify in a manner that will maximize the properties and avoid the formation of defects, such as shrinkage voids, gas porosity, and trapped inclusions.

BASIC REQUIREMENTS OF CASTING PROCESSES

Six basic steps are present in most casting processes:

1. A container must be produced with a *mold cavity*, having the desired shape and size, with due allowance for shrinkage of the solidifying material. Any geometrical feature desired in the finished casting must be present in the cavity. The mold material must provide the desired detail and also withstand the high temperatures and not contaminate the molten material that it will contain. In some processes, a new mold is prepared for each casting (single-use molds) while in other processes the mold is made from a material that can withstand repeated use, such as metal or graphite. The **multiple-use molds** tend to be quite costly and are generally employed with products where large quantities are desired. The more economical **single-use molds** are usually preferred for the production of smaller quantities, but may be required when casting the higher melting-temperature materials.
2. A *melting process* must be capable of providing molten material at the proper temperature, in the desired quantity, with acceptable quality, and at a reasonable cost.
3. A *pouring technique* must be devised to introduce the molten metal into the mold. Provision should be made for the escape of all air or gases present in the cavity prior to pouring, as well as those generated by the introduction of the hot metal. The molten material must be free to fill the cavity, producing a high-quality casting that is fully dense and free of defects.
4. The *solidification process* should be properly designed and controlled. Castings should be designed so that solidification and solidification shrinkage can occur without producing internal porosity or voids. In addition, the molds should not provide excessive restraint to the shrinkage that accompanies cooling, a feature that may cause the casting to crack when it is still hot and its strength is low.
5. It must be possible to remove the casting from the mold (i.e., *mold removal*). With single-use molds that are broken apart and destroyed after each casting, mold

removal presents no serious difficulty. With multiple-use molds, however, the removal of a complex-shaped casting may be a major design problem.

6. Various *cleaning, finishing, and inspection* operations may be required after the casting is removed from the mold. Extraneous material is usually attached where the metal entered the cavity, excess material may be present along mold parting lines, and mold material may adhere to the casting surface. All of these must be removed from the finished casting.

Each of the six steps will be considered in more detail as we move through the chapter. The fundamentals of solidification, pattern design, gating, and risering will all be developed. Various defects will also be considered, together with their causes and cures.

11.3 CASTING TERMINOLOGY

Before we proceed to the process fundamentals, it is helpful to first become familiar with a bit of casting vocabulary. Figure 11-2 shows a two-part mold, its cross section, and a variety of features or components that are present in a typical casting process. To produce a casting, we begin by constructing a **pattern,** an approximate duplicate of the

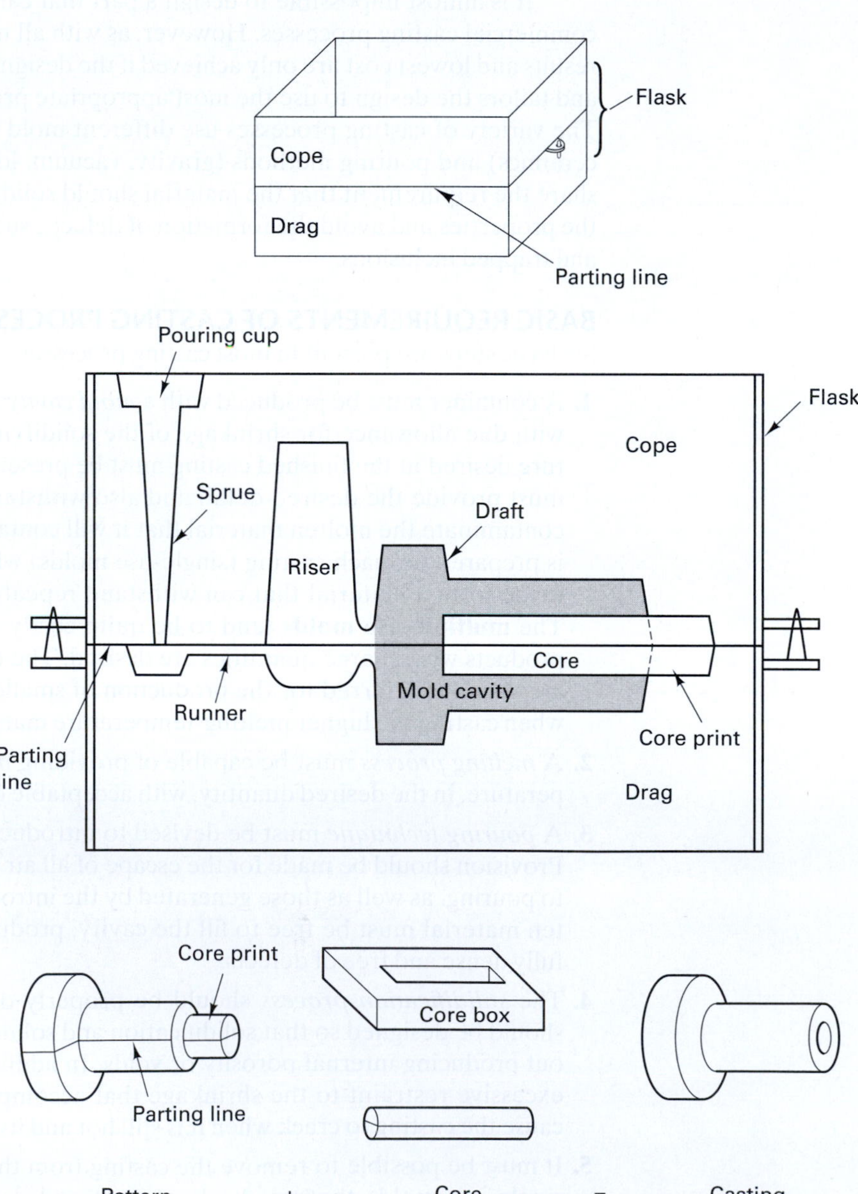

FIGURE 11-2 Cross section of a typical two-part sand mold, indicating various mold components and terminology.

final casting. **Molding material** will then be packed around the pattern and the pattern is removed to create all or part of the mold cavity. The rigid metal or wood frame that holds the molding aggregate is called a **flask.** In a horizontally parted two-part mold, the top half of the pattern, flask, mold, or core is called the **cope.** The bottom half of any of these features is called the **drag.** A **core** is a sand (or metal) shape that is inserted into a mold to produce the internal features of a casting, such as holes or passages. Cores are produced in wood, metal or plastic tooling, known as **core boxes.** A **core print** is a feature that is added to a pattern, core, or mold and is used to locate and support a core within the mold. The mold material and the cores then combine to produce a completed **mold cavity,** a shaped hole into which the molten metal is poured and solidified to produce the desired casting. A **riser** is an additional void in the mold that also fills with molten metal. Its purpose is to provide a reservoir of additional liquid that can flow into the mold cavity to compensate for any shrinkage that occurs during solidification. By designing so the riser contains the last material to solidify, shrinkage voids should be located in the riser and not the final casting.

The network of connected channels used to deliver the molten metal to the mold cavity is known as the **gating system.** The **pouring cup** (or pouring basin) is the portion of the gating system that receives the molten metal from the pouring vessel and controls its delivery to the rest of the mold. From the pouring cup, the metal travels down a **sprue** (the vertical portion of the gating system), then along horizontal channels, called **runners,** and finally through controlled entrances, or **gates,** into the mold cavity. Additional channels, known as **vents,** may be included in a mold or core to provide an escape for the gases that are originally present in the mold or are generated during the pour. (These and other features of a gating system will be discussed later in the chapter and are illustrated in Figure 11-9.)

The **parting line** or **parting surface** is the interface that separates the cope and drag halves of a mold, flask, or pattern and also the halves of a core in some core-making processes. **Draft** is the term used to describe the taper on a pattern or casting that permits it to be withdrawn from the mold. The draft usually expands toward the parting line. Finally, the term *casting* is used to describe both the process and the product when molten metal is poured and solidified in a mold.

11.4 THE SOLIDIFICATION PROCESS

Casting is a **solidification** process where the molten material is poured into a mold and then allowed to freeze into the desired final shape. Many of the structural features that ultimately control product properties are set during solidification. Furthermore, many casting defects, such as **gas porosity** and **solidification shrinkage,** are also solidification phenomena, and they can be reduced or eliminated by controlling the solidification process.

Solidification is a two-stage, nucleation and growth, process, and it is important to control both of these stages. **Nucleation** occurs when stable particles of solid form from within the molten liquid. When a material is at a temperature below its melting point, the solid state has a lower energy than the liquid. As solidification occurs, internal energy is released. At the same time, however, interface surfaces must be created between the new solid and the parent liquid. Formation of these surfaces requires energy. In order for nucleation to occur, there must be a net reduction or release of energy. As a result, nucleation generally begins at a temperature somewhat below the equilibrium melting point (the temperature where the internal energies of the liquid and solid are equal). The difference between the melting point and the actual temperature of nucleation is known as the amount of **undercooling.**

If nucleation can occur on some form of existing surface, it no longer requires the creation of a full, surrounding interface, and the required energy is reduced. Such surfaces are usually present in the form of mold or container walls or as solid impurity particles contained within the molten liquid. When ice cubes are formed in a tray, the initial solid forms on the walls of the container. The same phenomena can be expected with metals and other engineering materials.

Each nucleation event produces a crystal or grain in the final casting. Because fine-grained materials (many small grains) possess enhanced mechanical properties, efforts may be made to promote nucleation. Particles of existing solid may be introduced into the liquid before it is poured into the mold. These particles provide the surfaces required for nucleation and promote the formation of a uniform, fine-grained product. This practice of introducing solid particles is known as **inoculation** or **grain refinement.**

The second stage in the solidification process is **growth,** which occurs as the heat of fusion is extracted from the liquid material. The direction, rate, and type of growth can be controlled by the way in which this heat is removed. **Directional solidification,** in which the solidification interface sweeps continuously through the material, can be used to ensure the production of a sound casting. The molten material on the liquid side of the interface can flow into the mold to continuously compensate for the shrinkage that occurs as the material changes from liquid to solid. The relative rates of nucleation and growth control the size and shape of the resulting crystals. Faster rates of cooling generally produce products with finer grain size and superior mechanical properties.

COOLING CURVES

Cooling curves, such as those introduced in Chapter 5, can be one of the most useful tools for studying the solidification process. By inserting thermocouples into a casting and recording the temperature versus time, one can obtain valuable insight into what is happening in the various regions.

Figure 11-3 shows a typical cooling curve for a pure or eutectic-composition material (one with a distinct melting point) and is useful for depicting many of the features and terms related to solidification. The **pouring temperature** is the temperature of the liquid metal when it first enters the mold. **Superheat** is the difference between the pouring temperature and the freezing temperature of the material. Most metals are poured at temperatures of 100 to 200°C (200 to 400°F) above the temperature where solid begins to form. The higher the superheat, the more time is given for the material to flow into the intricate details of the mold cavity before it begins to freeze. The **cooling rate** is the rate at which the liquid or solid is cooling and can be viewed as the slope of the cooling curve at any given point. The **thermal arrest** is the plateau in the cooling curve that occurs during the solidification of a material with fixed melting point. At this temperature, the energy or heat being removed from the mold comes from the latent heat of fusion that is being released during the solidification process. The time from the start of pouring to the end of solidification is known as the **total solidification time.** The time from the start of solidification to the end of solidification is the **local solidification time.**

If the metal or alloy does not have a distinct melting point, such as the copper–nickel alloy shown in Figure 11-4, solidification will occur over a range of temperatures. The **liquidus** temperature is the lowest temperature where the material is all liquid, and

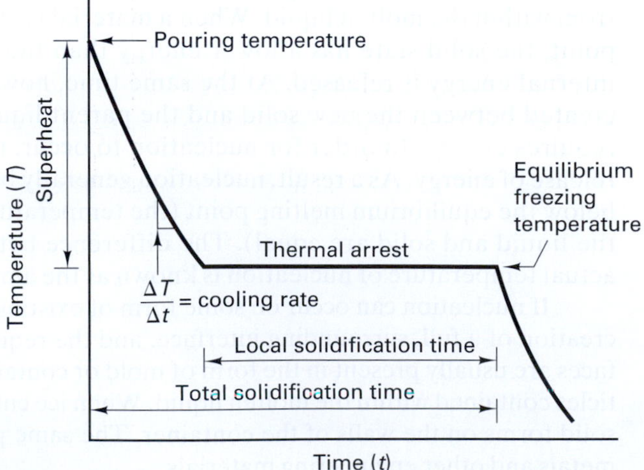

FIGURE 11-3 Cooling curve for a pure metal or eutectic-composition alloy (metals with a distinct freezing point), indicating major features related to solidification.

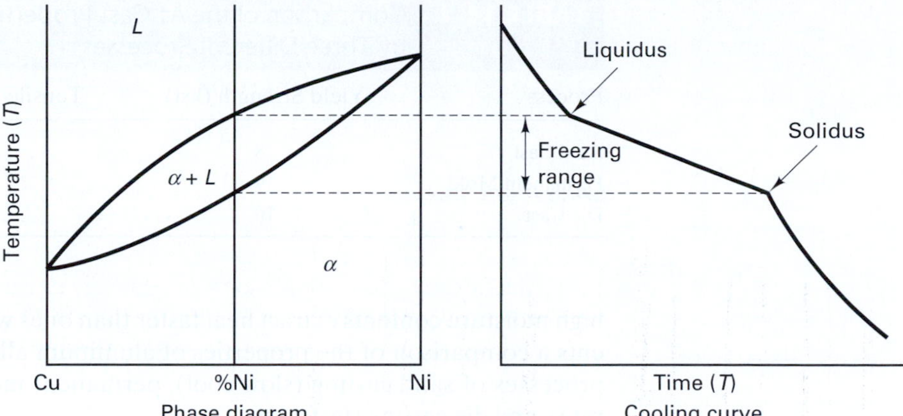

FIGURE 11-4 Phase diagram and companion cooling curve for an alloy with a freezing range. The slope changes indicate the onset and termination of solidification.

the **solidus** temperature is the highest temperature where it is all solid. The region between the liquidus and solidus temperatures is known as the **freezing range.** The onset and termination of solidification appear as slope changes in the cooling curve. The actual form of a cooling curve will depend on the type of material being poured, the nature of the nucleation process, and the rate and means of heat removal from the mold. By analyzing experimental cooling curves, we can gain valuable insight into both the casting process and the cast product. Fast cooling rates and short solidification times generally lead to finer structures and improved mechanical properties.

PREDICTION OF SOLIDIFICATION TIME: CHVORINOV'S RULE

The amount of heat that must be removed from a casting to cause it to solidify depends on both the amount of superheating and the volume of metal in the casting. Conversely, the ability to remove heat from a casting is directly related to the amount of exposed surface area through which the heat can be extracted and the environment surrounding the molten material (i.e., the mold and mold surroundings). These observations are reflected in **Chvorinov's rule,**[1] which states that the total solidification time, t_s, can be computed by:

$$t_s = B\,(V/A)^n \quad \text{where } n = 1.5 \text{ to } 2.0 \tag{11-1}$$

The total solidification time, t_s, is the time from pouring to the completion of solidification; V is the volume of the casting; A is the surface area through which heat is extracted; and B is the **mold constant.** The mold constant, B, incorporates the characteristics of the metal being cast (heat capacity and heat of fusion), the mold material (heat capacity and thermal conductivity), the mold thickness, initial mold temperature, and the amount of superheat.

Test specimens can be cast to determine the value of B for a given mold material, casting material, and condition of casting. This value can then be used to compute the solidification times for other castings made under the same conditions. Because a riser and casting both lie within the same mold and fill with the same metal under the same conditions, Chvorinov's rule can be used to compare the solidification times of each, and thereby ensure that the riser will solidify after the casting. This condition is absolutely essential if the liquid within the riser is to effectively feed the casting and compensate for solidification shrinkage. Aspects of riser design, including the use of Chvorinov's rule, will be developed later in this chapter.

Different cooling rates and solidification times can produce substantial variation in the structure and properties of the resulting casting. Die casting, for example, uses water-cooled metal molds, and the faster cooling produces higher-strength products than sand casting, where the mold material is more thermally insulating. Even variations in the type and condition of sand can produce different cooling rates. Sands with

[1] N. Chvorinov, "Theory of Casting Solidification", *Giesserei*, vol. 27 (1940), pp. 177–180, 201–208, 222–225.

TABLE 11-1 Comparison of the As-Cast Properties of Alloy 443 Aluminum Cast by Three Different Processes

Process	Yield Strength (ksi)	Tensile Strength (ksi)	Elongation (%)
Sand Cast	8	19	8
Permanent Mold	9	23	10
Die Cast	16	33	9

high moisture contents extract heat faster than ones with low moisture. Table 11-1 presents a comparison of the properties of aluminum alloy 443 cast by the three different processes of sand casting (slow cool), permanent mold casting (intermediate cooling rate), and die casting (fast cool).

THE CAST STRUCTURE

The products that result when molten metal is poured into a mold and permitted to solidify may have as many as three distinct regions or zones. The rapid nucleation that occurs when molten metal contacts the cold mold walls results in the production of a **chill zone,** a narrow band of randomly oriented crystals on the surface of a casting. As additional heat is removed, the grains of the chill zone begin to grow inward, and the rate of heat extraction and solidification decreases. Because most crystals have directions of rapid growth, a selection process begins. Crystals with rapid-growth direction perpendicular to the casting surface grow fast and shut off adjacent grains whose rapid-growth direction is at some intersecting angle.

The favorably oriented crystals continue to grow, producing the long, thin columnar grains of a **columnar zone.** The properties of this region are highly directional because the selection process has converted the purely random structure of the surface into one of parallel crystals of similar orientation. Figure 11-5 shows a cast structure containing both chill and columnar zones.

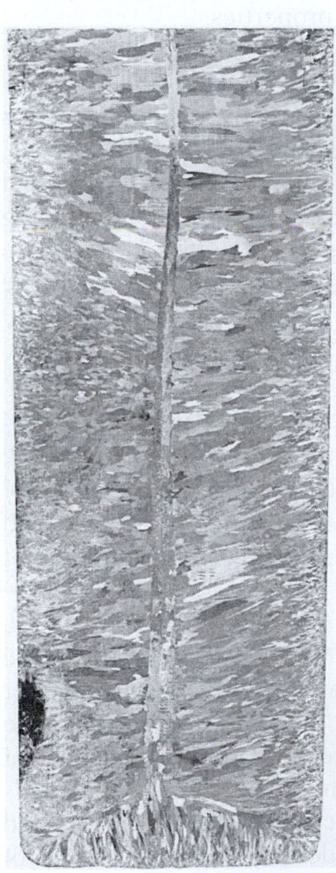

FIGURE 11-5 Internal structure of a cast metal bar showing the chill zone at the periphery, columnar grains growing toward the center, and a central shrinkage cavity. *(Courtesy Ronald Kohser)*

In many materials, new crystals then nucleate in the interior of the casting and grow to produce another region of spherical, randomly oriented crystals, known as the **equiaxed zone.** Low pouring temperatures, alloy additions, and the addition of inoculants can be used to promote the formation of this region, whose isotropic properties (uniform in all directions) are far more desirable than those of columnar grains.

MOLTEN METAL PROBLEMS

Castings begin with molten metal, and there are a number of chemical reactions that can occur between molten metal and its surroundings. These reactions and their products can often lead to defects in the final casting. For example, oxygen and molten metal can react to produce metal oxides (a nonmetallic or ceramic material), which can then be carried with the molten metal during the pouring and filling of the mold. Known as **dross** or **slag,** this material can become trapped in the casting and impair surface finish, machinability, and mechanical properties. Material eroded from the linings of furnaces and pouring ladles, and loose sand particles from the mold surfaces, can also contribute nonmetallic components to the casting.

Dross and slag can be controlled by using special precautions during melting and pouring and by good mold design. Lower pouring temperatures or superheat slows the rate of dross-forming reactions. Fluxes can be used to cover and protect molten metal during melting, or the melting and pouring can be performed under a vacuum or protective atmosphere. Measures can be taken to agglomerate the dross and cause it to float to the surface of the metal, where it can be skimmed off prior to pouring. Special ladles can be used which extract metal from beneath the surface, such as those depicted in Figure 11-6. Gating systems can be designed to trap any dross, sand or eroded mold material and keep it from flowing into the mold cavity. In addition, ceramic **filters** can be inserted into the feeder channels of the mold. These filters are available in a variety of shapes, sizes, and materials.

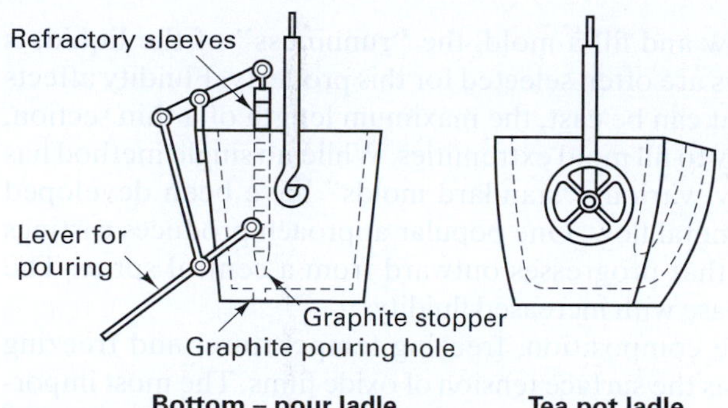

FIGURE 11-6 Two types of ladles used to pour castings. Note how each extracts molten material from the bottom, avoiding transfer of the impure material from the top of the molten pool.

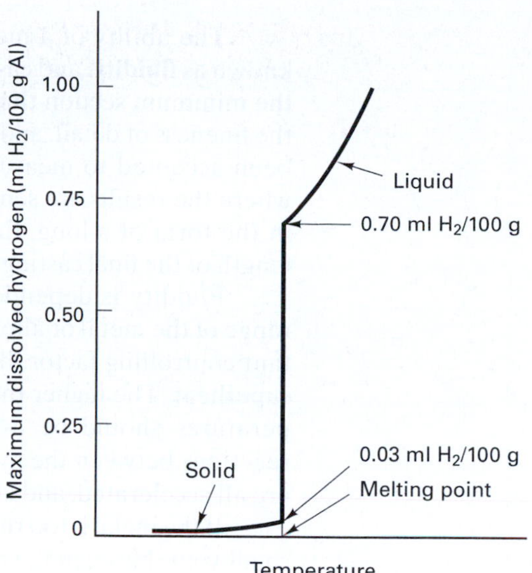

FIGURE 11-7 The maximum solubility of hydrogen in aluminum as a function of temperature.

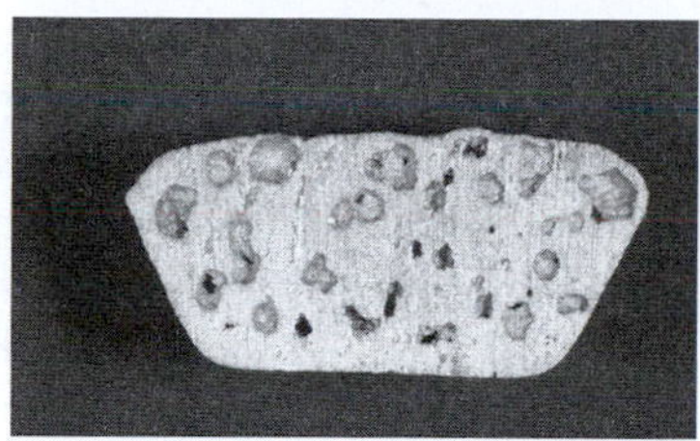

FIGURE 11-8 Demonstration casting made from aluminum that has been saturated in dissolved hydrogen. Note the extensive gas porosity. *(Courtesy Ronald Kohser)*

Liquid metals can also contain significant amounts of dissolved gas. When these materials solidify, the solid structure cannot accommodate the gas, and the rejected atoms form bubbles or gas porosity within the casting. Figure 11-7 shows the maximum solubility of hydrogen in aluminum as a function of temperature. Note the substantial decrease that occurs as the material goes from liquid to solid. Figure 11-8 shows a small demonstration casting that has been made from aluminum that has been saturated with dissolved hydrogen.

Several techniques can be used to prevent or minimize the formation of gas porosity. One approach is to prevent the gas from initially dissolving in the molten metal. Melting can be performed under vacuum, in an environment of low-solubility gases, or under a protective flux that excludes contact with the air. Superheat temperatures can be kept low to minimize gas solubility. In addition, careful handling and pouring can do much to control the flow of molten metal and minimize the turbulence that brings air and molten metal into contact.

Another approach is to remove the gas from the molten metal before it is poured into castings. **Vacuum degassing** sprays the molten metal through a low-pressure environment. Spraying creates a large amount of surface area, and the amount of dissolved gas is reduced as the material seeks to establish equilibrium with its new surroundings. (See a discussion of Sievert's law in any basic chemistry text.) Passing small bubbles of inert or reactive gas through the melt, known as **gas flushing,** can also be effective. In seeking equilibrium, the dissolved gases enter the flushing gas and are carried away. Bubbles of nitrogen or chlorine, for example, are particularly effective in removing hydrogen from molten aluminum. Ultrasonic vibrations, alone or with an assist gas, have been shown to be quite effective in degassing aluminum alloys.

The dissolved gas can also be reacted with something to produce a low-density compound, which then floats to the surface and can be removed with the dross or slag. Oxygen can be removed from copper by the addition of phosphorus. Steels can be deoxidized with addition of aluminum or silicon. The resulting phosphorus, aluminum, or silicon oxides are then removed by skimming, or are left on the top of the container as the remaining high-quality metal is extracted from beneath the surface.

FLUIDITY AND POURING TEMPERATURE

When molten metal is poured to produce a casting, it should first *flow* into all regions of the mold cavity and then *freeze* into this new shape. It is vitally important that these two functions occur in the proper sequence. If the metal begins to freeze before it has completely filled the mold, defects known as **misruns** and **cold shuts** are produced.

The ability of a metal to flow and fill a mold, the "runniness" of the liquid, is known as **fluidity,** and casting alloys are often selected for this property. Fluidity affects the minimum section thickness that can be cast, the maximum length of a thin section, the fineness of detail, and the ability to fill mold extremities. While no single method has been accepted to measure fluidity, various "standard molds" have been developed where the results are sensitive to metal flow. One popular approach produces castings in the form of a long, thin spiral that progresses outward from a central sprue. The length of the final casting will increase with increased fluidity.

Fluidity is dependent on the composition, freezing temperature, and freezing range of the metal or alloy, as well as the surface tension of oxide films. The most important controlling factor, however, is usually the pouring temperature or the amount of superheat. The higher the pouring temperature, the higher the fluidity. Excessive temperatures should be avoided, however. At high pouring temperatures, chemical reactions between the metal and the mold, and the metal and its pouring atmosphere, are all accelerated, and larger amounts of gas can be dissolved.

If the metal is too runny, it may not only fill the mold cavity but may also flow into the small voids between the particles that compose a sand mold. The surface of the resulting casting then contains small particles of embedded sand, a defect known as **penetration.**

THE ROLE OF THE GATING SYSTEM

When molten metal is poured into a mold, the gating system conveys the material and delivers it to all sections of the mold cavity. The speed or rate of metal movement is important as well as the amount of cooling that occurs while it is flowing. Slow filling and high loss of heat can result in misruns and cold shuts. Rapid rates of filling, on the other hand, can produce erosion of the gating system and mold cavity and might result in the entrapment of mold material in the final casting. It is imperative that the cross-sectional areas of the various channels be selected to regulate flow. The shape and length of the channels affect the amount of temperature loss. When heat loss is to be minimized, short channels with round or square cross sections (minimum surface area) are the most desirable. The gates are usually attached to the thickest or heaviest sections of a casting to control shrinkage and to the bottom of the casting to minimize turbulence and splashing. For large castings, multiple gates and runners may be used to introduce metal to more than one point of the mold cavity.

Gating systems should be designed to minimize **turbulent flow,** which tends to promote absorption of gases, oxidation of the metal, and erosion of the mold. Figure 11-9 shows a typical gating system for a mold with a horizontal parting line and can be used to identify some of the key components that can be optimized to promote the smooth flow of molten metal. Short sprues are desirable because they minimize the

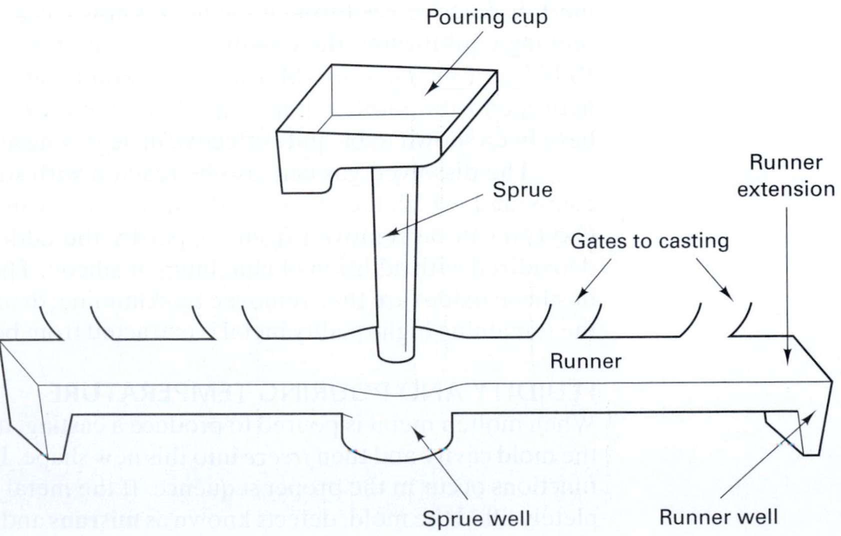

FIGURE 11-9 Typical gating system for a horizontal-parting-plane mold, showing key components involved in controlling the flow of metal into the mold cavity.

distance that the metal must fall when entering the mold and the kinetic energy that the metal acquires during that fall. Rectangular pouring cups prevent the formation of a vortex or spiraling funnel, which tends to suck gas and oxides into the sprue. Tapered sprues also prevent vortex formation. A large **sprue well** can be used to dissipate the kinetic energy of the falling stream and prevent splashing and turbulence as the metal makes the turn into the runner.

The **choke,** or smallest cross-sectional area in the gating system, serves to control the rate of metal flow. If the choke is located near the base of the sprue, flow through the runners and gates is slowed and flow is rather smooth. If the choke is moved to the gates, the metal might enter the mold cavity with a fountain effect, an extremely turbulent mode of flow, but the small connecting area would enable easier separation of the casting and gating system.

Gating systems can also be designed to trap dross and sand particles and keep them from entering the mold cavity. Given sufficient time, the lower-density contaminants will rise to the top of the molten metal. Long, flat runners can be beneficial (but these promote cooling of the metal), as well as gates that exit from the lower portion of the runners. Because the first metal to enter the mold is most likely to contain the foreign matter (dross from the top of the pouring ladle and loose particles washed from the walls of the gating system), **runner extensions** and **runner wells** (see Figure 11-9) can be used to catch and trap this first metal and keep it from entering the mold cavity. These features are particularly effective with aluminum castings because aluminum oxide has approximately the same density as molten aluminum.

Screens or ceramic filters of various shapes, sizes, and materials can also be inserted into the gating system to trap foreign material. Wire mesh can often be used with the nonferrous metals, but ceramic materials are generally required for irons and steel. Figure 11-10 shows several ceramic filters and depicts the two basic types—extruded and foam. The pores on the extruded ceramics are uniform in size and shape and provide parallel channels. The foams contain interconnected pores of various size and orientation, forcing the material to change direction as it negotiates its passage

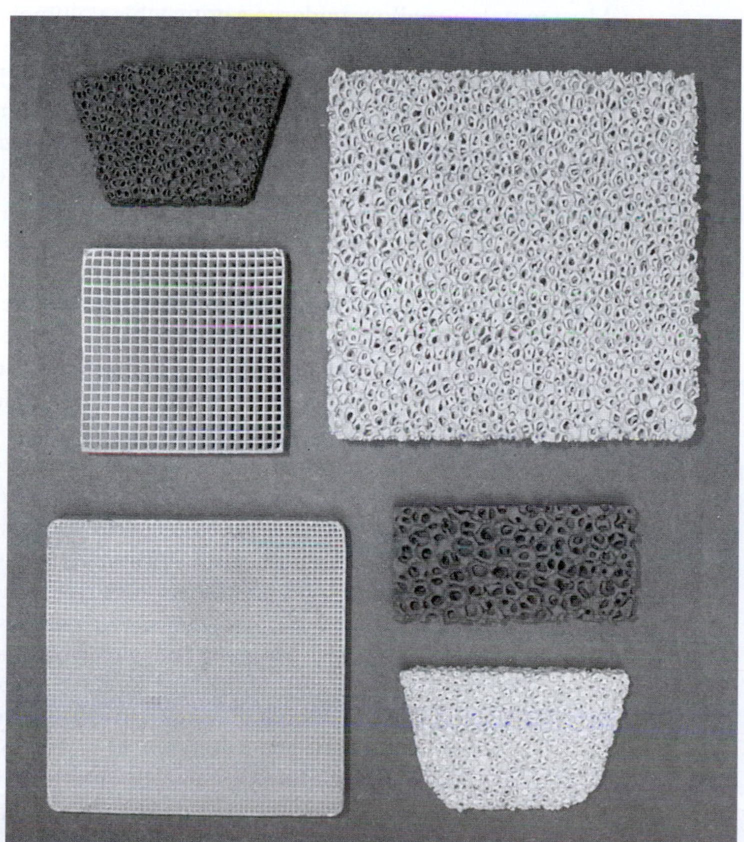

FIGURE 11-10 Various types of ceramic filters that may be inserted into the gating systems of metal castings. *(Courtesy Ronald Kohser)*

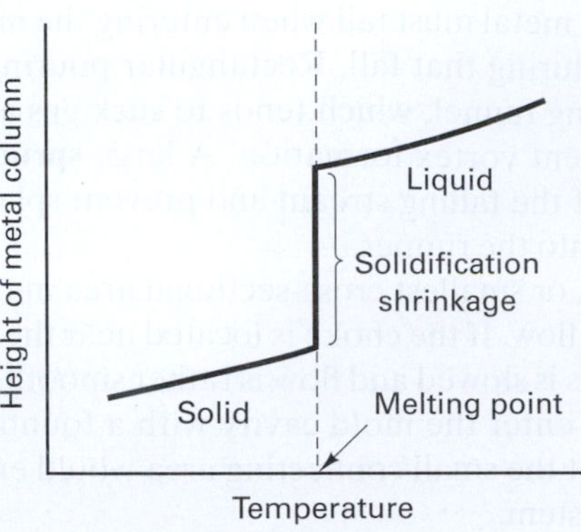

FIGURE 11-11 Dimensional changes experienced by a metal column as the material cools from a superheated liquid to a room-temperature solid. Note the significant shrinkage that occurs upon solidification.

through the filter. Contaminant removal can be by either particle entrapment or by a wetting action, whereby the nonmetallic contaminant adheres to the filter surface while the metal does not wet and flows freely through.

Because these devices can also restrict the fluid velocity, streamline the fluid flow, or reduce turbulence, proper placement is an important consideration. To ensure removal of both dross and eroded sand, the filter should be as close to the mold cavity as possible, but because a filter can also act as the choke, it may be positioned at other locations, such as the base of the pouring cup, base of the sprue, or in one or more of the runners.

The specific details of a gating system often vary with the metal being cast. Turbulent-sensitive metals (such as aluminum and magnesium), and alloys with low melting points, generally employ gating systems that concentrate on eliminating turbulence and trapping dross. Turbulent-insensitive alloys (such as steel, cast iron, and most copper alloys), and alloys with a high melting point, generally use short, open gating systems that provide for quick filling of the mold cavity.

SOLIDIFICATION SHRINKAGE

Most metals and alloys undergo a noticeable volumetric contraction once they enter the mold cavity and begin to cool. Figure 11-11 shows the typical changes experienced by a metal column as the material goes from superheated liquid to room-temperature solid. There are three principal stages of **shrinkage:** (1) *shrinkage of the liquid* as it cools to the temperature where solidification begins, (2) *solidification shrinkage* as the liquid turns into solid, and (3) *solid metal contraction* as the solidified material cools to room temperature.

The amount of liquid metal contraction depends on the coefficient of thermal contraction (a property of the metal being cast) and the amount of superheat. Liquid contraction is rarely a problem, however, because the metal in the gating system continues to flow into the mold cavity as the liquid already in the cavity cools and contracts.

As the metal changes state from liquid to crystalline solid, the new atomic arrangement is usually more efficient, and significant amounts of shrinkage can occur. The actual amount of shrinkage varies from alloy to alloy, as shown in Table 11-2. As indicated in that table, not all metals contract upon solidification. Some actually expand, such as gray cast iron, where low-density graphite flakes form as part of the solid structure.

When solidification shrinkage does occur, however, it is important to control the form and location of the resulting void. Metals and alloys with short freezing ranges, such as pure metals and eutectic alloys, tend to form large cavities or pipes. These can be avoided by designing the casting to have directional solidification where freezing begins farthest away from the feed gate or riser and moves progressively toward it. As the metal solidifies and shrinks, the shrinkage void is continually filled with additional liquid metal. When the flow of additional liquid is exhausted and solidification is complete, we hope that the final shrinkage void is located external to the desired casting in either the riser or the gating system.

TABLE 11-2 Solidification Shrinkage of Some Common Engineering Metals (Expressed in Percent)

Aluminum	6.6
Copper	4.9
Magnesium	4.0
Zinc	3.7
Low-carbon steel	2.5–3.0
High-carbon steel	4.0
White cast iron	4.0–5.5
Gray cast iron	−1.9

Alloys with large freezing ranges have a period of time when the material is in a slushy (liquid plus solid) condition. As the material cools between the liquidus and solidus, the relative amount of solid increases and tends to trap small, isolated pockets of liquid. It is almost impossible for additional liquid to feed into these locations, and the resultant casting tends to contain small but numerous shrinkage pores dispersed throughout. This type of shrinkage is far more difficult to prevent by means of gating and risering, and a porous product may be inevitable. If a gas- or liquid-tight product is required, these castings may need to be impregnated (the pores filled with a resinous material or lower-melting-temperature metal) in a subsequent operation. Castings with dispersed porosity tend to have poor ductility, toughness, and fatigue life.

After solidification is complete, the casting will contract further as it cools to room temperature. This solid metal contraction is often called *patternmaker's contraction* because compensation for these dimensional changes should be made when the mold cavity or pattern is designed. Examples of these compensations will be provided later in this chapter. Concern arises, however, when the casting is produced in a rigid mold, such as the metal molds used in die casting. If the mold provides constraint during the time of contraction, tensile forces can be generated within the hot, weak casting, and cracking can occur **(hot tears).** It is often desirable, therefore, to eject the hot castings as soon as solidification is complete.

RISERS AND RISER DESIGN

Risers are added reservoirs designed to fill with liquid metal, which is then fed to the casting as a means of compensating for solidification shrinkage. To effectively perform this function, the risers must solidify after the casting. If the reverse were true, liquid metal would flow from the casting toward the solidifying riser and the casting shrinkage would be even greater. Hence, castings should be designed to produce directional solidification that sweeps from the extremities of the mold cavity toward the riser. In this way, the riser can continuously feed molten metal and will compensate for the solidification shrinkage of the entire mold cavity. Figure 11-12 shows a three-level step block cast in aluminum with and without a riser. Note that the riser is positioned so directional solidification moves from thin to thick and the shrinkage void is moved from

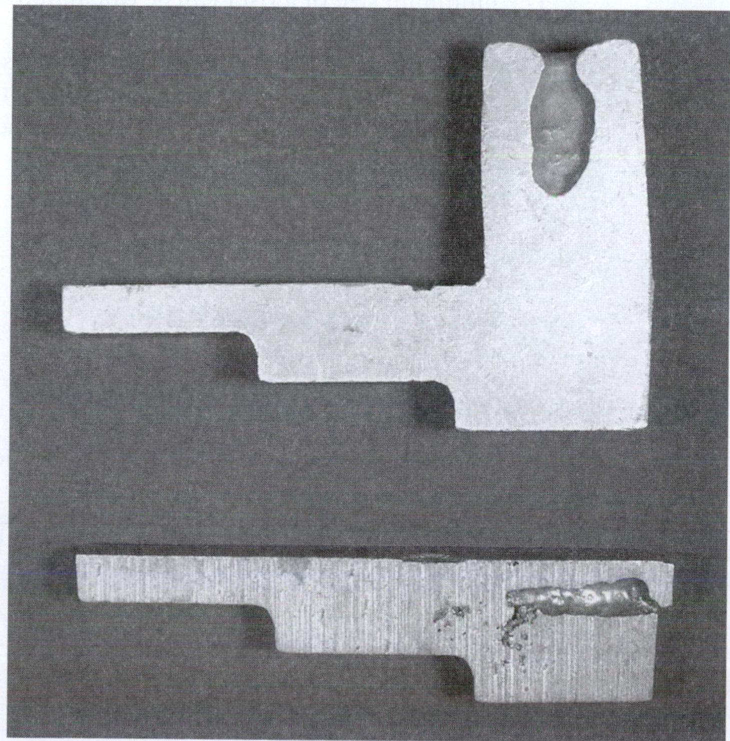

FIGURE 11-12 A three-tier step-block aluminum casting made with (top) and without (bottom) a riser. Note how the riser has moved the shrinkage void external to the desired casting. *(Courtesy Ronald Kohser)*

the casting to the riser. If a single directional solidification is not possible, multiple risers may be required, with various sections of the casting each solidifying toward their respective riser.

The risers should also be designed to conserve metal. If we define the **yield** of a casting as the casting weight divided by the total weight of metal poured (complete gating system, risers, and casting), it is clear that there is a motivation to make the risers as small as possible, yet still able to perform their task. This is usually done through proper consideration of riser size, shape, and location, as well as the type of connection between the riser and casting.

A good shape for a riser would be one that has a long freezing time. According to Chvorinov's rule, this would favor a shape with small surface area per unit volume. While a sphere would make the most efficient riser, this shape presents considerable difficulty to both patternmaker and moldmaker. The most popular shape for a riser, therefore, is a cylinder, where the height-to-diameter ratio is varied depending on the nature of the alloy being cast, the location of the riser, the size of the flask, and other variables. A 1-to-1 height-to-diameter ratio is generally considered to be ideal.

Risers should be located so that directional solidification occurs from the extremities of the mold cavity back toward the riser. Because the thickest regions of a casting will be the last to freeze, risers should feed directly into these locations. Various types of risers are possible. A **top riser** is one that sits on top of a casting. Because of their location, top risers have shorter feeding distances and occupy less space within the flask. They give the designer more freedom for the layout of the pattern and gating system. **Side risers** are located adjacent to the mold cavity, displaced horizontally along the parting line. Figure 11-13 depicts both a top and a side riser. If the riser is contained entirely within the mold, it is known as a **blind riser.** If it is open to the atmosphere, it is called an **open riser.** Blind risers are usually larger than open risers because of the additional heat loss that occurs where the top of the riser is in contact with mold material.

Live risers (also known as *hot risers*) receive the last hot metal that enters the mold and generally do so at a time when the metal in the mold cavity has already begun to cool and solidify. Thus, they can be smaller than **dead** (or *cold*) **risers,** which fill with metal that has already flowed through the mold cavity. As shown in Figure 11-13, top risers are almost always dead risers. Risers that are part of the gating system are generally live risers.

The minimum size of a riser can be calculated from Chvorinov's rule by setting the total solidification time for the riser to be greater than the total solidification time for the casting. Because both cavities receive the same metal and are in the same mold, the mold constant, B, will be the same for both regions. Assuming that $n = 2$, and a safe difference in solidification time is 25% (the riser takes 25% longer to solidify than the casting), we can write this condition as

$$t_{\text{riser}} = 1.25\, t_{\text{casting}} \tag{11-2}$$

or

$$(V/A)^2_{\text{riser}} = 1.25\, (V/A)^2_{\text{casting}} \tag{11-3}$$

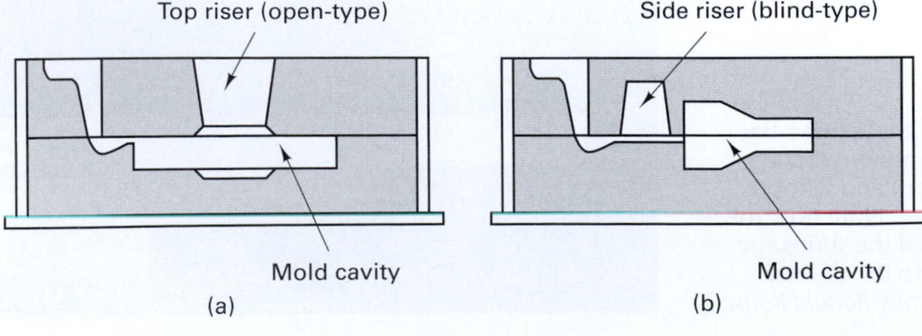

FIGURE 11-13 Schematic of a sand casting mold, showing (a) an open-type top riser and (b) a blind-type side riser (right). The side riser is a live riser, receiving the last hot metal to enter the mold. The top riser is a dead riser, receiving metal that has flowed through the mold cavity.

Calculation of the riser size then requires selection of a riser geometry, which is generally cylindrical. For a cylinder of diameter D and height H, the volume and surface area can be written as:

$$V = \pi D^2 H/4$$
$$A = \pi DH + 2(\pi D^2/4)$$

Selecting a specific height-to-diameter ratio for the riser then enables equation 11-3 to be written as a simple expression with one unknown, D. The volume-to-area ratio for the casting is computed for its particular geometry, and equation 11-3 can then be solved to provide the size of the required riser. One should note that if the riser and casting share a surface, as with a blind top riser, the area of the common surface should be subtracted from both components because it will not be a surface of heat loss to either. It should also be noted that there are a number of methods to calculate riser size. The Chvorinov's rule method is the only one presented here.

A final aspect of riser design is the connection between the riser and the casting. Because the riser must ultimately be separated from the casting, it is desirable that the connection area be as small as possible. On the other hand, the connection area must be sufficiently large so that the link does not freeze before solidification of the casting is complete. If the risers are placed close to the casting with relatively short connections, the mold material surrounding the link will receive heat from both the casting and the riser. It should heat rapidly and remain hot throughout the cast, thereby preventing solidification of the metal in the channel.

RISERING AIDS

Various methods have been developed to assist the risers in performing their job. Some are intended to promote directional solidification, while others seek to reduce the number and size of the risers, thereby increasing the yield of a casting. These techniques generally work by either speeding the solidification of the casting **(chills)** or retarding the solidification of the riser (**sleeves** or toppings).

External chills are masses of high-heat-capacity, high-thermal-conductivity material (such as steel, iron, graphite, or copper) that are placed in the mold, adjacent to the casting, to absorb heat and accelerate the cooling of various regions. Chills can effectively promote directional solidification or increase the effective feeding distance of a riser. They can also be used to reduce the number of risers required for a casting. External chills are frequently covered with a protective wash, silica flour, or other refractory material to prevent bonding with the casting.

Internal chills are pieces of metal that are placed within the mold cavity to absorb heat and promote more rapid solidification. When the molten metal of the pour surrounds the chill, it absorbs heat as it seeks to come to equilibrium with its surroundings. Internal chills ultimately become part of the final casting, so they must be made from an alloy that is the same or compatible with the alloy being cast.

The cooling of risers can be slowed by methods that include (1) switching from a blind riser to an open riser, (2) placing **insulating sleeves** around the riser, and (3) surrounding the sides or top of the riser with exothermic material that supplies added heat to just the riser segment of the mold. The objective of these techniques is generally to reduce the riser size rather than promote directional solidification.

It is important to note that risers are not always necessary or functional. For alloys with large freezing ranges, risers would not be particularly effective, and one generally accepts the fine, dispersed porosity that results. For processes such as die casting, low-pressure permanent molding, and centrifugal casting, the positive pressures associated with the process provide the feeding action that is required to compensate for solidification shrinkage.

11.5 Patterns

Casting processes can be divided into two basic categories: (1) those for which a new mold must be created for each casting (the **expendable-mold processes**), and (2) those that employ a permanent, **reusable mold.** Most of the expendable mold processes begin with some form of reusable pattern—a duplicate of the part to be cast, modified dimensionally to reflect both the casting process and the material being cast. Patterns can be made from wood, metal, foam or plastic, with urethane now being the material of choice for nearly half of all casting patterns.

The dimensional modifications that are incorporated into a pattern are called **allowances,** and the most important of these is the **shrinkage allowance.** Following solidification, a casting continues to contract as it cools to room temperature, the amount of this contraction being as much as 2%, or $\frac{1}{4}$ in./ft. To produce the desired final dimensions, the pattern (which sets the dimensions upon solidification) must be slightly larger than the room-temperature casting. The exact amount of this shrinkage compensation depends on the metal that is being cast, and can be estimated by the equation:

$$\Delta \text{length} = \text{length}\, \alpha\, \Delta T, \quad (11\text{-}4)$$

where α is the coefficient of thermal expansion and ΔT is the difference between the freezing temperature and room temperature. Typical allowances for some common engineering metals are:

Cast iron	0.8–1.0%
Steel	1.5–2.0%
Aluminum	1.0–1.3%
Magnesium	1.0–1.3%
Brass	1.5%

Shrinkage allowances are often incorporated into a pattern through use of special **shrink rules**—measuring devices that are larger than a standard rule by an appropriate shrink allowance. For example, a shrink rule for brass would designate 1 ft as a length that is actually 1 ft $\frac{3}{16}$ in., because the anticipated 1.5% shrinkage will reduce the length by $\frac{3}{16}$ in. A complete pattern made to shrink rule dimensions will produce a proper size casting after cooling.

Caution should be exercised when using shrink rule compensations, however, because thermal contraction may not be the only factor affecting the final dimensions. The various phase transformations discussed in Chapter 5 are often accompanied by significant dimensional expansions or contractions. Examples include eutectoid reactions, martensitic reactions, and graphitization.

In many casting processes, mold material is formed around the pattern and the pattern is then extracted to create the mold cavity. To facilitate pattern removal, molds are often made in two or more sections that separate along mating surfaces called the parting line or parting plane. A flat parting line is usually preferred, but the casting design or molding practice may dictate the use of irregular or multiple parting surfaces. In general, the best parting line will be a flat plane that allows for proper metal flow into the mold cavity, requires the fewest cores and molding steps, and provides adequate core support and venting.

If the pattern contains surfaces that are perpendicular to the parting line (parallel to the direction of pattern withdrawal), friction between the pattern and the mold material as well as any horizontal movement of the pattern during extraction could induce damage to the mold. This damage could be particularly severe at the corners where the mold cavity intersects the parting surface. Such extraction damage can be minimized by incorporating a slight taper, or draft, on all pattern surfaces that are parallel to the direction of withdrawal. A slight withdrawal of the pattern will free it from the mold material on all surfaces, and it can then be further removed without damage to the mold. Figure 11-14 illustrates the use of draft to facilitate pattern removal.

The size and shape of the pattern, the depth of the mold cavity, the method used to withdraw the pattern, the pattern material, the mold material, and the molding procedure all influence the actual amount of draft required. Draft is seldom less than 1 degree or $\frac{1}{8}$ in./ft., with a minimum taper of about $\frac{1}{16}$ in. over the length of any surface. Because draft allowances increase the size of a pattern (and thus the size and weight of a casting),

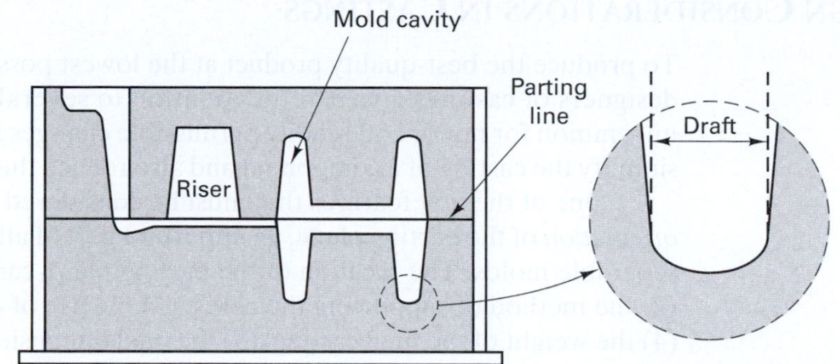

FIGURE 11-14 Two-part mold showing the parting line and the incorporation of a draft allowance on vertical surfaces.

it is generally desirable to keep them to the minimum that will permit satisfactory pattern removal. Molding procedures that produce higher strength molds and the use of mechanical pattern withdrawal can often enable reductions in draft allowances. By reducing the taper, casting weight and the amount of subsequent machining can both be reduced.

When smooth machined surfaces are required, it may be necessary to add an additional **machining allowance,** or **finish allowance,** to the pattern. The amount of this allowance depends to a great extent on the casting process and the mold material. Ordinary sand castings have rougher surfaces than those of shell-mold castings. Die castings have smooth surfaces that may require little or no metal removal, and the surfaces of investment castings are even smoother. It is also important to consider the location of the desired machining and the presence of other allowances, because the draft allowance may provide part or all of the extra metal needed for machining.

Some casting shapes require yet an additional allowance for **distortion.** Consider a U-shaped section where the arms are restrained by the mold at a time when the base of the U is shrinking. The result will be a final casting with outwardly sloping arms. If the design is modified to have the arms originally slope inward, the subsequent distortion will produce the desired final shape. Distortion depends greatly on the particular configuration of the casting, and casting designers must use experience and judgment to provide an appropriate distortion allowance.

Figure 11-15 illustrates the manner in which the various allowances are incorporated into a casting pattern. Similar allowances are applied to the cores that create the holes or interior passages of a casting.

If a casting is to be made in a multi-use metal mold, all of the "pattern allowances" discussed earlier should be incorporated into the machined cavity. The dimensions of this cavity will further change, however, as sequential casts raise the mold temperature to a steady-state level. An additional correction should be added to compensate for this effect.

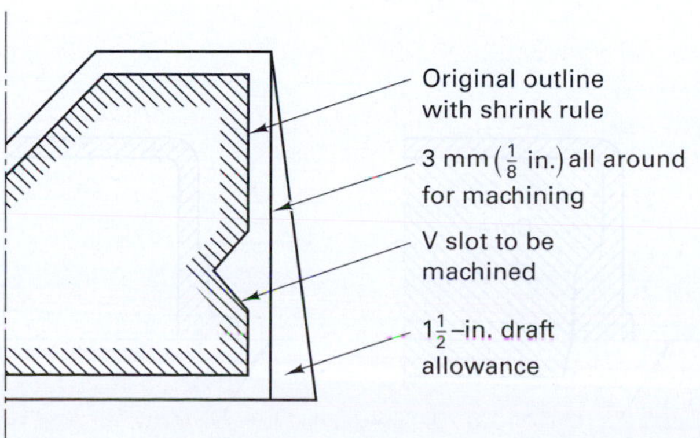

FIGURE 11-15 Various allowances incorporated into a casting pattern.

11.6 DESIGN CONSIDERATIONS IN CASTINGS

To produce the best-quality product at the lowest possible cost, it is important that the designers of castings give careful attention to several process requirements. It is not uncommon for minor and readily permissible changes in design to greatly facilitate and simplify the casting of a component and also reduce the number and severity of defects.

One of the first features that must be considered by a designer is the *location* and *orientation* of the parting plane, an important part of all processes that use segmented or separable molds. The location of the parting plane can affect (1) the number of cores, (2) the method of supporting the cores, (3) the use of effective and economical gating, (4) the weight of the final casting, (5) the final dimensional accuracy, and (6) the ease of molding.

In general, it is desirable to minimize the use of cores. A change in the location or orientation of the parting plane can often assist in this objective. The change illustrated in Figure 11-16 not only eliminates the need for a core but can also reduce the weight of the casting by eliminating the need for draft. Figure 11-17 shows another example of how a core can be eliminated by a simple design change. Figure 11-18 shows six different parting line arrangements for the casting of a simple ring. Arrangements (a) through (e) provide flat and parallel side faces with draft tapers on the inner and outer diameters. The (f) alternative requires the use of a core but might be preferred if draft cannot be tolerated on the inner and outer diameter surfaces. This figure also shows that simply noting the desired shape and the need to provide sufficient draft can provide considerable design freedom. Because mold closure may not always be consistent, consideration should also be given to the fact that dimensions across the parting plane are subject to greater variation than those that lie entirely within a given segment of the mold.

Controlling the solidification process is of prime importance in obtaining quality castings, and this control is also related to design. Those portions of a casting that have a high ratio of surface area to volume will experience more rapid cooling and will be stronger and harder than the other regions. Thicker or heavier sections will cool more slowly and may contain shrinkage cavities and porosity, or have weaker, large grain-size structures.

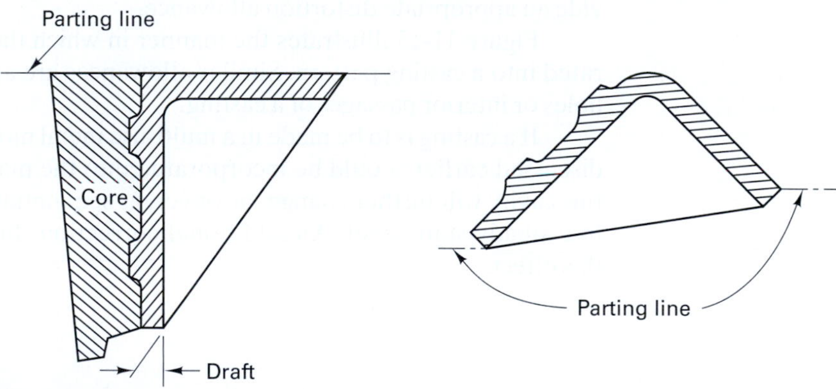

FIGURE 11-16 Elimination of a core by changing the location or orientation of the parting plane.

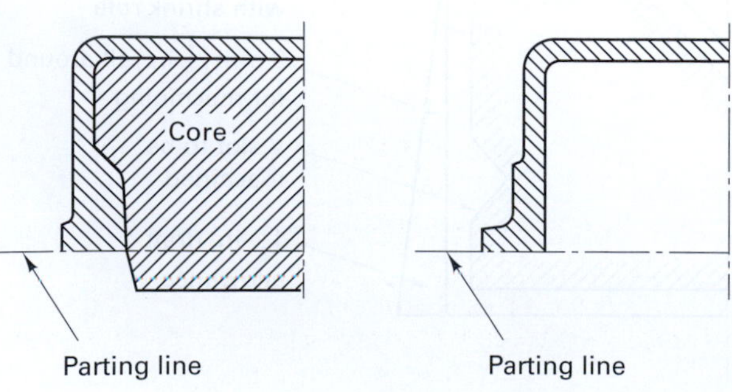

FIGURE 11-17 Elimination of a dry-sand core by a change in part design.

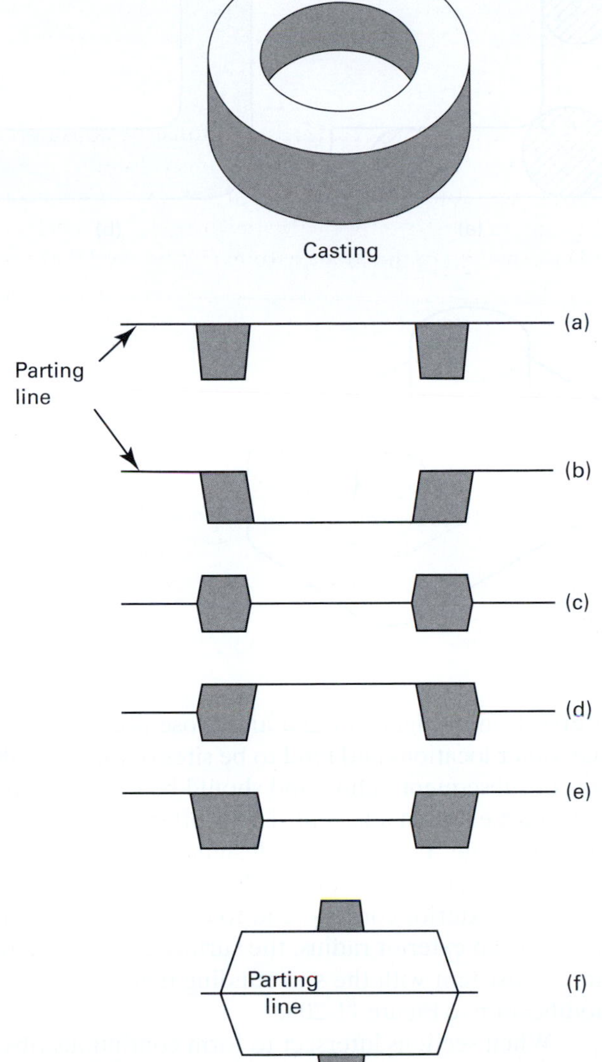

FIGURE 11-18 Multiple options to cast a simple ring with draft to the parting line. Evaluate the six options with respect to the following possible concerns: (1) flat and parallel side surfaces, (2) flat and parallel inner and outer diameters, (3) amount of material that must be removed if no tapers are allowed on any surfaces, and (4) possibility for nonuniform wall thickness if one of the mold segments is shifted with respect to the other.

Ideally, a casting should have uniform thickness at all locations. Instead of thicker sections, ribs or other geometric features can often be used to impart additional strength while maintaining uniform wall thickness. When the section thickness must change, it is best if these changes are gradual, as indicated in the recommendations of Figure 11-19.

When sections of castings intersect, as in Figure 11-20a, two problems can arise. The first of these is **stress concentration.** Generous **fillets** (inside radii) at all interior corners can better distribute stresses and help to minimize potential problems, including shrinkage cracks. If the fillets are excessive, however, the additional material can augment the second problem, known as **hot spots.** Thick sections, like those at the

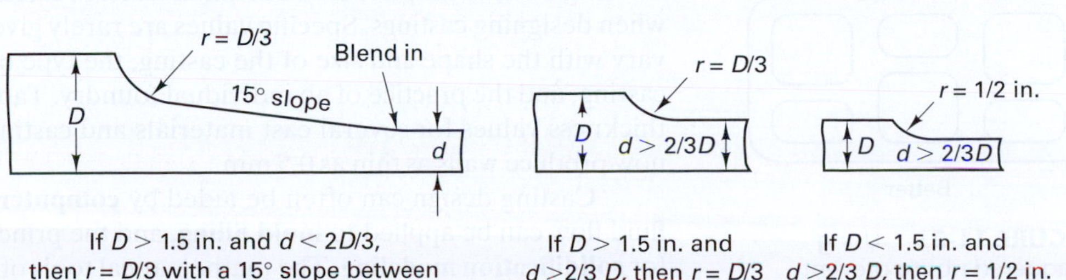

FIGURE 11-19 Typical guidelines for section change transitions in castings.

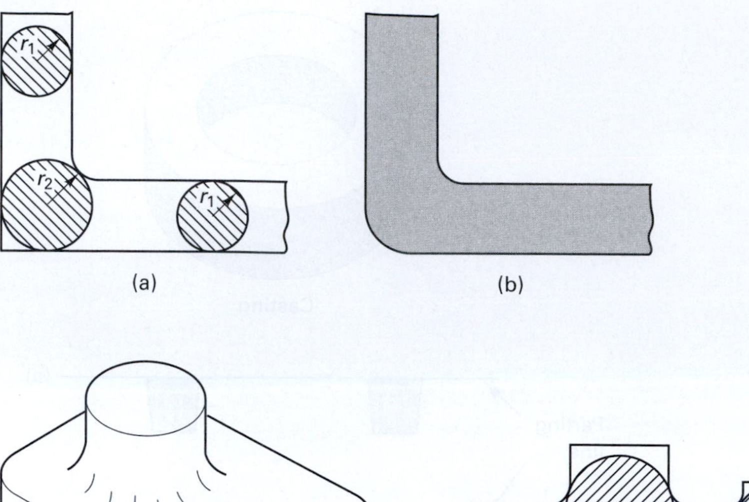

FIGURE 11-20 (a) The "hot spot" at section r_2 is caused by intersecting sections. (b) An interior fillet and exterior radius leads to more uniform thickness and more uniform cooling.

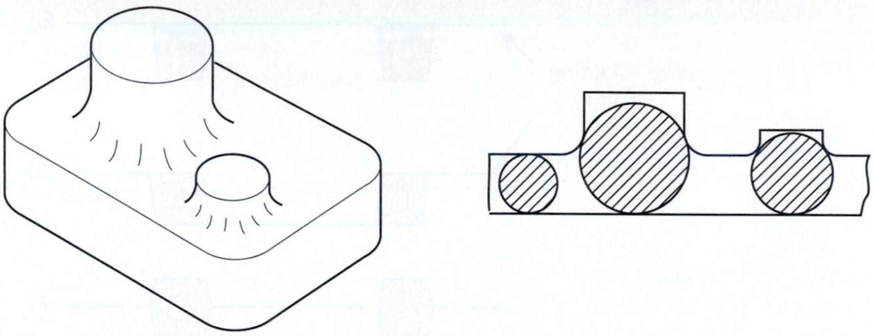

FIGURE 11-21 Hot spots often result from intersecting sections of various thickness.

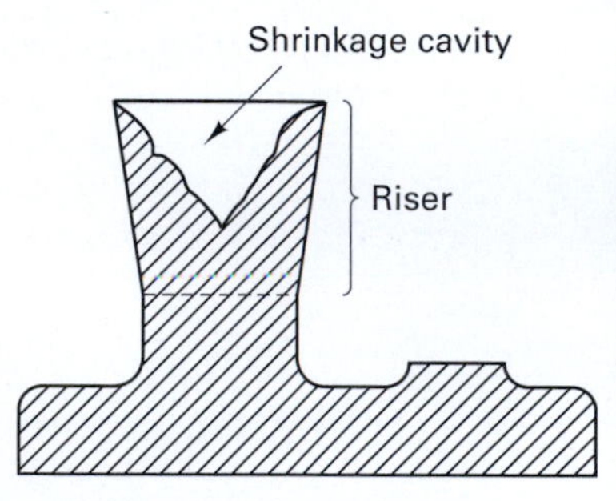

FIGURE 11-22 Attached risers can move the shrinkage cavity external to the actual casting.

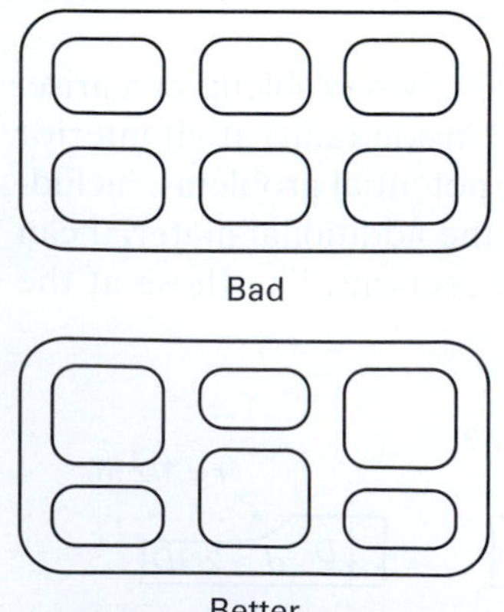

FIGURE 11-23 Using staggered ribs to prevent cracking during cooling.

intersection in Figure 11-20a and those illustrated in Figure 11-21, cool more slowly than other locations and tend to be sites of localized shrinkage. Shrinkage voids can be sites of subsequent failure and should be prevented if at all possible. Where thick sections must exist, an adjacent riser is often used to feed the section during solidification and shrinkage. If the riser is designed properly, the shrinkage cavity will lie totally within the riser, as illustrated in Figure 11-22, and will be removed when the riser is cut off. Sharp exterior corners tend to cool faster than the other sections of a casting. By providing an exterior radius, the surface area can be reduced and cooling slowed to be more consistent with the surrounding material. Figure 11-20b shows a recommended modification to Figure 11-20a.

When sections intersect to form continuous ribs, like those in Figure 11-23, contraction occurs in opposite directions as each of the arms cool and shrink and cracking may occur at the intersections, which are also local hot spots. By staggering the ribs, as shown in the second portion of Figure 11-23, the negative effects of thermal contraction and hot spots can be minimized.

The location of the parting line may also be an appearance consideration. A small amount of fin, or **flash,** is often present at the parting line, and when the flash is removed (or left in place if it is small enough), a line of surface imperfection results. If the location is in the middle of a flat surface, it will be clearly visible in the product. If the parting line can be moved to coincide with a corner, however, the associated "defect" will go largely unnoticed.

Thin-walled castings are often desired because of their reduced weight, but thin walls can often present manufacturing problems related to mold filling (premature freezing before complete fill). Minimum section thickness should always be considered when designing castings. Specific values are rarely given, however, because they tend to vary with the shape and size of the casting, the type of metal being cast, the method of casting, and the practice of an individual foundry. Table 11-3 presents typical minimum thickness values for several cast materials and casting processes. Zinc die casting can now produce walls as thin as 0.5 mm.

Casting design can often be aided by **computer simulation.** The mathematics of fluid flow can be applied to **mold filling,** and the principles of heat transfer can be used for **solidification modeling.** The mathematical tools of finite element or finite difference calculations can be coupled with the use of high-speed computers to permit beneficial

TABLE 11-3 Typical Minimum Section Thickness Values for Various Engineering Metals and Casting Processes

Casting Method	Minimum Section Thickness (mm.)		
	Aluminum	Magnesium	Steel
Sand Casting	3.18	3.96	4.75
Permanent Mold	2.36	3.18	—
Die Cast	1.57	2.36	—
Investment Cast	1.57	1.57	2.36
Plaster Mold	2.03	—	—

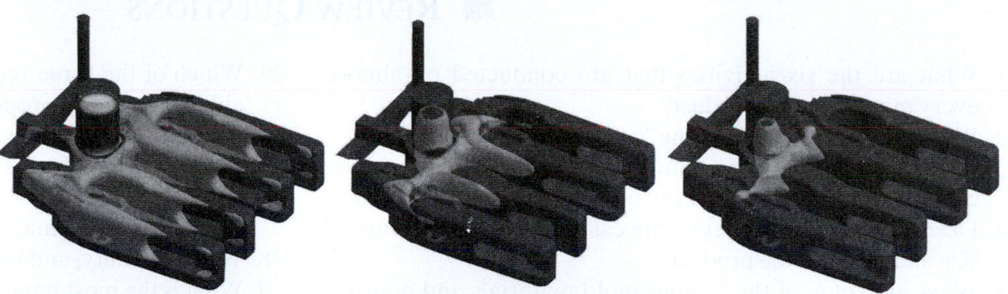

FIGURE 11-24 Computer model showing the progressive solidification of a cast steel mining shovel adapter. Molten metal appears light in the figures. As time passes (left to right), the material directionally solidifies back toward the riser at the left side of the casting. *(Copyright 2009, Giesseri-Verlag GmbH, Düsseldorf, Germany)*

design changes before the manufacture of patterns or molds. The computer model in Figure 11-24 shows the progressive solidification of a cast steel mining shovel adapter. Note the directional solidification back toward the riser at the left.

11.7 The Casting Industry

The U.S. metal casting industry ships more than 14 million pounds of castings every year, valued at more than $18 billion, with gray iron, ductile iron, aluminum alloys, and copper-base metals comprising the major portion. Metal castings form primary components in agricultural implements; construction equipment; mining equipment; valves and fittings; metalworking machinery; power tools; pumps and compressors; railroad equipment; power transmission equipment; and heating, refrigeration, and air conditioning equipment. Ductile iron pipe is a mainstay for conveying pressurized fluids, and household appliances and electronics all utilize metal castings.

Key Words

additive manufacturing
allowance
blind riser
casting
chill
chill zone
choke
Chvorinov's rule
cold shut
columnar zone
computer simulation
consolidation
cooling curve
cooling rate
cope
core
core box
core print
dead riser
deformation
direct digital manufacturing
directional solidification
distortion
draft
drag
dross
equiaxed zone
expendable-mold process
external chill
fillet
filters
finish allowance
flash
flask
fluidity
freezing range
gas flushing
gas porosity
gate
gating system
grain refinement
growth
hot spot
hot tears
inoculation
insulating sleeve
internal chill
liquidus
live riser
local solidification time
machining
machining allowance
material removal
materials processing
misruns

mold cavity
mold constant
mold filling
molding material
multiple-use mold
nucleation
open riser
parting line
(parting surface)
pattern
penetration
pouring cup
pouring temperature
powder metallurgy
reusable mold
riser
runner
runner extension
runner well
shrink rule
shrinkage
shrinkage allowance
side riser
single-use mold
slag
sleeves
solidification
solidification modeling
solidification shrinkage
solidus
sprue
sprue well
stress concentration
superheat
thermal arrest
top riser
total solidification time
turbulent flow
undercooling
vacuum degassing
vent
yield

REVIEW QUESTIONS

1. What are the six activities that are conducted on almost every manufactured product?
2. What is "materials processing"?
3. What are the five basic families of shape-production processes? Cite one advantage and one limitation of each family.
4. Describe the capabilities of the casting process in terms of size and shape of the product.
5. What are some of the various mold materials and pouring methods used in casting?
6. How might the desired production quantity influence the selection of a single-use or multiple-use molding process?
7. Why is it important to provide a means of venting gases from the mold cavity?
8. What types of problem or defect can occur if the mold material provides too much restraint to the solidifying and cooling metal?
9. What is a casting pattern? Flask? Core? Mold cavity? Riser?
10. What are some of the components that combine to make up the gating system of a mold?
11. What is a parting line or parting surface?
12. What is draft, and why is it used?
13. What are the two stages of solidification, and what occurs during each?
14. Why is it that most solidification does not begin until the temperature falls somewhat below the equilibrium melting temperature (i.e., undercooling is required)?
15. Why might it be desirable to promote nucleation in a casting through inoculation or grain refinement processes?
16. Heterogeneous nucleation begins at preferred sites within a mold. What are some probable sites for heterogeneous nucleation?
17. Why might directional solidification be desirable in the production of a cast product?
18. Describe some of the key features observed in the cooling curve of a pure metal.
19. What is superheat?
20. What is the freezing range for a metal or alloy?
21. Discuss the roles of casting volume and surface area as they relate to the total solidification time and Chvorinov's rule.
22. What characteristics of a specific casting process are incorporated into the mold constant, B, of Chvorinov's rule?
23. What is the correlation between cooling rate and final properties of a casting?
24. What is the chill zone of a casting, and why does it form?
25. Which of the three regions of a cast structure is least desirable? Why are its properties highly directional?
26. What is dross or slag, and how can it be prevented from becoming part of a finished casting?
27. What are some of the possible approaches that can be taken to prevent the formation of gas porosity in a metal casting?
28. What is fluidity, and how can it be measured?
29. What is the most important factor controlling the fluidity of a casting alloy?
30. What is a misrun or cold shut, and what causes them to form?
31. What defect can form in sand castings if the pouring temperature is too high and fluidity is too great?
32. Why is it important to design the geometry of the gating system to control the rate of metal flow as it travels from the pouring cup into the mold cavity?
33. What are some of the undesirable consequences that could result from turbulence of the metal in the gating system and mold cavity?
34. What is a choke, and how does its placement affect metal flow?
35. What features can be incorporated into the gating system to aid in trapping dross and loose mold material that is flowing with the molten metal?
36. What features of the metal being cast tend to influence whether the gating system is designed to minimize turbulence and reduce dross or to promote rapid filling to minimize temperature loss?
37. What are the three stages of contraction or shrinkage as a liquid is converted into a finished casting?
38. Why is it more difficult to prevent shrinkage voids from forming in metals or alloys with large freezing ranges?
39. What type of flaws or defects form during the cooling of an already-solidified casting?
40. Why is it desirable to design a casting to have directional solidification sweeping from the extremities of a mold toward a riser?
41. Based on Chvorinov's rule, what would be an ideal shape for a casting riser? A desirable shape from a practical perspective?
42. What is yield, and how does it relate to the number and size of the specified risers?
43. Define the following riser-related terms: top riser, side riser, open riser, blind riser, live riser, and dead riser.
44. What assumptions were made when using Chvorinov's rule to calculate the size of a riser in the manner presented in the text? Why is the mold constant, B, not involved in the calculations?

45. Discuss aspects relating to the connection between a riser and the casting.
46. What is the purpose of a chill? Of an insulating sleeve? Of exothermic material?
47. What types of modifications or allowances are generally incorporated into a casting pattern?
48. What is a shrink rule, and how does it work?
49. What is the purpose of a draft or taper on pattern surfaces?
50. Why is it desirable to make the pattern allowances as small as possible?
51. What are some of the features of the casting process that are directly related to the location of the parting plane?
52. What are hot spots, and what sort of design features cause them to form?
53. What are some appearance considerations in parting line location?
54. What determines the minimum section thickness for various casting materials and processes?
55. How can computer simulation be used in the design of successful castings?

■ Problems

1. Using Chvorinov's rule as presented in the text with $n = 2$, calculate the dimensions of an effective riser for a casting that is a 2 in. × 4 in. × 6 in. rectangular plate. Assume that the casting and riser are not connected, except through a gate and runner, and that the riser is a cylinder of height/diameter ratio $H/D = 1.5$. The finished casting is what fraction of the combined weight of the riser and casting?
2. Reposition the riser in Problem 1 so that it sits directly on top of the flat rectangle, with its bottom circular surface being part of the surface of the casting, and recompute the size and yield fraction. Which approach is more efficient?
3. A rectangular casting having the dimensions 3 in. × 5 in. × 10 in. solidifies completely in 11.5 minutes. Using $n = 2$ in Chvorinov's rule, calculate the mold constant, B. Then compute the solidification time of a 0.5 in. × 8 in. × 8 in. casting poured under the same conditions.
4. Figure 11-A shows the wall profile of a cast iron coupling, and the shrinkage porosity that was observed in the cast product. If the interior profile must be maintained, suggest possible changes to the design that would enable the casting of a defect-free product.
5. Investigate various experimental techniques to evaluate molten metal fluidity.

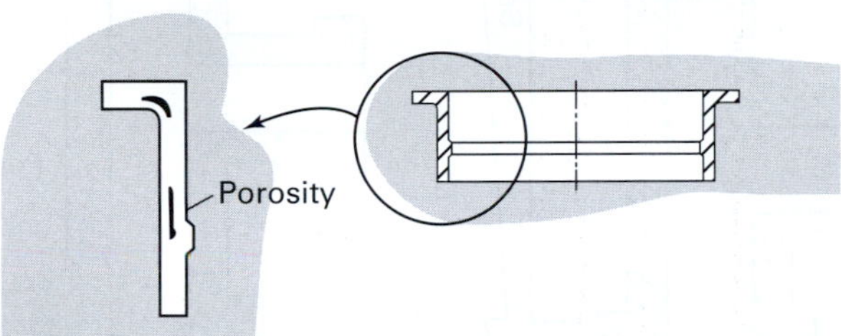

FIGURE 11-A

Chapter 11 CASE STUDY

The Cast Oil-Field Fitting

A cast iron, T-type fitting is being produced for the oil drilling industry, using an air-set or no-bake sand for both the mold and the core. A silica sand has been used in combination with a catalyzed alkyd-oil/urethane binder. Figure CS-11 shows a cross section of the mold with the core in place (part a), and a cross section of the finished casting (part b). The final casting contains several significant defects. Gas bubbles are observed in the bottom section of the horizontal tee. A penetration defect is observed near the bottom of the inside diameter, and there is an enlargement of the casting at location *C*.

1. What is the most likely source of the gas bubbles? Why are they present only at the location noted? What might you recommend as a solution?
2. What factors may have caused the penetration defect? Why is the defect present on the inside of the casting, but not on the outside? Why is the defect near the bottom of the casting, but not near the top?
3. What factors led to the enlargement of the casting at point *C*? What would you recommend to correct this problem?
4. Another producer has noted penetration defects on all surfaces of his castings, both interior and exterior. What would be some possible causes? What could you recommend as possible cures?
5. Could these molds and cores be reclaimed (i.e., recycled) after breakout? Discuss.

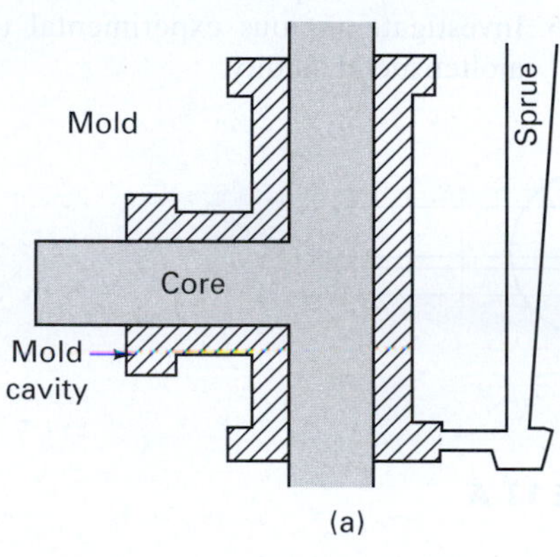

(a)

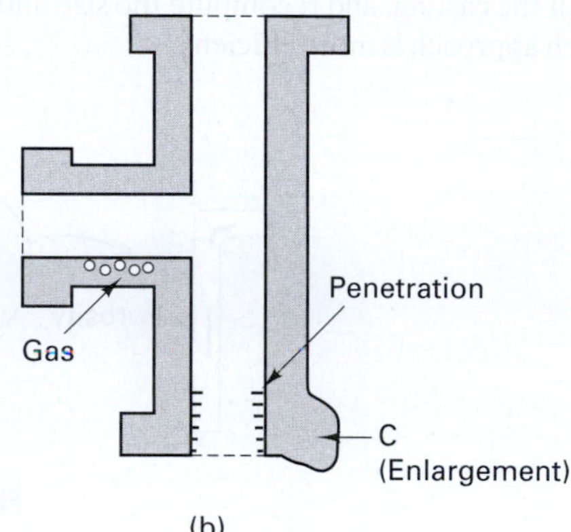

(b)

CHAPTER 12

EXPENDABLE-MOLD CASTING PROCESSES

12.1 INTRODUCTION

The versatility of metal casting is made possible by a number of distinctly different processes, each with its own set of characteristic advantages and benefits. Selection of the best process requires a familiarization with the various options and capabilities as well as an understanding of the needs of the specific product. Some factors to be considered include the desired dimensional precision and surface quality, the number of castings to be produced, the type of pattern and core box that will be needed, the cost of making the required mold or die, and restrictions imposed by the selected material.

As we begin to survey the various casting processes, it is helpful to have some form of process classification. One approach focuses on the molds and patterns and utilizes the following three categories:

1. Single-use molds with multiple-use patterns.
2. Single-use molds with single-use patterns.
3. Multiple-use molds.

Categories 1 and 2 are often combined under the more general heading of **expendable-mold casting processes** and will be presented in this chapter. Sand, plaster, ceramics, or other refractory materials are combined with binders to form the mold. Those processes where a mold can be used multiple times (Category 3) will be presented in Chapter 13. The multiple-use molds are usually made from metal.

Because the casting processes are primarily used to produce metal products, the emphasis of the casting chapters will be on metal casting. The metals most frequently cast are iron, steel, stainless steel, aluminum alloys, brass, bronze and other copper alloys, magnesium alloys, certain zinc alloys, and nickel-based superalloys. Among these, cast iron and aluminum are the most common, primarily because of their low cost, good fluidity, adaptability to a variety of processes, and the wide range of product properties that are available. The processes used to fabricate products from polymers, ceramics (including glass), and composites, including casting processes, will be discussed in Chapter 15.

12.2 SAND CASTING

Sand casting is by far the most common and possibly the most versatile of the casting processes, accounting for more than 90% of all metal castings. Granular refractory material (such as silica, zircon, olivine, or chromite sand) is mixed with small amounts of other materials, such as clay and water, and is then packed around a pattern that has the shape of the desired casting. Because the grains can pack into thin sections and can be economically used in large quantities, products spanning a wide range of sizes and detail can be made by this method. If the pattern is to be removed before pouring, the

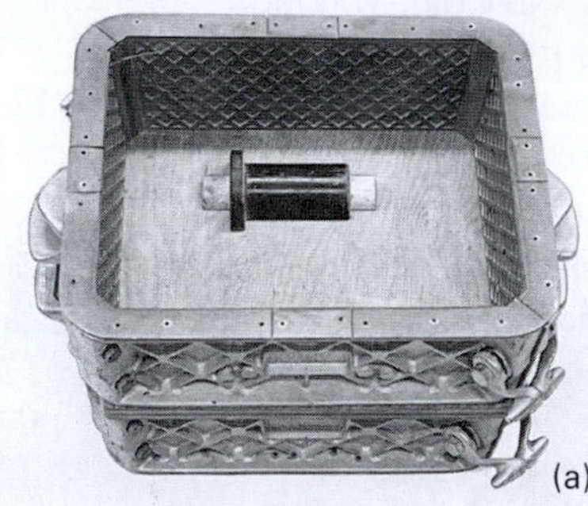

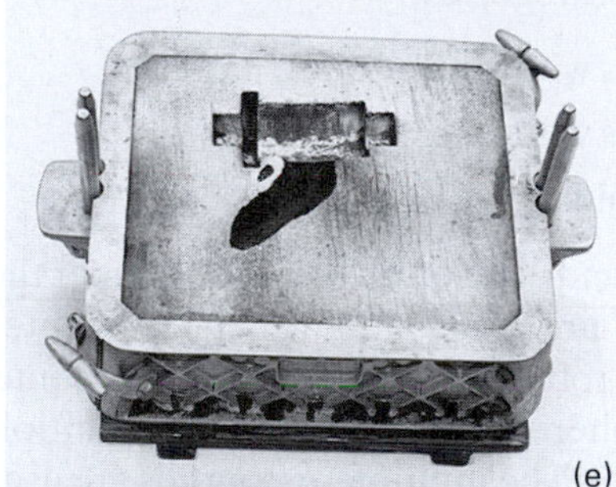

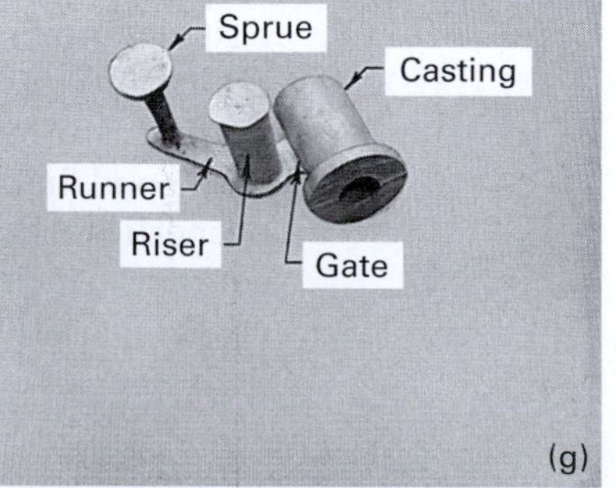

FIGURE 12-1 Sequential steps in making a sand casting. (a) A pattern board is placed between the bottom (drag) and top (cope) halves of a flask, with the bottom side up. (b) Sand is then packed into the bottom or drag half of the mold. (c) A bottom board is positioned on top of the packed sand, and the mold is turned over, showing the top (cope) half of pattern with sprue and riser pins in place. (d) The upper or cope half of the mold is then packed with sand. (e) The mold is opened, the pattern board is drawn (removed), and the runner and gate are cut into the bottom parting surface of the sand. (é) The parting surface of the upper or cope half of the mold is also shown with the pattern and pins removed. (f) The core is positioned, the pattern board is removed, the mold is reassembled, and molten metal is poured through the sprue. (g) The contents are shaken from the flask and the metal segment is separated from the sand, ready for further processing. *(E. Paul DeGarmo)*

mold is usually made in two or more segments. An opening called a **sprue** is cut from the top of the mold through the sand and connected to a system of horizontal channels called **runners.** The molten metal is poured down the sprue hole, flows through the runners, and enters the mold cavity through one or more openings, called **gates.** Gravity flow is the most common means of introducing the metal into the mold. The metal is allowed to solidify, and the mold is then broken to permit removal of the finished casting. Because the mold is destroyed in product removal, a new mold must be made for each casting. Figure 12-1 shows the essential steps and basic components of a sand casting process. A two-part cope and drag mold is illustrated, and the casting incorporates both a core and a riser (discussed in Chapter 11).

PATTERNS AND PATTERN MATERIALS

The first step in making a sand casting is the design and construction of a **pattern.** This is a duplicate of the part to be cast, modified in accordance with the requirements of the casting process, the metal being cast, and the particular molding technique that is being used. Selection of the pattern material is determined by the number of castings to be made, the size and shape of the casting, the desired dimensional precision, and the molding process. Wood patterns are relatively easy to make and are frequently used when small quantities of castings are required. Wood, however, is not very dimensionally stable. It may warp or swell with changes in humidity, and it tends to wear with repeated use. Metal patterns are more expensive but are more stable and durable. Hard plastics, such as urethanes, offer another alternative, and are often preferred with processes that use strong, organically bonded sands that tend to stick to other pattern materials. In the full-mold and lost-foam processes, expanded polystyrene (EPS) is used, and investment casting uses patterns made from wax. In the latter processes, both the pattern and the mold are single-use, each being destroyed when a casting is produced.

TYPES OF PATTERNS

Many types of patterns are used in the foundry industry, with selection being based on the number of duplicate castings required and the complexity of the part.

FIGURE 12-2 Single-piece pattern for a pinion gear. *(E. Paul DeGarmo)*

One-piece or *solid,* **patterns,** such as the one shown in Figure 12-2, are the simplest and often the least expensive type. They are essentially a duplicate of the part to be cast, modified only by the various allowances discussed in Chapter 11 and by the possible addition of core prints. One-piece patterns are relatively cheap to construct, but the subsequent molding process is usually slow. As a result, they are generally used when the shape is relatively simple and the number of duplicate castings is rather small.

If the one-piece pattern is simple in shape and contains a flat surface, it can be placed directly on a **follow board.** The entire mold cavity will be created in one segment of the mold, with the follow board forming the parting surface. If the parting plane is to be more centrally located, special follow boards are produced with inset cavities that position the one-piece pattern at the correct depth for the parting line. Figure 12-3 illustrates this technique, where the follow board again forms the parting surface.

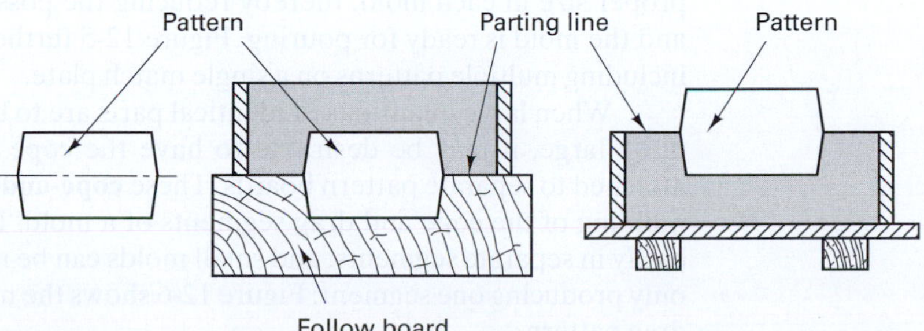

FIGURE 12-3 Method of using a follow board to position a single-piece pattern and locate a parting surface. The final figure shows the flask of the previous operation (the drag segment) inverted in preparation for construction of the upper portion of the mold (cope segment).

FIGURE 12-4 Split pattern, showing the two sections together and separated. The light-colored portions are core prints. *(E. Paul DeGarmo)*

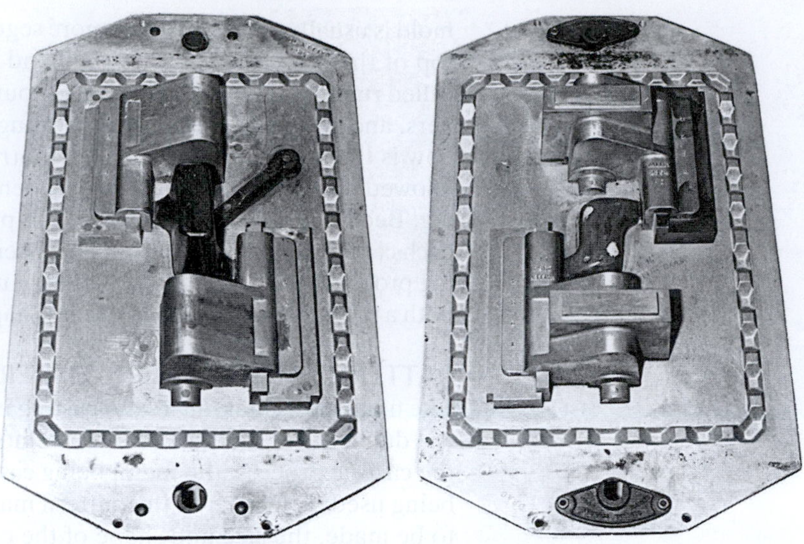

FIGURE 12-5 Match-plate pattern used to produce two identical parts in a single flask: (left) cope side; (right) drag side. (*Note:* The views are opposite sides of a single pattern board.) *(E. Paul DeGarmo)*

Split patterns are used when moderate quantities of a casting are desired. The pattern is divided into two segments along what will become the parting plane of the mold. The bottom segment of the pattern is positioned in the **drag** portion of a **flask,** and the bottom segment of the mold is produced. This portion of the flask is then inverted, and the upper segment of the pattern and flask are attached. Tapered pins in the cope half of the pattern align with holes in the drag segment to assure proper positioning. Mold material is then packed around the full pattern to form the upper segment **(cope)** of the mold. The two segments of the flask are separated, and the pattern pieces are removed to produce the mold cavity. Sprues and runners are cut and the mold is then reassembled, ready for pour. Figure 12-4 shows a split pattern that also contains several core prints (lighter color).

Match-plate patterns, like the one shown in Figure 12-5, further simplify the process and can be coupled with modern molding machines to produce large quantities of duplicate molds. The cope and drag segments of a split pattern are permanently fastened to opposite sides of a wood or metal **match plate.** The match plate is positioned between the upper and lower segments of a flask using holes that align with pins on one of the flask segments. Mold material is then packed on both sides of the match plate to form the cope and drag segments of a two-part mold. The mold sections are then separated and the match-plate pattern is removed. The segments are then reassembled with the pins and guide holes, ensuring that the cavities in the cope and drag are in proper alignment. The necessary gates, runners, and risers are usually incorporated on the match plate, as well. This guarantees that these features will be uniform and of the proper size in each mold, thereby reducing the possibility of defects. The sprue is cut and the mold is ready for pouring. Figure 12-5 further illustrates a common practice of including multiple patterns on a single match plate.

When large quantities of identical parts are to be produced, or when the casting is quite large, it may be desirable to have the cope and drag halves of split patterns attached to separate pattern boards. These **cope-and-drag patterns** enable independent molding of the cope and drag segments of a mold. Large molds can be handled more easily in separate segments, and small molds can be made at a faster rate if a machine is only producing one segment. Figure 12-6 shows the mating pieces of a typical cope-and-drag pattern.

When the geometry of the product is such that a one-piece or two-piece pattern could not be removed from the molding sand, a **loose-piece pattern** can sometimes be developed. Separate pieces are joined to a primary pattern segment by beveled grooves

FIGURE 12-6 Cope-and-drag pattern for producing two heavy parts: (left) cope section; (right) drag section. (*Note:* These are two separate pattern boards.) *(E. Paul DeGarmo)*

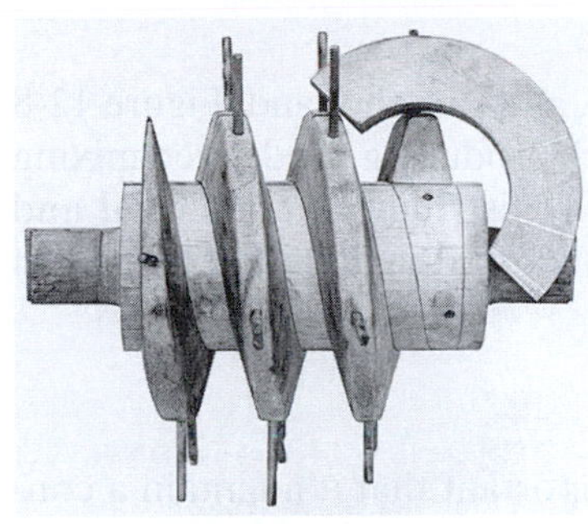

FIGURE 12-7 Loose-piece pattern for molding a large worm gear. After sufficient sand has been packed around the pattern to hold the pieces in position, the wooden pins are withdrawn. The mold is then completed, after which the pieces of the pattern can be removed in a designated sequence. *(E. Paul DeGarmo)*

or pins (Figure 12-7). After molding, the primary segment of the pattern is withdrawn. The hole that is created then permits the remaining segments to be sequentially extracted. Loose-piece patterns are expensive. They require careful maintenance, slow the molding process, and increase molding costs. They do, however, enable the sand casting of complex shapes that would otherwise require the full-mold, lost-foam, or investment processes.

SANDS AND SAND CONDITIONING

The sand used to make molds must be carefully prepared if it is to provide satisfactory and uniform results. Ordinary silica (SiO_2), zircon, olivine, or chromite sands are compounded with additives to meet four requirements:

1. **Refractoriness:** the ability to withstand high temperatures without melting, fracture or deterioration.
2. **Cohesiveness** (also referred to as *bond*): the ability to retain a given shape when packed into a mold.
3. **Permeability:** the ability of mold cavity, mold, and core gases to escape through the sand.
4. **Collapsibility:** the ability to accommodate metal shrinkage after solidification and provide for easy removal of the casting through mold disintegration **(shakeout)**.

Refractoriness is provided by the basic nature of the sand. Cohesiveness, bond, or strength is obtained by coating the sand grains with clays, such as bentonite, kaolinite, or illite, that become cohesive when moistened. Permeability is a function of the size of the sand particles, the amount and type of clay or bonding agent, the moisture content, and the compacting pressure. Collapsibility is sometimes enhanced by adding cereals or other organic materials, such as cellulose, that burn out when they come in contact with the hot metal. The combustion of these materials reduces both the volume and strength of the restraining sand.

Good molding sand always represents a compromise among competing factors. The size of the sand particles, the amount of bonding agent (such as clay), the moisture content, and the organic additives are all selected to obtain an acceptable compromise of the four basic requirements. The overall composition must be carefully controlled to assure satisfactory and consistent results. Because molding material is often reclaimed and recycled, the temperature of the mold during pouring and solidification is also important. If organic materials have been incorporated into the mix to provide collapsibility, a portion will burn during the pour. Adjustments will be necessary, and ultimately some or all of the mold material may have to be discarded and replaced with new.

A typical **green-sand** mixture contains about 88% silica sand, 9% clay, and 3% water. To achieve good molding, it is important for each grain of sand to be coated uniformly with the proper amount of additive agents. This is achieved by putting the

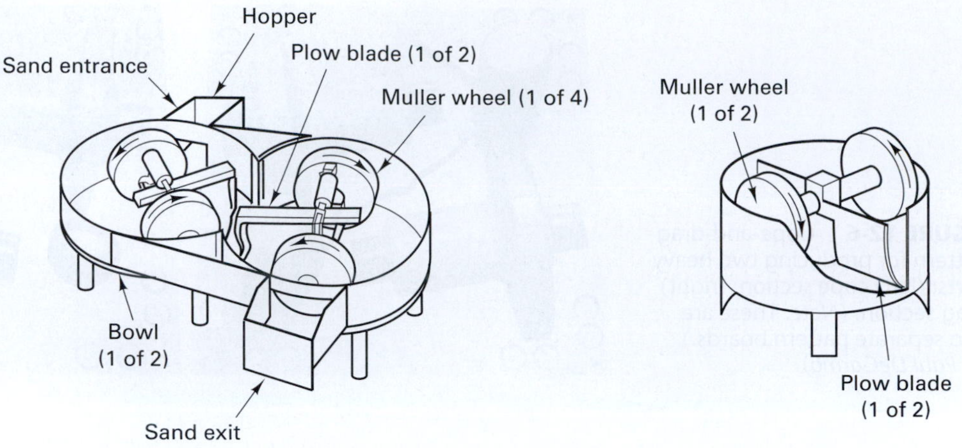

FIGURE 12-8 Schematic diagram of a continuous (left) and batch-type (right) sand muller. Plow blades move and loosen the sand and the muller wheels compress and mix the components. *(Courtesy of ASM International, Materials Park, OH)*

ingredients through a **muller,** a device that kneads, rolls, and stirs the sand. Figure 12-8 shows both a continuous and batch-type muller, with each producing the desired mixing through the use of rotating blades that lift, fluff, and redistribute the material and wheels that compress and squeeze. After mixing, the sand is often discharged through an aerator, which fluffs it for further handling.

SAND TESTING

If a foundry is to produce high-quality products, it is important that it maintain a consistent quality in its molding sand. The sand itself can be characterized by grain size, grain shape, surface smoothness, density, and contaminants. Blended molding sand can be characterized by moisture content, clay content, and compactibility. Key properties of compacted sand or finished molds include **mold hardness,** *permeability,* and *strength.* Standard tests and procedures have been developed to evaluate many of these properties.

Grain size can be determined by shaking a known amount of clean, dry sand downward through a set of 11 standard screens or sieves of decreasing mesh size. After shaking for 15 min, the amount of material remaining on each sieve is weighed, and these weights are used to compute an AFS (American Foundry Society) grain fineness number.

Moisture content can be determined by a special device that measures the electrical conductivity of a small sample of compressed sand. A more direct method is to measure the weight lost by a 50-g sample after it has been subjected to a temperature of about 110°C (230°F) for sufficient time to drive off all the water.

Clay content is determined by washing the clay from a 50-g sample of molding sand, using water that contains sufficient sodium hydroxide to make it alkaline. Several cycles of agitation and washing may be required to fully remove the clay. The remaining sand is then dried and weighed to determine the amount of clay removed from the original sample.

Permeability and strength tests are conducted on compacted sands, using a **standard rammed specimen.** An amount of sand is first placed into a 2-in.-diameter steel tube. A 14-lb weight is then dropped on it three times from a height of 2 in., and the height of the resulting specimen must be within $\frac{1}{32}$ in. of a targeted 2-in. height.

Permeability is a measure of how easily gases can pass through the narrow voids between the sand grains. Air in the mold before pouring, plus the steam that is produced when the hot metal contacts the moisture in the sand, along with various combustion gases, must all be allowed to escape, rather than prevent mold filling or be trapped in the casting as porosity or blow holes. During the permeability test, shown schematically in Figure 12-9, a sample tube containing the standard rammed specimen is subjected to an air pressure of 10 g/cm^2. By means of either a flow rate determination or a measurement of the steady-state pressure between the orifice and the sand specimen, an

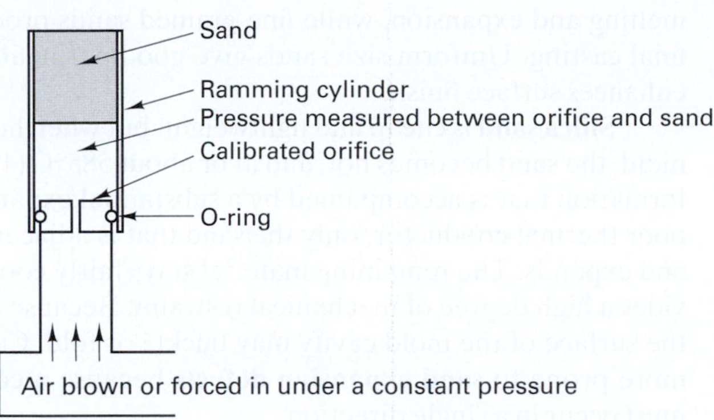

FIGURE 12-9 Schematic of a permeability tester in operation. A standard sample in a metal sleeve is sealed by an O-ring onto the top of the unit while air is passed through the sand. *(Courtesy of Dietert Foundry Testing Equipment Inc., Detroit, MI)*

AFS permeability number[1] can be computed. Most test devices are now calibrated to provide a direct readout of this number.

All molding material must have sufficient strength to retain the integrity of the mold cavity while the mold is being handled between molding and pouring. The mold material must also withstand the erosion of the liquid metal as it flows into the mold, and the pressures induced by a column of molten metal. The **compressive strength** of the sand (also referred to as **green compressive strength**) is a measure of the mold strength at this stage of processing. It is determined by removing the rammed specimen from the compacting tube and placing it in a mechanical testing device. A compressive load is then applied until the specimen breaks, which usually occurs in the range of 10 to 30 psi (0.07 to 0.2 MPa). If there is too little moisture in the sand, the grains will be poorly bonded and strength will be poor. If there is excess moisture, the extra water acts as a lubricant and strength is again poor. In between, there is a condition of maximum strength with an optimum water content that will vary with the content of other materials in the mix. A similar optimum also applies to permeability, because unwetted clay blocks vent passages, as does excess water. Sand coated with a uniform thin film of moist clay provides the best molding properties. A ratio of one part water to three parts clay (by weight) is often a good starting point.

The **hardness** of compacted sand can give additional insight into the strength and permeability characteristics of a mold. Hardness can be determined by the resistance of the sand to the penetration of a 0.2-in.(5.08-mm)-diameter spring-loaded steel ball. A typical test instrument is shown in Figure 12-10.

Compactibility is determined by sifting loose sand into a steel cylinder, leveling off the column, striking it three times with the standard weight (as in making a standard rammed specimen), and then measuring the final height. The *percent compactibility* is the change in height divided by the original height times 100%. This value can often be correlated with the moisture content of the sand, where a compactibility of around 45% indicates a proper level of moisture. A low compactibility is usually associated with too little moisture.

FIGURE 12-10 Sand mold hardness tester. *(Courtesy of Dietert Foundry Testing Equipment Inc., Detroit, MI)*

SAND PROPERTIES AND SAND-RELATED DEFECTS

The characteristics of the sand granules themselves can be very influential in determining the properties of foundry molding material. Round grains give good permeability and minimize the amount of clay required because of their low surface area. Angular sands give better green strength because of the mechanical interlocking of the grains. Large grains provide good permeability and better resistance to high-temperature

[1] The AFS permeability number is defined as

$$AFS\ number = (V \times H)/(P \times A \times T)$$

where V is the volume of air (2000 cm^3), H is the height of the specimen (5.08 cm), P is the pressure (10 g/cm^2), A is the cross section area of the specimen (20.268 cm^2), and T is the time in seconds to pass a flow of 2000 cm^3 of air through the specimen. Substituting each of the above constants, the permeability number becomes equal to $3000.2/T$.

melting and expansion, while fine-grained sands produce a better surface finish on the final casting. Uniform size sands give good permeability, while a distribution of sizes enhances surface finish.

Silica sand is cheap and lightweight, but when hot metal is poured into a silica sand mold, the sand becomes hot, and at or about 585°C (1085°F) it undergoes a phase transformation that is accompanied by a substantial expansion in volume. Because sand is a poor thermal conductor, only the sand that is adjacent to the mold cavity becomes hot and expands. The remaining material stays fairly cool, does not expand, and often provides a high degree of mechanical restraint. Because of this uneven heating, the sand at the surface of the mold cavity may buckle or fold. Castings with large, flat surfaces are more prone to **sand expansion defects** because a considerable amount of expansion must occur in a single direction.

Sand expansion defects can be minimized in a number of ways. Certain particle geometries permit the sand grains to slide over one another, thereby relieving the expansion stresses. Excess clay can be added to absorb the expansion, or volatile additives, such as cellulose, can be added to the mix. When the casting is poured, the cellulose burns, creating voids that can accommodate the sand expansion. Another alternative is the use of olivine or zircon sand in place of silica. Because these sands do not undergo phase transformations upon heating, their expansion is only about one-half that of silica sand. Unfortunately, these sands are much more expensive and heavier in weight than the more commonly used silica.

Trapped or evolved gas can create gas-related **voids** or **blows** in finished castings. The most common causes are low sand permeability (often associated with angular, fine, or wide-size distribution sands, fine sand additives, and overcompaction) and large amounts of evolved gas due to high mold-material moisture or excessive amounts of volatiles. If adjustments to the mold composition are not sufficient to eliminate the voids, vent passages may have to be cut into the mold, a procedure that may add significantly to the mold-making cost.

Molten metal can also penetrate between the sand grains, causing the mold material to become embedded in the surface of the casting. This defect, known as **penetration,** can be the result of high pouring temperatures (excess fluidity); high metal pressure (possibly due to excessive cope height or pouring from too high an elevation above the mold); or the use of high permeability sands with coarse, uniform particles. Fine-grained materials, such as silica flour (fine powdered sand), can be blended in to fill the voids, but this reduces permeability and increases the likelihood of both gas and expansion defects.

Hot tears or **cracks** can form in castings made from metals or alloys with large amounts of solidification shrinkage. As the metal contracts during solidification and cooling to room temperature, it may find itself restrained by a strong mold or core. Tensile stresses can develop while the metal is still partially liquid, or fully solidified, but still hot and weak. If these stresses become great enough, the casting will crack or tear. Hot tears are often attributed to poor mold collapsibility. Additives, such as cellulose, can be used to improve the collapsibility of sand molds.

Table 12-1 summarizes the many desirable properties of a sand-based molding material.

TABLE 12-1 Desirable Properties in Sand-Based Molding Materials

1. Is inexpensive in bulk quantities
2. Retains properties through transportation and storage
3. Uniformly fills a flask or container
4. Can be compacted or set by simple methods.
5. Has sufficient elasticity to remain undamaged during pattern withdrawal
6. Can withstand high temperatures and maintains its dimensions until the metal has solidified
7. Is sufficiently permeable to allow the escape of gases
8. Is sufficiently dense to prevent metal penetration
9. Is sufficiently cohesive to prevent wash-out of mold material into the pour stream
10. Is chemically inert to the metal being cast
11. Can yield to solidification and thermal shrinkage, thereby preventing hot tears and cracks
12. Has good collapsibility to permit easy removal and separation of the casting
13. Can be recycled

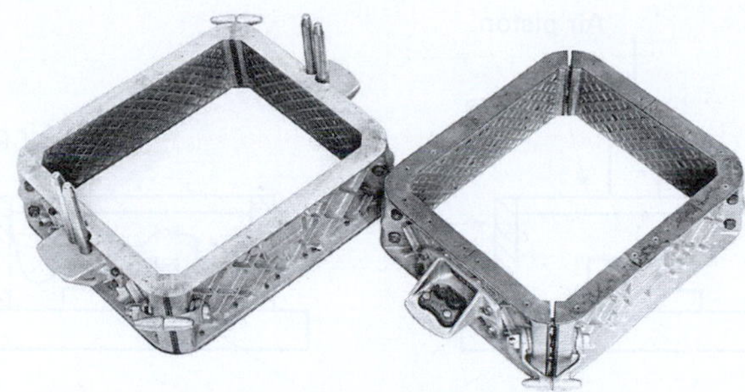

FIGURE 12-11 Bottom and top halves of a snap flask: (left) drag segment in closed position; (right) cope segment with latches opened for easy removal from the mold. *(E. Paul DeGarmo)*

THE MAKING OF SAND MOLDS

Molding usually begins with a pattern, like the match-plate pattern discussed earlier, and a flask. The flasks may be straight-walled containers with guide pins or removable jackets, and they are generally constructed of lightweight aluminum or magnesium. Figure 12-11 shows a *snap flask*, so named because it is designed to snap open for easy removal after the mold material has been packed in place.

When only a few castings are to be made, **hand ramming** is often the preferred method of packing sand to make a sand mold. Hand ramming, however, is slow, labor intensive, and usually results in nonuniform compaction. For normal production, sand molds are generally made using specially designed molding machines. The various methods differ in the type of flask, the way the sand is packed within the flask, whether mechanical assistance is provided to turn or handle the mold, and whether a flask is even required. In all cases, however, the molding machines greatly reduce the labor and required skill, and lead to castings with good dimensional accuracy and consistency.

Several techniques have been developed to pack the mixed sand (mold material) into the flask. A **sand slinger** uses a rotating impeller to fling or throw sand against the pattern. The slinger is manipulated to progressively deposit compacted sand into the mold. Sand slinging is a common method of achieving uniform sand compaction when making large molds and large castings.

In a method known as **jolting,** a flask is positioned over a pattern; filled with sand; and the pattern, flask, and sand are then lifted and dropped several times, as shown in Figure 12-12. The weight and kinetic energy of the sand produces optimum packing at the bottom of the mass, directly around the pattern. Jolting machines can be used on the first half of a match-plate pattern or on both halves of a cope-and-drag operation.

Squeezing machines use an air-operated squeeze head, a flexible diaphragm, or small individually activated squeeze heads to compact the sand. The squeezing motion provides firm packing adjacent to the squeeze head, with density diminishing as you move farther into the mold. Figure 12-13 illustrates the squeezing process, and

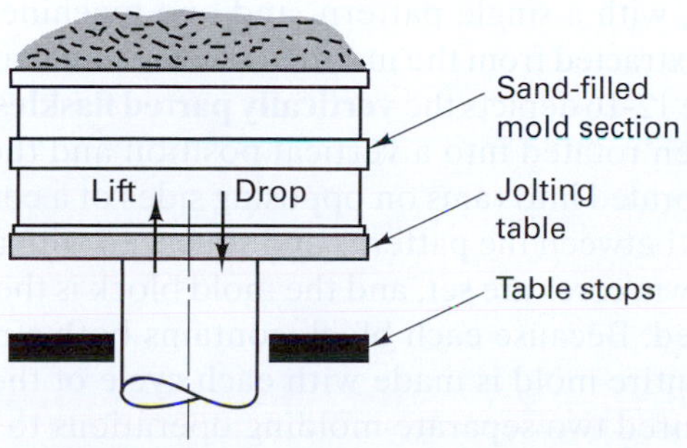

FIGURE 12-12 Jolting a mold section. (*Note*: The pattern is on the bottom where the greatest packing is expected.)

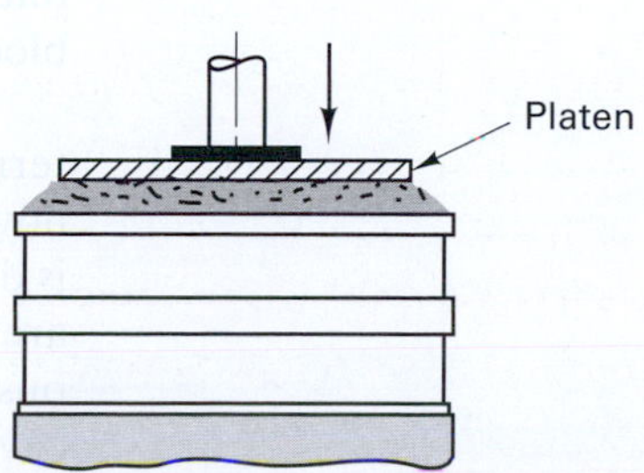

FIGURE 12-13 Squeezing a sand-filled mold section. While the pattern is on the bottom, the highest packing will be directly under the squeeze head.

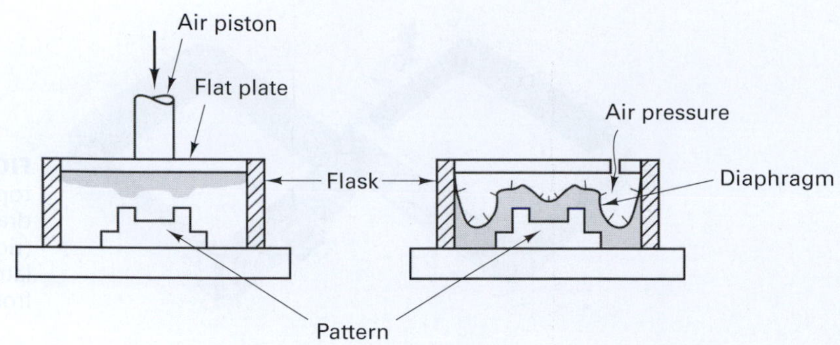

FIGURE 12-14 Schematic diagram showing relative sand densities obtained by flat-plate squeezing where all areas get vertically compressed by the same amount of movement (left) and by flexible-diaphragm squeezing where all areas flow to the same resisting pressure (right).

Figure 12-14 compares the density achieved by squeezing with a flat plate and squeezing with a flexible diaphragm.

In match-plate molding, a combination of jolting and squeezing is often used to produce a more uniform density throughout the mold. The match-plate pattern is positioned between the cope and drag sections of a flask, and the assembly is placed drag side up on the molding machine. A parting compound is sprinkled on the pattern, and the drag section of the flask is filled with mixed sand. The entire assembly is then jolted a specified number of times to pack the sand around the drag side of the pattern. A squeeze head is then swung into place, and pressure is applied to complete the drag portion of the mold. The entire flask is then inverted and a squeezing operation is performed to compact loose sand in the cope segment. (*Note:* Jolting is not performed here because it might cause the already-compacted sand to break free of the inverted drag section of the pattern!) Because the drag segment sees both jolting and squeezing, while the cope is only squeezed, the pattern side with the greatest detail is generally placed in the drag. If the cope and drag segments of a mold are made on separate machines (using separate cope-and-drag patterns), the combination of jolting and squeezing can be performed on each segment of the mold.

The sprue hole is most often cut by hand, with this operation being performed before removal of the pattern to prevent loose sand from falling into the mold cavity. The pouring basin may also be hand cut, or it may be shaped by a protruding segment on the squeeze board. The gates and runners are usually included on the pattern.

The pattern board is removed, and the segments of the mold are reassembled ready for pour. Heavy metal weights are often placed on top of the molds to prevent the cope section from rising and "floating" when the hydrostatic pressure of the molten metal presses upward. The weights are left in place until solidification is complete, and they are then moved to other molds.

For mass-production molding, a number of automatic mold-making methods have been developed. These include **automatic match-plate molding,** automatic cope-and-drag molding, and methods that produce some form of stacked segments. Figure 12-15 shows the production sequence for one of the variations of automatic match-plate molding, where the sand is introduced into the cope-and-drag mold segments from the side and then vertically compressed. The two-part cope-and-drag mold is produced in one station, with a single pattern, and one machine squeeze cycle. The compressed blocks are extracted from the molding machine and are poured in a flaskless condition.

Figure 12-16 depicts the **vertically parted flaskless molding** process, where the pattern has been rotated into a vertical position and the cope-and-drag impressions are now incorporated into rams on opposing sides of a compaction machine. Molding sand is deposited between the patterns and squeezed with a horizontal motion. The patterns are withdrawn, cores are set, and the mold block is then joined to those that were previously molded. Because each block contains both a right-hand cavity and a left-hand cavity, an entire mold is made with each cycle of the machine. (*Note:* Previous techniques required two separate molding operations to produce the individual cope and drag segments of a two-part mold.) A vertical gating system is usually included on one side of the pattern, and the assembled molds are usually poured individually. If a common horizontal runner is used to connect multiple mold segments, the method is known

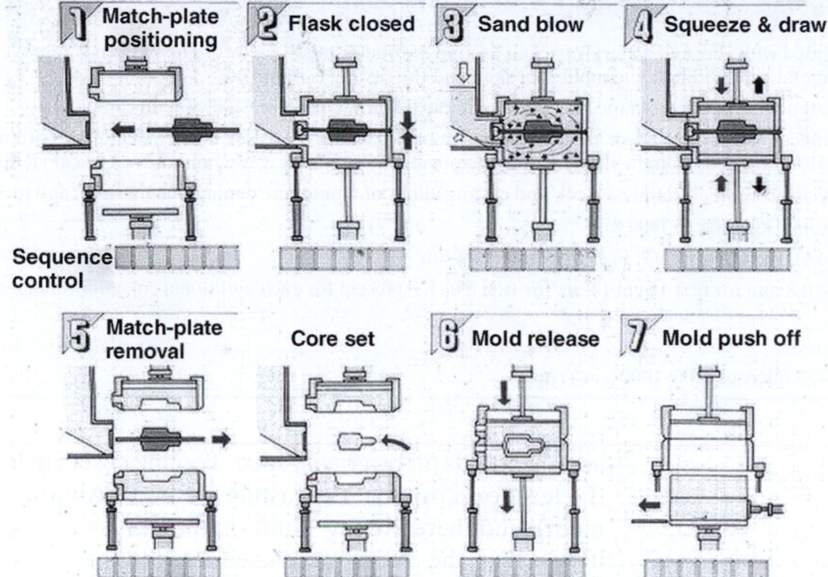

FIGURE 12-15 Activity sequence for automatic match-plate molding. Green sand is blown from the side and compressed vertically. The final mold is ejected from the flask and poured in a flaskless condition. *(Copyright 2000 American Foundry Society)*

as the **H-process.** Because metal cools as it travels through long runners, the individual cavities of the H-process often fill with different temperature metal. To ensure product uniformity, most producers reject the H-process and prefer to pour their vertically parted molds individually.

In **stack molding,** sections containing a cope impression on the bottom and a drag impression on the top are piled vertically on top of one another. Metal is poured down a common vertical sprue, which is connected to horizontal gating systems at each of the parting planes. Pneumatic rammers can provide additional tamping.

For molds that are too large to be made by either hand ramming or by one of the previous molding processes, large flasks can be placed directly on the foundry floor. Various types of mechanical aids, such as a sand slinger, can then be used to add and pack the sand. Pneumatic rammers can provide additional tamping. Even larger molds can be constructed in sunken pits. Because of the size, complexity, and need for strength, pit molds are often created by assembling smaller sections of baked or dried sand. Added binders may be required to provide the strength required for these large molds.

GREEN-SAND, DRY-SAND, AND SKIN-DRIED MOLDS

Green-sand casting (where the term *green* implies that the mold material has not been fired or cured) is the most widely used process for casting both ferrous and nonferrous metals. The mold material is composed of sand blended with clay, water, and additives,

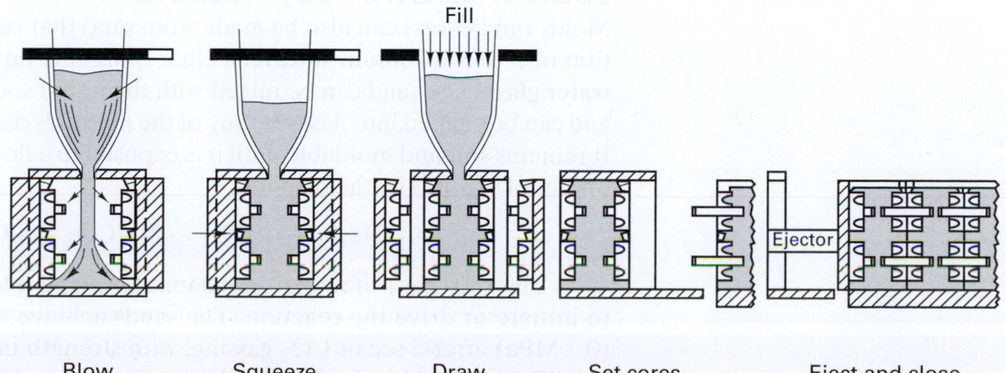

FIGURE 12-16 Vertically parted flaskless molding with inset cores. Note how one mold block now contains both the cope and drag impressions.

TABLE 12-2 Green-Sand Casting

Process: Sand, bonded with clay and water, is packed around a wood or metal pattern. The pattern is removed, and molten metal is poured into the cavity. When the metal has solidified, the mold is broken and the casting is removed.

Advantages: Almost no limit on size, shape, weight, or complexity; low cost; almost any metal can be cast.

Limitations: Tolerances and surface finish are poorer than in other casting processes; some machining is often required; relatively slow production rate; a parting line and draft are needed to facilitate pattern removal; due to sprues, gates, and risers, typical yields range from 50% to 85%.

Common metals: Cast iron, steel, stainless steel, and casting alloys of aluminum, copper, magnesium, and nickel.

Size limits: 30 g to 3000 kg (1 oz to 7000 lb).

Thickness limits: As thin as 0.25 cm ($\frac{3}{32}$ in.), with no maximum.

Typical tolerances: 0.8 mm for first 15 cm ($\frac{1}{32}$ in. for first 6 in.), 0.003 cm for each additional cm; additional increment for dimensions across the parting line

Draft allowances: 1–3°.

Surface finish: 2.5–25 microns (100–1000 μin.) rms.

and the molds fill by gravity feed. Tooling costs are low, and the entire process is one of the least expensive of the casting methods. Almost any metal can be cast (except titanium), and there are few limits on the size, shape, weight, and complexity of the products. Over the years, green-sand casting has evolved from a manually intensive operation to a mechanized and automated system capable of producing more than 300 molds per hour. As a result, it can be economically applied to both small and large production runs.

Design limitations are usually related to the rough surface finish and poor dimensional accuracy—and the resulting need for finish machining. Still other problems can be attributed to the low strength of the mold material and the moisture that is present in the clay-and-water binder. Table 12-2 provides a process summary for green-sand casting, and Figure 12-17 shows a variety of parts that have been produced in aluminum.

Some of the problems associated with the green-sand process can be reduced if the mold is heated to a temperature between 150 and 300°C (300 to 575°F), and baked until most of the moisture is driven off. This drying strengthens the mold and reduces the volume of gas generated when the hot metal enters the cavity. **Dry-sand molds** are very durable and may be stored for a relatively long period of time. They are not very popular, however, because of the long time required for drying, the added cost of that operation, and the availability of alternative processes. An attractive compromise may be the production of a **skin-dried mold,** drying only the sand that is adjacent to the mold cavity. Torches are often used to perform the drying, and the water is usually removed to a depth of about 13 mm ($\frac{1}{2}$-in.).

The molds used for casting of large steel parts are almost always skin dried, because the pouring temperatures for steel are significantly higher than those for cast iron. These molds may also be given a high-silica wash prior to drying to increase the refractoriness of the surface, or the more thermally stable zircon sand may be used as a facing. Additional binders, such as molasses, linseed oil, or corn flour, can be added to the facing sand to enhance the strength of the skin-dried segment.

SODIUM SILICATE—CO_2 MOLDING

Molds (and cores) can also be made from sand that receives its strength from the addition of 3 to 6% **sodium silicate,** a clear inorganic liquid binder, commonly known as **water glass.** The sand can be mixed with the liquid sodium silicate in a standard muller and can be packed into flasks by any of the methods discussed previously in this chapter. It remains soft and moldable until it is exposed to a flow of CO_2 gas. It then hardens in a matter of seconds by the reaction:

$$Na_2SiO_3 + CO_2 \rightarrow Na_2CO_3 + SiO_2 \text{ (colloidal)}$$

The CO_2 gas is nontoxic, nonflammable, and odorless, and no heating is required to initiate or drive the reaction. The sands achieve a tensile strength of about 40 psi (0.3 MPa) after 5 sec of CO_2 gassing, with strength increasing to 100 to 200 psi (0.7 to 1.4 MPa) after 24 hr of aging. Unlike most other sands, however, the heating that occurs

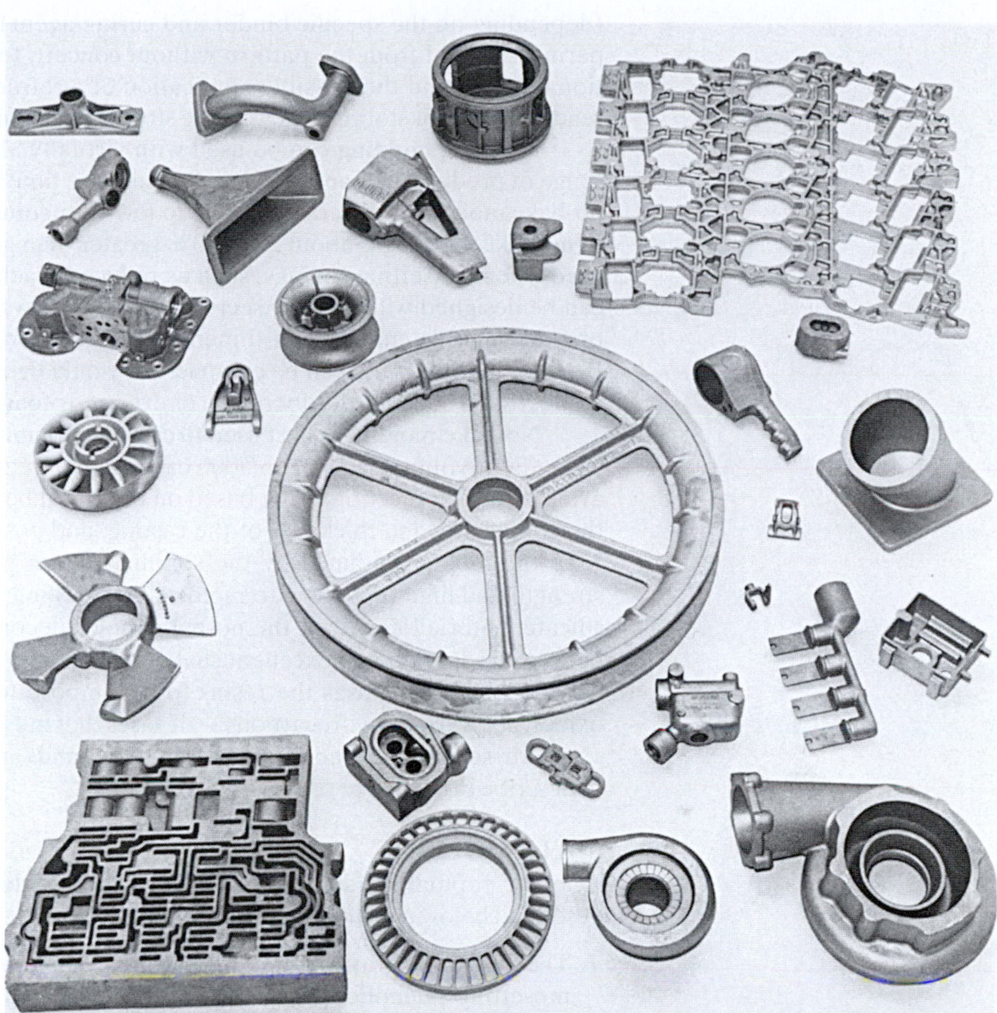

FIGURE 12-17 A variety of sand cast aluminum parts. *(Courtesy Bodine Aluminum, Inc., St. Louis, MO)*

as a result of the metal pour further increases the strength of the material (a phenomenon similar to the firing of a ceramic material). As a result, the sodium silicate sands have extremely poor collapsibility, making shakeout and core removal quite difficult. Additives that will burn out during the pour are frequently used to enhance the collapsibility of sodium silicate molds. Care must also be taken to prevent the carbon dioxide in the air from hardening the premixed sand before the mold-making process is complete. As a result, the **bench life** of sodium silicate sands (available time between mixing and the completion of molding) is quite short.

A modification of this process can be used when certain portions of a mold require better accuracy, thinner sections, or deeper draws than can be achieved with ordinary molding sand. Sand mixed with sodium silicate is packed around a special metal pattern to a thickness of about $2\frac{1}{2}$ cm (1 in.), followed by regular molding sand as a backing material. After the sand is fully compacted, CO_2 is introduced through vents in the metal pattern. The adjacent sand is further hardened, and the pattern can be withdrawn with less possibility of damage to the mold.

NO-BAKE, AIR-SET, OR CHEMICALLY BONDED SANDS

An alternative to the sodium silicate-CO_2 process involves room-temperature chemical reactions that can occur between organic or inorganic resin binders and liquid curing agents or catalysts. The two or more components are mixed with sand just prior to the molding operation, and the curing reactions begin immediately. The molds (or cores) are then made in a reasonably rapid fashion, because the mix remains workable for only a short period of time. After a few minutes to a few hours at room temperature (depending on the specific binder and curing agent), the sands harden sufficiently to permit removal from the pattern without concern for distortion. After time for additional curing and the possible application of a refractory coating, the molds are then ready for pour or storage (they can be stored almost indefinitely).

No-bake molding can be used with virtually all engineering metals over a wide range of product sizes and weights. Because the time for mold curing slows production, no-bake molding is generally limited to low-to-medium production quantities. The cost of no-bake molding is about 20 to 30% greater than green sand, so no-bake is generally used where offsetting savings, such as reduced machining, can be achieved. Products can be designed with thinner sections, deeper draws, and smaller draft, and the rigid, bricklike molds enable high dimensional precision, along with good surface finish. Because no-bake sand can be compacted by only light vibrations, patterns can often be made from wood, plastic, fiberglass, or even styrofoam, thereby reducing pattern cost.

No-bake molding is best used to cast components with higher complexities in low-to medium-volume runs (1 to 5000 castings per year). A wide variety of systems are available, with selection being based on the metal being poured, the cure time desired, the complexity and thickness of the casting, and possible desire for sand reclamation. Like the molds produced by the sodium silicate process, no-bake offers good hot strength and high resistance to mold-related casting defects. In contrast to the sodium silicate material, however, the no-bake molds decompose readily after the metal has been poured, providing excellent shakeout characteristics. Permeability must be good, because the heat causes the resins to decompose to hydrogen, water vapor, carbon oxides, and various hydrocarbons—all gases that must be vented.

Air-set molding and **chemically bonded sands** are other terms that have been used to describe the no-bake process.

SHELL MOLDING

Another popular sand casting process is **shell molding,** the basic steps of which are described below and are illustrated in Figure 12-18.

1. The individual grains of fine silica sand are first precoated with a thin layer of thermosetting phenolic resin and heat-sensitive liquid catalyst. This material is then dumped, blown, or shot onto a metal pattern (usually some form of cast iron) that has been preheated to a temperature between 230 and 340°C (450 and 650°F). During a period of sustained contact, heat from the pattern partially cures (polymerizes

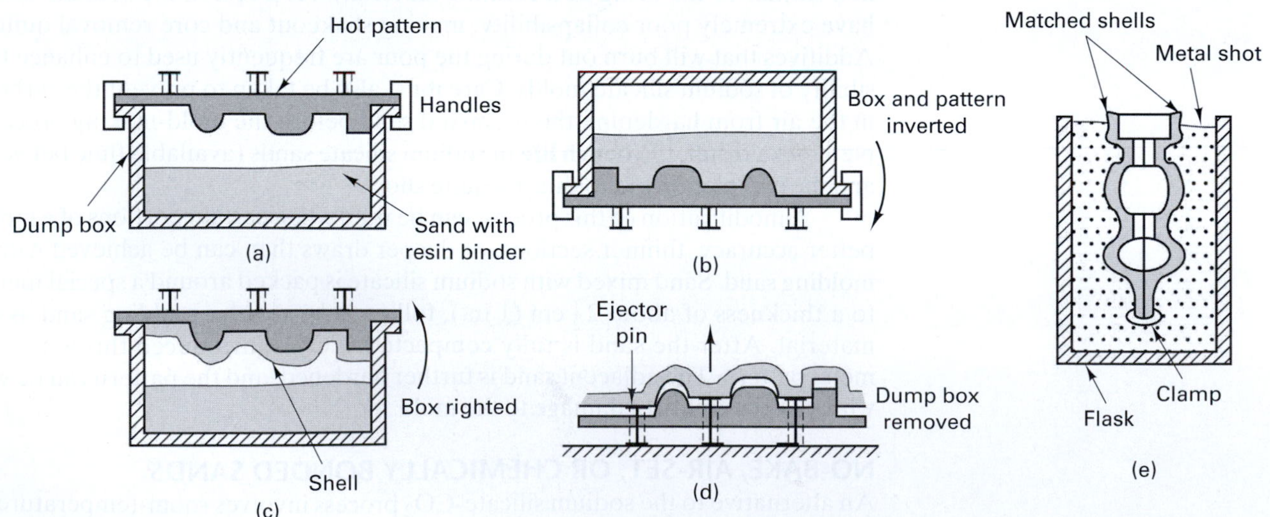

FIGURE 12-18 Schematic of the dump-box version of shell molding. (a) A heated pattern is placed over a dump box containing granules of resin-coated sand. (b) The box is inverted and the heat forms a partially-cured shell around the pattern. (c) The box is righted, the top is removed, and the pattern and partially-cured sand is placed in an oven to further cure the shell. (d) The shell is stripped from the pattern. (e) Matched shells are then joined and supported in a flask ready for pouring.

and cross-links) a layer of material. This forms a strong, solid-bonded region adjacent to the pattern. The actual thickness of cured material depends on the pattern temperature and the time of contact but typically ranges between 10 and 20 mm (0.4 and 0.8 in.).

2. The pattern and sand mixture are then inverted, allowing the excess (uncured) sand to drop free. Only the layer of partially cured material remains adhered to the pattern.
3. The pattern with adhering shell is then placed in an oven where additional heating completes the curing process.
4. The hardened shell, with tensile strength between 350 and 450 psi (2.4 and 3.1 MPa), is then stripped from the pattern.
5. Two or more shells are then clamped or glued together with a thermoset adhesive to produce a mold, which may be poured immediately or stored almost indefinitely.
6. To provide extra support during the pour, shell molds are often placed in a pouring jacket and surrounded with metal shot, sand, or gravel.

Because the shell is formed and partially cured around a metal pattern, the process offers excellent dimensional accuracy. Tolerances of 0.08 to 0.13 mm (0.003 to 0.005 in.) are quite common. Shell-mold sand is typically finer than ordinary foundry sand and, in combination with the plastic resin, enables fine detail and a very smooth casting surface. Cleaning, machining, and other finishing costs can be significantly reduced, and the mold process offers an excellent level of product consistency.

Figure 12-19 shows a set of metal patterns, the two shells before clamping, and the resulting shell-mold casting. Machines for making shell molds vary from simple ones for small operations, to large, completely automated devices for mass production. The cost

FIGURE 12-19 (Top) Two halves of a shell-mold pattern. (Bottom) The two shells before clamping, and the final shell-mold casting with attached pouring basin, runner and riser. *(Courtesey Roberts Sinto Corp./ Shalco Systems, Lansing, MI)*

TABLE 12-3 Shell-Mold Casting

Process: Sand coated with a thermosetting plastic resin is dropped onto a heated metal pattern, which cures the resin. The shell segments are stripped from the pattern and assembled. When the poured metal solidifies, the shell is broken away from the finished casting.

Advantages: Faster production rate than sand molding, high dimensional accuracy with smooth surfaces. Thin shells and hollow cores enable use of more expensive molding materials.

Limitations: Requires expensive metal patterns. Plastic resin adds to cost; part size is limited.

Common metals: Cast irons and casting alloys of aluminum and copper.

Size limits: 30 g (1 oz) minimum; usually less than 10 kg (25 lb); mold area usually less than 0.3 m^2 (500 in^2).

Thickness limits: Minimums range from 0.15 to 0.6 cm ($\frac{1}{16}$ to $\frac{1}{16}$ in.), depending on material.

Typical tolerances: Approximately 0.005 cm/cm or in/in.

Draft allowance: $\frac{1}{4}$ or $\frac{1}{2}$ degree.

Surface finish: $\frac{1}{3}$ – 4.0 microns (50–150 μin.) in.) rms.

of a metal pattern is often rather high, and its design must include the gate and runner system, because these cannot be cut after molding. Large amounts of expensive binder are required, but the amount of material actually used to form a thin shell is not that great. High productivity, low labor costs, smooth surfaces, and a level of precision that reduces the amount of subsequent machining all combine to make the process economical for even moderate quantities. The thin shell provides for the easy escape of gases that evolve during the pour, and the volume of evolved gas is rather low because of the absence of moisture in the mold material. When the shell becomes hot, some of the resin binder burns out, providing excellent collapsibility and shakeout characteristics. In addition, both the molding sand and completed shells can be stored for indefinite periods of time. Table 12-3 summarizes the features of shell molding.

OTHER SAND-BASED MOLDING METHODS

Over the years, a variety of processes have been proposed to overcome some of the limitations or difficulties of the more traditional sand-based molding methods. While few have become commercially significant, several are included here to illustrate the nature of these efforts.

In the **V-process** or **vacuum molding,** a vacuum performs the role of the sand binder. Figure 12-20 depicts the production sequence, which begins by draping a thin sheet of heat-softened plastic (similar to Saran-wrap) over a special vented pattern, which is often made of urethane or other plastic. A vacuum is applied within the pattern, drawing the sheet tight to its surface. A special vacuum flask is then placed over

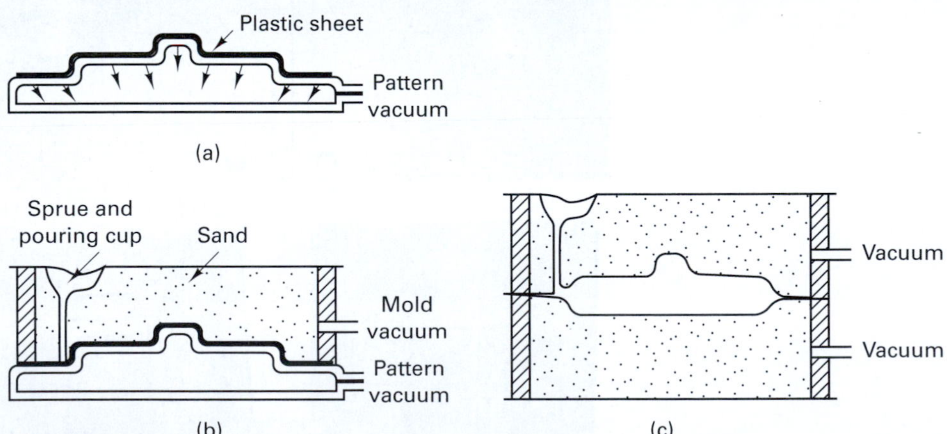

FIGURE 12-20 Schematic of the V-process or vacuum molding. (a) A vacuum is pulled on a pattern, drawing a heated shrink-wrap plastic sheet tightly against it. (b) A vacuum flask is placed over the pattern and filled with dry unbonded sand; a pouring basin and sprue are formed; the remaining sand is leveled; a second heated plastic sheet is placed on top; and a mold vacuum is drawn to compact the sand and hold the shape. (c) With the mold vacuum being maintained, the pattern vacuum is then broken and the pattern is withdrawn. The cope and drag segments are assembled, and the molten metal is poured.

the pattern; the flask is filled with vibrated dry, unbonded sand; a sprue and pouring cup are formed; and a second sheet of heated plastic is placed over the mold. A vacuum is then drawn on the flask itself, compacting the sand to provide the necessary strength and hardness. The pattern vacuum is released, and the pattern is then withdrawn. The other segment of the two-part cope-and-drag mold is made in a similar fashion, and the mold halves are assembled to produce a plastic-lined cavity. The mold is then poured with a vacuum of 300 to 600 torr being maintained in both the cope and drag segments of the flask. During the pour, the thin plastic film melts and vaporizes and is replaced immediately by metal, allowing the vacuum to continue holding the sand in shape until the casting has cooled and solidified. When the vacuum is released, the sand reverts to its loose, unbonded state and falls away from the casting.

Products can be made with zero draft and walls can be as thin as 3 mm (0.125 in.) over large areas. There is no pattern wear because the sand never touches the pattern. With the vacuum serving as the binder, there is a total absence of moisture-related defects; binder cost is eliminated; and the loose, dry sand is completely and directly reusable. With no clay, water, or other binder to impair permeability, finer sands can be used, resulting in better surface finish in the resulting castings. With no burning binders (only the thin plastic sheets are burned), few fumes are generated during the pouring operation. Shakeout characteristics are exceptional, because the mold collapses when the vacuum is released. Unfortunately, the process is relatively slow because of the additional steps and the time required to pull a sufficient vacuum. The V-process is used primarily for the production of prototype, frequently modified, or low- to medium-volume parts (more than 10 but less than 15,000).

In the **Eff-set process,** wet sand with just enough clay to prevent mold collapse is packed around a pattern. The pattern is removed and the surface of the mold is sprayed with liquid nitrogen. The ice that forms serves as the binder, and the molten metal is poured into the mold while the surface is in its frozen condition. This process offers low binder cost and excellent shakeout, but is not being used in a commercial operation.

12.3 CORES AND CORE MAKING

Casting processes are unique in their ability to easily incorporate complex internal cavities or reentrant sections. To produce these features, however, it is often necessary to use **cores** as part of the mold. Figure 12-21 shows an example of a product that uses multiple cores to produce the various cylinders, cooling passages, and other internal features. While cores constitute an added cost, they significantly expand the capabilities of the process.

Cores can often be used to improve casting design and optimize processes. Consider the simple belt pulley shown schematically in Figure 12-22. Various methods of fabrication are suggested in the four sketches, beginning with the casting of a solid form and the subsequent machining of the through-hole for the drive shaft. A large volume of metal would have to be removed by a secondary machining process. A more economical approach would be to make the pulley with a cast-in hole. In Figure 12-22b each half of the pattern includes a tapered hole, which fills with the same green sand being used for the remainder of mold. These protruding sections are an integral part of the mold, but they are also known as **green-sand cores.** Green-sand cores have a relatively low strength. If the protrusions are long or narrow, it might be difficult to withdraw the pattern without breaking them, or they may not have enough strength to even support their own weight. For long cores, a considerable amount of machining may still be required to remove the draft that must be provided on the pattern. In addition, green sand cores are not an option for more complex shapes where it is often impossible to withdraw the pattern.

Dry-sand cores can overcome some of the cited difficulties. These cores are produced separate from the remainder of the mold, and are then inserted into core prints that hold them in position. The sketches in Figure 12-22c and Figure 12-22d show dry-sand cores in the vertical and horizontal positions. Dry-sand cores can be made in a number of ways. In each, the sand, mixed with some form of binder, is packed into a

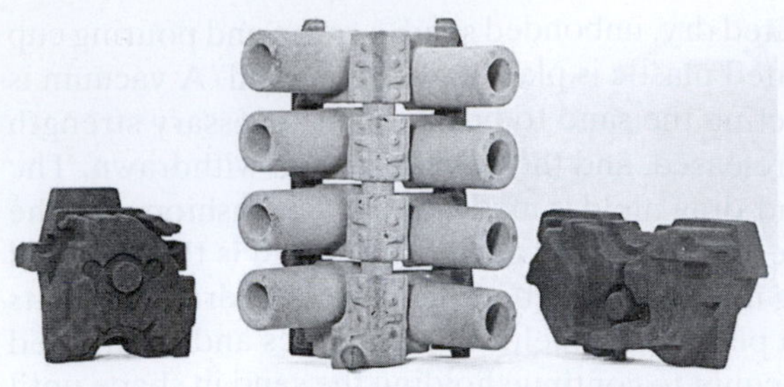

FIGURE 12-21 V-8 engine block (bottom center) and the five dry-sand cores that are used in the construction of its mold. *(Courtesy General Motors Corporation, Detroit, MI)*

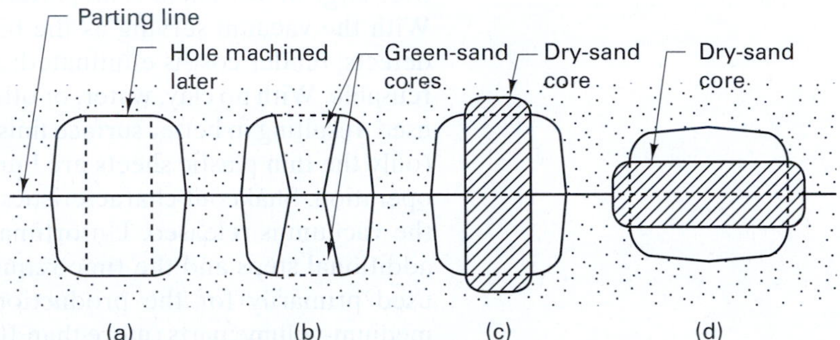

FIGURE 12-22 Four methods of making a hole in a cast pulley. Three involve the use of a core.

wood or metal core box that contains a cavity of the desired shape. A **dump core box** such as the one shown in Figure 12-23 offers the simplest approach. Sand is packed into the cavity and scraped level with the top surface (which acts like the parting line in a traditional mold). A wood or metal plate is then placed over the top of the box, and the box is inverted and lifted, leaving the molded sand segment resting on the plate. After baking or hardening, the various core segments are assembled with hot-melt glue or some other bonding agent. Rough spots along the parting line are removed with files or sanding belts, and the final core may be given a thin coating to provide a smoother surface or greater resistance to heat. Graphite, silica, or mica can be sprayed or brushed onto the surface.

Single-piece cores can be made in a **split core box.** Two halves of a core box are clamped together, with an opening in one or both ends through which sand is introduced and rammed. After the sand is compacted, the halves of the box are separated to

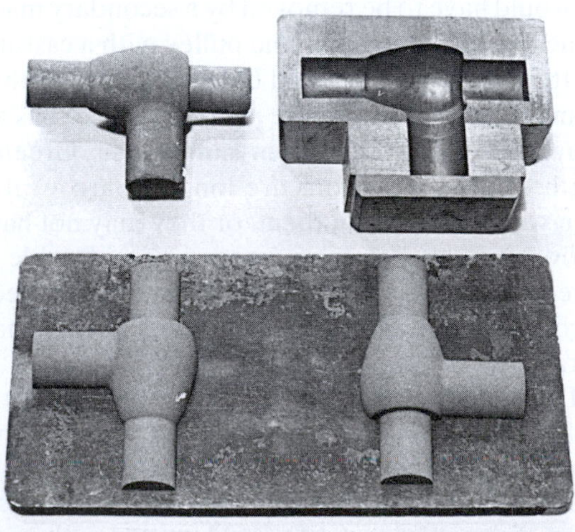

FIGURE 12-23 (Upper right) A dump-type core box; (bottom) two core halves ready for baking; (upper left) a completed core made by gluing two opposing halves together. *(E. Paul DeGarmo)*

permit removal of the core. Cores with a uniform cross section can be formed by a core-extruding machine and cut to the desired length as the product emerges. The individual cores are then placed in core supports for subsequent hardening. More complex cores can be made in core-blowing machines that use separating dies and receive the sand in a manner similar to injection molding or diecasting.

Cores are frequently the most fragile part of a mold assembly. To provide the necessary strength, the various core-making processes utilize a number of special binders. In the **core-oil process,** sand is blended with about 1% vegetable or synthetic oil, along with 2 to 4% water and about 1% cereal or clay to help develop green strength (i.e., to help retain the shape prior to curing). The wet sand is blown or rammed into a relatively simple core box at room temperature. The fragile uncured cores are then gently transferred to flat plates or special supports and placed in convection ovens at 200 to 260°C (400 to 500°F) for curing. The heat causes the binder to cross-link or polymerize, producing a strong organic bond between the grains of sand. While the process is simple and the materials are inexpensive, the dimensional accuracy of the resultant cores is often difficult to maintain.

In the **hot-box method,** sand blended with a liquid thermosetting binder and catalyst is packed into a core box that has been heated to around 230°C (450°F). When the sand is heated, the initial stages of curing begin within 10 to 30 sec. After this brief period, the core can be removed from the pattern and will hold its shape during subsequent handling. For some materials, the cure completes through an exothermic curing reaction. For others, further baking is required to complete the process.

In the preceding methods, cores must be handled in an uncured or partially cured state, and breakage or distortion is not uncommon. Processes that produce finished cores while still in the core box, and do not require heating operations, would appear to offer distinct advantages.

In the **cold-box** or **gas-hardened processes,** resin-coated sand is first blown into a room-temperature core box, which can be made from wood, metal, or even plastic. The box is sealed, and a gas or vaporized catalyst is then passed through the permeable sand to polymerize the resin binder. In a variation of the process, hollow cores are produced by introducing small amounts of curing gas through holes in the core box pattern, with the uncured sand in the center being dumped and reused. Unfortunately, the required gases tend to be either toxic (an amine gas with phenolic urethane resin) or odorous (SO_2 with furan or acrylic), making special handling of both incoming and exhaust gas a process requirement. The cold-box process is most attractive for large numbers of small-sized products. Cold-box cores can be stored for lengthy periods of time prior to use **(shelf life).**

Room-temperature cores can also be made with the air-set or no-bake sands. These systems eliminate the gassing operation of the cold-box process through the use of a reactive organic resin and a curing catalyst. As discussed previously, there is only a brief period of time to form the core once the components have been mixed. Shell molding is another core-making alternative, producing hollow cores with excellent strength and permeability.

Selecting the actual method of core production is usually based on a number of considerations, including production quantity, production rate, required precision, required surface finish, and the metal being poured. Certain metals may be sensitive to gases that are emitted from the cores when they come into contact with the hot metal. Other materials with low pouring temperatures may not break down the binder sufficiently to provide collapsibility and easy removal from the final casting.

To function properly, casting cores must have the following characteristics:

1. Sufficient strength before hardening if they will be handled in the "green" condition.
2. Sufficient hardness and strength after hardening to withstand handling and the forces of the casting process. As metal fills the mold, most cores want to "float." The cores must be strong enough to resist the induced stresses, and the supports must be sufficient to hold them in place. Flowing metal can also cause surface erosion. Compressive strength should be between 100 and 500 psi (0.7 and 3.5 MPa).

3. A smooth surface.
4. Minimum generation of gases when heated by the pour.
5. Adequate permeability to permit the escape of gas. Because cores are largely surrounded by molten metal, the gases must escape through the core.
6. Adequate refractoriness. Being surrounded by hot metal, cores can become quite a bit hotter than the adjacent mold material. They should not melt or adhere to the casting.
7. Collapsibility. After pouring, the cores must be weak enough to permit the casting to shrink as it cools, thereby preventing cracking. In addition, the cores must be easily removed from the interior of the finished product via shakeout.

Various techniques have been developed to enhance the natural properties of cores and core materials. Additional strength can be imparted by the addition of internal wires or rods. Collapsibility can be enhanced by producing hollow cores or by placing a material such as straw in the center. Hollow cores may be used to provide for the escape of trapped or evolved gases. Vent holes can be formed by pushing small wires into the core, and coke or cinders are sometimes placed in the center of large cores to enhance venting.

Because the gases must be expelled from the casting, and the core material itself must be removed to produce the desired hole or cavity, the cores must be connected to the outer surfaces of the mold cavity. Recesses at these connection points, known as **core prints,** are used to support the cores and hold them in proper position during mold filling. The dry-sand cores in Figure 12-22c and 12-22d are supported by core prints.

If the cores do not pass completely through the casting where they can be supported on both ends, a single core print may not be able to provide adequate support. Additional measures may also be necessary to support the weight of large cores or keep lighter ones from becoming buoyant as the molten metal fills the cavity. Small metal supports, called **chaplets,** can be placed between cores and the surfaces of a mold cavity, as illustrated in Figure 12-24. Because the chaplets are positioned within the mold cavity, they become an integral part of the finished casting. Chaplets should, therefore, be of the same, or at least comparable, composition as the material being poured. They should be large enough that they do not completely melt and permit the core to move, but small enough that their surface melts and fuses with the metal being cast. Because chaplets are one more source of possible defects and may become a location of weakness in the finished casting, efforts are generally made to minimize their use.

Additional sections of mold material can also be used to produce castings with reentrant angles. Figure 12-25 depicts a round pulley with a recessed groove around its perimeter. By using a third segment of flask, called a **cheek,** and adding a second parting

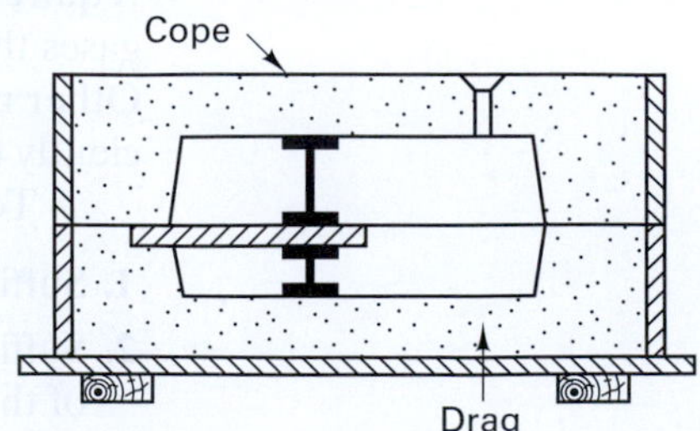

FIGURE 12-24 (Left) Typical chaplets. (Right) Method of supporting a core by use of chaplets (relative size of the chaplets is exaggerated). *(E. Paul DeGarmo)*

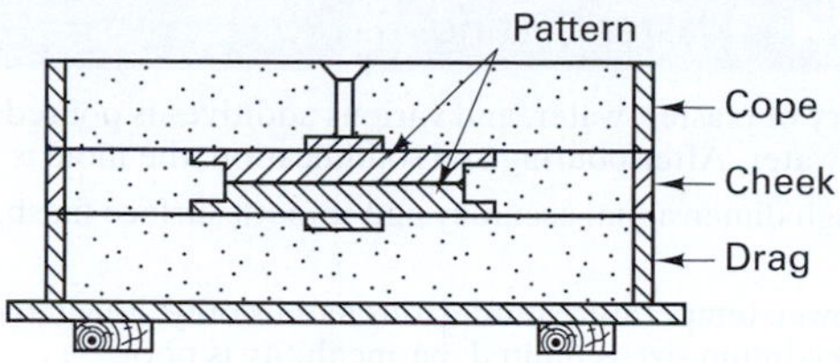

FIGURE 12-25 Method of making a reentrant angle or inset section by using a three-piece flask.

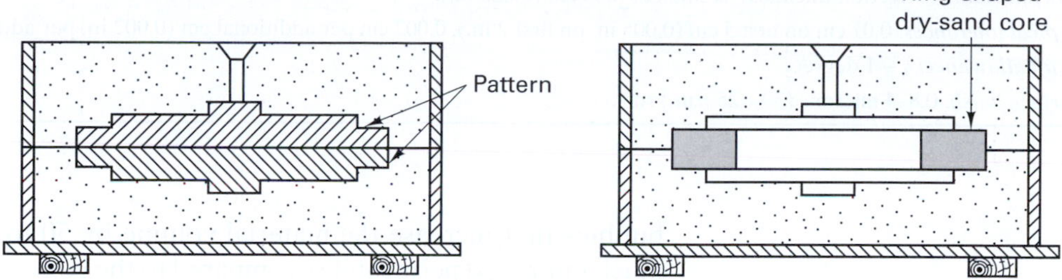

FIGURE 12-26 Molding an inset section using a dry-sand core.

plane, the entire mold can be made by conventional green-sand molding around withdrawable patterns. While additional molding operations are required, this may be an attractive approach for small production runs.

If we want to produce a large number of identical pulleys, rapid machine molding of a simple green-sand mold might be preferred. As shown in Figure 12-26, the pattern would be modified to include a seat for an inserted ring-shaped core. Molding time is reduced at the expense of a core box and a separate core-making operation.

12.4 OTHER EXPENDABLE-MOLD PROCESSES WITH MULTIPLE-USE PATTERNS

PLASTER MOLD CASTING

In **plaster molding,** the mold material is plaster of paris (also known as calcium sulfate or gypsum), combined with various additives to improve green strength, dry strength, permeability, and castability. Talc or magnesium oxide can be added to prevent cracking and reduce the setting time. Lime or cement helps to reduce expansion during baking. Glass fibers can be added to improve strength, and sand can be used as a filler.

The mold material is first mixed with water, and the creamy slurry is then poured over a metal pattern (wood patterns tend to warp or swell) and allowed to set. Hydration of the plaster produces a hard mold that can be easily stripped from the pattern. (*Note:* Flexible rubber patterns can be used when complex angular surfaces or reentrant angles are required. The plaster is strong enough to retain its shape during pattern removal.) The plaster mold is then baked to remove excess water, assembled, and poured.

With metal patterns and plaster mold material, surface finish and dimensional accuracy are both excellent. Cooling is slow because the plaster has low heat capacity and low thermal conductivity. The poured metal stays hot and can flow into thin sections and replicate fine detail, which can often reduce machining cost. Unfortunately, plaster casting is limited to the lower-melting-temperature nonferrous alloys (such as aluminum, copper, magnesium, and zinc). At the high temperatures of ferrous metal casting, the plaster would first undergo a phase transformation and then melt, and the water of hydration can cause the mold to explode. Table 12-4 summarizes the features of plaster mold casting.

The **Antioch process** is a variation of plaster mold casting where the mold material is comprised of 50% plaster and 50% sand, mixed with water. An autoclave process is used to prepare the molds, which offer improved permeability and reduced solidification time. The addition of a foaming agent to a plaster–water mix can add fine air

TABLE 12-4 Plaster Casting

Process: A slurry of plaster, water, and various additives is poured over a pattern and allowed to set. The pattern is removed, and the mold is baked to remove excess water. After pouring and solidification, the mold is broken and the casting is removed.

Advantages: High dimensional accuracy and smooth surface finish; can reproduce thin sections and intricate detail to make net- or near-net-shaped parts.

Limitations: Lower-temperature nonferrous metals only; long molding time restricts production volume or requires multiple patterns; mold material is not reusable; maximum size is limited; permeability is poor.

Common metals: Primarily aluminum and copper.

Size limits: As small as 30 g (1 oz) but usually less than 7 kg (15 lb).

Thickness limits: Section thickness as small as 0.06 cm (0.025 in.).

Typical tolerances: 0.01 cm on first 5 cm (0.005 in. on first 2 in.), 0.002 cm per additional cm (0.002 in. per additional in.)

Draft allowance: $\frac{1}{2}-1$ degree.

Surface finish: 0.8–4 microns (30–125 μin.) rms.

bubbles that increase the material volume by 50 to 100%. The resulting molds have much improved permeability compared to the conventional process.

CERAMIC MOLD CASTING

Ceramic mold casting (summarized in Table 12-5) is similar to plaster mold casting, except that the mold is now made from a ceramic material that can withstand the higher-melting-temperature metals. Much like the plaster process, ceramic molding can produce thin sections, fine detail, and smooth surfaces, thereby eliminating a considerable amount of finish machining. These advantages, however, must be weighed against the greater cost of the mold material. For large molds, the ceramic can be used to produce a facing around the pattern, which is then backed up by a less-expensive material such as reusable fireclay.

One of the most popular of the ceramic molding techniques is the **Shaw process.** A reusable pattern is positioned inside a slightly tapered flask, and a slurrylike mixture of refractory aggregate, hydrolyzed ethyl silicate, alcohol, and a gelling agent is poured on top. This mixture sets to a rubbery state that permits removal of both the pattern and the flask. The mold surface is then ignited with a torch. Most of the volatiles are consumed during the "burn-off," and a three-dimensional network of microscopic cracks (microcrazing) forms in the ceramic. The gaps are small enough to prevent metal penetration but large enough to provide venting of air and gas (permeability) and to accommodate both the thermal expansion of the ceramic particles during the pour and the subsequent shrinkage of the solidified metal. A baking operation then removes all of the remaining volatiles, making the mold hard and rigid. Ceramic molds are often preheated prior to pouring to ensure proper filling and to control the solidification characteristics of the metal.

TABLE 12-5 Ceramic Mold Casting

Process: Stable ceramic powders are combined with binders and gelling agents to produce the mold material.

Advantages: Intricate detail, close tolerances, and smooth finish.

Limitations: Mold material is costly and not reusable.

Common metals: Ferrous and high-temperature nonferrous metals are most common; can also be used with alloys of aluminum, copper, magnesium, titanium, and zinc.

Size limits: 100 grams to several thousand kilograms (several ounces to several tons).

Thickness limits: As thin as 0.06 cm (0.025 in.); no maximum.

Typical tolerances: 0.01 cm on the first 2.5 cm (0.005 in. on the first in.), 0.003 cm per each additional cm (0.003 in. per each additional in.).

Draft allowances: 1° preferred.

Surface finish: 2–4 microns (75–150 μin.) rms.

Like plaster molding, the ceramic molding process can effectively produce small-size castings in small to medium quantities.

EXPENDABLE GRAPHITE MOLDS

For metals such as titanium, which tend to react with many of the more common mold materials, powdered **graphite** can be combined with additives, such as cement, starch, and water and compacted around a pattern. After "setting," the pattern is removed and the mold is fired at 1000°C (1800°F) to consolidate the graphite. The casting is poured, and the mold is broken to remove the product.

RUBBER-MOLD CASTING

Artificial elastomers can also be compounded in liquid form and poured over a pattern to produce a semi-rigid mold. These molds are sufficiently flexible to permit stripping from an intricate shape or patterns with reverse-taper surfaces. Unfortunately, rubber molds are generally limited to small castings and low melting- point materials. The wax patterns used in investment casting are often made by **rubber-mold casting,** as are small quantities of finished parts made from plastics or metals which can be poured at temperatures below 250°C (500°F). Molds for plaster casting can also be made by pouring the plaster into reverse-pattern rubber molds.

12.5 EXPENDABLE-MOLD PROCESSES USING SINGLE-USE PATTERNS

INVESTMENT CASTING

Investment casting is actually a very old process—it was used in ancient China and Egypt and more recently performed by dentists and jewelers for a number of years. It was not until the end of World War II, however, that it attained a significant degree of industrial importance. Products such as rocket components and jet engine turbine blades required the fabrication of high-precision complex shapes from high-melting-point metals that are not easily machined. Investment casting offers almost unlimited freedom in both the complexity of shapes and the types of materials that can be cast, and millions of investment castings are now produced each year.

Investment casting uses the same type of molding aggregate as the ceramic molding process, and typically involves the following sequential steps:

1. *Produce a master pattern* i.e. a modified replica of the desired product made from metal, wood, plastic, or some other easily worked material.
2. *From the master pattern, produce a master die.* This can be made from low-melting-point metal, steel, or possibly even wood. If a low-melting-point metal is used, the die might be cast directly from the master pattern. Rubber molds can also be made directly from the master pattern. Steel dies are often machined directly, eliminating the need for step 1.
3. *Produce wax patterns.* Patterns are then made by pouring molten wax into the master die, or by injecting it under pressure (injection molding), and allowing it to harden. Release agents, such as silicone sprays are used to assist in pattern removal. Polystyrene plastic is an alternate pattern material and may be preferred for producing thin and complex surfaces where its higher strength and greater durability are desired. If cores are required, they can generally be made from soluble wax or ceramic. The soluble wax cores are dissolved out of the patterns prior to further processing, while the ceramic cores remain and are not removed until after solidification of the metal casting.
4. *Assemble the wax patterns onto a common wax sprue.* Using heated tools and melted wax, a number of wax patterns can be attached to a central sprue and runner system to create a pattern cluster, or a **tree.** If the product is sufficiently complex that its pattern could not be withdrawn from a single master die, the pattern may be made in pieces and assembled prior to attachment.

5. *Coat the cluster or tree with a thin layer of investment material.* This step is usually accomplished by dipping into a watery slurry of finely ground refractory material. A thin but very smooth layer of investment material is deposited onto the wax pattern, ensuring a smooth surface and good detail in the final product.
6. *Form additional investment around the coated cluster.* After the initial layer has dried, the cluster can be redipped, but this time the sand or other refractory aggregate is rained over the wet surface, a process called stuccoing. After drying, this process is repeated until the investment coating has the desired thickness (typically 5 to 15 mm [3/16 to 5/8 in.] with up to eight layers). As an alternative, the single-dipped cluster can be placed upside down in a flask and liquid investment material poured around it. The flask is then vibrated to remove entrapped air and ensure that the investment material now surrounds all surfaces of the cluster.
7. *Allow the investment to fully harden.*
8. *Remove the wax pattern from the mold by melting or dissolving.* Molds or trees are generally placed upside down in an oven, where the wax can melt and run out, and any subsequent residue vaporizes. This step is the most distinctive feature of the process because it enables a complex pattern to be removed from a single-piece mold. Extremely complex shapes can be readily cast. (*Note:* In the early years of the process, only small parts were cast, and when the molds were placed in the oven, the molten wax was absorbed into the porous investment. Because the wax "disappeared," the process was called the **lost-wax process,** and the name is still used.) Figure 9-7, in the chapter on ceramic materials, shows an investment casting mold at this stage of the process with part of the wall cut away to reveal the inner cavities.
9. *Heat the mold in preparation for pouring.* Heating to 550 to 1100°C (1000 to 2000°F) ensures complete removal of the mold wax, cures the mold to give added strength, and allows the molten metal to retain its heat and flow more readily into all of the thin sections and details. Mold heating also gives better dimensional control because the mold and the metal can shrink together during cooling.
10. *Pour the molten metal.* While gravity pouring is the simplest, other methods may be used to ensure complete filling of the mold. When complex, thin sections are involved, mold filling may be assisted by positive air pressure, evacuation of the air from the mold, or some form of centrifugal process.
11. *Remove the solidified casting from the mold.* After solidification, techniques such as mechanical chipping or vibration, high-pressure water jet, or sand blasting are used to break the mold and remove the mold material from the metal casting. Cutoff and finishing completes the process.

Figure 12-27 depicts the investment procedure where the investment material fills the entire flask, and Figure 12-28 shows the shell-investment method. Table 12-6 summarizes the features of investment casting.

Compared to other methods of casting, investment casting is a complex process and tends to be rather expensive. However, its unique advantages can often justify its use, and many of the steps can be easily automated. Extremely complex shapes can be cast as a single piece. Thin sections, down to 0.40 mm (0.015 in.), can be produced. No draft is required, and no parting line is present. Excellent dimensional precision can be achieved in combination with very smooth as-cast surfaces. Machining can often be completely eliminated or greatly reduced. When machining is required, allowances of as little as 0.4 to 1 mm (0.015 to 0.040 in.) are usually ample. These capabilities are especially attractive when making products from the high-melting-temperature, difficult-to-machine metals that cannot be cast with plaster- or metal-mold processes.

While most investment castings are less than 10 cm (4 in.) in size and weigh less than $\frac{1}{2}$ kg (1 lb), castings up to 1 m (36 in.) and 45 kg (100 lb) have been produced. Products ranging from stainless steel or titanium golf club heads to superalloy turbine

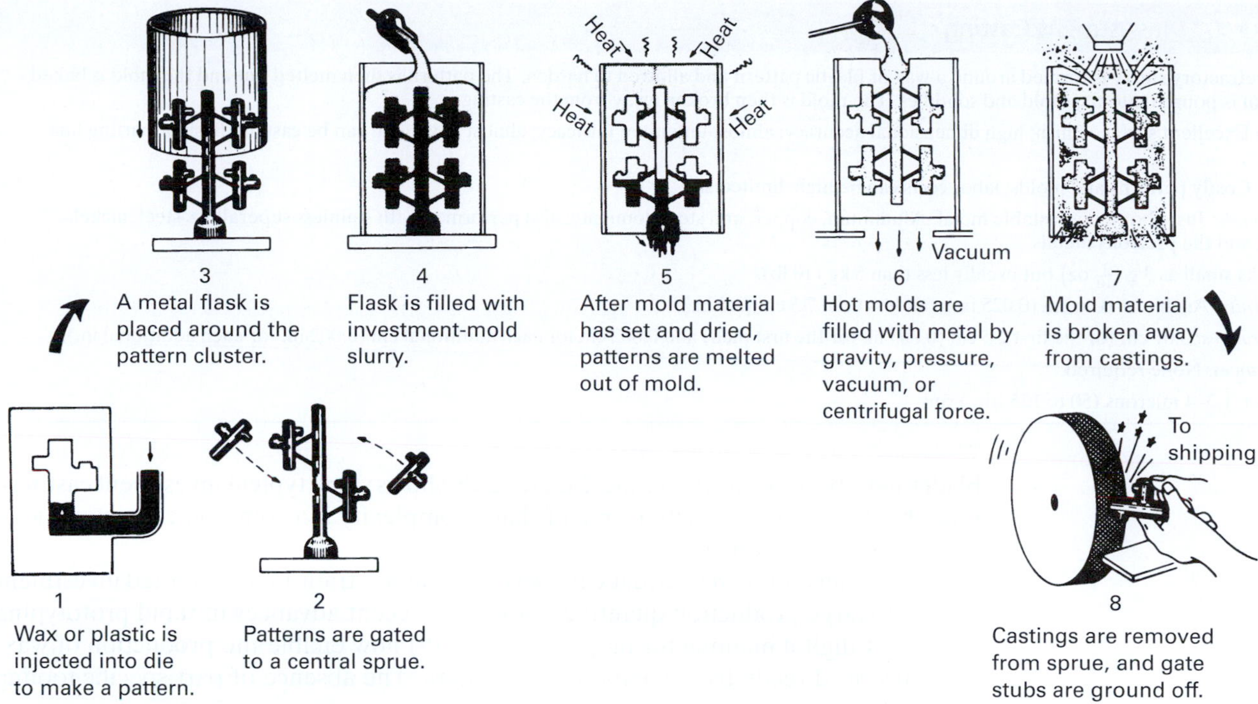

FIGURE 12-27 Investment casting steps for the flask-cast method. *(Courtesy of Investment Casting Institute, Dallas, TX)*

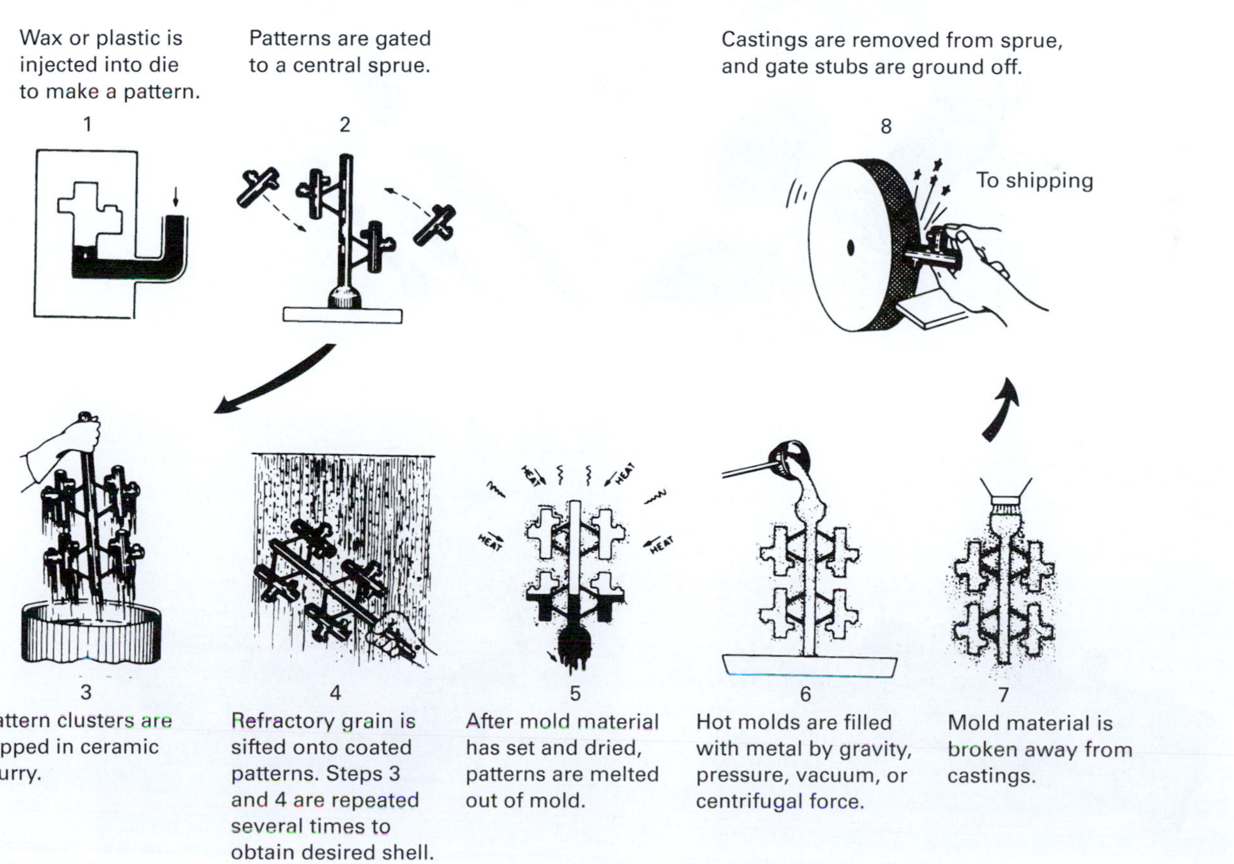

FIGURE 12-28 Investment casting steps for the shell-casting procedure. *(Courtesy of Investment Casting Institute, Dallas, TX)*

TABLE 12-6 Investment Casting

Process: A refractory slurry is formed around a wax or plastic pattern and allowed to harden. The pattern is then melted out and the mold is baked. Molten metal is poured into the mold and solidifies. The mold is then broken away from the casting.

Advantages: Excellent surface finish; high dimensional accuracy; almost unlimited intricacy; almost any metal can be cast; no flash or parting line concerns.

Limitations: Costly patterns and molds; labor costs can be high; limited size.

Common metals: Just about any castable metal. Aluminum, copper, and steel dominate; also performed with stainless superalloys steel, nickel, magnesium, and the precious metals.

Size limits: As small as 3 g ($\frac{1}{10}$ oz) but usually less than 5 kg (10 lb).

Thickness limits: As thin as 0.06 cm (0.025 in.), but less than 7.5 cm (3.0 in.).

Typical tolerances: 0.01 cm for the first 2.5 cm (0.005 in. for the first inch) and 0.002 cm for each additional cm (0.002 in. for each additional in.).

Draft allowances: None required.

Surface finish: 1.3–4 microns (50 to 125 μin.) rms.

blades have become quite routine. Figure 12-29 shows some typical investment castings. One should note that a high degree of shape complexity is a common characteristic of investment cast products.

The high cost of dies to make the wax patterns has traditionally limited investment casting to large production quantities. However, recent advances in rapid prototyping and direct-digital manufacturing (see Chapter 19) now enable the production of wax-like patterns directly from computer design data. The absence of part-specific tooling

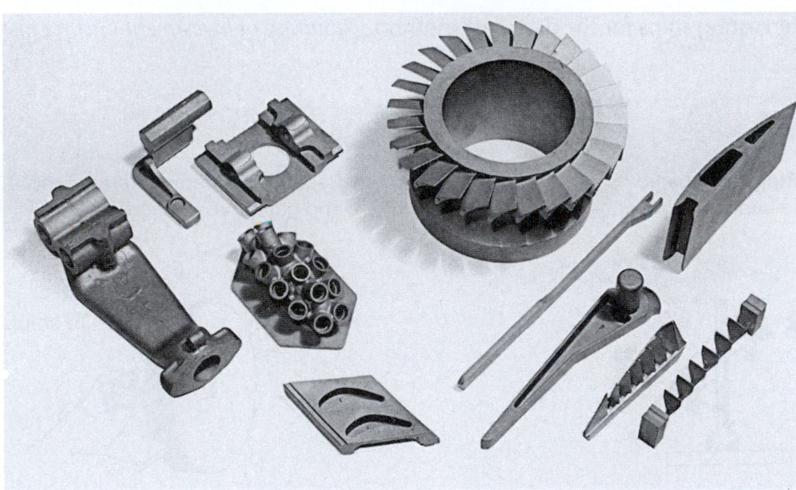

(a)

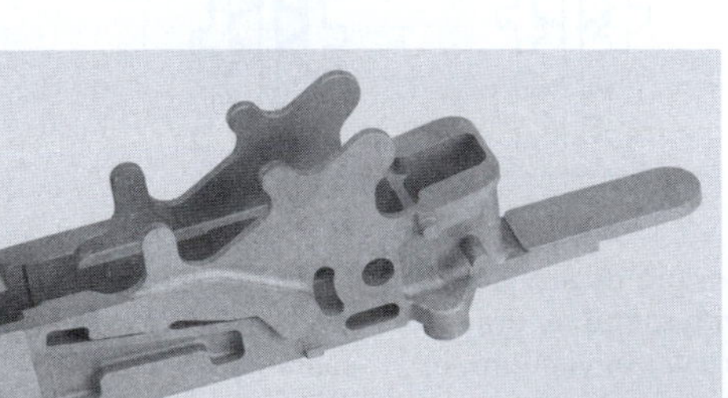

(b)

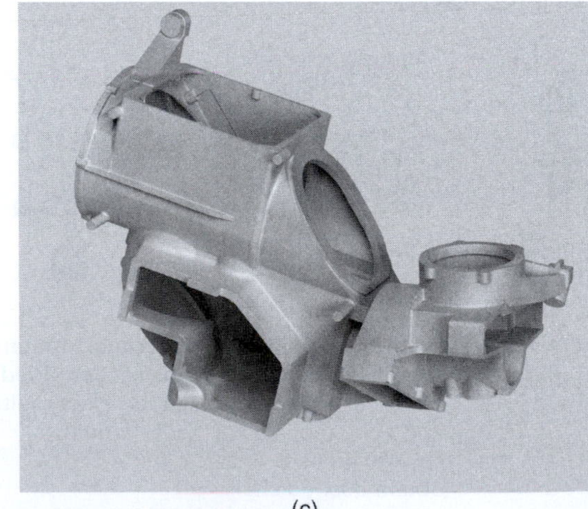

(c)

FIGURE 12-29 Typical parts produced by investment casting: (a) nickel and cobalt superalloys; (b) copper alloy; (c) aluminum alloy. *[(a) Haynes International Inc.; (b and c) O'Fallon Casting]*

now enables the economical casting of one-of-a-kind or small quantity products using the investment methods. The majority of investment castings now fall within the range of 100 to 10,000 pieces per year.

Another attractive feature of direct-digital patternmaking is the opportunity to produce wax patterns that have solid exteriors but honeycomb-like interiors with a high percentage of open volume. There is less material to melt out and considerably less volume expansion during the wax-removal operation. A common problem during wax removal is expansion-induced **shell cracking,** which destroys the shell and results in loss of the time and expense of reaching that stage of manufacture. A polyurethane foam has also been developed for investment cast patterns, with claims of enhanced dimensional stability, simpler pattern removal, and significantly reduced shell cracking.

COUNTER-GRAVITY INVESTMENT CASTING

Counter-gravity investment casting turns the pouring process upside down. In one variation of the process, a ceramic shell mold is placed in an open-bottom chamber with the sprue end down. The open end of the sprue is lowered into a pool of molten metal, and the bottom of the chamber is set against a seal. A vacuum is then induced within the chamber. As the air is withdrawn, the vacuum draws metal up through the central sprue and into the mold. The castings are allowed to solidify, the vacuum is released, and any unsolidified metal in the sprue and gating flows back into the melt. In another variation, a low-pressure inert gas is used to push the molten metal upward into the mold. This approach is discussed in more detail and is also illustrated in the section on low-pressure permanent-mold casting in Chapter 13.

The counter-gravity processes have a number of distinct advantages. Because the molten metal is withdrawn from below the surface of its ladle, it is generally free of slag and dross, and has a very low level of inclusions. The vacuum or low-pressure filling allows the metal to flow with little turbulence, further enhancing metal quality. The reduction in metallic inclusions improves machinability and enables mechanical properties to approach or equal those of wrought material. Because the molten metal is taken directly from the furnace or larger holding vessel, the temperature of the metal can be more consistently maintained from mold to mold.

Because the gating system does not need to control turbulence, simpler gating systems can be used, reducing the amount of metal that does not become product. In the counter-gravity process, between 60 and 95% of the withdrawn metal becomes cast product, compared to a 15 to 50% level for gravity-poured castings. The pressure differential enables metal to flow into thinner sections, and lower "pouring" temperatures can be used, resulting in improved grain structure and better surface finish.

EVAPORATIVE PATTERN (FULL-MOLD AND LOST-FOAM) CASTING

Several limitations are common to most of the casting processes that have been presented. Some form of pattern is usually required, and this pattern may be costly to design and fabricate. Pattern costs may be hard to justify, especially when the number of identical castings is rather small or the part is extremely complex. In addition, reusable patterns must be withdrawn from the mold, and this withdrawal often requires some form of design modification or compromise, division into multiple pieces, or special molding procedures. Investment casting overcomes the withdrawal limitations through single-use patterns that can be removed by melting and vaporization. Unfortunately, investment casting has its own set of limitations, including a large number of individual operations and the need to remove the investment material from the finished casting.

In the **evaporative pattern process,** the pattern is made of **expanded polystyrene (EPS),** or expanded polymethylmethacrylate (EPMMA), and remains in the mold. During the pour, the heat of the molten metal melts and burns the polymer, and the molten metal fills the space that was previously occupied by the pattern.

When small quantities are required, patterns can be cut by hand or machined from pieces of foamed polystyrene (a material similar to that used in Styrofoam drinking cups). This material is extremely light in weight and can be cut by a number of methods, including tools as simple as an electrically heated wire. Preformed material in the form of a pouring basin, sprue, runner segments, and risers can be attached with hot-melt glue to form a complete gating and pattern assembly. Small products can be assembled into clusters or trees, similar to investment casting.

When producing larger quantities of identical parts, a metal mold or die is generally used to mass produce the evaporative patterns. Hard beads of polystyrene are first preexpanded and stabilized. The preexpanded beads are then injected into a heated metal die or mold, usually made from aluminum. A steam cycle causes them to further expand, fill the die, and fuse, after which they are cooled in the mold. The resulting pattern, a replica of the product to be cast, consists of about 2.5% polymer and 97.5% air. Pattern dies can be quite complex and large quantities of patterns can be accurately and rapidly produced. When size or complexity is great, or geometry prevents easy removal, the pattern can be divided into multiple segments, or slices, which are then assembled by hot-melt gluing. The ideal glue should be strong, fast setting, and produce a minimum amount of gas when it decomposes or combusts.

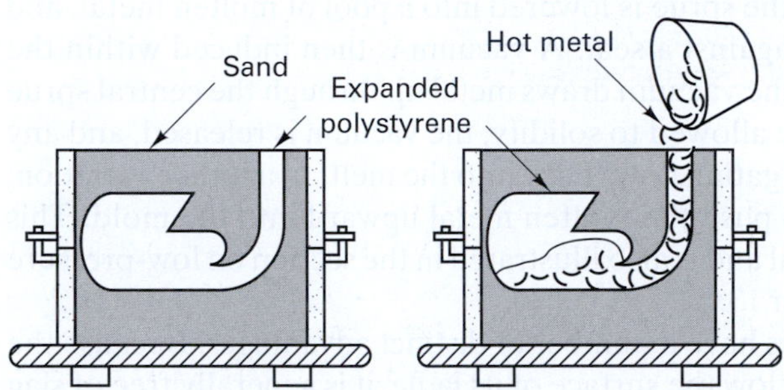

FIGURE 12-30 Schematic of the full-mold process. (Left) An uncoated expanded polystyrene pattern is surrounded by bonded sand to produce a mold. (Right) Hot metal progressively vaporizes the expanded polystyrene pattern and fills the resulting cavity.

After a polystyrene gating system is attached to the polystyrene pattern, there are several options for the completion of the mold. In the **full-mold process,** shown schematically in Figure 12-30, green sand or some type of chemically-bonded (no-bake) sand is compacted around the pattern and gating system, taking care not to crush or distort it. The mold is then poured like a conventional sand-mold casting.

In **lost-foam casting,** depicted schematically in Figure 12-31, the polystyrene assembly is first dipped into a water-based ceramic that wets both external and internal surfaces and dries to form a thin refractory coating. The coating must be thin enough and sufficiently permeable to permit the escape of the molten and gaseous pattern material while preventing metal penetration. At the same time, it must be thick enough and rigid enough to support the foam pattern and prevent mold collapse during pouring. The pattern assembly is then suspended in a one-piece flask and surrounded by fine, unbonded, low-thermal-expansion sand. Vibration ensures that the sand compacts around the pattern and fills all cavities and passages. During the pour, the molten metal melts, vaporizes, and ultimately replaces the expanded polystyrene, while the coating isolates the metal from the loose, unbonded sand and provides a smooth surface finish. After the casting has cooled and solidified, the loose sand is then dumped from the flask, freeing the casting and attached gating system. The backup sand can then be reused, provided the coating residue is removed and the organic condensates are periodically burned off. Figure 12-32 shows the Styrofoam pattern and finished casting of a five-cylinder engine block.

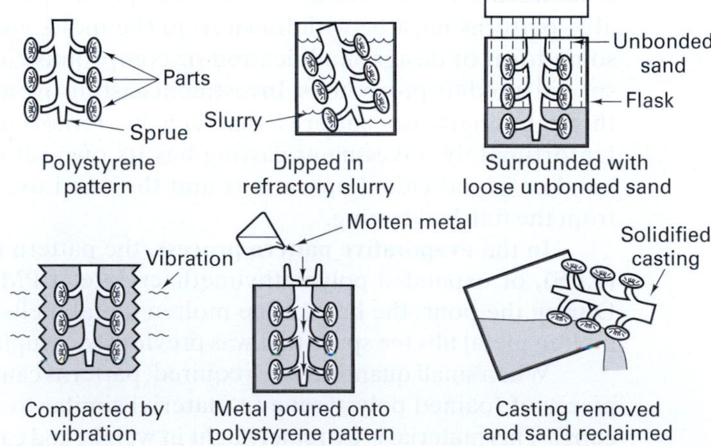

FIGURE 12-31 Schematic of the lost-foam casting process. In this process, the polystyrene pattern is dipped in a ceramic slurry, and the coated pattern is then surrounded with loose, unbonded sand.

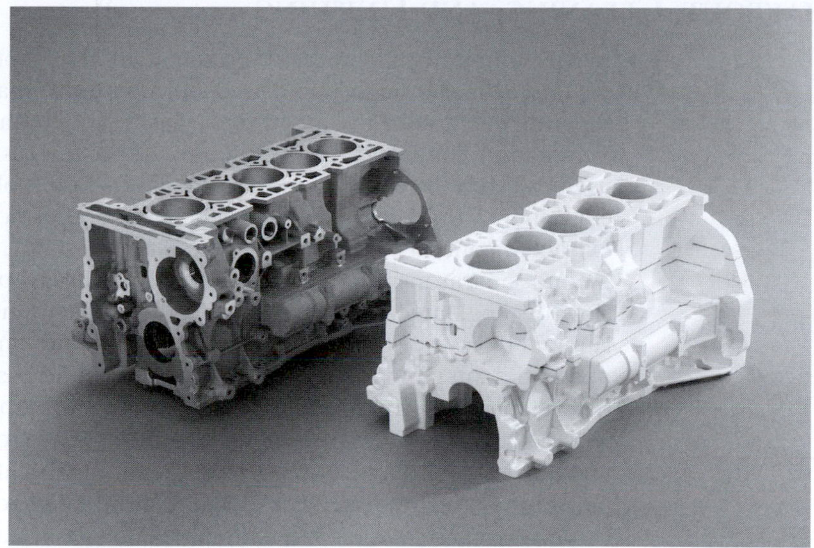

FIGURE 12-32 The Styrofoam pattern and the finished casting of a five-cylinder engine block produced by lost foam casting. *(Courtesy General Motors Corporation, Detroit, MI)*

The full-mold and lost-foam processes can produce castings of any size in both ferrous and nonferrous metals. Because the pattern need not be withdrawn, no draft is required in the design. Complex patterns can be produced to make shapes that would ordinarily require multiple cores, loose-piece patterns, or extensive finish machining. Component integration is a major attraction—multicomponent assemblies can often be replaced by a single casting. Because of the high precision and smooth surface finish, machining and finishing operations can often be reduced or totally eliminated. Fragile or complex-geometry cores are no longer required, and the absence of parting lines eliminates the need to remove associated lines or fins on the metal casting.

Unlike gravity-pour casting, where the mold cavity fills from the bottom, the evaporative-pattern molds fill by the progressive melting and volatilization of the foam. As the molten metal progresses through the pattern, it loses heat. As a result, the material farthest from the gate is the coolest, and solidification tends to proceed in a directional manner back toward the gate. For many castings, risers are not required. Metal yield (product weight versus the weight of poured metal) tends to be rather high. For these and other reasons, evaporative-pattern casting has grown rapidly in popularity and use. Table 12-7 summarizes the process and its capabilities.

In **pressurized lost-foam casting,** the mold is placed in a pressure vessel, which closes within 5 sec of pouring and a pressure of approximately 10 atm is applied during solidification. The applied pressure serves to reduce porosity by more than an order of magnitude over that experienced in normal sand casting. Accompanying the decrease in porosity is a significant increase in high-cycle fatigue strength.

TABLE 12-7 Lost-Foam Casting

Process: A pattern containing a sprue, runners, and risers is made from single or multiple pieces of foamed plastic, such as polystyrene. It is dipped in a ceramic material, dried, and positioned in a flask, where it is surrounded by loose sand. Molten metal is poured directly onto the pattern, which vaporizes and is vented through the sand.

Advantages: Almost no limits on shape and size; most metals can be cast; no draft is required and no flash is present (no parting lines).

Limitations: Pattern cost can be high for small quantities; patterns are easily damaged or distorted because of their low strength.

Common metals: Aluminum, iron, steel, and nickel alloys; also performed with copper and stainless steel.

Size limits: 0.5 kg to several thousand kg (1 lb to several tons).

Thickness limits: As small as 2.5 mm (0.1 in.) with no upper limit.

Typical tolerances: 0.003 cm/cm (0.003 in./in.) or less.

Draft allowance: None required.

Surface finish: 2.5–25 microns (100–1000 μin.) rms.

12.6 SHAKEOUT, CLEANING, AND FINISHING

In each of the casting processes presented in this chapter, the final step involves removing the castings from the molds and mold material. Shakeout operations are designed to separate the molds and sand from the flasks (i.e., containers), separate the castings from the molding sand, and remove the cores from the castings. Punchout machines can be used to force the entire contents of a flask (both molding sand and casting) from the container. Vibratory machines, which can operate on either the entire flasks or the extracted contents, are available in a range of styles, sizes, and vibratory frequencies. Rotary separators remove the sand from castings by placing the mold contents inside a slow-turning, large-diameter, rotating drum. The tumbling action breaks the gates and runners from the castings, crushes lumps of sand, and extracts the cores. Because of possible damage to lightweight or thin-sectioned castings, rotary tumbling is usually restricted to cast iron, steel, and brass castings of reasonable thickness.

Processes such as blast cleaning can be used to remove adhering sand, oxide scale, and parting line burrs. Compressed air or centrifugal force is used to propel abrasive particles against the surfaces of the casting. The propelled media can be metal shot (usually iron or steel), fine aluminum oxide, glass beads, or naturally occurring quartz or silica. The blasting action may be combined with some form of tumbling or robotic manipulation to expose the various surfaces. Additional finishing operations may include grinding, trimming, or various forms of machining.

12.7 SUMMARY

Liquids have a characteristic property that they assume the shape of their container. A number of processes have been developed to create shaped containers and then utilize liquid fluidity and subsequent solidification to produce desired shapes. Each process has its unique set of capabilities, advantages, and limitations, and the selection of the best method for a given application requires an understanding of all possible options. This chapter has presented processes that produce castings with a single-use (expendable) mold. The following chapter will supplement this knowledge with a survey of multiple-use mold processes.

KEY WORDS

AFS permeability number
air-set molding
Antioch process
automatic match-plate molding
bench life
blows
ceramic mold casting
chaplets
cheek
chemically bonded sands
clay content
cohesiveness
cold-box process
collapsibility
compactibility
compressive strength
cope
cope-and-drag pattern
core
core-oil process
core prints
counter-gravity investment casting
cracks
drag
dry-sand cores
dry-sand mold
dump core box
Eff-set process
evaporative pattern process
expanded polystyrene (EPS)
expendable-mold casting process
flask
follow board
full-mold process
gas-hardened process
gates
grain size
graphite mold
green compressive strength
green sand
green-sand cores
H-process
hand ramming
hardness
hot-box method
hot tears
investment casting
jolting
loose-piece pattern
lost-foam casting
lost-wax process
match-plate
match-plate pattern
mold hardness
moisture content
muller
no-bake molding
one-piece pattern
pattern
penetration
permeability
plaster molding
pressurized lost-foam casting
refractoriness
rubber-mold casting
runner
sand casting
sand expansion defects
sand slinger
shakeout
Shaw process
shelf life
shell cracking
shell molding
silica sand
single-piece cores
skin-dried mold
sodium silicate
squeezing
split-core box
split pattern
sprue
stack molding
standard rammed specimen
trees
V-process
vacuum molding
vertically parted flaskless molding
voids
water glass

Review Questions

1. What are some of the factors that influence the selection of a specific casting process as a means of making a product?
2. What are the three basic categories of casting processes when classified by molds and patterns?
3. What metals are frequently cast into products?
4. What features combine to make cast iron and aluminum the most common cast metals?
5. Which type of casting is the most common and most versatile?
6. What is a casting pattern?
7. What are some of the materials used in making casting patterns? What features should be considered when selecting a pattern material?
8. What is the simplest and least expensive type of casting pattern?
9. What is a match plate, and how does it aid molding?
10. How is a cope-and-drag pattern different from a match-plate pattern? When might this be attractive?
11. For what types of products might a loose-piece pattern be required?
12. What are the four primary requirements of a molding sand?
13. In what ways might a molding sand be a compromise material?
14. What is a muller, and what function does it perform?
15. What are some of the properties or characteristics of foundry sands that can be evaluated by standard tests?
16. What is a standard rammed specimen for evaluating foundry sands, and how is it produced?
17. What is permeability, and why is it important in molding sands?
18. How does the ratio of water to clay affect the compressive strength of green sand?
19. How is the hardness of a molding sand determined?
20. How does the size and shape of the sand grains relate to molding sand properties?
21. What are the attractive features of silica sand?
22. What is a sand expansion defect, and what is its cause?
23. How can sand expansion defects be minimized?
24. What causes "blows" to form in a casting, and what can be done to minimize their occurrence?
25. What features can cause the penetration of molten metal between the grains of the molding sand?
26. What are hot tears, and what can cause them to form?
27. Describe the distribution of sand density after compaction by sand slinging, jolting, squeezing, and a jolt–squeeze combination.
28. How can the use of vertically parted flaskless molding reduce the number of mold sections required to produce a series of castings?
29. What concern has limited the acceptance of the H-process?
30. What is stack molding?
31. How might extremely large molds be made?
32. What is green sand?
33. What are some of the limitations or problems associated with green sand as a mold material?
34. What restricts the use of dry sand molding?
35. What is a skin-dried mold?
36. What are some of the advantages and limitations of the sodium silicate–CO_2 process?
37. What is the primary feature of no-bake sands?
38. What is bench life? How does it relate to the sodium silicate–CO_2 process? No-bake? What may determine the bench life of green sand?
39. What material serves as the binder in the shell-molding process, and how is it cured?
40. Why do shell molds have excellent permeability and collapsibility?
41. What is the sand binder in the V-process? The Eff-set process?
42. What are a few of the attractive features of the V-process?
43. What types of geometric features might require the use of cores?
44. What is the primary limitation of green-sand cores?
45. What is the sand binder in the core-oil process, and how is it cured?
46. What is the binder in the hot-box core-making process?
47. What is the primary attraction of the cold-box core-making process? The primary negative feature?
48. What is shelf life? How is it different from bench life?
49. What is an attractive feature of shell-molded cores?
50. Why is it common for greater permeability, collapsibility, and refractoriness to be required of cores than for the base molding sand?
51. How can the properties of cores be enhanced beyond those offered by the various core materials?
52. What is the role of chaplets, and why is it important that they not completely melt during the pouring and solidification of a casting?
53. Why are plaster molds only suitable for the lower-melting-temperature nonferrous metals and alloys?
54. What is the primary performance difference between plaster and ceramic molds?
55. For what materials might a graphite mold be required?
56. What materials are used to produce the expendable patterns for investment casting?
57. Describe the progressive construction of an investment casting mold.
58. Why are investment casting molds generally preheated prior to pouring?
59. Why are investment castings sometimes called "lost-wax" castings?
60. What are some of the attractive features of investment casting?
61. What recent development has made one-of-a-kind or small quantities of investment castings an economic feasibility?
62. How can hollowlike investment casting patterns reduce the possibility of costly shell cracking?
63. What are some of the advantages of counter-gravity investment casting over the conventional gravity pour approach?
64. What are some of the benefits of not having to remove the pattern from the mold (as in investment casting, full-mold casting, and lost-foam casting)?
65. What are some of the ways by which expanded polystyrene patterns can be made?
66. Although both use expanded polystyrene as a pattern, what is the primary difference between full-mold and lost-foam casting?
67. What are some of the attractive features of the evaporative pattern processes?
68. What are some of the objectives of a shakeout operation?
69. How might castings be cleaned after shakeout?

PROBLEMS

1. While cores increase the cost of castings, they also provide a number of distinct advantages. The most significant is the ability to produce complex internal passages. They can also enable the production of difficult external features, such as undercuts, or allow the production of zero draft walls. Cores can reduce or eliminate additional machining, reduce the weight of a casting, and reduce or eliminate the need for multipiece assembly. Answer the following questions about cores.
   a. The cores themselves must be produced, and generally have to be removed from core boxes or molds. What geometric limitations might this impose? How might these limitations be overcome?
   b. Cores must be positioned and supported within a mold. Discuss some of the limitations associated with core positioning and orientation. Consider the weight of a core, prevention of core fracture, minimization of core deflection, and possible buoyancy.
   c. Because cores are internal to the casting, adequate venting is necessary to eliminate or minimize porosity problems. Discuss possible features to aid in venting.
   d. How might core behavior vary with different materials being cast—steel versus aluminum, for example?
   e. Discuss several of the reasons cores may be made from a different material than the molding material used in the primary mold.
   f. Core removal is another design concern. Discuss how several different core-making processes might perform in the area of removal. What are some ways to assist or facilitate core removal?

www.wiley.com/go/global/degarmo

Chapter 12 CASE STUDY

Movable and Fixed Jaw Pieces for a Heavy-Duty Bench Vise

The figure presents a sketch of a standard bench vise and a cut-away sketch of the movable and fixed jaw pieces of a heavy-duty vise that might see use in vocational schools, factories, and machine shops. The vise is intended to have a rated maximum clamping force of 15 tons. The slide of the moving jaw has been designed to be a 2-in. box channel. The jaw width is 5 in., the maximum jaw opening is 6 in., and the depth of the throat is 4 in. The designer has elected to use replaceable, serrated jawsand suggests that the material used for the receiving jaw pieces have a yield strength in excess of 35 ksi, with at least 15% elongation in a uniaxial tensile test (to ensure that an overload or hammer impact would not produce brittle fracture).

1. Determine some possible combinations of materials and process that could fabricate the desired shapes with the required properties. Of the alternatives presented, which would you prefer, and why?
2. Would the components require some form of subsequent heat treatment? Consider the possibilities of stress relief, homogenization, or the establishment of desired final properties. What would you recommend?
3. One of your colleagues has suggested that the slides be finished with a coat of paint. Do you think a surface treatment is necessary or desirable for your selected material and process? If so, what would you recommend? If not, defend your recommendation.

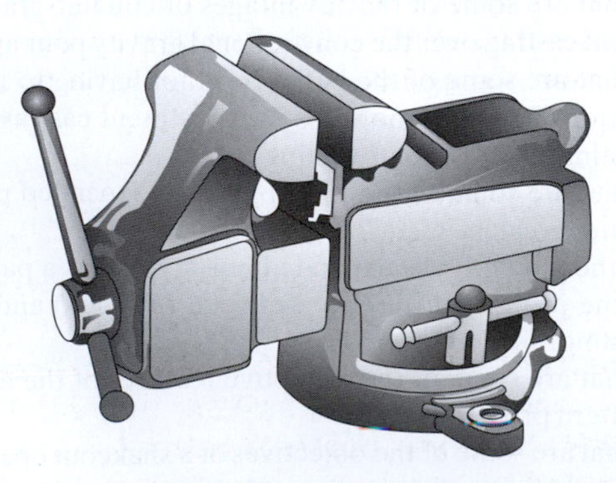

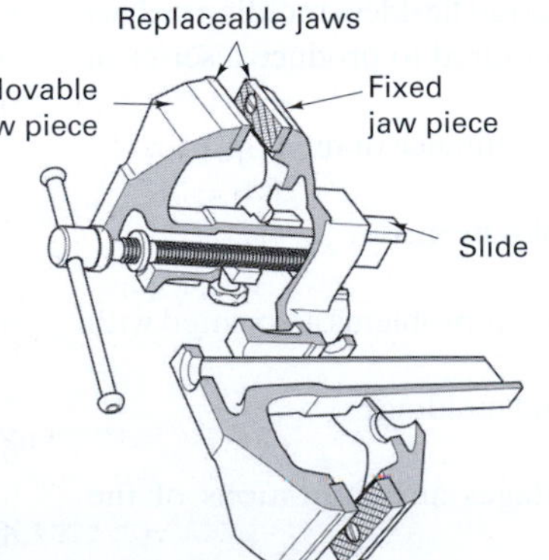

Note: Shaded surfaces have been produced by cross-sectional cuts

(*Courtesy Ronald Kohser*)

CHAPTER 13

Multiple-Use-Mold Casting Processes

13.1 Introduction

In each of the expendable-mold casting processes discussed in Chapter 12, a separate mold had to be created for each pour. Variations in mold consistency, mold strength, moisture content, pattern removal, and other factors contribute to dimensional and property variation from casting to casting. In addition, the need to create and then destroy a separate mold for each pour results in rather low production rates.

The multiple-use-mold casting processes overcome many of these limitations, but they, in turn, have their own assets and liabilities. Because the molds are generally made from metal, many of the processes are restricted to casting the lower-melting-point nonferrous metals and alloys. Part size is often limited, and the dies or molds can be rather costly.

13.2 Permanent-Mold Casting

In the **permanent-mold casting** process, also called *gravity die casting*, a reusable mold is machined from gray cast iron, alloy cast iron, steel, bronze, graphite, or other material. The molds are usually made in segments, which are often hinged to permit rapid and accurate opening and closing. After preheating, a refractory or other material coating is applied to the preheated mold, and the mold is clamped shut. Molten metal is then poured into the pouring basin and flows through the feeding system into the mold cavity by simple gravity flow. After solidification, the mold is opened and the product is removed. Because the heat from the previous cast is usually sufficient to maintain mold temperature, the process can be immediately repeated, with a single refractory coating serving for several pouring cycles. Aluminum-, magnesium-, zinc-, lead-, and copper-based alloys are the metals most frequently cast, along with gray cast iron. If graphite is used as the mold material, iron and steel castings can also be produced.

Numerous advantages can be cited for the permanent-mold process. Near-net shapes can be produced that require little finish machining. The mold is reusable, and a good surface finish is obtained if the mold is in good condition. Dimensions are consistent from part to part, and dimensional accuracy can often be held to within 0.25 mm (0.010 in.). Directional solidification can be achieved through good design or can be promoted by selectively heating or chilling various portions of the mold or by varying the thickness of the mold wall. The result is usually a sound, defect-free casting with good mechanical properties. The faster cooling rates of the metal mold produce a finer grain structure and higher-strength products than would result from a sand casting process. Because permanent mold products have lower levels of porosity compared to die castings, they can often acquire enhanced properties through heat treatment. Cores of expendable sand, plaster, or retractable metal can be used to increase the complexity of the casting, and multiple cavities can often be included in a single mold. When sand cores are used, the process is often called **semipermanent-mold casting.**

On the negative side, the process is generally limited to the lower-melting-point alloys, and high mold costs can make low production runs prohibitively expensive. The useful life of a mold is generally set by molten metal erosion or thermal fatigue. When making products of steel or cast iron, mold life can be extremely short. For the lower-temperature metals, one can usually expect somewhere between 10,000 and 120,000 cycles. The actual mold life will depend on the following:

1. *Alloy being cast*. The higher the melting point, the shorter the mold life.
2. *Mold material*. Gray cast iron has about the best resistance to thermal fatigue and machines easily. Thus, it is used most frequently for permanent molds.
3. *Pouring temperature*. Higher pouring temperatures reduce mold life, increase shrinkage problems, and induce longer cycle times.
4. *Mold temperature*. If the temperature is too low, one can expect misruns and large temperature differences in the mold. If the temperature is too high, excessive cycle times result and mold erosion is aggravated
5. *Mold configuration*. Differences in section sizes of either the mold or the casting can produce temperature differences within the mold and reduce its life.

The permanent molds contain the mold cavity, pouring basin, sprue, runners, risers, gates, possible core supports, alignment pins, and some form of ejection system. The molds are usually heated at the beginning of a run and continuous operation then maintains the mold at a fairly uniform elevated temperature. This minimizes the degree of thermal fatigue, facilitates metal flow, and controls the cooling rate of the metal being cast. Because the mold temperature rises when a casting is produced, it may be necessary to provide a mold-cooling delay before the cycle is repeated. Refractory washes or graphite coatings can be applied to the mold walls to protect the die surface, control or direct the cooling, prevent the casting from sticking, improve surface finish, and prolong the mold life by minimizing thermal shock and fatigue. When pouring cast iron, an acetylene torch is often used to apply a coating of carbon black to the mold.

Because the molds are not permeable, special provision must be made for **venting.** This is usually accomplished through the slight cracks between mold halves or by very small vent holes that permit the escape of trapped air but not the passage of molten metal. While turbulence-related defects are also more common with the gravity pour, these can often be minimized by reducing the free-fall height at the beginning of the pouring process and using ceramic-foam filters to smooth the flow of the molten metal. Because gravity is the only means of inducing metal flow, risers must still be employed to compensate for solidification shrinkage, and with the necessary sprues and runners, yields are generally less than 60%.

Mold complexity is often restricted because the rigid cavity offers no collapsibility to compensate for the solid-state shrinkage of the casting. As a best alternative, it is common practice to open the mold and remove the casting immediately after solidification. This prevents the formation of hot tears that may form if the product is restrained during the shrinkage that occurs during cooldown to room temperature.

For permanent-mold casting, high-volume production is usually required to justify the high cost of the metal molds. Automated machines can be used to coat the mold, pour the metal, and remove the casting. Figure 13-1 shows an example of automotive pistons that have been mass-produced using the permanent mold process, which is summarized in Table 13-1.

SLUSH CASTING

Hollow castings can be produced by a variant of permanent-mold casting known as **slush casting.** Hot metal is poured into the metal mold and is allowed to cool until a shell

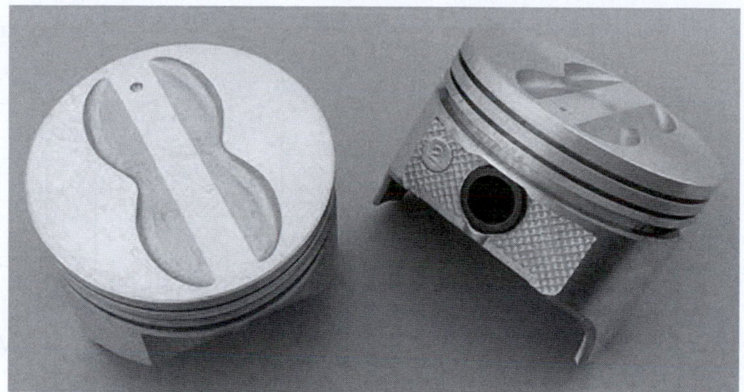

FIGURE 13-1 An example of automotive pistons that have been mass–produced by the millons using permanent mold casting. *(Courtesy of United Engine & Machine Co., Carson City, NV)*

TABLE 13-1 Permanent-Mold Casting

Process: Mold cavities are machined into mating metal die blocks, which are then preheated and clamped together. Molten metal is then poured into the mold and enters the cavity by gravity flow. After solidification, the mold is opened and the casting is removed.

Advantages: Good surface finish and dimensional accuracy; metal mold gives rapid cooling and fine-grain structure; multiple-use molds (up to 120,000 uses); metal cores or collapsible sand cores can be used.

Limitations: High initial mold cost; shape, size, and complexity are limited; yield rate rarely exceeds 60%, but runners and risers can be directly recycled; mold life is very limited with high-melting-point metals such as steel.

Common metals: Alloys of aluminum, magnesium, and copper are most frequently cast; irons and steels can be cast into graphite molds; alloys of lead, tin, and zinc are also cast.

Size limits: 100 grams to 75 kilograms (several ounces to 150 pounds).

Thickness limits: Minimum depends on material but generally greater than 3 mm ($\frac{1}{8}$ in.); maximum thickness about 50 mm (2.0 in.).

Geometric limits: The need to extract the part from a rigid mold may limit certain geometric features. Uniform section thickness is desirable.

Typical tolerances: 0.4. mm for the first 2.5. cm (0.015 in. for the first inch) and 0.02 mm for each additional centimeter (0.002 in. for each additional inch); 0.25mm (0.01 in.) added if the dimension crosses a parting line.

Draft allowance: 2°–3°.

Surface finish: 5.0 to 12.0 μm (200–400 μin.) rms.

of desired thickness has formed. The mold is then inverted and the remaining liquid is poured out. The resulting casting is a hollow shape with good surface detail but variable wall thickness. Common applications include the casting of ornamental objects such as candlesticks, lamp bases, and statuary from the low-melting-temperature metals.

TILT–POUR, LOW-PRESSURE, AND VACUUM PERMANENT-MOLD CASTING

Gravity pouring is the oldest, simplest and most traditional form of permanent-mold casting. In a variation known as **tilt–pour permanent-mold casting,** the mold is rotated into a horizontal position and a cup attached to the mold is filled with molten metal. The mold is then tilted upward, enabling the molten metal to flow through the gating system and into the mold cavity with reduced freefall and far less turbulence than when the metal is poured vertically down a sprue. Figure 13-2 shows a tilt–pour permanent mold in its vertical position where the pouring cup extends to become the sprue and riser.

In low-pressure and vacuum permanent-mold casting, the mold is turned upside down and positioned above a sealed airtight chamber that contains a crucible of molten metal. A small pressure difference then causes the molten metal to flow upward through a connecting tube into the die cavity. In **low-pressure permanent-mold (LPPM) casting,** illustrated in Figure 13-3, a low-pressure gas (3 to 15 psi) is introduced into a sealed chamber, driving molten metal up through a refractory fill tube and into the gating system or cavity of a metal mold. This metal is exceptionally clean because it flows from the center of the melt and is fed directly into the mold (a distance of about 10 cm, or 3 to 4 in.),

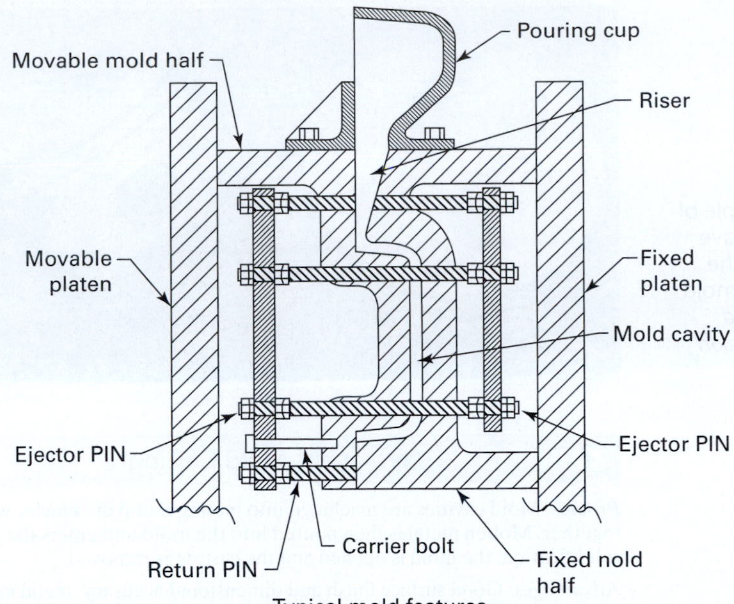

FIGURE 13-2 Schematic of the tilt–pour permanent-mold process, showing the mold assembly in its tilted or vertical position. *(Courtesy of C. M. H. Manufacturing, Lubbock, TX)*

never passing through the atmosphere. Product quality is further enhanced by the nonturbulent mold filling that helps to minimize gas porosity and dross formation.

Through design and cooling, the products directionally solidify from the top down. The molten metal in the pressurized fill tube acts as a riser to continually feed the casting during solidification. When solidification is complete, the pressure is released and the unused metal in the feed tube simply drops back into the crucible. The reuse of this metal, coupled with the absence of additional risers, leads to yields that are often greater than 85%.

Nearly all low-pressure permanent-mold castings are made from aluminum or magnesium, but some copper-base alloys can also be used. Mechanical properties are

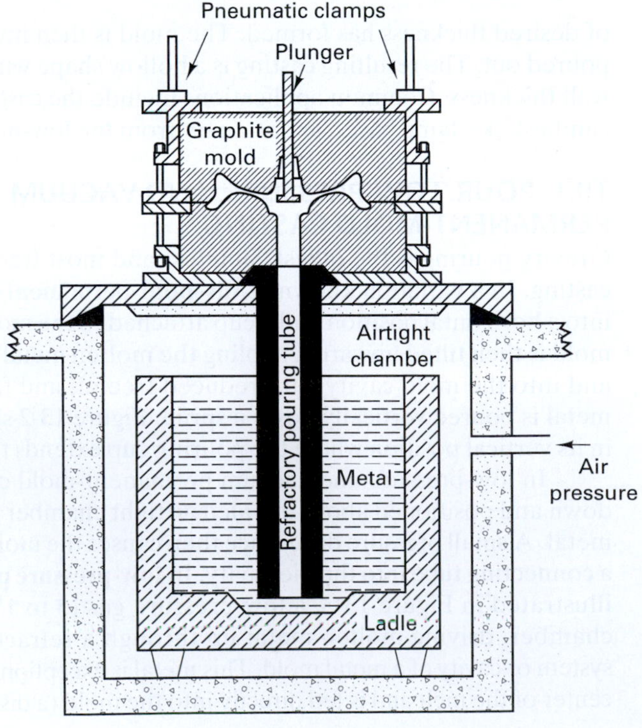

FIGURE 13-3 Schematic of the low-pressure permanent-mold process. *(Courtesy of Amsted Industries, Chicago, IL)*

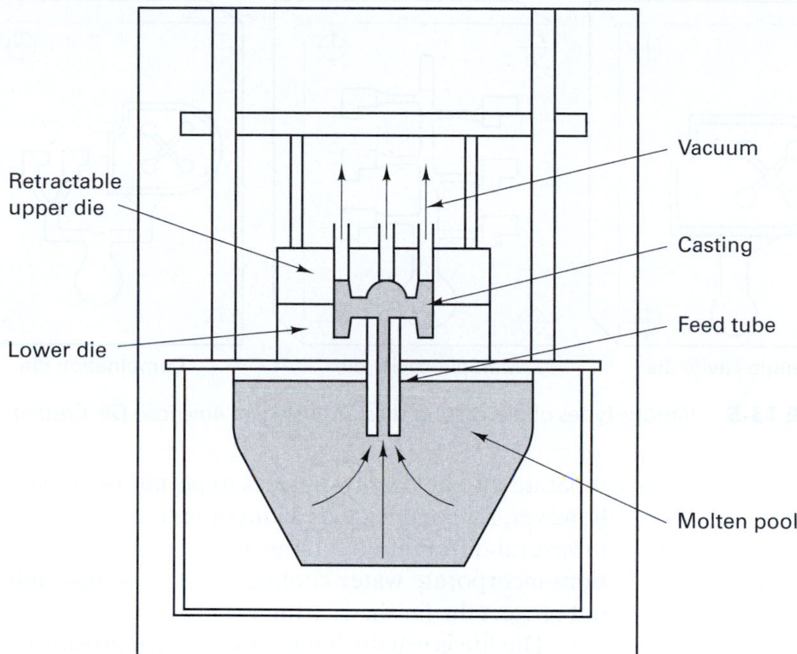

FIGURE 13-4 Schematic illustration of vacuum permanent-mold casting. Note the similarities to the low-pressure process.

typically about 5% better than those of conventional permanent-mold castings. Cycle times are somewhat longer, however, than those of conventional permanent molding.

Figure 13-4 depicts a similar variation. **Vacuum permanent-mold casting** is where a vacuum is drawn on the die assembly and atmospheric pressure in the chamber forces the metal upward. All of the benefits and features of the low-pressure process are retained, including the subsurface extraction of molten metal from the melt, the bottom feed to the mold, the low metal turbulence during filling, the self-risering action, and the downward directional solidification. Thin-walled castings can be produced with high metal yield and excellent surface quality. Because of the vacuum, the cleanliness of the metal and the dissolved gas content are superior to that of the low-pressure process. Final castings typically range from 0.2 to 5 kg (0.4 to 10 lb), and have mechanical properties that are even better than those of the low-pressure permanent-mold products.

13.3 DIE CASTING

In the **die-casting** process—or, more specifically, *pressure die casting*—molten metal is forced into metal molds under pressures of several thousand pounds per square inch (tens of MPa) and held under high pressure during solidification. Because of the combination of metal molds or dies and high pressure, fine sections and excellent detail can be achieved, together with long mold life. Most die castings are made from nonferrous metals and alloys, with special zinc-, copper-, magnesium-, and aluminum-based alloys having been designed to produce excellent properties when die cast. Ferrous-metal die castings are possible but are generally considered to be uncommon. Production rates are high, the products exhibit good strength, shapes can be quite intricate, and dimensional precision and surface qualities are excellent. There is almost a complete elimination of subsequent machining. Most die castings can be classified as small- to medium-sized parts, but the size and weight of die castings are continually increasing. Parts can now be made with weights up to 10 kg (20 lb) and dimensions as large as 600 mm (24 in.).

Die temperatures are usually maintained about 150° to 250°C (300° to 500°F) below the solidus temperature of the metal being cast in order to promote rapid freezing. Because cast iron cannot withstand the high casting pressures, die-casting dies are usually made from hardened hot-work tool steels and are typically quite expensive. As shown in Figure 13-5, the dies may be relatively simple, containing only one or two mold cavities, or they may be complex, containing multiple cavities of the same or different products or all of the components for a multipiece assembly. The rigid dies must

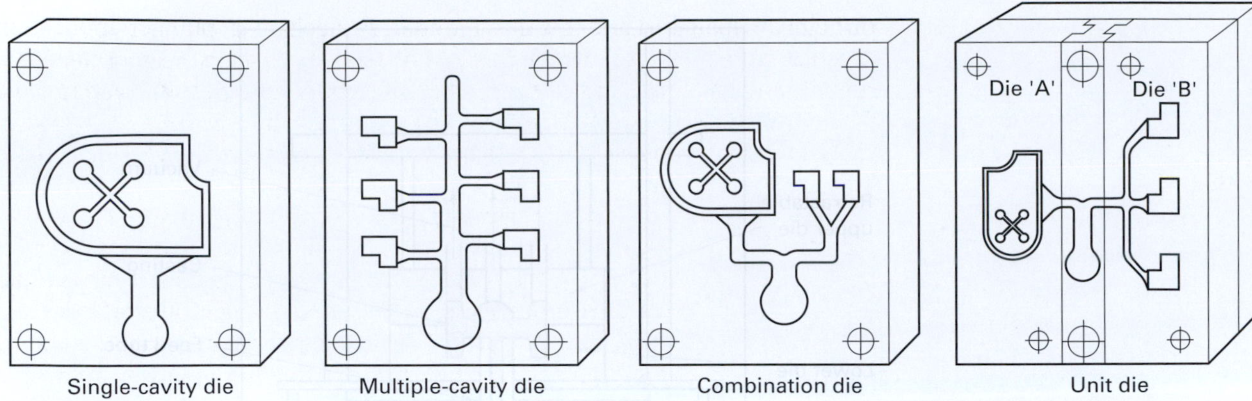

FIGURE 13-5 Various types of die-casting dies. *(Courtesy of American Die Casting Institute Inc., Des Plaines, IL)*

separate into at least two pieces to permit removal of the casting. It is not uncommon, however, for complex die castings to require multiple-segment dies that open and close in several different directions. Die complexity is further increased as the various sections incorporate water-cooling passages, **retractable cores,** and moving pins to knock out or eject the finished casting.

Die life is usually limited by wear (or erosion), which is strongly dependent on the temperature of the molten metal. Surface cracking can also occur in response to the large number of heating and cooling cycles that are experienced by the die surfaces. If the rate of temperature change is the dominant feature, the problem is called **heat checking.** If the number of cycles is the primary cause, the problem is called **thermal fatigue.** Typical die life is 100,000 shots, and production runs of 10,000 pieces or more are generally required to justify die set-up.

In the basic die-casting process, water-cooled dies are first lubricated and clamped tightly together. Molten metal is then injected under high pressure. Because high injection pressures cause turbulence and air entrapment, the specified values of pressure and the time and duration of application vary considerably. The pressure need not be constant, and there has been a trend toward the use of larger gates and lower injection pressures, followed by the application of higher pressure after the mold has completely filled and the metal has started to solidify. By reducing turbulence and solidifying under high pressure, both the porosity and inclusion content of the finished casting are reduced. After solidification is complete, the pressure is released, the dies separate, and ejector pins extract the finished casting along with its attached runners and sprues.

There are two basic types of die-casting machines. Figure 13-6 schematically illustrates the **hot-chamber,** or **gooseneck,** variety. A gooseneck chamber is partially submerged in a reservoir of molten metal. With the plunger raised, molten metal flows

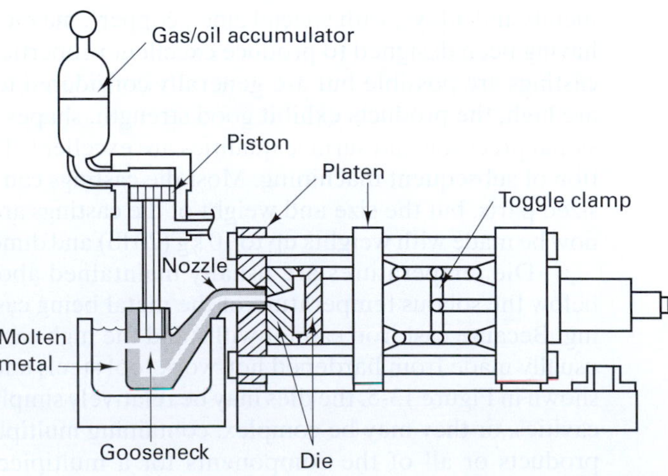

FIGURE 13-6 Principal components of a hot-chamber die-casting machine. *(Adapted from* Metals Handbook, *9th ed., Vol. 15, p. 287, ASM International, Materials Park, OH)*

through an open port and fills the chamber. A mechanical plunger applies a pressure of about 20 MPa (3000 psi), which forces the metal up through the gooseneck, through the runners and gates, and into the die, where it rapidly solidifies. Retraction of the plunger then allows the gooseneck to refill as the casting is being ejected, and the cycle repeats at speeds ranging from 100 shots per hour for large castings to more than 1000 shots per hour for small products.

Hot-chamber die-casting machines offer fast cycling times (set by the ability of the water-cooled dies to cool and solidify the metal) and the added advantage that the molten metal is injected from the same chamber in which it is melted (i.e., there is no handling or transfer of molten metal and minimal heat loss). Unfortunately, the hot-chamber design cannot be used for the higher-melting-point metals, and it is unattractive for aluminum because the molten aluminum tends to pick up some iron during the extended time of contact with the casting equipment. Hot-chamber machines, therefore, see primary use with zinc-, tin-, and lead-based alloys.

Zinc die castings can also be made by a process known as **heated-manifold direct-injection die casting** (also known as *direct-injection die casting* or *runnerless die casting*). The molten zinc is forced through a heated manifold and then through heated mini-nozzles directly into the die cavity. This approach totally eliminates the need for sprues, gates, and runners. Scrap is reduced, energy is conserved (less molten metal per shot and no need to provide excess heat to compensate for cooling in the gating system), and product quality is increased. Existing die-casting machines can be converted through the addition of a heated manifold and modification of the various dies.

Cold-chamber machines are usually employed for the die casting of materials that are not suitable for the hot-chamber design. These include alloys of aluminum, magnesium, and copper; high-aluminum zin;, and even some ferrous alloys. As illustrated in Figure 13-7, metal that has been melted in a separate furnace is transported to the die-casting machine, where a measured quantity is transferred to an unheated shot chamber (or injection cylinder) and subsequently driven into the die by a hydraulic or mechanical plunger under pressures of around 70 MPa (10,000 psi) The pressure is then maintained or increased until solidification is complete. Because molten metal must be transferred to the chamber for each shot, the cold-chamber process has a longer operating cycle compared to hot-chamber machines. Nevertheless, productivity is still high.

In all variations of the process, die-casting dies fill with metal so fast that there is little time for the air in the runner system and mold cavity to escape, and the metal molds offer no permeability. The air can become trapped and cause a variety of defects, including blowholes, porosity, and misruns. To minimize these defects, it is crucial that the dies be properly vented, usually by wide, thin (0.13 mm or 0.005 in.) vents positioned along the parting line. Proper positioning is a must, because all of the air must escape before the molten metal contacts the vents. The long thin slots allow the escape of gas but promote rapid freezing of the metal and a plugging of the hole. The metal that solidifies in the vents must be trimmed off after the casting has been ejected. This can be done with special trimming dies that also serve to remove the sprues and runners.

Risers are not used in the die-casting process because the high injection pressures ensure the continuous feed of molten metal from the gating system into the casting. The

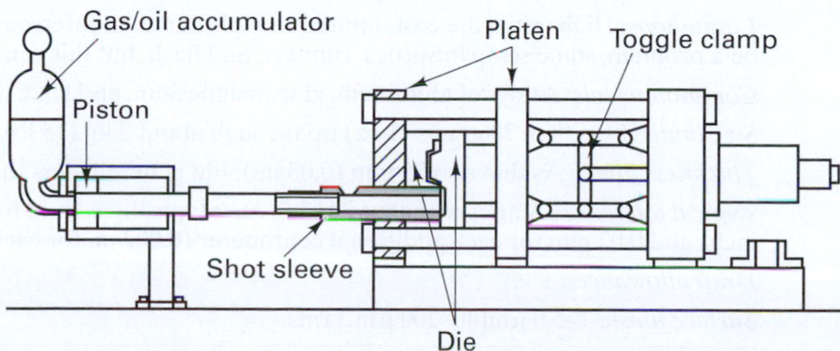

FIGURE 13-7 Principal components of a cold-chamber die-casting machine. *(Adapted from* Metals Handbook, *9th ed., Vol. 15, p. 287, ASM International, Materials Park, OH)*

porosity that is often found in die castings is not shrinkage porosity but is more likely to be the result of either entrapped air or the turbulent mode of die filling. This porosity tends to be confined to the interior of castings, and its formation can often be minimized by smooth metal flow, good venting, and proper application of pressure. The rapidly solidified surface is usually harder and stronger than the slower-cooled interior and is usually sound and suitable for plating or decorative applications.

Sand cores cannot be used in die casting because the high pressures and flow rates cause the cores to either disintegrate or have excessive metal penetration. As a result, metal cores are required, and provisions must be made for their retraction, usually before the die is opened for removal of the casting. As with all mating segments and moving components, a close fit must be maintained to prevented the pressurized metal from flowing into the gap. Loose core pieces (also metal) can also be positioned into the die at the beginning of each cycle and then removed from the casting after its ejection. This procedure permits more complex shapes to be cast, such as holes with internal threads, but production rate is slowed and costs increase.

Cast-in **inserts** can also be incorporated in the die-casting process. Examples include pre-threaded bosses, electrical heating elements, threaded studs, and high-strength bearing surfaces. These high-temperature components are positioned in the die before the lower-melting-temperature metal is injected. Suitable recesses must be provided in the die for positioning and support, and the casting cycle tends to be slowed by the additional operations.

Table 13-2 summarizes the key features of the die-casting process. Attractive aspects include smooth surfaces and excellent dimensional accuracy. For aluminum-, magnesium-, zinc-, and copper-based alloys, linear tolerances of 3 mm/m (0.003 in./in.) are not uncommon. Thinner sections can be cast than with either sand or permanent mold casting. The minimum section thickness and draft vary with the type of metal, with typical values as follows:

Metal	Minimum Section	Minimum Draft
Aluminum alloys	0.89 mm (0.035 in.)	1:100 (0.010 in./in.)
Brass and bronze	1.27 mm (0.050 in.)	1:80 (0.015 in./in.)
Magnesium alloys	1.27 mm (0.050 in.)	1:100 (0.010 in./in.)
Zinc alloys	0.63 mm (0.025 in.)	1:200 (0.005 in./in.)

Because of the precision and finish, most die castings require no finish machining except for the removal of excess metal fin, or *flash*, around the parting line and the possible drilling or tapping of holes. Production rates are high, and a set of dies can produce many thousands of castings without significant change in dimensions. While die casting

TABLE 13-2 Die Casting

Process: Molten metal is injected into closed metal dies under pressures ranging from 10 to 175 MPa (1500–25,000 psi). Pressure is maintained during solidification, after which the dies separate and the casting is ejected along with its attached sprues and runners. Cores must be simple and retractable and take the form of moving metal segments.

Advantages: Extremely smooth surfaces and excellent dimensional accuracy; rapid production rate; product tensile strengths as high as 415 Mpa (60 ksi).

Limitations: High initial die cost; limited to high-fluidity nonferrous metals; part size is limited; porosity may be a problem; some scrap in sprues, runners, and flash, but this can be directly recycled.

Common metals: Alloys of aluminum, zinc, magnesium, and lead; also possible with alloys of copper and tin.

Size limits: Less than 30 grams (1 oz) up through about 7 kg (15 lb) most common.

Thickness limits: As thin as 0.75 mm (0.03 in.), but generally less than 13 mm ($\frac{1}{2}$ in.).

Typical tolerances: Varies with metal being cast; typically 0.1mm for the first 2.5 cm (0.005 in. for the first inch) and 0.02 mm for each additional centimeter (0.002 in. for each additional inch).

Draft allowances: 1°–3°.

Surface finish: 1–5.0 μm(40–200 μin.) rms.

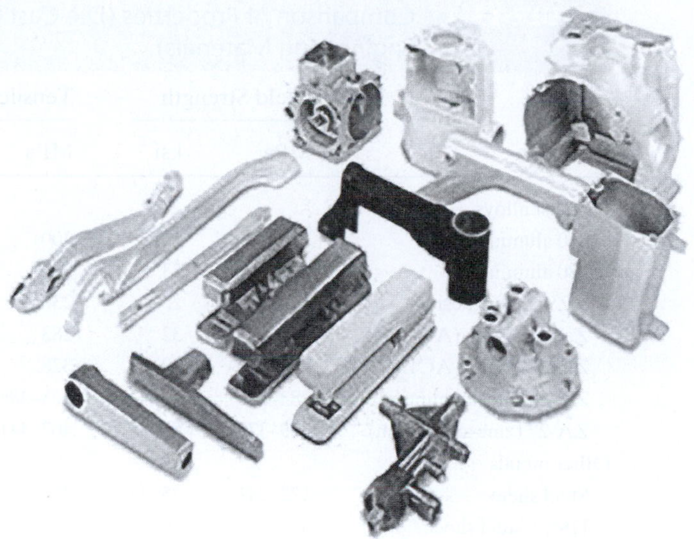

FIGURE 13-8 A variety of die cast products: zinc commercial door handles and executive desk staplers; magnesium housings for pneumatic staplers and nailers; and aluminum aircraft food tray arms. *(Courtesy of Chicago White Metal Casting, Inc., Bensenville, IL)*

is most economical for large production volumes, quantities as low as 2000 can be justified if extensive secondary machining or surface finishing can be eliminated.

Thin-wall zinc die castings are now considered to be significant competitors to plastic injection moldings. The die castings are stronger, stiffer, more dimensionally stable, and more heat resistant. In addition, the metal parts are more resistant to ultraviolet radiation, weathering, and stress cracking when exposed to various reagents. The metal casting also provides an excellent base for a wide variety of decorative platings and coatings.

Figure 13-8 presents a variety of aluminum and zinc die castings. Table 13-3 compares the key features of the four dominant families of die casting alloys, and Table 13-4 compares the mechanical properties of various die cast alloys with the properties of other engineering materials.

13.4 SQUEEZE CASTING AND SEMISOLID CASTING

Squeeze casting and semisolid casting are methods that enable the production of high-quality, near-net-shape, thin-walled parts with good surface finish and dimensional precision and properties that approach those of forgings. Both processes can be viewed as derivatives of conventional high-pressure die casting, because they employ tool steel dies and apply high pressure during solidification. While the majority of applications involve alloys of aluminum, each of the processes has been successfully applied to magnesium, zinc, copper, and a limited number of ferrous alloys.

TABLE 13-3 Key Properties of the Four Major Families of Die-Cast Metal

Metal	Key Properties
Aluminum	Lowest cost per unit volume; second lightest to magnesium; highest rigidity; good machinability, electrical conductivity, and heat-transfer characteristics.
Magnesium	Lowest density, faster production than aluminum since hot-chamber cast, highest strength-to-weight ratio, good vibration damping, best machinability, can provide electromagnetic shielding.
Zinc	Attractive for small parts; tooling lasts 3–5 times longer than for aluminum; heaviest of the die-castable metals but can be cast with thin walls for possible weight savings; good impact strength, machinability, electrical conductivity, and thermal conductivity.
Zinc–Aluminum	Highest yield and tensile strength, lighter than conventional zinc alloys, good machinability.

TABLE 13-4 Comparison of Properties (Die Cast Metals versus Other Engineering Materials)

Material	Yield Strength MPa	Yield Strength ksi	Tensile Strength MPa	Tensile Strength ksi	Elastic Modulus GPa	Elastic Modulus 10^6 psi
Die-cast alloys						
360 aluminum	170	25	300	44	71	10.3
380 aluminum	160	23	320	46	71	10.3
AZ91D magnesium	160	23	230	34	45	6.5
Zamak 3 zinc (AG40A)	221	32	283	41	—	—
Zamak 5 zinc (AC41A)	269	39	328	48	—	—
ZA-8 (zinc–aluminum)	283–296	41–43	365–386	53–56	85	12.4
ZA-27 (zinc–aluminum)	359–379	52–55	407–441	59–64	78	11.3
Other metals						
Steel sheet	172–241	25–35	276	40	203	29.5
HSLA steel sheet	414	60	414	60	203	29.5
Powdered iron	483	70	—	—	120–134	17.5–19.5
Plastics						
ABS	—	—	55	8	7	1.0
Polycarbonate	—	—	62	9	7	1.0
Nylon 6[a]	—	—	152	22	10	1.5
PET[a]	—	—	145	21	14	2.0

[a] 30% glass reinforced.

In the **squeeze casting**[1] process, molten metal is introduced into the die cavity of a metal mold, using large gate areas and slow metal velocities to avoid turbulence. When the cavity has filled, high pressure (20 to 175 MPa, or 3,000 to 25,000 psi) is then applied and maintained during the subsequent solidification. Parts must be designed to directionally solidify toward the gates, and the gates must be sufficiently large that they freeze after solidification in the cavity, thereby allowing the pressurized runner to feed additional metal to compensate for shrinkage.

Intricate shapes can be produced at lower pressures than would normally be required for hot or cold forging. Both retractable and disposable cores can be used to create holes and internal passages. Gas and shrinkage porosity are substantially reduced, and mechanical properties are enhanced. While the squeeze casting process is most commonly applied to aluminum and magnesium castings, it has also been adapted to the production of metal–matrix composites where the pressurized metal is forced around or through foamed or fiber reinforcements that have been positioned in the mold.

For most alloy compositions, there is a range of temperatures where liquid and solid coexist, and several techniques have been developed to produce shapes from this **semisolid** material. In the **rheocasting** process, molten metal is cooled to the semisolid state with constant stirring. The stirring or shearing action breaks up the dendrites, producing a slurry of rounded particles of solid in a liquid melt. This slurry, with about a 30% solid content, can be readily shaped by high-pressure injection into metal dies. Because the slurry contains no superheat and is already partially solidified, it freezes quickly.

In the **thixocasting** variation, there is no handling of molten metal. The material is first subjected to special processing (stirring during solidification as in rheocasting) to produce solid blocks or bars with a nondendritic structure. The solid material is then cut to prescribed length and reheated to a semisolid state where the material is about

[1] The term *squeeze casting* has also been applied to another process that should, more appropriately, go by its other name, *liquid-metal forging*. Molten metal is poured into the bottom half of a preheated die set. As the metal starts to solidify, the upper die closes and applies pressure during the further solidification. The pressure is considerably less than that required for conventional forging, and parts of great detail can be produced. Cores can be used to produce holes and recesses. The process has been applied to both ferrous and nonferrous metals, but aluminum and magnesium are the most common materials.

40% liquid and 60% solid. In this condition, the **thixotropic material** can be handled like a solid, but it flows like a liquid when agitated or squeezed. It is then mechanically transferred to the shot chamber of a cold-chamber die casting machine and injected under pressure. In a variation of the process, solid metal granules or pellets are fed into a barrel chamber, where a rotating screw shears and advances the material through heating zones that raise the temperature to the semisolid region. When a sufficient volume of thixotropic material has accumulated at the end of the barrel, a shot system drives it into the die or mold at velocities of 1 to 2.5 m/sec (40 to 100 in/sec). The injection system of this process is a combination of the screw feed used in plastic injection molding and the plunger used in conventional die casting.

In all of the semisolid casting processes, the absence of turbulent flow during the casting operation minimizes gas pickup and entrapment. Because the material is already partially solid, the lower injection temperatures and reduced solidification time act to extend tool life. The prior solidification coupled with further solidification under pressure results in a significant reduction in solidification shrinkage and related porosity. The minimization of porosity enables the use of high-temperature heat treatments, such as the T6 solution treatment and artificial aging of aluminum, to further enhance strength. Because the thixocasting process does not use molten metal, both wrought and cast alloys have been successfully shaped. Walls have been produced with thickness as low as 0.2 mm (0.01 in.).

13.5 CENTRIFUGAL CASTING

The inertial forces of rotation or spinning are used to distribute the molten metal into the mold cavity or cavities in the **centrifugal casting** processes, a category that includes true centrifugal casting, semicentrifugal casting, and centrifuging. In **true centrifugal casting,** a dry-sand, graphite, or metal mold is rotated about either a horizontal or vertical axis at speeds of 300 to 3000 rpm. As the molten metal is introduced, it is flung to the surface of the mold, where it solidifies into some form of hollow product. The exterior

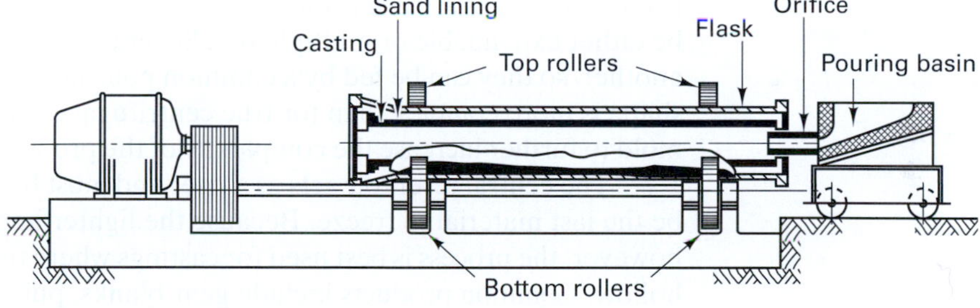

FIGURE 13-9 Schematic representation of a horizontal centrifugal casting machine. *(Courtesy of American Cast Iron Pipe Company, Birmingham, AL)*

profile is usually round (as with pipes, tubes, and gun barrels), but hexagons and other symmetrical shapes are also possible.

No core or mold surface is needed to shape the interior, which will always have a round profile because the molten metal is uniformly distributed by the centrifugal forces. When rotation is about the horizontal axis, as illustrated in Figure 13-9, the inner surface is always cylindrical. If the mold is oriented vertically, as in Figure 13-10, gravitational forces cause the inner surface to become parabolic, with the exact shape being a function of the speed of rotation. Wall thickness can be controlled by varying the amount of metal that is introduced into the mold.

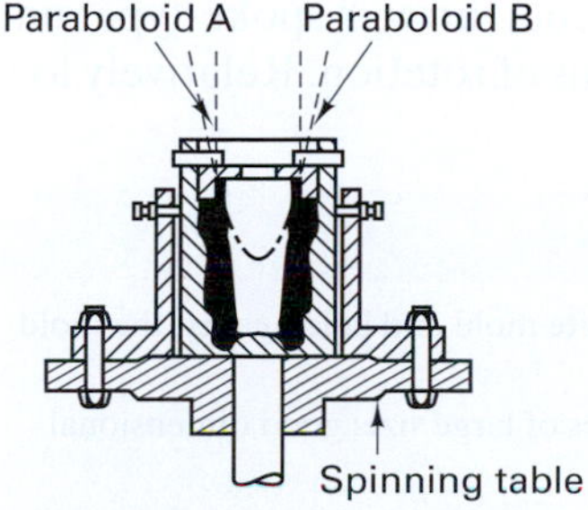

FIGURE 13-10 Vertical centrifugal casting, showing the effect of rotational speed on the shape of the inner surface. Paraboloid A results from fast spinning, whereas slower spinning will produce paraboloid B.

During the rotation, the metal is forced against the outer walls of the mold with considerable force, and solidification begins at the outer surface. Centrifugal force continues to feed molten metal as solidification progresses inward. Because the process compensates for shrinkage, no risers are required. The final product has a strong, dense exterior with all of the lighter impurities (including dross and pieces of the refractory mold coating) collecting on the inner surface of the casting. This surface is often left in the final casting, but for some products it may be removed by a light boring operation.

Products can have outside diameters ranging from 7.5 cm. to 1.4 m (3 to 55 in.) and wall thickness up to 25 cm (10 in.). Pipe (up to 12 m, or 40 ft, in length), pressure vessels,

FIGURE 13-11 Brass and bronze bushings that have been produced by centrifugal casting. *(Courtesy Machine Rebuilding Services, Milford, CT)*

cylinder liners, brake drums, the starting material for bearing rings, and all of the parts illustrated in Figure 13-11 can be manufactured by centrifugal casting. The equipment is rather specialized and can be quite expensive for large castings. The permanent molds can also be expensive, but they offer a long service life, especially when coated with some form of refractory dust or wash. Because no sprues, gates, or risers are required, yields can be greater than 90%. Composite products can also be made by the centrifugal casting of a second material on the inside surface of an already cast product. Table 13-5 summarizes the features of the centrifugal casting process.

In **semicentrifugal casting** (Figure 13-12) the centrifugal force assists the flow of metal from a central reservoir to the extremities of a rotating symmetrical mold. The rotational speeds are usually lower than for true centrifugal casting, and the molds may be either expendable or multiple use. Several molds may also be stacked on top of one another, so they can be fed by a common pouring basin and sprue. In general, the mold shape is more complex than for true centrifugal casting, and cores can be placed in the mold to further increase the complexity of the product.

The central reservoir acts as a riser and must be large enough to ensure that it will be the last material to freeze. Because the lighter impurities concentrate in the center, however, the process is best used for castings where the central region will ultimately be hollow. Common products include gear blanks, pulley sheaves, wheels, impellers, and electric motor rotors.

Centrifuging, or *centrifuge centrifugal casting* (Figure 13-13) uses centrifugal action to force metal from a central pouring reservoir or sprue, through spoke-type runners, into separate mold cavities that are offset from the axis of rotation. Relatively low

TABLE 13-5 Centrifugal Casting

Process: Molten metal is introduced into a rotating sand, metal, or graphite mold and held against the mold wall by centrifugal force until it is solidified.

Advantages: Can produce a wide range of cylindrical parts, including ones of large size; good dimensional accuracy, soundness, and cleanliness.

Limitations: Shape is limited; spinning equipment can be expensive.

Common metals: Iron; steel; stainless steel; and alloys of aluminum, copper, and nickel.

Size limits: Up to 3 m (10 ft) in diameter and 15 m (50 ft) in length.

Thickness limits: Wall thickness 2.5 to 125 mm (0.1–5 in.).

Typical tolerances: O.D. to within 2.5 mm (0.1 in.); I.D. to about 4 mm (0.15 in.).

Draft allowance: 10 mm/m ($\frac{1}{8}$ in./ft).

Surface finish: 2.5–12.5 μm (100–500 μin.) rms.

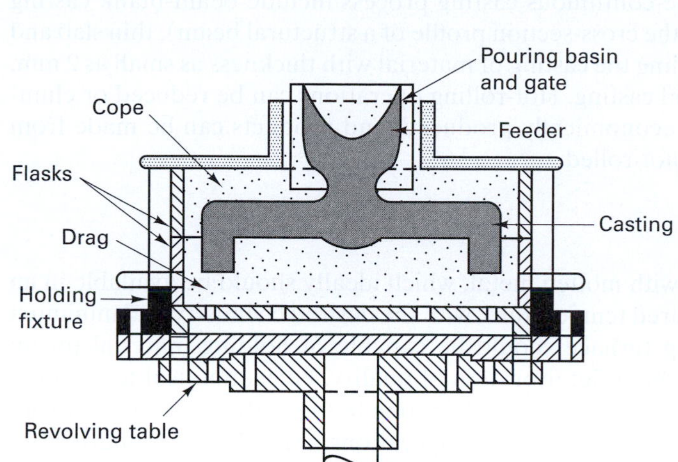

FIGURE 13-12 Schematic of a semicentrifugal casting process.

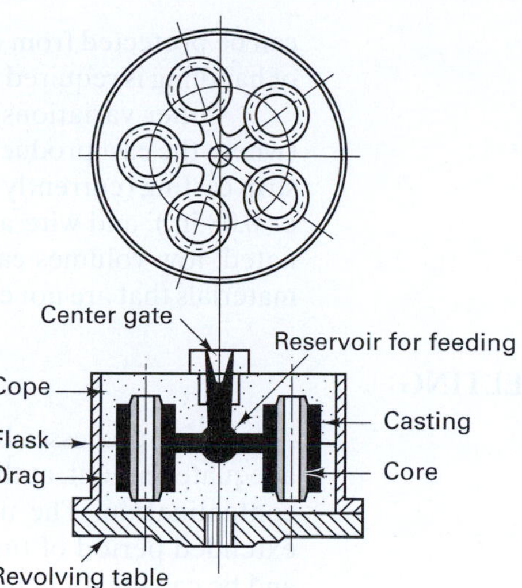

FIGURE 13-13 Schematic of a centrifuging process. Metal is poured into the central pouring sprue and spun into the various mold cavities. *(Courtesy of American Cast Iron Pipe Company, Birmingham, AL)*

rotational speeds are required to produce sound castings with thin walls and intricate shapes. Centrifuging is often used to assist in the pouring of multiple-product investment casting trees.

Centrifuging can also be used to drive pewter, zinc, or wax into spinning rubber molds to produce products with close tolerances, smooth surfaces, and excellent detail. These can be finished products or the low-melting-point patterns that are subsequently assembled to form the "trees" for investment casting.

13.6 CONTINUOUS CASTING

As discussed in Chapter 7 and depicted in Figure 7-3, **continuous casting** is usually employed in the solidification of basic shapes that become the feedstock for deformation processes such as rolling and forging. By producing a special mold, continuous casting can also be used to produce long lengths of complex cross-section product, such as the one depicted in Figure 13-14. Because each product is simply a cutoff section of the continuous strand, a single mold is all that is required to produce a large number of pieces. Near-net-shape continuous casting can significantly reduce the number of intermediate steps in the manufacture of a product, simultaneously reducing material losses, energy consumption and required manpower. Quality is high as well, because the metal

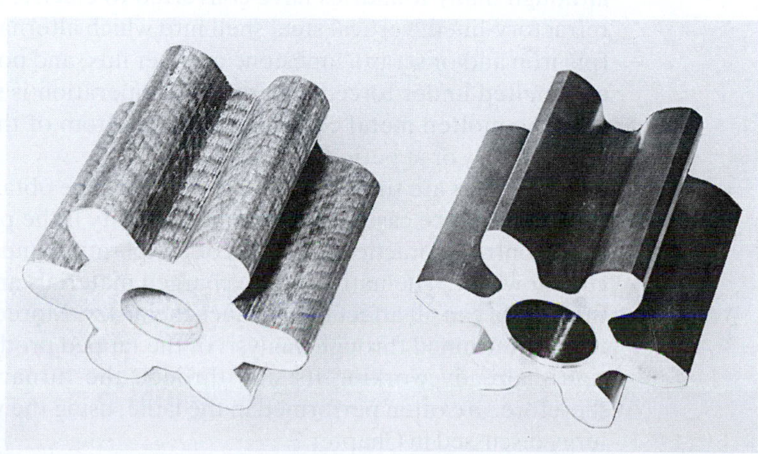

FIGURE 13-14 Gear produced by continuous casting: (left) as-cast material; (right) after machining. *(Courtesy of ASARCO, Tuscon, AZ)*

can be protected from contamination during melting and pouring, and only a minimum of handling is required.

Other variations of the continuous casting process include beam-blank casting (where the cast product has the cross-section profile of a structural beam), thin slab and strip casting (currently enabling the casting of material with thickness as small as 2 mm, or 0.08 in.), and wire and rod casting. Hot-rolling operations can be reduced or eliminated, low volumes can be economically produced, and products can be made from materials that are not easily hot-rolled.

13.7 MELTING

All casting processes begin with molten metal, which ideally should be available in an adequate amount, at the desired temperature, with the desired chemistry and minimum contamination. The melting furnace should be capable of holding material for an extended period of time without deterioration of quality, be economical to operate, and be capable of being operated without contributing to the pollution of the environment. Except for experimental or very small operations, virtually all foundries use cupolas, air furnaces (also known as direct fuel-fired furnaces), electric-arc furnaces, electric resistance furnaces, or electric-induction furnaces. In locations such as primary steel mills, molten metal may be taken directly from a steelmaking furnace and poured into casting molds. This practice is usually reserved for exceptionally large castings. For small operations, gas-fired crucible furnaces are also common, but these have rather limited capacities.

Selection of the most appropriate melting method depends on such factors as (1) the temperature needed to melt and superheat the metal, (2) the alloy being melted and the form of available charge material, (3) the desired melting rate or the desired quantity of molten metal, (4) the desired quality of the metal, (5) the availability and cost of various fuels, (6) the variety of metals or alloys to be melted, (7) whether melting is to be batch or continuous, (8) the required level of emission control, and (9) the various capital and operating costs.

The feedstock entering the melting furnace may take several forms. While prealloyed ingot may be purchased for remelt, it is not uncommon for the starting material to be a mix of commercially pure primary metal and commercial scrap, along with recycled gates, runners, sprues, and risers, as well as defective castings. The chemistry can be adjusted through alloy additions in the form of either pure materials or master alloys that are high in a particular element but are designed to have a lower melting point than the pure material and a density that allows for good mixing. Preheating the metal being charged is another common practice, and it can increase the melting rate of a furnace by as much as 30%.

CUPOLAS

A significant amount of gray, nodular, and white cast iron is still melted in **cupolas**, although many foundries have converted to electric induction furnaces. A cupola is a refractory-lined, vertical steel shell into which alternating layers of coke (carbon), iron (pig iron and/or scrap), limestone or other flux, and possible alloy additions are charged and melted under forced air draft. The operation is similar to that of a blast furnace, with the molten metal collecting at the bottom of the cupola to be tapped off either continuously or at periodic intervals.

Cupolas are simple and economical, can be obtained in a wide range of capacities, and can produce cast iron of excellent quality if the proper raw materials are used and good control is practiced. Control of temperature and chemistry can be somewhat difficult, however. The nature of the charged materials and the reactions that occur within the cupola can all affect the product chemistry. Moreover, by the time the final chemistry is determined through analysis of the tapped product, a substantial charge of material is already working its way through the furnace. Final chemistry adjustments, therefore, are often performed in the ladle, using the various techniques of ladle metallurgy discussed in Chapter 7.

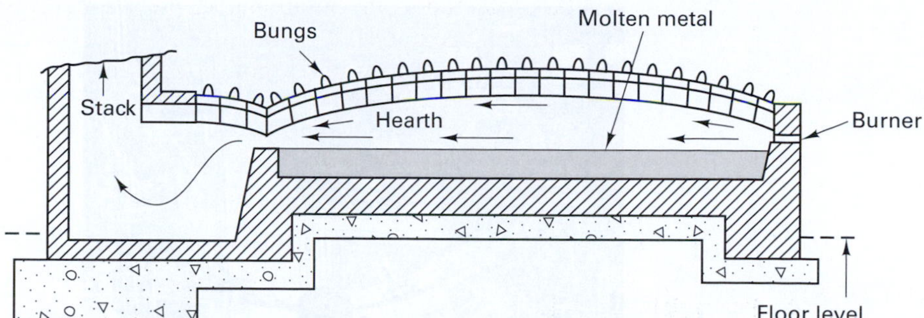

FIGURE 13-15 Cross section of a direct fuel-fired furnace. Hot combustion gases pass across the surface of a molten metal pool.

Various methods can be used to increase the melting rate and improve the economy of a cupola operation. In a hot-blast cupola, the stack gases are put through a heat exchanger to preheat the incoming air. Oxygen-enriched blasts can also be used to increase the temperature and accelerate the rate of melting. Plasma torches can be employed to melt the iron scrap. With typical enhancements, the melting rate of a continuously operating cupola can be quite high, such that production of 120 tons of hot metal per hour is not uncommon.

INDIRECT FUEL-FIRED FURNACES (OR CRUCIBLE FURNACES)

Small batches of nonferrous metal are often melted in **indirect fuel-fired furnaces** that are essentially crucibles or holding pots whose outer surface is heated by an external flame. The containment crucibles are generally made from clay and graphite, silicon carbide, cast iron, or steel. Stirring action, temperature control, and chemistry control are often poor, and furnace size and melting rate are limited. Nevertheless, these furnaces do offer low capital and operating cost. Better control of temperature and chemistry can be obtained, however, if the crucible furnaces are heated by electrical resistance heating.

DIRECT FUEL-FIRED FURNACES OR REVERBERATORY FURNACES

Direct fuel-fired furnaces, also known as **reverberatory furnaces,** are similar to small open-hearth furnaces but are less sophisticated. As illustrated in Figure 13-15, a fuel-fired flame passes directly over the pool of molten metal, with heat being transferred to the metal through both radiant heating from the refractory roof and walls and convective heating from the hot gases. Capacity is significantly greater than the crucible furnace, but the operation is still limited to the batch melting of nonferrous metals and the holding of cast iron that has been previously melted in a cupola. The rate of heating and melting and the temperature and composition of the molten metal are all easily controlled.

ARC FURNACES

Arc furnaces are the preferred method of melting in many foundries because of the (1) rapid melting rates, (2) ability to hold the molten metal for any desired period of time, and (3) greater ease of incorporating pollution control equipment. The basic features and operating cycle of a *direct-arc furnace* can be described with the aid of Figure 13-16. The top of the wide, shallow unit is first lifted or swung aside to permit the introduction of charge material. The top is then repositioned, and the electrodes are lowered to create an arc between the electrodes and the metal charge. The path of the heating current is usually through one electrode, across an arc to the metal charge, through the metal charge, and back through another arc to another electrode.

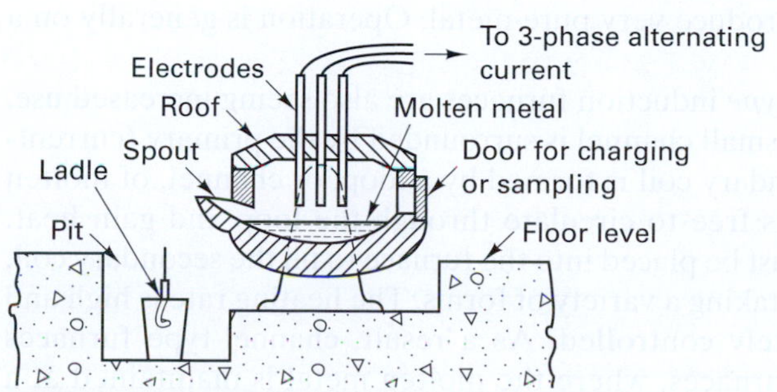

FIGURE 13-16 Schematic diagram of a three-phase electric-arc furnace.

Fluxing materials are usually added to create a protective slag over the pool of molten metal. Reactions between the slag and the metal serve to further remove impurities and

FIGURE 13-17 Electric-arc furnace, tilted for pouring. *(Courtesy Lectromelt Corporation, Pittsburgh, PA)*

are efficient because of the large interface area and the fact that the slag is as hot as the metal. Because the metal is covered and can be maintained at a given temperature for long periods of time, arc furnaces can be used to produce high-quality metal of almost any desired composition. They are available in sizes up to about 200 tons (but capacities of 25 tons or less are most common), and up to 50 tons per hour can be melted conveniently in batch operations. Arc furnaces are generally used with ferrous alloys—especially steel—and provide good mixing and homogeneity to the molten bath. Unfortunately, the noise and level of particle emissions can be rather high, and the consumption of electrodes, refractories, and power results in high operating costs. Figure 13-17 shows the pouring of an electric arc furnace. Note the still-glowing electrodes at the top of the furnace.

INDUCTION FURNACES

Because of their very rapid melting rates and the relative ease of controlling pollution, electric **induction furnaces** have become another popular means of melting metal. There are two basic types of induction furnaces. The *high-frequency*, or *coreless*, units, shown schematically in Figure 13-18, consist of a crucible surrounded by a water-cooled coil of copper tubing. A high-frequency electrical current passes through the coil, creating an alternating magnetic field. The varying magnetic field induces secondary electrical currents in the metal being melted, which bring about a rapid rate of heating.

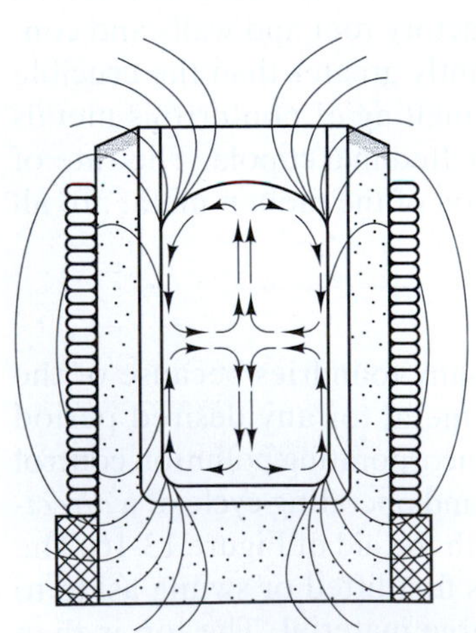

FIGURE 13-18 Schematic showing the basic principle of a coreless induction furnace.

Coreless induction furnaces are used for virtually all common alloys, with the maximum temperature being limited only by the refractory and the ability to insulate against heat loss. They provide good control of temperature and composition and are available in a range of capacities up to about 65 tons. Because there is no contamination from the heat source, they produce very pure metal. Operation is generally on a batch basis.

Low-frequency or *channel-type* induction furnaces are also seeing increased use. As shown in Figure 13-19, only a small channel is surrounded by the primary (current-carrying or heating) coil. A secondary coil is formed by a loop, or channel, of molten metal, and all the liquid metal is free to circulate through the loop and gain heat. To start, enough molten metal must be placed into the furnace to fill the secondary coil, with the remainder of the charge taking a variety of forms. The heating rate is high and the temperature can be accurately controlled. As a result, channel-type furnaces are often preferred as holding furnaces, where the molten metal is maintained at a constant temperature for an extended period of time. Capacities can be quite large, up to about 250 tons.

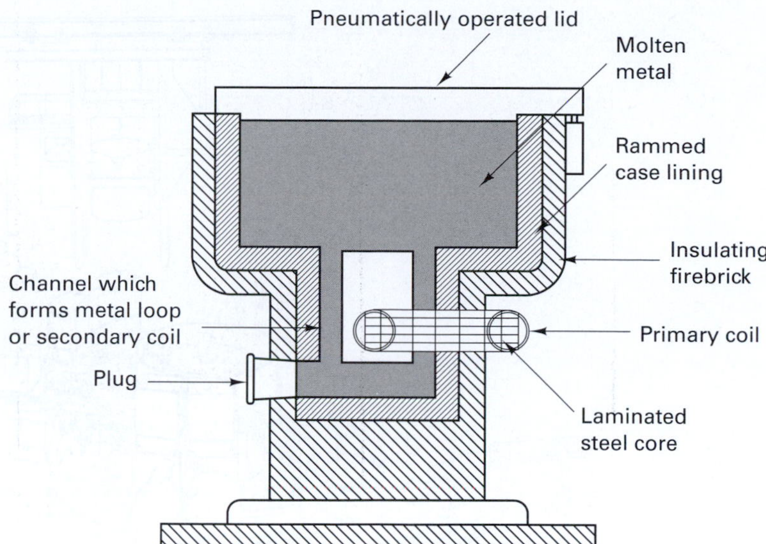

FIGURE 13-19 Cross section showing the principle of the low-frequency or channel-type induction furnace.

13.8 POURING PRACTICE

Some type of pouring device, or ladle, is usually required to transfer the metal from the melting furnace to the molds. The primary considerations for this operation are (1) to maintain the metal at the proper temperature for pouring and (2) to ensure that only high-quality metal is introduced into the molds. The specific type of **pouring ladle** is determined largely by the size and number of castings to be poured. In small foundries, a handheld, shank-type ladle is used for manual pouring. In larger foundries, either bottom-pour or teapot-type ladles are used, like the ones illustrated in Figure 11-6. These are often used in conjunction with a conveyor line that moves the molds past the pouring station. Because metal is extracted from beneath the surface, slag and other impurities that float on top of the melt are not permitted to enter the mold. To prevent the formation of new oxide inclusions, some form of protective shroud is often used to isolate the molten metal as it transfers from the ladle to the mold.

High-volume, mass-production foundries often use automatic pouring systems, like the one shown in Figure 13-20. Molten metal is transferred from a main melting furnace to a holding furnace. A programmed amount of molten metal is further transferred into individual pouring ladles and is then poured into the corresponding molds as they traverse by the pouring station. Laser-based control units position the pouring ladle over the sprue and control the flow rate into the pouring cup and fill height in the mold.

13.9 CLEANING, FINISHING, AND HEAT TREATING OF CASTINGS

CLEANING AND FINISHING

After solidification and removal from the mold, most castings require some additional cleaning and finishing. Specific operations may include all or several of the following:

1. Removing cores.
2. Removing gates and risers.
3. Removing fins, flash, and rough spots from the surface.
4. Cleaning the surface.
5. Repairing any defects.

Because cleaning and finishing operations can often comprise a considerable portion of the final cost of a cast component, consideration should be given to their minimization when designing the product and selecting the specific method of casting.

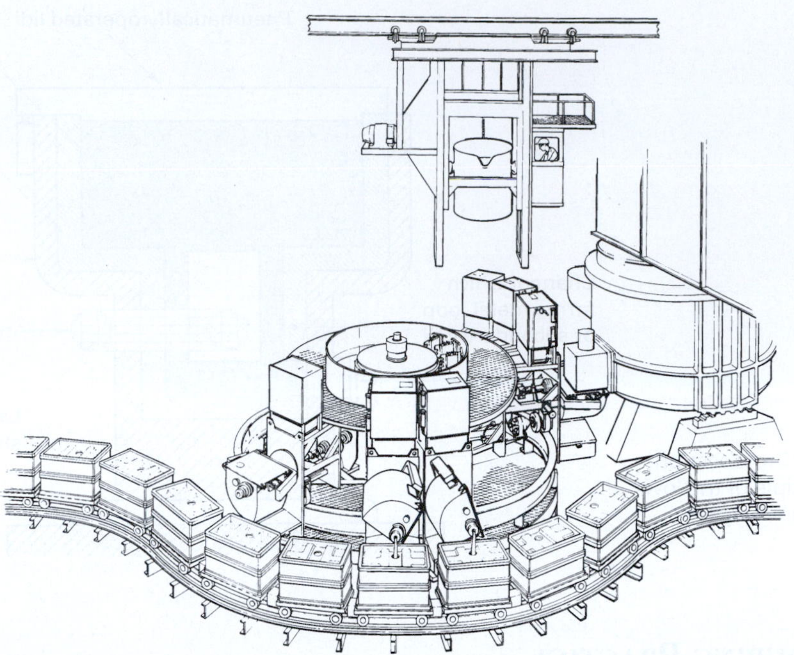

FIGURE 13-20 Automatic pouring of molds on a conveyor line. *(Courtesy of Roberts Sinto Corporation, Lansing, MI)*

In addition, consideration should also be directed toward the possibility of automating the cleaning and finishing.

Sand cores can usually be removed by mechanical shaking. At times, however, they must be removed by chemically dissolving the core binder. On small castings, sprues, runners, gates, and risers can sometimes be knocked off. For larger castings, a cutting operation is usually required. Most nonferrous metals and cast irons can be cut with an abrasive cutoff wheel, power hacksaw, or band saw. Steel castings frequently require an oxyacetylene torch. Plasma arc cutting can also be used. For high-volume castings, a trim press can be used to shear off the gates, riser connections and parting-line material. After the initial removal, the remaining segments of gates and excess material along parting lines are generally removed by abrasive grinding.

The specific method of cleaning often depends on the size and complexity of the casting. After the gates and risers have been removed, small castings are often tumbled in barrels to remove fins, flash, and sand that may have adhered to the surface. Tumbling may also be used to remove cores and, in some cases, gates and risers. Metal shot or abrasive material is often added to the barrel to aid in the cleaning. Conveyors can be used to pass larger castings through special cleaning chambers, where they are subjected to blasts of abrasive or cleaning material. Extremely large castings usually require manual finishing, using pneumatic chisels, portable grinders, and manually directed blast hoses.

While defect-free castings are always desired, flaws such as cracks, voids, and laps are not uncommon. In some cases, especially when the part is large and the production quantity is small, it may be more attractive to repair the part rather than change the pattern, die, or process. If the material is weldable, repairs are often made by removing the defective region (usually by chipping or grinding) and filling the created void with deposited weld metal. Porosity that is at or connected to free surfaces can be filled with resinous material, such as polyester, by a process known as **impregnation.** If the pores are filled with a lower-melting-point metal, the process becomes **infiltration.** (See Chapter 18 for a further discussion of these processes.)

HEAT TREATMENT AND INSPECTION OF CASTINGS

Heat treatment is an attractive means of altering properties while retaining the shape of the product. Steel castings are frequently given a full anneal to reduce the hardness and brittleness of rapidly cooled thin sections and to reduce the internal stresses that result from uneven cooling. Nonferrous castings are often heat treated to provide chemical homogenization or stress relief, as well as to prepare them for subsequent machining.

For final properties, virtually all of the treatments discussed in Chapter 6 can be applied. Ferrous-metal castings often undergo a quench-and-temper treatment, and many nonferrous castings are age hardened to impart additional strength. The variety of heat treatments is largely responsible for the wide range of properties and characteristics available in cast metal products.

Virtually all of the nondestructive **inspection techniques** can be applied to cast metal products. X-ray radiography, liquid penetrant inspection, and magnetic particle inspection are extremely common.

13.10 AUTOMATION IN FOUNDRY OPERATIONS

Many of the operations that are performed in a foundry are ideally suited for robotic automation because they tend to be dirty, dangerous, or dull. Robots can dry molds, coat cores, vent molds, and clean or lubricate dies. They can tend stationary, cyclic equipment, such as die-casting machines, and if the machines are properly grouped, one robot can often service two or three machines. In the investment casting process, robots can be used to dip the wax patterns into refractory slurry and produce the desired molds. In a similar manner, robots have been used in the full-mold and lost-foam processes to dip the Styrofoam patterns in their refractory coating and hang them on conveyors to dry. In a fully automated lost-foam operation, robots could be used to position the pattern, fill the flask with sand, pour the metal, and use a torch to remove the sprue. In the finishing room, robots can be equipped with plasma cutters or torches to remove sprues, gates, and runners. They can perform grinding and blasting operations, as well as various functions involved in the heat treatment of castings.

13.11 PROCESS SELECTION

As shown in the individual process summaries that have been included throughout Chapters 12 and 13, each of the casting processes has a characteristic set of capabilities, assets, and limitations. The requirements of a particular product (such as size, complexity, required dimensional precision, desired surface finish, total quantity to be made, and desired rate of production) often limit the number of processes that should be considered as production candidates. Further selection is usually based on cost.

Some aspects of product cost, such as the cost of the material and the energy required to melt it, are somewhat independent of the specific process. The cost of other features—such as patterns, molds, dies, melting and pouring equipment, scrap material, cleaning, inspection, and all related labor—can vary markedly and be quite dependent on the process. For example, pattern and mold costs for sand casting are quite a bit less than the cost of die-casting dies. Die casting, on the other hand, offers high production rates and a high degree of automation. When a small quantity of parts is desired, the cost of the die or tooling must be distributed over the total number of parts, and unit cost (or cost per casting) is high. When the total quantity is large, the tooling cost is distributed over many parts, and the cost per piece decreases.

Figure 13-21 shows the relationship between unit cost and production quantity for a product that can be made by both sand and die casting. Sand casting is an

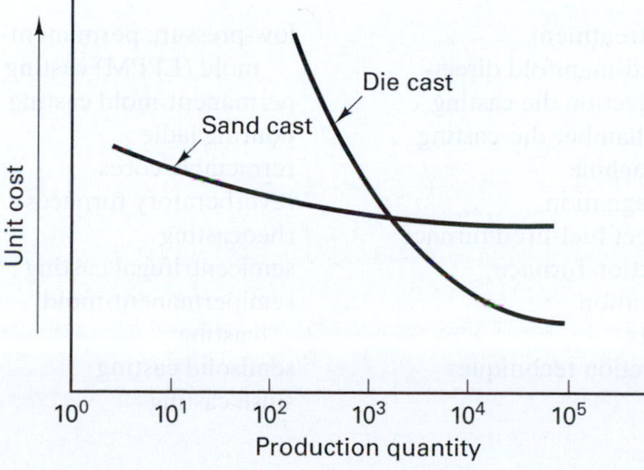

FIGURE 13-21 Typical unit cost of castings comparing sand casting and die casting. Note how the large cost of a die-casting die diminishes as it is spread over a larger quantity of parts.

TABLE 13-6 Comparison of Casting Processes

Property or Characteristic	Green-Sand Casting	Chemically Bonded Sand (Shell, Sodium Silicate, Air-Set)	Ceramic Mold and Investment Casting	Permanent-Mold Casting	Die Casting
Relative cost for small quantity	Lowest	Medium high	Medium	High	Highest
Relative cost for large quantity	Low	Medium high	Highest	Low	Lowest
Thinnest section (inches)	$\frac{1}{10}$	$\frac{1}{10}$	$\frac{1}{16}$	$\frac{1}{8}$	$\frac{1}{32}$
Dimensional precision (+/− in inches)	0.01–0.03	0.005–0.015	0.01–0.02	0.01–0.05	0.001–0.015
Relative surface finish	Fair to good	Good	Very good	Good	Best
Ease of casting complex shape	Fair to good	Good	Best	Fair	Good
Ease of changing design while in production	Best	Fair	Fair	Poor	Poorest
Castable metals	Unlimited	Unlimited	Unlimited	Low-melting-point metals	Low-melting-point metals

expendable mold process. Because an individual mold is required for each pour, increasing quantity does not lead to a significant drop in unit cost. Die casting involves a multiple-use mold, and the cost of the die can be distributed over the total number of parts. As shown in the figure, sand casting is often less expensive for small production runs, and processes such as die casting are preferred for large quantities. One should note that while the die-casting curve in Figure 13-21 is a smooth line, it is not uncommon for an actual curve to contain abrupt discontinuities. If the lifetime of a set of tooling is 50,000 casts, the cost *per part* for 45,000 pieces, using one set of tooling, would actually be less than for 60,000 pieces, because the latter would require a second set of dies.

In most cases, multiple processes are reasonable candidates for production, and the curves for all of the options should be included. The final selection is often based on a combination of economic, technical, and management considerations.

Table 13-6 presents a comparison of casting processes, including green-sand casting, chemically bonded sand molds (shell, sodium silicate, and air-set), ceramic mold and investment casting, permanent-mold casting, and die casting. The processes are compared on the basis of cost for both small and large quantities, thinnest section, dimensional precision, surface finish, ease of casting a complex shape, ease of changing the design while in production, and range of castable materials.

KEY WORDS

arc furnace
centrifugal casting
centrifuging
cold-chamber die-casting machine
continuous casting
cupola
die casting
direct fuel-fired furnace
gooseneck die-casting machine
heat checking
heat treatment
heated-manifold direct-injection die casting
hot-chamber die-casting machine
impregnation
indirect fuel-fired furnace
induction furnace
infiltration
inserts
inspection techniques
low-pressure permanent-mold (LPPM) casting
permanent-mold casting
pouring ladle
retractable cores
reverberatory furnaces
rheocasting
semicentrifugal casting
semipermanent-mold casting
semisolid casting
slush casting
squeeze casting
thermal fatigue
thixocasting
thixotropic material
tilt-pour permanent-mold casting
true centrifugal casting
vacuum permanent-mold casting
venting

REVIEW QUESTIONS

1. What are some of the major disadvantages of the expendable-mold casting processes?
2. What are some possible limitations of multiple-use molds?
3. What are some common mold materials for permanent-mold casting? What are some of the metals more commonly cast?
4. Describe some of the process advantages of permanent-mold casting.
5. Why do permanent-mold castings generally have higher strength than sand castings made from the same material?
6. Why might low production runs be unattractive for permanent-mold casting?
7. What features affect the life of a permanent mold?
8. How is venting provided in the permanent-mold process?
9. Why are permanent-mold castings generally removed from the mold immediately after solidification has been completed?
10. What types of products would be possible candidates for manufacture by slush casting?
11. What is the benefit of the tilt–pour version of permanent-mold casting?
12. How does low-pressure permanent-mold casting differ from the traditional gravity-pour process?
13. What are some of the attractive features of the low-pressure permanent-mold process?
14. What are some additional advantages of vacuum permanent-mold casting over the low-pressure process?
15. Contrast the feeding pressures on the molten metal in low-pressure permanent molding and die casting.
16. What are the most common die-cast materials?
17. Contrast the materials used to make dies for gravity-pour permanent-mold casting and die casting. Why is there a notable difference?
18. By what mechanisms do die-casting dies typically fail?
19. Why might it be advantageous to vary the pressure on the molten metal during the die-casting cycle?
20. For what types of materials would a hot-chamber die-casting machine be appropriate?
21. What is the benefit of heated-manifold direct-injection die casting?
22. What metals are routinely cast with cold-chamber die-casting machines?
23. How does the air in the mold cavity escape in the die-casting process?
24. Are risers employed in die casting? Can sand cores be used?
25. What is the most likely source of the porosity observed in die castings?
26. Give some examples of cast-in inserts.
27. What are some of the attractive features of die casting compared to alternative casting methods?
28. When might low quantities be justified for the die-casting process?
29. In what ways might a thin-walled zinc die casting be more attractive than a plastic injection molding?
30. Describe the squeeze-casting process.
31. What is a thixotropic material? How does it provide an attractive alternative to squeeze casting or rheocasting?
32. What are some of the attractive features of semisolid casting?
33. Contrast the structure and properties of the outer and inner surfaces of a centrifugal casting.
34. What are the key differences among true centrifugal casting, semicentrifugal casting, and centrifuging?
35. What are some common products that can be cast by the rotating mold casting processes?
36. How can continuous casting be used in the direct production of products?
37. What are some of the factors that influence the selection of a furnace type or melt procedure in a casting operation?
38. What are some of the possible feedstock materials that may be put in foundry melt furnaces?
39. What types of metals are commonly melted in cupolas?
40. What are some of the ways that the melting rate of a cupola can be increased?
41. What are some of the pros and cons of indirect fuel-fired furnaces?
42. What are some of the attractive features of arc furnaces in foundry applications?
43. Why are channel induction furnaces attractive for metal-holding applications where molten metal must be held at a specified temperature for long periods of time?
44. What are the primary functions of a pouring operation?
45. What are some of the typical cleaning and finishing operations that are performed on castings?
46. What are some common ways to remove cores from castings? To remove sprues, runners, gates, and risers?
47. What are some of the methods used to clean and finish castings?
48. How might defective castings be repaired to permit successful use in their intended applications?
49. What are some of the heat treatments that are applied to metal castings?
50. What are some of the ways that industrial robots can be employed in metal-casting operations?
51. Describe some of the features that affect the cost of a cast product. Why might the cost vary significantly with the quantity to be produced?
52. What are some of the key factors that should be considered when selecting a casting process?

PROBLEMS

1. Attractive properties for casting alloys include low melting points, high fluidity (or runniness), and high as-solidified strength. Wrought alloys (those fabricated as solids by processes such as rolling, forging, and extrusion) are best if they possess low yield strength, high ductility, and good strain-hardening characteristics. Reflecting these differences, the Aluminum Association uses distinctly different designation systems—a four-digit number for wrought alloys and a three-digit number for cast. Aluminum casting alloys are often further classified as "recommended for sand casting," "recommended for permanent mold casting," and "recommended for die casting." Why might different material properties be

required for casting into the three different mold materials—sand, cast iron, and water-cooled tool steel? What features or characteristics would be desired for each? Why?

2. Small, intricate shaped products can often be made by a variety of processes, including die casting, automated machining (screw machines), powder metallurgy, injection molding, and direct-digital manufacture. What features would favor manufacture by die casting over the alternative or competing processes?

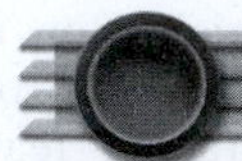

www.wiley.com/go/global/degarmo

Chapter 13 CASE STUDY

Baseplate for a Household Steam Iron

The item depicted in the figure is the baseplate of a high-quality household steam iron. It is rated for operation at up to 1200 watts and is designed to provide both steady steam and burst of steam features. Incorporated into the design is an embedded electrical resistance heating "horseshoe" that must be thermally coupled to the baseplate but must also remain electrically insulated. (This component often takes the form of a resistance heating wire, surrounded by ceramic insulation, all encased in a metal tube.) The steam emerges through a number of small vent holes in the base, each about $\frac{1}{16}$-in. in diameter. There are about a dozen larger threaded recesses, about $\frac{1}{8}$-in. in diameter, that are used in assembling the various components.

1. Discuss the various features that this component must possess in order to function in an adequate fashion. Consider strength, impact resistance, wear resistance, thermal conductivity, corrosion resistance, weight, and other factors.
2. Based on the required properties, what material or materials would appear to be strong candidates?
3. What are some possible means of producing the desired shape? What are the major advantages of the various methods you propose?
4. For each of your proposed shape-producing alternatives, answer the following: Could the heating element assembly be incorporated during initial manufacture? Could all of the design features (holes, webs, and recesses) be incorporated in the initial manufacturing operation? For what features might secondary processing be required? How would you recommend that they be produced?
5. One manufacturer has replaced the embedded tubular heating element with a square-cross-section element that is pressed into a receiving groove in the iron baseplate. Discuss this alternative in terms of manufacturing ease as well as product performance and quality.
6. Some commercial irons have baseplates for which the bottom surfaces have been finished by a simple buff and polish, while others have a Teflon coating or have been anodized. If your desire is to produce a product of the highest quality, what form of surface finishing would you recommend?

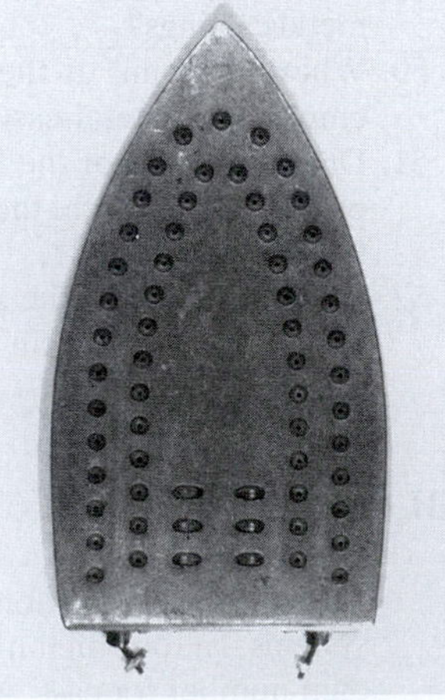

Baseplate of a steam iron: (left) from bottom; (right) from top.
(Ron Kohser)

CHAPTER 14

FABRICATION OF PLASTICS, CERAMICS, AND COMPOSITES

14.1 INTRODUCTION

In Chapters 7, 8 and 9, **plastics, ceramics,** and **composites** were shown to be substantially different from metals in both structure and properties. While the specific material will still be selected for its ability to provide the required properties, and the fabrication processes for their ability to produce the desired shape in an economical and practical manner, it is reasonable to expect that there will be significant differences in product design, material selection, and fabrication.

Plastics, ceramics, and composites tend to be used closer to their design limits, and many of the fabrication processes convert the raw material into a finished product in a single operation. Large, complex shapes can often be formed as a single unit, eliminating the need for multipart assembly operations. Materials in these classes can often provide integral and variable color, and the processes used to manufacture the shape can frequently produce the desired finish and precision. Finishing operations are often unnecessary—an attractive feature because altering the final dimensions or surface would be both difficult and costly for some of these materials. The joining and fastening operations also tend to be different from those used with metals.

As with metals, the desired properties are often affected by the processes used to produce the shape. The fabrication of an acceptable product, therefore, involves the selection of both an appropriate material and a companion method of processing, such that the resulting combination provides the desired shape, properties, precision, and finish.

14.2 FABRICATION OF PLASTICS

The manufacture of a successful plastic product requires satisfying the various mechanical and physical property requirements through the use of the most economical resin or **compound** that will perform satisfactorily, coupled with a manufacturing process that is compatible with both the part design and the selected material.

Chapter 9 presented material about the wide variety of plastics or polymers that are currently used as engineering materials. As we move our attention to the fabrication of parts and shapes, we find that there are also a number of processes from which to choose. Determination of the preferred method depends on the desired size, shape, and quantity, as well as whether the polymer is a *thermoplastic, thermoset*, or *elastomer*. **Thermoplastic polymers** can be heated to produce either a soft, formable solid or a liquid. The material can then be cast, injected into a mold, or forced into or through dies to produce a desired shape. **Thermosetting polymers** have far fewer options, because once the polymerization has occurred, the framework structure is established, and no further deformation can occur. Thus, the polymerization reaction must take place during the shape-forming operation. **Elastomers** are sufficiently unique that they will be treated in a separate section of this chapter.

Casting, blow molding, compression molding, transfer molding, cold molding, injection molding, reaction injection molding, extrusion, thermoforming, rotational molding, and *foam molding* are all processes that are used to shape polymers. Each has its distinct set of advantages and limitations that relate to part design, compatible materials, and production cost. To make optimum selections, we must become familiar with the shape capabilities of a process as well as how the process affects the properties of the material.

CASTING

Casting is the simplest of the shape forming processes because no fillers are used and no pressure is required. While not all plastics can be cast, there are a number of castable thermoplastics, including acrylics, nylons, urethanes, and PVC plastisols. The thermoplastic polymer is simply melted, and the liquid is poured into a container having the shape of the desired part. Several variations of the process have been developed. Small products can be cast directly into shaped molds. Plate glass can be used as a mold to cast individual pieces of thick plastic sheet. Continuous sheets and films can be produced by injecting the liquid polymer between two moving belts of highly polished stainless steel, the width and thickness being set by resilient gasket strips on either end of the gap. Thin sheets can be made by ejecting molten liquid from a gap-slot die onto a temperature-controlled chill roll. The molten plastic can also be spun against a rotating mold wall (centrifugal casting) to produce hollow or tubular shapes.

Some thermosets (such as phenolics, polyesters, epoxies, silicones, and urethanes) can also be cast, as well as any resin that will polymerize at low temperatures and atmospheric pressure. Because of the need for curing, the casting of thermoset resins usually involves additional processing, often some form of heating while in the mold. Figure 14-1 depicts a process where a steel pattern is dipped into molten lead, withdrawn, and allowed to cool. A thin lead sheath is produced when the pattern is removed, and this becomes the mold for the plastic resin. Curing occurs, either at room temperature or by heating for long times at temperatures in the range of 65 to 95°C (150 to 200°F). After curing, the product is removed, and the lead sheaths can be reused.

Because cast plastics contain no fillers, they have a distinctly lustrous appearance, and a wide range of transparent and translucent colors are available. Because the product is shaped as a liquid, fiber or particulate reinforcement can be easily incorporated. The process is relatively inexpensive because of the comparative lack of costly dies,

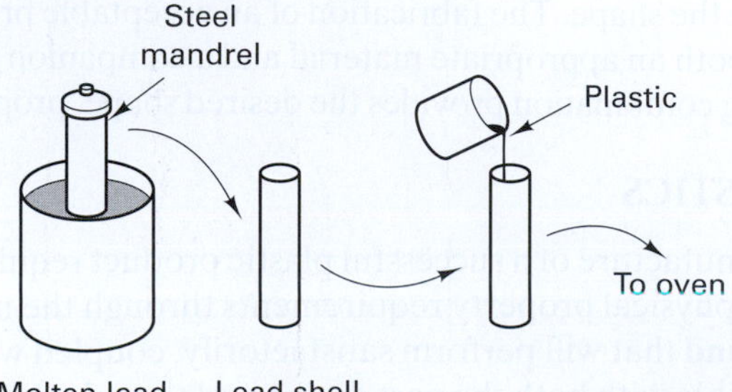

FIGURE 14-1 Steps in the casting of thermoset plastic parts using a lead shell mold.

equipment, and controls. Typical products include sheets, plates, films, rods, and tubes, as well as small objects, such as jewelry, ornamental shapes, gears, and lenses. While dimensional precision can be quite high, quality problems can occur because of inadequate mixing, air entrapment, gas evolution, and shrinkage. Hollow parts can be made by allowing the external surfaces to cool or cure, then pouring the remaining liquid back out of the mold, a variation of the slush casting process discussed in Chapter 13.

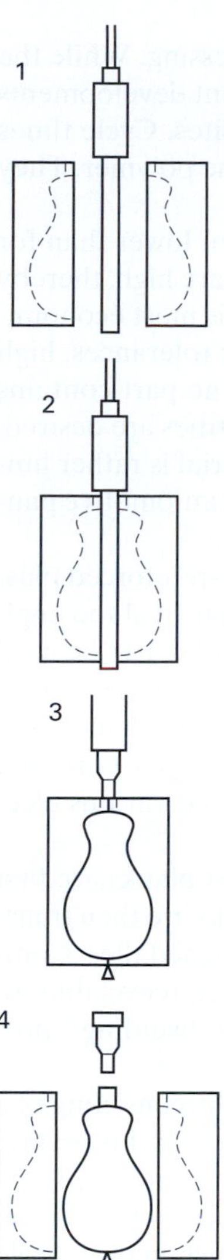

FIGURE 14-2 Steps in blow molding plastic parts: (1) a tube of heated plastic is placed in the open mold; (2) the mold closes over the tube, simultaneously sealing the bottom; (3) air expands the tube against the sides of the mold; and (4) after sufficient cooling, the mold opens to release the product.

BLOW MOLDING

A variety of **blow molding** processes have been developed, the most common being used to convert thermoplastic polyethylene, polyvinyl chloride (PVC), polypropylene, and polyether ether ketone PEEK resins into bottles and other seamless, hollow-shape containers. A solid-bottom, hollow-tube preform, known as a **parison,** is made from heated plastic by either extrusion or injection molding. The heated preform is then positioned between the halves of a split mold, the mold closes, and the preform is expanded against the mold by air or gas pressure. The mold is then cooled, the halves separated, and the product is removed. Any flash is then trimmed for direct recycling. Figure 14-2 depicts a form of this process where the starting material is a simple tube and the solid bottom is created by the pinching action of die closure. Blow molding has recently expanded to include engineering thermoplastics and has been used to produce products as diverse as automotive fuel tanks, seat backs, ductwork, and bumper beams.

Variations of blow molding have been designed to provide both axial and radial expansion of the plastic (for enhanced strength) as well as to produce multilayered products. In one process, a sheet of heated plastic is placed between the upper and lower cavities, the lower one having the shape of the product. Both cavities are then pressurized to 2 to 4 MPa (300 to 600 psi) with a nonreactive gas such as argon. When the pressure in the lower segment is then vented, the gas in the upper segment "blows" the material into the lower die cavity.

Because the thermoplastics must be cooled before removal from the mold, the molds for blow molding must contain the desired cavity as well as a cooling system, venting system, and other design features. The mold material must provide thermal conductivity and durability while being inexpensive and compatible with the resins being processed. Beryllium, copper, aluminum, tool steels, and stainless steels are all popular mold materials.

COMPRESSION MOLDING OR HOT-COMPRESSION MOLDING

Compression molding is one of the most widely used molding processes for thermosetting polymers. As illustrated schematically in Figure 14-3, a premeasured amount of solid granules, powders, or pellets, or preformed tablets of *unpolymerized* plastic (often preheated) are first introduced into an open, heated cavity. A heated plunger then descends to close the cavity and apply pressure, typically in the range of 10 to 150 MPa (1500 to 20,000 psi). As the material melts and becomes fluid, it is driven into all portions of the cavity. The heat and pressure are maintained until the material has "set" (i.e., cured or polymerized). The mold is then opened and the part is removed. A wide variety of heating systems and mold materials are used, and multiple cavities can

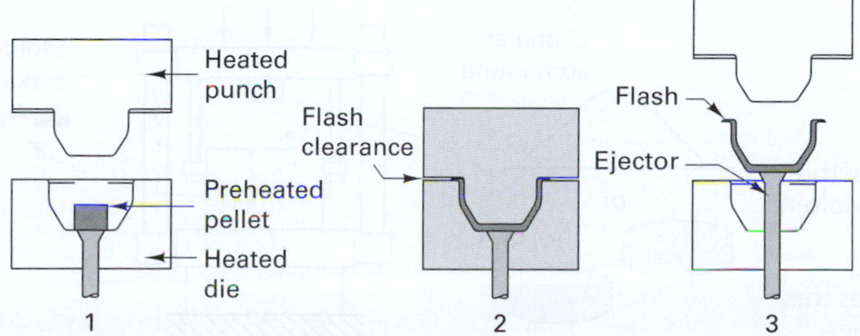

FIGURE 14-3 The hot-compression molding process: (1) solid granules or a preform pellet is placed in a heated die; (2) a heated punch descends and applies pressure; and (3) after curing (thermosets) or cooling (thermoplastics), the mold is opened and the part is removed.

be placed within a mold to produce more than one part in a single pressing. While the process has been used primarily with the thermosetting polymers, recent developments permit the shaping of thermoplastics and various reinforced composites. Cycle times are set by the rate of heat transfer, and the reaction or curing rate of the polymer. They typically range from under 1 min to as much as 20 min or more.

The tool and machinery costs for compression molding are often lower than for competing processes, and the dimensional precision and surface finish are high, thereby reducing or eliminating secondary operations. Compression molding is most economical when it is applied to small production runs of parts requiring close tolerances, high impact strength, and low mold shrinkage. It is a poor choice when the part contains thick sections (the cure times become quite long), or when large quantities are desired. Most products have relatively simple shapes because the flow of material is rather limited. Typical compression-molded parts include gaskets, seals, exterior automotive panels, aircraft fairings, and a wide variety of interior panels.

More recently, compression molding has been used to form fiber-reinforced plastics, both thermoplastics and thermosets, into parts with properties that rival the engineering metals. In the thermoset family, polyesters, epoxies, and phenolics can be used as the base of fiber-containing sheet-molding compound, bulk-molding compound, or sprayed-up reinforcement mats. These are introduced into the mold and shaped and cured in the normal manner. Cycle times range from about 1 to 5 min per part, and typical products include wash basins, bathtubs, equipment housings, and various electrical components.

If the starting material is a fiber-containing thermoplastic, precut blanks are first heated in an infrared oven to produce a soft, pliable material. The blanks are then transferred to the press, where they are shaped and cooled in specially designed dies. Compared to the thermosets, cycle times are reduced and the scrap is often recyclable. In addition, the products can be joined or assembled using the thermal "welding" processes applied to plastics.

Compression molding equipment is usually rather simple, typically consisting of a hydraulic or pneumatic press with parallel platens that apply the heat and pressure. Pressing areas range from 15 cm^2 (6 in.^{2}) to as much as 2.5 m^2 (8 ft^2), and the force capacities range from 6 to 9000 metric tons. The molds are usually made of tool steel and are polished or chrome plated to improve material flow and product quality. Mold temperatures typically run between 150 and 200°C (300 and 400°F) but can go as high as 650°C (1200°F). They are heated by a variety of means, including electric heaters, steam, oil, and gas.

TRANSFER MOLDING

Transfer molding is sometimes used to reduce the turbulence and uneven flow that can result from the high pressures of hot-compression molding. As shown in Figure 14-4, the *unpolymerized* raw material is now placed in a plunger cavity, where it is heated until molten. The plunger then descends, forcing the molten plastic through channels or runners into adjoining die cavities. Temperature and pressure are maintained until the thermosetting resin has completely cured. To shorten the cycle and extend the lifetime of the cavity, plunger, runner, and gates, the charge material may be preheated before being placed in the plunger cavity.

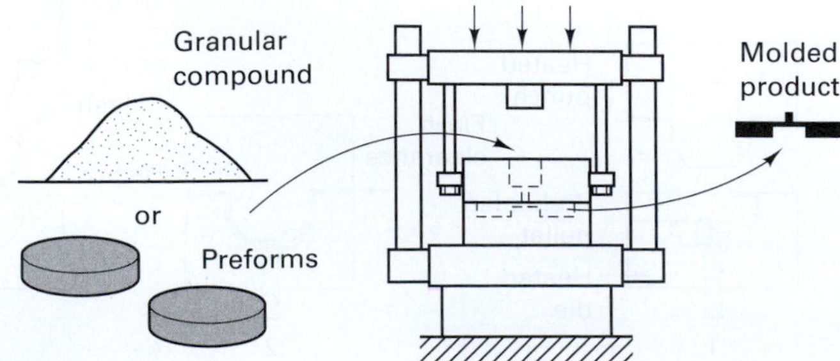

FIGURE 14-4 Diagram of the transfer molding process. Molten or softened material is first formed in the upper heated cavity. A plunger then drives the material into an adjacent die.

Because the material enters the die cavities as a liquid, there is little pressure until the cavity is completely filled. Thin sections, excellent detail, and good tolerances and finish are all characteristics of the process. In addition, inserts can be incorporated into the products of transfer molding. They are simply positioned within the cavity and maintained in place as the liquid resin is introduced around them.

Transfer molding is attractive for producing small to medium-sized parts with relatively complex shapes. It combines elements of both compression molding and injection molding (to be discussed) and enables some of the advantages of injection molding to be utilized with thermosetting polymers. The thermosetting resins can be reinforced with fillers—such as cellulose, glass, silica, alumina, or mica—to improve the mechanical or electrical properties and reduce shrinkage or warping. The main limitation is the loss of material, because resin left in the pot or well, sprue, and runners also cures and must now be discarded. Common products include electrical switchgear and wiring devices, parts of household appliances that require heat resistance, structural parts that require hardness and rigidity under load, under-hood automotive parts, and parts that require good resistance to chemical attack.

COLD MOLDING

In **cold molding,** the uncured thermosetting material is pressed to shape while cold and is then removed from the mold and cured in a separate oven. While the process is faster and more economical, the resulting products generally lack good surface finish and dimensional precision.

INJECTION MOLDING

Injection molding is the most widely used process for the high-volume production of relatively complex thermoplastic parts, and is often considered to be the polymer equivalent to metal die casting. Figure 14-5 illustrates one approach to the process, where granules of raw material are fed by gravity from a hopper into a cavity that lies ahead of a moving plunger. As the plunger advances, the material is forced through a preheating chamber and on through a torpedo section, where it is mixed, melted, and superheated. The superheated material is then driven through a nozzle that seats against a mold. Other types of injection units control the flow of material and generate the injection pressure with screws that have both rotational and axial movements, or combinations of screws and plungers. Alternative methods of heating the material include heated barrels and the shearing action as material moves through the screws.

Sprues and runners then channel the molten material into one or more closed-die cavities. Because the dies remain cool, the plastic solidifies almost as soon as the mold is filled. Premature solidification would cause defective parts, so the superheated material must be rapidly forced into the mold cavities by pressures in the range 35 to 140 MPa (5 to 20 ksi), which are maintained during solidification. The mold halves must clamp tightly together during molding and then be easily separated for part ejection. Impact forces should be minimized during die closure because they can adversely affect die life. Various types of clamping designs have been developed, including toggle, hydraulic, and hydro-mechanical.

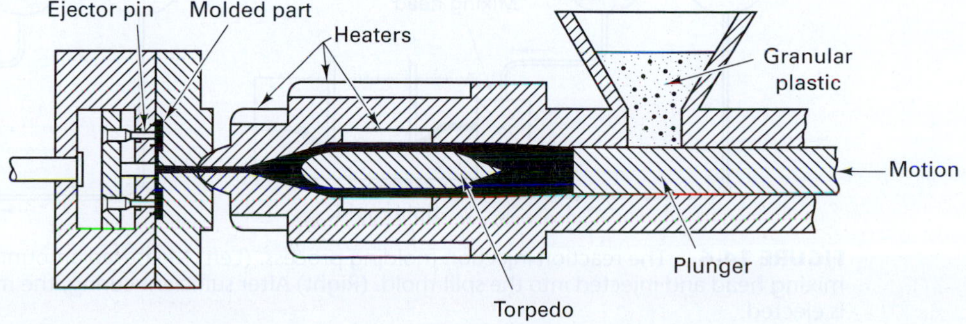

FIGURE 14-5 Schematic diagram of the injection molding process. A moving plunger advances material through a heating region (in this case, through a heated manifold and over a heated torpedo) and further through runners into a mold where the molten thermoplastic cools and solidifies.

Control systems coordinate all of the functions of the process, including the time required for cooling within the mold. By heating the material for the next part as the mold is separating for part ejection, a molding cycle can generally be completed in 10 to 30 s. The process is quite similar to the die casting of molten metal, and the result is usually a finished product needing no further work before assembly or use. Part size can be as small as 50 g (2 oz) or a large as 25 kg (>50 lb).

Some injection molding machines incorporate a hot runner distribution system to transfer the material from the injection nozzle to the mold cavities. If the runners are cold, the material in the runner solidifies with each cycle and needs to be ejected and reprocessed or disposed. With hot runners, the thermoplastic material is maintained in a liquid state until it reaches the gate. The material in the runners can be used in the subsequent shot, thereby reducing shot size and cycle time because less material must be heated. Quality is improved because all material enters the mold at the same temperature, recycled sprues and runners are not incorporated into the charge, and there is less turbulence because pressurized material is not injected into empty runners. Hot runners do add an additional degree of complexity to the design, operation, and control of the system, so the additional cost must be weighed against the benefits cited.

Injection molding can also be applied to the thermosetting materials, but the process must be modified to provide the temperature, pressure, and time required for curing. The injection chamber is now at a significantly lower temperature, and the mold is heated. The time in the heated mold must be sufficient to complete the curing process, typically between 20 s and 2 min. The relatively long cycle times are the major deterrent to the injection molding of the thermosets. Speed-up alternatives include completing the cure outside the mold and having multiple dies that can be fed with one injector.

REACTION INJECTION MOLDING

Figure 14-6 depicts the **reaction injection molding** process, in which two or more liquid reactants are metered into a unit where they are intimately mixed by the impingement of liquid streams that have been pressurized to a value between 13 and 20 MPa (2000 and 3000 psi). The combined material flows through a pressure-reducing chamber and exits the mix-head directly into a mold. An exothermic chemical reaction takes place between the two components, resulting in thermoset polymerization. Because no heating is required, the production rates are set primarily by the curing time of the polymer, which may be less than 1 min or up to 10 min. Molds are made from steel, aluminum, or nickel shell, with selection being made on the basis of number of parts to be made and the desired quality. The molds are generally clamped in low-tonnage presses.

Currently, the dominant materials for reaction injection molding are polyurethanes, polyamides, and composites containing short fibers or flakes. Properties can span a wide range, depending on the combination and percentage of base chemicals and the additives that are used. Different formulations can result in elastomeric, or

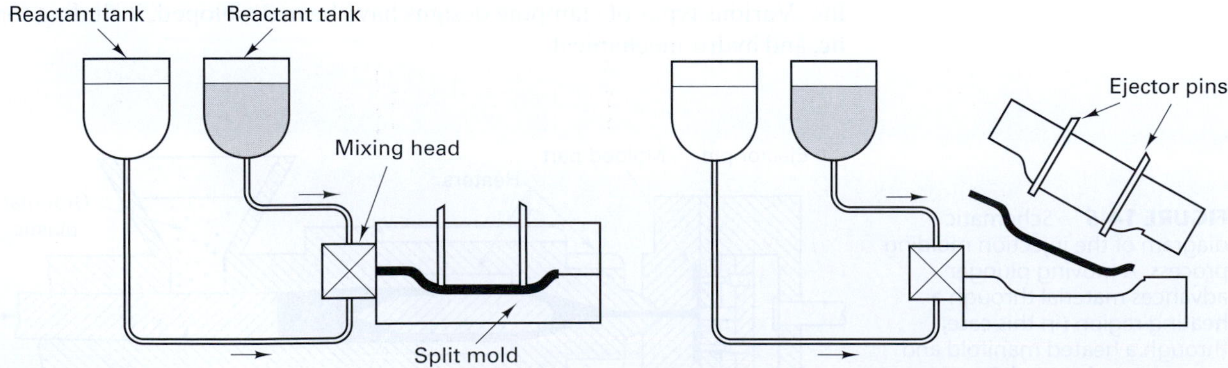

FIGURE 14-6 The reaction injection molding process. (Left) Measured amounts of reactants are combined in the mixing head and injected into the split mold. (Right) After sufficient curing, the mold is opened and the component is ejected.

flexible, structural foam (foam core with a hard, solid outer skin); solid (no foam core); or composite products. Part size can range from $\frac{1}{2}$ to 50 kg (1 to 100 lb), shapes can be quite complex (with variable wall thickness), and surface finish is excellent. Automotive applications include steering wheels, airbag covers, instrument panels, door panels, armrests, headliners, and center consoles, as well as body panels, bumpers, and wheel covers. Rigid polyurethanes are also used in such products as computer housings, household refrigerators, water skis, hot-water heaters, and picnic coolers.

From a manufacturing perspective, reaction injection molding has a number of attractive features. The low processing temperatures and low injection pressures make the process attractive for molding large parts, and the large size can often enable parts consolidation. Thermoset parts can generally be fabricated with less energy than the injection molding of thermoplastics, with similar cycle times and a similar degree of automation. The metering and mixing equipment and related controls tend to be quite sophisticated and costly, but the lower temperatures and pressures enable the use of cheaper molds, which can be quite large.

EXTRUSION

Long plastic products with uniform cross sections can be readily produced by the **extrusion** process depicted in Figure 14-7. Thermoplastic pellets or powders are fed through a hopper into the barrel chamber of a screw extruder. A rotating screw propels the material through a preheating section, where it is heated, homogenized, and compressed, and then forces it through a heated die and onto a conveyor belt. To preserve its newly imparted shape, the material is cooled and hardened by jets of air or sprays of water. It continues to cool as it passes along the belt and is then either cut into lengths or coiled, depending on whether the material is rigid or flexible and the desires of the customer.

The process is continuous and provides a cheap and rapid method of molding. As shown in Figure 14-8, common production shapes include a wide variety of constant cross-section profiles, including solid shapes (such as window and trim molding), hollow shapes (such as tubes and pipes), coated wires and cables, sheets and films, and small-diameter fibers or filaments. Thermoplastic foam shapes can also be produced.

By using a narrow slit as the die opening, sheet (thicknesses between 0.5 and 12.5 mm or 0.02 to 0.5 in) and film (thicknesses below 0.5 mm or 0.02 in.) can be produced in widths up to 3 m (10 ft). The extruded material is rapidly cooled by either wrapping around chilled rolls or direct immersion into a cooling bath. A combination of extrusion and blowing has been used in the manufacture of thin plastic bags, like those that are used as kitchen or bathroom trashcan liners. This sequence begins with the extrusion of a thin-walled plastic tube through an open-ended metal die. Air flowing through the center of the die causes the diameter of the tube to expand substantially as it emerges from the die constraint. Air jets around the circumference of the expanded tube then cool the thin plastic material, after which it is passed through pinch rollers. The flattened tube is then periodically seam welded to form the bottom of the bag, perforated for tearing, and wound on a roll for easy dispensing. The product can also be cut on both edges to produce strips of thin film.

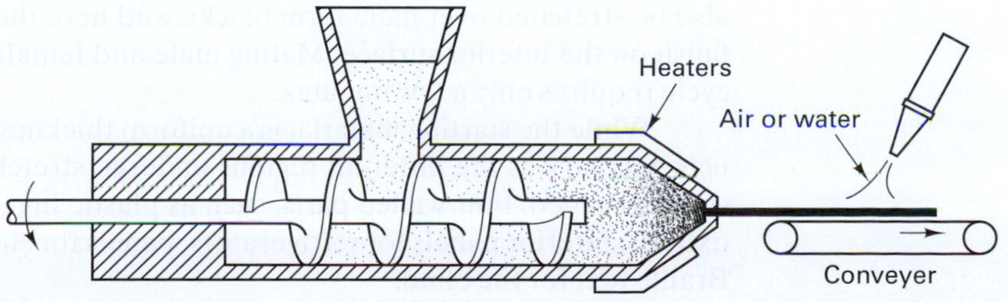

FIGURE 14-7 A screw extruder producing thermoplastic product. Some units may have a changeable die at the exit to permit production of different shaped parts.

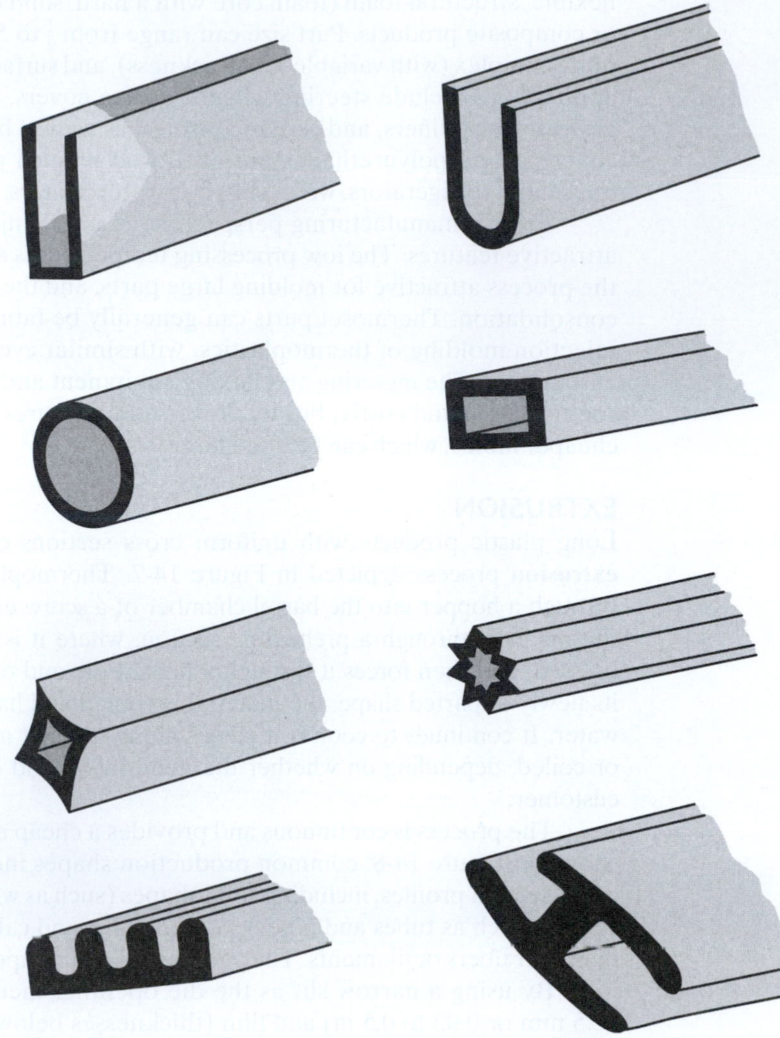

FIGURE 14-8 Typical shapes of polymer extrusions.

THERMOFORMING

In the **thermoforming** process, thermoplastic sheet material, as either discrete sheets or in a continuous roll, is heated to a working temperature and then formed into desirable shapes. If continuous material is used, it is usually heated by passing through an oven or other heating device. The material emerges over a male or female mold and is formed by the application of vacuum, pressure, or another mechanical tool. Cooling occurs upon contact with the mold, and the product hardens in its new shape. After sufficient cooling, the part is removed from the mold and trimmed, and the unused strip material is diverted for recycling.

Figure 14-9 shows the process using a female mold cavity and a discrete sheet of material. Here, the material is placed directly over the die or pattern and is heated in place, often by infrared radiant heaters. Either pressure or vacuum (sometimes both) are then applied, causing the material to draw into the cavity. The female die imparts both the dimensions and finish or texture to the exterior surface. The sheet material can also be stretched over male form blocks, and here the tooling controls dimensions and finish on the interior surface. Mating male and female dies can also be used. An entire cycle requires only a few minutes.

While the starting material is a uniform thickness sheet, the thickness of the products will vary as the different regions undergo stretching. Typical products tend to be simple-shaped, thin-walled parts, such as plastic luggage, plastic trays, panels for light fixtures, interior panels for refrigerators, panels for shower enclosures, or even pages of Braille text for the blind.

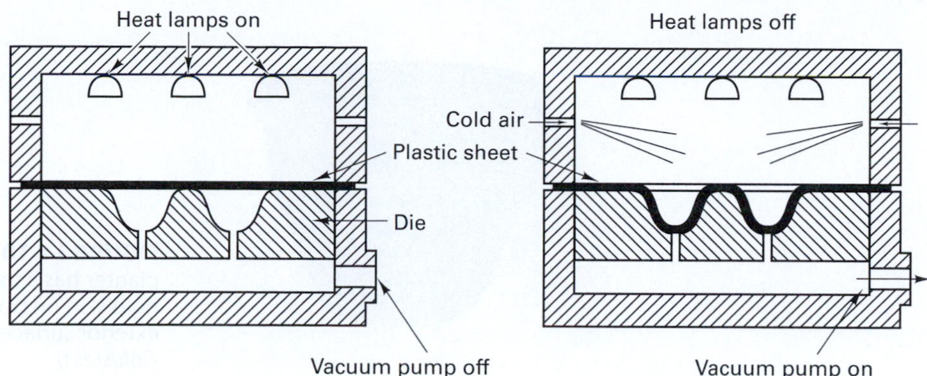

FIGURE 14-9 A type of thermoforming where thermoplastic sheets are shaped using a combination of heat and vacuum.

ROTATIONAL MOLDING

Rotational molding can be used to produce hollow, seamless products of a wide variety of sizes and shapes, including storage tanks, bins and refuse containers, small swimming pools, boat and canoe hulls, luggage, footballs, helmets, garbage cans, portable outhouses, septic tanks, and numerous automotive and truck panels and parts—generally parts with more complex external geometries, larger size, or lower production quantities than those produced by blow molding. Common materials include polyethylene, polypropylene, acrylonitrile butadiene styrene (ABS), and high-impact polystyrene.

The process begins with a closed mold or cavity that has been filled with a premeasured amount of thermoplastic powder or liquid. The molds are either preheated or placed in a heated oven and are then rotated simultaneously about two perpendicular axes. Other designs rotate the mold about one axis while tilting or rocking about another. In either case, the resin melts and is distributed, by gravity, to produce a uniform-thickness coating over all of the surfaces of the mold. The mold is then transferred to a cooling chamber, where the motion is continued and air or water is used to slowly drop the temperature. After the material has solidified, the mold is opened and the hollow product is removed. All of the starting material is used in the product; no scrap is generated. The lightweight rotational molds are frequently made from cast aluminum, but sheet metal is often used for larger parts, and electroformed or vapor-formed nickel is used when fine detail is to be reproduced. Production times are long compared to the other processes, as much as 10 min or more per cycle.

FOAM MOLDING

Foamed plastic products have become an important and widely used form of polymer offering light weight, good thermal insulation, and good energy absorption. In **foam molding,** a foaming agent is mixed with the plastic resin and releases gas or volatilizes when the material is heated during molding. The materials expand to 2 to 50 times their original size, resulting in products with densities ranging from 32 to 640 g/L (2 to 40 lb/ft^3). **Open-cell foams** have interconnected pores that permit the permeability of gas or liquid. **Closed-cell foams** have the property of being gas- or liquid-tight.

Both rigid and flexible foams have been produced using both thermoplastic and thermosetting materials. The rigid type is useful for structural applications (including housings for computers and business machines), packaging, and shipping containers; as patterns for the full-mold and lost-foam casting processes (see Chapter 12); and for injection into the interiors of thin-skinned metal components, such as aircraft fins and stabilizers. Flexible foams are used primarily for cushioning and padding.

Parts are typically produced by variations of the previously described processes. Large sheets of foamed insulation board can be produced by extruding foaming material. Thinner foamed sheets can be further thermoformed to create products such as the cushioning egg cartons. Prefoamed polystyrene beads can be expanded and fused to

FIGURE 14-10 This garden planter has a foamed core and rigid skin on both the interior and exterior surfaces. *(Photo by Rachel Goldstein)*

create the various Styrofoam products, including drinking cups, disposable plates, and the expendable patterns for lost-foam casting. Reactive ingredients can be combined and projected through a spray gun, where polymerization and foaming occur simultaneously to create a rigid insulating foam. By introducing foaming material into the interior of a cold mold, parts can be produced with a solid outer skin and a rigid foam core. The rapid cooling against the cold mold surfaces produces a rigid skin that can be as thick as 2 mm (0.08 in.). Dual-structure products can also be produced by injecting the foaming polymer into a mold that contains a hollow, partially formed product. Figure 14-10 shows a common product with a foamed core and a rigid skin.

OTHER PLASTIC-FORMING PROCESSES

In the **calendering** process (described in more detail in Section 14-3), a mass of dough-like thermoplastic is forced between and over two or more counter-rotating rolls to produce thin sheets or films of polymer, which are then cooled to induce hardening. Product thicknesses generally range between 0.3 and 1.0 mm (0.01 to .006 in.), but can be reduced further to as low as 0.05 mm (0.002 in.) by subsequent stretching. Embossed designs can be incorporated into the rolls to produce products with textures or patterns.

Conventional **drawing** can be used to produce fibers, and **rolling** can be performed to change the shape of thermoplastic extrusions. In addition to changing the product dimensions, these processes can also serve to induce crystallization or to produce a preferred orientation to the thermoplastic polymer chains.

Filaments, fibers, and yarns can be produced by **spinning,** a modified form of extrusion in which molten thermoplastic polymer is forced through a die containing many small holes called a **spinneret.** The emerging filaments are then drawn to further decrease their diameter and are air-cooled. Additional operations further elongate the fiber, aligning the structure and increasing the tensile strength. Most of the polymeric fibers—which include polyesters, nylons, rayons, and acrylics—are used in the textile industry. When multistrand yarns or cables are desired, the extrusion dies can rotate or spin to produce the necessary twists and wraps.

MACHINING OF PLASTICS

Plastics can be milled, sawed, drilled, and threaded much like metals, but their properties are so variable that it is impossible to give descriptions that would be correct for all types. Instead, let us first consider some of the general characteristics of plastics that affect their machinability. Because plastics tend to be poor thermal conductors, most of the heat generated during chip formation remains near the cut interface and is not conducted into the material or carried away in the chips. Thermoplastics tend to soften and swell, and they occasionally bind or clog the cutting tool. Considerable elastic flexing can also occur, and this couples with material softening to reduce the precision of final dimensions. Because the thermosetting polymers have higher rigidity and reduced softening, they generally machine to greater dimensional precision.

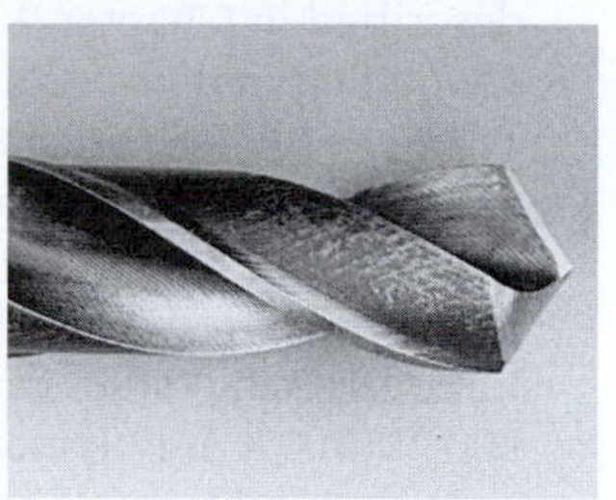

FIGURE 14-11 Straight-flute drill (left) and "dubbed" drill (right) used for drilling plastics. *(E. Paul DeGarmo)*

The high temperatures that develop at the point of cutting also cause the tools to run very hot, and they may fail more rapidly than when cutting metal. Carbide tools may be preferred over high-speed tool steels if the cuts are of moderate duration or if high-speed cutting is performed. Coolants can often be used advantageously if they do not discolor the plastic or induce gumming. Water, soluble oil and water, and weak solutions of sodium silicate have been used effectively.

The tools that are used to machine plastics should also be kept sharp at all times. Drilling is best done by means of straight-flute drills or by "dubbing" the cutting edge of a regular twist drill to produce a zero rake angle. These configurations are shown in Figure 14-11. Rotary files, saws, and milling cutters should be run at high speeds to improve cooling but with the feed carefully adjusted to avoid clogging the cutter.

Laser machining may be an attractive alternative to mechanical cutting. By vaporizing the material instead of forming chips, precise cuts can be achieved. Minute holes can be drilled, such as those in the nozzles of aerosol cans. Abrasive materials, such as filled and laminated plastics can be machined in a manner that also eliminates the fine machining dust that is often considered to be a health hazard.

FINISHING AND ASSEMBLY OPERATIONS

Polymeric materials frequently offer the possibility of integral color, and the as-formed surface is often adequate for final use. Some of the finishing processes that can be applied to plastics include printing, hot stamping, vacuum metallizing, electroplating, and painting. Chapter 34 presents the processes of surface finishing and surface engineering.

Thermoplastic polymers can often be joined by heating the relevant surfaces or regions. The joining heat can be provided by a stream of hot gases, applied through a tool like a soldering iron or generated by ultrasonic vibrations. The welding techniques that are applied to plastics are presented in Chapter 32. Adhesive bonding, another popular means of joining plastic, is presented in Chapter 33. Because of the low modulus of elasticity, plastics can also be easily flexed, and **snap-fits** are another popular means of assembling plastic components. Because of the softness of some polymeric materials, self-tapping screws can also be used.

DESIGNING FOR FABRICATION

The primary objective of any manufacturing activity is the production of satisfactory components or products, in the necessary quantity, and at the desired rate of production. This activity begins with the selection of an appropriate material or materials. When polymers are selected as the material of construction, it is usually as a result of one or more of their somewhat unique properties, which include light weight, corrosion resistance, good thermal and electrical insulation, ease of fabrication and the possibility of integral color. While these properties are indeed attractive, one should also be aware of the more common limitations, such as strength and stiffness values that are lower than the engineering metals, high amounts of thermal expansion, creep under mechanical load, softening or burning at elevated temperatures, poor dimensional stability, and the deterioration of properties with age or exposure to sunlight or other forms of radiation.

The basic properties and characteristics of polymeric materials have been described in Chapter 9. One should note, however, that property evaluation tests are conducted under specific test conditions. Polymers are often speed-sensitive materials. While a standard tensile test may show a polymer to have a moderately high strength value, a reduction in loading rate by two or three orders of magnitude may reduce this strength by as much as 80%. Conversely, an increase in loading rate may double or triple tensile strength. Polymers can also be extremely sensitive to changes in temperature. Strength values can vary by a factor of 10 over a temperature range of as little as 200°F (100°C). Polymeric materials should be selected with full consideration to the specific conditions of temperature, loads and load rates, and operating environments that will be encountered.

The second area of manufacturing concern is selecting the process or processes to be used in producing the shape and establishing the desired properties. Each of the wide variety of fabrication processes has distinct advantages and limitations, and efforts should be made to utilize their unique features. Once a process has been selected, the production of quality products further requires an awareness of all of the various aspects of that process. For example, consider a molding process in which a liquid or semifluid polymer is introduced into a mold cavity and allowed to harden. The proper amount of material must be introduced and caused to flow in such a way as to completely fill the cavity. Air that originally occupied the cavity needs to be vented and removed. Shrinkage will occur during solidification and/or cooling and may not occur in a uniform manner. Heat transfer must be provided to control the cooling and/or solidification. Finally, a means must be provided for part removal or ejection from the mold. Surface finish and appearance, the resultant engineering properties, and the ultimate cost of production are all dependent on good design and proper execution of the molding process. Product properties can be significantly affected by such factors as melt temperature, direction of flow, pressure during molding, thermal degradation, and cooling rate.

In all molded products, it is important to provide adequate fillets between adjacent sections to ensure smooth flow of the polymer into all sections of the mold and to eliminate stress concentrations at sharp interior corners. These fillets also make the mold less expensive to produce and reduce the danger of mold fracture during use. Even the exterior edges of the product should be rounded where possible. A radius of 0.25 to 0.40 mm (0.010 to 0.015 in.) is scarcely noticeable but will do much to prevent an edge from chipping. Sharp corners should also be avoided in products that will be used for electrical applications, because they tend to increase voltage gradients, which can lead to product failure.

Wall or section thickness is also very important because the hardening or curing time of a polymer is determined by the thickest section. If possible, sections should be kept nearly uniform in thickness because nonuniformity can lead to serious warpage and dimensional control problems. As a general rule, one should use the minimum thickness that will provide satisfactory end-use performance. The specific value will be determined primarily by the size of the part and, to some extent, the process and the type of polymer being used. Recommended minimum thicknesses for molded polymers are as follows:

Small parts	1.25 mm (0.050 in.)
Average-sized parts	2.15 mm (0.085 in.)
Large parts	3.20 mm (0.125 in.)

Thick corners should also be avoided because they can lead to gas pockets, undercuring, or cracking. When extra strength is needed in a corner, it can usually be provided by incorporating ribs into the design.

Economical production is also facilitated by appropriate dimensional tolerances. A minimum tolerance of 0.08 mm (0.003 in.) should be allowed in directions that are parallel to the parting line of a mold or contained within a mold segment. In directions that cross a parting surface, a minimum tolerance of 0.25 mm (0.010 in.) is desirable.

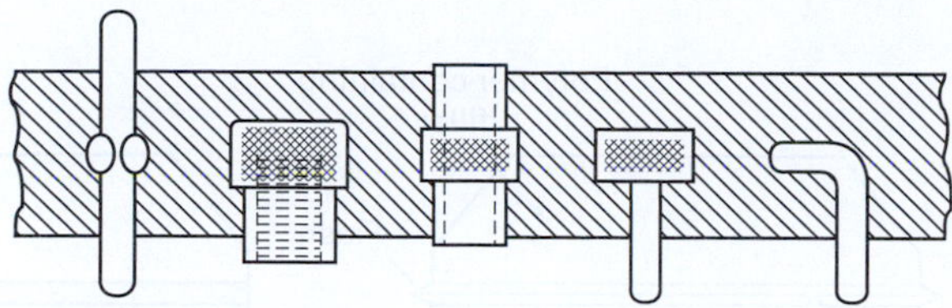

FIGURE 14-12 Typical metal inserts used to provide threaded cavities, holes, and alignment pins in plastic parts.

In both cases, increasing these values by about 50% can simultaneously reduce both manufacturing difficulty as well as cost.

Because most molds are reusable, careful attention should also be given to the removal of the part. Rigid-metal molds should be designed so that they can be easily opened and closed. A small amount of unidirectional taper should be provided to facilitate part withdrawal. Undercuts should be avoided whenever possible because they will prevent part removal unless additional mold sections are used. These must move independently of the major segments of the mold, adding to the costs of mold production and maintenance and slowing the rate of production.

INSERTS

Metal **inserts,** usually of brass or steel, are often incorporated into plastic products to provide enhanced performance or unique features. Because molded threads are difficult to produce, machined threads require additional processing, and both types tend to chip or deform, threaded metal inserts are frequently used when assemblies require considerable strength or when frequent disassembly and reassembly is anticipated. Figure 14-12 depicts one form of threaded insert, along with other types that provide pins or holes for alignment or mounting.

The successful use of inserts requires careful attention to design because they are generally held in place by only a mechanical bond that must resist both rotation and pullout. Knurling or grooving is often required to provide suitable sites for gripping. A medium or coarse knurl is usually adequate to resist torsional loads and moderate axial forces. Circumferential grooves are excellent for axial loads but offer little resistance to torsional rotation. Axial grooves resist rotation, but do little to prevent pullout. Other means of anchoring include bending, splitting, notching, and swaging. Headed parts with noncircular heads may be used as formed. Combinations of notches, grooves, and shoulders are also common. Figure 14-13 depicts some of these common means of insert attachment.

If an insert is to act as a boss for mounting or serve as an electrical terminal, it should protrude slightly above the surface of the plastic. This permits a firm connection to be made without creating an axial load that would tend to pull the insert from its surroundings. If the insert serves to hold two mating parts together or align them, it should be flush with the surface. In this way, the parts can be held together snugly without danger of loosening the insert. In all cases, the wall thickness of the surrounding plastic must be sufficient to support any load that may be transmitted through the insert. For small inserts, the wall thickness should be at least half the diameter of the insert. For inserts larger than 13 mm ($\frac{1}{2}$ in.) in diameter, the wall thickness should be at least 6.5 mm ($\frac{1}{4}$ in.).

DESIGN FACTORS RELATED TO FINISHING

Because plastics are frequently used where consumer acceptance is of great importance, special attention should be given to finish and appearance. In many cases, plastic parts can be designed to require very little finishing or decorative treatment. For small

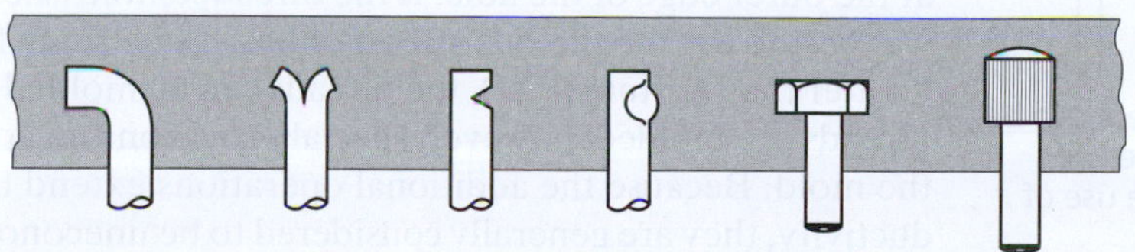

FIGURE 14-13 Various ways of anchoring metal inserts in plastic parts (left to right): bending, splitting, notching, swaging, noncircular head, and grooves and shoulders. Knurling is depicted in Figure 14-12.

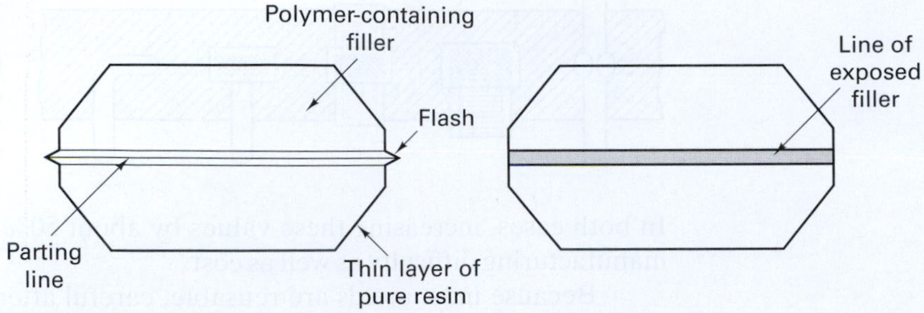

FIGURE 14-14 Trimming the flash from a plastic part ruptures the thin layer of pure resin along the parting line and creates a line of exposed filler.

parts, fins and rough spots can often be removed by barrel tumbling with suitable abrasives or polishing agents. Smoothing and polishing occur in the same operation.

By etching the surfaces of a mold, decorations or letters can be produced that protrude approximately 0.01 mm (0.004 in.) above the surface of the plastic. When higher relief is required, the mold can be engraved, but this adds significantly to mold cost. Whenever possible, depressed letters or designs should be avoided. These features, when transferred to the mold, become raised above the surrounding surface. Mold making then requires a considerable amount of intricate machining as the surrounding material must be cut away from the design or letters. When recessed features are absolutely required, manufacturing cost can be reduced if they can be incorporated into a small area that is originally raised above the primary surface.

When designing plastic parts, a prime objective is often the elimination of secondary machining, especially on surfaces that would be exposed to the customer. Even when fillers are used (as they are in most plastics), the surfaces of molded parts have a thin film of pure resin. This film provides the high luster that is characteristic of polymeric products. Machining cuts through the surface, exposing the underlying filler. The result is a poor appearance, as well as a site for the absorption of moisture.

One location that frequently requires machining is the parting line that is produced where the mold segments come together. Because perfect mating is difficult to achieve, a small fin, or **flash,** is usually produced around the part perimeter, as illustrated in Figure 14-14. When the flash is trimmed off, the resulting line of exposed filler may be objectionable. By locating the parting line along a sharp corner, it is easier to maintain satisfactory mating of the mold sections, and the exposed filler that is created by flash removal will be confined to a corner, where it is less noticeable.

Because plastics have a low modulus of elasticity, large flat areas are not rigid and should be avoided whenever possible. Ribbing or doming, like that illustrated in Figure 14-15, can be used to provide the required stiffness. In addition, flat surfaces tend to reveal flow marks from the molding operation, as well as scratches that occur during handling or service. External ribbing then serves the dual function of increasing strength and rigidity while masking any surface flaws. Dimpled or textured surfaces can also be used to provide a pleasing appearance and conceal scratches.

Holes that are formed by pins protruding from the mold often require special consideration. During the mold closure and filling stages of compression molding, these pins can be subjected to considerable bending. When they are supported only at one end, the length should not exceed twice the diameter. In processes with reduced filling pressures, the length can be as much as five times the diameter without excessive problems.

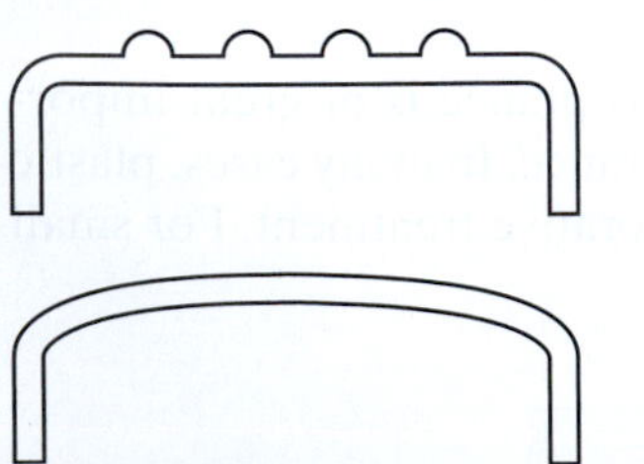

FIGURE 14-15 Stiffness can be imparted to large surfaces of plastic parts through the use of ribbing or doming.

Holes that are to be threaded or used to receive self-tapping screws should be countersunk. This not only assists in starting the tap or screw but also reduces chipping at the outer edge of the hole. If the threaded hole is less than 6.5 mm ($\frac{1}{4}$ in.) in diameter, it is best to cut the threads after molding, using some form of thread tap. For diameters greater than 6.5 mm ($\frac{1}{4}$ in.), the threads can be molded or an insert should be used. If the threads are molded, however, special provisions must be made to remove the part from the mold. Because the additional operations extend the molding time and reduce productivity, they are generally considered to be uneconomical.

14.3 PROCESSING OF RUBBER AND ELASTOMERS

Rubber and elastomeric products can be produced by a variety of fabrication processes. Relatively thin parts with uniform wall thickness, such as boots, gloves, and fairings, are often made by some form of **dipping.** A master form is first produced, usually from some type of metal. This form is then immersed into a liquid preparation or compound (usually based on natural rubber or latex, neoprene, or silicone), then removed and allowed to dry. With each dip, a certain amount of the liquid adheres to the surface, with repeated dips being used to produce a final desired thickness. After vulcanization, usually in steam, the products are stripped from the molds.

The dipping process can be accelerated by using electrostatic charges. A negative charge is introduced to the latex particles, and the form or mold receives a positive charge, either through an applied voltage or by a coagulant coating that releases positive ions when dipped into the solution. The attraction and neutralization of the opposite charges causes the elastomeric particles to be deposited on the form at a faster rate and in thicker layers than the basic process. With electrostatic deposition, many products can be made in a single immersion.

When the parts are thicker or are complex in shape, the first step is the compounding of elastomeric resin, vulcanizers, fillers, antioxidants, accelerators, and pigments. This is usually done in some form of mixer, which blends the components to form a homogeneous mass. Adaptations of the processes previously discussed for plastics are frequently used to produce the desired shapes. Injection, compression, and transfer molding are used, along with special techniques for foaming. Urethanes and silicones can also be directly cast to shape.

Rubber compounds can be made into sheets using calenders, like that shown in Figure 14-16. A warm mass of compound is fed into the gap between rotating rolls and the emerging product is typically 0.3 to 1 mm (0.01 to 0.40 in.) in thickness. The sheet coming from the calender is often rolled with a fabric between the layers to prevent the material from sticking to itself. Three- or four-roll calenders can also be used to place a rubber or elastomer covering over cord or woven fabric. In the three-roll geometry, only one side of the fabric is coated in a single pass. The four-roll arrangement, shown in Figure 14-17, enables both sides to be coated simultaneously. Rubber-coated fabrics can also be produced by dipping the fabric into a rubber solution, spraying the solution onto the fabric, or skimming a thick solution of rubber compound and solvent onto the cloth and then driving off the solvent. The resultant material is used in products such as tires, conveyor belts, inflatable rafts, and raincoats.

Rubber products such as inner tubes, garden hoses, tubing, and strip moldings can be produced by the extrusion process. The compounded elastomer is forced through a die by a screw device similar to that described for plastics.

(a)

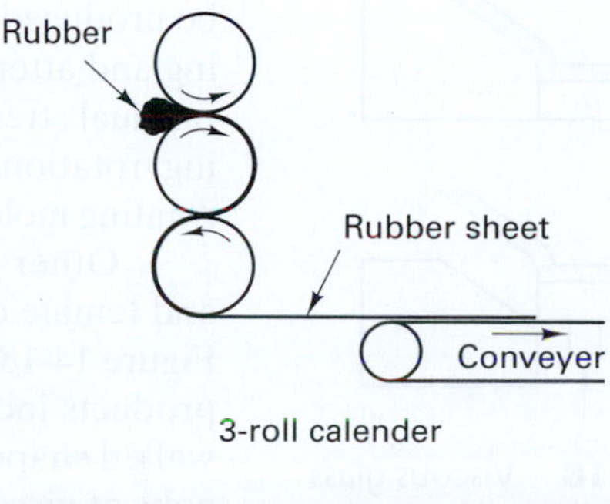

(b)

FIGURE 14-16 (a) Three-roll calender used for producing rubber or plastic sheet. (b) Schematic diagram showing the method of making sheets of rubber with a three-roll calender. *[(a) (Courtesy of Farrel-Birmingham Company, Inc. Ansonia, CT)]*

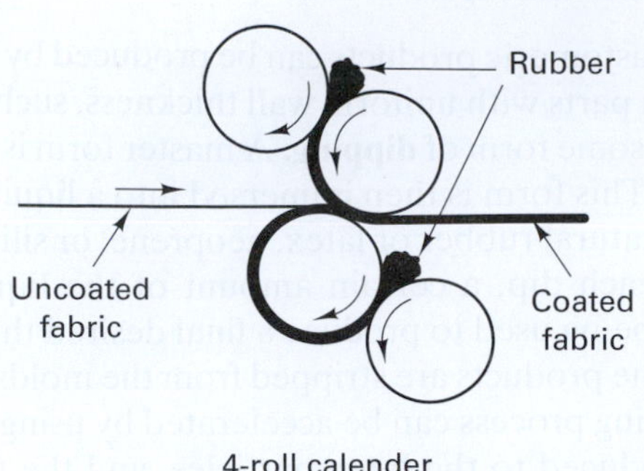

FIGURE 14-17 Arrangement of the rolls, fabric, and coating material for coating both sides of a fabric in a four-roll calender.

Rubber or artificial elastomers can also be bonded to metal, such as brass or steel, using a variety of polymeric **adhesives**. Only moderate pressures and temperatures are required to obtain excellent adhesion.

Vulcanization or **cross-linking** is a critical step in all of the rubber processing sequences because it imparts the desired stiffness and strength while retaining the elastic behavior. Sulfur and other cross-linking materials are blended into the starting material. In molding operations, the mold is simply maintained at elevated temperature for sufficient time to impart curing. Other processes perform the vulcanization after the part has been shaped. In batch processing, the parts are heated in either an autoclave (a pressure vessel containing hot steam) or an air-atmosphere furnace or oven. Continuous products can be vulvanized by passing them through hot-air tunnels or over or through heated rolls.

14.4 PROCESSING OF CERAMICS

The fabrication processes applied to ceramic materials generally fall into two distinct classes, based on the properties of the material. **Glasses** can be manufactured into useful articles by first heating the material to produce a molten or viscous state, shaping the material by means of **viscous flow,** and then cooling the material to produce a solid product. **Crystalline ceramics** have a characteristically brittle behavior and are normally manufactured into useful components by pressing moist aggregates or powder into a shape, followed by drying, and then bonding by one of a variety of mechanisms, which include chemical reaction, **vitrification** (cementing with a liquefied material), and **sintering** (solid-state diffusion).

FABRICATION TECHNIQUES FOR GLASSES

Glass is generally shaped at elevated temperatures where the viscosity can be controlled. A number of the processes begin with material in the liquid or molten condition at temperatures between 1000 and 1200°C (1850 and 2200°F). Sheet and plate glass is formed by processes such as extruding through a narrow slit, rolling through water-cooled rolls, or floating on a bath of molten tin. Glass shapes, such as larger lenses, can be produced by pouring the molten material directly into a mold. The cooling rate during and after solidification is then controlled (usually as slow as possible) to minimize residual stresses and the tendency for cracking. In a process similar to centrifugal casting, rotationally symmetrical parts can be produced by introducing molten glass into a rotating mold. Constant-cross-section shapes can be made by extrusion.

Other glass-forming processes begin with viscous masses and use mating male and female die members to press the material into the desired shape, as illustrated in Figure 14-18. This **pressing** process is similar to the closed-die forging of metal. Typical products include dishes and bake ware. The millions of bottles, jars, and other thin-walled shapcs are made by a process similar to the blow molding of plastics. Hollow gobs of viscous material are expanded against the outside of heated dies in a manner similar to that illustrated in Figure 14-19.

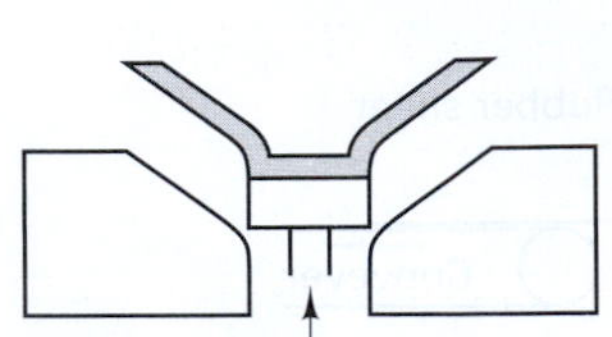

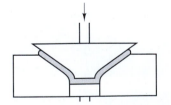

FIGURE 14-18 Viscous glass can be easily shaped by mating male and female die members.

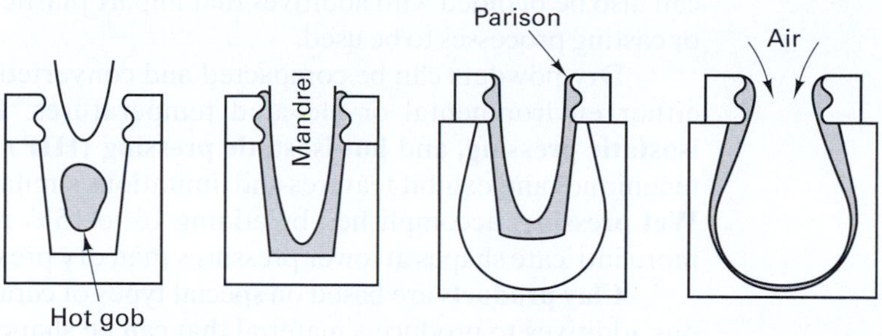

FIGURE 14-19 Thin-walled glass shapes can be produced by a combination of pressing and blow molding.

Glass fibers have been used for applications as diverse as thermal insulation, air and fluid filtration, sound damping, reinforcing polymers, weaving into cloth and fabric, and fiber-optic communication. When short length and random orientation are permissible, the fibers can be made by pouring molten glass into a rotating chamber with multiple small orifices around its exterior. Long, continuous fibers are generally made by extruding through a multi-orifice die or pulling through a heated plate with multiple small orifices. Various coatings are often applied to the fibers to lubricate and protect the surface.

Special heat treatments are often applied to glass products before their manufacture is complete. The most common of these is **annealing,** performed to relieve the unfavorable residual stresses that form during shaping and subsequent cooling. The glass product is heated to an elevated temperature (typically around 500°C or 900°F) and held for a period of time, then slowly cooled to prevent the formation of new residual stresses.

Beneficial residual stresses can be imparted by a process known as **tempering.** Glass, generally in the form of precut sheets or plates, is heated to a temperature above that of annealing and held to create a uniform temperature through the thickness. The surfaces are then rapidly cooled, usually with jets of air. These regions then cool and contract, and the softer interior flows to conform. Subsequently, the interior cools and attempts to contract but is restrained by the already cold surface, placing the surface in compression. The resulting product, called **tempered glass,** is stronger and more fracture resistant, because cracks tend to initiate on free surfaces and these are now compressed. When tempered glass fails, the internal tension is released, causing the material to shatter into numerous small fragments that are less likely to injure an individual.

Similar results can be achieved by **chemical tempering.** The glass is heated in a bath of molten salt (potassium nitrate, sodium nitrate, or potassium sulfate), and the resulting diffusion and ion exchange (where larger atoms replace smaller surface atoms) creates the residual compression. Because of the nature of the process, more complex shapes can be tempered.

Glass-ceramics—materials that are part crystalline and part glass—are formed by another heat treatment process. Products are first fabricated into shape as a glass, and are then subjected to a **devitrification** heat treatment that controls the nucleation and growth of the more-stable, lower-energy (i.e., equilibrium structure) crystalline component. Because of the dual structure, the final properties include good strength and toughness, along with low thermal expansion. Typical products include cookware (such as the white CorningWare products), dishes (Corelle), ceramic stove tops, and materials used in electrical and computer components.

FABRICATION OF CRYSTALLINE CERAMICS

Crystalline ceramics are hard, brittle materials with high melting points. As a result, they cannot be formed by techniques requiring either plasticity (i.e., forming methods) or melting (i.e., casting methods). Instead, these materials are generally processed in the solid state by techniques that utilize particles or aggregates that have been produced by crushing or grinding and resemble those used in powder metallurgy. The particles can also be blended with additives that impart plasticity or flow and enable the forming or casting processes to be used.

Dry powders can be compacted and converted into useful shapes by pressing at either environmental or elevated temperatures. **Dry pressing** with rigid tooling, **isostatic pressing,** and **hot-isostatic pressing (HIP)** with flexible molds are common techniques and exhibit features and limitations similar to those discussed in Chapter 18. **Wet pressing,** accomplished by adding 10 to 15% moisture, can be used to produce more intricate shapes at lower pressures than dry pressing.

Clay products are based on special types of ceramics blended with water and various additives to produce a material that can be shaped by most of the traditional forming methods. Plastic forming can also be applied to other ceramics if the ceramic particles are combined with additives that impart plasticity when subjected to pressure and heat. Extrusion can be used to produce products with constant cross sections.

Injection molding was discussed earlier in this chapter as a means of forming plastics, and metal injection molding (MIM) will be presented in Chapter 18 as a way of producing small, complex-shaped metal parts. A form of injection molding can also be used to form complex, three-dimensional shapes from ceramic materials. Ceramic powder is mixed with polymer material, and, after heating to 125 to 150°C (250 to 300°F), the mix is injected into an aluminum die under pressures on the order of 30 to100 MPa (5 to 15 ksi). The product cools, and after about 30 s, it is hard enough to permit ejection from the die. The additive materials are then removed by thermal, solvent, catalytic, or wicking methods, and the remaining ceramic is fused together by a firing operation. As with metal injection molding, the die shapes a part that is considerably oversized, and controlled shrinkage during additive removal and firing produces the final dimensions. The major dimensions of most injection-molded ceramic parts are less than 10 cm (4 in.); the wall thickness is less than 6 mm ($\frac{1}{4}$ in.); and tolerances are on the order of 1% or 0. mm (0.005 in.), whichever is greater. Most parts are made from the oxide ceramics, such as alumina or zirconia, but the process has also been used with silicon carbide and silicon nitride.

Several casting processes can be used to produce ceramic shapes beginning with a pourable slurry that strengthens by partial removal of the liquid or the gellation, polymerization, or crystallization of a matrix phase. In the **slip casting** process, ceramic powder is mixed with a liquid to form a **slip** or slurry, which is then cast into a mold containing very fine pores. Capillary action pulls the liquid from the slurry, allowing the ceramic particles to arrange into a "green" body with sufficient strength for subsequent handling. Pressure applied to the slurry, vacuum applied to the mold, or centrifugal pressure can all aid in liquid removal. Hollow shapes can be produced by pouring out the remaining slurry once a desired thickness of solid has formed on the mold walls. Slip casting has been used to produce a variety of porcelain products, including bathroom fixtures, fine china and dinnerware, and ceramic products for the chemical industry.

In the **tape casting** or **doctor-blading** process, a controlled-thickness film of slurry is formed on a substrate. Evaporation of the liquid during controlled drying produces a thin, flexible, rubbery tape or sheet that has smooth surfaces and uniform thickness. These products are widely used in the multilayer construction of electronic circuits and capacitors.

In other casting-type processes, slurries containing bonding agents can be used to produce cast-in-place products, such as furnace linings or dental fillings. When mixed with a sticky binder, the material can be blown through a pipe to apply ceramic coatings or build up refractory linings.

The numerous variations of **sol-gel processing** can be used to produce ceramic films and coatings, fibers, and bulk shapes. These processes begin with a solution or colloidal dispersion (sol), which undergoes a molecular polymerization to produce a gel, which is then dried. This approach offers higher purity and homogeneity, lower firing temperatures, and finer grain size—at the expense of higher raw-material cost, large volume shrinkages during processing, and longer processing times.

Table 14-1 summarizes some of the primary processes used to fabricate shapes from crystalline ceramics.

TABLE 14-1 Processes Used to Form Products from Crystalline Ceramics

Process	Starting material	Advantages	Limitations
Dry axial pressing	Dry powder	Low cost; can be automated	Limited cross sections; density gradients
Isostatic processing	Dry powder	Uniform density; variable cross sections; can be automated	Long cycle times; small number of products per cycle
Slip casting	Slurry	Large sizes; complex shapes; low tooling cost	Long cycle times; labor-intensive
Injection molding	Ceramic–plastic blend	Complex cross sections; fast; can be automated; high volume	Binder must be removed; high tool cost
Forming processes (e.g., extrusion)	Ceramic–binder blend	Low cost; variable shapes (such as long lengths)	Binder must be removed; particles oriented by flow
Clay products	Clay, water, and additives	Easily shaped by forming methods; wide range of size and shape	Requires controlled drying

PRODUCING STRENGTH IN PARTICULATE CERAMICS

Each of the processes just described can be used to produce useful shapes from ceramic materials, but the products are largely just shapes of packed particles with strength levels similar to those of aspirin tablets. At this stage, the ceramic parts are said to be in their **green condition.** Useful mechanical strength generally requires a subsequent heating operation, known as **firing** or sintering. Slurry-type materials must first be dried in a manner that is designed to control dimensional changes (shrinkage) and minimize stresses, distortion, and cracking. The material is then heated to temperatures between 0.5 and 0.9 times the absolute melting point, where diffusion processes act to fuse the particles together and impart the desired mechanical and physical properties. As the bonds form between the particles, the interparticle pores shrink in size, density increases, and the overall part shrinks in size. The sintering temperature and sintering time are selected to control the resulting grain size, pore size, and pore shape. In some firing operations, surface melting **(liquid-phase sintering)** or component reactions **(reaction sintering)** can produce a substantial amount of liquid material *(vitrification).* The liquid then flows to produce a glassy bond between the ceramic particles and either solidifies as a glass or crystallizes.

Pressure and elevated temperature can be combined in the **hot pressing** and *hot-isostatic pressing* operations. **Microwave sintering** and **spark-plasma sintering** are recent additions to the sintering options. These processes, along with more detail on the sintering process, will be presented in Sections 18.8, 18.9, and 18.10 in the chapter on powder metallurgy.

Cementation is an alternative method of producing strength that does not require elevated temperature. A liquid binder material is used to coat the ceramic particles, and a subsequent chemical reaction converts the liquid to a solid, forming strong, rigid bonds.

Prototypes or small production quantities of ceramic products have been made by the **laser sintering** of ceramic powders. Successive layers of material are fused together by the laser sintering (or laser melting) of thin layers of heat-fusible powder. For ceramic parts, the powder particles are actually coated with a very thin thermoplastic polymer binder. The laser then acts on the polymer coating to produce the bond. After the laser bonding, the parts then undergo conventional debinding and sintering to about 55 to 65% of theoretical density. Isostatic pressing prior to sintering can raise the final density to 90 to 99% of ideal.

MACHINING OF CERAMICS

Most ceramic materials are brittle, and the techniques used to cut metals will generally produce uncontrolled or catastrophic cracks. In addition, ceramics are typically hard materials. Because ceramics are often used as abrasives or coatings on cutting tools, the tools needed to cut them have to be even harder.

Direct production to the desired final shape is clearly the most attractive alternative, but there are times when a material removal operation is necessary. This machining can be performed either before or after the final firing. Before firing, the material tends to be rather weak and fragile. While fracture is always a concern, a more significant consideration might be the dimensional changes that will occur upon subsequent firing. Shrinkage may be as much as 30%, so it may be difficult to achieve or maintain close final tolerances. For this reason, machining before firing, known as **green machining,** is usually rough machining designed to reduce the amount of finishing that will be required after firing.

When machining is performed after firing, the processes are generally ones that might be considered nonconventional. Grinding, lapping, and polishing with diamond abrasives; drilling with diamond-tipped tooling; cutting with diamond saws; ultrasonic machining; laser and electron-beam machining; water-jet machining; and chemical etching have all been used. When mechanical forces are applied, material support is quite critical (because ceramic materials are almost always brittle). Because of the hardness of the ceramic, the tools must be quite rigid. Selection and use of coolants are also important issues.

Materials producers have developed "machinable" ceramics that lend themselves to precision shaping by more traditional machining operations. It should be noted, however, that these are indeed special materials and not characteristic of ceramics as a whole.

JOINING OF CERAMICS

When we consider joining operations, the unique properties of ceramics once again introduce fabrication limitations. Brittle ceramics cannot be joined by fusion welding or deformation bonding, and threaded assemblies should be avoided whenever possible. Therefore, most joining utilizes some form of adhesive bonding, brazing, diffusion bonding, or special cements. Even with these methods, the stresses that develop on the surfaces can lead to premature failure. As a result, most ceramic products are designed to be monolithic (single-piece) structures rather than multipart assemblies.

DESIGN OF CERAMIC COMPONENTS

Because ceramics are brittle materials, special care should be taken to minimize bending and tensile loading as well as design stress raisers. Sharp corners and edges should be avoided where possible. Outside corners should be chamfered to reduce the possibility of edge chips. Inside corners should have fillets of sufficient radius to minimize crack initiation. Undercuts are difficult to produce and should be avoided. Specifications should generally use the largest possible tolerances, because these can often be met with products in the as-fired condition. Extremely precise dimensions usually require hand grinding, and costs can escalate significantly. In addition, consideration should be given to surface finish requirements, because grinding, polishing, and lapping operations can increase production cost substantially.

14.5 FABRICATION OF COMPOSITE MATERIALS

As shown in Chapter 9, composite materials can be designed to offer a number of attractive properties. In some market areas, such as aerospace and sporting goods, their acceptance and growth have been phenomenal. Use can only occur, however, if the material can be produced in useful shapes at an acceptable cost and rate of production. Many of the manufacturing processes designed for composites are slow, and some require considerable amounts of hand labor. There is often a degree of variability between nominally identical products, and inspection and quality control methods are not as well developed as for other materials. While these limitations may be acceptable for certain applications, they often restrict the use of composites for high-volume, mass-produced items. Faster production speeds, increased use of automation, reduced variability, and integrated quality control continue to be important issues in the expanded use of composite materials.

In Chapter 9, composite materials were classified by their basic geometry as particulate, laminar, and fiber-reinforced. Because the fabrication processes are often unique to a specific type of composite, they will also be grouped in the same manner.

FABRICATION OF PARTICULATE COMPOSITES

Particulate composites usually consist of discrete particles dispersed in a ductile, fracture-resistant polymer or metal matrix. They offer isotropic properties and ease of fabrication relative to their fiber-reinforced counterparts. Their fabrication rarely requires processes unique to composite materials. Instead, the particles are simply dispersed in the matrix by introduction into a liquid melt or slurry or by blending the various components as solids, using powder metallurgy methods. Subsequent processing generally follows the conventional methods of casting or forming or utilizes the various techniques of powder metallurgy. These processes have been presented elsewhere in the text and will not be repeated here.

Recent developments include the successful blending of reinforcement particles into the highly viscous slurries of rheocast material, the semisolid mixtures that are viscous when agitated but retain their shape when static. In the **sinter-forge** process, metal and ceramic powders (such as SiC and aluminum) are blended, compacted at room temperature, sintered, and then hot forged to produce near-net-shape parts at near-full density. Particle-reinforceed composites have also been produced by spray forming multicomponent feeds.

FABRICATION OF LAMINAR COMPOSITES

Laminar composites include coatings and protective surfaces, claddings, bimetallics, laminates, and a host of other materials as previously presented in Chapter 9. Their production generally involves processes designed to form a high-quality bond between distinct layers of different materials. When the layers are metallic, as in claddings and bimetallics, the bonds are usually formed by one of the solid-state welding processes, such as roll bonding, ultrasonic welding, diffusion bonding, and explosive welding (all described in Chapter 31).

In the **roll bonding** process, sheets of the various materials are passed simultaneously through the rolls of a conventional rolling mill. If the amount of deformation is great enough, surface oxides and contaminants are broken up and dispersed as fresh metal surfaces are created, metal-to-metal contact is established, and the two surfaces become joined by a solid-state bond. U.S. coinage (dimes and quarters) is a common example of a roll-bonded material.

Explosive bonding is another means of joining layers of metal. A sheet of explosive material progressively detonates above the layers to be joined, causing a pressure wave to sweep across the interface. A small open angle is maintained between the two surfaces. As the pressure wave propagates, any surface films are liquefied or scarfed off, and are jetted out the open interface. Clean metal surfaces are then forced together at high pressures, forming a solid-state bond with a characteristically wavy configuration at the interface. Large areas, wide plates (too wide to roll bond conveniently), and dissimilar materials with large differences in mechanical properties are attractive candidates for explosive bonding.

Diffusion bonding can be used to join a number of dissimilar metals and even metals to ceramics. Both metallic and nonmetallic materials, essentially any solid to any other solid, can be joined by **adhesive bonding,** described in Chapter 33. Plywood is an excellent example. By gluing the layers at various orientations, the directional effects of wood grain can be minimized within the plane of the sheet. Later in this chapter, we will discuss the lamination of fiber-reinforced polymer matrix composites where each ply is a fiber-containing or woven layer. Films of unpolymerized resin are created between the layers. Pressing at elevated temperature cures the resin and completes the bond. In a manner similar to adhesive bonding, layers of metal can also be joined by brazing to form composites that can withstand moderate elevated temperatures.

In **sandwich structures,** such as corrugated cardboard or the honeycomb shown in Figure 14-20, thin layers of facing material are bonded, usually by adhesive, to a lightweight filler material. Special fabrication methods may be employed to produce the foam, corrugated, or honeycomb filler.

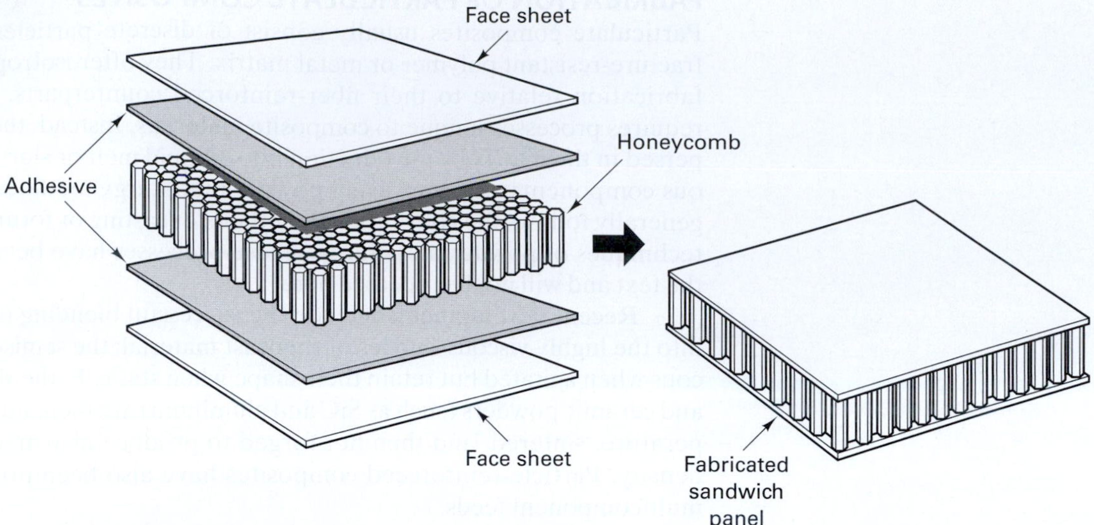

FIGURE 14-20 Fabrication of a honeycomb sandwich structure using adhesive bonding to join the facing sheets to the lightweight honeycomb filler. *(Courtesy of ASM International, Materials Park, OH)*

FABRICATION OF FIBER-REINFORCED COMPOSITES

In the fiber-reinforced composites, the matrix and fiber reinforcement combine to provide a system that offers properties not attainable by the individual components acting alone. The fiber reinforcement produces a significant increase in strength and stiffness, while the matrix functions as a binder, transfers the stresses, imparts toughness and fracture resistance, and provides protection against abrasion and environmental effects.

A number of processes have been developed to produce and shape the fiber-reinforced composites, with key differences relating to the orientation of the fibers, the length of continuous filaments, and the geometry of the final product. Each process seeks to embed the **fibers** in a selected **matrix** with the proper alignment and spacing necessary to produce the desired properties. Discontinuous fibers can be combined with a matrix to provide either a random or a preferred orientation. Continuous fibers are normally aligned in a unidirectional fashion in rods or tapes, woven into fabric layers, wound around a mandrel, or woven into a three-dimensional shape.

Some of the fiber-reinforced processes are identical to those previously described for unreinforced plastics: compression, transfer and injection molding, extrusion, rotational molding, and thermoforming. Others are standard processes with simple modifications, such as reinforced reaction injection molding and resin transfer molding. Still others are specific to fiber-reinforced composites, such as hand lay-up, spray-up, vacuum-bag molding, pressure-bag molding, autoclave molding, filament winding, and pultrusion.

Production of Reinforcing Fibers. A number of processes have been developed to produce the various types of reinforcement fibers described previously in Section 9.5. Metallic fibers, glass fibers, and many polymeric fibers (including the popular Kevlar) are produced by variations of conventional wire drawing and extrusion. Boron, carbon, and ceramic fibers such as silicon carbide are too brittle to be produced by the deformation methods. Boron fibers are produced by chemical vapor deposition around a tungsten filament. Carbon (graphite) fibers can be made by carbonizing (decomposing) an organic material that is more easily formed to the fiber shape. Natural fibers of flax, hemp, and sisal are also being used.

The individual fine filaments are often bundled into **yarns** (twisted assemblies of filaments), **tows** (untwisted assemblies of fibers), and **rovings** (untwisted assemblies of yarns or tows), and these can be further woven into cloth. Fibers can also be chopped into short lengths, usually 12 mm ($\frac{1}{2}$ in.) or less, for incorporation into **mats** or the various sheet or bulk molding compounds where the fibers assume a random orientation.

Processes Designed to Combine Fibers and a Matrix. A variety of processes have been developed to combine the fiber and the matrix into a unified material suitable for further processing. If the matrix material can be liquefied and the temperature is not harmful to the fibers, casting-type processes can be an attractive means of coating the reinforcement. The pouring of concrete around steel reinforcing rod is a macroscopic example of this approach. In the case of the fiber-reinforced plastics and metals, the liquid can be introduced between the fibers by means of **capillary action, vacuum infiltration,** or **pressure casting.** In a modification of **centrifugal casting,** resin is introduced into the center of a rotating mold where it is uniformly forced against and into the reinforcing material. Yet another alternative is to draw the fibers through a bath of molten material and combine them into aligned bundles before the liquid solidifies.

Mats are sheets of nonwoven, randomly oriented fibers similar to felt, where the fibers are held together by a matrix-material binder and possibly a carrier fabric. **Prepregs,** or pre-impregnated reinforcements, are sheets of unidirectional fibers or woven fabric that have been infiltrated with a matrix material. When the matrix is a polymeric material, the resin in the prepreg or mat is usually only partially cured. Later fabrication then involves the stacking of layers and the application of heat and pressure to further cure the resin and bond the layers to produce a continuous solid matrix. Prepreg layers can be stacked in various orientations to provide desired directional properties.

Individual **filaments** can be coated with a matrix material by drawing through a molten bath, plasma spraying, vapor deposition, electrodeposition, or other techniques. The coated fibers can then be used, either individually or in various assemblies. They can also be wound around a mandrel with a specified spacing, and then cut to produce **tapes** that contain continuous, unidirectionally aligned filaments. These tapes are generally one fiber diameter in thickness and can be up to 1.2 m (48 in.) wide.

When the temperatures of the molten matrix become objectionable or potentially damaging to the fiber, bonding between the fiber and the matrix can often be achieved through lower-temperature diffusion or deformation processing (hot pressing or rolling). A common arrangement is to position aligned or woven fibers between sheets of matrix material in foil form. Loosely woven fibers can also be infiltrated with a particulate matrix, which is then compacted at high pressures and sintered to produce a continuous solid.

Sheet-molding compounds (SMCs) are composed of chopped fibers (usually glass in lengths of 12.5 to 50 mm or $\frac{1}{2}$ to 2 in.) and partially cured thermoset resin, along with fillers, pigments, catalysts, thickeners, and other additives, in sheets approximately 2.5 to 5 mm (0.1 to 0.2 in.) thick. With strengths in the range of 35 to 70 MPa (5 to 10 ksi) and the ability to be press-formed in heated dies, these materials offer a feasible alternative to sheet metal in applications where light weight, corrosion resistance, and integral color are attractive features.

After initial compounding and a few days of curing, sheet-molding compounds generally take on the consistency of leather, making them easy to handle and mold. When they are placed in a heated mold, the viscosity is quickly reduced and the material flows easily under pressures of about 7 MPa (1000 psi). The elevated temperatures accelerate the chemical reactions, and final curing can often be completed in less than 60 s. As an added benefit, sheet-molding compounds can be easily recycled. One possible disadvantage, however, is that polymer flow may orient the reinforcing fibers, making the final orientation nonrandom and difficult to predict and control.

When the chopped fibers and thermosetting resin are combined to produce sheets up to 50 mm (2 in.) thick, or billets for compression or injection molding, the material is known as **thick-molding compound (TMC).** Because of the increased dimensions, handling costs are reduced and an increased amount of filler can be incorporated.

Bulk-molding compounds (BMCs) are fiber-reinforced, thermoset, molding materials, where short fibers (2 to 12 mm, or 0.1 to 0.5 in.) are distributed in random orientation. The starting material usually has the consistency of putty or modeling clay, although pellets and granules are also possible. The final shape is usually produced by compression molding in heated dies, but transfer molding and injection molding are other possibilities.

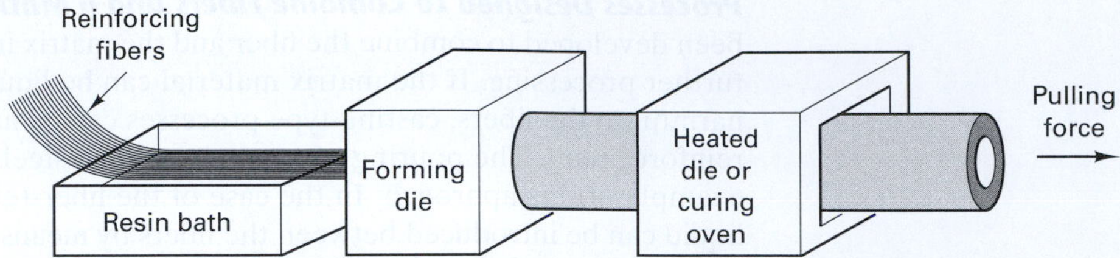

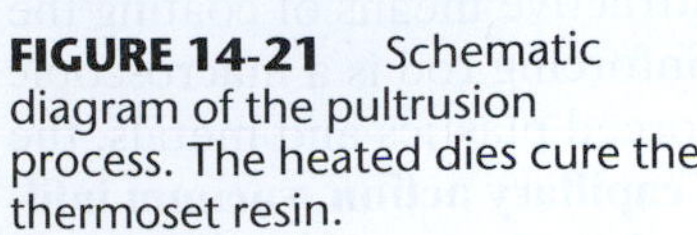
FIGURE 14-21 Schematic diagram of the pultrusion process. The heated dies cure the thermoset resin.

Fabrication of Final Shapes from Fiber-Reinforced Composites. A number of processes have been developed for the production of finished products from fiber-reinforced material. Many are simply extensions or adaptations of processes that are used to shape the matrix materials (usually metals or polymers). Others are unique to the family of fiber-reinforced composites. The dominant techniques will be discussed individually in the sections that follow.

PULTRUSION **Pultrusion** is a continuous process that is used to produce long lengths of uniform cross section, simple-complexity, shapes, such as round, rectangular, tubular, plate, sheet, and structural products. As shown in Figure 14-21, bundles of continuous fiber are first drawn through a bath of thermoset polymer resin, followed by dies to begin shaping the product. This preshaped material is then pulled through a long-length heated die (1 to 1.5 m, or 3 to 5 ft), which completes the shaping and cures the resin. When it emerges from the final die, the product is cooled by air or water and then cut to length. Some products, such as structural shapes, are complete at this stage, while others are further fabricated into products such as fishing poles, golf club shafts, and ski poles. Extremely high strengths and stiffnesses are possible because the reinforcement can be as much as 75% of the final structure. Tensile strengths of 210 MPa (30 ksi) and elastic modulus of 17 GPa (2.5×10^6 psi) are coupled with densities about 20% that of steel or 60% that of aluminum. Cross sections can be as much as 1.5 m (60 in.) wide and 0.3 m (12 in.) thick.

The pultrusion process can be modified if the desired products have curvatures (such as leaf springs) or require variations in cross section (such as hammer handles). Product emerging from the shaping dies (but before curing) is fed into heated molds that complete the forming of the more-complex shape and cure the resin while it is held in that shape. This modification is known as **pulforming.**

FILAMENT WINDING AND TAPE LAYING In the **filament winding** process, resin-coated (or resin-impregnated) continuous filaments, bundles, or tapes made from fibers of glass, graphite, boron, Kevlar (aramid), or similar materials can be used to produce cylinders, spheres, cones, and other container-type shapes with exceptional strength-to-weight ratios. The filaments are wound over a rotating form or mandrel, using longitudinal, circumferential, or helical patterns, or a combination of these, designed to take advantage of their highly directional strength properties. By adjusting the density of the filaments in various locations and selecting the orientation of the wraps, products can be designed to have strength where needed and lighter weight in less critical regions. After winding, the part and mandrel are placed in an oven for curing, after which the product is removed from the mandrel or form. In some cases, part removal requires the use of an inflatable/deflatable mandrel, segmented mandrels, or mandrels made from soluble materials, such as salts or plaster. In the final product, the matrix, often an epoxy-type polymer, binds the structure together and transmits the stresses to the fibers.

Figure 14-22 shows a large tank being produced by filament winding. Products such as pressure tanks, airplane fuselage segments, and rocket motor casings can be made in virtually any size, some as large as 6 m (20 ft) in diameter and 20 m (65 ft) long. Smaller parts include helicopter rotor blades, baseball bats, and light poles. Moderate production quantities are feasible, and because the process can be highly mechanized, uniform quality can be maintained. A new form block is all that is required to produce a new size or design. Because the tooling is so inexpensive, the process offers tremendous potential for cost savings and flexibility. With advancements in computer software, and

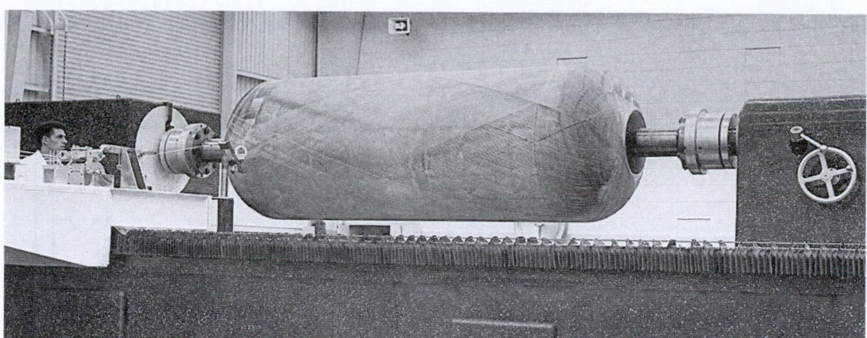

FIGURE 14-22 A large tank being made by filament winding. *(Courtesy of Rohr Inc./Goodrich)*

multi-axis computer-controlled equipment, parts no longer need to be axisymmetric. Filament-wound products can now be made with changing surfaces, nonsymmetric cross sections, and compound curvatures.

Lamination and Lamination-Type Processes In the **lamination** processes, prepregs, mats, or tapes are stacked, often in varying orientation, to produce a desired thickness and cured under pressure and heat. The resulting products possess unusually high strength as a result of the integral fiber reinforcement. Because the surface is a thin layer of pure resin, laminates usually possess a smooth, attractive appearance. If the resin is transparent, the fiber material is visible and can impart a variety of decorative effects. Other decorative laminates use a separate patterned face sheet that is bonded to the laminate structure.

Laminated materials can be produced as sheets, tubes, and rods. Flat sheets can be made using the method illustrated in Figure 14-23. Prepreg sheets or reinforcement sheets saturated in resin are stacked and then compressed under pressures on the order of 7 MPa (1000 psi). Figure 14-24 depicts the technique used to produce rods or tubes. For tubing, the impregnated stock is wound around a mandrel of the desired internal diameter. Solid rods are made by using a small-diameter mandrel, which is removed prior to curing, or by wrapping the material tightly about itself. Sheet laminating can also be a continuous process in which multiple reinforcement sheets are passed through a resin bath and then through squeeze rolls. In all of the preceding cases, the final operation is a curing, usually involving elevated temperature and possibly applied pressure. Because of their excellent strength properties, plastic laminates find a wide variety of uses. Some sheets can be easily blanked and punched. Gears machined from thick laminated sheets have unusually quiet operating characteristics when matched with metal gears.

Many laminated products are not flat but contain relatively simple curves and contours. Manufacturing processes that require zero to moderate pressures and relatively low curing temperatures can be performed where the only required tooling is often a female mold or male form block that can be made from metal, hardwood, or

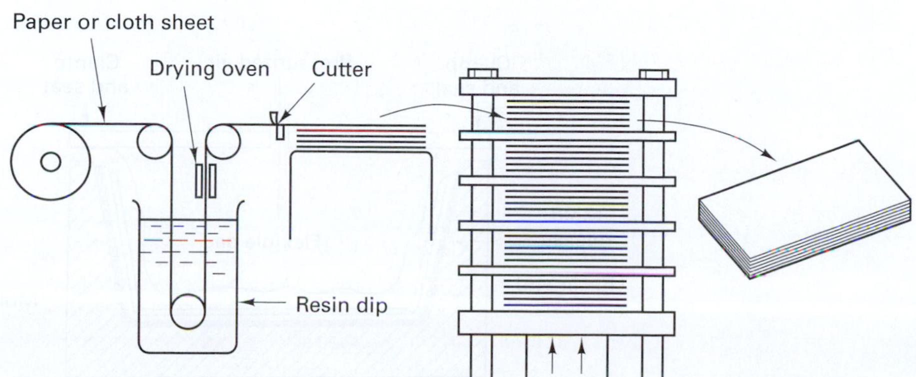

FIGURE 14-23 Method of producing multiple sheets of laminated plastic material.

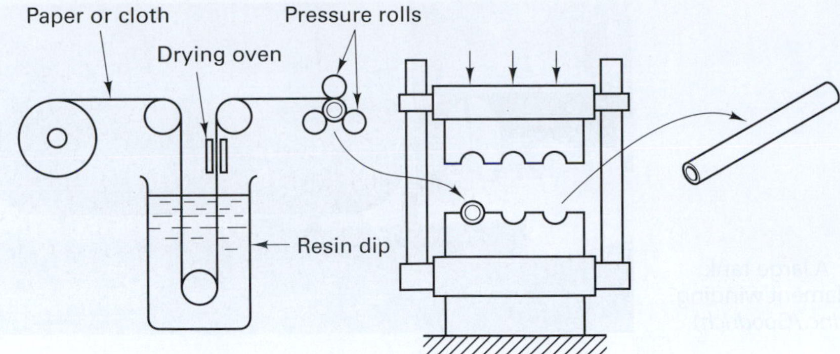

FIGURE 14-24 Method of producing laminated plastic tubing. In the final operation, the rolled tubes are cured by being held in heated tooling.

even particle board. The layers of prepreg or resin-dipped fabric are stacked in various orientations until the desired thickness is obtained. Care must be taken to avoid the entrapment of air bubbles and ensure that no impurities (such as oil, dirt, or other contaminants) are introduced between the layers. In **pressure-bag molding,** depicted in Figure 14-25, a flexible membrane is positioned over the female mold cavity and is pressurized to force the individual plies together and drive out entrapped air and excess resin. Pressures usually range from 0.2 to 0.4 MPa (30 to 50 psi) but can be as high as 2 MPa (250 psi). This pressing is coupled with room- or low-temperature curing. Pressure-bag molding has been used to produce extremely large components, such as the skins of military aircraft and other aerospace panels, automobile and truck body panels, large air deflectors for tractor-trailers, boat bodies, and similar products. In the **vacuum-bag molding** process, the entire assembly (mold and material) is placed below, depicted in Figure 14-26; a vacuum is pulled beneath the nonadhering, flexible bag; and the contained air is evacuated. Pressure from the outside air forces the laminate against the mold while the resin is cured. While curing may occur at room temperature, moderately elevated temperatures may also be used.

Higher heats and pressures are used when parts are cured in an **autoclave.** The supporting molds or vacuum-bagged lay-ups are placed inside a heated pressure vessel where curing occurs under elevated temperatures and pressures in the range 0.4 to 0.7 MPa (50 to 100 psi). Denser, void-free moldings are produced, and the properties can be further enhanced through the use of matrix resins that require higher-temperature cures. The size of the autoclave limits the size of the product.

When the quality demands are not as great, the reinforcement-to-resin ratio is not exceptionally high, and only one surface needs to be finished to high quality, the pressing operations can often be eliminated. In a process known as **hand lay-up** or **open-mold processing,** depicted in Figure 14-27, successive layers of pliable resin-coated cloth are simply placed in an open mold or draped over a form. Squeegees or rollers are used to manually ensure good contact and remove any entrapped air, and the assembly is then allowed to cure, generally at room temperature. If prepreg layers are not used, a layer of

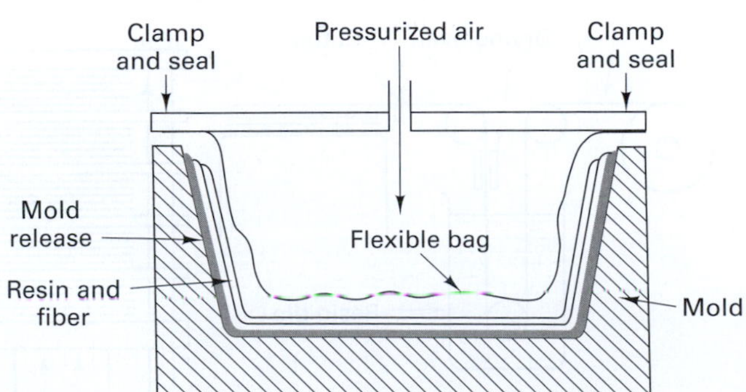

FIGURE 14-25 Schematic of the pressure-bag process.

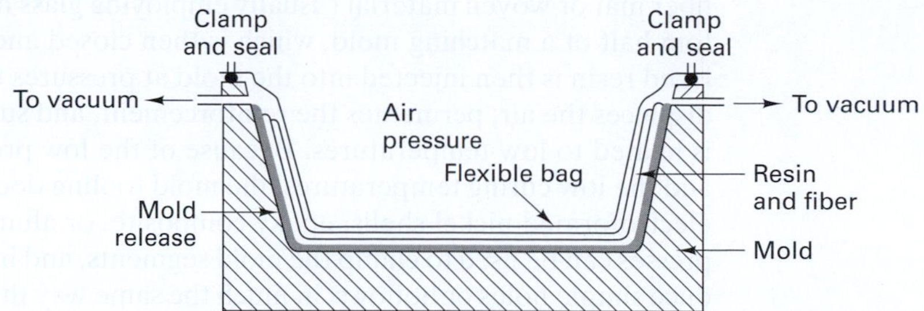

FIGURE 14-26 Schematic of the vacuum-bag process.

mat, cloth, or woven roving can be put in place, and a layer of resin brushed, sprayed, or poured on. This process can then be repeated to build the desired thickness.

While the hand lay-up process is slow and labor intensive, and has part-to-part and operator-to-operator variability, the tooling costs are sufficiently low that single items or small quantities become economically feasible. Molds or forms can be made from wood, plaster, plastics, aluminum, or steel, so design changes and the associated tool modifications are rather inexpensive, and manufacturing lead time can be quite short. In addition, large parts can be produced as single units, significantly reducing the amount of assembly, and various types of reinforcement can be incorporated into a single product, expanding design options. A high-quality surface can be produced by applying a pigmented gel coat to the mold before the lay-up.

The open-mold lay-up process can be automated by using a programmed tape-laying machine that deposits strips of prepreg tape, typically 75 mm (3 in.) wide. Unlike the filament-winding approach where the tape is deposited in a continuous strip, this adaptation uses a machine that deposits and cuts finite-length strips, each following a programmed three-dimensional contour, to build up the desired number of plies. While labor and time are reduced and quality and repeatability are improved, the desired quantity must be sufficient to justify the time and expense of programming the operation.

When production quantities are large, the quality needs to be high, part complexity is increased, and all surfaces need to have a smooth finish, matched metal dies mounted in a press can be used in place of the above techniques. This **closed-molding process** is essentially a modification of polymer compression molding. Sheet-molding compound, bulk-molding compound, or preformed mat is placed between the dies, and heat and pressure are applied. Temperatures typically range from 110 to 160°C (225 to 325°F), coupled with pressures from 1 to 7 MPa (250 to 1000 psi). With heated dies, the thermoset resin cures during the compression operation, with cycles repeating every 1 to 5 min. Because all surfaces now lie within the closed mold, precise tolerances can be maintained in any direction, including thickness. Part size is limited by the size of the molds.

Resin-transfer molding (RTM) is a low-pressure process that is intermediate to the slow, labor-intensive lay-up processes and the faster compression molding or injection molding processes, which generally require more expensive tooling. Continuous

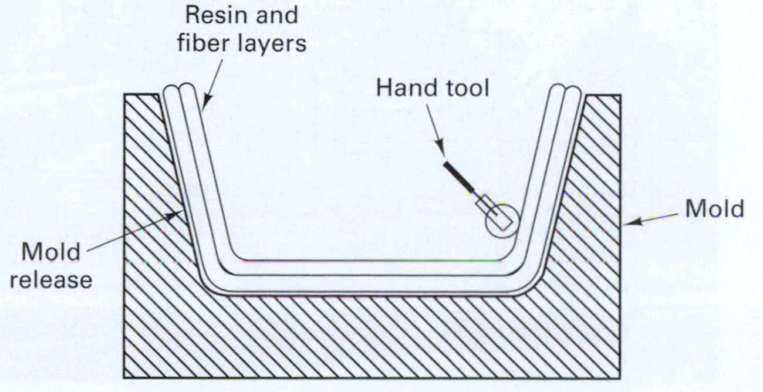

FIGURE 14-27 Schematic of the hand lay-up lamination process.

fiber mat or woven material (usually employing glass fiber) is positioned dry in the bottom half of a matching mold, which is then closed and clamped. A low-viscosity catalyzed resin is then injected into the mold at pressures up to 2.1 MPa (300 psi), where it displaces the air, permeates the reinforcement, and subsequently cures when the mold is heated to low temperatures. Because of the low pressures employed in the process and the low curing temperatures, the mold tooling does not need to be steel but can be electroformed nickel shells, epoxy composite, or aluminum. In addition, low-capacity presses can be used to clamp the mold segments, and inflatable bags can be used to produce simple holes or hollows, in much the same way that cores are used in conventional casting. By curing in the mold, the autoclave operation is eliminated, reducing both cost and production time. The resulting products can have excellent surfaces on both sides, because both mold surfaces can be precoated with a pigmented gel. Large parts can often be made as a single unit with a relatively low capital investment. Cycle times range from a few minutes to a few hours, depending on the part size and the resin system being used. The aerodynamic hood and fender assembly for Ford Motor Company's AeroMax heavy-duty truck (Figure 14-28) is an example of a large resin-transfer molding.

Resin-transfer molding uses a two-sided mold (upper and lower) with the resin being pumped in under pressure. **Vacuum-assisted resin-transfer molding (VARTM)** uses a single tool and a vacuum bag that surrounds the tool and dry fiber material, allowing the vacuum to draw the resin around the fibers. Large parts can be made by this process, originally used to produce large yacht hulls, but now being used extensively in the aerospace industry.

SPRAY MOLDING When continuous or woven fibers are not required to produce the desired properties, sheet-type parts can be produced by mixing chopped fibers, fillers, and catalyzed liquid resin and spraying the combination into or onto a mold form, as shown in Figure 14-29. To facilitate **spray molding,** the fiber content is limited to about 35%, significantly less than the laminated prepreg or mat approach, and the fibers assume a random orientation. Rollers or squeegees are used to remove entrapped air and work the resin into the reinforcement. Room-temperature curing is usually preferred, but elevated temperatures are sometimes used to accelerate the cure. As with the hand lay-up process, an initial gel coat can be used to produce a smooth, pigmented surface.

Because the fibers are no longer long and oriented, the resultant mechanical properties are not as great as with the lamination processes. Some typical products include bathtubs, shower stalls, furniture, large architectural panels, and truck-bed liners and toppers.

FIGURE 14-28 Aerodynamic styling and smooth surfaces characterize the hood and fender of Ford Motor Company's AeroMax truck. This one-piece panel was produced as a resin transfer molding by Rockwell International. *(From* Advanced Materials & Processes *magazine, Jan. 1991, Vol. 139, Issue 2, a publication of ASM International®. All rights reserved. www.asminternational.org)*

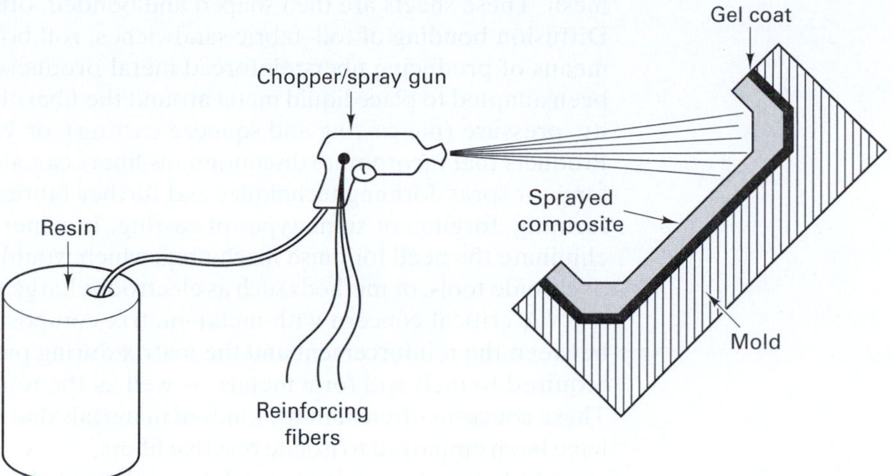

FIGURE 14-29 Schematic diagram of the spray forming of chopped-fiber-reinforced polymeric composite.

SHEET STAMPING Thermoplastic sheets that have been reinforced with nonwoven fiber can often be heated and press-formed in a manner similar to conventional sheet metal forming. Precut blanks are heated and placed between the halves of a matched metal mold that is mounted in a vertical press. Ribs, bosses, and contours can be formed in parts with essentially uniform thickness. Because the parts must be cooled prior to removal from the mold, cycle times range from 25 to 50 s for most parts.

INJECTION MOLDING The injection molding of fiber-reinforced plastics is a process that competes with metal die castings, and offers comparable properties at considerably reduced weight. In its simplest form, chopped or continuous fibers are placed in a mold cavity that is then closed and injected with resin. If the resin is a thermoplastic, it is heated and injected into a cold mold. Thermosetting resins are injected cold into a heated mold. In a variation known as *reaction injection molding,* reactive components are mixed just prior to injection and begin their cure, immediately following mixing. When reinforcing fibers are introduced into the mixture, the process becomes **reinforced reaction injection molding.**

In other variations, chopped fibers, up to 6 mm ($\frac{1}{4}$ in.) in length, are premixed with the heated thermoplastic (often nylon) prior to injection or the feedstock is discrete pellets that have been manufactured by slicing continuous-fiber pultruded rods. The benefits of fiber reinforcement (compared to conventional plastic molding) include increased rigidity and impact strength, reduced possibility of brittle failure during impact, better dimensional stability at elevated temperatures and in humid environments, improved abrasion resistance, and better surface finish due to the reduced dimensional contraction and absence of related sink marks. The molding process is quite rapid, and the final parts can be both precise and complex.

BRAIDING, THREE-DIMENSIONAL KNITTING, AND THREE-DIMENSIONAL WEAVING The primary causes of failure in lamination-type composites are interlaminar cracking and delamination (layer-separation) upon impact. To overcome these problems, the high-strength reinforcing fibers can also be interwoven into three-dimensional preforms by processes that include weaving, **braiding,** and stitching through the thickness of stacked two-dimensional preforms. Resin is then injected into the assembly and the resultant product is cured for use. Complex shapes can be produced with the fiber orientations selected for optimum properties. Computers can be used to design and control the weaving, making the process less expensive than many of the more labor-intensive techniques.

FABRICATION OF FIBER-REINFORCED METAL–MATRIX COMPOSITES

Continuous-fiber **metal–matrix composites** can be produced by variations of filament winding, extrusion, and pultrusion. Fiber-reinforced sheets can be produced by electroplating, plasma spray deposition coating, or vapor deposition of metal onto a fabric or mesh. These sheets are then shaped and bonded, often by some form of hot pressing. Diffusion bonding of foil–fabric sandwiches, roll bonding, and coextrusion are other means of producing fiber-reinforced metal products. Various casting processes have been adapted to place liquid metal around the fibers by means of capillary action, gravity, pressure (die casting and squeeze casting), or vacuum (counter-gravity casting). Products that incorporate discontinuous fibers can also be produced by powder metallurgy or spray-forming techniques and further fabricated by hot pressing, superplastic forming, forging, or some types of casting. In general, efforts are made to reduce or eliminate the need for finish machining, which would often require the use of diamond or carbide tools, or methods such as electrodischarge machining (EDM).

A critical concern with metal–matrix composites is the possibility of reactions between the reinforcement and the matrix during processing at the high temperatures required to melt and form metals, as well as the temperatures of subsequent service. These concerns often limit the kinds of materials that can be combined. Barrier coatings have been employed to isolate reactive fibers.

In terms of properties, graphite-reinforced aluminum has been shown to be twice as stiff as steel and one-third to one-fourth the weight, with practically zero thermal expansion. Aluminum reinforced with silicon carbide exhibits increased strength (tension, compression, and shear at both room and elevated temperature), as well as increased hardness, fatigue strength, and elastic modulus. Thermal creep and thermal expansion are both reduced, but ductility, thermal conductivity, and electrical conductivity are also decreased. Magnesium, copper, and titanium alloys, as well as the superalloys, have also been used as the matrix in fiber-reinforced metal–matrix composites.

FABRICATION OF FIBER-REINFORCED CERAMIC–MATRIX COMPOSITES

Unlike polymeric– or metal–matrix composites, where failures originate in or along the reinforcement fibers, **ceramic–matrix composites** often fail due to flaws in the matrix. If the reinforcement is bonded strongly to the matrix, a matrix crack might propagate right through the fibers. To impart toughness to the assembly, it is often desirable to promote a weak bond between the fiber and matrix. Cracks are redirected along the fiber–matrix interface rather than through the fiber and the remaining matrix.

The matrix materials and reinforcement fibers for ceramic–ceramic composites have been discussed in Chapter 9, along with some of the unique property combinations that can be achieved. Fabrication techniques are often quite different from the other composite families. One approach is to pass the fibers or mats through a slurry mixture that contains the matrix material. The impregnated material is then dried, assembled, and fired. Other techniques include the chemical vapor deposition or chemical vapor infiltration of a coated fiber base, where the coating serves to weaken an otherwise strong bond. Silicon nitride matrices can be formed by reaction bonding. The reinforcing fibers are dispersed in silicon powder, which is then reacted with nitrogen. Hot-pressing techniques can also be used with the various ceramic matrices. When the matrix is a glass, the heated material behaves much like a polymer, and the processing methods are often similar to those used for polymer–matrix composites.

SECONDARY PROCESSING AND FINISHING OF FIBER-REINFORCED COMPOSITES

The various fiber-reinforced composites can often be processed further with conventional equipment (sawed, drilled, routed, tapped, threaded, turned, milled, sanded, and sheared), but special considerations should be exercised because composites are not uniform materials. Cutting some materials may be like cutting multilayer cloth, and precautions should be used to prevent the formation of splinters and cracks as well as frayed or delaminated edges. Sharp tools, high speeds, and low feeds are generally required. Cutting debris should be removed quickly to prevent the cutters from becoming clogged.

In addition, many of the reinforcing fibers are extremely abrasive and quickly dull most conventional cutting tools. Carbide, diamond-coated carbide, or polycrystalline diamond tooling is generally required to achieve realistic tool life. Abrasive slurrys can be used in conjunction with rigid tooling to ensure the production of smooth surfaces.

Lasers and water-jets are alternative cutting tools. Lasers, however, can burn or carbonize the material or produce undesirable heat-affected zones. Water-jets can create moisture problems with some plastic resins, and pressurized water can cause delaminations, but the low heat and light cutting force are attractive characteristics. Elastic deflections are minimized during the cut. Parts can often be held in place by simple vacuum cups, and water-jets also minimize the generation of dust, which may be toxic.

When fiber-reinforced materials must be joined, the major concern is the lack of continuity of the fibers in the joint area. Thermoplastics can be softened and welded by applying pressure with heated tools, combining pressure and ultrasonic vibration, or using pressure and induction heating. Thermoset materials generally require the use of mechanical joints or adhesives, with each method having its characteristic advantages and limitations. Metal–matrix composites are often brazed.

Key Words

adhesives
adhesive bonding
annealing
autoclave
blow molding
braiding
bulk-molding compound (BMC)
capillary action
compound
calendering
casting
cementation
centrifugal casting
ceramics
ceramic–matrix composites
chemical tempering
clay products
closed-cell foam
closed-molding process
cold molding
composites
compression molding
cross-linking
crystalline ceramics
devitrification
diffusion bonding
dipping
doctor-blading
drawing
dry pressing
elastomer
explosive bonding
extrusion
fibers
filaments
filament winding
firing
flash
foam molding
glass
glass-ceramic
green condition
green machining
hand lay-up
hot-isostatic pressing (HIP)
hot pressing
injection molding
inserts
isostatic pressing
lamination
laser sintering
liquid-phase sintering
matrix
mats
metal–matrix composites
microwave sintering
open-cell foam
open-mold processing
parison
plastics
plastic forming
prepregs
pressing
pressure-bag molding
pressure casting
pulforming
pultrusion
reaction injection molding
reaction sintering
reinforced reaction injection molding
resin-transfer molding (RTM)
roll bonding
rolling
rotational molding
rovings
sandwich structures
sheet-molding compound (SMC)
sinter-forge
sintering
slip
slip casting
snap-fit
sol-gel processing
spark-plasma sintering
spinneret
spinning
spray molding
tape casting
tapes
tempered glass
tempering
thermoforming
thermoplastic polymer
thermosetting polymer
thick-molding compound (TMC)
tows
transfer molding
vacuum-assisted resin-transfer molding (VARTM)
vacuum-bag molding
vacuum infiltration
viscous flow
vitrification
vulcanization
wet processing
yarns

Review Questions

1. Why are the fabrication processes applied to plastics, ceramics, and composites often different from those applied to metals? What are some of the key differences?
2. How does the fabrication of a thermoplastic polymer differ from the processing of a thermosetting polymer?
3. What are some of the ways that plastic sheet, plate, and tubing can be cast?
4. Why do cast plastic resins typically have a lustrous appearance?
5. What types of polymers are most commonly blow molded?
6. Why do blow-molding molds typically contain a cooling system?
7. For what types of parts and production volumes would compression molding be an appropriate process?
8. What are typical mold temperatures for compression molding? What is the most common mold material?
9. What are some of the attractive features of the transfer molding process? Some process negatives?
10. Cold molding is faster and more economical than other types of molding. What limits its use?
11. What is the most widely used process for the fabrication of thermoplastic materials (in terms of number of parts produced)?
12. In what ways is injection molding of plastic similar to the die casting of metal?
13. What is the benefit of a hot runner distribution system in plastic injection molding?

14. Why is the cycle time for the injection molding of thermosetting polymers significantly longer than that for the thermoplastics?
15. How are the individual components mixed in the reaction injection molding process?
16. What are some of the attractive consequences of the low temperatures and low pressures of the reaction injection molding process?
17. What are some of the typical production shapes that are produced by the extrusion of plastics?
18. How can the extrusion process be used to produce polymeric sheet and film? To produce thin plastic bags?
19. For what types of materials and products might thermoforming be considered to be attractive?
20. What types of products are produced by rotational molding?
21. What is the difference between open-cell and closed-cell foamed plastics?
22. What are some typical applications for rigid-type foamed plastics? For flexible foams?
23. What types of products are produced by calendering?
24. What types of products are produced by the spinning process?
25. What are some of the general properties of plastics that affect their machinability?
26. What are some of the attractive features of laser machining when applied to plastics?
27. What property of plastics is responsible for making snap-fit assembly a popular alternative for plastic products?
28. What are some of the attractive properties of plastics that favor their selection? What are some of the common limitations?
29. What are some cautions regarding the use of standard test data when designing with plastics?
30. What are some of the design concerns when specifying and setting up a plastic molding process?
31. When designing with plastics, why should adequate fillets be included between adjacent sections of a mold? What is a major benefit of rounding exterior corners?
32. Why is it most desirable to have uniform wall thickness in plastic products?
33. Why are product dimensions less precise when they cross a mold parting line?
34. Why might threaded inserts be preferred over other means of producing threaded holes in a plastic component?
35. What are some of the ways in which metal inserts are held in place in a plastic part?
36. When designing a decorative surface (design or lettering) on a plastic product, why is it desirable that the details be raised on the product rather than depressed?
37. Why does locating a parting line on a sharp corner make that feature less noticeable?
38. What is the benefit of countersinking holes that are to be threaded or used for self-tapping screws?
39. What types of products can be produced from elastomeric materials using the dipping process?
40. What process or equipment is used to form rubber compounds into sheets?
41. At what stage during the manufacture of an elastomeric product should vulcanization or cross-linking be induced?
42. What are the two basic classes of ceramic materials, and how does their processing differ?
43. What are some of the most common processes used to shape glass?
44. How are glass fibers produced?
45. What are some of the special heat treatment operations performed on glass products?
46. What are glass-ceramics? How are they produced?
47. What are some of the techniques for compacting crystalline ceramic powders?
48. Describe the differences between the injection molding of plastics and the injection molding of ceramics.
49. What is the difference between slip casting and tape casting?
50. What is the purpose of the firing or sintering operations in the processing of crystalline ceramic products?
51. What is the attractive feature of liquid-phase sintering or reaction sintering?
52. How does cementation differ from sintering?
53. What are the benefits and limitations of machining ceramic materials before firing versus after firing?
54. What are some of the nonconventional methods used to machine ceramics?
55. Why are joining operations usually avoided when fabricating products from ceramic materials?
56. Discuss some of the design guidelines that relate to the production of parts from ceramic material.
57. Why are the processes used to fabricate particulate composites essentially the same as those used for conventional material?
58. How are metals and ceramics combined in the sinter-forge process?
59. What are some of the processes that can be used to produce a high-quality bond between the layers of a laminar composite?
60. List several fabrication processes for fiber-reinforced products that are essentially the same as for unreinforced plastics. List several that are unique to reinforced materials.
61. What types of materials are used as reinforcing fibers in fiber-reinforced composites?
62. What are some of the forms in which reinforcement fibers appear in composite materials?
63. What are some of the ways that liquefied matrix material can be introduced between the fibers of a fiber-reinforced material?
64. What is a prepreg?
65. What are sheet-molding compounds (SMC)? Bulk-molding compounds (BMC)?
66. In what way is pultrusion similar to wire drawing?
67. How are the products of pulforming different from those of pultrusion?
68. What are some typical products that are made by filament winding?
69. What are some of the various molding processes that can be used to shape products from laminated sheets of woven fibers?
70. What are the benefits of using an autoclave instead of room-temperature and low-pressure curing?
71. For what conditions might hand lay-up be a preferred process?
72. How can an automated tape-lay-up machine produce a laminated-ply fiber-reinforced composite?
73. What form of reinforcing fibers can be incorporated in the spray molding process?

74. What is the major benefit of three-dimensional fiber reinforcement?
75. Describe some of the ways in which a metal matrix can be introduced into a fiber-reinforced composite.
76. Why might it be desirable to have a weak bond between a reinforcing fiber and a ceramic–matrix material?
77. Discuss some of the concerns when cutting or machining fiber-reinforced composites.
78. What is the major concern when considering the joining of fiber-reinforced composites?

PROBLEMS

1. Consider some of the more prominent sporting goods that are fabricated from composite materials, such as skis, snowboards, tennis rackets, golf club shafts, bicycle frames, and body panels for racing cars. For two specific products, identify composite materials that are currently being used and the companion shape-producing fabrication methods.
2. Figure 14-A depicts the handles of two large wrenches: a ratchet wrench and a pipe wrench. These components are traditionally forged from a ferrous alloy or made from a cast steel or cast iron. For various reasons, alternative materials may be desired. The ratchet wrench is quite long, and reduced weight may be a reasonable desire. Both of these tools could be used in areas, such as a gas leak, where a nonsparking safety tool would be required. Current specifications for the ratchet handle call for a yield strength in excess of 50 ksi and a minimum of 2% elongation in all directions to ensure prevention of brittle fracture. The pipe wrench most likely has similar requirements.
   a. Could a plastic or composite material be used to make a quality product with these additional properties? (*Note:* Metal jaw inserts can be used in the pipe wrench, enabling the other components to be considered as separate pieces.)
   b. If so, how would you propose to manufacture the new handles?
3. Tires are the dominant product of the rubber industry, accounting for nearly three-fourths of all rubber material.
   a. What are some of the functions of a vehicle tire?
   b. Modern tires are built from multiple plies of material, where a ply is a layer of rubber-coated cord or fabric. What are some of the different constructions that can be made from these plies?
   c. Tire construction generally consists of three steps: manufacturing the individual components (plies and belts, filler strips, inner lining, tread strips, sidewall, and others), assembling these components in the desired sequence, and then molding and curing to produce an integral piece. At what stage is the tread design imparted to the tire?
   d. Tires are generally cured in the mold. Investigate the methods, temperatures, and times of heating.
4. Rubber hoses often have to have different properties on the inside and outside surfaces, as well as some form of strengthening reinforcement. The inside surface comes in contact with the material flowing through the hose, while the outer surface must resist environmental conditions. The reinforcement is generally applied at the interface between the inner and outer materials. How might such a product be manufactured?
5. Lavatory wash basins (bathroom sinks) have been successfully made from a variety of engineering materials, including cast iron, steel, stainless steel, ceramics, and polymers (such as melamine). Your company, Diversified Household Products Inc., is considering a possible entrance into this market

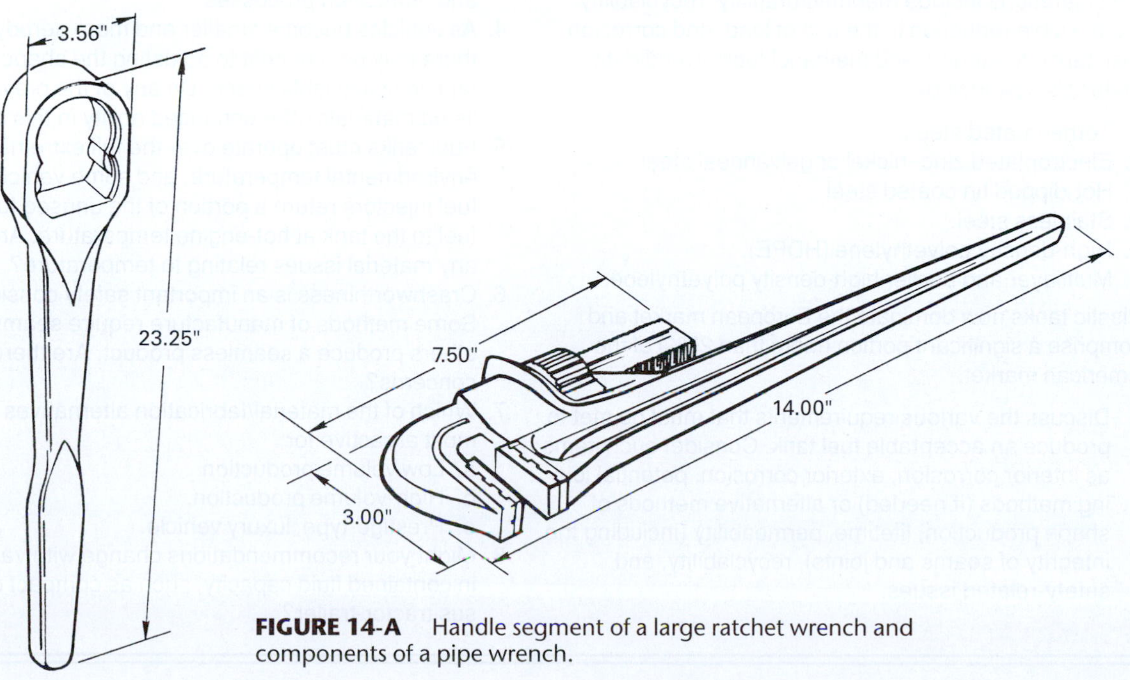

FIGURE 14-A Handle segment of a large ratchet wrench and components of a pipe wrench.

and has assigned you the tasks of assessing the competition and recommending the "best" approach toward producing this product.

a. For each of the materials (or families of materials), describe the material properties that are attractive for a wash basin application. What are the primary limitations or disadvantages?
b. For each of the materials (or families of materials), describe possible means of fabricating lavatory wash basins. Consider sheet metal forming, casting, molding, joining, and other types of fabrication processes. If multiple options exist, which one do you consider to be most attractive? Comment on the attractive features of the proposed system (materials and process) as well as the relative quality and cost.
c. Wash basins generally require a surface that is nonporous and stain resistant, scratch resistant, corrosion resistant, and attractive (and possibly available in a variety of colors). One approach to providing these properties on a steel or cast iron substrate is a coating of porcelain enamel. For each of the systems discussed in Problem 2, discuss the need for additional surface treatment. What type of treatment would you recommend?
d. Most sinks contain an overflow feature that diverts excess water to the drain at a location beneath the stoppered basin. Discuss how this feature can be incorporated into each of your material-process manufacturing systems.
e. If your company were to consider producing lavatory wash basins on a competitive basis, which of the alternative manufacturing systems (material and manufacturing process) would you recommend? What features make it the most attractive?

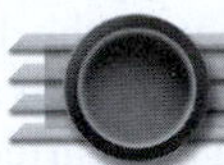

www.wiley.com/go/global/degarmo

Chapter 14 CASE STUDY

Automotive and Light Truck Fuel Tanks

For many years, seam-welded, terne-coated (8% tin–lead coating) steel sheet has been the standard material for automotive fuel tanks. Several factors, however, have driven a reevaluation of this material—including the desire for increased lifetime, weight reduction, enhanced safety, and reduced cost, as well as the pressures of recent legislation that include permeability and emission standards. Additional considerations include manufacturability, recyclability, the possible reduction in the use of lead, and corrosion resistance to ethanol and methanol fuels. Candidate materials now include:

a. Terne-coated steel.
b. Electroplated zinc–nickel or galvanneal steel.
c. Hot-dipped tin coated steel.
d. Stainless steel.
e. High-density polyethylene (HDPE).
f. Multilayer and barrier high-density polyethylene.

Plastic tanks now dominate the European market and comprise a significant portion (more than 25%) of the American market.

1. Discuss the various requirements that must be met to produce an acceptable fuel tank. Consider such items as interior corrosion, exterior corrosion, potential joining methods (if needed) or alternative methods of shape production, lifetime, permeability (including the integrity of seams and joints), recyclability, and safety-related issues.
2. For each of the materials listed, briefly discuss the advantages and disadvantages, considering the various factors that have been identified.
3. Other components are often incorporated in the fuel tank assembly, including the filler tube; level control; baffles; a fuel-level sending unit; and various seals, fittings, and tubes. Discuss the integration and compatibility of these features with the various materials and fabrication processes.
4. As vehicles become smaller and more aerodynamic, there may be a benefit to adjusting the shape of the tank to fit available space. Do any of the previously listed materials offer enhanced ability in this area?
5. Fuel tanks must operate over the full extremes of environmental temperature, and some vehicles with fuel injectors return a portion of the unused fuel-pump fuel to the tank at hot-engine temperature. Are there any material issues relating to temperature?
6. Crashworthiness is an important safety consideration. Some methods of manufacture require seams, while others produce a seamless product. Are there related concerns?
7. Which of the material/fabrication alternatives might be most attractive for:
   a. Low volume production.
   b. High-volume production.
   c. Prestige-type, luxury vehicle.
8. Might your recommendations change with variations in contained fluid capacity, such as compact car versus tractor-trailer?

CHAPTER 15

FUNDAMENTALS OF METAL FORMING

15.1 INTRODUCTION

Chapters 11 through 14 have already presented a variety of methods for producing a desired shape from an engineering material. Each of those methods had its characteristic set of capabilities, advantages, and limitations. If we are to select the best method to make a given product, however, we must have a reasonable understanding of the entire spectrum of available techniques for shape production and their related features.

The next several chapters will expand our study of shape production methods by considering the family of **deformation processes.** These processes have been designed to exploit a remarkable property of some engineering materials (most notably metals) known as **plasticity,** the ability to flow as solids without deterioration of their properties. Because all processing is done in the solid state, there is no need to handle molten material or deal with the complexities of solidification. Because the material is simply moved (or rearranged) to produce the shape, as opposed to the cutting away of unwanted regions, the amount of waste can be substantially reduced. Unfortunately, the forces required are often high. Machinery and tooling can be quite expensive, and large production quantities may be necessary to justify the approach.

The overall usefulness of metals is largely due to the ease of fabrication into useful shapes. Nearly all metal products undergo metal deformation at some stage of their manufacture. Through the process of rolling, cast ingots, strands, and slabs are reduced in size and converted into basic forms such as sheets, rods, plates, or structural shapes, such as I-beams. These forms may then undergo further deformation to produce wire, or the myriad of finished products formed by processes such as forging, extrusion, sheet metal forming, and others. The deformation may be **bulk flow** in three dimensions, simple **shearing,** simple or compound **bending,** or complex combinations of these. The stresses producing these deformations can be tension, compression, shear, or any of the other varieties included in Table 15-1. Table 15-2 depicts a wide variety of specific processes and identifies the primary state of stress responsible for the deformation. For most of these processes, a wide range of speeds, temperatures, tolerances, surface finishes, and amounts of deformation are possible.

TABLE 15-1 Classification of States of Stress

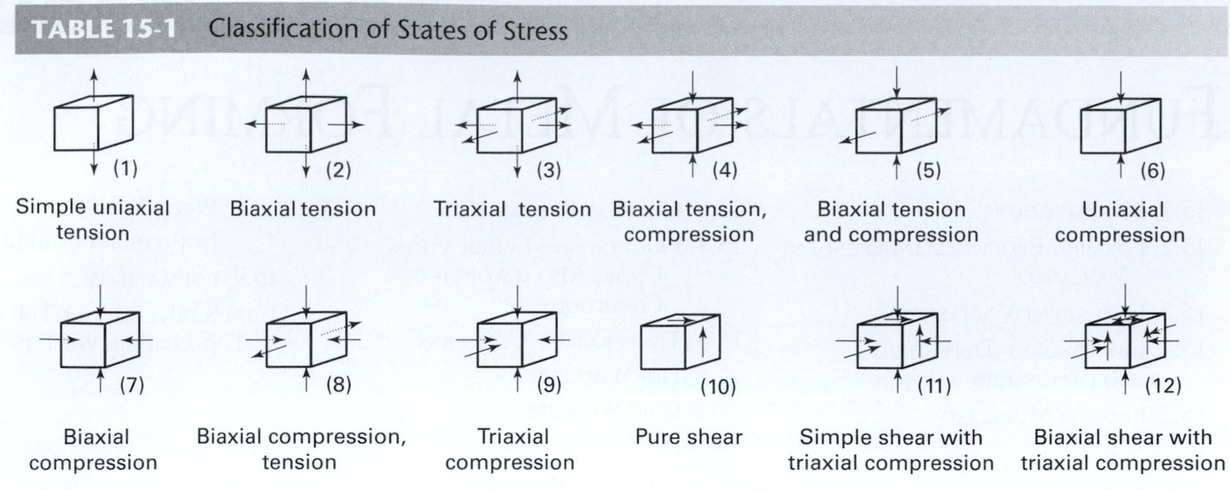

15.2 FORMING PROCESSES: INDEPENDENT VARIABLES

Forming processes tend to be complex systems consisting of independent variables, dependent variables, and independent-dependent interrelations. **Independent variables** are those aspects of a process over which the engineer or operator has direct control, and they are generally selected or specified during setup. Consider some of the independent variables in a typical forming process:

1. *Starting material.* When specifying the starting material, we may define not only the chemistry of that material, but also its condition (as-cast, annealed, as-previously hot rolled, etc.). In so doing, we define the initial properties and characteristics. These may be chosen entirely for ease of fabrication, or they may be restricted by the desire to achieve the required final properties upon completion of the deformation process.
2. *Starting geometry of the workpiece.* The starting geometry may be dictated by previous processing, or it may be selected from a variety of available shapes. Economic considerations often influence this decision.
3. *Tool or die geometry.* This is an area of major significance and has many aspects, such as the diameter and profile of a rolling mill roll, the bend radius in a sheet-forming operation, the die angle in wire drawing or extrusion, and the cavity details when forging. Because the **tooling** will induce and control the metal flow as the material goes from starting shape to finished product, success or failure of a process often depends on tool geometry.
4. *Lubrication.* It is not uncommon for friction between the tool and the workpiece to account for more than 50% of the power supplied to a deformation process. In addition to reducing friction, lubricants can also act as coolants, thermal barriers, corrosion inhibitors, and parting compounds. Hence, their selection is an important aspect in the success of a forming operation. Specification includes type of lubricant, amount to be applied, and the method of application.
5. *Starting temperature.* Because material properties can vary greatly with temperature, temperature selection and control are often key to the success or failure of a metal-forming operation. Specification of starting temperatures may include the temperatures of both the workpiece and the tooling.
6. *Speed of operation.* Most deformation processing equipment can be operated over a range of speeds. Because speed can directly influence the forces required for deformation (see Figure 3-32), the lubricant effectiveness, and the time available for heat transfer, its selection affects far more than just the production rate.
7. *Amount of deformation.* While some processes control this variable through the design of tooling, others, such as rolling, may permit its adjustment at the discretion of the operator.

TABLE 15-2 Common Forming Operations and Their Stress States

Process	Schematic Diagram	State of Stress in Main Part During Forming[a]
Rolling		7
Forging		9
Extrusion		9
Shear spinning		12
Tube spinning		9
Swaging or kneading		7
Deep drawing		In flange of blank, 5 In wall of cup, 1
Wire and tube drawing	(a) (b)	8
Stretching		2
Straight bending		At bend, 2 and 7
Contoured flanging	(a) Convex	At outer flange, 6 At bend, 2 and 7
	(a) Concave	At outer flange, 1 At bend, 2 and 7

[a] Numbers correspond to those in parentheses in Table 15-1.

15.3 DEPENDENT VARIABLES

After the independent variables have been specified, the process then determines the nature and values of a second set of features, known as **dependent variables,** which, in essence, are the consequences of the independent variable selection. Examples of dependent variables include the following:

1. *Force or power requirements*. A certain amount of force or power is required to convert a selected material from a starting shape to a final shape, with a specified lubricant, tooling geometry, speed, and starting temperature. A change in any of the independent variables will result in a change in the required force or power, but the effect is indirect. We cannot directly specify the force or power; we can only specify the independent variables and then experience the consequences of that selection.

   It is extremely important, however, that we be able to predict the forces or powers that will be required for any forming operation. Without a reasonable estimate, we would be unable to specify the equipment for the process, select appropriate tool or die materials, compare various die designs or deformation methods, or ultimately optimize the process.
2. *Material properties of the product*. While we can easily specify the properties of the starting material, the combined effects of deformation and the temperatures experienced during forming will certainly change them. The starting properties of the material may be of interest to the manufacturer, but the customer is far more concerned with receiving the desired final shape with the desired final properties. It is important to know, therefore, how the initial properties will be altered by the shape-producing process.
3. *Exit (or final) temperature*. Deformation generates heat within the material. Hot workpieces cool when in contact with colder tooling. Lubricants can break down or decompose when overheated or may react with the workpiece. The properties of an engineering material can be altered by both the mechanical and thermal aspects of a deformation process. Therefore, if we are to control a process and produce quality products, it is important to know and control the temperature of the material throughout the deformation. (*Note:* The fact that temperature may vary from location to location within the product further adds to the complexity of this variable.)
4. *Surface finish and precision*. The **surface finish** and **dimensional precision** of the resultant product depend on the specific details of the forming process.
5. *Nature of the material flow*. In deformation processes, dies or tooling generally exert forces or pressures and control the movement of the external surfaces of the workpiece. While the objective of an operation is the production of a desired shape, the internal flow of material may actually be of equal importance. As will be shown later in this chapter, product properties can be significantly affected by the details of material flow, and that flow depends on all aspects of a process. Customer satisfaction requires not only the production of a desired geometric shape, but also that the shape possesses the right set of companion properties, without any surface or internal defects.

15.4 INDEPENDENT–DEPENDENT RELATIONSHIPS

Figure 15-1 serves to illustrate the major problem facing metal-forming personnel. On the left side are the *independent* variables—*those aspects of the process for which control is direct and immediate*. On the right side are the *dependent* variables—*those aspects for which control is entirely indirect*. Unfortunately, it is the dependent variables that we want to control, but their values are determined by the process, as complex consequences of the independent variable selection. If we want to change a dependent variable, we must determine which independent variable (or combination of independent variables) is to be changed, in what manner, and by how much. To make appropriate decisions, therefore, it is important for us to develop an understanding of the independent variable–dependent variable interrelations.

Independent variables	Links	Dependent variables
Starting material		Force or power requirements
Starting geometry	-Experience-	Product properties
Tool geometry		Exit temperature
Lubrication	-Experiment-	Surface finish
Starting temperature		Dimensional precision
Speed of deformation	-Modeling-	Material flow details
Amount of deformation		

FIGURE 15-1 Schematic representation of a metalforming system showing independent variables, dependent variables, and the various means of linking the two.

Understanding the links between independent and dependent variables is truly the most important area of knowledge for a person in metalforming. Unfortunately, this knowledge is often difficult to obtain. Metalforming processes are complex systems composed of the material being deformed, the tooling performing the deformation, lubrication at surfaces and interfaces, and various other process parameters such as temperature and speed. The number of different forming processes (and variations thereof) is quite large. In addition, different materials often behave differently in the same process, and there are multitudes of available lubricants. Some processes are sufficiently complex that they may have 15 or more interacting independent variables.

We can gain information on the interdependencies of independent and dependent variables in three distinct ways:

1. *Experience*. Unfortunately, this generally requires long-time exposure to a process and is often limited to the specific materials, equipment, and products encountered during past contact. Younger employees may not have the experience necessary to solve production problems. Moreover, a single change in an area such as material, temperature, speed, or lubricant may make the bulk of past experience irrelevant.
2. *Experiment*. While possibly the least likely to be in error, direct experiment can be both time consuming and costly. Size and speed of deformation are often reduced when conducting laboratory studies. Unfortunately, lubricant performance and heat transfer behave differently at different speeds and sizes, and their effects are generally altered. The most valid experiment, therefore, is one conducted under full-size and full-speed production conditions—generally too costly to consider to any great degree. While laboratory experiments can provide valuable insight, caution should be exercised when extrapolating lab-scale results to more realistic production conditions.
3. *Process modeling*. Here, the process is approached through high-speed computing and one or more mathematical models. Numerical values are selected for the various independent variables, and the models are used to compute predictions for the dependent outcomes. Most techniques rely on the applied theory of plasticity with various simplifying assumptions. Alternatives vary from crude, first-order approximations to sophisticated, computer-based methods, such as finite element analysis. Various models may incorporate strain hardening, thermal softening, heat transfer, and other phenomena. Solutions may be algebraic relations that describe the process and reveal trends and relations between the variables, or simply numerical values based on the specific input features.

15.5 PROCESS MODELING

Metalforming simulations using the finite element modeling method became common in the 1980s but generally required high-power minicomputers or engineering workstations. By the mid-1990s, the rapid increase in computing power made it possible to model complex processes on desktop personal computers. With the continued expansion of computing power and speed, process simulations are now quick, inexpensive, and quite accurate. As a result, modeling is being used in all areas of manufacturing, including part design, manufacturing process design, heat-treatment and surface-treatment optimization, and others. Models can predict how a material will respond to a rolling process, fill a forging die, flow through an extrusion die, or solidify in a casting. Entire heat treatments can be simulated, including cooling rates in various quenchants. Models can even predict the strain distribution, residual stresses, microstructure, and final properties at all locations within a product.

Advanced simulation techniques can provide a clear and thorough understanding of a process, eliminating costly trial-and-error development cycles. Product design and manufacturing methods can be optimized for quality and reliability, while reducing production costs and minimizing lead times. When coupled with appropriate sensors, the same models can be used to determine the type of adjustments needed to provide online process control. Process models can also serve as laboratory tools to explore new ideas or new products. New employees can become familiar with what works and what doesn't in a quick and inexpensive manner.

It is important to note, however, that the accuracy of any model can be no better than that of the input variables. For example, when modeling a metal-forming operation, the mechanical properties of the deforming material (i.e., yield strength, ductility, etc.) must be known for the specific conditions of temperature, strain (amount of prior deformation), and strain rate (speed of deformation) being considered. The mathematical descriptions of material behavior as a function of the process conditions are known as **constitutive relations.** The development of such relationships is not an easy task, however, because the same material may respond differently to the same conditions if its microstructure is different. A 1040 steel that has been annealed (ferrite and pearlite) will not have the same properties as a quenched and tempered (tempered martensite) steel of the same chemistry. Microstructure and its effects on properties are difficult to describe in quantitative terms that can be input to a model.

Another rather elusive variable is the **friction** between the tool and the workpiece. Studies have shown that friction depends on contact pressure, contact area, lubricant, speed, and the surface finish and mechanical properties of the two contacting materials. We know that these parameters often vary from location to location and also change with time during a process, but many models tend to describe friction with a single variable of constant magnitude. Any variations with time and location are simply ignored in favor of mathematical simplicity or because of a lack of any better information.

At first glance, problems such as those just discussed appear to be a significant barrier to the use of mathematical models. It should be noted, however, that the same difficulties apply to the person trying to document, characterize, and extrapolate the results of experience or experiments. **Process modeling** often reveals features that might otherwise go unnoticed and can be quite useful when attempting to prevent or eliminate defects, optimize performance, or extend a process into a previously unknown area.

15.6 GENERAL PARAMETERS

While much metalforming knowledge is specific to a given process, there are certain features that are common to all processes, and these will be presented here.

It is extremely important to characterize the *material being deformed*. What is its strength or resistance to deformation at the relevant conditions of temperature, speed of deformation, and amount of prior straining? What are the formability limits and conditions of anticipated fracture? What is the effect of temperature or variations in temperature? To what extent does the material strain-harden? What are the recrystallization kinetics? Will the material react with various environments or lubricants? These and many other questions must be answered to assess the suitability of a material to a given deformation process. Because the properties of engineering materials vary widely, the details will not be presented at this time. The reader is referred to the various chapters on engineering materials, as well as the more in- depth references cited in Chapter 10.

Another general parameter is the *speed of deformation* and the various related effects. Some rate-sensitive materials may shatter or crack if impacted, but will deform plastically when subjected to slow-speed loadings. Other materials appear to be stronger when deformed at higher speeds. For these *speed-sensitive materials*, more energy is needed to produce the same result if we wish to do it faster, and stronger tools may be required. Mechanical data obtained at slow strain rates (as in tensile tests) may be totally useless if the deformation process operates at a significantly greater rate of deformation. **Speed sensitivity** is also greatest when the material is at elevated temperature, a condition that is frequently encountered in metalforming operations. The selection of hammer or press for the hot forging of a small product may well depend on the speed sensitivity of the material being forged.

In addition to the changes in mechanical properties, faster deformation speeds tend to promote improved lubricant efficiency. Faster speeds also reduce the time for heat transfer and cooling. During hot working, workpieces stay hotter, and less heat is transferred to the tools.

Other general parameters include *friction and lubrication* and *temperature*. Both of these are of sufficient importance that they will be discussed in some detail.

15.7 FRICTION AND LUBRICATION UNDER METALWORKING CONDITIONS

In metalforming, high forces or high pressures are applied through tools to induce the deformation of a material. Because of the relative motion between the workpiece and the tool, an important consideration in metal deformation processes is the friction that exists at this interface. For some processes, more than 50% of the input energy is spent in overcoming friction. Changes in lubrication can alter the mode of material flow during forming, create or eliminate defects, alter the surface finish and dimensional precision of the product, and modify product properties. Production rates, tool design, tool wear, and process optimization all depend on the ability to determine and control friction between the tool and workpiece.

In most cases, we want to economically reduce the effects of friction. However, some deformation processes, such as rolling, can only operate when sufficient friction is present. Regardless of the process, friction effects are hard to measure. As previously noted, the specific friction conditions depend on a number of variables, including contact area, contact pressure, surface finish, speed, lubricant, and temperature. Because of the many variables, the effects of friction are extremely difficult to scale down for laboratory testing or to extrapolate from laboratory tests to production conditions.

It should be noted that friction under metalworking conditions is significantly different from the friction encountered in most mechanical devices. The friction conditions of gears, bearings, journals, and similar components generally involve (1) two surfaces of similar material and similar strength, (2) experiencing elastic loads such that neither body undergoes permanent change in shape, (3) with wear-in cycles that produce surface compatibility, and (4) with operating temperatures in the low-to-moderate range. In contrast, metalforming operations involve a hard, nondeforming tool interacting with a soft workpiece at pressures sufficient to cause plastic flow in the weaker material. Only a single pass is involved as the tool and workpiece interact, the workpiece is often at elevated temperature, and the contact area is frequently changing as the workpiece deforms.

Figure 15-2 shows the typical relationship between frictional resistance and contact pressure. For light, elastic loads, friction is directly proportional to the applied pressure, with the proportionality constant, m, being known as the **coefficient of friction** or, more specifically, the *Coulomb coefficient of friction.* At high pressures, friction becomes independent of contact pressure and is more closely related to the strength of the weaker material.

An understanding of these results can be obtained from modern friction theory, whose primary premise is that "flat surfaces are not flat" but have some degree of roughness. When two irregular surfaces interact, sufficient contact is established to support the applied load. At the lightest of loads, only three points of contact may be necessary to support a plane. As the load is increased, the contacting points deform and the

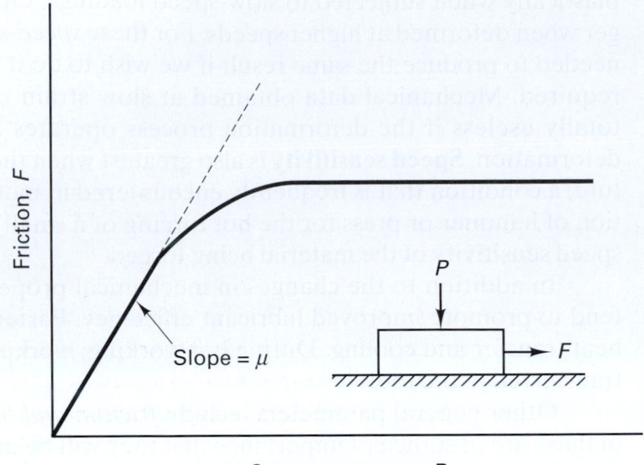

FIGURE 15-2 The effect of contact pressure on the frictional resistance between two surfaces.

contact area increases, initially in a linear fashion. As the load continues to increase, more area comes into contact. Finally, at some high value of load, there is full contact between the surfaces. Additional loads can no longer bring additional area into contact, and friction can now be described by a constant, independent of pressure.

Friction is the resistance to sliding along an interface. From a mechanistic viewpoint, this resistance can be attributed to **abrasion,** the force necessary to plow the peaks of a harder material through a softer one, and/or **adhesion,** the force necessary to rip apart microscopic weldments that form between the two materials. Because the weldment tears generally occur in the weaker of the two materials, it is reasonable to assume that the resistance attributed to both features would be proportional to the strength of the weaker material and also to the actual area of metal-to-metal contact. Thus, the curve depicted in Figure 15-2 could also be viewed as a plot of actual contact area at the interface versus contact pressure. Unfortunately, Figure 15-2, and the associated theory, applies only to unlubricated metal-to-metal contact. The addition of a lubricant, as well as any variation in its type or amount, can significantly alter the frictional response.

Surface deterioration or **wear** is another phenomenon that is directly related to friction. Because the workpiece only interacts with the tooling during a single forming operation, any wear experienced by the workpiece is usually not objectionable. In fact, a shiny, fresh-metal surface produced by wear is often viewed as desirable. Manufacturers whose processes retain most or all of the original dull finish may be accused of selling old or substandard products. Wear on the tooling, however, is quite the reverse. Tooling is expensive and it is expected to shape many workpieces. Tooling wear will generally result in a change of workpiece dimensions. Tolerance control will be lost, and at some point the tools will have to be replaced. Other consequences of tool wear include increased frictional resistance (increased required power and decreased process efficiency), poor surface finish on the product, and loss of production during tool changes.

Lubrication is a key to success in many metalforming operations. While lubricants are generally selected for their ability to reduce friction and suppress tool wear, secondary considerations may include the ability to act as a thermal barrier (keeping heat in the workpiece and away from the tooling), the ability to act as a coolant (removing heat from the tools), and the ability to retard corrosion if left on the formed product. Other influencing factors include ease of application and removal; lack of toxicity, odor, and flammability; reactivity or lack of reactivity with material surfaces; adaptability over a useful range of pressure, temperature, and velocity; surface wetting characteristics; cost; availability; and the ability to flow or thin and still function as a lubricant. Lubricant selection is further complicated by the fact that lubricant performance may change with any change in the interface conditions. The exact response is often dependent on such factors as the finish of both surfaces, the area of contact, the applied load, the speed, the temperature, and the amount of lubricant.

The ability to select an appropriate lubricant can be a critical factor in determining whether a process is successful or unsuccessful, efficient or inefficient. Considerable effort, therefore, has been directed to the study of friction and lubrication, a subject known as **tribology,** as it applies to both general metalworking conditions and specific metalforming processes. There are thousands of lubricant chemistries available, and most can be further modified by various additives. All of them, however, can be grouped into several general types. *Straight oils* are petroleum-based compounds that are easy to apply and offer good corrosion protection, but they can be difficult to clean. *Water-soluble oils* (emulsions) are easier to clean but compromise on corrosion and long-term product stability. *Synthetic lubricants* are water based and free of petroleum. *Semisynthetics* are a blend of water, petroleum oils, and emulsifiers. *Dry-film lubricants* are often composed of soaps or polymers that are applied in aqueous form and dried before the forming operation. *Chemically bonded agents* include plated coatings or products that form by chemical reaction with the workpiece.

Extreme pressure additives form a chemical film that bonds to both the workpiece and the tooling, enhancing lubrication under high pressure and high temperature. Boundary additives (including fats and solids such as graphite) attach themselves to the metal surfaces and provide cushioning or separation under high pressure. Other additives can help to promote **hydrodynamic lubrication.** If the combination of lubricant and the speed of relative motion is sufficient to create a lubricant layer thick enough to prevent mechanical contact between the tool and the workpiece, the forces and power required may decrease by as much as 30 to 40%, and tool wear becomes almost nonexistent.

A substantial information base has been developed that can aid in optimizing the use of lubricants in metalworking.

15.8 TEMPERATURE CONCERNS

In metalworking operations, workpiece temperature can be one of the most important process variables. The role of temperature in altering the properties of a material has been discussed in Chapter 3. In general, an increase in temperature brings about a decrease in strength, an increase in ductility, and a decrease in the rate of strain hardening—all effects that would tend to promote ease of deformation.

Forming processes tend to be classified as hot working, cold working, or warm working based on both the temperature and the material being formed. In hot working, the deformation is performed under conditions of temperature and strain rate where recrystallization occurs simultaneously with the deformation. To achieve this, the temperature of deformation is usually in excess of 0.6 times the melting point of the material on an absolute temperature scale (Kelvin or Rankine). Cold working is deformation under conditions where the recovery processes are not active. Here the working temperatures are usually less than 0.3 times the workpiece melting temperature. Warm working is deformation under the conditions of transition (i.e., a working temperature between 0.3 and 0.6 times the melting point).

HOT WORKING

Hot working is defined as the plastic deformation of metals at a temperature above the recrystallization temperature. It is important to note, however, that the recrystallization temperature varies greatly with different materials. Tin is near hot-working conditions at room temperature; steels require temperatures near 1100°C (2000°F); and tungsten does not enter the hot-working regime until about 2200°C (4000°F). Thus, the term *hot working* does not necessarily correlate with high or elevated temperature, although such is usually the case.

As shown in Figures 3-30 and 3-31, elevated temperatures bring about a decrease in the yield strength of a metal and an increase in ductility. At the temperatures of hot working, recrystallization eliminates the effects of strain hardening, so there is no significant increase in yield strength or hardness or corresponding decrease in ductility. The true stress–true strain curve is essentially flat once we exceed the yield point, and deformation can be used to drastically alter the shape of a metal without fear of fracture, and without the requirement of excessively high forces. In addition, the elevated temperatures promote diffusion that can remove or reduce chemical inhomogeneities, pores can be welded shut or reduced in size during the deformation; and the metallurgical structure can often be altered through recrystallization to improve the final properties. An added benefit is observed for steels, where hot working involves the deformation of the weak, ductile, face-centered-cubic austenite structure, which then cools and transforms to the stronger body-centered-cubic ferrite, or much stronger nonequilibrium structures, such as martensite.

From a negative perspective, the high temperatures of hot working may promote undesirable reactions between the metal and its surroundings. Tolerances are poorer due to thermal contractions, and warping or distortion can occur due to nonuniform cooling. The metallurgical structure may also be nonuniform, because the final grain size depends on the amount of deformation, the temperature of the last deformation/recrystallization, the cooling history after the deformation, and other factors—all of which may vary throughout a workpiece.

While recrystallization sets the minimum temperature for hot working, the upper limit for hot working is usually determined by factors such as excess oxidation, grain growth, or undesirable phase transformations. To keep the forming forces as low as possible and enable hot deformation to be performed for a reasonable amount of time, the starting temperature of the workpiece is usually set at or near the highest temperature for hot working.

Structure and Property Modification by Hot Working. When metals solidify into the large sections that are typical of ingots or continuously cast slabs or strands, coarse structures tend to form with a certain amount of chemical segregation. The size of the grains is usually not uniform, and undesirable grain shapes can be quite common, such as the columnar grains that have been revealed in Figure 15-3. Small gas cavities or shrinkage porosity can also form during solidification.

If a cast metal is reheated without prior deformation, it will simply experience grain growth and the accompanying deterioration in engineering properties. However, if the metal first experiences a sufficient amount of deformation, the distorted structure will be rapidly replaced by new strain-free grains. This **recrystallization** is then followed by (1) grain growth, (2) additional deformation and recrystallization, or (3) a drop in temperature that will terminate diffusion and "freeze in" the recrystallized structure. The structure in the final product is that formed by the last recrystallization and the thermal history that follows. By replacing the initial structure with a new one consisting of fine, spherical-shaped grains, it is possible to produce an increase not only in strength, but also in ductility and toughness—a somewhat universal enhancement of properties.

Engineering properties can also be improved through the reorientation of inclusions or impurity particles that are present within the metal. With normal melting and cooling, many impurities tend to locate along grain boundary interfaces. If these are

FIGURE 15-3 Cross section of a 4-in.-diameter cast copper bar polished and etched to show the as-cast grain structure. *(Courtesy Ronald Kohser)*

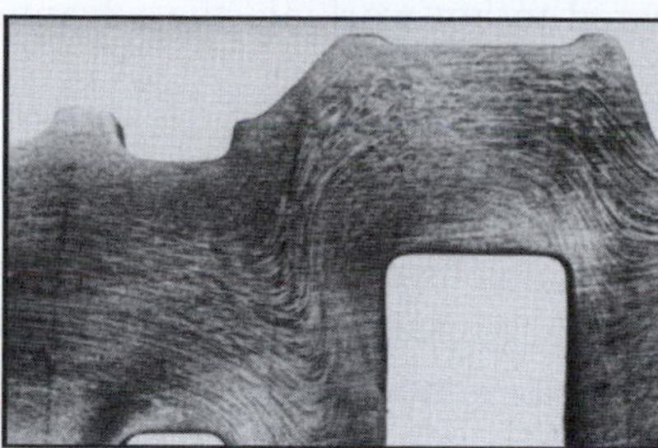

FIGURE 15-4 Etched surface of a sectioned crankshaft showing the grain flow pattern that gives forged parts enhanced strength, fracture resistance, and fatigue life. *(Courtesy of the Forging Industry Association, Cleveland, OH)*

unfavorably oriented or intersect surfaces, they can initiate a crack or assist its propagation. When a metal is plastically deformed, the impurities tend to flow along with the base metal, or fracture into rows of fragments **(stringers)** that are aligned in the direction of working. These nonmetallic impurities do not recrystallize with the base metal but retain their distorted shape and orientation. The product exhibits an **oriented** or **flow structure,** like the one shown in Figure 15-4, and final properties tend to exhibit directional variation. Through proper design of the deformation, impurities can often be reoriented into a "crack-arrestor" configuration, where they are perpendicular to the direction of crack propagation. In Figure 15-4, the flow lines are parallel to the external surfaces at all high-stress locations. This crankshaft will exhibit excellent fracture and fatigue resistance.

Figure 15-5 schematically compares a machined thread and a rolled thread in a threaded fastener. If the axial defects in the starting wire or rod are reoriented to be parallel to the thread profile, the rolled thread will offer improved strength and fracture resistance.

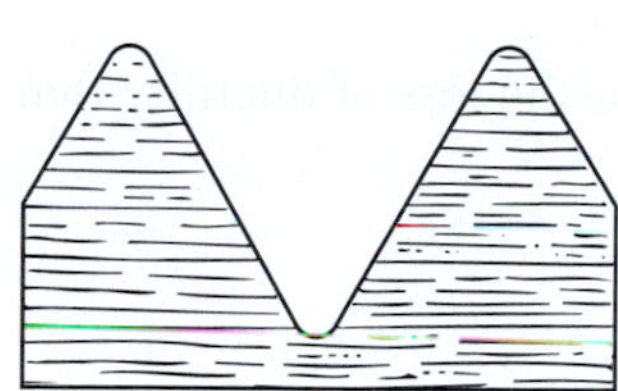

(a)

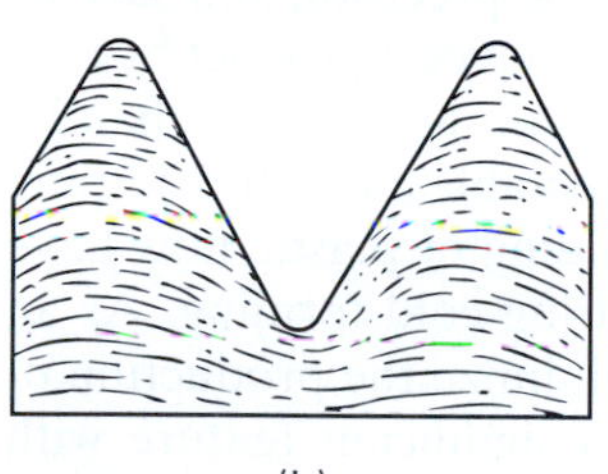

(b)

FIGURE 15-5 Schematic comparison of the grain flow in (a) a machined thread and (b) a rolled thread. The rolling operation further deforms the axial structure produced by the previous wire- or rod-forming operations, while machining simply cuts through it.

Temperature Variations. The success or failure of a hot deformation process often depends on the ability to control the temperatures within the workpiece. More than 90% of the energy imparted to a deforming workpiece will be converted into heat. If the deformation process is sufficiently rapid, the temperature of the workpiece may actually increase. More common, however, is the cooling of the workpiece in its lower-temperature environment. Heat is lost through the workpiece surfaces, with the majority of the loss occurring where the workpiece is in direct contact with lower-temperature tooling. Nonuniform temperatures are produced, and flow of the hotter, weaker interior may well result in cracking of the colder, less ductile surfaces. Thin sections cool faster than thick sections, and this may further complicate the flow behavior.

To minimize problems, it is desirable to keep the workpiece temperatures as uniform as possible. Heated dies can reduce the rate of heat transfer, but die life tends to be compromised. For example, dies are frequently heated to 325 to 450°C (600 to 850°F) when used in the hot forming of steel. Tolerances could be improved and contact times could be increased if the tool temperatures could be raised to 550 to 650°C (1000 to 1200°F), but tool life drops so rapidly that these conditions become quite unattractive.

A final concern is the cooldown from the temperatures of hot working. Nonuniform cooling can introduce significant amounts of **residual stress** in hot-worked products. Associated with these stresses may be warping or distortion, and possible cracking.

COLD WORKING

The plastic deformation of metals below the recrystallization temperature is known as **cold working.** Here, the deformation is usually performed at room temperature, but mildly elevated temperatures may be used to provide increased ductility and reduced strength. From a manufacturing viewpoint, cold working has a number of distinct advantages, and the various cold-working processes have become quite prominent. Recent advances have expanded the capabilities, and a trend toward increased cold working appears likely to continue.

When compared to hot working, the advantages of cold working include the following:

1. No heating is required.
2. Better surface finish is obtained.
3. Superior dimensional control is achieved because the tooling sets dimensions at room temperature. As a result, little, if any, secondary machining is required.
4. Products possess better reproducibility and interchangeability.
5. Strength, fatigue, and wear properties are all improved through strain hardening.
6. Directional properties can be imparted.
7. Contamination problems are minimized.

Some disadvantages associated with cold-working processes include:

1. Higher forces are required to initiate and complete the deformation.
2. Heavier and more powerful equipment and stronger tooling are required.
3. Less ductility is available.
4. Metal surfaces must be clean and scale-free.
5. Intermediate anneals may be required to compensate for the loss of ductility that accompanies strain hardening.
6. The imparted directional properties may be detrimental.
7. Undesirable residual stresses may be produced.

The strength levels induced by **strain hardening** are often comparable to those produced by the strengthening heat treatments. Even when the precision and surface finish of cold working are not required, it may be cheaper to produce a product by cold working a less expensive alloy (achieving the strength by strain hardening) than by heat-treating parts that have been hot formed from a heat-treatable alloy. In addition, better and more ductile metals and an improved understanding of plastic flow have done much to reduce the difficulties often experienced during cold forming. As an added benefit, most cold-working processes eliminate or minimize the production of waste material and the need for subsequent machining—a significant feature with today's emphasis on conservation and materials recycling.

Because the cold-forming processes require powerful equipment and product-specific tools or dies, they are best suited for large-volume production of precision parts where the quantity of products can justify the cost of the equipment and tooling. Considerable effort has been devoted to developing and improving cold-forming machinery along with methods to enable these processes to be economically attractive for modest production quantities. By grouping products made from the same starting material and using quick-change tooling, cold-forming processes can often be adapted to small-quantity or just-in-time manufacture.

Metal Properties and Cold Working. The suitability of a metal for cold working is determined primarily by its tensile properties, and these are a direct consequence of its metallurgical structure. Cold working then alters that structure, thereby altering the tensile properties of the resulting product. It is important for both the incoming and outgoing properties to be considered when selecting metals that are to be processed by cold working.

Figure 15-6 presents the true stress–true strain curves for both a low- and high-carbon steel. Focusing on the low-carbon material, we note that plastic deformation cannot occur until the strain exceeds the strain associated with the elastic limit, point a on the stress-strain curve. Plastic deformation then continues until the strain reaches the value x_4, where the metal ruptures. From the viewpoint of cold working, two features are significant: (1) the magnitude of the yield-point stress, which determines the force required to initiate permanent deformation, and (2) the extent of the strain region from x_1 to x_4, which indicates the amount of plastic deformation (or ductility) that can

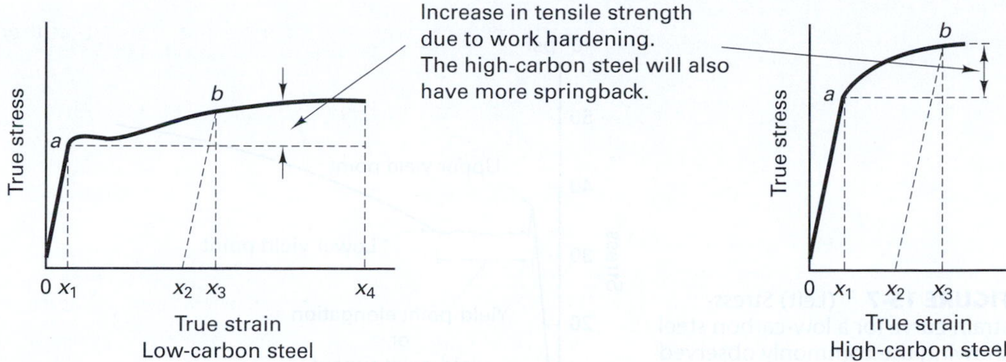

FIGURE 15-6 Use of true stress–true strain diagrams to assess the suitability of two metals for cold working. *(Courtesy Ronald Kohser)*

be achieved without fracture. If a considerable amount of deformation is desired, a material like the low-carbon steel is more desirable than the high-carbon variety. Greater ductility would be available, and less force would be required to initiate and continue the deformation. The curve on the right, however, has a higher strain-hardening coefficient (see Chapter 3 for discussion). If strain hardening is being used to impart strength, this material would have a greater increase in strength for the same amount of cold work. In addition, the material on the right would be more attractive for shearing operations and may be easier to machine (see Chapter 20).

Springback is another cold-working phenomenon that can be explained with the aid of a stress-strain diagram. When a metal is deformed by the application of a load, part of the resulting deformation is elastic. For example, if a metal is stretched to point x_1 on either of the curves of Figure 15-6 and the load is removed, it will return to its original size and shape because all of the deformation is elastic. If, on the other hand, the metal is stretched by an amount x_3, corresponding to point b on the stress-strain curve, the total strain is made up of two parts: a portion that is elastic and another that is plastic. When the deforming load is removed, the stress relaxation will follow line bx_2, and the final strain will only be x_2. The decrease in strain, $x_3 - x_2$, is known as **elastic springback.**

In cold-working processes, springback can be extremely important. If a desired size is to be achieved, the deformation must be extended beyond that point by an amount equal to the springback. Because different materials have different elastic moduli, the amount of springback from a given load will change from one material to another. A substitution in material, therefore, may well require adjustments in the forming process. Fortunately, springback is a predictable phenomenon, and most difficulties can be prevented by proper design procedures.

Initial and Final Properties in a Cold-Working Process. The quality of the starting material is often key to the success or failure of a cold-working operation. To obtain a good surface finish and maintain dimensional precision, the starting material must be clean and free of oxide or scale that might cause abrasion and damage to the dies or rolls. Scale can be removed by **pickling,** a process in which the metal is dipped in acid and then washed. In addition, sheet metal and plate are sometimes given a light cold rolling prior to the major deformation. The rolling operation not only ensures uniform starting thickness, but also produces a smooth starting surface.

The light cold-rolling pass can also serve to remove the **yield-point phenomenon** and the associated problems of nonuniform deformation and surface irregularities in the product. Figure 15-7 presents an expansion of the left-hand region of Figure 3-6 or the low-carbon portion of Figure 15-6. After loading to the upper yield point, the material exhibits a **yield-point runout,** wherein the material can strain up to several percent with no additional force being required. Consider a piece of sheet metal that is to be formed into an automotive body panel. If a segment of that panel were to receive a total stretch less than the magnitude of the yield-point runout, it would be induced by a stress equal to the yield-point stress. Because the stress is constant in the runout region, the

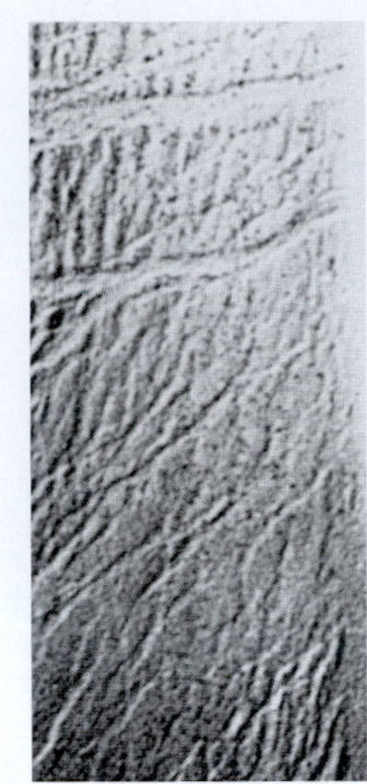

FIGURE 15-7 (Left) Stress-strain curve for a low-carbon steel showing the commonly observed yield-point runout. (Right) Luders bands or stretcher strains that form when this material is stretched to an amount less than the yield-point runout. *(Courtesy Ronald Kohser)*

material is free to not deform at all, to deform the entire amount of the yield-point runout, or to select some point in between. It is not uncommon for some regions to deform the entire amount and thin correspondingly, while adjacent regions resist deformation and retain the original thickness. The resulting ridges and valleys, shown in Figure 15-7, are referred to as **Luders bands** or **stretcher strains** and are very difficult to remove or conceal. By first cold rolling the material to a strain near or past the yield-point runout, all subsequent forming occurs in a region where a well-defined strain corresponds to each value of stress. If the body panel were shaped from prerolled material, the deformation and thinning would be uniform throughout the piece.

Figure 15-8 shows how the mechanical properties of pure copper are affected by cold working. Individual tensile tests were conducted on specimens that had experienced progressively greater amounts of cold work.[1] As the graph shows, yield strength and tensile strength increase significantly with increased deformation. Hardness is not presented on the graph, but it generally follows tensile strength. Elongation, along with

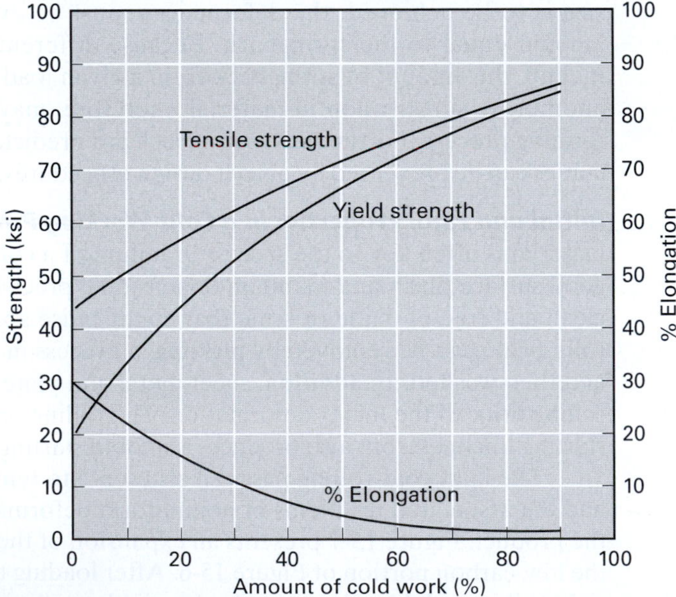

FIGURE 15-8 Mechanical properties of pure copper as a function of the amount of cold work (expressed in percent).

[1] A block of copper was rolled in increments to progressively reduced thickness, the decrease in height being compensated by an increase in length with width remaining constant. The amount of cold work can then be computed as: Percent cold work $= [(A_o - A_f)/A_o] \times 100\%$, where A_o is the original cross-sectional area and A_f is the cross-sectional area at each of the computed increments.

reduction in area, electrical conductivity, and corrosion resistance, will decline. Because the ductility decreases, the amount of cold working is generally limited by the onset of fracture.

In order to maximize the starting ductility, an **annealing** heat treatment is often applied to a metal prior to cold working. If the required amount of deformation exceeds the fracture limit, one or more **intermediate anneals** may be necessary to restore ductility (set the amount of cold work back to zero). If the desired final properties require a specific amount of cold work (as in Problem 1 at the end of this chapter), the last anneal can be judiciously positioned in the deformation cycle. In this way, the desired shape can be produced along with the mechanical properties that accompany the amount of cold work imparted after the final anneal. In all annealing operations, care should be exercised to control the grain size of the resulting material. Grain sizes that are too large or too small can both be detrimental.

Like hot working, cold working also produces an anisotropic structure—one whose properties vary with direction. Here, the **anisotropy** is related to the distorted crystal structure and is not simply a function of the nonmetallic inclusions. Also associated with cold working is the generation of residual stresses. While anisotropy and residual stresses can be beneficial, they can also be quite harmful. Because they occur as a consequence of cold working, their effect on performance should always be considered.

WARM WORKING

Deformation produced at temperatures intermediate to hot and cold forming is known as **warm working.** Compared to cold working, warm working offers the advantages of reduced loads on the tooling and equipment, increased material ductility, and a possible reduction in the number of anneals due to a reduction in the amount of strain hardening. The use of higher forming temperatures can often expand the range of materials and geometries that can be formed by a given process or piece of equipment. High-carbon steels may be formed without a spheroidization treatment.

Compared to hot working, the lower temperatures of warm working produce less scaling and decarburization and enable production of products with better dimensional precision and smoother surfaces. Finish machining is reduced and less material is converted into scrap. Because of the finer structures and the presence of some strain hardening, the as-formed properties may be adequate for many applications, enabling the elimination of final heat-treatment operations. The warm regime generally requires less energy than hot working due to the decreased energy in heating the workpiece (lower temperature), energy saved through higher precision (less material being heated), and the possible elimination of postforming heat treatments. Although the tools must exert 25 to 60% higher forces, they last longer because there is less thermal shock and thermal fatigue.

When energy was cheap, metal forming was usually conducted in either the hot- or cold-working regimes, and warm working was largely ignored. Even today, material behavior is less well characterized for the warm-working temperatures (the warm-working temperatures for steel are between 550 and 800°C, or 1000 and 1500°F). Lubricants have not been as fully developed for the warm-working temperatures and pressures, and die design technology is not as well established. Nevertheless, the pressures of energy and material conservation, coupled with the other cited benefits, strongly favor the continued development of warm working. Cold forming is still the preferred method for fabricating small components, but warm forming is considered to be attractive for larger parts (up to about 10 lb) and steels with more than 0.35% carbon and/or high alloy content.

Hot working and warm forming are usually applied to bulk-forming processes, like forging and extrusion. For sheet material, the surface-to-volume ratio is sufficiently large that the workpiece can rapidly lose its heat. As the major auto manufacturers seek to increase fuel efficiency, there has been significant interest in aluminum sheet as a replacement for steel. Unfortunately, the formability of high-strength aluminum is much lower than similar-strength low-carbon steels. If the steel is simply replaced with

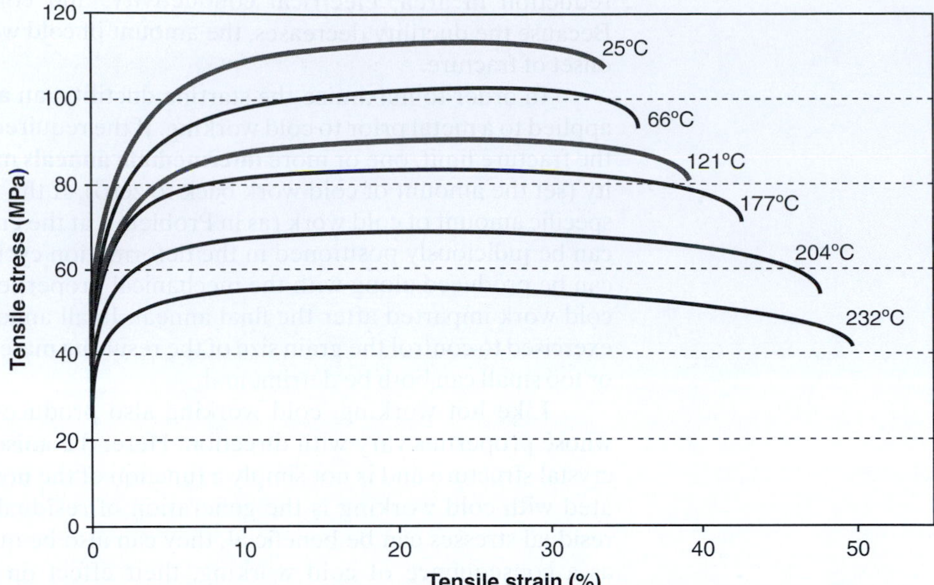

FIGURE 15-9 Increasing the temperature increases the formability of aluminum, decreasing the strength and increasing the ductility. *(Adapted from data provided by Interlaken Technology Corporation, Chaska, MN)*

aluminum, and the design and tooling remain unchanged, fracture often occurs in the more heavily worked regions. If the material, die, and blank holder are all heated to 200 to 300°C (400 to 575°F), however, the aluminum sheet shows a significant increase in formability and a decrease in elastic springback. Figure 15-9 shows how the properties of aluminum change with temperature in the warm-working regime. By heating the material, satisfactory parts can generally be produced.

ISOTHERMAL FORMING

Figure 15-10 shows the relationship between yield strength (or forging pressure) and temperature for several engineering metals. The 1020 and 4340 steels show a moderate increase in strength with decreasing temperature. In contrast, the strength of the titanium alloy (open circles) and the A-286 nickel-based superalloy (solid circles) show a much stronger variation. Within the range of typical hot-working temperatures, cooling of as little as 100°C (200°F) could result in a doubling of strength. During hot forming, cooling surfaces surround a hotter interior. Any variation in strength can result in nonuniform deformation and cracking of the less ductile surface.

To successfully deform such temperature-sensitive materials, deformation may have to be performed under **isothermal-forming** (constant temperature) conditions. The dies or tooling must be heated to the same temperature as the workpiece, sacrificing die

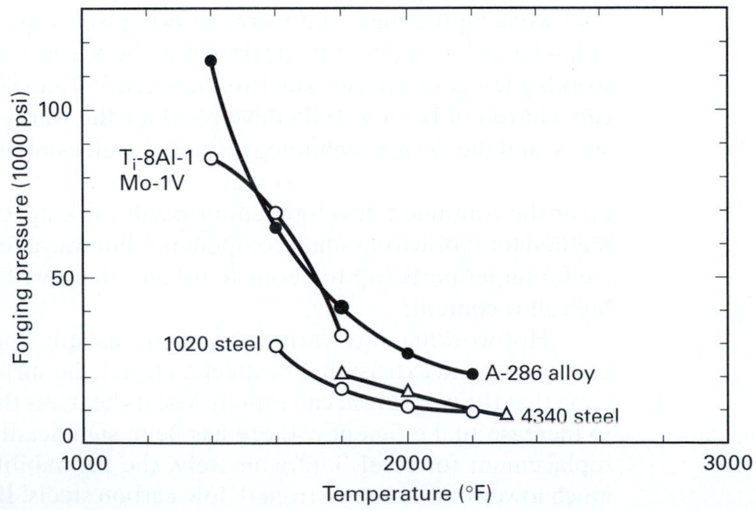

FIGURE 15-10 Yield strength of various materials (indicated by pressure required to forge a standard specimen) as a function of temperature. Materials with steep curves require isothermal forming. *(From "A Study of Forging Variables," ML-TDR-64-95, March 1964; courtesy of Battelle Columbus Laboratories, Columbus, OH)*

life for product quality. Deformation speeds must be slowed so that any heat generated by deformation can be removed in a manner that would maintain a uniform and constant temperature. Inert atmospheres may be required because of the long times at elevated temperature. Although such methods are indeed costly, they are often the only means of producing satisfactory products from certain materials. Because of the uniform temperatures and slow deformation speeds, isothermally formed components generally exhibit close tolerances, low residual stresses, and fairly uniform metal flow.

15.9 FORMABILITY

As discussed in Section 3-5, the word **formability,** while used to describe the ability of a material to be deformed, is really an extremely complex and rather nebulous concept. There is no universal formability scale. A material that responds well to one forming process and one set of process conditions, may fail miserably when used in another process or under a different set of process conditions. All of the independent variables—part design, die design, lubrication, speed of deformation, material temperature, die or tooling temperature, and even the specific operator—can have a significant effect on the observed formability.

KEY WORDS

abrasion
adhesion
anisotropy
annealing
bending
bulk flow
coefficient of friction
cold working
constitutive relation
deformation processes
dependent variable
dimensional precision
elastic springback
flow structure
formability
friction
hot working
hydrodynamic lubrication
independent variable
intermediate anneal
isothermal forming
lubrication
Luders bands
oriented structure
pickling
plasticity
process modeling
recrystallization
residual stresses
shearing
speed sensitivity
springback
strain hardening
stretcher strains
stringers
surface finish
tooling
tribology
warm working
wear
yield-point phenomenon
yield-point runout

REVIEW QUESTIONS

1. What is plasticity?
2. What are some of the general assets of the metal deformation processes? Some general liabilities?
3. Why might large production quantities be necessary to justify metal deformation as a means of manufacture?
4. What is an independent variable in a metalforming process?
5. What are some considerations regarding selection of the starting material for a forming process?
6. What is the significance of tool and die geometry in designing a successful metalforming process?
7. Why is lubrication often a major concern in metalforming?
8. What are some of the secondary effects that may occur when the speed of a metalforming process is varied?
9. What is a dependent variable in a metalforming process?
10. Why is it important to be able to predict the forces or powers required to perform specific forming processes?
11. Why is it important to control the final properties of a deformation product?
12. Why is it important to know and control the thermal history of a metal as it undergoes deformation?
13. Why is it important to control not only the external shape of a formed product, but also its internal flow?
14. Why is it often difficult to determine the specific relationships between independent and dependent variables?
15. What are the three distinct ways of determining the interrelation of independent and dependent variables?
16. What features limit the value of laboratory experiments in modeling metalforming processes?
17. What features have contributed to the expanded use of process modeling?
18. What are some of the uses or applications of process models?
19. What is a constitutive relation for an engineering material?
20. What features may limit the accuracy of a mathematical model?
21. What simplifying assumptions are often made regarding friction between the tool and workpiece?
22. What type of information about the material being deformed may be particularly significant to a metalforming engineer?
23. How might a material's performance vary with changes in the speed of deformation?
24. Why is friction such an important parameter in metalworking operations?
25. Why are friction effects in metalworking difficult to scale down for laboratory testing or scale up from laboratory conditions to production conditions?
26. How might friction be mathematically described under conditions of light, elastic loading? Under conditions of heavy loads, sufficient to induce plastic deformation?

27. What are several ways in which the friction conditions during metalworking differ from the friction conditions found in most mechanical equipment?
28. According to modern friction theory, frictional resistance can be attributed to what two physical phenomena?
29. Discuss the significance of wear in metalforming: (a) wear on the workpiece and (b) wear on the tooling.
30. Lubricants are often selected for properties in addition to their ability to reduce friction. What are some of these additional properties?
31. What is tribology?
32. What are some of the common types of metalforming lubricant additives?
33. What are some of the benefits of hydrodynamic lubrication?
34. If the temperature of a material is increased, what changes in properties might occur that would promote the ease of deformation?
35. Define the various regimes of cold working, warm working, and hot working in terms of the melting point of the material being formed.
36. What is an acceptable definition of hot working? Is a specific temperature involved?
37. What are some of the attractive manufacturing and metallurgical features of hot-working processes?
38. What are some of the negative aspects of hot working?
39. What factors generally set the upper temperature limit for hot working?
40. How can hot working be used to improve the grain structure of a metal?
41. If the deformed grains recrystallize during hot working, how can the process impart an oriented or flow structure (and directionally dependent properties)?
42. Why might a rolled thread offer improved strength and fracture resistance compared to a machined thread?
43. Why are heated dies or tools often employed in hot-working processes?
44. What generally restricts the upper temperature to which dies or tooling is heated?
45. What is the primary cause of residual stresses in hot-worked products?
46. Compared to hot working, what are some of the advantages of cold-working processes?
47. What are some of the disadvantages of cold-forming processes?
48. How could cold working be used to reduce the cost of a moderate- to high-strength product?
49. Why are cold forming processes best suited for large-volume production of precision parts?
50. How can the tensile test properties of a metal be used to assess its suitability for cold forming?
51. Why is elastic springback an important consideration in cold-forming processes?
52. What are Luders bands or stretcher strains, and what causes them to form? How can they be eliminated?
53. What engineering properties are likely to decline during the cold working of a metal?
54. How can the selective placement of the final intermediate anneal be used to establish desired final properties in a cold-formed product?
55. Is the anisotropy induced by cold working an asset or a liability? What about the residual stresses?
56. What are some of the advantages of warm forming compared to cold forming? Compared to hot forming?
57. What material feature is considered to be the driving force for isothermal forming?
58. Why is isothermal forming considerably more expensive than conventional hot forming?
59. Why is there no standard test to assess or quantify formability?

■ PROBLEMS

1. Copper is being reduced from a hot-rolled $\frac{3}{8}$-in.-diameter rod to a final diameter of 0.100 in. by wire drawing through a series of dies. The final wire should have a yield strength in excess of 50,000 psi and an elongation greater than 10%. Use Figure 15-8 to determine a desirable amount of final cold work. Compute the placement of the last intermediate anneal so that the final product has both the desired size and the desired properties.
2. a. List and discuss the various economic factors that should be considered when evaluating a possible switch from cold forming to warm forming.
   b. Repeat part a for a possible conversion from hot forming to warm forming.
3. An advertisement for automobile spark plugs has cited the superiority of rolled threads over machined threads. Figure 15-5 shows such a comparison for hot forming, where the deformation process reorients flaws and defects without significantly changing the structure and properties of the metal. The spark plug threads, however, were formed by cold rolling. Do the same benefits apply? Discuss the assets and liabilities of the cold rolling of threads compared to thread formation by conventional machining.
4. Computer modeling of metal deformation processes is a powerful and extremely useful tool. At the same time, there are several areas of limitation that can significantly compromise or even invalidate the final results. Consider each of the following areas of limitation, investigating what is currently being used or what current options are available:
   a. *A mathematical description of material behavior (a constitutive equation).* In almost all cases, some simplification of actual flow behavior is assumed. For accurate modeling, flow behavior should be known and mathematically characterized as a function of strain, strain rate, and temperature.
   b. *Interfacial friction between the tooling and the workpiece.* How is this being modeled? Does it consider the effects of surface finish, sliding velocity, interface temperature, and numerous other factors? As the process commences, lubricants may thin or be wiped from surfaces, forces and pressures change, temperatures change, and surface roughness or texture is modified. Does the model reflect any of these changes? Some models assign a single value to friction over the entire contact surface. This value may also remain constant throughout the entire operation.

c. *Assignment of boundary conditions.* Often, the mathematical solutions must conform to assigned features, such as defined motions or stresses at specified surfaces. The boundary conditions have a profound effect on the results that are calculated. Poor choices or choices made so as to facilitate easy analysis can often produce misleading or erroneous results.

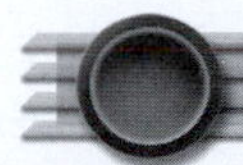

www.wiley.com/go/global/degarmo

Chapter 15 CASE STUDY

Interior Tub of a Top-Loading Washing Machine

The interior tub of a washing machine is the container that holds the clothes during the washing and rinsing cycles, but it also contains the perforations that permit removal of the water by draining and spinning. The component will see mechanical loadings from the weight of the clothes and water and also the dynamic action of spinning water-laden fabrics. There will be exposure to a wide range of water quality, as well as the full spectrum of soaps, detergents, bleaches, and other laundry additives. The surfaces should also be resistant to the impact and abrasion of buttons, zippers, and snaps.

This part has traditionally been manufactured by the sequential deep drawing, perforating and trimming of metal sheet, followed by some form of surface-coating treatment. For a long time, the standard material was "enameling iron"—a steel sheet with less than 0.03% carbon—which was then coated with a fired porcelain enamel. Due to the difficulties of producing ultra-low-carbon material in today's steelmaking operations, enameling iron became increasingly scarce, and manufacturers were forced to substitute the lowest-carbon, most readily available material—namely, 1008 steel. This substitution further required modification of the enameling process to prevent defects and blistering from CO evolution.

Your employer is presently manufacturing these tubs from 1008 steel sheet with a subsequent coating of fired porcelain enamel. Your marketing staff, however, reports that consumers tend to view a stainless steel tub to be of higher quality. As a result, your supervisor has asked you to evaluate the merits of converting to this material. You must first familiarize yourself with the current product (the base material, the forming process, and the porcelain enameling), and then determine what might be involved in converting to stainless steel. Consider the following specific questions:

1. What are the obvious pros and cons of the current product and process? Where would you expect most problems to occur in the current manufacturing process? Which aspects of fabrication are likely to be the most costly?
2. What would be the pros and cons of converting to stainless steel? In what ways would the product be superior? Are there any assets or liabilities associated with product fabrication from stainless material?
3. Which stainless steel would you recommend? Begin by considering the basic types (ferritic, austenitic, and martensitic), and then refine your selection to a specific alloy if possible. Discuss the rationale for your selection.
4. Because deep drawing is a metal-deformation process, we could use cold working (strain hardening) as a strengthening mechanism (some types of stainless steels strengthen considerably during deformation). Would you find this to be attractive, or would you prefer to use a recrystallization anneal after drawing and prior to use? Why? If you elect to use cold work, might you want to at least perform a stress-relief heat treatment prior to use? Could this be done and still preserve the deformation strengthening?
5. In deep drawing, the deformation is not uniform (increasing as we move up the sidewalls of the container), and the bottom of the tub simply retains the properties of the starting sheet. In order to ensure a minimum amount of strength at all locations, it may be desirable to begin the drawing with a partially cold-rolled sheet. Do you find this suggestion to be desirable? Why or why not?
6. After drawing and perforating, the residual drawing lubricant is removed from the part. Would any additional surface treatment be required? What would be your recommendation?

CHAPTER 16

BULK-FORMING PROCESSES

16.1 INTRODUCTION

The shaping of metal by deformation is as old as recorded history. The Bible, in the fourth chapter of Genesis, introduces Tubal-cain and cites his ability as a worker of metal. While we have no description of his equipment, it is well established that metal forging was practiced long before written records. Processes such as rolling and wire drawing were common in the Middle Ages and probably date back much further. In North America, by 1680 the Saugus Iron Works near Boston had an operating drop forge, rolling mill, and slitting mill.

Although the basic concepts of many forming processes have remained largely unchanged throughout history, the details and equipment have evolved considerably. Manual processes were converted to machine processes during the industrial revolution. The machinery then became bigger, faster, and more powerful. Water wheel power was replaced by steam and then by electricity. More recently, computer-controlled, automated operations have become the norm.

16.2 CLASSIFICATION OF DEFORMATION PROCESSES

A wide variety of processes have been developed to mechanically shape material, and a number of classification methods have been proposed. One approach divides the processes into primary and secondary. *Primary processes* reduce a cast material into intermediate shapes, such as slabs, plates, or billets. *Secondary processes* further convert these shapes into finished or semifinished products. Unfortunately, some processes clearly fit both categories, depending on the particular product being made.

In Chapter 15, we discussed the temperature of deformation and presented the various regimes based on the temperature of the workpiece. These included cold working, warm working, hot forming, and isothermal deformation. This classification has also become somewhat blurred, especially with the increased emphasis on energy conservation. Processes that were traditionally performed hot are now being performed cold, and cold-forming processes can often be enhanced by some degree of heating. Warm working has experienced considerable growth.

Chapters 16 and 17 utilize a division that focuses on the size and shape of the workpiece and how that size and shape is changed. **Bulk deformation processes** are those where the thicknesses or cross sections are reduced or shapes are significantly changed. Because the volume of the material remains constant, changes in one dimension require proportionate changes in others. Thus, the enveloping surface area changes significantly, usually increasing as the product lengthens or the shape becomes more complex. The bulk-forming operations can be performed in all of the temperature regimes. Common processes include rolling, forging; extrusion; cold forming; and wire, rod, and tube drawing.

In contrast, **sheet-forming operations** involve the deformation of a material where the thickness and surface area remain relatively constant. Common processes include shearing or blanking, bending, and deep drawing. Because of the large surface-to-volume ratio, sheet material tends to lose heat rapidly, and most sheet-forming operations are performed cold.

Even this division is not without confusion, however. Coining, for example, begins with sheet material but alters the thickness in a complex manner that is essentially bulk deformation. The bulk deformation processes will be presented here in Chapter 16. Sheet-forming processes can be found in Chapter 17.

16.3 BULK DEFORMATION PROCESSES

The bulk deformation processes that will be presented in this chapter include:

1. Rolling.
2. Forging.
3. Extrusion.
4. Wire, rod, and tube drawing.
5. Cold forming, cold forging, and impact extrusion.
6. Piercing.
7. Other squeezing processes.

These processes can be further divided in several ways. One grouping separates the processes by focusing on the size and shape of the deforming region. In continuous flow processes, such as forging, the size and shape are continually changing, and process analysis must reflect this change. In processes such as rolling or wire drawing, material moves through the deforming region, but the size and shape of that region remains unchanged. Some form of steady-state analysis can often be applied.

In all of the bulk forming processes, the primary deformation stress is compression. This may be applied directly by tools or dies that squeeze the workpiece or indirectly as in wire drawing, where the workpiece is pulled in tension but the resisting die generates compression in the region undergoing deformation.

16.4 ROLLING

Rolling operations reduce the thickness or change the cross section of a material through compressive forces exerted by rolls. As shown in Figure 16-1, rolling is often the first process that is used to convert material into a finished wrought product. Thick starting stock can be rolled into blooms, billets, or slabs, or these shapes can be obtained directly from continuous casting. A **bloom** has a square or rectangular cross section, with a thickness greater than 15 cm (6 in.) and a width no greater than twice the thickness. A **billet** is usually smaller than a bloom and has a square or circular cross section. Billets are usually produced by some form of deformation process, such as rolling or extrusion. A **slab** is a rectangular solid where the width is greater than twice the thickness. Slabs can be further rolled to produce **plate, sheet,** and **strip.** Plates have

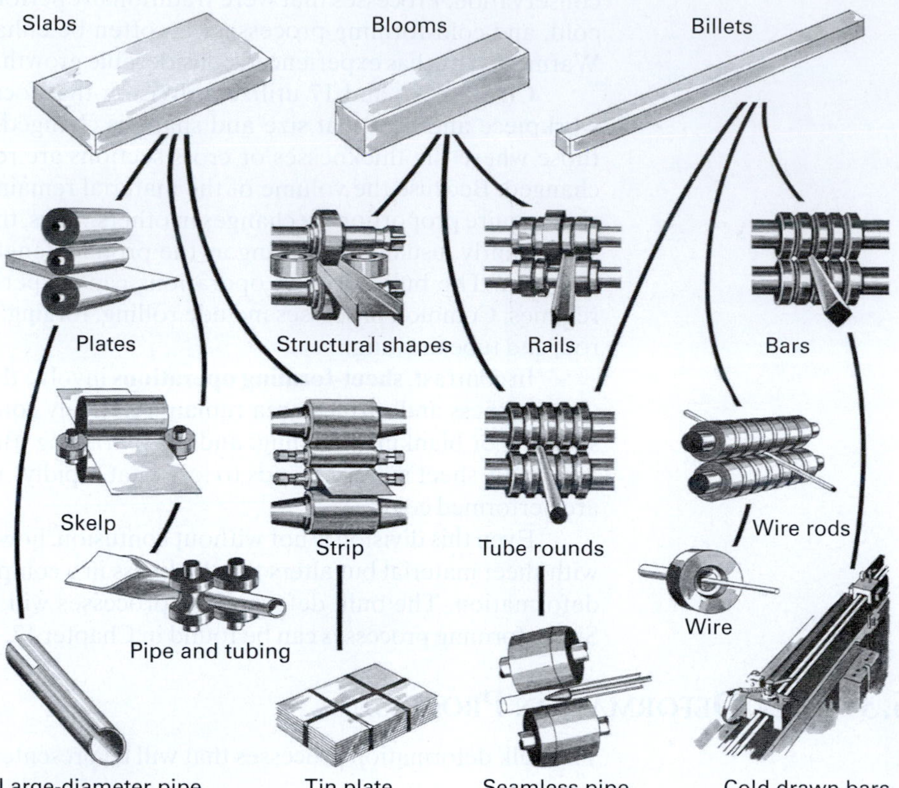

FIGURE 16-1 Flow chart for the production of various finished and semifinished steel shapes. Note the abundance of rolling operations. *(Courtesy of American Iron and Steel Institute, Washington, DC)*

thickness greater than 6 mm ($\frac{1}{4}$ in.), while sheet and strip ranges from 6 mm to 0.1 mm ($\frac{1}{4}$ in. to 0.004 in.).

These hot-rolled products often form the starting material for subsequent processes, such as cold forming or machining. Sheet and strip can be fabricated into products or further cold rolled into thinner, stronger material, or even into foil (thicknesses less than 0.1 mm). Blooms and billets can be further rolled into finished products, such as **structural shapes** or railroad rail, or they can be processed into semifinished shapes, such as **bar, rod, tube,** or **pipe.**

From a tonnage viewpoint, rolling is clearly predominant among all manufacturing processes, with approximately 90% of all metal products experiencing at least one rolling operation. Rolling equipment and rolling practices are sufficiently advanced that standardized, uniform-quality products can be produced at relatively low cost. Because shaped rolls are both massive and costly, shaped products are only available in standard forms and sizes where there is sufficient demand to permit economical production.

BASIC ROLLING PROCESS

In the basic rolling process, shown in Figure 16-2, metal is passed between two rolls that rotate in opposite directions, the gap between the rolls being somewhat less than the thickness of the entering metal. Because the rolls rotate with a surface velocity that exceeds the speed of the incoming metal, friction along the contact interface acts to propel the metal forward. The metal is then squeezed and elongates to compensate for the decrease in thickness or cross-sectional area. The amount of deformation that can be achieved in a single pass between a given pair of rolls depends on the friction conditions along the interface. If too much is demanded, the rolls cannot advance the material and simply skid over its surface. If too little deformation is taken, the operation will be successful, but the additional passes required to produce a given part will increase the cost of production.

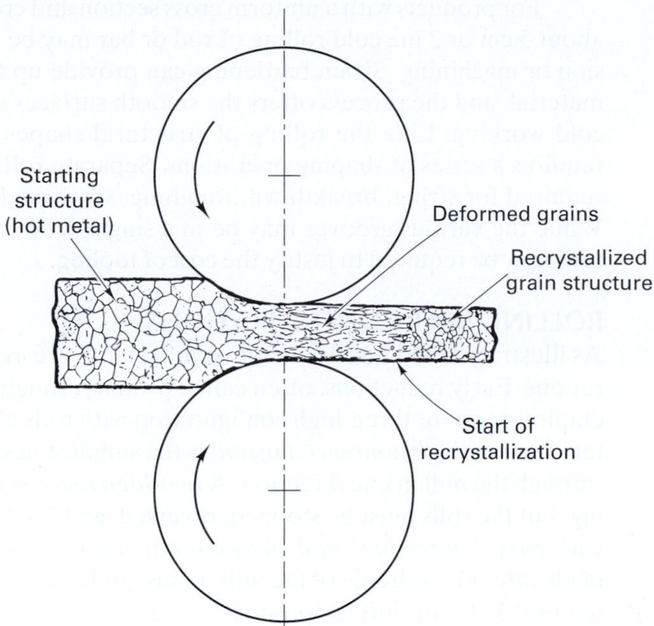

FIGURE 16-2 Schematic representation of the hot-rolling process, showing the deformation and recrystallization of the metal being rolled.

HOT ROLLING AND COLD ROLLING

In hot rolling, as with all hot-working processes, temperature control is required for success. The starting material should be heated to a uniform elevated temperature. If the temperature is not uniform, the subsequent deformation will not be uniform. Consider a piece being reheated for rolling. If the soaking time is insufficient, the hotter exterior will flow in preference to the cooler, stronger interior. Conversely, if a uniform-temperature material is allowed to cool prior to working or has cooled during previous working operations, the cooler surfaces will tend to resist deformation. Cracking and tearing of the surface may result as the hotter, weaker interior tries to deform.

It is not uncommon for high-volume producers to begin with continuous-cast feedstock. The cooling from solidification is controlled so as to enable direct insertion into a hot-rolling operation without additional handling or reheating. For smaller operations or secondary processing, the starting material is often a room-temperature solid, such as an ingot, slab, or bloom. This material must first be brought to the desired rolling temperature, usually in gas- or oil-fired soaking pits or furnaces. For plain-carbon and low-alloy steels, the soaking temperature is usually about 1200°C (2200°F). For smaller cross sections, induction coils may be used to heat the material prior to rolling.

Hot-rolling operations are usually terminated when the temperature falls to about 50 to 100°C (100 to 200°F) above the recrystallization temperature of the material being rolled. Such a **finishing temperature** ensures the production of a uniform fine grain size and prevents the possibility of unwanted strain hardening. If additional deformation is required, a period of reheating will be necessary to reestablish desirable hot-working conditions.

Cold rolling can be used to produce sheet, strip, bar, and rod products with extremely smooth surfaces and accurate dimensions. Cold-rolled sheet and strip can be obtained in various conditions, including *skin-rolled*, *quarter-hard*, *half-hard*, and *full-hard*. Skin-rolled metal is subjected to only a 0.5 to 1% reduction to produce a smooth surface and uniform thickness and to remove or reduce the yield-point phenomenon (i.e., prevent formation of Luders bands upon further forming). This material is well suited for subsequent cold-working operations where good ductility is required. Quarter-hard, half-hard, and full-hard sheet and strip experience greater amounts of cold reduction, up to 50%. Their yield points are higher, properties have become directional, and ductility has decreased. Quarter-hard steel can be bent back on itself across the grain without breaking. Half-hard and full-hard can be bent back 90 and 45 degrees, respectively, about a radius equal to the material thickness.

For products with a uniform cross section and cross-sectional dimensions less than about 5 cm or 2 in., cold rolling of rod or bar may be an attractive alternative to extrusion or machining. Strain hardening can provide up to 20% additional strength to the material, and the process offers the smooth surfaces and high dimensional precision of cold working. Like the rolling of structural shapes, however, the process generally requires a series of shaping operations. Separate roll passes (and roll grooves) may be required for sizing, breakdown, roughing, semiroughing, semifinishing, and finishing. While the various grooves may be in a single set of rolls, a minimum order of several tons may be required to justify the cost of tooling.

ROLLING MILL CONFIGURATIONS

As illustrated in Figure 16-3, rolling mill stands are available in a variety of roll configurations. Early reductions, often called primary, roughing, or breakdown passes, usually employ a two- or three-high configuration with rolls 60 to 140 cm (24 to 55 in.) in diameter. The *two-high nonreversing mill* is the simplest design, but the material can only pass through the mill in one direction. A *two-high reversing mill* permits back-and-forth rolling, but the rolls must be stopped, reversed, and brought back to rolling speed between each pass. A *three-high mill* eliminates the need for roll reversal but requires some form of elevator on each side of the mill to raise or lower the material and mechanical manipulators to turn or shift the product between passes.

As shown in Figure 16-4, smaller-diameter rolls produce less length of contact for a given reduction and therefore require lower force and less energy to produce a given change in shape. The smaller cross section, however, provides reduced stiffness, and the rolls are prone to flex elastically because they are supported on the ends and pressed apart by the metal passing through the middle (a condition known as three-point bending). *Four-high* and *cluster* arrangements use backup rolls to support the smaller work rolls. These configurations are used in the hot rolling of wide plate and sheets, and in cold rolling, where even small deflections in the roll would result in an unacceptable variation in product thickness. Foil is almost always rolled on **cluster mills** because the small thickness requires small-diameter rolls. In a cluster mill, the roll in contact with the work can be as small as 6 mm ($\frac{1}{4}$ in.) in diameter. To counter the need for even smaller rolls, some foils are produced by **pack rolling,** a process where two or more layers of metal are rolled simultaneously as a means of providing a thicker input material. Household aluminum foil is usually rolled as a double sheet, as evidenced by the one shiny side (in contact with the roll) and one dull side (in contact with the other piece of foil).

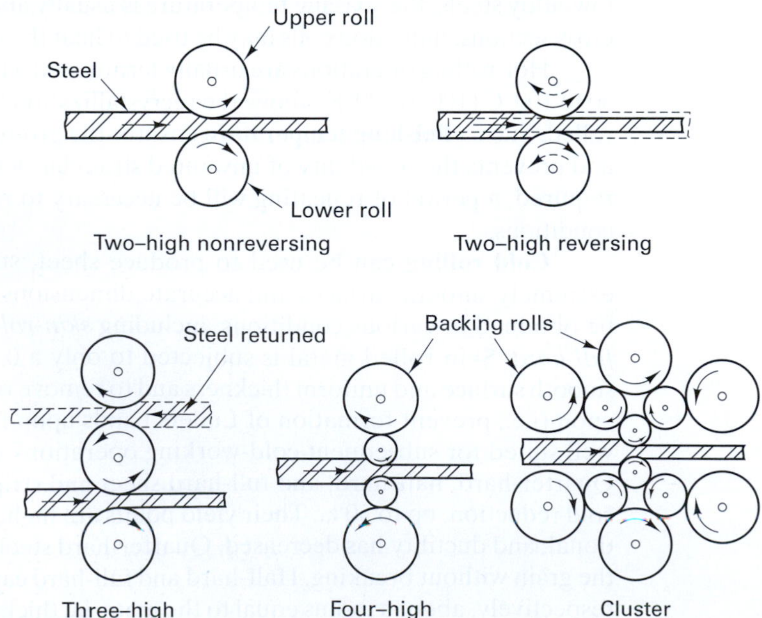

FIGURE 16-3 Various roll configurations used in rolling operations.

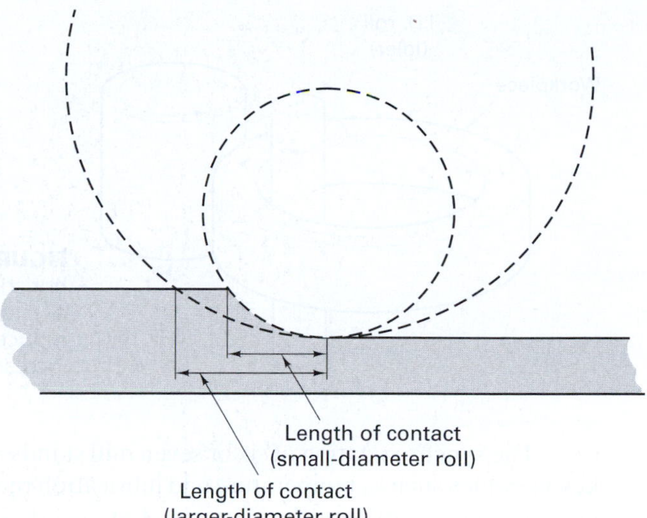

FIGURE 16-4 The effect of roll diameter on the length of contact for a given reduction.

In the rolling of nonflat or shaped products, such as structural shapes and railroad rail, the sets of rolls contain contoured grooves that sequentially form the desired shape, reduce the cross-sectional area, and control the metal flow. Figure 16-5 shows some typical roll-pass sequences used in the production of structural shapes. Length increases as the cross-section is reduced.

CONTINUOUS (OR TANDEM) ROLLING MILLS

When the volume of a product justifies the investment, rolling may be performed on a *continuous* or *tandem rolling mill*. Billets, blooms, or slabs are heated and fed through an integrated series of nonreversing rolling mill stands. Continuous mills for the hot rolling of steel strip, for example, often consist of a roughing train of approximately four four-high mill stands and a finishing train of six or seven additional four-high stands. In a continuous structural mill, the rolls in each stand contain only one set of shaped grooves, in contrast to the multigrooved rolls used when the product is produced by back-and-forth passes through a single stand.

If a single piece of material is in multiple rolling stations at the same time, it is imperative that the same volume pass through each stand in the same amount of time. If the cross section is reduced, speed must be increased proportionately. Therefore, as a material is reduced in size, the rolls of each successive stand must turn faster than those of the preceding one. If a subsequent stand is running too slowly, material will accumulate between stands. If the demand for incoming material exceeds the output of the previous stand, the material is placed in tension and may tear or rupture.

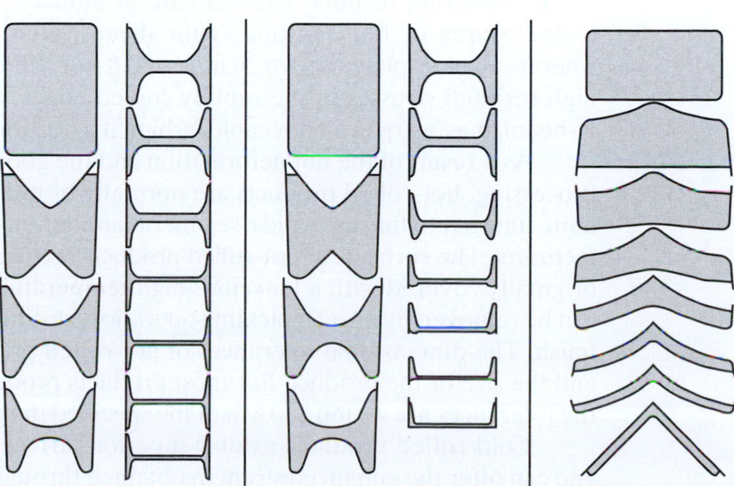

FIGURE 16-5 Typical roll-pass sequences used in producing structural shapes.

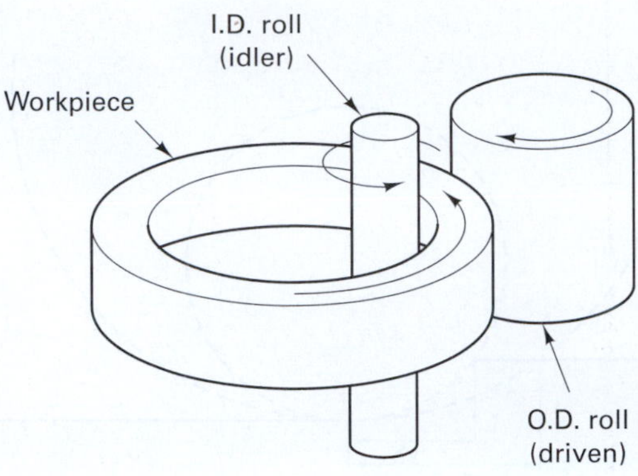

FIGURE 16-6 Schematic of a horizontal ring rolling operation. As the thickness of the ring is reduced, its diameter will increase.

The synchronization of six or seven mill stands is not an easy task, especially when key variables such as temperature and lubrication may vary during a single run, and the product may be exiting the final stand at speeds in excess of 110 km/hr (70 mph). Computer control is basic to successful rolling, and modern mills are equipped with numerous sensors to provide the needed information. When continuous casting units feed directly into continuous rolling mills, the time lapse from final solidification to finished rolled product is often a matter of a few minutes.

RING ROLLING

Ring rolling is a special rolling process where one roll is placed through the hole of a thick-walled ring, and a second roll presses in from the outside (Figure 16-6). As the rolls squeeze and rotate, the wall thickness is reduced and the diameter of the ring increases. Shaped rolls can be used to produce a wide variety of cross-section profiles. The resulting seamless rings have a circumferential grain orientation and find application in products such as rockets, turbines, airplanes, pipelines, and pressure vessels. Diameters can be as large as 8 m (25 ft) with face heights as great as 2 m (80 in.).

THREAD ROLLING

Thread rolling is a deformation alternative to the cutting of threads, and is illustrated in Figure 15-5.

CHARACTERISTICS, QUALITY, AND PRECISION OF ROLLED PRODUCTS

Because hot-rolled products are formed and finished above their recrystallization temperature, they have little directionality in their properties and are relatively free of deformation-induced residual stresses. These characteristics may vary, however, depending on the thickness of the product and the presence of complex sections. Nonmetallic inclusions do not recrystallize, so they may impart some degree of directionality. In addition, residual stresses can be induced by nonuniform cooling from the temperatures of hot working. Thin sheets often show directional characteristics, whereas thicker plate (above 20 mm, or 0.8 in.) will usually have very little. Because of high residual stresses in the rapidly cooled edges, a complex shape, such as an I- or H-beam, may warp in a noticeable fashion if a portion of one flange is cut away.

As a result of the hot deformation and the good control that is maintained during processing, hot-rolled products are normally of uniform and dependable quality. It is quite unusual to find any voids, seams, or laminations when produced by reliable manufacturers. The surfaces of hot-rolled products are usually a bit rough, however, and are originally covered with a tenacious high-temperature oxide, known as **mill scale.** This can be removed by an acid pickling operation, resulting in a surprisingly smooth surface finish. The dimensional tolerances of hot-rolled products vary with the kind of metal and the size of the product. For most products produced in reasonably large tonnages, the tolerances are within 2 to 5% of the specified dimension (either height or width).

Cold-rolled products exhibit superior surface finish and dimensional precision and can offer the enhanced strength obtained through strain hardening.

FLATNESS CONTROL AND ROLLING DEFECTS

If rolling a flat product with uniform thickness, the gap between the rolls must be a uniform one. Attaining such an objective, however, may be difficult. Consider the upper roll in a set that is rolling sheet or plate. As shown in Figure 16-7, the material presses upward in the middle of the roll, while the roll is held in place by bearings that are mounted on either end and are supported in the mill frame. The roll, therefore, is loaded in three-point bending and tends to flex in a manner that produces a thicker center and thinner edge. Because the thicker center will not lengthen as much as the thinner edge, the result is often a product with either wavy edge or fractured center.

If the rolls are always used to reduce the same material at the same temperature by the same amount, the forces and deflections can be predicted, and the roll can be designed to have a specified profile. If a **crowned,** or barrel-shaped, **roll** is subjected to the designed load, it will deflect into flatness, as illustrated in Figure 16-8. If the applied load is not of the designed magnitude, however, the resulting profile will not be flat and defects may result. If the correction is insufficient, for example, wavy edges or center fractures may still occur. If the correction is excessive, the center becomes thinner and longer, and the result may be a wavy center or cracking of the edges.

Because roll deflections are proportional to the forces applied to the rolls, product flatness can also be improved by measures that reduce these forces. If possible, friction could be reduced, smaller-diameter rolls could be used, and smaller reductions could be employed. Heating the workpiece generally makes it weaker, so increased workpiece temperature will also reduce the force on the rolls. Horizontal tensions can be applied to the piece as it is being rolled (strip tension in sheet metal rolling). Because these tensions combine with the vertical compression to deform the piece (stretching while squeezing), the roll forces and associated deflections are less. Other techniques to improve flatness include an increase in the elastic modulus of the rolls themselves through material selection or to provide some form of backup support to oppose deflection, as with the four-high and cluster mill configurations.

Successful rolling requires the balancing of many factors relating to the material being rolled, the variables of the rolling process, and lubrication between the workpiece and the rolls. Common defects include the nonuniform thickness previously discussed, dimensional variations caused by changes in workpiece temperature, surface flaws (such as rolled-in scale and roll marks), laps, seams, and various types of distortions.

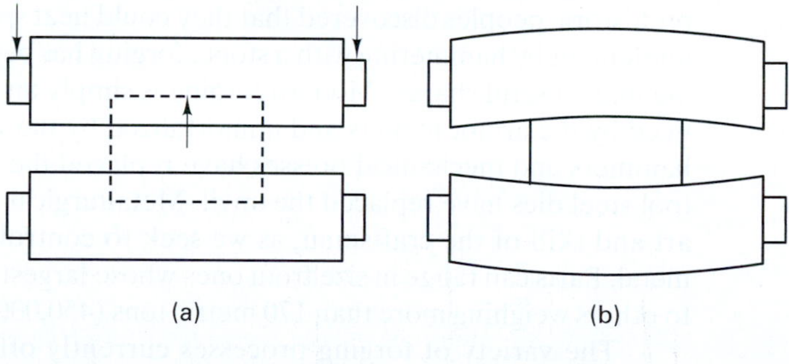

FIGURE 16-7 (a) Loading on a rolling mill roll. The top roll is pressed upward in the center while being supported on the ends. (b) The elastic response to the three-point bending.

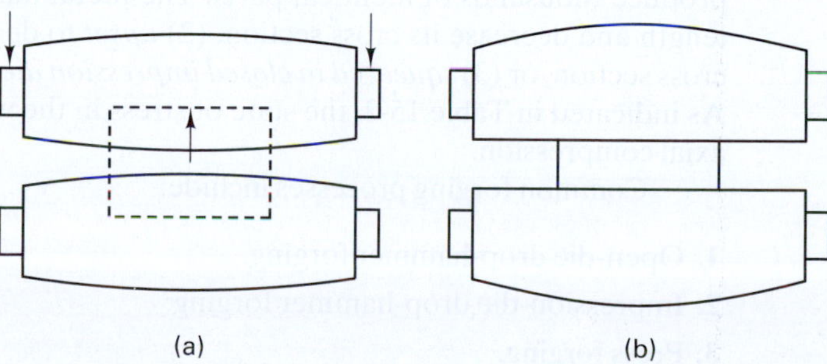

FIGURE 16-8 Use of a "crowned" roll to compensate for roll flexure. When the roll flexes in three-point bending, the crowned roll flexes into flatness.

THERMOMECHANICAL PROCESSING AND CONTROLLED ROLLING

As with most deformation processes, rolling is generally considered to be a way of changing the shape of a material. While heat may be used to reduce forces and promote plasticity, the thermal processes that produce or control product properties (heat treatments) are usually performed as subsequent operations. **Thermomechanical processing,** of which **controlled rolling** is an example, consists of integrating deformation and thermal processing into a single process that will produce not only the desired shape, but also the desired properties, such as strength and toughness. The heat for the property modification is the same heat used in the rolling operation, and subsequent heat treatment becomes unnecessary.

A successful thermomechanical operation begins with process design. The starting material must be specified and the composition closely maintained. Then a time–temperature–deformation system must be developed to achieve the desired objective. Possible goals include production of a uniform fine grain size; controlling the nature, size, and distribution of the various transformation products (such as ferrite, pearlite, bainite, and martensite in steels); controlling the reactions that produce solid solution strengthening or precipitation hardening; and producing a desired level of toughness. Starting structure (controlled by composition and prior thermal treatments), deformation details, temperature during the various stages of deformation, and the conditions of cooldown from the working temperature must all be specified and controlled. Moreover, the attainment of uniform properties requires uniform temperatures and deformations throughout the product. Computer-controlled facilities are an absolute necessity if thermomechanical processing is to be successfully performed.

Possible benefits of thermomechanical processing include improved product properties; substantial energy savings (by eliminating subsequent heat treatment); and the possible substitution of a cheaper, less-alloyed metal for a highly alloyed one that responds to heat treatment.

16.5 FORGING

Forging is a term applied to a family of processes that induce plastic deformation through localized compressive forces applied through dies. The equipment can take the form of hammers, presses, or special forging machines. While the deformation can be performed in all temperature regimes (hot, cold, warm, or isothermal), most forging is done with workpieces above the recrystallization temperature.

Forging is clearly the oldest known metalworking process. From the days when prehistoric peoples discovered that they could heat sponge iron and beat it into a useful implement by hammering with a stone, forging has been an effective method of producing many useful shapes. Modern forging is simply an extension of the ancient art practiced by the armor makers and immortalized by the village blacksmith. High-powered hammers and mechanical presses have replaced the strong arm and the hammer, and tool steel dies have replaced the anvil. Metallurgical knowledge has supplemented the art and skill of the craftsman, as we seek to control the heating and handling of the metal. Parts can range in size from ones whose largest dimension is less than 2 cm (1 in.) to others weighing more than 170 metric tons (450,000 lb).

The variety of forging processes currently offers a wide range of capabilities. A single piece can be economically fashioned by some methods, while others can mass-produce thousands of identical parts. The metal may be (1) *drawn out* to increase its length and decrease its cross section, (2) *upset* to decrease the length and increase the cross section, or (3) *squeezed in closed impression dies* to produce multidirectional flow. As indicated in Table 15-2, the state of stress in the work is primarily uniaxial or multiaxial compression.

Common forging processes include:

1. Open-die drop-hammer forging.
2. Impression-die drop-hammer forging.
3. Press forging.

4. Orbital forging.
5. Upset forging.
6. Automatic hot forging.
7. Roll forging.
8. Swaging.
9. Net-shape and near-net-shape.

OPEN-DIE HAMMER FORGING

In concept, **open-die hammer forging** is the same type of forging done by the blacksmith of old, but massive mechanical equipment is now used to impart the repeated blows. The metal is first heated to the proper temperature using a furnace or electrical induction heating. An impact is then delivered by some type of mechanical hammer. The simplest industrial hammer is a **gravity drop hammer,** where a free-falling ram strikes the workpiece and the energy of the blow is varied by adjusting the height of the drop. Most forging hammers now employ some form of energy augmentation, however, where pressurized air, steam, or hydraulic fluids are used to raise and propel the hammer. Higher striking velocities are achieved, with more control of striking force, easier automation, and the ability to shape pieces up to several tons. **Computer-controlled hammers** can provide blows of differing impact speed (energy) for different products or for each of the various stages of a given operation. Their use can greatly increase the efficiency of the process and also minimize the amount of noise and vibration, which are the most common outlets for the excess energy not absorbed in the deformation of the workpiece. Figure 16-9 shows a large double-frame hammer along with a labeled schematic of a hammer operation.

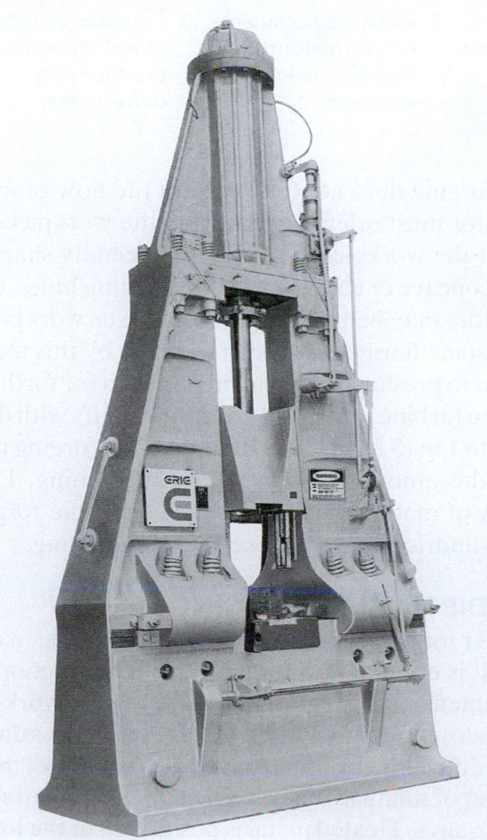

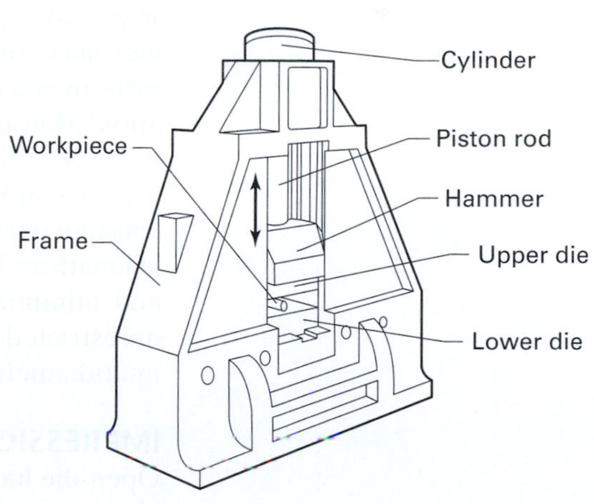

FIGURE 16-9 (Left) Double-frame drop hammer. (Right) Schematic diagram of a forging hammer. *(Courtesy of Erie Press Systems, Erie, PA)*

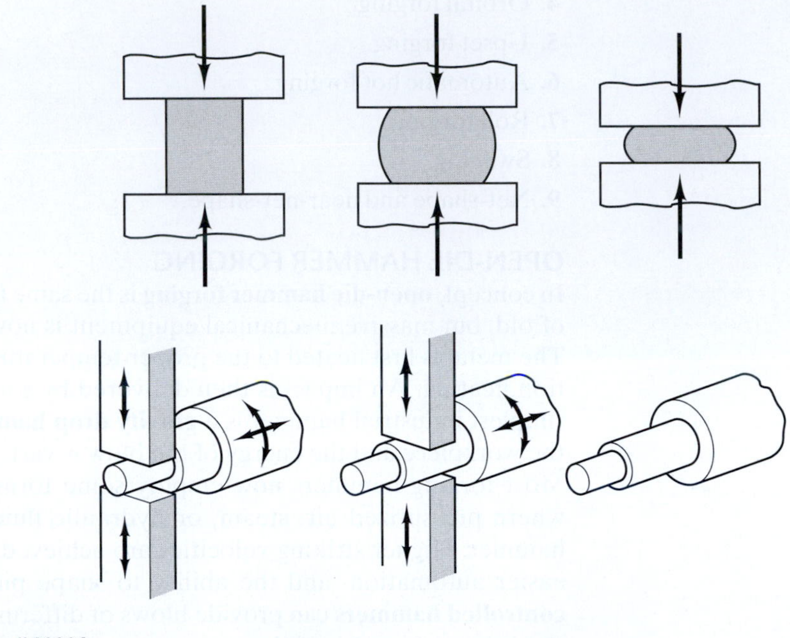

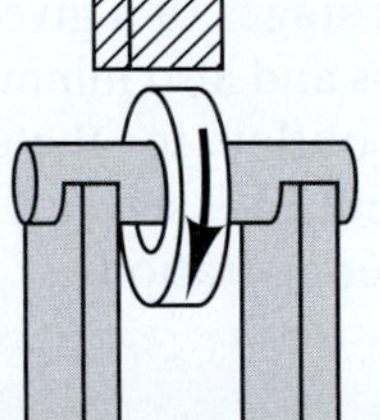

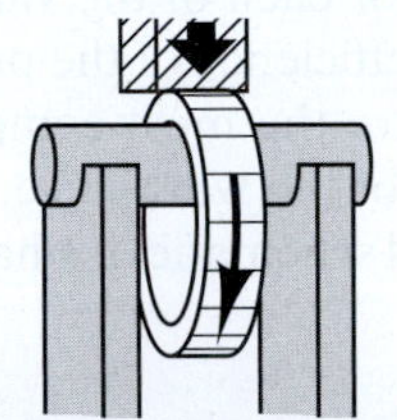

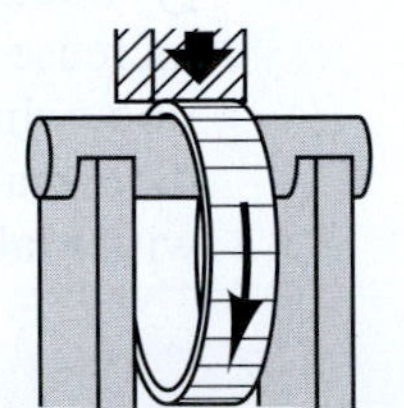

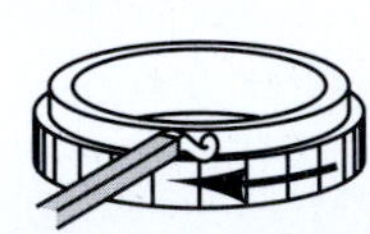
1 Preform mounted on saddle/mandrel.
2 Metal displacement– reduce preform wall thickness to increase diameter.
3 Progressive reduction of wall thickness to produce ring dimensions.
4 Machining to near net shape.

FIGURE 16-10 (Top) Illustration of the unrestrained flow of material in open-die forging. Note the barrel shape that forms due to friction between the die and material. (Middle) Open-die forging of a multidiameter shaft. (Bottom) Forging of a seamless ring by the open-die method. *(Courtesy of Forging Industry Association, Cleveland, OH)*

Open-die forging does not fully control the flow of metal. To obtain the desired shape, the operator must orient and position the workpiece between blows. The hammer may contact the workpiece directly, or specially shaped tools can be inserted to assist in making concave or convex surfaces, forming holes, or performing a cutoff operation. Manipulators may be used to position larger workpieces, which may weigh several tons. While some finished parts can be made by this technique, open-die forging is usually employed to preshape metal in preparation for further operations. For example, consider parts like turbine rotors and generator shafts with dimensions up to 20 m (70 ft) in length and up to 1 m (3 ft) in diameter. Open-die forging induces oriented plastic flow and minimizes the amount of subsequent machining. Figure 16-10 illustrates the unrestricted flow of material and shows how open-die forging can be used to shape a multidiameter cylindrical shaft and a seamless metal ring.

IMPRESSION-DIE HAMMER FORGING

Open-die hammer forging (or *smith forging,* as it has been called) is a simple and flexible process, but it is not practical for large-scale production. It is a slow operation, and the shape and dimensional precision of the resulting workpiece are dependent on the skill of the operator. As shown in Figure 16-11, **impression-die** or **closed-die forging** overcomes these difficulties by using shaped dies to control the flow of metal. Figure 16-12 shows a typical set of multicavity dies. The upper piece attaches to the hammer and the lower piece to the anvil. Heated metal is positioned in the lower cavity and struck one or more blows by the upper die. The hammering causes the metal to flow and completely fill the die cavity. Excess metal is squeezed out along the parting line to form a **flash** around the periphery of the cavity. This material cools rapidly; increases in strength;

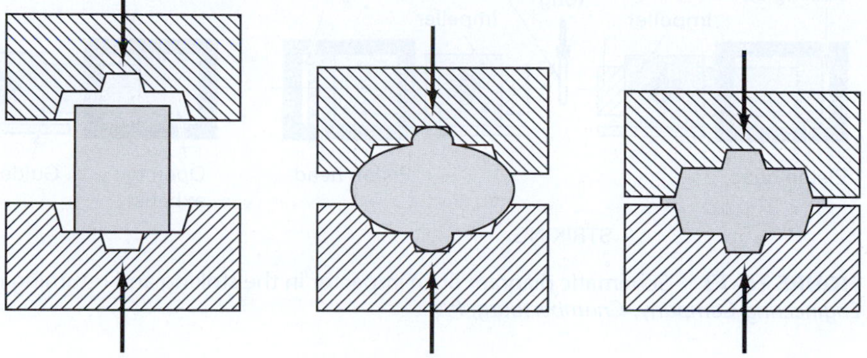

FIGURE 16-11 Schematic of the impression-die forging process showing partial die filling and the beginning of flash formation in the center sketch, and the final shape with flash in the right-hand sketch.

and, by resisting deformation, effectively blocks the formation of additional flash. By trapping material within the die, the flash then ensures the filling of all of the cavity details. The flash is ultimately trimmed from the part in a final forging operation.

In **flashless forging,** also known as *true closed-die forging,* the metal is deformed in a cavity that provides total confinement. Accurate workpiece sizing is required because complete filling of the cavity must be ensured with no excess material. Accurate workpiece positioning is also necessary, along with good die design and control of lubrication. The major advantage of this approach is the elimination of the scrap generated during flash formation, an amount that is often between 20 and 45% of the starting material.

Most conventional forgings are impression-die with flash. They are produced in dies with a series of cavities where one or more blows of the hammer are used for each step in the sequence. The first impression is often an *edging*, *fullering*, or *bending*

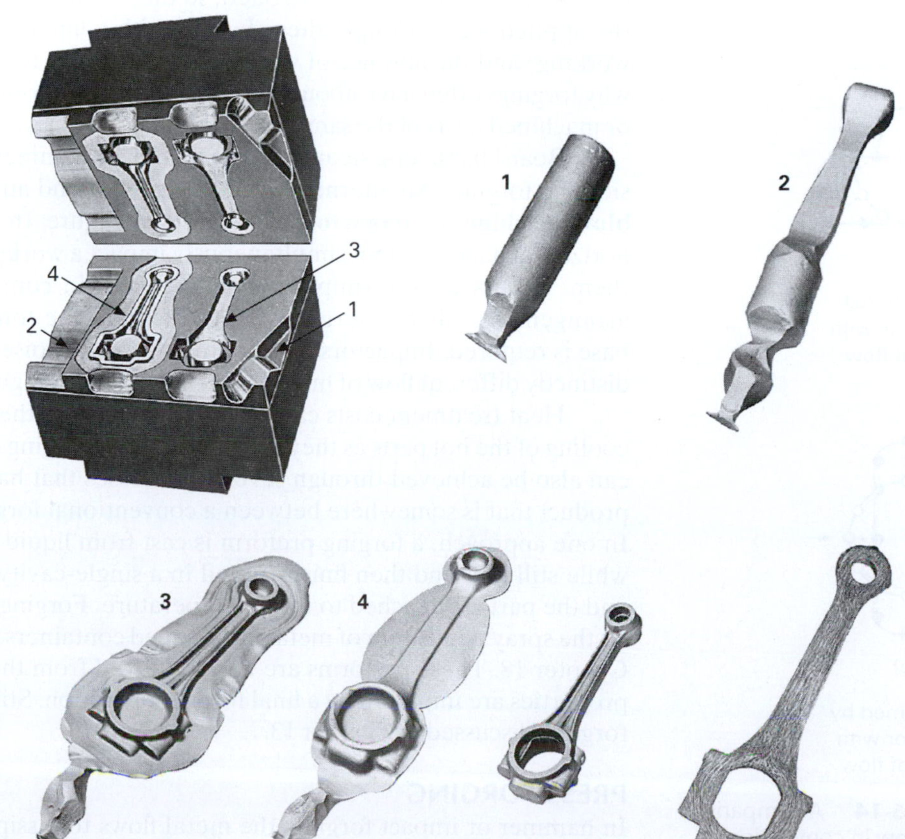

FIGURE 16-12 Impression drop-forging dies and the product resulting from each impression. The flash is trimmed from the finished connecting rod in a separate trimming die. The sectional view shows the grain flow resulting from the forging process. *(Courtesy of the Forging Industry Association, Cleveland, OH)*

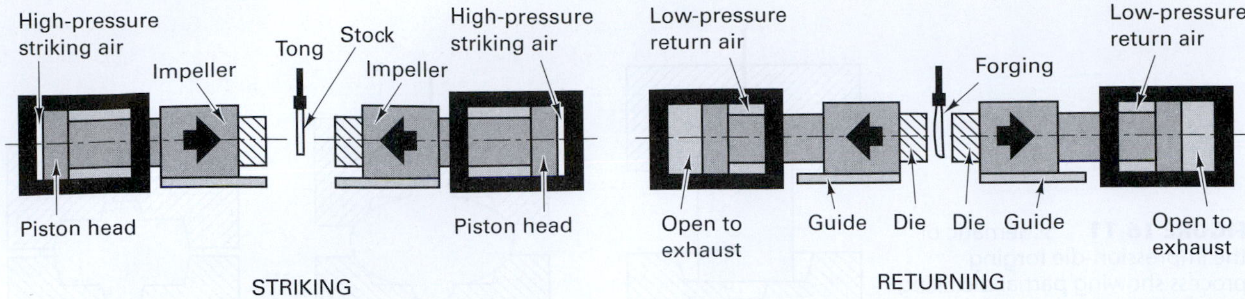

FIGURE 16-13 Schematic diagram of an impactor in the striking and returning modes. *(Courtesy of Chambersburg Engineering Company, Chambersburg, PA)*

impression to distribute the metal roughly in accordance with the requirements of the later cavities. Edging gathers material into a region, while fullering moves material away. Intermediate impressions are for **blocking** the metal to approximately its final shape, with generous corner and fillet radii. For small production lots, the cost of further cavities may not be justified, and the blocker-type forgings are simply finished by machining. More often, the final shape and size are imparted by an additional forging operation in a *final* or **finisher impression,** after which the flash is trimmed from the part. Figure 16-12 shows an example of these steps and the shape of the part at the conclusion of each. Because every part is shaped in the same die cavities, each mass-produced part is a close duplicate of all the others.

Conventional closed-die forging begins with a simple hot-rolled shape and utilizes reheating and working to progressively convert it into a more complex geometry. The shape of the various cavities controls the flow of material, and the flow, in turn, imparts the oriented structure discussed in Chapter 15. Grain flow that follows the external contour of the component is in the crack-arrestor orientation, improving strength, ductility, and resistance to impact and fatigue. Through forging, the size and shape of various cross sections can also be controlled, so the metal can be distributed as needed to resist the applied loads. Couple these factors with a fine recrystallized grain structure (hot working) and the absence of voids (compressive forming stresses), and it is easy to see why forgings often have about 20% higher strength-to-weight ratios compared with cast or machined parts of the same material.

Board hammers, steam hammers, and air hammers have all been used in impression die forging. An alternative to the hammer and anvil arrangement is the **counterblow machine,** or **impactor,** illustrated in Figure 16-13. These machines have two horizontal hammers that simultaneously impact a workpiece that is positioned between them. Excess energy simply becomes recoil, in contrast to the hammer and anvil arrangement, where energy is lost to the machine foundation, and a heavy machine base is required. Impactors also operate with less noise and less vibration and produce distinctly different flow of material, as illustrated in Figure 16-14.

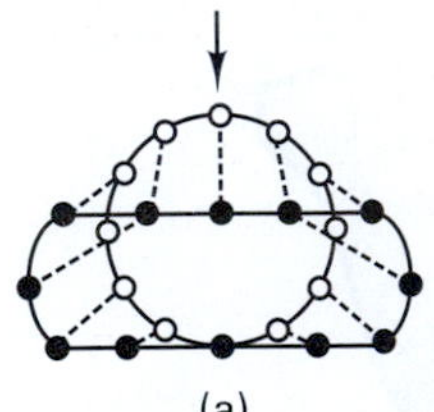

(a)

Conventional forged disk with paths of flow

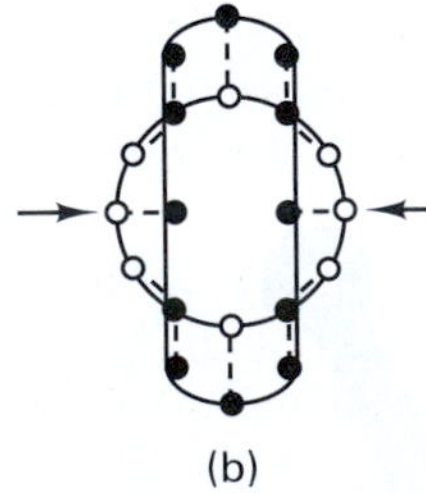

(b)

Disk formed by impactor with paths of flow

FIGURE 16-14 A comparison of metal flow in conventional forging and impacting.

Heat treatment costs can often be reduced by the direct quenching or controlled cooling of the hot parts as they emerge from the forging operation. Energy conservation can also be achieved through several processes that have been designed to produce a product that is somewhere between a conventional forging and a conventional casting. In one approach, a forging preform is cast from liquid metal, removed from the mold while still hot, and then finish-forged in a single-cavity die. The flash is then trimmed and the part is quenched to room temperature. Forging preforms can also be produced by the spray deposition of metal into shaped containers, as described in Section 18-11 in Chapter 18. These preforms are then removed from the mold, and the final shape and properties are imparted by a final forging operation. Still another approach is semisolid forging, discussed in Chapter 13.

PRESS FORGING

In hammer or impact forging, the metal flows to dissipate the energy imparted in the hammer–workpiece collision. Speeds are high, so the forming time is short. Contact times under load are on the order of milliseconds. There is little time for heat transfer and cooling of the workpiece, and the adiabatic heating that occurs during deformation helps to minimize chilling. It is possible, however, that all of the energy can be dissipated by deformation of just the surface of the metal (coupled with additional absorption by the anvil and foundation), and the interior of the workpiece remains essentially undeformed. Consider the deformation of a metal wood-splitting wedge after it has been struck repeatedly by a sledge hammer. The top is usually "mushroomed," while the remainder retains the original geometry and taper.

If large pieces or thick products are to be formed, **press forging** may be required. The deformation is now analyzed in terms of forces or pressures (rather than energy), and the slower squeezing action penetrates completely through the metal, producing a more uniform deformation and flow. New problems can arise, however, because of the longer time of contact between the dies and the workpiece. As the surface of the workpiece cools, it becomes stronger and less ductile, and it may crack if deformation is continued. Heated dies are generally used to reduce heat loss, promote surface flow, and enable the production of finer details and closer tolerances. Periodic reheating of the workpiece may also be required. If the dies are heated to the same temperature as the workpiece, and pressing proceeds at a slow rate, **isothermal forging** can be used to produce near-net-shape components with uniform microstructure and mechanical properties.

Forging presses are of two basic types, mechanical and hydraulic, and are usually quite massive. **Mechanical presses** use cams, cranks, or toggles to produce a preset and reproducible stroke. Because of their mechanical drives, different forces are available at the various stroke positions. Production presses are quite fast, capable of up to 50 strokes per minute, and are available in capacities ranging from 300 to 18,000 tons (3 to 160 MN). **Hydraulic presses** move in response to fluid pressure in a piston and are generally slower, more massive, and more costly to operate. On the positive side, hydraulic presses are much more flexible and can have greater capacity. Because motion is in response to the flow of pressurized drive fluids, hydraulic presses can be programmed to have different strokes for different operations and even different speeds within a stroke. Presses can be used to perform all types of forging, including open-die and impression-die. Impression-die press forgings usually require less draft than hammer forgings and have higher dimensional accuracy. In addition, press forgings can often be completed in a single closing of the dies as opposed to the multiple blows of a hammer. Machines with capacities up to 50,000 tons (445 MN) are currently in operation in the United States.

A third type of press is the **screw press,** which in many ways acts like a hammer. A large flywheel stores a predetermined amount of energy. This energy is then transmitted to a vertical screw, which drives a descending ram. Downward motion stops when all of the energy from the flywheel has been dissipated.

Additional information about the various types of presses and drive mechanisms can be found in the closing section of Chapter 17.

DESIGN OF IMPRESSION-DIE FORGINGS AND ASSOCIATED TOOLING

The geometrical possibilities for impression-die forging are quite numerous, with complex shapes like connecting rods, crankshafts, wrenches and gears being commonly produced. Figure 16-15 shows a forged-and-machined steel automotive crankshaft that significantly outperformed similar components made of austempered ductile cast iron. Parts typically range from under 1.4 kg (3 lb) up to about 340 kg (750 lb), with the major dimension being between 7 and 20 in. (20 and 50 cm). Steels; stainless steels; and alloys of aluminum, copper, and nickel can all be forged with fair to excellent results.

The forging dies are usually made of high-alloy or tool steel and can be expensive to design and construct. Impact resistance, wear resistance, strength at elevated temperature, and the ability to withstand cycles of rapid heating and cooling must all be outstanding. In addition, considerable care is required to produce and maintain a smooth and accurate cavity and parting plane. Better and more economical results will be obtained if the following rules are observed:

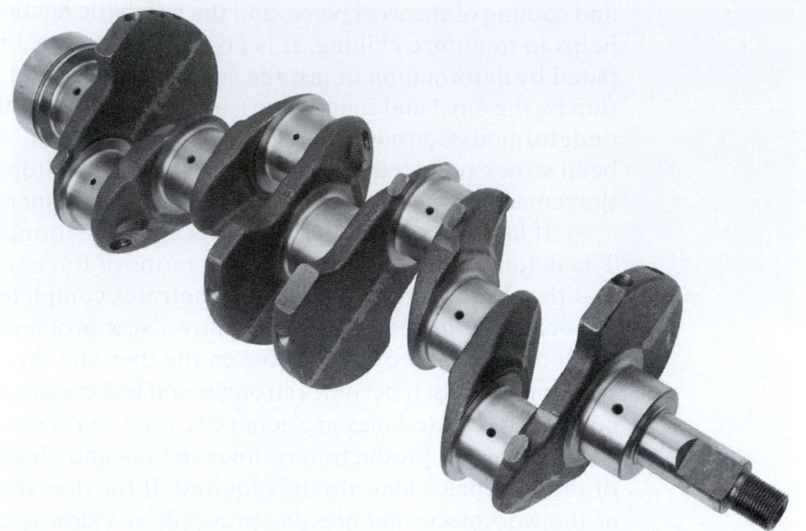

FIGURE 16-15 A forged and machined automobile engine crankshaft. Forged steel crankshafts provide superior performance, compared to those of ductile cast iron. *(© Sergiy Goruppa/iStockphoto)*

1. The dies should part along a single, flat plane if at all possible. If not, the parting plane should follow the contour of the part.
2. The parting surface should be a plane through the center of the forging and not near an upper or lower edge.
3. Adequate draft should be provided—at least 3 degrees for aluminum and 5 to 7 degrees for steel.
4. Generous fillets and radii should be provided.
5. Ribs should be low and wide.
6. The various sections should be balanced to avoid extreme differences in metal flow.
7. Full advantage should be taken of fiber flow lines.
8. Dimensional tolerances should not be closer than necessary.

The various design details—such as the number of intermediate steps, the shape of each, the amount of excess metal required to ensure die filling, and the dimensions of the flash at each ste— are often a matter of experience. Each component is a new design entity and brings its own unique challenges. Computer-aided design has made notable advances, however, and the development and accessibility of high-speed, immense-memory computers have enabled the accurate modeling of many complex shapes.

Good dimensional accuracy is a characteristic of impression-die forging. With reasonable care, the dimensions for steel products can be maintained within the tolerances of 0.50 to 0.75 mm (0.02 to 0.03 in.). It should be noted, however, that the dimensions across the parting plane are affected by closure of the dies and are, therefore, dependent on die wear and the thickness of the final flash. Dimensions contained entirely within a single die segment can be maintained at a significantly greater level of accuracy. Surface finish values range from 80 to 300 μin. Draft angles can sometimes be reduced, occasionally approaching zero, but this is not recommended for general practice.

Selection of a lubricant is also critical to successful forging. The lubricant not only affects the friction and wear and associated metal flow, but it may also be expected to act as a thermal barrier (restricting heat flow from the workpiece to the dies) and a parting compound (preventing the part from sticking in the cavities).

ORBITAL FORGING

Uniform-radius disk-shaped or conical products (such as gear blanks or bevel gears) can be formed by a process known as **orbital forging,** depicted in Figure 16-16. A conical-faced upper die is tilted or inclined so a portion of the cone is in contact with the

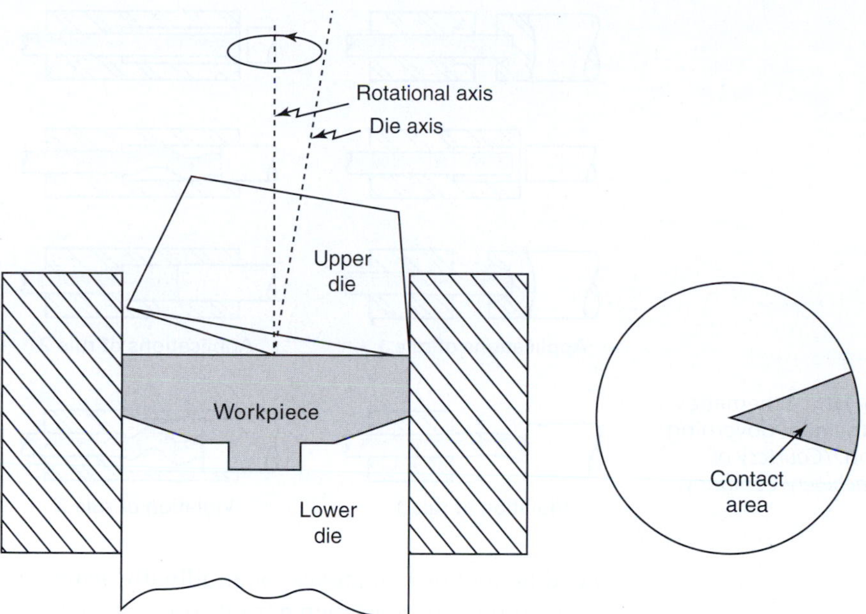

FIGURE 16-16 Schematic depiction of orbital forging.

workpiece. The upper die then orbits about a central axis, sweeping the contact area around the workpiece circumference. As the upper die is orbiting, it then descends or the lower die is raised to produce a continuous deformation that is similar to a combined rolling and forging—kneading the material into the lower die, much like a rolling pin moving across a lump of dough. Parts are generally formed with 10 to 20 orbits of the upper die.

The forming force is reduced due to the small area of contact between the tool and the workpiece and is further reduced due to the improved lubrication that results from the higher speed of movement between the orbiting tool and the workpiece. Smaller presses can be used, reducing cost and space, and tool life is extended. Strength, hardness, and surface finish can all be improved by cold working.

UPSET FORGING

Upset forging involves increasing the diameter of a material by compressing its length. Because of its use with a multitude of fasteners, it is the most widely used of all forging processes when evaluated in terms of the number of pieces produced. Parts can be upset-forged both hot and cold, with the operation generally being performed on special high-speed machines. The forging motion is usually horizontal and the workpiece is rapidly moved from station to station. While most operations start with wire or rod, some machines can upset bars up to 25 cm (10 in.) in diameter.

Upset forging generally employs split dies that contain multiple positions or cavities, as seen in the typical die set of Figure 16-17. The dies separate enough for the bar to advance between them and move into position. They are then clamped together, and a

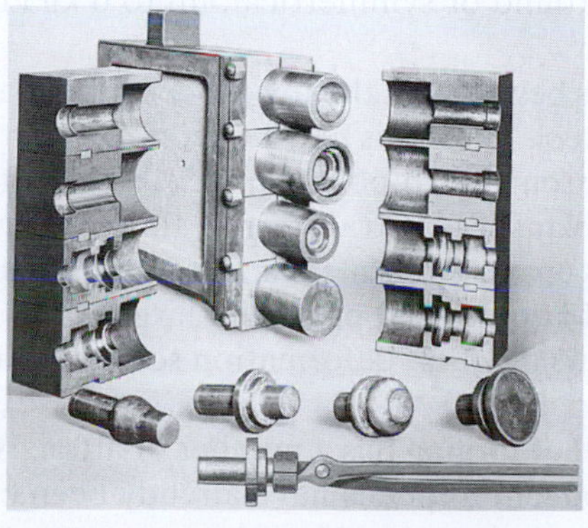

FIGURE 16-17 Set of upset forging dies and punches. The product resulting from each of the four positions is shown along the bottom. *(Courtesy of Ajax Manufacturing Company Euclid, OH)*

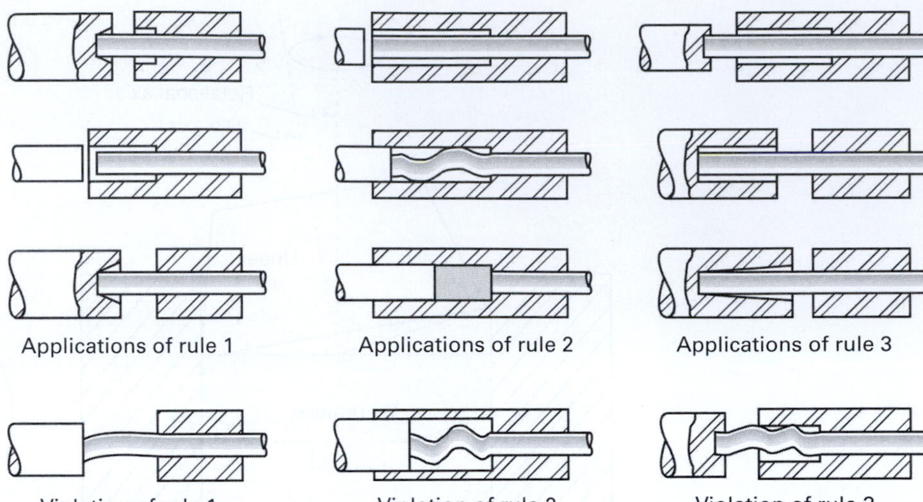

FIGURE 16-18 Schematics illustrating the rules governing upset forging. *(Courtesy of National Machinery Company, Tiffin, OH)*

heading tool or ram moves longitudinally against the bar, upsetting it into the cavity. Separation of the dies then permits transfer to the next position or removal of the product. If a new piece is started with each die separation, and an operation is performed in each cavity simultaneously, a finished product can be made with each cycle of the machine. By including a shearing operation as the initial piece moves into position, the process can operate with continuous coil or long-length rod as its incoming feedstock.

Upset-forging machines are often used to form heads on bolts and other fasteners and to shape valves, couplings, and many other small components. The upset region can be on the end or central portion of the workpiece, and the final diameter may be up to three times the original. The following three rules, illustrated in Figure 16-18, should be followed when designing parts that are to be upset forged:

1. The length of unsupported metal that can be gathered or upset in one blow without injurious buckling should be limited to three times the diameter of the bar.
2. Lengths of stock greater than three times the diameter may be upset successfully provided that the increase in diameter is not more than one times the diameter of the bar.
3. In an upset requiring stock length greater than three times the diameter of the bar, and where the increase in diameter is not more than one times the diameter of the bar (the conditions of rule 2), the length of unsupported metal beyond the face of the die must not exceed the diameter of the bar.

AUTOMATIC HOT FORGING

Several equipment manufacturers now offer highly automated upset equipment in which mill-length steel bars (typically 7-m, or 24-ft long) are fed into one end at room temperature and hot-forged products emerge from the other end at rates of up to 180 parts per minute (ppm—or 86,400 parts per eight-hour shift). These parts can be solid or hollow, round or symmetrical, up to 6 kg (12 lb) in weight, and up to 18 cm (7 in.) in diameter.

The process begins with the lowest-cost steel bar stock: hot-rolled and air-cooled carbon or alloy steel. The bar is first heated to 1200 to 1300°C (2200 to 2350°F) in under 60 s as it passes through high-power induction coils. It is then descaled by rolls, sheared into individual blanks, and transferred through several successive forming stages, during which it is upset, preformed, final forged, and pierced (if necessary). Small parts can be produced at up to 180 ppm, and larger parts at rates on the order of 90 ppm. Figure 16-19 shows a typical deformation sequence and a variety of hot-forged ferrous products.

Automatic hot forging has a number of attractive features. Low-cost input material and high production speeds have already been cited. Minimum labor is required,

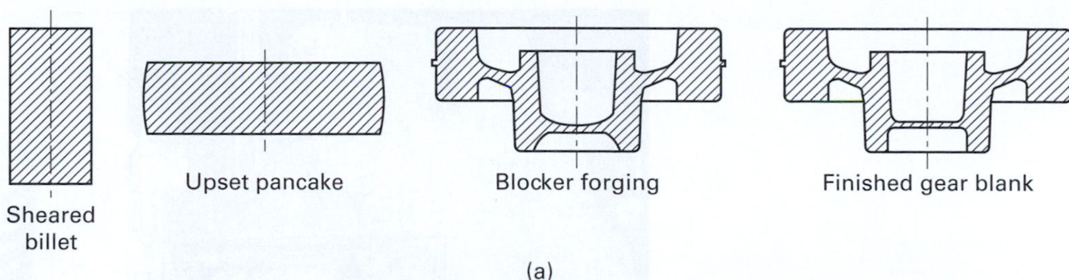

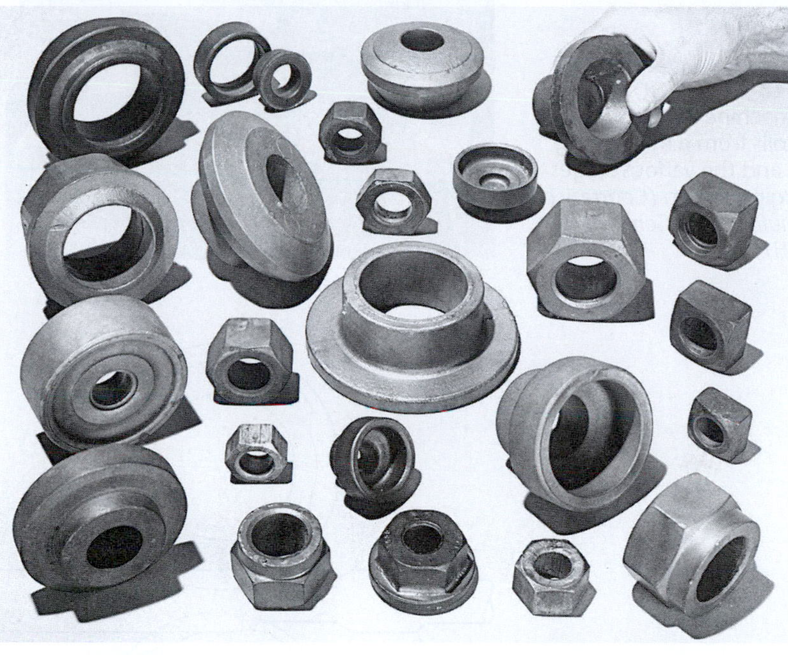

FIGURE 16-19 (a) Typical four-step sequence to produce a spur-gear forging by automatic hot forging. The sheared billet is progressively shaped into an upset pancake, blocker forging, and finished gear blank. (b) Samples of ferrous parts produced by automatic hot forging at rates between 90 and 180 parts per minute. *(Courtesy of National Machinery Company, Tiffin, OH)*

and because no flash is produced, material usage can be as much as 20 to 30% greater than with conventional forging. With a consistent finishing temperature near 1050°C (1900°F), an air cool can often produce a structure suitable for machining, eliminating the need for an additional anneal or normalizing treatment. Tolerances are generally within 0.3 mm (0.012 in.), surfaces are clean, and draft angles need only be 0.5 to 1 degree (as opposed to the conventional 3 to 5 degrees). Tool life is nearly double that of conventional forging because the contact times are only on the order of $\frac{6}{100}$ of a second.

Automatic hot formers can also be coupled with high-rate, cold-forming operations. Preform shapes can be hot formed at rates that approach 180 ppm. These products can then be cold formed to final shape on machines that operate at speeds near 90 ppm. The benefits of the combined operations include high-volume production at low cost coupled with the precision, surface finish, and strain hardening that are characteristic of a cold-finished material.

To justify an automatic hot-forging operation, however, large quantities of a given product must be required. A single production line may well require an initial investment in excess of $10 million.

ROLL FORGING

In **roll forging,** round or flat bar stock is reduced in thickness and increased in length to produce such products as axles, tapered levers, and leaf springs. As illustrated in Figure 16-20, roll forging is performed on machines that have two cylindrical or semi-cylindrical rolls, each containing one or more shaped grooves. A heated bar is inserted between the rolls. When the bar encounters a stop, the rolls rotate, and the bar is progressively shaped as it is rolled out toward the operator. The piece is then transferred to the next set of grooves (or rotated and reinserted in the same groove) and the process

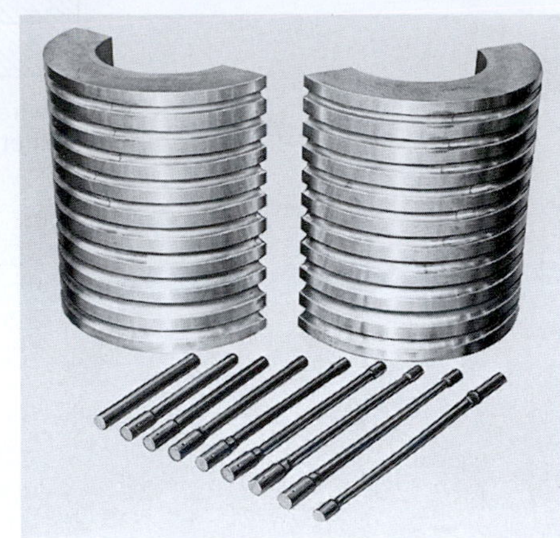

FIGURE 16-20 (Left) Roll-forging machine in operation. (Right) Rolls from a roll-forging machine and the various stages in roll forging a part. *(Courtesy of Ajax Manufacturing Company, Euclid, OH)*

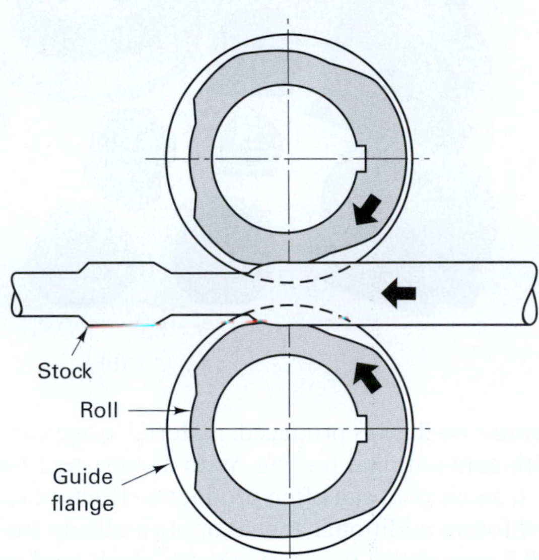

FIGURE 16-21 Schematic of the roll-forging process showing the two shaped rolls and the stock being formed. *(Courtesy of Forging Industry Association, Cleveland, OH)*

repeats until the desired size and shape is produced. Figure 16-20 also shows a set of rolls and the product formed by each set of grooves. Figure 16-21 shows the cross section of one set of grooves and a piece being formed. In most cases there is no flash, and the oriented structure imparts favorable forging-type properties.

SWAGING

Swaging (also known as *rotary swaging* or *radial forging*) uses external hammering to reduce the diameter or produce tapers or points on round bars or tubes. Figure 16-22 shows a typical swaging machine, and Figure 16-23 shows a schematic of its internal components. The dies, located in the center of the apparatus, consist of two blocks of hardened tool steel. They combine to form a central hole that generally has a conical input transitioning to a cylinder. An external motor drives a large, massive flywheel, which is connected to the central spindle of the machine. High-speed rotation of the central unit generates centrifugal force, which causes the matching die segments and backing blocks to separate. As the spindle rotates, the backing blocks are driven into opposing rollers that have been mounted in a massive machine housing. To pass beneath the rollers, the backer blocks must squeeze the dies tightly together. Once the assembly clears the rollers, the dies once again separate and the cycle repeats, generating between 1000 and 3000 blows per minute.

FIGURE 16-22 Tube being reduced in a rotary swaging machine. *(Courtesy of the Timken Company, Canton, OH)*

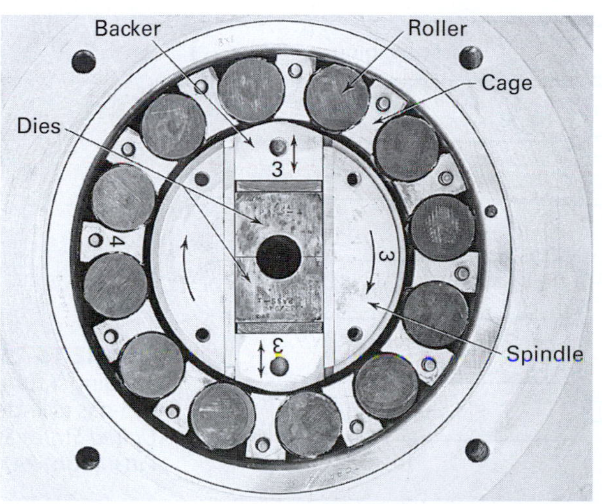

FIGURE 16-23 Basic components and motions of a rotary swaging machine. (*Note:* The cover plate has been removed to reveal the interior workings.) *(Courtesy of the Timken Company, Canton OH)*

With the machine in motion, the operator simply inserts a rod or tube between the dies and advances it during the periods of die separation. Because the dies rotate, the repeated blows are delivered from various angles, reducing the diameter and increasing the length. Because the rotating spindle is usually hollow, the workpiece can be passed completely through the machine, or withdrawn after a preset length has been reduced.

Swaging operations can also be used to form tubular products with internal cavities of constant cross section. A shaped mandrel is inserted into a thick-walled tube (or hollow-end workpiece), and the metal is collapsed around it to simultaneously shape and size both the interior and exterior of the product. Swaging over a mandrel can be used to form parts with internal gears, splines, recesses, or sockets. Figure 16-24 shows a variety of swaged products, many of which contain shaped holes.

The term *swaging* has also been applied to a process where material is forced into a confining die to reduce its diameter. This process is usually performed on heated material. Figure 16-25 shows a hot swaging sequence being used to form the end of a pressurized gas cylinder.

NET-SHAPE AND NEAR-NET-SHAPE FORGING

As much as 80% of the cost of a forged gear can be incurred during the machining operations that follow forging, and a finished aerospace wing spar may contain as little as 4% of the original billet (the remaining 96% being lost as scrap in the forging and subsequent machining operations). To minimize both the expense and waste, considerable

FIGURE 16-24 A variety of swaged parts, some with internal details. *(Courtesy of Cincinnati Milacron, Inc., Cincinnati, OH)*

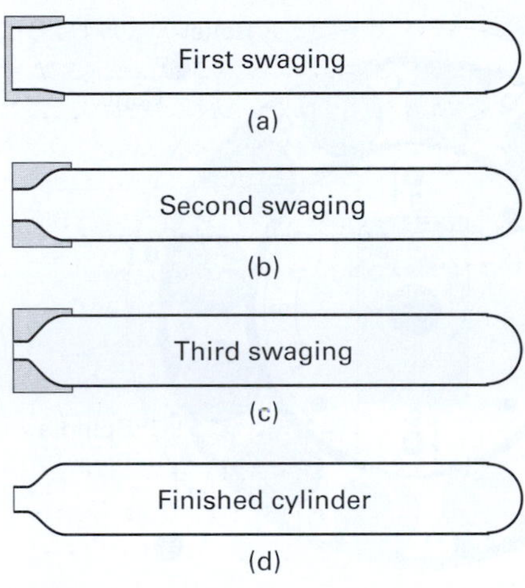

FIGURE 16-25 Steps in swaging a tube to form the neck of a gas cylinder. *(Courtesy of United States Steel Corporation, Pittsburgh, PA)*

effort has been made to develop processes that can form parts close enough to final dimensions that little or no final machining is required. These are known as **net-shape** or **near-net-shape forging** and may also be referred to as **precision forging.** Cost savings often result from the reduction or elimination of secondary machining (and the associated handling, positioning, and fixturing), the companion reduction in scrap, and an overall decrease in the amount of energy required to produce the product.

Precision or near-net-shape forgings can now be produced with draft angles of less than 1 degree (or even zero draft). Complex shapes can be forged with such close tolerances that little or no finish machining is required. Because the design and implementation of net-shape processing can be rather expensive, application is usually reserved for parts where a significant cost reduction can be achieved.

16.6 EXTRUSION

In the **extrusion** process, metal is compressed and forced to flow through a suitably shaped die to form a product with reduced but constant cross section. Although extrusion may be performed either hot or cold, hot extrusion is commonly employed for many metals to reduce the forces required, eliminate cold-working effects, and reduce directional properties. Basically, the extrusion process is like squeezing toothpaste out of a tube. In the case of metals, a common arrangement is to have a heated billet placed inside a confining chamber. A ram advances from one end, causing the billet to first upset and conform to the confining chamber. As the ram continues to advance,

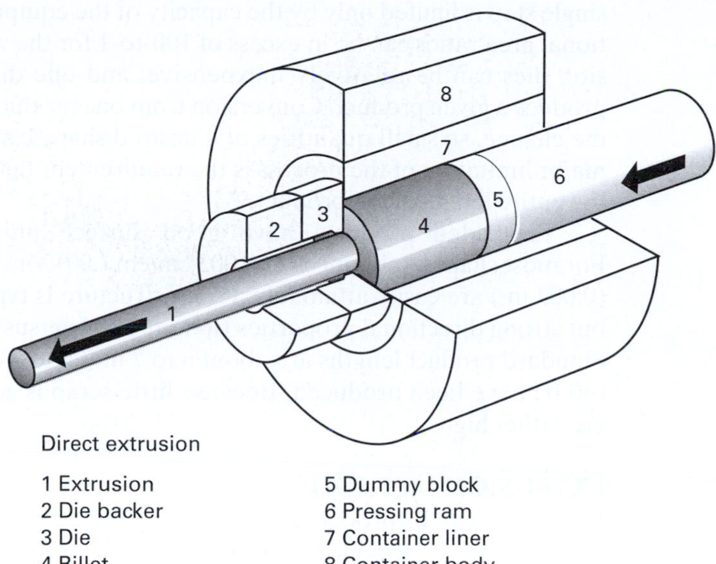

FIGURE 16-26 Direct extrusion schematic showing the various equipment components. *(Courtesy of Danieli Wean United, Cranberry Township, PA)*

the pressure builds until the material flows plastically through the die and *extrudes*, as depicted in Figure 16-26. The stress state within the material is one of triaxial compression.

Aluminum, magnesium, copper, lead, and alloys of these metals are commonly extruded, taking advantage of the relatively low yield strengths and low hot-working temperatures. Steels, stainless steels, nickel-based alloys, and titanium are far more difficult to extrude. Their yield strengths are high, and the metals tend to weld to the walls of the die and confining chamber under the required conditions of temperature and pressure. With the development and use of phosphate-based and molten glass lubricants, however, hot extrusions can be routinely produced from these high-strength, high-temperature metals. These lubricants are able to withstand the required temperatures and adhere to the billet, flowing and thinning in a way that prevents metal-to-metal contact throughout the process.

As shown in the left-hand segment of Figure 16-27, almost any cross-sectional shape can be extruded from the nonferrous metals. Size limitations are few because presses are now available that can extrude any shape that can be enclosed within a circle 75 cm (30 in.) in diameter. In the case of steels and the other high-strength metals, the shapes and sizes are a bit more limited, but, as the right-hand segment of Figure 16-27 shows, considerable freedom still exists.

Extrusion has a number of attractive features. Many shapes can be produced as extrusions that are not possible by rolling, such as ones containing reentrant angles or longitudinal holes. No draft is required, so extrusions can offer savings in both metal and weight. Because the deformation is compressive, the amount of reduction in a

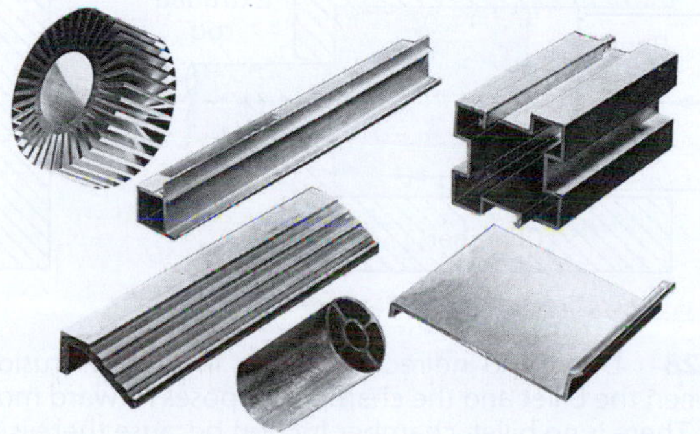
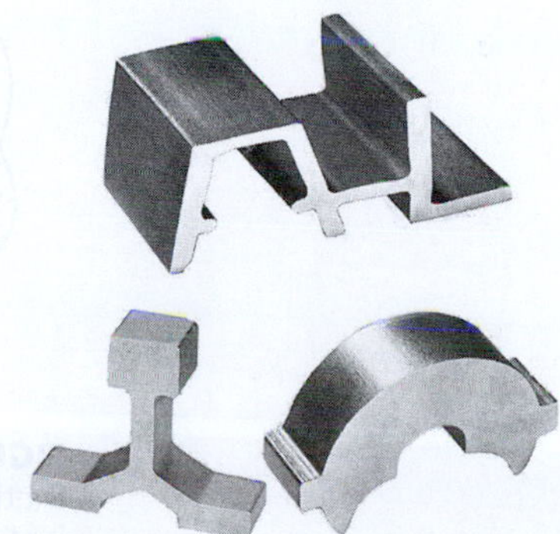

FIGURE 16-27 Typical shapes produced by extrusion. (Left) Aluminum products. (Right) Steel products. *[(Left) Alcoa Fastening Systems; (Right) Courtesy of Allegheny Ludlum Steel Corporation, Pittsburgh, PA]*

single step is limited only by the capacity of the equipment. Billet-to-product cross-sectional area ratios can be in excess of 100-to-1 for the weaker metals. In addition, extrusion dies can be relatively inexpensive, and one die may be all that is required to produce a given product. Conversion from one product to another requires only a single die change, so small quantities of a desired shape can be produced economically. The major limitation of the process is the requirement that the cross section be uniform for the entire length of the product.

Extruded products have good surface finish and dimensional precision. For most shapes, tolerances of 0.003 cm/cm (.0012 in./in.) with a minimum of 0.075 mm (0.003 in.) are easily attainable. Grain structure is typical of other hot-worked metals, but strong directional properties (longitudinal versus transverse) are usually observed. Standard product lengths are about 6 to 7 m (20 to 24 ft), but lengths in excess of 12 m (40 ft) have been produced. Because little scrap is generated, billet-to-product yields are rather high.

EXTRUSION METHODS

Extrusions can be produced by various techniques and equipment configurations. Hot extrusion is usually done by either the direct or indirect method, both of which are illustrated in Figure 16-28. In **direct extrusion,** a solid ram drives the entire billet to and through a stationary die and must provide additional power to overcome the frictional resistance between the surface of the moving billet and the confining chamber. With **indirect extrusion,** also called *reverse, backward,* or *inverted extrusion,* a hollow ram pushes the die back through a stationary, confined billet. Because there is no relative motion, friction between the billet and the chamber is eliminated. The required force is lower, and longer billets can be used with no penalty in power or efficiency.

Figure 16-29 shows the ram force versus ram position curves for both direct and indirect extrusion. The areas below the lines have units of Newton-meters or foot-pounds and are, therefore, proportional to the work required to produce the part. The area between the two curves is the work required to overcome the billet–chamber friction during direct extrusion—an amount that can be saved by converting to indirect extrusion. Unfortunately, the added complexity of the indirect process (applying force through a hollow ram, extracting the product through the hollow, and removing residual billet material at the end of the stroke) serves to increase the purchase price and maintenance cost of the required equipment.

With either process, the speeds of hot extrusion are usually rather fast, so as to minimize the cooling of the billet within the chamber. Extruded products can emerge at rates up to 300 m/min (1000 ft/min). The extrusion speed may be restricted, however, by the large amounts of heat that are generated by the massive deformation and the associated rise in temperature. Sensors are often used to monitor the temperature of the emerging product and feed this information back to a control system. For materials whose properties are not sensitive to strain rate, ram speed may be maintained at the highest level that will keep the product temperature below some predetermined value.

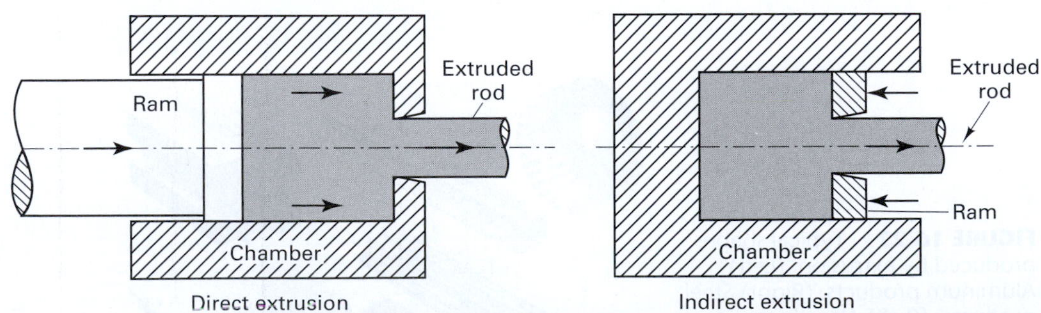

FIGURE 16-28 Direct and indirect extrusion. In direct extrusion, the ram and billet both move and friction between the billet and the chamber opposes forward motion. For indirect extrusion, the billet is stationary. There is no billet–chamber friction because there is no relative motion.

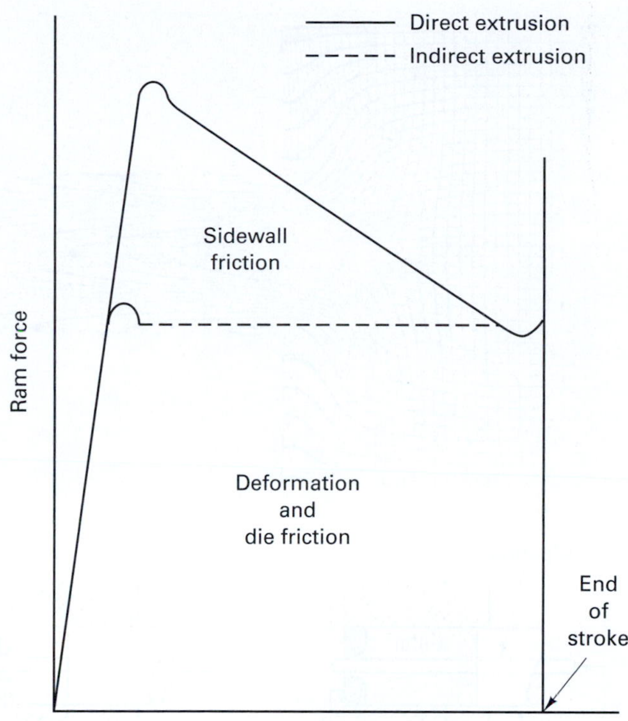

FIGURE 16-29 Diagram of the ram force versus ram position for both direct and indirect extrusion of the same product. The area under the curve corresponds to the amount of work (force × distance) performed. The difference between the two curves is attributed to billet–chamber friction.

Lubrication is another important area of concern. If the reduction ratio (cross section of billet to cross section of product) is 100, the product will be 100 times longer than the starting billet. If the product has a complex cross section, its perimeter can be significantly greater than a circle of equivalent area. Because the surface area of the product is the length times the perimeter, this value can be more than an order of magnitude greater than the surface area of the original billet. A lubricant that is applied to the starting piece must thin considerably as the material passes through the die and is converted to product. An acceptable lubricant is expected to reduce friction and act as a barrier to heat transfer at all stages of the process.

METAL FLOW IN EXTRUSION

The flow of metal during extrusion is often complex, and some care must be exercised to prevent surface cracks, interior cracks, and other flow-related defects. Metal near the center of the chamber can often pass through the die with little distortion, while metal near the surface undergoes considerable shearing. In direct extrusion, friction between the forward-moving billet and both the stationary chamber and die serves to further impede surface flow. The result is often a deformation pattern similar to the one shown in Figure 16-30. If the surface regions of the billet undergo excessive cooling, surface deformation is further impeded, often leading to the formation of surface cracks. If quality is to be maintained, process control must be exercised in the areas of design, lubrication, extrusion speed, and temperature.

EXTRUSION OF HOLLOW SHAPES

Hollow shapes, and shapes with multiple longitudinal cavities, can be extruded by several methods. For tubular products, the stationary or moving **mandrel** processes of Figure 16-31 are quite common. The die forms the outer profile, while the mandrel shapes and sizes the interior.

For products with multiple or more complex cavities, a **spider-mandrel die** (also known as a *porthole, bridge,* or *torpedo die*) may be required. As illustrated in Figure 16-32, metal flows around the arms of a "spider," and a further reduction then forces the material back together. Because the metal is never exposed to

FIGURE 16-30 Grid pattern showing the metal flow in a direct extrusion. The billet was sectioned, and the grid pattern was engraved prior to extrusion. *(Courtesy Ronald Kohser)*

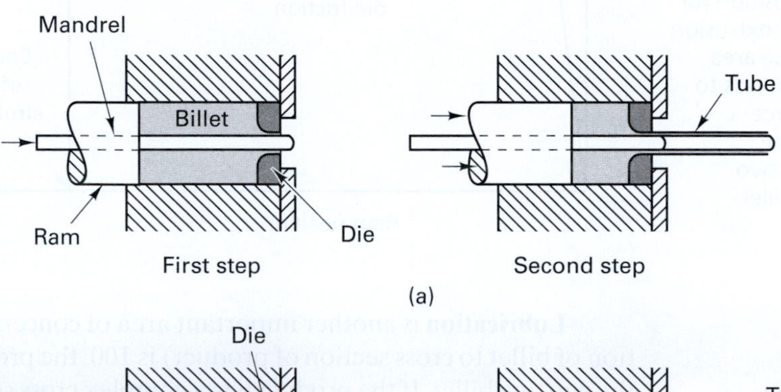

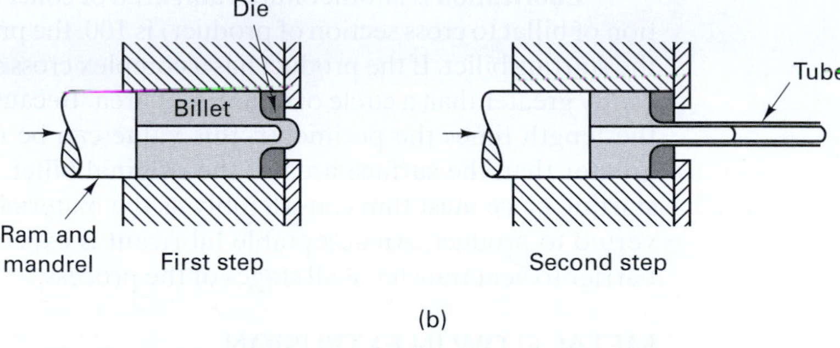

FIGURE 16-31 Two methods of extruding hollow shapes using internal mandrels. (a) The mandrel and ram have independent motions; (b) they move as a single unit.

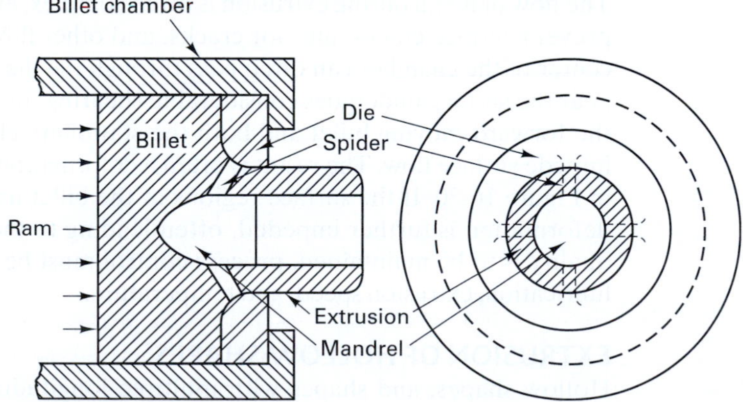

FIGURE 16-32 Hot extrusion of a hollow shape using a spider-mandrel die. Note the four arms connecting the external die and the central mandrel.

contamination, perfect welds result. Unfortunately, lubricants cannot be used because they will contaminate the surfaces to be welded. Thus, the process is limited to materials that can be extruded without lubrication and can also be easily pressure welded.

Because additional tooling is required, hollow extrusions will obviously cost more than solid ones, but a wide variety of continuous cross-section shapes can be produced that cannot be made economically by any other process.

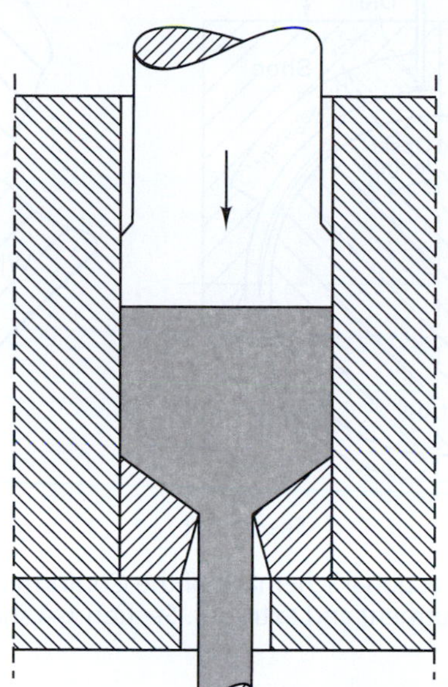

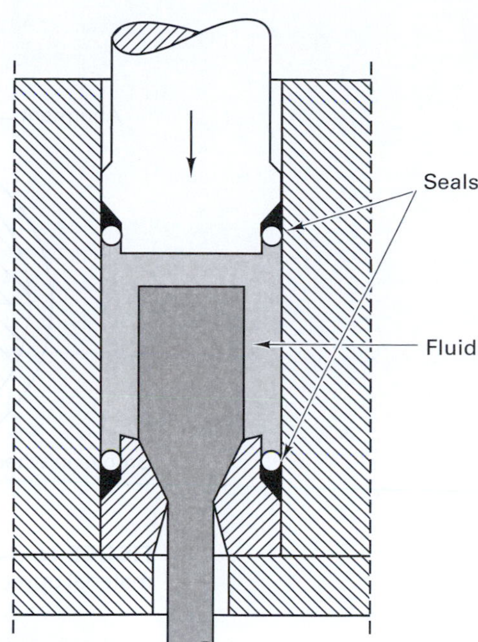

FIGURE 16-33 Comparison of conventional (left) and hydrostatic (right) extrusion. Note the addition of the pressurizing fluid and the O-ring and miter-ring seals on both the die and ram.

HYDROSTATIC EXTRUSION

Another type of extrusion, known as **hydrostatic extrusion,** is illustrated schematically in Figure 16-33. Here, high-pressure fluid surrounds the workpiece and applies the force necessary to extrude it through the die. The product emerges into either atmospheric pressure or a lower-pressure, fluid-filled chamber. The process resembles direct extrusion, but the pressurized fluid surrounding the billet prevents any upsetting. Because the billet does not come into contact with the surrounding chamber, billet–chamber friction is eliminated. In addition, the pressurized fluid can also emerge between the billet and the die, acting in the form of a lubricant.

While the efficiency can be significantly greater than most other extrusion processes, there are problems related to the fluid and the associated high pressures (which typically range between 900 and 1700 MPa, or 125 to 250 ksi). Temperatures are limited because the fluid acts as a heat sink, and many of the pressurizing fluids (typically light hydrocarbons and oils) burn or decompose at moderately low temperatures. Seals must be designed to contain the pressurized fluid without leaking, and measures must be taken to prevent the complete ejection of the product, often referred to as **blowout.** Because of these features, hydrostatic extrusion is usually employed only where the process offers unique advantages that cannot be duplicated by the more conventional methods.

Pressure-to-pressure extrusion is one of those unique capabilities. In this variant, the product emerges from an extremely high-pressure chamber into a second pressurized chamber. In effect, the metal deformation is performed in a highly compressed environment. Crack formation begins with void formation, void growth and void coalescence. Because voids are suppressed in a compressed environment, the result is a phenomenon known as **pressure-induced ductility.** Relatively brittle materials such as molybdenum, beryllium, tungsten, and various intermetallic compounds can be plastically deformed without fracture, and materials with limited ductility become highly formable. Products can be made that could not be otherwise produced, and materials can be considered that would have been rejected because of their limited ductility at room temperature and atmospheric pressure.

CONTINUOUS EXTRUSION

Conventional extrusion is a discontinuous process, converting finite-length billets into finite-length products. If the pushing force could be applied to the periphery of the feedstock, rather than the back, continuous feedstock could be converted into continuous

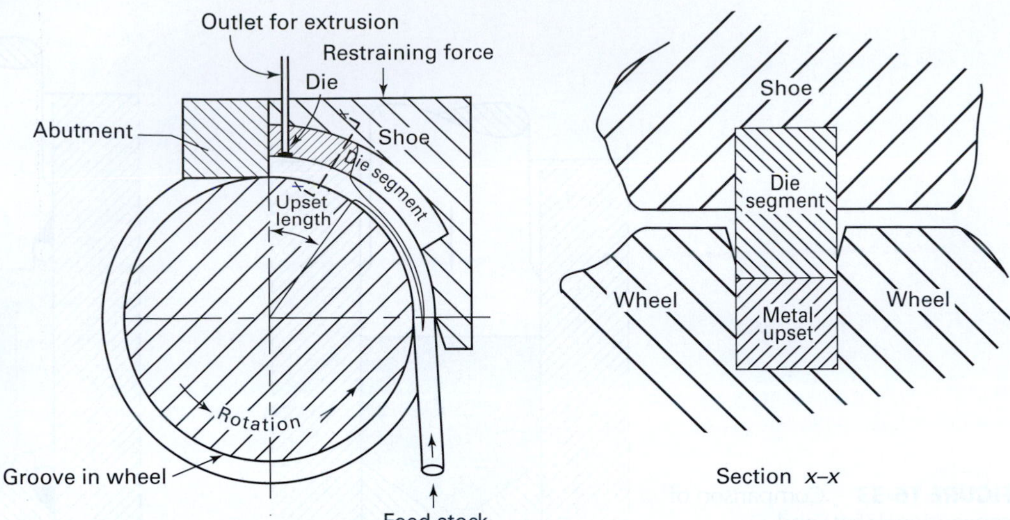

FIGURE 16-34 Cross-sectional schematic of the Conform continuous extrusion process. The material upsets at the abutment and extrudes. Section *x–x* shows the material in the shoe.

product, and the process could become one of **continuous extrusion.** The first continuous extrusion of solid metal feedstock was performed in 1970. Since then, a number of techniques have been proposed with varying degrees of success. In terms of commercial application, the most significant is probably the **Conform process,** illustrated schematically in Figure 16-34. Continuous feedstock is inserted into a grooved wheel and is driven by surface friction into a chamber created by a mating die segment. Upon impacting a protruding abutment, the material upsets to conform to the chamber, and the increased wall contact further increases the driving friction. Upsetting continues until the pressure reaches a value sufficient to extrude the material through a die opening that has been provided in either the shoe or abutment. At this point, the rate of material entering the machine equals the rate of product emerging, and a steady-state continuous process is established.

Because surface friction is the propulsion force, the feedstock can take a variety of forms, including solid rod, metal powder, punchouts from other forming operations, or chips from machining. Metallic and nonmetallic powders can be intimately mixed and co-extruded. Rapidly solidified material can be extruded without exposure to the elevated temperatures that would harm the properties. Polymeric materials and even fiber-reinforced plastics have been successfully extruded. The most common feed, however, is coiled aluminum or copper rod.

Continuous extrusion complements and competes with wire drawing and shape rolling as a means of producing nonferrous products with small, but uniform, cross sections. It is particularly attractive for complex profiles and cross sections that contain one or more holes. Because extrusion operations can perform massive reductions through a single die, one Conform operation can produce an amount of deformation equivalent to 10 conventional drawing or cold-rolling passes. In addition, sufficient heat can be generated by the deformation that the product will emerge in an annealed condition, ready for further processing without intermediate heat treatment.

16.7 WIRE, ROD, AND TUBE DRAWING

Wire, rod and tube **drawing** operations reduce the cross section of a material by pulling it through a die. In many ways, the processes are similar to extrusion, but the applied stresses are now tensile, pulling on the product rather than pushing on the workpiece. **Rod** or **bar drawing,** illustrated schematically in Figure 16-35, is probably the simplest of these operations. One end of a rod is reduced or pointed, so that it can pass through a die of somewhat smaller cross section. The protruding material is then placed in grips and pulled in tension, drawing the remainder of the rod through the die. The rods

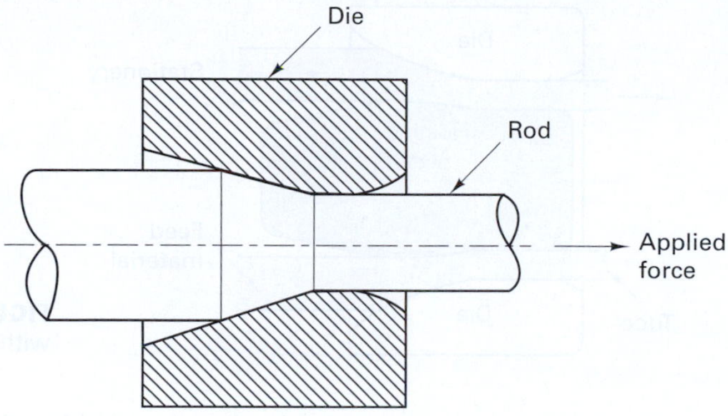

FIGURE 16-35 Schematic diagram of the rod- or bar-drawing process.

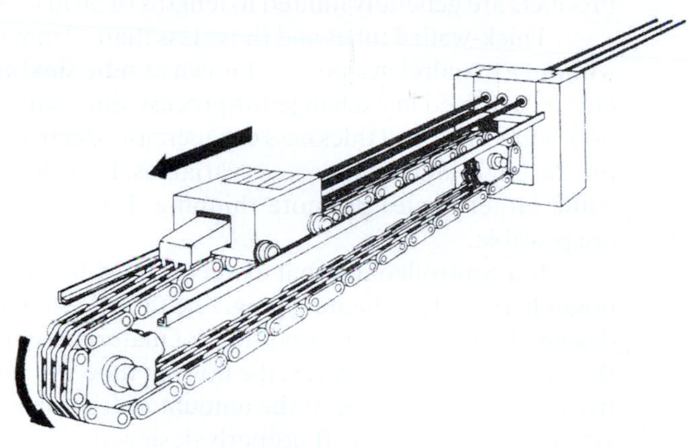

FIGURE 16-36 Diagram of a chain-driven multiple-die-draw bench used to produce finite lengths of straight rod or tube. *(Courtesy of Danieli Wean United, Cranberry Township, PA)*

reduce in section, elongate, and become stronger (strain harden). Because the product cannot be readily bent or coiled, straight-pull **draw benches** are generally employed with finite-length feedstock. Hydraulic cylinders can be used to provide the pull for short-length products, while chain drives, as depicted in Figure 16-36, can be used to draw products up to 30 m (100 ft) in length.

The reduction in area is usually restricted to between 20 and 50%, because higher values require higher pulling forces that may exceed the tensile strength of the reduced product. To produce a desired size or shape, multiple draws may be required through a series of progressively smaller dies. Intermediate anneals may also be required to restore ductility and enable further deformation.

Tube drawing can be used to produce high-quality tubing where the product requires the smooth surfaces, thin walls, accurate dimensions, and added strength (from the strain hardening) that are characteristic of cold forming. Internal mandrels are often used to control the inside diameter of tubes, which range from about 12 to 250 mm (0.5 to 10 in.) in diameter. As shown in Figure 16-37, these mandrels are

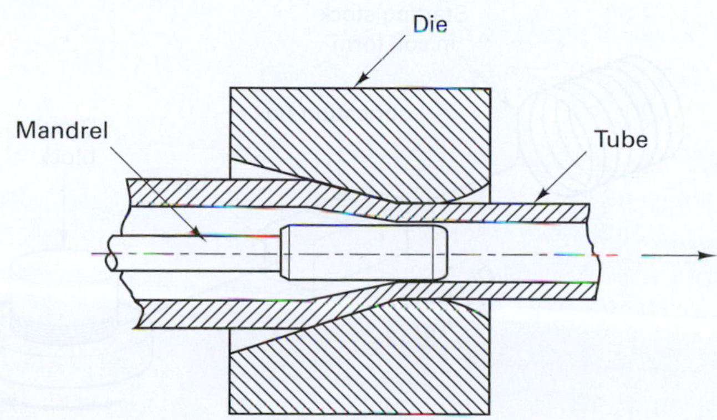

FIGURE 16-37 Cold-drawing smaller tubing from larger tubing. The die sets the outer dimension while the stationary mandrel sizes the inner diameter.

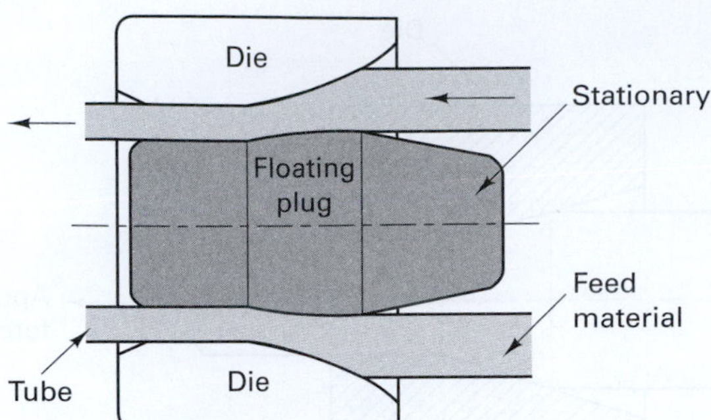

FIGURE 16-38 Tube drawing with a floating plug.

inserted through the incoming stock and are held in place during the drawing operation. Products are generally limited to lengths of 30 m (100 ft) or less.

Thick-walled tubes and those less than 12 mm (0.5 in.) in diameter are often drawn without a mandrel in a process known as **tube sinking**. Precise control of the inner diameter is sacrificed in exchange for process simplicity and the ability to draw long lengths of product. The wall thickness can increase, decrease, or remain the same, depending on the die angle and other process variables. Low die angles tend to favor wall thickening, while larger angles promote thinning. Because a mandrel is not used, long lengths are possible.

If a controlled internal diameter must be produced in a long-length product, it is possible to utilize a **floating plug,** like the one shown in Figure 16-38. This plug must be designed for the specific conditions of material, reduction, and friction. If the friction on the plug surface is too great, the flowing tube will pull it too far forward, pinching off or fracturing the tube wall. If the amount of friction is insufficient, the plug will chatter or vibrate within the tube. If properly designed, the floating plug will assume a stable position within the die and size the internal diameter, while the external die shapes and sizes the outside of the tube.

The drawing of bar stock can also be used to make products with shaped cross sections. By using cold drawing instead of hot extrusion, the material emerges with precise dimensions and excellent surface finish. Inexpensive materials strengthened by strain hardening can often replace stronger alloys or ones that would require additional heat treatment. Small parts with complex but constant cross sections can be economically made by sectioning long lengths of cold-drawn shaped bars to produce the individual products. Steels, copper alloys, and aluminum alloys have all been cold drawn into shaped bars. For steel, the cross-sectional area is usually less than 4 cm^2 (0.6 in.^{2}) and the largest cross-sectional dimension is generally less than 3 cm (1.25 in.).

Wire drawing is essentially the same process as bar drawing except that it involves smaller-diameter material. Because the material can now be coiled, the process can be conducted in a somewhat continuous manner on rotating draw blocks, like the one illustrated schematically in Figure 16-39. Wire drawing usually begins with large coils of

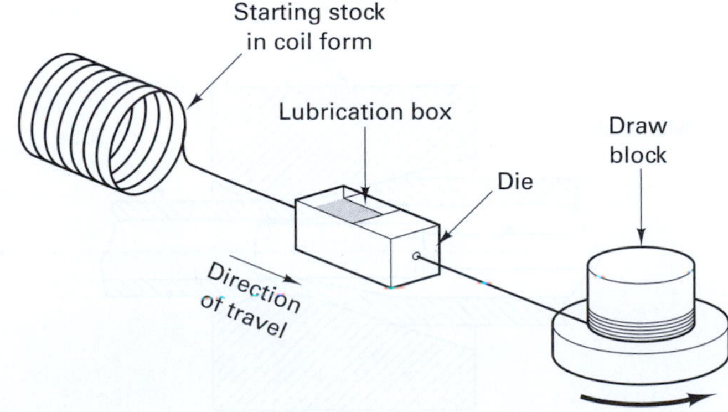

FIGURE 16-39 Schematic of wire drawing with a rotating draw block. The rotating motor on the draw block provides a continuous pull on the incoming wire.

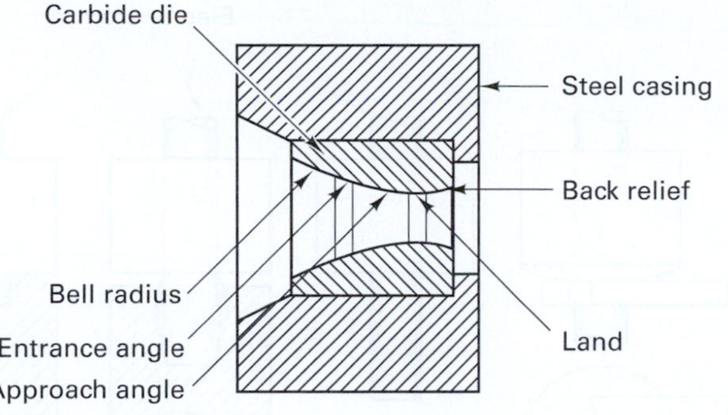

FIGURE 16-40 Cross section through a typical carbide wire drawing die showing the characteristic regions of the contour.

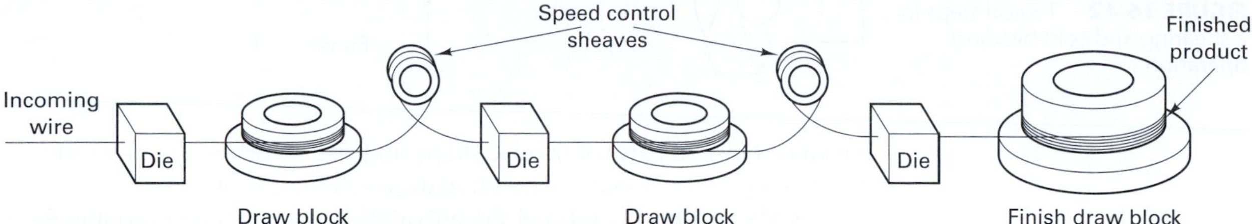

FIGURE 16-41 Schematic of a multistation synchronized wire drawing machine. To prevent accumulation or breakage, it is necessary to ensure that the same volume of material passes through each station in a given time. The loops around the sheaves between the stations use wire tensions and feedback electronics to provide the necessary speed control.

hot-rolled rod stock approximately 9 mm ($\frac{3}{8}$ in.) in diameter.[1] After descaling or other forms of surface preparation, one end of the coil is pointed, fed through a die, gripped, and the drawing process begins.

Wire dies generally have a configuration similar to the one shown in Figure 16-40. The contact regions are usually made of wear-resistant tungsten carbide or polycrystalline, manufactured diamond. Single-crystal diamonds can be used for the drawing of very fine wire, and wear-resistant and low-friction coatings can be applied to the various die material substrates. Lubrication boxes often precede the individual dies to help reduce friction drag and wear of the dies.

Because the tensile load is applied to the already reduced product, the amount of reduction is severely limited. Multiple draws are usually required to affect any significant change in size. To convert hot-rolled rod stock to the fine wire that is used in household telephone lines requires passes through as many as 20 or 30 individual dies. To minimize handling and labor, these operations are usually performed on tandem machines, like the one shown schematically in Figure 16-41. Between 3 and 12 dies are mounted in a single machine, and the material moves continuously from one station to another in a synchronized manner that prevents any localized accumulation or tension that might induce fracture.

After passing through all the dies in a tandem machine, the material usually requires an intermediate anneal before it can be subjected to further deformation. By controlling the placement of the last anneal so the final product has a selected amount of cold work, wires can be made with a wide range of strengths (or tempers). [See Problem 1 at the end of Chapter 15.] When maximum ductility and conductivity is desired, the wire should be annealed in controlled atmosphere furnaces after the final draw.

16.8 COLD FORMING, COLD FORGING, AND IMPACT EXTRUSION

Large quantities of products are now being made by **cold forming,** a family of processes in which slugs of material are squeezed into or extruded from shaped die cavities to

[1] This is approximately the smallest diameter material that can be produced by rolling with grooved rolls. With smaller diameter grooves, the concentrated stresses are likely to crack the rolls.

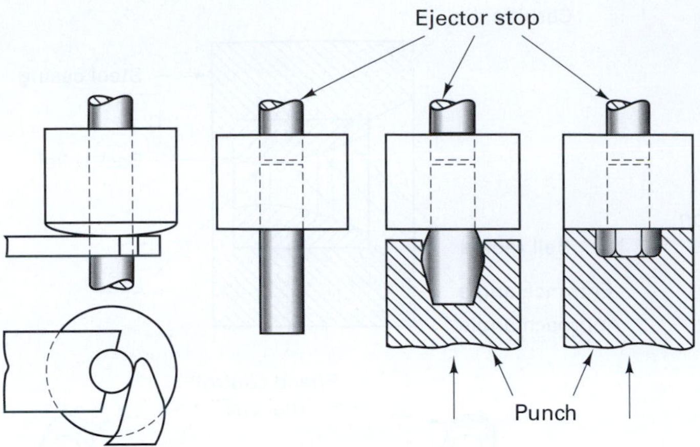

FIGURE 16-42 Typical steps in a shearing and cold-heading operation.

produce finished parts of precise shape and size. Workpiece temperature varies from room temperature to several hundred degrees Fahrenheit.

Cold heading is a form of the previously discussed upset forging. As illustrated in Figure 16-42, it is used for making enlarged sections on the ends of rod or wire, such as the heads of nails, bolts, rivets, or other fasteners. Two variations of the process are common. In the first, a piece of rod is sheared to a preset length and then transferred to a holder-ejector assembly. Heading punches then strike one or more blows on the exposed end to perform the upsetting. If intermediate shapes are required, the piece is transferred from station to station, or the various heading punches sequentially rotate into position. When the heading is completed, the ejector stop advances and expels the product. In the second variation, a continuous rod (or wire) is fed forward to produce a preset extension, clamped, and the head is formed. The rod is then advanced to a second preset length and sheared, and the cycle repeats. This procedure is particularly attractive for producing nails, because the point can be formed in the shearing or cutoff operation. Enlarged sections can also be produced at locations other than the ends of a rod or wire, in the manner illustrated in Figure 16-43.

While cold heading generally produces symmetrical parts, the expanded regions can also be square, hexagonal, or even offset. Production speeds tend to vary with the diameter of the incoming material. When the blanks are less than 6 mm (0.25 in.) in diameter, speeds of 400 to 600 ppm are typical. For larger diameters, the speeds may reduce to 40 to 100 ppm. Alloys of aluminum and copper have excellent formability, while mild steel and stainless steel are rated fair to good. Alloy steels are a bit more difficult because of their higher strength and lower ductility.

A variety of extrusion operations, commonly called **impact extrusion,** can also be incorporated into cold forming. In these processes, a metal slug of predetermined size is positioned in a die cavity, where it is struck a single blow by a rapidly moving punch. The metal may flow forward through the die, backward around the punch, or in a combination mode. Figure 16-44 illustrates the forward and backward variations, using both open and closed dies. In **forward extrusion,** the diameter is decreased while the length increases. **Backward extrusion** shapes hollow parts with a solid bottom. The punch

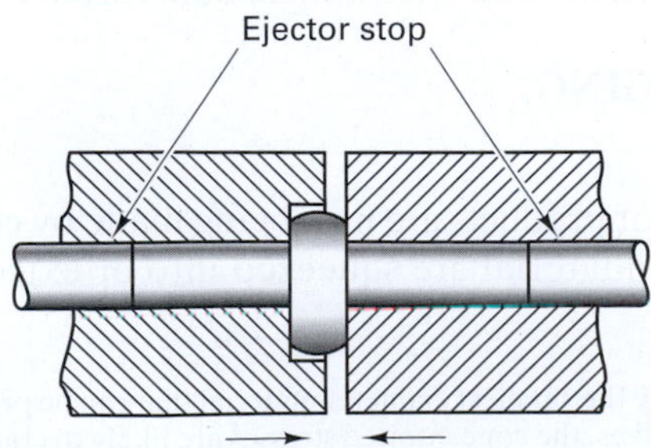

FIGURE 16-43 Method of upsetting the center portion of a rod. The stock is supported in both dies during upsetting.

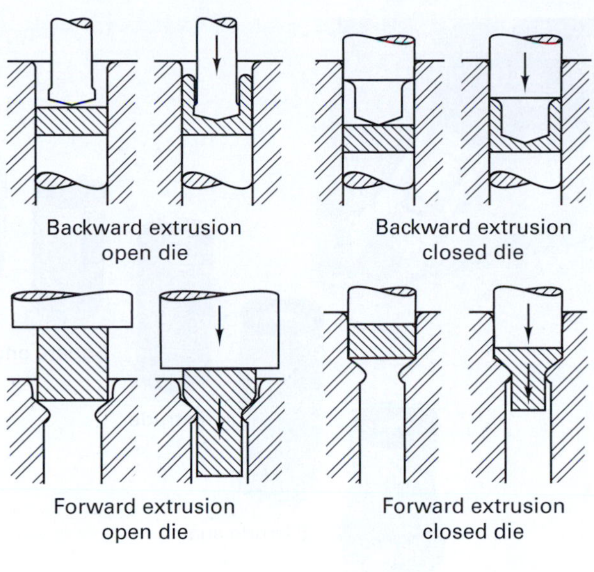

FIGURE 16-44 Backward and forward extrusion with open and closed dies.

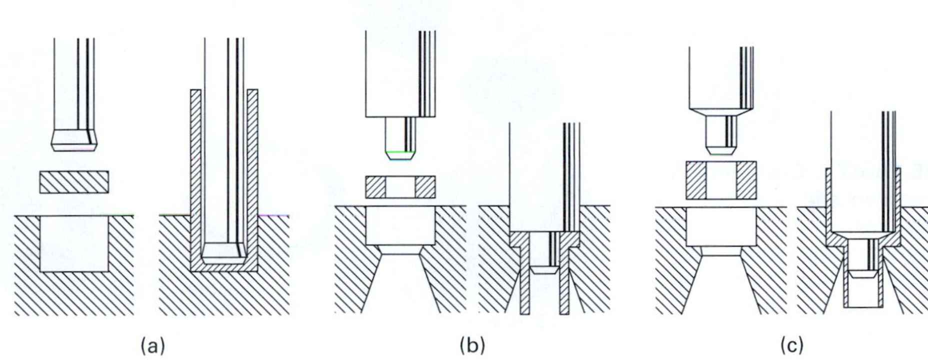

FIGURE 16-45 (a) Reverse, (b) forward, and (c) combined forms of cold extrusion. *(Courtesy of The Aluminum Association, Arlington, VA)*

controls the inside shape, while the die shapes the exterior. The wall thickness is determined by the clearance between the punch and die, and the bottom thickness is set by the stop position of the punch. Figure 16-45 provides additional schematics of forward, backward and combination impacting. Typical production speeds range from 20 to 60 strokes per minute.

The impact extrusion processes were first used to shape low-strength metals such as lead, tin, zinc, and aluminum into products such as collapsible tubes for toothpaste, medications, and other creams; small "cans" for shielding electronic components; zinc cases for flashlight batteries; and larger cans for food and beverages. In recent years, impact extrusion has expanded to the forming of mild steel parts, where it is often used in combination with cold heading, as in the example of Figure 16-46. When heading alone is used, there is a definite limit to the ratio of the head and stock diameters (as presented in Figure 16-18 and related discussion). The combination of forward extrusion and cold heading overcomes this limitation by using an intermediate starting

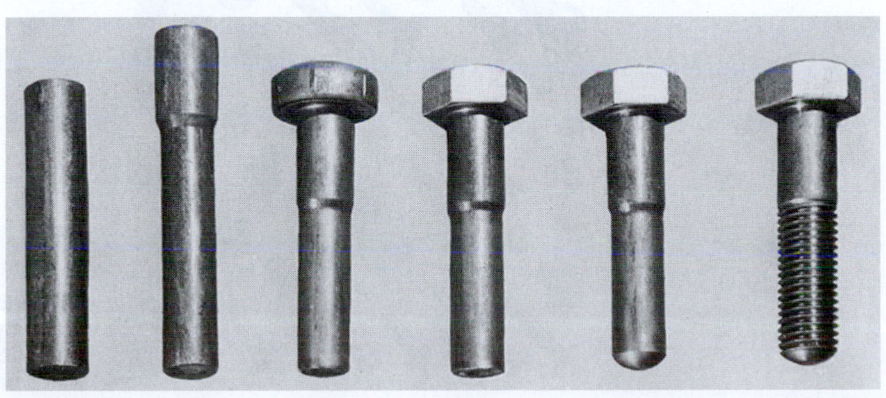

FIGURE 16-46 Steps in the forming of a bolt by cold extrusion, cold heading, and thread rolling. *(Courtesy of National Machinery Co., Tiffin, OH)*

FIGURE 16-47 Cold-forming sequence involving cutoff, squaring, two extrusions, an upset, and a trimming operation. Also shown are the finished part and the trimmed scrap. *(Courtesy of National Machinery Co., Tiffin, OH)*

diameter. The shank portion is reduced by forward extrusion while upsetting is used to increase the diameter of the head.

By using various types of dies and combining high-speed operations such as heading, upsetting, extrusion, piercing, bending, coining, thread rolling, and knurling, a wide variety of relatively complex parts can be cold formed to close tolerances. Figure 16-47 illustrates an operation that incorporates two extrusions, a central upset, and a final operation to shape and trim that upset. Figure 16-48 presents an array of upset and

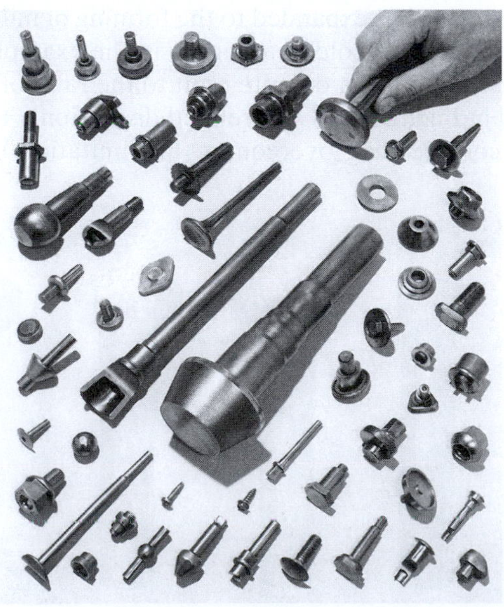

FIGURE 16-48 Typical parts made by upsetting and related operations. *(Courtesy of National Machinery Co., Tiffin, OH)*

FIGURE 16-49 Manufacture of a spark plug body: (left) by machining from hexagonal bar stock; (right) by cold forming. Note the reduction in waste. *(Courtesy of National Machinery Co., Tiffin, OH)*

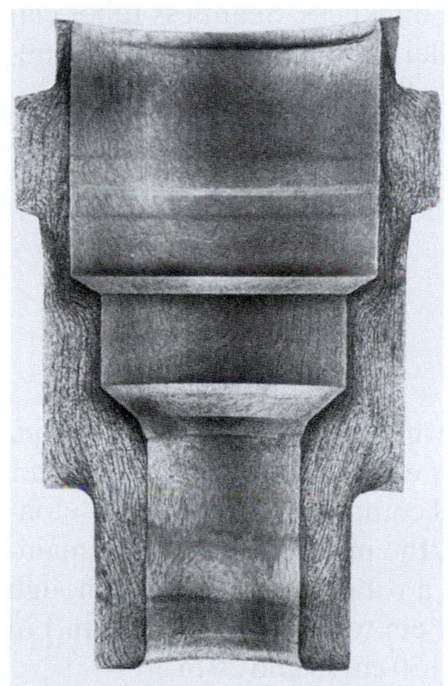

FIGURE 16-50 Section of the cold-formed spark-plug body of Figure 16.49, etched to reveal the flow lines. The cold-formed structure produces an 18% increase in strength over the machined product. *(Courtesy of National Machinery Co., Tiffin, OH)*

extruded products. The larger parts are generally hot formed and machined, while the smaller ones are cold formed.

Because cold forming is a chipless manufacturing process, producing parts by deformation that would otherwise be machined from bar stock or hot forgings, the material is used more efficiently and waste is reduced. Figure 16-49 compares the manufacture of a spark-plug body by machining from hexagonal bar stock with manufacture by cold forming. Material is saved, machining time and cost are reduced, and the product is stronger, due to cold work, and tougher, as illustrated by the flow lines revealed in Figure 16-50. By converting from screw machining to cold forming, a manufacturer of cruise-control housings was able to reduce material usage by 65%, while simultaneously increasing production rate by a factor of 5.

While cold forming is generally associated with the manufacture of small, symmetrical parts from the weaker nonferrous metals, the process is now used extensively on steel and stainless steel, with parts up to 45 kg (100 lb) in weight and 18 cm (7 in.) in diameter. At the small end of the scale, microformers are now cold forming extremely small electronic components with dimensional accuracies within 0.005 mm (0.0002 in.).

Cold-formed shapes are usually axisymmetric or those with relatively small departures from symmetry. Production rates are high; dimensional tolerances and surface finish are excellent; and there are no draft angles, parting lines, or flash to trim off. There is almost no material waste, and a considerable amount of machining can often be eliminated when used in place of alternate processes. Strain hardening can provide additional strength (up to 70% stronger than machined parts), and favorable grain flow can enhance toughness and fatigue life. As a result, parts can often be made smaller or thinner or from lower-cost materials. Unfortunately, the cost of the required tooling, coupled with the high production speed, generally requires large-volume production—typically in excess of 50,000 parts per year.

16.9 PIERCING

Thick-walled **seamless tubing** can be made by **rotary piercing,** a process illustrated in Figure 16-51. A heated billet is fed longitudinally into the gap between two large, convex-tapered rolls. These rolls are rotated in the same direction, but the axes of the rolls are offset from the axis of the billet by about 6 degrees, one to the right and the other to the left. The clearance between the rolls is preset at a value less than the diameter of the incoming billet. As the billet is caught by the rolls, it is simultaneously rotated and driven forward. The reduced clearance between the rolls forces the billet to deform into a rotating ellipse. As shown in the right-hand segment of Figure 16-51, rotation of the elliptical section causes the metal to shear about the major axis. A crack tends to form down the center axis of the billet, and the cracked material is then forced over a pointed mandrel that enlarges and shapes the opening to create a seamless tube. The result is a short length of thick-walled seamless tubing, which can then be passed

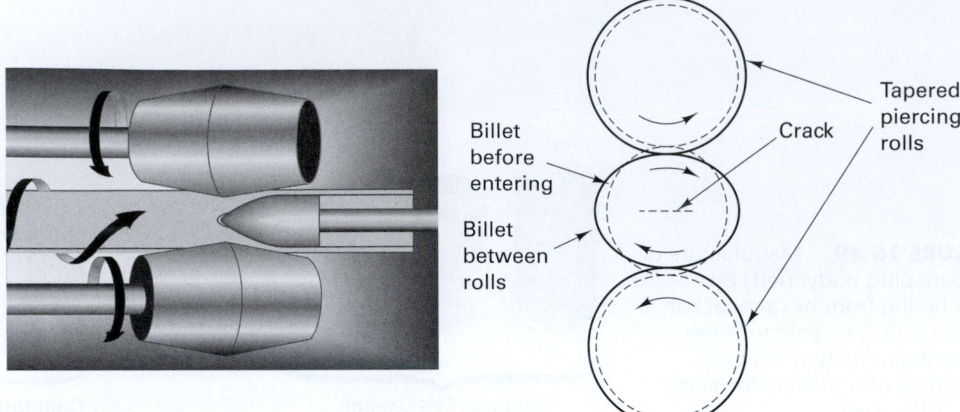

FIGURE 16-51 (Left) Principle of the Mannesmann process of producing seamless tubing. (Right) Mechanism of crack formation in the Mannesmann process.

through sizing rolls to reduce the diameter and/or wall thickness. Seamless tubes can also be expanded in diameter by passing them over an enlarging mandrel. As the diameter and circumference increase, the walls correspondingly thin.

The **Mannesmann mills** commonly used in hot piercing can be used to produce tubing up to 300 mm (12 in.) in diameter. Larger-diameter tubes can be produced on **Stiefel mills,** which use the same principle but replace the convex rolls of the Mannesmann mill with larger-diameter conical disks.

16.10 OTHER SQUEEZING PROCESSES

ROLL EXTRUSION

Thin-walled cylinders can be produced from thicker-wall material by **roll extrusion.** In the variant depicted in Figure 16-52a, internal rollers expand the internal diameter as they squeeze the rotating material against an external confining ring. The tube elongates as the wall thickness is reduced. In Figure 16-52b, the internal diameter is maintained as external rollers squeeze the material against a rotating mandrel. Although the process has been used to produce cylinders from 2 cm to 4 m (0.75 to 156 in.) in diameter, most products have diameters between 7.5 and 50 cm (3 and 20 in.).

SIZING

Sizing involves squeezing all or selected regions of forgings, ductile castings, or powder metallurgy products to achieve a prescribed thickness or enhanced dimensional

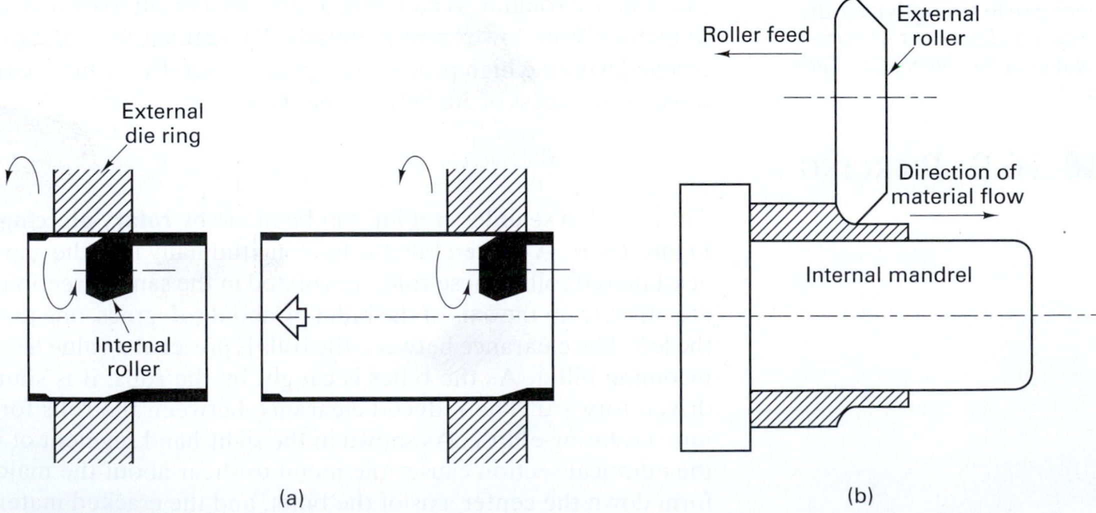

FIGURE 16-52 The roll extrusion process: (a) with internal rollers expanding the inner diameter; (b) with external rollers reducing the outer diameter.

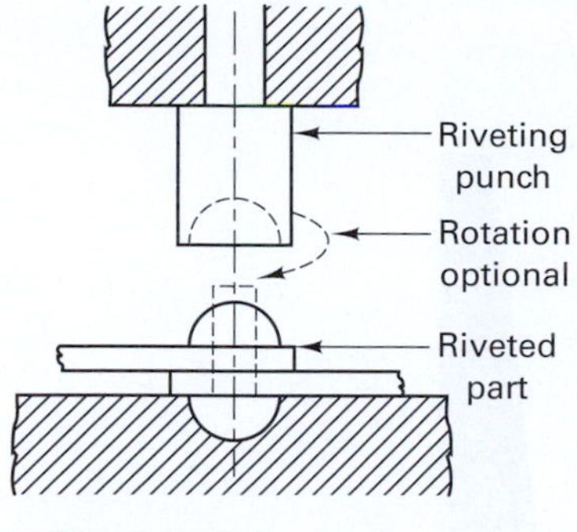

FIGURE 16-53 Joining components by riveting.

precision. By incorporating sizing, designers can make the initial tolerances of a part more liberal, enabling the use of less costly production methods. Those dimensions that must be precise are then set by one or more sizing operations that are usually performed on simple, mechanically driven presses.

RIVETING

In **riveting,** an expanded head is formed on the shank end of a fastener to permanently join sheets or plates of material. Although riveting is usually done hot in structural applications, it is almost always done cold in manufacturing. Where there is access to both sides of the work, the method illustrated in Figure 16-53 is commonly used. The shaped punch may be driven by a press or contained in a special, handheld riveting hammer. When a press is used, the rivet is usually headed in a single squeezing action, although the heading punch may also rotate so as to shape the head in a progressive manner, an approach known as *orbital forming*. Special riveting machines, like those used in aircraft assembly, can punch the hole, place the rivet in position, and perform the heading operation—all in about 1 sec.

It is often desirable to use riveting in situations where there is access to only one side of the assembly. Figure 16-54 shows two types of special rivets that can be used for

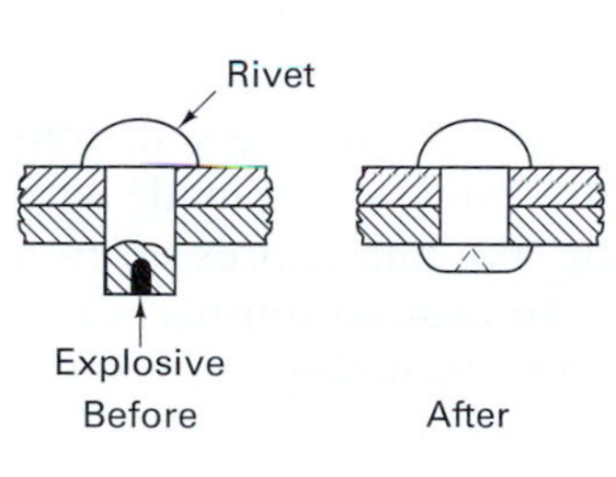

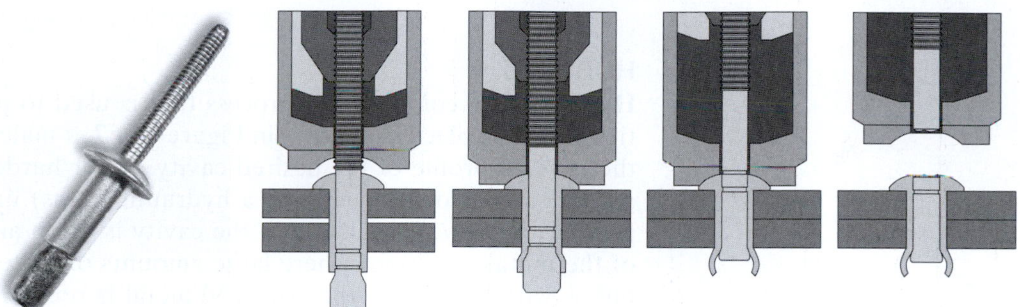

FIGURE 16-54 Rivets for use in "blind" riveting: (left) explosive type; (center) shank-type pull-up; (right) installation sequence for the shank-type pull-up rivet. *(Alcoa Fastening Systems, Magna-Lok®)*

one-side-access applications. The shank on the "blind" side of an **explosive rivet** expands upon detonation to form a retaining head when a heated tool is touched against the exposed segment. In the pull type, or **pop rivet,** a pull-up pin is used to expand a tubular shank. After performing its function, the pull pin breaks or is cut off flush with the head.

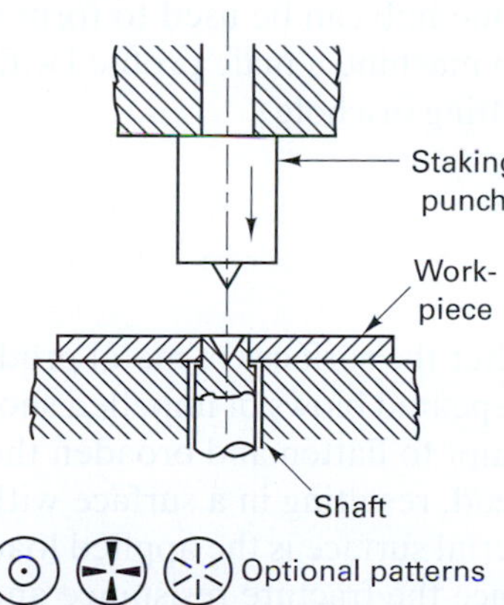

FIGURE 16-55 Permanently attaching a shaft to a plate by staking.

STAKING

Staking is a method of permanently joining parts together when a segment of one part protrudes through a hole in the other. As shown in Figure 16-55, a shaped punch is driven into the exposed end of the protruding piece. The deformation causes radial expansion, mechanically locking the two pieces together. Because the tooling is simple and the operation can be completed with a single stroke of a press, staking is a convenient and economical method of fastening when permanence is desired and the appearance of the punch mark is not objectionable. Figure 16-55 includes some of the decorative punch designs that are commonly used.

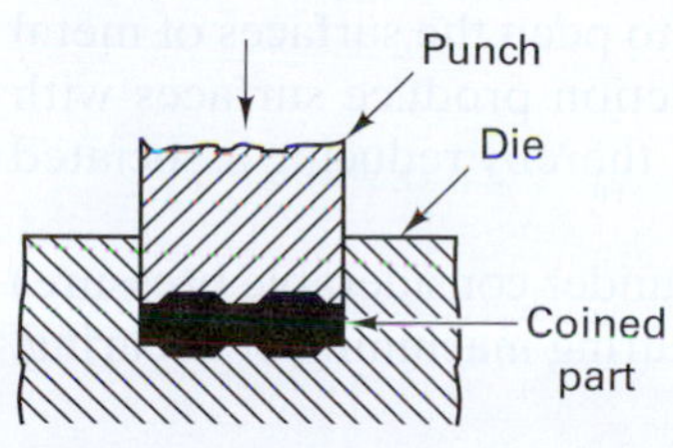

FIGURE 16-56 The coining process.

COINING

The term **coining** refers to the cold squeezing of metal while all of the surfaces are confined within a set of dies. The process, illustrated schematically in Figure 16-56, is used to produce coins, medals, and other products where exact size and fine detail are required and where thickness varies about a well-defined average. Because of the total confinement (there is no possibility for excess metal to escape from the die), the input material must be accurately sized to avoid breakage of the dies or press. Coining pressures may be as high as 1400 MPa (200,000 psi).

FIGURE 16-57 Hubbing a die block in a hydraulic press. Inset shows close-up of the hardened hub and the impression in the die block. The die block is contained in a reinforcing ring. The upper surface of the die block is then machined flat to remove the bulged metal. *(E. Paul DeGarmo)*

HUBBING

Hubbing[2] is a cold-working process that is used to plastically form recessed cavities in a workpiece. As shown in Figure 16-57, a male hub (or master) is made with the reverse profile of the desired cavity. After hardening, the hub is pressed into an annealed block (usually by a hydraulic press) until the desired impression is produced. (*Note:* Production of the cavity is often aided by machining away some of the metal in regions where large amounts of material would be displaced.) The hub is withdrawn, and the displaced metal is removed by a facing-type machining cut. The workpiece, which now contains the desired cavity, is then hardened by heat treatment.

Hubbing is often more economical than die sinking (machining the cavity)—especially when multiple impressions are to be produced. One hub can be used to form a number of identical cavities, and it is generally easier to machine a male profile (with exposed surfaces) than a female cavity (where you are cutting in a hole).

16.11 SURFACE IMPROVEMENT BY DEFORMATION PROCESSING

Deformation processes can also be used to improve or alter the surfaces of metal products. **Peening** is the mechanical working of surfaces by repeated blows of impelled shot or a round-nose tool. The highly localized impacts attempt to flatten and broaden the metal surface, but the underlying material restricts spread, resulting in a surface with residual compression. Because the net loading on a material surface is the applied load minus the residual compression, peening tends to enhance the fracture resistance and fatigue life of tensile-loaded components. For this reason, shot impellers are frequently used to peen shafting, crankshafts, connecting rods, gear teeth, and other cyclic-loaded components.

Manual or pneumatic hammers are frequently used to peen the surfaces of metal weldments. Solidification shrinkage and thermal contraction produce surfaces with residual tension. Peening can reduce or cancel this effect, thereby reducing associated distortion and preventing cracking.

Burnishing involves rubbing a smooth, hard object (under considerable pressure) over the minute surface irregularities that are produced during machining or shearing.

[2] This process should not be confused with "hobbing," a machining process used for cutting gears.

FIGURE 16-58 Tools for roller burnishing. (a) Tool for internal diameter burnishing; (b) Tool for the burnishing of outer diameters The burnishing rolers move inward or outward by means of an adjustable taper. *(Courtesy Monaghan & Associates, Inc., Dayton, OH)*

The edges of sheet metal stampings can be burnished by pushing the stamped parts through a slightly tapered die having its entrance end a little larger than the workpiece and its exit slightly smaller. As the part rubs along the sides of the die, the pressure is sufficient to smooth the slightly rough edges that are characteristic of a blanking operation (see Figure 17-2).

Roller burnishing, illustrated in Figure 16-58, can be used to improve the size and finish of internal and external cylindrical and conical surfaces. The hardened rolls of a burnishing tool press against the surface and deform the protrusions to a more-nearly-flat geometry. The resulting surfaces possess improved wear and fatigue resistance, because they have been cold worked and are now in residual compression.

■ KEY WORDS

automatic hot forging
backward extrusion
bar
bar drawing
billet
blocking
bloom
blowout
bulk deformation processes
burnishing
closed-die forging
cluster mill
coining
cold forming
cold heading
cold rolling
computer-controlled hammer
Conform process
continuous extrusion
controlled rolling
counterblow machine
crowned roll
direct extrusion
draw bench
drawing
explosive rivet
extrusion
finisher impression
finishing temperature
flash
flashless forging
floating plug
forging
forward extrusion
gravity-drop hammer
hubbing
hydraulic press
hydrostatic extrusion
impact extrusion
impactor
impression-die forging
indirect extrusion
isothermal forging
lubrication
mandrel
Mannesmann mill
mechanical press
mill scale
near-net-shape forging
net-shape forging
open-die hammer forging
orbital forging
pack rolling
peening
pipe
plate
pop rivet
precision forging
press forging
pressure-induced ductility
pressure-to-pressure extrusion
ring rolling
riveting
rod
rod drawing
roll extrusion
roll forging
roller burnishing
rolling
rotary piercing
seamless tubing
screw press
sheet
sheet-forming operations
sizing
slab
spider-mandrel die
staking
Stiefel mill
strip
structural shape
swaging
thermomechanical processing
thread rolling
tube
tube drawing
tube sinking
upset forging
wire drawing

■ REVIEW QUESTIONS

1. Briefly describe the evolution of forming equipment from ancient to modern.
2. What are some of the possible means of classifying metal deformation processes?
3. How are bulk deformation processes different from sheet-forming operations? Describe each.
4. What are some common bulk deformation processes? Sheet-forming operations?
5. Why might the method of analysis be different for a process like forging and a process like rolling?
6. What are some of the common terms applied to the various sizes and shapes of rolled products?
7. Why are hot-rolled shaped products generally limited to standard forms and sizes?
8. Why is it undesirable to minimize friction between the workpiece and tooling in a rolling operation?
9. Why is it desirable to have uniform temperature when hot rolling a material?
10. Why is it important to control the finishing temperature of a hot-rolling operation?
11. What are some of the attractive attributes of cold rolling?
12. What are some of the possible conditions to which cold-rolled material can be purchased?
13. Discuss the relative advantages and typical uses of two-high rolling mills with large-diameter rolls, three-high mills, and four-high mills.
14. Roll flexing can result in nonuniform thickness in the rolled product. How do four-high and cluster mills minimized this flexing?
15. Why is foil almost always rolled on a cluster mill?
16. Why is speed synchronization of the various rolls so vitally important in a continuous or multiple-stand rolling mill?
17. What types of products are produced by ring rolling?
18. Explain how hot-rolled products can have directional properties and residual stresses.
19. Discuss the problems in maintaining uniform thickness in a rolled product and some of the associated defects.
20. Why is a "crowned" roll always designed for a specific operation on a specific material?
21. How might the addition of horizontal tensions act to improve the thickness uniformity of rolled products?
22. What is thermomechanical processing, and what are some of its possible advantages?
23. Provide a concise description of the forging process.
24. What are some of the types of flow that can occur in forging operations?
25. Why are steam or air hammers more attractive than gravity-drop hammers for hammer forging?
26. What are some of the attractive features of computer-controlled forging hammers?
27. What is the difference between open-die and impression-die forging?
28. Why is open-die forging not a practical technique for large-scale production of identical products?
29. What additional controls must be exercised to perform flashless forging?
30. What is a blocker impression in a forging sequence?
31. What attractive features are offered by counterblow forging equipment, or impactors?
32. For what types of forging products or conditions might a press be preferred over a hammer?
33. Why are heated dies generally employed in hot-press forging operations?
34. Describe some of the primary differences among hammers, mechanical presses, and hydraulic presses.
35. Why are different tolerances usually applied to dimensions contained within a single die cavity and dimensions across the parting plane?
36. What are some of the roles played by lubricants in forging operations?
37. What are some of the attractive features of orbital forging?
38. What types of product geometry can be produced by orbital forging?
39. What is upset forging?
40. What are some of the typical products produced by upset-forging operations?
41. What types of products can be produced by automatic hot forging?
42. What are some of the attractive features of automatic hot forging? What is a major limitation?
43. How does roll forging differ from a conventional rolling operation?
44. What is swaging? What kind of products are produced?
45. How can the swaging process impart different sizes and shapes to an interior cavity and the exterior of a product?
46. What are some possible objectives of near-net-shape forging?
47. Provide a concise definition of extrusion.
48. What metals can be shaped by extrusion?
49. What are some of the attractive features of the extrusion process?
50. What is the primary shape limitation of the extrusion process?
51. What is the primary benefit of indirect extrusion?
52. What property of a lubricant is critical in extrusion that might not be required for processes such as forging?
53. What types of products are made using a spider-mandrel die? Why can lubricants not be used in spider-mandrel extrusion?
54. What are some of the unique capabilities and special limitations of hydrostatic extrusion?
55. What is the unique capability provided by pressure-to-pressure hydrostatic extrusion?
56. How is the feedstock pushed through the die in continuous extrusion processes?
57. What types of feedstock can be used in continuous extrusion other than the conventional solid rod?
58. Why are rods generally drawn on draw benches, while wire is drawn on draw block machines?
59. Why is the reduction in area significantly restricted during wire, rod, and tube drawing?
60. What is the difference between tube drawing and tube sinking?
61. For what types of products might a floating plug be employed?
62. What are the size limitations for the cold drawing of complex cross-section steel products?
63. What types of materials are used for wire-drawing dies?
64. What types of products are produced by cold heading?

65. What is impact extrusion, and what variations exist?
66. If a product contains a large-diameter head and a small-diameter shank, how can the processes of cold extrusion and cold heading be combined to save metal?
67. What are some of the attractive properties or characteristics of cold-forming operations?
68. How might cold forming be used to substantially reduce material waste?
69. What processes can be used to produce seamless pipe or tubing?
70. What type of products can be made by the roll-extrusion process?
71. What types of rivets can be used when there is access to only one side of a joint?
72. How is coining different from a process known as embossing?
73. Why might hubbing be an attractive way to produce a number of identical die cavities?
74. How might a peening operation increase the fracture resistance of a product?
75. What is burnishing?

PROBLEMS

1. Some snack foods, such as rectangular corn chips, are often formed by a rolling-type operation and are subject to the same types of defects common to rolled sheet and strip. Obtain a bag of such a snack and examine the chips to identify examples of rolling-related defects such as those discussed in Section 16-4 and shown in Figure 16-A.

FIGURE 16-A Some typical defects that occur during rolling: wavy edges, edge cracking, and center cracking.

2. Consider the extrusion of a cylindrical billet, and compute the following.
   a. Assume the starting billet to have a length of 0.3 m and a diameter of 15 cm. This is extruded into a cylindrical product that is 3 cm in diameter and 7.5 m long (a reduction ratio of 25). Neglecting the areas on the two ends, compute the ratio between the product surface area (wrap around cylinder) and the surface area of the starting billet.
   b. How would this ratio change if the product were a square with the same cross-sectional area as that of the 3-cm-diameter circle?
   c. Consider a cylinder-to-cylinder extrusion with a reduction ratio of R. Derive a general expression of the relative surface areas of product to billet as a function of R. (*Hint:* Start with a cylinder with length and diameter both equal to 1 unit. Because the final area will be $1/R$ times the original, the final length will be R units, and the final diameter will be proportional to $1/\sqrt{R}$).
   d. If the final product had a more complex cross-sectional shape than a cylinder., would the final area be greater than or less than that computed in part c?
   e. Relate your answers to parts a–d to a consideration of lubrication during large-reduction extrusion operations.
3. The force required to compress a cylindrical solid between flat parallel dies (see Figure 16-10) has been estimated (by a theory of plasticity analysis) to be

$$\text{Force} = \pi R^2 \sigma_o \frac{1 + 2mR}{3\sqrt{3}T}$$

where:

R = radius of the cylinder
T = thickness of the cylinder
σ_o = yield strength of the material
m = friction factor (between 0 and 1 where 0 is frictionless and 1 is complete sticking)

An engineering student is attempting to impress his date by demonstrating some of the neat aspects of metal forming. He places a shiny penny between the platens of a 60,000-lb capacity press and proceeds to apply pressure. Assume that the coin has a $\frac{3}{4}$-in. diameter and is $\frac{1}{16}$ in. thick. The yield strength is estimated as 50,000 psi, and because no lubricant is applied, friction is that of complete sticking, or $m = 1.0$.
   a. Compute the force required to induce plastic deformation.
   b. If this force is greater than the capacity of the press (60,000 lb), compute the pressure when the full-capacity force of 60,000 lb is applied.
   c. If the press surfaces are made from thick plates of steel with a yield strength of 120,000 psi, describe the results of the demonstration.
   d. A simple model of forging force uses the equation:

$$\text{Force} = K\sigma_o A$$

where:

K = a dimensionless multiplying factor
σ_o = yield strength of the material
A = projected area of the forging

K is assigned a value of 3–5 for simple shapes without flash, 5\–8 for simple shapes with flash, and 8–12 for complex shapes with flash. Consider the two equations for forging force and discuss their similarities and differences.
4. Mathematical analysis of the rolling of flat strip reveals that the roll-separation force (the squeezing force required to deform the strip) is directly proportional to the term:

$$1 + \frac{K_1 mR}{t_{av}}$$

where:

K_1 = geometric constant
m = friction factor
L = length of contact
t_{av} = average thickness of the strip in the roll bite

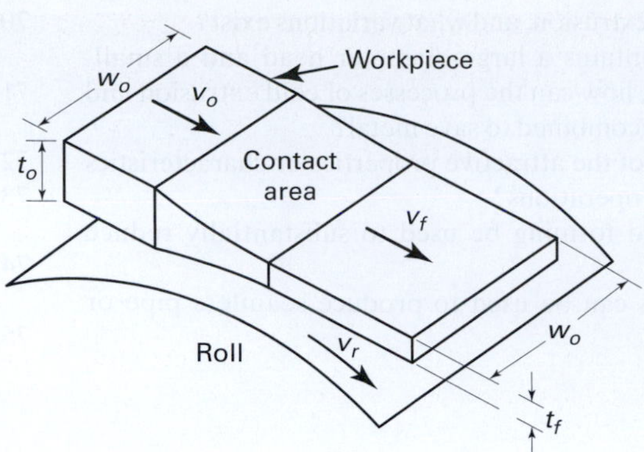

FIGURE 16-B Strip rolling where the width of the strip remains unchanged. The lines across the workpiece identify the area of contact with the rolls. The top roll has been removed for ease of visualization.

Because L is proportional to the roll radius, R, K_1L can be replaced by K_2R so the force becomes proportional to the term:

$$1 + \frac{K_2mR}{t_{av}}$$

If K_2 and m are both positive numbers, how will the roll-separation force change as the strip becomes thinner? How can this effect be minimized? Relate your observations to the types of rolling mills used for various thicknesses of product.

5. In Figure 16-29, the vertical axis is force and the horizontal axis is position. If force is measured in pounds and position in feet, the area under the curves has units of foot-pound, and is a measure of the work performed. If the area under the indirect extrusion curve is proportional to the work required to extrude a product without billet–chamber frictional resistance, how could the relative regions of the direct extrusion curve be used to determine a crude measure of the mechanical "efficiency" of direct extrusion?
6. Compare the forming processes of wire drawing, conventional extrusion, and continuous extrusion with respect to continuity, reduction in area possible in a single operation, possible materials, speeds, typical temperatures, and other important processing variables.
7. Figure 16-B shows the rolling of a wide, thin strip where the width remains constant as thickness is reduced. Material enters the mill at a rate equal to $t_ow_ov_o$ and exits at a rate of $t_fw_ov_f$. Because material cannot be created or destroyed, these rates must be equal, and the w_o terms will cancel. As a result, v_f is equal to $(t_o/t_f)v_o$. The material enters at velocity v_o and accelerates to velocity v_f as the material passes through the mill. For stable rolling, the velocity of the roll surface, v_r, which is a constant, must be a value between v_o and v_f. For these conditions, describe the relative sliding between the strip and the rolls as the strip moves through the region of contact.

www.wiley.com/go/global/degarmo

Chapter 16 CASE STUDY

Handle and Body of a Large Ratchet Wrench

The figure shows the handle and body segment of a relatively large ratchet wrench, such as those used with conventional socket sets. The design specifications require a material with a minimum yield strength of 50,000 psi and an elongation of at least 2% in all directions. Additional consideration should be given to weight minimization (because of the relatively large size of the wrench), corrosion resistance (due to storage and use environments), machinability (if finish machining is required), and appearance.

1. Based on the size and shape of the product, describe several methods that could be used to produce the component. For each method, briefly discuss the relative pros and cons.
2. What types of engineering materials might be able to meet the requirements? What would be the pros and cons of each general family?
3. For each of the shape generation methods in part 1, select an appropriate material from the alternatives discussed in part 2, making sure that the process and material are compatible. (*Note:* Casting alloys should be matched with casting processes, wrought alloys with forming processes, etc.)
4. Which of the combinations do you feel would be the "best" solution to the problem? Why?
5. For this system, outline the specific steps that would be required to produce the part from reasonable starting material.
6. For your proposed solution, would any additional heat treatment or surface treatment be required or recommended? If so, what would you suggest?
7. How might you modify your proposed solution for each of the following additional conditions?
   a. An adequate tool, but made at the lowest possible cost.
   b. A "top-of-the line" tool where high cost is perfectly acceptable.
   c. The tool is to be marketed as a "safety tool" that could be used in areas of gas leaks where a spark might be fatal—the material must be "nonpyrophoric."

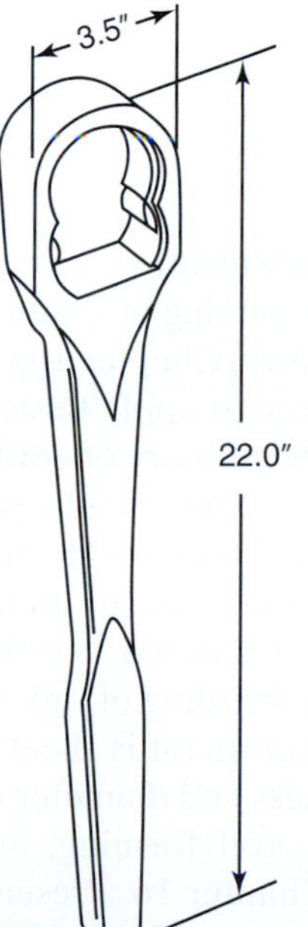

CHAPTER 17

SHEET-FORMING PROCESSES

17.1 INTRODUCTION

The various classification schemes for metal deformation processes have been presented at the beginning of Chapter 16, with the indication that our text will be grouping by bulk processes (Chapter 16) and sheet processes (Chapter 17). Bulk forming uses heavy machinery to apply three-dimensional stresses, and most of the processes are considered to be primary operations. Sheet metal processes, on the other hand, generally involve plane stress loadings and lower forces than bulk forming. Almost all sheet metal forming is considered to be secondary processing.

The classification into **bulk** and **sheet** is far from distinct, however. Some processes can be considered as either, depending on the size, shape, or thickness of the workpiece. The bending of rod or bar is often considered to be bulk forming, while the bending of sheet metal is sheet forming. Tube bending can be either, depending on the wall thickness and diameter of the tube. Similar areas of confusion can be found in deep drawing, roll forming, and other processes. The squeezing processes were described in Chapter 16. Presented here will be the processes that involve *shearing*, *bending*, and *drawing*. Table 17-1 lists some of the processes that fit these categories.

17.2 SHEARING OPERATIONS

Shearing is the mechanical cutting of materials without the formation of chips or the use of burning or melting. It is often used to prepare materials between 0.025 and 20 mm (0.001 and 0.8 in.) in thickness for subsequent operations, and its success helps to ensure the accuracy and precision of the finished product. When the two cutting blades are straight, the process is called *shearing*. When the blades are curved, the processes have special names, such as *blanking*, *piercing*, *notching*, and *trimming*. In terms of tool design and material behavior, however, all are shearing-type operations.

TABLE 17-1 Classification of the Nonsqueezing Metal-Forming Operations

Shearing	Bending	Drawing and Stretching
1. Simple shearing	1. Angle bending	1. Spinning
2. Slitting	2. Roll bending	2. Shear forming or flow turning
3. Piercing	3. Draw bending	3. Stretch forming
4. Blanking	4. Compression bending	4. Deep drawing and shallow drawing
5. Fineblanking	5. Press bending	5. Rubber-tool forming
6. Lancing	6. Tube bending	6. Sheet hydroforming
7. Notching	7. Roll forming	7. Tube hydroforming
8. Nibbling	8. Seaming	8. Hot drawing
9. Shaving	9. Flanging	9. High-energy-rate forming
10. Trimming	10. Straightening	10. Ironing
11. Cutoff		11. Embossing
12. Dinking		12. Superplastic sheet forming

A simple type of shearing operation is illustrated in Figure 17-1. As the punch (or upper blade) pushes on the workpiece, the metal responds by flowing plastically into the die (or over the lower blade). Because the clearance between the two tools is small, usually between 5 and 20% of the thickness of the metal being cut, the deformation occurs as highly localized shear. As the punch pushes downward on the metal, the material flows into the die, with the opposite surface bulging slightly. An instability arises when the penetration is between 15 and 60% of the metal thickness, the actual amount depending on the strength and ductility of the material. The applied stress exceeds the shear strength of the remaining material, and the metal tears or ruptures through the rest of its thickness, linking the cutting edges of the punch and die to produce an inwardly inclined fracture and a ragged edge or burr. As shown in Figure 17-2, the two

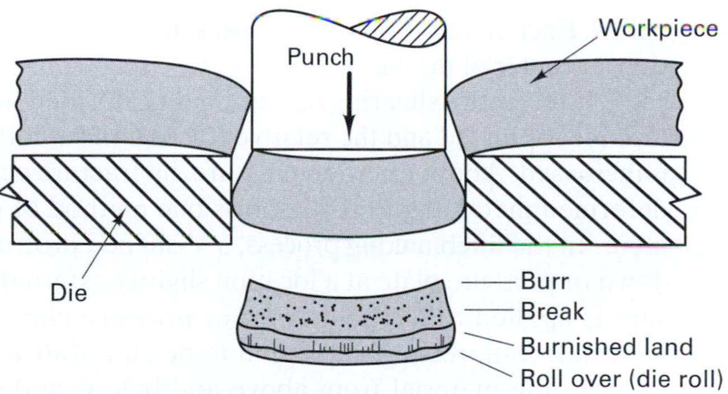

FIGURE 17-1 Simple blanking with a punch and die.

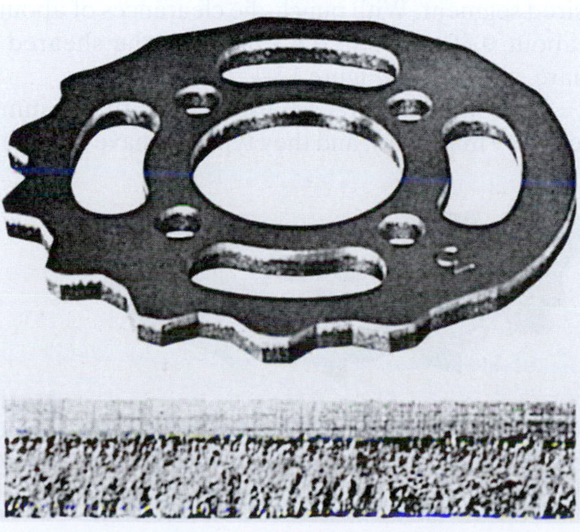

FIGURE 17-2 (Top) Conventionally sheared surface showing the distinct regions of deformation and fracture. (Bottom) Magnified view of the sheared edge. *(Courtesy of Feintool Equipment Corp., Cincinatti, OH)*

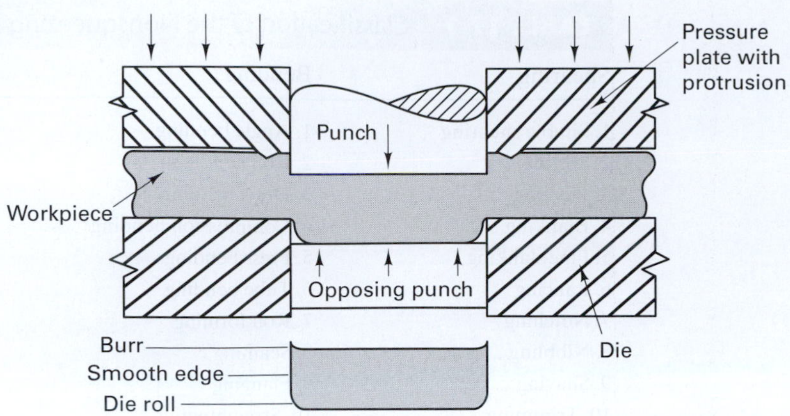

FIGURE 17-3 Fineblanking method of obtaining a smooth edge in shearing by using a shaped pressure plate to put the metal into localized compression and a punch and opposing punch descending in unison. The resulting product shows die roll and burr formation on opposite sides and a smooth edge surface.

distinct stages of the shearing process—deformation (producing the smooth burnished band) and fracture—are often visible on the edges of sheared parts.

Because of the normal inhomogeneities in a metal and the possibility of nonuniform clearance between the shear blades, the final shearing does not occur in a uniform manner. Fracture and tearing begin at the weakest point and proceed progressively or intermittently to the next-weakest location. This usually results in a rough and ragged edge that, combined with possible microcracks and work hardening of the sheared edge, can adversely affect subsequent forming processes.

Changing the clearance between the punch and the die can greatly change the condition of the cut edge. If the punch and die (or upper and lower shearing blades) have proper alignment and clearance, and are maintained in good condition, sheared edges can be produced that have sufficient smoothness to permit use without further finishing. The quality of the sheared edge can often be improved by clamping the starting stock firmly against the die (from above) and restraining the movement of the sheared piece by a plunger or rubber die cushion that applies opposing pressure from below. Each of these measures causes the shearing to take place more uniformly around the perimeter of the cut.

If the entire shearing operation is performed in a compressive environment, fracture is suppressed and the relative fraction of smooth edge (produced by deformation) is increased. Above a certain pressure, no fracture occurs and the entire edge is smooth, deformed metal. Figure 17-3 shows one method of producing a compressive environment. In the **fineblanking** process, a V-shaped protrusion is incorporated into the hold-down or pressure plate at a location slightly external to the contour of the cut. As pressure is applied to the hold-down or pressure plate, the protrusion is driven into the material, compressing the region to be cut. Matching upper and lower punches then squeeze the material from above and below, and descend in unison, extracting the desired segment. With punch-die clearances of about $\frac{1}{10}$, those of conventional blanking or about 0.5% of material thickness, the sheared edges are now both smooth and square, as shown in Figure 17-4.

Fineblanked parts are usually less than 6 mm ($\frac{1}{4}$ in.) in thickness, but can be as thick as 19 mm ($\frac{3}{4}$ in.), and they typically have complex-shaped perimeters. Dimensional

FIGURE 17-4 Fineblanked surface of the same component shown in Figure 17-2. *(Courtesy of Feintool Equipment Corp., Cincinatti, OH)*

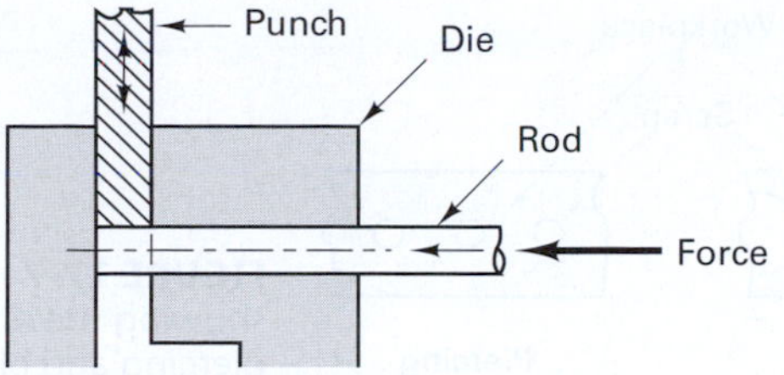

Punch
Die
Rod
Force

FIGURE 17-5 Method of smooth shearing a rod by putting it into compression during shearing.

accuracy is often within 0.05 mm (0.002 in.), and holes, slots, bends, and semipierced projections can be incorporated as part of the fineblanking operation. Holes and web sections can be smaller than the material thickness. Because the parts are pressed from both sides during the shearing, excellent flatness is maintained. Secondary edge finishing can often be eliminated, and the work hardening that occurs during the shearing process enhances wear resistance.

In fineblanking, however, a triple-action mechanical or hydraulic press is generally required. The fineblanking force is about 40% greater than conventional blanking of the same contour, and the extra material required for the impinging protrusion often forces a greater separation between nested parts. The added costs must be offset by a reduction in secondary operations, such as drilling, reaming, grinding, or other machining.

Similar results have been observed for the **GRIPflow** process (EBway Corporation), another process for blanking under pressure. Figure 17-5 illustrates yet another means of shearing under compression. Bar stock is pressed against the closed end of a feed hole, placing the stock in a state of compression. A transverse punch then shears the material into smooth-surface, burr-free slugs, ready for further processing.

SIMPLE SHEARING

When sheets of metal are to be sheared along a straight line, **squaring shears,** like the one shown in Figure 17-6, are frequently used. As the upper ram descends, a clamping bar or set of clamping fingers presses the sheet of metal against the machine table to hold it firmly in position. A moving blade then comes down across a fixed blade and shears the metal. On larger shears, the moving blade is often set at an angle or "rocks" as it descends, so the cut is made in a progressive fashion from one side of the material to the other, much like a pair of household scissors. This action significantly reduces the amount of cutting force required, replacing a high force–short stroke operation with one of low force and longer stroke.

The upper blade may also be inclined about 0.5 to 2.5 degrees with respect to the lower blade and set to descend along this line of inclination. While squareness and edge quality may be compromised, this action helps to ensure that the sheared material does not become wedged between the blades.

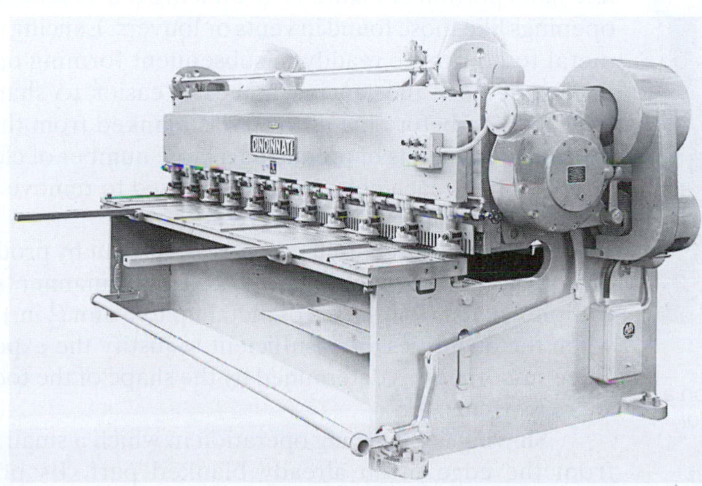

FIGURE 17-6 A 3-m (10-ft) power shear for 6.5 mm ($\frac{1}{4}$ in.) steel. *(Courtesy of Cincinnati Incorporated, Cincinnati, OH)*

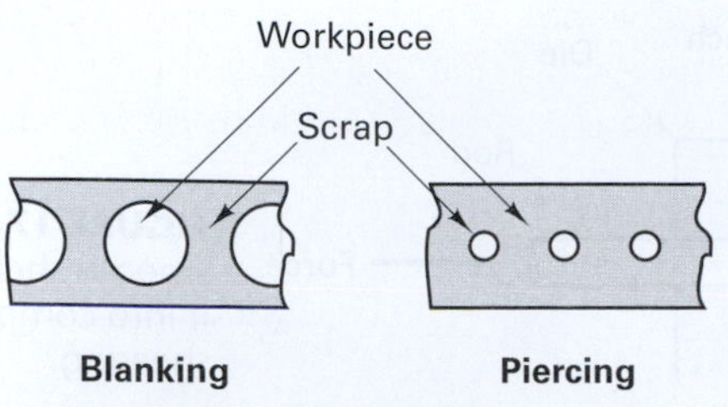

FIGURE 17-7 Schematic showing the difference between piercing and blanking.

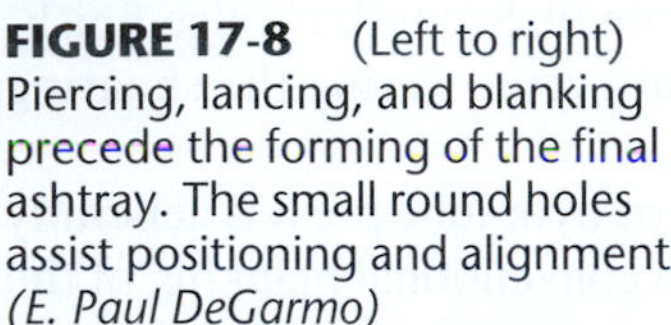

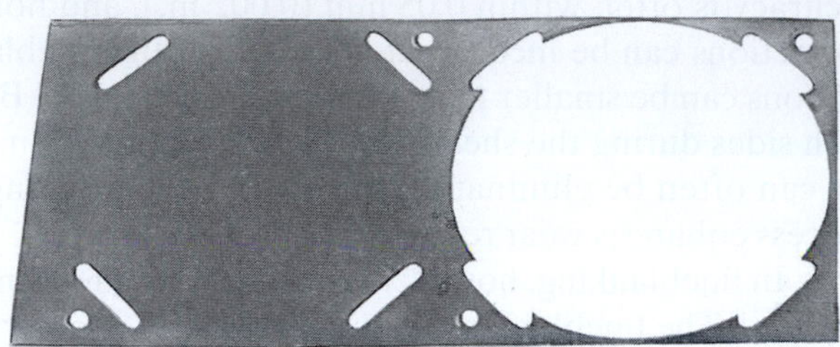

FIGURE 17-8 (Left to right) Piercing, lancing, and blanking precede the forming of the final ashtray. The small round holes assist positioning and alignment. *(E. Paul DeGarmo)*

SLITTING

Slitting is the lengthwise shearing process used to cut coils of sheet metal into several rolls of narrower width. Here, the shearing blades take the form of cylindrical rolls with circumferential mating grooves. The raised ribs of one roll match the recessed grooves on the other. The process is now continuous and can be performed rapidly and economically. Moreover because the distance between adjacent shearing edges is fixed, the resultant **strips** have accurate and constant width, more consistent than that obtained from alternative cutting processes.

PIERCING AND BLANKING

Piercing and **blanking** are shearing operations where a part is removed from sheet material by forcing a shaped punch through the sheet and into a shaped die. Any two-dimensional shape can be produced, with one surface having a slightly rounded edge and the other containing a slight burr. Because both processes involve the same basic cutting action, the primary difference is one of definition. Figure 17-7 shows that in blanking, the piece being punched out becomes the workpiece. In piercing, the punch-out is the scrap and the remaining strip is the workpiece. The term *piercing* can also be used to describe the formation of a hole by a pointed punch where no metal is removed from the sheet (like driving a nail through sheet metal). Piercing and blanking are usually done on some form of mechanical press.

Several variations of piercing and blanking are known by specific names. **Lancing** is a piercing operation that forms either a line cut (slit) or hole, like those shown in the left-hand portion of Figure 17-8. Lancing can be combined with bending to form tabs or openings like those found in vents or louvers. Lancing is also used to permit the adjacent metal to flow more readily in subsequent forming operations. In the case illustrated in Figure 17-8, the lancing makes it easier to shape the recessed grooves, which were formed before the ashtray was blanked from the strip stock and shallow drawn. **Perforating** consists of producing a large number of closely spaced holes where a slug is removed from each hole. **Notching** is used to remove segments from along the edge of an existing product.

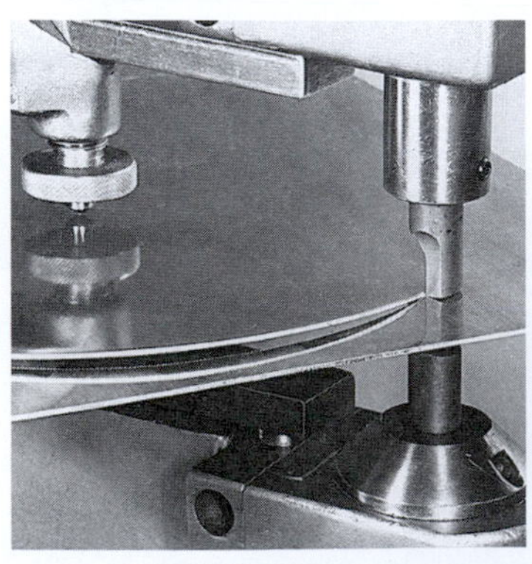

FIGURE 17-9 Shearing operation being performed on a nibbling machine. *(Courtesy of Pacific Press Technologies, Mt. Carmel, IL)*

In **nibbling,** a contour is progressively cut by producing a series of overlapping slits or notches, as shown in Figure 17-9. In this manner, simple tools can be used to cut a complex shape from sheets of metal up to 6 mm ($\frac{1}{4}$ in.) thick. The process is widely used when the quantities are insufficient to justify the expense of a dedicated blanking die. Edge smoothness is determined by the shape of the tooling and the degree of overlap in successive cuts.

Shaving is a finishing operation in which a small amount of metal is sheared away from the edge of an already blanked part. Its primary use is to obtain greater

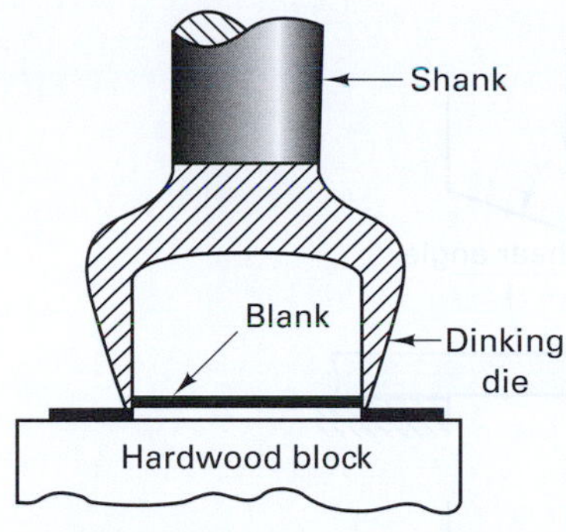

FIGURE 17-10 The dinking process.

dimensional accuracy, but it may also be employed to produce a squared or smoother edge. Because only a small amount of metal is removed, the punches and dies must be made with very little clearance. Blanked parts, such as small gears, can be shaved to produce dimensional accuracies within 0.025 mm (0.001 in.).

In a **cutoff** operation, a punch and die are used to separate a stamping or other product from a strip of stock. The contour of the cutoff frequently completes the periphery of the workpiece. Cutoff operations are quite common in progressive die sequences, like several to be presented shortly.

Dinking is a modified shearing operation that is used to blank shapes from low-strength materials, such as rubber, fiber, or cloth. As illustrated in Figure 17-10, the shank of a die is either struck with a hammer or mallet or the entire die is driven downward by some form of mechanical press.

TOOLS AND DIES FOR PIERCING AND BLANKING

As shown in Figure 17-11, the basic components of a piercing and blanking die set are a **punch,** a **die,** and a **stripper plate,** which is attached above the die to keep the strip material from ascending with the retracting punch. The position of the stripper plate and the size of its hole should be such that it does not interfere with either the horizontal motion of the strip as it feeds into position or the vertical motion of the punch.

Theoretically, the punch should fit within the die with a uniform clearance that approaches zero. On its downward stroke, it should not enter the die but should stop just as its base aligns with the top surface of the die. In general practice, the clearance is between 5 to 7% of the stock thickness and the punch enters slightly into the die cavity.

If the face of the punch is normal to the axis of motion, the entire perimeter is cut simultaneously. The maximum cutting force can then be computed as:

$$\text{Perimeter length} \times \text{Thickness} \times \text{Material shear stress}^{1}$$

By tilting the punch face on angle, a feature known as **shear** or **rake angle,** the cutting force can be reduced substantially. As shown in Figure 17-12, the periphery is now cut in a progressive fashion, similar to the action of a pair of scissors or the opening of a "pop-top" beverage can. Variation in the shear angle controls the amount of cut that is made at any given time and the total stroke that is necessary to complete the operation. Adding shear increases the stroke, but reduces the force, providing an attractive way to cut thicker or stronger material on an existing piece of equipment.

Punches and dies should also be in proper alignment so that a uniform clearance is maintained around the entire periphery. The die is usually attached to the bolster plate of the press, which, in turn, is attached to the main press frame. The punch is attached to the movable ram, enabling motion in and out of the die with each stroke of the

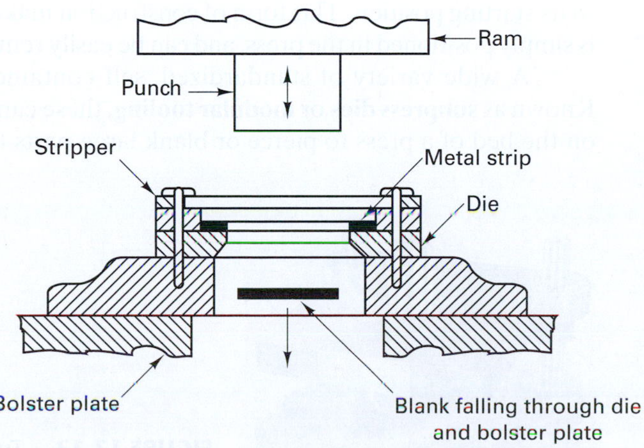

FIGURE 17-11 The basic components of piercing and blanking dies.

[1] Tensile strength is often substituted for shear strength to provide a margin of safety when designing tooling or specifying machinery.

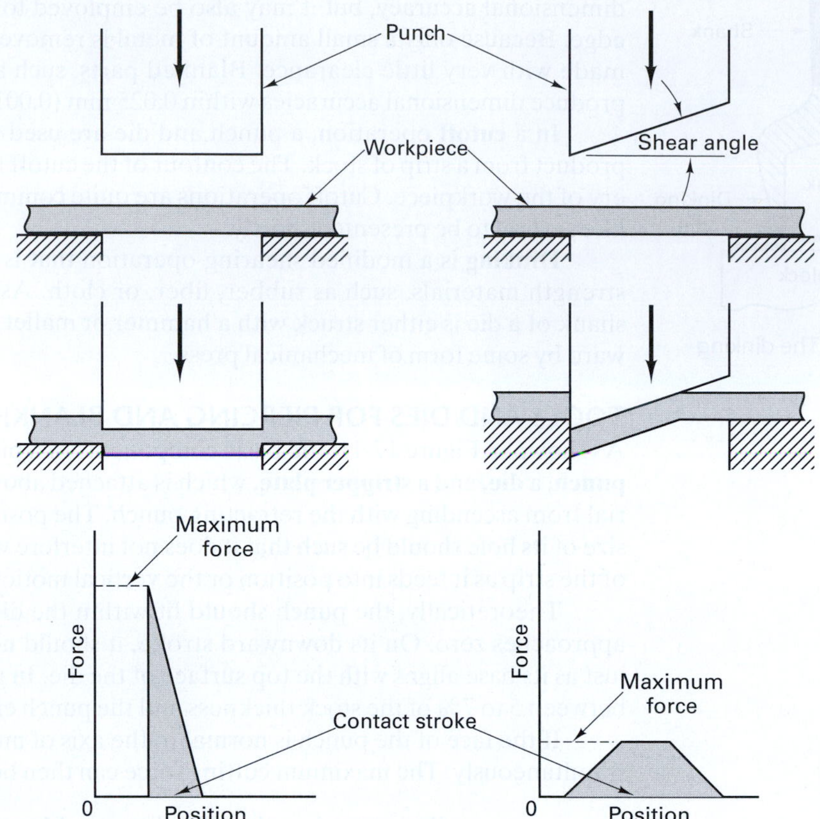

FIGURE 17-12 Blanking with a square-faced punch (left) and one containing angular shear (right). Note the difference in maximum force and contact stroke. The total work (the area under the curve) is the same for both processes.

press. Punches and dies can also be mounted on a separate **punch holder** and **die shoe,** like the one shown in Figure 17-13, to create an **independent die set.** The holder and shoe are permanently aligned and guided by two or more guide pins. By aligning a punch and die, and fastening them to the die set, an entire unit can be inserted into a press without having to set or check the tool alignment. This can significantly reduce the amount of production time lost during tool change. Moreover, when a given punch and die are no longer needed, they can be removed and new tools attached to the shoe and holder assembly.

In most cases, the punch holder attaches directly to the ram of the press, and ram motion acts to both raise and lower the punch. On smaller die sets, springs can be incorporated to provide the upward motion. The ram simply pushes on the top of the punch holder, forcing it downward. When the ram retracts, the springs cause the punch to return to its starting position. This form of construction makes the die set fully self-contained. It is simply positioned in the press, and can be easily removed, thereby reducing setup time.

A wide variety of standardized, self-contained die sets have been developed. Known as **subpress dies** or **modular tooling,** these can often be assembled and combined on the bed of a press to pierce or blank large parts that would otherwise require large

FIGURE 17-13 Typical die set having two alignment guideposts. *(Courtesy of Danly IEM, Ithaca, MI)*

and costly complex die sets. A single downward motion of the press activates each of the subpress dies, producing a variety of holes and slots, each in the proper relation to one another.

Punches and dies are usually made from low-distortion or air-hardenable tool steel so they can be hardened after machining with minimal warpage. The die profile is maintained for a depth of about 3 mm ($\frac{1}{8}$ in.) from the upper face, beyond which an angular clearance or back relief is generally provided (see Figure 17-11) to reduce friction between the part and the die and to permit the part to fall freely from the die after being sheared. The 3-mm depth provides adequate strength and sufficient metal so that the shearing edge can be resharpened by grinding a few thousandths of an inch from the face of the die.

Dies can be made as a single piece, or they can be made in component sections that are assembled on the punch holder and die shoe. The component approach simplifies production and enables the replacement of single sections in the event of wear or fracture. Substantial savings can often be achieved by modifying the design of parts to enable the use of standard die components. A further advantage of this approach is that when the die set is no longer needed, the components can be removed and used to construct tooling for another product.

When the cut periphery is composed of simple lines, and the material being cut is either soft metal or other soft material (such as plastics, wood, cork, felt, fabrics, and cardboard), **steel-rule,** or *cookie-cutter,* **dies** can often be used. The cutting die is fashioned from hardened steel strips, known as *steel rule,* that are mounted on edge in grooves that have been machined in the upper die block. The mating piece of tooling can be either a flat piece of hardwood or steel, a male shape that conforms to the part profile (such that the protruding strips descend around it), or a set of matching grooves into which the upper die can descend. Rubber pads are usually inserted between the strips to replace the stripper plate. During the compression stroke, the rubber compresses and allows the cutting action to proceed. As the ram ascends, the rubber then expands to push the blank free of the steel-rule cavity. Steel-rule dies are usually less expensive to construct than solid dies and are quite attractive for producing small quantities of parts.

Many parts require multiple cutting-type operations, and it is often desirable to produce a completed part with each cycle of a press. Several types of dies have been designed to accomplish this task. For simplicity, their operations are discussed in terms of manufacturing simple, flat washers from a continuous strip of metal.

The **progressive die** set, depicted in Figure 17-14, is the simpler of the two types. Basically, it consists of two or more sets of punches and dies mounted in tandem. Strip stock is fed into the first die, where a hole is pierced as the ram descends. When the ram raises, the stock advances and the pierced hole is positioned under the blanking punch.

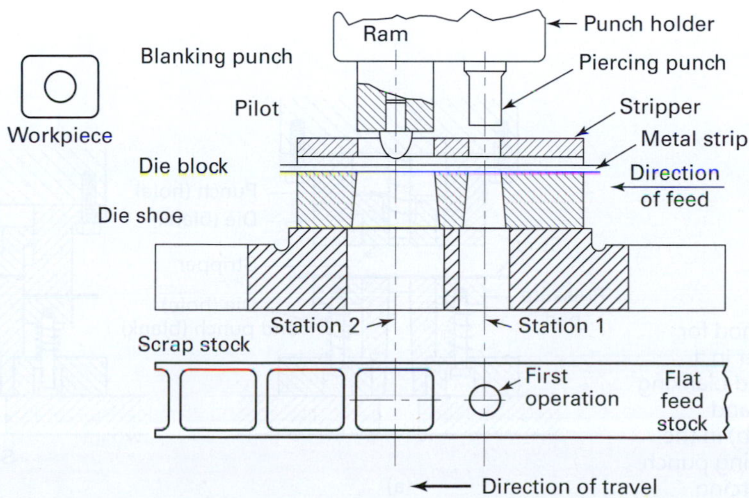

FIGURE 17-14 Progressive piercing and blanking die for making a square washer. Note that the punches are of different length.

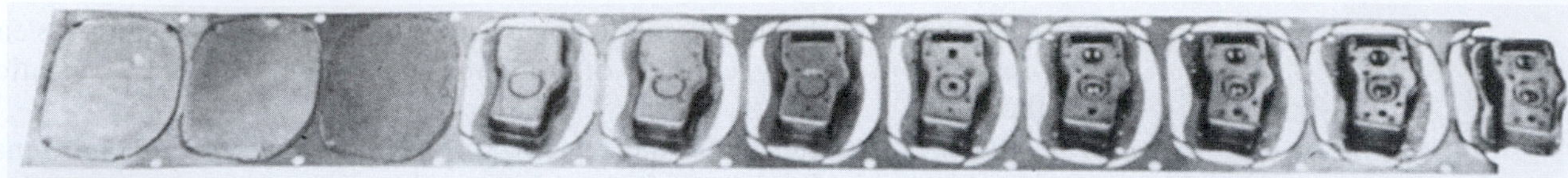

FIGURE 17-15 The various stages of an 11-station progressive die. *(Courtesy of the Minster Machine Company, Minster, OH)*

Upon the second descent, a pilot on the bottom of the blanking punch enters the hole that was pierced on the previous stroke to ensure accurate alignment. Further descent of the punch blanks the completed washer from the strip, and, at the same time, the first punch pierces the hole for the next washer. As the process continues, a finished part is completed with each stroke of the press.

Progressive dies can be used for many combinations of piercing, blanking, forming, lancing, and drawing, as shown by the examples in Figures 17-8 and 17-15. They are relatively simple to construct and are economical to maintain and repair because a defective punch or die does not require replacement of the entire die set. The material moves through the operations in the form of a continuous strip, advancing a preset distance with each cycle of the press. As the products are shaped, they remain attached to the strip or carrier until a final cutoff operation. While the attachment may restrict some of the forming operations and prevents part reorientation between steps, it also enables the quick and accurate positioning of material in each of the die segments.

If individual parts are mechanically moved from operation to operation within a single press, the dies are known as **transfer dies.** Part handling must operate in harmony with the press motions to move, orient, and position the pieces as they travel through the die.

In **compound dies** like the one shown schematically in Figure 17-16, piercing and blanking, or other combinations of operations, occur sequentially during a single stroke of the ram. Dies of this type are usually more expensive to construct and are more susceptible to breakage, but they generally offer more precise alignment of the sequential operations.

If many holes of varying sizes and shapes are to be placed in sheet components, numerically controlled **turret-type punch presses** may be specified. In these machines, as many as 60 separate punches and dies are contained within a turret that can quickly be rotated to provide the specific tooling required for an operation. Between operations, the workpiece is repositioned through numerically controlled movements of the worktable. This type of machine is particularly attractive when a variety of materials and thicknesses (0.4 to 8.0 mm, or 0.015 to 0.3 in.) are being processed. Still greater flexibility can be achieved by single machines that combine punch pressing with laser, plasma, or water-jet cutting.

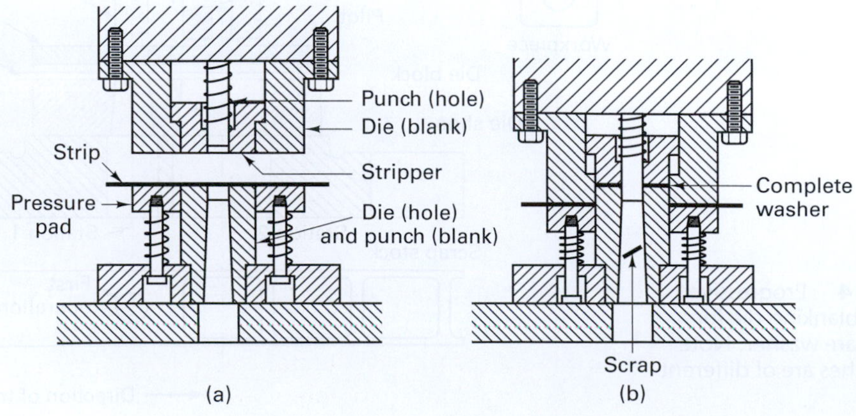

FIGURE 17-16 Method for making a simple washer in a compound piercing and blanking die. Part is blanked (a) and subsequently pierced (b) in the same stroke. The blanking punch contains the die for piercing.

DESIGN FOR PIERCING AND BLANKING

The construction, operation, and maintenance of piercing and blanking dies can be greatly facilitated if designers of the parts to be fabricated keep a few simple rules in mind:

1. Diameters of pierced holes should not be less than the thickness of the metal, with a minimum of 0.3 mm (0.025 in.). Smaller holes can be made, but with difficulty.
2. The minimum distance between holes, or between a hole and the edge of the stock, should be at least equal to the metal thickness.
3. The width of any projection or slot should be at least one times the metal thickness and never less than 2.5 mm ($\frac{3}{32}$ in.).
4. Keep tolerances as large as possible. Tolerances below about 0.075 mm (0.003 in.) will require shaving.
5. Arrange the pattern of parts on the strip to minimize scrap.

17.3 BENDING

Bending is the plastic deformation of metals about a linear axis with little or no change in the surface area. Multiple bends can be made simultaneously, but to be classified as true bending, and treatable by simple bending theory, each axis must be linear and independent of the others. If multiple bends are made with a single die, the process is often called **forming.** When the axes of deformation are not linear or are not independent, the processes are known as **drawing** and/or **stretching,** and these operations will be treated later in the chapter.

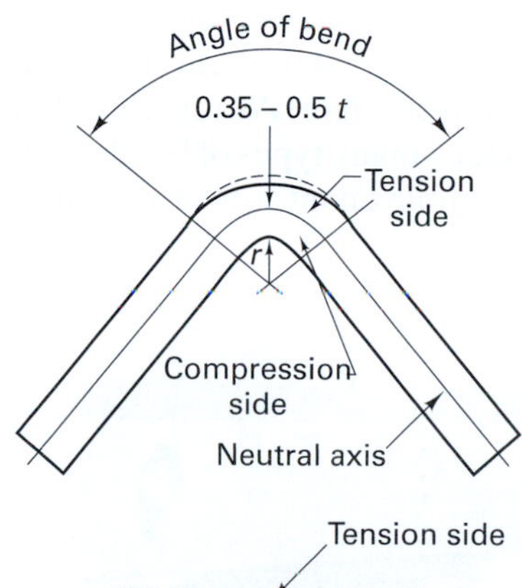

FIGURE 17-17 (Top) Nature of a bend in sheet metal showing tension on the outside and compression on the inside. (Bottom) The upper portion of the bend region, viewed from the side, shows how the center portion will thin more than the edges.

As shown in Figure 17-17, simple bending causes the metal on the outside to be stretched while that on the inside is compressed. The location that is neither stretched nor compressed is known as the **neutral axis** of the bend. Because the yield strength of metals in compression is somewhat higher than the yield strength in tension, the metal on the outer side yields first, and the neutral axis is displaced from the midpoint of the material. The neutral axis is generally located between one-third and one-half of the way from the inner surface, depending on the bend radius and the material being bent. Because of this lack of symmetry and the dominance of tensile deformation, the metal is generally thinned at the bend. In a linear bend, thinning is greatest in the center of the sheet and less near the free edges where inward movement can provide some compensation.

On the inner side of a bend, the compressive stresses can induce upsetting and a companion thickening of material. While this thickening somewhat offsets the thinning of the outer section, the upsetting can also produce an outward movement of the free edges. This contraction of the tensile segment and expansion of the compression segment can produce significant distortion of the edge surfaces that terminate a linear bend. This distortion is particularly pronounced when bends are produced across the width of thick but narrow plates.

Still another consequence of the combined tension and compression is the elastic recovery that occurs when the bending load is removed. The stretched region contracts and the compressed region expands, resulting in a small amount of "unbending," known as **springback.** To produce a product with a specified angle, the metal must be overbent by an amount equal to the subsequent springback. The actual amount of springback will vary with a number of factors, including the type of material and material thickness. Springback is typically about 0.5 degree for softer metals, 1 degree for steels, and as much as 3 degrees for stainless steel. Because the amount of springback increases with an increase in material strength, the newer advanced high-strength steels have been shown to have springback amounts as great as eight times that observed with annealed low-carbon steel.

ANGLE BENDING (BAR FOLDER AND PRESS BRAKE)

Machines like the **bar folder,** shown in Figure 17-18, can be used to make angle bends up to 150 degrees in sheet metal under 1.5 mm ($\frac{1}{16}$ in.) thick. The workpiece is inserted under the folding leaf and aligned in the proper position. Raising the handle then

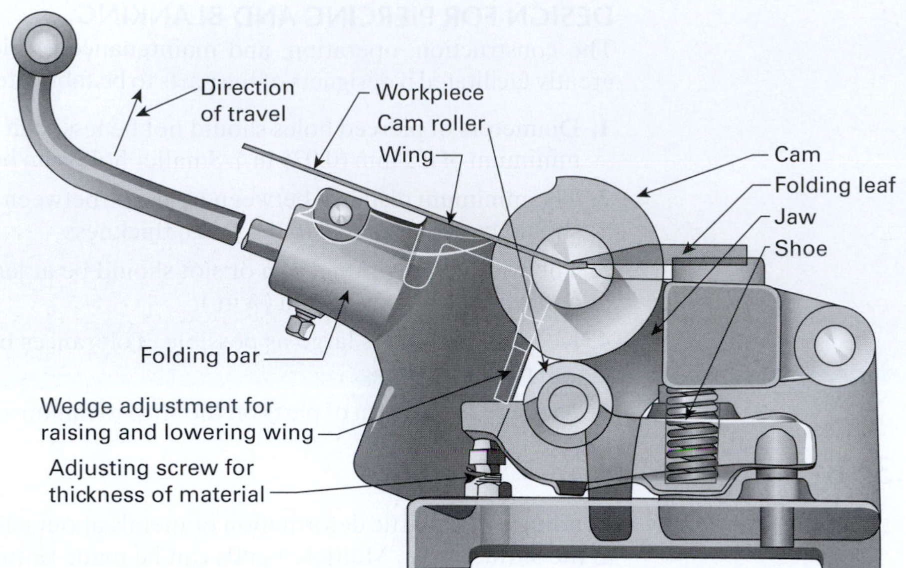

FIGURE 17-18 Phantom section of a bar folder, showing position and operation of internal components.

actuates a cam, causing the leaf to clamp the sheet. Further movement of the handle bends the metal to the desired angle. These manually operated machines can be used to produce linear bends up to about 3.5 m (12 ft) in length.

Bends in heavier sheet or more complex bends in thin material are generally made on **press brakes,** like the one shown in Figure 17-19. These are mechanical or hydraulic presses with a long, narrow bed and short strokes. The metal is bent between interchangeable dies that are attached to both the bed and the ram. As illustrated in Figures 17-19 and 17-20, different dies can be used to produce many types of bends. The metal can be repositioned between strokes to produce complex contours or repeated

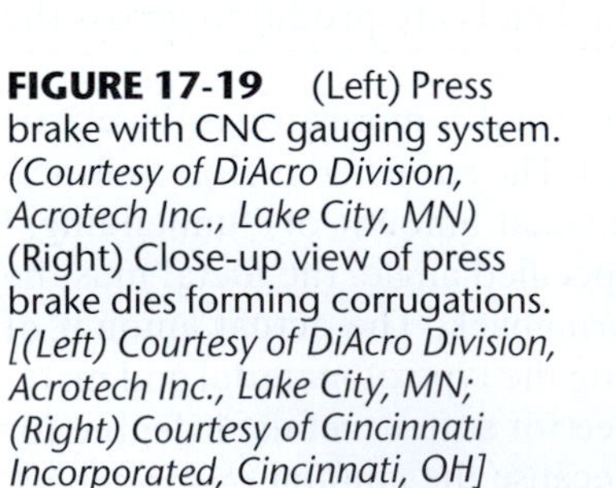
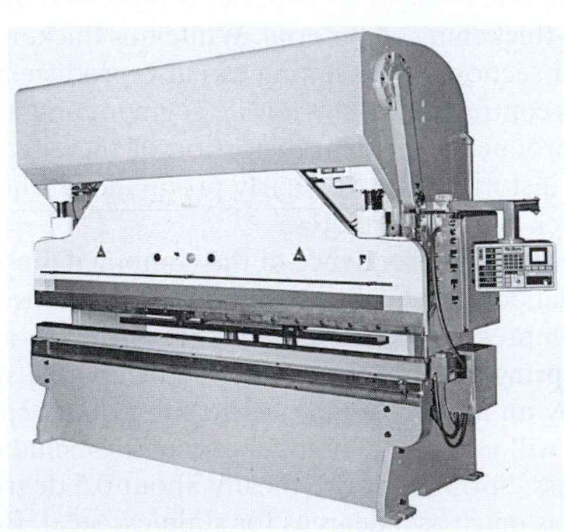

FIGURE 17-19 (Left) Press brake with CNC gauging system. *(Courtesy of DiAcro Division, Acrotech Inc., Lake City, MN)* (Right) Close-up view of press brake dies forming corrugations. *[(Left) Courtesy of DiAcro Division, Acrotech Inc., Lake City, MN; (Right) Courtesy of Cincinnati Incorporated, Cincinnati, OH]*

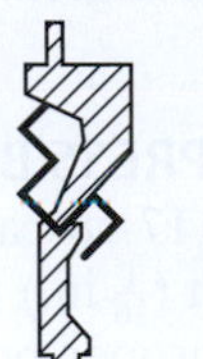
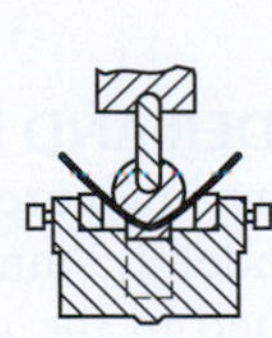
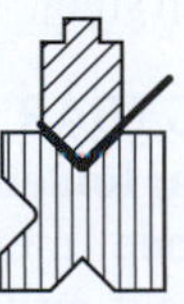
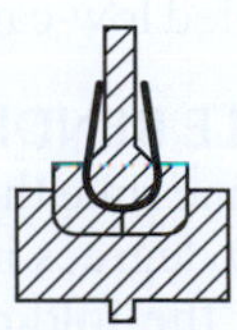

FIGURE 17-20 Press brake dies can form a variety of angles and contours. *(Courtesy of Cincinnati Incorporated, Cincinnati, OH)*

FIGURE 17-21 Dies and operations used in the press brake forming of a roll bead. *(Courtesy of Cincinnati Incorporated, Cincinnati, OH)*

bends, such as corrugations. Figure 17-21 shows how a roll bead can be formed with repeated strokes, repositioning, and multiple sets of tooling. Seaming, embossing, punching, and other operations can also be performed with press brakes, but these operations can usually be done more efficiently on other types of equipment.

The tools and support structures on a press brake are often loaded in the same three-point bending discussed in Chapter 16 for rolling mill rolls. Elastic deflections can cause a variety of bend deviations and defects, and a number of means have been developed to overcome these problems.

DESIGN FOR BENDING

Several factors must be considered when designing parts that are to be shaped by bending. One of the primary concerns is determining the smallest bend radius that can be formed without metal cracking (i.e., the **minimum bend radius**). This value is dependent on both the ductility of the metal (as measured by the percent reduction in area observed in a standard tensile test) and the thickness of the material being bent. Figure 17-22 shows how the ratio of the minimum bend radius, R, to the thickness of the material, t, varies with material ductility. As this plot reveals, an extremely ductile material is required to produce a bend with radius less than the thickness of the metal. If possible, bends should be designed with large bend radii. This permits easier forming and allows the designer to select from a wider variety of engineering materials.

If the punch radius is large and the bend angle is shallow, large amounts of springback are often encountered. The sharper the bend, the more likely the surfaces will be stressed beyond the yield point. Less severe bends have large amounts of elastically stressed material and large amounts of springback. In general, when the bend radius is greater than four times the material thickness, the tooling or process must provide springback compensation.

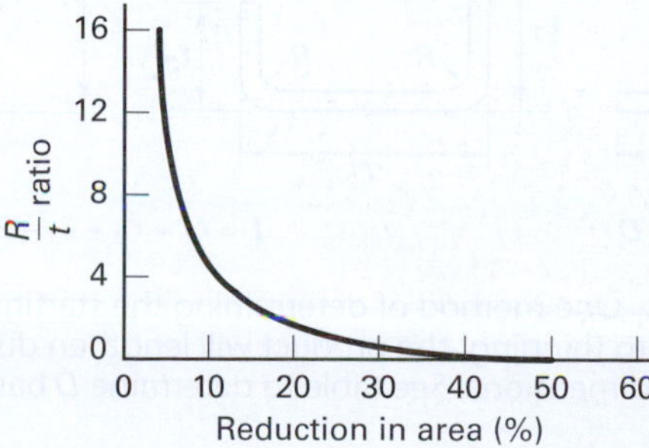

FIGURE 17-22 Relationship between the minimum bend radius, (relative to thickness) and the ductility of the metal being bent (as measured by the reduction in area in a uniaxial tensile test).

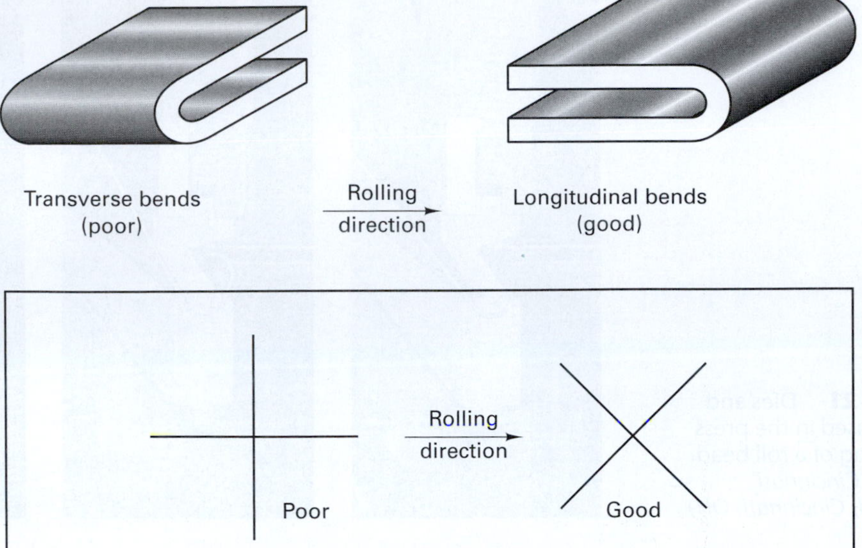

FIGURE 17-23 Bends should be made with the bend axis perpendicular to the rolling direction. When intersecting bends are made, both should be at angle to the rolling direction, as shown.

If the metal has experienced previous cold work or has marked directional properties, these features should be considered when designing the bending operation. Whenever possible, it is best to make the bend axis perpendicular to the direction of previous working, as shown in the upper portion of Figure 17-23. The explanation for this recommendation has little to do with the grain structure of the metal but is more closely related to the mechanical loading applied to the weak, oriented inclusions. Cracks can easily start along tensile-loaded inclusions and propagate to full cracking of the bend. If intersecting or perpendicular bends are required, it is often best to place each at an angle to the rolling direction, as shown in the lower portion of Figure 17-23, rather than have one longitudinal and one transverse.

Another design concern is determining the dimensions of a flat blank that will produce a bent part of the desired precision. As discussed earlier in the chapter, metal tends to thin and lengthen when it is bent. The amount of lengthening is a function of both the stock thickness and the bend radius. Figure 17-24 illustrates one method that has been found to give satisfactory results for determining the blank length for bent products. In addition, the minimum length of any protruding leg should be at least equal to the bend radius plus 1.5 times the thickness of the metal.

Whenever possible, the tolerance on bent parts should not be less than 0.8 mm ($\frac{1}{32}$ in). Bends of 90 degrees or greater should not be specified without first determining whether the material and bending method will permit them. Parts with multiple bends should be designed with most (or preferably all) of them to be of the same bend radius. This will reduce setup time and tooling costs. Consideration should also be given to providing regions for adequate clamping or support during manufacture. Bending near the

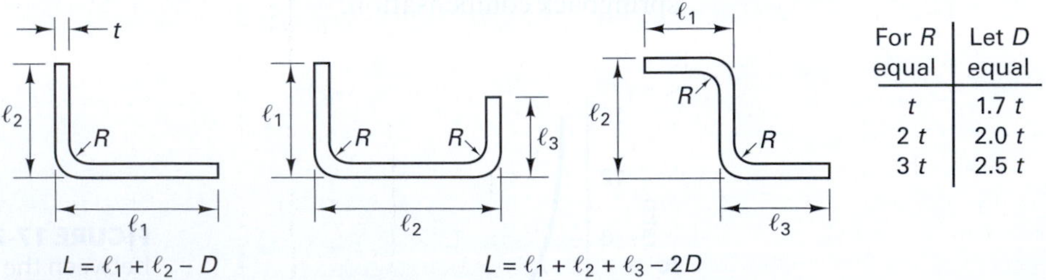

For R equal	Let D equal
t	1.7 t
2 t	2.0 t
3 t	2.5 t

FIGURE 17-24 One method of determining the starting blank size (L) for several bending operations. Due to thinning, the product will lengthen during forming; l_1, l_2, and l_3 and are the desired product dimensions. See table to determine D based on size of radius R, where t is the stock thickness.

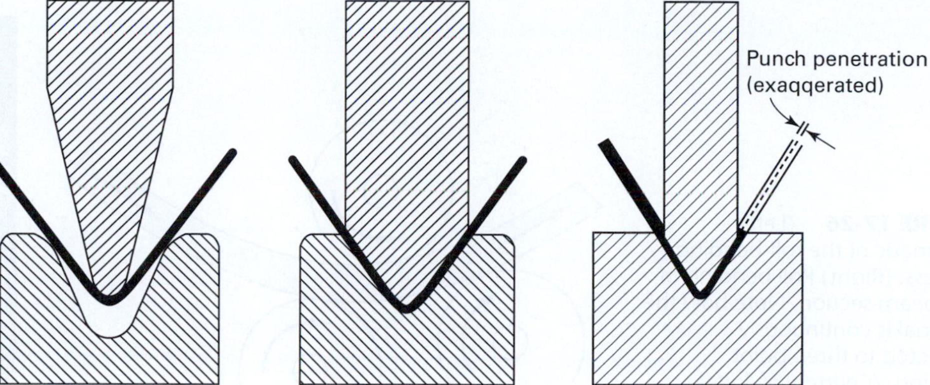

FIGURE 17-25 Comparison of air-bend (left), bottoming (center), and coining (right) press brake dies. With the air-bend die, the amount of bend is controlled by the bottoming position of the upper die.

edge of a material will distort the edge. If an undistorted edge is required, additional material must be included and a trimming operation performed after bending.

AIR-BEND, BOTTOMING, AND COINING DIES

Yet another design decision is the use of air-bend, bottoming, or coining dies. As shown in Figure 17-25, **bottoming dies** contact and compress the full area within the tooling. The angle of the resulting bend is set by the geometry of the tooling, adjusted for subsequent springback, and the inside bend radius is that machined on the nose of the punch. Bottoming dies are designed for a specific material and material thickness, and they form bends of a single configuration. If the results are outside specifications, or the material is changed and produces a different amount of springback, the geometry of the tooling will have to be modified. Once the geometry of the tool is successfully set, however, reproducibility of the bend geometry is excellent (usually within 0.25 degree), provided there is consistency within the size and properties of the material being bent.

In contrast, **air-bend dies** produce the desired geometry by simple three-point bending. Because the resulting angle is controlled by the bottoming position of the upper die, a single set of tooling can produce a range of bend geometries from 180 degrees through the included angle of the die. Air bending can also accommodate a variety of materials in a range of material gages and requires the least force of the three options (bottoming, air bending, and coining). Product reproducibility depends on the ability to control the stroke of the press. Adaptive control and on-the-fly corrections are frequently used with air-bend tooling and can generally produce consistent bends within 0.5-degree accuracy.

If bottoming dies continue to move beyond the full-contact position, the thickness of the bent material is reduced. Because of the extensive plastic deformation, the operation becomes one of **coining** (or bottoming with penetration). Springback can be significantly reduced, and more consistent results can be achieved with materials having variation in structure and thickness. Unfortunately, the loading on both the press and the tools is typically 5 to 10 times greater than with bottoming or air bending.

Reproducible-stroke mechanical presses are generally used for bottom bending and coining, while adjustable-stroke hydraulic presses are preferred for air bending.

Yet another approach is to replace the bottom die with a urethane (rubber) pad, which is usually mounted in a steel retainer. The urethane acts like a bladder of hydraulic fluid, maintaining full contact with the workpiece as the bend forms from the centerline outward. The pad, and sheet material, follows the punch geometry. Because both the sheet material and the urethane are deforming, bending with a urethane tool requires more force than air bending but less than bottoming with tool-steel tooling. Reinforcing bars and relief holes can be designed into the urethane tool to control its deformation. Prepainted materials, perforated sheets, and textured materials can all be bent with minimum marring and scratching. Depending on the range of materials and bend profiles, a single urethane pad may be able to act as a universal bottom die. The use of urethane tooling in blanking, drawing, and bulging will be discussed later in this chapter.

FIGURE 17-26 (Left) Schematic of the roll-bending process. (Right) Roll bending of an I-beam section. Note how the material is continuously subjected to three-point bending. *(Courtesy of Buffalo Machines, Inc., formerly known as Buffalo Forge, Lockport, NY)*

ROLL BENDING

Roll bending is a continuous form of three-point bending where plates, sheets, beams, pipe, and even rolled shapes and extrusions are bent to a desired curvature using forming rolls. As shown in Figure 17-26, roll bending machines usually have three rolls in the form of a triangle. The two lower rolls are driven and the position of the upper roll is adjusted to control the degree of curvature in the product. The rolls on the machine pictured in Figure 17-26 are supported on only one end. When wider material is being formed, the longer rolls often require support on both ends. The support frame on one end may be swung clear, however, to permit the removal of closed circular shapes or partially rolled product. Because of the variety of applications, roll bending machines are available in a wide range of sizes, some being capable of bending plate up to 25 cm (10 in.) thick.

DRAW BENDING, COMPRESSION BENDING, AND PRESS BENDING

Bending machines can also utilize clamps and pressure tools to bend material against a form block. In **draw bending,** illustrated in Figure 17-27, the workpiece is clamped against a bending form and the entire assembly is rotated to draw the workpiece along a stationary pressure tool. In **compression bending,** also illustrated in Figure 17-27, the bending form remains stationary and the pressure tool moves along the surface of the workpiece.

Press bending, also shown in Figure 17-27, utilizes a downward descending bend die, which pushes into the center of material that is supported on either side by wing dies. As the ram descends, the wing dies pivot up, bending the material around the form on the ram. The flexibility of each of the above processes is somewhat limited because a certain length of the product must be used for clamping.

TUBE BENDING

Quite often the material being bent is a tube or pipe **(tube bending),** and this geometry presents additional problems. Key parameters are the outer diameter of the tube, the

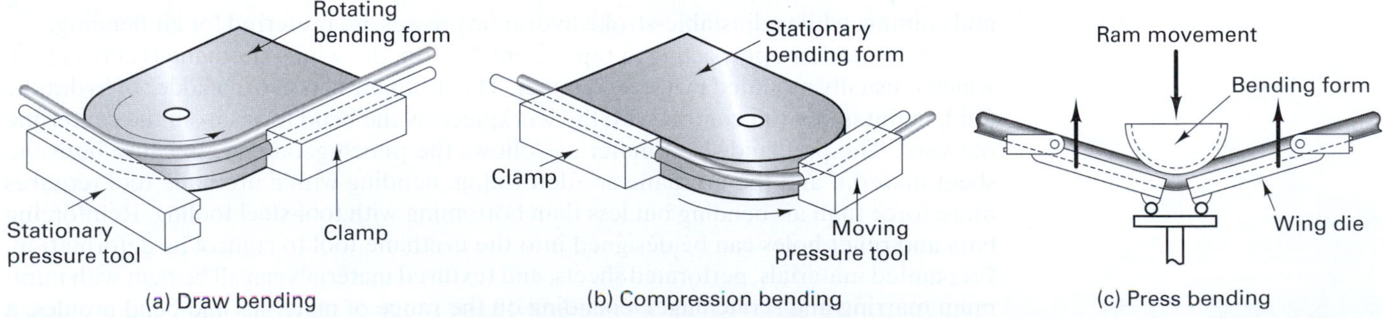

FIGURE 17-27 (a) Draw bending, in which the form block rotates; (b) compression bending, in which a moving tool compresses the workpiece against a stationary form, (c) press bending, where the press ram moves the bending form.

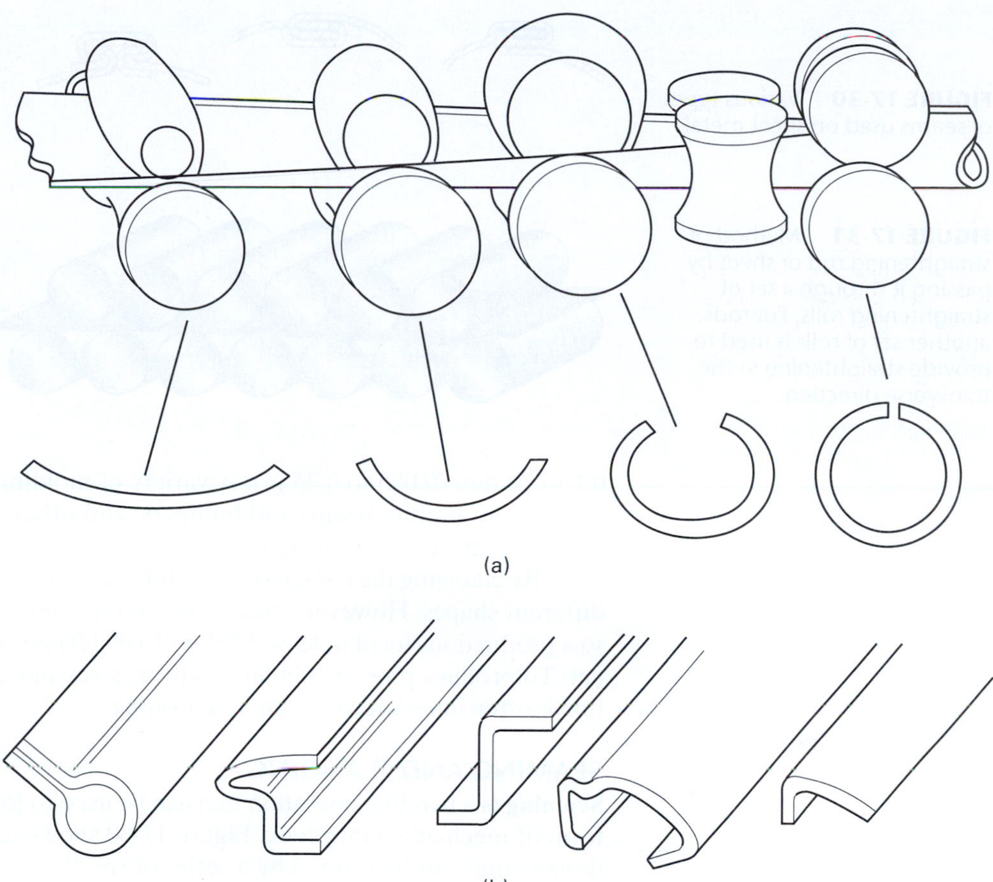

FIGURE 17-28 (a) Schematic representation of the cold roll-forming process being used to convert sheet or plate into tube. (b) Some typical shapes produced by roll forming.

wall thickness, and the radius of the bend. Small-diameter, thick-wall tubes usually present little difficulty. As the outer diameter increases, the wall thickness decreases, or the bend radius becomes smaller; the outside of the tube tends to pull to the center, flattening the tube; and the inside surface may wrinkle. For many years, a common method of overcoming these problems was to pack the tube with wet sand, produce the bend, and then remove the sand from the interior. Flexible mandrels have now replaced the sand, and are currently available in a wide variety of styles and sizes.

ROLL FORMING

The continuous **roll forming** of flat strip into complex sections has become a highly developed forming technique that competes directly with press brake forming, extrusion, and stamping. As shown in Figures 17-28 and 17-29, the process involves the progressive bending of metal strip as it passes through a series of forming rolls at speeds up to 80 m/min (270 ft/min). Only bending takes place, and all bends are parallel to one another. The thickness of the starting material is preserved, except for thinning at the bend radii.

Any material that can be bent can be roll formed—including cold-rolled, hot-rolled, polished, prepainted, coated, and plated metals—in thicknesses ranging from

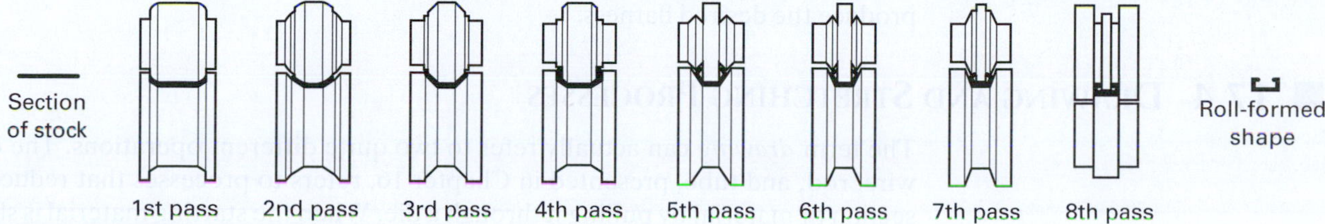

FIGURE 17-29 Eight-roll sequence for the roll forming of a box channel. *(Courtesy of the Aluminum Association, Washington, DC)*

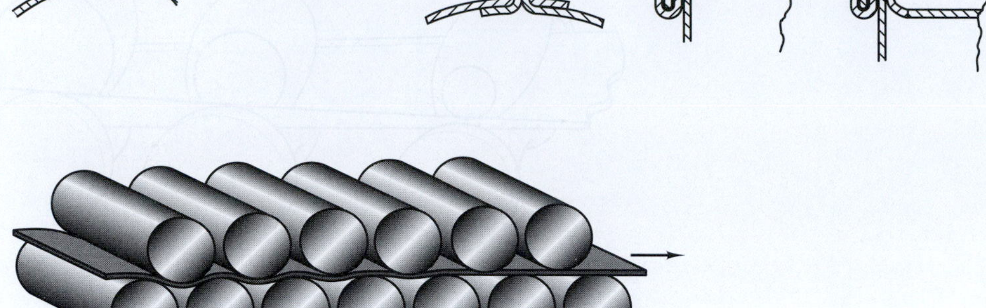

FIGURE 17-30 Various types of seams used on sheet metal.

FIGURE 17-31 Method of straightening rod or sheet by passing it through a set of straightening rolls. For rods, another set of rolls is used to provide straightening in the transverse direction.

0.1 to 20 mm (0.005 to 0.75 in). A variety of moldings, channeling, gutters and downspouts, automobile beams and bumpers, and other shapes of uniform wall thickness and uniform cross section are now being formed.

By changing the rolls, a single roll-forming machine can produce a wide variety of different shapes. However, changeover, setup, and adjustment may take several hours, so a production run of at least 3000 m (10,000 ft) is usually required for any given product. To produce pipe or tubular products, a resistance welding unit or seaming operation is often integrated with the roll forming.

SEAMING AND FLANGING

Seaming is a bending operation that can be used to join the ends of sheet metal in some form of mechanical interlock. Figure 17-30 shows several of the more common seam designs that can be formed by a series of small rollers. Seaming machines range from small hand-operated types to large automatic units capable of producing hundreds of seams per minute. Common products include cans, pails, drums, and other similar containers.

Flanging involves a rolling operation on sheet metal in essentially the same manner as seams. In many cases, however, the forming of both flanges and seams is a drawing operation because the bending can occur along a curved axis.

STRAIGHTENING, FLATTENING, OR LEVELING

The objective of **straightening, flattening,** or **leveling** is the opposite of bending, and these operations are often performed before subsequent forming to ensure the use of flat or straight material that is reasonably free of residual stresses. **Roll straightening** or **roller leveling,** illustrated in Figure 17-31, subjects the material to a series of reverse bends. The rod, sheet, or wire is passed through a series of alternating upper and lower rolls with progressively decreased offsets. This configuration produces deep bends at the entry, decreasing to little or no bend at the exit. As the material is bent, first up and then down, the surfaces are stressed beyond their elastic limit, replacing any permanent set with a flat or straight profile. Tension applied along the length of the product can help induce the required deformation.

Sheet material can also be straightened by a process called **stretcher leveling.** Finite-length sheets are gripped mechanically and stretched beyond the elastic limit to produce the desired flatness.

17.4 DRAWING AND STRETCHING PROCESSES

The term *drawing* can actually refer to two quite different operations. The drawing of wire, rod, and tube, presented in Chapter 16, refers to processes that reduce the cross section of material by pulling it through a die. When the starting material is sheet, drawing refers to a family of operations where plastic flow occurs over a curved axis, and the flat sheet is formed into a recessed, three-dimensional part with a depth more than

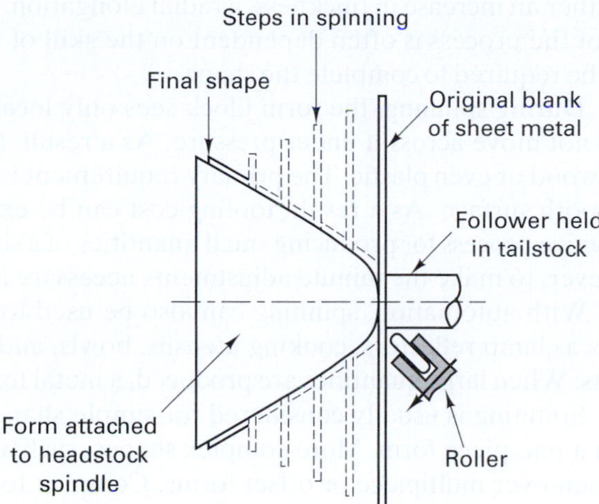

FIGURE 17-32 Progressive stages in the spinning of a sheet metal product.

several times the thickness of the metal. These operations can be used to produce a wide range of shapes, from small cups to large automobile and aerospace panels.

Drawing and stretching are also different. In drawing, the metal flows inward as it is forced to change shape. This deformation allows the thickness of the material to remain relatively constant or even thicken. Stretching operations restrict the compensating movement, and the increase in surface area is accompanied by a decrease in material thickness. The loads during stretching are tensile loads, and the material is at a greater risk of failing during stretching than during drawing.

SPINNING

Spinning is a cold-forming operation where a rotating disk of sheet metal is progressively shaped over a male form, or mandrel, to produce rotationally symmetrical shapes, such as cones, hemispheres, cylinders, bells, and parabolas. A form block possessing the shape of the desired part is attached to a rotating spindle, such as the drive section of a simple lathe. A disk of metal (or preformed shape) is centered on the small end of the form and held in place by a pressure pad. As the disk and form rotate, localized pressure is applied through a round-ended wooden or metal tool or small roller that traverses the entire surface of the part, causing it to flow progressively against the form. Figure 17-32 depicts the progressive operation, and Figure 17-33 shows a part at two stages of forming. Because the final diameter of the final part is less than that of the starting disk, the circumferential length decreases. This decrease must be compensated

FIGURE 17-33 An array of metal parts produced by spinning. *(Photo Courtesy of ACME Metal Spinning, Minneapolis, MN)*

by either an increase in thickness, a radial elongation, or circumferential buckling. Control of the process is often dependent on the skill of the operator, and multiple passes may be required to complete the shape.

During spinning, the form block sees only localized compression, and the metal does not move across it under pressure. As a result, form blocks can often be made of hardwood or even plastic. The primary requirement is simply to replicate the shape with a smooth surface. As a result, tooling cost can be extremely low, making spinning an attractive process for producing small quantities of a single part. Skilled labor is required, however, to make the minute adjustments necessary for the production of quality products. With automation, spinning can also be used to mass-produce such high-volume items as lamp reflectors, cooking utensils, bowls, and the bells of some musical instruments. When large quantities are produced, a metal form block is generally preferred.

Spinning is usually considered for simple shapes that can be directly withdrawn from a one-piece form. More complex shapes, such as those with reentrant angles, can be spun over multipiece or offset forms. Complex form blocks can also be made from frozen water, which is melted out of the product after spinning. Any ductile material with more than 2% elongation can be formed by metal spinning, and workpiece sizes can be as large as 6 m (20 ft) in diameter.

A recent modification of metal spinning is **laser-assisted metal spinning.** A laser beam is focused directly ahead of the roller to heat and soften the material. Materials such as titanium, nickel-based alloys, and austenitic stainless steels can be spun with less difficulty and the operation can be performed with less force and power.

SHEAR FORMING OR FLOW TURNING

Cones, hemispheres, and similar shapes are often formed by **shear forming** or **flow turning,** a modification of the spinning process in which each element of the blank maintains its distance from the axis of rotation and the spinning flange remains vertical throughout the process. The roller provides highly localized pressure that plastically deforms and cold-works the metal. Because there is no circumferential shrinkage, the metal flow is entirely by shear, and no compensating stretch has to occur. As shown in Figure 17-34, the wall thickness of the product, t_c, will vary with the angle of the particular region according to the relationship: $t_c = t_b \sin \alpha$, where t_b is the thickness of the starting blank. If $a < 30°$, it may be necessary to complete the forming in two stages with an intermediate anneal in between. Reductions in wall thickness as high as 8:1 are possible, but the limit is usually set at about 5:1 or 80%.

Conical shapes are usually shear formed by the direct process depicted in Figure 17-34. The bottom of the product is held against the face of the form block or mandrel, while the material being formed moves in the same direction as the roller. Products can also be formed by a reverse process, like that illustrated in Figure 17-35. By

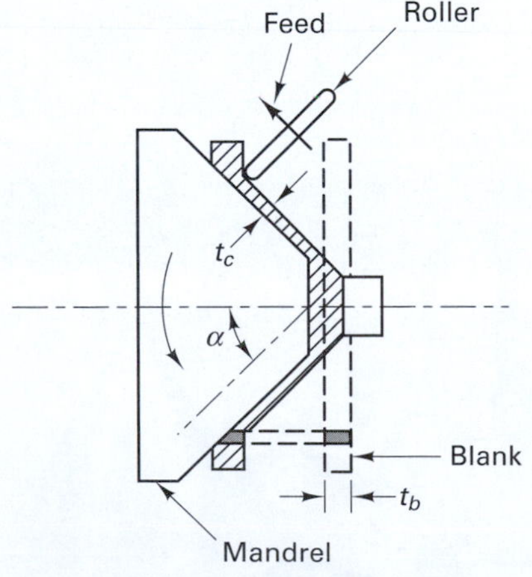

FIGURE 17-34 Schematic representation of the basic shear-forming process.

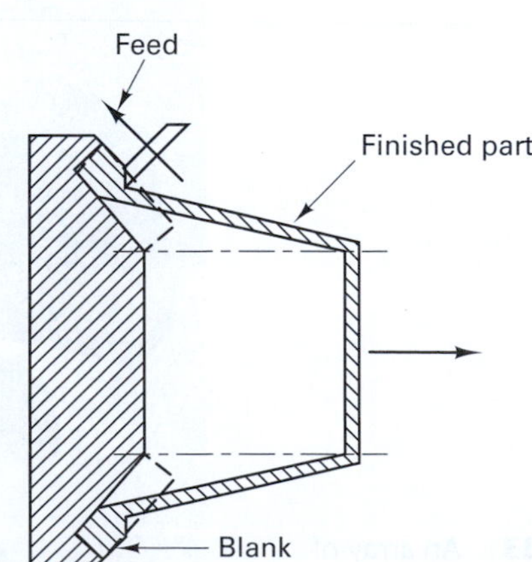

FIGURE 17-35 Forming a conical part by reverse shear forming.

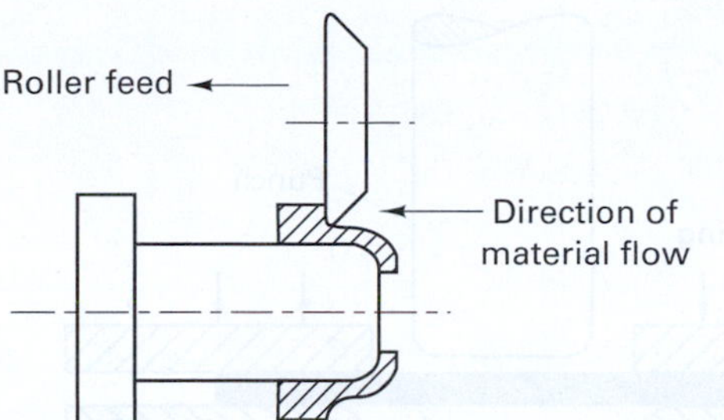

FIGURE 17-36 Shear forming a cylinder by the direct process.

controlling the position and feed of the forming rollers, the reverse process can be used to shape concave, convex, or conical parts without a matching form block or mandrel. Cylinders can be shear formed by both the direct and reverse processes. As shown in Figure 17-36, the direct process restricts the length of the product to the length of the mandrel. No schematic is provided for the reverse process because it is essentially the same as the roll extrusion process, depicted in Figure 16-52.

STRETCH FORMING

Stretch forming, illustrated in principle in Figure 17-37, is an attractive means of producing large sheet metal parts in low or limited quantities. A sheet of metal is gripped by two or more sets of jaws that stretch it and wrap it around a single form block. Various combinations of stretching, wrapping, and block movements can be employed, depending on the shape of the part.

Because most of the deformation is induced by the tensile stretching, the forces on the form block are far less than those normally encountered in bending or forming. Consequently, there is very little springback, and the workpiece conforms very closely to the shape of the tool. Because stretching accompanies bending or wrapping, wrinkles are pulled out before they occur. Because the forces are so low, the form blocks can often be made of wood, low-melting-point metal, or even plastic.

Stretch forming, or *stretch-wrap forming* as it is often called, is quite popular in the aircraft industry and is frequently used to form aluminum and stainless steel into cowlings, wing tips, scoops, and other large panels. Low-carbon steel can be stretch formed to produce large panels for the automotive and truck industry. If mating male and female dies are used to shape the metal while it is being stretched, the process is known as *stretch-draw forming*.

DEEP DRAWING AND SHALLOW DRAWING

The forming of solid-bottom cylindrical or rectangular containers from metal sheet is one of the most widely used manufacturing processes. When the depth of the product is less than its diameter (or the smallest dimension of its opening), the process is considered to be **shallow drawing.** If the depth is greater than the diameter, it is known as **deep drawing.**

Consider the simple operation of converting a circular disk of sheet metal into a flat-bottom cylindrical cup. Figure 17-38 shows the blank positioned over a die opening and a circular punch descending to pull or draw the material into the die cavity. The material beneath the punch remains largely unaffected and simply becomes the bottom of the cup. The cup wall is formed by pulling the remainder of the disk inward and over the radius of the die, as shown in Figure 17-39. As the material is pulled inward, its circumference decreases. Because the volume of material must remain constant, the

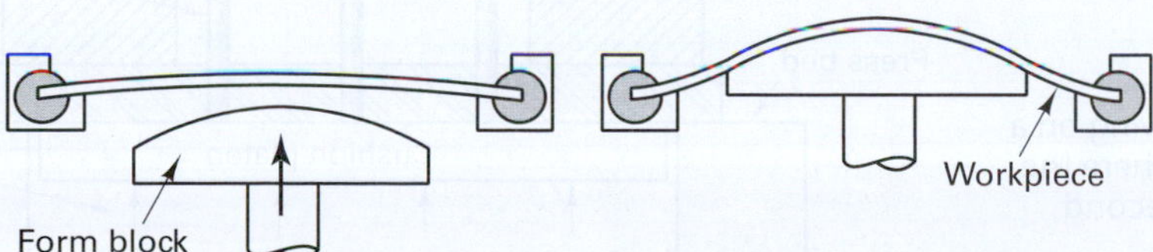

FIGURE 17-37 Schematic of a stretch-forming operation.

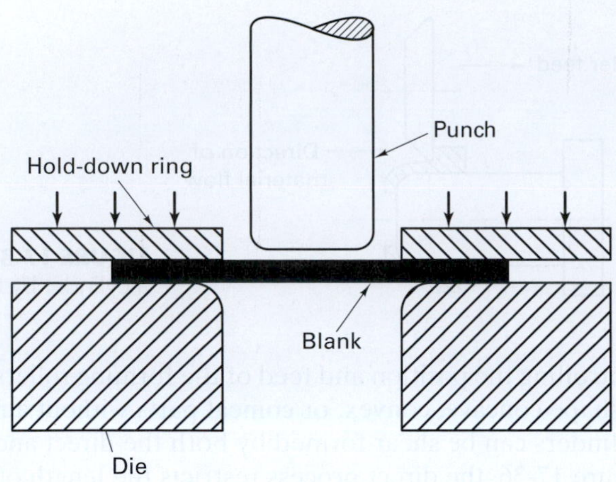

FIGURE 17-38 Schematic of the deep-drawing process.

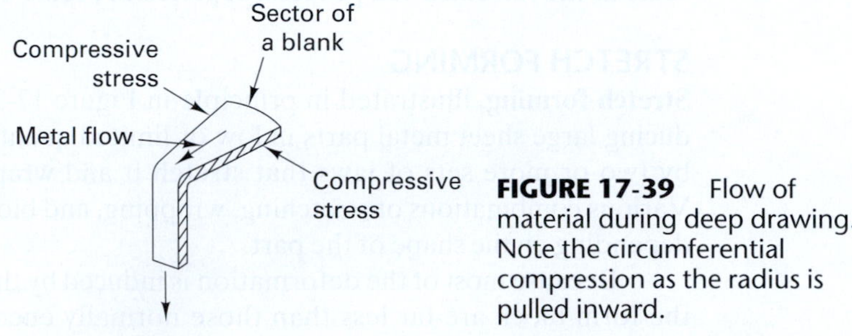

FIGURE 17-39 Flow of material during deep drawing. Note the circumferential compression as the radius is pulled inward.

decrease in circumferential dimension must be compensated by an increase in another dimension, such as thickness or radial length. Because the material is thin, an alternative is to relieve the circumferential compression by buckling or wrinkling. Wrinkle formation can be suppressed, however, by compressing the sheet between the die and a blankholder or hold-down ring during the forming (as shown in Figure 17-38).

In single-action presses, where there is only one movement that is available, springs or air pressure are often used to clamp the metal between the die and pressure ring. When multiple-actions are available (two or more independent motions), as shown in Figure 17-40, the hold-down force can now be applied in a manner that is

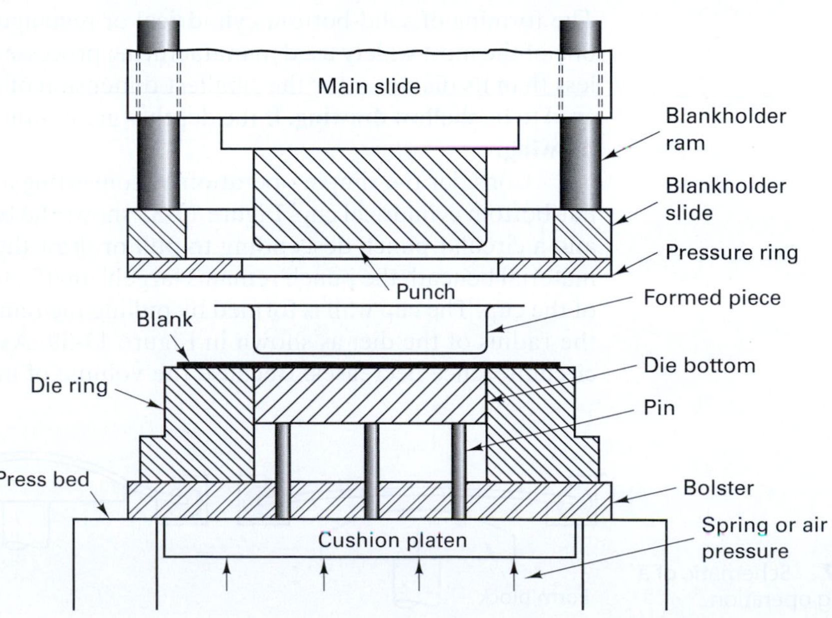

FIGURE 17-40 Drawing on a double-action press, where the blankholder uses the second press action.

independent of the punch position. This restraining force can also be varied during the drawing operation. For this reason, multiple-action presses are usually specified for the drawing of more complex parts, while single-action presses can be used for the simpler operations.

Key variables in the deep drawing process include the blank diameter and the punch diameter (which combine to determine the **draw ratio** and the height of the side walls), the die radius, the punch radius, the clearance between the punch and the die, the thickness of the blank, the surface finish of the die face and blankholder, lubrication, and the hold-down pressure. The punch-nose radius cannot be too small or the punch will cut or pierce the blank instead of pulling it into the die. If the punch radius is too large, the material will stretch and thin as it passes over the punch as opposed to pulling in material from the flange. If the die radius is too small, tearing may occur as the material bends over the tight radius. Excessive die radius, however, may lead to wrinkling in the unsupported region between the punch face and die face.

Once a process has been designed and the tooling manufactured, the primary variable for process adjustment is the **hold-down pressure** or **blankholder force.** If the force or pressure is too low, wrinkling may occur at the start of the stroke. If it is too high, there is too much restraint, and stretching and thinning of the material will occur. At the extreme, the descending punch will simply tear the disk or some portion of the already-formed cup wall.

When drawing a shallow cup, there is little change in circumference, and a small area is being confined by the blankholder. As a result, the tendency to wrinkle or tear is low. As cup depth increases, there is an increased tendency for forming both of the defects. In a similar manner, thin material is more likely to wrinkle or tear than thick material. Figure 17-41 summarizes the effects of cup depth, blank thickness, and blankholder force or pressure. Note that for thin materials and deep draws, a defect-free product may not be possible in a single operation, as the defects simply transition from wrinkling, to wrinkle plus tear, to just tearing as blankholder force is increased. To produce a defect-free product, the draw ratio (blank diameter/cup diameter) is often limited to values less than about 2.2.

The limitations of wrinkling and tearing (and the limiting draw ratio) can often be overcome by using multiple operations. Figure 17-42 shows two alternatives for converting drawn parts into deeper cups. In the **forward redraw** option, the material undergoes reverse bending as it flows into the die. In **reverse redrawing,** the starting cup is placed over a tubular die, and the punch acts to turn it inside out. Because all bending is in one direction, reverse redraws can produce greater changes in diameter. Because there is more bending and straightening and the material has been previously

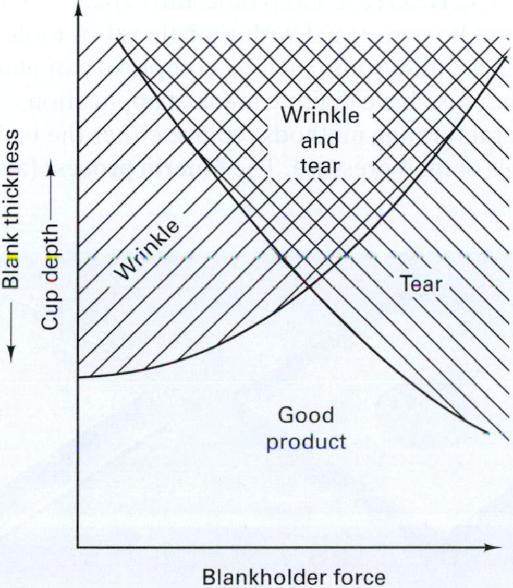

FIGURE 17-41 Defect formation in deep drawing as a function of blankholder force, blank thickness, and cup depth.

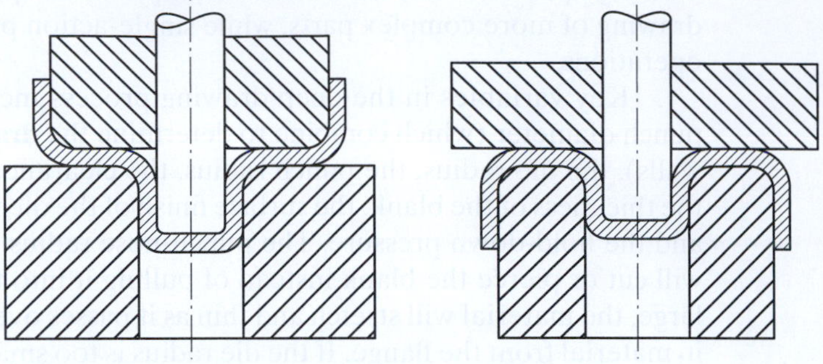

FIGURE 17-42 Cup redrawing to further reduce diameter and increase wall height. (Left) forward redraw; (right) reverse redraw.

worked, redraws are always performed with successively lower-diameter reductions than initial draws.

When the part geometry becomes more complex, as with rectangular or asymmetric parts, it is best if the surface area and thickness of the material can remain relatively constant. Different regions may need to be differentially constrained. One technique that can produce variable constraint is the use of **draw beads**—vertical projections and matching grooves in the die and blankholder. The added force of bending and unbending as the material flows over the draw bead restricts the flow of material. The degree of constraint can be varied by adjusting the height, shape, and size of the bead and bead cavity. An alternative approach is to apply different blankholder pressures at different locations.

Because of prior rolling and other metallurgical and process features, the flow of sheet metal is generally not uniform, even in the simplest drawing operation. Excess material may be required to ensure final dimensions, and **trimming** may be required to establish both the size and uniformity of the final part. Figure 17-43 shows a shallow-drawn part before and after trimming. Trimming obviously adds to the production cost because it not only converts some of the starting material to scrap and but also adds another operation to the manufacturing process—one that must be performed on an already-produced shape.

FORMING WITH RUBBER TOOLING OR FLUID PRESSURE

Blanking and drawing operations usually require mating male and female, or upper and lower die sets, and process setup requires that the various components be properly positioned and aligned. When the amount of deformation is great, multiple operations may be required, each with its own set of dedicated tooling, and intermediate anneals may also be necessary. Numerous processes have been developed that seek to (1) reduce tooling cost, (2) decrease setup time and expense, or (3) extend the amount of deformation that can be performed with a single set of tools. Although most of these methods have distinct limitations, such as complexity of shape or types of metal that can be formed, they also have definite areas of application.

Several forming methods replace either the male or female member of the die set with rubber or fluid pressure. The **Guerin process** (also known as *rubber-die forming*) is

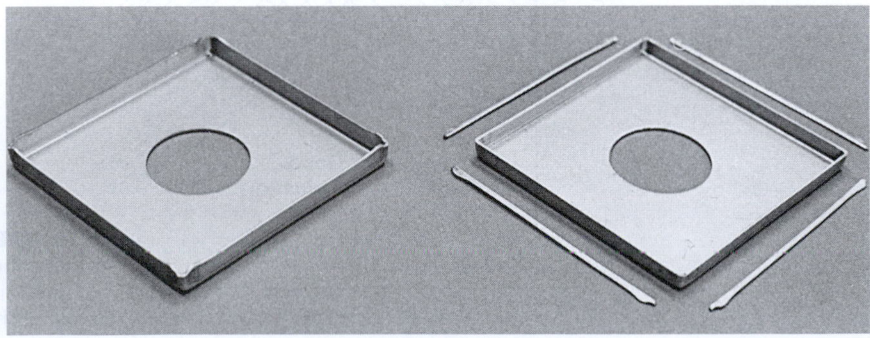

FIGURE 17-43 Pierced, blanked, and drawn part before and after trimming. *(E. Paul DeGarmo)*

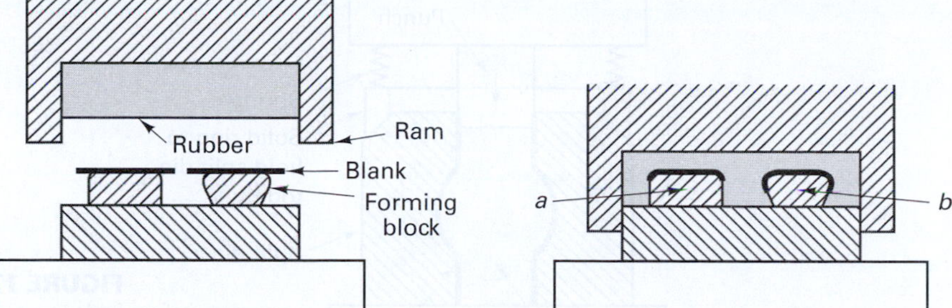

FIGURE 17-44 The Guerin process for forming sheet metal products.

depicted in Figure 17-44. It is based on the phenomenon that rubber of the proper consistency, *when totally confined,* acts as a fluid and transmits pressure uniformly in all directions. Blanks of sheet metal are placed on top of form blocks, which can be made of wood, Bakelite, polyurethane, epoxy, or low-melting-point metal. The upper ram contains a pad of rubber 20 to 25 cm (8 to 10 in.) thick mounted within a steel container. As the ram descends, the rubber pad becomes confined and transmits force to the metal, causing it to bend to the desired shape. Because no female die is used and inexpensive form blocks replace the male die, the total tooling cost is quite low. There are no mating tools to align, process flexibility is quite high (different shapes can even be formed at the same time), wear on the material and tooling is low, and the surface quality of the workpiece is easily maintained a feature that makes the process attractive for forming prepainted or specially coated sheet. When reentrant sections are produced (as with product *b* in Figure 17-44), it must be possible to slide the parts lengthwise from the form blocks or to disassemble a multipiece form from within the product.

The Guerin process was developed by the aircraft industry, where the production of small numbers of duplicate parts clearly favors the low cost of tooling. It can be used on aluminum sheet up to 3 mm ($\frac{1}{8}$ in.) thick and on stainless steel up to 1.5 mm ($\frac{1}{16}$ in.). Magnesium sheet can also be formed if it is heated and shaped over heated form blocks.

Most of the forming done with the Guerin process is multiple-axis bending, but some shallow drawing can also be performed. The process can also be used to pierce or blank thin gages of aluminum, as illustrated in Figure 17-45. For this application the blanking blocks are shaped the same as the desired workpiece, with a sharp face, or edge, of hardened steel. Round-edge supporting blocks are positioned a short distance from the blanking blocks to support the scrap skeleton and permit the metal to bend away from the sheared edges.

In **bulging,** fluid or rubber transmits the pressure required to expand a metal blank or tube outward against a split female mold or die. For simple shapes, **rubber tooling** can be inserted, compressed, and then easily removed, as shown in Figure 17-46. For complicated shapes, fluid pressure may be required to form the bulge. More complex equipment is required because pressurized seals must be formed and maintained, while still enabling the easy insertion and removal of material that is required for mass production.

SHEET HYDROFORMING

Sheet hydroforming is really a family of processes in which a rubber bladder backed by fluid pressure, or pockets of pressurized liquid, replaces either the solid punch or female

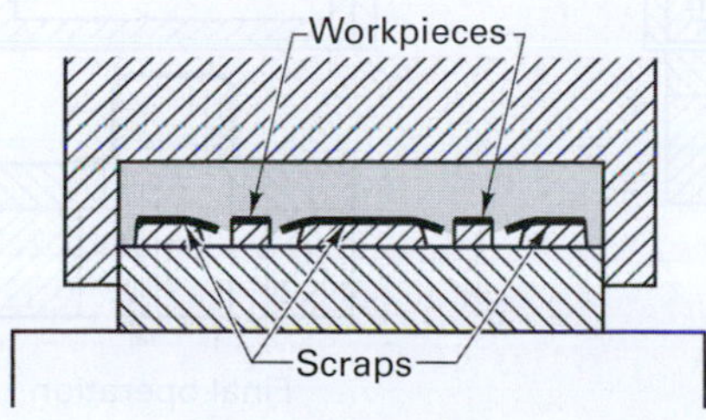

FIGURE 17-45 Method of blanking sheet metal using the Guerin process.

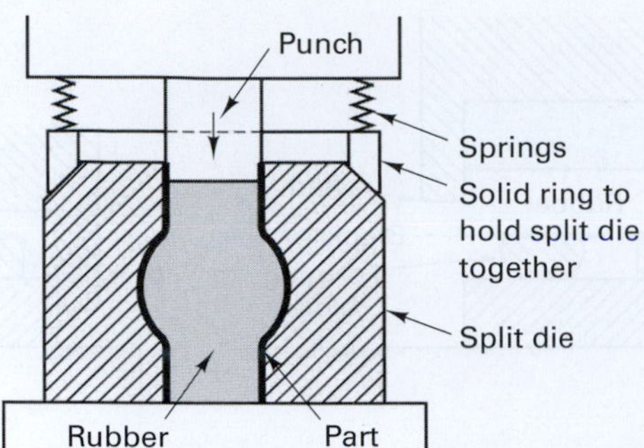

FIGURE 17-46 Method of bulging tubes with rubber tooling.

die of the traditional tool set. In a variant known as *high-pressure flexible-die forming*, or **flexforming,** the rubber pad of the Guerin process is replaced by a flexible rubber diaphragm backed by controlled hydraulic pressure at values between 140 and 200 MPa (20,000 and 30,000 psi). As illustrated in Figure 17-47, the solid punch drives the sheet into the resisting bladder, whose pressure is adjusted throughout the stroke.

In a variation of this process that shares similarity to stretch forming, the sheet is first clamped against the opening and the punch is retracted downward. The fluid is then pressurized, causing the workpiece to balloon downward toward the punch. Because the pressure is uniformly distributed over the workpiece, the sheet is uniformly stretched and uniformly thinned. The punch then moves upward, causing the prestretched metal to conform to the punch profile.

The flexible membrane can also be used to replace the hardened male punch, as shown in Figure 17-48. The ballooning action now causes the material to descend and conform fully to the female die, which may be made of epoxy or other low cost material. **Parallel-plate hydroforming,** or **pillow forming,** extends the process to the simultaneous production of upper and lower contours. As shown in Figure 17-49, two sheet metal blanks are laser welded around their periphery or are firmly clamped between upper and lower dies. Pressurized fluid is then injected between the sheets, simultaneously forming both upper and lower profiles. This may be a more attractive means of producing complex

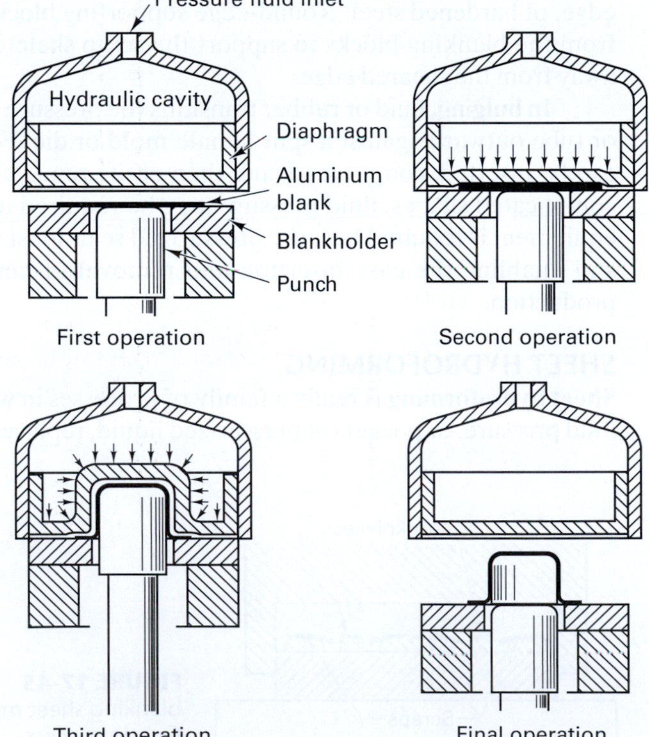

FIGURE 17-47 High-pressure flexible-die forming, showing (1) the blank in place with no pressure in the cavity, (2) press closed and cavity pressurized, (3) ram advanced with cavity maintaining fluid pressure, and (4) pressure released and ram retracted. *(Courtesy of Aluminum Association, Washington, DC)*

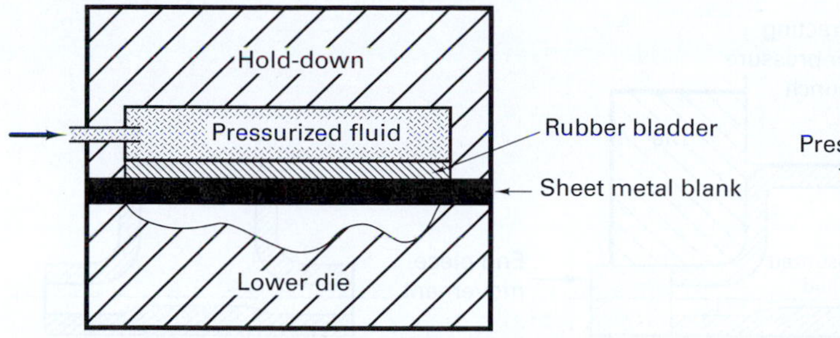

FIGURE 17-48 One form of sheet hydroforming.

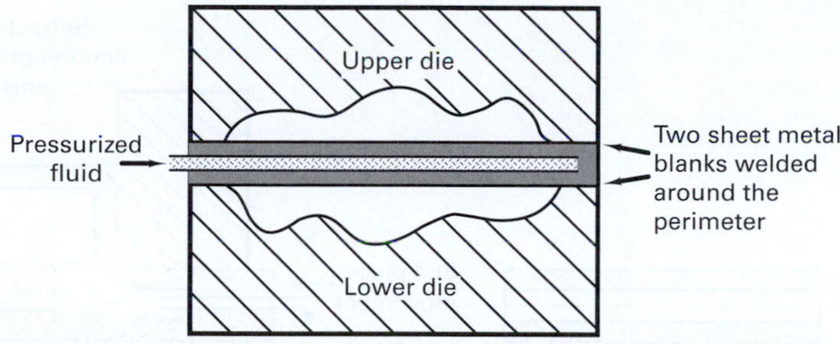

FIGURE 17-49 Two-sheet hydroforming or pillow-forming.

sheet metal containers because the manufacturer no longer has to cope with the problems of aligning and welding separately formed upper and lower pieces.

While the most attractive feature of sheet hydroforming is probably the reduced cost of tooling, there are other positive attributes. Materials exhibit greater formability because a more uniform distribution of strain is produced. Deeper parts can be formed without fracture (drawing limits are generally about 1.5 times those of conventional deep drawing), and complex shapes that require multiple operations can often be formed in a single pressing. Surface finish is excellent, and part dimensions are more accurate and more consistent.

Because the cycle times for sheet hydroforming (20 to 30 s) are slow compared to mechanical presses, conventional deep drawing is preferred when the draw depth is shallow and the part is not complex. The reduced tool costs of hydroforming make the process attractive for prototype manufacturing and low-volume production (up to about 10,000 identical parts), such as that encountered in the aerospace industry. Sheet hydroforming has also attracted attention within the automotive community because of its ability to not only produce low-volume parts in an economical manner, but also to successfully shape lower-formability materials, such as alloyed aluminum sheet and high-strength steels. Warm hydroforming has further expanded this capability.

TUBE HYDROFORMING

Tube hydroforming, illustrated in Figure 17-50, has emerged as a significant process for manufacturing strong, lightweight, tubular components that frequently replace an assembly of welded stampings. As shown in Figure 17-51, current automotive parts include engine cradles, frame rails, roof headers, radiator supports, and exhaust components.

In elementary terms, a tubular blank, either straight or preshaped, is placed in an encapsulating die, and the ends are sealed. A fluid is then introduced through one of the end plugs, achieving sufficient pressure to expand the material to the shape of the die. At the same time, the end closures may move inward to help compensate or overcome the thinning that would otherwise accompany radial expansion. In actual operation, the process may use combinations or even sequences of internal pressure, axial motions, and even external counter-pressure applied to bulging regions, to control the flow and final thickness of the material. Product length is currently limited by the ability of end movements to create axial displacements within the die.

In *low-pressure tube hydroforming* (pressures up to 35 MPa, or 5000 psi), the tube is first filled with fluid and then the dies are closed around the tube. The primary purpose of the fluid is to act as a liquid mandrel that prevents collapsing as the tube is bent to the contour of the die. While the cross-section shape can be changed, the shapes must be simple, corner radii must be large, and there must be minimal expansion of the tube diameter. In *high-pressure tube hydroforming,* an internal pressure between 100 and 700 MPa (15,000 and 100,000 psi) is used to expand the diameter of the tube, forming tight corner radii and significantly altered cross sections. *Pressure-sequence hydroforming* begins by applying low internal pressure as the die is closing. This supports the inside wall of the tube and allows it to conform to the cavity. When the die is fully closed, high pressure is then applied to complete the forming of the tube walls.

Another modification of tube hydroforming uses *pulsing or vibrating pressure* (also called hammering action) to reduce the frictional drag as the tube moves across

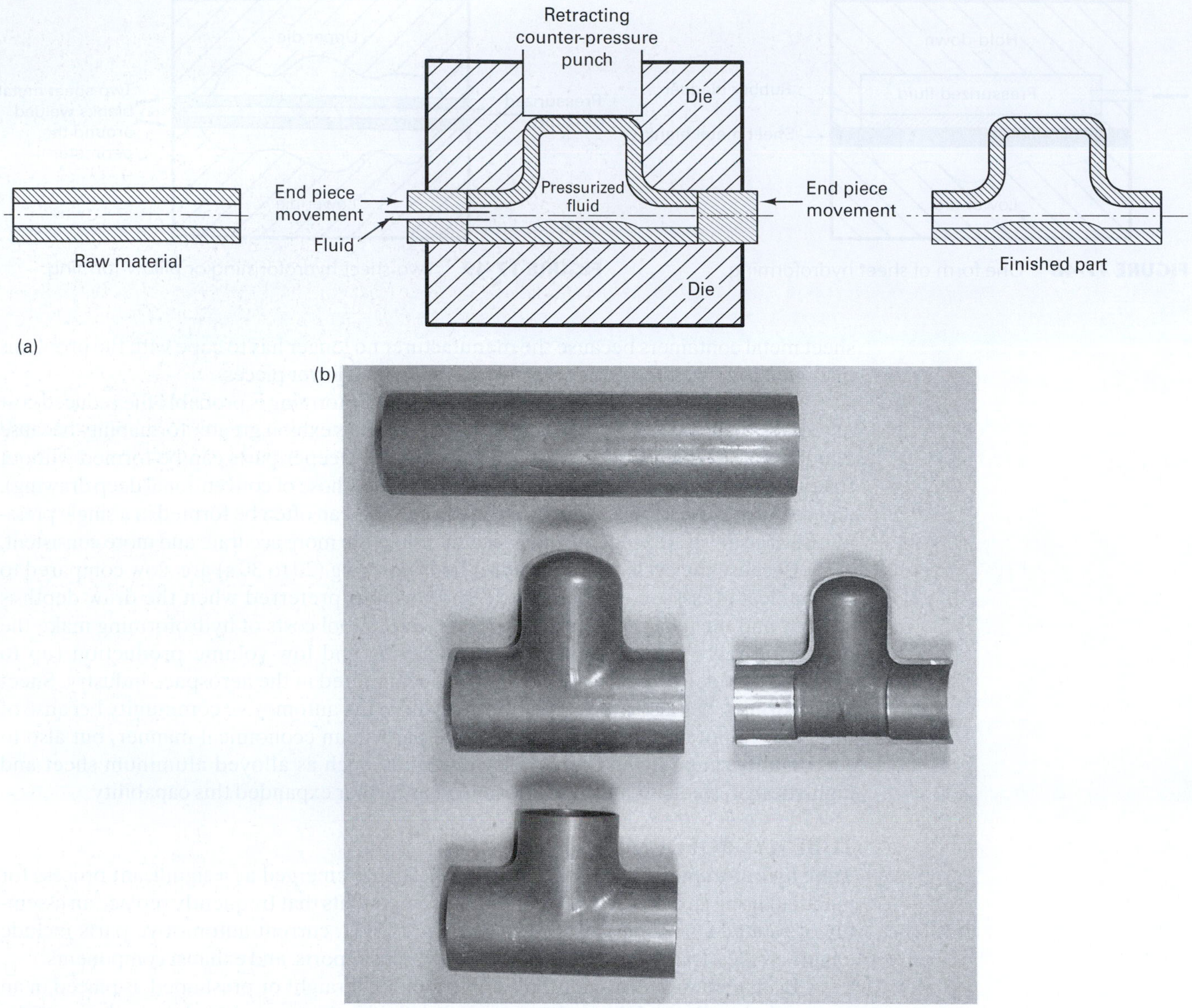

Retracting
counter-pressure
punch
Die
End piece
movement
Pressurized
fluid
End piece
movement
Fluid
Raw material
Die
Finished part
(a)
(b)

FIGURE 17-50 Tube hydroforming. (a) Process schematic. (b) Actual copper product. Note the inward movement of the tube ends and the nonuniform wall thickness of the nonsymmetric product. *(Courtesy Ronald Kohser)*

Roof headers
Instrument
panel supports
Radiator
supports
Frame rails
Engine
cradles
Upper frame
rails

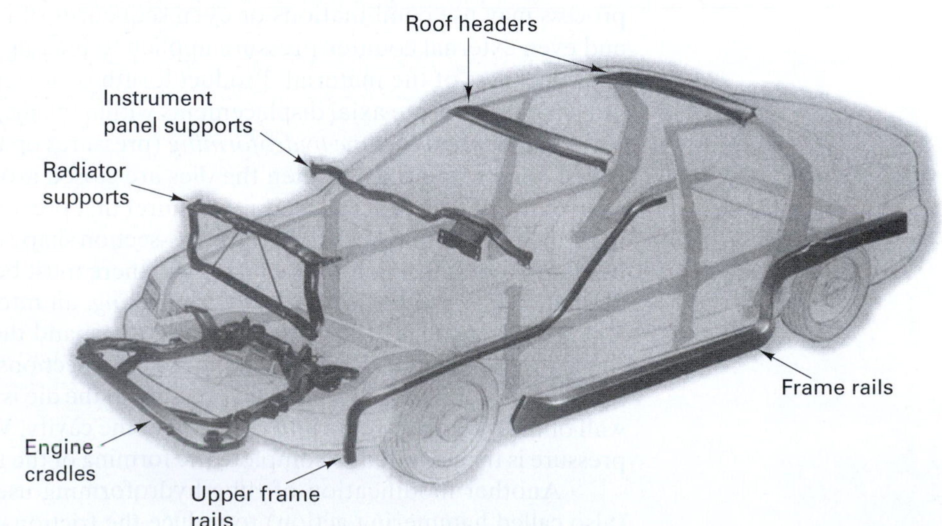

FIGURE 17-51 Use of hydroformed tubes in automotive applications. *(Courtesy of* MetalForming *magazine, a publication of PMA Services, Inc., for the Precision Metalforming Association, Independence, OH; www .metalformingmagazine.com)*

the dies. By oscillating the pressure between two preset values, the tube can move more freely, increasing the amount of allowable deformation and reducing the amount of thinning in regions of surface expansion. The pulsation frequency is generally between 1 and 3 cycles per second (Hz).

Attractive features of tube hydroforming include the ability to use lightweight, high-strength materials; the increase in strength that results from strain hardening; and the ability to utilize optimized designs with varying thickness or varying cross section. Smooth transitions can be produced among round, oval, or rectangular cross sections, and the cross section can change along a designed curvature. Stamped-and-welded assemblies can often be replaced by one-piece components with improved structural strength and rigidity, and secondary operations can often be reduced. Parts can be produced from materials as diverse as brass, stainless steel, carbon steel, and aluminum. Disadvantages include the long cycle time (low production rate) and relatively high cost of tooling and process setup.

An emerging alternative to tube hydroforming is **hot-metal-gas forming.** In this process, a straight or preformed tube is heated to forming temperature by induction heating and then pressurized internally with inert gas to drive the material outward to the enclosed die cavity. By using high temperatures, the parts can be elongated or stretched to greater degree than with the cold- or warm-forming fluid methods. The forming pressures are typically an order of magnitude lower than tube hydroforming, and the formed parts can be quenched directly from the high-temperature forming operation. Expectations for the process include faster speed, lower cost and greater flexibility compared to tube hydroforming. Lower formability metals, such as ferritic stainless steels, may be used in place of the higher formability, but often more expensive types, like the austenitic grades.

HOT DRAWING OPERATIONS

Because sheet material has a large surface area and small thickness, it cools rapidly in a lower-temperature environment. For this reason, most sheet forming is performed at room or mildly elevated temperature. Cold drawing uses relatively thin metal, changes the thickness very little or not at all, and produces parts in a wide variety of shapes. In contrast, hot drawing is used for forming relatively thick-walled parts of simple geometries, and the material thickness may change significantly during the operation.

As shown in Figure 17-52, hot drawing operations are extremely similar to previously discussed processes. The upper-left schematic shows a simple disk-to-cup drawing operation without a hold-down. While the increased thickness of the material acts to resist wrinkling, the height of the cup wall is still restricted by defect formation. When smaller diameter and higher wall height are desired, redraws, like that depicted in the upper-right schematic, can be used. The lower schematic of Figure 17-52 illustrates an alternative where the cup is pushed through a series of dies with a single punch.

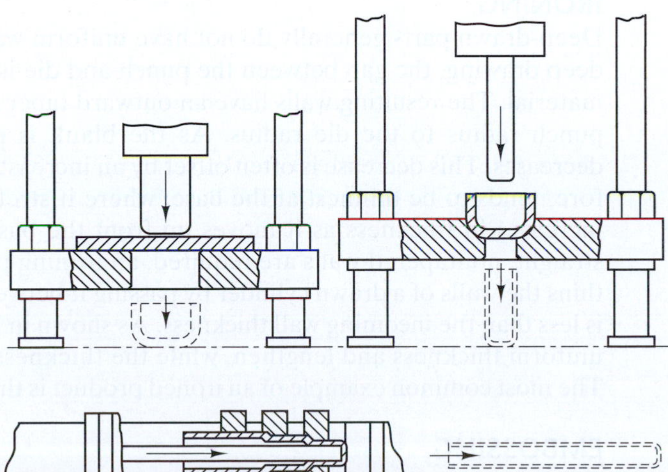

FIGURE 17-52 Methods of hot drawing a cup-shaped part. (Upper left) First draw. (Upper right) Redraw operation. (Lower) Multiple-die drawing. *(Courtesy of United States Steel Corp., Pittsburgh, PA)*

If the drawn products are designed to utilize part of the original disk as a flange around the top of the cup, the punch does not push the material completely through the die but descends to a predetermined depth and then retracts. The partially drawn cup is then ejected upward, and the perimeter of the remaining flange is trimmed to the desired size and shape.

Another hot-forming or hot-stamping process is being applied to the more-difficult-to-deform advanced high-strength steels. The material blank is first heated to between 900° and 950°C (1650° and 1750°F) to create an austenite microstructure. It is then transferred to a water-cooled (chilled) die, where it is immediately formed while in a soft, weak and very ductile condition. The formed part is then held in the die, where it rapidly cools, forming a martensitic structure with strengths in the range of 1025 to 1375 MPa (150 to 200 ksi). The entire forming process takes approximately 20 sec, and produces high-strength, complex-shaped stampings in a single step with virtually no springback because the part is stress-relieved as it forms. Secondary operations cannot be performed and the emerging part must be trimmed with a laser because the material is now too hard to be formed or trimmed using traditional methods.

HIGH-ENERGY-RATE FORMING

A number of methods have been developed to form metals through the application of large amounts of energy in a very short time (high strain rate). These are known as **high-energy-rate forming** processes and often go by the abbreviation **HERF.** Many metals tend to deform more readily under the ultra-rapid load application rates used in these processes. As a consequence, HERF makes it possible to form large workpieces and difficult-to-form metals with less expensive equipment and cheaper tooling than would otherwise be required. HERF processes also produce less springback. This is probably associated with two factors: (1) the high compressive stresses that are created and (2) the elastic deformation of the die produced by the ultrahigh pressure.

High-energy-release rates can be obtained by five distinct methods:

1. Underwater explosions.
2. Underwater spark discharge (electro-hydraulic techniques).
3. Pneumatic-mechanical means.
4. Internal combustion of gaseous mixtures.
5. The use of rapidly formed magnetic fields (electromagnetic techniques).

Specific processes were developed around each of these approaches and attracted considerable attention during the 1960s and 1970s when they were used to produce one-of-a-kind and small quantities of parts for the space program. They are playing a relatively insignificant role in current manufacturing technology.

IRONING

Deep-drawn parts generally do not have uniform wall thickness and vertical walls. In deep drawing, the gap between the punch and die is greater than the thickness of the material. The resulting walls have an outward taper as they progress upward from the punch radius to the die radius. As the blank is pulled inward, its circumference decreases. This decrease is often offset by an increase in thickness. The cup wall, therefore, tends to be thinnest at the base, where it stretches around the punch face, and increases in thickness as it moves up from the base. If uniform wall thickness and straight, nontapered walls are required, an ironing process may be necessary. **Ironing** thins the walls of a drawn cylinder by passing it between a punch and die where the gap is less than the incoming wall thickness. As shown in Figure 17-53, the walls reduce to a uniform thickness and lengthen, while the thickness of the base remains unchanged. The most common example of an ironed product is the thin-walled beverage can.

EMBOSSING

Embossing, shown in Figure 17-54, is a press-working process in which raised lettering or other designs are impressed in sheet material while the material thickness remains

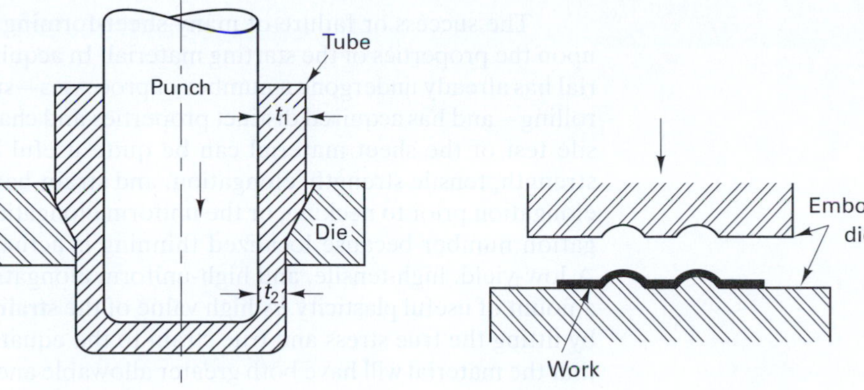

FIGURE 17-53 The ironing process. **FIGURE 17-54** Embossing.

largely unchanged. Basically, it is a very shallow drawing operation where the depth of the draw is limited to one to three times the thickness of the metal. A common example of an embossed product is the patterned or textured industrial stair tread.

SUPERPLASTIC SHEET FORMING

Conventional metals and alloys typically exhibit tensile elongations in the range of 10 to 30%. By producing sheet materials with ultra-fine grain size, and performing the deformation at low strain rates and elevated temperatures, elongations can exceed 100% and may be as high as 2000 to 3000%. This **superplastic forming** behavior can be used to mold material into large, complex-shaped products with compound curves. Deep or complex shapes can be made as single-piece, single-operation pressings rather than multistep conventional pressings or multipiece assemblies.

At the elevated temperatures required to promote superplasticity (about 900°C for titanium, between 450 and 520°C for aluminum, or generally above half of the melting point of the material on an absolute scale), the strength of material is sufficiently low that many of the superplastic forming techniques are adaptations of processes used to form thermoplastics (discussed in Chapter 14). The tooling does not have to be exceptionally strong, so form blocks can often be used in place of die sets. In thermoforming, a vacuum or pneumatic pressure causes the sheet to conform to a heated male or female die. Blow forming, vacuum forming, deep drawing, and combined superplastic forming and diffusion bonding are other possibilities. Precision is excellent, and fine details or surface textures can be reproduced accurately. Springback and residual stresses are almost nonexistent, and the products have a fine, uniform grain size.

The major limitation to superplastic forming is the low forming rate that is required to maintain superplastic behavior. Cycle times may range from 2 min to as much as 2 hr per part, compared to the several seconds that is typical of conventional presswork. As a result, applications tend to be limited to low-volume products such as those common to the aerospace industry. By making the products larger and eliminating assembly operations, the weight of products can often be reduced, there are fewer fastener holes to initiate fatigue cracks, tooling and fabrication costs are reduced, and there is a shorter production lead time.

PROPERTIES OF SHEET MATERIAL

A wide variety of materials have been used in sheet-forming operations, including hot- and cold-rolled steel, stainless steel, copper alloys, magnesium alloys, aluminum alloys, and even some types of plastics. These materials are available in a range of tempers (amounts of prior cold working) and surface finishes. Sheet material can also have a variety of platings or coatings or even be a clad or laminated composite. With the desire to reduce the weight of automotive vehicles, sheet formers are now being asked to shape parts from thinner-gauge, higher-strength steels and more-difficult-to-deform aluminum alloys. Some of the advanced high-strength steels undergo significant metallurgical and property changes during deformation.

The success or failure of many sheet-forming operations is strongly dependent upon the properties of the starting material. In acquiring the sheet geometry, the material has already undergone a number of processes—such as casting, hot rolling, and cold rolling—and has acquired distinct properties and characteristics. A simple uniaxial tensile test of the sheet material can be quite useful by providing values for the yield strength, tensile strength, elongation, and strain-hardening exponent. The amount of elongation prior to necking, or the uniform elongation, is probably a more useful elongation number because localized thinning is actually a form of sheet metal failure. A low-yield, high-tensile, and high-uniform elongation all combine to indicate a large amount of useful plasticity. A high value of the **strain-hardening exponent, *n*** [obtained by fitting the true stress and true strain to the equation: Stress $= K(\text{strain})^n$] indicates that the material will have both greater allowable and more uniform stretch.

In a uniaxial tensile test, the sheet is stretched in one direction and is permitted to contract and compensate in both width and thickness. Many sheet-forming operations subject the material to stretching in more than one direction, however. When the material is in biaxial tension, as in deep drawing or the various stretch-forming operations, all of the elongation must be compensated by a decrease in thickness. The strain to fracture is typically about one-third of the value observed in a uniaxial test. As a result, some form of biaxial stretching test may be preferred to assess formability, such as a dome-height or hydraulic bulge test.

Sheet metal is often quite anisotropic—with properties varying with direction or orientation. A useful assessment of this variation can be obtained through the **plastic strain ratio, *R*,** defined in ASTM specification E517. During a uniaxial tensile test, as the length increases, the width and thickness both reduce. The *R*-value is simply the ratio of the width strain to the thickness strain. Materials with values greater than 1 tend to compensate for stretching by flow within the sheet, and resist the thinning that leads to fracture. Hence materials with high *R*-values have good formability. Sheet materials can also have directional variations within the plane of the sheet. These are best evaluated by performing four uniaxial tensile tests: one cut longitudinally (generally along the direction of prior rolling), one transverse (perpendicular to the first), and one along each of the 45-degree axes. Values are then computed for the average **normal anisotropy** (the sum of the four values of *R* divided by 4) and the **planar anisotropy, ΔR** [computed as $(R_{\text{longitudinal}} + R_{\text{transverse}} - R_{45\ \text{right}} - R_{45\ \text{left}})/2$). Because a ΔR of zero means that the material is uniform in all directions within the plane of the sheet, an ideal drawing material would have a high value of *R* and a low value of ΔR. Unfortunately, the two values tend to be coupled, such that high formability is often accompanied by directional variations within the plane of the sheet. Due to these variations, discs being drawn into cylindrical cups will have variations in the final wall height.

Sheet-forming properties have also been shown to change significantly with variations in both temperature and speed of deformation, or strain rate. Low-carbon steels generally become stronger when deformed at higher speeds, while many aluminum alloys weaken and become more prone to failure during the operation.

DESIGN AIDS FOR SHEET METAL FORMING

The majority of sheet metal failures occur due to thinning or fracture, and both are the result of excessive deformation in a given region. A quick and economical means of evaluating the severity of deformation in a formed part is to use **strain analysis** coupled with a **forming limit diagram.** A pattern or grid, such as the one on the left side of Figure 17-55, is placed on the surface of a sheet by scribing, printing, or etching. The circles generally have diameters between 2.5 and 5 mm (0.1 to 0.2 in.) to enable detection of point-to-point variations in strain distribution. During deformation, the circles convert into ellipses, and the distorted pattern can be measured and evaluated. Regions where the enclosed area has expanded are locations of sheet thinning and possible failure. Regions where the enclosed area has decreased are locations of sheet thickening and may be sites of possible buckling or wrinkles.

Using the ellipses on the deformed grid, the major strains (strain in the direction of the largest radius or diameter) and the associated minor strains (strain 90 degrees

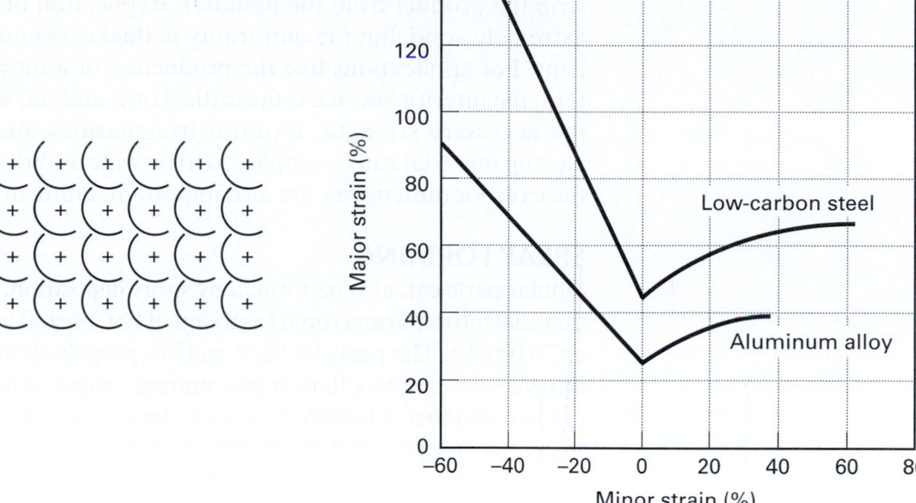

FIGURE 17-55 (Left) Typical pattern for sheet metal deformation analysis. (Right) Forming limit diagram used to determine whether a metal can be shaped without risk of fracture. Fracture is expected when strains fall above the lines.

from the major) can be determined for a variety of locations. These values can then be plotted on a forming-limit diagram such as the one shown on the right side of Figure 17-55. If both major and minor strains are positive (right-hand side of the diagram), the deformation is known as *stretching,* and the sheet metal will definitely decrease in thickness. If the minor strain is negative, the contraction may partially or wholly compensate for any positive stretching in the major direction. This combination of tension and compression is known as *drawing,* and the thickness of the material may decrease, increase, or stay the same, depending on the relative magnitudes of the two strains. Regions where both strains are negative do not appear on the diagram because its purpose is to reveal locations of possible fracture, and fractures only occur in a tensile environment.

Those strains that fall above the forming limit line indicate regions of probable fracture. Possible corrective actions include modification of the lubricant, change in the die design, or variation in the clamping or hold-down pressure. Strain analysis can also be used to determine the best orientation of blanks relative to the rolling direction, assist in the design of dies for complex-shaped products, or compare the effectiveness of various lubricants. Because of the large surface-to-volume ratio in sheet material, lubrication can play a key role in process success or failure.

Dies are now designed using solid modeling, providing a three-dimensional visualization that often eliminates confusion and interpretation errors. Forming simulation software often eliminates the need for prototype tooling and testing. Fractures, splits, wrinkles, and excessive thinning appear in the simulated part just as they would if an actual die had been built. Features such as blank size, hold-down pressures, and draw bead details can all be determined prior to die construction. Simulation software can also calculate the approximate springback that might be expected, permitting consideration of design changes prior to actual fabrication.

17.5 ALTERNATIVE METHODS OF PRODUCING SHEET-TYPE PRODUCTS

ELECTROFORMING

Several manufacturing processes have been developed to produce sheet-type products by directly depositing metal onto preshaped forms or mandrels. In a process known as **electroforming,** the metal is deposited by plating. Nickel, iron, copper, or silver can be deposited in thicknesses up to 16 mm ($\frac{5}{8}$ in.). When the desired thickness has been attained, plating is stopped and the product is stripped from the mandrel.

A wide variety of sizes and shapes can be made by electroforming, and the fabrication of a product requires only a single pattern or mandrel. Low production quantities can be made in an economical fashion, with the principal limitation being the need to strip the product from the mandrel. Replication of the contact surface and profile are extremely good, but the uniformity of thickness and external profile may present problems. For applications like the production of multiple molds from a single master pattern, the interior surface is the critical one, and the wall thickness serves only to provide the necessary strength. Exterior irregularities are not critical, and various types of backup material can be employed to provide additional support. For applications where the exterior dimensions are also important, uniform deposition is required.

SPRAY FORMING

Similar parts can also be formed by **spray deposition.** One approach is to inject powdered material into a plasma torch (a stream of hot ionized gas with temperatures up to 11,000°C, or 20,000°F). The particles melt and are propelled onto a shaped form or mandrel. Upon impact, the droplets flatten and undergo rapid solidification to produce a dense, fine-grained product. Multiple layers can be deposited to build up a desired size, shape, and thickness. Because of the high adhesion, the mandrel or form is often removed by machining or chemical etching. Most applications of plasma spray forming involve the fabrication of specialized products from difficult-to-form or ultra-high-melting-point materials.

The **Osprey process**, described in Chapter 18 (Section 18-11), can also be adapted to produce thin, spray-formed products. Here, molten metal flows through a nozzle, where it is atomized and carried by high-velocity nitrogen jets. The semisolid particles are propelled toward a target, where they impact and complete their solidification. Tubes, plates, and simple forms can be produced from a variety of materials. Layered structures can also be produced by sequenced deposition.

17.6 PIPE MANUFACTURE

Large quantities of steel pipe are made by two processes that use the hot forming of steel strip coupled with deformation welding of the free edges. Both of these processes—**butt welding** of pipe and **lap welding** of pipe—utilize steel in the form of **skelp,** which is long strips with specified width, thickness, and edge configurations. Because the skelp was produced by hot rolling, and the welding process produces further deformation and recrystallization, pipe welded by these processes tends to be very uniform in structure and properties.

BUTT-WELDED PIPE

In the manufacture of butt-welded pipe, steel skelp is heated to a specified hot-working temperature by passing it through a tunnel-type furnace. Upon exiting the furnace, the skelp is pulled through forming rolls that shape it into a cylinder and bring the free ends into contact. The pressure exerted between the opposite edges of the skelp is sufficient to upset the metal and produce a welded seam. Additional sets of rollers then size and shape the pipe, and it is cut to standard, preset lengths. Product diameters range from 3 mm ($\frac{1}{8}$ in.) to 75 mm (3 in.), and speeds can approach 150 m/min (500 ft/min).

LAP-WELDED PIPE

The lap-welding process for making pipe differs from the butt-welding technique in that the skelp now has beveled edges and the rolls form the weld by forcing the overlapped edges down against a supporting mandrel. This process is used primarily for larger sizes of pipe, with diameters from about 50 mm (2 in.) to 400 mm (14 in.). Because the product is driven over an internal mandrel, product length is limited to about 6 to 7 m (20 to 25 ft).

17.7 PRESSES

CLASSIFICATION OF PRESSES

The primary tool for performing most of the sheet-forming operations discussed in this chapter, and many of the bulk-forming operations presented in Chapter 16, is some form of press, and successful manufacture often depends on using the right kind of

TABLE 17-2 Classification of the Drive Mechanisms of Commercial Presses

Manual	Mechanical	Hydraulic
Kick presses	Crank	Single slide
	Single	Multiple slide
	Double	
	Eccentric	
	Cam	
	Knuckle joint	
	Toggle	
	Screw	
	Rack and pinion	

equipment. When selecting a press for a given application, consideration should be given to the capacity required, the type of power (manual, mechanical, or hydraulic), the number of slides or drives, the type of drive, the stroke length for each drive, the type of frame or construction, and the speed of operation. Table 17-2 lists some of the major types of presses and groups them by the type of drive.

Manually operated presses such as foot-operated or **kick presses** are generally used for very light work such as shearing small sheets and thin material.

Moving toward larger equipment, we find that **mechanical presses** tend to provide fast motion and positive control of displacement. Once built, however, the flexibility of a mechanical press is limited, because the length of the stroke is set by the design of the drive. The available force usually varies with position, so mechanical presses are preferred for operations that require the maximum pressure near the bottom of the stroke, such as cutting, shallow forming, drawing (up to about 10 cm) and progressive and transfer die operations. Typical capacities range up to about 9000 metric tons.

Figure 17-56 depicts some of the basic types of mechanical press drive mechanisms. *Crank-driven presses* are the most common type because of their simplicity. They are used for most piercing and blanking operations and for simple drawing. Double-crank presses offer a means of actuating blank holders or operating multiple-action dies. *Eccentric* or *cam drives* are used where the ram stroke is rather short. Cam

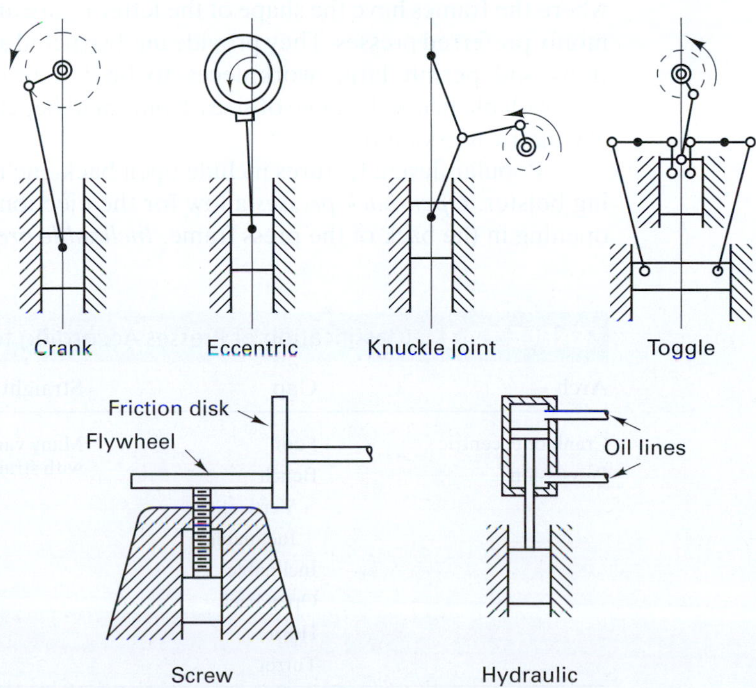

FIGURE 17-56 Schematic representation of the various types of press drive mechanisms.

action can also provide a dwell at the bottom of the stroke and is often the preferred method of actuating the blank holder in deep-drawing processes. *Knuckle-joint drives* provide a very high mechanical advantage along with fast action. They are often preferred for coining, sizing, and Guerin forming. *Toggle mechanisms* are used principally in drawing presses to actuate the blank holder, and *screw-type drives* offer great mechanical advantage coupled with an action that resembles a drop hammer (but slower and with less impact). For this reason, screw presses have become quite popular in the forging industry.

In contrast to the mechanical presses, **hydraulic presses** produce motion as the result of piston movement. The stroke can be programmed to any length within the limits of the cylinder (which may be as long as 250 cm, or 100 in.), reducing or eliminating excessive ram movement. Forces and pressures are more accurately controlled, and full pressure is available throughout the entire stroke. A built-in pressure relief valve provides overload protection to both the press and any inserted tooling. Moving parts are few in number, and most remain fully lubricated, being immersed in the pressurizing oil. Speeds can be varied within the stroke or remain constant during an operation, and the return stroke can be programmed for fast reset. Because position is varied through fluid displacement, the reproducibility of position will have greater variation than a mechanical press, but the noise level will be considerably less.

Hydraulic presses are available in capacities exceeding 50,000 metric tons and are preferred for operations requiring a steady pressure throughout a substantial stroke (such as deep drawing), operations requiring wide variation in stroke length, and operations requiring high or widely variable forces. In general, hydraulic presses tend to be slower than the mechanical variety, but some are available that can provide up to 600 strokes per minute in a high-speed blanking operation. By using multiple hydraulic cylinders, programmed loads can be applied to the main ram, while a separate force and timing are used on the blank holder.

TYPES OF PRESS FRAME

As shown in Table 17-3, presses should also be selected with consideration for the type of frame. Frame design often imposes limitations on the size and type of work that can be accommodated, how that work is fed and unloaded, the overall stiffness of the machine, and the time required to change dies.

Presses that have their frames in the shape of an arch *(arch-frame presses)* are seldom used today, except with screw drives for coining operations. **Gap-frame presses,** where the frames have the shape of the letter *C*, are among the most versatile and commonly preferred presses. They provide unobstructed access to the dies from three directions and permit large workpieces to be fed into the press. Gap-frame presses are available in a wide range of sizes, from small bench types of about 1 metric ton up to 300 metric tons or more.

Popular design features include open back, inclinability, adjustable bed, and sliding bolster. *Open-back presses* allow for the ejection of products or scrap through an opening in the back of the press frame. *Inclinable presses* can be tilted so that ejection

TABLE 17-3 Classification of Presses According to Type of Frame

Arch	Gap	Straight Sided
Crank or eccentric	Foot	Many variations but all with straight-sided frames
Percussion	Bench	
	Vertical	
	Inclinable	
	Inclinable	
	Open back	
	Horn	
	Turret	

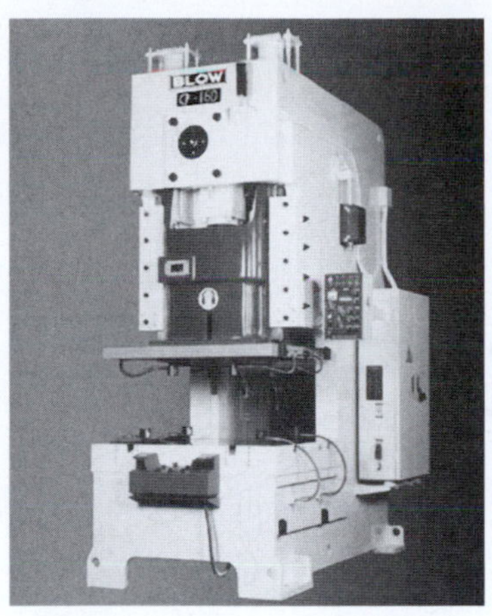

FIGURE 17-57 Gap-frame press. *(Courtesy of Blow Press, Guelph, Ontario, Canada)*

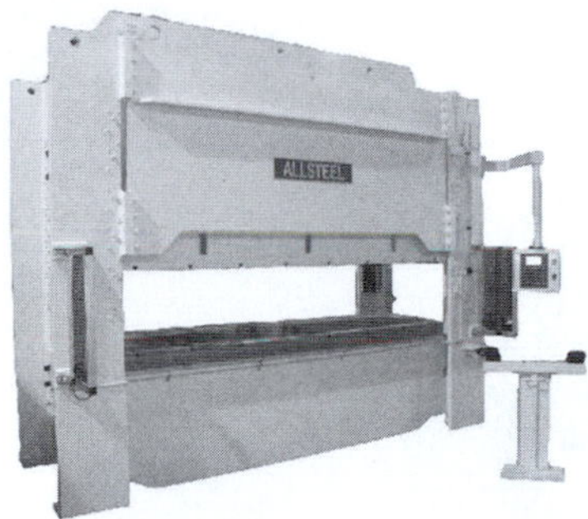

FIGURE 17-58 A hydro-mechanical straight side press. *(Courtesy Allsteel, Salaberry-de-Valleyfield, Canada)*

can be assisted by gravity or compressed air jets. As a result of these features, open-back inclinable (OBI) presses are the most common form of gap-frame press. The addition of an *adjustable bed* allows the base of the machine to raise or lower to accommodate different workpieces. A *sliding bolster* permits a second die to be set up on the press while another is in operation. Die changeover then requires only a few minutes to unclamp the punch segment of the active die, move the second die set into position, clamp the new punch to the press ram, and resume operation. Figure 17-57 shows an open-back, inclinable gap-frame press with a sliding bolster.

A *horn press* is a special type of gap-frame press where a heavy cylindrical shaft or "horn" appears in place of the usual bed. Curved or cylindrical workpieces can be placed over the horn for such operations as seaming, punching, and riveting. On some presses, both a horn and a bed are provided, with provision for swinging the horn aside when not needed.

Turret presses are especially useful in the production of sheet metal parts with numerous holes or slots that vary in size and shape. They usually employ a modified gap-frame structure and add upper and lower turrets that carry a number of punches and dies. The two turrets are geared together so that any desired tool set can be quickly rotated into position.

Straight-sided presses have frames that consist of a crown, two uprights, a base or bed, and one or more moving slides. Accessibility is generally from the front and rear, but openings are often provided in the side uprights to permit feeding and unloading of workpieces. Straight-sided presses are available in a wide variety of sizes and designs and are the preferred design for most hydraulic, large-capacity, or specialized mechanical-drive presses. As an added benefit, elastic deflections tend to be uniform across the working surface, as opposed to the angular deflections that are typical of the gap-frame design. Figure 17-58 shows a typical straight-sided press.

SPECIAL TYPES OF PRESSES

Presses have also been designed to perform specific types of operations. **Transfer presses** have a long moving slide that enables multiple operations to be performed simultaneously in a single machine. Multiple die sets are mounted side-by-side along the slide. After the completion of each stroke, a continuous strip is advanced or individual workpieces are transferred to the next station by a mechanism like the one shown in Figure 17-59. Transfer presses can be used to perform blanking, piercing, forming, trimming, drawing, flanging, embossing, and coining. Figure 17-60 illustrates the production of a part that incorporates a variety of these operations.

By using a single machine to perform multiple operations, transfer presses offer high production rates, high flexibility, and reduced costs (attributed to the reduced labor, floor space, energy, and maintenance). Because production is usually between 500 and 1500 parts per hour, these machines are usually restricted to operations where 4000 or more identical parts are required daily, each involving three or more separate operations. A total production run of 30,000 or more identical parts is generally desired between major changes in tooling. As a result, transfer presses are used primarily in

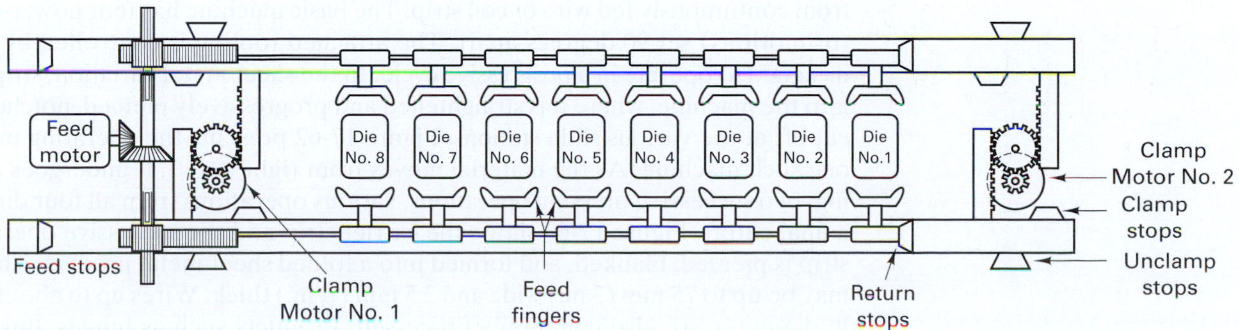

FIGURE 17-59 Schematic showing the arrangement of dies and the transfer mechanism used in transfer presses. *(Courtesy of Verson Allsteel Press Company, Chicago, IL)*

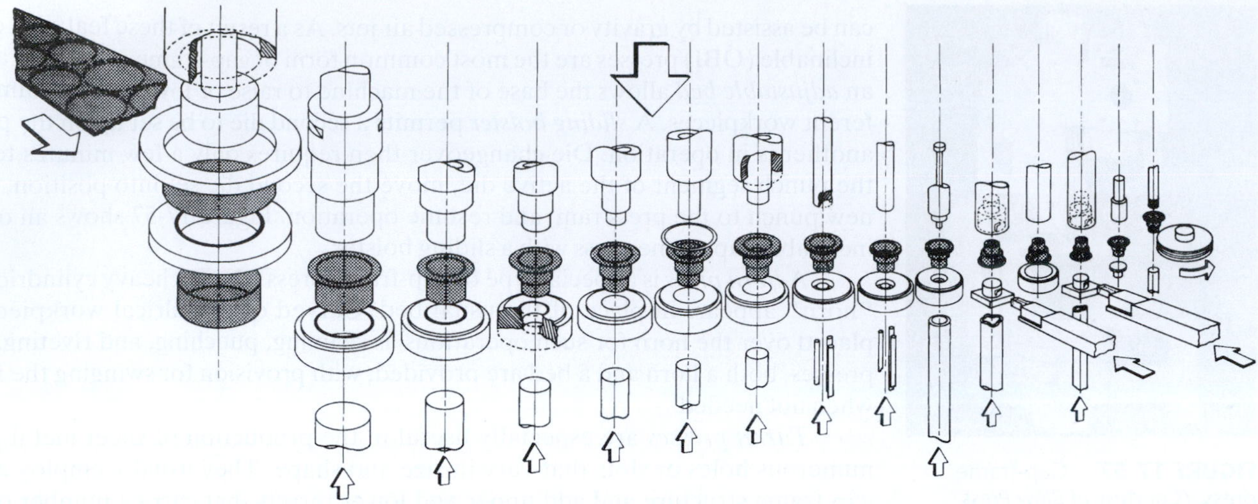

FIGURE 17-60 Various operations can be performed during the production of stamped and drawn parts on a transfer press. *(Courtesy of U.S. Baird Corporation, Stratford, CT)*

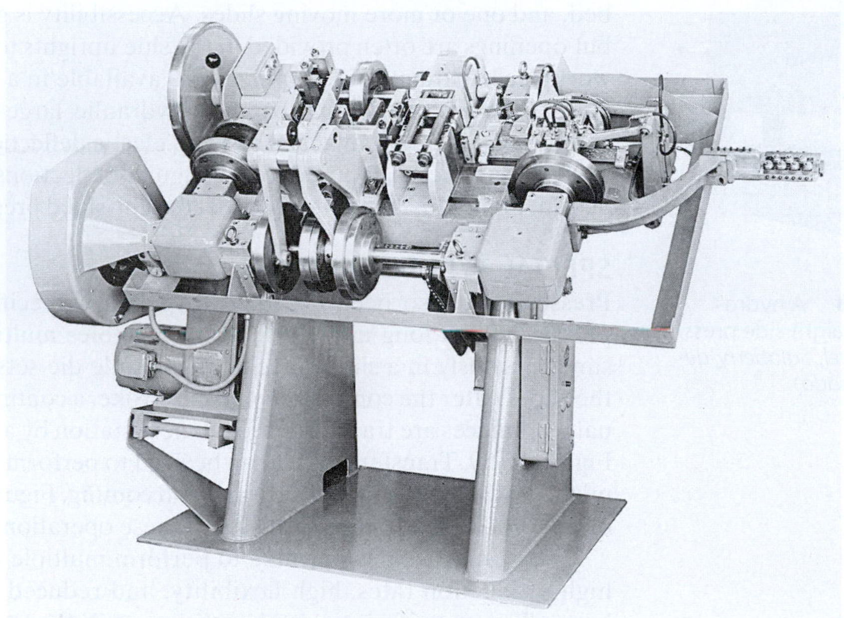

FIGURE 17-61 Multislide machine with guards and covers removed. *(The Baird Machinery Corporation, Thomaston, CT 06787)*

industries such as automotive and appliances, where large numbers of identical products are being produced.

Four-slide or **multislide machines,** like the one shown in Figure 17-61 are extremely versatile presses that are designed to produce small, intricately shaped parts from continuously fed wire or coil strip. The basic machine has four power-driven slides (or motions) set 90 degrees apart. The attached tooling is controlled by cams and is designed to operate in a progressive cycle. In the sheet metal variation, strip stock is fed into the machine, where it is straightened and progressively pierced, notched, bent, and cut off at the various slide stations. Figure 17-62 presents the operating mechanism of one such machine. As the material moves from right to left, it undergoes a straightening, two successive pressing operations, various operations from all four directions, and a final cutoff. Figure 17-63 shows the carrier strip and the successive operations as flat strip is pierced, blanked, and formed into a folded sheet metal product. The strip stock may be up to 75 mm (3 in.) wide and 2.5 mm ($\frac{3}{32}$ in.) thick. Wires up to about 3 mm ($\frac{1}{8}$ in.) in diameter are also commonly processed. Products such as hinges, links, clips, and razor blades can be formed at very high rates approaching 15,000 pieces per hour. Setup times are long, so large production runs are preferred.

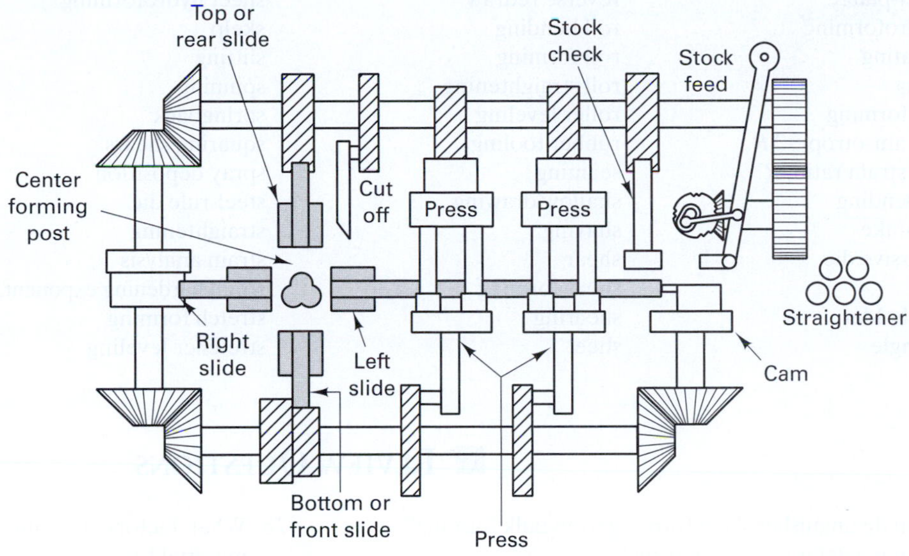

FIGURE 17-62 Schematic of the operating mechanism of a multislide machine. The material enters on the right and progresses toward the left as operations are performed. *(Courtesy of U.S. Baird Corporation, Stratford, CT)*

FIGURE 17-63 Example of the piercing, blanking, and forming operations performed on a multislide machine. *(The Baird Machinery Corporation, Thomaston, CT 06787)*

PRESS-FEEDING DEVICES

Although hand feeding may still be used in some press operations, operator safety and the desire to increase productivity have motivated a strong shift to feeding by some form of mechanical device. When continuous strip is used, it can be fed automatically by double-roll feeds mounted on the side of the press. Discrete products can be moved and positioned in a wide variety of ways. Dial-feed mechanisms enable an operator to insert workpieces into the front holes of a rotating dial, which then indexes with each stroke of the press to move the parts progressively into proper position between the punch and die. Lightweight parts can be fed by suction-cup mechanisms, vibratory-bed feeders, and similar devices. Robots are frequently used to place parts into presses and remove them after forming.

■ KEY WORDS

air-bend die
bar folder
bending
blankholder force
blanking
bottoming die
bulging
bulk
butt welding
coining
compound die
compression bending
cutoff
deep drawing
die
die shoe
dinking
draw bead
draw bending
draw ratio
drawing
electroforming
embossing
fineblanking
flanging
flattening
flexforming
flow turning
forming
forming limit diagram
forward redraw
gap-frame press
GRIPflow
Guerin process
high-energy-rate forming (HERF)
hold-down pressure
hot-metal-gas forming
hydraulic press
independent die set
ironing
kick press
lancing
lap welding
laser-assisted metal spinning
leveling
mechanical press
minimum bend radius
modular tooling
multislide machine
neutral axis
nibbling
normal anisotropy
notching
Osprey process

parallel-plate hydroforming
perforating
piercing
pillow forming
planar anisotropy, ΔR
plastic strain ratio, R
press bending
press brake
progressive die
punch
punch holder
rake angle
reverse redraw
roll bending
roll forming
roll straightening
roller leveling
rubber tooling
seaming
shallow drawing
shaving
shear
shear forming
shearing
sheet
sheet hydroforming
skelp
slitting
spinning
springback
squaring shears
spray deposition
steel-rule die
straightening
strain analysis
strain hardening exponent, n
stretch forming
stretcher leveling
stretching
strip
stripper plate
subpress die
superplastic forming
transfer die
transfer press
trimming
tube bending
tube hydroforming
turret-type punch press

REVIEW QUESTIONS

1. What distinguishes sheet forming from bulk forming?
2. What is a definition of shearing?
3. Why are sheared or blanked edges generally not smooth? What are the various regions on a sheared edge?
4. What measures can be employed to improve the quality of a sheared edge?
5. How does fineblanking create shearing in a compressive environment?
6. Why are fineblanking presses more complex than those used in conventional blanking?
7. Why might a long shearing cut be made in a progressive fashion like cutting with scissors?
8. What is a slitting operation?
9. What are the differences between piercing and blanking?
10. What are some types of blanking or piercing operations that have come to acquire specific names?
11. What are the basic components of a piercing or blanking die set?
12. What is the purpose of having a shear angle or rake angle on a punch?
13. Why is it important that a blanking punch and die be in proper alignment?
14. What is the major benefit of mounting punches and dies on independent die sets?
15. What is the major benefit of assembling a complex die set from standard subpress dies?
16. What is the benefit of making dies as a multipiece assembly?
17. What is a "steel-rule die," and for what types of designs and materials is it used?
18. What is a progressive die set?
19. What is the difference between progressive dies and transfer dies?
20. How do compound dies differ from progressive dies?
21. What is the attractive feature of a turret-type punch press?
22. When making bends in sheet metal, what is the distinction among bending, forming, and drawing?
23. What are the stress states on the exterior surface and interior surface of a bend?
24. Why does a metal usually become thinner in the region of a bend?
25. What is springback, and why is it a concern during bending?
26. Why is springback an increased concern for the newer advanced high-strength steels?
27. What types of operations can be performed on a press brake?
28. What factors determine the minimum bend radius for a material?
29. If a right-angle bend is to be made in a cold-rolled sheet, should it be made with the bend lying along or perpendicular to the direction of previous rolling?
30. From a manufacturing viewpoint, why is it desirable for all bends in a product (or component) to have the same radius?
31. What is the difference between air-bend and bottoming dies? Which is more flexible? Which produces more reproducible bends?
32. What is the primary benefit of incorporating a coining action in bottom bending? The primary negative feature?
33. Why are different types of presses (mechanical and hydraulic) used in bottom bending and air bending?
34. What is the benefit of using a urethane (rubber) bottom die in bending operations?
35. What type of products are produced by roll bending?
36. What is the role of the form block in draw bending and compression bending?
37. How can flattening or wrinkling be prevented when bending a tube?
38. What type of product geometry can be produced by cold-roll forming?
39. Is the roll forming process appropriate for making short lengths of specialized products?
40. What are some methods for straightening or flattening rod or sheet?
41. What two distinctly different metal forming processes use the term "drawing"?
42. What is the difference between drawing and stretching?
43. Why is the tooling cost for a spinning operation relatively low?
44. What is the benefit of the laser assist in laser-assisted metal spinning?
45. How is shear forming different from spinning?
46. For what types of products would stretch forming be an appropriate manufacturing technique?
47. What is the distinction between shallow drawing and deep drawing?
48. What is the function of the pressure ring or hold-down in a deep-drawing operation?
49. What are the key variables in a deep-drawing operation?
50. Explain why thin material may be difficult to draw into a defect-free cup.

51. How can redraw operations be used to produce a taller, smaller-diameter cup than can be produced in a single, deep-drawing operation?
52. What are draw beads, and what function do they perform?
53. Why is a trimming operation often included in a deep-drawing manufacturing sequence?
54. How does the Guerin process reduce the cost of tooling in a drawing operation?
55. How can fluid pressure or rubber tooling be used to perform bulging?
56. What is sheet hydroforming?
57. What are the various forms of sheet hydroforming?
58. What explanation can be given for the greater formability observed during sheet hydroforming? How is this feature being used in the automotive industry?
59. What is the purpose of the inward movement of the end plugs during tube hydroforming?
60. What is the difference between low-pressure tube hydroforming and high-pressure tube hydroforming in terms of both pressures and the nature of the deformations produced?
61. What is the advantage of hot-metal-gas-forming over conventional tube hydroforming?
62. What are some of the basic methods that have been used to achieve the high energy-release rates needed in the HERF processes?
63. Why is springback rather minimal in high-energy-rate forming?
64. What are some well-known products that have been produced by processes including ironing? By embossing?
65. What material and process conditions are associated with superplastic forming?
66. What is the major limitation of the superplastic forming of sheet metal? What are some of the more attractive features?
67. What properties from a uniaxial tensile test can be used to assess sheet metal formability?
68. How is the formability in biaxial tension different from that in uniaxial tension?
69. What techniques can be used to assess "normal anisotropy" and "planar anisotropy" in sheet material?
70. How can strain analysis be used to determine locations of possible defects or failure in sheet metal components?
71. What is a forming limit diagram?
72. Explain the key difference between the right- and left-hand sections of a forming limit diagram. These sections correspond to stretching and drawing.
73. Describe two alternative methods of producing complex-shaped thin products without requiring sheet metal deformation techniques.
74. What two hot-forming operations can be used to produce pipe from steel strip?
75. What features should be considered when selecting a press for a given application?
76. What are the primary assets and limitations of mechanical press drives? Of hydraulic drives?
77. What are some of the common types of press frames?
78. What is the purpose of inclining or tilting a press?
79. Describe how multiple operations are performed simultaneously in a transfer press.
80. What types of products are produced on a four-slide or multislide machine?

■ Problems

1. The maximum punch force in blanking can be estimated by the equation:

$$\text{Force} = StL$$

where:

S = the material shear strength
t = the sheet thickness, and
L = the total length of sheared edge (circumference or perimeter)

   a. How would this number change if the rake angle is equal to a $1t$ change across the width or diameter of the part being sheared?
   b. How would this number change if the rake were increased to a $3t$ change across the width or diameter?
2. Consider the various means of producing tubular products, such as extrusion, seam welding, butt-welding during forming, piercing, and the various drawing operations. Describe the advantages, limitations, and typical applications of each.
3. Tube and sheet hydroforming have been undergoing rapid growth. Investigate current uses for these processes in automotive and other fields.
4. What are some of the techniques for minimizing the amount of springback in sheet forming operations?
5. Select a forming process from either Chapter 16 or 17, and investigate the residual stresses that typically accompany or result from that process. If they are considered to be detrimental, how could these residual stresses be reduced or removed without damaging or deteriorating the product?
6. In most products, the painting or surface finishing operations are usually performed on individual pieces after the shape has been produced. An alternative approach is to paint, plate, or surface finish the sheet material while it is in the form of continuous coil (coil-coated sheet material), and then fabricate the shape in a manner so as to retain the integrity of the surface. Discuss some of the pros and cons of this approach, such as uncoated sheared edges, special handling or modified tooling, etc.

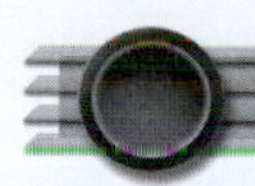

www.wiley.com/go/global/degarmo

Chapter 17 CASE STUDY

Automotive Body Panels

Automotive body panels have been a prime target for weight reduction through the use of thinner material, higher-strength steels, and lightweight alternative materials, such as aluminum and polymeric sheet-molding compound. Associated with each candidate material is a related manufacturing or fabrication system. In order to select the "best" material, we need to evaluate both the material and the related processing with respect to a number of factors, including:

a. Mechanical Properties
   - Elastic modulus
   - Yield strength
   - Tensile strength
   - Ductility (total elongation or uniform elongation)
   - Fatigue resistance
   - Fracture resistance

b. Weight

c. Corrosion resistance

d. Cost of material

e. Methods and ease of manufacture
   - Formability
   - Ease and methods of assembly

f. Surface finishing
   - Paint or other alternative

g. Fabrication costs

h. Fabrication speeds/rate of production

i. Cost of equipment, tooling, energy, labor

j. Dent resistance, ease of repair, and cost of repair

k. Recyclability

l. Other

Body panels were traditionally made from drawing-quality, low-carbon steel (such as AISI 1008 or 1010), formed by traditional sheet metal forming, and assembled by spot welding. Some panels contained hemmed edges for both appearance and rigidity.

1. Reflecting on the key factors listed, discuss the pros and cons of the following four materials:
   a. The conventional low-carbon steel.
   b. Advanced high-strength steel.
   c. Aluminum.
   d. Polymeric sheet molding compound (resin, fiber, and filler).
2. The traditional low-carbon steels have largely been replaced by thinner gauge high-strength steels. What fabrication difficulties were encountered during this transition? How was the fabrication system changed to accommodate the change in material?
3. If aluminum is the material of choice, what mechanism will be used to provide the necessary strength (cold working, age hardening, etc.)? If an age-hardening heat treatment is specified, will it be performed before or after shape production, or possibly integrated into the production such as: solution treat, shape the product, then age to desired strength? (*Note:* The times and temperatures of paint curing are not sufficient to age-harden aluminum alloys.)
4. Polymeric materials have characteristically low rigidity. How might the desired rigidity be provided in a polymeric-based automotive body panel?
5. If adhesive bonding is specified as a replacement for the traditional spot welding as a means of joining or assembly, discuss some of the considerations relating to the curing process or the time and conditions necessary to produce full-strength joints.
6. Which of the material/process options do you feel would have the:
   a. Lowest cost?
   b. Highest cost?
   c. Lightest weight?
   d. Fastest production rate?
   e. Slowest production rate?
7. Which of the material/process options in Part 6 do you feel would be best for:
   a. A high-volume, low-cost vehicle?
   b. A limited-quantity, more-costly, high-performance vehicle?

CHAPTER 18

POWDER METALLURGY

18.1 INTRODUCTION

Powder metallurgy is the name given to the process by which fine powdered materials are blended, pressed into a desired shape (compacted), and then heated (sintered) in a controlled atmosphere to bond the contacting surfaces of the particles and establish desired properties. The process, commonly designated as **P/M,** readily lends itself to the mass production of small, intricate parts of high precision, often eliminating the need for additional machining or finishing. There is little material waste, unusual materials or mixtures can be utilized, and controlled degrees of porosity or permeability can be produced. Major areas of application tend to be those for which the P/M process has strong economical advantage (compared to machined components, castings, or forgings) or where the desired properties and characteristics would be difficult to obtain by any other method (such as products made from tungsten, molybdenum or tungsten carbide, porous bearings, filters, and various magnetic components). Because of its level of manufacturing maturity, powder metallurgy should actually be considered as a possible means of manufacture for any part where the geometry and production quantity are appropriate.

While a crude form of iron powder metallurgy existed in Egypt as early as 3000 BC, and the ancient Incas made jewelry and other artifacts from precious metal powders, mass manufacturing of P/M products did not begin until the mid- or late 19th century. At this time, powder metallurgy was used to produce copper coins and medallions, platinum ingots, lead printing type, and tungsten wires (the primary material for light bulb filaments). By the 1920s, the tips of tungsten carbide cutting-tools and nonferrous bushings were being produced. Self-lubricating bearings and metallic filters were other early products.

A period of rapid technological development occurred after World War II, based primarily on automotive applications, and iron and steel replaced copper as the dominant P/M material. Aerospace and nuclear developments created accelerated demand for refractory and reactive metals, materials for which powder processing is quite attractive. Full-density products emerged in the 1960s, and high-performance superalloy components, such as aircraft turbine engine parts, were a highlight of the 1970s. Developments in the 1980s and 1990s included the commercialization of rapidly solidified and amorphous powders, powder forging, warm compacting, and P/M injection molding. Even newer technologies include submicron and nanophase powders; direct powder rolling; high-velocity and ultra-high-pressure compaction; and high-temperature, plasma- and microwave-sintering. With products now incorporating ceramics, ceramic fibers, and intermetallic compounds, even the term *powder metallurgy* fails to encompass the breadth of the industry, and some have come to prefer **particulate processing.**

Recent years have been ones of rapid growth for the P/M industry. From 1960 to 1980, the consumption of iron powder increased 10-fold. A similar increase occurred between 1980 and 1990, the exponential growth continued through the 1990s, and demand has stabilized over the past decade. While most products are still under 50 mm (2 in.) in size, some have been produced with weights up to 45 kg (100 lb) with linear dimensions up to 500 mm (20 in.).

Automotive applications now account for nearly 70% of the powder metallurgy market. In 1990, the average U.S. automobile contained about 10 kg (21 lb) of P/M parts. By 1995, the amount had increased to more than 13.6 kg (30 lb), then to 16.3 kg (36 lb) in 2000, and by 2005 it was at 19.5 kg (nearly 45 lb). One version of the 2008 Cadillac CTS and STS V-6 engine contains nearly 32 lb of P/M parts. More than 500 million P/M forged connecting rods have been made for cars produced in the United States, Europe, and Japan. Automatic transmissions often contain between 15 and 25 lb of P/M components. Commercial aircraft engines contain between 1500 and 4500 lb of P/M superalloy components.

Other areas where powder metallurgy products are used extensively include household appliances, recreational equipment, hand and power tools, hardware items, office equipment, industrial motors, and hydraulics. Areas of rapid growth include aerospace applications, advanced composites, electronic components, magnetic materials, metalworking tools, and a variety of biomedical and dental applications. Iron and low-alloy steels now account for 85% of all P/M usage, with copper and copper-based powders comprising about 7%. Stainless steel, high-strength and high-alloy steels, and aluminum and aluminum alloys are other high-volume materials. Titanium, magnesium, refractory metals, particulate composites, and intermetallics are seeing increased use.

18.2 THE BASIC PROCESS

The powder metallurgy process generally consists of four basic steps: (1) powder manufacture, (2) mixing or blending, (3) compacting, and (4) sintering. Compaction is generally performed at room temperature, and the elevated temperature process of sintering is usually conducted at atmospheric pressure. Optional secondary processing often follows to obtain special properties or enhanced precision. Figure 18-1 presents a simplified flowchart of the conventional die-compaction P/M process.

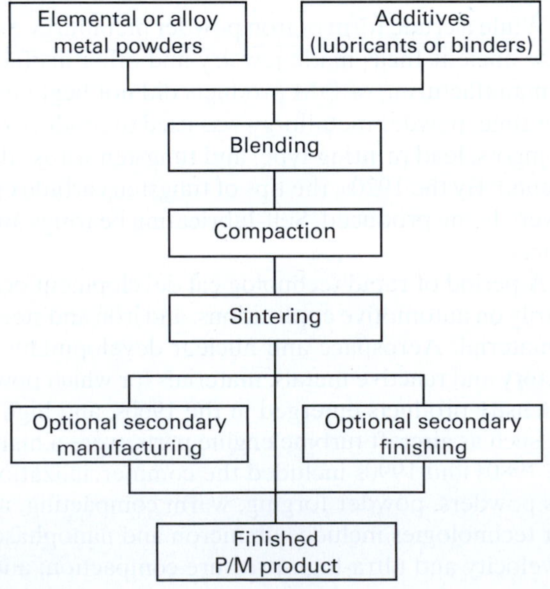

FIGURE 18.1 Simplified flow chart of the basic powder metallurgy process.

18.3 POWDER MANUFACTURE

The properties of powder metallurgy products are highly dependent on the characteristics of the starting powders. Some important properties and characteristics include **chemistry** and **purity, particle size, size distribution, particle shape,** and the **surface texture** of the particles. Several processes can be used to produce powdered material, with each imparting distinct properties and characteristics to the powder and hence to the final product.

More than 80% of all commercial powder is produced by some form of melt **atomization,** where liquid material is fragmented into small droplets that cool and solidify into particles before they come into contact with each other or with a solid surface. Various methods have been used to form the droplets, several of which are illustrated in Figure 18-2. Part a illustrates **gas atomization,** where jets of high pressure gas (usually nitrogen argon or helium) strike a stream of liquid metal as it emerges from an orifice. Pressurized liquid (usually water) can replace the pressurized gas, converting the process to **liquid atomization** or **water atomization.** In part b, an electric arc impinges on a rapidly rotating electrode. Centrifugal force causes the molten droplets to fly from the surface of the electrode and freeze in flight. Particle size is very uniform and can be varied by changing the speed of rotation.

Regardless of the specific process, atomization is an extremely useful means of producing **prealloyed powders.** By starting with an alloyed melt or prealloyed electrode, each powder particle has the desired alloy composition. Powders of aluminum alloys, copper alloys, stainless steel, nickel-based alloys (such as Monel), titanium alloys, cobalt-based alloys, and various low-alloy steels have all been commercially produced. The size, shape, and surface texture of the powder particles varies, depending on such process features as the velocity and media of the atomizing jets or the speed of electrode rotation, the starting temperature of the liquid (which affects the time that surface tension can act on the individual droplets prior to solidification), and the environment provided for cooling. When cooling is slow (such as in gas atomization) and surface tension is high, smooth-surface spheres can form before solidification. With the more rapid cooling of water atomization, irregular shapes tend to be produced.

Other methods of powder manufacture include:

1. *Chemical reduction of particulate compounds* (generally crushed oxides or ores). The powders that result from these solid-state reactions are usually soft, irregular in shape, and spongy in texture. Powder purity depends on the purity of the starting materials. A large amount of iron powder is produced by reducing iron ore or rolling mill scale.
2. *Electrolytic deposition* from solutions or fused salts with process conditions favoring the production of a spongy or powdery deposit that does not adhere to the cathode. Purity is generally high, but the energy required is also high. Therefore, electrolysis is

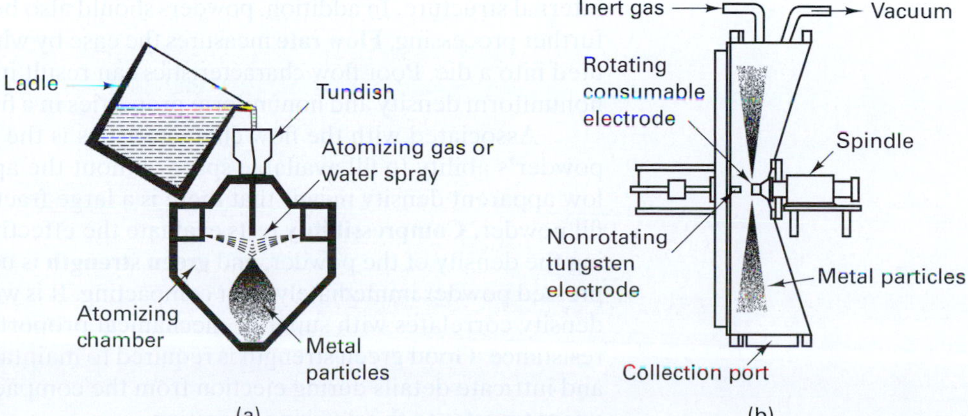

FIGURE 18.2 Two methods for producing metal powders: (a) melt atomization; (b) atomization from a rotating consumable electrode.

usually restricted to the production of high-value powders, such as high-conductivity copper.

3. *Pulverization* or *grinding* of brittle materials (comminution).
4. *Thermal decomposition of particulate hydrides or carbonyls.* Iron and nickel powders are produced by carbonyl decomposition, resulting in small, spherical particles.
5. *Precipitation from solution.*
6. *Condensation of metal vapors.*

Almost any metal, metal alloy, or nonmetal (ceramic, polymer, or wax or graphite lubricant) can be converted into powder form by one or more of the powder production methods. Some methods can produce only elemental powder (often of high purity), while others can produce prealloyed particles. Alloying can also be achieved mechanically by processes that cause elemental powders to successively adhere and break apart. Material is transferred as traces of one particle are left on the other. Unusual compositions can be produced that are not possible with conventional melting.

Prior to further processing, powders may also undergo further operations, such as drying or heat treatment. The usual objective of heat treatment is to weaken the material and make it more responsive to compaction.

18.4 Microcrystalline and Amorphous Material Produced by Rapid Cooling

Increasing the cooling rate of an atomized liquid can result in the formation of an ultra-fine or microcrystalline grain size. In these materials, a large percentage of the atoms are located in grain boundary regions, giving unusual properties (such as high diffusivity), expanded alloy possibilities, and good formability. If the cooling rate approaches or exceeds 10^6 °C/s, metals can solidify without becoming crystalline. These **amorphous,** or glassy, **metals** can also exhibit unusual or unique properties, which include high strength, improved corrosion resistance, and reduced energy to induce and reverse a magnetization. Amorphous metal transformer cores lose 60 to 70% less energy in magnetization than conventional silicon steels. As a result, it is estimated that more than half of all new power distribution transformers purchased in the United States will utilize amorphous metal cores.

Production of amorphous material, however, requires immensely high cooling rates and hence ultra-small dimensions. Atomization with rapid cooling and the "splat quenching" of a metal stream onto a cool surface to produce a continuous ribbon are two prominent methods. Because much of the ribbon material is further fragmented into powder, powder metallurgy is the primary means of fabricating useful products from microcrystalline and amorphous material.

18.5 Powder Testing and Evaluation

Key properties of powdered material include bulk chemistry, surface chemistry, particle size and size distribution, particle shape and shape distribution, surface texture, and internal structure. In addition, powders should also be evaluated for their suitability for further processing. **Flow rate** measures the ease by which powder can be fed and distributed into a die. Poor flow characteristics can result in nonuniform die filling, as well as nonuniform density and nonuniform properties in a final product.

Associated with the flow characteristics is the **apparent density,** a measure of a powder's ability to fill available space without the application of external pressure. A low apparent density means that there is a large fraction of unfilled space in the loose-fill powder. **Compressibility** tests evaluate the effectiveness of applied pressure in raising the density of the powder, and **green strength** is used to describe the strength of the pressed powder immediately after compacting. It is well established that higher product density correlates with superior mechanical properties, such as strength and fracture resistance. Good green strength is required to maintain smooth surfaces, sharp corners, and intricate details during ejection from the compacting die or tooling and the subsequent transfer to the sintering operation.

The overall objective is often to achieve a useful balance of the key properties. The smooth-surface spheres produced by gas atomization, for example, tend to pour and flow well, but the compacts have extremely low green strength, disintegrating easily during handling. The irregular particles of water-atomized powder have better compressibility and green strength, but poorer flow characteristics. The sponge iron powders produced by chemical reduction of iron oxide are extremely porous and have highly irregular, extremely rough surfaces. They have poor flow characteristics and low compacted density, but green strength is quite high. Thus, the same material can have widely different performance characteristics, depending on the specifics of powder manufacture.

18.6 POWDER MIXING AND BLENDING

It is rare that a single powder will possess all of the characteristics desired in a given process and product. Most likely, the starting material will be a mixture of various grades or sizes of powder, or powders of different compositions, along with additions of **lubricants** or **binders.**

In powder products, the final chemistry is often obtained by combining pure metal or nonmetal powders, rather than starting with a prealloyed material. To produce a uniform chemistry and structure in a blended powder product, sufficient diffusion must occur during the sintering operation. Unique **composites** can also be produced, such as the distribution of an immiscible reinforcement material in a matrix or the combination of metals and nonmetals in a single product such as a tungsten carbide–cobalt matrix cutting tool for high-temperature service.

Some powders, such as graphite, can even play a dual role, serving as a lubricant during compaction and a source of carbon as it alloys with iron during sintering to produce steel. Lubricants such as graphite or stearic acid improve the flow characteristics and compressibility at the expense of reduced green strength. Binders produce the reverse effect. Because most lubricants or binders are not wanted in the final product, they are removed (volatilized or burned off) in the early stages of sintering, leaving holes that are reduced in size or closed during subsequent heating.

Blending or **mixing** operations can be done either dry or wet, where water or other solvent is used to enhance particle mobility, reduce dust formation, and lessen explosion hazards. Large lots of powder can be homogenized with respect to both chemistry and distribution of components, sizes, and shapes. Quantities up to 16,000 kg (35,000 lb) have been blended in single lots to ensure uniform behavior during processing and large-run production of a consistent product.

18.7 COMPACTING

One of the most critical steps in the P/M process is **compaction.** Loose powder is compressed and densified into a shape known as a *green compact,* usually at room temperature. High product density and the uniformity of that density throughout the compact are generally desired characteristics. In addition, the mechanical interlocking and cold welding of the particles should provide sufficient green strength for in-process handling and transport to the sintering furnace.

Most compacting is done with mechanical presses and rigid tools, but hydraulic and hybrid (combinations of mechanical, hydraulic, and pneumatic) presses can also be used. Figure 18-3 shows a typical mechanical press for compacting powders and a removable set of compaction tooling. The removable die sets allow the time-consuming alignment and synchronization of tool movements to be set up while the press is producing parts with another die set. Compacting pressures generally range between 3 and 120 ton/in.^{2} depending on material and application (see Table 18-1), with the range of 10 to 50 ton/in.^{2} being the most common. While most P/M presses have total capacities of less than 100 tons, increasing numbers are being purchased with higher capacity. Because of pressures and press capacity, powder metallurgy products are often limited to pressing areas of less than 10 in.^{2}, but larger parts have become more common. Some P/M presses now have capacities up to 3000 tons and are capable of pressing areas up to 100 in.^{2} When even larger products are desired, compaction can be performed by dynamic methods, such as use of

FIGURE 18.3 An 880-ton compacting press, capable of compacting multi-level powder metallurgy products. A second die set can be seen at the lower right side of the press, enabling quick change of the compaction tooling. *(Courtesy Cincinnati Incorporated, Cincinnati, OH)*

an explosively induced shock wave. Metal-forming processes—such as rolling, forging, extrusion, and swagging—have also been adapted for powder compaction.

Figure 18-4 shows the typical compaction sequence for a mechanical press. With the bottom punch in its fully raised position, a feed shoe moves into position over the die. The feed shoe is an inverted container filled with powder, connected to a large powder container by a flexible feed tube. With the feed shoe in position, the bottom punch descends to a preset fill depth, and the shoe retracts, with its edges leveling the powder. The upper punch then descends and compacts the powder as it penetrates the die. The upper punch retracts and the bottom punch then rises to eject the green compact. As the die shoe advances for the next cycle, its forward edge clears the compacted product from the press, and the cycle repeats.

During uniaxial or one-direction compaction, the powder particles move primarily in the direction of the applied force. Because the loose-fill dimensions are two to two-and-a-half times the pressed dimensions, the amount of particle travel in the pressing direction can be substantial. The amount of lateral flow, however, is quite limited. In fact, it is rare to find a particle in the compacted product that has moved more than three particle diameters off of its original axis of pressing. Thus, the powder does not flow like a liquid but simply compresses until an equal and opposing force is created. This opposing force is probably a combination of (1) resistance by the bottom punch and (2) friction between the particles and the die surfaces. Densification occurs by particle movement, as well as plastic deformation of the individual particles.

TABLE 18-1 Typical Compaction Pressures for Various Applications

	Compaction Pressures	
Application	tons/in.^{2}	Mpa
Porous metals and filters	3–5	40–70
Refractory metals and carbides	5–15	70–200
Porous bearings	10–25	146–350
Machine parts (medium-density iron & steel)	20–50	275–690
High-density copper and aluminum part	18–20	250–275
High-density iron and steel parts	50–120	690–1650

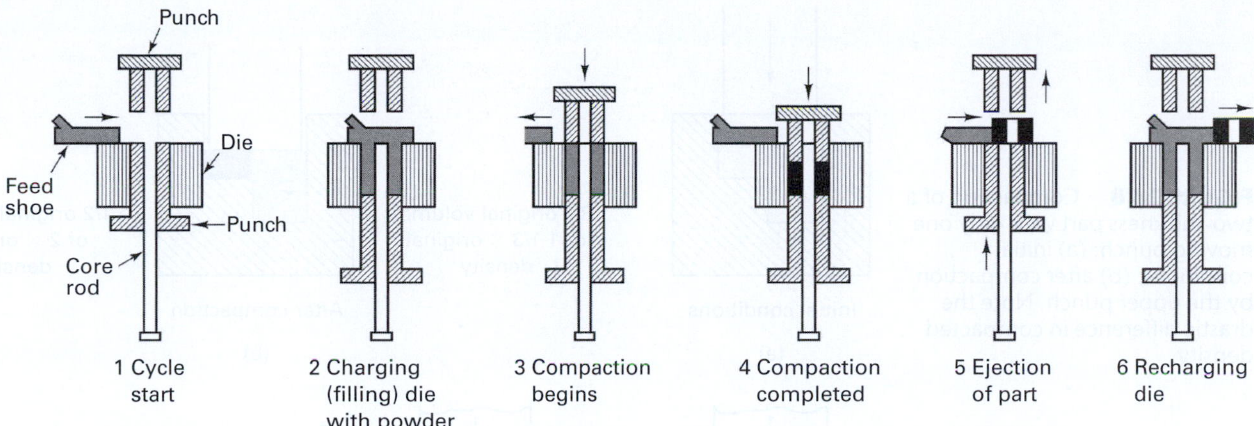

FIGURE 18.4 Typical compaction sequence for a single-level part, showing the functions of the feed shoe, die, core rod, and upper and lower punches. Loose powder is shaded; compacted powder is solid black.

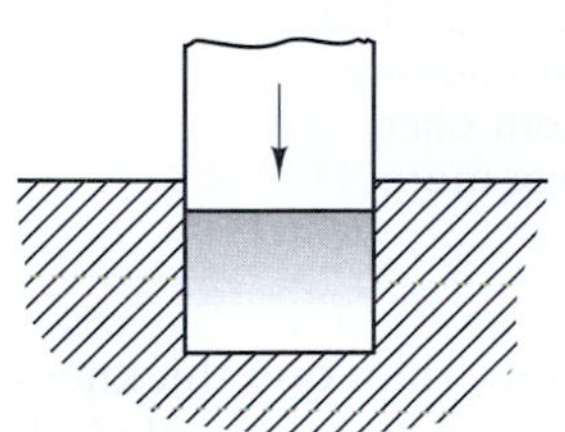

FIGURE 18.5 Compaction with a single moving punch, showing the resultant nonuniform density (shaded), highest where particle movement is the greatest.

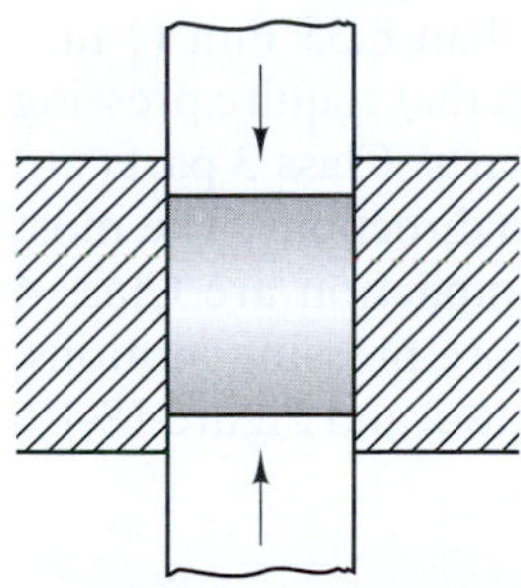

FIGURE 18.6 Density distribution obtained with a double-acting press and two moving punches. Note the increased uniformity compared to Figure 18-5. Thicker parts can be effectively compacted.

As illustrated in Figure 18-5,when the pressure is applied by only one punch, maximum density occurs below the punch and decreases as one moves down the column. It is very difficult to transmit uniform pressures and produce uniform density throughout a compact, especially when the thickness is large. By use of a double-action press, where pressing movements occur from both top and bottom (Figure 18-6), thicker products can be compacted to a more uniform density. Because sidewall friction is a key factor in compaction, the resulting density shows a strong dependence on both the thickness and width of the part being pressed. For uniform compaction, the ratio of thickness/width should be kept below 2.0 whenever possible. When the ratio exceeds 2.0, the products tend to exhibit considerable variation in density.

As shown in Figure 18-7, the average density of the compact depends on the amount of pressure that is applied, with the specific response being strongly dependent upon the characteristics of the powder being compressed (its size, shape, surface texture, mechanical properties, etc.). The final density may be reported as either an absolute density in units such as grams per cubic centimeter, or as a percentage of the pore-free or theoretical density. The difference between this percentage and 100% corresponds to the amount of void space remaining within the compact.

Figure 18-8 shows that a single displacement will produce different degrees of compaction in different thicknesses of powder. It is impossible, therefore, for a single punch

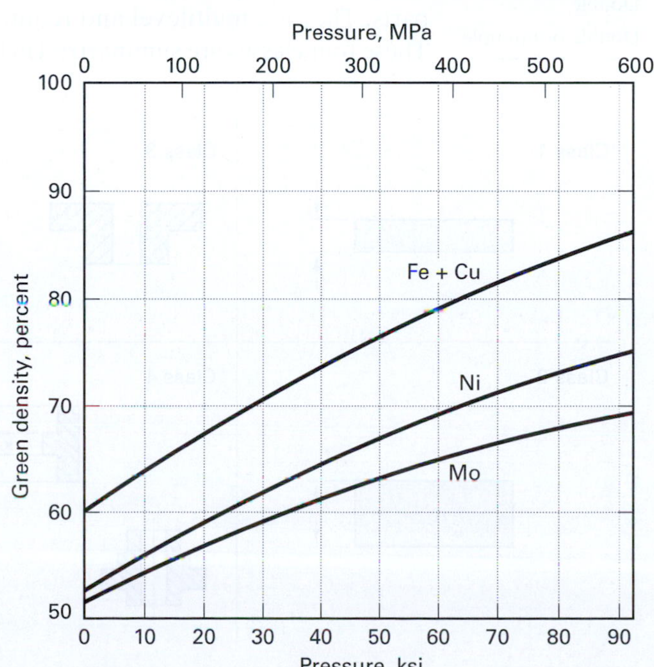

FIGURE 18.7 Effect of compacting pressure on green density (the density after compaction but before sintering). Separate curves are for several commercial powders.

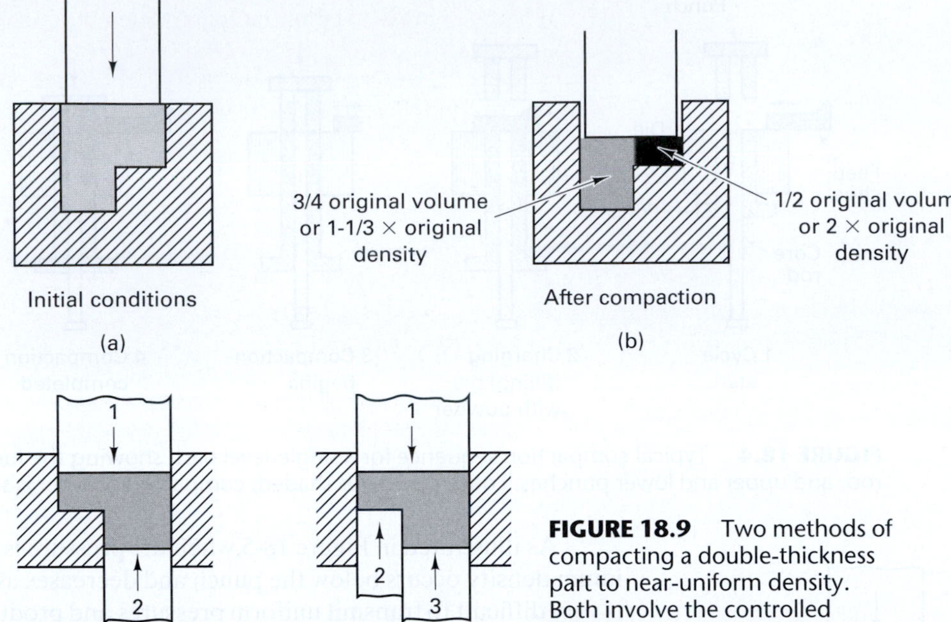

FIGURE 18.8 Compaction of a two-thickness part with only one moving punch: (a) initial conditions; (b) after compaction by the upper punch. Note the drastic difference in compacted density.

FIGURE 18.9 Two methods of compacting a double-thickness part to near-uniform density. Both involve the controlled movement of two or more punches.

to produce uniform density in a multi-thickness part. When more than one thickness is required, more complicated presses or compaction methods must be employed. Figure 18-9 illustrates two methods of compacting a dual-thickness part. By providing different amounts of motion to the various punches and synchronizing these movements to provide simultaneous compaction, a uniform-density product can be produced.

Because the complexity of the part dictates the complexity of equipment, powder metallurgy components have been grouped into classes. Class 1 components are the simplest and easiest to compact. They are thin, single-level parts that can be pressed with a force from one direction. The thickness is generally less than 6.35 mm ($\frac{1}{4}$ in.). Class 2 parts are single-level parts of any thickness that require pressing from two directions. These are usually thicker parts. Class 3 parts are double-level parts that require pressing from two directions. The most complex of the parts produced by rigid die compaction are Class 4 parts. They are multilevel and require two or more pressing motions. These four classes are summarized in both Table 18-2 and Figure 18-10.

TABLE 18-2 Features That Define the Various Classes of Press-and-Sinter P/M Parts

Class	Levels	Press Actions
1	1	Single
2	1	Double
3	2	Double
4	More than 2	Double or multiple

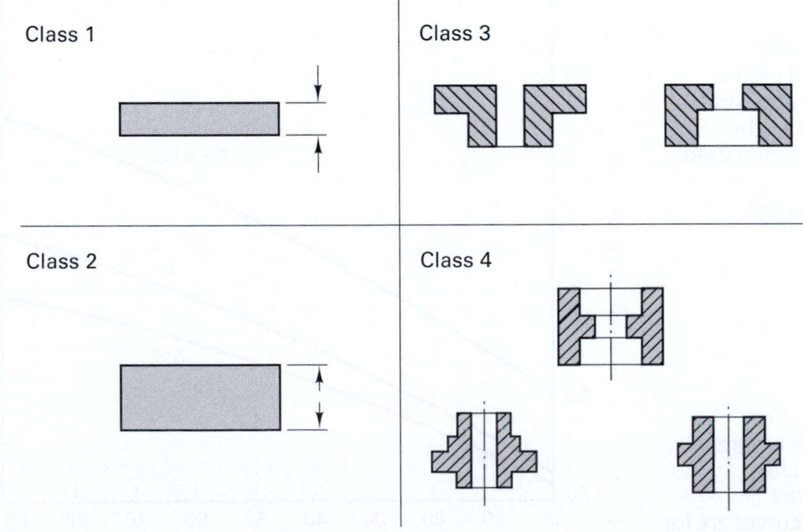

FIGURE 18.10 Sample geometries of the four basic classes of press-and-sinter powder metallurgy parts. Note the increased pressing complexity that would be required as class increases.

If a large part with complex shape is desired, the powder is generally encapsulated in a flexible mold, usually made of rubber or some elastomeric material, which is then immersed in a pressurized gas or liquid. This process is known as **isostatic** *(uniform-pressure)* **compaction.** Because the pressure is applied in all directions, lower compaction pressures produce densities higher than conventional punch-and-die compaction. Production rates are extremely low, but parts with weights up to several hundred pounds have been effectively compacted.

Warm compaction emerged as a common practice in the 1990s. By preheating the powder prior to pressing, the metal is softened and responds better to the applied pressures. The better compaction results in improved properties, both in the as-compacted state and after final processing.

Compaction can also be enhanced by increasing the amount of lubricant in the powder. This reduces the friction between the powder and the die wall, improves the transmission of pressure through the powder, and makes it easier to eject the compact from the die. If too much lubricant is used, however, the green strength may be reduced to the point where it is insufficient for part ejection and handling, or the final properties may become unacceptable.

While pressing rates vary widely, small mechanical presses can typically compact up to 100 pieces (or parts) per minute (ppm). By means of bulk movement of particles, deformation of individual particles, and particle fracture or fragmentation, mechanical compaction can raise the density of loose powder to about 80% of an equivalent cast or forged metal. Sufficient strength can be imparted to retain the shape and permit a reasonable amount of careful handling. In addition, the compaction process sets both the nature and distribution of the porosity remaining in the product.

Because powder particles tend to be somewhat abrasive and high pressures are involved during compaction, wear of the tool components is a major concern. Consequently, compaction tools are usually made of hardened tool steel. For particularly abrasive powders, or for high-volume production, cemented carbides may be employed. Die surfaces should be highly polished and the dies should be heavy enough to withstand the high pressing pressures. Lubricants are also used to reduce die wear.

18.8 SINTERING

In the **sintering** operation, the pressed-powder compacts are heated in a controlled atmosphere environment to a temperature below the melting point but high enough to permit solid-state diffusion and are held for sufficient time to permit bonding of the particles. Most metals are sintered at temperatures of 70 to 80% of their melting point, while certain refractory materials may require temperatures near 90%. Table 18-3 presents a summary of some common sintering temperatures. When the product is composed of more than one material, the sintering temperature may be above the melting temperature of one or more

TABLE 18-3 Typical Sintering Temperatures for Some Common Metals and Materials

	Sintering Temperature	
Metal	°C	°F
Aluminum alloys	590–620	1095–1150
Brass	850–950	1550–1750
Copper	750–1000	1400–1850
Iron/steel	1100–1200	2000–2200
Stainless steel	1200–1280	2200–2350
Cemented carbides	1350–1450	2450–2650
Molybdenum	1600–1700	2900–3100
Tungsten	2200–2300	4000–4200
Various ceramics	1400–2100	2550–3800

components. The lower-melting-point materials then melt and flow into the voids between the remaining particles, and the process becomes **liquid-phase sintering.** In **activated sintering,** a small amount of additive is used to increase the rate of diffusion.

Most sintering operations involve three stages, and many sintering furnaces employ three corresponding zones. The first operation, the *preheat, burn-off,* or *purge,* is designed to combust any air, volatilize and remove lubricants or binders that would interfere with good bonding, and slowly raise the temperature of the compacts in a controlled manner. Rapid heating would produce high internal pressure from air entrapped in closed pores or volatilizing lubricants and would result in swelling or fracture of the compacts. When the compacts contain appreciable quantities of volatile materials, their removal creates additional *porosity* and *permeability* within the pressed shape. The manufacture of products such as metal filters is designed to take advantage of this feature. When the products are load-bearing components, however, high amounts of porosity are undesirable, and the amount of volatilizing lubricant is kept to an optimized minimum. The second, or *high-temperature,* stage is where the desired solid-state diffusion and bonding between the powder particles take place. As the material seeks to lower its surface energy, atoms move toward the points of contact between the particles. The areas of contact become larger, and the part becomes a solid mass with small pores of various sizes and shapes. The mechanical bonds of compaction become true metallurgical bonds. The time in this stage must be sufficient to produce the desired density and final properties, usually varying from 10 min to several hours. Finally, a *cooling period* is required to lower the temperature of the products while retaining them in a controlled atmosphere. This feature serves to prevent oxidation that would occur upon direct discharge into air, as well as possible thermal shock from rapid cooling. Both batch and continuous furnaces are used for sintering.

All three stages of sintering must be conducted in the oxygen-free conditions of a vacuum or **protective atmosphere.** This is critical because the compacted shapes typically have 10 to 25% residual porosity, and some of the internal voids are connected to exposed surfaces. At elevated temperatures, rapid oxidation would occur and significantly impair the quality of interparticle bonding. **Reducing atmospheres,** commonly based on hydrogen, dissociated ammonia, or cracked hydrocarbons, are preferred because they can reduce any oxide already present on the particle surfaces and combust harmful gases that are liberated during the sintering. **Inert gases** cannot reduce existing oxides but will prevent the formation of any additional contaminants. **Vacuum sintering** is frequently employed with stainless steel, titanium, and the refractory metals. **Nitrogen atmospheres** are also common.

During the sintering operation, a number of changes occur in the compact. Metallurgical bonds form between the powder particles as a result of solid-state atomic diffusion, and strength, ductility, toughness, and electrical and thermal conductivities all increase. If different chemistry powders were blended, interdiffusion promotes the formation of alloys or intermetallic phases. As the lubricant is removed and the pores reduce in size, there will be a concurrent increase in density and contraction in product dimensions. To meet final tolerances, the dimensional shrinkage will have to be compensated through the design of oversized compaction dies. During sintering, not all of the porosity is removed, however. Conventional pressed-and-sintered P/M products generally contain between 5 and 25% residual porosity.

Sinter brazing is a process in which two or more separate pieces are joined by brazing while they are also being sintered. The individual pieces are compacted separately and are assembled with the braze metal positioned so it will flow into the joint. When the assembly is heated for sintering, the braze metal melts and flows between the joint surfaces to create the bond. As sintering continues, much of the braze metal diffuses into the surrounding metal, producing a final joint that is often stronger than the materials being joined.

18.9 RECENT ADVANCES IN SINTERING

Because product properties improve with increases in density, various techniques have been developed to produce higher-density components. One way of achieving this while using the conventional "press-and-sinter" approach is to increase the temperature of sintering. While **high-temperature sintering** may seem easily attainable, for iron and steel components, the increased temperatures generally require significant changes in furnace design and materials. The higher product densities, however, can often enable the use of less costly materials, such as chromium- or silicon-alloyed steels in place of the nickel- or molybdenum-steels.

Sinter hardening integrates a strengthening heat-treatment directly into the sintering operation. At the temperatures of sintering, iron and steel parts are austenite, and a rapid cool can produce the stronger, nonequilibrium microstructures. In place of the usual slow cool under protective atmosphere, parts can undergo rapid convective cooling, and some may experience an oil quench.

Microwave sintering has recently moved from the research laboratory to full-scale production. Unlike convection heating, where heat is transmitted through external surfaces, microwaves interact with the entire volume of material, uniformly heating the whole part, thereby reducing both processing time and energy consumption to as little as 20% of traditional processing. The use of microwaves in the sintering of ceramic materials has been very successful and has demonstrated the overall viability of the process. More recently, the technique has been extended to metal powders, where it has been successfully used with ferrous alloys, tungsten, and other metals. In bulk metals, the microwaves induce surface eddy currents and skin heating but are otherwise reflected. Powdered metals, however, heat up well due to the large amounts of surface area and poor electrical connectivity. Surface oxides, moisture, and other surface contaminants aid in the initial heating. Once the metal becomes conductive and interparticle connectivity improves, however, microwave heating becomes less effective.

Microwave heating can also be extremely useful in reducing the time required for removal of binders and lubricants. In traditional processing, heat is conducted from the surface of the part into the interior. If the part is heated too quickly, volatilization can create gas pockets that cause volume expansion or even cracking. If the temperature is too high, the surface can densify, trapping binder or lubricant in the interior. Optimal removal can often be achieved by combining conventional heating with a microwave assist that heats the lubricant or binder from the inside and drives it to the surface.

Compaction and sintering are combined in **spark-plasma sintering,** where axial force compaction is coupled with high-frequency, high-amperage, low-voltage pulses of direct current that is applied through the punches. Spark discharges occur in the gaps between particles while electrical resistance heating occurs at points of particle contact. Some surface melting is observed, and the heating of the particles combined with the axial pressure causes the particles to deform, further aiding densification. No binders are required, and full density can be achieved with both metal and ceramic powders. Sintering can be achieved at lower overall temperatures, and processing time can be greatly reduced.

18.10 HOT-ISOSTATIC PRESSING

In conventional press-and-sinter powder metallurgy, the pressing or compaction is usually performed at room temperature, and the sintering at atmospheric pressure. **Hot-isostatic pressing (HIP)** combines powder compaction and sintering into a single operation by using gas-pressure squeezing at elevated temperature. While this may seem to be an improvement over the two-step approach, it should be noted that heated powders may need to be "protected" or isolated from harmful environments, and the pressurizing media must be prevented from entering the voids between the particles. One approach to hot-isostatic pressing begins by sealing the powder in a flexible, air-tight, evacuated container, which is then subjected to a high-temperature, high-pressure environment. Conditions for processing irons and steels involve pressures around 70 to 100 MPa (10,000 to 15,000 psi), coupled with temperatures in the neighborhood of 1250°C (2300°F). For the nickel-based superalloys, refractory metals, and ceramic powders, the equipment must be capable of 310 MPa (45,000 psi) and 1500°C (2750°F). Multiple pieces, totaling up to several tons, can now be processed in a single cycle that typically lasts several hours.

After processing, the products emerge at full density with uniform, isotropic properties that are often superior to those of other processes. Near-net shapes are possible, thereby reducing material waste and costly machining operations. Because the powder is totally isolated and compaction and sintering occur simultaneously, the process is attractive for reactive or brittle materials, such as beryllium, uranium, zirconium, and titanium. Difficult-to-compact materials, such as superalloys, tool steels, and stainless steels, can be readily processed. Because die compaction is not required, large parts are now possible, and shapes can be produced that would be impossible to eject from rigid compaction dies. Hot-isostatic pressing has also been employed to densify existing parts (such as those that have been conventionally pressed and sintered), heal internal porosity in castings, and seal internal cracks in a variety of products. The elimination or reduction of defects yields startling improvements in strength, toughness, fatigue resistance, and creep life.

Several aspects of the HIP process make it expensive and unattractive for high-volume production. The first is the high cost of **canning** the powder in a flexible isolating medium that can resist the subsequent temperatures and pressures, and then later removing this material from the product **(decanning).** Sheet metal containers are most common, but glass and even ceramic molds have been used. The second problem involves the relatively long time for the HIP cycle. While the development of advanced cooling methods has reduced cycle times from 24 hr to 6 to 8 hr, production is still limited to several loads a day, and the number of parts per load is limited by the ability to produce and maintain uniform temperature throughout the pressure chamber.

The **sinter-HIP process** and **pressure-assisted sintering** are techniques that have been developed to produce full-density powder products without the expense of canning and decanning. Conventionally compacted P/M parts are placed in a pressurizable chamber and sintered (heated) under vacuum for a time that is sufficient to seal the surface and isolate all internal porosity. (*Note:* This generally requires achieving a density greater than 92 to 95%.) While maintaining the elevated temperature, the vacuum is broken and high pressure is then applied for the remainder of the process. The sealed surface produced during the vacuum sintering acts as an isolating can during the high-pressure stage. Because these processes start with as-compacted powder parts, they eliminate the additional heating and cooling cycle that would be required if parts were first sintered in the conventional manner and then subjected to the HIP process for further densification.

18.11 OTHER TECHNIQUES TO PRODUCE HIGH-DENSITY POWDER METALLURGY PRODUCTS

High-temperature metal deformation processes can also be used to produce high-density P/M parts. Sheets of sintered powder (produced by roll compaction and sintering) can be reduced in thickness and further densified by hot rolling in the process depicted in Figure 18-11. Rods, wires, and small billets can be produced by the hot extrusion of encapsulated powder or pressed-and-sintered slugs. Forging can be applied to form complex shapes from canned powder or simple-shaped sintered preforms. By using powdered material, these processes offer the combined benefits of powder metallurgy and the respective forming process, such as the production of fabricated shapes with uniform fine grain size, uniform chemistry, or unusual alloy composition.

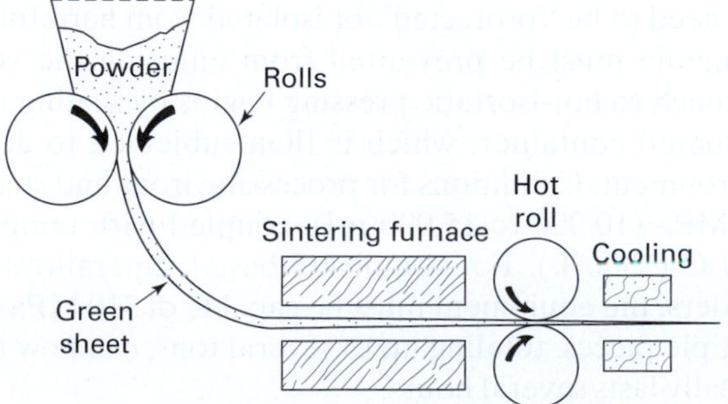

FIGURE 18.11 One method of producing continuous sheet products from powdered feedstock.

The **Ceracon process** is another method of raising the density of conventional pressed-and-sintered P/M products without requiring encapsulation or canning. A heated preform is surrounded by hot granular material, usually smooth-surface ceramic particles. When the assembly is then compacted in a conventional hydraulic press, the granular material transmits a somewhat uniform pressure. Encapsulation is not required because the pressurizing medium is not capable of entering pores in the material. When the pressure cycle is complete, the part and the pressurizing medium separate freely, and the pressure-transmitting granules are reheated and reused.

Yet another means of producing a high-density shape from fine particles is **in-situ compaction** or **spray forming** (also known as the **Osprey process**). Consider an atomizer similar to that of Figure 18-2a, in which jets of inert or harmless gas (nitrogen or carbon dioxide) propel molten droplets down into a collecting container. If the droplets solidify before impact, the container fills with loose powder. If the droplets remain liquid during their flight, the container fills with molten metal, which then solidifies into a conventional casting. However, if the cooling of the droplets is controlled so that they are semi-solid (and computers can provide the necessary process control), they act as "slush balls" and flatten upon impact. The remaining freezing occurs quickly, and the resultant product is a uniform chemistry, fine-grain-size, high-density (in excess of 98%) solid. Depending on the shape of the collecting container, the spray-formed product can be a finished part, a strip or plate, a deposited coating, or a preform for subsequent operations, such as forging. Both ferrous and nonferrous products can be produced with deposition rates as high as 200 kg/min (400 lb/min). Unique composites can be produced by the simultaneous deposition of two or more materials, injecting secondary particles into the stream, and promoting in-stream reactions.

18.12 METAL INJECTION MOLDING OR POWDER INJECTION MOLDING

For many years, injection molding has been used to produce small, complex-shaped components from plastic. A thermoplastic resin is heated to impart the necessary degree of fluidity and is then pressure injected into a die, where it cools and hardens. Die casting is a similar process for metals but is restricted to alloys with relatively low melting temperature, such as lead-, zinc-, aluminum-, and copper-based materials. Small, complex-shaped products of the higher-melting-point metals are generally made by more costly processes, which include investment casting, machining directly from metal stock, or conventional powder metallurgy. **Metal injection molding (MIM),** also called **powder injection molding (PIM),** is a rather recent extension of conventional powder metallurgy that combines the shape forming capability of plastics, the precision of die casting, and the materials flexibility of powder metallurgy.

Because powdered material does not flow like a fluid, complex shapes are produced by first combining ultra-fine (usually in the range of 3 to 20 mm) spherical-shaped metal, ceramic, or carbide powder with a low-molecular-weight thermoplastic or wax material in a mix that is typically 60 vol% powder. This mixture is frequently produced in the form of pellets or granules, which become the feedstock for the injection process. After heating to a pastelike consistency (about 260°C, or 500°F), the material is injected into a heated mold cavity under sufficient pressure (about 70 MPa, or 10,000 psi) to ensure die filling. After cooling and ejection, the binder material is removed by one of a variety of processes that include solvent extraction, controlled heating to above the volatilization temperature, or heating in the presence of a catalyst that breaks the binder down into removable products. Removing the binder is currently the most expensive and time-consuming part of the process. Heating rates, temperatures, and debinding times must be carefully controlled and adjusted for part thickness. The parts then undergo conventional sintering, where any remaining binder is first removed, and the diffusion processes then set the final properties of the product. During sintering, MIM parts typically shrink 15 to 25%, and the density increases from about 60% up to as much as 99% of ideal. (*Note:* Because MIM parts are molded without density

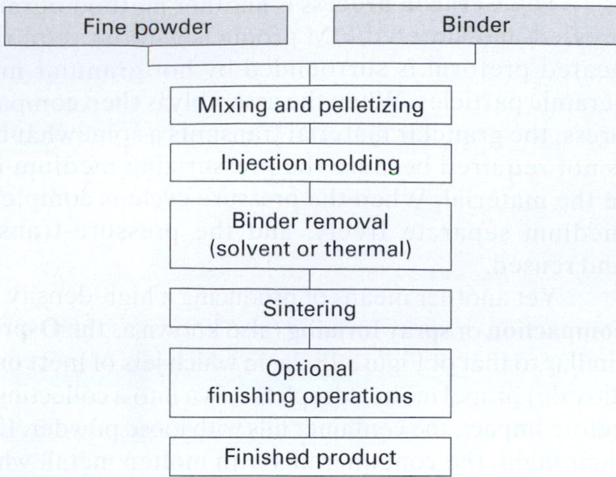

FIGURE 18.12 Flow chart of the metal injection molding process (MIM) used to produce small, intricate-shaped parts from metal powder.

variations, the subsequent shrinkage tends to be both uniform and repeatable.) Secondary processes may take the form of surface cleaning or finishing, plating, machining, or heat treating. The high final density enables the secondary processes to be conducted in the same manner as for wrought materials. Figure 18-12 summarizes the full sequence of activities, and Table 18-4 provides a summary comparison of conventional powder metallurgy and MIM.

While the size of conventional P/M products is generally limited by press capacity, the size of P/M injection moldings is more limited by economics (cost of the fine powders) and binder removal. The best candidates for P/M injection molding are complex-shaped parts (generally too complex to compact by conventional powder metallurgy) with thicknesses of less than $\frac{1}{4}$ in., weights under 20 g (20 oz), and made from a metal that cannot be economically die cast. MIM parts compete with and frequently replace machined components or investment castings. Section thicknesses as small as 0.25 mm (0.010 in.) are possible because of the fineness of the powder. As a general rule, the smaller the part and the greater the complexity, the more likely MIM will be an attractive alternative to machining, casting, stamping, cold forming, or traditional powder metallurgy. Figure 18-13 shows a variety of MIM products.

Medium to large production volumes (more than 2000 to 5000 identical parts) are generally required to justify the cost of die design and manufacture. The relatively high final density (95 to 99% compared to 75 to 90% for conventional P/M parts), the uniformity of that density, the close tolerances (0.3 to 0.5%), and excellent surface finish (about 125 μ-in.) all combine to make the process attractive for many applications. Parts can be made from a wide selection of metal alloys, including steels, stainless steel, tool steel, brass, copper, titanium, tungsten, nickel-based superalloys, ceramics, and many specialty materials. The final properties are superior to those of conventional powder metallurgy and are generally close to those of wrought or cast equivalents.

TABLE 18-4 Comparison of Conventional Powder Metallurgy and Metal Injection Molding

Feature	P/M	MIM
Particle size	20–250 μm	<20 μm
Particle response	Deform plastically	Undeformed
Porosity (% nonmetal)	10–20%	30–40%
Amount of binder/lubricant	0.5–2%	30–40%
Homogeneity of green part	Nonhomogeneous	Homogeneous
Final sintered density	<92%	>96%

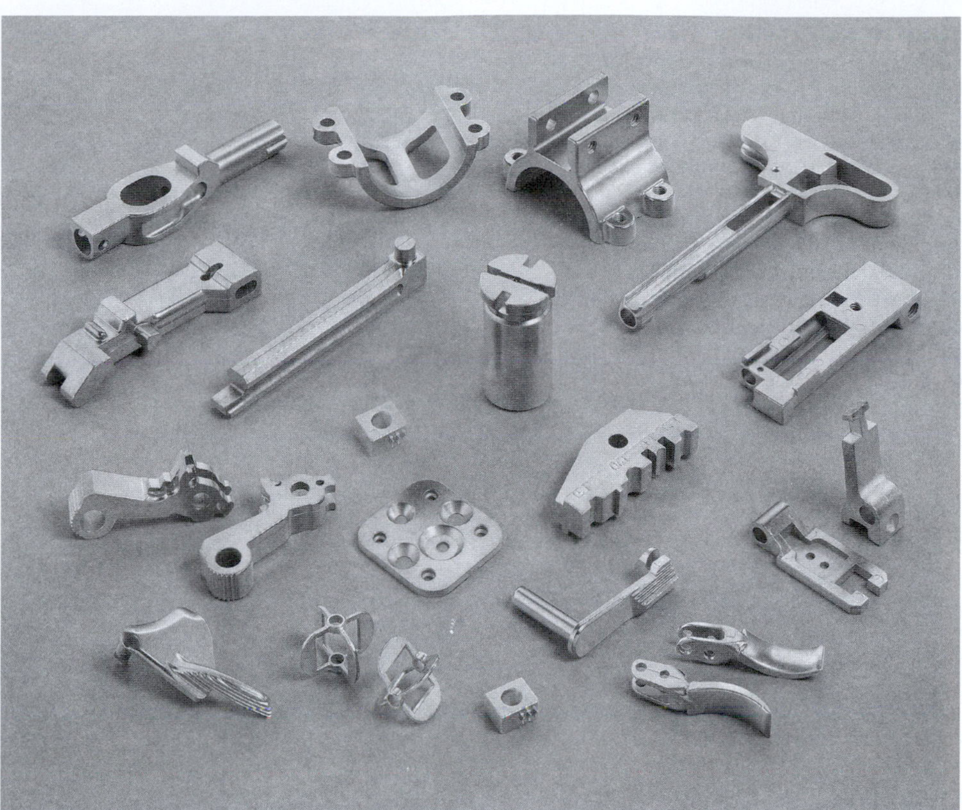

FIGURE 18.13 Metal injection molding (MIM) is ideal for producing small, complex parts. *(Courtesy of Megamet Solid Metals, Inc., St. Louis, MO)*

18.13 SECONDARY OPERATIONS

Powder metallurgy products are often ready to use when they emerge from the sintering furnace. Many P/M products, however, utilize one or more secondary operations to provide enhanced precision, improved properties, or special characteristics.

During sintering, product dimensions shrink due to densification. In addition, warping or distortion may occur during nonuniform cooldown from elevated temperatures. As a result, a second, room-temperature pressing operation, known as **repressing, coining,** or **sizing,** may be required to restore or improve dimensional precision. The part is placed in a die and subjected to pressures equal to or greater than the initial pressing pressure. A small amount of plastic flow takes place, resulting in high dimensional accuracy, sharp detail, and improved surface finish. The associated cold working and increase in part density may combine to increase part strength by 25 to 50%. (*Note:* Because of the shrinkage that occurs during sintering, repressing cannot be performed with the same set of tooling that was used for the original powder compaction.)

If massive metal deformation takes place in the second pressing, the operation is known as **P/M forging.** Conventional press-and-sinter powder metallurgy is used to produce a preform, which is one forging operation removed from the finished shape. The normal forging sequence of billet or bloom production, shearing, reheating, and sequential deformation is replaced by the manufacture of a comparatively simple-shaped powder metallurgy preform followed by a single hot-forging operation. The forging stage produces a more-complex shape, adds precision, provides the benefits of metal flow, and increases the density (often up to 99%). The increase in density is accompanied by a significant improvement in mechanical properties, which are often equivalent or superior to those of wrought materials. While protective atmospheres or coatings are required to prevent oxidation of the powder perform during heating and hot forging, the added cost and operations can often be offset by a significant reduction in scrap or waste. (By controlling preform weight to within 0.5%, flash-free forging can often be performed.) Forged products can benefit from the improved properties of powder metallurgy, such as the absence of segregation, the uniform fine grain size, and the

FIGURE 18.14 Comparison of conventional forging and the forging of a powder metallurgy preform to produce a gear blank (or gear). Moving left to right, the top sequence shows the sheared stock, upset section, forged blank, and exterior and interior scrap associated with conventional forging. The finished gear is generally machined from the blank with additional generation of scrap. The bottom pieces are the powder metallurgy preform and forged gear produced entirely without scrap by P/M forging. *(Courtesy of GKN Sinter Metals, Auburn Hills, MI)*

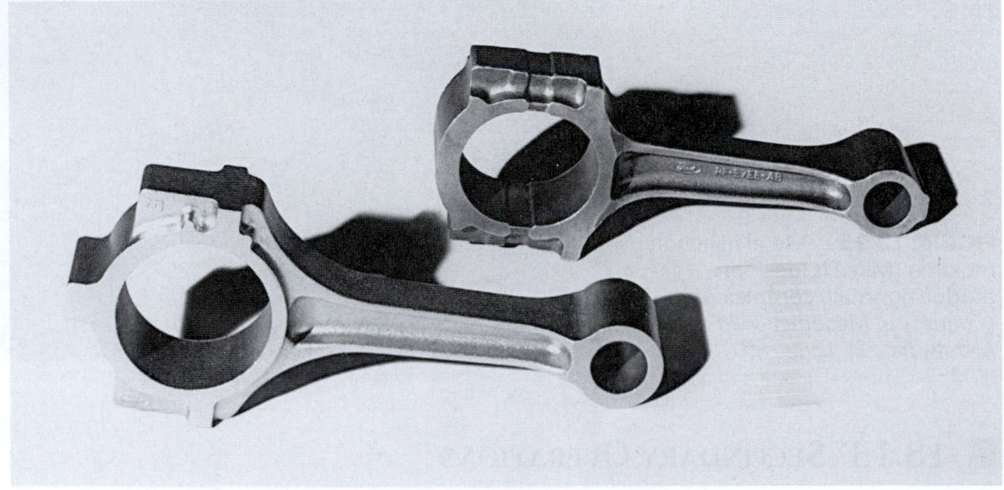

FIGURE 18.15 P/M forged connecting rods have been produced by the millions. *(Photos Courtesy of Metal Powder Industries Federation, Princeton, NJ 08540)*

use of novel alloys or unique composites. The conventional powder metallurgy process can be expanded to larger size products with increased complexity. The tolerance requirements of cams, splines, and gears can often be met without subsequent machining. Figure 18-14 illustrates the reduction in scrap by comparing the same part made by conventional forging and the P/M forge approach. P/M forged connecting rods, like those shown in Figure 18-15, currently account for more than 60% of all connecting rods used in North America and are typical of the high volume steel parts currently being produced.

Impregnation and infiltration are secondary processes that utilize the interconnected porosity or permeability of low-density P/M products. **Impregnation** refers to the forcing of oil or other liquid, such as a polymeric resin, into the porous network. This can be done by immersing the part in a bath and applying pressure or by a combination vacuum-pressure process. The most common application is that of oil-impregnated bearings. After impregnation, the bearing material will contain from 10 to 40 vol% oil, which will provide lubrication over an extended lifetime of operation. In a similar manner, P/M parts can be impregnated with fluorocarbon resin (such as Teflon) to produce products offering a combination of high strength and low friction.

When the presence of pores is undesirable, P/M products may be subjected to metal **infiltration.** In this process, a molten metal or alloy with a melting point lower than the P/M constituent flows into the interconnected pores of the product under pressure or by capillary action. Steel parts are often infiltrated with copper, for example. After infiltration, the engineering properties—such as strength and toughness—are improved to a level where they are generally comparable to those of solid metal products. Infiltration can also be used to seal pores prior to plating, improve machinability or corrosion resistance, or make the components gas- or liquid-tight. Additional heating after infiltration can cause interdiffusion between the infiltrant and base metal, further enhancing mechanical properties.

Powder metallurgy products can also be subjected to the conventional finishing operations of *heat treatment, machining,* and *surface treatment*. If the part is of high density (<10% porosity) or has been metal impregnated, conventional processing can often be employed. Special precautions must be taken, however, when processing low-density P/M products. During heat treatment, protective atmospheres must again be used, and certain liquid quenchants (such as water or brine) should be avoided. Speeds and feeds must be adjusted when machining, and care should be taken to avoid pickup of lubricant or coolant. In general, P/M products should be machined using sharp tools, light cuts, and high feed rates. When a large amount of machining is required, special machinability-enhancing additions may be incorporated into the initial powder blend. Nearly all common methods of surface finishing can be applied to P/M products, including platings and coatings, diffusion treatments, surface hardening, and steam treatment (which is used to produce a hard, corrosion-resistant oxide on ferrous parts). As with the other secondary processes, some process modifications may be required if the part has a reasonable amount of porosity or permeability. Because most parts are small and are produced in large quantity, barrel tumbling is another common means of cleaning, deburring, and surface modification.

18.14 PROPERTIES OF POWDER METALLURGY PRODUCTS

Because the properties of powder metallurgy products depend on so many variables—type and size of powder, amount and type of lubricant, pressing pressure, sintering temperature and time, finishing treatments, and so on—it is difficult to provide generalized information. Products can range all the way from low-density, highly porous parts with tensile strengths as low as 70 MPa (10 ksi)up to high-density pieces with tensile strengths of 1250 MPa (180 ksi) or greater.

Many of the properties of P/M parts are closely related to **final density,** reported as either straight mass per unit volume or as relative density (the ratio of the P/M part density to that of a pore-free equivalent). P/M parts with less than 75% relative density are considered to be low density. Parts with more than 90% relative density are high-density products. **Porosity** is the percentage of void volume. A medium-density P/M product, therefore, will have between 10 and 25% porosity. This porosity can be a network of interconnected voids extending from the surface, or a multitude of isolated holes. **Permeability** is the ability of a part to pass fluids or gas, a measure of the interconnectedness of the porosity. Permeable products can filter materials, diffuse the flow of liquids or gases, regulate flow or pressure drops in lines, or act as flame arrestors by cooling gases below combustion temperatures.

As shown in Figure 18-16, most mechanical properties of P/M products exhibit a strong dependence on product density, with the fracture-limited properties of

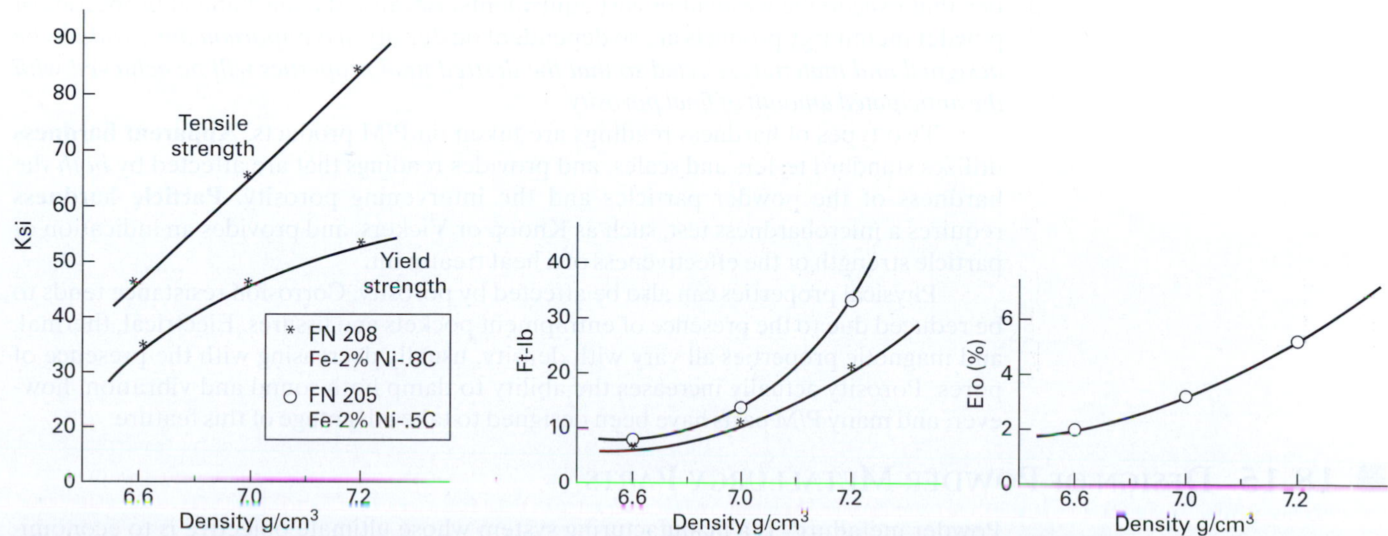

FIGURE 18.16 Mechanical properties versus as-sintered density for two iron-based powders. Properties depicted include yield strength, tensile strength, Charpy impact energy (shown in foot-pounds), and percent elongation in a 1-in. gage length.

TABLE 18-5 Comparison of Properties of Powder Metallurgy Materials and Equivalent Wrought Metals

Material[a]	Form and Composition	Condition[b]	Percent of Theoretical Density	Tensile Strength 10^3 psi	Tensile Strength Mpa	Elongation in 2 in. (%)
Iron	Wrought	HR	—	48	331	30
	P/M—49% Fe min	As sintered	89	30	207	9
	P/M—99% Fe min	As sintered	94	40	276	15
Steel	Wrought AISI 1025	HR	—	85	586	25
	P/M—0.25% C. 99.75% Fe	As sintered	84	34	234	2
Stainless steel	Wrought type 303	Annealed	—	90	621	50
	P/M type 303	As sintered	82	52	358	2
Aluminum	Wrought 2014	T6	—	70	483	20
	P/M 201 AB	T6	94	48	331	2
	Wrought 6061	T6	—	45	310	15
	P/M 601 AB	T6	94	36.5	252	2
Copper	Wrought OFHC	Annealed	—	34	234	50
	P/M copper	As sintered	89	23	159	8
		Repressed	96	35	241	18
Brass	Wrought 260	Annealed	—	44	303	65
	P/M 70% Cu-30% Zn	As sintered	89	37	255	26

[a] Equivalent wrought metal shown for comparison.
[b] HR. hot rolled: 16 age hardened.
Note how porosity diminishes mechanical performance.

toughness, ductility, and fatigue life being more sensitive than strength and hardness. The voids in the P/M part act as stress concentrators and assist in starting and propagating fractures. The yield strength of P/M products made from the weaker metals is often equivalent to the same material in wrought form. If higher-strength materials are used or the fracture-related tensile strength is specified, the properties of the P/M product tend to fall below those of wrought equivalents by varying but usually substantial amounts. Table 18-5 shows the properties of a few powder metallurgy materials compared with those of wrought material of similar composition. When larger presses or processes such as P/M forging or hot-isostatic pressing are used to produce higher density, the strength of the P/M products approaches that of the wrought material. If the processing results in full density with fine grain size, P/M parts can actually have properties that exceed the wrought or cast equivalents. Because the mechanical properties of powder metallurgy products are so dependent on density, *it is important that products be designed and materials selected so that the desired final properties will be achieved with the anticipated amount of final porosity*.

Two types of hardness readings are taken on P/M products. **Apparent hardness** utilizes standard testers and scales, and provides readings that are affected by *both* the hardness of the powder particles and the intervening porosity. **Particle hardness** requires a microhardness test, such as Knoop or Vickers, and provides an indication of particle strength or the effectiveness of a heat treatment.

Physical properties can also be affected by porosity. Corrosion resistance tends to be reduced due to the presence of entrapment pockets and fissures. Electrical, thermal, and magnetic properties all vary with density, usually decreasing with the presence of pores. Porosity actually increases the ability to damp both sound and vibration, however, and many P/M parts have been designed to take advantage of this feature.

18.15 DESIGN OF POWDER METALLURGY PARTS

Powder metallurgy is a manufacturing system whose ultimate objective is to economically produce products for specific engineering applications. Success begins with good design and follows with good material and proper processing. In designing parts that are to be made by powder metallurgy, it must be remembered that P/M is a special manufacturing process and provision should be made for its unique factors. Products that are converted from other manufacturing processes without modification in design rarely perform as well as parts designed specifically for manufacture by powder metallurgy. Some basic rules for the design of P/M parts are as follows:

1. The shape of the part must permit ejection from the die. Sidewall surfaces should be parallel to the direction of pressing. Holes or recesses should have uniform cross section with axes and sidewalls parallel to the direction of punch travel.
2. The shape of the part should be such that powder is not required to flow into small cavities such as thin walls, narrow splines, or sharp corners.
3. The shape of the part should permit the construction of strong tooling.
4. The thickness of the part should be within the range for which P/M parts can be adequately compacted.
5. The part should be designed with as few changes in section thickness as possible.
6. Parts can be designed to take advantage of the fact that certain forms and properties can be produced by powder metallurgy that are impossible, impractical, or uneconomical to obtain by any other method.
7. The design should be consistent with available equipment. Pressing areas should match press capability, and the number of thicknesses should be consistent with the number of available press actions.
8. Consideration should also be made for product tolerances. Higher precision and repeatability are observed for dimensions in the radial direction (set by the die) than for those in the axial or pressing direction (set by punch movement).
9. Finally, design should consider and compensate for the dimensional changes that will occur after pressing, such as the shrinkage that occurs during sintering.

The ideal powder metallurgy part, therefore, has a uniform cross section and a single thickness that is small compared to the cross-sectional width or diameter. More complex shapes are indeed possible, but it should be remembered that uniform strength and properties require uniform density. Holes that are parallel to the direction of pressing are easily accommodated. Holes at angles to this direction, however, must be made by secondary processing. Multiple-stepped diameters, reentrant holes, grooves, and undercuts should be eliminated whenever possible. Abrupt changes in section, narrow deep flutes, and internal angles without generous fillets should also be avoided. Straight serrations can be readily molded, but diamond knurls cannot. Punches should be designed to eliminate sharp points or thin sections that could easily wear or fracture. Figure 18-17 illustrates some of these design recommendations and restrictions.

18.16 POWDER METALLURGY PRODUCTS

The products that are commonly produced by powder metallurgy can generally be classified into six groups.

1. *Porous or permeable products, such as bearings, filters, and pressure or flow regulators*. Oil-impregnated bearings, made from either iron or copper alloys, constitute a large volume of P/M products. They are widely used in home appliance and automotive applications because they require no lubrication or maintenance during their service life. P/M filters can be made with pores of almost any size, some as small as 0.0025 mm (0.0001 in.). Unlike many alternative filters, powder metallurgy filters can withstand the conditions of elevated temperature, high applied stresses, and corrosive environments.
2. *Products of complex shapes that would require considerable machining when made by other processes*. Because of the dimensional accuracy and fine surface finish that are characteristic of the P/M process, many parts require no further processing, and others require only a small amount of finish machining. Tolerances can generally be

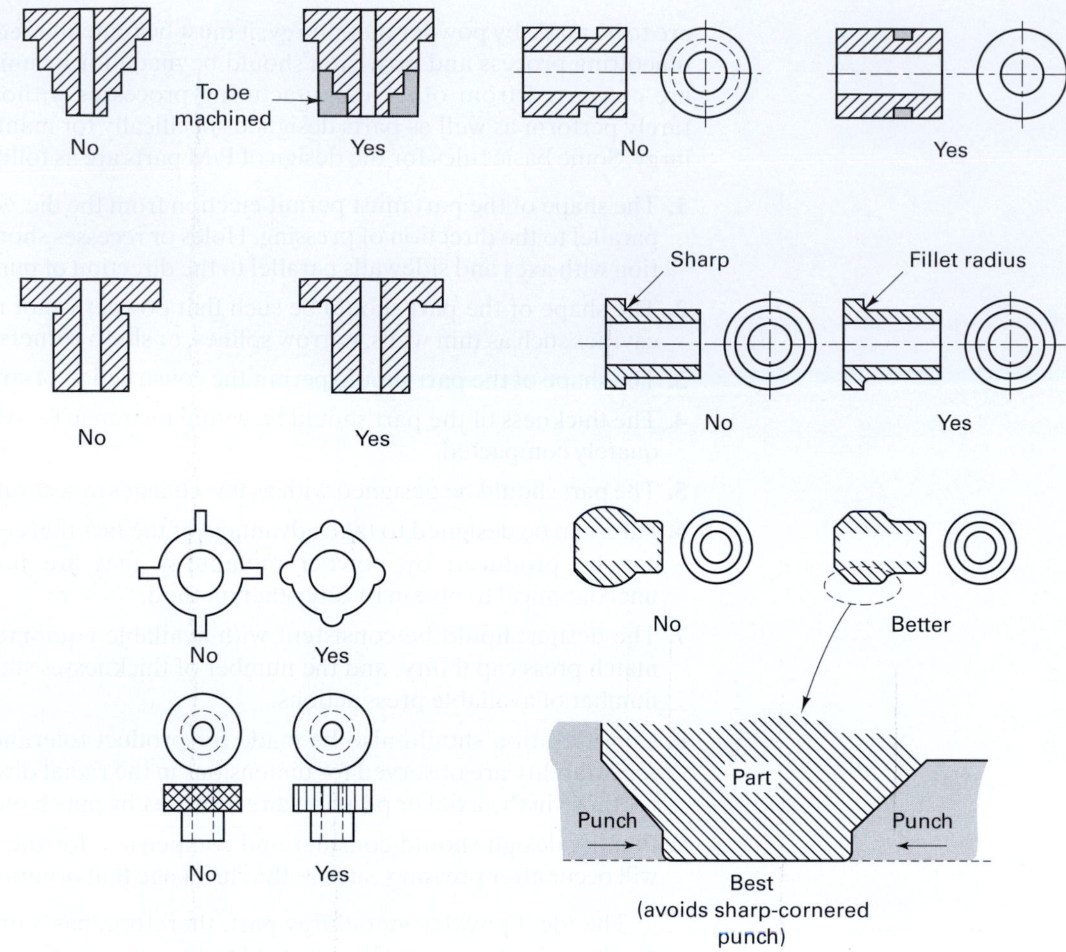

FIGURE 18.17 Examples of poor and good design features for powder metallurgy products. Recommendations are based on ease of pressing, design of tooling, uniformity of properties, and ultimate performance.

held to within 0.1 mm (0.005 in.). Large numbers of small gears are currently being made by the powder metallurgy process. Other complex shapes, such as pawls, cams, and small activating levers, can be made quite economically.

3. *Products made from materials that are difficult to machine or materials with high melting points.* Some of the first modern uses of powder metallurgy were the production of tungsten lamp filaments and tungsten carbide cutting tools.

4. *Products where the combined properties of two or more metals (or metals and nonmetals) are desired.* This unique capability of the powder metallurgy process is applied to a number of products. In the electrical industry, copper and graphite are frequently combined in applications like motor or generator brushes, where copper provides the current-carrying capacity and graphite provides lubrication. Bearings have been made of graphite combined with iron or copper or from mixtures of two metals, such as tin and copper, where the harder material provides wear resistance and the softer material deforms in a way that better distributes the load. Electrical contacts often combine copper or silver with tungsten, nickel, or molybdenum. Here, the copper or silver provides high conductivity, while the high melting temperature material provides resistance to fusion when the contacts experience arcing and subsequent closure.

5. *Products where the powder metallurgy process produces clearly superior properties.* The development of processes that produce full density has resulted in P/M products that are clearly superior to those produced by competing techniques. In areas of critical importance such as aerospace applications, the additional

FIGURE 18.18 Typical parts produced by the powder metallurgy process. *(Courtesy of PTX-Pentronix, Inc.)*

cost of the processing may be justified by the enhancement of properties. Some of the premium grades of tool steel are full-density P/M products. As another example, consider the production of P/M magnets. A magnetic field can be used to align particles prior to sintering, resulting in a product with extremely high flux density.

6. *Products where the powder metallurgy process offers definite economic advantage.* The process advantages described in the next section often make powder metallurgy the most economical among competing ways to produce a part.

Figure 18-18 shows an array of typical powder metallurgy products.

18.17 ADVANTAGES AND DISADVANTAGES OF POWDER METALLURGY

Like all other manufacturing processes, powder metallurgy has distinct advantages and disadvantages that should be considered if the technique is to be employed economically and successfully. Among the important advantages are these:

1. *Elimination or reduction of machining.* The **dimensional accuracy** and **surface finish** of P/M products are such that subsequent machining operations can be totally eliminated for many applications. If unusual dimensional accuracy is required, simple coining or sizing operations can often give accuracies equivalent to those of most production machining. Reduced machining is especially attractive for difficult-to-machine materials.
2. *High production rates.* All steps in the P/M process are simple and readily automated. Labor requirements are low, and product uniformity and reproducibility are among the highest in manufacturing.
3. *Complex shapes can be produced.* Subject to the limitations discussed previously, complex shapes can be produced, such as combination gears, cams, and internal keys. It is often possible to produce parts by powder metallurgy that cannot be economically machined or cast.
4. *Wide variations in compositions are possible.* Parts of very high purity can be produced. Metals and ceramics can be intimately mixed. Immiscible materials can be combined, and solubility limits can be exceeded. Compositions are available that are virtually impossible with any other process. In most cases, the chemical homogeneity of the product exceeds that of all competing techniques.

5. *Wide variations in properties are available.* Products can range from low-density parts with controlled permeability to high-density parts with properties that equal or exceed those of equivalent wrought counterparts. Damping of noise and vibration can be tailored into a P/M product. Magnetic properties, wear properties, friction characteristics, and others can all be designed to match the needs of a specific application.
6. *Scrap is eliminated or reduced.* Powder metallurgy is the only common manufacturing process in which so little material is wasted. In casting, machining, and press forming, the scrap can often exceed 50% of the starting material. In contrast, more than 97% of the starting material typically appears in finished P/M parts. This is particularly important where expensive materials are involved and may make it possible to use more costly materials without increasing the overall cost of the product. An example of such a product would be the rare-earth magnets.

The major disadvantages of the powder metallurgy process are these:

1. *Inferior strength properties.* Because of the residual porosity, powder metallurgy parts generally have mechanical properties that are inferior to wrought or cast products of the same material. Their use may be limited when high stresses are involved. The required strength and fracture resistance, however, can often be obtained by using different materials or by employing alternate or secondary processing techniques that are unique to powder metallurgy.
2. *Relatively high tooling cost.* Because of the high pressures and severe abrasion involved in the process, the P/M dies must be made of expensive materials and be relatively massive. Because of the need for part-specific tooling, production quantities of less than 10,000 identical parts are normally not practical.
3. *High material cost.* On a unit weight basis, powdered metals are considerably more expensive than wrought or cast stock. However, the absence of scrap and the elimination of machining can often offset the higher cost of the starting material. In addition, powder metallurgy is usually employed for rather small parts where the material cost per part is not very great.
4. *Size and shape limitations.* The powder metallurgy process is simply not feasible for many shapes. Parts must be able to be ejected from the die. The thickness/diameter (or thickness/width) ratio is limited. Thin vertical sections are difficult, and the overall size must be within the capacity of available presses. Few parts exceed 150 cm^2 (25 $in.^2$) in pressing area.
5. *Dimensions change during sintering.* While the actual amount depends on a variety of factors—including as-pressed density, sintering temperature, and sintering time—it can often be predicted and controlled.
6. *Density variations produce property variations.* Any nonuniform product density that is produced during compacting generally results in property variations throughout the part. For some products, these variations may be unacceptable.
7. *Health and safety hazards.* Many metals, such as aluminum, titanium, magnesium, and iron, are pyrophoric—they can ignite or explode when in particle form with large surface-to-volume ratios. Fine particles can also remain airborne for long times and can be inhaled by workers. To minimize the health and safety hazards, the handling of metal powders frequently requires the use of inert atmospheres, dry boxes, and hoods, as well as special cleanliness of the working environment.

18.18 PROCESS SUMMARY

For many years, powder metallurgy products carried the stigma of "low strength" or "inferior mechanical properties." This label was largely the result of comparisons where "identical" parts were made of the same material, but by various methods of manufacture. In essence, the size, shape, *and material* were all specified. In such a

TABLE 18-6 Comparison of Four Powder Processing Methods

Characteristic	Conventional Press and Sinter	Metal Injection Molding (MIM)	Hol-Isostatic Pressing (HIP)	P/M Forging
Size of workpiece	Intermediate <5 pounds	Smallest <1/4 pounds	Largest 1–1000 pounds	Intermediate <5 pounds
Shape complexity	Good	Excellent	Very good	Good
Production rate	Excellent	Good	Pour	Excellent
Production quantity	>5000	>5000	1–1000	>10,000
Dimensional precision	Excellent ±0.001 in./in.	Good ±0.003 in./in.	Poor 0.020 in./in.	Very good ±0.0015 in./in.
Density	Fair	Very good	Excellent	Excellent
Mechanical properties	80–90% of wrought	90–95% of wrought	Greater than wrought	Equal to wrought
Cost	Low $0.50–5.00/lb	Intermediate $1.00–10.00/lb	High >$100.00/lb	Somewhat low $1.00–5.00/lb

comparison, any product with 10 to 25% residual porosity would naturally be inferior to a fully dense product made by casting, forming, or machining processes. Unfortunately, it is this type of comparison that is frequently made when converting an existing design or existing part to P/M manufacture.

A far more valid comparison can be obtained by specifying size, shape, *and desired mechanical properties*. Each process can then be optimized by the selection of *both* material and process conditions. Powder metallurgy can use its unique materials, such as iron–copper blends for which there are no cast or wrought equivalents. The P/M products can be designed to provide the targeted properties while containing the typical amounts of residual porosity. Because all products will then possess the targeted mechanical properties, process comparison can then be based on economic factors, such as total production cost. On this basis, powder metallurgy has emerged as a significant manufacturing process, and its products no longer carry the stigma of "inferior mechanical properties."

Table 18-6 summarizes some of the important manufacturing features of four powder processing methods. Note the variations in product size, production rate, production quantity, mechanical properties, and cost.

■ Key Words

activated sintering
amorphous metals
apparent density
apparent hardness
atomization
binder
blending
canning
Ceracon process
chemistry
coining
compaction
composites
compressibility
decanning
dimensional accuracy
final density
flow rate
gas atomization
green strength
high-temperature sintering
hot-isostatic pressing (HIP)
impregnation
inert gases
infiltration
in-situ compaction
isostatic compaction
liquid atomization
liquid-phase sintering
lubricant
metal injection molding (MIM)
microwave sintering
mixing
nitrogen atmospheres
Osprey process
P/M forging
particle hardness
particle shape
particle size
particulate processing
permeability
porosity
powder injection molding (PIM)
powder metallurgy (P/M)
prealloyed powder
pressure-assisted sintering
protective atmosphere
purity
reducing atmospheres
repressing
sinter brazing
sinter hardening
sinter-HIP process
sintering
size distribution
sizing
spark-plasma sintering
spray forming
surface finish
surface texture
vacuum sintering
warm compaction
water atomization

■ REVIEW QUESTIONS

1. What type of product would be considered to be a prospect for powder metallurgy manufacture?
2. What were some of the earliest powder metallurgy products?
3. What are some of the newest technologies in powder metallurgy?
4. What are some of the primary market areas for P/M products?
5. Which metal family currently dominates the powder metallurgy market?
6. What are the four basic steps that are usually involved in making products by powder metallurgy?
7. What are some of the important properties and characteristics of metal powders to be used in powder metallurgy?
8. What is the most common method of producing metal powders?
9. What are some of the other techniques that can be employed to produce particulate material?
10. Which of the powder manufacturing processes are likely to be restricted to the production of elemental (unalloyed) metal particles?
11. Why might the powdered material be heat treated prior to compaction?
12. What are some of the unique properties of amorphous metals?
13. Why is powder metallurgy a key process in producing products from amorphous or rapidly solidified material?
14. Why is flow rate an important powder characterization property?
15. What is apparent density, and how is it related to the final density of a P/M product?
16. What is green strength, and why is it important to the manufacture of high-quality P/M products?
17. How do the various powder properties relate to the method of powder manufacture?
18. What are some of the objectives of powder mixing or blending?
19. How does the addition of a lubricant affect compressibility? Green strength?
20. How might the use of a graphite lubricant be fundamentally different from the use of wax or stearates?
21. What types of composite materials can be produced through powder metallurgy?
22. Why might mixing or blending be performed under wet conditions?
23. What are some of the objectives of the compaction operation?
24. What is the benefit of a removable die set in a P/M compaction press?
25. What limits the cross-sectional area of most P/M parts to several square inches or less?
26. Describe the movement of powder particles during uniaxial compaction.
27. For what conditions might a double-action pressing be more attractive than compaction with a single, moving punch?
28. How is the density of a P/M product typically reported?
29. Why is it more difficult to compact a multiple-thickness part?
30. Describe the four classes of conventional powder metallurgy products.
31. What is isostatic compaction? For what product shapes might it be preferred?
32. What is the benefit of warm compaction?
33. What is a reasonable compacted density? How much residual porosity is still present?
34. What types of materials are used in compaction tooling?
35. How do the common sintering temperatures compare to material melting points?
36. What is liquid-phase sintering? Activated sintering?
37. What are the three stages associated with most P/M sintering operations?
38. Why is it necessary to raise the temperature of P/M compacts slowly to the temperature of sintering?
39. Why is a protective atmosphere required during sintering? During the cooldown period?
40. What types of atmospheres are used during sintering?
41. What are some of the changes that occur to the compact during sintering?
42. What is the purpose of the sinter brazing process?
43. What are some benefits of high-temperature sintering? Sinter hardening? Microwave sintering?
44. How can microwave heating help in the removal of binders and lubricants?
45. Describe the process of spark-plasma sintering.
46. The combined heating and pressing of powder would seem to be an improvement over separate operations. What features act as deterrents to this approach?
47. What are some of the attractive properties of hot-isostatic pressed products?
48. What is canning and decanning, and how do these operations relate to the HIP process?
49. What is the attractive feature of the sinter-HIP and pressure-assisted sintering processes?
50. What are some of the other methods that can produce high-density P/M products?
51. Describe the spray-forming process and the unique feature that enables production of high-density, fine-grain-size products.
52. How is the injection molding of powdered material similar to the injection molding of plastic or polymeric products?
53. In the MIM process, what is done to enable metal powder to flow like a fluid under pressure?
54. How is the metal powder used in metal injection molding (MIM) different from the metal powder used in a conventional press-and-sinter production?
55. What are some of the ways that the binder can be removed from metal injection molded parts?
56. Why are MIM products injection molded to sizes that are considerably larger than the desired product?
57. For what types of parts is P/M injection molding an attractive manufacturing process?
58. How does the final density of a MIM product compare to a press-and-sinter P/M part?
59. What is the purpose of repressing, coining, or sizing operations?
60. Why can the original compaction tooling not be used to perform repressing?
61. What is the major difference between repressing and P/M forging?
62. What is the difference between impregnation and infiltration? How are they similar?
63. Why might different conditions be required for the heat treatment, machining, or surface treatment of a powder metallurgy product?

64. The properties of P/M products are strongly tied to density. Which properties show the strongest dependence?
65. What is apparent hardness? How is it different from particle hardness?
66. How do the physical properties of P/M products vary with density?
67. What advice would you want to give to a person who is planning to convert the manufacture of a component from die casting to powder metallurgy?
68. What is the shape of an "ideal" powder metallurgy product?
69. What are some P/M products that have been intentionally designed to use the porosity or permeability features of the process?
70. Give an example of a product where two or more materials are mixed to produce a composite P/M product with a unique set of properties.
71. What are the primary assets of the powder metallurgy method of parts manufacture?
72. Why is finish machining such an expensive component in parts manufacture?
73. Describe some of the materials that can be made into P/M parts that could not be used for processes such as casting and forming.
74. Why is P/M not attractive for parts with low production quantities?
75. What features of the P/M process often compensate for the higher cost of the starting material?
76. Why is it so important to achieve uniform density in a P/M product?
77. How might you respond to the criticism that P/M parts have inferior properties?

Problems

1. When specifying the starting material for casting processes, the primary variables are chemistry and purity. Any structural features of the starting material will be erased by the melting. For forming processes, the material remains in the solid state, so the principal concerns relating to the starting material are chemistry and purity, ductility, yield strength, strain-hardening characteristics, grain size, and so on. What are some of the characteristics that should be specified for the starting powder to ensure the success of a powder metallurgy process? In what ways are these similar or different from those mentioned for casting and forming processes?
2. Various techniques are used to "measure" the size of small particles. One method evaluates the volume of fluid that a particle displaces and computes a *volume diameter*—the diameter of a sphere having the same volume as the particle. Sedimentation methods measure the terminal settling velocity and compute a *Stokes diameter*—the diameter of a sphere having the same density that would have the same settling velocity. Surface area methods determine the total surface for a given mass of powder. The *surface diameter* is the size of spheres that would have the same surface area as that measured. Direct observation methods can be used to determine linear dimensions, projected areas, and projected particle perimeters, all of which can be adjusted to the diameter of a spherical particle that would produce the same result (*projected area diameter* and *perimeter diameter*). *Sieve diameter* is the minimum width of a square aperture that would permit passage of the particle. Select a simple geometrical shape (such as a cube or rectangular solid), and compute several of the cited diameters. Note the variation in results for the different methods.
3. In conventional powder metallurgy manufacture, the material is compacted with applied pressure at room temperature and then sintered by elevated temperature at atmospheric pressure. With P/M hot pressing, the loose powder is subjected to pressure while it is also at elevated temperature. It would appear, therefore, that hot pressing could produce a finished part in a single operation and would be a more economical and attractive manufacturing process. What features have been overlooked in this argument that would tend to favor the press-and-sinter sequence for conventional manufacture?
4. Investigate the method(s) used to produce tungsten incandescent lamp filaments. How does the method used today compare to the method developed by Coolidge in the late 1800s?

Chapter 18 CASE STUDY

Steering Gear for a Riding Lawn Mower/ Garden Tractor

The photo depicts the steering gear for a riding lawn mower/garden tractor (where the photo is approximately half of actual size). The center hole mates with the spline of the steering shaft to properly position the gear teeth on the outer sector. The two round holes have been included simply to reduce the weight of the product, and the "stops" on either side of the bottom surface serve to limit the turning radius of the tractor. Dimensions in both tooth positioning and tooth profile, and the surface finish in this area should be sufficient to hold backlash with the mating gear to within .10 mm (0.004 in). Design parameters have specified a minimum tensile strength of 586 MPa (85,000 psi), and a minimum hardness on the gear teeth of Rockwell B 80 (equivalent to about 150 Brinell or zero on the Rockwell C scale—i.e., rather low!). Assume the quantity to be produced is sufficient to justify most manufacturing processes.

1. Briefly discuss the properties and characteristics that this piece must possess to function properly, and discuss the important fabrication requirements.
2. Based on the size, shape, and reasonable precision of the component, identify and describe several fabrication methods that could be used to produce the part.
3. Identify several material families that could be used to meet the specified requirements.
4. Using your answers to Question 3, present material-process combinations that would be viable options to produce this item.
5. Which of your combinations in Question 4 do you feel is the "best" solution? Why?
6. For your "best" solution of Question 5 select a specific metal, alloy, or other material, and justify your selection.
7. Outline the fabrication steps that would be necessary to produce the desired shape from reasonable starting material.
8. Would some form of final heat treatment be required to establish the desired properties? If so, what do you recommend?
9. Would any form of surface treatment be required to establish the desired properties? If so, what do you recommend?

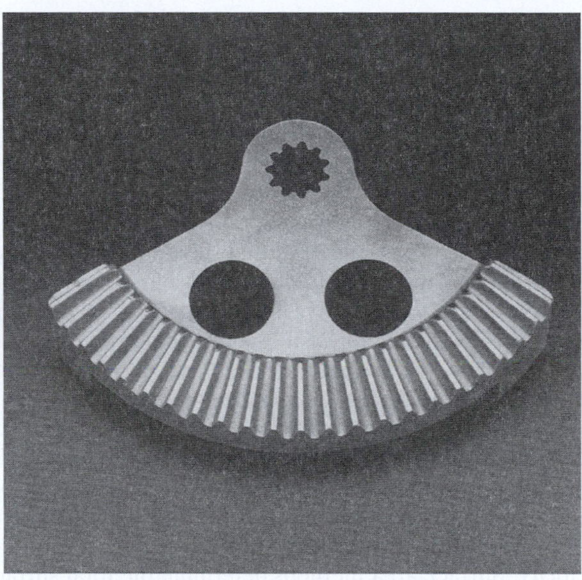

Steering Gear for a Riding Mower/Lawn Tractor. *(Photos Courtesy of Metal Powder Industries Federation, Princeton, NJ 08540)*

CHAPTER 19

ADDITIVE PROCESSES: RAPID PROTOTYPING AND DIRECT-DIGITAL MANUFACTURING

19.1 INTRODUCTION

The beginning of Chapter 11 introduced the five families of processes by which useful shaped products are made. Liquids assume the shape of their container, and **casting processes** produce shaped containers, fill them with liquid, and allow the material to solidify while being held in that shape. **Deformation processes** exploit the plasticity of certain materials and use mechanical forces to rearrange solids into a more desirable shape. **Material removal processes** begin with an oversized solid and progressively remove unwanted segments to create the desired product. Discrete pieces of material are joined together by the **consolidation processes.** Each of these families, as well as the various processes within them, has distinct assets and limitations. Deformation processes, for example, frequently require strong, part-specific tooling, and the cost of this tooling is usually distributed over a large number of identical products. In contrast, basic machine tools, in the hands of a skilled operator, can produce millions of different shapes, one-of-a-kind if desired, and with extremely high dimensional precision. The material must be clamped or fixture for each of the material removal operations, and the removed material, often in the form of cuttings, turnings or chips, must be discarded or recycled, producing added cost. Liquids can flow around inserts, enabling castings to easily incorporate internal holes and passages, but most liquids also shrink when they solidify, causing significant concern for the casting processes.

This chapter will present a relatively new grouping of processes that create a desired shape by the incremental addition of material in a layer-by-layer fashion, and are known as **additive processes.** Because the earliest application was the manufacture of prototype products that could be produced with extremely short lead times, the term *rapid prototyping* was commonly applied. As design concepts are transitioned from engineering drawings or computer-aided design models to working products, **prototypes** are often produced. A prototype provides a physical representation of the product that can be used to assess a designer's success in achieving a part's form, fit, and function. Prototypes often have high unit cost because all of the required tooling applies to only one or a small number of parts. In addition, delays related to the fabrication of specialized tooling (e.g., patterns or molds for casting, dies for forming, or fixtures for machining) often led to long lead times for prototype manufacture and evaluation. Based on the prototype, design changes may be required, which would require another iteration of tooling. Cheaper and faster prototype production can be an attractive competitive edge.

It was not long after the development of the computer that its capabilities were put to use in manufacturing, and these applications have continued to grow to include computer numerical controlled (CNC) machines, industrial robots, and computer-aided design/computer-aided manufacturing processes (CAD/CAM) to name just a few. CNC machining processes are now common throughout the manufacturing industry. They begin with a three-dimensional computer model that is used to generate tool paths that progressively subtract or remove material. The additive processes begin with the same three-dimensional computer model, which is then sectioned into a large number of individual layers. Liquids, powders, plasticized extrusionsor sheets of material are then used to deposit the layers and build the products that are now available in plastics, ceramics, metals, and even composites. The additive processes are a much newer addition to the manufacturing inventory, with the first patent being granted in March 1986.

As the additive manufacturing processes matured and new concepts were introduced, the array of applications also expanded and now include: (1) **rapid prototyping (RP);** (2) the production of scale models to enhance visualization or for purposes such as wind-tunnel testing; (3) **rapid tooling (RT)**—the production of tooling to be used in another process, such as foundry patterns, cores or molds, or jigs and fixtures for machining or joining processes; and (4) **direct-digital manufacturing (DDM)**—the manufacture of finished products directly from a computer file, with no intervening tooling. With no required tooling, one-of-a-kind and small-batch production may be quite economical, provided material cost, build time, surface finish, and resulting precision are all considered to be acceptable.

The past 25 yr have seen extensive activity, and a number of new approaches have been developed. Along with the evolving technology, the terminology has also evolved. Because some of the original machines were quite compact[1] and produced small products, they were often classified as **desktop manufacturing.** Other general terms include **additive manufacturing, free-form fabrication, rapid manufacturing,** and **layered manufacturing.**

19.2 RAPID PROTOTYPING AND DIRECT-DIGITAL MANUFACTURING

Rapid prototyping and direct-digital manufacturing are additive processes, building the shape progressively through the accumulation of thin layers. They are free-form fabrication processes, capable of producing any required geometry, providing that the dimensions fit within the working envelope of the equipment. They are also **tool-less manufacturing** processes, able to make any product geometry without requiring part-specific tooling, such as dies, molds, or fixtures. Complex, free-form surfaces and contours can be produced as quickly as simple planar and cylindrical features. When the geometry is simple, subtractive machining processes, using general-purpose tools, will usually produce a quicker and cheaper product. As the shape complexity increases, the additive processes begin to excel.

As shown in Figure 19-1, the first step in all additive processes is **preprocessing,** the conversion of the three-dimensional **computer-aided design (CAD)** into the set of

[1] The original 10 × 10 × 10 in. build chambers were selected based on an industry investigation of the sizes of typical injection-molded parts. 3-D Systems performed the study and were the first to market with a stereolithography machine with this size build chamber. It then became the default standard for other entry-level or first-generation process platforms.

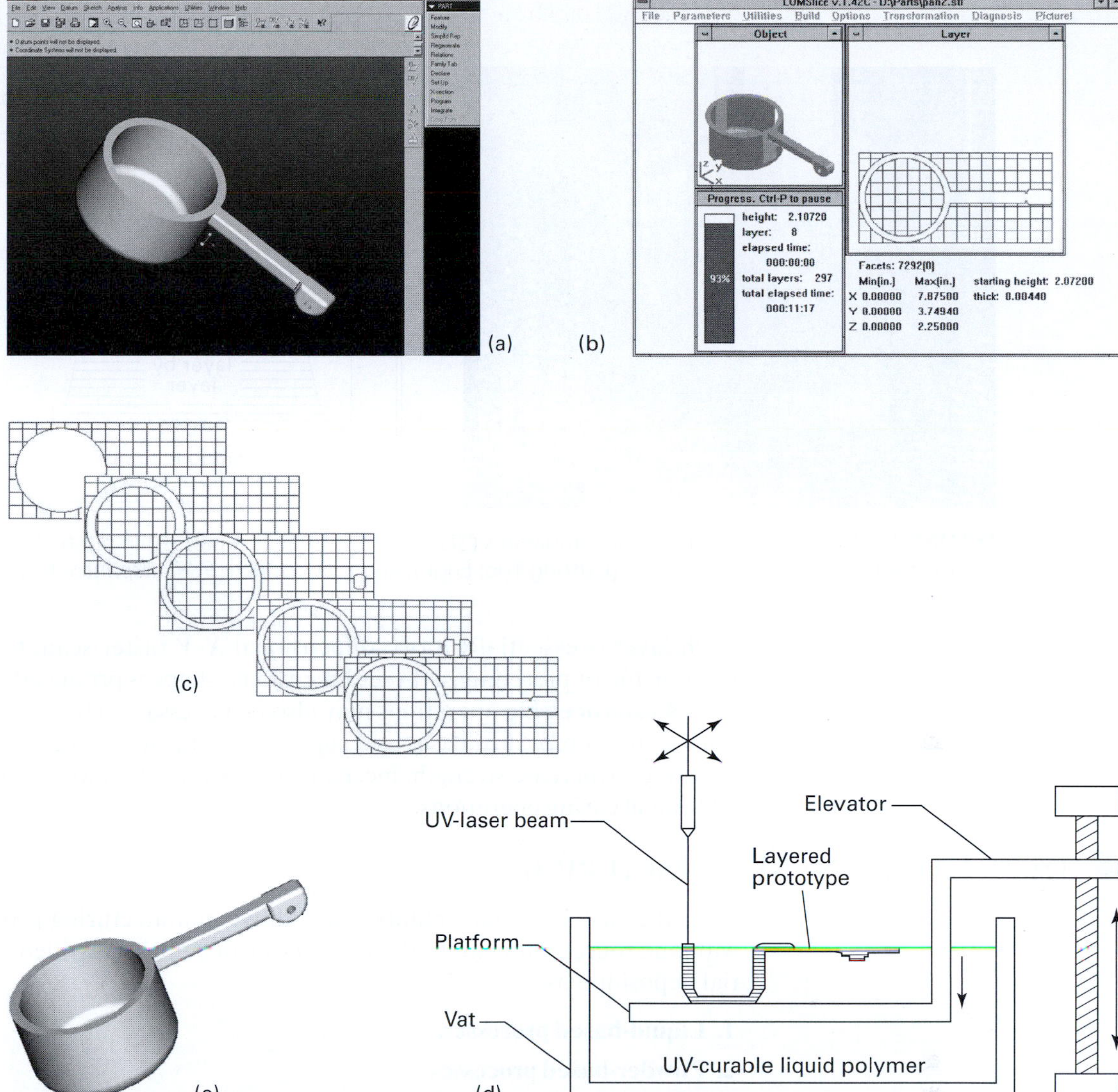

FIGURE 19-1 Conceptual framework for additive processes: (a) development of a virtual model in CAD; (b) model is converted to STL file format and loaded into CAM software; (c) CAM software slices the model to generate tool paths for the laser; (d) product is produced layer-by-layer in machine tool; (e) final prototype or product.

instructions that will be used to control and direct the manufacturing tool. Using **computer-aided manufacturing (CAM)** software, the part is mathematically sectioned into a succession of layers, starting at the bottom and moving up. **Bases** and **supports** can be added to the part during this stage to facilitate separation from the machine tool after fabrication and to support free-standing geometric features that would distort during the process, such as cantilevers. Each layer is then reduced to a set of **tool paths** that will guide the deposition of energy or material during fabrication. In many ways, the computerized preprocessing for additive manufacturing is similar to the reduction of CAD geometries to the series of tool paths used in the subtractive process of **computer numerical control (CNC)** machining.

Figures 19-1 and 19-2 both illustrate the manufacture of an additive product, using the process of stereolithography (to be presented in more detail later) as an example. For any given layer, the laser follows the programmed tool path, converting the light-curable resin (liquid photopolymer) into polymerized solid. When the layer is completed, the build platform descends one layer of thickness, and uncured material flows across the surface. The laser then follows the path for the next successive layer, bonding it to the previously cured solid. The cycle then repeats, building the part layer-by-layer.

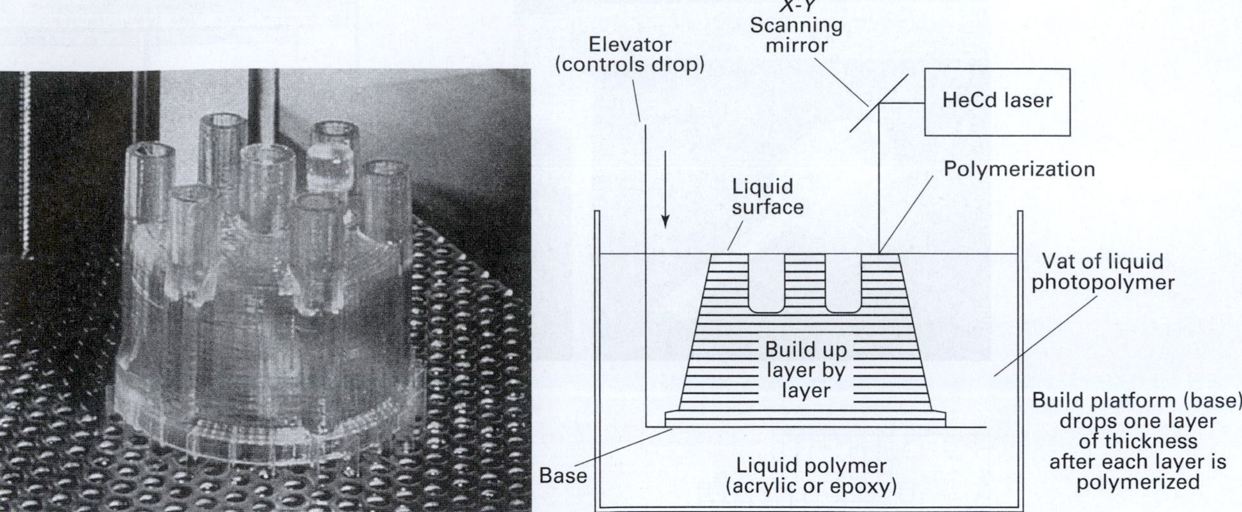

FIGURE 19-2 The stereolithography apparatus (SLA) (right) can build a part (b) in plastic layer-by-layer, using a laser to polymerize liquid photopolymer. *(Cutting Tool Engineering, December 1989. Reprinted with Permission.)*

Each layer is essentially a two-dimensional *X-Y* raster scan, but by processing each layer on top of preceding, a three-dimensional shape is produced.

Postprocessing operations may also be necessary. These may include such operations as the removal of bases and supports, removal of excess uncured resin, firing or **sintering** to increase strength, mechanical finishing to remove stairstepping artifacts, or additional curing operations.

19.3 LAYERWISE MANUFACTURING

Over the past 25 years, a number of additive manufacturing processes have emerged, with one source citing more than 40. They can be best classified by the nature of material deposition as:

1. **Liquid-based processes.**
2. **Powder-based processes.**
3. **Deposition-based processes** (or solid-based).

These families and the key members within them will be presented in later sections of this chapter.

The numerous processes have attained different levels of development and process maturity. They typically have low material addition rates compared to the material removal rates in machining, poorer dimensional accuracy and surface finish, and smaller work envelopes. The types and range of materials are often limited, and material cost is typically higher. The primary advantage is the ability to produce complex geometries with a minimal lead time and no part-specific tooling. Each individual process has its own set of advantages and limitations, but each builds the product in a layerwise fashion, and they share common features and a common terminology.

COMPUTER-AIDED DESIGN (CAD) TO COMPUTER-AIDED MANUFACTURING (CAM)

Because each of the additive manufacturing processes is highly automated, computer planning and control software is essential to their operation. The manufacturing sequence begins with preprocessing, the conversion of the CAD solid model into the data needed to drive and control the additive machine tool. The specialized CAM software is often unique to each process, but generally begins by converting the geometry of the part into *stereolithography* or **STL files,** beginning with a process known as **tesselation.** The desired part is approximated as a multifaceted, water-tight surface, model by converting its surface into a network of interconnected triangles, often numbering in the millions. For each triangle, the data consist of its three vertices and a normal vector pointing to the interior of the product.

The STL format was developed by 3D Systems Inc. for the first additive process, stereolithography, in the late 1980s, and it has become the default standard for conversion. All of the major CAD modeling software programs now permit exportation of data in the STL format. When the manufacturing objective is to replicate an existing product, as with **reverse engineering,** input (point cloud) data from devices such as laser trackers, laser scanners, or coordinate measuring machines can be used to create the initial database.

The process-specific CAM software then slices the tessalated solid representation into a series of cross-sectional layers, with a thickness specified by the process. Each layer is then reduced to a series of tool paths that will be used to guide the deposition of material or the application of energy used to form that specific layer. Some processes begin by outlining the perimeter of the solid regions and then filling in the area with techniques, such as **rastoring,** wherein a two-dimensional area is reduced to a series of specified-width and spaced lines.

In summary, the CAD-to-CAM conversion begins with a three-dimensional solid that is then sliced into a series of two-dimensional layers, each of which is further reduced into a series of linear tool paths.

THICKNESS CONTROL, STAIRSTEPPING, AND SURFACE FINISH

When a product is produced by creating thousands of stacked layers, it is important to control the thickness of the individual layers. Some processes have excellent control and permit fabrication by simple deposition of layer upon layer. Others may require intermediate attention, such as a surface milling after each deposition, or an intermediate measurement of build height followed by a modification of slice data to compensate for any inaccuracies. The maximum thickness of a given layer is typically set by the depth of solid that can be formed by the deposition process, but considerations of surface finish and dimensional accuracy may well warrant a reduction in layer thickness and an increase in the number of layers.

The various additive manufacturing processes produce products with a range of surface finishes and surface textures, but all suffer from a phenomenon known as **stairstepping,** illustrated in Figure 19-3. All surfaces not perpendicular to the slice plane (i.e., curved surfaces, rounded corners, and tapered sides) are approximated by the edges of stacked, uniform-thickness layers. Surface geometry, therefore, will tend to be rougher in the z-direction than around the perimeter of the x-y planes. Surfaces that have shallow inclination angles with respect to the x-y plane will have more obvious stairstepping artifacts because the distance between the edges of the z-height layers will be greater. If finish is critical in the z-direction, it can be improved by reorienting the part so that this surface is closer to being perpendicular to the build plane or by reducing the layer thickness during the build cycle. As shown in Figure 19-4, however, decreasing the layer thickness often results in an increase in the time required to build the part. Because stairstepping is often more pronounced on the downward-facing surfaces of a part, reorientation of critical surfaces may also yield an improvement.

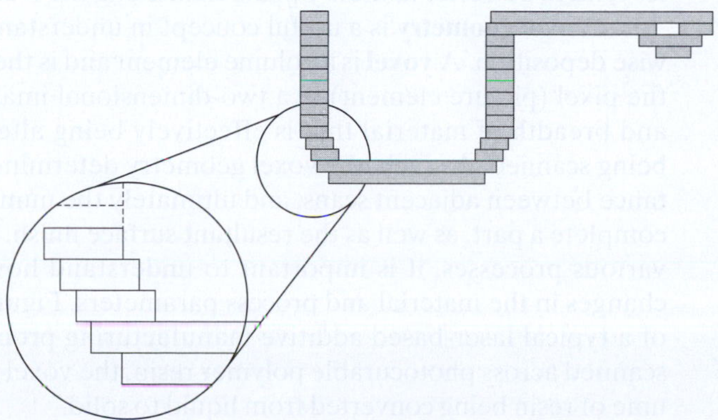

FIGURE 19-3 Building by layers results in stairstepping of curved surfaces and rounded corners.

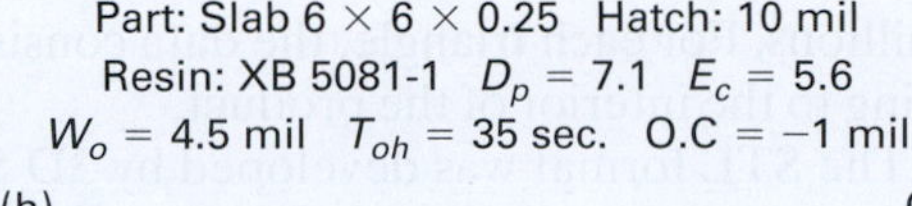

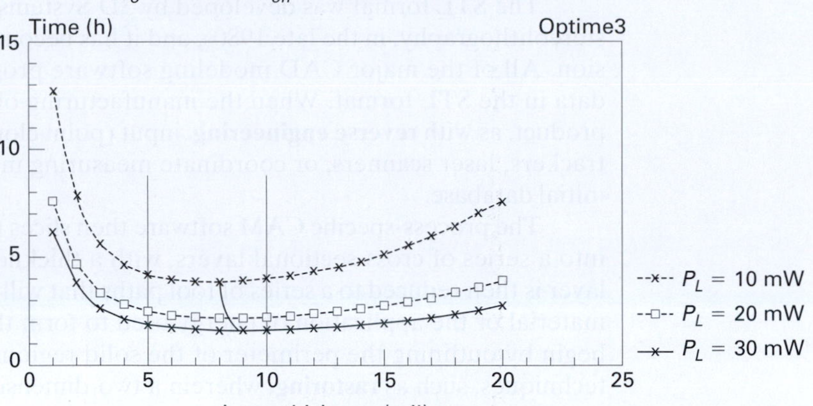

FIGURE 19-4 Build time versus layer thickness for various laser power levels. *(From P. F. Jacobs,* Rapid Prototyping and Manufacturing: Fundamentals of Stereolithography, *SME, Dearborn, MI, 1992)*

In general, the surface finish of the liquid-based processes is better than that of the powder-based approaches. For certain applications, such as the manufacture of investment casting patterns or scale models for wind tunnel testing, the as-deposited surface finish may not be adequate. Secondary grinding, sanding, or polishing may be required, with these operations adding to both the time and the cost.

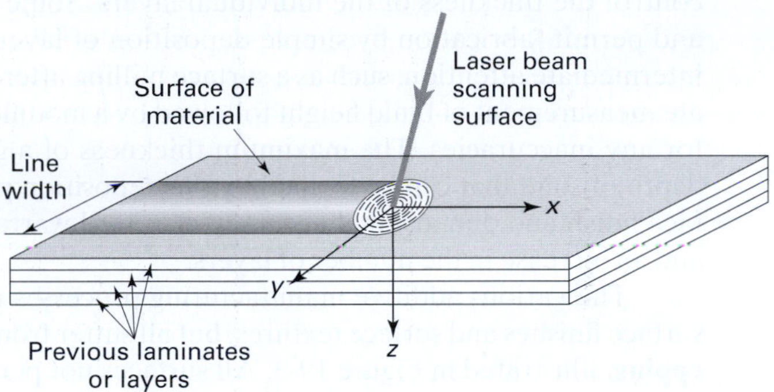

FIGURE 19-5 Schematic of a generic laser-based additive process showing the traditional coordinate system.

VOXEL GEOMETRY

Figure 19-5 shows a generic laser-beam scanning process and presents the coordinate system that has become common for all additive manufacturing techniques. The *x-y* plane is oriented parallel to the material layers and the material surface. The *z*-axis is perpendicular to the material layers. For processes that involve scanning, the *x-y* plane contains both the direction of scan travel and the width of the scanned line. The *x-y-z* coordinate system is useful when considering the anisotropy of various material properties, including both yield strength and surface texture. In general, the material properties tend to be better in the *x-y* plane than along the *z*-axis.

Voxel geometry is a useful concept in understanding the scanning-types of layer-wise deposition. A **voxel** is a volume element and is the three-dimensional equivalent of the pixel (picture element) in a two-dimensional image. It describes the depth, width, and breadth of material that is effectively being altered by the energy or deposition being scanned. As such, the voxel geometry determines the thickness of layers, the distance between adjacent scans, and ultimately the number of layers and scans needed to complete a part, as well as the resultant surface finish. To understand the capabilities of various processes, it is important to understand how voxel geometry is affected by changes in the material and process parameters. Figure 19-6 shows the voxel geometry of a typical laser-based additive manufacturing process. When an ultraviolet laser is scanned across photocurable polymer resin, the voxel geometry corresponds to the volume of resin being converted from liquid to solid.

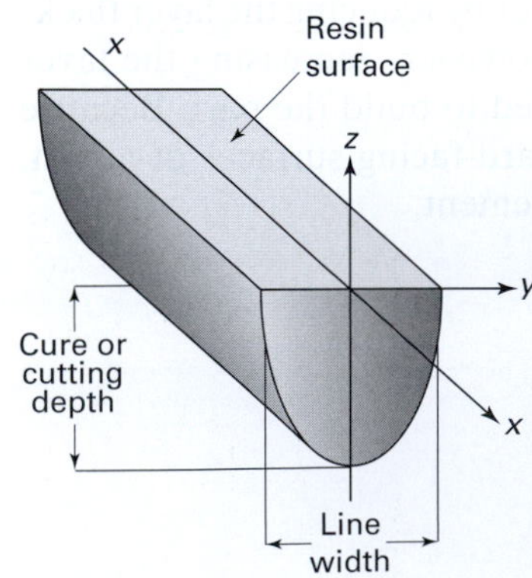

FIGURE 19-6 In stereolithography, the laser produces a cure line of photopolymer, where the width, depth, area, and profile are determined by the voxel.

DIMENSIONAL PRECISION

It is very difficult to set established dimensional tolerances for the various additive processes. There are various mechanisms by which the products are created, including photocuring, layer stacking, chemical consolidation, and thermal consolidation. Some have excellent dimensional tolerances, while others are poor. In addition, the specific operating conditions can greatly affect dimensional accuracy. A product fabricated under one set of processing conditions may have a different overall accuracy than one built under a different set of conditions. Part geometry, build rate, preprocessing, postprocessing, type of material, and even ambient temperature and humidity may combine to affect the accuracy of additive manufacturing products.

Additional problems arise due to the phase changes that occur during most of the additive processes. Material frequently goes from liquid to solid or from solid to liquid and then back to solid. Each of these changes is accompanied by a change in density, with the material shrinking upon solidification. These dimensional changes occur in an anisotropic manner as material is deposited layer by layer and voxel by voxel.

When layers are produced by scanning techniques, it is common for the initial scans to trace the boundaries of the particular cross section, followed by scans across the outlined areas. This filling in of the cross-sectional area is known as **hatching.** If the material did not experience shrinkage, hatching could occur in a straightforward manner, such as a simple sweep across the area. When the amount of shrinkage is large, alternative methods of hatching may be used to minimize the effects of shrinkage on material properties and dimensional accuracy. For example, the space between scan lines may be increased to retain small pockets of uncured resin that is then cured during postprocessing. Because many of the product uses require only geometric surfaces of the correct size and shape, hollow products are often permissible. In this case, hatching is modified to create internal honeycomb structures, where the uncured resins can be drained prior to final postprocessing.

On a more macroscopic scale, the interior temperatures of a solid part tend to exceed the surface temperatures during the build process, leading to the creation of residual stresses within the material as it cools and possible part warpage. Because the additive processes are generally used with complex geometries, and complex geometries cool in complex ways, the resultant residual stresses and associated dimensional changes are often quite complex.

SUPPORTS

Some of the additive manufacturing processes require additional preprocessing and postprocessing because of the need for supports. As shown in Figure 19-7, supports are extra material used to support segments of material layers that do not have solid material underneath. Supports may be required when the geometry contains segments that are cantilevered out from the main body, as with the panhandle in Figure 19-7. Supports may also be required to retain the position of segments that are initially unattached during the build, as with the "island" in the same figure. In the metallic processes, the supports can also be used to control shrinkage and distortion and to provide thermally conductive paths for the removal of the thermal energy provided by the laser or electron beam. The supports must be designed in to the part before processing and must be removed after the build, possibly negating some of the advantages of tool-less processing.

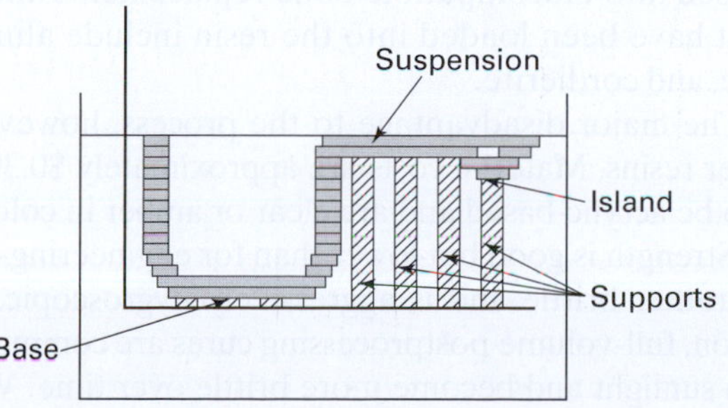

FIGURE 19-7 Many of the additive manufacturing processes require the use of bases and supports.

BUILD TIME

A major asset of the additive manufacturing processes is the significant reduction in the time required to produce a product. The **build time** for these processes consists of three major components: preprocessing, fabrication, and postprocessing. Preprocessing involves the creation of the CAM software necessary to drive the machinery, and the required time has been reduced considerably by advances in the capacity and speed of computers. Fabrication is often the largest component of both time and cost, with variation depending on the specific process, part size, part complexity, and batch size. All of the additive processes permit the manufacture of multiple products within a working envelope, allowing for possible reduction in both time and cost. Postprocessing involves cleaning and finishing the product after fabrication with details varying greatly depending on the process being used.

19.4 LIQUID-BASED PROCESSES

STEREOLITHOGRAPHY APPARATUS (SLA)

The first of the additive manufacturing processes, **stereolithography,** already presented in Figures 19-1 and 19-2, was made possible by a series of technological advances, including: high-speed computers, computer-aided design (CAD), precise motion control, UV lasers, and **photocurable polymers.** Like all of the additive manufacturing processes, stereolithography begins by converting a three-dimensional CAD solid model into a series of thin cross-sectional layers, or slices, each being between 0.05 and 0.15 mm (0.002 in and 0.006 in.) in thickness. Using the tool path data from the CAM software, the process then draws the cross section onto the surface of liquid photocurable epoxy-based resin with a HeCd or argon ion UV laser. The small, but intense spot of UV light causes the resin to locally solidify (polymerize) wherever it is scanned. When a layer is complete, an elevator within a vat of liquid photopolymer descends one layer of thickness, allowing uncured resin to wash across the surface. A mechanical wiper then sweeps across the surface, leveling the resin and removing excess material. The uniform layer of liquid resin is then ready for the next curing scan. By repeating these steps over and over, the desired three-dimensional geometry is created with typical build times ranging from an hour to more than a day.

The completed part is then removed from the resin bath and excess partially cured resin is removed from surfaces with the use of a solvent. Supports are mechanically removed while they are still soft, and a postcuring operation in an ultraviolet oven is used to achieve more thorough polymerization and enhanced material properties.

Advantages of a **stereolithography apparatus (SLA)** include high dimensional accuracy and good surface finish. Tolerances can be a tight as 0.002 cm/cm (in./in.). Part volumes are typically within an envelope of 0.5 × 0.5 × 0.75 m (19 × 19 × 30 in.), but one machine can produce products up to 1.5 m (59 in.) in length. Larger parts can be produced as joined segments by breaking the original CAD model and adding joint features that may include dowel holes and gaps for adhesives. Wax investment casting patterns can be replaced by hollow SLA shells supported by an internal lattice network as illustrated in Figure 19-8 that shows an SLA investment casting pattern and the resulting aluminum/silicon carbide (metal–matrix composite) casting. Ceramic products can be produced by loading nano- or microsized ceramic powders into the photocure resin with stirring steps to keep them in suspension. Hydroxyapatite can be processed into biocompatible bone replacements and implants. Other ceramic materials that have been loaded into the resin include alumina, zirconia, aluminum nitride, mullite, and cordierite.

The major disadvantage to the process, however, is the use of expensive photopolymer resins. Material costs are approximately \$0.30/cm^3 or \$5.00/in.^{3} The basic resins tend to be acrylic-based and are clear or amber in color (can be dyed but remain translucent). Strength is good but lower than for engineering-grade resins. In addition, the material is rather brittle and is aggressively hygroscopic. Because it is transparent to UV radiation, full-volume postprocessing cures are common, but the material will continue to cure in sunlight and become more brittle over time. When opaque products are desired,

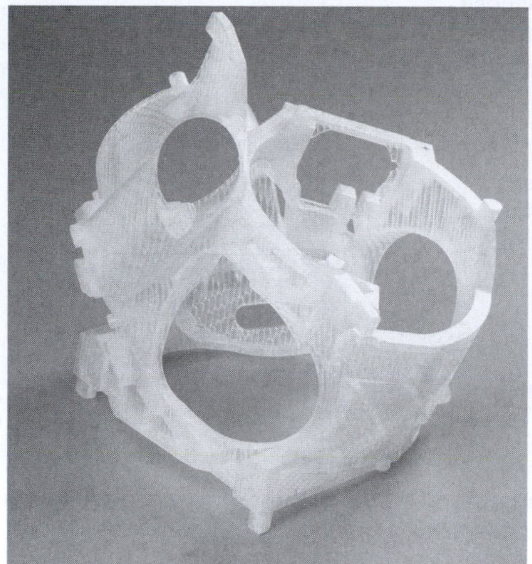

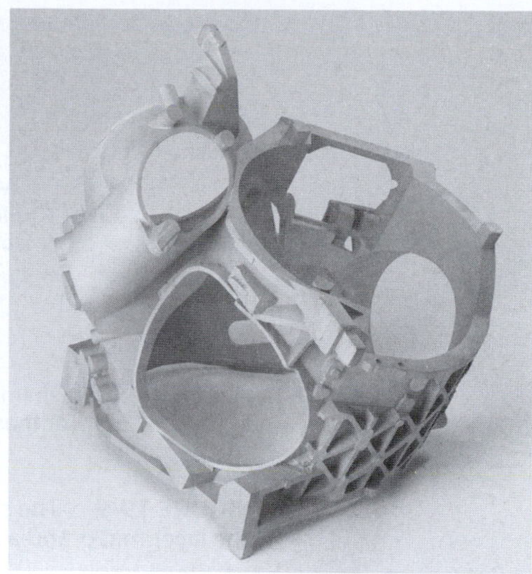

FIGURE 19-8 (Left) A honeycombed pattern (solid surface with honeycomb interior) produced by stereolithography is used to produce an investment casting (Right) of particle-reinforced aluminum. *(Courtesy O'Fallon Casting)*

the resins can be loaded with ceramic materials to produce blue, flesh, and gray colors. Toughness and strength are improved, but ductility is still quite limited. Unfortunately, many of the resins have toxicity concerns ranging from possible skin irritation through being possible carcinogens. Because of the necessary safety precautions, SLA is generally not performed in an office environment. In addition, the process may not be attractive for parts with multiple free-hanging segments, because each will require support.

Stereolithography was the first of the additive processes to be commercialized, and it continues to be the most-used and most-studied of the techniques. Additional scanned laser polymerization (SLP) processes have been developed by modifying certain features of the SLA technique, each having their own advantages and limitations. In the **masked-lamp descending-platform** approach, photomasks are created for each layer and are positioned above the resin surface. A single photoexposure forms the entire layer as opposed to the serial laser scan, thereby decreasing the required build time. In the **ascending surface** approach, the base and partially built solid remain fixed and a premeasured amount of uncured resin is added to the chamber to create each new layer. A final approach inverts the process, placing the laser beneath a transparent window. As the part is built, the partial product is raised by incremental amounts, allowing a new layer of resin to flow between the window and product.

SOLID GROUND CURING (SGC)

Solid ground curing (SGC) cures layer upon layer of photopolymer with a photomask and high-intensity UV lamp similar to the masked-lamp descending-platform method mentioned in the previous section, but the details are somewhat different. As illustrated in Figure 19-9, fabrication of a layer begins with the creation of a photomask. As shown in the mask plotter cycle segment of the figure, this process is similar to the one used in photocopiers. A pattern of static charge is applied to a plate of glass, based on the control data from the slice file. Black powder, or toner, is electrostatically attracted to the charged areas of the glass. The resulting photomask is then used to selectively expose a layer of photopolymer to high-intensity ultraviolet light. The glass plate is then erased by removing the charge and powder and a new photomask is created for the next layer.

Once a layer of photopolymer has been exposed under the photomask, it is then further processed through a model grower cycle. Any unexposed excess resin is first removed from the layer by wiping and/or vacuuming. Then, a layer of liquid water-soluble wax is applied and solidified to fill any voids left by removing the unexposed resin. Depending on the ability to control layer thickness, which often depends on the liquid viscosities, a face-milling operation may be performed to reduce the entire layer of wax and cured photopolymer to a specific thickness. The process then repeats for the next slice, each layer adhering to the previous one, until the object is finished.

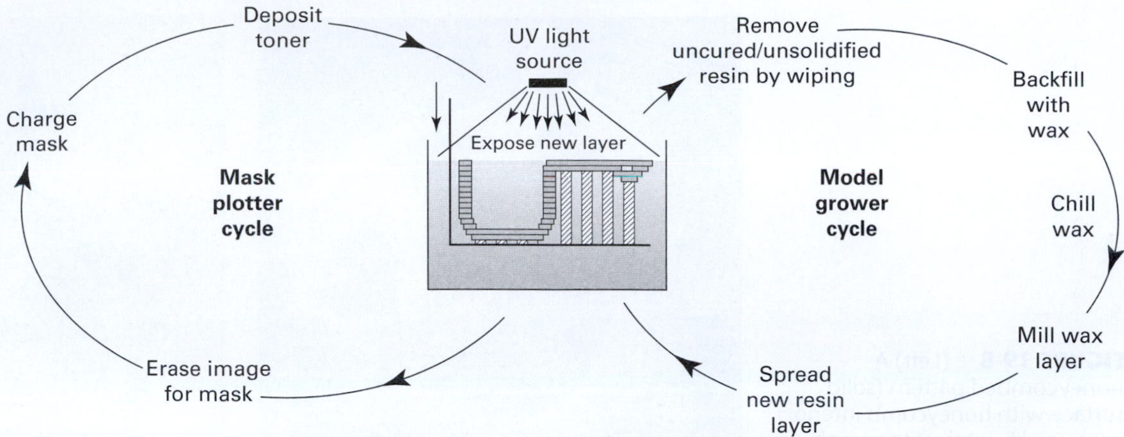

FIGURE 19-9 The solid ground curing process simultaneously operates a mask plotter cycle to produce the layer masks and a model grower cycle to deposit the layers.

When the object is removed from the machine, it is essentially a large block of wax with embedded product. The wax can be removed by melting or rinsing with hot water or left on the part for shipping or security purposes and simply removed by the end user.

Dimensional accuracy with the SGC process tends to be good. Claimed accuracy for the process is 0.1% up to a maximum of 0.5 μm (0.02 in.). Because of the milling operation, the z-axis accuracy of the SGC process tends to be better than for other additive manufacturing processes. In addition to producing a precise, uniform layer thickness, the milling operation is also performed to promote adhesion between layers by roughening the surface.

An advantage of the SGC process is that the time to build every layer is the same, independent of either part geometry or the number of parts being built. Therefore, build time can be predicted quite accurately by multiplying the time per layer by the number of layers. Once a mask is developed, layers can be photopolymerized within 3 s. However, because of the additional steps involved, the process cycle time is about 1 min per layer. Because the polymerized resin is surrounded by solidified wax, there is no need for supports. All of the work envelope can be dedicated to the fabrication of parts, and multiple parts can be closely nested in a single build, resulting in one of the highest part production rates of the additive manufacturing processes.

The primary disadvantage of the process is its complexity and the large number of process steps, leading to larger, heavier, and more expensive equipment. Maintenance issues, coupled with the need for full-time skilled operators, has led to the disfavor of this approach.

INKJET DEPOSITION (ID) OR DROPLET DEPOSITION MANUFACTURING (DDM)

Using layer-sectioning software and tool path geometries similar to previous processes, the **inkjet deposition (ID),** or **droplet deposition manufacturing,** systems create the individual layers by selectively depositing molten material (wax, thermoplastic polymers, or low-melting-point metals) onto a substrate as a series of uniformly spaced, uniform-size micro-droplets, often at rates of several thousand per second. As depicted in Figure 19-10, these droplets then adhere to the substrate, forming the new layer, and ultimately a solid mass.

When supports are needed, they are often deposited in a perforated pattern to facilitate easy removal. Different materials can be deposited for supports and product, and multi-axis robots can be used to deposit material at nonvertical angles, often eliminating the need for supports.

Ultrathin layers, on the order of 16 μm, provide exceptionally fine details and smooth surfaces. Dimensional accuracy is good, but milling of the intermediate layers may be required to improve precision in the z-direction. Build times tend to be long, however. To improve build time, multiple jets may be incorporated onto a single print head, and alternative hatching patterns may be used when solid products are not

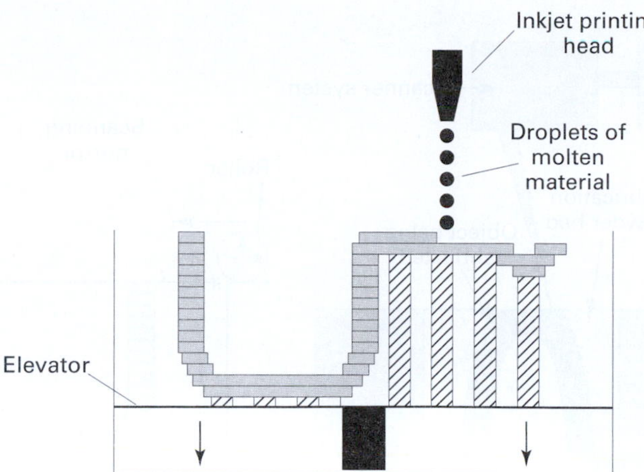

FIGURE 19-10 Inkjet deposition uses the ballistic particle concept to build solid objects.

required. The use of an inkjet to directly deposit molten material has become less viable as a commercial process in favor of the deposition of a binder onto powdered materials, a process known as three-dimensional printing that will be presented later in the chapter.

19.5 POWDER-BASED PROCESSES

SELECTIVE LASER SINTERING (SLS) AND SELECTIVE LASER MELTING SLM)[2]

Like most of the other processes, **selective laser sintering (SLS)** and **selective laser melting (SLM)** begin with the same part CAD file that is sliced in the STL format and a vertically positionable elevator. As depicted in Figure 19-11, the process begins by spreading a layer of heat-fusible powder across the part build chamber. A CO_2 or YAG laser is then scanned over the layer to selectively fuse together those areas defined by the the cross-section geometry. The interaction of the laser beam raises the powder temperature above the melting point, fusing the powder particles to one another and the layer below them. The unfused material remains in place as the support structure for subsequent layers. The elevator lowers, and another layer of powdered material is deposited, leveled, and laser sintered, with the process repeating until the part is complete. After processing, the part is removed from the build chamber, and the loose powder is removed for reuse.

Because the excess powder acts as a **natural support,** the SLS process does not require the design, fabrication, and removal of support structures. Thus, time and materials are not wasted in their build and removal. In addition, the unfused powder can be immediately reused, and optimal packing densities can be achieved within a build because additional geometries can be produced within the open volume of a part.

Selective laser sintering uses one of the broadest range of build materials, including plastics, waxes (such as those used in investment casting), metals, ceramics, and particulate composites. A variety of polymeric materials are available for the SLS process, including nylons, polyamides, polycarbonates, polyvinylchloride, elastomers, and acrylic styrene. (*Note:* These materials are less expensive than the photosensitive resins used in stereolithography and are also nontoxic.) Strength, toughness, and elongation values are among the best of the additive manufacturing processes and are often near-isotropic. Glass beads and other fillers can easily be dry-blended with the polymers to enhance stiffness and reduce shrinkage. Steels, stainless steels, tool steels, aluminum, titanium, and even superalloys have been used. Powder costs generally range between \$7 and \$90/kg (\$15 and \$200/lb), but because the unbonded powders can be reused,

[2] This terminology is a bit confusing. *Selective laser sintering* is a name given to plastic processes, and *selective laser melting* is given to metal processes. Melting actually occurs in both processes. In fact, the plastic materials stay liquid for the entire layer dwell time and through the next couple of layers. To achieve this, the process chamber is heated to a temperature close to the low end of the melting range. In contrast, the metals are liquid for only a short time after the laser melts the powder. The energy required to keep the metal liquid is much greater, and the chambers are not preheated.

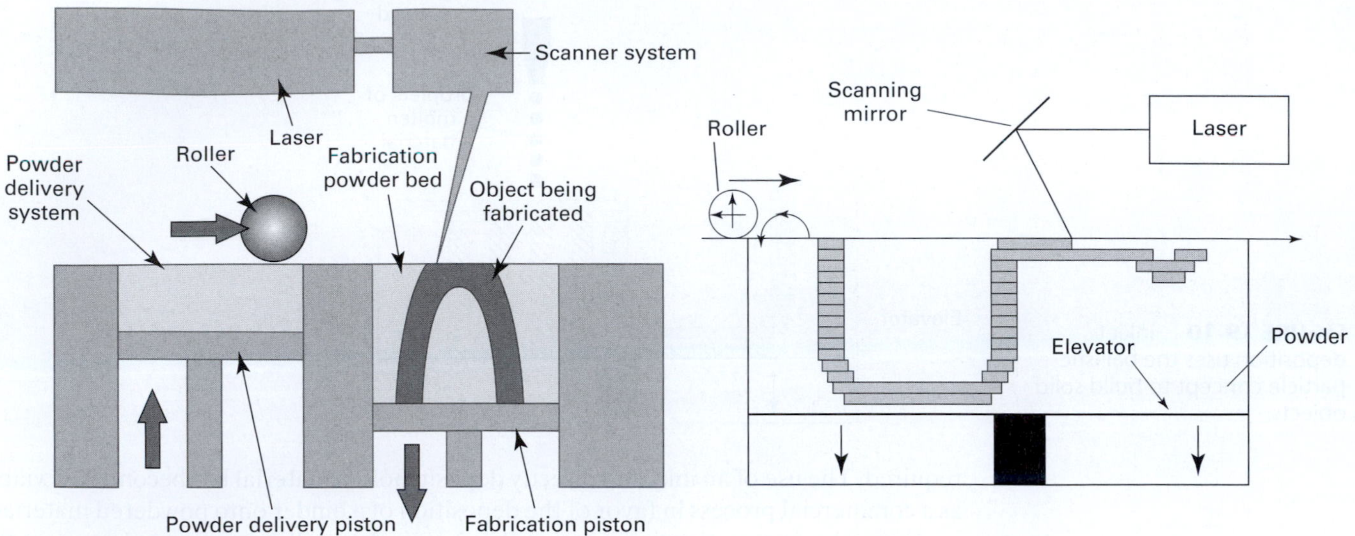

FIGURE 19-11 Selective laser sintering uses a laser to scan and sinter powdered material into solid shapes in a layer-by-layer manner.

FIGURE 19-12 Half of a silica sand casting mold produced by selective laser sintering. *(Courtesy of Mr. C. S. Huskamp, St. Louis, MO)*

material utilization is among the best of the additive processes. When the starting material is aluminum silicate sand or quartz sand that has been coated with phenolic resin, the process can be used to create laser-sintered sand cores and molds for metal casting, as depicted in Figure 19-12. Build speeds up to 2,500 cm^3/hr (150 $in.^3$/hr) can be achieved.

The primary mechanisms for binding particles are **fusion,** based on melting and resolidification, and the viscous flow of thermally softened (but not melted) surface material. For amorphous thermoplastics, such as polyvinyl chloride (PVC) or polycarbonate, the absorbed laser energy causes the powder particles to soften and bind to one another at their points of contact, producing a porous material. The porosity of these materials may be as much as 40% of the part volume. Semicrystalline materials, such as nylon, are locally melted and can essentially produce fully dense parts. These materials are preferred for the SLS process because they have the ability to exist in both the liquid and solid forms over a range of temperatures. Crystalline materials are not suitable because they become brittle due to the relatively long cooling times associated with the plastic SLS processes.

The bonding of metal and ceramic powders requires more energy input per unit time than the fusing of polymers. One approach to overcome this limitation is the use of metal and ceramic powders that have been coated with a thin layer of polymer binder. During the build, the laser heats the polymer coating, fusing it to the coatings on adjacent particles. The part is then transferred to a separate sintering operation, where the polymer binder is burned off, leaving a porous metal or ceramic product. The temperature is then raised to complete densification. In some cases, a full-density product is created by infiltrating a lower-melting temperature metal, such as bronze into tool steel, using capillary action to pull it into the loosely sintered product. Because of this ability to fabricate metals and ceramics, selective laser sintering has been among the first of the additive processes to offer direct fabrication of metal tooling for processes such as injection molding and die casting.

To minimize the required laser power and increase process speed, radiant heaters are often used to bring the powder to temperatures just below the fusing point. This also helps to decrease the material distortion that occurs during processing due to thermal residual stresses. Because of the high temperatures created by the laser, the build chamber is typically filled with nitrogen or argon to avoid oxygen contamination of the bonding surfaces and to reduce the possibility of combustion or explosion due to the high surface energy of powder particles.

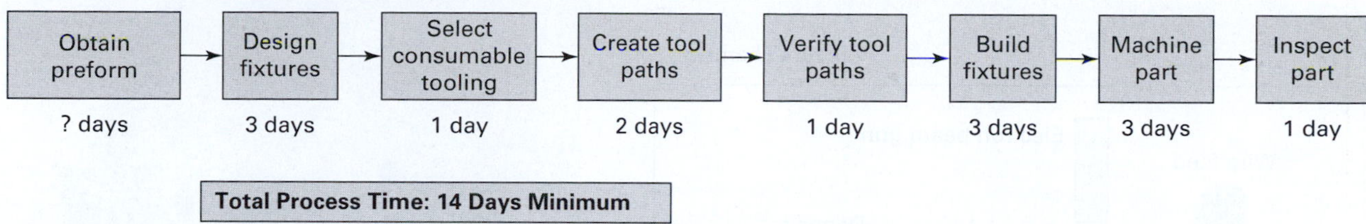

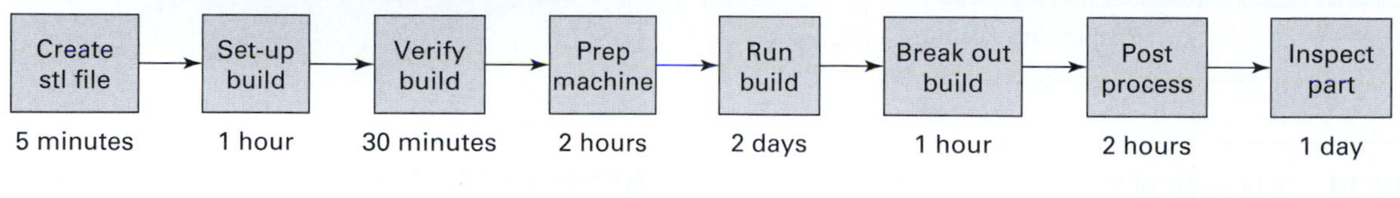

FIGURE 19-13 Comparison of metal prototype manufacture by CNC machining and direct-digital manufacture using selective laser sintering. Note the significant reduction in total process time. *(Courtesy of Mr. C. Huskamp, St. Louis, MO)*

Dimensional accuracy and surface roughness tend to be poorer in selective laser sintering than in other additive processes. Dimensional stability is largely a function of the volumetric shrinkage of materials, and the volumetric shrinkage associated with the melting and fusing of nylon and wax powders to fully dense materials can be quite large in comparison with other processes. As a result, trade-offs exist among the density, the mechanical properties, and the dimensional accuracy of these materials. Dimensions that require the greatest accuracy should be built in the *x-y* plane if possible. Material porosity is also the dominant source of surface roughness. Particles in the SLS process are approximately spherical and range in diameter from 50 μm (.002 in) to 125 μm (.005 in). When particles fuse only at their respective contact points, significant surface roughness can result. Mechanical properties tend to be very isotropic in both the thermoplastic materials and metals and are often equivalent to those produced by altenative processes. Aerospace components made from Ti-6Al-4V regularly achieve performance equivalent to parts produced by investment casting.

Figure 19-13 compares the required steps and associated time for the maqnufacture of a metal prototype by CNC machining and direct-digital manufacturing using selective laser sintering. Note the significant reduction in total process time.

ELECTRON BEAM MELTING (EBM)

Electron beam melting (EBM) is another powder-based, scanning-beam process, where the higher power of the electron beam can be used to produce full-density metal products from layers of metal powder. A vacuum is now required, however, to permit the electron beam to travel within the build chamber. The vacuum also provides a clean environment, leading to the production of superior-quality products with excellent mechanical properties. A high-speed diffuse beam is scanned over the top-most layer to preheat the material and reduce thermal stresses. The new powder layer is spread, and a high-power focused-beam scan then melts the selected regions. Build rates are on the order of 60 cm^3/hr (3.7 $in.^3$/hr).

Electron beam melting has been used to produce fully functional products, especially in areas where strength and elevated temperature performance are required. Materials must be electrically conductive but include reactive metals, such as titanium, and high-temperature metals, such as cobalt-chrome, tool steels, and the superalloys. Surface finish is typically rougher and the minimum feature size is larger than selective laser melting. This must be weighed against the higher mechanical performance, which is achieved by the lack of oxygen in the chamber (vacuum) and the high energy levels of

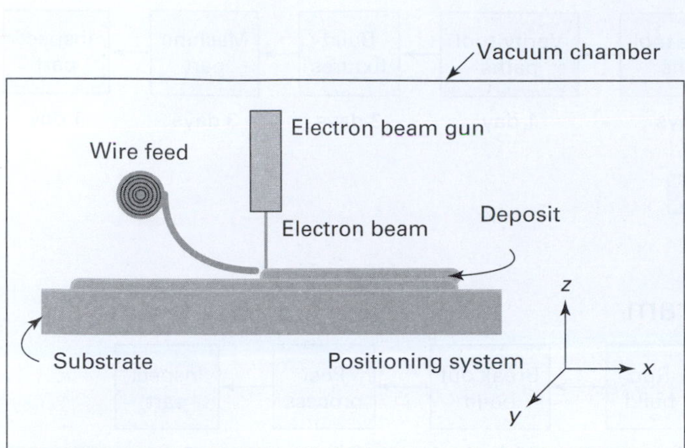

FIGURE 19-14 Schematic of the electron beam manufacturing process.

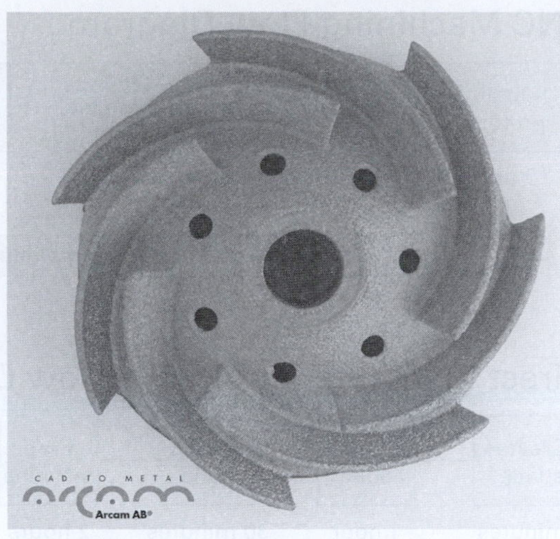

FIGURE 19-15 Small metal impeller made from titanium (Ti-6Al-4V). *(Arcam AB)*

the electron beam that fully consolidate the powder. Applications include injection molding dies, die-casting tooling, and stamping tools, as well as finished parts for direct service, such as aerospace parts and orthopedic implants. Figure 19-14 presents the sequence of electron beam melting manufacture, and Figure 19-15 shows a 120-mm (4.72-in.)-diameter, 30-mm (1.2-in.)-thick impeller made from titanium (Ti-6Al-4V).

THREE-DIMENSIONAL PRINTING

In yet another approach, inkjet printing has been combined with powdered materials by using the inkjet to deposit a liquid binder. As shown in the sequence of Figure 19-16, **three-dimensional printing (3DP),** also known as **selective inkjet binding,** begins by loosely depositing a thin layer of powder for processing. An inkjet printing head then scans the powder surface and selectively injects a binder material (such as polymeric resin or colloidal silica), joining the powder together in those areas defined by the cross-section geometry. As in selective laser sintering, the unbounded powder remains in place, surrounding the product and serving as supports.

The primary assets of three-dimensional printing are the variety of engineering materials that can be used (virtually any material that can be obtained as powder, including metals, ceramics, cermets and polymers) and the elimination of thermal residual stresses and their related problems because of the absence of thermal processing during the build. Inorganic binders, such as colloidal silica, can be used with ceramic powders. Rather than being evaporated away like organic binders, these inorganic binders fuse together with the ceramic powder and reduce the amount of volumetric

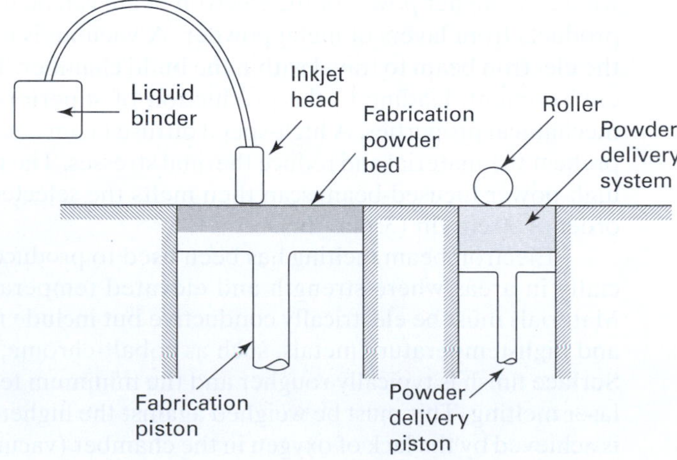

FIGURE 19-16 Three-dimensional printing builds components by selectively binding powdered material in a layer-by-layer manner. Metal parts are then sintered and possibly infiltrated to produce strong, high-density products.

FIGURE 19-17 A casting mold produced by binding sand using three-dimensional printing, and the aluminum casting that was made in the mold. *(Courtesy of Z Corporation)*

shrinkage. When strength is not required and low material cost is preferred, plaster powder can be used along with deposited drops of water where solidification is needed.

By incorporating dyes or pigments, colored binders and multiple printheads can be used to produce multicolored products without the need for paint or surface processing. This is often used for marketing prototyping, tailoring the prototype to the particular customer. Using different colors to differentiate the various components of an assembly can often enhance communication and understanding.

Three-dimensional printing offers some of the fastest build times of the additive processes. Unfortunately, the process usually requires multiple processing steps, including postprocessing in a sintering furnace to remove the binder (if needed) and to fuse and densify the product. Residual porosity tends to compromise material strength. For metal powders, such as stainless steel, postprocessing after sintering can also include infiltration of the metal matrix by a lower-melting-temperature metal (such as bronze) to increase densification.

Foundry molds and cores can be created directly from CAD data by replacing the powder material with quartz or specialty sands (or starch- or plaster-based powders for use with lower-melting-point metals) and injecting microdroplets of foundry-grade resin as a binder. The binder material, often as little as 2% of the product, hardens in contact with an activator that has been blended with the sand. Cores can also be printed with a hollow interior to allow for better venting. Figure 19-17 shows a printed mold and the resulting aluminum casting.

In another variation, radiation absorbing material is used in place of the binder. After each layer is printed, infrared light is flashed over the surface. The deposited material absorbs the energy, causing the selected material to sinter. By sintering as each layer is deposited, subsequent processing can often be eliminated.

SINTERMASK

Sintermask is a relatively new additive manufacturing process that melts layers of thermoplastic powders using masks, similar to those used in solid ground curing, and a single exposure flash of energy whose wavelength is set to maximize absorption into the powder (usually infrared). If the layer thickness is 0.12 μm and the desired product has a height of 25 cm, 2083 layers will be required. The average layer time for a sintermask machine is approximately 4 s, which compares to an average layer time of 30 s on selective laser sintering processes using the same material. Comparing the two processes, sintermask offers a 15-hr time savings. Creating a full layer in a single flash is quite efficient, and the use of thermoplastic powder, instead of photocuring resins is an additional cost saving.

19.6 DEPOSITION-BASED PROCESSES

FUSED DEPOSITION MODELING (FDM)

Fused deposition modeling (FDM) produces laminated three-dimensional objects through robotically guided extrusion of a commercial-grade thermoplastic material,

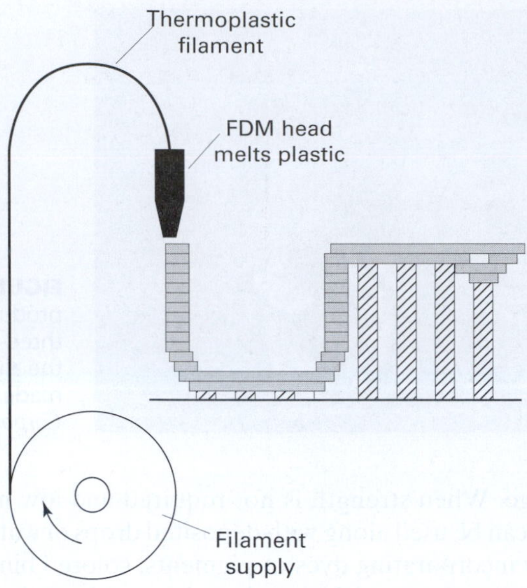

FIGURE 19-18 Schematic of fused deposition modeling. This process can be "office friendly" (i.e., small in size and using nontoxic materials).

as illustrated in Figure 19-18. A spool of amorphous thermoplastic filament is unwound and fed through a robotic extruding head where it is heated to a semi-liquid state. The emerging extrudate then builds the *x-y* plane, like building an object with a hot glue gun, first forming the contour and then filling in the areas by hatching. The material solidifies upon contact and bonds with the layer below. When the layer is completed, the build platform indexes down one layer in thickness (as fine as 0.05 μm, or 0.002 in., or as much as 0.75 μm, or 0.03 in.), and the process repeats. As a whole, the process resembles an *x-y* pen plotter, except that the plots are three dimensional. FDM requires the use of a base because the material does not cool fast enough to maintain rigidity and dimensionality during deposition. The base is a brittle material that can be easily separated from the finished product. If supports are needed, a dual-extrusion head is used, and a different material is used to create the supports. Depending on the material selected for this application, it may be easily removed from the primary build material by dissolving in a water-based solution or by breaking the support structures away from the part.

The build cycle for fused deposition modeling is much simpler than other additive manufacturing processes. As a result, the equipment is compact and relatively low cost, ideal for applications in a design engineering office environment. In addition, the process uses nontoxic thermoplastic materials, a requirement in office environments, and it does not require venting or postprocessing. As a desktop unit, the FDM process is ideal for concept-modeling applications early in product development cycles. Colors can be incorporated in the build material and are easily changed for the various regions of a product.

Disadvantages of the process include the limited selection of engineering materials, which includes machinable wax, investment-casting wax, polycarbonate, polyphenolsulfone, and acrylonitrile-butadiene-styrene (ABS). Most of these materials have poor mechanical properties due to the anisotropic behavior in the *z*-direction. While parts do not require postprocessing, the build time is the longest among the additive processes, and the material cost of the starting filaments is among the highest per unit volume. Common defects include gaps between contour perimeter traces and the rastored fill traces, voids between restored fill traces when the extrusion carriage makes a 180-degree direction reversal, and visible hatch lines due to the extrusion of round profile material. If the surface roughness or stairstepping effect is objectionable, a heated tool can be used to smooth the surface, the surface can be sanded, or a coating can be applied, provided the necessary tolerances can be maintained.

LASER-ENGINEERED NET SHAPING (LENS) AND DIRECT METAL DEPOSITION (DMD)

In a process known as **direct metal deposition (DMD),** lasers and powders have been combined to enable the direct fusing of uncoated metal powders into fully dense functional metal products. Powder delivery nozzles direct streams of metal powder at the focal point of a high-power neodymium: yttrium–aluminum–garnet (Nd:YAG) laser. The powder particles then melt and fuse with the underlying layer as the part is scanned on an *x-y* positioning stage beneath the laser. Upon completion of the layer, the laser and powder delivery nozzle move upward and begin patterning the next layer. By adjusting the focal point of the laser (as small as 0.2 μm to as large as 2.5 μm), the material can be deposited as either fine detail or bulk build. In **laser-engineered net shaping (LENS),** the high-power laser produces a tiny molten pool on the surface of a build, and a nozzle blows a small amount of metal powder into the pool to increase its volume. The process then repeats to complete the layer. Yet another alternative uses laser power to melt and fuse a solid wire with a flat cross-sectional profile. In essence, these processes are microscopic, computer-controlled versions of laser cladding performed in a controlled-atmosphere chamber, typically argon, to prevent undesirable oxidation. The major drawback to these processes is the long layer times that can lead to build times in weeks or months for large parts.

A variety of metals, alloys, and nonmetallic materials can be deposited—including titanium, tool steels, nickel and cobalt superalloys, and other high-performing materials—and the composition can be varied during a build to create products with composition gradients. These technologies were initially targeted for the rapid manufacture of metal molds for the injection molding of plastic parts but have expanded into the production of molds, dies, and finished products with full density, fine grain structure, and mechanical and metallurgical properties that are equivalent or superior to forged materials. By inputting both current and desired geometries, the process can also be used to add material to existing components to repair damage or restore worn products to useful life.

An electron beam equivalent has also been developed in which a metal wire is fed into a molten pool created by the beam. The beam focus and power output are adjustable, allowing for bulk deposition or fine detail. The vacuum environment allows for the deposition of reactive materials, and the process can be extended to accommodate refractory materials, including tantalum and tungsten.

LAMINATED-OBJECT MANUFACTURING (LOM)

Laminated-object manufacturing (LOM) involves the sequential bonding and patterning of solid material sheets or *laminae,* with a common process variation using sheets of paper with a thermally activated organic compound coating one side. As shown in Figure 19-19, this lamination-based process begins by delivering a sheet of paper onto the working surface either by roller or some other means. Next, a heated roller is passed over the surface to bond the paper sheet to the layer below it. Once bonded, a laser or cutter is guided to cut the perimeter of that particular cross section, cutting only the uppermost layer. These steps are then repeated until part the part is completed.

A unique characteristic of the lamination processes is the method for extracting the product after fabrication. Measures must be taken to remove the excess paper mass that accumulates around the periphery of the part. This is handled by drawing a consistent set of cross-hatches in the excess material regions during patterning. Over many layers, these cross-hatches form the boundaries of small blocks, which are later removed from around the finished part. On the upside, the lamination processes do not require the fabrication of any specialized support structures. The excess solid material provides a natural support of the part while it is being fabricated. Thus, complex cantilevered geometries can be fabricated as easily as any other structure.

Equipment and materials tend to be cheaper due to the simplicity of the process and reduced maintenance requirements. Large components can be readily produced with sizes up to about 0.5 × 0.5 × 0.75 m (20 × 20 × 30 in.). Because the material does

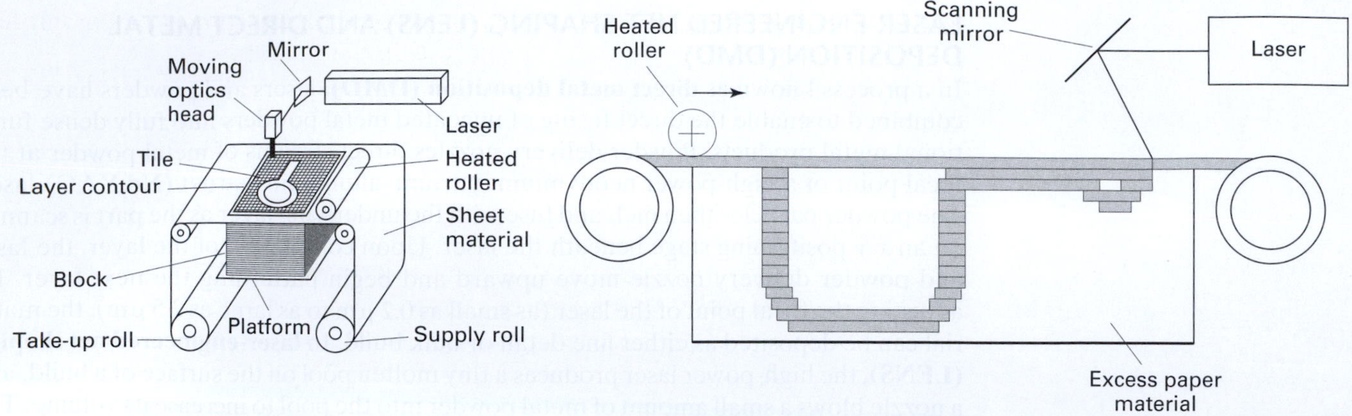

FIGURE 19-19 Schematic of the laminated object manufacturing process where solid sheets are used to create the layers.

not go through a phase change (e.g., liquid to solid), residual stresses are low, resulting in less warpage and better dimensional stability. Because paper is hygroscopic in nature, laminated parts must be surface treated with water-resistant resins to prevent swelling. Combining multiple parts into a single build is often difficult because the removal of excess material often prohibits parts from being spaced closely together. Layer thickness is fixed at the thickness of the lamina, which decreases the flexibility in build time and surface roughness. The *z*-axis dimensions tend to be worse in laminated processes than in the other additive processes.

With paper laminates, the properties of the finished part are similar to those of birch wood, so automotive patternmakers and modelmakers in woodworking shops find the process to be attractive. Other sheet materials have been introduced with limited success, including polyester and other plastics, as well as metals, ceramics, and fiber-reinforced materials. Fully dense ceramic products can be made by replacing the paper rolls in the process with ceramic tape (see tape casting in Chapter 14). Once the part is removed from the excess material, it goes through a debinding operation and is fired in a sintering oven to achieve full densification. Excellent material properties can be achieved because of the high green density of the ceramic tape. The cost of metal and ceramic laminations is rather high, however, due to the high cost of metal and ceramic tape combined with the potential for large amounts of material waste.

OTHER DEPOSITION-BASED PROCESSES

In a modification of the lamination approach, **ultrasonic consolidation** (i.e., ultrasonic welding) is used to bond layers of metal foil into full-density products. True metallurgical bonds are produced with modest pressure and temperatures that do not exceed half of the melting point. A periodic machining pass is incorporated to maintain flatness and dimensional tolerance. Features include the ability to produce extremely complex shapes, as well as laminate dissimilar metals, maintain low temperatures, and embed electronics and/or fibers.

The **Poly-jet conceptual model** uses the inkjet approach to deposit droplets of liquid photopolymer, which are immediately cured by exposure to ultraviolet light while a second system deposits noncurable support material. Layer thickness is approximately 20 μm (0.001 in.), and features as small as 0.04 μm (0.0015 in.) can be produced. Surface finish is among the best of the additive processes, but strength is lower than that obtained with commercial-grade polymers.

19.7 USES AND APPLICATIONS

In less than three decades, additive manufacturing processes have evolved from a method of making low-strength, three-dimensional, plastic design prototypes to a family of processes that are changing the way organizations design and manufacture products. Starting with a CAD drawing, computer tomography (CT scan), magnetic resonance imaging (MRI), or three-dimensional digitizing, the data can be quickly converted to a fully functional prototype, scale model, manufacturing tool, or finished product, facilitating considerable savings in both time and money.

PROTOTYPE MANUFACTURE:

A large number of additive products are used produce both **conceptual models** and **functional models** to assess aspects of form, fit, and function. Converting a design concept to a physical representation offers the following advantages:

1. Ability to evaluate concepts and seek input from others during the design phase.
2. Ability to detect and correct design errors before the costly tooling stage.
3. Ability to have multiple iterations to a design before manufacture, allowing comparison of different shapes and styles.
4. Ability to assess assembly concerns and ensure proper alignment and interaction with companion parts.
5. Can possibly allow for functional testing under real-life or simulated conditions, including wind-tunnel testing, assessment of fluid-flow through orifice designs, and evaluation of stress distribution and strength under load.
6. Ability to serve as communication models among management, design, manufacturing, sales and marketing, and potential customers.
7. Can serve as sales and marketing models or samples.
8. Can serve as a visual aid during bidding, tooling design, and other activities.
9. Often enables a significant reduction in product development time.

Figure 19-20 depicts an assembly of SLS (stereolithography) prototypes developed to assess form, fit and function. Figure 19-21 presents a collection of scale models that have been produced by additive manufacturing processes.

RAPID TOOLING

Rapid tooling (RT) is an excellent application for the additive manufacturing processes for several reasons:

1. Tooling is usually a low-production-volume or one-of-a-kind commodity.
2. Tooling typically has an associated long lead time.
3. Tooling typically requires the high cost of skilled labor.

FIGURE 19-20 Example of additively produced prototypes used to assess form, fit, and function. An assembly of components and SLS prototypes (including engine, exhaust headers, and active suspension) has been prepared for wind tunnel testing of an open-wheeled race car.

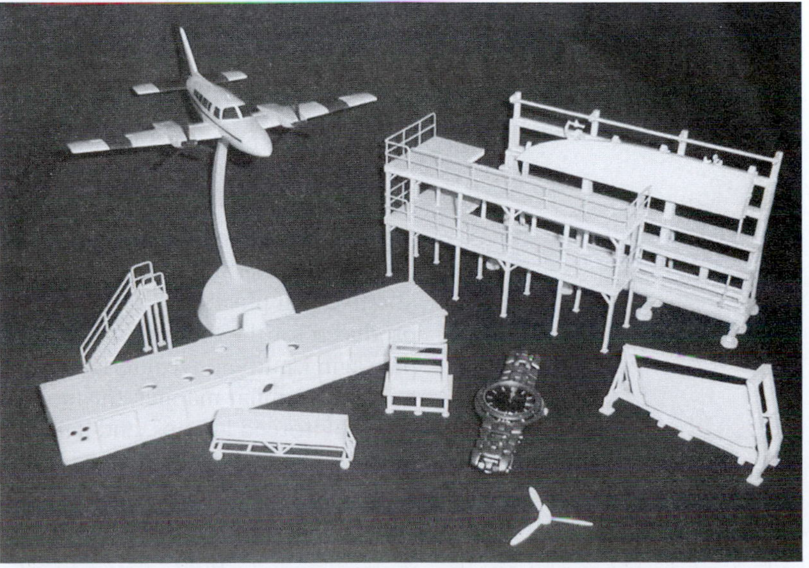

FIGURE 19-21 Assembly of small-scale models produced directly from CAD data by additive manufacturing processes. *(Courtesy of Mr. C. S. Huskamp, St. Louis, MO)*

4. The high cost of tooling often requires its use in producing a large number of identical products.
5. Tooling is a requirement of most manufacturing processes, so there are a number of potential applications.
6. Tooling is usually specific to one or a few part geometries.

As a rapid source of cheap yet effective tooling, additive manufacturing technology can lower fixed tooling costs. Manufacturers may be able to justify die casting, stamping, or injection molding runs of a few hundred or a few thousand parts or produce those parts with a much shorter lead time.

The simplest of the rapid tooling approaches is the direct production of patterns or molds used for metal casting. Most of the additive techniques can be used to produce patterns for the sand-casting processes, where the direct conversion of design to pattern can significantly reduce lead time while requiring no change in standard foundry practices or procedures. Patterns must be abrasion resistant and able to withstand the ramming forces of sand compaction, and many additive process materials are adequate. The choice of CNC machining or additive manufacturing will depend on a number of factors, including size of the casting, geometric complexity, number to be produced, and required precision. The additive manufacturing techniques that employ sand and binder can be used to directly produce both molds and cores. The cores can be hollow, providing excellent venting and permeability. Several examples of this application have been presented in earlier figures in the chapter.

Investment casting (or lost wax casting) is frequently used to manufacture intricate parts that are difficult, if not impossible, to machine, forge, or cast by other methods. Designs often include internal passages and ports (as in valve bodies), curved surfaces (as in the vanes of impellers), and internal cooling channels (like those in turbine blades). Each casting requires an expendable wax or thermoplastic pattern, a replica of the desired shape, usually produced in an injection mold. Many of the additive manufacturing processes, including stereolithography (SLA), selective laser sintering (SLS) and fused deposition modeling (FDM), can now produce these expendable patterns directly from design data, bypassing the costly and time-consuming manufacture of injection mold tooling. There is no increase in cost as part complexity increases. Hollow patterns, or patterns with a honeycomb or scaffold interior, can be produced, significantly reducing the possibility of shell cracking when the pattern material is melted out of the ceramic shell.

Die-casting dies can be produced with internal cooling channels that are positioned for optimum performance. Molds for plastic injection molding can be economically produced, and these molds can be produced in a modular form that reduces cost and enables simple replacement of worn or broken segments. Rubber or epoxy molds for vacuum casting of polymers or spin casting of low-temperature metal alloys can be produced by pouring the mold material over a pattern that has been produced by additive manufacturing and allowing it to cure.

Molds have been produced to perform the blow molding of prototype polyethylene and PET containers. Mold manufacturing time was reduced from several weeks to less than 5 days, and cost to less than half of a typical aluminum tool. The only significant change to the process was an increase in cooling time due to the lower thermal conductivity of the additive manufacturing polymer. Both male and female molds and form tools have been produced for the thermoforming of thermoplastic sheets into parts that include consumer packaging, automotive interior and body panels, aerospace panels and ducts, and even boat hulls. In a similar manner, the molds and form tools have been used in the manufacture of fiber-reinforced lay-ups, and lightweight, honeycombed or semihollow additive products have also been used as enclosed cores for fiber-reinforced products.

More durable metal molds and dies can be produced by first creating a pattern using one of the additive processes. A metal deposition operation, such as thermal spray or electroforming, is then employed to coat the pattern and build a metal shell of the desired thickness. The resulting metal shell is then separated from the mandrel and attached to a hollow frame that is reinforced with a backing material such as a metal-reinforced epoxy or chemically bonded ceramic. Production runs in excess of 5000 have been made on a plastics compression molding machine using a pure nickel shell backed with chemically bonded ceramic.

There are also a number of manufacturing tools that are used within the manufacturing system, including jigs, fixtures, templates, and gauges. They are used to align, assemble, clamp, hold, test, and calibrate components and subassemblies. Much like any other manufactured product, they too go through the operations of design, ducumentation, and production and have the same concerns of lead time and cost. If a manufacturing tool is damaged or lost, a replacement can be made quickly, with little effort and expense.

DIRECT-DIGITAL MANUFACTURE

With the numerous additive manufacturing processes and the developments that have occurred over the past several decades, what was once considered to be the fabrication of low-strength, plastic prototypes has expanded to include the direct production of metal, ceramic, and polymer components, many with properties equivalent to their standard-production-process equivalents. By producing finished parts direct from digital data, eliminating all tooling, a number of applications have emerged, including:

1. The production of small batch sizes that could not be produced economically by traditional methods that required dedicated tooling.
2. The production of parts with intricate geometries or internal features that could not be made without extensive assembly operations.
3. The production of one-of-a-kind custom or personalized products, such as bone replacements, dental implants, orthodontics, custom-fit hearing aids, and other items that must be tailored to a specific individual. Figure 19-22 shows such an application where CT-scan data from a skull injury were used to create a custom titanium patch using the electron beam melting process.
4. The production of replacement parts on demand, enabling reduced inventory costs.
5. The repair or reworking of worn, distorted, or damaged parts, restoring to original dimensions and mechanical properties comparable to the original product, thereby enabling reuse as opposed to costly replacement. This is especially important when replacement parts are not available.
6. Manufacture of products from hard-to-fabricate or hard-to-machine materials.
7. The manufacture of products within pressing time constraints.

Designing for direct-digital manufacture is different from traditional design. Much of traditional design is design for manufacture. For example, castings must consider the various features of solidification and shrinkage, with preference for uniform wall thickness and parting lines and draft incorporated to facilitate pattern removal or part extraction. Consideration must be given to core placement and support and aspects such as minimum wall thickness, gate placement, risering, and others. Machining operations require accessibility of the cutting tool and chip removal. Forging and the other deformation processes

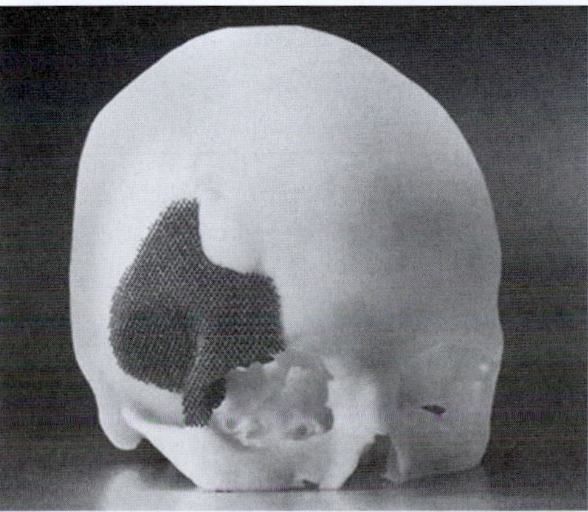

FIGURE 19-22 Titanium skull patch produced by the electron beam melting process beginning with data obtained from a CT scan of an injured person. This is an example of a one-of-a-kind, or custom, product. *(Arcam AB)*

have similar considerations. A part designed for forging would not make an ideal casting, and a part designed for casting would not make an ideal forging.

Direct-digital manufacture breaks the rules. For the additive manufacturing processes, the design process should focus on function. Because cost and time are no longer a function of complexity, parts can be as intricate, complex, or detailed as they need to be. Internal features and/or sculptured surfaces do not add to cost. Continued iterations and refinements are encouraged. There is no penalty for changes late in the product development cycle. Design changes can be made or variations of a product can be offered even after production has started. There is no need to simplify, because additional features do not increase cost. Wall thickness can vary as needed, and weight can be reduced by making some walls or features with an internal lattice or even completely hollow. Part consolidation should be considered to eliminate assembly operations, provide better dimensional precision, and reduce parts inventory. Conversely, because there is no cost of part-specific tooling, breaking a product into subcomponents to better facilitate serviceability or reduce replacement costs does not add to the manufacturing expense.

19.8 PROS, CONS, AND CURRENT AND FUTURE TRENDS

Additive manufacturing has grown from a concept to a diverse and fully developed manufacturing family in less than three decades. Table 19-1 summarizes the primary features of some of the more popular processes. What began as a means of producing plastic models directly from computer data has expanded to the production of a multitude of components, including direct-usage products. Available materials have expanded from low-strength photopolymers to now include engineering plastics and elastomers, ceramics, metals, and even composites. Properties include good tensile and flexural strength, good impact resistance, and reduced anisotropy. Metals have moved to full density, with properties equivalent to wrought and cast products, and now range from low-strength, low-melting-temperature metals through titanium, tool steels, and superalloys. Candidate applications are most attractive when complexity is high, lead times are short, and volumes are low (often as low as a single part).

The range of available materials is still limited, however, especially within certain processes, and the mechanical properties are often less than those of similar manufacturing materials. Raw material costs are often high, and the time required to produce each part is too great to enable the manufacture of large production runs. The long-term durability of products, especially as it relates to fatigue and fracture, has not been proven.

Current trends include:

1. The development of high productivity machines with large build volumes for large corporate customers and service providers.
2. The development of small, low-cost machines, aimed at desktop users and educational institutions.
3. The development of machines for targeted applications, such as dental, jewelry, etc.

TABLE 19-1 Comparison of Additive Processes

Process	Starting Material	Layering Mechanism	Possible Materials
Stereolithography (SLA)	Liquid photopolymer	Polymer curing by UV light	Photopolymers
Inkjet deposition (ID)	Molten liquid	Solidification of droplets	Thermoplastic polymers, waxes, low melting-point metals
Selective laser sintering (SLS)	Powders	Sintering or melting	Thermoplastic polymers, waxes, metals, binder coated sands
Electron beam melting (EBM)	Powders	Melting	Metals (titanium, tool steels, superalloys)
Three-dimensional printing	Powders	Deposited molten binder	Metals, ceramics, polymers, cermets, sand
Fused deposition modeling (FDM)	Semi-liquid polymer	Solidification of extruded strand	Thermoplastic polymers
Laminated-object manufacturing (LOM)	Solid sheets	Fusing of sheet material	Paper, polymers, ceramic

TABLE 19-2 Primary Uses of Additive Manufacturing, 2009 (in decreasing order of use)

1. Functional models
2. Direct part production
3. Assess fit and function
4. Visual aids
5. Patterns for prototype testing
6. Presentation models
7. Patterns for metal casting
8. Tooling components
9. Other

Data taken from Wohlers Report—2009, Wohlers Associates Inc., Fort Collins, CO

4. Integration of optical scanning to better facilitate reverse engineering, part replication, and repair and restoration.
5. Increased use in the direct-digital manufacturing of complex-shaped, low-volume products.
6. Expansion of candidate materials.
7. Improvements in dimensional precision and surface finish.
8. Reduction in cost.

Future trends include:

1. The ability to produce parts on demand from a computer data bank with relatively short waiting time.
2. Ability to mix materials during the build process to produce tailored properties and functional gradients.
3. Products with integral and multiple colors.
4. Functional composite materials:
   - Incorporation of graphite for thermal conductivity.
   - Copper and aluminum microspheres.
5. Multifunctional parts through embedded components—such as conductors, insulators, microfluidics, photovoltaics, or sensors directly on or inside a thermoplastic part.
6. Three-dimensional biomaterials for skin grafts, bone repair, and functional implants.

Table 19-2 lists the primary applications of the additive manufacturing processes in decreasing order of use. This tabulation is compiled annually, and direct part production has risen from near bottom to second ranking, with a prediction that it will soon become the largest single use.

19.9 Economic Considerations

In 2009, there were 29 additive manufacturing system manufacturers, 65 service providers, and more than 5000 users and customers. When deciding to use additive manufacturing processes, several questions must be answered, including whether to buy manufacturing equipment or to outsource work to vendors. If purchasing equipment, a company must first decide which additive manufacturing machine to buy. If outsourcing, the company must determine which vendor to choose. In addition, the company must recognize and prepare for the effect that additive manufacturing technology will have on the organization.

It is important to first define the purpose for using the additive manufacturing technology. There are many different applications, including concept modeling, form/fit verification, marketing demonstrations, functional testing, prototype tooling, production tooling, and direct manufacture of finished products. Comparisons must be made between the defined requirements and the equipment capabilities. Full-scale models of an engine block, for example, can only be fabricated on three or four of the additive manufacturing processes that have been presented because of constraints in the size of the work envelope. Other factors to be considered include material properties, dimensional tolerances, surface finish, build time, and dimensional stability over time. In general, material removal processes will produce better dimensional accuracy, surface finish, and material properties, while the additive manufacturing processes provide expanded geometric complexity and shorter lead time. Lead times depend on the amount of specialized fixturing and cutting tools required for machining, which depends on the geometric complexity of the part. Generally, lead times are better on machining processes for simple geometries and are better on additive processes for complex geometries.

Another important consideration is the batch size of products to be fabricated on the machine. Nesting parts within the build cycle can often make the difference between economical and expensive additive manufacturing processing. For example, the lamination processes generally do not permit large batch sizes, but they are the fastest and cheapest alternative when building large individual parts.

Finally, an evaluation of costs and benefits can help make the final decision regarding whether to buy or outsource additive manufacturing technology. To buy an additive manufacturing machine tool, several costs must be factored into its procurement, including equipment costs, support technology costs (including CAD systems, cleaning equipment, postcuring devices, etc.), freight and installation costs (including any building modifications), and personnel costs (including any new hires or training). Beyond procurement are the costs required for its operation, including material costs, maintenance costs (including service contracts, preventive maintenance, equipment downtime, etc.), facility costs (including floor space, utilities, etc.), and labor costs.

Experienced personnel are very important when operating some additive manufacturing processes. For example, stereolithography has more than 30 input variables, which require fine adjustment in order to make good parts. Typically, one person is required for each manufacturing machine. Site preparation can also be a factor. Certain additive systems use toxic resins that require high-capacity venting, or cleaning solvents which must be changed frequently. Many additive manufacturing systems require additional room for postprocessing equipment, including cleaning and postcuring. In all, the personnel, maintenance, training, and site preparation costs must be anticipated in order for the additive manufacturing system to be profitable.

Benefits resulting from the use of additive manufacturing equipment, in addition to those previously cited, may include:

1. Reduction of current operating costs (including labor, quality, purchasing, etc.).
2. Improved sales and marketing due to the ability to respond to customer requests for bids with actual models.
3. Improved product development, including improved customer satisfaction and improved product manufacturability.
4. Improved process development as lead times and costs are reduced, leading to more iterations.
5. No required operator observation during the build.

These benefits must be weighed against the time required to learn how to make high-quality, dimensionally accurate parts. If the company has not had experience with CAD solid modeling prior to its involvement with additive manufacturing technology, the learning time will be greater.

Prior to purchase of additive manufacturing equipment, it is generally a good idea to first test a particular technology by using the services of a job shop or service bureau. Small and medium-sized companies may find it preferable to use service bureaus rather than incur the costs of purchasing an additive manufacturing machine.

Key Words

additive manufacturing
additive processes
ascending surface
bases
build time
casting processes
computer-aided design (CAD)
computer-aided manufacturing (CAM)
computer numerical control (CNC)
conceptual models
consolidation processes
deposition-based processes
desktop manufacturing
direct metal deposition (DMD)
direct-digital manufacturing (DDM)
droplet deposition manufacturing
electron beam melting (EBM)
free-form fabrication
functional models
fused deposition modeling (FDM)
fusion
hatching
inkjet deposition (ID)
laminated-object manufacturing (LOM)
laser engineered net shaping (LENS)
layered manufacturing
liquid-based processes
masked-lamp descending platform
material removal processes
natural support
photocurable polymer
poly-jet conceptual models
postprocessing
powder-based processes
preprocessing
prototype
rapid manufacturing
rapid prototyping (RP)
rapid tooling (RT)
rastoring
reverse engineering
selective inkjet binding
selective laser melting (SLM)
selective laser sintering (SLS)
sintering
sintermask
solid ground curing (SGC)
stairstepping
stereolithography (STL)
stereolithography apparatus (SLA)
STL files
supports
tesselation
three-dimensional printing (3DP)
tool-less manufacturing
tool paths
ultrasonic consolidation
voxel
voxel geometry

Review Questions

1. What are the four traditional families of shape-production processes?
2. What is the common feature of the additive processes?
3. Why are the additive manufacturing processes often referred to as rapid prototyping?
4. Why is traditional prototyping considered to be costly? Time-consuming?
5. What are some of the manufacturing changes or new approaches that have been enabled by the computer?
6. When was additive manufacturing first introduced as a manufacturing process?
7. What are the four areas of application for the additive manufacturing processes?
8. What are some of the other terms that have been applied to additive manufacturing? What is the basis of these terms?
9. What are the benefits of "free-form" and "tool-less"?
10. What is involved in preprocessing?
11. What is the role of bases and supports?
12. What are some possible postprocessing operations?
13. What are the three families of layerwise manufacturing processes?
14. What is the key, somewhat universal, advantage of the additive manufacturing processes?
15. How do the capabilities of RP processes in general compare with those of material removal processes such as turning?
16. What does the term *tessellation* mean, and what is it used for?
17. What is reverse engineering, and how are the input data obtained?
18. How is geometric data presented in the STL format?
19. What is rastoring?
20. What are some of the ways of controlling the thickness of a multilayered part?
21. What is stairstepping?
22. Describe the coordinate system that has become common for all of the additive processes.
23. What is a voxel? Why is it significant in the additive processes?
24. What are some of the factors that affect the dimensional accuracy in additive processes?
25. What is hatching, and how might it be affected by dimensional changes, such as those occurring during phase transformations?
26. Of the three components of build time (preprocessing, fabrication, and postprocessing), which typically requires the greatest amount of time?
27. What technical advances were necessary for the stereolithography process?
28. How are the various layers produced during stereolithography?
29. What postprocessing is required for SLA products?
30. What are some of the advantages of stereolithography? Disadvantages?
31. How are the various layers produced during solid ground curing (SGC)?
32. How are the two build materials used in the solid ground curing (SGC) process?
33. Why can the packing density of products in the SGC process be significantly higher than in the SLP processes?
34. What are some of the disadvantages of solid ground curing (SGC)?
35. How are the various layers produced during inkjet deposition (ID)?
36. What are the advantages of inkjet deposition (ID) processes? Disadvantages?
37. How are the various layers produced during selective laser sintering (SLS)?
38. What is the difference between selective laser sintering and selective laser melting?
39. What acts as a support structure during selective laser sintering (SLS)?
40. What are some of the build materials that can be used in selective laser sintering (SLS)?

41. How is the processing of metal and ceramic powder different from the processing of thermoplastic powder in selective laser sintering (SLS)?
42. What are some of the advantages of using an electron beam as opposed to a laser beam?
43. How are the various layers produced during three-dimensional printing (or selective inkjet binding)?
44. What types of materials can be used in three-dimensional printing (3DP)?
45. What type of postprocessing is required for three-dimensional printing processes?
46. How can foundry molds and cores be produced by three-dimensional printing?
47. How are the various layers produced during the sintermask process?
48. How are the various layers produced during fused deposition modeling (FDM)?
49. What are some of the advantages and disadvantages of fused deposition modeling (FDM)?
50. How are the various layers produced during laser engineered net shaping (LENS) and direct metal deposition (DMD)?
51. What types of products are produced by laser engineered net shaping (LENS) and direct metal deposition (DMD)?
52. How are the various layers produced during laminated-object manufacturing (LOM)?
53. How are products extracted after a lamination-process build?
54. How are the various layers produced during ultrasonic consolidation?
55. What are some of the unique features or capabilities of ultrasonic consolidation?
56. What are some of the advantages of a physical prototype or model?
57. What are some of the attractive features of being able to produce rapid tooling?
58. How can the additive manufacturing processes be employed in metal casting?
59. How can additive manufacturing be coupled with thermal spray or electroforming to create shaped metal shells?
60. What are some of the benefits of creating jigs, fixtures, templates, and gauges by additive manufacturing techniques?
61. What are some of the areas where direct-digital manufacture has been applied?
62. How is design for direct-digital manufacture different from traditional design?
63. What are some of the current trends within additive manufacturing?
64. What are some of the future trends within additive manufacturing?
65. What are some of the considerations when deciding whether to buy additive manufacturing equipment or outsource work to vendors?

■ PROBLEMS

1. a. Which of the additive manufacturing processes can produce full-density metal products from the higher-strength, higher-melting-temperature metals (such as titanium, tool steels, and nickel superalloys)?
   b. Which of the additive manufacturing processes can produce parts from ceramic materials?
   c. Which of the additive manufacturing processes can directly produce foundry cores and molds?
2. Perform a Web survey of additive manufacturing service providers, and determine which of the various processes appear to be in most common use. Would your answer differ if it were to be focused on prototype products? Rapid tooling? Direct-digital manufacturing?
3. Describe several biomaterials-type applications where the manufacture of products tailored to a specific individual would be attractive and desirable.
4. The ability to restore worn parts to original dimensions or repair damaged components has been cited in the chapter. For what types of products would this ability be most attractive?
5. Looking to the future, people have cited the ability to build parts on demand from computer files and the potential of significantly reducing inventory of spare parts in areas as diverse as automotive, aerospace, and military. What do you see as potentially limiting concerns to this future projection?

CHAPTER 20

FUNDAMENTALS OF MACHINING/ ORTHOGONAL MACHINING

20.1 INTRODUCTION

Machining is the process of removing unwanted material from a workpiece in the form of chips. If the workpiece is metal, the process is often called **metal cutting** or *metal removal*. U.S. industries annually spend well over $100 billion to perform metal removal operations because the vast majority of manufactured products require machining at some stage in their production, ranging from relatively rough or nonprecision work, such as cleanup of castings or forgings, to high-precision work involving tolerances of 0.0001 in. or less and high-quality finishes. Thus, machining undoubtedly is the most important of the basic manufacturing processes.

Beginning with the work of F. W. Taylor at Midvale Steel in the 1880s, the process has been the object of considerable research and experimentation that have led to improved understanding of the nature of both the process itself and the surfaces produced by it. While this research effort led to marked improvements in machining productivity, the complexity of the process has resulted in slow progress in obtaining a complete theory of chip formation. What makes this process so unique and difficult to analyze?

- Prior work-hardening of the material greatly affects the process.
- Different materials deform differently.
- The process is asymmetrical and unconstrained, bounded only by the cutting tool.
- The level of strain is very large.
- The strain rate is very high.
- The process is sensitive to variations in tool geometry, tool material, tool wear, temperature, environment (cutting fluids), and process dynamics (chatter and vibration).

The objective of this chapter is to put all this in perspective for the practicing engineer.

20.2 FUNDAMENTALS

The process of metal cutting is complex because it has such a wide variety of inputs (which are outlined in Figure 20-1):

- The machine tool selected to perform the process.
- The cutting tool selected (geometry and material).
- The properties and parameters of the workpiece.

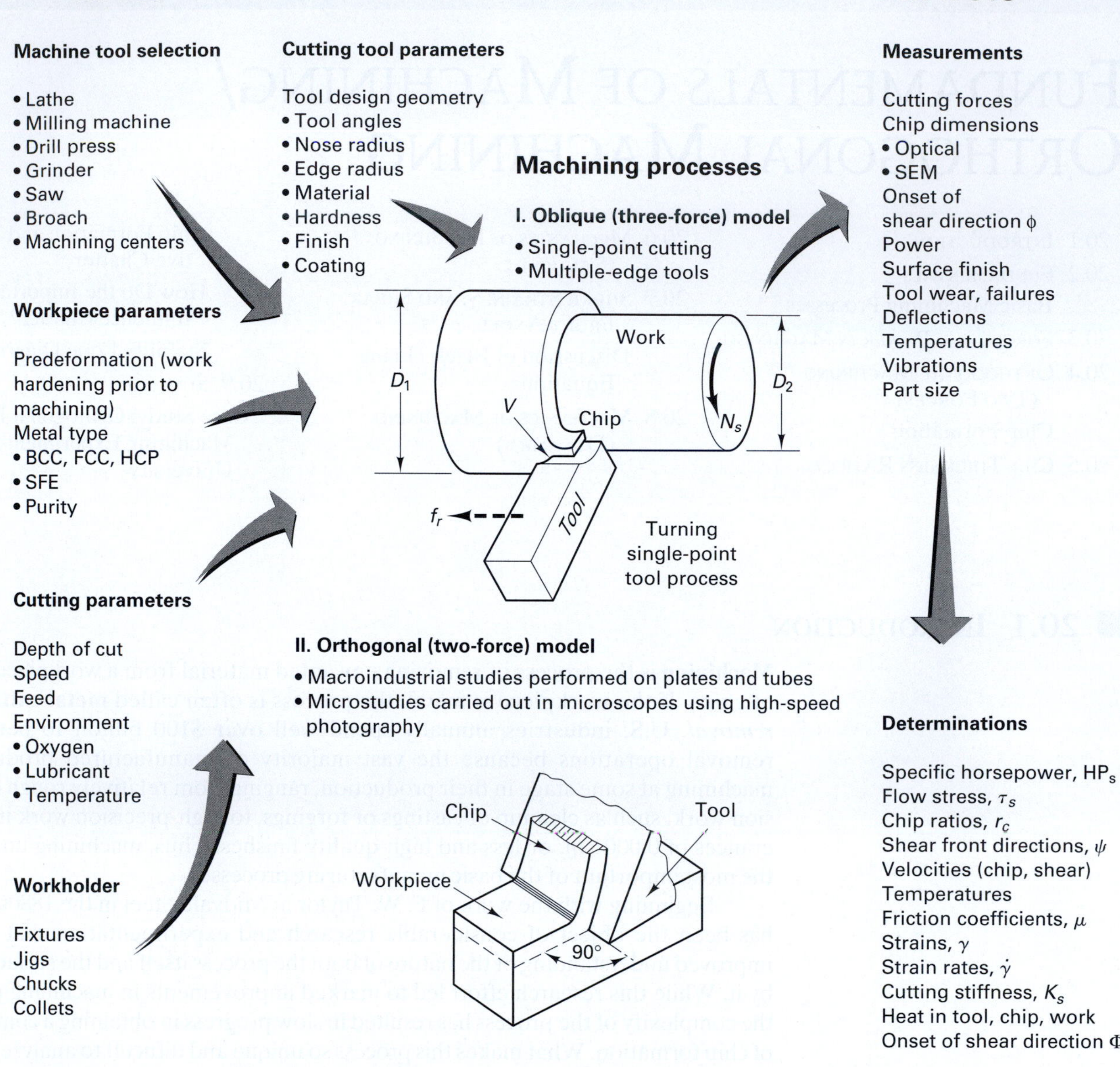

FIGURE 20-1 The fundamental inputs and outputs to machining processes.

- The cutting parameters selected (speed, feed, depth of cut).
- The workpiece holding devices or fixtures or jigs.

As we can see from Figure 20-1, the wide variety of inputs creates a host of outputs, most of which are critical to satisfactory performance of the component and product.

There are seven basic chip formation processes (see Figure 20-2): **turning, milling, drilling, sawing, broaching, shaping** *(planing)*, and **grinding** (also called *abrasive machining*), discussed in Chapters 22, 23, 24, and 26. Chapter 27 describes workholding devices that go into the machine tools and hold the work with respect to the cutting tools. Chapter 21 will provide additional insights into the selection of the cutting tools. Usually the workpiece material is determined by the design engineer to meet the functional requirements of the part in service. The manufacturing engineer will be called on to do the process planning. Therefore the manufacturing engineer (MfE) will have to select the cutting-tool materials and workholder parameters and then cutting parameters based on that work material decision. Let us begin with the assumption that the workpiece material has been selected as part of the design. To make the component, you decided to use a high-speed steel cutting tool for a turning operation (see Figure 20-3).

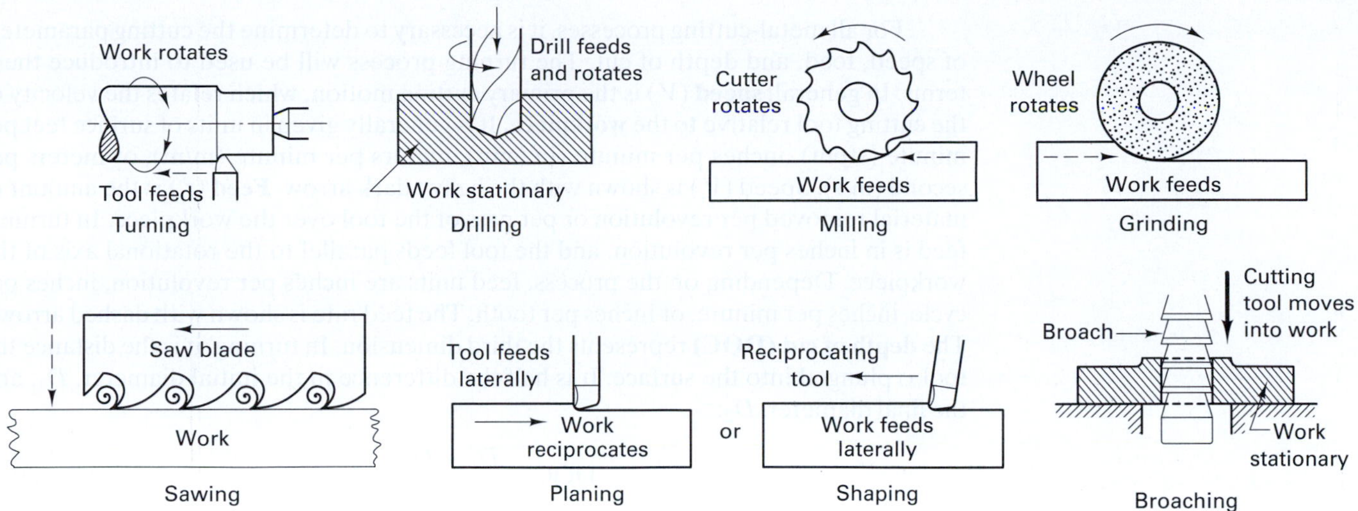

FIGURE 20-2 The basic machining processes used in chip formation are widely varied.

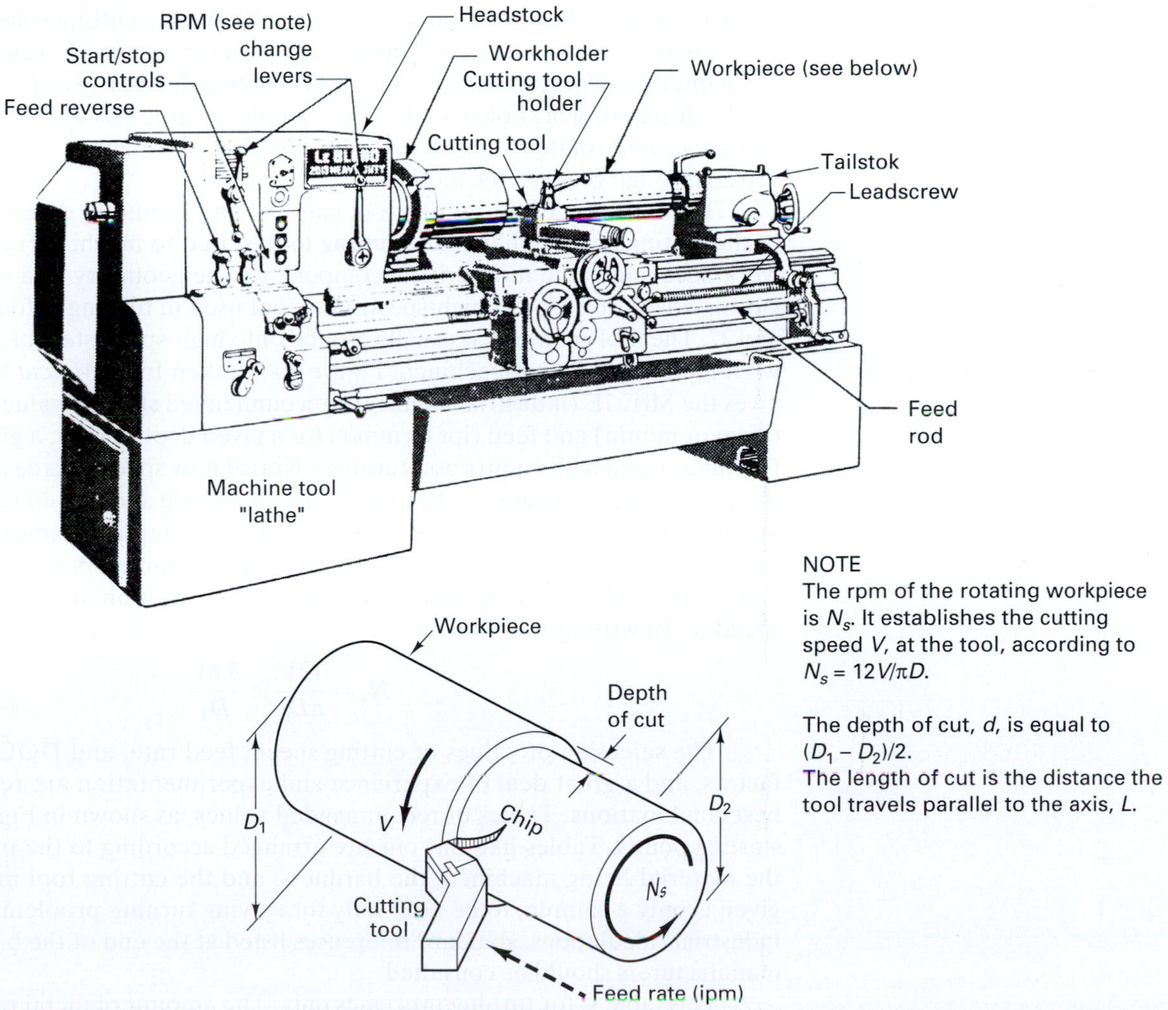

NOTE
The rpm of the rotating workpiece is N_s. It establishes the cutting speed V, at the tool, according to $N_s = 12V/\pi D$.

The depth of cut, d, is equal to $(D_1 - D_2)/2$.
The length of cut is the distance the tool travels parallel to the axis, L.

FIGURE 20-3 Turning a cylindrical workpiece on a lathe requires you to select the cutting speed, feed, and depth of cut.

For all metal-cutting processes, it is necessary to determine the cutting parameters of speed, feed, and depth of cut. The turning process will be used to introduce these terms. In general, **speed** (V) is the primary cutting motion, which relates the velocity of the cutting tool relative to the **workpiece.** It is generally given in units of surface feet per minute (sfpm), inches per minute (in./min), meters per minute (m/m), or meters per second (m/s). Speed (V) is shown with the heavy dark arrow. **Feed** (f_r) is the amount of material removed per revolution or per pass of the tool over the workpiece. In turning, feed is in inches per revolution, and the tool feeds parallel to the rotational axis of the workpiece. Depending on the process, feed units are inches per revolution, inches per cycle, inches per minute, or inches per tooth. The feed rate is shown with dashed arrows. The **depth of cut (DOC)** represents the third dimension. In turning, it is the distance the tool is plunged into the surface. It is half the difference in the initial diameter, D_1, and the final diameter, D_2:

$$\text{DOC} = \frac{D_1 - D_2}{2} = d \tag{20-1}$$

The selection of the cutting speed V determines the surface speed of the rotating part that is related to the outer diameter of the workpiece.

$$V = \frac{\pi D_1 N_s}{12} \tag{20-2}$$

where D_1 is in inches, V is speed in surface feet per minute, and N_s is the revolutions per minute (rpm) of the workpiece. The operator inputs the desired revolutions per minute (rpm) of the spindle into the lathe to produce the desired cutting velocity.

Figure 20-3 shows a typical **machine tool** for the turning process, a lathe, discussed in Chapter 22. Workpieces are held in **workholding devices**. (See Chapter 27 for details on the design of workholders.) In this example, a three-jaw chuck is used to hold the workpiece and rotate it against the tool. The chuck is attached to the spindle, which is driven through gears by the motor.

The selection of the cutting tool material and geometry determines the selection of the cutting speed. That is, the **cutting tool** is used to machine (i.e., make chips) the workpiece and is the most critical component. The geometry of a single point (single cutting edge) of a typical high-speed steel tool used in turning is found in Chapters 21 and 22. The tool geometry is usually ground onto high-speed steel blanks, depending on what material is being machined. Figure 20-4, taken from *Metcut's Data Handbook*, gives the MfE/IE (industrial engineer)–recommended starting values for cutting speed (sfpm or m/min) and feed (ipr or mm/r) for a given depth of cut, a given work material (hardness), and a given process (turning). Notice how speed decreases as DOC or feed increases. Cutting speeds can be increased with carbide and coated-carbide tool materials. Thus, we see that to process different metals, the input parameters to the machine tools must be determined. For the lathe, the input parameters are DOC, the feed rate, and the rpm value of the spindle. The rpm value depends on the selection of the cutting speed, V. Rewriting equation for N_s:

$$N_s = \frac{12V}{\pi D_1} \cong \frac{3.8V}{D_1} \tag{20-3}$$

The selection of values of cutting speed, feed rate, and DOC depend on many factors, and a great deal of experience and experimentation are required to find the best combinations. Tables of recommended values, as shown in Figure 20-4 are good starting points. Tables like this one are arranged according to the process being used, the material being machined, the hardness, and the cutting tool material. The table given is only a sample, to be used only for solving turning problems in the book. For industrial calculations, standard references listed at the end of the book or cutting tool manufacturers should be consulted.

This table is for turning processes only. The amount of metal removed per pass is called the **metal removal rate,** and it depends on V, f_r, and DOC. In practice, roughing cuts are heavier than finishing cuts in terms of DOC and feed and are run at a lower

Turning, Single Point and Box Tools

Material	Hard-ness	Condition	Depth of Cut*	High Speed Steel Tool: Speed	High Speed Steel Tool: Feed	High Speed Steel Tool: Tool Material	Carbide Tool, Uncoated: Speed Brazed	Carbide Tool, Uncoated: Speed Index-able	Carbide Tool, Uncoated: Feed	Carbide Tool, Uncoated: Tool Material Grade	Carbide Tool, Coated: Speed	Carbide Tool, Coated: Feed	Carbide Tool, Coated: Tool Material Grade
	Bhn		in / mm	fpm / m/min	ipr / mm/r	AISI / ISO	fpm / m/min	fpm / m/min	ipr / mm/r	C / ISO	fpm / m/min	ipr / mm/r	C / ISO
1. FREE MACHINING		Hot Rolled,	.040	160	.008	M2, M3	500	610	.007	C-7	925	.007	CC-7
CARBON STEELS,		Normalized,	.150	125	.015	M2, M3	390	480	.020	C-6	600	.015	CC-6
WROUGHT (cont.)		Annealed,	.300	100	.020	M2, M3	310	375	.030	C-6	500	.020	CC-6
Medium Carbon	225	Cold Drawn	.625	80	.030	M2, M3	.240	290	.040	C-6	—	—	—
Leaded	to	or	1	49	.20	S4, S5	150	185	.18	P10	280	.18	CP10
(cont.)	275	Quenched	4	38	.40	S4, S5	120	145	.50	P20	185	.40	CP20
(materials listed		and	8	30	.50	S4, S5	95	115	.75	P30	150	.50	CP30
on preceding page)		Tempered	16	24	.75	S4, S5	73	88	1.0	P40	—	—	—
		Hot	.040	135	.007	T15, M42†	460	545	.007	C-7	825	.007	CC-7
		Rolled,	.150	105	.015	T15, M42†	350	425	.020	C-6	525	.015	CC-6
		Normalized,	.300	85	.020	T15, M42†	275	380	.030	C-6	425	020	CC-6
	275	Annealed	.625	—	—	—	—	—	—	—	—	—	—
	to	or	1	41	.18	S9, S11†	140	165	.18	P10	250	.18	CP10
	325	Quenched	4	32	.40	S9, S11†	105	130	.50	P20	160	.40	CP20
		and	8	26	.50	S9, S11†	84	100	.75	P30	130	.50	CP30
		Tempered	16	—	—	—	—	—	—	—	—	—	—
			.040	100	.007	T15, M42†	390	480	.007	C-7	725	.007	CC-7
			.150	80	.015	T15, M42†	300	375	.020	C-6	475	.015	CC-6
	325	Quenched	.300	65	.020	T15, M42†	230	290	.030	C-6	375	.020	CC-6
	to	and	.625	—	—	—	—	—	—	—	—	—	—
	375	Tempered	1	30	.18	S9, S11†	120	145	.18	P10	220	.18	CP10
			4	24	.40	S9, S11†	90	115	.50	P20	145	.40	CP20
			8	20	.50	S9, S11†	70	88	.75	P30	115	.50	CP30
			16	—	—	—	—	—	—	—	—	—	—
			.040	70	.007	T15, M42†	325	400	.007	C-7	600	.007	CC-7
			.150	55	.015	T15, M42†	250	310	.020	C-6	400	.015	CC-6
	375	Quenched	.300	45	.020	T15, M42†	200	240	.030	C-6	325	.020	CC-6
	to	and	.625	—	—	—	—	—	—	—	—	—	—
	425	Tempered	1	21	.18	S9, S11†	100	120	.18	P10	185	.18	CP10
			4	17	.40	S9, S11†	76	95	.50	P20	120	.40	CP20
			8	14	.50	S9, S11†	60	73	.75	P30	100	.50	CP30
			16	—	—	—	—	—	—	—	—	—	—
2. CARBON STEELS,			.040	185	.007	M2, M3	535	700	.007	C-7	1050	.007	CC-7
WROUGHT		Hot	.150	145	.015	M2, M3	435	540	.020	C-6	700	.015	CC-6
Low Carbon		Rolled,	.300	115	.020	M2, M3	340	420	.030	C-6	550	.020	CC-6
	85	Normalized,	.625	90	.030	M2, M3	265	330	.040	C-6	—	—	—
1005 1010 1020	to	Annealed	1	56	.18	S4, S5	165	215	.18	P10	320	.18	CP10
1006 1012 1023	125	or Cold	4	44	.40	S4, S5	135	165	.50	P20	215	.40	CP20
1008 1015 1025		Drawn	8	35	.50	S4, S5	105	130	.75	P30	170	.50	CP30
1009 1017			16	27	.75	S4, S5	81	100	1.0	P40	—	—	—
			.040	150	.007	M2, M3	485	640	.007	C-7	950	.007	CC-7
		Hot	.150	125	.015	M2, M3	410	500	.020	C-6	625	.015	CC-6
		Rolled,	.300	100	.020	M2, M3	320	390	.030	C-6	500	.020	CC-6
	125	Normalized,	.625	80	.030	M2, M3	245	305	.040	C-6	—	—	—
	to	Annealed	1	46	.18	S4, S5	150	195	.18	P10	290	.18	CP10
	175	or Cold	4	38	.40	S4, S5	125	150	.50	P20	190	.40	CP20
		Drawn	8	30	.50	S4, S5	100	120	.75	P30	150	.50	CP30
			16	24	.75	S4, S5	75	95	1.0	P40	—	—	—
			.040	145	.007	M2, M3	460	570	.007	C-7	850	.007	CC-7
		Hot	.150	115	.015	M2, M3	385	450	.020	C-6	550	.015	CC-6
		Rolled,	.300	95	.020	M2, M3	300	350	.030	C-6	450	.020	CC-6
	175	Normalized,	.625	75	.030	M2, M3	235	265	.040	C-6	—	—	—
	to	Annealed	1	44	.18	S4, S5	140	175	.18	P10	260	.18	CP10
	225	or Cold	4	35	.40	S4, S5	115	135	.50	P20	170	.40	CP20
		Drawn	8	29	.50	S4, S5	90	105	.75	P30	135	.50	CP30
			16	23	.75	S4, S5	72	81	1.0	P40	—	—	—
			.040	125	.007	M2, M3	410	510	.007	C-7	750	.007	CC-7
			.150	95	.015	M2, M3	360	400	.020	C-6	500	.015	CC-6
		Annealed	.300	75	.020	M2, M3	285	315	.030	C-6	400	.020	CC-6
	225	or	.625	60	.030	M2, M3	220	240	.040	C-6	—	—	—
	to	Cold	1	38	.18	S4, S5	125	155	.18	P10	230	.18	CP10
	275	Drawn	4	29	.40	S4, S5	110	120	.50	P20	150	.40	CP20
			8	23	.50	S4, S5	87	95	.75	P30	120	.50	CP30
			16	18	.75	S4, S5	67	73	1.0	P40	—	—	—

See section 15.1 for Tool Geometry.
*Caution: Check Horsepower requirements on heavier depths of cut.

See section 16 for Cutting Fluid Recommendations.
† Any premium HSS (T15, M33, M41–M47) or (S9, S10, S11, S12).

FIGURE 20-4 Examples of a table for selection of speed and feed for turning. *(Source: Metcut's Machinability Data Handbook)*

surface speed. Note that this table provides recommendations of V and f_r in both English and metric units based on the DOC needed to perform the job. Table values are usually conservative and should be considered starting points for determining the operational parameters for a process.

Once cutting speed (V) has been selected, equation 20-3 is used to determine the spindle rpm, N_s. The speed and feed can be used with the DOC to estimate the metal removal rate for the process, or MRR. For turning, the MRR is

$$\text{MRR} \cong 12V f_r d \tag{20-4}$$

This is an approximate equation for MRR. For turning, MRR values can range from 0.1 to 600 in.^{3}/min. The MRR can be used to estimate the horsepower needed to perform a cut, as will be shown later. For most processes, the MRR equation can be viewed as the volume of metal removed divided by the time needed to remove it:

$$\text{MRR} = \frac{\text{Volume of cut}}{T_m}$$

where T_m is the cutting time in minutes. For turning, the cutting time depends on the length of cut L divided by the rate of traverse of the cutting tool past the rotating workpiece $f_r N_s$, as shown in Figure 20-5. Therefore,

$$T_m = \frac{L + \text{Allowance}}{f_r N_s} \tag{20-5}$$

An allowance is usually added to the L term to allow for the tool to enter and exit the cut.

BASIC MACHINING PROCESSES

Turning is an example of a single-point tool process, as is shaping. Milling and drilling are examples of multiple-point tool processes. Figures 20-5 through 20-9 show schematics of the basic processes. Speed (V) is shown in these figures with a dark heavy

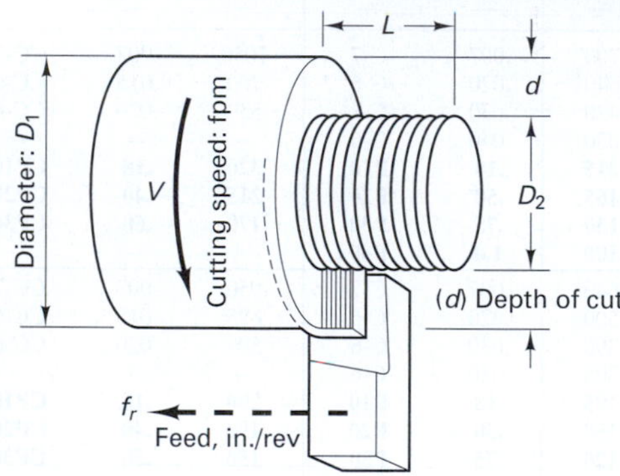

Turning

Speed, stated in surface feet per minute (sfpm), is the peripheral speed at the cutting edge. Feed per revolution in turning is a linear motion of the tool parallel to the rotating axis of the workpiece. The depth of cut reflects the third dimension.

L = length of cut

$T_m = \frac{L + A}{f_r N_s}$

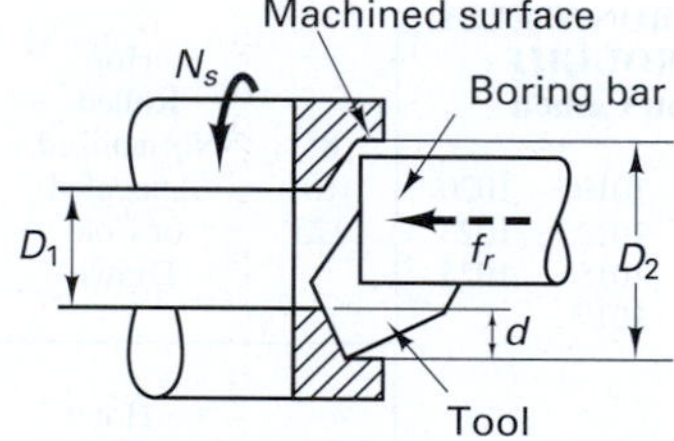

Boring

Enlarging hole of diameter D_1 to diameter D_2. Boring can be done with multiple cutting tools. Feed in inches per revolution, f_r.

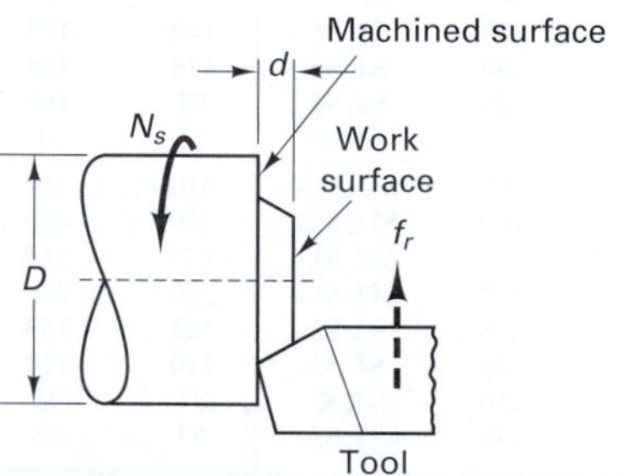

Facing

Tool feeds to center of workpiece so $L = D/2$. The cutting speed is decreasing as the tool approaches the center of the workpiece.

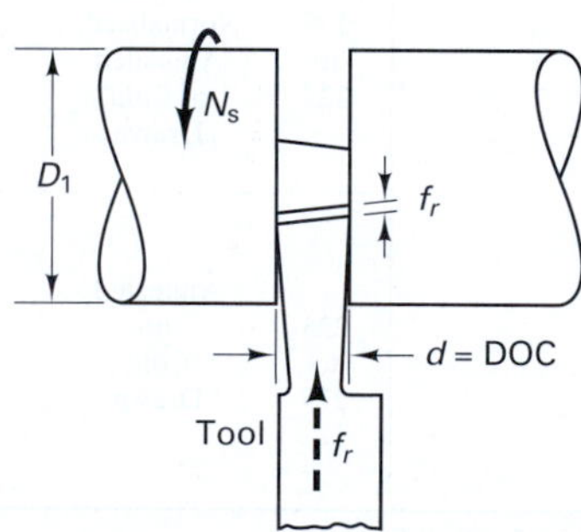

Grooving, parting, or cutoff

Tool feed perpendicular to the axis of rotation. The width of the tool produces the depth of cut (DOC).

FIGURE 20-5 Relationship of speed, feed, and depth of cut in turning, boring, facing, and cutoff operations typically done on a lathe.

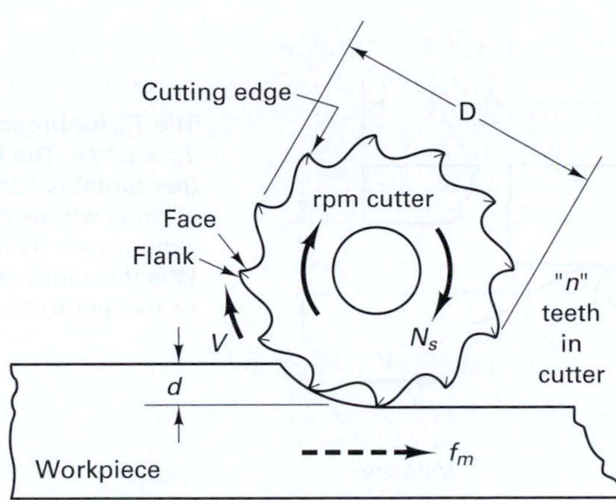

Slab milling – multiple tooth

Slab milling is usually performed on a horizontal milling machine. Equations for T_m and MRR derived in Chapter 25.

The tool rotates at rpm N_s. The workpiece translates past the cutter at feed rate f_m, the table feed. The length of cut, L, is the length of workpiece plus allowance, L_A,

$$L_A = \sqrt{\frac{D^2}{4} - \left(\frac{D}{2} - d\right)^2} = \sqrt{d(D-d)} \text{ inches}$$

$$T_m = (L + L_A)/f_m$$

The MRR = Wdf_m where W = width of the cut and d = depth of cut.

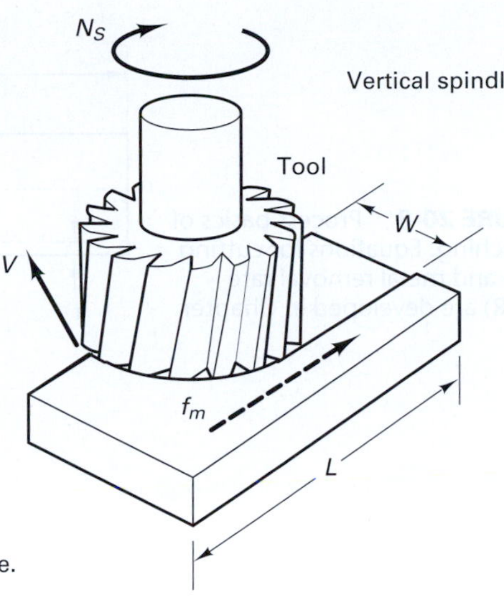

Face milling
Multiple-tooth cutting

Given a selected cutting speed V and a feed per tooth f_t, the rpm of the cutter is $N_s = 12V/\pi D$ for a cutting of diameter D. The table feed rate is $f_m = f_t nN_s$ for a cutter with n teeth.

The cutting time, $T_m = (L + L_A + L_o)/f_m$

where $L_o = L_A = \sqrt{W(D-W)}$ for $W < D/2$

or $L_o = L_A = D/2$ for $W \geqslant D/2$.

The MRR = Wdf_m where d = depth of cut.

FIGURE 20-6 Basics of milling processes (slab, face, and end milling), including equations for cutting time and metal removal rate (MRR).

arrow. Feed (f) is the amount of material removed per pass of the tool over the workpiece and is shown as a dashed arrow.

Table 20-1 provides a summary of the equations for T_m and MRR for these basic processes. These equations are commonly referred to as **shop equations** and are as fundamental as the processes themselves, so the student should be as familiar with them as with the basic processes. If one keeps track of the units and visualizes the process, the equations are, for the most part, straightforward.

In addition to turning, other operations can be performed on the lathe. For example, as shown in Figure 20-5, a flat surface on the rotating part can be produced by facing

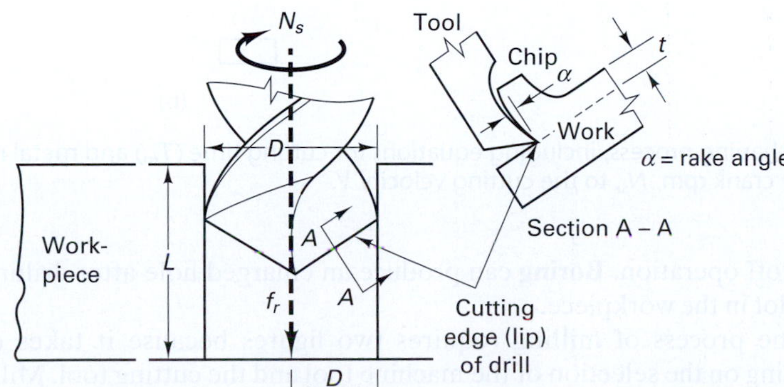

Drilling multiple-edge tool

Select cutting speed V, fpm and feed, f_r, in./rev. Select drill.
D = diameter of the drill which rotates 2 cutting edges at rpm N_s. V = velocity of outer edge of the lip of the drill.
$N_s = 12V/\pi D$. T_m = cutting time = $(L + A)/f_rN_s$ where f_r is the feed rate in in. per rev. The allowance $A = D/2$.
The MRR = $(\pi D^2/4)f_rN_s$ in.^{3}/min which is approximately $3DVf_r$.

FIGURE 20-7 Basics of the drilling (hole-making) processes, including equations for cutting time and metal removal rate (MRR).

L_W
Work
t
d
t
Tool
V
The T_m for broaching is $T_m = L/12V$. The MRR (per tooth) is $12tWV$ in.^{3}/min where V = cutting velocity in fpm, W is the width of cut, t = rise per tooth.

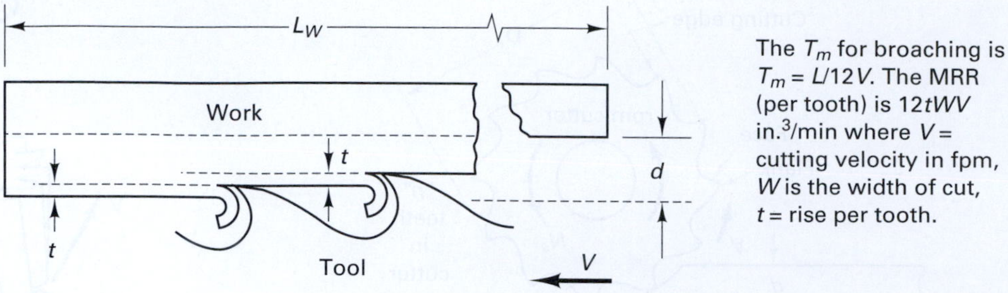

FIGURE 20-8 Process basics of broaching. Equations for cutting time and metal removal rate (MRR) are developed in Chapter 26.

Feed, in./stroke
Depth of cut (d)
V
L
f_c Feed
W
(a)
The tool cuts at velocity V with a return velocity of V_R dictated by the rpm of the crank, N_s. The cutting speed $V = (l + A)N_s/12R_s$ where R_s = stroke ratio = 200°/360° and the length of stroke is $l = L$ + ALLOW. The tool feed is f_c inches per stroke.
$T_m = W/N_s f_c$
MRR = $LdN_s f_c$ in^3/min
Velocity
V
Velocity diagram
Cut
0
Displacement
V_R
Ram
V
Advance
200°
N_s rpm of crank
Return
160°
Shaper (quick return) mechanism for driving tool past work.
(b)

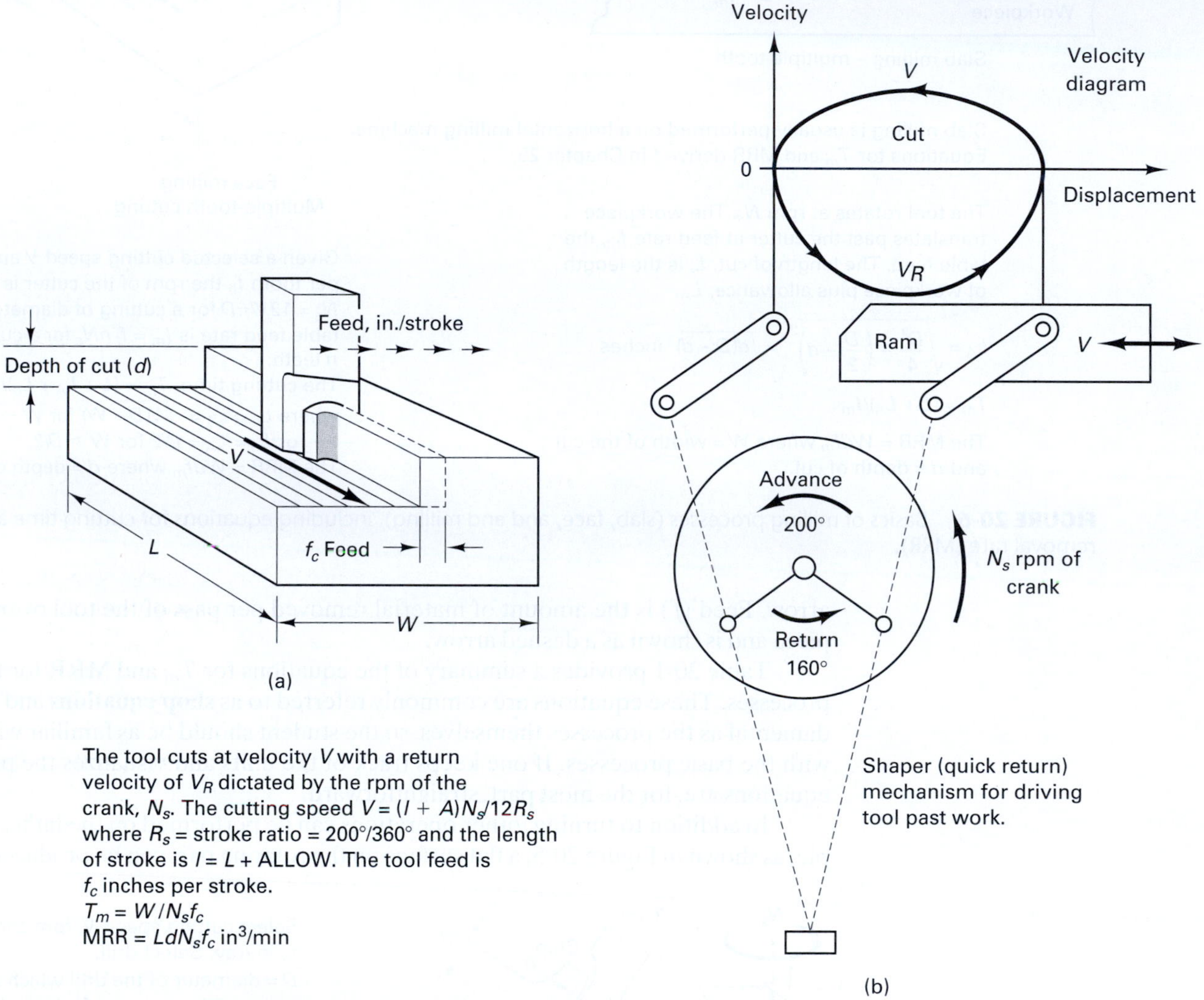

FIGURE 20-9 (a) Basics of the shaping process, including equations for cutting time (T_m) and metal removal rate (MRR). (b) The relationship of the crank rpm, N_s, to the cutting velocity V.

or a cutoff operation. **Boring** can produce an enlarged hole after drilling and grooving puts a slot in the workpiece.

The process of milling requires two figures because it takes different forms depending on the selection of the machine tool and the cutting tool. Milling, a multiple-tooth process, has two feeds: the amount of metal an individual tooth removes, called the feed per tooth, f_t, and the rate at which the table translates past the rotating tool, called the table feed rate, f_m, in inches per minute. It is calculated from

$$f_m = f_t n N_s \tag{20-6}$$

TABLE 20-1 Shop Formulas for Turning, Milling, Drilling, and Broaching (English Units)

Parameter	Turning	Milling	Drilling	Broaching
Cutting speed, fpm	$V = 0.262 \times D_1 \times \text{rpm}$	$V = 0.262 \times D_m \times \text{rpm}$	$V = 0.262 \times D_d \times \text{rpm}$	V
Revolutions per minute, N_s	$\text{rpm} = 3.82 \times V_c/D_1$	$\text{rpm} = 3.82 \times V_c/D_m$	$\text{rpm} = 3.82 \times V_c/D_d$	—
Feed rate, in./min	$f_m = f_r \times \text{rpm}$	$f_m = f_r \times \text{rpm}$	$f_m = f_r \times \text{rpm}$	—
Feed per rev tooth pass, in./rev	f_r	f_t	f_r	—
Cutting time, min, T_m	$T_m = L/f_m$	$T_m = L/f_m$	$T_m = L/f_m$	$T_m = L/12V$
Rate of metal removal, in.^{3}/min	$\text{MRR} = 12 \times d \times f_r \times V_c$	$\text{MRR} = w \times d \times f_m$	$\text{MRR} = \pi D^2 d/4 \times f_m$	$\text{MRR} = 12 \times w \times d \times V$
Horsepower required at spindle	$\text{hp} = \text{MRR} \times \text{HP}_s$	$\text{hp} = \text{MRR} \times \text{HP}_s$	$\text{hp} = \text{MRR} \times \text{HP}_s$	—
Horsepower required at motor	$\text{hp}_m = \text{MRR} \times \text{HP}_s/E$	$\text{hp}_m = \text{MRR} \times \text{HP}_s/E$	$\text{hp}_m = \text{MRR} \times \text{HP}_s/E$	$\text{hp}_m = \text{MRR} \times \text{HP}_s/E$
Torque at spindle	$t_s = 63{,}030\ \text{hp/rpm}$	$t_s = 63{,}030\ \text{hp/rpm}$	$t_s = 63{,}030\ \text{hp/rpm}$	—
Symbols	D_I = Diameter of workpiece in turning, inches D_m = Diameter of milling cutter, inches D_d = Diameter of drill, inches d = Depth of cut, inches E = Efficiency of spindle drive f_m = Feed rate, inches per minute f_r = Feed, inches per revolution f_t = Feed, inches per tooth hp_m = Horsepower at motor MRR = Metal removal rate, in.^{3}/min		hp = horsepower at spindle L = Length of cut, inches n = Number of teeth in cutter HP_s = Unit power, horsepower per cubic inch per minute, specific horsepower N_s = Revolution per minute of work or cutter t_s = Torque at spindle, inch-pound T_m = Cutting time, minutes V = Cutting speed, feet per minute w = Width of cut, inches	

Values for specific horsepower (unit power) are given in Table 20-3.

where n is the number of teeth in a cutter and N_s is the rpm value of the cutter. Just as was shown for turning, standard tables of speeds and feeds for milling provide values for the recommended cutting speeds and feeds per tooth, f_t.

Figure 20-10 provides an overview of the basic machining processes in terms of typical machine tools that can generate cylindrical surfaces. Figure 20-11 provides an overview of the basic processes that can generate flat surfaces. Table 20-2 provides a summary on typical sizes (minimum–maximum), the production rates (part/hour), tolerances (precision or repeatability), and surface finish (roughness). Milling has pretty much replaced shaping and planing, although gear shaping is still a viable process. Milling combined with other rotational multiple-edge tool processes (drilling or reaming) is often performed in machining centers rather than on milling machines. The turret lathe has been replaced by CNC turning centers with multiple turrets in many factories, discussed in Chapters 22 and 25.

20.3 FORCES AND POWER IN MACHINING

Most of the cutting operations process described to this point are examples of oblique, or three-force, cutting. The cutting force system in a conventional, oblique-chip formation process is shown schematically in Figure 20-12. Oblique cutting has three components:

1. F_c: Primary cutting force acting in the direction of the cutting velocity vector. This force is generally the largest force and accounts for 99% of the power required by the process.
2. F_f: Feed force acting in the direction of the tool feed. This force is usually about 50% of F_c but accounts for only a small percentage of the power required because feed rates are usually small compared to cutting speeds.
3. F_r: radial or thrust force acting perpendicular to the machined surface. This force is typically about 50% of F_r and contributes very little to power requirements because velocity in the radial direction is negligible. Figure 20-12 shows the general

Operation	Block diagram	Most commonly used machines	Machines less frequently used	Machines seldom used
Turning	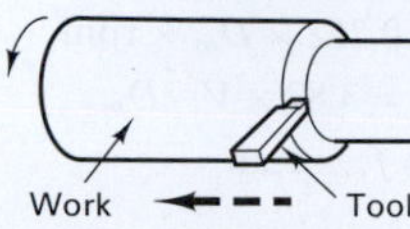	Lathe NC lathe Machining center	Boring mill	Turret lathe
Grinding	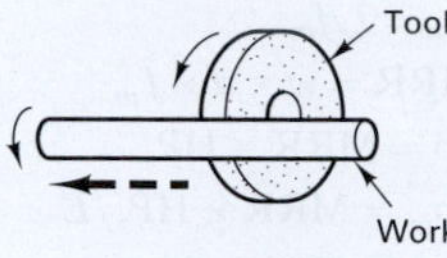	Cylindrical grinder		Lathe (with special attachment)
Sawing (of plates and sheets)	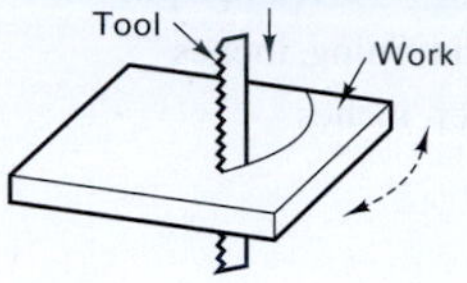	Contour or band saw	Laser Flame cutting Plasma arc Water jet	
Drilling	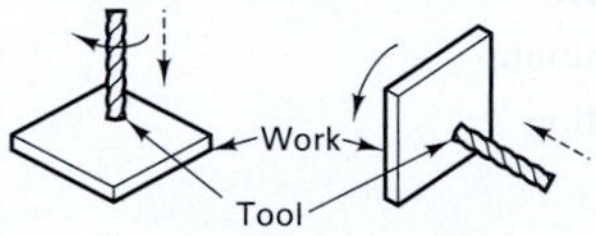	Drill press Machining center (nc) Vert. milling machine	Lathe Horizontal boring machine	Horizontal milling machine Boring mill
Boring	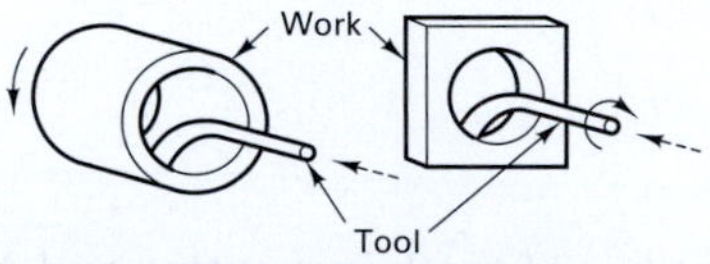	Lathe Boring mill Horizontal boring machine Machining center		Milling machine Drill press
Reaming	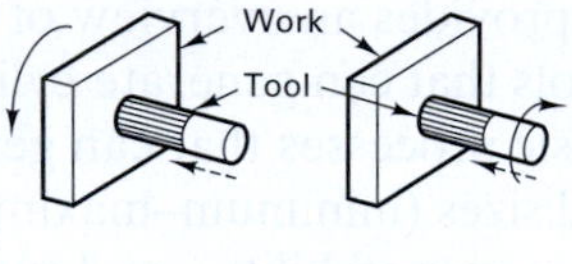	Lathe Drill press Boring mill Horizontal boring machine Machining center	Milling machine	
Grinding	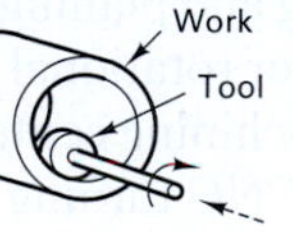	Cylindrical grinder		Lathe (with special attachment)
Sawing	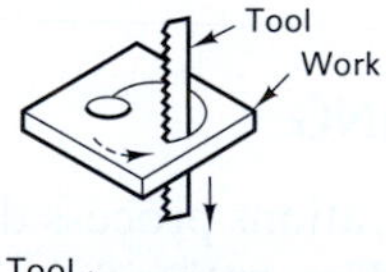	Contour or band saw		
Broaching	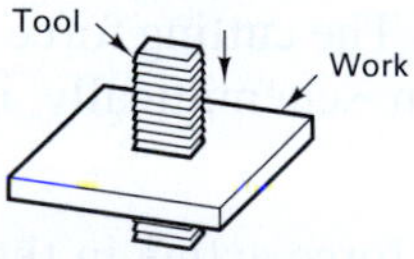	Broaching machine	Arbor press (keyway broaching)	

FIGURE 20-10 Operations and machines used for machining cylindrical surfaces.

relationship between these forces and changes in speed, feed, and depth of cut. Note that these figures cannot be used to determine forces for a specific process.

The power required for cutting is

$$P = F_c V (\text{ft-lb/min}) \tag{20-7}$$

Operation	Block diagram	Most commonly used machines	Machines less frequently used	Machines seldom used
Facing	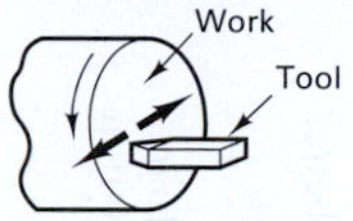 Work, Tool	Lathe	Boring mill	
Broaching	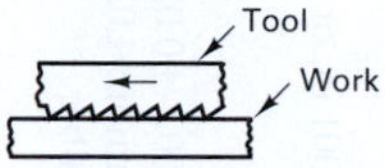 Tool, Work	Broaching machine		Turret broach
Grinding	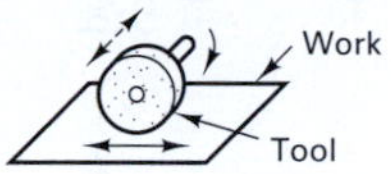 Work, Tool	Surface grinder		Lathe (with special attachment)
Sawing	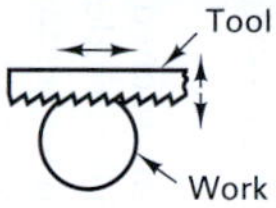 Tool, Work	Band saw Cutoff saw Hack saw	Contour saw	
Shaping	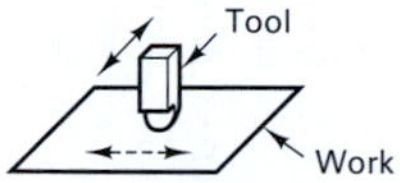 Tool, Work	Horizontal shaper	Vertical shaper	
Planing	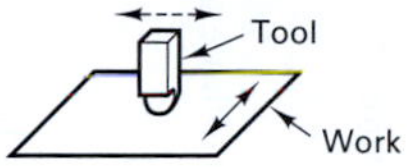 Tool, Work	Planer		
Milling	slab milling: 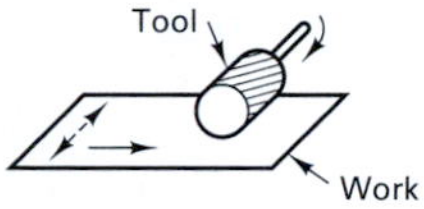 Tool, Work	Milling machine	Lathe with special milling tools	
	face milling: 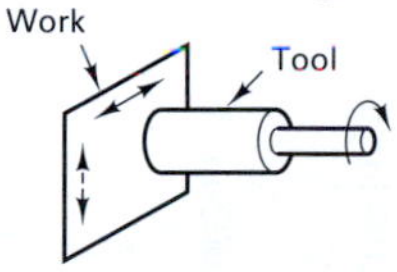 Work, Tool	Milling machine Machining center	Lathe with special milling tools	Drill press (light cuts)

FIGURE 20-11 Operations and machines used to generate flat surfaces.

The horsepower at the spindle of the machine is therefore

$$\text{hp} = \frac{F_c V}{33,000} \tag{20-8}$$

In metal cutting, a very useful parameter is called the unit, or specific, horsepower, HP_s, which is defined as

$$\text{HP}_s = \frac{\text{hp}}{\text{MRR}} \,(\text{hp/in}^3/\text{min}) \tag{20-9}$$

In turning, for example, where $\text{MRR} \cong 12V f_r d$, then

$$\text{HP}_s = \frac{F_c}{396,000 f_r d} \tag{20-10}$$

Thus, this term represents the approximate power needed at the spindle to remove a cubic inch of metal per minute.

Values for **specific horsepower, $\mathbf{HP_s}$,** which is also called unit power, are given in Table 20-3. These values are obtained through orthogonal metal-cutting experiments described later in this chapter.

Specific power can be used in a number of ways. First, it can be used to estimate the motor horsepower required to perform a machining operation for a given material.

TABLE 20-2 Basic Machining Process

Applicable Process	Raw Material Form	Size: Maximum	Size: Minimum	Typical Production Rate	Material Choice	Typical Tolerance	Typical Surface Roughness
Turning (engine lathes)	Cylinders preforms, castings forgings	78 in. dia. × 73 in. long	$\frac{1}{64}$ in. typical	1–10 parts/hour	All ferrous and nonferrous material considered machinable	±0.002 in. on dia. common; ±0.001 in. obtainable	125–250
Turning (CNC)	Bar, rod, tube preforms	36 in. dia. × 93 in. long	$\frac{1}{64}$ in. dia.	1–2 parts/minute to 1–4 parts/hour	Any material with good machinability rating	±0.001 in. on dia. where needed; ±0.0005 in. possible	63 or better
Turning (automatic screw machine)	Bar, rod	Generally 2 in. dia. × 6 in. long	$\frac{1}{16}$ in. dia. and less, weight less than 1 ounce	10–30 parts/minute	Any material with good machinability rating ±0.001 to ±0.003 in.	±0.0005 in. possible ±0.001 to ±0.003 in. common	63 average
Turning (Swiss automatic machining)	Rod	Collets adapt to $\frac{1}{2}$ in. dia.	Collets adapt to less than $\frac{1}{2}$ in.	12–30 parts/minute	Any material with good machinability rating	±0.0002 in. to ±0.001 in. common	63 and better
Boring (vertical)	Casting, preforms	98 in. × 72 in.	2 in. × 12 in.	2–20 hours/piece	All ferrous and nonferrous	±0.0005 in.	90–250
Milling	Bar, plate, rod, tube	4–6 ft long	Limited usually by ability to hold part	1–100 parts/hour	Any material with good machinability rating	±0.0005 in. possible; ±0.001 in. common	63–250
Hobbing (milling gears)	Blanks, preforms, rods	10-ft-dia. gears 14-in. face width	0.100 in. dia.	1 part/minute	Any material with good machinability rating	±0.001 in. or better	63
Drilling	Plate, bar, preforms	$3\frac{1}{2}$-in.-dia.drills (1-in.-dia. normal)	0.002-in. drill dia.	2–20 second/hole after setup	Any unhardened material; carbides needed for some case-hardened parts	±0.002–±0.010 in. common; ±0.001 in. possible	63–250
Sawing	Bar, plate, sheet	2-in. armor plate $\frac{1}{2}$ in. is preferred)	0.010 in. thick	3–30 parts/hour	Any nonhardened material	±0.015 in. possible	250–1000
Broaching	Tube, rod, bar, plate	74 in. long	1 in.	300–400 parts/minute	Any material with good machinability rating	±0.0005–±0.001 in.	32–125
Grinding	Plate, rod, bars	36 in. wide × 7 in. dia.	0.020 in. dia.	1–1000 pieces/hour	Nearly all metallic materials plus many nonmetallic	0.0001 in. and less	16
Shaping	Bar, plate, casting	3 ft × 6 ft	Limited usually by ability to hold part	1–4 parts/hour	Low-to medium-carbon steels and nonferrous metals best; no hardened parts	±0.001–±0.002 in. (larger parts) ±0.0001–±0.0005 in. (small–medium parts)	63–250
Planing	Bar, plate, casting	42 ft wide × 18 ft high × 76 ft long	Parts too large for shaper work	1 part/hour	Low- to medium-carbon steels or nonferrous materials best	±0.001–±0.005 in.	63–125
Gear shaping	Blanks	120-in.-dia. gears 6-in. face widh	1 in. dia.	1–60 parts/hour	Any material with good machinability rating	±0.001 in. or better at 200 D.P. to 0.0065 in. at 30 D.P.	63

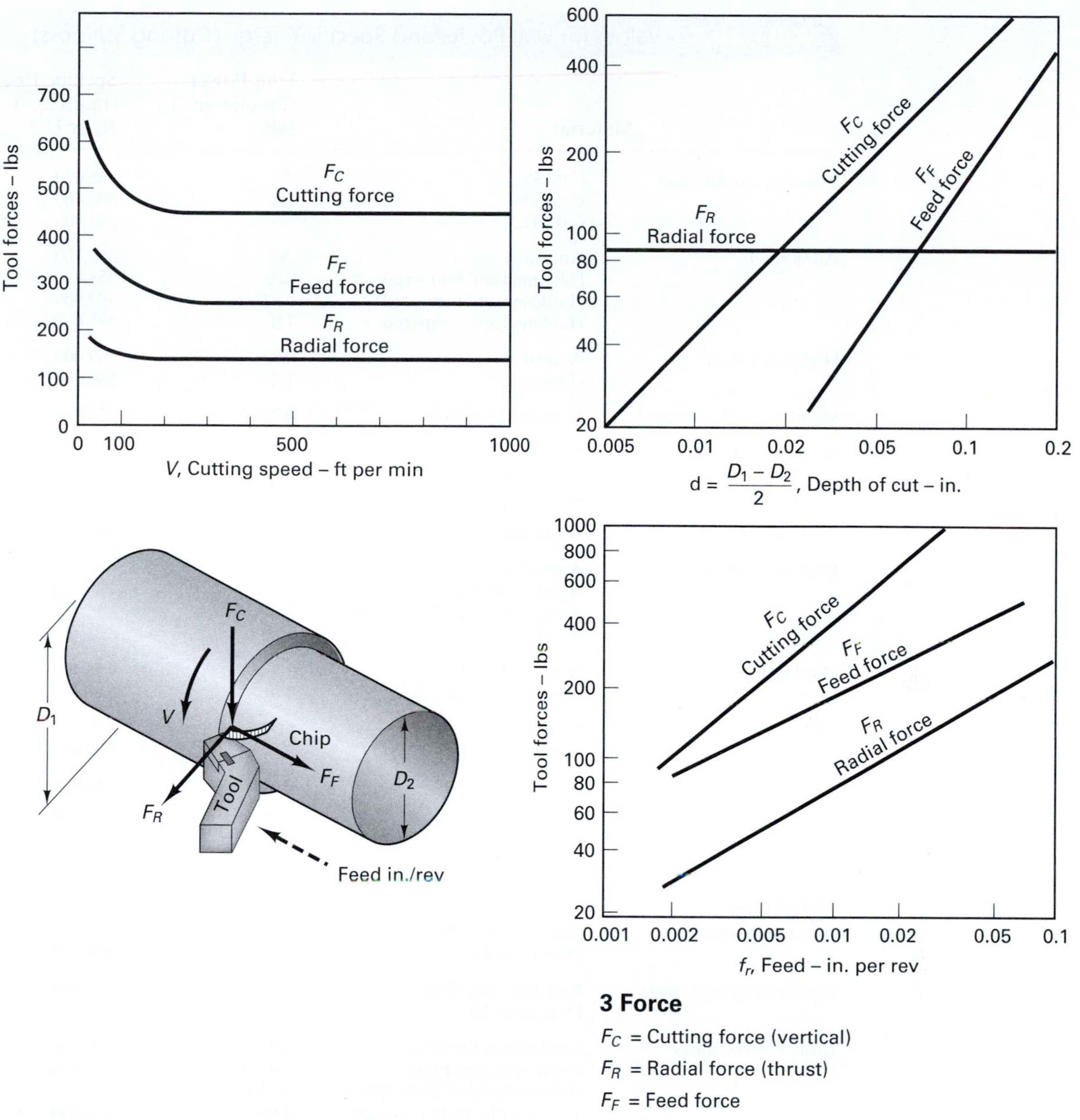

FIGURE 20-12 Oblique (three-force) machining has three measurable components of forces acting on the tool. The forces vary with speed, depth of cut, and feed.

HP_s values from the table are multiplied by the approximate MRR for the process. The motor horsepower HP_m, is then

$$HP_m = \frac{HP_s \times MRR \times CF}{E} \tag{20-11}$$

where E is the efficiency of the machine. The E factor accounts for the power needed to overcome friction and inertia in the machine and drive moving parts. Usually, 80% is used. Usually the maximum MRR is used in this calculation. Correction factors (CFs) may also be used to account for variations in cutting speed, feed, and rake angle. There is usually a tool wear correction factor of 1.25, used to account for the fact that dull tools use more power than sharp tools.

The primary cutting force F_c can be roughly estimated according to

$$F_c \cong \frac{HP_s \times MRR \times 33{,}000}{V} \tag{20-12}$$

TABLE 20-3 Values for Unit Power and Specific Energy (Cutting Stiffness)

Material		Unit Power (hp-min. in.^{3}) HP_s	Specific Energy (in.-lb/in.^{3}) K_s or U	Hardness Brinell HB
Nonalloy carbon steel	C 0.15%	.58	268,000	125
	C 0.35%	.58	302,400	150
	C.0.60%	.75	324,800	200
Alloy steel	Annealed	.50	302,400	180
	Hardened and tempered	0.83	358,400	275
	Hardened and tempered	0.87	392,000	300
	Hardened and tempered	1.0	425,000	350
High-alloy steel	Annealed	0.83	369,000	200
	Hardened	1.2	560,000	325
Stainless steel, annealed	Martensitic/ferritic	0.75	324,800	200
Steel castings	Nonalloy	0.62	257,000	180
	Low-alloy	0.67	302,000	200
	High-alloy	0.80	336,000	225
Stainless steel, annealed	Austenitic	0.73	369,600	180
Heat-resistant alloys	Annealed	0.78		200
	Aged—Iron based	—		280
	Annealed—Nickel or cobalt	1.10		250
	Aged	1.20		350
Hard steel	Hardened steel	1.4	638,400	$55HR_c$
	Manganese steel 12%	1.0	515,200	250
Malleable iron	Ferritic	0.42	156,800	130
	Pearlitic	—	257,600	230
Cast iron, low tensile		0.62	156,800	180
Cast iron, high tensile		0.80	212,800	260
Nodular SG iron	Ferritic	0.55	156,800	160
	Pearlitic	0.76	257,600	250
Chilled cast iron		—	492,800	400
Aluminum alloys	Non-heat-treatable	.25	67,200	60
	Heat-treatable	.33	100,800	100
Aluminum alloys (cast)	Non-heat-treatable	.25	112,000	75
	Heat-treatable	.33	123,200	90
Bronze-brass alloys	Lead alloys, Pb>1%	.25	100,800	110
	Brass, cartridge brass	1.8–2.0	112,000	90
	Bronze and lead-free copper	0.33–0.83		
	Includes Electrolytic copper	0.90	246,400	100
Zinc alloy	Diecast	0.25	—	—
Titanium		.034	250,275	

Values assume normal feed ranges and sharp tools. Multiply values by 1.25 for a dull tool.
Calculation of unit power (HP_s)
$HP = F_c V/33000$
$HP_s = HP/MRR$ Where
$MRR = 12Vtw$ for tube turning
$HP_s = F_c V/12Vtw \times 33000 = F_c/tw \times 396000$
Calculation of specific energy (U)
$U = F_c V/Vtw = F_c/tw$ for tube turning

This type of estimate of the major force F_c is useful in analysis of deflection and vibration problems in machining and in the proper design of workholding devices, because these devices must be able to resist movement and deflection of the part during the process.

In general, increasing the speed, the feed, or the depth of cut will increase the power requirement. Doubling the speed doubles the horsepower directly. Doubling the feed or the depth of cut doubles the cutting force, F_c. In general, increasing the speed does not increase the cutting force, F_c, a surprising experimental result. However, speed has a strong effect on tool life because most of the input energy is converted into heat, which raises the temperature of the chip, the work, and the tool—to the latter's detriment. Tool life (or tool death) is discussed in Chapter 22.

Equation 20-12 can be used to estimate the maximum depth of cut, d, for a process as limited by the available power.

$$d_{\max} = \frac{\mathrm{HP_m} \times E}{12\mathrm{HP_s}\, VF_r(CF)} \tag{20-13}$$

Another handbook value useful in chatter or vibration calculations is **cutting stiffness, K_s.** In this text, the term *specific energy, U,* will be used interchangeably with cutting stiffness, K_s.

It is interesting to compute the total specific energy in the process and determine how it is distributed between the primary shear and the secondary shear that occurs at the interface between the chip and the tool. It is safe to assume that the majority of the input energy is consumed by these two regions.

Therefore,

$$U = U_s + U_f \tag{20-14}$$

where specific energy (also called cutting stiffness) is

$$U = \frac{F_c V}{V f_r d} = \frac{F_c}{f_r d} = K_s(\text{turning}) \tag{20-15}$$

Usually, 30 to 40% of the total energy goes into friction and 60 to 70% into the shear process.

Typical values for U are given in Table 20-3. This is experimental data developed by the orthogonal machining experiment described in the next section.

In the metal-cutting process, the energy put into the process ($F_c V$) is largely converted into heat, elevating the temperatures of the chips, the tool, and the workpiece—all of which, along with the environment (the cutting fluids) act as heat sinks. In the next chapter, the effect of cutting speed on temperature will be discussed as a prime cause of tool wear and tool failure. The data that appear in tables like Table 20-1 for recommended speeds are usually based on the expected tool life of 30 to 60 min and are developed experimentally.

20.4 Orthogonal Machining (Two Forces)

Orthogonal machining (OM) is carried out mostly in research laboratories in order to better understand this complex process. In OM, the tool geometry is simplified from the three-dimensional (oblique) geometry, to a two-force geometry, as shown in Figure 20-1.

Using this simplified tool geometry, metals can be cut to test machining mechanics and theory and develop machining values for specific power and energy given in Table 20-2 and 20-3. There are basically three orthogonal machining setups (as shown in Figure 20-13):

1. Orthogonal plate machining (OPT), or machining a plate in a milling machine—low-speed cutting.
2. Orthogonal tube turning (OTT), or end-cutting a tube wall in a turning setup—medium-speed ranges.
3. Orthogonal disk machining (ODM), or end-cutting a plate with tool feeding in a facing direction—high-speed cutting.

In **oblique machining,** as in shaping, drilling, and single-point turning, the cutting edge and the cutting motion are not perpendicular to each other. In the orthogonal case, the cutting velocity vector and the cutting edge are perpendicular. The OPM low-speed plate machining is shown in more detail in Figure 20-14, using a modified horizontal milling machine where the table traverse provides the cutting speed. The tool is

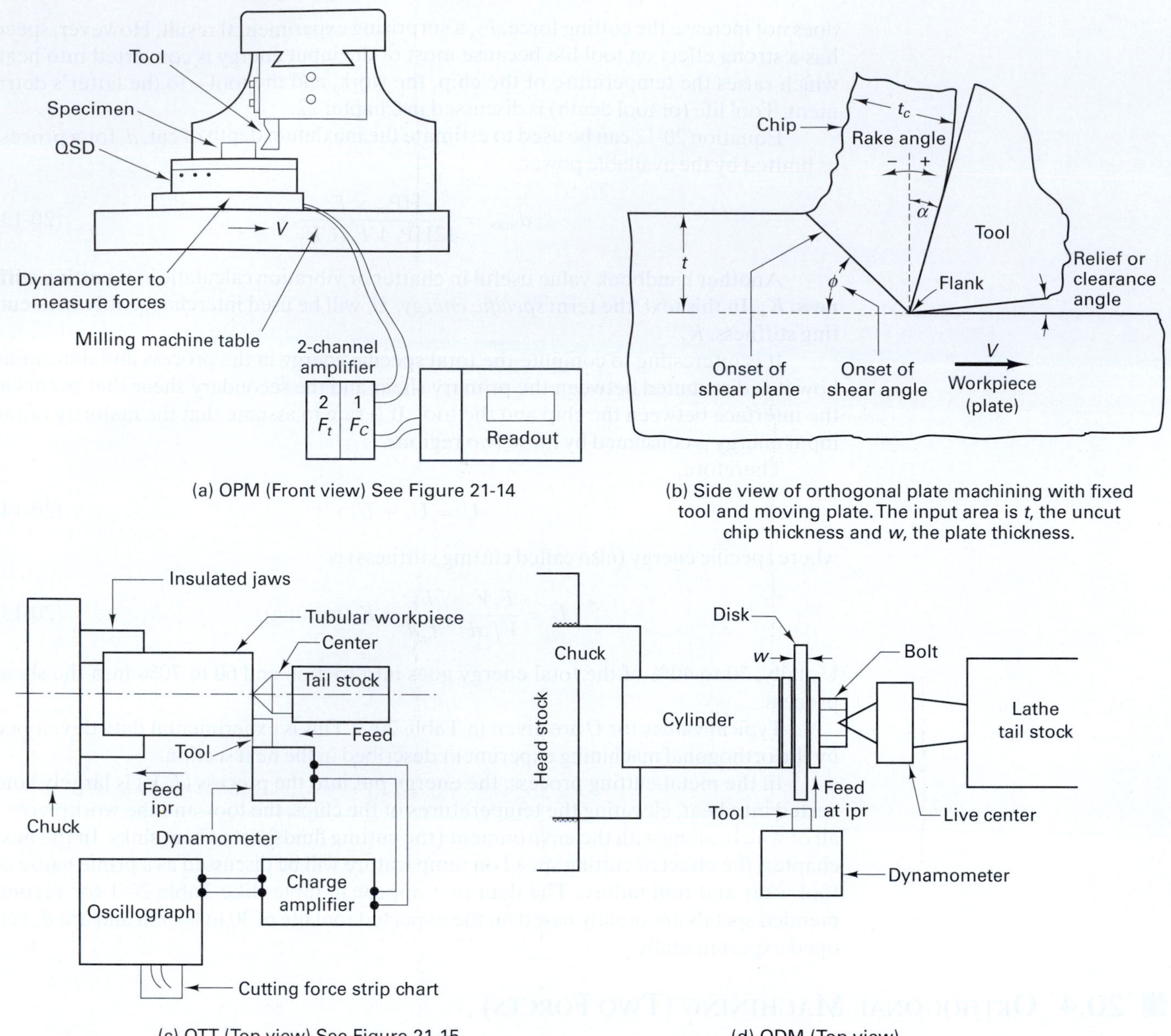

FIGURE 20-13 Three ways to perform orthogonal machining. (a) Orthogonal plate machining on a horizontal milling machine, good for low-speed cutting. (b) Orthogonal plate machining enlarged view (see Figure 20-14). (c) Orthogonal tube turning on a lathe; high speed cutting (see Figure 20-15). (d) Orthogonal disk machining on a lathe; very high-speed machining with tool feeding (ipr) in the facing direction.

mounted in a tool holder in the overarm. Low-speed orthogonal plate machining uses a flat plate setup in a milling machine. The workpiece moves past the tool at velocity V. The feed of the work vertically up into the tool is now called t, the uncut chip thickness. The DOC is the width of the plate, w. The cutting edge of the tool is perpendicular to the direction of motion V. The angle that the tool makes with respect to a vertical from the workpiece is called the **back rake angle, α.** A positive angle is shown in the schematic. The chip is formed by **shearing.** The **onset of shear** occurs at a low boundary deformed by angle ϕ with respect to the horizontal. This model is sufficient to allow us to consider the behavior of the work material during chip formation, the influence of the most critical elements of the tool geometry (the edge radius of the cutting tool and the back rake angle, α), and the interactions that occur between the tool and the freshly generated surfaces of the chip against the rake face and the new surface as rubbed by the flank of the tool.

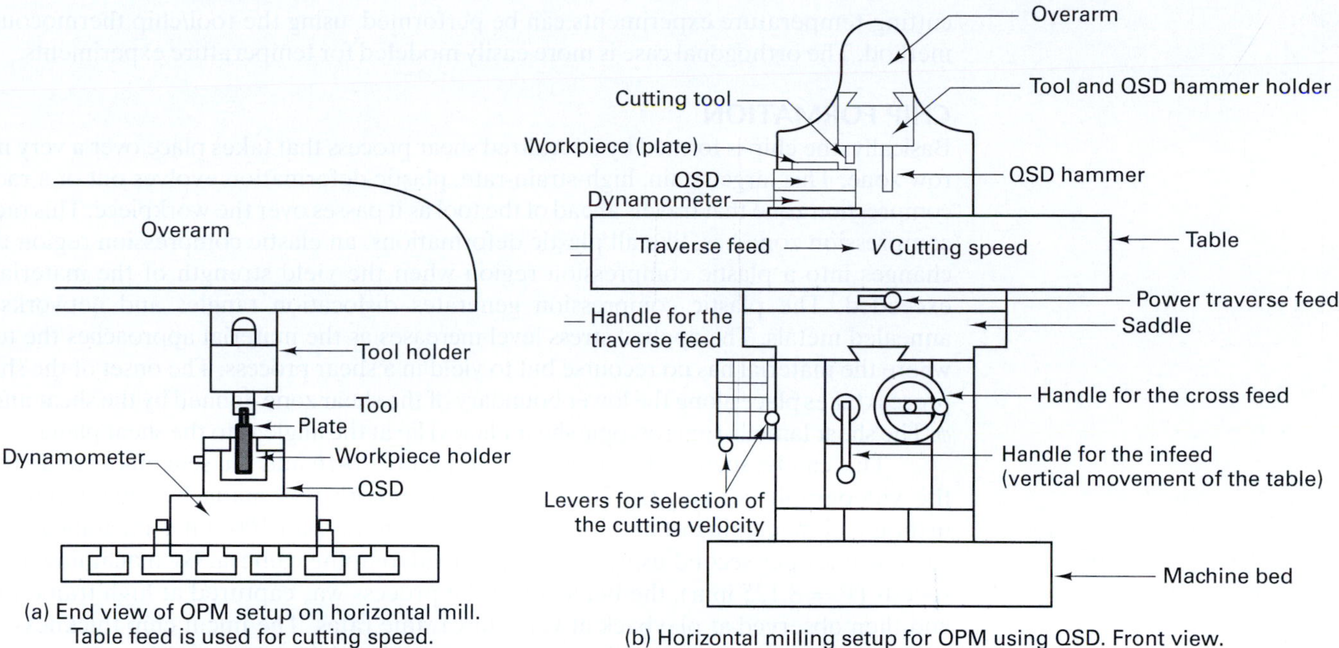

(a) End view of OPM setup on horizontal mill. Table feed is used for cutting speed.

(b) Horizontal milling setup for OPM using QSD. Front view.

FIGURE 20-14 Schematics of the orthogonal plate machining setup on a horizontal milling machine, using a quick-stop device (QSD) and dynamometer. The table feed mechanism is used to provide a low speed cut.

As shown in Figure 20-13 and 20-15, orthogonal machining can be performed in lathes at higher speeds. OTT is done on solid cylinders that have had a groove machined on the end to form a tube wall, w, or a tubular workpiece can be used. The tubular workpieces can be mounted in a lathe and normal cutting speeds developed for the machining experiment. This setup has the advantage of being very easy to modify so that

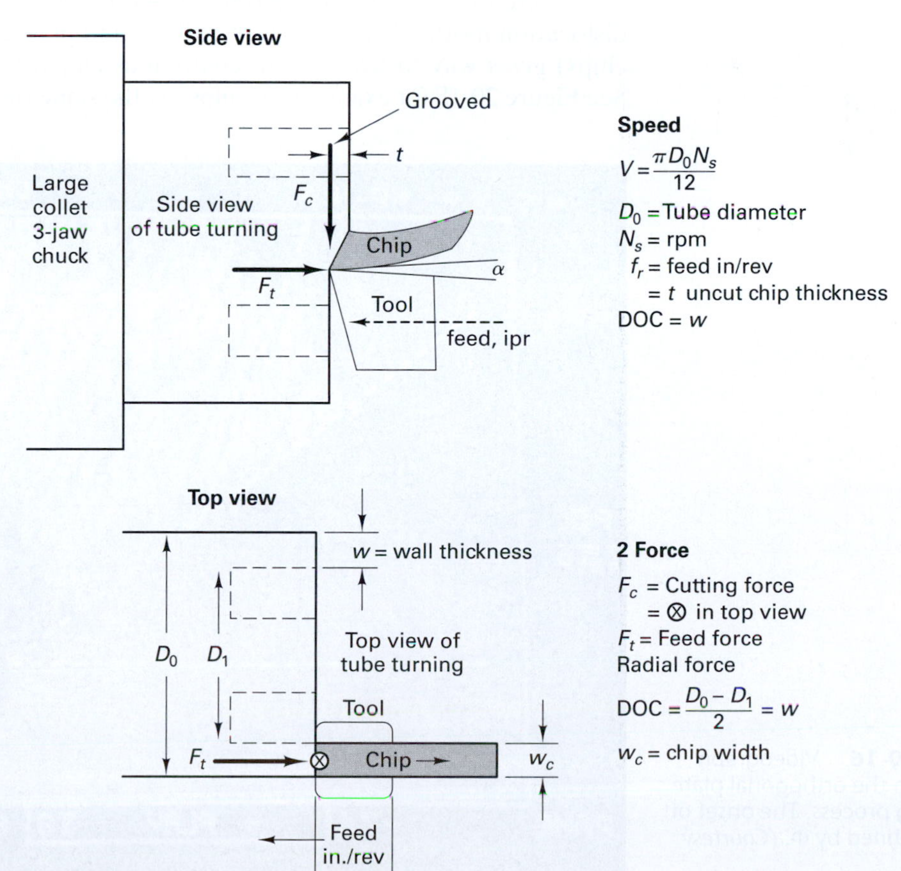

FIGURE 20-15 Orthogonal tube turning (OTT) produces a two-force cutting operation at speeds equivalent to those used in most oblique machining operations. The slight difference in cutting speed between the inside and outside edge of the chip can be neglected.

cutting-temperature experiments can be performed, using the tool/chip thermocouple method. The orthogonal case is more easily modeled for temperature experiments.

CHIP FORMATION

Basically, the chip is formed by a localized shear process that takes place over a very narrow zone. This large-strain, high-strain-rate, plastic deformation evolves out of a radial compression zone that travels ahead of the tool as it passes over the workpiece. This radial compression zone has, like all plastic deformations, an elastic compression region that changes into a plastic compression region when the yield strength of the material is exceeded. The plastic compression generates dislocation tangles and networks in annealed metals. The applied stress level increases as the material approaches the tool, where the material has no recourse but to yield in a shear process. The onset of the shear process takes place along the lower boundary of the shear zone defined by the shear angle, ϕ. The shear lamella (microscopic shear planes) lie at the angle ψ to the shear plane.

This can be seen in the videograph in Figure 20-16 and the schematic made from the videograph (see Figure 20-17). The videograph was made by videotaping the orthogonal machining of an aluminum plate at more than 100 times magnification at 1,000 frames per second using a high-speed videotaping camera. By machining at low speeds ($V = 8.125$ ipm), the behavior of the process was captured at high frame rates and then observed at playback at very slow frame rates. The uncut chip thickness was $t = 0.020$. The termination of the shear process as defined by the upper boundary cannot be observed in the still videograph but can easily be seen in the videos. Videographic experiments showed how the onset of shear angle, Φ, increases with metal hardness. This directly agrees with the material behavior observed in tensile/compression testing—that yield (and ultimate) strength increases with hardness. In steels the correlation is so good, hardness tests are used to estimate ultimate tensile strength. So in tensile testing, we observe that the onset of plastic deformation (yielding) is delayed by increased hardness or increased dislocation density. Correspondingly, in metal cutting, we observe that the onset of shear (to form the chip) is delayed by increased hardness (so ϕ increases directly with hardness). As the material being machined gets harder, dislocation motion becomes more difficult and plastic deformation (with continuous chips) gives way to fracture (discontinuous chips) just as it does in tensile testing. See Figure 20-18 for examples of chips. At the same time, the angle Φ is increasing, the

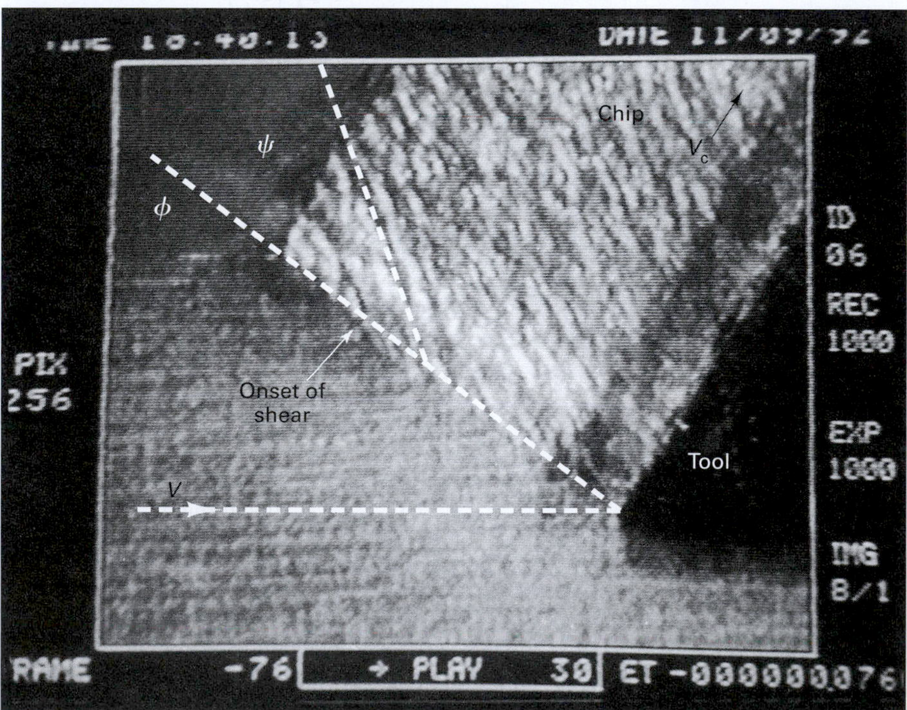

FIGURE 20-16 Videograph made from the orthogonal plate machining process. The onset of shear is defined by Φ. *(Courtesy J T. Black)*

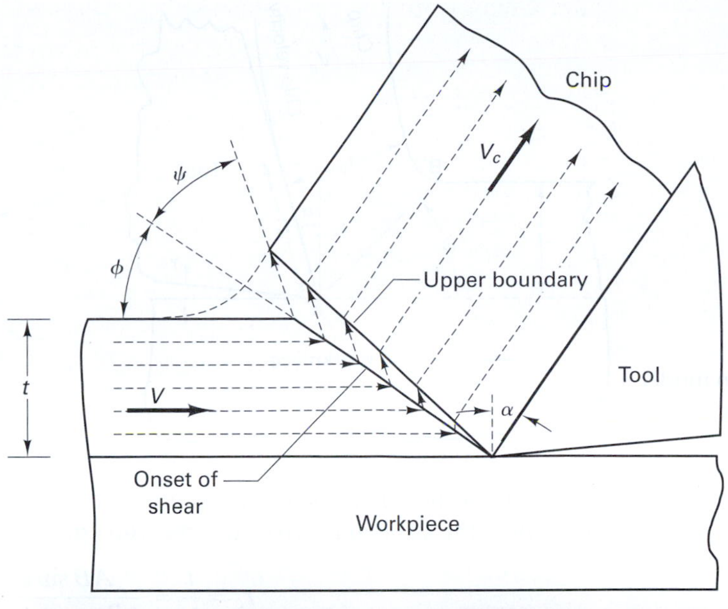

FIGURE 20-17 Schematic representation of the material flow, that is, the chip-forming shear process. Here, ϕ defines the onset of shear or lower boundary and ψ defines the direction of slip due to dislocation movement.

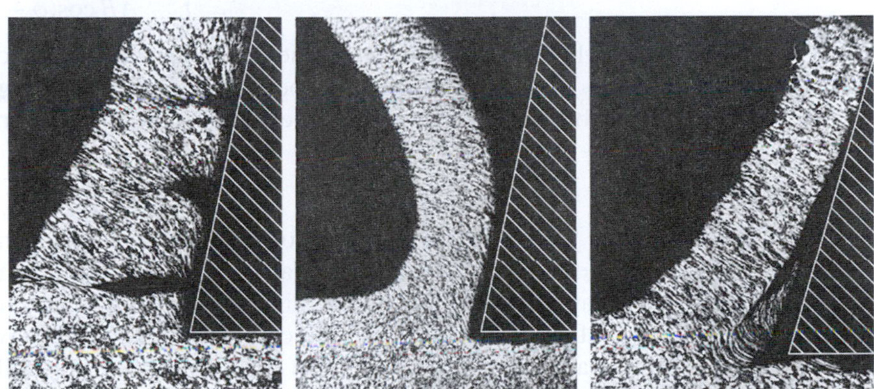

FIGURE 20-18 Three characteristic types of chips. (Left to right) Discontinuous, continuous, and continuous with built-up edge. Chip samples produced by quick-stop technique. *(Courtesy of Eugene Merchant at Cincinnati Milacron, Inc., Ohio)*

angle ψ is decreasing with increased hardness. Over the entire spectrum of hardness, from dead soft to full hard, Φ and ψ are related according to

$$\phi + \varphi = 45^\circ + \frac{\alpha}{2}.$$

If the work material has hard second-phase particles dispersed in it, they can act as barriers to the shear front dislocations, which cannot penetrate the particle. The dislocations create voids around the particles. If there are enough particles of the right size and shape, the chip will fracture through the shear zone, forming segmented chips. **Free-machining steels,** which have small percentages of hard second-phase particles added to them, use this metallurgical phenomenon to break up the chips for easier chip handling.

20.5 CHIP THICKNESS RATIO, r_c

For the purpose of modeling chip formation, assume that the shear process takes place on a single narrow plane, shown in Figure 20-19 as A–B rather than on the set of shear fronts that actually comprise a narrow shear zone. Further, assume that the tool's cutting edge is perfectly sharp and no contact is being made between the flank of the tool and the new surface. The workpiece passes the tool with velocity V, the cutting speed. The uncut chip thickness is t. Ignoring the compression deformation, chips having thickness t_c are formed by the shear process. The chip has velocity V_c. The shear process then has velocity V_s and occurs at the onset of shear angle ϕ. The tool geometry is given by the back rake angle α and the clearance angle γ. The velocity triangle for V, V_c, and V_s is also shown (see Figure 20-19). The chip makes contact with the rake face of the tool over length l_c. The workpiece is a plate thickness of w.

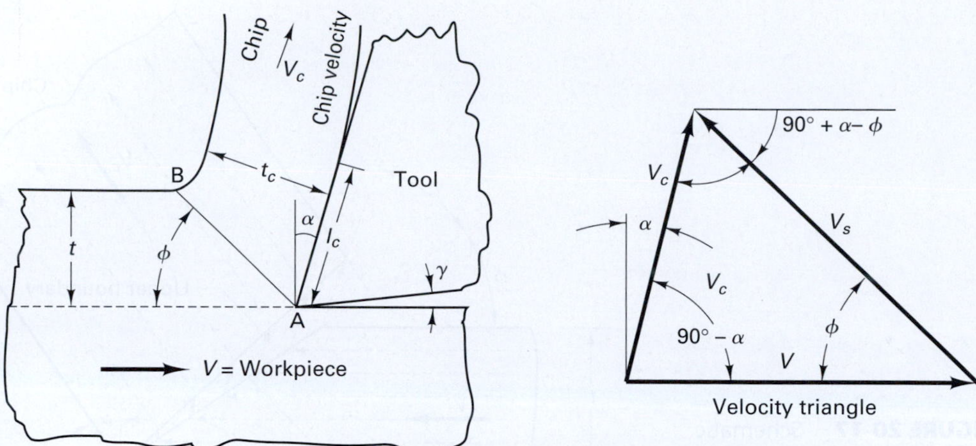

FIGURE 20-19 The orthogonal chip formation model is used to define the chip thickness ratio $r_c = t/t_c$.

From orthogonal machining experiments, the chip thickness is measured and used to compute the shear angle from the **chip thickness ratio, r_c,** defined as t/t_c:

$$r_c = \frac{t}{t_c} = \frac{AB \sin \phi}{AB \cos(\phi - \alpha)} \tag{20-16}$$

where AB is the length of the shear plane from the tool tip to the free surface.

Equation 20-16 may be solved for the **shear angle, ϕ,** as a function of the measurable chip thickness ratio by expanding the cosine term and simplifying:

$$\tan \phi = \frac{r_c \cos \alpha}{1 - r_c \sin \alpha} \tag{20-17}$$

There are numerous other ways to measure chip ratios and obtain shear angles both during (dynamically) and after (statically) the cutting process. For example, the ratio of the length of the chip L_c to the length of the cut L can be used to determine r_c. Many researchers use the chip compression ratio, which is the reciprocal of r_c, as a parameter. The shear angle can be measured statically by instantaneously interrupting the cut through the use of **quick-stop devices (QSDs).** These devices disengage the cutting tool from the workpiece while cutting is in progress, leaving the chip attached to the workpiece. Optical and scanning electron microscopy is then used to observe the direction of shear. Figure 20-14 shows a QSD on a plate machining setup, and Figure 20-18 was made using a quick-stop device.

Dynamically, high-speed motion pictures and high-speed videographic systems have also been used to observe the process at frame rates as high as 30,000 frames per second. Figure 20-16 is a high-speed videograph.

Machining stages have been built that allow the process to be performed inside a scanning electron microscope and recorded on videotapes for high-resolution, high-magnification examination of the deformation process. Using sophisticated electronics and slow-motion playback, this technique can be used to measure the **shear velocity.** The vector sum of V_s and V_c equals V.

For consistency of volume, we observe that

$$r_c = \frac{t}{t_c} = \frac{\sin \phi}{\cos(\phi - \alpha)} = \frac{V_c}{V} \tag{20-18}$$

indicating that the chip ratio (and therefore the onset of shear angle) can be determined dynamically if a reliable means to measure V_c can be found.

The ratio of V_s to V is

$$\frac{V_s}{V} = \frac{\cos \alpha}{\cos(\phi - \alpha)} \tag{20-19}$$

These velocities are important in power calculations, heat and temperature calculations, and vibration analysis associated with chatter in chip formation.

20.6 MECHANICS OF MACHINING (STATICS)

Orthogonal machining has been defined as a two-force system. Consider Figure 20-20, which shows a free-body diagram of a chip that has been separated at a shear plane. It is assumed that the resultant force, R, acting on the back of the chip is equal and opposite to the resultant force, R', acting on the shear plane. The resultant R is composed of the **friction force,** F, and the normal force N acting on the tool–chip interface contact area. The resultant force, R', is composed of a **shear force,** F_s, and normal force, F_n, acting on the shear plane area, A_s. Because neither of these two sets of forces can usually be measured, a third set is needed, which can be measured using a dynamometer (force transducer) mounted either in the workholder or the tool holder. Note that this set has resultant R, which is equal in magnitude to all the other resultant forces in the diagram. The resultant force, R, is composed of a **cutting force,** F_c, and a tangential (normal) force, F_t. Now it is necessary to express the desired forces (F_s, F_n, F, N) in terms of the measured dynamometer components, F_c and F_t, and appropriate angles. To do this, a circular force diagram is developed in which all six forces are collected in the same force circle (Figure 20-21). The only symbol in this figure as yet undefined is β, which is the angle between the normal force N and the resultant R. It is called friction angle β and is used to describe the friction coefficient μ on the tool–chip interface area, which is defined as F/N so that

$$\beta = \tan^{-1}\mu = \tan^{-1}\frac{F}{N} \tag{20-20}$$

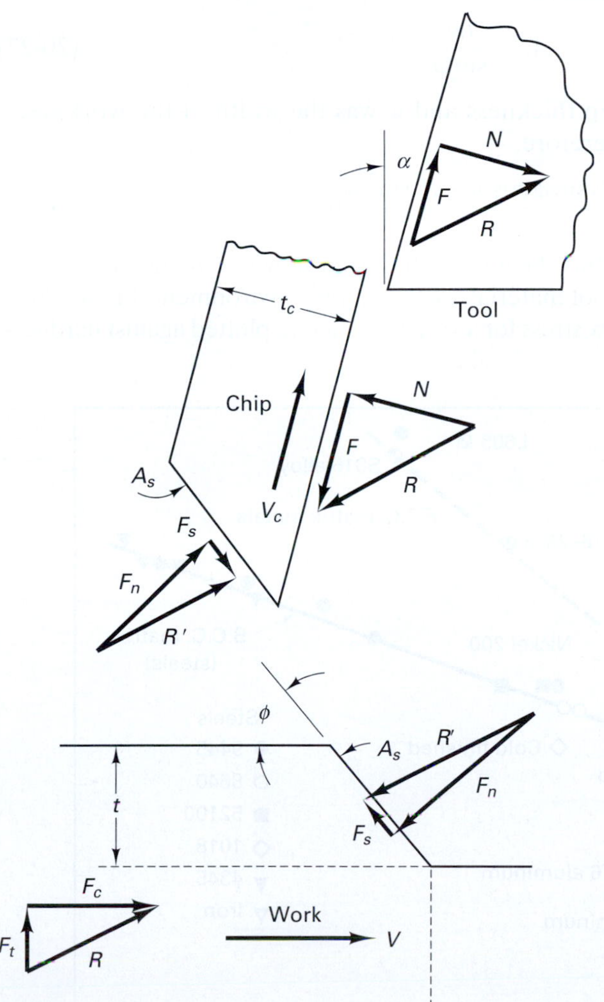

FIGURE 20-20 Free-body diagram of orthogonal chip formation process, showing equilibrium condition between resultant forces R and R'.

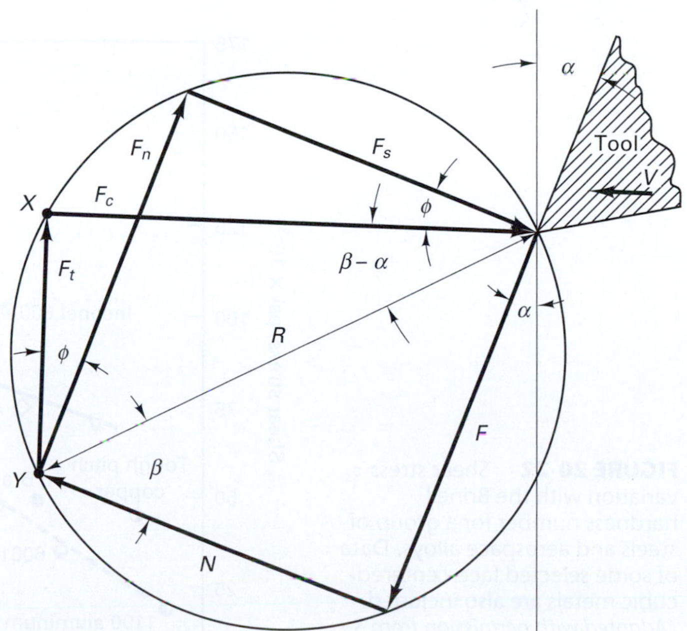

FIGURE 20-21 Merchant's circular force diagram used to derive equations for F_s, F_n, F_t, and N as functions of F_c, F_t, ϕ, α, and β.

The friction force F and its normal N can be shown to be

$$F = F_c \sin\alpha + F_t \cos\alpha \tag{20-21}$$

$$N = F_c \cos\alpha - F_t \sin\alpha \tag{20-22}$$

and the resultant R is

$$R = \sqrt{F_c^2 + F_t^2} \tag{20-23}$$

Notice that in the special situation where the back rake angle α is zero then, $F = F_t$ and $N = F_c$, so that in this orientation, the friction force and its normal can be directly measured by the dynamometer.

The forces parallel and perpendicular to the shear plane can be shown from the circular force diagram (Figure 20-21) to be

$$F_s = F_c \cos\phi - F_t \sin\phi \tag{20-24}$$

$$F_n = F_c \sin\phi + F_t \cos\phi \tag{20-25}$$

F_s is of particular interest, because it is used to compute the shear stress on the shear plane. This shear stress is defined as

$$\tau_s = \frac{F_s}{A_s} \tag{20-26}$$

where A_s is the area of the shear plane, as

$$A_s = \frac{tw}{\sin\phi} \tag{20-27}$$

recalling that t was the uncut chip thickness and w was the width of the workpiece. The **shear stress (flow stress)** is, therefore,

$$\tau_s = \frac{F_c \sin\phi\cos\phi - F_t \sin^2\phi}{tw} \text{ psi} \tag{20-28}$$

For a given polycrystalline metal, this shear stress has been shown to be not sensitive to variations in cutting parameters, tool material, or the cutting environment. Figure 20-22 gives some typical values for the flow stress for a variety of metals, plotted against hardness.

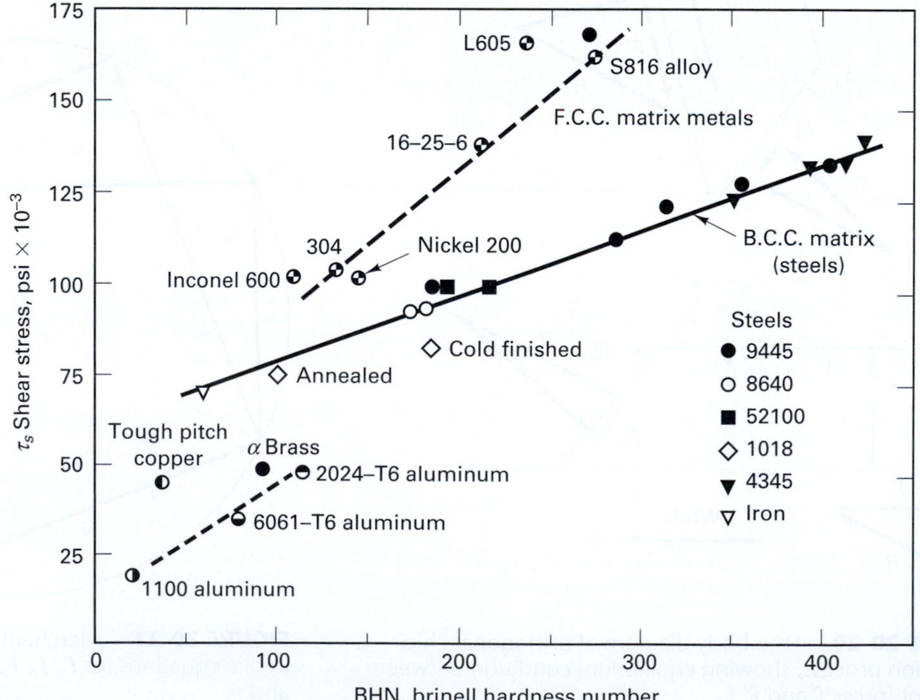

FIGURE 20-22 Shear stress τ_s variation with the Brinell hardness number for a group of steels and aerospace alloys. Data of some selected face-centered-cubic metals are also included. *(Adapted with permission from S. Ramalingham and K. J. Trigger,* Advances in Machine Tool Design and Research, *1971, Pergamon Press)*

Using orthogonal machining setups, values for HP_s as given in Table 20-3, can be computed from equation 20-10. K_s can be computed from

$$K_s = F_c/tw \quad (20\text{-}29)$$

Specific horsepower is related to and correlates well with shear stress for a given metal. Unit power is sensitive to material properties (e.g., hardness), rake angle, depth of cut, and feed, whereas τ_s is sensitive to material properties only.

20.7 SHEAR STRAIN, γ, AND SHEAR FRONT ANGLE, φ

Merchant's shear strain model[1] is cited in almost every textbook for manufacturing processes. It is widely used in theoretical studies and industrial practice. Merchant's model is based on the "stack of cards" model shown in Figure 20-23a. In this model, the material shears on a single plane defined by Φ, in the direction of the shear plane, so the model combines ψ and Φ into Φ, coupling them in any following analysis, Thus, Merchant's shear strain equation is:

$$\gamma = \cos\alpha/\sin\phi\cos(\phi - \alpha)$$

In the Black-Huang model,[2] the onset of shear begins at the lower boundary, defined by Φ with the material shearing in the direction of ψ relative to this lower boundary.

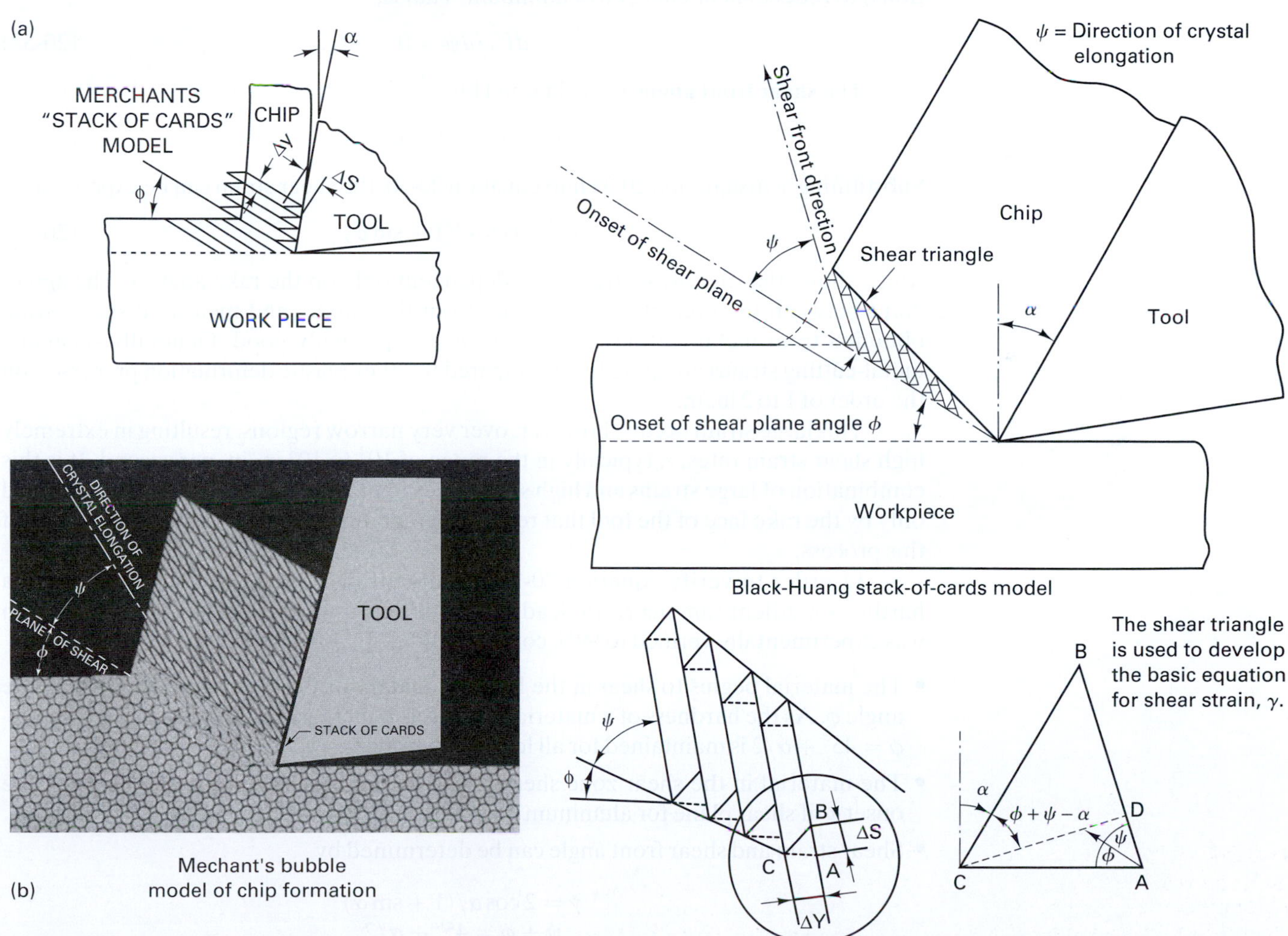

FIGURE 20-23 The Black-Huang "stack-of-cards" model for calculating shear strain in metal cutting is based on Merchant's bubble model for chip formation, shown on the left.

[1] M. E. Merchant, *Journal of Applied Physics,* 16, 267(a) and 318(b) (1945).

[2] J T. Black and J. M. Huang, "Shear Strain Model in Machining," *Manufacturing Science and Energy,* MED-Vol 2-1 (1995), p. 81.

Using Merchant's chip formation bubble model shown in Figure 20-23b verified by videographic images like that in Figure 20-16 show that a new "stack-of-cards" model can be developed as shown in Figure 20-23c on the right. From the shear triangle, an equation for strain, γ, is

$$\gamma = \cos\alpha/[\sin(\phi + \varphi)\cos(\phi + \varphi - \alpha)] \tag{20-30}$$

where

$\phi =$ the angle of the onset of the shear plane
$\psi =$ the shear front angle or direction of crystal elongation

Using the available machining data, for a given metal of some hardness, ψ is observed to decrease, reach a minimum, and rise again for all rake angles.

The minimum energy principle has been used in many scientific fields such including physics, metalforming processes, and machining processes. Applied to metal cutting, the specific shear energy (shear energy/volume) equals shear stress times shear strain:

$$U_s = \tau_s \times \gamma \tag{20-31}$$

The minimum energy principle is used here, where ψ will take on values (shear directions) to reduce shear energy to a minimum. That is,

$$dU_s/d\psi = 0 \tag{20-32}$$

The **shear front angle, φ,** is obtained by

$$\psi = 45^\circ - \phi + \alpha/2 \tag{20-33}$$

Substituting ψ in equation 20-34 into equation 20-30, the shear strain can be expressed as

$$\gamma = 2\cos\alpha/(1 + \sin\alpha) \tag{20-34}$$

which shows that the **shear strain, γ,** is dependent only on the rake angle α. The agreement between the predicted shear strain from this model and measured shear strain obtained from metal-cutting experiments is exceptionally good. Generally speaking, metal-cutting strains are quite large compared to other plastic deformation processes, on the order of 1 to 2 in./in.

This large strain occurs, however, over very narrow regions, resulting in extremely high shear strain rates, ε, typically in the range of 10^4 to 10^8 in./in. per second. It is this combination of large strains and high strain rates operating within a process constrained only by the rake face of the tool that results in great difficulties in theoretical analysis of this process.

In order to verify equation 20-33, metal-cutting experiments in copper with a hardness gradient ranging from dead soft to full hard were performed. The equation was experimentally verified to 99% confidence!

- The material begins to shear at the lower boundary of the shear zone, defined by the angle ϕ. As the hardness of a material increases, ϕ increases while ψ decreases, so $\psi + \phi = 45^\circ + \alpha/2$ is maintained for all levels of hardness.
- The material in the shear zone shears at an inclination angle ψ to the plane of the onset ϕ of shear plane for aluminum and steel.
- Shear strain and shear front angle can be determined by

$$\gamma = 2\cos\alpha/(1 + \sin\alpha)$$
$$\psi + \phi = 45^\circ + \alpha/2$$

where ϕ and ψ vary with hardness.

DISCUSSION OF BLACK-HUANG EQUATION

What has been done in the Black-Huang analysis is to separate the plastic deformation process (in chip formation) in two pieces, one angle defining the boundary and another defining the shear direction. The latter parameter, ψ, correctly obeys the minimum energy principle while ϕ is determined by mechanics, the state of the material (hardness), and the tool/chip interface boundary. Merchant's error was that his model combined ψ and ϕ into ϕ, so he coupled them and could not separate them.

Work hardness prior to machining is the most important factor because it controls the onset of shear. Just as in tensile testing, the onset of shear is delayed by increased hardness, so ϕ increases with hardness, as does τ_s, HP_s, and K_s. Highly ductile materials permit extensive plastic deformation of the chip during cutting, which increases heat generation and temperature, and also result in longer, "continuous" chips that remain in contact longer with the tool face, thus causing more frictional heat. Chips of this type are severely deformed and have a characteristic curl. On the other hand, materials that are already heavily work hardened or brittle, such as gray cast iron, lack the ductility necessary for appreciable plastic deformation. Consequently, the compressed material ahead of the tool fails in brittle fracture, sometimes along the shear front, producing small fragments. Such chips are termed *discontinuous* or *segmented*.

A variation of the continuous chip, often encountered in machining ductile materials, is associated with a **built-up edge (BUE)** formation on the cutting tool. The local high temperature and extreme pressure in the cutting zone cause the work material to adhere or pressure weld to the cutting edge of the tool forming the built-up edge, rather like a dead metal zone in the extrusion process. Although this material protects the cutting edge from wear, it modifies the geometry of the tool. BUEs are not stable and will break off periodically, adhering to the chip or passing under the tool and remaining on the machined surface. Built-up edge formation can be eliminated or minimized by reducing the depth of cut, altering the cutting speed, using positive rake tools, applying a coolant, or changing cutting tool materials.

20.8 MECHANICS OF MACHINING (DYNAMICS)

Machining is a dynamic process of large strain and high strain rates. All the process variables are dependent variables. The process is intrinsically a closed-loop interactive process as shown in Figure 20-24.

Starting at the top, inputs to the processes (speed, feed, depth of cut) determine the chip load on the tool. The chip load determines the cutting forces (magnitude and direction) (usually elastic), which alters the chip load on the tool. The altered chip load produces new forces. The cycle repeats, producing chatter and vibration.

Remember that plastic deformation is always preceded by elastic deformation. The elastic deflection behaves like a big spring. The mechanism by which a process dissipates energy is called **chatter** or **vibration.** In machining, it has long been observed in practice that rotational speed may greatly influence process stability and chatter. Experienced operators commonly listen to machining noise and interactively modify the speed when optimizing a specific application. In addition, experience demonstrates that the performance of a particular tool may vary significantly based on the machine tool employed and other characteristics such as the workpiece, fixture holder, and the like. Today more than ever, the manufacturing industry is more competitive and responsive, characterized by both high-volume and small-batch production, seeking economies of scale. High productivity is achieved by increased machine and tooling capabilities along with the elimination of all non-value-added activities. Few companies can afford lengthy trial-and-error approaches to machining-process optimization or additional processes to treat the effect of chatter.

In metal cutting, chatter is a self-excited vibration that is caused by the closed-loop force-displacement response of the machining process. The process-induced variations in the cutting force may be caused by changes in the cutting velocity, chip cross section (area), tool–chip interface friction, built-up edge, workpiece variation, or, most commonly, process modulation resulting in regeneration of vibration. When more energy is input into the dynamic machining system than can be dissipated by mechanical work, damping, and friction, equilibrium (the state of minimum potential energy) is sought by the machining system through the generation of chatter vibration.

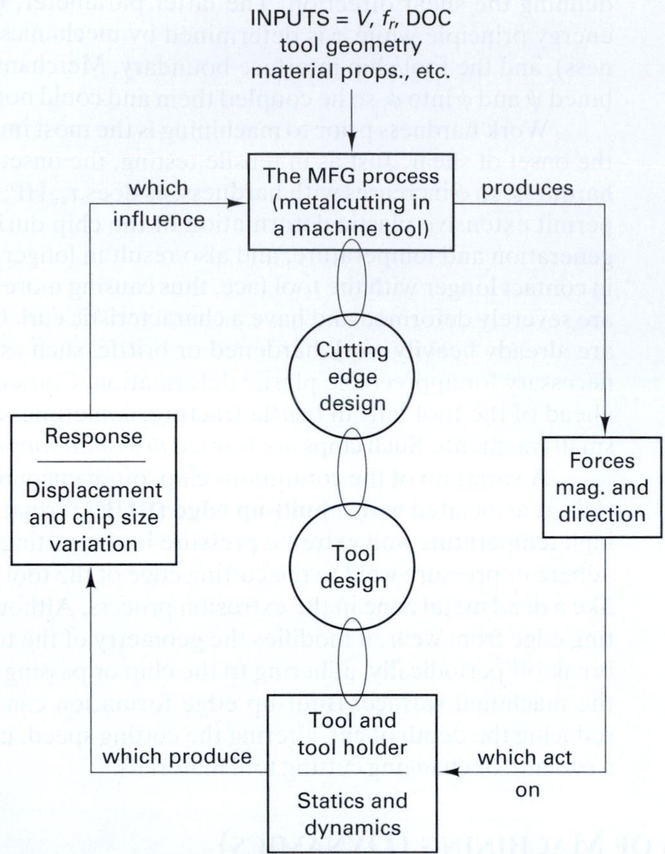

FIGURE 20-24 Machining dynamics is a closed-loop interactive process that creates a force-displacement response.

The proper classification of the type of vibration is the first step in identifying and solving the cause of unwanted vibration (see Figure 20-25):

- **Free vibration** is the response to any initial condition or sudden change. The amplitude of the vibration decreases with time and occurs at the natural frequency of the system. Interrupted machining is an example that often appears as lines or shadows following a surface discontinuity.
- **Forced vibration** is the response to a periodic (repeating with time) input. The response and input occur at the same frequency. The amplitude of the vibration remains constant for set input conditions and is linearly related to speed. Unbalance, misalignment, tooth impacts, and resonance of rotation systems are the most common examples.
- **Self-excited vibration** is the periodic response of the system to a constant input. The vibration may grow in amplitude (become unstable) and occurs near the natural frequency of the system regardless of the input. Chatter due to the regeneration of waviness in the machined surface is the most common metal cutting example.

How do we know chatter exists? Listen and look! Chatter is characterized by the following:

1. There is a sudden onset of vibration (a screech or buzz or whine) that rapidly increases in amplitude until a maximum threshold (saturation) is reached.
2. The frequency of chatter remains very close to a natural frequency (critical frequency) of the machining system and changes little with variation of process parameters. The largest force-displacement response occurs at resonance and therefore the greatest energy dissipation.
3. Chatter often results in unacceptable surface finish, exhibited by a helical or angular pattern (pearled or fish scaled) superimposed over normal feed marks.

- **Free Vibration** The response to an initial condition or sudden change. The amplitude of the vibration decreases with time and occurs at the natural frequency of the system often produced by interrupted machining. Often appears as lines or shadows following a surface discontinuity.

K C Free vibration M tool Workpiece

- **Forced Vibration** The response to a periodic (repeating with time) input. The response and input occur at the same frequency. The amplitude of the vibration remains constant for a set input condition and is nonlinearly related to speed. Unbalance, misalignment, tooth impacts, and resonance of rotating systems are the most common examples.

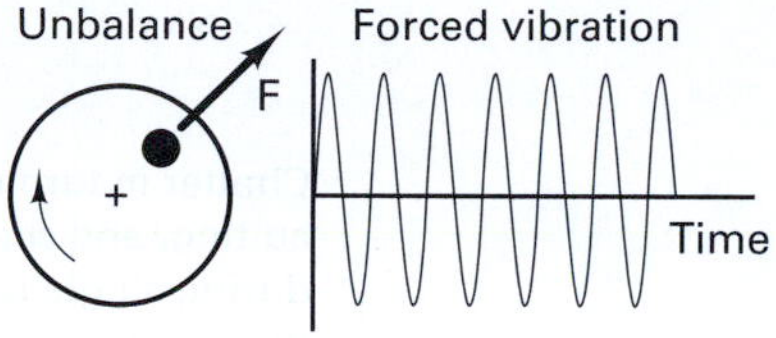

- **Self-Excited Vibration** The periodic response of the system to a constant input. The vibration may grow in amplitude (unstable) and occurs near the natural frequency of the system regardless of the input. Chatter due to the regeneration of surface waviness is the most common metal cutting example.

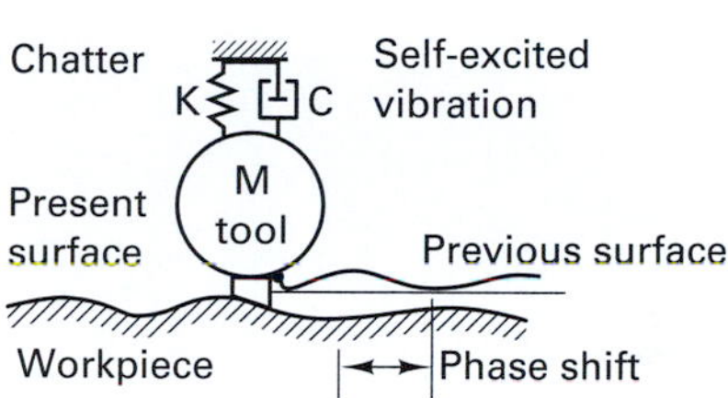

FIGURE 20-25 There are three types of vibration in machining.

4. Visible surface undulations are found in the feed direction and corresponding wavy or serrated chips with variable thickness.

Figure 20-26 shows some typical examples of chatter visible in the surface finish marks.

There are several important factors that influence the stability of a machining process:

- Cutting stiffness of the workpiece material (related to the machinability), K_s.
- Cutting-process parameters (speed, feed, DOC, total width of chip).
- Cutter geometry (rake and clearance angles, edge prep, insert size and shape).
- Dynamic characteristics of the machining process (tooling, machine tool, fixture, and workpiece).

Cutting stiffness, K_s, is closely aligned with flow stress but simpler to calculate in that ϕ is not used. Like flow stress, cutting stiffness can be viewed as a material property of the workpiece, dependent on hardness.

CHIP FORMATION AND REGENERATIVE CHATTER

In machining, the chip is formed due to the shearing of the workpiece material over the chip area (A = Thickness × Width = $t \times w$), which results in a cutting force, F_c. The magnitude of the resulting cutting force is predominantly determined by the material cutting stiffness, K_s, and the chip area, such that $F_c = K_s \times t \times w$. The direction of the cutting force, F_c, is influenced mainly by the geometries of the rake and clearance angles as well as the edge prep.

Machining operations require an overlap of cutting paths that generate the machined surface (see Figure 20-27). In single-point operations, the overlap of cutting paths does not occur until one complete revolution. In milling or drilling, overlap occurs in a fraction of a revolution, depending on the number of cutting edges on the tool.

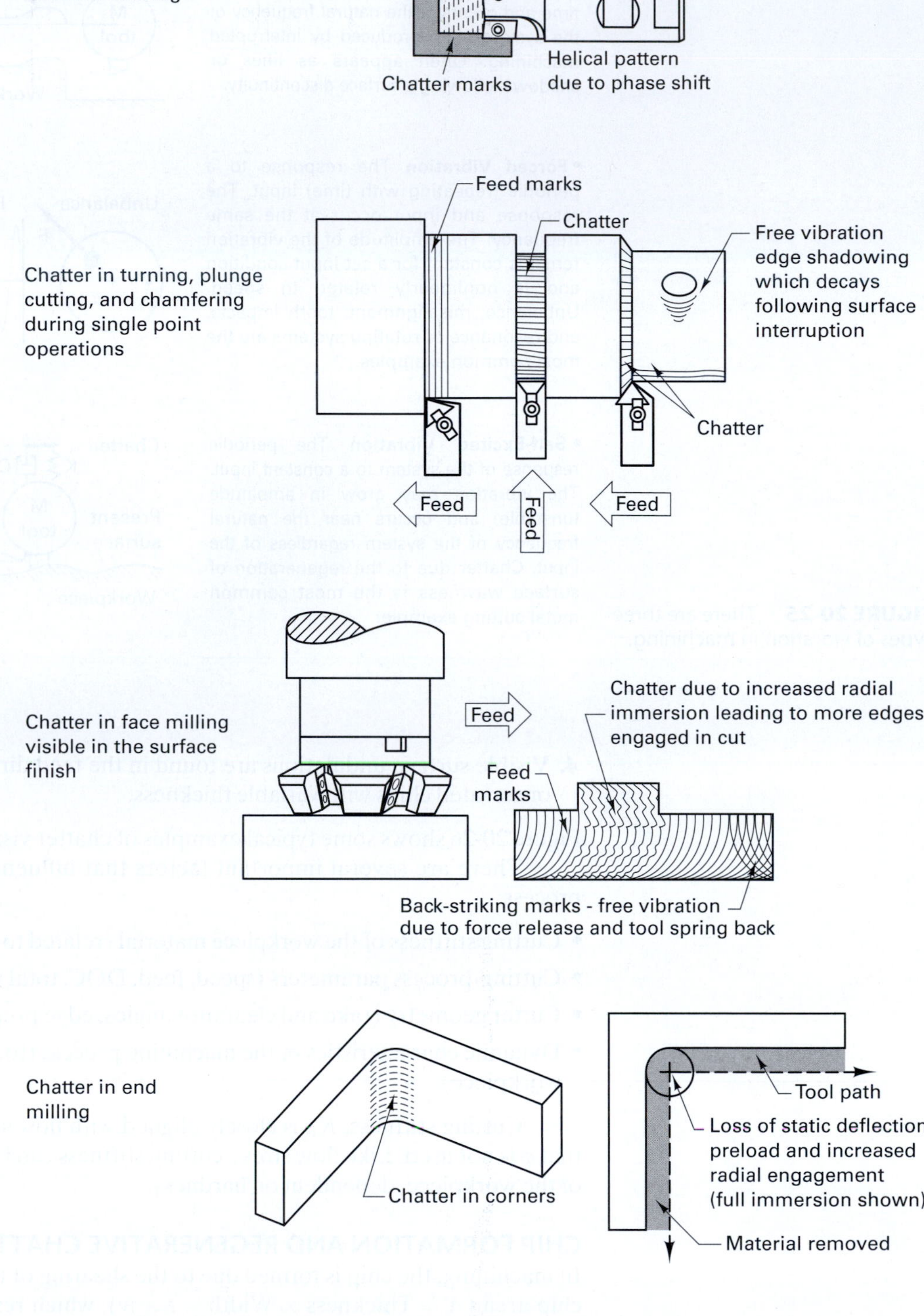

FIGURE 20-26 Some examples of chatter that are visible on the surfaces of the workpiece.

The cutting force causes a relative displacement, X, between the tool and workpiece, which affects the uncut chip thickness, t, and, in turn, the cutting force. This coupled relationship between displacement in the Y-direction (modulation direction) and the resulting cutting force forms a closed-loop response system. The modulation direction is normal to the surface defining the chip thickness.

A phase shift, ε, between subsequent overlapping surfaces results in a variable chip thickness and modulation of the displacement, causing chatter vibration. The phase shift between overlapping cutting paths is responsible for producing chatter.

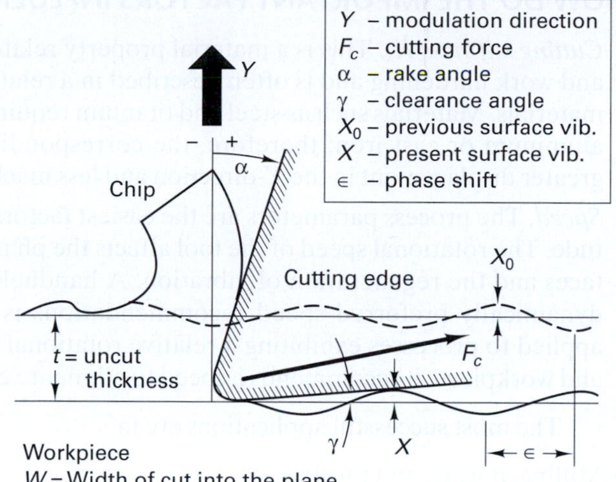

FIGURE 20-27 When the overlapping cuts get out of phase with each other, a variable chip thickness is produced, resulting in a change in F_c on the tool or workpiece.

However, there is a preferred speed that corresponds to a phase-locked condition ($\varepsilon = 0$) that results in a constant chip thickness, t. A constant chip thickness results in a steady cutting force and the elimination of the feedback mechanism responsible for **regenerative chatter.** This what the operators are trying to achieve when they vary cutting speeds (see Figure 20-28).

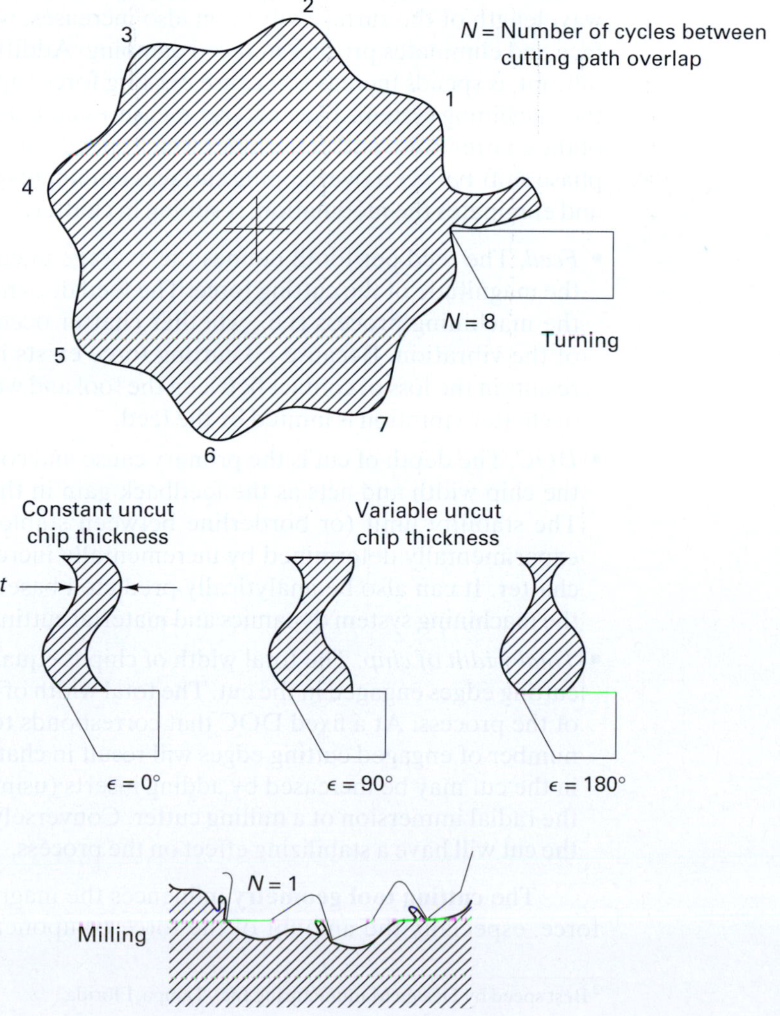

FIGURE 20-28 Regenerative chatter in turning and milling produced by variable uncut chip thickness.

HOW DO THE IMPORTANT FACTORS INFLUENCE CHATTER?

- *Cutting stiffness* K_s. This is a material property related to shear flow stress, hardness, and work hardening and is often described in a relative sense of the machinability of materials. Materials such as steel and titanium require much greater shear forces than aluminum or cast iron; therefore, the corresponding larger cutting forces lead to greater displacement in the Y-direction and less machining stability.
- *Speed*. The process parameters are the easiest factors to change chatter and its amplitude. The rotational speed of the tool affects the phase shift between overlapping surfaces and the regeneration of vibration. A handheld speed analyzer[3] that produces dynamically preferred speed recommendations is commercially available. When applied to processes exhibiting a relative rotational motion between the cutting tool and workpiece, it recommends a speed to eliminate chatter.

The most successful applications are in

- Milling, boring, and turning.
- Multipoint tools.
- Machining aluminum and cast iron.
- High-speed machining.
- Thin-chip, high-speed die machining.

At slow speeds (relative to the vibration frequency) process stability is mainly due to increased frictional losses occurring between the tool clearance angle (defined by γ) and the present surface vibration, X. This interference and friction dissipates energy in the form of heat and is called **process damping.** As machining speeds are increased, the wavelength of the surface vibration also increases, which reduces the slope of the surface and eliminates process-induced damping. Additionally, chatter becomes more significant as speeds increase, because existing forces approach the natural frequencies of the machining system. The analyzer measures and identifies the vibrational frequencies of the chatter noise and determines which speeds will most closely result in $\varepsilon = 0$. A zero phase shift between overlapping surfaces eliminates the variation of the chip thickness and eliminates the modulation, resulting in chatter.

- *Feed*. The feed per tooth defines the average uncut chip thickness, t, and influences the magnitude of the cutting force. The feed does not greatly influence the stability of the machining process (i.e., whether chatter occurs) but does control the severity of the vibration. Because no cutting force exists if the vibration in the Y-direction results in the loss of contact between the tool and workpiece, the maximum amplitude of chatter vibration is limited by the feed.
- *DOC*. The depth of cut is the primary cause and control of chatter. The DOC defines the chip width and acts as the feedback gain in the closed-loop machining process. The stability limit (or borderline between stable machining and chatter) may be experimentally determined by incrementally increasing the DOC until the onset of chatter. It can also be analytically predicted based on a thorough understanding of the machining system dynamics and material cutting stiffness.
- *Total width of chip*. The total width of chip is equal to the DOC times the number of cutting edges engaged in the cut. The total width of cut directly influences the stability of the process. At a fixed DOC that corresponds to the stability limit, increasing the number of engaged cutting edges will result in chatter. The number of engaged teeth in the cut may be increased by adding inserts (using a fine-pitch cutter) or increasing the radial immersion of a milling cutter. Conversely, reducing the number of edges in the cut will have a stabilizing effect on the process.

The **cutting tool geometry** influences the magnitude and direction of the cutting force, especially the amount of the force component in the modulation direction Y.

[3] Best speed by Design Manufacturing Inc., Tampa, Florida.

A greater projection of force in the Y-direction results in increased displacement and vibration normal to the surface, leading to potential chatter.

- As the *back rake angle* α increases (becomes more positive), Φ becomes larger (see Equation 20-33); the length of the onset of shear plane decreases, which reduces the magnitude of the cutting force, F_c. A more positive rake also directs the cutting force to be more tangential and reduces the force component in the Y-direction. In general, a more positive cutting geometry increases process stability, especially at higher speeds. An insufficient feed compared to the edge radius results in less efficient machining, greater tool deflection, and poorer machining stability.
- A *reduced clearance angle* γ, which increases the frictional contact between the tool and workpiece, may produce process damping. The stabilizing effect is due to energy dissipation in the form of heat, which potentially decreases tool life and may thermally distort the workpiece or increase the heat-affected zone in the workpiece. The initial wear of a new cutting edge may have a stabilizing effect on chatter.
- The *size* (nose radius), *shape* (diamond, triangular, square, round), and *lead angle* of the insert all influence the chip area shape and the corresponding Y-direction (Figure 20-29). In milling, the feed direction is transverse to the tool axis (i.e., radial), and the DOC is defined by the axial immersion of the tool. In milling, as the lead angle of the cut increases or the shape of the insert becomes rounded (same effect as a large nose radius compared to a small DOC), the Y-orientation is directed away from the more flexible radial tool direction and toward the stiffer axial direction. The orientation of the modulation direction Y toward a dynamically more rigid direction results in decreased vibrational response and therefore greater process stability (less tendency for chatter). In boring, the feed direction is axial, and the DOC is determined by the radial direction. Therefore, in boring, a reduced lead angle or a less round (and smaller nose radius compared to the DOC) insert maintains a more axial (stiffer tool direction) orientation of Y, leading to greater stability.

Because the stability of the machining process is a direct result of the dynamic *force-displacement characteristics* between the tool and workpiece, all components of the machining system (tool, spindle, workpiece, fixture, machine tool) may, to varying degrees, influence chatter. Maximizing the **dynamics** (the product of the static stiffness and damping) of the machining system leads to increased process stability. For

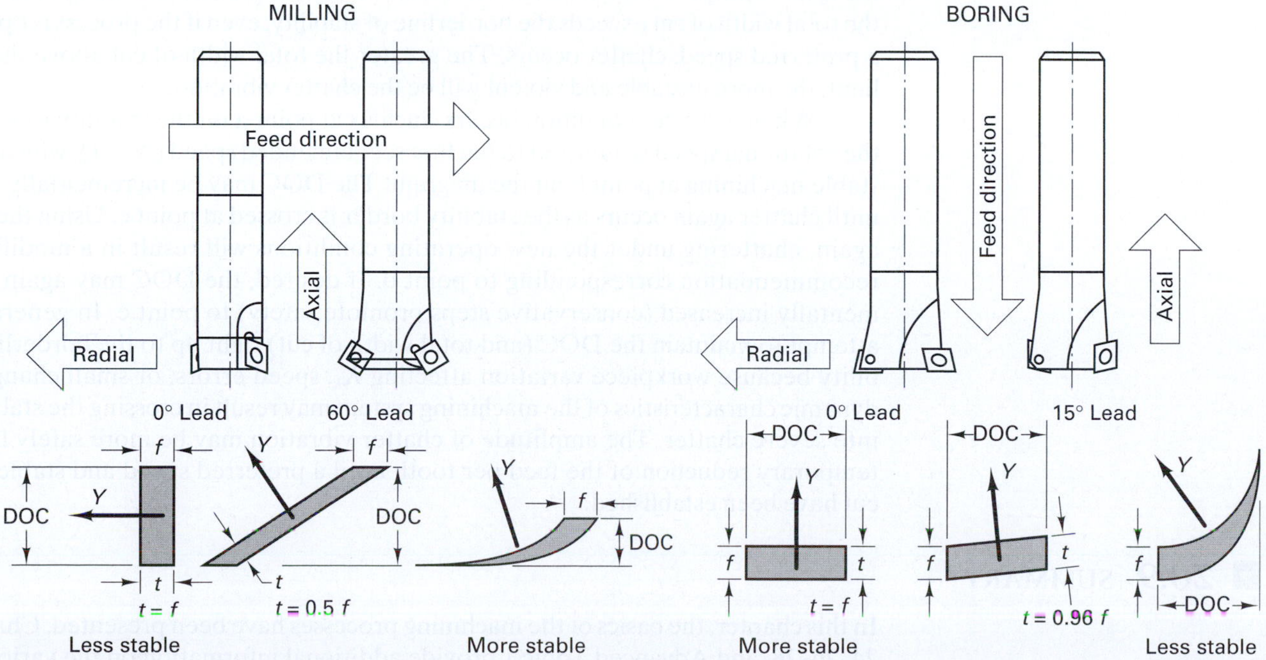

FIGURE 20-29 Milling and boring operations can be made more stable by correct selection of insert geometry.

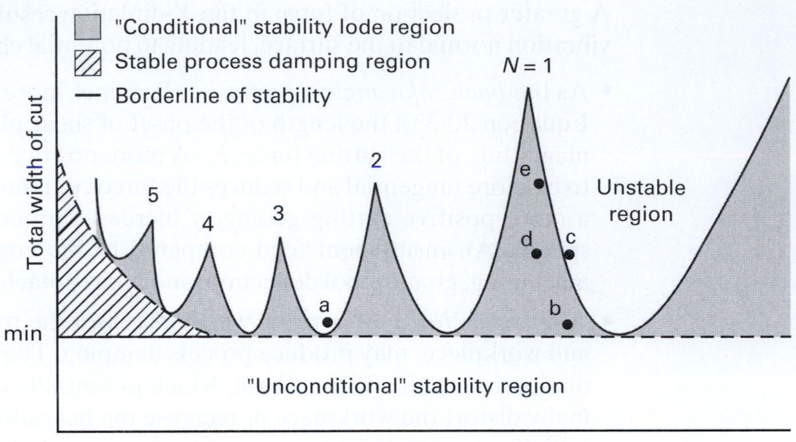

FIGURE 20-30 Dynamic analysis of the cutting process produces a stability lobe diagram, which defines speeds that produce stable and unstable cutting conditions.

example, the static stiffness of an overhung (cantilever) tool with circular cross section varies nonlinearly with the diameter D and the unsupported length L. Machining stability is increased by having the tool with the largest possible diameter with the minimum overhand. The frequency of chatter occurs near the most flexible vibrational mode of the machining system.

STABILITY LOBE DIAGRAM

A **stability lobe diagram** (Figure 20-30) relates the total width of cut that can be machined to the rotational speed of the tool with a specified number of cutting edges. If the total width of cut is maintained below a minimum level (although this may be of limited practical value for some machining systems), then the process stability exhibits speed independence or "unconditional" stability. At slow speeds, increased stability may be achieved within the process damping region. The "conditional" stability lobe regions allow increased total width of cut (DOC × Number of edges engaged in the cut) at dynamically preferred speeds at which the phase shift ε between overlapping or consecutive cutting paths approaches zero. The stability lobe number N indicates complete cycles of vibration that exist between overlapping surfaces. As can be seen by the diagram, the higher speeds correspond to lower lobe numbers and provide the greatest potential increase in the total width of cut and material removal rate (due to greater lobe height and width). If the total width of cut exceeds the borderline of stability, even if the process is operating at a preferred speed, chatter occurs. The greater the total width of cut above the stability limit, the more unstable and violent will be the chatter vibration.

When a chatter condition occurs, such as at point a on the stability lobe diagram, the rotational speed is adjusted to the first recommended speed ($N = 1$), which results in stable machining at point b on the diagram. The DOC may be incrementally increased until chatter again occurs as the stability border is crossed at point c. Using the analyzer again, chattering under the new operating conditions will result in a modified speed recommendation corresponding to point d. If desired, the DOC may again be incrementally increased (conservative steps promote safety) to point e. In general, do not attempt to maintain the DOC (and total width of cut) right up to the borderline of stability because workpiece variation affecting K_s, speed errors, or small changes in the dynamic characteristics of the machining system may result in crossing the stability limit into severe chatter. The amplitude of chatter vibration may be more safely limited by temporary reduction of the feed per tooth until a preferred speed and stable depth of cut have been established.

20.9 SUMMARY

In this chapter, the basics of the machining processes have been presented. Chapters 22–24, 26–28, and Advanced Topic 1 provide additional information on the various operations and machine tools. Modern machine tools can often perform several basic operations. As described in Chapter 25, machining centers operated by computer numerical controls are now widely used. These chapters on basic processes must be carefully studied. The relationship between the basic processes and the machine tools that can be used to perform these processes has been presented. Most of the machines described in these chapters will be of the A(1) to A(3) levels of automation. Machining centers, which are programmable machines, are A(4). Machining centers have automatic tool-change capability and are usually capable of milling, drilling, boring, reaming, tapping (hole threading), and other minor machining processes. For particular machines, you will need to become familiar with new terminology, but in general, all machining processes will need inputs concerning revolutions per minute (given that you selected the cutting speed), feeds, and depths of cut. Note that the same process can be performed on different machine tools. There are many ways to produce flat surfaces, internal and external cylindrical surfaces, and special geometries in parts. Generally, the quantity to be made is the driving factor in the selection of processes.

This chapter also introduced orthogonal machining, a two-force machining process used to experimentally determine values for specific horsepower, cutting stiffness, and flow stress for machining and to investigate the theory of the process.

As noted previously, the properties of the work material are important in chip formation. High-strength materials produce larger cutting forces than materials of lower strength, causing greater tool and work deflection, increased friction, heat generation, and operating temperatures. The structure and composition also influence metal cutting. Hard or abrasive constituents, such as carbides in steel, accelerate tool wear. This chapter introduced the student to the Black-Huang model, which decouples the onset of shear angle, ϕ, from the direction of shear angle, ψ. The model holds over all levels of hardness.

Finally, this chapter introduced the student to the dynamics of machining and the reality of chatter and vibration.

■ KEY WORDS

back rake angle, α
boring
broaching
built-up edge (BUE)
chatter
chip thickness ratio, r_c
chip velocity
cutting force
cutting stiffness, K_s
cutting tool
cutting tool geometry
depth of cut (DOC)
drilling
dynamics
feed
flow stress
free vibration
free-machining steel
forced vibration
friction force
grinding
machine tool
machining
metal cutting
metal removal rate
milling
oblique machining
onset of shear
orthogonal machining (OM)
quick-stop device (QSD)
process damping
regenerative chatter
sawing
self-excited vibration
shaping
shearing
shear angles
shear force
shear front angle, φ
shear stress
shear strain, γ
shear velocity
shop equations
specific horsepower, HP_s
speed
stability lobe diagram
turning
vibration
workholding device
workpiece

■ REVIEW QUESTIONS

1. Why has the metal-cutting process resisted theoretical solution for so many years?
2. What variables must be considered in understanding a machining process?
3. Which of the seven basic chip formation processes are single point, and which are multiple point?
4. How is feed related to speed in the machining operations called turning?
5. Before you select speed and feed for a machining operation, what have you had to decide? (*Hint:* See Figure 20-4.)
6. Milling has two feeds. What are they, and which one is an input parameter to the machine tool?
7. What is the fundamental mechanism of chip formation?
8. What is the difference between oblique machining and orthogonal machining?
9. What are the implications of Figure 20-16, given that this videograph was made at a very low cutting speed?
10. Note that the units for the approximate equation for MRR for turning are not correct. When is the approximate equation not very good (yields a large error in MRR values)?

11. For orthogonal machining, the cutting edge radius is assumed to be small compared to the uncut chip thickness. Why?
12. How do the magnitude of the strain and strain rate values of metal cutting compare to those of tensile testing?
13. Why is titanium such a difficult metal to machine? (Note its high value of HP_s).
14. Explain why you get segmented or discontinuous chips when you machine cast iron.
15. Why is metal cutting shear stress such an important parameter?
16. Which of the three cutting forces in oblique cutting consumes most of the power?
17. How is the energy in a machining process typically consumed?
18. Where does the energy consumed in metal cutting ultimately go?
19. What are two ways of estimating the primary cutting force F_c.
20. What is the relationship between speed and temperature in metal cutting tool materials?
21. Why does the cutting force F_c increase with increased feed or DOC?
22. Why doesn't the cutting force F_c increase with increased speed V?
23. Why doesn't the radial force F_r, increase with DOC?
24. How does the selection of the machining parameters (speed, feed, DOC) influence chatter?
25. Suppose you had a machining operation (boring) running perfectly and you changed work materials. All of a sudden, you are getting lots of chatter. Why?
26. Make a sketch like that shown in Figure 20-1 with w, t, and V and show F_c and F_t.
27. Show how you would do near orthogonal machining in a turning operation like that shown in Figure 20-3. (See Figure 20-5).
28. Explain how or why you can or cannot do orthogonal machining on a shaper or planer.
29. What process and material combination would yield the largest cutting force, based on the data in Table 20-3 or 20-4.
30. What is meant by the statement that machining dynamics is a closed-loop interactive process?

PROBLEMS

1. For a turning operation, you have selected a high-speed steel (HSS) tool and turning a hot rolled free machining steel, BHN = 300. Your depth of cut will be 0.150 in. The diameter of the workpiece is 1.00 in.
   a. What speed and feed would you select for this job?
   b. Using a speed of 105 sfpm and a feed of 0.015, calculate the spindle rpm for this operation.
   c. Calculate the metal removal rate.
   d. Calculate the cutting time for the operation with a length of cut of 4 in. and 0.10-in. allowance.
2. For a slab milling operation using a 5-in.-diameter, 11-tooth cutter (see Figure 20-6), the feed per tooth is 0.005 in./tooth with a cutting speed of 100 sfpm (HSS steel). Calculate the rpm of the cutter and the feed rate (f_m) of the table, then calculate the metal removal rate, MRR, where the width of the block being machined is 2 in. and the depth of cut is 0.25 in. Calculate the time to machine (T_m) a 6-in.-long block of metal with this setup. Suppose you switched to a coated-carbide tool, so you increase the cutting speed to 400 sfpm. Now recalculate the machining time (T_m) with all the other parameters the same.
3. The power required to machine metal is related to the cutting force (F_c) and the cutting speed. For Problem 1, estimate cutting force F_c for this turning operation. (*Hint:* You have to estimate a value of HPs for this material.)
4. In order to drill a hole in the material described in Problem 1 using an HSS drill, you have to select a cutting speed and a feed rate. Using a speed of 105 sfpm for the HSS drill, calculate the rpm for a $\frac{3}{4}$-in.-diameter drill and the MRR if the feed rate is 0.008 inches per revolution.
5. Explain how the constant 33,000 in equation 20-8 is obtained.
6. Explain how the constant 396,000 in equation 20-10 is obtained.
7. Suppose you have the following data obtained from a metal-cutting experiment (orthogonal machining). Compute the shear angle, the shear stress, the specific energy, the shear strain, and the coefficient of friction at the tool–chip interface. How do your HP_s and τ_s values compare with the values found in Table 20–3?
8. For the data in Problem 7, determine the specific shear energy and the specific friction energy.
9. Derive equations for F and N using the circular force diagram. (*Hint:* Make a copy of the diagram. Extend a line from point X intersecting force F perpendicularly. Extend a line from point Y intersecting the previous line perpendicularly. Find the angle α made by these constructions.)
10. Derive equations for F_s and F_n using the circular force diagram. (*Hint:* Construct a line through X parallel to vector F_n. Extend vector F_s to intersect this line. Construct a line from X perpendicular to F_n. Construct a line through point Y perpendicular to the line through X.)
11. For the data in Problem 7, calculate the shear strain and compare it to $1/r_c$. Comment on the comparison $1/r_c = t_c/t = L/L_c$ assuming that $W = W_c$.
12. A manufacturing engineer needs an estimate of the cutting force F_c to estimate the loss of accuracy of a machining process due to deflection. The material being machined is Inconel 600 with a BHN value of 100.The cutting speed was 250 fpm, the feed was 0.020 in./rev, and the depth of cut was 0.250 in. The chip from the process measured 0.080 in. thick. Estimate the cutting force F_c assuming that $F_t = F_c/2$.
13. Using Figure 20-4 for input data, determine the maximum and minimum MRR values for rough machining (turning) a 1020 carbon steel with a BHN value of 200. Repeat for finish machining assuming a DOC value equal to 10% of the roughing DOC.
14. Estimate the horsepower needed to remove metal at 550 in.^{3}/min with a feed of 0.005 in./rev at a DOC value of 0.675 in. The cutting force F_c was measured at 10,000 lb. Comment on these values.
15. For a turning process, the horsepower required was 24 hp. The metal removal rate was 550 in.^{3}/min. Estimate the specific horsepower and compare to published values for 1020 steel at 200 BHN.

TABLE 7-A

Data										
Run Number	F_c	F_t	Feed ipr ×1/1000	Chip Ratio r_c	ϕ	τ_s	U	HP_s	γ	μ
1	330	295	4.89	0.331						
2	308	280	4.89	0.381						
3	410	330	7.35	0.426						
4	420	340	7.35	0.426						
5	510	350	9.81	0.458						
6	540	395	9.81	0.453						

Table 7-A has machining data for 1020 steel, as-received, in air with a K3H carbide tool, orthogonally (tube cutting on lathe) with tube OD = 2.875. The cutting speed was 530 fpm. The tube wall thickness was 0.200 in. The back rake angle was zero for all cuts. Only the feed rate was changed. Fill in the blanks.

Chapter 20 CASE STUDY

Orthogonal Plate Machining Experiment at Auburn University

Jeremy has just been to his ISNY3800 metal-cutting lab where he learned about orthogonal machining.

This lab provided him with a hands-on-experience in material cutting analysis. The experiment varied cutting speed and uncut chip thickness (UCT), two of the most important cutting parameters in orthogonal metal cutting.

An orthogonal plate machining setup was used in the experiment. A horizontal milling machine equipped with a dynamometer was used to perform the experiment. The material for the experiment was cartridge brass. Eighteen runs were performed using two levels of speed, three levels of UCT, and three levels of positive rake angles. A toolmakers microscope were used to measure the thickness of the chip (t_c). The cutting and thrust forces (F_c and F_t) were measured with a dynamometer. The data were recorded in Table CS-20a.

Make a table with headings for r_c, ϕ, F, N, β, μ, F_s, τ_s, HP_s, and specific energy (U). Complete the table and then discuss (using plots) the effect (on the forces and other calculated values) of the changing level in the input parameters.

Jeremy's professor had the student use a designed experiment so that it can be statistically analyzed. Show how this can be done.

TABLE CS20A The Data for OPM Experiment

Run	α	t_o	V	F_c	F_t	t_c
1	10	0.005	HI	92	18	0.019
2	10	0.005	LO	62	11	0.015
3	10	0.010	HI	140	35	0.020
4	10	0.010	LO	128	25	0.025
5	10	0.020	HI	270	45	0.044
6	10	0.020	LO	200	25	0.040
7	20	0.005	HI	48	4	0.011
8	20	0.005	LO	54	4	0.012
9	20	0.010	HI	110	15	0.027
10	20	0.010	LO	106	12	0.029
11	20	0.020	HI	251	30	0.041
12	20	0.020	LO	249	25	0.057
13	30	0.005	HI	55	4	0.011
14	30	0.005	LO	640	4	0.010
15	30	0.010	HI	98	14	0.028
16	30	0.010	LO	106	3	0.027
17	30	0.020	HI	170	22	0.051
18	30	0.020	LO	192	5	0.040

$V = 16$ in./min
$V = 2$ in./min

CHAPTER 21

Cutting Tools for Machining

21.1 Introduction

Success in **metal cutting** depends on the selection of the proper cutting tool (material and geometry) for a given work material. A wide range of **cutting tool materials** are available with a variety of properties, performance capabilities, and costs. These include high-carbon steels and low-/medium-alloy steels, high-speed steels, cast cobalt alloys, cemented carbides, cast carbides, coated carbides, coated high-speed steels, ceramics, cermets, whisker-reinforced ceramics, sialons, sintered polycrystalline **cubic boron nitride (CBN),** sintered polycrystalline diamond, and single-crystal natural diamond. Figure 21-1 shows some of these common tool materials ranked by the cutting speeds used to machine a unit volume of steel materials, assuming equal tool lives. As the speed (feed rate and DOC) increases, so does the metal removal rate. The time required to remove a given unit volume of material therefore decreases. Notice the five-fold increase in speed that the AL_2O_3-coated carbide has over the WC/Co tool (250 → 1200 sfpm). Today, approximately 85% of carbide tools are coated, almost exclusively by the **chemical vapor deposition (CVD)** process. The cutting tool (material and

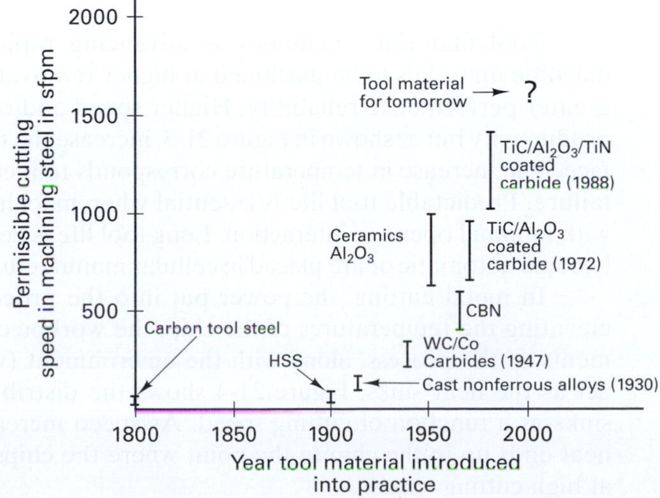

FIGURE 21-1 Improvements in cutting tool materials have led to significant increases in cutting speeds (and productivity) over the years.

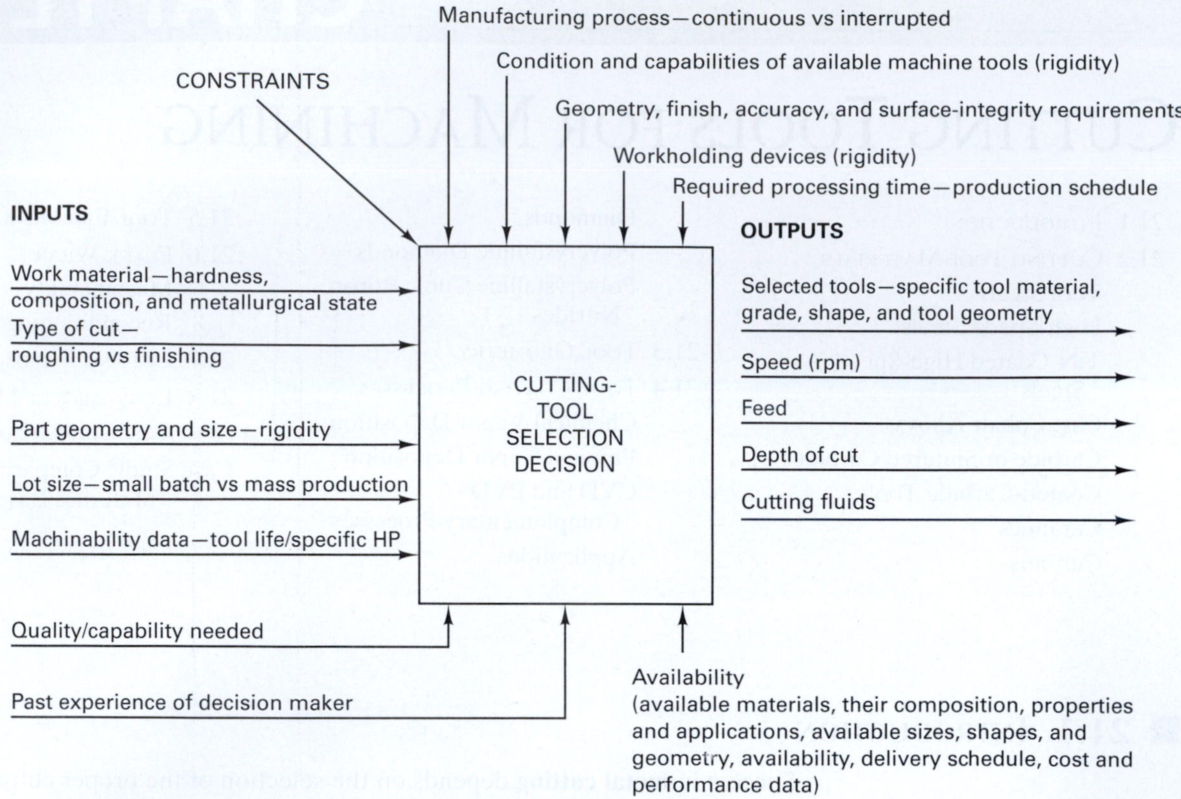

FIGURE 21-2 The selection of the cutting tool material and geometry followed by the selection of cutting conditions for a given application depends on many variables.

geometry) is the most critical aspect of the machining process. Clearly, the cutting tool material, cutting parameters, and tool geometry selected directly influence the productivity of the machining operation. Figure 21-2 outlines the input variables that influence the tool material selection decision. The elements that influence the decision are:

- Work material characteristics, hardness, chemical and metallurgical state.
- Part characteristics (geometry, accuracy, finish, and surface-integrity requirements).
- Machine tool characteristics, including the workholders (adequate rigidity with high horsepower, and wide speed and feed ranges).
- Support systems (operator's ability, sensors, controls, method of lubrication, and chip removal).

Tool material technology is advancing rapidly, enabling many difficult-to-machine materials to be machined at higher removal rates and/or cutting speeds with greater performance reliability. Higher speed and/or removal rates usually improve productivity but as shown in Figure 21-3, increase the temperature at the tool/chip interface. The increase in temperature corresponds to increase in tool wear leading to tool failure. Predictable tool life is essential when machine tools are computer controlled with minimal operator interaction. Long tool life is desirable especially when machines become automatic or are placed in cellular manufacturing systems.

In metal cutting, the power put into the process is largely converted to heat, elevating the temperatures of the chip, the workpiece, and the tool. These three elements of the process, along with the environment (which includes the cutting fluid), act as the heat sinks. Figure 21-4 shows the distribution of the heat to these three sinks as a function of cutting speed. As speed increases, a greater percentage of the heat ends up in the chip to the point where the chips can be cherry red or even burn at high cutting speeds.

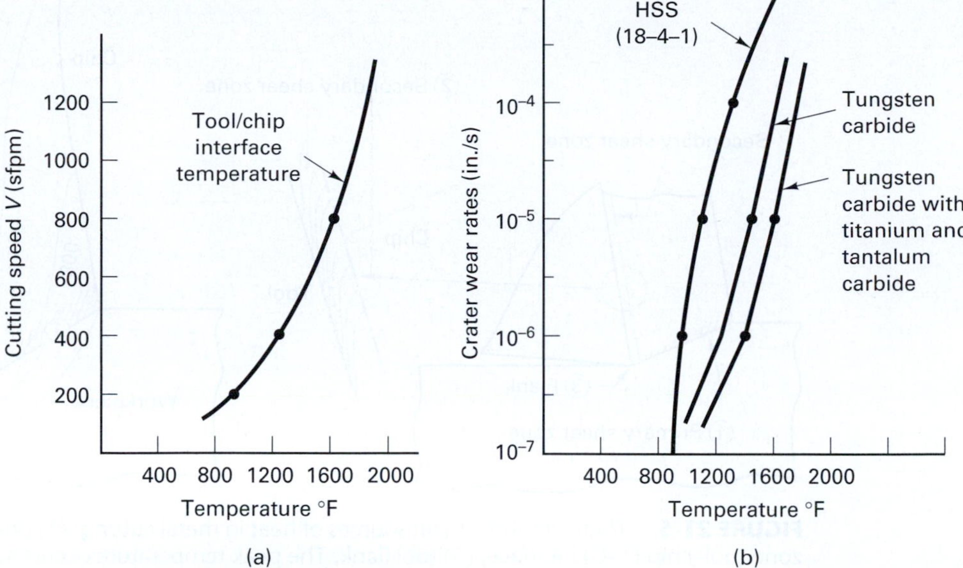

FIGURE 21-3 The typical relationship of temperature at the tool–chip interface to cutting speed shows a rapid increase. Correspondingly, the tool wears at the interface rapidly with increased temperature, often created by increased speed.

There are three main sources of heat. Listed in order of their heat-generating capacity, they are shown in Figure 21-5:

1. The shear front itself, where plastic deformation results in the major heat source. Most of this heat stays in the chip.
2. The tool/chip interface contact region, where additional plastic deformation takes place in the chip and considerable heat is generated due to sliding friction. This heat goes into the chip and the tool.
3. The flank of the tool, where the freshly produced workpiece surface rubs the tool.

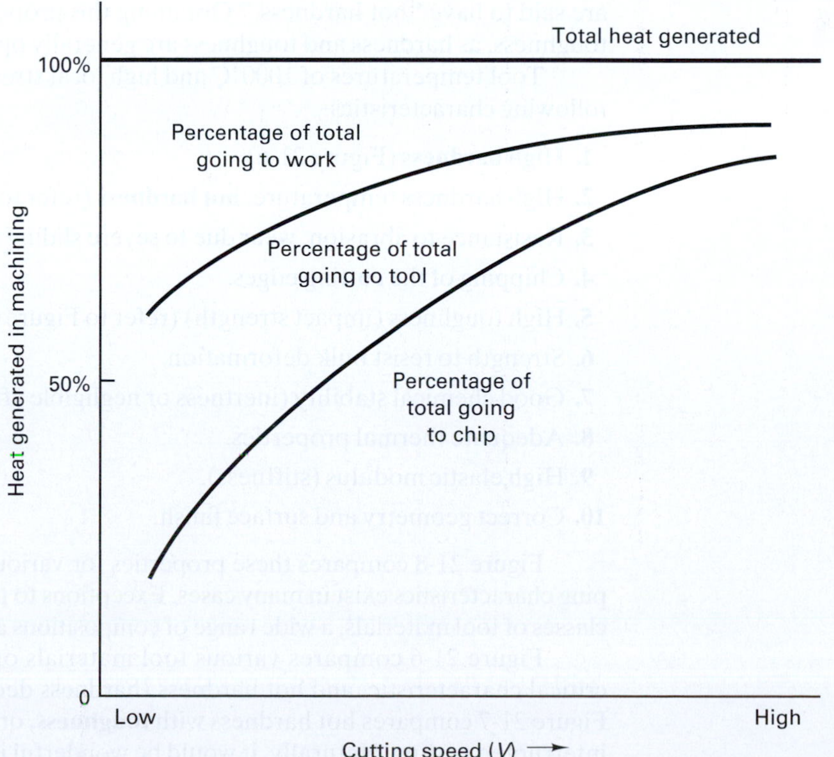

FIGURE 21-4 Distribution of heat generated in machining to the chip, tool, and workpiece. Heat going to the environment is not shown. Figure based on the work of A. O. Schmidt.

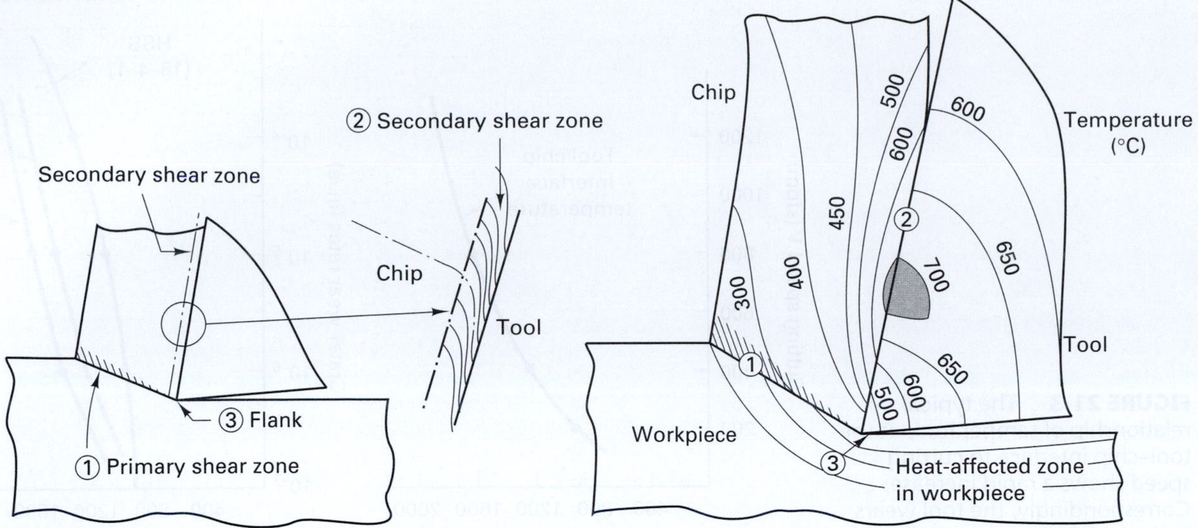

FIGURE 21-5 There are three main sources of heat in metal cutting: (1) primary shear zone; (2) secondary shear zone tool–chip (T–C) interface; (3) tool flank. The peak temperature occurs at the center of the interface, in the shaded region.

There have been numerous experimental techniques developed to measure cutting temperatures and some excellent theoretical analyses of this "moving" multiple-heat-source problem. Space does not permit us to explore this problem in depth. Figure 21-3a shows the effect of cutting speed on the tool–chip interface temperature. The rate of wear of the tool at the interface can be shown to be directly related to temperature (see Figure 21-3b). Because cutting forces are concentrated in small areas near the cutting edge, these forces produce large pressures. The tool material must be hard (to resist wear) and tough (to resist cracking and chipping). Tools used in interrupted cutting, such as milling, must be able to resist impact loading as well. Tool materials must sustain their hardness at elevated temperatures. The challenge to manufacturers of cutting tools has always been to find materials that satisfy these severe conditions. Cutting tool materials that do not lose hardness at the high temperatures associated with high speeds are said to have "hot hardness." Obtaining this property usually requires a trade-off in toughness, as hardness and toughness are generally opposing properties.

Tool temperatures of 1000°C and high local stresses require that the tool have the following characteristics:

1. High **hardness** (Figure 21-6).
2. High hardness temperature, **hot hardness** (refer to Figure 21-6).
3. Resistance to abrasion, wear due to severe sliding friction.
4. Chipping of the cutting edges.
5. High toughness (impact strength) (refer to Figure 21-7).
6. Strength to resist bulk deformation.
7. Good chemical stability (inertness or negligible affinity with the work material).
8. Adequate thermal properties.
9. High elastic modulus (stiffness).
10. Correct geometry and surface finish.

Figure 21-8 compares these properties for various cutting tool materials. Overlapping characteristics exist in many cases. Exceptions to the rule are very common. In many classes of tool materials, a wide range of compositions and properties are obtainable.

Figure 21-6 compares various tool materials on the basis of hardness, the most critical characteristic, and hot hardness (hardness decreases slowly with temperature). Figure 21-7 compares hot hardness with **toughness,** or the ability to take impacts during interrupted cutting. Naturally, it would be wonderful if these materials were also easy to

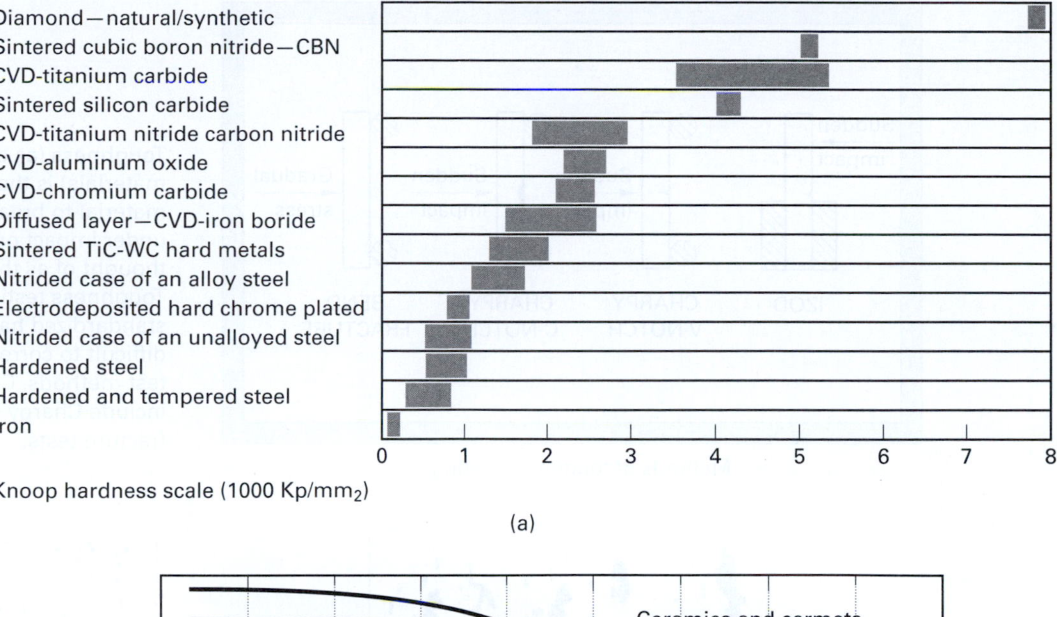

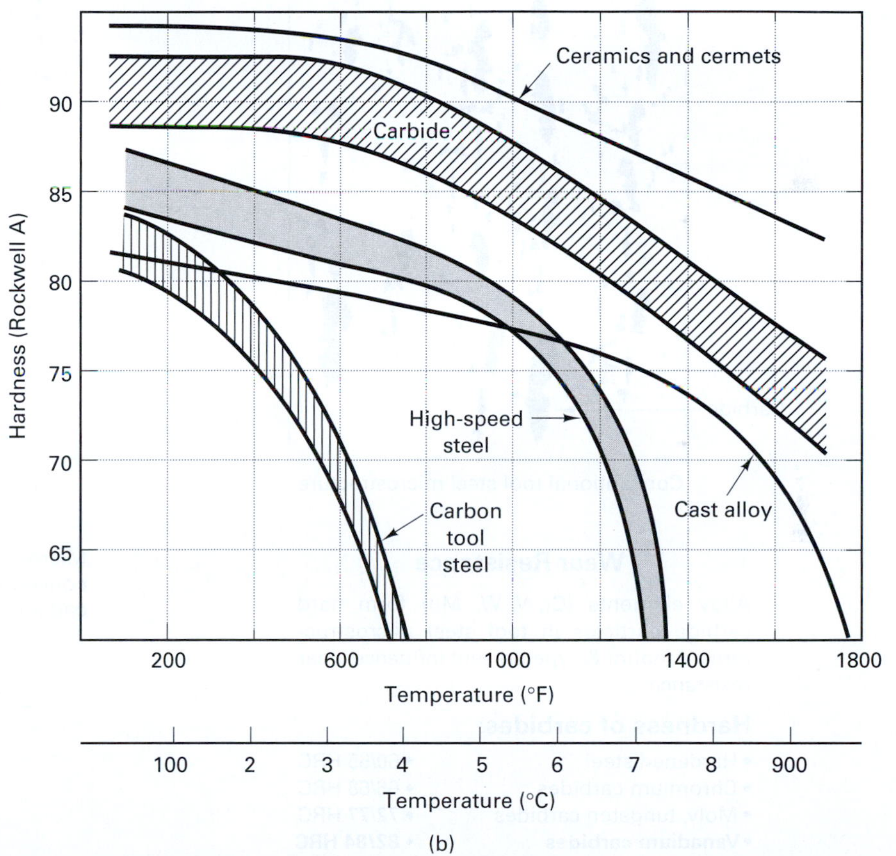

FIGURE 21-6 (a) Hardness of cutting materials and (b) decreasing hardness with increasing temperature, called hot hardness. Some materials display a more rapid drop in hardness above some temperatures. *(From* Metal Cutting Principles, *2nd ed. Courtesy of Ingersoll Cutting Tool Company)*

fabricate, readily available, and inexpensive, because cutting tools are routinely replaced, but this is not usually the case. Obviously, many of the requirements conflict, and therefore, tool selection will always require trade-offs.

21.2 Cutting Tool Materials

In nearly all machining operations, cutting speed and feed are limited by the capability of the tool material. Speeds and feeds must be kept low enough to provide for an

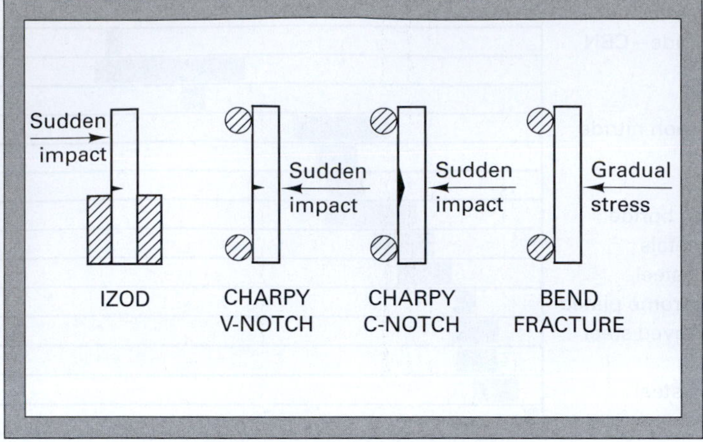

Methods of toughness testing

Toughness

Toughness (as considered for tooling materials) is the relative resistance of a material to breakage, chipping, or cracking under impact or stress. Toughness may be thought of as the opposite of brittleness. Toughness testing is not the same as standardized hardness testing. It may be difficult to correlate the results of different test methods. Common toughness tests include Charpy impact tests and bend fracture tests.

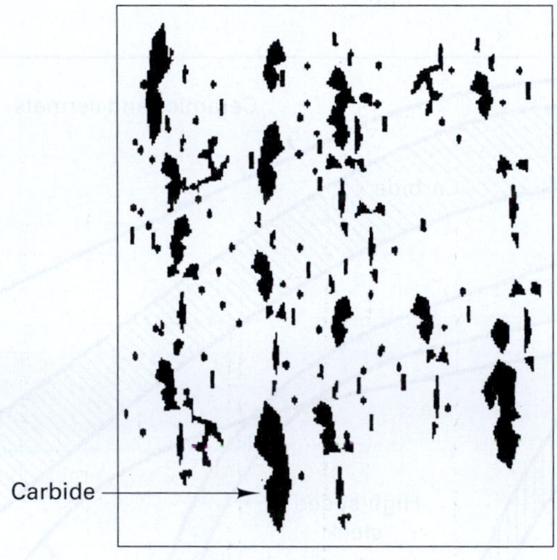

Conventional tool steel microstructure

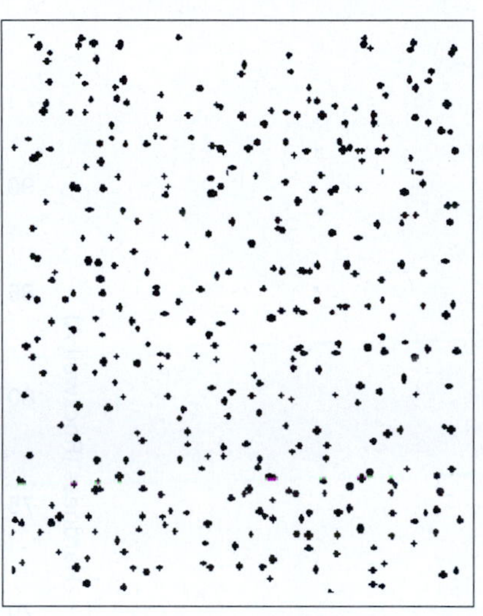

P/M tool steels microstructure

Microstructure of P/M tool steel versus conventional tool steels shows the fine carbide distribution, uniformly distributed.

Wear Resistance

Alloy elements (Cr, V, W, Mo) form hard carbide particles in tool steel microstructures. Amount & type present influence wear resistance.

Hardness of carbides:

• Hardened steel	• 60/65 HRC
• Chromium carbides	• 66/68 HRC
• Moly, tungsten carbides	• 72/77 HRC
• **Vanadium carbides**	• **82/84 HRC**

FIGURE 21-7 The most important properties of tool steels are:
1. Hardness—resistance to deforming and flattening.
2. Toughness—resistance to breakage and chipping.
3. Wear resistance—resistance to abrasion and erosion.

acceptable tool life. If not, the time lost changing tools may outweigh the productivity gains from increased cutting speed. Coated high-speed steel (HSS) and uncoated and coated carbides are currently the most extensively used tool materials.

Coated tools cost only about 15 to 20% more than uncoated tools, so a modest improvement in performance can justify the added cost. About 15 to 20% of all tool steels are coated, mostly by the **physical vapor deposition (PVD)** processes. Diamond and CBN are used for applications in which, despite higher cost, their use is justified.

FIGURE 21-8 Salient properties of cutting tool materials.[a]

	Carbon and Low-/Medium-Alloy Steels	High-Speed Steels	Sintered Cemented Carbides	Coated HSS	Coated Carbides	Ceramics	Polycrystalline CBN	Diamond
Toughness	▶ Decreasing ▶							
Hot hardness	▶ Increasing ▶							
Impact strength	◀ Decreasing ◀							
Wear resistance	▶ Increasing ▶							
Chipping resistance	◀ Decreasing ◀							
Cutting speed	▶ Increasing ▶							
Depth of cut	Light to medium	Light to heavy	Light to heavy	Light to heavy	Light to heavy	Light to heavy	Light to heavy	Very light for single-crystal diamond
Finish obtainable	Rough	Rough	Good	Good	Good	Very good	Very good	Excellent
Method of manufacture	Wrought	Wrought cast, HIP sintering	Cold pressing and sintering, PM	PVD[b] after forming	CVD[c]	Cold pressing and sintering or HIP sintering	High-pressure–high-temperature sintering	High-pressure–high-temperature sintering
Fabrication	Machining and grinding	Machining and grinding	Grinding	Machining and grinding, coating	Grinding before coating	Grinding	Grinding and polishing	Grinding and polishing
Thermal shock resistance	▶ Increasing ▶							
Tool material cost	▶ Increasing ▶							

[a]Overlapping characteristics exist in many cases. Exceptions to the rule are very common. In many classes of tool materials, a wide range of compositions and properties are obtainable.
[b]Physical vapor deposition.
[c]Chemical vapor deposition.

Cast cobalt alloys are being phased out because of the high raw-material cost and the increasing availability of alternate tool materials. New ceramic materials called *cermets* (ceramic material in a metal binder) are having a significant impact on future manufacturing productivity.

Tool requirements for other processes that use noncontacting tools, as in electrodischarge machining (EDM) and electrochemical machining (ECM), or *no tools at all* (as in laser machining), are discussed in Chapter 28. Grinding abrasives will be discussed in Chapter 26.

TOOL STEELS

Carbon steels and low-/medium-alloy steels, called **tool steels,** were once the most common cutting tool materials. Plain-carbon steels of 0.90 to 1.30% carbon when hardened and tempered have good hardness and strength and adequate toughness and can be given a keen cutting edge. However, tool steels lose hardness at temperatures above 400°F because of tempering and have largely been replaced by other materials for metal cutting.

The most important properties for tool steels are hardness, hot hardness, and toughness. Low-/medium-alloy steels have alloying elements such as molybdenum and chromium, which improve hardenability, and tungsten and molybdenum, which improve wear resistance. These tool materials also lose their hardness rapidly when heated to about their tempering temperature of 300° to 650°F, and they have limited abrasion resistance. Consequently, low-/medium-alloy steels are used in relatively inexpensive cutting tools (e.g., drills, taps, dies, reamers, broaches, and chasers) for certain low-speed cutting applications when the heat generated is not high enough to reduce their hardness significantly. High-speed steels, cemented carbides, and coated tools are also used extensively to make these kinds of cutting tools. Although more expensive, they have longer tool life and improved performance. These steels greatly benefit from **powder metallurgy (P/M)** manufacturing due to uniformly distributed carbides.

HIGH-SPEED STEELS

First introduced in 1900 by F. W. Taylor and Mansel White, high-alloy steel is superior to tool steel in that it retains its cutting ability at temperatures up to 1100°F, exhibiting good "red hardness." Compared with tool steel, it can operate at about double or triple cutting speeds to about 100 sfpm with equal life, resulting in its name: **high-speed steel,** often abbreviated **HSS.**

Today's high-speed steels contain significant amounts of tungsten, molybdenum, cobalt, vanadium, and chromium besides iron and carbon. Tungsten, molybdenum, chromium, and cobalt in the ferrite as a solid solution provide strengthening of the matrix beyond the tempering temperature, thus increasing the hot hardness. Vanadium, along with tungsten, molybdenum, and chromium, improves hardness (R_c 65–70) and wear resistance. Extensive solid solutioning of the matrix also ensures good hardenability of these steels.

Although many formulations are used, a typical composition is that of the 18-4-1 type (tungsten 18%, chromium 4%, vanadium 1 %), called T1. Comparable performance can also be obtained by the substitution of approximately 8% molybdenum for the tungsten, referred to as a tungsten equivalent (W_{eq}). High-speed steel is still widely used for drills and many types of general-purpose milling cutters and in single-point tools used in general machining. For high-production machining, it has been replaced almost completely by carbides, coated carbides, and coated HSS.

HSS main strengths are as follows:

- Great toughness—superior transverse rupture strength.
- Easily fabricated.
- Best for sever applications where complex tool geometry is needed (gear cutters, taps, drills, reamers, dies).

High-speed steel tools are fabricated by three methods: cast, wrought, and sintered (using the powder metallurgy technique). Improper processing of cast and wrought products can result in carbide segregation, formation of large carbide particles and significant variation of carbide size, and nonuniform distribution of carbides in the matrix. The material will be difficult to grind to shape and will cause wide fluctuations of properties, inconsistent tool performance, distortion, and cracking.

To overcome some of these problems, a powder metallurgy technique has been developed that uses the hot-isostatic pressing (HIP) process on atomized, prealloyed tool steel mixtures. Because the various constituents of the P/M alloys are "locked" in place by the compacting procedure, the end product is a more homogeneous alloy, Figure 21-7. P/M high-speed steel cutting tools exhibit better grindability, greater toughness, better wear resistance, and higher red (or hot) hardness; they also perform more consistently. They are about double the cost of regular HSS.

TiN-COATED HIGH-SPEED STEELS

Coated high-speed steel provides significant improvements in cutting speeds, with increases of 10 to 20% being typical. First introduced in 1980 for gear cutters (hobs) and in 1981 for drills, TiN-coated HSS tools have demonstrated their ability to more than pay for the extra cost of the coating process.

In addition to hobs, gear-shaper cutters, and drills, HSS tooling coated by TiN now includes reamers, taps, chasers, spade-drill blades, broaches, bandsaw and circular saw blades, insert tooling, form tools, end mills, and an assortment of other milling cutters.

Physical vapor deposition has proved to be the most viable process for coating HSS, primarily because it is a relatively low-temperature process that does not exceed the tempering point of HSS. Therefore, no subsequent heat treatment of the cutting tool is required. Films 0.0001 to 0.0002 in. thick adhere well and withstand minor elastic, plastic, and thermal loads. Thicker coatings tend to fracture under the typical thermomechanical stresses of machining.

There are many variations to the PVD process, as outlined in Table 21-1. The process usually depends on gas pressure and is performed in a vacuum chamber. PVD processes are carried out with the workpieces heated to temperatures in the range of 400 to 900°F. Substrate heating enhances coating adhesion and film structure.

Because surface pretreatment is critical in PVD processing, tools to be coated are subject to a vigorous cleaning process. Precleaning methods typically involve degreasing, ultrasonic cleaning, and Freon drying. Deburring, honing, and more active cleaning methods are also used.

The main advantage of TiN-coated HSS tooling is reduced tool wear. Less tool wear results in less stock removal during tool regrinding, thus allowing individual tools to be reground more times. For example, a TiN hob can cut 300 gears per sharpening; the uncoated tool would cut only 75 parts per sharpening. Therefore, the cost per gear is reduced from 20 to 2¢. Naturally, reduced tool wear means longer tool life.

Higher hardness, with typical values for the thin coatings, "equivalent" to R_c 80–85, as compared to R_c 65–70 for hardened HSS, means reduced abrasion wear. Relative inertness (i.e., TiN does not react significantly with most workpiece materials) results in greater tool life through a reduction in adhesion. TiN coatings have a low coefficient of friction. This can produce an increase in the shear angle, which in turn reduces the cutting forces, spindle power, and heat generated by the deformation processes. PVD coatings generally fail in high-stress applications such as cold extrusion, piercing, roughing, and high-speed machining.

CAST COBALT ALLOYS

Cast cobalt alloys, popularly known as **stellite tools,** are cobalt-rich, chromium–tungsten–carbon cast alloys having properties and applications in the intermediate range between high-speed steel and cemented carbides. Although comparable in room-temperature hardness to high-speed steel tools, cast cobalt alloy tools retain their hardness to a much higher temperature. Consequently, they can be used at higher cutting speeds (25% higher) than HSS tools. Cast cobalt alloys are hard as cast and cannot be softened or heat treated.

Cast cobalt alloys contain a primary phase of cobalt-rich solid solution strengthened by chromium and tungsten and dispersion hardened by complex hard, refractory

TABLE 21-1 Surface Treatments for Cutting Tools

Process	Method	Hardness[a] and Depth	Advantages	Limitations
Black oxide	HSS cutting tools are oxidized in a steam atmosphere at 1000°F	No change in prior steel hardness	Prevents built-up edge formations in machining of steel.	Strictly for HSS tools.
Nitriding case hardening	Steel surface is coated with nitride layer by use of cyanide salt at 900° to 1600°F, or ammonia, gas. or N_2 ions.	To 72 R_c; Case depth: 0.0001 to 0.100 in.	High production rates with bulk handling. High surfcice hardness. Diffuses into the steel surfaces. Simulates strain hardening.	Can only be applied to steel. Process has embrittling effect because of greater hardness. Post-heat treatment needed for some allovs.
Electrolytic electroplating	The part is the cathode in a chromic acid solution; anode is lead. Hard chronic plating is the most common process for wear resistance.	70-72 R_c 0.0002 to 0.100 in.	Low friction coefficient, antigalling Corrosion resistance. High hardness.	Moderate production: pieces must be fixtured. Part must be very clean. Coating does: not diffuse into surface, which can affect impact properties.
Vapor deposition chemical vapor deposition (CVD)	Deposition of coating material by chemical reactions in the gaseous phase. Reactive gases replace a protective atmosphere in a vacuum chamber, At temperatures of 1800° to 1200°F, a thin diffusion zone is created between the base metal and the coating.	To 84 R_c; 0.0002 to 0.0004 in.	Large quantities per batch. Short reaction times reduce substrate stresses. Excellent adhesion, recommended for forming tools. Multiple coatings can be applied (TiN-TiC. Al_2O_3) Line-of-sight not a problem.	High temperatures can affect substrate metallurgy, requiring post-heat treatment, which can cause dimensional distortion (except when coating sintered carbides). Necessary to reduce effects, of hydrogen chloride on material properties, such as impact strength. Usually not diffused. Tolerances of + 0.001 required for HSS tools.
Physical vapor desposition (PVD sputtering)	Plasma is generated in a vacuum chamber by ion bombardment to dislodge particles from a target made of the coating material. Metal is evaporated and is condensed or attracted to substrate surfaces.	To 84 R_c; To 0.0002 in. thick	A useful experimental procedure for developing wear surfaces. Can coat substrates with metals, alloys, compounds, and refractories. Applicable for all tooling.	Not a high-production method. Requires care in cleaning. Usually not diffused
PVD (electron beam)	A plasma is generated in vacuum by evaporation from a molten pool that is heated by an electron-beam gun.	To 84 R_c To 0.0002 in. thick	Can coat reasonable quantities per batch cycle. Coating materials, are metals, compounds, alloys, and refractories. Substrate metallurgy is preserved. Very good adhesion. Fine particle deposition. Applicable for all tooling.	Parts require fixturing and orientation in line-of-sight process. Ultra-cleanliness required.
PVD/ARC	Titanium is evaporated in a vacuum and reacted with nitrogen Gas. Resulting titanium nitride plasma is ionized and electrically attracted to the substrate surface. A high-energy process with. multiple plasma guns.	To 85 R_c; To 0.0002 in. thick.	Process at 900°F preserves substrate metallurgy. Excellent coating adhesion. Controllable deposition of grain size and growth. Dimensions, surface finish, and sharp edges are preserved Can coat all high-speed steels without distortion.	Parts must be fixtured for line-of-sight process. Parts must be very clean. No by-products formed in reaction. Usually only minor diffusion.

[a] Rockwell hardness values above 68 are estimates.

carbides of tungsten and chromium. Other elements added include vanadium, boron, nickel, and tantalum. The casting provides a tough core and elongated grains normal to the surface. The structure is not, however, homogeneous.

Tools of cast cobalt alloys are generally cast to shape and finished to size by grinding. They are available only in simple shapes, such as single-point tools and saw blades, because of limitations in the casting process and the expense involved in the final shaping (grinding). The high cost of fabrication is primarily due to the high hardness of the material in the as-cast condition. Materials that can be machined with this tool material include plain-carbon steels, alloy steels, nonferrous alloys, and cast iron.

Cast cobalt alloys are currently being phased out for cutting-tool applications because of increasing costs, shortages of strategic raw materials (cobalt, tungsten, and chromium), and the development of other, superior tool materials at lower cost.

CARBIDE OR SINTERED CARBIDES

Carbide cutting tool inserts are traditionally divided into two primary groups:

1. Straight tungsten grades, which are used for machining cast irons, austenitic stainless steel, and nonferrous and nonmetallic materials.

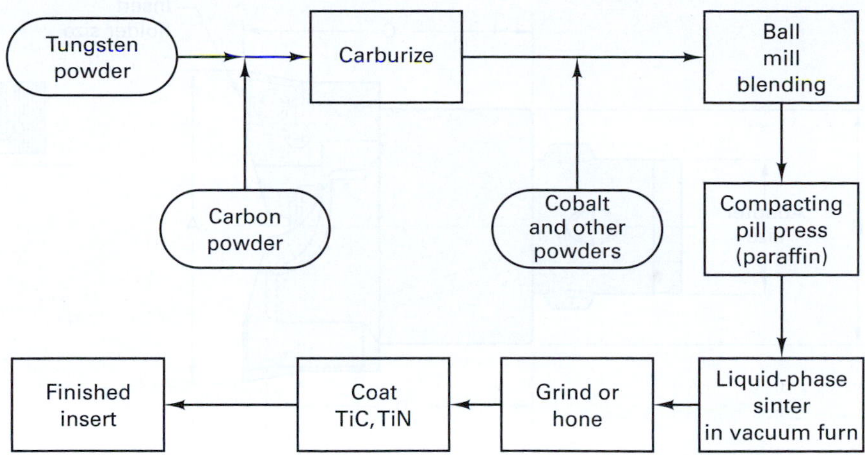

Tungsten is carburized in a high-temperature furnace, mixed with cobalt and blended in large ball mills. After ball milling, the powder is screened and dried. Paraffin is added to hold the mixture together for compacting. Carbide inserts are compacted using a pill press. The compacted powder is sintered in a high-temperature vacuum furnace. The solid cobalt dissolves some tungsten carbide, then melts and fills the space between adjacent tungsten carbide grains. As the mixture is cooled, most of the dissolved tungsten carbide precipitates onto the surface of existing grains. After cooling, inserts are finish ground and honed or used in the pressed condition.

FIGURE 21-9 P/M process for making cemented carbide insert tools.

2. Grades containing major amounts of titanium, tantalum, and or columbium carbides, which are used for machining ferritic workpieces. There are also the titanium carbide grades, which are used for finishing and semifinishing ferrous alloys.

The classification of carbide insert grades employs a C-classification system in the United States and ISO P and M classification system in Europe and Japan. These classifications are based on application, rather than composition or properties. Each cutting-tool vendor can provide proprietary grades and recommended applications.

Carbides, which are nonferrous alloys, are also called **sintered** (or cemented) **carbides** because they are manufactured by powder metallurgy techniques. The P/M process is outlined in Figure 21-9. See Chapter 18 for details on powder metallurgy processes. These materials became popular during World War II because they afforded a four- or fivefold increase in cutting speeds. The early versions had tungsten carbide as the major constituent, with a cobalt binder in amounts of 3 to 13 %. Most carbide tools in use today are either straight WC or multicarbides of W–Ti or W–Ti–Ta, depending on the work material to be machined. Cobalt is the binder. These tool materials are much harder, are chemically more stable, have better hot hardness, have high stiffness, have lower friction, and operate at higher cutting speeds than HSS. They are more brittle and more expensive and use strategic metals (tungsten, tantalum, cobalt) more extensively.

Cemented carbide tool materials based on TiC have been developed primarily for auto industry applications using predominantly nickel and molybdenum as a binder. These are used for higher-speed (>1000 ft/min) finish machining of steels and some malleable cast irons.

Cemented carbide tools are available in insert form in many different shapes: squares, triangles, diamonds, and rounds. They can be either brazed or mechanically clamped onto the tool shank. Mechanical clamping (Figure 21-10) is more popular because when one edge or corner becomes dull, the insert is rotated or turned over to expose a new cutting edge. Mechanical inserts can be purchased in the as-pressed state, or the insert can be ground to closer tolerances. Naturally, precision-ground inserts cost more. Any part tolerance less than ±0.003 normally cannot be manufactured without radial adjustment of the cutting tool, even with ground inserts. If no radial adjustment is performed, precision-ground inserts should be used only when the part tolerance is between ±0.006 and ±0.003. Pressed inserts have an application advantage because the

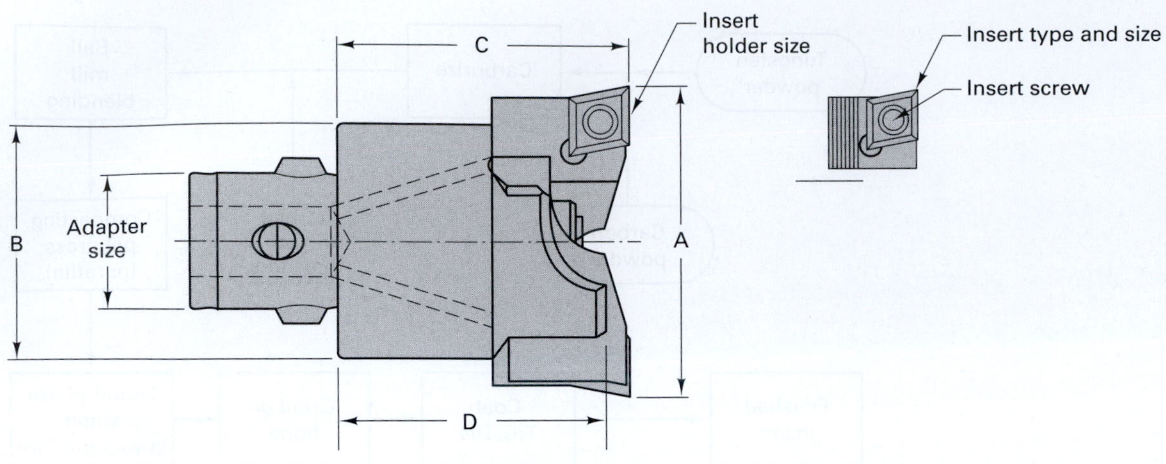

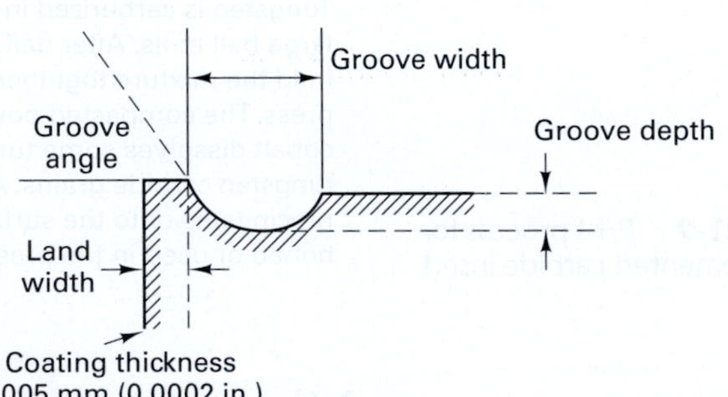

FIGURE 21-10 Boring head with carbide insert cutting tools. These inserts have a chip groove that can cause the chips to curl tightly and break into small, easily disposed lengths.

cutting edge is unground and thus does not leave grinding marks on the part after machining. Ground inserts can break under heavy cutting loads because the grinding marks on the insert produce stress concentrations that result in brittle fracture. Diamond grinding is used to finish carbide tools. Abusive grinding can lead to thermal cracks and premature (early) failure of the tool. Brazed tools have the carbide insert brazed to the steel tool shank. These tools will have a more accurate geometry than the mechanical insert tools, but they are more expensive. Because cemented carbide tools are relatively brittle, a 90-degree corner angle at the cutting edge is desired. To strengthen the edge and prevent edge chipping, it is rounded off by honing, or an appropriate chamfer or a negative land (a T-land) on the rake face is provided. The preparation of the cutting edge can affect tool life. The sharper the edge (smaller edge radius), the more likely the edge is to chip or break. Increasing the edge radius will increase the cutting forces, so a trade-off is required. Typical edge radius values are 0.001 to 0.003 in.

A **chip groove** (see Figure 21-10) with a positive rake angle at the tool tip may also be used to reduce cutting forces without reducing the overall strength of the insert significantly. The groove also breaks up the chips by causing them to curl tightly, thus making disposal easier.

For very low-speed cutting operations, the chips tend to weld to the tool face and cause subsequent microchipping of the cutting edge. Cutting speeds for carbides are generally in the range of 150 to 600 ft/min. Higher speeds (>1000 ft/min) are recommended for certain less-difficult-to-machine materials (such as aluminum alloys) and much lower speeds (100 ft/min) for more difficult-to-machine materials (such as titanium alloys). In interrupted cutting applications, it is important to prevent edge chipping by choosing the appropriate cutter geometry and cutter position with respect to the workpiece. For interrupted cutting, finer grain size and higher cobalt content improve toughness in straight WC–Co grades.

After use, carbide inserts (called disposable or throwaway inserts) are generally recycled in order to reclaim the tantalum, WC, and cobalt. This recycling not only conserves strategic materials but also reduces costs. A new trend is to regrind these tools for future use where the actual size of the insert is not of critical concern.

COATED-CARBIDE TOOLS

Beginning in 1969 with TiC-coated WC, **coated tools** became the norm in the metalworking industry because coating can consistently improve tool life 200 or 300% or more. In cutting tools, material requirements at the surface of the tool need to be abrasion resistant, hard, and chemically inert to prevent the tool and the work material from interacting chemically with each other during cutting. A thin, chemically stable, hard refractory coating accomplishes this objective. The bulk of the tool is a tough, shock-resistant carbide that can withstand high-temperature plastic deformation and resist breakage. The result is a composite tool as shown in Figure 21-11. The coatings must be fine grained, free of binders and porosity. Naturally, the coatings must be metallurgically bonded to the substrate. Interface coatings are graded to match the properties of the coating and the substrate. The coatings must be thick enough to prolong tool life but thin enough to prevent brittleness.

Coatings should have a low coefficient of friction so the chips do not adhere to the rake face. Coating materials include single coatings of TiC, TiN, Al_2O_3, HfN, or HfC. Multiple coatings are used, with each layer imparting its own characteristic to the tool. Successful coating combinations include TiN/TiC/TiCN/TiN and TiC/Al_2O_3/TiN.

Chemical vapor deposition is used to obtain coated carbides. The coatings are formed by chemical reactions that take place only on or near the substrate. Like electroplating, CVD is a process in which the deposit is built up atom by atom. It is, therefore, capable of producing deposits of maximum density and of closely reproducing fine detail on the substrate surface.

Control of critical variables such as temperature, gas concentration, and flow pattern is required to ensure adhesion of the coating to the substrate. The coating-to-substrate adhesion must be better for cutting tool inserts than for most other coatings applications to survive the cutting pressure and temperature conditions without flaking off. Grain size and shape are controlled by varying temperature and/or pressure.

The purpose of multiple coatings is to tailor the coating thickness for prolonged tool life. Multiple coatings allow a stronger metallurgical bond between the coating and the substrate and provide a variety of protection processes for machining different work materials, thus offering a more general-purpose tool material grade. A very thin final coat of TiN coating (in microns, or μm) can effectively reduce crater formation on the tool face by one to two orders of magnitude relative to uncoated tools.

Coated inserts of carbides are finding wide acceptance in many metalcutting applications. Coated tools have two or three times the wear resistance of the best uncoated tools with the same breakage resistance. This results in a 50 to 100% increase in speed for the same tool life. Because most coated inserts cover a broader application range, fewer grades are needed; therefore, inventory costs are lower. Aluminum oxide coatings have demonstrated excellent crater wear resistance by providing a chemical diffusion reaction barrier at the tool–chip interface, permitting a 90% increase in cutting speeds in machining some steels.

Coated-carbide tools have progressed to the place where in the United States about 80 to 90% of the carbide tools used in metalworking are coated.

CERAMICS

Ceramics are made of pure **aluminum oxide,** Al_2O_3, or Al_2O_3 used as a metallic binder. Using P/M, very fine particles are formed into cutting tips under a pressure of 267 to 386 MPa (20 to 28 ton/in.^{2}) and sintered at about 1000°C (1800°F). Unlike the case with ordinary ceramics, sintering occurs without a vitreous phase.

Ceramics are usually in the form of disposable tips. They can be operated at two to three times the cutting speeds of tungsten carbide. They almost completely resist cratering, run with no coolant, and have about the same tool life at their higher speeds as tungsten carbide does at lower speeds. As shown in Table 21-2, ceramics are usually as hard as carbides but are more brittle (lower bend strength) and therefore require more rigid tool holders and machine tools in order to take advantage of their capabilities. Their hardness and chemical inertness make ceramics a good material for high-speed finishing and/or high-removal-rate machining applications of superalloys, hard-chill

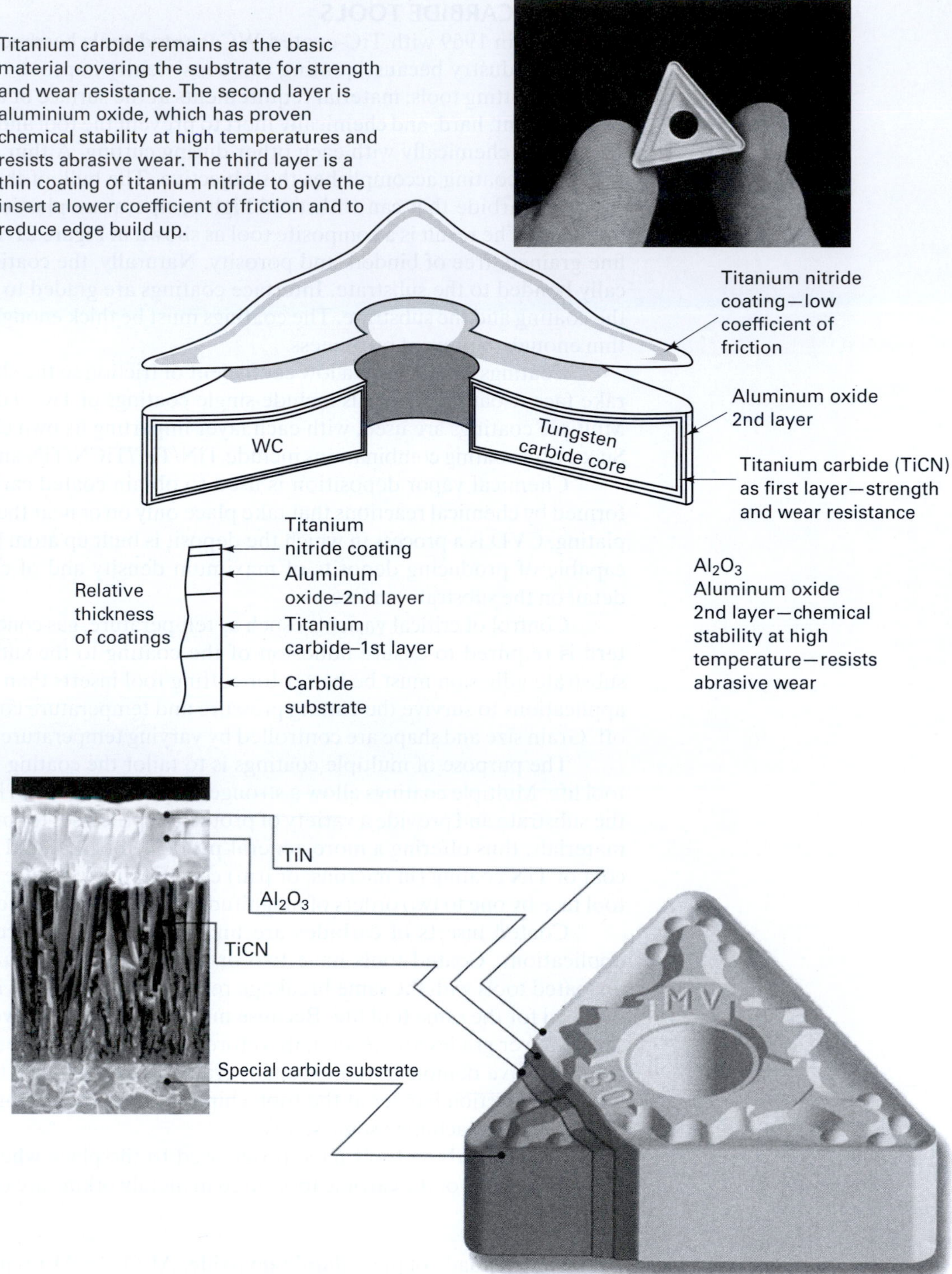

FIGURE 21-11 Triple-coated carbide tools provide resistance to wear and plastic deformation in machining of steel, abrasive wear in cast iron, and built-up edge formation. *(Courtesy J T. Black)*

cast iron, and high-strength steels. Because ceramics have poor thermal and mechanical shock resistance, interrupted cuts and interrupted application of coolants can lead to premature tool failure. Edge chipping is usually the dominant mode of tool failure. Ceramics are not suitable for aluminum, titanium, and other materials that react chemically with alumina-based ceramics. Recently, whisker-reinforced ceramic materials that have greater transverse rupture strength have been developed. The whiskers are made from silicon carbide.

TABLE 21-2 Properties of Cutting Tool Materials Compared for Carbides, Ceramics, HSS, and Cast Cobalt[a]

	Hardness Rockwell A or C	Transverse Rupture (bend) Strength ($\times 10^3$ psi)	Compressive Strength ($\times 10^3$ psi)	Modulus of Elasticity (e) ($\times 10^6$ psi)
Carbide C1–C4	90–95 R_A	250–320	750–860	89–93
Carbide C5–C8	91–93 R_A	100–250	710–840	66–81
High-speed steel	86 R_A	600	600–650	30
Ceramic (oxide)	92–94 R_A	100–125	400–650	50–60
Cast cobalt	46–62 R_C	80–120	220–335	40

[a] Exact properties depend on materials, grain size, bonder content, volume.

CERMETS

Cermets are a class of tool materials often used for finishing processes. Cermets are ceramic TiC, nickel, cobalt, and tantalum nitrides. TiN and other carbides are used for binders. Cermets have superior wear resistance, longer tool life, and can operate at higher cutting speeds with superior wear resistance. Cermets have higher hot hardness and oxidation resistance than cemented carbides. The better finish imparted by a cermet is due to its low level of chemical reaction with iron [less cratering and **built-up edge (BUE)** formation]. Compared to carbide, the cermet has less toughness, lower thermal conductivity, and greater thermal expansion, so thermal cracking can be a problem during interrupted cuts.

Cermets are usually cold pressed, and proper processing techniques are required to prevent insert cracking. New cermets are designed to resist thermal shocking during milling by using high nitrogen content in the titanium carbonitride phase (produces finer grain size) and adding WC and TaC to improve shock resistance. PVD-coated cermets have the wear resistance of cermets and the toughness range of a coated carbide, and they perform well with a coolant.

Figure 21-12 shows a comparison of speed feed coverage of typical cermets compared to ceramics, carbides, and coated carbides. The values illustrate that cermets can clearly cover a wide range of important metal-cutting applications.

DIAMONDS

Diamond is the hardest material known. Industrial diamonds are now available in the form of polycrystalline compacts, which are finding industrial application in the machining of aluminum, bronze, and plastics, greatly reducing the cutting forces as compared to carbides. Diamond machining is done at high speeds, with fine feeds for finishing, and produces excellent finishes. Recently, *single-crystal* diamonds, with a cutting-edge radius of 100 Å or less, have been used for precision machining of large mirrors. However, single-crystal diamonds have been used for years to machine brass watch faces, thus eliminating polishing. They have also been used to slice biological materials into thin films for viewing in transmission electron microscopes. (This process, known as *ultramicrotomy,* is one of the few industrial versions of orthogonal machining in common practice.)

The salient features of diamond tools include high hardness; good thermal conductivity; the ability to form a sharp edge of cleavage (single-crystal, natural diamond); very low friction; nonadherence to most materials; the ability to maintain a sharp edge for a long period of time, especially in machining soft materials such as copper and aluminum; and good wear resistance.

To be weighed against these advantages are some shortcomings, which include a tendency to interact chemically with elements of Group IVB to Group VIII of the periodic table. In addition, diamond wears rapidly when machining or grinding mild steel. It wears less rapidly with high-carbon alloy steels than with low-carbon steel and has occasionally machined gray cast iron (which has high carbon content) with long life.

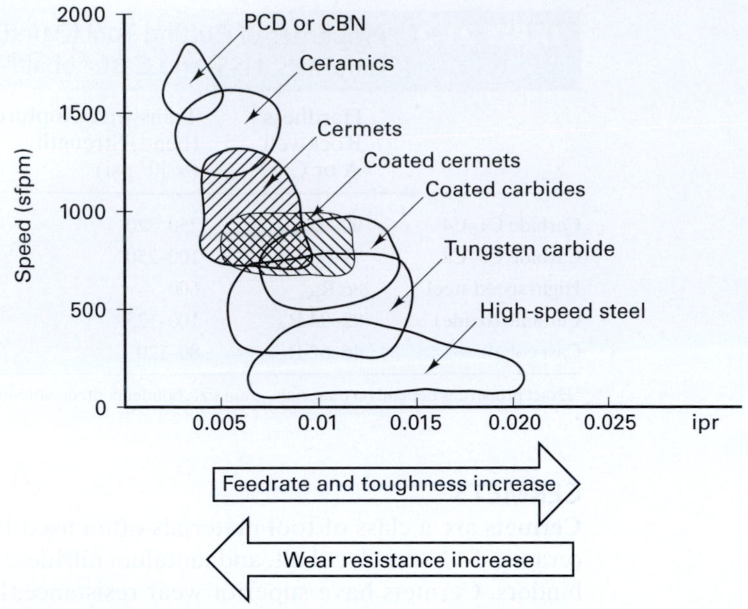

Tool Material Group	General Applications	Versus Cermet
PCD (polycrystal diamond)	High-speed machining of aluminum alloys, nonferrous metals, and nonmetals.	Cermets can machine same materials, but at lower speeds and significantly less cost per corner.
CBN (cubic boron nitride)	Hard workpieces and high-speed machining on cast irons.	Cermets cannot machine the harder workpieces that CBN can. Cermets cannot machine cast iron at the speeds CBN can. The cost per corner of cermets is significantly less.
Ceramics (cold press)	High-speed turning and grooving of steels and cast iron.	Cermets are more versatile and less expensive than cold press ceramics but cannot run at the higher speeds.
Ceramics (hot press)	Turning and grooving of hard workpieces; high-speed finish machining of steels and irons.	Cermets cannot machine the harder workpieces or run at the same speeds on steels and irons but are more versatile and less expensive.
Ceramics (silicon nitride)	Rough and semirough machining of cast irons in turning and milling applications at high speeds and under unfavorable conditions.	Cermets cannot machine cast iron at the high speeds of silicon nitride ceramics, but in moderate-speed applications cermets may be more cost effective.
Coated carbide	General-purpose machining of steels, stainless steels, cast iron, etc.	Cermets can run at higher cutting speeds and provide better tool life at less cost for semiroughing to finishing applications.
Carbides	Tough material for lower-speed applications on various materials.	Cermets can run at higher speeds, provide better surface finishes and longer tool life for semiroughing to finishing applications.

FIGURE 21-12 Comparison of cermets with various cutting tool materials.

Diamond has a tendency to revert at high temperatures (700°C) to graphite and/or to oxidize in air. Diamond is very brittle and is difficult and costly to shape into cutting tools—the process for doing the latter being a tightly held industry practice.

POLYCRYSTALLINE DIAMONDS

The limited supply of, increasing demand for, and high cost of natural diamonds led to the ultra-high-pressure (50 Kbar), high-temperature (1500°C) synthesis of diamond from graphite at the General Electric Company in the mid-1950s and the subsequent development of *polycrystalline* sintered diamond tools in the late 1960s.

Polycrystalline diamond (PCD) tools consist of a thin layer (0.5 to 1.5 mm) of fine-grain-size diamond particles sintered together and metallurgically bonded to a cemented carbide substrate. A high-temperature/high-pressure process, using conditions close to those used for the initial synthesis of diamond, is needed. Fine diamond powder (1 to 30 μm) is first packed on a support base of cemented carbide in the press. At the appropriate sintering conditions of pressure and temperature in the diamond stable region, complete consolidation and extensive diamond-to-diamond bonding take place. Laser cutting followed by grinding is used to shape, size, and accuracy finish PCD tools, as outlined in Figure 21-13. The cemented carbide provides the necessary elastic support for the hard and brittle diamond layer above it. The main advantages of sintered polycrystalline tools over natural single-crystal tools are better quality, greater toughness, and improved wear resistance, resulting from the random orientation of the diamond grains and the lack of large cleavage planes. Diamond tools offer dramatic

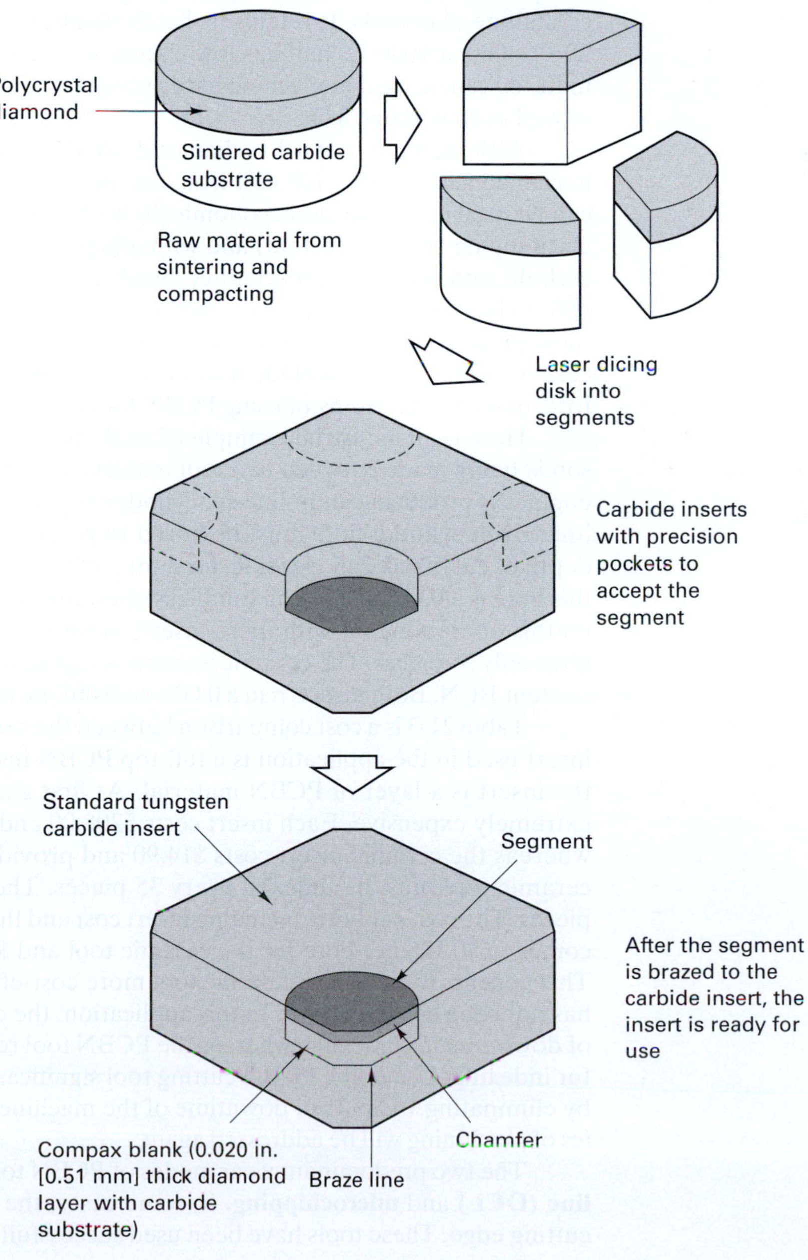

FIGURE 21-13 The process to make polycrystalline diamond tools uses carbides and diamond inserts and are restricted to simple geometries.

performance improvements over carbides. Tool life is often greatly improved, as is control over part size, finish, and surface integrity.

Positive rake tooling is recommended for the vast majority of diamond tooling applications. If BUE formation is a problem, increasing cutting speed and using more positive rake angles may eliminate it. If edge breakage and chipping are problems, the feed rate can be reduced. Coolants are not generally used in diamond machining unless, as in the machining of plastics, it is necessary to reduce airborne dust particles. Diamond tools can be reground.

There is much commercial interest in being able to coat HSS and carbides directly with diamond, but getting the diamond coating to adhere reliably has been difficult. Diamond-coated inserts would deliver roughly the same performance as PCD tooling when cutting nonferrous materials but could be given more complex geometries and chip breakers while reducing the cost per cutting edge.

POLYCRYSTALLINE CUBIC BORON NITRIDES

Polycrystalline cubic boron nitride (PCBN) is a man-made tool material widely used in the automotive industry for machining hardened steels and superalloys. It is made in a compact form for tools by a process quite similar to that used for sintered polycrystalline diamonds. It retains its hardness at elevated temperatures (Knoop 4700 at 20°C, 4000 at 1000°C) and has low chemical reactivity at the tool–chip interface. This material can be used to machine hard aerospace materials like Inconel 718 and René 95 as well as chilled cast iron.

Although not as hard as diamond, PCBN is less reactive with such materials as hardened steels, hard-chill cast iron, and nickel- and cobalt-based superalloys. PCBN can be used efficiently and economically to machine these difficult-to-machine materials at higher speeds (fivefold) and with a higher removal rate (fivefold) than cemented carbide, and with superior accuracy, finish, and surface integrity. PCBN tools are available in basically the same sizes and shapes as sintered diamond and are made by the same process. The cost of an insert is somewhat higher than either cemented carbide or ceramic tools, but the tool life may be five to seven times that of a ceramic tool. Therefore, to see the economy of using PCBN tools, it is necessary to consider all the factors.

Here is an industrial example of analysis of tooling economics, where a comparison is being made between two tool materials (insert tools). A manufacturer of diesel engines is producing an in-line six-cylinder engine block that is machined on a transfer line. Each cylinder hole must be bored to accept a sleeve liner. This operation has a depth of cut of 0.062 in. per side, for a total of 0.125 in. stock removal. The tolerance on this bore is ±0.001 in. and the spindle is operating at 2000 sfpm. Ceramic inserts are used on this operation, but with these inserts, wear was severe enough to require indexing after only 35 pieces. The ceramic insert was replaced with PCBN inserts made of a high-content BCN. Both inserts had a 0.001- to 0.002-in. radius hone for edge preparation.

Table 21-3 is a cost comparison between the ceramic and PCBN insert. The PCBN insert used in the application is a full-top PCBN insert, meaning that the entire top of the insert is a layer of PCBN material. At first glance the PCBN tool appears to be extremely expensive. Each insert costs $208.00 and provides only three usable edges, whereas the ceramic insert costs $14.90 and provides six usable edges. However, the ceramic tool must be indexed every 35 pieces. The PCBN tool is indexed every 500 pieces. The cost per bore, including insert cost and the cost of labor to perform indexing, comes to $0.125 per bore for the ceramic tool and $0.142 per bore for the PCBN tool. This appears to make the ceramic tool more cost-effective, but downtime for indexing has not been accounted for. In this application, the ceramic insert required 10.75 hours of downtime for indexing, whereas the PCBN tool required only 0.75 hour of downtime for indexing. Use of the PCBN cutting tool significantly reduces the total cost per piece by eliminating 10 hours of downtime of the machine. Later in this chapter the economics of machining will be addressed again.

The two predominant wear modes of PCBN tools are notching at the **depth-of-cut line (DCL)** and **microchipping.** In some cases, the tool will exhibit flank wear of the cutting edge. These tools have been used successfully for heavy interrupted cutting and

TABLE 21-3 Cost Comparison for Machining Liner Bores in 1500 Engine Blocks[a]

	Ceramic TNG-433	PCBN BTNG-433
Cost per insert	$14.90	$208.00
Edges per insert	6	3
Cost per edge	$2.48	$69.33
Time per index (6 tools)	0.25 hr	0.25 hr
Cost per index at $45 per hour	$11.25	$11.25
Indexes per 1500 blocks	43	3
Indexing cost (indexes × $11.25)	$483.75	$33.75
Insert cost for 6 spindles	$638.34	$1248.00
Labor and tool cost	$1122.09	$1281.00
Cost per bore	$0.125	$0.142
Total number of tool changes	43	3
Downtime for 1500 blocks	× 0.25 hr 10.75 hr	× 0.25 hr 0.75 hr

[a] To see the economy of using PCBN cutting tools, it is important to consider all factors of the operation, especially downtime for tool changing.

TABLE 21-4 Suggested Application of Four (4) Cutting Tool Materials to Workpiece Materials

	Applicable Tool Material			
Workplace Material	Carbide-Coated Carbide	Ceramic, Cermet	Cubic Boron Nitride	Diamond Compacts
Cast irons		uninterrupted finishing cuts		
carbon steels	X	X		
Alloy steels				
alloy cast iron	X	X	X	
Aluminium, brass	X	X		X
High-silicon aluminium	X			X
Nickel-based	X	X	X	
Titanium	X			
Plastic composites	X		X	

for milling white cast iron and hardened steels using negative lands and honed cutting edges. See Table 21-4 for suggested applications of CBN and diamonds along with carbides and ceramics.

Because diamond and PCBN are extremely hard but brittle materials, new demands are being placed on the machine tools and on machining practice in order to take full advantage of the potential of these tool materials. These demands include:

- Use of more rigid machine tools and machining practices involving gentle entry and exit of the cut in order to prevent microchipping.
- Use of high-precision machine tools, because these tools are capable of producing high finish and accuracy.
- Use of machine tools with higher power, because these tools are capable of higher metal removal rates and faster spindle speeds.

21.3 TOOL GEOMETRY

Selecting tool geometry is a critical part of selecting the cutting tool. Tool geometry can be very complex. Figure 21-14 shows the cutting-tool geometry for a single-point tool (HSS) used in turning in oblique (three forces) machining. The **back rake angle** affects

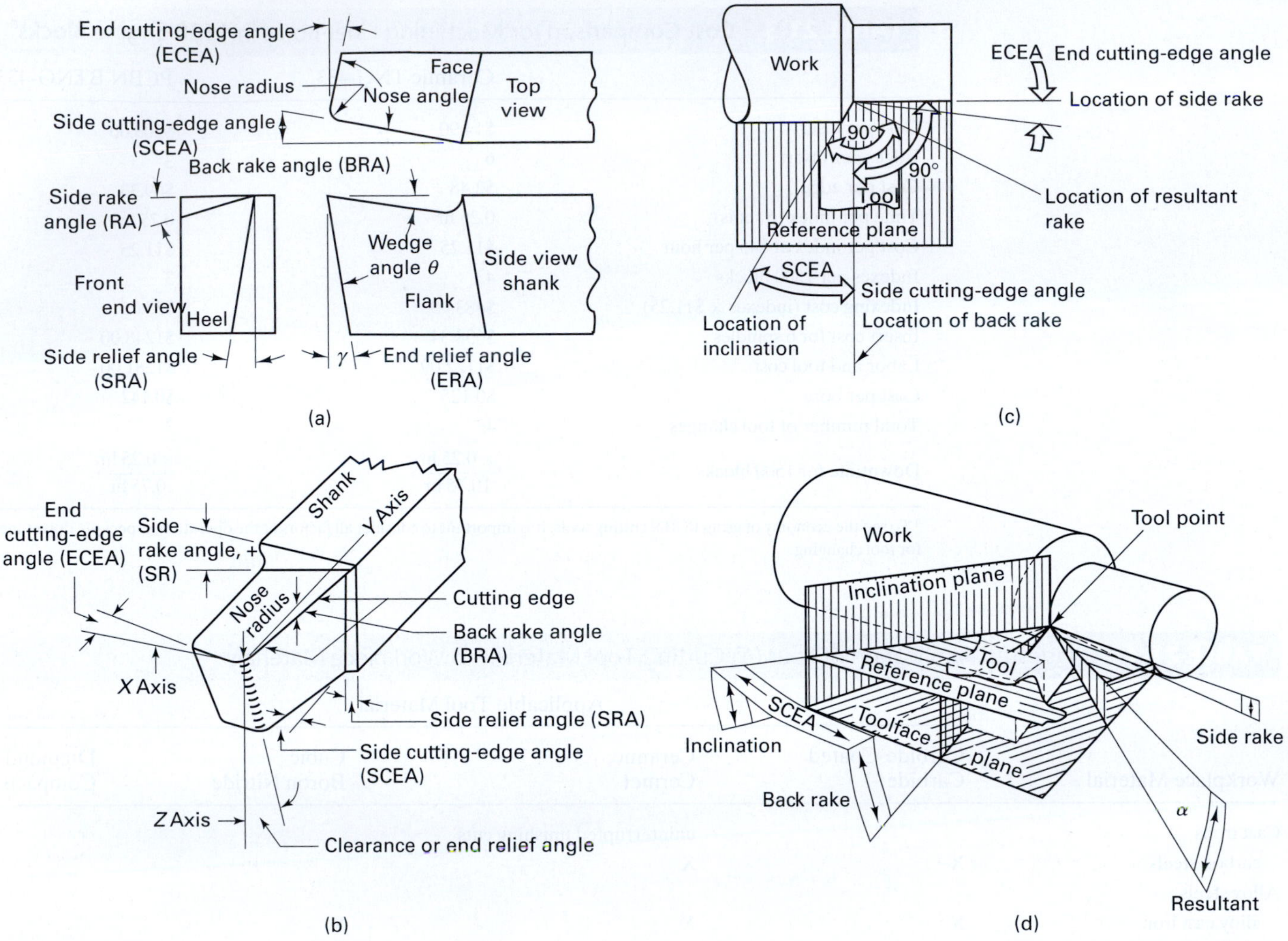

FIGURE 21-14 Standard terminology to describe the geometry of single-point tools: (a) three dimensional views of tool, (b) oblique view of tool from cutting edge, (c) top view of turning with single-point tool, (d) oblique view from shank end of single-point turning tool.

the ability of the tool to shear the work material and form the chip. It can be positive or negative. Positive rake angles reduce the cutting forces, resulting in smaller deflections of the workpiece, tool holder, and machine. In machining hard work materials, the back rake angle must be small, even negative for carbide and diamond tools. Generally speaking, the higher the hardness of the workpiece, the smaller the back rake angle. For high-speed steels, back rake angle is normally chosen in the positive range, depending on the type of tool (turning, planing, end milling, face milling, drilling, etc.) and the work material.

For carbide tools, inserts for different work materials and tool holders can be supplied with several standard values of back rake angle: −6 to +6 degrees. The side rake angle and the back rake angle combine to form the *effective rake angle.* This is also called the *true* rake angle or *resultant* rake angle of the tool.

True rake inclination of a cutting tool has a major effect in determining the amount of chip compression and the onset of shear angle, ϕ. A small rake angle causes high compression, tool forces, and friction, resulting in a thick, highly deformed, hot chip. Increasing the back or side rake angles reduces the compression, the forces, and the friction, yielding a thinner, less deformed, and cooler chip. In general, the power consumption is reduced by approximately 1% for each 1-degree change in alpha (α). The end relief angle is called gamma (γ). Unfortunately, it is difficult to take much advantage of the desirable effects of larger positive rake angles because they are offset by the reduced strength of the cutting tool, due to the reduced tool section, and by its greatly reduced capacity to conduct heat away from the cutting edge.

To provide greater strength at the cutting edge and better heat conductivity, zero or negative rake angles are commonly employed on carbide, ceramic, polydiamond, and PCBN cutting tools. These materials tend to be brittle, but their ability to hold their superior hardness at high temperatures results in their selection for high-speed and continuous machining operations. While the negative rake angle increases tool forces, it keeps the tool in compression and provides added support to the cutting edge. This is particularly important in making intermittent cuts, as in milling, and in absorbing the impact during the initial engagement of the tool and work.

The wedge angle, Θ, determines the strength of the tool and its capacity to conduct heat and depends on the values of α and γ. The relief angles mainly affect the tool life and the surface quality of the workpiece. To reduce the deflections of the tool and the workpiece and to provide good surface quality, larger relief values are required. For high-speed steel, relief angles in the range of 5 to 10 degrees are normal, with smaller values being for the harder work materials. For carbides, the relief angles are lower to give added strength to the tool.

The side and end cutting-edge angles define the nose angle and characterize the tool design. The nose radius has a major influence on surface finish. Increasing the nose radius usually decreases tool wear and improves surface finish.

Tool nomenclature varies with different cutting tools, manufacturers, and users. Many terms are still not standard because of this variety. The most common tool terms will be used in later chapters to describe specific cutting tools.

The introduction of coated tools has spurred the development of improved tool geometries. Specifically, **low-force groove (LFG)** geometries have been developed that reduce the total energy consumed and break up the chips into shorter segments. These grooves effectively increase the rake angle, which increases the shear angle and lowers the cutting force and power. This means that higher cutting speeds or lower cutting temperatures (and better tool lives) are possible.

As a chip breaker, the groove deflects the chip at a sharp angle and causes it to break into short pieces that are easier to remove and are not as likely to become tangled in the machine and possibly cause injury to personnel. This is particularly important on high-speed, mass-production machines.

The shapes of cutting tools used for various operations and materials are compromises, resulting from experience and research so as to provide good overall performance. For coated tools, edge strength is an important consideration. A thin coat enables the edge to retain high strength, but a thicker coat exhibits better wear resistance. Normally, tools for turning have a coating thickness of 6 to 12 μm. Edge strength is higher for multilayer coated tools. The radius of the edge should be 0.0005 to 0.005 in.

21.4 TOOL-COATING PROCESSES

The two most effective coating processes for improving the life and performance of tools are the chemical vapor deposition and physical vapor deposition of **titanium nitride (TiN)** and **titanium carbide (TiC).** The selection of the *cutting materials for cutting tools* depends on what property you are seeking. If you want

Oxidation and corrosion resistance; high-temperature stability	select	Al_2O_3, TiN, TiC
Crater resistance	select	Al_2O_3, TiN, TiC
Hardness and edge retention	select	TiC, TiN, Al_2O_3
Abrasion resistance and flank wear	select	Al_2O_3, TiN, TiC
Low coefficient of friction and high lubricity	select	TiN, Al_2O_3, TiC
Fine grain size	select	TiN, TiC, Al_2O_3

The CVD process, used to deposit a protective coating onto carbide inserts, has been benefiting the metal removal industry for many years and is now being applied with equal success to steel. The PVD processes have quickly become the preferred TiN coating processes for high-speed steel and carbide-tipped cutting tools.

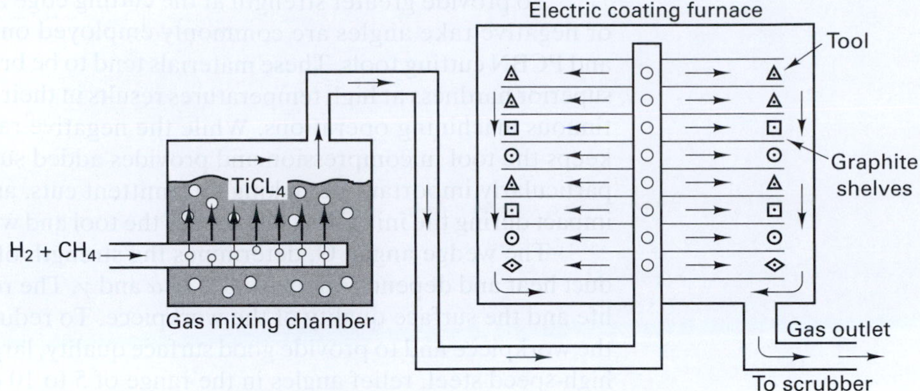

FIGURE 21-15 Chemical vapor deposition (CVD) is used to apply layers (TiC, TiN, etc.) to carbide cutting tools.

CHEMICAL VAPOR DEPOSITION

Chemical vapor deposition is an atmosphere-controlled process carried out at temperatures in the range of 950 to 1050°C (1740 to 1920°F). Figure 21-15 shows a schematic of the CVD process.

Cleaned tools ready to be coated are staged on precoated graphite work trays (shelves) and loaded onto a central gas distribution column (tree). The tree loaded with parts to be coated is placed inside the retort of the CVD reactor. The tools are heated under an inert atmosphere until the coating temperature is reached. The coating cycle is initiated by the introduction of titanium tetrachloride ($TiCL_4$), hydrogen, and methane (CH_4) into the reactor. $TiCL_4$ is a vapor and is transported into the reactor via a hydrogen carrier gas; CH_4 is introduced directly. The chemical reaction for the formation of TiC is:

$$TiCl_4 + CH_4 \rightarrow TiC + 4HCl \quad (21\text{-}1)$$

To form titanium nitride, a nitrogen–hydrogen gas mixture is substituted for methane. The chemical reaction for TiN is:

$$2TiCl_4 + 2H_2 + N_2 \rightarrow 2TiN + 4HCl \quad (21\text{-}2)$$

PHYSICAL VAPOR DEPOSITION

The simplest form of PVD is evaporation, where the substrate is coated by condensation of a metal vapor. The vapor is formed from a source material called the *charge*, which is heated to a temperature less than 1000°C. PVD methods currently being used include reactive sputtering, reactive ion plating, low-voltage electron-beam evaporation, triode high-voltage electron-beam evaporation, cathodic evaporation, and arc evaporation. In each of the methods, the TiN coating is formed by reacting free titanium ions with nitrogen away from the surface of the tool and relying on a physical means to transport the coating onto the tool surface.

All of these PVD processes share the following common features:

1. The coating takes place inside a vacuum chamber under a hard vacuum with the workpiece heated to 200 to 405°C (400 to 900°F).
2. Before coating, all parts are given a final cleaning inside the chamber to remove oxides and improve coating adhesion.
3. The coating temperature is relatively low (for cutting and forming tools), typically about 450°C (842°F).
4. The metal source is vaporized in an inert gas atmosphere (usually argon), and the metal atoms react with gas to form the coating. Nitrogen is the reactive gas for nitrides, and methane or acetylene (along with nitrogen) is used for carbides.
5. All four are ion-assisted deposition processes. The ion bombardment compresses the atoms on the growing film, yielding a dense, well-adhered coating.

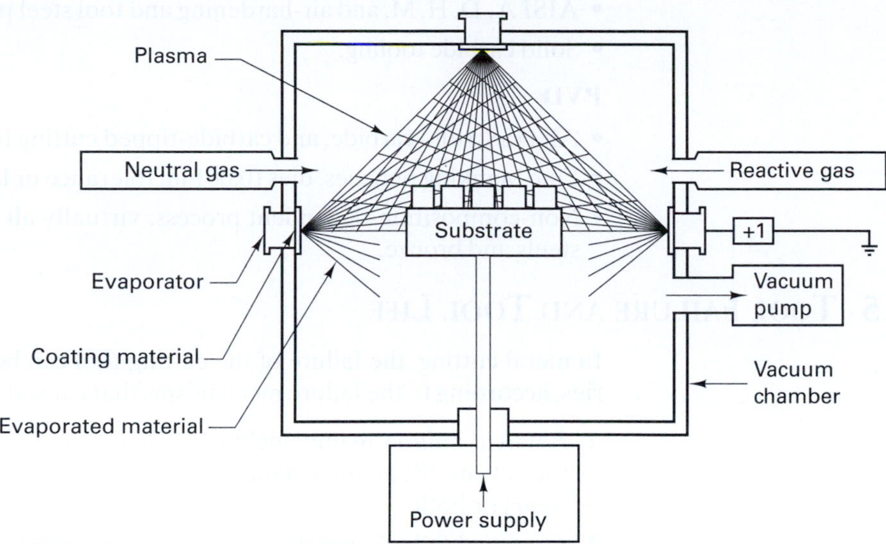

FIGURE 21-16 Schematic of physical vapor deposition (PVC) arc evaporation process.

A typical cycle time for the coating of functional tools, including heat-up and cool down, is about 6 hr.

In Figure 21-16, PVD arc evaporation is shown. The plasma sources are from several arc evaporators located on the sides and top of the vacuum chamber. Each evaporator generates plasma from multiple arc spots. In this way, a highly localized electrical arc discharge causes minute evaporation of the material of the cathode, and a self-sustaining arc is produced that generates a high-energy and concentrated plasma.

The kinetic energy of deposition is much greater than that found in any other PVD method. During coating, this energy is of the order of 150 eV and more. Therefore, the plasma is highly reactive and the greater percentage of the vapor is atomic and ionized.

Coating temperatures can be selected and controlled so that metallurgy is preserved. This enables a coating of a wide variety of sintered carbide tools—for example, brazed tools and solid carbide tools such as drills, end mills, form tools, and inserts. The PVD arc evaporation process will preserve substrate metallurgy, surface finish, edge sharpness, geometrical straightness, and dimensions.

CVD AND PVD—COMPLEMENTARY PROCESSES

CVD and PVD are complementary coating processes. The differences between the two processes and resultant coatings dictate which coating process to use on different tools.

Because CVD is done at higher temperatures, the adhesion of these coatings tends to be superior to a PVD–CVD-deposited coating. CVD coatings are normally deposited thicker than PVD coatings (6 to 9 μm for CVD, 1 to 3 μm for PVD). See Table 21-4.

With CVD multiple coatings, layers may be readily deposited, but the tooling materials are restricted. CVD coated tools must be heat treated after coating. This limits the application to loosely toleranced tools. However, the CVD process, being a gaseous process, results in a tool that is coated uniformly all over; this includes blind slots and blind holes.

Because PVD is mainly a line-of-sight process, all surfaces of the part to be coated may be masked. PVD also requires fixturing of each part in order to affect the substrate bias.

APPLICATIONS

Applications for the two different processes are as follows:

CVD

- Loosely toleranced tooling.
- Piercing and blanking punches, trim dies, phillips punches, upsetting punches.

- AISI A, D, H, M, and air-hardening and tool steel parts.
- Solid carbide tooling.

PVD

- All HSS, solid carbide, and carbide-tipped cutting tools.
- Fine blanking punches, dies (0.001 in. tolerance or less).
- Non-composition-dependent process; virtually all tooling materials, including mold steels and bronze.

21.5 TOOL FAILURE AND TOOL LIFE

In metal cutting, the failure of the cutting tool can be classified into two broad categories, according to the failure mechanisms that caused the tool to die (or fail):

1. *Physical failures* mainly include gradual tool wear on the flank(s) of the tool below the cutting edge (called **flank wear**) or wear on the rake face of the tool (called **crater wear**) or both.
2. *Chemical failures*, which include wear on the rake face of the tool (crater wear) are rapid, usually unpredictable, and often catastrophic failures resulting from abrupt, premature death of a tool.

Other modes of failure are outlined in Figure 21-17. The selection of failure criteria is also widely varied. Figure 21-17 also shows a sketch of a "worn" tool, showing crater wear and flank wear, along with wear of the tool nose radius and an outer-diameter groove (the DCL groove). Tools also fail by edge chipping and edge fracture.

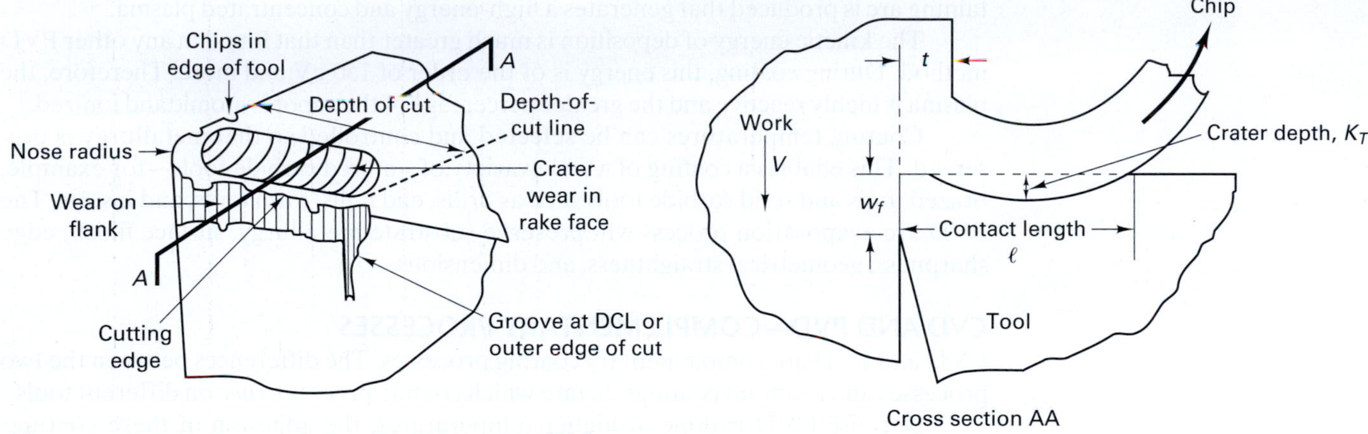

No.	Failure		Cause
1-3	Flank wear		Due to the abrasive effect of hard grains contained in the work material
4-5	Groove		Due to wear at the DCL or outer edge of the cut
6	Chipping	Physical	Fine chips caused by high-pressure cutting, chatter, vibration, etc.
7	Partial fracture		Due to the mechanical impact when an excessive force is applied to the cutting edge
8	Crater wear		Carbide particles are removed due to degradation of tool performances and chemical reactions at high temperature
9	Deformation	Chemical	The cutting edge is deformed due to its softening at high temperature
10	Thermal crack		Thermal fatigue in the heating and cooling cycle with interrupted cutting
1	Built-up edge		A portion of the workpiece material adheres to the insert cutting edge

FIGURE 21-17 Tools can fail in many ways. Tool wear during oblique cutting can occur on the flank or the rake face; t = uncut chip thickness; k_t = crater depth; w_f = flank wear land length; DCL = depth-of-cut line.

As the tool wears, its geometry changes. This geometry change will influence the cutting forces, the power being consumed, the surface finish obtained, the dimensional accuracy, and even the dynamic stability of the process. Worn tools are duller, creating greater cutting forces and often resulting in chatter in processes that otherwise are usually relatively free of vibration. The actual wear mechanisms active in this high-temperature environment are abrasion, adhesion, diffusion, or chemical interactions. It appears that in metal cutting, any or all of these mechanisms may be operative at a given time in a given process.

Tool failure by plastic deformation, brittle fracture, fatigue fracture, or edge chipping can be unpredictable. Moreover, it is difficult to predict which mechanism will dominate and result in a tool failure in a particular situation. What can be said is that tools, like people, die (or fail) from a great variety of causes under widely varying conditions. Therefore, **tool life** should be treated as a random variable, or probabilistically, not as a deterministic quantity.

21.6 FLANK WEAR

During machining, the tool is performing in a hostile environment in which high-contact stresses and high temperatures are commonplace; therefore, tool wear is always an unavoidable consequence. At lower speeds and temperatures, the tool most commonly wears on the flank. Suppose that the tool wear experiment were to be repeated 15 times without changing any of the input parameters. The result would look like Figure 21-18, which depicts the variable nature of tool wear and shows why tool wear must be treated as a random variable. In Figure 21-18 the average time is denoted as μ_T and the standard deviation as σ_T, where the wear limit criterion was 0.025 in. At a given time during the test, 35 min, the tool displayed flank wear ranging from 0.013 to 0.021 in, with an average of $\mu_w = 0.00175$ in. with standard deviation σ_w 0.001 in.

In Figure 21-19 four characteristic tool wear curves (average values) are shown for four different cutting speeds, V_1 through V_4; V_1 is the fastest cutting speed and therefore generates the fastest wear rates. Such curves often have three general regions, as shown in the figure. The central region is a steady-state region (or the region of secondary wear). This is the normal operating region for the tool. Such curves are typical for both

w_f values for general life determination (for cemented carbides)

Width of Wear (in.)	Applications
0.008	Finish cutting of nonferrous alloys, fine and light cut, etc.
0.016	Cutting of special steels
0.028	Normal cutting of cast irons, steels, etc.
0.040–0.050	Rough cutting of common cast irons

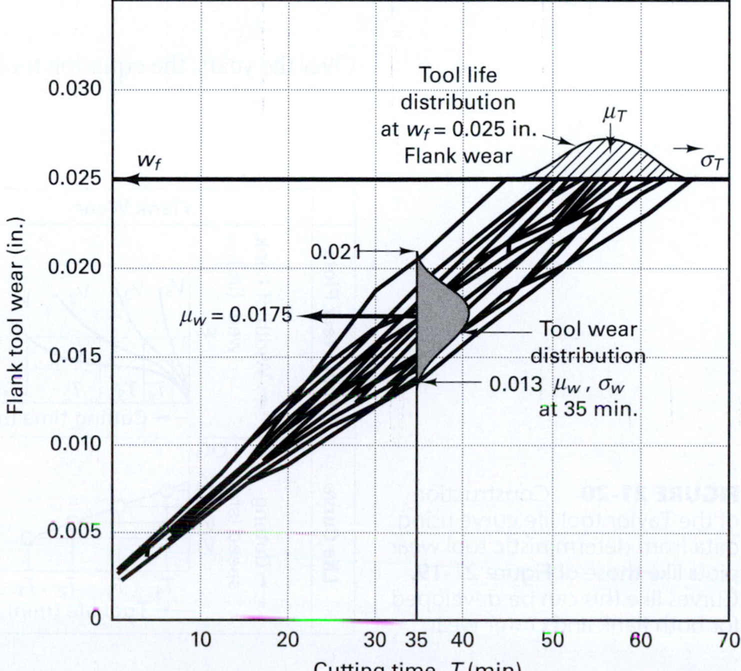

FIGURE 21-18 Tool wear on the flank displays a random nature, as does tool life. w_f = flank wear limit value.

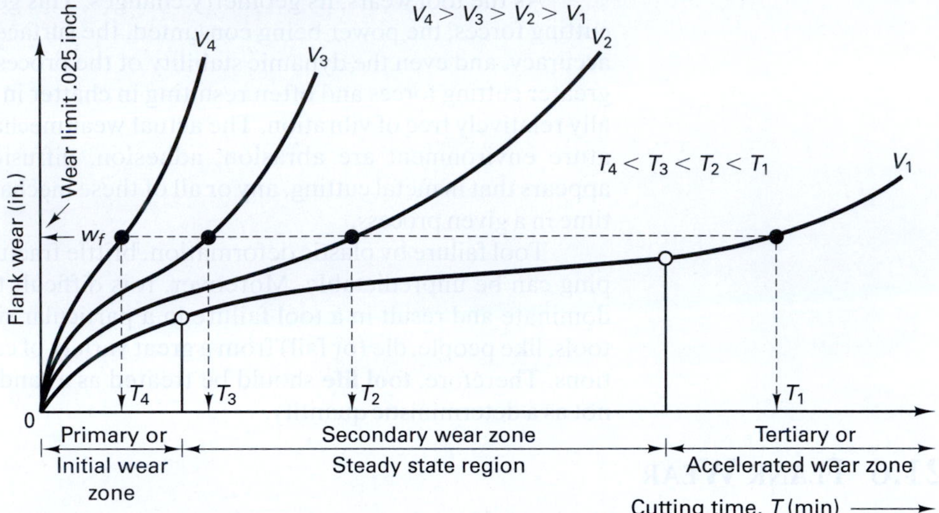

FIGURE 21-19 Typical tool wear curves for flank wear at different velocities. The initial wear is very fast, then it evens out to a more gradual pattern until the limit is reached; after that, the wear substantially increases.

flank wear and crater wear. When the amount of wear reaches the value w_f, the permissible tool wear on the flank, the tool is said to be "worn out." The value w_f is typically set at 0.025 to 0.030 in. for flank wear for high-speed steels and 0.008 to 0.050 for carbides, depending on the application. For crater wear, the depth of the crater, k_b, is used to determine tool failure.

Using the empirical tool wear data shown in Figure 21-19, which used the values of T (time in minutes) associated with V (cutting speed) for a given amount of tool wear, w_f (see the dashed-line construction), Figure 21-20 was developed. When V and T are plotted on log-log scales, a linear relationship appears, described by the equation

$$VT^n = \text{Constant} = C \tag{21-3}$$

This equation is called the Taylor tool life equation because in 1907, F. W Taylor published his now-famous paper, "On the Art of Cutting Metals," in *ASME Transactions,* wherein tool life (T) was related to cutting speed (V) and feed (f). This equation had the form[1]

$$T = \frac{\text{Constant}}{f_x V_y} \tag{21-4}$$

Over the years, the equation took the more widely published form

$$VT^n = C$$

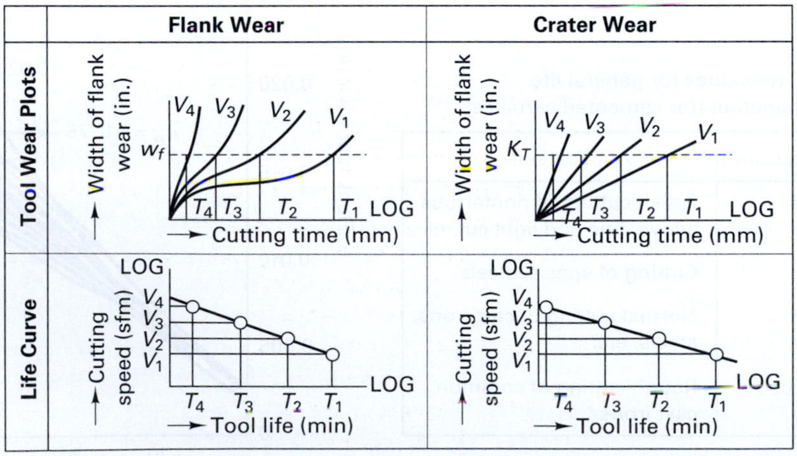

FIGURE 21-20 Construction of the Taylor tool life curve using data from deterministic tool wear plots like those of Figure 21-19. Curves like this can be developed for both flank and crater wear.

[1] Carl Barth, who was Taylor's mathematical genius, is generally thought to be the author of these formulations along with early versions of slide rules.

TABLE 21-5 Tool Life Information for Various Materials and Conditions

Source	Tool Material	Geometry	Workpiece Material	Size of Cut (in.)		Cutting Fluid	$VT^n = C$	
				Depth	Feed		n	C
1	High-carbon steel	8.14, 6.6, 6.15, 3/64	Yellow brass (.60 Cu,	.050	.0255	Dry	.081	242
			40 Zn, 85 Ni, .006 Pb)	.100	.0127	Dry	.096	299
1	High-carbon steel	8.14, 6.6, 6.15, 3/64	Bronze (.9 Cu, .1.5n)	.050	.0255	Dry	.086	190
				.100	.0127	Dry	.111	232
1	HSS-18-4-1	8.14, 6.6, 6.15, 3/64	Cast iron 160 Bhn	.050	.0255	Dry	.101	172
			Cast iron. Nickel, 164 Bhn	.050	.0255	Dry	.111	186
			Cast iron. Ni-Cr, 207 Bhn	.050	.0255	Dry	.088	102
1	HSS-18-4-1	8.14, 6.6, 6.15, 3/64	Stell, SAE B1113 C.D.	.050	.0127	Dry	.080	260
			Stell, SAE B1112 C.D.	.050	.0127	Dry	.105	225
			Stell, SAE B1120 C.D.	.050	.0127	Dry	.100	270
			Stell, SAE B1120 + Pb C.D.	.050	.0127	Dry	.060	290
			Stell, SAE B1035 C.D.	.050	.0127	Dry	.110	130
			Stell, SAE B1035 + Pb C.D.	.050	.0127	Dry	.110	147
1	HSS-18-4-1	8.14, 6.6, 6.15, 3/64	Stell, SAE 1045 CD.	.100	.0127	Dry	.110	192
		8.14, 6.6, 6.13, 3/64	Stell, SAE 2340 185 Bhn	.100	.0125	Dry	.147	143
		8.14, 6.6, 6.15, 3/64	Stell, SAE 2345 198 Bhn	.050	.0255	Dry	.105	126
		8.14, 6.6, 6.15, 3/64	Stell, SAE 3140 190 Bhn	.100	.0125	Dry	.160	178
1	HSS-18-4-1	8.14, 6.6, 6.15, 3/64	Stell, SAE 4350 363 Bhn	.0125	.0127	Dry	.080	181
			Stell, SAE 4350 363 Bhn	.0125	.0255	Dry	.125	146
			Stell, SAE 4350 363 Bhn	.0250	.0255	Dry	.125	95
			Stell, SAE 4350 353 Bhn	.100	.0127	Dry	.110	78
			Stell, SAE 4350 363 Bhn	.100	.0255	Dry	.110	46
1	HSS-18-4-1	8.14, 6.6, 6.15, 3/64	Stell, SAE 4140 230 Bhn	.050	.0127	Dry	.180	190
			Stell, SAE 4140 271 Bhn	.050	.0127	Dry	.180	159
			Stell, SAE 6140 240 Bhn	.050	.0127	Dry	.150	197
1	HSS-18-4-1	8.22, 6.6, 6.15, 3/64	Monel metal 215 Bhn	.100	.0127	Dry	.084	170
				.150	.0255	Dry	.074	127
				.100	.0127	Em	.080	185
				.100	.0127	SMO	.105	189
1	Stellite 2400	0.0, 6.6, 6.0, 3/32	Steel. SAE 3240 annealed	.187	.031	Dry	.190	215
				.125	.031	Dry	.190	240
				.062	.031	Dry	.190	270
				.031	.031	Dry	.190	310
1	Stellite No. 3	0.0, 6.6, 6.0, 3/32	Cast iron 200 Bhn	.062	0.31	Dry	.150	205
1	Carbide (T 64)	6.12, 5.5, 10.45	Steel. SAE 1040 annealed	.062	.025	Dry	.156	800
			Steel. SAE 1060 annealed	.125	.025	Dry	.167	660
			Steel. SAE 1060 annealed	.187	.025	Dry	.167	615
			Steel. SAE 1060 annealed	.250	.025	Dry	.167	560
			Steel. SAE 1060 annealed	.062	.021	Dry	.167	880
			Steel. SAE 1060 annealed	.062	.042	Dry	.164	510
			Steel. SAE 1060 annealed	.062	.062	Dry	.162	400
			Steel. SAE 2340 annealed	.062	.025	Dry	.162	630
2	Ceramic	not available	AISI 4150	.160	.016	Dry	.400	2000
			AISI 4150	.160	.016	Dry	.200	620

Sources: 1- *Fundamentals of Tool Design*. ASTME. A. R. Koneeny, W. J. Potthoff; 2 - *Theory of Metal Cutting*, P. N. Black

where n is an exponent that depends mostly on tool material but is affected by work material, cutting conditions, and environment and C is a constant that depends on all the input parameters, including feed. Table 21-5 provides some data on Taylor tool life constants.

Figure 21-21 shows typical tool life curves for one tool material and three work materials. Notice that all three plots have about the same slope, n. Typical values for n

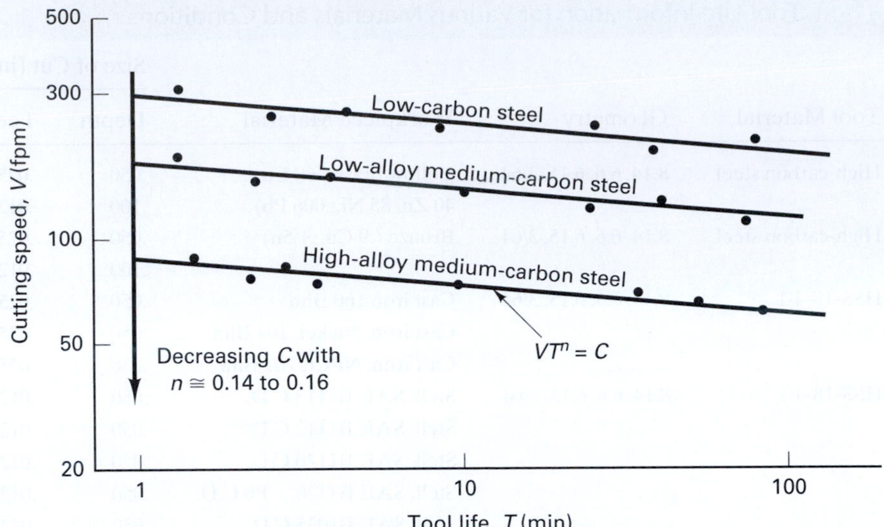

FIGURE 21-21 Log-log tool life plots for three steel work materials cut with HSS tool material.

are 0.14 to 0.16 for HSS, 0.21 to 0.25 for uncoated carbides, 0.30 for TiC inserts, 0.33 for poly-diamonds, 0.35 for TiN inserts, and 0.40 for ceramic-coated inserts.

It takes a great deal of experimental effort to obtain the constants for the Taylor equation because each combination of tool and work material will have different constants. Note that for a tool life of 1 min, $C = V$, or the cutting speed that yields about 1 min of tool life for this tool.

A great deal of research has gone into developing more sophisticated versions of the Taylor equation, wherein constants for other input parameters (typically feed, depth of cut, and work material hardness) are experimentally determined. For example,

$$VT^n F^m d^p = K' \tag{21-5}$$

where n, m, and p are exponents and K' is a constant. Equations of this form are also deterministic and determined empirically.

The problem has been approached probabilistically in the following way. Because T depends on speed, feed, materials, and so on, one writes

$$T = \frac{K^{1/n}}{V^{1/n}} = \frac{K}{V^m} \tag{21-6}$$

where K is now a random variable that represents the effects of all unmeasured factors and is an input variable.

The sources of tool life variability include factors such as:

1. Variation in work material hardness (from part to part and within a part).
2. Variability in cutting tool materials, geometry, and preparation.
3. Vibrations in machine tool, including rigidity of work and tool-holding devices.
4. Changing surface characteristics of workpieces.

The examination of the data from a large number of tool life studies in which a variety of steels were machined shows that regardless of the tool material or process, tool life distributions are usually log normal and typically have a large standard deviation. As shown in Figure 21-22, tool life distributions have a large coefficient of variation, which means that tool life is not very predictable.

Other criteria can be used to define tool death in addition to wear limits:

- When surface finish deteriorates unacceptably.
- When workpiece dimension is out of tolerance.

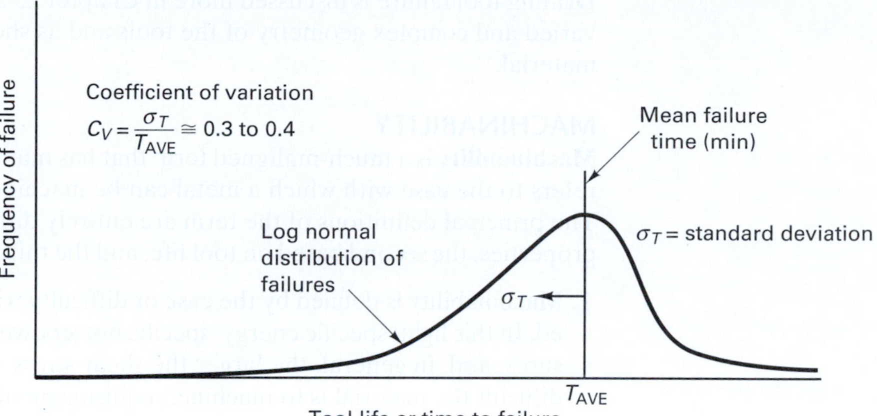

FIGURE 21-22 Tool life viewed as a random variable has a log normal distribution with a large coefficient of variation.

- When power consumption or cutting forces increase to a limit.
- Sparking or chip discoloration and disfigurement.
- Cutting time or component quantity.

In automated processes, it is very beneficial to be able to monitor the tool wear online so that the tool can be replaced prior to failure, wherein defective products may also result. The feed force has been shown to be a good, indirect measure of tool wear. That is, as the tool wears and dulls, the feed force increases more than the cutting force increases.

Once criteria for failure have been established, tool life is that time elapsed between start and finish of the cut, in minutes. Other ways to express tool life, other than time, include:

1. Volume of metal removed between regrinds or replacement of tool.
2. Number of pieces machined per tool.
3. Number of holes drilled with a given tool (see Figure 21-23).

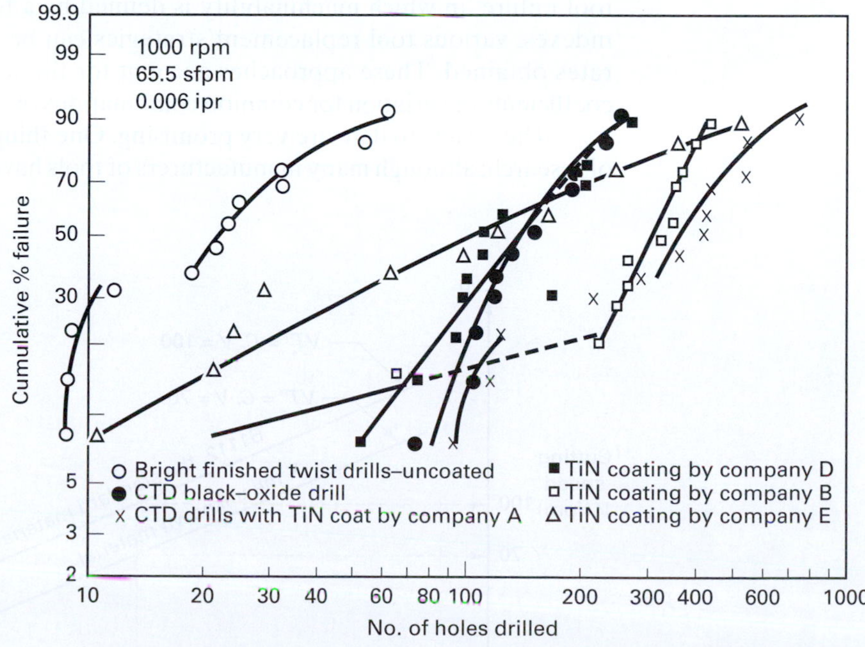

Drill performance based on the number of holes drilled with 1/4-in.-diameter drills in T–1 structural steel.

FIGURE 21-23 Tool life test data for various coated drills. TiN-coated HSS drills outperform uncoated drills. Life based on the number of holes drilled before drill failure.

Drilling tool failure is discussed more in Chapter 23 and is very complex because of the varied and complex geometry of the tools and as shown here in Figure 21-23, the tool material.

MACHINABILITY

Machinability is a much-maligned term that has many different meanings but generally refers to the ease with which a metal can be machined to an acceptable surface finish. The principal definitions of the term are entirely different—the first based on material properties, the second based on tool life, and the third based on cutting speed.

1. Machinability is defined by the ease or difficulty with which the metal can be machined. In this light, specific energy, specific horsepower, and shear stress are used as measures, and, in general, the larger the shear stress or specific power values, the more difficult the material is to machine, requiring greater forces and lower speeds. In this definition, the material is the key.
2. Machinability is defined by the relative cutting speed for a given tool life while cutting some material, compared to a standard material cut with the same tool material. As shown in Figure 21-24, tool life curves are used to develop machinability ratings. For example, in steels, the material chosen for the standard material was B1112 steel, which has a tool life of 60 min at a cutting speed of 100 sfpm. Material X has a 70% rating, which implies that steel X has a cutting speed of 70% of B1112 for equal tool life. Note that this definition assumes that the tool fails when machining material X by whatever mechanism dominated the tool failure when machining the B1112. There is no guarantee that this will be the case. ISO standard 3685 has machinability index numbers based on 30 min of tool life with flank wear of 0.33 mm.
3. Cutting speed is measured by the maximum speed at which a tool can provide satisfactory performance for a specified time under specified conditions. (See ASTM standard E 618-81, "Evaluating Machining Performance of Ferrous Metals Using an Automatic Screw Bar Machine.")
4. Other definitions of machinability are based on the ease of removal of the chips (chip disposal), the quality of the surface finish of the part itself, the dimensional stability of the process, or the cost to remove a given volume of metal.

Further definitions are being developed based on the probabilistic nature of the tool failure, in which machinability is defined by a tool reliability index. Using such indexes, various tool replacement strategies can be examined and optimum cutting rates obtained. These approaches account for the tool life variability by developing coefficients of variation for common combinations of cutting tools and work materials.

The results to date are very promising. One thing is clear, however, from this sort of research: although many manufacturers of tools have worked at developing materials

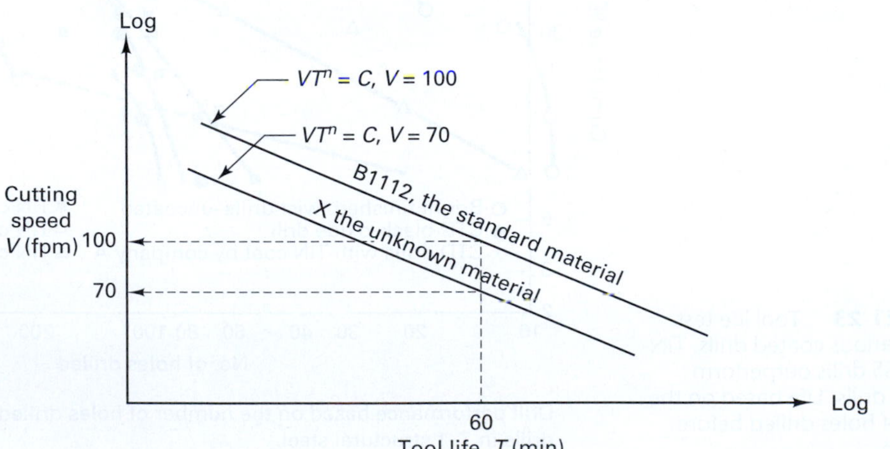

FIGURE 21-24 Machinability ratings defined by deterministic tool life curves.

that have greater tool life at higher speeds, few have worked to develop tools that have less variability in tool life at all speeds. The reduction in variability is fundamental to achieving smaller coefficients of variation, which typically are of the order of 0.3 to 0.4. This means that a tool with a 100-min average tool life has a standard deviation of 30 to 40 min, so there is a good probability that the tool will fail early. In automated equipment, where early, unpredicted tool failures are extremely costly, reduction of the tool life variability will pay great benefits in improved productivity and reduced costs.

RECONDITIONING CUTTING TOOLS

In the reconditioning of tools by sharpening and recoating, care must be taken in grinding the tool's surfaces. The following guidelines should be observed:

1. Resharpen to original tool geometry specifications. Restoring the original tool geometry will help the tool achieve consistent results on subsequent uses. Computer numerical control (CNC) grinding machines for tool resharpening have made it easier to restore a tool's original geometry.
2. Grind cutting edges and surfaces to a fine finish. Rough finishes left by poor and abusive regrinding hinder the performance of resharpened tools. For coated tools, tops of ridges left by rough grinding will break away in early tool use, leaving uncoated and unprotected surfaces that will cause premature tool failure.
3. Remove all burrs on resharpened cutting edges. If a tool with a burr is coated, premature failure can occur because the burr will break away in the first cut, leaving an uncoated surface exposed to wear.
4. Avoid resharpening practices that overheat and burn or melt (called *glazing over*) the tool surfaces, because this will cause problems in coating adhesion. Polishing or wire brushing of tools causes similar problems.

The cost of each recoating is about one-fifth the cost of purchasing a new tool. By recoating, the tooling cost per workpiece can be cut by between 20 and 30%, depending on the number of parts being machined.

21.7 CUTTING FLUIDS

From the day that Frederick W. Taylor demonstrated that a heavy stream of water flowing directly on the cutting process allowed the cutting speeds to be doubled or tripled, **cutting fluids** have flourished in use and variety and have been employed in virtually every machining process. The cutting fluid acts primarily as a coolant and secondly as a lubricant, reducing the friction effects at the tool/chip interface and the work flank regions. The cutting fluids also carry away the chips and provide friction (and force) reductions in regions where the bodies of the tools rub against the workpiece. Thus, in processes such as drilling, sawing, tapping, and reaming, portions of the tool apart from the cutting edges come in contact with the work, and these (sliding friction) contacts greatly increase the power needed to perform the process, unless properly lubricated.

The reduction in temperature greatly aids in retaining the hardness of the tool, thereby extending the tool life or permitting increased cutting speed with equal tool life. In addition, the removal of heat from the cutting zone reduces thermal distortion of the work and permits better dimensional control. Coolant effectiveness is closely related to the thermal capacity and conductivity of the fluid used. Water is very effective in this respect but presents a rust hazard to both the work and tools and also is ineffective as a lubricant. Oils offer less effective coolant capacity but do not cause rust and have some lubricant value. In practice, straight cutting oils or emulsion combinations of oil and water or wax and water are frequently used. Various chemicals can also be added to serve as wetting agents or detergents, rust inhibitors, or polarizing agents to promote formation of a protective oil film on the work. The extent to which the flow of a cutting fluid washes the very hot chips away from the cutting area is an important factor in heat removal. Thus, the application of a coolant should be copious and of some velocity.

TABLE 21-6 Cutting Fluid Contaminants

Category	Contaminants	Effects
Solids	Metallic fines, chips Grease and sludge Debris and trash	Scratch product's surface Plug coolant lines Produce wear on tools and machines
Tramp fluids	Hydraulic oils (coolant) Water (oils)	Decrease cooling efficiency Cause smoking Clog paper filters Grow bacteria faster
Biologicals (coolants)	Bacteria Fungi Mold	Acidity coolant **Break down** emulsions Cause rancidity, dermatitis Require toxic biocides

The possibility of a cutting fluid providing lubrication between the chip and the tool face is an attractive one. An effective lubricant can modify the process, perhaps producing a cooler chip, discouraging the formation of a built-up edge on the tool, and promoting improved surface finish. However, the extreme pressure at the tool/chip interface and the rapid movement of the chip away from the cutting edge make it virtually impossible to maintain a conventional hydrodynamic lubricating film at the tool/chip interface. Consequently, any lubrication action is associated primarily with the formation of solid chemical compounds of low shear strength on the freshly cut chip face, thereby reducing tool/chip shear forces or friction. For example, carbon tetrachloride is very effective in reducing friction in machining several different metals and yet would hardly be classified as a good lubricant in the usual sense. Chemically active compounds, such as chlorinated or sulfurized oils, can be added to cutting fluids to achieve such a lubrication effect. Extreme-pressure lubricants are especially valuable in severe operations, such as internal threading (tapping), where the extensive tool-work contact results in high friction with limited access for a fluid. In addition to functional effectiveness as coolant and lubricant, cutting fluids should be stable in use and storage, noncorrosive to work and machines, and nontoxic to operating personnel. The cutting fluid should also be restorable by using a closed recycling system that will purify the used coolant and cutting oils. Cutting fluids become contaminated in three ways (Table 21-6). All these contaminants can be eliminated by filtering, hydrocycloning, pasteurizing, and centrifuging. Coolant restoration eliminates 99% of the cost of disposal and 80% or more of new fluid purchases. See Figure 21-25 for a schematic of a coolant recycling system.

21.8 ECONOMICS OF MACHINING

The cutting speed has such a great influence on the tool life compared to the feed or the depth of cut that it greatly influences the overall economics of the machining process. For a given combination of work material and tool material, a 50% increase in speed results in a 90% decrease in tool life, while a 50% increase in feed results in a 60% decrease in tool life. A 50% increase in depth of cut produces only a 15% decrease in tool life. Therefore, in limited-horsepower situations, depth of cut and then feed should be maximized, while speed is held constant and horsepower consumed is maintained within limits. As cutting speed is increased, the machining time decreases, but the tools wear out faster and must be changed more often. In terms of costs, the situation is as shown in Figure 21-26, which shows the effect of cutting speed on the cost per piece.

The total cost per operation is comprised of four individual costs: machining costs, tool costs, tool-changing costs, and handling costs. The machining cost is observed to decrease with increasing cutting speed because the cutting time decreases. Cutting time is proportional to the machining costs. Both the tool costs and the tool-changing costs increase with increases in cutting speeds. The handling costs are independent of cutting speed. Adding up each of the individual costs results in a total unit cost curve that is

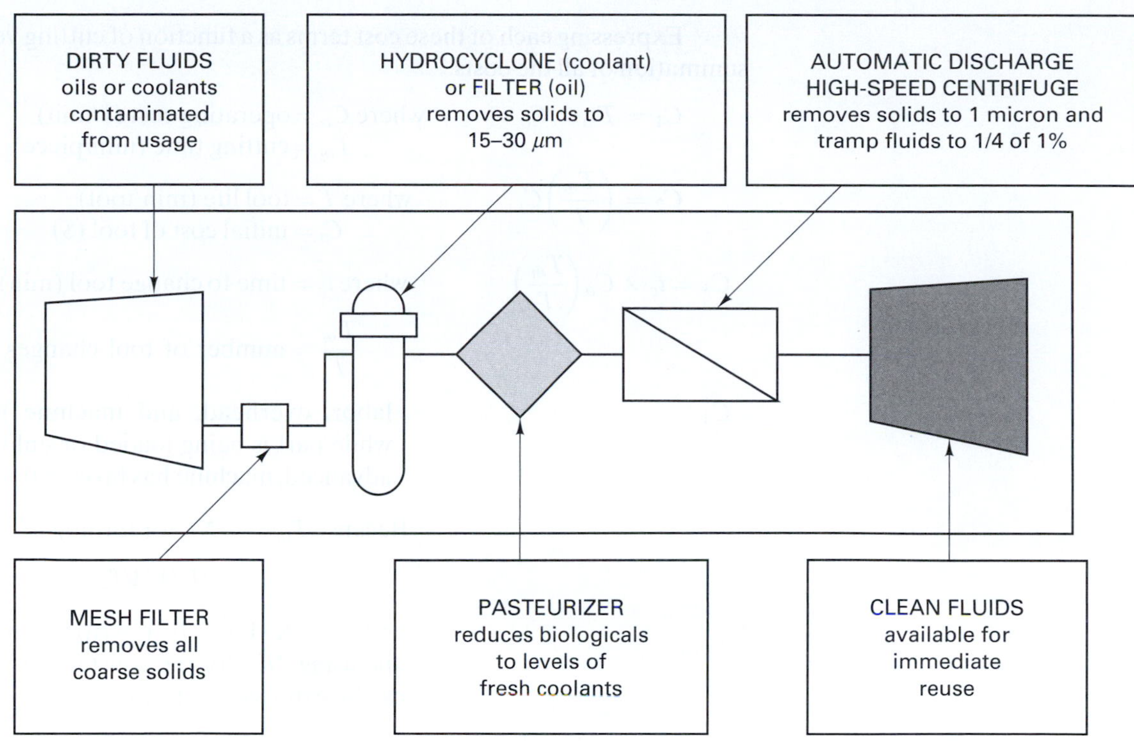

FIGURE 21-25 A well-designed recycling system for coolants will return more than 99% of the fluid for reuse.

observed to go through a minimum point. For a turning operation, the total cost per piece C equals

$$C = C_1 + C_2 + C_3 + C_4$$
$$= \text{Machining cost} + \text{Tooling cost} + \text{Tool-changing cost} + \text{Handling cost per piece} \quad (21\text{-}7)$$

Note: This "C" is not the same "C" used in the Taylor Tool life equation. In this analysis, that "C" will be called "K."

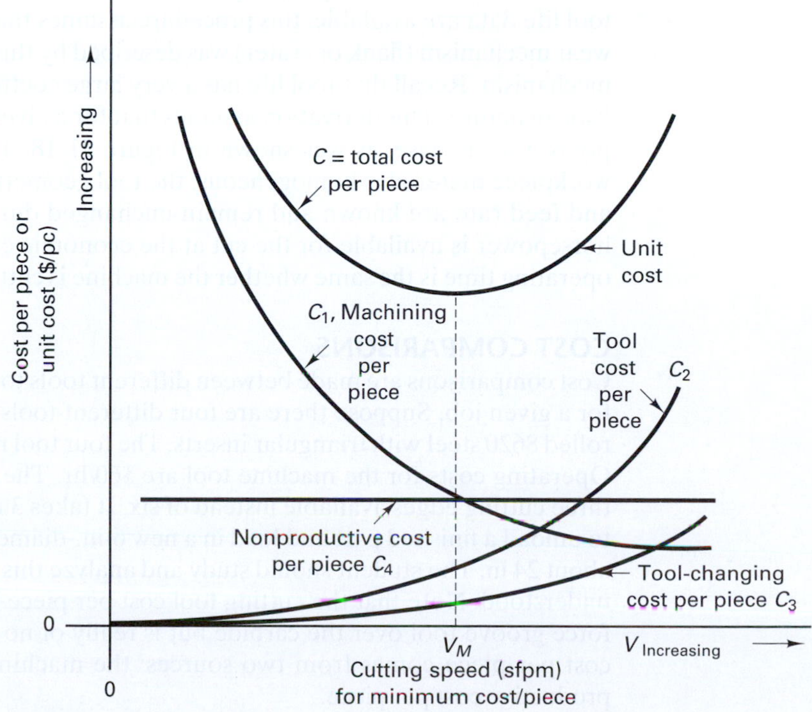

FIGURE 21-26 Cost per unit for a machining process versus cutting speed. Note that the "C" in this figure and related equations is not the same "C" used in the Taylor tool life (equation 21-3).

Expressing each of these cost terms as a function of cutting velocity will permit the summation of all the costs.

$C_1 = T_m \times C_o$ where C_o = operating cost ($/min)
T_m = cutting time (min/piece)

$C_2 = \left(\frac{T_m}{T}\right) C_t$ where T = tool life (min/tool)
C_t = initial cost of tool ($)

$C_3 = t_c \times C_o \left(\frac{T_m}{T}\right)$ where t_c = time to change tool (min)

$\frac{T_m}{T}$ = number of tool changes per piece

C_4 labor, overhead, and machine tool costs consumed while part is being loaded or unloaded, tools are being advanced, machine has broken down, and so on.

Because $T_m = L/Nf_r$ for turning

$= \pi DL/12Vf_r$

and $T = (K/V)^{1/n}$, by rewriting equation 21-3, and using "K" for the constant "C", the cost per unit, C, can be expressed in terms of V:

$$C = \frac{L\pi DC_o}{12Vf_r} + \frac{C_t V^{1/n}}{K^{1/n}} + \frac{t_c C_o V^{1/n}}{K^{1/n}} + C_4 \qquad (21\text{-}8)$$

To find the minimum, take $dc/dV = 0$ and solve for V:

$$V_m = K\left[\frac{n}{1-n} \bullet \frac{C_o}{C_o\, t_t + C_t}\right]^n \qquad (21\text{-}9)$$

Thus, V_m represents a cutting speed that will minimize the cost per unit, as depicted in Figure 21-26. However, a word of caution here is appropriate. Note that this derivation was totally dependent upon the Taylor tool life equation. Such data may not be available because they are expensive and time consuming to obtain. Even when the tool life data are available, this procedure assumes that the tool fails only by whichever wear mechanism (flank or crater) was described by this equation and by no other failure mechanism. Recall that tool life has a very large coefficient of variation and is probabilistic in nature. This derivation assumes that for a given V, there is one T—and this simply is not the case, as was shown in Figure 21-18. The model also assumes that the workpiece material is homogeneous, the tool geometry is preselected, the depth of cut and feed rate are known and remain unchanged during the entire process, sufficient horsepower is available for the cut at the economic cutting conditions, and the cost of operating time is the same whether the machine is cutting or not cutting.

COST COMPARISONS

Cost comparisons are made between different tools to decide which tool material to use for a given job. Suppose there are four different tools that can be used for turning hot-rolled 8620 steel with triangular inserts. The four tool materials are shown in Table 21-7. Operating costs for the machine tool are $60/hr. The low-force groove insert has only three cutting edges available instead of six. It takes 3 min to change inserts and 0.5 min to unload a finished part and load in a new 6-in.-diameter bar stock. The length of cut is about 24 in. The student should study and analyze this table carefully so that each line is understood. Note that the cutting tool cost per piece is three times higher for the low-force groove tool over the carbide but is really of no consequence, because the major cost per piece comes from two sources: the machining cost per piece and the non-productive cost per piece.

TABLE 21-7 Cost Comparison of Four Tool Materials, Based on Equal Tool Life of 40 Pieces per Cutting Edge

	Uncoated	TiC-Coated	Al_2O_3-Coated	Al_2O_3 LFG
Cutting speed (surface ft/min)	400	640	1100	1320
Feed (in./rev)	0.020	0.02	0.024	0.028
Cutting edges available per insert	6	6	6	3
Cost of an insert ($/insert)	4.80	5.52	6.72	6.72
Tool life (pieces/cutting edge)	192	108	60	40
Tool-change time per piece (min)	0.075	0.075	0.075	0.075
Nonproductive cost per piece ($/pc)	0.50	0.50	0.50	0.50
Machining time per piece (min/pc)	4.8	2.7	1.50	1.00
Machining cost per piece ($/unit)	4.8	2.7	1.5	1.00
Tool-change cost per piece ($/pc)	0.08	0.08	0.08	0.08
Cutting tool cost per piece ($/pc)	0.02	0.02	0.03	0.06
Total cost per piece ($/pc)	5.40	3.30	2.11	1.64
Production rate (pieces/hr)	11	18	29	38
Improvement in productivity based on pieces/hr (%)	0	64	164	245

Source: Data from T. E. Hale et al., "High Productivity Approaches to Metal Removal," *Materials Technology*, Spring 1980, p. 25.

KEY WORDS

aluminum oxide
back rake angle
BUE (built-up edge)
carbides
cast cobalt alloy
ceramics
cermets
chemical vapor deposition (CVD)
chip groove
coated tools
crater wear
cubic boron nitride (CBN)
cutting fluids
cutting tool materials
depth-of-cut line (DCL)
diamonds
flank wear
hardness
high-speed steel (HSS)
hot hardness
low-force groove (LFG)
machinability
metal cutting
microchipping
physical vapor deposition (PVD)
polycrystalline cubic boron nitride (PCBN)
polycrystalline diamond (PCD)
powder metallurgy (P/M)
sintered carbides
stellite tools
tool life
tool steels
titanium carbide (TiC)
titanium nitride (TiN)

REVIEW QUESTIONS

1. For metal-cutting tools, what is the most important material property (i.e., the most critical characteristic)? Why?
2. What is hot hardness compared to hardness?
3. What is impact strength, and how is it measured?
4. Why is impact strength an important property in cutting tools?
5. Is a cemented carbide tool made by a powder metallurgy method?
6. What are the primary considerations in tool selection?
7. What is the general strategy behind coated tools?
8. What is a cermet?
9. How is a CBN tool manufactured?
10. F. W. Taylor was one of the discoverers of high-speed steel. What else is he well known for?
11. What casting process do you think was used to fabricate cast cobalt alloys?
12. Discuss the constraints in the selection of a cutting tool.
13. What does *cemented* mean in the manufacture of carbides?
14. What advantage do ground carbide inserts have over pressed carbide inserts?
15. What is a chip groove?
16. What is the DCL?
17. Suppose you made four beams out of carbide, HSS, ceramic, and cobalt. The beams are identical in size and shape, differing only in material. Which beam would do each of the following?
    a. Deflect the most, assuming the same load.
    b. Resist penetration the most.
    c. Bend the farthest without breaking.
    d. Support the greatest compressive load.
18. Multiple coats or layers are put on the carbide base for what different purposes?
19. What tool material would you recommend for machining a titanium aircraft part?
20. What makes the process that makes TiC coatings for tools a problem? See equation 21-1.
21. Why does a TiN-coated tool consume less power than an uncoated HSS under exactly the same cutting conditions?
22. For what work material are CBN tools more commonly used, and why?
23. Why is CBN better for machining steel than diamond?
24. What is the typical coefficient of variation for tool life data, and why is this a problem?
25. What is meant by the statement "Tool life is a random variable"?

26. The typical value of a coefficient of variation in metal-cutting tool life distributions is 0.3. How could it be reduced?
27. Machinability is defined in many ways. Explain how a rating is obtained.
28. What are the chief functions of cutting fluids?
29. How are CVD tools manufactured?
30. Why is the PVD process used to coat HSS tools?
31. Why is there no universal cutting tool material?
32. What is an 18-4-1 HSS composed of?
33. Over the years, tool materials have been developed that have allowed significant increases in MRR. Nevertheless, HSS is still widely used. Under what conditions might HSS be the material of choice?
34. Why is the rigidity of the machine tool an important consideration in the selection of the cutting tool material?
35. Explain how it can be that the tool wears when it may be four times as hard as the work material.
36. What is a honed edge on a cutting tool and why is it done?

PROBLEMS

1. Figure 21-A gives data for cutting speed and tool life. Determine the constants for the Taylor tool life equation for these data. What do you think the tool material might have been?

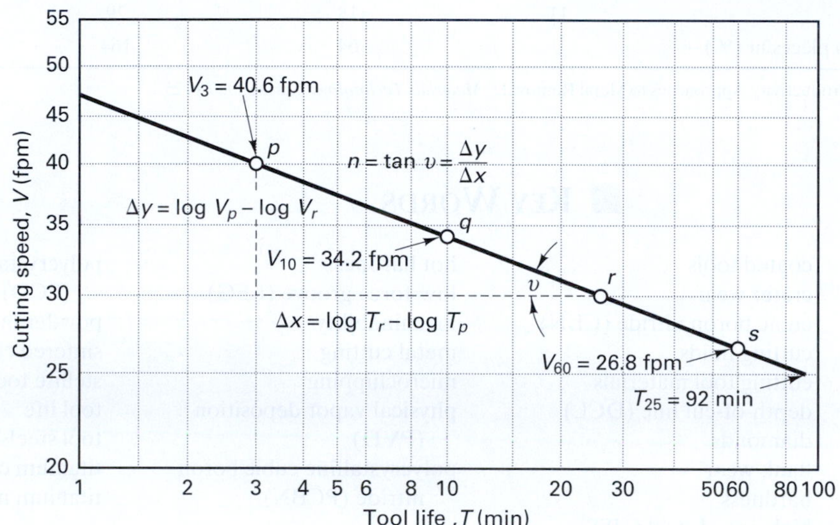

FIGURE 21-A

2. Suppose you have a turning operation using a tool with a zero back rake and 5-degree end relief. The insert flank has a wear land on it of 0.020 in. How much has the diameter of the workpiece grown (increased) due to this flank wear, assuming the tool has not been reset to compensate for the flank wear?

3. In Figure 21-B, a single-point tool is shown. Identify points A through G using tool nomenclature.

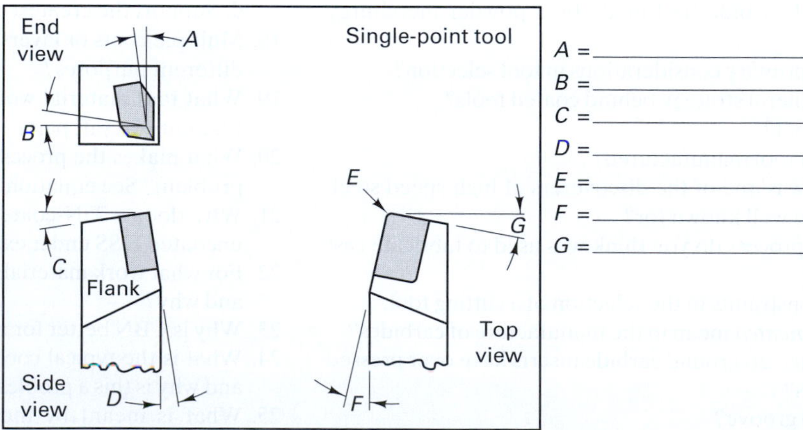

FIGURE 21-B

4. The following data have been obtained for machining an Si-Al alloy:

Workpiece Material	Tool Material	Cutting Speed (m/min) for Tool Life (m/min) of 20 min	30 min	60 min
Sand casting	Diamond polycrystal	731	642	514
Permanent-mold casting	Diamond polycrystal	591	517	411
PMC with flood cooling	Diamond polycrystal	608	554	472
Sand casting	WC-K-20	175	161	139

Compute the C and n values for the Taylor tool life equation. How do these n values compare to the typical values?

5. In Figure 21-C, the insert at the top is set with a 0-degree side cutting-edge angle. The insert at the bottom is set so that the edge contact length is increased from a 0.250-in. depth of cut to 0.289 in. The feed was 0.010 ipr.
   a. Determine the side cutting-edge angle for the offset tool.
   b. What is the uncut chip thickness in the offset position?
   c. What effect will this have on the forces and the process?

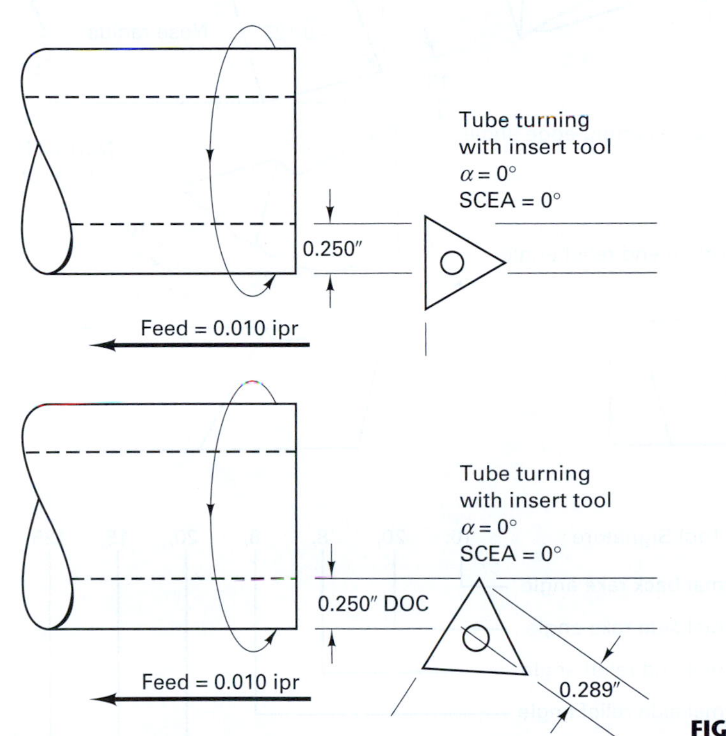

FIGURE 21-C

6. Tool cost is often used as the major criterion for justifying tool selection. Either silicon nitride or PCBN insert tips can be used to machine (bore) a cylinder block on a transfer line at a rate of 312,000 part/yr (material: gray cast iron). The operation requires 12 inserts (2 per tool), as six bores are machined simultaneously. The machine was run at 2600 sfpm with a feed of 0.014 in. at 0.005-in. DOC for finishing. Here are some additional data:

	SiN	PCBN
Tips in use per part	12	12
Tool life (parts per tool)	200	4700
Cost per tip	$1.25	$28.50

   a. Which tool material would you recommend?
   b. On what basis?

7. A 2-in.-diameter bar of steel was turned at 284 rpm, and tool failure occurred in 10 min. The speed was changed to 132 rpm, and the tool failed in 30 min of cutting. Assume that a straight-line relationship exists. What cutting speed should be used to obtain a 60-min tool life of V_{60}?
8. Table 21-7 shows a cost comparison for four tool materials. Show how the data in this table were generated.
9. Problem 6 provided data stating the cutting speed for this job was 2600 sfpm. Use equation 21-9 to verify that speed.
10. The outside diameter of a roll for a steel (AISI 1015) rolling mill is to be turned. In the final pass, Starting diameter = 26.25 in.

and Length = 48.0 in. The cutting conditions will be Feed = 0.0100 in./rev and Depth of cut = 0.125 in. A cemented carbide cutting tool is to be used, and the parameters of the Taylor tool life equation for this setup are $n = 0.25$ and $C = 1300$. It is desirable to operate at a cutting speed such that the tool will not need to be changed during the cut. Determine the cutting speed that will make the tool life equal to the time required to complete this turning operation. (Problem suggested by Groover, *Fundamentals of Modern Manufacturing: Materials, Processes, and Systems*, 2nd ed., John Wiley & Sons, 2002.)

11. Using data from Problems 8 and 10, estimate the necessary horsepower for the machine tool to make this cut.
12. Figure 21-B shows a sketch of a single-point tool and its associated tool signature. Put the signature from the tool in Figure 21-B in the same order as shown in Figure 21-D. Which tool would produce the larger F_c, given that both are cutting at the same V, f_r, and DOC in the same material?

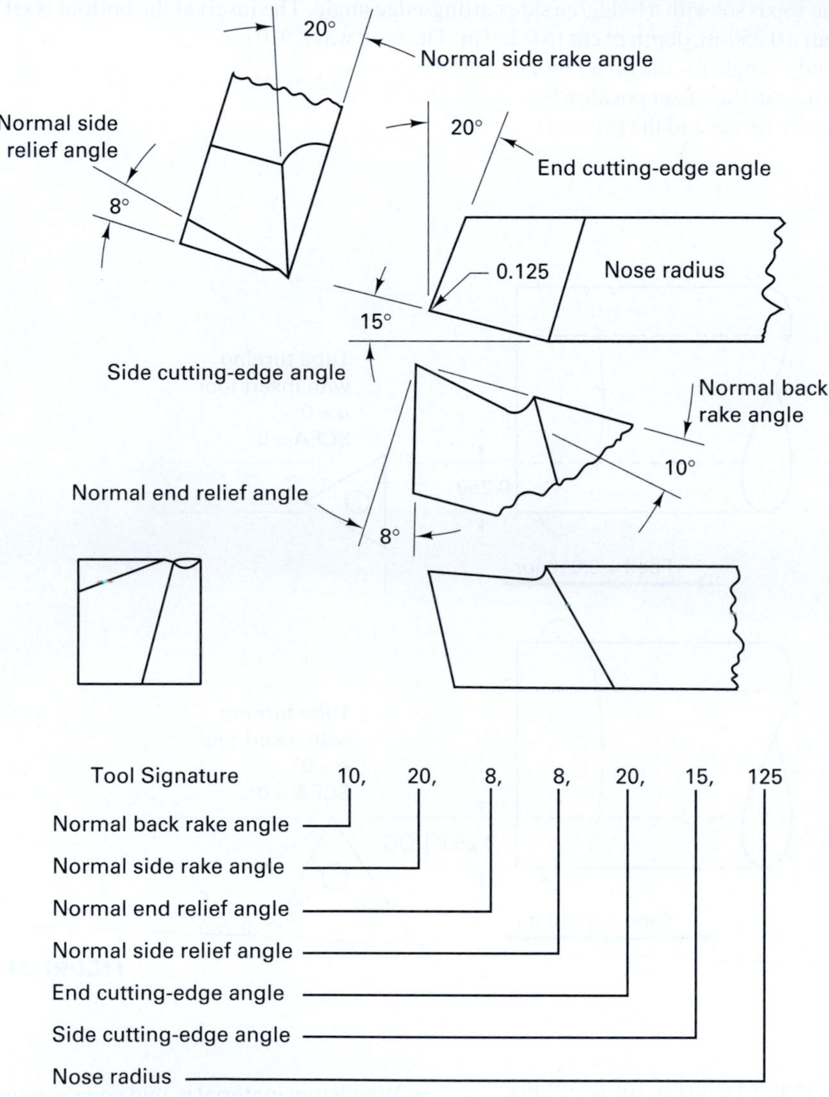

FIGURE 21-D

13. What are the various ways a cutting tool can fail and what can be done to remedy this? See Figures 21-17 and 21-E.

	Failure		Basic Remedy
Edge Failure	Excessive flank wear 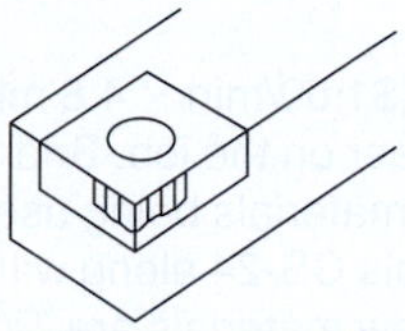	Tool material Cutting conditions	• Use a more wear-resistant grade carbide ⟶ {coated, cermet} • Decrease speed
	Excessive crater wear 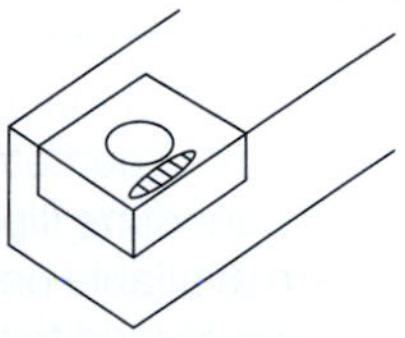	Tool material Tool design Cutting conditions	• Use a more wear-resistant grade carbide ⟶ {coated, cermet} • Enlarge the rake angle • Select the correct chip breaker • Decrease speed • Reduce the depth of cut and feed
	Cutting-edge chipping 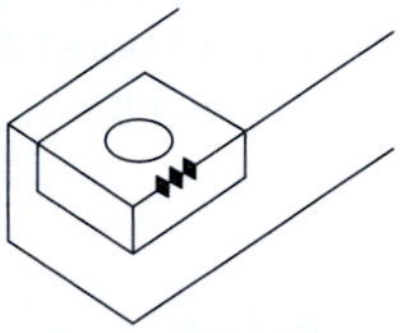	Tool material Tool design Cutting conditions	• Use tougher grades If carbide, (AC2000 ⟶ AC3000) • If built-up edge occurs, change to a less susceptible grade (cermet) • Reinforce the cutting edge (honing) • Reduce the rake angle • Increase speed (if caused by edge build-up)
	Partial fracture of cutting edges 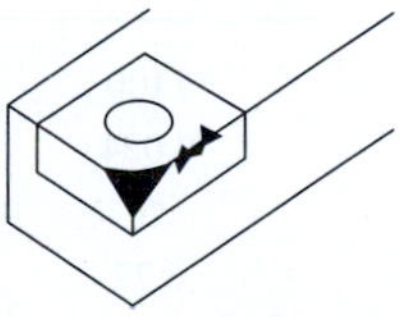	Tool material Tool design Cutting conditions	• Use tougher grades if carbide, (AC2000 ⟶ AC3000) • Use holder with a large approach angle • Use larger shank-size holder • Reduce the depth of cut and feed
	Built-up edge	Tool material Cutting conditions	• Change to a grade that is adhesion resistant • Increase the cutting speed and feed • Use cutting fluids
	Plastic deformation	Tool material Cutting conditions	• Change to highly thermal-resistant grades • Reduce the cutting speed and feed

FIGURE 21-E

www.wiley.com/go/global/degarmo

Chapter 21 CASE STUDY

Comparing Tool Materials Based on Tool Life

The Zachary Milling Company is trying to decide on what kind of inserts to use in their milling cutters. These milling cutters often provide a marked productivity improvement compared to conventional HDD drills, particularly when they are combined with coated insert tools. The company is trying to determine which type of insert to use in the drill for machining some hot-rolled 8620 steel shafts using triangular inserts. The operating cost of the machine tool is $60/hr. It takes about 3 min to change the inserts and about 30 s to unload a finished part and load a new part in the machine. The company is currently using uncoated inserts at the following operating conditions: 400 sfpm and 0.020 ipr. These speeds and feeds resulted in each cutting edge producing about 40 pieces before the tool's cutting edge became dull. The tool was then indexed. Because it was a triangular tool, each tool had six cutting edges available before it had to be replaced. At these speeds and feeds, the milling time was 4.8 min and the production rate was 11 part/hr, while the machining cost per piece was $4.80 ($1.00/min × 4.8 min/pc). The manufacturing engineer on the job, Brian Graney, has found three new tool materials being used in face mills. They are listed in Table CS-24 along with the data for the uncoated tool. The new materials are TiC-coated carbide, Al_2O_3 coated carbide, and a ceramic-coated insert with a single-side, low-force groove geometry. The expected cutting conditions, speeds, and feeds are given in the table along with Brian's estimates of the production rates in pieces per hour for each of these new tool materials. The low-force groove geometry tools can only be used three times—they cannot be flipped over—so only three cutting edges are available per insert before it has to be replaced. Brian has argued that even though the ceramic-coated inserts cost more, they result in a lower cost per piece, considering all the costs. Determine the machining cost per piece, the tool changing cost per piece, and the tool cost per piece that make up the total cost per piece, and verify Brian's belief that these coated tools will provide some cost savings.

TABLE CS-21 Cost Comparison of Four Tool Materials, Based on Equal Tool Life

	Uncoated	TiC-Coated	Al_2O_3-coated	Al_2O_3 LFG
Cutting speed (surface ft/min)	400	640	1100	1320
Feed (in/rev)	0.020	0.02	0.024	0.028
Cutting edges available per insert	6	6	6	3
Cost of an insert ($/insert)	4.80	5.52	6.72	6.72
Tool life (pieces/cutting edge)	192	108	60	40
Tool change time per piece (min)	0.075	0.075	0.075	0.075
Nonproductive cost per piece ($/pc)	0.50	0.50	0.50	0.50
Machining time per piece (min/pc)	4.8	2.7	1.50	1.00
Machining cost per piece ($/pc)	4.80			
Tool change cost per piece ($/pc)	0.08			
Cutting tool cost per piece ($/pc)	0.02			
Total cost per piece ($)	5.40			
Production rate (pieces/hr)	11	18	29	38
Improvement in productivity based on pieces/hr (%)	0	64	164	245

Source: Data from T. E. Hale et al., "High Productivity Approaches to Metal Removal," *Materials Technology*, Spring 1980, p. 25.

CHAPTER 22

TURNING AND BORING PROCESSES

22.1 INTRODUCTION

Turning is the process of machining external cylindrical and conical surfaces. It is usually performed on a machine tool called a lathe. An engine lathe is shown in Figure 22-1, using a cutting tool. The workpiece is held in a workholder as indicated in Figure 22-2, relatively simple work and tool movements are involved in turning a cylindrical surface. The workpiece is rotated and the single-point cutting tool is fed longitudinally into the workpiece and then travels parallel to the axis of workpiece rotation, reducing the diameter by the **depth of cut (DOC).** The tool feeds at a rate, f_r, cutting at a speed, V, which is determined by the revolutions per minute (rpm) and the diameter of the workpiece, according to

$$V = \frac{\pi D_1 N_s}{12} \tag{22-1}$$

If the tool is fed at an angle to the axis of rotation, an external conical surface results. This is called **taper turning** (see Figure 22-3). If the tool is fed to the axis of

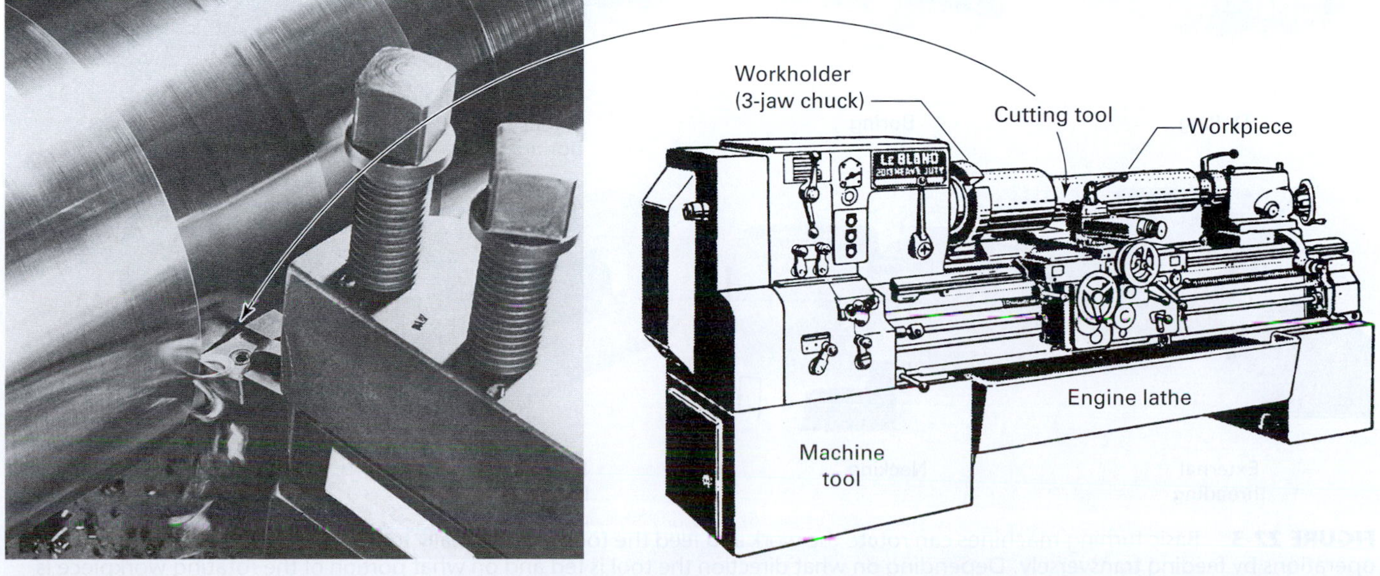

FIGURE 22-1 Schematic of a standard engine lathe performing a turning operation, with the cutting tool shown in inset. *(Courtesy J T. Black)*

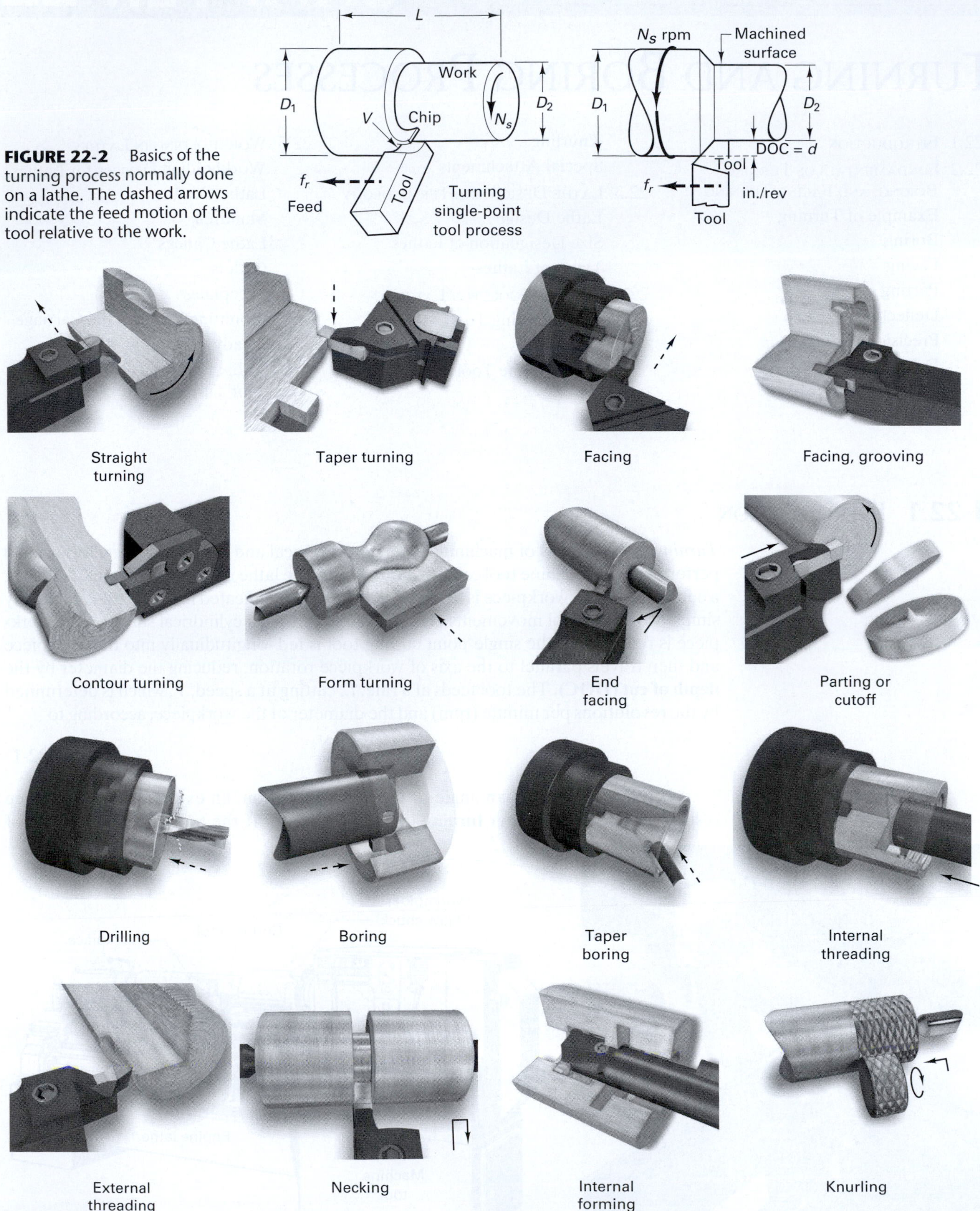

FIGURE 22-2 Basics of the turning process normally done on a lathe. The dashed arrows indicate the feed motion of the tool relative to the work.

FIGURE 22-3 Basic turning machines can rotate the work and feed the tool longitudinally for turning and can perform other operations by feeding transversely. Depending on what direction the tool is fed and on what portion of the rotating workpiece is being machined, the operations have different names. The dashed arrows indicate the tool feed motion relative to the workpiece. *(Courtesy J T. Black)*

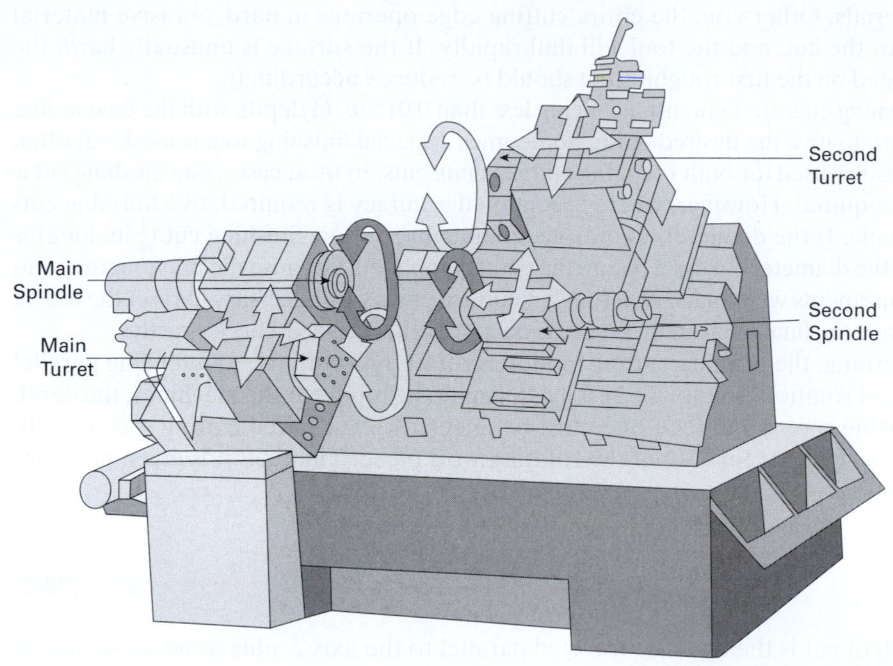

FIGURE 22-4 NC twin-spindle horizontal turning centers use coaxial spindles so work can be transferred from the main spindle to the second spindle.

rotation, using a tool that is wider than the depth of the cut, the operation is called *facing,* and a flat surface is produced on the end of the cylinder.

By using a tool having a specific form or shape and feeding it radially or inward against the work, external cylindrical, conical, and irregular surfaces of limited length can also be turned. The shape of the resulting surface is determined by the shape and size of the cutting tool. Such machining is called **form turning** (see Figure 22-3). If the tool is fed all the way to the axis of the workpiece, it will be cut in two. This is called *parting* or *cutoff,* and a simple, thin tool is used. A similar tool is used for *necking* or *partial cutoff.*

Boring is a variation of turning. Essentially, boring is internal turning. Boring can use single-point cutting tools to produce internal cylindrical or conical surfaces. It does not create the hole but, rather, machines or opens the hole up to a specific size. Boring can be done on most machine tools that can do turning. However, boring also can be done using a rotating tool with the workpiece remaining stationary. Also, specialized machine tools have been developed that will do boring, drilling, and reaming but will not do turning. Other operations, like *threading* and *knurling,* can be done on machines used for turning. In addition, drilling, reaming, and tapping can be done on the rotation axis of the work.

In recent years, turning centers have been developed that use turrets to hold multiple-edge rotary tools in powered heads. Some new machine tools feature two opposing spindles with automatic transfer from one to the other and two turrets of tools (see Figure 22-4).

22.2 FUNDAMENTALS OF TURNING, BORING, AND FACING TURNING

Turning constitutes the majority of lathe work. The cutting forces, resulting from feeding the tool from right to left, should be directed toward the headstock to force the workpiece against the workholder and thus provide better work support.

If good finish and accurate size are desired, one or more roughing cuts are usually followed by one or more finishing cuts. Roughing cuts may be as heavy as proper chip thickness, cutting dynamics, tool life, lathe horsepower, and the workpiece permit. Large depths of cut and smaller feeds are preferred to the reverse procedure, because fewer cuts are required and less time is lost in reversing the carriage and resetting the tool for the following cut.

On workpieces that have a hard surface, such as castings or hot-rolled materials containing mill scale, the initial roughing cut should be deep enough to penetrate the hard materials. Otherwise, the entire cutting edge operates in hard, abrasive material throughout the cut, and the tool will dull rapidly. If the surface is unusually hard, the cutting speed on the first roughing cut should be reduced accordingly.

Finishing cuts are light, usually being less than 0.015 in. in depth, with the feed as fine as necessary to give the desired finish. Sometimes a special finishing tool is used, but often the same tool is used for both roughing and finishing cuts. In most cases, one finishing cut is all that is required. However, where exceptional accuracy is required, two finishing cuts may be made. If the diameter is controlled manually, a short finishing cut ($\frac{1}{4}$ in. long) is made and the diameter checked before the cut is completed. Because the previous micrometer measurements were made on a rougher surface, it may be necessary to reset the tool in order to have the final measurement, made on a smoother surface, check exactly.

In turning, the primary cutting motion is rotational, with the tool feeding parallel to the axis of rotation (Figure 22-2). To determine the inputs to the machines, the depth of cut, cutting speed, and feed rate must be selected. The desired cutting speed establishes the necessary rpm (N_s) of the rotating workpiece. The feed f_r is given in inches per revolution (ipr).

The depth of cut is d, where

$$d = \text{DOC} = \frac{D_1 - D_2}{2} \text{ in.} \tag{22-2}$$

The length of cut is the distance traveled parallel to the axis L plus some allowance or overrun A to allow the tool to enter and/or exit the cut.

The inputs to the turning process are determined as follows. Based on the material being machined and the cutting tool material being used, the engineer selects the cutting speed, V, in feet per minute, the feed (f_r), and the depth of cut. The rpm value for the machine tool can be determined by

$$N_s = \frac{12V}{\pi D_1} \tag{22-3}$$

(using the larger diameter), where the factor of 12 is used to convert feet to inches. The cutting time is

$$T_m = \frac{L + A}{f_r N_s} \tag{22-4}$$

where A is overrun allowance and f_r is the selected feed in inches per revolution.

EXAMPLE OF TURNING

The 1.78-in.-diameter steel bar is to be turned down to a 1.10-in. diameter on a standard engine lathe. The overall length of the bar is 18.750 in., and the region to be turned is 16.50 in. The part is made from cold-drawn free-machining steel (this means the chips breakup nicely) with a BHN of 250. Because you want to take the bar from 1.78 to 1.10, you have a total depth of cut, d, of 0.34 in. (0.68/2). You decide you want to make two cuts, a roughing pass and a finishing pass. Rough at $d = 0.300$ and finish with $d = 0.04$ in. Looking at the table in Figure 20-4, for selecting speed and feed, you select $V = 100$ fpm and feed $= 0.020$ ipr because you have decided to use high-speed steel cutting tools.

The bar is held in a chuck with a feed through the hole in the spindle and is supported on the right end with a live center. The ends of the bar have been center drilled. Allowance should be 0.50 in. for approach (no overtravel). Allow 1.0 min to reset the tool after the first cut. To determine the inputs to the lathe, we calculate the spindle rpm:

$$N_s = 12V/\pi D_1 = 12 \times 100/3.14 \times 1.78 = 214$$

But your lathe does not have this particular rpm, so you select the closest rpm, which is 200. You don't need any further calculations for lathe inputs as you input the feed in ipr directly.

The time to make the cut is

$$T_m = \frac{L + ALL}{f_r \times N_s} = \frac{16.50 + 0.50}{0.020 \times 200} = 4.25 \text{ min}$$

You could reduce this time by changing to a coated carbide tool that would allow you to increase the cutting speed to 925 sfpm (see the table in Figure 20-4). The time for the second cut will be different if you change the feed and/or the speed to improve the surface finish. Again from a table of speeds and feeds, you select a speed of 0.925 and a feed rate of 0.007, so

$N_s = 12 \times 925/3.14 \times 1.10 = 3213$ with 3200 rpm the closest value:

$$T_m = \frac{L + ALL}{0.007 \times 3200} = 0.75 \text{ min}$$

The **metal removal rate (MRR)** is

$$\text{MRR} = \frac{\text{Volume removed}}{\text{Time}} = \frac{(\pi D_1^2 - \pi D_2^2)L}{4L/f_r N}$$

(omitting the allowance term). By rearranging and substituting N_s, an exact expression for MRR is obtained:

$$\text{MRR} = 12Vf_r \frac{(D_1^2 - D_2^2)}{4D_1} \tag{22-5}$$

Rewriting the last term,

$$\frac{(D_1^2 - D_2^2)}{4D_1} = \frac{D_1 - D_2}{2} \times \frac{D_1 - D_2}{2D_1}$$

Therefore, because

$$d = \frac{(D_1 - D_2)}{2} \text{ and } \frac{D_1 - D_2}{2D_1} \cong 1 \text{ for small } d$$

then,

$$\text{MRR} \cong 12\, Vf_r d \text{ in.}^3/\text{min} \tag{22-6}$$

Note that equation 22-6 is an approximate equation that assumes that the depth of cut d is small compared to the uncut diameter D_1.

BORING

Boring always involves the enlarging of an existing hole, which may have been made by drilling or may be the result of a core in a casting. An equally important and concurrent purpose of boring may be to make the hole concentric with the axis of rotation of the workpiece and thus correct any eccentricity that may have resulted from the drill starting or drifting off the centerline. Concentricity is an important attribute of bored holes.

When boring is done in a lathe, the work usually is held in a chuck or on a faceplate. Holes may be bored straight, tapered, with threads, or to irregular contours. Figure 22-5a shows the relationship of the tool and the workpiece for boring. Think

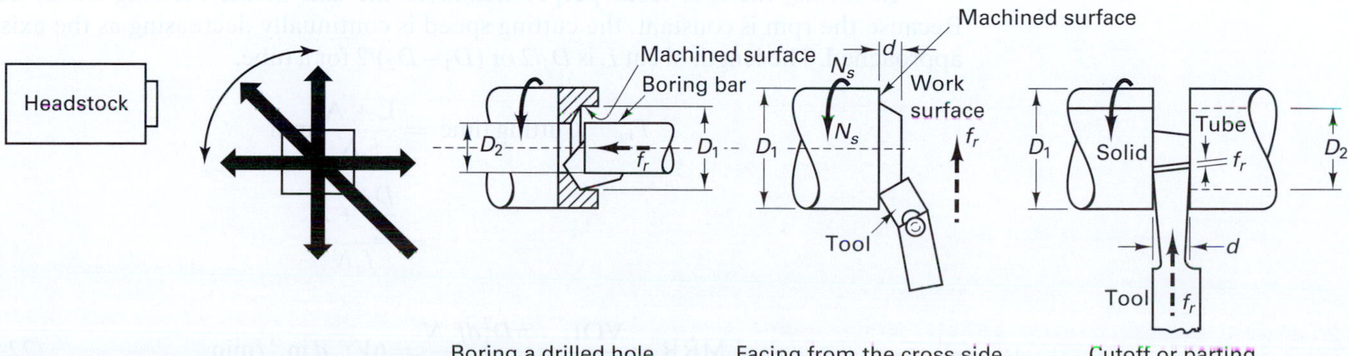

FIGURE 22-5 Basic movement of boring, facing, and cutoff in a lathe, where cutting is performed by one-single-point cutting tool at a time and the tool can be fed in any direction.

of boring as internal turning while feeding the tool parallel to the rotation axis of the workpiece, with two important differences.

First, the relief and clearance angles on the tool should be larger and the tool overhang (length to diameter) must be considered with regard to stability and deflection problems.

As before V and f_r are selected. For a cut of length L, the cutting time is

$$T_m = \frac{L + A}{f_r N_s} \tag{22-7}$$

where $N_s = 12V/\pi D_1$ for D_1, the diameter of bore, and A, the overrun allowance. The metal removal rate is

$$\text{MRR} = \frac{L(\pi D_1^2 - \pi D_2^2)/4}{L/f_r N}$$

where D_2 is the original hole diameter;

$$\text{MRR} \cong 12Vf_r d \tag{22-8}$$

(omitting allowance term), where d is the depth of cut.

In most respects, the same principles are used for boring as for turning. Again, the tool should be set exactly at the same height as the axis of rotation. Larger end clearance angles help to prevent the heel of the tool from rubbing on the inner surface of the hole. Secondly, because the tool overhang will be greater, smaller feeds and depths of cut may be selected to reduce the cutting forces that cause tool vibration and chatter. In some cases, the boring bar may be made of tungsten carbide because of this material's greater stiffness.

FACING

Facing is the producing of a flat surface as the result of the tool being fed across the end of the rotating workpiece, as shown in Figure 22-5b. Unless the work is held on a mandrel, if both ends of the work are to be faced, it must be turned end for end after the first end is completed and the facing operation repeated.

The cutting speed should be determined from the largest diameter of the surface to be faced. Facing may be done either from the outside inward or from the center outward. In either case, the point of the tool must be set exactly at the height of the center of rotation. Because the cutting force tends to push the tool away from the work, it is usually desirable to clamp the carriage to the lathe bed during each facing cut to prevent it from moving slightly and thus producing a surface that is not flat.

In the facing of castings or other materials that have a hard surface, the depth of the first cut should be sufficient to penetrate the hard material to avoid excessive tool wear.

In facing, the tool feeds perpendicular to the axis of the rotating workpiece. Because the rpm is constant, the cutting speed is continually decreasing as the axis is approached. The length of cut L is $D_1/2$ or $(D_1 - D_2)/2$ for a tube.

$$T_m = \text{Cutting time} = \frac{\text{L} + \text{A}}{f_r N} \text{ min}$$

$$= \frac{\frac{D_1}{2} + A}{f_r N}$$

$$\text{MRR} = \frac{\text{VOL}}{T_m} = \frac{\pi D_1^2 d f_r N}{4L} = 6Vf_r d \text{ in.}^3/\text{min} \tag{22-9}$$

where d is the depth of cut and $L = D_1/2$ is the length of cut.

PARTING

Parting is the operation by which one section of a workpiece is severed from the remainder by means of a cutoff tool, as shown in Figure 22-5c. Because parting tools are quite thin and must have considerable overhang, this process is more difficult to perform accurately. The tool should be set exactly at the height of the axis of rotation, be kept sharp, have proper clearance angles, and be fed into the workpiece at a proper and uniform feed rate.

In parting or **cutoff** work, the tool is fed (plunged) perpendicular to the rotational axis, as it was in facing. The length of cut for solid bars is $D_1/2$. For tubes,

$$L = \frac{D_1 - D_2}{2}$$

In cutoff operations, the width of the tool is d in inches, the width of the cutoff operation. The equations for T_m and MRR are then basically the same as for facing.

DEFLECTION

In boring, facing, and cutoff operations, the speeds, feeds, and depth of cut selected are generally less than those recommended for straight turning because of the large overhang of the tool often needed to complete the cuts. Recall the basic equation for deflection of a cantilever beam, modifying for machining,

$$\delta = \frac{Pl^3}{3EI} = \frac{F_c l^3}{3EI} \tag{22-10}$$

In equation 22-10, l represents the overhang of the tool, which greatly affects the deflection, so it should be minimized whenever possible. In equation 22-10,

E = modulus of elasticity
I = moment of inertia of cross section of tool
$P = F_c$ = applied load or cutting force

where

$I = \pi D_1^2/64$ solid round bar
$I = \pi(D_1^4 - (D_2^4))/64$ bar with hole
D_1 = diameter of tube or bar
D_2 = inside diameter of the tube

Deflection is proportional to the fourth power of the boring bar diameter and the third power of the bar overhang. Select the largest bar diameter and minimize the overhand. Use carbide shank boring bars ($E \cong 80{,}000{,}000$ psi), and select tool geometries that direct cutting forces into the feed direction to minimize chatter. The reduction of the feed or depth of cut reduces the forces operating on the tools. The cutting speed usually controls the occurrence of chatter and vibration. See the dynamics of machining discussion in Chapter 20.

Any imbalance in the cutting forces will deflect the tool to the side, resulting in loss of accuracy in cutoff lengths. At the outset, the forces will be balanced if there is no side rake on the tool. As the cutoff tool reaches the axis of the rotating part, the tool will be deflected away from the spindle, resulting in a change in the length of the part.

PRECISION BORING

Sometimes bored holes are slightly bell mouthed because the tool deflects out of the work as it progresses into the hole. This often occurs in castings and forgings where the holes have draft angles so that the depth of cut increases as the tool progresses down the bore. This problem can usually be corrected by repeating the cut with the same tool setting; however, the total cutting time for the part is increased. Alternately, a more robust setup can be used. Large holes may be precision bored using the setup shown in

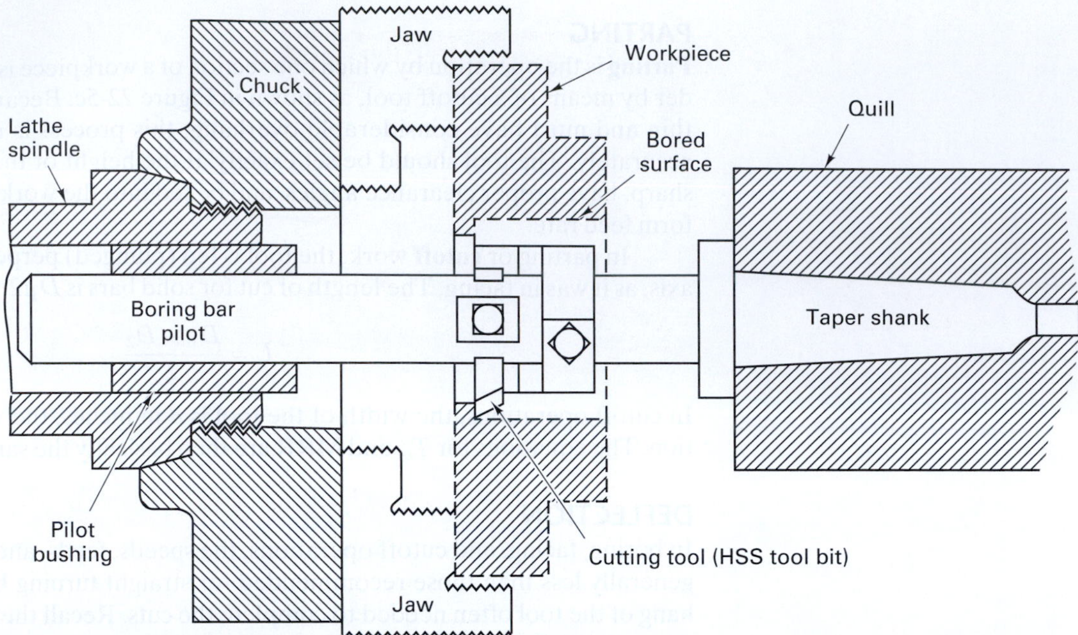

FIGURE 22-6 Pilot boring bar mounted in tailstock of lathe for precision boring large hole in casting. The size of the hole is controlled by the rotation diameter of the cutting tool.

Figure 22-6, where a pilot bushing is placed in the spindle to mate with the hardened ground pilot of the boring bar. This setup eliminates the cantilever problems common to boring.

Because the rotational relationship between the work and the tool is a simple one and is employed on several types of machine tools—such as lathes, drilling machines, and milling machines—boring is very frequently done on such machines. However, several machine tools have been developed primarily for boring, especially in cases involving large workpieces or for large-volume boring of smaller parts. Such machines as these are also capable of performing other operations, such as milling and turning. Because boring frequently follows drilling, many boring machines also can do drilling, permitting both operations to be done with a single setup of the work.

DRILLING

Drilling, discussed in detail in Chapter 23, can be done on lathes with the drill mounted in the tailstock quill of engine lathes or the turret on turret lathes and fed against a rotating workpiece. Straight-shank drills can be held in Jacobs chucks, or drills with taper shanks mounted directly in the quill hole can drill holes online (center of rotation). Drills can also be mounted in the turrets of modern turret centers and fed automatically on the rotational axis of the workpiece or off-axis with power heads. It also is possible to drill on a lathe with the drill bit mounted and rotated in the spindle while the work remains stationary, supported on the tailstock or the carriage of the lathe.

The usual speeds used for drilling should be selected for lathe work. Because the feed may be manually controlled, care must be exercised, particularly in drilling small holes. Coolants should be used where required. In drilling deep holes, the drill should be withdrawn occasionally to clear chips from the hole and to aid in getting coolant to the cutting edges. This is called *peck drilling*. See Chapter 23 for further discussion on drilling.

REAMING

Reaming on a lathe involves no special precautions. Reamers are held in the tailstock quill, taper-shank types being mounted directly and straight-shank types by means of a drill chuck. Rose-chucking reamers are usually used (see Chapter 23). Fluted-chucking reamers may also be used, but these should be held in some type of holder that will permit the reamer to float (i.e., have some compliance) in the hole and conform to the geometry created by the boring process.

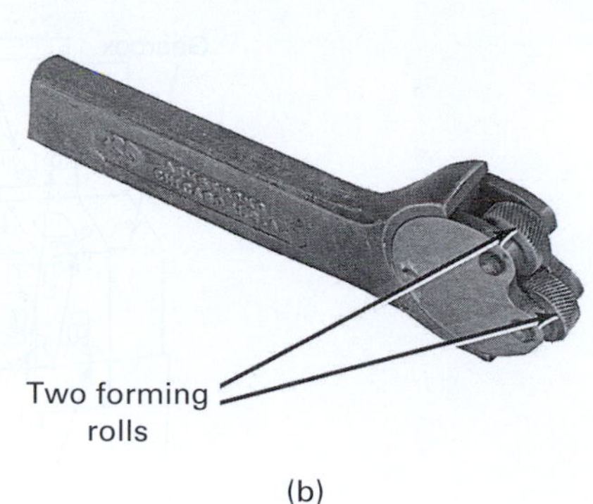

(a) (b)

FIGURE 22-7 (a) Knurling in a lathe, using a forming-type tool, and showing the resulting pattern on the workpiece; (b) knurling tool with forming rolls. *(Courtesy Armstong Industrial Hand Tools)*

KNURLING

Knurling produces a regularly shaped, roughened surface on a workpiece. Although knurling also can be done on other machine tools, even on flat surfaces, in most cases it is done on external cylindrical surfaces using lathes. Knurling is a chipless, cold-forming process. See Figures 22-7 and 22-10 for examples. The two hardened rolls are pressed against the rotating workpiece with sufficient force to cause a slight outward and lateral displacement of the metal to form the knurl in a raised, diamond pattern. Another type of knurling tool produces the knurled pattern by cutting chips. Because it involves less pressure and thus does not tend to bend the workpiece, this method is often preferred for workpieces of small diameter and for use on automatic or semiautomatic machines.

SPECIAL ATTACHMENTS

For engine lathes, taper turning and **milling** can be done on a lathe but require special attachments. The *milling attachment* is a special vise that attaches to the cross slide to hold work. The milling cutter is mounted and rotated by the spindle. The work is fed by means of the cross-slide screw. *Tool-post grinders* are often used to permit grinding to be done on a lathe. Taper turning will be discussed later. *Duplicating attachments* are available that, guided by a template, will automatically control the tool movements for turning irregularly shaped parts. In some cases, the first piece, produced in the normal manner, may serve as the template for duplicate parts. To a large extent, duplicating lathes using templates have been replaced by numerically controlled lathes and milling is done with power tools in numerical control turret lathes.

22.3 LATHE DESIGN AND TERMINOLOGY

Knowing the terminology of a machine tool is fundamental to understanding how it performs the basic processes, how the workholding devices are interchanged, and how the cutting tools are mounted and interfaced to the work. **Lathes** are machine tools designed primarily to do turning, facing, and boring. Very little turning is done on other types of machine tools, and none can do it with equal facility. Lathes also can do facing, drilling, and reaming. Modern turning centers permit milling and drilling operations using live (also called powered) spindles in multiple-tool turrets, so their versatility permits multiple operations to be done with a single setup of the workpiece. Consequently, the lathe is probably the most common machine tool, along with milling machines.

Lathes in various forms have existed for more than 2000 years, but modern lathes date from about 1797, when Henry Maudsley developed one with a leadscrew, providing controlled, mechanical feed of the tool. This ingenious Englishman also developed a change-gear system that could connect the motions of the spindle and leadscrew and thus enable threads to be cut.

LATHE DESIGN

The essential components of an **engine lathe** (Figure 22-8) are the bed, headstock assembly (which includes the spindle), tailstock assembly, carriage assembly,

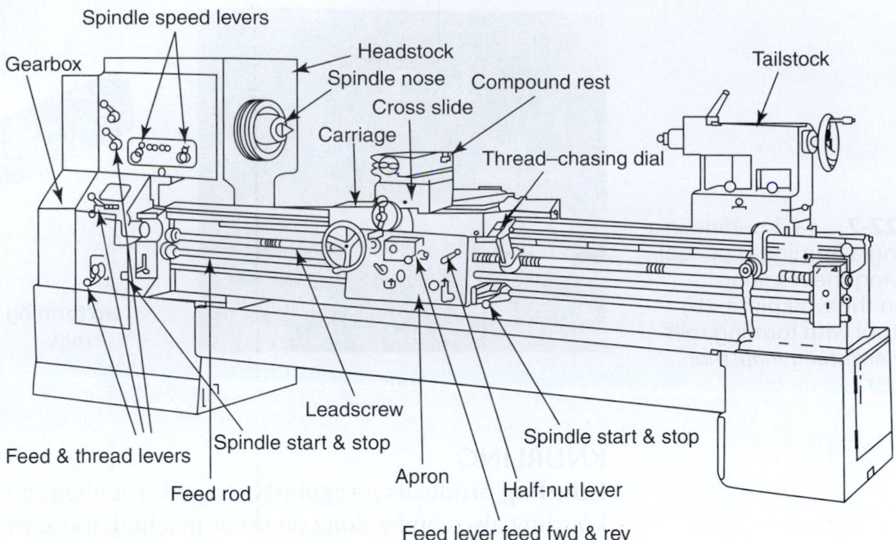

FIGURE 22-8 Schematic of an engine lathe, a versatile machine tool that is easy to set up and operate and is intended for short-run jobs.

quick-change gearbox, and the leadscrew and feed rod. The **bed** is the base and backbone of a lathe. The bed is usually made of well-normalized or aged gray or nodular cast iron and provides a heavy, rigid frame on which all the other basic components are mounted. Two sets of parallel, longitudinal **ways,** inner and outer, are contained on the bed. On modern lathes, the ways are surface hardened and precision machined, and care should be taken to ensure that the ways are not damaged. Any inaccuracy in them usually means that the accuracy of the entire lathe is destroyed.

The **headstock,** mounted in a fixed position on the inner ways, provides powered means to rotate the work at various rpm values. Essentially, it consists of a hollow **spindle,** mounted in accurate bearings, and a set of transmission gears—similar to a truck transmission—through which the spindle can be rotated at a number of speeds. Most lathes provide from 8 to 18 choices of rpm. On modern lathes all the rpm rates can be obtained merely by moving from two to four levers or the lathe has a continuously variable spindle rpm using electrical or mechanical drives.

The accuracy of a lathe is greatly dependent on the spindle. It carries the workholders and is mounted in heavy bearings, usually preloaded tapered roller or ball types. The spindle has a hole extending through its length through which long bar stock can be fed. The size of this hole is an important dimension of a lathe because it determines the maximum size of bar stock that can be machined when the materials must be fed through the spindle.

The spindle protrudes from the gearbox and contains means for mounting various types of workholding devices (chucks, face and dog plates, collets). Power is supplied to the spindle from an electric motor through a V-belt or silent-chain drive. Most modern lathes have motors of from 5 to 25 hp to provide adequate power for carbide and ceramic tools cutting hard materials at high cutting speeds.

For the classic engine lathe, the **tailstock** assembly consists, essentially, of three parts. A lower casting fits on the inner ways of the bed, can slide longitudinally, and can be clamped in any desired location. An upper casting fits on the lower one and can be moved transversely upon it, on some type of keyed ways, to permit aligning the tailstock and headstock spindles (for turning tapers). The third major component of the assembly is the tailstock **quill.** This is a hollow steel cylinder, usually about 2 to 3 in. diameter, that can be moved longitudinally in and out of the upper casting by means of a handwheel and screw. The open end of the quill hole has a Morse taper. Cutting tools or a **lathe center** are held in the quill. A graduated scale is usually engraved on the outside of the quill to aid in controlling its motion in and out of the upper casting. A locking device permits clamping the quill in any desired position. In recent years, dual-spindle numerical control (NC) turning centers have emerged, where a subspindle replaces the tailstock assembly. Parts can be automatically transferred from the spindle to the subspindle for turning the back end of the part. See Advanced Topic 1 for more on NC machining and turning centers.

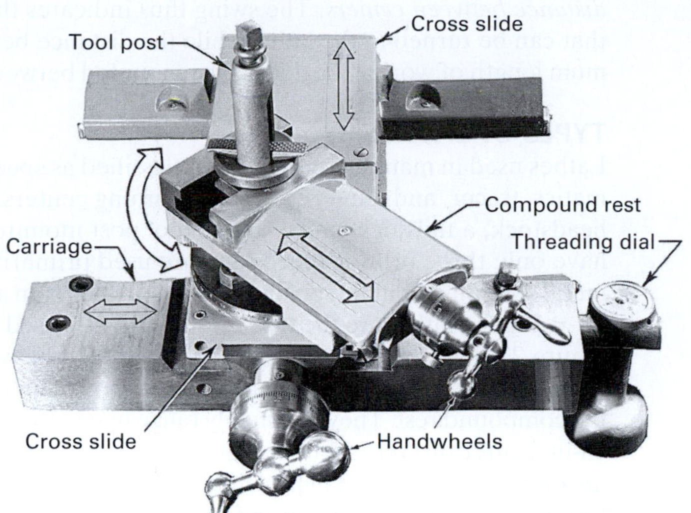

FIGURE 22-9 The cutting tools for lathe work are held in the tool post on the compound rest, which can translate and swivel. *(Courtesy J T. Black)*

The **carriage assembly,** together with the apron, provides the means for mounting and moving cutting tools. The **carriage,** a relatively flat H-shaped casting, rides on the outer set of ways on the bed. The **cross slide** is mounted on the carriage and can be moved by means of a feed screw that is controlled by a small handwheel and a graduated dial. The cross slide thus provides a means for moving the lathe tool in the facing or cutoff direction.

On most lathes, the tool post is mounted on a **compound rest** (see Figure 22-9). The compound rest can rotate and translate with respect to the cross slide, permitting further positioning of the tool with respect to the work. The **apron,** attached to the front of the carriage, has the controls for providing manual and powered motion for the carriage and powered motion for the cross slide. The carriage is moved parallel to the ways by turning a handwheel on the front of the apron, which is geared to a pinion on the back side. This pinion engages a rack that is attached beneath the upper front edge of the bed in an inverted position.

Powered movement of the carriage and cross slide is provided by a rotating **feed rod.** The feed rod, which contains a keyway, passes through the two reversing bevel pinions and is keyed to them. Either pinion can be activated by means of the feed reverse lever, thus providing "forward" or "reverse" power to the carriage. Suitable clutches connect either the rack pinion or the cross-slide screw to provide longitudinal motion of the carriage or transverse motion of the cross slide.

For cutting threads, a **leadscrew** is used. When a friction clutch is used to drive the carriage, motion through the leadscrew is by a direct, mechanical connection between the apron and the leadscrew. A **split nut** is closed around the leadscrew by means of a lever on the front of the apron directly driving the carriage without any slippage.

Modern lathes have **quick-change gearboxes,** driven by the spindle, that connect the feed rod and leadscrew. Thus, when the spindle turns a revolution, the tool (mounted on the carriage) translates (longitudinally or transversely) a specific distance in inches—that is, inches per revolution (ipr). This revolutions per minute, rpm or N_s, times the feed, f_r, gives the feed rate, f, in inches per minute (ipm) that the tool is moving. In this way, the calculations for turning rpm and feed in ipr are "mechanically related." Typical lathes may provide as many as 48 feeds, ranging from 0.002 to 0.118 in. (0.05 to 3 mm) per revolution of the spindle, and, through the leadscrew, leads from to 92 threads per inch.

SIZE DESIGNATION OF LATHES

The size of a lathe is designated by two dimensions. The first is known as the **swing.** This is the maximum diameter of work that can be rotated on a lathe. Swing is approximately twice the distance between the line connecting the lathe centers and the nearest point on the ways. The maximum diameter of a workpiece that can be mounted between centers is somewhat less than the swing diameter because the workpiece must clear the carriage assembly as well as the ways. The second size dimension is the *maximum* *distance between centers*. The swing thus indicates the maximum workpiece diameter that can be turned in the lathe, while the distance between centers indicates the maximum length of workpiece that can be mounted between centers.

TYPES OF LATHES

Lathes used in manufacturing can be classified as speed, engine, toolroom, turret, automatics, tracer, and numerical control turning centers. Speed lathes usually have only a headstock, a tailstock, and a simple tool post mounted on a light bed. They ordinarily have only three or four speeds and are used primarily for wood turning, polishing, or metal spinning. Spindle speeds up to about 4000 rpm are common.

Engine lathes are the type most frequently used in manufacturing. Figure 22-1 and Figure 22-8 are examples of this type. They are heavy-duty machine tools with all the components described previously and have power drive for all tool movements except on the compound rest. They commonly range in size from 12- to 24-in. swing and from 24- to 48-in. center distances, but swings up to 50 in. and center distances up to 12 ft are not uncommon. Very large engine lathes (36- to 60-ft-long beds) are therefore capable of performing roughing cuts in iron and steel at depths of cut of $\frac{1}{2}$ to 2 in. and at cutting speeds at 50 to 200 sfpm with WC tools run at 0.010 to 0.100 in./rev. To perform such heavy cuts requires rigidity in the machine tool, the cutting tools, the workholder, and the workpiece (using steady rests and other supports) and large horsepower (50 to 100 hp).

Most engine lathes are equipped with chip pans and a built-in coolant circulating system. Smaller engine lathes, with swings usually not greater than 13 in., also are available in *bench type*, designed for the bed to be mounted on a bench or table.

Toolroom lathes have somewhat greater accuracy and, usually, a wider range of speeds and feeds than ordinary engine lathes. Designed to have greater versatility to meet the requirements of tool and die work, they often have a continuously variable spindle speed range and shorter beds than ordinary engine lathes of comparable swing because they are generally used for machining relatively small parts. They may be either bench or pedestal type.

Several types of special-purpose lathes are made to accommodate specific types of work. On a *gap-bed lathe*, for example, a section of the bed, adjacent to the headstock, can be removed to permit work of unusually large diameter to be swung. Another example is the *wheel lathe*, which is designed to permit the turning of railroad-car wheel-and-axle assemblies.

Figure 22-10 shows a large vertical turning lathe machining a large steel casting. Vertical lathes are an excellent alternative to large horizontal CNC lathes.

FIGURE 22-10 This large vertical turning center is used for turning large circular parts rotated under vertically mounted tools. (*© Alvis Upitis/SuperStock*)

Gravity-aided seating of large/heavy workpieces allows a high degree of process repeatability. A smaller footprint, lower initial cost, and increased productivity are all advantages when compared to traditional horizontal lathes.

Although engine lathes are versatile and very useful, the time required for changing and setting tools and for making measurements on the workpiece is often a large percentage of the cycle time. Often, the actual chip-production time is less than 30% of the total cycle time. Methods to reduce setup and tool-change time are now well known, reducing setups to minutes and unload/load steps to seconds. The placement of single-cycle machine tools into interim or lean manufacturing cells will increase the productivity of the workers because they can run more than one machine. Turret lathes, screw machines, and other types of semiautomatic and automatic lathes have been highly developed and are widely used in manufacturing as another means to improve cutting productivity.

Turret Lathes. The basic components of a **turret lathe** are depicted in Figure 22-11. Basically, a longitudinally feedable, hexagon turret replaces the tailstock. The turret, on which six tools can be mounted, can be rotated about a vertical axis to bring each tool into operating position, and the entire unit can be translated parallel to the ways, either manually or by power, to provide feed for the tools. When the turret assembly is backed away from the spindle by means of a capstan wheel, the turret indexes automatically at the end of its movement, thus bringing each of the six tools into operating position in sequence.

The square turret on the cross slide can be rotated manually about a vertical axis to bring each of the four tools into operating position. On most machines, the turret

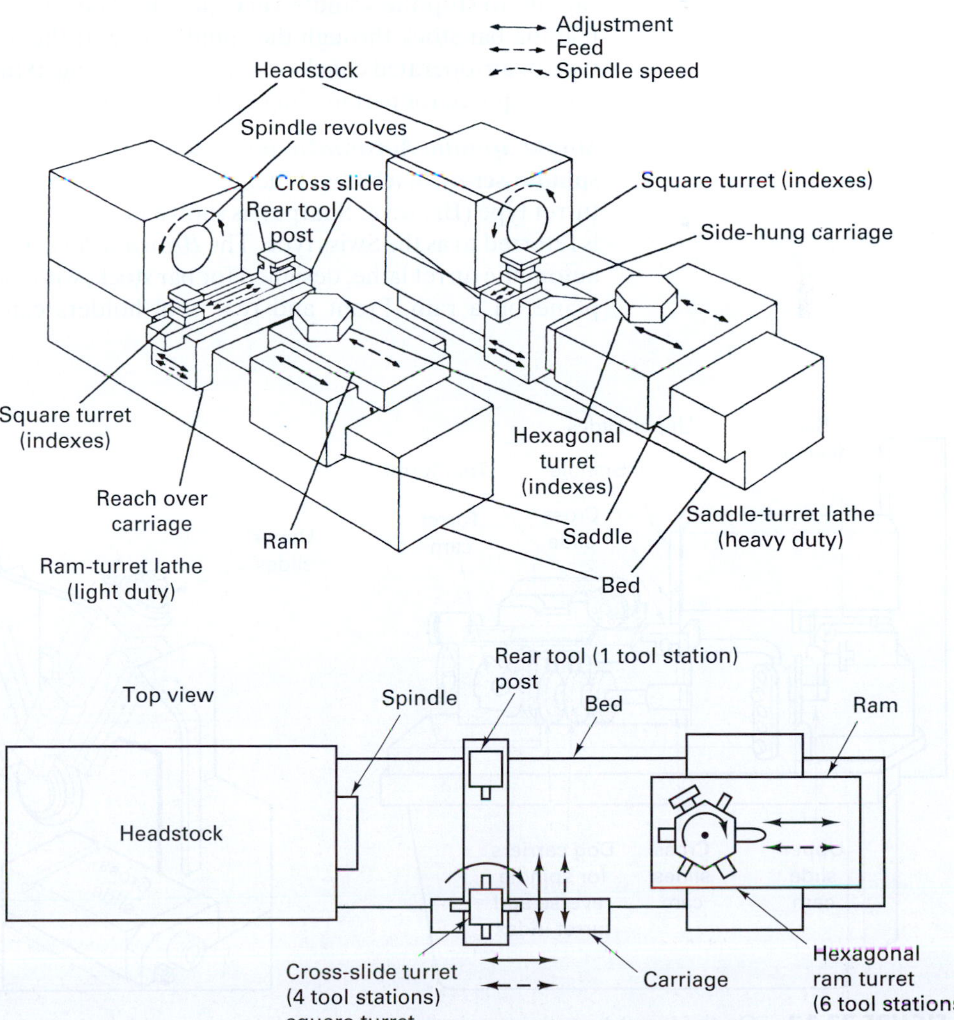

FIGURE 22-11 Block diagrams of ram- and saddle-turret lathe.

can be moved transversely, either manually or by power, by means of the cross slide, and longitudinally through power or manual operation of the carriage. In most cases, a fixed tool holder also is added to the back end of the cross slide; this often carries a parting tool.

Through these basic features of a turret lathe, a number of tools can be set up on the machine and then quickly be brought successively into working position so that a complete part can be machined without the necessity for further adjusting, changing tools, or making measurements.

The two basic types of turret lathes are the ram-type turret lathe and the saddle-type turret lathe. In the *ram-type turret lathe*, the ram and turret are moved up to the cutting position by means of the capstan wheel, and the power feed is then engaged. As the ram is moved toward the headstock, the turret is automatically locked into position so that rigid tool support is obtained. Rotary stopscrews control the forward travel of the ram, one stop being provided for each face on the turret. The proper stop is brought into operating position automatically when the turret is indexed. A similar set of stops is usually provided to limit movement of the cross slide. The *saddle-type turret lathe* provides a more rugged mounting for the hexagon turret than can be obtained by the ram-type mounting. In saddle-type lathes, the main turret is mounted directly on the saddle, and the entire saddle and turret assembly reciprocates. Larger turret lathes usually have this type of mounting. However, because the saddle-turret assembly is rather heavy, this type of mounting provides less rapid turret reciprocation. When such lathes are used with heavy tooling for making heavy or multiple cuts, a **pilot arm** attached to the headstock engages a pilot hole attached to one or more faces of the turret to give additional rigidity. Turret-lathe headstocks can shift rapidly between spindle speeds and brake rapidly to stop the spindle very quickly. They also have automatic stock-feeding for feeding bar stock through the spindle hole. If the work is to be held in a chuck, some type of air-operated chuck or a special clamping fixture is often employed to reduce the time required for part loading and unloading.

Single-Spindle Automatic Screw Machines. There are two common types of **single-spindle screw machines.** One, an American development and commonly called the turret type (Brown & Sharpe), is shown in Figure 22-12. The other is of Swiss origin and is referred to as the Swiss type. The *Brown & Sharpe screw machine* is essentially a small automatic turret lathe, designed for bar stock, with the main turret mounted in a vertical plane on a ram. Front and rear tool holders can be mounted on the cross slide.

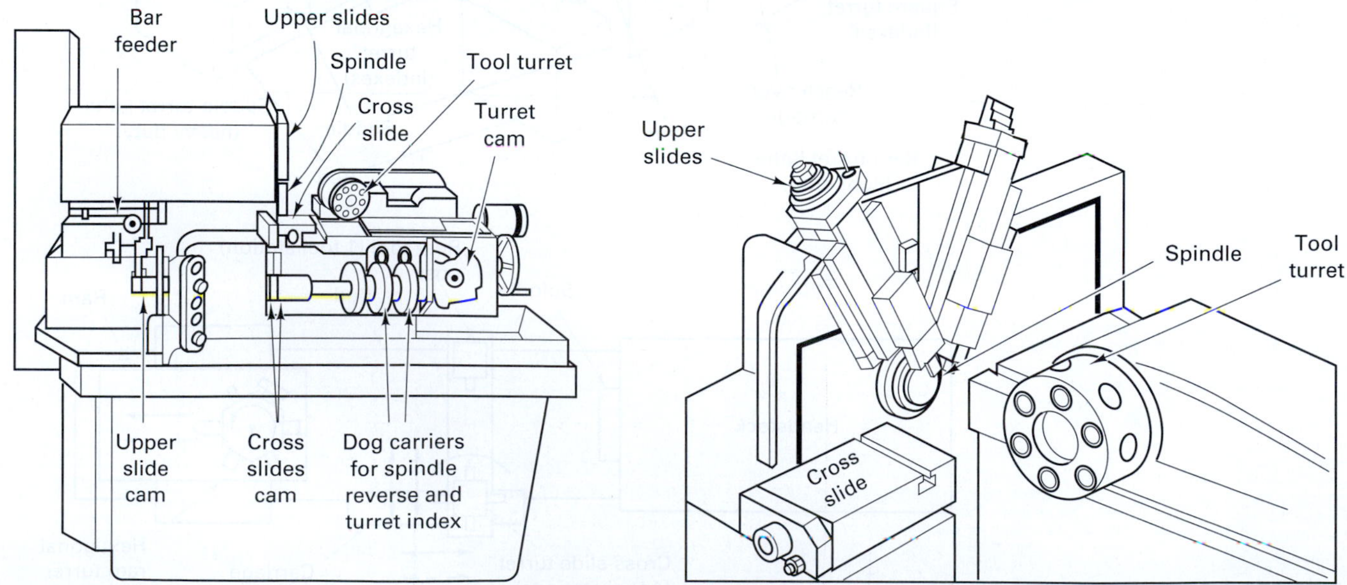

FIGURE 22-12 On the turret-type single-spindle automatic, the tools must take turns to make cuts.

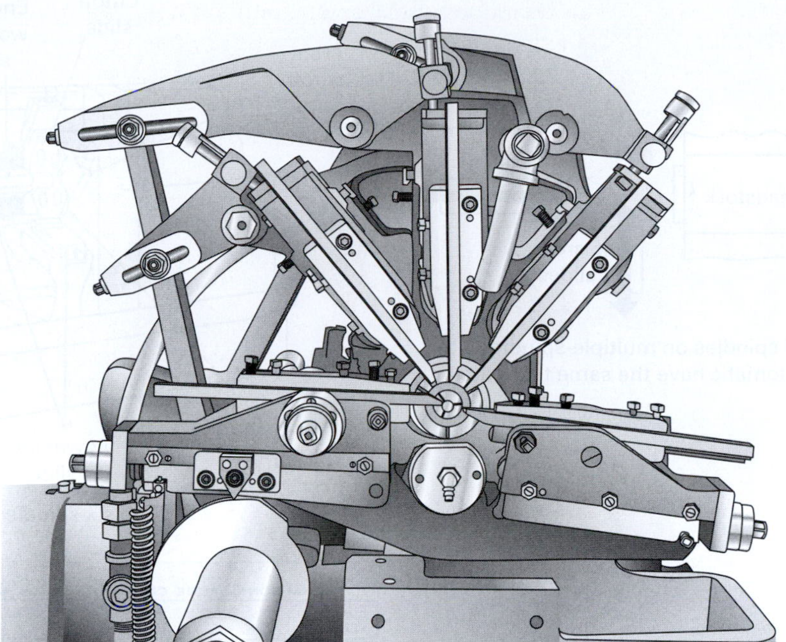

FIGURE 22-13 Close-up view of a Swiss-type screw machine, showing the tooling and radial tool sides, actuated by rocker arms controlled by a disk cam, shown in lower left. *(Courtesy J T. Black)*

All motions of the turret, cross slide, spindle, chuck, and stock-feed mechanism are controlled by cams. The turret cam is essentially a program that defines the movement of the turret during a cycle. These machines are usually equipped with an automatic rod-feeding magazine that feeds a new length of bar stock into the collet (the workholding device) as soon as one rod is completely used.

Often, screw machines of the Brown & Sharpe type are equipped with a transfer or "picking" attachment. This device picks up the workpiece from the spindle as it is cut off and carries it to a position where a secondary operation is performed by a small, auxiliary power head. In this manner, screwdriver slots are put in screw heads, small flats are milled parallel with the axis of the workpiece, or holes are drilled normal to the axis.

On the *Swiss-type automatic screw machine*, the cutting tools are held and moved in radial slides (Figure 22-13). Disk cams move the tools into cutting position and provide feed into the work in a radial direction only; they provide any required longitudinal feed by reciprocating the headstock.

Most machining on Swiss-type screw machines is done with single-point cutting tools. Because they are located close to the spindle collet, the workpiece is not subjected to much deflection. Consequently, these machines are particularly well suited for machining very small parts.

Both types of single-spindle screw machines can produce work to close tolerances, the Swiss-type probably being somewhat superior for very small work. Tolerances of 0.0002 to 0.0005 in. are not uncommon. The time required for setting up the machine is usually an hour or two and can be much less. One person can tend many machines, once they are properly tooled. They have short cycle times, frequently less than 30 s/piece.

Multiple-Spindle Automatic Screw Machines. Single-spindle screw machines utilize only one or two tooling positions at any given time. Thus, the total cycle time per workpiece is the sum of the individual machining and tool-positioning times. On **multiple-spindle screw machines,** sufficient spindles—usually four, six, or eight—are provided so that many tools can cut simultaneously. Thus, the cycle time per piece is equal to the maximum cutting time of a single tool position plus the time required to index the spindles from one position to the next.

The two distinctive features of multiple-spindle screw machines are shown in Figure 22-14. First, the six spindles are carried in a rotatable drum that indexes in order to bring each spindle into a different working position. Second, a nonrotating tool slide contains the same number of tool holders as there are spindles and thus provides and

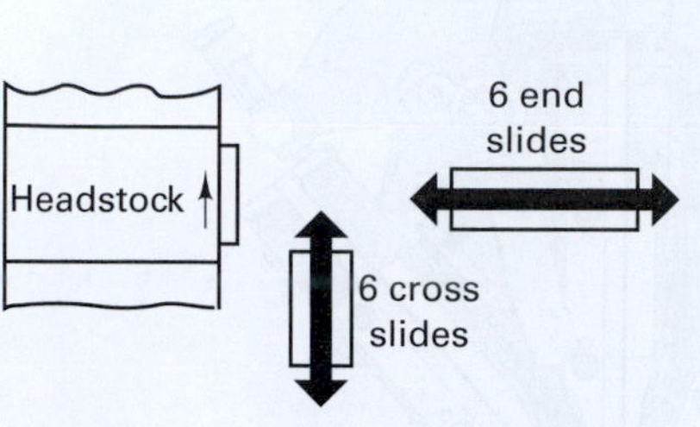

All spindles on multiple-spindle automatic have the same tool path

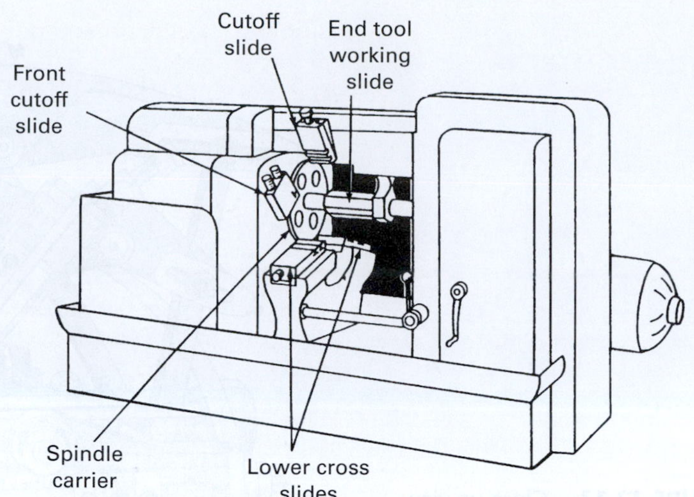

The six-spindle automatic

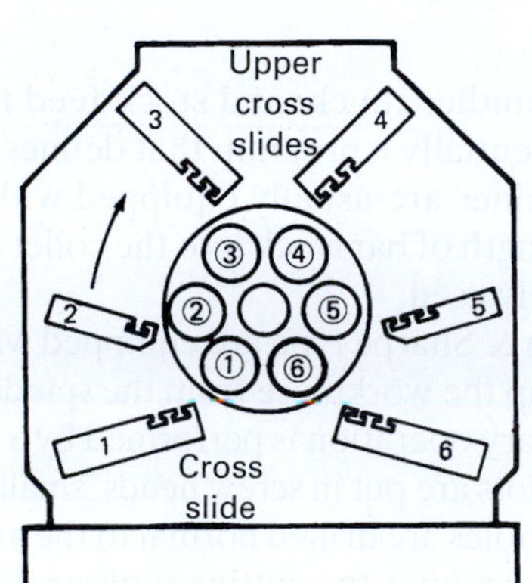

Spindle arrangement for six-spindle automatic. The barstock is usually fed to a stop at position 6. The cutoff position is the one preceding the bar feed position.

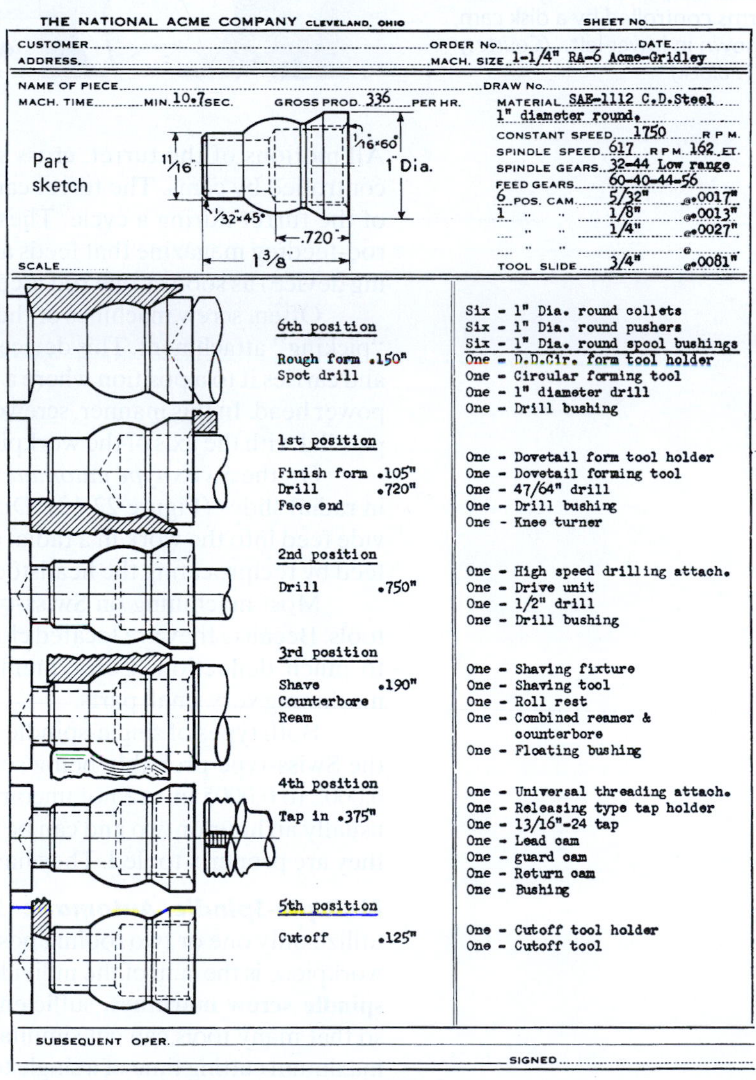

THE NATIONAL ACME COMPANY CLEVELAND, OHIO SHEET NO.

CUSTOMER ORDER No. DATE

ADDRESS MACH. SIZE 1-1/4" RA-6 Acme-Gridley

NAME OF PIECE DRAW No.

MACH. TIME MIN. 10.7 SEC. GROSS PROD. 336 PER HR.

Part sketch: 11/16"; 1/16×60°; 1" Dia.; 1/32×45°; .720"; 1 3/8"

SCALE

MATERIAL SAE-1112 C.D.Steel 1" diameter round.

CONSTANT SPEED 1750 R P M.

SPINDLE SPEED 617 R P M. 162 FT.

SPINDLE GEARS 32-44 Low range

FEED GEARS 60-40-44-56

6 POS. CAM 5/32" @.0017"

1 " " 1/8" @.0013"

" " 1/4" @.0027"

" " @

TOOL SLIDE 3/4" @.0081"

Six - 1" Dia. round collets
Six - 1" Dia. round pushers
Six - 1" Dia. round spool bushings

6th position

Rough form .150"
Spot drill

One - D.D.Cir. form tool holder
One - Circular forming tool
One - 1" diameter drill
One - Drill bushing

1st position

Finish form .105"
Drill .720"
Face end

One - Dovetail form tool holder
One - Dovetail forming tool
One - 47/64" drill
One - Drill bushing
One - Knee turner

2nd position

Drill .750"

One - High speed drilling attach.
One - Drive unit
One - 1/2" drill
One - Drill bushing

3rd position

Shave .190"
Counterbore
Ream

One - Shaving fixture
One - Shaving tool
One - Roll rest
One - Combined reamer & counterbore
One - Floating bushing

4th position

Tap in .375"

One - Universal threading attach.
One - Releasing type tap holder
One - 13/16"-24 tap
One - Lead cam
One - guard cam
One - Return cam
One - Bushing

5th position

Cutoff .125"

One - Cutoff tool holder
One - Cutoff tool

SUBSEQUENT OPER.

SIGNED

Tooling sheet for making a part on a six-spindle.

FIGURE 22-14 The multiple-spindle, automatic screw machine makes all cuts simultaneously and then performs the noncutting functions (tool withdrawal, index, bar feed) at high speed.

positions a cutting tool (or tools) for each spindle. Tools are fed by longitudinal reciprocating motion. Most machines have a cross slide at each spindle position so that an additional tool can be fed from the side for facing, grooving, knurling, beveling, and cutoff operations. All motions are controlled automatically.

Starting with the sixth position, follow the sequence of processing steps on the tooling sheet for making a part shown in Figure 22-14. With a tool position available on the end tool slide for each spindle (except for a stock-feed stop at position 6), when the slide moves forward, these tools cut essentially simultaneously. At the same time, the tools in the cross slides move inward and make their cuts. When the forward cutting motion of the end tool slide is completed, it moves away from the work, accompanied by the outward movement of the radial slides. The spindles are indexed one position, by rotation of the spindle carrier, to position each part for the next operation to be performed. At spindle position 5, finished pieces are cut off. Bar stock 1 in. in diameter is fed to correct length for the beginning of the next operation. Thus, a piece is completed each time the tool slide moves forward and back. Multiple-spindle screw machines are made in a considerable range of sizes, determined by the diameter of the stock that can be accommodated in the spindles. There may be four, five, six, or eight spindles. The operating cycle of the end tool slide is determined by the operation that requires the longest time.

Once a multiple-spindle screw machine is set up, it requires only that the bar stock feed rack be supplied and the finished products checked periodically to make sure that they are within desired tolerances. One operator usually services many machines.

Most multiple-spindle screw machines use cams to control the motions. Setting up the cams and the tooling for a given job may require from 2 to 20 hr. However, once such a machine is set up, the processing time per part is very short. Often, a piece may be completed every 10 s. Typically, a minimum of 2000 to 5000 parts are required in a lot to justify setting up and tooling a multiple-spindle automatic screw machine. The precision of multiple-spindle screw machines is good, but seldom as good as that of single-spindle machines. However, tolerances from 0.0005 to 0.001 in. on the diameter are typical.

22.4 CUTTING TOOLS FOR LATHES

LATHE CUTTING TOOLS

Most lathe operations are done using single-point **cutting tools,** such as the classic tool designs shown in Figure 22-15. On right-hand turning (and left-hand turning) and facing tools, the cutting usually takes place on the side of the tool; therefore, the side rake angle is of primary importance, particularly when deep cuts are being made. On the round-nose turning tools, cutoff tools, finishing tools, and some threading tools, cutting takes place on or near the tip of the tool, and the back rake is therefore of importance. Such tools are used with relatively light depths of cut.

Because tool materials are expensive, it is desirable to use as little as possible. At the same time, it is essential that the cutting tool be supported in a strong, rigid manner to minimize deflection and possible vibration. Consequently, lathe tools are supported in various types of heavy, forged steel tool holders. The high-speed steel (HSS) tool bit should be clamped in the tool holder with minimum overhang; otherwise, tool chatter and a poor surface finish may result.

In the use of carbide, ceramic, or coated carbides for higher speed cutting, throwaway inserts are used that can be purchased in a great variety of shapes, geometrics (nose radius, tool angles, and groove geometry), and sizes (see Figure 22-16 for some examples).

When lathes are incorporated into lean manufacturing cells, requiring that different operations be performed and the time required for changing and setting tools may constitute as much as 50% of the total cycle time. Quick-change tool holders (Figure 22-17) can reduce the manual tool-changing time. The individual tools, preset in their holders, can be interchanged in the special tool post in a few seconds. With some systems, a second tool may be set in the tool post while a cut is being made with the first tool and can then be brought into proper position by rotating the post.

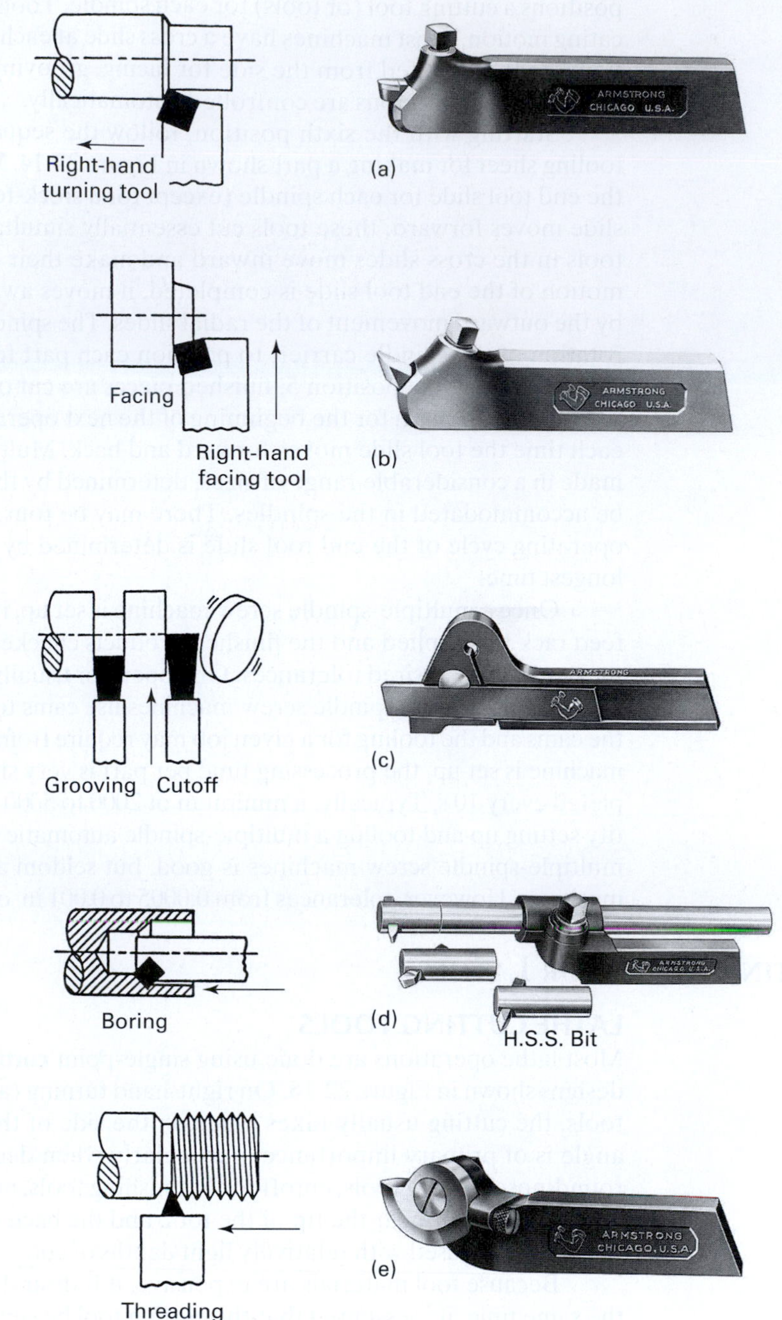

FIGURE 22-15 Common types of forged tool holders: (a) right-hand turning, (b) facing, (c) grooving cutoff, (d) boring, (e) threading. *(Courtesy of Armstrong Industrial Hand Tools)*

In lathe work, the nose of the tool should be set exactly at the same height as the axis of rotation of the work. However, because any setting below the axis causes the work to tend to "climb" up on the tool, most machinists set their tools a few thousandths of an inch above the axis, except for cutoff, threading, and some facing operations.

FORM TOOLS

In Figure 22-15, the use of form tools was shown in automatic lathe work. Form tools are made by grinding the inverse of the desired work contour into a block of HSS or tool steel. A **threading** tool is often a form tool. Although form tools are relatively expensive to manufacture, it is possible to machine a fairly complex surface with a single inward feeding of one tool. For mass-production work, adjustable form tools of either flat or rotary types, such as are shown in Figure 22-18, are used. These are expensive to make initially but can be resharpened by merely grinding a small amount off the face and then raising or rotating the cutting edge to the correct position.

Insert shape	Available cutting edges	Typical insert holder
Round	4–10 on a side 8–20 total	15° Square insert
80°/100° diamond	4 on a side 8 total	
Square	4 on a side 8 total	0° Triangular insert
Triangle	3 on a side 6 total	
55° diamond	2 on a side 4 total	35° diamond 5°
35° diamond	2 on a side 4 total	

FIGURE 22-16 Typical insert shapes, available cutting edges per insert, and insert holders for throwaway insert cutting tools. *(Adapted from* Turning Handbook of High Efficiency Metal Cutting, *Courtesy Armstong Industrial Hand Tools)*

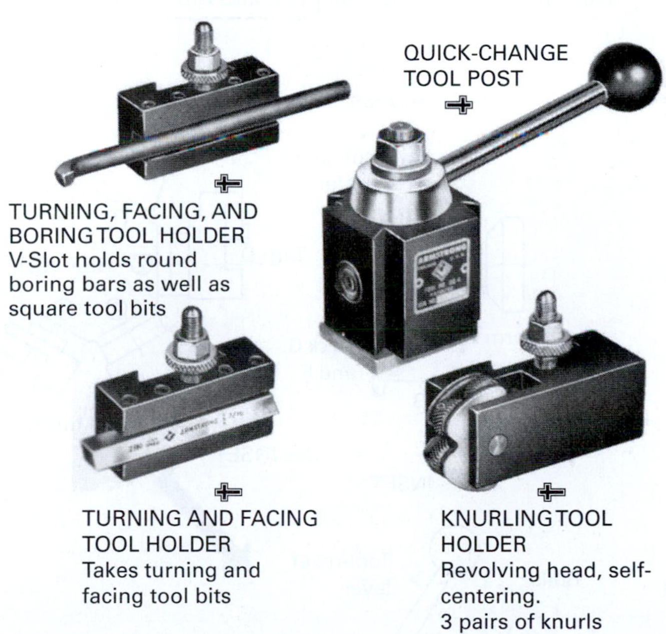

FIGURE 22-17 Quick-change tool post and accompanying tool holders. *(Courtesy Armstong Industrial Hand Tools)*

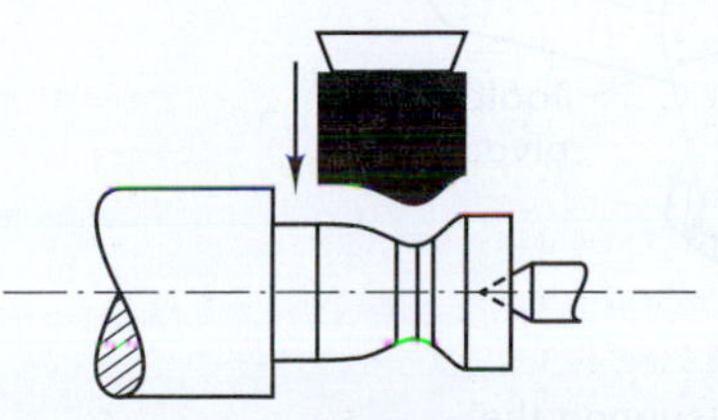

FIGURE 22-18 Circular and block types of form tools.

The use of form tools is limited by the difficulty of grinding adequate rake angles for all points along the cutting edge. A rigid setup is needed to resist the large cutting forces that develop with these tools. Light feeds with sharp, coated HSS tools are used on multiple-spindle automatics, turret lathes, and transfer line machines.

TURRET-LATHE TOOLS

In turret lathes, the work is generally held in collets and the correct amount of bar stock is fed into the machine to make one part. The tools are arranged in sequence at the tool stations with depths of cut all preset. The following factors should be considered when setting up a turret lathe.

1. *Setup time:* time required to install and set the tooling and set the stops. Standard tool holders and tools should be used as much as possible to minimize setup time. Setup time can be greatly reduced by eliminating adjustment in the setup.
2. *Workholding time:* time to load and unload parts and/or stock.
3. *Machine-controlling time:* time required to manipulate the turrets. Can be reduced by combining operations where possible. Dependent on the sequence of operations established by the design of the setup.
4. *Cutting time:* time during which chips are being produced. Should be as short as is economically practical and represent the greatest percentage of the total cycle time possible.
5. *Cost:* cost of the tool, setup labor cost, lathe operator labor cost, and the number of pieces to be made.

There are essentially 11 tooling stations, as shown in Figure 22-19, with six in the turret, four in the indexable tool post, and one in the rear tool post. The tooling is more rugged

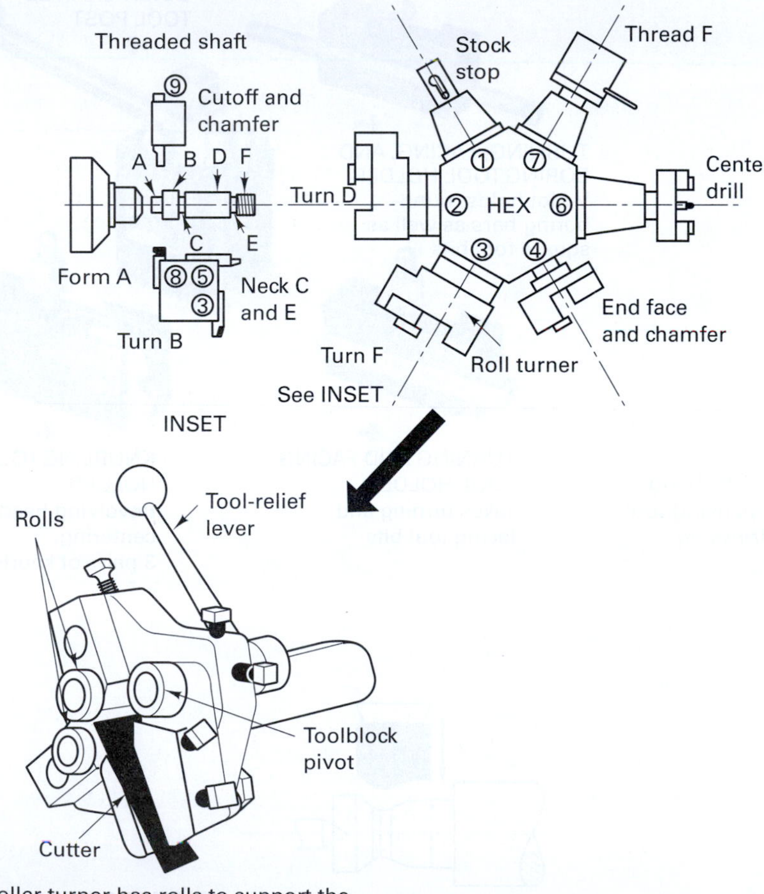

FIGURE 22-19 Turret-lathe tooling setup for producing part shown. Numbers in circles indicate the sequence of the operations from 1 to 9. The letters A through F refer to the surfaces being machined. Operation 3 is a combined operation. The roll turner is turning surface F, while tool 3 on the square post is turning surface B. The first operation stops the stock at the right length. The last operation cuts the finished bar off and puts a chamfer on the bar, which will next be advanced to the stock stop.

in turret lathes because heavy, simultaneous cuts are often made. Tools mounted in the hex turret that are used for turning are often equipped with pressure rollers set on the opposite side of the rotating workpiece from the tool to counter the cutting forces.

The setup times are usually 1 or 2 hr, so turret lathes are most economical in producing lots too large for engine lathes but too small for automatic screw machines or automatic lathes. In recent years, much of this work has been assumed by numerical control lathes, turning centers, or lean manufacturing cells. For example, the component (threaded shaft) shown in Figure 22-19 could have also been made on an NC turret lathe with some savings in cycle time or in a manufacturing cell with further savings in cycle time.

22.5 WORKHOLDING IN LATHES

WORKHOLDING DEVICES FOR LATHES

Five methods are commonly used for supporting workpieces in lathes:

1. Held between centers.
2. Held in a chuck.
3. Held in a collet.
4. Mounted on a faceplate.
5. Mounted on the carriage.

In the first four of these methods, the workpiece is rotated during machining. In the fifth method, which is not used extensively, the tool rotates while the workpiece is fed into the tool.

A general discussion of workholding devices is found in Chapter 27, and the student involved in designing working devices should study the reference materials under "Tool Design." For lathes, **workholding** is a matter of selecting from standard tooling.

LATHE CENTERS

Workpieces that are relatively long with respect to their diameters are usually held between centers (see Figure 22-20). Two lathe centers are used, one in the spindle hole and the other in the hole in the tailstock quill. Two types are used, called *dead* and *live*. Dead centers are *solid*, that is, made of hardened steel with a Morse taper on one end so that it will fit into the spindle hole. The other end is ground to a taper. Sometimes the tip of this taper is made of tungsten carbide to provide better wear resistance. Before a center is placed in position, the spindle hole should be carefully wiped clean. The

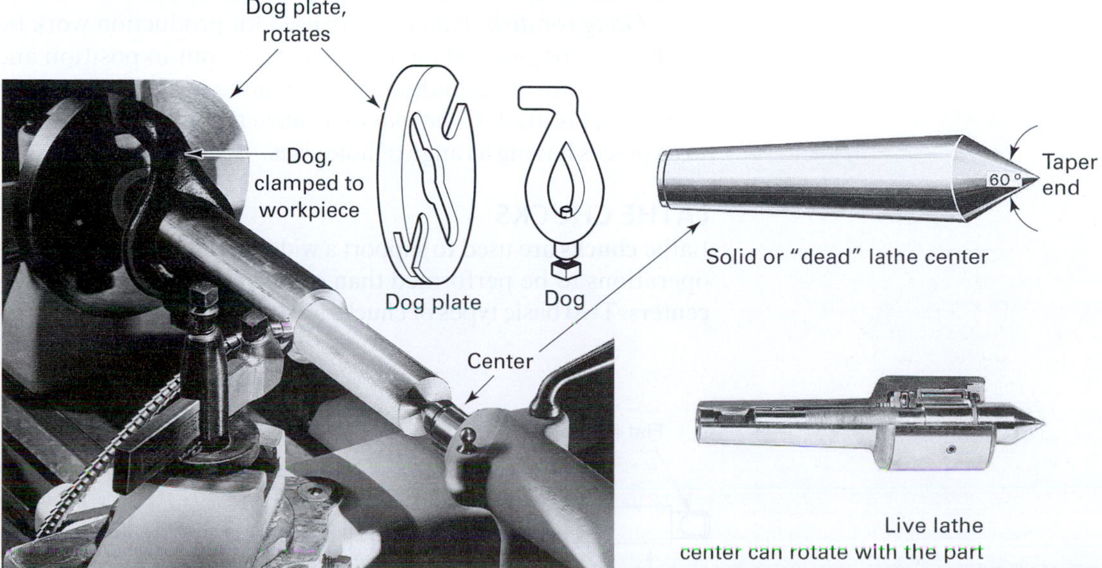

FIGURE 22-20 Work being turned between centers in a lathe, showing the use of a dog and dog plate. *(Courtesy of South Bend Lathe)*

presence of foreign material will prevent the center from seating properly, and it will not be aligned accurately.

A mechanical connection must be provided between the spindle and the workpiece to provide rotation. This is accomplished by a **lathe dog** and **dog plate.** The dog is clamped to the work. The tail of the dog enters a slot in the dog plate, which is attached to the lathe spindle in the same manner as a lathe chuck. For work that has a finished surface, a piece of soft metal, such as copper or aluminum, can be placed between the work and the dog setscrew clamp to avoid marring. Live centers are designed so that the end that fits into the workpiece is mounted on ball or roller bearings. It is free to rotate. No lubrication is required. Live centers may not be as accurate as the solid type and therefore are not often used for precision work.

Before a workpiece can be mounted between lathe centers, a center hole must be drilled in each end. This is typically done in a drill press or on the lathe with a tool held in the rear turret. A combination center drill and countersink ordinarily is used, with care taken that the center hole is deep enough so that it will not be machined away in any facing operation and yet is not drilled to the full depth of the tapered portion of the center drill (see Chapter 23).

Because the work and the center of the headstock end rotate together, no lubricant is needed in the center hole at this end. The center in the tailstock quill does not rotate; adequate lubrication must be provided. A mixture of white lead and oil is often used. Failure to provide proper lubrication at all times will result in scoring of the workpiece center hole and the center, and inaccuracy and serious damage may occur. Live centers are often used in the tailstock to overcome these problems.

The workpiece must rotate freely, yet no looseness should exist. Looseness will usually be manifested in chattering of the workpiece during cutting. The setting of the centers should be checked after cutting for a short time. Heating and thermal expansion of the workpiece will reduce the clearances in the setup.

MANDRELS

Workpieces that must be machined on both ends or are disk-shaped are often mounted on **mandrels** for turning between centers. Three common types of mandrels are shown in Figure 22-21. *Solid mandrels* usually vary from 4 to 12 in. in length and are accurately ground with a 1:2000 taper (0.006 in./ft). After the workpiece is drilled and/or bored, it is pressed on the mandrel. The mandrel should be mounted between centers so that the cutting force tends to tighten the work on the mandrel taper. Solid mandrels permit the work to be machined on both ends as well as on the cylindrical surface. They are available in stock sizes but can be made to any desired size.

Gang (or disk) mandrels are used for production work because the workpieces do not have to be pressed on and thus can be put in position and removed more rapidly. However, only the cylindrical surface of the workpiece can be machined when this type of mandrel is used. *Cone mandrels* have the advantage that they can be used to center workpieces having a range of hole sizes.

LATHE CHUCKS

Lathe **chucks** are used to support a wider variety of workpiece shapes and to permit more operations to be performed than can be accomplished when the work is held between centers. Two basic types of chucks are used, three-jaw and four-jaw (Figure 22-22).

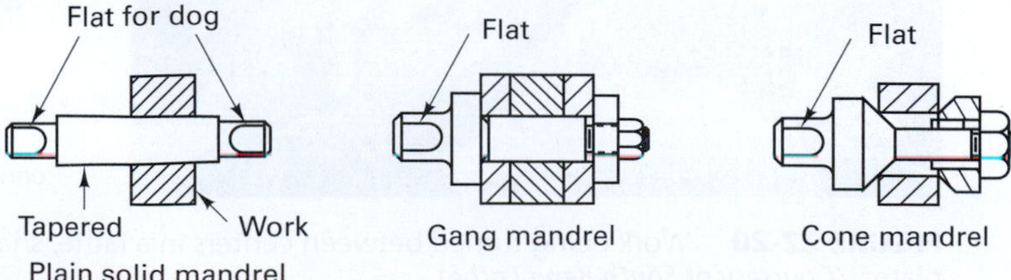

FIGURE 22-21 Three types of mandrels, which are mounted between centers for lathe work.

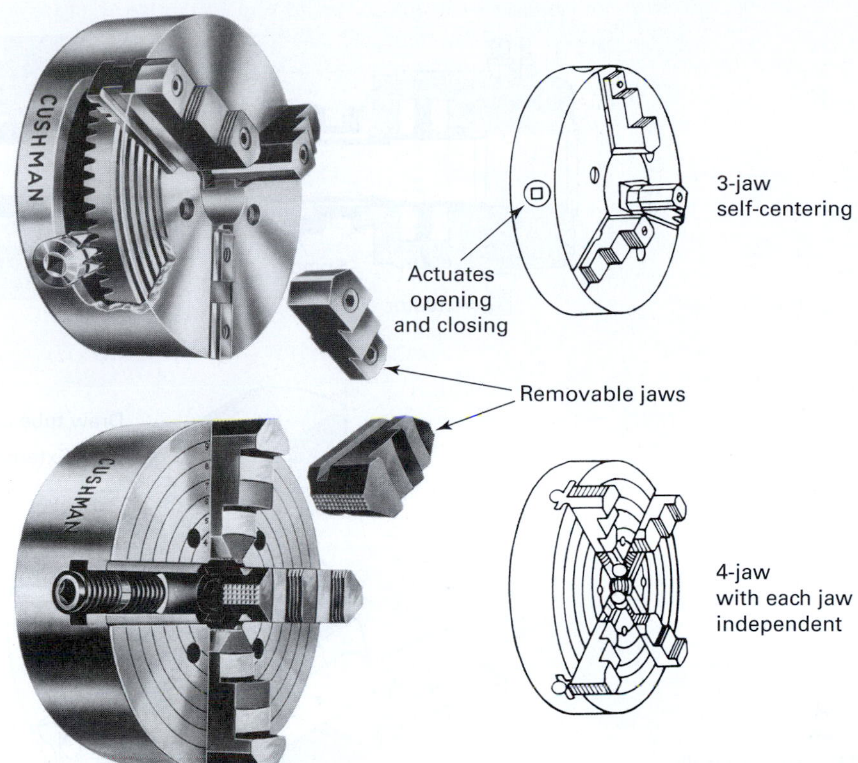

FIGURE 22-22 The jaws on chucks for lathes (four-jaw independent or three-jaw self-centering) can be removed and reversed.

Three-jaw self-centering chucks are used for work that has a round or hexagonal cross section. The three jaws are moved inward or outward simultaneously by the rotation of a spiral cam, which is operated by means of a special wrench through a bevel gear. If they are not abused, these chucks will provide automatic centering to within about 0.001 in. However, they can be damaged through use and will then be considerably less accurate.

Each jaw in a *four-jaw independent chuck* can be moved inward and outward independent of the others by means of a chuck wrench. Thus, they can be used to support a wide variety of work shapes. A series of concentric circles engraved on the chuck face aid in adjusting the jaws to fit a given workpiece. Four-jaw chucks are heavier and more rugged than the three-jaw type, and because undue pressure on one jaw does not destroy the accuracy of the chuck, they should be used for all heavy work. The jaws on both three- and four-jaw chucks can be reversed to facilitate gripping either the inside or the outside of workpieces.

Combination four-jaw chucks are available in which each jaw can be moved independently or can be moved simultaneously by means of a spiral cam. *Two-jaw chucks* are also available. For mass-production work, special chucks are often used in which the jaws are actuated by air or hydraulic pressure, permitting very rapid clamping of the work. See Figure 22-23 for a schematic. The rapid exchange of tooling is a key manufacturing strategy in manufacturing cells. Chuck jaw sets are dedicated and customized for specific parts. The first time a chuck jaw set is used, each jaw is marked with the number of the jaw slot where it was installed and an index mark that corresponds with the alignment of the jaw serrations and the first tooth on the chuck master jaw. The jaws can now be reinstalled on the chuck exactly where they were bored. The adjustability of the chuck body lets the operator dial in part concentrically without resetting the jaw.

COLLETS

Collets are used to hold smooth cold-rolled bar stock or machined workpieces more accurately than with regular chucks. As shown in Figure 22-24, collets are relatively thin tubular steel bushings that are split into three longitudinal segments over about two-thirds of their length. At the split end, the smooth internal surface is shaped to fit the

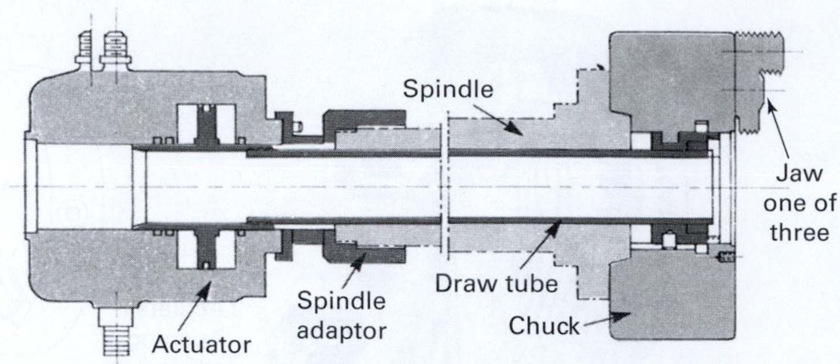

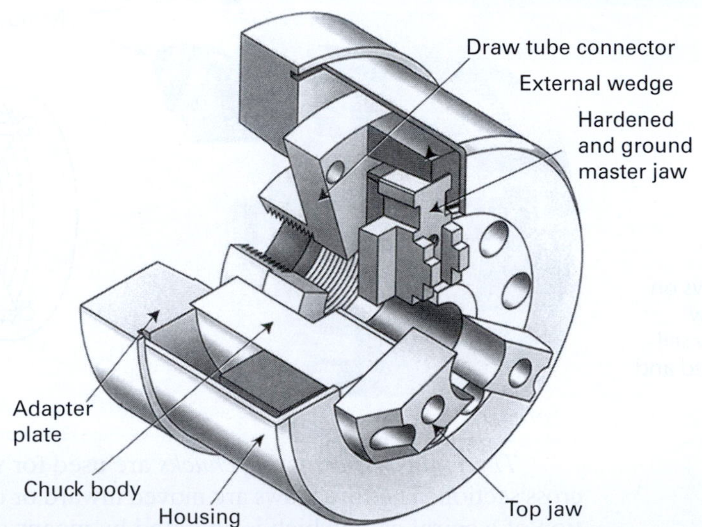

FIGURE 22-23 Hydraulically actuated through-hole three-jaw power chuck shown in section view to left and in the spindle of the lathe above connected to the actuator.

piece of stock that is to be held. The external surface of the collet is a taper that mates with an internal taper of a collet sleeve that fits into the lathe spindle. When the collet is pulled inward into the spindle (by means of the draw bar), the action of the two mating tapers squeezes the collet segments together, causing them to grip the workpiece.

Collets are made to fit a variety of symmetrical shapes. If the stock surface is smooth and accurate, good collets will provide very accurate centering, with runout less than 0.0005 in. However, the work should be no more than 0.002 in. larger or 0.005 in.

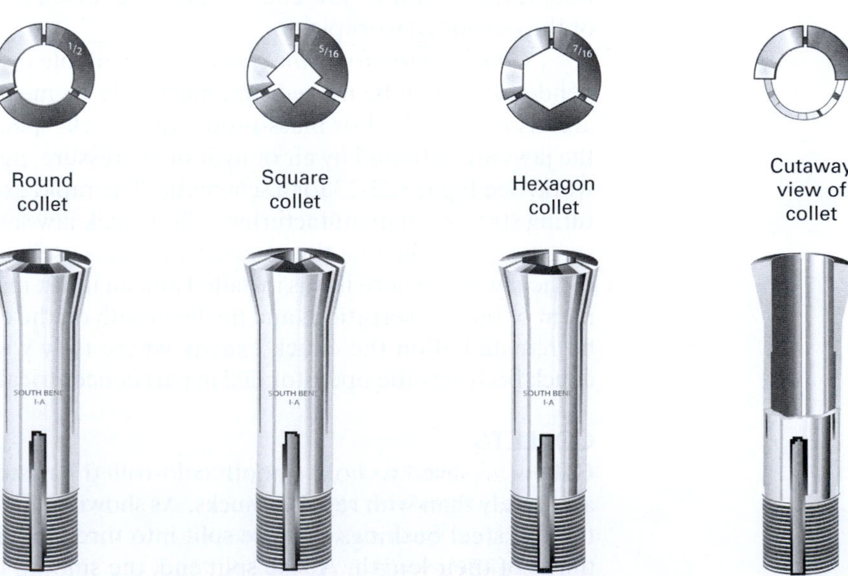

FIGURE 22-24 Several types of lathe collets. *(Courtesy of South Bend Lathe)*

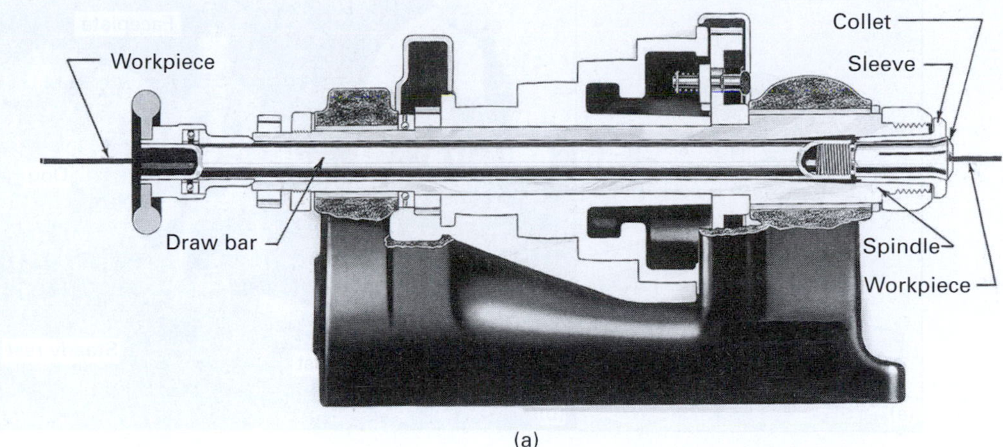

(a)

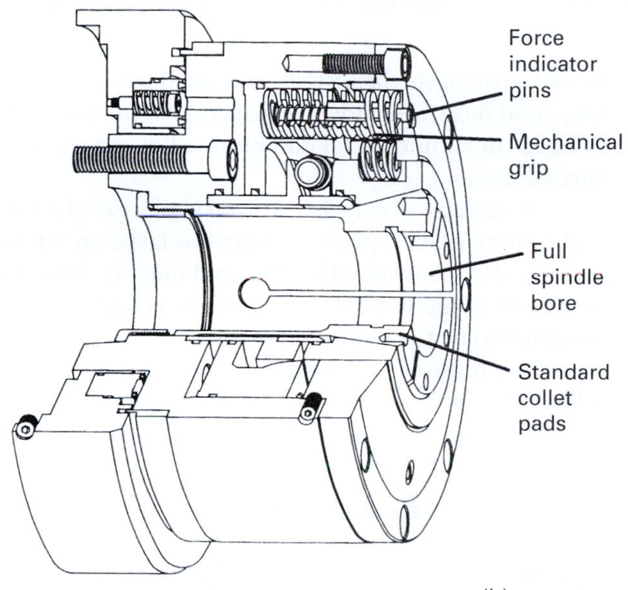

(b)

FIGURE 22-25 (a) Method of using a draw-in collet in lathe spindle. (b) Schematic of a collet chuck in which the clamping force can be adjusted. *[(a) Courtesy of South Bend Lathe]*

smaller than the nominal size of the collet. Consequently, collets are used only on drill-rod, cold-rolled, extruded, or previously machined stock.

Collets that can open automatically and feed bar stock forward to a stop mechanism are commonly used on automatic lathes and turret lathes. An example of a collet chuck is shown in Figure 22-25. Another type of collet similar to a Jacobs drill chuck has a greater size range than ordinary collets; therefore, fewer are required.

FACEPLATES

Faceplates are used to support irregularly shaped work that cannot be gripped easily in chucks or collets. The work can be bolted or clamped directly on the faceplate or can be supported on an auxiliary fixture that is attached to the faceplate. The latter procedure is time saving when identical pieces are to be machined.

MOUNTING WORK ON THE CARRIAGE

When no other means are available, boring is occasionally done on a lathe by mounting the work on the carriage, with the boring bar mounted between centers and driven by means of a dog.

STEADY AND FOLLOW RESTS

If one attempts to turn a long, slender piece between centers, the radial force exerted by the cutting tool or the weight of the workpiece itself may cause it to be deflected out of line. **Steady rests** and **follow rests** (Figure 22-26) provide a means for supporting such

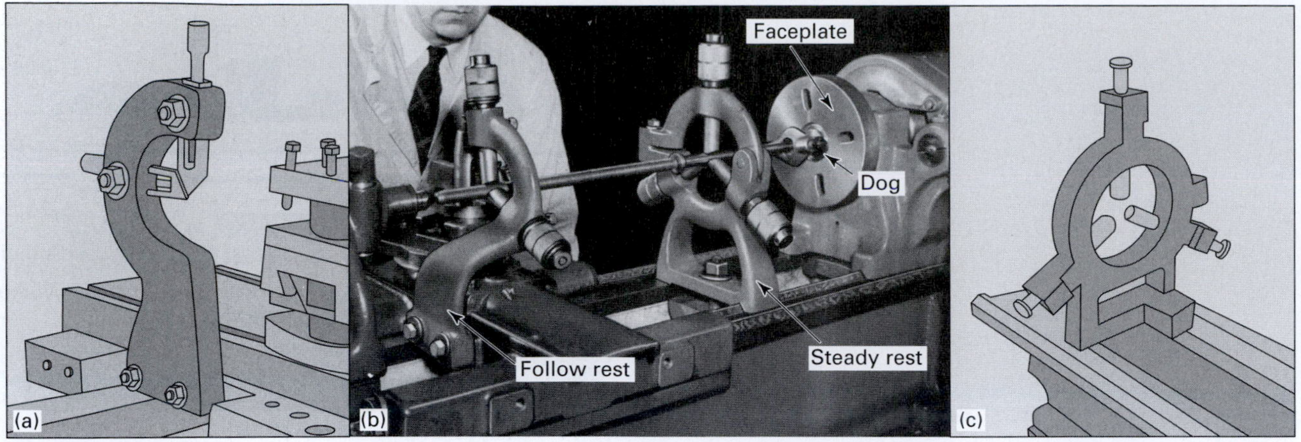

FIGURE 22-26 Cutting a thread on a long, slender workpiece, using a follow rest (left) and a steady rest (right) on an engine lathe. Note the use of a dog and faceplate to drive the workpiece. *((b) Courtesy of South Bend Lathe)*

work between the headstock and the tailstock. The steady rest is clamped to the lathe ways and has three movable fingers that are adjusted to contact the work and align it. A light cut should be taken before adjusting the fingers to provide a smooth contact-surface area.

A steady rest can also be used in place of the tailstock as a means of supporting the end of long pieces, pieces having too large an internal hole to permit using a regular dead center, or work where the end must be open for boring. In such cases, the headstock end of the work must be held in a chuck to prevent longitudinal movement. Tool feed should be toward the headstock.

The follow rest is bolted to the lathe carriage. It has two contact fingers that are adjusted to bear against the workpiece, opposite the cutting tool, in order to prevent the work from being deflected away from the cutting tool by the cutting forces.

Key Words

apron
bed
boring
carriage
carriage assembly
chucks
collets
compound rest
cross slide
cutoff
cutting tools
deflection
depth of cut (DOC)
dog plate
drilling
engine lathe
faceplates
facing
feed rod
follow rest
form turning
headstock
knurling
lathes
lathe centers
lathe dog
leadscrews
mandrels
metal removal rate (MRR)
milling
multiple-spindle screw machine
parting
pilot arm
quick-change gearbox
quill
reaming
single-spindle screw machine
spindle
split nut
steady rest
swing
tailstock
taper turning
threading
turning
turret lathe
ways
workholding

Review Questions

1. How is the tool–work relationship in turning different from that in facing?
2. What different kinds of surfaces can be produced by turning versus facing?
3. How does form turning differ from ordinary turning?
4. What is the basic difference between facing and a cutoff operation?
5. Which machining operations shown in Figure 22-3 do not form a chip?
6. Why is it difficult to make heavy cuts if a form turning tool is complex in shape?
7. Show how equation 22-6 is an approximate equation.
8. Why is the spindle of the lathe hollow?
9. What function does a lathe carriage have?
10. Why is feed specified for a boring operation typically less than that specified for turning if the MRR equations are the same?
11. What function is provided by the leadscrew on a lathe that is not provided by the feed rod?

12. How can work be held and supported in a lathe?
13. How is a workpiece that is mounted between centers on a lathe driven (rotated)?
14. What will happen to the workpiece when turned, if held between centers, and the centers are not exactly in line?
15. Why is it not advisable to hold hot-rolled steel stock in a collet?
16. How does a steady rest differ from a follow rest?
17. What are the advantages and disadvantages of a four-jaw independent chuck versus a three-jaw chuck?
18. Why should the distance the cutting tool overhands from the tool holder be minimized?
19. What is the difference between a ram- and a saddle-turret lathe?
20. How can a tapered part be turned on a lathe?
21. Why might it be desirable to use a heavy depth of cut and a light feed at a given speed in turning rather than the opposite?
22. If the rpm for a facing cut (assuming given work and tool materials) is being held constant, what is happening during the cut to the speed? To the feed?
23. Why is it usually necessary to take relatively light feeds and depths of cut when boring on a lathe?
24. How does the corner radius of the tool influence the surface roughness?
25. What effect does a BUE have on the diameter of the workpiece in turning?
26. How does the multiple-spindle screw machine differ from the single-spindle machine?
27. Why does boring ensure concentricity between the hole axis and the axis of rotation of the workpiece (for boring tool), whereas drilling does not?
28. Why are vertical spindle machines better suited for machining large workpieces than horizontal lathes?
29. In which figures in this chapter is a workpiece being held in a three-jaw chuck?
30. How is the workpiece in Figure 22-13 being held?
31. In which figures in this chapter is a dead center shown?
32. In which figures in this chapter is a live center shown?
33. In which figures in this chapter showing setups do you find the following being used as a workholding device?
    a. Three-jaw chuck
    b. Collet
    c. Faceplate
    d. Four-jaw chuck
34. How many form tools are being utilized in the process shown in Figure 22-14 to machine the part?
35. From the information given in Figure 22-14, start with a piece of round bar stock and show how it progresses, operation by operation, into a finished part—a threaded shaft.

PROBLEMS

1. Select the speed, feed, and depth of cut for turning wrought, low-carbon steel (hardness of 200 BHN) on a lathe with AISI tool material of HSS M2 or M3. (*Hint:* Refer back to Chapter 20 for recommended parameters.)
2. Calculate the rpm (N_s) to run the spindle on a lathe to generate a cutting speed of 80 sfpm if the outside diameter of the workpiece is 8 in.
3. The lathe has rpm settings of 20, 30, and 35 that are near the calculated value. What rpm would you pick, and what will actual cutting speed be?
4. Calculate the cutting time if the length of cut is 24 in., the feed rate is 0.030 ipr, and the cutting speed is 80 fpm. The allowance is 0.5 in and the diameter is 8 in.
5. Calculate the metal removal rate for machining at a speed of 80 fpm, feed of 0.030 ipr, at a depth of 0.625 in. Use any data from Problem 4 that you need.
6. If the cutting force is 11,135 lb, calculate the horsepower that a process operating at a speed of 80 fpm is going to use.
7. Explain how you would estimate the cutting force for a turning operation if you do not have a dynamometer.
8. Determine the speed, feed, and depth of cut when boring wrought, low-carbon steel (hardness of 120 BHN) on a lathe with AISI tool material of HSS M2 or M3.
9. At a speed of 90 fpm, feed of 0.030 ipr, and depth at 0.625, calculate the input parameters for a lathe if the diameter to be machined (bored) is 6 in. Find the rpm (N_s) to run the spindle to generate the cutting speed.
10. Calculate the cutting time for a 4-in. length of cut, given that the feed rate is 0.030 ipr at a speed of 90 fpm.
11. For a boring operation at $V = 90$ sfpm, $fr = 0.030$ ipr, and $d = 0.625$ in., calculate the MMR.
12. Calculate the horsepower that a process is going to use if the cutting force is 563 lb and the speed is 90 fpm.
13. Explain how you would estimate the cutting force for this boring operation if you do not have a dynamometer.
14. A cutting speed of 100 sfpm has been selected for a turning cut. The workpiece is 8 in. (203.2 mm) long and a feed of 0.020 in. (0.51 mm) per revolution is used. Using cutting speeds and feeds, calculate the machining time to cut off the bar.
15. The following data apply for machining a part on a turret lathe and on an engine lathe:

	Engine Lathe	Turret Lathe
Times, in minutes, to machine part	30 min	5 min
Cost of special tooling	0	$300
Time to set up the machine	30 min	3 hr
Labor rates	$8/hr	$8/hr
Machine rates (overhead)	$10/hr	$12/hr

One of the jobs for the engineer is to estimate the run time for a sequence of machining processes. For example:
    a. How many pieces would have to be made for the cost of the engine lathe to just equal the cost of the turret lathe? This is the BEQ.
    b. What is the cost per unit at the BEQ?
16. A finish cut for a length of 10 in. on a diameter of 3 in. is to be taken in 1020 steel with a speed of 100 fpm and a feed of 0.005 ipr. What is the machining time?
17. A workpiece 10 in. in diameter is to be faced down to a diameter of 2 in. on the right end. The CNC lathe (see Chapter 39)

controls the spindle speed and maintains the cutting speed at 100 fpm throughout the cut by changing the rpm. What should be the time for the cut? Now suppose the spindle rpm for the workpiece is set to give a speed of 100 fpm for the 10-in. diameter and is not changed during the cut. What is the machining time for the cut now? The feed rate is 0.005 ipr.

18. A hole 89 mm in diameter is to be drilled and bored through a piece of 1340 steel that is 200 mm long, using a horizontal boring, drilling, and milling machine. High-speed tools will be used. The sequence of operations will be center drilling; drilling with an 18-mm drill followed by a 76-mm drill; then boring to size in one cut, using a feed of 0.50 mm/rev. Drilling feeds will be 0.25 mm/rev for the smaller drill and 0.64 mm/rev for the larger drill. The center drilling operation requires 0.5 min. To set or change any given tool and set the proper machine speed and feed requires 1 min. Select the initial cutting speeds, and compute the total time required for doing the job. (Neglect setup time for the fixture.) This is often referred to as the run time or the cycle time. (*Hint:* Check Chapter 20 for recommended speeds for turning.)

Chapter 22 CASE STUDY

Estimating the Machining Time for Turning

As the plant manufacturing engineer at BRC Inc., Jay Strom has been called into the production department to provide an expert opinion on a machining problem. Unfortunately, the only tool or instrument available at the time is a 1-in. micrometer.

Katrin Zachary, the production manager would like to know the minimum time required to machine a large forging. The 8-ft-long forging is to be turned down from an original diameter of 10 in. to a final diameter of 6 in. The forging has a BHN of 300 to 400. The turning is to be performed on a heavy-duty lathe, which is equipped with a 50 hp motor and a continuously variable speed drive on the spindle. The work will be held between centers, and the overall efficiency of the lathe has been determined to be 75%. (See Chapter 21.)

The forging (or log) is made from medium-carbon, 4345 alloy steel. The steel manufacturer, some basic experimentation, and established knowledge of the product and its manufacture have provided the following information:

1. A tool-life equation developed for the most suitable type of tool material at a feed of 0.020 ipr and a rake angle of $\alpha = 10$ degrees. The equation $VT_n = C$ generally fits the data, with $V =$ cutting speed and $T =$ the time in minutes to tool failure. Two test cuts were run, one at $V = 60$ sfpm, where $T = 100$ min, and another at $V = 85$ sfpm, where $T = 10$ min.
2. According to the vendor, the dynamic shear strength of the material is on the order of 125,000 psi.
3. Jay decides to make two test cuts at the standard feed of 0.020 ipr. He assumes that the chip thickness ratio varies almost linearly between the speeds of 20 and 80 fpm, the values being 0.4 at the speed of 20 fpm and 0.6 at 80 fpm. The chip thickness values were determined by micrometer measurement in order to determine the value of r_c.
4. The machined forging (log) will be used as a roller in a newspaper press and must be precisely machined. If the log deflects during the cutting more than 0.005 in., the roll will end up barrel-shaped after final grinding and polishing.

How should Jay proceed to estimate the minimum time required to machine this forging, assuming that one finishing pass will be needed when the log has been reduced to 6 in. in diameter? The deflection due to cutting forces must be kept below 0.005 in. at the mid-log location.

You can assume that $F_c \times 0.5 = F_f$ and $F_f \times 0.5 = F_R$ and that F_R causes the deflection.

CHAPTER 23

DRILLING AND RELATED HOLE-MAKING PROCESSES

23.1 INTRODUCTION

In manufacturing it is probable that more holes are produced than any other shape, and a large proportion of these are made by **drilling**. Of all the machining processes performed, drilling makes up about 25%. Consequently, drilling is a very important process. Although drilling appears to be a relatively simple process, it is really a complex process. Most drilling is done with a cutting tool called a *twist drill* that has two cutting edges, or *lips,* as shown in Figure 23-1. The twist drill has the most the most common drill geometry. The cutting edges are at the end of a relatively flexible tool. Cutting action takes place inside the workpiece. The only exit for the chips is the hole that is mostly filled by the drill. Friction between the margin and the hole wall produces heat that is additional to that due to chip formation. The counterflow of the chips in the flutes

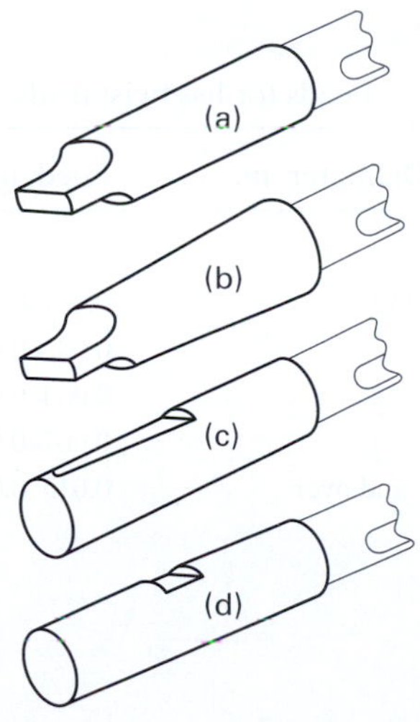

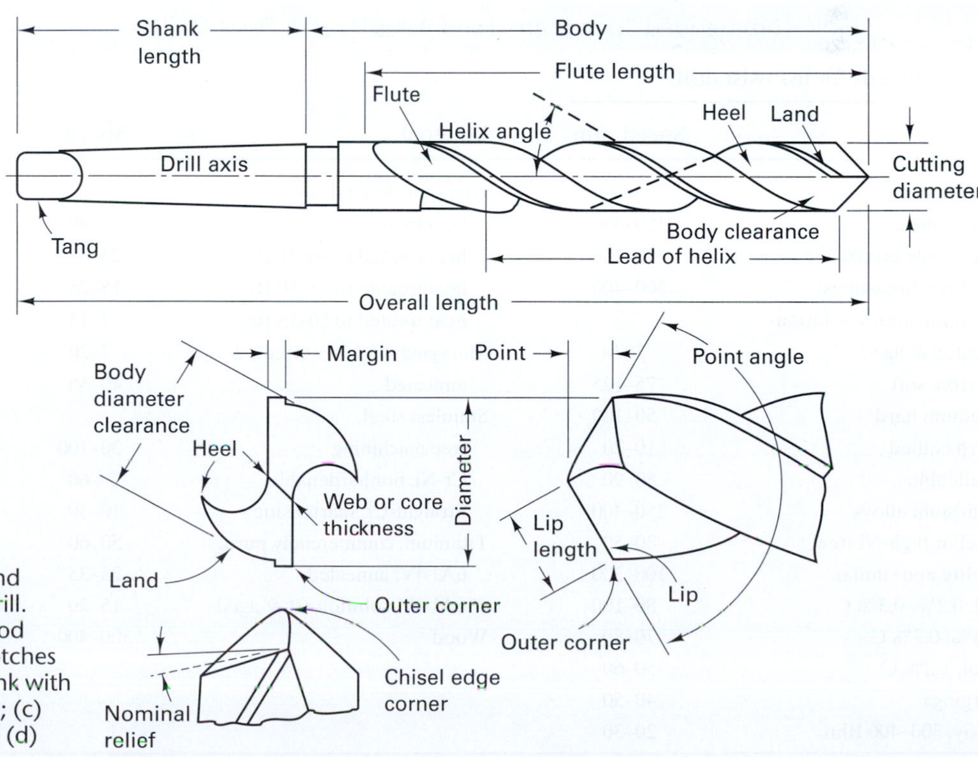

FIGURE 23-1 Nomenclature and geometry of conventional twist drill. Shank style depends on the method used to hold the drill. Tangs or notches prevent slippage. (a) Straight shank with tang; (b) tapered shank with tang; (c) straight shank with whistle notch; (d) straight shank with flat notch.

makes lubrication and cooling difficult. There are four major actions taking place at the point of a drill:

1. A small hole is formed by the web—chips are not cut here in the normal sense.
2. Chips are formed by the rotating lips.
3. Chips are removed from the hole by the screw action of the helical flutes.
4. The drill is guided by lands or margins that rub against the walls of the hole.

In recent years, new drill-point geometries and TiN coatings have resulted in improved hole accuracy, longer life, self-centering action, and increased feed-rate capabilities. However, the great majority of drills manufactured are twist drills. One estimate has U.S. manufacturing companies consuming 250 million twist drills per year.

When high-speed steel (HSS) drills wear out, the drill can be reground to restore its original geometry. However, if regrinding is not done properly, the original drill geometry may be lost, and so will drill accuracy and precision. Drill performance also depends on the machine tool used for drilling, the workholding device, the drill holder, and the surface of the workpiece. Poor surface conditions (sand pockets and/or chilled hard spots on castings, or hard oxide scale on hot-rolled metal) can accelerate early tool failure and degrade the hole-drilling process.

23.2 FUNDAMENTALS OF THE DRILLING PROCESS

The process of drilling creates two chips. A conventional two-flute drill, with drill of diameter D, has two principal cutting edges rotating at an rpm rate of N and feeding axially. Using a table like Table 23-1, the starting cutting speeds and feed rates are obtained, depending on the type of drill, the tool material and the material being machines. The rpm of the drill is established by the selected cutting speed

$$N_s = \frac{12V}{\pi D} \tag{23-1}$$

where V is in surface feet per minute and D is in inches (mm), using equation (23-1). This equation assumes that V is the cutting speed at the outer corner of the cutting lip

TABLE 23-1 Recommended Speeds and Feed Rates for HSS Twist Drills

Speeds for hss twist drills				Feeds for hss twist drills	
Material	Speed, sfm	Material	Speed	Diameter, in.	Feed, ipr
Aluminum alloys	200–300	High-tensile steel,			
Brass, bronze	150–300	heat treated to 35–40 Rc	30–40	Under $\frac{1}{8}$	0.001–0.002
High-tensile bronze	70–150	heat treated to 40–45 Rc	25–35	$\frac{1}{8}-\frac{1}{4}$	0.002–0.004
Zinc-base diecastings	300–400	heat treated to 45–50 Rc	15–25	$\frac{1}{4}-\frac{1}{2}$	0.004–0.007
High-temp alloys, solution-		heat treated to 50–55 Rc	7–15	$\frac{1}{2}-1$	0.007–0.015
treated & aged	7–20	Maraging steel, heat treated	7–20	1 and over	0.015–0.025
Cast iron, soft	75–125	annealed	40–55		
medium hard	50–100	Stainless steel,			
hard chilled	10–20	free machining	30–100		
malleable	80–90	Cr-Ni, nonhardenable	20–60		
Magnesium alloys	250–400	Straight-Cr, martensitic	10–30		
Monel or high-Ni steel	30–50	Titanium, commercially pure	50–60		
Bakelite and similar	100–300	6Al-4V, annealed	25–35		
Steel, 0.2%–0.3% C	80–100	6Al-4V, solution-trt & aged	15–20		
0.4%–0.5% C	70–80	Wood	300–400		
tool, 1.2% C	50–60				
forgings	40–50				
alloy, 300–400 Bhn	20–30				

Source: Cleveland Twist Drill

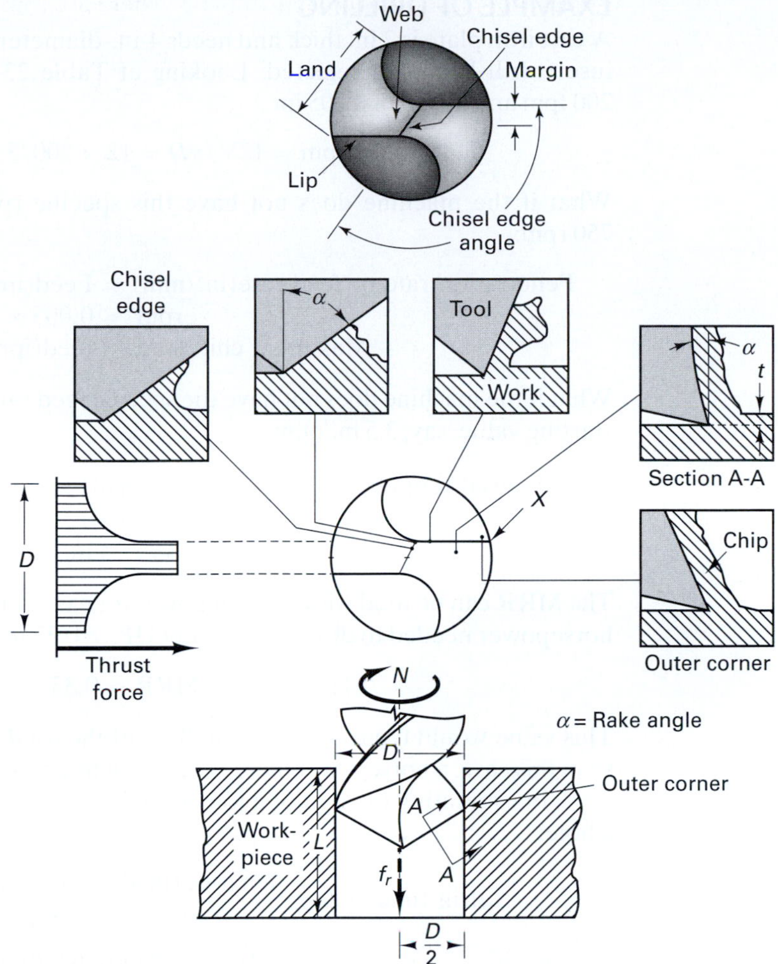

FIGURE 23-2 Conventional drill geometry viewed from the point showing how the rake angle varies from the chisel edge to the outer corner along the lip. The thrust force increases as the web is approached.

(point X in Figure 23-2). The velocity is very small near the center of the chisel end of the drill.

The feed, f_r, is given in inches per revolution (ipr). The depth of cut in drilling is equal to half the feed rate, or $t = f_r/2$ (see section A–A in Figure 23-2). The feed rate in inches per minute (ipm), f_m, is $f_r N_s$. The length of cut in drilling equals the depth of the hole, L, plus an allowance for approach and for the tip of drill, usually $A = D/2$. Table 23-1 gives some typical values for speed and feed rates and for carbide indexable insert drills, a type of drill shown later in the chapter.

For drilling, cutting time is given in equation 23-2:

$$T_m = \frac{(L + A)}{f_r N_s} = \frac{L + A}{f_m} \tag{23-2}$$

The metal removal rate is

$$\begin{aligned} \text{MRR} &= \frac{\text{Volume}}{T_m} \\ &= \frac{\pi D^2 L/4}{L/f_r N_s} \text{(omitting allowances)} \end{aligned} \tag{23-3}$$

which reduces to

$$\text{MRR} = (\pi D^2/4) f_r N_s \text{ in.}^3 \tag{23-4}$$

Substituting for N with equation 23-1, we obtain an approximate form

$$\text{MRR} \cong 3DVf_r \tag{23-5}$$

EXAMPLE OF DRILLING

A cast iron plate is 2 in. thick and needs 4-in.-diameter holes drilled in it. An indexable-insert drill has been selected. Looking at Table 23-1, we select a cutting speed of 200 fpm and a feed of 0.005 ipr.

$$\text{Spindle rpm} = 12V/\pi D = 12 \times 200/3.14 \times 1 = 764 \text{ rpm}$$

What if the machine does not have this specific rpm? Pick the closest value: say, 750 rpm:

$$\begin{aligned}\text{Penetration rate or feed rate(in./min)} &= \text{Feed(ipr)}\\ \text{rpm} &= 0.005 \times 750 = 3.75 \text{ in./min}\\ \text{Maximum chip load} &= \text{feed(ipr)}/2 = 0.005/2 = 0.0025 \text{ in./rev}\end{aligned}$$

What if the machine does not have the specific feed rate? Pick the next lowest value as a starting value, say, 3.5 in./min:

$$\begin{aligned}\text{Material removal rate (in.}^3\text{/min)} &= (\pi/4) \times (D)^2 \times \text{Feed(ipr)} \times \text{rpm}\\ &= (\pi/4)\, 1^2 \times \text{Feed rate} = 3.14/4 \times 1^2 \times 3.50\\ &= 2.75 \text{ in.}^3\text{/min}\end{aligned}$$

The MRR can be used with the unit power for cast iron (see Chapter 20) to estimate the horsepower needed to drill the hole. Let $HP_s = 0.33$ for this cast iron:

$$HP = HP_s \times MRR = 0.33 \times 2.75 = 0.90$$

This value would typically represent 80% of the total motor horsepower (HP) needed, so in this case, a horsepower motor greater than 1.5 or 2 would be sufficient.

In estimating the cost of a job, it is often necessary to determine the time to drill a hole:

$$\text{drill time/hole} = \frac{\text{length drilled} + \text{allowance}}{\text{feed rate(in./min)}} + \frac{\text{rapid traverse length of withdrawal}}{\text{rapid traverse rate}} + \text{prorated downtime to change drills per hole}$$

The last term prorated downtime is

$$\frac{\text{drill change downtime}}{\text{holes drilled per drill (tool life)}}$$

And the cost/hole is

$$\text{Drilling time/holes} \times (\text{Labor} + \text{Machine rate}) + \text{Prorated cost of drill/hole}$$

23.3 TYPES OF DRILLS

The most common types of drills are **twist drills.** These have three basic parts: the *body*, the *point*, and the *shank*, shown in Figures 23-1 and 23-2. The body contains two or more spiral or helical grooves, called **flutes,** separated by **lands.** To reduce the friction between the drill and the hole, each land is reduced in diameter except at the leading edge, leaving a narrow margin of full diameter to aid in supporting and guiding the drill and thus aiding in obtaining an accurate hole. The lands terminate in the point, with the leading edge of each land forming a cutting edge. The flutes serve as channels through which the chips are withdrawn from the hole and coolant gets to the cutting edges. Although most drills have two flutes, some, as shown in Figure 23-3, have three, and some have only one.

The principal rake angles behind the cutting edges are formed by the relation of the flute **helix angle** to the work. This means that the rake angle of a drill varies along

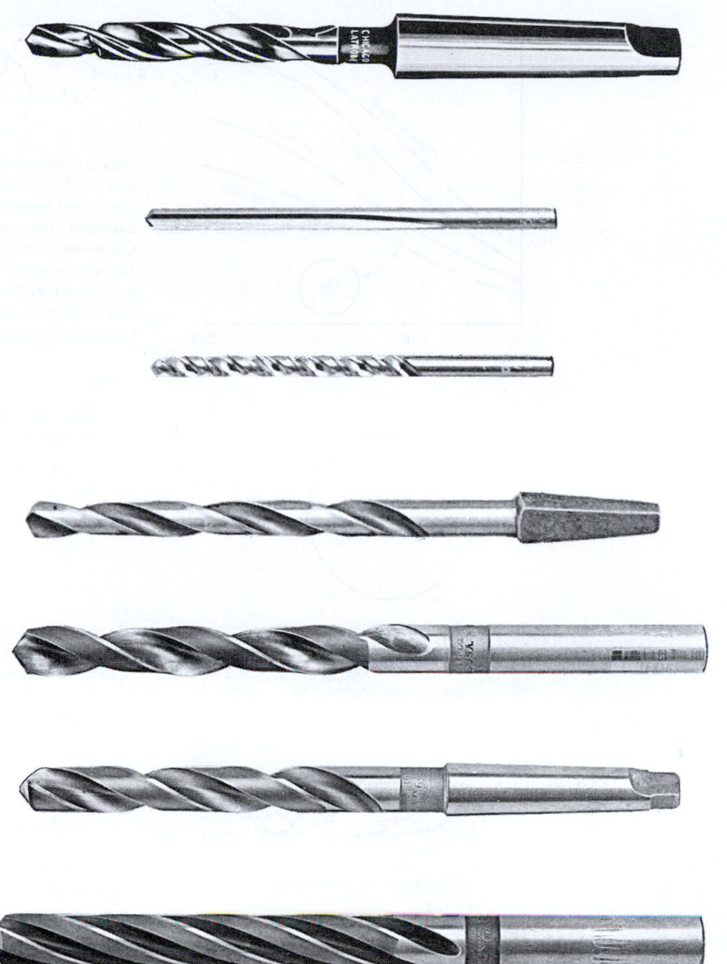

FIGURE 23-3 Types of twist drills and shanks. Bottom to top: straight-shank, three-flute core drill; straight-shank; taper-shank; bit-shank; straight-shank, high-helix angle; straight-shank, straight-flute; taper-shank, subland drill. *(Courtesy J T. Black)*

the cutting edges (or **lips**), being negative close to the point and equal to the helix angle out at the lip. Because the helix angle is built into the twist drill, the primary rake angle cannot be changed by normal grinding. The helix angle of most drills is 24 degrees, but drills with larger helix angles—often more than 30 degrees—are used for materials that can be drilled very rapidly, resulting in a large volume of chips. Helix angles ranging from 0 to 20 degrees are used for soft materials, such as plastics and copper. Straight-flute drills (zero helix and rake angles) are also used for drilling thin sheets of soft materials. It is possible to change the rake angle adjacent to the cutting edge by a special grinding procedure called **dubbing.**

The cone-shaped point on a drill contains the cutting edges and the various clearance angles. This cone angle affects the direction of flow of the chips across the tool face and into the flute. The 118-degree cone angle that is used most often has been found to provide good cutting conditions and reasonable tool life when drilling mild steel, thus making it suitable for much general-purpose drilling. Smaller cone angles—from 90 to 118 degrees—are sometimes used for drilling more brittle materials, such as gray cast iron and magnesium alloys. Cone angles from 118 to 135 degrees are often used for the more ductile materials, such as aluminum alloys. Cone angles less than 90 degrees are frequently used for drilling plastics. Many methods of grinding drills have been developed that produce point angles other than 118 degrees.

The drill produces a **thrust force,** *T*, and a **torque,** *M*. Drill torque increases with feed (in./rev) and drill diameter, while the thrust force is influenced greatly by the web or chisel end design, as shown in Figure 23-4.

The relatively thin **web** between the flutes forms a metal column or backbone. If a plain conical point is ground on the drill, the intersection of the web and the cone

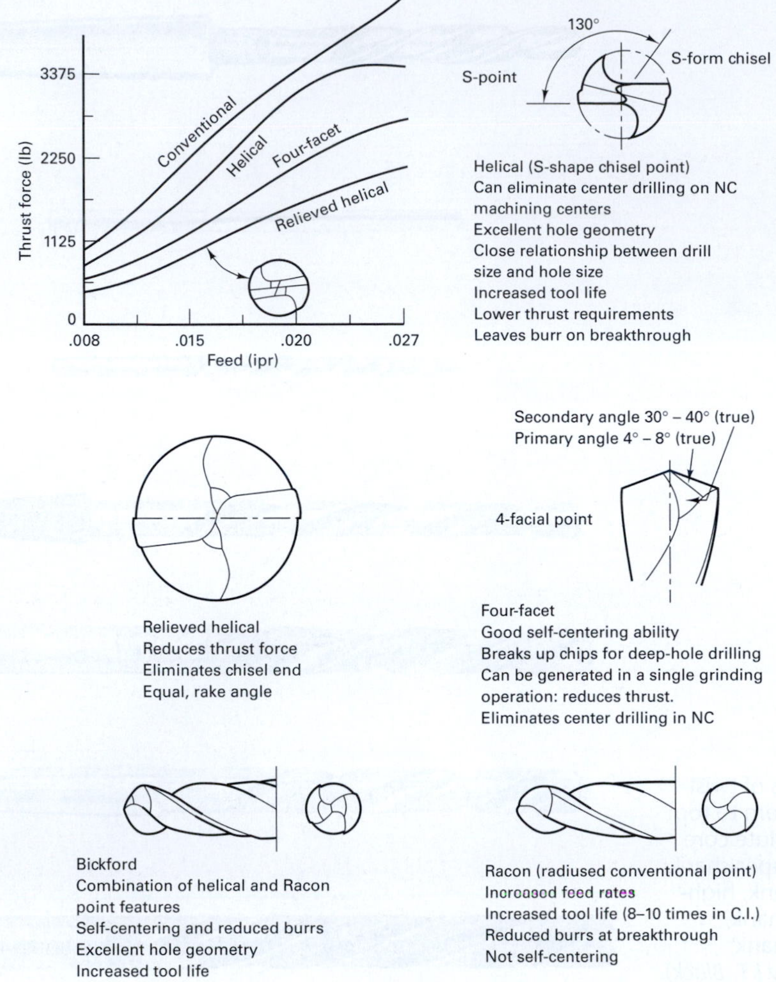

FIGURE 23-4 As the drill advances, it produces a thrust force. Variations in the drill-point geometry are aimed at reducing the thrust force.

produces a straight-line **chisel end,** which can be seen in the end view of Figure 23-2. The chisel point, which also must act as a cutting edge, forms a 56-degree negative rake angle with the conical surface. Such a large negative rake angle does not cut efficiently, causing excessive deformation of the metal. This results in high thrust forces and excessive heat being developed at the point. In addition, the cutting speed at the drill center is low, approaching zero. As a consequence, drill failure on a standard drill occurs both at the center, where the cutting speed is lowest, and at the outer tips of the cutting edges, where the speed is highest.

When the rotating, straight-line chisel point comes in contact with the workpiece, it has a tendency to slide or "walk" along the surface, thus moving the drill away from the desired location. The conventional point drill, when used on machining centers or high-speed automatics, will require additional supporting operations like center drilling, burr removal, and tool change—all of which increase total production time and reduce productivity.

Many special methods of grinding drill points have been developed to eliminate or minimize the difficulties caused by the chisel point and to obtain better cutting action and tool life (see Figure 23-4 for some examples).

The center core or slot-point drill shown in Figure 23-5 has twin carbide tips brazed on a steel shank and a hole (or slot) in the center. The work material in the slot is not machined but, rather, fractured away. The center core drill has a self-centering action and greatly relieves the thrust force produced by the chisel edge of conventional twist drills. This drill operates at about 30 to 50% less thrust than that of conventional drills. All rake angles of the cutting edge are positive, which further reduces the cutting force.

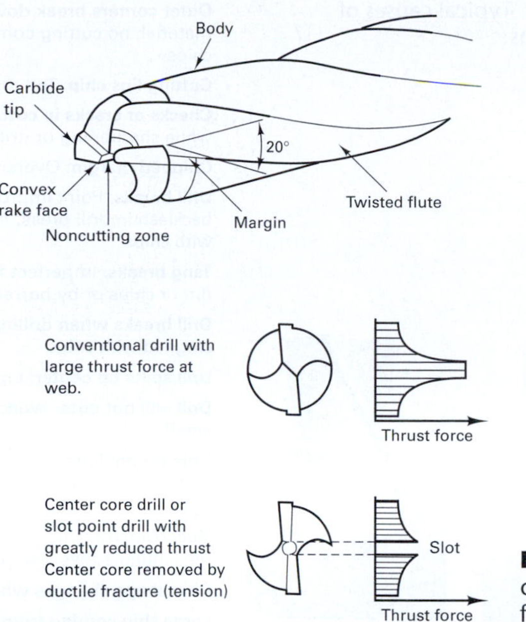

FIGURE 23-5 Center core drills can greatly reduce the thrust force.

The conventional point also has a tendency to produce a burr on the exit side of a hole. Some type of chip breaker is often incorporated into drills. One procedure is to grind a small groove in the rake face, parallel with and a short distance back from the cutting edge. Drills with a special chip-breaker rib as an integral part of the flute are available. The rib interrupts the flow of the chip, causing it to break into short lengths.

The split-point drill is a form of **web thinning** to shorten the chisel edge. This design reduces thrust and allows for higher feed rates. Web thinning uses a narrow grinding wheel to remove a portion of the web near the point of the drill. Such methods have had varying degrees of success, and they require special drill-grinding equipment.

Also shown in Figure 23-4 is a four-facet self-centering point that works well in tougher materials. The facets refer to the number of edges on the clearance surfaces exposed to the cutting action. The self-centering drill lasts longer and saves machining time on numerical control (NC) centers as they can eliminate the need for center drills. A common aspect in drill-point terminology is total indicator runout (TIR). This is a measure of the cutting lips' relative side-to-side accuracy. The original drill point produced by the manufacturer lasts only until the first regrind; thereafter, performance and life depend on the quality of regrind. Proper regrinding (reconditioning) of a drill is a complex and important operation. If satisfactory cutting and hole size are to be achieved, it is essential that the point angle, lip clearance, lip length, and web thinning be correct. As illustrated in Figure 23-6, incorrect sharpening often results in unbalanced cutting forces at the tip, causing misalignment and oversized holes. Drills, even small drills, should always be machine ground, never hand ground. Drill grinders, often computer controlled, should be used to ensure exact reproduction of the geometry established by the manufacturer of the drill. This is extremely important when drills are used on mass-production or numerically controlled machines. Companies invest huge sums in NC machining centers but overlook the value of a top-quality drill-grinding machine.

Drill shanks are made in several types. The two most common types are the straight and the taper. **Straight-shank** drills are usually used for sizes up to $\frac{3}{8}$-in. diameter and must be held in some type of drill chuck. **Taper shanks** are available on larger drills and are common on drills above 1 in. Morse tapers are used on taper-shank drills, ranging from a number 1 taper to a number 6.

Taper-shank drills are held in a female taper in the end of the machine tool spindle. If the taper on the drill is different from the spindle taper, adapter sleeves are available. The taper assures the drill's being accurately centered in the spindle. The **tang** at

FIGURE 23-6 Typical causes of drilling problems.

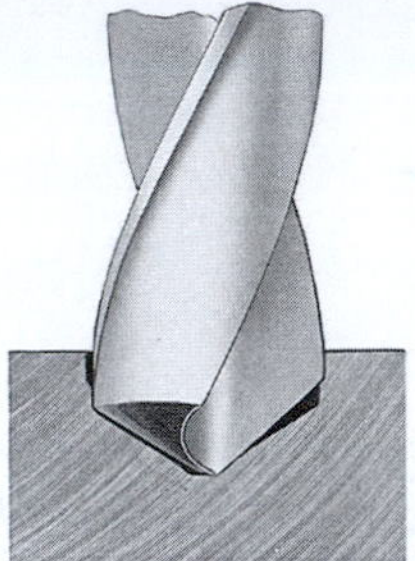

(a) Angle unequal

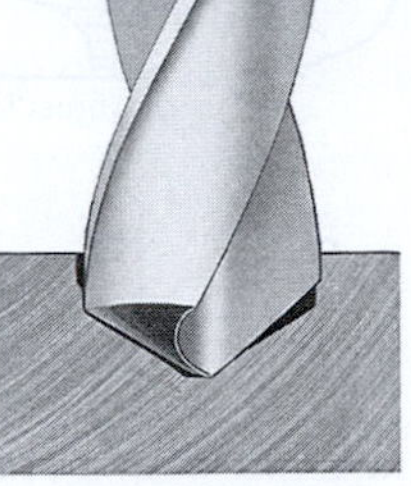

(b) Length unequal

Outer corners break down: Cutting speed too high; hard spots in material; no cutting compound at drill point; flutes clogged with chips

Cutting lips chip: Too much feed; lip relief too great

Checks or cracks in cutting lips: Overheated or too quickly cooled while sharpening or drilling

Chipped margin: Oversize jig bushing

Drill breaks: Point improperly ground; feed too heavy; spring or backlash in drill press, fixture, or work; drill is dull; flutes clogged with chips

Tang breaks: Imperfect fit between taper shank and socket caused by dirt or chips or by burred or badly worn sockets

Drill breaks when drilling brass or wood: Wrong type drill; flutes clogged with chips

Drill spilts up center: Lip relief too small; too much feed

Drill will not enter work: Drill is dull; web too heavy; lip relief too small

Hole rough: Point improperly ground or dull; no cutting compounds at drill point; improper cutting compound; feed too great; fixture not rigid

Hole oversize: Unequal angle of the cutting edges; unequal length of the cutting edges; see part (a)

Chip shape changes while drilling: Dull drill or cutting lips chipped

Large chip coming from one flute, small chip from the other: Point improperly ground, one lip doing all the cutting

the end of the taper shank fits loosely in a slot at the end of the tapered hole in the spindle. The drill may be loosened for removal by driving a metal wedge, called a **drift,** through a hole in the side of the spindle and against the end of the tang. It also acts as a safety device to prevent the drill from rotating in the spindle hole under heavy loads. However, if the tapers on the drill and in the spindle are proper, no slipping should occur. The driving force to the drill is carried by the friction between the two tapered members. Standard drills are available in four size series, the size indicating the diameter of the drill body:

- *Millimeter series:* 0.01- to 0.50-mm increments, according to size, in diameters from 0.015 mm.
- *Numerical series:* no. 80 to no. 1 (0.0135 to 0.228 in.).
- *Lettered series: A* to *Z* (0.234 to 0.413 in.).
- *Factional series:* to 4 in. (and over) by 64ths.

TiN coating of conventional drills greatly improves drilling performance. The increase in tool life of TiN-coated drills over uncoated drills in machining steel is more than 200 to 1000%.

DEPTH-TO-DIAMETER RATIO

The depth of the hole to be drilled divided by the diameter of the drill is the depth-to-diameter ratio. Most machinists consider a ratio of 3:1 to be deep-hole drilling, after which hole accuracy (location) drilling speed and tool life will be reduced. The bores of rifle barrels were once drilled using conventional drills. Today, **deep-hole drills,** or **gundrills,** are used when deep holes are to be drilled.

The oldest of these deep-hole techniques is gundrilling. The original gundrills were very likely half-round drills, drilled axially with a coolant hole to deliver cutting fluids to the cutting edge (see Figure 23-7). Modern gundrills typically consist of an alloy-steel-tubing shank with a solid carbide or carbide-edged tip brazed or mechanically fixed to it. Guide pads following the cutting edge by about 90 to 180 degrees are also standard.

The gundrill is a single-lipped tool, and its major feature is the delivery of coolant through the tool at extremely high pressures—typically from 300 to 1800 psi, depending on diameter—to force chips back down the flute. Successful application of a gundrill

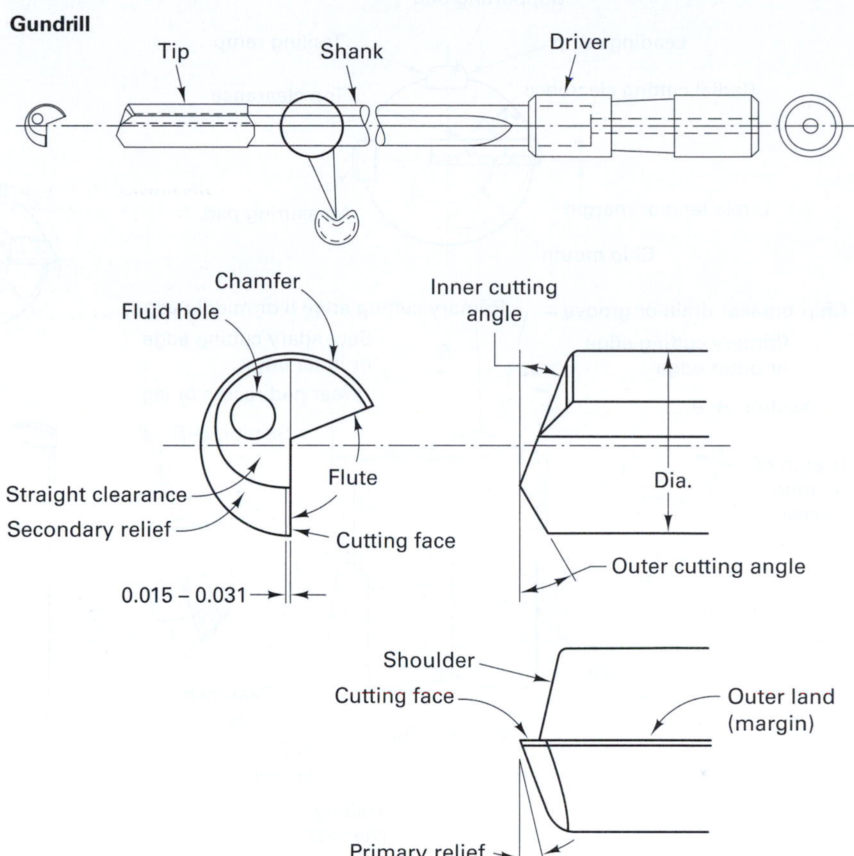

FIGURE 23-7 The gundrill geometry is very different from that of conventional drills.

depends almost entirely on the formation of small chips that can be effectively evacuated by the flow of cutting fluids.

Standard gundrills are made in diameters from 0.0078 in. (2 mm) to 2 in. or more. Depth-to-diameter ratios of 100:1 or more are possible.

In gundrilling tolerances for diameters of drilled holes under 1 in. can be held to 0.0005-in. total tolerance, and, should not exceed 0.001 in. over all. According to one source, "roundness accuracies of 0.00008 inch can be attained." Because of the burnishing effect of the guide pads, excellent surface finishes can be produced.

Hole straightness is affected by a number of variables, such as diameter, depth, uniformity of workpiece material, condition of the machine, sharpness of the gundrill, feeds and speeds used, and the specific technique used (rotation of the tool, of the work, or both), but deviation should not exceed about 0.002 in. TIR in a 4-in. depth at any diameter, and it can be held to 0.002 in./ft.

Basic setup for a gundrilling operation, which is generally horizontal, requires a drill bushing very close to the work entry surface and may involve rotating the work or the tool, or both. Best concentricity and straightness are achieved by the work and the tool rotating in opposite directions.

Other deep-hole drills are called **BTA** (Boring Trepanning Association) **drills** and **ejector drills.** A deep hole is one in which the length (or depth) of the hole is three or more times the diameter. Coolants can be fed internally through these drills to the cutting edges. See Figure 23-8 for schematic of an ejector drill and the machine tool used for gundrilling. The coolants flush the chips out the flutes. The special design of these drills reduces the tendency of the drill to drift, thus producing a more accurately aligned hole. The typical BTA deep-hole drilling tools are designed for single-lip end cutting of a hole in a single pass. Solid deep-hole drills have alloy-steel shanks with a carbide-edged tip that is fixed to it mechanically. The cutting edge cuts through the center on one side of the hole, leaving no area of material to be extruded. The cutting is done by

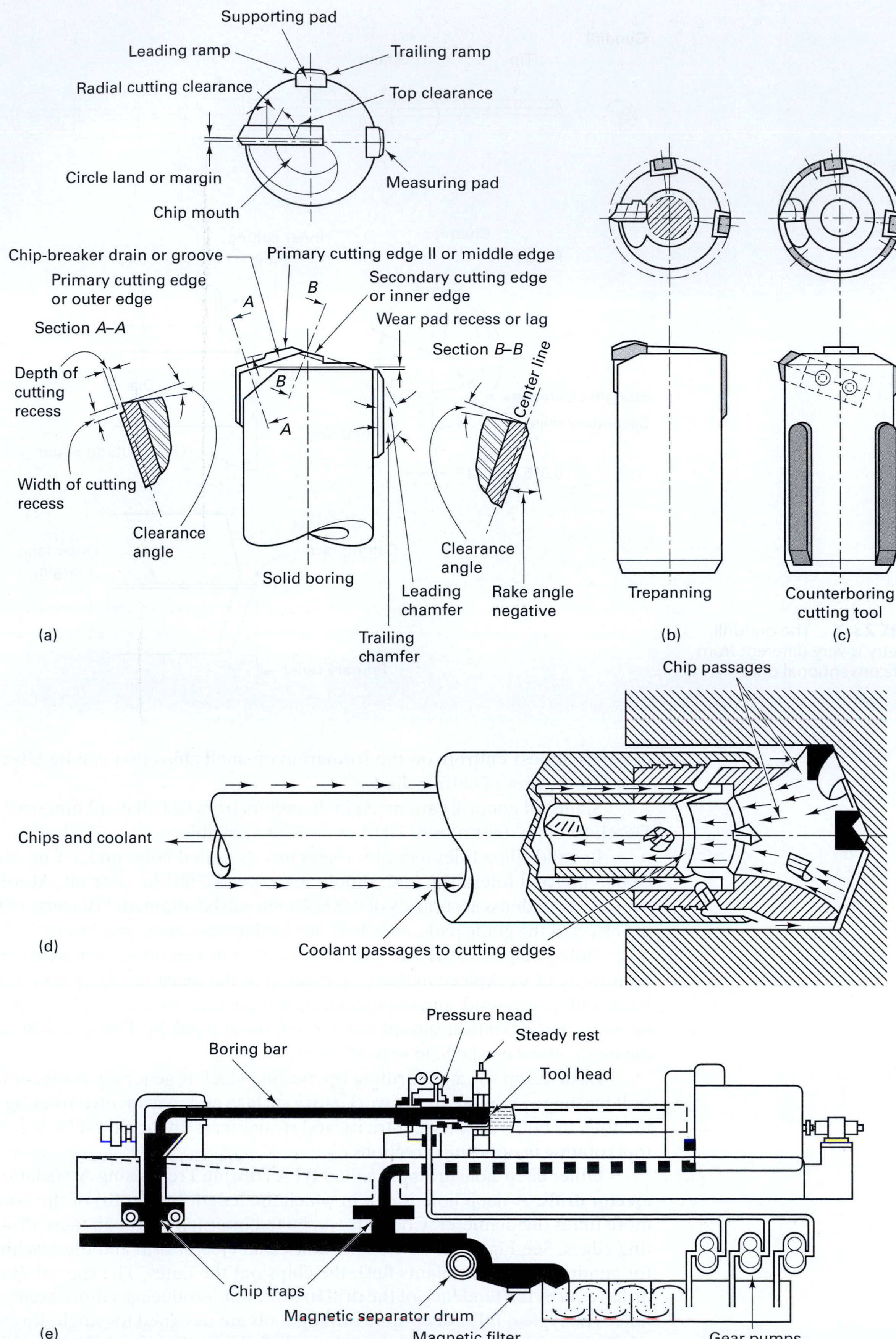

FIGURE 23-8 BTA drills for (a) boring, (b) trepanning, (c) counterboring, (d) deep-hole drilling with ejector drill, (e) horizontal deep-hole-drilling machine. *(S. Azad and S. Chandeashekar,* Mechanical Engineering, *September 1985, pp. 62, 63)*

TABLE 23-2 Drilling Processes Compared

	Twist Drill	Pivot (micro)	Spade (inserted blade)	Indexable-Insert Drill	Gundrill	BTA System	Ejector Drill	Trepanning
Diameter, in.	0.020–2	0.001–0.020	1–6	$\frac{5}{8}$–3	0.078–1	$\frac{7}{16}$–8	$\frac{3}{4}$–$2\frac{1}{2}$	$1\frac{3}{4}$–10
Typical range	0.0059	<0.0001	$\frac{5}{8}$ spec, 1	$\frac{5}{8}$	0.039	$\frac{3}{4}$	$\frac{3}{4}$	$1\frac{3}{4}$
Min	$3\frac{1}{2}$ std	$\frac{1}{8}$	std	3	$2\frac{1}{2}$	12	7	>24
Max	6 spec		18					
Depth/Diameter Ratio								
Min. practical	No min	No min	No min	<1	1	1	1	10
Common max[a]	5–10	3–10	>40 (horiz)	2–3	100	100	50	100
Ultimate	>50	20	10 (vert) >100 (horz)	—	200	>100	>50	>100

[a] Maximum depth/diameter ratios in this table are estimates of what can be achieved with special attention and under ideal conditions Equality of tolerances should not be assumed for the different process.

the outer and inner cutting angles, which meet at a point. Theoretically, the depth of the hole has no limit, but practically, it is restricted by the torsional rigidity of the shank.

Gundrills have a single-lip cutting action. Bearing areas and lifting forces generated by the coolant pressure counteract the radial and tangential loads. The single-lip construction forces the edge to cut in a true circular pattern. The tip thus follows the direction of its own axis. The **trepanning gundrill** leaves a solid core.

Hole straightness is affected by variables such as diameter, depth, uniformity of the workpiece material, condition of the machine, sharpness of cutting edges, feeds and speeds used, and whether the tool or the workpiece is rotated or counter-rotated.

Two-flute drills are available that have holes extending throughout the length of each land to permit coolant to be supplied, under pressure, to the point adjacent to each cutting edge. These are helpful in providing cooling and also in promoting chip removal from the hole in drilling to moderate depths. They require special fittings through which the coolant can be supplied to the rotating drill, and they are used primarily on automatic and semiautomatic machines. See Table 23-2 for comparison of drilling processes.

Larger holes in thin material may be made with a **hole cutter** (Figure 23-9), where the large hole is produced by the thin-walled, multiple-tooth cutter with saw teeth and the metal hole with a twist drill. Hole cutters are often called **hole saws**.

When starting to drill a hole, a drill can deflect rather easily because of the "walking" action of the chisel point. Hole location accuracy is lost. Consequently, to ensure that a hole is started accurately, a **center drill** (Figure 23-10) is used prior to a regular

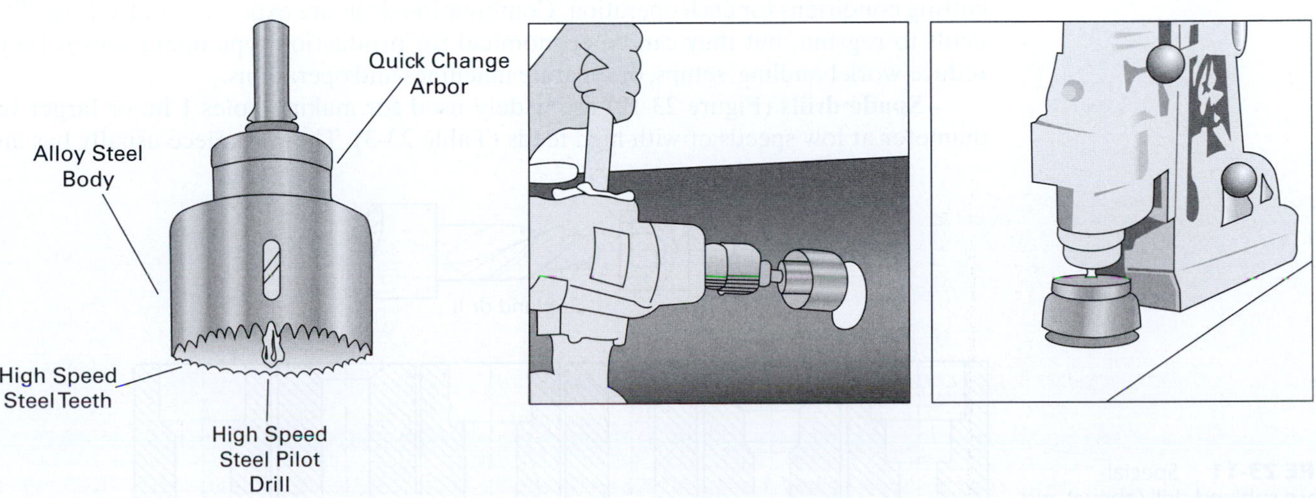

FIGURE 23-9 High Speed Edge Hole Saw can cut holes $\frac{9}{16}$ in. to 6 in. in diameter in any machinable material up to $1\frac{1}{8}$ in. thick. They can be used in portable electric or drill presses.

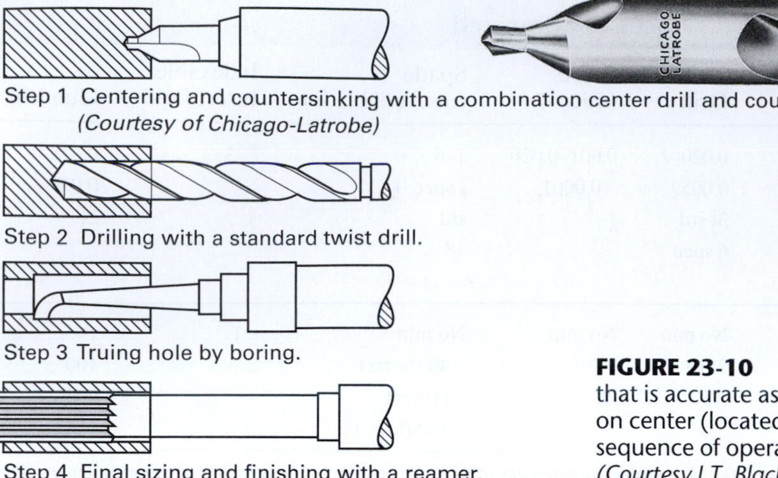

CHICAGO LATROBE
Step 1 Centering and countersinking with a combination center drill and countersink.
(Courtesy of Chicago-Latrobe)
Step 2 Drilling with a standard twist drill.
Step 3 Truing hole by boring.
Step 4 Final sizing and finishing with a reamer.

FIGURE 23-10 To obtain a hole that is accurate as to size and aligned on center (located), this four-step sequence of operations is usual. *(Courtesy J T. Black)*

chisel-point twist drill. The center drill and countersink tool have a short, straight drill section extending beyond a 60-degree taper portion. The heavy, short body provides rigidity so that a hole can be started with little possibility of tool deflection. The hole should be drilled only partway up on the tapered section of countersink. The conical portion of the hole serves to guide the drill being used to make the main hole. Combination center drills are made in four sizes to provide a starting hole of proper size for any drill. If the drill is sufficiently large in diameter, or if it is sufficiently short, satisfactory accuracy may often be obtained without center drilling. Special drill holders are available that permit drills to be held with only a very short length protruding.

Because of its flexibility and endpoint geometry, a drill may start or drift off centerline during drilling. The use of a center (start) drill will help to ensure that a drill will start drilling at the desired location. Nonhomogeneities in the workpiece and imperfect drill geometries may also cause the hole to be oversize or off-line. For accuracy, it is necessary to follow center drilling and drilling by boring and reaming. Boring corrects the hole alignment, and reaming brings the hole to accurate size and improves the surface finish.

Special **combination drills** can drill two or more diameters, or drill and countersink and/or counterbore, in a single operation (Figure 23-11). Countersinking and counterboring usually follow drilling. These operations are described in more detail later in this chapter. A **step drill** has a single set of flutes and is ground to two or more diameters. **Subland drills** have a separate set of flutes on a single body for each diameter or operation; they provide better chip flow, and the cutting edges can be ground to give proper cutting conditions for each operation. Combination drills are expensive and may be difficult to regrind, but they can be economical for production-type operations if they reduce work handling, setups, or separate machines and operations.

Spade drills (Figure 23-12) are widely used for making holes 1 in. or larger in diameter at low speeds or with high feeds (Table 23-3). The workpiece usually has an

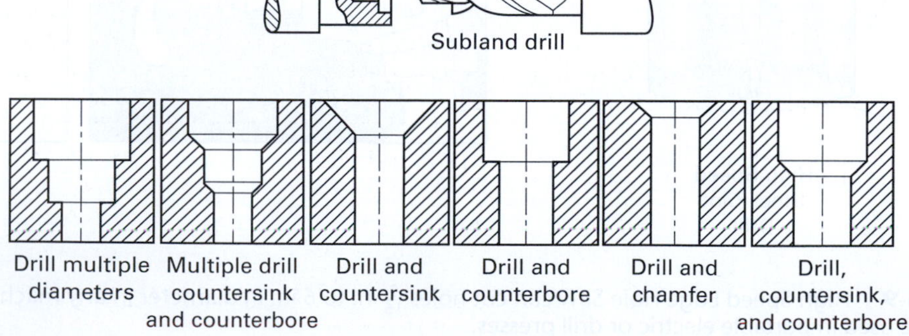

FIGURE 23-11 Special-purpose subland drill (above) and some of the operations possible with other combination drills (below).

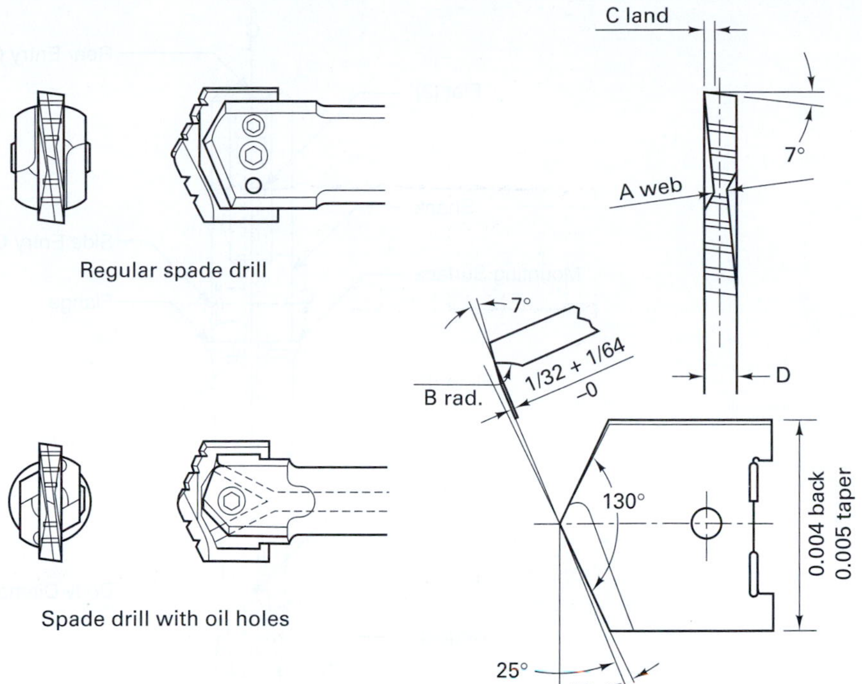

FIGURE 23-12 (Top) Regular spade drill; (middle) spade drill with oil holes; (bottom) spade drill geometry, nomenclature.

existing hole, but a spade drill can drill deep holes in solids or stacked materials. Spade drills are less expensive because the long supporting bar can be made of ordinary steel. The drill point can be ground with a minimum chisel point. The main body can be made more rigid because no flutes are required, and it can have a central hole through which a fluid can be circulated to aid in cooling and in chip removal. The cutting blade is easier to sharpen; only the blades need to be TiN-coated.

TABLE 23-3 Recommended Surface Speeds and Feeds for High-Speed Steel Spade Drills for Various Materials

Material	Surface Speed (ft per min)	Material	Speed
Mild machinery steel 0.2 and 0.3 carbon	65–100	Cast iron, medium hard	55–100
Steel, annealed 0.4 to 0.5 carbon	55–80	Cast iron, hard, chilled	25–40
Tool steel, 1.2 carbon	45–60	Malleable iron	79–90
Steel forging	35–50	Brass and bronze, ordinary	200–300
Alloy steel	45–70	Bronze, high tensile	70–150
Stainless steel, free machining	50–70	Monel metal	35–50
Stainless steel, hard	25–40	Aluminum and its alloys	200–300
Cast iron, soft	80–150	Magnesium and its alloys	250–400

Feed Rates for Spade Drilling (inches per revolution)

Drill Size (inches)	Cast Iron Malleable Iron Brass Bronze	Medium Steel Stainless Steel Monel Metal Drop-Forged Alloys Tool Steel (annealed)	Tough Steel Drop Forging Aluminum
1 to $1\frac{1}{4}$	0.010–0.020	0.008–0.014	0.006–0.012
$1\frac{1}{4}$ to $\frac{3}{4}$	0.010–0.024	0.008–0.018	0.008–0.017
$1\frac{3}{4}$ to $2\frac{1}{2}$	0.010–0.030	0.010–0.024	0.010–0.017
$2\frac{1}{2}$ to 4	0.012–0.032	0.012–0.030	0.010–0.017
4 to 6	0.012–0.032	0.010–0.024	0.008–0.017

Source: Waukesha Cutting Tools, Inc.

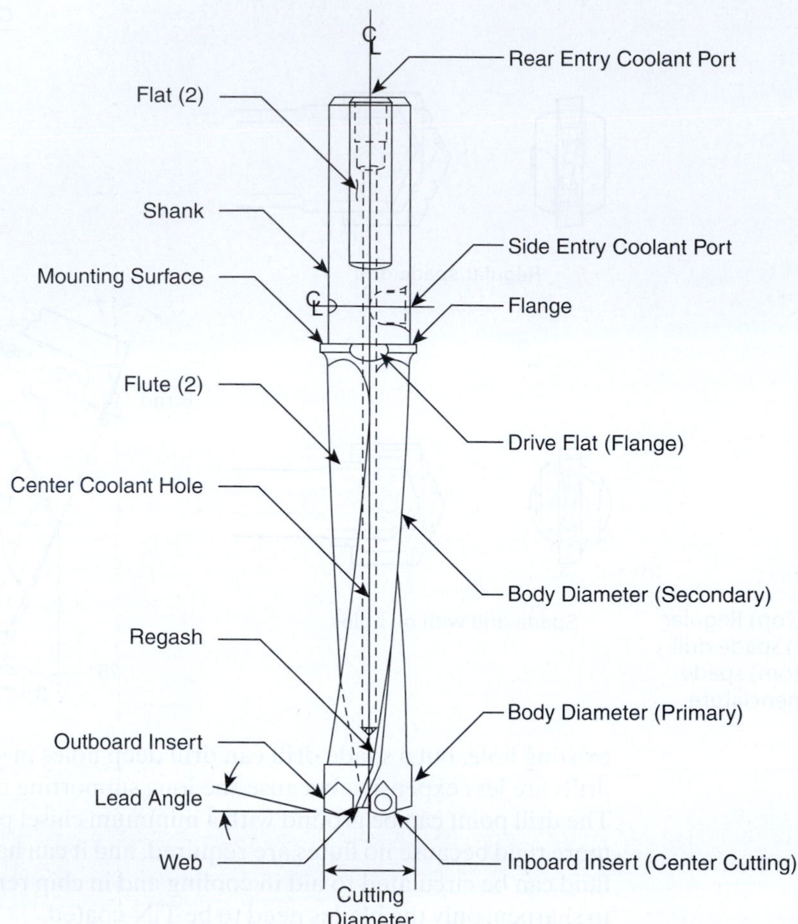

FIGURE 23-13 Design of an indexable insert drill with 2 inserts, the most common style. *(Indexable Drill Applications, T. Benjamin,* Machining Technology, *Vol. 8, No. 2, SME, 1997)*

Spade drills are often used to machine a shallow locating cone for a subsequent smaller drill and at the same time to provide a small bevel around the hole to facilitate later tapping or assembly operations. Such a bevel also frequently eliminates the need for deburring. This practice is particularly useful on mass-production and numerically controlled machines.

Carbide-tipped drills and drills with indexable inserts are also available (see Figure 23-13) with one- and two-piece inserts for drilling shallow holes in solid workpieces. **Indexable insert drills** can produce a hole four times faster than a spade drill because they run at high speeds/low feeds and are really more of a boring operation than a drilling process.

However, to use indexable drills, you must have an extremely rigid machine tool and setup, adequate horsepower, and lots of cutting fluid. Indexable drills are roughing tools generating hole tolerances of and surface finish of 250 rpm or greater. The tool is designed for the inboard insert to cut past the centerline of the tool so the inboard tool is positioned radially below the center. See Table 23-4 for an indexable drilling troubleshooting guide.

A high-pressure, pulsating coolant system can generate pressures up to 300 psi and works well with indexable drilling. It can have disadvantages, however. High pressure with pulsating action can decrease chip control and cause drill deflection. A high-pressure coolant stream can flatten chips at the point of forming and forces them into the cut, causing recutting, insert chipping, and poor hole finishes. The pressure can force chips between the drill body and hole diameter, wrapping them around the drill. Friction then will weld the chips to the tool body or hole.

The diameter of the hole and the length/diameter ratio usually determine what kind of drill to use. Figure 23-14 explores how drill selection depends on the depth of the hole and the diameter of the drill: Section A shows the drilling areas of relatively shallow holes and small diameters. About half of all the drilling process falls within the

TABLE 23-4 Indexable Drilling Troubleshooting Guide[a]

Problem	Source	Solution
Insert chipping or breakage[b]	Off-center drill, caused by misalignment	Maintain proper alignment.
		Concentricity not to exceed ± 0.005 TIR.
	Improper seating of tool in tool holder, spindle, or turret	Check tool shank and socket for nicks and dirt. Check parting line between tool shank and socket with feeler gage. Check to see if tool is locked tightly.
	Deflection because of too much overhand and Jack of rigidity	With indicator, check if tool can be moved by hand. Check if tool can be held shorter.
	Improper seating of inserts in pockets	Clean pockets whenever indexing or changing inserts.
		Check pockets for nicks and burrs.
		Check if inserts rest completely on pocket bottoms.
	Damaged insert screws	Check head and thread for nicks and burns. Do not overtorque screws.
	Improper speeds and feeds	Check recommended guidelines for given materials.
	Insufficient coolant supply	Check coolant flow.
	Improper carbide grade in inboard station	Recommend straight grade for multiple-insert drills.
Grooving on back stroke; drill body rubbing hole wall; over- or undersize holes	Off-center drill	Maintain proper alignment and concentricity.
		Check bottom of hole or disk for center stub.
	Deflection	Check setup rigidity. Check speed and feed guidelines.
Poor hole surface finish	Vibrations	Check setup and part rigidity. Check seat in spindle or tool holder.
		Check speeds and feeds.
	Insufficient coolant pressure and volume	Increase coolant pressure and flow. Is coolant flow constant?
		Make sure coolant reaches inserts at all times.
	Recutting chips, causing drill to jump	Increase coolant flow. Add coolant grooves.
	Poor chip control; chips trapped in hole	Mostly speed or feed.
	Chatter	Mostly feed rate.
Very short, thick, flat chips	Feed rate too high in relation to cutting speed	Lower feed or increase speed.
Long and stringy chips	Feed rate too low in relation to cutting speed	Increase feed rate or decrease speed.
		Use dimple inserts
Unable to loosen insert locking screws	Seized threads, caused by coolant or heat	Apply water and heat-resistant lubricant to threads.

[a] *Source:* "Fundamentals of Indexable Drilling," K.L. Anderson, *Machining Technology*, vol. 2, no. 3, 1991.
[b] If constant chipping occurs, especially on an inner insert, and conditions are optimum, try an uncoated-carbide insert or a grade with higher transverse rapture strength.

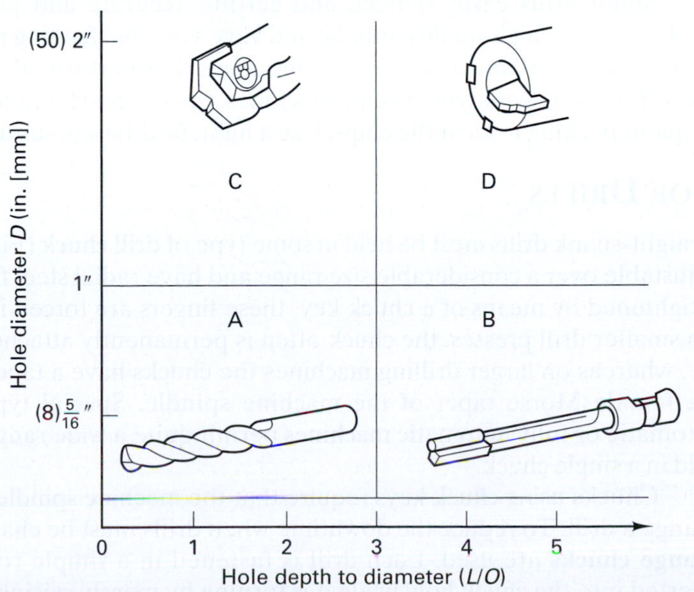

Sector	Typical drill types
A	Twist drill (HSS) Center core drill
B	Twist drill Gundrill BTA Ejector drill
C	Twist drill Indexable insert drill Spade drill Center core drill
D	BTA Ejector drill

FIGURE 23-14 Drill selection depends on hole diameter and hole depth.

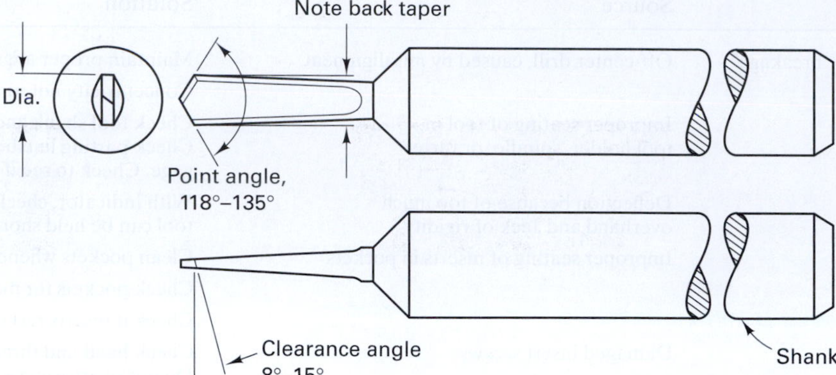

FIGURE 23-15 Pivot microdrill for drilling very-small-diameter holes.

category of this section. It is the section for which the majority of the work is done by twist drills and a very few cemented carbide drills. Section B is the drilling of deep holes for which cemented carbide gundrills are used. Section C is that of shallow holes having large diameters, for which spade drills are used. Section D is that of deep holes having large diameters, for which BTA tools are used.

MICRODRILLING

As the term suggests, **microdrilling** involves very-small-diameter cutting tools, including drills, end mills, routers, and other special tools. Drills from 0.002 in. (0.05 mm) and mills to 0.005 in. in diameter are used to produce geometries involving dimensions at which many workpiece materials no longer exhibit uniformity and homogeneity. Grain borders, inclusions, alloy or carbide segregates, and microscopic voids are problems in microdrilling, where holes of 0.02 to 0.0001 in. have been drilled using pivot drills, as shown in Figure 23-15.

Pivot drills are two-lipped (two-fluted), end-cutting tools of relatively simple geometry. Web thickness tapers toward the point, and a generous back-taper is incorporated. For softer workpiece materials, point angles are typically 118 degrees and lip clearance is 15 degrees. For steels and general use in harder metals, 135-degree points and 8-degree clearance are recommended. The chisel edge is similar to that of a twist drill. Pivot drills are made of tungsten-alloy tool steel in standard sizes from 0.0001 to 0.125 in. and of sintered tungsten carbide from 0.001 to 0.125 in.

Small drills easily deflect, and getting accurate and precise holes requires a machine with a high-quality spindle and very sensitive feeding pressure. In the medical components field, much of this machining work is performed on computer numerical control (CNC) Swiss-type turning machines. Speeds and feeds are greatly reduced with frequent pecking to clear the chips. Use a light, lard-based, sulfurized cutting oil.

23.4 TOOL HOLDERS FOR DRILLS

Straight-shank drills must be held in some type of drill **chuck** (Figure 23-16). Chucks are adjustable over a considerable size range and have radial steel fingers. When the chuck is tightened by means of a chuck key, these fingers are forced inward against the drill. On smaller drill presses, the chuck often is permanently attached to the machine spindle, whereas on larger drilling machines the chucks have a tapered shank that fits into the female Morse taper of the machine spindle. Special types of chucks in semiautomatic or fully automatic machines permit quite a wide range of sizes of drills to be held in a single chuck.

Chucks using chuck keys require that the machine spindle be stopped in order to change a drill. To reduce the downtime when drills must be changed frequently, **quick-change chucks** are used. Each drill is fastened in a simple round collet that can be inserted into the chuck hole while it is turning by merely raising and lowering a ring on the chuck body. With the use of this type of chuck, center drills, drills, counterbores,

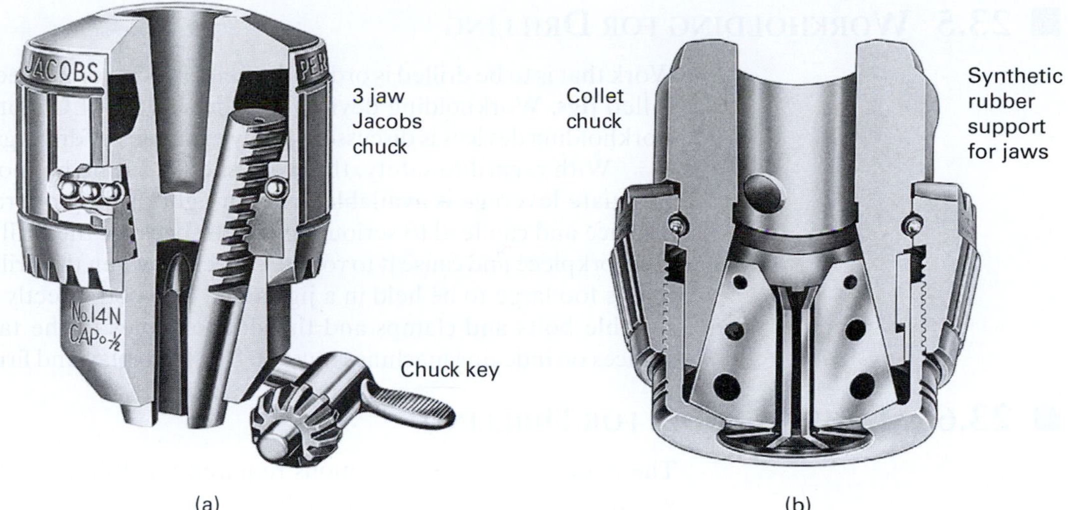
JACOBS
3 jaw
Jacobs
chuck
No.14N
CAPo-½
Chuck key
Collet
chuck
Synthetic
rubber
support
for jaws
(a)
(b)

FIGURE 23-16 Two of the most commonly used types of drill chucks are the three-jaw Jacobs chuck (above) and the collet chuck with synthetic rubber support for jaws. *(Image provided by Jacobs Chuck, Apex Tool Group, Sparks, MD)*

reamers, and so on, can be manually changed in quick succession. For carbide drills, collet-type holders with thrust bearings are recommended (Figure 23-17). For drills using an internal coolant supply, a very rigid chuck with either an inducer or through-spindle coolant source is recommended.

Conventional holders such as keyless chucks cannot be used because the gripping strength is limited. Collet holders should be cleaned periodically with oil to remove small chips.

The entire flute length must protrude from the chuck. At maximum hole depth, the length of flute protruding from the hole must be at least one to one-and-a-half times the drill diameter. Radial runout at the drill tip must not exceed 0.001 in.

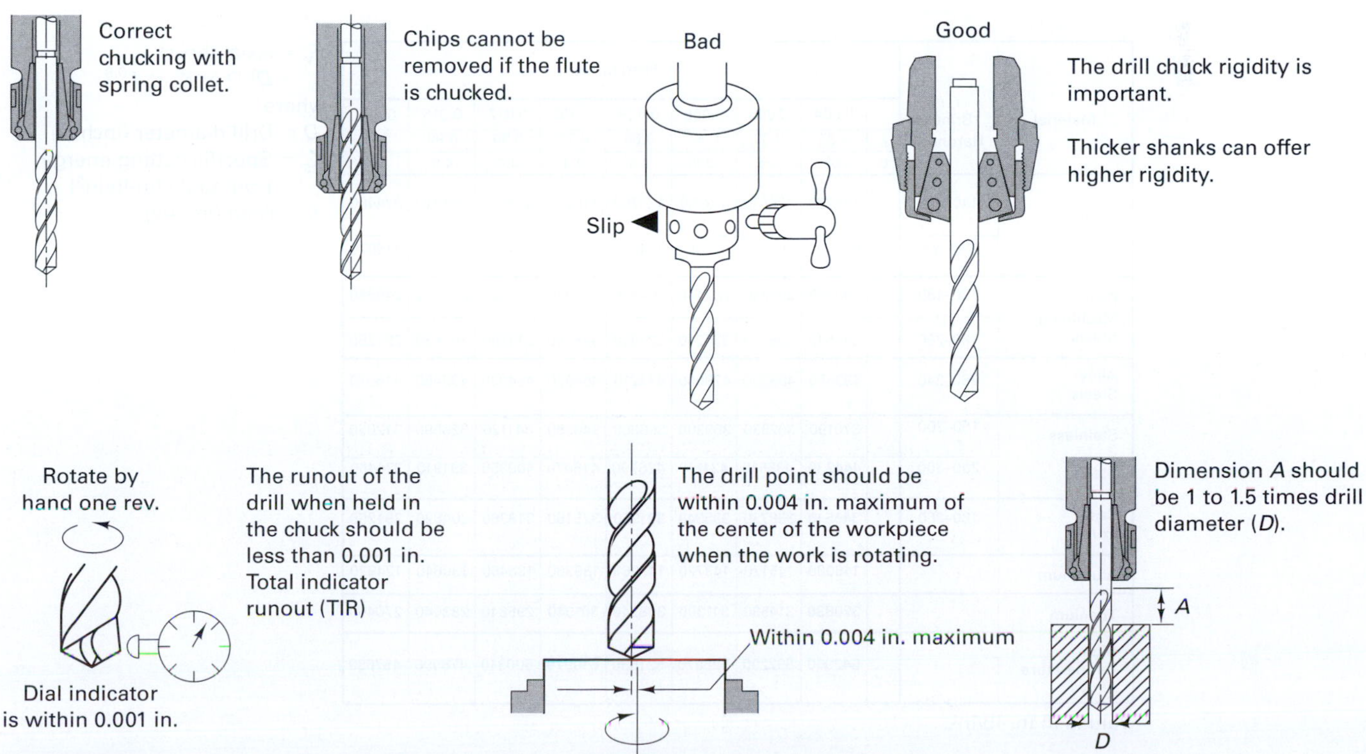
Correct chucking with spring collet.
Chips cannot be removed if the flute is chucked.
Bad
Slip
Good
The drill chuck rigidity is important.
Thicker shanks can offer higher rigidity.
Rotate by hand one rev.
Dial indicator is within 0.001 in.
The runout of the drill when held in the chuck should be less than 0.001 in. Total indicator runout (TIR)
The drill point should be within 0.004 in. maximum of the center of the workpiece when the work is rotating.
Within 0.004 in. maximum
Dimension A should be 1 to 1.5 times drill diameter (D).
A
D

FIGURE 23-17 Here are some suggestions for correct chucking of carbide drills.

23.5 WORKHOLDING FOR DRILLING

Work that is to be drilled is ordinarily held in a vise or in specially designed workholders called **jigs.** Workholding devices are the subject of Chapter 27, where the design of workholding devices is discussed. Many examples of drill jigs are shown.

With regard to safety, the work should not be held on the table by hand unless adequate leverage is available, even in light drilling operations. This is a dangerous practice and can lead to serious accidents, because the drill has a tendency to catch on the workpiece and cause it to rotate, especially when the drill exits the workpiece. Work that is too large to be held in a jig can be clamped directly to the machine table using suitable bolts and clamps and the slots or holes in the table. Jigs and workholding devices on indexing machines must be free from play and firmly seated.

23.6 MACHINE TOOLS FOR DRILLING

The basic work and tool motions required for drilling—relative rotation between the workpiece and the tool, with relative longitudinal feeding—occur in a wide variety of machine tools. Thus, drilling can be done on a variety of machine tools such as lathes, horizontal and vertical milling machines, boring machines, and machining centers. This section will focus on those machines that are designed, constructed, and used primarily for drilling.

First, the machine tools must have sufficient power (torque) and thrust to perform the cut. It is the task of the engineer to select the correct machine or select the cutting parameters (speed and feed) based on the type and size of the drill, drill material, and the work material (hardness). Because of the complex geometry of the drill, empirical equations are widely used. Figure 23-18 shows the type of information provided by cutting tool manufacturers to calculate (estimate) thrust in drilling. Specific cutting force values K_s are given in Figure 23-18, while empirical constants X and Y are obtained from cutting tool manufacturers. Much of these data have been developed for high-speed-steel tools. When using solid carbide tools, rigid machines such as machining centers or NC turning machines are recommended, whereas a radial drilling machine is not recommended (not rigid enough).

Material	Brinell Hardness		Feed (ipr)							
			0.004	0.005	0.006	0.008	0.010	0.012	0.016	0.020
		E	0.35	0.39	0.47	0.60	0.70	0.80	0.90	1.08
		C	3.3	3.5	3.8	4.4	4.4	4.4	4.0	3.6
Plain-Carbon Steel	140–220		444230	435510	431150	426790	418070	409350	391910	374460
	220–300		493590	483900	479060	474210	464520	454830	435450	416070
Free-Machining Steels	120–180		296150	290340	287440	284530	278710	272900	261270	249640
	180–260		345510	338730	335340	331950	325160	318380	304820	291250
Alloy Steels	260–340		493590	483900	479060	474210	464520	454830	435450	416070
Stainless Steels	150–200		370190	362930	359300	355660	348390	341120	326590	312050
	200–300		444230	435510	431150	426790	418070	409350	391910	374460
Cast Iron	180–250		345510	338730	335340	331950	325160	318380	304820	291250
Aluminum			148080	145170	143720	142260	139360	136450	130640	124820
Titanium			320830	314530	311390	308240	301940	295640	283040	270450
High-Temperature Alloys			542950	532290	526970	521630	510970	500310	478990	457680

Values in in.-lb/in².

F_v = Axial thrust
$= D^{1.15} \times K_s \times f_r^{0.8}$
where
D = Drill diameter (inches)
K_s = Specific cutting energy from table (in-lb/in²)
f_r = Feed (in./rev)

FIGURE 23-18 Estimating the thrust force, F_v, in drilling using, K_s values in in-lb/in². *(Waukesha Cutting Tools)*

Rigidity is especially important in avoiding chatter. A lack of rigidity in the cutting tool, the workpiece workholding device or the machine tool permits the affected members to deflect due to the cutting forces. Conditions for chatter are discussed in Chapter 20. The cutting lips can have a hammering action against the work. Using the shortest tool possible can help.

In addition, backlash in the feed mechanism should be kept at a minimum to reduce strain on the drill when it breaks through the bottom of the hole.

The common name for the machine tool used for drilling is the **drill press.** Drill presses consist of a *base*, a *column* that supports a *powerhead*, a *spindle*, and a *worktable*. On small machines, the base rests on a workbench, whereas on larger machines it rests on the floor (Figure 23-19). The column may be either round or of box-type construction—the latter being used on larger, heavy-duty machines, except in radial types. The powerhead contains an electric motor and means for driving the spindle in rotation at several speeds. On small drilling machines, this may be accomplished by shifting a belt on a step-cone pulley, but on larger machines a geared transmission is used.

The heart of any drilling machine is its **spindle.** In order to drill satisfactorily, the spindle must rotate accurately and also resist whatever side forces result from the drilling process. In virtually all machines, the spindle rotates in preloaded ball or taper-roller bearings. In addition to powered rotation, provision is made so that the spindle can be moved axially to feed the drill into the work. On small machines, the spindle is fed by hand, using the handles extending from the capstan wheel; on larger machines, power feed is provided. Except for some small bench types, the spindle contains a hole with a Morse taper in its lower end into which taper-shank drills or drill chucks can be inserted.

The worktables on drilling machines may be moved up and down on the column to accommodate work of various sizes. On round-column machines, the table can usually be rotated out of the way so that workpieces can be mounted directly on the base. On some box-column machines, the table is mounted on a subbase so that it can be moved in two directions in a horizontal plane by means of feed screws.

Figure 23-19 shows examples of common types of drilling machines used in production environments. Drilling machines are usually classified as bench, upright with single spindle, turret or NC turret, gang, multispindle, deep-hole, and transfer.

With bench drill presses, holes up to $\frac{1}{2}$ in. in diameter can be drilled. The same type of machine can be obtained with a long column so that it can stand on the floor. The size of bench and upright drilling machines is designated by *twice* the distance from the centerline of the spindle to the nearest point on the column, this being an indication of the maximum size of the work that can be drilled in the machines. For example, a 15-in. drill press will permit a hole to be drilled at the center of a workpiece 15 in. in diameter.

Sensitive drilling machines are essentially smaller, plain, bench-type machines with more accurate spindles and bearings. They are capable of operating at higher speeds, up to 30,000 rpm. Very sensitive hand-operated feeding mechanisms are provided for use in drilling small holes. Such machines are used for tool and die work and for drilling very small holes, often less than a few thousandths of an inch in diameter, when high spindle speeds are necessary to obtain proper cutting speed and sensitive feel to provide delicate feeding to avoid breakage of the very small drills.

Upright drilling machines usually have spindle speed ranges from 60 to 3500 rpm and power feed rates, from 4 to 12 steps, from about 0.004 to 0.025 in./rev. Most modern machines use a single-speed motor and a geared transmission to provide the range of speeds and feeds. The feed clutch disengages automatically when the spindle reaches a preset depth.

Worktables on most upright drilling machines contain holes and slots for use in clamping work and nearly always have a channel around the edges to collect cutting fluid, when it is used. On box-column machines, the table is mounted on vertical ways on the front of the column and can be raised or lowered by means of a crank-operated elevating screw.

In mass production, **gang-drilling machines** are often used when several related operations—such as drilling holes of different sizes, reaming, or counterboring—must be done on a single part. These consist essentially of several independent columns, heads, and spindles mounted on a common base and having a single table. The work can be slid

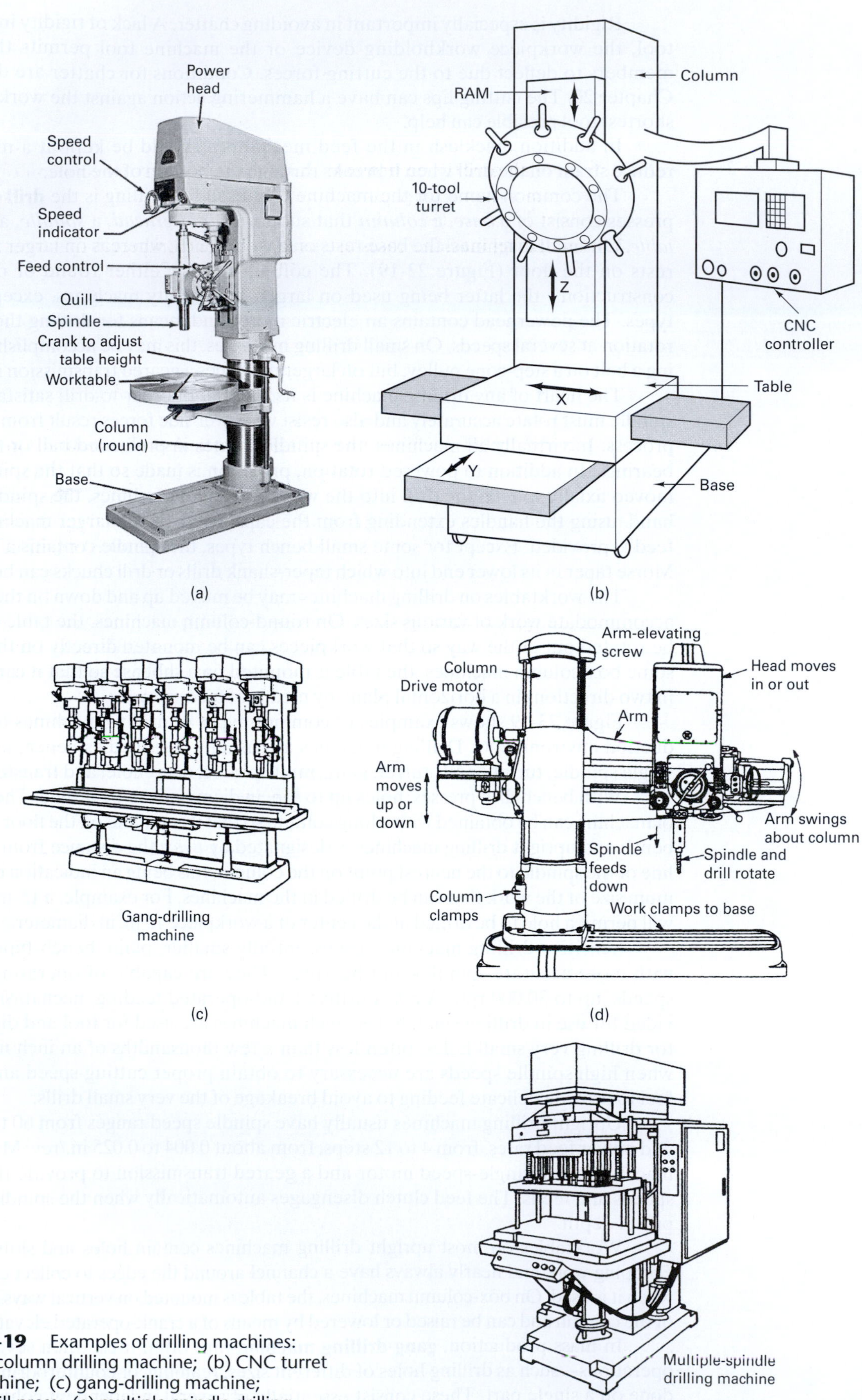

FIGURE 23-19 Examples of drilling machines: (a) upright column drilling machine; (b) CNC turret drilling machine; (c) gang-drilling machine; (d) radial drill press; (e) multiple-spindle drilling machine. *(Courtesy J T. Black)*

into position for the operation at each spindle. They are available with or without power feed. One or several operators may be used. This machine would be an example of a simple small cell, except the machines are usually not single-cycle automatics.

Turret-type, upright drilling machines are used when a series of holes of different sizes, or a series of operations (such as center drilling, drilling, reaming, and spot facing), must be done repeatedly in succession. The selected tools are mounted in the turret. Each tool can quickly be brought into position merely by rotation of the turret. These machines automatically provide individual feed rates for each spindle and are often numerically controlled.

Radial drilling machine tools are used on large workpieces that cannot be easily handled manually. As shown in Figure 23-19, these machines have a large, heavy, round, vertical column supported on a large base. The column supports a radial arm that can be raised and lowered by power and rotated over the base. The spindle head, with its speed- and feed-changing mechanism, is mounted on the radial arm. It can be moved horizontally to any desired position on the arm. Thus, the spindle can be quickly positioned properly for drilling holes at any point on a large workpiece mounted either on the base of the machine or even sitting on the floor.

Plain radial drilling machines provide only a vertical spindle motion. On *semiuniversal machines,* the spindle head can be pivoted at an angle to a vertical plane. On *universal machines,* the radial arm is rotated about a horizontal axis to permit drilling at any angle.

Radial drilling machines are designated by the radius of the largest disk in which a center hole can be drilled when the spindle head is at its outermost position. Sizes from 3 to 12 ft are available. Radial drilling machines have a wide range of speeds and feeds, can do boring, and include provisions for tapping internal threading.

Multiple-spindle drilling machines (Figure 23-19) are mass-production machines with as many as 50 spindles driven by a single power head and fed simultaneously into the work. Figure 23-20 shows an adjustable multiple-spindle head that can be mounted on a regular single spindle-drill press. Figure 23-20 shows the methods of driving and positioning the spindles, which permit them to be adjusted so that holes can be drilled at any location within the overall capacity of the head. Special drill jigs are often designed and built for each job to provide accurate guidance for each drill. Although such machines and workholders are quite costly, they can be cost-justified when the quantity to be produced will justify the setup cost and the cost of the jig. Reducing setup on these machines is difficult. Numerically controlled drill presses other than turret drill presses are not common because drilling and all its related processes can be done on vertical or horizontal NC machining centers equipped with automatic tool changers (see Chapter 25).

Special machines are used for drilling long (deep) holes, such as are found in rifle barrels, connecting rods, and long spindles. High cutting speeds, very light feeds, and a copious flow of cutting fluid ensure rapid chip removal. Adequate support for the long, slender drills is required. In most cases horizontal machines are used. The work is rotated in a chuck with steady rests providing support along its length, as required. The drill does not rotate and is fed into the work. Vertical machines are also available for shorter workpieces. Notice the similarity between this process and boring.

23.7 CUTTING FLUIDS FOR DRILLING

For shallow holes, the general rules relating to cutting fluids, as given in Chapter 21, are applicable. When the depth of the hole exceeds one diameter, it is desirable to increase the lubricating quality of the fluid because of the rubbing between the drill margins and the wall of the hole. The effectiveness of a cutting fluid as a coolant is quite variable in drilling. While the rapid exit of the chips is a primary factor in heat removal, this action also tends to restrict entry of the cutting fluid. This is of particular importance in drilling materials that have poor heat conductivity. Recommendations for cutting fluids for drilling are given in Table 23-5.

If the hole depth exceeds two or three diameters, it is usually advantageous to withdraw the drill each time it has drilled about one diameter of depth to clear chips from the hole. Some machines are equipped to provide this "pecking" action automatically.

Adjustable drill head
Spindle: 6 Production: 50 pieces

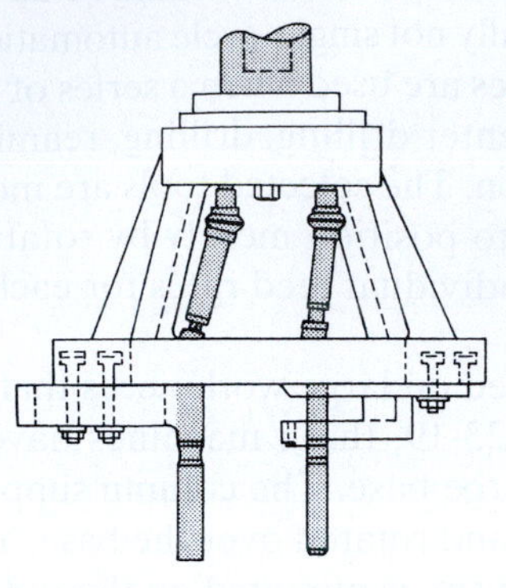

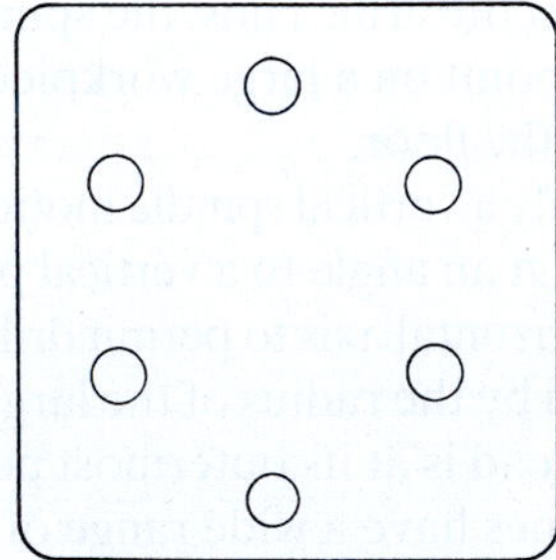

An adjustable drill head should be considered for low-production jobs. However, many short-run jobs such as this would be required to justify a multiple-spindle head.

Geared drill head
Spindle: 8 Production: 80,000 pieces

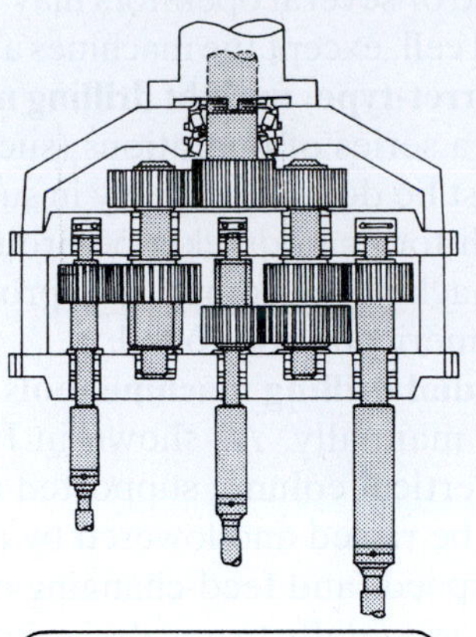

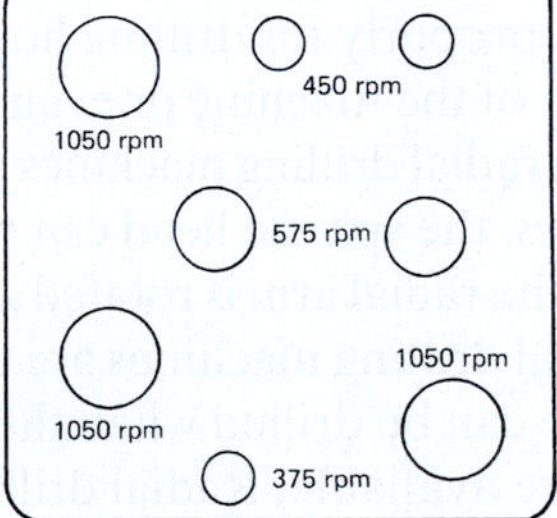

A geared drill head is most appropriate in this situation, where there is a large difference in sizes and a high daily production.

Gearless drill head
Spindle: 16 Production: 30,000 pieces

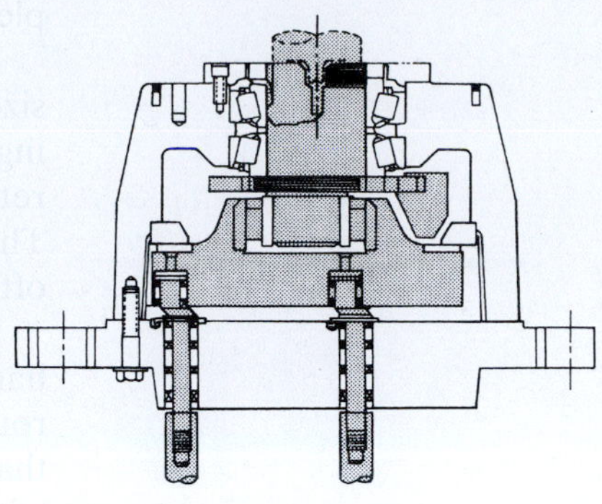

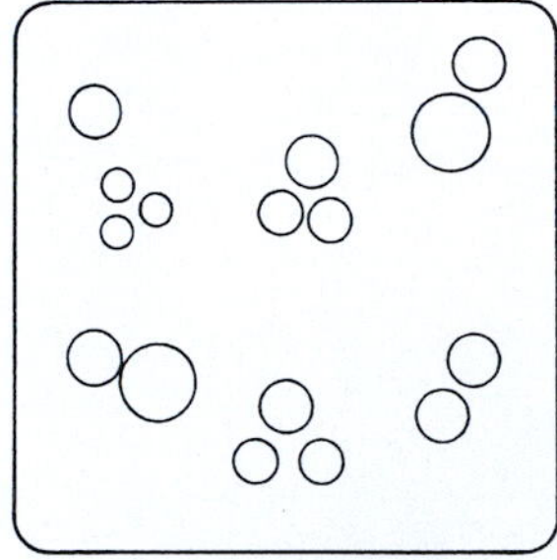

Only a gearless head can perform this operation in one pass, due to the close proximity of the spindle centers.

FIGURE 23-20 Three basic types of multiple-spindle drill heads: (left) adjustable; (middle) geared; (right) gearless. *(Courtesy of Zagar Incorporated)*

Where cooling is desired, the fluid should be applied copiously. For severe conditions, drills containing coolant holes have a considerable advantage. Not only is the fluid supplied near the cutting edges, but the coolant flow aids in flushing the chips from the hole. Where feasible, drilling horizontally has distinct advantages over drilling vertically downward.

TABLE 23-5 Cutting Fluids for Drilling

Work Material	Cutting Fluid
Aluminum and its alloys	Soluble oil, kerosene, and lard-oil compounds; light, nonviscous neutral oil; kerosene and soluble oil mixtures
Brass	Dry or a soluble oil; kerosene and lard-oil compounds; light, nonviscous neutral oil
Copper	Soluble oil, strained lard oil, oleic-acid compounds
Cast iron	Dry or with a jet of compressed air for cooling
Malleable iron	Soluble oil, nonviscous neutral oil
Monel metal	Soluble oil, sulfurized mineral oil
Stainless steel	Soluble oil, sulfurized mineral oil
Steel, ordinary	Soluble oil, sulfurized oil, high extreme-pressure-value mineral oil
Steel, very hard	Soluble oil, sulfurized oil, turpentine
Wrought iron	Soluble oil, sulfurized oil, mineral-animal oil compound

- Neat oil can be used effectively with the solid carbide drills for low-speed drilling (up to 130 sfpm)
- If the work surface becomes hard or blue in color, decrease the rpm and use neat oil.
- For heavy-duty cutting, emulsion-type oil containing some extreme pressure additive is recommended.
- A volume of 3.0 gal/min at a pressure of 37–62 lb/in.^{2} is recommended.
- A double stream supply of fluid is recommended.

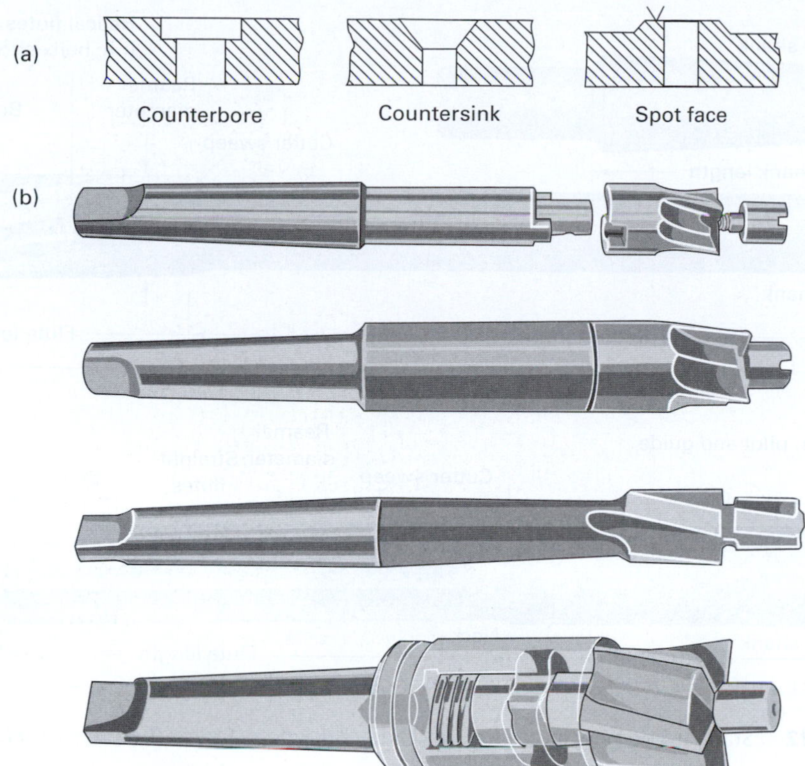

FIGURE 23-21 (a) Surfaces produced by counterboring, countersinking, and spot facing. (b) Counterboring tools: (bottom to top) interchangeable counterbore; solid, taper-shank counterbore with integral pilot; replaceable counterbore and pilot; replaceable counterbore, disassembled. *(Courtesy of Ex-Cell-O Corporation and Chicago Latrobe Twist Drill Works)*

23.8 COUNTERBORING, COUNTERSINKING, AND SPOT FACING

Drilling is often followed by *counterboring, countersinking,* or *spot facing.* As shown in Figure 23-21, each provides a bearing surface at one end of a drilled hole. They are usually done with a special tool having from three to six cutting edges.

Counterboring provides an enlarged cylindrical hole with a flat bottom so that a bolt head, or a nut, will have a smooth bearing surface that is normal to the axis of the hole; the depth may be sufficient so that the entire bolt head or nut will be below the surface of the part. The pilot on the end of the tool fits into the drilled hole and helps to ensure concentricity with the original hole. Two or more diameters may be produced in a single counterboring operation. Counterboring also can be done with a single-point tool, although this method ordinarily is used only on large holes and essentially is a boring operation. Some counterboring tools are shown in Figure 23-21b.

Countersinking makes a beveled section at the end of a drilled hole to provide a proper seat for a flat-head screw or rivet. The most common angles are 60, 82, and 90 degrees. Countersinking tools are similar to counterboring tools except that the cutting edges are elements of a cone, and they usually do not have a pilot because the bevel of the tool causes them to be self-centering.

Spot facing is done to provide a smooth bearing area on an otherwise rough surface at the opening of a hole and normal to its axis. Machining is limited to the minimum depth that will provide a smooth, uniform surface. Spot faces thus are somewhat easier and more economical to produce than counterbores. They are usually made with a multiedged end-cutting tool that does not have a pilot, although counterboring tools are frequently used.

23.9 REAMING

Reaming removes a small amount of material from the surface of holes. It is done for two purposes: to bring holes to a more exact size and to improve the finish of an existing

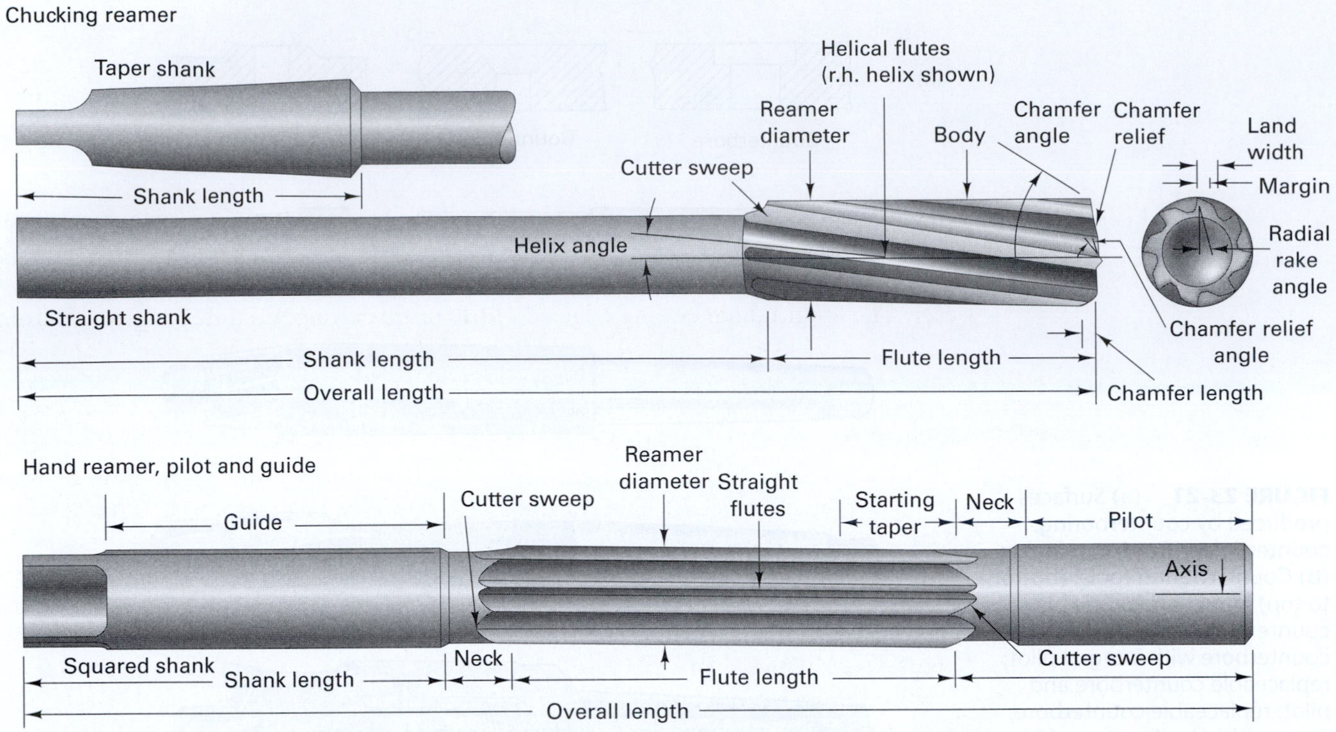

Chucking reamer
Taper shank
Shank length
Straight shank
Shank length
Overall length
Helical flutes (r.h. helix shown)
Reamer diameter
Cutter sweep
Helix angle
Body
Chamfer angle
Chamfer relief
Land width
Margin
Radial rake angle
Chamfer relief angle
Flute length
Chamfer length

Hand reamer, pilot and guide
Guide
Cutter sweep
Reamer diameter
Straight flutes
Starting taper
Neck
Pilot
Axis
Squared shank
Shank length
Neck
Flute length
Cutter sweep
Overall length

FIGURE 23-22 Standard nomenclature for hand and chucking reamers. *(Courtesy J T. Black)*

hole. Multiedge cutting tools are used, as shown in Figure 23-22. No special machines are built for reaming. The same machine that was employed for drilling the hole can be used for reaming by changing the cutting tool.

To obtain proper results, only a minimum amount of materials should be left for removal by reaming. As little as 0.005 in. is desirable, and in no case should the amount exceed 0.015 in. A properly reamed hole will be within 0.001 in. of correct size and have a fine finish.

The principal types of reamers are shown in Figures 23-22 and 23-23. **Hand reamers** are intended to be turned and fed by hand and to remove only a few thousandths of

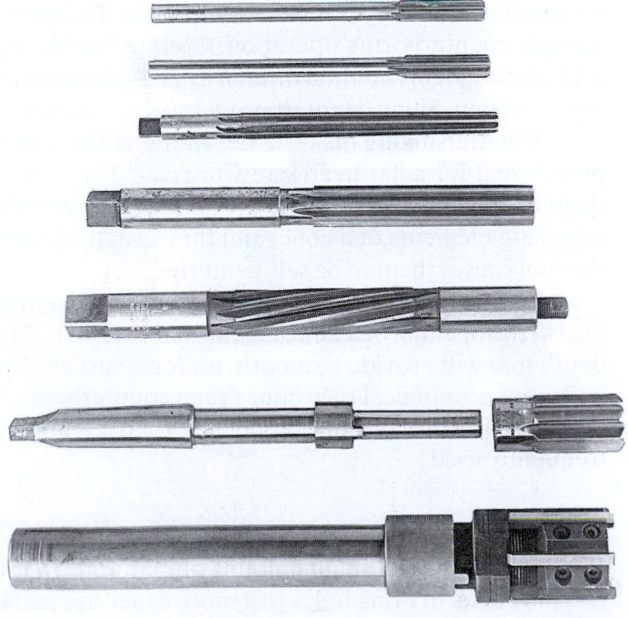

FIGURE 23-23 Types of reamers (top to bottom): straight-fluted rose reamer, straight-fluted chucking reamer, straight-fluted taper reamer, straight-fluted hand reamer, expansion reamer, shell reamer, adjustable insert-blade reamer. *(Courtesy J T. Black)*

an inch of metal. They have a straight shank with a square tang for a wrench. They can have straight or spiral flutes and can be solid or expandable. The teeth have relief along their edges and thus may cut along their entire length. However, the reamer is tapered from 0.005 to 0.010 in. in the first third of its length to assist in starting it in the hole, and most of the cutting therefore takes place in this portion.

Machine or **chucking reamers** are for use with various machine tools at slow speeds. The best feed is usually two to three times the drilling feed. Machine reamers have chamfers on the front end of the cutting edges. The chamfer causes the reamer to seat firmly and concentrically in the drilled hole, allowing the reamer to cut at full diameter. The longitudinal cutting edges do little or no cutting. Chamfer angles are usually 45 degrees. Reamers have straight or tapered shanks and straight or spiral flutes. **Rose-chucking reamers** are ground cylindrical and have no relief behind the outer edges of the teeth. All cutting is done on the beveled ends of the teeth. **Fluted-chucking reamers,** on the other hand, have relief behind the edges of the teeth as well as beveled ends. They can, therefore, cut on all portions of the teeth. Their flutes are relatively short, and they are intended for light finishing cuts. For best results they should not be held rigidly but permitted to float and be aligned by the hole.

Shell reamers often are used for larger sizes in order to save cutting tool material. The shell, made of tool steel for smaller sizes and with carbide edges for larger sizes or for mass-production work, is held on an arbor that is made of ordinary steel. One arbor may be used with any number of shells. Only the shell is subject to wear and needs to be replaced when worn. They may be ground as rose or fluted reamers.

Expansion reamers can be adjusted over a few thousandths of an inch to compensate for wear or to permit some variation in hole size to be obtained. They are available in both hand and machine types.

Adjustable reamers have cutting edges in the form of blades that are locked in a body. The blades can be adjusted over a greater range than expansion reamers. This permits adjustment for size and to compensate for regrinding. When the blades become too small from regrinding, they can be replaced. Both tool steel and carbide blades are used.

Taper reamers are used for finishing holes to an exact taper. They may have up to eight straight or spiral flutes. Standard tapers, such as Morse, Jarno, or Brown & Sharpe, come in sets of two. The **roughing reamer** has nicks along the cutting edge to break up the heavy chips that result as a cylindrical hole is cut to a taper. The **finishing reamer** has smooth cutting edges.

REAMING PRACTICE

If the material to be removed is free cutting, reamers of fairly light construction will give satisfactory results. However, if the material is hard, then tough, solid-type reamers are recommended, even for fairly large holes.

To meet quality requirements, including both finish and accuracy (tolerances on diameter, roundness, straightness, and absence of bell-mouth at ends of holes), reamers must have adequate support for the cutting edges, and reamer deflection must be minimal. Reaming speed is usually two-thirds the speed for drilling the same materials. However, for close tolerances and fine finish, speeds should be slower.

Feeds are usually much higher than those for drilling and depend on material. A feed of between 0.0015 and 0.004 in. per flute is recommended as a starting point. Use the highest feed that will still produce the required finish and accuracy. Recommended cutting fluids are the same as those for drilling. Reamers, like drills, should not be allowed to become dull. The chamfer must be reground long before it exhibits excessive wear. Sharpening is usually restricted to the starting taper or chamfer. Each flute must be ground exactly even, or the tool will cut oversize.

Reamers tend to chatter when not held securely, when the work or workholder is loose, or when the reamer is not properly ground. Irregularly spaced teeth may help reduce chatter. Other cures for chatter in reaming are to reduce the speed, vary the feed rate, chamfer the hole opening, use a piloted reamer, reduce the relief angle on the chamfer, or change cutting fluid. Any misalignment between the workpiece and the reamer will cause chatter and improper reaming.

KEY WORDS

adjustable reamer
BTA drills
center core drill
chisel end
chuck
chuck reamer
combination drill
counterboring
countersinking
deep-hole drills
drift
drill press
drilling
dubbing
ejector drills
expansion reamer
finishing reamer
flute
fluted-chucking reamer
gang-drilling machine
gundrill
gundrill trepanning
hand reamer
helix angle
hole cutter
hole saw
indexable insert drill
jig
lands
lip
machine reamer
microdrilling
multiple-spindle drilling machine
quick-change chuck
radial drilling machine
reaming
rose-chucking reamer
roughing reamer
shell reamer
spade drill
spindle
spot facing
step drill
straight shank
subland drill
tang
tape reamer
taper shank
thrust force
trepanning torque
turret-type, upright drilling machine
twist drill
web
web thinning

REVIEW QUESTIONS

1. What functions are performed by the flutes on a standard twist drill?
2. What determines the rake angle of a drill? See Figure 23-2.
3. Basically, what determines what helix angle a drill should have?
4. When a large-diameter hole is to be drilled, why is a smaller-diameter hole often drilled first?
5. Equation 23-4 for the MRR for drilling can be thought of as ___ × ___, where $f_r N_s$ is the feed rate of the drill bit.
6. Are the recommended surface speeds for spade drills given in Table 23-3 typically higher or lower than those recommended for twist drills? How about the feeds? Why?
7. What can happen when an improperly ground drill is used to drill a hole?
8. Why are most drilled holes oversize with respect to the nominally specified diameter?
9. What are the two primary functions of a combination center drill?
10. What is the function of the margins on a twist drill?
11. What factors tend to cause a drill to "drift" off the centerline of a hole?
12. The drills shown in Figure 23-13 have coolant passages in the flutes. What is the purpose of these holes?
13. In drilling, the deeper the hole, the greater the torque. Why?
14. Why do cutting fluids for drilling usually have more lubricating qualities than those for most other machining operations?
15. How does a gang-drilling machine differ from a multiple-spindle drilling machine?
16. How does a multiple-spindle drilling machine differ from a numerical control (NC) drilling machine with a tool changer that would hold all the drills found in the multiple-spindle machine? See Chapter 25 for discussion on NC.
17. How does the thrust force vary with feed? Why?
18. Holding the workpiece by hand when drilling is not a good idea. Why?
19. What is the rationale behind the operation sequence shown in Figure 23-10?
20. In terms of thrust, what is unusual about the slot-point drill compared to other drills?
21. What is the purpose of spot facing?
22. How does the purpose of counterboring differ from that of spot facing?
23. What are the primary purposes of reaming?
24. What are the advantages of shell reamers?
25. A drill that operated satisfactorily for drilling cast iron gave very short life when used for drilling a plastic. What might be the reason for this?
26. What precautionary procedures should be used when drilling a deep, vertical hole in mild steel when using an ordinary twist drill?
27. What is the advantage of a spade drill? Is it really a drill?
28. What is a "pecking" action in drilling?
29. Why does drill feed increase with drill size?
30. Suppose you specified a drilling feed rate that was too large. What kinds of problems do you think this might cause? See Figure 23-6 and Table 23-4 for help.

PROBLEMS

1. Suppose you wanted to drill a 1.5-in.-diameter hole through a piece of 1020 cold-rolled steel that is 2 in. thick, using an indexable insert drill. What values of feed and cutting speed will you specify, along with an appropriate allowance. Is this the correct tool? What other drill types could be used?
2. How much time will be required to drill the hole in Problem 1 using the insert drill?
3. What is the metal removal rate when a 1.5-in.-diameter hole, 2 in. deep, is drilled in 1020 steel at a cutting speed of 200 fpm with a feed of 0.010 ipr? What is the cutting time?
4. If the specific horsepower for the steel in Problem 3 is 0.9, what horsepower would be required, assuming 80% efficiency in the machine tool?

5. If the specific power of an AISI 1020 steel of 0.9, and 80% of the output of the 1.0-kW motor of a drilling machine is available at the tool, what is the maximum feed that can be used in drilling a 1-in.-diameter hole with a carbide drill? (Use the cutting speed suggested in Problem 3.)
6. Show how the approximate equation 23-5 for MRR in drilling was obtained. What assumption was needed?
7. A workpiece must have 10 holes finished in it. Manual layout time is $\frac{1}{2}$ hr/piece. To drill and ream all the holes requires 1 hr on the machine for each piece, not counting layout or setup. The labor rate is $10/hr and the machine rate is $20/hr. If a jig is used, the labor cost to lay out each piece can be saved. Both methods give the same-quality product, but this jig saves 40 min in processing time on the machine. How large a lot justifies the use of a jig that costs $150 to make (labor and materials)?
8. A part has two holes located for drilling by manual layout. If a drill jig is used, 0.5 min in processing time is saved for each piece. The labor rate is $9/hr. The overhead rate on the labor saved is 100%. Setup time is no more with than without the jig. The combined rate for interest, insurance, taxes, and maintenance is 35%. The cost of the jig is $500.
   a. How many pieces must be made in one lot to make the jig worthwhile?
   b. How many pieces must be made on the jig in one lot each month to earn the cost of the jig in 2 yr?
9. Manufacturer's charts will help determine the best feed and speed to run the drills. For example, a 1.5-in hole is to be drilled in 4140 steel annealed to BHN 275. For the spade drill, speed is 80 sfpm; feed, 0.009 ipr; and spindle rotation, 204 rpm. For the indexable insert drill, speed is 358 sfpm; feed, 0.007 ipr; and spindle rotation, 891 rpm. Typically, an indexable insert drill can produce a hole four times faster than a spade drill but may cost (with inserts) 50 to 75% more than the equivalent spade blade and holder. For making only a couple of holes, the extra cost is not usually justified. Determine the number of holes needed to justify the extra cost of the indexable insert drill. Some additional cost data are given in the accompanying table.

   Ignore tool life and assume that the blades and the indexable drills make about the same number of holes. (Why is this a reasonable assumption?) The holes are 3 in. deep, with no allowance needed. Cost of drills:

Spade Drill	Indexable-Insert Drill
$139.00 holder	$273.00 drill
+21.90 per blade	+12.80 per two inserts
$160.90	$285.80

   Assume for this example that a machine rate of $45/hr includes the cost of labor and machine burden.
10. Assume that you are drilling eight holes, equally spaced in a bolt-hole circle. That is, there would be holes at 12, 3, 6, and 9 o'clock and four more holes equally spaced between them. The diameter of the bolt hole circle is 6 in. The designer says that the holes must be 45 degrees ± 1 degree from each other around the circle.
    a. Compute the tolerance between hole centers.
    b. Do you think a typical multiple-spindle drill setup could be used to make this bolt circle—using eight drills all at once? Why or why not?
    c. Do you think that the use of a jig may help improve the situation?
    d. Do you think a CNC drilling process could do the holes best?
11. A part with seven holes can be machined on a numerically controlled turret drill press in 3 min (estimated time based on similar parts). The rate on the CNC machine for labor is $34/hr. Currently, the part is being machined on a gang drill press with a special jig in 10 min/piece. The jig for the gang drill costs $300; the combined rate for depreciation, interest, insurance, and taxes is 135%; and the hourly rate for the gang drill and operator is $16. Setup time is about the same for both machines. For how many pieces is it economical to switch to the CNC?
12. It is estimated that a jig for machining a part with three holes costs $400 and with it the operation takes 15 min/part. The operation can be done without a jig on a numerically controlled drill press in 5 min. Assume that any other conditions are the same as in Problem 11. How many pieces are needed to cost-justify the use of a jig?

Chapter 23 CASE STUDY

Bolt-down Leg on a Casting

Larry Cornelius and Marcelo Veras are consulting engineers and have just received the drawing shown in Figure CS-23. This is one of four legs on a casting made by the the Lorac Warpug Company. These legs are used to attach the device to the floor. The section drawing to the right shows the typical loading to which the leg is subjected. The company is currently drilling the bolt hole and then counterboring the land, but manufacturing has experienced some difficulty in machining the four holes. They report a lot of drill breakage. Quality control reports that distances between the four holes are frequently too large. Sales has recently reported that a substantial number of in-service failures have occurred with these legs. Larry and Marcelo have obtained a sketch from sales showing where the legs typically fail. This casting is manufactured from gray cast iron using the sand casting process.

1. What machining difficulties should Larry and Marcelo suspect this leg to have?
2. Why were the distances between the holes too large?
3. What should Larry and Marcelo recommend for solving these problems in the future in terms of materials, design, and manufacturing methods?
4. What should Larry and Marcelo recommend be done with the units in the field to stop the failures?

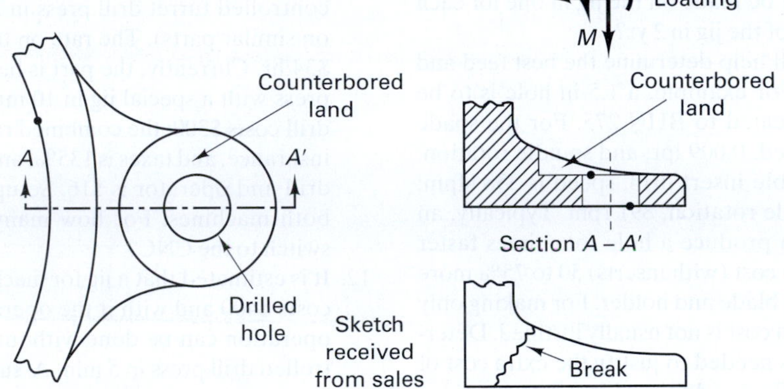

FIGURE CS-23 Shows the design of one of four legs on a casting made by the Cruftsmobile Company.

CHAPTER 24

MILLING

24.1 INTRODUCTION

Milling is a basic machining process by which a surface is generated by progressive chip removal. The workpiece is fed into a rotating cutting tool. Sometimes the workpiece remains stationary, and the cutter is fed to the work. In nearly all cases, a multiple-tooth cutter is used so that the material removal rate is high. Often the desired surface is obtained in a single pass of the cutter or work and, because very good surface finish can be obtained, milling is particularly well suited and widely used for mass-production work. Many types of milling machines are used, ranging from relatively simple and versatile machines that are used for general-purpose machining in job shops and tool and die work (these are NC or CNC machines) to highly specialized machines for mass production. Unquestionably, more flat surfaces are produced by milling than by any other machining process.

The cutting tool used in milling is known as a **milling cutter.** Equally spaced peripheral teeth will intermittently engage and machine the workpiece. This is called **interrupted cutting.** The workpieces are typically held in fixtures, as described in Chapter 27.

24.2 FUNDAMENTALS OF MILLING PROCESSES

Milling operations can be classified into broad categories called peripheral milling, end milling, and face milling. Each has many variations. In **peripheral milling,** the surface is generated by teeth located on the periphery of the cutter body (Figure 24-1). The surface is parallel with the axis of rotation of the cutter. Both flat and formed surfaces can be produced by this method, the cross section of the resulting surface corresponding to the axial contour of the cutter. This process, often called **slab milling,** is usually performed on horizontal spindle milling machines. In slab milling, the tool rotates (mills) at some rpm (N_s) while the work feeds past the tool at a table feed rate, f_m, in inches per minute (ipm), which depends on the feed per tooth, f_t.

As in the other processes, the cutting speed, V, and feed per tooth are selected by the engineer or the machine tool operator. As before, these variables depend on the work material, the tool material, and the specific process. The cutting velocity is that which occurs at the cutting edges of the teeth in the milling center. The rpm of the spindle is determined from the surface cutting speed, where D is the cutter diameter in inches according to

$$N_s = \frac{12V}{\pi D} \tag{24-1}$$

The depth of cut, called DOC or d in Figure 24-1, is simply the distance between the old and new machined surface.

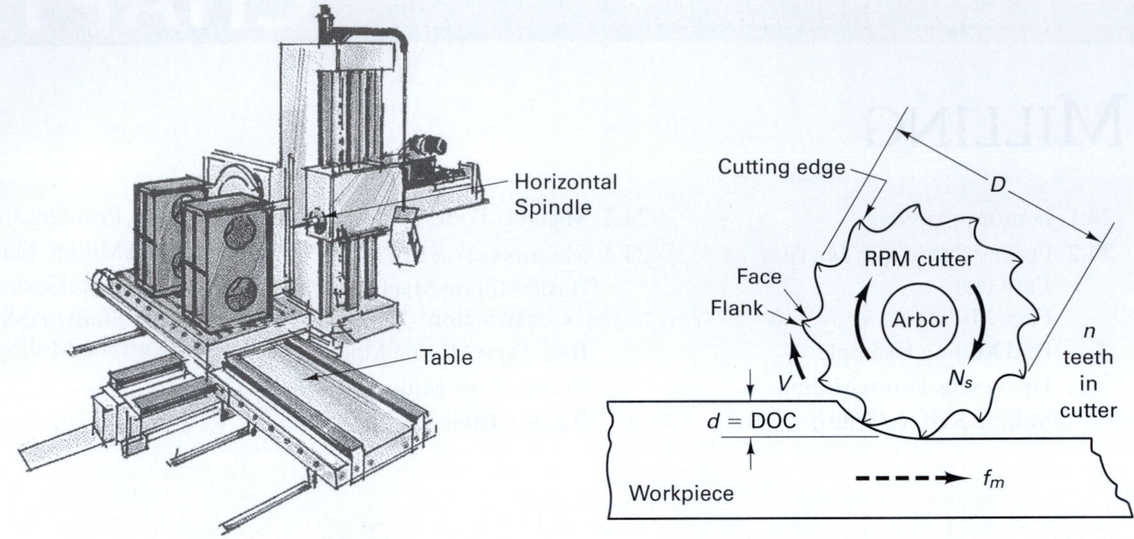

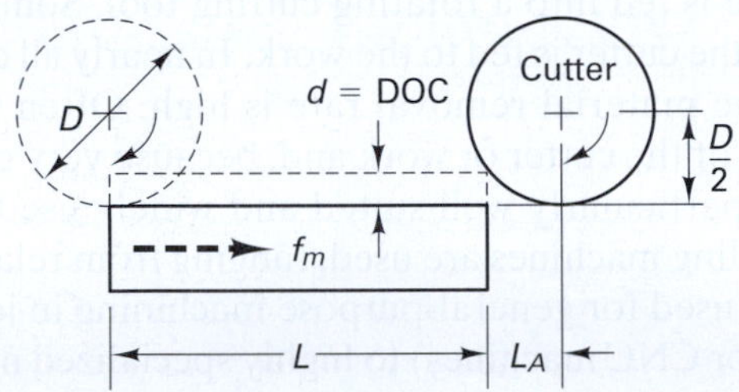

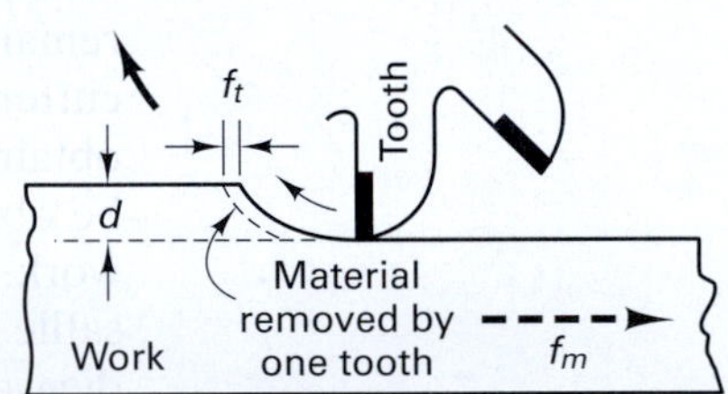

FIGURE 24-1 Peripheral milling can be performed on a horizontal-spindle milling machine. The cutter rotates at rpm N_s, removing metal at cutting speed, V. The allowance for starting and finishing the cut depends on the cutter diameter and depth of cut, d. The feed per tooth, f_t, and cutting speed are selected by the operator or process planner. (*Courtesy J T. Black*)

The width of cut is the width of the cutter or the work, in inches, and is given the symbol W. The length of the cut, L, is the length of the work plus some allowance, L_A, for approach and overtravel. The feed of the table, f_m, in inches per minute, is related to the amount of metal each tooth removes during a revolution, the feed per tooth, f_t, according to

$$f_m = f_t N_s n \tag{24-2}$$

where n is the number of teeth in the cutter (teeth rev.).

The **cutting time** is

$$T_m = \frac{L + L_A}{f_m} \tag{24-3}$$

The length of approach is

$$L_A = \sqrt{\frac{D^2}{4} - \left(\frac{D}{2} - \text{DOC}\right)^2} = \sqrt{d(D - d)} \tag{24-4}$$

The **metal removal rate (MRR)** is

$$\text{MRR} = \frac{\text{Volume}}{T_m} = \frac{LWd}{T_m} = Wf_m d \text{ in.}^3/\text{min} \tag{24-5}$$

ignoring L_A. Values for f_t are given in Table 24-1, along with recommended cutting speeds in feet per minute.

TABLE 24-1 Suggested Starting Feeds and Speeds Using High-Speed Steel and Carbide Cutters[a]

		Carbide Cutters						High-speed-Steel Cutters					
Material	Feed (in./tooth) Speed (fpm)	Face Mills	Slab Mills	End Mills	Full and Half-Side Mills	Saws	Form Mills	Face Mills	Slab Mills	End Mills	Full and Half-Side Mills	Saws	Form Mills
Malleable iron	Feed per tooth	.005–.015	.005–.015	.005–.010	.005–.010	.003–.004	.005–.010	.005–.015	.005–.015	.003–.015	.006–.012	.003–.006	.005–.010
Soft/hard	Speed, fpm	200–300	200–300	200–350	200–300	200–350	175–275	60–100	60–90	60–100	60–100	60–100	60–80
Cast steel	Feed per tooth	.008–.015	.005–.015	.003–.010	.005–.010	.002–.004	.005–.010	.010–.015	.010–.015	.005–.010	.005–.010	.002–.005	.008–.012
Soft/hard	Speed, fpm	150–350	150–350	150–350	150–350	150–300	150–300	40–60	40–60	40–60	40–60	40–60	40–60
100–150	Feed per tooth	.010–.015	.008–.015	.005–.010	.008–.012	.003–.006	.004–.010	.015–.030	.008–.015	.003–.010	.010–.020	.003–.006	.008–.010
BHN steel	Speed, fpm	450–800	450–600	450–600	450–800	350–600	350–600	80–130	80–130	80–140	80–130	70–100	70–100
150–250	Feed per tooth	.010–.015	.008–.015	.005–.010	.007–.012	.003–.006	.004–.010	.010–.020	.008–.015	.003–.010	.010–.015	.003–.006	.006–.010
BHN steel	Speed, fpm	300–450	300–450	300–450	300–450	300–450	300–450	50–70	50–70	60–80	50–70	50–70	50–70
250–350	Feed per tooth	.008–.015	.007–.012	.005–.010	.005–.012	.002–.005	.003–.008	.005–.010	.005–.010	.003–.010	.005–.010	.002–.005	.005–.010
BHN steel	Speed, fpm	180–300	150–300	150–300	160–300	150–300	150–300	35–60	35–50	40–60	35–50	35–50	35–50
350–450	Feed per tooth	.008–.015	.007–.012	.004–.008	.005–.012	.001–.004	.003–.008	.003–.008	.005–.008	.003–.101	.003–.008	.001–.004	.003–.008
BHN steel	Speed, fpm	125–180	100–150	100–150	125–180	100–150	100–150	20–35	20–35	20–40	20–35	20–35	20–35
Cast iron, hard	Feed per tooth	.005–.010	.005–.010	.003–.008	.003–.010	.002–.003	.005–.010	.005–.012	.005–.010	.003–.008	.005–.010	.002–.004	.005–.010
BHN 180–225	Speed, fpm	125–200	100–175	125–200	125–200	125–200	100–175	40–60	35–50	40–60	40–60	35–60	35–50
Cast iron, medium	Feed per tooth	.008–.015	.008–.015	.005–.010	.005–.012	.003–.004	.006–.012	.010–.020	.008–.015	.003–.010	.008–.015	.003–.005	.008–.012
BHN 180–225	Speed, fpm	200–275	175–250	200–275	200–275	200–250	175–250	60–80	50–70	60–90	60–80	60–70	50–60
Cast iron, soft	Feed per tooth	.015–.025	.010–.020	.005–.012	.008–.015	.003–.004	.008–.015	.015–.030	.010–.025	.004–.010	.010–.020	.002–.005	.010–.015
BHN 150–180	Speed, fpm	275–400	250–350	275–400	275–400	250–350	250–350	80–120	70–110	80–120	80–120	70–110	60–80
Bronze	Feed per tooth	.010–.020	.010–.020	.005–.010	.008–.012	.003–.004	.008–.015	.010–.025	.008–.020	.003–.010	.008–.015	.003–.005	.008–.015
Soft/hard	Speed, fpm	300–100	300–800	300–1000	300–1000	300–1000	200–800	50–225	50–200	50–250	50–225	50–250	50–200
Brass	Feed per tooth	.010–.020	.010–.020	.005–.010	.008–.012	.003–.004	.008–.015	.010–.025	.008–.020	.005–.015	.008–.015	.003–.005	.008–.015
Soft/hard	Speed, fpm	500–1500	500–1500	500–1500	500–1500	500–1500	500–1500	150–300	100–300	150–350	150–350	150–300	100–300
Aluminum alloy	Feed per tooth	.010–.040	.010–.030	.003–.015	.008–.025	.003–.006	.008–.015	.010–.040	.015–.040	.015–.040	.010–.030	.004–.008	.010–.020
Soft/hard	Speed, fpm	2000 UP	2000 UP	2000 UP	2000 UP	2000 UP	2000 UP	300–1200	300–1200	300–1200	300–1200	300–1000	300–1200

[a] Generally lower end of range used for inserted blade cutters, higher end of range for indexable insert cutters.

FACE MILLING

In face milling and end milling, the generated surface is at right angles to the cutter axis (Figure 24-2). Most of the cutting is done by the peripheral portions of the teeth, with the face portions providing some finishing action. **Face milling** is done on both horizontal- and vertical-spindle machines.

The tool rotates (face mills) at some rpm (N_s) while the work feeds past the tool. The rpm is related to the surface cutting speed, V, and the cutting tool diameter, D, according to equation 24-1. The depth of cut is d, in inches, as shown in Figure 24-2b. The width of cut is W, in inches, and may be width of the workpiece or width of the cutter, depending on the setup. The length of cut is the length of the workpiece, L, plus an allowance, L_A, for approach and overtravel, L_O, in inches. The feed rate of the table, f_m, in inches per minute, is related to the amount of metal each tooth removes during a pass over the work, called the feed per tooth, f_t,. As before $f_m = f_t N_s n$, where the number of teeth in the cutter is n. The cutting time is

$$T_m = \frac{L + L_A + L_o}{f_m} \text{min} \tag{24-6}$$

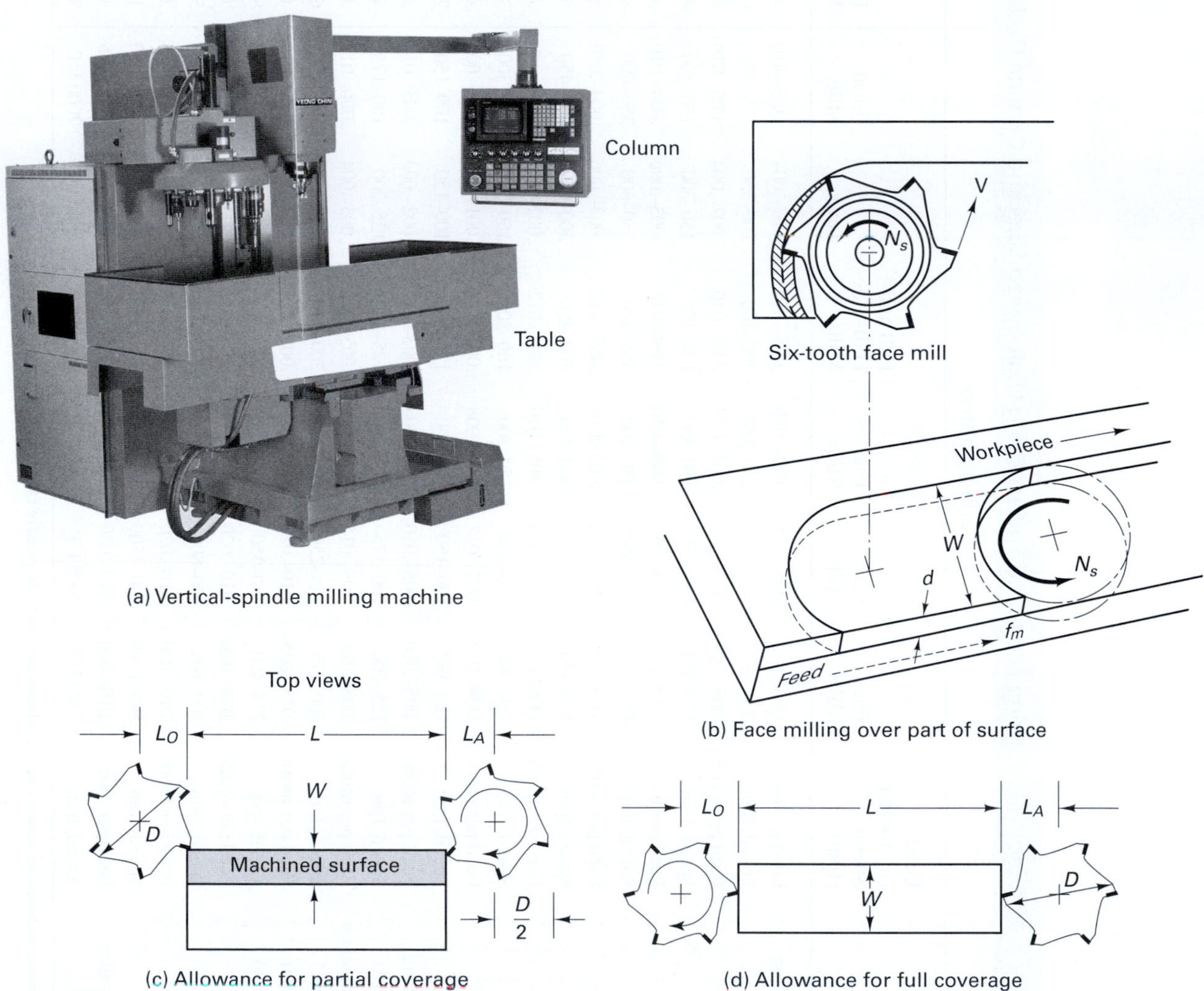

FIGURE 24-2 Face milling is often performed on a vertical spindle milling machine using a multiple-tooth cutter (n = 6 teeth) rotating N_s at rpm to produce cutting speed, V. The workpiece feeds at rate f_m, in inches per minute past the tool. The allowance depends on the tool diameter and the width of cut. (*Courtesy J T. Black*)

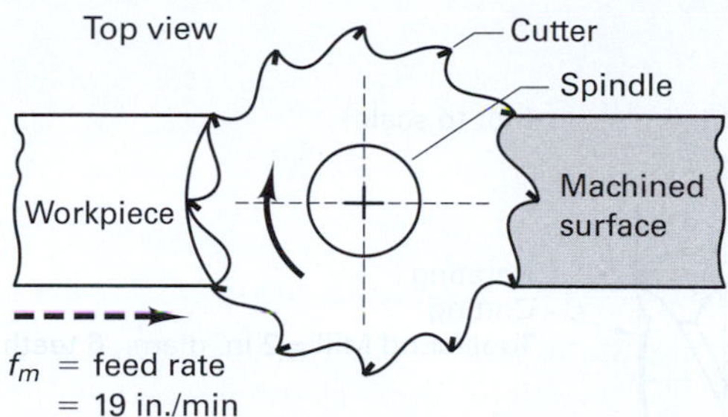

FIGURE 24-3 Face milling viewed from above with vertical spindle-machine.

The metal removal rate is

$$\text{MRR} - \frac{\text{Volume}}{T_m} = \frac{LWd}{T_m} = f_m W d \text{ in.}^3/\text{min}$$

When calculating the MRR, ignore L_O and L_A. The length of approach is usually equal to the length of overtravel, which usually equals *D/2* in. For a setup where the tool does not completely pass over the workpiece,

$$L_o = L_A = \sqrt{W(D - W)} \text{ for } W < \frac{D}{2} \quad (24\text{-}7)$$

$$L_o = L_A = \frac{D}{2} \text{ for } W \geq \frac{D}{2} \quad (24\text{-}8)$$

Here is an example of a face milling calculation. A 4-in.-diameter, six-tooth face mill is selected, using carbide inserts (Figure 24-3). The material being machined is low-alloy steel, annealed. Using cutting data recommendations, the cutting speed chosen is 400 sfpm with a feed of 0.008 in./tooth at a *d* of 0.12 in. Determining rpm at the spindle,

$$N_s = \frac{12V}{\pi D} = \frac{12 \times 400}{3.14 \times 4} = 392 \text{ rpm}$$

Determining the feed rate of the table, $fm = nN_s f_t$,

$$f_m = 0.008 \times 6 \times 392 = 19 \text{ in./min}$$

If slab or side milling were being performed, with the same parameters selected as given earlier, the setup would be different but the spindle rpm and table feed rate the same. The cutting time would be different because the allowances for face milling are greater than for slab milling. In milling, power consumption is usually the limiting factor. A thick chip is more power efficient than a thin chip.

END MILLING EXAMPLE

End milling is a very common operation performed on both vertical- and horizontal-spindle milling machines or machining centers. Figure 24-4 shows a vertical spindle end milling process, cutting a step in the workpiece. This cutter can cut on both the sides and ends of the tool. If you were performing this operation on a block of metal (for example, 430F stainless steel), you (the manufacturing engineer) would select a specific machine tool. You would have to determine how many passes (rough and finish cuts) were needed to produce the geometry specified in the design. Why? The number of passes determines the total cutting time for the job.

Using a vertical-spindle milling machine, an end mill can produce a step in the workpiece. In Figure 24-4, an end mill with six teeth on a 2-in. diameter is used to cut a step in 430F stainless. The *d* (depth of cut) is 0.375 in., and the depth of immersion

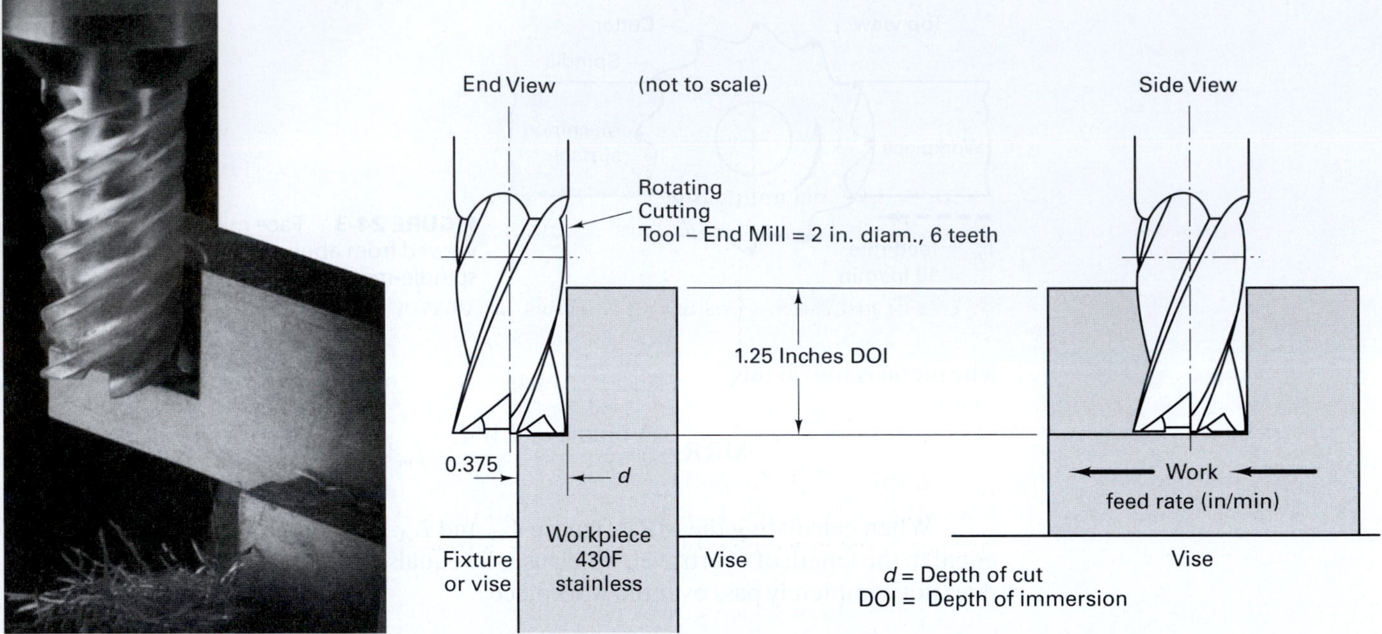

FIGURE 24-4 End milling a step feature in a block using a flat-bottomed, end mill cutter in a vertical spindle-milling machine. On left, photo. In middle, end view, table moving the block into the cutter. On right, side view, workpiece feeding right to left into tool. *(Courtesy J T. Black)*

(DOI) is 1.25 in. The tool deflects due to the cutting forces, so the cut needs to be made at full immersion. The engineer should check to see if there is enough power for a full DOC. Can the step be cut in one pass, or will multiple cuts be necessary? The vertical milling machine tool available has a 5-hp motor with 80% efficiency. The specific horsepower for 430F stainless is 1.3 hp/in.^{3}/min.

The maximum amount of material that can be removed per pass is usually limited by the available power. Using the horsepower equation from Chapter 20,

$$\text{hp} = \text{HP}_s \times \text{MRR} = \text{HP}_s \times f_m W D = \text{HP}_s f_m \times \text{DOI} \times d \tag{24-9}$$

Select $f_t = 0.005$ ipt and $V = 250$ fpm from Table 24-1. Calculate the spindle rpm:

$$N_s = \frac{12 \times 250}{3.14 \times 2} = 477 \text{ rpm of cutter}$$

Next, assuming the machine tool has this rpm available, calculate the table feed rate:

$$f_m = f_t \times n \times N_s = 0.005 \times 6 \times 477 = 14.31 \text{ in./min}$$

But the actual table feed rates for the selected machine are 11 in./min or 16 in./min, so, being conservative, select

$$f_m = \text{table feed rate} = 11.00 \text{ in./min}$$

Next, assuming 80% of the available power is used for cutting, calculate the depth of cut from equation 24-9:

$$d = \text{DOC} \cong \frac{5 \times 0.8}{1.3 \times 11.00 \times 1.25} \cong 0.225 \text{ in. maximum}$$

Therefore, two passes are needed because (0.375/0.225 = 1.6):

$$0.375 - 0.225 = 0.150 \text{ in. second pass DOC}$$

$$\begin{aligned} \text{2 passes: DOC} &= 0.225 \text{ rough cut} \\ \text{DOC} &= \underline{0.150} \text{ finish cut} \\ & 0.375 \text{ total DOC} \end{aligned}$$

Note that for $d = 0.150$, the feed per tooth would be only slightly increased to 0.0051 ipt:

$$f_t = \frac{0.5 \times 0.8}{1.3 \times 6 \times 477 \times 0.150 \times 1.25} = 0.0051 \text{ in./tooth}$$

You may want to change f_t to improve the surface finish. With a smaller feed per tooth, a better surface finish is usually obtained. However, there are other factors to consider, like machining time.

In general (for face, slab, or end milling), if machine power is lacking, the following actions may help.

1. Use a cutter with a positive rake as this can be more efficient than one with a negative rake.
2. Use a cutter with a coarser pitch (fewer teeth).
3. Use a smaller cutter and take several passes (reduce d or DOI).

UP VERSUS DOWN MILLING

One of the subtle aspects of milling concerns the direction of rotation of the cutter with respect to the movement of the workpiece. Surfaces can be generated by two distinctly different methods (Figure 24-5). **Up milling** is the traditional way to mill and is also called **conventional milling.** The cutter rotates against the direction of feed of the workpiece. In **climb** or **down milling,** the cutter rotation is in the same direction as the feed rate. The method of chip formation is completely different in the two cases.

In up milling, the chip is very thin at the beginning, where the tooth first contacts the work; then it increases in thickness, becoming a maximum where the tooth leaves the work. The cutter tends to push the work along and lift it upward from the table. This action tends to eliminate any effect of looseness in the feed screw and nut of the milling machine table and results in a smooth cut. However, the action tends to loosen the work from the fixture. Therefore, greater clamping forces must be employed, with the danger of deflecting the part. In addition, the smoothness of the generated surface depends greatly on the sharpness of the cutting edges. In up milling, chips can be carried into the newly machined surface, causing the surface finish to be poorer (rougher) than in down milling and causing damage to the insert.

In down milling, maximum chip thickness occurs close to the point at which the tooth contacts the work. Because the relative motion tends to pull the workpiece into the cutter, any possibility of looseness in the table feed screw must be eliminated if down milling is to be used. It should never be attempted on machines that are not

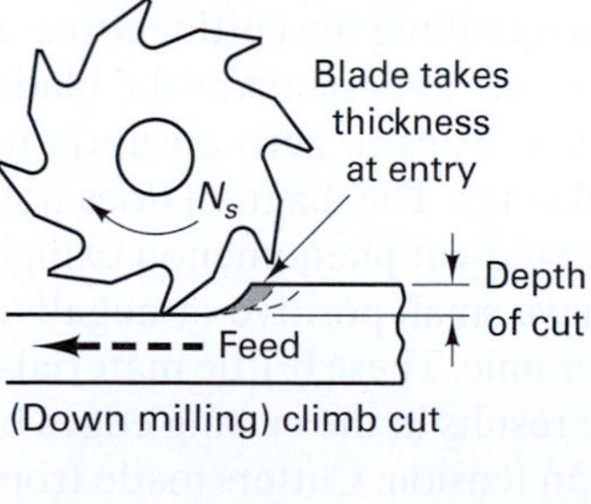

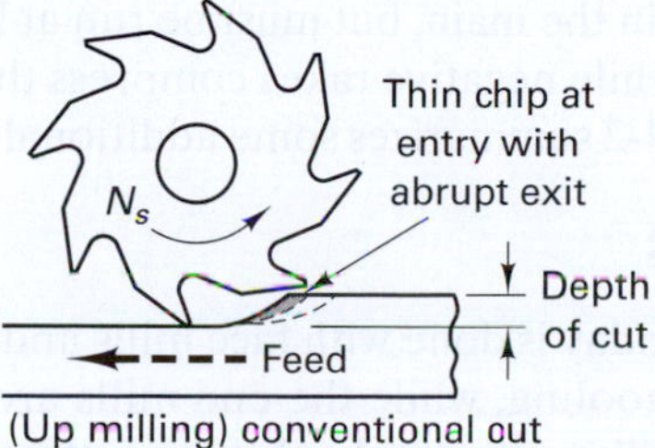

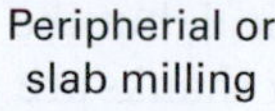

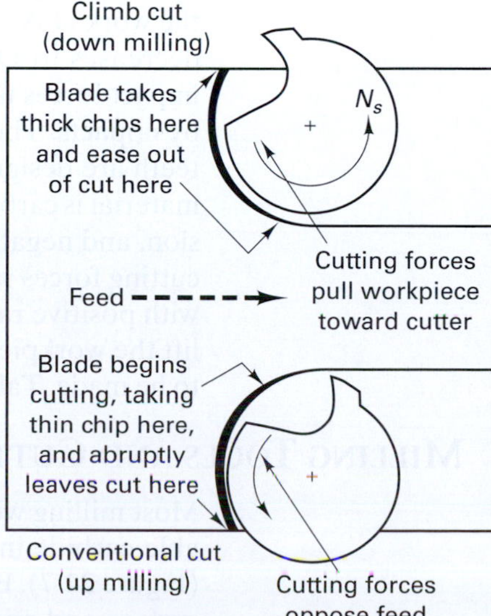

FIGURE 24-5 Climb cut or down milling versus conventional cut or up milling for slab or face or end milling.

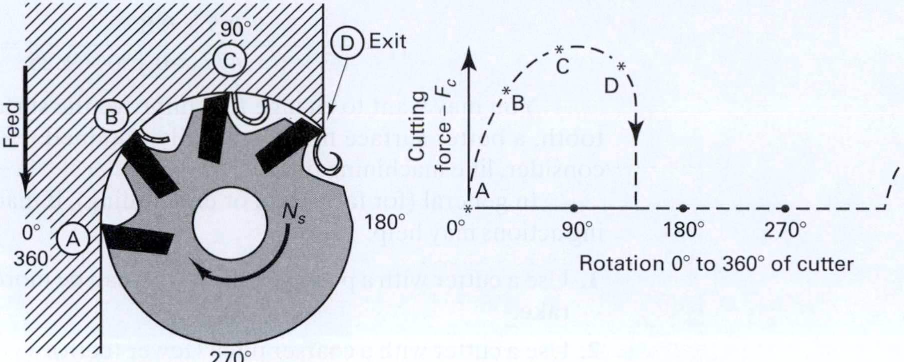

FIGURE 24-6 Conventional face milling (left) with cutting force diagram for F_c (right) showing the interrupted nature of the process. *(From* Metal Cutting Principles, *2nd ed., Ingersoll Cutting Tool Company)*

designed for this type of milling. Virtually all modern milling machines are capable of down milling, and it is a most favorable application for carbide cutting edges. Because the material yields in approximately a tangential direction at the end of the tooth engagement, there is less tendency (than when up milling is used) for the machined surface to show toothmarks, and the cutting process is smoother, with less chatter. Another advantage of down milling is that the cutting force tends to hold the work against the machine table, permitting lower clamping forces. However, the fact that the cutter teeth strike against the surface of the work at the beginning of each chip can be a disadvantage if the workpiece has a hard surface, as castings sometimes do. This may cause the teeth to dull rapidly. Metals that readily work-harden should be down milled, and many toolmakers recommend that down milling should always be the first choice.

MILLING SURFACE FINISH

The average surface finishes that can be expected on free-machining materials range from 60 to 150 μin. Conditions exist, however, that can produce wide variations on either side of these ranges. For example, some inserts are designed with wiper flats (short parallel surface behind the tool tip). If the feed per revolution (Feed per tooth $\times$ Number of teeth) of the cutter is smaller than the length of the wiper flat (the land on the tool), then the surface finish on the workpiece will be generated by the highest insert. In finishing cuts, keeping the depth of cut small will limit the axial cutting force, reducing vibrations and producing a superior finish. See Chapter 34 for discussions on measuring surface finish.

Milling is an interrupted cutting process. The individual teeth enter and leave the cut and subject the tool to impact loading, cyclic heating, and cycle cutting forces. As shown in Figure 24-6, in upmilling the cutting force, F_c, builds rapidly as the tool enters the work at A and progresses to B, peaks as the blade crosses the direction of feed at C, decreases to D, and then drops to zero abruptly upon exit. Down milling produces impact forces upon tool entry. The diagram does not indicate the impulse loads caused by impacts. The interrupted-cut phenomenon explains in large part why milling cutter teeth are designed to have small positive or negative rakes, particularly when the tool material is carbide or ceramic. These brittle materials tend to be very strong in compression, and negative rake results in the cutting edges being placed in compression by the cutting forces rather than tension. Cutters made from high-speed steel (HSS) are made with positive rakes, in the main, but must be run at lower speeds. Positive rake tends to lift the workpiece, while negative rakes compress the workpiece and allow heavier cuts to be made. Table 24-2 summarizes some additional milling problems.

■ 24.3 MILLING TOOLS AND CUTTERS

Most milling work today is done with face mills and end mills. The face mills use indexable carbide insert tooling, while the end mills are either solid HSS or insert tooling (Figure 24-7). Basically, *mills* are shank-type cutters having teeth on the circumferential surface and one end. They can thus be used for facing, profiling, and end milling. The teeth may be either straight or helical, but the latter is more common. Small end mills have straight shanks, whereas taper shanks are used on larger sizes (Figure 24-8).

TABLE 24-2 Probable Causes of Milling Problems

Problem	Probable Cause	Cures
Chatter (vibration)	**1.** Lack of rigidity in machine, fixtures, arbor, or workpiece **2.** Cutting load too great **3.** Dull cutter **4.** Poor lubrication **5.** Straight-tooth cutter **6.** Radial relief too great **7.** Rubbing, insufficient clearance	Use larger arbors. Change rpm (cutting speed). Decrease feed per tooth or number of teeth in contact with work. Sharpen or replace inserts. Flood coolant. Use helical cutter. Check tool angles.
Loss of accuracy (cannot hold size)	**1.** High cutting load causing deflection **2.** Chip packing, between teeth **3.** Chips not cleaned away before mounting new piece of work	Decrease number of teeth in contact with work or feed per tooth. Adjust cutting fluid to wash chips out of teeth.
Cutter rapidly dulls	**1.** Cutting load too great **2.** Insufficient coolant	Decrease feed per tooth or number of teeth in contact. Add blending oil to coolant.
Poor surface finish	**1.** Feed too high **2.** Tool dull **3.** Speed too low **4.** Not enough cutter teeth	Check to see if all teeth are set at same height.
Cutter digs in (hogs into work)	**1.** Radial relief too great **2.** Rake angle too large **3.** Improper speed	Check to see that workpiece is not deflecting and is securely clamped.
Work bumishing	**1.** Cut is too light **2.** Tool edge worn **3.** Insufficient radial relief **4.** Land too wide	Enlarge feed per tooth. Sharpen cutter.
Cutter burns	**1.** Not enough lubricant **2.** Speed too high	Add sulfur-based oil. Reduce cutting speed. Flood coolant.
Teeth breaking	**1.** Feed too high **2.** Depth of cut too large	Decrease feed per tooth. Use cutter with more teeth. Reduce table feed rate.

Adapted from *Cutting Tool Engineering*, October 1990, p. 90, by Peter Liebhold, museum specialist, Division of Engineering and Industry, the Smithsonian Institute, Washington, DC.

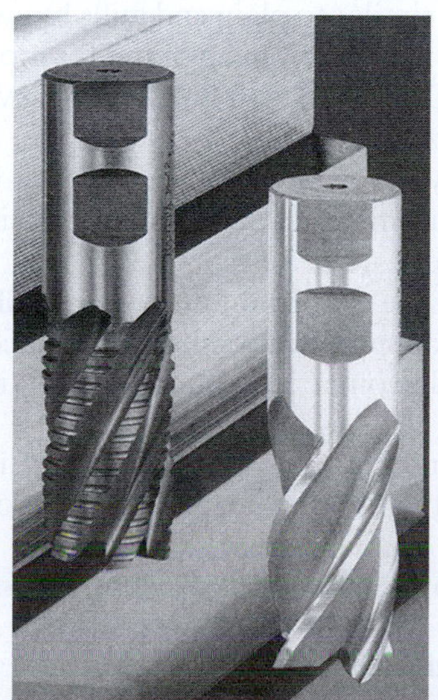

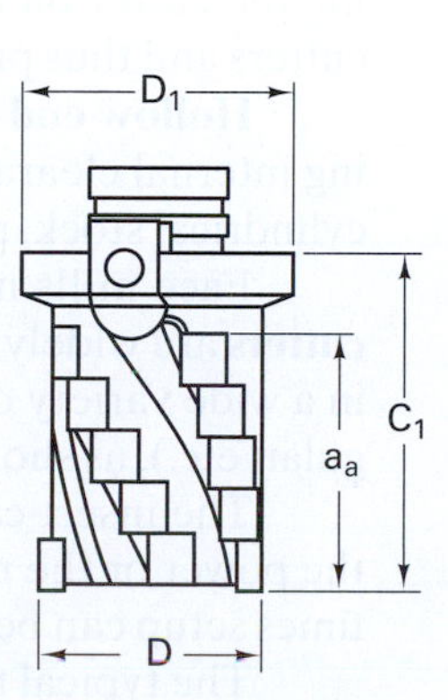

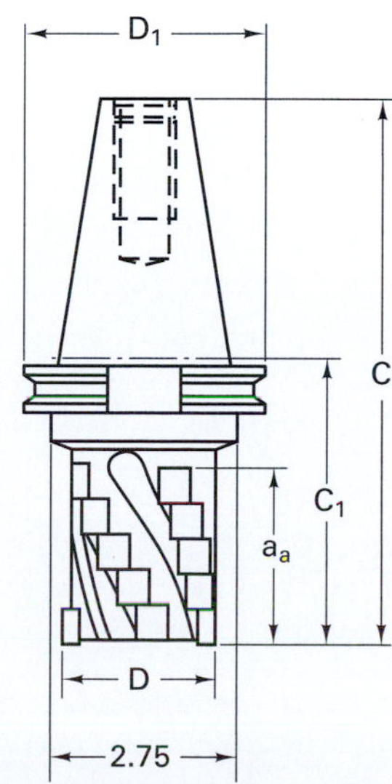

FIGURE 24-7 Solid end mills are often coated. Insert tooling end mills come in a variety of sizes and are mounted on taper shanks. (*Courtesy J T. Black*)

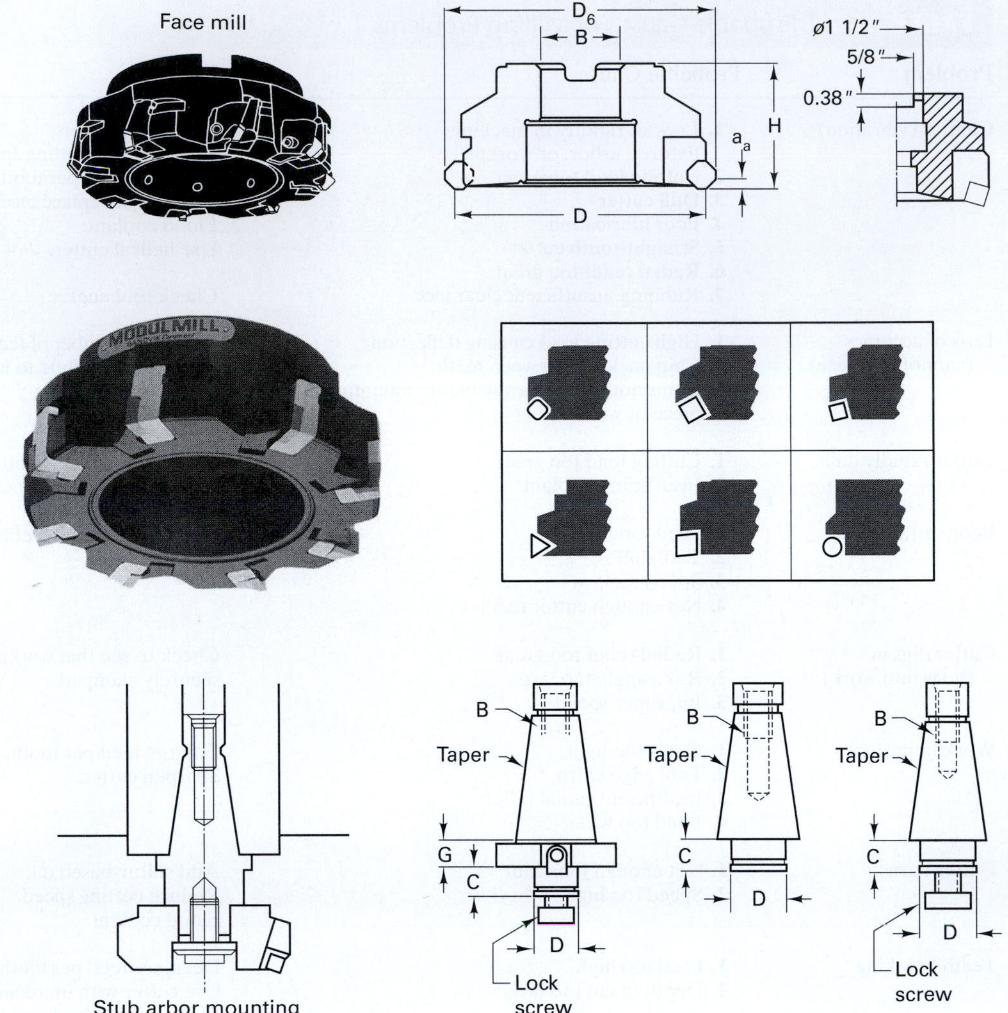

FIGURE 24-8 Face mills come in many different designs using many different insert geometries and different mounting arbors.

Plain end mills have multiple teeth that extend only about halfway toward the center on the end. They are used in milling slots, profiling, and facing narrow surfaces. **Two-lip mills** have two straight or helical teeth that extend to the center. Thus, they may be sunk into material, like a drill, and then fed lengthwise to form a groove, a slot, or a pocket.

Shell end mills are solid multiple-tooth cutters, similar to plain end mills but without a shank. The center of the face is recessed to receive a screw head or nut for mounting the cutter on a separate shank or a stub arbor. One shank can hold any of several cutters and thus provides great economy for larger-sized end mills.

Hollow end mills are tubular in cross section, with teeth only on the end but having internal clearance. They are used primarily on automatic screw machines for sizing cylindrical stock, producing a short cylindrical surface of accurate diameter.

Face mills have a center hole so that they can be arbor mounted. **Face-milling cutters** are widely used in both horizontal- and vertical-spindle machine tools and come in a wide variety of sizes (diameters and heights) and geometries (round, square, triangular, etc.), as shown in Figure 24-8.

The insert can usually be indexed four times and must be well supported. Either the power or the rigidity of the machine tool will be the limiting factor, although sometimes setup can be the limiting factor.

The typical tooth geometry for a slab mill is shown in Figure 24-9. Here, a positive rake angle is shown.

Another common type of **arbor-mounted milling cutter** is called a *side mill* because it cuts on the ends and sides of the cutters. Figure 24-10 shows the geometry of a staggered-tooth side-milling cutter.

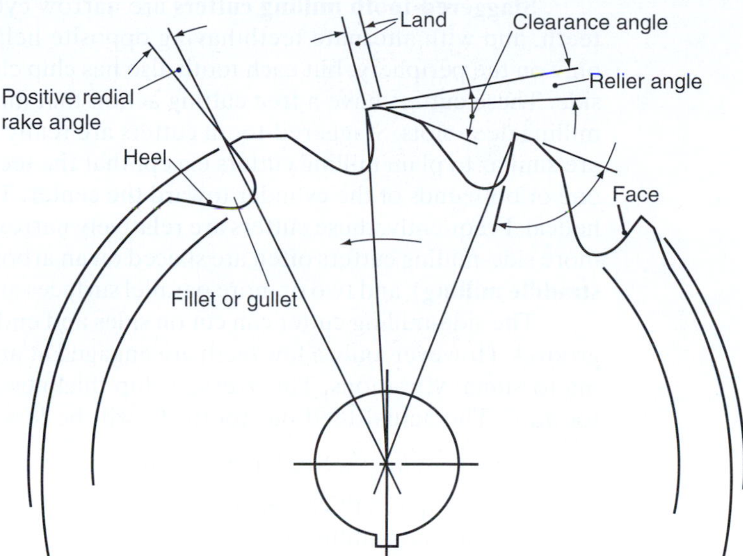

FIGURE 24-9 The geometry of a slab milling cutter showing the main cutting angles.

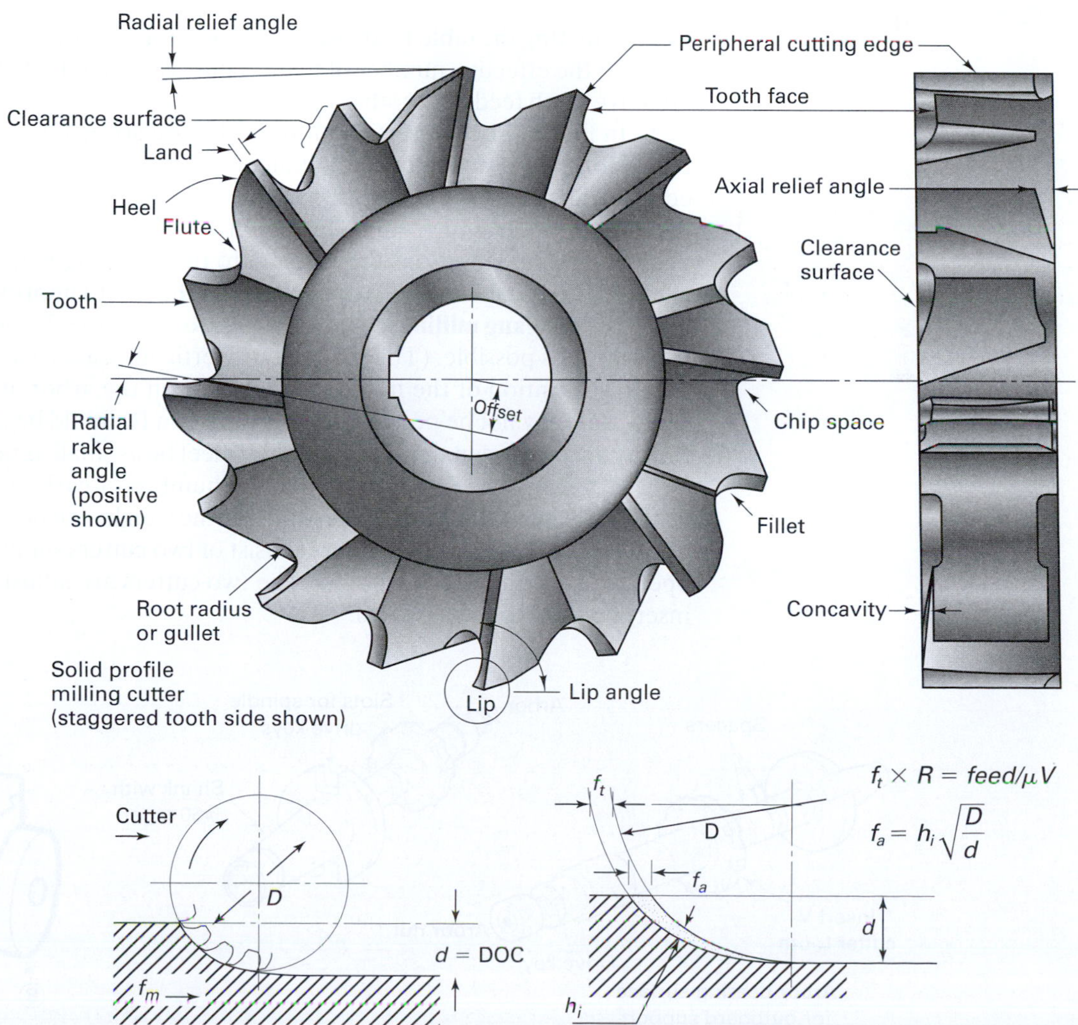

FIGURE 24-10 The staggered tooth side mill can cut on both the ends and sides of the cutter so it can cut slots or grooves.

Staggered-tooth milling cutters are narrow cylindrical cutters having staggered teeth, and with alternate teeth having opposite helix angles. They are ground to cut only on the periphery, but each tooth also has chip clearance ground on the protruding side. These cutters have a free cutting action that makes them particularly effective in milling deep slots. Staggered-tooth cutters are really special **side-milling cutters,** which are similar to plain milling cutters except that the teeth extend radially part way across one or both ends of the cylinder toward the center. The teeth may be either straight or helical. Frequently, these cutters are relatively narrow, being disk-like in shape. Two or more side-milling cutters often are spaced on an arbor to straddle the workpiece (called **straddle milling**), and two or more parallel surfaces are machined at once.

The side-milling cutter can cut on sides and ends of the teeth, so it can cut slots or grooves. However, only a few teeth are engaged at any one point in time, causing heaving torsional vibrations. The average chip thickness, h_i, will be less than the feed per tooth, f_t. The actual feed per tooth, f_a, will be less than feed per tooth selected, F_t, according to $f_a = h_i \sqrt{\frac{D}{d}}$. See Figure 24-10.

For example, a thickness (h_i) of 0.004 in. corresponds to 0.012 in. feed per tooth in most side- and face-milling operations.

If the radial depth of cut, d, is very small compared to the cutter diameter, D, use this formula:

$$\text{Feed per tooth} = ft = 0.004\sqrt{\frac{\text{D}}{\text{d}}}\,(\text{ipt})$$

For calculating the table feed, use half the number of inserts in a full side and face mill to arrive at the effective number of teeth. Thus, Table feed rate (ipm) = rpm × Number of effective teeth feed per tooth.

In Figure 24-11, insert-tooth side mills are arranged in a gang-milling setup to cut three slots in the workpiece simultaneously. Thus, the desired part geometry is repeatedly produced by the setup as the position of the cutters is fixed. However, in side- and face-milling operations, only a few teeth are engaged at any point in time, resulting in heavy torsional vibrations detrimental to the resulting machined product. A flywheel can solve this problem and in many cases be the key to improved productivity.

For the **gang milling** as shown in Figure 24-11, the diameter of the flywheel should be as large as possible. (The moment of inertia increases with the square of the radius.) The best position of the flywheel is inboard on the arbor at A, but depending on the setup, this may not be possible, so then position B should be chosen. It is important that the distance between the cutters and flywheel be as small as possible.

A flywheel can be built up from a number of carbon steel disks, each having a center hole and keyway to fit the arbor, so the weight can be easily varied.

Interlocking slotting cutters consist of two cutters similar to side mills but made to operate as a unit for milling slots. The two cutters are adjusted to the desired width by inserting shims between them.

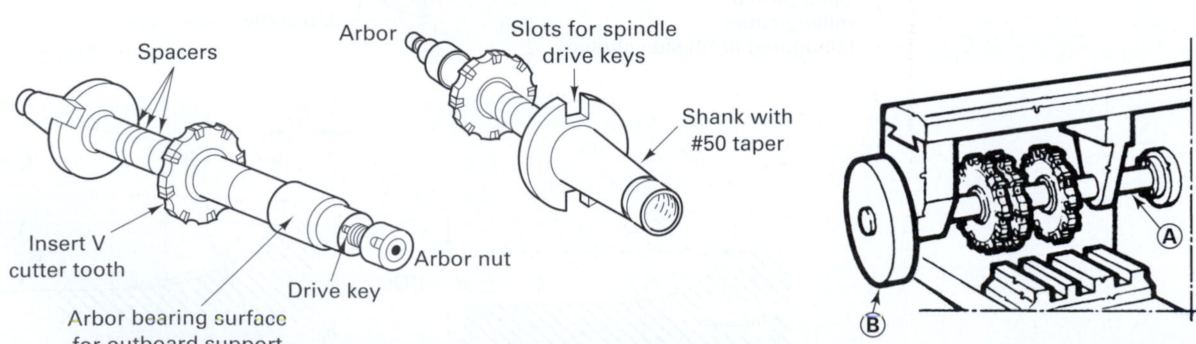

FIGURE 24-11 Arbor (two views) used on a horizontal-spindle milling machine on left. On right, a gang-milling setup showing three side-milling cutters mounted on an arbor (A) with an outboard flywheel (B).

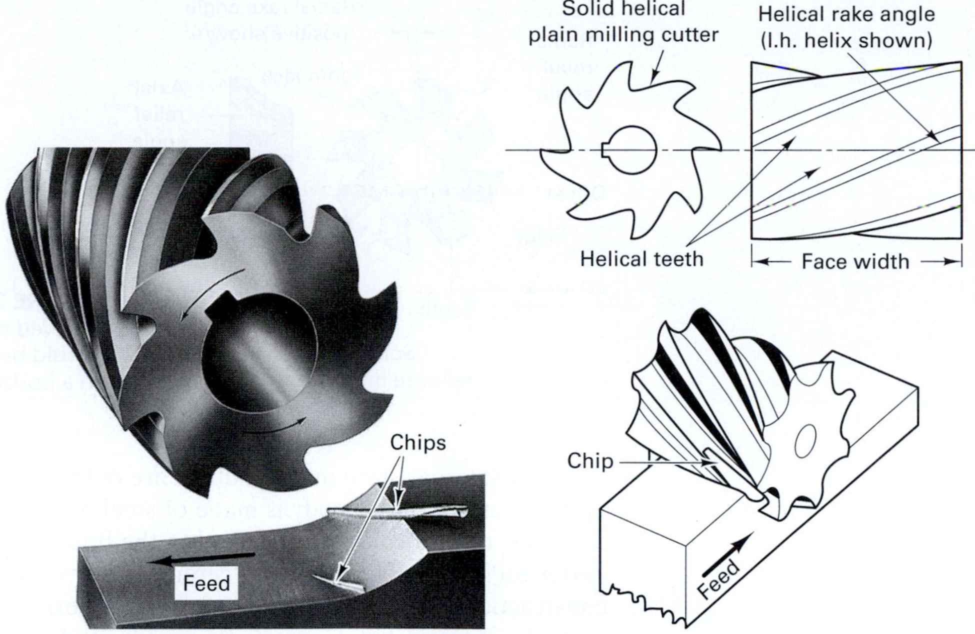

FIGURE 24-12 The chips are formed progressively by the teeth of a plain helical-tooth milling cutter during up milling.

Slitting saws are thin, plain milling cutters, usually from $\frac{1}{32}$ to $\frac{3}{16}$ in. thick, which have their sides slightly "dished" to provide clearance and prevent binding. They usually have more teeth per unit of diameter than ordinary plain milling cutters and are used for milling deep narrow slots and cutting-off operations.

Another method of classification for face and end mill cutters relates to the direction of rotation. A **right-hand cutter** must rotate counterclockwise when viewed from the front end of the machine spindle. Similarly, a **left-hand cutter** must rotate clockwise. All other cutters can be reversed on the arbor to change them from one hand to the other. Positive rake angles are used on general-purpose HSS milling cutters. Negative rake angles are commonly used on carbide- and ceramic-tipped cutters employed in mass-production milling in order to obtain the greater strength and cooling capacity. TiN coating of these tools is quite common, resulting in significant increases in tool life.

Plain milling cutters used for plain or slab milling have straight or helical teeth on the periphery and are used for milling flat surfaces. **Helical mills** (Figure 24-12) engage the work gradually, and usually more than one tooth cuts at a given time. This reduces shock and chattering tendencies and promotes a smoother surface. Consequently, this type of cutter usually is preferred over one with straight teeth.

Angle milling cutters are made in two types: single angle and double angle. Angle cutters are used for milling slots of various angles or for milling the edges of workpieces to a desired angle. *Single-angle cutters* have teeth on the conical surface, usually at an angle of 45 to 60 degrees to the plane face. *Double-angle cutters* have V-shaped teeth, with both conical surfaces at an angle to the end faces but not necessarily at the same angle. The V-angle usually is 45, 60, or 90 degrees.

Form milling cutters have the teeth ground to a special shape—usually an irregular contour—to produce a surface having a desired transverse contour. They must be sharpened by grinding only the tooth face, thereby retaining the original contour as long as the plane of the face remains unchanged with respect to the axis of rotation. Convex, concave, corner-rounding, and gear-tooth cutters are common examples (Figure 24-13). Solid HSS cutters of simple shape and reasonably small size are usually more economical in initial cost than inserted-blade cutters. However, inserted-blade cutters may be lowest in overall cost on large production jobs.

Form-relieved cutters can be cost effective where intricately shaped cuts are needed. Solid or carbide insert tool cutters may need large volumes to be cost-justified by high-production requirements.

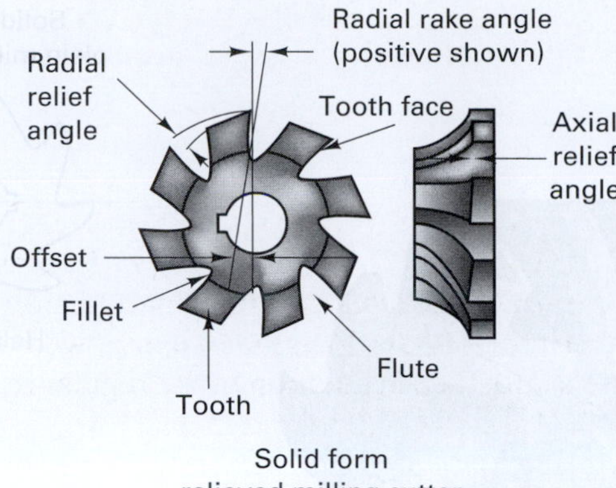

FIGURE 24-13 Solid form relieved milling cutter, which would be mounted on an arbor in a horizontal milling machine.

Most larger-sized milling cutters are of the insert-tooth type. In such **insert-tooth milling cutters,** cutter body is made of steel, with the teeth made of high-speed steel, carbides, or TiN carbides, fastened to the body by various methods. An insert-tooth cutter uses indexable carbide or ceramic inserts, as shown in Figure 24-9. This type of construction reduces the amount of costly material that is required and can be used for any type of cutter, but it is most often used with face mills.

T-slot cutters are integral-shank cutters with teeth on the periphery and *both* sides. They are used for milling the wide groove of a T-slot. To use them, the vertical groove must first be made with a slotting mill or an end mill to provide clearance for the shank. Because the T-slot cutter cuts on five surfaces simultaneously, it must be fed with care.

Woodruff keyseat cutters are made for the single purpose of milling the semi-cylindrical seats required in shafts for Woodruff keys. They come in standard sizes corresponding to Woodruff key sizes. Those less than 2 in. in diameter have integral shanks; the larger sizes may be arbor mounted.

Occasionally, **fly cutters** may be used for face milling or boring. Both operations may be done with a single tool at one setup. A single-point cutting tool is attached to a special shank, usually with provision for adjusting the effective radius of the cutting tool with respect to the axis of rotation. The cutting edge can be made in any desired shape and, because it is a single-point tool, is very easy to grind.

24.4 MACHINES FOR MILLING

The four most common types of manually controlled milling machines are listed here in order of increasing power (and therefore metal removal capability):

1. Ram-type milling machines.
2. Column-and-knee-type milling machines:
   a. Horizontal spindle.
   b. Vertical spindle.
3. Fixed-bed-type milling machines.
4. Planer-type milling machines.

Milling machines whose motions are electronically controlled are listed in order of increasing production capacity and decreasing flexibility:

1. Manual data input milling machines.
2. Programmable CNC milling machines.
3. Machining centers (tool changer and pallet exchange capability).
4. Flexible manufacturing cell and flexible manufacturing system.
5. Transfer lines.

BASIC MILLING MACHINE CONSTRUCTION

Most basic milling machines are of **column-and-knee construction,** employing the components and motions shown in Figure 24-14. The column, mounted on the base, is the main supporting frame for all the other parts and contains the spindle with its driving mechanism. This construction provides controlled motion of the worktable in three mutually perpendicular directions: (1) through the *knee*, moving vertically on ways on the front of the column; (2) through the *saddle*, moving transversely on ways on the knee; and (3) through the *table*, moving longitudinally on ways on the saddle. All these motions can be imparted by either manual or powered means. In most cases, a powered rapid traverse is provided in addition to the regular feed rates for use in setting up work and in returning the table at the end of a cut.

The **ram-type milling machine** is one of the most versatile and popular milling machines, using the knee-and-column design. Ram-type machines have a head equipped with a motor-stopped pulley and belt drive as well as a spindle. The ram, mounted on horizontal ways at the top of the column, supports the head and permits positioning of the spindle with respect to the table. Ram-type milling machines are normally 10 hp or less and are suitable for light-duty milling, drilling, reaming, and so on (Figure 24-14).

Milling machines having only the three mutually perpendicular table motions are called **plain column-and-knee milling machines.** These are available with both horizontal and vertical spindles (Figure 24-14). On the older, horizontal-spindle-type machines, an adjustable overarm provides an outboard bearing support for the end of the cutter arbor, which is shown in Figure 24-11 and 24-14. These machines are well suited for slab, side, or straddle milling.

In some vertical-spindle machines, the spindle can be fed up and down, either by power or by hand. Vertical-spindle machines are especially well suited for face and end milling operations. They also are very useful for drilling and boring, particularly where holes must be spaced accurately in a horizontal plane, because of the controlled table motion.

Turret-type column-and-knee milling machines have dual heads that can be swiveled about a horizontal axis on the end of a horizontally adjustable ram. This permits milling to be done horizontally, vertically, or at any angle. This added flexibility is advantageous when a variety of work has to be done, as in tool and die or experimental shops. They are available with either plain or universal tables.

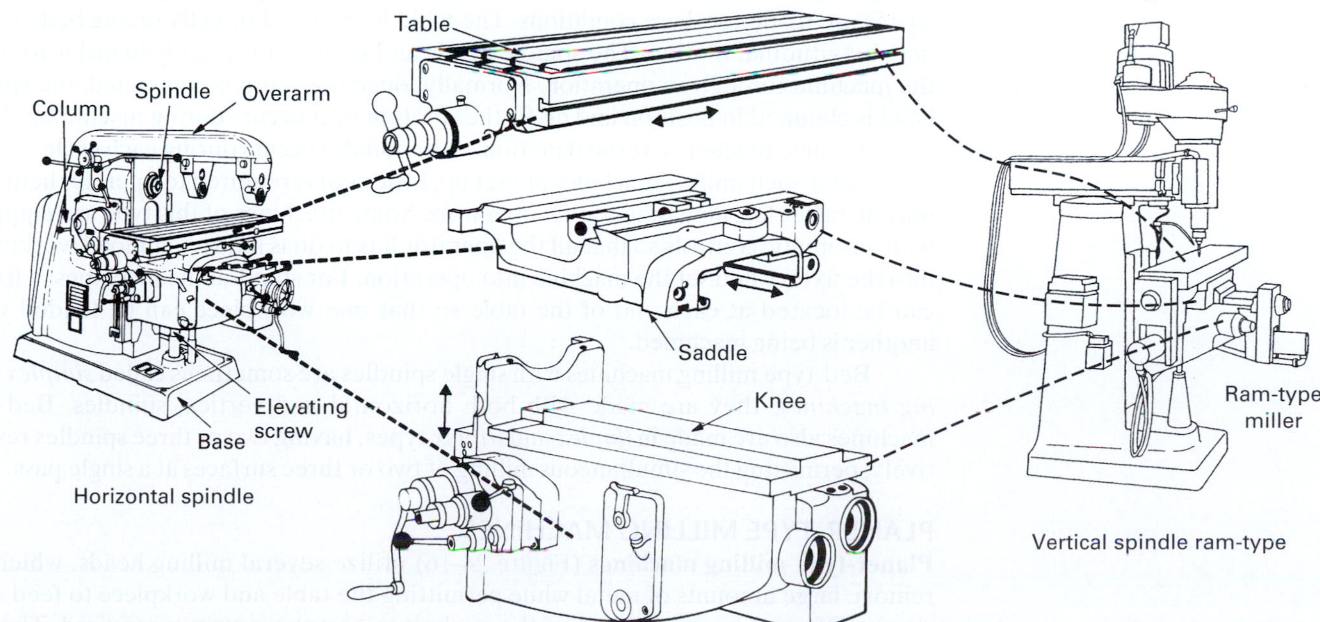

FIGURE 24-14 Major components of a plain column-and-knee-type milling machine, which can have horizontal spindle (shown on the left) or a turret type machine with a vertical spindle (shown on the right). The workpiece and workholder on the table can be translated in *X, Y,* and *Z* directions with respect to the tool.

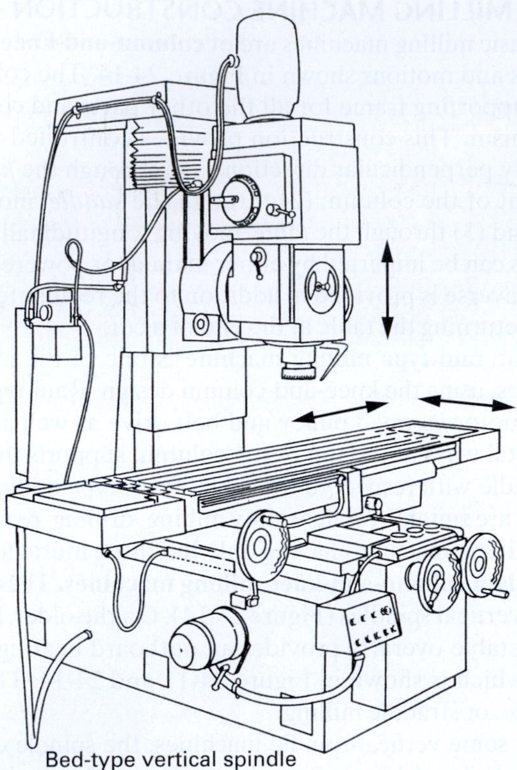

FIGURE 24-15 Bed-type vertical-spindle heavy-duty production machine tools for milling usually have three axes of motion.

Universal column-and-knee milling machines differ from plain column-and-knee machines in that the table is mounted on a housing that can be swiveled in a horizontal plane, thereby increasing its flexibility. Helices, as found in twist drills, milling cutters, and helical gear teeth, can be milled on universal machines.

BED-TYPE MILLING MACHINES

In production manufacturing operations, ruggedness and the capability of making heavy cuts are of more importance than versatility. **Bed-type milling machines** (Figure 24-15) are made for these conditions. The table is mounted directly on the bed and has only longitudinal motion. The spindle head can be moved vertically in order to set up the machine for a given operation. Normally, once the setup is completed, the spindle head is clamped in position and no further motion of it occurs during machining. However, on some machines, vertical motion of the spindle occurs during each cycle.

After such milling machines are set up, little skill is required to operate them, permitting faster learning time for the operators. Some machines of this type are equipped with automatic controls so that all the operator has to do is load and unload workpieces into the fixture and set the machine into operation. For stand-alone machines, a fixture can be located at each end of the table so that one workpiece can be loaded while another is being machined.

Bed-type milling machines with single spindles are sometimes called *simplex milling machines;* they are made with both horizontal and vertical spindles. Bed-type machines also are made in *duplex* and *triplex* types, having two or three spindles respectively, permitting the simultaneous milling of two or three surfaces at a single pass.

PLANER-TYPE MILLING MACHINES

Planer-type milling machines (Figure 24-16) utilize several milling heads, which can remove large amounts of metal while permitting the table and workpiece to feed quite slowly. Often only a single pass of the workpiece past the cutters is required. Through the use of different types of milling heads and cutters, a wide variety of surfaces can be machined with a single setup of the workpiece. This is an advantage when heavy workpieces are involved.

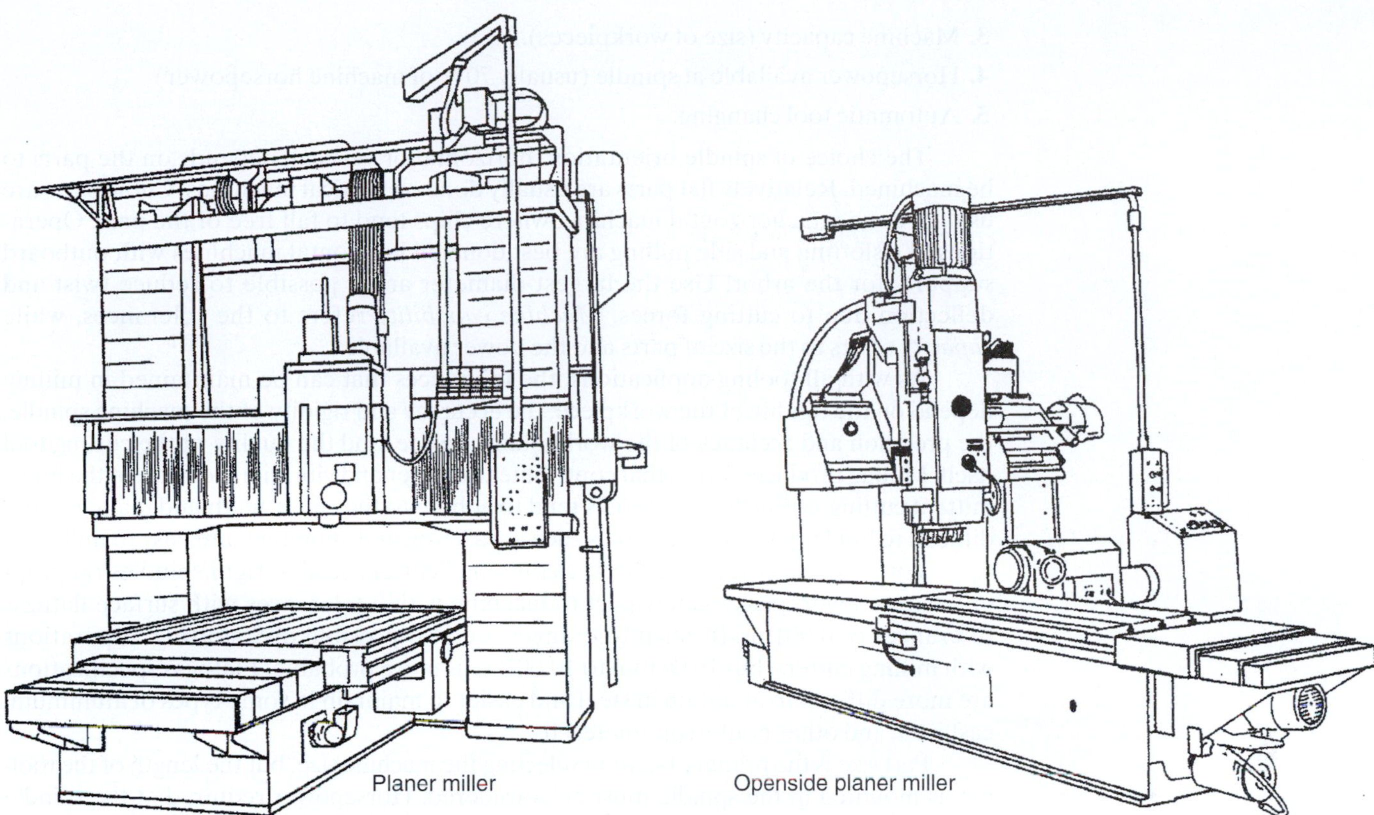

FIGURE 24-16 Large planer-type milling machines shares its basic design with that of a planer with the planing tool replaced by a powered milling head (not shown). (American Machinist Special Report to Fundamentals of Milling, *J Jablonowski, Feb, 1978*)

ROTARY-TABLE MILLING MACHINES

Some types of face milling in mass-production manufacturing are often done on **rotary-table milling machines.** Roughing and finishing cuts can be made in succession as the workpieces are moved past the several milling cutters while held in fixtures on the rotating table. The operator can load and unload the work without stopping the machine.

PROFILERS AND DUPLICATORS

Milling machines that can duplicate external or internal geometries in two dimensions are called **profilers** or tracer-controlled machines. A tracing probe follows a two-dimensional pattern or template and, through electronic or hydraulic air-actuated mechanisms, controls the cutting spindles in two mutually perpendicular directions.

All hydraulic tracers work basically the same way, in that they utilize a stylus connected to a precision servomechanism for each axis of control. The servos are connected to hydraulic actuators on the machine slides. As the stylus traces a template, the servos control the motion of the slides so that the milling cutter duplicates the template shape onto the workpiece.

Duplicators produce forms in three dimensions and are widely used to machine molds and dies. Sometimes these machines are called **die-sinking machines.** They are used extensively in the aerospace industry to machine parts from wrought plate or bar stock as substitutes for forgings when the small number of parts required would make the cost of forging dies uneconomical. Many of these kinds of jobs are now done on NC and CNC-type machines.

MILLING MACHINE SELECTION

When purchasing or using a milling machine, consider the following issues:

1. Spindle orientation and rpm.
2. Machine capability (accuracy and precision).

3. Machine capacity (size of workpieces).
4. Horsepower available at spindle (usually 70% of machine horsepower).
5. Automatic tool changing.

The choice of spindle orientation, horizontal or vertical, depends on the parts to be machined. Relatively flat parts are usually done on vertical machines. Cubic parts are usually done on a horizontal machine, where chips tend to fall free of the part. Operations like slotting and side milling are best done on horizontal machines with outboard supports for the arbor. Use the largest-diameter arbor possible to reduce twist and deflection due to cutting forces. *Machine capability* refers to the tolerances, while *capacity* refers to the size of parts and the power available.

As with all tooling applications, the tolerances that can be maintained in milling depend on the rigidity of the workpiece, the accuracy and rigidity of the machine spindle, the precision and accuracy of the workholding device, and the quality of the cutting tool itself. Milling produces forces that contribute to chatter and vibration because of the intermittent cutting action. Soft materials tend to adhere to the cutter teeth and make it more difficult to hold tolerances. Materials such as cast iron and aluminum are easy to mill.

Within these criteria, properly maintained cutters used in rigid spindles on properly fixtured workpieces can expect to machine within tolerances with surface flatness tolerances of 0.001 in./ft. Such tolerances are also possible on "slotting" operations with milling cutters, but 10.001 in. to 10.002 in. is more probable. Flatness specifications are more difficult to maintain in steel and easier to maintain in some types of aluminum, cast iron, and other nonferrous material.

Part size is the primary factor in selecting the machine size, but the length of the tooling as mounted in the spindle must be considered. Horsepower required at the spindle depends on the MRR and the materials (unit horsepower, HP_s). Remember, coated inserts allow the MRR (the cutting speed) to be increased and available power may be exceeded.

Finally, the capacity of the tool changers on machining centers is limited by the number, size, and weight of the tools—especially if large-diameter tools are being employed. These often have to be stored in every other space in the storage mechanism.

ACCESSORIES FOR MILLING MACHINES

The usefulness of ordinary milling machines can be greatly extended by employing various accessories or attachments. Here are some examples.

A horizontal milling machine can be equipped with a vertical milling attachment to permit vertical milling to be done. Ordinarily, heavy cuts cannot be made with such an attachment.

The **universal milling attachment** (Figure 24-17) is similar to the vertical attachment but can be swiveled about both the axis of the milling machine spindle and a second, perpendicular axis to permit milling to be done at any angle.

The **universal dividing head** is by far the most widely used milling machine accessory, providing a means for holding and indexing work through any desired arc of

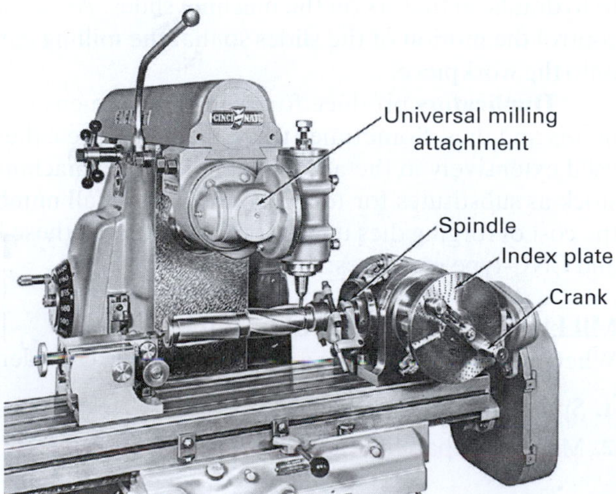

FIGURE 24-17 End milling a helical groove on a horizontal-spindle milling machine using a universal dividing head and a universal milling attachment. *(Courtesy of Cincinnati Milacron, Inc., Cincinnati, OH)*

rotation. The work may be mounted between centers (Figure 24-17) or held in a chuck that is mounted in the spindle hole of the dividing head. The spindle can be tiled from about 5 degrees below horizontal to beyond the vertical position.

Basically, a dividing head is a rugged, accurate, 40:1 worm-gear reduction unit. The spindle of the dividing head is rotated one revolution by turning the input crank 40 turns. An index plate mounted beneath the crank contains a number of holes, arranged in concentric circles and equally spaced, with each circle having a different number of holes. A plunger pin on the crank handle can be adjusted to engage the holes of any circle. This permits the crank to be turned an accurate, fractional part of a complete circle as represented by the increment between any two holes of a given circle on the index plate. Utilizing the 40:1 gear ratio and the proper hole circle on the index plate, the spindle can be rotated a precise amount by the application of either of the following rules:

$$\text{Number of turns of crank} = \frac{40}{\text{Cuts per revolution of work}}$$

$$\text{Holes to be indexed} = \frac{40 \times \text{Holes index circle}}{\text{Cuts per revolution of work}}$$

If the first rule is used, an index circle must be selected that has the proper number of holes to be divisible by the denominator of any resulting fractional portion of a turn of the crank. In using the second rule, the number of holes in the index circle must be such that the numerator of the fraction is an even multiple of the denominator. For example, if 24 cuts are to be taken about the circumference of a workpiece, the number of turns of the crank required would be $1\frac{2}{3}$. An index circle having 12 holes could be used with one full turn plus eight additional holes. The second rule would give the same result. Adjustable **sector arms** are provided on the index plate that can be set to a desired number of holes, less than a full turn, so that fractional turns can be made readily without the necessity for counting holes each time. Dividing heads are made having ratios other than 40:1.The ratio should be checked before using.

Because each full turn of the crank on a standard dividing head represents 360/40, or 9 degrees of rotation of the spindle, indexing to a fraction of a degree can be obtained. Indexing can be done in three ways. **Plain indexing** is done solely by the use of the 40:1 ratio in the dividing head. In **compound indexing,** the index plate is moved forward or backward a number of hole spaces each time the crank handle is advanced. For **differential indexing,** the spindle and the index plate are connected by suitable gearing so that as the spindle is turned by means of the crank, the index plate is rotated a proportional amount.

The dividing head can also be connected to the feed screw of the milling machine table by means of gearing. This procedure is used to provide a definite rotation of the workpiece with respect to the longitudinal movement of the table, as in cutting helical gears.

■ Key Words

angle milling cutter
arbor-mounted milling cutter
bed-type milling machine
climb (down) milling
column-and-knee milling machine
compound indexing
conventional (up) milling
cutting time
die-sinking machine
differential indexing
down (climb) milling
duplicator
end milling
face milling
face milling cutter
fly cutter
form milling cutter
gang milling
helical mill
hollow end mill
interlocking slotting cutter
insert-tooth milling cutter
interrupted cutting
left-hand cutter
machining center
metal removal rate (MRR)
milling
milling cutters
milling machines
peripheral milling
plain column-and-knee milling machine
plain end mill
plain indexing
plain milling cutter
planer-type milling machine
profiler
ram-type milling machine
right-hand cutter
rotary-table milling machine
sector arm
shell end mill
side-milling cutter
slab milling
slitting saw
staggered-tooth milling cutter
straddle milling
T-slot cutter
turret-type column-and-knee milling machine
two-lip mill
universal column-and-knee milling machine
universal dividing head
universal milling attachment
up (conventional) milling
Woodruff keyseat cutter

REVIEW QUESTIONS

1. Suppose you wanted to machine a cast iron with BHN of 275. The process to be used is face milling and an HSS cutter is going to be used. What feed and speed values would you select?
2. Calculate the spindle rpm and table feed (ipm) for a face-milling machine after the speed and feed per tooth are selected. Use Figure 24-3 for reference.
3. Why must the number of teeth on the cutter be known when calculating milling machine table feed, in in./min?
4. Why is the question of up or down milling more critical in horizontal slab milling than in vertical-spindle (end or face) milling?
5. For producing flat surfaces in mass-production machining, how does face milling basically differ from peripheral milling?
6. Milling has a higher metal removal rate than planing. Why?
7. Which type of milling (up or down) is being done in Figures 24-1b, 24-1d, and 24-2b?
8. Why does down milling dull the cutter more rapidly than up milling when machining sand castings?
9. What parameters do you need to specify in order to calculate MRR in milling?
10. In Figure 24-2b, the tool material is carbide. What would you change in the process?
11. What is the advantage of a helical-tooth cutter over a straight-tooth cutter for slab milling?
12. What would the cutting force diagram for F_c look like if the cutter were performing climb milling?
13. Could the stub arbor–mounted face mill shown in Figure 24-9 be used to machine a T-slot? Why or why not?
14. In a typical solid arbor milling cutter shown in Figure 24-10, why are the teeth staggered? (Review the discussion of dynamics in Chapter 20.)
15. Make some sketches to show how you would you set up a plain column-and-knee milling machine to make it suitable for milling the top and sides of a large block.
16. Make some sketches to show how you would set up a horizontal milling machine to cut both sides of a block of metal simultaneously.
17. Explain how controlled movements of the work in three mutually perpendicular directions are obtained in column-and-knee-type milling machines.
18. What is the basic principle of a universal dividing head?
19. What is the purpose of the hole-circle plate on a universal dividing head?

PROBLEMS

1. You have selected a feed per tooth and a cutting speed for a face milling process, using Table 24-1. Reasonable values for feed and speed are 0.010 in. per tooth and 200 sfpm. The cutter is 8 in. in diameter, as shown in Figure 24-9. Compute the input values for the machine tool.
2. How much time will be required for a milling machine to face mill an AISI 1020 steel surface BHN 150, that is 12 in. long and 5 in. wide, using a 6-in.-diameter, eight-tooth tungsten carbide inserted-tooth face mill cutter? Select values of feed per tooth and cutting speed from Table 24-1.
3. If the depth of cut is 0.35 in., what is the metal removal rate in Problem 2?
4. Estimate the power required for the operation of Problem 3. Do not forget to consider Figure 24-7.
5. Examine the pinion shown in Chapter 1. The slot on the left end must be produced by machining. Provide a process plan (a description [sketch] of how the part would set up in the machine for machining the slot and the details regarding cutting tools, such as material, sizes, and so on). Specify (select) the type of milling machine, the cutting parameters, and any other information needed to make this component.
6. A gray cast iron surface 6 in. wide and 18 in. long may be machined on either a vertical milling machine, using an 8-in.-diameter face mill having eight inserted HSS teeth, or on a horizontal milling machine using an HSS slab mill with eight teeth on a 4-in. diameter. Which machine has the faster cutting time?
7. An operation is to be performed to machine three grooves on a number of parts shown in Figure 24-11. Setup time is 40 min on a shaper (not shown) and 30 min on the horizontal milling machine. The direct time to machine each piece on the shaper is 14 min and on the miller is 6 min. Labor costs $10/hr. The charge for the use of the shaper is $10/hr and for the milling machine $20/hr. What is the breakeven quantity, below which the shaper is more economical than the mill?
8. In Figure 24-12, the feed is 0.006 in. per tooth. The cutter is rotating at an rpm that will produce the desired surface cutting speed of 125 sfpm. The cutter diameter is 3 in. The depth of cut is 0.5 in. The block is 2 in. wide.
   a. What is the feed rate, in inches per minute, of the milling machine table?
   b. What is the MRR for this situation?
   c. What is horsepower (HP) consumed by this process, assuming an 80% efficiency and a HP_s value for this material of 1.8?
9. Suppose you want to do the job described in Problem 6 by slab milling. You have selected a 6-in.-diameter cutter with eight TiN-coated carbide teeth. The cutting speed will be 500 sfpm and the feed per tooth will be 0.010 in. per tooth. Determine the input parameters for the machine (rpm of arbor and table feed), then calculate the T_m and MRR. Compare these answers with what you got for slab milling the block with HSS teeth.
10. The Bridgeport vertical-spindle milling machine is perhaps the single most popular machine tool. Virtually every factory (or shop) that does machining has one or more of these type machines. Go to your nearest machine shop and find a Bridgeport, make a sketch to show how it works, and explain what makes it so popular.

Chapter 24 CASE STUDY

HSS versus Tungsten Carbide Milling

The KC Machine works, which does job shop machining, has received an order to make 40 duplicate pieces, made of AISI 4140 steel, which will require 1 hr/piece of actual cutting time if a high-speed-steel (M1) milling cutter is used. Abigail Langley, a new machinist, says the cutting time could be reduced significantly if the company would purchase a suitable tungsten carbide milling cutter. Brandon Kannan, the foreman for the milling area, says he does not believe that Abigail's estimate is realistic, and he is not going to spend $450 (the current price from the vendor) of the company's money on a carbide cutter that probably would not be used again. The machine hour rate, including labor for the shop is $40/hr. Abigail and Brandon have come to you, the manufacturing engineer (MfE) of the plant, for a decision on whether or not to buy the cutter, which is readily available from a local supplier.

What factors should you consider in this situation? How much faster could the carbide cutter cut compared to the HSS cutter? See Table 24-1, in Chapter NaN. Based on your best guess as to the savings in actual cutting time per piece, who do you think is correct: Abigail or Brandon?

TABLE CS-24 Representative Cutting Data

Material						Forces	
Work	Tool	Back Rake (deg.)	Feed (ipt)	Width (in.)	Velocity (fpm)	Cutting (lb)	Thrust (lb)
AISI4140	HSS	0	0.0104	0.100	100	360	190
AISI4140	Carbide	0	0.011	0.15	540	540	156
AISI4145	Carbide	0	0.015	0.25	560	1190	560

CHAPTER 25

NC/CNC Processes and Adaptive Control: A(4) and A(5) Levels of Automation

25.1 Introduction

The first **numerically controlled (NC)** machine tool was developed in 1952 at the Massachusetts Institute of Technology (MIT). It had three-axis positional feedback control and is generally recognized as the first NC machine tool. By 1958, the first NC **machining center** was being marketed by Kearney and Trecker. A machining center was a compilation of many machine tools capable of performing many processes (milling, drilling, tapping, and boring), as shown in Figure 25-1. This NC machine had automatic capability. Almost from the start, computers were needed to help program these machines. Within 10 years, NC machine tools had become **computer numerical control (CNC)** machine tools with onboard microprocessors and could be programmed directly.

With the advent of the NC type of machine (and, more recently, programmable robots), two types of **automation** were defined. **Hard** or **fixed automation** is exemplified by transfer machines or automatic screw machines controlled by a mechanical cam. **Flexible** or **programmable automation** is typified by CNC machines or robots that can be taught or programmed externally by means of computers. The control is in computer software rather than mechanical hardware.

Most NC machines are A(4) level and employ the concept of **feedback** control, where some aspect (usually **position**) of the process is measured using a detection device **(sensor).** This information is fed back to an electronic comparator, housed in the **machine control unit (MCU),** which makes comparisons with the desired position. If the output and input are not equal, an error signal is generated and the table is adjusted to reduce the error.

Using a milling machine table as an example, Figure 25-2 shows the difference between **open-loop machine** and a **closed-loop machine,** with feedback provided on the location of the table and the part with respect to the axis of the spindle of the cutting tool. Three position control schemes are shown.

25.2 Basic Principles of Numerical Control

NC uses a processing language to control the movement of the cutting tool or workpiece or both. The programs contain information about the machine tool and cutting-tool geometry, the part dimensions (from rough to finish size), and the machining parameters (speeds and feeds and depth of cut). Thus, NC machines can duplicate consecutive parts, and a part made at a later date will be the same as one made today. Repeatability

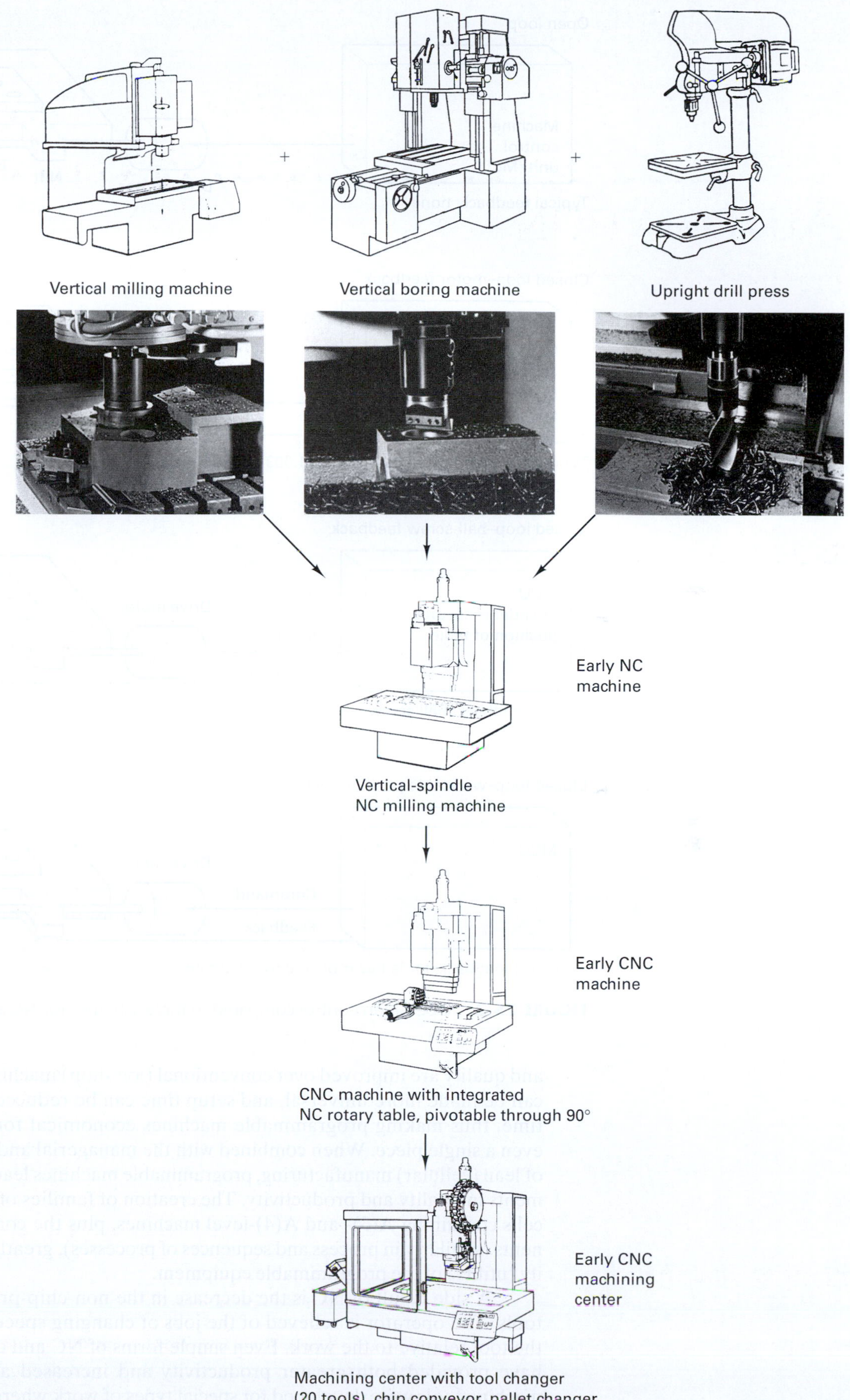

FIGURE 25-1 Early NC machine tools were controlled by paper tape. Soon, onboard computers were added, followed by tool changers and pallet changers. *(Courtesy J T. Black)*

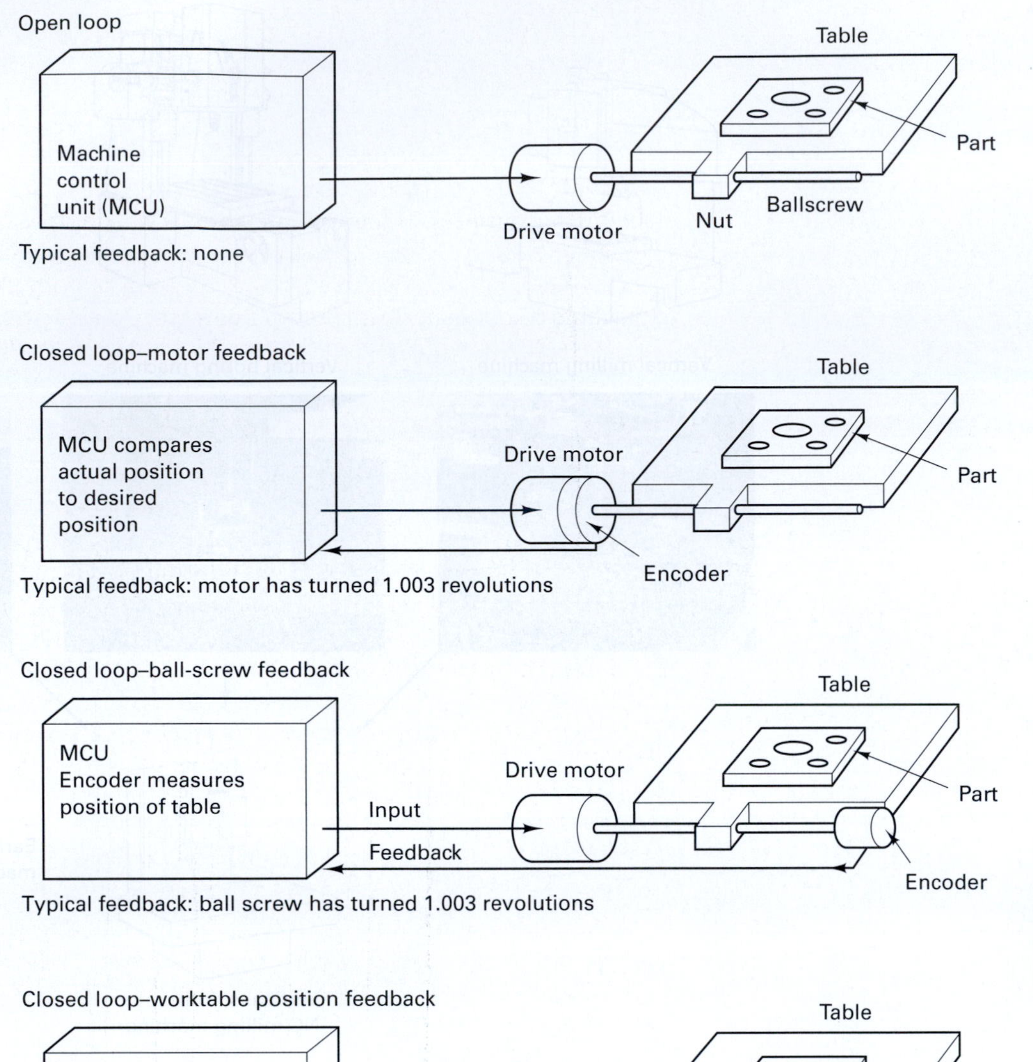

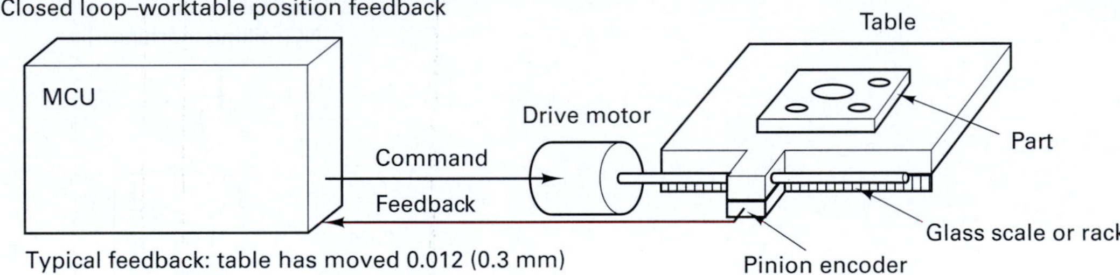

FIGURE 25-2 Open-loop control compared to three schemes for NC and CNC closed-loop.

and quality are improved over conventional (job shop) machines. Workholding devices can be made more universal, and setup time can be reduced, along with tool-change time, thus making programmable machines economical for producing small lots or even a single piece. When combined with the managerial and organizational strategies of lean (cellular) manufacturing, programmable machines lead to tremendous improvements in quality and productivity. The creation of families of parts made in machining cells containing A(3)- and A(4)-level machines, plus the compatibility of the components (similarly in process and sequences of processes), greatly enhances the productivity (utility) of the programmable equipment.

A side result of NC is the decrease in the non-chip-producing time of machine tools. The operator is relieved of the jobs of changing speeds and feeds and locating the tool relative to the work. Even simple forms of NC and digital readout equipment have provided both greater productivity and increased accuracy. Most early NC machine tools were developed for special types of work where accuracies of as much as 0.00005 in. were required, and many NC machines are built to provide accuracies of at least 0.0001 in., regardless of whether or not it is needed. This forced many machine tool builders to redesign their machines and improve their quality because the operator was not available to compensate for the machine (positioning) error. While most NC machine tools today will provide greater accuracy than is required for most jobs, the trend is toward greater accuracy and precision (i.e., better quality) but at no increase in cost. Therefore, CNC/NC machines will continue to be the very backbone of the machine tool business. Hopefully, all builders will one day arrive at a common language so that if one can program one machine, one can program them all.

As the name implies, *numerical control* is a method of controlling the motion of machine components by means of numbers or coded instructions. Assume that three 1-in. holes in the part shown in Figure 25-3 are to be drilled and bored on a vertical-spindle machine. The centers of these holes must be located relative to each other and with respect to the left-hand edge (x direction) and the bottom edge (y direction) of the workpiece. The depth of the hole will be controlled by the z- (or w-) axis. For this part, this is the zero reference point for the part.

The holes will be produced by center drilling, hole drilling, boring, reaming, and counterboring (five tool changes). If this were done conventionally or in manned cells, three or four different machines might be required. On the NC machining center, all the tools are held in a tool magazine and are changed automatically. The movements of the table are controlled by servo mechanisms for each direction. The machine tool has a zero point. The accurate positioning of the cutting tool with respect to the work is established by these zero points. The center of the table or a point along the edge of the traverse range is commonly used. The workpiece is positioned on the table with respect to this zero point on the table. For our example, the lower left-hand corner of the part is placed on the table 12 in. in the $x+$ direction, 4 in. in the $y+$ direction, and 2 in. in the $z+$ direction with respect to the machine zero point.

A software program is written that instructs the table to move with respect to the axis of the spindle to bring the holes to the correct location for machining. The machine shown in Figure 25-3 is called a five-axis machine because it has five movements (shown by the dark arrows) under numerical control. No fixture is shown in this example, but one would typically be used to obtain quick and repeatable location of the part on the table.

HOW CNC MACHINES WORK

Controlling a machine tool using variable input via a computer program, is known as numerical control and is defined by the Electronic Industries Association (EIA) as "a system in which actions are controlled by direct insertion of a numerical data at some point (the measured data is call the parts program). The system must automatically interpret at least some portion of the data."

NC and CNC machines can be subdivided into two types, shown in Figure 25-4. In **point-to-point machines,** the tool path is not controlled but tools can be moved in straight lines or parallel traverses at desired table feed rates, but only one axis drive is operated at a time. **Contouring** permits two or three axes to be controlled simultaneously, permitting two- or three-dimensional geometries to be generated. Another term for contouring is **continuous path.** Most milling and turning machining centers have contouring capability and are closed loop. The point-to-point machines can be open loop rather than closed loop.

The x-axis of the three-axis vertical spindle CNC machine tool shown in Figure 25-3 will be used to explain how the closed-loop positional control works. Figure 25-5 shows a schematic of the table control system.

Traditionally, NC machine tool has a machine control unit (MCU). The MCU is further divided into two elements: the **data-processing unit (DPU)** and the **control-loops unit (CLU).** The DPU processes the coded data that are read from the tape or some other input medium that it gives to the CPU, specifically the position of each axis, its direction of motion feed, and its auxiliary-function control signals. The CLU operates the drive mechanisms of the machine.

The CNC control system uses a resolver or encoder to provide axis-position feedback to the MCU, see Figure 25-6. Closed-loop control requires a transducer or sensing

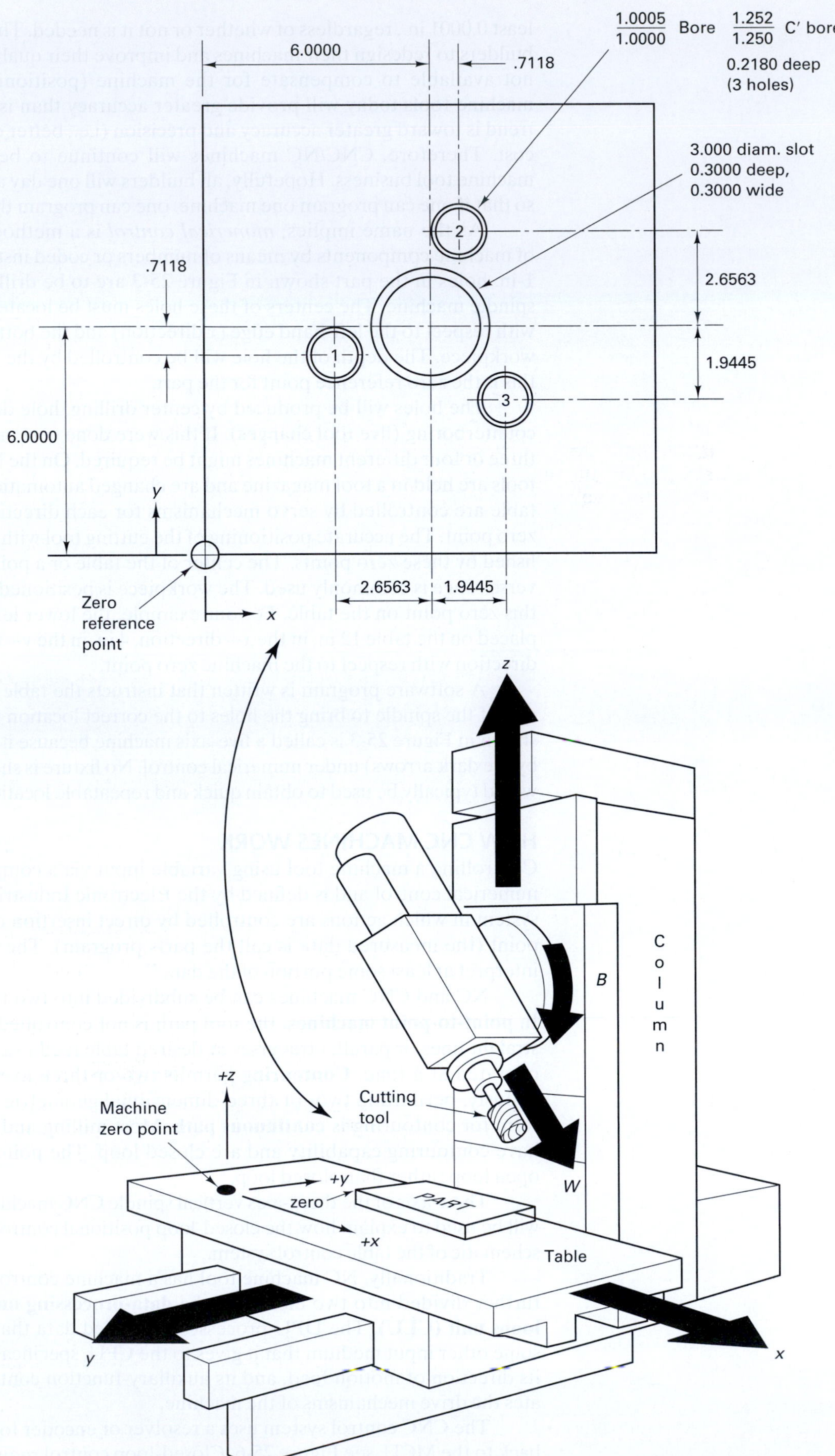

FIGURE 25-3 The part (above) to be machined on the NC machine (below) has a zero reference point. The machine also has a zero reference point.

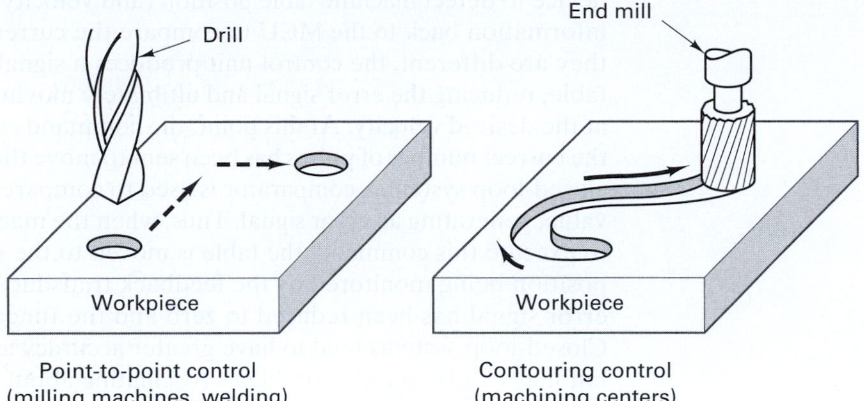

FIGURE 25-4 NC and CNC systems are subdivided into two basic categories: point-to-point controls and contouring controls.

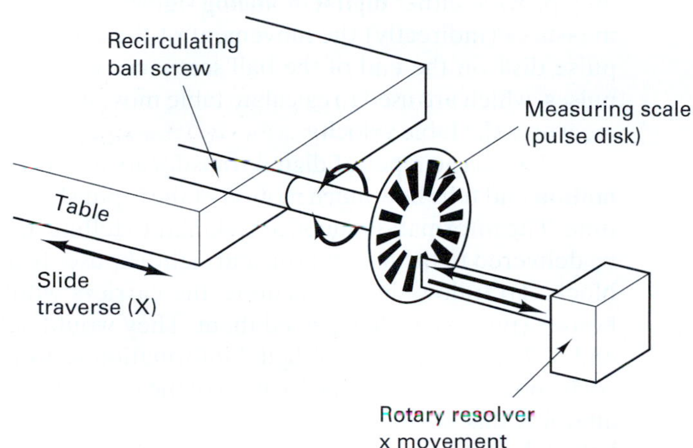

FIGURE 25-5 The table of the CNC is located using a resolver or encorder attached to the ball screw. (See Figure 25-6.)

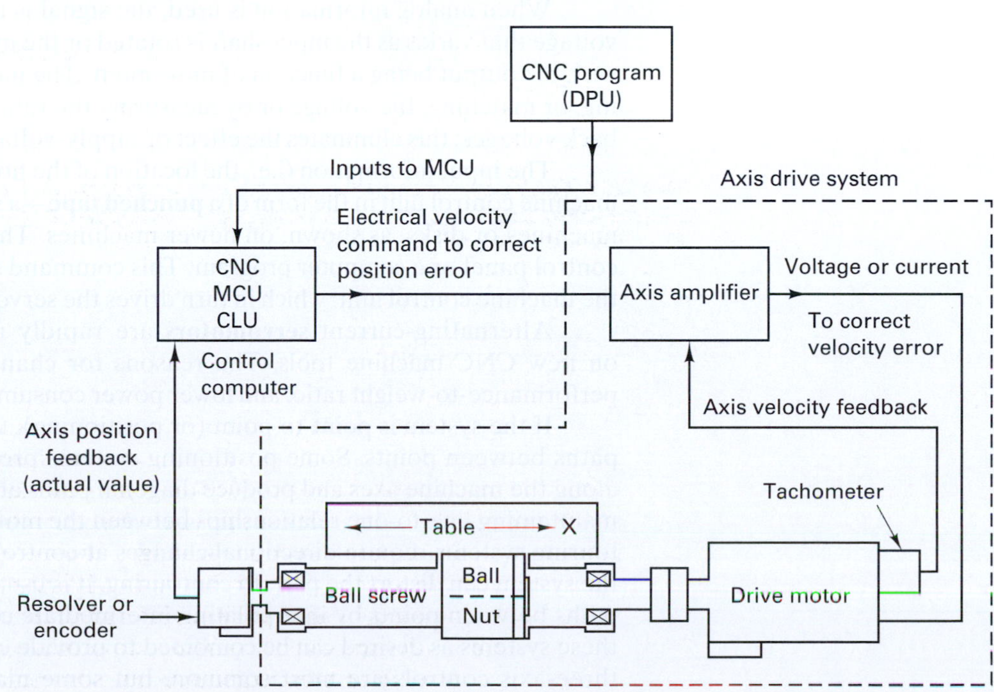

FIGURE 25-6 The table of the CNC machine is translated with a ball screw mechanism, and its location is detected with a resolver. The schematic shows how the table is located with respect to the spindle axis of the machine tool.

device to detect machine table position (and velocity for contouring) and transmit that information back to the MCU to compare the current status with the desired state. If they are different, the control unit produces a signal to the drive motors to move the table, reducing the error signal and ultimately moving the table to the desired position at the desired velocity. At this point, the command counter reaches zero, meaning that the correct number of pulses has been sent to move the table to the desired position. In a closed-loop system, a comparator is used to compare feedback pulses with the original value, generating an error signal. Thus, when the machine control unit receives a signal to execute this command, the table is moved to the specified location, with the actual position being monitored by the feedback transducer. Table motion ceases when the error signal has been reduced to zero and the function (drilling a hole) takes place. Closed-loop systems tend to have greater accuracy and respond faster to input signals but may exhibit stability problems (oscillating about a desired value instead of achieving it) not found in open-loop machines.

For most NC controls, the feedback signals are supplied by transducers actuated either by the feed screw or by the actual movement of the component. The transducers may provide either **digital** or **analog signals** (information). The resolver in Figure 25-6 measures (indirectly) the movement of the table by the rotation of the ball screw. A pulse disk on the end of the ball screw converts analog movement back into digital pulses, which are used to calculate table movement. The tachometer on the drive motor measures the table velocity.

Two basic types of digital transducers are used. One supplies **incremental information** and tells how much motion of the input shaft or table has occurred since the last time. The information supplied is similar to telling newspaper carriers that papers are to be delivered to the first, fourth, and eighth houses from a given corner on one side of the block. To follow the instructions, the carriers would need a means of counting the houses (pulses) as they passed them. They would deliver papers as they counted 1, 4, and 8. The second type of digital information is **absolute** in character, with each pulse corresponding to a specific location of the machine components. To continue the carrier analogy, this would correspond to telling the carriers to deliver papers to the houses having house numbers 2400, 2406, and 2414. In this case, it would only be necessary for the carriers (machine component) to be able to read the house numbers (addresses) and stop to deliver a paper when arriving at a proper address. This **address system** is a common one in numerical control systems because it provides absolute location information relative to a **machine zero point.**

When analog information is used, the signal is usually in the form of an electric voltage that varies as the input shaft is rotated or the machine component is moved, the variable output being a function of movement. The movement is evaluated by measuring, or matching, the voltage or by measuring the ratios between the applied and feedback voltages; this eliminates the effect of supply-voltage variations.

The input information (i.e., the location of the holes) is given in binary form to the machine control unit in the form of a **punched tape**—a magnetic tape in a cassette on old machines or disks, as shown, on newer machines. The data are input directly via the control panel or a computer program. This command signal is converted into pulses by the machine control unit, which in turn drives the servomotor or stepper motor.

Alternating-current **servomotors** are rapidly replacing direct-current motors on new CNC machine tools. The reasons for change are better reliability, better performance-to-weight ratio, and lower power consumption.

If the system is point-to-point (or positioning), the control system disregards the paths between points. Some positioning systems provide for control of straight cuts along the machine axes and produce diagonal paths at a 45-degree angle to the axes by maintaining one-to-one relationships between the motions of perpendicular axes. Contouring systems require directional changes at controlled velocity. Any lost motion in the system can distort the part. In contouring, it is usually necessary to control multiple paths between points by interpolating intermediate coordinate positions. As many of these systems as desired can be combined to provide control in several axes—two- and three-axis controls are most common, but some machines have as many as seven.

In many, conversion to either English or metric measurement is available merely by throwing a switch.

The components required for such a numerical control system are now standardized items of hardware. In most cases, the drive motor is electric, but hydraulic systems are also used. They are usually capable of moving the machine elements, such as tables, at high rates of speed, up to 200 in./min being common. Thus, exact positioning can be achieved more rapidly than by manual means. The transducer can be placed on the drive motor or connected directly to the leadscrew, with special precautions being taken, such as the use of extra-large screws and ball nuts, to avoid backlash and to ensure accuracy. In other systems, the encoder is attached to the machine table, providing direct measurement of the table position. Various degrees of accuracy are obtainable. Guaranteed positioning accuracies of 0.001 or 0.0001 in. are common, but greater accuracies can be obtained at higher cost. Most NC systems are built into the machines, but they can be retrofitted to some machine tools.

Most modern NC machine can change the tools automatically, change the speeds and feeds as needed for different operations, position the work relative to the tools, control the cutter path and velocity, reposition the tool rapidly between operations, and start and stop the sequence as needed. Tables and tools are positioned using recirculating ball-screw drives (see Figure 25-7), or linear accelerators, which greatly reduce the backlash in the drive systems, helping to eliminate problems of servoloop oscillation and machine instability. Using such hardware, NC machines are manufactured with greater accuracy and repeatability and more rapid table movements than is possible for conventional machines.

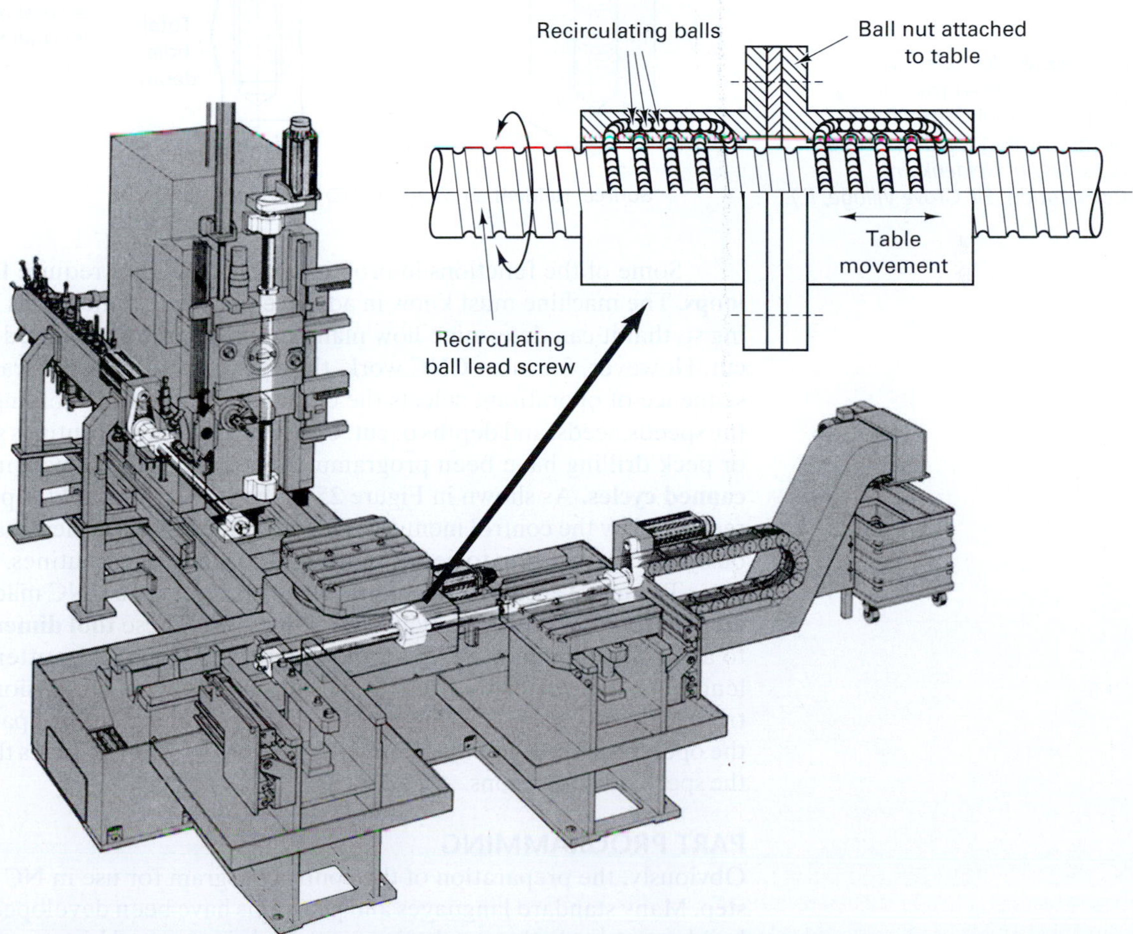

FIGURE 25-7 The ball leadscrew shown in detail provides great accuracy and position to NC and CNC machine tools.

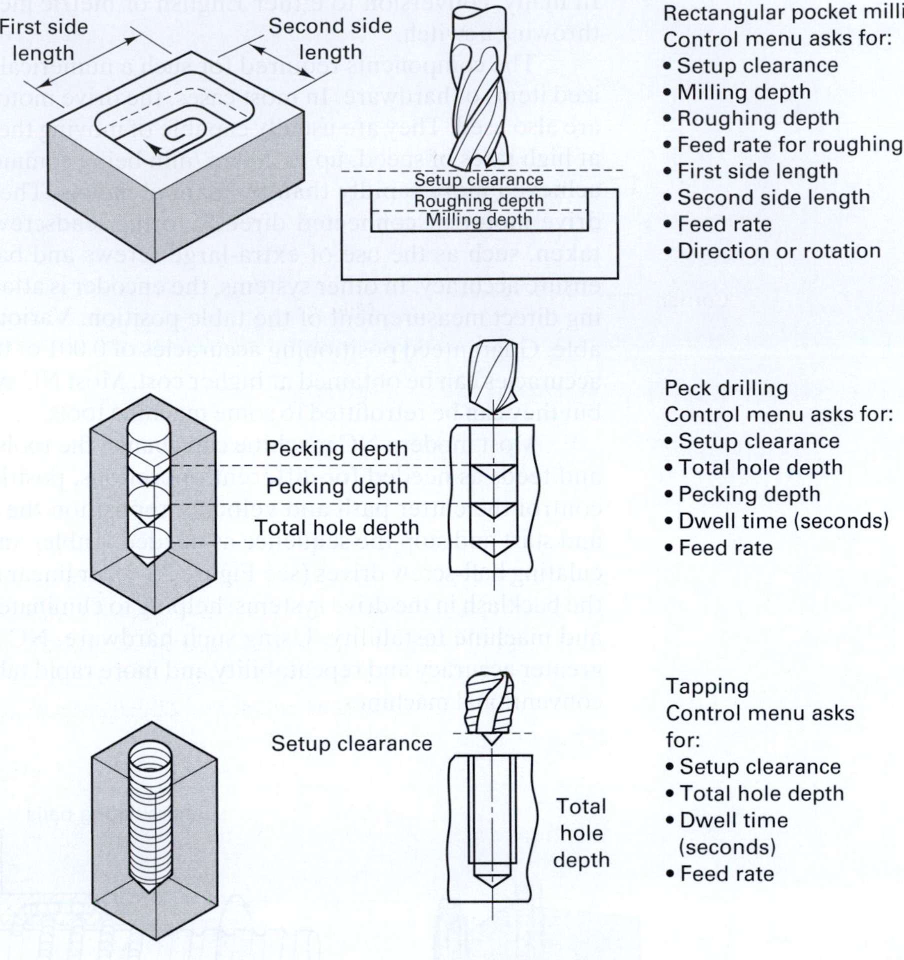

FIGURE 25-8 Canned or preprogrammed machining routines greatly simplify programming CNC machines. *(Courtesy of Heidenkain Corporation, Elk Grove Village, IL)*

Some of the functions in programmable machines require **feedforward** or preset **loops.** The machine must know in advance the rough dimensions of a casting or a forging so that it can determine how many roughing cuts are needed prior to the finishing cut. However, for most CNC work, the operator (or part programmer) still plans the sequence of operations; selects the cutting tools and workholding devices; and selects the speeds, feeds, and depths of cut. Common machining routines such as pocket milling or peck drilling have been programmed into many CNC machines. These are called **canned cycles.** As shown in Figure 25-8, the operator merely supplies the information requested by the control menu, and the machine fills in the necessary data for the previously written program to perform the desired machining routines.

To ensure accurate machining of a workpiece on a CNC machine, the control system has to *know* certain dimensions of the tools. These **tool dimensions** are referenced to a fixed **setting point** on the tool holder. For the milling cutter, the dimensions are length, L, and cutter radius. For the turning tool, the dimensions are length, L, and transverse overhang (see Figure 25-9). These dimensions are part of the information the operator inputs into the control panel. The program assumes that the tools will have the specified dimensions.

PART PROGRAMMING

Obviously, the preparation of the control program for use in NC machines is a critical step. Many standard languages and programs have been developed by the machine tool builders, and only the very basic aspect can be presented here.

The basic steps can be illustrated by reference to the part shown in Figure 25-10. The first step is to modify the part drawing to establish the zero reference axes: the x and

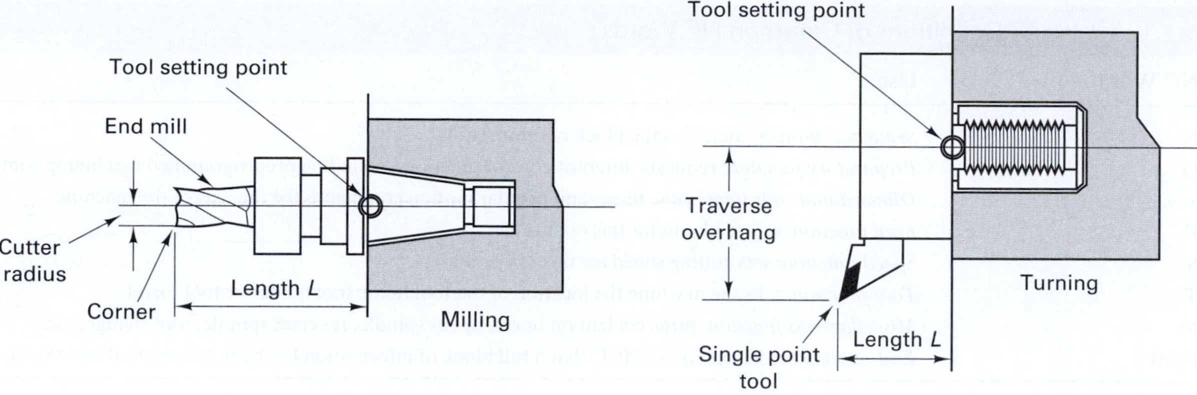

FIGURE 25-9 The location of the corner of the end mill (left) or the tip of a single-point tool (right) must be known with respect to the tool setting points so that tool dimensions are accurately set.

y directions. The **zero reference point** is the lower left-hand corner of the part. The hole labeled 1 is located +3.3437 in the *x* direction and ±5.2882 in the *y* direction, with respect to its zero point. The part should be dimensioned with respect to its zero reference point. This may require redrawing and redimensioning the part. Obviously, this step can be avoided if the original drawing is made in the desired form or if a CAD design is used. The setup instructions given to the operator establish the position of the workpiece properly on the machine table with respect to the axis of the spindle or the machine zero reference point.

The second step is to develop a **part program.** The program (1) defines the sequence of operations required to fabricate the part; (2) gives the *x*, *y*, and *z* coordinate positions of the operations; (3) specifies the spindle traverse that determines the depth

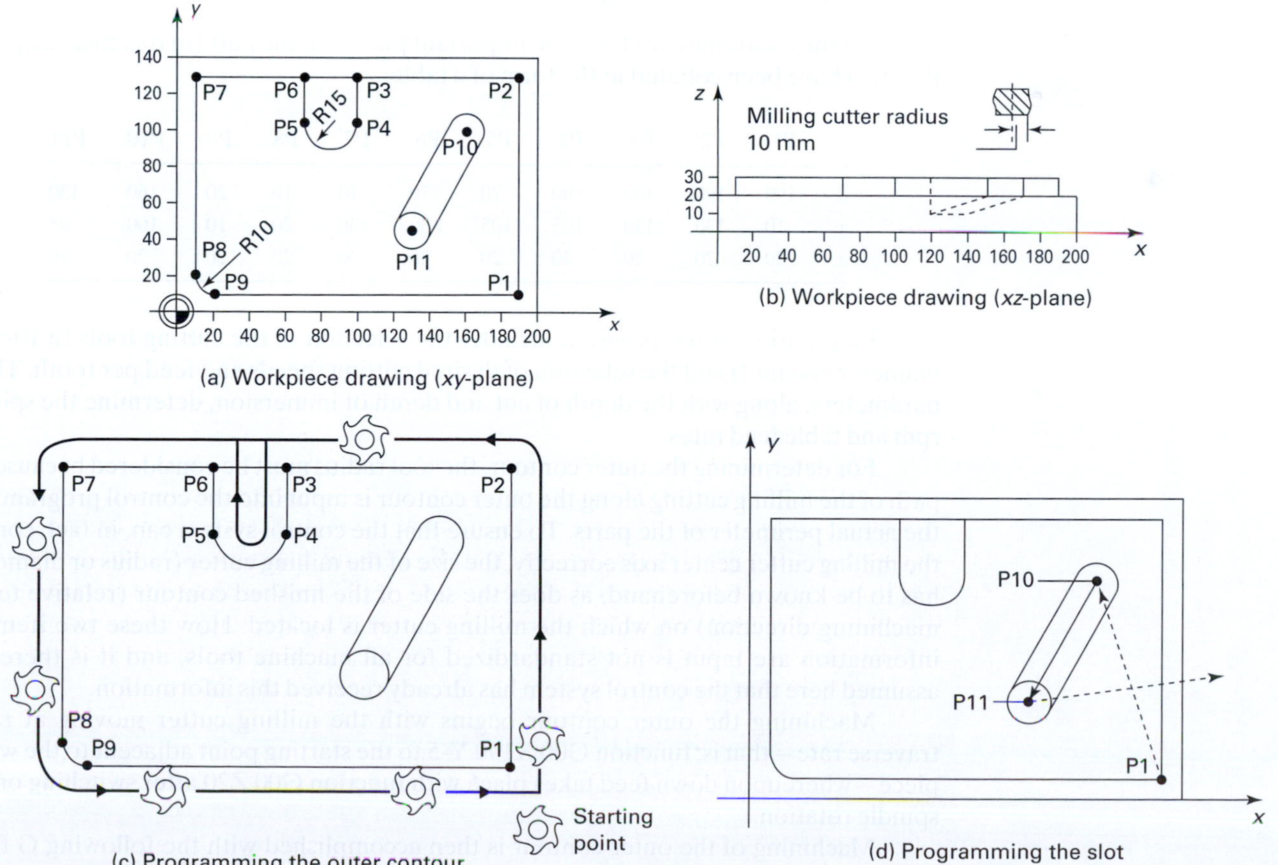

FIGURE 25-10 Example of programming a part in a vertical-spindle NC machine.

TABLE 25-1 Definitions of Common NC Words

NC Word	Use
N	*Sequence number:* identifies the block of information
G	*Preparatory function:* requests different control functions, including preprogrammed machining routines
x, y, z, b	*Dimensional coordinate data:* linear and angular motion commands for the axis of the machine
F	*Feed function:* sets feed rate for this operation
S	*Speed function:* sets cutting speed for this operation
T	*Tool function:* tells the machine the location of the tool in the tool holder or tool turret
M	*Miscellaneous function:* turns coolant on or off, opens spindle, reverses spindle, tool change, etc.
EOB	*End of block:* indicates to the MCU that a full block of information has been transmitted and the block can be executed

of the cut, the spindle speed, and the feed rate; and (4) determines whether the same tool can continue the next operation or whether a tool change is required. These items are specified by code symbols or NC words (see Table 25-1). The NC words are put together in a specified order to define a **block** of information needed to execute an operation. By convention, the data are usually arranged in blocks in the sequential order shown in the table.

Before starting with programming tool movements, the table movements and the workpiece drawing are studied. As a first step, the workpiece zero point is established. The workpiece drawing, two views, is redone into a coordinate system. In Figure 25-3 the workpiece zero point is located at the bottom left-hand corner of the workpiece drawing. The three axes of the machine coordinate system is established in relation to the workpiece as follows:

- The x- and y-axes are used to show the workpiece geometry in the XY-plane.
- The z-axis shows the down feed or depth.

All the coordinates of the most important points on the part (in this case the points P1–P11) have been collated in the form of a table:

	P1	P2	P3	P4	P5	P6	P7	P8	P9	P10	P11
x	190	190	100	100	70	70	10	10	20	160	130
y	10	130	130	105	105	130	130	20	10	100	45
z	20	20	20	20	20	20	20	20	20	20	10

Preparation of the program includes the selection of the cutting tools (a 10-mm-diameter end mill) and the selection of desired cutting speeds and feed per tooth. These parameters, along with the depth of cut and depth of immersion, determine the spindle rpm and table feed rates.

For determining the outer contour, the tool radius must be considered because the path of the milling cutting along the outer contour is input into the control program, not the actual perimeter of the parts. To ensure that the control system can, in fact, control the milling cutter center axis correctly, the size of the milling cutter (radius or diameter) has to be known beforehand, as does the side of the finished contour (relative to the machining direction) on which the milling cutter is located. How these two items of information are input is not standardized for all machine tools, and it is therefore assumed here that the control system has already received this information.

Machining the outer contour begins with the milling cutter moving at rapid traverse rate—that is, function G00 X190 Y-5 to the starting point adjacent to the workpiece—whereupon down feed takes place with function G00 Z20 after switching on the spindle rotation.

Machining of the outer contour is then accomplished with the following G functions: The supplementary function F produces the necessary feed rate in millimeters per minute. The downfeed depth is taken up by the milling cutter at the starting point.

G01	Y130	F200			Straight line from starting point to P2
G01	X100				Straight line from P2 to P3
G01	X105	F150			Straight line from P3 to P4
G02	X70	Y105	R15		Radial arc, clockwise, with 15 radius
G01	Y130	F200			Straight line from P5 to P6
G01	X10				Straight line from P6 to P7
G01	Y20				Straight line from P7 to P8
G03	X20	Y10	R10	F150	Radial arc, counterclockwise with 10 radius
G01	X190	F200			Straight line from P9 to P1

For milling the slot, the milling cutter is first retracted by G01 Z35 from the workpiece to point P1 shown in Figure 25-10d.

Because the width of the slot is equal to the cutter diameter, the path of the cutter is easy to program. The tool radius compensation feature is switches off (by G40) so the following traverse instructions refer to the cutter center line axis.

Milling of the slot is accomplished by the following instructions:

G00	X160	X100		Rapid traverse to point P10
G01	Z20	F150		Down feed at point P10
G01	X130	Y45	Z10	Straight line from P10 to P11
G01	Z35	F200		Retraction from workpiece
G00	X300	Y300		Rapid traverse away from workpiece

After the program has been written, it is verified, which means the steps in making the part are graphically simulated. The **verification** step can use the computer monitor, which simulates the part being made by tracing all the toolwork paths as they would occur on the machine tool. Sometimes a sample part is machined in plastic or machinable wax for checking the part specifications from the drawing against the real part.

The program can be directly entered into the control panel of a CNC machine. Usually, longer programs are entered by tape or disk and short programs are entered manually. The inputs may be in binary code (in older machines) or in Arabic numbers with verbal commands that the machine understands.

Historically, punched tape was used to control the movements. The idea came from the old player-piano roll, which was a form of tape control, and punched cards had been used for many years for controlling complicated weaving and business machines. Thus, tape control of machine tools was an extension of an existing basic concept in which holes, representing information that has been punched into the tape, were read by sensing devices and used to actuate the devices that control various electrical or mechanical mechanisms.

Over the years, four basic types of tape format have been used for NC input to communicate dimensional and nondimensional information: **fixed-sequential format, block-address format, tab-sequential format,** and **word-address format.** Most new NC or CNC systems use the word-address format, which allows the words to be presented in any order and is the most flexible.

Today, CNC machines permit the user to program the interface between the machine tool and the control, greatly reducing the number of machining system components and interconnections. Current CNCs have extensive self-diagnostics and performance-monitoring systems. The greatest advances in CNC technology, however, are in part programming, where easy-to-use, menu-driven software makes programming almost as simple as setting up the machine manually. With older NC machines, the operator could override the program when necessary but could not reprogram the machine unless a new program was written. The CNC machine has the capability of reading a program into its computer memory, and the program can be modified at the machine like any other computer program.

On CNC machines, the machine tool operator may perform all the programming steps right at the console of the machine, programming the processing steps for the part directly into the computer memory. The program can be saved by having the machine print out a copy of the program, which can be used later for reorders of the same part. Features such as program edit, canned routines, program storage, diagnostics, constant surface speed, and tape punch are common on today's CNC machines.

As more and more design work is done on the computer (CAD) using databases and software that are compatible to the machine tools, there will be less dependence on tape for program storage and more utilization of floppy disks, hard disks, and other typical computer storage means. For example, a machining cell composed of NC machine tools designed for a family of 10 component parts is able to make the 10 different parts without needing retooling or refixturing, but it will still have 10 different programs for these parts for each machine. If the programs are stored in the computer, they can be readily accessed, but if they are stored on disk, delays will occur in dumping the different programs into or out of the control computer.

INTERPOLATION AND CUTTER OFFSET IN NUMERICAL CONTROL

In milling machines, the centerline of the cutter is offset from the desired surface by the radius of the cutter. The path that the cutter needs to take to generate the desired geometry is not simply the perimeter or profile of the part. Thus, **cutter offset** programs must be included in the software. Obviously, contour machining (with cutter offset and interpolation) requires that complex information be entered into the computer because the number of straight-line or curved segments may be quite large. Manual programming of the tape can be quite laborious. Computer programming can translate simple commands into the complex information required by the machine.

The required curves and contours of the part are generated approximately by a series of very short straight lines or segments of some type of regular curves, such as hyperbolas. This is called **interpolation.** The program fed to the machine is arranged to approximate the required curve within the desired accuracy. Figure 25-11 illustrates how a desired straight-line or curved surface can be approximated by means of short segments. Interpolation refers to the fact that curved surfaces as generated by machine

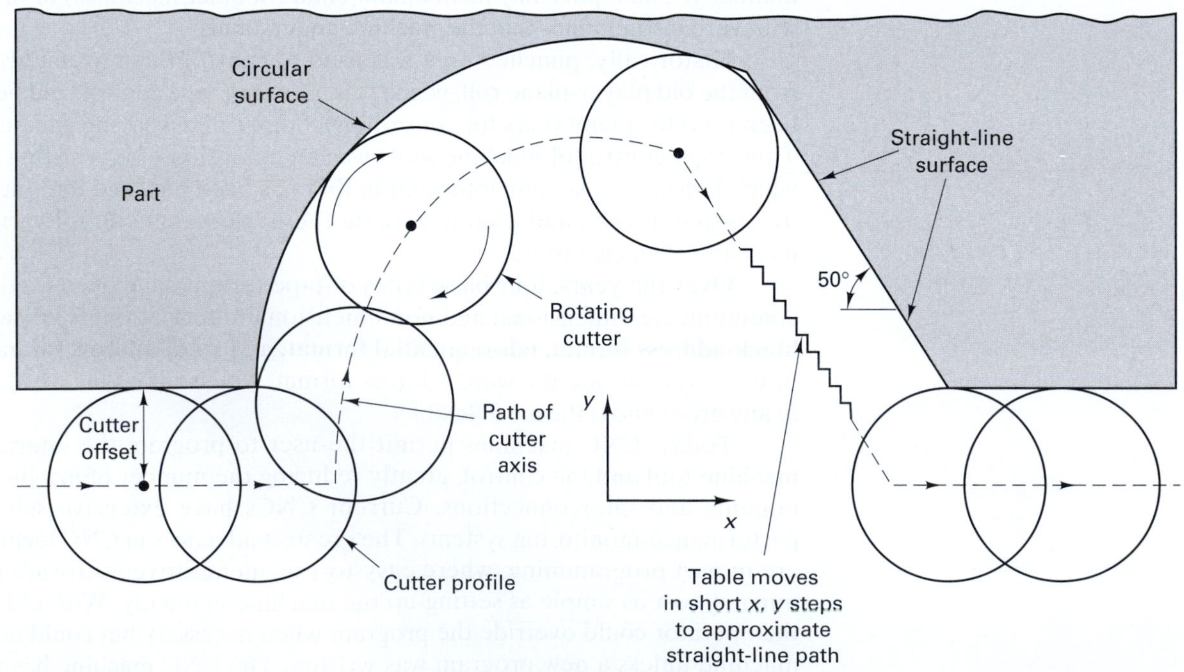

FIGURE 25-11 Two classic problems in NC programming are the determination of cutter offset and interpolation of cutter parts.

tools must be approximated by a series of very short, straight-line movements in the x, y, and z directions. The length of the segment must be varied in accordance with the deviation permitted. Most NC machine tools with contouring capability will produce a surface that is within 0.001 in. of the one desired, and many will provide considerably better performance. Most contouring machines have either two- or three-axis capability, but five-axis capability is readily available.

BASICS OF INTERPOLATION

For numerically controlled machine tools, the program to control the relative tool/work motions is expressed by a sequence of numbers. In addition, the numerical program also contains other commands to control spindle speeds, tool changes, coolant cycles, and other tasks. Since the first model of a NC milling machine was demonstrated at MIT in 1952, this level of automation has undergone tremendous development.

Interpolation in machine tools is discussed using Figure 25-12. The discussion is limited to two dimensions, even though today's machines can involve three, four, or five axes of control. Starting with the part drawing and the associated geometry of the machining process—contour end milling—the path of the center of the cutter is offset from the desired geometry of the workpiece by the radius of the cutter. The numerical controlled machine has a capability known as **cutter radius compensation,** which automatically generates the correct path of the cutter. In this example, the path consists of three straight sections and one curved section, called **segments.** The engineer develops a process plan that includes the details of the geometry of the four sections and the tool path strategy. For example, whether up milling or down milling are used

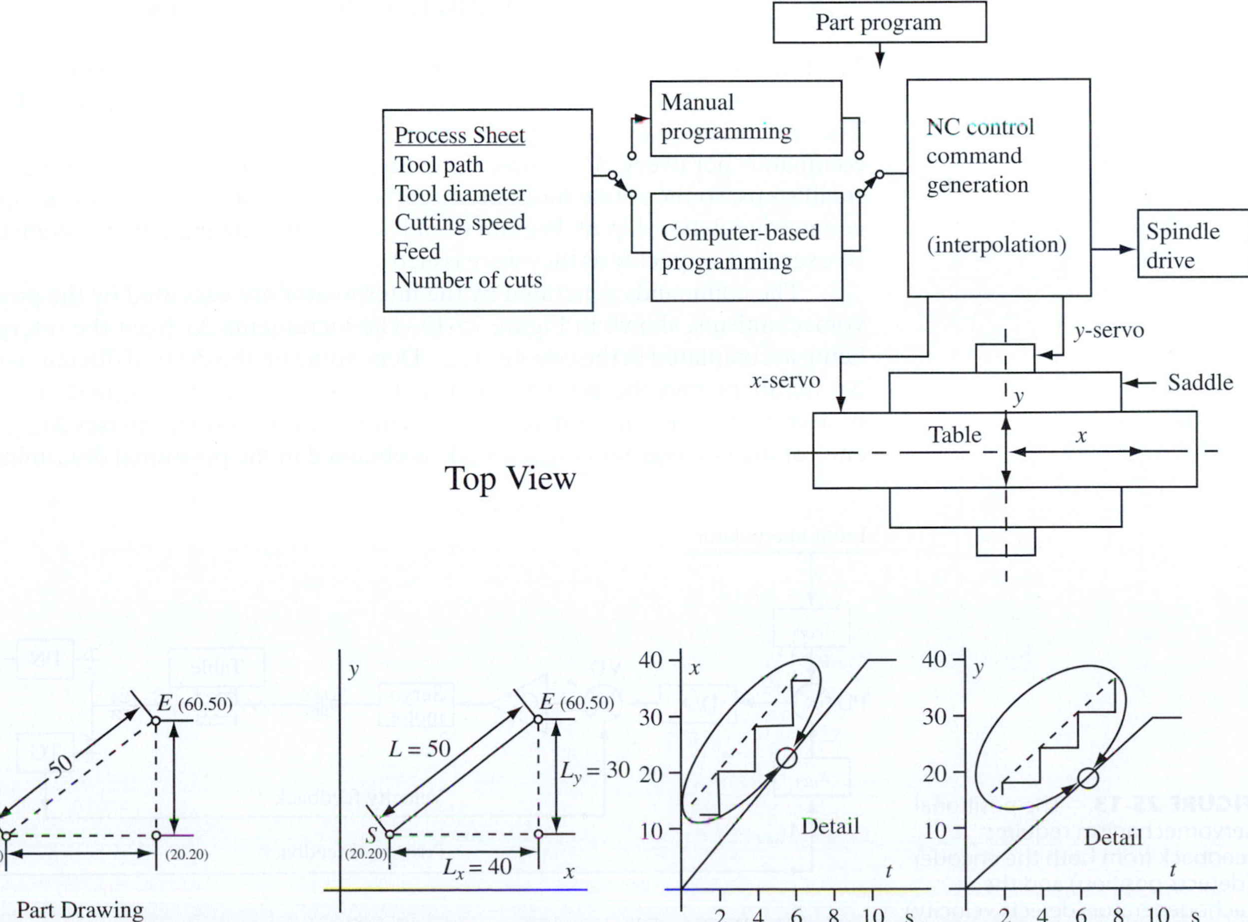

FIGURE 25-12 The process sheet has the information on the part needed to develop the part program, either manually or computer-aided. The 50-in. cut requires linear interpolation to move the cutter from point S to point E.

(see Chapter 24 for discussion) will determine the direction of the cutter path with respect to the part. Also, the cut may be made in one or more passes, depending on the form and dimensions of the raw material and on the data describing the machining operations, such as spindle speed and feed rate (which depend on the work material and the cutting-tool material). These data are entered into the control program. The control program for an X-Y table will consist of Δx and Δy commands for the desired tool path issued at regular time intervals; typically, intervals of 1 to 5 msec are used. Most NC machine tools have linear and circular interpolation capability. The commanded path increments go to the positional x and y servomechanisms of the machine, which include motors and transmission devices driving the saddle and the table of the machine. Referring again to Figure 25-12, the command for linear interpolation from the starting point S, where $x_s = 20$ and $y_s = 20$, to the end point E, with $x_e = 60$ and $y_e = 50$, are dimensioned in millimeters or inches. Assume the time of the command generation cycle is $\Delta t = 0.005$ s and the commanded feed rate (velocity of motion) $f_n = 5$ mm/s. The total motions are $L_s = 40$ mm and x and $L_y = 30$ mm for y; the length of the motion between points S and E is $L = 50$ mm.

The cutter is moving in small x and y steps to produce the 50 mm cut.

The control as well as the positional servos work digitally with numbers that are a **basic-length unit (BLU).** Assume a basic length unit is BLU = 0.001 mm.

The total time for this motion is $T = L/f = 50$ mm/(5 mm/s) = 10 s; the component velocities are $v_x = 40$ mm/10 s = 4 mm/s, and $v_y = 30$ mm/10 s = 3 mm/s. Correspondingly, the commands will be issued every 5 ms in increments of $\Delta x\ \Delta y$ represented by the following numbers:

$$\Delta x = 4000\ \text{BLU/s} \times 0.005\ \text{s} = 20\ (\text{BLU})$$
$$\Delta y = 3000\ \text{BLU/s} \times 0.005\ \text{s} = 15\ (\text{BLU})$$

The total numbers issued will be counted up to 40,000 in x and 30,000 in y.

The detail of motion commands for x and y versus time are shown in Figure 25-12. The increments $\Delta x = 20$ and $\Delta y = 15$ are issued discretely, as "chunks" of the travel commands per every $\Delta t = 5$ ms. In reality, however, the servos cannot execute these small steps, so the actual motions are rather smooth along both axes. In any case, the geometric relationship of diagram would hold even if the incremental commands could be executed as quickly as they were issued.

The commands generated by the interpolator are executed by the positional servomechanisms, shown in Figure 25-13. The increments Δx from the interpolator are being accumulated in the counter x_{com}. Depending on the desired direction of the table, Δx numbers may be positive or negative, so x_{com} is correspondingly increasing or decreasing. Another counter, x_{act}, accumulates the feedback pulses Δx_{act}. The difference of the two registers ($x_{com} - x_{act}$), as obtained in the positional discriminator PD, is

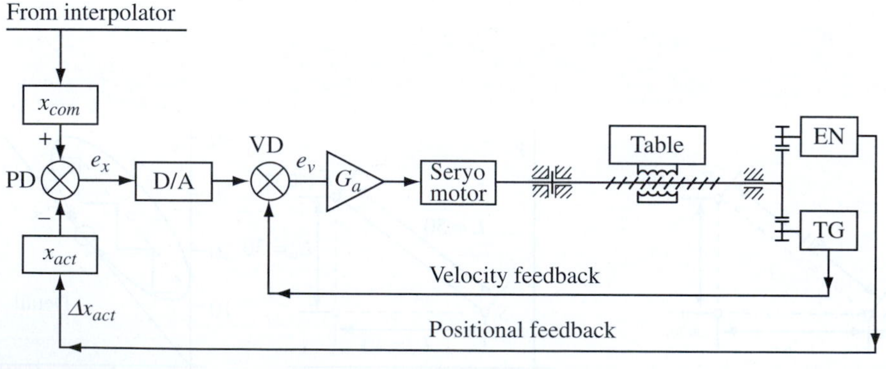

FIGURE 25-13 The positional servomechanism requires feedback from both the encoder (detects position) and the tachogenerator (defects velocity) to drive the servomotor that moves the table to the desired location.

the error of position e_x, and it is interpreted by the servo as a command to move the table. These functions may be executed as a software routine in the microprocessor. The error e_x is then output through a digital/analog converter and becomes the control signal for the velocity servo. In the velocity discriminator, *VD*, this command signal is compared with the **tachogenerator** feedback, and the velocity error, e_y, is fed into the power amplifier with gain, G_a, and further into the servomotor. The servomotor drives the table of the machine through a leadscrew and nut transmission of the recirculating ball type, shown in Figure 25-7. The nut consists of two halves preloaded axially against each other so as to eliminate any backlash. High-precision gears transmit the rotation of the leadscrew to a tachogenerator, *TG*, for velocity feedback and also to an encoder, *EN*, for positional feedback (see Figure 25-5). The encoder emits one pulse for every BLU of the table motion. Correspondingly, the positional feedback has the form of a train of pulses; their frequency corresponds to the velocity of the motion, and their total number to the total distance traveled.

ADAPTIVE CONTROL: A(5) LEVEL

Adaptive control (A/C) can deal with the problems caused by variations in the size of the uncut workpiece, which may be a casting or a forging. The NC program was prepared with assumptions about the amount of material to be removed. The program may have to adapt to variations in the size—that is, the depth of cut. Also, NC programs are prepared under certain assumptions regarding the tool-wear rate for the cutting speeds that were selected, but the actual wear rate may be different from that which was assumed. Tool wear can also change the depth of cut. Consequently, the cutting speed may have to be modified to reduce the tool-wear rate. Other phenomena that are difficult to predict, such as chatter vibrations, may occur, and the cutting conditions need to be changed to stabilize the cut.

In all these instances, it is possible to use sensors to measure the significant parameters of the actual cutting process and to change the NC program—that is, change the feed rate, f_r, the spindle rpm or the depth of cut so as to improve the cutting process. This is shown schematically in Figure 25-14, which depicts as an example a pocket end-milling operation (see Figure 25-6). The process parameters to be measured may be the spindle torque on the spindle, the cutting force, vibrations, and tool wear (by measuring the change in size of the part). This information is fed back into the CNC **controller,** which contains in its software the corresponding adaptive control strategies in the form of algorithms to modify cutting conditions. A simple A/C system can be conceived wherein the actual cutting force, F_{act}, is measured and compared with the nominal force, F_{nom}, which the cutter can safely maintain. Their relative difference establishes the force error, e_f. For $F_{act} < F_{nom}$, e_f is positive, and for $F_{act} > F_{nom}$, e_f is negative. Changing the feed rate of the table can eliminate the force error. So if $F_{act} > F_{nom}$, it is necessary to reduce the feed rate, f_r. (See Chapter 20 for relation between feed rate and cutting force.) The CNC controller will correspondingly start changing the feed rate of the commanded travel x_{com} until $F_{act} = F_{nom}$. In this way, the cutter will move rapidly where there is little material to cut and slowly when the depth of cut is large. In the design of

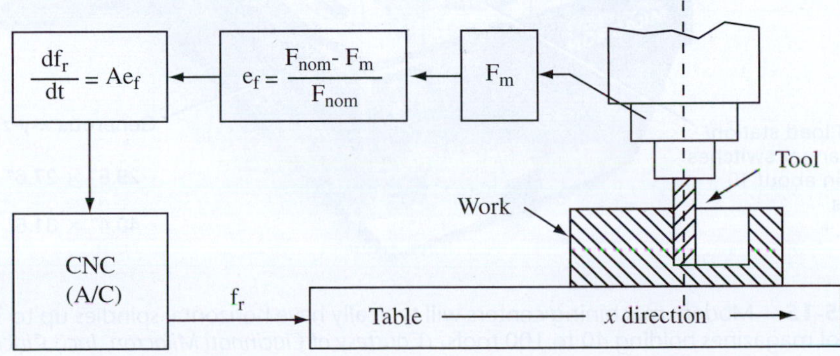

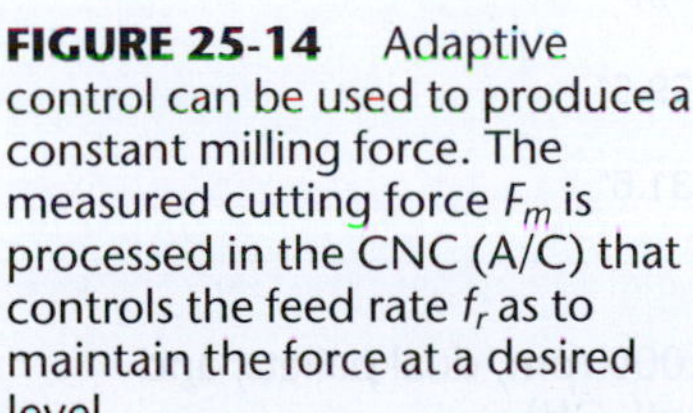
FIGURE 25-14 Adaptive control can be used to produce a constant milling force. The measured cutting force F_m is processed in the CNC (A/C) that controls the feed rate f_r as to maintain the force at a desired level.

NC control systems with adaptive control, it is necessary to solve a number of problems that are mainly due to changes in the dept of cut. Thus, very few machines have adaptive control programs in their software.

25.3 MACHINING CENTER FEATURES AND TRENDS

Computer numerical control is used on a wide variety of machine tools. These range from single-spindle drilling machines, which often have only two-axis control and can be obtained for about $10,000, to machining centers, such as shown in Figure 25-15. The machining center can do drilling, boring, milling, tapping, and so on, with four- or five-axis control. It can automatically select and change 40 to 180 preset tools. The table can move left/right and in/out, and the spindle can move up/down and in/out, with positioning accuracy in the range 0.00012 in. with repeatability to ±0.00004 in. over 40 in. of travel. Beyond accuracy and repeatability and the number of controlled axes, CNC machines are also categorized by spindle speed, horsepower, and the size of the workpiece. The accuracy and repeatability (precision) result from a combination of factors including the resolution of the control instrumentation and the accuracy of the hardware. The control resolution is the minimum length distinguishable by the control unit. It is called the basic-length unit (BLU) and is mainly a factor of the axis transducer and the quality of the leadscrew or linear translator. There are many sources of error in a machine, including wear in the machine sliding elements, machine tool assembly errors, spindle runout (wear), and leadscrew backlash.

Tool deflection due to the cutting forces produces dimensional error and chatter marks. **Thermal error,** caused by the thermal expansion of machine elements, is not uniform and is normally the greatest source of machine error. Methods used to remove

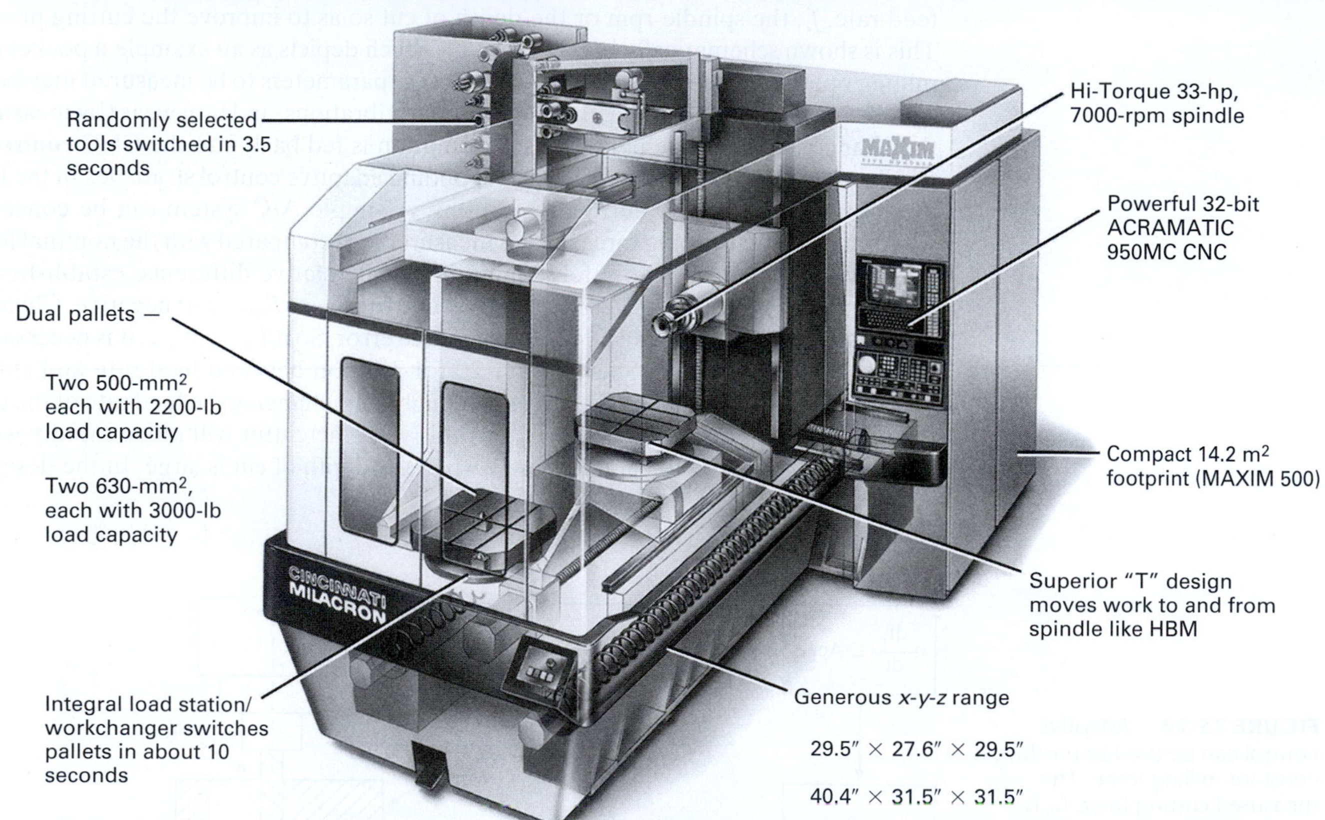

FIGURE 25-15 Modern machining centers will typically have horizontal spindles up to 15,000 rpm, dual pallets, and cutting-tool magazines holding 40 to 100 tools. *(Courtesy of Cincinnati Milacron, Inc., Cincinnati, OH)*

heat from a machine include cutting fluids, locating drive motors away from the center of a machine, reducing friction from the ways and bearings, and spray cooling control element of the machine.

Most modern machining centers have automatic tool-change and automatic work-transfer capability, so that workpieces can be loaded and unloaded while machining is in process. Such a machine can cost more than $200,000. Between these extremes are numerous machine tools that do less varied work than the highly sophisticated machining centers but that combine high output and minimum setup time with remarkable flexibility (large number of tool motions provided).

CNC TURNING CENTERS

The modern lathe is called a **turning center** and has CNC control and tools mounted in turrets on slanted beds. The tailstock has been replaced by a live, powered spindle and chuck. On some lathes, the concept of automatic tool changing has been implemented. The tools are held on a rotating tool magazine, and a gantry-type tool changer is used to change the tools. Each magazine holds one type of cutting tool. This is an example of the trend of providing greater versatility along with high productivity in lathes. The versatility is being further increased by combining both rotary-work and rotary-tool operations—turning and milling in a single machine. Live, powered, or driven tools replace regular tools in the turrets and perform milling and drilling operations when the spindle is stopped (see Figure 25-16).

OTHER NC MACHINES

Numerical control has been applied to a wide variety of other production processes. NC turret punches with *X-Y* control on the table, CNC wire EDM machines, laser and water-jet abrasive machining, flame cutters, and many other machines are readily available. Some new trends are being observed in the development of machining centers,

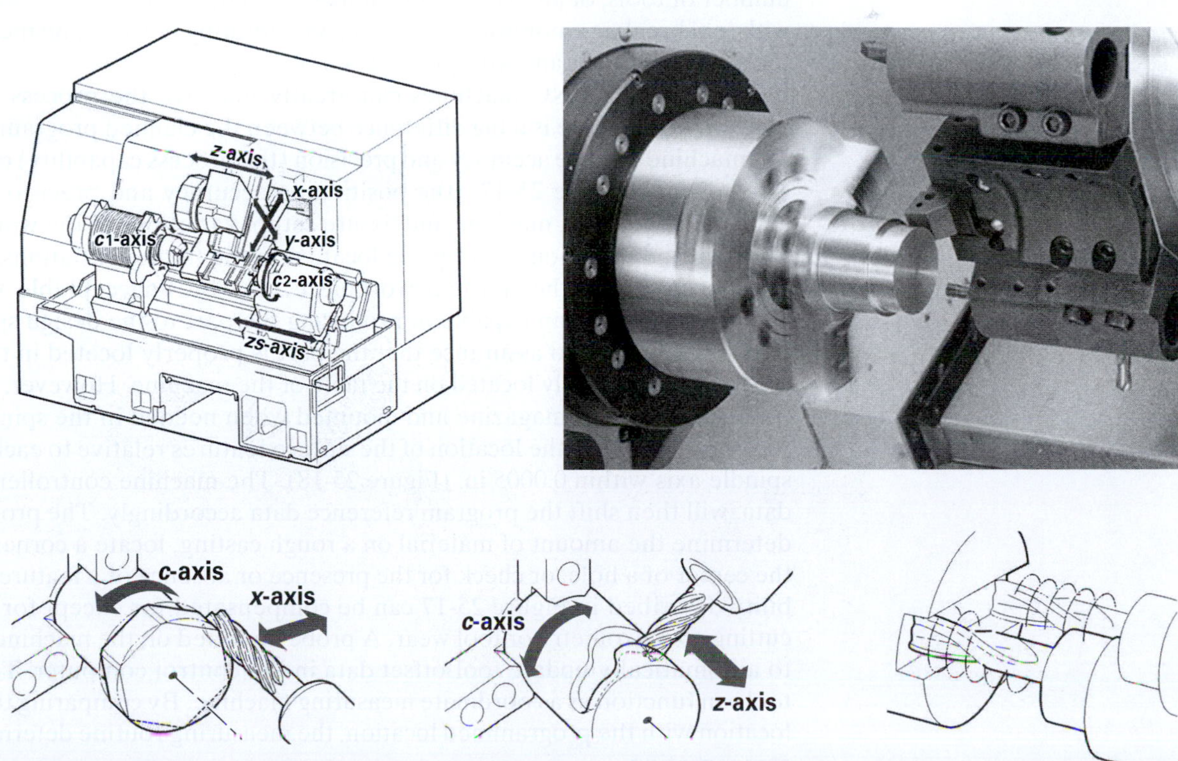

FIGURE 25-16 This CNC turning center has a multiple-axis capability with two spindles and a 12-tool turret with *x*, *y*, and *z* control as well as axis control of the spindles. *(Courtesy J T. Black)*

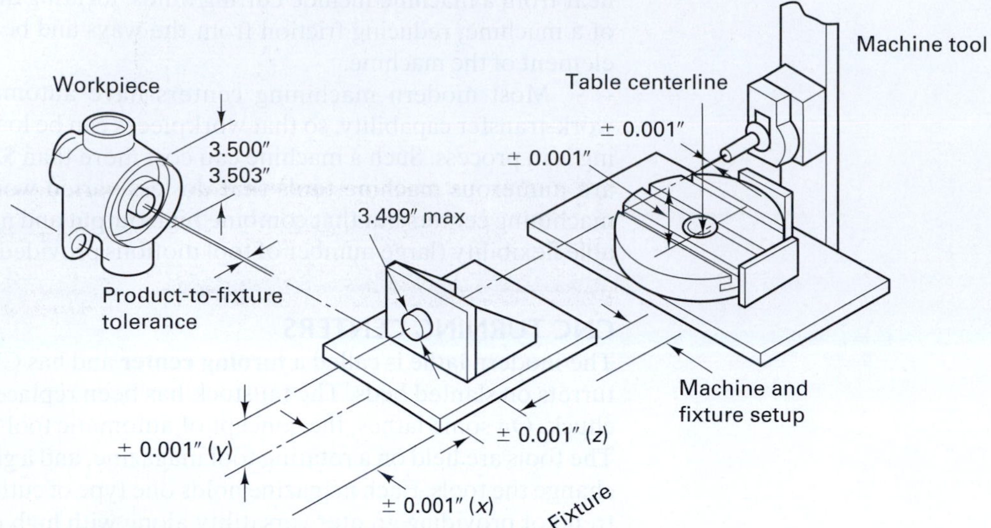

FIGURE 25-17 Process capability in NC machines is affected by many factors.

such as smaller, compact machining centers with higher spindle speeds. Machines with four- and five-axis capability are readily available. Modern machining centers have contributed significantly to improved productivity in many companies. They have eliminated the time lost in moving workpieces from machine to machine and the time needed for workpiece loading and unloading for separate operations. In addition, they have minimized the time lost in changing tools, carrying out gaging operations, and aligning workpieces on the machine.

The latest generation of machining centers is aimed at further improving utilization by reducing the time when machines are stopped, either during pauses in a shift or between shifts. Delays are caused by tool breakage, unforeseen tool wear, a limited number of tools, or an inadequate number of available workpieces. Machines are fitted with tool breakage monitors, tool-wear compensating devices, and means for increasing the number of tools and workpieces available.

Probes on CNC machines can greatly improve the process capability of the machine tool. There is a big difference between the claimed program resolution for an NC machine and the accuracy and precision (the process capability) of the actual parts. As shown in Figure 25-17, true positioning accuracy and precision are affected by machine alignment, machine and fixture setup, variations in the workholding device, raw material variations, workpiece location in the fixture variations, and cutting-tool tolerances. Thus, the finished workpiece may be unacceptable even though the machine is more than capable of producing the part to the design specifications. The part program has no assurance that the part is properly located in the fixture or that the fixture is properly located on the table of the machine. However, a probe—carried in the tool storage magazine and mounted when needed in the spindle like a cutting tool—can establish the location of the surface features relative to each other and to the spindle axis within 0.0005 in. (Figure 25-18). The machine controller, using the probe data, will then shift the program reference data accordingly. The probe can be used to determine the amount of material on a rough casting, locate a corner of a part, define the center of a hole, or check for the presence or absence of a feature. All of the variability described in Figure 25-17 can be compensated for except for variations in the cutting-tool geometry or tool wear. A probe mounted on the machine tool can be used to automatically update tool-offset data in the control computer. Thus, the machine tool can function as a coordinate measuring machine. By comparing the actual touched location with the programmed location, the measuring routine determines appropriate compensation.

Modern machining centers are equipped with automatic **pallet changers** and workpiece loading and unloading devices. Robots are increasingly being applied for

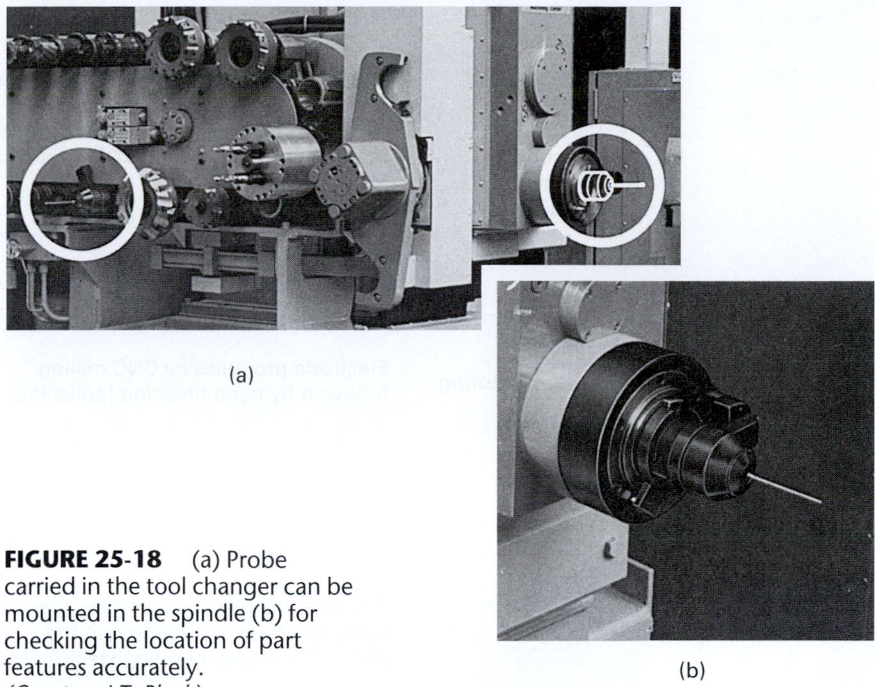

FIGURE 25-18 (a) Probe carried in the tool changer can be mounted in the spindle (b) for checking the location of part features accurately. *(Courtesy J T. Black)*

workpiece handling in machine groups, including manufacturing cells. In some cases, the robot is also used for tool-changing functions.

25.4 ULTRA-HIGH-SPEED MACHINING CENTERS

Anyone involved in the product development process knows that the longest lead-time path from a new-product (like a car) design to a finished product coming off the final assembly line always includes the die-making process. For example, in the automotive industry, each new car design may have many die sets for forged and sheet metal parts. Machining of the dies is a key process in the conventional die-making process (see Figure 25-19), where the major steps are milling, electronic discharge machining (EDM), polishing, die assembly, tryout, and modification.

Ultra-high-speed machining centers (UHSMCs) can shorten the die-making process lead time. The new machining centers have highly accurate, high-speed spindles capable of 30,000 to 50,000 rpm, cutting feed rates of 60 m/min, and cutting feed accelerations of 9.8 m/s^2. The machine requires a spindle with high stiffness utilizing ceramic ball bearings, with a constant-pressure preload mechanism and jet lubrication for superior performance (see Figure 25-20).

Robust machining centers of this sort are capable of very high metal removal rates, particularly in materials like aluminum. However, before investing in a **high-speed machining center (HSMC),** many issues must be addressed. The new machine will usually require the purchase of new cutting tools and tool holders. At the higher rpms and feed rates, smaller-diameter tools are easier to balance. Tungsten carbide is the primary tool material. Make sure the insert retention mechanism is adequate in case of a failure. Other major problems will include chatter (see Chapter 20) and removal of the large volume of chips.

UHSMCs require exceptionally accurate, stiff spindles using ceramic ball bearings, raising the possible rpm as compared to regular ball bearings, air bearings, or magnetic bearings. Synchronized sets of ball screws are used to feed the tool in the x and y directions to reduce the errors caused by distortion in the frame during rapid acceleration of the tool. These machines have usually four or five axes under control and represent the pinnacle of machine tool development at this time.

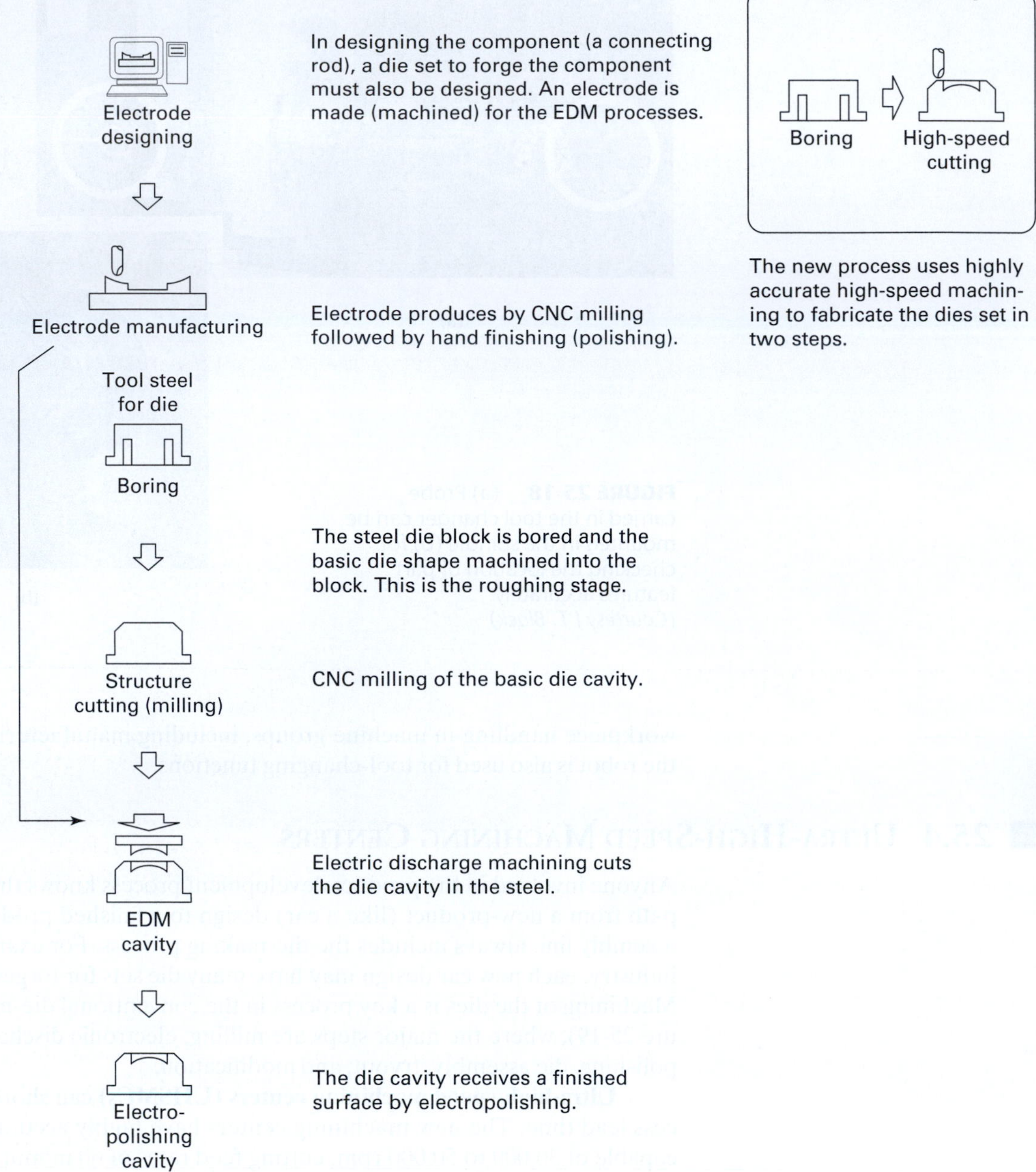
Electrode designing
In designing the component (a connecting rod), a die set to forge the component must also be designed. An electrode is made (machined) for the EDM processes.
High-speed cutting
Boring
High-speed cutting
The new process uses highly accurate high-speed machining to fabricate the dies set in two steps.
Electrode manufacturing
Electrode produces by CNC milling followed by hand finishing (polishing).
Tool steel for die
Boring
The steel die block is bored and the basic die shape machined into the block. This is the roughing stage.
Structure cutting (milling)
CNC milling of the basic die cavity.
EDM cavity
Electric discharge machining cuts the die cavity in the steel.
Electro-polishing cavity
The die cavity receives a finished surface by electropolishing.

FIGURE 25-19 The sequence of operations to manufacture dies for forging processes is shown on the left. Using ultra-high-speed machining centers reduces the sequence to two steps. (See Figure 25-20 for UHSMC details.)

25.5 SUMMARY

Computer numerical control machines are critical elements of the advanced manufacturing technologies available for today's factories, as computer-integrated manufacturing (CIM) becomes closer to reality. Computer technology abounds: computer-aided design; computer-aided manufacturing with NC, CNC, or A/C along with DNC systems computer-aided process planning; computer-aided testing and inspection (CATI); artificial intelligence; smart robots; and much more. Of course, any company can buy computers, robots, and other pieces of automation hardware and software. The secret to manufacturing success lies in the design of a simple, unique manufacturing system that can achieve superior quality at low cost with on-time delivery and still be flexible. Flexibility means the system can readily adapt to changes in the customer demand

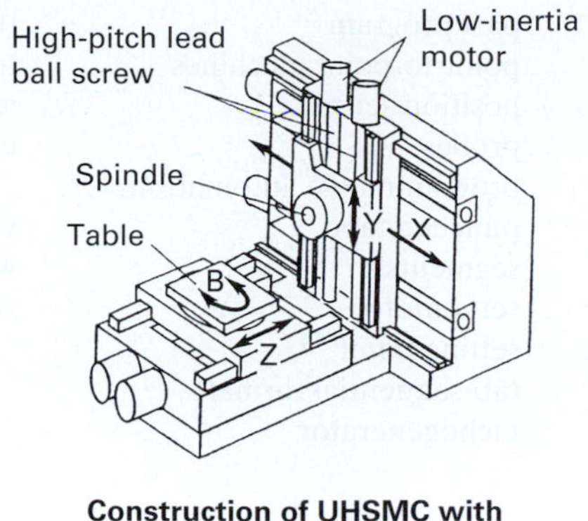

Construction of UHSMC with Horizontal Spindle

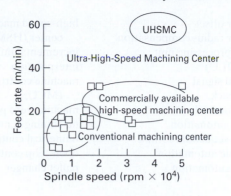

Capabilities of Machining Centers

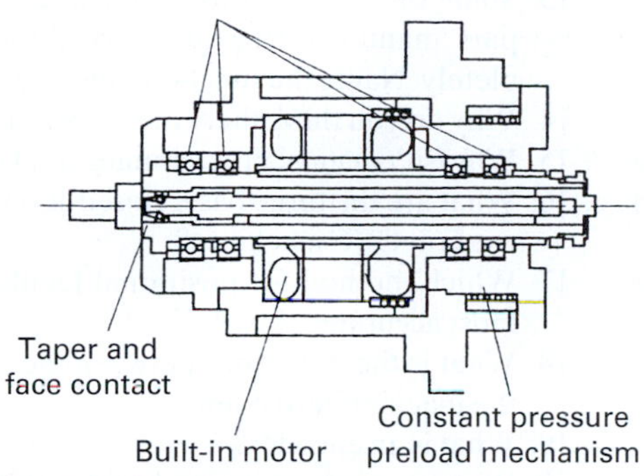

Construction of Spindle

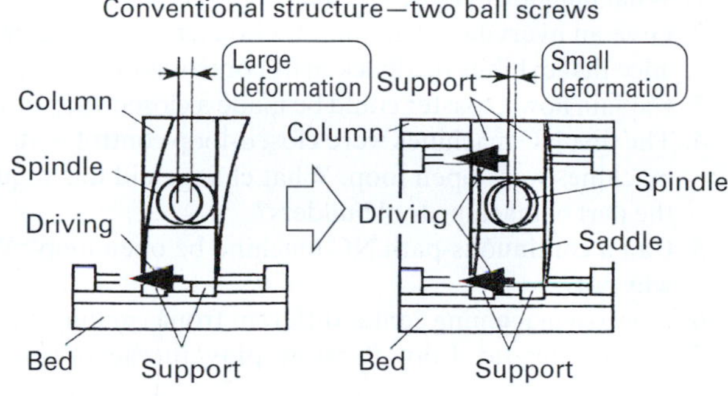

Feed Mechanism of Conventional (x-axis) MC vs UHSMC

FIGURE 25-20 Ultra-high-speed machining centers (UHSMCs) are being developed with ceramic ball bearings in the spindles, synchronized ball screws on the x-axis to reduce distortion (due to inertia) in the moving components. *(From "Development of Ultra High Speed Machining Center",* Toyota Technical Review, *vol. 49, No. 1, September 1999)*

(both volume and mix) while quickly implementing engineer design changes. Changing how people work in a manufacturing system means you have to change the manufacturing system design, but change is always difficult to implement.

No better example of this can be found than the Toyota Motor Company. Led by their vice president for manufacturing, Taiichi Ohno, who conceived, developed, and implemented Toyota's unique manufacturing system, this company has emerged as the world leader in car production. The implementation of this system has saved many companies (e.g., Harley Davidson) and carried many others to the top position in their industry. Recent surveys show that more than 60% of all manufacturing industries are implementing some version of the Toyota Production System. The Toyota system is unique and as revolutionary today as the American Armory System (job shops) and the Ford system (flow shops) were in their day. This new system is now being called *lean production* (to contrast it to mass production). Toyota does not use the term *CIM* because the computer is only a tool used in their system, a manufacturing system design that recognizes people as the most flexible element. Today's manufacturing engineer is a lean engineer, fully knowledgeable regarding this new system.

■ Key Words

absolute information
adaptive control (A/C)
address system
analog signal
automation
basic-length unit (BLU)
block
block-address format
canned cycles
closed-loop control
computer numerical control (CNC)
continuous path
contouring
controller
control-loop unit (CLU)

cutter offset
cutter radius compensation
data-processing unit (DLU)
digital signal
feedback
feedforward
fixed automation
fixed-sequential format
flexible automation
hard automation
high-speed machining center (HSMC)
incremental information
interpolation
machine control unit (MCU)
machine zero point
machining center
numerical control (NC)
open-loop control
pallet changer
part program
point-to-point machines
position sensor
probes
programmable automation
punched tape
segments
servomotor
setting point
tab-sequential format
tachogenerator
therma lerror
tool dimensions
turning center
ultra-high-speed machining centers (UHSMCs)
verification
word-address format
zero reference point

REVIEW QUESTIONS

1. What human attribute is replaced by an NC machine?
2. Give an everyday example of a household device or appliance that exhibits feedback in its control system.
3. Explain how a toaster could be made a closed-loop device.
4. The first NC machines were closed-loop control. Later some machines were open loop. What change did this require on the part of machine tool builders?
5. Can a continuous-path NC machine be open loop? Why or why not?
6. How is a machining center different from a milling machine?
7. What role did John Parsons play in the development of NC?
8. Many manufacturers have purchased large machining centers tied to an ASRS (automated storage and retrieval system). Go on the Internet to find a video of an ASRS, and explain how such a system works.
9. How does feedforward differ from feedback in a process control system?
10. Why was it necessary for machine tool builders to improve the leadscrews on their machines when they made them into NC machines?
11. Explain what is meant by interpolation in NC programming.
12. Explain the problem of cutter offset in NC programming by making a sketch showing an end mill cutting a perimeter on a square plate.
13. Some of the functions performed by the operator in piece-part manufacturing are very difficult to automate completely. Name the functions and explain why.
14. Why do you think there are no NC shapers?
15. Why isn't manual programming used for continuous-path NC?
16. What are the three basic closed-loop feedback A(4) schemes used on CNC machines?
17. Which method for positional feedback do you think is the most accurate? Why?
18. What is the difference between the zero reference point and the machine zero point?
19. What is an encoder?
20. What is a peck-drilling subroutine for a CNC machine?
21. What is pocket milling, and what kind of milling cutter is usually used to perform it?
22. How are probes used in CNC machines to improve process capability?
23. What are G words used for in NC?
24. What structural major changes are introduced into a UHSMCs?
25. What process steps are eliminated by the UHSMC in die production?
26. Draw a large circle and imagine you are (the tool) moving from point A to B on a semicircular segment. What is different about the increments and velocities here compared to linear interpolation?

PROBLEMS

1. What are the x and y dimensions for the center position of holes 1, 2, and 3 in the part shown in Figure 25-3?
2. Configurations obtained from continuous-path machining are the result of a series of straight-line, parabolic-span, or higher-order curves. The degree to which curved surfaces correspond to their design depends on how many lines or spans are used. Four equal chords in a circle describe a square (Figure 25-A). Six make a hexagon. As the number of sides increases, the lines themselves come closer to a perfect circle. The number of lines needed is determined by a maximum tolerance allowed between the design of the curved section and the actual chord programmed. Call this the dimension T. The program for a parabolic-span control unit requires enough spans for any deviation to stay within an acceptable tolerance. For a tolerance of $T = 0.001$ in., how long should the span be for a curve with a 5-in. radius? Assume that the arc is part of a circle. What is the span angle here, in degrees?
3. In Problem 2, suppose that the acceptable tolerance, T, was 0.0001 in. Determine the span angle.

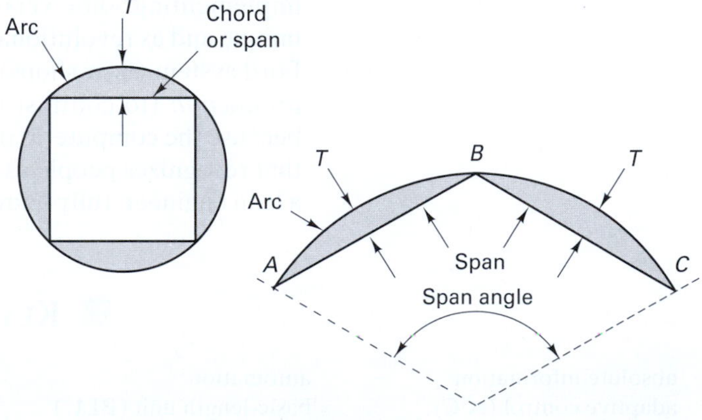

FIGURE 25-A

4. Suppose that the plate shown in Figure 25-B was to be profile milled around the periphery with a 1-in.-diameter end milling cutter. The dashed line is the cutter path. The programmer must calculate an offset path to allow for cutter diameter. Because the programmed points are followed by the cutter centerline and the profile is made at the tool's periphery, the programmer called for a $\frac{1}{2}$-in. cutter offset. Working with computer assistance, the programmer would describe the part profile to be machined and specify the cutter. The computer would generate the cutter path. Complete the following table to specify the cutter path, starting with the origin at the zero reference point. Move the tool around the plate counterclockwise.

Programmed Point Locations

PT	*x*	*y*
1		
2		
3		
4		
5		
1		
5.		

5. Suppose that surface finish is very important for the profile milling job described in Problem 4. Thus down milling is going to be used. Rewrite the NC program points to accommodate this requirement. Show the new path on a sketch such as Figure 25-B. (Up versus down milling is discussed in Chapter 24.)

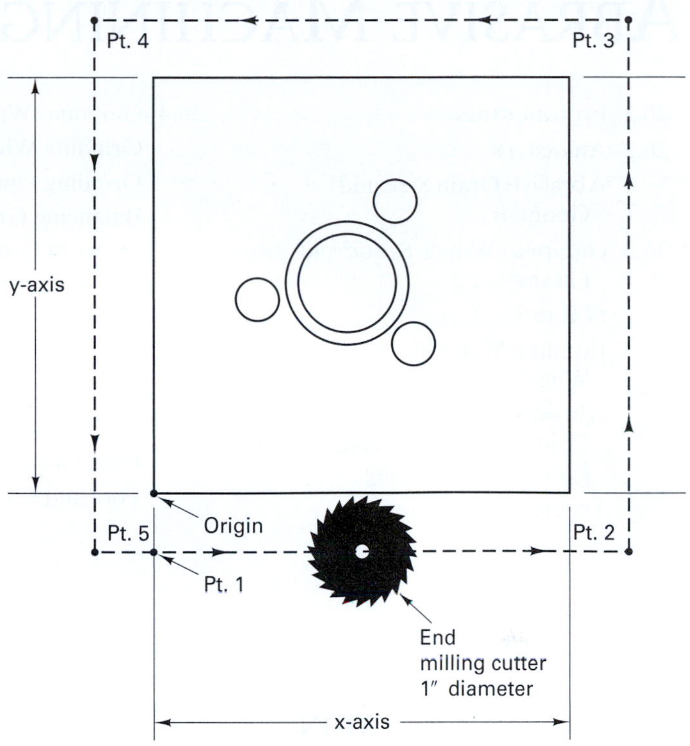

FIGURE 25-B

CHAPTER 26

ABRASIVE MACHINING PROCESSES

26.1 INTRODUCTION

Abrasive machining is a material removal process that involves the interaction of abrasive grits with the workpiece at high cutting speeds and shallow penetration depths. The chips that are formed resemble those formed by other machining processes. Unquestionably, abrasive machining is the oldest of the basic machining processes. Museums abound with examples of utensils, tools, and weapons that ancient peoples produced by rubbing hard stones against softer materials to abrade away unwanted portions, leaving desired shapes. For centuries, only natural abrasives were available for grinding, while other more modern basic machining processes were developed using superior cutting materials. However, the development of manufactured abrasives and a better fundamental understanding of the abrasive machining process have resulted in placing abrasive machining and its variations among the most important of all the basic machining processes.

The results that can be obtained by abrasive machining range from the finest and smoothest surfaces produced by any machining process, in which very little material is removed, to rough, coarse surfaces that accompany high material removal rates. The abrasive particles may be (1) free; (2) mounted in resin on a belt (called **coated product**); or, most commonly (3) close packed into wheels or stones, with abrasive grits held together by bonding material (called **bonded product** or a grinding wheel). Figure 26-1 shows a surface grinding process using a grinding wheel. The depth of cut d is determined by the infeed and is usually very small, 0.002 to 0.005 in., so the arc of contact (and the chips) is small. The table reciprocates back and forth beneath the rotating wheel. The work feeds into the wheel in the cross-feed direction. After the work is clear of the wheel, the wheel is lowered and another pass is made, again removing a couple of thousandths of inches of metal. The metal removal process is basically the same in all abrasive machining processes but with important differences due to spacing of active grains (grains in contact with work) and the rigidity and degree of fixation of the grains. Table 26-1 summarizes the primary abrasive processes. The term *abrasive machining* applied to one particular form of the grinding process is unfortunate, because all these process are machining with abrasives.

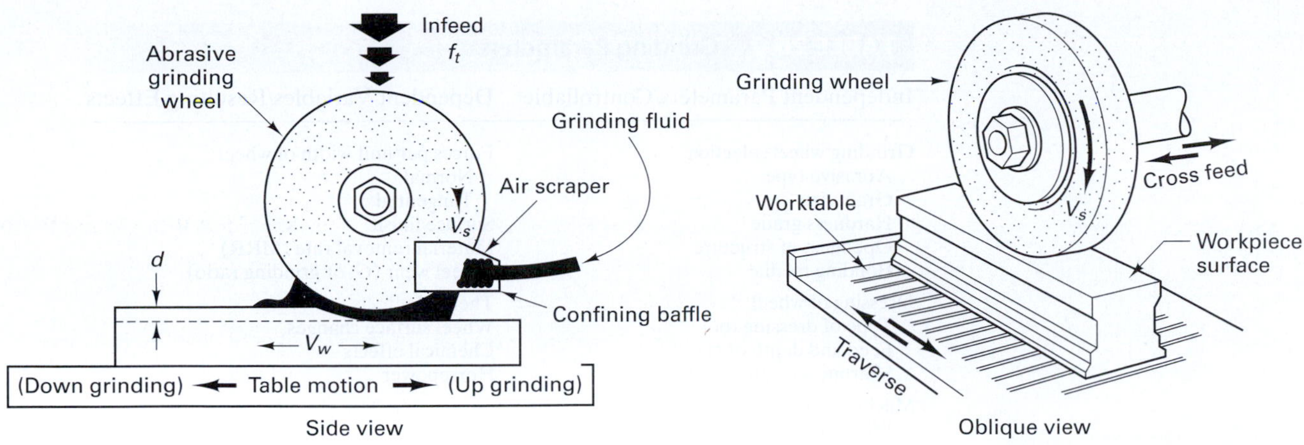

FIGURE 26-1 Schematic of surface grinding, showing infeed and cross-feed motions along with cutting speeds, V_S, and workpiece velocity, V_W.

Compared to machining, abrasive machining processes have three unique characteristics. First, each cutting edge is very small, and many of these edges can cut simultaneously. When suitable machine tools are employed, very fine cuts are possible, and fine surfaces and close dimensional control can be obtained. Second, because extremely hard abrasive grits, including diamonds, are employed as cutting tool materials, very hard materials, such as hardened steel, glass, carbides, and ceramics, can readily be machined. As a result, the abrasive machining processes are not only important as manufacturing processes, they are indeed essential. Many of our modern products, such as modern machine tools, automobiles, space vehicles, and aircraft, could not be manufactured without these processes. Third, in grinding, you have no control over the actual tool geometry (rake angles, cutting edge radius) or all the cutting parameters (depth of cut). As a result of these parameters and variables, grinding is a complex process.

To get a handle on the complexity, Table 26-2 presents the primary grinding parameters, grouped by their independence or dependence. Independent variables are those that are controllable (by the machine operator) while the dependent variables are the resultant effects of those inputs. Not listed in the table is workpiece hardness, which has a significant effect on all the resulting effects. Workpiece hardness will be an input factor but it is not usually controllable.

26.2 ABRASIVES

An **abrasive** is a hard material that can cut or abrade other substances. Natural abrasives have existed from the earliest times. For example, sandstone was used by ancient

TABLE 26-1 Abrasive Machining Processes

Process	Particle Mounting	Features
Grinding	Bonded	Uses wheels, accurate sizing, finishing, low MRR; can be done at high speeds (>12,000 sfpm)
Creep feed grinding	Bonded open, soft	Uses wheels with long cutting arc, very slow feed rate, and large depth of cut
Abrasive machining	Bonded	High MRR, to obtain desired shapes and approximate sizes
Snagging	Bonded belted	High MRR, rough rapid technique to clean up and deburr castings, forgings
Honing	Bonded	"Stones" containing fine abrasives; primarily a hole-finishing process
Lapping	Free	Fine particles embedded in soft metal or cloth; primarily a surface-finishing process
Abrasive water-jet	Free in jet	Water jets with velocities up to 3000 sfpm* carry abrasive particles (silica and garnet)
Ultrasonic	Free in liquid	Vibrating tool impacts abrasives at high velocity
Abrasive flow	Free in gel	Abrasives in gel flow over surface-edge finishing
Abrasive jet	Free in	A focused jet of abrasives in an inert gas at high velocity

* sfpm = surface feet per minute; this is the cutting velocity.

TABLE 26-2 Grinding Parameters

Independent Parameters/Controllable	Dependent Variables/Resulting Effects
Grinding wheel selection	Forces per unit width of wheel
Abrasive type	Normal
Grain size	Tangential
Hardness grade	Surface finish
Openness of structure	Material removal rate (MRR)
Bonding media	Wheel wear (*G*, or grinding ratio)
Dressing of wheel	Thermal effects
Type of dressing tool	Wheel surface changes
Feed and depth of cut	Chemical effects
Sharpness of dressing tool	Horsepower
Machine settings	
Wheel speed	
Infeed rate (depth of cut)	
Cross-feed rate	
Workpiece speed	
Rigidity of setup	
Type and quality of machine	
Grinding fluid	
Type	
Cleanliness	
Method of application	

peoples to sharpen tools and weapons. Early grinding wheels were cut from slabs of sandstone, but because they were not uniform in structure throughout, they wore unevenly and did not produce consistent results. **Emery,** a mixture of alumina (Al_2O_3) and magnetite (Fe_3O_4), is another natural abrasive still in use today and is used on coated paper and cloth (emery paper). **Corundum** (natural Al_2O_3) and diamonds are other naturally occurring abrasive materials. Today, the only natural abrasives that have commercial importance are quartz, sand, garnets, and diamonds. For example, **quartz** is used primarily in coated abrasives and in air blasting, but artificial abrasives are also making inroads in these applications. The development of artificial abrasives having known uniform properties has permitted abrasive processes to become precision manufacturing processes.

Hardness, the ability to resist penetration, is the key property for an abrasive. Table 26-3 lists the primary abrasives and their approximate Knoop hardness (kg/mm^2). The particles must be able to decompose at elevated temperatures. Two other properties are significant in abrasive grits—attrition and friability. **Attrition** refers to the abrasive wear action of the grits resulting in dulled edges, grit flattening,

TABLE 26-3 Knoop Hardness Values for Common Abrasives

Abrasive Material	Year of Discovery	Hardness (Knoop)	Temperature of Decomposition in Oxygen (°C)	Comments and Uses
Quartz	?	320		Sand blasting
Aluminum oxide	1893	1600–2100	1700–2400	Softer and tougher than silicon carbide; used on steel, iron, brass, silicon
Carbide	1891	2200–2800	1500–2000	Used for brass, bronze, aluminum, and stainless and cast iron
Borazon [cubic boron nitride stainless (CBN)]	1957	4200–5400	1200–1400	For grinding hard, tough tool steels, stainless steel, cobalt and nickel based, superalloys, and hard coatings
Diamond (synthetic)	1955	6000–9000	700–800	Used to grind nonferrous materials, tungsten carbide, and ceramics

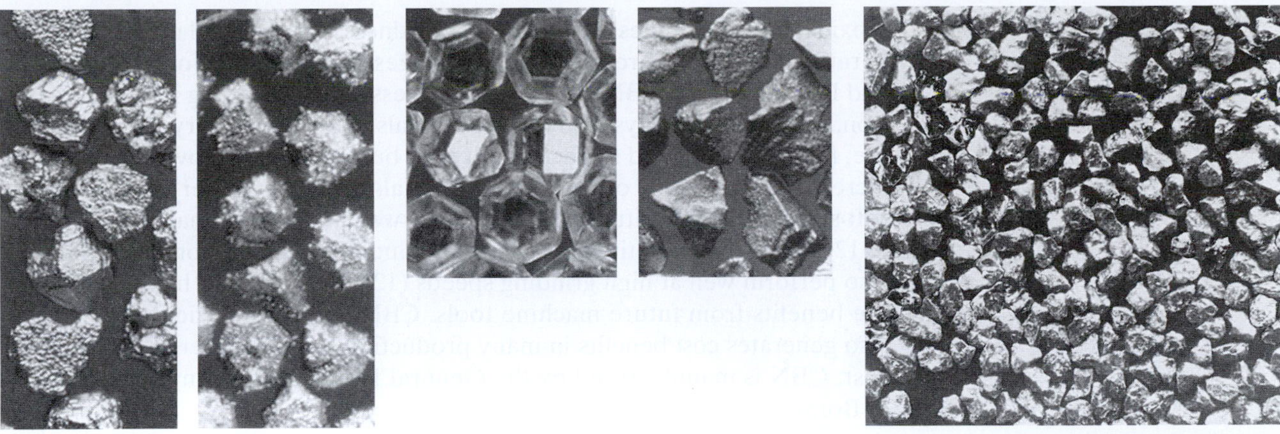

FIGURE 26-2 Loose abrasive grains at high magnification, showing their irregular, sharp cutting edges. *(Courtesy of Norton Abrasives/Saint Gobain)*

and wheel glazing. **Friability** refers to the fracture of the grits and is the opposite of toughness. In grinding, it is important that grits be able to fracture to expose new, sharp edges.

Artificial abrasives date from 1891, when E. G. Acheson, while attempting to produce precious gems, discovered how to make **silicon carbide** (SiC). Silicon carbide is made by charging an electric furnace with silica sand, petroleum coke, salt, and sawdust. By passing large amounts of current through the charge, a temperature of greater than 4000°F is maintained for several hours, and a solid mass of silicon carbide crystals results. After the furnace has cooled, the mass of crystals is removed, crushed, and graded (sorted) into various desired sizes. As can be seen in Figure 26-2, the resulting grits, or grains, are irregular in shape, with cutting edges having every possible rake angle. Silicon carbide crystals are very hard (Knoop 2480), friable, and rather brittle. This limits their use. Silicon carbide is sold under the trade names Carborundum and Crystolon.

Aluminum oxide (Al_2O_3) is the most widely used artificial abrasive. Also produced in an arc furnace from bauxite, iron filings, and small amounts of coke, it contains aluminum hydroxide, ferric oxide, silica, and other impurities. The mass of aluminum oxide that is formed is crushed, and the particles are graded to size. Common trade names for aluminum oxide abrasives are Alundum and Aloxite. Although aluminum oxide is softer (Knoop 2100) than silicon carbide, it is considerably tougher. Consequently, it is a better general-purpose abrasive.

Diamonds are the hardest of all materials. Those that are used for abrasives are either natural, off-color stones (called **garnets**) that are not suitable for gems, or small, synthetic stones that are produced specifically for abrasive purposes. Manufactured stones appear to be somewhat more friable and thus tend to cut faster and cooler. They do not perform as satisfactorily in metal-bonded wheels. Diamond abrasive wheels are used extensively for sharpening carbide and ceramic cutting tools. Diamonds also are used for truing and dressing other types of abrasive wheels. Diamonds are usually used only when cheaper abrasives will not produce the desired results. Garnets are used primarily in the form of very finely crushed and graded powders for fine polishing.

Cubic boron nitride (CBN) is not found in nature. It is produced by a combination of intensive heat and pressure in the presence of a catalyst. CBN is extremely hard, registering at 4700 on the Knoop scale. It is the second-hardest substance created by nature or manufactured and is often referred to, along with diamonds, as a superabrasive. Hardness, however, is not everything.

CBN far surpasses diamond in the important characteristic of thermal resistance. At temperatures of 650°C, at which diamond may begin to revert to plain carbon dioxide, CBN continues to maintain its hardness and chemical integrity. When the temperature of 1400°C is reached, CBN changes from its cubic form to a hexagonal form and loses hardness. CBN can be used successfully in grinding iron, steel, and alloys of iron, nickel-based alloys, and other materials. CBN works very effectively (long wheel life, high *G* ratio, good surface quality, no burn or chatter, low scrap rate, and overall increase in parts/shift) on hardened materials (R_c 50 or higher). It can also be used for soft steel in selected situations. CBN does well at conventional grinding speeds (6,000 to 12,000 ft/min), resulting in lower total grinding in conventional equipment. CBN can also perform well at high grinding speeds (12,000 ft/min and higher) and will enhance the benefits from future machine tools. CBN can solve difficult-to-grind jobs, but it also generates cost benefits in many production grinding operations despite its higher cost. CBN is manufactured by the General Electric Company under the trade name of Borazon.

ABRASIVE GRAIN SIZE AND GEOMETRY

To enhance the process capability of grinding, abrasive grains are sorted into sizes by mechanical sieving machines. The number of openings per linear inch in a sieve (or screen) through which most of the particles of a particular size can pass determines the grain size (Figure 26-3).

A no. 24 grit would pass through a standard screen having 24 openings per inch but would not pass through one having 30 openings per inch. These numbers have since been specified in terms of millimeters and micrometers (see ANSI B74.12 for details). Commercial practice commonly designates grain sizes from 4 to 24, inclusive, as *coarse;* 30 to 60, inclusive, as *medium;* and 70 to 600, inclusive, as *fine*. Grains smaller than 220 are usually termed *powders*. Silicon carbide is obtainable in grit sizes ranging from 2 to 240 and aluminum oxide in sizes from 4 to 240. Superabrasive grit sizes normally range from 120 grit for CBN to 400 grit for diamond. Sizes from 240 to 600 are designated as *flour* sizes. These are used primarily for lapping, or in fine-honing stones for fine finishing tasks.

The grain size is closely related to the surface finish and metal removal rate. In grinding wheels and belts, coarse grains cut faster (higher removal rate) while fine grains provide better finish, as shown in Figure 26-4.

The grain diameter can be estimated from the screen number (*S*), which corresponds to the number of openings per inch. The mean diameter of the grain (*g*) is related to the screen number by $g \cong 0.7/S$.

Regardless of the size of the grain, only a small percentage (2 to 5%) of the surface of the grain is operative at any one time. That is, the depth of cut for an individual grain (the actual feed per grit) with respect to the grain diameter is very small. Thus, the chips are small. As the grain diameter decreases, the number of active grains per unit area increases and the cuts become finer because grain size is the controlling factor for surface finish (roughness). Of course, the MRR also decreases.

The grain shape is also important, because it determines the tool geometry—that is, the back rake angle and the clearance angle at the cutting edge of the grit

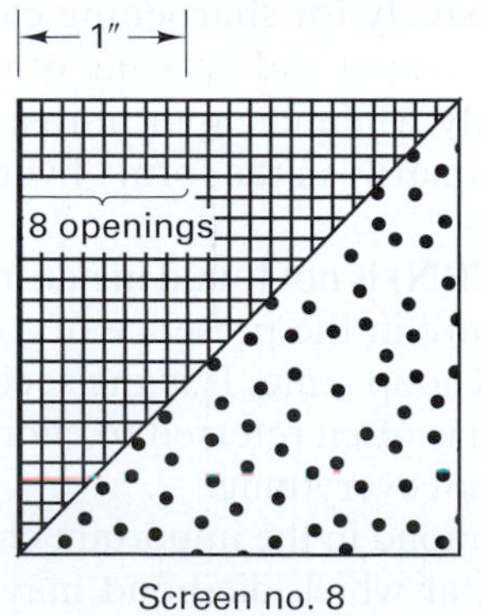

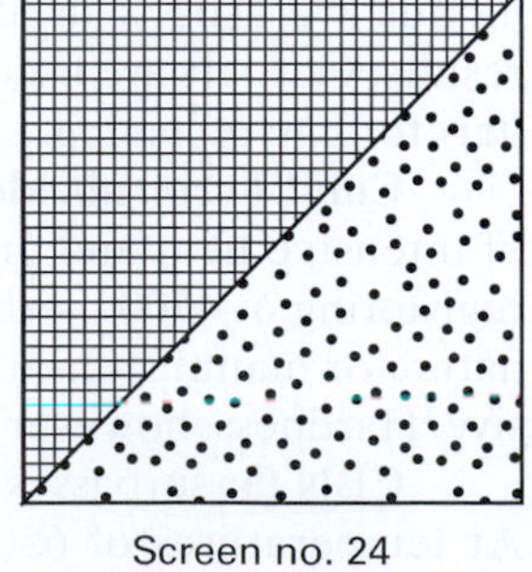

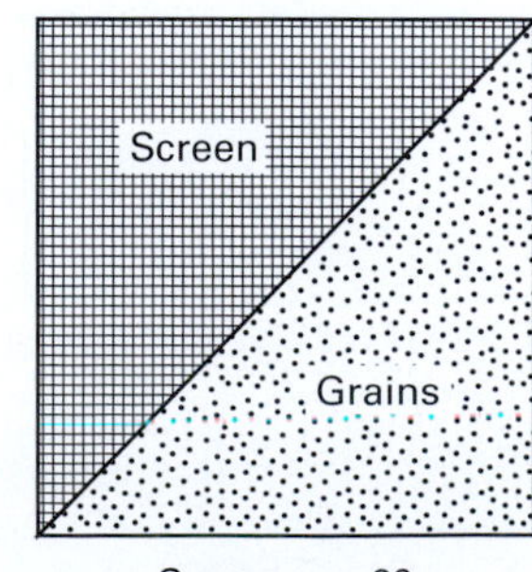

FIGURE 26-3 Typical screens for sifting abrasives into sizes. The larger the screen number (of opening per linear inch), the smaller the grain size. *(Courtesy of Carborundum Company)*

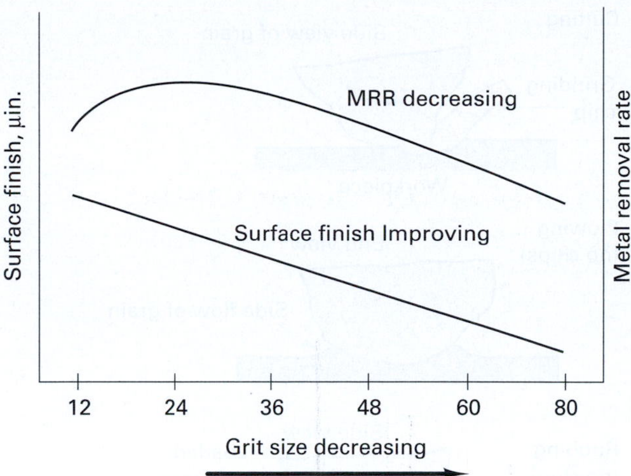

FIGURE 26-4 MRR and surface finish versus grit size.

(Figure 26-5). In the figure, γ is the clearance angle, θ is the wedge angle, and α is the rake angle. The cavities between the grits provide space for the chips, as shown in Figure 26-6. The volume of the cavities must be greater than the volume of the chips generated during the cut.

Obviously, there is no specific rake angle but rather a distribution of angles. Thus a grinding wheel can present to the surface rake angles in the range of +45 to −60 degrees or greater. Grits with large negative rake angles or rounded cutting edges do not form chips but will rub or *plow* a groove in the surface (Figure 26-7). Thus abrasive machining is a mixture of **cutting, plowing,** and **rubbing,** with the percentage of each being highly dependent on the geometry of the grit. As the grits are continuously abraded, fractured, or dislodged from the bond, new grits are exposed and the mixture of cutting, plowing, and rubbing is changing continuously. A high percentage of the

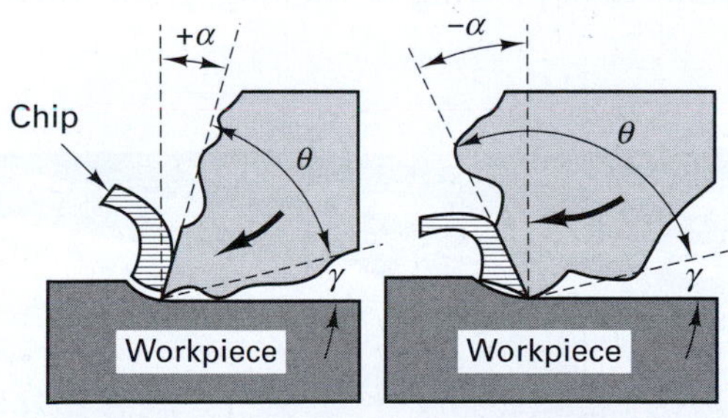

FIGURE 26-5 The rake angle of abrasive particles can be positive, zero, or negative.

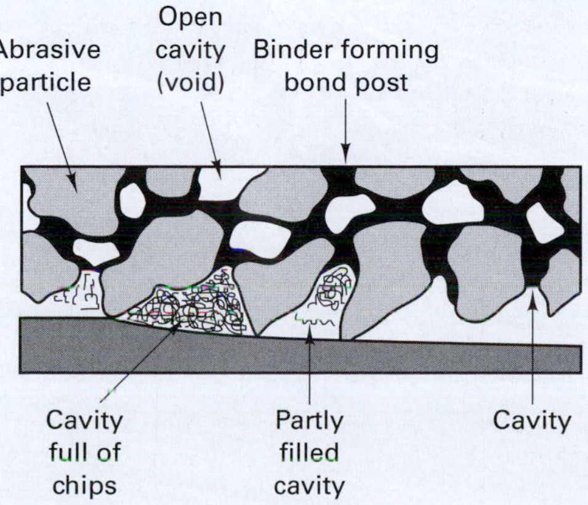

FIGURE 26-6 The cavities or voids between the grains must be large enough to hold all the chips during the cut.

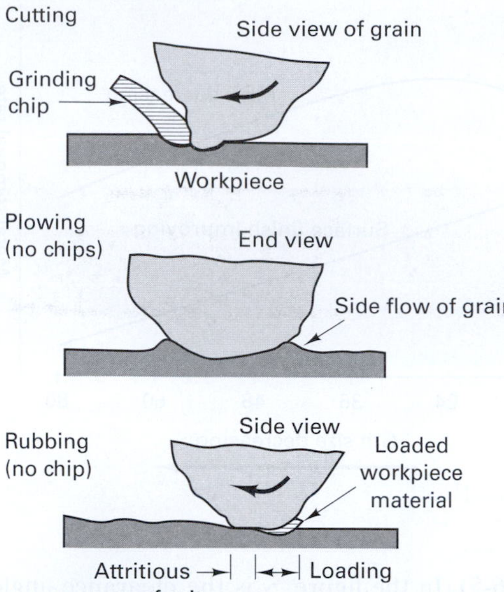

FIGURE 26-7 The grits interact with the surface in three ways: cutting, plowing, and rubbing.

energy used for rubbing and plowing goes into the workpiece, but when chips are found, 95 to 98% of the energy (the heat) goes into the chip. Figure 26-8 shows a scanning electron microscope (SEM) micrograph of a ground surface with a plowing track.

In grinding, the chips are small but are formed by the same basic mechanism of compression and shear as discussed in Chapter 20 for regular metal cutting. Figure 26-9 shows steel chips from a grinding process at high magnification. They show the same structure as chips from other machining processes. Chips flying in the air from a grinding process often have sufficient heat energy to burn or melt in the atmosphere. Sparks observed during grinding steel with no cutting fluid are really burning chips.

The feeds and depths of cut in grinding are small while the cutting speeds are high, resulting in high specific horsepower numbers. Because cutting is obviously more efficient than plowing or rubbing, grain fracture and grain pullout are natural

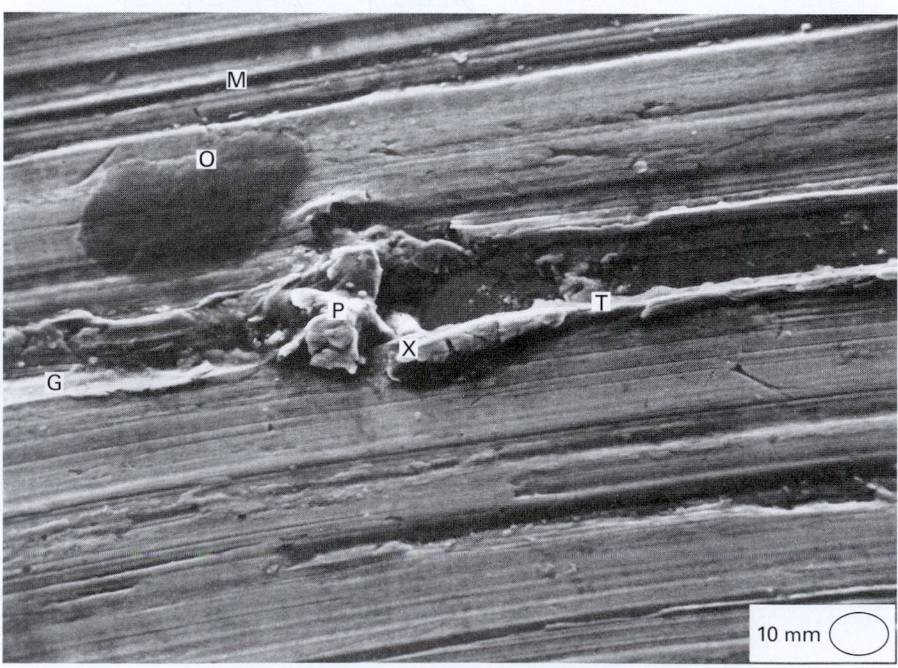

FIGURE 26-8 SEM micrograph of a ground steel surface showing a plowed track (T) in the middle and a machined track (M) above. The grit fractured, leaving a portion of the grit in the surface (X), a prow formation (P), and a groove (G) where the fractured portion was pushed farther across the surface. The area marked (O) is an oil deposit. *(Courtesy J T. Black)*

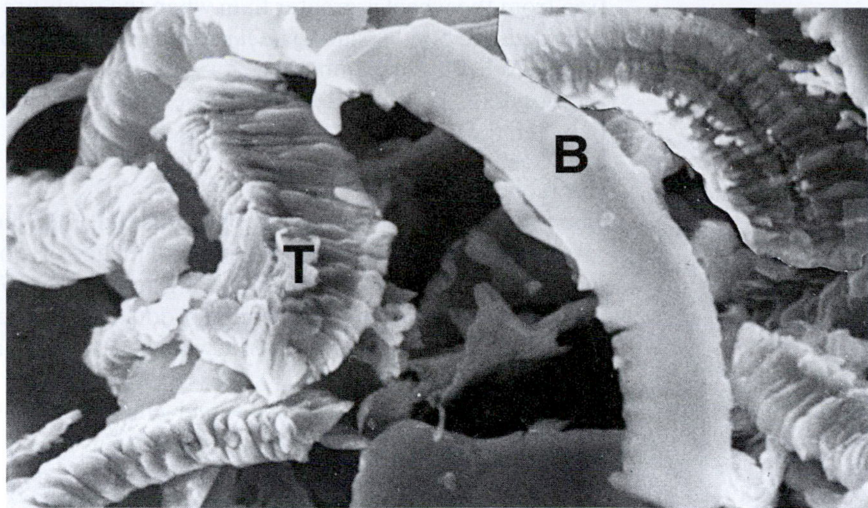

FIGURE 26-9 SEM micrograph of stainless steel chips from a grinding process. The tops (T) of the chips have the typical shear-front-lamella structure while the bottoms (B) are smooth where they slide over the grit 4800×. *(Courtesy J T. Black)*

phenomena used to keep the grains sharp. As the grains become dull, cutting forces increase, and there is an increased tendency for the grains to fracture or break free from the bonding material.

26.3 GRINDING WHEEL STRUCTURE AND GRADE

Grinding, wherein the abrasives are bonded together into a wheel, is the most common abrasive machining process. The performance of grinding wheels is greatly affected by the bonding material and the spatial arrangement of the particles' grits.

The spacing of the abrasive particles with respect to each other is called **structure.** Close-packed grains have dense structure; open structure means widely spaced grains. Open-structure wheels have larger chip cavities but fewer cutting edges per unit area (Figure 26-10a).

The fracturing of the grits is controlled by the bond strength, which is known as the **grade.** Thus, grade is a measure of how strongly the grains are held in the wheel. It is really dependent on two factors: the strength of the bonding materials and the amount of the bonding agent connecting the grains. The latter factor is illustrated in Figure 26-10b. Abrasive wheels are really porous. The grains are held together with "posts" of bonding material. If these posts are large in cross section, the force required to break a grain

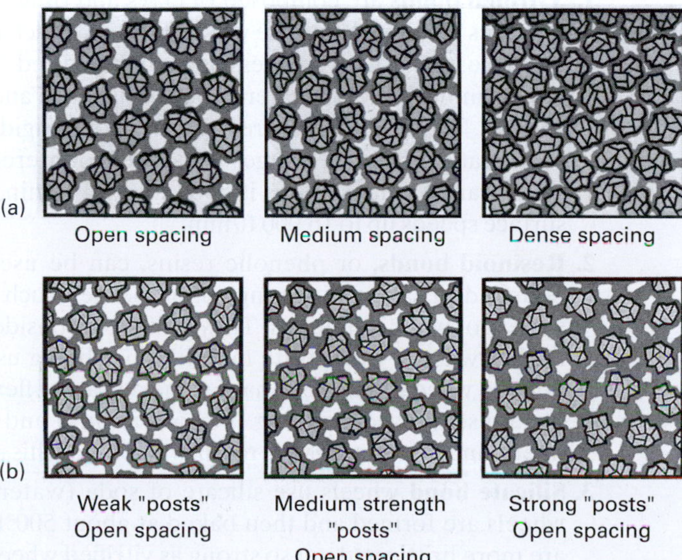

FIGURE 26-10 Meaning of terms *structure* and *grade* for grinding wheels. (a) The structure of a grinding wheel depends on the spacing of the grits. (b) The grade of a grinding wheel depends on the amount of bonding agent (posts) holding abrasive grains in the wheel.

free from the wheel is greater than when the posts are small. If a high dislodging force is required, the bond is said to be *hard*. If only a small force is required, the bond is said to be *soft*. Wheels are commonly referred to as hard or soft, referring to the net strength of the bond, resulting from both the strength of the bonding material and its disposition between the grains.

G RATIO

The loss of grains from the wheel means that the wheel is changing size. The grinding ratio, or ***G* ratio,** is defined as the cubic inches of stock removed divided by the cubic inches of wheel lost. In conventional grinding, the *G* ratio is in the range 20:1 to 80:1. The *G* ratio is a measure of grinding production and reflects the amount of work a wheel can do during its useful life. As the wheel loses material, it must be reset or repositioned to maintain workpiece size.

A typical vitrified grinding wheel will consist of 50 vol% abrasive particles, 10 vol% bond, and 40 vol% cavities; that is, the wheels have porosity. The manner in which the wheel performs is influenced by the following factors:

1. The mean force required to dislodge a grain from the surface (the grade of the wheel).
2. The cavity size and distribution of the porosity (the structure).
3. The mean spacing of active grains in the wheel surface (grain size and structure).
4. The properties of the grain (hardness, attrition, and friability).
5. The geometry of the cutting edges of the grains (rake angles and cutting-edge radius compared to depth of cut).
6. The process parameters (speeds, feeds, cutting fluids) and type of grinding (surface, or cylindrical).

It is easy to see why grinding is a complex process, difficult to control.

BONDING MATERIALS FOR GRINDING WHEELS

Bonding material is a very important factor to be considered in selecting a grinding wheel. It determines the strength of the wheel, thus establishing the maximum operating speed. It determines the elastic behavior or deflection of the grits in the wheel during grinding. The wheel can be hard or rigid, or it can be flexible. Finally, the bond determines the force required to dislodge an abrasive particle from the wheel and thus plays a major role in the cutting action. Bond materials are formulated so that the ratio of bond wear matches the rate of wear of the abrasive grits. Bonding materials in common use are the following:

1. **Vitrified bonds** are composed of clays and other ceramic substances. The abrasive particles are mixed with the wet clays so that each grain is coated. Wheels are formed from the mix, usually by pressing, and then dried. They are then fired in a kiln, which results in the bonding material becoming hard and strong, having properties similar to glass. Vitrified wheels are porous, strong, rigid, and unaffected by oils, water, or temperature over the ranges usually encountered in metal cutting. The operating speed range in most cases is 5500 to 6500 ft/min, but some wheels now operate at surface speeds up to 16,000 ft/min.
2. **Resinoid bonds,** or phenolic resins, can be used. Because plastics can be compounded to have a wide range of properties, such wheels can be obtained to cover a variety of work conditions. They have, to a considerable extent, replaced shellac and rubber wheels. Composite materials are being used in rubber-bonded or resinoid-bonded wheels that are to have some degree of flexibility or are to receive considerable abuse and side loading. Various natural and synthetic fabrics and fibers, glass fibers, and nonferrous wire mesh are used for this purpose.
3. **Silicate bond** wheels use silicate of soda (waterglass) as the bond material. The wheels are formed and then baked at about 500°F for a day or more. Because they are more brittle and not so strong as vitrified wheels, the abrasive grains are released

more readily. Consequently, they machine at lower surface temperatures than vitrified wheels and are useful in grinding tools when heat must be kept to a minimum.

4. **Shellac-bonded** wheels are made by mixing the abrasive grains with shellac in a heated mixture, pressing or rolling into the desired shapes, and baking for several hours at about 300°F. This type of bond is used primarily for strong, thin wheels having some elasticity. They tend to produce a high polish and thus have been used in grinding such parts as camshafts and mill rolls.
5. **Rubber bonding** is used to produce wheels that can operate at high speeds but must have a considerable degree of flexibility so as to resist side thrust. Rubber, sulfur, and other vulcanizing agents are mixed with the abrasive grains. The mixture is then rolled out into sheets of the desired thickness, and the wheels are cut from these sheets and vulcanized. Rubber-bonded wheels can be operated at speeds up to 16,000 ft/min. They are commonly used for snagging work in foundries and for thin cutoff wheels.
6. **Superabrasive bond** wheels are either electroplated (single layer of superabrasive plated to outside diameter of a steel blank) or a thin-segmented drum of vitrified CBN surrounds a steel core. The steel core provides dimensional accuracy, and the replaceable segments provide durability, homogeneity, and repeatability while increasing wheel life. The latter type of wheels can use resin, metal, or vitrified bonding. Selection of bond grade and structure (also called abrasive concentration) is critical.

For the electroplated wheels, nickel is used to attach a single layer of CBN (or diamond) to the outside diameter (OD) of an accurately ground or turned steel blank. For the vitrified wheel, superabrasives are mixed with bonding media and molded (or preformed and sintered) into segments or a ring. The ring is mounted on a split steel body. Porosity is varied (to alter structure) by varying preform pressure or by using "pore-forming" additives to the bond material that are vaporized during the sintering cycle. The steel-cored segmented design can rotate at 40,000 sfpm (200 m/s) whereas a plain vitrified wheel may burst at 20,000 fpm.

ABRASIVE MACHINING VERSUS CONVENTIONAL GRINDING VERSUS LOW-STRESS GRINDING

The condition wherein very rapid metal removal can be achieved by grinding is the one to which some have applied the term *abrasive machining*. The metal removal rates are compared with, or exceed, those obtainable by milling or turning or broaching, and the size tolerances are comparable. It is obviously just a special type of grinding, using abrasive grains as cutting tools, as do all other types of abrasive machining. Abrasive grinding done in an aggressive way can produce sufficient localized plastic deformation and heat in the surface so as to develop tensile residual stresses, layers of overtempered martensite (in steels), and even microcracks, because this process is quite abusive. See Figure 26-11 for a discussion of residual stresses produced by various surface-grinding processes.

Conventional grinding can be replaced by procedures that develop lower surface stresses when service failures due to fatigue or stress corrosion are possible. This is accomplished by employing softer grades of grinding wheels, reducing the grinding speeds and infeed rates, using chemically active cutting fluids (e.g., highly sulfurized oil or KNO_2 in water), as outlined in the table of grinding conditions in Figure 26-11. These procedures may require the addition of a variable-speed drive to the grinding machine. Generally, only about 0.005 to 0.010 in. of surface stock needs to be finish ground in this way, as the depth of the surface damage due to conventional grinding or abusive grinding is 0.005 to 0.007 in. High-strength steels, high-temperature nickel, and cobalt-based alloys and titanium alloys are particularly sensitive to surface deformation and cracking problems from grinding. Other postprocessing processes, such as polishing, honing, and chemical milling plus peening, can be used to remove the deformed layers in critically stressed parts. It is strongly recommended, however, that testing programs be used

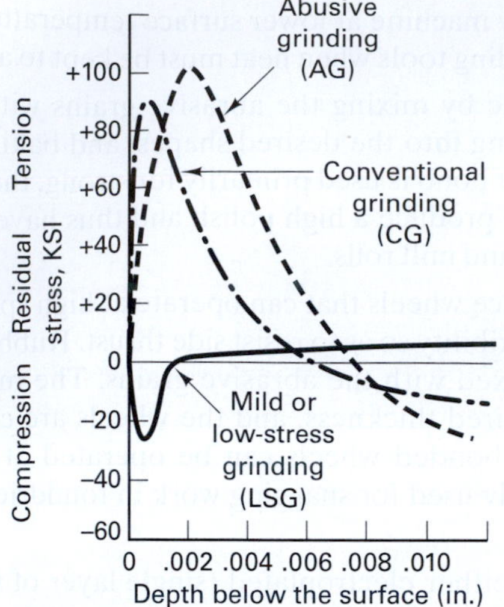

Grinding conditions	Abusive AG	Conventional CG	Low-stress LSG
Wheel	A46MV	A46KV	A46HV or A60IV
Wheel speed ft/min	6,000–18,000	4,500–6,500	2500–3000
Down feed in./pass	.002–.004	.001–.003	.0002–.005
Cross feed in./pass	.004–.060	.040–.060	.040–.060
Table speed ft/min	40–100	40–100	40–100
Fluid	Dry	Sol oil (1 : 20)	Sulfurized oil

FIGURE 26-11 Typical residual stress distributions produced by surface grinding with different grinding conditions for abusive, conventional, and low-stress grinding. Material is 4340 steel. *(From M. Field and M. P. Kahles, "Surface Integrity in Grinding," in* New Developments in Grinding, *Carnegie-Mellon University Press, Pittsburgh, 1972, p. 666)*

along with service experience on critical parts before these procedures are employed in production.

In the casting and forging industries, the term often used for abrasive machining is *snagging*. **Snagging** is a type of rough manual grinding that is done to remove fins, gates, risers, and rough spots from castings or flash from forgings, preparatory to further machining. The primary objective is to remove substantial amounts of metal rapidly without much regard for accuracy, so this is a form of abrasive machining except that pedestal-type or **swing grinders** ordinarily are used. Portable electric or hand air grinders are also used for this purpose and for miscellaneous grinding in connection with welding.

TRUING AND DRESSING

Grinding wheels lose their geometry during use. **Truing** restores the original shape. A single-point diamond tool can be used to *true* the wheel while fracturing abrasive grains to expose new grains and new cutting edges on worn, glazed grains (Figure 26-12). Truing can also be accomplished by grinding the grinding wheel with a controlled-path or

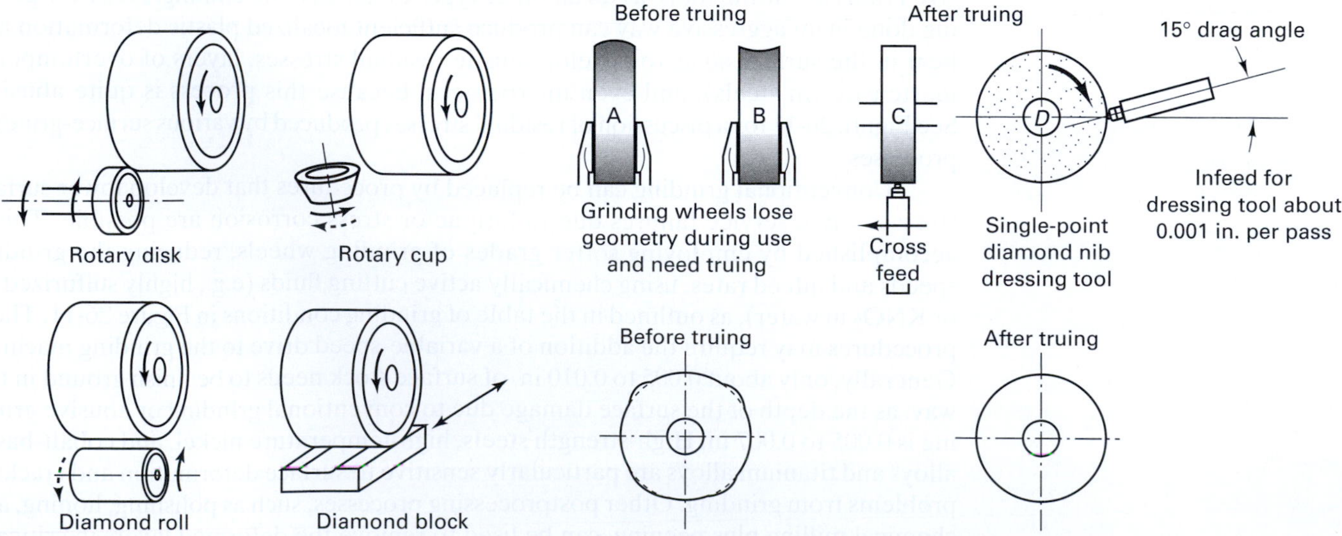

FIGURE 26-12 Truing methods for restoring grinding geometry include nibs, rolls, disks, cups, and blocks.

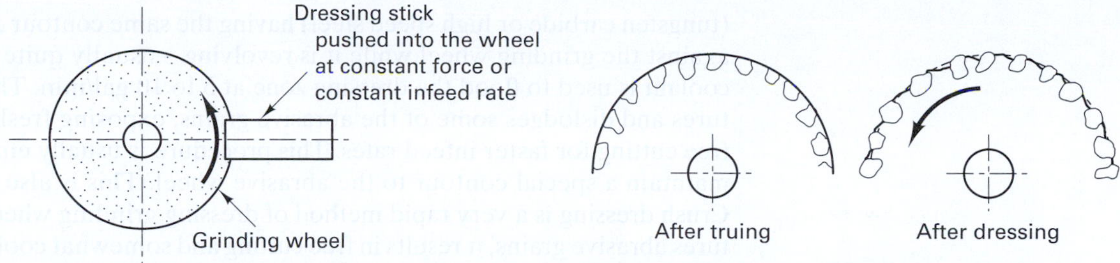

FIGURE 26-13 Schematic arrangement of stick dressing versus truing.

powered rotary device using conventional abrasive wheels. The precision in generating a trued wheel surface by these methods is poorer than by the method described earlier.

As the wheel is used, there is a tendency for the wheel to become *loaded* (metal chips become lodged in the cavities between the grains). Also, the grains dull or glaze (grits wear, flatten, and polish). Unless the wheel is cleaned and sharpened (or dressed), the wheel will not cut as well and will tend to plow and rub more. Figure 26-13 shows an arrangement for stick **dressing** a grinding wheel. The dulled grains cause the cutting forces on the grains to increase, ideally resulting in the grains' fracturing or being pulled out of the bond, thus providing a continuous exposure of sharp cutting edges. Such a continuous action ordinarily will not occur for light feeds and depths of cut. For heavier cuts, grinding wheels do become somewhat self-dressing, but the workpiece may become overheated and turn a bluish temper color (this is called **burn**) before the wheel reaches a fully dressed condition. A burned surface, the consequence of an oxide layer formation, results in the scrapping of several workpieces before parts of good quality are ground.

Resin-bonded wheels can be trued by grinding with hard ceramics such as tungsten carbide. The procedure for truing and dressing a CBN wheel in a surface grinder might be as follows: use 0.0002-in. downfeed per pass and cross feed slightly more than half the wheel thickness at moderate table speeds. The wheel speed is the same as the grinding speed. The grinding power will gradually increase, as the wheel is getting dull, while being trued. When the power exceeds normal power drawn during workpiece grinding, stop the truing operation. Dress the wheel face open using a J-grade stick, with abrasive one grit size smaller than CBN. Continue the truing. Repeat this cycle until the wheel is completely trued.

Modern grinding machines are equipped so that the wheel can be dressed and/or trued continuously or intermittently while grinding continues. A common way to do this is by **crush dressing** (Figure 26-14). Crush dressing consists of forcing a hard roll

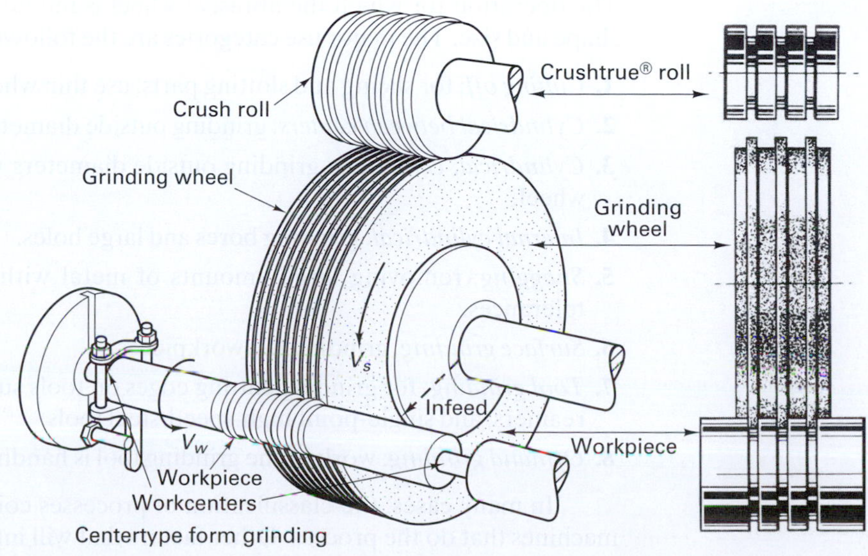

FIGURE 26-14 Continuous crush roll dressing and truing of a grinding wheel (form—truing and dressing throughout the process rather than between cycles) performing plunge-cut grinding on a cylinder held between centers.

(tungsten carbide or high speed steel) having the same contour as the part to be ground against the grinding wheel while it is revolving—usually quite slowly. A water-based coolant is used to flood the dressing zone at 5 to 10 gal/min. The crushing action fractures and dislodges some of the abrasive grains, exposing fresh sharp edges, allowing free cutting for faster infeed rates. This procedure is usually employed to produce and maintain a special contour to the abrasive wheel. This is also called wheel profiling. Crush dressing is a very rapid method of dressing grinding wheels, and because it fractures abrasive grains, it results in free cutting and somewhat cooler grinding. The resulting surfaces may be slightly rougher than when diamond dressing is used.

■ 26.4 GRINDING WHEEL IDENTIFICATION

Most grinding wheels are identified by a standard marking system that has been established by the American National Standards Institute (ANSI). This system is illustrated and explained in Figure 26-15. The first and last symbols in the marking are left to the discretion of the manufacturer.

GRINDING WHEEL GEOMETRY

The shape and size of the wheel are critical selection factors. Obviously, the shape must permit proper contact between the wheel and all of the surface that must be ground. Grinding wheel shapes have been standardized, and eight of the most commonly used types are shown in Figure 26-16. Types 1, 2, and 5 are used primarily for grinding external or internal cylindrical surfaces and for plain surface grinding. Type 2 can be mounted for grinding on either the periphery or the side of the wheel. Type 4 is used with tapered safety flanges so that if the wheel breaks during rough grinding, such as snagging, these flanges will prevent the pieces of the wheel from flying and causing damage. Type 6, the straight cup, is used primarily for surface grinding but can also be used for certain types of offhand grinding. The flaring-cup type of wheel is used for tool grinding. Dish-type wheels are used for grinding tools and saws.

Type 1, the straight grinding wheels, can be obtained with a variety of standard faces. Some of these are shown in Figure 26-17.

The size of the wheel to be used is determined primarily by the spindle rpm values available on the grinding machine and the proper cutting speed for the wheel, as dictated by the type of bond. For most grinding operations the cutting speed is about 2500 to 6500 ft/min. Different types and grades of bond often justify considerable deviation from these speeds. For certain types of work using special wheels and machines, as in thread grinding and "abrasive machining," much higher speeds are used.

GRINDING OPERATIONS

The operation for which the abrasive wheel is intended will also influence the wheel shape and size. The major use categories are the following:

1. *Cutting off*: for slicing and slotting parts; use thin wheel, organic bond.
2. *Cylindrical between centers*: grinding outside diameters of cylindrical workpieces.
3. *Cylindrical, centerless*: grinding outside diameters with work rotated by regulating wheel.
4. *Internal cylindrical*: grinding bores and large holes.
5. *Snagging*: removing large amounts of metal without regard to surface finish or tolerances.
6. *Surface grinding*: grinding flat workpieces.
7. *Tool grinding*: for grinding cutting edges on tools such as drills, milling cutters, taps, reamers, and single-point high-speed-steel tools.
8. *Offhand grinding*: work or the grinding tool is handheld.

In many cases, the classification of processes coincides with the classification of machines that do the process. Other factors that will influence the choice of wheel to be

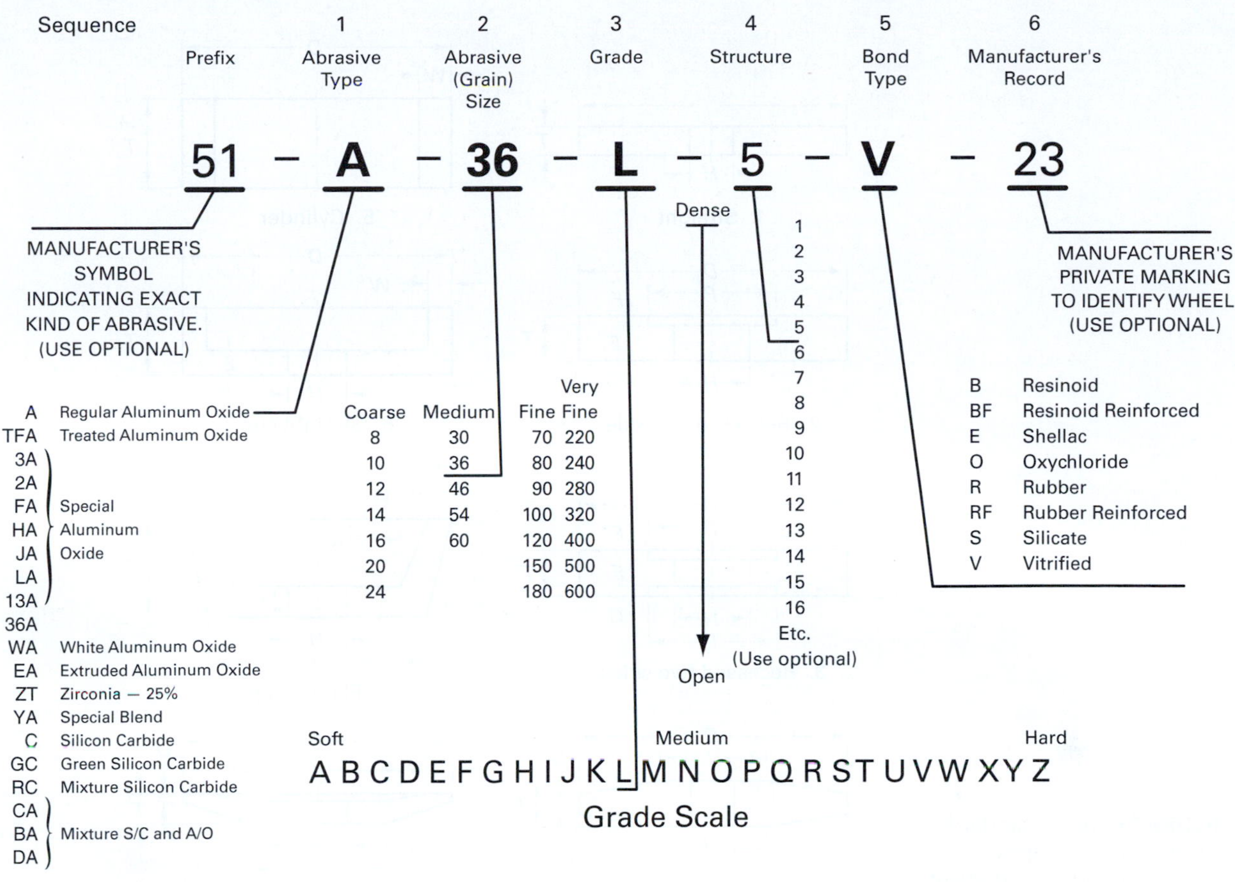

Standard bonded-abrasive wheel-marching system (ANSI *Standard* B74.13-1977)

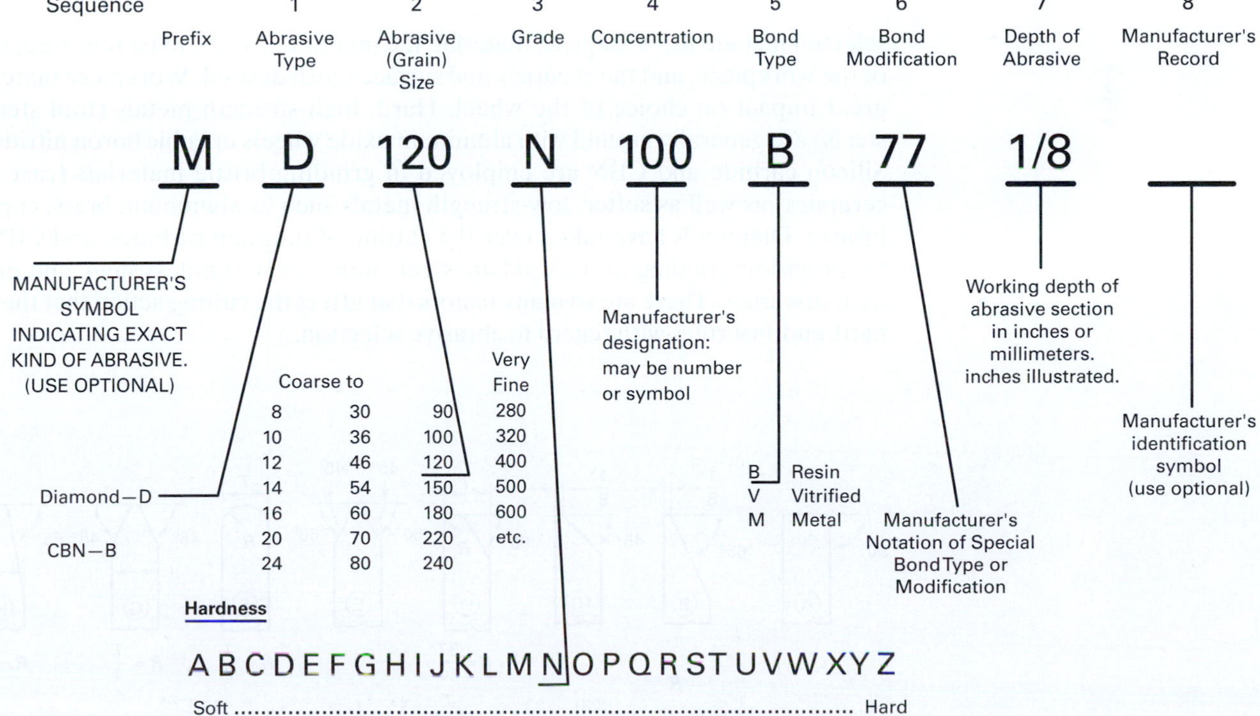

Wheel-marching system for diamond and cubic boron nitride wheels (ANSI *Standard* B74.13-1977).

FIGURE 26-15 Standard marking systems for grinding wheels (ANSI standard B74. 13-1977).

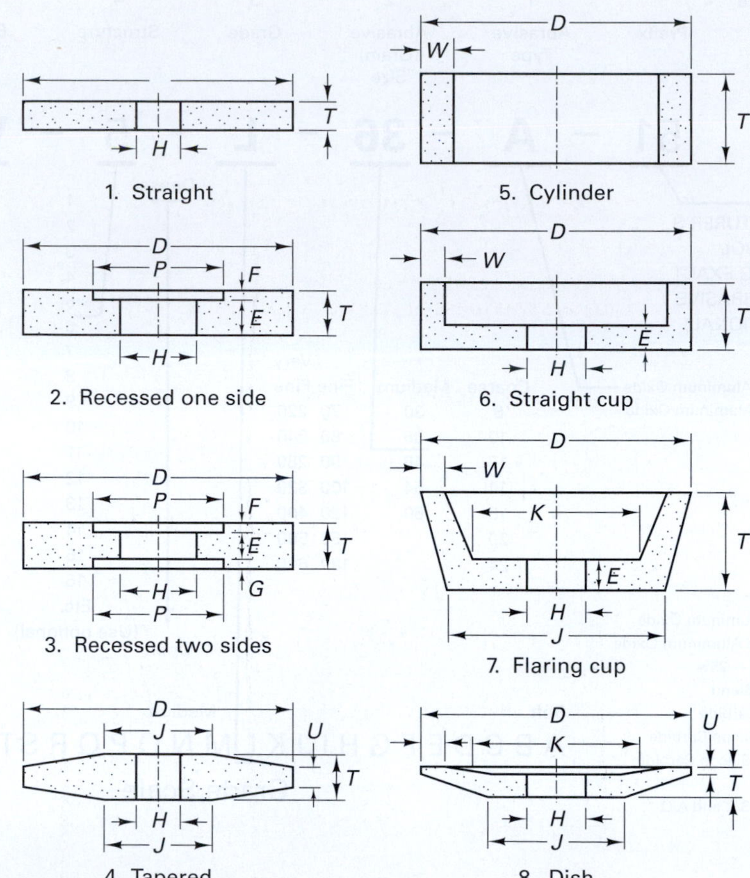

FIGURE 26-16 Standard grinding wheel shapes commonly used. *(Courtesy of Carborundum Company)*

selected include the workpiece material, the amount of stock to be removed, the shape of the workpiece, and the accuracy and surface finish desired. Workpiece material has a great impact on choice of the wheel. Hard, high-strength metals (tool steels, alloy steels) are generally ground with aluminum oxide wheels or cubic boron nitride wheels. Silicon carbide and CBN are employed in grinding brittle materials (cast iron and ceramics) as well as softer, low-strength metals such as aluminum, brass, copper, and bronze. Diamonds have taken over the cutting of tungsten carbides, and CBN is used for precision grinding of tool and die steel, alloy steels, stainless steel, and other very hard materials. There are so many factors that affect the cutting action that there are no hard-and-fast rules with regard to abrasive selection.

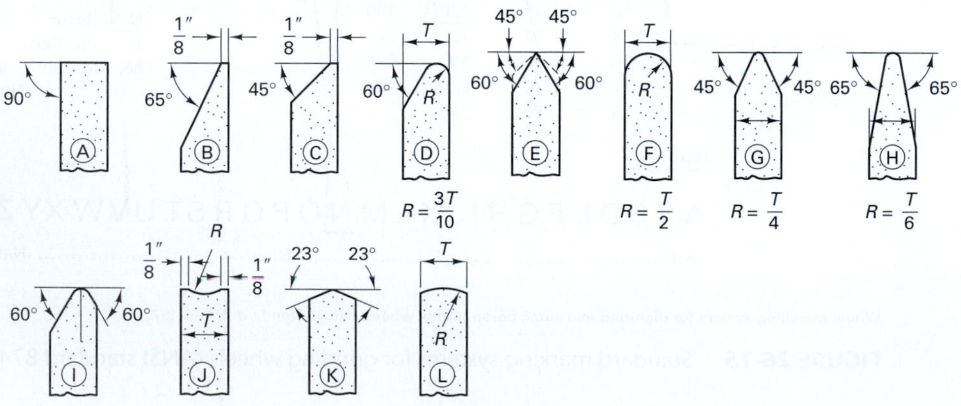

FIGURE 26-17 Standard face contours for straight grinding wheels. *(Courtesy of Carborundum Company)*

Selection of grain size is determined by whether coarse or fine cutting and finish are desired. Coarse grains take larger depths of cut and cut more rapidly. Hard wheels with fine grains leave smaller tracks and therefore are usually selected for finishing cuts. If there is a tendency for the work material to load the wheel, larger grains with a more open structure may be used for finishing.

BALANCING GRINDING WHEELS

Because of the high rotation speeds involved, grinding wheels must never be used unless they are in good balance. A slight imbalance will produce vibrations that will cause waviness in the work surface. It may cause a wheel to break, with the probability of serious damage and injury. The wheel should be mounted with proper bushings so that it fits snugly on the spindle of the machine. Rings of blotting paper should be placed between the wheel and the flanges to ensure that the clamping pressure is evenly distributed. Most grinding wheels will run in good balance if they are mounted properly and trued. Most machines have provision for compensating for a small amount of wheel imbalance by attaching weights to one mounting flange. Some have provision for semiautomatic balancing with weights that are permanently attached to the machine spindle.

SAFETY IN GRINDING

Because the rotational speeds are quite high, and the strength of grinding wheels is usually much less than that of the materials being ground, serious accidents occur much too frequently in connection with the use of grinding wheels. Virtually all such accidents could be avoided and are due to one or a combination of four causes. First, grinding wheels are occasionally operated at unsafe and improper speeds. All grinding wheels are clearly marked with the maximum rpm value at which they should be rotated. They are all tested to considerably above the designated rpm and are safe at the specified speed *unless abused*. *They should never, under any condition, be operated above the rated speed*. Second, a very common form of abuse, frequently accidental, is dropping the wheel or striking it against a hard object. This can cause a crack (which may not be readily visible), resulting in subsequent failure of the wheel while rotating at high speed under load. If a wheel is dropped or struck against a hard object, it should be discarded and never used unless tested at above the rated speed in a properly designed test stand. A third common cause of grinding wheel failure is improper use, such as grinding against the side of a wheel that was designed for grinding only on its periphery. The fourth and most common cause of injury from grinding is the absence of a proper safety guard over the wheel and/or over the eyes or face of the operator. The frequency with which operators will remove safety guards from grinding equipment or fail to use safety goggles or face shields is amazing and inexcusable.

USE OF CUTTING FLUIDS IN GRINDING

Because grinding involves cutting, the selection and use of a cutting fluid is governed by the basic principles discussed in Chapter 21. If a fluid is used, it should be applied in sufficient quantities and in a manner that will ensure that the chips are washed away, not trapped between the wheel and the work. This is of particular importance in grinding horizontal surfaces. In hardened steel, the use of a fluid can help to prevent fine microcracks that result from highly localized heating. The air scraper shown in Figure 26-18 permits the cutting fluid (lubricant) to get onto the face of the wheel. Metal air scrapers disrupt the airflow. Upper and lower nozzles cool the grinding zone, while a high-pressure scrubber helps deter loading of the wheel.

Much snagging and off-hand grinding is done dry. On some types of material, dry grinding produces a better finish than can be obtained by wet grinding.

Grinding fluids strongly influence the performance of CBN wheels. Straight, sulfurized, or sulfochlorinated oils can enhance performance considerably when used with straight oils.

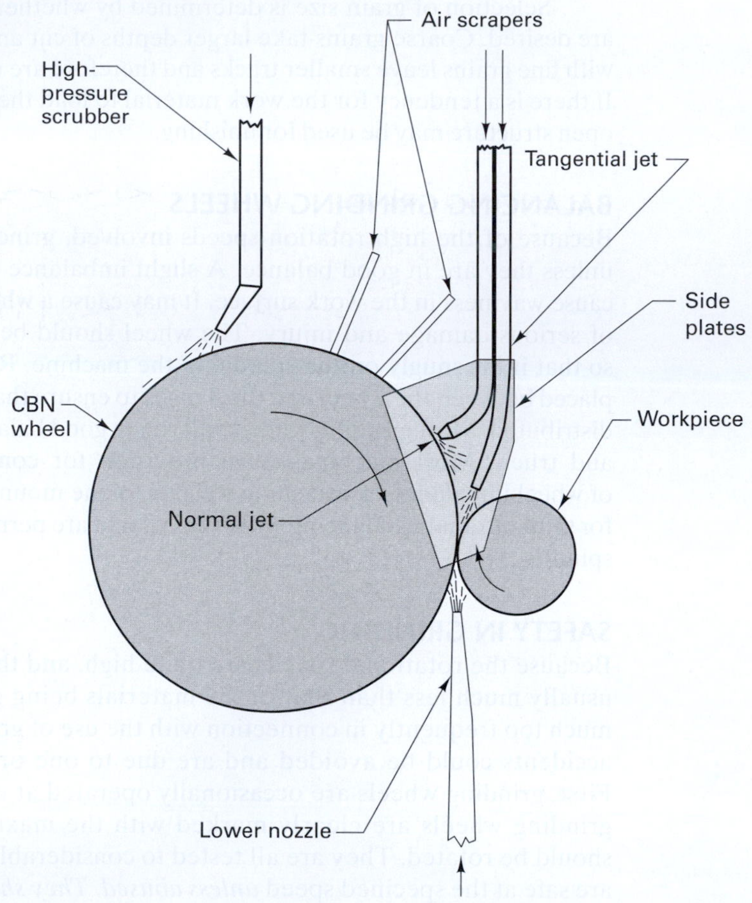

FIGURE 26-18 Coolant delivery system for optimum CBN grinding. *(M. P. Hitchiner, "Production Grinding with CBN," Machining Technology, Vol. 2, no. 2, 1991)*

26.5 GRINDING MACHINES

Grinding machines commonly are classified according to the type of surface they produce. Table 26-4 presents such a classification, with further subdivision to indicate characteristic features of different types of machines within each classification. Grinding on

TABLE 26-4 Classification of Grinding Machines

Type of Machine	Type of Surface	Specific Types or Features
Cylindrical external	External surface on rotating, usually cylindrical parts	Work rotated between centers Centerless Centerless Chucking Tool post Crankshaft, cam, etc.
Cylindrical internal	Internal diameters of holes	Chucking Planetary (work stationary) Centerless
Surface conventional	Flat surfaces	Reciprocating table or rotating table Horizontal or vertical spindle
Creep feed	Deep slots, profiles in hard steels, carbides, and ceramics using CBN and diamond	Rigid, chatter-free, creep feed rate Continuous dressing Heavy coolant flows NC or CNC control Variable speed wheel
Tool grinders	Tool angles and geometries	Universal Special
Other	Special or any of the above	Disk, contour, thread, flexible shaft, swing frame, snag, pedestal, bench

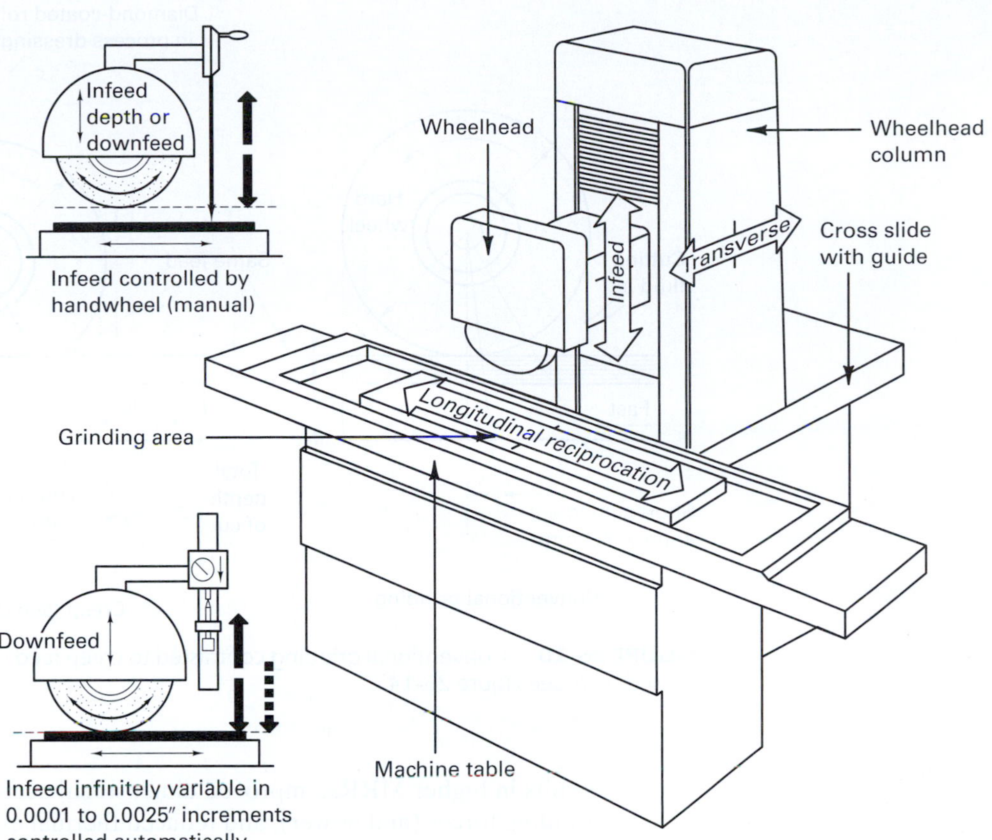

FIGURE 26-19 Horizontal-spindle surface grinder, with insets showing movements of wheelhead.

all machines is done in three ways. In the first, the depth of cut (d_t) is obtained by **infeed**—moving the wheel down into the work or the work up into the wheel (Figure 26-19). The desired surface is then produced by traversing the wheel across (cross feed) the workpiece, or vice versa. In the second method, known as **plunge-cut grinding,** the basic movement is of the wheel being fed radially into the work while the latter revolves on centers. It is similar to form cutting on a lathe; usually a formed grinding wheel is used (Figure 26-14). In the third method, the work is fed very slowly past the wheel and the total downfeed or depth (d) is accomplished in a single pass (Figure 26-20). This is called **creep feed grinding (CFG).** (Table 26-5 compares CFG to conventional and high-speed grinding for CBN applications.)

The CFG method, often done in the surface grinding mode, is markedly different from conventional surface grinding. The depth of cut is increased 1,000 to 10,000 times, and the work feed ratio is decreased in the same proportion; hence the name *creep feed grinding*. The long arc of contact between the wheel and the work increases the cutting forces and the power required. Therefore, the machine tools to perform this type of grinding must be specially designed with high static and dynamic stability, stick-slip-free ways, adequate damping, increased horsepower, infinitely variable spindle speed, variable but extremely consistent table feed (especially in the low ranges), high-pressure cooling systems, integrated devices for dressing the grinding wheels, and specially designed (soft with open structure) grinding wheels. The process is mainly applied when grinding deep slots with straight parallel sides or when grinding complex profiles in difficult-to-grind materials. The process is capable of producing extreme precision at relatively high metal removal rates (MRRs). Because the process can operate at relatively low surface temperatures, the surface integrity of the metals being ground is good.

However, in CFG, the grinding wheels must maintain their initial profile much longer, so continuous dressing is used that is form-truing and dressing the grinding wheel throughout the process rather than between cycles. Continuous crush dressing

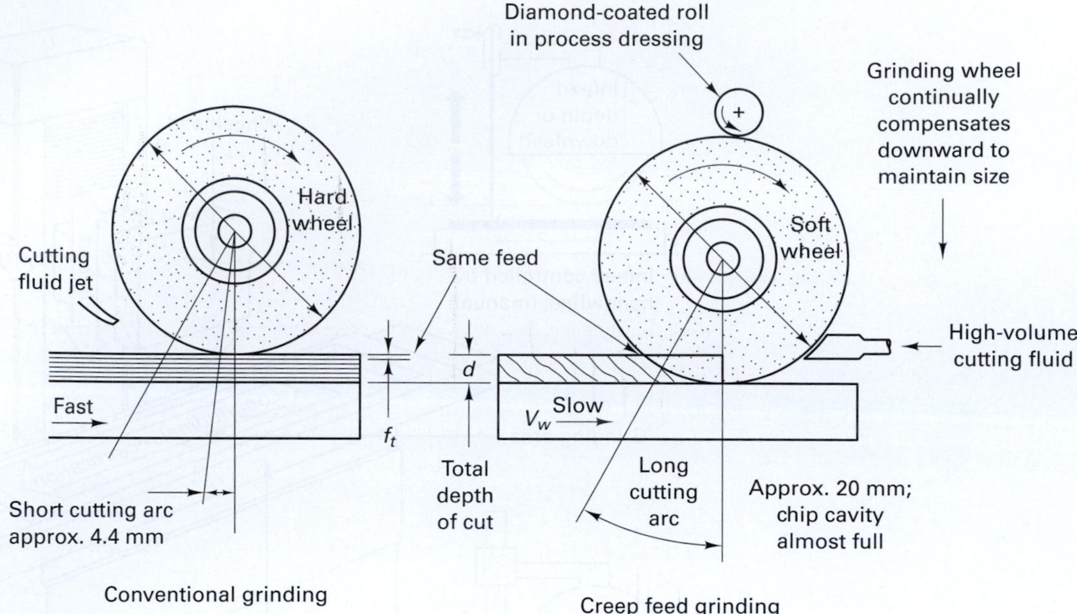

FIGURE 26-20 Conventional grinding contrasted to creep feed grinding. Note that crush roll dressing is used here; see Figure 26-14.

results in higher MRRs, improved dimensional accuracy and form tolerance, reduced grinding forces (and power), and reduced thermal effects while sacrificing wheel wear. Creep feed grinding eliminates preparatory operations such as milling or broaching, because profiles are ground into the solid workpiece. This can result in significant savings in unit part costs.

Grinding machines that are used for precision work have certain important characteristics that permit them to produce parts having close dimensional tolerances. They are constructed very accurately, with heavy, rigid frames to ensure permanency of alignment. Rotating parts are accurately balanced to avoid vibration. Spindles are mounted in very accurate bearings, usually of the preloaded ball-bearing type. Controls are provided so that all movements that determine dimensions of the workpiece can be made with accuracy—usually to 0.001 or 0.00001 in.

The abrasive dust that results from grinding must be prevented from entering between moving parts. All ways and bearings must be fully covered or protected by seals. If this is not done, the abrasive dust between moving parts becomes embedded in the softer of the two, causing it to act as lap and abrade the harder of the two surfaces, resulting in permanent loss of accuracy.

These special characteristics add considerably to the cost of these machines and require that they be operated by trained personnel. Production-type grinders are more fully automated and have higher metal removal rates and excellent dimensional accuracy. Fine surface finish can be obtained very economically.

TABLE 26-5 Starting Conditions for CBN Grinding

Grinding Variable	Conventional Grinding	Creep Feed Grinding	High-speed Grinding
Wheel speed (fpm)	5500–9500 versus 4500–6500 vitrified	5000–9000 versus 3000–5000	12000–2500
Table speed (fpm)	80–150	0.5–5	5–20
Feed (f_t) in./pass	0.0005–0.0015	0.100–0.250	250–500
Grinding fluids	10% heavy-duty soluble oil or 3–5% light-duty soluble for light feeds	Sulfurized or sulfochlorinated straight grinding oil applied at 80 to 100 gal/min at 100 psi or more	

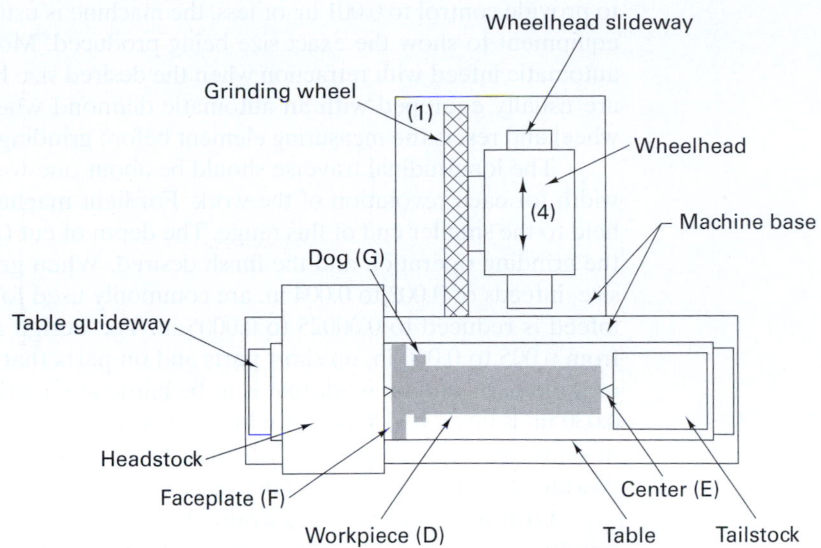

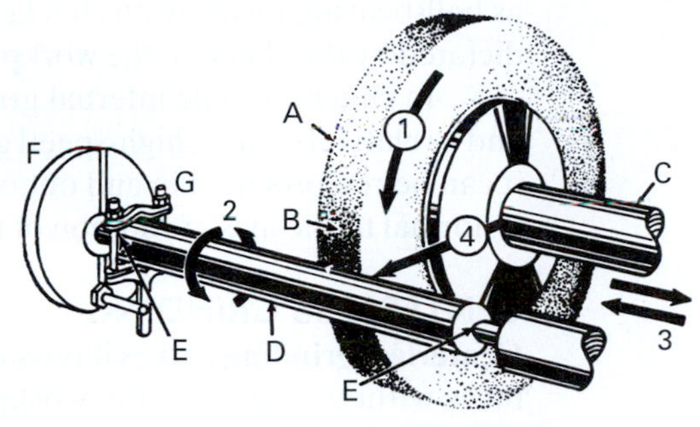

FIGURE 26-21 Cylindrical grinding between centers. A = edge of wheel, B = face of wheel, C = shaft for wheel, D = workpiece, E = centers, F = faceplate, G = dog.

CYLINDRICAL GRINDING

Center-type cylindrical grinding is commonly used for producing external cylindrical surfaces. Figures 26-14 and Figure 26-21 show the basic principles and motions of this process. The grinding wheel revolves at an ordinary cutting speed, and the workpiece rotates on centers at a much slower speed, usually from 75 to 125 ft/min. The grinding wheel and the workpiece move in opposite directions at their point of contact. The depth of cut is determined by infeed of the wheel or workpiece. Because this motion also determines the finished diameter of the workpiece, accurate control of this movement is required. Provision is made to traverse the workpiece with the wheel, or the work can be reciprocated past the wheel. In very large grinders, the wheel is reciprocated because of the massiveness of the work. For form or plunge grinding, the detail of the wheel is maintained by periodic crush roll dressing.

A plain center-type cylindrical grinder is shown in Figure 26-21. On this type the work is mounted between headstock and tailstock centers. Solid dead centers are always used in the tailstock, and provision is usually made so that the headstock center can be operated either dead or alive. High-precision work is usually ground with a dead headstock center, because this eliminates any possibility that the workpiece will run out of round due to any eccentricity in the headstock.

The table assembly can be reciprocated—in most cases, by using a hydraulic drive. The speed can be varied, and the length of the movement can be controlled by means of adjustable trip dogs.

Infeed is provided by movement of the wheelhead at right angles to the longitudinal axis of the table. The spindle is driven by an electric motor that is also mounted on the wheelhead. If the infeed movement is controlled manually by some type of vernier drive to provide control to 0.001 in. or less, the machine is usually equipped with digital readout equipment to show the exact size being produced. Most production-type grinders have automatic infeed with retraction when the desired size has been obtained. Such machines are usually equipped with an automatic diamond wheel-truing device that dresses the wheel and resets the measuring element before grinding is started on each piece.

The longitudinal traverse should be about one-fourth to three-fourths of the wheel width for each revolution of the work. For light machines and fine finishes, it should be held to the smaller end of this range. The depth of cut (infeed) varies with the purpose of the grinding operation and the finish desired. When grinding is done to obtain accurate size, infeeds of 0.002 to 0.004 in. are commonly used for roughing cuts. For finishing, the infeed is reduced to 0.00025 to 0.0005 in. The design allowance for grinding should be from 0.005 to 0.010 in. on short parts and on parts that are not to be hardened. On long or large parts and on work that is to be hardened, a grinding allowance of from 0.015 to 0.030 in. is desirable. When grinding is used primarily for metal removal (called abrasive machining), infeeds are much higher, 0.020 to 0.040 in. being common. Continuous downfeed is often used, with rates up to 0.100 in./min being common.

Grinding machines are available in which the workpiece is held in a chuck for grinding both external and internal cylindrical surfaces. **Chucking-type external grinders** are production-type machines for use in rapid grinding of relatively short parts, such as ball-bearing races. Both chucks and collets are used for holding the work, the means dictated by the shape of the workpiece and rapid loading and removal.

In chucking-type internal grinding machines, the chuck-held workpiece revolves, and a relatively small, high-speed grinding wheel is rotated on a spindle arranged so that it can be reciprocated in and out of the workpiece. Infeed movement of the wheelhead is normal to the axis of rotation of the work (Figure 26-21).

CENTERLESS GRINDING

Centerless grinding makes it possible to grind both external and internal cylindrical surfaces without requiring the workpiece to be mounted between centers or in a chuck. This eliminates the requirement of center holes in some workpieces and the necessity for mounting the workpiece, thereby reducing the cycle time.

The principle of centerless *external* grinding is illustrated in Figure 26-22. Two wheels are used. The larger one operates at regular grinding speeds and does the actual

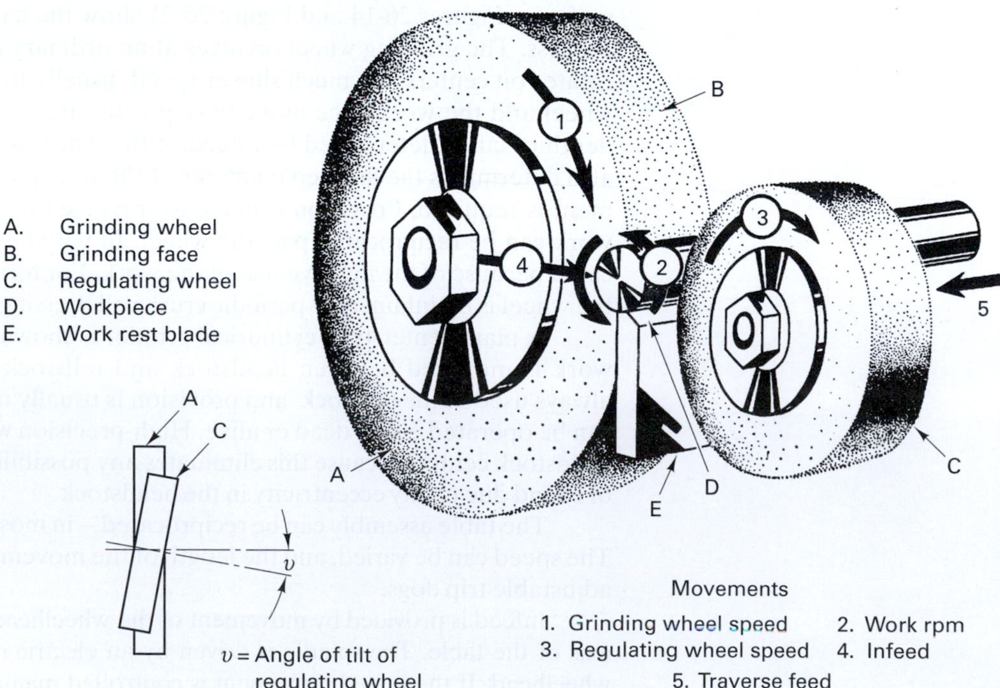

FIGURE 26-22 Centerless grinding showing the relationship among the grinding wheel, the regulating wheel, and the workpiece in centerless method. *(Courtesy of Carborundum Company)*

grinding. The smaller wheel is the **regulating wheel.** It is mounted at an angle to the plane of the grinding wheel. Revolving at a much slower surface speed—usually 50 to 200 ft/min—the regulating wheel controls the rotation and longitudinal motion of the workpiece and is a usually a plastic- or rubber-bonded wheel with a fairly wide face.

The workpiece is held against the work-rest blade by the cutting forces exerted by the grinding wheel and rotates at approximately the same surface speed as that of the regulating wheel. This axial feed is calculated approximately by the equation

$$F = ND \sin \phi \tag{26-1}$$

where

F = feed (mm/min or in./min)
D = diameter of the regulating wheel (mm or in.)
N = revolutions per minute of the regulating wheel
ϕ = angle of inclination of the regulating wheel

Centerless grinding has several important advantages:

1. It is very rapid; infeed centerless grinding is almost continuous.
2. Very little skill is required of the operator.
3. It can often be made automatic (single-cycle automatic).
4. Where the cutting occurs, the work is fully supported by the work rest and the regulating wheel. This permits heavy cuts to be made.
5. Because there is no distortion of the workpiece, accurate size control is easily achieved.
6. Large grinding wheels can be used, thereby minimizing wheel wear.

Thus, centerless grinding is ideally suited to certain types of mass-production operations. The major disadvantages are as follows:

1. Special machines are required that can do no other type of work.
2. The work must be round—no flats, such as keyways, can be present.
3. Its use on work having more than one diameter or on curved parts is limited.
4. In grinding tubes, there is no guarantee that the OD and inside diameter (ID) are concentric.

Special centerless grinding machines are available for grinding balls and tapered workpieces. The centerless grinding principle can also be applied to internal grinding, but the external surface of the cylinder must be finished accurately before the internal operation is started. However, it ensures that the internal and external surfaces will be concentric. The operation is easily mechanized for many applications.

SURFACE GRINDING MACHINES

Surface grinding machines are used primarily to grind flat surfaces. However formed, irregular surfaces can be produced on some types of surface grinders by use of a formed wheel. There are four basic types of surface grinding machines, differing in the movement of their tables and the orientation of the grinding wheel spindles (Figure 26-23):

1. Horizontal spindle and reciprocating table.
2. Vertical spindle and reciprocating table.
3. Horizontal spindle and rotary table.
4. Vertical spindle and rotary table.

The most common type of surface grinding machine has a reciprocating table and horizontal spindle (Figures 26-19). The table can be reciprocated longitudinally either by handwheel or by hydraulic power. The wheelhead is given transverse (cross-feed) motion at the end of each table motion, again either by handwheel or by hydraulic

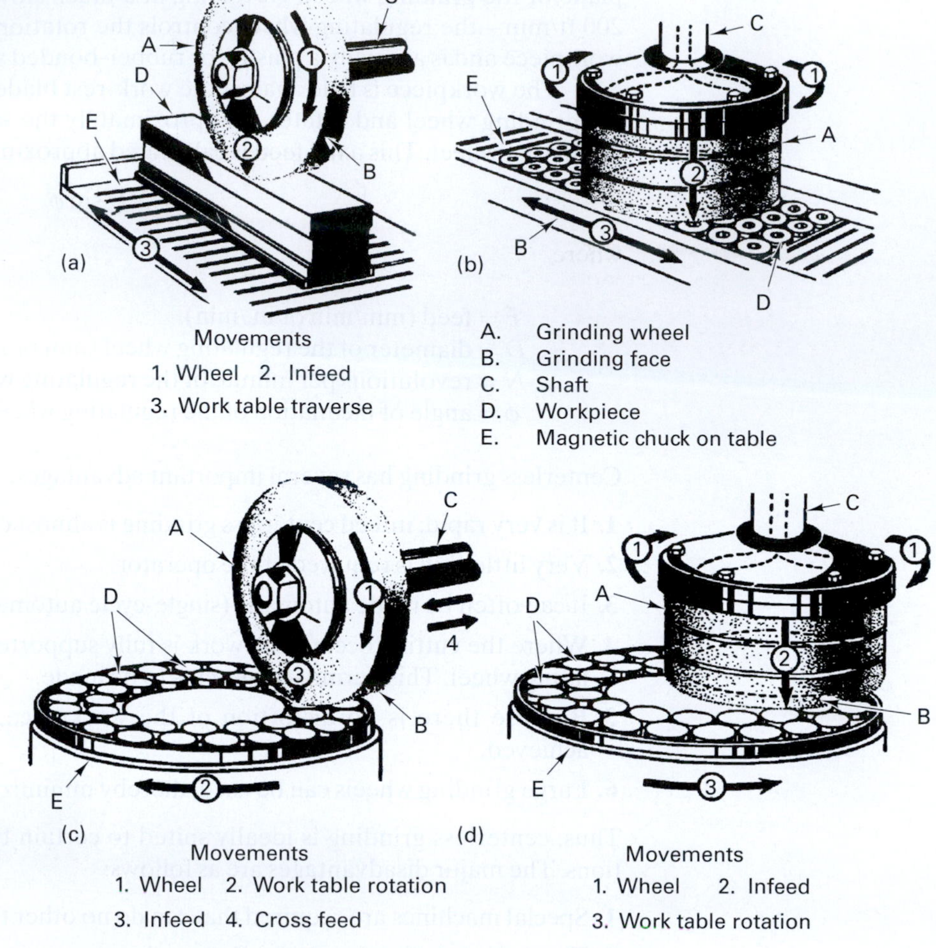

FIGURE 26-23 Surface grinding: (a) horizontal surface grinding and reciprocating table; (b) vertical spindle with reciprocating table; (c) and (d) both horizontal- and vertical-spindle machines can have rotary tables. *(Courtesy of Carborundum Company)*

power feed. Both the longitudinal and transverse motions can be controlled by limit switches. Infeed or downfeed on such grinders is controlled by handwheels or automatically. The size of such machines is determined by the size of the surface that can be ground.

In using such machines, the wheel should overtravel the work at both ends of the table reciprocation, so as to prevent the wheel from grinding in one spot while the table is being reversed. The transverse or cross-feed motion should be one-fourth to three-fourths of the wheel width between each stroke.

Vertical-spindle reciprocating-table surface grinders differ basically from those with horizontal spindles only in that their spindles are vertical and that the wheel diameter must exceed the width of the surface to be ground. Usually, no transverse motion of either the table or the wheelhead is provided. Such machines can produce very flat surfaces.

Rotary-table surface grinders can have either vertical or horizontal spindles, but those with horizontal spindles are limited in the type of work they will accommodate and therefore are not used to a great extent. Vertical-spindle rotary-table surface grinders are primarily production-type machines. They frequently have two or more grinding heads, and therefore, both rough grinding and finish grinding are accomplished in one rotation of the workpiece. The work can be held either on a magnetic chuck or in special fixtures attached to the table.

By using special rotary feeding mechanisms, machines of this type often are made automatic. Parts are dumped on the rotary feeding table and fed automatically onto workholding devices and moved past the grinding wheels. After they pass the last grinding head, they are automatically unloaded.

DISK-GRINDING MACHINES

Disk grinders have relatively large side-mounted abrasive disks. The work is held against one side of the disk for grinding. Both single- and double-disk grinders are used; in the latter type, the work is passed between the two disks and is ground on both sides simultaneously. On these machines, the work is always held and fed automatically. On small, single-disk grinders, the work can be held and fed by hand while resting on a supporting table. Although manual disk grinding is not very precise, flat surfaces can be obtained quite rapidly with little or no tooling cost. On specialized, production-type machines, excellent accuracy can be obtained very economically.

TOOL AND CUTTER GRINDERS

Simple, single-point tools are often sharpened by hand on bench or pedestal grinders **(off-hand grinding).** More complex tools, such as milling cutters, reamers, hobs, and single-point tools for production-type operations require more sophisticated grinding machines, commonly called **universal tool and cutter grinders.** These machines are similar to small universal cylindrical center-type grinders, but they differ in four important respects:

1. The headstock is not motorized.
2. The headstock can be swiveled about a horizontal as well as a vertical axis.
3. The wheelhead can be raised and lowered and can be swiveled through at 360-degree rotation about a vertical axis.
4. All table motions are manual. No power feeds being provided.

Specific rake and clearance angles must be created, often repeatedly, on a given tool or on duplicate tools. Tool and cutter grinders have a high degree of flexibility built into them so that the required relationships between the tool and the grinding wheel can be established for almost any type of tool. Although setting up such a grinder is quite complicated and requires a highly skilled worker, after the setup is made for a particular job, the actual grinding is accomplished rather easily. Figure 26-24 shows several typical setups on a tool and cutter grinder.

Hand-ground cutting tools are not accurate enough for automated machining processes. Many numerically controlled (NC) machine tools have been sold on the premise that they can position work to very close tolerances—within ±0.0001 to 0.0002 in.—only to have the initial workpieces produced by those machines out of tolerance by as much as 0.015 to 0.020 in. In most instances, the culprit was a poorly ground tool. For example, a twist drill with a point ground 0.005 in. off-center can "walk" as much as 0.015 in., thus causing poor hole location. Many companies are turning to computer numerical control (CNC) grinders to handle the regrinding of their cutting tools. A six-axis CNC grinder is

FIGURE 26-24 Three typical setups for grinding single- and multiple-edge tools on a universal tool and cutter grinder. (a) Single-point tool is held in a device that permits all possible angles to be ground. (b) Edges of a large hand reamer are being ground. (c) Milling cutter is sharpened with a cupped grinding wheel. *(Courtesy J T. Black)*

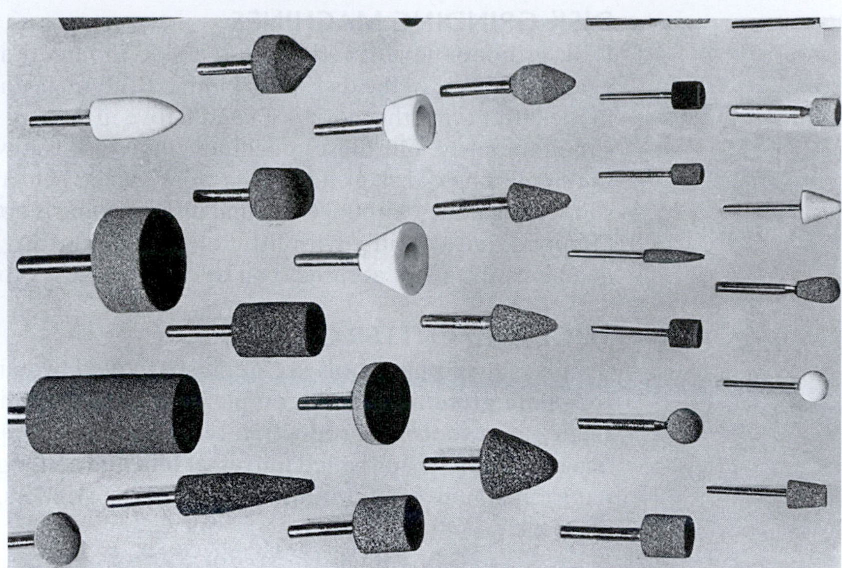

FIGURE 26-25 Examples of mounted abrasive wheels and points. *(Courtesy of Norton Abrasives/Saint Gobain)*

capable of restoring the proper tool angles (rake and clearance), concentricity, cutting edges, and dimensional size.

MOUNTED WHEELS AND POINTS

Mounted wheels and points are small grinding wheels of various shapes that are permanently attached to metal shanks that can be inserted in the chucks of portable, high-speed electric or air motors. They are operated at speeds up to 100,000 rpm, depending on their diameters, and are used primarily for deburring and finishing in mold and die work. Several types are shown in Figure 26-25.

COATED ABRASIVES

Coated abrasives are being used increasingly in finishing both metal and nonmetal products. These are made by gluing abrasive grains onto a cloth or paper backing (Figure 26-26). Synthetic abrasives—aluminum oxide, silicon carbide, aluminum, zirconia, CBN, and diamond—are used most commonly, but some natural abrasives—sand, flint, garnet, and emery—also are employed. Various types of glues are utilized to attach the abrasive grains to the backing, usually compounded to allow the finished product to have some flexibility.

Coated abrasives are available in sheets, rolls, endless belts, and disks of various sizes. Some of the available forms are shown in Figure 26-26. Although the cutting action of coated abrasives basically is the same as with grinding wheels, there is one major difference: they have little tendency to be self-sharpened when dull grains are pulled from the backing. Consequently, when the abrasive particles become dull or the belt loaded, the belt must be replaced. Finer grades result in finer first cuts but slower material removal rates. This versatile process is now widely used for rapid stock removal as well as fine surface finishing.

26.6 HONING

Honing is a stock-removal process that uses fine abrasive stones to remove very small amounts of metal. Cutting speed is much lower than that of grinding. The process is used to size and finish bored holes, remove common errors left by boring (taper, waviness, and tool marks), or remove the tool marks left by grinding. The amount of metal removed is typically about 0.005 in. or less. Although honing is occasionally done by hand, as in finishing the face of a cutting tool, it usually is done with special equipment. Most honing is done on internal cylindrical surfaces, such as automobile cylinder walls. The honing stones are usually held in a honing head, with the stones being held against

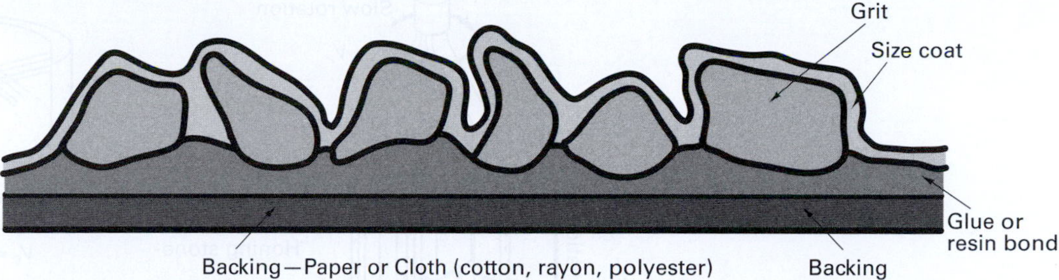

Grit Size—grade

vs	Approx.	Finish (rms)
24	300	μin.
36	250	"
50	140	"
80	125	"
120	60-80	"
150	40-60	"

Bonds

Name	Make coat	Size coat	Backing
Glue bond	Glue	Glue	Non WP
Modified glue	Mod. glue	Mod. glue	"
Resin over glue	Glue	Resin	"
Resin over resin	Resin	Resin	"
Waterproof	Resin	Resin	WP

WP = waterproof

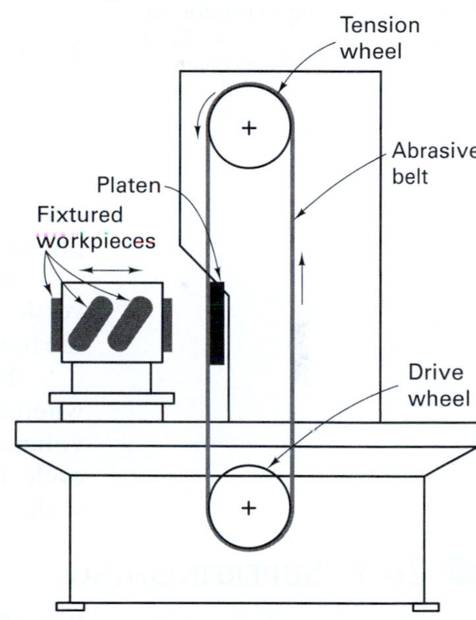

FIGURE 26-26 Belt composition for coated abrasives (top). Platen grinder (right) and examples of belts and disks for abrasive machining. *(Courtesy J T. Black)*

the work with controlled light pressure. The honing head is not guided externally but, instead, *floats* in the hole, being guided by the work surface (Figure 26-27).

The stones are given a complex motion so as to prevent a single grit from repeating its path over the work surface. Rotation is combined with an oscillatory axial motion. For external and flat surfaces, varying oscillatory motions are used. The length of the motions should be such that the stones extend beyond the work surface at the end. A cutting fluid is used in virtually all honing operations. The critical process parameters are rotational speed, V_r, oscillation speed, V_o, the length and position of stroke, and the honing stick pressure. Note that V_c and the inclination angle are both products of V_o and V_r.

Virtually all honing is done with stones made by bonding together various fine artificial abrasives. **Honing stones** differ from grinding wheels in that additional materials, such as sulfur, resin, or wax, are often added to the bonding agent to modify the

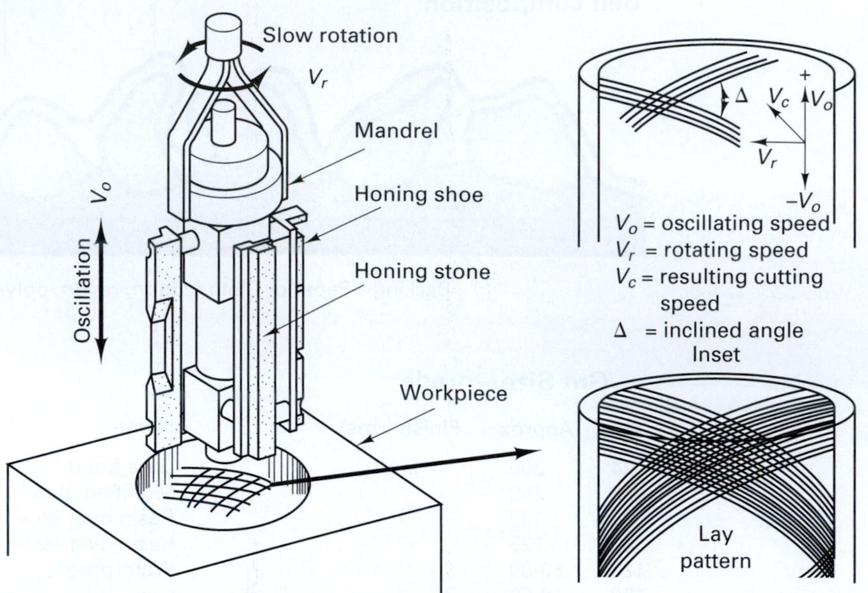

FIGURE 26-27 Schematic of honing head showing the manner in which the stones are held. The rotary and oscillatory motions combine to produce a cross-hatched lay pattern. Typical values for V_c and P_s are given below.

For:	Honing Parameters	Conventional Abrasives	Diamonds	CBN
High MRR	V_c(m/min)	20–30	40–70	35–90
	P_s(N/min^2)	1–2	2–8	2–4
Best-quality service	V_c(m/min)	5–30	40–70	20–60
	P_s(N/min^2)	0.5–1.5	1.0–3.0	1.0–2.0

cutting action. The abrasive grains range in size from 80 to 600 grit. The stones are equally spaced about the periphery of the tool. Reference values for V_c and honing stick pressure, P_s, for various abrasives are shown in Figure 26-27.

Single- and multiple-spindle honing machines are available in both horizontal and vertical types. Some are equipped with special sensitive measuring devices that collapse the honing head when the desired size has been reached.

For honing single, small, internal cylindrical surfaces, a procedure is often used wherein the workpiece is manually held and reciprocated over a rotating hone. If the volume of work is sufficient, honing is a fairly inexpensive process. A complete honing cycle, including loading and unloading the work, is often less than 1 min. Size control within 0.0003 in. is achieved routinely.

26.7 SUPERFINISHING

Superfinishing is a variation of honing that is typically used on flat surfaces. The process is:

1. Very light, controlled pressure, 10 to 40 psi.
2. Rapid (more than 400 cycles per minute), short strokes—less than $\frac{1}{4}$ in.
3. Stroke paths controlled so that a single grit never traverses the same path twice.
4. Copious amounts of low-viscosity lubricant-coolant flooded over the work surface.

This procedure, illustrated in Figure 26-28, results in surfaces of very uniform, repeatable smoothness.

Superfinishing is based on the phenomenon that a lubricant of a given viscosity will establish and maintain a separating, lubricating film between two mating surfaces if their roughness does not exceed a certain value and if a certain critical pressure, holding them apart, is not exceeded. Consequently, as the minute peaks on a surface are cut away by the honing stone, applied with a controlled pressure, a certain degree of smoothness is achieved. The lubricant establishes a continuous film between the stone and the workpiece and separates them so that no further cutting action occurs. Thus, with a given pressure, lubricant, and honing stone, each workpiece is honed to the same degree of smoothness.

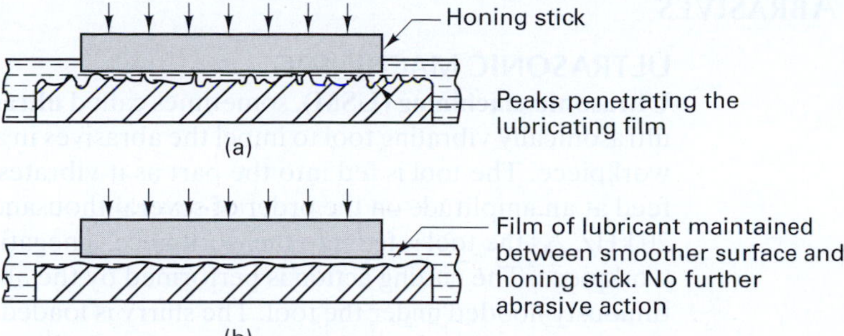

FIGURE 26-28 In superfinishing and honing, a film of lubricant is established between the work and the abrasive stone as the work becomes smoother.

Superfinishing is applied to both cylindrical and plane surfaces. The amount of metal removed usually is less than 0.002 in., most of it being the peaks of the surface roughness. Copious amounts of lubricant-coolant maintain the work at a uniform temperature and wash away all abraded metal particles to prevent scratching.

LAPPING

Lapping is an abrasive surface finishing process wherein fine abrasive particles are *charged* (caused to become embedded) into a soft material, called a **lap.** The material of the lap may range from cloth to cast iron or copper, but it is always softer than the material to be finished, being only a holder for the hard abrasive particles. Lapping is applied to both metals and nonmetals.

As the charged lap is rubbed against a surface, the abrasive particles in the surface of the lap remove small amounts of material from the surface to be machined. Thus, the abrasive does the cutting, and the soft lap is not worn away because the abrasive particles become embedded in its surface instead of moving across it. This action always occurs when two materials rub together in the presence of a fine abrasive: the softer one forms a lap, and the harder one is abraded away.

In lapping, the abrasive is usually carried between the lap and the work surface in some sort of a vehicle, such as grease, oil, or water. The abrasive particles are from 120 grit up to the finest powder sizes. As a result, only very small amounts of metal are removed, usually considerably less than 0.001 in. Because it is such a slow metal removing process, lapping is used only to remove scratch marks left by grinding or honing or to obtain very flat or smooth surfaces, such as are required on gage blocks or for liquid-tight seals where high pressures are involved.

Materials of almost any hardness can be lapped. However, it is difficult to lap soft materials because the abrasive tends to become embedded. The most common lap material is fine-grained cast iron. Copper is used quite often and is the common material for lapping diamonds. For lapping hardened metals for metallographic examination, cloth laps are used.

Lapping can be done either by hand or by special machines. In hand lapping, the lap is flat, similar to a surface plate. Grooves are usually cut across the surface of a lap to collect the excess abrasive and chips. The work is moved across the surface of the lap, using an irregular, rotary motion, and is turned frequently to obtain a uniform cutting action.

In lapping machines for obtaining flat surfaces, workpieces are placed loosely in holders and are held against the rotating lap by means of floating heads. The holders, rotating slowly, move the workpieces in an irregular path. When two parallel surfaces are to be produced, two laps may be employed—one rotating below and the other above the workpieces.

Various types of lapping machines are available for lapping round surfaces. A special type of centerless lapping machine is used for lapping small cylindrical parts, such as piston pins and ball-bearing races.

Because the demand for surfaces having only a few micrometers of roughness on hardened materials has become quite common, the use of lapping has increased greatly. However, it is a very slow method of removing metal, obviously costly compared with other methods, and should not be specified unless such a surface is absolutely necessary.

26.8 FREE ABRASIVES

ULTRASONIC MACHINING

Ultrasonic machining (USM), sometimes called **ultrasonic impact grinding,** employs an ultrasonically vibrating tool to impel the abrasives in a slurry at high velocity against the workpiece. The tool is fed into the part as it vibrates along an axis parallel to the tool feed at an amplitude on the order of several thousandths of an inch and a frequency of 20 kHz. As the tool is fed into the workpiece, a negative of the tool is machined into the workpiece. The cutting action is performed by the abrasives in the slurry, which is continuously flooded under the tool. The slurry is loaded up to 60 wt% with abrasive particles. Lighter abrasive loadings are used to facilitate the flow of the slurry for deep drilling (up to 2 in. deep). Boron carbide, aluminum oxide, and silicon carbide are the most commonly used abrasives in grit sizes ranging from 400 to 2000. The amplitude of the vibration should be set approximately to the size of the grit. The process can use shaped tools to cut virtually any material but is most effective on materials with hardnesses greater than R_C 40, including brittle and nonconductive materials such as glass. Figure 26-29 shows a simple schematic of this process.

USM uses piezoelectric or magnetostrictive transducers to impart high-frequency vibrations to the tool holder and tool. Abrasive particles in the slurry are accelerated to great speed by the vibrating tool. The tool materials are usually brass, carbide, mild steel, or tool steel and will vary in tool wear depending on their hardness. Wear ratios (workpiece material removed versus tool material lost) from 1:1 (for tool steel) to 100:1 (for glass) are possible. Because of the high number of cyclic loads, the tool must be strong enough to resist fatigue failure.

The cut will be oversize by about twice the size of the abrasive particles being used, and holes will be tapered, usually limiting the hole depth-to-diameter ratio to about 3:1. Surface roughness is controlled by the size of the abrasive particles (finer finish with smaller particles). Holes, slots, or shaped cavities can be readily eroded in any hard material—conductive or nonconductive, metallic, ceramic, or composite. Advantages of the process include that it is one of the few machining methods capable of machining glass. Also, it is the safest machining method. Skin is impervious to the process because of its ductility. High-pitched noise can be a problem due to secondary

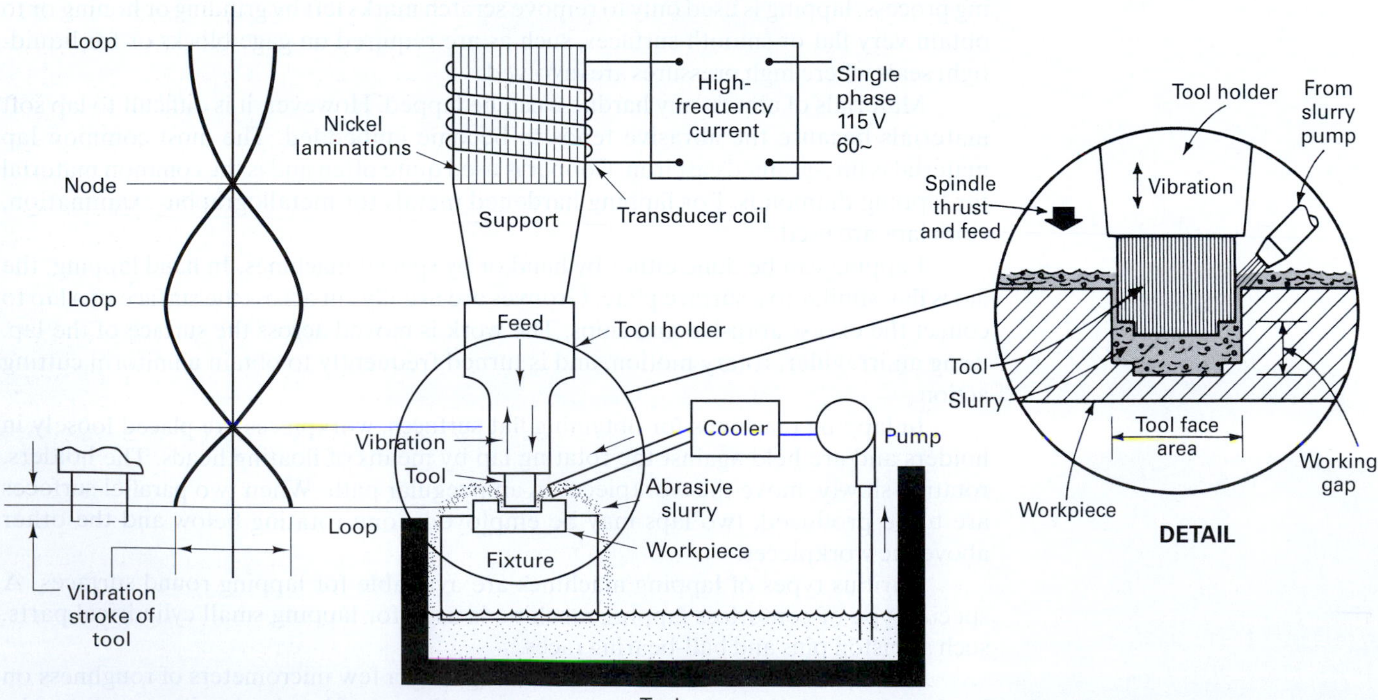

FIGURE 26-29 Sinking a hole in a workpiece with an ultrasonically vibrating tool driving an abrasive slurry.

vibrations. In addition to machining, ultrasonic energy has also been employed for coining, lapping, deburring, and broaching. Plastics can be welded using ultrasonic energy.

WATER-JET CUTTING AND ABRASIVE WATER-JET MACHINING

Water-jet cutting (WJC), also known as **water-jet machining** or **hydrodynamic machining,** uses a high-velocity fluid jet impinging on the workpiece to perform a slitting operation (Figure 26-30). Water is ejected from a nozzle orifice at high pressure (up to 60,000 psi). The jet is typically 0.003 to 0.020 in. in diameter and exits the orifice at velocities up to 3000 ft/s. Key process parameters include water pressure, orifice diameter, water flow rate, and working distance (distance between the workpiece and the nozzle).

Nozzle materials include synthetic sapphire, due to its machinability and resistance to wear. Tool life on the order of several hundred hours is typical. Mechanisms for tool failure include chipping from contaminants or constriction due to mineral deposits. This emphasizes the need for high levels of filtration prior to pressure intensification. In the past, long-chain polymers were added to the water to make the jet more coherent (i.e., not come out of the jet dispersed). However, with proper nozzle design, a tight, coherent water-jet may be produced without additives.

The advantages of WJC include the ability to cut materials without burning or crushing the material being cut. Figure 26-30 shows a comparison (end view) of cutting corrugated boxboard with a mechanical knife and with WJC. The mechanism for material removal is simply the impinging pressure of the water exceeding the compressive strength of the material. This limits the materials that can be cut by the process to leather, plastics, and other soft nonmetals, which is the major disadvantage of the process. Alternative fluids (alcohol, glycerine, cooking oils) have been used in processing meats, baked goods, and frozen foods. Other disadvantages include that the process is noisy and requires operators to have hearing protection.

The majority of the metalworking applications for water-jet cutting require the addition of abrasives. This process is known as **abrasive water-jet cutting (AWC).** A full range of materials, including metals, plastics, rubber, glass, ceramics, and composites, can be machined by AWC. Cutting feed rates vary from 20 in./min for acoustic tile to 50 in./min for epoxies and 500 in./min for paper products.

Abrasives are added to the water-jet in a mixing chamber on the downstream side of the water-jet orifice. A single, central water-jet with side feeding of abrasives into a mixing chamber is shown in Figure 26-30. In the mixing chamber, the momentum of the water is transferred to the abrasive particles, and the water and particles are forced out through the AWC nozzle orifice, also called the mixing tube. This design can be made quite compact; however, it also experiences rapid wear in the mixing tube. An alternate configuration is to feed the abrasives from the center of the nozzle with a converging set of angled water-jets imparting momentum to the abrasives. This nozzle design produces better mixing of the water and abrasives as well as increased nozzle life. The inside diameter of the mixing tube is normally from 0.04 to 0.125 in. in diameter. These tubes are normally made of carbide.

Generally, the kerf of the cut is about 0.001 in. greater than the nozzle orifice. AWC requires control of additional process parameters over water-jet machining, including abrasive material (density, hardness, shape), abrasive size or grit, abrasive flow rate (pounds per minute), abrasive feed mechanism (pressurized or suction), and AWC nozzle (design, orifice diameter, and material). Typical AWC systems operate under the following conditions: water pressures of 30,000 to 50,000 psi, water orifice diameters from 0.01 to 0.022 in., and working distances of 0.02 to 0.06 in. Working distances are much smaller than in WJC to minimize the dispersion of the abrasive water-jet prior to entering the material. Abrasive materials used include garnet, silica, silicon carbide, or aluminum oxide. Abrasive grit sizes range from 60 to 120 and abrasive flow rates from 0.5 to 3 lb/min. For many applications, the AWC tool is combined with a CNC-controlled X–Y table, which permits contouring and surface engraving.

AWC can be used to cut any material through the appropriate choice of the abrasive, water-jet pressure, and feed rate. Table 26-6 gives cutting speeds for various metals. The ability of the abrasive water-jet to cut through thick materials (up to 8 in.) is

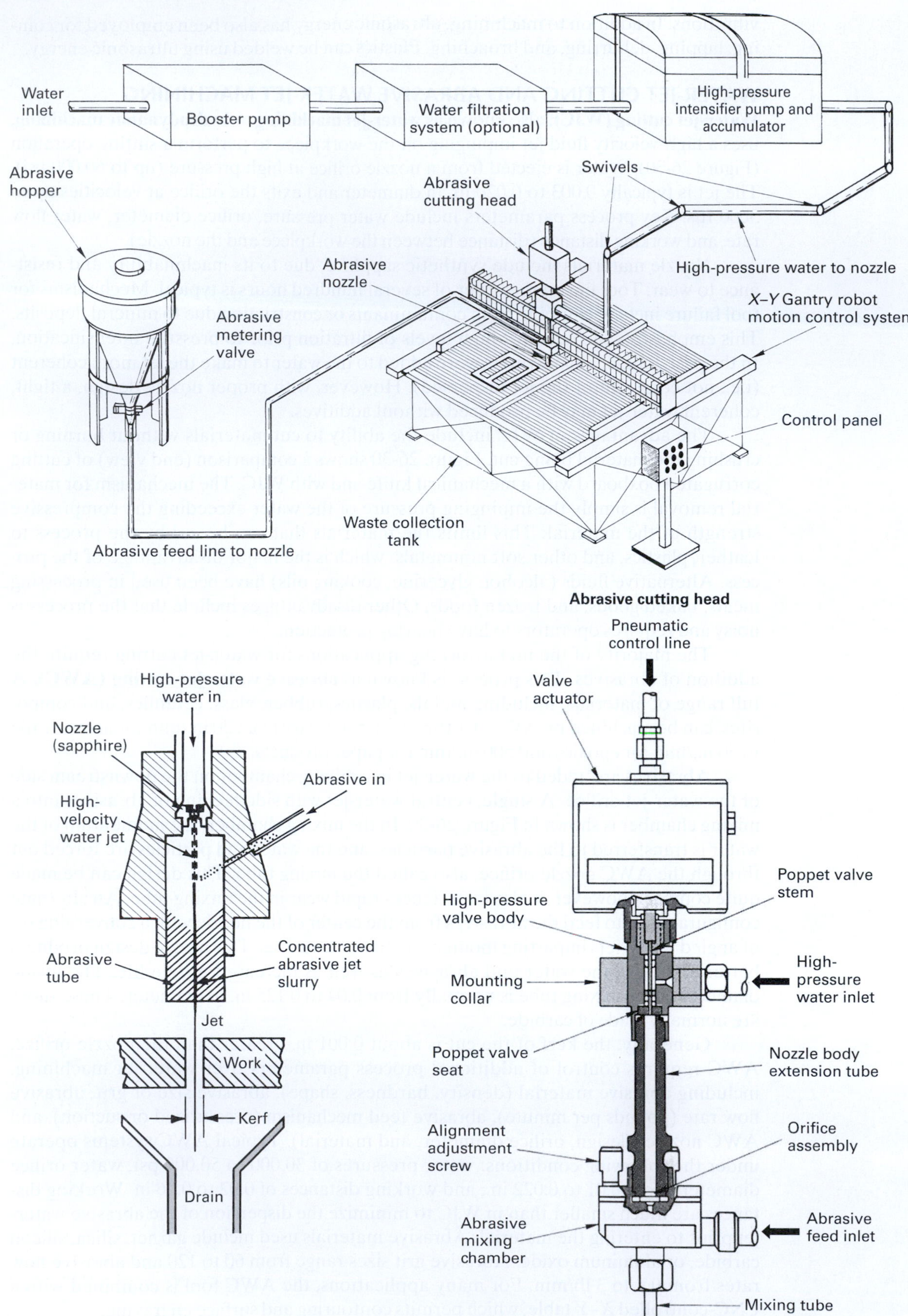

FIGURE 26-30 Schematic of hydrodynamic jet machining. The intensifier elevates the fluid to the desired nozzle pressure, while the accumulator smoothes out the pulses in the fluid jet. Schematic of an abrasive water-jet machining nozzle is shown on the right.

TABLE 26-6 Typical Values for Through-cutting Speeds for Simple Water-Jet and Abrasive Water-Jet of Machining Metals and Nonmetals.

Cutting speeds with abrasive waterjet

Material	Thickness (in.)	Nozzle speed (in./min)	Edge quality (comments)
Aluminum	0.130	20–40	good
Aluminum tube	0.220	50	burred
Aluminum casting	0.400	15	
Aluminum	0.500	6–10	
Aluminum	3.0	0.5–5	
Aluminum	4.0	0.2–2	
Brass	0.125	18–20	good or small burr
Brass	0.500	4–5	
Brass	0.75	0.75–3	striations at 1 +
Bronze	1.100	1.0	good
Copper	0.125	22	good
Copper-nickel	0.125	12–14	fair edge
Copper-nickel	2.0	1.5–4.0	fair edge
Lead	0.25	10–50	good to striated
Lead	2.0	3–8	slower = better
Magnesium	0.375	5–15	good
Armor plate	0.200	1.5–15	good
Carbon steel	0.250	10–12	good
Carbon steel	0.750	4–8	good to bad edge
Carbon steel	3.0	0.4	good w. sm. nozzle
4130 carbon steel	0.5	3.0	
Mild steel	7.5	0.017–0.05	
High-strength steel	3.0	0.38	
Cast iron	1.5	1.0	good edge
Stainless steel	0.1	10–15	good to striated
Stainless steel	0.25	4–12	good to striated
Stainless steel	1.0	1.0	65–150 RMS
15–5 PH stainless	4.0	0.3	striated
Inconel 718	1.25	0.5–1.0	good
Inconel	0.250	8–12	good to striated
Inconel	2–2.5	0.2	good to fair
Titanium	0.025–0.050	5–50	good
Titanium	0.500	1–5	65–150 RMS
Titanium	2.0	0.5–1.0	125 RMS
Tool steel	0.250	3–15	125 RMS
Tool steel	1.0	2–5	
Nonmetals			
Acrylic	0.375	15–50	good to fair
C-glass	0.125	100–200	shape dependent
Carbon/carbon comp.	0.125	50–75	good
Carbon/carbon comp.	0.500	10–20	good
Epoxy/glass composite	0.125	100–250	good
Fiberglass	0.100	150–300	good
Fiberglass	0.250	100–150	good
Glass (plate)	0.063	40–150	good
Glass (plate)	0.75	10–20	125 RMS
Graphite/epoxy	0.250	15–70	good to practical
Graphite/epoxy	1.0	3–5	good
Kevlar (steel reinf.)	0.125	30–50	good
Kevlar	0.375/0.580	10–25	good
Kevlar	1.0	3–5	good
Lexan	0.5	10	good
Phenolic	0.25–0.50	10–15	good
Plexiglass	0.175/0.50	25	
Rubber belting	0.300	200	good
Ceramic matrix composites			
Toughened zirconia	0.250	1.5	
SiC fiber in SiC	0.125	1.5	
Al_2O_3/CoCrAly (60%/40%)	0.125	2	
SiC./TiB_2 (15%)	0.250	0.35	
Metal matrix composites			
Mg/B_4C (15%)	0.125	35	fair
Al/SiC (15%)	0.500	8–12	good to fair
Al/Al_2O_3 (15%)	0.250	15–20	good to fair

Table 2. Cutting speeds with simple waterjet

Material	Thickness (in.)	Nozzle speed (in./min)	Edge quality (comments)
ABS plastic	0.087	20–50	100% separation
Aluminum	0.050	2–5	burr
Cardboard	0.055	240–600	slits very well
Delrin	0.500	2–5	good to stringers
Fiberglass	0.100	40–150	good to raggy
Formica	0.040	1450	
Graphite composite	0.060	25	
Kevlar	0.040–0.250	50–53	fair, some furring
Lead	0.125	10	good, slight burr
Plexiglass	0.118	30–35	fair
Printed circuit bd	0.050–0.125	50–5	good
PVC	0.250	10–20	good to fair
Rubber	0.050	2400–3600	good
Vinyl	0.040	2000–2400	good
Wood	0.125	40	fair

Comment on these tables: In trying to provide data on waterjet and abrasive waterjet cutting we have collected material from diverse sources. But we must note that most of the data presented is not from uniform tests. Also, note that in many cases data was largely absent on such parameters as pump horsepower, waterjet pressure, abrasive-particle rate of flow or type or size, and standoff distance. So these cutting rates vary widely in value—from laboratory control to shop floor ballpark estimates. Many of the top speeds cited either represent cuts made to illustrate speed alone, without regard to surface quality, or may reflect data from machines with very high power output. (American Machinist, *October 1989*.)

attributed to the reentrainment of abrasive particles in the jet by the workpiece material. AWC is particularly suited for composites because the cutting rates are reasonable and they do not delaminate the layered material. In particular, AWC is used in the airplane industry to cut carbon-fiber composite sections of the airplane after autoclaving.

ABRASIVE JET MACHINING

One of the least expensive of the nontraditional processes is **abrasive jet machining (AJM).** AJM removes material by a focused jet of abrasives and is similar in many respects to AWC, with the exception that momentum is transferred to the abrasive particles by a jet of inert gas. Abrasive velocities on the order of 1000 ft/s are possible with AJM. The small mass of the abrasive particles produces a microscale chipping action on the workpiece material. This makes AJM ideal for processing hard, brittle materials, including glass, silicon, tungsten, and ceramics. It is not compatible with soft, elastic materials.

Key process parameters include working distance, abrasive flow rate, gas pressure, and abrasive type. Working distance and feed rate are controlled by hand. If necessary, a hard mask can be placed on the workpiece to control dimensions. Abrasives are typically smaller than those used in AWC. Abrasives are typically not recycled, since the abrasives are cheap and are used only on the order of several hundred grams per hour. To minimize particulate contamination of the work environment, a dust-collection hood should be used in concert with the AJM system.

26.9 DESIGN CONSIDERATIONS IN GRINDING

Almost any shape and size of work can be finished on modern grinding equipment, including flat surfaces, straight or tapered cylinders, irregular external and internal surfaces, cams, anti-friction-bearing races, threads, and gears. For example, the most accurate threads are formed from solid cylindrical blanks on special thread-grinding machines. Gears that must operate without play are hardened and then finish ground to close tolerances. Two important design recommendations are to reduce the area to be ground and to keep all surfaces that are to be ground in the same or parallel planes (Figure 26-31). This is an example of **design for manufacturing (DFM).**

Original design of base plate

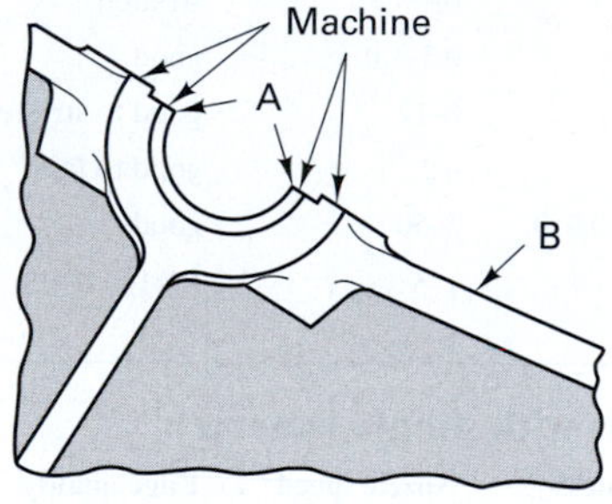

Original design of crankshaft bearing bracket

Redesigned to reduce weight and grinding time

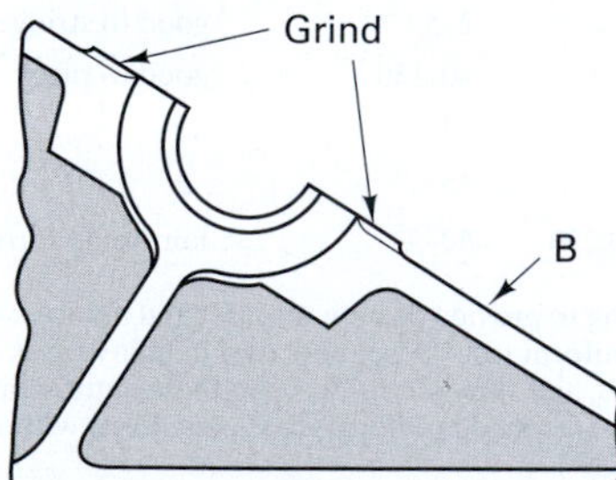

Redesign eliminated shoulders and made part suitable for grinding in single setup

FIGURE 26-31 Reducing area to be ground and keeping all surface to be ground in the same or parallel planes are two important design recommendations. *(From* Machine Design, *June 1, 1972, p. 87)*

Abrasive machining can remove scale as well as parent metal. Large allowances of material, needed to permit conventional metal-cutting tools to cut below hard or abrasive inclusions, are not necessary for abrasive machining. An allowance of 0.015 in. is adequate, assuming, of course, that the part is not warped or out of round. This small allowance requirement results in savings in machining time, in material (often 60% less metal is removed), and in shipping of unfinished parts.

Key Words

abrasive
abrasive machining
abrasive jet machining (AJM)
abrasive water-jet cutting (AWC)
aluminum oxide
attrition
bonded product
burn
centerless grinding
center-type cylindrical grinding
chucking-type external grinding
coated abrasive
coated product
corundum
creep feed grinding (CFG)
crush dressing
cubic boron nitride (CBN)
cutting
design for manufacturing (DFM)
diamond
disk grinder
dressing
emery
friability
G ratio
garnet
grade
grinding
hardness
honing
honing stone
hydrodynamic machining
infeed
lap
lapping
off-hand grinding
plowing
plunge-cut grinding
quartz
regulating wheel
resinoid bond
rubber bond
rubbing
shellac bond
silicate bond
silicon carbide
snagging
structure
superadhesive bond
superfinishing
surface grinding
swing grinder
truing
ultrasonic impact grinding
ultrasonic machining (USM)
universal tool and cutter grinder
vitrified bond
water-jet cutting (WJC)
water-jet machining

Review Questions

1. What are machining processes that use abrasive particles for cutting tools called?
2. What is attrition in an abrasive grit?
3. Why is friability an important grit property?
4. Explain the relationship between grit size and surface finish.
5. Why is aluminum oxide used more frequently than silicon carbide as an abrasive?
6. Why is CBN superior to silicon carbide as an abrasive in some applications?
7. What materials commonly are used as bonding agents in grinding wheels?
8. Why is the grade of a bond in a grinding wheel important?
9. How does grade differ from structure in a grinding wheel?
10. What is crush dressing?
11. How does loading differ from glazing?
12. What is meant by the statement that grinding is a mixture of processes?
13. What is accomplished in dressing a grinding wheel?
14. How does abrasive machining differ from ordinary grinding?
15. What is a grinding ratio or *G* ratio?
16. How is the feed of the workpiece controlled in centerless grinding?
17. Why is grain spacing important in grinding wheels?
18. Why should a cutting fluid be used in copious quantities when doing wet grinding?
19. How does plunge-cut grinding compare to cylindrical grinding?
20. If grinding machines are placed among other machine tools, what precautions must be taken?
21. What is the purpose of low-stress grinding?
22. How is low-stress grinding done compared to conventional grinding?
23. The number of grains per square inch that actively contact and cut a surface decreases with increasing grain diameter. Why is this so?
24. Why are centerless grinders so popular in industry compared to center-type grinders?
25. Explain how an SEM micrograph is made. Check the Internet or the library to find the answer.
26. Why are vacuum chucks and magnetic chucks widely used in surface grinding but not in milling?
27. How does creep feed grinding differ from conventional surface grinding?
28. Why does a lap not wear, even though it is softer than the material being lapped?
29. How do honing stones differ from grinding wheels?
30. What is meant by "charging" a lap?
31. Why is a honing head permitted to float in a hole that has been bored?
32. How does a coated abrasive differ from an abrasive wheel?
33. Figure out why the bottoms of chips shown in Figure 26-9 are so smooth. The magnification of the micrograph is 4800B. How thick are these chips?
34. What is the inclined angle in honing, and what determines it?
35. What are the common causes of grinding accidents?
36. What other machine tool does a surface grinder resemble?
37. Figure 26-11 showed residual stress distributions produced by surface grinding. What is a residual stress?
38. In grinding, what is infeed versus cross feed?
39. One of the problems with water-jet cutting is that the process is very noisy. Why?
40. In AWC, what keeps the abrasive jet from machining the orifice?

PROBLEMS

1. Perhaps you have observed the following wear phenomena: A set of marble or wooden steps shows wear on the treads in the regions where people step when they climb (or descend) the steps. The higher up the steps, the less the wear on the tread. Given that soles of shoes (leather, rubber) are far softer than marble or granite, explain:
   a. Why and how the steps wear.
   b. Why the lower steps are more worn than the upper steps.
2. Explain why it is that a small particle of a material can be used to abrade a surface made of the same material (i.e., why does the small particle act harder or stronger than the bulk material)?
3. In grinding, both the wheel and workpiece are moving (or rotating). Using the data in Figure 26-11 and assuming that you are doing surface grinding (see Figure 26-1), what are some typical MRR values? How do these compare to MRR values for other machining processes, such as milling? What is the significance of this?

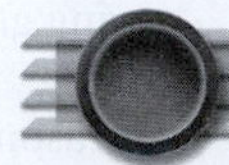

www.wiley.com/go/global/degarmo

Chapter 26 CASE STUDY

Process Planning for the MfE

At Lulu's WarEagle company, Figure CS-26A shows a part design for a small flange to be made out of 1020 steel. Prepare a sequence of operations or process plan to make this part. Note that you need to machine the top, bottom, and all the holes. Specify the machines and the tools you would use, assuming you select bar stock as your raw material. The term "Co bore" in the drawing means counter bore. See Chapter 2 for an example of a process plan. Assume the lot size here is 1200 part/yr made in lots of 120 every month.

Now assume that the designer provides you a design like Figure CS-26B and calls for the flange to be made from cast iron. The casting will probably have a large hole in the center cored during the casting process. What casting process will you use? Prepare another process plan for the cast part. Will the sequence of operations be the same? What about the cost per unit? Which will be lower?

1. The part drawings failed to provide a critical specification in the design. What is it?
2. The drawing failed to show the final geometry (in the top view) correctly. Redraw the part to show these corrections.
3. As the process engineer, how would you advise the design engineer he screwed up—twice?

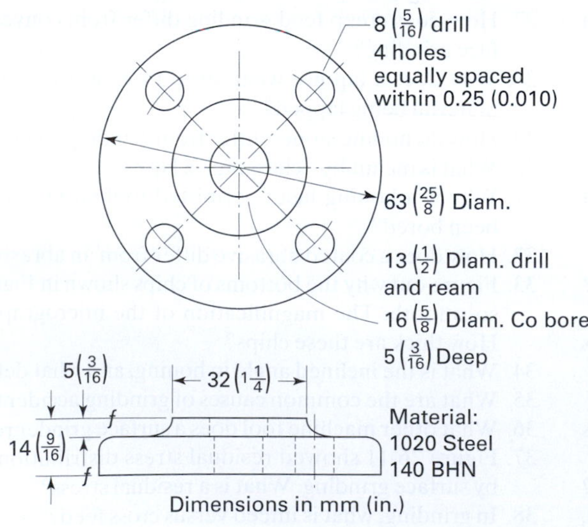

Figure CS 26a

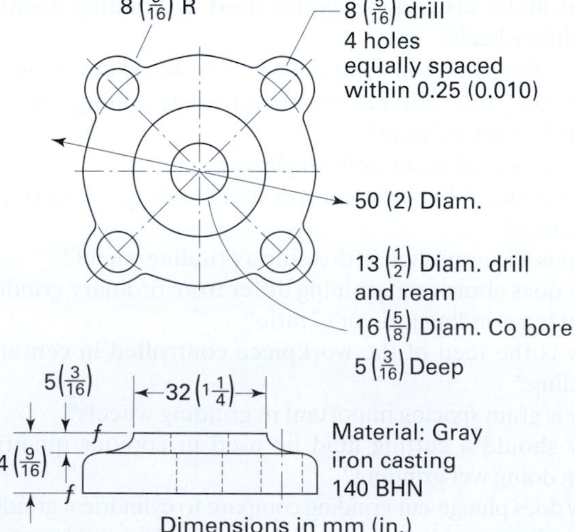

Figure CS 26b

FIGURE CS-26 Two different designs of a flange. The design on the right suggested by Doyle et al. in 3rd edition of *Manufacturing Process and Materials for Engineers*, Prentice-Hall.

CHAPTER 27

Workholding Devices for Machine Tools

27.1 Introduction

Workholding devices, often called *jigs* and *fixtures,* are critical components in the manufacturing of interchangeable parts. **Workholders** hold and locate the work in the machine tool with respect to the cutting tool. For example, Figure 27-1 shows a machining center with a fixture that holds the workpiece in the correct location with respect to the cutting tools. For many machine tools, jigs and fixtures hold the workpieces while providing location with respect to the cutting tools. With workholders, process accuracy and precision (repeatability) can be achieved that otherwise would be impossible with a given combination of cutting tools and machine tools. In this chapter, workholding devices (jigs and fixtures) will be considered as important production tools or adjuncts, with primary attention being directed toward their functional characteristics, their relationship to the machine tools, and the manufacturing processes.

In recent years, workholding devices have become more flexible; that is, they are able to (1) hold more than one part and (2) be quickly changed over for different parts. Workholders that can be quickly exchanged are critical elements in lean manufacturing cells, where components are made in families of parts (groups of parts of similar design). Further, being able to change from one device to another quickly to accommodate different parts means smaller lot sizes can be run, which reduces inventory levels in plants. In computer numerical control (CNC) machines (described in Chapter 25), the workholders can be automatic. For example in Figure 27-2, the CNC turning center has two chucks and the workpiece can be automatically transferred from one to the other, so the powered tools on the turrets can work on both ends of the part. These flexibility requirements add significantly to the complexity of conventional jig and fixture design. Let's begin with a discussion of the basics of jig and fixture design.

27.2 Conventional Fixture Design

In the conventional method of fixture design, tool designers rely on their experience and intuition to design simple, single-purpose fixtures for specific machining operations, often using a trial-and-error method until the workholders perform satisfactorily. Of course, these designers should calculate the clamping forces or stress distributions in the fixturing elements to make sure that the loads will not deform the fixtures or the workpieces elastically or plastically. In the design of the workholding devices, two primary functions must be considered: locating and clamping. **Locating** refers to orienting

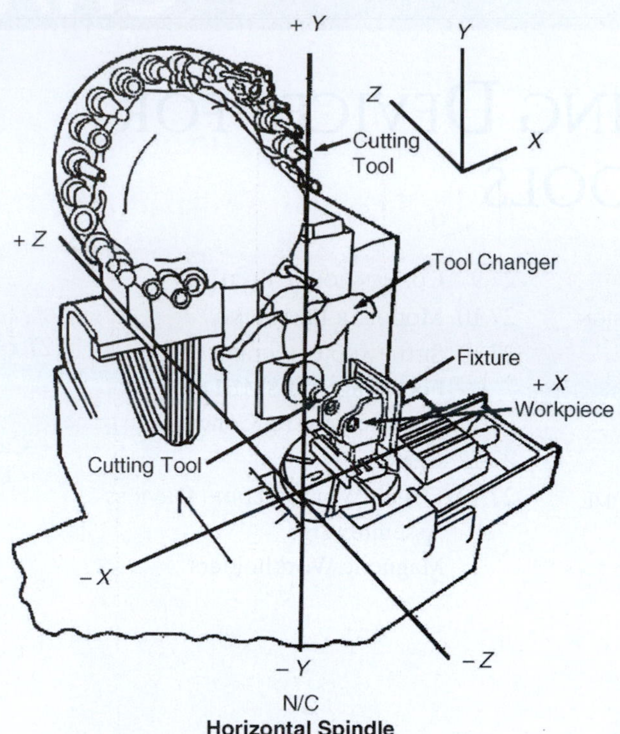

FIGURE 27-1 An example of a fixture in a milling machine holding a part.

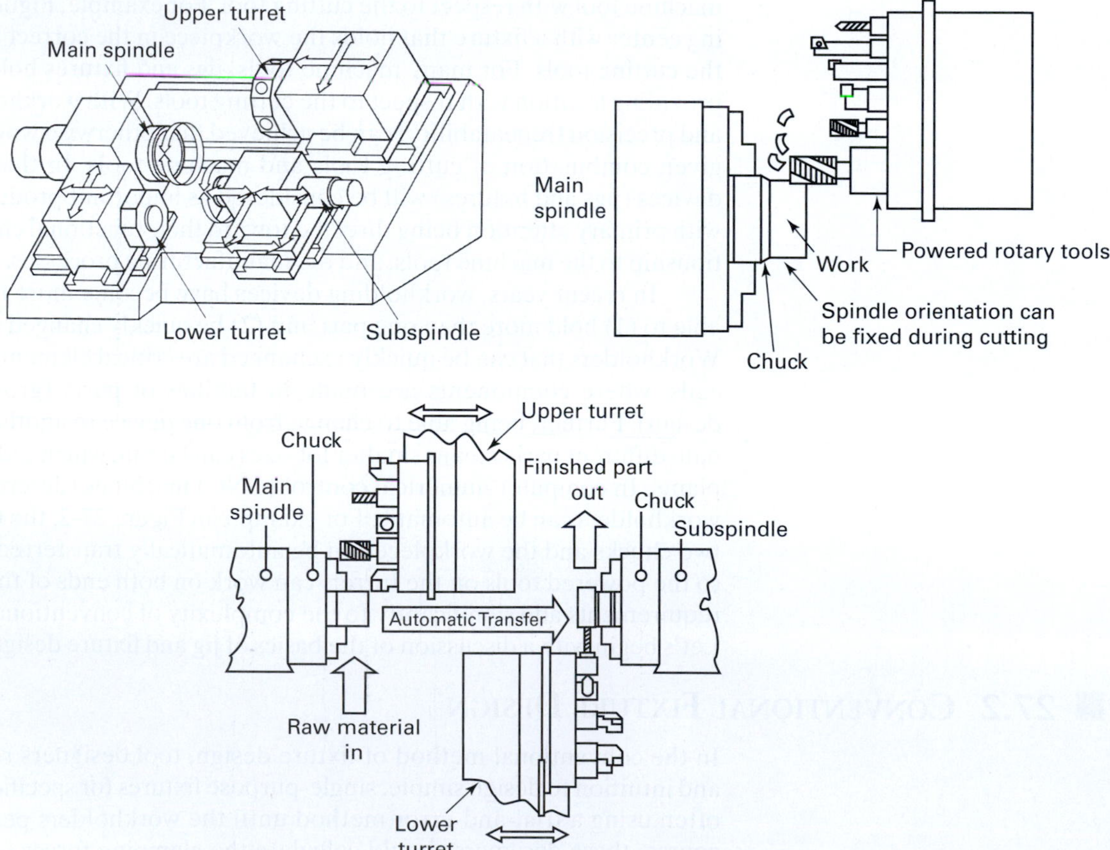

FIGURE 27-2 CNC turning center with two chucks, turrets for cutting tools, and C-axis control for the main spindle. The C-axis control, on the spindle, can stop it in any orientation so the powered tools can operate on the workpiece.

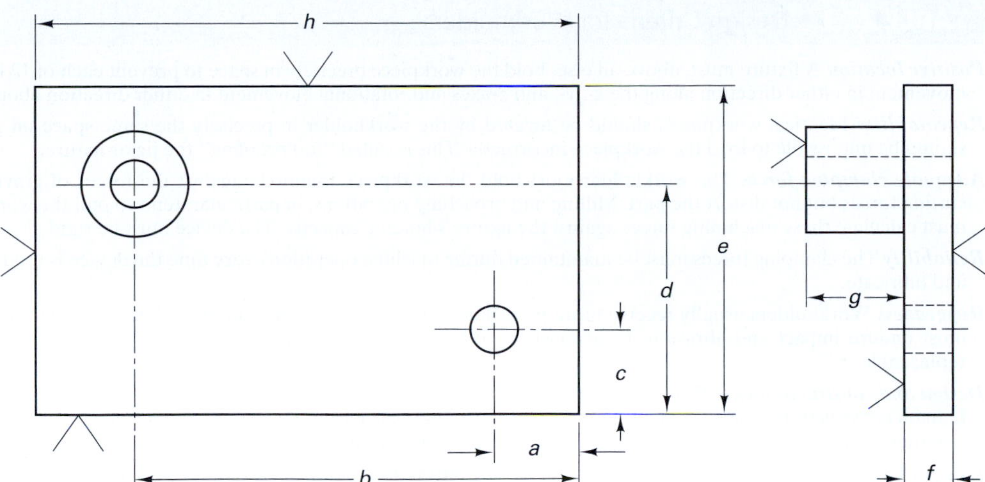

FIGURE 27-3 Drawing of a plate showing locating dimensions *(a, b, c, d)* versus sizing dimensions *(e, f, g, h)*.

and positioning the part in the machine tool with respect to the cutting tools to achieve the required specifications. **Clamping** refers to holding or maintaining the part in that location during the cutting operations (resisting the cutting forces).

Jigs and **fixtures** are specially designed and built workholding devices that hold the work during machining or assembly operations. In addition, a jig determines a location dimension that is produced by machining or fastening. For example, location dimensions determine the position of a hole on a plate (Figure 27-3). Consider the subject of dimensioning as used in drafting practice. Dimensions are of two types: size and location. **Size dimensions** denote the size of geometrical shapes—holes, cubes, parallelepipeds, and so on—of which objects are composed. **Location dimensions,** on the other hand, determine the position or location of these geometrical shapes *with respect to each other*. Thus *a* and *c* in Figure 27-3 are location dimensions, whereas *e* and *g* are size dimensions. With location dimensions in mind, one can precisely define a jig as follows: *a jig is a special workholding device that, through built-in features, determines location dimensions that are produced by machining or fastening operations*. The key requirement of a jig is that it determine a location dimension. Thus, jigs accomplish layout by means of their design.

In order to establish location dimensions, jigs may do a number of other things. They frequently guide tools, as in drill jigs, and thus determine the location of a component geometrical shape. However, they do not always guide tools. In the case of welding jigs, component parts are held (located) in a desired relationship with respect to each other while an unguided tool accomplishes the fastening. The guiding of a tool is not a necessary requirement of a jig.

Similarly, jigs usually hold the work that is to be machined, fastened, or assembled. However, in certain cases, the work actually supports the jig. Thus, although a jig *may* incidentally perform other functions, the basic requirement is that, through qualities that are built into it, certain critical dimensions of the workpiece are determined.

A fixture is defined as *a special workholding device that holds work during machining or assembly operations and establishes size dimensions*. The key characteristic is that it is a *special* workholding device, designed and constructed for a particular part or shape. A general-purpose device, such as a chuck in a lathe or a clamp on a milling machine table, is usually not considered to be a fixture. Thus a fixture has as its specific objective the facilitating of **setup,** or making the part holding easier. Because many jigs hold the work while determining critical location dimensions, they usually meet all the requirements of a fixture. Alternatively, many fixtures are used in numerical control (NC) machines holding parts where holes are located and drilled according to a program. So the strict definition of jigs and fixtures has been blurred by the changes in technology.

In designing workholders, the designer must consider whether the part is a casting, forging, or bar stock. With castings and forgings, variations in shape and size must be accommodated in the design, and usually a machining operation is required to establish a reference surface (called the **datum surface**) to aid initial fixturing.

TABLE 27-1 Design Criteria for Workholders

Positive location A fixture must, above all else, hold the workpiece precisely in space to prevent each of 12 kinds of degrees of freedom—linear movement in either direction along the x-, y-, and z-axes and rotational movement in either direction about each axis.

Repeatability Identical workpieces should be located by the workholder in precisely the same space on repeated loading and unloading cycles. It should be impossible to load the workpiece incorrectly. This is called "foolproofing" the jig or fixture.

Adequate clamping forces The workholder must hold the workpiece immobile against the forces of gravity, centrifugal forces, inertial forces, and cutting forces but not distort the part. Milling and broaching operations, in particular, tend to pull the workpiece out of the fixture, and the designer must calculate these machining forces against the fixture's holding capacity. The device must be rigid.

Reliability The clamping forces must be maintained during machine operation every time the device is used. The mechanism must be easy to maintain and lubricate.

Ruggedness Workholders usually receive more punishment during the loading and unloading cycle than during the machining operation. The device must endure impact and abrasion for at least the life of the job. Elements of a device that are subject to damage and wear should be easily replaceable.

Design and construction ease Workholders should use standard elements as much as possible to allow the engineer to concentrate on function rather than on construction details. Modular fixtures epitomize this design rule as the entire workholder is made from standard elements, permitting a bolt-together approach for substantial time and cost savings over custom workholders.

Low profile Workholder elements must be clear of the cutting-tool path. Designing lugs on the part for clamping can simplify the fixture and allow proper tool clearance.

Workpiece accommodation Surface contours of castings or forgings vary from one part to the next. The device should tolerate these variations without sacrificing positive location or other design objectives.

Ergonomics and safety Clamps should be selected and positioned to eliminate pinch points and facilitate ease of operation. The workholder elements should not obstruct the loading or unloading of workpieces. In manual operations, the operator should not have to reach past the tool to load or unload parts. A rule sometimes used is that the operator can repeatedly exert a force of 30 to 40 lb to open or close a clamp, but greater forces than this can cause ergonomic problems.

Freedom from part distortion Parts being machined can be distorted by gravity, the machining forces, or the clamping forces. Once clamped into the device, the part must be unstressed or, at least, undistorted. Otherwise, the newly machined surfaces take on any distortions caused by the clamping forces.

Flexibility The workholding device can locate and restrain more than one type (design) of part. Many different schemes are being proposed to provide workholder flexibility. Modular vise fixturing, programmable clamps using air-activated plungers, part encapsulation with a low-melting-point alloy, and NC clamping machines are some of the more recently developed systems. Despite their flexibilities, these clamping systems have some significant drawbacks. They are expensive, and the individual systems may not integrate well into individual machine tools. (See the discussions of intermediate jig concept and group jigs for additional thoughts on flexibility.)

Even in parts cut from bar stock, allowances must be made for inaccuracies and irregularities produced by the cutoff operations.

Table 27-1 provides a summary of design criteria for workholders for careful to review. Obviously, it is impossible to meet all these design criteria for workholders. Compromise is inevitable. Still, it is useful to know the optimal design objectives to illustrate the positioning, holding, and supporting functions that fixtures must fulfill.

27.3 TOOL DESIGN STEPS

The classical design of a workholder (e.g., a drill jig) involves the following steps:

1. Review the design criteria for workholders in Table 27-1.
2. Analyze the drawing of the workpiece and determine (visualize) the machining operations required to machine it. Note the critical (size and location) dimensions and tolerances. This is called the sequence of operations.
3. Determine the orientations of the workpiece in relation to the cutting tools and the movements of the tools and tables.
4. Perform an analysis to estimate the magnitude and direction of the cutting forces (see Chapter 20).
5. Study the standard devices available for workholders and for the clamping functions. Can an off-the-shelf device be modified? What standard elements can be used?
6. Form a mental picture of the workpiece in position in the workholder in the machine tool with the cutting tools performing the required operation(s). See Figure 27-4 and chapters on machining for more examples.
7. Make a three-dimensional sketch of the workpiece in the workholder in its required position to determine the location of all the elements: clamps, locator buttons, bushings, and so on. Use the 3-2-1 location principle discussed next.

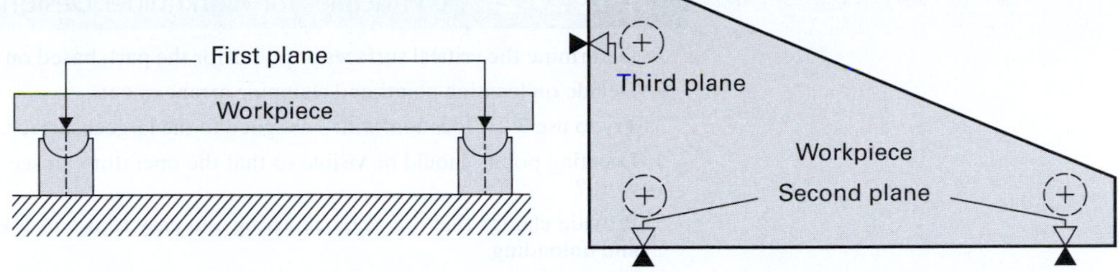

FIGURE 27-4 Workpiece location is based on the 3-2-1 principle. Three points will define a base surface, two points in a vertical plane will establish an end reference, and one point in a third plane will positively locate most parts.

8. Make a sketch of the workholder and workpiece in the machine tool to show the orientation of these elements with respect to the cutting tool in the machine tool.

3-2-1 LOCATION PRINCIPLE

After determining the orientation of the workpiece in the workholder, the next step is to locate it in that position. This location is also used for all similar workpieces. The tool designer must select or design locating devices (supports) *that ensure that every workpiece placed in the device occupies the same position with respect to the cutting tools.* Thus, when the machining operation is performed, the workpieces are processed identically. This is, of course, the key to making interchangeable parts. In locating the workpiece, the basic **3-2-1 principle** of location is used (Figure 27-4). For positive location, the fixture must position the workpiece in each of three perpendicular planes. Positioning processes can vary greatly, but workholder design always begins by defining the first plane of reference with three points. The first plane is supported by three flattened balls that swivel in sockets, providing a self-adapting surface. Once the object is defined in a single plane, supported at three points (like a three-legged stool on a floor), a second plane can be assigned that is perpendicular to the first. To do this, the object is brought up against any two points in the second plane. To continue the example, the stool is slid along the floor until two legs touch a wall.

A third plane, perpendicular to each of the other two, is then defined by designating one point on it. As long as an object is in contact with three points on the first plane, two points on the second, and a single point on the third, it is positively located in space. The location points within each plane should be selected as far apart as possible for maximum stability.

In practice, it is often necessary to support a workpiece on more points than this 3-2-1 formula dictates. The machining of a large rectangular plate, for example, typically requires support at four or more points. However, any extra points must be established carefully to support the workpiece in a plane defined by three—and only three—points.

Appropriate clamping devices are selected so that the clamping forces hold the workpiece in the proper location and resist the effects of the cutting forces, centrifugal forces, and vibrations. If possible, the machining forces should act into the location points, not into the clamps, so that smaller clamps can be used. In reality, the worker often determines clamping force when loading the part into the workholder. Fixtures are usually fastened to the table of the machine tool. Although used primarily on milling and broaching machines, fixtures are also designed and used to hold workpieces for various operations on most of the standard machine tools and machining centers. Black's 20 principles for fixture design are given in Table 27-2.

■ 27.4 CLAMPING CONSIDERATIONS

Clamping of the work is closely related to support of the work. Any clamping, of course, induces some stresses into the part that can cause some distortion of the workpiece, usually elastic. If this distortion is measurable, it will cause some inaccuracy in final dimensions of the part, as illustrated in an exaggerated manner in Figure 27-5. The obvious solution is to spread the clamping forces over a sufficient area to reduce the stresses to a level that will not produce appreciable distortion. The clamping forces should direct the work against the points of location and work support. Clamped surfaces often

TABLE 27-2 20 Principles for Workholder Design

1. Determine the critical surfaces or points for the part, based on the design.
2. Decide on locating points and clamping arrangements.
3. Try to use 3-2-1 location, with 3 assigned to the largest surface. Additional points should be adjustable.
4. Locating points should be visible so that the operator can see if they are clean. Can they be replaced if worn?
5. Provide clamps that are as quick acting and easy to use as is economically justifiable for rapid loading and unloading.
6. Clamps should not require undue effort by the operator to close or to open, nor should they harm hands and fingers during use.
7. Clamps should be integral parts of device. Avoid loose parts that can get lost.
8. Avoid complicated clamping arrangements or combinations that can wear out or malfunction. Keep it simple.
9. Locate clamps opposite locaters (if possible) to avoid deflection/distortion during machining and spring-back afterward.
10. Take the thrust of the cutting forces on the locaters (if possible), not on the clamps.
11. Arrange the workholder so that the workpiece can easily be loaded and unloaded from the device.
12. Design the workholder so that the part can be loaded only in the correct manner (mistake-proof) and in such a way that the location can be found quickly (visually).
13. Consistent with strength and rigidity, make the workholder as light as possible.
14. Provide ample room for chip clearance and chip extraction.
15. Provide accessibility for cleaning.
16. Provide for entrance and exit of cutting fluids (which may carry off chips) if one is to be used.
17. Provide four feet on all movable workholders.
18. Provide hold-down lugs on all fixed workholders.
19. Provide keys to align fixtures on machine tables so fixtures can be replaced in exactly the same position.
20. Never sacrifice safety for production.

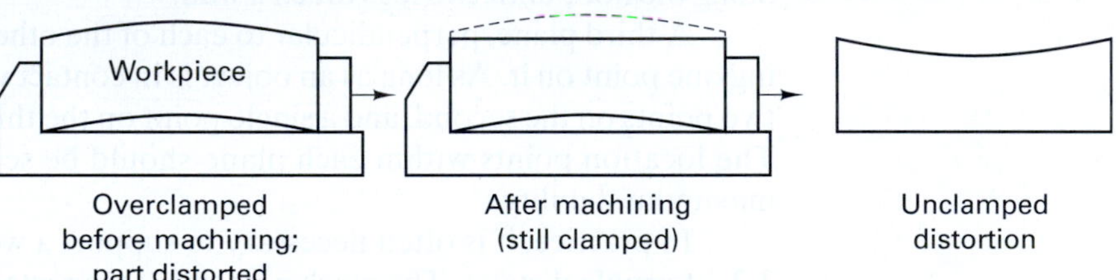

FIGURE 27-5 Exaggerated illustration of the manner in which excessive clamping forces can affect the final dimensions of a workpiece.

have some irregularities that may produce force components in an undesired direction. Consequently, clamping forces should be applied in directions that will ensure that the work will remain in the desired position.

Whenever possible, jigs and fixtures should be designed so that the forces induced by the cutting process act to hold the workpiece in position against the supports. These forces are predictable, and proper utilization of them can materially aid in reducing the magnitude of the clamping stresses required. In addition to locating the work properly, the stops or work-supporting areas must be arranged so as to provide adequate support against the cutting forces. As shown in Figure 27-6a, having the cutting force act against a fixed portion of the jig or fixture and not against a movable section permits lower clamping forces to be used. Figure 27-6b illustrates the principle of keeping the points of clamping as nearly as possible in line with the action forces of the cutting tool so as to reduce their tendency to pull the work from the clamping jaws. Compliance with this principle results both in lower clamping stresses and less massive clamping devices. Don't forget that down milling produces different forces than up milling. The location points should be as far apart as possible but positioned so as not to allow the cutting force to distort the work, as shown in Figure 27-6c. The cutting forces may distort the work, with resulting inaccuracy or broken tools. These design suggestions materially reduce vibration and chatter during the cutting process.

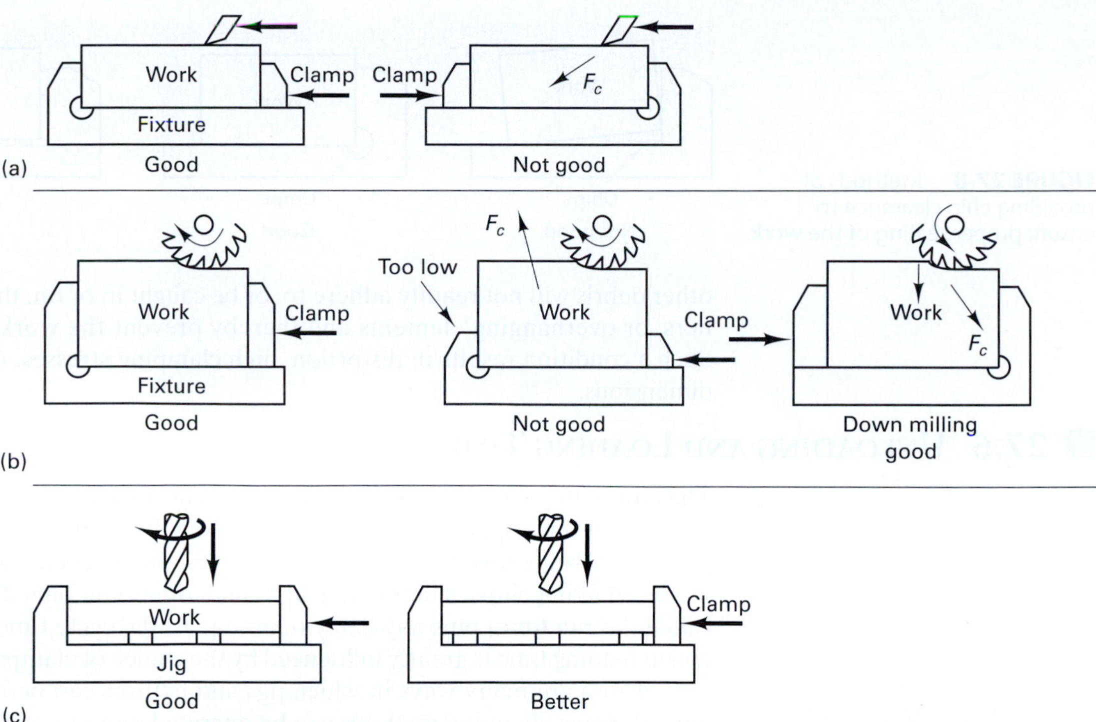

FIGURE 27-6 In (a) and (b), proper work support to resist the forces imposed by cutting tools is demonstrated. In (c), three buttons form a triangle for the work to rest on.

In the job shop, try to perform as many operations as are possible with each clamping of the workpiece. This has both physical and economic aspects. Because some stresses result from each clamping, with the possibility of accompanying distortion, greater accuracy is achieved if multiple operations are performed with each clamping. From the economic viewpoint, if the number of jigs or fixtures is reduced, less capital will be required and less time will be spent handling the workpiece loading and unloading. In the lean shop, simple, fast-acting fixtures are used with rapid tooling exchange.

27.5 Chip Disposal

When jigs or fixtures are used in connection with chip-making operations, adequate provision must be made for the easy removal of the chips. This is essential for several reasons. First, if chips become packed around the tool, heat will not be carried away and tool life can be decreased. Figure 27-7 illustrates how insufficient clearance between the end of a drill bushing and the workpiece can prevent the chips from escaping, whereas too much clearance may not provide accurate drill guidance and can result in broken drills.

A second reason chips must be removed is so that they do not interfere with the proper seating of the work in the jig or fixture (Figure 27-8). Even though chips and dirt always have to be cleaned from the locating and supporting surfaces by a worker or by automatic means, such as an air blast, the design details should be such that chips and

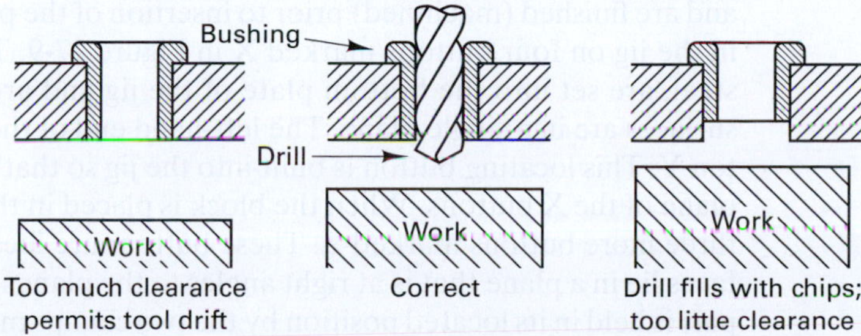

FIGURE 27-7 Proper clearance between drill bushing and tool of workpiece is important.

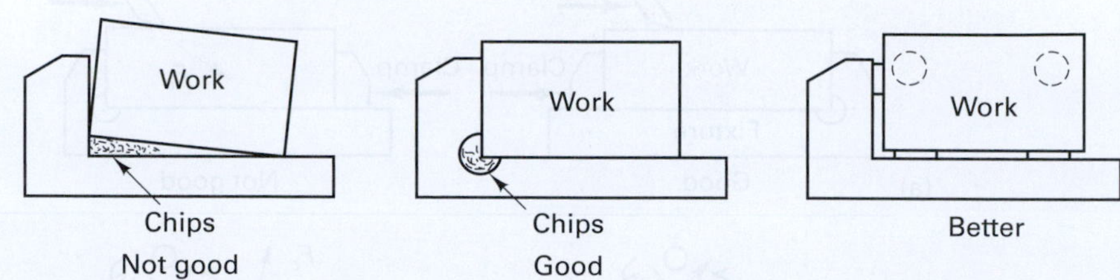

FIGURE 27-8 Methods of providing chip clearance to ensure proper seating of the work.

other debris will not readily adhere to, or be caught in or on, the locating surfaces, corners, or overhanging elements and thereby prevent the work from seating properly. Such a condition results in distortion, high clamping stresses, and incorrect workpiece dimensions.

27.6 UNLOADING AND LOADING TIME

The cost of the workholders must be justified by the quantities of production involved, and their primary purpose is to increase productivity and quality. While work is being put into or being taken out of jigs and fixtures, the machines with which they are used are not making chips. The loading and unloading time plus the machining time (also called the **run time**) plus any delay times equals the cycle time for a part. The loading and unloading time is greatly influenced by the choice of clamps.

There are many ways in which jigs and fixtures can be made easier to load and unload. Some clamping methods can be operated more readily than others. For example, in the drill jig shown in Figure 27-9, a **knurled clamping screw** on the right side is used to hold the part against the buttons at the left end of the jig. To clamp or unclamp the block in this direction requires several motions. On the other hand, a **cam latch** is used to close the jig and hold the workpiece against the rear locating buttons. This type of latch can be operated with a single motion.

Certainly, the device should be designed so that the part cannot be loaded incorrectly. Defect prevention is often accomplished by the clamping device so that a part loaded improperly cannot be clamped. Ease of operation of workholders not only directly increases the productivity of such equipment but also results indirectly in better quality and fewer lost-time accidents.

The workholder is as critical as the machine tool and the cutting tool to the final quality of the part. The use of the workholder eliminates manual layout of the desired features of the part on the raw material. Manual layout requires a highly skilled worker and is very time consuming. The workholder permits a less skilled person to achieve quality and repeatable production with far greater efficiency.

27.7 EXAMPLE OF JIG DESIGN

Several principles of work location and tool guidance are illustrated in Figure 27-9. The two mounting holes in the base of the bearing block are to be located and drilled. The dimensions *A*, *B*, and *C* are determined by the jig. While it is not specified on the drawing, there is one other location dimension that must be controlled. The axes of the mounting holes must be at right angles to the bottom surface of the block, so the bottom must be machined (milled) prior to this drilling step.

The way in which the part dimensions are obtained in the finished workpiece is as follows. The surfaces marked with a large carat V are reference (or location) surfaces and are finished (machined) prior to insertion of the part into the drill jig. The part rests in the jig on four buttons marked X in Figure 27-9. These buttons, made of hardened steel, are set into the bottom plate of the jig and are accurately ground so that their surfaces are in a single plane. The left-hand end of the part is held against another button Y. This locating button is built into the jig so that its surface is at right angles to the plane of the X buttons. When the block is placed in the jig, its rear surface rests against three more buttons marked Z. These buttons are located and ground so that their surfaces lie in a plane that is at right angles to the planes of both the X and Y buttons. The part is held in its located position by the two clamps marked C.

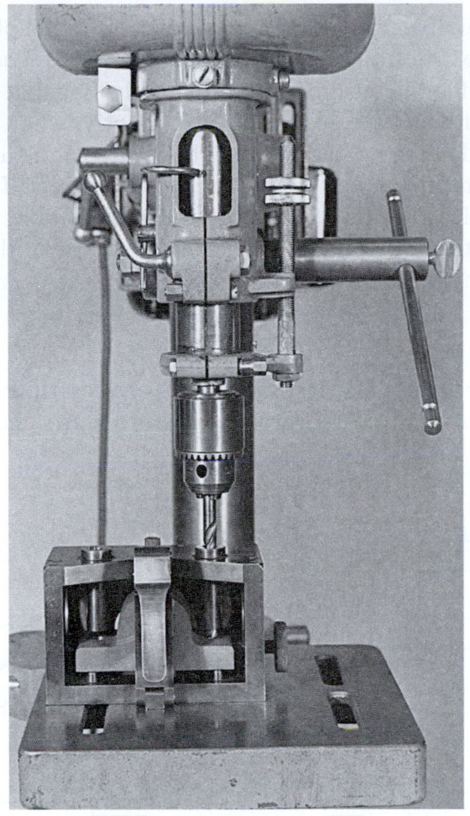

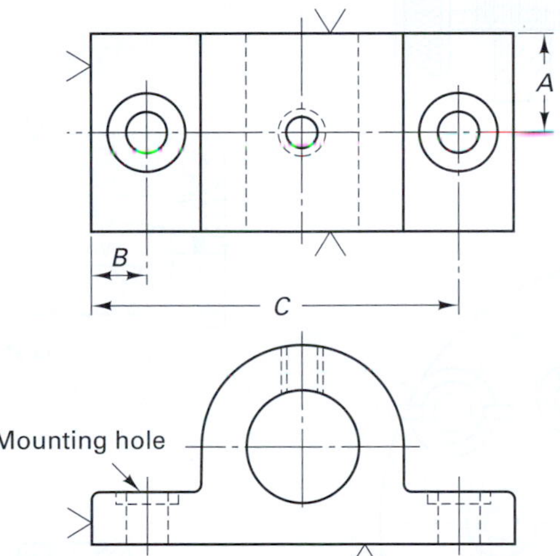

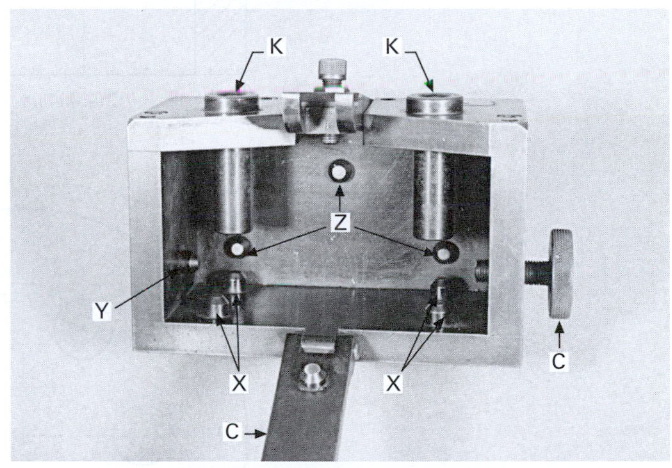

FIGURE 27-9 (Lower left) Part to be drilled; (lower right) box drill jig for drilling two holes; (upper left) jig in drill press; (upper right) drill being guided by drill bushing. *(Courtesy J T. Black)*

The use of four buttons on the bottom of this jig (X buttons) appears not to adhere to the 3-2-1 principle stated previously. However, although only two X buttons would have been required for complete location, the use of only two buttons would not have provided adequate support during drilling. The thrust from the drills would have dislodged the part from the locators. Thus the 3-2-1 principle is a *minimum* concept and often must be exceeded.

To ensure that the mounting holes are drilled in their proper locations, the drill must be located and then guided during the drilling process. This is accomplished by the two drill bushings marked K. Such drill bushings are accurately made of hardened steel, with their inner and outer cylindrical surfaces concentric. The inner diameter is made slightly larger than the drill—usually 0.0005 to 0.002 in.—so that the drill can turn freely but not shift appreciably. The bushings are accurately mounted in the upper plate of the jig and positioned so that their axes are exactly perpendicular to the plane of the X buttons, at a distance A from the Z buttons and at distances B and C, respectively, from the plane of the Y button. Note that the bushings are sufficiently long that the drill is guided close to the surface where it will start drilling. Consequently, when the workpiece is properly placed and clamped in the jig, the drill will be located and guided by the bushings so that the critical dimensions on the workpiece will be correct. The right hole will be drilled in a vertical-spindle drill press (not running), and then the jig will be shifted (manually by the operator) to the right, and the left hole will be drilled. The box construction is rigid but open for chip removal.

27.8 TYPES OF JIGS

Jigs are made in several basic forms and carry names that are descriptive of their general configurations or predominant features. Several of these are illustrated in Figure 27-10.

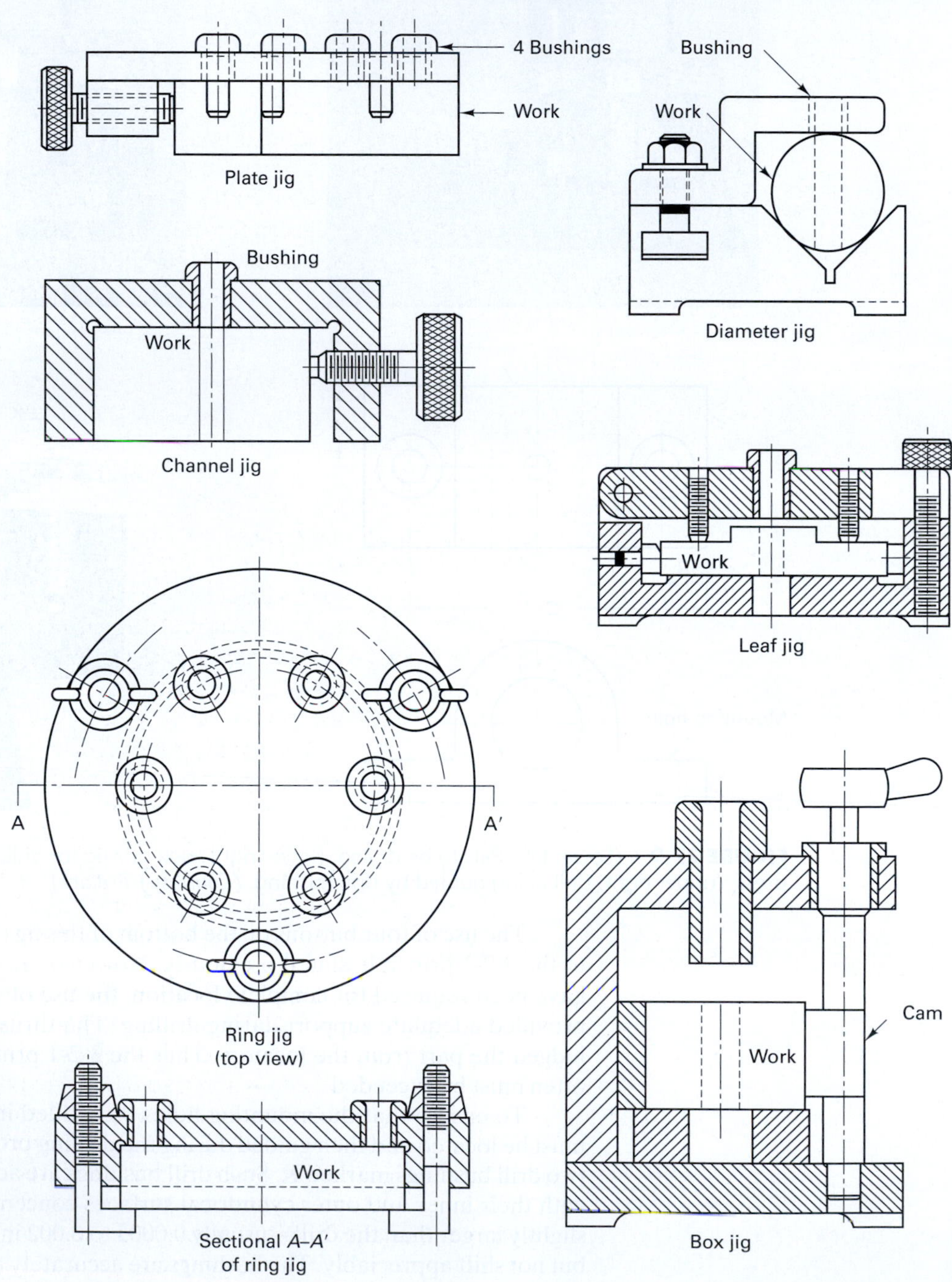

FIGURE 27-10 Examples of some common types of workholders—jigs.

A **plate jig** is one of the simplest types, consisting only of a plate that contains the drill bushings and a simple means of clamping the work in the jig or the jig to the work. In the latter case, wherein the jig is clamped to the work, the device is sometimes called a **clamp-on jig.** Such jigs are frequently used on large parts, where it is necessary to drill one or more holes that must be spaced accurately with respect to each other, or to a corner of the part, but that need not have an exact relationship with other portions of the work.

Channel jigs also are simple and derive their name from the cross-sectional shape of the main member. They can be used only with parts having fairly simple shapes.

Ring jigs are used only for drilling round parts, such as pipe flanges. The clamping force must be sufficient to prevent the part from rotating in the jig. **Diameter jigs** provide a means of locating a drilled hole exactly on a diameter of a cylindrical or spherical piece.

Leaf jigs derive their name from the hinged leaf or cover that can be swung open to permit the workpiece to be inserted and then closed to clamp the work in position. Drill bushings may be located in the leaf as well as in the body of the jig to permit locating and drilling holes on more than one side of the workpiece. Such jigs are called **rollover jigs** or **tumble jigs** when they require turning to permit drilling from more than one side.

Box jigs are very common, deriving their name from their boxlike construction. They have four fixed sides a bottom and a hinged cover or leaf, which opens to permit loading the workpiece, and a cam that locks the workpiece in place. Usually, the drill bushings are located in the fixed sides to ensure retention of their accuracy. The fixed sides of the box are usually fastened by means of dowel pins and screws so that they can be taken apart and reassembled without loss of accuracy. Because of their more complex construction, box jigs are costly, but their inherent accuracy and strength can be justified when there is sufficient volume of production. They have two obvious disadvantages: (1) it is usually more difficult to put work into them than into simpler types, and (2) there is a greater tendency for chips to accumulate within them. Figure 27-9 shows a box-type jig.

Because jigs must be constructed very accurately and be made sufficiently rugged so as to maintain their accuracy despite the use (and abuse) to which they inevitably are subjected, they are expensive. Consequently, several methods have been devised to aid in lowering the cost of manufacturing jigs. One way to reduce this cost is to use simple, standardized plate and clamping mechanisms called **universal jigs** (Figure 27-11). These can easily be equipped with suitable locating buttons and drill bushings to construct a jig for a particular job. Such universal jigs are available in a variety of configurations and sizes, and because they can be produced in quantities, their cost is relatively low. However, the variety of work that can be accommodated by such jigs obviously is limited. While the drill bushing should be spaced far enough from the work to allow chips to escape without entering the bushing, when drilling into an angled surface, the bushing should be very close. Once the drill has penetrated to at least one-half of the drill diameter, the bushing should be retracted to provide chip clearance. Design of the drill jig must not obstruct coolant flow to where it is needed. Bushing length should be 1.75 to 2.5 times the drill diameter.

27.9 CONVENTIONAL FIXTURES

Many examples of **conventional fixtures** have appeared in the text. Production milling, broaching, and boring processes as performed on NC machines, conventional equipment, or machining centers routinely use fixtures to locate and hold the part properly with respect to the cutting tools on the machine tool. Like cutting tools, tooling for workholding is sold separately and is not usually supplied by the machine tool builder. Traditionally, beginning with Eli Whitney, manufacturers have designed and built custom-made, dedicated fixtures. Because of the pressure of shorter production runs and smaller lot sizes, many companies are turning to modified fixturing approaches where the fixture can be constructed quickly or manufacturing cells where families of parts are made in group fixtures.

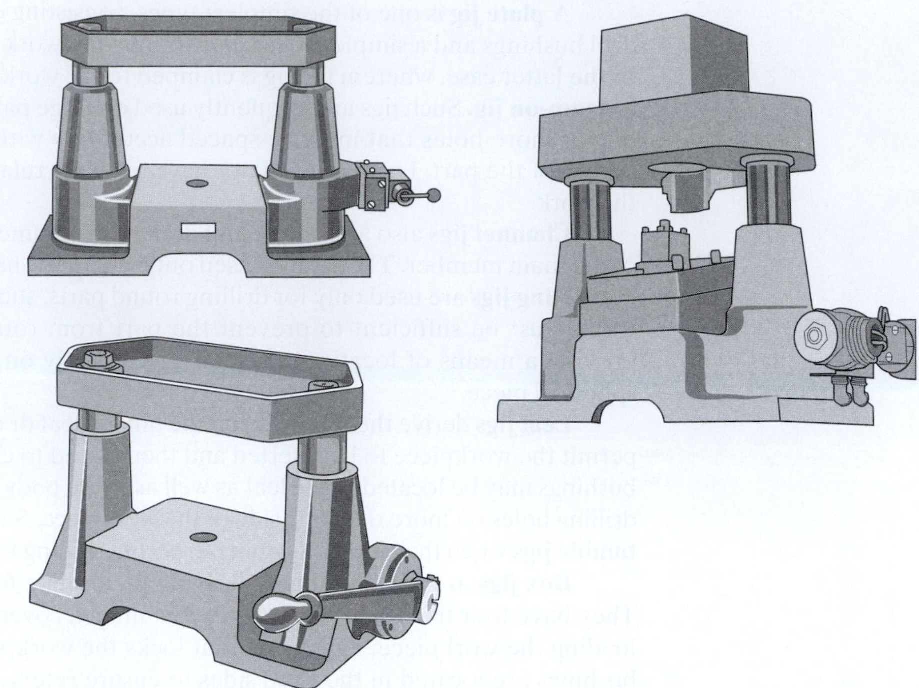

FIGURE 27-11 Two types of universal jigs are manual (bottom) and power-actuated (center). A completed jig (on the top) made from unit right below.

Perhaps the most common fixture uses the **vise** as its base element. Figure 27-12 shows a schematic and photo of a typical commercially available vise that can be adapted for use as a fixture. As shown, the vise jaws are readily modified to conform to the 3-2-1 location principle and provide adequate clamping forces for almost every machining operation. Four vises can be mounted on a subplate for rapid insertion and location in the machine, or four vices can be mounted on a tombstone for milling parts in a CNC machine.

The chucks used in lathes are really general-purpose fixtures for rotational parts. Newer chuck designs have greatly improved their flexibility (the range of diameters the chuck can accommodate in a given setup and speed of setup). Figure 27-13 shows a complete change of top jaws for a three-jaw chuck being done in less than 5 min. The normal time for this part of the setup might exceed 15 min. New quick-change insert top jaws may even snap in by hand with no jaw nuts, keys, screws, or tools. Jaws that can be exchanged by robots can also be designed.

Most producers of chucks use some variation of equation 27-1 to compute the maximum rpm rate at which the chuck can run:

$$S_m^2 = \frac{F_m}{3 \times \left(2.84 \times 10^{-5}\right) \times W \times D} \qquad (27\text{-}1)$$

where

S_m = maximum rpm value at which gripping force $= \frac{1}{3} F_m$
F_m = maximum rate gripping force, at rest (lb)
W = combined weight of jaws (lb)
D = distance from spindle centerline to center of jaw mass (in.)

Thus, with this equation, a 10-in. power chuck with a published rating F_m of 13,200 lb would retain one-third of its initial gripping force at 2507 rpm. (Check this calculation using $W = 8$ lb, $D = 3.1$ in.) The higher the rpm value, the greater the centrifugal force factor. This is an important factor in high-speed machining operations in which the part is rotating.

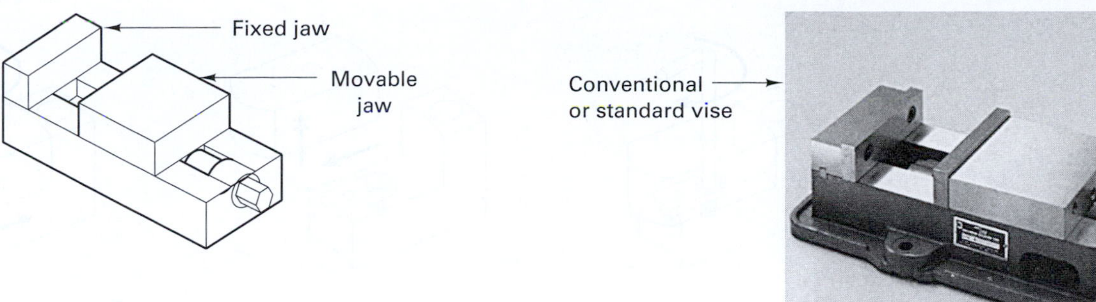

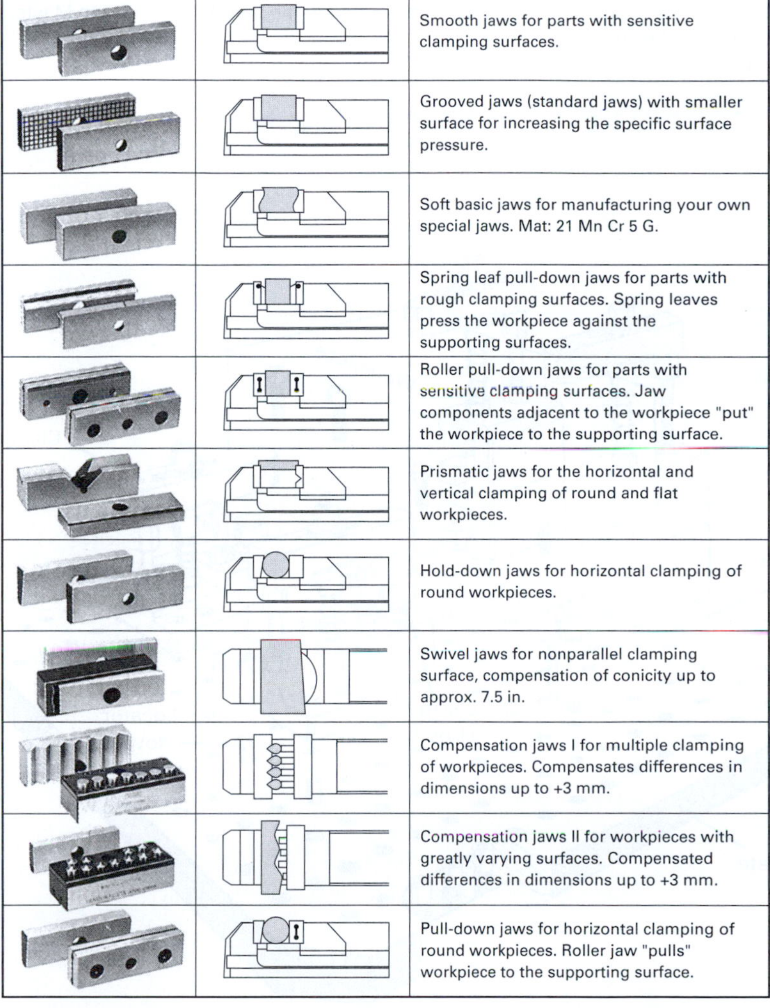

		Smooth jaws for parts with sensitive clamping surfaces.
		Grooved jaws (standard jaws) with smaller surface for increasing the specific surface pressure.
		Soft basic jaws for manufacturing your own special jaws. Mat: 21 Mn Cr 5 G.
		Spring leaf pull-down jaws for parts with rough clamping surfaces. Spring leaves press the workpiece against the supporting surfaces.
		Roller pull-down jaws for parts with sensitive clamping surfaces. Jaw components adjacent to the workpiece "put" the workpiece to the supporting surface.
		Prismatic jaws for the horizontal and vertical clamping of round and flat workpieces.
		Hold-down jaws for horizontal clamping of round workpieces.
		Swivel jaws for nonparallel clamping surface, compensation of conicity up to approx. 7.5 in.
		Compensation jaws I for multiple clamping of workpieces. Compensates differences in dimensions up to +3 mm.
		Compensation jaws II for workpieces with greatly varying surfaces. Compensated differences in dimensions up to +3 mm.
		Pull-down jaws for horizontal clamping of round workpieces. Roller jaw "pulls" workpiece to the supporting surface.

FIGURE 27-12 The conventional or standard vise (top left and right) can be modified with removable jaw plates to adapt to different part geometries. These vices can be integrated into milling fixtures (right middle and bottom). *(Courtesy J T. Black)*

27.10 Modular Fixturing

Modular fixtures have all the same design criteria as those of conventional fixtures, plus one more—**flexibility** or **versatility**. Modular fixture elements must be useful for a variety of machining applications and easily adaptable to different workpiece geometries. Individual fixture designs can be photographed or entered into a computer-aided design (CAD) library for future reference. After the job is done, the fixture itself can be dismantled and the elements returned to the toolroom. The erector-set approach uses either T-slot or dowel-pin designs. Figure 27-14 examples of modular fixturing. The designs begin with baseplates. Elements for locating and clamping are added to the subplate. Rectangular, square, and round are the typical patterns for the subplates. Also

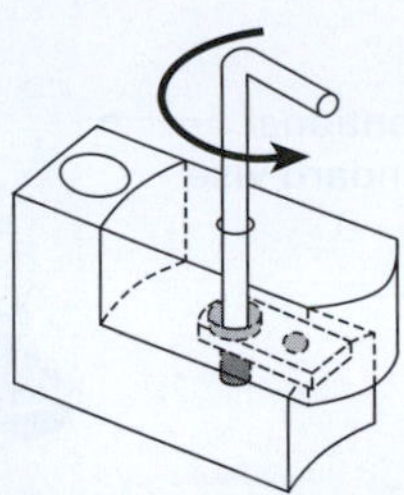

1 Insert allon wrench into hole on top jaw insert turn 1 to 2 revolutions to loosen locking bolt.

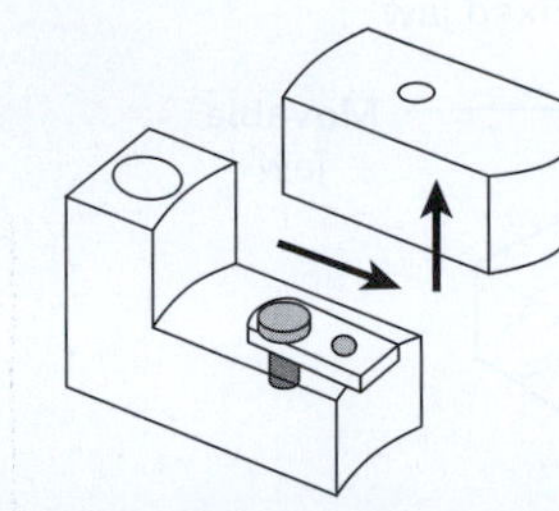

2 Slide top jaw insert forward then up to remove the insert.

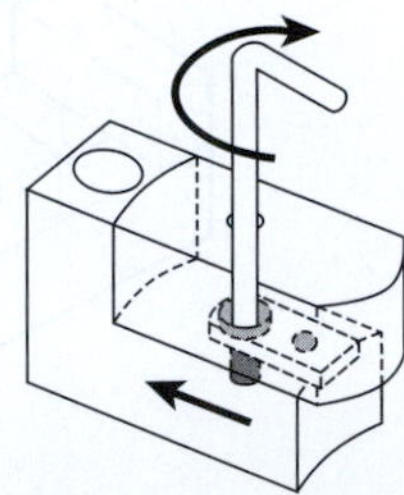

3 Place new top jaw insert on Master jaw then slide backwards and tighten.

FIGURE 27-13 Quick-changing of the top jaws on a three-jaw chuck.

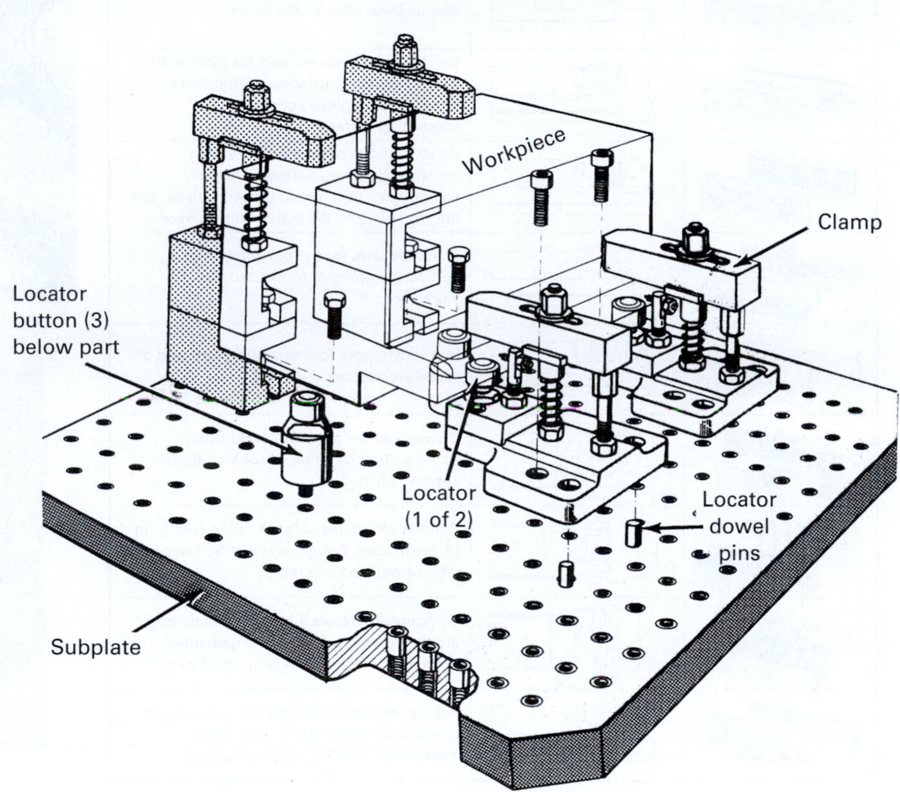

FIGURE 27-14 Modular fixturing begins with a subplate (grid base) and adds locators and clamps.

shown are the typical components for modular fixturing systems used for mounting points, locators, attachments, and so on. The standard elements needed to construct the fixture include riser blocks, vee blocks, angle plates, cubes, box parallels, and the like. Smaller elements such as locator pins, supports, pads, and clamps are added to the subplate on the larger structural elements. Mechanical clamping devices are shown, but power-assisted clamps are available. The base and fixturing elements are made to tolerances of ±0.0002 to 0.0004 in. in flatness, parallelism, and size. Figure 27-15 shows a part in a dedicated fixture compared to a modular fixture. The dedicated fixture represents a capital investment that must be absorbed by the job and must be maintained after the job is complete. The modular fixture is disassembled and the elements reused later in fixtures for other parts. Modular fixtures are commonly used for prototype tooling and small-batch production runs. They are being incorporated more frequently into regular production as users gain confidence in this approach.

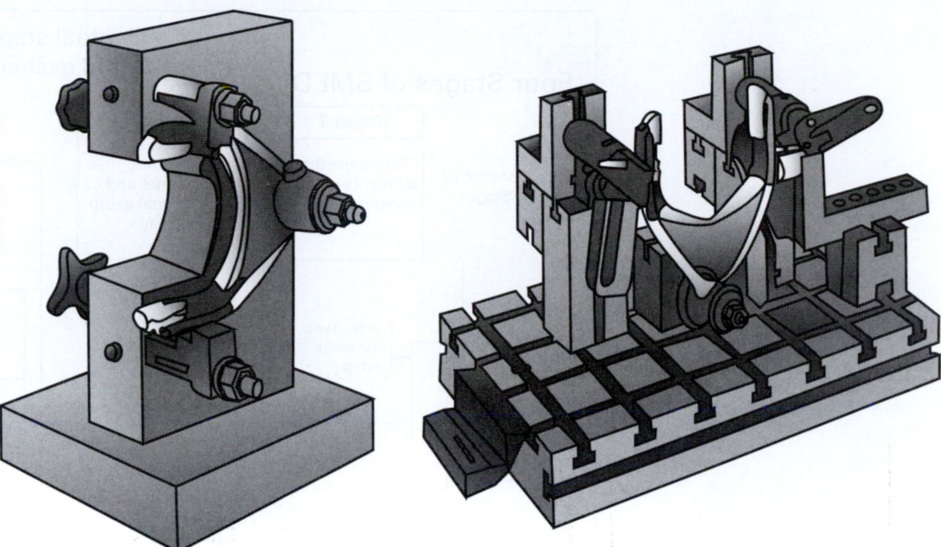

FIGURE 27-15 Dedicated fixture on the left versus modular fixture on the right.

■ 27.11 SETUP AND CHANGEOVER

Every part coming out of the workholder should be the same, resulting in interchangeable parts. But what about the first part? Does it meet specifications? What about the initial setup of the workholder into the machine tool? In many cases, the setup operation takes hours and the machine is not producing anything during this time. Rapid exchange of workholding devices is a key technique in modern manufacturing systems. The reduction in setup times permits shorter production runs (smaller lot sizes). Do not confuse initial setup (of workholders) with part loading and unloading or tool changing. The idea is to do setup quickly and to get the first part out of the process as a good part, with no adjustment of the machine, the tooling, or the workholder. Quick tool (and die) exchange is a critical component in the strategy for the manufacturing cells discussed in Chapter 36. The basic methodology is called **SMED,** an acronym for **single-minute-exchange-of-dies** (or workholding devices). It was developed by an industrial engineer consulting with the Toyota Motor Company. The basics of the SMED methodology to improve tool and die exchanges are discussed next.

FOUR STAGES OF SMED

The basic stages of a SMED setup time reduction program are outlined in Figure 27-16. The four stages are as follows (note that the setup is broken down into short elements and activities that consume the most time):

1. Determine the existing method, using videotaping.
2. Separate the internal elements from the external elements, where *external setup* refers to all operations (such as transporting workholders to storage or to the machine) that can be conducted while a machine is in operation or running and *internal setup* refers to all operations (such as mounting or removing fixtures) that can be performed only when a machine is stopped.
3. Shift the internal elements to external elements.
4. Improve all elemental operations. Reduce or eliminate internal elements by continuously improving setup. Apply methods analysis and practice doing setups, thus eliminating adjustments and abolishing the setup itself.

Operational analysis, using motion and time study, can be used to determine the current setup procedure. The usual objective is to improve work methods, eliminate all unnecessary motions, and arrange the necessary motions into the best sequence.

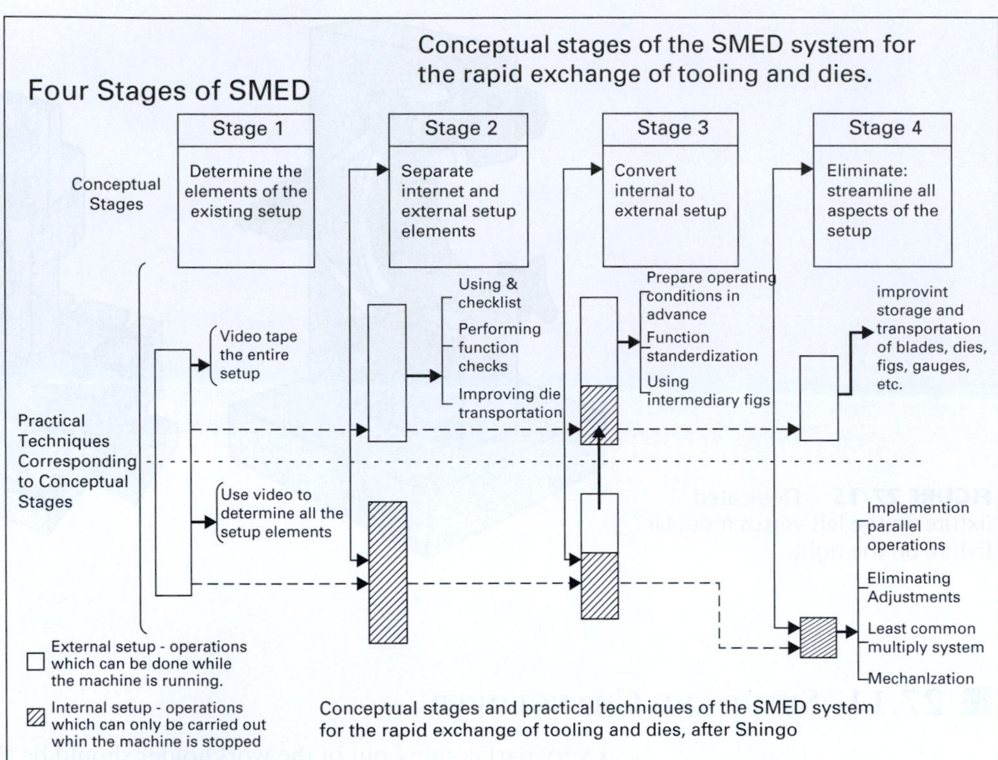

FIGURE 27-16 The conceptual stages of the SMED system for rapid exchange of tooling.

Problem-solving techniques can be applied separately to each particular activity to achieve the lowest possible time.

Figure 27-17 shows the outcome of a SMED analysis for a drilling operation on a group or family of four components. The machine tool would be placed in a manufacturing cell after these modifications, reducing setup time to 5 to 10 s.

Another approach to rapid setup is shown in Figure 27-18, where instead of five different jigs, a master jig is made (also called a **group jig**) for a family of similar components and then a set of adapters is made that customizes the jig for each part in the part family. This concept of group jigs and fixtures originated from the group technology (GT) concept for master jigs as a method to form cellular manufacturing systems by determining a family of parts where an imaginary part, called the composite part, is designed that has all the key features of all the parts in the family. In other words, the composite part is an envelope, the shape of which encompasses the shapes of all the parts in the family. The theory is that if the tooling is designed for the composite part, any part that fits within the envelope could be machined without any tooling changes. This part is used for designing the workholder. The workholding devices should be able to accommodate all the parts within the parts family. For manufacturing cells, the workholders will also have to compensate for variation in cutting forces, centrifugal forces, and so on. Group workholders are designed to accept every part-family member, with or without adapters, that accommodate minor part variations.

INTERMEDIATE JIG CONCEPT

One way to achieve rapid fixture exchange is to employ the **intermediate jig concept.** This means that the workholding devices are designed so that they all appear the same to the machine tool but different to the parts. This usually requires construction of intermediate jig or fixture plates to which the jig or fixture is attached. The jigs or fixtures are all different, but the plates are all identical.

The cassette tape for a VCR is an example of an intermediate workholder. To the VCR, every cassette appears to be the same and can be quickly loaded and unloaded with one handling—that is, one touch. From the outside, every tape appears to be the same,

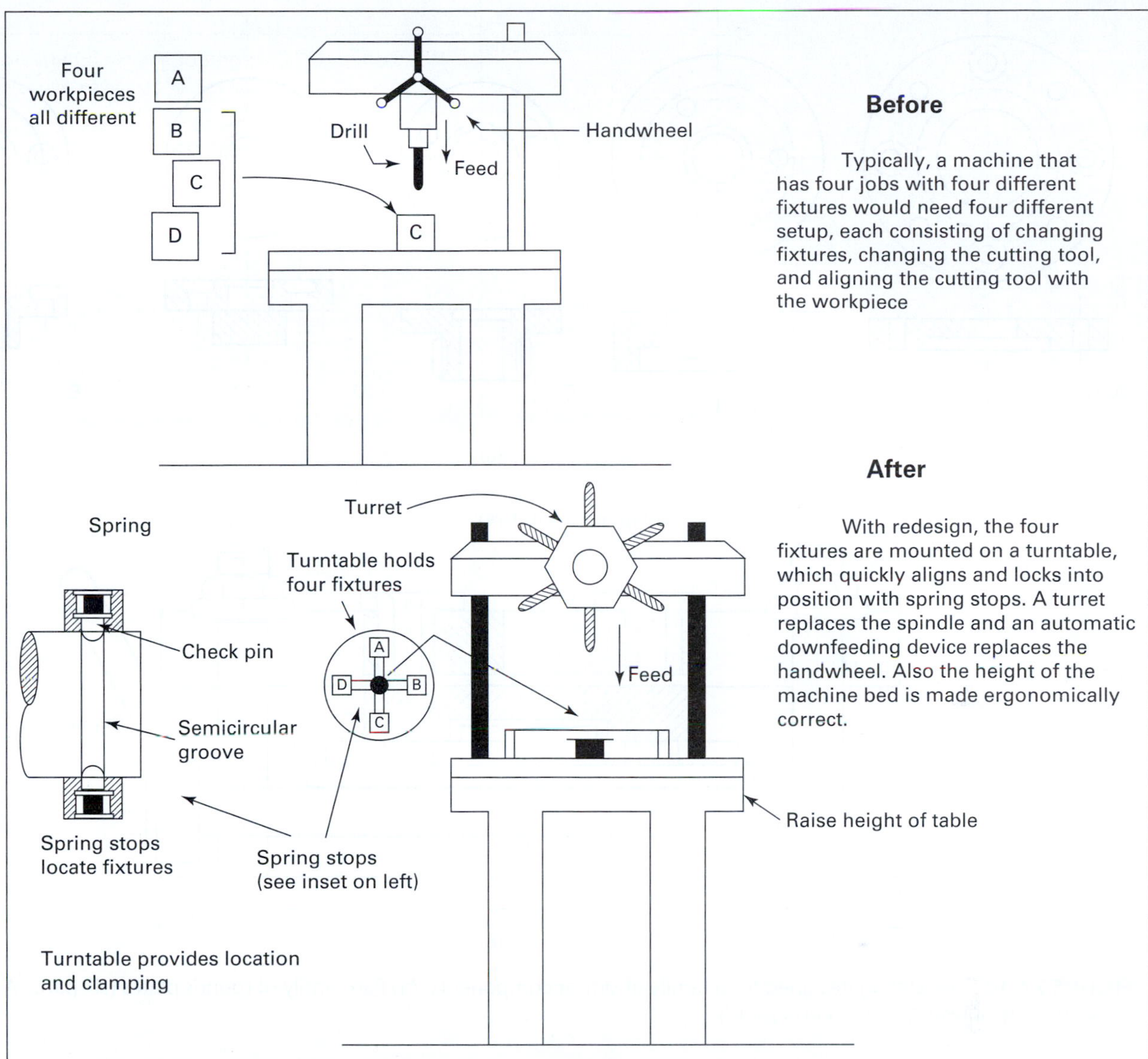
Four workpieces all different
A
B
C
D
Drill
Feed
Handwheel
C
Before
Typically, a machine that has four jobs with four different fixtures would need four different setup, each consisting of changing fixtures, changing the cutting tool, and aligning the cutting tool with the workpiece
After
Spring
Turret
Turntable holds four fixtures
With redesign, the four fixtures are mounted on a turntable, which quickly aligns and locks into position with spring stops. A turret replaces the spindle and an automatic downfeeding device replaces the handwheel. Also the height of the machine bed is made ergonomically correct.
Check pin
Feed
Semicircular groove
Raise height of table
Spring stops locate fixtures
Spring stops (see inset on left)
Turntable provides location and clamping

FIGURE 27-17 Machine tools can be modified to reduce setup time.

but on the inside, every tape is different. If you think about the workholding devices in terms of the intermediate jig concept, you can quickly achieve one-touch setups.

Figure 27-19 shows an example of the intermediate jig concept, applied to lathes and chucks. An adapter or intermediate fixture is bolted to the lathe's spindle and is a permanent part of the machine tool. The intermediate fixture will accept mating chucks that have been preset for the workpiece prior to insertion. Different chuck designs mount interchangeably on the common actuator. This method greatly reduces setup time and permits the operator to perform chuck maintenance and retooling (setup) while the machine is running. The chucks can be exchanged automatically.

Quick-change fixtures for CNC milling machines and machining centers (using the intermediate jig concept) are now available commercially.

■ 27.12 CLAMPS

When designing a jig or fixture, there are many choices to be made regarding the clamps. Manual clamps, which include screw, strap, swing, edge, cam, toggle, and C-clamps, each with certain strengths and weaknesses, are usually cheaper but slower. Figure 27-20 shows typical types of clamps that are used in fixtures. The **strap clamp** comes in many forms and sizes and is simple, low cost, and flexible. The force can be

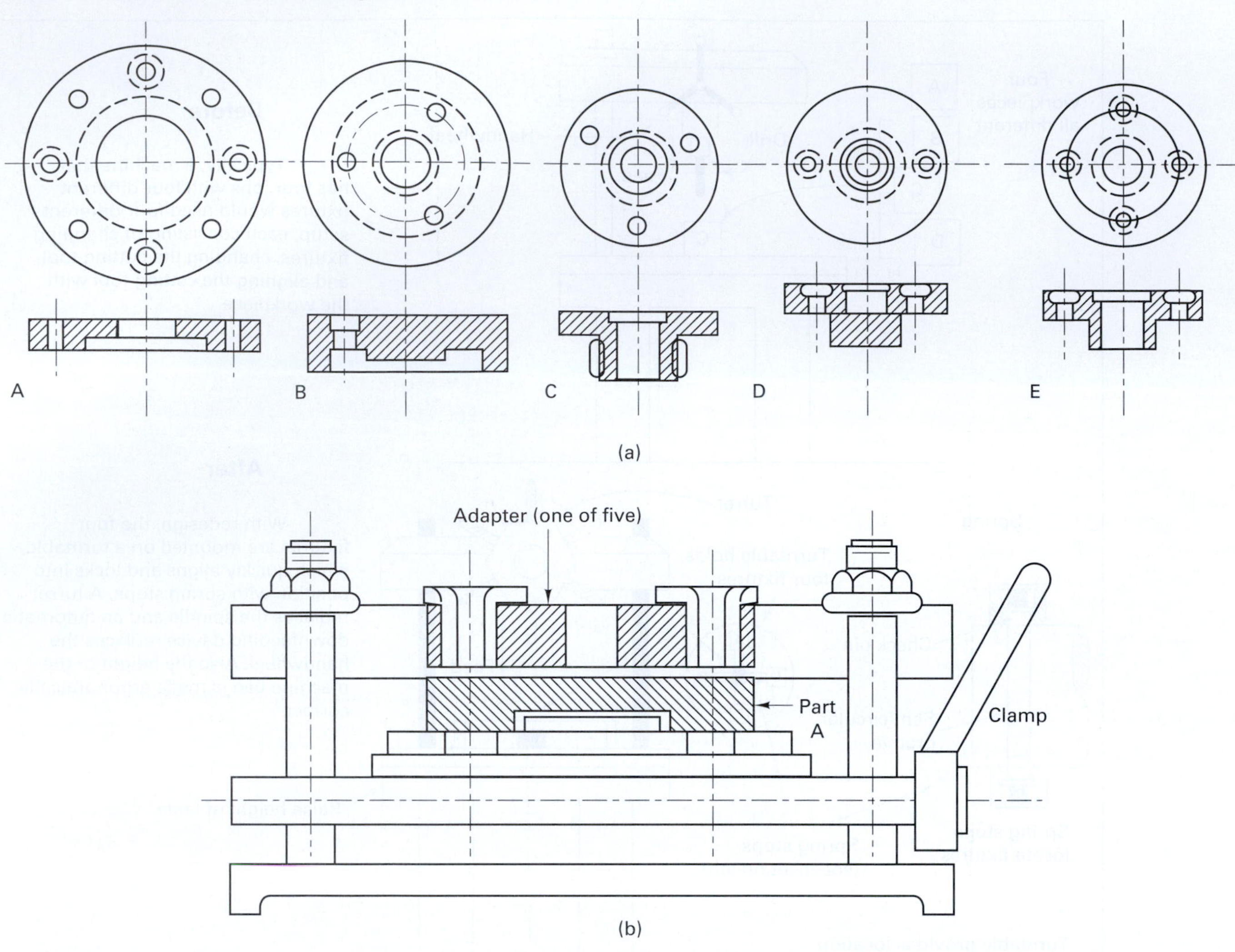

FIGURE 27-18 Master jig designed for a family of similar components. (a) Part family of rounds plates (six parts, A–F); (b) group jig for drilling, showing adapter and part A.

applied by a hand knob, a cam, or a wrench turning down a nut. A conventional **toggle clamp** accommodates only small thickness variation from part to part yet provides an excellent, consistent clamping force.

Power-actuated clamps (shown in Figure 27-21) provide more consistent clamping forces than do manual clamps, especially in applications that promote operator fatigue. The higher cost must be weighed against the capability for consistent and repeatable operation, automatic adjustment of holding forces, remote actuations, and automating sequencing of clamping actions. **Extending clamps** operate in a manner similar to that of a manual clamp-strap assembly. They extend forward horizontally, then clamp down. **Edge clamps** have a very low profile. They clamp down and forward simultaneously.

27.13 OTHER WORKHOLDING DEVICES

ASSEMBLY JIGS

Because **assembly jigs** usually must provide for the introduction of several component parts and the use of some type of fastening equipment, such as welding or riveting, they commonly are of the open-frame type. Such jigs are widely used in automobile body welding and aircraft assembly. Large jigs of the type are shown in Figure 27-22 are used for the assembly and usually feature automatic clips. This jig is constructed mainly of reinforced concrete.

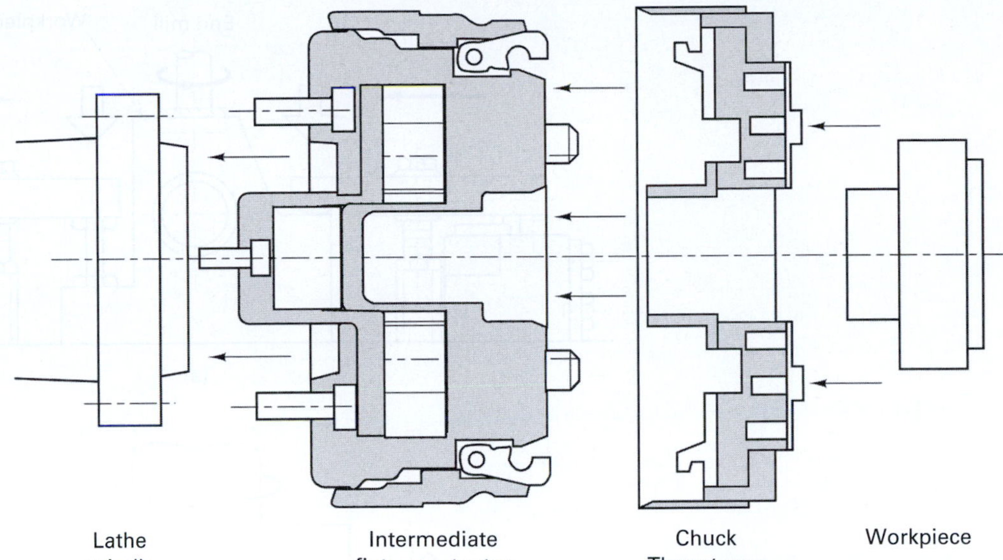

FIGURE 27-19 Example of the intermediate jig concept applied to lathe chucks. The actuator is mounted on the lathe and can quickly adapt to three different chuck types. *(Courtesy ITW Workholding)*

MAGNETIC WORKHOLDERS

Because of the light cuts and low cutting forces, workpieces can be held in a different manner on surface grinders than on other machine tools. **Magnetic chucks** are used for ferromagnetic materials. To obtain high accuracy, it is desirable to reduce clamping

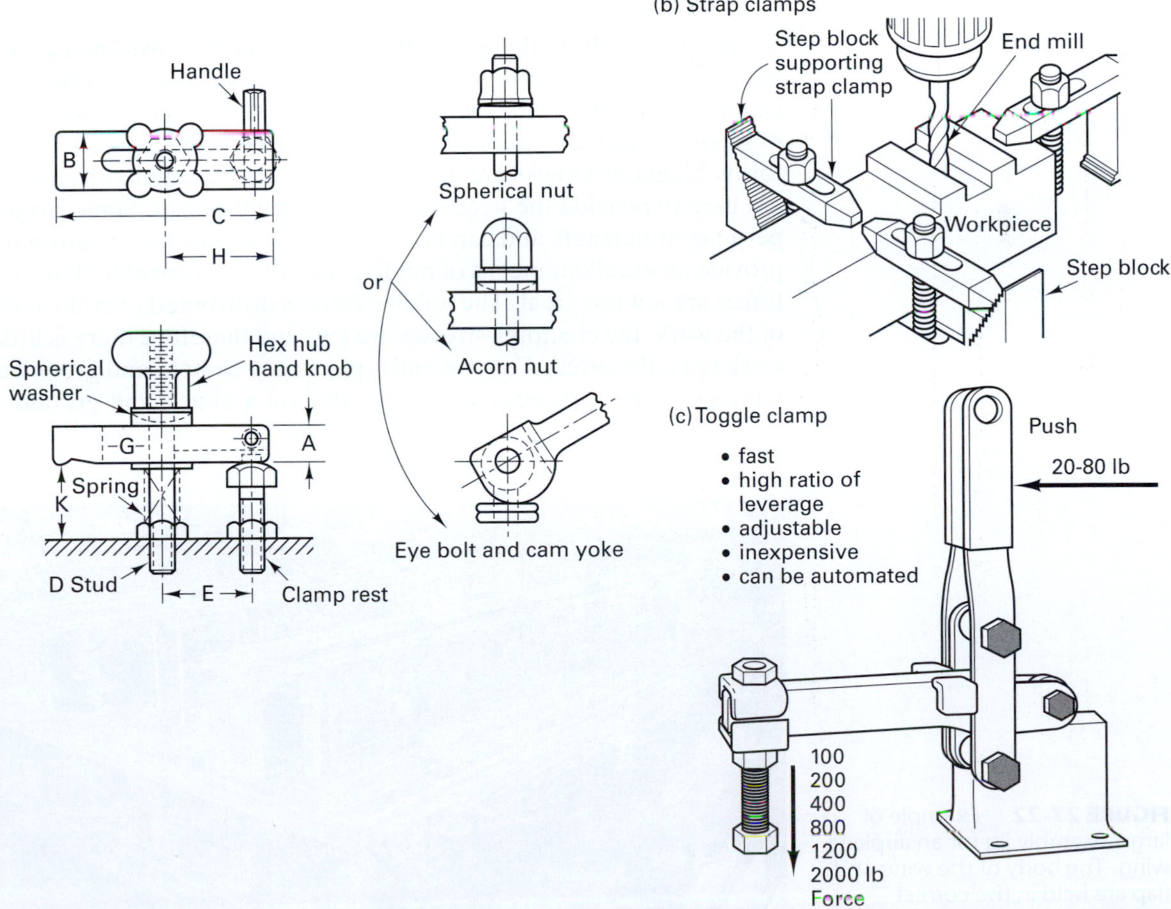

FIGURE 27-20 Examples of basic types of clamps used for workholding. The clamp elements come in a wide variety of sizes.

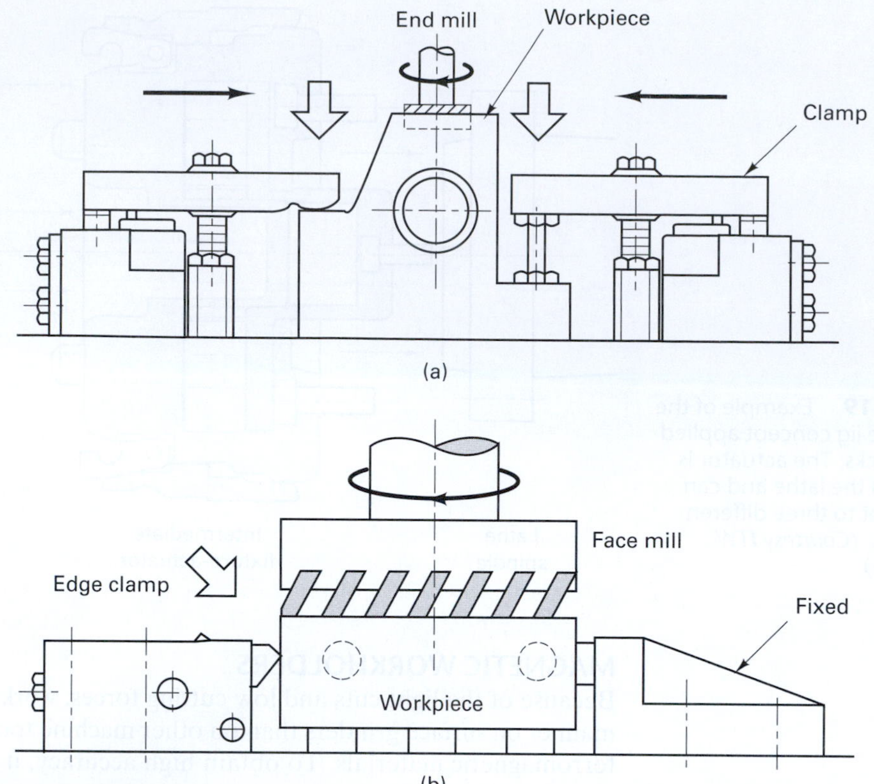

FIGURE 27-21 Examples of power-clamping devices: (a) extending clamp; (b) edge clamp.

forces and distribute them over the entire area of the workpiece. Also, grinding is frequently done on quite thin or relatively delicate workpieces, which would be difficult to clamp by normal methods. In addition, there is often the problem of grinding a number of small, duplicate workpieces. Magnetic chucks solve all these problems very satisfactorily. Magnetic chucks are available in disk or rectangular shapes. Dry-disk rectifiers are used to provide the necessary direct-current power. Some magnetic chucks utilize permanent magnets and can be tilted so that angles can be ground. Magnetic chucks provide an excellent means of holding workpieces, provided that the cutting or inertial forces are not too great. The holding force is distributed over the entire contact surface of the work, the clamping stresses are low, and therefore there is little tendency for the work to be distorted. Consequently, pieces can be held and ground accurately. Also, a number of small pieces can be mounted on a chuck and ground at the same time.

FIGURE 27-22 Example of large assembly jig for an airplane wing. The body of the wing and flap are held in the correct location with each other and then the flap is mechanically attached. *(Courtesy J T. Black)*

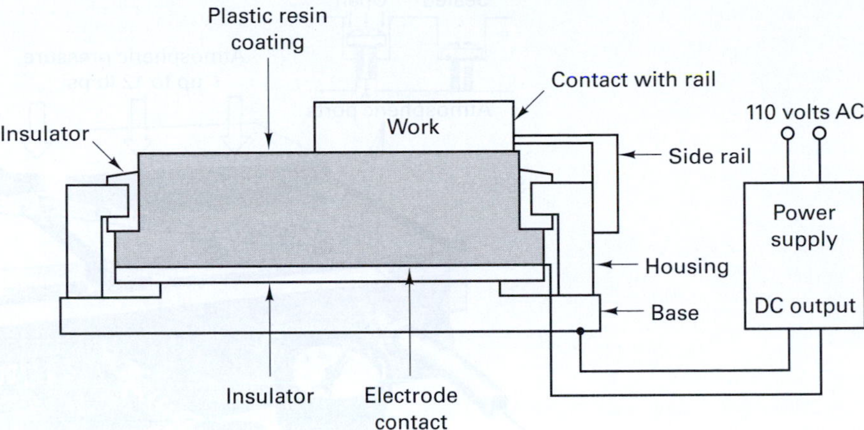

FIGURE 27-23 Principle of electrostatic chuck.

Magnetic chucks provide great part-to-part repeatability because the holding power from one part to the next is the same. Initial setup is usually fast, simple, and relatively inexpensive. Parts loading and unloading are also relatively easy.

It often is necessary to demagnetize work that has been held on a magnetic chuck. Some electrically powered chucks provide satisfactory demagnetization by reversing the direct current briefly when the power is shut off.

ELECTROSTATIC WORKHOLDERS

Magnetic chucks can be used only with ferromagnetic materials. **Electrostatic chucks** can be used with any electrically conductive material. This principle (Figure 27-23) directs that work be held by mutually attracting electrostatic fields in the chuck and the workpiece. These provide a holding force of up to 20 psi (21,000 Pa). Nonmetal parts can usually be held if they are flashed (i.e., coated) with a thin layer of metal. These chucks have the added advantage of not inducing residual magnetism in the work.

VACUUM CHUCKS

Vacuum chucks are also available. In one type, illustrated in Figure 27-24, the holes in the workplate are connected to a vacuum pump and can be opened or closed by means of valve screws. The valves are opened in the area on which the work is to rest. The other type has a porous plate on which the work rests. The workpiece and plate are covered with a polyethylene sheet. When the vacuum is turned on, the film forms around the workpiece, covering and sealing the holes not covered by the workpiece and thus producing a seal. The film covering the workpiece is removed or the first cut removes the film covering the workpiece. Vacuum chucks have the advantage that they can be used on both nonmetals and metals and can provide an easily variable force. Magnetic, electrostatic, and vacuum chucks are used for some light milling and turning operations.

As shown in Figure 27-24, T-slots are provided on milling machine tables so that workpieces can be clamped directly to the table. More often various workholding devices, called vices or fixtures, are utilized. Smaller workpieces are usually held in a vise mounted on the table. Fixtures designed to specifically hold a part in the correct location with respect to the tool are used for larger volumes. Fixtures reduce the time it takes to put the part in the machine and assure repeatable location with respect to the cutting tools. Fixtures provide clamping forces that counteract the cutting forces.

27.14 ECONOMIC JUSTIFICATION OF JIGS AND FIXTURES

As discussed previously, workholders are expensive, even when designed and constructed by using standard components. Obviously, their cost is a part of the total cost of production, and one must determine whether they can be justified economically by the savings in labor and machine cost and improvements in quality that will result from their use. Often it is only through the use of such devices that the design specifications

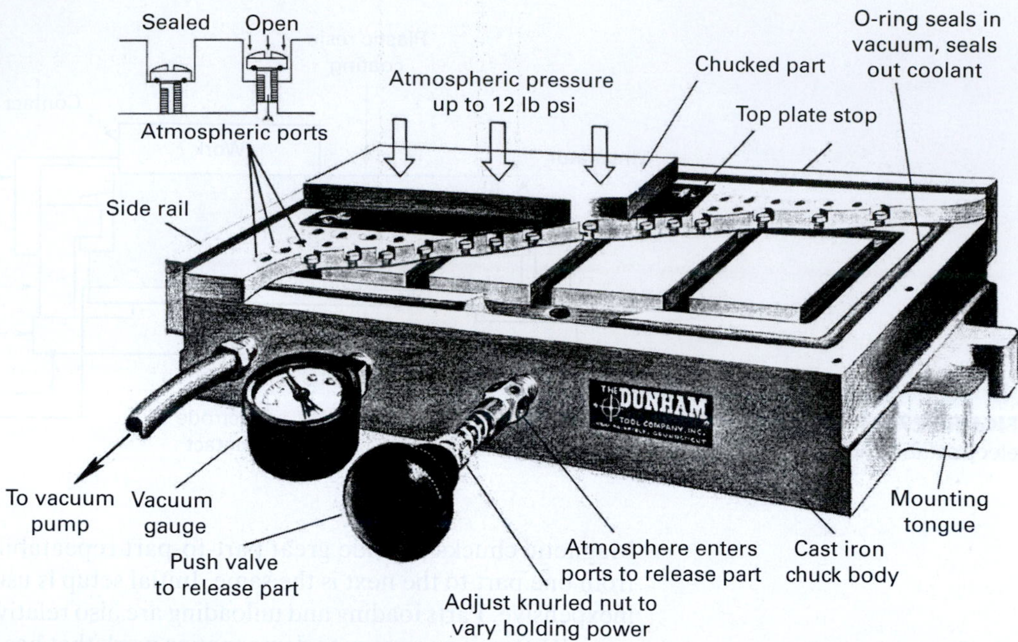

FIGURE 27-24 Cutaway view of a vacuum chuck. *(Courtesy of Dunham Tool Company, Inc.)*

can be met and sustained from part to part. To determine the economic justification of any special tooling, the following factors must be considered:

1. The cost of the tooling.
2. Interest or profit charges on the tooling cost.
3. The savings resulting from the use of the tooling, which can result from reduced cycle times or improved quality or lower-cost labor.
4. The savings in machine cost due to increased productivity.
5. The number of units that will be produced using the tooling.

The economic relationship between these factors can be expressed in the following manner:

$$\text{Savings per piece (exclusive of tooling costs)} \geq \text{Additional cost per piece}$$

$$\left\{\begin{array}{c}\text{Total cost per piece}\\ \text{without tooling}\end{array}\right\} - \left\{\begin{array}{c}\text{Total cost per piece}\\ \text{using tooling}\\ \text{(exclusive of tooling cost)}\end{array}\right\} \geq \text{Tooling cost per piece}$$

$$\overbrace{\begin{array}{c}\text{Labor cost}\\ \text{per piece}\\ \text{without}\\ \text{tooling}\end{array} + \begin{array}{c}\text{Machine and}\\ \text{overhead cost}\\ \text{per piece}\\ \text{without tooling}\end{array}} - \overbrace{\begin{array}{c}\text{Labor cost}\\ \text{per piece}\\ \text{with}\\ \text{tooling}\end{array} + \begin{array}{c}\text{Machine and}\\ \text{overhead cost}\\ \text{per piece}\\ \text{with tooling}\end{array}} \geq \overbrace{\begin{array}{c}\text{Cost}\\ \text{of}\\ \text{tooling}\end{array} + \begin{array}{c}\text{Interest on}\\ \text{tooling cost}\end{array}}$$

$$[(R)(t) + (R_m)(t)] - [(R_t)(t_t) + (R_m)(t_t)] \geq \frac{C_t + (C_t/2)(n)(i)}{N} \tag{27-2}$$

where

R = labor rate per hour, without tooling
R_t = labor rate per hour, using tooling
t = hours per piece, without tooling
t_t = hours per piece, using tooling
R_m = machine cost per hour, including all overhead
C_t = cost of the special tooling

n = number of years tooling will be used
i = interest rate (or what invested capital is worth)
N = number of pieces that will be produced with the tooling

Equation 27-2 can be expressed in a simpler form:

$$(R + R_m)t - (R_t + R_m)t_t \geq \frac{C_t}{N}\left(1 + \frac{n \times i}{2}\right) \tag{27-3}$$

This equation assumes straight-line depreciation and computes interest on the average amount of capital invested throughout the life of the tooling.[1] When the time over which the tooling is to be used is less than 1 yr, companies often do not include an interest cost. If this factor is neglected, the right-hand term of equation 27-3 reduces to C_t/N.

The equations assume that the material cost will be the same regardless of whether special tooling is used. This is not always true. Although these equations are not completely accurate for all cases, they are satisfactory for determining tooling justification in most cases, because the life of tooling for machine tools seldom exceeds 5 yr and often does not exceed 2 yr. The equation does not include the cost of poor quality. This can be included by estimating the decrease in the number of defective parts when the workholder is used versus when it is not used.

JIG COST EXAMPLE

The following example illustrates the use of equation 27-3 to determine tooling justification for a dedicated jig. In drilling a series of holes on a radial drill, the use of a drill jig will reduce the time from $\frac{1}{2}$ hr/piece to 15 min/piece. If a jig is not used, a machinist—whose hourly rate is \$18/hr—must be used. If the jig is used, the job can be done by a machinist whose rate is \$12/hr. The hourly rate for the radial drill is \$32/hr.

The cost of making the jig would include \$350 for design, \$150 for material, and 50 hr of toolmaker's labor, which is charged at the rate of \$22/hr to include all machine and overhead costs in the toolmaking department. Investment capital is worth 16% to the company. It is estimated that the jig would last 3 yr and that it would be used for the production of 300 parts over this period. Is the jig justified?

The cost of the jig, C_t, is estimated to be

$$C_t = \$350 + \$150 + \$150 + \$22 \times 50\text{ hr} = \$1,600$$

Substituting the values given in equation 27-3, we find:

$$(18 + 32)0.5 - (12 + 32)0.25 \geq \$\frac{1600}{300}\left(1 + \frac{3 \times 0.16}{2}\right)$$

$$\text{or } 14.00 \cong 13.76$$

So this jig is not justified based on cost savings.

One could also ask how many parts would have to be produced with the jig to break even (i.e., increased costs just equal savings). By omitting the value 300 in the preceding solution and solving for N, it is found that at least 1627 pieces would have to be produced annually with the jig for it to break even.

This analysis assumes that the time (of the people and machines) saved by the use of the special tooling can be used for other productive work. If this is not the case, the cost analysis should be altered to take this important fact into account. Otherwise, the tooling justification may be substantially in error.

The application of group technology, NC machines, and lean manufacturing techniques may eliminate the need for designing and building a new jig or fixture every time a new part is designed. New measures of manufacturing productivity that include terms for quality and flexibility are being developed.

[1] For the use of more sophisticated economic analysis see C.S. Park, *Contemporary Engineering Economics*, 2nd ed., New York, Wiley, 1999.

KEY WORDS

3-2-1 principle
assembly jig
box jig
cam latch
channel jigs
clamping
clamp-on jig
conventional fixtures
datum surface
diameter jig
edge clamp
electrostatic chuck
extending clamp
fixture
flexibility
group jig
intermediate jig concept
jig
knurled clamping screw
leaf jig
locating
location dimensions
magnetic chuck
modular fixtures
plate jig
ring jig
rollover jig
run time
setup
size dimensions
SMED (single-minute-exchange-of-dies)
strap clamp
toggle clamp
tumble jig
universal jig
vacuum chuck
versatility
vice
workholder

REVIEW QUESTIONS

1. What are the two primary functions of a workholding device?
2. What distinguishes a jig from a fixture?
3. An early treatise defined a jig as "a device that holds the work and guides a tool." Why was this definition incorrect?
4. Does an ordinary vise qualify as a fixture? Why or why not?
5. What basic criteria should be considered in designing jigs and fixtures?
6. In any part drawing, what are the critical surfaces of a part (i.e., what makes a part surface critical)? (This question requires an understanding of basic part drawings.)
7. What difficulties can result from not keeping clamping stresses low in designing jigs and fixtures?
8. Explain the 3-2-1 concept for workpiece location in a workholder on a machine tool.
9. Which of the basic design principles relating to jigs and fixtures would most likely be in conflict with the 3-2-1 location concept?
10. What are two reasons for not having drill bushings actually touching the workpiece? How many of the designs shown in this chapter violate this rule? It is not uncommon to have conflicts and trade-offs in fixture design situations.
11. Why does the use of down milling often make it easier to design a milling fixture than if up milling were used?
12. Name another example of the intermediate workholder concept aside from the video cassette.
13. A large assembly jig for an airplane-wing component gave difficulty when it rested on four-point support. The assembled wing components were not consistent in shape. It was satisfactory when only three supporting points were used. Why?
14. Explain why the use of a given fixture may not be economical when used with one machine tool but may be economical when used in conjunction with another machine tool.
15. What are rollover jigs, and what advantages do they offer?
16. In the clamps shown in Figure 27-20, what is the purpose of the spherical washer?
17. What are other common types of clamps?
18. What is the purpose of dimensioning the strap-clamp assembly in Figure 27-20 with letters?
19. Figure 27-9 showed the part sitting on locator buttons. Why not have the part rest on the flat plate?
20. In Figure 27-9, why aren't there three points put on the x-plane, two points on the z-plane, and one point on the y-plane?
21. Which set of locators in Figure 27-9 establishes the A dimension on the part?
22. What are the four stages of the SMED methodology for setup time reduction?
23. How are external setup tasks different from internal setup tasks?
24. How would the intermediate jig concept be used in the master jig shown in Figure 27-18?

PROBLEMS

1. Develop a sequence of operations to produce the part shown in Figure 27-A, from a casting. The part is shown in a drilling fixture that is going into a NC milling machine.
2. Using the following values, determine the number of pieces that would have to be made to justify the use of a jig costing \$3000.
   $R = \$5.75$
   $R_t = \$4.50$
   $t_t = 1\frac{1}{4}$
   $t = 2\frac{1}{2}$
   $I = 10\%$
   $R_m = \$4.50$
   $N = 3$
3. Suppose in the sample problem at the end of this chapter that modular fixturing is used, which reduces the toolmaker's labor to 4 hr and the design cost to \$100 (4 hr at \$25/hr), and that the material cost (modular elements) for the subplate structural elements, clamps, and so on was \$300. What is the breakeven quantity for a modular fixture? (*Note:* The modular fixture is used for the job, then disassembled and returned to the toolroom. The parts are reused in other workholders.)
4. Suppose the fixture Figure 27-14 could be improved by replacing the strap clamp with powered swing clamps. See Figure 27-B. Many things are needed to be able to cost justify the improvement in the fixture. This problem requires that the engineer estimate or determine the following:

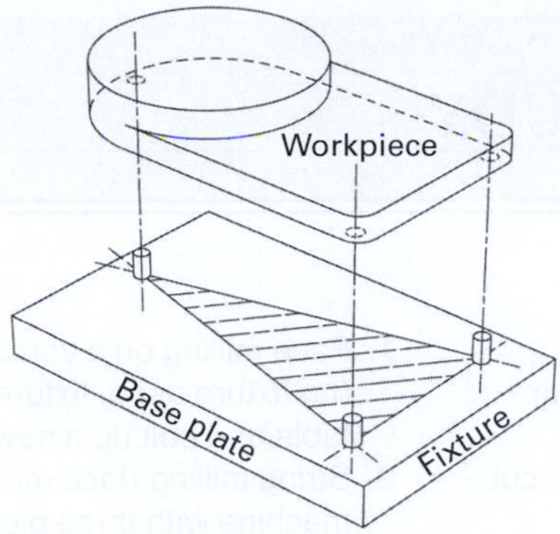

Three-point support location in any one plane

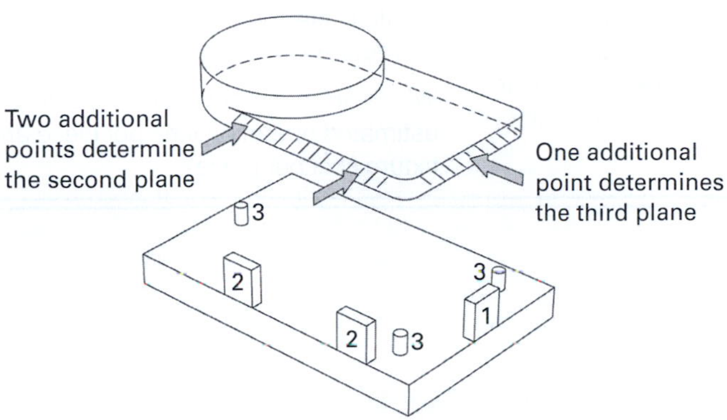

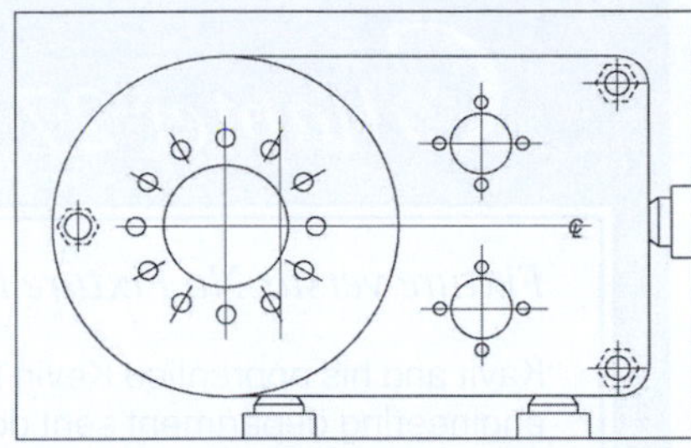

Workpiece in fixture—top view

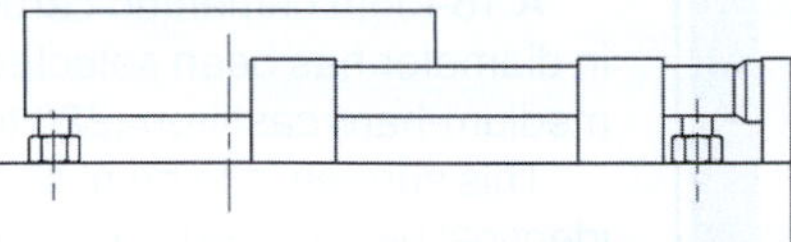

Fixture base—side view of workpiece

Clamp close–near support locations

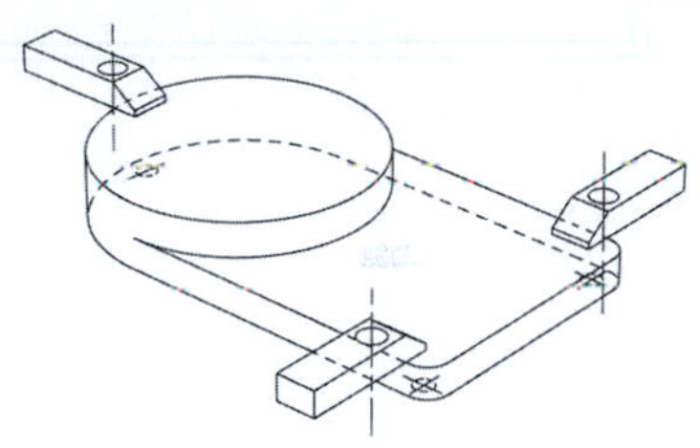

FIGURE 27-A

4a. How much time is saved with powered swing clamps (in the loading and unloading) cycle?

4b. How much does a swing clamp cost?

4c. How much will it cost to modify the existing fixture? Currently, for this job, the machining cycle time to mill the block is 3 min., the unload/load time is 90 s, the operator is getting $12/hr, and the machine cost is $30/hr.

4d. Are the swing clamps justified if they reduce the unload/load time to 30s?

5. Notice that the holes in the part in Figure 27-9 need to be countersunk after they are drilled. How can the jig be designed to put the countersinks on the mounting holes while the part is in the jig, or would this operation be done afterward?

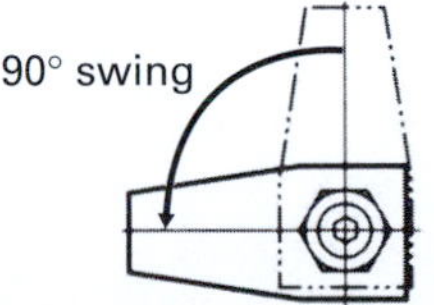

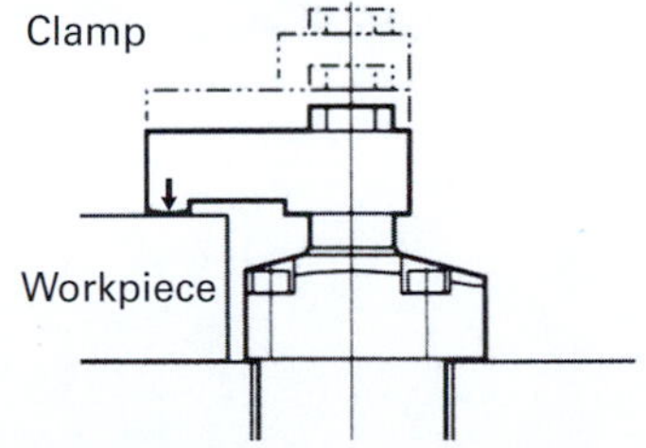

FIGURE 27-B

www.wiley.com/go/global/degarmo

Chapter 27 CASE STUDY

Fixture versus No Fixture in Milling

Kavit and his apprentice Kevin from the design engineering department sent down to manufacturing engineering a drawing that calls for a surface $4\frac{1}{2}$ in. wide × 10 in. long to be rough milled with a depth of cut of 0.30 in.

A 16-tooth cemented carbide face mill 150 mm (6 in.) in diameter has been selected for the job. The material is medium-hard cast iron (220 to 260 BHN).

This surface is to be milled on a large number of identical pieces (500 is the size of the first order of parts). The estimated time to unload and load a piece in a fixture is 0.30 min. Here are three possible choices of machine setup you could use to do this job.

1. Face milling on a vertical-spindle milling machine (no fixture used, fixture, 6 min to remove part from table and bolt up a new part).
2. String milling (face milling) on a vertical milling machine with three pieces 0.6 in. apart in a fixture.
3. Index base milling (0.15 min required to index the base) on a vertical single-spindle mill where the operator is unloading/loading the index table while the alternate piece is being machined (2 fixtures required).

Which arrangement should be selected based on your estimated operation time per piece and your estimated fixture cost per piece?

CHAPTER 28

NONTRADITIONAL MANUFACTURING PROCESSES

28.1 INTRODUCTION

Many material removal processes have been developed since World War II to address problems that can't be handled with conventional, that is, "traditional chip-forming" machining processes. The processes described in this chapter are often called **non-traditional machining (NTM)** processes, and have the following advantages:

- Complex geometries beyond simple planar or cylindrical features can be machined.
- Parts with extreme surface-finish and tight tolerance requirements can be obtained.
- Delicate components that cannot withstand large cutting forces can be machined.
- Parts can be machined without producing burrs or inducing residual stresses.
- Brittle materials or materials with very high hardness can be easily machined.
- Microelectronic or integrated circuits can be mass-produced.

NTM processes can often be divided into four groups based upon the material removal mechanism: See Table 28-1.

1. *Chemical.* Chemical reaction between a liquid reagent and the workpiece results in etching.
2. *Electrochemical.* An electrolytic reaction at the workpiece surface is responsible for material removal.
3. *Thermal.* High temperatures in very localized regions evaporate materials.
4. *Mechanical.* High-velocity abrasives or liquids like abrasive jet machining, ultra sonic machining and water-jet machining are used to remove material (see Chapter 26).

Machining processes that involve **chip** formation have a number of inherent limitations. Large amounts of energy are expended to produce unwanted chips that must be removed and discarded. Much of the machining energy ends up as undesirable heat that often produces problems of distortion and surface cracking. Cutting forces require that the workpiece be held, which can also lead to distortion. Unwanted distortion, residual stresses, and burrs caused by the machining process often require further processing. Finally, some geometries are too delicate to machine, while others are too complex. When examining these processes, be aware that conventional end milling (see Chapter 24) has these typical machining parameters:

- Feed rate—25 to 5000 mm/min (5 to 200 in./min).
- Surface finish—1.5 to 3.75 μm (60 to 150 μin.) Arithmetic average (AA).
- Dimensional accuracy—0.025 to 0.05 mm (0.001 to 0.002 in.).
- Workpiece/feature size—61 cm × 61 cm (25 in. × 24 in.); 2.5 cm (1 in.) deep.

TABLE 28-1 Summary of NTM Processes[‡]

Process	Typical Penetration or Feed Rate, Mm/m (ipm)	Typical Surface Finish AA, μm (μin.)	Typical Accuracy, Mm (in.)	Typical Workpiece or Feature Size, cm (in.)	Comments
Chemical—see Chapter 41 on the web					
Chemical milling (cut-and-peel)	0.013 to 0.076 (0.0005 to 0.003)	1.6 to 6.35; as low as 0.2 (63 to 250; 8)	greater than 0.127 (0.005)	as large as 365 × 1524 (144 × 600); up to 1.27 (0.5) thick	No burrs; no surface stresses; tooling cost low
Photochemical machining (see Chapter 41 on the web)	as above	as above	0.025 to 0.05 (0.001 to 0.002)	30 × 30 (12 × 12); up to 0.15 (0.06) thick	Limited to thin material; burr-free blanking of brittle material; tooling cost low; used in microelectronics
Electrochemical					
Electrochemical machining (ECM)	2.5 to 12.7 (0.1–0.5)	0.4 to 1.6 (16–63)	0.013 to 0.13 (0.0005–0.005); 0.05 (0.002) in cavities	30 × 30 (12 × 12); 5 (2) deep	Stress-free, burr-free metal removal in hard-to-machine metals; tool design expensive; disposal of wastes a problem; MRR independent of hardness; deep cuts will have tapered walls
Electrostream drilling	1.5 to 3 (0.06–0.12)	0.25 to 1.6 (10–63)	0.025 (0.001) or 5% of hole dia.	up to 0.5 (0.2) thick	Charged high-velocity stream of electrolyte; hole diameters down to 0.127 mm (0.005 in.); 40:1-hole aspect ratios possible
Shaped-tube electrolytic machining (STEM)	as above	0.8 to 3.1 (32–125)	0.025 to 0.125 (0.001–.005)	routinely up to 127 (5) thick	Special form of ECM using conductive tube with insulated surface and acidic electrolyte; 300:1-hole aspect ratios; hole diameters down to 0.5 mm (0.02 in.)
Thermal					
Electrical discharge machining (EDM)	up to 0.5 (0.02)	0.8 to 2.7 (32–105)	0.013 to 0.05 (0.0005–.002)	up to 200 × 200 (79 × 79); 5 (2) deep	Widely used and disseminated; dies expensive; cuts any conductive material regardless of hardness; forms recast layer
Electron-beam machining (EBM)	30 to 1500 (1.2–60)	0.8 to 6.35 (32–250)	0.005 to 0.025 (0.0002–0.001)	0.025 to 0.63 (0.01–0.25) thick	Capable of micromachining thin materials; hole sizes down to 0.05 mm (0.002 in.); 100:1 hole aspect ratios; requires high vacuum
Laser-beam machining (LBM)	100 to 2500 (4–100)	0.8 to 6.35 (32–250)	0.013 to 0.13 (0.0005–0.005)	up to 2.5 (1) thick	Capable of drilling holes down to 0.127 mm (0.005 in.) at 20:1 aspect ratio in seconds; has heat-affected zone and recast layers which may require removal
Plasma arc cutting (PAC)	250 to 5000 (10–200)	0.6 to 12.7 (25–500)	0.5 to 3.2 (0.02–0.125)	up to 15 (6) thick	Clean rapid cuts and profiles in almost all plates; 5° to 10° taper; cheaper capital equipment
Precision PAC	as above	as above	0.25 (0.01)	up to 1.5 (0.625) thick	Special form of PAC limited to thin sheets of material; straighter, smaller kerf
Wire EDM	100 to 250 (4–10)	0.8 to 1.6; as low as 0.38 (32–64; 15)	0.0025 to 0.1 (0.0001–0.004)	as large as 100 × 160 (40 × 64); up to 45 (18) thick	Special form of EDM using traveling wire; cuts straight narrow kerfs; wire diameters as small as 0.05 mm (0.002 in.); CNC machines permit complex geometries
Mechanical—see Chapter 26 on Abrasive Machining					
Abrasive jet machining	76 (3)	0.25 to 1.27 (10–50)	0.12 (0.005)	up to 0.15 (0.06) thick	Used for cutting brittle materials; produces tapers; inexpensive to implement; can cut up to 6.3 mm (0.25 in.) thick glass
Abrasive waterjet machining	15 to 450 (0.6–18)	2.0 to 6.35 (80–250)	0.13 to 0.38 (0.005–0.015)	up to 20 (8) thick	Use in glass, titanium, composites, nonmetals, and heat-sensitive or brittle materials; produces tapered walls in deep cuts; no burrs

TABLE 28-1 Continued

Ultrasonic machining (impact grinding)	0.5 to 3.8 (0.02–.15)	0.4 to 1.6; as low as 0.15 (16–63; 6)	0.013 to 0.025 (0.0005–.001)	up to 100 cm^2 (16 $in.^2$)	Most effective in hard materials, $R_C >$ 40; tool wear and taper limit hole aspect ratio at 2.5:1
Water-jet machining	250 to 200,000 (10–7900) soft materials	1.27 to 1.9 (50–100)	0.13 to 0.38 (0.005–0.015)	up to 2.5 (1) thick	Used on leather, plastics, and other non-metals; pressures of 60,000 psi and jet velocity of up to 3000 ft/sec

‡ Mechanical process discussed in Chapter 26.

In comparison, NTM processes typically have lower feed rates and require more power consumption when compared to machining. However, some processes permit batch processing, which increases the overall throughput of these processes and enables them to compete with machining. A major advantage of some NTM processes is that feed rate is independent of the material being processed. As a result, these processes are often used for difficult-to-machine materials. NTM processes typically have better accuracy and surface finish, with the ability of some processes to machine larger feature sizes at lower capital costs. In most applications, NTM requires part-specific tooling, while general-purpose cutting and workholding tools make machining very flexible. There are numerous hybrid forms of all these processes, developed for special applications, but only the main NTM processes are described here due to space limitations.

28.2 CHEMICAL MACHINING PROCESSES

CHEMICAL MACHINING

Chemical machining (CHM) is the simplest and oldest of the chipless machining processes. The use of CHM dates back 4500 yr to the Egyptians, who used it to etch jewelry. In modern practice, it is applied to parts ranging from very small microelectronic circuits to very large engravings up to 15 m (50 ft) long. Typically, metals are chemically machined, although methods do exist for etching ceramics and even glass.

In CHM, material is removed from a workpiece by selectively exposing it to a chemical reagent or **etchant.** The mechanism for metal removal is the chemical reaction between the etchant and the workpiece, resulting in dissolution of the workpiece. One means for accomplishing CHM is called **gel milling**, where the etchant is applied to the workpiece in gel form. However, the most common method of CHM involves covering selected areas of the workpiece with a **maskant** (or etch resist) and imparting the remaining exposed surfaces of the workpiece to the etchant. The general material removal steps for CHM are shown in Figure 28-1 and summarized here:

1. *Cleaning.* Contaminants on the surface of the workpiece are removed to prepare for application of the maskant and permit uniform **etching.** This may include degreasing, rinsing, and/or pickling.
2. *Masking.* If selective etching is desired, an etch-resistant maskant is applied and selected areas of the workpiece are exposed through the maskant in preparation for etching.
3. *Etching.* The part is either immersed in an etchant or an etchant is continuously sprayed onto the surface of the workpiece. The chemical reaction is halted by rinsing.
4. *Stripping.* The maskant is removed from the workpiece, and the surface is cleaned and desmutted as necessary.

Lateral dimensions in CHM are controlled in large part by the patterned maskant. Masking can be performed in one of several ways depending on the level of precision required in CHM. The simplest method of applying a maskant is the **cut-and-peel method.** In this procedure, the maskant material—typically neoprene, polyvinyl chloride, or polyethylene—is applied to the entire surface of the workpiece by dipping or spraying. Once the coating dries, it is then selectively removed in those areas where etching is desired by scribing the maskant with a knife and peeling away the unwanted portions. When volume permits, scribing templates may be used to improve accuracy.

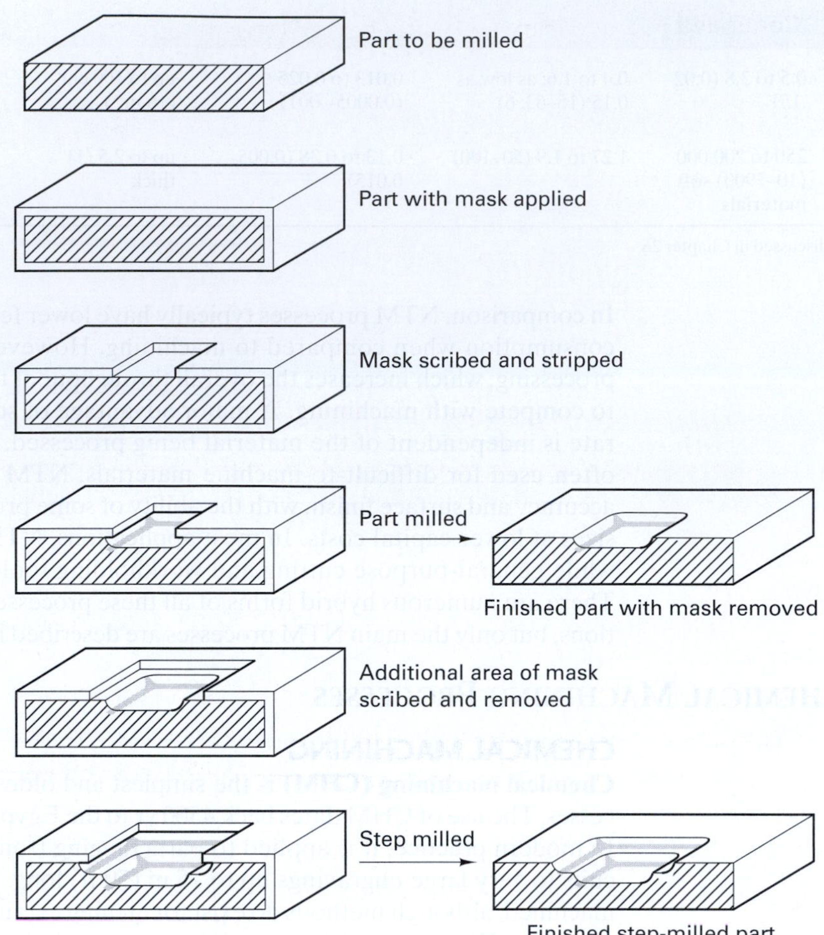

FIGURE 28-1 Steps required to produce a stepped contour by chemical machining.

Cut-and-peel coatings are thick, ranging from 0.025 to 0.13 mm (0.001 to 0.005 in.). Because of this thickness, the maskant can withstand exposure to the etchant for the extended periods of time necessary to remove large volumes of material. This technique is generally preferred when the workpiece is not flat or is very large, or for low-volume work where the development of screens or phototools necessary for other masking methods is not justified. The scribe-and-peel method for **stepped machining** is shown in Figure 28-1.

Another method used to apply maskants is **screen printing,** which involves the use of traditional silk-screening technology. The method applies the maskant through a mask made from a fine silk mesh or stainless steel screen. Masks are typically formed by application, exposure, and development of a light-sensitive emulsion on top of the screen. The screen is pressed against the surface of the workpiece, and the maskant is rolled on. Screen printing is good for high-volume, low-precision applications with tolerances typically in the 0.05 to 0.18 mm (0.002 to 0.007 in.) range. Etch depth is limited to about 1.5 mm (0.06 in.) by the thickness of the maskant, typically on the order of 0.05 mm (0.002 in.).

Etch rates in CHM are very slow compared with other nontraditional machining processes. However, etching proceeds on all exposed surfaces simultaneously, which significantly increases the overall material removal rate on large parts. The etch rate in CHM is directly proportional to the etchant concentration directly adjacent to the area being machined. For parts machined by immersion, the uniformity of the etchant concentration within the bath can be improved by agitation. If the bath is not agitated properly, several defect conditions can result, as shown in Figure 28-2. **Islands,** or isolated high spots, can be the result of improper agitation on large parts. Islands can also be formed due to inadequate cleaning or inhomogeneity with the work material.

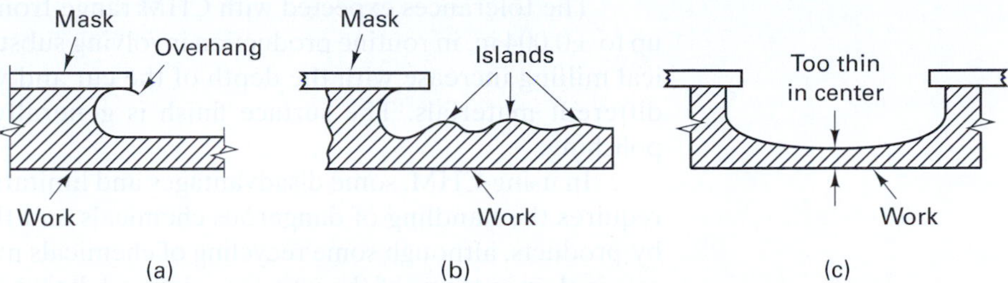

FIGURE 28-2 Typical chemical milling defects: (a) overhang: deep cuts with improper agitation; (b) islands: isolated high spots from dirt, residual maskant, or work material inhomogeneity; (c) dishing: thinning in center due to improper agitation or stacking of parts in tank.

Single-sided, blind etching of the part is called **chemical milling** or, when the photoresist method of applying maskants is used, **photochemical milling.** Chemical milling is so named because its earliest use was for replacing mechanical milling on large components. Chemical milling is often used to remove weight on aircraft components, as shown in Figure 28-3. Through-etching of the workpiece is called **chemical** (or **photochemical**) **blanking**. The process competes with blanking, laser cutting, and electrical discharge machining (EDM) for through-cutting of thin material sheets. Chemical blanking is typically performed using double-sided etching to increase production rates and minimize taper on the etched walls of the feature. A key requirement for chemical blanking is registration of top-side and bottom-side screens or phototools during masking. Because of the precision required, chemical blanking is not performed with the cut-and-peel method of masking.

Advantages and Disadvantages of Chemical Machining. Chemical machining has a number of distinct advantages when compared with other machining and forming processes. Except for the preparation of the artwork and phototool, screen, or scribing template, the process is relatively simple, does not require highly skilled labor, induces no stress or cold working in the metal, and can be applied to almost any metal—aluminum, magnesium, titanium, and steel being the most common. Large areas can be machined; tanks for parts up to 12 ft × 50 ft and spray lines up to 10 ft wide are available. Machining can be done on parts of virtually any shape. Thin sections, such as honeycomb, can be machined because there are no mechanical forces involved.

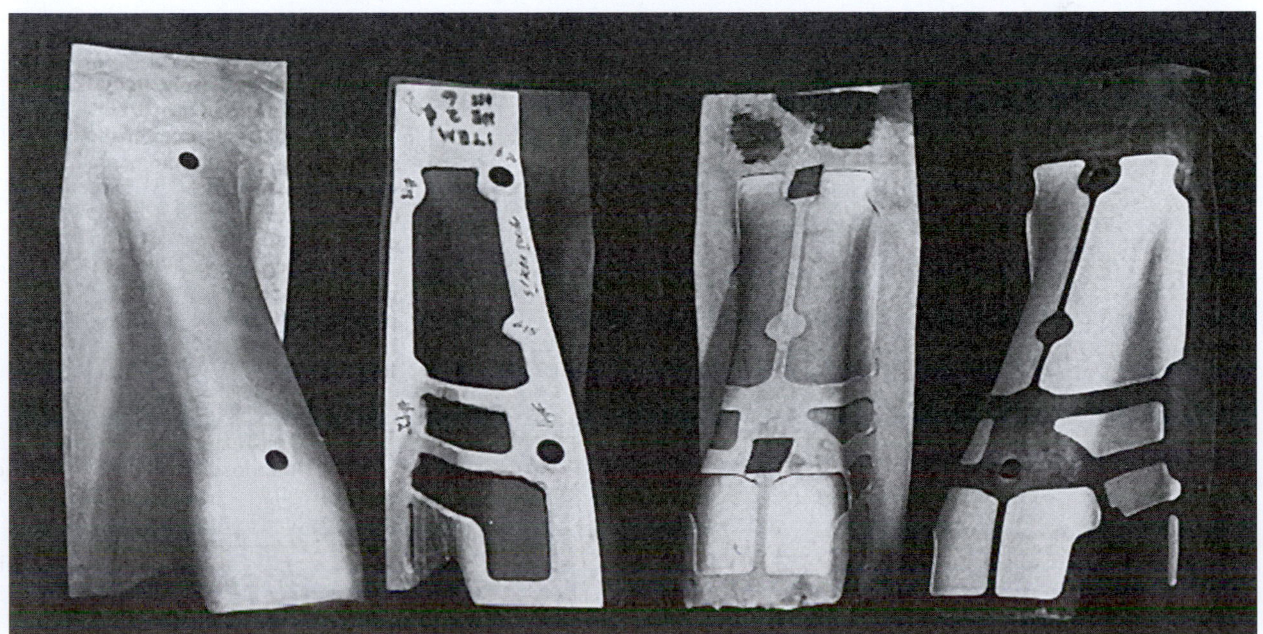

FIGURE 28-3 Iconel 718 aircraft engine parts. These sheet metal parts for a jet fighter engine are chemically milled to remove weight. (Left) As-formed workpiece; (middle left) workpiece coated with liquid rubber, fiberglass scribing template in place; (middle right) scribed workpiece; (right) finished part. About 0.035 in. of stock is removed from the 0.070-in.-thick workpieces. Tolerances are held to ±0.004 in. *(Courtsey J T. Black)*

The tolerances expected with CHM range from ±0.0005 in. on small etch depths up to ±0.004 in. in routine production involving substantial depths. Tolerances in chemical milling increase with the depth of the cut and with faster etch rates and vary for different materials. The surface finish is generally good to excellent for chemical polishing.

In using CHM, some disadvantages and limitations should be kept in mind. CHM requires the handling of dangerous chemicals and the disposal of potentially harmful by-products, although some recycling of chemicals may be possible. The metal removal rate is slow in terms of the unit area exposed, being about 0.2 to 0.04 lb/min per square foot exposed in the case of steel. However, because large areas can be exposed all at once, the overall removal rate may compare favorably with other metal removal processes, particularly when the work material is not machinable or the workpiece is thin and fragile, unable to sustain large cutting forces.

Photochemical Machining. Figure 28-4 shows the specific steps that are involved when **photochemical machining (PCM)** is performed with the use of **photoresists**:

1. Clean the workpiece.
2. Coat the workpiece with a photoresist, usually by hot-roller lamination of dry-film photoresists, on both sides, although liquid photoresists may also be applied by dipping, flowing, rolling, or electrophoresis (i.e., migration of charged molecules in the presence of an electric field). For liquid photoresists, the coating is heated in an oven to remove solvents.
3. Prepare the artwork. A drawing of the workpiece is made on a computer-aided design (CAD) system.
4. Develop the phototool. The CAD file is used to derive a photographic negative of the workpiece. Several methods may be used. Typically, the CAD drawing is downloaded to a laser-imaging system that exposes the desired image directly onto photographic (e.g., silver halide) film. In the past, oversized artwork was used to increase the accuracy of the phototool through photographic reduction of the artwork.
5. Expose the photoresist. Bring the phototool in contact with the workpiece, using a vacuum frame to ensure good contact, and expose the workpiece to intense ultraviolet (UV) light.

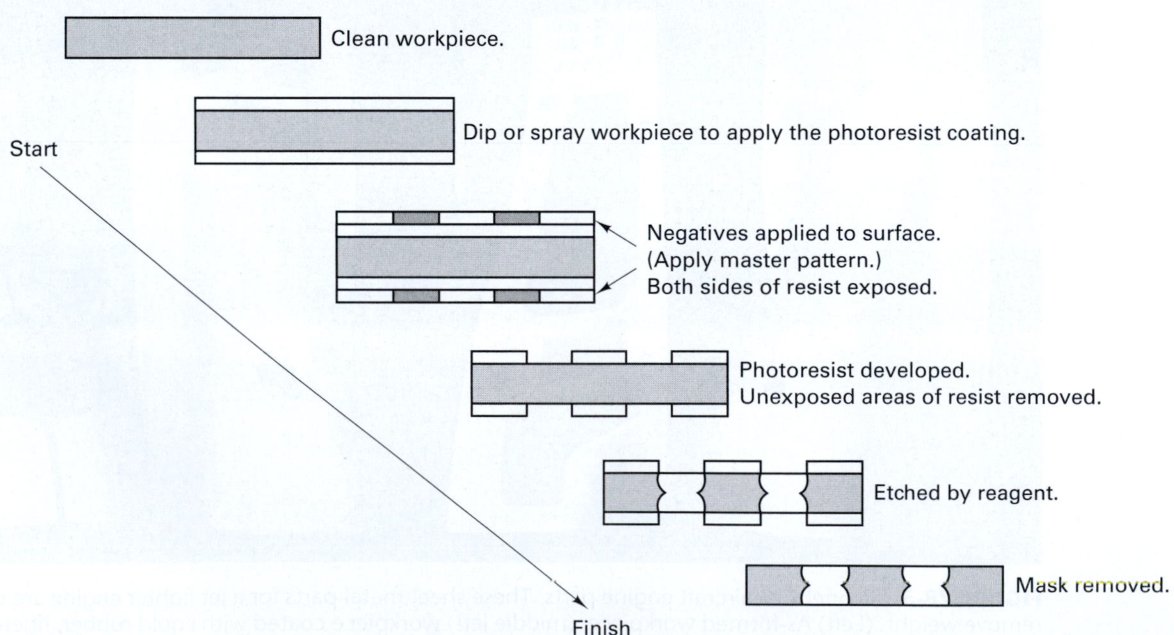

FIGURE 28-4 Basic steps in photochemical machining (PCM).

6. Develop the photoresist. Exposure of the photoresist to intense UV light alters the chemistry of the photoresist, making it more resistant to dissolution in certain solvents. By placing the exposed maskant in the proper solvent, the unexposed areas of the resist are removed, exposing the underlying material for etching. All residue is rinsed away.
7. Spray the workpiece with (or immerse it in) the reagent.
8. Remove the remaining maskant.

PCM has been widely used for the production of small, complex parts, such as printed circuit boards, and very thin parts that are too small or too thin to be blanked or milled by ordinary sheet metal forming or machining operations, respectively. Refinements to the PCM process are used in microeletronics fabrication (see Chapter 35).

Design Factors in Chemical Machining. When designing parts that are to be made by chemical machining, several unique factors related to the process must be kept in mind. First, if artwork is used, dimensional variations can occur through size changes in the artwork or phototool film due to temperature and humidity changes. These can be controlled or eliminated by putting the artwork on thicker polyester films or glass, by controlling the temperature or humidity in the artwork and phototool production areas, or by using a direct-write laser-imaging system.

The second item that must be considered is the **etch factor,** sometimes referred to as **etch radius,** which describes the undercutting of the maskant. The etchant acts isotropically on whatever surface is exposed. Areas that are exposed longer will have more metal removed from them. Consequently, as the depth of the etch increases, there is a tendency to undercut or etch under the maskant. The etch factor, E, in chemical machining is defined as

$$E = \frac{U}{d} \qquad (28\text{-}1)$$

where d is the depth of cut and U is the undercut as defined in Figure 28-5. In photochemical machining, the term **anisotropy** (sometimes referred to as etch factor) is used to describe the directionality of the cut. Anisotropy, A, of a material-etchant interaction in photochemical machining is defined as

$$A = \frac{d}{U} \qquad (28\text{-}2)$$

which is the inverse of equation 28-1. In many electrical and electronic products, anisotropies much greater than 1 are desirable in order to permit greater densities of electrical and electronic components and wires.

An allowance for the etch factor must be taken into account in designing the part and the artwork or scribing template. In the case of chemical milling, the width of the opening in the maskant must be reduced by an amount sufficient to compensate for the undercut under both sides of the maskant:

$$W_m = W_f - (E \times d) \qquad (28\text{-}3)$$

where W_f is the final desired width of the cut. This allowance is considered a minimum allowance; it has been found that results will vary based on etching conditions, and actual etch allowances will have to be somewhat greater and adapted to specific conditions.

In double-sided chemical blanking, a sharp edge remains along the line at which breakthrough occurs. Because such an edge is usually objectionable, etching ordinarily is continued to produce nearly straight sidewalls, as shown in Figure 28-5g. In order to achieve nearly straight sidewalls, it is typical to allow the process to continue for the amount of time necessary to do a through-cut from one side.

The anisotropy of the cut defines the maximum limit for aspect ratio (i.e., ratio of depth to width) of the cut, which is a measure of how deep and narrow a cut can be made. In chemical blanking, by using a double-sided etch, the maximum aspect ratio of the cut

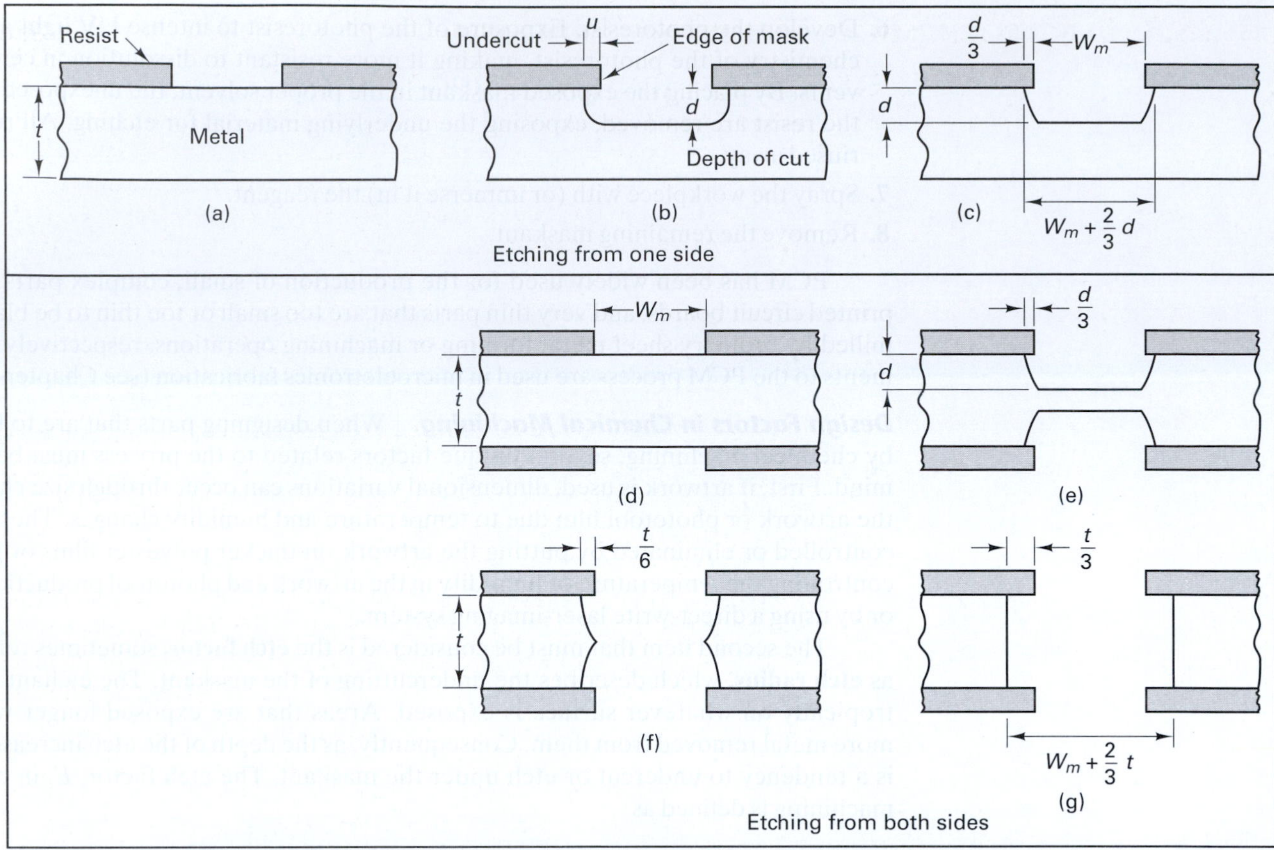

FIGURE 28-5 Undercutting of the mask or resist is defined by the etch factor, which must be accounted for in designing the part using the artwork on the scribing template.

may be effectively doubled if the process is stopped at breakthrough (Figure 28-5f). This is the effective maximum aspect ratio that may be achieved in CHM. Consequently, it is difficult to produce deep, narrow cuts in materials using CHM. Table 28-2 shows some etch rates and etch factors for common metal and etchant combinations in CHM.

The soundness and homogeneity of the metal are very important. Wrought materials should be uniformly heat treated and stress relieved prior to processing. Although chemical machining induces no stresses, it may release existing residual stresses in the metal and thus cause warpage. Castings can be chemically machined provided that they are not porous and have uniform grain size. Lack of the latter can cause nonuniform etching rates, producing islands. Because of the different grain structures that exist near welds, weldments usually are not suitable for chemical machining. Preferential etching due to **intergranular attack** can also be an issue in CHM.

TABLE 28-2 Etch Rates and Etch Factors for Some Common Metal–Etchant Combinations in CHM

Metal	Preferred Etchant	Penetration Rate (mm/mm)	Etch Factor *E*
Aluminum	$FeCl_3$	0.025	1.7:1
Copper	$FeCl_3$	0.05	2.7:1
Nickel alloys	$FeCl_3$	0.018	2.0:1
Phosphor-bronze	Chromic acid	0.013	2.0:1
Silver	$FeNO_3$	0.02	1.5:1
Titanium	HF	0.25	2.0:1
Tool steel	HNO_3	0.018	1.5:1

Source: G. F. Benedict, *Nontraditional Manufacturing Processes*, Marcel Dekker, New York, 1987, p. 200.

28.3 ELECTROCHEMICAL MACHINING PROCESSES

ELECTROCHEMICAL MACHINING

Electrochemical machining, commonly designated **ECM,** removes material by anodic dissolution with a rapidly flowing electrolyte. The process is shown schematically in Figure 28-6. It is basically a deplating process in which the tool is the cathode and the workpiece is the anode; both must be electrically conductive. The electrolyte, which can be pumped rapidly through or around the tool, sweeps away any heat and waste product (sludge) given off during the reaction. The sludge is captured and removed from the electrolyte through filtration. The shape of the cavity is defined by the tool, which is advanced by means of a servomechanism that controls the gap between the electrodes (i.e., the interelectrode gap) to a range from 0.003 to 0.03 in. (0.01 in. typical). The tool advances into the work at a constant feed rate, or penetration rate, that matches the deplating rate of the workpiece. The electrolyte is a highly conductive solution of inorganic salt—usually NaCl, KCl, and $NaNO_3$ (or other proprietary mixtures)—and is operated at about 75 to 150°F with flow rates ranging from 50 to 200 ft/sec. The temperature of the electrolyte is maintained through appropriate temperature controls. Tools are usually made of copper or brass and sometimes stainless steel.

The behavior of the ECM process is governed by the laws of electrolysis (the use of electrical current to bring about chemical change). Faraday's first law of electrolysis states that the amount of chemical change (material removed) during electrolysis is proportional to the charge (number of electrons) passed. This can be expressed mathematically as

$$n = K \times I \times t \tag{28-4}$$

where n is the theoretical number of moles of material removed; I is the applied current, which is assumed to stay constant; and t is the time over which the current is applied. The term K is a proportionality constant, that is, inversely proportional to the valence of the workpiece material.

To convert equation 28-4 into a useful expression for estimating the theoretical material removal rate, the number of moles, n, can be converted into a material volume by multiplying through the atomic weight, A_w, and dividing by the density, p, of the

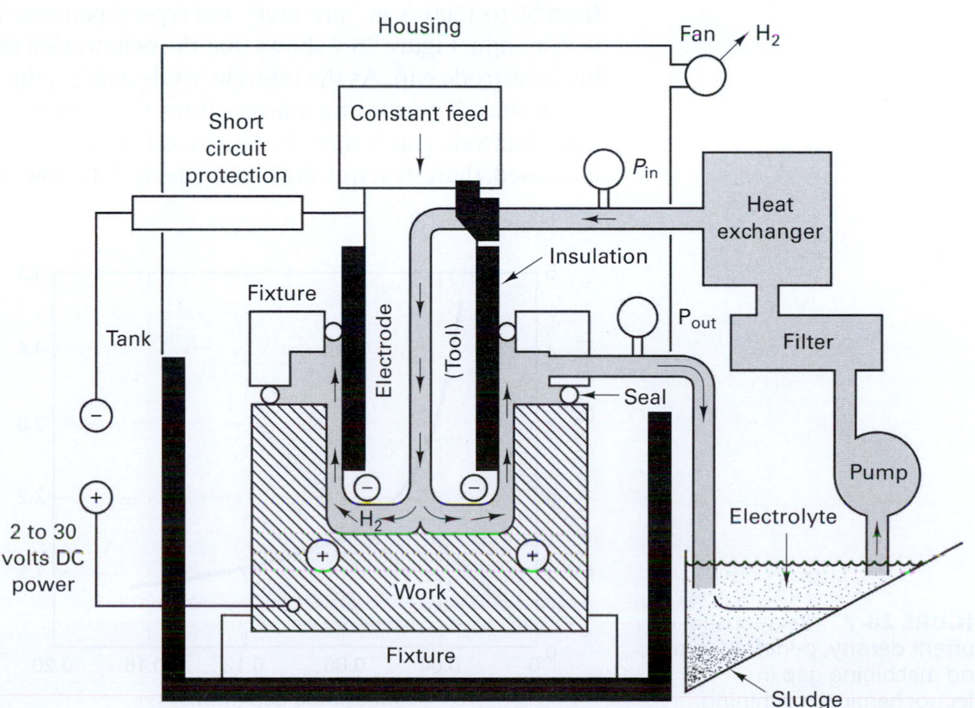

FIGURE 28-6 Schematic of electrochemical machining process (ECM).

TABLE 28-3 Material Removal Rates for ECM of Alloys Assuming 100% Current Efficiency

Alloy	Theoretical Removal Rates for 1000 Amperes per Square Inch	
	in.^{3}/min	cm^3/min
4340 steel	0.133	2.18
17-4 PH	0.123	2.02
A-286	0.117	1.92
M252	0.110	1.80
Rene 41	0.108	1.77
Udimet 500	0.110	1.80
Udimet 700	0.108	1.77
L605	0.107	1.75

Note: Rates listed were calculated using Faraday's law and valences as follows:

Aluminum	3	Copper	2	Silicon	0
Carbon	0	Iron	4	Titanium	4
Columbium	3	Manganese	3	Tungsten	6
Cobalt	2	Molybdenum	4	Vanadium	5
Chromium	3	Nickel	2		

workpiece material. Further, by dividing through by the time, t, an expression for the volumetric material removal rate is obtained,

$$\text{MRR} = \left(\frac{K \times A_w}{p}\right) \times I = \text{MRR}_s \times I \qquad (28\text{-}5)$$

where MRR_s is a proportionality constant called the specific material removal rate based on the valence, atomic weight, and density of the workpiece material. In suitable applications, material removal rates on the order of 0.1 in.^{3}/min per 1000 A can be expected. Table 28-3 shows the specific material removal rate for some different materials.

An estimate for the theoretical feed rate, f_t, or penetration rate, in ECM can be made by dividing equation 28-5 by the area of workpiece material, A, exposed to the ECM tool at the interelectrode gap. This yields:

$$f_r = \text{MRR}_s \times \frac{I}{A} = \text{MRR}_s \times J \qquad (28\text{-}6)$$

Equations 28-4, 28-5 and 28-6 assume that the efficiency, E, for metal removal by electrolysis is 100%, though normally it is in the range of 90 to 100% depending on the current, voltage, and other operating conditions. Therefore, equations 28-5 and 28-6 may be modified as follows:

$$\text{MRR}_s = E \times \text{MRR}_i \qquad (28\text{-}7)$$

$$f_r = E \times f_i \qquad (28\text{-}8)$$

where MRR and f_r are the actual material removal rate and actual feed rate, respectively.

As shown in equation 28-6, the penetration rate in ECM is primarily a function of the current density, J, and the physical properties of the material. Current densities from 50 to 1500 A/in.^{2} are used, and typical penetration rates in metals are from 0.02 to 0.75 in./min. Figure 28-7 shows that the penetration rate in ECM is also a function of the interelectrode gap. As the interelectrode gap is reduced, the resistance across the gap is also reduced, permitting a larger flow of electrons. One disadvantage of reducing the interelectrode gap is that the likelihood of plating anodic material onto the cathode is increased, thereby requiring higher electrolyte flow rates.

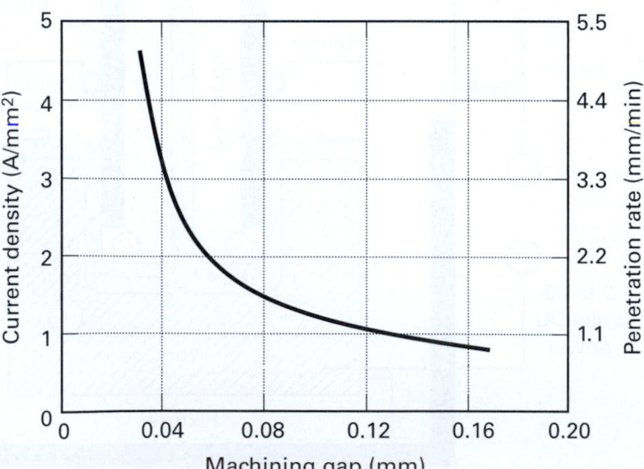

FIGURE 28-7 Relationship of current density, penetration rate, and machining gap in electrochemical machining.

The importance of these implications is that the penetration rate, while a function of current density and interelectrode gap, is not directly affected by the hardness or toughness of the work material. This makes ECM advantageous for the machining of certain high-strength materials with poor machinability. Penetration rates up to 0.1 in./min are obtained routinely in Waspalloy, a very hard metal alloy.

Pulsed-current ECM (PECM) has recently shown the potential to improve accuracies and surface finish in traditional ECM. In PECM, high-current densities (>100 A/cm^2) are pulsed on for durations on the order of 1 ms and pulsed off for intervals on the order of 10 ms. The relaxation (pulse-off) interval permits reaction by-products to be removed from the interelectrode gap at low electrolyte flow rates without electrolytic deposition on the ECM tool. As a result, high-current densities can be used at small interelectrode distances, improving both removal rates and precision. Because PECM allows lower electrolyte flow rates, it has been proposed to remove recast layers from the surface of dies produced by electrical discharge machining (EDM). Some efforts are being made to integrate this technology into EDM machines.

Pulsed currents have also been used in **electrochemical micromachining (EMM).** In one variation of the process, photolithographic masks are used to concentrate material removal in selective areas of thin films. In addition to providing smoother surfaces (on the order of electropolished surfaces) than those found in traditional ECM, the high-current density (~100 A/cm^2) results in less taper in etching profiles and more uniform material removal across the workpiece, which are important for shrinking feature sizes in micromachining.

Electrochemical polishing is a modification of the ECM process that operates essentially the same as ECM, but with a much slower penetration rate. Current density is lowered, which greatly reduces the material removal rate and produces a fine finish on the order of 10 min. Electrochemical polishing must be differentiated from **electropolishing**, which operates without the use of a part-specific hard tool.

Electrochemical Hole Machining. Several electrochemical hole-drilling processes have been developed for drilling small holes with high aspect (depth-to-diameter) ratios in difficult-to-machine materials. In **electrostream drilling,** large numbers of holes (more than 50) can be simultaneously gang drilled in nickel and cobalt alloys with diameters down to 0.005 in. at aspect ratios of 50:1. Machining is performed by a high-velocity stream of charged, acidic electrolyte ejected from the capillary end of a drawn glass tube (the tool). An internal electrode (e.g., a small titanium wire) is fed into the large end of the tube and placed close to the capillary. The electrode is used to charge the electrolyte, which is pumped through the tube. An acidic electrolyte is used so that the sludge by-product goes into solution instead of clogging flow. Voltages used in the process are 10 to 20 times higher than that of typical ECM processes. This technique was originally developed to drill high-incident-angle, cooling holes in turbine blades for jet engines.

A second process, known as the **shaped-tube electrolytic machining (STEM)** process, was also created in response to unique challenges presented in the jet engine industry. See Figure 28-8 for a schematic. Like the electrostream process, STEM is also capable of gang drilling small holes in difficult-to-machine materials. However, the STEM process is generally not capable of drilling holes smaller than about 0.02 in. STEM is capable of making shaped holes with aspect ratios as high as 300:1. Holes up to 24 in. in depth have been drilled. Like the electrostream process, it uses an acidic electrolyte to minimize clogging due to sludge buildup. The major differences between the STEM process and the electrostream process are the reduced voltage levels (5 to 10 Vdc) and the special electrodes, which are long, straight, metallic tubes coated with an insulator. The insulator helps to eliminate taper by constraining the electrolytic action between the bottom of the tool and the workpiece. Titanium is often used for its ability to resist acids. The electrolyte is pressure-fed through the tube and returns through the gap (0.001 to 0.002 in.) between the insulated tube wall and the hole wall. Electrolyte concentrations may include up to 10% sulfuric acid. Lower concentrations may be used to increase tool life. Table 28-4 compares various NTM processes used in holemaking.

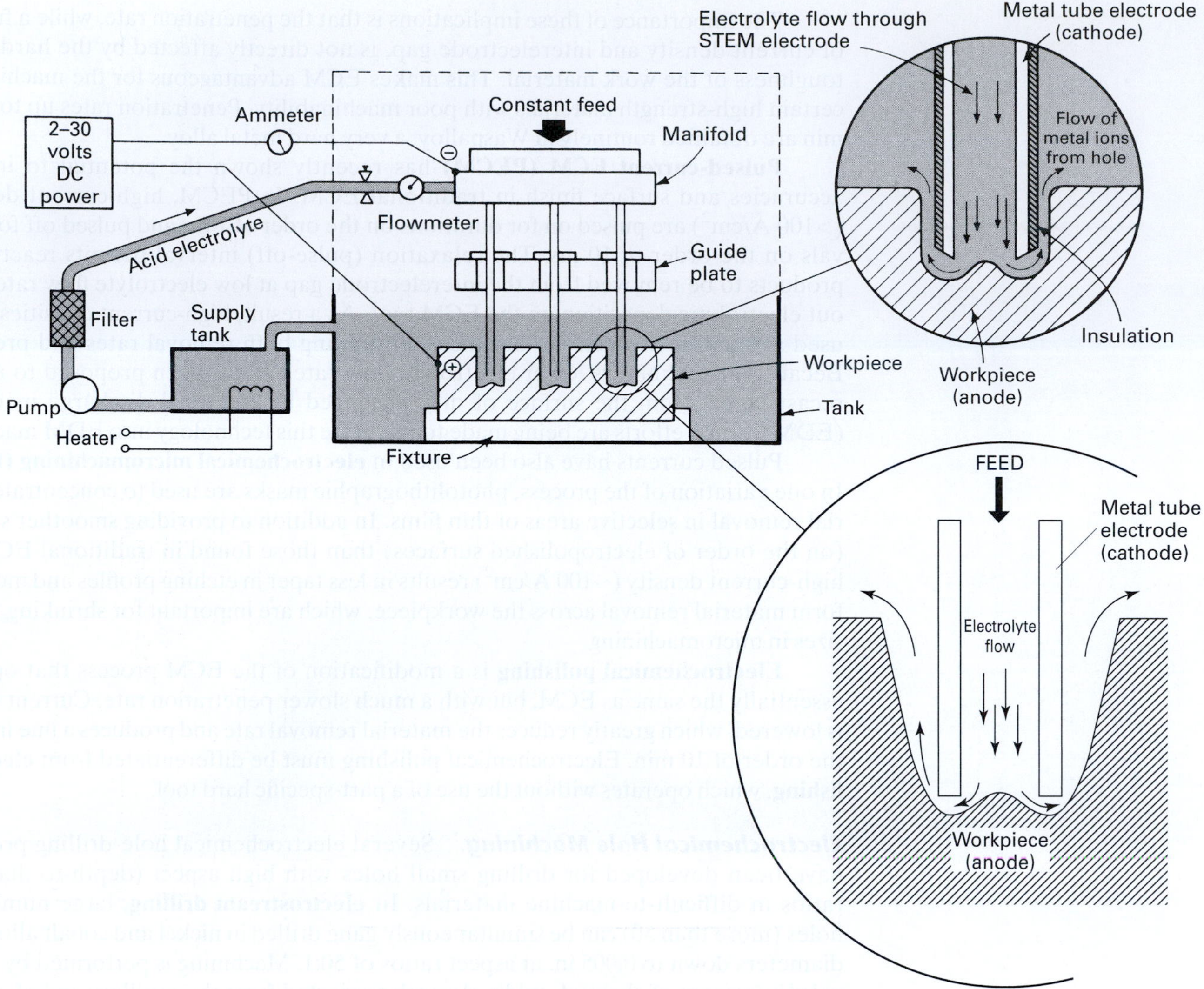

FIGURE 28-8 The shaped-tube electrolytic machining (STEM) cell process is a specialized ECM technique for drilling small holes using a metal tube electrode or metal tube electrode with dielectric coating.

Electrochemical Grinding. **Electrochemical grinding,** commonly designated **ECG,** is a low-voltage, high-current variant of ECM in which the tool cathode is a rotating, metal-bonded, diamond grit grinding wheel. The setup shown in Figure 28-9 for grinding a cutting tool is typical. As the electric current flows through the electrolyte between the workpiece and the wheel, some surface metal is removed electrochemically and some is changed to a metal oxide, which is ground away by the abrasives. As the oxide film is removed, new surface metal is exposed to electrolytic action. Most of the material removal is electrochemical, with only about 10% of the material being removed by grinding. The metal removal rate (MRR) is dependent on many variables. Table 28-5 gives some typical values for the MRRs for various metals.

The wheels used in ECG must be electrically conductive abrasive wheels. For most metals, resin-bonded aluminum oxide wheels are recommended. The resin bond is loaded with copper to provide for negligible electrical resistance. The wheels are dressable, using a variety of wheel-dressing measures. Once dressed, the wheels can then be used for precision form-grinding operations.

The abrasive particles are hard, nonconducting materials such as aluminum oxide, diamond, or borazon (cubic boron nitride). In addition to increasing the efficiency of the process, the abrasives act as an insulating spacer, maintaining a separation of from 0.0005 to 0.003 in. (0.012 to 0.05 mm) between the electrodes. A dead short would result if the insulating particles were absent. The particles also serve to grind away any

TABLE 28-4 Various Nontraditional Production Holemaking Processes

Process	ECM	ES	STEM	EDM	LBM	EBM
Types of material	Conductive	Conductive	Conductive	Conductive	Any	Any
Hole size (d), in.						
Typical	$\frac{1}{4}$ and up	0.005–0.035	0.030–0.100	0.005–$\frac{1}{4}$	0.005–0.050	0.001–0.040
Minimum	$\frac{1}{8}$	0.002	0.020	0.000,35	0.000,075	0.0008
Max practical	3**	0.040	0.250	$\frac{1}{4}$**	0.060**	0.050**
Hole depth (l), in.						
Common max*	5	0.750	5	0.125	0.2	0.10
Ultimate*	12	1.0	36	2.5	0.7	0.3
Aspect ratio (l:d)						
Typical	8:1	16:1	16:1	10:1	16:1	6:1
Maximum*	20:1	40:1	300:1	20:1	75:1	100:1
Practical multiple drilling	100	100	100	200	2	1
Typical penetrating or cutting rate	0.3 ipm	0.060 ipm	0.060 ipm	0.030 ipm	<1 sec	0.010 ips
Typical tolerance, ±	0.002 in.	5% d	10% d	0.0005 in.	5 to 20% d	5 to 10% d
Finish, μin. AA	16–63	10–63	32–125	63–125	32–250	32–250
Surface integrity	No residual stress; polished surfaces; no burrs	No residual stress; no burrs	No residual stress; no burrs	Heat-affected surface; no burr	Heat-affected surface	Heat-affected surface
Production rates achieved					10 holes per sec	5000 holes per sec
Special attributes	Contoured and irregular shapes; no burrs	High integrity; no burrs	Shaped and multiangled; no burrs	Irregular piercings or contours; no burrs; delicate components	Rapidly-adjustable	Rapid positioning

PCM — Photochemical machining
USM — Ultrasonic machining
*Maximum depth and aspect ratios are estimates of what can be achieved with special attention, special tools, and the best conditions
**Not including trepanning or cavity sinking

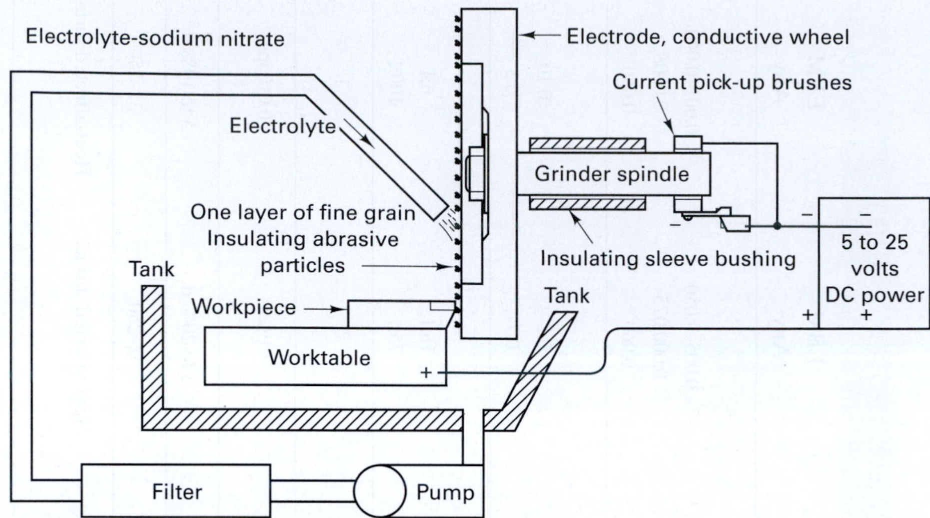

FIGURE 28-9 Equipment setup and electrical circuit for electrochemical grinding.

TABLE 28-5 Metal Removal Rates for ECG for Various Metals

Metal	Valency	Density		Metal Removal Rate at 1000 A		
		lb/in^3	g/cm^3	lb/hr	in^3/min	cm^3/min
Aluminum	3	0.098	2.67	0.74	0.126	2.06
Beryllium	2	0.067	1.85	0.37	0.092	1.50
Chromium	2	0.260	7.19	2.14	0.137	2.25
	3			1.43	0.092	1.51
	6			0.71	0.046	0.75
Cobalt	2	0.322	8.85	2.42	0.125	2.05
	3			1.62	0.084	1.38
Niobium (columbium)	3	0.310	8.57	2.55	0.132	2.16
	4			1.92	0.103	1.69
	5			1.53	0.082	1.34
Copper	1	0.324	8.96	5.22	0.268	4.39
	2			2.61	0.134	2.20
Iron	2	0.284	7.86	2.30	0.135	2.21
	3			1.53	0.090	1.47
Magnesium	2	0.063	1.74	1.00	0.265	4.34
Manganese	2	0.270	7.43	2.26	0.139	2.28
	4			1.13	0.070	1.15
	7			0.65	0.040	0.66
Molybdenum	3	0.369	10.22	2.63	0.119	1.95
	4			1.97	0.090	1.47
	6			1.32	0.060	0.98
Nickel	2	0.322	8.90	2.41	0.129	2.11
	3			1.61	0.083	1.36
Silicon	4	0.084	2.33	0.58	0.114	1.87
Silver	1	0.379	10.49	8.87	0.390	6.39
Tin	2	0.264	7.30	4.88	0.308	5.05
	4			2.44	0.154	2.52
Titanium	3	0.163	4.51	1.31	0.134	1.65
	4			0.99	0.101	
Tungsten	6	0.697	19.3	2.52	0.060	0.98
	8			1.89	1.89	0.74
Uranium	4	0.689	19.1	4.90	0.117	1.92
	6			3.27	0.078	1.29
Vanadium	3	0.220	6.1	1.40	0.106	1.74
	5			0.84	0.064	1.05
Zinc	2	0.258	7.13	2.69	0.174	2.85

Source: 1985 SCTE Conference Proceedings, ASM, Metals Park, OH, 1986.

passivating oxide that should form during electrolysis. The process is used for shaping and sharpening carbide cutting tools, which cause high wear rates on expensive diamond wheels in normal grinding. ECG greatly reduces wheel wear. Fragile parts (e.g., honeycomb structures), surgical needles, and tips of assembled turbine blades have also been ECG-processed successfully. The lack of heat damage, burrs, and residual stresses is very beneficial, particularly when coupled with MRRs that are competitive with conventional grinding and far less wheel wear.

Electrochemical Deburring. **Electrochemical deburring** is a deburring process that works on the principle that electrolysis is accelerated in areas with small interelectrode gaps and prevented in areas with insulation between electrodes. The cathodic tool in electrochemical deburring is stationary and generally shaped as a negative of the workpiece to focus the electrolysis on the region of the workpiece where burrs are to be removed. Portions of the tool not used for deburring are coated with insulation to prevent the electrolytic reaction. The process does not require a feed mechanism. A fixture made of insulating material is used to position the workpiece with respect to the cathodic tool. Because of the small amount of material removed, electrochemical deburring generally requires short cycle times under one minute.

Design Factors in Electrochemical Machining. In general, current densities tend to concentrate at sharp edges or features and, therefore, produce rounded corners. Therefore, ECM processes should not be used to create sharp corners or pockets in the workpiece, although sharp corners are possible on **through-hole** features. Due to the need for an interelectrode gap, the actual feature size will be larger than the tool size. The overcut on the side of the tool is normally on the order of 0.005 in. Taper should be expected in all features due to electrolytic reaction between the side of the tool and the workpiece. Depending on the tool design, taper can be held to 0.001 in./in. For micromachining applications, insulator material may be placed on the side of the tool to prevent the taper effects.

Control of the electrolyte flow can be difficult in parts with irregular shapes. Changes in electrolyte concentration due to varying flow patterns can change the local resistance across the interelectrode gap, resulting in local variations in removal rates and tolerances. In addition, high electrolytic flow rates can cause erosion of workpiece features. Therefore, complex geometries requiring tortuous interelectrode flow paths are discouraged.

Generally, no detrimental effects to the properties of the workpiece material are expected when using ECM. However, the lack of compressive residual stresses imparted to the surface of the workpiece during operation can have a negative impact on the fatigue resistance of the part when compared with mechanically machined parts. Further, if parameters are set improperly, the process may favor electrolysis at the grain boundaries, resulting in intergranular attack, which may also have a negative effect on the fatigue resistance of the part. If possible, *shot peening* or some similar process may be used to impart compressive stresses to the surface of the workpiece, thereby improving its fatigue resistance. See Chapter 34 for additional discussion on fatigue.

Advantages and Disadvantages of Electrochemical Machining. ECM is well suited for the machining of complex two-dimensional shapes into delicate or difficult-to-machine geometries made from poorly machinable but conductive materials. The principal tooling cost is for the preparation of the tool electrode, which can be time consuming and costly, requiring several cut-and-try efforts for complex shapes, because it is difficult to predict the precise final geometry due to variable current densities produced by certain electrode geometries (e.g., corners) or electrolyte flow variations. There is no tool wear during actual cutting, which suggests that the process becomes more economical with increasing volume. The process produces a stress-free surface, which can be advantageous, especially for small, thin parts. The ability to cut a large area simultaneously (as in chemical and electrical discharge machining) makes the production of small parts very productive. However, as in chemical machining, ECM requires the disposal of environmentally harmful by-products.

■ 28.4 ELECTRICAL DISCHARGE MACHINING

As early as the 1700s, Benjamin Franklin wrote of witnessing the removal of metal by electrical sparks. However, it was not until the 1940s that development of the **electrical discharge machining (EDM)** process, also known as *electric* or *electrodischarge machining*, began in earnest. Its initial use was to remove broken taps from metal holes, but today EDM is a widely used machining process.

EDM processes remove metal by discharging electric current from a pulsating DC power supply across a thin interelectrode gap between the tool and the workpiece. See Figure 28-10 for a schematic. The gap is filled by a dielectric fluid, which becomes locally ionized at the point where the interelectrode gap is the narrowest—generally, where a high point on the workpiece comes close to a high point on the tool. The ionization of the dielectric fluid creates a conduction path in which a spark is produced. The spark produces a tiny crater in the workpiece by melting and vaporization, and consequently tiny, spherical "chips" are produced by resolidification of the melted quantity of workpiece material. Bubbles from discharge gases are also produced. In addition to machining the workpiece, the high temperatures created by the spark also melt or vaporize the tool, creating tool wear. The dielectric fluid is pumped through the interelectrode gap and flushes out the chips and bubbles while confining the sparks. Once the highest point on the workpiece is removed, a subsequent spark is created between the tool and the next highest point, and so the process proceeds into the workpiece. Literally hundreds of thousands of sparks may be generated per second. This material removal mechanism is described as **spark erosion**.

Two different types of EDM exist based on the shape of the tool electrode used. In **ram EDM,** also known as *die-sinking EDM* or simply *EDM*, the electrode is a die in the shape of the negative of the cavity to be produced in a bulk material. By feeding the die into the workpiece, the shape of the die is machined into the workpiece.

In **wire EDM**, also known as **electrical discharge wire cutting,** the electrode is a wire used for cutting through-cut features, driving the workpiece with a computer numerical controlled (CNC) table. See Figure 28-11 for a schematic and Chapter 25 for a discussion of CNC.

Wire EDM uses a continuously moving conductive wire as the tool electrode.The tensioned wire of copper, brass, tungsten, or molybdenum is used only once, traveling

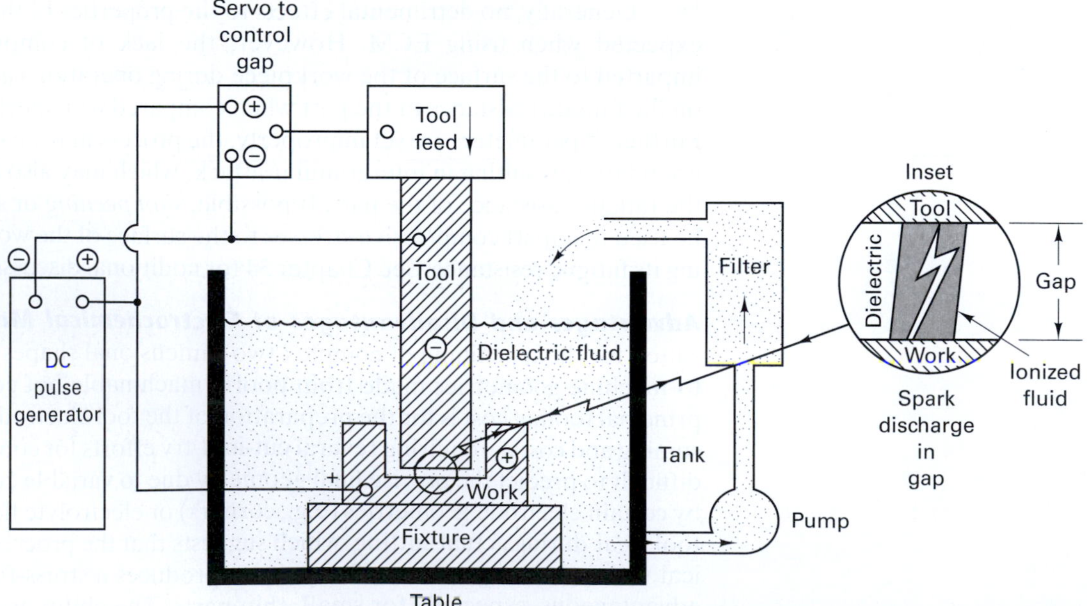

FIGURE 28-10 EDM or spark erosion machining of metal, using high-frequency spark discharges in a dielectric, between the shaped tool (cathode) and the work (anode). The table can make *X–Y* movements.

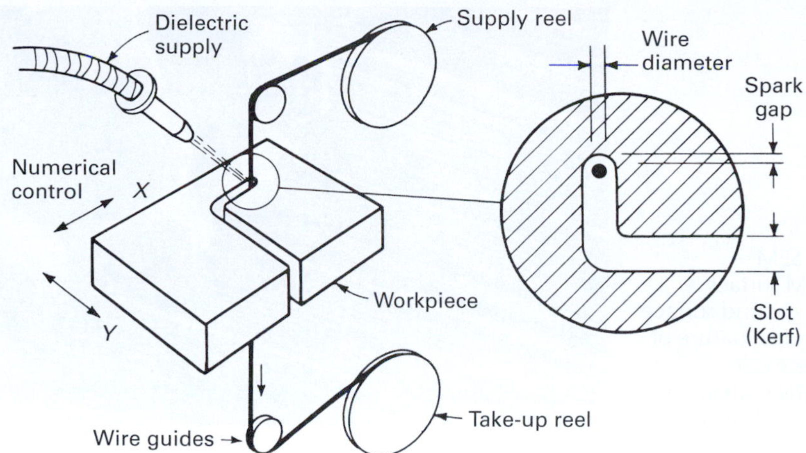

FIGURE 28-11 Schematic of equipment for wire EDM using a moving wire electrode.

from a take-off spool to a take-up spool while being "guided" to produce a straight, narrow kerf in plates up to 3 in. thick. The wire diameter ranges from 0.002 to 0.01 in., with positioning accuracy up to ±0.00002 in. in machines with numerical control (NC) or tracer control. The dielectric is usually deionized water because of its low viscosity. This process is widely used for the manufacture of punches, dies, and stripper plates, with modern machines capable of routinely cutting die relief, intricate openings, tight radius contours, and corners. See Figure 28-12 for some examples.

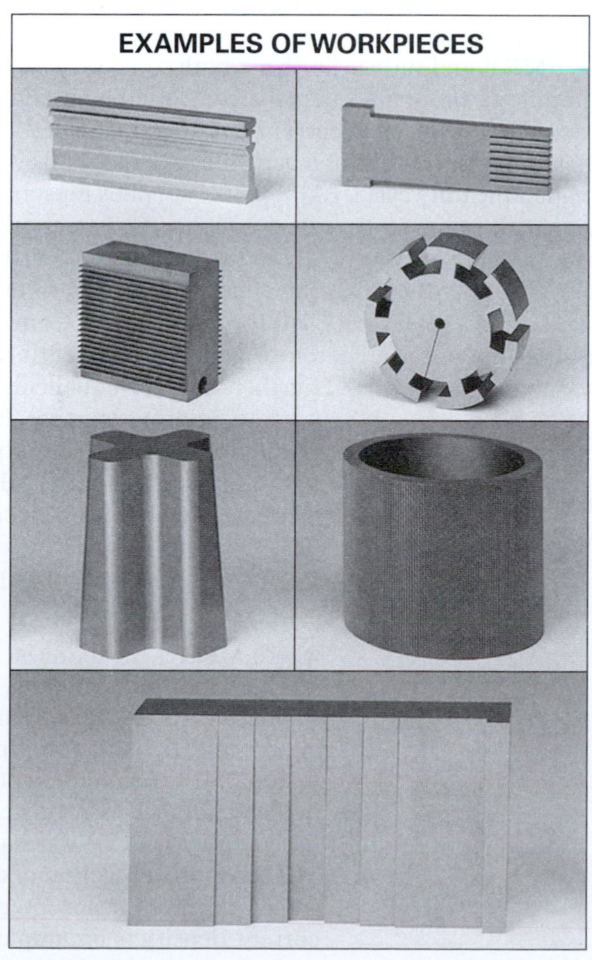

FIGURE 28-12 Examples of wire EDM workpieces made on an NC machine (Hatachi). *(Courtsey J T. Black)*

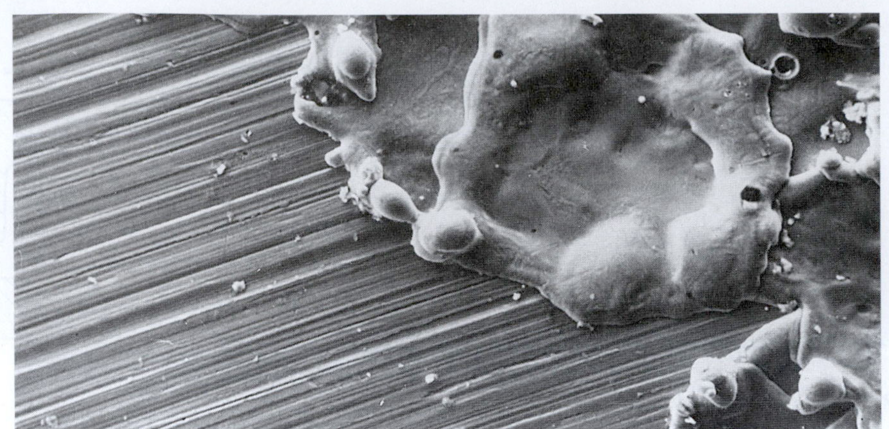

FIGURE 28-13 SEM micrograph of EDM surface (right) on top of a ground surface in steel. The spheroidal nature of debris on the surface is in evidence around the craters (300×). *(Courtsey J T. Black)*

EDM processes are slow compared to more conventional methods of machining, and they produce a matte surface finish composed of many small craters. While feed rates in EDM are slow, EDM processes can still compete with conventional machining in producing complex geometries, particularly in hardened tool materials. As a result, one of the biggest applications of EDM processes is tool and die making. Another drawback of EDM is the formation of a recast or remelt layer on the surface and a **heat-affected zone (HAZ)** below the surface of the workpiece. Figure 28-13 shows a scanning electron micrograph of a recast layer on top of a ground surface. Note the small spheres in the lower-right corner attached to the surface, representing chips that did not escape the surface. Below the recast layer is a heat-affected zone on the order of 0.001 in. thick. The effect of the recast layer and heat-affected zone is poor surface finish as well as poor surface integrity and poor fatigue strength.

MRR and surface finish are both controlled by the spark energy. In modern EDM equipment, the spark energy is controlled by a DC power supply. The power supply works by pulsing the current on and off at certain frequencies (between 10 and 500 kHz). The on-time as a percentage of the total cycle time (inverse of the frequency) is called the **duty cycle.** EDM power supplies must be able to control the pulse voltage, current, duration, duty cycle, frequency, and electrode polarity. The power supply controls the spark energy mainly by two parameters: current on-time and discharge current.

Figure 28-14 shows the effect of current on-time and discharge current on crater size. Larger craters are good for high MRRs. Conversely, small craters are good for finishing operations. Therefore, generally, higher duty cycles and lower frequencies are used to maximize MRR. Further, higher frequencies and lower discharge currents are used to improve surface finish while reducing the MRR. Higher frequencies generally cause increased tool wear.

An estimate for the MRR in ram EDM for a given workpiece material has been given in Weller:

$$\text{MRR} = \frac{C \times I}{T_m^{1,23}} \qquad (28\text{-}9)$$

where MRR is the material removal rate (in in.^{3}/min, or cm^3/min); C is a proportionality constant equal to 5.08 in British standard units (39.86 in SI units); I is the discharge current (in amps); and T_m is the melting temperature of the workpiece material, °F (°C). Melting points of selected metals are shown in Table 28-6. While this equation shows that the MRR is based mainly on the discharge current, it is recognized that MRR is more a function of current density. For graphite electrodes, a generally accepted rule of thumb for the maximum current density is 65 amp/cm^2 of surface area. Given these

TABLE 28-6 Melting Temperatures for Selected EDM Workpiece Materials

Metal	Melting Temperature, Metal °F (°C)
Aluminum	1200 (660)
Carbon steel	2500 (1371)
Cobalt	2696 (1480)
Copper	1980 (1082)
Manganese	2300 (1260)
Molybdenum	4757 (2625)
Nickel	2651 (1455)
Titanium	3308 (1820)
Tungsten	6098 (3370)

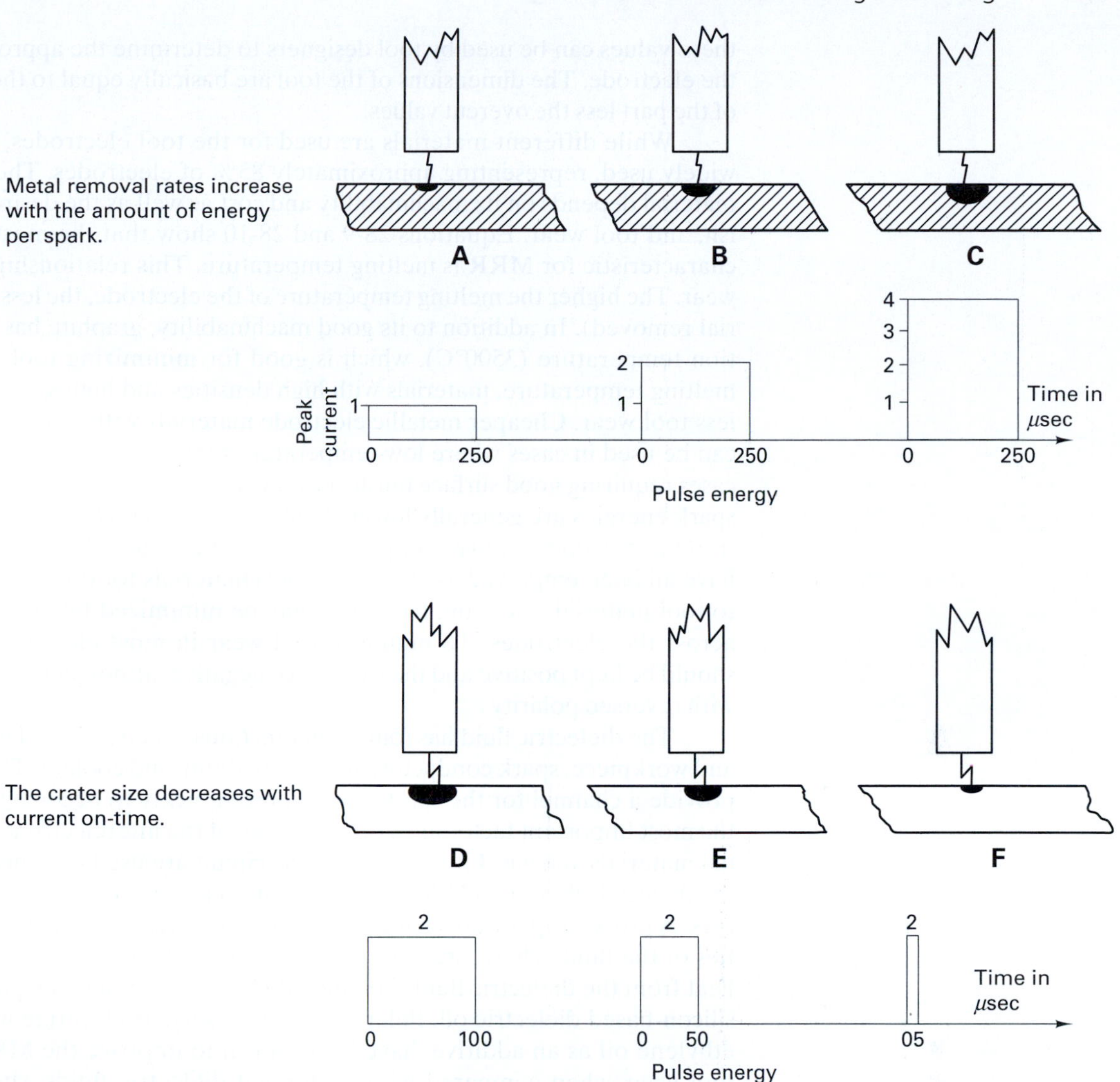

FIGURE 28-14 The principles of metal removal for EDM.

guidelines, an expression for the maximum MRR, MRR_{max}, can be derived:

$$MRR_{max} = \frac{C_{max} \times A_s}{T_m^{1.23}} \tag{28-10}$$

where C_{max} is a proportionality constant equal to 51.2 in British standard units (2591 in SI units) and is the bottom-facing surface area of the electrode (in in.^{2}, or cm^2).This formula suggests that the MRR in EDM is also a function of the tooling geometry. The larger and more highly contoured the electrode surface, the slower the MRR at the same discharge current.

EDM OVERCUT

The size of the cavity cut by the EDM tool will be larger than the tool. That is, the distance between the surface of the electrode and the surface of the workpiece represents the **overcut** and is constrained by the minimum interelectrode distance necessary for a spark, which is essentially constant over all areas of the electrode, regardless of size or shape. Typical overcut values range from 0.0005 to 0.02 in. Overcut depends on the gap voltage plus the chip size, which varies with the amperage. EDM equipment manufacturers publish overcut charts for the different power supplies for their machines, and these values can be used by tool designers to determine the appropriate dimensions of the electrode. The dimensions of the tool are basically equal to the desired dimensions of the part less the overcut values.

While different materials are used for the tool electrodes, graphite is the most widely used, representing approximately 85% of electrodes. The choice of electrode material depends on its machinability and cost as well as the desired MRR, surface finish, and tool wear. Equations 28-9 and 28-10 show that the most important material characteristic for MRR is melting temperature. This relationship also applies to tool wear. The higher the melting temperature of the electrode, the less tool wear (i.e., material removed). In addition to its good machinability, graphite has a very high sublimation temperature (3500°C), which is good for minimizing tool wear. In addition to melting temperature, materials with high densities and high specific heats tend toward less tool wear. Cheaper metallic electrode materials with lower melting temperatures can be used in cases where low-temperature metals are to be machined. And in some cases requiring good surface finish, a case can be made for metallic electrodes, because spark energies are generally lower, resulting in reduced temperatures in the interelectrode gap. Copper, brass, copper–tungsten, aluminum, 70Zn–30Sn, and other alloys have all been employed as electrode (tool) materials for different reasons. In addition to tool material selection, tool wear may be minimized by using the proper polarity across the electrodes. To minimize tool wear in most electrode materials, the tool should be kept positive and the workpiece negative, although larger MRRs are possible with reversed polarity.

The dielectric fluid has four main functions: electrical insulation between the tool and workpiece, spark conductor, flushing medium, and coolant. The fluid must ionize to provide a channel for the spark and deionize quickly to become an insulator. Perhaps the most important factor in EDM is flushing of the interelectrode gap to remove residual materials and gas. Filters in the fluidic circuit are used to remove these wastes from the dielectric fluid. In addition, the fluid must carry away the heat produced in the process. A gross temperature change in the dielectric fluid significantly changes the properties of the fluid. Therefore, a heat exchanger is added to the fluidic circuit to remove heat from the dielectric fluid. Common dielectric fluids include paraffin, kerosene, and silicon-based dielectric oil. Polar compounds, such as glycerine water (90:10) with triethylene oil as an additive, have been shown to improve the MRR and decrease the tool wear when compared with traditional dielectric fluids, such as kerosene. The dielectric materials must be safe for inhalation and skin contact because operators are in constant in contact with the fluid.

Advantages and Disadvantages of EDM. EDM is applicable to all materials that are fairly good electrical conductors, including metals, alloys, and most carbides. The hardness, toughness, or brittleness of the material imposes no limitations. EDM provides a relatively simple method for making holes and pockets of any desired cross section in materials that are too hard or too brittle to be machined by most other methods. The process leaves no burrs on the edges. About 80 to 90% of the EDM work performed in the world is in the manufacture of tool and die sets for injection molding, forging, stamping, and extrusions. The absence of almost all mechanical forces makes it possible to EDM fragile or delicate parts without distortion. EDM has been used in micromachining to make feature sizes as small as 0.0004 in. (0.01 mm).

On most materials, the process produces a thin, hard recast surface, which may be an advantage or a disadvantage, depending on the use. When the workpiece material is one that tends to be brittle at room temperature, the surface may contain fine cracks caused by the thermally induced stresses. Consequently, some other finishing process is often used subsequent to EDM to remove the thin recast and heat-affected layers, particularly if the product will be fatigued. Fumes, resulting from the bubbles produced during spark erosion, are given off during the EDM process. Fumes can be toxic when electrical discharge machining boron carbide, titanium boride, and beryllium, posing a significant safety issue, although machining of these materials is hazardous in many other processes as well.

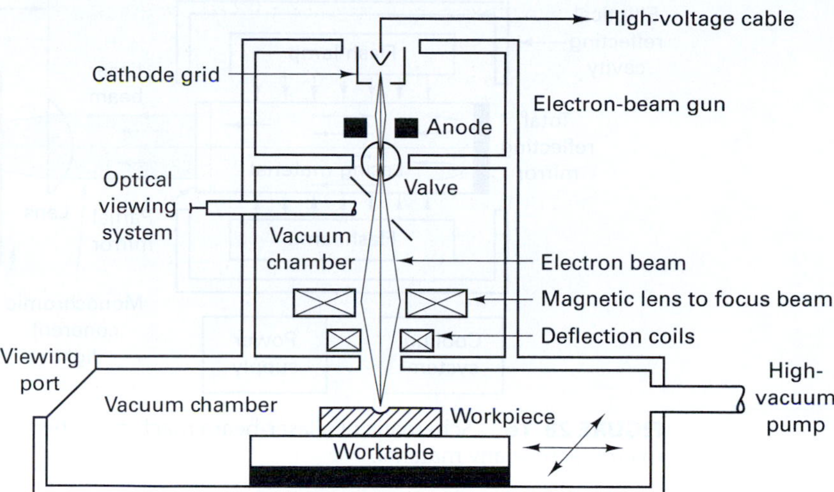

FIGURE 28-15 Electron-beam machining uses a high-energy electron beam (10^9 W/in.^{2}).

ELECTRON AND ION MACHINING

As a metals-processing tool, the electron beam is used mainly for welding, to some extent for surface hardening, and occasionally for cutting (mainly drilling). **Electron-beam machining (EBM)** is a thermal process that uses a beam of high-energy electrons focused on the workpiece to melt and vaporize metal. This process shown in Figure 28-15 is performed in a vacuum chamber. The electron beam is produced in the electron gun (also under vacuum) by thermionic emission. In its simplest form, a filament (tungsten or lanthanum-hexaboride) is heated to temperatures in excess of 2000°C, where a stream (beam) of electrons (more than 1 billion per second) is emitted from the tip of the filament. Electrostatic optics are used to focus and direct the beam. The desired beam path can be programmed with a computer to produce any desired pattern in the workpiece. The diameter of the beam is on the order of 0.0005 to 0.001 in., and holes or narrow slits with depth-to-width ratios of 100:1 can be "machined" with great precision in any material. The interaction of the beam with the surface produces dangerous X-rays; therefore, electromagnetic shielding of the process is necessary. The layer of recast material and the depth of the heat damage are very small. For micromachining applications, MRRs can exceed that of EDM or ECM. Typical tolerances are about 10% of the hole diameter or slot width. These machines require high voltages (50 to 200 kV) to accelerate the electrons to speeds of 0.5 to 0.8 the speed of light and should be operated by fully trained personnel.

Ion-beam machining (IBM) is a nanoscale (10^{-9}) machining technology used in the microelectronics industry to cleave defective wafers for characterization and failure analysis. IBM uses a focused ion beam created by thermionic emission similar to EBM to machine features as small as 50 nm. The ion beam may be focused down to a 50-nm diameter and is focused and positioned by an electrostatic optics column. Current densities up to 5 A/cm^2 and voltages between 4 and 150 kV provide ion energies up to 300 A/cm^2/keV. Target substrates as large as 7 in. $\times$ 7 in. $\times \frac{1}{4}$ in. thick can be processed. Chapter 36 has additional discussion on micro- and nanoprocesses.

LASER-BEAM MACHINING

Laser stands for "light amplification by stimulated emission of radiation." Lasers were invented five decades ago and are used everywhere. Semiconductor lasers are found inside laptops and DVD players and are used to ring up purchase at the store. For manufacturing processes, like **laser-beam machining (LBM),** an intensely focused, coherent stream of light called a laser is used to vaporize or chemically *ablate* materials. A schematic of the LBM process is shown in Figure 28-16. Lasers are also used for joining (welding, crazing, soldering), heat-treating materials (see Chapters 6), inspection (see Chapter 37 and Advanced Topic 2 for additional discussion), and free-form

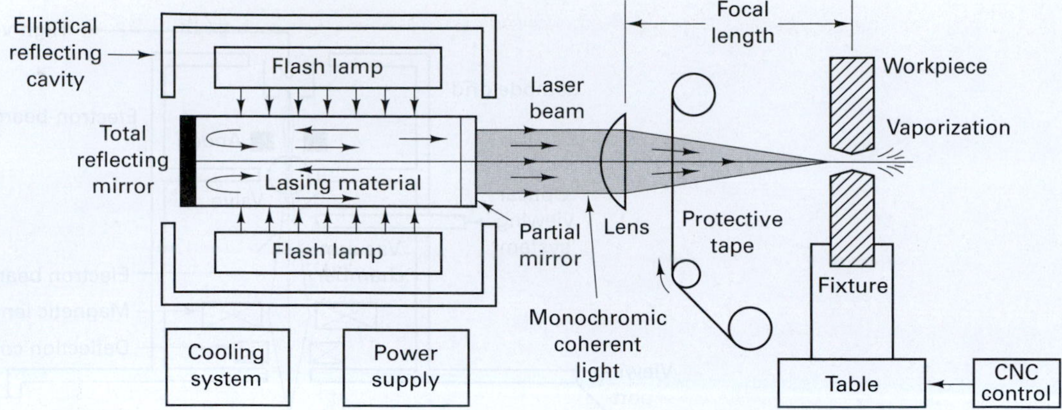

FIGURE 28-16 Schematic of a laser-beam machine, a thermal NTM process that can micromachine any material.

manufacturing (see Chapter 19). Power density and interaction time are the basic parameters in laser applications, as shown in Figure 28-17. Drilling requires higher-power densities and shorter interaction times compared to most other applications.

The material removal mechanism in LBM is dependent on the wavelength of the laser used. At UV wavelengths (i.e., between about 200 and 400 nm), the material removal mechanism in polymers (for example) is generally thermal evaporation. Below 400 nm, polymeric material is typically removed by chemical ablation. In *ablation*, the chemical bonds between atoms are broken by the excess amount of laser energy absorbed by the valence electrons in the material. The advantage of *chemical* ablation is that because it is not a thermal process, it does not result in a heat-affected zone.

Here is how the laser works. Laser light is produced within a laser cavity, which is a highly reflective cavity containing a laser rod and a high-intensity light source, or laser lamp. The light source is used to "pump up" the laser rod, which includes atoms of a lasing media that is capable of absorbing the particular wavelength of light produced by the light source. When an atom of lasing media is struck by a photon of light, it becomes energized. When a second photon strikes the energized atom, the atom gives off two photons of identical wavelength, moving in the same direction and with the same phase. This process is called **stimulated emission.** As the two photons now stimulate further emission from other energized atoms, a cascading of stimulated emission ensues. To increase the number of stimulated emissions, the laser rod has mirrors on both ends that are precisely parallel to one another. Only photons moving perpendicular to these

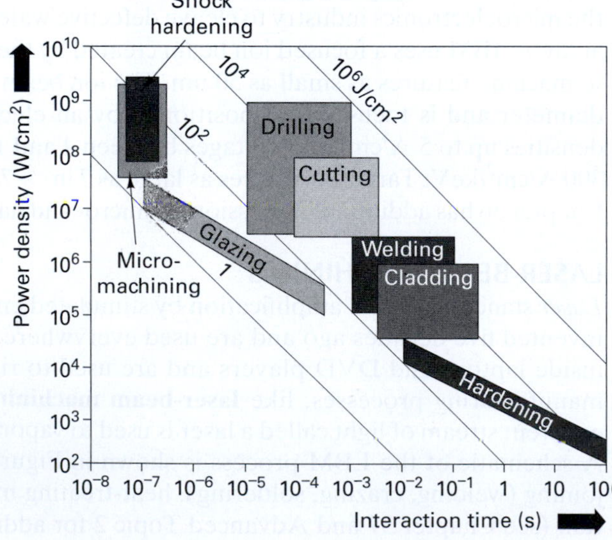

FIGURE 28-17 Power densities and interaction times in laser processing vary with the application.

TABLE 28-7 Commercial Lasers Available for Machining, Welding, and Trimming

Laser Type	Wavelength (μm)	Mode of Operation	Power (W)	Pulses per Second	Pulse Length of Time	Application	Comments
Argon	0.4880 0.5145	Pulse	20 peak; 0.005 average	60	50 μs	Scribing thin films	Power low
Ruby	0.6943	Pulse	2×10^5 peak	Low (5 to 10)	0.2–7 ms	Large material removal in one pulse, drilling diamond dies, spot welding	Often uneconomical
Nd-glass	1.06	Pulse	2×10^6 peak	Low (0.2)	0.5–10 ms	Large material removal in one pulse	Often uneconomical
Nd-YAG[a]	1.06	Continuous	1000	—	—	Welding	Compact, economical at low powers
Nd-YAG	1.06	Repetitively Q-switched	3×10^5 peak; 30 average	1–24,000	50–250 ns	Resistor trimming, electronic	Compact and economical
Nd-YAG	1.06	Pulse	400	300	0.5–7 ms	Spot weld, drill	
CO_2^b	10.6	Continuous	15,000	—	—	Cutting organic materials, oxygen-assisted metal	Very bulky at high powers
CO_2	10.6	Repetitively Q-switched	75,000 peak 1.5 average	400	50–200 ns	Resistor trimming	Bulky but economical
CO_2	10.6	Pulse	100 average	100	100 μs and up	Welding, hole production, cutting	Bulky but economical
KrF (excimer)	0.248	Pulse	Up to several kW	100–1000	15–45 ns	Micromachining, industrial materials processing, and laser annealing	Short wavelength, high energy, and high average power
XeCl (excimer)	0.308	Pulse	200	300	40–50 ns		

Source: Modified from J. F. Ready, Selecting a laser for material working, *Laser Focus* (March 1970), p. 40.
[a] Neodymium-yttrium aluminum garnet.
[b] CO_2 plus He plus N_2 mixture.

two mirrors stay within the laser rod, causing additional stimulated emission. One of the mirrors is partially transmissive and permits some percent of the laser energy to escape the cavity. The energy leaving the laser rod is the laser beam.

Table 28-7 lists some commercially available lasers for material processing. The most common industrial laser is the CO_2 laser. The CO_2 laser is a gas laser that uses a tube of helium and carbon dioxide as the laser rod. Output is in the far-infrared range (10.6 *jiva*), and the power can be up to 10 kW. Nd:YAG lasers are called solid-state lasers. The laser rod in these lasers is a solid crystal of yttrium, aluminum, and garnet that has been doped with neodymium atoms (the lasing media). The output wavelength is in the near-infrared range (1064 nm), and power up to 500 W is common. For micromachining applications, modifications can be made to a 50-W Nd:YAG laser to output at one-half (532 nm), one-third (355 nm), and one-fourth (266 nm) of this wavelength, with roughly an order of magnitude reduction in power from 1064 to 532 to 266 nm. More recently, gas lasers, called **excimer** (from excited *dimer)* **lasers,** have been developed with laser rods consisting of excited complex molecules (usually noble gas halides) called dimers. Excimer lasers are pulsed lasers that output in the near and deep UV range at powers up to 100 W. Excimer lasers are significantly more expensive to purchase and operate than CO_2 or Nd:YAG lasers.

Lasers produce highly collimated, coherent (in-phase) light, which, when focused to a small diameter, produces high-power densities that are good for machining. It is generally accepted that in order to evaporate materials, infrared power densities in excess of 10^5 W/mm^2 are needed. For CO_2 lasers, these levels are directly achievable. However, in Nd:YAG lasers, these high power conditions would significantly decrease the life of the laser lamp. Therefore, Nd:YAG lasers make use of a Q-switch, which breaks up the continuous light stream into a series of higher power pulses. Pulsed Nd:

YAG lasers may have peak powers in excess of 30 kW at 1064 nm and peak power densities in excess of 2×10^7 W/mm^2, which are easily enough for metal sublimation. With this magnitude of power density, the thermal effects of LBM are minimal.

Applications of LBM are widely varied. Most CO_2 industrial laser CNC machining centers can focus the beam down to a diameter of about 0.005 in. Applications for these systems range from cigarette paper cutting to drilling microholes in turbine engine blades to cutting steel plate for chain saw blades. For printed circuit board and chip-scale packaging applications, Nd:YAG, excimer, and now pulsed lasers are being used to drill holes down to 0.001 in. in milliseconds in polyimide or polyester films. However, this drilling is limited to rather thin stock (0.01 in.), as the cutting speed drops off rapidly with penetration into the material. Hole depth-to-diameter ratios of 10:1 are common, and hole geometry is irregular. Recast and heat-affected zones exist adjacent to the cuts. Deep UV excimer and Nd:YAG lasers are ideal for micromachining applications. Deep UV lasers use primarily a chemical ablation mechanism for material removal. In addition to providing the potential for eliminating thermal effects in machining, deep UV lasers (having lower wavelengths) may be focused to tighter diameters than lasers with higher wavelengths. While less powerful than infrared lasers, deep UV lasers (because of chemical ablation mechanisms) experience much greater energy efficiencies in cutting. Holes as small as 60 μin. (1.5 μm) in diameter have been machined in thin-film materials with excimer lasers.

In LBM, the wavelength used to process the material is mainly determined by the optical characteristics (reflectivity, absorptivity, and transmissivity) of the workpiece. Not all materials can be machined by all lasers. Protective eyeware is necessary when working around laser equipment because of the potential damage to eyesight from either direct or scattered laser light.

PLASMA ARC CUTTING

Plasma arc cutting (PAC) uses a superheated stream of electrically ionized gas to melt and remove material (Figure 28-18). The 20,000° to 50,000°F plasma is created inside a water-cooled nozzle by electrically ionizing a suitable gas such as nitrogen, hydrogen,

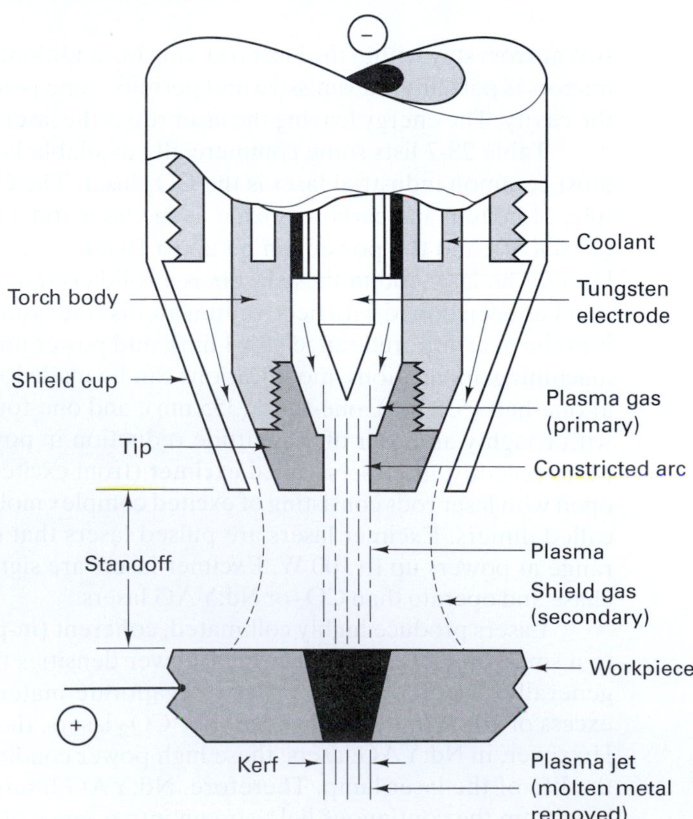

FIGURE 28-18 Plasma arc machining or cutting.

argon, or mixtures of these gases. The process can be used on almost any conductive metal. The plasma arc is a mixture of free electrons, positively charged ions, and neutral atoms. The arc is initiated in a confined gas-filled chamber by a high-frequency spark. The high-voltage, DC power sustains the arc, which exits from the nozzle at near-sonic velocity. The workpiece is electrically positive. The high-velocity gases melt and blow away the molten metal "chips." Dual-flow torches use a secondary gas or water shield to assist in blowing the molten metal out of the kerf, giving a cleaner cut. The process may be performed underwater, using a large tank to hold the plates being cut. The water assists in confining the arc and reducing smoke. The main advantage of PAC is speed. Mild steel $\frac{1}{4}$ in. thick can be cut at 125 in./min. Speed decreases with thickness. Greater nozzle life and faster cutting speeds accompany the use of water-injection-type torches. Control of nozzle stand-off from the workpiece is important. One electrode size can be used to machine a wide variety of materials and thicknesses by suitable adjustments to the power level, gas type, gas flow rate, traverse speed, and flame angle. PAC is sometimes called plasma-beam machining.

PAC can machine exotic materials at high rates. Profile cutting of metals, particularly of stainless steel and aluminum, has been the most prominent commercial application. However, mild steel, alloy steel, titanium, bronze, and most metals can be cut cleanly and rapidly. Multiple-torch cuts are possible on programmed or tracer-controlled cutting tables on plates up to 6 in. thick in stainless steel. Smooth cuts free from contaminants are a PAC advantage. Well-attached dross on the underside of the cut can be a problem, and there will be a heat-affected zone. The depth of the HAZ is a function of the metal, its thickness, and the cutting speed. Surface heat treatment and metal joining are beginning to use the plasma torch. See Chapter 30 for more discussion on plasma arc processes.

Some of the drawbacks of the PAC process include poor tolerances, tapered cuts, and double arcing, leading to premature wear on the nozzle. **Precision PAC,** also called **high-definition plasma** and **fine plasma cutting,** uses a special nozzle, where either a high-flow vortex or a magnetic field causes the plasma to spin rapidly and stabilizes the plasma pressure (Figure 28-19). The fast-spinning plasma results in a finely defined beam that cuts a narrow kerf with a perpendicular edge. Precision PAC also reduces the problems of HAZ and dross on the bottom of parts.

THERMAL DEBURRING

Thermal deburring, also known as the **thermal energy method,** has been developed for the removal of burrs and fins by exposing the workpiece to hot corrosive gases for a short period of time, typically on the order of a few milliseconds (Figure 28-20). The hot gases are formed by combusting (with a spark plug) explosive mixtures of oxygen and fuel (e.g., natural gas) in a chamber holding the workpieces. A thermal shock wave,

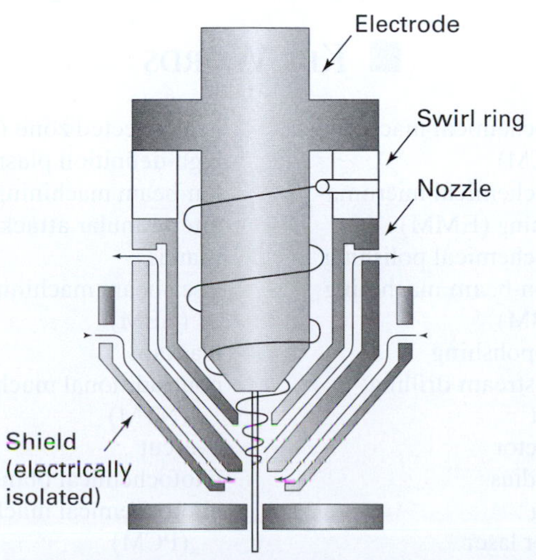

FIGURE 28-19 Precision plasma cutting nozzle.

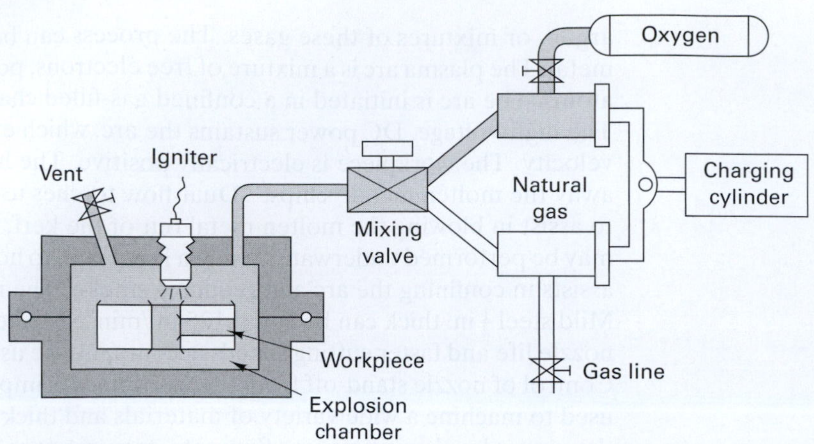

FIGURE 28-20 Thermochemical machining process for the removal of burrs and fins.

moving at Mach 8 (2700 m/s, or 6000 mph) with temperatures up to 3300°C (6000°F), vaporizes the burr in about 25 ms. Because of the intense heat, the burrs and fins are unable to dissipate the heat fast enough to the surrounding workpiece and, consequently, sublimate. The workpiece remains unaffected and relatively cool because of its low surface-to-mass ratio and the short exposure time. Small amounts of metal are removed from all exposed surfaces, and this must be permissible if the process is to be used. Consequently, while large burrs and fins can be removed with this process, the procedure can usually be used only for removing small burrs. Care must be taken with parts with thin cross sections. Maximum burr thickness should be about the thinnest feature on the workpiece.

Thermal deburring will remove burrs or fins from a wide range of materials, but it is particularly effective with materials of low thermal conductivity, which easily oxidize. It will deburr thermosetting plastics, but not thermoplastic materials. Any workpiece of modest size requiring manual deburring or flash removal should be considered a candidate for thermal deburring. Die castings, gears, valves, rifle bolts, and similar small parts are deburred readily, including blind, internal, and intersecting holes in inaccessible locations. Carburetor parts are processed in automated equipment. The major advantage of thermal deburring is that fine burrs are removed much more quickly and cheaply than if they were removed by hand. Uniformity of results and greater quality assurance over hand deburring are also advantages of thermal deburring. One outcome of thermal deburring is that the part is coated with a fine oxide dust that can be removed easily by solvents. Capital costs can be several hundred thousand dollars, and the maximum workpiece dimensions are on the order of 250 mm (10 in.) by 690 mm (27 in.). See Chapter 34 for additional discussions on deburring.

KEY WORDS

ablation
anisotropy
chemical blanking
chemical machining (CHM)
chemical milling
chip
cut-and-peel method
duty cycle
electrical discharge machining (EDM)
electrical discharge wire cutting
electrochemical deburring
electrochemical grinding (ECG)
electrochemical machining (ECM)
electrochemical micromachining (EMM)
electrochemical polishing
electron-beam machining (EBM)
electropolishing
electrostream drilling
etchant
etch factor
etch radius
etching
excimer laser
fine cutting plasma
gel milling
heat-affected zone (HAZ)
high-definition plasma
ion-beam machining (IBM)
intergranular attack
islands
laser-beam machining (LBM)
maskant
nontraditional machining (NTM)
overcut
photochemical blanking
photochemical machining (PCM)
photochemical milling
photoresists
plasma arc cutting (PAC)
plasma etching
precision PAC
pulsed-current ECM (PECM)
ram EDM
screen printing
shaped-tube electrolytic machining (STEM)
simulated emission
spark erosion
stepped machining
thermal deburring
thermal energy method
through-hole
wire EDM

■ Review Questions

1. How do the MRRs for most NTM processes compare to conventional metal cutting?
2. What are the steps in chemical machining using photosensitive resists?
3. In chemical machining, should the etchant be applied by spraying or immersion?
4. What are the advantages of chemical blanking over regular blanking using punch and die methods?
5. How are multiple depths of cut (steps) produced by chemical machining?
6. Would it be feasible to produce a groove 2 mm wide and 3 mm deep by chemical machining?
7. A drawing calls for making a groove 23 mm wide and 3 mm deep by chemical machining. What should be the width of the opening in the maskant?
8. Could an ordinary steel weldment be chemically machined? Why or why not?
9. How could you produce a tapered section by chemical machining?
10. What is the principal application of thermochemical machining?
11. How is ECM related to chemical machining?
12. What effect does work material hardness have on the metal removal rate in ECM?
13. What is the principal cause of tool wear in ECM?
14. Would electrochemical grinding be a suitable process for sharpening ceramic tools? Why or why not?
15. Upon what factors does the metal removal rate depend in ECM?
16. Why is the tool insulated in the ECM schematic?
17. What is the nature of the surface obtained by electrodischarge machining (EDM)?
18. What is the principal advantage of using a moving wire electrode in electrodischarge machining?
19. What effect would increasing the voltage have on the metal removal rate in electrodischarge machining? Why?
20. If a metal part is quite brittle and the part will be subjected to repeated tensile loads, would you select ECM or electrodischarge machining for making it? Why or why not?
21. If you had to make several holes in a large number of delicate parts, would you prefer ECM, EDM, EBM, or LBM? Why?
22. What process would you recommend to make many small holes in a very hard alloy where the holes will be used for cooling and venting?
23. Explain (using a little physics and metallurgy) why the "chips" in a thermal process like EDM are often hollow spheres.
24. What is undercutting?
25. What are some possible defects that can result from underetching? From overetching?
26. What are some other uses for the laser other than machining?
27. How does the laser produce coherent light (stimulated emission)?
28. What is ablation?
29. What is an excimer?
30. In Figure 28-16, what is the protective tape protecting?
31. Why is the EBM process done in a vacuum?

Chapter 28 CASE STUDY

Vented Cap Screws

The machine shop at the Fly Space Laboratories received an order from their engineering department for 10,000 of the vented cap screws shown in Figure CS-28A. Joseph, the machine shop foreman, returned the order to engineering, stating that there was no practical way to make these other than by hand. Carol, the design engineer, insisted that the vent slot, as designed, was essential to ensure no pressure buildup around the threads of the screw body in the intended application and that she knew they could be made because she had seen such cap screws.

After some discussions, Julia submitted a sketch of a redesigned part, Figure CS-28B, to manufacturing, now specifying that a vent hole be drilled parallel to the axis of the cap at location D. The small hole, with cross-sectional area equal to the previous 0.062 in. × 0.031 in. slot, would be drilled $\frac{1}{2}$ in. deep through the cap to connect to the longitudinal slot. Carol said that this would eliminate the right-angle slot geometry.

1. What problems will be incurred in the manufacture of the vented cap screw, as initially designed? As the manufacturing engineer, what do you think of this design? Why is the slot width 0.031 in., and how would you make it?
2. What is the major problem with the redesigned cap screw? What should the drill diameter be? How could it be made?
3. As the manufacturing engineer, working with the design engineer, can you suggest an alternative redesign that would be more practical to manufacture? Show a sketch of your redesigned cap screw and explain how you would make this new design.

Do you think you have a good solution to this case study? Send it to blackjt@eng.auburn.edu. Maybe your design will make the next edition instead of the Carol design.

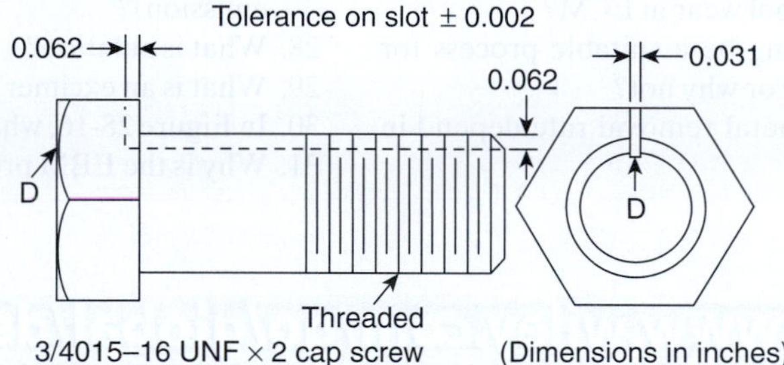

CHAPTER 29

FUNDAMENTALS OF JOINING

29.1 Introduction to Consolidation Processes

Large-size products, products with a high degree of shape complexity, or products with a wide variation in required properties are often manufactured as joined assemblies of two or more component pieces. These pieces may be smaller and therefore easier to handle, simpler shapes that are easier to manufacture, or segments that have been made from different materials. Assembly is an important part of the manufacturing process, and a wide variety of **consolidation processes** have been developed to meet the various needs.

Each of the methods has its own distinctive characteristics, strengths, and weaknesses. The metallurgical processes of welding, brazing, and soldering are usually used to join metals and often involve the solidification of molten material. The use of discrete fasteners (such as bolts and nuts, screws, and rivets) requires the creation of aligned holes and produces stress localization. While the holes may affect performance, disassembly and reassembly can often be performed with relative ease. Adhesive bonding has grown with new developments in polymeric materials and is being used extensively in automotive and aircraft production. Any material can be joined to any other material, and the low-temperature joining is particularly attractive for composite materials. Production rates are often low, however, because of the time required for the adhesive to develop full strength. Lesser-known joining techniques include shrink fits, slots and tabs, and a wide variety of other mechanical methods. From a technical viewpoint, powder metallurgy is another consolidation process, because the end product is built up by the joining of a multitude of individual particles.

Welding is the dominant method of joining in manufacturing, and a large fraction of products would have to be drastically modified if welding were not available. Our survey of consolidation or joining processes, therefore, will begin with a spectrum of techniques known by the generic term of **welding**—the permanent joining of two materials, usually metals, by **coalescence,** which is induced by a combination of temperature, pressure, and metallurgical conditions. The particular combination of these variables can range from high temperature with no pressure to high pressure with no increase in temperature. Because welding can be accomplished under such a wide variety of conditions, a number of different processes have been developed.

Coalescence requires sufficient proximity and activity between the atoms of the two pieces so as to create the formation of common crystals. The ideal metallurgical bond, for which there would be no noticeable or detectable interface, would require (1) perfectly smooth, flat, or matching surfaces; (2) surfaces that are clean and free from oxides, absorbed gases, grease, and other contaminants; and (3) the joining of single crystals with identical crystallographic structure and orientation alignment. These conditions would be difficult to obtain under laboratory conditions and are virtually impossible to achieve in normal production. Consequently, the various joining methods have been designed to overcome or compensate for the common deficiencies.

Surface roughness can be overcome either by force, causing plastic deformation and flattening of the high points, or by melting the two surfaces so that fusion occurs. The various processes also employ different approaches to cleaning the metal surfaces prior to welding and preventing further oxidation or contamination during the joining process. In solid-state welding, contaminated layers are generally removed by mechanical or chemical cleaning prior to welding or by causing sufficient metal flow along the interface so that new surface is created and existing impurities are displaced from the joint. In **fusion welding,** where molten material is produced and high temperatures accelerate the reactions between the metal and its surroundings, contaminants are often removed from the pool of molten metal through the use of fluxing agents. If welding is performed in a vacuum, the contaminants are removed much more easily, and coalescence is easier to achieve. In the vacuum of outer space, mating parts may weld under extremely light loads, even when welding is not intended.

When the process requires heat, the structure of the metal may be significantly altered. Melting and resolidification will certainly change structure. Even when no melting occurs, the heating and cooling of the welding process can affect the metallurgical structure and quality of both the weld and the adjacent material. Because many of the changes are detrimental, the possible consequences of heating and cooling should be a major consideration when selecting a joining process.

To produce a high-quality weld, we will need (1) a source of satisfactory heat and/or pressure; (2) a means of protecting or cleaning the metals to be joined; and (3) caution to avoid, or compensate for, harmful metallurgical effects. These aspects will be developed in the sections that follow.

29.2 Classification of Welding and Thermal Cutting Processes

Wherever possible, this text will utilize the nomenclature of the American Welding Society (AWS). The various welding processes have been classified in the manner presented in Figure 29-1, and letter symbols have been assigned to facilitate process designation. The variety of processes provide multiple ways of achieving coalescence, and

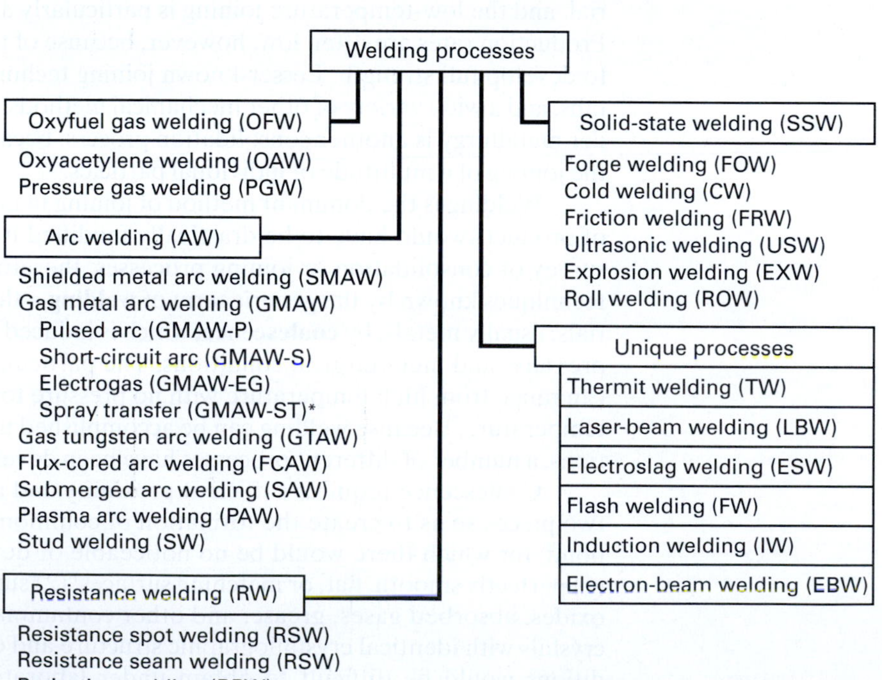

FIGURE 29-1 Classification of common welding processes along with their AWS (American Welding Society) designations.

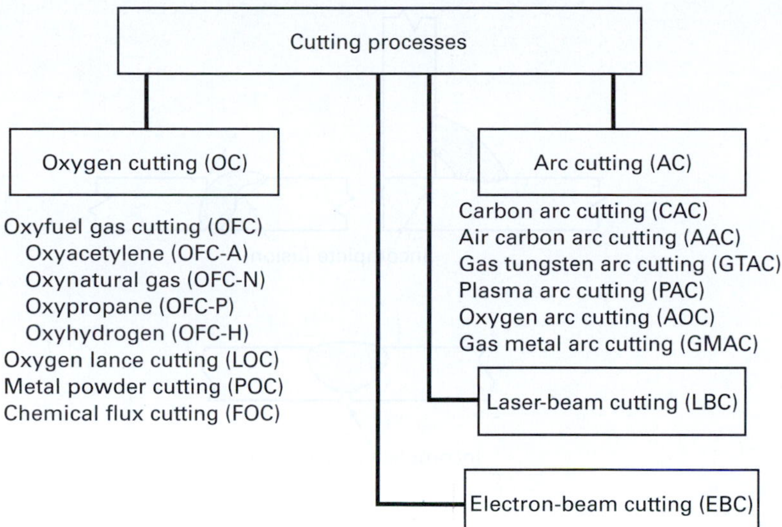

FIGURE 29-2 Classification of thermal cutting processes along with their AWS (American Welding Society) designations.

make it possible to produce effective and economical welds in nearly all metals and combinations of metals. Chapter 30 will present the gas and arc welding processes; Chapter 31 will cover resistance and solid-state welding; and Chapter 32 will present a variety of other processes, including brazing and soldering.

For many years, welding equipment (such as oxyfuel torches and electric arc units) has also been used to cut metal sheets and plates. Developed originally for salvage and repair work, then used for preparing plates for welding, this type of equipment is now widely used to cut sheets and plates into desired shapes for a variety of uses and operations. Laser and electron-beam equipment can now cut both metals and nonmetals at speeds up to 25 m/min (1000 in./min), with accuracies of up to 0.25 mm (0.01 in.). Figure 29-2 summarizes the commonly used **thermal cutting** processes and provides their AWS designations. Because cutting is often an adaptation of welding, the welding and cutting capabilities will be presented together as the individual processes are discussed.

29.3 SOME COMMON CONCERNS

Many of the problems that are inherent to welding and joining can be avoided by selecting the proper process with further consideration of both general and process-specific characteristics and requirements. Proper design of the joint is extremely critical. Heating, melting, and resolidification can produce drastic changes in the properties of base and filler materials. **Weld metal** properties can also be changed by dilution of the filler by melted base metal, vaporization of various alloy elements, and gas–metal reactions.

Various types of weld defects can also be produced. These include cracks in various forms, cavities (both gas and shrinkage), inclusions (slag, flux, and oxides), **incomplete fusion** between the weld and base metals, **incomplete penetration** (insufficient weld depth), unacceptable weld shape or contour, arc strikes, spatter, undesirable metallurgical changes (aging, grain growth, or transformations), and excessive distortion. Figure 29-3 depicts several of these defects.

29.4 TYPES OF FUSION WELDS AND TYPES OF JOINTS

Figure 29-4 illustrates four basic types of fusion welds. **Bead welds,** or **surfacing welds,** are made directly onto a flat surface and therefore require no edge preparation. Because the penetration depth is limited, bead welds are used primarily for joining thin sheets of metal, building up surfaces, and depositing hard-facing (wear-resistant) materials.

Groove welds are used when full-thickness strength is desired on thicker material. Some sort of edge preparation is required to form a groove between the abutting edges. V, double-V (top and bottom), U, and J (one-sided V) configurations are most common

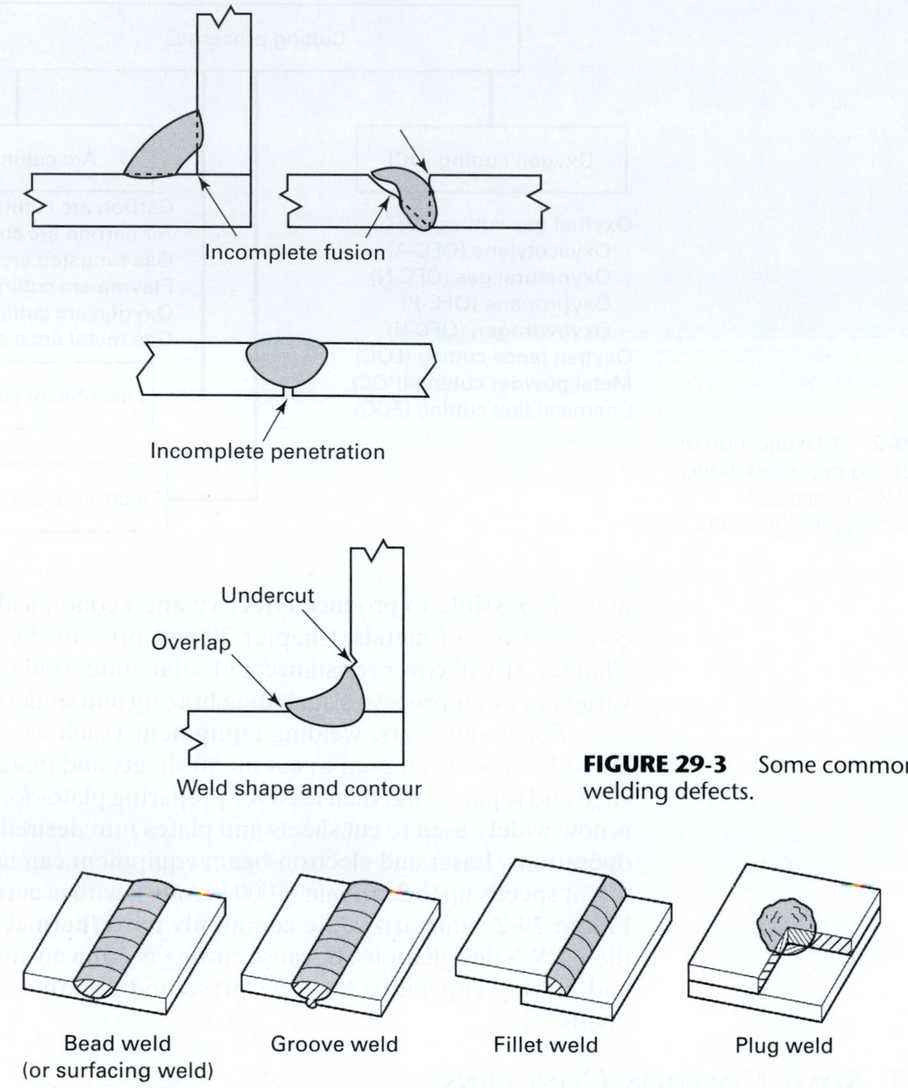

FIGURE 29-3 Some common welding defects.

FIGURE 29-4 Four basic types of fusion welds.

and are often produced by oxyacetylene flame cutting. The specific type of groove usually depends on the thickness of the joint, the welding process to be employed, and the position of the work. The objective is to obtain a sound weld throughout the full thickness with a minimum amount of additional weld metal. If possible, single-pass welding is preferred, but multiple passes may be required, depending on the thickness of the material and the welding process being used. As shown in Figure 29-5, special consumable **inserts** can be used to ensure proper spacing between the mating edges and good quality in the root pass. These inserts are particularly useful in pipeline welding and other applications where welding must be performed from only one side of the work.

Fillet welds are used for tee, lap, and corner joints, and require no special edge preparation. The size of the fillet is measured by the leg of the largest 45-degree right triangle that can be inscribed within the contour of the weld cross section. This is shown in Figure 29-6, which also depicts the proper shape for fillet welds to avoid excess metal deposition and reduce stress concentration.

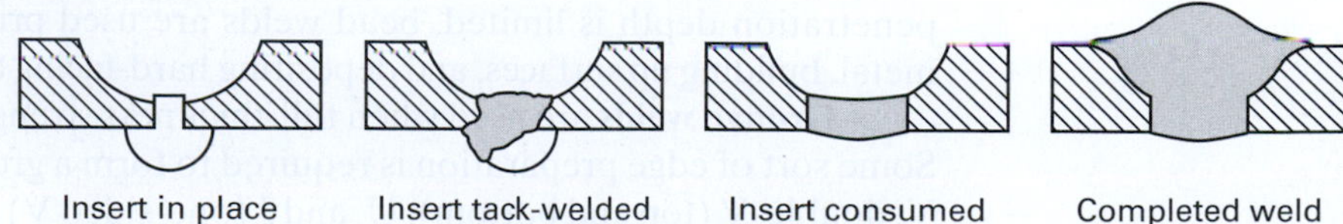

FIGURE 29-5 The use of a consumable backup insert in making a fusion weld. *(Courtesy Arcos Industries, Mount Carmel, PA)*

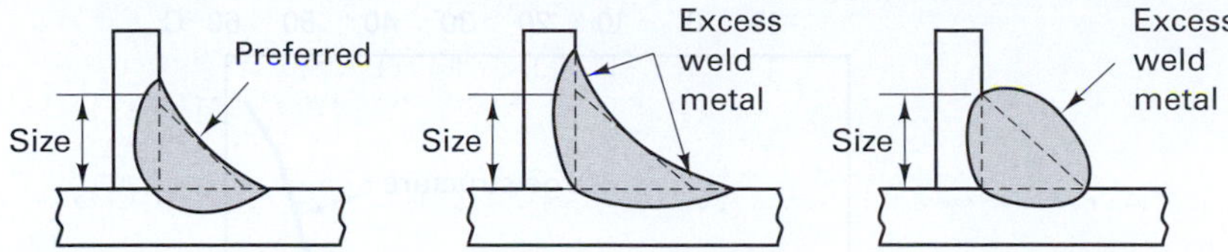

FIGURE 29-6 Preferred shape and the method of measuring the size of fillet welds.

Plug welds attach one part on top of another and are often used to replace rivets or bolts. A hole is made in the top plate and welding is started at the bottom of this hole.

Figure 29-7 shows five basic types of joints **(joint configurations)** that can be made with the use of bead, groove, and fillet welds, and Figure 29-8 shows some of the methods to construct these joints. In selecting the type of joint to be used, a primary consideration should be the type of loading that will be applied. A large portion of what are erroneously called "welding failures" can more accurately be attributed to inadequate consideration of loading. Cost and accessibility for welding are other important factors when specifying joint design but should be viewed as secondary to loading. Cost is affected by the amount of required edge preparation, the amount of weld metal that must be deposited, the type of equipment that must be used, and the speed and ease with which the welding can be accomplished.

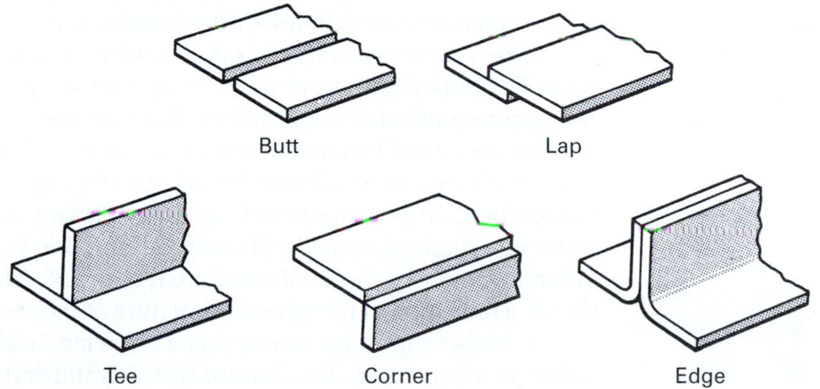

FIGURE 29-7 Five basic joint designs for fusion welding.

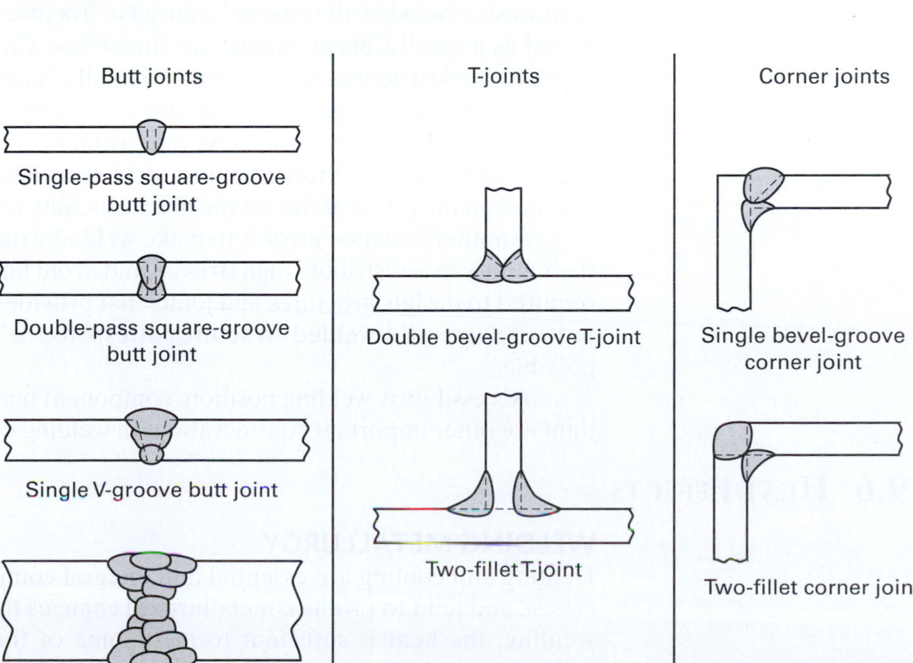

FIGURE 29-8 Various weld procedures used to produce welded joints. *(Courtesy Republic Steel Corporation, Youngstown, OH)*

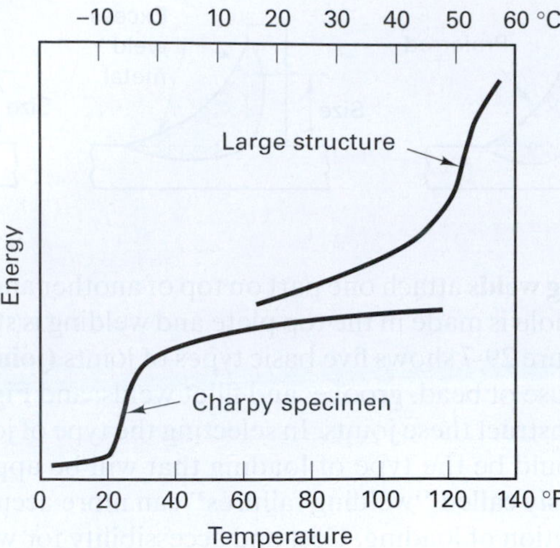

FIGURE 29-9 Effect of size on the transition temperature and energy-absorbing ability of a certain steel. While the larger structure absorbs more energy because of its size, it becomes brittle at a much higher temperature.

29.5 DESIGN CONSIDERATIONS

Welding is a unique process that cannot be directly substituted for other methods of joining without proper consideration of its particular characteristics and requirements. Unfortunately, welding is also easy and convenient, and the considerations of proper design and implementation are often overlooked.

One very important fact is that welding produces **monolithic,** or one-piece, structures. When two pieces are welded together, they become one continuous piece. This can cause significant complications. For example, a crack in one piece of a multipiece structure may not be catastrophic, because it will seldom progress beyond the single piece in which it occurs. However, when a large structure—such as a ship hull, pipeline, storage tank, or pressure vessel—consists of many pieces welded together, a crack that starts in a single plate or weld can propagate for a great distance and cause complete failure. Obviously, this kind of failure is not the fault of the welding process itself but is simply a reflection of the monolithic nature of the product.

It is also important to note that a given material in small pieces may not behave as it does in a larger size. This feature is clearly illustrated in Figure 29-9, which shows the relationship between the energy required to fracture and temperature for the same steel tested as a small Charpy impact specimen (see Chapter 3 and Figure 3-20) and as a large, welded structure. In the form of a small Charpy bar, the material exhibits ductile behavior and good energy absorption at temperatures down to 4°C (25°F). When welded into a large structure, however, brittle behavior is observed at temperatures as high as 43°C (110°F). More than one welded structure has failed because the designer overlooked the effect of size on the notch-ductility of metal.

Another common error is to make welded structures too rigid, thereby restricting their ability to redistribute high stresses and avoid failure. Considerable thought may be required to design structures and joints that provide sufficient flexibility, but the multitude of successful welded structures attests to the fact that such designs are indeed possible.

Accessibility, welding position, component match-up, and the specific nature of a joint are other important considerations in welding design.

29.6 HEAT EFFECTS

WELDING METALLURGY

Heating and cooling are essential and integral components of almost all welding processes, and tend to produce metallurgical changes that are often undesirable. In fusion welding, the heat is sufficient to melt some of the **base metal** (the material being welded), and this is often followed by a rapid cooling. Thermal effects tend to be most pronounced for this type of welding, but also exist to a lesser degree in processes where

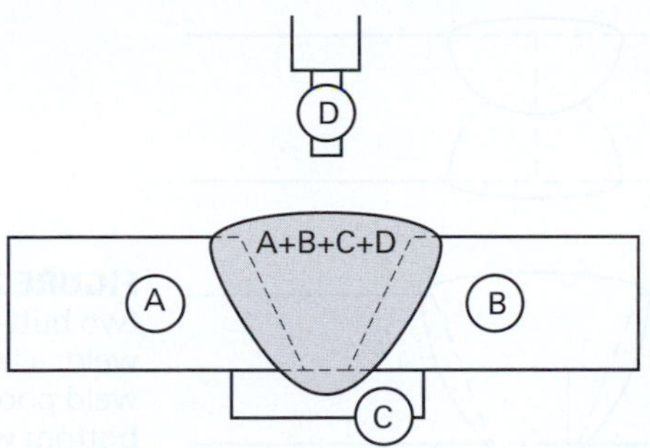

FIGURE 29-10 Schematic of a butt weld between a plate of metal A and a plate of metal B, with a backing plate of metal C and filler of metal D. The resulting weld nugget becomes a complex alloy of all four metals.

the heating–cooling cycle is less severe. If the thermal effects are properly considered, adverse results can usually be avoided or minimized, and excellent service performance can be obtained. If they are overlooked, however, the results can be disastrous.

Because such a wide range of metals are welded and a variety of processes are used, welding metallurgy is an extensive subject, and the material presented here serves only as an introduction. In fusion welding, a pool of molten metal is created, with the molten metal coming from either the parent plate alone **(autogenous welding)** or a mixture of parent and filler material. Figure 29-10 shows a butt weld between plates of material A and material B. A backing strip of material C is used with filler metal of material D. In this situation, the molten pool is actually a complex alloy of all four materials. The molten material is held in place by a metal "mold" formed by the surrounding solids. Because the molten pool is usually small compared to the surrounding metal, fusion welding can often be viewed as *a small metal casting in a large metal mold*. The resultant structure and its properties can be best understood by first analyzing the casting and then considering the effects of the associated heat on the adjacent base material.

Figure 29-11 shows a typical microstructure produced by a fusion weld. In the center of the weld is a region composed of metal that has solidified from the molten state. The material in this **weld pool,** or **fusion zone,** is actually a mixture of parent metal and electrode or filler metal, with the ratio depending upon the particular process, the type of joint, and the edge preparation. Figure 29-12 compares two butt weld designs where the weld pool in the upper design would contain a large percentage of base metal and the weld pool in the lower design would be largely filler material. The metal in the fusion zone is cast material with a microstructure reflecting the cooling rate of the weld.

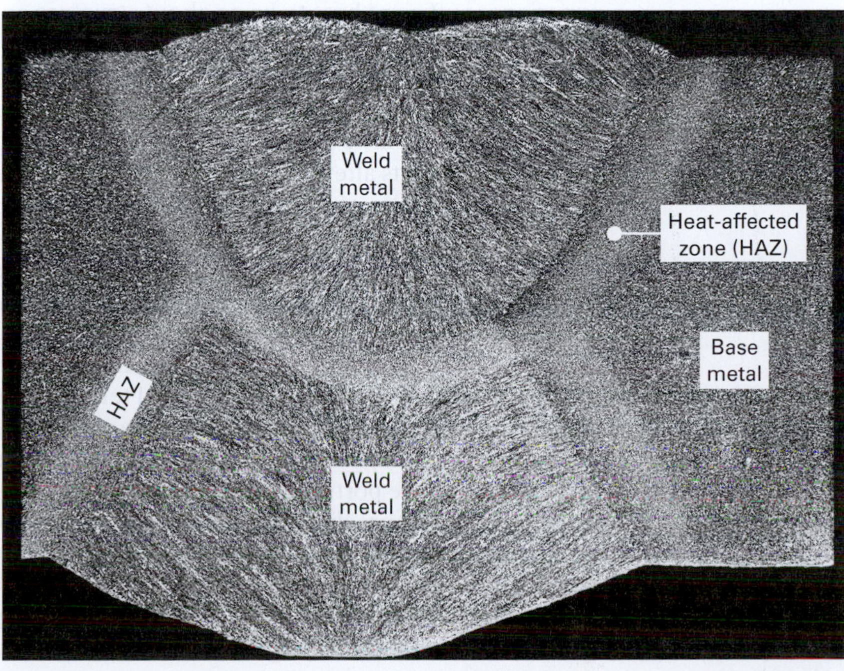

FIGURE 29-11 Grain structure and various zones in a fusion weld. *(Courtesy Ronald Kohser)*

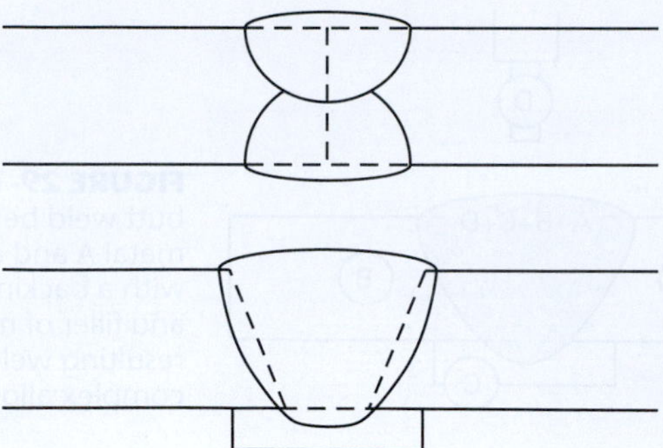

FIGURE 29-12 Comparison of two butt-weld designs. In the top weld, a large percentage of the weld pool is base metal. In the bottom weld, most of the weld pool is filler metal.

This region cannot be expected to have the same properties and characteristics as the wrought material being welded because their processing histories and resulting structures are usually different. Adequate mechanical properties, therefore, can only be achieved by selecting filler rods or electrodes, which have properties *in their as-deposited condition* that equal or exceed those of the wrought parent metal. It is not uncommon, therefore, for the filler metal to have a different chemistry than the metal being welded. The grain structure in the fusion zone may be fine or coarse, equiaxed or dendritic, depending on the type and volume of weld metal and the rate of cooling, but most electrode and filler rod compositions tend to produce fine, equiaxed grains. The matching or exceeding of base metal strength, in the as-solidified condition, is the basis for several AWS specifications for electrodes and filler rods.

The pool of molten metal created by fusion welding is prone to all of the problems and defects associated with metal casting, such as gas porosity, inclusions, blowholes, cracks, and shrinkage. Because the amount of molten metal is usually small compared to the total mass of the workpiece, rapid solidification and rapid cooling of the solidified metal are quite common. Associated with these conditions may be the entrapment of dissolved gases, chemical segregation, grain-size variation, grain shape problems, and orientation effects.

Adjacent to the fusion zone, and wholly within the base material, is the ever-present and generally undesirable **heat-affected zone (HAZ).** In this region, the parent metal has not melted but has been subjected to elevated temperatures for a brief period of time. Because the temperature and its duration vary widely with location, fusion welding might be more appropriately described as "a metal casting in a metal mold, coupled with an abnormal and widely varying heat treatment." The adjacent metal may experience sufficient heat to bring about structure and property changes, such as phase transformations, recrystallization, grain growth, precipitation or precipitate coarsening, embrittlement, or even cracking. The variation in thermal history can produce a variety of microstructures and a range of properties. In steels, the structures can range from hard, brittle martensite all the way through coarse pearlite and ferrite.

Because of its altered structure, the heat-affected zone may no longer possess the desirable properties of the parent material, and because it was not melted, it cannot assume the properties of the solidified weld metal. Consequently, this is often the weakest area in the as-welded joint. Except where there are obvious defects in the weld deposit, most welding failures originate in the heat-affected zone. This region extends outward from the weld to the location where the base metal has experienced too little heat to be affected or altered by the welding process. Figure 29-13 presents a schematic of a fusion weld in steel and uses standard terminology for the various regions and interfaces. Part of the heat-affected zone has been heated above the A_1 transformation temperature and could assume a totally new structure through phase transformation. The lower temperature portion of the heat-affected zone (peak temperatures below the A_1 value) can experience diffusion-induced changes within the original structure.

Because of the melting, solidification, and exposure to a range of high temperatures, the structure and properties of welds can be extremely complex and varied. Through proper concern, however, associated problems can often be reduced or totally eliminated. Consideration should first be given to the thermal characteristics of the various processes.

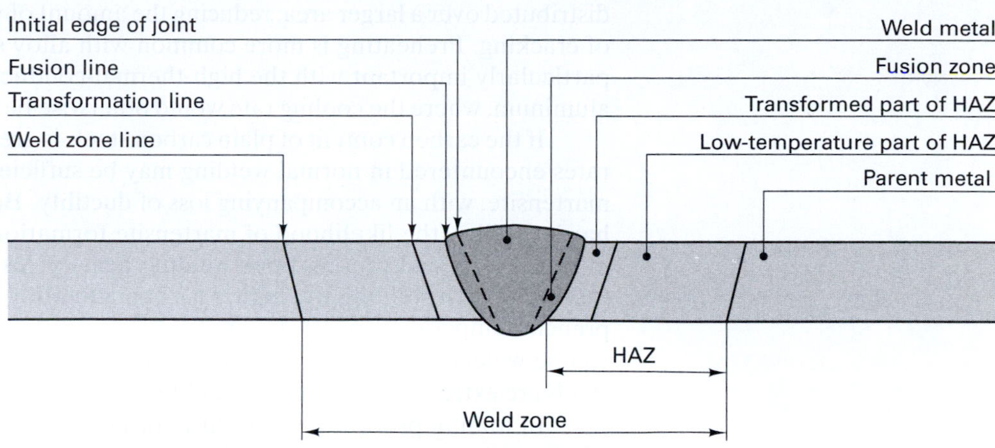

FIGURE 29-13 Schematic of a fusion weld in steel, presenting proper terminology for the various regions and interfaces. Part of the heat-affected zone has been heated above the transformation temperature and will form a new structure upon cooling. The remaining segment of the heat-affected zone experiences heat alteration of the initial structure. *(Courtesy Sandvik AB, Sandviken, Sweden)*

TABLE 29-1 Classification of Common Welding Processes by Rate of Heat Input

Low Rate of Heat Input	High Rate of Heat Input
Oxyfuel welding (OFW)	Plasma arc welding (PAW)
Electroslag welding (ESW)	Electron-beam welding (EBW)
Flash welding (FW)	Laser welding (LBW)
	Spot and seam resistance welding (RW)
	Percussion welding
Moderate Rate of Heat Input	
Shielded metal arc welding (SMAW)	
Flux cored arc welding (FCAW)	
Gas metal arc welding (GMAW)	
Submerged arc welding (SAW)	
Gas tungsten arc welding (GTAW)	

Table 29-1 classifies some of the more common welding processes with regard to their **rate of heat input.** Processes with low rates of heat input (slow heating) tend to produce large total heat content within the metal, slow cooling rates, large heat-affected zones, and resultant structures with lower strength and hardness, but higher ductility. High-heat-input processes, on the other hand, have low total heats, fast cooling rates, and small heat-affected zones. The size of the heat-affected zone will also increase with increased starting temperature, decreased welding speed, increased thermal conductivity of the base metal, and a decrease in base metal thickness. Weld geometry is also important, with fillet welds producing smaller heat-affected zones than butt welds.

If the as-welded properties are unacceptable, the entire welded assembly might be heat-treated after welding. Structure variations can be reduced or eliminated, but the results are restricted to those that can be produced through heat treatment. The structures and properties associated with cold working, for example, could not be achieved. In addition, problems may be encountered in trying to achieve controlled heating and cooling within the large, complex-shaped structures commonly produced by welding. Moreover, furnaces, quench tanks, and related equipment may not be available to handle the full size of welded assemblies.

An alternative technique to reduce microstructural variation, or the sharpness of that variation, is to preheat either the entire base metal or material at least 10 cm (4 in.) on either side of the joint just prior to welding. This heating serves to reduce the cooling rate of both the weld deposit and the immediately adjacent metal in the heat-affected zone. The slower cooling produces a softer, more ductile structure and provides more time for the out-diffusion of harmful dissolved hydrogen. The welding stresses are distributed over a larger area, reducing the amount of weld distortion and the possibility of cracking. Preheating is more common with alloy steels and thicker sections and is particularly important with the high-thermal-conductivity metals, such as copper and aluminum, where the cooling rate would otherwise be extremely rapid.

If the carbon content of plain carbon steels is greater than about 0.3%, the cooling rates encountered in normal welding may be sufficient to produce hard, untempered martensite, with an accompanying loss of ductility. Because alloy steels possess higher hardenability, the likelihood of martensite formation will be even greater with these materials. Special pre- and postwelding heat cycles (**preheat** and **postheat**) may be required when welding the higher carbon and alloy steels. For plain carbon steels, a preheat temperature of 100 to 200°C (200 to 400°F) is usually adequate. Because they can be welded without the need for preheating or postheating, low-carbon, low-alloy steels are extremely attractive for welding applications.

In joining processes where little or no melting occurs, considerable pressure is often applied to the heated metal (as in forge or resistance welding). The weld region experiences deformation, and the resultant structure exhibits the characteristics of a wrought material.

Because steel is the primary metal that is welded, our discussion of metallurgical effects has largely focused on steel. It should be noted, however, that other metals also exhibit heat-related changes in their structure and properties. The exact effects of the heating and cooling associated with welding will depend on the specific transformations and structural changes that can occur within the materials being joined.

THERMAL EFFECTS IN BRAZING AND SOLDERING

In brazing and soldering, there is no melting of the base metal, but the joint still contains a region of solidified liquid and heat-affected sections within the base material. For these processes, however, another thermal effect may become quite significant. The base and filler metals are usually of radically different chemistries, and the elevated temperatures of joining also promote interdiffusion. Intermetallic phases can form at interfaces and alter the properties of the joint. If present in small amounts, they can enhance bonding and provide strength reinforcement. Most **intermetallic compounds,** however, are extremely brittle. Too much intermetallic material can result in the formation of continuous layers with significant loss of both strength and ductility.

THERMAL-INDUCED RESIDUAL STRESSES

Another effect of heating and cooling is the introduction of **residual stresses.** In welding, these may be of two types and are most pronounced in fusion welding, where the greatest amount of heating occurs. Their effects can be observed in the form of dimensional changes, distortion, and cracking.

Residual welding stresses are the result of restraint to thermal expansion and contraction by the pieces being welded. Consider a rectangular bar of metal that is uniformly heated and cooled. When heated, the material expands and becomes larger in length, width, and thickness. Upon cooling, the material contracts, and each dimension returns to its original value. Now insert the bar between rigid restraints so that lengthwise expansion cannot take place and repeat the thermal cycle. Upon heating, all of the expansion is restricted to the width and thickness, but the contraction upon cooling will still occur uniformly. The resulting rectangle will be shorter, thicker, and wider than the original specimen.

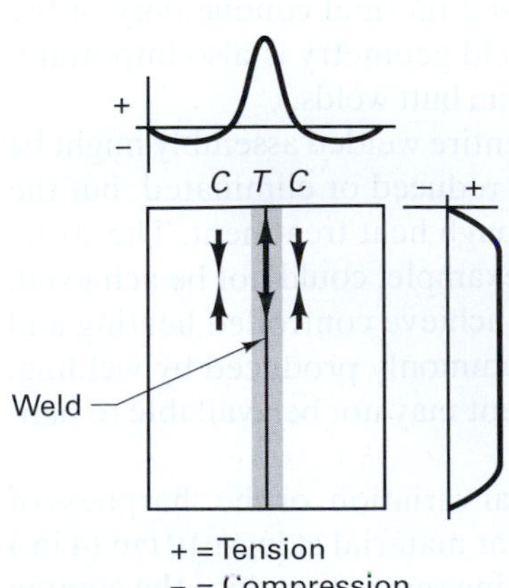

FIGURE 29-14 Schematic of the longitudinal residual stresses in a fusion-welded butt joint.

Now consider a weld being made between two flat plates, as illustrated in Figure 29-14. As the weld is produced, the liquid region conforms to the shape of the "mold," and the adjacent material becomes hot and expands (the heat-affected zone). The molten pool can absorb expansion of the plate perpendicular to the weld line, but expansion along the length of the weld tends to be restrained by the nearby plate material that has remained cooler and stronger. This resistance or restraint is often sufficient to induce deformation of the hot, weak, and thermally expanding heat-affected zone, which now becomes thicker instead of longer.

After the weld pool solidifies, both the weld metal and adjacent heat-affected region cool and contract. The surrounding metal, however, resists this contraction,

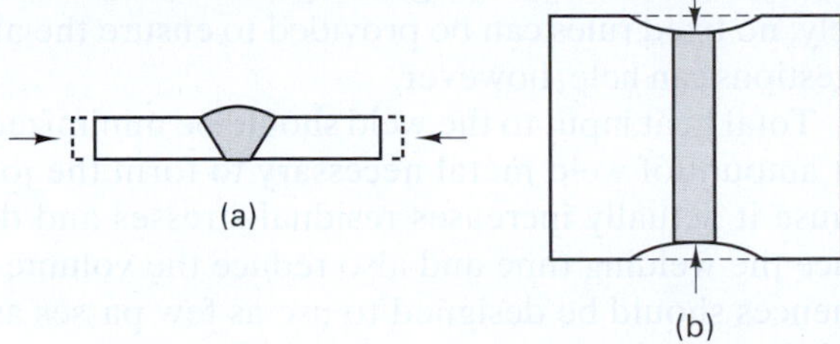

FIGURE 29-15 Shrinkage of a typical butt weld in the transverse (a) and longitudinal (b) directions as the material responds to the induced stresses. Note that restricting transverse motion will place the entire weld in transverse tension.

forcing the weld region remain in a "stretched" condition, known as residual tension (region *T*). The cooling weld, in turn, exerts forces on the adjacent material, producing regions of residual compression (region *C*). While the net force must remain at zero (in keeping with the equilibrium laws of physics and mechanics), the localized tensions and compressions can be substantial. As Figure 29-14 depicts, a high longitudinal residual tension is observed in the weld metal, which becomes longitudinal compression and then returns to zero as one moves away from the weld centerline. The magnitude of the residual tension will be relatively uniform along the weld line, except at the ends where the stresses can be relieved by a pulling in of the edges.

During cooling, the thermal contractions occur both parallel (longitudinal) and perpendicular (transverse) to the weld line. Lateral movement of the material being welded can often compensate for the transverse contractions. The width of the welded assembly simply becomes less than that of the positioned components at the time of welding. Figure 29-15a depicts this reduction in width, while Figure 29-15b illustrates the longitudinal contractions that generate the complex stresses of Figure 29-14.

Components being joined during fabrication typically have considerable freedom of movement, but welds made on nearly completed structures or repair welds often join components that are somewhat restrained. If the welded plates in Figure 29-15a are restrained from horizontal movement, *additional stresses will be induced*. These residual stresses are known as **reaction stresses**, and they can cause cracking of the hot weld or heat-affected material or can contribute to failure during subsequent use. Their magnitude will be an inverse function of the length between the weld joint and the point of restraint and can be as high as the yield strength of the parent metal (because yielding would occur to relieve any higher stresses).

EFFECTS OF THERMAL STRESSES

Distortion or warping can easily result from the nonuniform temperatures and thermal stresses induced by welding. Figure 29-16 depicts the some of the distortions that can

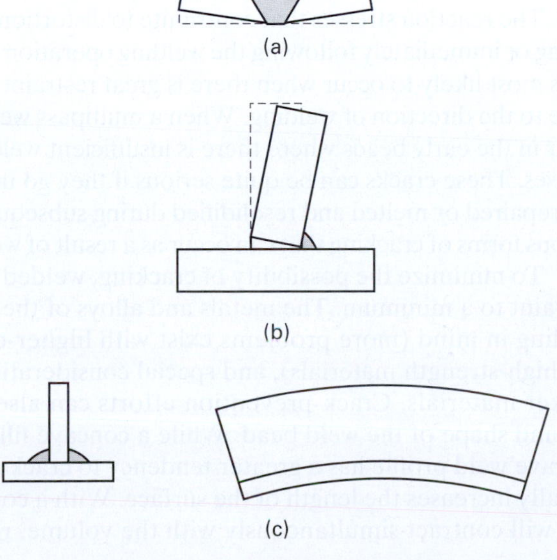

FIGURE 29-16 Distortions or warpage that may occur as a result of welding operations: (a) V-groove butt weld where the top of the joint contracts more than the bottom; (b) one-side fillet weld in a T-joint; (c) two-fillet weld T-joint with a high vertical web.

occur during various welding configurations. Because the causing conditions can vary widely, no fixed rules can be provided to ensure the absence of warping. The following suggestions can help, however.

Total heat input to the weld should be minimized. Welds should be made with the least amount of weld metal necessary to form the joint. Overwelding is not an asset, because it actually increases residual stresses and distortion. Faster welding speeds reduce the welding time and also reduce the volume of metal that is heated. Welding sequences should be designed to use as few passes as possible, and the base material should be permitted to have a high freedom of movement. When constructing a multi-weld assembly, it is beneficial to weld toward the point of greatest freedom, such as from the center to the edge.

The initial components can also be oriented out of position, so that the subsequent distortion will move them to the desired final shape. Another common procedure is to completely restrain the components during welding, thereby forcing some plastic flow in the joint and surrounding material. This procedure is used most effectively on small weldments where the reaction stresses will not be high enough to cause cracking.

Still another procedure is to balance the resulting thermal stresses by depositing the weld metal in a specified pattern, such as short lengths along a joint (intermittent or stitch welding) or on alternating sides of a plate (such as a double-V weld on thick plate). Warping can also be reduced by the use of **peening.** As the weld bead surface is hammered with the peening tool or material, the metal is flattened and tries to spread. Being held back by the underlying material, the surface becomes compressed or squeezed. Surface rolling of the weld-bead area can have the same effect. In both processes, the compressive stresses induced by the surface deformation serve to offset the tensile stresses induced by welding.

Residual stresses should not have a harmful effect on the strength of weldments, except in the presence of notches or in very rigid structures where no plastic flow can occur. These two conditions should not exist if the welds have been properly designed and proper workmanship has been employed. Unfortunately, it is easy to inadvertently join heavy sections and produce rigid configurations that will not permit the small amounts of elastic or plastic movement required to reduce highly concentrated stresses. In addition, geometric notches, such as sharp interior corners, are often incorporated into welded structures. Other harmful "notches"—such as gas pockets, rough beads, porosity, and arc "strikes"—can serve as initiation sites for weld failures. These can generally be avoided by proper welding procedures, good workmanship, and adequate supervision and inspection.

The residual stresses of welding can also cause additional distortion when subsequent machining removes metal and upsets the stress equilibrium. For this reason, welded assemblies that are to undergo subsequent machining are frequently given a **stress-relief** heat treatment prior to that operation.

The reaction stresses that contribute to distortion are often associated with cracking during or immediately following the welding operation (as the weld is cooling). This cracking is most likely to occur when there is great restraint to the shrinkage that occurs transverse to the direction of welding. When a multipass weld is being made, cracking tends to occur in the early beads where there is insufficient weld metal to withstand the shrinkage stresses. These cracks can be quite serious if they go undetected and are not chipped out and repaired or melted and resolidified during subsequent passes. Figure 29-17 shows the various forms of cracking that can occur as a result of welding.

To minimize the possibility of cracking, welded joints should be designed to keep restraint to a minimum. The metals and alloys of the structure should be selected with welding in mind (more problems exist with higher-carbon steels, higher-alloy steels, and high-strength materials), and special consideration should be given when welding thicker materials. Crack-prevention efforts can also include maintaining the proper size and shape of the weld bead. While a concave fillet is desirable when machining, a concave weld profile has a greater tendency to crack upon cooling because contraction actually increases the length of the surface. With a convex profile, the length of the surface will contract simultaneously with the volume, reducing the possibility of surface

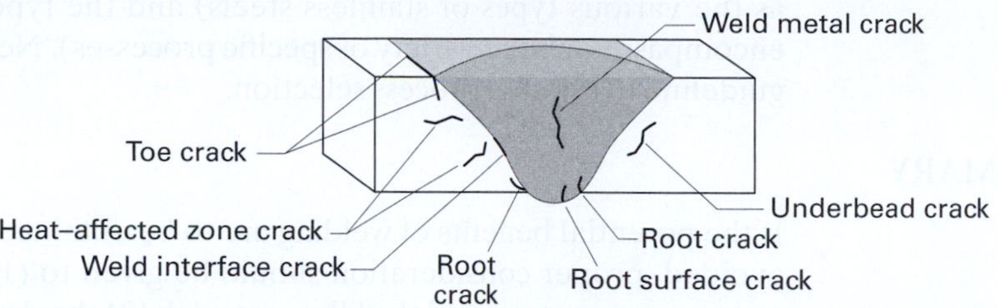

FIGURE 29-17 Various types and locations of cracking that can occur as a result of welding.

tension and cracking. Weld beads with high penetration (high depth/width ratio) are also more prone to cracking.

Still other methods to suppress cracking focus on reducing the stresses by making the cooling more uniform or relaxing the stresses by promoting plasticity in the metals being welded. The metals to be welded may be preheated and additional heat may be applied between the welding passes to retard cooling. Some welding codes also require the inclusion of a thermal stress relief (postweld heat treatment) after welding but prior to use. Hydrogen dissolved in the molten weld metal can also induce cracking. Slower welding and cooling or a post-weld heating will allow any hydrogen to escape, and the use of low-hydrogen electrodes and low-moisture fluxes will reduce the likelihood of hydrogen being present. Welding stresses can also be relieved through a postweld vibratory stress relief.

29.7 WELDABILITY OR JOINABILITY

It is important to note that not all joining processes are compatible with all engineering materials. While the terms **weldability** or **joinability** imply a reliable measure of a material's ability to be welded or joined, they are actually quite nebulous. One process might produce excellent results when applied to a given material, whereas another may produce a dismal failure. Within a given process, the quality of results may vary greatly with variations in the process parameters, such as electrode material, shielding gas, welding speed, and cooling rate.

Table 29-2 presents the compatibility of the various joining processes with some of the major classes of engineering materials. In each case, the process is classified as recommended (*R*), commonly performed (*C*), performed with some difficulty (*D*), seldom used (*S*), and not used (*N*). It should be noted, however, that the classifications are generalizations, and exceptions often exist within both the family of materials (such

TABLE 29-2 Weldability or Joinability of Various Engineering Materials[a]

Material	Arc Welding	Oxyacetylene Welding	Electron-Beam Welding	Resistance Welding	Brazing	Soldering	Adhesive Bonding
Cast iron	C	R	N	S	D	N	C
Carbon and low-alloy steel	R	R	C	R	R	D	C
Stainless steel	R	C	C	R	R	C	C
Aluminum and magnesium	C	C	C	C	C	S	R
Copper and copper alloys	C	C	C	C	R	R	C
Nickel and nickel alloys	R	C	C	R	R	C	C
Titanium	C	N	C	C	D	S	C
Lead and zinc	C	C	N	D	N	R	R
Thermoplastics	Heated tool R	Hot gas R	N	Induction C	N	N	C
Thermosets	N	N	N	N	N	N	C
Elastomers	N	N	N	N	N	N	R
Ceramics	N	S	C	N	N	N	R
Dissimilar metals	D	D	C	D	D/C	R	R

[a] R, recommended (easily performed with excellent results); C, commonly performed; D, difficult; S, seldom used; N, not used.

as the various types of stainless steels) and the types of processes (arc welding here encompasses a large variety of specific processes). Nevertheless, the table can serve as a guideline to assist in process selection.

29.8 SUMMARY

If the potential benefits of welding are to be obtained and harmful side effects are to be avoided, proper consideration should be given to (1) the selection of the process, the process parameters, and the filler material; (2) the design of the joint; and (3) the effects of heating and cooling on both the weld and parent material. Joint design should consider manufacturability, durability, fatigue resistance, corrosion resistance, and safety. Welding metallurgy helps determine the structure and properties across the joint and the need for additional thermal treatments. Further attention may be required to control or minimize residual stresses and distortion.

Parallel considerations apply to brazing and soldering operations, with additional attention to the effects of interdiffusion between the filler and base metals. Flame and arc-cutting operations involve localized heating and cooling, and they also create altered structures in the heat-affected zone. Because many cut products undergo further welding or machining, however, the regions of undesirable structure may not be retained in the final product.

KEY WORDS

autogenous weld
base metal
bead weld (or surfacing weld)
coalescence
consolidation processes
distortion
fillet weld
fusion welding
fusion zone
groove weld
heat-affected zone (HAZ)
incomplete fusion
incomplete penetration
insert
intermetallic compound
joinability
joint configuration
monolithic
peening
plug weld
postheat
preheat
rate of heat input
reaction stresses
residual stresses
residual welding stresses
stress relief
surfacing weld
thermal cutting
welding
weldability
weld metal
weld pool

REVIEW QUESTIONS

1. What types of design features favor manufacture as a joined assembly?
2. What types of manufacturing processes fall under the classification of consolidation processes?
3. Define *welding*.
4. What conditions are required to produce an ideal metallurgical bond?
5. What are some of the ways in which welding processes compensate for the inability to meet the conditions of an ideal bond?
6. What are some possible problems associated with the high temperatures that are commonly used in welding?
7. What are the three primary aspects required to produce a high-quality weld?
8. How are welding processes identified by the American Welding Society?
9. What is thermal cutting?
10. What are some of the common types of weld defects?
11. What are the four basic types of fusion welds?
12. What is the role of an insert in welding?
13. What types of weld joints commonly employ fillet welds?
14. What are the five basic joint types for fusion welding?
15. What are some of the factors that influence the cost of making a weldment?
16. Why is it important to consider welded products as monolithic structures?
17. How does the fracture resistance and temperature sensitivity of a steel vary with changes in material thickness?
18. How might excessive rigidity actually be a liability in a welded structure?
19. What is autogenous welding?
20. In what way is the weld-pool segment of a fusion weld like a small metal casting?
21. Why is it possible for the fusion zone to have a chemistry that is different from the filler metal?
22. Why is it not uncommon for the selected filler metal to have a chemical composition that is different from the material being welded?
23. What are some of the defects or problems that can occur in the molten metal region of a fusion weld?
24. Why can the resulting material properties vary widely within welding heat-affected zones?
25. What are some of the structure and property modifications that can occur in welding heat-affected zones?
26. Why do most welding failures occur in the heat-affected zone?
27. What are some of the characteristics and consequences of welding with processes that have low rates of heat input?

28. What process features can increase the size of the heat-affected zone?
29. What are some of the difficulties or limitations encountered in heat-treating large, complex structures after welding?
30. What is the purpose of pre- and postheating in welding operations?
31. What heat-related metallurgical effects may produce adverse results when brazing or soldering?
32. What causes weld-induced residual stresses?
33. What are some of the undesirable consequences of residual stresses?
34. What is the cause of reaction-type residual stresses?
35. How are reaction stresses affected by the distance between the weld and the point of fixed constraint?
36. What are some of the techniques that can reduce the amount of distortion in a welded structure?
37. Under what conditions might residual stresses have a harmful effect on load-bearing abilities?
38. In what ways might welding create geometric notches in a welded structure?
39. Why might a welded structure warp if the structure is machined after welding?
40. What are some of the techniques that can be employed to reduce the likelihood of cracking in a welded structure?
41. Why are the terms *weldability* and *joinability* somewhat nebulous?

Problems

1. Through the 1940s, the hulls of ocean-going freighters were constructed by riveting plates of steel together. When the defense efforts of World War II demanded accelerated production of freighters to supply U.S. troops overseas, construction of the hulls was converted to welding. The resulting Liberty Ships proved quite successful but also drew considerable attention when minor or moderate impacts (usually under low-temperature conditions) produced cracks of lengths sufficient to scuttle the ship, often up to 15 m (50 ft) or more. Because the material was essentially the same and the only significant process change had been the conversion from riveting to welding, the welding process was blamed for the failures.
   a. Is this a fair assessment?
   b. What do you think may have contributed to the problem?
   c. What evidence might you want to gather to support your beliefs?
2. Two pieces of AISI 1025 steel are being shielded-metal-arc welded with E6012 electrodes. Some difficulty is being experienced with cracking in the weld beads and in the heat-affected zones. What possible corrective measures might you suggest?
3. Figure 29-A schematically depicts the design of a go-cart frame with cross bars and seat support. The assembly is to be constructed from hot-rolled, low-carbon, box-channel material with miter, butt, and fillet welds at the 12 numbered joints. Due to the solidification shrinkage and subsequent thermal contraction of the joint material, the welds are best made when one or more of the sections are unrestrained. If the structure is too rigid at the time of welding, the associated dimensional changes are restricted, causing the generation of residual stresses that can lead to distortion, cracking, or tears.
   a. Consider the 12 welds in the proposed structure and recommend a welding sequence that would minimize the possibility of hot tears and cracks due to the welding of a restrained joint.
   b. Your company is developing a computer-assisted design program. Suggest one or more rules that may be programmed to aid in the selection of an acceptable weld sequence.

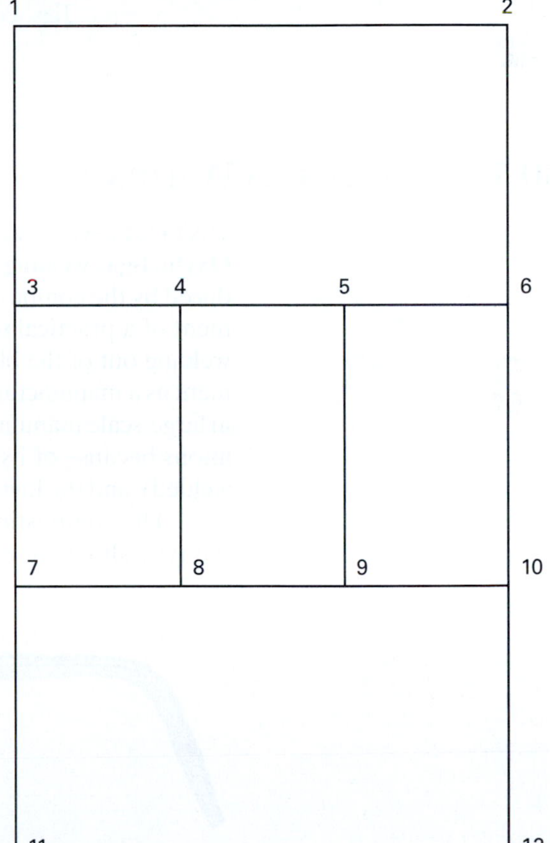

FIGURE 29-A

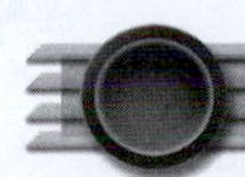

CHAPTER 30

GAS FLAME AND ARC PROCESSES

30.1 OXYFUEL-GAS WELDING

OXYFUEL-GAS WELDING PROCESSES

Oxyfuel-gas welding (OFW) refers to a group of welding processes that use the flame produced by the combustion of a fuel gas and oxygen as the source of heat. It was the development of a practical **torch** to burn acetylene and oxygen, shortly after 1900, that brought welding out of the blacksmith's shop, demonstrated its potential, and started its development as a manufacturing process. Other processes have largely replaced gas-flame welding in large-scale manufacturing, but the process is still popular for small-scale and repair operations because of its portability, versatility (most ferrous and nonferrous metals can be welded), and the low capital investment required. Acetylene is still the principal fuel gas.

The combustion of oxygen and **acetylene** (C_2H_2) by means of a welding torch of the type shown in Figure 30-1 produces a temperature of about 3250°C (5850°F) in a

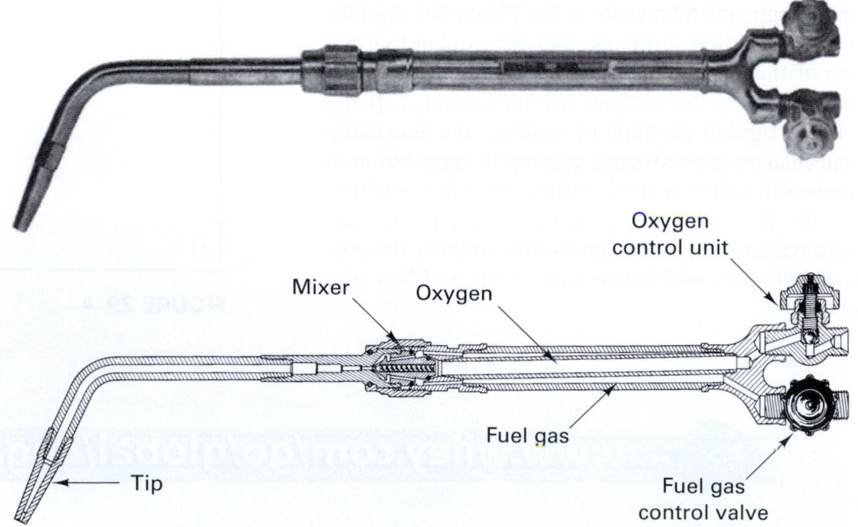

FIGURE 30-1 Typical oxyacetylene welding torch and cross-sectional schematic. *(Courtesy of Thermadyne Industries, Inc., St. Louis, MO)*

two-stage reaction. In the first stage, the supplied oxygen and acetylene react to produce carbon monoxide and hydrogen:

$$C_2H_2 + O_2 \rightarrow 2CO + H_2 + \text{Heat}$$

This reaction occurs near the tip of the torch and generates intense heat. The second stage of the reaction involves the combustion of the CO and H_2 and occurs just beyond the first combustion zone. The specific reactions of the second stage are:

$$2CO + O_2 \rightarrow 2CO_2 + \text{Heat}$$

$$H_2 + \tfrac{1}{2}O_2 \rightarrow H_2O + \text{Heat}$$

The oxygen for these secondary reactions is generally obtained from the surrounding atmosphere.

The two-stage combustion process produces a flame having two distinct regions. As shown in Figure 30-2, the maximum temperature occurs near the end of the inner cone, where the first stage of combustion is complete. Most welding should be performed with the torch positioned so that this point of maximum temperature is just above the metal being welded. The outer envelope of the flame serves to preheat the metal and, at the same time, provides shielding from oxidation, because oxygen from the surrounding air is consumed in the secondary combustion.

Three different types of flames can be obtained by varying the oxygen-to-acetylene (or oxygen-to-fuel gas) ratio. If the ratio is between 1:1 and 1.15:1, all reactions are carried to completion and a **neutral flame** is produced. Most welding is done with a neutral flame, because it will have the least chemical effect on the heated metal.

A higher ratio, such as 1.5:1, produces an **oxidizing flame,** which is hotter than the neutral flame (about 3600°C, or 6000°F) but similar in appearance. Such flames are used when welding copper and copper alloys but are generally considered harmful when welding steel because the excess oxygen reacts with the carbon in the steel, lowering the carbon in the region around the weld.

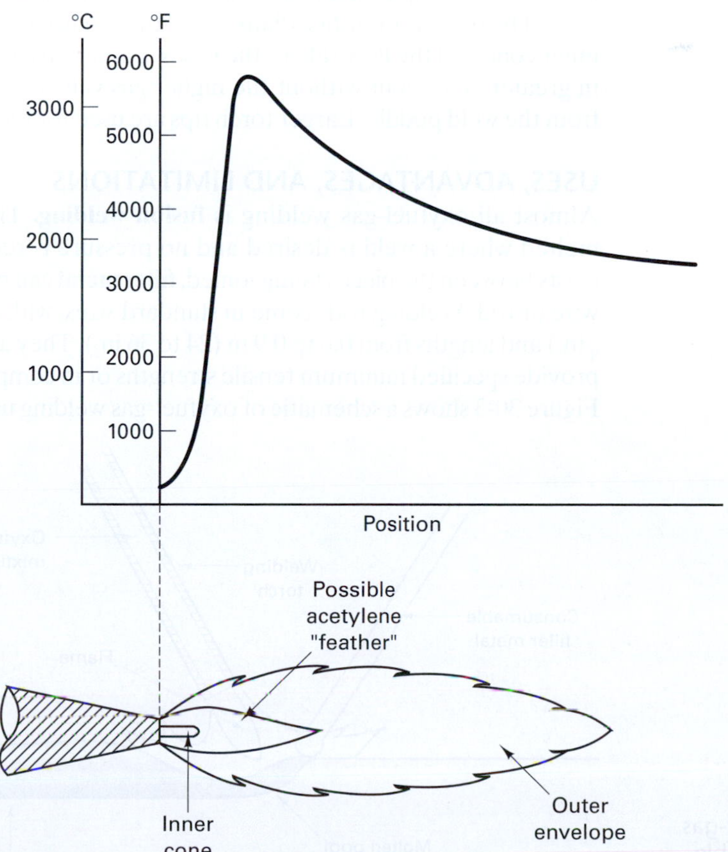

FIGURE 30-2 Typical oxyacetylene flame and the associated temperature distribution.

Excess fuel, on the other hand, produces a **carburizing flame.** The excess fuel decomposes to carbon and hydrogen, and the flame temperature is not as great (about 3050°C, or 5500°F). Flames with a slight excess of fuel are reducing flames. No carburization occurs, but the metal is well protected from oxidation. Flames of this type are used in welding Monel (a nickel–copper alloy), high-carbon steels, and some alloy steels, and for applying some types of hard-facing material.

For welding purposes, oxygen is usually supplied from pressurized tanks in a relatively pure form, but, in rare cases, air can also be used. The acetylene is usually obtained in portable storage tanks that hold up to 8.5 m^3 (300 ft^3) at 1.7 MPa (250 psi) pressure. Because acetylene is not safe when stored as a gas at pressures above 0.1 MPa (15 psi), it is usually dissolved in acetone. The storage cylinders are filled with a porous filler. Acetone is absorbed into the voids in the filler material and serves as a medium for dissolving the acetylene.

Acetylene is the hottest and most versatile of the fuel gases. Alternative fuels include propane, propylene, and stabilized **methyl-acetylene-propadiene,** best known by the trade name of **MAPP** gas. MAPP is the second-hottest gas, with a flame temperature between 2875 and 3000°C (5200 and 5400°F). *Propylene* is actually a generic name for a variety of mixed gases, often consisting of propane and ethylene or other hotter-burning chemicals. The flame temperature is usually between 2650 and 2925°C (4800 and 5300°F). Propane has a flame temperature between 2480 and 2540°C (4500 and 4600°F). While flame temperature is slightly lower, these gases can be safely stored in ordinary pressure tanks. Three to four times as much gas can be stored in a given volume, and cost per cubic foot can be less than acetylene. Butane, natural gas, and hydrogen have also been used in combination with air or oxygen for brazing and to weld the low-melting-temperature, nonferrous metals. They are generally not suited to the ferrous metals because the flame atmosphere is oxidizing and the heat output is too low.

The pressures used in gas-flame welding range from 0.006 to 0.1 MPa (1 to 15 psi) and are controlled by pressure regulators on each tank. Because mixtures of acetylene and oxygen or air are highly explosive, precautions must be taken to avoid mixing the gases improperly or by accident. All acetylene fittings have left-hand threads, while those for oxygen are equipped with right-hand threads. This prevents improper connections.

The tip size, or orifice diameter of the torch, can be varied to control the shape of the inner cone and the flow rate of the gases. Larger tips permit greater flow of gases, resulting in greater heat input without the higher gas velocities that might blow the molten metal from the weld puddle. Larger torch tips are used for the welding of thicker metal.

USES, ADVANTAGES, AND LIMITATIONS

Almost all oxyfuel-gas welding is **fusion welding.** The metals to be joined are simply melted where a weld is desired and no pressure is required. Because a slight gap often exists between the pieces being joined, **filler metal** can be added in the form of a solid metal wire or rod. Welding rods come in standard sizes, with diameters from 1.5 to 9.5 mm ($\frac{1}{16}$ to $\frac{3}{8}$ in.) and lengths from 0.6 to 0.9 m (24 to 36 in.). They are available in standard grades that provide specified minimum tensile strengths or in compositions that match the base metal. Figure 30-3 shows a schematic of oxyfuel-gas welding using a consumable welding rod.

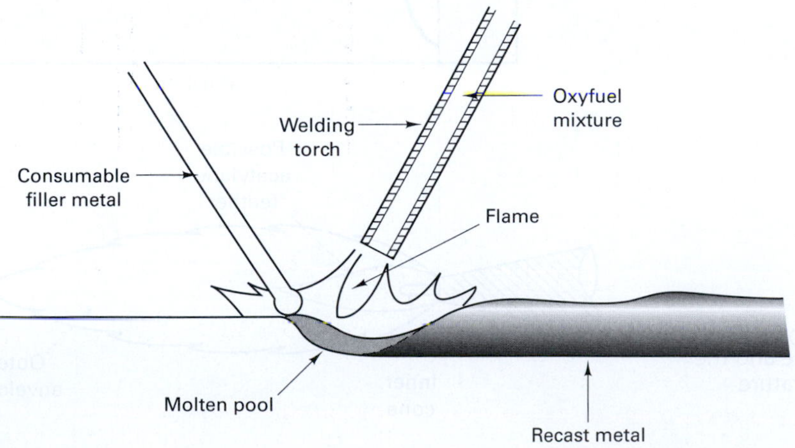

FIGURE 30-3 Oxyfuel-gas welding with a consumable welding rod.

TABLE 30-1 Process Summary: Oxyfuel Gas Welding (OFW)

Heat source	Fuel gas—oxygen combustion
Protection	Gases produced by combustion
Electrode	None
Material joined	Best for steel and other ferrous metals
Rate of heat input	Low
Weld profile (Depth/Width)	$\frac{1}{3}$
Max. penetration	3 mm
Assets	Cheap, simple equipment, portable, versatile
Limitations	Large HAZ, slow

To promote the formation of a better bond, **fluxes** may be used to clean the surfaces and remove contaminating oxide. In addition, the gaseous shield produced by vaporizing flux can prevent further oxidation during the welding process, and the slag produced by solidifying flux can protect the weld pool as it cools. Flux can be added as a powder, the welding rod can be dipped in a flux paste, or the rods can be precoated.

The oxyfuel-gas welding (OFW) processes can produce good-quality welds if proper caution is exercised. Welding can be performed in all positions, the temperature of the work can be easily controlled, and the puddle is visible to the welder. However, exposure of the heated and molten metal to the various gases in the flame and atmosphere makes it difficult to prevent contamination. Because the heat source is not concentrated, heating is rather slow. As the heat spreads, a large volume of metal is heated, and distortion is likely to occur. The thickness of the material being joined is usually less than 6.5 mm ($\frac{1}{4}$ in.). In production applications, therefore, the flame-welding processes have largely been replaced by arc welding. Nevertheless, flame welding is still quite common in fieldwork, in maintenance and repairs, and in fabricating small quantities of specialized products.

Oxyfuel equipment is quite portable, relatively inexpensive, and extremely versatile. A single set of equipment can be used for welding, brazing, and soldering, and as a heat source for bending, forming, straightening, and hardening. With the modifications to be discussed shortly, it can also perform flame cutting. Table 30-1 summarizes some of the key features of oxyfuel-gas welding. Table 30-2 shows its compatibility with some common engineering materials.

PRESSURE GAS WELDING

Pressure gas welding (PGW) is a process that uses equipment similar to the oxyfuel-gas process to produce butt joints between the ends of objects such as pipe and railroad rail. The ends are heated with a gas flame to a temperature below the melting point, and the

TABLE 30-2 Engineering Materials and Their Compatibility with Oxyfuel Welding

Material	Oxyfuel Welding Recommendation
Cast iron	Recommended with cast iron filler rods; braze welding recommended if there are no corrosion objections
Carbon and low-alloy steels	Recommended for low-carbon and low-alloy steels, using rods of the same material; more difficult for higher carbon
Stainless steel	Common for thinner material; more difficult for thicker
Aluminum and magnesium*	Common for aluminum thinner than 1 in.; difficult for magnesium alloys
Copper and copper alloys*	Common for most alloys; more difficult for some types of bronzes
Nickel and nickel alloys	Common for nickel, Monels, and Inconels
Titanium	Not recommended
Lead and zinc	Recommended
Thermoplastics, thermosets, and elastomers	Hot-gas welding used for thermoplastics, not used with thermosets and elastomers
Ceramics and glass	Seldom used with ceramics, but common with glass
Dissimilar metals	Difficult; best if melting points are within 50°F; concern for galvanic corrosion
Metals to nonmetals	Not recommended
Dissimilar nonmetals	Difficult

* Due to the high thermal conductivity, a large volume of metal may reach the melting temperature and it may be difficult to control the size of the weld pool.

soft metal is then forced together under pressure. Pressure gas welding, therefore, is actually a form of solid-state welding where the gas flame simply softens the metal and coalescence is produced by pressure.

30.2 OXYGEN TORCH CUTTING

PROCESSES

Oxyfuel-gas cutting (OFC), commonly called *flame cutting*, is the most common **thermal cutting** process. In some cases the metal is merely melted by the flame of the oxyfuel-gas torch and is blown away to form a gap, or **kerf,** as illustrated in Figure 30-4. When ferrous metal is cut, however, the process becomes one where the iron actually burns (or oxidizes) at high temperatures according to one or more of the following reactions:

$$Fe + O \rightarrow FeO + Heat$$

$$3Fe + 2O_2 \rightarrow Fe_3O_4 + Heat$$

$$4Fe + 3O_2 \rightarrow 2Fe_2O_3 + Heat$$

Because these reactions do not occur until the metal is above 815°C (1500°F), the oxyfuel flame is first used to raise the metal to the temperature where burning can be initiated. Then a stream of pure oxygen is added to the torch (or the oxygen content of the oxyfuel mixture is increased) to oxidize the iron. The liquid iron oxide and any unoxidized molten iron are then expelled from the joint by the kinetic energy of the oxygen-gas stream. Because of the low rate of heat input and the need for preheating ahead of the cut, oxyfuel cutting produces a relatively large heat-affected zone and associated distortion compared to competing techniques. Therefore, the process is best used where the edge finish or tolerance is not critical and the edge material will either be subsequently welded or removed by machining. Cutting speeds are relatively slow, but the low cost of both the required equipment and its operation make the process attractive for many applications.

Theoretically, the heat supplied by the oxidation will be sufficient to keep the cut progressing, but additional heat is often necessary to compensate for losses to the atmosphere and the surrounding metal. If the workpiece is already hot from other processing, such as solidification or hot working, no supplemental heating is required, and a supply of oxygen through a small pipe is all that is needed to initiate and continue a cut. This is known as **oxygen lance cutting (LOC).** A workpiece temperature of about 1200°C (2200°F) is required to sustain continuous cutting.

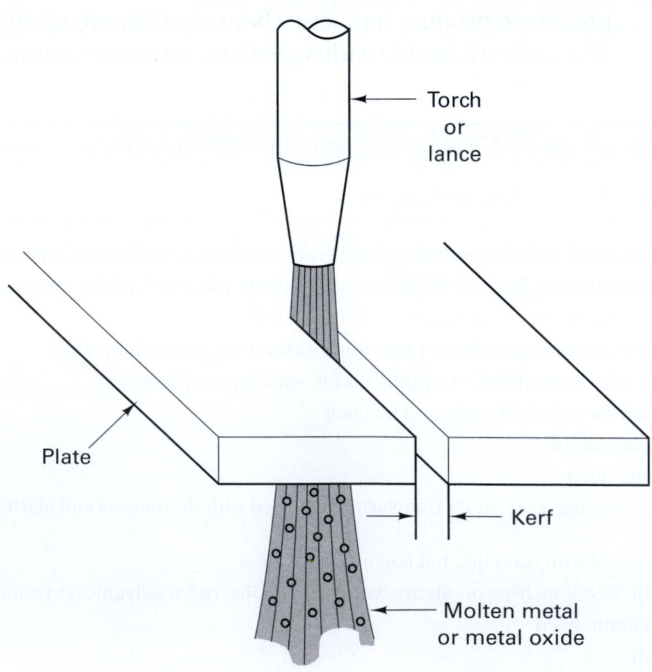

FIGURE 30-4 Flame cutting of a metal plate.

Oxyfuel-gas cutting works best on metals that oxidize readily but do not have high thermal conductivities. Carbon and low-alloy steels can be readily cut in thicknesses from 5 mm to in excess of 75 cm (30 in.). Stainless steels contain oxidation-resistant ingredients and are difficult to cut, as are aluminum and copper alloys.

FUEL GASES FOR OXYFUEL-GAS CUTTING

Acetylene is by far the most common fuel used in oxyfuel-gas cutting, and the process is often referred to as **oxyacetylene cutting (OFC-A)**. Figure 30-5 shows a typical cutting torch. The tip contains a circular array of small holes through which the oxygen–acetylene mixture is supplied to form the heating flame. A larger hole in the center supplies a stream of oxygen and is controlled by a lever valve. The rapid flow of the cutting oxygen not only oxidizes the hot metal, but also blows the formed oxides from the cut.

If the torch is adjusted and manipulated properly, it is possible to produce a relatively smooth cut. Cut quality, however, depends on careful selection of the process variables, including preheat conditions, oxygen flow rate, and cutting speed. Oxygen purities greater than 99.5% are required for the most efficient cutting. If the purity drops to 98.5%, cutting speed will be reduced by 15%, oxygen consumption will increase by 25%, and the quality of the cut will diminish.

Cutting torches are often manipulated manually. However, when the process is applied to manufacturing, the desired path is usually controlled by mechanical or programmable means. Specialized equipment has been designed to produce straight cuts in flat stock and square-cut ends on pipe. The marriage of computer numerical control (CNC) machines and cutting torches has also proven to be quite popular. This approach, along with the use of robot-mounted torches, provides great flexibility along with good precision and control.

Fuel gases other than acetylene can also be used for oxyfuel-gas cutting, the most common being *natural gas (OFC-N)* and *propane (OFC-P)*. While their flame temperatures are lower than acetylene, their use is generally a matter of economics and gas availability. For certain special work, *hydrogen* can also be used *(OFC-H)*.

STACK CUTTING

When a modest number of duplicate parts are to be cut from thin sheet, but not enough to justify the cost of a blanking die, **stack cutting** may be the answer. The sheets should be flat, smooth, and free of scale, and they should be stacked and clamped together tightly so that there are no intervening gaps that could interrupt uniform oxidation or

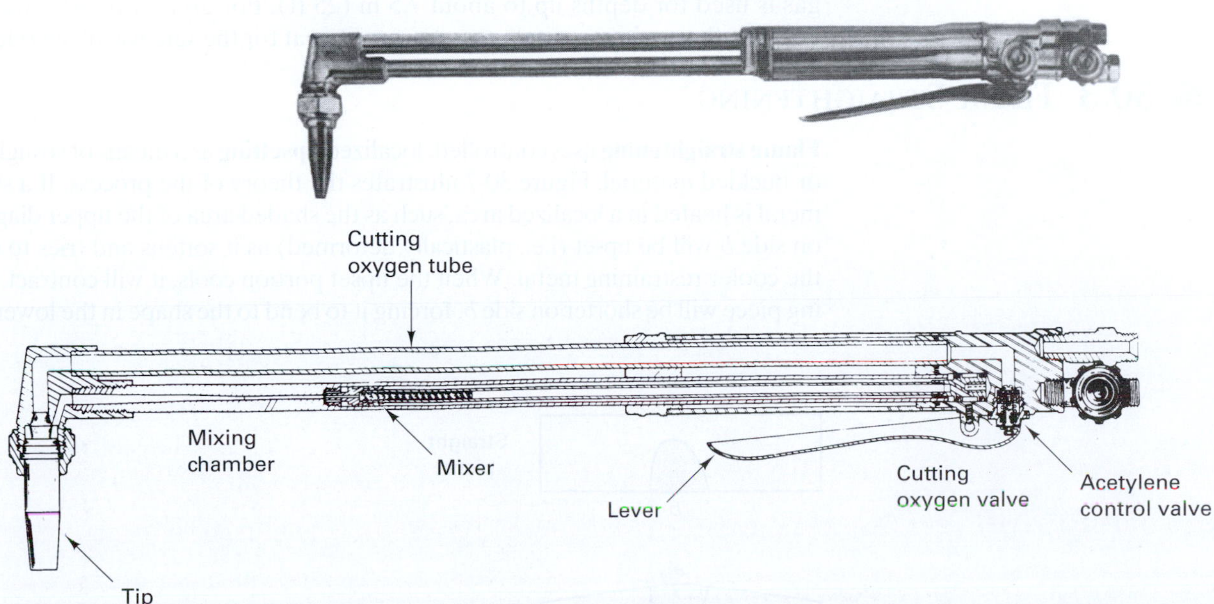

FIGURE 30-5 Oxyacetylene cutting torch and cross-sectional schematic. *(Courtesy of Thermadyne Industries, Inc., St. Louis, MO)*

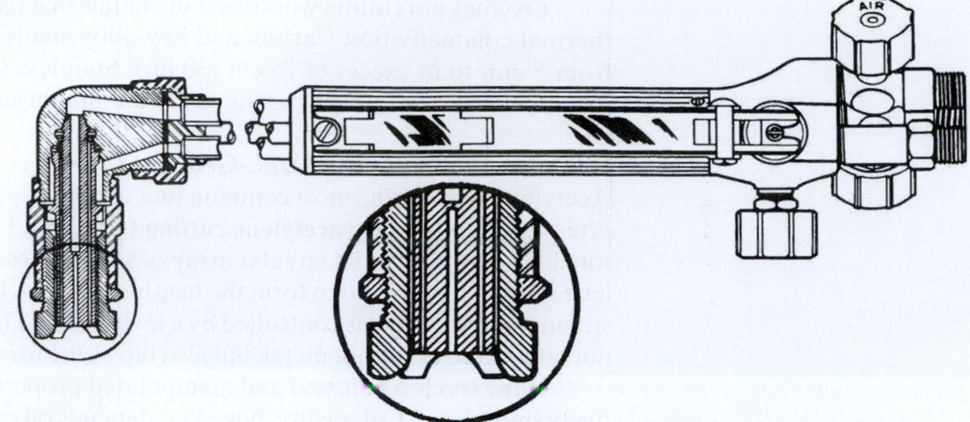

FIGURE 30-6 Underwater cutting torch. Note the extra set of gas openings in the nozzle to permit the flow of compressed air and the extra control valve. *(Courtesy of Bastian-Blessing Company, Chicago, IL)*

permit slag or molten metal to be entrapped. The resulting cut, however, will be less accurate than one produced by a blanking die.

METAL POWDER CUTTING, CHEMICAL FLUX CUTTING, AND OTHER THERMAL METHODS

When cutting hard-to-cut materials, modified torch techniques may be required. Metal powder cutting (POC) injects iron or aluminum powder into the flame to raise its cutting temperature. Chemical flux cutting (FOC) adds a fine stream of special flux to the cutting oxygen to increase the fluidity of the high-melting-point oxides. Both of these methods, however, have largely been replaced by plasma arc cutting (PAC), which is discussed as an extension of plasma arc welding to be presented later in this chapter. Laser- and electron-beam cutting will be presented with their welding parallels in a future chapter.

UNDERWATER TORCH CUTTING

The thermal cutting of materials underwater presents a special challenge. A specially designed torch, like the one shown in Figure 30-6, is used to cut steel. An auxiliary skirt surrounds the main tip, and an additional set of gas passages conducts a flow of compressed air that provides secondary oxygen for the oxyacetylene flame and expels water from the zone where the burning of metal occurs. The torch is either ignited in the usual manner before descent or by an electric spark device after being submerged. Acetylene gas is used for depths up to about 7.5 m (25 ft). For greater depths, hydrogen is used because the environmental pressure is too great for the safe use of acetylene.

30.3 FLAME STRAIGHTENING

Flame straightening uses controlled, localized **upsetting** as a means of straightening warped or buckled material. Figure 30-7 illustrates the theory of the process. If a straight piece of metal is heated in a localized area, such as the shaded area of the upper diagram, the metal on side *b* will be upset (i.e., plastically deformed) as it softens and tries to expand against the cooler restraining metal. When the upset portion cools, it will contract, and the resulting piece will be shorter on side *b*, forcing it to bend to the shape in the lower diagram.

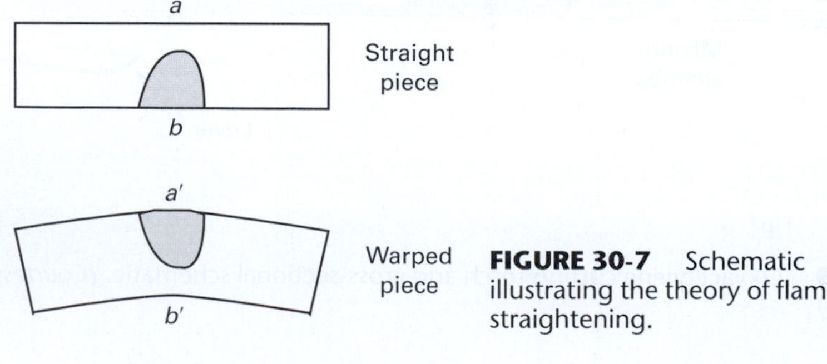

FIGURE 30-7 Schematic illustrating the theory of flame straightening.

If the starting material is bent or warped, as in the lower segment of Figure 30-7, the upper surface can be heated. Upsetting and subsequent thermal contraction will shorten the upper surface at a', bringing the plate back to a straight or flat configuration. This type of procedure can be used to restore structures that have been bent in an accident, such as automobile frames.

A similar process can be used to flatten metal plates that have become dished. Localized spots about 50 mm (2 in.) in diameter are quickly heated to the upsetting temperature while the surrounding metal remains cool. Cool water is then sprayed onto the plate, and the contraction of the upset spot brings the buckle into an improved degree of flatness. To remove large buckles, the process may have to be repeated at several spots within the buckled area.

Several cautions should be noted. When straightening steel, consideration should be given to the possible phase transformations that could occur during the heating and cooling. Because rapid cooling is used and martensite may form, a subsequent tempering operation may be required. In addition, one should also consider the residual stresses that are induced and their effect on subsequent cracking, stress-corrosion cracking, and other modes of failure. The effects of phase transformations and residual stresses were discussed more fully in Chapter 29.

Also, flame straightening should not be attempted with thin material. For the process to work, the metal adjacent to the heated area must have sufficient strength and rigidity to induce upsetting. If the material is too thin, localized heating and cooling will simply transfer the buckle from one area to another.

30.4 Arc Welding

With the development of commercial electricity in the late 19th century, it was soon recognized that an **arc** between two electrodes was a concentrated heat source that could produce temperatures approaching 4000°C (7000°F). As early as 1881, various attempts were made to use an arc as the heat source for fusion welding. A carbon rod was selected as one **electrode** and the metal workpiece became the other. Figure 30-8 depicts the basic electrical circuit. If needed, filler metal was provided by a metallic wire or rod that was independently fed into the arc. As the process developed, the filler metal replaced the carbon rod as the upper electrode. The metal wire not only carried the welding current, but as it melted in the arc, it also supplied the necessary filler.

The results of these early efforts were extremely uncertain. Because of the instability of the arc, a great amount of skill was required to maintain it, and contamination of the weld resulted from the exposure of hot metal to the atmosphere. There was little or no understanding of the metallurgical effects and requirements of arc welding. Consequently, while the great potential was recognized, very little use was made of the process until after World War I. Shielded metal electrodes were developed around 1920. These electrodes enhanced the stability of the arc by shielding it from the atmosphere and provided a fluxing action to the molten pool. The major problems of arc welding were overcome, and the process began to expand rapidly. It is estimated that 90% of all industrial welding is now performed with arc welding.

All **arc-welding** processes employ the basic circuit depicted in Figure 30-8. Welding currents vary from 1 to 4000 A (amps), with the range from 100 to 1000 being most typical. Voltages are generally in the range of 20 to 50 V. If direct current is used and the

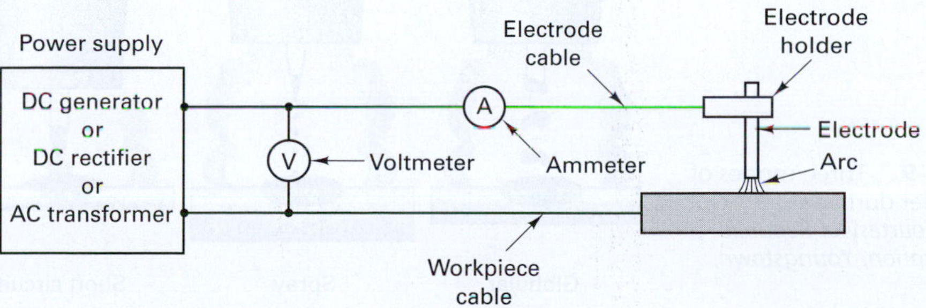

FIGURE 30-8 The basic electrical circuit for arc welding.

electrode is made negative, the condition is known as **straight polarity** (SPDC) or **DCEN,** for **direct-current electrode-negative.** Electrons are attracted to the positive workpiece, while ionized atoms in the arc column are accelerated toward the negative electrode. Because the ions are far more massive than the electrons, the heat of the arc is more concentrated at the electrode. DCEN processes are characterized by fast melting of the electrode (high metal deposition rates) and a shallow molten pool on the workpiece (weld penetration). If the work is made negative and the electrode positive, the condition is known as **reverse polarity** (RPDC) or **DCEP,** for **direct-current electrode-positive.** The positive ions impinge on the workpiece, breaking up any oxide films and giving deeper penetration. Metal deposition rate is lower, however. Sinusoidal **alternating current** provides a 50–50 average of the preceding two modes and is a popular alternative to the direct current conditions. **Variable polarity** power supplies also alternate between DCEP and DCEN conditions, using rectangular waveforms to vary the fraction of time in each mode, as well as the frequency of switching. With these power supplies, weld characteristics can now be varied over a continuous range between DCEN and DCEP conditions.

In one group of arc-welding processes, the electrode is consumed **(consumable-electrode processes)** and thus supplies the metal needed to fill the joint. Consumable electrodes have a melting temperature below the temperature of the arc. Small droplets are melted from the end of the electrode and pass to the workpiece. The size of these droplets varies greatly, and the transfer mechanism depends on the type of electrode, welding current, and other process parameters. Figure 30-9 depicts metal transfer by the globular, spray, and short-circuit transfer modes. As the electrode melts, the arc length and the electrical resistance of the arc path will vary. To maintain a stable arc and satisfactory welding conditions, the electrode must be moved toward the work at a controlled rate. Manual arc welding is almost always performed with shielded (covered) electrodes, where the coating provides molten slag or isolating gas that protects the hot weld metal from oxidation and contamination. Continuous bare-metal wire can be used as the electrode in automatic or semiautomatic arc welding, but this is always in conjunction with some form of shielding and arc-stabilizing medium and automatic feed control devices that maintain the proper arc length.

The second group of arc-welding processes employs a tungsten (or carbon) electrode, which is not consumed by the arc, except by relatively slow vaporization. In these **nonconsumable-electrode processes,** a separate metal wire is required to supply the filler metal.

Because of the wide variety of processes available, arc welding has become a widely used means of joining material. Each process and application, however, requires the selection or specification of the welding voltage, welding current, arc polarity (straight polarity, reversed polarity, or alternating), arc length, welding speed (how fast the electrode is moved across the workpiece), arc atmosphere, electrode or filler material, and flux. Filler materials must be selected to match the base metal with respect to properties and/or alloy content (chemistry). For many of the processes, the quality of the weld also depends on the skill of the operator. Automation and robotics are reducing this dependence, but the selection and training of welding personnel are still of great importance.

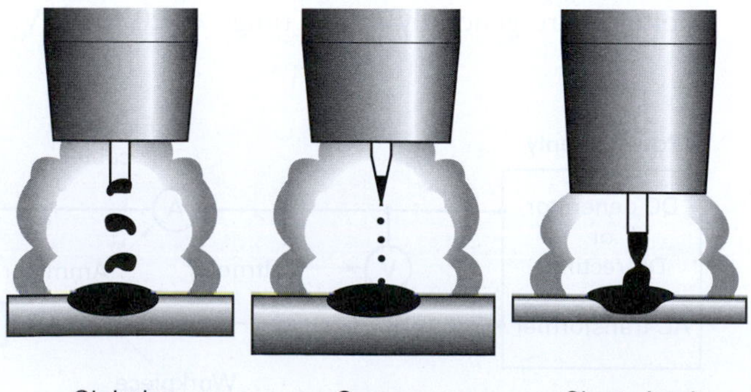

FIGURE 30-9 Three modes of metal transfer during arc welding. *(Courtesy of Republic Steel Corporation, Youngstown, OH)*

30.5 CONSUMABLE-ELECTRODE ARC WELDING

Four processes make up the bulk of consumable-electrode arc welding:

1. Shielded metal arc welding (SMAW).
2. Flux-cored arc welding (FCAW).
3. Gas metal arc welding (GMAW).
4. Submerged arc welding (SAW).

These processes all have a medium rate of heat input and produce a fusion zone whose depth is approximately equal to its width. Because the fusion zone is composed of metal from both of the pieces being joined plus melted filler (i.e., electrode), the electrode must be of the same material as that being welded. These processes cannot be used to join dissimilar metals or ceramics.

SHIELDED METAL ARC WELDING

Shielded metal arc welding (SMAW), also called **stick welding** or covered-electrode welding, is among the most widely used welding processes because of its versatility and because it requires only low-cost equipment. The key to the process is a finite-length electrode that consists of metal wire, usually from 1.5 to 6.5 mm ($\frac{1}{16}$ to $\frac{1}{4}$ in.) in diameter and 20 to 45 cm (8 to 18 in.) in length. Surrounding the wire is a bonded coating containing chemical components that add a number of desirable characteristics, including all or many of the following:

1. Vaporize to provide a protective atmosphere (a gas shield around the arc and pool of molten metal).
2. Provide ionizing elements to help stabilize the arc, reduce weld metal spatter, and increase efficiency of deposition.
3. Act as a *flux* to deoxidize and remove impurities from the molten metal.
4. Provide a protective **slag** coating to accumulate impurities, prevent oxidation, and slow the cooling of the weld metal.
5. Add alloying elements that often enhance the ductility or strength of the weld.
6. Add additional filler metal.
7. Affect arc **penetration** (the depth of melting in the workpiece).
8. Influence the shape of the weld bead.

The coated electrodes are classified by the tensile strength of the deposited weld metal, the welding position in which they may be used, the preferred type of current and polarity (if direct current), and the type of coating. For mild-steel electrodes, a four- or five-digit system of designation has been adopted by the American Welding Society (AWS classification A5.1) and is presented in Figure 30-10. As an example, type E7016 is a low-alloy steel electrode that will provide a deposit with a minimum tensile strength of 485 MPa (70,000 psi) in the non-stress-relieved condition; it can be used in all positions, with either alternating or reverse-polarity direct current; and it has a low-hydrogen plus potassium coating. To assist in identification, all electrodes are marked with colors in accordance with a standard established by the National Electrical Manufacturers Association. Electrode selection consists of determining the electrode coating, coating thickness, electrode composition, and electrode diameter. The current type and polarity are matched to the electrode.

A variety of electrode coatings have been developed and can be classified as cellulosic, rutile (titanium oxide), and basic. The cellulosic and rutile coatings contain variable amounts of SiO_2, TiO_2, FeO, MgO, Na_2O, and volatile matter. Upon decomposition, the volatile matter may release hydrogen, which can dissolve in the weld metal and lead to embrittlement or cracking in the joint. The basic coatings contain large amounts of calcium carbonate (limestone) and calcium fluoride (fluorspar) and produce low hydrogen weld metal. Because many of the electrode coatings can absorb moisture, and this is another source of undesirable hydrogen, the coated low-hydrogen electrodes are often baked at temperatures between 200 and 300°C (400 and 600°F) just prior to use and stored in an oven at 110 to 150°C (225 to 300°F).

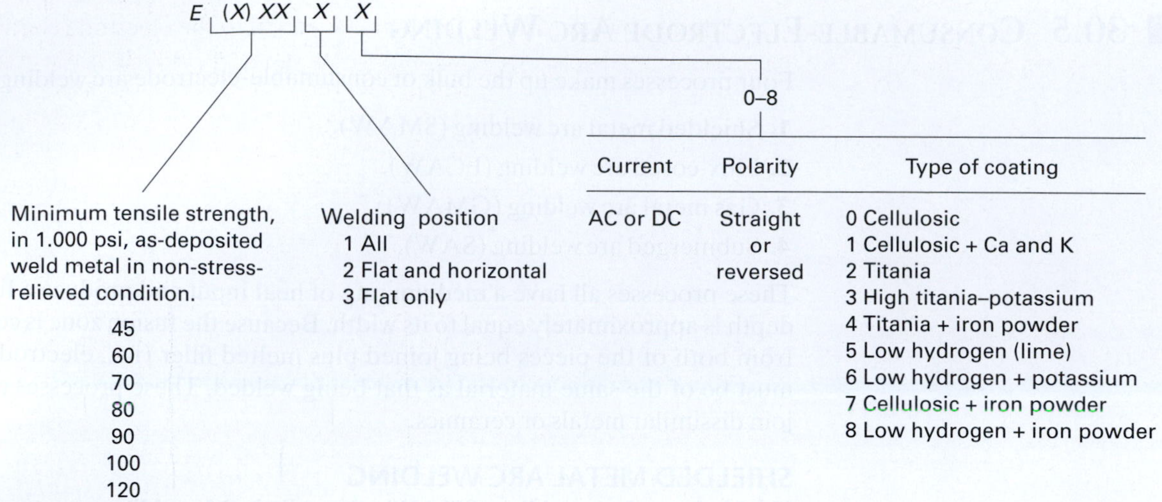

FIGURE 30-10 Designation system for arc-welding electrodes.

To initiate a weld, the operator briefly touches the tip of the electrode to the workpiece and quickly raises it to a distance that will maintain a stable arc. The intense heat quickly melts the tip of the electrode wire, the coating, and portions of the adjacent base metal. As part of the electrode coating melts and vaporizes, it forms a protective atmosphere of CO, CO_2, and other gases that stabilizes the arc and protects the molten and hot metal from contamination. Other coating components surround the metal droplets with a layer of liquid flux and slag. The fluxing constituents unite with any impurities in the molten metal and float them to the surface to be entrapped in the slag coating that forms over the weld. The slag coating then protects the cooling metal from oxidation and slows the cooling rate to prevent the formation of hard, brittle structures. The glassy slag is easily chipped from the weld when it has cooled. Figure 30-11 illustrates the shielded metal arc welding process, and Figure 30-12 provides a schematic of metal deposition from a shielded electrode.

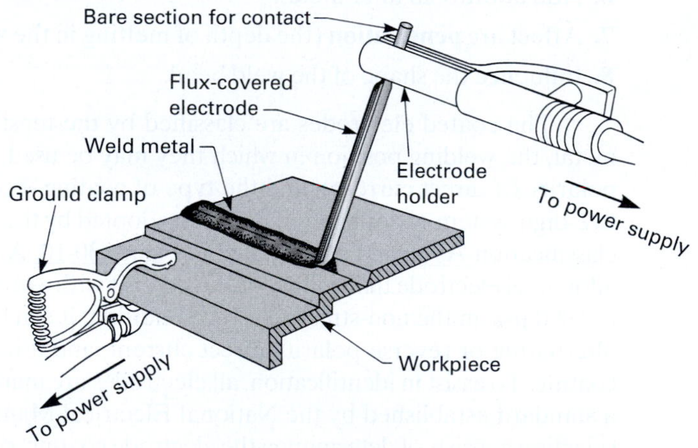

FIGURE 30-11 A shielded metal arc welding (SMAW) system.

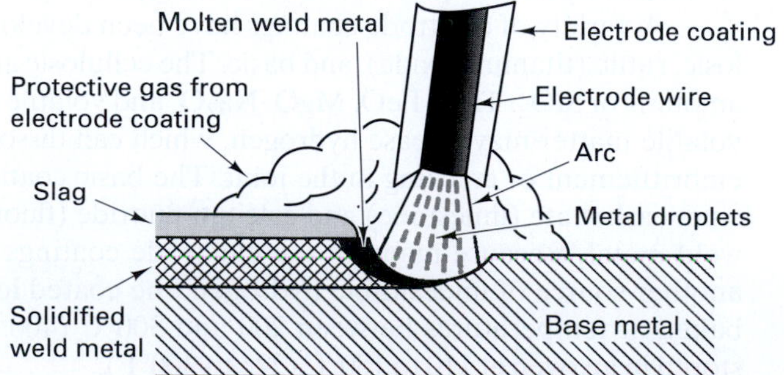

FIGURE 30-12 Schematic diagram of shielded metal arc welding (SMAW). *(Courtesy of American Iron and Steel Institute, Washington, DC)*

Metal powder (usually iron) can be added to the electrode coating to significantly increase the amount of weld metal that can be deposited with a given size electrode wire and current—but with a noticeable decrease in penetration depth. Alloy elements can also be incorporated into the coating to adjust the chemistry of the weld. Special contact or drag electrodes utilize coatings that are designed to melt more slowly than the filler wire. If these electrodes are tracked along the surface of the work, the faster melting center wire will be recessed by the proper length to maintain a stable arc.

Because electrical contact must be maintained with the center wire, SMAW electrodes are finite-length "sticks." Stick length is limited because the current must be supplied near the arc, or the electrode will tend to overheat (by electrical resistance heating) and ruin the coating. Overheating also restricts the weld currents to values below 300 A (generally about 40 A/mm of electrode diameter). As a result, the arc temperatures are somewhat low, and penetration is generally less than 5 mm ($\frac{3}{16}$ in.). Welding of material thicker than 5 mm will require multiple passes, and the slag coating must be removed between each pass.

The shielded metal arc process is best used for welding ferrous metals: carbon steels, alloy steels, stainless steels, and cast irons can all be welded. DCEP conditions are used to obtain the deepest possible penetration, with alternate modes being employed when welding thin sheet. The mode of metal transfer is either globular or short circuit.

Shielded metal arc welding is a simple, inexpensive, and versatile process, requiring only a power supply, power cables, electrode holder, and a small variety of electrodes. The equipment is portable and can even be powered by gasoline or diesel generators. Therefore, it is a popular process in job shops and is used extensively in repair operations. The electrode provides and regulates its own flux, and there is less sensitivity to wind and drafts than in the gas-shielded processes. Welds can be made in all positions. Unfortunately, the process is discontinuous (only short lengths of weld can be produced before a new electrode needs to be inserted into the holder), produces shallow welds, and requires slag removal after each welding pass. Table 30-3 presents a process summary for shielded metal arc welding.

FLUX-CORED ARC WELDING

Flux-cored arc welding (FCAW) overcomes some of the limitations of the shielded metal arc process by moving the powdered flux to the interior of a continuous tubular electrode (Figure 30-13). When the arc is established, the vaporizing flux again produces a protective atmosphere and also forms a slag layer over the weld pool that will require subsequent removal. Alloy additions (metal powders) can be blended into the flux to create a wide variety of filler metal chemistries. Compared to the stick electrodes of the shielded metal arc process, the flux-cored electrode is both continuous and less bulky because binders are no longer required to hold the flux in place.

The continuous electrode is fed automatically through a welding gun, with electrical contact being maintained through the bare-metal exterior of the wire at a position near the exit of the gun. Overheating of the electrode is no longer a problem, and welding currents can be increased to about 500 A. The higher heat input increases penetration depth to about 1 cm ($\frac{3}{8}$ in.). The process is best used for welding steels, and welds

TABLE 30-3 Process Summary: Shielded Metal Arc Welding (SMAW)

Heat source	Electric arc
Protection	Slag from flux and gas from vaporized coating material
Electrode	Discontinuous, consumable
Material joined	Best for steel
Rate of heat input	Medium
Weld profile (D/W)	1
Current	<300 amps
Max. penetration	3–6 mm
Assets	Cheap, simple equipment
Limitations	Discontinuous, shallow welds; requires slag removal

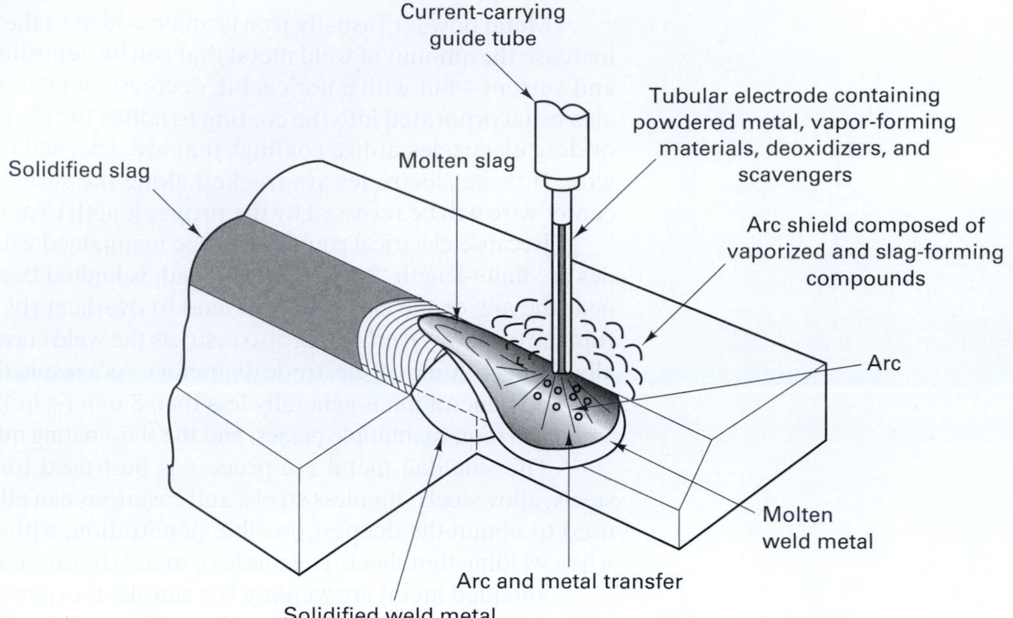

FIGURE 30-13 The flux-cored arc welding (FCAW) process. *(Courtesy of American Welding Society, New York, NY)*

can be made in all positions. Direct-current electrode-positive (DCEP) conditions are almost always used for the enhanced penetration. High deposition rates are possible, but the equipment cost is greater than SMAW because of the need for a controlled wire feeder and a more costly power supply. Good ventilation is required to remove the fumes generated by the vaporizing flux.

In the basic flux-cored arc welding process, the shielding gas is provided by the vaporization of flux components (self-shielding electrodes). Better protection and cleaner welds can be produced by a process variation that combines the flux with a flow of externally supplied shielding gas, such as CO_2.

Table 30-4 presents a process summary of flux-cored arc welding.

GAS METAL ARC WELDING

If the supplemental shielding gas flowing through the torch (described earlier) becomes the primary protection for the arc and molten metal, there is no longer a need for the volatilizing flux. The consumable electrode can now become a continuous, solid, uncoated metal wire or a continuous hollow tube with powdered alloy additions in the center, known as a metal-cored electrode. The resulting process, shown in Figure 30-14, was formerly called **metal inert-gas (MIG) welding,** but it is now known as **gas metal arc welding (GMAW).** The arc is still maintained between the workpiece and the automatically fed bare-wire electrode, which continues to provide the necessary filler metal. Electrode diameters range from 0.6 to 6.4 mm (0.02 to 0.25 in.) The welding current, penetration depth, and process cost are all similar to the flux-cored process.

TABLE 30-4 Process Summary: Flux-Cored Arc Welding (FCAW)

Heat source	Electric arc
Protection	Slag and gas from flux (optional secondary gas shield)
Electrode	Continuous, consumable
Material joined	Best for steel
Rate of heat input	Medium
Weld profile (D/W)	1
Current	<500 amps
Max. penetration	6–10 mm
Assets	Continuous electrode
Limitations	Requires slag removal

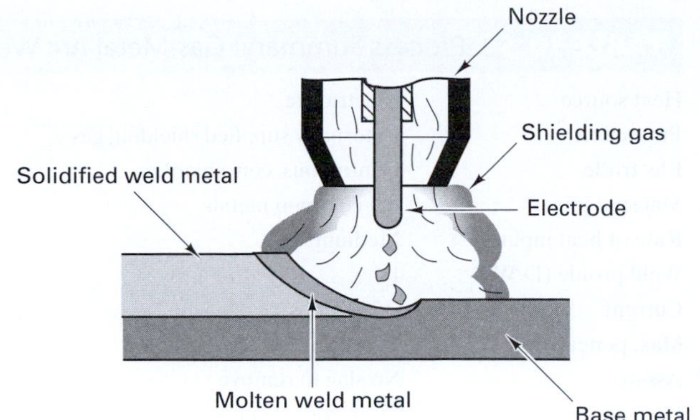

FIGURE 30-14 Schematic diagram of gas metal arc welding (GMAW). *(Courtesy of American Iron and Steel Institute, Washington, DC)*

Because shielding is provided by the flow of gas, and fluxing and slag-forming agents are no longer required, the gas metal arc process can be applied to all metals. Carbon and stainless steels and alloys of aluminum, magnesium, copper, and nickel are the most common. Argon, helium, and mixtures of the two are the primary shielding gasses. When welding steel, some O_2 or CO_2 is usually added to improve the arc stability and reduce weld spatter. The cheaper CO_2 can also be used alone when welding steel, provided that a deoxidizing electrode wire is employed. Nitrogen and hydrogen may also be added to modify arc characteristics. Because these shielding gases only provide protection and do not remove existing contamination, starting cleanliness is critical to the production of a good weld.

The specific shielding gases can have considerable effect on the stability of the arc; the metal transfer from the electrode to the work; and also the heat transfer behavior, penetration, and tendency for undercutting (weld pool extending laterally beneath the surface of the base metal). Helium produces the hottest arc and deepest penetration. Argon is intermediate, and CO_2 yields the lowest arc temperatures and shallowest penetrations. Because argon is heavier than air, it tends to blanket the weld area, enabling the use of low gas flow rates. The lighter-than-air helium generally requires higher flow rates than either argon or carbon dioxide.

Electronic controls can be used to alter the welding current, enabling further control of the metal transfer mechanism, shown previously in Figure 30-9. **Short-circuit transfer** *(GMAW-S)* is promoted by the lowest currents and voltages (14 to 21 V) and the use of CO_2 shield gas. The advancing electrode (or molten metal on its tip) makes direct contact with the weld pool, and the short circuit causes a rapid rise in current. Big molten globs form on the tip of the electrode and then separate, forming a gap between the electrode and workpiece. This gap reinitiates a brief period of arcing, but the rate of electrode advancement exceeds the rate of melting in the arc, and another short circuit occurs. The power conditions oscillate between arcing and short circuiting at a rate of 20 to 200 cycles per second. Short-circuit transfer is preferred when joining thin materials and can be used in all welding positions.

If the voltage and amperage are increased, the mode becomes one of **globular transfer.** The electrode melts from the heat of the arc, and metal drops form with a diameter equal to or greater than the diameter of the electrode wire. Gravity and electromagnetic forces then transfer the drops to the workpiece at a rate of several per second. Because gravity plays a role in metal transfer, there is a definite limitation on the positions of welding. This is the least desirable mode of transfer because the arc is loud and erratic, with lots of splashing or spatter. **Spray transfer** *(GMAW-ST)* occurs with even higher currents and voltages (25 to 32 V and about 200 A), argon gas shielding, and DCEP conditions. Small droplets emerge from a pointed electrode at a rate of hundreds per second. Because of their small size and the greater electromagnetic effects, the droplets are easily propelled across the arc in any direction, irrespective of the effects of gravity. Spray transfer is accompanied by deep penetration and low spatter. The biggest problem with out-of-position welding may be keeping the rather large molten weld pool in place until it solidifies.

Pulsed spray transfer *(GMAW-P)* was developed in the 1960s to overcome some of the limitations of conventional spray transfer. In this mode, a low welding current is first used to create a molten globule on the end of the filler wire. A burst of high current then "explodes" the globule and transfers the metal across the arc in the form of a spray.

TABLE 30-5 Process Summary: Gas-Metal Arc Welding (GMAW)

Heat source	Electric arc
Protection	Externally supplied shielding gas
Electrode	Continuous, consumable
Material joined	All common metals
Rate of heat input	Medium
Weld profile (D/W)	1
Current	<500 amps
Max. penetration	6–10 mm
Assets	No slag to remove
Limitations	More costly equipment than SMAW or FCAW

By alternating low and high currents at a rate of 60 to 600 times per second, the filler metal is transferred in a succession of rapid bursts, similar to the emissions of a rapidly squeezed aerosol atomizer. With the pulsed form of deposition, the weld pool cools between the periods of molten metal deposition. There is less heat input to the weld, and the weld temperatures and size of the heat-affected zone are reduced. Thinner material can be welded, distortion is reduced, workpiece discoloration is minimized, heat-sensitive parts can be welded, high-conductivity metals can be joined, electrode life is extended, electrode cooling techniques may not be required, and fine microstructures are produced in the weld pool. Welds can be made in all positions, and the use of pulsed power lowers spattering and improves the safety of the process. The high speed of the process is attractive for productivity, and the energy or power required to produce a weld is lower than with other methods (reduced cost). Controls can be adjusted to alter the shape of the weld pool and vary the penetration.

In general, the gas metal arc process is fast and economical and currently accounts for more than half of all weld metal deposition. There is no frequent change of electrodes as with the shielded metal arc process. No flux is required, and no slag forms over the weld. Thus, multiple-pass welds can be made without the need for intermediate cleaning. The process can be readily automated, and the lightweight, compact welding unit lends itself to robotic manipulation. A DCEP arc is generally used because of its deep penetration, spray transfer, and the ability to produce smooth welds with good profile. Process variables include type of current, current magnitude, shielding gas, electrode diameter, electrode composition, electrode stickout (extension beyond the gun), welding speed, welding voltage, and arc length. Table 30-5 provides a process summary for gas metal arc welding.

Several process modifications have recently emerged for GMAW. While nearly all GMAW welding is performed in the direct-current electrode-positive mode because of the features cited earlier, newer power supplies allow allow a variable combination of DCEP and DCEN. Weld penetration can be varied, enabling this *variable-polarity GMAW* to weld thin steel and aluminum sheet metal and tubing. In a process modification known as *advanced gas metal arc welding (AGMAW)*, a second power source is used to preheat the filler wire before it emerges from the welding torch. Less arc heating is needed to produce a weld, so less base metal is melted, producing less dilution of the filler metal and less penetration. *Twin-wire GMAW* increases the deposition rate by feeding two wires through the same torch.

Another recent modification is the use of flat electrode wire, typically having a rectangular cross section of about 4 mm × 0.5 mm (0.15 in. × 0.02 in.). By having a larger surface area participating in the arc, deposition rate is similar to a two-wire feed with only a single wire delivery system. The arc is also asymmetric. Orienting the wire perpendicular to the weld seam produces a wide, shallow weld pool, suitable for bridging gaps and often eliminating the need to weave during deposition. A narrower, deeper weld pool results when the wire is parallel to the weld. Varying the angle between parallel and perpendicular generates a spectrum of weld pool geometries.

SUBMERGED ARC WELDING

No shielding gas is used in the **submerged arc welding (SAW)** process, depicted in Figure 30-15. Instead, a thick layer of granular flux is deposited just ahead of a solid bare-wire

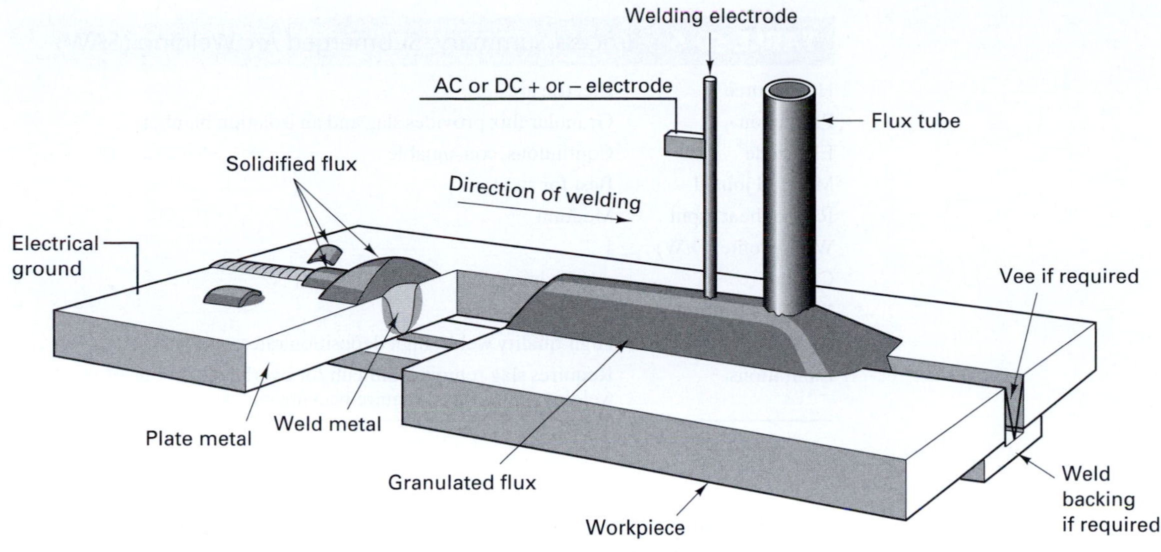

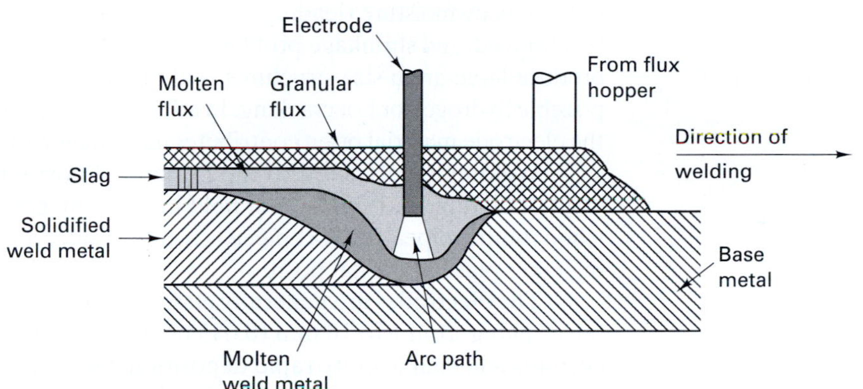

FIGURE 30-15 (Top) Basic features of submerged arc welding (SAW). *(Courtesy of Linde Division, Union Carbide Corporation, Houston, TX)* (Bottom) Cutaway schematic of submerged arc welding. *(Courtesy of American Iron and Steel Institute, Washington, DC)*

consumable electrode, and the arc is maintained beneath the blanket of flux with only a few small flames being visible. A portion of the flux melts and acts to remove impurities from the rather large pool of molten metal, while the unmelted excess provides additional shielding. The molten flux solidifies into a glasslike covering over the weld. This layer, along with the flux that is not melted, provides good thermal insulation. The slow cooling of the weld metal helps to produce soft, ductile welds. Upon further cooling, the solidified flux cracks loose from the weld (due to the differential thermal contraction) and is easily removed. The unmelted granular flux is recovered by a vacuum system and reused.

Submerged arc welding is most suitable for making flat-butt or fillet welds in low-carbon steels (0.3% carbon). With some preheat and postheat precautions, medium-carbon and alloy steels and some cast irons, stainless steels, copper alloys, and nickel alloys can also be welded. The process is not recommended for high-carbon steels, tool steels, aluminum, magnesium, titanium, lead, or zinc. The reasons for this incompatibility are somewhat varied, including the unavailability of suitable fluxes, reactivity at high temperatures, and low sublimation temperatures.

Because the arc is totally submerged, high welding currents can be used (600 to 2000 A). High welding speeds, high deposition rates, deep penetration, and high cleanliness (due to the flux action) are all characteristic of submerged arc welding. A welding speed of 0.75 m/min (25 in./min) in 2.5-cm (1-in.)-thick steel plate is typical. Single-pass welds can be made with penetrations up to 2.5 cm (1.0 in.), and greater thicknesses can be joined by multiple passes. Because the metal is deposited in fewer passes than with alternative processes, there is less possibility of entrapped slag or voids, and weld quality is further enhanced. For even higher deposition rates, multiple electrode wires can be employed (up to five wires feeding one weld pool). Deposition rates of more than 50 kg/hr (100 lb/hr) have been reported.

TABLE 30-6 Process Summary: Submerged Arc Welding (SAW)

Heat source	Electric arc
Protection	Granular flux provides slag and an isolation blanket
Electrode	Continuous, consumable
Material joined	Best for steel
Rate of heat input	Medium
Weld profile (D/W)	1
Current	<1000 amps
Max. penetration	25 mm
Assets	High-quality welds, high deposition rates
Limitations	Requires slag removal, difficult for overhead and out-of-position welding, joints often require backing plates

The submerged arc process is applied almost exclusively to ferrous metals. Other limitations to the process include the need for extensive flux handling, possible contamination of the flux by moisture (leading to porosity in the weld), the large volume of slag that must be removed, and shrinkage problems due to the large weld pool. The high heat inputs can produce large-grain-size structures, and the slow cooling rate may enable segregation and possible hydrogen or hot cracking. In addition, chemical control is quite important, because the electrode material often contributes more than 70% of the molten weld region.

Welding is restricted to the horizontal position because the large, high-fluidity weld pool, flux, and slag are all held in place by gravity. For circumferential joints, as when joining pipe, the workpiece is rotated under a fixed welding head so that all welding takes place in the flat position.

The electrodes are generally classified by composition and are available in diameters ranging from 1 to 10 mm (0.045 to .375 in.). The larger electrodes can carry higher currents and enable more rapid deposition, but penetration is shallower. The welding of alloy steels can be performed in several ways: solid wire electrodes of the desired alloy, plain carbon electrodes with the alloy additions being incorporated into the flux, or tubular metal electrodes with the alloy additions in the hollow core. Various fluxes are also available and are selected for compatibility with the weld metal. All are designed to have low melting temperatures, good fluidity, and brittleness after cooling.

In a modification of the submerged arc process known as **bulk welding,** iron powder is first deposited into the joint (ahead of the flux) as a means of increasing deposition rate. A single weld pass can then produce enough filler metal to be equivalent to seven or eight conventional submerged arc passes.

Table 30-6 provides a summary of the submerged arc welding process.

STUD WELDING

Stud welding (SW) is an arc-welding process used to attach studs, screws, pins, or other fasteners to a metal surface. A special gun is used, such as the one shown in Figure 30-16. The inserted stud acts as an electrode, and a DC arc is established between the end of the stud and the workpiece. Welding currents vary from 200 to 2400 A, and weld times from 0.1 to 1.5 s, depending on the diameter of the fastener and the materials being joined. After a small amount of metal is melted, the two pieces are brought together under light pressure and allowed to solidify. Automatic equipment controls the arc, its duration, and the application of pressure to the stud.

Figure 30-17 shows some of the wide variety of studs that are specially made for this process. Many contain a recessed end that is filled with flux. A ceramic ferrule, such as the one shown in the center photo of Figure 30-17, may be placed over the end of the stud before it is positioned in the gun. During the arc, the ferrule serves to concentrate the heat and isolate the hot metal from the atmosphere. It also confines the molten or softened metal and shapes it around the base of the stud, as shown in the photo on the right of Figure 30-17. After the weld has cooled, the brittle ceramic is broken free and

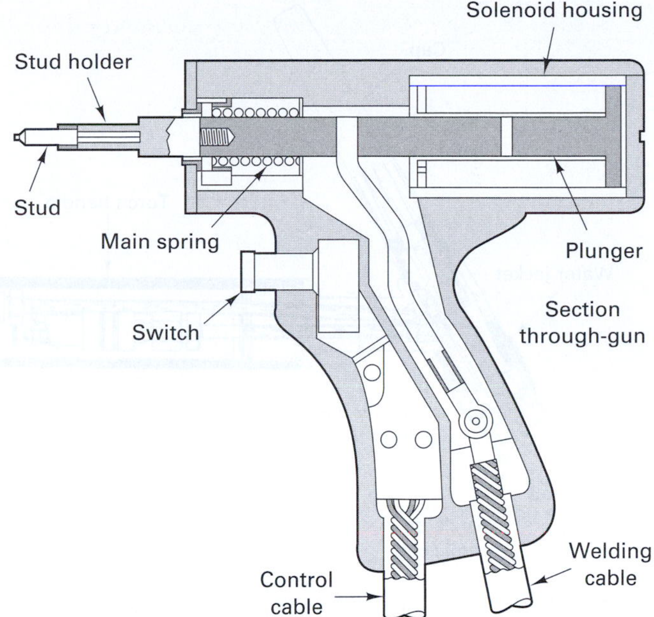

FIGURE 30-16 Diagram of a stud welding gun. *(Courtesy of American Machinist)*

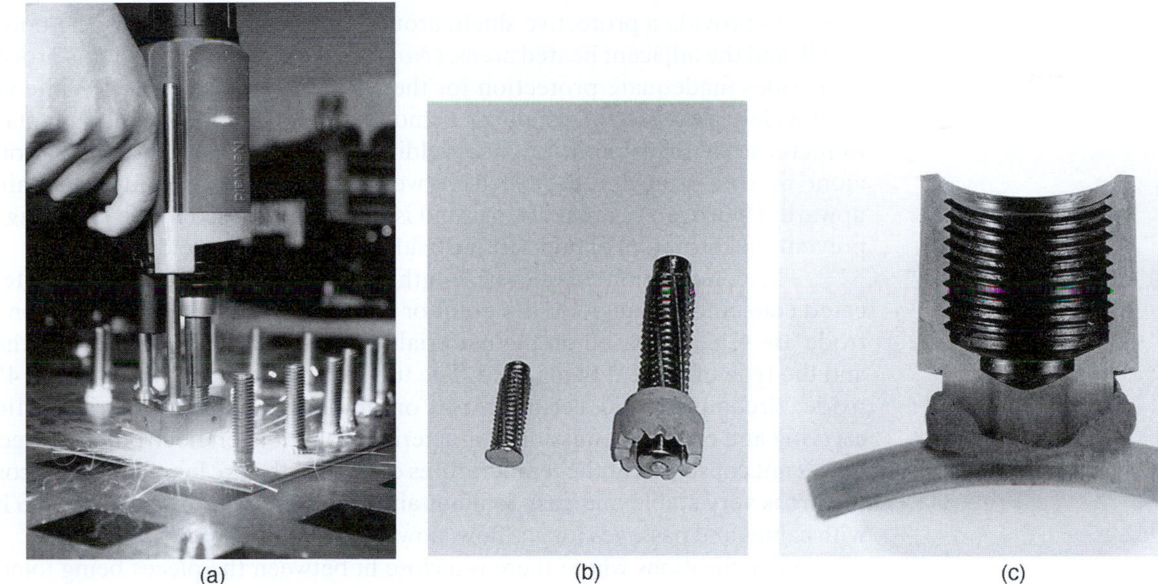

FIGURE 30-17 (a) Threaded studs being welded utilizing the drawn arc welding process; (b) (left) Flanged stud (right) stud with flux and ceramic ferrule; (c) cross section of an internally-tapped stud weld. Note that the fusion zone is comprised of metal from both the stud and the plate *(Courtesy of Nelson Stud Welding Co, Elyria, OH)*

removed. Because burn-off or melting reduces the length of the stud, the original dimensions should be selected to compensate.

Stud welding requires almost no skill on the part of the operator. Once the stud and ferrule are placed in the gun and the gun positioned on the work, all the operator has to do is pull the trigger. The cycle is executed automatically and takes less than 1 s. Thus, the process is well suited to manufacturing and can be used to eliminate the drilling and tapping of many special holes. Production-type stud welders can produce more than 1000 welds per hour.

30.6 NONCONSUMABLE-ELECTRODE ARC WELDING

GAS TUNGSTEN ARC WELDING

Gas tungsten arc welding (GTAW) produces very high quality welds, and was formerly known as **tungsten inert-gas (TIG) welding**, or **heliarc welding** when helium was the shielding gas. A nonconsumable tungsten electrode provides the arc but not the filler metal. Inert gas (argon, helium, or a mixture of them) flows through the electrode

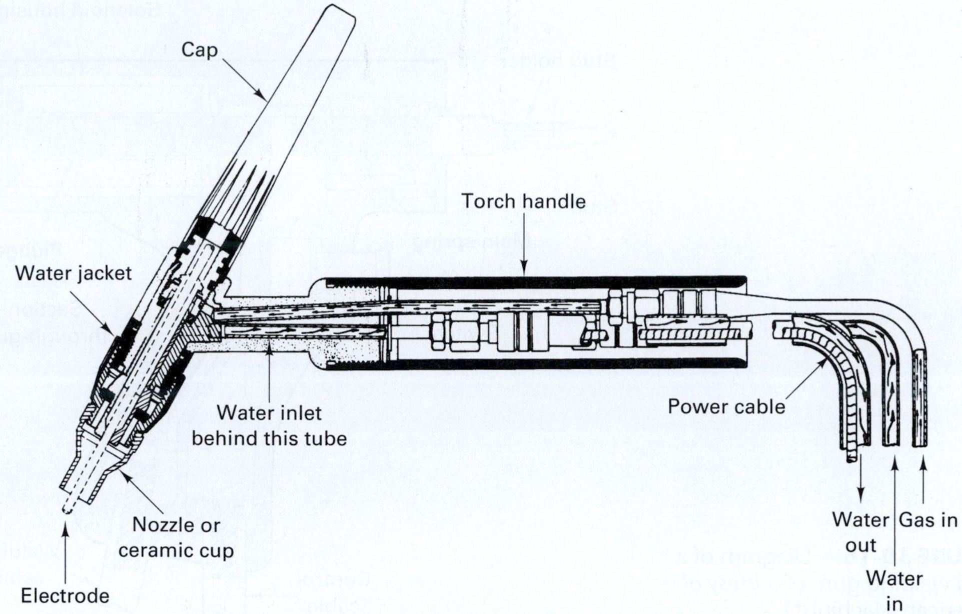

FIGURE 30-18 Welding torch used in nonconsumable electrode, gas tungsten arc welding (GTAW), showing feed lines for power, cooling water, and inert gas flow. *(Courtesy of Linde Division, Union Carbide Corporation, Houston, TX)*

holder to provide a protective shield around the electrode, the arc, the pool of molten metal, and the adjacent heated areas. (*Note:* CO_2 cannot be used in this process because it provides inadequate protection for the hot tungsten electrode.) While argon is the most widely used gas and produces a smoother, more stable arc, helium may be added to increase the heat input (higher welding speeds and deeper penetration). Helium alone may be preferred for overhead welding because it is lighter than air and flows upward. Hydrogen is a reactive gas and is sometimes added to the shielding mixture to prevent the formation of undesirable oxides in the molten weld metal.

The composition, diameter, length, and tip geometry (balled, pointed, or truncated cone and the angle of the point or cone) of the tungsten or tungsten alloy electrode are selected based on the material being welded, the thickness of the material, and the type of current being used. The tungsten is often alloyed with 1 to 4% thorium oxide, zirconium oxide, cerium oxide, or lanthanum oxide to provide better current-carrying and electron-emission characteristics and longer electrode life. Because tungsten is not consumed at the temperatures of the arc, the arc length remains constant, and the arc is very stable and easy to maintain. Figure 30-18 shows a typical GTAW torch with cables and passages for gas flow, power, and cooling water.

In applications where there is a close fit between the pieces being joined, no filler metal may be needed. When filler metal is required, it is usually supplied as a separate rod or wire as illustrated in Figure 30-19. The filler metal is generally selected to match the

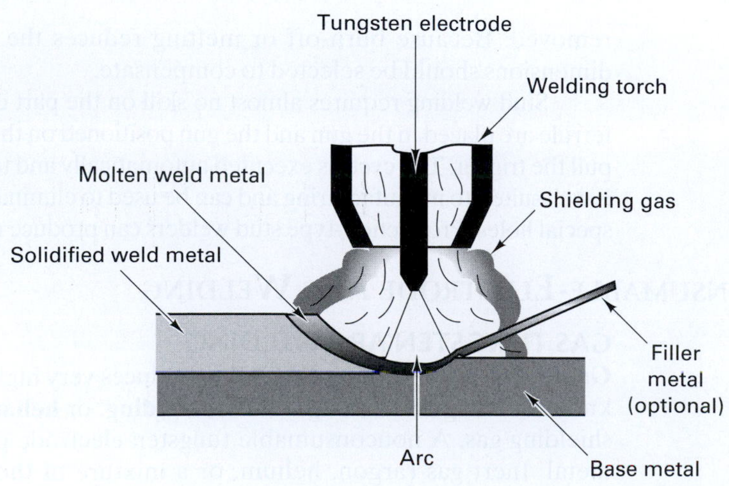

FIGURE 30-19 Diagram of gas tungsten arc welding (GTAW). *(Courtesy of American Iron and Steel Institute, Washington, DC)*

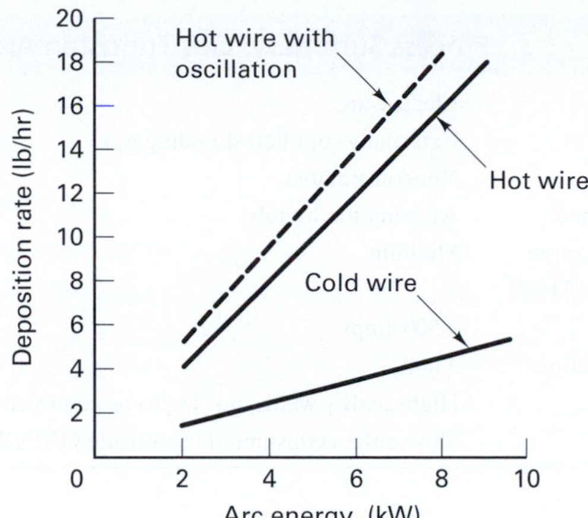

FIGURE 30-20 Comparison of the metal deposition rates in GTAW with cold, hot, and oscillating-hot filler wire. *(Courtesy of* Welding Journal*)*

chemistry and/or tensile strength of the metal being welded. When high deposition rates are desired, a separate resistance heating circuit can be provided to preheat the filler wire. As shown in Figure 30-20, the deposition rate of heated wire can be several times that of a cold wire. By oscillating the filler wire from side to side while making a weld pass, the deposition rate can be further increased. The hot-wire process is not practical when welding copper or aluminum, however, because it is difficult to preheat the low-resistivity filler wire.

With skilled operators, gas tungsten arc welding can produce high-quality welds that are very clean and scarcely visible. Because no flux is employed, no special cleaning or slag removal is required. However, the surfaces to be welded must be clean and free of oil, grease, paint, and rust because the inert gas does not provide any cleaning or fluxing action. It is also important to control the arc length throughout the process. Because the arc is somewhat bell-shaped, decreasing the stand-off distance will decrease the melt and heat-affected widths on the workpiece. However, if the hot tungsten electrode comes into contact with the workpiece or molten pool, it will contaminate the electrode.

Most common engineering metals and alloys can be welded by this process, and the use of inert gas makes it particularly attractive for the reactive metals, such as aluminum, magnesium, and titanium, as well as the high-temperature superalloys and refractory metals. Maximum penetration is obtained with direct-current electrode-negative conditions, although alternating current may be specified to break up surface oxides (as when welding aluminum). DCEP or reverse-polarity conditions provide oxide break-up but with low penetration, and are used only when welding thin pieces where the shallow penetration is desirable. Weld currents should be kept low in the DCEP configuration because this mode tends to melt the tungsten electrode. Weld voltage is typically 20 to 40 V, and weld current varies from less than 125 A for DCEP to 1000 A for DCEN. A high-frequency, high-voltage, alternating current is often superimposed on the regular AC or DC welding current to make it easier to start and maintain the arc. The *pulsed arc gas tungsten arc welding (GTAW-P)* modification offers all of the advantages previously cited for pulsed gas metal arc, including reduced bead width, increased penetration, reduced heat input, smaller heat-affected zones, increased weld travel speeds, the ability to weld thinner materials, and better overall weld quality. Key variables include the peak amperage, background amperage, pulse frequency, and percent of time at peak.

GTAW costs more than SMAW and is slower than GMAW. However, it produces a high-quality weld in a very wide range of thicknesses, positions, and geometries. The process has a medium rate of heat input, and the welds have a depth that is approximately equal to the width. The materials being welded are generally thinner than 6.5 mm ($\frac{1}{4}$ in.). Table 30-7 provides a process summary.

GAS TUNGSTEN ARC SPOT WELDING

A variation of gas tungsten arc welding can be used to produce **spot welds** between two pieces of metal where access is limited to one side of the joint or where thin sheet is being attached to heavier material. The basic procedure is illustrated in Figures 30-21. A modified tungsten inert-gas gun is used with a vented nozzle on the end. The nozzle is

TABLE 30-7 Process Summary: Gas Tungsten Arc Welding (GTAW)

Heat source	Electric arc
Protection	Externally supplied shielding gas
Electrode	Nonconsumable
Material joined	All common metals
Rate of heat input	Medium
Weld profile (D/W)	1
Current	<500 amps
Max. penetration	3 mm
Assets	High-quality welds, no slag to be removed
Limitations	Slower than consumable electrode GMAW

pressed firmly against the material, holding the pieces in reasonably good contact. (The workpieces must be sufficiently rigid to sustain the contact pressure.) Inert gas, usually argon or helium, flows through the nozzle to provide a shielding atmosphere. Automatic controls then advance the electrode to initiate the arc and retract it to the correct distance for stabilized arcing. The duration of arcing is timed automatically to produce an acceptable spot weld. The depth and size of the weld nugget are controlled by the amperage, time, and type of shielding gas.

In arc spot welding, the weld nugget begins to form at the surface where the gun makes contact. This is in contrast to the more standard resistance spot-welding methods, where the weld nugget forms at the interface between the two members. Each technique has its characteristic advantages and disadvantages.

PLASMA ARC WELDING

In **plasma arc welding (PAW),** the arc is maintained between a nonconsumable electrode and either the welding gun **(nontransferred arc)** or the workpiece **(transferred arc)** as illustrated in Figure 30-22. The nonconsumable tungsten electrode is set back within the "torch" in such a way as to force the arc to pass through or be contained within a small-diameter nozzle. An inert gas (usually argon) is forced through this constricted arc, where it is heated to a high temperature and forms a hot, fast-moving **plasma.** The emerging gas then transfers its heat to the workpiece and melts the metal. This flow is called the **orifice gas.** A second flow of inert gas surrounds the plasma column and provides shielding to the weld pool. When filler metal is needed, it is provided by an external feed.

Figure 30-23 presents a comparison of the nonconstricted arc of the GTAW process and the constricted arc of plasma arc welding and shows the differences in

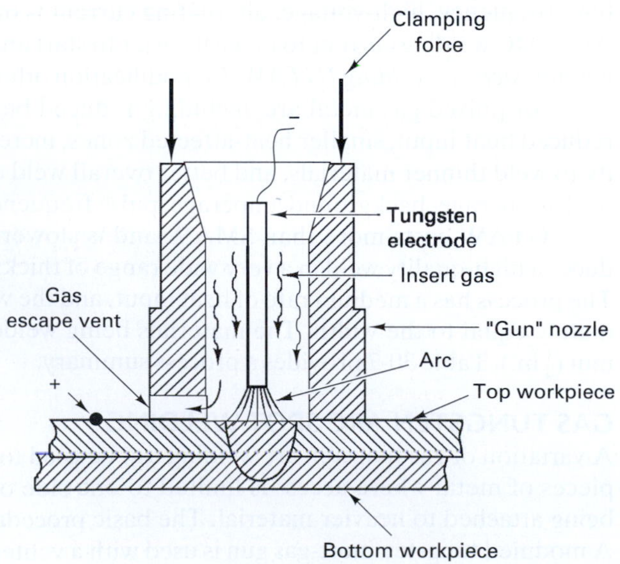

FIGURE 30-21 Process schematic of spot welding by the inert-gas shielded tungsten arc process.

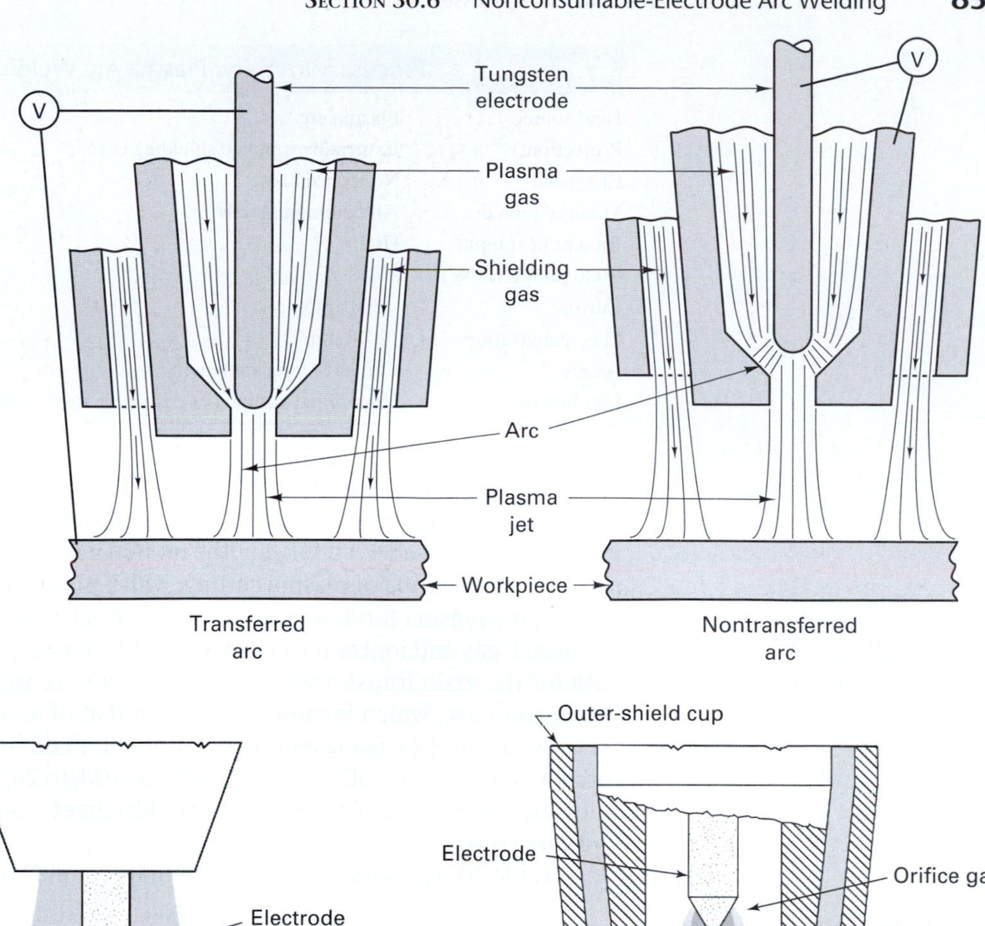

FIGURE 30-22 Two types of plasma arc torches: (left) transferred arc; (right) nontransferred arc.

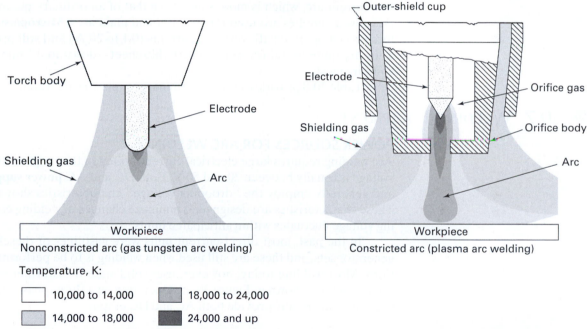

Torch body
Electrode
Shielding gas
Arc
Workpiece
Nonconstricted arc (gas tungsten arc welding)

Outer-shield cup
Electrode
Orifice gas
Shielding gas
Orifice body
Arc
Workpiece
Constricted arc (plasma arc welding)

Temperature, K:

10,000 to 14,000
14,000 to 18,000
18,000 to 24,000
24,000 and up

FIGURE 30-23 Comparison of the nonconstricted arc of gas tungsten arc welding and the constricted arc of the plasma arc process. Note the level and distribution of temperature. *(Courtesy ASM International, Materials Park, OH)*

temperature distribution. Plasma arc welding is characterized by a high rate of heat input and temperatures on the order of 16,500°C (30,000°F). This in turn offers fast welding speeds, narrow welds with deep penetration (a depth-to-width ratio of about 3:1), a narrow heat-affected zone, reduced distortion, and a process that is insensitive to variations in arc length because the plasma column is cylindrical. Welds can be made in all positions, and nearly all metals and alloys can be welded.

With a low-pressure plasma and currents between 20 and 100 A, the metal simply melts and flows into the joint. This condition is similar to the gas tungsten arc process, but with higher penetration and greater tolerance to surface contamination. At higher pressures and currents in excess of 100 A, a **keyhole effect** occurs in which the plasma gas creates a hole completely through the sheet (up to 10 mm, or $\frac{3}{8}$ in. thick) that is

TABLE 30-8	Process Summary: Plasma Arc Welding (PAW)
Heat source	Plasma arc
Protection	Externally supplied shielding gas
Electrode	Nonconsumable
Material joined	All common metals
Rate of heat input	High
Weld profile (D/W)	3
Current	<500 amps
Max. penetration	12–18 mm
Assets	Can have long arc length
Limitations	High initial equipment cost, large torches may limit accessibility

surrounded by molten metal. As the torch is moved, liquid metal flows to fill the keyhole. The keyhole condition offers deep penetration and high welding speeds. If the gas pressure is increased even further, the molten metal is expelled from the region, and the process becomes one of plasma cutting, which is discussed later in this chapter.

Many plasma torches employ a small, nontransferred arc within the torch to heat the orifice gas and ionize it (a pilot arc). The ionized gas then forms a good conductive path for the main transferred arc. This dual-arc technique permits instant ignition of a low-current arc, which is more stable than that of an ordinary plasma torch. Separate DC power supplies are used for the pilot and main arcs. **Microplasma,** or **needle arc,** torches can operate with very low currents (0.1 to 20 A) and still produce stable arcs. They are quite useful for welding very thin sheet—down to 0.1 mm (0.004 in.)—wire and wire mesh.

Table 30-8 provides a process summary for plasma arc welding.

30.7 WELDING EQUIPMENT

POWER SOURCES FOR ARC WELDING

Arc welding requires large electrical currents, often in the range of 100 to 1000 A. The voltage is usually between 20 and 50 V. Both DC and AC **power supplies** are available and generally employ the "drooping voltage" characteristics shown in Figure 30-24. These characteristics are designed to minimize changes in welding current as the welding voltage fluctuates within anticipated limits.

In the past, most direct-current units were gasoline- or diesel-powered motor-generator sets, and these are still used when welding is to be performed in remote locations. Most welding today, however, uses solid-state transformer-rectifier machines, such as the one shown in Figure 30-25. Operating on a three-phase electrical line, these machines can usually provide both AC and DC output.

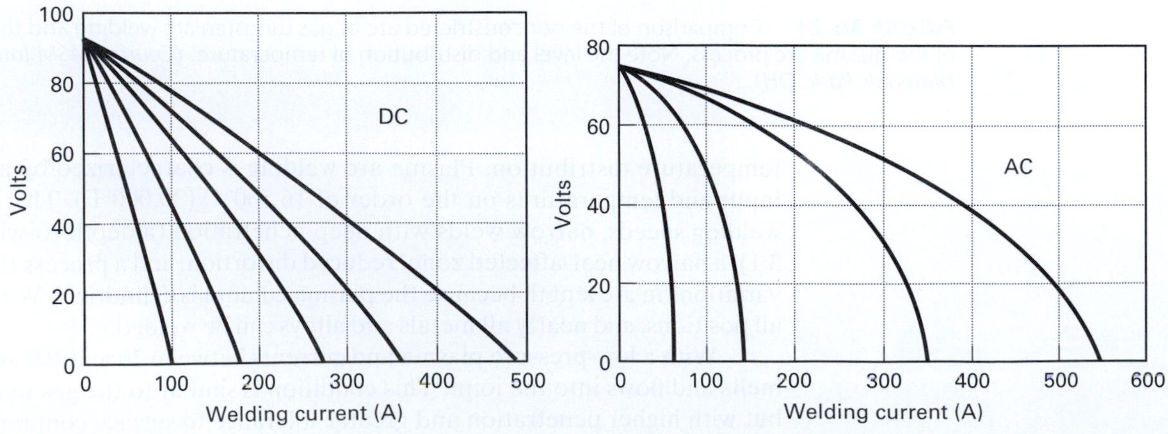

FIGURE 30-24 Drooping-voltage characteristics of typical arc-welding power supplies: (left) direct current; (right) alternating current.

FIGURE 30-25 Rectifier-type AC and DC welding power supply. *(Photo used with permission of The Lincoln Electric Company, Cleveland, OH)*

If only AC welding is to be performed, relatively simple transformer-type power supplies can be used to convert the high-voltage, low-amperage primary power into the low-voltage, high-amperage power needed for welding. These are usually single-phase devices with low power factors. When multiple machines are to be operated, as in a production shop, they are often connected to the various phases of a three-phase supply to help balance the load.

Inverter-based power supplies, introduced in the 1980s, provide great flexibility. Through solid-state electronics, these AC machines can quickly modify the shape and frequency of the pulse waveform or momentarily change the power output. The square wave technology currently being employed provides improved arc starts and more stable arcs. The percentage of time in electrode-negative or electrode-positive can be adjusted along with the amperages in each of those conditions. Increasing the electrode-negative portion narrows the weld bead and increases penetration. Reducing the electrode-negative portion widens the bead, decreases penetration, and produces greater cleaning action to remove adhered oxides. Output frequency can be varied from 20 to 400 Hz, with higher frequencies giving a tighter, more focused arc and a narrower weld bead. Through feedback and logic control, the power supply can actually adjust to compensate for changes in a number of process variables.

JIGS, POSITIONERS, AND ROBOTS

Jigs or fixtures (also called **positioners**) are frequently used to hold the work in production welding. By positioning and manipulating the workpiece, the welding operations can often be performed in a more favorable orientation. Parts can also be mounted on numerically controlled (NC) tables that position them with respect to the welding tool.

Industrial robots have replaced humans for many welding applications. They can operate in hostile environments and are capable of producing high-quality welds in a repetitive mode.

30.8 ARC CUTTING

While oxygen torch cutting was discussed earlier in this chapter, and laser cutting will be covered in a future chapter, there are a number of arc-cutting methods. Virtually all metals can be cut by some form of electric arc. In these processes, the material is melted by the intense heat of the arc and then permitted, or forced, to flow away from the region of the slit or notch (kerf). Most of the techniques are simply adaptations of the arc-welding procedures discussed in this chapter. Each has its inherent characteristics and capabilities, including tolerance, thickness capability, kerf width, edge squareness, size of the heat-affected zone, and cost. Selection depends on factors such as tolerance requirements, the subsequent processes that will be performed on the cut part, and the end use of the product. The ideal cut would have a low bevel angle (less than 1 degree), no rounding of the top edge, no dross on the bottom edge, minimal heat-affected zone, and a smooth cut face.

CARBON ARC AND SHIELDED METAL ARC CUTTING

The carbon arc cutting (CAC) and shielded metal arc cutting (SMAC) methods use the arc from a carbon or shielded metal arc electrode to melt the metal, which is then removed from the cut by gravity or the force of the arc itself. These processes are generally limited to small shops, garages, and homes, where there is limited investment in equipment.

AIR CARBON ARC CUTTING

In air carbon arc cutting (AAC), the arc is again maintained between a carbon electrode and the workpiece, but high-velocity jets of air are directed at the molten metal from holes in the electrode holder. While there is some oxidation, the primary function of the air is to blow the molten material from the cut. Air carbon arc cutting is particularly effective for cutting cast iron and preparing steel plates for welding. Speeds up to 0.6 m/min are possible, but the process is quite noisy, and hot metal particles tend to be blown over a substantial area.

OXYGEN ARC CUTTING

In oxygen arc cutting (AOC), an electric arc and a stream of oxygen are combined to make the cut. The electrode is a coated ferrous-metal tube. The coated metal serves to establish a stable arc, while oxygen flows through the bore and is directed on the area of incandescence. With easily oxidized metals, such as steel, the arc preheats the base metal, which then reacts with oxygen, becomes liquefied, and is expelled by the oxygen stream.

GAS METAL ARC CUTTING

If the wire feed rate and other variables of gas metal arc welding (GMAW) are adjusted so that the electrode penetrates completely through the workpiece, cutting rather than welding will occur, and the process becomes gas metal arc cutting (GMAC). The wire feed rate controls the quality of the cut, and the voltage determines the width of the slit or kerf.

GAS TUNGSTEN ARC CUTTING

Gas tungsten arc cutting (GTAC) employs the same basic circuit and shielding gas as used in gas tungsten arc welding, with a high-velocity jet of gas added to expel the molten metal.

PLASMA ARC CUTTING

The torches used in plasma arc cutting (PAC) produce the highest temperatures available from any practical source. With the nontransferred type of torch, the arc column is completely within the nozzle, and a temperature of about 16,500°C (30,000°F) is obtained. With the transfer-type torch, the arc is maintained between the electrode and the workpiece, and temperatures can be as high as 33,000°C (60,000°F). Ionized gases flowing at these temperatures and near supersonic speeds are capable of cutting virtually any electrically conductive material simply by melting it and blowing it away from the cut. Materials up to 75 mm (3 in.) thick can be cut with plasma torches.

Early efforts to employ this technique showed that the speed, versatility, and operating cost were far superior to those of the oxyfuel cutting methods. However, the early systems could not constrict the arc sufficiently to produce the quality of cut needed to meet the demands of manufacturing. Therefore, plasma arc cutting was generally limited to those materials that could not be cut by the oxidation type of cutting techniques. In the 1970s, swirling of the plasma gas combined with radial impingement of water on the arc was found to produce the desired constriction. A steam curtain forms at the plasma–water interface, shielding the plasma; shrinking its diameter; and creating an intense, highly focused column of heat that liquefies metal and ejects it out the bottom of the column. Water-injected torches can now cut virtually any metal in any position. Magnetic fields have also been used to constrict the arc and can produce high-quality cuts without the need for water impingement.

Compared to oxyfuel cutting, plasma cutting is more economical (cost per cut is a fraction of oxyfuel), more versatile (can cut all metals as easily as mild steel), and much faster (typically, five to eight times faster than oxyfuel). Cutting speeds up to 7.5 m/min (25 ft/min) have been obtained in 6-mm ($\frac{1}{4}$-in.)-thick aluminum, and up to 2.5 m/min (8 ft/min) in 12.5-mm ($\frac{1}{2}$-in.)-thick steel. The combination of the extremely high temperatures and jet-like action of the plasma produces narrow kerfs and remarkably smooth surfaces, nearly as smooth as can be obtained by sawing. Plasma-cut surfaces are often within 2 degrees of vertical, and surface oxidation is nearly eliminated by the cooling effect of the water spray. In addition, the heat-affected zone in the metal is only one-third to one-fourth as large as that produced by oxyfuel cutting, and a preheat cycle is not required in the cutting of steel. Heat-related distortion is extremely small.

Transferred-arc torches are usually used for cutting metals, while the nontransferred type are employed with the low-conductivity nonmetals. Ordinary air or inexpensive nitrogen can be used as the plasma gas for the cutting of all types of metal. Oxygen plasma systems were introduced in the 1980s and are used on carbon and low-alloy steel products with thicknesses ranging from 2 to 32 mm (up to $1\frac{1}{4}$ in.). As with the oxyfuel cutting systems, the oxygen gas reacts with the iron to provide additional heat and produces a better-quality cut at higher speeds. When cutting thick sections (greater than 12 mm or $\frac{1}{2}$ in.), stainless steels, or nonferrous metals, an argon–hydrogen mixture may be preferred to provide a hotter, deeper-penetrating arc. A secondary flow of shielding gas (nitrogen, air, or carbon dioxide) may be used to help cool the torch, blow the molten metal away, shield the arc, and prevent oxidation of the cut surface. The arc-constricting water flow can also serve as a shielding medium.

During the 1990s, *high-density*, or **precision plasma,** systems began to appear. Various designs are used to restrict the orifice (i.e., superconstrict the plasma), producing vertical edges (less than 1-degree taper), close tolerances (one-third that of conventional), and dross-free plasma cutting of thin materials. The lower-amperage torches (10 to 100 A) are limited to cutting carbon and low-alloy steels less than 16 mm ($\frac{5}{8}$ in.) thick and higher-performance metals (such as stainless and high-strength steels, nickel alloys, titanium, and aluminum) less than 12 mm ($\frac{1}{2}$ in.) thick. The cutting speeds are slower than conventional plasma cutting, but there is no change in the size of the heat-affected zone. **Pulsed plasma arc cutting,** another recent development, can reduce heat input to the workpiece while producing kerfs that are 50% narrower and cleaner edges on the cuts.

Combining a plasma torch with CNC manipulation can provide fast, clean, and accurate cutting, like that shown in Figure 30-26. Because plasma torches work under water, the cutting table may be submerged as a means of reducing noise, air pollution, dust, and arc glare (dyes are placed in the water). Plasma arc torches can also be incorporated into punch presses to provide a manufacturing machine with outstanding flexibility in producing cut and punched products from a variety of materials.

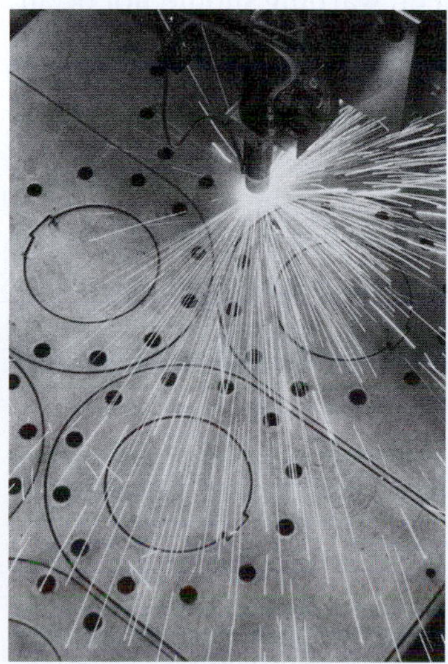

FIGURE 30-26 Cutting sheet metal with a plasma torch. *(Roger Tully/Getty Images, Inc.)*

Plasma arc cutting is also suitable for robot application. A single robot system can be used for both cutting and welding of intricate shapes and contours. Water constriction of the manipulated arc is a problem, however, making it important to select the right process parameters and type of gas for the particular application.

Table 30-9 compares the features of oxyfuel cutting, plasma arc cutting, and laser cutting.

TABLE 30-9 Cutting Process Comparison: Oxyfuel, Plasma Arc, and Laser

Feature	Oxyfuel Cutting	Plasma Arc Cutting	Laser Cutting
Preferred Materials	Carbon steel and titanium	All electrically conductive metals	Metal, plastic, wood, textiles
Quality of Cut	Average	Similar to oxyfuel almost as good as laser on thin material	Good quality—best for plate material $>\frac{1}{2}$ in. thick
Thickness Range			
1. Steel	$\frac{3}{16}$ to unlimited	26 ga. to 3 in.	Foil to 1 in.
2. Stainless	not used	26 ga. to 5 in.	20 ga to $\frac{3}{4}$ in.
3. Aluminum	not used	22 ga. to 6 in.	20 ga. to $\frac{3}{4}$ in.
Cutting Speed or Time	Long preheat is required	Fast cutting	Slower than plasma, but faster than oxyfuel

30.9 METALLURGICAL AND HEAT EFFECTS IN THERMAL CUTTING

When used for cutting, the flame and arc processes expose materials to high localized temperatures and can produce harmful metallurgical effects. If the cut edges will be subsequently welded, or if they will be removed by machining, there is little cause for concern. When the edges are retained in the finished product, however, consideration should be given to the effects of cutting heat and their interaction with the applied loads. In some cases, additional steps may be required to avoid or overcome harmful consequences.

For carbon steels with less than 0.25% carbon, thermal cutting does not produce serious metallurgical effects. However, in steels of higher carbon content, the metallurgical changes can be quite significant, and preheating and/or postheating may be required. For alloy steels, additional consideration should be given to the effects of the various alloy elements.

Because of the low rate of heat input, oxyacetylene cutting will produce a rather large **heat-affected zone (HAZ).** The arc-cutting methods produce intermediate effects that are quite similar to those of arc welding. Plasma arc cutting is so rapid, and the heat is so localized, that the original properties of a metal are only modified within 1.5 mm (0.06 in.) of the cut.

All of the thermal cutting processes produce some **residual stresses,** with the cut surface generally in tension. Except in the case of thin sheet, warping should not occur. However, if subsequent machining removes only a portion of the cut surface, or does not penetrate to a sufficient depth, the resulting imbalance in residual stresses can induce distortion. It may be necessary to remove all cut surfaces to a substantial depth to ensure good dimensional stability.

Thermal cutting can also introduce geometrical features into the edge. All flame- or arc-cut edges are rough to varying degrees and thus contain notches that can act as stress raisers and reduce the endurance or fracture strength. If cut edges are to be subjected to high or repeated tensile stresses, the cut surfaces and the heat-affected zone should be removed by machining or at least subjected to a stress-relief heat treatment.

KEY WORDS

acetylene
alternating current
arc
arc welding
bulk welding
carburizing flame
consumable-electrode process
direct-current electrode-negative (DCEN)
direct-current electrode-positive (DCEP)
electrode
filler metal
flame straightening
flux
flux-cored arc welding (FCAW)
fusion welding
gas metal arc welding (GMAW)
gas tungsten arc welding (GTAW)
globular transfer

heat-affected zone (HAZ)
heliarc welding
jig
kerf
keyhole effect
methyl-acetylene-propadiene (MAPP)
microplasma
metal inert-gas (MIG) welding
needle arc
neutral flame
nonconsumable-electrode process
nontransferred arc
orifice gas
oxidizing flame
oxyacetylene cutting (OFC-A)
oxyfuel-gas cutting (OFC)
oxyfuel-gas welding (OFW)
oxygen lance cutting (LOC)
penetration
plasma
plasma arc welding (PAW)
positioner
power supply
precision plasma
pressure gas welding (PGW)
pulsed plasma arc cutting
pulsed spray transfer
residual stresses
reverse polarity
shielded metal arc welding (SMAW)
short-circuit transfer
slag
spot weld
spray transfer
stack cutting
stick welding
straight polarity
stud welding (SW)
submerged arc welding (SAW)
thermal cutting
tungsten inert-gas (TIG) welding
torch
transferred arc
upsetting
variable polarity

■ REVIEW QUESTIONS

1. Why does an oxyfuel-gas welding torch usually have a flame with two distinct regions?
2. What is the location of the maximum temperature in an oxyacetylene flame?
3. What function or functions are performed by the outer zone of the welding flame?
4. What three types of flames can be produced by varying the oxygen-to-fuel ratio?
5. Which type of oxyfuel flame is most commonly used?
6. What are some of the alternative fuels (other than acetylene) for oxyfuel-gas welding? What attractive features might they offer?
7. Why might a welder want to change the tip size (or orifice diameter) in an oxyacetylene torch?
8. What is filler metal, and why might it be needed to produce a joint?
9. What is the role of a welding flux?
10. Oxyfuel-gas welding has a low rate of heat input. What are some of the adverse features that result from the slow rate of heating?
11. What are some of the more attractive features of the oxyfuel-gas process?
12. How does pressure gas welding differ from the oxyfuel-gas process?
13. In what way does the torch cutting of ferrous metals differ from cutting nonoxidizing metals?
14. Why might it be possible to use only an oxygen lance to cut hot steel strands as they emerge from a continuous casting operation?
15. How does an oxyacetylene cutting torch differ from an oxyacetylene welding torch?
16. What are some of the ways in which cutting torches can be manipulated?
17. When might stack cutting be an attractive process?
18. What modification must be incorporated into a cutting torch to permit it to cut metal under water?
19. If a curved plate is to be straightened by flame straightening, should the heat be applied to the longer or shorter surface of the arc? Why?
20. Why does the flame-straightening process not work for thin sheets of metal?
21. What sorts of problems plagued early attempts to develop arc welding?
22. What are the three basic types of current and polarity that are used in arc welding?
23. What is the difference between a consumable and nonconsumable electrode? For which processes does a filler metal have to be added by a separate mechanism?
24. What are the three types of metal transfer that can occur during arc welding?
25. What are some of the process variables that must be specified when setting up an arc-welding process?
26. What are the four primary consumable-electrode arc-welding processes?
27. What are some general properties of the consumable-electrode arc-welding processes?
28. What are some of the functions of the electrode coatings used in shielded metal arc welding?
29. How are welding electrodes commonly classified, and what information does the designation usually provide?
30. Why are shielded metal arc electrodes often baked just prior to welding?
31. What is the function of the slag coating that forms over a shielded metal arc weld?
32. What benefit can be obtained by placing iron powder in the coating of shielded metal arc electrodes that will be used to weld ferrous metals?
33. Why are shielded metal arc electrodes generally limited in length, forcing the process to be one of intermittent operation?
34. Why is the shielded metal arc welding process limited to low welding currents and shallow penetration?
35. What are some of the attractive features of the shielded metal arc welding process?
36. What is the advantage of placing the flux in the center of an electrode (flux-cored arc welding) as opposed to a coating on the outside (shielded metal arc welding)?
37. What feature enables the welding current in FCAW to be higher than in SMAW?
38. What are some of the advantages of gas metal arc welding compared to the shielded metal arc process?
39. Describe the relative performance of argon, helium, and carbon dioxide gases in creating a high-temperature arc and promoting weld penetration.
40. For what welding conditions would short-circuit transfer be preferred?

41. What is the least desired mode of metal transfer?
42. Which of the metal transfer mechanisms is most used in arc welding?
43. Describe the metal transfer that occurs during pulsed arc gas metal arc welding.
44. What are some of the benefits that can be obtained by the reduced heating of the pulsed arc process?
45. What are some of the primary process variables in the gas metal arc welding process?
46. What is the attractive feature of advanced gas metal arc welding?
47. What benefits can be gained by using a rectangular cross-section electrode wire as opposed to a round one?
48. What are some of the functions of the flux in submerged arc welding?
49. What are some of the attractive features of submerged arc welding? Major limitations?
50. What is the primary goal or objective in bulk welding?
51. What is the primary objective of stud welding?
52. What is the function of the ceramic ferrule placed over the end of the stud in stud welding?
53. What is the current (or proper) designation for MIG welding? TIG welding? Heliarc welding?
54. What types of shielding gases are used in the gas tungsten arc process?
55. What are some of the features that must be specified for the tungsten electrode?
56. What can be done to increase the rate of filler metal deposition during gas tungsten arc welding?
57. What are some of the attractive features of gas tungsten arc welding?
58. For the GTAW process, what are the attractive features of the DCEN polarity? DCEP?
59. How are the spot welds produced by gas tungsten arc spot welding different from those made by conventional resistance spot welding?
60. How is the heating of the workpiece during plasma arc welding different from the heating in other arc welding techniques?
61. What are the two different gas flows in plasma arc welding?
62. What are some of the attractive features of plasma arc welding?
63. What is the keyhole effect in plasma arc welding?
64. What is the primary difference between plasma arc welding and plasma arc cutting?
65. What are the attractive features or benefits of an inverter-based power supply?
66. What are jigs, fixtures, or positioners, and how are they used in welding?
67. What is the kerf in thermal cutting operations?
68. What is the purpose of the oxygen in oxygen arc cutting?
69. Why is plasma arc cutting an attractive way of cutting high-melting-point materials?
70. What techniques can be used to constrict the arc in plasma arc cutting, producing a narrower, more controlled cut?
71. Compared to oxyfuel cutting, what are some of the attractive features of the plasma technique?
72. How can a nontransferred arc plasma torch be used to cut low-conductivity nonmetals?
73. What is the attractive feature of pulsed plasma arc cutting?
74. Describe the relative size of the heat-affected zone for the various cutting processes: oxyfuel, arc, and plasma.
75. Why might the residual stresses induced during cutting operations be objectionable?
76. Why might it be wise to machine away the thermally cut edge and heat-affected zone of metal that will be used as a stressed machine part?

www.wiley.com/go/global/degarmo

Chapter 30 CASE STUDY

Bicycle Frame Construction and Repair

As a new employee in a bicycle shop, customers frequently seek your advice on a number of matters. What type of bicycle is best for them? What material is the bicycle made from? How has it been manufactured? And can it be repaired when it has been damaged? One of the key components of a bicycle is the frame—the backbone of the bike—which provides the necessary strength and stiffness. Early frames were made of wood or cast iron, which transitioned to steel tubing; then to thinner-wall and higher-strength steel tubing; and now to aluminum, titanium, carbon-fiber, and other exotic materials. In addition to strength and stiffness, customers also want their bike to offer light weight and a smooth ride—properties that often must be compromised with the original strength and stiffness.

Steel is the most common bike frame material. Carbon steel is the cheapest but also results in one of the heavier bike frames. Chrome-molybdenum steel (alloy steel) can provide the necessary strength with thinner tubing and has become a popular frame material. Aluminum is widely used to provide light weight and adequate strength, coupled with good corrosion resistance. Its fatigue resistance is poor, however. Titanium offers the strength of steel with a weight between steel and aluminum and good corrosion resistanc but at a significantly higher cost. Carbon fiber can produce one of the lightest frames, but the composite is brittle and the design must accommodate this feature.

Bent or broken frames appear to be a common problem, and you learn that inexpensive bicycles generally have frames made from welded low- or medium-carbon steel tubing. This material can be heated to straighten and is easily repair-welded using a wide variety of welding techniques. As you move to the more costly, lightweight, or high-performance models, you learn that repair is generally not quite as simple.

1. In one case, the frame of a high-quality, lightweight bicycle had been fabricated from aluminum alloy tubing, with joints made by adhesive bonding of the tubes to connectors that incorporate either internal lugs or external sleeves. When the frame fractured near one of the joints, the owner contacted a local auto body shop and requested that a repair be made using conventional gas tungsten arc welding. The welder was familiar with the welding of aluminum, and the repair seemed to be of good quality. Shortly thereafter, however, the frame broke again. This time the fracture was adjacent to the repair weld and the characteristics of the break were different. While the first fracture was somewhat brittle in nature, the second appeared to be more ductile, with evidence of metal flow prior to fracture. Because the second fracture occurred in the tube material and not the weld itself, the welder felt that the failure was not related to the attempted repair and the material in the tubing must be defective.
   a. If the material had been cold-drawn aluminum tubing (i.e., strain hardened), explain what may have occurred during the repair. What is the probable cause of the second fracture? Was the weld in any way defective? Was the second failure related to the welding repair?
   b. If the tubing had been strengthened by an age-hardening heat treatment, could the same results have occurred? Explain.
   c. Is there a better means of repairing the original fracture? What would have been your recommendation?
2. Magnesium, while not as strong as steel or aluminum, is the lightest weight engineering metal. Would magnesium be an appropriate material for bicycle frame construction? If so, how would the frames be assembled?
3. If the bicycle frame deflects, motion of the cyclist and related energy can be wasted. Therefore, a rigid frame may be quite desirable. Beryllium is an extremely rigid, lightweight metal. Could it be used as a material for bicycle frames? How would you fabricate the tubing and assemble the frame? How would cost compare to steel or aluminum?
4. Titanium offers the strength of heat-treated steel at approximately half of the weight. How would you propose to join the tubular segments of a titanium bicycle frame?
5. Composite materials can be used to produce tailored sets of properties. Fiber-reinforced composites can have extremely high rigidity in the direction of fiber orientation, coupled with extremely light weight.
   a. If you were to assemble a composite frame using fiber-reinforced tubing, such as graphite fiber reinforced epoxy, how would you join the assembly?
   b. The strength and stiffness of carbon fiber can be extremely directional. Maximum strength can be provided in directions of maximum stress. Bicycle frames, however, often experience a variety of stresses in different directions. Consider the properties in directions transverse to the fiber orientation. Are they equivalent to alternate materials or would they be compromised?
6. Premium quality racing bikes (such as Tour De France models) have used one-piece carbon fiber composite frames (i.e., no joints at all!). What is the benefit of such a design?
7. Numerous joints appear in bicycle frames, some of which may involve dissimilar materials. Consider each of the following possibilities and discuss possible joining methods and the properties of each.
   a. Aluminum to aluminum.
   b. Carbon fiber-to-aluminum lugs.
   c. Carbon fiber-to-carbon fiber.
8. Some adhesive-bonded joints may require a thermal treatment to complete the cure. Welded or brazed frames may benefit from a stress-relief heat treatment, or a complete reheat treatment to establish a uniform set of properties throughout the joined assembly. What materials would benefit from a subsequent heat treatment? What concerns might you have regarding these processes?

CHAPTER 31

RESISTANCE- AND SOLID-STATE WELDING PROCESSES

31.1 INTRODUCTION

As indicated in the lists of Figures 29-1 and 29-2, there are a number of welding and cutting processes that utilize heat sources other than oxyfuel flames and electric arcs, and some use no heat source at all. We begin this chapter with a group of processes that use electrical resistance heating to form the joint. The second group of processes in this chapter, known as **solid-state welding** processes, create joints without any melting of the workpiece or filler material.

31.2 THEORY OF RESISTANCE WELDING

In **resistance welding,** heat and pressure are combined to induce coalescence. **Electrodes** are placed in contact with the material, and electrical resistance heating is used to raise the temperature of the workpieces and the interface between them. The same electrodes that supply the current also apply the pressure, which is usually varied throughout the weld cycle. A certain amount of pressure is applied initially to hold the workpieces in contact and thereby control the electrical resistance at the interface. When the proper temperature has been attained, the pressure is increased to induce coalescence. Because pressure is utilized, coalescence occurs at a lower temperature than that required for oxyfuel gas or arc welding. In fact, melting of the base metal does not occur in many resistance-welding operations. Resistance-welding processes might well be considered as a form of solid-state welding, although they are not officially classified as such by the American Welding Society.

In some resistance-welding processes, additional pressure is applied immediately after coalescence to provide a certain amount of forging action. Accompanying the deformation is a certain amount of grain refinement. Additional heating can also be employed after welding to provide tempering and/or stress relief.

The required temperature can often be attained, and coalescence can be achieved, in a few seconds or less. Resistance welding, therefore, is a very rapid and economical process, extremely well suited to automated manufacturing. No filler metal is required, and the tight contact maintained between the workpieces excludes air and eliminates the need for fluxes or shielding gases.

HEATING

The heat for resistance welding is obtained by passing a large electrical current through the workpieces for a short period of time. The amount of heat input can be determined by the basic relationship:

$$H = I^2 R t$$

where:

H = total heat input in joules
I = current in amperes
R = electrical resistance of the circuit in ohms
t = length of time during which current is flowing in seconds

It is important to note that the workpieces actually form part of the electrical circuit, as illustrated in Figure 31-1, and that the total resistance between the electrodes consists of three distinct components:

1. The bulk resistance of the electrodes and workpieces—the upper electrode, upper workpiece, lower workpiece, and lower electrode.
2. The contact resistance between the electrodes and the workpieces—between the upper electrode and upper workpiece and the lower electrode and lower workpiece.
3. The resistance between the surfaces to be joined, known as the **faying surfaces.**

Because the maximum amount of heat is generated at the point of maximum resistance, it is desirable for this to be the location where the weld is to be made. Therefore, it is essential to keep components 1 and 2 as low as possible with respect to resistance 3. The bulk resistance of the electrodes is always quite low, and that of the workpieces is determined by the type and thickness of the metal being joined. Because of the large areas involved and the relatively high electrical conductivity of most metals, the workpiece resistances are usually much less than the contact or interface values. Resistance 2 (the resistance between the electrodes and workpieces) can be minimized by using electrode materials that are excellent electrical conductors and by controlling the size and shape of the electrodes and the applied pressure. Any change in the pressure between the electrodes and workpieces, however, also affects the contact between the faying surfaces. Therefore, the ability to control the electrode-to-work resistance by pressure variation is quite limited.

The final resistance, that between the faying surfaces, is a function of (1) the quality (surface finish or roughness) of the surfaces; (2) the presence of nonconductive scale, dirt, or other contaminants; (3) the pressure; and (4) the contact area. These factors must all be controlled if uniform resistance welds are to be produced.

As indicated in Figure 31-2, the objective of resistance welding is to bring both of the faying surfaces to the proper temperature, while simultaneously keeping the remaining material and the electrodes relatively cool. Water cooling is usually used to keep the electrode temperature low and thereby extend their useful life.

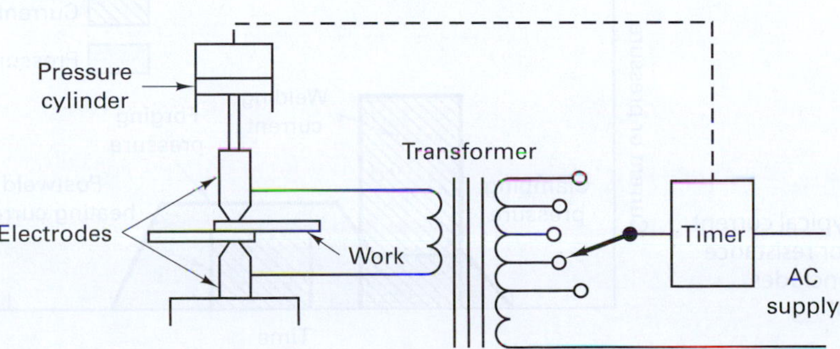

FIGURE 31-1 The basic resistance-welding circuit.

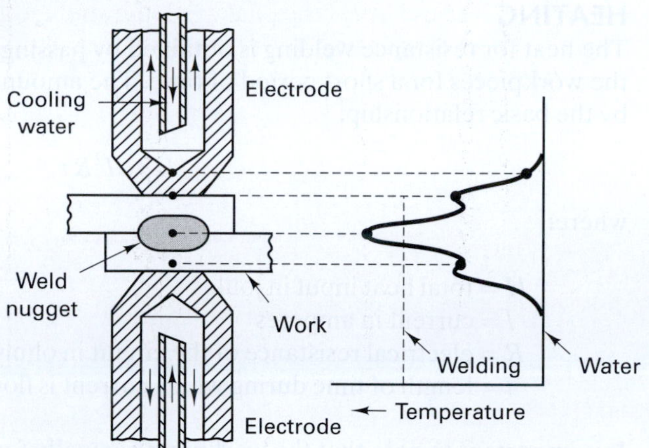

FIGURE 31-2 The desired temperature distribution across the electrodes and workpieces during resistance welding.

PRESSURE

Because the applied pressure promotes a forging action, resistance welds can be produced at lower temperatures than welds made by other processes. Controlling both the magnitude and timing of the pressure, however, is very important. If too little pressure is used, the contact resistance will be high and surface burning or pitting of the electrodes may result. If excessive pressure is applied, molten or softened metal may be expelled from between the faying surfaces, or the electrodes may indent the softened workpiece. Ideally, moderate pressure should be applied to hold the workpieces in place and establish proper resistance at the interface prior to and during the passage of the welding current. The pressure should then be increased considerably just as the proper welding temperature is attained. This completes the coalescence and forges the weld to produce a fine-grained structure.

On small, foot-operated machines, only a single spring-controlled pressure is used. On larger, production-type welders, the pressure is generally applied through controllable air or hydraulic cylinders.

CURRENT AND CURRENT CONTROL

While surface conditions and pressure are important variables, the temperature achieved during resistance welding is primarily determined by the magnitude and duration of the welding current. The various resistances change as current flows and the material heats. The bulk resistances of metal increase as temperature rises, and the contact resistances decrease as the metal softens and pressure improves the contact. Because the best conditions are the initial ones, high currents and short time intervals are generally preferred. The weld location can attain the desired temperature while minimizing the amount of heat generated in or dissipated to the adjacent material.

In production-type welders, the magnitude, duration, and timing of both current and pressure can be programmed to follow specified cycles. Figure 31-3 shows a relatively simple cycle for a resistance weld that includes both forging and postheating

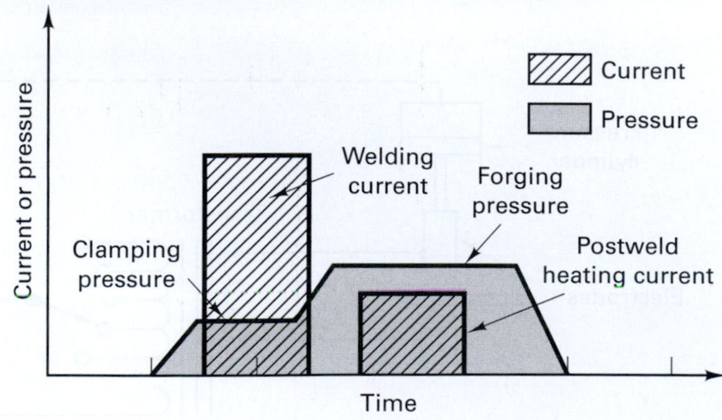

FIGURE 31-3 A typical current and pressure cycle for resistance welding. This cycle includes forging and postheating operations.

operations. The quality of the final weld, therefore, often depends more on the development of a proper schedule and the subsequent setup, adjustment, and maintenance of equipment than it does on operator skill.

POWER SUPPLY

Because the overall resistance in the welding circuits can be quite low, high currents are generally required to produce a resistance weld. Power transformers convert the high-voltage, low-current line power to the high-current (up to 100,000 A), low-voltage (0.5 to 10 V) power required for welding. Smaller machines may utilize single-phase circuitry, but the larger units generally operate on three-phase power. Many resistance welders use DC welding current, obtained through solid-state rectification of the three-phase power. These machines reduce the current demand per phase, give a balanced load, and produce excellent welds.

Newer power supplies can also adapt to changes in the process conditions. The resistance changes as the initial contact interface is replaced with a weld. The resistance can also change from one weld to the next with variations in part geometry, surface conditions, and part alignment. Over time, the electrodes can mushroom, changing the area of contact, and the temperature of the electrodes can vary. These and other changes can be compensated by adjustments in voltage, current, or duration of current flow.

31.3 RESISTANCE-WELDING PROCESSES

RESISTANCE SPOT WELDING

Resistance spot welding (RSW) is the simplest and most widely used form of resistance welding, providing a fast, economical means of joining overlapped materials that will not require subsequent disassembly. Even with all of the advances in technology, resistance spot welding is still the dominant method for joining sheet material, and the average steel-bodied automobile contains between 2000 and 5000 spot welds. Figure 31-4 presents a schematic of the process. Overlapped metal sheets are positioned between water-cooled electrodes, which have reduced areas at the tips to produce welds that are usually from 1.5 to 13 mm ($\frac{1}{16}$ to $\frac{1}{2}$ in.) in diameter. The electrodes close on the work, and the controlled cycle of pressure and current is applied to produce a weld at the metal interface. The electrodes are then opened, and the work is removed.

A satisfactory spot weld, like the one shown in Figure 31-5, consists of a **nugget** of coalesced metal formed between the faying surfaces. There should be little indentation of the metal under the electrodes. As shown in Figure 31-6, the strength of the weld should be such that in a tensile or tear test, the weld will remain intact while failure occurs in the heat-affected zone surrounding the nugget. Sound spot welds can be

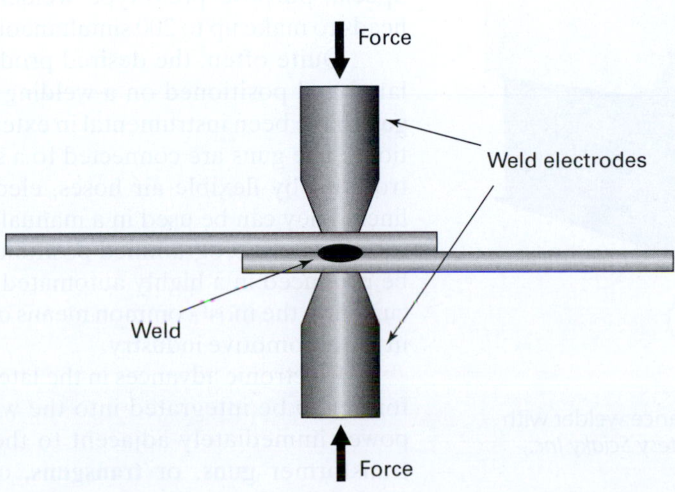

FIGURE 31-4 The arrangement of the electrodes and workpieces in resistance spot welding.

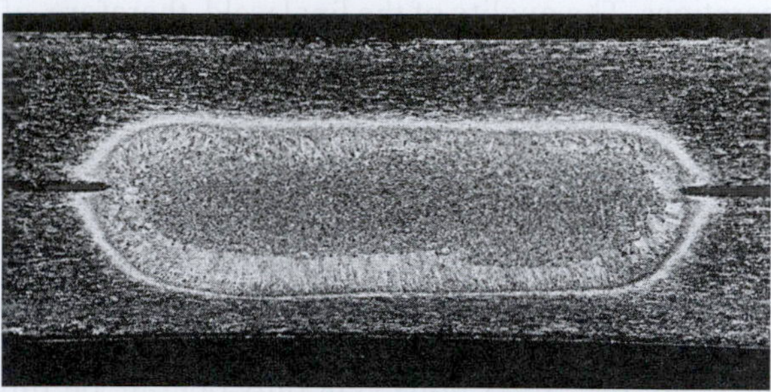

FIGURE 31-5 A spot-weld nugget between two sheets of 1.3-mm (0.05-in.) aluminum alloy. The nugget is not symmetrical because the radius of the upper electrode was greater than that of the lower electrode. *(Courtesy Locheed Martin Corporation, Bethesda, MD)*

FIGURE 31-6 Tear test of a satisfactory spot weld, showing how failure occurs outside of the weld. *(E. Paul DeGarmo)*

obtained with excellent consistency if proper current density and timing, electrode shape, electrode pressure, and surface conditions are maintained.

Spot-Welding Equipment. A variety of spot-welding equipment is available to meet the needs of production operations. For light-production work where complex current-pressure cycles are not required, a simple **rocker-arm machine** is often used. The lower electrode arm is stationary, while the upper electrode, mounted on a pivot arm, is brought down into contact with the work by means of a spring-loaded foot pedal. Rocker-arm machines are available with throat depths up to about 1.2 m (48 in.) and transformer capacities up to 50 kVa. They are used primarily on steel.

FIGURE 31-7 Three-phase, air-operated, press-type resistance welder with microprocessor control. *(Courtesy Sciaky Inc., Chicago, IL.)*

Larger spot welders, and those used at high production rates, are generally of the press type, as shown in Figure 31-7. On these machines, the movable electrode is controlled by an air or hydraulic cylinder, and complex pressure cycles can be programmed. Capacities up to 500 kVa with a 1.5-m (60-in.) throat depth are quite common. Special-purpose press-type welders can employ multiple welding heads to make up to 200 simultaneous spot welds in less than 60 s.

Quite often, the desired products are too large to be manipulated and positioned on a welding machine. Portable **spot-welding guns** have been instrumental in extending the process to such applications. The guns are connected to a stationary power supply and control unit by flexible air hoses, electrical cables, and water-cooling lines. They can be used in a manual fashion or installed on industrial robots where programmed positioning enables quality spot welds to be produced in a highly automated fashion. Robotic spot welding is currently the most common means of joining sheet metal components in the automotive industry.

Electronic advances in the late 1980s enabled the welding transformer to be integrated into the welding gun. By transforming the power immediately adjacent to the area of use, the small integral transformer guns, or **transguns,** offer reduced power losses and enhanced process efficiency. However, if accurate positioning is required in an articulated system like an industrial robot, the added weight of the integral transformer may become a disadvantage. Servomotors have also been incorporated into a variety of spot-welding machines to control the electrode positioning, speed of closure, level of applied torque or pressure, and the rate at which the load is applied.

Electrodes. Resistance spot-welding electrodes must conduct the welding current to the work, set the current density at the weld location, apply force, and help dissipate heat during the noncurrent portions of the welding cycle. Electrical and thermal conductivity properties are important considerations for electrode selection. Hot compressive strength must be sufficient to resist electrode deformation during the application of pressure. In addition, the electrode should not melt under welding conditions and should be of a composition that does not alloy with the material being welded—a phenomenon that promotes sticking or galling and electrode wear. The Resistance Welder Manufacturers Alliance (RWMA) has standardized various electrode geometries (size and shape) and has approved a variety of electrode materials, including copper-base alloys, refractory metals, and refractory–metal composites. A straight electrode with a 6.5-mm ($\frac{1}{4}$-in.)-diameter flat face suits most sheet metal welding applications. A dome-shaped nose may be preferred for applications where alignment may be difficult. Offset electrodes may be used for difficult-to-reach welds.

Spot-Weldable Metals and Geometries. While steel is clearly the most common metal that is spot welded, one of the greatest advantages of the process is that virtually all of the commercial metals can be joined, and most of them can be joined to each other. In only a few cases do the welds tend to be brittle. Table 31-1 shows some of the many combinations of metals that can be successfully spot welded.

While spot welding is primarily used to join wrought sheet material, other forms of metal can also be welded. Sheets can be attached to rolled shapes and steel castings, as well as some types of nonferrous die castings. Most metals require no special preparation, except to be sure that the surface is free of corrosion and is not badly pitted. For best results, aluminum and magnesium should be cleaned immediately prior to welding by some form of mechanical or chemical technique. Metals that have high electrical conductivity require clean surfaces to ensure that the electrode-to-metal resistance is

TABLE 31-1 Metal Combinations That Can Be Spot Welded

Metal	Aluminum	Brass	Copper	Galvanized Iron	Iron (Wrought)	Monel	Nichrome	Nickel	Nickel Silver	Steel	Tin Plate	Zinc
Aluminum	x										x	x
Brass		x	x	x	x	x	x	x	x	x	x	x
Copper		x	x	x	x	x	x	x	x	x	x	x
Galvanized Iron		x	x	x	x	x	x	x	x	x	x	
Iron (Wrought)		x	x	x	x	x	x	x	x	x	x	
Monel		x	x	x	x	x	x	x	x	x	x	
Nichrome		x	x	x	x	x	x	x	x	x	x	
Nickel		x	x	x	x	x	x	x	x	x	x	
Nickel Silver		x	x	x	x	x	x	x	x	x	x	
Steel		x	x	x	x	x	x	x	x	x	x	
Tin Plate	x	x	x	x	x	x	x	x	x	x		
Zinc	x	x	x									x

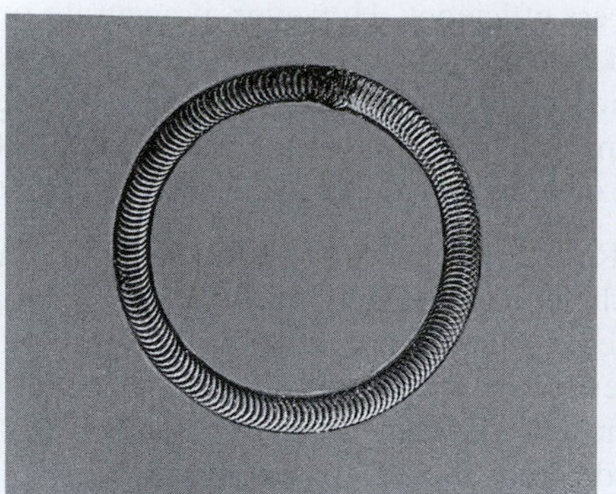

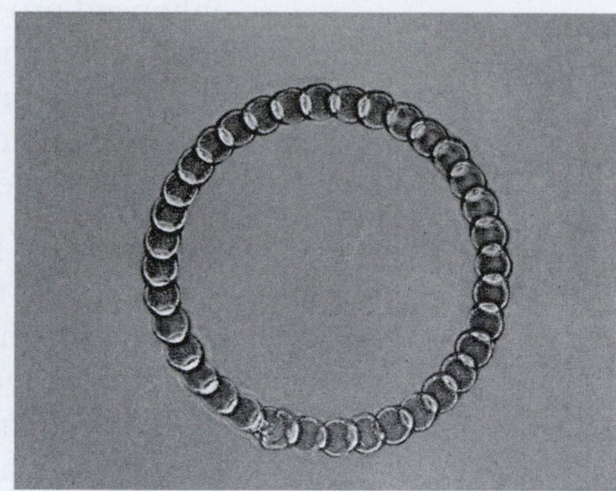

FIGURE 31-8 Seam welds made with overlapping spots of varied spacing. *(Courtesy Taylor-Winfield Corporation, Youngstown, OH)*

low enough for adequate temperature to be developed within the metal itself. Silver and copper are especially difficult to weld because of their high thermal conductivity. Higher welding currents coupled with water cooling of the surrounding material may be required if adequate welding temperatures are to be obtained.

When the two pieces being joined are of the same thickness, the practical limit for spot welding is about 3 mm ($\frac{1}{8}$ in.) for each sheet. Sheets of differing thickness can also be joined, and thin pieces can be attached to material that is considerably thicker than 3 mm. When metals of different thickness or different conductivity are to be welded, however, a larger electrode or one with higher conductivity is often used against the thicker or higher-resistance material to ensure that both workpieces will be brought to the desired temperature in a simultaneous fashion.

RESISTANCE SEAM WELDING

Resistance seam welding (RSEW) require two distinctly different processes. In the first process, sheet metal segments are joined to produce gas- or liquid-tight vessels, such as gas tanks, mufflers, and simple heat exchangers. The weld is made between overlapping sheets of metal, and the seam is simply a series of overlapping spot welds, like those shown in Figure 31-8. The basic equipment is the same as for spot welding, except that the electrodes now assume the form of rotating disks, like those shown schematically in Figure 31-9. As the metal passes between the electrodes, timed pulses of current form the overlapping welds. The timing of the welds and the movement of the work are controlled to ensure that the welds overlap and the workpieces do not get too hot. The welding current is usually a bit higher than in conventional spot welding (to compensate for the short circuit of the adjacent weld), and the workpiece is often cooled by a flow of air or water. In a variation of the process, a continuous current is passed through the rotating electrodes to produce a continuous seam. This form of seam welding is best suited for thin materials, but metals up to 6 mm ($\frac{1}{4}$ in.) can be joined. A typical welding speed is about 2 m/min for thin sheet.

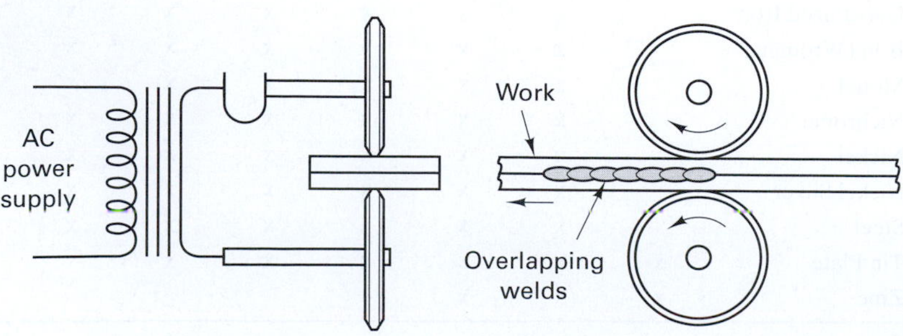

FIGURE 31-9 Schematic representation of the seam welding process.

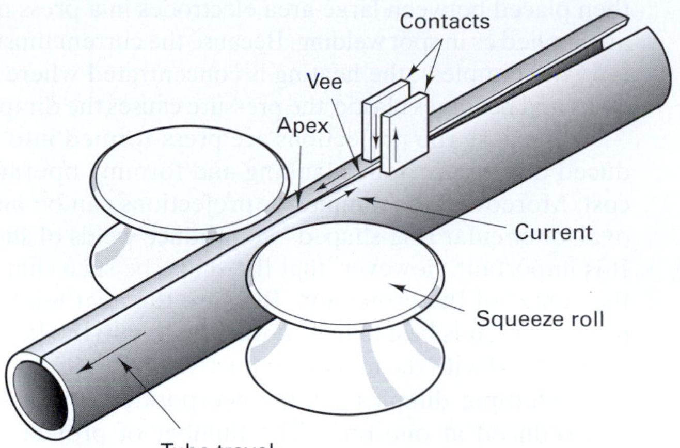

FIGURE 31-10 Using high-frequency AC current to produce a resistance seam weld in butt-welded tubing. Arrows from the contacts indicate the path of the high-frequency current.

The second type of resistance seam welding, known as **resistance butt welding,** is used to produce butt welds between thicker metal plates. The electrical resistance of the abutting metals is still used to generate heat, but high-frequency current (up to 450 kHz) is now employed. At high frequencies, current flows only on or near the surface of conductive material—the higher the frequency, the lower the penetration depth. (*Note:* This is similar to the results obtained in the parallel process of high-frequency induction heating.) When the abutting surfaces attain the desired temperature, they are pressed together to form a weld.

Resistance butt welding is used extensively in the manufacture of pipes and tubes, as illustrated in Figure 31-10, but the process is also used to construct simple structural shapes from sections of plate. Material from 0.1 mm (0.004 in.) to more than 20 mm ($\frac{3}{4}$ in.) in thickness can be welded at speeds up to 80 m/min (250 fpm). The combination of high-frequency current and high welding speed produces a very narrow heat-affected zone. Almost any type or combination of metal can be welded, including difficult dissimilar metals and the high-conductivity metals, such as aluminum and copper.

PROJECTION WELDING

In a mass-production operation, conventional spot welding is plagued by two significant limitations. Because the small electrodes provide both the high currents and the required pressure, the electrodes generally require frequent attention to maintain their geometry. In addition, the process is designed to produce only one spot weld at a time. When increased strength is required, multiple welds are often needed, and this means multiple operations. **Projection welding (RPW)** provides a means of overcoming these limitations.

Figure 31-11 illustrates the principle of projection welding. A dimple is **embossed** into one of the workpieces at the location where a weld is desired. The two workpieces are

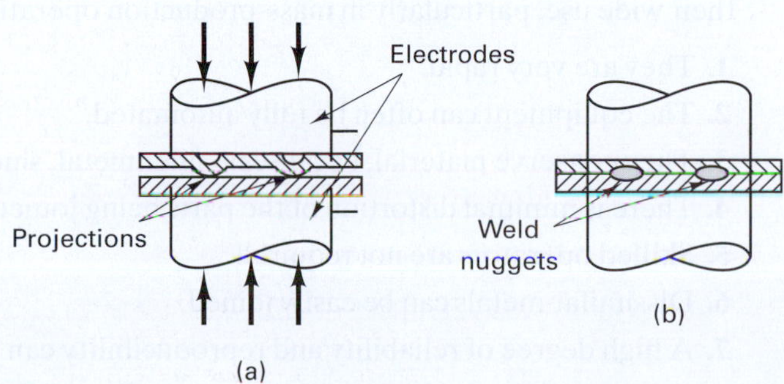

FIGURE 31-11 Principle of projection welding (a) prior to application of current and pressure and (b) after formation of the welds.

then placed between large-area electrodes in a press machine, and pressure and current are applied as in spot welding. Because the current must flow through the points of contact (i.e., the dimples), the heating is concentrated where the weld is desired. As the metal heats and becomes plastic, the pressure causes the dimple to flatten and form a weld.

Because the projections are press-formed into the sheet, they can often be produced during previous blanking and forming operations with virtually no additional cost. Moreover, the dimples or projections can be made in almost any shape—round, oval, or circular ring-shaped—to produce welds of shapes that optimize a given design. It is important, however, that the shape be such that the weld will form outward from the center of the projection. Because the heat tends to develop in the piece with the projections, it is best to incorporate them into the heavier of the two pieces to be joined or the metal with the higher conductivity.

Multiple dimples can be incorporated into a sheet, enabling multiple welds to be produced at one time. The number of projections is limited only by the ability of the machine to provide the required current and pressure and the need to uniformly distribute both. If more than three projections are to be made at one time, however, the height of all projections should be uniform to ensure uniform contact and heating.

An attractive feature of projection welding is the fact that it does not require special equipment. Conventional spot-welding machines can be converted to projection welding simply by changing the size and shape of the electrodes. In addition, projection welding leaves no indentation mark on the exterior surfaces, a definite advantage over spot welding when good surface appearance is required.

In a variation of the process, projection welding can also be used to attach bolts and nuts to other metal parts. Contact is made at a projection that has been machined or forged onto the bolt or nut. Current is applied, and the pieces are pressed together to form a weld. In a variation known as **capacitor-discharge stud welding** a burst of current from an electrostatic storage system melts the projection and the pieces are pushed together, all within a time of 6 to 10 ms.

RESISTANCE WELDS INVOLVING TUBING

Brazing and welding have been the traditional means of joining tube to sheet or tube to tube. A process known as **deformation resistance welding** now offers an alternative approach that can produce quality joints within 2 to 3 s between both similar and dissimilar materials. High currents heat the metal surfaces to the point of softening, and they are then rapidly pressed together to create a solid-state joint through the deformation and displacement of material at the weld interface. To maintain an open inner diameter, the tubing is often upset bulged or flared at the location of the weld. Applications include heat exchangers, automotive exhaust systems and tubular space frames, fluid handling systems, and tubular structural assemblies.

31.4 ADVANTAGES AND LIMITATIONS OF RESISTANCE WELDING

The resistance-welding processes have a number of distinct advantages that account for their wide use, particularly in mass-production operations:

1. They are very rapid.
2. The equipment can often be fully automated.
3. They conserve material, because no filler metal, shielding gases, or flux is required.
4. There is minimal distortion of the parts being joined.
5. Skilled operators are not required.
6. Dissimilar metals can be easily joined.
7. A high degree of reliability and reproducibility can be achieved.

TABLE 31-2 Process Summary: Resistance Welding

Heat source	Electrical resistance heating with high current
Protection	None; isolation of weld site is adequate
Material joined	All common metals (steel, aluminum, and copper)
Rate of heat input	High
Weld profile (D/W)	Does not apply
Maximum penetration	Does not apply
Assets	High speed; Small HAZ; no flux, filler metal, or shielding gas required; adaptable to mass production
Limitations	Equipment is more expensive than arc welding; welds are weaker than arc welds; requires access to both sides of a joint

The primary limitations of resistance welding include:

1. The equipment has a high initial cost.
2. There are limitations to the thickness of material that can be joined (generally less than 6 mm, or $\frac{1}{4}$ in.), and the type of joints that can be made (mostly **lap joints**). Lap joints tend to add weight and material.
3. Access to both sides of the joint is usually required to apply the proper electrode force or pressure.
4. Skilled maintenance personnel are required to service the control equipment.
5. For some materials, the surfaces must receive special preparation prior to welding.

The resistance-welding processes are among the most common techniques for high-volume joining. The rapid heat inputs, short welding times, and rapid quenching by both the base metal and the electrodes can produce extremely high cooling rates in and around the weld. While these conditions can be quite attractive for most nonferrous metals, untempered martensite can form in steels containing more than 0.15% carbon. For these materials, some form of postweld heating is generally required to eliminate possible brittleness.

Table 31-2 provides a process summary for resistance welding.

31.5 SOLID-STATE WELDING PROCESSES

FORGE WELDING

Being the most ancient of the welding processes, **forge welding (FOW)** has both historical and practical value: it helps illustrate how and why the modern welding practices were developed. The armor makers of ancient times occupied positions of prominence in their society, largely because of their ability to join pieces of metal into single, strong products. The village blacksmith was a more recent master of forge welding. With his hammer and anvil, coupled with skill and training, he could create a wide variety of useful shapes from metal.

Using a charcoal forge, the blacksmith heated the pieces to be welded to a practical forging temperature and then prepared the ends by hammering so that they could be properly fitted together. The ends were then reheated and dipped into a borax flux. Heating was continued until the blacksmith judged (by color) that the workpieces were at the proper temperature for welding. They were then withdrawn from the heat and either struck on the anvil or hit by the hammer to remove any loose scale or impurities. The ends to be joined were then overlapped on the anvil and hammered to the degree necessary to produce an acceptable weld.

As the two pieces reduced in thickness, they spread in width, resulting in the creation of new, fresh, uncontaminated metal surface. As these surfaces were being created, the hammer blows also provided the necessary pressure to produce instant coalescence. Thus, by the correct combination of heat and deformation, a competent blacksmith could produce joints that might be every bit as strong as the original metal.

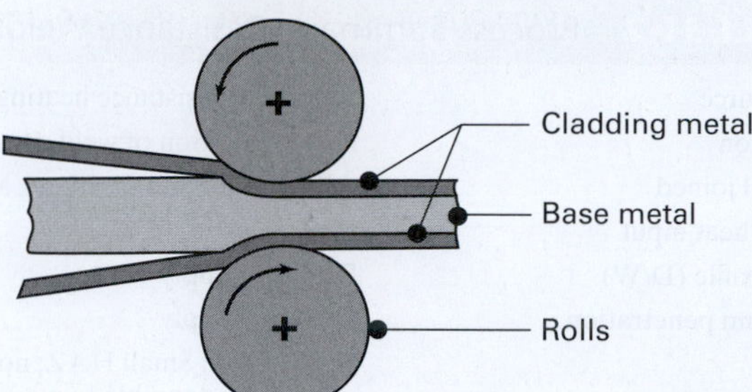

FIGURE 31-12 Schematic representation of the roll bonding process.

However, because of the crudeness of the heat source, the uncertainty of temperature, and the difficulty in maintaining metal cleanliness, a great amount of skill was required, and the results were highly variable.

FORGE-SEAM WELDING

Although forge welding has largely been replaced by other joining methods, a large amount of **forge-seam welding** is still used in the manufacture of pipe. In this process, a heated strip of steel is first formed into a cylinder, and the edges are simply pressed together in either a lap or a butt configuration. Welding is the result of pressure and deformation when the metal is pulled through a conical welding bell or passed between welding rolls.

COLD WELDING

Cold welding is a variation of forge welding that uses no heating but produces metallurgical bonds by means of room-temperature plastic deformation. The surfaces to be joined are first cleaned and placed in contact. They are then subjected to high localized pressure, sufficient to cause about 30 to 50% cold work. While some heating will occur due to the severe deformation, the primary factor in producing coalescence is the high pressure acting on newly formed surface material. The cold-welding process is generally confined to the joining of small parts made from soft, ductile metal, like those encountered in various electrical connections.

ROLL WELDING OR ROLL BONDING

In the **roll-welding (ROW)** or **roll-bonding** process, shown schematically in Figure 31-12, two or more sheets or plates of metal are joined by passing them simultaneously through a rolling mill. As the materials are reduced in thickness, the length and/or width must increase to compensate. The newly created uncontaminated interfaces are pressed together by the rolls, and coalescence is produced. Roll bonding can be performed either hot or cold and can be used to join either similar or dissimilar metals (such as the Alclad aluminums—a skin of high-corrosion-resistance aluminum over a core of high-strength aluminum—or conventional steel with a stainless steel cladding). The resulting bond can be quite strong, as evidenced by the roll-bonded "sandwich" material used in the production of various U.S. coins.

By precoating select portions of one interface surface with a material that prevents bonding, the roll-bonding process can be used to produce sheets that have both bonded and nonbonded areas. Subsequent heating in an oven or furnace can cause the no-bond coating to volatilize. The resulting pressure expands the no-bond regions, producing flow paths for gases or liquids. A common example of this technique is in the manufacture of refrigerator freezer panels, where inexpensive sheet metal is used to produce structural panels that also serve to conduct the coolant.

FRICTION WELDING: ROTATIONAL FRICTION WELDING, INERTIA WELDING, AND LINEAR FRICTION WELDING

In **friction welding (FRW)**, the heat required to produce the joint is generated by mechanical friction at the interface between a moving workpiece and a stationary

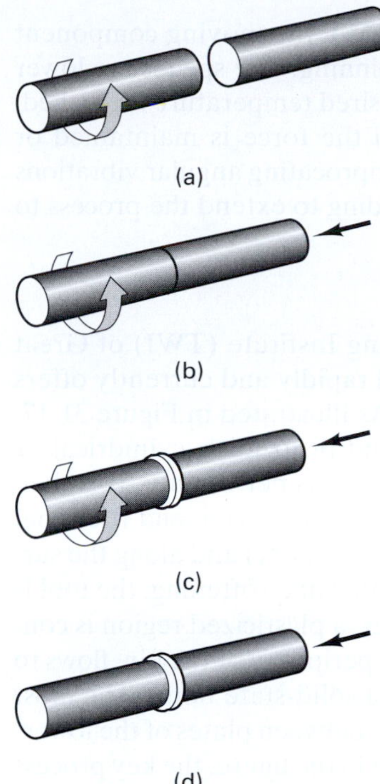

FIGURE 31-13 Sequence for making a friction weld. (a) Components with square surfaces are inserted into a machine where one part is rotated and the other is held stationary. (b) The components are pushed together with a low axial pressure to clean and prepare the surfaces. (c) The pressure is increased, causing an increase in temperature, softening, and possibly some melting. (d) Rotation is stopped and the pressure is increased rapidly, creating a forged joint with external flash.

component. The pieces to be joined are first prepared to have smooth, square-cut surfaces. As shown in Figure 31-13, one piece is mounted in a motor-driven chuck or collet and rotated against a stationary piece at high speed. A low contact pressure may be applied initially to permit cleaning of the surfaces by a burnishing action. The pressure is then increased, and contact friction quickly generates enough heat to soften both components and raise the abutting surfaces to the welding temperature. As soon as this temperature is reached, rotation is stopped and the pressure is further increased to complete the weld. The softened metal is squeezed out to the edges of the joint, forming a **flash,** which can be removed by subsequent machining. Clean, uncontaminated material is left on the interface, and the force creates a "forged" structure in the joint. Friction welding has been used to join steel bars up to 20 cm (8 in.) in diameter and tubes of even larger diameter. The process is also ideal for welding dissimilar metals with very different melting temperatures and physical properties, such as copper to aluminum, titanium to copper, and nickel alloys to steel. Figure 31-14 shows a schematic of the equipment required for friction welding.

Inertia welding is a modification of friction welding where the moving piece is attached to a rotating flywheel (Figure 31-15). The flywheel is brought to a specified rotational speed, storing a predetermined amount of kinetic energy, and is then separated from the driving motor. The rotating and stationary components are then pressed together, and the kinetic energy of the flywheel is converted into frictional heat at the interface between the two pieces. The weld is formed when the flywheel stops its motion and the pieces are firmly pressed together. Because the conditions of inertia welding are easily duplicated, welds of extremely consistent quality can be produced, and the process can be readily automated.

With inertia welding, the time required to form a weld can be very short, often on the order of several seconds. Because of the high rate of heat input and the limited time for heat to flow away from the joint, both the weld and heat-affected zones are usually very narrow. Oxides and other surface impurities tend to be displaced radially into the upset flash, which is generally removed after welding. Because virtually all of the energy is converted to heat, the process is very efficient. No material is melted, so joints can be formed with a wide variety of metals or combinations of metals, including some not normally considered compatible, such as aluminum to steel. Thermoplastic polymers can also be joined. Graphite-bearing cast irons, free-machining metals, and some bearing materials must be excluded because the graphite, lead, or free-machining additive smears across the surface, reducing the friction heating and preventing good solid-state bonding. One, or preferably both, of the components must be sufficiently ductile (when hot) to permit deformation during the forging stage. Grain size tends to be refined during the hot deformation, so the strength of the weld is about the same as that of the base metal. In addition, the friction processes are environmentally attractive because no smoke, fumes, or gases are generated, and no fluxes are required.

Because of the rotational motion, both of the above processes require that one of the components have rotational symmetry. They are used primarily to join round bars or tubes of the same size or connect bars or tubes to flat surfaces as shown in the examples of Figure 31-16.

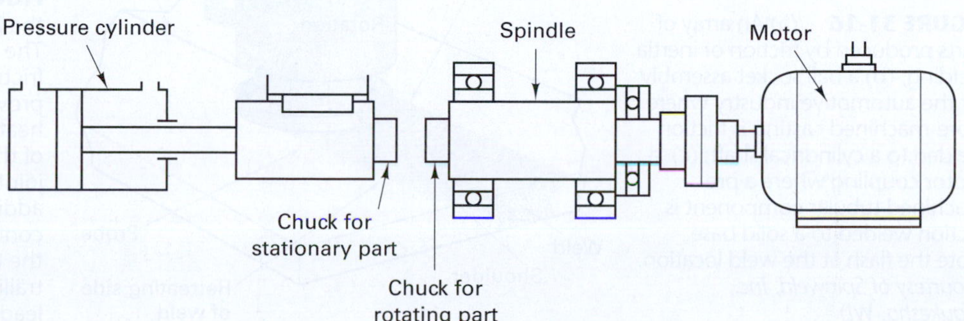

FIGURE 31-14 Schematic of the equipment used for friction welding. *(Courtesy of* Materials Engineering*)*

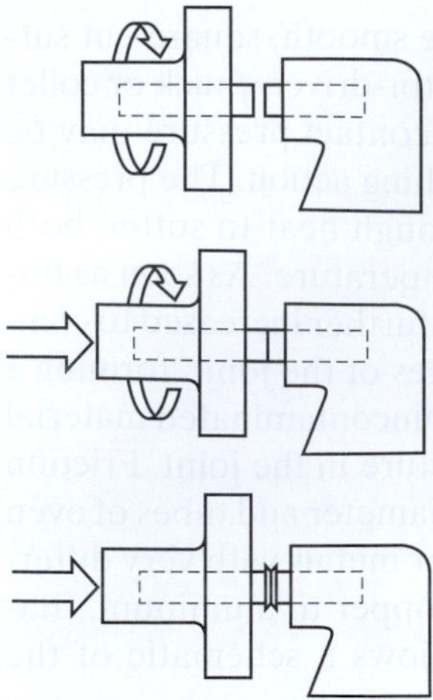

FIGURE 31-15 Schematic representation of the various steps in inertia welding. The rotating part is now attached to a large flywheel.

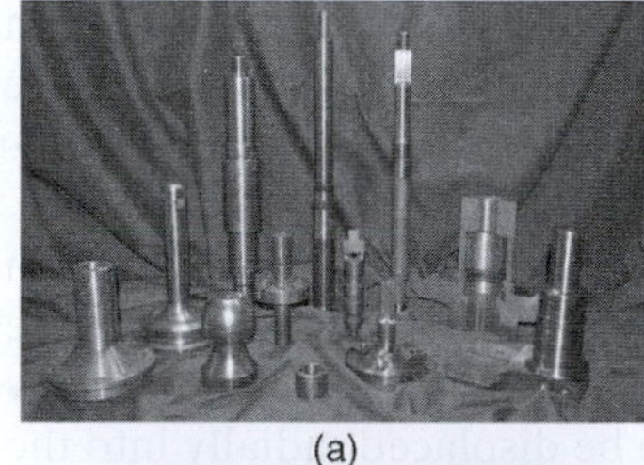

(a)

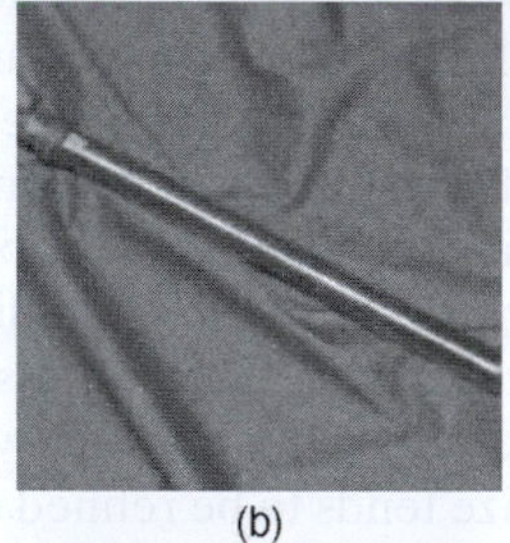

(b)

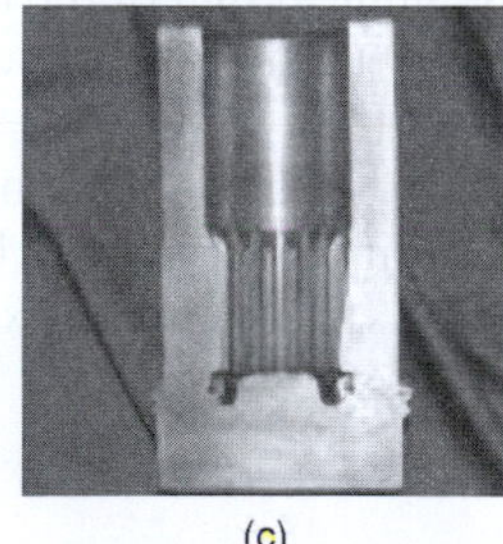

(c)

FIGURE 31-16 (a) An array of parts produced by friction or inertia welding; (b) a ball socket assembly for the automotive industry where a pre-machined casting is friction welded to a cylindrical shaft; (c) a motor coupling where a pre-machined tubular component is friction welded to a solid base. Note the flash at the weld location. *(Courtesy of Spinweld, Inc., Waukesha, WI)*

In a variation known as **linear friction welding (LFW),** the moving component oscillates laterally (parallel to the interface) instead of spinning. The speeds are lower and the pieces are kept in constant pressure. When the desired temperature is reached, the pieces are brought into the desired alignment, and the force is maintained or increased to consolidate the joint. Orbital motions and reciprocating angular vibrations are other movements that have been used in friction welding to extend the process to additional joint geometries.

FRICTION-STIR WELDING

A relatively new process, first performed by The Welding Institute (TWI) of Great Britain in 1991, **friction-stir welding (FSW)** has matured rapidly and currently offers significant benefits compared to conventional methods. As illustrated in Figure 31-17, a nonconsumable welding tool containing a shoulder and protruding cylindrical or tapered probe or pin is rotated at several hundred revolutions per minute. It is then lowered into the interface between pieces of rigidly clamped material, and frictional heat is generated along the top surface (under the rotating shoulder) and along the surfaces of the rotating probe. After a period of time for heating and softening, the tool is driven along the material interface. As the probe traverses, a plasticized region is continually created. This softened material is swept along the periphery of the pin, flows to the back of the advancing probe, and coalesces to form a solid-state bond. The most common application is the formation of butt welds, usually between plates of the lower-melting-point metals or thermoplastic polymers. As shown in the figure, the key process variables include probe geometry (diameter, depth, and profile), shoulder diameter, rotation speed, downward force, travel speed and possible tilt to the tool.

Weld quality is excellent. The extensive plastic deformation creates a refined grain structure with no entrapped oxides or gas porosity. The top photo in Figure 31-18 shows a friction-stir weld in aluminum plate, viewed from the top, and the bottom photo cross section shows the structural changes induced by a friction-stir pass through an aluminum alloy casting. Note the significant refinement in structure. As a result, the strength, ductility, fatigue life, and toughness of the resulting weld are all quite good. Welds in aircraft aluminum are 30 to 50% stronger than those formed by arc welding. Because no material is melted, both wrought and cast alloys can be joined, and they can be joined to each other. No filler material or shielding gas is required, and the process is environmentally friendly (no fumes, weld spatter, or arc glare). Because of the high-energy efficiency, total heat input and associated distortion and shrinkage are all low. Joint preparation is minimal, and surface oxides need not be removed. Welding can be performed in any position and requires access to only one side of the plate. Gaps up to 10% of the material thickness can be accommodated with no reduction in weld quality or performance. Weld speed, however, is slower than most fusion processes. Table 31-3 summarizes some of the attractive features of the friction-stir process.

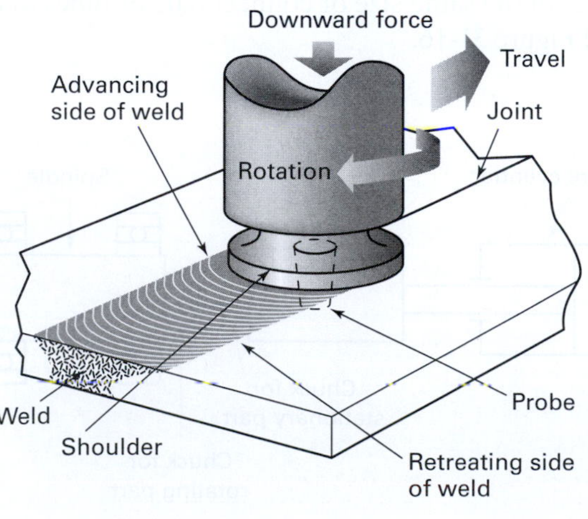

FIGURE 31-17 Schematic of the friction-stir welding process. The rotating probe generates frictional heat, while the shoulder provides additional friction heating and prevents expulsion of the softened material from the joint. (*Note*: To provide additional forging action and confine the softened material, the tool may be tilted so the trailing edge is lower than the leading segment.)

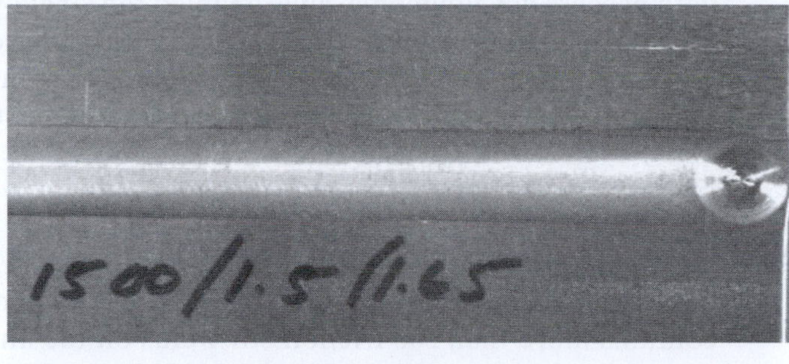

(a)

(b)

FIGURE 31-18 (a) Top surface of a friction-stir weld joining 1.5-mm- and 1.65-mm-thick aluminum sheets with 1500-rpm pin rotation. The welding tool has traversed left-to-right and has retracted at the right side of the photo. (b) Metallurgical cross section through an alloy 356 aluminum casting that has been modified by friction-stir processing. *(Courtesy Ronald Kohser)*

Friction-stir welding has been used to weld nearly all of the wrought aluminum alloys, including some that are classified as "unweldable" by fusion processes (the 2xxx and 7xxx series). Aluminum plates up to 75 mm (3 in.) thick have been successfully welded from a single side in a single pass. Copper, lead, magnesium, and zinc have all been welded with relative ease. High-strength and high-hardness tools, along

TABLE 31-3 Attractive Features of Friction-Stir Welding

Metallurgical Benefits

- Excellent weld quality.
- Applicable to a wide range of materials, including both wrought and cast alloys, as well as some "nonweldable" by fusion methods.
- Solid-state process.
- Low distortion of the workpiece.
- High joint strength and good fatigue properties.
- No loss of alloy elements.
- Fine microstructure.
- No cracking or porosity.
- Low shrinkage.

Environmental Benefits

- No shielding gas is required.
- No surface cleaning is required.
- No solvent degreasing is used.
- No fumes, gases, or smoke are produced.
- Postweld finishing is often unnecessary.
- No arc glare or reflected laser beams.

Energy Benefits

- Welds produced with far less energy than other processes.
- Weight reduction in aircraft, automobiles, and ships is possible.

Process Features

- Nonconsumable tool.
- No filler metal.
- Can tolerate small alignment and surface imperfections.
- Thin oxide layers are acceptable.
- Can be performed in all positions.

with process developments, have enabled the single-pass welding of titanium, ferritic, high-strength and stainless steels, and the nickel-base superalloys in thicknesses up to 25 mm. (1 in.). Various thermoplastics—including polyethylene, polypropylene, nylon, polycarbonate, and ABS—have also been successfully welded, with several exhibiting as-welded strengths exceeding 95% of the tensile strength of the base material. Different alloys of the same base metal can be welded, and wrought plate or forgings can be joined to castings.

As an indication of process capability, the Eclipse 500 very-light jet—a six-passenger, short-hop air taxi built by Eclipse Aviation of Albuquerque, New Mexico—contains more than 260 friction-stir welds on the aluminum fuselage and wing panel assemblies with a combined length of 135 m (nearly 450 ft). These welds replaced more than 7000 rivets and other fasteners—with a significant reduction in cost—and produced a structure whose fatigue life is equal to or greater than a riveted assembly.

The friction-stir process has also been modified to produce spot welds, and thin sheet material can now be welded. A bobbin-type tool, with shoulders that rub against both top and bottom surfaces, is fed from the side, eliminating the need for downward pressure. A hybrid laser–friction-stir process has also been developed, where a laser is used to preheat the metal ahead of the stir tool. Because the spinning tool no longer needs to generate all of the required heat, the load on the tool can be reduced, producing a significant increase in tool life. In addition, the hybrid process enhances the ability of friction-stir to weld the high-strength or high-melting-temperature materials.

In an extension known as **friction-stir processing,** the thermomechanical features of friction-stir have been used for purposes other than creating a joint. By tracking the stir tool through the material with overlapping passes, ultra-fine grain size (<10 mm) can be produced that enables superplastic forming at comparatively high strain rates. While superplasticity is usually limited to thin sheets, thicker material can now be made superplastic, and by stirring only selected locations, large parts can be made from less-expensive conventional materials, with enhanced formability in only the needed locations. In other applications, key surfaces of large castings can be enhanced by the passage of the probe. The cast structure, with possible microporosity and segregation, is replaced by a fine, homogeneous, wrought microstructure. Strength, ductility, corrosion resistance, and fatigue resistance are all improved. Fusion welds can be stirred to replace the cast structure with a fine, worked structure, removing any weld defects and enhancing properties. Reinforcement particles can be stirred into a material to create a particle-reinforced composite surface on a standard alloy substrate. Unique composition powder products can be brought to full density with attractive strength and ductility. In a modification known as **friction-stir channeling,** a slight upward helix is incorporated onto the surface of the rotating probe. As the probe rotates, material is displaced upward, creating a continuous subsurface channel that can be used for purposes such as the flow of cooling water.

ULTRASONIC WELDING

In **ultrasonic welding (USW),** coalescence is produced by the localized application of high-frequency (10,000 to 200,000 Hz with 20,000 Hz being most common) shear vibrations of low amplitude (usually <100 μm or 0.004 in.) to surfaces that are held together under rather light normal pressure. Although there is some heating at the faying surfaces, the interface temperature rarely exceeds one-half of the melting point of the material on an absolute-temperature scale. Instead, it appears that the rapid reversals of stress along the contact interface deform, shear, and flatten the surface asperities and break up and disperse the oxide films and surface contaminants, allowing the clean metal surfaces to coalesce into a high-strength bond.

Figure 31-19 depicts the basic components of the ultrasonic welding process. The ultrasonic transducer is essentially the same as that employed in ultrasonic machining. It is coupled to a force-application system that contains a welding tip on one end, either

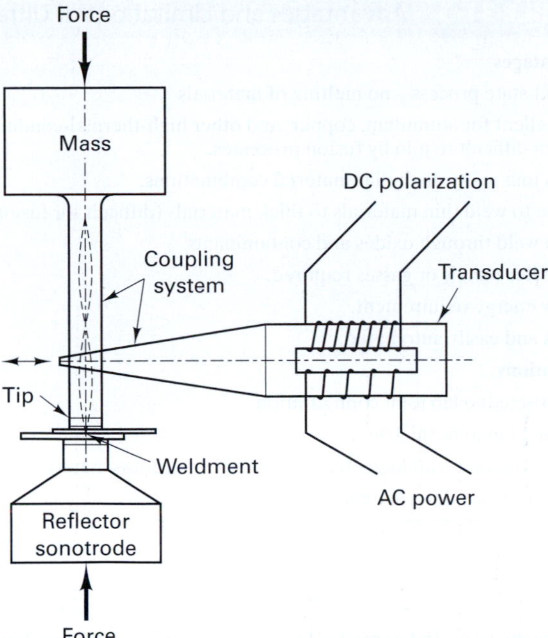

FIGURE 31-19 Diagram of the equipment used in ultrasonic welding.

stationary for spot welds or rotating for seams. The pieces to be welded are placed between this tip and a reflecting anvil, thereby concentrating the vibratory energy to a relatively small area, typically about 40 mm^2 (0.06 in^2). Seam welds can be produced by rolling an ultrasonically vibrated disc over the workpieces.

Ultrasonic welding is restricted to the lap joint welding of thin materials—sheet, foil, and wire—or the attaching of thin sheets to heavier structural members. The maximum thickness is about 2.5 mm (0.1 in.) for aluminum and 1.0 mm (0.04 in.) for harder metals. As indicated in Table 31-4, the process is particularly attractive because of the number of metals and dissimilar metal combinations that can be joined. It is even possible to bond metals to nonmetals, such as aluminum to ceramics or glass. Because the temperatures are low and no arcing or current flow is involved, the process is often preferred for heat-sensitive electronic

TABLE 31-4 Metal Combinations Weldable by Ultrasonic Welding

Metal	Aluminum	Copper	Germanium	Gold	Molybdenum	Nickel	Platinum	Silicon	Steel	Zirconium
Aluminum	x	x	x	x	x	x	x	x	x	x
Copper		x		x		x	x		x	x
Germanium			x	x		x	x	x		
Gold				x		x	x	x		
Molybdenum					x	x			x	x
Nickel						x	x		x	x
Platinum							x		x	
Silicon										
Steel									x	x
Zirconium										x

TABLE 31-5 Advantages and Limitations of Ultrasonic Welding

Advantages
Solid-state process—no melting of materials
Excellent for aluminum, copper, and other high-thermal-conductivity materials that are often difficult to join by fusion processes.
Can join many dissimilar material combinations.
Able to weld thin materials to thick materials (difficult for fusion processes).
Can weld through oxides and contaminants.
No filler metals or gasses required.
Low energy requirement.
Fast and easily automated.
Limitations
Restricted to lap joint configuration.
Limited in joint thickness.
More difficult with high-strength and high-hardness materials.
Material may deform under the tooling.

components. Intermetallic compounds seldom form, and there is no contamination of the weld or surrounding area. The equipment is simple and reliable, and only moderate skill is required of the operator. Surface preparation is less than for most competing processes (such as resistance welding), and less energy is needed to produce a weld. Typical applications include joining the dissimilar metals in bimetallics, making microcircuit electrical contacts, welding refractory or reactive metals, bonding ultra-thin metal, and encapsulating explosives or chemicals.

Ultrasonic welding has also been used to produce spot and seam welds on thin plastics and to seal foil or plastic envelopes and pouches. Compared to joining methods that employ solvents or adhesives, the ultrasonic method is considerably faster and results in products with cleaner surfaces.

Table 31-5 summarizes some of the advantages and limitations of ultrasonic welding.

DIFFUSION WELDING

Diffusion welding (DFW) or **diffusion bonding** occurs when properly prepared surfaces are maintained in contact under sufficient pressure and time at elevated temperature. In contrast to the deformation welding methods, plastic flow is limited and the principal bonding mechanism is atomic diffusion. Clean, flat, well-prepared surfaces can produce an interface that can be viewed as a planar grain boundary with intervening voids and impurities. Under low pressure (enough to produce firm contact) and elevated temperature (usually between 0.5 and 0.8 times the melting temperature on an absolute temperature scale), atomic diffusion will provide the necessary void shrinkage and grain boundary migration to form a metallurgical bond. The required time can range from 1 min to more than 1 hr, depending on the materials being bonded and the process conditions. Most diffusion bonding is performed in a vacuum or an inert gas atmosphere.

Diffusion bonding is capable of joining a wide range of metal and ceramic materials, as well as composite material combinations, in both small and large sizes. Furnaces with inert or protective atmospheres can be used to produce high-quality joints with the reactive metals—such as titanium, beryllium, and zirconium—and the high-temperature refractory metals. Because the bonding process is quite slow, multiple parts are generally loaded into a furnace, or the application is restricted to low-volume production. The quality of a diffusion weld depends on the surface condition of the materials, temperature, time at temperature, and pressure.

Modifications of the basic process often use intermediate material layers, which can either promote diffusion or prevent the formation of undesirable intermetallic compounds. Intermediate layers are often used when joining dissimilar materials, especially metals to ceramics, where they aid the bonding process and help to modify the stress distribution in the bonded product. A dissimilar material layer can also be designed to melt at a lower temperature than the materials being joined, forming a temporary liquid that significantly accelerates the rate of atom movement. While being held at the bonding temperature, the atoms in the liquid subsequently diffuse into the adjoining solids, producing a strong, solid-state bond.

Because the conditions for the diffusion bonding of titanium and titanium alloys coincide with the conditions required for superplastic forming, aerospace companies have combined the two processes into a single operation known as **superplastic forming/diffusion bonding.** The products are generally sandwich-type sheet metal structures with the bonds being used to create internal reinforcing elements.

EXPLOSIVE WELDING

Explosive welding (EXW) is typically used to bond flat surfaces of widely different materials (often with very different crystal structures), particularly when large areas are involved. As shown in Figure 31-20, the bottom sheet or plate is positioned on a rigid base or anvil, and the top sheet is inclined to it with a small open angle between the surfaces to be joined. An explosive material, usually in the form of a sheet, is placed on top of the two layers of metal and detonated in a progressive fashion, beginning where the surfaces touch. A compressive stress wave, on the order of thousands of megapascals (hundreds of thousands of pounds per square inch), sweeps across the surface of the plates. Surface films are liquefied or scarfed off the metals and are jetted out of the interface. The clean metal surfaces are then thrust together under high contact pressure. The result is a low-temperature weld with an interface configuration consisting of a series of interlocking ripples. Because the bond strength is quite high (equal to or greater than the strength of the weaker material), explosively clad plates can be subjected to a wide variety of subsequent processing, including further reduction in thickness by rolling. Because it is a solid-state welding process, numerous combinations of dissimilar metals can be joined.

Explosive welding is best suited to large products and exotic material combinations. A common application is the cladding of corrosion-resistant metal to heavier plates of base metal. The claddings have been aluminum, titanium, zirconium, copper, and stainless steel, and the substrate is usually some form of steel. Dimensions are often measured in meters or tens-of-feet. Process assets include the flexibility in product size, the wide variety of material combinations, and the quick delivery of a finished product.

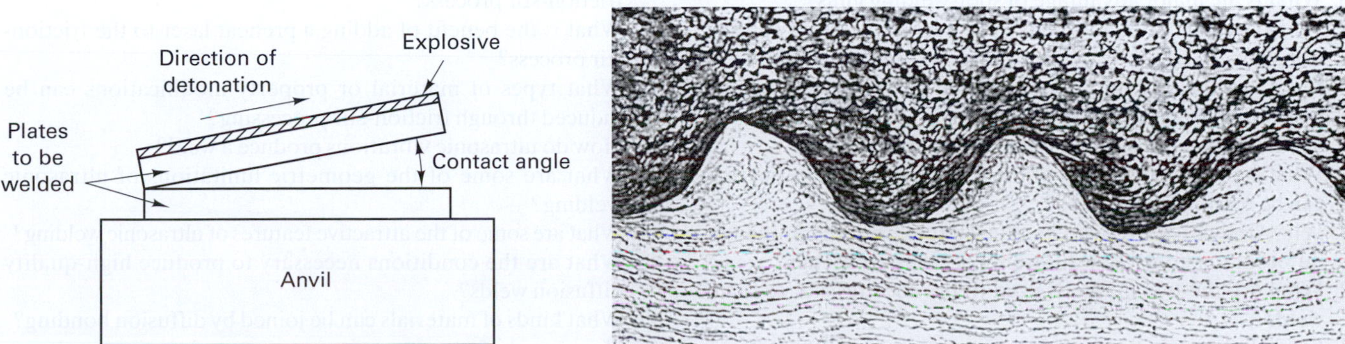

FIGURE 31-20 (Left) Schematic of the explosive welding process. (Right) Explosive weld between mild steel and stainless steel, showing the characteristic wavy interface. *(Courtesy Ronald Kohser)*

■ KEY WORDS

capacitor-discharge stud welding
cold welding
deformation resistance welding
diffusion welding (or bonding) (DFW)
electrodes
embossing
explosive welding (EXW)
faying surfaces
flash
forge welding (FOW)
forge-seam welding
friction welding (FRW)
friction-stir channeling
friction-stir processing
friction-stir welding (FSW)
inertia welding
lap joint
linear friction welding (LFW)
nugget
projection welding (RPW)
resistance butt welding
resistance seam welding (RSEW)
resistance spot welding (RSW)
resistance welding
rocker-arm machine
roll welding (or bonding) (ROW)
solid-state welding
spot-welding gun
superplastic forming/diffusion bonding
transgun
ultrasonic welding (USW)

■ REVIEW QUESTIONS

1. What are the two primary functions of the electrodes in resistance welding?
2. What are the two major roles of the applied pressure in resistance welding?
3. Why might resistance welding be considered as a form of solid-state welding?
4. Why is there no need for fluxes or shielding gases in resistance welding?
5. Based on the heat input equation, which term is most significant in providing heat: current, resistance, or time?
6. What are the three components that contribute to the total resistance between the electrodes?
7. What measures can be taken to reduce the resistance between the electrodes and the workpieces?
8. What factors control the resistance between the faying surfaces?
9. What are the possible consequences of too little pressure during the resistance-welding cycle? Too much pressure?
10. What is the ideal sequence for pressure application during resistance welding?
11. Why do the resistance-welding conditions become less favorable as the material heats and softens?
12. What magnitude of current may be used to produce resistance welds?
13. What is the simplest and most widely used form of resistance welding?
14. What is the typical size of a spot-weld nugget?
15. What are the two basic types of stationary spot-welding machines?
16. What is the major advantage of spot-welding guns?
17. What are the pros and cons of a resistance spot-welding transgun?
18. What are some of the properties that must be possessed by resistance-welding electrodes?
19. What is the most common metal that is spot welded?
20. What is the practical limit of the thicknesses of material that can be readily spot welded?
21. What design features can be altered to permit the joining of different thicknesses or different conductivity metals?
22. What are the two methods used to produce resistance seam welds?
23. For what products would resistance butt welding be a common approach?
24. What two limitations of spot welding can be overcome by using the projection approach?
25. What limits the number of projection welds that can be formed in a single operation?
26. What are some of the attractive features of resistance welding when viewed from a manufacturing standpoint?
27. What are some of the primary limitations to the use of resistance welding?
28. What type of metallurgical problem might be encountered when spot welding medium- or high-carbon steels?
29. What were some of the limitations that made the forge welds of a blacksmith somewhat variable in terms of quality?
30. What features promote coalescence in cold welding?
31. Describe how the roll-bonding process can be used to fabricate products that contain pressure-tight, fluid-flow channels that once required the use of metal tubing.
32. How is inertia welding similar to friction welding? Different from friction welding?
33. How are surface impurities removed in the friction- and inertia-welding processes?
34. Why are inertia welds of more consistent quality than friction welds?
35. What are some of the geometric limitations of friction and inertia welding?
36. How does linear friction welding differ from friction welding?
37. How does friction-stir welding differ from friction welding?
38. What are the primary process variables in friction-stir welding?
39. What are some of the attractive features of friction-stir welding?
40. What are some of the materials that have been welded by the friction-stir process?
41. What is the benefit of adding a preheat laser to the friction-stir process?
42. What types of material or property modifications can be induced through friction-stir processing?
43. How do ultrasonic vibrations produce a weld?
44. What are some of the geometric limitations of ultrasonic welding?
45. What are some of the attractive features of ultrasonic welding?
46. What are the conditions necessary to produce high-quality diffusion welds?
47. What kinds of materials can be joined by diffusion bonding?
48. How are surface contaminants removed during explosive welding?
49. If the interface of a weld is viewed in cross section, what is the distinctive geometric feature of an explosive weld?

PROBLEMS

1. Many advanced engineering products, as well as composite materials, require the joining of dissimilar materials. Select several of the processes discussed in this chapter, and investigate the capability of the process to join dissimilar materials and the associated limitations.

2. Friction-stir processing is an interesting extension of friction-stir welding. Can you identify other examples of where a welding or joining process is currently being used for purposes other than those for which it was initially developed?

www.wiley.com/go/global/degarmo

Chapter 31 CASE STUDY

Manufacture of an Automobile Muffler

An automobile muffler is the device for reducing the amount of noise emitted from the exhaust of an internal combustion engine vehicle. Typically installed lengthwise underneath the vehicle, the muffler is simply a container with a set of tubes with holes in them, as shown in the accompanying figure. The tubes act to create reflected waves that interact with one another to cause destructive interference, reducing or canceling the intensity of the sound.

Because of their location, the exterior of mufflers is exposed to a variety of environments, including rain; salt spray (winter roads); and dirt, dust, and debris. The interior sees the flow of exhaust gases, along with moisture condensation. Both the interior and exterior surfaces, therefore, see corrosive environments.

Mufflers have been made from a variety of materials. Uncoated low-carbon steel is probably the cheapest and easiest to fabricate, but it clearly has a limited lifetime due to corrosion. Various corrosion-resistant or corrosion-reducing coatings have been applied. Zinc coatings (galvanized) provide a sacrificial anode, and have been applied in a variety of means, including hot-dip immersion, electroplate, and others. Barrier-type aluminum coatings have also been applied by various means. Stainless steel (most commonly, type 304 austenitic stainless steel sheet) has also been used. Because of the nature of the design and the need to provide corrosion protection to the interior surfaces, the coatings are usually applied to sheet metal before the fabrication of the muffler.

Various types of joints are involved as the tubes, end caps, and wrap-around body are joined to produce the muffler, including sheet metal seams and tube-to-sheet fillets.

1. Consider a muffler being made entirely from uncoated low-carbon steel sheet and tubing. Describe its weldability and the welding techniques you would recommend for its manufacture.
2. Rework Question 1 using type 304 austenitic stainless steel as the base material.
3. If a zinc coating were applied to the low-carbon steel, how would this affect your answers to Question 1? What changes might be required in order to produce a satisfactory muffler?
4. An alternate material might be "one-side galvanized," where the low-carbon steel is dipped in molten zinc and the majority of the liquid is removed from one side by compressed air jets. The material is then subjected to an acid immersion to produce a zinc-free surface on one side and a zinc-coated surface on the other. Which surface would be most weldable? Would you want to put the zinc-coated surface to the inside or outside of your muffler? How might your welding design have to be modified?
5. If the applied coating is changed to aluminum, how would you assemble the components of a muffler?
6. Can you suggest any alternative that might provide improved manufacturing ease? Consider both material and design.

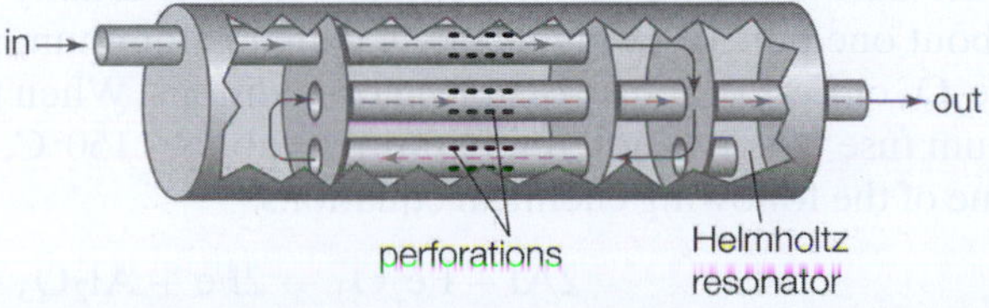

Cut-away of an automobile muffler. (© *Universal Images Group Limited/Alamy)*

CHAPTER 32

OTHER WELDING PROCESSES, BRAZING, AND SOLDERING

32.1 INTRODUCTION

We have already surveyed gas-flame and arc welding (Chapter 30), as well as resistance and solid-state joining processes (Chapter 31). Other processes within the realm of welding include some that are quite old (thermit welding) and others that are relatively new (laser and electron beam). These and several others will be presented here, along with a brief section devoted to the application of welding and welding-related processes to surfacing and thermal spray coating.

There are also many joining or assembly operations where welding may not be the best choice. Perhaps the heat of welding is objectionable, the materials possess poor weldability, welding is too expensive, or the joint involves thin or dissimilar materials. In such cases, low-temperature joining methods may be preferred. These include brazing, soldering, adhesive bonding, and the use of mechanical fasteners. Brazing and soldering will be explored in this chapter, while adhesive bonding and mechanical fasteners are deferred to Chapter 33.

32.2 OTHER WELDING AND CUTTING PROCESSES

THERMIT WELDING

Thermit welding (TW) is an extremely old process in which superheated molten metal and slag are produced from an exothermic chemical reaction between a metal oxide and a metallic reducing agent. The name **thermit** usually refers to a mechanical mixture of about one part (by weight) finely divided aluminum and three parts iron oxide (either Fe_2O_3 or Fe_3O_4), plus possible alloy additions. When this mixture is ignited by a magnesium fuse (the ignition temperature is about 1150°C, or 2100°F), it reacts according to one of the following chemical equations:

$$2Al + Fe_2O_3 \rightarrow 2Fe + Al_2O_3 + Heat$$

$$8Al + 3Fe_3O_4 \rightarrow 9Fe + 4Al_2O_3 + Heat$$

The temperature rises to more than 2750°C (5000°F) in about 30 s, superheating the molten iron, which then flows by gravity into a prepared joint, providing both heat and filler metal. Runners and risers must be provided, as in a casting, to channel the molten metal and compensate for solidification shrinkage.

Steels and cast irons can be welded using the process previously described. Copper, brass, and bronze can be joined using a starting mixture of copper oxide and aluminum. Nickel, chromium, and manganese oxides have also been used in the thermit welding of more exotic metals.

To a large degree, thermit welding has been replaced by alternative methods. Nevertheless, it is still effective and can be used to produce economical, high-quality welds in thick sections of material, particularly in remote locations or where more sophisticated welding equipment is not available. The field repair of large steel castings that have broken or cracked is one such application.

ELECTROSLAG WELDING

Electroslag welding (ESW), depicted in Figure 32-1, is a very effective process for welding thick sections of steel plate. There is no arc involved (except to start the weld), so the process is entirely different from submerged arc welding, and the electrical resistance of the metal being welded plays no part in producing the heat. Instead, heat is derived from the passage of electrical current through a pool of electrically conductive liquid slag. Resistance heating raises the temperature of the slag to around 1750°C (3200°F). The molten slag then melts the edges of the pieces that are being joined, as well as continuously fed solid or flux-cored electrodes. Multiple electrodes are often used to provide an adequate supply of filler metal and maintain the molten pool. Under normal operating conditions, there is a 65-mm (2.5-in.)-deep layer of molten slag, which serves to protect and cleanse the underlying 12- to 20-mm ($\frac{1}{2}$- to $\frac{3}{4}$-in.)-deep pool of molten metal. These liquids are confined to the region between the materials being joined by means of sliding, water-cooled **molding plates** that are usually made of copper. As the weld metal solidifies at the bottom of the pool, the molding plates move upward at a rate that is typically between 12 and 40 mm/min ($\frac{1}{2}$ to $1\frac{1}{2}$ in./min), a relatively slow welding rate.

Because a vertical joint provides the easiest geometry for maintaining a deep slag bath, the process is used most frequently in this configuration. Circumferential joints

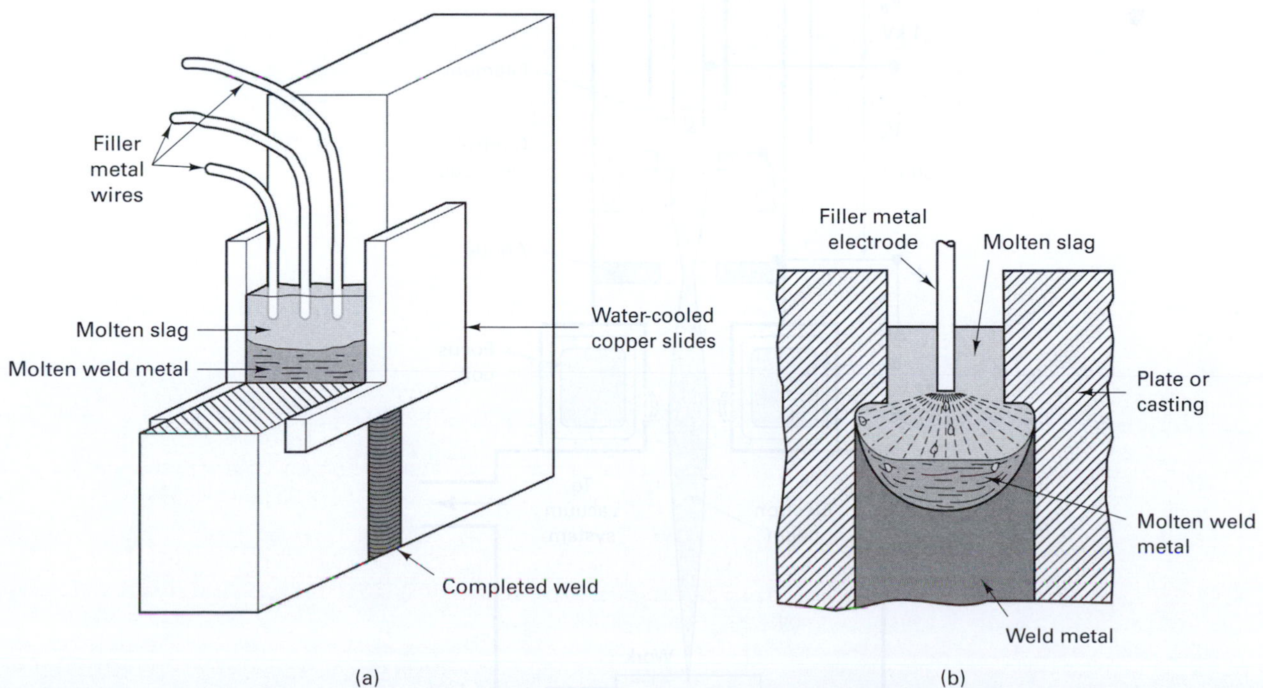

FIGURE 32-1 (a) Arrangement of equipment and workpieces for making a vertical weld by the electroslag process. (b) Cross section of an electroslag weld, looking through the water-cooled copper slide.

can also be produced in large pipe by using special, curved slag-holder plates and rotating the pipe to maintain the welding area in a vertical position.

Because large amounts of weld metal and heat can be supplied, electroslag welding is the best of all the welding processes for making welds in thick plates. The thickness of the plates can vary from 25 to 900 mm (1 to 38 in.), and the length of the weld (amount of vertical travel) is almost unlimited. Edge preparation is minimal, requiring only squared edges separated by 25 to 35 mm (1 to $1\frac{3}{8}$ in.). Applications have included building construction, shipbuilding, machine manufacture, heavy pressure vessels, and the joining of large castings and forgings.

Solidification control is vitally important to obtaining a good electroslag weld because the large molten pool and slow cooling tends to produce a coarse grain structure. Cracking tendencies can be suppressed by adjusting the current, voltage, slag depth, number of electrodes, and electrode extension to produce a wide shallow pool of molten metal. A large heat-affected zone and extensive grain growth are common features of the process. While these are undesirable metallurgical features, the long thermal cycle does serve to minimize residual stresses, distortion, and cracking in the heat-affected zone. Subsequent heat treatment of the welded structure, usually by normalizing, may be necessary if good fracture toughness is required.

ELECTRON-BEAM WELDING

In the **electron-beam welding (EBW)** process, the metal to be welded is heated by the impingement of a beam of high-velocity electrons. Originally developed for obtaining ultra-high-purity welds in reactive and refractory metals, the unique qualities of the process have led to a much wider range of applications.

Figure 32-2 presents the electron optical system. An electric current heats a tungsten filament to about 2200°C (4000°F), causing it to emit a stream of electrons by

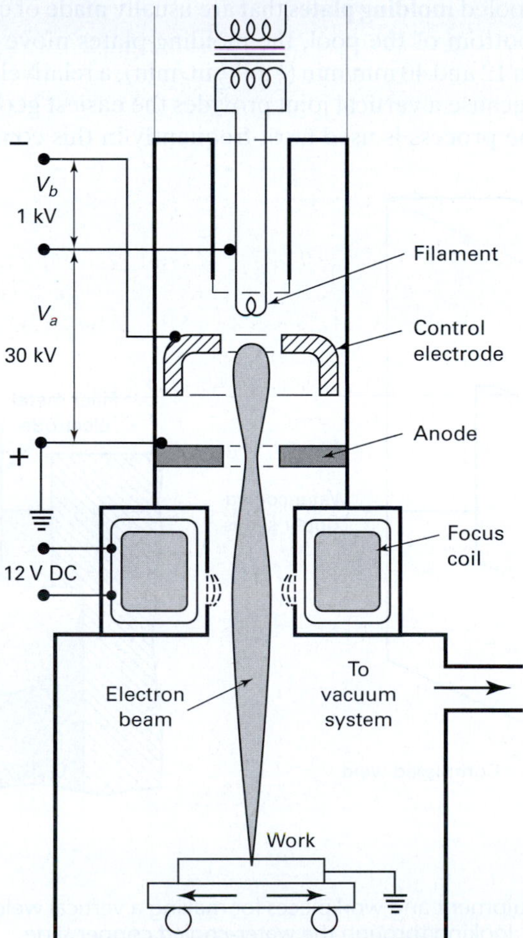

FIGURE 32-2 Schematic of the electron-beam welding process.

thermal emission. By means of a control grid, accelerating voltage, and focusing coils, these electrons are collected into a concentrated beam, accelerated, and directed to a focused spot between 0.8 and 3.2 mm. ($\frac{1}{32}$ to $\frac{1}{8}$ in.) in diameter. Because electrons accelerated at 150 kV achieve speeds nearly two-thirds the speed of light, the electron beam is concentrated energy, capable of producing temperatures in excess of 1 million degrees Celsius when its kinetic energy is converted to heat. Because the beam is composed of charged particles, it can be positioned and moved by electromagnetic lenses. Unfortunately, the electrons cannot travel well through air. To be effective as a welding heat source, the beam must be generated and focused in a very high vacuum, typically at pressures of 0.01 Pa (1×10^{-4} mm Hg) or less.

In many operations, the workpiece must also be enclosed in the high-vacuum chamber, with provision for positioning and manipulation. The vacuum then ensures degasification and decontamination of the molten weld metal, and welds of very high quality are obtained. The size of the vacuum chamber, however, tends to impose serious limitations on the size of the workpiece that can be accommodated, and the need to break and reestablish the high vacuum as pieces are inserted and removed places a considerable restriction on productivity. As a consequence, electron-beam welding machines have been developed that operate at pressures considerably higher than those required for beam generation. Some permit the workpiece to remain outside the vacuum chamber, with the beam emerging through a small orifice in the vacuum chamber to strike an adjacent surface. High-capacity vacuum pumps are required to compensate for the leakage through the orifice. While these machines offer more production freedom, they do produce shallower, wider welds because the beam loses energy and diffuses as the pressure increases.

Two distinct ranges of accelerating voltage are generally employed in electron beam welding. High-voltage equipment operates between 60 and 150 kV and produces a smaller spot size and greater penetration than does the lower-voltage type, which uses from 10 to 50 kV. Because of their high electron velocities, the high-voltage units emit considerable quantities of harmful X-rays and thus require expensive shielding and indirect viewing systems for observing the work. The X-rays produced by the low-voltage machines are sufficiently soft that the walls of the vacuum chamber absorb them, and the parts can be viewed directly through viewing ports.

Almost any metal can be welded by the electron-beam process, including those that are difficult to weld by other methods, such as zirconium, beryllium, and tungsten. Dissimilar metals, including those with extremely different melting points, can also be readily welded, because the intense beam will melt both metals simultaneously. Electron beam welds typically exhibit a narrow profile and remarkable penetrations like those shown in Figure 32-3. The high power and heat concentrations can produce fusion zones with depth-to-width ratios up to 25:1. This is coupled with low total heat input, low distortion, and a very narrow heat-affected zone. Heat-sensitive materials can often be welded without damage to the base metal. Deep welds can be made in a single pass. High welding

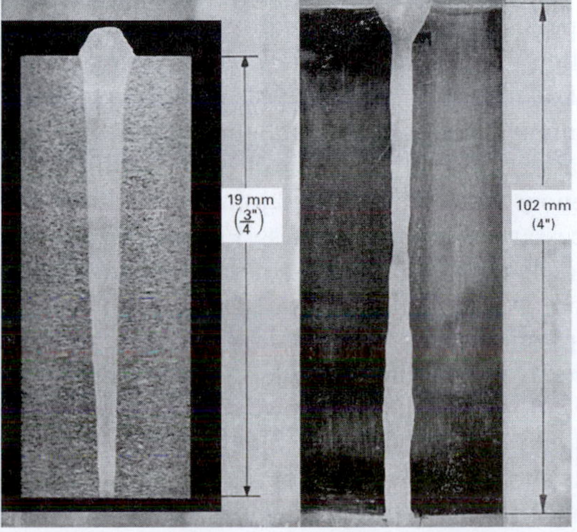

FIGURE 32-3 (Left to right) Electron-beam welds in 19-mm-thick 7079 aluminum and 102-mm-thick stainless steel. *(Courtesy of Hamilton Standard Division of United Technologies Corporation, Hartford, CT)*

TABLE 32-1 Process Summary: Electron-Beam Welding (EBW)

Heat source	High-energy electron beam
Protection	Vacuum
Electrode	None
Material joined	All common metals
Rate of heat input	High
Weld Profile (D/W)	20
Maximum penetration	175 mm (7 in.)
Assets	High precision; high quality; deep and narrow welds; small HAZ, low distortion; fast welding speed; beam is easily positioned and deflected; no filler metal, flux, or shielding gas required
Limitations	Beam and target must be within a vacuum; very expensive equipment; work piece size is limited by the vacuum chamber; significant edge preparation and alignment required; requires safety protection from X-ray and visible radiation

speeds are common; no shielding gas, flux, or filler metal is required; the process can be performed in all positions; and preheat or postheat is generally unnecessary.

On the negative side, the equipment is quite expensive, and extensive joint preparation is required. Because of the deep and narrow weld profile, joints must be straight and precisely aligned over the entire length of the weld. Machining and fixturing tolerances are often quite demanding. The vacuum requirements tend to limit production rate, and the size of the vacuum chamber may restrict the size of workpiece that can be welded.

The electron-beam process is best employed where welds of extremely high quality are required or where other processes will not produce the desired results. Electron beam welds often exhibit joint strength 15 to 25% greater than arc welds in the same material. The unique capabilities have resulted in its routine use in a number of applications, particularly in the automotive and aerospace industries. Table 32-1 provides a process summary for electron-beam welding.

LASER-BEAM WELDING

Laser beams can be used as a heat source for welding, hole making, cutting, cladding, and heat treating a wide variety of engineering metals. When used for **laser-beam welding (LBW),** the beam of coherent light can be focused to a diameter of 0.1 to 1.0 mm (0.004 to 0.04 in.), providing a power density in excess of 10^6 W/mm^2. The high-intensity beam can be used to simply melt the material at the joint, but more often, it produces a very narrow column of vaporized metal (a "keyhole") with a surrounding liquid pool. As the beam traverses, the liquid flows into the joint to produce a weld with depth-to-width ratio generally greater than 5:1. Because of the narrow weld pool geometry, high travel speed of the beam (typically several meters per minute), and low total heat input, the molten metal solidifies quickly, producing a very thin heat-affected zone and little thermal distortion. Finishing costs are quite low. Because welds require only one-side access, many different joint configurations are possible.

FIGURE 32-4 Laser butt weld of 3-mm (0.125-in.) stainless steel, made at 1.5 m/min with a 1250-W laser. *(Courtesy of Coherent, Inc., Santa Clara, CA)*

Laser-beam welding is most effective for simple fusion welds without filler metal **(autogenous welds),** but careful joint preparation is required to produce the narrow gap and necessary level of cleanliness. Filler metal can be added if the gap is excessive, and a low-velocity flow of inert gas (generally helium or argon) may be used to protect the weld pool from oxidation. Carbon steel, stainless steel, aluminum, titanium, nickel alloys and thermoplastics can all be welded with laser techniques. Some of these materials are shiny and reflective, and a light-absorbing surface treatment may be required prior to the welding operation. Figure 32-4 shows a typical laser beam weld and Table 32-2 provides a process summary for laser-beam welding.

As shown in Figure 32-5, laser-beam welding and electron-beam welding both offer some of the highest power densities of the welding processes. The well-collimated

TABLE 32-2 Process Summary: Laser-Beam Welding (LBW)

Heat source	Laser light
Protection	None, or externally supplied gas
Electrode	None
Material joined	All common metals
Rate of heat input	High
Weld profile (D/W)	5
Maximum penetration	25 mm (1 in.)
Assets	High heat-transfer efficiency; high rate of heat input but low total heat input; can weld any location that is light-accessible; small HAZ; low distortion; can accurately focus the beam with light optics; high welding speed easily automated
Limitations	Possible problems with reflectivity of some metals; good positioning and fit-up required

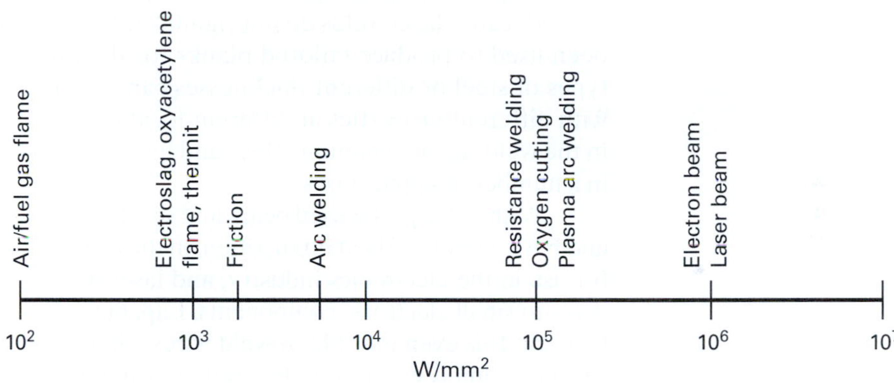

FIGURE 32-5 Comparison of the power densities of various welding processes. The high-power densities of the electron-beam and laser-beam welding processes enable the production of deep, narrow welds with small heat-affected zones. Welds can be made quickly and at high travel speeds.

beam of intense laser energy can produce deep penetration welds that are similar to electron-beam welds, but the laser-beam technique offers several distinct advantages:

1. The beam can be transmitted through air (i.e., a vacuum environment is not required). There is no physical contact between the welding equipment and the workpiece. The originating laser can be a considerable distance removed.
2. No harmful X-rays are generated.
3. The laser beam can be easily shaped, directed, and focused with both transmission and reflective optics (lenses and mirrors), and some beams can be transmitted through fiber-optic cables.
4. The only restriction on weld location is optical accessibility. Welds can be made in difficult-to-reach places, and materials can be joined within transparent containers, such as inside a vacuum tube.

A laser welding system consists of an industrial laser, a means of guiding and focusing the beam, and a means of positioning and manipulating the parts to be welded. The traditional equipment for laser welding is a CO_2 laser with a power output range of 1.5 to 10 kW (or even higher). Because CO_2 lasers emit light with a far-infrared wavelength, they require mirror systems or special optical materials to focus and position the beam. They can be used to weld steel up to 25 mm (1 in.) thick at speeds ranging from 1 to 20 m/min (3 to 65 ft/min). Nd:YAG (neodymium: yttrium–aluminum–garnet) lasers, are more limited in power and capability, but operate in a near-infrared wavelength that can utilize conventional glass lenses or delivery by flexible fiber-optic cable (as much as 3000 W of energy can be transmitted up to 150 m through a 0.6-mm-diameter fiber!). Recent alternatives include various high-power, higher-efficiency fiber-delivery lasers, including ytterbium fiber lasers and Nd:YAG disk lasers. Another process option is the use of pulsed power. Pulsing can enable high-power, deep-penetration bursts from an otherwise lower-power laser. By controlling peak power, pulse duration and pulse shape, the equipment can be tailored to the product and heat-input can be reduced and controlled.

The industrial lasers lend themselves to automation and robotic manipulation. A single Nd:YAG laser with a fiber-optic system can distribute its beam to multiple workstations in either a simultaneous (distributed power) or time-sharing (full-power) fashion. Multiple welds can be made on a single part, thereby speeding production, or the power can be distributed to individual stations as much as 100 m (300 ft) from the laser. By using fiber-optic cables, laser energy can be piped directly to the end of a robot arm. This eliminates the need to mount and maneuver a heavy, bulky laser and, by reducing weight, enhances the speed and accuracy of both positioning and manipulation. Cutting, drilling, welding, and heat treating can all be performed with the same unit, and multiple axes of motion can provide a high degree of mobility and accessibility.

The equipment cost for a CO_2 or Nd:YAG laser-beam welding system is quite high, but this cost can be somewhat offset by the faster welding speeds, the ability to weld without filler metal, and low distortion, which enables a reduction in postweld straightening and machining. Caution should be used with such equipment, however, because reflected or scattered laser beams can be quite dangerous, even at great distances from the welding site. Eye protection is a must.

Because laser welds do not significantly reduce sheet metal formability, they have been used to produce tailored blanks for the production of sheet products. Different types of steel or different thicknesses can be joined to produce single-piece products with different properties at different locations. Laser welding has made great progress in the welding of aluminum alloys and has replaced gas tungsten-arc welding or riveting in a number of applications.

With a sharply focused beam and short exposure times, laser welds can be very small and have a low total heat input, often on the order of 0.1 to 10 J. These conditions are ideal for use in the electronics industry, and laser welding is frequently used to connect lead wires to small electronic components. Lap, butt, tee, and cross-wire configurations can all be used. It is even possible to weld wires without removing the polyurethane insulation. The laser simply evaporates the insulation and completes the weld with the internal wire.

Lasers have also been used in **hybrid processes** that combine laser welding with arc welding (GMAW, GTAW, or PAW), with both operating in one process zone and producing one weld pool. Hybrid laser-arc welding (HLAW) combines the deep penetration, low distortion, and high-welding speed features of laser welding with the wider pool, gap-bridging capability of arc welding. The resulting weld pool is wide and shallow at the surface, transitioning to deep and narrow. In addition to the unique and flexible weld pool geometry, another benefit is the enhanced arc stabilization provided by the material that the laser evaporates. Laser power can be reduced from that required for lasers operating alone, and welds can be made faster than with just the arc-welding processes. The shielding gas from the arc-welding process protects the entire weld pool, and the added filler metal from the arc welding process adds mass to the weld pool, slowing cooling, thereby helping to retard weld cracking.

LASER-BEAM CUTTING

Cutting small holes, narrow slots, or closely spaced patterns in a variety of materials, or producing small quantities of complex-contoured sheet or plate, is another widely used application of industrial lasers. **Laser-beam cutting (LBC)** begins by "drilling" a hole through the material and then moving the beam along a programmed path. As shown in Figure 32-6, the intense heat from the laser is used to melt and/or evaporate the material being cut. A stream of **assist gas** blows the molten metal through the cut, cools the workpiece, minimizes the heat-affected zone, and may participate in a combustion reaction with the material being cut.

Oxygen is the usual gas for cutting mild steel. The laser heats the metal to a temperature where the iron and oxygen combine in an exothermic reaction. The molten iron and iron oxide have a low viscosity and are easily blown away by the flow of assist gas. In this exothermic cutting process, the assist gas actually contributes additional heat, up to 50% of the energy required to cut the material. High cutting speeds are possible, the speed being limited by the rate of material burning. Nitrogen is used with stainless steel and aluminum, and, because of its high reactivity, titanium requires an

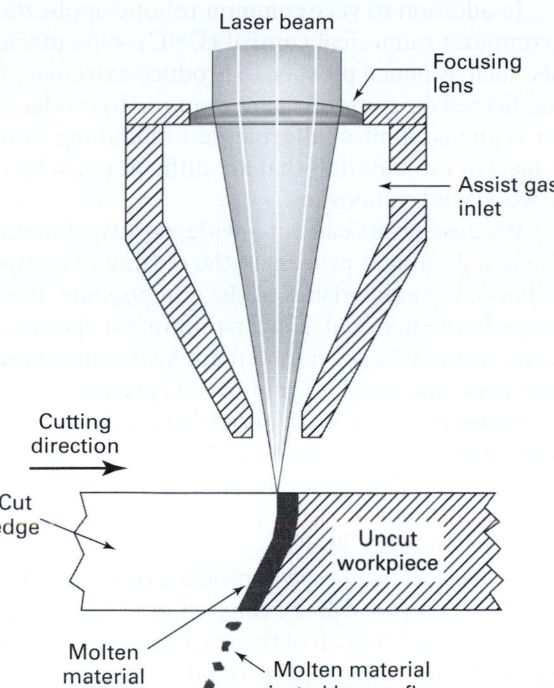

FIGURE 32-6 Schematic of laser-beam cutting. The laser provides the heat, and the flow of assist gas propels the molten droplets from the cut.

inert gas, such as argon. Inert gas or compressed air is generally used to cut a wide range of nonmetallic materials. The latter processes are ones of endothermic cutting, because the gas actually absorbs energy as it is heated. Cutting speed is set by the rate at which the laser can melt and/or vaporize material. **Exothermic cutting** produces an oxidized edge, while **endothermic cutting** (also called *clean cutting*) results in oxide-free surfaces.

Because of its features and recent advancements, laser cutting has replaced a number of plasma and oxy-fuel cutting operations, as well as some mechanical cutting processes Clean, accurate, square-edged cuts are characteristic, and the **kerf** (typically as small as 0.25 mm, or 0.01 in.) and heat-affected zone are narrower than with any other thermal cutting process. Because laser cutting is noncontact, no clamping or fixturing is required, as is common with mechanical cutting. Virtually any material can be cut, and no postcut finishing is required in many applications, even though the process does produce a thin recast surface.

Originally developed for the cutting of thin sheet, laser cutting is rapidly moving into heavier gauges and plate material. A 5000-W laser can produce the following cuts:

$\frac{1}{2}$-in. mild steel at 70 inches per minute (ipm)
$\frac{3}{4}$-in. mild steel at 42 ipm
$1\frac{1}{8}$-in. mild steel at 23 ipm
$\frac{3}{8}$-in. aluminum at 50 ipm
$\frac{1}{2}$-in. aluminum at 26 ipm

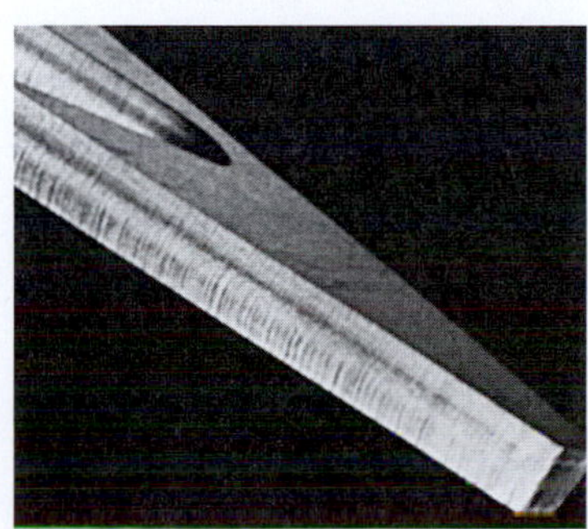

FIGURE 32-7 Surface of 6-mm-thick carbon steel cut with a 1250-W laser at 1.8 m/min. *(Courtesy of Coherent, Inc., Santa Clara, CA)*

On thin steel sheet, and with some nonmetals, cutting speeds can be in excess of 25 m/min or 1000 ipm (80 ft/min), and positioning accuracy can be within 0.025 mm or 0.001 in. Figure 32-7 shows the edge of 6-mm (0.25-in.)-thick carbon steel, laser cut at 1.8 m/min with a 1250-W laser.

Carbon dioxide and Nd:YAG lasers have been used in both continuous and pulsed modes. While cutting speed depends on the material being cut and its thickness, it is greatest in the continuous mode that is preferred for straight and mildly contoured cuts. The pulsed mode is preferred for thin materials and enables tight corners and intricate details to be cut without excessive burning. Metal plates as well as a variety of nonmetals can be cut in thicknesses up to 30 mm ($1\frac{1}{4}$ in.). Cutting temperatures can be in excess of 11,000°C (20,000°F).

In addition to very common robotic applications, lasers have also been mounted on computer numerical control (CNC)–type machines, or combined with traditional tools, such as punch presses, to produce extremely flexible hybrid equipment. Because no dedicated dies or tooling are required to produce a cut and there is no setup time, the laser is an economical alternative to blanking or nibbling for prototype or short-run products or for materials that are difficult to cut by conventional methods, such as plastics, wood, and composites.

Because lasers can cut a wide variety of metals and nonmetals, laser cutting has become a dominant process in the cutting of composite materials. The more uniform the thermal characteristics of the components, the better the cut and the less thermal damage to the material. Kevlar-reinforced epoxy cuts easiest and gives a narrow heat-affected zone. Glass-reinforced epoxy is more difficult because of the greater thermal differences, and graphite-reinforced epoxy is even worse because of the high dissociation temperature and thermal conductivity of the graphite. By the time the graphite has absorbed sufficient cutting heat, the epoxy matrix will have decomposed to a significant depth. The use of lasers for machining is discussed further in Chapter 28.

LASER SPOT WELDING

Lasers have also been used to produce spot welds in a manner that offers unique advantages when compared to the conventional resistance methods. A small clamping force is applied to assure contact of the workpieces, and a fine-focused beam then scans the area of the weld. Welding is performed in the keyhole mode, where the laser produces a small hole through the molten puddle. As the beam is moved, molten metal flows into the hole and solidifies, forming a fusion-type nugget.

Laser spot welding can be performed with access to only one side of the joint. It is a noncontact process and produces no indentations. No electrodes are involved, so electrode wear is no longer a production problem. Weld quality is independent of material resistance, surface resistance, and electrode condition, and no water cooling is required. The total heat input is low, so the heat-affected zone is small. Speed of welding and strength of the resulting joint are comparable to resistance spot welds.

FLASH WELDING

Flash welding (FW) is a process used to produce butt welds between similar or dissimilar metals in solid or tubular form. The two pieces of metal are first secured in current-carrying grips and lightly touched together. An electric current may be passed through the joint to provide optional preheat, after which the pieces are withdrawn slightly. An intense flashing arc forms across the gap, which melts the material on both surfaces. The pieces are then forced together under high pressure (on the order of 70 MPa, or 10,000 psi), expelling the liquid and oxides, and upsetting the softened metal. The electric current is turned off, and the force is maintained until solidification is complete. If desired, the upset portion can then be removed by machining. Figure 32-8 shows a schematic of the flash welding process, including both the equipment setup and the completed weld.

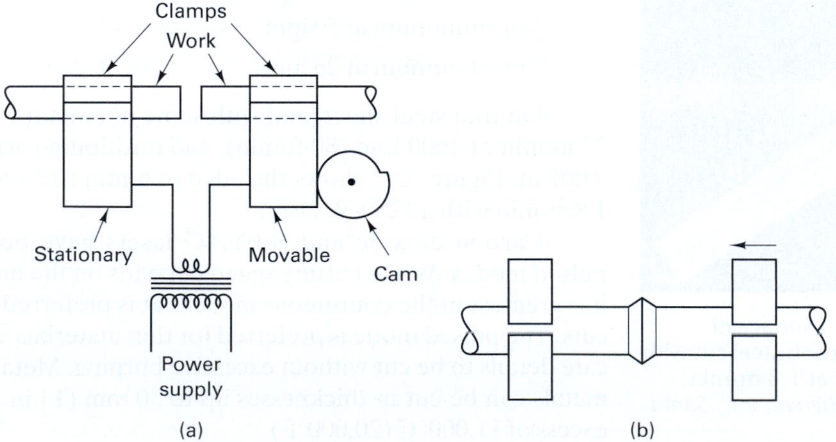

FIGURE 32-8 Schematic of the flash-welding process: (a) equipment and setup; (b) completed weld.

To produce a high-quality weld, it is important that the initial surfaces be flat and parallel so that the flashing is even across the area to be joined. The flashing action must be continued long enough to melt the interface and also soften the adjacent metal. Sufficient plastic deformation must occur during the upsetting to transfer the impurities and contaminants outward into the flash. The equipment required is generally large and expensive, but excellent welds can be made at high production rates.

Percussion welding (PEW) is a similar process, in which a rapid discharge of stored energy produces a brief period of arcing, which is followed by the rapid application of force to expel the molten metal and produce the joint. In percussion welding, the duration of the arc is on the order of 1 to 10 ms. The heat is intense but highly concentrated. Only a small amount of molten metal is produced, little or no upsetting occurs at the joint, and the heat-affected zone is quite small. Application is generally restricted to the butt welding of bar or tubing where heat damage is a major concern.

Upset welding (UW) is also similar to flash welding, but there is no period of arcing. The equipment and geometries are similar, but the heating is achieved through electrical resistance. The parts are clamped in the machine, pressure is applied, and high current is passed through the joint. When the abutting surfaces have been heated to a suitable forging temperature, the current is stopped and an upsetting force is applied to produce coalescence. The initial conditions of interface flatness, finish, and alignment must create uniform contact if a good quality weld is to be produced.

32.3 SURFACE MODIFICATION BY WELDING-RELATED PROCESSES

SURFACING (INCLUDING HARDFACING)

Surfacing or **thermal cladding** is the process of depositing a layer of weld metal on the surface or edge of a different composition base material. The usual objectives are to obtain improved resistance to wear, abrasion, heat, or chemical attack without having to make the entire piece from an expensive material, one that is difficult to fabricate, or one that would not possess the desired bulk properties. Because the deposited surfaces are generally harder than the base metal, the process is often called **hardfacing.** This is not always true, however, for in some cases a softer metal (such as bronze) is applied to a harder base material.

Surfacing Materials. The materials most commonly used for surfacing include (1) carbon and low-alloy steels; (2) high-alloy steels and irons; (3) cobalt-based alloys; (4) nickel-based alloys, such as Monel, Nichrome, and Hastelloy; (5) copper-based alloys; (6) stainless steels; and (7) ceramic and refractory carbides, oxides, borides, silicides, and similar compounds.

Surfacing Methods and Applications. Because some of the base metal melts during the deposition, surfacing is actually a variation of fusion welding and can be performed by nearly all of the gas-flame or arc-welding techniques, including oxyfuel gas, shielded metal arc, gas metal arc, gas tungsten arc, submerged arc, and plasma arc. Arc welding is frequently used for the deposition of high-melting-point alloys. Submerged arc welding is used when large areas are to be surfaced or a large amount of surfacing material is to be applied. The plasma arc process further extends the process capabilities because of its extreme temperatures. To obtain true fusion of the surfacing material, a transferred arc is used and the surfacing material is injected in the form of a powder. If a nontransferred arc is used, only a mechanical bond is produced, and the process becomes a form of **metallizing.** Lasers are also used in surfacing or thermal cladding operations where they melt powders that have been deposited on the substrate surface.

THERMAL SPRAY COATING OR METALLIZING

The **thermal spray** processes offer a means of applying a coating of high-performance material (metals, alloys, ceramics, intermetallics, cermets, carbides, or even plastics) to more economical and more easily fabricated base metals. A wire or rod of the coating material is fed into a gas flame or arc, where it melts and becomes atomized by a stream of gas, such as argon, nitrogen, combustion gases, or compressed air. The gas stream propels the 0.01- to 0.05-mm (0.0004- to 0.002-in.)-diameter molten particles toward the target surface, where they impact ("splat"), cool, and bond. Very little heat is transferred to the

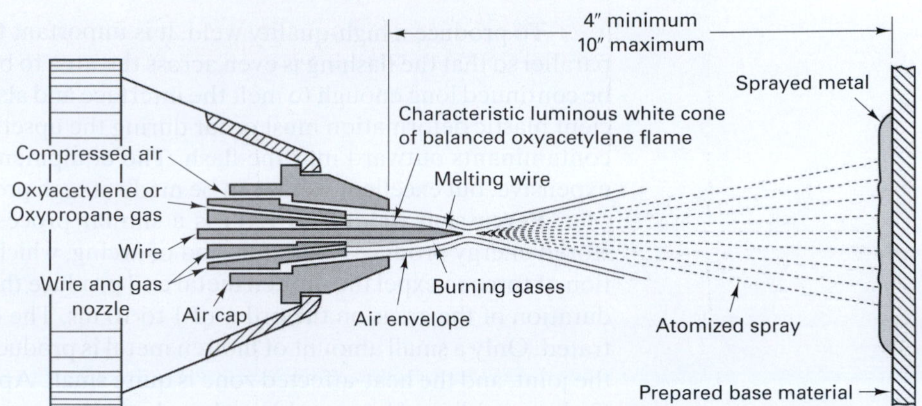

FIGURE 32-9 Schematic of an oxyacetylene metal-spraying gun. *(Courtesy of Sulzer Metco, Winterthur, Switzerland)*

substrate, whose peak temperatures generally range from 100 to 250°C (200 to 500°F). As a result, thermal spraying does not induce undesirable metallurgical changes or excessive distortion, and coatings can be applied to thin or delicate targets or to heat-sensitive materials such as plastics. The applied coating can range in thickness from 0.1 to 12 mm (0.004 to 0.5 in.), and there is no limit to the size of the workpiece that can be coated.

Several of the thermal spray processes use adaptations of oxyfuel welding equipment. Figure 32-9 shows a schematic of an oxyacetylene metal spraying gun designed to utilize a solid wire feed. The flame melts the wire and a flow of compressed air disintegrates the molten material and propels it to the workpiece. An alternative type of oxyfuel gun uses material in the form of powder, which is gravity or pressure fed into the flame, where it is melted and carried by the flame gas onto the target. The powder feed permits the deposition of material that would be difficult to fabricate into wire, such as cermets, oxides, and carbides. In addition, the droplet size is controlled by the size of the powder, not by the factors that control atomization.

The lower temperatures and lower particle velocities of the oxyfuel deposition methods result in coatings with high porosity and low cohesive strength. By modifying the process to produce a supersonic stream of hot gas, an adaptation known as high-velocity oxyfuel spraying (HVOF), the particles now impact with high kinetic energies, and the resulting coating is dense and well bonded.

The simplest of the electric arc methods is wire arc or electric arc spraying. In the twin-wire arc spray process, two oppositely charged electrode wires are fed through a gun, meeting at the tip, where they form an arc. A stream of atomizing gas flows through the gun, stripping off the molten metal to produce a high-velocity spray. Because all of the input energy is used to melt the metal, this process is extremely energy efficient, and deposition rates are higher than for most other processes.

Plasma spray metallizing, illustrated in Figure 32-10, is a more sophisticated technique. A plasma-forming gas serves as both the heat source and propelling agent for the

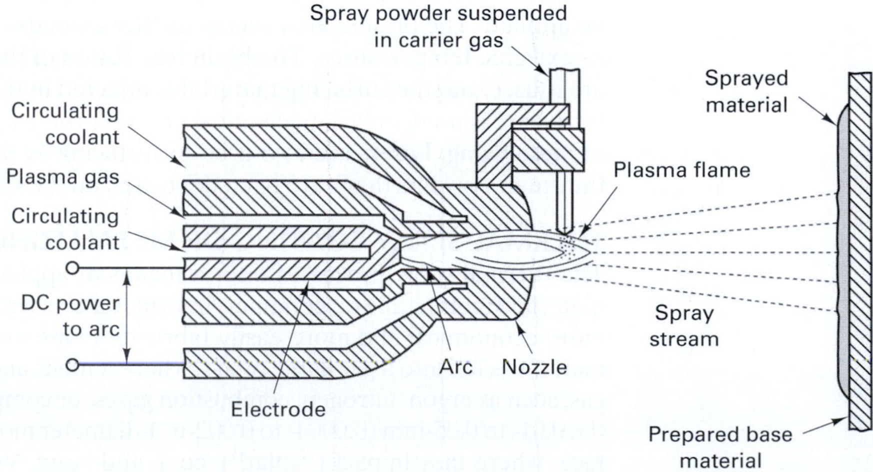

FIGURE 32-10 Diagram of a plasma-arc spray gun. *(Courtesy of Sulzer Metco, Winterhur, Switzerland)*

TABLE 32-3 Comparison of Five Thermal Spray Deposition Techniques

Method	Heat: Source	Heat: Temperature (°C)	Deposited Materials	Particle Impact Velocity (m/sec)	Adhesion Strength	Maximum Spray Rate (kg/hr)
Flame spray						
Wire	Oxyfuel	3000	Metals	180	Medium	9
Powder	Oxyfuel	3000	Metals, ceramics, plastics	30	Low	7
High-velocity oxyfuel (HVOF)	Oxyfuel	3100	Metals, carbides	600–1000	Very high	14
Wire arc	DC arc	5500	Metals only	250	High	16
Plasma spray	DC arc	5500 to 16,500	All	250–1200	High to Very high	5-25

coating material, which is usually fed in the form of powder. The molten particles attain high velocity and therefore produce a dense, strongly bonded coating. Because temperatures can reach 16,500°C (30,000°F), plasma spraying can be used to deposit materials with extremely high melting points. Metals, alloys, ceramics, carbides, cermets, intermetallics, and plastic-based powders have all been successfully deposited.

While thermal spraying or metallizing is similar to surfacing and is often applied for the same reasons, the coatings are usually thinner and the process is more suitable for irregular surfaces or heat-sensitive substrates. The deposition guns can be either handheld or machine driven. A stand-off distance of 0.15 to 0.25 m (6 to 10 in.) is usually maintained between the spray nozzle and the workpiece. Table 32-3 compares the features of five methods of thermal spray deposition.

Surface Preparation for Metallizing. Unlike surfacing, metallizing does not melt the base metal. Adhesion is entirely mechanical, so it is essential that the base metal be prepared in a way that promotes good mechanical interlocking. The target material must first be clean and free of dirt, moisture, oil, and other contaminates. The surface is then roughened by one of a variety of methods to create minute crevices that can anchor the solidifying particles. Grit blasting with a sharp, abrasive grit is the most common technique, and a surface roughness of 2.5 to 7.5 μm is adequate for most applications.

Characteristics and Applications of Sprayed Metals. During deposition, the atomized, molten, or semimolten particles mix with air and then cool rapidly upon impact with the base metal. The resultant coatings consist of bonded particles that span a range of size, shape, and degree of melting. Some particles become oxidized, and interparticle voids can become entrapped. Compared to conventional wrought material, the coatings are harder, more porous (0.1 to 15% porosity), and more brittle. Thermal spray coatings add little, if any, additional strength to a part, because the strength of the porous coating is usually between one-third and one-half of its normal wrought strength. Applications, therefore, generally look to the coating to provide resistance to heat, wear, erosion, and/or corrosion or to restore worn parts to original dimensions and specifications. Some typical applications include:

1. *Protective coatings.* Zinc and aluminum are sprayed on iron and steel to provide corrosion resistance—a process that can extend the lifetime of bridges, buildings, and other infrastructure items. Compared to electroplating or hot-dip immersion, there is no size limit, coating thickness can be varied from location to location as needed, and the coating can be applied on site. The interior surfaces of power boilers can be coated with high-chromium alloys to extend wall life by providing both heat-resistance and corrosion resistance.

2. *Building up worn surfaces*. Worn parts may be salvaged or their life extended by adding new metal to the depleted regions. The repair and restoration of aircraft engine components is probably the largest single use of thermal spraying.
3. *Hard surfacing*. Although metal spraying should not be compared to hardfacing deposits that are applied by welding techniques, it can be used when thin coatings are considered to be adequate. Typical applications might include automobile cylinder liners and piston rings; thread guides in textile plants; and critical parts within pumps, bearings, and seals.
4. *Applying coatings of expensive metals*. Metal spraying provides a simple method for applying thin coatings of noble metals to surfaces where conventional plating would not be economical.
5. *Electrical properties*. Because metal can be deposited on almost any surface, thermal spraying can be used to apply a conductive surface to an otherwise poor conductor or nonconductor. Copper, aluminum, or silver is frequently sprayed on glass or plastics for this purpose. Conversely, sprayed alumina (Al_2O_3) can be used to impart insulating or dielectric properties.
6. *Reflecting surfaces*. Aluminum, sprayed on the back of glass by a special fusion process, makes an excellent mirror.
7. *Decorative effects*. One of the earliest and still important uses of metal spraying was to obtain decorative effects. Because sprayed metal can be treated in a variety of ways, such as buffed, wire brushed, or left in the as-sprayed condition, it is frequently specified for finishing manufactured products and architectural materials.
8. *Tailored surface characteristics*. Porous coatings of cobalt or titanium alloys, or certain ceramic materials, have been applied to medical implants to help promote adhesion and in-growth of bone and tissue.

32.4 BRAZING

In brazing and soldering, the surfaces to be joined are first cleaned and the components assembled or fixtured; a low-melting-point nonferrous metal is then melted, drawn into the space between the two solids by capillary action, and allowed to solidify. **Brazing** is the permanent joining of similar or dissimilar metals or ceramics (or composites based on those two materials) through the use of heat and a filler metal whose melting temperature (actually, liquidus temperature) is above 450°C (840°F),[1] but below the melting point (or solidus temperature) of the materials being joined. The brazing process is different from welding in a number of ways:

1. The *composition* (or chemistry) of the brazing alloy is significantly different from that of the base metal.
2. The *strength* of the brazing alloy is usually lower than that of the base metal.
3. The *melting point* of the brazing alloy is lower than that of the base metal, so none of the base metal is melted.
4. Bonding requires **capillary action** to distribute the filler metal between the closely fitting surfaces of the joint. The specific flow depends on the viscosity of the liquid, the geometry of the joint, and surface wetting characteristics.

Because of these differences, the brazing process has several distinct advantages:

1. A wide range of metallic and nonmetallic materials can be brazed. The process is ideally suited for joining dissimilar materials, such as ferrous metal to nonferrous metal, cast metal to wrought metal, metals with widely different melting points, or even metal to ceramic. Fiber- and dispersion-strengthened composites can be joined.
2. Because less heating is required than for welding, the process can be performed quickly and economically.

[1] This temperature is an arbitrary one, selected to distinguish brazing from soldering.

3. The lower temperatures reduce problems associated with heat-affected zones (or other material property alteration), warping, and distortion. Thinner and more complex assemblies can be joined successfully. Thin sections can be joined to thick. Metal as thin as 0.01 mm (0.0004 in.) and as thick as 150 mm (6 in.) can be brazed.
4. Assembly tolerances are closer than for most welding processes, and joint appearance is usually quite neat.
5. Brazing is highly adaptable to automation and performs well when mass-producing complex or delicate assemblies. Complex products requiring multiple joints can be brazed in several steps using filler metals with progressively lower melting temperatures.
6. A strong permanent joint is formed.

Successful brazing or soldering requires that the parts have relatively good fit-up (i.e., small joint clearances) to promote capillary flow of the filler metal. The parts must be thoroughly cleaned prior to joining, and many parts will require flux removal after joining. It is also important to remember that any subsequent heating of the assembly can cause inadvertent melting of the braze metal, thereby weakening or destroying the joint.

Another concern with brazed joints is their enhanced susceptibility to corrosion. Because the **filler metal** is of different composition from the materials being joined, the brazed joint is actually a localized galvanic corrosion cell. Corrosion problems can often be minimized, however, by proper selection of the filler metal.

NATURE AND STRENGTH OF BRAZED JOINTS

Brazing, like welding, forms a strong metallurgical bond at the interfaces. Clean surfaces, proper clearance, good wetting, and good fluidity will all enhance the bonding. The strength of the resulting joint can be quite high, certainly higher than the strength of the brazing alloy and often greater than the strength of the metal being brazed. Attainment of a high-strength joint, however, requires optimum processing and design.

Of all of the factors contributing to joint strength, **joint clearance** is the most important. If the joint is too tight, it may be difficult for the braze metal to flow into the gap (leaving unfilled voids), and flux may not be able to escape (remaining in locations that should be filled with braze material). There must be sufficient clearance for the braze metal to wet the joint and flow into it under the force of capillary action. As the gap is increased beyond an optimum value, however, the joint strength decreases rapidly, dropping off to the strength of the braze metal itself. If the gap becomes too great, the capillary forces may be unable to draw the material into the joint or hold it in place during solidification. Figure 32-11 shows the tensile strength of a butt-joint braze as a function of joint clearance.

Proper clearance can vary considerably, depending primarily on the type of braze metal being used. The ideal clearance is usually between 0.01 and 0.04 mm (0.0005 and 0.0015 in.), an "easy-slip" fit. A press fit can even be acceptable if fluxes are not used and surface roughness is sufficient to ensure adequate flow of the filler metal into the joint. Clearances up to 0.075 mm (0.003 in.) can be accommodated with a more sluggish filler

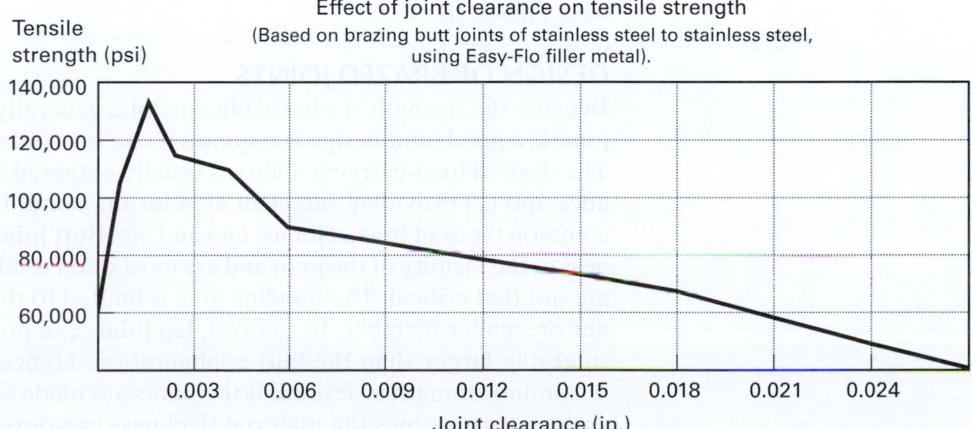

FIGURE 32-11 Typical variation of tensile strength with clearance in a butt-joint braze. *(Courtesy of Handy & Harman, Rye, NY)*

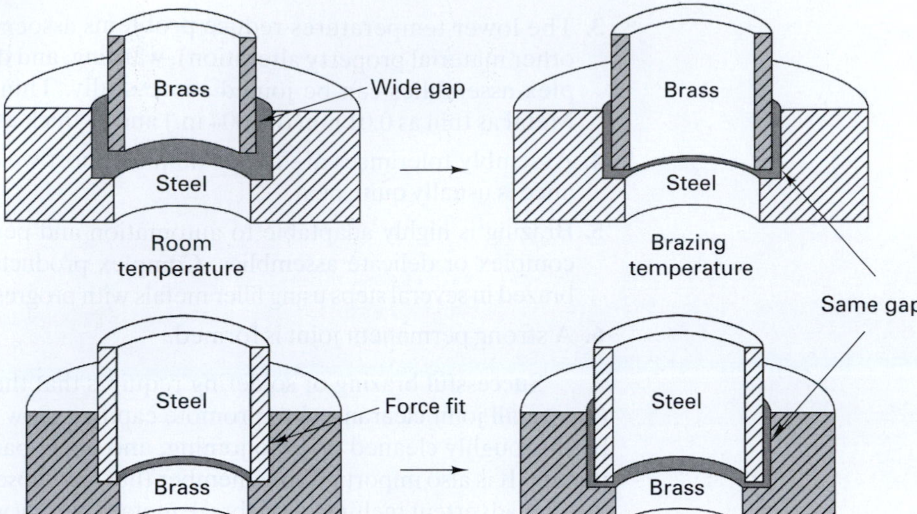

FIGURE 32-12 When brazing dissimilar metals, the initial joint clearance should be adjusted for the different thermal expansions (here, brass expands more than steel). Proper brazing clearances should exist at the temperature where the filler metal flows.

metal, such as nickel. When clearances range between 0.075 and 0.13 mm (0.003 and 0.005 in.), however, acceptable brazing becomes somewhat difficult, and joints with gaps in excess of 0.13 mm (0.005 in.) are almost impossible to braze. It should be noted that the specified gap should be maintained over the entire braze area—braze surfaces should be parallel.

It is also important to recognize that the dimensions cited above are the clearances that should exist *at the temperature of the brazing process*. Any effects of thermal expansion should be compensated when specifying the dimensions of the starting components. This is particularly significant when dissimilar materials are to be joined: here, the joint clearance will change as one material expands at a faster rate than the other. Consider a joint between brass and steel, like the one depicted in Figure 32-12. Brass expands more than steel when temperature is increased. Therefore, if the insert tube is the brass component, the initial fit should be somewhat loose. The brass will expand more than the steel as the temperature is increased, and at the brazing temperature, the gap will assume the desired dimensions. Conversely, if a steel tube is to be inserted into a brass receiver, an initial force fit may be required because the interface will widen as the brass expands more than the steel. Problems can also occur when the reverse dimensional changes occur during cooldown. Significant residual stresses can form in the new joint, and tensile stresses can induce cracking.

Wettability is a strong function of the surface tensions between the braze metal and the base alloy. Generally, the wettability is good when the surfaces are clean and the two metals can form intermediate diffused alloys. Sometimes the wettability can be improved, as is done when steel is tin plated to accept a lead–tin solder, or plated with nickel or copper to enhance brazing. **Fluidity** is a measure of the flow characteristics of the molten braze metal and is a function of the metal, its temperature, surface cleanliness, and clearance.

DESIGN OF BRAZED JOINTS

Because the strength of a braze filler metal is generally less than that of the metals being joined, a good joint design is required if one is to obtain adequate mechanical strength. The desired load-carrying ability is usually obtained by (1) assuring proper joint clearance and (2) providing sufficient area for the bond. Figure 32-13 depicts the two most common types of brazed joints: *butt* and *lap*. **Butt joints** do not require additional thickness in the vicinity of the joint and are most often used where the strength requirements are not that critical. The bonding area is limited to the cross-sectional area of the thinner or smaller member. In contrast, **lap joints** can provide bonding areas that are considerably larger than the butt configuration. Hence, they are often preferred when maximum strength is desired. If the joints are made very carefully, a lap of one to one-and-a-quarter times the material thickness can develop strength equal to that of the

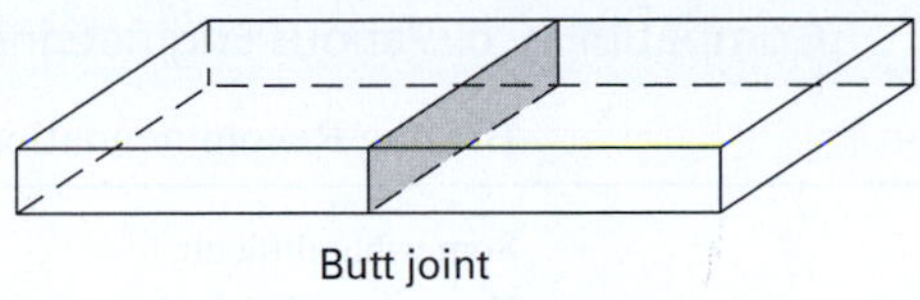

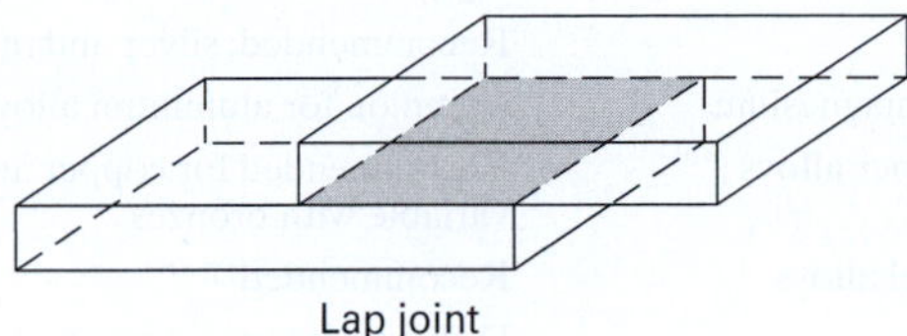

FIGURE 32-13 The two most common types of braze joints are butt and lap. Butt offers uniform thickness across the joint, whereas lap offers greater bonding area and higher strength.

parent metal. For joints that are made by routine production, it is best to use a lap of three to six times the material thickness to ensure that failure will occur in the base metal, not in the brazed joint.

Variations of the two basic joint designs include the *butt-lap* and *scarf* configurations, shown in Figure 32-14. The butt-lap design is an attempt to combine the advantage of a uniform thickness with a large bonding area and companion high strength. Unfortunately, it also requires a higher degree of joint preparation. The scarf joint maintains uniform thickness and increases bonding area by tilting the butt joint interface. Careful joint preparation and component alignment is required to maintain the desired clearance dimensions throughout the length of the joint. Figure 32-14 shows relatively simple butt, lap, butt-lap, and scarf joints for both flat and tubular parts. Figure 32-15 shows some common brazing designs for a variety of joint configurations.

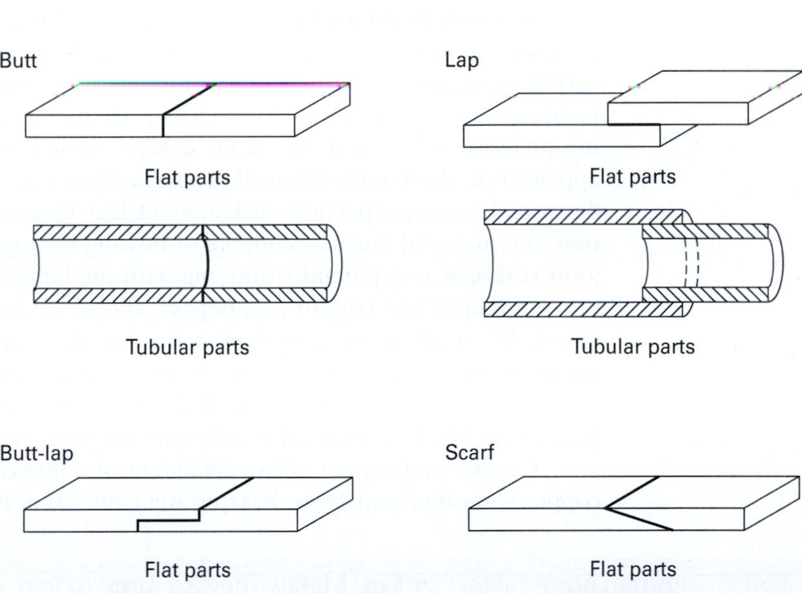

FIGURE 32-14 Variations of the butt and lap configurations include the butt-lap and scarf. The four types are shown for both flat and tubular parts.

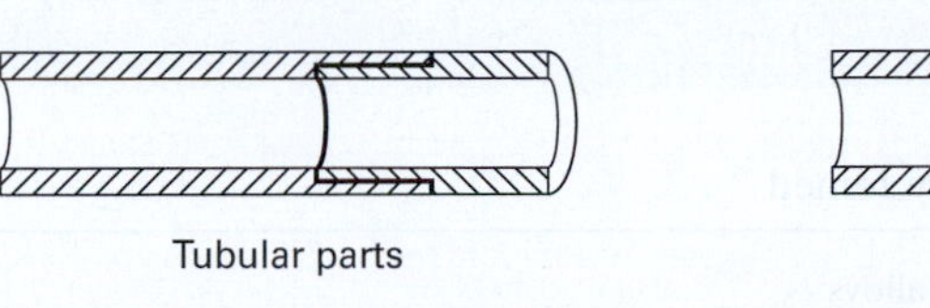

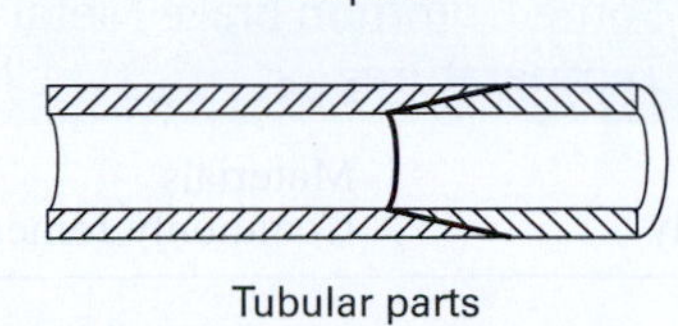

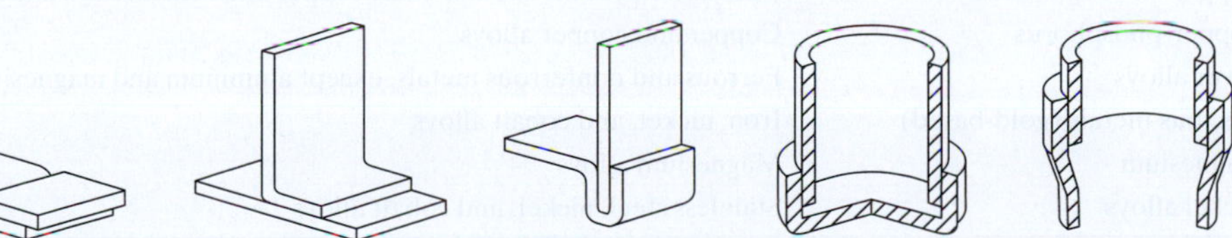

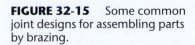

FIGURE 32-15 Some common joint designs for assembling parts by brazing.

TABLE 32-4 Compatibility of Various Engineering Materials with Brazing

Material	Brazing Recommendation
Cast iron	Somewhat difficult
Carbon and low-alloy steels	Recommended for low- and medium-carbon materials; difficult for high-carbon materials; seldom used for heat-treated alloy steels
Stainless steel	Recommended; silver and nickel brazing alloys are preferred
Aluminum and magnesium	Common for aluminum alloys and some alloys of magnesium
Copper and copper alloys	Recommended for copper and high-copper brasses; somewhat variable with bronzes
Nickel and nickel alloys	Recommended
Titanium	Difficult, not recommended
Lead and zinc	Not recommended
Thermoplastics, thermosets, and elastomers	Not recommended
Ceramics and glass	Not recommended
Dissimilar metals	Recommended, but may be difficult, depending on degree of dissimilarity
Metals to nonmetals	Not recommended
Dissimilar nonmetals	Not recommended

The materials being brazed also need to be considered when designing a brazed joint. Table 32-4 summarizes the compatibility of various engineering materials with the brazing process.

FILLER METALS

The filler metal used in brazing can be any metal that melts between 450°C (840°F) and the melting point of the material being joined. Actual selection, however, considers a variety of factors, including compatibility with the base materials, brazing-temperature restrictions, restrictions due to service or subsequent processing temperatures, the brazing process to be used, the joint design, anticipated service environment, desired appearance, desired mechanical properties (such as strength, ductility, and toughness), desired physical properties (such as electrical, magnetic, or thermal), and cost. In addition, the material must be capable of flowing through small capillaries, "wetting" the joint surfaces, and partially alloying with the base metals. The most commonly used brazing metals are copper and copper alloys, silver and silver alloys, and aluminum alloys. Many of the brazing metals are based on eutectic reactions (See Chapter 5) where the material melts at a single temperature that is lower than the melting points of the individual metals in the alloy. Table 32-5 presents some common braze metal families, the metals they are used to join, and the typical brazing temperatures.

Copper and *copper alloys* are the most commonly used braze metals. Unalloyed copper is used primarily for brazing steel and other high-melting-point materials, such

TABLE 32-5 Some Common Braze Metal Families, Metals They Are Used to Join, and Typical Brazing Temperatures

Braze Metal Family	Materials Commonly Joined	Typical Brazing Temperature (°C)
Aluminum-silicon	Aluminum alloys	565–620
Copper and copper alloys	Various ferrous metals as well as copper and nickel alloys and stainless steel	925–1150
Copper-phosphorus	Copper and copper alloys	700–925
Silver alloys	Ferrous and nonferrous metals, except aluminum and magnesium	620–980
Precious metals (gold-based)	Iron, nickel, and cobalt alloys	900–1100
Magnesium	Magnesium alloys	595–620
Nickel alloys	Stainless steel, nickel, and cobalt alloys	925–1200

as high-speed steel and tungsten carbide. Its melting point is rather high (about 1100°C, or 2000°F), and tight fitting joints are required (gaps less than 0.075 mm, or 0.003 in.). Copper–zinc alloys offer lower melting points and are used extensively for brazing steel, cast irons, and copper. Copper–phosphorus alloys are used for the fluxless brazing of copper because the phosphorus can reduce the copper oxide film. These alloys should not be used with ferrous or nickel-based materials, however, because these metals form brittle compounds with phosphorus and the resulting joints may be brittle. A copper–nickel–titanium alloy can be used to braze titanium and some of its alloys. Manganese bronzes can also be used as filler metal in brazing operations.

Pure silver can be used for brazing titanium. **Silver solders** (alloys based on silver and copper) have brazing temperatures significantly below that of pure copper and are used in joining steels, copper, brass, and nickel. While silver and silver alloys are expensive, only a small amount is required to make a joint, so the cost per joint is still low.

Aluminum–silicon alloys, containing between 6 and 12% silicon, are used for brazing aluminum and other aluminum alloys. By using a braze metal that is similar to the base metal, the possibility of galvanic corrosion is reduced. These brazing alloys, however, have melting points of about 610°C (1130°F), and the melting temperature of commonly brazed aluminum alloys, such as 3003, is around 670°C (1290°F). Therefore, control of the brazing temperature is critical. In brazing aluminum, proper fluxing action, surface cleaning, and/or the use of a controlled atmosphere or vacuum environment is required to ensure adequate flow of the braze metal.

Nickel- and cobalt-based alloys are attractive for joining assemblies that will be subjected to elevated-temperature service conditions and/or extremely corrosive environments. The service temperature for brazed assemblies can be as high as 1200°C (2200°F). Gold and palladium alloys offer outstanding oxidation and corrosion resistance, as well as good electrical and thermal conductivity. Magnesium alloys can be used to braze other types of magnesium.

A variety of brazing alloys are currently available in the form of amorphous foils, formed by cooling metal at extremely rapid rates. These foils are extremely thin (0.04 mm, or 0.0015 in. being typical) and exhibit excellent ductility and flexibility, even when they are made from alloys whose crystalline form is quite brittle. Shaped inserts can be cut or stamped from the foil and positioned in the joint region. Because the braze material is fully dense, no shrinkage or movement occurs during the brazing operation.

One amorphous alloy—composed of nickel, chromium, iron, and boron—is used to produce assemblies that can withstand high temperatures. When the filler metal is liquid (during the brazing operation), the boron diffuses into the base metal, raising the melting point of the remaining filler. The brazed assembly can then be reheated to temperatures above the melting point of the original braze alloy, and the brazed joint will not melt.

FLUXES

In a normal atmosphere, the heat required to melt the brazing alloy would also cause the formation of surface oxides that oppose the wetting of the surface and subsequent bonding. **Brazing fluxes,** therefore, play an important part in the process by (1) dissolving oxides that may have formed on the surfaces prior to heating, (2) preventing the formation of new oxides during heating, and (3) lowering the surface tension between the molten brazing metal and the surfaces to be joined, thereby promoting the flow of the molten material into the joint. Ideally, the **flux** will melt and become active at a temperature below the solidus of the filler metal, yet remain active throughout the entire range of temperatures encountered while making the braze.

Surface cleanliness is one of the most significant factors affecting the quality and uniformity of brazed joints. Although fluxes can dissolve modest amounts of oxides, *they are not cleaners.* Before a flux is applied, dirt, grease, oil, rust, and heat-treat scale should be removed from the surfaces that are to be brazed. Cleaning operations can involve water- or solvent-based techniques; high-temperature burn-off of oils, greases, and fuel residues; acid pickling; grit blasting with selected media; other mechanical methods; or exposure to high-temperature reducing atmospheres. The less cleaning the flux has to do, the more effective it will be during the brazing operation. Because the presence of surface graphite impairs wetting, cast iron materials often require special treatment. Graphite removal by chemical etching may be required before cast iron can be brazed.

Brazing fluxes usually take the form of chemical compounds in which the most common ingredients are borates, fused borax, fluoroborates, fluorides, chlorides, acids, alkalies, wetting agents, and water. The particular flux should be selected for compatibility with the base metal being brazed and the particular process being used. Paste fluxes are utilized for furnace, induction, and dip brazing, and they are usually applied by brushing. Either paste or powdered fluxes can be used with the torch-brazing process, where application is usually achieved by dipping the heated end of the filler wire into the flux material.

APPLYING THE BRAZE METAL

The brazing filler metal can be applied to joints in several ways. The oldest method (and still a common technique when torch brazing) uses brazing metal in the form of a rod or wire. The joint area is first heated to a temperature high enough to melt the braze alloy and ensure that it remains molten while flowing into the joint. The torch is then used to melt the braze metal, and capillary action draws it into the prepared gap.

This method of braze metal application requires considerable labor, and care must be taken to ensure that the filler metal has flowed into the inner portions of the joint. To avoid these difficulties, the braze metal is often inserted into the joint prior to heating, usually in the form of wires, foils, shims, powders, or preformed rings, washers, disks, or slugs. Rings or shims can also be fitted into internal grooves in the joint before the parts are assembled.

When using preloaded joints, care must be exercised to ensure that the filler metal is not drawn away from the intended surface by the capillary action of another surface of contact. Capillary action will always pull the molten braze metal into the smallest clearance, regardless of whether that was the intended location. In addition, the flow of filler metal must not be cut off by inadequate clearances or the presence of entrapped or escaping air. Fillets and grooves within the joint can also act as reservoirs and trap the filler metal.

Yet another approach is to precoat one or both of the surfaces to be joined with the brazing alloy. Simply placing the materials in contact and heating forms the desired bond. By having the braze material already in place over the full area of contact, the joining operation does not have to rely on capillary action and metal flow. More complex assemblies can be produced than with conventional methods, and the thickness of the braze material is precisely controlled to provide maximum strength to the joint.

All of the components must maintain fixed positions during the brazing operation, and some form of restraint or fixturing is often required. Alignment and clearances can often be maintained by tack welding, riveting, staking, expanding or flaring, swaging, knurling, or dimpling. Shims, wires, ribbons, and screens can also be employed to assist in locating pieces or maintaining fit. For more complex components, special brazing **jigs and fixtures** are often used to hold the components during the heating. When these are used, however, it is necessary to provide springs that will compensate for thermal expansion, particularly when dissimilar metals are being joined.

HEATING METHODS USED IN BRAZING

Because molten metal tends to flow toward the location of highest temperature, it is important that the heat sources used in brazing control both the temperature and the uniformity of that temperature throughout the joint. In specifying the heating method, a number of factors should be considered, including the size and shape of the parts being brazed, the type of material being joined, and the desired quantity and rate of production.

A common source of heat for brazing is a gas-flame torch. In the **torch-brazing** procedure, oxyacetylene, oxyhydrogen, or another gas-flame combination can be used. Most repair brazing is done in this manner because of its flexibility and simplicity, but the process is also widely used in production applications where specially shaped torches speed the heating and reduce the amount of skill required. Local heating permits the retention of most of the original material strength and enables large components to be joined with little or no distortion. The major drawbacks are the difficulty in controlling the temperature and maintaining uniformity of heating, as well as meeting

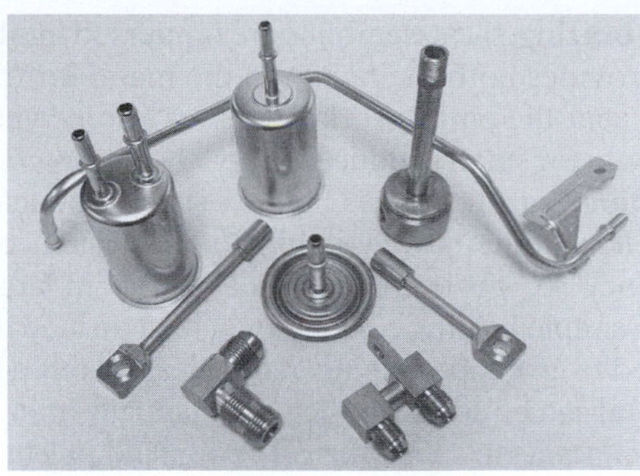

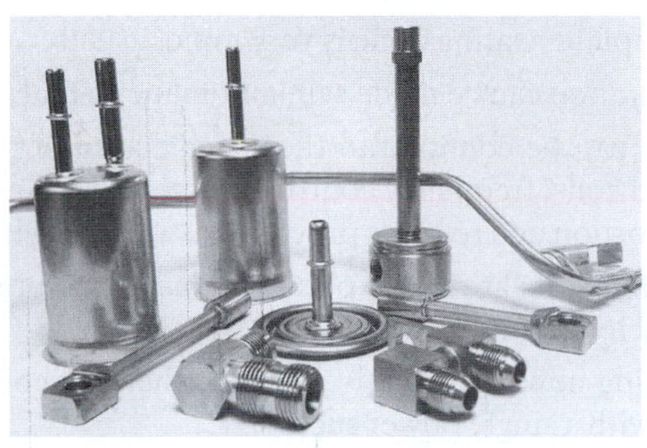

FIGURE 32-16 Two views of an array of furnace-brazed assemblies. *(www.franklinbrazing.com)*

the cost of skilled labor. Because the heating is performed in air, a protective flux is usually required, and the flux residue must be removed after brazing.

If the flux and filler metal can be preloaded into the joints and the part can endure uniform heating, a number of assemblies can be brazed simultaneously in controlled-atmosphere or vacuum furnaces, a process known as **furnace brazing.** If the components are not likely to maintain their alignment, brazing jigs or fixtures must be used. Fortunately, for most assemblies that are to be furnace brazed, a light press fit is usually sufficient to maintain alignment. Figure 32-16 shows some typical furnace-brazed assemblies.

Because excellent control of the furnace temperature is possible and no skilled labor is required, furnace brazing is particularly well suited for mass-production operations, with either batch- or continuous-type furnaces being used. Furnace brazing heats the entire assembly in a uniform manner and therefore produces less warpage and distortion than processes that employ localized heating. Extremely complex assemblies can be produced, with multiple joints being formed in a single heating.

A variety of furnace atmospheres can be utilized to reduce oxide films and prevent both the base and filler metals from oxidizing during the brazing operation. A chemical flux may no longer be needed, and the parts emerge clean and free of contaminants. When reactive materials are to be joined or the joint must meet the highest of standards, a vacuum furnace is frequently used.

A third type of heating is **salt-bath brazing,** where the parts are preheated and then dipped in a bath of molten salt that is maintained at a temperature slightly above the melting point of the brazing metal. This process offers three distinct advantages:

1. The salt bath acts as the brazing flux, preventing oxidation and enhancing wettability.
2. The work heats very rapidly because it is in complete contact with the heating medium.
3. Temperature can be accurately controlled so thin pieces can be attached to thicker pieces without danger of overheating. This last feature makes the process well suited for brazing aluminum, where precise temperature control is often required.

In salt-bath brazing, the parts must be held in jigs or fixtures (or be prefastened in some manner), and the brazing metal must be preloaded into the joints. To ensure that the bath remains at the desired temperature during the immersion process, its volume must be substantially larger than that of the assemblies to be brazed.

In **dip brazing,** the assemblies are immersed in a bath of molten brazing metal. The bath thus provides both the heat and the metal for the joint. Because the braze metal will usually coat the entire workpiece, it is a somewhat wasteful process and is usually employed only for small products.

Induction brazing utilizes high-frequency induction currents as the source of heat and is therefore limited to the joining of electrically conductive materials. A variety of high-frequency AC power supplies is available in large and small capacities. These are coupled to a simple heating coil designed to fit around the joint. The heating coils are generally formed from copper tubing and typically carry a supply of cooling water. Although the filler metal can be added to the joint manually after it is heated, the usual practice is to use preloaded joints to speed the operation and produce more uniform bonds. Induction brazing offers the following advantages, which account for its extensive use:

1. The complete heating cycle is very rapid, usually only a few seconds in duration.
2. The operation can be made semiautomatic so that only semiskilled labor is required.
3. Heating can be confined to the specific area of the joint through use of specially designed coils, frequency control, and short heating times. This minimizes softening and distortion and reduces problems associated with scale and discoloration.
4. Uniform results are easily obtained due to the precise control of both heating rate and final temperature.
5. By making new, and relatively simple heating coils, a wide variety of work can be brazed with a single power supply.

Resistance brazing can be used to produce relatively simple joints in metals with high electrical conductivity. The parts to be joined are pressed between two electrodes and a current is passed through. Unlike resistance welding, the carbon or graphite electrodes provide most of the resistance in resistance brazing, and the heating of the joint is primarily by conduction from the hot electrodes.

Infrared heat lamps, lasers, and electron beams can also be used to provide the heat required for brazing. Recent studies have also shown microwave energy to be an efficient heat source. Silicon carbide plates are positioned around the joint and are heated by the microwaves. Heat is then transferred to the joint by radiation.

FLUX REMOVAL AND OTHER POSTBRAZE OPERATIONS

Because most brazing fluxes are corrosive, the flux residue should be removed from the work as soon as brazing is completed. Rapid and complete flux removal is particularly important in the case of aluminum, where chlorides can be particularly detrimental. Fortunately, many brazing fluxes are water soluble, and an immersion in a hot-water tank for a few minutes will often provide satisfactory results. Blasting with grit or sand is another effective method of flux removal, but this procedure may not be attractive if a good surface finish is to be maintained. Fortunately, such drastic treatment is seldom necessary.

Other postbraze operations may include heat treating, cleaning, and inspection. A visual examination is probably the simplest of the inspection techniques and is most effective when both sides of a brazed joint are accessible for examination. A proof test can be performed by subjecting the joint to loads in excess of those expected during service. Leak tests or pressure tests can ensure gas- or liquid-tightness. Cracks and other flaws can be detected by dye-penetrant, magnetic particle, ultrasonic, or radiographic examination. Destructive forms of evaluation include peel tests, tension or shear tests, and metallographic examination.

FLUXLESS BRAZING

Because the application and removal of brazing flux involves significant costs, particularly where complex joints and assemblies are involved, a large amount of work has been devoted to the development of procedures where a flux is not required—that is, **fluxless brazing.** Controlled furnace atmospheres can make a flux unnecessary by reducing existing oxides and preventing the formation of new ones. Vacuum furnaces can also

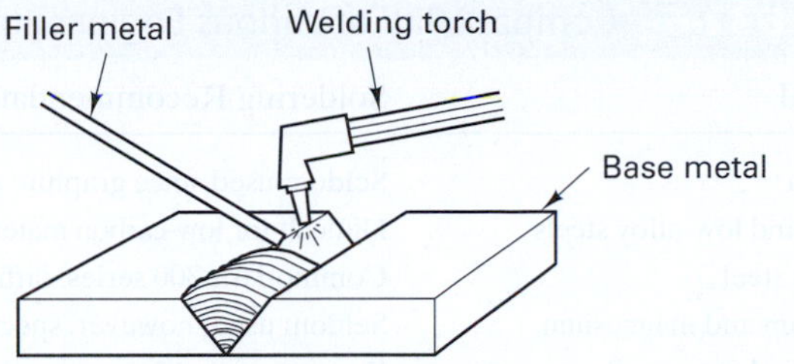

FIGURE 32-17 Schematic of the braze-welding process.

be used to create and preserve clean brazing surfaces. Special brazing metals have been developed with alloy additions, such as phosphorus, that can also fulfill the role of a flux.

BRAZE WELDING

Braze welding differs from straight brazing in that capillary action is not required to distribute the filler metal. Here, the molten filler is simply deposited by gravity, as in oxyacetylene gas welding. Because relatively low temperatures are required and warping is minimized, braze welding is very effective for the repair of steel products and ferrous castings. It is also attractive for joining cast irons because the low heat does not alter the graphite shape, and the process does not require good wetting characteristics. Strength is determined by the braze metal being used and the amount applied. Considerable buildup may be required if full strength is to be restored to the repaired part.

Braze welding is almost always done with an oxyacetylene torch. The surfaces are first "tinned" with a thin coating of the brazing metal, and the remainder of the filler metal is then added. Figure 32-17 shows a schematic of braze welding.

32.5 SOLDERING

Soldering is a brazing-type operation where the filler metal has a melting temperature (or liquidus temperature if the alloy has a freezing range) below 450°C (840°F). It is typically used for joining thin metals, connecting electronic components, joining metals while avoiding exposure to high elevated temperatures, and filling surface flaws and defects. The process generally involves six important steps: (1) design of an acceptable joint; (2) selection of the correct solder for the job; (3) selection of the proper type of flux; (4) cleaning the surfaces to be joined; (5) application of flux, solder, and sufficient heat to allow the molten solder to fill the joint by capillary action and solidify; and (6) removal of the flux residue.

DESIGN AND STRENGTH OF SOLDERED JOINTS

Soldering can be used to join a wide variety of sizes, shapes, and thicknesses and is employed extensively to provide electrical coupling or gas- or liquid-tight seals. While

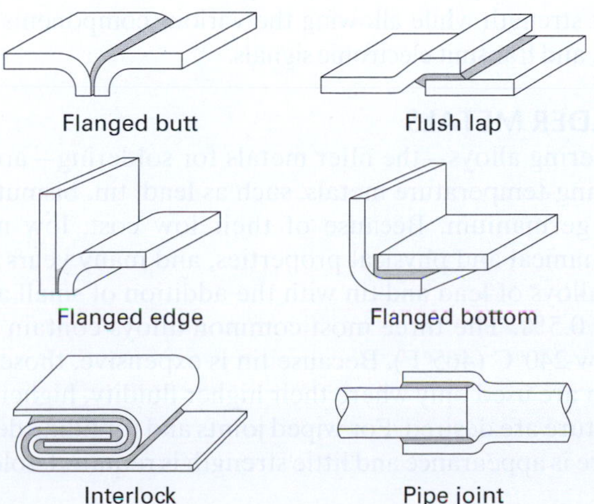

FIGURE 32-18 Some common designs for soldered joints.

TABLE 32-6 Compatibility of Various Engineering Materials with Soldering

Material	Soldering Recommendation
Cast iron	Seldom used since graphite and silicon inhibit bonding
Carbon and low-alloy steels	Difficult for low-carbon materials; seldom used for high-carbon materials
Stainless steel	Common for 300 series; difficult for 400 series
Aluminum and magnesium	Seldom used; however, special solders are available
Copper and copper alloys	Recommended for copper, brass, and bronze
Nickel and nickel alloys	Commonly performed using high-tin solders
Titanium	Seldom used
Lead and zinc	Recommended, but must use low-melting-temperature solders
Thermoplastics, thermosets, and elastomers	Not recommended
Ceramics and glass	Not recommended
Dissimilar metals	Recommended, but with consideration for galvanic corrosion
Metals to nonmetals	Not recommended
Dissimilar nometals	Not recommended

the low joining temperatures are attractive for heat-sensitive materials, soldered joints seldom develop shear strengths in excess of 1.75 MPa (250 psi). Consequently, if appreciable strength is required, soldered joints should be avoided, the contact area should be large, or some form of mechanical joint, such as a rolled-seam lock, should be made prior to soldering. Butt joints should never be used, and designs where peeling action is possible should be avoided. Figure 32-18 shows some of the more common solder joint designs, including lap, flanged butt, and interlock.

As with brazing, there is an optimal clearance for best performance. For typical solder joints, a clearance of 0.025 to 0.13 mm (0.001 to 0.005 in.) provides for capillary flow of the solder, expulsion of the flux, and reasonable joint strength. The parts should be held firmly so that no movement can occur until the solder has cooled to well below the solidification temperature. Otherwise, the resulting joint may contain cracks and have very little strength.

METALS TO BE JOINED

Table 32-6 summarizes the compatibility of soldering with a variety of engineering materials. Copper, silver, gold, and tin, as well as steels plated with these metals, are all easily soldered. Because aluminum has a strong, adherent oxide that makes soldering difficult, special fluxes and modified techniques may be required. Adequate joints are indeed possible, however, as evidenced by the large number of soldered aluminum radiators currently in automotive use.

Soldering is used extensively in electronic assemblies where the joints provide sufficient strength while allowing the various components to expand and contract, dissipate heat, and transmit electronic signals.

SOLDER METALS

Soldering alloys—the filler metals for soldering—are generally combinations of low-melting-temperature metals, such as lead, tin, bismuth, indium, cadmium, silver, gold, and germanium. Because of their low cost, low melting temperature, acceptable mechanical and physical properties, and many years of use, the most common solders are alloys of lead and tin with the addition of small amounts of antimony, usually less than 0.5%. The three most common alloys contain 60, 50, and 40% tin and all melt below 240°C (465°F). Because tin is expensive, those alloys having higher proportions of tin are used only where their higher fluidity, higher strength, and lower melting temperature are desired. For wiped joints and for filling dents and seams, where the primary desire is appearance and little strength is required, solders containing only 10 to 20% tin are preferred. Joints made with the 5% tin alloy require higher temperatures to produce but will withstand service temperatures as high as 150°C (300°F).

Other soldering alloys may be specified for special purposes or where environmental or health concerns dictate the use of lead-free joints. Lead and lead compounds can be quite toxic. Since 1988, the use of lead-containing solders in drinking water lines has been prohibited in the United States, and concern has been expresses regarding other applications and industries. Japan and the European Union have banned the use of lead-containing solders in electronic equipment. If substitute solders are to be acceptable, however, they should not only be harmless to the environment, but should also exhibit desirable characteristics in the areas of melting temperature, wettability, electrical and thermal conductivity, thermal-expansion coefficient, mechanical strength, ductility, creep resistance, thermal fatigue resistance, corrosion resistance, manufacturability, and cost. At present, none of the **lead-free solders** meet all of these requirements, and most are deficient in more than one area. Compatible fluxes must also be identified, and assembly methods may need to be modified.

Most of the alternative solders have been proposed from other eutectic alloy systems. Tin–antimony and tin–copper alloys are useful in electrical applications and have good strength and creep resistance but high melting points. Bismuth alloys have very low melting points and good fluidity, but suffer from poor wettability. Indium alloys offer low melting points, ductility that is retained even at cryogenic temperatures, and rapid creep that allows joints between dissimilar metals to adjust to changes in temperature without generating internal stresses. Tin–indium alloys have been used for metal-to-glass and glass-to-glass joints. They have very low melting points and good wettability, but they are expensive and can be somewhat brittle. Aluminum is often soldered with tin–zinc, cadmium–zinc, or aluminum–zinc alloys. Tin–silver and tin–gold offer possibilities when a somewhat higher melting point is desired (typically above 205°C, or 400°F) coupled with good mechanical strength and creep resistance, but both systems are limited by the high cost of their components. Lead–silver and cadmium–silver alloys can also be used for higher-temperature service.

TABLE 32-7 Some Common Solders and Their Properties

Composition (wt %)	Freezing Temperature (°C) Liquidus	Solidus	Range	Applications
Lead-tin solders				
98 Pb–2 Sn	322	316	6	Side seams in three-piece can
90 Pb–10 Sn	302	268	34	Coating and joining metals
80 Pb–20 Sn	277	183	94	Filling and seaming auto bodies
70 Pb–30 Sn	255	183	72	Torch soldering
60 Pb–40 Sn	238	183	55	Wiping solder, radiator cores, heater units
50 Pb–50 Sn	216	183	33	General purpose
40 Pb–60 Sn	190	183	7	Electronic (low temperature)
Silver solders				
97.5 Pb–1 Sn–1.5 Ag	308	308	0	Higher–temperature service
36 Pb–62 Sn–2 Ag	189	179	10	Electrical
96 Sn–4 Ag	221	221	0	Electrical
Other alloys				
45 Pb–55 Bi	124	124	0	Low temperature
43 Sn–57 Bi	138	138	0	Low temperature
95 Sn–5 Sb	240	234	6	Electrical
50 Sn–50 In	125	117	8	Metal-to-glass
37.5 Pb–25 In–37.5 Sn	138	138	0	Low temperature
95.5 Sn–3.9 Ag–0.6 Cu	217	217	0	Electrical
91.8 Sn–3.4 Ag–4.8 Bi	213	211	2	Electrical (must be lead-free)

The three-component tin–silver–copper system has emerged as the predominant lead-free solder for electrical and electronics applications. Typical compositions include 3 to 4% silver and 0.5 to 0.8% copper, with the remainder being tin. Other ternary (three-component) systems showing promise include tin–silver–bismuth and bismuth–indium–zinc.

Like the filler metal used in brazing and braze welding, solders are available as wire and paste, as well as in a variety of standard and special preshaped forms. Table 32-7 presents some of the more common solder alloys with their melting properties and typical applications.

SOLDERING FLUXES

As in brazing, soldering requires that the metal surfaces be clean and free of oxide so that the solder can wet the surfaces and be drawn into the joint to produce an effective bond. Soldering fluxes are used to remove surface oxides and prevent oxide formation during the soldering process, but it is essential that dirt, oil, and grease be removed before the flux is applied. This precleaning or surface preparation can be performed by a variety of chemical or mechanical means, including solvent or alkaline degreasers, acid immersion (pickling), grit blasting, sanding, wire brushing, and other mechanical abrasion techniques.

Soldering fluxes are generally classified as **corrosive** or **noncorrosive.** The most common noncorrosive flux is **rosin** (the residue after distilling turpentine) dissolved in alcohol. Rosin fluxes are suitable for making joints to copper and brass and to tin-, cadmium-, and silver-plated surfaces, provided that the surfaces have been adequately cleaned prior to soldering. Aniline phosphate is a more active noncorrosive flux, but it has limited use because it emits toxic gases when heated. The wide variety of corrosive fluxes provide enhanced cleaning action, but require complete removal after the soldering operation to prevent corrosion problems during service.

THE SOLDERING OPERATION

Soldering requires a source of sufficient heat and a means of transferring it to the metals being joined. Any method of heating that is suitable for brazing can be used for soldering, but furnace and salt-bath heating are seldom used. Most hand soldering is still done with soldering irons or small oxyfuel or air-fuel (acetylene, propane, butane or MAPP) torches. Induction heating is used when large numbers of identical parts are to be soldered. For low-melting-point solders, infrared heat sources can also be employed. The joints can be preloaded with solder, or the filler metal can be supplied from a wire. The particular method of heating usually dictates which procedure is used.

Wave soldering, depicted in Figure 32-19, is a process used to solder wire ends, such as the multiple connectors that protrude through holes in electronic circuit boards. Molten solder is pumped upward through a submerged nozzle to create a wave or crest in a pool of molten metal. The circuit boards are then passed across this wave at a height where each of the pins sees contact with the molten metal. Wetting and capillary action pulls solder into each joint, and numerous connections are made as each board passes across the wave.

In **vapor-phase soldering,** a product with prepositioned solder is passed through a chamber containing hot, saturated vapors, which condense on the cooler product, transferring the heat of vaporization. The result is rapid and uniform heating, with excellent

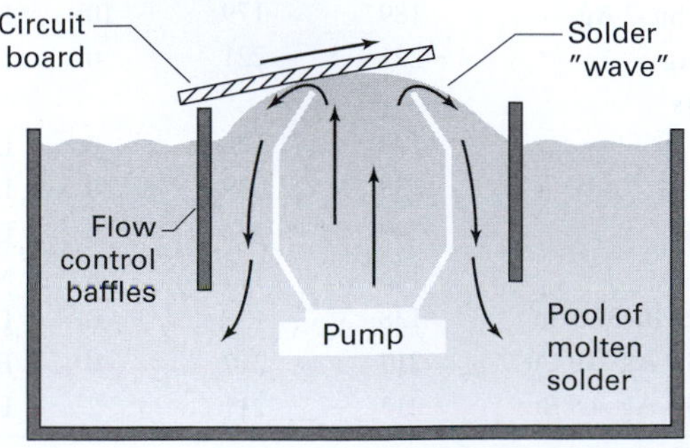

FIGURE 32-19 Schematic of wave soldering.

temperature control, combined with the possibility of an oxygen-free environment. The soldering temperature is linked to the boiling point of the fluid, with current materials operating in the range of 100 to 265°C (212 to 510°F). The vapor-phase process has also been used to cure epoxies and stress-relieve metals, but its primary application is the soldering of surface-mounted components to substrate materials. Because of the precise temperature control, multipass soldering is possible, using up to three different solder compositions with three different melting temperatures. Because the solder is pre-positioned, this process is also known as **vapor-phase reflow soldering.**

Dip soldering, where the entire piece is immersed in molten metal, has been used to produce automobile radiators and "tinned" coatings.

FLUX REMOVAL

After soldering, the flux residues should be removed from the finished joints, either to prevent corrosion or for the sake of appearance. Flux removal is rarely difficult, provided that the type of solvent in the flux is known. Water-soluble fluxes can be removed with hot water and a brush. Alcohol will remove most rosin fluxes. However, when the flux contains some form of grease, as in most paste fluxes, a grease solvent must be used, followed by a hot water rinse. In the past, solvents containing chlorofluorocarbons (CFCs) were the cleaners of choice, but because they have been implicated in the depletion of atmospheric ozone, an alternative means of flux removal should be employed or the process converted to fluxless soldering.

FLUXLESS SOLDERING

Several **fluxless soldering** techniques have been developed using controlled atmospheres (such as hydrogen plasma), thermomechanical surface activation (such as plasma gas impingement), or protective coatings that prevent oxide formation and enhance wetting. Additional successes have been reported with both laser and ultrasonic soldering.

Key Words

assist gas
autogenous weld
braze welding
brazing
brazing fluxes
butt joint
capillary action
corrosive flux
dip brazing
dip soldering
electron-beam welding (EBW)
electroslag welding (ESW)
endothermic cutting
exothermic cutting
filler metal
flash welding (FW)
fluidity
flux
fluxless brazing
fluxless soldering
furnace brazing
hardfacing
hybrid processes
induction brazing
jigs and fixtures
joint clearance
kerf
lap joint
laser spot welding
laser-beam cutting (LBC)
laser-beam welding (LBW)
lead-free solder
metallizing
molding plates
noncorrosive flux
percussion welding (PEW)
resistance brazing
rosin
salt-bath brazing
silver solder
soldering
surfacing
thermal cladding
thermal spray
thermit
thermit welding (TW)
torch brazing
upset welding (UW)
vapor-phase reflow soldering
vapor-phase soldering
wave soldering
wettability

Review Questions

1. What are some of the lower-temperature methods of joining?
2. In what ways is a thermit weld similar to the production of a casting?
3. What is the source of the welding heat in thermit welding?
4. For what types of applications might thermit welding be attractive?
5. What is the source of the welding heat in electroslag welding?
6. What are some of the various functions of the slag in electroslag welding?
7. Electroslag welding would be most attractive for the joining of what types of geometries and thicknesses?
8. What is the source of heat in electron-beam welding?
9. Why is a high vacuum required in the electron-beam chamber of an electron-beam welding machine?

10. What types of production limitations are imposed by the high-vacuum requirements of electron-beam welding? What compromises are made when welding is performed on pieces outside the vacuum chamber?
11. What are the major assets and negative features of high-voltage electron-beam welding equipment?
12. What are some of the attractive features of electron-beam welding? Negative features?
13. What is unique about the fusion zone geometry of electron-beam welds?
14. What might be necessary to permit the laser welding of shiny or reflective materials?
15. What are some of the ways in which laser-beam welding is more attractive than electron-beam welding?
16. Which type of laser light can be transmitted through fiber-optic cable?
17. What are some of the attractive features of a fiber-optic laser coupled with robotic manipulation?
18. Why is laser-beam welding an attractive process for producing tailored blanks for sheet metal forming? For use on small electronic components?
19. What are the attractive properties of hybrid processes that combine laser and arc welding?
20. What is the function of the assist gas in laser-beam cutting?
21. How do the cut edges differ with endothermic laser cutting and exothermic laser cutting?
22. What are the attractive features of laser-beam cutting compared to plasma and oxyfuel processes?
23. What features have made lasers a common means of cutting composite materials?
24. What are some of the attractive features of laser spot welding?
25. In the flash-welding process, why is it important to have a sufficient duration of arcing and sufficient amount of upsetting?
26. Percussion welding and upset welding have geometries that are similar to flash welding. How are these processes different?
27. What are some common objectives of surfacing operations?
28. What types of materials are applied by surfacing methods?
29. What are some of the primary methods by which surfacing materials can be deposited onto a metal substrate?
30. What are some of the techniques that can be used to apply a thermal spray coating?
31. How is thermal spraying similar to surfacing? How is it different?
32. Why is surface preparation such a critical feature of metallizing?
33. What are some of the more common applications of sprayed coatings?
34. Provide a reasonable definition of brazing?
35. What are some key differences between brazing and fusion welding?
36. Why is brazing an attractive process for joining dissimilar materials?
37. What advantages can be gained by the lower temperatures of the brazing process?
38. Why do brazed joints have an enhanced susceptibility to corrosion?
39. What is the most important factor contributing to the strength of a brazed joint?
40. How does capillary action relate to joint clearance?
41. Why is it necessary to adjust the initial room-temperature clearance of a joint between two significantly dissimilar metals?
42. What is wettability? Fluidity? How do each relate to brazing?
43. What are the two most common types of brazed joints and the attractive features of each?
44. How do the butt-lap and scarf joint configurations enhance or improve the conventional butt design?
45. What are some important considerations when selecting a brazing alloy?
46. What are some of the most commonly used brazing metals?
47. What special measures should be taken when brazing aluminum?
48. What are the three primary functions of a brazing flux?
49. Why is it important to preclean brazing surfaces before applying the flux?
50. In what ways might braze metal be preloaded into joints?
51. What is the purpose of brazing jigs and fixtures?
52. What is the primary attraction of furnace-brazing operations?
53. Why might reducing atmospheres or a vacuum be employed during furnace brazing operations?
54. What are some of the attractive features of salt-bath brazing?
55. Why is dip brazing usually restricted to use with small parts?
56. What are some of the attractive features of induction brazing?
57. Why is flux removal a necessary part of many brazing operations?
58. What benefits can be achieved through fluxless brazing?
59. How does braze welding differ from traditional brazing?
60. What is the primary difference between brazing and soldering?
61. Why is soldering unattractive if a high-strength joint is desired?
62. For many years, the most common solders were alloys of what two base metals?
63. What is driving the conversion to lead-free solders?
64. What are some of the difficulties encountered when attempting a conversion to lead-free solder?
65. What are the two basic families of soldering flux?
66. What are some of the more common heat sources for producing a soldered joint?
67. Why is wave soldering attractive for making the multiple connections of circuit boards?

PROBLEMS

1. A common problem with brazed or soldered joints is galvanic corrosion, because the joint usually involves dissimilar metals in direct metal-to-metal electrical contact.
   a. For each of the various solder or braze joints described here, determine which material will act as the corroding anode.
      i. Two pieces of low-carbon steel being brazed with a copper-base brazing alloy.

Chapter 32 CASE STUDY

Impeller of a Pharmaceutical Company Industrial Shredder/Disposal

The impeller of a large industrial disposal unit has been manufactured by brazing rectangular pieces of cobalt-bonded tungsten carbide into recesses that have been machined in a circular plate of type 316 stainless steel, as illustrated in the accompanying figure. A copper-based brazing alloy has been used in the joint. In service, the impeller rotates at approximately 700 rpm, shredding outdated and off-chemistry capsules and pills (as well as other possible contaminants), eliminating all identification, thereby enabling a safe and permissible disposal. Water is used to clean and flush the disposal unit.

After five weeks of use, the impeller must be removed. Several pieces of carbide have broken free and are chewing up the unit. In addition, craters have formed in the baseplate around the recesses where the carbide has broken free and around the remaining carbide inserts.

1. Evaluate the appropriateness of the original design. Consider the base materials of stainless steel with carbide inserts. Is brazing an appropriate method of joining? Is a copper-base brazing alloy an appropriate selection?
2. What do you suspect to be the cause of the problem? What data or information would you need to confirm it?
3. How would you alter the design, materials, or method of fabrication to produce an acceptable product?

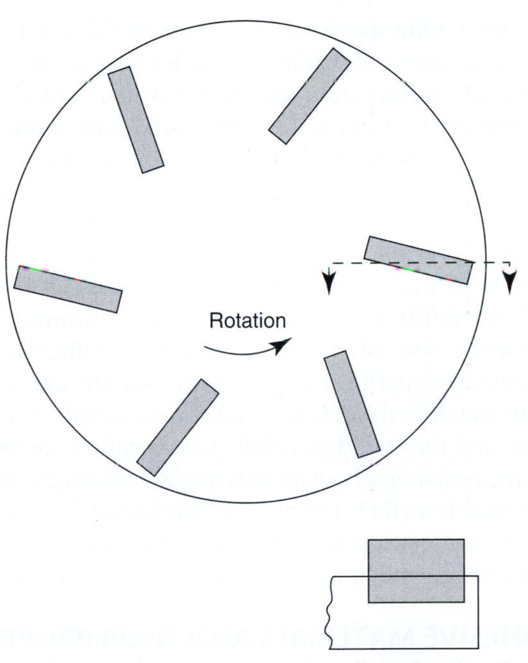

CHAPTER 33

ADHESIVE BONDING, MECHANICAL FASTENING, AND JOINING OF NONMETALS

33.1 ADHESIVE BONDING

The ideal **adhesive** bonds to any material, needs no surface preparation, cures rapidly, and maintains a high bond strength under all operating conditions. It also does not exist. However, tremendous advances have been made in the development of adhesives that are stronger, easier to use, less costly, and more reliable than many of the alternative methods of joining. From early applications, such as plywood, the use of structural adhesives has grown rapidly. Adhesives are everywhere—in construction, packaging, furniture, appliances, electronics, bookbinding, product assembly, and even medical and dental applications. They are used to bond metals, ceramics, glass, plastics, rubbers, composite materials, woods, and even a variety of roofing materials. Even such quality- and durability-conscious fields as the automotive and aircraft industries now make extensive use of adhesive bonding. Adhesives in the automotive industry have advanced from the attaching of interior and exterior trim to the joining of major components (such as door, hood, and trunk assemblies) and the installation of the nonmoving front and rear windows. Adhesive bonding has become the preferred means of assembly for polymeric body panels made from sheet-molding compounds and reaction-injection-molded (RIM) materials. Moreover, because adhesive bonding has the ability to bond such a wide variety of materials, its use has grown significantly with the ever-expanding applications of plastics and composites.

ADHESIVE MATERIALS AND THEIR PROPERTIES

In **adhesive bonding,** a nonmetallic material (the adhesive) is used to fill the gap and create a joint between two surfaces. The actual adhesives span a wide range of material types and forms, including **thermoplastic** resins, **thermosetting** resins, artificial **elastomers,** and even some ceramics. They can be applied as drops, beads, pellets, tapes, or coatings (films) and are available in the form of liquids, pastes, gels, and solids. **Curing** can be induced by the use of heat, radiation or light (photoinitiation), moisture, activators, catalysts, multiple-component reactions, or combinations thereof. Applications can be full load bearing (structural adhesives), light-duty holding or fixturing, or simply sealing (the forming of liquid- or gas-tight joints). With such a wide range of possibilities, the selection of the best adhesive for the task at hand can often be quite challenging.

The **structural adhesives** are selected for their ability to effectively transmit load across the joint, and include epoxies, cyanoacrylates, anaerobics, acrylics, urethanes, silicones, high-temperature adhesives, and hot melts. Both strength and rigidity may be important, and the bond must be able to be stressed to a high percentage of its maximum load for extended periods of time without failure.

Consider some of the more important families of adhesives:

1. **Epoxies.** The thermosetting epoxies are the oldest, most common, and most diverse of the adhesive systems, and can be used to join most engineering materials, including metal, glass, and ceramic. They are strong, versatile adhesives that can be designed to offer high adhesion, good tensile and shear strength, toughness, high rigidity, creep resistance, easy curing with little shrinkage, good chemical resistance, and tolerance to elevated temperatures. Various epoxies can be used over a temperature range from –50 to +250°C (–60 to +500°F). After curing at room temperature, shear strengths can be as high as 35 to 70 MPa (5,000 to 10,000 psi).

   Single-component epoxies use heat as the curing agent. Most epoxies, however, are two-component blends involving a resin and a curing agent, plus possible additives such as accelerators, plasticizers, and fillers that serve to enhance cure rate, flexibility, peel resistance, impact resistance, or other characteristics. Heat may again be required to drive or accelerate the cure.

   Low peel strength and poor flexibility limit epoxy adhesives, and the bond strength can be sensitive to moisture and surface contamination. Epoxies are often brittle at low temperatures, and the rate of curing is comparatively slow. Sufficient strength for structural applications is generally achieved in 8 to 12 hr, with full strength often requiring two to seven days.

2. **Cyanoacrylates.** These are liquid monomers that polymerize when spread into a thin film between two surfaces. Trace amounts of moisture on the surfaces promote curing at amazing speeds, often in as little as 2 s. Thus, the cyanoacrylates offer a one-component adhesive system that cures at room temperature with no external impetus. Commonly known as **superglues,** this family of adhesives is now available in the form of liquids and gels of varying viscosity, toughened versions designed to overcome brittleness, and even nonfrosting varieties.

   The cyanoacrylates provide excellent tensile strength, fast curing, and good shelf life and adhere well to most commercial plastics, metals, and rubbers. They are limited by their high cost, poor peel strength, and brittleness. Bond properties are poor at elevated temperatures and effective curing requires good component fit (gaps must be smaller than 0.25 mm, or 0.010 in.).

3. **Anaerobics.** These one-component, thermosetting, polyester acrylics remain liquid when exposed to air. When confined to small spaces and shut off from oxygen, as in a joint to be bonded or along the threads of an inserted fastener, the polymer becomes unstable. In the presence of iron or copper, it polymerizes into a bonding-type resin, without the need for elevated temperature. Curing can occur across gaps as large as 1 mm (0.04 in.). Additives can reduce odor, flammability, and toxicity and can speed the curing operation. Slow-curing anaerobics require 6 to 24 hr to attain useful strength. With selected additives and heat, however, curing can be reduced to as little as 5 min.

   The anaerobics are extremely versatile and can bond almost anything, including oily surfaces. The joints resist vibrations and offer good sealing to moisture and other environmental influences. Unfortunately, they are somewhat brittle and are limited to service temperatures below 150°C (300°F).

4. **Acrylics.** The acrylic-based adhesives offer good strength, toughness, and versatility, and they are able to bond a variety of materials, including plastics, metals, ceramics, and composites and even oily or dirty surfaces. Most involve application systems where a catalyst primer (curing agent) is applied to one of the surfaces to be joined and the adhesive is applied to the other. The pretreated parts can be stored separately for weeks without damage. Upon assembly, the components react to produce a strong bond at room temperature. Heat can often accelerate the curing, and at least one variety cures with ultraviolet light. In comparison to other varieties of adhesives, the acrylics offer strengths comparable to the epoxies, flexible bonds, good resistance to water and humidity, and the added advantages of room temperature curing and a no-mix application system. Major limitations include poor strength at high temperatures, flammability, and an unpleasant odor when still uncured.

5. **Urethanes.** Urethane adhesives are a large and diverse family of polymers that are generally targeted for applications that involve temperatures below 65°C (150°F) and components that require great flexibility. Both one-part thermoplastic and two-part thermosetting systems are available. Urethanes cure quickly to handling strength but are slow to reach the full-cure condition. Two minutes for handling with 24 hr to complete cure is common at room temperature.

   Compared to other structural adhesives, the urethanes offer good flexibility and toughness, even at low operating temperatures. They are somewhat sensitive to moisture, degrade in many chemical environments, and can involve toxic components or curing products.

6. **Silicones.** The silicone thermosets cure from the moisture in the air or adsorbed moisture from the surfaces being joined. They form low-strength structural joints and are usually selected when considerable amounts of expansion and contraction are expected in the joint; flexibility is required (as in sheet metal parts); or good gasket, gap-filling, or sealing properties are necessary. Metals, glass, paper, plastics, and rubbers can all be joined. The adhesives are relatively expensive, and curing is slow, but the bonds that are produced can resist moisture, hot water, oxidation, and weathering, and they retain their flexibility at low temperature.

7. **High-temperature adhesives.** When strength must be retained at temperatures in excess of 300°C (500°F), high-temperature structural adhesives should be specified. These include epoxy phenolics, modified silicones or phenolics, polyamides, and some ceramics. High cost and long cure times are the major limitations for these adhesives, which see primary application in the aerospace industry.

8. **Hot melt adhesives.** Hot-melt adhesives can be used to bond dissimilar substrates, such as plastics, rubber, metals, ceramics, glass, wood, and fibrous materials like paper, fabric, and leather. They can produce permanent or temporary bonds, seal gaps, and plug holes. While generally not considered to be true structural adhesives, the hot melts are being used increasingly to transmit loads, especially in composite material assemblies. The joints can withstand exposure to vibration, shock, humidity, and numerous chemicals and offer the added features of sound deadening and vibration damping.

   Most hot-melt adhesives are thermoplastic resins that are solid at room temperature, but melt abruptly when heated into the range of 100 to 150°C (200 to 300°F). They are usually applied as heated liquids (between 160 and 180°C, or 320 and 355°F) and form a bond as the molten adhesive cools and resolidifies. Another method of application is to position the adhesive in the joint prior to operations such as the paint-bake process in automobile manufacture. During the baking, the adhesive melts, flows into seams and crevices, and seals against the entry of corrosive moisture. These adhesives contain no solvents and do not need time to cure or dry. Hot melts achieve more than 80% of their bond strength within seconds of solidification, but they do soften and creep when subsequently exposed to elevated temperatures and can become brittle when cold.

   The traditional hot melts do not cross-link or form three-dimensional network structures, but retain their linear thermoplastic structure throughout their history. As a result, they are characterized by poor strength, poor heat resistance, and the tendency to creep under load. A relatively new class of material, the **reactive hot melts,** overcomes many of these limitations, positioning the hot melts as true structural adhesives. These materials are applied as liquids at elevated temperature, cool to solids at room temperature, but then react (often with moisture) to form a cross-linked or three-dimensional network thermoset polymer with enhanced performance properties. They melt at lower temperatures than the conventional hot melts and can bond to many different surfaces. Their tensile strengths range from 14 to 24 MPa (2000 to 3500 psi), with elongations between 290 and 750%. They can also endure higher service temperatures than their conventional counterparts (100°C, or 212°F, for long time and 125°C, or 260°F, for intermittent exposure).

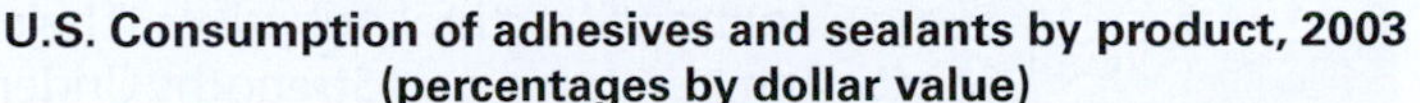

U.S. Consumption of adhesives and sealants by product, 2003
(percentages by dollar value)

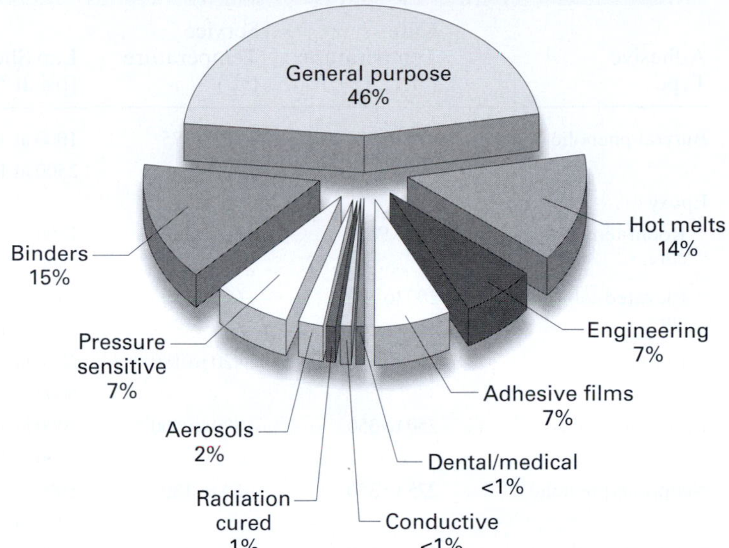

FIGURE 33-1 Distribution among the common types of adhesives and sealants. *(Reprinted with permission from* The Rauch Guide to the US Adhesives & Sealants Industry, *5th ed., 2006, Grey House Publishing, Millerton, NY)*

Additives also play a large part in the success of industrial adhesives. They can impart or enhance properties like toughness, joint durability, moisture resistance, adhesion, and flame retardance. Rheological additives and plasticizers control viscosity and flow. Adhesives must penetrate the surfaces to be bonded but not flow in an uncontrollable fashion. Pigments, antioxidants, and ultraviolet light stabilizers impart still other properties. Fillers and extenders provide bulk and reduce cost.

Figure 33-1 provides the distribution of various types of adhesive and sealant products for a recent year, and Figure 33-2 classifies adhesives by end-use markets. Table 33-1 lists some popular structural adhesives along with their service and curing temperatures and expected strengths. Table 33-2 presents the advantages and disadvantages of various curing processes.

NONSTRUCTURAL AND SPECIAL ADHESIVES

There are a number of other types of adhesives whose limited load-bearing capabilities place them in a nonstructural classification. Nevertheless, they still play roles in

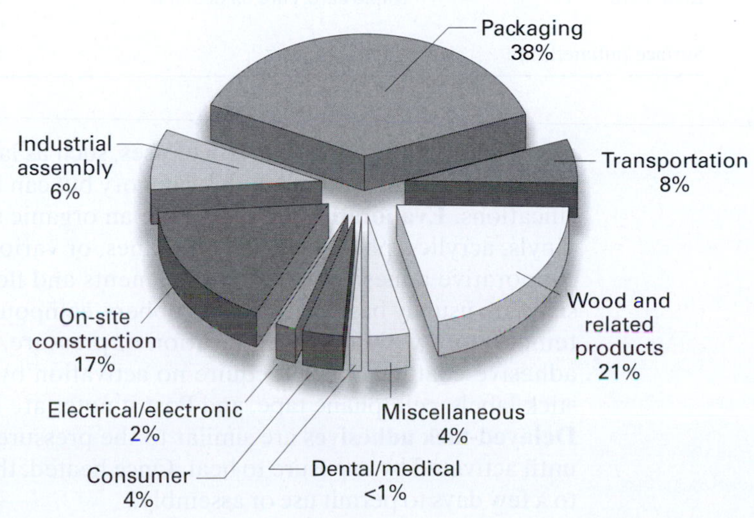

FIGURE 33-2 Distribution of adhesives and sealants by end-use areas. *(Reprinted with permission from* The Rauch Guide to the US Adhesives & Sealants Industry, *5th ed., 2006, Grey House Publishing, Millerton, NY)*

TABLE 33-1 Some Common Structural Adhesives, Their Cure Temperatures, Maximum Service Temperatures, and Strengths Under Various Types of Loading

Adhesive Type	Cure Temperature (°F)	Service Temperature (°F)	Lap Shear Strength (psi at °F)[a]	Peel Strength at Room Temperature (lb/in.)
Butyral-phenolic	275 to 350	−60 to 175	1000 at 175 2500 at RT	10
Epoxy				
Room-temperature cure	60 to 90	−60 to 180	1500 at 180 2500 at RT	4
Elevated-temperature cure	200 to 350	−60 to 350	1500 at 350 2500 at RT	5
Epoxy-nylon	250 to 350	−420 to 180	2000 at 180 6000 at RT	70
Epoxy-phenolic	250 to 350	−420 to 500	1000 at 175 2500 at RT	10
Neoprene-phenolic	275 to 350	−60 to 180	1000 at 180 2000 at RT	15
Nitrile-phenolic	275 to 350	−60 to 250	2000 at 250 4000 at RT	60
Polyimide	550 to 650	−420 to 1000	1000 at 1000 2500 at RT	3
Urethane	75 to 250	−420 to 175	1000 at 175 2500 at RT	50

[a] RT, room temperature.

TABLE 33-2 Advantages and Disadvantages of Various Structural Adhesive Curing Processes

Curing Process	Advantages	Disadvantages
Mixing reactive components	Good shelf life, unlimited depth of cure, accelerated with heat	High processing costs, mix ratio critical to performance
Anaerobic cure	Single-component adhesive, good shelf life	Poor depth of cure, require primer on many surfaces, sensitive to surface contaminants
Heat cure	Unlimited depth of cure, heat can aid adhesion	Expenses for oven energy cost, heat can adversely affect some substrates
Moisture cure	Room-temperature process, one component, no curing equipment required	Long cure cycles (12–72 hr), minimum % humidity required, limited depth of cure
Light cure	Rapid cure, cure on demand	Expenses for UV light source, limited depth of cure, most allow light to reach bond
Surface-initiated cure	Rapid cure	Poor depth of cure

manufacturing through a variety of uses, such as labeling and packaging. The hot-melt adhesives are often placed in this category but can be used for applications in both classifications. **Evaporative adhesives** use an organic solvent or water base, coupled with vinyls, acrylics, phenolics, polyurethanes, or various types of rubbers. Some common evaporative adhesives are rubber cements and floor waxes. **Pressure-sensitive adhesives** are usually based on various rubbers, compounded with additives to bond at room temperature with a brief application of pressure. No cure is involved, and the tacky adhesive-coated surfaces require no activation by water, solvents, or heat. Peel-and-stick labels, cellophane tape, and Post-it notes are examples of this group of adhesives. **Delayed-tack adhesives** are similar to the pressure sensitive systems but are nontacky until activated by exposure to heat. Once heated, they remain tacky for several minutes to a few days to permit use or assembly.

While most adhesives are electrical and thermal insulators, **conductive adhesives** can be produced by incorporating selected fillers, such as silver, copper, aluminum, nickel, and gold in the form of flakes or powder. These conductive materials must be added in sufficient quantity so as to be in physical contact with one another—an amount that often compromises flexibility and adhesion. The resin must then provide bonding between the particles and between the adhesive and the substrates being joined. Metal particles can also be used to provide thermal conductivity. When thermal conductivity is desired without electrical conductivity, certain ceramic fillers can be used, including aluminum oxide, beryllium oxide, boron nitride, and silica.

Still another group of commercial adhesives are those designed to cure by exposure to radiation, such as visible, infrared, or ultraviolet light; microwaves; or electron beams. These **radiation-curing adhesives** offer rapid conversion from liquid to solid at room temperature and a curing mechanism that occurs throughout, rather than progressing from exposed surfaces (as with the competing low-temperature air or moisture cures). Current applications include a wide variety of dental amalgams that can fill cavities or seal surfaces while matching the color of the remaining tooth. In the manufacturing realm, heat-sensitive materials can be effectively bonded, and the rapid cure time significantly reduces the need for fixturing.

DESIGN CONSIDERATIONS

The structural adhesives have been used for a wide range of applications in fields as diverse as automotive, aerospace, appliances, biomedical, electronics, construction, machinery, and sporting goods. Proper selection and use, however, requires consideration of a number of factors:

1. What materials are being joined? What are their surface finishes, hardnesses, and porosities? Will the thermal expansions or contractions be different?
2. How will the joined assembly be used? What type of joint is proposed, what will be the bond area, and what will be the applied stresses? How much strength is required? Will there be mechanical vibration, acoustical vibration, or impacts?
3. What temperatures might be required to affect the cure, and what temperatures might be encountered during service? Consideration should be given to the highest temperature, lowest temperature, rates of temperature change, frequency of change, duration of exposure to extremes, the properties required at the various conditions, and differential expansions or contractions.
4. Will there be subsequent exposure to solvents, water or humidity, fuels or oils, light, ultraviolet radiation, acid solutions, or general weathering?
5. What is the desired level of flexibility or stiffness? How much toughness is required?
6. Over what length of time is stability desired? What portion of this time will be under load?
7. Is appearance important?
8. How will the adhesive be applied? What equipment, labor, and skill are required?
9. Are their restrictions relating to storage or shelf life? Cure time? Disposal? Recyclability?
10. What will it cost?

Because there is such a large difference in bonding area, adhesive-bonded joints are often classified as either continuous surface or core-to-face. In **continuous-surface bonds,** both of the adhering surfaces are relatively large and are of the same size and shape. **Core-to-face bonds** have one **adherend** area that is very small compared to the other, like when the edges of lightweight honeycomb core structures are bonded to the face sheets (see Figure 14-20).

A major design consideration for both types is the nature of the stresses that the joint will experience. As shown in Figure 33-3, applied stresses can subject the joint to **tension, compression, shear, cleavage,** and **peel.** While it may be tempting to use joint

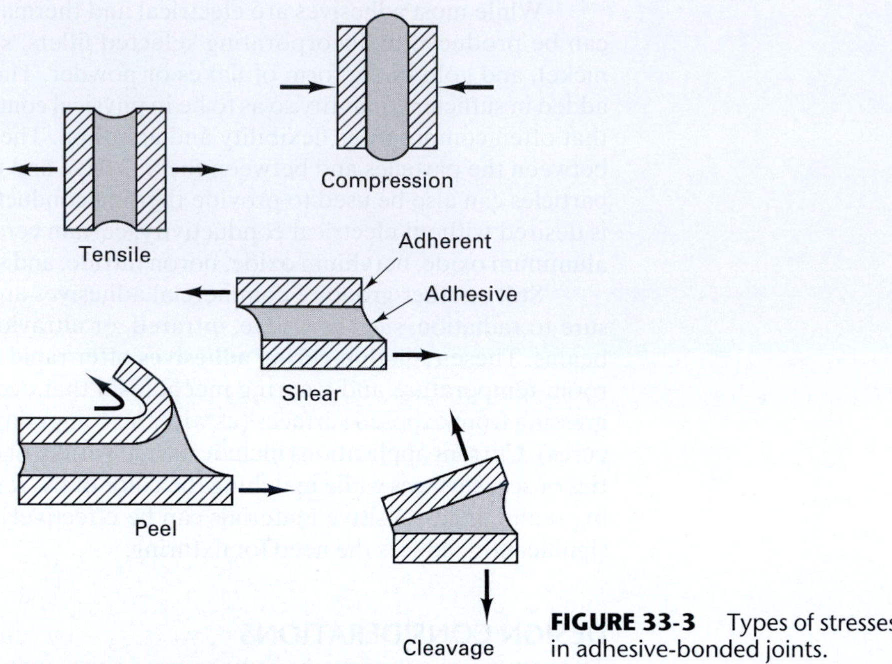

FIGURE 33-3 Types of stresses in adhesive-bonded joints.

designs intended for other methods of fastening, adhesives require specially designed joints to optimize their properties. Most of the structural adhesives are significantly weaker in peel and cleavage, where the stress is concentrated on only a very small area of the total bond, than they are in shear or tension. Therefore, adhesively bonded joints should be designed so as much of the stress as possible is in shear, tension, or compression, where all of the bonded area shares equally in bearing the load. The shear strengths of structural adhesives range from 14 to 40 MPa (2000 to 6000 psi) at room temperature, while the tensile strengths are only 4 to 8 MPa (600 to 1200 psi). The best adhesive-bonded joints, therefore, will be those that are designed to utilize the superior shear strengths. Creep, vibration and associated fatigue, thermal shock, and mechanical shocks can all induce additional stresses. When vibration or shock loading is expected, the elastomeric adhesives are quite attractive because they can provide valuable damping. Rigid adhesives are better for shear loadings, while flexible adhesives are better in peel.

Figure 33-4 shows some commonly used joint designs and indicates their relative effectiveness. The butt joint is unsatisfactory because it offers only a minimum of bond surface area and little resistance to cleavage. Useful strength is generally obtained by

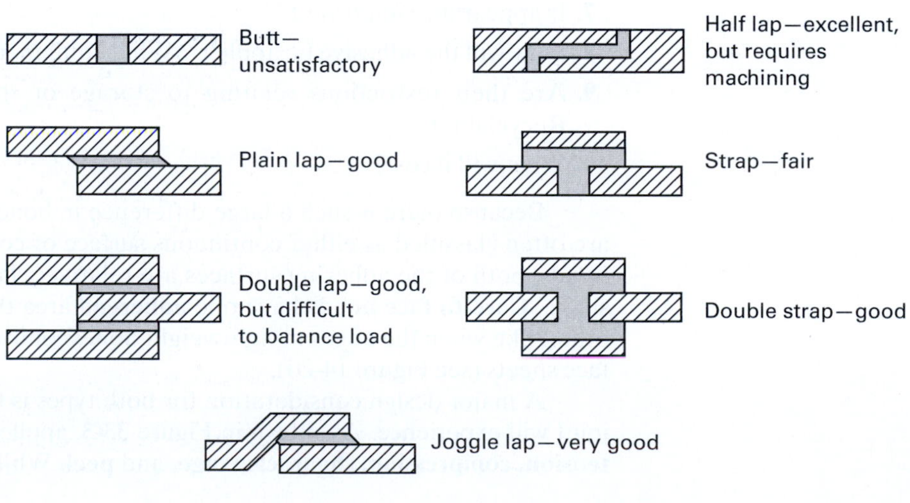

FIGURE 33-4 Possible designs of adhesive-bonded joints and a rating of their performance in service.

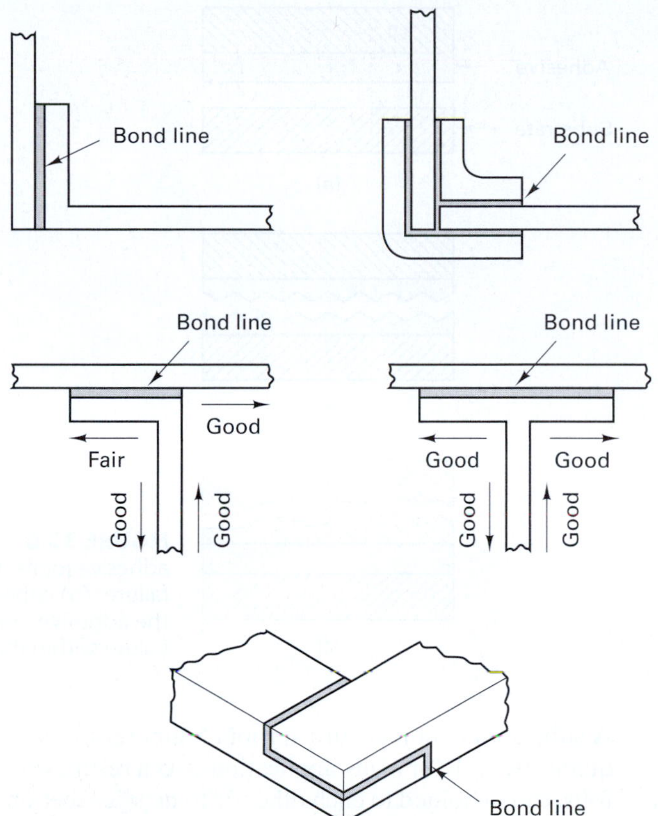

FIGURE 33-5 Adhesively bonded corner and angle joint designs.

increasing the bond area through the addition of straps or the conversion to some form of lap design. The scarf joint, shown previously in Figure 32-14, is also used when uniform thickness is required. Figure 33-5 shows some recommended designs for corner and angle joints. Adhesives can also be used in combination with welding, brazing, or mechanical fasteners. Spot welds or rivets can provide additional strength or simply prevent movement of the components when the adhesive is not fully cured or is softened by exposure to elevated temperature.

To obtain satisfactory and consistent quality in adhesive-bonded joints, it is essential that the surfaces be properly prepared. Procedures vary widely, but frequently include cleaning of the surfaces to be joined. Contaminants, such as oil, grease, rust, scale, or even mold-release agents, must be removed to ensure adequate wetting of the surfaces by the adhesive. Solvent or vapor cleaning is usually adequate. Chemical alteration of the surface to form a new intermediate layer, chemical etching, steam cleaning, or abrasive techniques may also be employed to further enhance wetting and bonding. While thick or loose oxide films are detrimental to adhesive bonding, a thin porous oxide or surface primer can often provide surface roughness and enhance adhesion.

The destructive testing of adhesive joints, or the examination of joint failures, can reveal much about the effectiveness of an adhesive system. If failure occurs by separation at the adhesive-substrate interface, as shown in Figure 33-6a, it is indicative of a bonding or adhesion problem. If the failure lies entirely within the adhesive as in Figure 33-6b, then the bonding with the substrate is adequate, but the strength of the adhesive may need to be enhanced. Finally, if failure occurs within the substrate materials, as in Figure 33-6c the joint is good, and failure is unrelated to the adhesive bonding operation.

ADVANTAGES AND LIMITATIONS

Adhesive bonding has many obvious advantages. Almost any material or combination of materials can be joined in a wide variety of sizes, shapes, and thicknesses. For most adhesives, the curing temperatures are low, seldom exceeding 180°C (350°F).

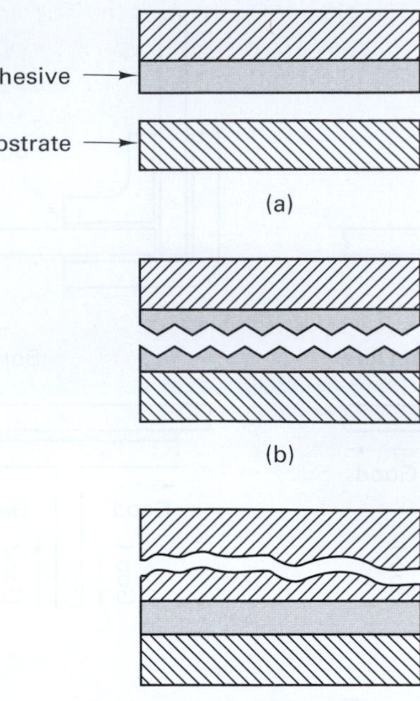

FIGURE 33-6 Failure modes of adhesive joints: (a) adhesive failure, (b) cohesive failure within the adhesive, and (c) cohesive failure within the substrate.

A substantial number cure at room temperature or slightly above and can provide adequate strength for many applications. As a result, very thin or delicate materials, such as foils, can be joined to each other or to heavier sections. Heat-sensitive materials can be joined without damage, and heat-affected zones are not present in the product. When joining dissimilar materials that experience subsequent changes in temperature, the adhesive often provides a bond that can tolerate the stresses of differential expansion and contraction. Industrial adhesives offer useful properties in the temperature range from −40 to +170°C (−40 to +340°F).

Because adhesives bond the entire joint area, good load distribution and fatigue resistance are obtained, and stress concentrations (such as those observed with screws, rivets, and spot welds) are avoided. Because of the high extension and recovery properties of flexible adhesives, the fatigue resistance can be up to 20 times that of riveted or spot-welded assemblies. The large contact areas that are usually employed provide a total joint strength that compares favorably with alternative methods of joining or attachment. Shear strengths of industrial adhesives can exceed 27.5 MPa (4000 psi), and additives can be incorporated to enhance strength, increase flexibility, or provide resistance to various environments.

Adhesives are generally inexpensive and frequently weigh less than the fasteners needed to produce a comparable-strength joint. In addition, an adhesive can also provide thermal and electrical insulation; act as a damper to noise, shock, and vibration; stop a propagating crack; and provide protection against galvanic corrosion when dissimilar metals are joined. By providing both a joint and a seal against moisture, gases, and fluids, adhesive-bonded assemblies often offer improved corrosion resistance throughout their useful lifetime. When used to bond polymers or polymer–matrix composites, the adhesive can be selected from the same polymer family to ensure good compatibility.

From a manufacturing viewpoint, the formation of a joint does not require the capillary-induced flow of material, as in brazing and soldering. The bonding adhesive is applied directly to the surfaces, and the joint is then formed by the application of heat and/or pressure. Most adhesives can be applied quickly, and useful strengths are achieved in a short period of time. Some curing mechanisms take as little as 2 to 3 s! Surface preparation may be reduced because bonding can occur with an oxide film in place, and rough surfaces are actually beneficial because of the increased contact area. Tolerances are less critical because the adhesives are more forgiving than alternative methods of bonding. Adhesives can compensate for dimensional irregularities by filling in small gaps and locations of poor part fit. The adhesives are often invisible; exposed surfaces are not defaced; smooth contours are not disturbed; and holes do not have to be made, as with rivets or bolts. These factors contribute to reduced manufacturing costs, which can be further reduced through the elimination of the mechanical fasteners and the absence of highly skilled labor. Bonding can often be achieved at locations that would prevent the access of many types of welding apparatus. Robotic dispensing systems can often be utilized.

The major disadvantages of adhesive bonding are the following:

1. There is no universal adhesive. Selection of the proper adhesive is often complicated by the wide variety of available options.
2. Most industrial adhesives are not stable above 180°C (350°F). Oxidation reactions are accelerated, thermoplastics can soften and melt, and thermosets decompose. While some adhesives can be used up to 260°C (500°F), elevated temperatures are usually a cause for concern.
3. Some adhesives shrink significantly during curing.
4. High-strength adhesives are often brittle (poor impact properties). Resilient ones often creep. Some become brittle when exposed to low temperatures.
5. Surface preparation and cleanliness, adhesive preparation, and curing can be critical if good and consistent results are to be obtained. Some adhesives are quite sensitive to the presence of grease, oil, or moisture on the surfaces to be joined. Surface roughness and wetting characteristics must be controlled.
6. Assembly times may be greater than for alternative methods depending on the curing mechanism. Elevated temperatures or pressure may be required as well as specialized fixtures.
7. It is difficult to determine the quality of an adhesive-bonded joint by traditional nondestructive techniques, although some inspection methods have been developed that give good results for certain types of joints.
8. Some adhesives contain objectionable chemicals or solvents or produce them upon curing.
9. Many structural adhesives deteriorate under certain operating conditions. Environments that may be particularly hostile include heat, ultraviolet light, ozone, acid rain (low pH), water and humidity, salt, and numerous solvents. Thus, long-term durability and reliability may be questioned, and life expectancy is hard to predict.
10. Adhesively bonded joints cannot be readily disassembled.

Nevertheless, the extensive and successful use of adhesive bonding provides ample evidence that these limitations can be overcome if adequate quality-control procedures are adopted and followed.

WELD BONDING

Welding and adhesive bonding are two very different methods of joining, and are typically considered to be competing technologies. When used together, however, in a process called **weld bonding,** they often combine in a way that compliments one another and cancels each other's negative features. Resistance, laser, friction-stir, or ultrasonic welding can be combined with adhesives to join steel and aluminum. The adhesives distribute the load over large areas, while the welds provide high peel resistance. The welds can also act as fixtures, holding the pieces in position or alignment while the adhesives cure. In some combinations, the welds are made through the adhesives, while others employ a staggered or patterned approach.

33.2 Mechanical Fastening

INTRODUCTION AND METHODS

Mechanical fastening includes a wide variety of techniques and fasteners designed to suit the individual requirements of a multitude of joints and assemblies. Included within this family are integral fasteners, threaded discrete fasteners (which includes screws, bolts, studs, and inserts), nonthreaded discrete fasteners (such as rivets, pins, retaining rings, nails, staples, and wire stitches), special-purpose fasteners (such as the quick-release and tamper-resistant types), shrink and expansion fits, press fits, seams, and others. Selection of the specific fastener or fastening method depends primarily on the materials to be joined, the function of the joint, strength and reliability requirements, weight limitations, dimensions of the components, and environmental conditions. Other considerations include cost, installation equipment and accessibility, appearance, and the need or desire for disassembly. When disassembly and reassembly are desired (as for parts replacement, maintenance, or repair), threaded fasteners, snap-fits, or other fasteners that can be removed quickly and easily should be specified. Such fasteners should not have a tendency to loosen after installation, however. If disassembly is not necessary, permanent fasteners are often preferred, or threaded fasteners can be coupled with anaerobic adhesives that cure to full strength at room temperature and "lock" the fastener in place.

A mechanical joint acquires its strength through either mechanical interlocking or interference as a result of a clamping force. No fusion or adhesion of the surfaces is required. The fasteners and fastening processes should be selected to provide the required strength and properties in view of the nature and magnitude of subsequent loading. Consider the possibility of vibrations and/or cyclic stresses that might promote loosening over a period of time. The added weight of fasteners may be a significant factor in certain applications, such as those in the aerospace and automotive industries. The need to withstand corrosive environments, operate at high or low temperatures, or face other severe conditions may be additional constraints to the selection of fasteners or fastening processes.

The effectiveness of a mechanical fastener often depends on (1) the material of the fastener, (2) the fastener design (including the load-bearing area of the head), (3) hole preparation, and (4) the installation procedure. The general desire is to achieve a uniform load transfer, a minimum of stress concentration, and uniformity of installation torque or interference fit. Various means are available for achieving these goals, as described in the following paragraphs.

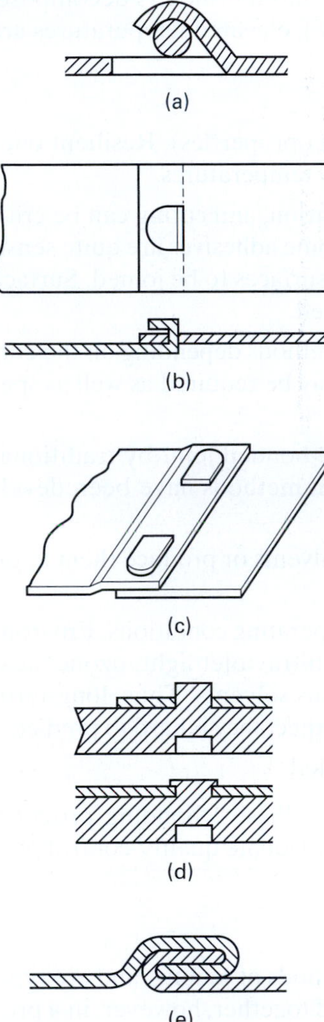

FIGURE 33-7 Several types of integral fasteners: (a) lanced tab to fasten wires or cables to sheet or plate; (b) and (c) assembly through folded tabs and slots for different types of loading; (d) use of a flattened embossed protrusion; (e) single-lock seam.

Integral fasteners are formed areas of a component that interfere or interlock with other components of the assembly and are most commonly found in sheet metal products. Examples include lanced or shear-formed tabs, extruded hole flanges, embossed protrusions, edge seams, and crimps. Figure 33-7 shows some of these techniques, each of which involve some form of metal shearing and/or forming. The common beverage can includes several of these joints—an edge seam to join the top of the can to the body (as in Figure 33-7e) and an embossed protrusion that is subsequently flattened to attach the opener-tab (as in Figure 33-7d).

Discrete fasteners, like those illustrated in Figure 33-8, are separate pieces whose function is to join the primary components. These include bolts and nuts (with accessory washers, etc.), screws, nails, rivets, quick-release fasteners, staples, and wire stitches. More than 150 billion discrete fasteners are consumed annually in the United States, with a variety so immense that the major challenge is usually selection of an appropriate, and hopefully optimum, fastener for the task at hand. Fastener selection is further complicated by inconsistent nomenclature and identification schemes. Some fasteners are identified by their specific product or application, while others are classified by the material from which they are made, their size, their shape, their strength, or primary operational features.[1] The commercial availability of such a wide range of standard and special types, sizes, materials, strengths, and finishes virtually ensures that an appropriate fastener can be found for most all joining needs. Discrete fasteners are easy to install, remove, and replace. In addition, most standard varieties are interchangeable.

[1] Discussion of the primary terms used in identifying discrete fasteners can be found in ANSI Standard B18.12. Both the Society of Automotive Engineers (SAE) and the American Society of Testing and Materials (ASTM) have formalized various "grades" of bolts and identified them by markings that are stamped on the bolt head. These grades often relate to allowable stress and temperatures of operation.

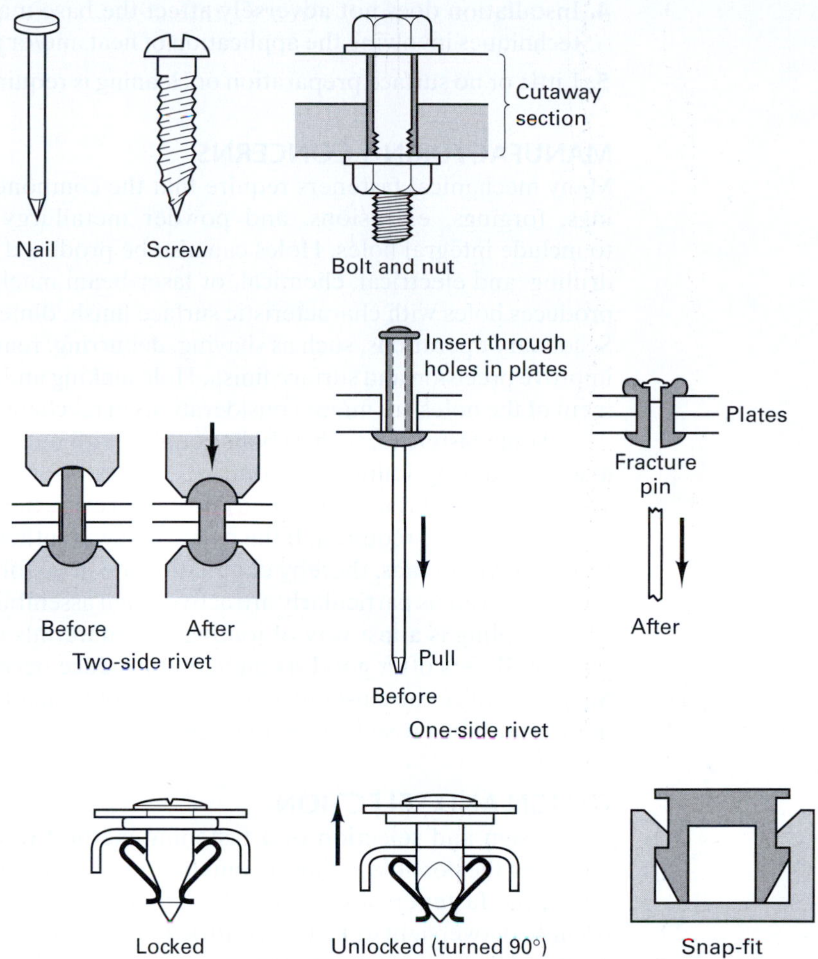

FIGURE 33-8 Various types of discrete fasteners, including a nail, screw, nut-and-bolt, two-side-access supported rivet, one-side-access blind rivet, quick-release fastener, and snap-fit.

Steel is the most common material due to its high strength and low cost. Various finishes and coatings can be applied to withstand a multitude of service conditions.

Shrink and **expansion fits** form another major class of mechanical joining. Here, a dimensional change is introduced to one or both of the components by heating or cooling (heating one part only, heating one and cooling the other, or cooling one). Assembly is then performed, and a strong interference fit is established when the temperatures return to uniformity. Joint strength can be exceptionally high. This technique can also be used to apply a corrosion-resistant cladding or lining to a less costly, but corrosion-prone, bulk material.

Press fits are similar to shrink and expansion fits, but the results are obtained through mechanical force instead of differential temperatures.

REASONS FOR SELECTION

Mechanical fastening offers a number of attractive features:

1. They are easy to disassemble and reassemble. The threaded fasteners are noteworthy for this feature, and semipermanent fasteners (such as rivets) can be drilled out for a major disassembly.
2. They can be used to join similar or different materials in a wide variety of sizes, shapes, and joint designs, including some—such as hinges and slides—that permit limited motion between the components.
3. Manufacturing cost is low. The fasteners are usually small-formed components that cost little compared to the components being joined. They are readily available in a variety of mass-produced sizes.

4. Installation does not adversely affect the base materials, as is often the case with techniques involving the application of heat and/or pressure.
5. Little or no surface preparation or cleaning is required.

MANUFACTURING CONCERNS

Many mechanical fasteners require that the components contain aligned holes. Castings, forgings, extrusions, and powder metallurgy components can be designed to include integral holes. Holes can also be produced by such techniques as punching, drilling, and electrical, chemical, or laser-beam machining. Each of these techniques produces holes with characteristic surface finish, dimensional features, and properties. Secondary operations, such as shaving, deburring, reaming, and honing, can be used to improve precision and surface finish. Hole making and the proper positioning and alignment of the holes are major considerations in mechanical fastening.

Some fasteners, such as bolts coupled with nuts, require access to both sides of an assembly during joining. In contrast, screws offer one-side joining. If a bolt can be inserted into a threaded (tapped) hole, however, the nut can be eliminated, and only one-side access is required. If the bolt or screw is sufficiently hard, the fastener can often form its own threads, thereby eliminating the need for a threaded receptacle. This self-tapping feature is particularly attractive when assembling plastic products.

Stapling is a fast way of joining thin materials and does not require prior hole-making. Rivets offer good strength, but produce permanent or semipermanent joints. Snap-fits utilize the elasticity of one of the components, but the necessary elastic deformation must be possible without fracture.

DESIGN AND SELECTION

The design and selection of a fastening method requires numerous considerations, including the possible means of joint failure. When a product is assembled with fastened joints, the fasteners are extremely vulnerable sites. Mechanical joints generally fail because of oversight or lack of control in one of four areas: (1) the design of the fastener itself and the manufacturing techniques used to make it, (2) the material from which the fastener is made, (3) joint design, or (4) the means and details of installation. Fasteners may have insufficient strength or corrosion resistance or may be subject to stress corrosion cracking or hydrogen embrittlement. They may be unable to withstand the temperature extremes (both high and low) experienced by the final assembly. Metal fasteners provide electrical conductivity between the components, and an inappropriate choice of fastener or component material can cause severe galvanic corrosion. Nonmetallic fasteners (such as threaded nylon) can be used for low-strength applications where corrosion is a concern, but creep under load is a concern for these materials. Because mechanical fasteners only join at discrete points, gases or liquids can easily penetrate the joint area and further aggravate conditions.

Many failures are the result of poor joint preparation or improper fastener installation. A high percentage of the cracks in aircraft structures originate at fastener holes, and fatigue of fasteners is the largest single cause of fastener failure. Installation frequently imparts too much or too little preload (too tight or too loose). The joint surfaces may not be flat or parallel, and the area under the fastener head may be insufficient to bear the load. Vibrational loosening enhances fastener fatigue. The details of joint design should further consider stress distribution because much of the load will be concentrated on the fasteners (in contrast to the previously discussed adhesive joints that distribute the load uniformly over the entire joint area).

Nearly all fastener failures can be avoided by proper design and fastener selection. Consideration should be given to the operating environment, required strength, and magnitude and frequency of vibration. Fastener design should incorporate a shank-to-head fillet whenever possible. Rolled threads can be specified for their superior strength and fracture resistance. Corrosion-resistant coatings can be employed for enhanced performance. Joint design should seek to avoid such features as offset or oversized holes. Proper installation and tightening are critical to good performance. Standard sizes, shapes, and grades should be used whenever possible with as little variety as is absolutely necessary.

33.3 JOINING OF PLASTICS

Mechanical fasteners, adhesives, and welding processes can all be employed to form joints between engineering plastics. Fasteners are quick and are suitable for most materials, but they may be expensive to use, they generally do not provide leak-tight joints, and the localized stresses may cause them to pull free of the polymeric material. Threaded metal inserts may have to be incorporated into the plastic components to receive the fasteners, further increasing the product cost. Adhesives can provide excellent properties and fully sound joints, but they are often difficult to handle and relatively slow to cure. In addition, considerable attention is required in the areas of joint preparation and surface cleanliness. In a modification of adhesive bonding, solvents may be used to soften surfaces, which are then pressed together to form a bond. Welding can be used to produce bonded joints with mechanical properties that approach those of the parent material. Unfortunately, only the *thermoplastic polymers* can be welded because these materials can be melted or softened by heat without degradation and good bonds can be formed with the subsequent application of pressure. The *thermosetting polymers* do not soften with heat, tending only to char or burn, and must be joined by alternative methods, such as mechanical fasteners, adhesives, snap-fits, or possible co-curing (placing the components together and curing while in contact).

Because the thermoplastics soften and melt at such low temperatures, the heat required to weld these materials is significantly less than that required in the welding of metals. The processes used to weld plastics can be divided into two groups: (1) those that utilize mechanical movement and friction to generate heat, such as ultrasonic welding, spin welding, and vibration welding, and (2) those that involve external heat sources, such as hot-plate welding, hot-gas welding, and resistive and inductive implant welding. In both groups, it is important to control the rate of heating. Plastics have low thermal conductivity, and it is easy to induce burning, charring or other material degradation before softening has occurred to the desired depth.

Ultrasonic welding of plastics uses high-frequency mechanical vibrations to create the bond. Parts are held together under pressure and are subjected to ultrasonic vibrations (20 to 40 kH frequency and 10 to 100 μm amplitude) perpendicular to the area of contact. The high-frequency vibrations are transmitted through the workpiece to the joint area where they generate sufficient heat through friction to produce a high-quality weld in a period of 0.5 to 1.5 s. The process can be readily automated, and produces fast, strong, clean and reliable welds. The tools are expensive, however, and large production runs are generally required. Ultrasonic welding is usually restricted to small components where relative movement is restricted and weld lengths do not to exceed a few centimeters.

In **vibration welding,** or **linear friction welding,** relative movement between the two parts is again used to generate the heat, but the direction of movement is now parallel to the interface and aligned with the longest dimension of the joint. The vibration amplitudes are significantly larger than in ultrasonic welding (1 to 4 mm), and the frequencies are considerably less (on the order of 100 to 240 Hz). When molten material is produced, the vibration is stopped, parts are aligned, and the weld region is allowed to cool and solidify. The entire process takes about 1 to 5 s. Long-length, complex joints can be produced at rather high production rates. Nearly all thermoplastics can be joined, independent of whether their prior processing was by injection molding, extrusion, blow molding, thermoforming, foaming, or stamping.

The **friction welding** of plastics (also called **spin welding**) is similar to vibration welding, except the relative motion is now continuous and rotational. The process is essentially the same as the friction welding of metals, but melting now occurs at the joint interface. High-quality welds are produced with good reproducibility, and little end preparation is required. The major limitations are that at least one of the components must exhibit circular symmetry and the axis of rotation must be perpendicular to the

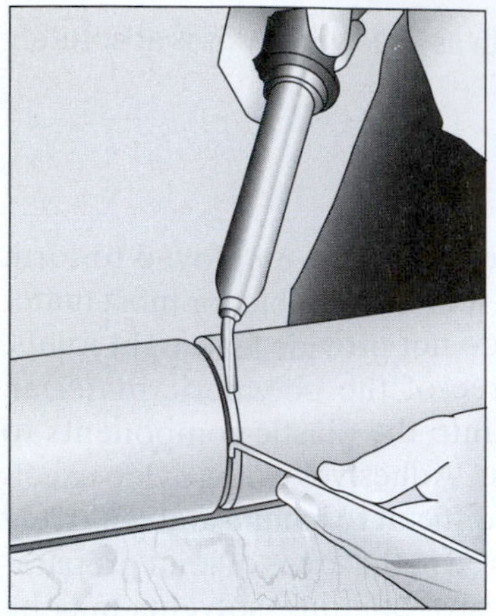

FIGURE 33-9 Using a hot-gas torch to make a weld in plastic pipe.

mating surface. Weld strengths vary from 50 to 95% of the parent material in bonds of the same plastic. Joints between dissimilar materials generally have poorer strengths.

Another friction process, **friction-stir welding** (described in Chapter 31), can be used to produce butt welds between plates of thermoplastic material.

Hot-plate welding uses an external heat source and is probably the simplest of the mass-production techniques used to join the thermoplastic polymers. The parts to be joined are held in fixtures and pressed against the opposite sides of an electrically heated tool. Contact is maintained until the surfaces have melted and the adjacent material has softened to a specified distance from the interface. The parts separate, the tool is removed, and the two prepared surfaces are pressed together and allowed to cool. Contaminated surface material is usually displaced into a flash region. Weld times are comparatively slow, ranging from 10 s to several minutes. The joint strength can be equal to that of the parent material, but the joint design is usually limited to a square-butt configuration, like that encountered when joining sections of plastic pipe. If the bond interface has a nonflat profile, shaped heating tools can be employed. Heated-tool welding can also be used to produce lap seams between flexible plastic sheets. Rollers apply pressure after the material has passed over a heater.

The **hot-gas welding** of plastics is similar to the oxyacetylene welding of metals, and V-groove or fillet welds are the most common joint configuration. A gas (usually compressed air or nitrogen) is heated by an electric coil as it passes through a welding gun, like the one shown in Figure 33-9. The hot-gas stream emerges from a nozzle at 200 to 400°C (400 to 750°F) and impinges on the joint area. Thin rods of thermoplastic material (the same material as the parts being joined) are heated along with the work-piece and are then forced into the softened joint area, providing both the filler material and the pressure needed to produce coalescence. Because this process is usually slow and the results are generally dependent on operator skill, it is seldom used in production applications. It is, however, a popular process for the repair of thermoplastic materials.

Extrusion welding is an established technique for joining polyethylene and polypropylene when the fabrications are large and the materials being joined are thick. A flow of hot gas is used to heat the substrate, as in the hot-gas process, but the external filler material rod is replaced by a continuously extruded stream of fully molten polymer that emerges from the weld tool as it moves along the joint. (*Note:* This is similar to the electrode feed in the consumable-electrode arc welding processes presented in Chapter 30.) When welding thick materials, the hot-gas process is often used to produce tack welds and then to create a single pass along the base to ensure full root penetration. Extrusion welding is then used to fill the remainder of the joint, using the same thermoplastic material as the substrate.

Implant welding processes involve positioning metal inserts between the pieces to be joined and then heating them by means of electrical resistance or induction. The resistance method requires a continuous current-carrying path, and the implants are often wire, braid, or mesh. As the implant heats, the surrounding thermoplastic softens and melts, and the subsequent application of pressure forms a weld. In the induction welding approach, a high-frequency AC coil is placed in proximity to the implants and the eddy-currents induced in the metal produce the desired heating. Because the welds form only in the vicinity of the implants, the process resembles spot welding and produces joints that are considerably weaker than those formed by processes that bond the entire contact area. When bonding is desired over larger areas, tapes, rods, or gaskets of thermoplastic material can be laced with iron oxide or metal particles, which then provide heat across the entire interface and simultaneously provide filler material.

A variation of implant welding has been used to successfully bond polymer–matrix carbon-fiber composite materials. A single ply of carbon fiber is positioned between the materials to be bonded, and provides sufficient conductivity to generate the necessary heat. In a process that is still under development, **microwave energy** is being used to weld thermoplastics. Most thermoplastics do not experience a temperature rise when irradiated with microwaves, but the insertion of a microwave-susceptible implant and the application of pressure after heating can be used to create an acceptable weld. Because the waves penetrate the entire component, complex three-dimensional joints can be produced. As with all forms of implant welding, the inserted material becomes an integral part of the final assembly.

The welding of thermoplastics using **infrared radiation welding** is a relative newcomer to the list of processes. One approach is essentially a noncontact version of hot-plate welding, where the hot plate is heated to higher temperatures than in the contact method. The parts to be welded are brought into close proximity to the heated plate (typically within 0.2 mm, or 0.01 in.) where they heat by radiation and convection. When large surface areas are to be joined, a bank of infrared lamps may replace the heated plate. Because the heating is performed without contact, there is less possibility of contamination entering the joint.

Lasers have also been used to weld thermoplastics. The radiation from a CO_2 laser is readily absorbed by plastics, and is used in a variation known as **direct laser welding.** Materials are heated from the outer surface, and depth of heating is limited, thereby restricting the process to the joining of thin films. The radiation from Nd:YAG, fiber, and diode lasers is easily transmitted through plastics. In **transmission laser welding,** one of the pieces transmits the laser light while the other absorbs its energy (concentrating the heat at the joint region), or an opaque surface coating is applied at the joint interface. Thicker materials, up to 10 mm (0.4 in.) can be welded. The attractive features of laser welding include controllable beam power and the ability to precisely focus the beam. Because the process is noncontact, it is quite attractive for hygienic applications, such as medical devices and food packaging, as well as applications where cleanliness is critical, like microelectronics.

All of the welding techniques discussed above use thermal energy or heating to achieve coalescence. In **solvent welding,** a solvent is applied to both surfaces to temporarily dissolve the polymer at room temperature. The solvent then permeates through the polymer and into the environment, leaving a solid material in the joint area. Probably the most common use of this technique is the joining of PVC (polyvinylchloride) pipe and fittings in plumbing-type applications.

The most common method of joining plastics to one another, and joining metal to plastic is not welding, however, but **mechanical fasteners.** Here, it is important that the plastic be able to withstand the strain of fastener insertion and the localized stresses around the fastener. Conventional machine screws are rarely used, except with extremely strong plastics. Instead, there are a number of fasteners designed specifically for use with plastics. Threaded fasteners work best with thick sections. Self-tapping, thread-cutting screws are used on hard plastics, and thread-forming screws are used with softer materials. If the joint is to undergo disassembly and reassembly, threaded metal inserts may be incorporated into the part to receive the fasteners. If recycling of the plastic product is desired, welding may be preferred over mechanical fasteners because only similar materials are involved and fasteners do not need to be removed or separated.

Because of the low elastic modulus of plastic materials, snap-fit assemblies are often an attractive alternative to the use of fasteners. Access panels to the battery compartments of electronic devices are a classic example. In addition, adhesives, described previously in this chapter, provide an attractive means of joining plastics. Because adhesives are polymeric materials, the joint material can be selected for compatibility with the material being joined.

■ 33.4 JOINING OF CERAMICS AND GLASS

The properties of ceramic materials are significantly different from the engineering metals, and these differences restrict or limit the processes that can be used for joining. High melting temperatures can be a significant deterrent to fusion welding. More significant, however, are the effects of low thermal conductivity and brittleness. Heating and cooling will likely result in nonuniform temperatures, and the thermally induced stresses are likely to result in cracking or fracture. The lack of useful ductility virtually eliminates any form of deformation bonding. Mechanical fasteners, and their associated threads and holes, create high concentrated stresses, and these stresses often lead to material fracture. As a result, most ceramic materials are joined by some form of adhesive bonding, brazing, diffusion or sinter bonding, or ceramic cements.

Adhesives and cements are probably the most common methods of joining ceramics to ceramics, ceramics to glasses, and ceramics to metals and other materials. The inserted material (polymer adhesive, glass or glass–ceramic frit, or ceramic cement or mortar) will bond to the surfaces and bridge what are often radically different compositions and structures.

Brazing and soldering use a low-melting-point metal or lower-melting ceramic as the intermediate material. Some materials, such as indium solders, directly wet ceramic surfaces. Other standard braze alloys can be made "active" by adding elements such as titanium, hafnium, or zirconium that promote wetting of the ceramic. To promote adhesion and bonding with nonwetting alloys, it may be necessary to first coat the ceramic with some form of metallized or deposited layer. These coatings bond to the ceramic, and the braze or solder material bonds to the coating.

Sinter bonding is a means of joining ceramic materials during their initial production. As the component pieces are held together and co-fired, diffusion bonds form across the interface while similar bonds form within the components. This process is best performed when the components are of identical material or materials with similar composition and structure. Various intermediate materials have been used to assist the joining of dissimilar ceramics.

The joining of glass is a much easier operation. Heating softens the two materials, which are then pressed together and cooled. A wide variety of heating methods are used, depending on the size, shape, and quantity of components to be joined.

33.5 JOINING OF COMPOSITES

Joining processes are often used to bond the various components within composite materials. Examples might include the bonding of honeycomb cores to the face materials in a honeycomb sandwich structure or the bonding of the individual layers in a laminate. At various times, however, we may need to join composite materials to other materials or other composites.

The joining of composite materials can be an extremely complex subject, especially when considering the variety of composites and the fact that the joint interface is likely to be a distinct disruption to the continuity of structure and properties. Particulate composites may have the least structural difference at an interface. Laminar composites will certainly behave differently if the joint surface is a core (or multilayer) surface or a single-material exposed face. Fiber-reinforced composites, regardless of the type of fiber and fiber configuration, will certainly lack fiber continuity across the joint.

The usual joining methods tend to be those used with the matrix of the composite. Metal–matrix composites can be welded, brazed, or soldered, or joined with screws or bolts or any of the other techniques applied to metals. The techniques for polymer–matrix and ceramic–matrix follow those for plastics and ceramics, as discussed previously in this chapter.

Theoretically, all composites can be adhesively bonded, but there may be limits set by the applied stresses, operating temperatures, or size of the workpiece (because many adhesives require a thermal cure). When working with polymer–matrix composites, the adhesive is often selected to match or be compatible with the matrix polymer. In many cases, however, the adhesive is being asked to bond to already-cured polymer surfaces.

KEY WORDS

acrylics
adherend
adhesive
adhesive bonding
anaerobics
cleavage
compression
conductive adhesives
continuous-surface bonds
core-to-face bonds
curing
cyanoacrylates
delayed-tack adhesives
direct laser welding
discrete fasteners
elastomer
epoxies
evaporative adhesives
expansion fit
extrusion welding
friction-stir welding
friction welding
high-temperature adhesives
hot-gas welding
hot-melt adhesives
hot-plate welding
implant welding

infrared radiation welding
integral fasteners
linear friction welding
mechanical fasteners
mechanical fastening
microwave energy
peel
press fit
pressure-sensitive adhesives
radiation-curing adhesive
reactive hot melts
shear
shrink fit
silicones
sinter bonding
solvent welding
spin welding
structural adhesive
superglue
tension
thermoplastic
thermosetting
transmission laser welding
ultrasonic welding
urethanes
vibration welding
weld bonding

REVIEW QUESTIONS

1. What would be some of the characteristics of an ideal adhesive?
2. What are some of the newer applications that have helped promote increased use of adhesive bonding?
3. What are some of the types of materials that have been used as industrial adhesives?
4. What are some of the ways in which adhesives can be cured?
5. What is a structural adhesive?
6. Characterize the temperature range over which epoxies might be used, typical values of shear strength, and commonly observed curing times.
7. What promotes the curing of cyanoacrylates? Of anaerobics?
8. Urethane adhesives might be favored when what property is desired?
9. What features or characteristics might favor the selection of a silicone adhesive?
10. What are some common applications of hot-melt adhesives?
11. What features or properties are provided or enhanced by adhesive additives?
12. What are some types of nonstructural or special adhesives.
13. How can polymeric adhesives be made electrically or thermally conductive?
14. What types of radiation can be used with radiation-curing adhesives?
15. What are some of the temperature considerations that should be made when selecting an adhesive?
16. What are some of the environmental conditions that might reduce the performance or lifetime of a structural adhesive?
17. What is the difference between a continuous-surface and core-to-face bond?
18. Why is it desirable for adhesive joints to be designed so the adhesive is loaded in shear, tension, or compression?
19. Why are butt joints unattractive for adhesive bonding?
20. What types of joints provide large bonding areas?
21. What are some common techniques by which surfaces are prepared for adhesive bonding?
22. How can destructive testing and the examination of the failed joint provide useful information about the effectiveness of an adhesive system?
23. Why are structural adhesives an attractive means of joining dissimilar metals or materials? Different sizes or thicknesses?
24. Why might an adhesive joint provide enhanced corrosion resistance compared to alternative joining methods?
25. What are some of the other attractive properties of structural adhesives?
26. In what ways might a structural adhesive offer manufacturing ease or reduced manufacturing cost?
27. In view of the relatively low strengths of the structural adhesives, how can adhesively bonded joints attain strengths comparable to other methods of joining?
28. Why are adhesive joints unattractive for applications that involve exposure to elevated temperature?
29. Describe some of the ways that structural adhesives might deteriorate over time.
30. What is weld bonding?
31. What factors would influence the selection of a specific type of mechanical fastener or fastening method?
32. What types of fasteners are attractive if the application requires the ability to disassemble and reassemble the product?
33. What factors determine the overall effectiveness of a mechanical fastener?
34. What is an integral fastener? Provide an example.
35. What are some of the primary types of discrete fasteners?
36. How are press fits similar to shrink or expansion fits? How do they differ?
37. What are some of the major assets of mechanical fasteners?
38. What are some of the ways that fastener holes can be made in manufactured products?
39. What is the benefit of a self-tapping fastener?
40. What are some of the common causes for failure of mechanically fastened joints?
41. From a manufacturing viewpoint, why is it desirable to use standard fasteners and minimize the variety of fasteners within a given product?
42. What are some of the ways that plastics can be joined?
43. Why can the thermoplastic polymers be welded, but not the thermosetting varieties?
44. Describe several of the plastic joining processes that use mechanical movement or friction to generate the required heat.
45. What are some of the external heat sources that can be used in the welding of plastic materials?
46. How can metal inserts be used to produce welds in plastics?
47. What is the difference between direct laser welding and transmission laser welding? What type of laser is used with each?
48. Give an everyday example of solvent welding.
49. What material property enables snap-fits to be a common means of connecting plastic parts?
50. Why are the crystalline ceramic materials particularly difficult to join?
51. What are some of the inserted or bridging materials that can be used to join ceramics?
52. What is sinter bonding, and how does it join ceramic materials?
53. When materials are joined, we create interfaces between the various components. Describe the structural features that might result at the interface when we join (a) particulate composites, (b) laminar composites, and (c) fiber-reinforced composites.

PROBLEMS

1. Some automakers are using adhesives and sealants that cure under the same conditions used for the paint-bake operation. Determine the conditions used for paint-bake, and identify some adhesives and sealants that could be used. What are some of the pros and cons of such an integration?
2. A contractor has installed aluminum siding on a house with steel nails. Use the galvanic series to evaluate the corrosion properties of this assembly. (*Note:* The aluminum is exposed to air, so it should be considered to be in its passive condition.) What do you expect will be the outcome of this fastener selection? Can you recommend a better alternative?
3. Mechanical fasteners are an attractive means of joining composite materials because they avoid exposing the composite to heat and/or high pressure. Assume that the composite is a polymer-based fiber-reinforced material with either uniaxial or woven fibers. For this particular system, what are some possible fastener-related problems? Consider joint preparation, assembly, and possible service failure.
4. The heat-resisting tiles on the U.S. Space Shuttle are made from heat-resisting ceramics. Determine the method or methods used to attach them to the structure. What difficulties or problems have been encountered relating to this bonding?
5. The bicycle frames used by riders in recent Tour de France races have been single-piece fiber-reinforced composites. What difficulties or property compromises might be associated with a fabrication method that uses joints? What methods might be available to join the carbon-fiber–epoxy-composite materials commonly used in these bicycles?
6. The processes described for the joining of plastics focused almost exclusively on the thermoplastic polymers. What types of joining techniques could be applied to thermosetting polymers? To elastomeric polymers? To ceramic materials?

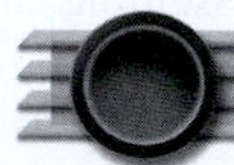

www.wiley.com/go/global/degarmo

Chapter 33 CASE STUDY

Golf Club Heads with Insert

You are employed by a small manufacturer of sporting goods equipment, and your design team has recently proposed a new line of high-performance golf clubs. While the club head of the irons is to be a "standard" AISI 431 martensitic stainless steel investment casting, the striking face will incorporate a metal insert, as shown in Figure CS-33. This insert will be produced by powder metallurgy and will consist of a copper-based alloy laced with presized particles of tungsten carbide. After a mild acid etching of the copper matrix, the carbide particles will protrude sufficiently to better grip the surface of the ball, imparting an enhanced amount of backspin to better control the "bite" of the ball upon landing. Because its purpose is to modify the striking face, the insert is rather thin, about 1.5 to 3 mm (1/16 to 1/8 in.). It is important that the insert be incorporated into the club face in a manner that does not dampen the impact or compromise the "feel" of the club.

FIGURE CS-33 Golf Club Head with Striking-Face Insert. (*Courtesy Ronald Kohser*)

1. You must devise a means of incorporating the proposed insert into the face of the club. What are some possible means of joining or bonding the dissimilar materials? What are the advantages and limitations of each of your alternatives? What would your recommend?
2. An additional joint occurs where the club head is attached to the shaft. If a graphite fiber reinforced epoxy is being considered for the shaft, how might it be attached to the stainless steel head? If the composite shaft were selected, which joining method would you recommend? If the shafts were metal, what other methods might be possible?
3. For production simplicity, it might be preferable to use the same joining procedure at all locations. In view of your answers to Questions 1 and 2, does this appear to be a possibility for this product?
4. If the bonding process and resulting interface proves to be problematic, there may be other ways to produce a raised-carbide surface on a stainless steel golf club face. Consider processes such as thermal spray, friction-stir to embed particles, and others. What do you see as the advantages and limitations for each of these alternatives, considering both manufacturing and performance? Would you expect them to be cheaper or more expensive than the proposed insert?
5. If quality and performance were the primary objective, which of the preceding options would you recommend? Why?
6. If cost minimization were to become important as you seek an edge over competitors, what would be your recommendation, and why?

CHAPTER 34

SURFACE ENGINEERING

34.1 INTRODUCTION

Surface engineering is a multidisciplinary activity intended to tailor the properties of the surfaces of manufactured components so that their function and serviceability can be improved. Processes include solidification treatments such as hot-dip coatings, weld-overlay coatings, and thermal spray surfaces; deposition surface treatments such as electrodeposition, chemical vapor deposition, and physical vapor deposition; and heat treatment coatings such as diffusion coatings and surface hardening. Electroplating means the electrodeposition of an adherent metallic coating onto an object that serves as the cathode in an electrochemical reaction. The resulting surface provides wear resistance, corrosion resistance, high-temperature resistance, or electrical properties different from those in the bulk material.

Many manufacturing processes influence surface properties, which in turn may significantly affect the way the component functions in service. The demands for greater strength and longer life in components often depend on changes in the surface properties rather than the bulk properties. These changes may be mechanical, thermal, chemical, and/or physical and therefore are difficult to describe in general terms.

Many metal-cutting processes specified by the manufacturing engineer to produce a specific geometry can often have the effect of producing alterations in the surface material of the component, which, in turn, produces changes in performance.

SURFACE INTEGRITY

The term **surface integrity** was coined by Field and Kahles in 1964 in reference to the nature of the surface condition that is produced by the manufacturing process. If we view the process as having five main components (workpiece, tool, machine tool, environment, and process variables), we see that surface properties can be altered by all of these parameters (see Table 34-1) by producing the following:

- High temperatures involved in the machining process.
- Plastic deformation of the work material (residual stress).

TABLE 34-1 Characteristics of Manufacturing Processes That Affect Surface Integrity

Workpiece–Tool–Machine–Environment–Process Variables	
Workpiece characteristics	**Tool characteristics**
Geometry	Tool body
Shape	Type of tool
Dimensions	Size
Material	Shape
Type	Number of cutting edges
Route of manufacture	Cutting edge
Mechanical properties	Shape (angles)
Elastic constants	Nose geometry/topography
Plastic constants	Microgeometry
Physical properties	Wear
Melting point	Material
Thermal diffusivity, conductivity, capacity	Type
Coefficient of thermal expansion	Coating
Phase transformations	Type
Chemical properties	Thickness
Chemical composition	Number and kind of layers
Chemical affinity to tool material and environment	Mechanical properties
Metallurgical properties	Elastic constants
Structure	Plastic properties
Grain size	Physical properties
Hardness	Thermal diffusivity, conductivity, capacity
	Coefficient of thermal expansion
	Chemical properties
Environment characteristics	Chemical composition
Type of medium (gas, fluid, mist)	Chemical affinity to tool material
Lubricity	Metallurgical properties
Cooling ability	Structure
Flow rate	Grain size
Temperature	
Chemical composition	**Process variables**
	Speed
Machine tool characteristics	Feed
Error motions	Depth of cut

Source: Advanced Manufacturing Engineering, Vol. 1, July 1989.

- Surface geometry (roughness, waviness, cracks, distortion).
- Chemical reactions, particularly between the tool and the workpiece.

More specifically, surface integrity refers to the impaired or enhanced surface condition of a component or specimen that influences its performance in service. Surface integrity has two aspects: topography characteristics and surface-layer characteristics. **Topography** is made up of surface roughness, waviness, errors of form, and flaws (Figure 34-1).

A typical roughness profile includes the peaks and valleys that are considered separately from waviness. Flaws also add to texture but should be measured independent of it. Changes in the surface layer, as a result of processing, include plastic deformation, **residual stresses,** cracks, and other metallurgical changes (hardness, overaging, phase changes, recrystallization, intergranular attack, and hydrogen embrittlement). The surface layer will always contain local surface deformation due to any machining passes.

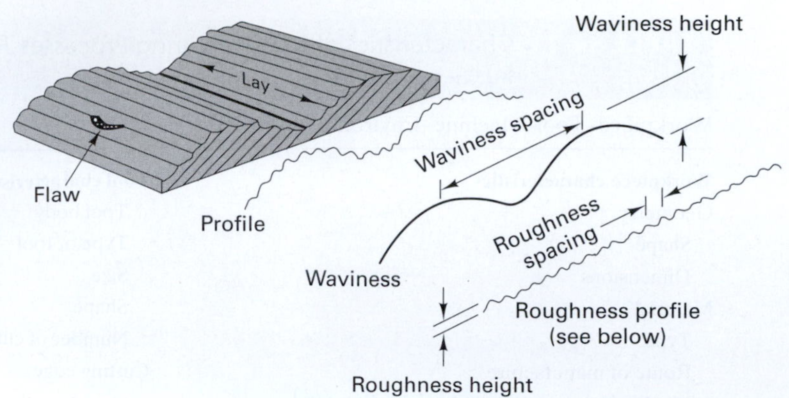

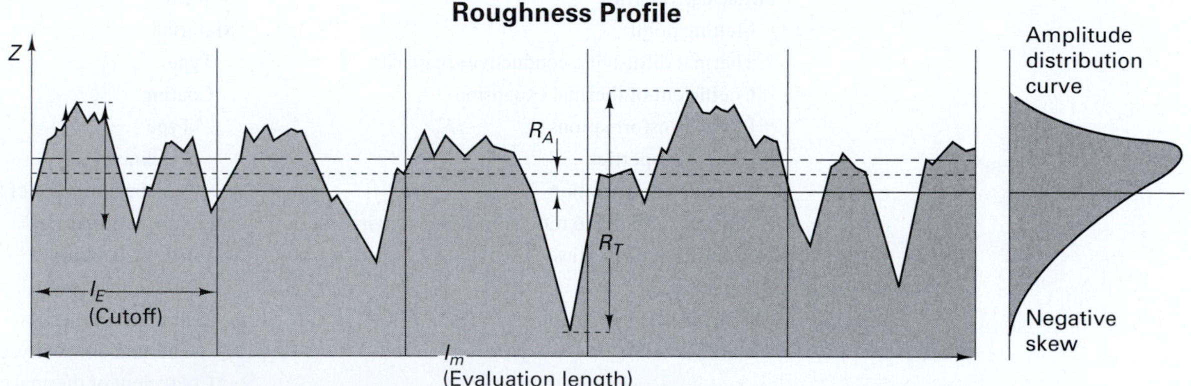

R_T = Maximum roughness depth (peak to valley) along l_m

R_A = Arithmetic roughness average

FIGURE 34-1 Machining processes produce surface flaws, waviness, and roughness that can influence the performance of the component. A roughness profile as produced by a stylus profile machine is shown with parameters.

The material removal processes generate a wide variety of surfaces textures, generally referred to as **surface finish.** The cutting processes leave a wide variety of surface patterns on the materials. **Lay** is the term used to designate the direction of the predominant surface pattern produced by the machining process. In addition, certain other terms and symbols have been developed and standardized for specifying the surface quality. The most important terms are *surface roughness*, *waviness*, and *lay* (Figure 34-2). **Roughness** refers to the finely spaced surface irregularities. It results from machining operations in the case of machined surfaces. **Waviness** is surface irregularity of greater spacing than in roughness. It may be the result of warping, vibration, or the work being deflected during machining.

A variety of instruments are available for measuring surface roughness and surface profiles. The majority of these devices use a diamond stylus that is moved at a constant rate across the surface, perpendicular to the lay pattern. The rise and fall of the stylus is detected electronically [often by a linear-variable differential transformer (LVDT)], is amplified and recorded on a strip-chart, or is processed electronically to produce average or root-mean-square readings for a meter (Figure 34-3).The unit containing the stylus and the driving motor may be handheld or supported by skids that ride on the workpiece or some other supporting surface.

Roughness is measured by the height of the irregularities with respect to an average line. These measurements are usually expressed in micrometers or microinches.

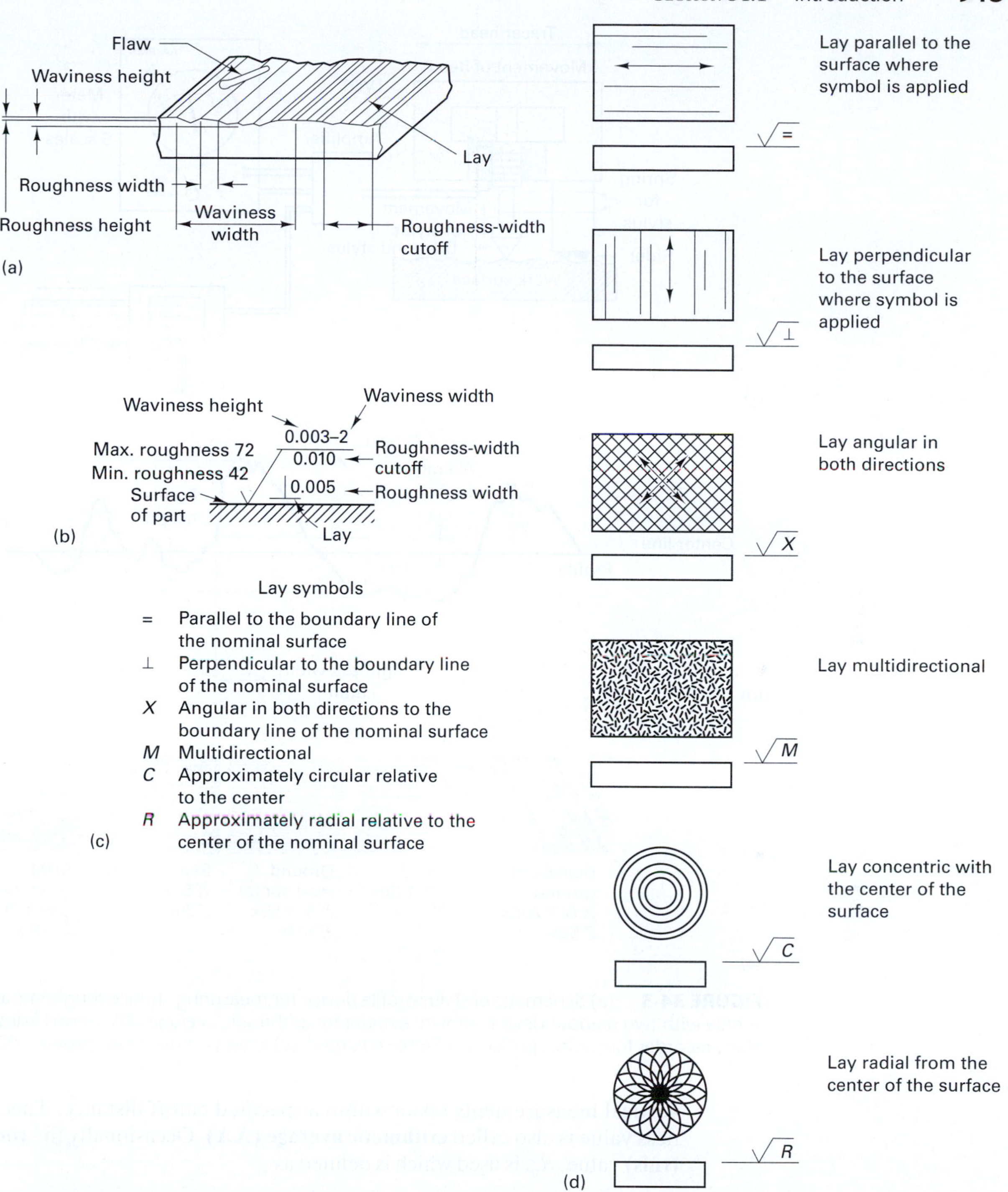

FIGURE 34-2 (a) Terminology used in specifying and measuring surface quality; (b) symbols used on drawing by part designers, with definitions of symbols; (c) lay symbols; (d) lay symbols applied on drawings.

In most cases, the arithmetic average, R_A, is used. In terms of the measurements, the R_A would be as follows:

$$R_A = \frac{\sum_{i=1}^{n} y_i}{n} \tag{34-1}$$

Cutoff refers to the sampling length used for the calculation of the roughness height. When it is not specified, a value of 0.030 in. (0.8 mm) is assumed. In the previous equation, y_i is a vertical distance from the centerline and n is the total number of

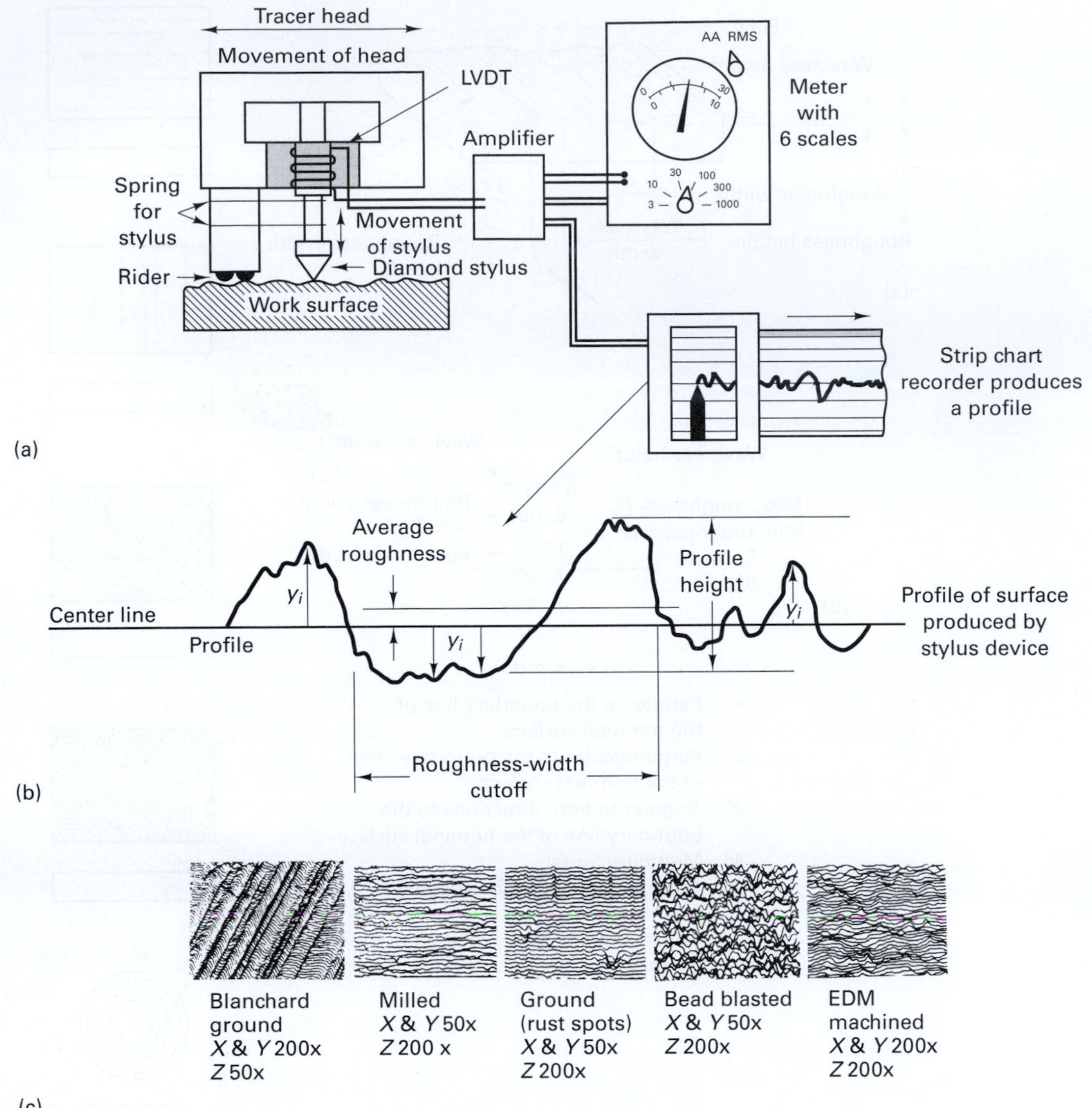

FIGURE 34-3 (a) Schematic of stylus profile device for measuring surface roughness and surface profile with two readout devices shown: a meter for arithmetic average (AA) or rms values and a strip chart recorder for surface profile. (b) Profile enlarged. (c) Examples of surface profiles. *(Courtesy J T. Black)*

vertical measurements taken within a specified cutoff distance. This average roughness value is also called **arithmetic average (AA)**. Occasionally the **root-mean-square (rms)** value, R_q, is used which is defined as

$$R_q = \text{rms} = \sqrt{\frac{\sum_{i=1}^{n} y_i^2}{n}} \tag{34-2}$$

The resolution of stylus profile devices is determined by the radius or the diameter of the tip of the stylus. When the magnitude of the geometric features begins to approach the magnitude of the tip of the stylus, great caution should be used in interpreting the output from these devices. As a case in point, Figure 34-4 shows a scanning electron micrograph of a face-milled surface on which has been superimposed (photographically) a scanning electron micrograph of the tip of a diamond stylus (tip radius of 0.0005 in.). Both micrographs have the same final magnification. Surface flaws of the same general size as the roughness created by the machining process are difficult to resolve with the stylus-type device, where both these features are about the same size as the stylus tip.

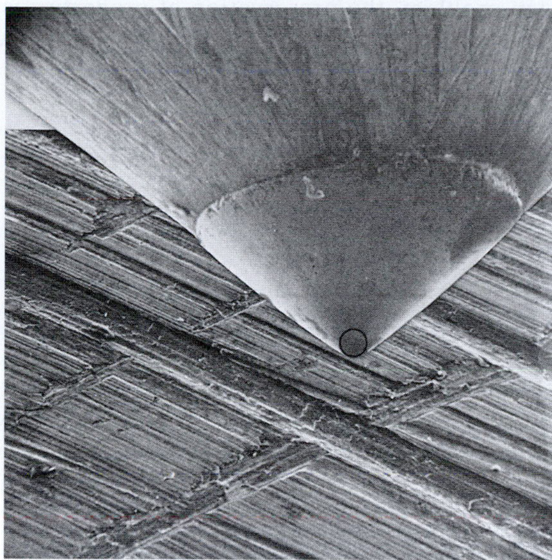

FIGURE 34-4 Typical machined steel surface as created by face milling and examined in the SEM. A micrograph (same magnification) of a 0.00005-in. stylus tip has been superimposed at the top. *(Courtesy J T. Black)*

This example points out the difference between *resolution* and *detection.* Stylus tracing devices can often *detect* the presence of a surface crack, step, or ridge on the part but cannot *resolve* the geometry of the defect when the defect is of the same order of magnitude as the stylus tip or smaller.

Another problem with these devices is that they produce a reading (a line on the chart) where the stylus tip is not touching the surface, as is demonstrated in Figure 34-5a, which shows the *S* from the word *TRUST* on a U.S. dime. The scanning electron microscope (SEM) micrograph was made after the topographical map of Figure 34-5b had been made. Both figures are at about the same magnification. The tracks produced by the stylus tip are easily seen in the micrograph. Notice the difference between the features shown in the micrograph and the trace, indicating that the stylus tip was not in contact with the surface many times during its passage over the surface (left no track in the surface), yet the trace itself is continuous.

Surface integrity has become the subject of intense interest because the traditional, nontraditional, and posttreatment methods used to manufacture hardware can change the material's properties. Although the consequence of these changes becomes

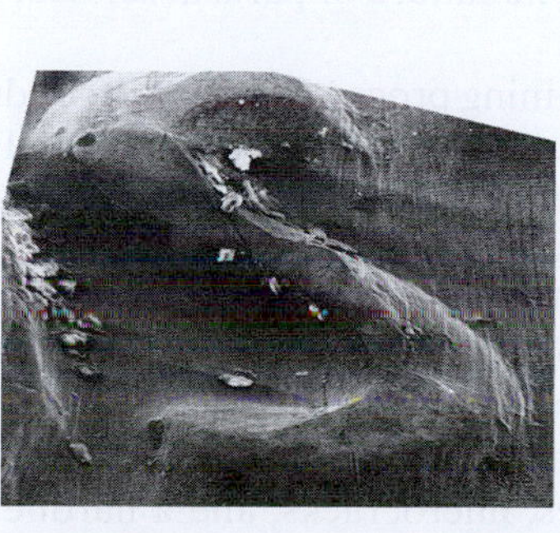

(a)

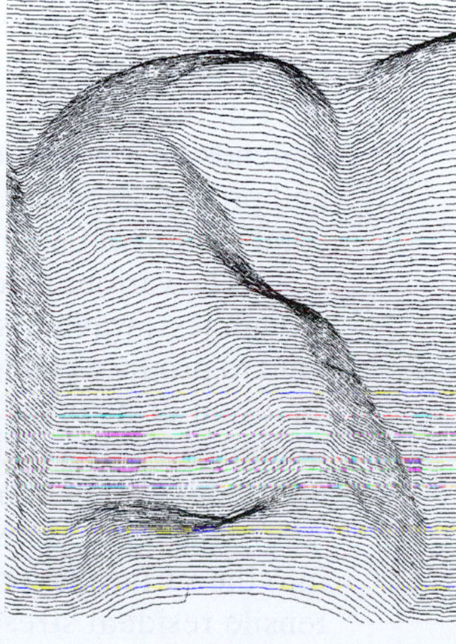

(b)

FIGURE 34-5 (a) SEM micrograph of a U.S. dime, showing the *S* in the word *TRUST* after the region has been traced by a stylus-type machine. (b) Topographical map of the S region of the word *TRUST* from a U.S. dime (compare to part a). *(Courtesy J T. Black)*

a design problem, the preservation of properties is a manufacturing consideration. Designs that require a high degree of surface integrity are the ones that display the following qualities:

- Are highly stressed.
- Employ low safety factors.
- Operate in severe environments.
- Must have prime reliability.
- Have a high surface areas–to–volume ratio.
- Are made with alloys that are sensitive to processing.

Surface integrity should be a joint concern of manufacturing and engineering. Manufacturing must balance cost and producibility with design requirements. It bears repeating to say that engineering must design components with knowledge of manufacturing processes. A reduction in fatigue life resulting from processing can be reversed with a posttreatment. This is another example of design for manufacturing.

The range of surface roughnesses that are typically produced by various manufacturing processes is indicated in Figure 34-6, which is a very general picture of typical ranges associated with these processes. However, one can usually count on its being more expensive to generate a fine finish (low roughness). To aid designers, metal samples with various levels of surface roughness are available.

All of the processes used to manufacture components are important if their effects are present in the finished part. It is convenient to divide processes that are used to manufacture parts into three categories: traditional, nontraditional, and finishing treatments. In **traditional processes,** the tool contacts the workpiece. Examples are grinding, milling, and turning. These material removal processes will inflict damage to the surface if improper parameters are used. Examples of improper parameters are dull tools, excessive infeed, inadequate coolant, and improper grinding wheel hardness. The **nontraditional processes** have intrinsic characteristics that, even if well controlled, will change the surface. In these processes, the workpiece does not touch the tool. Electrochemical machining (ECM), electrical discharge machining (EDM), laser machining, chemical milling, ultrasonic grinding, and abrasive water-jet machining are examples of these kinds of processes. Such methods can leave stress-free surfaces, remelted layers, and excessive surface roughness. **Finishing treatments** can be used to negate or remove the impact of both traditional and nontraditional processes as well as provide good surface finish. For example, residual tensile stresses can be removed by shot peening or roller burnishing. Chemical milling can remove the recast layer left by EDM.

SURFACE PROPERTIES AND PRODUCT PERFORMANCE

It is important to understand that the various manufacturing and surface-finishing processes each impart distinct properties to the materials that will influence the performance of the product. The achievement of satisfactory product performance obviously depends on a good design, high-quality manufacturing (including surface treatment), and proper assembly. The failure of parts in service, however, is usually the result of a combination of factors.

The various machining processes will each produce characteristic surface textures (roughness, waviness, and lay) on the workpieces. In addition, the various processes tend to produce changes in the chemical, physical, mechanical, and metallurgical properties on or near the surfaces that are created. For the most part, these changes are limited to a depth of 0.005 to 0.050 in. below the surface. The effects can be beneficial or detrimental, depending on the process, material, and function of the product.

Machining processes (both chip-forming and chipless) induce plastic deformation into the surface layer, as shown in Figure 34-7. The cut surfaces are generally left with tensile residual stresses, microcracks, and a hardness that is different from the bulk material.

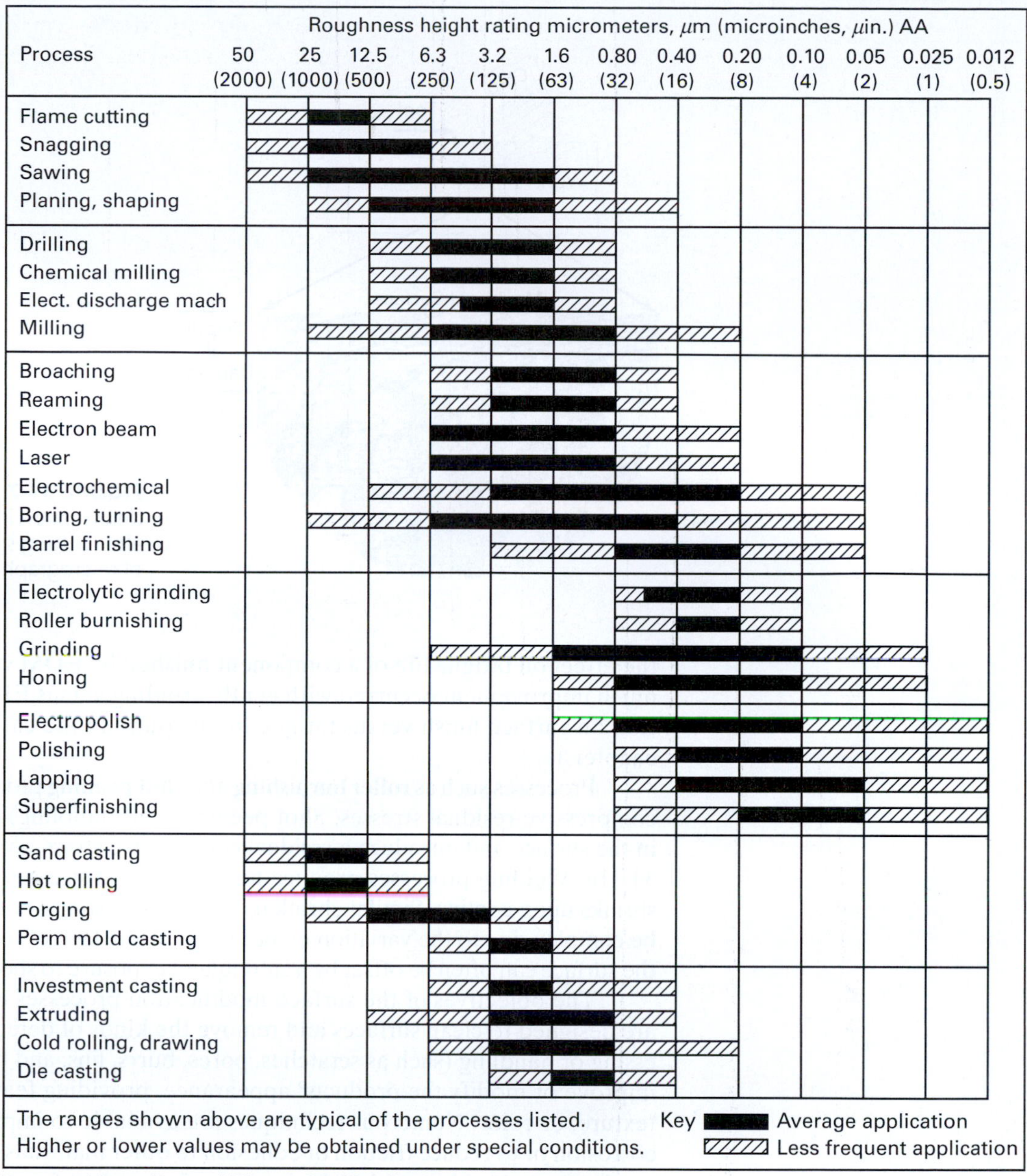

Extracted from General Motors Drafting Standards, June 1973 revision

FIGURE 34-6 Comparison of surface roughness produced by common production processes. *(Courtesy of* American Machinist*)*

Consider Figure 34-8, which shows the depth of "surface damage" due to machining as a function of the rake angle of the tool. To increase the cutting speed (and thereby increase the rate of production), an engineer might change from a high-speed tool steel cutter with a large rake angle (such as 30 degrees) to a carbide tool with a zero rake. While the resulting surface finish may be similar, the depth of surface damage is doubled. The increase in cutting speed may increase the heat going into the surface. If sufficient heat is generated, phase transformations can occur in the surface and subsurface regions. Failures may occur in service, whereas previous parts had performed quite admirably.

Processes such as EDM and laser machining leave a layer of hard, recast metal on the surface that usually contains microcracks. Ground surfaces can have either residual tension or residual compression, depending on the mix between chip formation and plowing or rubbing during the grinding operation. Figure 34-9 shows

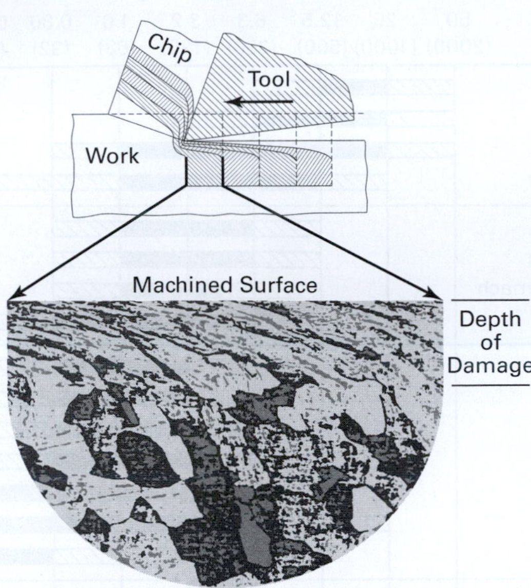

FIGURE 34-7 Plastic deformation in the surface layer after cutting shown in a drawing of micrograph at 120x.

the effect on fatigue life of a component finished by EDM versus grinding. Look how much improvement occurred with gentle grinding versus EDM. (Additional information on surface finish versus fatigue life is found in the case study at the end of this chapter.)

Processes such as **roller burnishing** and **shot peening** produce a smooth surface with compressive residual stresses. Shot peening (and tumbling) can increase the hardness in the surface and introduce a residual compressive stress, as shown in Figure 34-10 and 34-11a. Welding processes produce tensile residual stresses as the deposited material shrinks upon cooling. Similar shrinkage occurs in castings, but the resulting stresses may be complex due to the variation of shrinkage or the lack of restraint. Tensile stresses on the surface can often be offset by a subsequent exposure to shot peening or tumbling.

The objectives of the surface-modification processes can be quite varied. Some are designed to clean surfaces and remove the kinds of defects that occur during processing or handling (such as scratches, pores, burrs, fins, and blemishes). Others further improve or modify the products' appearance, providing features such as smoothness, texture, or color. Numerous techniques are available to improve resistance to wear or corrosion or to reduce friction or adhesion to other materials. Scarce or costly materials can be conserved by making the interior of a product from a cheaper, more common material and then coating or plating the product surface.

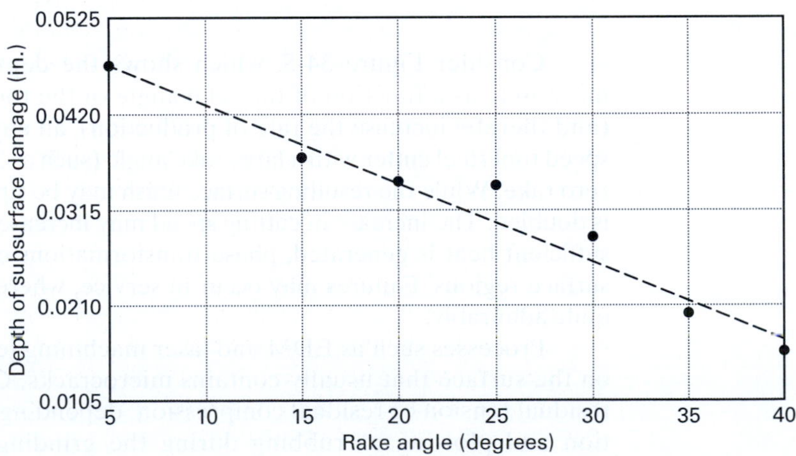

FIGURE 34-8 The depth of damage to the surface of a machined part increases with decreasing rake angle of the cutting tool.

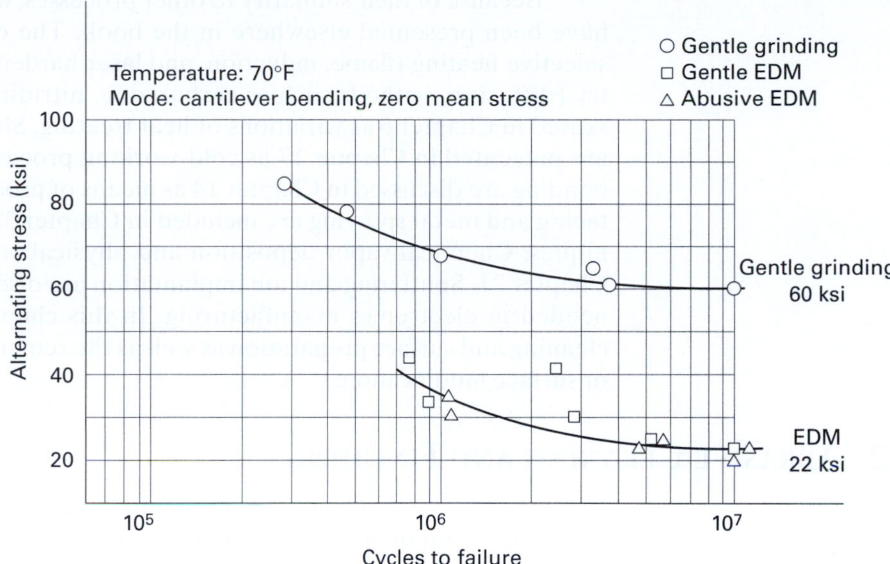

FIGURE 34-9 Fatigue strength of Inconel 718 components after surface finishing by grinding or EDM. *(Field and Kahles, 1971)*

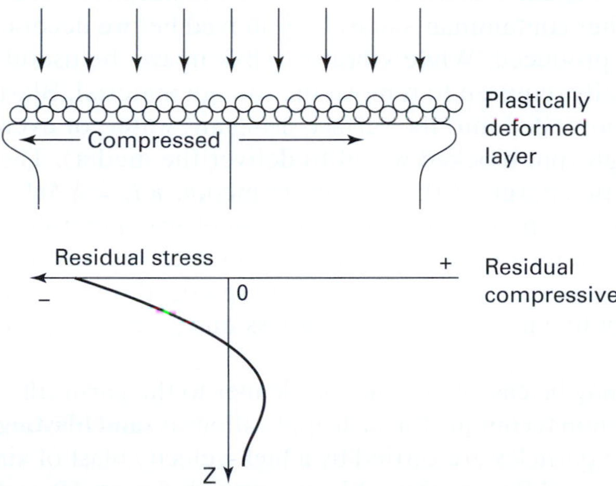

FIGURE 34-10 Mechanism for formation of residual compressive stresses in surface by cold plastic deformation (shot peening).

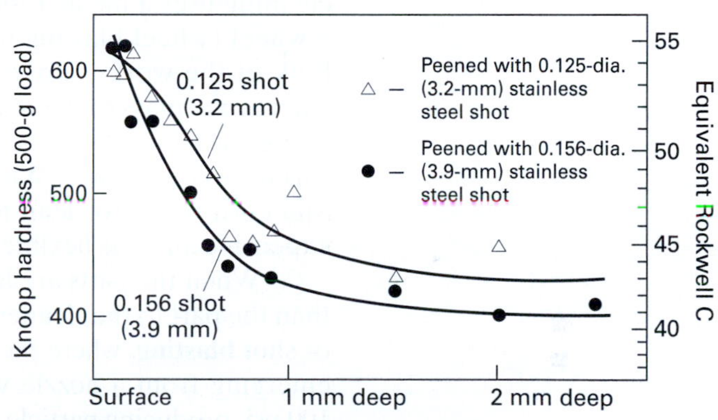

FIGURE 34-11 Hardness increased in surface due to shot peening.

As with all other processes, surface treatment requires time, labor, equipment, and material handling—and all of these have an associated cost. Efficiencies can be realized through process optimization and the integration of surface treatment into the entire manufacturing system. Design modifications can often facilitate automated or bulk finishing, eliminating the need for labor-intensive or single-part operations. Process selection should further consider the size of the part, the shape of the part, the quantity to be processed, the temperatures required for processing, the temperatures encountered during subsequent use, and any dimensional changes that might occur due to the surface treatment. Through knowledge of the available processes and their relative advantages and limitations, finishing costs can often be reduced or eliminated while maintaining or improving the quality of the product.

In addition to the preceding, the field of surface finishing has recently undergone another significant change. Many chemicals that were once "standard" to the field—such as cyanide, cadmium, chromium, and chlorinated solvents—have now come under strict government regulation. Wastewater treatment and waste disposal have also become significant concerns. As a result, processes may have to be modified or replacement processes may have to be used.

Because of their similarity to other processes, many surface finishing techniques have been presented elsewhere in the book. The **case hardening** techniques, both selective heating (flame, induction, and laser hardening) and altered surface chemistry (diffusion methods such as carburizing, nitriding, and carbonitriding), are presented in Chapter 6 as variations of heat treating. Shot peening and roller burnishing are presented in Chapter 17 as cold-working processes. Roll bonding and explosive bonding are discussed in Chapter 14 as means of producing laminar composites. Hard facing and metal spraying are included in Chapter 32 as adaptations of welding techniques. Chemical vapor deposition and physical vapor deposition are discussed in Chapter 21. Sputtering and ion implantation are discussed in Chapter 35 as processes needed in electronics manufacturing. In this chapter, we focus on techniques for cleaning and surface preparation as well as the remaining methods of surface finishing or surface modification.

34.2 ABRASIVE CLEANING AND FINISHING

It is not uncommon for the various manufacturing processes to produce certain types of surface contamination. Sand from the molds and cores used in casting often adheres to product surfaces. **Scale** (metal oxide) can be produced whenever metal is processed at elevated temperatures. **Oxides** such as rust can form if material is stored between operations. These and other contaminants must be removed before decorative or protective surfaces can be produced. While vibratory shaking can be useful, some form of **blast cleaning** is usually required to remove the foreign material. Blast cleaning uses a media (abrasive) propelled into the surface using air, water, or even a wheel (wheel blasting uses a high-rpm blocked wheel to deliver the media). The bulk of the work is done by kinetic energy of the impacting media; $KE = \frac{1}{2} MV^2$, where m = mass of the median and v = the velocity. Abrasives, steel grit, metal shot, fine glass shot, plastic beads, and even CO_2 are mechanically impelled against the surface to be cleaned. When sand is used, it should be clean, sharp-edged silica sand. Steel grit tends to clean more rapidly and generates much less dust, but it is more expensive and less flexible.

When the parts are large, it may be easier to bring the cleaner to the part rather than the part to the cleaner. A common technique for such applications is **sand blasting** or **shot blasting,** where the abrasive particles are carried by a high-velocity blast of air emerging from a nozzle with about a $\frac{3}{8}$-in. opening. Air pressures between 60 and 100 psi, producing particle speeds of 400 mph, are common when cleaning ferrous metals, and 10 to 60 psi is common for nonferrous metals. The abrasive may be sand or shot, or materials such as walnut shells, dry-ice pellets, or even baking soda. Pressurized water can also be used as a carrier medium.

When production quantities are large or the parts are small, the operation can be conducted in an enclosed hood, with the parts traveling past stationary nozzles. For large parts or small quantities, the blast may be delivered manually. Protective clothing and breathing apparatus must be provided and precautions taken to control the spread of the resulting dust. The process may even require a dedicated room or booth that is equipped with integrated air pollution control devices.

From a manufacturing perspective, these processes are limited to surfaces that can be reached by the moving abrasive (line-of-sight) and cannot be used when sharp edges or corners must be maintained (because the abrasive tends to round the edges).

BARREL FINISHING OR TUMBLING

Barrel finishing or **tumbling** is an effective means of finishing large numbers of small parts. In the Middle Ages, wooden casks were filled with abrasive stones and metal parts and were rolled about until the desired finish was obtained. Today, modifications of this technique can be used to deburr, radius (remove sharp corner on part), descale, remove rust, polish, brighten, surface-harden, or prepare parts for further finishing or assembly. The amount of stock removal can vary from as little as 0.0001 to as much as 0.005 in.

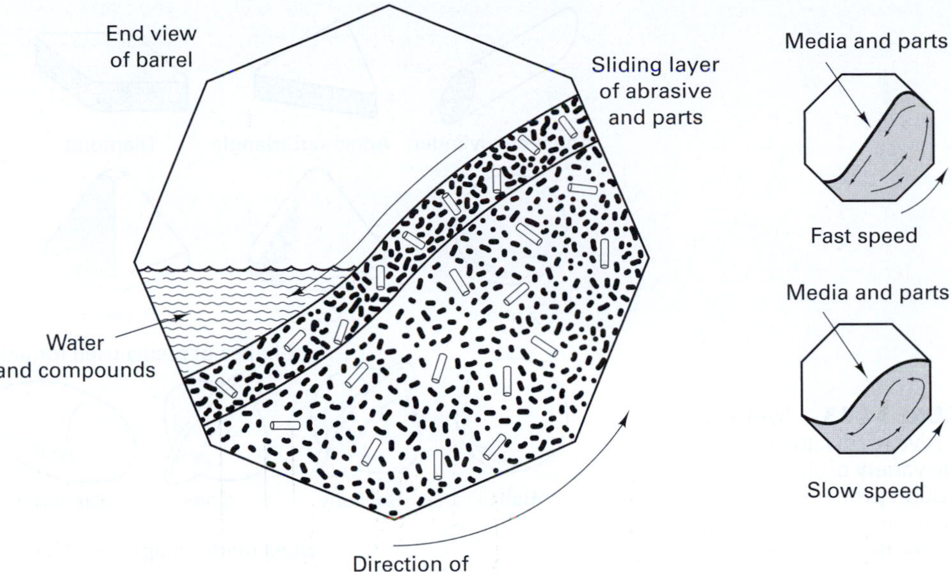

FIGURE 34-12 Schematic of the flow of material in tumbling or barrel finishing. The parts and media mass typically account for 50 to 60% of capacity.

In the typical operation, the parts are loaded into a special barrel or drum until a predetermined level is reached. Occasionally, no other additions are made, and the parts are simply tumbled against one another. In most cases, however, additional media of metal slugs or abrasives (such as sand, granite chips, slag, or ceramic pellets) are added. Rotation of the barrel causes the material to rise until gravity causes the uppermost layer to cascade downward in a "landslide" movement, as depicted in Figure 34-12. The sliding produces abrasive cutting that can effectively remove fins, flash, scale, and adhered sand. Because only a small portion of the load is exposed to the abrasive action, long times may be required to process the entire contents.

Increasing the speed of rotation adds centrifugal forces that cause the material to rise higher in the barrel. The enhanced action can often accelerate the process, provided that the speed is not so great as to destroy the cascading action and that the additional action does not damage the workpiece. By a suitable selection of abrasives, filler, barrel size, ratio of workpieces to abrasive, fill level, and speed, a wide range of parts can be tumbled successfully. Delicate parts may have to be attached to racks within the barrel to reduce their movement while permitting the media to flow around them.

Natural and synthetic abrasives are available in a wide range of sizes and shapes, including those depicted in Figure 34-13, that enable the finishing of complex parts with irregular openings. The various media are often mixed in a given load so that some will reach into all sections and corners to be cleaned.

Tumbling is usually done dry, but it can also be performed with an aqueous solution in the barrel. Chemical compounds can be added to the media to assist in cleaning, or descaling, or to provide features such as rust inhibition. Support equipment usually assists with loading and unloading the barrels as well as, with the separation of the workpieces from the abrasive media. The latter operation often uses mesh screens with selected size openings.

Barrel tumbling can be a very inexpensive way to finish large quantities of small parts and produce rounded edges and corners. Unfortunately, the abrasive action occurs on all surfaces and cannot be limited to selected areas. The cycle time is often long, and the process can be quite noisy.

In the **barrel burnishing** process, no cutting action is desired. Instead, the parts are tumbled against themselves or with media such as steel balls, shot, rounded-end pins, or ballcones. If the original material is free of visible scratches and pits, the combination of peening and rubbing will reduce minute irregularities and produce a smooth, uniform surface.

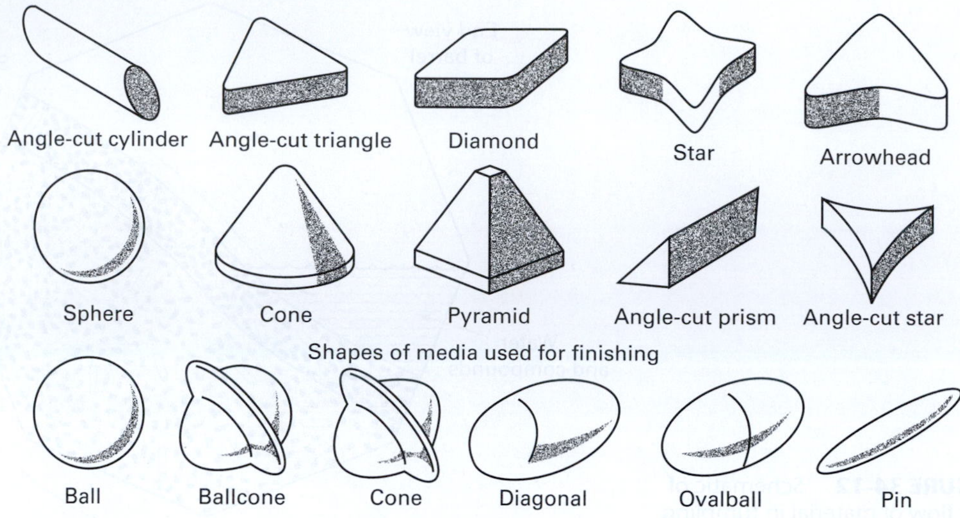

FIGURE 34-13 Synthetic abrasive media are available in a wide variety of sizes and shapes. Through proper selection, the media can be tailored to the product being cleaned.

Barrel burnishing is normally done wet, using a solution of water and lubricating or cleaning agents, such as soap or cream of tartar. Because the rubbing action between the work and the media is very important, the barrel should not be loaded more than half full, and the volume ratio of media to work should be about 2:1 so the workpieces rub against the media, not each other. The speed of rotation should be set to maintain the cascading action and not fling the workpieces free of the tumbling mass.

Centrifugal barrel tumbling places the tumbling barrel at the end of a rotating arm. This adds centrifugal force to the weight of the parts in the barrel and can accelerate the process by as much as 25 to 50 times.

In **spindle finishing,** the workpieces are attached to rotating shafts, and the assembly is immersed in media moving in a direction opposite to part rotation. This process is commonly applied to cylindrical parts and avoids the impingement of workpieces on one another. The abrasive action is accelerated, but time is required for fixturing and removal of the parts.

VIBRATORY FINISHING

Vibratory finishing is a versatile process widely used for deburring, radiusing, descaling, burnishing, cleaning, brightening, and fine finishing. In contrast to the barrel processing, vibratory finishing is performed in open containers. As illustrated in Figure 34-14, tubs or bowls are loaded with workpieces and media and are vibrated at frequencies between 900 and 3600 cycle/min. The specific frequency and amplitude are determined by the size, shape, weight, and material of the pan, as well as the media and compound. Because the entire load is under constant agitation, cycle times are less than with barrel operations. The process is less noisy and is easily controlled and automated. In addition, the open tubs allow for direct observation during the process, which can also deburr or smooth internal recesses or holes.

MEDIA

The success of any of the mass-finishing processes depends greatly on **media** selection and the ratio of media to parts, as presented in Table 34-2. The media may prevent the parts from impinging upon one another as they simultaneously clean and finish. Fillers—such as scrap punchings, minerals, leather scraps, and sawdust—are often added to provide additional bulk and cushioning.

Natural abrasives include slag, cinders, sand, corundum, granite chips, limestone, and hardwood shapes, such as pegs, cylinders, and cubes. Synthetic media typically

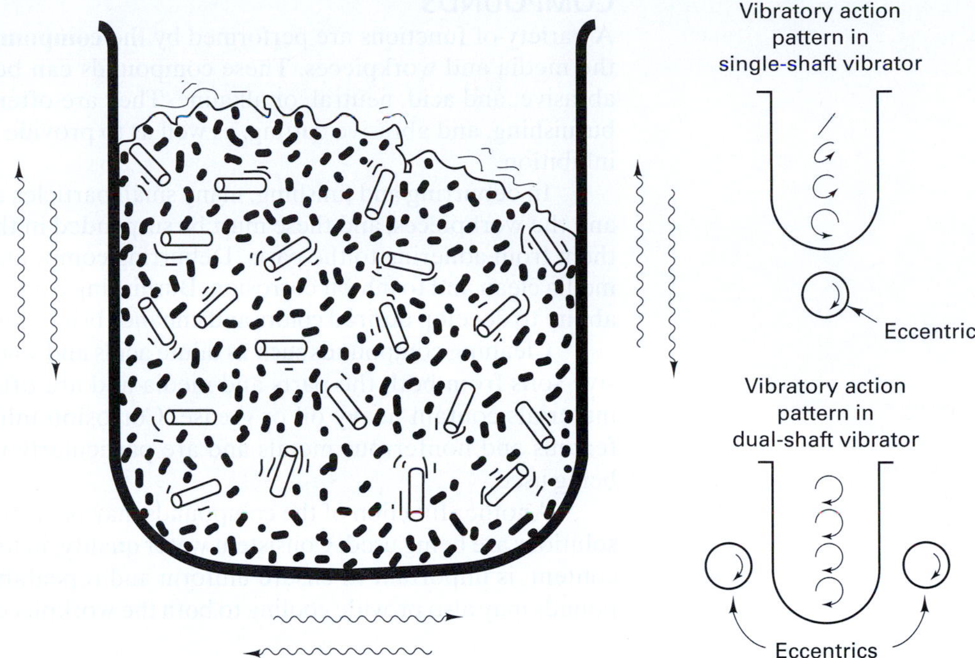

FIGURE 34-14 Schematic of a vibratory-finishing tub loaded with parts and media. The single eccentric shaft drive provides maximum motion at the bottom, which decreases as one moves upward. The dual-shaft design produces more uniform motion of the tub and reduces processing time.

contain 50 to 70 wt% of abrasives, such as alumina (Al_2O_3), emery, flint, and silicon carbide. This material is embedded in a matrix of ceramic, polyester, or resin plastic, which is softer than the abrasive and erodes, allowing the exposed abrasive to perform the work. The synthetics are generally produced by some form of casting operation, so their sizes and shapes are consistent and reproducible (as opposed to the random sizes and shapes of the natural media). Steel media with no added abrasive are frequently specified for burnishing and light deburring.

Media selection should also be correlated with part geometry, because the abrasives should be able to contact all critical surfaces without becoming lodged in recesses or holes. This requirement has resulted in a wide variety of sizes and shapes, including those presented in Figure 34-13. The different abrasives, sizes, and shapes can be selected or combined to perform tasks ranging from light deburring with a very fine finish to heavy cutting with a rough surface.

TABLE 34-2 Typical Media-to-Part Ratios for Mass Finishing

Media/Part Ratio by Volume	Typical Application
0:1	Part-on-part processing or burr removal without media
1:1	Produces very rough surfaces and is suitable for parts in which part-on-part damage is not a problem
2:1	Somewhat less severe part-on-part damage, but more action from less media
3:1	May be acceptable for very small parts and very small media. Part-on-part contact is likely on larger and heavier parts
4:1	In general, a good average ratio for many parts; a good ratio for evaluating a new deburring process
5:1	Better for nonferrous parts subject to part-on-part damage
6:1	Suitable for nonferrous parts, especially preplate surfaces on zinc parts with resin-bonded media
8:1	For improved preplate surfaces with resin-bonded media
10:1	Produces very fine finishes

Source: American Machinist, August 1983.

COMPOUNDS

A variety of functions are performed by the **compounds** that are added in addition to the media and workpieces. These compounds can be liquid or dry, abrasive or non-abrasive, and acid, neutral, or alkaline. They are often designed to assist in deburring, burnishing, and abrasive cutting, as well as to provide cleaning, descaling, or corrosion inhibition.

In deburring and finishing, many small particles are abraded from both the media and the workpieces, and these must be suspended in the compound solution to prevent them from adhering to the parts. Deburring compounds also act to keep the parts and media clean and to inhibit corrosion. Burnishing compounds are often selected for their ability to develop desired colors and enhance brightness.

Cleaning compounds such as dilute acids and soaps are designed to remove excessive soils from both the parts and media and are often specified when the incoming materials contain heavy oil or grease. Corrosion inhibitors can be selected for both ferrous and nonferrous metals and are particularly important when steel media are being used.

Another function of the compounds may be to condition the water when aqueous solutions are being used. Consistent water quality, in terms of "hardness" and metal ion content, is important to ensure uniform and repeatable finishing results. Liquid compounds may also provide cooling to both the workpieces and the media.

SUMMARY OF MASS-FINISHING METHODS

The barrel and vibratory finishing processes are really quite simple and economical and can process large numbers of parts in a batch procedure. Soft, nonferrous parts can be finished in as little as 10 min, while the harder steels may require 2 hr or more. Sometimes the operations are sequenced, using progressively finer abrasives. Figure 34-15 shows a variety of parts before and after the mass-finishing operation, using the triangular abrasive shown with each component.

Despite the high volume and apparent success, these processes may still be as much art as science. The key factors of workpiece, equipment, media, and compound are all interrelated, and the effect of changes can be quite complex. Media, equipment, and compounds are often selected by trial and error, with various approaches being tested until the desired result is achieved. Even then, maintenance of consistent results may still be difficult.

BELT SANDING

In the **belt sanding** operation, the workpieces are held against a moving abrasive belt until the desired degree of finish is obtained. This is really a surface grinding process. Because of the movement of the belt, the resulting surface contains a series of parallel scratches with a texture set by the grit of the belt. When smooth surfaces are desired, a series of belts may be employed, with progressively finer grits.

The ideal geometry for belt sanding is a flat surface, for the belt can be passed over a flat table where the workpiece can be held firmly against it. Belt sanding is frequently a hand operation and is therefore quite labor intensive. Furthermore, it is difficult to

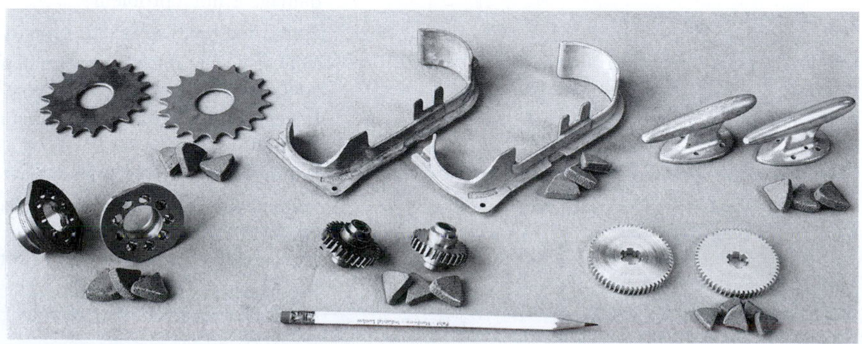

FIGURE 34-15 A variety of parts before and after barrel finishing with triangular-shaped media. *(Courtesy of Norton Abrasives/Saint Gobain)*

apply when the geometry includes recesses or interior corners. As a result, belt sanding is usually employed when the number of parts is small and the geometry is relatively simple (see Chapter 26).

WIRE BRUSHING

High-speed rotary **wire brushing** is sometimes used to clean surfaces and can also impart some small degree of material removal or smoothing. The resulting surface consists of a series of uniform curved scratches. For many applications, this *may* be an acceptable final finish. If not, the scratches can easily be removed by barrel finishing or buffing.

Wire brushing is often performed by hand application of a small workpiece to the brush or the brush to a larger workpiece. Automatic machines can also be used where the parts are moved past a series of rotating brushes. In another modification, the brushes are replaced with plastic or fiber wheels that are loaded with abrasive.

BUFFING

Buffing is a polishing operation in which the workpiece is brought into contact with a revolving cloth wheel that has been charged with a fine abrasive, such as polishing rouge. The "wheels," which are made of disks of linen, cotton, broadcloth, or canvas, achieve the desired degree of firmness through the amount of stitching used to fasten the layers of cloth together. When the operation calls for very soft polishing or polishing into interior corners, the stitching may be totally omitted, the centrifugal force of the wheel rotation being sufficient to keep the layers in the proper position. Various types of polishing compounds are also available, with many consisting of ferric oxide particles in some form of hinder or carrier.

The buffing operation is very similar to the lapping process that was discussed in Chapter 26. In buffing, however, the abrasive removes only minute amounts of metal from the workpiece. Fine scratch marks can be eliminated and oxide tarnish can be removed. A smooth, reflective surface is produced. When soft metals are buffed, a small amount of metal flow may occur, which further helps reduce high spots and produce a high polish.

In manual buffing, the workpiece is held against the rotating wheel and manipulated to provide contact with all critical surfaces. Once again, the labor costs can be quite extensive. If the workpieces are not too complex, semiautomatic machines can be used, where the workpieces are held in fixtures and move past a series of individual buffing wheels. By designing the part with buffing in mind, good results can be obtained quite economically.

ELECTROPOLISHING

Electropolishing is the reverse of *electroplating* (discussed later in this chapter) because material is removed from the surface rather than being deposited. A DC electrolytic circuit is constructed with the workpiece as the anode. As current is applied, material is stripped from the surface, with material removal occurring preferentially from any raised location. Unfortunately, it is not economical to remove more than about 0.001 in. of material from any surface. However, if the initial surface is sufficiently smooth (less than 8 in. rms), and the grain size is small, the result will be a smooth polish with irregularities of less than 2 μin.—a mirrorlike finish.

Electropolishing was originally used to prepare metallurgical specimens for examination under the microscope. It was later adopted as a means of polishing stainless steel sheets and other stainless products. It is particularly useful for polishing irregular shapes that would be difficult to buff.

■ 34.3 CHEMICAL CLEANING

Chemical cleaning operations are effective means of removing oil, dirt, scale, or other foreign material that may adhere to the surface of a product, as a preparation for subsequent painting or plating. Because of environmental, health, and safety concerns, however, many processes that were once the industrial standard have now been eliminated or substantially modified. While the major concern with the mechanical methods has usually been airborne particles, the chemical methods often require the disposal of spent or contaminated solutions, and they occasionally use hazardous, toxic, or environmentally unfriendly materials. Chlorofluorocarbons (CFCs) and carbon tetrachloride, for example, have been identified as ozone-depleting chemicals and have been phased out of commercial use. Process changes to comply with added regulations can significantly shift process economics. Manufacturers must now ask themselves if a part really has to be cleaned, what soils have to be removed, how clean the surfaces have to be, and how much they are willing to pay to accomplish that goal. Selection of the cleaning method will depend on cost of the equipment, power, cleaning materials, maintenance and labor, plus the cost of recycling and disposal of materials. Specific processes will depend on the quantity of parts to be processed (part per hour), part configuration, part material, desired surface finish, temperature of the process, and flexibility. Manufacturers want machines they can integrate with manufacturing cells so changes in products can be quickly handled.

ALKALINE CLEANING

Alkaline cleaning is basically the "soap and water" approach to parts cleaning and is a commonly used method for removing a wide variety of soils (including oils, grease, wax, fine particles of metal, and dirt) from the surfaces of metals. The cleaners are usually complex solutions of alkaline salts, additives to enhance cleaning or surface modification, and surfactants or soaps that are selected to reduce surface tension and displace, emulsify, and disperse the insoluble soils. The actual cleaning occurs as a result of one or more of the following mechanisms: (1) saponification, the chemical reaction of fats and other organic compounds with the alkaline salts; (2) displacement, where soil particles are lifted from the surface; (3) dispersion or emulsification of insoluble liquids; and (4) dissolution of metal oxides.

Alkaline cleaners can be applied by immersion or spraying, and they are usually heated to accelerate the cleaning action. The cleaning is then followed by a water rinse to remove all residue of the cleaning solution as well as to flush away some small amounts of remaining soil. A drying operation may also be required because the aqueous cleaners do not evaporate quickly, and some form of corrosion inhibitor (or rust preventer) may be required, depending on subsequent use.

Environmental issues relating to alkaline cleaning include (1) reducing or eliminating phosphate effluent, (2) reducing toxicity and increasing biodegradability, and (3) recycling the cleaners to extend their life and reduce the volume of discard.

SOLVENT CLEANING

In **solvent cleaning,** oils, grease, fats, and other surface contaminants are removed by dissolving them in organic solvents derived from coal or petroleum, usually at room temperature. The common solvents include petroleum distillates (such as kerosene, naphtha, and mineral spirits); chlorinated hydrocarbons (such as methylene chloride and trichloroethylene); and liquids such as acetone, benzene, toluene, and the various alcohols. Small parts are generally cleaned by immersion, with or without assisting agitation, or by spraying. Products that are too large to immerse can be cleaned by spraying or wiping. The process is quite simple, and capital equipment costs are rather low. Drying is usually accomplished by simple evaporation.

Solvent cleaning is an attractive means of cleaning large parts, heat-sensitive products, materials that might react with alkaline solutions (such as aluminum, lead, and zinc), and products with organic contaminants (such as soldering flux or marking crayon). Virtually all common industrial metals can be cleaned, and the size and shape of the workpiece are rarely a limitation. Insoluble contaminants—such as metal oxides, sand, scale, and the inorganic fluxes used in welding, brazing, and soldering—cannot be removed by solvents. In addition, resoiling can occur as the solvent becomes contaminated. As a result, solvent cleaning is often used for preliminary cleaning.

Many of the common solvents have been restricted because of health, safety, and environmental concerns. Fire and excessive exposure are common hazards. Adequate ventilation is critical. Workers should use respiratory devices to prevent inhalation of vapors and wear protective clothing to minimize direct contact with skin. In addition, solvent wastes are often considered to be hazardous materials and may be subject to high disposal cost.

VAPOR DEGREASING

In **vapor degreasing,** the vapors of a chlorinated or fluorinated solvent are used to remove oil, grease, and wax from metal products. A nonflammable solvent, such as trichloroethylene, is heated to its boiling point, and the parts to be cleaned are suspended in its vapors. The vapor condenses on the work and washes the soluble contaminants back into the liquid solvent. Although the bath becomes dirty, the contaminants rarely volatilize at the boiling temperature of the solvent. Therefore, vapor degreasing tends to be more effective than cold solvent cleaning because the surfaces always come into contact with clean solvent. Because the surfaces become heated by the condensing solvent, they dry almost instantly when they are withdrawn from the vapor.

Vapor degreasing is a rapid, flexible process that has almost no visible effect on the surface being cleaned. It can be applied to all common industrial metals, but the solvents may attack rubber, plastics, and organic dyes that might be present in product assemblies. A major limitation is the inability to remove insoluble soils, forcing the process to be coupled with another technique, such as mechanical or alkaline cleaning. Because hot solvent is present in the system, the process is often accelerated by coupling the vapor cleaning with an immersion or spray using the hot liquid.

Unfortunately, environmental issues have forced the almost complete demise of the process. While the vapor degreasing solvents are chemically stable, have low toxicity, are nonflammable, evaporate quickly, and can be recovered for reuse, the CFC materials have been identified as ozone-depleting compounds and have essentially been banned from use. Solvents that can be used in the same process, or in a replacement process that offers the necessary cleaning qualities, include chlorinated solvents (methylene chloride, perchloroethylene, and trichloroethylene); most manufacturers have converted to some form of water-based process using alkaline, neutral, or acid cleaners or to a process using chlorine-free, hydrocarbon-based solvents. Sealed chamber machines use non-**volatile organic compounds** (**VOCs**) nonchlorinated solvents that are continuously recycled.

ULTRASONIC CLEANING

When high-quality cleaning is required for small parts, **ultrasonic cleaning** may be preferred. Here, the parts are suspended or placed in wire-mesh baskets that are then immersed in a liquid cleaning bath, often a water-based detergent. The bath contains an ultrasonic transducer that operates at a frequency that causes **cavitation** in the liquid. The bubbles that form and implode provide the majority of the cleaning action, and if gross dirt, grease, and oil are removed prior to the immersion, excellent results can usually be obtained in 60 to 200 s. Most systems operate at between 10 and 40 kHz. Because of the ability to use water-based solutions, ultrasonic cleaning has replaced many of the environmentally unfriendly solvent processes.

ACID PICKLING

In the **acid-pickling** process, metal parts are first cleaned to remove oils and other contaminants and then dipped into dilute acid solutions to remove oxides and dirt that are left on the surface by the previous processing operations. The most common solution is a 10% sulfuric acid bath at an elevated temperature between 150 and 185°F. Muriatic acid is also used, either cold or hot. As the temperature increases, the solutions can become more dilute.

After the parts are removed from the pickling bath, they should be rinsed to flush the acid residue from the surface and then dipped in an alkaline bath to prevent rusting. When it will not interfere with further processing, an immersion in a cold milk of lime solution is often used. Caution should be used to avoid overpickling because the acid attack can result in a roughened surface.

34.4 COATINGS

Each of the surface finishing methods previously presented has been a material removal process designed to clean, smooth, and otherwise reduce the size of the part. Many other techniques have been developed to add material to the surface of a part. If the material is deposited as a liquid or organic gas (or from a liquid or gas medium), the process is called **coating.** If the added material is a solid during deposition, the process is known as **cladding.**

PAINTING, WET OR LIQUID

Paints and **enamels** are by far the most widely used finish on manufactured products, and a great variety is available to meet the wide range of product requirements. Most of today's commercial paints are synthetic organic compounds that contain pigments and dry by polymerization or by a combination of polymerization and adsorption of oxygen. Water is the most common carrying vehicle for the pigments. Heat can be used to accelerate the drying, but many of the synthetic paints and enamels will dry in less than an hour without the use of additional heat. The older oil-based materials have a long drying time and require excessive environmental protection measures. For these reasons, they are seldom used in manufacturing applications.

Paints are used for a variety of reasons, usually to provide protection and decoration but also to fill or conceal surface irregularities, change the surface friction, or modify the light or heat absorption or radiation characteristics. Table 34-3 provides a list of some of the more commonly used organic finishes, along with their significant characteristics. *Nitrocellulose lacquers* consist of thermoplastic polymers dissolved in organic solvent. Although fast drying (by the evaporation of the solvent) and capable of producing very beautiful finishes, they are not sufficiently durable for most commercial applications. The *alkyds* are a general-purpose paint but are not adequate for hard-service applications. *Acrylic enamels* are widely used for automotive finishes and may require catalytic or oven curing. *Asphaltic paints*, solutions of asphalt in a solvent, are used extensively in the electrical industry, where resistance to corrosion is required and appearance is not of prime importance.

When considering a painted finish, the temptation is to focus on the outermost coat, to the exclusion of the underlayers. In reality, painting is a complex system that includes the substrate material, cleaning and other pretreatments (such as anodizing, phosphating, and various conversion coatings), priming, and possible intermediate layers. The method of application is another integral feature to be considered.

TABLE 34-3 Commonly Used Organic Finishes and Their Qualities

Material	Durability (Scale of 1–10)	Relative Cost (Scale of 1–10)	Characteristics
Nitrocellulose lacquers	1	2	Fast drying; low durability
Epoxy esters	1	2	Good chemical resistance
Akyd-amine	2	1	Versatile; low adhesion
Acrylic lacquers	4	1.7	Good color retention; low adhesion
Acrylic enamels	4	1.3	Good color retention; tough high baking temperature
Vinyl solutions	4	2	Flexible; good chemical resistance; low solids
Silicones	4–7	5	Good gloss retention; low flexibility
Flouropolymers	10	10	Excellent durability; difficult to apply

PAINT APPLICATION METHODS

In manufacturing, almost all painting is done by one of four methods: *dipping, hand spraying, automatic spraying*, or *electrostatic spraying or electrocoating*. In most cases, at least two coats are required. The first coat (or **prime coat**) serves to (1) ensure adhesion, (2) provide a leveling effect by filling in minor porosity and other surface blemishes, and (3) improve corrosion resistance and thus prevent later coatings from being dislodged in service. These properties are less easily attainable in the more highly pigmented paints that are used in the final coats to promote color and appearance. When using multiple coats, however, it is important that the carrying vehicles for the final coats do not unduly soften the underlayers.

Dipping is a simple and economical means of paint application when all surfaces of the part are to be coated. The products can be manually immersed into a paint bath or passed through the bath while on or attached to a conveyor. Dipping is attractive for applying prime coats and for painting small parts where spray painting would result in a significant waste due to overspray. Conversely, the process is unattractive where only some of the surfaces require painting or where a very thin, uniform coating would be adequate, as on automobile bodies. Other difficulties are associated with the tendency of paint to run, producing both a wavy surface and a final drop of paint attached to the lowest drip point. Good-quality dipping requires that the paint be stirred at all times and be of uniform viscosity.

Spray painting is probably the most widely used paint application process because of its versatility and the economy in the use of paint. In the conventional technique, the paint is atomized and transported by the flow of compressed air. In a variation known as **airless spraying,** mechanical pressure forces the paint through an orifice at pressures between 500 and 4500 psi. This provides sufficient velocity to produce atomization and also propel the particles to the workpiece. Because no air pressure is used for atomization, there is less spray loss (paint efficiency may be as high as 99%) and less generation of gaseous fumes.

Hand spraying is probably the most versatile means of application but can be quite costly in terms of labor and production time. When air or mechanical means provide the atomization, workers must exercise considerable skill to obtain the proper coverage without allowing the paint to "run" or "drape." Only a very thin film can be deposited at one time, usually less than 0.001 in. As a result, several coats may be required with intervening time for drying.

One means of applying thicker layers in a single application is known as **hot spraying.** Special solvents are used that reduce the viscosity of the material when heated. Upon atomization, the faster-evaporating solvents are removed, and the drop in temperature produces a more viscous, run-resistant material that can be deposited in thicker layers.

When producing large quantities of similar or identical parts, some form of **automatic spraying** system is usually employed. The simplest automatic equipment consists of some form of parts conveyor that transports the parts past a series of stationary spray heads. While the concept is simple, the results may be unsatisfactory. A large amount of paint is wasted, and it is difficult to get uniform coverage.

Industrial robots can be used to move the spray heads in a manner that mimics the movements of a human painter, maintaining uniform separation distance and minimizing waste. This is an excellent application for the robot because a monotonous and repetitious process can be performed with consistent results. In addition, use of a robot removes the human from an unpleasant, and possibly unhealthy, environment. Nowdays, cars are painted almost exclusively with robots.

Both manual and automatic spray painting can benefit from the use of **electrostatic deposition.** A DC electrostatic potential is applied between the atomizer and the workpiece. The atomized paint particles assume the same charge as the atomizer and are therefore repelled. The oppositely charged workpiece then attracts the particles, with the actual path of the particle being a combination of the kinetic trajectory and the electrostatic attraction. The higher the DC voltage, the greater the electrostatic attraction. Overspraying can be reduced by as much as 60 to 80%, as can the

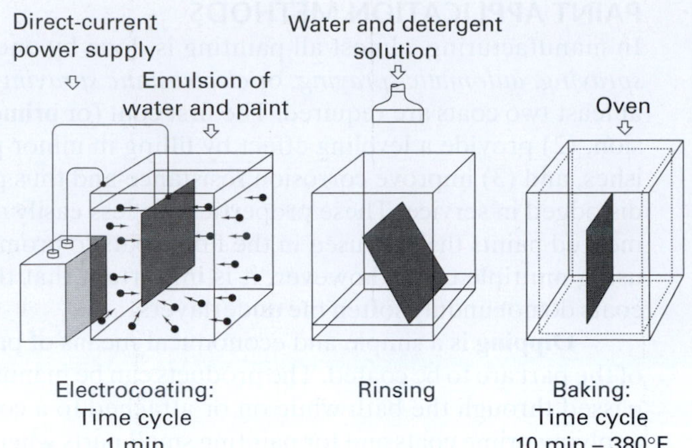

FIGURE 34-16 Basic steps in the electrocoating process.

generation of airborne particles and other emissions. Unfortunately, part edges and holes receive a heavier coating than flat surfaces due to the concentration of electrostatic lines of force on any sharp edge. Recessed areas will receive a reduced amount of paint, and a manual touch-up may be required using conventional spray techniques. Despite these limitations, electrostatic spraying is an extremely attractive means of painting complex-shaped products where the geometry would tend to create large amounts of overspray.

In an electrostatic variation of airless spraying, the paint is fed onto the surface of a rapidly rotating cone or disk that is also one electrode of the electrostatic circuit. Centrifugal force causes the thin film of paint to flow toward the edge, where charged particles are spun off without the need for air assist. The particles are then attracted to the workpiece, which serves as the other electrode of the electrostatic circuit. Because of the effectiveness of the centrifugal force, paints can be used with high-solids content, reducing the amount of volatile emissions and enabling a thicker layer to be deposited in a single application.

Electrocoating or **electrodeposition** applies paint in a manner similar to the electroplating of metals. As shown schematically in Figure 34-16, the paint particles are suspended in an aqueous solution and are given an electrostatic charge by applying a DC voltage between the tank (cathode) and the workpiece (anode). As the electrically conductive workpiece enters and passes through the tank, the paint particles are attracted to it and deposit on the surface, creating a uniform, thin coating that is more than 90% resin and pigment. When the coating reaches a desired thickness, determined by the bath conditions, no more paint is deposited. The workpiece is then removed from the tank, rinsed in a water spray, and baked at a time and temperature that depends on the particular type of paint. Baking of 10 to 20 min at 375°F is somewhat typical.

Electrocoating combines the economy of ordinary dip painting with the ability to produce thinner, more uniform coatings. The process is particularly attractive for applying the prime coat to complex structures, such as automobile bodies, where good corrosion resistance is a requirement. Hard-to-reach areas and recesses can be effectively coated. Because the solvent is water, no fire hazard exists (as with the use of many solvents), and air and water pollution are reduced significantly. In addition, the process can be readily adapted to conveyor-line production.

DRYING

Most paints and enamels used in manufacturing require from 2 to 24 hr to dry at normal room temperature. This time can be reduced to between 10 min and 1 hr if the temperature can be raised to between 275 and 450°F. As a result, elevated-temperature drying is often preferred. Parts can be batch processed in ovens or continuously passed through heated tunnels or under panels of infrared heat lamps.

TABLE 34-4 Thermosetting Powder Coatings (Dry Painting) Have a Wide Variety of Properties and Applications

Properties	Epoxy	Epoxy/Polyester Hybrid	TGIC Polyester	Polyester Urethane	Acrylic Urethane
Application thickness	0.5–20 mils[a]	0.5–10 mils	0.5–10 mils	0.5–10 mils	0.5–10 mils
Cure cycle (metal temperatures)[b]	450°F—3 min 250°F—30 min	450°F—3 min; 325°F—25 min	400°F—7 min; 310°F—20 min	400°F—7 min; 350°F—17 min	400°F—7 min; 360°F—25 min
Outdoor weatherability	Poor	Poor	Very good	Very good	Excellent
Pencil hardness	HB-5H	HB-2H	HB-2H	HB-3H	H-3H
Direct impact resistance, in lb[c]	80–160	80–160	80–160	80–160	20–60
Chemical resistance	Excellent	Very good Least expensive	Good	Good	Very good Most expensive
Cost (relative)	2	1	3	4	5
Applications	Furniture, cars, ovens, appliances	Water heaters, radiators, office furniture	Architectural aluminum, outdoor furniture, farm equipment	Car wheels/rims, playground equipment	Washing machines, refrigerators, ovens

[a] Thickness up to 150 mils can be applied via multiple coats in a fluidized bed.
[b] Time and temperature can be reduced, by utilizing accelerated curing mechanisms, while maintaining the same general properties.
[c] Tested at a coating thickness of 2.0 mils.

Elevated-temperature drying is rarely a problem with metal parts, but other materials can be damaged by exposure to the moderate temperatures. For example, when wood is heated, the gases, moisture, and residual sap are expanded and driven to the surface beneath the hardening paint. Small bubbles tend to form that roughen the surface, or break, producing small holes in the paint.

POWDER COATING

Powder coating is yet another variation of electrostatic spraying, but here the particles are solid rather than liquid. Several coats, such as primer and finish, can be applied and then followed by a single baking, in contrast to the baking after each coat that is required in the conventional spray processes. In addition, the overspray powder can often be collected and reused. While volatilized solvents are no longer a concern, operators must now address the possibility of powder explosion, as well as the health hazards of airborne particles.

Modern powder technology can produce a high-quality finish with superior surface properties and usually at a lower cost than liquid painting. Powder painting is more efficient in the use of materials (the overspray can be captured and reused) and lower energy requirements. The economic advantages must be weighed against the limitations of powder coating. Dry systems have a longer color change time than wet systems. The process is not good for large objects (massive tanks) or heat-sensitive objects. It is not easy to produce film thickness less than 1 mil (0.03 mm).

Table 34-4 provides details on powders that are used in powder coatings. Thermoplastics can also be used, but thermosetting powders are most common. The elements of a powder coating system are shown in Figure 34-17. The following aspects of the process must be considered:

- Types of guns—corona charged or tribo charged.
- Number of guns—depends on many factors, such as parts per hour, size of parts, line speed, and powder types.
- Color change time/frequency.
- Safety.
- Curing oven—coated parts put in ovens to melt, flow, and cure the powder.

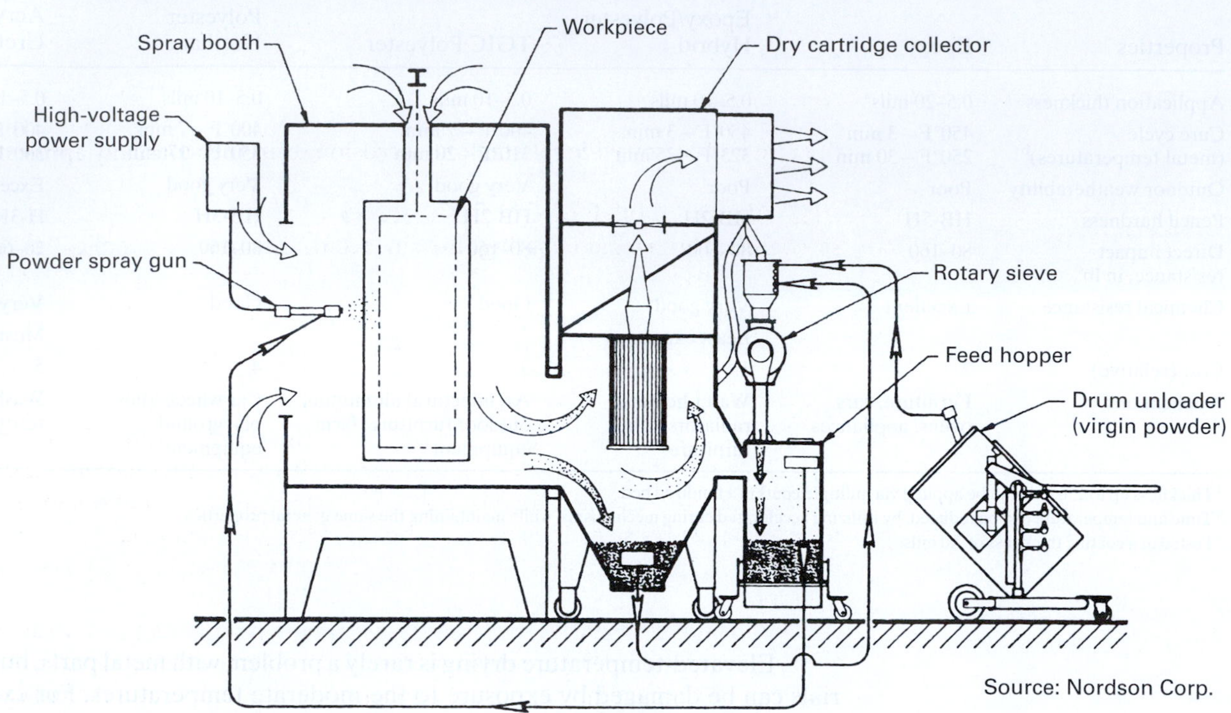

FIGURE 34-17 A schematic of a powder coating system. The wheels on the color modules permit it to be exchanged with a spare module to obtain the next color.

HOT-DIP COATINGS

Large quantities of metal products are given corrosion-resistant coatings by direct immersion into a bath of molten metal. The most common coating materials are zinc, tin, aluminum, and tene (an alloy of lead and tin).

Hot-dip galvanizing is the most widely used method of imparting corrosion resistance to steel. (The zinc acts as a sacrificial anode, protecting the underlying iron.) After the products, or sheets, have been cleaned to remove oil, grease, scale, and rust, they are fluxed by dipping into a solution of zinc ammonium chloride and dried. Next, the article is completely immersed in a bath of molten zinc. The zinc and iron react metallurgically to produce a coating that consists of a series of zinc–iron compounds and a surface layer of nearly pure zinc.

The coating thickness is usually specified in terms of weight per unit area. Values between 0.5 and 3.0 oz/ft^2 are typical, with the specific value depending on the time of immersion and speed of withdrawal. Thinner layers can be produced by incorporating some form of air-jet or mechanical wiping as the product is withdrawn. Because the corrosion resistance is provided through the sacrificial action of the zinc, the thin layers do not provide long-lasting protection. Extremely heavy coatings, on the other hand, may tend to crack and peel. The appearance of the coating can be varied through both the process conditions and alloy additions of tin, antimony, lead, and aluminum. When the coatings are properly applied, bending or forming can often follow galvanizing without damage to the integrity of the coating. Zinc-galvanized sheet can be heat treated with a zinc–iron alloy coating. The 10% iron content adds strength and makes for good corrosion and pitting/chipping resistance. In auto applications, galvannealing beats out pure zinc on several counts: spot weldability, pretreatability, and ease of painting. Electrogalvanized zinc–nickel coatings that contain 10 to 15% nickel can be used in thinner layers (5 to 6 μm) and are easier to form and spot weld.

The primary limitations to hot-dip galvanizing are the size of the product (which is limited to the size of the tank holding the molten zinc) and the "damage" that might occur when a metal is exposed to the temperatures of the molten material (approximately).

Tin coatings can also be applied by immersing in a bath of molten tin with a covering of flux material. Because of the high cost of tin and the relatively thick coatings applied by hot dipping, most tin coatings are now applied by electroplating. **Terne coating** utilizes an alloy of 15 to 20% tin and the remainder lead. This material is cheaper than tin and can provide satisfactory corrosion resistance for many applications.

CHEMICAL CONVERSION COATINGS

In **chemical conversion coating,** the surface of the metal is chemically treated to produce a nonmetallic, nonconductive surface that can impart a range of desirable properties. The most popular types of conversion coatings are chromate and phosphate. Aluminum, magnesium, zinc, and copper (as well as cadmium and silver) can all be treated by a **chromate** conversion process that usually involves immersion in a chemical bath. The surface of the metal is convened into a layer of complex chromium compounds that can impart colors ranging from bright clear through blue, yellow, brown, olive drab, and black. Most of the films are soft and gelatinous when they are formed but harden upon drying. They can be used to (1) impart exceptionally good corrosion resistance; (2) act as an intermediate bonding layer for paint, lacquer, or other organic finishes; or (3) provide specific colors by adding dyes to the coating when it is in its soft condition.

Phosphate coatings are formed by immersing metals (usually steel or zinc) in baths where metal phosphates (iron, zinc, and manganese phosphates are all common) have been dissolved in solutions of phosphoric acid. The resultant coatings can be used to precondition surfaces to receive and retain paint or enhance the subsequent bonding with rubber or plastic. In addition, phosphate coatings are usually rough and can provide an excellent surface for holding oils and lubricants. This feature can be used in manufacturing, where the coating holds the lubricants that assist in forming, or in the finished product, as with black-color bolts and fasteners, whose corrosion resistance is provided by a phosphate layer impregnated with wax or oil.

BLACKENING OR COLORING METALS

Many steel parts are treated to produce a black, iron oxide coating—a lustrous surface that is resistant to rusting when handled. Because this type of oxide forms at elevated temperatures, the parts are usually heated in some form of special environment, such as spent carburizing compound or special blackening salts.

Chemical solutions can also be used to blacken, blue, and even "brown" steels. Brown, black, and blue colors can also be imparted to tin, zinc, cadmium, and aluminum through chemical bath immersions or wipes. The surfaces of copper and brass can be made to be black, blue, green, or brown, with a full range of tints in between.

ELECTROPLATING

Large quantities of metal *and plastic* parts are electroplated to produce a metal coating that imparts corrosion or wear resistance, improves appearance (through color or luster), or increases the overall dimensions. Virtually all commercial metals can be plated, including aluminum, copper, brass, steel, and zinc-based die castings. Plastics can be electroplated, provided that they are first coated with an electrically conductive material.

Figure 34-18 depicts the typical **electroplating** process. A DC voltage is applied between the parts to be plated (which is made the cathode) and an anode material that is either the metal to be plated or an inert electrode. Both of these components are immersed in a conductive electrolyte, which may also contain dissolved salts of the metal to be plated as well as additions to increase or control conductivity. In response to the applied voltage, metal ions migrate to the cathode, lose their charge, and deposit on the surface. While the process is simple in its basic concept, the production of a high-quality plating requires selection and control of a number of variables, including the electrolyte and the concentrations of the various dissolved components, the

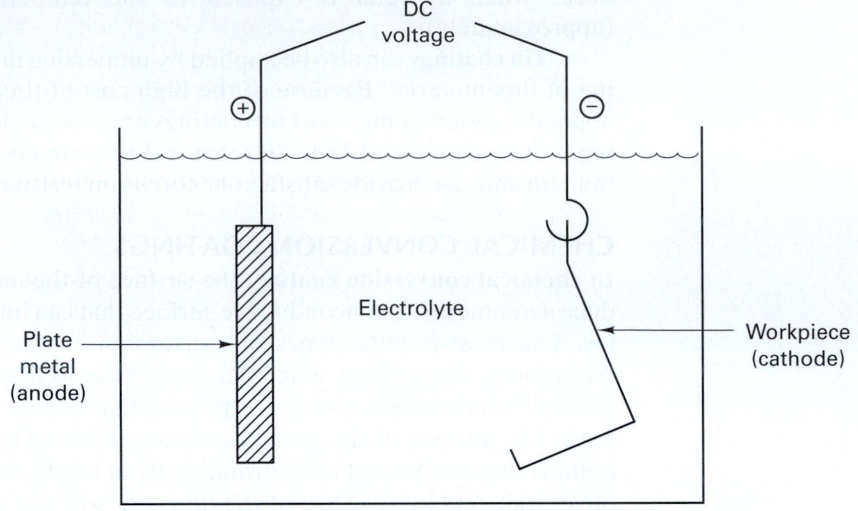

FIGURE 34-18 Basic circuit for an electroplating operation, showing the anode, cathode (workpiece), and electrolyte (conductive solution).

temperature of the bath, and the electrical voltage and current. The interrelation of these features adds to the complexity and makes process control an extremely challenging problem.

The surfaces to be plated must also be prepared properly if satisfactory results are to be obtained. Pinholes, scratches, and other surface defects must be removed if a smooth, lustrous finish is desired. Combinations of degreasing, cleaning, and pickling are used to ensure a chemically clean surface, one to which the plating material can adhere.

As shown in Figure 34-19, the plated metal tends to be preferentially attracted to corners and protrusions. This makes it particularly difficult to apply a uniform plating to irregular shapes, especially ones containing recesses, corners, and edges. Design features can be incorporated to promote plating uniformity, and improved results can often be obtained through the use of multiple spaced anodes or anodes whose shape resembles that of the workpiece.

The most common platings are zinc, chromium, nickel, copper, tin, gold, platinum, and silver. The electrogalvanized zinc platings are thinner than the hot-dip coatings and can be produced without subjecting the base metal to the elevated temperatures of

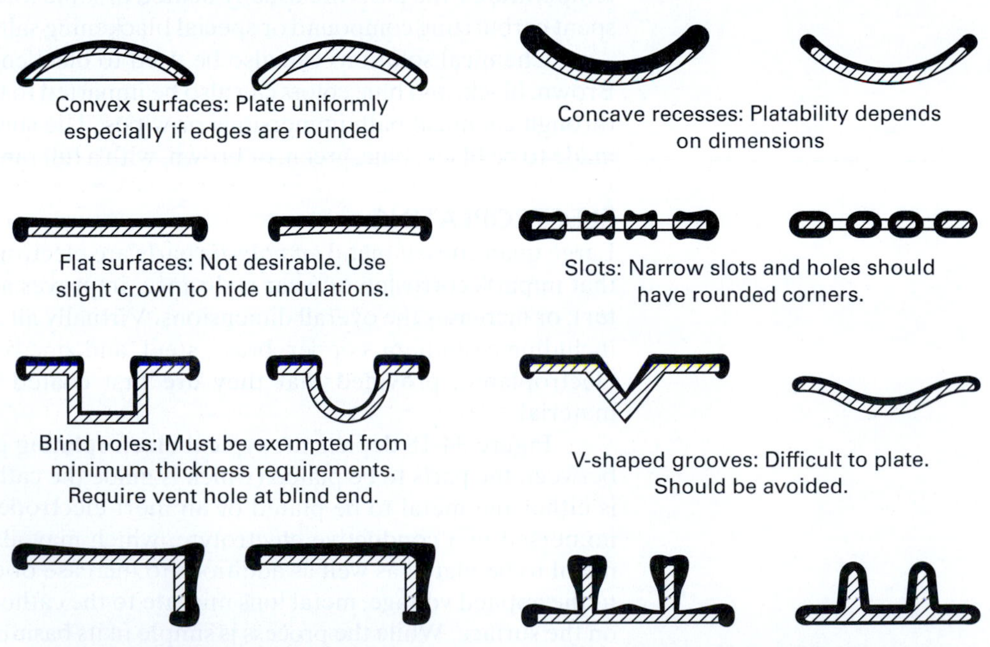

Convex surfaces: Plate uniformly
especially if edges are rounded
Concave recesses: Platability depends
on dimensions
Flat surfaces: Not desirable. Use
slight crown to hide undulations.
Slots: Narrow slots and holes should
have rounded corners.
Blind holes: Must be exempted from
minimum thickness requirements.
Require vent hole at blind end.
V-shaped grooves: Difficult to plate.
Should be avoided.
Sharply angled edges: Plating is thinner
in center areas. Round all areas.
Fins: Increase plating time and costs.
Reduce durability of finish.

FIGURE 34-19 Design recommendations for electroplating operations.

molten zinc. Nickel plating provides good corrosion resistance but is rather expensive and does not retain its lustrous appearance. Consequently, when lustrous appearance is desired, a chromium plate is usually specified. Chromium is seldom used alone, however. An initial layer of copper produces a leveling effect and makes it possible to reduce the thickness of the nickel layer that typically follows to less than 0.0006 in. The final layer of chromium then provides the attractive appearance. Gold, silver, and platinum platings are used in both the jewelry and electronics industries, where the thin layers impart the desired properties while conserving the precious metals.

Hard chromium plate, with Rockwell hardnesses between 66 and 70, can be used to build up worn parts to larger dimensions and to coat tools and other products that need reduced surface friction and good resistance to both wear and corrosion. Hard chrome coatings are always applied directly to the base material and are usually much thicker than the decorative treatments, typically ranging from 0.003 to 0.010 in. thick. Even thicker layers are used in applications such as diesel cylinder liners. Because hard chrome plate does not have a leveling effect, defects or roughness in the base surface will be amplified. If smooth surfaces are desired, subsequent grinding and polishing may be necessary.

Electroplating is frequently performed as a continuous process, where the individual parts to be plated are hung from conveyors. As they pass through the process, they are lowered into successive plating, washing, and fixing tanks. Ordinarily, only one type of workpiece is plated at a time, because the details of solutions, immersion times, and current densities are usually changed with changes in workpiece size and shape.

In the **electroforming** process, the coating becomes the final product. Metal is electroplated onto a mandrel (or mold) to a desired thickness and is then stripped free to produce small quantities of molds or other intricate-shaped sheet-metal-type products.

ANODIZING

Anodizing is an electrochemical process, somewhat the reverse of electroplating, that produces a conversion-type coating on aluminum that can improve corrosion and wear resistance and impart a variety of decorative effects. If the workpiece is made the anode of an electrolytic cell, instead of a plating layer being deposited on the surface, a reaction progresses inward, increasing the thickness of the hard hexagonal aluminum oxide crystals on the surface. The hardness depends on thickness, density, and porosity of the coating, which are controlled by the cycle time and applied currents along with the chemistry, concentration, and temperature of the electrolyte. The surface texture very nearly duplicates the prefinishing texture, so a buffing prefinish produces a smooth, lustrous coating, while sand blasting produces a grainy or satiny coating.

The flow diagram in Figure 34-20 shows the anodizing process. Coating thicknesses range from 0.1 to 0.25 mil. Note that the product dimensions will increase, however, because the aluminum oxide coating occupies about twice the volume of the metal from which it formed.

The nature of the developed coating is controlled by the electrolyte. If the oxide coating is not soluble in the anodizing solution, it will grow until the resistance of the oxide prevents current from flowing. The resultant coating, which is thin, nonporous, and nonconducting, is used in a variety of electrical applications.

If the oxide coating is slightly soluble in the anodizing solution, dissolution competes with oxide growth and a porous coating will be produced, where the pores provide for continued current flow to the metal surface. As the coating thickens, the growth rate decreases until it achieves steady state, where the growth rate is equal to the rate of dissolution. This condition is determined by the specific conditions of the process, including voltage, current density, electrolyte concentration, and electrolyte temperature. Sulfuric, chromic, oxalic, and phosphoric acids all produce electrolytes that dissolve oxide, with a sulfuric acid solution being the most common.

In a process variation known as **color anodizing,** a sulfuric acid bath is used to produce a layer of microscopically porous oxide that is transparent on pure aluminum and somewhat opaque on alloys. When this material is immersed in a dye solution,

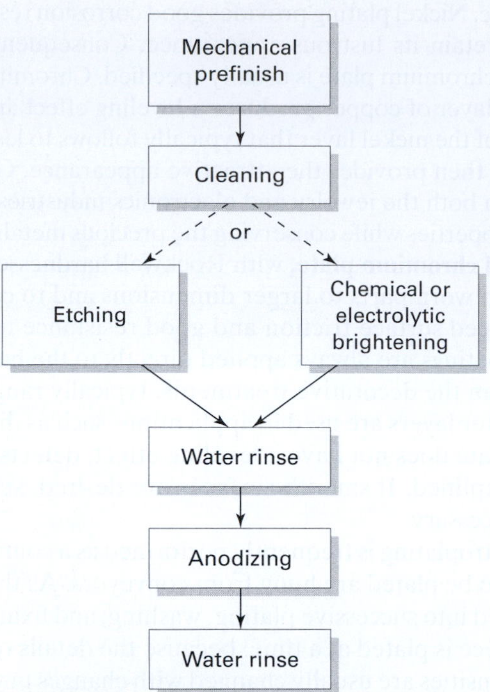

FIGURE 34-20 The anodizing process has many steps.

capillary action pulls the dye into the pores. The dye is then trapped in place by a sealing operation, usually performed simply by immersing the anodized metal in a bath of hot water. The aluminum oxide coating is converted to a monohydrate, with accompanying increase in volume. The pores close and become resistant to further staining or the leaching out of the dye.

While most people are familiar with the variety of colors in aluminum athletic goods, such as softball hats, the actual applications range from giftware through automotive trim to architectural use. Aluminum can be made to look like gold, copper, or brass, or it can take on a variety of colors with a combined metallic luster that cannot be duplicated by other methods.

If PTFE (Teflon) is introduced into the pores, coatings can be produced that couple high hardness and low friction. The porous oxide layer can also be used to enhance the adhesion of an additional layer of material, such as paint, or carry lubricant during a subsequent forming operation. Because the coating is integral to the part, subsequent operations can often be performed without destroying its integrity or reducing its protective qualities.

Anodizing can also be performed on other metals, such as magnesium, and the process is similar to the passivation of stainless steel.

ELECTROLESS PLATING

When using electroplating, it is almost impossible to obtain a uniform plating thickness on even moderately complex shapes: the platings cannot be applied to nonconductors, and a large amount of energy is required. For these reasons, a substantial effort has been directed toward the development of plating techniques that do not require an external source of electricity. These methods are known as **electroless** (or **autocatalytic**) **plating.** Considerable success has been achieved with nickel, but copper and cobalt, as well as some of the precious metals, can also be deposited.

In the electroless process, complex plating solutions (containing metal salts, reducing agents, complexing agents, pH adjusters, and stabilizers) are brought into contact with a substrate surface that acts as a catalyst or has been pretreated with catalytic material. The metallic ion in the plating solution is reduced to metal and deposits on the surface. Because the deposition is purely a chemical process, the coatings are uniform in thickness, independent of part geometry. Unfortunately, the rate of deposition is considerably slower than with electroplating.

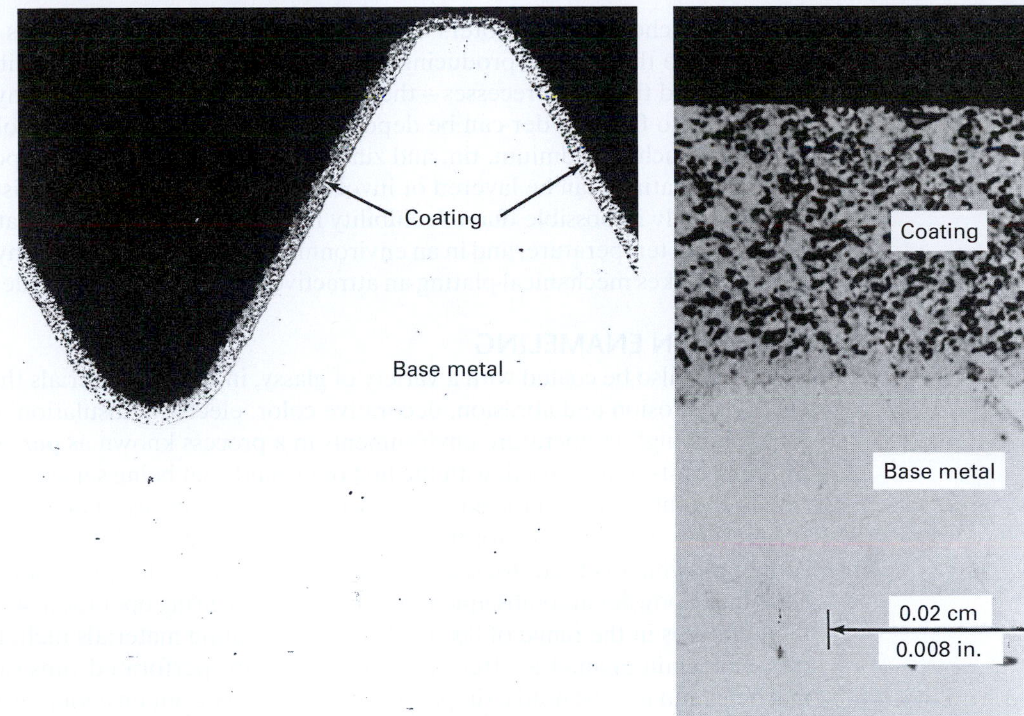

FIGURE 34-21 (Left) Photomicrograph of nickel carbide plating produced by electroless deposition. Notice the uniform thickness coating on the irregularly shaped product. (Right) High-magnification cross section through the coating. *(Images Courtesy of The L. S. Starrett, Co.)*

Probably the most popular of the electroless coatings is electroless nickel, and various methods exist for its deposition using both acid and alkaline solutions. The coatings offer good corrosion resistance, as well as hardnesses between Rockwell C 49 and 55. In addition, the hardness can be increased further to as high as Rockwell C 80 by subsequent heat treatment.

ELECTROLESS COMPOSITE PLATING

A very useful adaptation of the electroless process has been developed wherein minute particles are co-deposited along with the electroless metal to produce composite-material coatings; the process is called **electroless composite plating.** Finely divided solid particles, with diameters between 1 and 10 in., are added to the plating bath and deposit up to 50 vol% with the matrix. While it may appear that a large variety of materials could be co-deposited, commercial applications have largely been limited to diamond, silicon carbide, aluminum oxide, and Teflon (PTFE).

Figure 34-21 shows a deposit of silicon carbide particles in a nickel–alloy matrix, where the particles constitute about 25 percent/volume. The coating offers the same corrosion resistance as nickel, but the high hardness of the silicon carbide particles (about 4500 on the Vickers scale, where tungsten carbide is 1300 and hardened steel is about 900) contributes outstanding resistance to wear and abrasion. Because the deposition is electroless, the thickness of the coating is not affected by the shape of the part. Applications include the coating of plastic-molding dies, for use where the polymer resin contains significant amounts of abrasive filler.

MECHANICAL PLATING

Mechanical plating, also known as **peen plating** or **impact plating,** is an adaptation of barrel finishing in which coatings are produced by cold-welding soft, malleable metal powder onto the substrate. Numerous small products are first cleaned and may be given a thin galvanic coating of either copper or tin. They are then placed in a tumbling barrel, along with a water slurry of the metal powder to be plated, glass or ceramic tumbling media, and chemical promoters or accelerators. The media particles peen the metal powder onto the surface, producing uniform-thickness deposits (possibly a bit thinner on edges and thicker in recesses—the opposite of electroplating!). Any metal that can be made into fine powder can be deposited, but the best results are obtained for soft materials, such as cadmium, tin, and zinc. Because the material is deposited mechanically, the coatings can be layered or involve mixtures with bulk chemistries that would be chemically impossible due to solubility limits. The fact that the coatings are deposited at room temperature, and in an environment that does not induce hydrogen embrittlement, makes mechanical plating an attractive means of coating hardened steels.

PORCELAIN ENAMELING

Metals can also be coated with a variety of glassy, inorganic materials that impart resistance to corrosion and abrasion, decorative color, electrical insulation, or the ability to function in high-temperature environments in a process known as **porcelain enameling.** Multiple coats may be used, with the first or ground coat being selected to provide adhesion to the substrate and the cover coat to provide the surface characteristics. The material is usually applied in the form of a multicomponent suspension or slurry (by dipping or spraying), which is then dried and fired. An alternative dry process uses electrostatic spraying of powder and subsequent firing. During the firing operation, which may require temperatures in the range of 800 to 8000°F, the coating materials melt, flow, and resolidify. Porcelain enamel is often found on the inner, perforated tubs of many washing machines and may be used to impart the decorative exterior on cookpots and frying pans.

34.5 VAPORIZED METAL COATINGS

Vapor deposition processes can be classified into two main categories: **physical vapor deposition (PVD)** and **chemical vapor deposition (CVD).** While sometimes used as though it were a specific process, the term *PVD* applies to a group of processes in which the material to be deposited is carried physically to the surface of the workpiece. **Vacuum metallizing** and **sputtering** are key PVD processes, as are complex variations, such as ion plating. All are carried out in some form of vacuum, and most are line-of-sight processes in which the target surfaces must be positioned relative to the source. In contrast, the CVD processes deposit material through chemical reactions and generally require significantly higher temperatures. Tool steels treated by CVD may have to be heat treated again, while most PVD processes can be conducted below normal tempering temperatures. See Chapter 21 for additional discussions on PVD and CVD processes.

34.6 CLAD MATERIALS

Clad materials are actually a form of composite in which the components are joined as solids, using techniques such as roll bonding, explosive welding, and extrusion. The most common form is a laminate, where the surface layer provides properties such as corrosion resistance, wear resistance, electrical conductivity, thermal conductivity, or improved appearance, while the substrate layer provides strength or reduces overall cost. Alclad aluminum is a typical example. Here, surface layers of weaker but more corrosion-resistant single-phase aluminum alloys are applied to a base of high-strength but less corrosion-resistant, age-hardenable material. Aluminum-clad steel meets the same objective but with a heavier substrate, and stainless steel can be used to clad steels, reducing the need for nickel- and chromium-alloy additions throughout.

Wires and rods can also be made as claddings. Here, the surface layer often imparts conductivity, while the core provides strength or rigidity. Copper-clad steel rods that can be driven into the ground to provide electrical grounding for lightning rod systems are one example.

34.7 TEXTURED SURFACES

While technically not the result of a surface finishing process or operation, **textured surfaces** can be used to impart a number of desirable properties or characteristics. The types of textures that are often rolled onto the sheets used for refrigerator panels serve to conceal dirt, smudges, and fingerprints. Embossed or coined protrusions can enhance the grip of metal stair treads and walkways. Corrugations provide enhanced strength and rigidity. Still other textures can be used to modify the optical or acoustical characteristics of a material.

34.8 COIL-COATED SHEETS

Traditionally, sheet metal components, such as panels for appliance cabinets, have been fabricated from bare-metal sheets. Pans are blanked and shaped by the traditional metal-forming operations, and the shaped panels are then finished on an individual basis. This requires individual handling and the painting or plating of geometries that contain holes, bends, and contours. In addition, there is the time required to harden, dry, or cure the applied surface finish.

An alternative approach is to apply the finish to the sheet material after rolling but before coiling. Coatings can be applied continuously to one or both sides of the material while it is in the form of a flat sheet. Thus, the coiled material is effectively prefinished, and efforts need to be taken to protect the surface during the blanking and forming operations used to produce the final shape. Various paints have been applied successfully, as well as a full spectrum of metal coatings and platings. The sheared edges will not be coated, but if this feature can be tolerated, the additional measures to protect the surface may be an attractive alternative to the finishing of individual components. A second sequence that has some advantages takes the coils of steel that have been cut to length and stamps the holes and notches into them to create blanks. The blanks are pretreated, dried, powder coated, cured, and restacked. Then they are postformed to shape them into the back, side, and front panels of appliance cabinets.

The manufacturers call this **blank coating.** The coating thickness is about 1.5 mil $\pm$ 0.2 mil versus 2 mil $\pm$ 0.5 mil (less powder, better quality), and rusting at the corners of the holes is eliminated.

34.9 EDGE FINISHING AND BURRS

Burrs are the small, sometimes flexible projections of material that adhere to the edges of workpieces that are formed by machining, like the exit-side burrs formed in the milled slot of Figure 34-22. Dimensionally, they are typically only 0.003 in. thick and 0.001 to 0.005 in. in height, but if not removed, they can lead to assembly failures, short circuits, injuries to workers, or even fatigue failures.

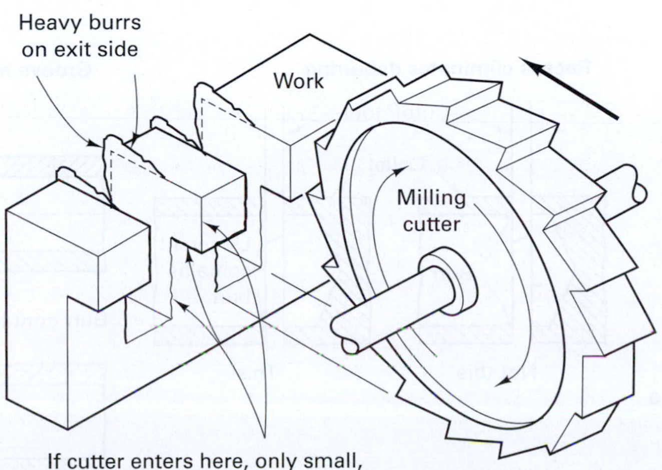

FIGURE 34-22 Schematic showing the formation of heavy burrs on the exit side of a milled slot. *(From L. X. Gillespie, American Machinist, November 1985)*

The most basic way to detect a burr is to run your finger or fingernail over the edges of the part. Probes and visual inspection techniques (microscopes) are used to find burrs as well.

A number of different processes have been used for **burr removal,** including some discussed previously in this chapter and others presented as special types of machining. These include grinding, chamfering, barrel tumbling, vibratory finishing, centrifugal and spindle finishing, abrasive-jet machining, water-jet cutting, wire brushing, belt sanding, chemical machining, electropolishing, buffing, electrochemical machining, filing, ultrasonic machining, and abrasive flow machining.

Other burr removal methods may be quite specialized, such as thermal-energy deburring, where parts are loaded into a chamber, which is then filled with a combustible gas mixture. When the gas is ignited, the short-duration wavefront heats the small burrs to as much as 6000°F, while the remainder of the workpiece rarely exceeds 300°F. The burrs are vaporized in less than 20 ms, including those in inaccessible or difficult-to-reach locations. Because the process does not use abrasive media, there is no change to any of the product dimensions. The product surfaces are rarely affected by the generated heat, and the cycle (including loading and unloading) can be repeated as many as 100 times an hour. Unfortunately, there is a thin recast layer and heat-affected zone that forms where the burrs were removed. This region is usually less than 0.001 in. thick, but it may be objectionable in hardened steels and highly stressed parts.

Of all of the burr removal methods, tumbling and vibratory finishing are usually the most economical, typically costing in the neighborhood of a few cents per part. Because most of the common methods also remove metal from exposed surfaces and produce a radius on all edges, it is important that the parts be designed for deburring. Table 34-5 provides a listing of the various deburring processes, as well as the edge radius, stock loss, and surface finish that would result from removal of a "typical burr" that is 0.003 in. thick.

By knowing how and where burrs are likely to *form,* the design engineer may be able to design parts to make the burrs easy to remove or even eliminate them. As shown in Figure 34-23, extra recesses or grooves can eliminate the need for deburring, because the burr produced by a cutoff tool or slot milling cutter will now lie below the surface. In this approach, one must determine whether it is cheaper to perform another machining operation (undercutting or grooving) or to remove the resulting burr.

Chamfers on sharp corners can also eliminate the need to deburr. The chamfering tool removes the large burrs formed by facing, turning, or boring and produces a relief for mating parts. The small burr formed during chamfering may be allowable or can easily be removed. Often, it may be preferable to give the manufacturer the freedom to use either a chamfer (produced by machining) or an edge radius (formed during the deburring operation) on all exposed corners or edges.

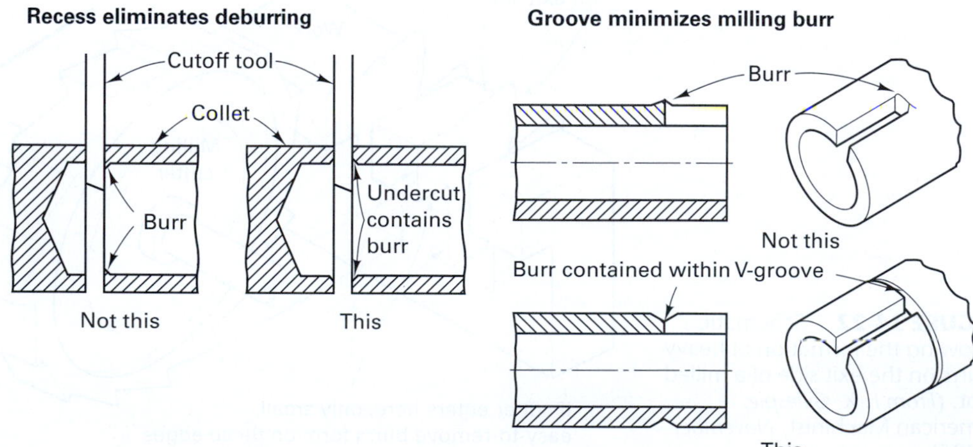

FIGURE 34-23 Designing extra recesses and grooves into a part may eliminate the need to deburr. *(From L. X. Gillespie,* American Machinist, *November 1985)*

TABLE 34-5 Recommended Allowances for Deburring Processes[a]

Process	Edge Radius, mm (in.)	Stock Loss, mm (in.)	Surface Finish, μmAA (μin.AA[b])
Barrel tumbling	0.08–0.5 (0.003–0.020)	0.0025 (0–0.001)	1.5–0.5 (60–20)
Vibratory deburring	0.08–0.5 (0.003–0.020)	0–0.025 (0–0.001)	1.8–0.9 (70–35)
Centrifugal barrel tumbling	0.08–0.5 (0.003–0.020)	0–0.025 (0–0.001)	1.8–0.5 (70–20)
Spindle finishing	0.08–0.5 (0.003–0.020)	0–0.025 (0–0.001)	1.8–0.5 (70–20)
Abrasive-jet deburring	0.08–0.25 (0.003–0.010)	0–0.05 (0–0.002)[c]	0.8–1.3 (30–50)
Water-jet deburring	0–0.13 (0–0.005)(p)	0(p)	
Liquid hone deburring	0–0.13 (0–0.005)	0–0.013 (0–0.0005)	
Abrasive-flow deburring	0.025–0.5 (0.001–0.020)	0.025–0.13 (0.001–0.005)[d]	1.8–0.5 (70–20)
Chemical deburring	0–0.5 (0–0.002)	0–0.025 (0–0.001)	1.3–0.5 (50–20)
Ultrasonic deburring	0–0.05 (0–0.002)	0–0.025 (0–0.001)	0.5–0.4 (20–15)
Electrochemical deburring	0.05–0.25 (0.002–0.010)	0.025–0.08 (0.001–0.003)[e]	
Electropolish deburring	0–0.25 (0–0.010)	0.025–0.08 (0.001–0.003)[e]	0.8–0.4 (30–15)
Thermal-energy deburring	0.05–0.5 (0.002–0.020)	0	1.5–1.3(p) (60–50)
Power brushing	0.08–0.5 (0.003–0.020)	0–0.013 (0–0.0005)	
Power sanding	0.08–0.8 (0.003–0.030)[f]	0.013–0.08 (0.0005–0.003)	1.0–08 (40–30)
Mechanical deburring	0.08–1.5 (0.003–0.060)		
Manual deburring	0.05–0.4 (0.002–0.015)[g]		

[a] Based on a burr 0.08 mm (0.003 in.) thick and 0.13 mm (0.005 in.) high in steel. Thinner burrs can generally be removed much more rapidly. Values shown are typical. Stock-loss values are for overall thickness or diameter. Location A implies that loss occurs over external surfaces, B that loss occurs over all surfaces, and C that loss occurs only near edge. (p) indicates best estimate.
[b] Values shown indicate typical before and after measurements in a deburring cycle.
[c] Abrasive is assumed to contact all surfaces.
[d] Stock loss occurs only at surfaces over which medium flows.
[e] Some additional stray etching occurs on some surfaces.
[f] Flat sanding produces a small burr and no radius.
[g] Chamfer is generally produced with a small burr.
Source: L. X. Gillespie, *American Machinist*, November 1985.

KEY WORDS

abrasive cleaning
acid pickling
airless spraying
alkaline cleaning
anodizing
arithmetic average (AA)
autocatalytic plating
automatic spraying
barrel burnishing
barrel finishing
belt sanding
blank coating
blast cleaning
buffing
burrs
burr removal
case hardening
cavitation
centrifugal barrel tumbling
chemical conversion coating
chemical cleaning
chemical vapor deposition (CVD)
chromate
cladding
coating
coil-coated sheets
color anodizing
compounds
cutoff
dipping
electrocoating
electrodeposition
electroforming
electroless composite plating
electroless plating
electroplating

electropolishing
electrostatic deposition
enamel
finishing processes
hand spraying
hard chromium plate
hot-dip galvanizing
hot spraying
impact plating
lay
mechanical cleaning
mechanical plating
media
nontraditional processes
oxides
paint
peen plating
phosphate coatings
physical vapor deposition (PVD)
porcelain enameling
powder coating
prime coat
residual stresses
roller burnishing
root-mean-square (rms)
roughness
sand blasting
scale
shot blasting
shot peening
solvent cleaning
spindle finishing
spray painting
sputtering
surface engineering
surface finish
surface integrity
surface roughness
terne coating
textured surfaces
tin coating
topography
traditional processes
tumbling
ultrasonic cleaning
vacuum metallizing
vapor degreasing
vibratory finishing
waviness
wire brushing

REVIEW QUESTIONS

1. Why are surface processes so important?
2. What are some of the factors that should be considered when selecting a surface-modification process?
3. How are the surface and its integrity altered by the process of metal cutting?
4. Two surfaces can have the same microinch roughness but be different in appearance. Explain.
5. What limits the resolution of a stylus-type surface-measuring device in finding profiles?
6. What is the general relationship between surface roughness and tolerance?
7. What is the relationship between tolerance and cost to produce the surface and/or tolerance?
8. What are some common abrasive media used in blasting or abrasive cleaning operations?
9. What types of quantities and part sizes are most attractive for barrel finishing operations?
10. Why might there be an optimum fill level in barrel finishing?
11. Using statistical methods, how might you find the optimum rotational speed?
12. Describe the primary differences between barrel finishing and vibratory finishing.
13. What are some of the possible functions of the compounds that are used in abrasive finishing operations?
14. What are some of the mechanisms of alkaline cleaning, and what types of soils can be removed?
15. What types of surface contaminants cannot be removed by solvent cleaning?
16. In view of its many attractive features, why has vapor degreasing become an unattractive process?
17. What is the primary type of surface contaminant removed by acid pickling?
18. What is the difference between coating and cladding operations?
19. What are some of the reasons that paints may be specified for manufactured items?
20. What are some of the functions of a prime coat in a painting operation? What features are desired in the final coat?
21. What produces atomization and propulsion in airless spraying?
22. What features make industrial robots attractive for spray-painting automobiles?
23. What are some of the attractive features of electrostatic spraying?
24. Why would it be difficult to apply electrostatic spray painting to products made from wood or plastic?
25. What are some of the metal coatings that can be applied by the hot-dip process?
26. What are the two most common types of chemical conversion coatings?
27. How can nonconductive materials such as plastic be coated by electroplating?
28. What are the attractive properties of hard chrome plate?
29. What are some of the common process variables in an electroplating cell?
30. Why is it difficult to mix parts of differing size and shape in an automated electroplating system?
31. How is electroforming different from electroplating?
32. When anodizing aluminum, what features determine the thickness of the resulting oxide when the oxide is not soluble in the electrolyte? When it is partially soluble?
33. What produces the various colors in the color anodizing process?
34. What are some of the attractive features of electroless plating?
35. What types of particulate composites can be deposited by electroless plating?
36. What is mechanical plating?
37. What are some of the attractive properties of a porcelain enamel coating?
38. How and why are burrs made by the milling process? See Figure 34-22.
39. What deburring processes are available that were not described in this chapter?

PROBLEMS

1. Fishermen are among the most superstitious people in the world, and their superstitions affect the type of equipment that they use. As a result, hook manufacturers generally offer their products in a wide range of colors and finishes. Your company manufactures a range of hooks from AISI 1080 carbon steel wire, forming them to precision shape (eye, bends,

barbs, and point) and then heat treating them by a quench-and-temper treatment.

Consider the size and shape of the product and the various properties that are required. The hooks must be strong enough to resist bending, but not so brittle that they might break. They must be corrosion-resistant to both fresh and salt water, and the desired appearance must be provided without fouling the point or the barbs. If the surface is applied before heat treatment, it must endure that process and maintain its appearance. If it is applied afterward, it cannot weaken or embrittle the hook.

a. Of the various surface-modification processes, which ones might be attractive for such an application? (*Note:* Make sure that the process is appropriate! For example, barrel plating would probably produce a hopelessly snarled mass of wires!)
b. For each of the possible processes, describe the advantages, limitations, possible colors or finishes, and relative cost.

2. Select one or more of the following products (as assigned by the instructor) and recommend a surface treatment or coating. Consider the appropriateness of the technique to the size, shape, quantity, and material. Cite the specific features that make the recommended treatment the most attractive. What, if any, are the primary limitations or production concerns?
   a. The exterior housing for the motor and drive unit of a chain saw that has been made as a magnesium–alloy die casting.
   b. Large quantities of steel bolts that are intended for use in outdoor construction. They have been fabricated from 4140 steel, have a shank diameter of $\frac{1}{2}$ in., and have been quenched and tempered to a final hardness of Rockwell C 45.
   c. The handle of a household utility knife (retractable-blade cutter) has been made as a two-part zinc die casting.
   d. The scoop portion of an inexpensive ice cream scoop that has been made as a zinc die casting.
   e. A decorative handle for a kitchen cabinet that is made as a zinc die casting.
   f. The exterior of an office filing cabinet that has been made from low-carbon steel sheet.
   g. A high-quality combination wrench (open-end and box-end) that has been forged from 4147 steel bar stock.
   h. The case of a moderately priced wristwatch that has been fabricated from yellow brass (to have a gold appearance).
   i. Tubular frame of a lightweight bicycle that has been made from age-hardened aluminum.
   j. The basket section of a grocery-store shopping cart that has been fabricated from welded steel mesh.
   k. The exterior of an automobile muffler to be fabricated from steel sheet. Describe how the coating treatment might best be integrated into the fabrication sequence.
   l. An inexpensive interior door knob that has been fabricated from deep-drawn cartridge brass sheet.
   m. High-quality steel sockets for a socket-wrench set. These have been forged from AISI 4145 bar stock and subsequently heat treated by a quench-and-temper process.
   n. Refrigerator door panels that have been fabricated from textured AISI 1010 steel sheet.
   o. The interior and exterior surfaces of a 1000-gal water storage tank that is fabricated by welding 5000 series aluminum plates. The water will be held at room temperature and is intended for human consumption.
   p. A standard office paper clip.
   q. A flashlight case that has been fabricated from deep-drawn yellow brass sheet.
   r. High-speed drill bits that have been fabricated from Ml tool steel.
   s. Injection-molded ABS plastic wheel covers for cars that are intended to look like chrome-plated metal.
   t. Inexpensive household scissors that have been cast from gray cast iron.
   u. The blade of a high-quality screwdriver that has been forged from AISI 1053 steel and quenched and tempered to Rockwell C 55.
   v. The exterior of high-quality, thick-walled cast aluminum cookware.
   w. A bathroom sink basin made from deep-drawn 1008 steel sheet.
   x. The body section of a child's toy wagon that has been deep-drawn from 1008 steel sheet.

Chapter 34 CASE STUDY

Dana Lynn's Fatigue Lesson

Dana Lynn has just come from her lab in the manufacturing processes course. The purpose of this lab was to introduce her to metal fatigue and its basic principles. Metal fatigue arises from the cyclic loading below the yield strength. It can be greatly influenced by the surface finish applied to the metal. Dr. Black, her instructor, said fatigue most likely accounts for 90% of all mechanical failures, so it is important for an engineer to understand how materials respond to fatigue conditions.

The procedure of this lab was for each student to finish the aluminum specimen with the emery paper, then load the specimen into the fatigue machine, see Figure CS-34A. Students were required to record the number of cycles that was needed to fracture the specimen. It was important that the student wipe the specimen clean and inspect the specimen for burrs, ridges, or flats before inserting it into the drive spindle. The presence of stress raisers can decrease the cumulative number of cycles needed to start a fatigue crack and cause it to fail, hence reducing the fatigue life and static strength of the metal. Small surface cracks, surface flows, or machining marks are examples of stress raisers. Therefore, it is important that one strive to eliminate stress raiser or surface flaws in the specimen that will be exposed to cyclic loadings. The experimental factors for this experiment were the surface finish (grit size of energy paper), applied stress level, and the direction of the surface finish (parallel or perpendicular to the specimen axis).

Dr. Black explained that repeated applications of stress can cause metals to fracture, even if all of the stresses are less than the yield tensile strength and less than the ultimate strength of the material. Surface conditions can heavily influence fatigue life because most fatigue cracks start at the surface of a metal. The fatigue data in Table CS-34 is to be analyzed using statistical experimental design techniques and summarized in the ANOVA table. Factor A was the applied load (50 and 5 lb), factor B was the treatment (fine and course emery paper), and factor C was the surface finish direction (horizontal and vertical).

TABLE CS-34 Data from 16 Fatigue Tests

	Horizontal		Vertical	
	Fine	Course	Fine	Course
50 lb-in.	167,000	126,700	102,200	88,600
	145,600	116,800	78,600	92,600
55 lb-in.	89,600	61,300	56,500	49,400
	98,800	59,800	63,200	41,200

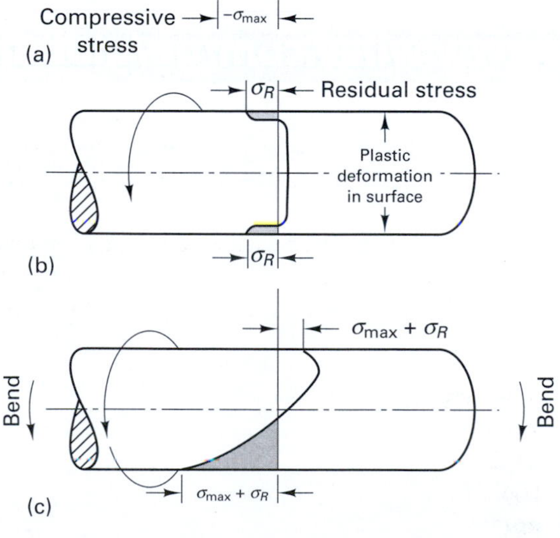

$+\sigma_{max}$
Tensile stress
Rotating
Bend
Bend
Compressive stress
$-\sigma_{max}$
(a)
σ_R
Residual stress
Plastic deformation in surface
σ_R
(b)
$\sigma_{max} + \sigma_R$
Bend
Bend
$\sigma_{max} + \sigma_R$
(c)
Rotating beam with bending results in cycles of tension and compression
Beam given roller burnish or shot peen to cold work surface and develop a residual compressive stress
Beam in service with reduced peak tensile stresses due to additive nature of applied and residual stresses

FIGURE CS-34A (Top) A cantilever-loaded (bent) rotating beam, showing the normal distribution of surface stresses (i.e., tension at the top and compression at the bottom). (Center) The residual stresses induced by roller burnishing or shot peening. (Bottom) Net stress pattern obtained when loading a surface-treated beam. The reduced magnitude of the tensile stresses contributes to increased fatigue life.

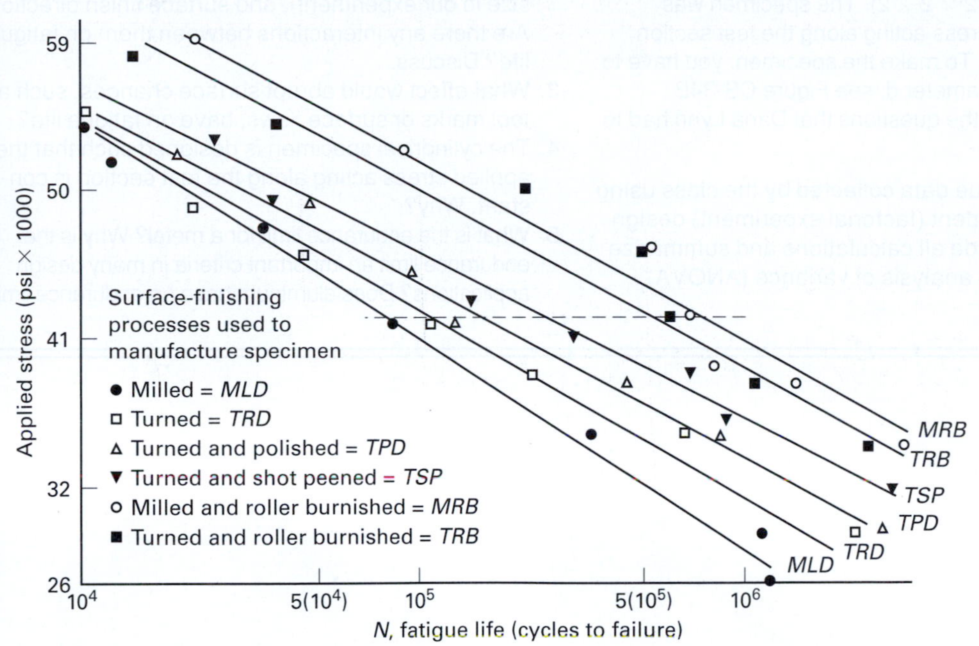

FIGURE CS-34B Fatigue life of rotating beam 2024-T4 aluminum specimens with a variety of surface-finishing operations. Note the enhanced performance that can be achieved by shot peening and roller burnishing.

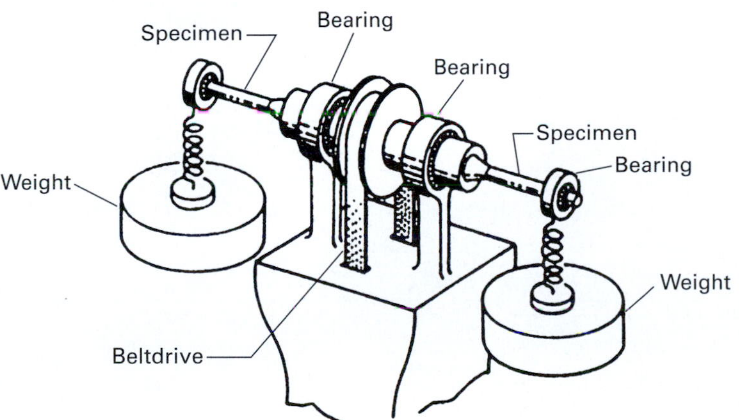

FIGURE CS-34C Rotary beam fatigue testing machine, used for cylindrical specimens.

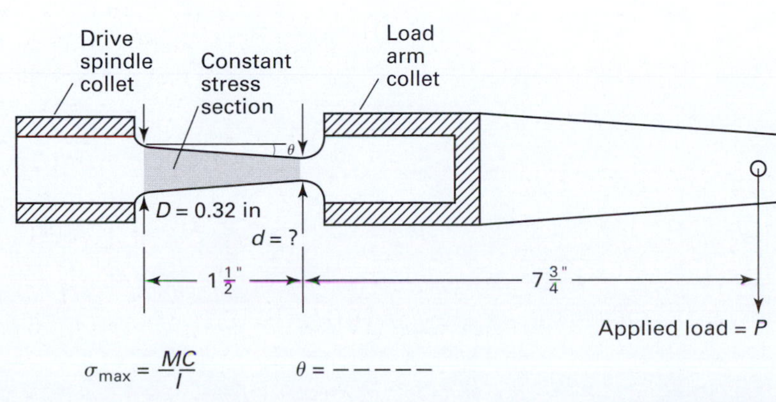

FIGURE CS-34D Cylindrical specimen with constant stress section. Determine the angle θ and the diameter d.

Continued on next page

The experiment was replicated so a total of 16 tests were run by the class (2 × 2 × 2 × 2). The specimen was tapered so that the stress acting along the test section (shaded) is constant. To make the specimen, you have to calculate θ and the diameter d, see Figure CS-34B.

Here are some of the questions that Dana Lynn had to answer:

1. Analyze the fatigue data collected by the class using statistical experiment (factorial experiment) design techniques. Include all calculations and summarize your results in an analysis of variance (ANOVA) table.
2. What are the effects of stress level, surface finish (grit size in our experiment), and surface finish direction? Are there any interactions between them on fatigue life? Discuss.
3. What effect would abrupt surface changes, such as tool marks or surface flaws, have on fatigue life?
4. The cylindrical specimen is designed such that the applied stress acting along the test section is constant. Why?
5. What is the endurance limit for a metal? Why is the endurance limit an important criteria in many design applications? Does aluminum have an endurance limit?

CHAPTER 35

MICROELECTRONIC MANUFACTURING AND ELECTRONIC ASSEMBLY

35.1 INTRODUCTION

Miniaturized microelectronic circuits are in common use today in wristwatches, portable CD players, cellular phones, home entertainment systems, fax machines, artificial hearts, military satellites, automotive fuel injection systems, and cardiac defibrillators, among others. Over the past three decades, the number of components per integrated circuit has increased from 2300 in 1971 to 42 million in 2001, while the number of calculations per second has increased 100,000 times, from 10,000 to more than a billion. The key to this progress has been the development of large-batch-size, semiconductor-processing methods coupled with miniaturization of electrical components and connectors. Unlike most other manufacturing processes in this book, semiconductor processes and other electrical and electronic manufacturing processes are concerned mainly with the manipulation of electrical properties rather than mechanical properties.

35.2 HOW ELECTRONIC PRODUCTS ARE MADE

The goal of all electronics is the processing and manipulation of electrical signals represented most fundamentally by the flow of electrons. A hierarchy for producing electronic products is illustrated in Figure 35-1. At the lowest level, microelectronic fabrication methods produce entire **integrated circuits (ICs)** of solid-state (no moving parts) components, complete with wiring and connections, on a single piece of semiconductor material. Arrays of ICs are produced on thin, round disks of semiconductor material called **wafers.** Once the semiconductor wafer has been processed, the finished wafer is sectioned into individual ICs, or **chips.** Next, these chips are individually housed within various types of IC packages for connection to other electronic components and protection from environmental elements. These IC packages, along with other discrete components (e.g., resistors, capacitors, etc.), are then combined together into even larger circuits on **printed circuit boards (PCBs).** This is sometimes referred to as **electronic assembly.** Electronic packages at this level are called **cards** or **printed wiring assemblies (PWAs).** Next, series of cards are combined on a **backpanel PCB,** also known as a **motherboard** or simply a **board.** This level of packaging is sometimes referred to as **card-on-board** packaging. Ultimately, card-on-board assemblies are put into housings and integrated with power supplies and other electronic peripherals through the use of cables to produce final commercial products.

In general, the lower the level of integration (i.e., the physically smaller the circuit and its components) within this hierarchy, the less expensive it is to produce in terms of cost per functional element. This is because, to some extent, the manufacturing cost per

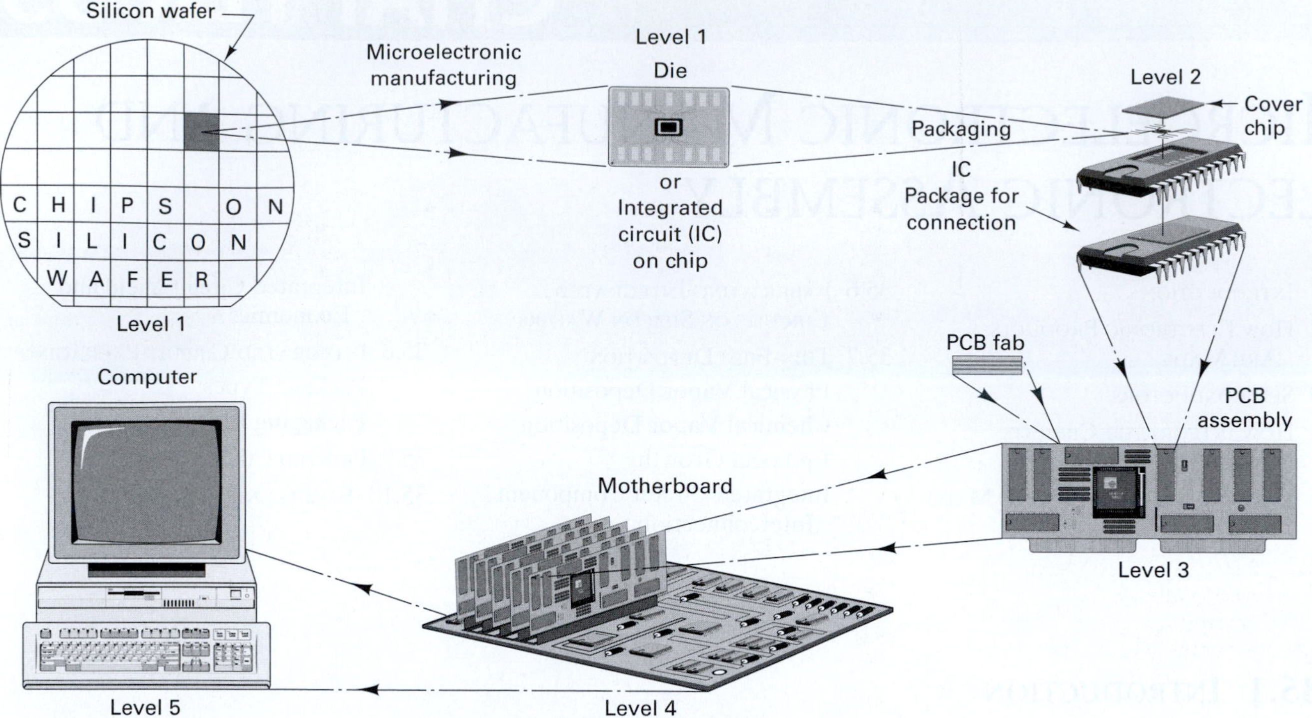

FIGURE 35-1 The hierarchy for producing electronic products has many levels. *(M. L. Minges,* Electronic Materials Handbook, Volume 1, Packaging, *Materials Park, OH: ASM International, 1989)*

IC is about the same regardless of how many components are packaged onto the chip. At the same time, the lower the level of integration is, the less flexibility in configuring the electronic system for different commercial applications. This balance between cost and flexibility is primarily what drives designers to implement circuits at various hierarchical levels.

35.3 SEMICONDUCTORS

Semiconductors—such as silicon, gallium arsenide, and germanium—are materials that can be made to be either electrically conducting or electrically insulating by changing the type and concentration of impurity atoms found within the material. Like metals, all semiconductors have crystalline microstructures exhibiting long-range order in the form of a **lattice.** However, unlike metals, semiconductor atoms are characterized as having half-filled valence shells, and so, when placed into a lattice, the semiconductor atoms form covalent bonds. Figure 35-2a shows a schematic of a lattice of covalently bonded silicon atoms.

At room temperature (25°C), silicon permits a small amount of electrical conductivity that is too small for most electronic applications. The electrical conductivity of semiconductors can be altered by inserting impurity atoms into the semiconductor lattice. The process of modifying the electrical properties of semiconductors by introducing impurity atoms is commonly referred to as **doping.** Figure 35-2b shows the same lattice as before with the middle silicon atom having been replaced by a phosphorous atom. Because phosphorous is a Column V element on the periodic chart, the phosphorous atom has one more valence electron than the surrounding silicon atoms. As such, the phosphorous atom is considered a **donor** of electrons to the silicon lattice and the phosphorus-doped semiconductor is now called an ***n*-type** (negatively charged type) **semiconductor.** *N*-type semiconductors have extra valence electrons, which are free to move about, providing increased electrical conductivity. Similarly, Figure 35-2c shows a third silicon lattice, this time with the middle silicon atom replaced by a boron atom (a Column III element). This lattice has a shortage of electrons represented as **electron holes** and is therefore termed a ***p*-type** (positively charged type) **semiconductor.**

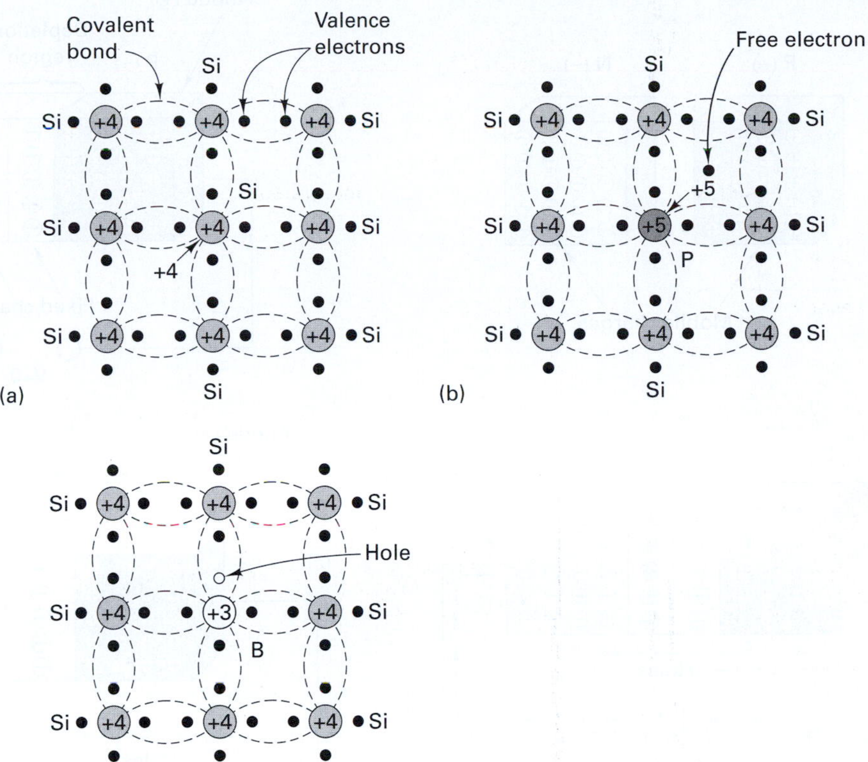

FIGURE 35-2 (a) Schematic of a lattice of silicon atoms; (b) doping with impurity atoms changes conductivity to *n*-type semiconductor or (c) to *p*-type, positively charged semiconductor. *(J. Millman,* Microelectronics: Digital and Analog Circuits and Systems, *New York: McGraw-Hill, 1979)*

Silicon is the most widely used semiconductor. It is plentiful and can be readily produced in single crystal form. Also, the native oxide, silicon dioxide, can be used both as a dieletric layer and a diffusion mask during processing.

35.4 How Integrated Circuits Are Made

The ability to selectively modify the electrical properties of semiconductors is the backbone of microelectronic manufacturing as shown in the following examples for producing IC components. A simple bipolar diode (allows current flow in only one direction) may be fabricated by forming two adjacent regions of *n*-type and *p*-type semiconductors whose electrical properties have been modified through the placement of impure, secondary atoms into the semiconductor lattice. At the interface of the regions, a so-called ***p-n* junction** is formed (Figure 35-3a). In the *p-n* junction, the excess electrons (from the *n*-type semiconductor) and holes (from *p*-type) recombine to form a **depletion region,** where all charge mobility (i.e., via electrons and holes) is effectively eliminated. While the recombination of holes and electrons fills out the valence shells in the lattice, an imbalance in charge exists, creating an electrostatic potential called the **barrier potential.** The barrier potential for a *p-n* junction in silicon is approximately 0.7 V. Application of a negative potential to the cathode and a positive potential to the anode (forward bias) at a level greater than the barrier potential of the *p-n* junction results in a flow of electrons (or holes) as shown in Figure 35-3b. In this state, the diode acts as a closed switch with very little electrical resistance. Application of a reverse bias (Figure 35-3c) causes the diode to act as an open switch with very high electrical resistance.

As shown in Figure 35-4, the manufacturing fabrication sequence for making a simple bipolar diode has many steps, beginning with the production of a silicon wafer from a predoped, single crystal ingot **(boule),** which is cut into wafers, lapped, and polished to produce silicon wafers. The wafers are placed in vacuum chambers, where an oxide layer is grown on the surface of the wafer to act as a mask during subsequent doping of the substrate. The oxide layer is patterned using **photolithography** in combination with **etching** (see Figure 35-5). Photolithography is used to produce a polymeric

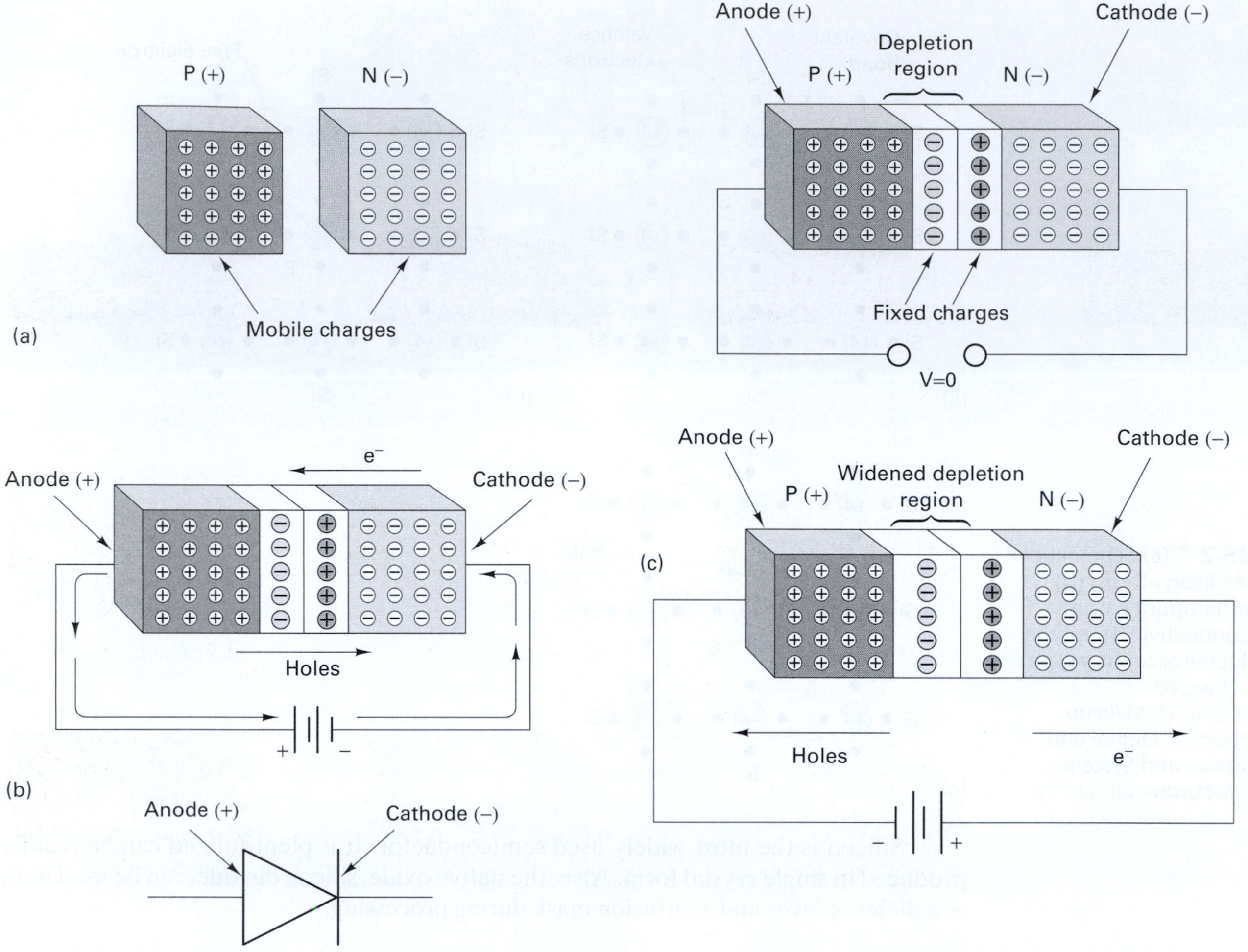

FIGURE 35-3 The diode is produced with a *p-n* junction or interface at (a) which allows electrons to flow at (b) or have high electrical resistance at (c) *(Texas Engineering Extension Service, Semiconductor Processing Overview, College Station, TX: The Texas A&M University System, 1996)*

mask over the oxide layer, which will allow only select areas of the oxide layer to be etched. After etching, the polymeric mask is removed from the silicon dioxide layer, and the *n*-type silicon is doped (by diffusion) with boron to produce a *p*-type region. After doping, the silicon dioxide mask is removed, and a second silicon dioxide layer is grown and patterned to establish openings in the silicon dioxide layer above the *n*-type and *p*-type regions. Next, a thin metal film is deposited on top of the silicon dioxide to provide an electrical pathway allowing the *p*-type and *n*-type regions of the diode to be connected to an external power supply. Photolithography and etching are used once again to pattern the thin film into leads and contact pads large enough for biasing the device. To protect the final integrated device from mechanical damage and moisture, a final passivation coating is added.

This example shows the production of a single IC component. Typically, multiple components and, further, multiple circuits are produced in parallel during IC fabrication. As shown in Table 35-1, IC fabrication has evolved from the original small-scale integration (SSI) architecture of the 1960s, with 2 to 50 electronic components per circuit, to the ultra-large-scale integration (ULSI) architectures of today, with tens of millions of components per circuit. The classification of ICs by scale of integration represents the successive advancement of semiconductor processing technologies to provide lower cost, higher-performance ICs. Each increase in the number of components represented a breakthrough in miniaturization technology (e.g., photolithography and clean rooms) that permitted the fabrication of smaller IC components with improved performance. To achieve lower cost, manufacturing

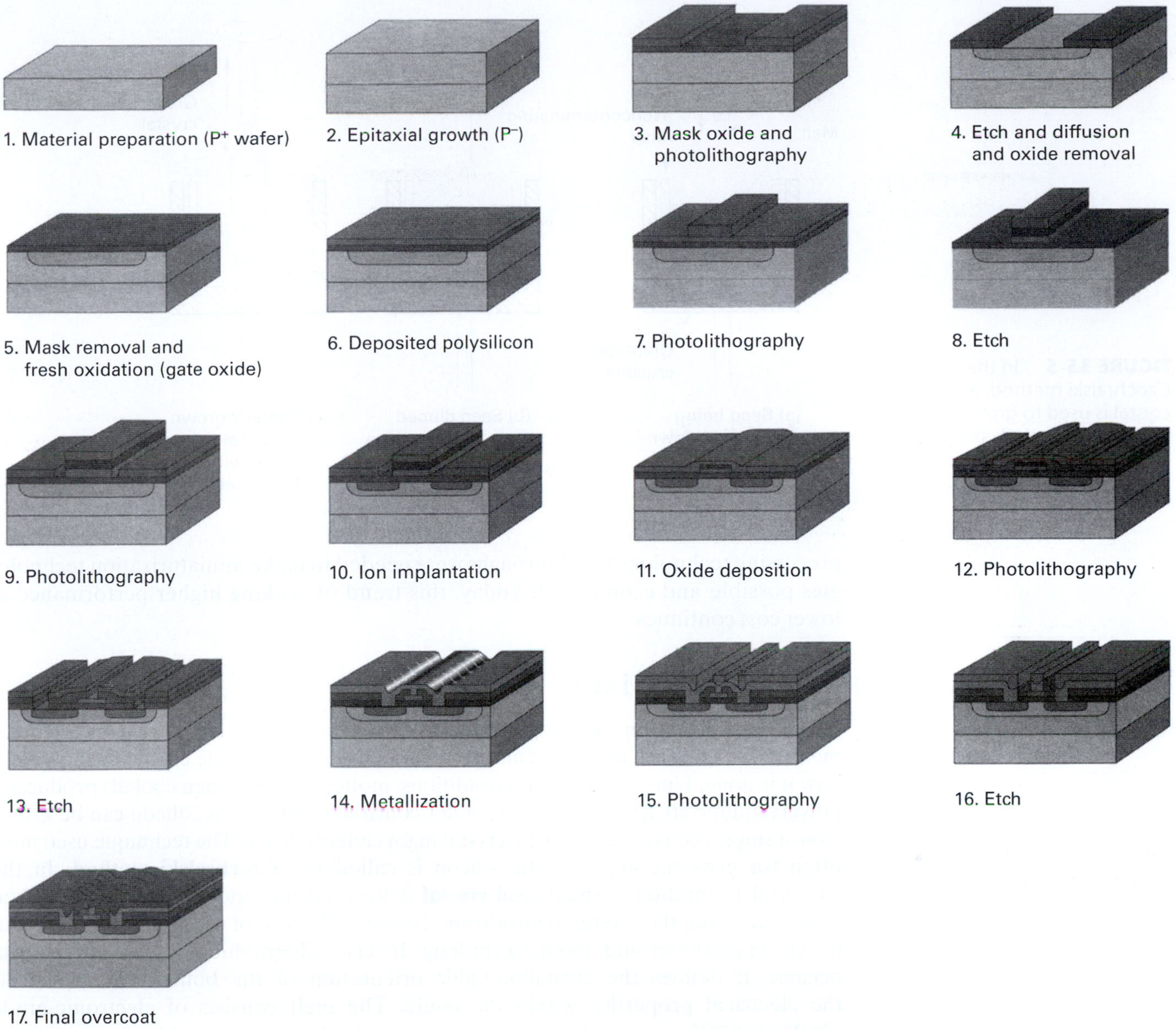

1. Material preparation (P+ wafer)
2. Epitaxial growth (P−)
3. Mask oxide and photolithography
4. Etch and diffusion and oxide removal
5. Mask removal and fresh oxidation (gate oxide)
6. Deposited polysilicon
7. Photolithography
8. Etch
9. Photolithography
10. Ion implantation
11. Oxide deposition
12. Photolithography
13. Etch
14. Metallization
15. Photolithography
16. Etch

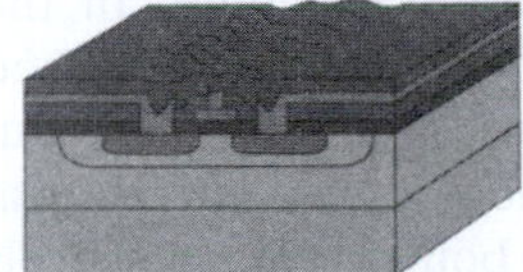

17. Final overcoat

FIGURE 35-4 The manufacture of a simple metal-oxide-semiconductor (MOS) field effect transistor device requires many steps as shown here. *Source:* Semiconductor Processor Overview, © 2008, Texas Engineering Extension Service. www.teex.org

TABLE 35-1 Classification in the Development of IC Architectures

Class	Number of Electrical Components per IC	Applications
SSI	2–50	Basic logic
MSI	50–5,000	Encoders, multiplexers, etc.
LSI	5,000–100,000	First generation microprocessors, memory ICs, early calculators, and electronic watches
VLSI	100,000–1,000,000	Integration of microprocessor, memory and I/O on single chip, digital signal processors, computer workstations and microcomputers
ULSI	Greater than 1,000,000	4–64 Mb memory ICs, latest microprocessors, advanced workstations and microcomputers

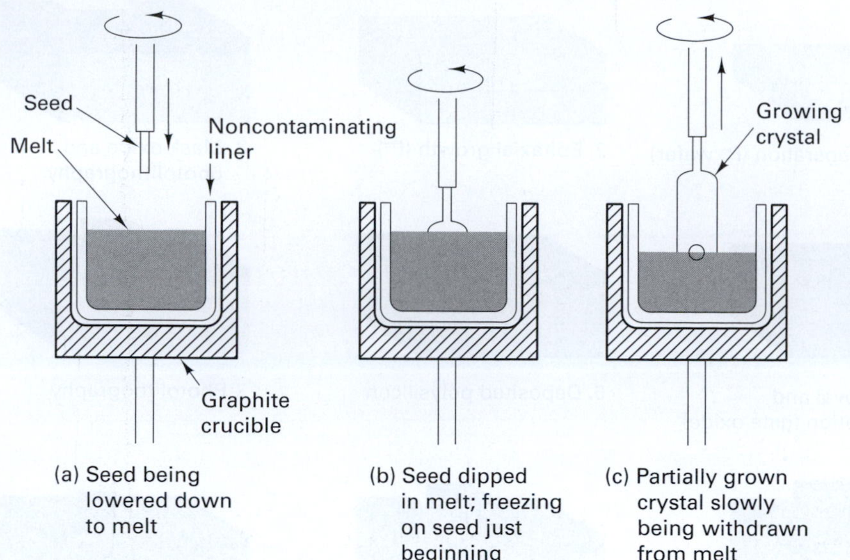

FIGURE 35-5 In the Czochralski method, a small seed crystal is used to grow large single crystals of silicon. *(MEMC Electronics International website, www.memc.com)*

processing technology breakthroughs were needed to make miniaturization technologies possible and economical. Today, this trend of seeking higher performance at lower cost continues.

35.5 HOW THE SILICON WAFER IS MADE

One of the key reasons that single crystal silicon is the most widely used semiconductor material is that it can be refined and grown economically in single crystal form. Here is how it is done. Under equilibrium conditions, molten silicon (when cooled) produces a polycrystalline structure. However, under controlled conditions, silicon can be grown from a single seed in a large single crystal ingot called a boule. The technique used most often for growing single crystal silicon is called the **Czochralski method.** In the Czochralski method, a small **seed crystal** is lowered into molten silicon and raised slowly, allowing the crystal to grow from the seed. The size of the seed crystal is about 0.5 cm in diameter and about 10 cm long. Its crystallographic orientation is critical because it defines the crystallographic orientation of the boule, which controls the electrical properties within the boule. The melt consists of electronic-grade (99.999999999% pure) polycrystalline silicon (polysilicon). If desirable, **dopant** may be added to the melt, although alloying complicates the crystal growth process. The silicon is melted in a fused silica crucible within a furnace chamber backfilled with an inert gas such as argon. The crucible is heated to approximately 1500°C and maintained at slightly above the melting point with a graphite resistance heater.

Once grown, the boule is characterized for resistivity and crystallographic defects. Table 35-2 provides a list of typical specifications for a silicon wafer. If acceptable, the unusable end portions of the boule are cut off, and the outside of the body is ground into a cylindrical ingot. For diameters less than 300 mm, **flats** are ground along the length of the ingot to denote crystal orientation and dopant type (Figure 35-6). The largest flat, called the primary flat, denotes the (011) plane. Flats are used to properly orient wafers during IC processing. In larger-diameter boules, notches are cut along the length of the boule to increase the surface area available for IC processing. Afterward, the boule is chemically etched to remove any damage imparted while grinding the flat.

TABLE 35-2 Typical Specifications for State-of-the-Art Silicon Wafer

Cleanliness (particle/cm^2)	<0.03
Oxygen concentration (cm^{-3})	Specified ± 3%
Carbon concentration (cm^{-3})	$< 1.5 \times 10^{17}$
Metal contaminants bulk (ppb)	< 0.001
Grown in dislocation (cm^{-2})	< 0.1
Oxidation induced stacking faults (cm^{-3})	< 3
Diameter (mm)	≥ 150
Thickness (μm)	625 or 675
Bow (μm)	10
Global flatness (μm)	3
Cost ($/cm^2)	0.2

Source: Campbell, S. A., *The Science and Engineering of Microelectronics*, Oxford: Oxford University Press, 2001.

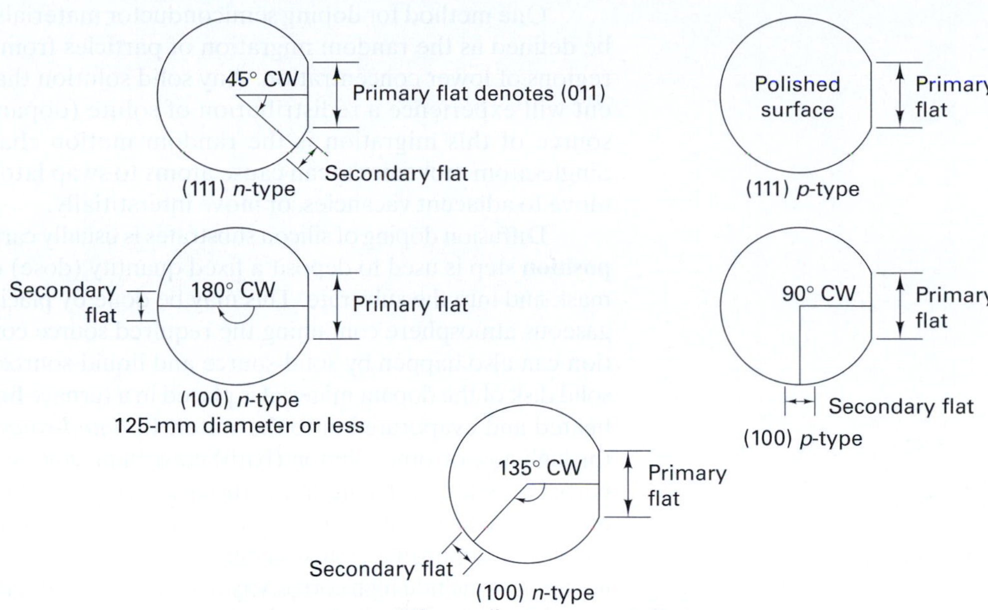

FIGURE 35-6 Flats are ground on the boule to denote the (011), (111), (100) planes, *n*-type and *p*-type materials. *(B. El-Kareh,* Fundamentals of Semiconductor Processing Technology, *Boston: Kluwer Academic Publishers, 1995)*

Next, the boule is sliced into wafers using a wire or diamond saw. Geometric concerns resulting from wafer slicing include flatness and bowing of the wafer. The wafers are typically ground on the edge because edge-rounded wafers handle better and have less mechanical damage during IC processing, and the pile-up of **photoresists** on the edge of the wafer during photolithography is minimized. Finally, a series of processing steps are needed to remove any sawing damage, including lapping, chemical etching, and polishing.

Single-crystal silicon, with few lattice imperfections, is necessary to produce the high yields required in IC processing. Several sources of crystalline defects exist during processing of the wafer, including contamination, improper pull rates, temperature gradients during pulling, and residual stress during wafer machining. Some methods exist for controlling crystalline defects during processing, such as by rotating the solidified boule and the melt in opposite directions during growth to minimize unbalanced growth caused by temperature gradients. However, not all defects can be avoided. To keep unwanted impurities and defects from diffusing into active regions of the wafer (i.e., where IC components are made), a strategy known as **gettering** is used. Gettering involves the use of hard-to-move crystalline defects in inactive regions of the wafer (i.e., away from where components will be made) to trap other impurities and defects that may otherwise diffuse into active regions thereby impairing device performance.

35.6 FABRICATING INTEGRATED CIRCUITS ON SILICON WAFERS

The first level of electronic manufacturing involves the manufacture of the ICs or chips. This is a complex process involving many steps, the sequence of which depends on the particular electrical device. The initial steps of doping by diffusion or ion implantation and oxidation are performed in large machines that manipulate the wafers in and out of various vacuum chambers in the correct sequence and duration.

Doping can be accomplished in bulk by alloying at the time of crystal formation. However, selective doping is required for IC production. Selective doping in most early IC devices involved thermal diffusion; more recently, as device dimensions have continued to shrink, ion implantation has become more suitable to better control the depth and concentration of the dopant atoms in the silicon wafer. The doped lateral geometry is primarily defined by the use of a low diffusivity mask (e.g., oxide mask) patterned by lithography methods (covered later in this chapter). The depth and concentration are controlled by the method of doping and its process parameters.

One method for doping semiconductor materials involves diffusion. **Diffusion** can be defined as the random migration of particles from regions of high concentration to regions of lower concentration. Any solid solution that contains a concentration gradient will experience a redistribution of solute (dopant) concentration over time. The source of this migration is the random motion characteristic of atoms above 0°K. Single-atom movements can cause atoms to swap lattice locations with adjacent atoms, move to adjacent vacancies, or move interstitially.

Diffusion doping of silicon substrates is usually carried out in two steps. First, a **predeposition** step is used to deposit a fixed quantity (dose) of dopant atoms through an oxide mask and into the substrate. This may be done by placing the wafer in a furnace having a gaseous atmosphere containing the required source concentration of dopant. Predeposition can also happen by solid-source and liquid-source doping. In *solid-source doping*, a solid disk of the dopant material is placed in a furnace **boat** adjacent to the wafer where it is heated and evaporated onto the wafer. In *liquid-source doping,* an inert gas is bubbled through an isotropic solution (bath) containing compounds with the desired dopant. The partial pressure of dopant in the furnace is controlled by the temperature of the bath, the pressure of the gas above the liquid, and the flow of other inert gases into the furnace. Liquid-source doping has gained significant acceptance because of improved purity levels. Disadvantages include high corrosivity and sensitivity to temperature changes in the bath.

Once the dose is deposited in the predeposition step, a **drive-in** step is used to redistribute the dose to achieve the proper depth and concentration. The drive-in step is performed in a vacuum oven without the presence of the dopant source. The advantage of thermal diffusion is that it is fast relative to other doping processes. The disadvantage is less control over the depth and concentration of dopant profiles.

As the overall size of IC devices has decreased, the required thickness of doped regions has also decreased, requiring greater control and precision of doped dimensions. Therefore, doping by thermal diffusion has been replaced by ion implantation within the current generation of IC devices. Ion implantation involves electrostatically accelerating a beam of ionized atoms or molecules toward the wafer surface, allowing the resultant kinetic energy to drive the particles into the substrate. Ion implantation has been found to control the amount of impurity and the depth of impurity penetration much better than thermal diffusion.

One disadvantage of ion implantation is that the kinetic energy of the ion particles damages the silicon substrate. The resulting lattice damage can significantly affect the electrical and chemical properties of the single crystal substrate. This damage can be minimized by annealing the substrates at temperatures up to 1000°C after ion implantation. However, annealing at these temperatures can create problems of its own, causing redistribution of dopant profiles within other previously processed regions of the device. To compensate, **rapid thermal processing** technologies have been developed to reduce the time the wafer is exposed to high temperature. In **rapid thermal annealing (RTA),** the wafer rests on quartz pins and is heated using a bank of high-intensity filament lamps. Problems with RTA include temperature measurement and thermal uniformity across the wafer. Excessive temperature gradients across the wafer can lead to plastic deformation in the wafer such as warpage and/or slip. RT technologies have been extended to include rapid thermal oxidation, chemical vapor deposition, and epitaxial growth, among others.

Under exposure to oxygen, a silicon surface oxidizes to form silicon dioxide, the same underlying chemical makeup of window glass. Silicon dioxide is an excellent dielectric material and so can be used as the "gate" dielectric in a MOSFET device or as an isolation layer between layers of metal wires that interconnect IC components. Thick oxides formed by thermal oxidation are generally used as masks during doping. The major objective in thermal oxidation is to create an oxide layer of uniform thickness. While silicon readily oxidizes at room temperature, deep penetration of the oxide into the single crystal is accelerated at high temperatures by thermal diffusion. From this standpoint, thermal oxidation is similar to diffusion doping. Other methods for producing thin-oxide layers (e.g., for device isolation) do exist and are briefly discussed in the section on deposition processes.

Though similar in some ways, atomic diffusion in oxidation is different than in doping. Compared with the number density of silicon (on the order of 5×10^{22} atom/cm^3), final dopant concentrations for active device regions within semiconductor substrates are small (on the order of 10^{17} atom/cm^3), whereas oxygen concentrations are of the same magnitude (10^{22} atom/cm^3). Due to the large oxygen concentrations and the stoichiometry of the reactions, about 46% of the silicon surface is "consumed" during oxidation. That is, for every $1\,\mu m$ of SiO_2 grown, about .046 μm of silicon is used.

The thermal oxidation of silicon is achieved by heating the substrate to temperatures typically in the range of 900 to 1200°C within an oxygen atmosphere. The atmosphere in the furnace can either contain pure oxygen **(dry oxidation)** or part oxygen and water vapor **(wet oxidation).** Initially, the growth of silicon dioxide is a surface reaction, and the growth rate depends on the reaction rate at the silicon surface. The chemical reaction for dry oxidation at the wafer surface is:

$$Si + O_2 \rightarrow SiO_2 \tag{35-1}$$

Thin gate oxides can be prepared with a very high uniformity over the wafer and from wafer to wafer using dry oxidation.

Because growth rates in wet oxidation are higher than in dry oxidation, wet oxidation is often used to grow thicker oxides. For thicker oxides, the arriving oxidant molecules must diffuse through the growing layer to get to the silicon surface in order to react. The primary reason for the faster growth rate in wet oxidation is because water vapor molecules are smaller than molecular oxygen and, therefore, diffuse more easily within silicon. One disadvantage of wet oxidation is that the oxide layer is not as dense. Therefore, wet oxidation is used in applications that are not subjected to electrical stress, such as for diffusion masks.

Techniques like diffusion and oxidation are used to modify the electrical properties of the silicon wafer. Additional techniques are needed to transfer the shape of the integrated circuit from the designer's workstation to the semiconductor wafer. In particular, lithography and etching are two intermediate steps necessary to pattern the silicon dioxide films formed as diffusion masks and as electrically insulating layers in components. In addition, these two steps are also needed to pattern the various conductive and insulating thin films necessary to fabricate and interconnect IC components.

Lithography is the process of transferring the geometric patterns of the IC design to a thin layer of polymer, called a **resist,** producing a **resist mask** on the surface of the silicon wafer. The purpose of the resist mask is to serve as a temporary barrier to etching or implantation, allowing for the selective patterning of **thin films** (e.g., thin films of deposited polysilicon, oxide, or metal for component fabrication, insulation, or interconnection) or the selective doping of semiconductor substrates underneath the resist in various steps of IC processing. Lithography is the most complicated, expensive, and critical process in mainstream microelectronic fabrication. A typical silicon IC device technology may involve 15 to 20 different lithography patterns, each with feature sizes, or **linewidths,** as small as 0.18 μm. Needless to say, the technologies needed to meet these requirements are expensive. In the early 1990s, using dynamic random access memory (DRAM) ICs as an example, lithography accounted for roughly one-third of the total fabrication cost.

Several different lithography methods exist including: (1) photolithography, (2) X-ray lithography, (3) electron-beam (e-beam) lithography, and (4) ion-beam lithography. The difference in the techniques is the source of ionizing radiation used to expose the resist. The first two methods involve the use of electromagnetic radiation, whereas the latter two involve particle radiation (i.e., an electron or ion beam). Lithography techniques based on electromagnetic radiation are **through-mask techniques,** requiring the use of a lithography mask to selectively pattern the resist, whereas techniques based on particle radiation are **direct-write techniques,** indicating that the particle beams scan the pattern onto the resist directly without the use of a mask. Photolithography is the most common method and will be discussed here.

In photolithography, UV sources of radiation are used to expose UV-sensitive materials called **photoresists,** or simply *resists.* The **photomasks,** sometimes called

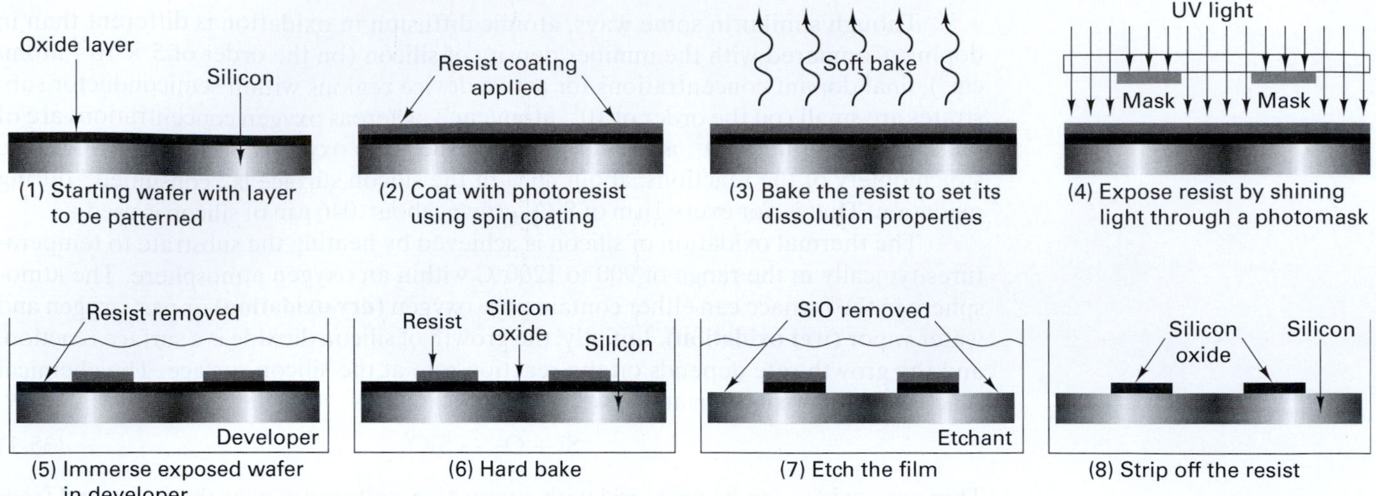

FIGURE 35-7 The process of making an IC using photolithography has many steps. *(S. A. Campbell,* The Science and Engineering of Microelectronic, *Oxford: Oxford University Press, 2001)*

reticles, are used to mask or screen parts of the surface from etching or doping processes. The photomask is a thin, high-optical-purity quartz plate onto which a thin film of opaque material (such as chromium) has been deposited and patterned (selectively etched). The photomask is patterned with the aid of computer-aided manufacturing (CAM) techniques using the original IC design data from computer-aided design (CAD) systems.

The photolithography process involves the sequence of steps shown in Figure 35-7. First, a liquid photoresist is applied to the surface of the silicon oxide layer over the silicon wafer. Typically, this is done with a process known as **spin coating.** In spin coating, centrifugal forces are used to produce a photoresist layer of uniform thickness. Next, the coated wafer is *soft baked* on a hot plate or in an oven. In this step, solvents used to reduce the viscosity of the photoresist during spin coating are evaporated, and adhesion between the wafer and the photoresist is improved. After soft bake, the photoresist is *exposed* using a photomask to transmit a pattern of electromagnetic radiation onto the surface of the photoresist. This step is performed using a machine called a **stepper,** because the lithographic pattern of the device is indexed or stepped across the wafer, subjecting it to repeated exposures—one for each chip you are making. Once the resist has been exposed, the wafer is *developed* in a chemical solvent. Development removes the unwanted resist materials, exposing the underlying material to be etched. Next, the resist is *hard baked* to remove any remaining solvents after development and to further toughen the remaining resist against downstream etching or implantation processes. Hard bakes generally take longer and are at slightly higher temperatures than soft bakes. Once the downstream etching or implantation has made use of the resist, a photoresist **stripping** step is necessary for removal of the resist.

During exposure, the UV radiation that is transmitted through the photomask selectively modifies the molecular weight of the polymer in desired regions. In *positive* photoresists, the UV radiation is responsible for decreasing the molecular weight in these regions by breaking molecular bonds. The lower molecular weight of these regions makes them more soluble in the chemical developer. In *negative* photoresists, the UV radiation increases the molecular weight of the exposed resist through cross-linking, making the exposed region more insoluble in the developer. These two types of photoresists are contrasted in Figure 35-8.

Obviously, the most important requirementof the photoresist is that it resists the downstream etching or implantation process. Other requirements important to the function of resists are their **resolution** and **sensitivity.** Resolution refers to the smallest linewidth that can be reproduced repeatably by the resist. The resolution of the resist is strongly a function of the source of ionizing radiation or the exposure machine tool used. Sensitivity refers to the amount of ionizing energy required to sufficiently modify

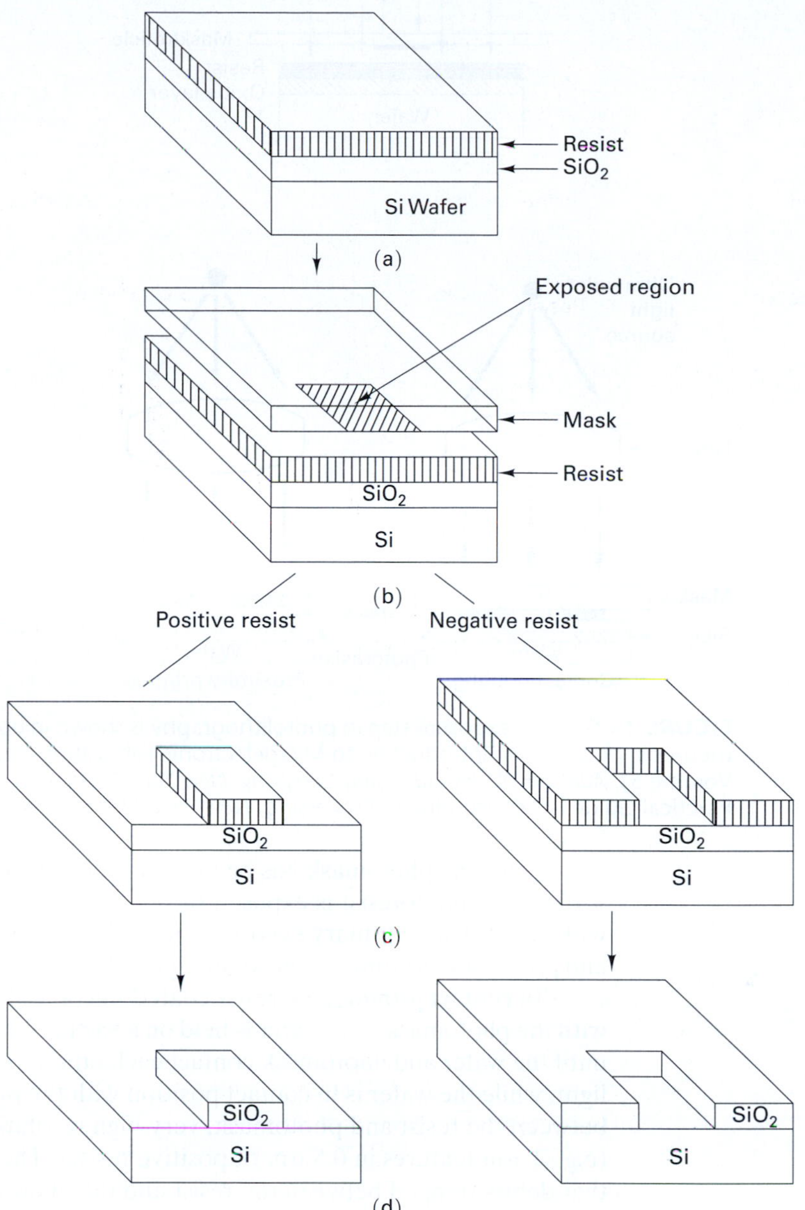

FIGURE 35-8 Photoresist material can be made soluble (positive) or insoluble (negative) to the developer. *(R. C. Jaeger, Introduction to Microelectronic Fabrication (Modular Series on Solid State Device Volume 5), New York: Addison-Wesley, 1990)*

the solubility of the resist. The more sensitive a resist is, the shorter the exposure cycle time and the greater the throughput. A final requirement of the resist is that it adheres to the substrate.

In the past, the most commonly used light source in photolithography was the mercury arc lamp. Its most useful wavelengths for photolithography occur at 436 and 365 nm (blue and UV light, respectively)—the so-called mercury *g*-line and *i*-line. As a result, *g*-line and *i*-line photoresists have long been used as standards within the IC industry. Negative photoresists were popular in the early history of IC processing because of their low cost and good adhesion, but positive photoresists are now most widely used because they offer better process control for small geometric features.

Schematic of the exposure step in photolithography are shown in Figure 35-9. Exposure begins with photomask alignment. The photomask is aligned with the wafer so that the pattern can be transferred onto the wafer surface. Each pattern after the first one requires photomask alignment to the previous pattern. For linewidths on the order of (current IC resolutions), misregistration errors as small as 6 nm can have detrimental effects on device performance. This registration requirement contributes to the high cost of lithography equipment.

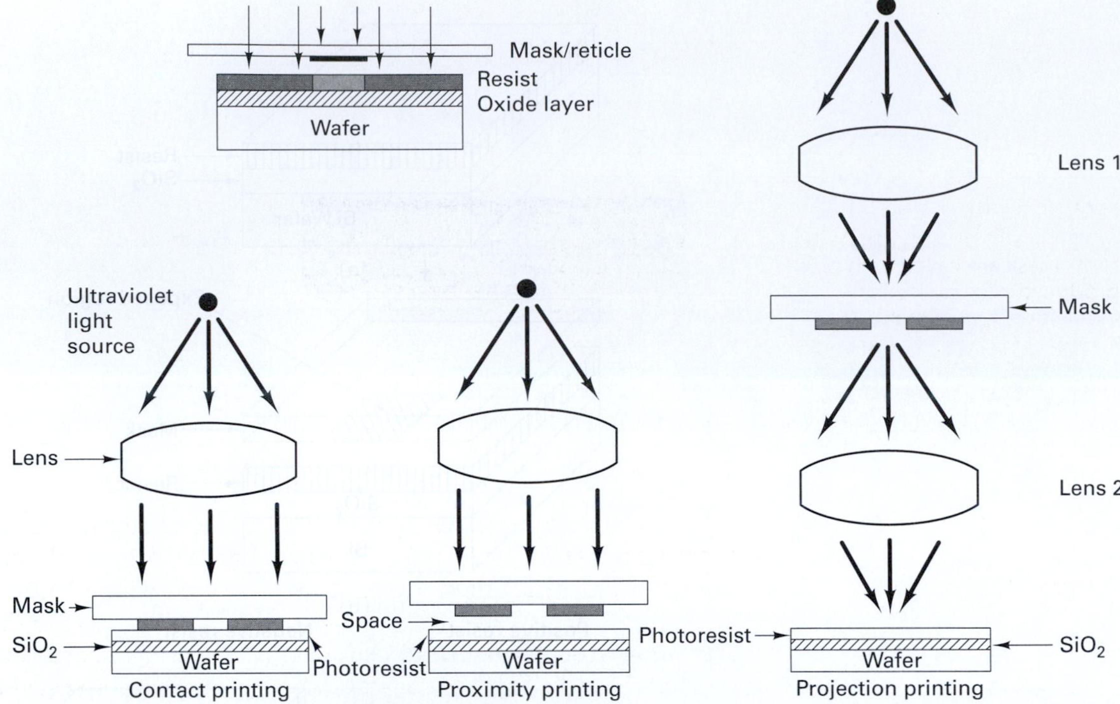

FIGURE 35-9 The exposure step in photolithography is shown in upper left with three primary exposure methods. *(R. C. Jaeger,* Introduction to Microelectronic Fabrication (Modular Series on Solid State Device Volume 5), *Addison-Wesley Publishing Company, New York, 1990; and P. V. Zant,* Microchip Fabrication: A Practical Guide to Semiconductor Processing, *New York: McGraw-Hill, 2000)*

Once the photomask has been accurately aligned with the pattern on the wafer's surface, the photoresist is exposed through the photomask with a high intensity ultraviolet light. Three primary exposure methods exist: contact printing, proximity printing, and projection printing, as shown in Figure 35-9.

In **contact printing,** the resist-coated silicon wafer is brought into physical contact with the photomask. The wafer is held on a vacuum chuck, and the whole assembly rises until the wafer and photomask contact each other. The photoresist is exposed with UV light, while the wafer is in contact position with the photomask. Because of the contact between the resist and photomask, very high resolution is possible in contact printing (e.g., 1-μm features in 0.5 μm of positive resist). The problem with contact printing is that debris trapped between the resist and the photomask can damage the photomask and cause defects in the resist mask.

Proximity printing is similar to contact printing except that a small gap, 1 to 25 μm wide, is maintained between the wafer and the photomask during exposure. This gap minimizes, but may not eliminate, resist mask damage entirely due to particles between the photomask and the wafer. Proximity printing offers higher throughput than the other methods but is limited in resolution. Approximately 2- to 4-μm resolution is possible with proximity printing.

Projection printing avoids photomask and resist-mask damage entirely. An image of the photomask is projected onto the resist-coated wafer, which can be many centimeters away. To achieve high resolution, only a small portion of the resist layer can be imaged—thus, the need to scan or *step* the small image over the surface of the wafer. Projection printers that step the photomask image over the wafer surface are called step-and-repeat systems, or steppers. Step-and-repeat projection printers are capable of submicron resolution.

After photolithography, the next step is the permanent removal of an underlying film or substrate by etching—by chemical or physical means or both. Typical materials etched during semiconductor processing include silicon dioxide to make diffusion masks, dielectric layers, and thin-metal films for device fabrication and interconnection. Typical etch rates in semiconductor processing are on the order of several hundred to several thousand Ångstrom per minute.

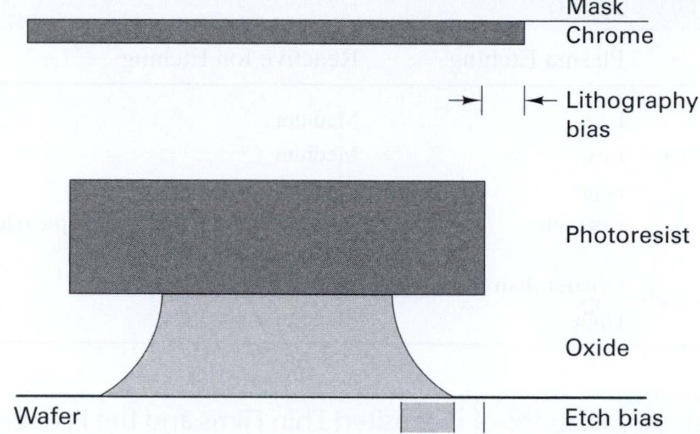

FIGURE 35-10 Deviations from the lateral dimensions in the resist mask are called etch bias, produced here by isotropic behavior of the etchant. *(S. A. Campbell,* The Science and Engineering of Microelectronic, *Oxford: Oxford University Press, 2001)*

The objective in etching is to produce the proper lateral dimensions of the IC in the target material while minimizing the removal of the mask and substrate (underlying the target) materials. Lateral dimensions in etching are controlled in large part by a resist (or perhaps an oxide) mask patterned as described in the lithography section. Deviations from the lateral dimensions in the resist mask are called **etch bias.**

Etching through the resist mask is typically accomplished by either wet chemical etching or dry plasma etching. **Wet etching** involves the immersion of the lithographically patterned wafer in a liquid etchant. The **etchant** removes material exposed through the resist mask, creating soluble by-products. A rinsing procedure is used to terminate the etching process. Critical parameters in wet-etching processes include immersion time, etchant concentration, and etchant temperature. Poor control of process parameters can cause underetching or overetching. Underetching of oxide masks may cause electrical opens in doped regions. In thin films, underetching can cause electrical shorts. Overetching results in etch bias due to undercutting of the mask. Undercutting is the lateral extent of the etch beneath the mask. Overetching can also cause damage to the properties of substrate materials or to the geometry of the resist mask, resulting in further bias.

One way to minimize damage in mask and substrate materials is to use an etchant with high **selectivity.** Selectivity of an etchant refers to the ratio of its etch rate in the target material to its etch rate in the mask or substrate material. As an example, hydrofluoric (HF) acid has a nearly infinite selectivity over silicon in the making of a diffusion mask. However, one disadvantage of using HF to etch is that the etch process is **isotropic,** meaning that it proceeds equally in all directions. As shown in Figure 35-10, isotropic etching results in an etch bias caused by undercutting.

Dry etching refers to those plasma-assisted etching techniques sometimes called gas-phase chemical etching. There are three main types of plasma-assisted etching, with the main difference being the gas pressure (vacuum) inside the plasma and, consequently, the kinetic energy generated by ions formed within the plasma. **Plasma etching** involves the use of a partially ionized gas (plasma) to chemically react with the target material surface, producing gaseous by-products.

At the opposite end of the dry-etching spectrum is **sputter etching** or **ion milling.** Sputter etching involves no chemical reaction with the target. Etching of target materials simply involves the physical removal of target atoms as electrostatically accelerated plasma ions slam into the target substrate. As such, sputter etching is the micromechanical equivalent of sandblasting. High-etch anisotropy is possible with sputter etching, meaning the etch is very directional with very little undercutting.

A cross between plasma etching and sputter etching is ion-assisted etching, better known as **reactive ion etching (RIE).** In RIE, plasma ions bombard the target material, creating physical damage, which increases the rate of chemical etching.

Table 35-3 compares the anisotropies, resist selectivities, and etch rates of the dry etching processes.

TABLE 35-3 Types of Dry Etching

	Plasma Etching	Reactive Ion Etching	Sputter Etching
Relative excitation energy	Low	Medium	High
Relative chance of radiation damage	Low	Medium	High
Relative selectivity	High	Medium	Low
Undercut, directionality	Isotropic	Directional from quasi-isotropic (slope) to anisotropic (vertical profile)	Highly anisotropic
Pressure (vacuum)	Greater than 100 mtorr	Approximately 100 mtorr	Less than 100 mtorr
Etch rate	High	Medium	Low

TABLE 35-4 Some Common Applications of Deposited Thin Films and the Processes Used to Make Them

	Process						
Function	VPE	MBE	APCVD	LPCVD	PECVD	Sputtering	Evaporation
Component Fabrication							
Growth of higher purity semiconductor for increased device performance	Single-crystal Si	Single-crystal GaAs					
Masking layer during oxidation			Si_3N_4				
Dopant source for diffusion				Doped polysilicon			
Metal and Dielectric Layers							
Dielectric layer for component fabrication			BPSG	SiO_2, SiO_3N_4			
Conduction path for component fabrication				Doped polysilicon			
Component Interconnection							
Dielectric layer for component interconnection			SiO_2, PSG		SiO_2, PSG	SiO_2, PSG	
Conduction path for component interconnection				W, TiN	TiN	Al, Cu	Al, Cu
IC Packaging							
Passivation of the IC after processing				SiO_2, PSG	SiO_2, Si_3N_4, PSG		

35.7 THIN-FILM DEPOSITION

In semiconductor processing, many thin layers of material must be deposited on top of the semiconductor substrate to build IC component features such as transistor gates and to interconnect IC components to form electrical circuits. These layers of material are often well below in thickness, and so the term *thin films* is used to describe them. Table 35-4 illustrates some typical ways that thin films are used in semiconductor processing. Various thin film deposition processes are necessary to accomplish these objectives. In general, deposition processes can be broken down into physical vapor deposition (PVD) and chemical vapor deposition (CVD) processes.

PHYSICAL VAPOR DEPOSITION

Physical vapor deposition (PVD) processes include both evaporation and sputtering. The simplest form of PVD is **evaporation.** In evaporation, the substrate is coated by condensation of a metal vapor.The vapor is formed from a source material called the **charge,** which is heated within a crucible in moderate vacuum (below 1 millitorr) at a high temperature (greater than 1000°C). Heat energy is provided by either electrical resistance or an electron beam. Electron-beam heating has the advantage of reducing contamination of the deposited thin film because it does not require crucible heating and, consequently, outgassing of the crucible.

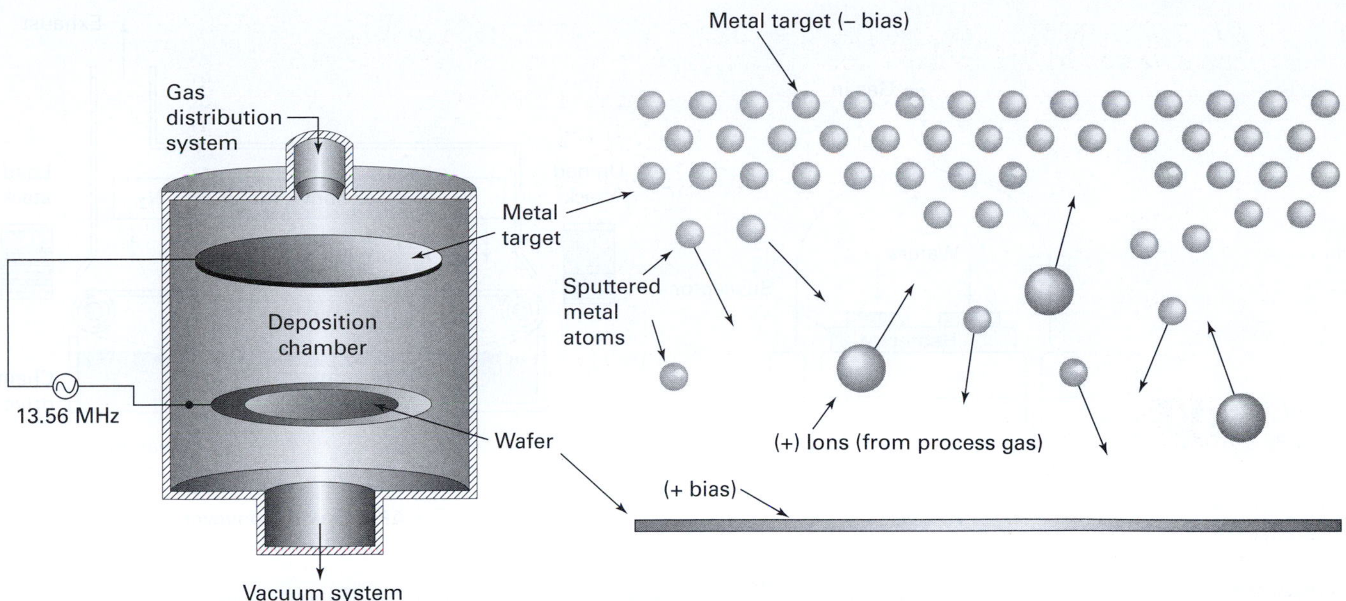

FIGURE 35-11 Sputtering is a PVD method for depositing thin films on microelectronic devices. *(Texas Engineering Extension Service, Semiconductor Processing Overview, College Station, TX: The Texas A&M University System, 1996)*

While early semiconductor technologies utilized evaporation, it is not used in mainstream processes today. The major disadvantage of evaporation is poor **step coverage.** Step coverage is important to avoid openings in wires that connect components during late-stage processing where the surface of the wafer can have severe topology as a result of the many deposition and etching steps. Step coverage has become even more important because the lateral dimensions of circuits continue to decrease with little change in vertical dimensions, resulting in device features with higher aspect ratios. Another disadvantage of evaporation is that it is limited primarily to the deposition of pure metals, although deposition of alloys can be accomplished with difficulty. Dielectrics and polysilicon cannot be deposited by evaporation. Finally, the uniformity of film thickness is hard to control over a large substrate.

An alternative PVD method for metal deposition is **sputtering.** The physics of sputtering are much the same as reactive ion etching. As shown in Figure 35-11, two electrodes are placed several centimeters apart in a low-pressure gas (typically, argon, at about 100 millitorr). A potential is placed across the electrodes forming a plasma. Plasma ions are accelerated toward the cathode on which is placed the charge material. At moderate ion energies, atoms and clusters of atoms are ejected from the charge material surface and accelerated toward the wafer.

One advantage of sputtering over evaporation is better step coverage due largely to greater transport energies leading to enhanced surface mobility of the atoms on the wafer. Further, sputtering can be performed on a wide range of materials including elemental metals, alloys, and dielectrics. For metals, a simple DC power source can be used to generate the plasma. For dielectrics, a radio frequency RF plasma is required. Due to its advantages, sputtering has replaced evaporation for most silicon-based technologies, although deposition rates for sputtering are lower than those for evaporation and require more expensive equipment.

CHEMICAL VAPOR DEPOSITION

Chemical vapor deposition (CVD) processes involve the growth of a thin film on a heated substrate by chemical reactions between the substrate and a gaseous compound containing reacting species. In general, CVD techniques provide the advantage of uniform step coverage and, therefore, have become the preferred deposition method for many materials. Figure 35-12 shows a simple configuration for an *atmospheric pressure CVD* (APCVD) system. The reactor consists of a tube with a heated susceptor on which the wafer rests. An inlet and outlet permits the flow of gasses over the surface of the

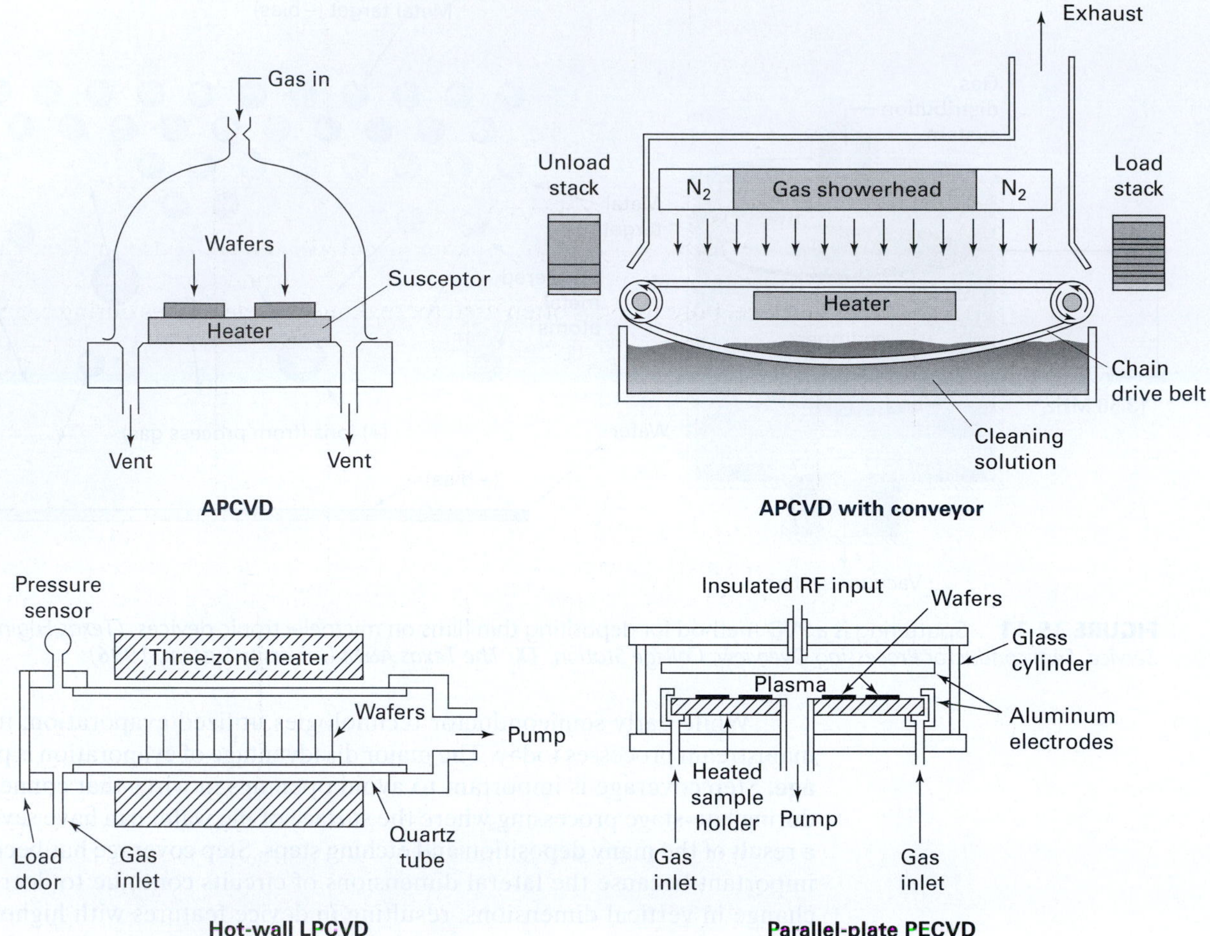

FIGURE 35-12 Chemical vapor deposition (CVD) processes include atmospheric pressure CVD (upper left); APCVD with conveyor (upper right); hot-wall, low-pressure CVD (lower left); cold-wall plasma enhanced CVD. *(S. A. Campbell,* The Science and Engineering of Microelectronic, *Oxford, U.K.: Oxford University Press, 2001; R. C. Jaeger,* Introduction to Microelectronic Fabrication (Modular Series on Solid State Device Volume 5), *New York: Addison-Wesley, 1990; and M. Madou,* Fundamentals of Microfabrication, *New York: CRC Press, 1997)*

wafer. A common thin film deposited in APCVD reactors is silicon dioxide used for passivating circuits. The chemical reaction for the deposition of silicon dioxide is:

$$SiH_4(g) + O_2(g) - (h) \rightarrow SiO_2(s) + 2H_2(g)$$

where the parenthetical entities represent gas, heat, and solid, respectively. In this reaction, silane and molecular oxygen enter the reactor at the inlet, silicon dioxide is deposited onto the wafer, and molecular hydrogen leaves the reactor at the outlet (along with any unused silane). Under proper conditions, this reaction takes place on the surface of the wafer at around 425°C.

APCVD can be performed at temperatures much lower than thermal oxidation, which has advantages in midstage processing of dielectrics. APCVD processes are also attractive because of high deposition rates and simple equipment design. To increase production, wafers can be conveyed through the APCVD reactor on a heated chain conveyor, fed one wafer at a time by multiwafer cassettes. In addition, APCVD can be used to deposit phosphorous-doped or phosphosilicate glass (PSG) otherwise known as **p-glass,** by adding phosphine to the reaction. P-glass can be used to smooth the wafer topology and getter wafer impurities, also during midstage processing. Problems with APCVD include impurities and poor control over film thickness.

Because CVD processes involve chemical reactions, one distinction from other processes involves the location of those reactions. Gas-phase (homogeneous) reactions resulting in solid particulates are generally undesirable because the ensuing particulate deposition produces thin films with poor morphology, increased contamination, and inconsistent properties. Chemical reactions on the wafer surface (heterogeneous reactions) result in the deposition of a solid, thin film with more uniform properties. In APCVD, homogeneous reactions are reduced by the introduction of diluent gases. However, greater control over gas-phase reactions can be obtained with the use of low gas pressures on the order of several hundred millitorr. This process is commonly called *low-pressure CVD* (LPCVD).

Most polycrystalline silicon, or **polysilicon,** is deposited through LPCVD. The chemical reaction for the deposition of polysilicon is carried out at 600°C on the wafer surface. Polysilicon is often used for making gate electrodes during component fabrication. While polysilicon can be deposited within an APCVD reactor, the uniformity of the film thickness is hard to control, which is problematic for gate electrodes. Therefore, polysilicon is normally deposited in an LPCVD reactor where uniformity is easier to control. Another benefit of LPCVD is that it consumes much less carrier gas, which reduces gas expense and handling.

To understand the reason for the improved process control in LPCVD reactors, it is important to differentiate CVD processes that are **reaction-rate limited** from those that are **mass-transport limited.** In reaction-rate limited processes, the deposition rate is controlled by the chemical reaction rate at the surface of the wafer. In contrast, mass-transport limited processes are controlled by the concentration of gases at the surface of the wafer. These two limits on process kinetics account for the major differences in CVD reactor designs.

Most LPCVD processes are reaction-rate limited. Because chemical reaction rates are heavily temperature dependent, thermal uniformity tends to be a design requirement for LPCVD reactors. However, at the low gas pressures in LPCVD reactors, there is more distance between molecules than in APCVD reactors, and consequently, there are fewer interactions between molecules. Therefore, it is more difficult to transfer energy between molecules and attain thermal equilibrium within LPCVD reactors. Because thermal equilibrium is hard to achieve within LPCVD reactors, most LPCVD reactors keep all surfaces within the reactor at the same temperature to minimize thermal gradients within the reactor. Because the walls of these reactors are heated, they are called hot-wall reactors. The ability to maintain thermal stability within hot-wall reactors is the reason for the improved process control of LPCVD reactors.

One disadvantage of hot-wall reactors is that the thin film is deposited along the walls of the reactor as well as on the wafer surface. Over time, these deposited films can flake off and contaminate the wafer surface. As a result, hot-wall LPCVD reactors must be dedicated to the growth of only one material, which reduces their flexibility. In some cases, cold-wall reactors can be used to reduce deposition on the walls. Cold-wall reactors have been used successfully to deposit tungsten for component interconnection, which has the advantage of reducing the size of metal contacts. However, cold-wall reactors do not permit the same level of temperature control and, therefore, do not permit the same level of deposition uniformity as hot-wall reactors.

In general, LPCVD reactors require higher capital expense due to their vacuum requirements and permit lower deposition rates than APCVD reactors. However, because LPCVD reactors typically are not mass-transport limited, wafers may be processed in higher densities within the reactor. Batch sizes in hot-wall LPCVD reactors may be as high as several hundred wafers, which more than makes up for the loss of deposition rate. With such large batch sizes, depletion of reacting species can cause variation in deposition rates from the front to the back of the reactor. This can be accommodated by setting up a temperature gradient from the front to the back of the reactor, resulting in higher chemical reaction rates in the back of the reactor.

In other CVD processes where the deposition rate is mass-transport limited, the major design requirement of reactors is to permit uniform transport of reactant gasses to all parts of the wafers. As a result, these reactors have geometries optimized for gas flow and have excellent gas flow controls. One example of this was shown in the conveyorized APCVD reactor in Figure 35-12. In such reactors, the natural convection of the reactant gas between the wafer surface and the cold walls of the reactor can cause circulation of the gas and make it difficult to control the concentration of the gas at the wafer surface. Consequently, this can make the deposition uniformity between wafers hard to control. These problems can be addressed with proper flow design of the reactor and appropriate parametric design.

Many integrated electronic features require the deposition accuracy of LPCVD. However, many LPCVD processes require high temperatures that are not compatible with late stage processing, because high temperatures will cause diffusion of previously deposited material layers. *Plasma-enhanced CVD* (PECVD) has been used effectively to lower the processing temperatures needed to sustain the necessary chemical reactions. One manifestation of a cold-wall PECVD reactor is shown in Figure 35-12. In this case, the wafer rests between electrodes through which an AC potential is applied at radio frequency.

One application of PECVD is in passivating the IC after processing, achieved by depositing silicon nitride as a durable, inert coating to protect the circuit from moisture and scratches. A chemical reaction for sustaining silicon nitride passivation is:

$$3SiH_4(g) + 4NH_3(g) - (h) \rightarrow Si_3N_4(s) + 12H_2(g)$$

An LPCVD reactor would drive this process at 900°C, but by substituting dichlorosilane, the temperature can be driven lower (700 to 900°C). PECVD can drive this reaction at 300 to 400°C. Both hot-wall and cold-wall PECVD reactors have been used for silicon nitride passivation.

Issues with PECVD include low deposition rates, poor throughput, poorer step coverage, and more impurities than LPCVD. Deposition uniformity can be a concern in hot-wall reactors due to gas depletion. Some effort has been made to improve the throughput of cold-wall reactors by permitting multiple process steps to be performed in a single vacuum chamber.

EPITAXIAL GROWTH

In some cases, it is desirable to deposit a thin film of single-crystal semiconductor material onto the silicon wafer prior to semiconductor processing. **Epitaxy** is the growth of a single crystal, thin film of identical crystallographic orientation as the surface on which it is grown. Epitaxy is used for the purpose of improving semiconductor properties or fabricating abrupt transitions between doped layers that would otherwise be hard to form by diffusion or ion implantation. Epitaxial layers, or epi-layers, are grown under tighter specifications than bulk single crystals, resulting in fewer crystal defects, higher purity, more uniform dopant distributions, and sharper transitions between doped layers. In early semiconductor processing, *n*-type epilayers were grown on top of *p*-type substrates for standard buried-collector bipolar processing. Due to the high temperatures involved (leading to the solid-state diffusion of dopants), current mainstream silicon manufacturing uses epitaxial deposition mainly for thick layers (1 to 10 μm) from which devices may be fabricated.

The mainstream method for silicon epitaxial deposition is **vapor-phase epitaxy (VPE).** VPE is an extension of LPCVD. One difference is that the VPE of single crystal silicon is performed at higher temperatures (1000°C) than the LPCVD of polysilicon. Because surface reaction rates are much faster at higher temperatures, VPE reactions tend to be mass-transport limited, and as a result, the reactor is designed to optimize gas flow to the wafer. Many other CVD techniques are able to produce semiconductor epitaxial growth, including ultra-high vacuum and laser, optical, and X-ray-assisted CVD. However, VPE is the mainstream epitaxial method used in silicon fabrications. Non-CVD methods (liquid phase, molecular beam, ion beam, and clustered ion beam epitaxy) have been found more important for depositing compound semiconductors such as galium arsonite (GaAs). CVD techniques such as metallorganic CVD are beginning to show promise for GaAs as well.

INTEGRATED CIRCUIT COMPONENT INTERCONNECTION

Up to this point, most of the discussion has focused on individual IC components. The components must be *interconnected* (connected together) with the use of thin-film

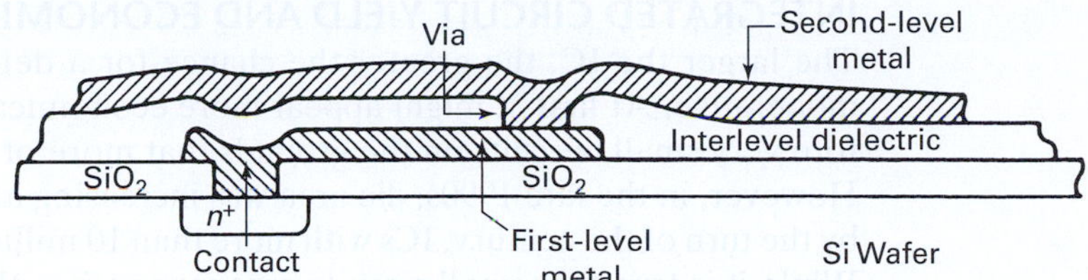

FIGURE 35-13 Schematic of a two-level metal interconnect structure typical of metallization process. *(R. C. Jaeger, Introduction to Microelectronic Fabrication (Modular Series on Solid State Device Volume 5), New York: Addison-Wesley, 1990)*

metal wires between thin-film dielectric layers. The deposition of thin film metals and dielectrics for component interconnection is called **metallization.**

As shown in Figure 35-13, the geometry of metallization involves metal wires, or **lines,** in between layers of dielectric. In metallization, metal deposition has conventionally been accomplished by PVD processes, whereas dielectric deposition has been accomplished by either PECVD or sputtering. In order to interconnect the IC components, access must be made between the first level of lines and the semiconductor. These access points are called **contacts.** To permit lines to cross over one another, access must be made between each layer of lines. These access points are called *vias* (discussed in more detail later). Holes for contacts and vias are produced by pattern transfer with photolithography and etching steps.

Modern ICs typically have between one and six metallization layers. During metallization, as each layer of metal and dielectric is deposited and etched on top of another, the topology of the wafer can become quite severe. If the topology of the wafer becomes too pronounced, the result can be shorts in metal wires due to poor step coverage during metal deposition. Further, as the feature resolution of photolithography exposure systems continues to improve, the depth of focus of these systems decreases, making photolithography on uneven topology difficult. To avoid these problems, interlayer dielectric layers must be *planarized* (made flat). As mentioned earlier, one method of reducing topology is to use p-glass as a dielectric interlayer, because it flows at a relatively low temperature and smooths out peaks and valleys. Another method is to perform an etch on the dielectric, which will referentially attack the high points on the surface. For devices with less than (transistor) gate thicknesses involving a large number of metallization layers, **chemical mechanical polishing (CMP)** has become the standard in planarization. CMP is similar to conventional mechanical polishing with a wet abrasive slurry, except that the slurry contains a chemical etchant as well. This process is discussed in more detail in Chapter 26.

Metals used for interconnection are those that exhibit low electrical resistance and good adhesion to dielectric insulating layers. Aluminum is a popular metal for interconnecting IC components. Small amounts of copper may be added to reduce the potential for **electromigration** effects in which the applied current to the device can induce the undesirable mass transport of metal atoms over time. As the device and wire sizes continue to decrease, electromigration, which can result in short or open circuits, is becoming a bigger problem. In addition to aluminum alloys, other alloys and pure metals such as tungsten, titanium, and copper are being considered for metallization for their ability to resist electromigration. Other key characteristics of deposited metal films include low film reflectivity (to reduce interference with optical alignment during photolithography) and low residual stress.

To finish component interconnection, a series of connectivity and functional tests called **wafer testing** are performed on each separate IC on the wafer, and the wafer is cut into individual ICs called **dies** or *chips*. The purpose of wafer testing is to eliminate any unnecessary packaging of defective ICs. At this stage, the wafer has an array of ICs that has been produced on it. Wafer testing is performed by computer-controlled probing equipment that introduces electrical signals into each IC by contacting each set of bonding pads on the wafer with needle-like probes. The multiprobe procedure involves indexing each IC under a probe head, which has a probe for each bonding pad. ICs that fail the test are marked with an ink dot and discarded. After testing, the wafer is diced into individual dies or chips. Wafer dicing is typically performed by **diamond sawing** to give clean edges with minimal damage.

INTEGRATED CIRCUIT YIELD AND ECONOMICS

The larger the IC, the greater the chance for a defect to appear and render the IC inoperative. At first. it might appear more economical to build very simple, and therefore very small, circuits on the grounds that more of them would likely be functional. However, in the late 1990s, die area was increasing at a rate of about 12% per year, so by the turn of the century, ICs with more than 10 million transistors had been produced. While it is true that small circuits are inexpensive, the cost of packaging, testing, and assembling the completed circuits into an electronic system must be taken into account. Once the ICs are separated into individual chips, each chip must be handled individually. From that point on, the cost of any processing is not spread over hundreds or thousands. Thus, packaging and testing costs often dominate the other production costs in the fabrication of ICs.

One way to improve the economics of microelectronic manufacturing is to increase wafer sizes. The key benefit from processing larger wafers is an increase in the percentage of usable area. Larger wafers have a smaller proportion of the area being affected by edge losses and wafer dicing. Since the mid-1980s, wafer diameters have increased threefold from 100 to 300 mm, which required the development of new equipment throughout the semiconductor manufacturing process. A second strategy for improving semiconductor economics involved increasing the number of chips per wafer by decreasing IC dimensions. IC dimensions have decreased more than 50-fold in the past 30 years. The smallest feature size in 1971 was 10 μm. By 2001, transistors with gate features as small as 0.18 μm were made. Again, the catalyst for this improvement was an investment in the process technology, in particular, photolithography.

Perhaps the most effective method of improving IC economics has been the improvement in **die yield.** Die yield improvement is much more desirable because considerable improvements in economics can be had without making large capital investments. The die yield depends on the **wafer yield** (the fraction of silicon wafers that started versus those that finished the process), which involves the **processing yield** (the fraction of good die per wafer), the **assembly yield** (the fraction of die that are packaged), and the **burn-in yield** (the fraction of packaged die that survives wafer testing). The largest contributors to lower yields are generally the wafer and processing yields. Wafer yields are driven by large-area defects, which might be the result of poor process control in deposition or etching that would eliminate the usefulness of the entire wafer. Processing yields are generally driven by point defects such as particle contamination, although large area defects can also affect processing yields.

A single, submicron dust particle trapped between the photoresist and reticle in a photolithographic step can cause a point defect that will result in the malfunction of an entire IC. As a result, all microelectronic manufacturing is conducted in **clean rooms,** where special clothing must be worn to prevent dust particle contamination of wafers being processed. The air is continuously filtered and recirculated using high-efficiency particulate-arresting (HEPA) and ultra-HEPA (ULPA) filters to keep the dust level at a minimum. Clean rooms are specified by their class of cleanliness with respect to federal standard 209D. Class 100,000 indicates that the filtration in the clean room limits the number of 0.5-μm diameter particles to 100,000/ft^3 volume. Wafers are commonly processed in Class 100 clean rooms.

35.8 INTEGRATED CIRCUIT PACKAGING

Several levels of packaging and assembly are necessary to integrate the IC chip with other electronic devices to make it part of a fully functional commercial or military product. IC packaging serves to distribute electronic signals and power as well as provide mechanical interfacing to test equipment and printed circuit boards (PCBs). In addition to this interconnection role, IC packages protect the delicate circuitry from mechanical stresses and electrostatic discharge during handling and corrosive environments during its operational life. Finally, because of the high density of the integrated circuits, dissipation of heat generated in the circuits has become more critical.

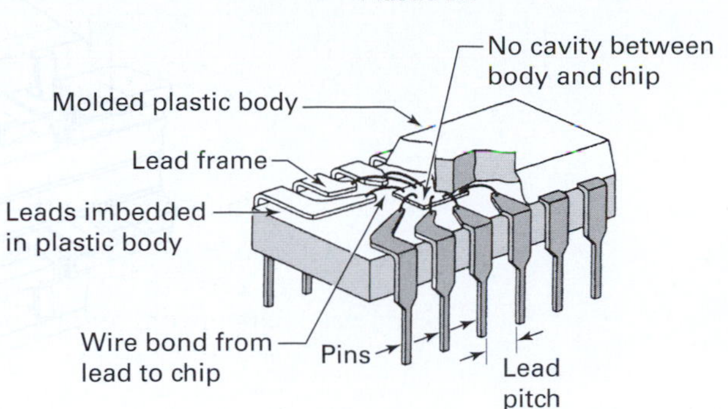

FIGURE 35-14 The dual-in-line package (DIP) has a lead frame and package body. The leads on the chips are connected to the pins. *(D. P. Seraphim, R. C. Lasky, and C.-Y. Li,* Principles of Electronic Packaging, *New York: McGraw-Hill, 1989)*

PACKAGE TYPES

ICs come in a variety of packages made from a variety of materials. Figure 35-14 shows a cutaway view of the most well-known IC chip package; the **dual in-line package (DIP)** refers to the two sets of in-line pins that go into holes in the PCB. The DIP, like all other IC packages, is made up of a lead frame and a package body. Typically composed of a copper alloy (sometimes with an aluminum coating), the lead frame provides electrical interface between the IC and the PCB. The DIP body is made from a low-cost epoxy, which facilitates mass production. In high-reliability applications (e.g., military), where hermetic (air-tight) sealing of the package is important, ceramic package bodies are used.

Generally, IC packages are grouped mainly based on the arrangement, shape, and quantity of leads. Lead pitch refers to the center-to-center distance between leads on an IC package. In conformance to standard-setting bodies, such as the Electronics Industries Association (EIA) in the United States and EIA Japan, lead pitches above 20 mils (0.02 in.) are measured in inches. Below 20 mils, lead pitches are measured in millimeters.

There are two methods by which components are connected to the circuit on the PCB. The DIP is the leading example of **through-hole (TH)** technology, also known as **pin-in-hole (PIH)** technology, where IC packages and discrete components are inserted into metal-plated holes in the PCB and soldered from the underside of the PCB. In **surface mount (SM)** technology, electronic components are placed onto solder paste pads that have been dispensed onto the surface of the PCB. Figure 35-15 shows the cross section of solder joints for typical SM- and TH-packaged components on a PCB.

SM packages are more cost effective in electronic assembly, and this SM technology has replaced a lot of the TH technology, but not entirely, because not all electronic components can be purchased in an SM package. SM packages are designed for automated production and allow for higher circuit board density than TH components. The manufacturing challenges associated with SM technology include weaker joint strength and solderabilty issues relating to lower in-process lead temperatures. Also, TH components have only one lead geometry, whereas SM components have many different designs. The key packaging families for TH technology are dual in-line packages (DIPs) and **pin grid arrays (PGAs).**

In SM technology, IC packages cannot be discussed separately from lead geometry. Lead geometries affect the electrical performance, size constraints on the PCB, and ease of assembly of the IC package. The most basic form of SM lead is the **butt lead,** or I-lead (see Figure 35-16). Butt leads are normally formed by clipping the leads on the TH component. This technique is sometimes used to convert an existing TH component to an SM component. Consequently, butt-leaded components do not typically save any space on the PCB. However, they can reduce costs by eliminating the need to perform TH soldering of the PCB after SM soldering. Butt-lead components tend to result in the lowest solder joint strengths, and therefore, reliability is an issue.

Gull-wing leads bend down and out, whereas **J-leads** bend down and in. Gull-wing leads allow for thinner package sizes and smaller leads, which is important for compact applications such as laptop computers. In addition, packages with gull-wing leads are compatible with most reflow soldering processes and have the ability to self-align

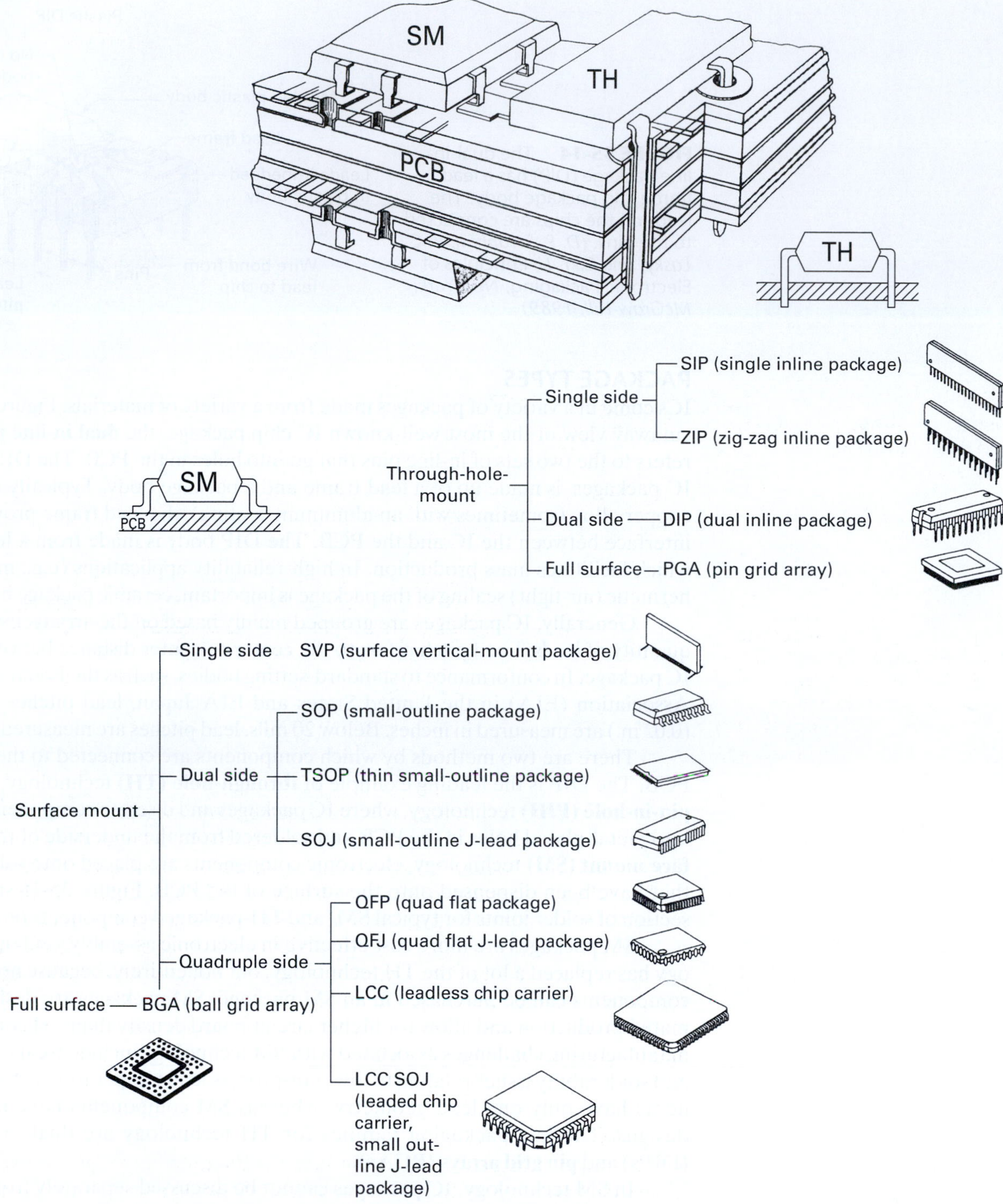

FIGURE 35-15 Here is a summary of the various types of packaging used for ICs. *(L. T. Manzione,* Plastic Packaging of Microelectronic Devices, *New York: Van Nostrand Reinhold, 1990)*

during reflow if they are slightly misoriented. Gull-wing leads are compatible with fine pitch packages, but inspection of solder joints is difficult in its final soldered configuration. Gull-wing leads are also susceptibile to lead damage and deviation from lead coplanarity. J-leads are sturdier than gull wings and stand up better in handling. The solder joint of J-leads face out, making inspection easier. J-leads have a higher profile than gull wings, which can be a disadvantage for compact applications. At the same time, this higher standoff makes postsolder cleaning easier. J-leads can be used for packages with between 20 and 84 leads.

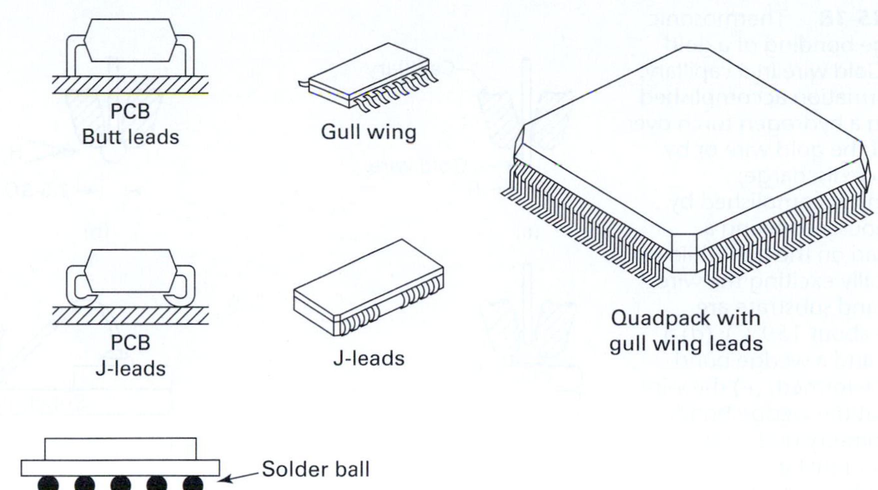

FIGURE 35-16 Basic lead geometries for surface mounted packages include butt leads, gull-wing leads, J-leads, solder balls, and plastic quad flat pack.

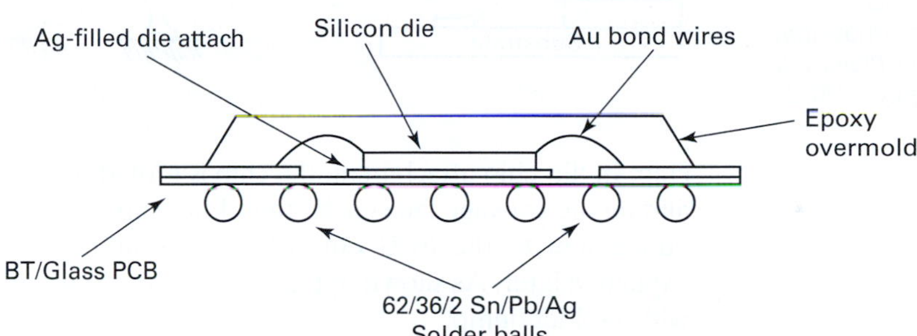

FIGURE 35-17 Ball grid arrays (BGAs) provide high numbers of connections for leads using solder balls arranged across the entire bottom of the package. *(R. Prasad,* Surface Mount Technology. Principles and Practice, *New York: Chapman & Hall, 1997, p. 493)*

Solder balls are increasingly being used to provide SM interconnection through **ball grid arrays (BGAs).** Figure 35-17 shows a BGA package. BGAs provide high lead density because the solder balls are arrayed across the entire bottom surface of the package. Lead counts on BGAs can go as high as 2400, with most in the 200 to 500 lead range. Because of their arrayed nature, BGAs do not need as fine of a pitch (40 to 50 mils) as **quad flat packages (QFPs),** which can help in electronic assembly yields. To further boost yield, the solder balls on BGAs have excellent self-aligning capability during reflow and require less coplanarity (6 to 8 mils) than other leads. The downside of BGAs is the difficulty associated with cleaning, inspection, and rework of solder joints and the lack of compatibility with some reflow methods because joints are out of sight beneath the package.

PACKAGING PROCESSES

The first step in IC packaging is to attach the die to the package. Die attachment techniques include **wire bonding, tape-automated bonding (TAB),** and **flip-chip** technology. In wire bonding, also known as **chip-and-wire attachment,** the chip is attached to the package with an adhesive, and a wire is attached to bonding pads on the chip and on the package. Gold wire as thin as 25 μm and aluminum wire as thin as 50 μm can be attached in wire bonding. As shown in Figure 35-18 for gold wire, the ball bond at the die pad is formed by melting the wire tip and compressing it against the die pad. After die pad bonding, the wire is then looped out and ultrasonically or thermosonically welded to the lead frame of the package. In ultrasonic welding, frictional energy, caused by placing the vibrating wire in contact with the lead frame, causes heating, melting, and coalescence of the two materials. Thermosonic welding is ultrasonic welding with the addition of heat.

In TAB attachment, a thin polymer tape carrying the lead circuitry (see Figure 35-19) is aligned with the die, and the leads are bonded under temperature and pressure to the IC

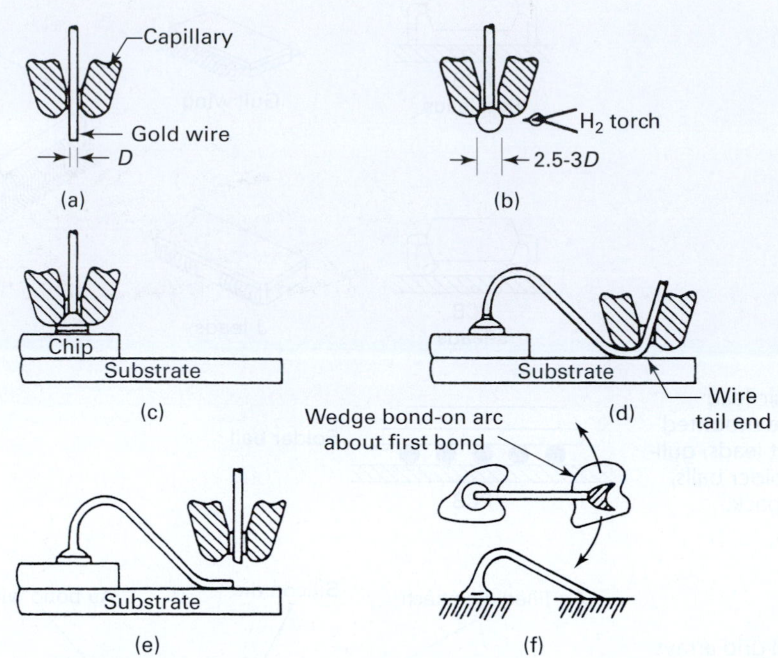

FIGURE 35-18 Thermosonic ball-wedge bonding of a gold wire. (a) Gold wire in a capillary; (b) ball formation accomplished by passing a hydrogen torch over the end of the gold wire or by capacitance discharge; (c) bonding accomplished by simultaneously applying a vertical load on the ball while ultrasonically exciting the wire (the chip and substrate are heated to about 150°C); (d) a wire loop and a wedge bond ready to be formed; (e) the wire is broken at the wedge bond; (f) the geometry of the ball-wedge bond that allows high-speed bonding. Because the wedge can be on an arc from the ball, the bond head or package table does not have to rotate to form the wedge bond. *(Semiconductor International magazine, May, Des Plaines, IL: Cahners Publishing Co., 1982)*

chip. In flip-chip attachment, the chip is turned over so that the bonding pads on the chip and on the package face each other. Flip-chip technology is more common for direct chip attachment to the PCB but is becoming more important for chip-scale packages, as explained later. As shown in Figure 35-20, flip-chips are normally attached to the package with a solder bump.

After die attachment, the package is sealed. Plastic packages are either premolded or postmolded. Premolded packages are sealed adhesively with a lid (Figure 35-21). Postmolded packages are sealed via a transfer molding or injection molding process (Figure 35-21). Prior to molding, the die is adhered and wire bonded to the lead frame, which is automatically inserted into the mold. The postmolding process is relatively harsh on the die and wire bonds and can cause major yield and reliability problems. To keep out environmental contaminates, ceramic packages are hermetically sealed by glass using either eutectic AuSi or silver-loaded glass adhesive technologies.

Once the package is sealed, leads are typically formed and may require a solder dip. BGA packages differ from the other packages in that the BGA is interconnected through the use of laminated substrates (plastic or ceramic) similar to PCB processing instead of through leadframes. In BGAs, the outermost layer of interconnection is covered with a solder mask, and openings in the solder mask allow for solder ball attachment during solder dipping.

A more advanced option for interconnecting ICs with PCBs is called **direct chip attachment (DCA),** also known as **chip-on-board** or **direct die mounting.** As suggested, DCA directly attaches the chip to the board using any of three die-attachment technologies mentioned previously. On paper, flip-chip technology has the greatest potential for DCA. However, one challenge associated with flip-chip technology is the coefficient of thermal expansion (CTE) mismatch between the chip and the PCB substrate, particularly as chip sizes increase and solder joint sizes decrease. As a result, the development of underfill encapsulants has become increasingly important for the reinforcement of the mechanical and thermal properties of flip-chip solder joints.

Disadvantages of DCA technologies include shipping and handling of the bare chip and the need for electronic assembly manufacturers to purchase die-attachment equipment. As a compromise, **chip scale packaging (CSP)** has been developed to help downstream processes take advantage of DCA technology. CSP is defined as any packaging that adds no more than 20% of additional board area to the chip. The micro BGA (MBGA) package shown in Figure 35-22 is one example of CSP.

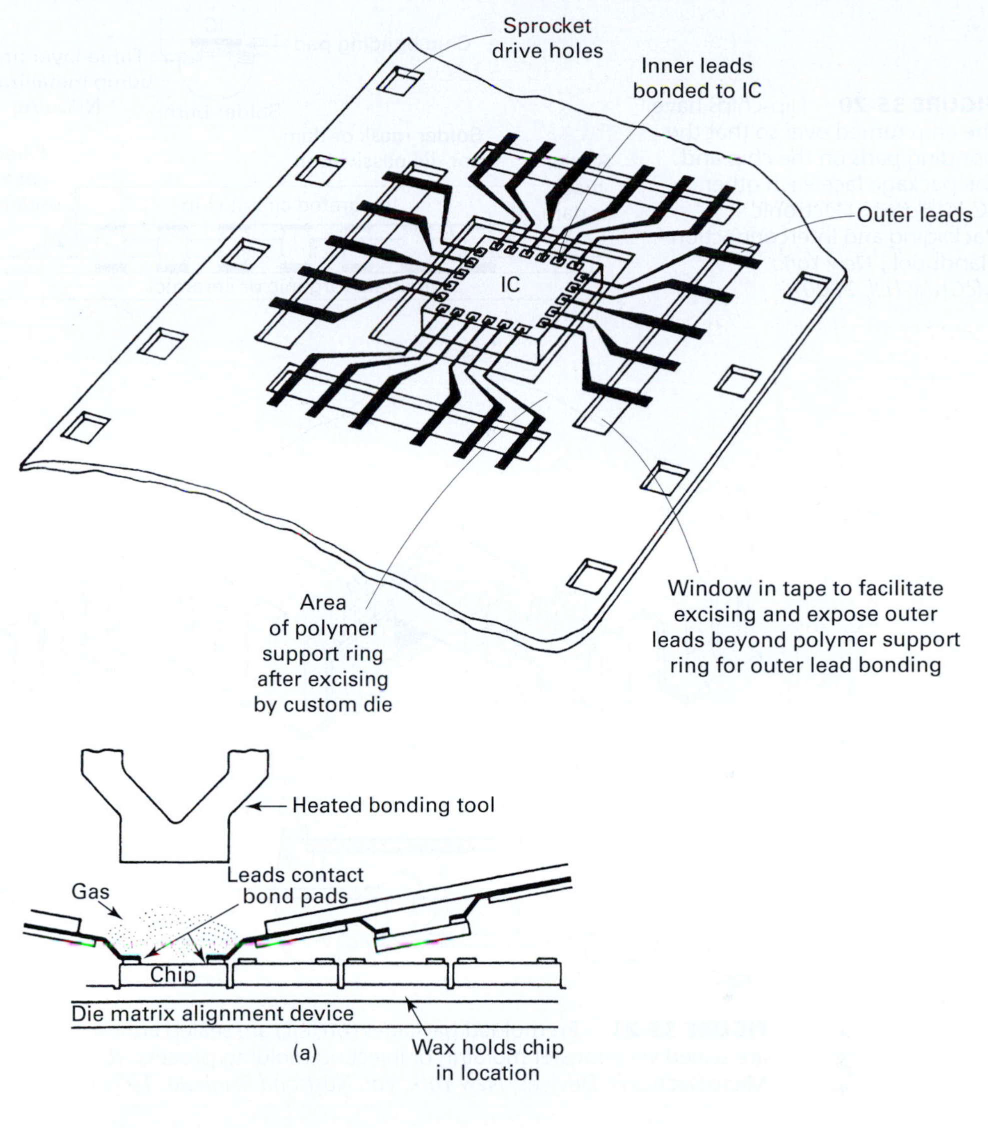

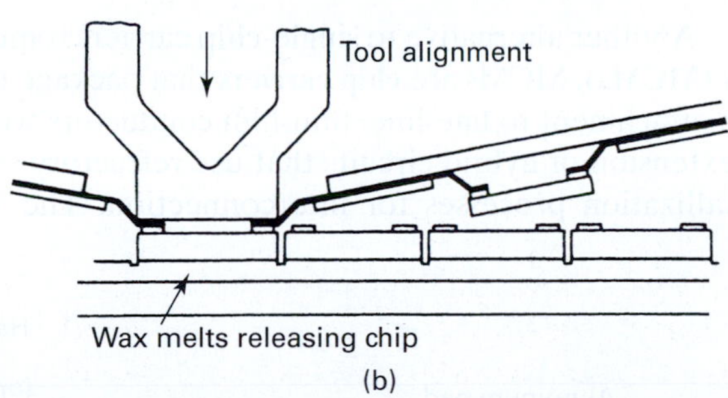

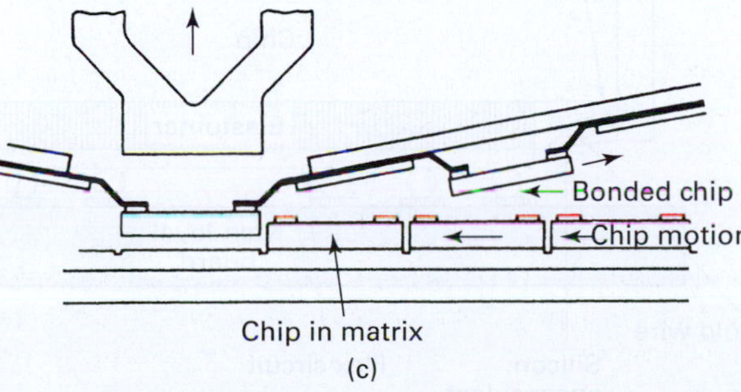

FIGURE 35-19 Tape-automated bonding (TAB) uses a polymer type to carry the leads to the chip for bonding. *(R. C. Jaeger,* Introduction to Microelectronic Fabrication (Modular Series on Solid State Device Volume 5), *New York: Addison-Wesley, 1990)*

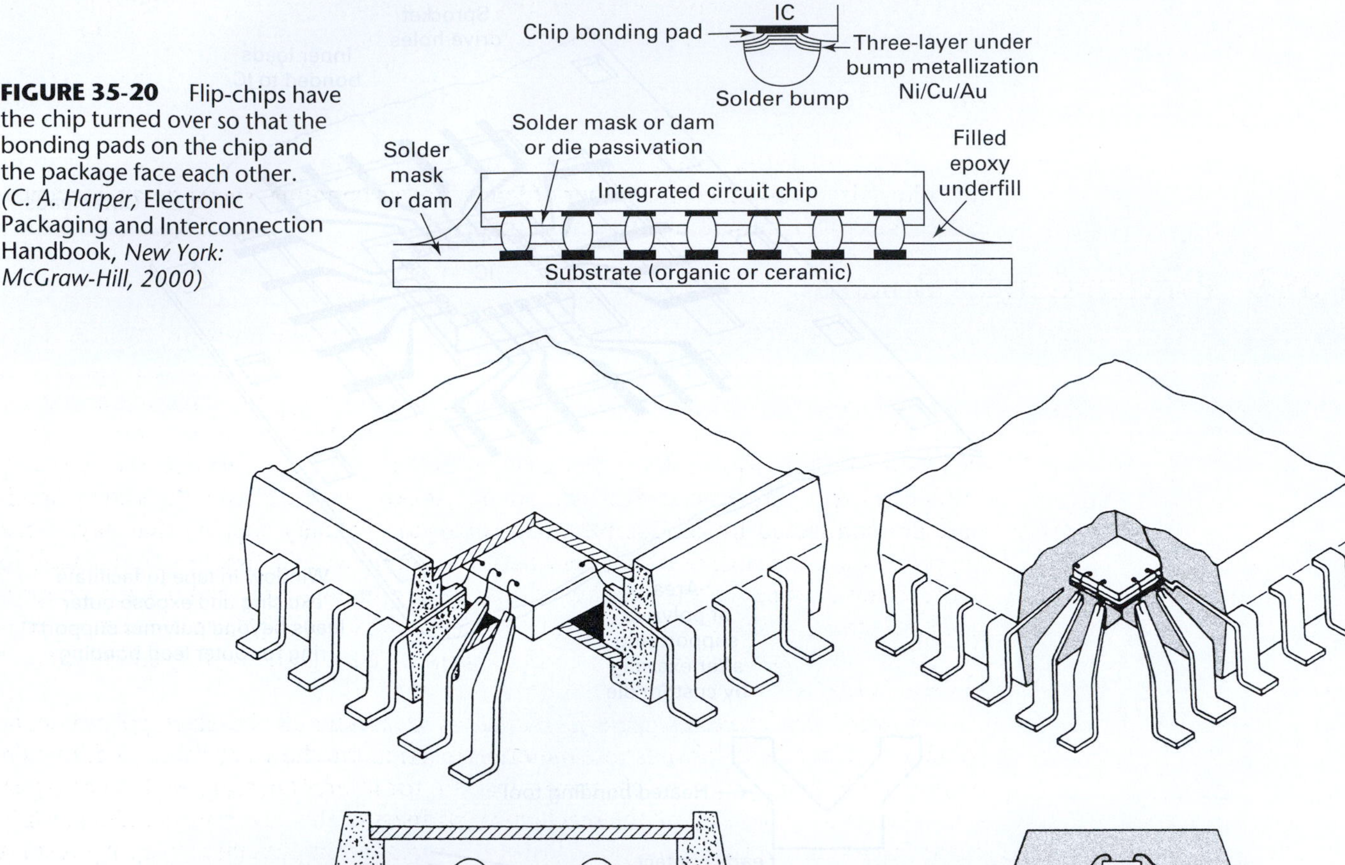

FIGURE 35-20 Flip-chips have the chip turned over so that the bonding pads on the chip and the package face each other. *(C. A. Harper,* Electronic Packaging and Interconnection Handbook, *New York: McGraw-Hill, 2000)*

FIGURE 35-21 Premolded packages (on left) are sealed adhesively with a lid while postmolded packages are sealed via atransfer molding or injecting molding process. *(L. T. Manzione,* Plastic Packaging of Microelectronic Devices, *New York: Van Nostrand Reinhold, 1990)*

Another alternative to single-chip carriers is multichip carriers or **multichip modules (MCMs).** MCMs are **chip carriers** that package more than one chip through direct chip attachment to fine-line, thin-film conductors within a ceramic carrier. MCMs are an extension of hybrid circuits that use refractory substrates and thick- and thin-film metallization processes for interconnection. The MCM is essentially a mini-PCB

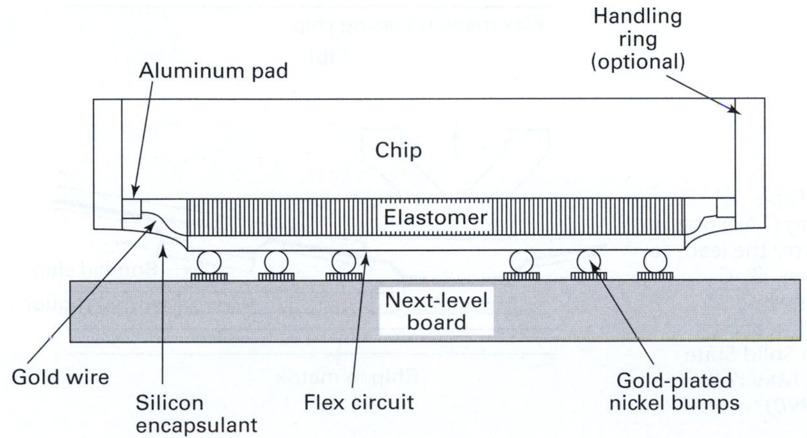

FIGURE 35-22 The Tessera micro BGA package is an example of chip scale packaging. *(R. Prasad,* Surface Mount Technology. Principles and Practice, *New York: Chapman & Hall, 1997, p. 493)*

with DCA interconnection between the chips and the board. Usually, the die is mounted in the MCM with flip-chip technology. The major advantage of MCMs is the reduction in electronic, single-path distance between ICs. The MCM replaces the typical die-wirebond-pin-board-pin-wirebond-die path with a much shorter die-bump-wire-bump-die path.

Selection of the final chip package for an IC device depends on several factors, including size, weight, cost, number of leads, power handling, signal delay, electrical noise, and cooling requirements, among others.

35.9 PRINTED CIRCUIT BOARDS

The printed circuit board (PCB), or printed wiring board, connects the IC with other components to produce a functional circuit. Specifically, a PCB is a laminated set of dielectric layers or laminates of bulk sheet materials that have metallic circuits that are used to interconnect the various packaged components. As shown in Figure 35-23, each PCB laminate comprises a base, tracks, and pads. The base material must be electrically insulating to provide support to all components making up the circuit. Pads on the laminate are connected by conductive tracks or traces (usually copper) that have been deposited onto the surface of the base. Screen printing was the first technology used to make circuits, hence the term *printed* circuits. Today, metal for traces and pads is deposited by electroless plating and electroplating. Surface mount (SM) components are connected to the PCB at pads (lands) or, in the case of through-hole (TH) technology, at insertion holes.

Typical base materials used may be epoxy-impregnated fiberglass, polyimide, or ceramic. Criteria used for substrate material selection are shown in Table 35-5. Epoxy-impregnated fiberglass is the cheapest substrate for interconnecting leaded packages. Fiberglass is used to increase the mechanical stiffness of the device for handling, while epoxy resin imparts better ductility. Prior to impregnation, the uncured epoxy resin is referred to as **A-stage.** The fiberglass is impregnated on a continuous line where A-stage resin infiltrates the fiberglass mat in a dip basin, and the soaked fabric passes through a set of rollers to control thickness and an oven where the resin is partially cured (Figure 35-24). The resulting glass-resin sheet is called **B-stage** or **prepreg.** Multiple prepregs are then pressed together between electroformed copper foil under precise heat and pressure conditions to form a copper-clad laminate. The fully cured glass-epoxy core is called **C-stage** material.

Many different types of epoxy-impregnated fiberglass exist, as shown in Table 35-6. FR-4 (flame retardant) and G-10 are the most popular PCB substrates in use today. Polyimide (without reinforcing fiberglass) is also now widely used in consumer products. PCBs using polyimide substrates are known as flexible printed circuits, or simply **flex circuits,** emphasizing the lack of rigidity. Flex circuits offer the advantages of reduced size and weight as well as the ability to route printed circuits around corners or other non-planar geometry. Polyester is also used as a flex circuit substrate, although polyimide is the most popular due to its high temperature stability.

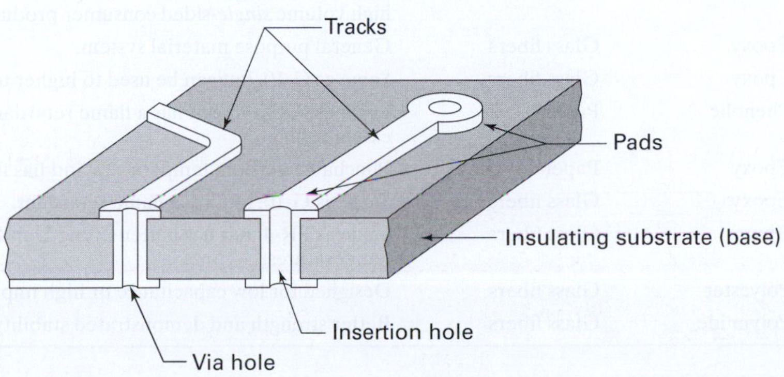

FIGURE 35-23 Double-sided PCB laminate has a base, tracks, and pads.

TABLE 35-5 Substrate Selection Criteria

	Material Properties								
Design Parameters	Transition Temperature	Coefficient of Thermal Expansion	Thermal Conductivity	Tensile Modulus	Flexural Modulus	Dirlectric Constant	Volume Resistivity	Surface Resistivity	Moisture Absorption
Temperature and Power Cycling	X	X	X	X					
Vibration				X	X				
Mechanical Shock				X	X				
Temperature and Humidity	X	X				X	X	X	X
Power Density	X		X						
Chip Carrier Size		X		X					
Circuit Density						X	X	X	
Circuit Speed						X	X	X	

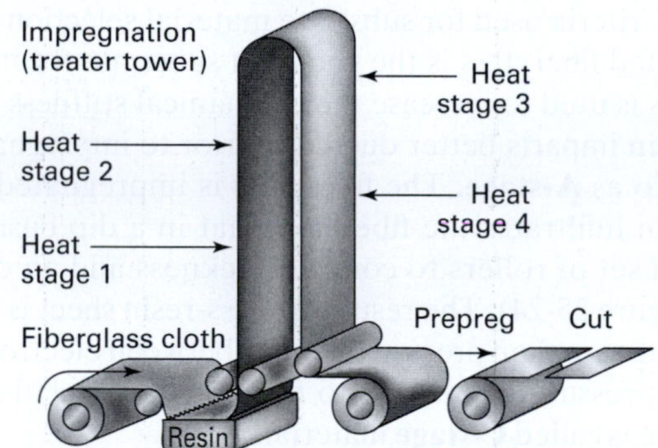

FIGURE 35-24 The PCB is made up of laminates of epoxy-impregnated fiberglass manufactured on a machine like this. *(D. P. Seraphim, R. C. Lasky, and C.-Y. Li,* Principles of Electronic Packaging, *New York: McGraw-Hill, 1989)*

TABLE 35-6 Laminate Materials Used in Printed Circuit Boards

Common Designation	Resin System	Base Material	Description
XXXP	Phenolic	Paper	Punchable at room temperature.
XXXPC	Phenolic	Paper	Punchable at or above room temperature. XXXP and XXXPC are widely used in high volume *single*-sided consumer products.
G-10	Epoxy	Glass fibers	General purpose material system.
G-11	Epoxy	Glass fibers	Same as G-10, but can be used to higher temperatures.
FR-2	Phenolic	Paper	Same as XXXPC, but has a flame retardant (FR) system that renders it self-extinguishing.
FR-3	Epoxy	Paper	Punchable at room temperature and has flame retardant.
FR-4	Epoxy	Glass fibers	Same as G-10, but has a flame retardant.
FR-5	Epoxy	Glass fibers	Same as FR-4, but has better strength and electrical properties at higher temperatures.
FR-6	Polyester	Glass fibers	Designed for low capacitance or high impact resistance; has flame retardant.
Polyimide	Polyimide	Glass fibers	Better strength and demonstrated stability to a higher temperature than FR-4.

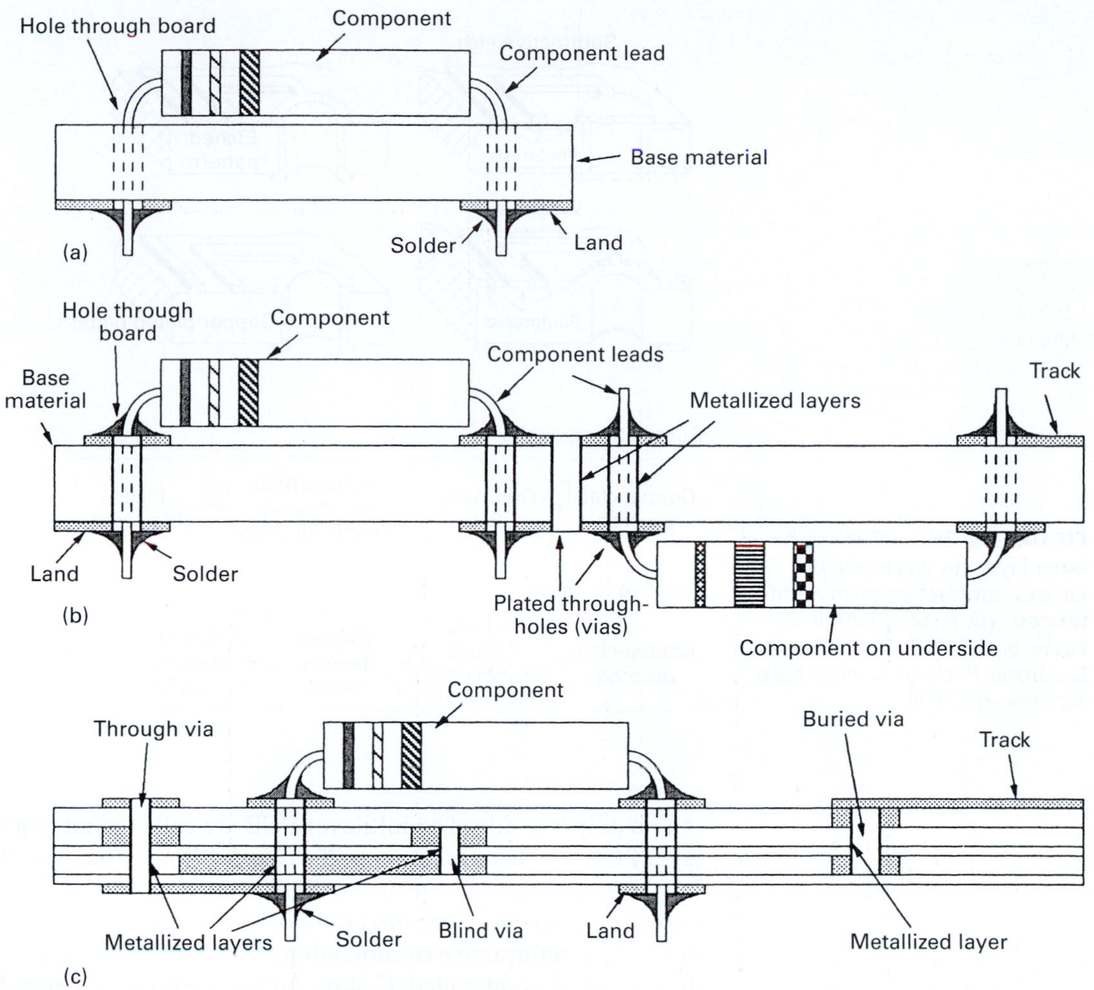

FIGURE 35-25 PCBs can be single-sided, double-sided, or multilayer. *(M. Judd and K. Brindley, Soldering in Electronics Assembly, Boston: Reed International Books, 1992)*

Ceramic substrates (typically alumina) are used primarily to minimize thermal stresses on joints in military applications that use leadless ceramic packages. In addition, ceramic substrates are used for hybrid circuits involving both semiconductor and thick-film components. **Thick film** refers to components that are screen printed as opposed to deposited by thin-film technology (i.e., evaporation, sputtering, or electroplating).

PCBs can be single-sided, double-sided, or multilayer. Single-sided PCBs simply have metallic circuits on one side of the laminate. Through-hole (TH), single-sided PCBs have insertion holes that extend through the board to the other side where TH components may be inserted into the board (Figure 35-25a). Surface mount (SM) components are simply mounted onto the pads on the same side as the circuit and do not require through-holes. Double-sided PCBs are used in cases where circuits must "jump," or cross over, one another. In this case, **via holes** (or simply **vias**) are needed to route the circuits over one another (Figure 35-25b). Vias are essentially metal-filled holes through the laminate material that connect a circuit on one side to the other. The metal inside of the via is electroplated. Vias that are also used as insertion holes are called **plated through-holes (PTHs).** As the number of packaged components on the board increases, the complexity of the circuits increases, giving rise to the need for multilayer PCBs in which multiple single- and double-sided boards are laminated together using prepreg. Vias that pass from an outermost track on one side of the board to the outermost track on the other side are called **through vias** (Figure 35-25c).Vias within a laminate core on the inside of a multilayer PCB are called **buried vias.** Vias that come

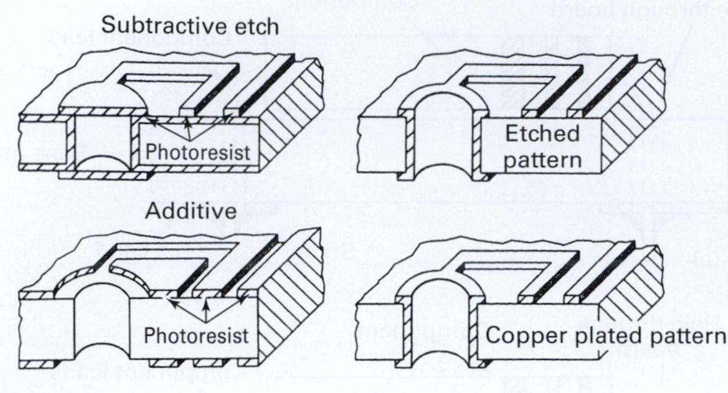

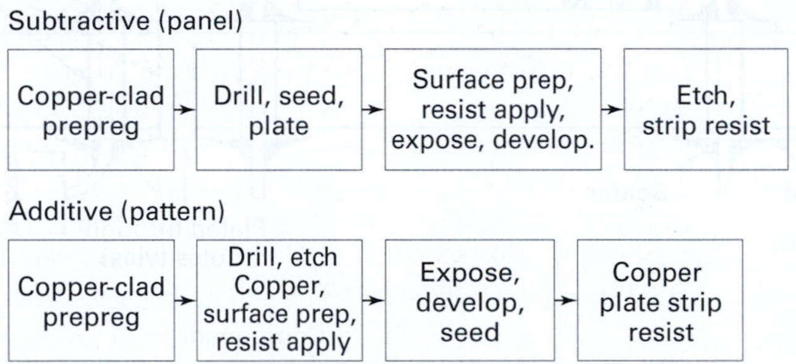

FIGURE 35-26 PCBs can have inner layer circuits made by either a subtractive or an additive process. *(D. P. Seraphim, R. C. Lasky, and C.-Y. Li,* Principles of Electronic Packaging, *New York: McGraw-Hill, 1989)*

out on only one side of a multilayer PCB are called **blind** or **partially buried vias.** Multilayer PCBs can have as many as 20 layers, although four to eight are more common.

Production of a multilayer PCB from a C-stage laminate begins with a process known as **inner-layer circuitization.** Inner-layer circuitization may be either subtractive or additive. **Subtractive circuitization** (Figure 35-26) for glass-epoxy PCBs begins with a double-sided, copper-clad, C-stage laminate known as a **panel,** which has been sheared to size. A film of dry photoresist is applied by hot roller to the copper surface. Next, the circuit pattern (traces and pads) is transferred to the photoresist by exposure through a reticle and chemical development of the photoresist in a photolithographic process similar to that used in semiconductor processing. The copper is then selectively etched through the resulting etch mask, and the resist is subsequently stripped from the laminate. Afterward, registration holes are drilled relative to locator marks, called **fiducials,** produced in the copper layer during the lithography and etching processes.

In **additive circuitization** (Figure 35-26), copper is selectively deposited instead of etched away. The process begins with a bare glass-epoxy laminate cut to size with registration holes. If necessary, via holes may be drilled. An etch mask is exposed and developed, exposing the underlying dielectric including all via holes. The exposed dielectric surfaces are **buttercoated,** or *seeded*, to permit electroless deposition by adsorbing a catalyst (usually palladium) from solution onto the surface of the dielectric. A thin layer of electroless copper is deposited on the seeded dielectric, followed by electroplating of thicker copper layers. Afterward, the resist is stripped. The additive process has the advantage of providing higher resolution for circuitry with finer lines and higher density but tends to be less economical.

The final multilayer board is produced by lamination of inner layers, drilling and preparation of via holes, and circuitization of outer layers. In lamination, as shown in Figure 35-27, a stack of inner layers is bonded together between B-stage prepreg of the appropriate shape and size under time, temperature, and pressure. Usually the inner layers are stacked up between copper layers on the top and bottom of the stack. Prior to lamination, the copper surfaces of the inner layers are oxidized to improve adhesion between layers. Alignment between layers is critical during lamination and is controlled with pins and registration holes. After lamination, excess resin that oozed out during bonding is sheared off. Next, via holes are drilled, deburred, and plated. After drilling,

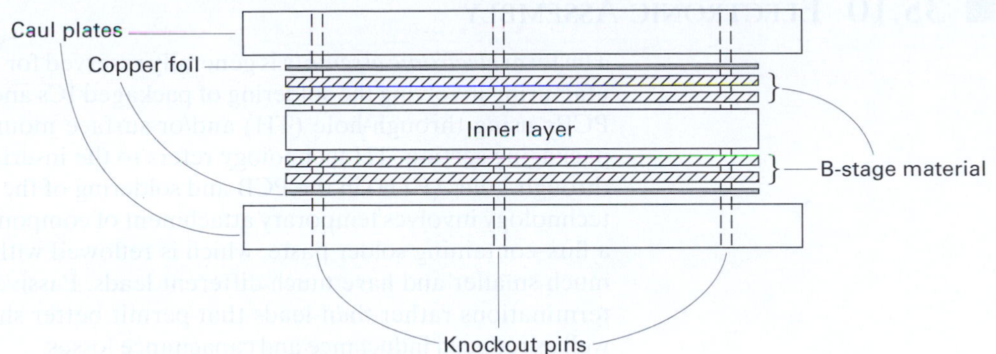

FIGURE 35-27 PCBs are often multilayers of laminations as shown here with a four-layer foil construction. *(M. L. Minges, Electronic Materials Handbook, Volume 1. Packaging, Materials Park, OH: ASM International, 1989)*

some epoxy smear may exist on the copper surface within the via, which will prevent electrical connection or weaken the mechanical integrity of the via. Therefore, an acidic solution is used to remove excess epoxy and actually to slightly etch back epoxy within the hole so that the copper layers protrude slightly, which improves the connection between copper layers within the via. Subsequently, the epoxy surface within the via is seeded, and a thin layer of electroless copper is deposited. **Outer-layer circuitization** involves the same method of dry resist patterning to produce a lithographic mask. One difference between inner- and outer-layer circuitization is that copper is electrolytically deposited onto exposed copper after masking to ensure good electrical contact between the inner and outer copper layers. To finish the PCB, the copper layer is etched, a photosensitive solder mask/encapsulant is applied, and the remaining exposed pads are "pre-tinned" with solder.

The manufacturing process for producing flex circuits is similar to the process just described. Flex circuits are typically not as complicated as glass–epoxy PCBs and so have fewer layers. The typical process makes use of a two-sided, copper-clad, polyimide film usually bonded with an epoxy resin. Encapsulation is performed with the use of flexible cover layers, which are either photosensitive or precut and bonded. After encapsulation, the final shape of the flex circuit is cut to size either by shearing (high volume) or by lasers and water jets (prototyping).

With the advent of CSP and MCM packaging technologies, the requirements for track and pad densities have increased. The practical limit of mechanical drilling is a diameter of about 200 μm. **Microvias** are via holes made by photo-imaging, laser ablation, and plasma etching that extend well below 200 μm. Microvias as small as 25 μm can be made that increase the density of pads and tracks eightfold over conventional mechanical drilling technologies. Advantages of such small vias include the elimination of bonding pads for direct trace-to-trace connections. Microvias can be used to produce **built-up multilayers** (Figure 35-28). Built-up multilayers are made by deposition and processing of one dielectric layer at a time, similar to IC interconnection. Connections between dielectric layers are made by drilling and electroplating into microvias. Circuit routing can be made extremely efficient, which results in optimally short signal lengths for high-performance applications.

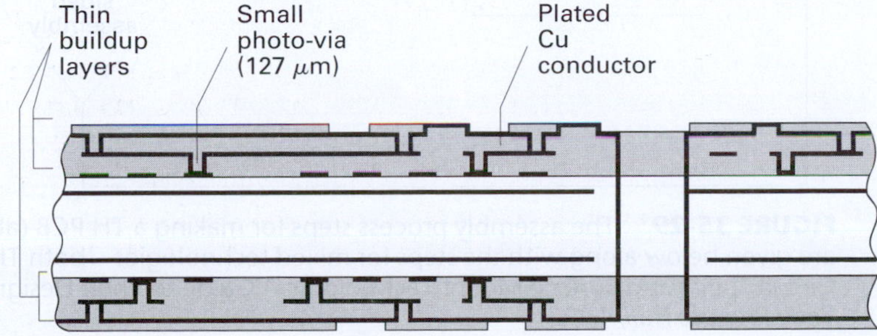

FIGURE 35-28 Microvias can be used to produce built-up microlayers. *(C. A. Harper, Electronic Packaging and Interconnection Handbook, New York: McGraw-Hill, 2000)*

35.10 ELECTRONIC ASSEMBLY

The term *electronic assembly* is generally reserved for the third level of electronics manufacturing involving the soldering of packaged ICs and other discrete components onto PCBs using through-hole (TH) and/or surface mount (SM). As explained in the IC packaging section, TH technology refers to the insertion of packaged leads into plated through-holes (PTH) in the PCB and soldering of the terminals from the backside. SM technology involves temporary attachment of components to the surface of the PCB via a flux-containing solder paste, which is reflowed within an oven. SM components are much smaller and have much different leads. Passive (non-IC) SM components have terminations rather than leads that permit better shock- and vibration-resistance as well as reduced inductance and capacitance losses.

The sequence of operations for SM and TH assembly is shown in Figure 35-29. Insertion can be performed either manually or with automatic insertion machines. After

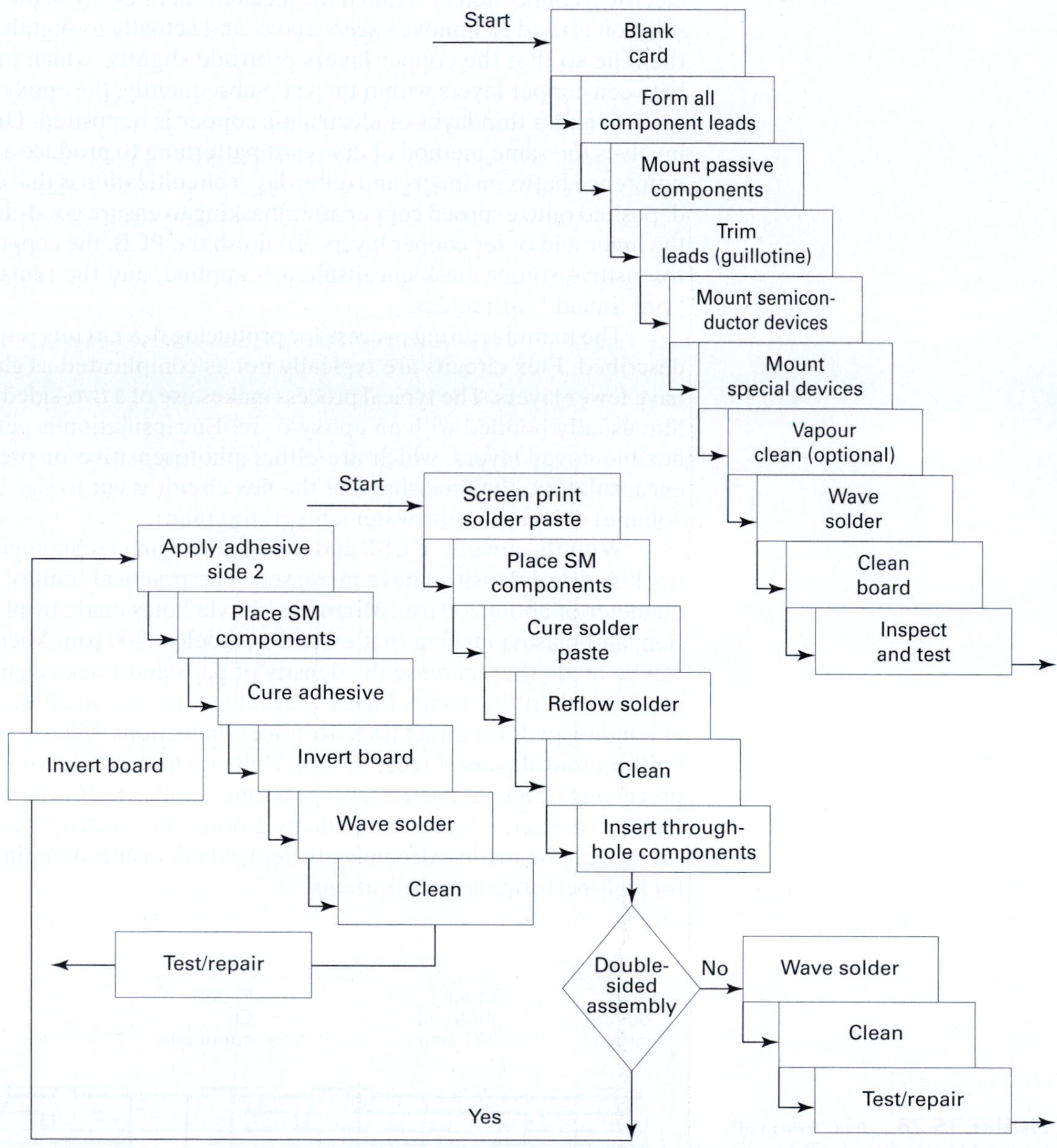

FIGURE 35-29 The assembly process steps for making a TH PCB (above). The steps for SM assembly are given below along with the steps for mixed technologies—both TH and SM. *(M. R. Haskard,* Electronic Circuit Cards and Surface Mount Technology: A Guide to Their Design, Assembly, and Application, *New York: Prentice Hall, 1992)*

insertion, leads are generally clinched and trimmed if necessary to avoid **bridging** between joints during soldering, which can cause electrical shorting of the circuit. Generally, soldering of TH components is performed automatically through a process known as **wave soldering.** Wave soldering involves the conveyance of a preheated and prefluxed PCB over a standing wave of solder created by pumping action. The combination of capillary action and pumping action permits flow of the solder from the underside of the board into the joint. Cleanliness of the PCB is critical for wetting of the lead and PTH. A high-pressure air jet is used to blow off excess solder from the underside of the board to prevent solder ridging. Postsolder cleaning of the board includes degreasing and defluxing.

One key consideration for TH solder joints is joint strength. The trade-off is the clearance between the insertion lead and insertion hole. As the clearance decreases, joint strength increases. However, with smaller clearances it is more difficult to insert pins in holes. Clearances on the order of 0.25 mm are typical. Another factor affecting joint strength involves the **clinching** of leads. Clinched lead joints are much stronger than unclenched joints. Because the mechanical strength of TH joints is generally superior to SM joints, large, heavy components are generally attached with TH technology.

SM assembly involves application of solder paste to the lands on the surface of the PCB, placement of SM components on top of this paste, and reflow of the solder paste within an oven. **Solder paste** consists of small spherical particles of solder less than a tenth of a millimeter in diameter together with flux and solvents used to dissolve the flux (imparting tackiness) and thicken the paste. At the time of application, the paste has the consistency of peanut butter and is applied by screening, stenciling, or dispensing. In screening and stenciling, a solder paste printer is used to apply solder paste through a mask (screen or stencil) by running a squeegee over the surface of the mask. The mask is typically held off of the surface by a distance on the order of 0.5 mm, known as the **snap-off distance** (see Figure 35-30). As the squeegee passes over the mask surface, the mask is pressed against the PCB, allowing contact between the paste and the lands on the board. After the squeegee has passed, the mask snaps back from the surface, leaving an island of solder paste on the PCB lands. Stencils are typically metal sheets or wire mesh that has been chemically etched using a lithographic process. Screens are typically formed by application, exposure, and development of a photosensitive emulsion on top of a wire mesh. Advantages of metal sheet stencils include longevity and multilevel (pads of varying thicknesses) printing, and the screens are cheaper to make. To decrease tooling costs during product development, pastes can also be dispensed without a mask through a syringe needle. Dispensing generally requires pastes with lower viscosity, which can lead to other problems including solder paste **slump** (spreading out of the solder pastes after application).

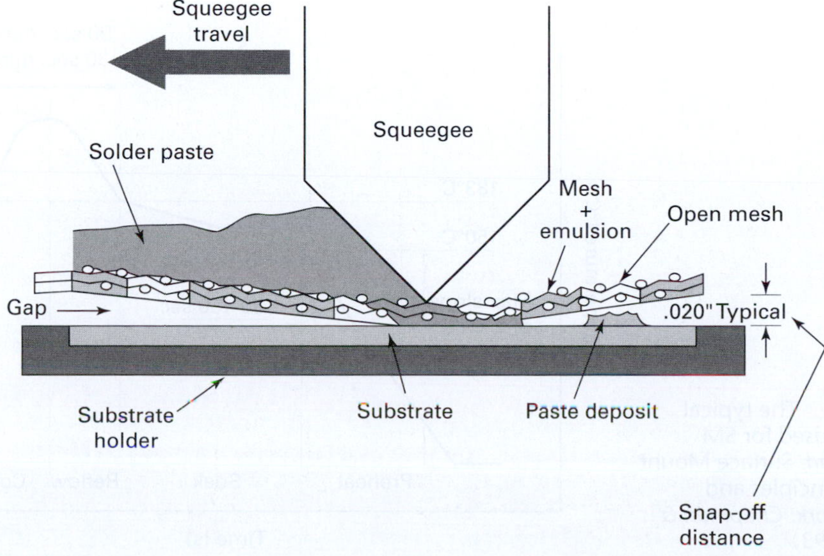

FIGURE 35-30 Schematic for applying solder paste on a substrate by squeegee in a screen printing process in SM technology. *(R. Prasad,* Surface Mount Technology. Principles and Practice, *New York: Chapman & Hall, 1997, p. 493)*

Once the solder paste is positioned on the board, a component placement machine, also known as a **pick-and-place** machine, is used to place the components onto the solder paste pads. The flux in the solder paste is tacky and holds positioned components in place until oven soldering. Components are fed to a robotic manipulator that has a vacuum chuck or a mechanical chuck, or both. Component feeders deliver components to the manipulator. Several types of **feeders** exist, including tape (or reel), bulk, tube (or stick), and waffle pack. The feeder system must be carefully selected based on the desired quantity per feeder, availability, part identification, component cost, inventory turns, and potential for damage during shipping and handling. Tape feeders are widely utilized and are most desirable for high-volume placement. Tube feeders are useful for smaller-volume assemblers, even though costs per component are higher. Waffle packs are flat-machined plates with inset pockets to hold various chips. In general, waffle packs increase the cost of assembly. However, some IC packages, like the bumperless, fine-pitch QFPs, require a high level of protection during handling to minimize lead damage, so this component requires the tape-feeding mechanism. Bulk feeding of IC components, through the use of a vibratory bowl, may be useful for prototyping environments.

The economics of SM technology are driven by component placement equipment, which determines the throughput of the SM line and is the source (at least partially) of most defects requiring rework. Further, placement equipment strongly influences startup costs because it may involve as much as 50% of the capital equipment cost in setting up a line. Key criteria in the selection of placement equipment include placement accuracy, placement rate, maximum PCB size, types and sizes of components, and maximum number of feeders, among others. In general, placement equipment has been classified as four discrete types: (1) high throughput, (2) high flexibility, (3) high flexibility and high throughput, and (4) low cost and low throughput with high flexibility. High-throughput placement machines are called **chip shooters.** Chip shooters are typically dedicated to the placement of passive (resistors, capacitors, etc.) and small active (IC) components and can place components at rates up to 60,000 components per hour with linear repeatability around 0.05 to 0.1 mm and rotational accuracy of 0.2 to 0.5 degree over a 350- × 450-mm area.

After components are placed, the PCB is placed in a **reflow oven,** where the solder paste melts, causing a fluxing action, which permits the melted solder to wet the leads and the PCB lands. To achieve this, the PCB must be exposed to an appropriate **thermal profile,** or **time–temperature curve,** as it passes through the oven. Figure 35-31 shows a common thermal profile for SM reflow. At a minimum, the thermal profile must include at least four zones. The first zone, called preheating, is used to drive off any nonflux volatiles within the paste. The second zone, the soak zone, is used to bring the entire assembly up to just below the reflow temperature of the paste. The third (reflow) zone

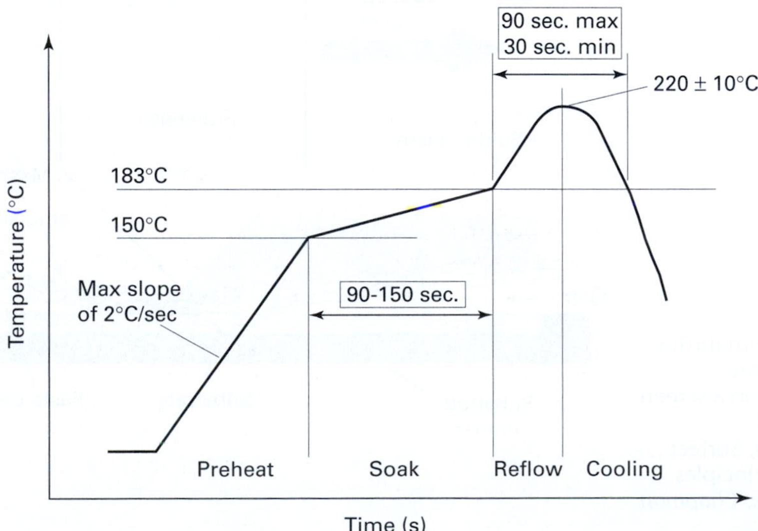

FIGURE 35-31 The typical thermal profile used for SM reflow. *(R. Prasad,* Surface Mount Technology. Principles and Practice, *New York: Chapman & Hall, 1997, p. 493)*

quickly raises the temperature of the solder paste above the reflow temperature allowing for fluxing and wetting of solder joints. The fourth zone cools the assembly permitting solidification. Reflow soldering is generally done in infrared (IR) reflow ovens, and heating involves both IR radiation as well as gas-forced convection. The minimum number of heating zones for a reflow oven must be three (the fourth is a cooling zone) but can contain as many as 20 to provide better control over the thermal profile.

An alternative to IR reflow soldering is **vapor-phase soldering,** or **condensation soldering,** involving the condensation of a hot perfluorocarbon vapor onto the assembly surface, releasing the latent heat of vaporization into the solder joints and substrates. By and large, this process has been replaced by IR reflow soldering due to improved process reliability and control. Other alternatives to IR reflow soldering include laser, hot bar, and hot belt reflow soldering.

For a variety of reasons, SM technology and through-hole insertion technology are mixed on the same PCB. Some components are not available in SM packages. Some components are large and require the added strength provided by TH solder joints. Some components require more heat dissipations than SM can accommodate. Thus, both methods will continue to be used in the future.

Key Words

additive circuitization
assembly yield
A-stage
backpanel PCB
ball grid array (BGA)
barrier potential
blind vias
board
boat
boule
bridging
B-stage
built-in multilayers
buried via
burn-in yield
butt lead
buttercoated
card-on-board
charge
chemical metal polishing (CMP)
chemical vapor deposition (CVD)
chip-and-wire attachment
chip carriers
chip-on-board
chip scale packaging (CSP)
chip shooter
chip
clean rooms
clinching
condensation soldering
contact printing
contacts
C-stage
Czochralski method
depletion region
diamond sawing
dies
die yield
diffusion
direct chip attachment (DCA)
direct die mounting
direct-write technique
donor
dopants
doping
drive-in
dry etching
dry oxidation
dual in-line packging (DIP)
electromigration
electronic assembly
electron holes
epitaxy
etch bias
etchant
etching
evaporation
feeders
fiducials
flats
flex circuits
flip-chip
gettering
gull-wing leads
inner-layer circuitization
integrated circuit (IC)
ion milling
isotropic
J-leads
lattice
lines
linewidths
lithography
mass-transport limited
metallization
microvias
motherboard
multichip module (MCM)
n-type semiconductors
outer-layer circuitization
panel
partially buried vias
p-glass
p-n junction
p-type semiconductors
photolithography
photomask
photoresists
physical vapor deposition (PVD)
pick-and-place
pin grid array (PGA)
pin-in-hole (PIH)
plasma etching
plated through-hole (PTH)
polishing
polysilicon
predisposition
prepreg
printed circuit board (PCB)
printed wirign assembly (PWA)
processing yield
projection printing
proximity printing
quad flat package (QFP)
rapid thermal annealing (RTA)
rapid thermal processing
reaction-rate limited
reactive ion etching (RIE)
reflow oven
resist
resist masks
resolution
reticles
seed crystal
selectivity
semiconductor
sensitivity
slump
snap-off distance
solder balls
solder paste
spin coating
sputter etching
sputtering
step coverage
stepper
stripping
subtractive circuitization
surface mount (SM)
tape-automated bonding (TAB)
thermal profile
thick film
thin film
through-hole (TH)
through-mask technique
through vias
time–temperature curve
vapor-phase epitaxy (VPE)
vapor-phase soldering
via holes (vias)
wafer testing
wafer yield
wafers
wave soldering
wet etching
wet oxidation
wire bonding

■ REVIEW QUESTIONS

1. What is the goal of the field of electronics?
2. What are the major advantages of integrated circuits over electronic assemblies? What is the advantage of electronic assemblies?
3. How many levels of electronic manufacturing exist? Name each level.
4. What is a semiconductor?
5. Name three common semiconductor materials.
6. What is meant by the term *doping?*
7. What is the difference between *n*-type and *p*-type semiconductors?
8. What are electron holes?
9. Give three reasons why silicon is the most popular semiconductor used today.
10. What is a *p-n* junction? What can it be used for?
11. What is the barrier potential of a *p-n* junction? Why does it exist?
12. List the sequence of steps necessary to produce a bipolar diode.
13. What is meant by the term ULSI? How is it different from VLSI?
14. In general, what technological breakthroughs were necessary to advance to each successive level of integration?
15. What is a silicon boule?
16. Why is it advantageous to add impurities during the formation of the single crystal?
17. Why are notches or flats cut or ground along the periphery of the single-crystal ingot?
18. Why are single-crystal ingots with diameters 300 mm and larger marked with a notch instead of a flat?
19. What are some geometric concerns involved with wafer production?
20. What does gettering mean in relation to wafer production?
21. Name three methods for doping a silicon wafer.
22. What are some advantages of ion implantation over thermal diffusion?
23. Why must ion-implanted substrates be annealed?
24. Why are rapid thermal processing technologies generally advantageous?
25. What are two ways in which silicon dioxide is commonly used in microelectronic manufacturing?
26. Give two reasons wet oxidation is better suited to making thicker oxides (e.g., diffusion masks) than dry oxidation.
27. How is the lateral geometry of the IC and its components patterned into microelectronic materials?
28. What is the most complicated, expensive, and critical step in microelectronics manufacturing?
29. In addition to photolithography, name three other lithographic methods that may be used for pattern transfer.
30. List the photolithographic steps necessary to produce a resist mask on a silicon substrate.
31. Of the two major classifications of photoresists, which type cross-links under exposure to electromagnetic energy of the proper wavelength, resulting in longer polymer chains?
32. List four requirements of a photoresist.
33. What were the two most important wavelengths for photolithography provided by the mercury arc lamp?
34. List the three types of exposure methods used in photolithography. Give an advantage of each.
35. What is the difference between wet and dry etching?
36. What is undercutting?
37. What are some possible defects that can result from underetching? From overetching?
38. List and describe two properties of etchants.
39. Compare the three dry etch processes, including a description of their etch mechanisms and advantages.
40. What are thin films? Why are they important to microelectronic manufacturing?
41. List two different types of physical vapor deposition, including the physical mechanisms and advantages of each.
42. List three different forms of chemical vapor deposition, and indicate in what application each might be used.
43. How are undesirable gas-phase reactions controlled within APCVD and LPCVD reactors?
44. What is the key difference in the reactor designs of APCVD and LPCVD processes?
45. What are the two types of LPCVD reactor designs? List the advantages and disadvantages of each.
46. What is the advantage of using PECVD processes?
47. What is epitaxy, and why is it important for microelectronic manufacturing?
48. In metallization, what is the difference between a contact and a via?
49. What is planarization, and why is it needed?
50. Give three methods for planarizing interconnect layers.
51. What is electromigration, and why is it a concern in IC processing?
52. What is the purpose of wafer testing?
53. What is meant by the term *chip?*
54. What drives the increase in component density and die area within microelectronic manufacturing?
55. Why are clean rooms so important to microelectronic processing?
56. What two subcomponents make up an IC package?
57. What are the advantages of surface mount technology and through-hole (or pin-in-hole) technology for attachment of IC packages and discrete electrical components to boards?
58. Name the two key classes of TH packages.
59. Name the four different types of SM lead geometries, and discuss the advantages of each.
60. List the key steps involved in conventional IC packaging.
61. List three techniques for attaching and electrically connecting dies to IC packages.
62. What is meant by direct chip attachment, and how does it differ from more conventional IC packaging? What are some disadvantages to direct chip attachment?
63. What are chip-scale packages, and how do they differ from direct chip attachment methods?
64. What are multichip modules, and why are they advantageous for IC packaging?
65. What is a printed circuit board (PCB)? What three elements does a PCB consist of?
66. Name three alternative materials used as dielectrics within PCBs, and give the major reason each is used.
67. What is the difference between plated through-holes and via holes?
68. What is the difference among blind, buried, and through vias?

69. List the four major process steps for producing a multilayer PCB.
70. List the two methods for inner layer circuitization, and discuss the physical process and advantages of each.
71. What are built-up mulitlayers and microvias? How are they different from laminates and vias?
72. List the four major steps for assembling a TH printed wiring assembly (only TH technology).
73. Why are leads trimmed and clinched after through-hole insertion?
74. List the four major steps for assembling an SM printed wiring assembly (only SM technology).
75. Name four methods for feeding components to robotic manipulators in pick-and-place robots. Discuss the application of each.
76. What step in the SM assembly process most strongly affects startup and operations costs for an SM assembly line? Why?
77. What four zones are necessary within an SM solder reflow thermal profile?

CHAPTER 36

MICRO/MESO/NANO FABRICATION PROCESSES

36.1 INTRODUCTION

The manufacture of products deals with the making of machines, structures, or processing equipment by casting, forming, machining, welding, and assembly. **Fabrication,** a synonym for **manufacturing,** can be classified into two main categories: **macro fabrication** and **micro fabrication.** The first one deals with the process of fabrication of structures/parts/products that are measurable and observable by the naked eye (greater than or equal to 1 mm in size), while the second category considers the miniature structures/parts/products which are not easily visible with the naked eye, having dimensions smaller than 1 mm (typically in the range of 1 mm to 999 mm). Although the term **nano fabrication** is currently quite popular, only polishing methods are routinely done at the true nano level of fabrication. Most fabrication occurs in between the micro and the nano extremes. This region of fabrication is now commonly being referred to as **meso fabrication** by the various trade industries. The two most common classification methods are to group the methods as either (1) additive processes or (2) material removal processes. These processes are often immediately recognizable as an extension of the previously discussed traditional and nontraditional fabrication methods.

THE SIZE EFFECT

The laws of physics, which were discussed for earlier additive processes, continue to hold at the micro/meso/nano level of fabrication. However, when metal is removed by machining of any type including abrasives, there is a substantial increase in the specific energy required with decreasing chip size. It is generally believed that this is because all metals contain dislocations and other defects (e.g., grain boundaries, missing and impure atoms, and substitutional defects). When the size of the material removed decreases, the probability of encountering a stress-reducing defect decreases. Because the shear stress and strain in the metal-cutting region is unusually high, discontinuous microcracks often form on the metal-cutting shear plane at the nano level. If the material being cut is very brittle or the compressive stress on the sheer plane is relatively low, these **microcracks** will grow into gross cracks, giving rise to discontinuous chip formation because the material can no longer absorb the damage of deformation through dislocation glide and slip. An alternative theory for the size effect in cutting is based on the premise that shear stress increases with increases in strain rate. When an attempt is made to apply this to metal cutting, it is assumed in the analysis that the von Mises' criterion applies on the shear plane. However, this is inconsistent with the experimental findings of Eugene Merchant. Until this difficulty is resolved, it should be assumed that the strain rate effect may be responsible for some portion of the size effect in metal cutting. The typical manufacturing student of today should understand that the art of metal cutting at extremely small geometries is a "target-rich" experimental field currently dominated by the empirical data produced by original equipment manufacturers. Such information remains largely proprietary and unpublished in the literature. Purchasing officers should clearly specify the delivery of machining data when purchasing materials or equipment for use at the micro/meso/nano region of fabrication.

36.2 ADDITIVE PROCESSES

In meso fabrication, **deposition** refers to the many different additives processes used to add a thin-film to one or both sides of a prepared substrate. Figure 36-1 summarizes the processes under the general groupings of *traditional, kinematics interaction, vapor deposition, direct deposition,* and *crystal growth.* Many of the processes have already been described as either electronics fabrication methods or nontraditional machining methods. Discussion in this section will be limited to a basic review of general topics and the introduction of a few new specific processes. Figure 36-2 summarizes some typical applications of the additive processes listed in Figure 36-1. Specific materials, deposition methods, and their typical applications within the electronics industry are summarized in Figure 36-3. (See also Chapter 35.)

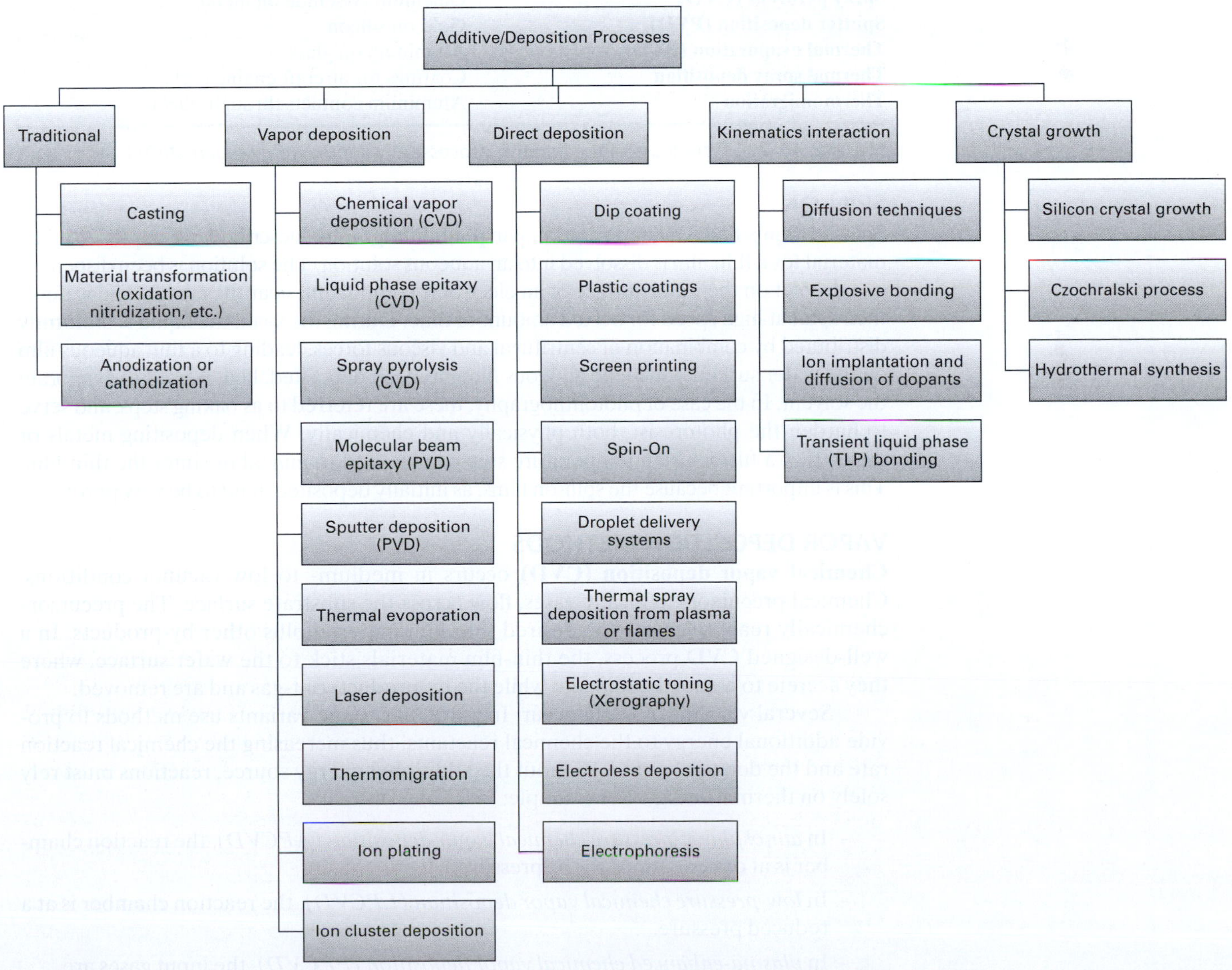

FIGURE 36-1 Summary listing of additive processes at the micro/meso/nano level of fabrication.

Additive	Application
Bonding techniques	7740 glass to silicon
Casting	Thick resist (10–1000 μm)
Chemical vapor deposition	Tungsten on metal
Dip coating	Wire-type ion selective electrodes
Droplet delivery systems	Epoxy, chemical sensor membranes
Electrochemical deposition	Copper on steel
Electroless deposition	Vias
Electrophoresis	Coating of insulation on heater wires
Electrostatic toning	Xerography
Ion cluster deposition	
Ion implantation and diffusion of dopants	Boron into silicon
Ion plating	Metal on insulators
Laser deposition	Superconductor compounds
Liquid phase epitaxy (CVD)	Galenium Arsenide
Material transformation	Growth of silicon dioxide (SiO_2) on silicon
Molecular beam epitaxy (PVD)	Galenium Arsenide
Plastic coatings	Electronic packages
Screen printing	Planar ion selective electrodes (ISEs)
Silicone Crystal growth	Primary process
Spin-on	Thin resist (0.1–2.0 μm)
Spray pyrolysis (CVD)	Galenium Arsenide on metal
Sputter deposition (PVD)	Gold on silicon
Thermal evaporation (PVD)	Aluminum on glass
Thermal spray deposition	Coatings for aircraft engine parts
Thermomigration	Aluminum contacts through silicon

FIGURE 36-2 Typical use of the additive processes. *(Adapted from Madou, 2001)*

SPIN-ON

Spin-on deposition is an important step in **photolithography** (described in Chapter 35). The material for a thin film is dissolved into an aqueous solution. The solution is been dispensed as a droplet on the wafer surface of an electronic component, near the center. The wafer is then spun at high speed for a fixed amount of time. During the span, the liquid is uniformly distributed by combination of centrifugal and viscous forces, leading to a thin aqueous film on the wafer surface. Once the aqueous film has been deposited, heat is used to evaporate the solvent. In the case of photolithography, these are referred to as baking steps, and serve to harden the photoresist, both physically and chemically. When depositing metals or dielectrics, a further high temperature step can be used to anneal or sinter the thin-film. This is important because the spin-on films, as initially deposited, tend to be very porous.

VAPOR DEPOSITION METHODS

Chemical vapor deposition (CVD) occurs in medium- to low-vacuum conditions. Chemical precursors, typically gases, flow across the substrate surface. The precursors chemically react to create the desired thin-film material plus other by-products. In a well-designed CVD process, the thin-film materials stick to the wafer surface, where they accrete to create the thin film while the by-products out-gas and are removed.

Several variants of CVD occur. In particular, some variants use methods to provide additional energy to the chemical reactants, thus increasing the chemical reaction rate and the deposition rate. Without the additional energy source, reactions must rely solely on thermal energy. For example:

- In *atmospheric-pressure chemical vapor deposition (APCVD),* the reaction chamber is at or near atmospheric pressure.
- In *low-pressure chemical vapor deposition (LPCVD),* the reaction chamber is at a reduced pressure.
- In *plasma-enhanced chemical vapor deposition (PECVD),* the input gases are turned into plasma by adding radio-frequency (RF) energy.

Material	Deposition Technique	Function
Organic thin films		
Hydrogel	Silk-screening	Internal electrolyte in chemical sensors
Photoresist	Spin-on	Masking, planarization
Polyimide	Spin-on	Electrical isolation, planarization, microstructures
Metal oxides		
Aluminum oxide	CVD, sputtering, anodization	Electrical isolation
Indium oxide	Sputtering	Semiconductor
Tantalum oxide	CVD, sputtering, anodization	Electrical isolation
Tin oxide (SnO_2)	Sputtering	Semiconductor in gas sensors
Zinc oxide	Sputtering	Electrical isolation, piezoelectric
Noncrystalline silicon compounds		
α-Si-H	CVD, sputtering, plasma CVD	Semiconductors
Polysilicone	CVD, sputtering, plasma CVD	Conductor, microstructures
Silicides	CVD, sputtering, plasma CVD Alloying of metal and silicon	Conductors
Metals (thin films)		
Silver	Evaporation, sputtering	Electrochemistry electrodes
Aluminum	CVD, sputtering, plasma CVD	Electrical interconnects (below 300°C)
Chromium	Evaporation, sputtering, electroplating	Electrical conduction, adhesion layer (10–100 nm)
Gold	Evaporation, sputtering, electroplating	Electrical interconnects (above 300°C) Optical reflection in the infrared
Iridium	Sputtering	Electrochemistry electrodes, biopotential measurements
Molybdenum	Splluttering	Electrical conduction
Platinum	Sputtering	Electrochemistry electrodes, biopotential measurements
Palladium	Sputtering	Electrical conduction, adhesion layer, electrochemistry electrodes, solder wetting layer
Tungsten	Sputtering	Electrical interconnects at higher temperatures
Titanium	Sputtering	Adhesion layer
Copper	Sputtering	Low resistivity interconnects
Alloys		
Al-Si-Cu	Evaporation, sputtering	Electrical conduction
Nichrome (NiCr)	Evaporation, sputtering	Thin-film laser-trimmed resistor
Permalloy[TM] (Ni_xFe_y)	Sputtering	Magnetoresistor, thermistor
TiNi (80)	Sputtering	Shape memory alloy
Chemically/physically modified silicon		
N/P type silicon	Implantation, diffusion, incorporation in the melt	Conduction modulation, etch stop
Porous silicon	Anodization	Electrical isolation, light emitting structures, porous junctions
Silicon dioxide	Thermal oxidation, sputtering, implantation, CVD, anodization	Electrical and thermal isolation, masking, encapsulation
Silicon nitride	Plasma CVD	Electrical and thermal isolation, masking, encapsulation

FIGURE 36-3 Integrated Circuits (IC) and Microelectromechanical Systems (MEMS) materials, deposition method, and typical applications. *(Adapted from Madou, 2001)*

- *Light-* or *laser-assisted chemical vapor deposition* uses photons to provide additional energy. The use of a laser also allows direct-write applications.

Physical vapor deposition (PVD) occurs in very low vacuums relative to chemical vapor deposition. In PVD systems, the density of gas molecules is so low that most

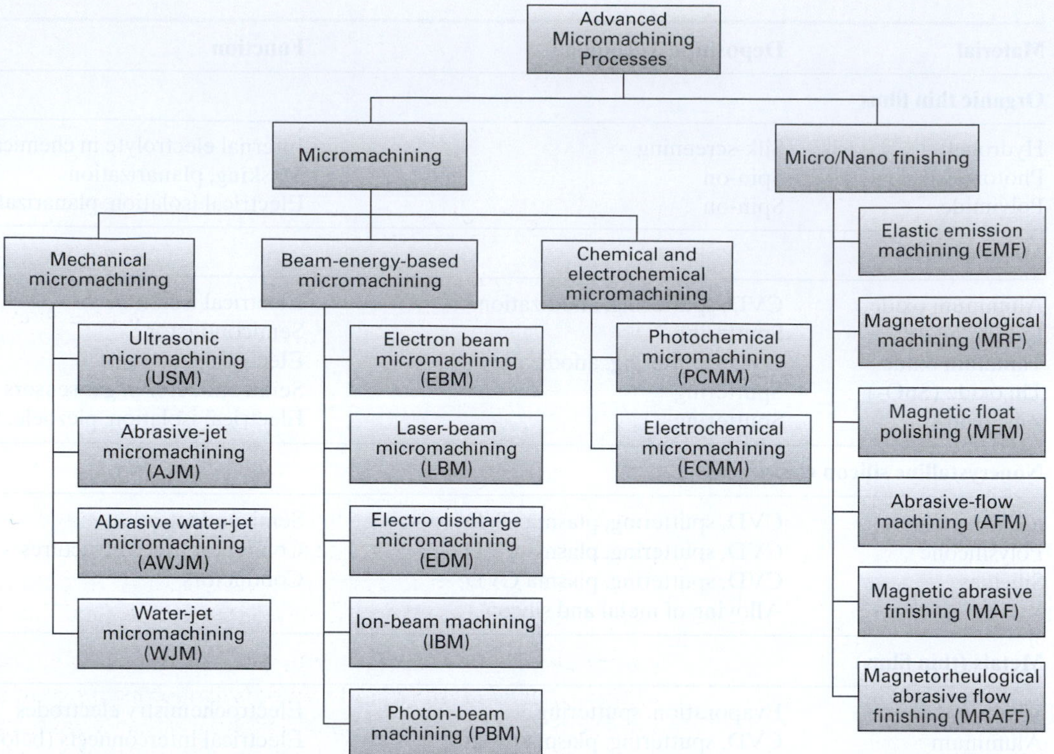

FIGURE 36-4 Summary of the material removal processes used for micro/meso/nano fabrication.

molecules can travel across the reaction chamber without interacting with any other molecules until they reached the target. This is called the **free molecular regime of space.**

- In **evaporation** systems, the source material is converted into a liquid, which then evaporates into the very low vacuum. The evaporated add-ons then move across the reaction chamber where they hopefully adhered to the substrate of the target.
- In a **sputtering** system, atoms are physically dislodged from the source material. Source material is often called the target.

The illustrated specifics of chemical vapor deposition (CVD) are explained in greater detail in Chapters 21 and 35 along with physical vapor deposition (PVD).

REMOVAL PROCESSES

Many of the traditional machining processes as well as the recently developed nontraditional machining processes have been adapted micro/meso/nano machining. Figure 36-4 organizes these methods into mechanical micromachining, beam-energy-based micromachining, chemical and electrochemical micromachining, and the finishing processes.

MECHANICAL MICROMACHINING

The **mechanical micromachining** processes have been discussed elsewhere as ultrasonic machining, water-jet machining, abrasive-jet machining (a form of sand blasting) and abrasive water-jet machining. The miniaturized techniques use the same principles but with more advanced control systems having a higher resolution of control movement.

Ultrasonic Micromachining. The **ultrasonic micromachining** process relies on the projection of very hard abrasive particles on the part to be machined, by use of a tool known as a **sonotrode,** vibrating at an ultrasonic frequency in excess of 20 kHz (Figure 36-5). The particles are normally conveyed by a fluid, water in most cases. Two different mechanisms are observed at the micro level:

1. Mechanical action of particles on the surface of the workpiece that hardens and causes brittleness in materials such as glass, ceramics, composites, courts, precious stones, and semiconductors.

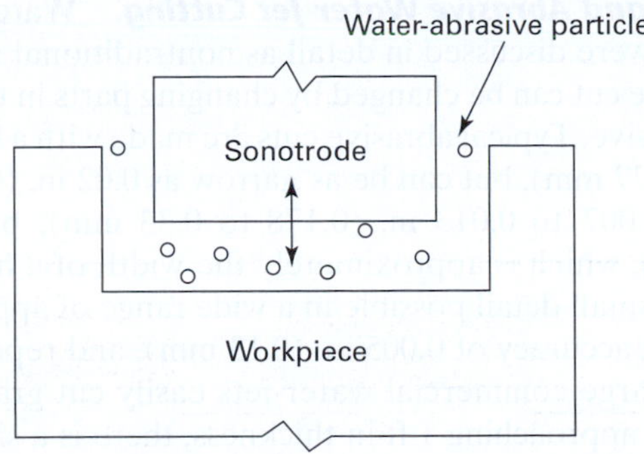

FIGURE 36-5 Geometry of ultrasonic machining, which uses a loose abrasive particle mixture. *(From McGeough, 2002)*

2. Cavitation erosion caused by rapid changes in pressure inside the fluid carrying the particles, which has a significant effect when machining fragile and porous materials such as graphite or porous ceramics.

Particles affect both the part and the sonotrode: material removal takes place at the workpiece; wear occurs on both the sonotrode and the particles. The industrial processes are therefore characterized by removal rate on the workpiece, sonotrode wear, and abrasive wear. Figure 36-6 summarizes the ultrasonic machinability of various materials.

Abrasive-Jet Machining. **Abrasive-jet machining,** also known as *abrasive microblasting, pencil blasting,* and *microabrasive blasting*, is an abrasive blasting process that uses scouring abrasives propelled by high-velocity gas to erode material from the workpiece. It is basically a refined version of traditional sand blasting. Propellant gases can be either air or an inert gas if the surface is naturally reactive. The units can be quite small (desktop) and relatively inexpensive. **Maskants** are frequently applied to the surface to prevent undesired overblasting. The main advantages are its flexibility, low heat production, and the ability to machine-hard and -brittle materials. Common blast media include aluminum oxide, silicon carbide and glass beads. Typical uses of abrasive jet machining include:

- Drill a 0.008-in. hole, without burrs, in the side of a hypodermic needle.
- Trim a 0.003-in. bubble from the edge of a gallium arsenide (GaAs) wafer.
- Precisely prepare dental molds for crowns.
- Dice electronic substrates without fracturing the margins.
- Remove oxidation from a priceless masterpiece without disturbing the original surface of the painting.

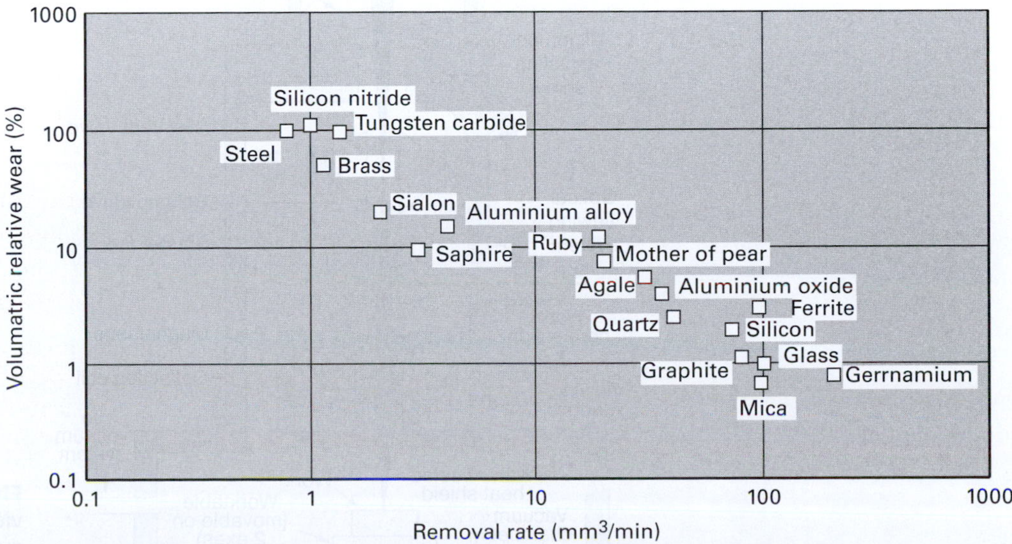

FIGURE 36-6 Ultrasonic machinability of various materials. *(From McGeough, 2002)*

Water-Jet and Abrasive Water-Jet Cutting. **Water-jet cutting** and **abrasive water-jet cutting** were discussed in detail as nontraditional machining methods. The **kerf,** or width, of the cut can be changed by changing parts in the nozzle, as well as the type and size of abrasive. Typical abrasive cuts are made with a kerf in the range of 0.04 to 0.05 in. (1.016 to 1.27 mm), but can be as narrow as 0.02 in. (0.508 mm). Nonabrasive cuts are normally 0.007"to 0.013 in. (0.178 to 0.33 mm), but can be as small as 0.003 in. (0.076 mm), which is approximately the width of a human hair. These small jets can make very small detail possible in a wide range of applications. Water-jets are capable of attaining accuracy of 0.005 in. (0.13 mm), and repeatability of 0.001 in. (0.03 mm). Although large commercial water-jets easily cut granite and steel at high speeds in thicknesses approaching 1 ft in thickness, there is a size effect as one miniaturizes the water-jet process due to backscatter/backsplash.

BEAM-ENERGY-BASED (DISSOLUTION) MICROMACHINING

Electron-Beam Machining. The operation of **electron-beam machining (EBM)** utilizes the the phenomenon of primary knock out when electrons are generated within a vacuum chamber (Figure 36-7) and then impinged upon a target surface. Electron-beam machining rates are usually evaluated in terms of the number of pulses required to evaporate a particular amount of material. As such, it is considered to be a dissolution method of machining. Some of the earliest work on the utilization of the electron beam for material removal can be attributed to Steigerwald, who designed a prototype machine as early as 1947. The components of an electron beam machining system are housed in a vacuum chamber, then evacuated to about 10^{-4} torr. The source of electrons or "electron gun" is basically a triode consisting of a cathode, a grid cup negatively biased with respect to the cathode to allow control, and an anode at ground potential. The cathode is usually made of a tungsten filament, which is heated to between 2500 and 3200°C to act as an electron emitter. After acceleration away from the electron emitter, the electrons are focused by the field so that they travel through an

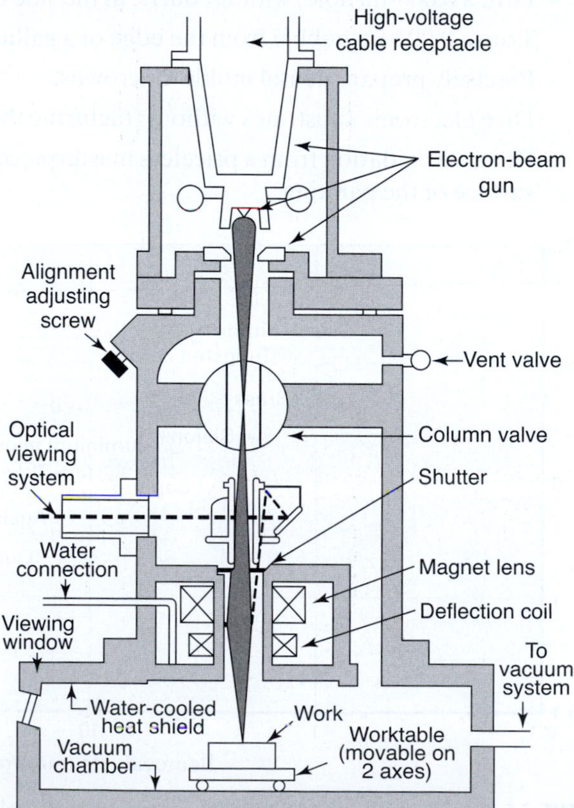

FIGURE 36-7 Cross-sectional view of an electron-beam micro cutter–welder. *(From Walker, 2004)*

aperture in the anode. On its exit from the anode cavity, the electron beam is refocused by a magnetic or electrostatic lens system. This lens system guides the beam toward the workpiece. The electrons maintain the velocity imparted by the accelerating voltage, until they strike to work the specimen, over a well-defined small area (typically 0.025 mm in diameter or less). There, the kinetic energy of electrons is rapidly translated into heat, causing a correspondingly rapid increase in the temperature of the workpiece, to well above its boiling point or plasma point, causing **dissolution** of the target. Material removal by evaporation of the solution then occurs. With power densities on the order of 2 MW/mm^2, virtually all engineering materials can be machined by this technique. Accurate manipulation of the workpiece coupled with precise control of the beam can yield a process that may be fully automated within a large vacuum chamber.

A strong attraction of EBM is a comparatively large depth-to-width ratio penetrated by the beam with applications of very fine hole drilling. The depth to which the electron beam can penetrate a material has received close attention in the literature, but the analytic expressions produced have often been extremely complicated. Kaczmarek derived expressions for the average depth of material removed by a single pulse. In practice, the number of pulses needed to produce a given hole depth is usually found to decrease with an increase in accelerating voltage. Kaczmarek also showed that for a fixed set of process conditions, the number of pulses required increases hyperbolically as the depth increases.

One must consider the limitations of surface finish and the material effects of a **heat-affected zone (HAZ)** when choosing an EBM method. The quality or surface roughness of the edges produced in EVM depends greatly on the type of material being machined. Local pitting of the surface is a common occurrence, the extent of which is influenced by the thermal properties of the workpiece and by the pulse energy or charge. Similarly, the surface layers of materials treated by EBM are affected by the high temperatures of the focused beam, frequently illustrated by white ring surrounding the hole. When using high-energy pulses, the heat-affected zone can be as much as 0.25 mm additional radius.

Although the size of the vacuum chamber can be a limiting factor in the conventional world of machining, quite the opposite is true in the world of micro/meso/nano fabrication, where the targets are frequently so small as to require magnification simply to be seen. Specialized scanning electron microscopes (SEMs) or transmission electron microscopes (TEMs) have been constructed to provide both the machining and necessary visual magnification for the operator. Common applications line the following main areas:

1. Drilling.
2. Sheet perforation.
3. Integrated circuit fabrication through pattern generation (i.e., electron beam milling).
4. Texturing.

Electron-beam machining is widely valued for its ability to drill holes of an exceptionally small radius to depths approaching 10 mm. This ability to drill holes of such a small radius rapidly over the surface of a foil at a rate of 10,000 to 100,000 holes per second also makes it unsurpassed for the production of sieves in manufacturing. Due to the complexity of the vacuum chamber and beam control system, prices range from $75,000 up to $2 million for an EBM machine capable of manufacturing materials on a large scale.

Laser-Beam Micromachining. **Laser-beam micromachining** is based on the interaction of laser light with solid matter. Two different phenomena have been identified for the removal of small amounts of material from the surface of the solid via laser micromachining: **pyrolithic** (thermal) and **photolithic** processes. In each case, short to ultra-short laser pulses are applied in order to remove small amounts of material in a controlled way.

Pyrolithic processes are based on a rapid thermal cycle: heating, melting, and (partly) evaporation of the heated volume. In the case of a photolithic process, the

TABLE 36-1 Laser Micromachining Applications

Ablation	Wire stripping	Soldering
Cutting	Marking	Trimming
Drilling	Welding	Hardening
Decoration	Structuring	Texturing

photon energy itself is sufficient for the direct breaking of the chemical bonds in a wide variety of materials. To date, it has been applied mostly on polymers using ultraviolet lasers in the wavelengths of 157 to 351 nm. Because the photon energy is converted directly into breaking chemical bonds, there is almost no thermal interaction with the product itself. The reaction products escape as gas or small particles. Table 36-1 summarizes current applications of the laser micromachining processes. With the development of new lasers such as *ultra-short-pulsed lasers* and *passively Q–switched microlasers,* new applications continue to arise. From the beginning of laser technology in the 1960s, a reduction in size by a factor of 2 every 7 yr has been observed. Although the cost of production equipment is growing much faster (three to five times every 7 yr), the cost of products is being reduced, owing to the higher production volumes achieved. As an example, the complicated optics of a step-and-repeat camera used in semiconductor production now costs well over $1 million.

The mechanism of laser-beam interaction with the material being removed dictates the frequency of the laser and the presence/absence of damage to the materials. If a suitable frequency can be found for the material to allow for photolithic removal, this is highly desirable—regardless of cost—because there is no heat damage to the target. If the material properties and laser frequency do not allow for this, then the designer must allow for absorption, heat conduction, melting, expansion and contraction, evaporation, and the presence of disruptive plumes.

Electrodischarge Micromachining. **Electrical discharge machining (EDM),** sometimes known as **spark erosion,** employs electrical energy to remove metal from the workpiece without touching it. As described in Chapter 28, a pulsating high-frequency electric current is applied between the tool point in the workpiece, causing sparks to arc and jump the gap. The electric arc generates a state change from solid immediately to plasma, thereby causing apparent dissolution of the material. Because no cutting forces are involved, light and delicate operations can be performed on thin workpieces and small targets. Plunge EDM (sometimes called Ram EDM because early machines looked like a small battering ram) uses a shaped electrode made from graphite or copper. The electrodes separated by nonconductive liquid and maintained at a close distance (about $\frac{1}{1000}$ in.) from the workpiece. A high DC voltage is pulsed to the electrode and jumps to the conductive workpiece. The resulting sparks dissolve the workpiece and result in a cavity, which is in the reverse shape of the electrode. A through-hole can also be drilled with the applying electrode. This process is also known as **die sinking** or **sinker electrical discharge machining. Wire electrical discharge machining** is similar to sinker electrical discharge machining except a small diameter wire is used as a traveling electrode (see Chapter 28).

As might be expected, commercial micro-EDM equipment has been produced by companies in Switzerland and Japan, where large knowledge centers of excellence in microtechnology and precision engineering exist. EDM has been applied on a wide variety of micro-parts from electrically conductive materials—including metals, alloys, sintered metals, cemented carbides, ceramics, and silicon. Sinker electrical discharge machining is often used to produce molds and dies that can themselves be utilized to manufacture other micro-parts from both conductive and nonconductive materials such as plastics.

Fine holes to be machined in conductive materials use a thin rod or terminated wire as an electrode, providing high aspect ratios (long axis to short axis of the hole) that are not achievable by other machining methods. The Swiss company Charmilles

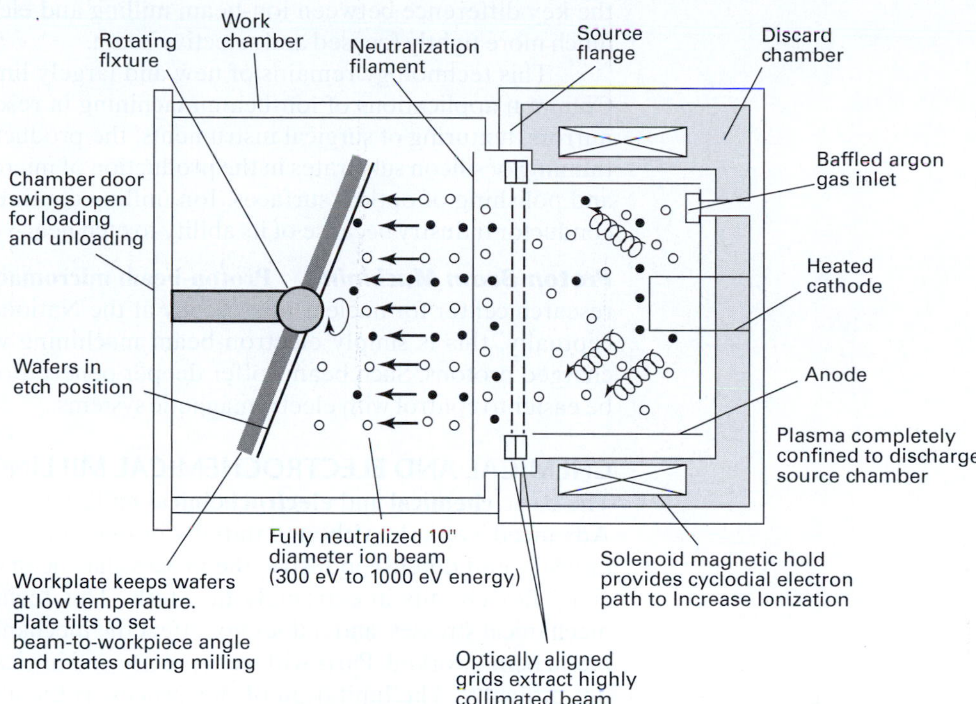

FIGURE 36-8 Typical construction of an ion-beam chamber. *(www.ionbeammilling.com)*

Technologies originally developed micro-EDM machines to manufacture parts for the inkjet printer industry, particularly the injection nozzles for bubble-jet color printers. Other applications of micro-electro-discharge machining include wheels, gears, micro-surgery forceps, and micro-scissors for plastic surgery. A limiting factor in the miniaturization of these techniques is the need for micro tools and dies to produce the next, smaller generation of wires and electrodes.

Ion-Beam Machining. Conceptually, **ion-beam machining** is similar to electron-beam machining because it also occurs in a vacuum chamber and an energy beam is directed at a target. However, ion-beam machining is more physically like sputtering than electron-beam machining. The ion-beam machine (Figure 36-8) has several main components:

1. A plasma source that generates the ions.
2. Extraction grids for removing the ions from the plasma and accelerating them toward the substrate (or target).
3. A table for holding/manipulating the target.

As indicated in Figure 36-8, a heated filament, usually tungsten, acts as a cathode, from which electrons are accelerated by means of a high-voltage greater than 1 kV toward the anode. Impingement upon the anode produces a cloud of ions, which drift toward the accelerating grid. Upon encountering the electrical fields of the grid, they accelerate rapidly and are directed by the shape of the grid and in the surrounding electrical coils toward the final target.

In the simplest of terms, ion-beam milling can be viewed as an atomic sandblaster. The grains of sand are replaced by submicron ion particles, which are then accelerated, bombarding the surface of the work mounted on a movable table inside the vacuum chamber. The workpiece is typically a wafer, substrate, or element that requires material removal. Alternatively, a selectively applied protective, photosensitive resistant may be applied to the workpiece prior to introduction into the island miller. The resistant protects the underlying material during the etching process that may last up to 8 or 12 hr, depending on the amount of material to be removed. Beams up to 15 in. in diameter can be applied over a large surface. This massive column of directed beam energy is the key difference between ion-beam milling and electron-beam milling, which has a much more tightly focused and selective beam.

This technology remains of new and largely limited to the research laboratory. Common applications of ion-beam machining in research include smoothing of laser mirrors, texturing of surgical instruments, the production of atomically clean surfaces, thinning of silicon substrates in the production of microprocessor chips, and the shaping and polishing of optical surfaces. Ion milling continues to be of interest to the semiconductor industry because of its ability to etch near-vertical walls in substrates.

Proton-Beam Machining. **Proton-beam micromachining** is being developed at the research center for nuclear microscopy at the National University of Singapore. Conceptually, this is simply electron-beam machining with a positive particle beam of charged protons. Such beams offer deeper penetration, faster removal rates, and may be easier to control with electromagnetic systems.

CHEMICAL AND ELECTROCHEMICAL MILLING

The basic **chemical** and **electrochemical milling** processes were explained in detail in Advanced Topic 1. Although initially designed to lighten airframes in the aerospace industry and defense industry, the process has been easily extended for creating specialty flex circuits in extremely thin foils. This technique avoids burrs and residual mechanical stresses, and it does not affect the mechanical or electrical properties of the metal being worked. Parts with very precise and intricate designs can be produced without difficulty. The limitation of this process is the ability of the operator to produce ever-finer meshes. Electron beams, abrasive blast jets, and proton beams have been applied in the production of increasingly intricate masks for the process.

FINISHING PROCESSES

All of the methods previously described in this chapter are capable of building shapes at the micro and meso levels. Some of them such as electron-beam machining have entered the nano level successfully. All of these processes, therefore, require **finishing processes,** which are capable of working at the atomic level of material removal. These finishing methods have historically been referred to as superfinishing methods in earlier literature. Some processes such as elastic emission machining work directly on the molecules, removing them from the workpiece surface, while other processes based on finishing by abrasives remove them in clusters. Most of the finishing processes are using abrasive particles either suspended in liquid or held by viscoelastic material, carbonyl iron particles, or magnetorheological fluids.

Typical comparisons to traditional grinding, honing, and lapping processes are summarized in the Table 36-2.

Elastic Emission Machining. If the material removal of a process can occur at the atomic level, then the finish generated can be close to the order of atomic dimensions

TABLE 36-2 Comparison of Traditional Finishing Results to Micro/Meso/Nano Results

Finishing Process	Workpiece	R_a value (nm)
Grinding	Various	25–6250
Honing	Various	25–1500
Lapping	Various	13–750
Abrasive flow machining	Hardened steel	50
Magnetic abrasive finishing	Stainless steel	7.6
Magnetic flow policy	Si_3N_4	4.0
Magnetorheological finishing	Flat BK7 glass	0.8
Elastic emission machining	Silicon	<0.5
Ion-beam machining	Cemented carbide	<0.1

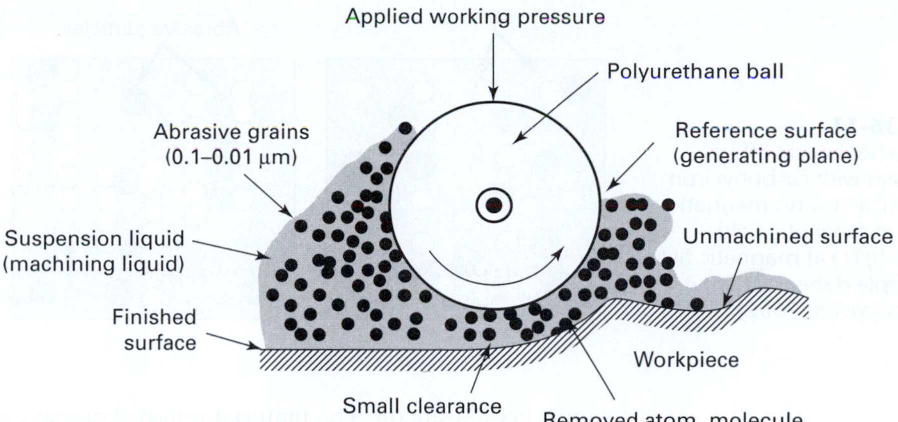

FIGURE 36-9 Rotating sphere and workpiece interface in EEM. *(From Mahalik, 2010)*

(0.2 to 0.4 nm). Using ultra-fine particles to collide with the workpiece surface, it may be possible to finish the surface by the atomic scale unit elastic fracture without plastic deformation (Mahalik, 2010). This new process is termed **elastic emission machining (EEM).** Mori (1987) established theoretically and experimentally that atomic scale fracture can be induced elastically and that the finished surface can be undisturbed crystallographically and physically. The rotating sphere and workpiece interface inelastic emission machining is illustrated in Figure 36-9.

In the EEM process, a polyurethane ball 56 mm in diameter is mounted on a shaft driven by a variable-speed motor (Figure 36-10). The axis of rotation is oriented and angled at approximately 45 degrees relative to the surface of the workpiece to be polished. The workpieces are submerged in the slurry of ZrO_2 or Al_2O_3 abrasive particles and water. The material removal rate for the workpiece was found to be linear with dwell time in a particular location, although it varied nonlinearly with concentration of abrasive particles in the slurry. The proposed mechanism the material removal due to slurry and workpiece interaction involves erosion of the surface atoms by the bombardment of abrasive particles without the introduction of dislocations. Surface roughness as low as 0.5 nm root-mean-square (rms) has been reported on glass and 1 nm rms on

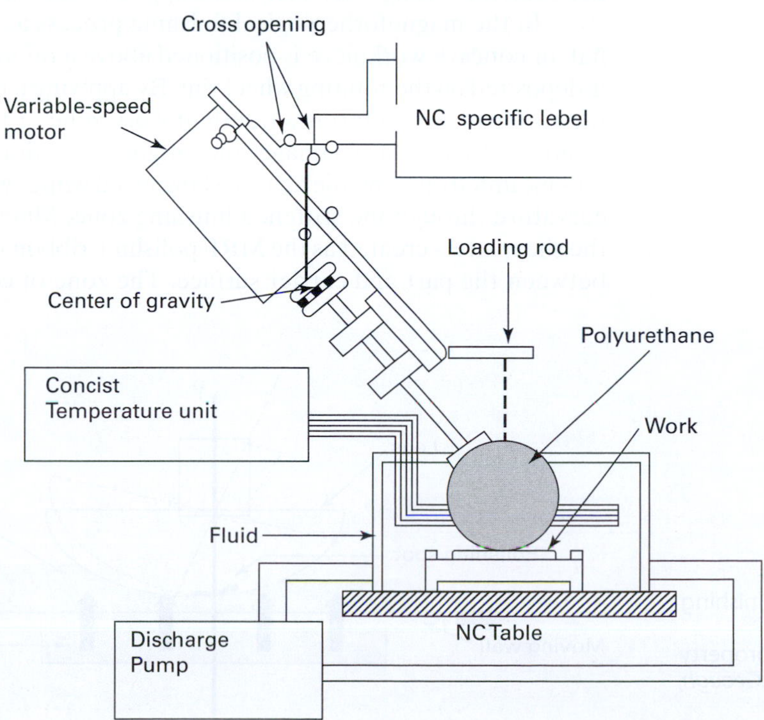

FIGURE 36-10 Schematic of EEM assembly used on numerically controlled machine. *(From Mor, 2001i)*

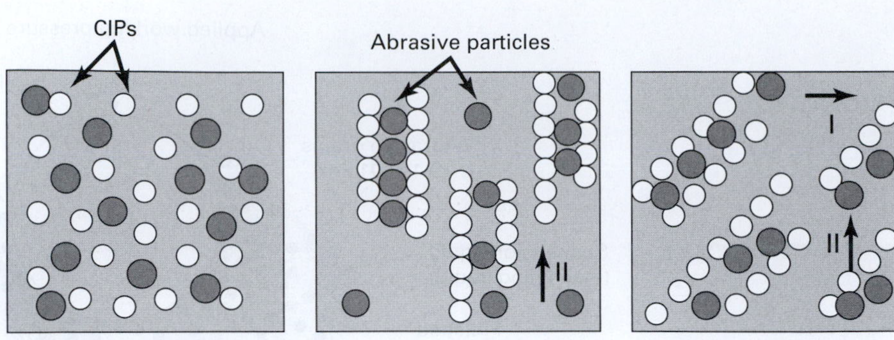

FIGURE 36-11 Magnetorheological effect: (a) MR fluid with carbonyl iron particles (CIPs) at no magnetic field; (b) at magnetic field strength, *H*; (c) at magnetic field, *H*, and applied shear strain, *g*. *(From McGeough, 2002)*

single crystal silicon. The material removal process is a surface energy phenomenon in which each abrasive particle removes a number of atoms after coming in contact with the surface. The type of abrasive used has been found to be critical to the removal efficiency.

Magnetorheological Machining. The **magnetorheological finishing (MRF)** process a relies on a unique "smart fluid" known as magnetorheological (MR) fluid. MR fluids are suspensions of micron-sized magnetizable particles, such as carbonyl iron, dispersed in a nonmagnetic carrier medium, such as silicone oil, mineral oil, or water. In the absence of a magnetic field, an ideal MR fluid exhibits Newtonian behavior. Such a Newtonian fluid has a linear stress versus strain rate curve, which passes through the origin. The slope of this line is the viscosity of the material. In common terms, the fluid will continue to flow, regardless of the forces acting on it. For example, water is Newtonian because it continues to exemplify fluid properties no matter how fast it is mixed. Contrast this with a non-Newtonian fluid, in which either stirring can leave a hole behind that gradually fills up over time or the fluid can climb the stirring rod because of shear thinning (the drop in viscosity causing it to flow more).

On the application of an external magnetic field to the MR fluid, a phenomenon known as the magnetorheological effect (shown in Figure 36-11) can take place. Because energy is required to form and rupture the chains, this microstructural transition is responsible for the onset of a large controllable finite yield stress. Figure 36-11c shows an increasing resistance to the applied shear strain, *g*, due to this yield stress.

In the magnetorheological finishing process as shown in Figure 36-12, a convex, flat, or concave workpiece is positioned above a reference surface. An MR fluid ribbon is deposited on the rotating wheel rim. By applying a magnetic field in the gap interface, the stiffened region forms a transient work zone or finishing spot. Surface smoothing, removal of subsurface damage, and fatigue correction are accomplished by rotating the lens mounted on a spindle at a constant speed while sweeping the lens about its radius of curvature through the stiffened finishing zone. Material removal takes place through the shear stress created as the MRF polishing ribbon is dragged into the converging gap between the part and carrier surface. The zone of contact is restricted to a spot that

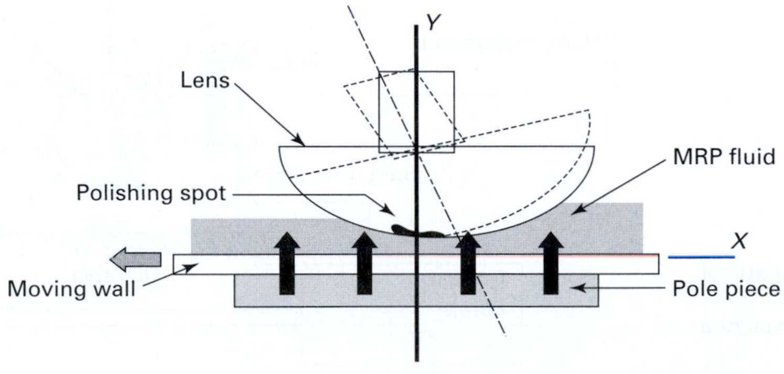

FIGURE 36-12 The magnetorheological finishing process using magnetorheological property (MRP) fluid. *(From McGeough, 2002)*

conforms perfectly to the topography of the part. Deterministic finishing of flats and spheres can be accomplished by mounting the part on a rotating spindle and sweeping it through the spot under computer control. The dwell time over the polishing spot determines the amount of material removal.

The MRF process has the following advantages over traditional lapping techniques:

- Compliance is adjustable through the magnetic field.
- Heat and debris are naturally carried away from the polishing zone.
- It does not load up as a grinding wheel.
- It is flexible and adapts to the shape of the part.

The computer-controlled MRF process has demonstrated the ability to produce a surface accuracy of the order of 10 to 100 nm peak-to-valley by overcoming many fundamental limitations inherent to traditional finishing techniques. MRF is regarded as the premier method for the high-precision finishing of optical equipment. Applications that use these high-precision lenses include medical equipment such as endoscopes, collision avoidance devices for transportation industries, scientific testing devices, and night vision equipment used by the military. Intercontinental ballistic missiles are equipped with a wide variety of high-precision lenses for navigation, target location, and other functions. Nano diamond-doped MR fluid removes edge chips, cracks, and scratches in sapphire for laser applications.

Magnetorheological Abrasive-Flow Finishing. In the MRF process, the medium is limited to only the magnetorheological fluid itself. When fine abrasive particles are dispersed in the MR fluid, then the process is said to be **magnetorheological abrasive-flow finishing (MRAFF).** Typically, carbonyl iron is used as the additive due to its magnetic susceptibility. This results in a higher material removal rate than the basic MRF process.

Magnetic Float Polishing. The applications of advanced ceramics are often limited because of their poor machinability and the difficulties involved in processing and useful shapes. They are extremely sensitive to surface defects resulting from grinding and polishing processes. Because fatigue failure of ceramics is driven by surface imperfections, it is of the utmost importance that the quality and finish of the elements of ceramic bearings the superior with minimal defects. A gentle and flexible polishing condition with a low level of control forces and the use of abrasive very slightly harder than the work material are required. **Magnetic float polishing** was developed to address this requirement. A typical instrument is shown in Figure 36-13.

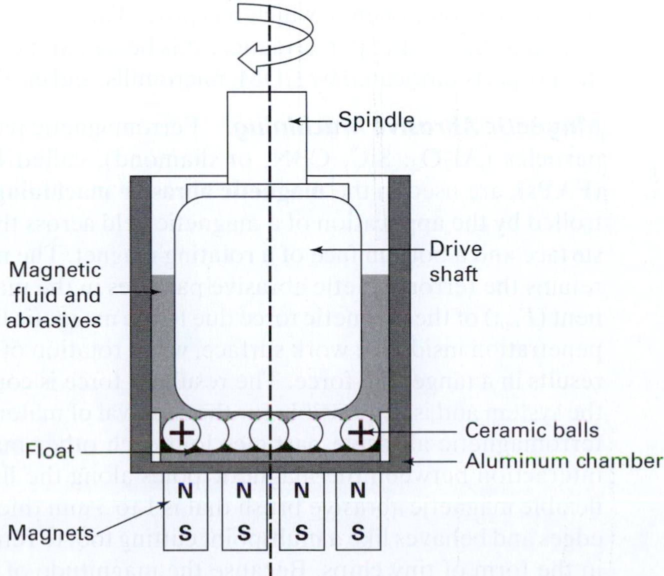

FIGURE 36-13 Schematic of the magnetic float polishing apparatus. *(From Jain, 2010)*

The magnetic float polishing technique is based on the ferro–hydrodynamic behavior of magnetic fluids that can levitate a nonmagnetic float and abrasive suspended in it by magnetic field. The levitation force applied by the abrasives is proportional to the magnetic field gradient and is extremely small and highly controllable. A magnetic fluid containing fine abrasive grains and extremely fine ferromagnetic particles in a carrier fluid such as water or kerosene fills the aluminum chamber. A bank of strong electromagnetics is arranged alternately north and south below the chamber. When a magnetic field is applied, the ferro-fluid is attracted downward toward the area of higher magnetic field, and an upward buoyant force is exerted on the nonmagnetic material to push it to the lower magnetic field. The buoyant force acts on the nonmagnetic body in the presence of magnetic fluid subject to the magnetic field. Although the abrasive grains, target (e.g., ceramic micro balls), and acrylic suspension now inside the chamber are of nonmagnetic materials, the magnetic buoyant force levitates all. The driveshaft is set downward to contact the balls and presses them downward to reach the desired normal force acting against the surfaces. The balls are polished by the relative motion between the balls and the abrasives under the influence of levitation force. In this polishing method, both a higher material removal rate and smoother surface finish are attained by a stronger magnetic field and finer abrasives.

Magnetic flow polishing and has been used to produce ceramic bearings and ultra-high-speed precision spindles of machine tools and jet turbines of aircraft. It is being investigated for use in micro/meso/nano surgical tools.

Abrasive-Flow Machining. **Abrasive-flow machining (AFM)** was identified in the early 1960s as a method to deburr, polish, and radius difficult-to-reach surfaces and edges by flowing abrasive-laden **viscoelastic polymer** over them. It uses two vertically opposed cylinders, which drive the abrasive medium back and forth through passages formed by the actual workpiece itself, and tooling, which guides the medium to the passages (Figure 36-14). Abrasive finishing occurs wherever the medium passes through the restrictive passages of the workpiece. Extrusion pressure, number of cycles, grit composition and type, and the fixture design are the process parameters that have the largest impact on AFM results. To formulate the AFM medium, the abrasive particles are blended into special viscoelastic polymers, which show a remarkable change in viscosity when forced to flow through restrictive passages. Inaccessible areas and complex internal passages can be finished economically and productively. It reduces surface roughness by 75 to 90% on cast, machined, or electrical discharge machined surfaces.

AFM offers simplicity, precision, consistency, and flexibility. Aerospace, medical components, electronics, automotive parts, and precision dies and molds manufacturing industries are extensively using AFM as a part of their manufacturing activities. Recently, AFM has been applied to the improvement in air and fluid flow for automotive engine components, which has proved an effective method for lowering emissions as well as increasing performance. It is being extended to the nano level as a means to deburr parts fabricated by EDM, micromills, and castings.

Magnetic Abrasive Machining. Ferromagnetic particles sintered with fine abrasive particles (Al_2O_3, SiC, CBN, or diamond), called **ferromagnetic abrasive particles (FAPs),** are used in the **magnetic abrasive machining** process. Finishing action is controlled by the application of a magnetic field across the gap between the workpiece top surface and a bottom face of a rotating magnet. The magnetic field acts as a binder and retains the ferromagnetic abrasive particles in the machining gap. The normal component (F_{mn}) of the magnetic force due to the magnetic field is responsible for the abrasive penetration inside the work surface, while rotation of the ferromagnetic abrasive brush results in a tangential force. The resultant force is considered to be the cutting force in the system and is responsible for the removal of material in the form of microchips. The ferromagnetic abrasive particles join each other magnetically due to dipole–dipole interaction between the magnetic poles along the lines of magnetic force, forming a flexible magnetic abrasive brush that is 1 to 3 mm thick. This brush has multiple cutting edges and behaves like a multipoint cutting tool to remove material from the workpiece in the form of tiny chips. Because the magnitude of the machine force caused by the

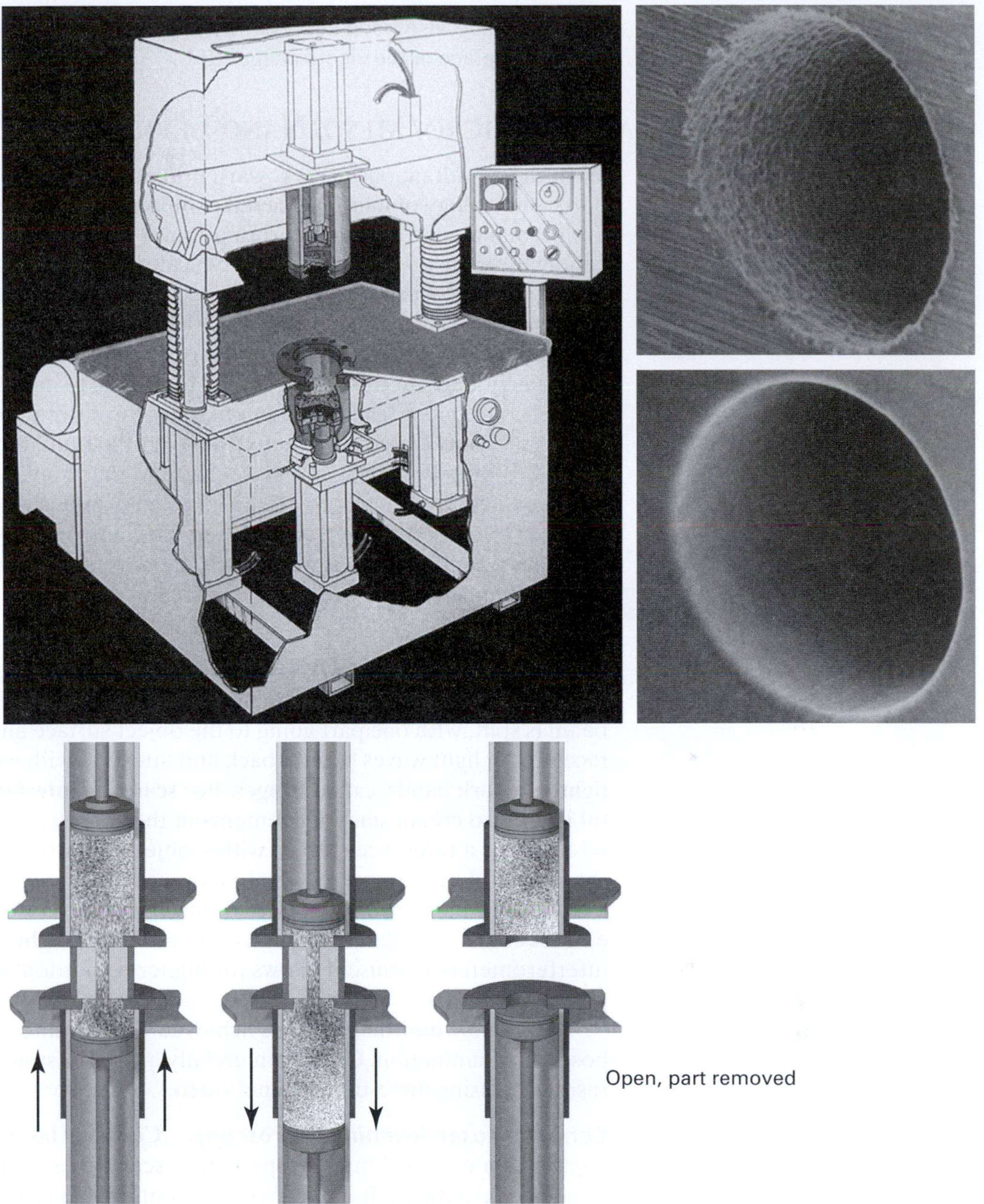

FIGURE 36-14 An abrasive flow machine produces the pressure necessary to force media through or across a workpiece. Abrasive action takes place wherever flow is restricted as the media is forced back and forth between two opposed media cylinders. The before and after surfaces are remarkably different as the above SEM micrographs show. *(Courtesy J T. Black)*

magnetic field is very low but controllable, a mirror-like surface finish (R_a value in the range of a nanometer) is easily obtained. The process is also used to perform operations such as polishing and the removal of thin oxide films from high-speed rotating shafts. Such shafts are commonly found cylindrical workpieces such as tubes, vacuum tubes, and sanitary tubes.

The material removal rate and finishing rate depend on the workpiece circumferential speed, working gap, and material hardness, as well as the size, type, and volume fraction of abrasives. Finishing of stainless steel rollers using magnetic abrasive finishing has obtained final R_a of 7.6 nm at an average finishing rate of 7.08 nm/s (Komanduri, 1996). Magnetic abrasive finishing produces mechanical electronic components with high accuracy and high surface finish having hardly any surface defects. Strong earth magnets have also been successfully applied for micro-deburring using a permanent magnet in place of the electromagnetic.

36.3 METROLOGY AT THE MICRO/MESO/NANO LEVEL

With the simultaneous drive toward **miniaturization** and global quality control standards, **metrology** of micromachined components becomes critical. Tools for metrology of micromachined components are primarily derived from the techniques of the semiconductor industry. These tools are typically limited to lateral dimensions and are insufficient for measuring the overall part geometry of micromachined mechanical components. The components can also have moving parts, and the dynamic behavior of the component may be critical to the operation, requiring direct observation. Therefore, the modern metrology tools need to be able to measure the dynamics of the miniaturized component. Table 36-3 summarizes the current methods being used by industry and researchers to accomplish metrology at the miniaturized level. Discussion of each of these methods in detail would require another book and is beyond the scope of this textbook. Karhade et al. provide an excellent summary description of these methods in Chapter 15 of the Jain reference. A very brief introduction is provided here for some of the more popular methods with which students may not already be familiar. Table 36-3 is an evolving list of the evolving field of nano metrology.

SCANNING WHITE LIGHT INTERFEROMETRY

An **interferometer** works on the principle of interference. Within the objective, a light beam is split, with one part going to the object surface and the other to a reference surface. These light waves bounce back and interfere with each other, forming a pattern of light and dark bands, called **fringes.** For **scanning interferometers,** a piezoelectric crystal is used to create small movements in the objective perpendicular to the surface of interest. As a reference surface within objective moves, the result of a combination of the reflected light varies. Several images are captured and then combined. Based on interference pattern, or fringes, and the wavelength of light employed, it is possible to extract coordinate data forming an image. **White light** is commonly used in scanning interferometers because it allows for higher resolution by comparing data from multiple wavelengths.

Dynamic metrology can be achieved with a white light interferometer using stroboscopic illumination. One commercially available system is capable of a 0.1-nm-scale resolution using three-dimensional video.

Confocal Laser Scanning Microscopy. **Confocal laser scanning microscopy (CLSM)** combines a confocal microscope with a scanning system in order to gather a three-dimensional data set. It differs from conventional microscopy in that it creates an image point by point. Also, because of the double-pinhole lens system employed, when the samples move out of the focal plane of the objectives, the light intensity at the detector decreases rapidly—in effect allowing the system to focus on single plane. A different plane can be imaged by moving the detection pinhole. When combined with a scanning system, the system has the ability to scan multiple times on different imaging planes, resulting in a three-dimensional data set. The system is unique in its ability to measure

TABLE 36-3 Metrology Methods at the Micro/Meso/Nano Manufacturing Level

Scanning electron microscope	Molecular measuring machine (M^3)
Computed tomography	Autofocusing probing
Scanning tunneling microscope	Atomic force microscopes
Optical microscopy	Digital volumetric imaging
Scanning white light microscopy	Confocal laser scanning microscopy
Scanning laser doppler vibrometry	Digital holographic microscopes
Fringe projection microscopy	Micro-coordinate measuring machines

steep slopes from overhead, almost achieving 90 degrees. This makes it a relatively inexpensive ($75,000) system capable of three-dimensional imaging of small electronics packages.

SCANNING PROBE MICROSCOPY

Scanning probe microscopes (SPMs) offer a high-resolution (sub-Angstrom) alternative to noncontact techniques. The two most widely used SPMs are the **scanning tunneling microscope (STM)** and the **atomic force microscope (AFM).** The older of the two technologies, the STM, uses a metallic probe brought into close proximity of the conductive surface so a small current flows between the probe and surface. The current is held constant by feedback control schemes, allowing the probe to track the height of the surface, generating a three-dimensional map. Sub-Angstrom resolution is attainable in the normal direction of the surface, and Angstrom-scale resolution is obtainable in the lateral direction of the surface.

Atomic force microscopy is a newer SPM technology; it retains the resolution of the STM but is not limited to conductive surfaces. The measurements of an AFM are performed with a sharp probe that collects a series of line scans across the surface of the part. The typography of the part is measured by bringing the probe close to the specimen and measuring the repulsive and attractive forces on the probe tip. An AFM is capable of working in both a contact and noncontact mode to collect surface data. In contact mode, the method of data acquisition is similar to a stylus profile device (Chapter 37), where the probe tip slides along the surface of the specimen to measure the relative height changes. In noncontact mode, the Van der Waal's forces between the probe tip and specimen are measured and converted to coordinate data. Having the advantage of not contacting the surface of the specimen and therefore eliminating tip erosion and vibration, this method has better resolution but is less stable than the than the sliding contact devices.

KEY WORDS

abrasive water-jet cutting
abrasive-flow machining (AFM)
abrasive-jet machining
atomic force microscope (AFM)
chemical milling
chemical vapor deposition (CVD)
confocal laser scanning microscopy (CLSM)
deposition
die sinking
dissolution
dynamic metrology
elastic emission machining (EEM)
electrical discharge machining (EDM)
electrochemical milling
electron-beam machining (EBM)
fabrication
ferromagnetic abrasive particle (FAP)
free molecular regime of space
fringes
heat-affected zone (HAZ)
interferometer
ion-beam machining
kerf
laser-beam micromachining
macro fabrication
magnetic abrasive machining
magnetic flow polishing
magnetorheological abrasive-flow finishing (MRAFF)
magnetorheological finishing (MRF)
manufacturing
maskants
mechanical micromachining
meso fabrication
metrology
micro fabrication
microcracks
miniaturization
nano fabrication
photlithic
photolithography
physical vapor deposition (PVD)
proton-beam micromachining
pyrolithic
scanning interferometer
scanning probe microscope (SPM)
scanning tunneling microscope (STM)
sinter electrical discharge machining
sonotrode
spark erosion
spin-on deposition
sputtering
ultrasonic micromachining
viscoelastic polymer
water-jet cutting
white light
wire electrical discharge machining

REVIEW QUESTIONS

1. In the spin-on deposition process, a droplet falls onto the target surface and then spreads as the subsurface rotates.
   a. What process variables are at play in this simple system?
   b. What are the possible defects in this type of deposition casting?
2. Which is the more expensive process, chemical vapor deposition or physical vapor deposition? Justify your answer.
3. What are the mechanisms for material removal in the ultrasonic machining process?
4. What are the advantages of abrasive micromachining?
5. Search of the World Wide Web for abrasive micromachining companies within your state. What uses are they advertising for the process within your state?
6. What is the limiting factor in applying water jet machining techniques at the micro/meso level?
7. Is there any commercially useful material which may not be machined with the electron beam method? What are the practical limitations of this method?
8. What is the mechanism which causes the formation of a heat affected zone in a material that has been subjected to electron beam machining?
9. Why is there no heat affected zone in laser beam machining as compared to electron beam machining?
10. In what ways is electrodischarge machining different than electron beam machining?
11. What are three common micro/meso applications of plunge EDM?
12. Ion beam machining differs from electron beam machining in what way?
13. Most of the true nano-machining processes occur during super finishing? What acts as the tool holder/carrier in most of these applications?
14. Compare the advantages and disadvantages of abrasive micromachining to abrasive flow machining.
15. Why is white light normally preferred for interferometers?
16. Atomic force microscopes are best used for what applications?

INDEX

Selected References for Additional Study

J T. Black and Ronald A. Kohser

HANDBOOKS AND GENERAL REFERENCES

ASM Engineered Materials Handbook Series, ASM, Materials Park, OH.

Vol. 1, Composites (1987)
Vol. 2, Engineering Plastics (1988)
Vol. 3, Adhesives and Sealants (1990)
Vol. 4, Ceramics and Glasses (1991)

ASM Engineered Materials Handbook, Desk Ed., ASM, Materials Park, OH (1995).

ASM Engineered Materials Reference Book, 2nd ed., ASM, Materials Park, OH (1993).

ASM Metals Reference Book. 3rd ed., ASM, Materials Park, OH (1993).

ASME Handbook, McGraw-Hill, New York.

Vol. 1, Metals Engineering: Design
Vol. 2, Metals Properties
Vol. 3, Engineering Tables
Vol. 4, Metals Engineering: Processes

ASTM Standards (multiple volumes, published annually), ASTM, Philadelphia, PA.

Machinery's Handbook, 28th ed., Industrial Press (2008).

Manufacturing Engineering Handbook, Hinauyu Geng, McGraw-Hill (2004).

Manufacturing Processes Reference Guide, R.H. Todd, D. K. Allen, and. F. Alting, Industrial Press (1994).

Mark's Standard Handbook for Mechanical Engineers, 11th ed., E. A. Avallone and T. Baumeister III (eds)., McGraw Hill (2006).

Materials Handbook, 14th ed., George S. Brady and Henry R. Clauser, McGraw-Hill (1996).

Materials Handbook: A Concise Reference, 2nd ed., F. Cardelli, Springer (2008).

Metals Handbook, 10th ed., ASM, Materials Park, OH.

Vol. 1, Properties and Selection: Irons, Steels and High-Performance Alloys (1990)
Vol. 2, Properties and Selection: Nonferrous Alloys and Special-Purpose Materials (1990)
Vol. 3, Alloy Phase Diagrams (1992)
Vol. 4, Heat Treating (1991)
Vol. 5, Surface Engineering (1994)
Vol. 6, Welding, Brazing and Soldering (1993)
Vol. 7, Powder Metal Technologies and Applications (1998)
Vol. 8, Mechanical Testing and Evaluation (2000)
Vol. 9, Metallography and Microstructures (2004)
Vol. 10, Materials Characterization (1986)
Vol. 11, Failure Analysis and Prevention (2002)
Vol. 12, Fractography (1987)
Vol. 13A, Corrosion: Fundamentals, Testing, and Protection (2003)
Vol. 13B, Corrosion: Materials (2005)
Vol. 13C, Corrosion: Environments and Industries (2006)
Vol. 14A, Metalworking: Bulk Forming (2005)
Vol. 14B, Metalworking: Sheet Forming (2006)
Vol. 15, Casting (2008)
Vol. 16, Machining (1989)
Vol. 17, Nondestructive Evaluation and Quality Control (1989)
Vol. 18, Friction, Lubrication and Wear Technology (1992)
Vol. 19, Fatigue & Fracture (1996)
Vol. 20, Materials Selection and Design (1997)
Vol. 21, Composites (2001)
Vol. 22A, Fundamentals of Modeling for Metals Processing (2009)
Vol. 22B, Metals Process Simulation (2010)

Metals Handbook, Desk Ed., 2nd ed., ASM, Materials Park, OH (1998).

MMPDS-02: Metallic Materials Properties Development and Standardization, Battelle Memorial Institute (2005) [NOTE: This is the successor to MIL-HDBK-5].

Vol. 1 Introduction and Guidelines
Vol. 2 Steel and Magnesium Alloys
Vol. 3 2000-6000 Series Aluminum Alloys
Vol. 4 7000 Series and Cast Aluminum Alloys
Vol. 5 Miscellaneous Alloys & Structural Joints

SAE Handbook, three-volume set. Society of Automotive Engineers, Warrendale, PA. (2005).

Smithells Metals Reference Book, 8th Ed., W.F. Gale and T.C. Totemcier, Butterworths (2004).

Tool and Manufacturing Engineers Handbook – Desk Edition, Society of Manufacturing Engineers, Dearborn, MI (1988).

Tool and Manufacturing Engineers Handbook, 4th ed., Society of Manufacturing Engineers, Dearborn, MI.

Vol. 1, Machining (1983)
Vol. 2, Forming (1984)
Vol. 3, Materials, Finishing and Coating 1985)
Vol. 4, Quality Control and Assembly (1987)
Vol. 5, Manufacturing Management (1987)
Vol. 6, Design for Manufacturability (1992)
Vol. 7, Continuous Improvement (1994)

Vol. 8, Plastic Part Manufacturing (1996)
Vol. 9, Material and Part Handling in Manufacturing (1997)

Woldman's Engineering Alloys, 9th ed., J. Frick, ASM, Materials Park, OH (2000).

BASIC AND GENERAL TEXTBOOKS IN MATERIALS AND PROCESSES

21st Century Manufacturing, Paul Kenneth Wright, Prentice Hall (2001).

Elements of Metallurgy and Engineering Alloys, Flake Campbell (ed.), ASM, Materials Park, OH (2008).

Engineering Materials, 3rd Ed, 2 volumes, M.F. Ashby and D.R.H. Jones, Butterworth-Heinemann (2005).

Engineering Materials, 8th ed., Kenneth Budinski and Michael Budinski, Prentice Hall (2005).

Foundations of Materials Science and Engineering, 5th Ed., Wm. F. Smith and J. Hashemi, McGraw Hill. (2010).

Fundamentals of Materials Science and Engineering: An Integrated Approach 3rd ed., W.D. Callister and D.G. Rethwisch, Wiley (2008).

Fundamentals of Manufacturing, 3rd Ed., Philip Rufe (ed.), Society of Manufacturing Engineers (2011).

Fundamentals of Modern Manufacturing, 4th ed., Mikell P. Groover, Wiley (2010).

Handbook of Manufacturing Processes: How Products, Components, and Materials are Made, James Bralla, Industrial Press (2007).

Introduction of Manufacturing Processes, 3rd ed., John A. Schey, McGraw-Hill (2000).

Introduction to Manufacturing Processes and Materials, Robert C. Creese, Marcel Dekker (1999).

Introduction to Materials Science for Engineers, 6th ed., James F. Shackelford, Prentice-Hall (2005).

Manufacturing Engineering and Technology, 6th ed., Serope Kalpakjian and Steven R. Schmid, Prentice-Hall (2010).

Manufacturing Processes, B.H. Amstead, P.F. Ostwald and M.L. Begeman, Wiley (1987)

Manufacturing Processes and Equipment, Jiri Tlusty, Prentice Hall, Upper Saddle River, NJ (2000).

Manufacturing Processes and Materials, 5th ed., Ahmad K. Elshennawy and Gamal Weheba, Society of Manufacturing Engineers (2011).

Manufacturing Processes and Systems, 9th ed., Phillip F. Ostwald and Jairo Muñoz, Wiley (1997).

Materials: Engineering, Science, Processing and Design 2nd ed., M. Ashby et al., Elsevier/Butterworth-Heinemann (2010).

Materials Science and Engineering: An Introduction, 8th ed., W.D. Callister and D. G. Rethwisch, Wiley (2010).

Mechanical Metallurgy, 3rd ed., G. E. Dieter, McGraw-Hill, New York (1986).

Principles of Materials Science and Engineering, 4th ed., W. F. Smith, McGraw-Hill (1996).

Processes and Design for Manufacturing, 2nd ed., Sherif D. El Wakil, PWS Publishing Company (1998).

Processes and Materials of Manufacture, 4th ed., Roy A. Lindberg, Allyn and Bacon (1990).

Structures and Properties of Engineering Alloys, 2nd ed., W. F. Smith, McGraw-Hill New York (1993).

The Science and Design of Engineering Materials, 2nd ed., J. P. Schaffer et al., McGraw-Hill (1999).

The Science and Engineering of Materials, 6th ed., D. R. Askeland, P. P. Fulay, and Wendelin Wright, Cengage Learning (2011).

FERROUS METALS

ASM Specialty Handbook - Carbon and Alloy Steels, ASM (1996).

ASM Specialty Handbook - Cast Irons, ASM (1996).

ASM Specialty Handbook - Stainless Steels, ASM (1994).

Cast Iron: Physical and Engineering Properties, H.T. Angus, Butterworths (1976).

Design Guidelines for the Selection and Use of Stainless Steel, Designers Handbook Series, Specialty Steel Industry of the United States, Washington, DC (1992).

Engineering Properties of Steel, ASM, Materials Park, OH (1982).

Handbook of Stainless Steels, Donald Peckner and I. M. Bernstein (eds.), McGraw-Hill. (1977).

Introduction to Stainless Steels, 3rd ed., J. Beddoes. and J.G. Parr, ASM (1999).

Metallurgy and Heat Treatment of Tool Steels, Robert Wilson, McGraw-Hill (1975).

Modern Steels and Their Properties (various editions), Bethlehem Steel Corporation, Bethlehem, PA.

Stainless Steel, R.A. Lula, ASM, Materials Park, OH (1985).

Steel Selection: A Guide for Improving Performance and Profits, R. F. Kern and M. E. Suess, Wiley-Interscience (1979).

Steels: Metallurgy & Applications, 3rd ed. D.T. Llewellyn, Butterworth-Heinemann (1999).

Steels: Heat Treatment and Processing Principles, G. Krauss, ASM, Materials Park, OH (1990).

Steels: Processing, Structure and Performance, George Krauss, ASM, Materials Park, OH (2005)

The Making, Shaping and Treating of Steel, 11th ed., Association for Iron and Steel Technology (1998).

The Tool Steel Guide, Jim Szumera, Industrial Press (2003).

Tool Steels, 5th ed., G.A. Roberts, G. Krauss and R. Kennedy, ASM, Materials Park, OH (1998).

Worldwide Guide to Equivalent Irons and Steels, 5th ed., ASM, Materials Park, OH (2006).

NONFERROUS METALS

Aluminum, ASM, Materials Park, OH (1967).
Vol. 1, Properties, Physical Metallurgy and Phase Diagrams
Vol. 2, Design and Application
Vol. 3, Fabrication and Finishing

Aluminum Standards and Data, The Aluminum Association, Washington, DC (bi-annual updates).

Aluminum: Properties and Physical Metallurgy, John E. Hatch (ed.), ASM, Materials Park, OH (1984).

ASM Specialty Handbook - Aluminum and Aluminum Alloys, ASM (1993).

ASM Specialty Handbook - Copper and Copper Alloys, ASM (2001).

ASM Specialty Handbook - Magnesium and Magnesium Alloys, ASM (1999).

ASM Specialty Handbook - Nickel, Cobalt, and Their Alloys, ASM (2000).

Beryllium Chemistry and Processing, Kenneth A. Walsh, ASM, Materials Park, OH (2009).

Engineering Properties of Zinc Alloys, International Lead Zinc Research Organization, New York (1980).

Introduction to Aluminum Alloys and Tempers, J.G. Kaufman, ASM (2000).

Materials Properties Handbook:Titanium Alloys, R. Boyer, E.W. Collings, and G. Welsch, ASM, Materials Park, OH (1994).

Properties of Aluminum Alloys: Fatigue Data and the Effects of Temperature, Product Form and Processing, J.G. Kaufman (ed.), ASM, Materials Park, OH (2008).

Properties of Aluminum Alloys:Tensile, Creep and Fatigue Data at High and Low Temperatures, J.G. Kaufman (ed.), ASM and Aluminum Association (1999).

Standards Handbook: Copper, Brass, and Bronze (7 volumes) Copper Development Association.

Superalloys: Alloying and Performance, Blaine Geddes, Hugo Leon and Xiao Huang, ASM, Materials Park, OH (2010)

Superalloys: A Technical Guide, 2nd ed., ASM (2002).

The Superalloys: Fundamentals and Applications, C. Reed, Cambridge University Press (2008).

Titanium, 2nd ed., G. Lutjering and J.C. Williams, Springer (2007).

Titanium: A Technical Guide, 2nd ed., M. Donachie (ed.), ASM, Materials Park, OH (2000).

Worldwide Guide to Equivalent Nonferrous Metals and Alloys, 4th Ed., ASM, Materials Park, OH (2001).

CERAMICS

Advanced Technical Ceramics, S. Somiya (ed.), Academic Press (1989).

Ceramic Manufacturing Practices and Technologies, B. Hiremath et al., American Ceramic Society (1997).

Ceramic Materials: Science and Engineering, C.B. Carter and M.G. Norton, Springer (2008).

Ceramic Processing, M.N. Rahaman, CRC Taylor& Francis, (2007).

Ceramics – Applications in Manufacturing, D.W. Richerson, Society of manufacturing Engineers (1989).

Ceramics Processing and Technology, A.G. King, Noyes Publishing (2001).

Fundamentals of Ceramics, M.W. Barsoum, CRC Taylor & Francis, (2003).

Handbook of Advanced Ceramics: Materials, Applications, Processing and Properties, S. Somiya et al., Academic Press (2003).

Handbook of Ceramics, Glasses and Diamonds, C.A. Harper (ed.), McGraw-Hill (2001).

Introduction to Ceramics, 2nd ed., W.D. Kingery, et al., Wiley (1995).

Modern Ceramic Engineering: Properties, Processing, and Use in Design, 3rd ed., D.W. Richerson, CRC Taylor & Francis (2006).

Physical Ceramics, Y-M Chiang et al., Wiley (1997).

Principles of Ceramics Processing, 2nd ed., J.S. Reed, Wiley (1995).

Structural Ceramics, M. Schwartz, McGraw-Hill (1994).
Vol 1: Manufacturing Process Fundamentals
Vol 2: Material Selection Fundamentals & Product Design
Vol 3: Mold Design & Construction Fundamentals
Vol 4: Manufacturing Startup and Management

PLASTICS

Characterization and Failure Analysis of Plastics, ASM, Materials Park, OH (2003).

Concise Polymeric Materials Encyclopedia, J.C. Salamone (ed.), CRC Press (1999).

Engineering Plastics and Composites, 2nd ed., W.A. Woishnis (ed.), ASM, Materials Park, OH (1993).

Engineering Plastics Handbook, J.M. Margolis, McGraw-Hill (2006).

Handbook of Elastomers, 2nd ed., A.K. Bhowmick and H.L. Stephens, CRC Press (2000).

Handbook of Plastics, Elastomers and Composites, 4th ed., C.A. Harper (ed.), McGraw-Hill (2003).

Handbook of Thermoset Plastics, S.H. Goodman, Noyes Publications (1986).

Introduction to Polymers, 3rd ed., R.J. Young and P. Lovell, CRC Taylor and Francis, (2008).

Modern Plastics Encyclopedia, published annually by Modern Plastics Magazine, a McGraw-Hill publication.

Modern Plastics Handbook, C.A. Harper, McGraw-Hill (2000).

Plastic Blow Molding Handbook, N. C. Lee (ed.), Van Nostrand Reinhold (1991).

Plastic Injection Molding – 4 volume set, Society of Manufacturing Engineers (1996-1999).

Plastic Materials and Processes, S. Schwartz and S. H. Goldman, Van Nostrand Reinhold, New York (1982).

Plastic Part Technology, E.A. Muccio, ASM, Materials Park, OH (1991).

Plastics: Materials and Processing, 3rd ed., A.B. Strong, Pearson Educational (2006).

Plastics Processing Technology, E.A. Muccio, ASM, Materials Park, OH (1994).

Plastics Engineering Handbook, 5th ed., M.L. Berins (ed.), Chapman & Hall (1994).

Plastics' Technology Handbook, 4th ed., M. Chanda and S. K. Roy (eds.), CRC Press (2006).

Polymer Handbook, 4th ed., J. Brandrup and E.E. Immergut eds., Wiley, 2004.

Polymer Processing, D.H. Morton-Jones, Chapman and Hall, London (2008).

Polymer Processing Principles and Design, D.G. Baird and D.I. Collias, Wiley (1998).

Principles of Polymer Engineering, 2nd ed., N.G. Crum et al., Oxford University Press (1997).

Principles of Polymer Processing, 2nd ed., Z. Tadmore and C.G. Gogos, Wiley (2006).

Textbook of Polymer Science, 3rd ed., F.W. Billmeyer Jr., Wiley (1984).

Thermoforming: Improving Process Performance, Stanley R. Rosen, Society of Manufacturing Engineers (2002).

COMPOSITES

Advanced Polymer Composites: Principles and Applications, B. J. Jang. ASM, Materials Park, OH (1993).

Ceramic Matrix Composites, R. Naslain and B. Harris, Elsevier Applied Science (1990).

Composite Materials Handbook, 2nd ed., M.M. Schwartz,. McGraw-Hill, New York (1995).

Composite Materials: Science and Engineering, 3rd ed., K.K. Chawla, Springer-Verlag (2008).

Composite Filament Winding, Stan Peters (ed.), ASM, Materials Park, OH (2011).

Composites Engineering Handbook, P.K. Mallick, Marcel Dekker (1997).

Composites Manufacturing: Materials, Products and Process Engineering, S.K. Mazumdar, CRC Press (2001).

Engineers Guide to Composite Materials, John W. Weeton (ed.), ASM, Materials Park, OH (1986).

Fabrication of Composite Materials—Source Book, M.M. Schwartz (ed.), ASM, Materials Park, OH (1985).

Fiber-Reinforced Composites: Materials, Manufacturing, and Designs, 3rd ed., P.K. Mallick, CRC Taylor & Francis (2007)

Fundamentals of Composite Manufacturing: Materials, Methods, and Applications2^{nd}. ed., A. Brent Strong, Society of Manufacturing Engineers (2008).

Fundamentals of Metal Matrix Composites, S. Suresh (ed.), Butterworth-Heinemann (1993).

Handbook of Ceramic Composites, N.P. Bansal (ed.), Springer (2005).

Handbook of Composites, George Lubin (ed.), Van Nostrand Reinhold, New York (1982).

Manufacturing Processes for Advanced Composites, E. Campbell, Elsevier (2004).

Metal Matrix Composites, N. Chawla and K.K. Chawla, Springer, (2006).

Structural Composite Materials, F.C. Campbell, ASM, Materials Park, OH. (2010).

ELEVATED TEMPERATURE APPLICATIONS

ASM Specialty Handbook - Heat-Resistant Materials, ASM (1997).

Engineer's Guide to High Temperature Materials, F. J. Clauss, Addison Wesley, Reading, MA (1969).

Fatigue and Durability of Metals at High Temperatures, S.S. Manson and G.R. Halford, ASM, Materials Park, OH (2009).

High Temperature Property Data: Ferrous Alloys, M. F. Rothman (ed.), ASM, Materials Park, OH (1987).

Properties of Refractory Metals, W.D. Wilkinson, Gordon and Breach (1969).

Source Book on Materials for Elevated Temperature Applications, ASM, Materials Park, OH (1979).

ELECTRONIC AND MAGNETIC APPLICATIONS

Advanced Electronic Packaging, 2nd ed., R.K. Ulrich and W.D. Brown, IEEE Press and Wiley (2006).

ASM Ready Reference: Electrical and Magnetic Properties of Materials, ASM (2000).

Electronic Materials Handbook, Vol. 1, Packaging, ASM (1989).

Electronic Materials & Processes Handbook, 3rd ed., C.A. Harper, McGraw-Hill (2009).

Electronic Packaging and Interconnection Handbook, 4th ed., C.A. Harper, McGraw-Hill, New York (2004).

Electronic Properties of Materials, 3rd ed., R.E. Hummel, Springer (2004).

Fundamentals of Semiconductor Manufacturing and Process Control, G.S. May and C.J. Spanos, Wiley (2006).

Fundamentals of Semiconductor Processing Technology, B. El-KarehKluwer Academic Publishers, Boston (1995).

Handbook of Electronics Manufacturing, 3rd ed., B.S. Matisoff, Chapman & Hall (1996).

Handbook of Photomask Manufacturing Technology, S. Rizvi, CRC Press, (2005).

Handbook of Semiconductor Technology, K.A. Jackson and W. Schroter, Wiley (2000).

Integrated Circuit Fabrication Technology, D. J. Elliott, McGraw-Hill, New York (1989).

Microchip Fabrication, 5th ed., P. Vanzant, McGraw-Hill (2005).

Principles of Electronic Packaging, D.P. Seraphim et al., McGraw-Hill (1989).

Printed Circuit Assembly Design, L. Marks and J Caterina, McGraw-Hill (2000).

Printed Circuits Handbook, 6th ed., C.F. Coombs Jr. (ed.), McGraw-Hill (2006).

Soldering in Electronics Assembly, M. Judd and K. Brindley, Reed International Books, Boston (1992).

Springer Handbook of Electronic and Photonic Materials, S. Kasap and P. Capper (Eds.), Springer (2006).

The Magnetic Properties of Materials, J. E. Thompson, CRC Press, Cleveland, OH (1968).

The Science and Engineering of Microelectronic Fabrication, S.A. Campbell, Oxford University Press (2001).

DESIGN

Application of Fracture Mechanics—for Selection of Metallic Structural Materials, J. E. Campbell, et al., ASM, Materials Park, OH (1982).

Atlas of Stress-Strain Curves, H. E. Boyer (ed.), ASM, Materials Park, OH (1986).

Atlas of Fatigue Curves, Howard E. Boyer (ed.), ASM, Materials Park, OH (1986).

Atlas of Stress Corrosion and Corrosion Fatigue Curves, A. J. McEvily, ASM, Materials Park, OH (1990).

Axiomatic Design, Nam P. Suh, Oxford University Press (2001).

Concurrent Design of Products and Processes, J. L. Nevins and Dan Whitney, McGraw-Hill (1989).

Die Design Handbook, 3rd ed., David A. Smith, Society of Manufactuirng Engineers (1990).

Engineering Design: A Materials and Processing Approach, 3rd ed., George Dieter, McGraw-Hill (2000).

Fatigue and Durability of Structural Materials, S.S. Manson and G.R. Halford, ASM, Materials Park, OH (2006).

Fatigue Data Book: Light Structural Alloys, ASM, Materials Park, OH (1995).

Fracture Mechanics: Fundamentals and Applications, T. L. Anderson, CRC Press (1991).

Fundamentals of Tool Design, 6th ed., J. Nee (ed.), Society of Manufacturing Engineers (2010).

Handbook of Design, Manufacturing and Automation, R.C. Dorrf and A. Kusiak (eds.), Wiley (1995).

Handbook of Product Design for Manufacturing—A Practical Guide for Low-Cost Production, James G. Bralla (ed.), McGraw-Hill, New York (1986).

Introduction to Engineering Design, Andrew Samuel and John Weir, Butterworth Heinemann, (1999).

*Material Selection in Mechanical Design*4th ed., Michael Ashby, Elsevier (2011).

Mechanical Assemblies, Daniel E. Whitney, Oxford University Press, (2004).

Probabilistic Approaches to Design, E. B. Haugen, Wiley & Sons, NY, (1968).

Product Design for Manufacture and Assembly, 2nd ed., G. Boothroyd, P. Dewhurst, W. Knight, Marcel Decker, (2002).

The Principles of Design, Nam P. Suh, Oxford University Press (1990).

The Principles of Material Selection for Engineering Design, Pat L. Mangonon, Prentice Hall (1999).

Tool Design, 3rd ed., C. Donaldson, G. LeCain, and V. C. Gould, Glencoe (1993).

CASTING

Aluminum Alloy Castings: Properties, Processes and Applications, J.G. Kaufman and E. L. Rooy, ASM and American Foundry Society (2004).

Cast Metals Handbook, 4th ed., American Foundryman's Society (1957).

Casting Design and Performance, ASM, Materials Park, OH (2009).

Castings, 2nd ed., J. Campbell, Butterworth-Heinemann (2003).

Foundry Technology, P.R. Beeley, Butterworths-Heinemann (2002).

Fundamentals of Metal Casting, R.A. Flinn, American Foundry Society (1987).

Gray and Ductile Iron Casting Handbook, Gray and Ductile Iron Founder's Society, Cleveland, OH (1971).

Investment Casting Handbook, Investment Casting Institute, Chicago, IL (1997).

Iron Castings Handbook, Charles F. Walton, Iron Castings Society (1981).

Metalcasting, C.W. Ammen, McGraw-Hill, (1999).

Principles of Metal Casting, 2nd ed., Heine, Loper and Rosenthal, McGraw-Hill (1967).

Steel Castings Handbook, 6th ed., Steel Founder's Society of America, OH (1995).

The NFFS (Non-Ferrous Founders Society) Guide to Aluminum Casting Design: Sand and Permanent Mold, NFFS (1994).

WELDING AND JOINING

Adhesive Bonding: Materials, Applications, and Technology, W. Brockman et al., Wiley (2009).

Adhesive Bonding: Science, Technology and Applications, R.S. Adams (ed.), CRC Taylor & Francis (2005).

Brazing Handbook, American Welding Society (1991).

Brazing, 2nd ed., M.M. Schwartz, ASM, Materials Park, OH (2003).

Ceramic Joining, M.M. Schwartz, ASM, Materials Park, OH (1990).

Design for Assembly, M. Andreasen et al., Springer-Verlag (1988).

Design of Welded Structures, James F. Lincoln Foundation.

Design of Weldments, James F. Lincoln Foundation.

Friction Stir Welding and Processing, R.S. Mishra and M.W. Mahoney (eds.), ASM, Materials Park, OH (2007).

Handbook of Adhesives, 3rd ed., Irving Skiest (ed.), Chapman & Hall (1990).

Handbook of Adhesives and Sealants, 2nd ed., E.M. Petrie, McGraw-Hill (2006).

Handbook of Plastics Joining: A Practical Guide, William Andrew Inc. (1996).

Industrial Brazing Practice, P. Roberts, CRC Press (2004).

Joining of Composite Matrix Materials, M.M. Schwartz, ASM, Materials Park, OH (1994).

Joining of Plastics: Handbook for Designers and Engineers, J. Rotheiser, Hanser Gardner (2004).

Joining Processes: Introduction to Brazing and Diffusion Bonding, M.G. Nicholas, Chapman & Hall (1998).

Lead-free Soldering, J. Bath (ed.), Springer (2007).

Mechanical Fastening, Joining, and Assembly, J.A. Speck, Marcel Dekker, (1997).

Metals Joining Manual, M.M. Schwartz, McGraw-Hill (1979).

Modern Welding Technology, 6th ed., H. B. Cary and S.C. Helzer, Pearson/Prentice Hall, NJ (2005).

Principles of Brazing, D.M. Jacobson and G. Humpston, ASM, Materials Park, OH (2005).

Principles of Soldering, G. Humpston and D.M. Jacobson, ASM, Materials Park, OH (2004).

Principles of Welding: Processes, Physics, Chemistry, and Metallurgy, R. W. Messler Jr., Wiley (1999).

Resistance Welding: Fundamentals and Applications, H. Zhang and J. Senkara, CRC Press (2005).

Soldering Manual, 2nd ed., American Welding Society (1978).

Solders and Soldering, 3rd ed., H.H. Manko, McGraw-Hill (1994).

Source Book on Brazing and Brazing Technology, ASM, Materials Park, OH (1980).

Source Book on Electron Beam and Laser Welding, ASM, Materials Park, OH (1980).

Standard Handbook of Fastening and Joining, 3rd ed., R.O. Parmley (ed.), McGraw-Hill (1997).

The Basics of Soldering, A. Rahn, Wiley (1993).

The Science and Practice of Welding, 10th ed., 2 vols., A.C. Davies, Cambridge University Press (1993).

Welding Encyclopedia, L.B. MacKenzie, 16th ed., revised and re-edited by T.B. Jefferson, Monticello Books, Morton Grove, IL (1968).

Welding: Fundamentals and Procedures, J. Galyen et al., Prentice-Hall (1991).

Welding Handbook, 9th ed., American Welding Society, New York (2007).

Welding Process Handbook, K. Weman, CRC Press (2003).

Welding: Principles and Applications, 6th ed., L.F. Jeffus, Cengage Learning (2007).

Welding Technology for Engineers, B. Raj et al.., Narosa Publishing House and ASM, Materials Park, OH (2006).

FABRICATION AND FORMING

Aluminum Extrusion Technology, P.K. Saha, ASM, Materials Park, OH (2000).

Cold and Hot Forging: Fundamentals and Applications, T. Altan et al.., ASM, Materials Park, OH (2005).

Cold Rolling of Steel, W.L. Roberts, Marcel Dekker (1978)

Die Design Handbook, 3rd ed., David A. Smith, Society of Manufacturing Engineers (1990).

Extrusion, 2nd ed., M. Bauser et al. (Eds.), ASM, Materials Park, OH (2006).

Extrusion, K. Laue and H. Stenger, ASM (1981).

Finite-Element Plasticity and Metalforming Analysts, G.W. Rowe, C. E. N. Sturgess, P. Hartley, and I. Pillingen, Cambridge University Press (1991).

Forging Handbook, T.G. Bryer (ed.), Forging Industry Association, Cleveland, OH and ASM (1985).

Fundamentals of Hydroforming, Harjinder Singh, Society of Manufacturing Engineers (2003).

Fundamentals of Metal Forming, R.H. Wagoner and J-L Chenot, Wiley (1997).

Fundamentals of Pressworking, David A. SmithSociety of Manufacturing Engineers (1994).

Fundamentals of Tool Design, 5th ed., David A. Smith and John Nee (Eds), Society of Manufacturing Engineers (2003).

Handbook of Fabrication Processes, O.D. Lascoe, ASM, Materials Park, OH (1988).

Handbook of Metal Forming, Kurt Lange (ed.), Society of Manufacturing Engineers (2006).

Handbook of Metalforming Processes, B. Avitzur, Wiley-Interscience, New York (1983).

Handbook of Workability and Process Design, G.E. Dieter et al., (eds.), ASM, Materials Park, OH (2003).

Hot Rolling of Steel, W.L. Roberts, Marcel Dekker (1983).

Mechanics of Sheet Metal Forming, DJ. Hu et. Al, Butterworth-Heinemann (2002).

Metal Forming: Fundamentals and Applications, T. Altan, S-I Oh, and H. Gegal, ASM, Materials Park, OH (1983).

Metal Forming: Mechanics and Metallurgy, 3rd ed., W.F. Hosford and R.M. Caddell, Cambridge University Press (2007).

Metal Forming Practice: Processes, Machines, Tools, H. Tschaetch, Springer (2007).

Metalforming and the Finite-Element Method, S. Kohavashi, S. Oh, and T. Altan, Oxford University Press (1989).

Metalworking Science and Engineering, Edward M. Mielnik, McGraw-Hill (1991).

Press Brake Technology, S. Benson, Society of Manufacturing Engineers (1997).

Progressive Dies: Principles and Practices of Design and Construction, Don Peterson (Ed.), Society of Manufacturing Engineers (1994).

Roll Forming Handbook, George T. Halmos, Taylor and Francis (2005)

Sheet Metal Forming, R. Pearce, Springer (2006).

Sheet Metal Forming Processes and Die Design, Vukota Boljanovic, Industrial Press (2004).

Steel Forgings: Design, Production, Selection, Testing, and Application, E. G. Nisbett, ASTM, West Conshohocken, PA (2005).

Techniques of Pressworking Sheet Metal, 2nd ed., D. F. Eary and E. A. Reed, Prentice Hall (1974).

The Metal Stamping Process, Jim Szumera, Industrial Press, (2003).

Tube Forming Processes: A Comprehensive Guide, Greg Miller, Society of Manufacturing Engineers (2002).

POWDER METALLURGY

A-Z of Powder Metallurgy, R. M. German, Elsevier Science (2006).

Fundamentals of Powder Metallurgy, Leander Pease and William West, Metal Powder Industries Federation (2002)

Handbook of Powder Metallurgy, H.H. Hausner, Chemical Publishing Co. (2007).

Injection Molding of Metals and Ceramics, Randall M. German and Animesh Bose, Metal Powder Industries Federation (1997).

Introduction to Powder Metallurgy, F. Thummler and R. Oberacker, The Institute of Materials (1994).

Powder Injection Molding, R.M. German, Metal Powder Industries Federation (1990).

Powder Injection Molding: Design & Applications, Randall M. German, Innovative Materials Solutions (2003).

Powder Metallurgy & Particulate Materials Processing, Randall M. German, Metal Powder Industries Federation (2006).

Powder Metallurgy Design Manual, 3rd ed., Metal Powder Industries Federation (1998).

Powder Metallurgy Science, 2nd Ed., R.M. German, Metal Powder Industries Federation (1994).

Powder Metallurgy, Principles and Applications, F.V. Lenel, Metal Powder Industries Federation, Princeton, NJ (1980).

Powder MetallurgyStainless Steels: Processing, Microstructure and Properties, E. Klar and P. Samal, ASM (2007).

MACHINING

Analysis of Material Removal Processes, W. R. DeVries, Springer-Verlag (1992).

Creep Feed Grinding, C. Andrew et al., Holt, Rinehart and Winston (1985).

Fundamentals of Machining and Machine Tools, 3rd ed., G. Boothroyd and Winston Knight, CRC Taylor & Francis (2006).

Gear Materials, Properties and Manufacture, J.R. Davis (ed.), ASM International (2006).

Grinding Technology, 2nd Ed., Changsheng Guo and Stephen Malkin, Industrial Press and SME (2007).

Grinding Technology: Theory and Applications of Machining with Abrasives, 2nd ed., S. Malkin, Industrial Press (2008).

Handbook of Machine Tools, 4 vols., M. Weck, Wiley (1984).

Handbook of Machining with Grinding Wheels, I.D. Marinescu et al., CRC Press (2006).

High-Speed Machining, Berthold Erdel, Society of Manufacturing Engineers (2003).

Machine Tools for High Performance Machining, L.N. Lopez and A. Lamikiz (eds.), Springer (2009).

Machine Tools Handbook, P.H. Joshi, McGraw-Hill (2008).

Machining and Metal Working Handbook, 3rd ed., Bah. Walsh and D. Cormier, McGraw-Hill, (2006).

Machining Data Handbook, 3rd ed., (2 volumes) Machinability Data Center, Cincinnati, OH (1989).

Machining Dynamics: Fundamentals, Applications and Practice, K. Cheng, Springer (2008).

Machining Fundamentals and Recent Advances, J.P. Davim (ed.), Springer (2008).

Machining of Plastics, Akira Kobayashi, Krieger Pub. Co. (1981).

Machining Technology: Machine Tools and Operations, H.A. Youssef and H. El-Hofy, CRC Press (2008).

McGraw-Hill Machining and Metalworking Handbook, 3rd ed., R.A. Walsh (2006).

Metal Cutting, 4th ed., E.M. Trent and P.K. Wright, Butterworth-Heinemann (2000).

Metal Cutting Mechanics, V.P. Astakhov, CRC Press (1998).

Metal Cutting Principles, 2nd ed, M. C. Shaw, Oxford University Press (2005).

Metal Cutting Theory and Practice, 2nd Ed., David A. Stephenson and John S. Agapiou, CRC/Taylor & Francis, (2006).

Metal Cutting Tool Handbook, 7th ed., Industrial Press, (1989).

Modern Grinding Process Technology, S.C. Salmon, McGraw-Hill (1992).

Modern Metal Cutting, AB Sandvik Coromant (1994).

Principles of Modern Grinding Technology, W. Rowe, Elsevier Applied Science (2009).

Technology of Machine Tools, 6th ed., S.F. Krar and A.F. Check, Macmillan/McGraw-Hill (2009).

Tribology of Metal Cutting, V. Astakhov, Elsevier (2007).

HEAT TREATMENT

Atlas of Isothermal and Cooling Transformation Diagrams, ASM, Materials Park, OH (1977).

Atlas of Isothermal and Cooling Transformation Diagrams for Engineering Steels, ASM, Materials Park, OH (1980).

Atlas of Time-Temperature Diagrams for Nonferrous Alloys, G. VanderVoort (ed.), ASM, Materials Park, OH (1991).

Atlas of Time-Temperature Diagrams for Irons and Steels, G. VanderVoort (ed.), ASM, Materials Park, OH (1991).

Handbook of Quenchants and Quenching Technology, G. E. Totten, C. E. Bates, and N. A. Clinton, ASM, Materials Park, OH (1992).

Heat Treater's Guide: Practices and Procedures for Irons and Steels, 2nd Ed., ASM, Materials Park, OH (1995).

Heat Treater's Guide: Practices and Procedures for Nonferrous Alloys, ASM (1996).

Heat Treatment, Structure and Properties of Nonferrous Alloys, Charles R. Brooks, ASM, Materials Park, OH (1982).

Heat Treating Data Book, 9th ed., Seco/Warwick Corporation (2006).

Practical Heat Treating, 2nd ed. J.L. Dossett and H.E. Boyer, ASM, Materials Park, OH (2006).

Practical Induction Heat Treating, R. E. Haimbaugh, ASM (2001).

Principles of Heat Treatment of Steel, G. Krauss, ASM, Materials Park, OH (1980).

Principles of Heat Treatment, M.A. Grossman and E. C. Bain, ASM, Materials Park, OH (1964).

Principles of the Heat Treatment Plain Carbon and Low Alloy Steels, C. R. Brooks, ASM, Materials Park, OH (1996).

Steel and Its Treatment, Bofors 2nd ed., Ed., G. E. Totten and M. A. H. Howes, Marcel Dekker (2006).

Steel Heat Treatment Handbook, 2nd ed., S.S. Babu and G.E. Totten, CRC Taylor & Francis (2006).

Surface Hardening of Steels: Understanding the Basics, J. R. Davis (ed.), ASM (2002).

SURFACES AND FINISHES

Advanced Machining Technology Handbook, James Brown, McGraw-Hill, (1998).

Chemical Vapor Deposition, J. Park, ASM International (2001).

Coating and Surface Treatment Systems for Metals: A Comprehensive Guide to Selection, J. Edwards, ASM International (1997).

Coatings and Coating Processes for Metals, J.H. Lindsay (ed.), ASM (1998).

Coatings Technology Handbook, 2nd ed., A.A. Tracton (ed.), CRC Taylor & Francis (2006).

Electroplating Engineering Handbook, 4th ed., Lawrence Durney (ed.), Van Nostrand-Reinhold (1984).

Guide to High Performance Powder Coating, Bob Utech, Society of Manufacturing Engineers (2002).

Handbook of Hard Coatings: Deposition Technologies, Properties and Applications, R.F. Bunshah (ed), Noyes (2001).

Handbook of Physical Vapor Deposition (PVD) Processing, D.M. Mattox, Noyes (1999).

Handbook of Surface Treatments and Coatings, ASME Press (2003).

Handbook of Thermal Spray Technology, J.R. Davis (ed.), Thermal Spray Society and ASM, Metals Park, OH (2004).

Industrial Painting and Powder Coating: Principles and Practice, 3rd ed., Norman Roobol, Hanser-Gardner Publications, (2005).

Mass Finishing Handbook, LaRoux Gillespie, Industrial Press and SME (2006).

Metal Cutting and High Speed Machining, D. Dudzinski, et al. (eds.), Kluwer Academic (2002).

Metal Finishing: Guidebook Directory, Metals and Plastics Publications, Inc., Westwood, NJ (published annually).

Metallizing of Plastics: Handbook of Theory and Practice, R. Suchentruck, Finishing Publications Ltd and ASM (1993).

Modern Electroplating, 4th ed., M. Schlesinger and M. Paunovic, Wiley (2001).

Properties of Electroplated Metals and Alloys, 2nd ed., Wm. H. Safranek, Amer. Electroplaters and Finishers Soc. (1986).

Surface Engineering for Corrosion and Wear Resistance, J. R. Davis (ed.), IOM Communications and ASM (2001).

Surface Engineering for Wear Resistance, K.G. Budinski, Prentice Hall (1988).

Surface Engineering of Metals: Principles, Equipment, Technologies, T. Burakowski and T. Wiershon, CRC Press (1998).

Surface Finishing Systems, George Rudski, ASM, Materials Park, OH (1983).

Surface Modification Technologies, T.S. Sudarshan (ed.), ASM International (1998).

The Surface Treatment and Finishing of Aluminum and Its Alloys, 6th ed., S. Wernick, R. Pinner and P. B. Sheasby, Finishing Publications Ltd. and ASM (2001).

CORROSION

An Introduction to Metallic Corrosion, 3rd ed., Ulick R. Evans, Edward Arnold Ltd. and ASM (1981).

Corrosion and Corrosion Control, 3rd ed., H.H. Uhlig and R. Winston Revie, Wiley-Interscience, New York (1985).

Corrosion and Its Control—An Introduction to the Subject, Atkinson and Van Droffelaar, National Association of Corrosion Engineers, Houston, TX (1982).

Corrosion Engineering, 3rd ed., M.G. Fontana, McGraw-Hill, New York (1986).

Corrosion Prevention by Protective Coatings, Charles G. Munger, National Association of Corrosion Engineers (1984).

Corrosion Resistance Tables, 4th ed., P.A. Schweitzer, Marcel Dekker (1995).

Corrosion, 2nd ed.—Vol. 1: Corrosion of Metals and Alloys and Vol. 2: Corrosion Control, L. L. Shrier (ed.).

Corrosion: Understanding the Basics, J.R. Davis (ed.), ASM (2000).

Encyclopedia of Corrosion Technology, 3rd ed., P.A. Schweitzer, Marcel dekker (2004).

Fundamentals of Electrochemical Corrosion, E. E. Stansbury and R. A. Buchanan, ASM (2000).

Handbook of Corrosion Data, 2nd ed., B. Craig and D. Anderson (eds.), ASM, Materials Park, OH (1995).

Material Selection for Corrosion Control, S. I. Chawla and R. K. Gupta, ASM, Materials Park, OH (1993).

NACE Corrosion Engineer's Reference Book, 2nd ed., R. S. Treseder (ed.), National Association of Corrosion Engineers (NACE) (1991).

NDT AND METROLOGY

Dimensioning and Tolerancing Handbook, P.J. Drake, McGraw-Hill (1999).

Geometric Dimensioning and Tolerancing for Mechanical Design, G. Cogorno, McGraw-Hill (2006).

Handbook of Dimensional Measurement, 4th ed., M. Curtis, Industrial Press (2007).

Handbook of Metrology, Brown and Sharpe (1992).

Hardness Testing, 2nd ed., H. Chandler (ed.), ASM (1999).

ISO System of Limits and Fits, General Tolerances and Deviations, American National Standards Institute.

Measurement of Geometrical Tolerances in Manufacturing, J.D. Meadows, Marcel Dekker (1998).

Nondestructive Testing Handbook, 2nd ed. (multiple volumes), American Society for Nondestructive Testing (1982–1985).

Nondestructive Testing, Louis Cartz, ASM, Materials Park, OH (1995).

Practical Non-Destructive Testing, 2nd ed., B. Raj, T. Jaykumar, and M. Thavasimuthu, Narosa Publishing and ASM (2002).

Principles of Measurement Systems, 4th ed., J.P. Bentley, Prentice Hall (2005).

Springer Handbook of Materials Measurement Methods, H. Czichos et al. (eds.), Springer (2006).

Tensile Testing, 2nd ed., J.R. Davis (Ed.), ASM, Materials Park, OH (2004).

The Testing of Engineering Materials, 4th ed., H. E Davis, G. E. Troxell, and G. F.W. Hauck, McGraw-Hill, New York (1982).

QUALITY CONTROL, TAGUCHI METHODS AND SIX SIGMA

A Primer on the Taguchi Method, Ranjit Roy, Society of Manufacturing Engineers (2010).

Creating Quality, William J. Kolarik, McGraw-Hill (1995).

Introduction to Quality Engineering, Genichi Taguchi, Kraus Int. Pub. (1986).

Introduction to Statistical Quality Control, 6th ed., D. C. Montgomery, Wiley (2008).

Juran's Quality Handbook, 5th ed., J.M. Juran and A.B. Godfrey (eds.), McGraw-Hill (2000).

Juran's Quality Planning & Analysis, 5th ed., F.M. Gryna, R.C.H. Chua and J.A. DeFeo, McGraw-Hill (2006).

Lean Six Sigma, M.L. George, McGraw-Hill (2002).

Process Quality Control, 2nd ed., E. R. Ott and E.G. Schilling, McGraw-Hill (1990).

Quality Control, 8th Ed., D.H. Besterfield, Prentice Hall (2008).

Quality Control and Industrial Statistics, 5th ed., Acheson J Duncan, Irwin, (1986).

Quality Control for Managers and Engineers, Elwood G. Kirkpatrick, Wiley (1970).

Quality Control Handbook, 4th ed., J.M. Juran and F. M. Gigna, McGraw-Hill (1988).

Quality Engineering Handbook, 2nd ed., T. Pyzdek and P. Keller, CRC Taylor & Francis, (2003).

Quality Engineering in Production Systems, G. Taguchi, E.A. ElSayed, and T. Hsiang, McGraw-Hill (1989).

Quality Engineering in Production Systems, Genichi Taguchi, Elsayed A. Edsayed, and Thomas C. Hsiang, McGraw-Hill (1989).

Quality Planning and Analysis, 3rd ed., Juran and Gryan, McGraw-Hill (1993).

Quality, Productivity, and Competitive Position, W. Edwards Deming, MIT Press, Boston (1982).

Reliability and Maintainability Management, Balbir S. Dhillon and Hans Reiche, Von Nostrand Reinhold (1985).

Statistical Quality Control, 6th ed., Eugene L. Grant and Richard S. Leavenworth, McGraw-Hill (1988).

Statistical Quality Design and Control, 2nd ed., R.E. DeVor, T. Chang, and J.W. Sutherland, Prentice Hall (2007).

Statistical Process Control and Quality Improvement, 5th ed., G.M. Smith, Prentice Hall (2004).

Taguchi Methods, A. Bendell, Springer (2007).

Taguchi Techniques for Quality Engineering, 2nd ed., P. J. Ross, McGraw-Hill (1996).

The Six Sigma Black Belt Handbook, McCarty et al.. (2005).

The Six Sigma Handbook, Thomas Pyzdek (ed.), McGraw-Hill, (2003).

The Six Sigma Performance Handbook, P. Gupta, McGraw-Hill (2005).

Total Quality Management—Text, Cases and Readings, Joel E. Ross, St. Lucie Press (1993).

Zero Quality Control: Source Inspection and the Poka-Yoke System, Shigeo Shingo, Productivity Press (1986).

FAILURE ANALYSIS AND PRODUCT LIABILITY

Analysis of Metallurgical Failures, V. J. Conangelo and F. A. Heiser, Wiley-Interscience (1974).

Case Histories in Failure Analysis, ASM, Materials Park, OH (1979).

Engineering Aspects of Product Liability, V. J. Conangelo and P. A. Thornton, ASM, Materials Park, OH (1981).

Failure Analysis of Engineering Structures: Methodology and Case Histories, V. Ramachandran et al., ASM, Materials Park, OH (2005).

Failure Analysis: Case Histories, F. K. Naumann, ASM, Materials Park, OH (1983).

Failure Analysis: The British Engine Technical Reports, F. K. Hutchings and P. M. Unterweiser, ASM, Materials Park, OH (1981).

Failure Analysis of Heat Treated Steel Components, L.C.F. Canale et al. (eds), ASM, Materials Park, OH (2008).

Failure of Materials in Mechanical Design, J.A. Collins, Wiley-Interscience (1981).

Fractography in Failure Analysis, American Society for Testing and Materials, Philadelphia, PA (1978).

Handbook of Case Histories in Failure Analysis: Vols. 1 and 2, K.A. Esaklul (ed.), ASM, Materials Park, OH (1992–93).

How to Organize and Run a Failure Investigation, Daniel P. Dennies, ASM, Materials Park, OH (2005).

Metal Failures: Mechanisms, Analysis, Prevention, A. J. McEvily (ed.), John Wiley & Sons, New York (2002).

Metallography in Failure Analysis, J. L. McCall and P. M. French, Plenum (1977).

Metallurgical Failure Analysis, C. R. Brooks and A. Choudhury, McGraw-Hill (1993).

Residual Stresses and Fatigue in Metals, J.O. Almen and P. H. Black, McGraw-Hill, New York (1963).

Stress Corrosion Cracking: Materials Performance and Evaluation, Russell Jones (ed.), ASM, Materials Park, OH (1992).

Tool and Die Failures Source Book, ASM, Materials Park, OH (1982).

Understanding How Components Fail, 2nd ed., Donald J. Wulpi, ASM, Materials Park, OH (1999).

Why Metals Fail, R.D. Barer and B. F. Peters, Gordon and Breach, New York (1970).

MANUFACTURING SYSTEMS ANALYSIS

Computational Intelligence in Design and Manufacturing, Andrew Kusiak, Wiley (2000).

Design and Analysis of Lean Production Systems, R.G. Askin and J.B. Goldberg, Wiley (2002).

Fundamentals of Systems Engineering, C. Jotin Kristy and Jamshid Mohammadi, Prentice-Hall (2001).

Manufacturing Systems Design and Analysis, 2nd ed., B. Wu, Chapman and Hall (1994).

Manufacturing Systems Engineering, S. B. Gershwin, PTR, Prentice Hall (1994).

Manufacturing Systems Engineering, 2nd ed., K. Hitomi, Taylor and Francis Ltd. (1996).

Modeling and Analysis of Manufacturing Systems, R.G. Askin and C. R. Standridge, Wiley (1993).

Performance Modeling of Automated Manufacturing Systems, N. Viswanadham and Y. Narahari, Prentice Hall (1992).

Stochastic Models of Manufacturing Systems, J.A. Buzacott and J. C. Shanthikumar, Prentice Hall (1993).

Systems Analysis and Modeling, Donald W. Boyd, Academic Press (2001).

MANUFACTURING ENGINEERING

Concurrent Design of Products and Processes, J. L. Nevins and D. E. Whitney, McGraw-Hill (1989).

Environmentally Conscious Manufacturing, Myer Kutz, John Wiley & Sons (2007).

Intelligent Manufacturing Systems, A. Kusiak, Prentice Hall (1990).

Manufacturing Intelligence, P. K. Wright and D. A. Bourne, Addison Wesley (1988).

ROBOTICS

Handbook of Industrial Robotics, 2nd ed., Shimon Nof (ed.), Wiley (1999).

Industrial Robotics: How to Implement the Right System for Your Plant, Andrew Glaser, Industrial Press (2008).

Industrial Robotics: Technology, Programming and Applications, M.P. Groover et al., McGraw-Hill (1986).

Industrial Robots: Computer Interfacing and Control, W. E. Snyder, Prentice Hall (1985).

Introduction to Robotics: Mechanics and Control, 3rd ed., J.J. Craig, Prentice-Hall (2004).

Introduction to Robot Technology, P. Coiffet and M. Chizouze, McGraw-Hill (1993).

Introduction to Robots in CIM Systems, J.A. Rehg, Prentice-Hall (2000).

Robot Analysis, L-W Tsai, Wiley (1999).

Robotics and Manufacturing Automation, 2nd ed., C. R. Asfahl, Wiley (1992).

AUTOMATION, CNC, CAM, CAD, FMS AND CIM

An Introduction to CNC Machining and Programming, D. Gibbs and T. M. Crandall, Industrial Press (1991).

Automatic Assembly, C. Boothroyd, C. Poli, and L. F. March, Marcel Dekker (1982).

Automation, Production Systems, and Computer-Integrated Manufacturing, 3rd ed., Mikell P. Groover, Prentice Hall (2008).

Computer-Aided Manufacturing, 3rd ed., T.C. Chang, R.A. Wysk and Hsu-Pin Wang, Prentice Hall (2006).

Computer-Integrated Design and Manufacturing, D.D. Bedworth, M. R. Henderson, and P. M. Wolfe, McGraw-Hill (1991).

Computer-Integrated Manufacturing, 3rd ed., J.A. Rehg and H.W. Kraebber, Prentice-Hall (2005).

Computer Control of Machines and Processes, J.G. Bollinger and N.A. Duffie, Addison-Wesley (1989).

Computer Numerical Control: Operation and Programming, 3rd ed., J. Stenerson and K.S. Curran, Prentice Hall (2006).

Engineering Robust Designs, J X. Wang, Prentice-Hall (2005).

Fundamentals of Computer-Integrated Manufacturing, A.L. Foston, et al., Prentice Hall (1991).

Getting Factory Automation Right (The First Time), Edwin H. Zimmerman, Society of Manufacturing Engineers (2001).

Handbook of Machine Vision, A. Hornberg, Wiley (2007).

Introduction to Computer Numerical Control, 4th ed., J.V. Valentino and J. Goldenberg, Regents/Prentice Hall (2008).

Industrial Automation, D.W. Pessen, Wiley (1989).

The Principles of Group Technology and Cellular Manufacturing Systems, H. Parsaei, et al., Wiley (2006).

COST ESTIMATING

Cost Estimator's Reference Manual, R.D. Steward and R.W. Wyskida, Wiley (1987).

Engineering Cost Estimating, 3rd ed., P. E. Ostwald, Prentice Hall (1992).

Realistic Cost Estimating for Manufacturing, 2nd ed., W. Winchell, Society of Manufacturing Engineers (1989).

NON-TRADITIONAL MACHINING/LASER MACHINING SOURCES

Advanced Machining Processes: Nontraditional and Hybrid Machining Processes, H. ElHofy, McGraw-Hill (2005).

Advanced Machining Processes of Metallic Materials: Theory, Modelling and Applications, W. Grzesik, Elsevier (2008).

Advanced Machining Technology Handbook, James Brown, McGraw-Hill, New York (1998).

Advanced Methods of Machining, J.A. McGeough, Chapman and Hall, New York (1988).

Advanced Machining Processes, H. El-Hofy, McGraw-Hill, (2005).

Electrical Discharge Machining, E.C. Jameson, Society of Manufacturing Engineers (2001).

Laser Fabrication and Machining of Metals, N.B. Dahotre and S.P. Harimkar, Springer (2007).

Laser Material Processing, W.M. Steen, Springer, London, England (1998).

Laser Materials Processing, L. Migliore, Marcel Dekker (1996).

Laser Machining, G. Chryssolouris and P. Sheng, Springer-Verlag (1991).

Non-Traditional Machining Handbook, 2nd ed., C. Sommer, Advance Publishing (2009).

Nontraditional Machining Processes, 2nd ed., E.J. Weller ed., Society of Manufacturing Engineers (1984).

Nontraditional Manufacturing Processes, G.F. Bennedict, Marcel Dekker (1987).

Principles of Abrasive Water Jet Machining, A.W. Momber and R. Kovacevic, Springer (1998).

The EDM Handbook, E. B. Guitrau and Hanser-Gardner, Cincinnati, OH (1997).

Theory and Practice of Electrochemical Machining, V.K. Jain and P.C. Pandey, Wiley (1993).

ADDITIVE PROCESSES: RAPID PROTOTYPING

Automated Fabrication: Improving Productivity in Manufacturing, M. Burns, Prentice Hall, Englewood Cliffs, NJ (1993).

Rapid Manufacturing: The Technologies and Applications of Rapid Prototyping and Rapid Tooling, D.T. Pham and S. S. Dimov, Springer-Verlag, London, UK (2001).

Rapid Prototyping, A. Gebhardt, Hanser Gardner, (2004).

Rapid Prototyping and Other RP&M Technologies: From Rapid Prototyping to Rapid Tooling, P. F. Jacobs, SME, Dearborn, MI (1996).

Rapid Prototyping: Principles and Applications, 2nd ed., C.C. Kai et al., World Scientific Publishing Company, Singapore (2003).

Rapid Prototyping: Principles and Applications, R.I. Noorani, Wiley, (2006).

Rapid Prototyping: Principles and Applications in Manufacturing, C.K. Chua and L.K. Fua, Wiley (1997).

Rapid Prototyping: Theory and Practice, A. Kamrani and E.A. Nasr (eds.), Springer (2007).

Rapid Tooling: Technologies and Industrial Applications, Peter D. Hilton and P. F. Jacobs, Marcel Dekker/CRC Press (2000).

Rapid Tooling: The Technologies and Applications of Rapid Prototyping and Rapid Tooling, D.T. Pham and S.S. Dimov, Springer-Verlag (2001).

Solid Freeform Fabrication, J.J. Beaman et al., Kluwer, (1997).

User's Guide to Rapid Prototyping, Todd Grimm, Society of Manufacturing Engineers, (2004).

Wohler's Report (published annually), Terry Wohlers, Wohlers Associates, Colorado.

MICROFABRICATION AND NANOTECHNOLOGY

An Introduction to Microelectromechanical Systems Engineering, 2nd ed., N. Maluf, and K. Williams Artech House (2004).

Applications of Microfabrication and Nanotechnology, M.J. Madou, CRC Press (2009).

Basics of Nanotechnology, 3rd ed., H-G Rubahn, Wiley-VCH, Germany (2008).

Fundamentals of Microfabrication, M.J. Madou, Boca Raton, FLCRC Press, (2002).

Fundamentals of Nanotechnology, G.L. Hornyak et al., CRC Taylor & Francis, (2009).

Handbook of Nanoscience, Engineering and Technology, W.A. Goddard III, et al., CRC Press, (2003).

Introduction to Micromaching, V. K., Jain., Ed. Alpha Sci., Kanpur, India (2010)

Introduction to Nanotechnology, C.P. Poole Jr., Wiley-Interscience, Wiley (2003).

Manufacturing Techniques for Microfabrication and Nanotechnology, M. Madou, CRC Taylor & Francis (2009).

MEMS/NEMS Handbook (5 vols.), C.T. Leondes, Springer (2007).

Micro and Nanomanufacturing, M.L. Jackson, et al., Springer (2007).

Microfabrication and Nanomanufacturing, Boca Raton, Fl. M. J. Jackson, CRC Taylor & Francis (2006).

Micromachining of Engineering Materials, J.A. McGeugh, Marcel Dekker, New York, (2002).

Micromanufacturing and Nanotechnology, N.P. Mahalik, New York, NY: Springer Publishing, (2010).

Micromechanical System Design, J.J. Allen, CRC Press (2006).

Nanofabrication Fundamentals and Applications, A.A. Tseng (ed.), World Scientific, Singapore, (2008).

Nanofuture: What's next for nanotechnology, J.S. Hall, Amherst, NY; Prometheus Books (2005).

Nanomaterials, S. Mitura, Elsevier (2000).

Nanomaterials Handbook, Y. Gogotsi, CRC Press (2006).

The MEMS Handbook, 2nd ed., M. Gad-el_Hak (ed.), CRC Press (2006).

Springer Handbook of Nanotechnology, 2nd ed., B. Bhushan (ed.), Springer (2006).

LEAN MANUFACTURING

A Revolution in Manufacturing: The SMED System, Shigeo Shingo, Productivity Press (1985).

A Study of The Toyota Production System, Shigeo Shingo, Productivity Press (1989).

Becoming Lean, Ed. J. Liker, Productivity Press, (2004).

Competitive Manufacturing Management, John Nicholas, McGraw-Hill, (1998).

Creating a Lean Culture: Tools to Sustain Lean Conversions, 2nd Ed., David Mann, Taylor & Francis (2010).

Design, Analysis and Control and Manufacturing Cells, PED-Vol. 53, J.T. Black, B. C. Jiang, and G. J. Wiens, ASME (1991).

Factory Physics, W.J. Hopp, and M.L. Spearman, McGraw-Hill (2001).

Handbook of Cellular Manufacturing Systems, Shahrukh A. Irani (ed.), Wiley (1999).

Inside the Mind of Toyota, Hino, Satuski, Productivity Press (2006).

Introduction to TPM: Total Productive Maintenance, Seiichi Nakajima, Productivity Press (1988).

Japanese Manufacturing Techniques: Nine Hidden Lessons in Simplicity, Richard J. Schonberger, Free Press (1982).

JIT Factory Revolution, H. Hirano and J T. Black, Productivity Press (1988).

Just-in-Time in American Manufacturing, William Duncan, SME, (1988).

Kaikaku, The Power and Magic of Lean, N. Bodek, PCS Press, (2004).

Kaizen for Quick Changeover, Kenichi Sekine and Keisuke Arai (translated by Bruce Talbot), Productivity Press (1992).

Kanban: Just-In-Time at Toyota, translation by D. J. Lu, Japan Management Association, Productivity Press (1986).

Kaizen Event Fieldbook: Foundation, Framework, and Standard Work for Effective Events, Mark Hamel, Society of Manufacturing Engineers (2010)

Lean Assembly, Michael Baudin, Productivity Press (2002).

Lean Enterprise Systems, Steve Bell, Wiley (2006).

Lean Manufacturing Implementation, D.P. Hobbs, APICS (2004).

Lean Manufacturing Systems and Cell Design, J T. Black and Steve L. Hunter, Society of Manufacturing Engineers (2003).

Lean Production: Implementing a World Class System, John Black, Industrial Press and SME (2008).

Lean Production Simplified, Pascal Dennis, (2002).

Lean Thinking, James Womack, Jones, Danie, Simon & Schuster, New York, (1996).

One-Piece Flow: Cell Design for Transforming the Production Process, Kenichi Sekine, Productivity Press (1990).

The Design of the Factory with A Future, J T. Black, McGraw-Hill (1991).

The Hitchhiker's Guide to Lean: Lessons from the Road, Jamie Flinchbaugh and Andy Carlino, Society of Manufacturing Engineers (2005).

The Lean Design Guidebook, Ronald Mascitelli, Technology Publications (2004).

The Machine That Changed The World, James P. Womack, Daniel T. Jones, and Daniel Roos, Harper Perennial (1991).

The Principles of Group Technology and Cellular Manufacturing, H. Parsai et al., Wiley (2006).

The Toyota Product Development System, J.M. Morgan and J.K. Liker, Productivity Press, (2006).

The Toyota Way: 14 Management Principles from the World's Greatest Manufacturer, Jeffrey Liker, McGraw-Hill, (2004).

Toyota Production System: Beyond Large-Scale Production, Taiichi Ohno, Productivity Press (1988).

Toyota Production System—An Integrated Approach to Just-In-Time, 3rd Ed., Yasuhiro Monden.Norcross, Georgia (1998).

World Class Manufacturing Casebook: Implementing JIT and TOC, Richard J. Schonberger, The Free Press (1987).

World Class Manufacturing, Richard J. Schonberger, The Free Press (1986).

World Class Manufacturing, The Next Decade, Richard Schonberger, The Free Press, (1996).

Acronyms

AC	Adaptive Control
AFM	Abrasive Flow Machining
AGVS	Automated Guided Vehicle System
AHSS	Advanced High-Strength Steel
AI	Artificial Intelligence
APT	Automatic Programming of Tools
AQL	Acceptable Quality Limit (or Level)
ASCII	American Standard Code
AS/RS	Automatic Storage/Retrieval System
ATE	Automatic Test Equipment
AWJM	Abrasive Water Jet Machining
BASIC	Beginner's All-Purpose Symbolic Instruction Code
BTA	boring trepanning association
BTRI	Behind the Tape Reader Interface
BUE	built-up edge
CAD	Computer-Aided Design
CAD/CAM	Computer-Aided Design/Computer-Aided Manufacturing
CAD/D	Computer-Aided Drafting and Design
CAE	Computer-Aided Engineering
CAM	Computer-Aided Manufacturing
CAPP	Computer-Aided Process Planning
CATI	Computer-Aided Testing and Inspection
CDC	Cutter Diameter Compensation
CHM	Chemical Machining
CIM	Computer-Integrated Manufacturing
CL	Center Line
CMM	Coordinate Measuring Machine
CMS	Cellular Manufacturing System
CNC	Computer Numerical Control
COBOL	Common Business Oriented Language
CPR	Capacity Resources Planning
CPU	Central Processing Unit *(Computer)*
CRT	Cathode Ray Tube
CVD	Chemical Vapor Deposition
DBM	Data-Base Management
CBN	cubic boron nitride
DBTT	Ductile-to-Brittle Transition Temperature
DCEN	Direct Current Electrode Negative
DCEP	Direct Current Electrode Positive
DDAS	Direct Data Acquisition System
DDC	Direct Digital Control
DDM	Direct Digital Manufacturing
DMD	Direct Metal Deposition
DNC	Digital (or Direct or Distributed) Numerical Control
DOC	depth of cut
DOI	depth of immersion
DOS	Disk Operating System
DP	Diametrical Pitch
DPRO	Digital Position Readout
DRO	Digital Readout
EAROM	Electrically-Alterable Read-Only Memory
EBCDIC	Extended Binary Coded Decimal Interchange Code
EBM	Electron Beam Machining *(EBW = Welding)* *(EBC = Cutting)*
ECM	Electrochemical Machining
ECO	engineering change order
EDM	Electrodischarge Machining *(EDG = Grinding)*
EMI	Electromagnetic Interface
EOB	End of Block
EOP	End of Program (workpiece)
EOT	End of Tape
EROM	Eraseable Read-Only Memory
ESW	Electroslag Welding
FATT	Fracture Appearance Transition Temperature
FCAW	Flux Cored Arc Welding
FDM	Fused Deposition Modeling
FEM	Finite-Element Method
FMC	Flexible Manufacturing Cell
FMS	Flexible Manufacturing System
FORTRAN	Formula Translation
FRN	Feed Rate Number
FSW	Friction Stir Welding
GMAW	Gas Metal Arc Welding
GT	Group Technology
GTAW	Gas Tungsten Arc Welding
HAZ	Heat Affected Zone
HERF	High Energy Rate Forming
HGVS	Human-Guided Vehicle System *(fork-lift with driver)*
HIP	Hot Isostatic Pressing
HP	horse power
HSLA	High-strength Low-alloy

HSS	high speed steel
IC	integrated circuit
ID	Inkjet Deposition
IGES	Initial Graphics Exchange System
IMPSs	Integrated Manufacturing Production Systems
I/O	Input/Output
IOCS	Input/Output Control System
JIT	Just-In-Time
LAN	Local Area Network
LASER	Light Amplification by Stimulated Emission of Radiation
LBM	Laser Beam Machining *(LBW = Welding)* *(LBC = Cutting)*
L-CMS	Linked-Cell Manufacturing System
LED	Light Emitting Diode
LE	lean engineering
LENS	Laser Engineered Net Shaping
LFG	low force grove
LOM	Laminated Object Manufacturing
LP	Lean Production
LSI	Large Scale Integration
MAP	Manufacturing Automation Protocol
MCU	Machine Control Unit
MDI	Manual Data Input
MIG	Metal-Inert Gas
MIM	Metal Injection Molding
MMFA	mixed model final assembly
MPS	Manufacturing Production System
mrp	Material Requirements Planning
MRPII	Manufacturing Resources Planning
MRR	metal removal rate
MSD	Manufacturing System Design
NC	Numerically Control
NDT	NonDestructive Testing *(NDE = Evaluation)* *(NDI = Inspection)*
OAW	Oxyacetylene Welding
OCR	Optical Character Recognition
OM	Orthogonal Machining
OPM	Orthogonal Plate Machining
OS	Operating System
OTT	Orthogonal Tube Turning
PAW	Plasma Arc Welding *(PAC = Cutting)* *(PAM = Machining)*
PC	process capability
PCB	Printed Circuit Board
PCN	polycrystalline cubic boron nitride
PD	Pitch Diameter
PDES	Product Design Exchange Specification
PIM	Powder Injection Molding
PLC	Programmable Logic Controller
POK	Production Ordering Kanban
POUS	point of use storage
PROM	Programmable Read-Only Memory
PS	Production System
P/M	Powder Metallurgy
PVD	Physical Vapor Deposition
QC	Quality Control
QMS	Quality Management System
RAM	Random Access Memory
RIM	Reaction Injection Molding
ROM	Read-Only Memory
RSW	Resistance Spot Welding
SAW	Submerged Arc Welding
SCA	Single Cycle Automatic
SEM	scanning electron microscope
SGC	Solid Ground Curing
SLA	Stereolithography Apparatus
SLM	Selective Laser Melting
SLS	Selective Laser Sintering
SMAW	Shielded Metal Arc Welding
SMED	single minute exchange of dies
SPC	Statistical Process Control
SPF	Single Piece Flow
SQC	Statistical Quality Control
TCM	Thermochemical Machining
TIG	Tungsten-Inert Gas
TIR	Total Indicator Readout
TPS	Toyota Production System
TQC	Total Quality Control
UHSMC	Ultra-high-speed machining center
USM	Ultrasonic Machining *(USW = Welding)*
VA	Value Analysis
WAN	Wide Area Network
WIP	Work-In-Progress (or Process)
WJM	Water Jet Machining
WLK	Withdrawl Kanban
YAG	Yttrium-Aluminum Garnet